T0181665

Springer Studium Mathematik – Master

Reihe herausgegeben von

Martin Aigner, Freie Universität Berlin, Berlin, Germany

Heike Faßbender, Technische Universität Braunschweig, Braunschweig, Germany

Barbara Gentz, Bielefeld, Germany

Daniel Grieser, Institut für Mathematik, Carl von Ossietzky Universität, Oldenburg, Germany

Peter Gritzmann, Zentrum Mathematik, Technische Universität München, Garching, Germany

Jürg Kramer, Institut für Mathematik, Humboldt-Universität zu Berlin, Berlin, Germany

Volker Mehrmann, Institut für Mathematik, TU Berlin, Berlin, Germany

Gisbert Wüstholz, ETH Zürich, Wermatswil, Switzerland

Die Reihe „Springer Studium Mathematik" richtet sich an Studierende aller mathematischen Studiengänge und an Studierende, die sich mit Mathematik in Verbindung mit einem anderen Studienfach intensiv beschäftigen, wie auch an Personen, die in der Anwendung oder der Vermittlung von Mathematik tätig sind. Sie bietet Studierenden während des gesamten Studiums einen schnellen Zugang zu den wichtigsten mathematischen Teilgebieten entsprechend den gängigen Modulen. Die Reihe vermittelt neben einer soliden Grundausbildung in Mathematik auch fachübergreifende Kompetenzen. Insbesondere im Bachelorstudium möchte die Reihe die Studierenden für die Prinzipien und Arbeitsweisen der Mathematik begeistern. Die Lehr- und Übungsbücher unterstützen bei der Klausurvorbereitung und enthalten neben vielen Beispielen und Übungsaufgaben auch Grundlagen und Hilfen, die beim Übergang von der Schule zur Hochschule am Anfang des Studiums benötigt werden. Weiter begleitet die Reihe die Studierenden im fortgeschrittenen Bachelorstudium und zu Beginn des Masterstudiums bei der Vertiefung und Spezialisierung in einzelnen mathematischen Gebieten mit den passenden Lehrbüchern. Für den Master in Mathematik stellt die Reihe zur fachlichen Expertise Bände zu weiterführenden Themen mit forschungsnahen Einblicken in die moderne Mathematik zur Verfügung. Die Bücher können dem Angebot der Hochschulen entsprechend auch in englischer Sprache abgefasst sein.

Diese Reihe wird in Zukunft systematisch weitergeführt in den Reihen:
Springer Studium Mathematik (Bachelor) - springer.com/series/16564
Springer Studium Mathematik (Master) - springer.com/series/16565

Ulrich Görtz · Torsten Wedhorn

Algebraic Geometry II: Cohomology of Schemes

With Examples and Exercises

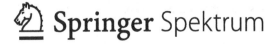
Springer Spektrum

Ulrich Görtz
Fakultät für Mathematik
Universität Duisburg-Essen
Essen, Germany

Torsten Wedhorn
Fachbereich Mathematik
TU Darmstadt
Darmstadt, Germany

ISSN 2509-9310   ISSN 2509-9329   (electronic)
Springer Studium Mathematik – Master
ISBN 978-3-658-43030-6   ISBN 978-3-658-43031-3   (eBook)
https://doi.org/10.1007/978-3-658-43031-3

This Springer Spektrum imprint is published by the registered company Springer Fachmedien Wiesbaden GmbH, part of Springer Nature.
The registered company address is: Abraham-Lincoln-Str. 46, 65189 Wiesbaden, Germany

Paper in this product is recyclable.

Contents

Introduction

Algebraic geometry is at the same time a very classical subject, and one that remains extraordinarily active to this day. There has been fascinating and spectacular progress in the last century, especially after Grothendieck had introduced his language of schemes which for instance allowed for a much easier treatment of families of varieties, and also immensely strengthened the connection between algebraic geometry and algebraic number theory. On the other hand, many intriguing questions remain open. Not surprisingly, the study of solutions of systems of polynomial equations, which is at the heart of algebraic geometry, is important in many mathematical subjects. In addition to the numerous applications of algebraic geometry in number theory and representation theory, say, there is also a close connection to mathematical physics, and over the last ten or twenty years advanced methods of algebraic geometry play an increasing role in solving problems arising in biology, chemistry, etc. Many of the results in modern algebraic geometry are deep in the sense that they rely on several layers of theory building upon each other. While it is particularly exciting when this large body of abstract machinery can be used to prove innocuous looking elementary statements, this also means that the subject is not easily accessible.

The aim of this book, which is a sequel to our introduction [GWI][O] to the theory of schemes, is to provide a systematic account of several key results and methods in algebraic geometry that were not discussed in [GWI][O], but are required knowledge for (almost) every advanced student and researcher in the field, at a level of generality that is suitable for using the theory in a more classical, geometric context, as well as in a more arithmetically oriented setting.

The content of the book can be grouped into three parts:

- Smooth morphisms (including the closely related notions of étale and unramified morphisms, and also locally complete intersections, and the beginnings of the theory of the étale topology and the étale fundamental group of a scheme), Chapters 17 – 20,

- cohomology of $\mathscr{O}_X$-modules and of quasi-coherent $\mathscr{O}_X$-modules (including finiteness results for cohomology of projective and proper schemes, the theorem of formal functions and some of its applications such as the Stein factorization, Grothendieck duality, and a rather long appendix on homological algebra), Chapters 21 – 25 and Appendix F, and

- as examples to illustrate how to put the technical machinery to use, and as an outlook on two more specific, fascinating topics in algebraic geometry, Chapter 26 on algebraic curves and Chapter 27 on abelian schemes.

The material in the first chapters, which center around the notion of smoothness, is of great importance in basically all areas of algebraic geometry, and is to a large extent "classical" in the sense that it can be found in [EGAIV][O] and [SGA1][O][X]. Some of the key results here are the infinitesimal lifting criterion for smoothness (and similarly for étale and for unramified morphisms), and the characterization of étale morphisms as flat and unramified morphisms. In Chapter 20 we explain the construction and basic properties of the étale fundamental group of a connected scheme.

© Springer Fachmedien Wiesbaden GmbH, ein Teil von Springer Nature 2023
U. Görtz und T. Wedhorn, *Algebraic Geometry II: Cohomology of Schemes*,
Springer Studium Mathematik – Master, https://doi.org/10.1007/978-3-658-43031-3_1

The bulk of the book is concerned with the theory of cohomology of (quasi-)coherent sheaves. It was our intention to present the topic in a way that makes full use of the powerful machinery of derived categories, including results such as (a suitable version of) the Brown representability theorem, which is used in Chapter 25 to construct a right adjoint of the derived direct image functor. Beyond the classical results by Serre, Grothendieck, Verdier and others, this approach relies on work of Neeman, Lipman and others. To set up the notation and to collect references for the results from homological algebra we use, we have included Appendix F.

Of course, other approaches, either more elementary or more advanced, are possible and have their merit. For instance using derived functors, but no derived categories, as in [Har3] $^\circ$ or in [Vak] $_\mathrm{X}$, or focusing on Čech cohomology as in [Liu] $^\circ$. In comparison to [Har1] $^\circ$, we work with the unbounded derived category wherever possible. We neither use ∞-categories, nor go into the more recent approach to Grothendieck duality using Clausen's and Scholze's theory of condensed mathematics.

The chapters on cohomology are roughly organized as follows. Chapter 21 contains general results on $\mathscr{O}_X$-modules on a ringed space X, their derived category, derived functors and relations between them. Starting from Chapter 22, we specialize to the case of schemes, and (mostly) to categories of quasi-coherent sheaves. We prove vanishing of higher cohomology of quasi-coherent modules on affine schemes and compare the derived category of quasi-coherent sheaves on a scheme X and the subcategory $D_{\mathrm{qcoh}}(X)$ of the derived category of $\mathscr{O}_X$-modules consisting of complexes with quasi-coherent cohomology. In Chapter 22 we also prove some important general results such as the (derived) projection formula and the Künneth isomorphism.

The next three chapters contain some of the cornerstones of the theory. We start with finiteness theorems for cohomology of coherent sheaves on proper schemes in Chapter 23. Building on this, we discuss intersection numbers and give an outlook on the famous theorem of Grothendieck-Riemann-Roch; however, we do not introduce Chow groups but rather choose an axiomatic approach and use the theorem that K-theory is the "initial multiplicative cohomology theory" (which we state without proof). In the final parts of the chapter, we develop the theory of cohomology and base change and apply it to Hilbert polynomials and the flattening stratification. The topic of Chapter 24 is the theorem on formal functions which relates, given a closed subscheme Z of a proper scheme X, the completion of the cohomology of X (with respect to the ideal corresponding to Z) to the cohomology of the "completion of X along Z", i.e., the limit of the cohomology groups of infinitesimal thickenings of Z. We base the proof of this fact on a detailed discussion of derived completion. The notion of formal scheme is not discussed in detail, however. As two important applications, we discuss the Stein factorization and Grothendieck's algebraization theorem. Next, in Chapter 25, we study Grothendieck duality for (complexes of) quasi-coherent sheaves. We first discuss that the derived direct image functor has a right adjoint (under mild conditions) and compute it for smooth and proper morphisms. We go on to construct the twisted inverse image functor, a variant which is better behaved for non-proper morphisms. At the end of the chapter we apply results on cohomology to the theory of algebraic surfaces.

The final two chapters serve to highlight some of the beautiful consequences of the theory that was developed before. In particular, we freely use all previous results. Nevertheless we have aimed at giving a somewhat coherent exposition of basics of the theory of algebraic curves over a field in Chapter 26, and of the foundations of the theory of abelian schemes in Chapter 27. Clearly, each of these topics could easily fill a whole (or several ...) books of its own, so it was impossible to give a comprehensive account. Thus we just hope that the material presented here will help the reader to get started. Probably the single most important theorem in the theory of algebraic curves, and in Chapter 26, is the theorem of Riemann-Roch which relates the degree of a line bundle to its cohomology. At this point of the book, it is an easy corollary of facts on the Euler characteristic and of Serre duality. We prove some consequences such as the formula of Riemann and Hurwitz, discuss curves of genus 0, of genus 1, and hyperelliptic curves, show the existence of the Harder-Narasimhan filtration of vector bundles on curves, and prove the Weil conjectures for curves over finite fields.

In Chapter 27 on abelian schemes we focused on studying abelian schemes over arbitrary base schemes. For instance we sketch Deligne's proof of the existence of the dual abelian scheme A^t of an abelian scheme A over an arbitrary base scheme. To this end we briefly introduce (but do not discuss in full detail) the notion of algebraic space, and use the representability of the relative Picard functor by an algebraic space without proof. We also prove a few other foundational results of the theory, for instance the Fourier-Mukai equivalence for abelian schemes or that every abelian scheme over a normal noetherian base scheme is projective.

Let us conclude this introduction by emphasizing that we had to make a choice what topics to present and how to present them. Our approach is certainly influenced by our mathematical background. There are many other excellent introductions to this beautiful and vast theory ([EGAInew] – [EGAIV] $^\circ$, [Har3] $^\circ$, [Vak]$_X$, [Sta], [Liu] $^\circ$, [Lu-DAG] – just to list a few), different in style and in choice of material. We engourage the reader to take advantage of this diversity of literature.

Leitfaden

The following diagram displays dependencies between the chapters. A double arrow indicates that the target chapter relies in an essential way on (some of) the results developed in the source chapter. A single arrow stands for a weaker dependency, in the sense that it might be possible to use some results of the source chapter as a black box in order to read the target chapter. The diagram is not meant to be complete in a formal sense; in particular, some references to "standard definitions outside Algebraic Geometry" given in previous chapters are not visible here.

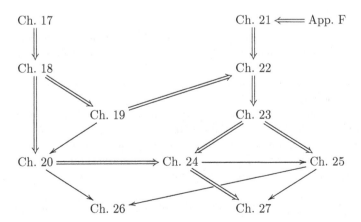

Corrigenda and addenda

Additions and corrections of the text will be posted on the web page `https://www.algebraic-geometry.de/` of this book. We encourage all readers to send us remarks and to give us feedback.

Acknowledgements

Almost none of the results presented here are new. The EGA volumes [EGAInew], [EGAII]$^\circ$, [EGAIII]$^\circ$, [EGAIV]$^\circ$ written by Dieudonné and Grothendieck, and their modern successor, the Stacks project [Sta] launched and largely written by de Jong, were of great importance for us when learning the topics discussed in this book. In addition to individual references in many places, some of the chapters have a short section at the end with pointers to the literature.

While we have devised many of the exercises ourselves, others are standard or folklore problems or exercises whose source is hard to track. For some of the more elaborate exercises, we gave references, but please notify us if there are places, in the exercises or elsewhere, where a reference would be appropriate but is currently missing.

We thank everybody who helped us writing this book, in discussions or via concrete feedback or comments, in particular:

Johannes Anschütz, Elmar Große-Klönne, Hasan Hasan, Tim Holzschuh, Moritz Kerz, Alex Küronya, Christopher Lang, Eike Lau, Jonas Lenz, Kin-Lok Li, Joseph Lipman, Martin Lüdtke, Catrin Mair, Lucas Mann, Timo Richarz, Benjamin Rosswinkel, Alexander Schmidt, Jakob Stix, Georg Tamme, Burt Totaro, Can Yaylali, Heer Zhao.

We would also like to use the occasion to make explicit here our thanks to Peng Du for his extensive list of remarks on the first volume, because his name is unfortunately missing in the list of acknowledgments in the preface to the second edition of that volume.

Notation

References to volume 1 are given by the number of the result in the second edition of that book, without the prefix [GWI]$^\circ$. They are "clickable" if you put the pdf file of volume 1 in the same folder as this file, under the name `GW1.pdf`.

We collect some general notation used throughout the book. By $\subseteq$ we denote an inclusion with equality allowed, and by $\subsetneq$ we denote a proper inclusion; by $\subset$ we denote an inclusion where we do not emphasize that equality must not hold, but where equality never occurs or would not make sense (e.g., $\mathfrak{m} \subset A$ a maximal ideal in a ring).

By Y^c we denote the complement of a subset Y of some bigger set. By $\overline{Y}$ we denote the closure of some subspace Y of a topological space. By convention, the empty topological space is not connected.

If R is a ring, then we denote by $M_{m \times n}(R)$ the additive group of $(m \times n)$-matrices over R, and by $GL_n(R)$ the group of invertible $(n \times n)$-matrices over R.

The letters $\mathbb{Z}$, $\mathbb{Q}$, $\mathbb{R}$, $\mathbb{C}$ denote the ring of integers and the fields of rational, real and complex numbers, respectively. By $\mathbb{N}$ we denote the set of natural numbers (including 0). Given a prime number p, we denote by $\mathbb{Z}_p$ the ring of p-adic integers and by $\mathbb{Q}_p$ its field of fractions.

We denote (projective) limits by lim, and colimits, i.e., inductive limits, by colim, see also Section (F.3).

17   Differentials

Content

- Differentials for rings and extensions of algebras
- Differentials for schemes
- The de Rham complex

In differential geometry the tangent bundle T_M of a manifold M and its dual $T_M^\vee$ play a central role. As both are vector bundles, one determines the other. Moreover, vector bundles over M are essentially the same as finite locally free modules over the structure sheaf of M, and the sections of the cotangent bundle over an open subset U form the module of differential forms $\Omega^1(U)$. In the presence of singularities, such as for non-smooth schemes over a field, it turns out that the module of differential forms is the more fundamental object. For a morphism $X \to S$ of schemes there is a suitable algebraic notion of differential forms, giving rise to a quasi-coherent $\mathscr{O}_X$-module $\Omega^1_{X/S}$ of differential 1-forms of X over S, called the sheaf of *Kähler differentials*. Its dual will be called the tangent sheaf $\mathscr{T}_{X/S}$ whose sections are the sections of the tangent bundle $T_{X/S}$.

If $X \to S$ is smooth, we will see in the next chapter that $\Omega^1_{X/S}$ is finite locally free (Corollary 18.58) and hence $\Omega^1_{X/S}$ and $\mathscr{T}_{X/S}$ determine each other. But in general the biduality homomorphism $\Omega^1_{X/S} \to (\Omega^1_{X/S})^{\vee\vee}$ is not an isomorphism and it is not possible to recover $\Omega^1_{X/S}$ from $\mathscr{T}_{X/S}$.

Taking exterior powers of $\Omega^1_{X/S}$, we obtain the sheaves of differential forms of higher order. Together they form the *de Rham complex of X over S* which we will introduce at the end of the chapter.

Differentials for rings and extensions of algebras

We start by defining the module of differentials $\Omega^1_{A/R}$ for a homomorphism of rings $R \to A$. It is an A-module which is characterized by the property that for every A-module M there is a functorial identification of $\mathrm{Hom}_A(\Omega^1_{A/R}, M)$ with the A-module of R-derivations of A with values in M.

(17.1) Derivations and Kähler differentials for rings.

Let R be a ring, let A be an R-algebra, and let M be an A-module.

Definition 17.1. *An R-derivation of A with values in M is a map $D: A \to M$ such that*
(a) *D is R-linear,*
(b) *$D(xy) = xD(y) + yD(x)$ for all $x, y \in A$ ("Leibniz rule").*
We denote by $\mathrm{Der}_R(A, M)$ the set of R-derivations from A to M.

© Springer Fachmedien Wiesbaden GmbH, ein Teil von Springer Nature 2023
U. Görtz und T. Wedhorn, *Algebraic Geometry II: Cohomology of Schemes*,
Springer Studium Mathematik – Master, https://doi.org/10.1007/978-3-658-43031-3_2

The sum of two R-derivations is again an R-derivation, and for $D \in \mathrm{Der}_R(A, M)$ and $a \in A$ the map $aD \colon x \mapsto aD(x)$ is again an R-derivation. This endows $\mathrm{Der}_R(A, M)$ with the structure of an A-module. For $M = A$ we write $\mathrm{Der}_R(A)$ instead of $\mathrm{Der}_R(A, A)$.

Remark 17.2.

(1) For $D \in \mathrm{Der}_R(A, M)$ the Leibniz rule implies $D(1) = D(1 \cdot 1) = D(1) + D(1)$ which shows

$$(17.1.1) \qquad\qquad D(\lambda \cdot 1) = 0, \qquad \text{for all } \lambda \in R.$$

(2) By induction the Leibniz rule implies

$$(17.1.2) \qquad\qquad D(x^n) = nx^{n-1}D(x), \qquad \text{for all } x \in A,\, n \geq 1.$$

(3) Finally, if $D \in \mathrm{Der}_R(A)$, another easy induction argument shows that

$$(17.1.3) \qquad D^n(xy) = \sum_{i=0}^{n} \binom{n}{i} D^i(x) D^{n-i}(y), \qquad \text{for all } x, y \in A,\, n \geq 1.$$

Example 17.3. Let R be a ring and let $A = R[T_1, \dots, T_n]$. Then the partial derivative $\partial/\partial T_i \colon A \to A$ is a derivation for all i.

The next two remarks establish a connection between derivations and liftings of R-algebra homomorphisms.

Remark 17.4. Let

$$
\begin{array}{c}
A \\
\downarrow \psi \\
P \xrightarrow{\;\pi\;} C
\end{array}
$$

be homomorphisms of R-algebras such that π is surjective and $I := \mathrm{Ker}(\pi)$ is an ideal of square zero. Thus $I = I/I^2$ is also a C-module and hence, via ψ, an A-module. Let L_ψ be the set of lifts of ψ, i.e., the set of R-algebra homomorphisms $\tilde{\psi} \colon A \to P$ with $\pi \circ \tilde{\psi} = \psi$.

We claim that

$$(17.1.4) \qquad\qquad \mathrm{Der}_R(A, I) \times L_\psi \to L_\psi, \qquad (D, \tilde{\psi}) \mapsto D + \tilde{\psi}$$

defines a simply transitive action of the additive group $\mathrm{Der}_R(A, I)$ on the set L_ψ. Note that L_ψ might be empty.

We first check that $D + \tilde{\psi} \in L_\psi$: For $m \in I$ and $p \in P$ one has $pm = \pi(p)m$, depending whether we consider I as ideal in P or as a C-module. For $D \in \mathrm{Der}_R(A, I)$ and $\tilde{\psi} \in L_\psi$ one has

$$(D + \tilde{\psi})(a)(D + \tilde{\psi})(b) = \psi(b)D(a) + \psi(a)D(b) + \tilde{\psi}(a)\tilde{\psi}(b)$$

for all $a, b \in A$. Here we use that $I^2 = 0$. Moreover, $\pi \circ (D + \tilde{\psi}) = \psi$.

The action (17.1.4) is simply transitive: Indeed if $\tilde{\psi}_1, \tilde{\psi}_2 \in L_\psi$, then $D := \tilde{\psi}_1 - \tilde{\psi}_2$ is an R-linear map $D \colon A \to I$ such that for $a, b \in A$

$$\psi(a)D(b) + \psi(b)D(a) = \tilde{\psi}_1(a)(\tilde{\psi}_1(b) - \tilde{\psi}_2(b)) + \tilde{\psi}_2(b)(\tilde{\psi}_1(a) - \tilde{\psi}_2(a)) = D(ab),$$

hence $D \in \mathrm{Der}_R(A, I)$.

Remark 17.5. We consider the following special case of Remark 17.4. Let $\psi\colon A \to C$ be an R-algebra and let N be a C-module. Define a C-algebra by

$$(17.1.5) \qquad D_C(N) := C \oplus N, \qquad (c,n) \cdot (c',n') := cc' + (cn' + c'n).$$

Then $\pi\colon D_C(N) \to C$, $(c,n) \mapsto c$ is a C-algebra homomorphism with kernel N of square zero. For $N = C$ we obtain $D_C(C) \cong C[T]/(T^2)$, the ring of dual numbers over C.

The composition of ψ with the homomorphism of C-algebras $C \to D_C(N)$, $c \mapsto (c,0)$ is an element of L_ψ. Applying Remark 17.4 we obtain a bijection

$$(17.1.6) \qquad \begin{aligned} \operatorname{Der}_R(A,N) &\xrightarrow{\sim} \{\, \tilde{\psi}\colon A \to D_C(N) \ R\text{-algebra homomorphism} \ ; \ \pi \circ \tilde{\psi} = \psi \,\}, \\ D &\mapsto (a \mapsto \psi(a) + D(a)). \end{aligned}$$

Applied to the special case $C = A$ and $\psi = \operatorname{id}_A$ we obtain

$$(17.1.7) \qquad \begin{aligned} \operatorname{Der}_R(A,N) &\xrightarrow{\sim} \{\, \psi\colon A \to D_A(N) \ R\text{-algebra homomorphism} \ ; \ \pi \circ \psi = \operatorname{id}_A \,\}, \\ D &\mapsto (a \mapsto a + D(a)). \end{aligned}$$

We will now show that there is an A-module $\Omega^1_{A/R}$ and an R-derivation $d\colon A \to \Omega_{A/R}$ such that for all A-modules the map

$$(17.1.8) \qquad \operatorname{Hom}_A(\Omega^1_{A/R}, M) \to \operatorname{Der}_R(A,M), \qquad u \mapsto u \circ d$$

is an isomorphism of A-modules. The bijection (17.1.8) is clearly functorial in M, thus $\Omega^1_{A/R}$ represents the covariant functor $M \mapsto \operatorname{Der}_R(A,M)$.

To construct $\Omega^1_{A/R}$, let I be the kernel of $\Delta^*\colon A \otimes_R A \to A$, $b_1 \otimes b_2 \mapsto b_1 b_2$, and set

$$(17.1.9) \qquad P^1_{A/R} := (A \otimes_R A)/I^2, \qquad \Omega^1_{A/R} := I/I^2.$$

Then $\Omega^1_{R/A}$ is the kernel of the R-algebra homomorphism $\pi\colon P^1_{A/R} \to A$ induced by Δ^*. It is an ideal of square zero in $P^1_{A/R}$ and so it is an A-module. With respect to this A-module structure an element $a \in A$ acts by multiplication by $a \otimes 1 \in A \otimes_R A$ (or equivalently by $1 \otimes a$ or by any element $\alpha \in A \otimes_R A$ such that $\Delta^*(\alpha) = a$).

The R-algebra homomorphisms

$$i_1, i_2\colon A \to P^1_{A/R}, \qquad i_1(a) := a \otimes 1 \bmod I^2, \quad i_2(a) := 1 \otimes a \bmod I^2$$

satisfy $\pi \circ i_1 = \pi \circ i_2 = \operatorname{id}_A$. Thus Remark 17.4 shows that

$$(17.1.10) \qquad d := d_{A/R}\colon A \to \Omega^1_{A/R}, \qquad b \mapsto i_2(a) - i_1(a)$$

is an R-derivation.

Proposition 17.6. *Let $X \subseteq A$ be a subset that generates A as an R-algebra. Then the A-module $\Omega^1_{A/R}$ is generated by $\{\, d(x) \ ; \ x \in X \,\}$.*

Proof. It suffices to show that I is generated as $A \otimes_R A$-module by the elements of the form $1 \otimes x - x \otimes 1$ for $x \in X$. But for all $a, a' \in A$ one has $a \otimes a' = aa' \otimes 1 + (a \otimes 1)(1 \otimes a' - a' \otimes 1)$. For $\sum_i a_i \otimes a'_i \in I$ one has $\sum_i a_i a'_i = 0$ by definition and therefore

$$\sum_i (a_i \otimes a_i') = \sum_i (a_i \otimes 1)(1 \otimes a_i' - a_i' \otimes 1).$$

This proves that I is generated by $1 \otimes a - a \otimes 1$ for $a \in A$. Moreover, if $a = xy$, then $1 \otimes a - a \otimes 1 = (1 \otimes y)(1 \otimes x - x \otimes 1) + (x \otimes 1)(1 \otimes y - y \otimes 1)$ and the claim follows by induction. $\qquad\square$

Proposition 17.7. *The map*

(17.1.11) $\alpha \colon \operatorname{Hom}_A(\Omega^1_{A/R}, M) \to \operatorname{Der}_R(A, M), \qquad u \mapsto u \circ d$

is an isomorphism of A-modules, functorial in M.

Proof. It is clear that α is A-linear and functorial in M. The injectivity of α follows from Proposition 17.6. For the surjectivity let $D \in \operatorname{Der}_R(A, M)$. By (17.1.7), D corresponds to a homomorphism $\psi_D \colon A \to D_A(N)$ of R-algebras. We obtain a commutative diagram of R-linear maps with exact rows

$$
\begin{array}{ccccccccc}
0 & \longrightarrow & \Omega^1_{A/R} & \longrightarrow & P^1_{A/R} & \longrightarrow & A & \longrightarrow & 0 \\
 & & \downarrow{\scriptstyle u_D} & & \downarrow{\scriptstyle a \otimes a' \mapsto a\psi_D(a')} & & \| & & \\
0 & \longrightarrow & M & \longrightarrow & D_A(M) & \longrightarrow & A & \longrightarrow & 0.
\end{array}
$$

Then $(u_D \circ d_{A/R})(a) = u(1 \otimes a - a \otimes 1) = \psi_D(a) - a = D(a)$ shows $\alpha(u_D) = D$. $\qquad\square$

Example 17.8. Let $A = R[(T_i)_{i \in I}]$ be a polynomial algebra. Then $\Omega^1_{A/R}$ is a free A-module with basis $(dT_i)_{i \in I}$.

Indeed, $\Omega^1_{A/R}$ is generated by $(dT_i)_{i \in I}$ (Proposition 17.6). From the definition of derivations, it follows that for every polynomial $f \in A$ one has

$$df = \sum_{i \in I} \frac{\partial f}{\partial T_i} dT_i.$$

Moreover $\frac{\partial}{\partial T_j} \in \operatorname{Der}_R(A)$ corresponds via the bijection (17.1.11) to an A-linear map $u_j \colon \Omega^1_{A/R} \to A$ such that $u_j(dT_i) = \delta_{ij}$. This shows that $(dT_i)_{i \in I}$ is also linearly independent.

(17.2) Extensions of algebras by modules.

In this section, we introduce certain "extension groups", which will later be useful when we investigate functoriality properties of derivations and differential forms.

Definition 17.9. *Let R be a ring, A an R-algebra, and M an A-module.*
(1) *An R-extension of A by M is an exact sequence of R-modules*

$$0 \to M \xrightarrow{\ j\ } E \xrightarrow{\ \pi\ } A \to 0,$$

where E is an R-algebra, π is a homomorphism of R-algebras, and one has

(17.2.1) $j(\pi(e)m) = ej(m) \qquad for\ all\ m \in M,\ e \in E.$

(2) *Two such extensions E and E' are called* equivalent, *if there exists an R-algebra homomorphism $u\colon E \to E'$ making the following diagram commutative*

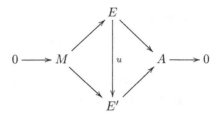

The Five Lemma shows that u as in (2) is automatically an isomorphism of R-algebras. The set of equivalence classes of R-extensions of A by M is denoted by $\mathrm{Ex}_R(A, M)$.

We think of an extension $0 \to M \xrightarrow{j} E \xrightarrow{\pi} A \to 0$ as a thickening E of A of order 1 with kernel M: Condition (17.2.1) implies that $j(M)$ is an ideal of E and

$$j(m)j(m') = j(\pi(j(m))m') = 0$$

shows that $j(M)$ is an ideal of square zero in E.

Remark 17.10. Let R be a ring, A an R-algebra, and M an A-module.
(1) The algebra $D_A(M)$ (17.1.5) is an extension of A by M when we define $\pi(a, n) = a$ and $j(m) = (0, m)$ for $m \in M$ and $a \in A$. It is called the *trivial extension*.

An extension $0 \to M \xrightarrow{j} E \xrightarrow{\pi} A \to 0$ is equivalent to the trivial extension if and only if there exists an R-algebra homomorphism $\iota\colon A \to E$ which is a section of π, i.e., such that $\pi \circ \iota = \mathrm{id}_A$.
(2) Let $0 \to M \xrightarrow{j} E \xrightarrow{\pi} A \to 0$ be an extension and let $\psi\colon C \to A$ be a homomorphism of R-algebras. Remark 17.4 shows that $\mathrm{Der}_R(C, M)$ acts simply transitively on the set of R-algebra homomorphisms $\tilde{\psi}\colon C \to E$ with $\pi \circ \tilde{\psi} = \psi$.

Remark 17.11. Let R be a ring, A an R-algebra, and M an A-module.
(1) $\mathrm{Ex}_R(A, M)$ is functorial in M: Let $u\colon M \to M'$ be a homomorphism of A-modules and let $0 \to M \xrightarrow{j} E \xrightarrow{\pi} A \to 0$ be in $\mathrm{Ex}_R(A, M)$. Then the pushout with u is an extension

$$0 \to M' \xrightarrow{j'} E \oplus^M M' \xrightarrow{\pi'} A \to 0$$

in $\mathrm{Ex}_R(A, M')$. Here $E \oplus^M M' := (E \oplus M')/\{\, (j(m), -u(m)) \; ; \; m \in M \,\}$ is the pushout in the category of R-modules, j' is given by $m' \mapsto (0, m')$ and π' is given by $(e, m') \mapsto \pi(e)$.
(2) Let $M = \prod_{i \in I} M_i$ be a product of A-modules. Then the projections $p_i\colon M \to M_i$ induce via functoriality a map

$$\mathrm{Ex}_R(A, M) \to \prod_{i \in I} \mathrm{Ex}_R(A, M_i).$$

This map is bijective: An inverse map is given by sending $(0 \to M_i \xrightarrow{j_i} E_i \xrightarrow{\pi_i} A \to 0)_{i \in I}$ in $\prod_{i \in I} \mathrm{Ex}_R(A, M_i)$ to $0 \to M \xrightarrow{j} E \xrightarrow{\pi} A \to 0$, where

$$E := \{\, (e_i)_{i \in I} \in \prod_{i \in I} E_i \; ; \; \forall i, j \in I : \pi_i(e_i) = \pi_j(e_j) \,\}$$

and where j is given by $\prod j_i$ and π is given by $(e_i) \mapsto \pi_i(e_i)$.

(3) Assertions (1) and (2) show that the addition and scalar multiplication on M yield the structure of an A-module on $\mathrm{Ex}_R(A, M)$ and that $M \mapsto \mathrm{Ex}_R(A, M)$ is an A-linear functor from the category of A-modules to itself. The zero element in $\mathrm{Ex}_R(A, M)$ is given by the trivial extension $D_A(M)$.

Proposition 17.12. *Let $\varphi\colon R \to A$, $\psi\colon A \to B$ be ring homomorphisms, and let M be a B-module. Then*

(17.2.2)
$$0 \to \mathrm{Der}_A(B, M) \xrightarrow{D \mapsto D} \mathrm{Der}_R(B, M) \xrightarrow{D \mapsto D \circ \psi} \mathrm{Der}_R(A, M)$$
$$\xrightarrow{\partial} \mathrm{Ex}_A(B, M) \xrightarrow{E \mapsto E} \mathrm{Ex}_R(B, M) \xrightarrow{E \mapsto E \times_B A} \mathrm{Ex}_R(A, M),$$

is an exact sequence of R-modules, where ∂ is given by

$$(D\colon A \to M) \mapsto (0 \to M \to E \to B \to 0)$$

with $E = D_B(M)$ with A-algebra structure $A \to B \oplus M$, $a \mapsto (\psi(a), D(a))$.

Proof. This is a straightforward calculation.   □

If A is a quotient of R we have the following description of $\mathrm{Ex}_R(A, -)$.

Proposition 17.13. *Let R be a ring, $I \subseteq R$ an ideal, $A := R/I$. Then for every A-module M there exists an isomorphism, functorial in M*

$$\Phi\colon \mathrm{Hom}_A(I/I^2, M) \xrightarrow{\sim} \mathrm{Ex}_R(A, M).$$

Proof. Consider the canonical exact sequence

$$0 \to I/I^2 \longrightarrow R/I^2 \longrightarrow A \to 0$$

which yields an element in $\mathrm{Ex}_R(A, I/I^2)$. For a homomorphism of A-modules $w\colon I/I^2 \to M$ we obtain by functoriality of $\mathrm{Ex}_R(A, -)$ an element $\Phi(w) \in \mathrm{Ex}_R(A, M)$.

Conversely, to $0 \to M \to E \xrightarrow{\pi} A \to 0$ in $\mathrm{Ex}_R(A, M)$ we attach the R-linear map $u\colon I \hookrightarrow R \to E$. Then $\pi \circ u = 0$ and we may consider u as an R-linear map $I \to M$, which induces an A-linear map $I/I^2 \to M$ because M is an ideal of square zero in E. This defines an inverse to Φ.   □

Differentials for sheaves on schemes

We now globalize the construction of the module of differentials to a morphism of schemes $X \to S$. In fact, the construction in the affine case (17.1.9) suggests that instead of gluing local $\Omega^1_{A/R}$'s for open affine $\mathrm{Spec}\, A \subseteq X$ and $\mathrm{Spec}\, R \subseteq S$ (which is one possible definition) we may define $\Omega^1_{X/S}$ as the "conormal sheaf" of the diagonal $X \to X \times_S X$. Hence we start with a short section on the conormal sheaf of an arbitrary immersion.

(17.3) Conormal sheaf of an immersion.

Let $i\colon Y \to X$ be an immersion of schemes. By definition (Definition 3.43) there exists an open subscheme U of X and a closed immersion $i_1\colon Y \to U$ such that i is the composition of i_1 followed by the inclusion $U \hookrightarrow X$. Let $\mathscr{I} \subseteq \mathscr{O}_U$ be the quasi-coherent ideal defining i_1. For any integer $n \geq 0$ the closed subscheme

$$Y_i^{(n)} := V(\mathscr{I}^{n+1})$$

of U is called the *n-th infinitesimal neighborhood of Y in X*. Moreover we call

(17.3.1) $$\mathscr{C}_{Y/X} := \mathscr{C}_i := \mathscr{I}/\mathscr{I}^2$$

the *conormal sheaf of i^1*. It is a quasi-coherent $\mathscr{O}_U$-module that is annihilated by $\mathscr{I}$. Thus we will consider it as a quasi-coherent $\mathscr{O}_Y$-module.

One obtains a sequence of closed immersions of subschemes of X,

$$Y = Y_i^{(0)} \hookrightarrow Y_i^{(1)} \hookrightarrow Y_i^{(2)} \hookrightarrow \cdots$$

each closed immersion defined by a quasi-coherent ideal of square 0. The immersion $Y_i^{(n-1)} \hookrightarrow Y_i^{(n)}$ is defined by the quasi-coherent ideal $\mathscr{I}^n/\mathscr{I}^{n+1}$ which we consider as a quasi-coherent $\mathscr{O}_Y$-module. In particular, $\mathscr{C}_i$ defines the closed immersion $Y \to Y_i^{(1)}$.

We have the following alternative description of $Y_i^{(n)}$ and of $\mathscr{C}_i$ which shows in particular that they do not depend on the choice of U. The homomorphism of sheaves of rings $i^\#\colon i^{-1}\mathscr{O}_X \to \mathscr{O}_Y$ on Y is surjective. If $\mathscr{J}$ denotes its kernel, then $\mathscr{J} = i_1^{-1}\mathscr{I}$. Hence $Y_i^{(n)}$ is the ringed space $(Y, i^{-1}\mathscr{O}_X/\mathscr{J}^{n+1})$ and $\mathscr{C}_i = \mathscr{J}/\mathscr{J}^2$ as $\mathscr{O}_Y$-modules.

Remark 17.14. The formation of $Y_i^{(n)}$ and of $\mathscr{C}_i$ is functorial in the following sense. Let

(17.3.2)
$$\begin{array}{ccc} Y' & \xrightarrow{g} & Y \\ {\scriptstyle i'}\downarrow & & \downarrow{\scriptstyle i} \\ X' & \xrightarrow{f} & X \end{array}$$

be a commutative diagram of scheme morphisms such that i and i' are immersions. Let $i = j \circ i_1$, where $i_1\colon Y \to U$ is a closed immersion, U is an open subscheme of X, and j is the inclusion $U \hookrightarrow X$. Let $j'\colon U' \hookrightarrow X'$ be an open subscheme such that $U' \subseteq f^{-1}(U)$ such that $i' = j' \circ i_1'$, where $i_1'\colon Y' \to U'$ is a closed immersion. Replacing X by U and X' by U' we may assume that i and i' are closed immersions.

Let $\mathscr{I}$ (resp. $\mathscr{I}'$) be the quasi-coherent ideal defining i (resp. i'). The commutativity of (17.3.2) implies that $f^*(\mathscr{I}^n) \to \mathscr{O}_{X'}$ factors through $(\mathscr{I}')^n$ and thus yields for all $n \geq 1$ a homomorphism of $\mathscr{O}_{X'}$-modules

$$u^{(n)}\colon f^*(\mathscr{I}^n) \to (\mathscr{I}')^n.$$

Hence f induces scheme morphisms

(17.3.3) $$g^{(n)}\colon Y_{i'}'^{(n)} \to Y_i^{(n)}$$

---

[1] It is denoted by $\mathscr{N}_{Y/X}$ in [EGAIV]^{O} (16.1.1). But we will denote its dual, the normal bundle, by $\mathscr{N}_i$, at least for quasi-regular immersions, see Definition 19.21.

and $u^{(1)}$ induces a homomorphism of $\mathscr{O}_{Y'}$-modules

(17.3.4) $$w := w_{i',i} \colon g^*(\mathscr{C}_i) \to \mathscr{C}_{i'}.$$

Remark 17.15. Assume that the diagram (17.3.2) is cartesian.
(1) Then $u^{(n)}$ is surjective. Indeed, as above one can assume that i and i' are closed immersions. As the diagram is cartesian, $\mathscr{I}' = f^{-1}(\mathscr{I})\mathscr{O}_{X'}$ and $u^{(n)}$ is the surjective homomorphism

(17.3.5) $$f^{-1}(\mathscr{I}^n) \otimes_{f^{-1}\mathscr{O}_X} \mathscr{O}_{X'} \to (\mathscr{I}')^n.$$

In particular $w \colon g^*(\mathscr{C}_i) \to \mathscr{C}_{i'}$ is surjective.
(2) If f is flat, then the description of $u^{(n)}$ in (17.3.5) shows that $u^{(n)}$ is an isomorphism for all $n \geq 1$. In particular, w is an isomorphism.
(3) Assume that there exist morphisms $p \colon X \to Y$ and $p' \colon X' \to Y'$ such that $p \circ i = \mathrm{id}_Y$, $p' \circ i' = \mathrm{id}_{Y'}$ and such that the following diagram is cartesian

$$\begin{array}{ccc} Y' & \xrightarrow{\ g\ } & Y \\ {\scriptstyle p'}\uparrow & & \uparrow{\scriptstyle p} \\ X' & \xrightarrow{\ f\ } & X. \end{array}$$

Then p (resp. p') makes $\mathscr{O}_{Y^{(n)}}$ (resp. $\mathscr{O}_{Y'^{(n)}}$) into a quasi-coherent $\mathscr{O}_Y$-algebra (resp. $\mathscr{O}_{Y'}$-algebra) endowed with augmentations defined by i (resp. i'). Thus we obtain a direct sum decomposition of $\mathscr{O}_Y$-modules $\mathscr{O}_{Y^{(n)}} = \mathscr{O}_Y \oplus \mathscr{I}/\mathscr{I}^{n+1}$; similarly for $\mathscr{O}_{Y'^{(n)}}$. Moreover, if $i^{(n)}$ (resp. $(i')^{(n)}$) denotes the immersion $Y^{(n)} \to X$ (resp. $Y'^{(n)} \to X$), then the diagram

$$\begin{array}{ccc} Y'^{(n)} & \xrightarrow{\ g^{(n)}\ } & Y^{(n)} \\ {\scriptstyle p'\circ(i')^{(n)}}\downarrow & & \downarrow{\scriptstyle p\circ i^{(n)}} \\ Y' & \xrightarrow{\ g\ } & Y \end{array}$$

is cartesian. Hence $g^*\mathscr{O}_{Y^{(n)}} \to \mathscr{O}_{Y'^{(n)}}$ is an isomorphism of $\mathscr{O}_{Y'}$-algebras, restricting to an isomorphism $g^*(\mathscr{I}/\mathscr{I}^{n+1}) \to \mathscr{I}'/\mathscr{I}'^{n+1}$ of $\mathscr{O}_{Y'}$-modules. For $n = 1$ this shows that (17.3.4) is an isomorphism

$$w \colon g^*(\mathscr{C}_i) \xrightarrow{\sim} \mathscr{C}_{i'}.$$

Remark 17.16. If i is an immersion locally of finite presentation (i.e., $\mathscr{I}$ is an $\mathscr{O}_U$-module of finite type (Proposition 10.35)), then $\mathscr{C}_i$ is a quasi-coherent $\mathscr{O}_Y$-module of finite type.

Proposition 17.17. *Let $i \colon Y \to X$ be an immersion locally of finite presentation, and let $y \in Y$. Then the following assertions are equivalent.*
(i) *There exists an open neighborhood V of y in Y such that $i_{|V}$ is an open immersion.*
(ii) *One has $(\mathscr{C}_i)_y = 0$.*

Proof. Clearly, (i) implies (ii). The question is local on X and we may assume that i is a closed immersion, $X = \operatorname{Spec} A$ is affine, $Y = V(I)$, where I is a finitely generated ideal of A. Thus $\mathscr{C}_i = \widetilde{I/I^2}$. As $\mathscr{C}_i$ is of finite type, its support is closed (Corollary 7.32). Hence (ii) implies that there exists an open neighborhood V of y in Y such that $\mathscr{C}_{i|V} = 0$. So we may assume that $I/I^2 = 0$. Nakayama's lemma then implies $I = 0$ because $I_y \subseteq \mathfrak{m}_y$ for all $y \in Y$. $\qquad\square$

(17.4) Derivations and Kähler differentials for schemes.

The following is the central definition of this chapter; the sheaf of Kähler differentials will also be of crucial importance in Chapter 18 where we study smooth morphisms in detail.

Definition 17.18. *Let $f\colon X \to S$ be a morphism of schemes and let $\Delta_f\colon X \to X \times_S X$ be the diagonal. Its conormal sheaf*

$$\Omega^1_f := \Omega^1_{X/S} := \mathscr{C}_{\Delta_f}$$

is called the $\mathscr{O}_X$-module of Kähler differentials (of X over S).

Remark 17.19. Let $f\colon X \to S$ be a morphism of schemes.
(1) If $S = \operatorname{Spec} R$ and $X = \operatorname{Spec} A$ are affine, then the construction of $\Omega^1_{A/R}$ (17.1.9) shows that $\Omega^1_{X/S}$ is the quasi-coherent $\mathscr{O}_X$-module attached to $\Omega^1_{A/R}$.
(2) Let $j\colon U \to X$ be a monomorphism of S-schemes. Then

$$
\begin{array}{ccc}
U & \xrightarrow{\;\;j\;\;} & X \\
{\scriptstyle \Delta_{U/S}}\big\downarrow & & \big\downarrow{\scriptstyle \Delta_{X/S}} \\
U \times_S U & \xrightarrow{\;j \times j\;} & X \times_S X
\end{array}
$$

is a cartesian diagram. If j is in addition flat, then $j^*(\Omega^1_{X/S}) = \Omega^1_{U/S}$ by Remark 17.15. In particular, if U is an open subscheme of X, we obtain $\Omega^1_{X/S|U} = \Omega^1_{U/S}$.
(3) It follows from (1) and (2) that $\Omega^1_{X/S}$ is a quasi-coherent $\mathscr{O}_X$-module for an arbitrary morphism $X \to S$ of schemes.
(4) Assume that $S = \operatorname{Spec} R$ and $X = \operatorname{Spec} A$ are affine and let $\varphi\colon R \to A$ be the ring homomorphism corresponding to f. Let $T \subseteq A$ be a multiplicative set. Then $\operatorname{Spec}(T^{-1}A) \to \operatorname{Spec} A$ is a flat monomorphism and hence (2) shows

(17.4.1) $$T^{-1}\Omega^1_{A/R} = \Omega^1_{T^{-1}A/R}.$$

If $U \subseteq R$ is a multiplicative set such that $f(U) \subseteq T$, then $T^{-1}A \otimes_R T^{-1}A = T^{-1}A \otimes_{U^{-1}R} T^{-1}A$ which shows $\Omega^1_{T^{-1}A/R} = \Omega^1_{T^{-1}A/U^{-1}R}$ and thus

(17.4.2) $$T^{-1}\Omega^1_{A/R} = \Omega^1_{T^{-1}A/U^{-1}R}.$$

In particular one obtains for an arbitrary scheme morphism $f\colon X \to S$ and for $x \in X$ an isomorphism of $\mathscr{O}_{X,x}$-modules

(17.4.3) $$(\Omega^1_{X/S})_x = \Omega^1_{\mathscr{O}_{X,x}/\mathscr{O}_{S,f(x)}}.$$

Remark 17.20. Let $f\colon X \to S$ be a monomorphism (e.g., if f is an immersion), then Δ_f is an isomorphism (Exercise 9.1) and therefore $\Omega^1_{X/S} = 0$.

We have the following finiteness properties of the quasi-coherent $\mathscr{O}_X$-module $\Omega^1_{X/S}$.

Remark 17.21. If $f\colon X \to S$ is a morphism locally of finite type, then the first projection $X \times_S X \to X$ is locally of finite type. Thus Δ_f is locally of finite presentation (Proposition 10.35 (3)) and $\Omega^1_{X/S}$ is of finite type (Remark 17.16). Alternatively, this follows from Proposition 17.6.

If S is in addition locally noetherian, then X is locally noetherian (Proposition 10.9) and hence $\Omega^1_{X/S}$ is coherent (Proposition 7.46).

In Corollary 17.34 below we will see that $\Omega^1_{X/S}$ is of finite presentation, if f is locally of finite presentation.

Remark 17.22. Let $f\colon X \to S$ be a morphism of schemes and let $e\colon S \to X$ be a section of f. Then e is an immersion by Example 9.12. Applying Remark 17.15 (3) to the cartesian diagram

$$
\begin{array}{ccc}
S & \xrightarrow{\;\;e\;\;} & X \\
{\scriptstyle f}\Big\uparrow{\scriptstyle e} & & \Big\uparrow{\scriptstyle \mathrm{pr}_1}{\scriptstyle \Delta_{X/S}} \\
X & \xrightarrow{(ef,\mathrm{id}_X)_S} & X \times_S X
\end{array}
$$

one obtains an isomorphism of $\mathscr{O}_S$-modules

$$(17.4.4) \qquad\qquad e^*(\Omega^1_{X/S}) \xrightarrow{\sim} \mathscr{C}_e.$$

Example 17.23. Let k be a field, X be a k-scheme and let $x \in X(k)$ be a k-rational point, considered as a closed point of X. Let $\mathfrak{m}_x \subset \mathscr{O}_{X,x}$ be the maximal ideal such that $\mathscr{O}_{X,x}/\mathfrak{m}_x = k$. Then (17.4.4) shows that

$$x^*(\Omega^1_{X/k}) = \mathscr{C}_x = \mathfrak{m}_x/\mathfrak{m}_x^2,$$

in other words, the absolute tangent space as defined in Section (6.2) is the dual space of $x^*(\Omega^1_{X/k}) = \mathscr{C}_x$. If X is locally noetherian (e.g., if X is locally of finite type over k), then $\mathscr{C}_x$ is finite-dimensional.

For affine schemes we saw in Proposition 17.7 that the module of differentials represents the functor of derivations. Globally, we have the results given by Proposition 17.27 and Corollary 17.28.

Definition and Remark 17.24. Let $f\colon X \to S$ be a morphism of schemes, and let $\mathscr{F}$ be an $\mathscr{O}_X$-module. An *S-derivation of X with values in $\mathscr{F}$* is an $f^{-1}(\mathscr{O}_S)$-linear homomorphism $D\colon \mathscr{O}_X \to \mathscr{F}$ satisfying the Leibniz rule

$$D_U(ab) = a D_U(b) + b D_U(a), \qquad U \subseteq X \text{ open}, \quad a, b \in \mathscr{O}_X(U).$$

The sum of two S-derivations and the product of an element in $\Gamma(X, \mathscr{O}_X)$ with a derivation are again derivations. This makes the set $\mathrm{Der}_S(\mathscr{O}_X, \mathscr{F})$ of all S-derivations of X with values in $\mathscr{F}$ a $\Gamma(X, \mathscr{O}_X)$-module.

For $D \in \mathrm{Der}_S(\mathscr{O}_X, \mathscr{F})$ and $U \subseteq X$ open, the restriction $D_{|U}$ is an S-derivation of U with values in $\mathscr{F}_{|U}$. We obtain an $\mathscr{O}_X$-module $\mathscr{D}er_S(\mathscr{O}_X, \mathscr{F})$ by setting

$$\Gamma(U, \mathscr{D}er_S(\mathscr{O}_X, \mathscr{F})) = \mathrm{Der}_S(\mathscr{O}_U, \mathscr{F}_{|U}), \qquad U \subseteq X \text{ open.}$$

Remark 17.25. Let $f\colon X \to S$ be a morphism of schemes. Remark 17.19 shows that there exists a unique S-derivation

$$(17.4.5) \qquad\qquad d := d_{X/S}\colon \mathscr{O}_X \to \Omega^1_{X/S}$$

such that for all open affine subschemes $\mathrm{Spec}\, R \subseteq S$ and $U = \mathrm{Spec}\, A \subseteq f^{-1}(\mathrm{Spec}\, R)$ the derivation $d_U\colon A = \Gamma(U, \mathscr{O}_X) \to \Omega^1_{A/R} = \Gamma(U, \Omega^1_{X/S})$ is the derivation $d_{A/R}$ defined in (17.1.10).

Remark 17.26. Let $f\colon X \to S$ be a morphism of schemes, and let $\mathscr{F}$ be an $\mathscr{O}_X$-module. Globalizing (17.1.5) we set $D_{\mathscr{O}_X}(\mathscr{F}) := \mathscr{O}_X \oplus \mathscr{F}$, endowed with the structure of an $\mathscr{O}_X$-algebra as in (17.1.5).

Assume now that $\mathscr{F}$ is quasi-coherent, then $D_{\mathscr{O}_X}(\mathscr{F})$ is a quasi-coherent $\mathscr{O}_X$-algebra and we set

$$(17.4.6) \qquad\qquad D_X(\mathscr{F}) := \mathrm{Spec}(D_{\mathscr{O}_X}(\mathscr{F}))$$

(see Section (11.2) for the definition of spectra of quasi-coherent $\mathscr{O}_X$-algebras). We obtain a contravariant functor $\mathscr{F} \mapsto D_X(\mathscr{F})$ from the category of quasi-coherent $\mathscr{O}_X$-modules to the category of X-schemes that are affine over X.

Let $\varepsilon\colon X \to D_X(\mathscr{F})$ be the morphism corresponding to the $\mathscr{O}_X$-algebra homomorphism $D_{\mathscr{O}_X}(\mathscr{F}) \to \mathscr{O}_X$, $(a, n) \mapsto a$ for $a \in \mathscr{O}_X(U)$, $n \in \mathscr{F}(U)$ and $U \subseteq X$ open. By restriction to the affine case, (17.1.7) yields a bijection, functorial in $\mathscr{F}$,

$$(17.4.7) \qquad \mathrm{Der}_S(\mathscr{O}_X, \mathscr{F}) \xrightarrow{\sim} \{\, f \in \mathrm{Hom}_S(D_X(\mathscr{F}), X)\;;\; f \circ \varepsilon = \mathrm{id}_X \,\}$$

If $S = \mathrm{Spec}\, R$ and $X = \mathrm{Spec}\, A$ are affine and $\mathscr{F} = \tilde{M}$ for an A-module M, then the description of derivations in terms of algebra homomorphisms in (17.4.7) shows that evaluation on global sections yields an isomorphism of A-modules

$$(17.4.8) \qquad \mathrm{Der}_S(\mathscr{O}_X, \mathscr{F}) \xrightarrow{\sim} \mathrm{Der}_R(A, M).$$

Restricting to the affine case and using Remark 17.19 (1) and (17.4.8) shows that there is the following global version of Proposition 17.7.

Proposition 17.27. *Let $f\colon X \to S$ be a morphism of schemes, and let $\mathscr{F}$ be a quasi-coherent $\mathscr{O}_X$-module. The map*

$$(17.4.9) \qquad \mathrm{Hom}_{\mathscr{O}_X}(\Omega^1_{X/S}, \mathscr{F}) \to \mathrm{Der}_S(\mathscr{O}_X, \mathscr{F}), \qquad u \mapsto u \circ d_{X/S},$$

is an isomorphism of $\Gamma(X, \mathscr{O}_X)$-modules, functorial in $\mathscr{F}$.

Corollary 17.28. *The morphism*

$$(17.4.10) \qquad \mathscr{H}om_{\mathscr{O}_X}(\Omega^1_{X/S}, \mathscr{F}) \to \mathscr{D}er_S(\mathscr{O}_X, \mathscr{F}), \qquad u \mapsto u \circ d_{X/S},$$

is a functorial isomorphism of $\mathscr{O}_X$-modules for every quasi-coherent $\mathscr{O}_X$-module $\mathscr{F}$.

In fact, one can show that (17.4.9) and hence (17.4.10) are isomorphisms for an arbitrary $\mathscr{O}_X$-module $\mathscr{F}$. We leave this as an exercise (Exercise 17.12).

Example 17.29. Let S be a scheme and let $\mathscr{E}$ be a quasi-coherent $\mathcal{O}_S$-module. Let $X :=$ $\mathbb{V}(\mathscr{E}) = \operatorname{Spec}(\operatorname{Sym}(\mathscr{E}))$ be the corresponding quasi-coherent bundle (Definition 11.2) and let $f\colon X \to S$ be the structure homomorphism. Then there is an $\mathcal{O}_X$-linear isomorphism

$$(17.4.11) \qquad f^*(\mathscr{E}) \overset{\sim}{\to} \Omega^1_{X/S}, \qquad f^*(a) \mapsto d_{X/S}(a),$$

where a is a local section of $\mathscr{E} \subseteq \operatorname{Sym}(\mathscr{E})$. Indeed, for every quasi-coherent $\mathcal{O}_X$-module $\mathscr{F}$ there are bijections, functorial in $\mathscr{F}$,

$$
\begin{aligned}
\operatorname{Hom}_{\mathcal{O}_X}(\Omega^1_{X/S}, \mathscr{F}) &\overset{(17.4.9)}{=} \operatorname{Der}_S(\mathcal{O}_X, \mathscr{F}) \\
&\overset{(17.4.7)}{=} \{\, f \in \operatorname{Hom}_S(D_X(\mathscr{F}), X) \;;\; f \circ \varepsilon = \operatorname{id}_X \,\} \\
&\overset{(11.3.3)}{=} \{\, u \in \operatorname{Hom}_{\mathcal{O}_X}(f^*\mathscr{E}, D_{\mathcal{O}_X}(\mathscr{F}) \;;\; \varepsilon^* \circ u = \operatorname{id}_{\mathcal{O}_X} \,\} \\
&= \operatorname{Hom}_{\mathcal{O}_X}(f^*\mathscr{E}, \mathscr{F}),
\end{aligned}
$$

and this bijection is induced by (17.4.11). In particular $\Omega^1_{\mathbb{A}^n_S/S} \cong \mathcal{O}^n_X$. This also follows by globalizing Example 17.8.

(17.5) Fundamental exact sequences for Kähler differentials.

Consider a commutative diagram of scheme morphisms

$$(17.5.1) \qquad
\begin{array}{ccc}
X' & \overset{h}{\longrightarrow} & X \\
{\scriptstyle f'}\downarrow & & \downarrow{\scriptstyle f} \\
S' & \overset{g}{\longrightarrow} & S.
\end{array}
$$

We obtain a commutative diagram

$$(17.5.2) \qquad
\begin{array}{ccc}
X' & \overset{h}{\longrightarrow} & X \\
{\scriptstyle \Delta_{X'/S'}}\downarrow & & \downarrow{\scriptstyle \Delta_{X/S}} \\
X' \times_{S'} X' & \longrightarrow & X \times_S X
\end{array}
$$

and thus by Remark 17.14 a homomorphism of $\mathcal{O}_{X'}$-modules

$$(17.5.3) \qquad u := u_{X'/S', X/S}\colon h^*\Omega^1_{X/S} \to \Omega^1_{X'/S'}$$

In the affine case ($S = \operatorname{Spec} R$, $S' = \operatorname{Spec} R'$, $X = \operatorname{Spec} A$, $X' = \operatorname{Spec} A'$ and $\psi\colon A \to A'$ corresponding to h) (17.5.3) is given by

$$(17.5.4) \qquad A' \otimes_A \Omega^1_{A/R} \to \Omega^1_{A'/R'}, \qquad a' \otimes d(a) \mapsto a'd(\psi(a)).$$

The homomorphism (17.5.3) satisfies an obvious transitivity property for a composition of commutative squares.

Proposition 17.30. *If the diagram (17.5.1) is cartesian, then the homomorphism $h^*\Omega^1_{X/S} \to \Omega^1_{X'/S'}$ in (17.5.3) is an isomorphism.*

Proof. This is Remark 17.15 (3) applied to $i = \Delta_{X/S}$, $i' = \Delta_{X'/S'}$, and the cartesian diagram

$$\begin{array}{ccc} X' & \xrightarrow{\ h\ } & X \\ \scriptstyle p' \uparrow & & \uparrow \scriptstyle p \\ X' \times_{S'} X' & \longrightarrow & X \times_S X, \end{array}$$

where p and p' are the first projections. $\qquad\square$

Now consider a commutative diagram of scheme morphisms

(17.5.5)

$$\begin{array}{ccc} X & \xrightarrow{\ f\ } & Y \\ & \searrow{\scriptstyle g} \quad \swarrow{\scriptstyle h} & \\ & S. & \end{array}$$

As special cases of (17.5.3) we obtain homomorphisms of $\mathscr{O}_X$-modules

(17.5.6)
$$v := v_{f/S} := u_{X/S,Y/S} \colon f^*\Omega^1_{Y/S} \to \Omega^1_{X/S},$$
$$w := w_{X/h} := u_{X/Y,X/S} \colon \Omega^1_{X/S} \to \Omega^1_{X/Y}.$$

In the affine case, $S = \operatorname{Spec} R$, $X = \operatorname{Spec} B$, $Y = \operatorname{Spec} A$, and f corresponding to $\varphi \colon A \to B$, these maps are given by

(17.5.7)
$$v \colon \Omega^1_{A/R} \otimes_A B \to \Omega^1_{B/R}, \qquad d(a) \otimes b \mapsto b\,d(\varphi(a)),$$
$$w \colon \Omega^1_{B/R} \to \Omega^1_{B/A}, \qquad d(b) \mapsto d(b).$$

Proposition 17.31. *The following sequence of $\mathscr{O}_X$-modules is exact.*

(17.5.8)
$$f^*\Omega^1_{Y/S} \xrightarrow{\ v\ } \Omega^1_{X/S} \xrightarrow{\ w\ } \Omega^1_{X/Y} \to 0.$$

Proof. This is a local question, hence we may assume that $S = \operatorname{Spec} R$, $Y = \operatorname{Spec} A$ and $X = \operatorname{Spec} B$ are affine. We have to show that $B \otimes_A \Omega^1_{A/R} \to \Omega^1_{B/R} \to \Omega^1_{B/A} \to 0$ is exact. But this sequence of B-modules is exact, if and only if it is exact after applying $\operatorname{Hom}_B(\cdot, N)$ for all B-modules N (this follows from the Yoneda Lemma, but can also easily be checked directly). Thus by Proposition 17.7 we have to show the exactness of the sequence

$$0 \to \operatorname{Der}_A(B, M) \xrightarrow{D \mapsto D} \operatorname{Der}_R(B, M) \xrightarrow{D \mapsto D \circ \psi} \operatorname{Der}_R(A, M),$$

where $\psi \colon A \to B$ is the homomorphism corresponding to f. This follows from Proposition 17.12. $\qquad\square$

Corollary 17.32. *Let $f \colon X \to S$, $g \colon Y \to S$ be morphisms of schemes, set $Z := X \times_S Y$ and let $p \colon Z \to X$ and $q \colon Z \to Y$ be the projections. Then $v_{p/S}$ and $v_{q/S}$ (17.5.6) yield an isomorphism of $\mathscr{O}_Z$-modules*

$$p^*\Omega^1_{X/S} \oplus q^*\Omega^1_{Y/S} \xrightarrow{\ \sim\ } \Omega^1_{X\times_S Y/S}.$$

Proof. Proposition 17.31 yields an exact sequence

(*) $$p^*\Omega^1_{X/S} \xrightarrow{v_{p/S}} \Omega^1_{Z/S} \xrightarrow{w_{Z/f}} \Omega^1_{Z/X} \to 0.$$

The composition of $w_{Z/g}\colon \Omega^1_{Z/S} \to \Omega^1_{Z/Y}$ followed by the isomorphism $\Omega^1_{Z/Y} \cong p^*\Omega^1_{X/S}$ (Proposition 17.30) is a left inverse of $v_{p/S}$. Hence $v_{p/S}$ is injective, and the sequence (*) splits. Using $\Omega^1_{Z/X} = q^*\Omega^1_{Y/S}$ (again by Proposition 17.30) we obtain $\Omega^1_{Z/S} \cong p^*\Omega^1_{X/S} \oplus q^*\Omega^1_{Y/S}$. $\qquad\square$

Let S be a scheme and let $i\colon Y \to X$ be a closed immersion of S-schemes defined by a quasi-coherent ideal $\mathscr{I}$ of $\mathscr{O}_X$. Then the restriction of $d_{X/S}\colon \mathscr{O}_X \to \Omega^1_{X/S}$ to $\mathscr{I}$ induces an $\mathscr{O}_Y$-linear map

(17.5.9) $$d\colon \mathscr{C}_i = \mathscr{I}/\mathscr{I}^2 \to i^*\Omega^1_{X/S}.$$

Indeed, we may assume that $S = \operatorname{Spec} R$, $X = \operatorname{Spec} A$, $Y = \operatorname{Spec} A/I$ are affine. Then $\delta := d_{A/R|I}\colon I \to \Omega^1_{A/R}/I\Omega^1_{A/R}$ maps I^2 to 0: For $a, b \in I$ one has $\delta(ab) = a\delta(b) + b\delta(a) = 0 \pmod{I\Omega^1_{A/R}}$. Moreover, δ is A-linear because for $a \in A$, $b \in I$ one has $\delta(ab) = a\delta(b) + b\delta(a) = a\delta(b)$.

Proposition 17.33. *The following sequence of $\mathscr{O}_Y$-modules is exact.*

(17.5.10) $$\mathscr{C}_i \xrightarrow{d} i^*\Omega^1_{X/S} \xrightarrow{v_{i/S}} \Omega^1_{Y/S} \to 0.$$

Proof. Again we may assume that $S = \operatorname{Spec} R$, $X = \operatorname{Spec} A$, $Y = \operatorname{Spec} B$, $B = A/I$ are affine as above. Let $\pi\colon A \to B$ the canonical projection. It suffices to show that (17.5.10) is exact after applying $\operatorname{Hom}_B(\,\cdot\,, N)$ for all B-modules, i.e., we have to show that

$$0 \longrightarrow \operatorname{Hom}_B(\Omega^1_{B/R}, N) \longrightarrow \operatorname{Hom}_B(B \otimes_A \Omega^1_{A/R}, N) \longrightarrow \operatorname{Hom}_B(B \otimes_A I, N)$$
$$\big\| \qquad\qquad\qquad\qquad \big\| \qquad\qquad\qquad\qquad \big\|$$
$$\operatorname{Der}_R(B, N) \qquad\qquad \operatorname{Der}_R(A, N) \qquad\qquad \operatorname{Hom}_A(I, N)$$

ist exact. Here $\operatorname{Der}_R(B, N) \to \operatorname{Der}_R(A, N)$ is given by $D \mapsto D \circ \pi$, and $\operatorname{Der}_R(A, N) \to \operatorname{Hom}_A(I, N)$ is given by $D' \mapsto D'_{|I}$. This follows from Proposition 17.12 using Proposition 17.13 and $\operatorname{Der}_A(B, N) = 0$. $\qquad\square$

Corollary 17.34. *Let $f\colon X \to S$ be locally of finite presentation. Then $\Omega^1_{X/S}$ is an $\mathscr{O}_X$-module of finite presentation.*

Proof. The question is local, thus we may assume that $S = \operatorname{Spec} R$ and $X = \operatorname{Spec} B$ with $B = R[T_1, \ldots, T_n]/I$, where I is a finitely generated ideal. Then (17.5.10) yields an exact sequence

$$I/I^2 \to B \otimes_{R[T_1, \ldots, T_n]} \Omega^1_{R[T_1, \ldots, T_n]/R} \to \Omega^1_{B/R} \to 0.$$

By Example 17.8 the middle term is a free B-module of rank n and I/I^2 is finitely generated by hypothesis. Thus $\Omega^1_{B/R}$ is of finite presentation. $\qquad\square$

Remark 17.35. Let R be a ring, and let A be an R-algebra. By choosing generators of A as R-algebra we may write $A = R[(T_i)_{i \in I}]/(f_j)_{j \in J}$. As $\Omega^1_{R[(T_i)_{i \in I}]/R}$ is freely generated by $(dT_i)_{i \in I}$, Proposition 17.33 shows that

$$\Omega^1_{A/R} \cong \left(\bigoplus_{i \in I} A\, dT_i \right)/N,$$

where N is generated by

$$\left\{ df_j = \sum_{i \in I} \frac{\partial f_j}{\partial T_i} dT_i\, ;\ j \in J \right\}$$

Example 17.36. Let R be a ring.
(1) Let $A = R[T]/(f)$ for some $f \in R[T]$. Then $\Omega^1_{A/R} \cong R[T]/(f, f')$, where f' is the formal derivative of f.
(2) Consider $A = R[T_1, T_2]/(T_1 T_2)$ (so Spec A is the "union of the coordinate axes in $\mathbb{A}^2_R$"). Then dT_1 and dT_2 are generators of $\Omega^1_{A/R}$ as an A-module with the single relation $T_1 dT_2 + T_2 dT_1 = 0$. Let $N \subseteq \Omega^1_{A/R}$ be the submodule generated by $T_1 dT_2 = -T_2 dT_1$. Then $T_1 N = T_2 N = 0$, and N is a free R-module of rank 1. In $\Omega^1_{A/R}/N$ one has $T_1 dT_2 = T_2 dT_1 = 0$ and we obtain an exact sequence

$$0 \to N \to \Omega^1_{A/R} \to \Omega^1_{R[T_1]/R} \oplus \Omega^1_{R[T_2]/R} \to 0.$$

Thus $\Omega^1_{A/R}$ is an extension of $\Omega^1_{R[T_1]/R} \oplus \Omega^1_{R[T_2]/R}$ by a torsion module.

Remark 17.37. Let R be a ring, let A be an R-algebra, let $B = A/I$ for an ideal $I \subseteq A$ and let $\pi \colon A \to B$ be the canonical projection. Combining the map d (17.5.9), the boundary map in the exact sequence (17.2.2), and Proposition 17.13 one obtains for all B-modules N a diagram

$$
\begin{array}{ccc}
\mathrm{Der}_R(A, N) & \xrightarrow{\ \partial\ } & \mathrm{Ex}_A(B, N) \\
\downarrow{\scriptstyle\cong} & & \downarrow{\scriptstyle\cong} \\
\mathrm{Hom}_A(\Omega^1_{A/R}, N) & \xrightarrow{(-)\circ d} & \mathrm{Hom}_B(I/I^2, N),
\end{array}
$$

where the vertical maps are isomorphisms. This diagram is commutative. In fact the definitions show that both compositions send an R-derivation $D \colon A \to N$ to the B-linear map $I/I^2 \to N$ induced by the A-linear map $I \to N$ that sends $a \in I$ to $D(a)$.

(17.6) Tangent bundles.

As mentioned in the introduction, we define the tangent bundle of an S-scheme X as the dual of the sheaf of differentials of X over S. See Remark 17.44 below for the connection with the notion of (relative) tangent space defined in Section (6.6).

Definition 17.38. *Let* $g \colon X \to S$ *be a morphism of schemes. We call the* $\mathcal{O}_X$-*module*

$$(17.6.1) \qquad \mathcal{T}_g := \mathcal{T}_{X/S} := \mathcal{H}om_{\mathcal{O}_X}(\Omega^1_{X/S}, \mathcal{O}_X) \stackrel{(17.4.10)}{=} \mathcal{D}er_S(\mathcal{O}_X, \mathcal{O}_X)$$

the tangent sheaf of g *or of* X *over* S.

Remark 17.39. If g is locally of finite presentation, $\Omega^1_{X/S}$ is of finite presentation and hence $\mathcal{T}_{X/S}$ is a quasi-coherent $\mathcal{O}_X$-module by Proposition 7.29.

Remark 17.40. Let $g\colon X \to S$ be a morphism of schemes. For $D_1, D_2 \in \mathrm{Der}_S(\mathcal{O}_X, \mathcal{O}_X)$ the bracket

$$(17.6.2) \qquad\qquad\qquad [D_1, D_2] := D_1 \circ D_2 - D_2 \circ D_1$$

is again an S-derivation of $\mathcal{O}_X$. It is easy to check that $[\cdot, \cdot]$ is $\Gamma(S, \mathcal{O}_S)$-bilinear (but not $\Gamma(X, \mathcal{O}_X)$-bilinear in general) and that it satisfies the Jacobi identity

$$[D_1, [D_2, D_3]] + [D_2, [D_3, D_1]] + [D_3, [D_1, D_2]] = 0.$$

Therefore $\mathrm{Der}_S(\mathcal{O}_X, \mathcal{O}_X)$ is a $\Gamma(S, \mathcal{O}_S)$-Lie algebra (see also Exercise 17.1). This structure is compatible with restrictions to open subsets. Thus $\mathcal{T}_{X/S}$ obtains the structure of an $g^{-1}(\mathcal{O}_S)$-Lie algebra.

Note that the tangent sheaf is in general not compatible with base change: For a cartesian diagram

$$\begin{array}{ccc} X \times_S T & \xrightarrow{\ p_1\ } & X \\ {\scriptstyle p_2}\downarrow & & \downarrow \\ T & \longrightarrow & S \end{array}$$

one obtains a canonical map

$$(17.6.3) \qquad\qquad\qquad p_1^* \mathcal{T}_{X/S} \longrightarrow \mathcal{T}_{X \times_S T/T}$$

as the composition

$$p_1^* \mathcal{T}_{X/S} = p_1^*((\Omega^1_{X/S})^\vee) \longrightarrow (p_1^* \Omega^1_{X/S})^\vee \xrightarrow{\ \sim\ } (\Omega^1_{X \times_S T/T})^\vee = \mathcal{T}_{X \times_S T/T},$$

where the second isomorphism is induced by the inverse of the isomorphism in Proposition 17.30. It is in general not an isomorphism because forming the dual is not compatible with pullback. If $\Omega^1_{X/S}$ is locally free of finite type (for instance if $X \to S$ is smooth, see Corollary 18.58 below), then (17.6.3) is an isomorphism (Proposition 7.7).

The notion of the tangent bundle is better behaved. To define it, recall that we attached in Section (11.3) to every quasi-coherent $\mathcal{O}_X$-module $\mathscr{E}$ an X-scheme $\mathbb{V}(\mathscr{E}) = \mathrm{Spec}(\mathrm{Sym}(\mathscr{E}))$ such that the sections of $\mathbb{V}(\mathscr{E})$ over some open $U \subseteq X$ are identified with $\Gamma(U, \mathscr{E}^\vee)$. If $\pi\colon \mathbb{V}(\mathscr{E}) \to X$ is the structure morphism, π is affine and $\pi_* \mathcal{O}_{\mathbb{V}(\mathscr{E})} \cong \mathrm{Sym}(\mathscr{E})$ is a graded quasi-coherent $\mathcal{O}_X$-algebra.

Definition 17.41. *Let* $g\colon X \to S$ *be a morphism of schemes. We call*

$$T_g := T_{X/S} := \mathbb{V}(\Omega^1_{X/S})$$

the tangent bundle *of* g *or of* X *over* S.

The tangent bundle of a morphism g is of finite type (resp. of finite presentation), if g is locally of finite type (resp. locally of finite presentation). As recalled above, for every open subscheme U of X there is an identification

$$\mathrm{Hom}_X(U, T_{X/S}) = \Gamma(U, \mathcal{T}_{X/S}).$$

Remark 17.42. The tangent bundle is functorial in the following sense. Let $f\colon X \to Y$ be a morphism of S-schemes. Then the homomorphism $f^*\Omega^1_{Y/S} \to \Omega^1_{X/S}$ (17.5.6) yields by functoriality of $\mathbb{V}(\ )$ a morphism of X-schemes $T_{X/S} \to \mathbb{V}(f^*\Omega^1_{Y/S}) = T_{Y/S} \times_Y X$, or equivalently a morphism $T_f\colon T_{X/S} \to T_{Y/S}$ making the diagram

(17.6.4)

$$
\begin{array}{ccc}
T_{X/S} & \xrightarrow{\ T_f\ } & T_{Y/S} \\
\downarrow & & \downarrow \\
X & \xrightarrow{\ f\ } & Y
\end{array}
$$

commutative.

The tangent bundle has also the following description. For every scheme Y we set $Y[\varepsilon] := D_Y(\mathcal{O}_Y) = \mathrm{Spec}\,\mathcal{O}_Y[T]/(T^2)$. We consider $Y[\varepsilon]$ as Y-scheme and denote by $\iota_Y\colon Y \to Y[\varepsilon]$ the section corresponding to $T \mapsto 0$.

Proposition 17.43. *Let $f\colon X \to S$ be a morphism of schemes. Then for all S-schemes Y there is a commutative diagram, functorial in Y,*

$$
\begin{array}{ccc}
\mathrm{Hom}_S(Y[\varepsilon], X) & \xrightarrow{\qquad \tau_Y \qquad} & \mathrm{Hom}_S(Y, T_{X/S}) \\
& \searrow_{X_S(\iota_Y)} \qquad \swarrow & \\
& \mathrm{Hom}_S(Y, X), &
\end{array}
$$

such that the horizontal map is bijective.

Proof. We have to construct an isomorphism τ of functors over $h_X\colon Y \mapsto \mathrm{Hom}_S(Y, X)$ which is the same as to construct an isomorphism of functors on the category of X-schemes. Thus we have to show that for all X-schemes $g\colon Y \to X$ there is a bijection, functorial in Y,

(17.6.5) $$\{\, \tilde{g} \in \mathrm{Hom}_S(Y[\varepsilon], X) \ ; \ \tilde{g} \circ \iota_Y = g \,\} \leftrightarrow \mathrm{Hom}_X(Y, T_{X/S}).$$

But by restriction to the affine case (17.1.6) (for $N = C$) implies that the left hand side can be identified functorially with

$$\mathrm{Der}_S(\mathcal{O}_X, g_*\mathcal{O}_Y) = \mathrm{Hom}_{\mathcal{O}_X}(\Omega^1_{X/S}, g_*\mathcal{O}_Y) = \mathrm{Hom}_{\mathcal{O}_Y}(g^*\Omega^1_{X/S}, \mathcal{O}_Y)$$

$$= \mathrm{Hom}_X(Y, T_{X/S}). \qquad \square$$

The proposition shows in particular that $T_{X/S}$ represents the functor

$$(\mathrm{Sch}/S)^{\mathrm{opp}} \to (\mathrm{Sets}), \qquad Y \mapsto \mathrm{Hom}_S(Y[\varepsilon], X).$$

Remark 17.44. In Section (6.6) we defined the relative tangent space $T_\xi(X/S)$ of $X \to S$ at a K-valued point $\xi\colon \mathrm{Spec}\,K \to X$, where K is field. It was defined as the left hand side of (17.6.5) with $Y = \mathrm{Spec}\,K$, and we obtain

$$T_\xi(X/S) = \mathrm{Hom}_X(\mathrm{Spec}\,K, T_{X/S}) = \mathrm{Hom}_K(\xi^*\Omega^1_{X/S}, K).$$

In other words, the relative tangent spaces are the K-valued points of the tangent bundle. If $X \to S$ is locally of finite type, this is a finite-dimensional K-vector space (because $\Omega^1_{X/S}$ is then of finite type).

More generally, we can define for every morphism of schemes $x \colon Z \to X$ the *tangent space of X over S at x* as the $\mathscr{O}_Z$-module

(17.6.6) $$T_x(X/S) := \mathscr{H}om_{\mathscr{O}_Z}(x^*\Omega^1_{X/S}, \mathscr{O}_Z).$$

Then we have

(17.6.7) $$\mathscr{T}_{X/S} = T_{\mathrm{id}_X}(X/S)$$

and

(17.6.8) $$\Gamma(Z, T_x(X/S)) = \mathrm{Hom}_{\mathscr{O}_Z}(x^*\Omega^1_{X/S}, \mathscr{O}_Z) = \mathrm{Hom}_X(Z, T_{X/S}).$$

Let $Y \to S$ be another S-scheme and let $f \colon X \to Y$ be a morphism of S-schemes. Then the homomorphism $f^*\Omega^1_{Y/S} \to \Omega^1_{X/S}$ (17.5.6) induces by functoriality an $\mathscr{O}_Z$-linear map

(17.6.9) $$T_x(X/S) \longrightarrow T_{f \circ x}(Y/S).$$

Given S schemes X and Y with Z-valued points $x \colon Z \to X$ and $y \colon Z \to Y$, Corollary 17.32 shows that one has an isomorphism

(17.6.10) $$T_{(x,y)}(X \times_S Y/S) \xrightarrow{\sim} T_x(X/S) \times T_y(Y/S)$$

of $\mathscr{O}_Z$-modules.

Remark 17.45. Let k be a field, let X and Y be k-schemes locally of finite type, and let $f \colon X \to Y$ be a proper k-morphism. By the functorial description of $T_{X/k}$, Proposition 12.94 may be reformulated as follows.

The morphism f is a closed immersion if and only if there exists an algebraically closed extension K of k such that $T_f(K) \colon T_{X/k}(K) \to T_{Y/k}(K)$ is injective.

(17.7) Differentials of Grassmannians and of projective bundles.

We now determine the sheaf of differentials of Grassmannians and thus in particular of projective bundles. Let S be a scheme, let $\mathscr{E}$ be a quasi-coherent $\mathscr{O}_S$-module, and let $e \geq 1$ be an integer. Recall that $X := \mathrm{Grass}^e(\mathscr{E})$ is the S-scheme parametrizing submodules $\mathscr{U}$ of $\mathscr{E}$ such that $\mathscr{E}/\mathscr{U}$ is locally free of rank e, see Section (8.6). Let $\pi \colon X \to S$ be the structure morphism. For $e = 1$, X is the projective bundle $\mathbb{P}(\mathscr{E})$. In particular, if $\mathscr{E} = (\mathscr{O}_S^{n+1})^\vee$, then $\mathbb{P}(\mathscr{E}) = \mathbb{P}_S^n$.

Recall that by definition of $X = \mathrm{Grass}^e(\mathscr{E})$ there is a universal surjection of $\mathscr{O}_X$-modules

(17.7.1) $$u \colon \pi^*(\mathscr{E}) \to \mathscr{Q} \to 0,$$

where $\mathscr{Q}$ is a locally free $\mathscr{O}_X$-module of rank e, such that for every S-scheme $f \colon T \to S$ the map

(17.7.2)
$$\mathrm{Hom}_S(T, X) \to \{\, \mathscr{U} \subseteq f^*\mathscr{E} \;;\; f^*\mathscr{E}/\mathscr{U} \text{ locally free of rank } e \,\},$$
$$h \mapsto \ker(h^*(u) \colon h^*(\mathscr{E}) \to h^*(\mathscr{Q}))$$

is a bijection.

Theorem 17.46. *Let S be a scheme, let $\mathscr{E}$ be a finite locally free $\mathscr{O}_S$-module and set $X = \mathrm{Grass}^e(\mathscr{E})$. Define $\mathscr{K} := \ker(u)$ so that there is an exact sequence $0 \to \mathscr{K} \to \pi^*\mathscr{E} \to \mathscr{Q} \to 0$ of finite locally free $\mathscr{O}_X$-modules. Then*

$$\Omega^1_{X/S} \cong \mathscr{H}om_{\mathscr{O}_X}(\mathscr{Q}, \mathscr{K})$$

Proof. We will show that the duals of these two $\mathscr{O}_S$-modules are isomorphic. This suffices because both $\mathscr{O}_X$-modules are finite locally free: This is clear for $\mathscr{H}om_{\mathscr{O}_X}(\mathscr{Q}, \mathscr{K})$ because $\mathscr{Q}$ is locally free of rank e by definition and $\mathscr{K}$ is finite locally free because $\mathscr{E}$ and $\mathscr{Q}$ are. Its dual is isomorphic to $(\mathscr{Q}^\vee \otimes \mathscr{K})^\vee \cong \mathscr{K}^\vee \otimes \mathscr{Q} \cong \mathscr{H}om_{\mathscr{O}_X}(\mathscr{K}, \mathscr{Q})$. To check that $\Omega^1_{X/S}$ is finite locally free[2], we can work locally on X and in particular on S. Hence we may assume that $\mathscr{E} = \mathscr{O}_S^n$. Then X has an open covering $(U_i)_i$ such that $U_i \cong \mathbb{A}_S^{e(n-e)}$ for all i (Corollary 8.15). Therefore $(\Omega^1_{X/S})|_{U_i}$ is a free $\mathscr{O}_{U_i}$-module of rank $e(n-e)$ (Example 17.29).

Let $f \colon \mathrm{Spec}\, R \to S$ be a morphism with affine domain and let $h \colon \mathrm{Spec}\, R \to X$ be an S-morphism. Then h corresponds via (17.7.2) to an R-submodule N of $M := \Gamma(\mathrm{Spec}\, R, f^*\mathscr{E})$ such that M/N is projective of rank e. Verbatim the same argument as in the calculation of the tangent space of the Grassmannian (Section (8.9)) shows that

$$\mathrm{Hom}_X(\mathrm{Spec}\, R[\varepsilon], X) = \mathrm{Hom}_R(N, M/N),$$

functorially in R. Therefore we obtain for $\mathrm{Spec}\, R = U \subseteq X$ open affine:

$$\begin{aligned}
\Gamma(U, (\Omega^1_{X/S})^\vee) &= \mathrm{Hom}_X(U, \mathbb{V}(\Omega^1_{X/S})) \\
&= \mathrm{Hom}_X(\mathrm{Spec}\, R[\varepsilon], X) \\
&= \mathrm{Hom}_R(N, M/N) \\
&= \mathrm{Hom}_{\mathscr{O}_U}(\mathscr{K}|_U, \mathscr{Q}|_U) = \Gamma(U, \mathscr{H}om_{\mathscr{O}_X}(\mathscr{K}, \mathscr{Q})),
\end{aligned}$$

where the first equality is (11.3.3) and the second equality holds by Proposition 17.43 $\square$

For $e = 1$ one has $\mathrm{Grass}^1(\mathscr{E}) = \mathbb{P}(\mathscr{E}) = \mathrm{Proj}(\mathrm{Sym}(\mathscr{E})) =: P$ and $\mathscr{Q} = \mathscr{O}_P(1)$ (Theorem 13.32). Hence we obtain the following description of the sheaf of differentials for projective bundles.

Corollary 17.47. *Let $\mathscr{E}$ be a finite locally free $\mathscr{O}_S$-module, $P = \mathbb{P}(\mathscr{E})$, $\pi \colon P \to S$ its structure morphism. Let $\mathscr{K}$ be the kernel of the universal quotient $\pi^*\mathscr{E} \to \mathscr{O}_P(1)$. Then*

$$\Omega^1_{P/S} \cong \mathscr{H}om_{\mathscr{O}_P}(\mathscr{O}_P(1), \mathscr{K}) = \mathscr{K}(-1).$$

In particular there is an exact sequence of locally free $\mathscr{O}_P$-modules

(17.7.3) $$0 \to \Omega^1_{P/S} \to \pi^*(\mathscr{E})(-1) \to \mathscr{O}_P \to 0,$$

called the Euler sequence.

Example 17.48. For $P = \mathbb{P}^1_S$ the universal quotient (17.7.1) is given by $\mathscr{O}_P^2 \to \mathscr{O}_P(1) \to 0$ and its kernel is $\mathscr{O}_P(-1)$. Therefore

$$\Omega^1_{P/S} \cong \mathscr{O}_P(-2).$$

---

[2] This also follows from Corollary 18.58 below since $X \to S$ is smooth as the argument here shows.

The de Rham complex

Our next goal is the definition of the de Rham complex. It is the exterior algebra of $\Omega^1_{X/S}$ together with an extension of $d_{X/S}$ which makes it into an "strictly graded commutative differential graded algebra". We start by recalling these notions.

(17.8) The exterior algebra.

Let R be a ring (commutative as usual).

Definition 17.49. *Let $L = \bigoplus_{p \in \mathbb{Z}} L_p$ be a $\mathbb{Z}$-graded not necessarily commutative R-algebra. Then L is called* graded commutative *if for all homogeneous elements $x, y \in L$ one has*

$$xy = (-1)^{\deg(x)\deg(y)} yx.$$

It is called strictly graded commutative *if in addition $x^2 = 0$ for $x \in L$ homogeneous of odd degree.*

Note that if 2 is not a zero-divisor in L, then every graded commutative graded algebra is automatically strictly graded commutative.

An important example is the *exterior algebra* of an R-module M. It is defined as $\bigwedge_R M := \oplus_{p \geq 0} \bigwedge^p_R(M)$. The R-module $\bigwedge^p(M)$ is generated by elements of the from $m_1 \wedge \cdots \wedge m_p$ for $m_1, \ldots, m_p \in M$ and the multiplication is given by

$$\bigwedge{}^p(M) \times \bigwedge{}^q(M) \longrightarrow \bigwedge{}^{p+q}(M),$$
$$(m_1 \wedge \cdots \wedge m_p) \wedge (m'_1 \wedge \cdots \wedge m'_q) := m_1 \wedge \cdots \wedge m_p \wedge m'_1 \wedge \cdots \wedge m'_q.$$

This is well defined and defines by bilinear extension the structure of a strictly graded commutative graded R-algebra on $\bigwedge_R M$.

We obtain a functor $M \mapsto \bigwedge_R M$ from the category of R-modules to the category of strictly graded commutative graded R-algebras. By [BouAI] ^{O} III, §7, 1, Rem. (1) we have the following adjointness property.

Proposition 17.50. *The functor $M \mapsto \bigwedge_R M$ is left adjoint to the functor that sends a strictly graded commutative graded R-algebra E to the R-module E_1.*

In other words, given any strictly graded commutative graded R-algebra E and any homomorphism $u \colon M \to E_1$ of R-modules, there is a unique extension of u to a homomorphism $\bigwedge_R M \to E$ of graded R-algebras.

All these notions have obvious globalizations. Let $(X, \mathscr{O}_X)$ be a ringed space. It is clear how to define the notion of a (strictly) graded commutative graded $\mathscr{O}_X$-algebra. If $\mathscr{F}$ is an $\mathscr{O}_X$-module we define the *exterior algebra*

$$\bigwedge \mathscr{F} := \bigwedge{}_{\mathscr{O}_X} \mathscr{F} := \bigoplus_{p \geq 0} \bigwedge{}^p_{\mathscr{O}_X} \mathscr{F}.$$

The formation of $\bigwedge \mathscr{F}$ is clearly functorial in $\mathscr{F}$. Moreover, the properties of $\bigwedge^p_{\mathscr{O}_X}(\mathscr{F})$ (see Section (7.20)) immediately imply the following properties of the exterior algebra.

Remark 17.51. Let X be a scheme and let $\mathscr{F}$ be quasi-coherent.
(1) $\bigwedge \mathscr{F}$ is a quasi-coherent $\mathscr{O}_X$-algebra.
(2) For every morphism $f \colon X' \to X$ of schemes there is an isomorphism $f^*(\bigwedge \mathscr{F}) \xrightarrow{\sim} \bigwedge(f^*\mathscr{F})$ of $\mathscr{O}_{X'}$-algebras, functorial in $\mathscr{F}$.
(3) If $X = \operatorname{Spec} A$ and $\mathscr{F} = \tilde{M}$ for an A-module M, then $\bigwedge \mathscr{F}$ is the quasi-coherent $\mathscr{O}_X$-algebra corresponding to the A-algebra $\bigwedge_A M$.

(17.9) Differential graded algebras.

The notion of differential graded algebra which we will define below combines the notions of graded algebra and of a derivation, giving the algebra the structure of a complex.

Definition 17.52. *Let R be a ring.*
(1) *Let L be a graded R-algebra (not necessarily commutative). A* graded R-derivation *of L is an R-linear map $d \colon L \to L$ homogeneous of degree 1 such that for all $x \in L_p$, $y \in L$ one has the graded Leibniz rule*

$$(17.9.1) \qquad d(xy) = (dx)y + (-1)^p x\, dy.$$

(2) *A* differential graded R-algebra *is a graded algebra L together with a graded R-derivation d of L such that $d \circ d = 0$.*
(3) *A differential graded R-algebra L is called* (strictly) graded commutative *if the underlying graded algebra is (strictly) graded commutative.*

Sometimes a graded R-derivation is instead defined to be of degree -1. We will always stick with the degree 1 version as defined above. In the literature, the term *differential graded algebra* is often abbreviated as *dga*.

Again we have the obvious globalizations. Let $f \colon (X, \mathscr{O}_X) \to (S, \mathscr{O}_S)$ be a morphism of ringed spaces and let $\mathscr{L}$ be a graded $\mathscr{O}_X$-algebra. An $f^{-1}(\mathscr{O}_S)$-linear map $d \colon \mathscr{L} \to \mathscr{L}$ is called a *graded $\mathscr{O}_S$-derivation of $\mathscr{L}$*, if it is homogeneous of degree 1 and if (17.9.1) holds for all local sections x of $\mathscr{L}_p$ and y of $\mathscr{L}$. A pair $(\mathscr{L}, d)$ consisting of a graded $\mathscr{O}_X$-algebra and a graded $\mathscr{O}_S$-derivation of $\mathscr{L}$ is called *differential graded $\mathscr{O}_S$-algebra over X* if $d \circ d = 0$. Such a differential graded $\mathscr{O}_S$-algebra over X is called (strictly) graded commutative if $xy = (-1)^{\deg(x)\deg(y)} yx$ holds for all local homogeneous sections x and y (and if $x^2 = 0$ for all local homogeneous sections x of odd degree).

(17.10) The de Rham complex.

We now come to the definition of the de Rham complex. Let $X \to S$ be a morphism of schemes, $p \geq 0$ an integer. The p-th exterior power (Section (7.20))

$$\Omega^p_f := \Omega^p_{X/S} := \bigwedge^p_{\mathscr{O}_X} \Omega^1_{X/S}$$

is called the *sheaf of p-differential forms for f* or *of X over S*. In particular $\Omega^0_{X/S} = \mathscr{O}_X$. Moreover, for an integer $p < 0$ we set $\Omega^p_{X/S} := 0$. We also denote by

$$\Omega^\bullet_{X/S} := \bigwedge_{\mathscr{O}_X} \Omega^1_{X/S} = \bigoplus_{p \in \mathbb{Z}} \Omega^p_{X/S}$$

the exterior algebra of $\Omega^1_{X/S}$.

To define the differentials of the de Rham complex we will use the following result which follows from [BouAI]^{O} III, §10.9, Prop. 14.

Lemma 17.53. *Let R be a ring and let A be a commutative R-algebra, let M be an A-module, and let $L := \bigwedge_A(M)$ be its exterior algebra. Let $d^0 \colon A \to M$ be an R-derivation and let $d^1 \colon M \to \bigwedge^2(M)$ be an R-linear map such that for all $a \in A$ and $m \in M$ we have*

$$(17.10.1) \qquad d^1(am) = ad^1(m) + d^0(a) \wedge m.$$

Then there exists a unique R-derivation $d \colon L \to L$ of degree 1 such that $d_{|L_0} = d^0$ and $d_{|L_1} = d^1$.

Proposition 17.54. *Let $f \colon X \to S$ be a morphism of schemes. There exists a unique graded S-derivation $d \colon \Omega^\bullet_{X/S} \to \Omega^\bullet_{X/S}$ of degree 1 such that*
(a) $d \circ d = 0$.
(b) *For $f \in \Gamma(U, \mathscr{O}_X)$, $U \subseteq X$ open, one has $d(f) = d_{X/S}(f) \in \Gamma(U, \Omega^1_{X/S})$.*
For local sections $b, a_1, \ldots, a_p$ of $\mathscr{O}_X$, the differential is given by

$$(17.10.2) \qquad d \colon \Omega^p_{X/S} \longrightarrow \Omega^{p+1}_{X/S}, \qquad d(bda_1 \wedge \cdots \wedge da_p) = db \wedge da_1 \wedge \cdots \wedge da_p.$$

Proof. The question is local on X, thus we may assume that $S = \operatorname{Spec} R$ and $X = \operatorname{Spec} A$ are affine. To show the uniqueness of d we remark that $d \circ d = 0$ shows that for all $b, a_1, \ldots, a_p \in A$ one has

$$d(bda_1 \wedge \cdots \wedge da_p) = db \wedge da_1 \wedge \cdots \wedge da_p.$$

As the A-module $\Omega^p_{A/R}$ is generated by elements of the form $da_1 \wedge \cdots \wedge da_p$, this proves the uniqueness of d.

To show the existence we set $d^0 := d_{A/R} \colon A \to \Omega^1_{A/R}$. We will construct an R-linear map $d^1 \colon \Omega^1_{A/R} \to \Omega^2_{A/R}$ such that $d^1 \circ d^0 = 0$ and such that

$$(17.10.3) \qquad d^1(a\omega) = d^0(a) \wedge \omega + ad^1(\omega)$$

for all $a \in A$, $\omega \in \Omega^1_{A/R}$. Then Lemma 17.53 ensures the existence of d and we have $d \circ d = 0$ because $d^1 \circ d^0 = 0$ and $\Omega^\bullet_{A/R}$ is generated as A-algebra by elements of the form $d^0(a)$ for $a \in A$.

Let I be the kernel of the multiplication $m \colon A \otimes_R A \to A$. Then $\Omega^1_{A/R} = I/I^2$ by definition. Consider the R-linear homomorphism

$$u \colon A \otimes_R A \to \Omega^2_{A/R}, \qquad u(a \otimes b) := d^0(b) \wedge d^0(a).$$

Let $a \in A$ and $\tau \in A \otimes_R A$. Writing τ as sum of elementary tensors, one easily sees that

$$(17.10.4) \qquad u((a \otimes 1 - 1 \otimes a)\tau) = d^0(m(\tau)) \wedge d^0(a).$$

As I is generated by elements of the form $a \otimes 1 - 1 \otimes a$ for $a \in A$ (proof of Proposition 17.6), this shows that $u(I^2) = 0$. Thus u defines by restriction to I an R-linear map $d^1 \colon I/I^2 \to \Omega^2_{A/R}$.

For $\tau = b \otimes 1$ with $b \in A$, (17.10.4) shows $d^1(bd^0(a)) = d^0(b) \wedge d^0(a)$. For $b = 1$ this shows $d^1 \circ d^0 = 0$. For $c \in A$ and $\omega = bd^0(a)$ it shows

$$d^1(c\omega) = d^0(cb) \wedge d^0(a) = d^0(c) \wedge \omega + cd^1(\omega),$$

As $\Omega^1_{A/R}$ is generated as R-module by elements of the form $bd^0(a)$, this implies (17.10.3). $\qquad \square$

Definition 17.55. *Let* $f\colon X \to S$ *be a morphism of schemes. The complex* $(\Omega^\bullet_{X/S}, d)$ *is called the* de Rham complex *of* X *over* S *or of* f. *Sections of* $\Omega^i_{X/S}$ *are called* differential forms of degree i.

If $S = \operatorname{Spec} R$ and $X = \operatorname{Spec} A$ are affine, taking global sections of $(\Omega^\bullet_{X/S}, d)$ one obtains a complex $(\Omega^\bullet_{A/R}, d)$ of A-modules with R-linear differential of degree 1.

The de Rham complex has the following universal property.

Proposition 17.56. *Let* $f\colon X \to S$ *be a morphism of schemes. Then the de Rham complex* $(\Omega^\bullet_{X/S}, d)$ *of* f *is an initial object in the category of strictly graded commutative differential graded quasi-coherent* $\mathcal{O}_S$-*algebras over* X.

In other words, given any strictly graded commutative differential graded quasi-coherent $\mathcal{O}_S$-algebra $(\mathscr{E}^\bullet, d)$ over X, there is a unique homomorphism $(\Omega^\bullet_{X/S}, d) \to (\mathscr{E}^\bullet, d)$ of differential graded $\mathcal{O}_S$-algebras over X.

Proof. We may assume that $S = \operatorname{Spec} R$ and $X = \operatorname{Spec} A$. Then we have to show that $(\Omega^\bullet_{A/R}, d)$ is the initial object in the category of strictly graded commutative graded A-algebras with an R-linear differential d of degree 1 with $d \circ d = 0$ satisfying the graded Leibniz rule (17.9.1). Let (E, d) such an object. The structure of E as a graded A-algebra yields a map $A \to E^0$ of commutative rings whose composition with $d\colon E^0 \to E^1$ is an R-linear derivation $A \to E^1$. By the universal property of $\Omega^1_{A/R}$, this derivation corresponds to an A-linear map $\Omega^1_{A/R} \to E^1$. By the universal property of the exterior algebra (Proposition 17.50), this linear map corresponds to a map of graded algebras $\varphi\colon \Omega^\bullet_{A/R} \to E$.

It remains to show that φ preserves the differentials. By construction we have $d(\varphi(a)) = \varphi(d(a))$ for $a \in A$, where on left hand side we consider a as element of E^0 via the ring map $A \to E^0$. But as an R-module $\Omega^p_{A/R}$ is generated by elements of the form $x = b\,da_1 \wedge \cdots \wedge da_p$ for $b, a_1, \ldots, a_p \in A$ and we have

$$\varphi(dx) = \varphi(db \wedge da_1 \wedge \cdots \wedge da_p) = \varphi(db)\varphi(da_1)\cdots\varphi(da_p)$$
$$= d(b)d(\varphi(a_1))\cdots d(\varphi(a_p)),$$
$$d(\varphi(x)) = d(b\varphi(da_1)\cdots\varphi(da_p)) = d(bd(\varphi(a_1))\cdots d(\varphi(a_p)))$$
$$= d(b)d(\varphi(a_1))\cdots d(\varphi(a_p)). \qquad \square$$

Example 17.57. Let R be a ring and let $A = R[T_1, \ldots, T_n]$. Then $\Omega^1_{A/R}$ is a free A-module with basis $(dT_1, \ldots, dT_n)$ (Example 17.8). For every subset $I = \{i_1, \ldots, i_p\}$ of $\{1, \ldots, n\}$ with $i_1 < \cdots < i_p$ we set $dT_I := dT_{i_1} \wedge \cdots \wedge dT_{i_p}$. Then if I runs through the subsets of $\{1, \ldots, n\}$ with cardinality p, the dT_I form a basis of $\Omega^p_{A/R}$. For all $f \in A$ one has by (17.10.2)

$$d(f\,dT_I) = df \wedge dT_I = \sum_{i \notin I} (-1)^{n(I,i)} \frac{\partial f}{\partial T_i} dT_{I \cup \{i\}},$$

where $n(I, i)$ denotes the number of elements $j \in I$ with $j < i$.

One can show (Exercise 19.7) that $\Omega^\bullet_{A/R}$ is exact in degrees > 0 if R is a $\mathbb{Q}$-algebra and deduce that $\Omega^\bullet_{X/S}$ is exact in degrees > 0 if S is a $\mathbb{Q}$-scheme and $X \to S$ is a smooth morphism (Exercise 20.28).

In Corollary 17.47 we have seen that $\Omega^1_{\mathbb{P}^n_S/S}$ is a locally free module of rank n. Hence $\Omega^n_{\mathbb{P}^n_S/S}$ is locally free of rank 1. For its calculation we will use the following general remark (see also Exercise 7.29).

Definition and Remark 17.58. Let X be a ringed space. If $\mathscr{E}$ is a finite locally free $\mathscr{O}_X$-module, we denote by $\mathrm{rk}(\mathscr{E})$ its rank, which is in general a locally constant function $X \to \mathbb{Z}_{\geq 0}$. Moreover, we call

$$\det(\mathscr{E}) := \bigwedge^{\mathrm{rk}(\mathscr{E})}(\mathscr{E})$$

its determinant, which is a line bundle. By this we mean that if $X_n \subseteq X$ is the open and closed subscheme of X where the rank of $\mathscr{E}$ is equal to $n \in \mathbb{Z}_{\geq 0}$, then $\det(\mathscr{E})$ is the unique line bundle on X such that $\det(\mathscr{E})_{|X_n} = \bigwedge^n \mathscr{E}_{|X_n}$.

For any short exact sequence

$$0 \longrightarrow \mathscr{E}' \longrightarrow \mathscr{E} \longrightarrow \mathscr{E}'' \longrightarrow 0$$

of vector bundles on X there exists a unique isomorphism

(17.10.5) $\delta\colon \det(\mathscr{E}') \otimes \det(\mathscr{E}'') \overset{\sim}{\longrightarrow} \det(\mathscr{E})$

that makes the following diagram commutative

$$
\begin{array}{ccc}
 & \bigwedge^{r'}\mathscr{E}' \otimes \bigwedge^{r''}\mathscr{E} & \\
{\scriptstyle\theta}\swarrow & & \searrow{\scriptstyle\eta} \\
\bigwedge^{r'}\mathscr{E}' \otimes \bigwedge^{r''}\mathscr{E}'' \xrightarrow{\hspace{2cm}\delta\hspace{2cm}} & & \bigwedge^{r'+r''}\mathscr{E},
\end{array}
$$

where $r' := \mathrm{rk}(\mathscr{E}')$, $r'' := \mathrm{rk}(\mathscr{E}'')$, and where θ is induced by functoriality from $\mathscr{E} \to \mathscr{E}''$ and where η is induced by functoriality $\mathscr{E}' \to \mathscr{E}$ and by the multiplication in the exterior algebra $\bigwedge \mathscr{E}$.

Indeed, as θ is surjective, the uniqueness is clear and we can work locally on X to see its existence. Therefore we may assume that $\mathscr{E}'$ and $\mathscr{E}''$ are free of constant rank and that we find a basis $e''_{r'+1}, \dots, e''_{r'+r''} \in \Gamma(X, \mathscr{E}'')$ that can be lifted to global sections $e_{r'+1}, \dots, e_{r'+r''}$ of $\mathscr{E}$. Denote by $e_1, \dots, e_{r'}$ images of a basis $e'_1, \dots, e'_{r'}$ of $\mathscr{E}'$. Then $\mathscr{E}$ is a free, and $e_1, \dots, e_{r'+r''} \in \Gamma(X, \mathscr{E})$ is a basis. Therefore, we can define δ by sending the single basis element $(e'_1 \wedge \cdots \wedge e'_{r'}) \otimes (e''_{r'+1} \wedge \cdots \wedge e''_{r'+r''})$ of $\det(\mathscr{E}') \otimes \det(\mathscr{E}'')$ to the basis element $e_1 \wedge \cdots \wedge e_{r'+r''}$.

Example 17.59. Let S be a scheme, let $\mathscr{E}$ be a locally free $\mathscr{O}_S$-module of constant rank $n + 1$, and let $P = \mathbb{P}(\mathscr{E})$ be the corresponding projective bundle over S. Denote by $\pi\colon P \to S$ the natural morphism.

Let $\mathscr{L} = \pi^* \det(\mathscr{E})$ be the pullback of the determinant of $\mathscr{E}$ (Remark 17.58). We then have

(17.10.6) $\Omega^n_{P/S} \cong \mathscr{L}(-n-1)$.

This follows by passing to the top exterior powers of the terms in the short exact sequence (17.7.3) using (17.10.5).

In particular, for $\mathscr{E} = \mathscr{O}_S^{n+1}$, we obtain that

(17.10.7) $\Omega^n_{\mathbb{P}^n_S/S} \cong \mathscr{O}_{\mathbb{P}^n_S}(-n-1)$.

Exercises

Exercise 17.1. Let $R \to A$ be a homomorphism of rings. For $D, D' \in \operatorname{Der}_R(A)$ set $[D, D'] := D \circ D' - D' \circ D$.
(1) Show that $[D, D'] \in \operatorname{Der}_R(A)$ and that $[\ ,\ ]$ endows $\operatorname{Der}_R(A)$ with the structure of an R-Lie algebra.
(2) Let $\mathfrak{g} \subseteq \operatorname{Der}_R(A)$ be an A-submodule generated by a set X. Show that $[\ ,\ ]$ preserves $\mathfrak{g}$ if and only if $[D, D] \in \mathfrak{g}$ for all $D \in X$.

Exercise 17.2. Let $Z \xrightarrow{\ i\ } X \xrightarrow{\ f\ } Y$ be morphisms of schemes such that i and $j := f \circ i$ are immersions. Show that there is a canonical exact sequence of $\mathscr{O}_Z$-modules

$$\mathscr{C}_j \longrightarrow \mathscr{C}_i \longrightarrow i^* \Omega^1_{X/Y} \longrightarrow 0.$$

Exercise 17.3. Let $R \to A$ be a ring homomorphism, $I \subseteq A$ an ideal, and let $\hat{A} = \lim_n A/I^n$ be the I-adic completion of A. Let $D \in \operatorname{Der}_R(A)$. Show that $D(I^n) \subseteq I^{n-1}$ for all integers $n \geq 1$. Deduce that D induces a derivation on $\hat{A}$.

Exercise 17.4. Let k be a field of characteristic zero, let K/k be a field extension, and let $x_1, \ldots, x_n \in K$. Show that $dx_1, \ldots, dx_n \in \Omega^1_{K/k}$ are K-linearly independent if and only if $x_1, \ldots, x_n$ are algebraically independent over k.
Give a counterexample to this assertion in characteristic > 0.

Exercise 17.5. Let $R \to A$ be a homomorphism of rings. For $D \in \operatorname{Der}_R(A)$ and for an integer $i \geq 1$ write $D^i = D \circ \cdots \circ D$ (i times). Let p be a prime number and assume that $pA = 0$.
(1) Let $D \in \operatorname{Der}_R(A)$. Show that $D^p \in \operatorname{Der}_R(A)$.
(2) Assume that A is an integral domain and let $0 \neq D \in \operatorname{Der}_R(A)$. Show that $a_0 + a_1 D + a_2 D^2 + \cdots + a_{p-1} D^{p-1}$ is a derivation ($a_i \in A$) if and only if $a_0 = a_2 = \cdots = a_{p-1} = 0$.
Hint: Show first that $1, D, D^2, \ldots, D^{p-1}$ are linearly independent over $\operatorname{Frac} A$.
(3) Show that an analogous assertion as in (2) does not hold for $A = \mathbb{F}_p[T]/(T^p)$.
(4) Now let A again be an arbitrary ring with $pA = 0$. Show the "Hochschild formula": For $a \in A$ and $D \in \operatorname{Der}_R(A)$ one has $(aD)^p = a^p D^p + (aD)^{p-1}(a) \cdot D$. In particular $(aD)^p$ is a linear combination of D and D^p.
Hint: Reduce to an identity over $\mathbb{F}_p[T_1, T_2, \ldots]$. Then use (2).

Exercise 17.6. Let k be a field of characteristic $p > 0$ and let K be a field extension of k such that $a^p \in k$ for every $a \in K$. Let V be a k-vector space. For every $D \in \operatorname{Der}_k(K)$ we obtain a map

$$D_V := \operatorname{id}_V \otimes D \colon V \otimes_k K \to V \otimes_k K.$$

Show that a sub K-vector space $U' \subseteq V \otimes_k K$ is of the from $U \otimes_k K$ for some sub K-vector space $U \subseteq V$ if and only if $D_V(U') \subseteq U'$ for all $D \in \operatorname{Der}_k(K)$.

Exercise 17.7. Let R be a ring, $n \geq 1$, and let $X \subset \mathbb{P}^n_R$ be a hypersurface of degree d, i.e. $X = V_+(f)$ for a non-zero homogeneous polynomial f of degree d. Show that the conormal sheaf of the inclusion $i \colon X \to \mathbb{P}^n_R$ is $i^* \mathscr{O}_{\mathbb{P}^n_R}(-d)$.

Exercise 17.8. Let R be a discrete valuation ring and let $\pi \in R$ be a uniformizer. Fix $i \in \mathbb{N}$ and let $A = R[X, Y]/(XY^i - \pi)$, a regular domain. For which i is $\Omega_{A/R}$ a locally free A-module? For which i is it torsion-free?

Exercise 17.9. Let k be a field of characteristic $\neq 2, 3$, let $a, b \in k$ such that $g(X) := X^3 + aX + b$ is separable, and let $E = V_+(Y^2 Z - g(X)) \subset \mathbb{P}_k^2$. Show that $\omega := \frac{dX}{Y} \in \Omega_{K(E)/k}^1$ actually lies in $\Gamma(E, \Omega_{E/k}^1) \subset \Omega_{K(E)/k}^1$, and is a nowhere vanishing global section of $\Omega_{E/k}^1$. Conclude that $\Omega_{E/k}^1$ is a free $\mathscr{O}_E$-module of rank 1.

Remark: The curve E is an elliptic curve, see Section (16.35) and Section (26.17).

Exercise 17.10. Let $K = k((T_i)_{i \in I})$ be a purely transcendental extension of a field k. Show that $(dT_i)_{i \in I}$ is a basis of $\Omega_{K/k}^1$.

Exercise 17.11. Let $X \to S$ be a morphism of schemes that is locally of finite presentation. Show that $\mathscr{D}er_S(\mathscr{O}_X, \mathscr{F})$ is a quasi-coherent $\mathscr{O}_X$-module for every quasi-coherent $\mathscr{O}_X$-module $\mathscr{F}$.

Exercise 17.12. Let $X \to S$ be a morphism of schemes. Show that for every $\mathscr{O}_X$-module $\mathscr{F}$ the morphism $\mathscr{H}om_{\mathscr{O}_X}(\Omega_{X/S}^1, \mathscr{F}) \to \mathscr{D}er_S(\mathscr{O}_X, \mathscr{F})$, $u \mapsto u \circ d_{X/S}$, is an isomorphism of $\mathscr{O}_X$-modules.

Exercise 17.13. Let p be a prime, let S be a scheme of characteristic p, let $X \to S$ be an S-scheme, and let $F_{X/S} \colon X \to X^{(p)}$ be the relative Frobenius morphism (Remark 4.24).
(1) Show that the canonical homomorphism $v_{F/S} \colon F^* \Omega_{X^{(p)}/S}^1 \to \Omega_{X/S}^1$ is zero.
(2) Show that $\Omega_{X/S}^1 \to \Omega_{X/X^{(p)}}^1$ is an isomorphism.

Exercise 17.14. Let $f \colon X \to S$ be a morphism of schemes and let $\mathscr{F}$ be a quasi-coherent $\mathscr{O}_X$-module. An $f^{-1}(\mathscr{O}_S)$-linear homomorphism $\nabla^0 \colon \mathscr{F} \to \mathscr{F} \otimes_{\mathscr{O}_X} \Omega_{X/S}^1$ is called a *connection* if

$$\nabla^0(am) = a\nabla^0(m) + m \otimes da, \qquad a \in \mathscr{O}_X(U), \ m \in \mathscr{F}(U), \ U \subseteq X \text{ open.}$$

In the sequel we will consider $\mathscr{F} \otimes \Omega_{X/S}^\bullet$ as a right module over the $\mathscr{O}_X$-algebra $\Omega_{X/S}^\bullet$.
(1) Show that there exists a unique $f^{-1}(\mathscr{O}_S)$-linear graded endomorphism ∇ of $\mathscr{F} \otimes \Omega_{X/S}^\bullet$ of degree 1 which extends ∇^0 in degree 0 and which satisfies the equality

$$\nabla(x\omega) = (\nabla x)\omega + (-1)^p x(d\omega)$$

for x a local section of $\mathscr{F} \otimes \Omega_{X/S}^p$, ω a local section of $\Omega_{X/S}^\bullet$.
(2) Show that the composition $\nabla \circ \nabla$ is $\Omega_{X/S}^\bullet$-linear.
(3) Set $R := \nabla^1 \circ \nabla^0 \colon \mathscr{F} \to \mathscr{F} \otimes \Omega^2$. Deduce that R is $\mathscr{O}_X$-linear and that

$$(\nabla \circ \nabla)(m \otimes \omega) = R(m)\omega$$

for all local sections m of $\mathscr{F}$ and ω of $\Omega_{X/S}^\bullet$.

The homomorphism R is called the *curvature of the connection* ∇^0. If $R = 0$, then the connection ∇^0 is called *flat* and the complex $(\mathscr{F} \otimes \Omega_{X/S}^\bullet, \nabla)$ is called the *de Rham complex* of $(\mathscr{F}, \nabla^0)$ *over* S.

18 Étale and smooth morphisms

Content

- Formally unramified, formally smooth and formally étale morphisms
- Unramified and étale morphisms
- Smooth morphisms

In Chapter 6 we defined a morphism of schemes $f\colon X \to S$ to be smooth if Zariski locally on S and on X one may write f as an open immersion followed by a morphism of the form $\operatorname{Spec} R[T_1,\dots,T_n]/(f_1,\dots,f_r) \to \operatorname{Spec} R$ such that the rank of the Jacobian matrix of the f_i is equal to r in each point $x \in X$. We called f étale if in addition we may choose $r = n$. As the term suggests, smoothness should be thought of as being "without relative singularities". This point of view is also supported by the fact that for a smooth surjective morphism $X \to S$ the singularities of X "are as bad as" the singularities of S (see Corollary 14.60 and Remark 14.61 for precise statements). Moreover, in (20.12) we will attach to each scheme X locally of finite type over $\mathbb{C}$ a complex analytic space X^{an}. Then X is smooth over $\operatorname{Spec}(\mathbb{C})$ if and only if X^{an} is a complex manifold. This follows from the theorem on inverse functions in complex analysis and justifies the definition of smoothness via the Jacobian matrix.

If $S = \operatorname{Spec} k$ for a field k, then we showed that a morphism locally of finite type $X \to \operatorname{Spec} k$ is smooth if and only if $X \otimes_k K$ is a regular for some (or, equivalently, for all) algebraically closed extension K of k (Corollary 6.32). Moreover we saw in Theorem 14.24 that every smooth morphism is flat. In this chapter we will prove that conversely every flat morphism with geometrically regular fibers is smooth (Theorem 18.56). That theorem will also show that if $f\colon X \to S$ is locally of finite presentation, then f is smooth if and only if for every S-morphism $\operatorname{Spec} R \to X$ (R some ring) and for every nilpotent ideal $I \subset R$ the map $X(R) \to X(R/I)$ is surjective.

This last property is called *formal smoothness* and can be expressed by properties of Kähler differentials (Proposition 18.18). We start by studying these properties.

Formally unramified, formally smooth and formally étale morphisms

Let T be a scheme and let T_0 be a closed subscheme. Given a morphism $X \to S$ of schemes it is an important question whether it is possible to extend (maybe even uniquely) an S-morphism $T_0 \to X$ to an S-morphism $T \to X$. In other words, we look at a commutative diagram of scheme morphisms

© Springer Fachmedien Wiesbaden GmbH, ein Teil von Springer Nature 2023
U. Görtz und T. Wedhorn, *Algebraic Geometry II: Cohomology of Schemes*,
Springer Studium Mathematik – Master, https://doi.org/10.1007/978-3-658-43031-3_3

$$\begin{array}{ccc} T_0 & \xrightarrow{\;a_0\;} & X \\ {\scriptstyle i}\downarrow & & \downarrow{\scriptstyle f} \\ T & \xrightarrow{\;a\;} & S, \end{array}$$

(18.0.1)

where i is a closed immersion, and we study the question whether there exists a morphism $b\colon T \to X$ that commutes with (18.0.1) by which we mean that $b \circ i = a_0$ and $f \circ b = a$.

In general there is probably no hope to answer this question, but if $T = \operatorname{Spec} R$ is affine and hence $T_0 = \operatorname{Spec} R/I$ for an ideal $I \subset R$, and if I is "not too big", then the situation is more manageable. Here we study this question if I is a nilpotent ideal, i.e., there exists some $n \geq 0$ such that $I^{n+1} = 0$. Then we think of $\operatorname{Spec} R$ as an infinitesimal thickening of $\operatorname{Spec} R/I$. Morphisms f for which there always exists a lift b (resp. a unique lift b) as above whenever I is nilpotent will be called formally smooth (resp. formally étale). As the definition suggests, we will see then in the following sections that these notions are linked to the question whether $X \to S$ is smooth or étale.

Later in Sections (20.3) (for local rings) and (20.4) in general, we will also consider the more general case if (R, I) is a henselian pair (see Definition 20.15 below) which is in particular the case if I is nilpotent (Proposition 20.17 below).

(18.1) Definition of formally unramified, formally smooth and formally étale morphisms.

Definition 18.1. *Let $i\colon Y \to X$ be a closed immersion defined by a quasi-coherent ideal $\mathscr{I} \subseteq \mathscr{O}_X$. Then i is called a* nil-immersion *if i is a homeomorphism. We call i a* thickening *if $\mathscr{I}$ is locally nilpotent (i.e., there exist an open covering $(U_i)_{i \in I}$ of X and for all i an $n_i \in \mathbb{N}$ such that $(\mathscr{I}_{|U_i})^{n_i+1} = 0$). For $n \in \mathbb{N}$, a thickening is called of order at most n if $\mathscr{I}^{n+1} = 0$.*

Remark 18.2. Let $i\colon Y \to X$ be a closed immersion and let $\mathscr{I}$ be the quasi-coherent ideal of $\mathscr{O}_X$ defining i.
(1) Then i is a nil-immersion if and only if $\mathscr{I} \subseteq \mathscr{N}_X$, where $\mathscr{N}_X$ denotes the nil-radical of $\mathscr{O}_X$, cf. Section (3.18).
(2) If X is quasi-compact and i is a thickening, then there exists an integer $n \geq 0$ such that $\mathscr{I}^{n+1} = 0$.
(3) Consider the situation locally on X, i.e. $X = \operatorname{Spec} A$ is affine. Then $Y \cong \operatorname{Spec} A/I$ for an ideal $I \subseteq A$ and i is given by the canonical ring homomorphism $A \to A/I$. Then i is a nil-immersion (resp. a thickening) if and only if every element of I is nilpotent (resp. there exists an $n \geq 0$ such that $I^{n+1} = 0$).
(4) Every thickening is a nil-immersion. Conversely, if i is a nil-immersion and $\mathscr{I}$ is of finite type (e.g., if X is locally noetherian), then i is a thickening.
(5) The n-th infinitesimal neighborhood of Y in X (17.3) is a thickening of order at most n of Y.

Definition 18.3. *A morphism $f\colon X \to S$ of schemes is called* formally unramified *(resp.* formally smooth, *resp.* formally étale*) if for every diagram of the form (18.0.1) with T affine and i a thickening of order at most 1 there exists at most one (resp. at least one, resp. a unique) morphism $b\colon T \to X$ commuting with (18.0.1).*

If R is a ring, an R-algebra A is called formally unramified *(resp.* formally smooth, *resp.* formally étale*) over R, if $\operatorname{Spec} A \to \operatorname{Spec} R$ has this property.*

Remark 18.4. Let $f\colon X \to S$ be a morphism of schemes.

(1) f is formally étale if and only if f is formally unramified and formally smooth.

(2) Let f be formally unramified (resp. formally smooth, resp. formally étale) and let i be an arbitrary thickening of affine schemes given by the projection $R \to R/I$ for an ideal I such that $I^{n+1} = 0$ for some $n \geq 0$. We factorize $R \to R/I$ into $R = R/I^{n+1} \to R/I^n \to \cdots \to R/I^2 \to R/I$ and note that $\mathrm{Ker}(R/I^{i+1} \to R/I^i) = I^i/I^{i+1}$ is an ideal of square zero in R/I^{i+1}. This shows that there exists at most one (resp. at least one, resp. a unique) morphism $b\colon T \to X$ commuting with (18.0.1).

(3) The fact that we may glue morphisms and the uniqueness properties shows that if f is formally unramified (resp. formally étale), then there exists at most one (resp. a unique) morphism $b\colon T \to X$ commuting with (18.0.1) if i is an arbitrary thickening of arbitrary schemes.

For formally smooth morphisms a similar property as in (3) does not hold. We will see more precisely in Section (18.3) below that the obstruction to glue local lifts to a global lift is an element in some cohomology group. See also Exercise 18.24.

The proof of the following proposition is easy and mostly formal and is therefore omitted.

Proposition 18.5.

(1) *Every monomorphism of schemes (in particular every immersion) is formally unramified. Every open immersion is formally étale.*

(2) *The properties "formally unramified", "formally smooth", and "formally étale" are stable under composition and stable under base change.*

We will also see (Corollary 18.8 and Proposition 18.12) that all the above properties are local on source and target.

(18.2) Formally unramified morphisms and differentials.

Proposition 18.6. *A morphism of schemes $f\colon X \to S$ is formally unramified if and only if $\Omega^1_{X/S} = 0$.*

Proof. Let $X^{(1)}$ be the first infinitesimal neighborhood of X in $X \times_S X$ such that $\Omega^1_{X/S}$ is the quasi-coherent ideal defining $X \hookrightarrow X^{(1)}$, considered as a quasi-coherent $\mathcal{O}_X$-module. Let $p_1, p_2\colon X^{(1)} \to X$ be the restrictions of the projections $X \times_S X \to X$. Then $p_1{}_{|X} = p_2{}_{|X}$.

Assume that f is formally unramified. We apply Remark 18.4 (2) to $T_0 = X$ and $T = X^{(1)}$ and see that $p_1 = p_2$. By Proposition 17.6, $\Omega^1_{X/S}$ is generated by local sections of the form $p_1^\flat(a) - p_2^\flat(a)$ for local sections a of $\mathcal{O}_X$. Hence $\Omega^1_{X/S} = 0$.

Conversely assume that $\Omega^1_{X/S} = 0$ and hence $X = X^{(1)}$. Let $b, b'\colon T \to X$ be morphisms commuting with (18.0.1), where $T_0 \to T$ is a thickening of order at most 1. They yield a morphism $(b, b')\colon T \to X \times_S X$. Then $b = b'$ if and only if (b, b') factors through the subscheme X of $X \times_S X$. By hypothesis, $(b, b')_{|T_0}$ factors through X. As T_0 is defined in T by an ideal of square zero, (b, b') factors through $X^{(1)} = X$. $\square$

Therefore the exact sequence (17.5.8) shows:

Corollary 18.7. *Let S be a scheme and let $f\colon X \to Y$ be a morphism of S-schemes. Then f is formally unramified if and only if the morphism $v\colon f^*\Omega^1_{Y/S} \to \Omega^1_{X/S}$ (17.5.6) is surjective.*

As Kähler differentials are compatible with passage to open subschemes on source (Remark 17.19 (2)) and target (Proposition 17.30) we also deduce from Proposition 18.6 the following result.

Corollary 18.8. *The property "formally unramified" is local on the source and local on the target (Section (4.9)).*

In other words, if $f\colon X \to Y$ is a morphism of schemes and $(U_i)_i$ and $(V_i)_i$ are open coverings of X and Y, respectively, such that $f(U_i) \subseteq V_i$ for all i, then f is formally unramified if and only if the restriction $U_i \to V_i$ of f is formally unramified for all i.

(18.3) Gluing local lifts.

For thickenings of order at most 1 we have the following tool to glue local liftings b. Suppose given a diagram (18.0.1) where $T_0 \to T$ is a thickening of order at most 1 defined by a quasi-coherent ideal $\mathscr{J}$ of $\mathscr{O}_T$ with $\mathscr{J}^2 = 0$. Let $\mathscr{L}_a$ be the sheaf of lifts of a, i.e., for $U \subseteq T$ open we define $\mathscr{L}_a(U)$ as the set of morphisms $b\colon U \to X$ such that $f \circ b = a_{|U}$ and $b \circ i_{|i^{-1}(U)} = a_0{}_{|i^{-1}(U)}$. As the underlying topological spaces of T and T_0 are equal, we can consider $\mathscr{L}_a$ also as a sheaf on T_0.

Lemma 18.9. *The sheaf of groups $\mathscr{G} := \mathscr{H}om_{\mathscr{O}_{T_0}}(a_0^*\Omega^1_{X/S}, i^*\mathscr{J})$ on T_0 acts simply transitively on the sheaf of sets $\mathscr{L}_a$ on T_0.*

If all schemes are affine, this is essentially Remark 17.4 and we will reduce to it.

Proof. For $U_0 \subseteq T_0$ open we have to define an action of $\mathscr{G}(U_0)$ on $\mathscr{L}_a(U_0)$, compatible with restrictions to smaller open subsets, and to show that this action is simply transitive. It suffices to do this for open subsets U_0 that run through a basis of the topology. As the underlying topological spaces of T and T_0 are the same, we may do so for U_0 that are of the form $i^{-1}(U)$ for an open affine subscheme $U \subseteq T$ such that $a(U)$ is contained in an open affine $W = \operatorname{Spec} R$ of S and that $a_0(U_0)$ is contained in an open affine $V = \operatorname{Spec} A$ of $f^{-1}(W)$. Let $U = \operatorname{Spec} C$ be such an open affine subset. Then $U_0 = i^{-1}(U) = \operatorname{Spec} C/I$ with $I^2 = 0$ and we are now given a commutative diagram of rings

$$
\begin{array}{ccc}
C/I & \longleftarrow & A \\
\uparrow & & \uparrow \\
C & \longleftarrow & R
\end{array}
$$

and we have defined in Remark 17.4 a simply transitive action of

$$\operatorname{Der}_R(A, I) = \operatorname{Hom}_A(\Omega^1_{A/R}, I) = \operatorname{Hom}_{C/I}(\Omega^1_{A/R} \otimes_A C/I, I) = \mathscr{G}(U_0)$$

on $\mathscr{L}_a(U_0)$ which is easily seen to be compatible with passing to principal open subsets of U. $\qquad\square$

Remark 18.10. With the previous notations assume that there exists an open covering $(U_i)_{i \in I}$ of T and for each $i \in I$ a lift $b_i \colon U_i \to X$ (i.e. $b \in \mathcal{L}_a(U_i)$). Then $\mathcal{L}_a$ is a $\mathcal{G}$-torsor and hence defines a class in $H^1(T_0, \mathcal{G})$ (Section (11.5)). This class is trivial if and only if $\mathcal{L}_a(T) \neq \emptyset$, i.e., if and only if there exists a global lift $b \colon T \to X$.

Proposition 18.11. *Suppose there exists an open covering $(U_i)_{i \in I}$ of T and for each $i \in I$ a lift $b_i \colon U_i \to X$. If*

$$(*) \qquad H^1(T_0, \mathcal{H}om_{\mathcal{O}_{T_0}}(a_0^* \Omega^1_{X/S}, i^* \mathcal{J})) = 0,$$

then there exists a global lift $b \colon T \to X$ such that $b_{|U_i} = b_i$ for all i. Moreover, (*) holds if T_0 is affine and $X \to S$ is locally of finite presentation.

Proof. If (*) holds, b exists by Remark 18.10. It remains to show the last assertion. As T_0 is affine, we have $H^1(T_0, \mathcal{F}) = 0$ for any quasi-coherent $\mathcal{O}_{T_0}$-module $\mathcal{F}$ (Theorem 12.35). Hence it suffices to show that $\mathcal{G} := \mathcal{H}om_{\mathcal{O}_{T_0}}(a_0^* \Omega^1_{X/S}, i^* \mathcal{J})$ is quasi-coherent.

But if $X \to S$ is locally of finite presentation, then $\Omega^1_{X/S}$ is of finite presentation (Corollary 17.34) and hence $a_0^* \Omega^1_{X/S}$ is an $\mathcal{O}_{T_0}$-module of finite presentation. Therefore $\mathcal{G}$ is quasi-coherent by Proposition 7.29. $\qquad \square$

Proposition 18.12. *The properties "formally smooth", and "formally étale" are local on the source and local on the target (Section (4.9)).*

We will show the result only for morphisms locally of finite presentation. The general result is sketched in Exercise 18.5 (or can be found in [Sta] 0D0F).

Proof. As we have already seen that "formally unramified" is local on source and target (Corollary 18.8) it suffices to show the claim for "formally smooth".

Let us show that the property is local on the source. Let $f \colon X \to Y$ be a morphism of schemes and let $(U_i)_i$ be an open covering of X. If f is formally smooth, then $f_{|U_i}$ is formally smooth because the inclusion $U_i \to X$ is formally étale and all properties are stable under composition.

Conversely, suppose that for all i the morphism $f_{|U_i}$ is formally smooth. Suppose we are given a thickening square (18.0.1), where T_0 is a closed subscheme defined by an ideal of square 0. In particular T and T_0 have the same underlying topological space. Set $V_{0,i} := a_0^{-1}(U_i)$ and let V_i be the open subscheme of T that has the same underlying topological space as $V_{0,i}$.

Choosing open affine coverings of the V_i we see that Zariski locally on T there exists a lifting of a to a morphism to X. As T is affine, we may apply Proposition 18.11 (by our additional assumption that f is locally of finite presentation) to see that there exists a morphism b that commutes with (18.0.1).

It remains to show that "formally smooth" is local on the target. Let $f \colon X \to Y$ be a morphism of schemes and let $(V_j)_j$ be an open covering of Y. If f is formally smooth, then so is its restriction $f^{-1}(V_j) \to V_j$ since "formally smooth" is stable by base change. Conversely, suppose that $f^{-1}(V_j) \to V_j$ is formally smooth for all j. As open immersions are formally étale, the composition $f^{-1}(V_j) \to V_j \to Y$ is formally smooth for all j. Hence f is formally smooth since we already have seen that the property is local on the source. $\qquad \square$

(18.4) Formally smooth resp. formally étale morphisms and differentials.

As the notions of being formally smooth and formally étale are local on the source and target (Proposition 18.12), it often suffices to consider the affine case. Recall that an R-algebra A is by definition formally smooth (resp. formally étale) if and only if for every ring C, every ideal $I \subset C$ with $I^2 = 0$ and every commutative diagram

(18.4.1)

$$
\begin{array}{ccc}
C/I & \longleftarrow & A \\
\uparrow & & \uparrow \\
C & \longleftarrow & R
\end{array}
$$

there exists a (resp. there exists a unique) homomorphism $A \to C$ commuting with the diagram.

Example 18.13. Let R be a ring. Then every polynomial algebra $A = R[(T_\lambda)_{\lambda \in \Lambda}]$ is formally smooth over R.

Indeed, in this case an R-algebra homomorphism $\bar\varphi \colon A \to C/I$ corresponds via $\bar\varphi \mapsto (\bar\varphi(T_\lambda))_{\lambda \in \Lambda}$ simply to a Λ-tuple $(\bar c_\lambda)_\lambda$ of elements in C/I, and we can lift $\bar\varphi$ to an R-algebra homomorphism $A \to C$ by choosing a lift $c_\lambda \in C$ of every $\bar c_\lambda$.

Remark 18.14. Let R be a ring and let A be an R-algebra of finite presentation. Then $A \cong R[T_1, \ldots, T_n]/(f_1, \ldots, f_r)$ for polynomials $f_i \in R[T_1, \ldots, T_n]$. Consider the diagram (18.4.1). The homomorphism $A \to C/I$ corresponds to elements $\bar c_1, \ldots, \bar c_n \in C/I$ such that $f_i(\bar c_1, \ldots, \bar c_n) = 0$ for all i, i.e., to a solution of the system of equations $f_1 = \cdots = f_r = 0$ in C/I. A homomorphism $A \to C$ commuting with (18.4.1) corresponds to a lift $c_1, \ldots, c_n \in C$ of this solution. Corollary 18.57 below will show that A is a formally smooth R-algebra, i.e., such a lift of solution always exists, if and only if $\operatorname{Spec} A \to \operatorname{Spec} R$ is a smooth morphism. By definition (Definition 6.14), this is for instance the case if the Jacobian matrix

$$
\left(\frac{\partial f_i}{\partial T_j}(x) \right)_{i,j} \in M_{r \times n}(\kappa(x))
$$

has rank r for all $x \in \operatorname{Spec} A$.

The following result relates the notion of extensions (17.2) and the property of being formally smooth.

Proposition 18.15. *Let R be a ring and let A be an R-algebra. Then the following assertions are equivalent.*
(i) *A is a formally smooth R-algebra.*
(ii) *$\operatorname{Ex}_R(A, M) = 0$ for every A-module M.*

Proof. *(i) $\Rightarrow$ (ii).* Let $0 \to M \xrightarrow{j} E \xrightarrow{\pi} A \to 0$ be in $\operatorname{Ex}_R(A, M)$. As A is formally smooth over R, id_A can be lifted to an A-algebra homomorphism $A \to E$. Therefore the extension is trivial.

(ii) $\Rightarrow$ (i). Consider the diagram (18.4.1), where C is an arbitrary ring and $I^2 = 0$. Then $0 \to I \to C \times_{C/I} A \to A \to 0$ is an element of $\operatorname{Ex}_R(A, I)$. Hence (ii) implies that there exists a homomorphism $s \colon A \to C \times_{C/I} A$ of R-algebras. Then the composition of s followed by the projection $C \times_{C/I} A \to C$ commutes with the diagram. $\qquad\square$

For field extensions, formal smoothness has the following description which we will not use in the sequel except in the exercises (see [Mat2] Theorem 26.9 for a proof).

Example 18.16. A field extension $K \supseteq k$ is formally smooth if and only if it is separable (in the sense of Definition B.91).

Once we have shown that a formally smooth morphism locally of finite presentation is smooth (Corollary 18.57), this result can be viewed, at least for finitely generated extensions K, as a variant of generic smoothness (Theorem 6.19 and Remark 6.20), see also Section (18.17) below.

We will often use the following easy remark in the sequel.

Remark 18.17. Let A be a ring (not necessarily commutative) and let

(*) $$ 0 \to M' \to M \to M'' \to 0 $$

be a sequence of left A-modules. Then (*) is exact and split if and only if for every left A-module N the induced sequence

(**) $$ 0 \to \operatorname{Hom}_A(M'', N) \to \operatorname{Hom}_A(M, N) \to \operatorname{Hom}_A(M', N) \to 0 $$

is exact.

Indeed, the map $M \to M''$ is a cokernel of $u \colon M' \to M$ if and only if for every left module N a linear map $v \colon M \to N$ factors through $M \to M''$ if and only if $v \circ u = 0$. Hence (*) is right exact if and only if (**) is left exact for all N. Moreover, it is clear that if (*) is exact and split, so is (**). Conversely, suppose that (**) is exact for all N. To see that $M' \to M$ is has a left inverse r, take $N = M'$ and choose a preimage r of $\operatorname{id}_{M'}$ in $\operatorname{Hom}_A(M, M')$.

Proposition 18.18. *Let $f \colon X \to Y$ be a morphism of S-schemes and consider the sequence (17.5.8)*

(18.4.2) $$ 0 \to f^* \Omega^1_{Y/S} \longrightarrow \Omega^1_{X/S} \longrightarrow \Omega^1_{X/Y} \to 0. $$

(1) *If f is formally smooth, then (18.4.2) is exact and Zariski locally split.*
(2) *If (18.4.2) is exact and Zariski locally split and X is formally smooth over S, then f is formally smooth.*

The proof will show that if S, X, and Y are affine and f is formally smooth, then (18.4.2) is split exact. One can also use Exercise 18.4 which shows that if $X = \operatorname{Spec} B$ and $Y = \operatorname{Spec} A$ are affine and f is formally smooth, then $\Omega^1_{B/A}$ is a projective B-module, which also implies that (18.4.2) is split exact.

Proof. We may assume that $S = \operatorname{Spec} R$, $X = \operatorname{Spec} B$, and $Y = \operatorname{Spec} A$ are affine. For a B-module M consider the exact sequence (17.2.2)

$$ 0 \to \operatorname{Der}_A(B, M) \to \operatorname{Der}_R(B, M) \to \operatorname{Der}_R(A, M) \to \operatorname{Ex}_A(B, M) \to \operatorname{Ex}_R(B, M). $$
$$ \| \qquad\qquad \| \qquad\qquad \| $$
$$ \operatorname{Hom}_B(\Omega^1_{B/A}, M) \quad \operatorname{Hom}_B(\Omega^1_{B/R}, M) \quad \operatorname{Hom}_B(\Omega^1_{A/R} \otimes_A B, M) $$

If f is formally smooth, then $\operatorname{Ex}_A(B, M) = 0$ (Proposition 18.15). Therefore (18.4.2) is exact and split by Remark 18.17. This shows (1).

To show (2) we may assume that (18.4.2) is exact and split. Then $\operatorname{Der}_R(B, M) \to \operatorname{Der}_R(A, M)$ is surjective. If moreover $\operatorname{Ex}_R(B, M) = 0$, then $\operatorname{Ex}_A(B, M) = 0$. This shows (2). $\qquad\square$

Corollary 18.19. *Let S be a scheme and $f\colon X \to Y$ a morphism of S-schemes.*
(1) If f is formally étale, the morphism $v\colon f^*\Omega^1_{Y/S} \to \Omega^1_{X/S}$ (17.5.6) is an isomorphism.
(2) If $v\colon f^*\Omega^1_{Y/S} \to \Omega^1_{X/S}$ is an isomorphism and X is formally smooth over S, then f is formally étale.

Proof. The assertions follow from Proposition 18.18 because f is formally étale if and only if f is formally smooth and $\Omega^1_{X/Y} = 0$ (Proposition 18.6). $\square$

Proposition 18.20. *Let $f\colon X \to S$ and $g\colon Z \to S$ be S-schemes and let $i\colon X \to Z$ be an immersion of S-schemes. Consider the sequence (17.5.10)*

$$(18.4.3) \qquad\qquad 0 \longrightarrow \mathscr{C}_i \longrightarrow i^*\Omega^1_{Z/S} \longrightarrow \Omega^1_{X/S} \longrightarrow 0.$$

(1) If f is formally smooth, then (18.4.3) is exact and Zariski locally split.
(2) If g is formally smooth and (18.4.3) is exact and Zariski locally split, then f is formally smooth.

The proof will show that if f is formally smooth, i is a closed immersion and S and Z are affine, then (18.4.3) is split.

Proof. Let U be an open subscheme of Z such that i factors through a closed immersion $X \to U$. By replacing Z by U we may assume that i is a closed immersion. Moreover we may assume that $S = \operatorname{Spec} R$, $Z = \operatorname{Spec} C$ and $X = \operatorname{Spec} A$ with $A = C/I$ for an ideal $I \subseteq C$. For every A-module M we have $\operatorname{Der}_C(A, M) = 0$. Hence the exact sequence (17.2.2) takes the form

$$0 \to \underset{\substack{\| \\ \operatorname{Hom}_A(\Omega^1_{A/R}, M)}}{\operatorname{Der}_R(A, M)} \to \underset{\substack{\| \\ \operatorname{Hom}_A(\Omega^1_{C/R} \otimes_C A, M)}}{\operatorname{Der}_R(C, M)} \to \underset{\substack{\| \\ \operatorname{Hom}_A(I/I^2, M)}}{\operatorname{Ex}_C(A, M)} \to \operatorname{Ex}_R(A, M) \to \operatorname{Ex}_R(C, M),$$

where the description of $\operatorname{Ex}_C(A, M)$ comes from Proposition 17.13 and where the map $\operatorname{Der}_R(C, M) \to \operatorname{Ex}_C(A, M)$ is induced by the map $I/I^2 \to \Omega^1_{C/R} \otimes_C A$ defining the first arrow of (18.4.3) by Remark 17.37.

If f is formally smooth, then $\operatorname{Ex}_R(A, M) = 0$ by Proposition 18.15. This shows (1) by Remark 18.17.

To show (2), we may assume that (18.4.3) is split exact, in particular $\operatorname{Der}_R(C, M) \to \operatorname{Ex}_C(A, M)$ is surjective. As g is formally smooth, we have $\operatorname{Ex}_R(C, M) = 0$. Therefore $\operatorname{Ex}_R(A, M) = 0$ for all A-modules M. This proves that f is formally smooth (again by Proposition 18.15). $\square$

Corollary 18.21. *Let S be a scheme and let $i\colon X \to Z$ be an immersion of S-schemes with conormal sheaf $\mathscr{C}_i$.*
(1) If X is formally étale over S, then the morphism $d\colon \mathscr{C}_i \to i^*\Omega^1_{Z/S}$ (17.5.9) is an isomorphism.
(2) If $d\colon \mathscr{C}_i \to i^*\Omega^1_{Z/S}$ (17.5.9) is an isomorphism and Z is formally smooth over S, then X is formally étale over S.

Proof. Assertion (1) follows from Proposition 18.20 because $\Omega^1_{X/S} = 0$ if X is formally étale over S (Proposition 18.6).

Let us show (2). If d is an isomorphism, the exact sequence (17.5.10) shows that $\Omega^1_{X/S} = 0$. Hence X is formally unramified over S by Proposition 18.6. Moreover Proposition 18.20 (2) shows that $X \to S$ is also formally smooth. Hence it is formally étale. $\square$

Unramified and étale morphisms

We will call a morphism unramified (resp. étale) if it is formally unramified (resp. formally étale) and satisfies certain finiteness conditions. The main result is a theorem describing the local shape of unramified and étale morphisms (Theorem 18.42). Its proof relies on Zariski's main theorem. This result will then in particular show (Theorem 18.44) that our new definition of étale is equivalent to the previous definition given in Section (6.8).

(18.5) Unramified morphisms.

Definition 18.22. *Let $f\colon X \to S$ be a morphism of schemes.*
(1) f *is called* unramified *if and only if it is formally unramified and locally of finite type.*
(2) *Let $x \in X$. Then f is called* unramified at x *if there exists $x \in U \subseteq X$ open such that $f_{|U}$ is unramified.*

Here we follow Raynaud [Ray2] [^o] and deviate from the definition of [EGAIV] [^o] (17.3.1), where unramified morphisms are defined to be formally unramified and locally of finite presentation, because for many results below (e.g., Proposition 18.29 or Theorem 18.42 and the following remark) this slightly weaker finiteness assumption suffices.

Remark 18.23. The permanence properties of "formally unramified" and "locally of finite type" imply:
(1) Every immersion is unramified.
(2) The property "unramified" is stable under composition, stable under base change, local on the source, and local on the target.

We now characterize unramified morphisms, becoming more and more global. We first consider unramified schemes over a field.

Proposition 18.24. *Let k be a field and let $f\colon X \to \operatorname{Spec} k$ be locally of finite type. Then the following assertions are equivalent.*
(i) f *is unramified.*
(ii) *The k-scheme X is isomorphic to $\coprod_{i \in I} \operatorname{Spec} k_i$ with I some index set and finite separable field extensions k_i of k.*
(iii) *There exists an algebraically closed extension K of k, such that the K-scheme $X \otimes_k K$ is isomorphic to $\coprod_{j \in J} \operatorname{Spec} K$ for some index set J.*
(iv) X *is a geometrically reduced k-scheme and $\dim(X) = 0$.*

Proof. We have "(ii) $\Leftrightarrow$ (iii)" by Proposition B.97 and "(ii) $\Leftrightarrow$ (iv)" by Proposition 5.49. It remains to prove that (iii) holds if and only if $\Omega^1_{X/k} = 0$ (Proposition 18.6). As $\Omega^1_{X/k} \otimes_k K \cong \Omega^1_{X \otimes_k K/K}$ (Proposition 17.30), we may assume that $k = K$ is algebraically closed. The condition is clearly necessary. Conversely, assume that for every closed point $x \in X$ one has $0 = \Omega^1_{X/k}(x) = \mathfrak{m}_x/\mathfrak{m}_x^2$ (Example 17.23). Then for all $x \in X$ closed one has $\mathfrak{m}_x = 0$ by Nakayama's lemma and hence $\mathcal{O}_{X,x} = k$. This shows that $\dim X = 0$ (Theorem 5.22) and implies (iii) by Proposition 5.11. $\qquad\square$

Generalizing the case of a field extension, a k-algebra A is called *separable* if $\operatorname{Spec} A$ is geometrically reduced over k. Using this language, Proposition 18.24 implies:

Remark 18.25. Let k be a field and let A be a k-algebra. Then the following assertions are equivalent.
(i) $\operatorname{Spec} A \to \operatorname{Spec} k$ is unramified.
(ii) A is a finite-dimensional separable k-algebra.
(iii) A is isomorphic to a finite product of finite separable field extensions of k.

Example 18.26. Let k be a field and let $X \to \operatorname{Spec} k$ be a proper geometrically reduced k-scheme. Then $\Gamma(X, \mathscr{O}_X)$ is a finite separable k-algebra.

Indeed, $\Gamma(X, \mathscr{O}_X)$ is a finite k-algebra by Theorem 12.65. For every field extension K of k one has $\Gamma(X_K, \mathscr{O}_{X_K}) = \Gamma(X, \mathscr{O}_X) \otimes_k K$ as the formation of global sections commutes with flat base change (Corollary 12.8). As X_K is reduced, it follows that $\Gamma(X, \mathscr{O}_X)$ is geometrically reduced.

Proposition 18.27. *Let $f \colon X \to S$ be locally of finite type, $x \in X$, $s := f(x)$. Recall that $\Omega^1_{X/S}(x) = \Omega^1_{X/S,x} \otimes_{\mathscr{O}_{X,x}} \kappa(x)$ denotes the fiber of $\Omega^1_{X/S}$ at x. The following assertions are equivalent.*
(i) *f is unramified at x.*
(ii) *$\Omega^1_{X/S}(x) = 0$.*
(iii) *For every local ring C, every ideal $I \subset C$ with $I^2 = 0$ and for every commutative diagram of local ring homomorphisms*

there exists at most one local ring homomorphism $\mathscr{O}_{X,x} \to C$ commuting with the diagram.
(iv) *$\mathscr{O}_{X,x}/\mathfrak{m}_s \mathscr{O}_{X,x}$ is a finite separable field extension of $\kappa(s)$.*

Proof. As $\Omega^1_{X/S}$ is quasi-coherent of finite type (Remark 17.21), we have $\Omega^1_{X/S}(x) = 0$ if and only if the stalk $(\Omega^1_{X/S})_x$ is zero by Nakayama's lemma. As the support of an $\mathscr{O}_X$-module of finite type is closed, this means that we find an open neighborhood $x \in U \subseteq X$ such that $\Omega^1_{X/S|U} = \Omega^1_{U/S} = 0$. Hence (i) and (ii) are equivalent by Proposition 18.6. Clearly, (i) implies (iii).

Next we show that (iii) implies (ii) without assuming that f is locally of finite type. We may assume that $X = \operatorname{Spec} A$ with $A = \mathscr{O}_{X,x}$ and $S = \operatorname{Spec} R$ with $R = \mathscr{O}_{S,s}$ (17.4.3). Let J be the kernel of the multiplication $A \otimes_R A \to A$. Applying (iii) to $C = (A \otimes_R A)/J^2$ and to the image I of J in C we see that the homomorphisms $\chi \colon A \to C$, $\chi(a) = a \otimes 1$ and $\chi' \colon A \to C$, $\chi(a) = 1 \otimes a$ are equal. This implies $\Omega^1_{A/R} = 0$ because $\Omega^1_{A/R}$ is generated by elements of the form $\chi(a) - \chi'(a)$ (Proposition 17.6).

For the equivalence of (iv) and (ii) we may assume that $S = \operatorname{Spec} k$, where k is a field. Hence the equivalence follows from Proposition 18.24. □

In particular we see that x is isolated in its fiber if f is unramified at x. Hence Zariski's main theorem (in the version of Corollary 12.79) shows the following corollary (using the notation of ramification index and of inertia index introduced in Section (12.5)).

Corollary 18.28. *A morphism of finite type* $h\colon X \to S$ *is unramified at* $x \in X$ *if and only if there exists* $x \in U \subseteq X$ *open such that* $h_{|U}$ *is quasi-finite and* $e_{x/s} = f'_{x/s} = 1$, *where* $s := h(x) \in S$.

From these local characterizations it is now easy to deduce the following global characterization of a morphism to be unramified.

Proposition 18.29. *Let* $f\colon X \to S$ *be locally of finite type. Then the following assertions are equivalent.*

(i) f *is unramified.*

(ii) $\Omega^1_{X/S} = 0$.

(iii) *The diagonal* $\Delta_{X/S}\colon X \to X \times_S X$ *is an open immersion.*

(iv) *For an arbitrary nil-immersion of schemes* $i\colon T_0 \to T$ *there exists at most one morphism* $b\colon T \to X$ *commuting with* (18.0.1).

(v) *For all* $s \in S$ *the fiber* X_s *of* f *at* s *is unramified over* $\kappa(s)$.

Therefore to check that a morphism locally of finite type is unramified can be done fiber by fiber, and for the fibers one has the characterizations from Proposition 18.24 to be unramified.

Proof. We have already seen "(i) $\Leftrightarrow$ (ii)", and "(ii) $\Leftrightarrow$ (iii)" holds by Proposition 17.17 because $\Omega^1_{X/S}$ is of finite type (Remark 17.21). The implication "(iv) $\Rightarrow$ (i)" is clear.

We show that (iii) implies (iv). Let $i\colon T_0 \to T$ be a nil-immersion and let $b, b'\colon T \to X$ be two morphisms commuting with (18.0.1). Then $(b, b')_{|T_0}\colon T_0 \to X \times_S X$ factors through $\Delta_{X/S}$. As T_0 and T have the same underlying topological space and as $\Delta_{X/S}$ is an open immersion, this implies that (b, b') factors through $\Delta_{X/S}$. Hence $b = b'$.

Finally, if f is unramified, then each fiber is unramified because "unramified" is stable under base change. This shows that (i) implies (v). The converse follows from the implication "(iv) $\Rightarrow$ (i)" of Proposition 18.27. $\qquad\square$

Remark 18.30. Consider a commutative diagram of schemes

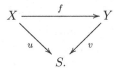

Assume that v is unramified. Then the diagonal $Y \to Y \times_S Y$ is an open immersion by Proposition 18.29.

(1) The cartesian diagram (9.1.4) shows that the graph Γ_f of f is an open immersion because it can be written as base change of the diagonal $\Delta_{Y/S}$. The same cartesian diagram shows that if $g\colon X \to Y$ is a second S-morphism, then the equalizer $\mathrm{Eq}(f, g)$ is an open subscheme of X.

(2) This can be used to show "cancellation" for properties **P** of scheme morphisms (in the sense of Appendix C). Assume that **P** is stable under composition and under base change. Moreover, we assume that every open immersion possesses **P**. Examples are the properties "flat", "(formally) unramified/smooth/étale", "open immersion", or "universally open". Then the same argument as in Remark 9.11 shows that if u possesses **P** and v is unramified, f also possesses **P** (simply write f as the composition of Γ_f followed by the projection $X \times_S Y \to Y$).

(3) Applying this to $u = \mathrm{id}_S$ and to the property "open immersion", we see that every section of an unramified morphism is an open immersion.

Another implication of this remark is the following.

Proposition 18.31. *Let S be a scheme, let $X \to S$ and $Y \to S$ be S-schemes, and let $g, f\colon X \to Y$ be morphisms of S-schemes. Suppose that $Y \to S$ is unramified and separated and that X is connected. If there exists a non-empty S-scheme T and a morphism $h\colon T \to Y$ such that $f \circ h = g \circ h$, then $f = g$.*

One usually applies this proposition to $T = \operatorname{Spec} K$, where K is some field extension of $\kappa(x)$ for some point $x \in X$, and to $T \to \operatorname{Spec} \kappa(x) \to X$ the canonical morphism. The proposition says that if f and g are equal in some K-valued point, then they are equal.

Proof. Since $Y \to S$ is unramified and separated, the diagonal $\Delta_{Y/S}\colon Y \to Y \times_S Y$ is an open and closed immersion. Consider the equalizer subscheme $\operatorname{Eq}(f, g)$ (Definition 9.1). As the immersion $\operatorname{Eq}(f, g) \to X$ is obtained by base change from $\Delta_{Y/S}$ (Proposition 9.3), $\operatorname{Eq}(f, g)$ is an open and closed subscheme of X. By hypothesis, $T \to X$ factors through $\operatorname{Eq}(f, g)$, so $\operatorname{Eq}(f, g)$ is non-empty. Hence $\operatorname{Eq}(f, g) = X$ because X is connected. This means $f = g$. $\square$

The characterization of unramified morphisms in Proposition 18.29 also allows us to prove the following descent property.

Corollary 18.32. *Let $f\colon X \to S$ be locally of finite type and let $g\colon S' \to S$ be a morphism of schemes. Let $x' \in X' := X \times_S S'$ and let $x \in X$ be the image of x'. Then f is unramified at x if and only if its base change $f'\colon X \times_S S' \to S'$ is unramified at x'.*

Proof. The condition is necessary since "unramified" is stable under base change. To see that it is sufficient by Proposition 18.29 it is enough show that if f' is unramified at x', then the fiber $f^{-1}(f(x))$ of f is unramified at x. Hence we can assume that $S = \operatorname{Spec} k$ and $S' = \operatorname{Spec} k'$ for a field extension $k \to k'$. In this case we can apply Proposition 18.24 (iii) by choosing an algebraically closed extension K of k'. $\square$

Remark 18.33. Using Corollary 18.32 it is not difficult to see that the property "unramified" is compatible with cofiltered limits of schemes in the following sense. In the situation of Section (10.13) 1.-3.,5., and 6., we assume that f_0 is locally of finite type (e.g., if X_0 and Y_0 are locally of finite type over S_0). Let $x \in X$ be a point and let x_λ be the image of x in X_λ. Then f is unramified at x if and only if there exists λ such that f_λ is unramified at x_λ.

Indeed, set $y = f(x)$ and $y_\lambda = f_\lambda(x_\lambda)$ be the image of y in Y_λ. Then $f^{-1}(y) = f_\lambda^{-1}(y_\lambda) \otimes_{\kappa(y_\lambda)} \kappa(y)$. By Proposition 18.27, it suffices to show that $f^{-1}(y)$ is unramified at x if and only if $f_\lambda^{-1}(y_\lambda)$ is unramified at x_λ. This follows from Corollary 18.32.

(18.6) Étale morphisms.

Definition 18.34. *A morphism of schemes $f\colon X \to S$ is called étale if f is formally étale and locally of finite presentation.*

Let $x \in X$. Then f is called étale at x if there exists $x \in U \subseteq X$ open such that $f_{|U}$ is étale.

An R-algebra A is called étale, if $\operatorname{Spec} A \to \operatorname{Spec} R$ is étale.

We already defined in Section (6.8) an étale morphism to be a smooth morphism of relative dimension 0. We will see in Theorem 18.44 that both notions are equivalent. Until then we will use the notion étale only in the sense of Definition 18.34.

Remark 18.35. Every open immersion is étale. The property "étale" is stable under composition, stable under base change, local on the source and local on the target. Indeed, all these permanencies hold for the properties "formally étale" and "locally of finite presentation".

In Corollary 18.43 we will see that it is compatible with filtered colimits and in Remark 18.46 we will see that "étale" is stable under faithfully flat descent.

Moreover, Remark 18.30 implies the following cancellation property.

Remark 18.36. Let S be a scheme and let $f\colon X \to Y$ be a morphism of S-schemes. If X is étale over S and Y is unramified over S, then f is étale.

(18.7) Local description of étale morphisms.

Our next goal is to show that locally all étale morphisms have a particularly simple form. We will use the following easy lemma.

Lemma 18.37. *Let A be a ring, $I \subseteq A$ a nil-ideal (i.e., every element of I is nilpotent). Let $u \in A$, u_0 its image in A/I. Then u is a unit in A if and only if u_0 is a unit in A_0.*

Proof. The condition is certainly necessary. Conversely, if $u_0 \in (A/I)^\times$ there exists $b \in A$ such that $ab = 1 + i$ with $i \in I$. But $1 + i$ is invertible with inverse $\sum_{n=0}^\infty (-1)^n i^n$, a finite sum because i is nilpotent. $\square$

Proposition 18.38. *Let R be a ring, $f \in R[T]$ a polynomial and set $A := R[T]/(f)$. Let $g \in R[T]$ such that the image of the formal derivative f' of f in A_g is invertible. Then A_g is an étale R-algebra.*

We will see below that locally on the source every étale morphism has this form with f monic. Hence we coin the following notion.

Definition 18.39. *An R-algebra of the form $(R[T]/(f))_g$, where $f, g \in R[T]$ such that f is monic and such that f' is invertible in $(R[T]/(f))_g$, is called a* standard étale algebra.

Proof. [of Proposition 18.38] Set $A := R[T]/(f)$. Clearly, A_g is an R-algebra of finite presentation. Hence it remains to prove that A_g is a formally étale R-algebra. Let C be an R-algebra, $I \subseteq C$ an ideal with $I^2 = 0$, and let $p\colon C \to C_0 := C/I$ be the canonical homomorphism. Let $u_0\colon A_g \to C_0$ be an R-algebra homomorphism.

Let $x \in A_g$ be the image of T. Let B an R-algebra. Then sending an R-algebra homomorphism $v\colon A_g \to B$ to $b := v(x)$ defines a bijection

(*) $$\operatorname{Hom}_{R\text{-Alg}}(A_g, B) \xrightarrow{\sim} \{\, b \in B \ ; \ f(b) = 0, g(b) \in B^\times \,\}.$$

By hypothesis we have $f'(b) \in B^\times$ for any such b.

Let $c_0 \in C_0$ be the element corresponding to u_0 and $\tilde{c} \in C$ be any element with $p(\tilde{c}) = c_0$. Then $g(\tilde{c}) \in C^\times$ by Lemma 18.37. It remains to show that there exist a unique $i \in I$ such that $f(\tilde{c} + i) = 0$. Then $\tilde{c} + i$ defines the unique lift $u\colon A_g \to C$ of u_0.

We have $f(\tilde{c}) \in I$ and for $i \in I$ we obtain as Taylor expansion

$$f(\tilde{c} + i) = f(\tilde{c}) + i f'(\tilde{c})$$

because $I^2 = 0$. As $f'(c_0) \in C_0^\times$, we have $f'(\tilde{c}) \in C^\times$ (again by Lemma 18.37) and hence there exists a unique $i \in I$ such that $f(\tilde{c} + i) = 0$. $\qquad\qquad\qquad\qquad\qquad\square$

Remark 18.40. It is also easy to see that standard étale algebras are smooth of relative dimension 0. For $f, g \in R[T]$ we have $(R[T]/(f))_g = R[T, U]/(f, gU - 1)$ and the Jacobian matrix is given by

$$\begin{pmatrix} f' & Ug' \\ 0 & g \end{pmatrix}.$$

This matrix is invertible over $(R[T]/(f))_g$ if f' is invertible in $(R[T]/(f))_g$.

To show that every étale morphism is locally given by a standard étale algebra, we start with the easy remark that being "standard étale" is compatible with filtered colimits of rings.

Remark 18.41. Let $(R_i)_{i \in I}$ be a filtered diagram of rings, R its colimit, let $i_0 \in I$, let A_{i_0} be an R_{i_0}-algebra of finite presentation and set $A_i := A_{i_0} \otimes_{R_{i_0}} R_i$ for $i \geq i_0$ and $A := A_{i_0} \otimes_{R_{i_0}} R$. If A is a standard étale R-algebra there exists $i \geq i_0$ such that A_i is a standard étale R_i-algebra.

Indeed, this is a standard limit argument. Let $A = (R[T]/(f))_g$ with f monic and f' invertible in A. Let $i \in I$ such that there exist $f_i, g_i \in R_i[T]$ whose images in $R[T]$ are f and g, respectively. After possibly enlarging i we may assume that f_i is monic and that f_i' is invertible in $(R_i[T]/(f_i))_{g_i}$. For all $j \geq i$ let f_j and g_j be the image of f_i and g_i in $R_j[T]$. We show that there exists a $j \geq i$ such that $A_j \cong (R_j[T]/(f_j))_{g_j}$.

For large $j \geq i$ we find $t_j \in A_j$ whose image in A is the image of T, and that under the map of R_j-algebras $R_j[T] \to A_j$, $T \mapsto t_j$, the polynomial f_j is sent to zero and the polynomial g_j is sent to a unit. We obtain a map φ_j of R_j-algebras of the standard étale R_j-algebra $(R_j[T]/(f_j))_{g_j}$ to A_j which is an isomorphism after base change to R. As A_j is generated by finitely many elements as an R_j-algebra, we may assume that φ_j is surjective, again after possibly enlarging j. As A_j is of finite presentation over R_j, the kernel $\mathrm{Ker}(\varphi_j)$ is a finitely generated ideal (Proposition B.11). Hence again after enlarging j we may assume that φ_j is also injective.

Theorem 18.42. *Let $f \colon X \to S$ be a morphism of schemes locally of finite presentation, let $x \in X$, $s := f(x)$. Then f is étale at x (resp. unramified at x) if and only if there exists an open affine neighborhood $V = \mathrm{Spec}\, R$ of s and an open affine neighborhood $U = \mathrm{Spec}\, A \subseteq f^{-1}(V)$ of x such that A is isomorphic to a standard étale R-algebra (resp. such that A is isomorphic to a quotient of a standard étale R algebra).*

The hypothesis that f is locally of finite presentation is in fact superfluous (in the étale case this holds by assumption and in the unramified case one can first reduce to a noetherian situation, see Exercises 18.12, 18.13).

The essential ingredient in the proof is Zariski's main theorem.

Proof. The condition is sufficient by Proposition 18.38 and because closed immersions are unramified. We prove the converse. We may assume that $X = \mathrm{Spec}\, A$ and $S = \mathrm{Spec}\, R$ are affine, and A is an R-algebra of finite presentation. Any localization by one element in a (quotient of a) standard étale algebra is again a (quotient of a) standard étale algebra. Hence if there exists an R-algebra B and elements $g \in A$ and $h \in B$ such that $g(x) \neq 0$ and $A_g \cong B_h$, we may replace A by B.

(I). We start by assuming that f is unramified at x. Let $\mathfrak{m} \subset R$ be the prime ideal corresponding to s. Writing $R_\mathfrak{m}$ as an inductive limit of R_h for $h \in R \setminus \mathfrak{m}$ we may assume that R is local with maximal ideal $\mathfrak{m}$ by Remark 18.41.

As f is unramified at x, the point x is isolated in its fiber (Proposition 18.24). Hence after passing to a principal open neighborhood of X we may assume that $A = \tilde{A}_g$ for some finite R-algebra $\tilde{A}$ and some $g \in \tilde{A}$ by a suitable version of Zariski's main theorem (Proposition 12.77).

Replacing A by $\tilde{A}$ we may assume that A is a finite R-algebra. Then a prime ideal of A is maximal if and only if it lies over $\mathfrak{m}$ and there are only finitely many maximal ideals in A.

(II). Next we show that we may assume that A is generated by one element. This follows from the fact that finite separable field extensions are generated by one element. Indeed, let $k = \kappa(s)$ be the residue field of R and let $\mathfrak{p} \subset A$ be the maximal ideal corresponding to x. As f is unramified at x, $K := A_\mathfrak{p}/\mathfrak{m}A_\mathfrak{p}$ is a finite separable field extension of k (Proposition 18.27) and hence generated by one element $\bar{a} \in K$ (Proposition B.98). Let $a \in A$ be a lift of $\bar{a}$ and define $C := R[a] \subseteq A$ and $\mathfrak{q} := \mathfrak{p} \cap C$. Then $\mathfrak{p}$ is the unique prime ideal over $\mathfrak{q}$. As A is finite over R, it is finite over C. Therefore $A_\mathfrak{p} = A \otimes_C C_\mathfrak{q}$ is finite over $C_\mathfrak{q}$. By the choice of a, the inclusion $C_\mathfrak{q} \hookrightarrow A_\mathfrak{p}$ is surjective modulo $\mathfrak{m}$. Therefore it is an isomorphism by Nakayama's lemma. As A and C are of finite presentation over R, there exists $g \in C \setminus \mathfrak{q}$ such that $C_g \xrightarrow{\sim} A_g$ (Proposition 10.52). Replacing A by C we may therefore assume that A is finite and generated by $a \in A$.

(III). Let $r := [K : k]$. Then Nakayama's lemma implies that $1, a, \dots, a^{r-1}$ generate the R-module A and hence $h(a) = 0$ for some monic polynomial $h \in R[T]$ of degree r. Hence we obtain a surjection $A' := R[T]/(h) \to A$ which modulo $\mathfrak{m}$ is an isomorphism $k[T]/(\bar{h}) \xrightarrow{\sim} K$. Let $\mathfrak{p}'$ be the inverse image of $\mathfrak{p}$ in A'. As A is unramified over R at $\mathfrak{p}$ and the property of being unramified can be checked on fibers (Proposition 18.27), A' is unramified at $\mathfrak{p}'$ and hence $(\Omega^1_{A'/R})_{\mathfrak{p}'} = 0$. As $\Omega^1_{A'/R} = A'/(h')$, where h' is the formal derivative of h (Example 17.36 (1)), the image of h' in $A'_{\mathfrak{p}'}$ is invertible. Hence there exists $g \in A' \setminus \mathfrak{p}'$ such that A'_g is standard étale. This proves the theorem if f is unramified at $\mathfrak{p}$.

(IV). Now assume that f is étale at $\mathfrak{p}$. We have already seen in Step (III) that we find a standard étale R-algebra B and surjective homomorphism of R-algebras $\varphi \colon B \to A$ which is modulo $\mathfrak{m}$ an isomorphism. Let $\mathfrak{q} := \varphi^{-1}(\mathfrak{p})$ and let $I := \operatorname{Ker}(\varphi)$. We will show that I is zero in a neighborhood of $\mathfrak{q}$. As A is of finite presentation, I is finitely generated (Proposition B.11) and hence it suffices to show that $I_\mathfrak{q} = 0$ or even $I_\mathfrak{q}/I_\mathfrak{q}^2 = (I/I^2)_\mathfrak{q} = 0$ by Nakayama's lemma.

As A is étale over R, there exists a unique R-algebra homomorphism $u \colon A \to B/I^2$ lifting the isomorphism $A \xrightarrow{\sim} B/I = (B/I^2)/(I/I^2)$. Then u is in particular a splitting of the exact sequence of R-modules

$$0 \to I/I^2 \to B/I^2 \xrightarrow{\bar{\varphi}} A \to 0.$$

Hence this exact sequence stays exact after tensoring with k. As $\bar{\varphi} \otimes_R \operatorname{id}_k$ is an isomorphism, this shows $I/I^2 \otimes_R k = 0$. Hence its base change to residue field of $\mathfrak{q}$ is also zero, i.e., $(I/I^2)_\mathfrak{q} \otimes_{B_\mathfrak{q}} \kappa(\mathfrak{q}) = 0$ and therefore $(I/I^2)_\mathfrak{q} = 0$ by Nakayama's lemma. $\qquad\square$

Corollary 18.43. *In the situation of Section* (10.13) *1.-3.,5., and 6., we assume that X_0 and Y_0 are locally of finite presentation over S_0, let $x \in X$ and let x_λ be its projection in X_λ. Then f is étale at x if and only if there exists λ such that f_λ is étale at x_λ.*

Proof. As being étale is stable under base change, the condition is sufficient. For the converse we may assume that $Y_0 = S_0$ because f_0 is locally of finite presentation (Proposition 10.35). We may assume that all schemes are affine and that X is standard étale over $Y = S$ by Theorem 18.42. Then the result is easy (Remark 18.41). $\qquad\square$

(18.8) Characterization of étale morphisms.

We now can characterize étale morphisms (in the sense of Definition 18.34) as smooth morphisms of relative dimension 0 (Definition 6.14, see also Section (18.10) below for further characterizations of smooth morphisms) and also as flat unramified morphisms.

Theorem 18.44. *Let $f\colon X \to S$ be locally of finite presentation. Then the following assertions are equivalent.*
(i) *f is étale.*
(ii) *f is flat and unramified.*
(iii) *f is smooth of relative dimension 0.*

Proof. (iii) $\Rightarrow$ (ii). If f is smooth, then f is flat (Theorem 14.24). Moreover, all fibers are geometrically reduced and of dimension 0 (Corollary 6.32). Hence all fibers of f are unramified (Proposition 18.24) and therefore f is unramified (Proposition 18.29).

(i) $\Rightarrow$ (iii). As the question is local on S and X, we may assume that $X = \operatorname{Spec} A$ and $S = \operatorname{Spec} R$, where A is a standard étale algebra over R. But a standard étale algebra is smooth of relative dimension 0 (Remark 18.40).

(ii) $\Rightarrow$ (i). Consider the usual diagram

$$
\begin{array}{ccc}
T_0 = \operatorname{Spec}(C/I) & \xrightarrow{\ a_0\ } & X \\
{\scriptstyle i}\downarrow & & \downarrow{\scriptstyle f} \\
T = \operatorname{Spec} C & \xrightarrow{\ a\ } & S,
\end{array}
$$

where I is an ideal of square zero. As f is unramified, we have to show that there exists $b\colon T \to X$ commuting with the diagram. This we may do locally on T (as we already know uniqueness of b). Hence after possibly shrinking T, we may assume that X and S are affine. Define affine schemes Z and Z_0 by the following diagram with cartesian squares

$$
\begin{array}{ccc}
Z_0 \longrightarrow Z \longrightarrow X \\
{\scriptstyle f_0'}\downarrow \quad {\scriptstyle f'}\downarrow \quad\ \ \downarrow{\scriptstyle f} \\
T_0 \xrightarrow{\ i\ } T \xrightarrow{\ a\ } S.
\end{array}
$$

Then f_0' has a section, namely $t_0 := (\operatorname{id}_{T_0}, a_0)\colon T_0 \to T_0 \times_S X = Z_0$. As f_0' is separated (as morphism between affine schemes), t_0 is a closed immersion (Example 9.12). As f_0' is unramified (being the base change of f), t_0 is an open immersion (Remark 18.30).

Therefore there exists an open and closed subscheme $W_0 \subseteq Z_0$ such that $f_{0|W_0}'\colon W_0 \to T_0$ is an isomorphism. As Z_0 and Z have the same underlying topological spaces, there exists a unique open and closed subscheme W of Z such that $W \times_Z Z_0 = W_0$.

Then $f'|_W \colon W \to T$ is a surjective closed immersion (because the corresponding homomorphisms of rings is surjective modulo I and hence surjective by Nakayama's lemma). Moreover it is flat and of finite presentation because f has these properties. Therefore $f'|_W \colon W \to T$ is an open immersion (Proposition 14.20) and hence an isomorphism. Therefore f' has a section $t \colon T \overset{\sim}{\to} W \hookrightarrow Z$ and we may take b as the composition of t followed by $Z \to X$. □

Corollary 18.45. *Let $f \colon X \to S$ be a morphism locally of finite presentation. Then f is étale if and only if it is flat and all fibers of f satisfy the equivalent conditions in Proposition 18.24.*

Proof. Combine Theorem 18.44 and Proposition 18.29. □

Remark 18.46. Theorem 18.44 implies easily that "étale" is stable under fpqc descent in the following sense. Let $f \colon X \to S$ be a morphism of schemes and let $g \colon S' \to S$ be a faithfully flat quasi-compact morphism. Then f is étale if and only if its base change $f' \colon X \times_S S' \to S'$ is étale.

Indeed, the condition is clearly necessary as "étale" is stable under base change. Conversely, "étale" is stable under faithfully flat descent as this is true for the properties "unramified" (Corollary 18.32), "flat" (Corollary 14.12) and "locally of finite presentation" (Proposition 14.53).

Example 18.47. Let R be a ring, $f \in R[T]$ a monic polynomial of degree $n \geq 1$. Set $A := R[T]/(f)$. Then A is a free R-module of rank n and in particular flat over R. Hence A is étale over R if and only if $\Omega^1_{A/R} = R[T]/(f, f') = 0$ (Proposition 18.29), i.e., if and only if $(f) + (f') = R[T]$. Moreover $(f) + (f') = R[T]$ if and only if for all maximal ideals $\mathfrak{m} \subset R$ the images of f and of f' in $(R/\mathfrak{m})[T]$ are prime to each other (because it suffices to show that $\Omega^1_{A/R}/\mathfrak{m}\Omega^1_{A/R} = \Omega^1_{(A/\mathfrak{m}A)/(R/\mathfrak{m})} = 0$ for all $\mathfrak{m}$ by Nakayama's lemma).

As a specific example, consider $\mu_{n,R}$, the group scheme of n-th roots of unity (n an integer ≥ 1) over R, i.e., $\mu_{n,R}(B) = \{\, b \in B \ ; \ b^n = 1 \,\}$. The underlying scheme of $\mu_{n,R}$ is $\operatorname{Spec} R[T]/(T^n - 1)$. Then the above discussion shows that $\mu_{n,R}$ is étale over R if and only if n is invertible in R.

In some cases, flatness is easy to see. For instance we get the following examples of étale morphisms using Theorem 14.128.

Corollary 18.48. *Let $f \colon X \to Y$ be a morphism of locally noetherian schemes which is locally of finite type, let $x \in X$, $y := f(x)$. Assume that the following conditions are satisfied.*
(a) $\dim(\mathcal{O}_{Y,y}) = \dim(\mathcal{O}_{X,x})$.
(b) $\mathcal{O}_{Y,y}$ is regular and $\mathcal{O}_{X,x}$ is Cohen-Macaulay.
(c) f is unramified at x.
Then f is étale at x.

It is often useful to express the fact that f is unramified in x by the triviality of ramification and inseparable inertia index $e_{x/y} = f'_{x/y} = 1$ (Corollary 18.28).

Proof. Replacing X by an open neighborhood of x we may assume that f is quasi-finite (Corollary 18.28). Then Conditions (a) and (b) imply that f is flat at x by Theorem 14.128. Therefore f is étale at x by Theorem 18.44. □

Smooth morphisms

We now examine the notion of smooth morphisms. The main results on smooth morphisms are Theorem 18.56 and its noetherian variant Theorem 18.63 which give several characterizations of smooth morphisms.

(18.9) Geometrically regular schemes.

One characterization of smoothness will be via flatness and geometric regularity of the fibers. We already touched upon this topic briefly in Section (6.12). Therefore we start our examination of smooth morphisms by studying geometrically regular schemes. We first recall the definition of "geometrically regular" (see also Exercise 5.20 and Exercise 6.19).

Definition 18.49. *Let k be a field, let X be a k-scheme locally of finite type, and let $x \in X$. Then X is called* geometrically regular *at x over k if for every field extension $K \supseteq k$ the local ring $\mathscr{O}_{X \otimes_k K, \bar{x}}$ is regular for all points $\bar{x} \in X \otimes_k K$ whose projection in X is x.*

The k-scheme X is called geometrically regular *over k if it is geometrically regular in all its points.*

In other words, X is geometrically regular if $X \otimes_k K$ is regular for all field extensions K of k. This definition is a special case of the Definition of a geometrically regular scheme given in Exercise 6.19. The exercise also shows that it suffices to check regularity of $X \otimes_k K$ for finite purely inseparable extensions K, see also Proposition 18.53 below for a similar result.

Remark 18.50. As in the discussion of geometric reducedness/irreducibility/connectedness in Section (5.13), the interesting question is whether a point stays regular after base change.

The converse is always true: Let K be a field extension of k and let $\bar{x}$ be a point of $X \otimes_k K$ lying over x. If $\mathscr{O}_{X \otimes_k K, \bar{x}}$ is regular, then $\mathscr{O}_{X,x}$ is regular. This follows from Proposition 14.59 because $X \otimes_k K \to X$ is faithfully flat.

Similarly, Proposition 14.59 also implies that if $X \to Y$ is a flat morphism of k-schemes locally of finite type, and X is geometrically regular at $x \in X$ over k, then Y is geometrically regular at $f(x)$ over k.

Lemma 18.51. *Let $(A_\lambda)_\lambda$ be a filtered inductive systems of regular rings. Assume that $A := \lim_{\longrightarrow} A_\lambda$ is noetherian. Then A is regular.*

Proof. We have to show that for all $\mathfrak{p} \in \operatorname{Spec} A$ the local ring $A_\mathfrak{p}$ is regular. As filtered colimits commute with localization and localization preserves flatness, we may replace A by $A_\mathfrak{p}$ and A_λ by $(A_\lambda)_{\mathfrak{p}_\lambda}$, where $\mathfrak{p}_\lambda$ is the inverse image of $\mathfrak{p}$ in A_λ. Hence we can assume that A and all A_λ are local and that the transition homomorphisms $\varphi_{\lambda\mu}$ are local.

Let $\mathfrak{m}_\lambda$ (resp. $\mathfrak{m}$) be the maximal ideal of A_λ (resp. of A). Then $\mathfrak{m}$ is the inductive limit of the $\mathfrak{m}_\lambda$. Recall that A is regular if and only if the canonical homomorphism $\sigma_A \colon \operatorname{Sym}_A(\mathfrak{m}/\mathfrak{m}^2) \to \bigoplus_{n \geq 0} \mathfrak{m}^n/\mathfrak{m}^{n+1}$ is an isomorphism (Proposition B.76). As the formation of symmetric algebra commutes with filtered colimits and as filtered colimits preserve exact sequences, the bijectivity of σ_{A_λ} implies the bijectivity of σ_A. □

Proposition 18.52. *Let k be a field, let X be a k-scheme locally of finite type, and let $x \in X$ such that $\mathscr{O}_{X,x}$ is regular. Let K be a separable extension of k. Then $\mathscr{O}_{X \otimes_k K, \bar{x}}$ is regular for all points $\bar{x}$ in $X \otimes_k K$ lying over x.*

Proof. As $\mathscr{O}_{X,x} \to \mathscr{O}_{X_K, \bar{x}}$ is a flat local homomorphism, $\mathscr{O}_{X_K, \bar{x}}$ is regular if and only if $\mathscr{O}_{X_K, \bar{x}}/\mathfrak{m}_x \mathscr{O}_{X_K, \bar{x}} = \kappa(x) \otimes_k K$ is regular (Proposition B.77 (5)). Hence it suffices to show that $L \otimes_k K$ is regular, where L is the finitely generated extension $\kappa(x)$ of k.

First note that $L \otimes_k K$ is noetherian. Indeed, L is the field of fractions of some finitely generated k-algebra B and hence $L \otimes_k K$ is the localization of the ring $B \otimes_k K$ which is noetherian because it is of finite type over K.

Writing K as filtered union of finitely generated subextensions K_λ we may assume that K is finitely generated by Lemma 18.51. If K is finitely generated, then K is a finite separable extension of a purely transcendental extension $k(\mathbf{T}) := k(T_1, \ldots, T_n)$ (Proposition B.97). Set $A := L \otimes_k k(T_1, \ldots, T_n)$. Then A is a localization of the regular ring $L \otimes_k k[T_1, \ldots, T_n] \cong L[T_1, \ldots, T_n]$ and therefore regular. As K is faithfully flat over $k(T_1, \ldots, T_n)$, $p \colon \operatorname{Spec} L \otimes_k K \to S := \operatorname{Spec} A$ is faithfully flat. Moreover, for every $s \in S$ the fiber $p^{-1}(s) = \operatorname{Spec}(\kappa(s) \otimes_{k(\mathbf{T})} K)$ is a product of finite separable field extensions of $\kappa(s)$, in particular $p^{-1}(s)$ is regular. Therefore $L \otimes_k K$ is regular by Proposition 14.59. $\square$

Proposition 18.53. *Let k be a field, let X be a k-scheme locally of finite type, and let $x \in X$. Then X is geometrically regular in x if and only if there exists a perfect field extension K of k and a point $\bar{x}$ of $X \otimes_k K$ lying over x such that $\mathscr{O}_{X \otimes_k K, \bar{x}}$ is regular.*

Proof. By Remark 18.50 it suffices to show that if k is perfect and $\mathscr{O}_{X,x}$ is regular, then X is geometrically regular in x. This follows from Proposition 18.52 because every extension of a perfect field is separable. $\square$

(18.10) Characterization of smooth morphisms.

Recall that a morphism of schemes $f \colon X \to S$ is called *smooth of relative dimension d at* $x \in X$, if there exist affine open neighborhoods U of x and $V = \operatorname{Spec} R$ of $f(x)$ such that $f(U) \subseteq V$, and an open immersion

$$j \colon U \hookrightarrow \operatorname{Spec} R[T_1, \ldots, T_n]/(f_1, \ldots, f_{n-d})$$

of R-schemes for suitable $n \geq d$ and $f_1, \ldots, f_{n-d}$, such that the *Jacobi criterion* holds:

$$(18.10.1) \qquad J_{f_1,\ldots,f_{n-d}}(x) = \left(\frac{\partial f_i}{\partial T_j}(x) \right)_{i,j} \in M_{(n-d) \times n}(\kappa(x)) \text{ has rank } n - d.$$

In fact, using the results of this chapter one can show that one may assume that j is an isomorphism (Exercise 18.21).

The morphism f is called *smooth at x* if it is smooth of some relative dimension d at x. Finally f is called *smooth (of relative dimension d)* if f is smooth (of relative dimension d) at all $x \in X$. If f is smooth of relative dimension d at x, then we set

$$(18.10.2) \qquad \dim_x(f) := d.$$

For a morphism $f \colon X \to S$ locally of finite presentation, the set

$$\{ x \in X \; ; \; f \text{ is smooth of relative dimension } d \text{ at } x \}$$

is an open subset of X. For a smooth morphism $f \colon X \to S$ the map $x \mapsto \dim_x(f)$ is locally constant. Smoothness has the usual permanence properties:

Remark 18.54. Clearly, all open immersions are smooth.

We have already seen in Proposition 6.15 that the property of being smooth is local on the source, local on the target, and stable under base change.

In Proposition 18.59 below we will see that smoothness is stable under composition, under faithfully flat descent, and under passage to cofiltered limits of schemes.

The Jacobi criterion can be expressed via differentials as follows.

Proposition 18.55. *Let R be a ring, $I \subseteq B := R[T_1, \ldots, T_n]$ a finitely generated ideal, $A := B/I$. Let $\mathfrak{p}$ be a prime ideal of A.*

Then the following assertions are equivalent.
(i) *A is smooth over R at $\mathfrak{p}$.*
(ii) *The localization of the sequence (17.5.10)*

$$0 \to I/I^2 \longrightarrow \Omega^1_{B/R} \otimes A \longrightarrow \Omega^1_{A/R} \to 0$$

at $\mathfrak{p}$ is exact and split.
(iii) *The $\kappa(\mathfrak{p})$-linear map*

$$I/I^2 \otimes_A \kappa(\mathfrak{p}) \to \Omega^1_{B/R} \otimes_B \kappa(\mathfrak{p})$$

induced by $I \to \Omega^1_{B/R}$, $b \mapsto d(b)$, is injective.

Proof. Recall that $\Omega^1_{B/R}$ is a free B-module with basis $dT_1, \ldots, dT_n$ (Example 17.8) and that the sequence in (ii) is always right exact (Proposition 17.33). Therefore the equivalence of (ii) and (iii) follows from Proposition 8.10.

Let $\mathfrak{q} \subset B$ be the unique (prime) ideal of B with $\mathfrak{q}/I = \mathfrak{p}$. Let $(\frac{\partial}{\partial T_i})_{1 \leq i \leq n}$ be the dual basis in $\mathrm{Der}_R(B, B) = \mathrm{Hom}_B(\Omega^1_{B/R}, B)$ of $(dT_i)_i$. As I is finitely generated, the smoothness of A at $\mathfrak{p}$ is equivalent to the existence of $f_1, \ldots, f_m \in I$ with $0 \leq m \leq n$ such that their images in $I_\mathfrak{q}$ generate $I_\mathfrak{q}$ and such that (after a possible renumbering of the indeterminates) we have $\det((\frac{\partial}{\partial T_i}, df_j))_{1 \leq i,j \leq m} \notin \mathfrak{q}$. But this is equivalent to the injectivity of the map $I/I^2 \otimes_A \kappa(\mathfrak{q}) = I \otimes_B \kappa(\mathfrak{q}) \to \Omega^1_{B/R} \otimes_B \kappa(\mathfrak{q})$ induced by d (again by Proposition 8.10). As we have $\kappa(\mathfrak{p}) = \kappa(\mathfrak{q})$, this proves the equivalence of (i) and (iii). $\square$

The same proof shows a similar equivalence if B is an arbitrary R-algebra such that $\Omega^1_{B/R}$ is a projective B-module of finite type. Then locally on $\mathrm{Spec}\, B$ one can choose $T_1, \ldots, T_n \in B$ such that $(dT_i)_i$ yields a basis of $\Omega^1_{B/R} \otimes_B \kappa(\mathfrak{p})$. This holds in particular if B is a smooth R-algebra (see Corollary 18.58 below). In fact one can show that a similar result holds if B is only formally smooth over R (see [BouAC10]$^\circ$ X, §7.9, Théorème 3).

We will now state the main theorem on smooth morphisms and deduce some corollaries. We will give further characterizations of smoothness in the situation where all involved schemes are locally noetherian in Theorem 18.63.

Theorem 18.56. *Let $f: X \to S$ be a morphism locally of finite presentation, let $x \in X$, and $s := f(x)$. Then the following assertions are equivalent.*
(i) *f is smooth at x.*
(ii) *There exist an open neighborhood U of x in X and a morphism $g: U \to \mathbb{A}^n_S$ such that g is étale in x and such that $f_{|U}$ is the composition*

$$U \xrightarrow{g} \mathbb{A}^n_S \longrightarrow S,$$

where the second morphism is the canonical morphism.

(iii) f *is flat in* x *and* $\dim_{\kappa(x)} \Omega^1_{X/S}(x) \leq \dim_x f^{-1}(f(x))$.
(iv) f *is flat in* x *and the* $\kappa(s)$*-scheme* $f^{-1}(f(x))$ *is geometrically regular at* x.
(v) *There exists* $x \in U \subseteq X$ *open such that* $f_{|U}$ *is formally smooth*.
Moreover, if f *is smooth of relative dimension* d *at* x, *then* $n = d$ *in (ii) and*

$$(18.10.3) \qquad d = \dim_{\kappa(x)} \Omega^1_{X/S}(x) = \dim_x f^{-1}(f(x)) = \dim_K T_\xi(X/S)$$

for every field K *and every* S*-morphism* $\xi \colon \operatorname{Spec} K \to X$ *with image* x.

Proof. Step I: (i) $\Rightarrow$ (iii)+(iv)+(18.10.3) and (iii) $\Leftrightarrow$ (iv).
We have already seen in Theorem 14.24 that a smooth morphism is flat. Hence we may assume that $S = \operatorname{Spec} k$ is a field.

We next claim that we may assume that $k = \kappa(x)$ by replacing X by $X \otimes_k \kappa(x)$. Indeed being smooth is stable under base change. Moreover, we have

$$\Omega^1_{X/S}(x) \cong T_x(X/k)^\vee = T_x(X \otimes_k \kappa(x)/\kappa(x))$$

by Remark 17.44 and we have $\dim_x(X) = \dim_x(X \otimes_k \kappa(x))$ (apply Proposition 5.38 to all components of X containing x). Finally, X is geometrically regular in x if and only if $X \otimes_k \kappa(x)$ is geometrically regular in x by Remark 18.50. This shows our claim.

But if $\kappa(x) = k$, then $\dim_x(X) = \dim \mathcal{O}_{X,x}$ (Proposition 5.26) and $\dim_k \Omega^1_{X/k}(x) = \dim T_x(X/k)$. Hence (i) implies (iv) by Corollary 6.32, and (iv) and (iii) are equivalent by Theorem 6.28. This also shows (18.10.3), where we also use the compatibility of the relative tangent base change with field extensions (Remark 6.12 (2)).

Step II: (iii)+(iv)+(18.10.3) $\Rightarrow$ (ii) with $n = \dim_{\kappa(x)} \Omega^1_{X/S}(x)$.
We may assume that $S = \operatorname{Spec} R$ and $X = \operatorname{Spec} A$ are affine. Set $n := \dim_{\kappa(x)} \Omega^1_{X/S}(x)$. After possibly localizing A we find $a_1, \ldots, a_n \in A$ such that the images of $da_1, \ldots, da_n \in \Omega^1_{A/R}$ in $\Omega^1_{A/R} \otimes_A \kappa(x)$ are a $\kappa(x)$-basis. Let $\psi \colon B := R[T_1, \ldots, T_n] \to A$ be the R-algebra homomorphism with $\psi(T_i) = a_i$. We will show that

$$g := \operatorname{Spec}(\psi) \colon X \to \mathbb{A}^n_R = \operatorname{Spec}(B)$$

is étale in x. The morphism g is of finite presentation by Proposition 10.35. By Theorem 18.44 it suffices to show that g is unramified and flat at x.

g *is unramified at* x. As f is locally of finite presentation, $\Omega^1_{A/R}$ is of finite presentation by Corollary 17.34. Moreover, we have an exact sequence (17.5.8)

$$\Omega^1_{B/R} \otimes_B A \xrightarrow{u} \Omega^1_{A/R} \longrightarrow \Omega^1_{A/B} \to 0,$$
$$dT_i \otimes 1 \longmapsto da_i.$$

As $da_1, \ldots, da_n$ generate $(\Omega^1_{A/R})_{\mathfrak{p}_x}$ (Nakayama's lemma), the homomorphism $u_{\mathfrak{p}_x}$ is surjective. Hence $(\Omega^1_{A/B})_{\mathfrak{p}_x} = 0$. Therefore g is unramified at x by Proposition 18.27.

g *is flat at* x. As X is flat over S in x, we can apply the fiber criterion for flatness (Theorem 14.25) and hence we may assume that $R = k$ is a field. As flatness can be checked after a faithfully flat base change, the same arguments as in Step I show that we may assume $\kappa(x) = k$. By (18.10.3) we then have $n = \dim(\mathcal{O}_{X,x}) = \dim(\mathcal{O}_{\mathbb{A}^n_k, g(x)})$. Therefore Corollary 18.48 shows that g is étale at x because both local rings are regular. In particular g is flat at x.

Step III: (ii) ⇒ (v).
We may assume that f is the composition of an étale morphism $g \colon X \to \mathbb{A}_S^n$ followed by the canonical morphism $\mathbb{A}_S^n \to S$. As being formally smooth is stable under base change, the implication follows because étale morphisms are formally smooth and $\mathbb{A}_S^n \to S$ is formally smooth by Example 18.13.

Step IV: (v) ⇒ (i).
Again we may assume that $S = \operatorname{Spec} R$ and $X = \operatorname{Spec} A$ are affine. As A is of finite presentation over R, we can write $A = R[T_1, \dots, T_n]/I$ for some finitely generated ideal I. By Proposition 18.55 it suffices to show that the sequence (17.5.10)

$$0 \to I/I^2 \longrightarrow \Omega^1_{R[T_1,\dots,T_n]/R} \otimes A \longrightarrow \Omega^1_{A/R} \to 0$$

is split and exact (as the middle is a free A-module, this implies that I/I^2 and $\Omega^1_{A/R}$ are finitely generated projective modules). This follows from Proposition 18.20. $\square$

We deduce the following global version of Theorem 18.56. The equivalence between (i) and (ii) is called the *infinitesimal lifting criterion* for smoothness.

Corollary 18.57. *Let $f \colon X \to S$ be a morphism of schemes. Then the following assertions are equivalent.*
(i) *f is smooth.*
(ii) *f is locally of finite presentation and formally smooth.*
(iii) *There exists an open covering $(U_i)_i$ of X and for all i a factorization of $f_{|U_i}$ of the form $U_i \to \mathbb{A}_S^d \to S$ with $U_i \to \mathbb{A}_S^d$ étale.*
(iv) *f is flat, locally of finite presentation, and all fibers of f are geometrically regular.*

Proof. All assertions are local on source and target (for (ii) use Proposition 18.12) and hence the equivalences follow from Theorem 18.56. $\square$

The following corollary shows in particular that the tangent bundle $T_{X/S}$ is a vector bundle of rank d over X if X is smooth of relative dimension d over S.

Corollary 18.58. *Let $X \to S$ be smooth of relative dimension d. Then $\Omega^1_{X/S}$ is a locally free $\mathcal{O}_X$-module of rank d.*

Proof. We may assume that there exists a factorization of f of the form $X \xrightarrow{g} \mathbb{A}_S^d \to S$ with g étale. Then $g^* \Omega^1_{\mathbb{A}_S^d/S} \to \Omega^1_{X/S}$ is an isomorphism by Corollary 18.19. As $\Omega^1_{\mathbb{A}_S^d/S}$ is free of rank d (Example 17.8), $\Omega^1_{X/S}$ is a free $\mathcal{O}_X$-module of rank d. $\square$

In particular, for a smooth S-scheme X of relative dimension d the top exterior power $\Omega^d_{X/S} = \det(\Omega^1_{X/S})$ (see Remark 17.58) is a line bundle on X; it is called the *canonical bundle* of X over S. Any divisor with associated line bundle $\Omega^d_{X/S}$ is called a *canonical divisor*. Using Theorem 18.56 the following permanence properties for smoothness are now easy to see.

Proposition 18.59. *The property of being smooth is stable under composition, stable under base change and stable under faithfully flat descent. More precisely:*
(1) *Let $f \colon X \to Y$, $g \colon Y \to Z$ be morphisms of schemes, let $x \in X$ and $y := f(x)$. If f is smooth at x and g is smooth at y, then $g \circ f$ is smooth at x with*

$$(18.10.4) \qquad\qquad \dim_x(g \circ f) = \dim_x(f) + \dim_y(g).$$

In (2) and (3), we consider a cartesian diagram of schemes

$$
\begin{array}{ccc}
X' & \xrightarrow{\;g'\;} & X \\
{\scriptstyle f'}\downarrow & \square & \downarrow{\scriptstyle f} \\
S' & \xrightarrow{\;g\;} & S,
\end{array}
$$

and let $x' \in X'$, $x := g'(x) \in X$.

(2) *If f is smooth of relative dimension d at x, then f' is smooth of relative dimension d at x'.*

(3) *If f' is smooth of relative dimension d at x' and g is flat at $f'(x')$, then f is smooth of relative dimension d at x.*

(4) *In the situation of Section (10.13) 1.-3., 5.,6., assume that X_0 and Y_0 are of finite presentation over S_0. Let $x \in X$ and let $x_\lambda \in X_\lambda$ be its image for all λ. Then $f\colon X \to Y$ is smooth at x if and only if it there exist a λ such that f_λ is smooth at x_λ.*

The stability under base change was also easy to see with our definition of smoothness given in Volume I (Proposition 6.15).

Proof. By Theorem 18.56 we know that a morphism is smooth if and only if it is formally smooth and locally of finite presentation, and these two properties are stable under composition. To show (18.10.4) we may assume, after shrinking X and Y, that f and g are smooth of relative dimension d and e, respectively. By Proposition 18.18 there is a locally split exact sequence $0 \to f^*\Omega^1_{Y/Z} \to \Omega^1_{X/Z} \to \Omega^1_{X/Y} \to 0$. By Corollary 18.58, $\Omega^1_{X/Y}$ (resp. $f^*\Omega^1_{Y/Z}$) is locally free of rank d (resp. e). Therefore $\Omega^1_{X/Z}$ is locally free of rank $d + e$. This proves (18.10.4).

To show (2) we use characterization (iii) of Theorem 18.56. As the formation of $\Omega^1_{X/S}$ is stable under base change $S' \to S$ (Proposition 17.30), we have

$$
\Omega^1_{X/S}(x) \otimes_{\kappa(x)} \kappa(x') \cong (g'^*\Omega^1_{X/S})(x') \cong \Omega^1_{X'/S'}(x')
$$

and in particular $\dim_{\kappa(x)} \Omega^1_{X/S}(x) = \dim_{\kappa(x')} \Omega^1_{X'/S'}(x')$. Moreover, Proposition 5.38 shows that $\dim_x f^{-1}(f(x)) = \dim_{x'} f'^{-1}(f'(x'))$.

Hence (2) follows because flatness is stable under base change. To see (3) we have to show that f is flat at x. To see this we may replace S by $\operatorname{Spec} \mathscr{O}_{S,f(x)}$ and S' by $\operatorname{Spec} \mathscr{O}_{S',s'}$. But then g is faithfully flat (Example B.18) and the flatness of f' in x' implies that f is flat in x because flatness is stable under faithfully flat descent.

Let us show (4). The hypotheses imply that f_0 is of finite presentation by Proposition 10.35 3. Hence their base changes f_λ and f are all of finite presentation. The condition is clearly sufficient since smoothness is stable under base change. Suppose that f is smooth at x. Choose an open quasi-compact neighborhood U of x such that there $f_{|U}$ can be factorized into $U \xrightarrow{g} \mathbb{A}^d_Y \to Y$ with g étale. By Theorem 10.57 there exists a λ and a quasi-compact open neighborhood U_λ of x_λ such that the preimage of U_λ in X is U. After possibly enlarging λ we also find a factorization of $f_{\lambda|U_\lambda}$ into a morphism $g_\lambda\colon U_\lambda \to \mathbb{A}^d_{Y_\lambda}$ followed by $\mathbb{A}^d_{Y_\lambda} \to Y_\lambda$ by Theorem 10.63. Again after possibly enlarging λ we may assume g_λ is étale by Corollary 18.43. Hence f_λ is smooth in x_λ. $\qquad\square$

To show (4) it is also possible to use the fact that morphisms locally of finite presentation are smooth if and only if they are flat with geometrically regular fibers. An argument similar as in the proof of Remark 18.33 then shows that if f is smooth in x then there exists λ such that f_λ has geometrically regular fiber in x_λ and it remains to prove a statement as in (4) for "smooth" replaced by "flat". This can be done, see [EGAIV]^O (11.2.6) and it gives even a slightly more general statement as one has only to assume that X_0 and Y_0 are locally of finite presentation over S_0, see [EGAIV]^O (17.7.8) for details. Together with Remark 18.33 this line of argument also gives a new proof of Corollary 18.43. Here we decided to circumvent the (difficult) theorem [EGAIV]^O (11.2.6). For yet another proof of (4) using the characterization of smoothness given in Proposition 18.55 see [Sta] 0C0C.

Proposition 18.60. *Let S be a scheme and let $f\colon X \to Y$ be a morphism of S-schemes locally of finite presentation over S. Let $x \in X$, $y := f(x)$.*
(1) If X is smooth at x over S and Y is unramified at y over S, then f is smooth at x.
(2) If X is smooth at x over S and f is flat at x, then Y is smooth at y over S.

Proof. To show (1) we may pass from Y and X to open neighborhoods of y and of x, respectively. Thus we may assume that X is smooth over S and that Y is unramified over S. Then Assertion (1) is a special case of Remark 18.30.

Let us show (2). As f is flat at x and X is flat at x over S, Y is flat over S by Corollary 14.27. Hence we may assume $S = \operatorname{Spec} k$ is a field. As X is smooth over k, X is geometrically regular at all $x \in X$, and as f is flat at x, Y is geometrically regular at y (Remark 18.50). Therefore Y is smooth over k. $\square$

We may express the fact that a scheme is smooth in terms of the functor it represents.

Remark 18.61. Let S be a scheme, and let us say that a ring C is an S-algebra if $\operatorname{Spec} C$ is given an S-scheme structure. Let X be an S-scheme, and let $h_X\colon (\mathrm{Sch}/S) \to (\mathrm{Sets})$ be the corresponding functor. As usual we write $h_X(C)$ instead of $h_X(T)$ if $T = \operatorname{Spec} C$ is an S-algebra. Then by definition X is formally unramified / formally smooth / formally étale over S if and only if $h_X(C) \to h_X(C/I)$ is injective / surjective / bijective for every S-algebra C and every nilpotent (or, equivalently, every square zero) ideal $I \subset C$.

Exercise 18.16 shows that if f is locally of finite presentation, then it suffices to consider only local rings C. Below in Theorem 18.63 we will see that if S is in addition locally noetherian, then it even suffices to consider local Artinian rings C.

Moreover, in Theorem 20.12 we will see that if X is smooth (resp. étale) over S, then $h_X(C) \to h_X(C/\mathfrak{m}_C)$ is surjective (resp. bijective) for every local henselian (see Definition 20.1 below) S-algebra C with maximal ideal $\mathfrak{m}_C$ (not necessarily nilpotent). In particular, this holds if C is a complete local ring (Example 20.3).

Finally, the property of being locally of finite presentation can also be expressed in terms of the functor h_X ([EGAIV]^O (8.14.2), see also Exercise 10.22): We call a functor $F\colon (\mathrm{Sch}/S)^{\mathrm{opp}} \to (\mathrm{Sets})$ *locally of finite presentation* if F is a sheaf for the Zariski topology (Section (8.3)) and if for all filtered system $(C_\lambda)_\lambda$ of S-algebras the canonical map

$$\operatorname{colim} F(\operatorname{Spec} C_\lambda) \to F(\operatorname{Spec}(\operatorname{colim} C_\lambda))$$

is bijective. Then an S-scheme X is locally of finite presentation over S if and only if the functor h_X is locally of finite presentation.

(18.11) Characterizations of smooth morphisms in the noetherian case.

If S is locally noetherian, there are further important characterizations of smooth morphisms $X \to S$. We use the following notion.

Definition 18.62. *A surjective ring homomorphism $\pi\colon C \to C_0$ of local Artinian rings is called* small *if $I = \ker(\pi)$ has length 1 as a C-module.*

In other words, $I = (\varepsilon)$ for some $0 \neq \varepsilon \in \mathfrak{m}_C$ with $\varepsilon\mathfrak{m}_C = 0$ (in particular $\varepsilon^2 = 0$).

Theorem 18.63. *Let S be a locally noetherian scheme, and let $f\colon X \to S$ be a morphism locally of finite type. Let $x \in X$ and $s := f(x)$. Then the following assertions are equivalent.*
(i) *f is smooth at x.*
(ii) *For all small surjections $C \to C_0$ of local Artinian rings and for all commutative diagrams of the form*

$$
\begin{array}{ccc}
\operatorname{Spec} C_0 & \xrightarrow{\ a_0\ } & X \\
\downarrow & & \downarrow{\scriptstyle f} \\
\operatorname{Spec} C & \xrightarrow{\ a\ } & S,
\end{array}
$$

such that a_0 has image $\{x\}$ and induces an isomorphism $\kappa(x) \xrightarrow{\sim} C/\mathfrak{m}_C$, there exists $b\colon \operatorname{Spec} C \to X$ commuting with the diagram.
If further $\kappa(s) = \kappa(x)$, then (i) and (ii) are equivalent to
(iii) *The $\widehat{\mathscr{O}}_{S,s}$-algebra $\widehat{\mathscr{O}}_{X,x}$ is isomorphic to the algebra $\widehat{\mathscr{O}}_{S,s}[\![T_1,\dots,T_d]\!]$ of formal power series.*
In this case the integer d in (iii) is the relative dimension of f in x.

In the proof of the theorem we will use the following lemma.

Lemma 18.64. *Let $R \to A$ and $R \to B$ be a local homomorphisms of local rings, and let $\varphi\colon A \to B$ be a local homomorphism of R-algebras. Suppose that φ induces an isomorphism on residue fields and a surjective map*

$$
\bar\varphi\colon \mathfrak{m}_A/(\mathfrak{m}_A^2 + \mathfrak{m}_R A) \to \mathfrak{m}_B/(\mathfrak{m}_B^2 + \mathfrak{m}_R B).
$$

Then the map $\hat\varphi\colon \hat A \to \hat B$ induced by φ on completions is surjective.

Proof. Let $\kappa_R = R/\mathfrak{m}_R$ be the residue field of R and similarly define κ_A and κ_B. The homomorphism φ yields a commutative diagram

$$
\begin{array}{ccccccc}
\mathfrak{m}_R/\mathfrak{m}_R^2 \otimes_{\kappa_R} \kappa_A & \longrightarrow & \mathfrak{m}_A/\mathfrak{m}_A^2 & \longrightarrow & \mathfrak{m}_A/(\mathfrak{m}_A^2 + \mathfrak{m}_R A) & \longrightarrow & 0 \\
\downarrow & & \downarrow & & \downarrow{\scriptstyle \bar\varphi} & & \\
\mathfrak{m}_R/\mathfrak{m}_R^2 \otimes_{\kappa_R} \kappa_B & \longrightarrow & \mathfrak{m}_B/\mathfrak{m}_B^2 & \longrightarrow & \mathfrak{m}_B/(\mathfrak{m}_B^2 + \mathfrak{m}_R B) & \longrightarrow & 0
\end{array}
$$

with exact rows. By hypothesis, the left vertical map is an isomorphism and $\bar\varphi$ is surjective. Hence the vertical map in the middle is surjective. Hence φ induces a surjective map $\operatorname{gr}^{\mathfrak{m}_A}(A) \to \operatorname{gr}^{\mathfrak{m}_B}(B)$ because $\mathfrak{m}_B/\mathfrak{m}_B^2$ generates $\operatorname{gr}^{\mathfrak{m}_B}(B)$ as a κ_B-algebra. Now Proposition B.49 implies that $\hat\varphi$ is surjective. $\qquad\square$

Proof. [of Theorem 18.63] All assertions of Theorem 18.63 are local on S and on X. Hence we can assume that $S = \operatorname{Spec} R$ for a noetherian ring R and $X = \operatorname{Spec} A$ for a finitely generated R-algebra A. If f is smooth at x, then we have already seen that f is formally smooth. Hence (i) implies (ii).

Step I: (iii) ⇒ (i) if $\kappa(x) = \kappa(s)$.
Theorem 18.56 shows that it suffices to show that f is flat at x and that the fiber X_s is smooth at x over $\kappa(s)$. As $\widehat{\mathscr{O}}_{S,s}$ is noetherian, $\widehat{\mathscr{O}}_{S,s}[\![T_1, \ldots, T_d]\!]$ is flat over $\widehat{\mathscr{O}}_{S,s}$ (Example B.47 and Proposition B.41 (1)). Hence (iii) implies that $\widehat{\mathscr{O}}_{X,x}$ is flat over $\widehat{\mathscr{O}}_{S,s}$. As the completion of a local noetherian ring C is faithfully flat over C, this implies that $\mathscr{O}_{X,x}$ is flat over $\mathscr{O}_{S,s}$ which means that f is flat at x.

Hence we may replace X by the fiber X_s and thus S by $\operatorname{Spec} \kappa(s)$. By hypothesis x is then $\kappa(s)$-rational. But in this case we have already shown in Theorem 6.28 that (iii) implies (i).

Step II: (ii) ⇒ (iii) if $\kappa(x) = \kappa(s)$.
Preliminary remark. If C is a local Artinian ring and $I \subseteq \mathfrak{m}_C$ is an ideal, then I is of finite length as a C-module (Proposition B.36). Hence there exists a sequence $0 = I_0 \subseteq I_1 \subseteq \cdots \subseteq I_l = I$ such that I_j/I_{j-1} has length 1. In other words, $C/I_{j-1} \to C/I_j$ is a small surjection of local Artinian rings. Hence we may assume that (ii) is satisfied for an arbitrary ideal $I \neq C$ of the local Artinian ring C.

Construction of $\psi\colon \widehat{\mathscr{O}}_{X,x} \to B := \widehat{\mathscr{O}}_{S,s}[\![T_1, \ldots, T_d]\!]$. Let $(\bar{t}_1, \ldots, \bar{t}_d)$ be a basis of the $\kappa(x)$-vector space $\Omega^1_{f^{-1}(s)/\kappa(s)}(x) = \mathfrak{m}_x/(\mathfrak{m}_x^2 + \mathfrak{m}_s \mathscr{O}_{X,x})$ and let $t_i \in \mathfrak{m}_x$ be a representative of $\bar{t}_i$. We have

$$B/(\mathfrak{m}_B^2 + \mathfrak{m}_s B) \cong \kappa(s) \oplus \kappa(s)T_1 \oplus \cdots \oplus \kappa(s)T_d,$$

$$\mathscr{O}_{X,x}/(\mathfrak{m}_x^2 + \mathfrak{m}_s \mathscr{O}_{X,x}) \cong \kappa(s) \oplus \kappa(s)\bar{t}_1 \oplus \cdots \oplus \kappa(s)\bar{t}_d,$$

where we use $\kappa(s) = \kappa(x)$ for the second isomorphism. Hence there exists a homomorphism of $\mathscr{O}_{S,s}$-algebras $\psi_0\colon \mathscr{O}_{X,x} \to B/(\mathfrak{m}_B^2 + \mathfrak{m}_s B)$ sending t_i to the images of the T_i. By the preliminary remark we can lift ψ_0 first to $\psi_1\colon \mathscr{O}_{X,x} \to B/\mathfrak{m}_B^2$ and then successively to a projective system $(\psi_n\colon \mathscr{O}_{X,x} \to B/\mathfrak{m}_B^{n+1})_{n \geq 1}$. By passing to the limit we obtain a continuous homomorphism $\mathscr{O}_{X,x} \to \varprojlim_n B/\mathfrak{m}_B^{n+1} = B$ and hence a continuous homomorphism

$$\psi\colon \widehat{\mathscr{O}}_{X,x} \to B.$$

Moreover, ψ is surjective by Lemma 18.64.

Construction of an inverse. As ψ is surjective, we find τ_i in the maximal ideal of $\widehat{\mathscr{O}}_{X,x}$ with $\psi(\tau_i) = T_i$ for all i. Let $\varphi\colon B \to \widehat{\mathscr{O}}_{X,x}$ be the unique continuous homomorphism with $\varphi(T_i) = \tau_i$ for all i. Then $(\psi \circ \varphi)(T_i) = T_i$ for all i and hence $\psi \circ \varphi = \operatorname{id}_B$. Hence φ induces an injective map $\bar{\varphi}\colon \mathfrak{m}_B/(\mathfrak{m}_B^2 + \mathfrak{m}_s B) \to \mathfrak{m}_x/(\mathfrak{m}_x^2 + \mathfrak{m}_s \mathscr{O}_{X,x})$ of $\kappa(s)$-vector spaces of the same dimension. Hence $\bar{\varphi}$ is surjective. Now we again apply Lemma 18.64 to obtain that φ is surjective. Hence ψ and φ are mutually inverse isomorphisms.

Step III: (ii) ⇒ (i).
It is enough to show that the proof of "(ii) ⇒ (i)" may be reduced to the case that $\kappa(s) = \kappa(x)$ (then we are done by Steps I and II). By Lemma 18.65 below there exists a faithfully flat local ring homomorphism $\mathscr{O}_{S,s} \to R'$ with R' noetherian such that $R'/\mathfrak{m}_{R'} = \kappa(x)$. Define X' by the cartesian diagram

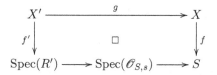

The points $x \in X$ and $s' := \mathfrak{m}_{R'} \in \operatorname{Spec} R'$ are both sent to $s \in S$. Hence $g^{-1}(x) \cap f'^{-1}(s') = \operatorname{Spec}(\kappa(x) \otimes_{\kappa(s)} \kappa(s'))$ (Lemma 4.28), and the multiplication $\kappa(x) \otimes_{\kappa(s)} \kappa(s') \to \kappa(x)$ corresponds to a point $x' \in X'$ with $f'(x') = s'$, $g(x') = x$ and $\kappa(x') = \kappa(s')$. As $\operatorname{Spec} R' \to S$ is flat, it suffices to show that f' is smooth at x' (Proposition 18.59). As $\kappa(x') = \kappa(s') = \kappa(x)$, the morphism f' still satisfies (ii). This gives the desired reduction to the case $\kappa(s) = \kappa(x)$. $\qquad\square$

Lemma 18.65. *Let R be a local ring with residue field k and let k' be a field extension of k. Then there exists a local ring R' with residue field k' and a faithfully flat ring homomorphism $\varphi \colon R \to R'$ such that $\varphi(\mathfrak{m}_R)R' = \mathfrak{m}_{R'}$.*

If $k' \supseteq k$ is finite, one can assume that R' is a finite free R-algebra.

If R is noetherian, one can assume that R' is a complete local noetherian ring.

In the proof of Theorem 18.63 we used the lemma only in the case where k' is a finitely generated extension of k, and we will prove the lemma only in this case. For the general case see [EGAInew] $\mathbf{0}_I$ (6.8.2) and (6.8.3).

Proof. [if k' is a finitely generated field extension of k] By induction on the number of generators of k' over k we may assume that k' is generated by one element over k.

We first assume in addition that k' is a finite extension of k, hence $k' \cong k[T]/(f)$ for some monic polynomial $f \in k[T]$. Choose a monic polynomial $F \in A[T]$ with image f in $k[T]$ and set $A' := A[T]/(F)$. Then A' is a finite free A-module. Let $I := \mathfrak{m}_A A[T] + (F) \subset A[T]$. Then $A[T]/I \cong k[T]/(f) = k'$. Hence $IA' = \mathfrak{m}_A A'$ is a maximal ideal of A' and this is the only prime ideal over $\mathfrak{m}_A$. As A' is a finite A-algebra, the maximal ideals of A' are exactly the prime ideals lying over $\mathfrak{m}_A$ (by the going up theorem, see Theorem B.56). Hence A' is local.

For a general monogeneous extension k' of k it thus suffices to consider the case of $k' = k(T)$ for a transcendental element $T \in k'$. But then we can set $A' := A[T]_{\mathfrak{p}}$, where $\mathfrak{p} \subset A[T]$ is the prime ideal of polynomials with coefficients in $\mathfrak{m}_A$.

In both cases we see that A' is noetherian if A is noetherian. By replacing A' by its completion we may then even assume that A' is complete. $\qquad\square$

As étale morphisms are smooth morphisms of relative dimension 0 we immediately obtain the following corollary.

Corollary 18.66. *Let S be a locally noetherian scheme and let $f \colon X \to S$ be a morphism locally of finite type over S, let $x \in X$, $s := f(x)$ such that $\kappa(s) = \kappa(x)$. Then f is étale at x if and only if the induced homomorphism $\widehat{\mathscr{O}}_{S,s} \to \widehat{\mathscr{O}}_{X,x}$ between the complete local rings is an isomorphism.*

(18.12) Smooth schemes over a field.

Using the results above we obtain the following criteria for smoothness over a field (see also Theorem 6.28). For a further characterization of smoothness over a field see Exercise 18.18.

Proposition 18.67. *Let k be a field, let X be a k-scheme locally of finite type, and let $x \in X$ be a point. Consider the following assertions.*

(i) *$\mathscr{O}_{X,x}$ is regular and $\kappa(x)$ is a separable extension of k.*

(ii) *X is smooth over k at x.*

(iii) *X is geometrically regular at x.*

(iv) *There exists a perfect field extension $K \supseteq k$ and a point $\bar{x} \in X \otimes_k K$ whose projection in X is x such that $\mathscr{O}_{X \otimes_k K, \bar{x}}$ is regular.*

(v) *$\Omega^1_{X/k,x}$ is a free $\mathscr{O}_{X,x}$-module of rank $\dim_x(X)$.*

(vi) *$\mathscr{O}_{X,x}$ is regular.*

Then the implications "(i) $\Rightarrow$ (ii) $\Leftrightarrow$ (iii) $\Leftrightarrow$ (iv) $\Leftrightarrow$ (v) $\Rightarrow$ (vi)" hold. If k is perfect, then all assertions are equivalent.

Proof. We have already seen "(ii) $\Leftrightarrow$ (iii)" (Theorem 18.56), "(v) $\Leftrightarrow$ (ii)" (Theorem 18.56 and Corollary 18.58), and "(iii) $\Leftrightarrow$ (iv)" (Proposition 18.53). Moreover, the implication "(iii) $\Rightarrow$ (vi)" is trivial. If k is perfect, then $\kappa(x)$ is automatically separable over k, hence (vi) implies (i) in this case.

If (i) holds, then Proposition 18.52 shows that $X \otimes_k \kappa(x)$ is still regular at the $\kappa(x)$-rational point $\bar{x}$ lying over x. But then $X \otimes_k \kappa(x)$ is smooth at $\bar{x}$ (Theorem 6.28) and hence X is smooth at x by faithfully flat descent (Proposition 18.59). $\square$

Corollary 18.68. *Let X be a k-scheme locally of finite type over a field k, and let $x \in X(k)$ be a k-valued point, considered as a closed point of X. Then the following assertions are equivalent*

(i) *X is smooth over k in x.*

(ii) *$\mathscr{O}_{X,x}$ is regular.*

(iii) *One has $T_x(X) = \dim_x(X)$.*

Proof. The equivalence of (i) and (ii) follows from Proposition 18.67. As $x \in X(k)$, $T_x(X) = (\mathfrak{m}_x/\mathfrak{m}_x^2)^{\vee}$ and $\dim_x(X) = \dim \mathscr{O}_{X,x}$. This shows the equivalence of (ii) and (iii). $\square$

In Assertion (v) of Proposition 18.67 it is in general important that $\Omega^1_{X/k,x}$ is a free $\mathscr{O}_{X,x}$-module of the correct rank, see Example 18.71 below. If the characteristic of k is 0, then the rank hypothesis is superfluous.

Proposition 18.69. *If $\operatorname{char}(k) = 0$, then a k-scheme X locally of finite type is smooth at $x \in X$ if and only if $\Omega^1_{X/k,x}$ is a free $\mathscr{O}_{X,x}$-module.*

Proof. The condition is necessary by Proposition 18.67. Let $R := \mathscr{O}_{X,x}$. Then $\Omega^1_{X/k,x} = \Omega^1_{R/k}$ (17.4.3). As formation of Kähler differentials commutes with base change and as we can check smoothness after the field extension $k \to \kappa(x)$, we may assume that the residue field of R is k. As k is perfect, it suffices to show that R is regular (Theorem 18.56). This follows from the following lemma. $\square$

Lemma 18.70. *Let k be a field of characteristic 0 and let R be a local noetherian k-algebra with residue field k. Let $n \geq 0$ be an integer. If $\Omega^1_{R/k}$ is a free R-module of rank n, then R is a regular ring of dimension n.*

Proof. Let $\mathfrak{m} \subset R$ be the maximal ideal. By Example 17.23, the map $f \mapsto df$ yields an isomorphism of k-vector spaces

(*)
$$\mathfrak{m}/\mathfrak{m}^2 \xrightarrow{\sim} \Omega^1_{R/k} \otimes_R k,$$

in particular $\dim_k(\mathfrak{m}/\mathfrak{m}^2) = n$. We now prove the lemma by induction on n.

If $n = 0$, then $\mathfrak{m} = 0$ by (*) and by Nakayama's lemma. Hence $R = k$. Now suppose $n > 0$. Then there exists $f \in \mathfrak{m} \setminus \mathfrak{m}^2$. By (*) multiplication by df induces an injective map $k \to \Omega^1_{R/k} \otimes_R k$. Therefore the submodule $\langle df \rangle$ of $\Omega^1_{R/k}$ generated by df is a direct summand (Proposition 8.10). Hence $\Omega^1_{(R/(f))/k} = \Omega^1_{R/k}/\langle df \rangle$ is free of rank $n - 1$. The induction hypothesis therefore shows that $R/(f)$ is regular of dimension $n - 1$. To deduce that R is regular it suffices to show that f is a nonzero divisor (Proposition G.19).

As $\langle df \rangle$ is a direct summand, we find a derivation $D \colon R \to R$ with $D(f) = 1$ and hence $D(f^r) = rf^{r-1}$ for all $r \geq 1$. Let $g \in R$ with $gf = 0$. By Proposition B.43 it suffices to show that $g \in (f^r)$ for all $r \geq 1$. As we have $0 = D(fg) = g + fD(g)$, we find $g \in (f)$. By induction we may assume that $g = hf^{r-1}$ for some $h \in R$. Then

$$0 = D(fg) = D(f^r h) = rf^{r-1}h + f^r D(h) = rg + f^r D(h).$$

Since we are in characteristic 0, this implies that $g \in (f^r)$. $\qquad\square$

The implications "(ii) $\Rightarrow$ (i)" and "(vi) $\Rightarrow$ (v)" of Proposition 18.67 do not hold over non-perfect fields. Neither does Proposition 18.69 hold in positive characteristic:

Example 18.71.
(1) Let k be a field that is not perfect, let p be its characteristic, and let $a \in k$ be an element that is not a p-th power of an element in k. Then $K := k[a^{1/p}] \cong k[T]/(T^p - a)$ is a purely inseparable field extension of degree p.

   Consider $X = \operatorname{Spec} K$. As the derivative of $T^p - a$ is zero, $\Omega^1_{K/k}$ is a 1-dimensional K-vector space (Example 17.36 (1)). Moreover $\mathcal{O}_{X,x} = K$ is regular, if x denotes the unique point of X. This example shows that in Proposition 18.67 Assertion (vi) does not imply Assertion (v) in general.

   Now consider $X = \mathbb{A}^1_k$ and let x be the closed point corresponding to the irreducible polynomial $T^p - a$. Then $\kappa(x) = k[a^{1/p}]$ is a non-separable extension of k. This shows that in Proposition 6.28 Assertion (ii) does not imply Assertion (i) in general.
(2) Let k be an arbitrary field of characteristic $p > 0$ and consider $X = \operatorname{Spec} A$ with $A = k[T]/(T^p)$, $x \in X$ its unique point. Then again $\Omega^1_{X/k}$ is free of rank 1, but $\mathcal{O}_{X,x} = A$ is not reduced and in particular not regular. Hence the freeness of $\Omega^1_{X/k}$ does not imply regularity of $\mathcal{O}_{X,x}$.
(3) Finally, Exercise 18.19 gives an example of a scheme X over a non-perfect field k and a point $x \in X$ such that $\mathcal{O}_{X,x}$ is regular and geometrically reduced, but $\Omega^1_{X/k,x}$ is not a free $\mathcal{O}_{X,x}$-module.

(18.13) Smooth morphisms and differentials.

If X and Y are schemes locally of finite presentation over a scheme S, then every S-morphism $X \to Y$ is locally of finite presentation (Proposition 10.35). Hence the two following results follow from the analogous assertions for formally smooth morphisms (Proposition 18.18 and Proposition 18.20) and the fact that a morphism is smooth if and only if it is locally of finite presentation and formally smooth (Corollary 18.57).

Corollary 18.72. *Let $f \colon X \to Y$ be a morphism of S-schemes locally of finite presentation.*

(1) *Let f be smooth, then the sequence* (17.5.8)

$$(18.13.1) \qquad\qquad 0 \longrightarrow f^*\Omega^1_{Y/S} \longrightarrow \Omega^1_{X/S} \longrightarrow \Omega^1_{X/Y} \longrightarrow 0$$

is exact and locally on X split.
(2) *Conversely, assume that X is smooth over S and that* (18.13.1) *is exact and locally split. Then f is smooth.*

Corollary 18.73. *Let $i\colon X \to Z$ be an immersion of S-schemes locally of finite presentation.*
(1) *If X is smooth over S, then the sequence* (17.5.10)

$$(18.13.2) \qquad\qquad 0 \longrightarrow \mathscr{C}_i \longrightarrow i^*\Omega^1_{Z/S} \longrightarrow \Omega^1_{X/S} \longrightarrow 0$$

is exact and locally split.
(2) *Conversely, assume that Z is smooth over S and that* (18.13.2) *is exact and locally split, then X is smooth over S.*

(18.14) Smooth and étale morphisms between smooth schemes.

Theorem 18.74. *Let $f\colon X \to Y$ be a morphism of S-schemes locally of finite presentation, let $x \in X$ and $y := f(x)$. Then the following assertions are equivalent.*
(i) *Y is smooth over S at y and f is smooth (resp. étale) at x.*
(ii) *X is smooth over S at x and f is smooth (resp. étale) at x.*
(iii) *X is smooth over S at x and the homomorphism*

$$\Omega^1_{Y/S}(y) \otimes_{\kappa(y)} \kappa(x) \to \Omega^1_{X/S}(x),$$

which is induced by $f^*\Omega^1_{Y/S} \to \Omega^1_{X/S}$, is injective (resp. bijective).
If f induces a bijection $\kappa(y) \to \kappa(x)$, then (i), (ii), and (iii) are equivalent to the following assertion.
(iv) *X is smooth over S at x and the map on relative tangent spaces $T_x(X/S) \to T_y(Y/S)$ induced by f is surjective (resp. bijective).*

Proof. Let us consider the smooth case first. As smoothness ist stable under composition, (i) implies (ii). The implication "(ii) $\Rightarrow$ (i)" follows from Proposition 18.60 (2).

Corollary 18.72 (1) shows that (ii) implies (iii). Conversely, if X is smooth at x, then $(\Omega^1_{X/S})_x$ is a free finitely generated $\mathscr{O}_{X,x}$-module (Corollary 18.58). Moreover, the hypothesis means that $u_x\colon f^*(\Omega^1_{Y/S})_x \to (\Omega^1_{X/S})_x$ is injective after passing to fibers in x. Therefore u_x has a left inverse (Proposition 8.10). Thus there exists an open neighborhood U of x such that the exact sequence $0 \to f^*(\Omega^1_{Y/S})_{|U} \to (\Omega^1_{X/S})_{|U} \to (\Omega^1_{X/Y})_{|U} \to 0$ splits. Hence Corollary 18.72 (2) shows that f is smooth at x.

Finally, if $\kappa(y) = \kappa(x)$, then $T_x(X/S) \to T_y(Y/S)$ is the dual of $\Omega^1_{Y/S}(y) \to \Omega^1_{X/S}(x)$. This proves the equivalence of (iii) and (iv) in this case.

The equivalences in the étale case are proved in the same way using Corollary 18.19 instead of Corollary 18.72 and using that a homomorphism of free finitely generated modules over a local ring is an isomorphism if and only if it is an isomorphism modulo the maximal ideal. $\qquad\square$

(18.15) Open immersions and étale morphisms.

Lemma 18.75. *Let $f: X \to S$ be a morphism locally of finite type. Then f is a monomorphism if and only if f is unramified and universally injective.*

Proof. The morphism f is a monomorphism, if and only if $\Delta_f: X \to X \times_S X$ is an isomorphism. This is equivalent to Δ_f being a surjective open immersion. But Δ_f is an open immersion if and only if f is unramified (Proposition 18.29) and Δ_f is surjective if and only if f is universally injective (Exercise 9.9). $\square$

Proposition 18.76. *Let $f: X \to S$ be a morphism of schemes. Then the following assertions are equivalent.*
(i) *f is an open immersion.*
(ii) *f is étale and universally injective.*
(iii) *f is a flat monomorphism locally of finite presentation.*

Proof. Clearly, (i) implies (iii). The equivalence "(iii) $\Leftrightarrow$ (ii)" follows from Lemma 18.75 and the fact that a morphism is étale if and only if it is flat, unramified and locally of finite presentation (Theorem 18.44).

It remains to show that (ii) and (iii) imply (i). The question is local on the target and, as f is injective, local on the source. Hence we may assume that X and S are affine. In particular f is in addition quasi-compact. As flat morphisms locally of finite presentation are open (Theorem 14.35), we may replace S by the open subscheme $f(S)$ and hence may assume that f is also surjective and therefore faithfully flat. The base change of f with f itself is the second projection $X \times_S X \to X$ which is an isomorphism because the diagonal Δ_f is an isomorphism. As the property of being an isomorphism is stable under descent by faithfully flat quasi-compact morphisms, this implies that f is an isomorphism. $\square$

In the proof of "(iii) $\Rightarrow$ (i)" the reduction to the case that f is quasi-compact is superfluous because the property of being an isomorphism is also stable under descent by faithfully flat morphisms locally of finite presentation (Exercise 14.9).

(18.16) Fibre criterion for smooth and étale morphisms.

We deduce for a number of properties that they can be checked on fibers in the presence of flatness.

Corollary 18.77. *Let $f: X \to Y$ be a morphism of S-schemes. For a point $s \in S$ let $f_s: X_s \to Y_s$ be the morphism induced by f on the fiber over s. Let X and Y be locally of finite presentation over S and let X be flat over S. Let $\mathbf{P}$ be one of the following properties.*
(a) *flat,*
(b) *smooth,*
(c) *étale,*
(d) *open immersion,*
(e) *isomorphism.*
Then f has property $\mathbf{P}$ if and only if f_s has property $\mathbf{P}$ for all $s \in S$.

This assertion holds also holds for the property "syntomic" defined below (see Corollary 19.54).

Proof. If X and Y are locally of finite presentation over S, then f is locally of finite presentation (Proposition 10.35). For the property "flat", the assertion is the fiber criterion of flatness (Theorem 14.25).

If f is flat, then we have seen that all other properties hold if and only if they hold for the fibers $f^{-1}(y) \to \mathrm{Spec}(\kappa(y))$ for all $y \in Y$: For "smooth" this is Theorem 18.56. For "étale" this holds because the flat morphism f is étale if and only if it is unramified (Theorem 18.44) and being unramified can be checked on fibers (Proposition 18.29). For "open immersion" this follows from Proposition 18.76 and the fact that a morphism is universally injective if and only if the induced extensions of residue fields are purely inseparable (Proposition 4.35). Finally, a morphism is an isomorphism if and only if it is a surjective open immersion, and "surjective" can of course also be checked on fibers.

As all properties are stable under base change, they hold for f_s for all $s \in S$ if they hold for f. Conversely, if they hold for all f_s, then they hold for all $s \in S$ and $y \in Y_s$ for the fiber $f_s^{-1}(y) = f^{-1}(y)$. Therefore they hold for all fibers of f and hence for f by the remarks above. $\square$

(18.17) Generic Smoothness.

For schemes locally of finite type over a field let us recall (and slightly refine) the following result of generic smoothness already obtained in Volume I.

Proposition 18.78. *Let k be a field and let X be a scheme locally of finite type over k. Then the following assertions are equivalent.*
(i) *There exists an open dense subscheme U_0 of X such that U_0 is smooth over k.*
(ii) *There exists an open dense subscheme U_1 of X such that U is geometrically reduced over k.*
(iii) *There exists an open dense subscheme U_2 of X such that U_2 is reduced and for every maximal point η of X the field extension $k \to \kappa(\eta)$ is separable.*

If k is perfect, then every field extension of k is separable. Hence in this case X has an open dense smooth subscheme if and only if it has an open dense open reduced subscheme.

Proof. Any smooth k-scheme is geometrically regular and in particular geometrically reduced. This shows that (i) implies (ii). The converse holds by Remark 6.20. The equivalence of (ii) and (iii) holds by Proposition 5.49. $\square$

We now generalize this result to morphisms of arbitrary schemes. The following result (see also Exercise 10.40) shows in particular that in characteristic 0 dominant morphisms of finite type between integral noetherian schemes are automatically smooth on an open dense subscheme of the source.

Proposition 18.79. *Let $f\colon X \to Y$ be a dominant morphisms of integral schemes that is locally of finite presentation. Assume that the extension of functions fields $K(Y) \subseteq K(X)$ is separable (e.g., if the field $K(Y)$ is of characteristic 0).*
(1) *Then the smooth locus of f is an open dense subscheme of X.*
(2) *Assume that X is regular and that $K(Y)$ is perfect. Then there exists a dense open subscheme $V \subseteq Y$ such that the restriction $f^{-1}(V) \to V$ of f is smooth.*

Recall that there is a similar result on generic flatness which only assumes that Y is integral (Theorem 10.84).

Proof. Let $\eta \in Y$ be the generic point. For $U = \operatorname{Spec} A \subseteq X$ open affine, $f^{-1}(\eta) \cap U$ is a localization of A. Hence $f^{-1}(\eta)$ is reduced.

(1). As $K(X)$ is a separable extension of $K(Y)$, $f^{-1}(\eta)$ is even geometrically reduced (Proposition 5.49). Hence by Remark 6.20 there exists an open dense subscheme $U \subseteq X$ such that $U \cap f^{-1}(\eta)$ is smooth over $K(Y)$. Therefore by replacing X by U we may assume that $f^{-1}(\eta)$ is smooth.

Now $K(Y) = \mathscr{O}_{Y,\eta}$ is the filtered colimit of $\Gamma(V, \mathscr{O}_Y)$, where V runs through the open affine neighborhoods of η. As smoothness is compatible with colimits, there exists $\eta \in V \subseteq Y$ open affine such that the restriction $f^{-1}(V) \to V$ is smooth.

(2). If X is regular, $f^{-1}(\eta)$ is regular because localizations of regular rings are again regular (Proposition B.77). Therefore $f^{-1}(\eta)$ is a smooth $K(Y)$-scheme because $K(Y)$ is perfect. Now we conclude as in the proof of (1). $\qquad\square$

Exercises

Exercise 18.1. Let A be a ring, $S \subseteq A$ a multiplicative set. Show that $S^{-1}A$ is a formally étale A-algebra.

Exercise 18.2. Let $f : X \to Y$ and $g : Y \to Z$ be morphisms of schemes.
(1) Suppose that g is formally unramified. Show that if $g \circ f$ is formally smooth (resp. formally étale), then so is f.
(2) Let g be formally étale. Show that $g \circ f$ is formally smooth (resp. formally unramified, resp. formally étale) if and only if so is f.

Exercise 18.3. Show that the property "formally unramified" is stable under faithfully flat descent.
Hint: Use that two morphisms are equal if they are equal after a faithfully flat quasi-compact base change.

Exercise 18.4. Let R be a ring and let A be a formally smooth R-algebra. Show that $\Omega^1_{A/R}$ is a projective A-module.

Exercise 18.5. Show that the property "formally smooth" is local on the source and local on the target.
Hint: If A is a ring, $X = \operatorname{Spec} A$, and M is a projective A-module, then show that $H^1(X, \mathscr{H}om_{\mathscr{O}_X}(\tilde{M}, \mathscr{F})) = 0$ for every quasi-coherent $\mathscr{O}_X$-module $\mathscr{F}$. Now use Exercise 18.4 and Remark 7.44.

Exercise 18.6. This exercise shows that formally étale morphisms are not necessarily flat. Let A be a ring and let $\mathfrak{a} \subseteq A$ be an ideal with $\mathfrak{a} = \mathfrak{a}^2$.
(1) Show that the projection $A \to A/\mathfrak{a}$ is formally étale.
(2) Give an example of a pair $(A, \mathfrak{a})$ as above such that $A \to A/\mathfrak{a}$ is not flat.
    Hint: One can choose as A the integral closure of the ring of p-adic integers $\mathbb{Z}_p$ in an algebraic closure of $\mathbb{Q}_p$ and $\mathfrak{a}$ its maximal ideal.

Exercise 18.7. Let $k \to K$ be a field extension.
(1) Let $\operatorname{char}(k) = 0$. Show that K is formally unramified over k if and only if K is an algebraic extension of k.

(2) Let $\operatorname{char}(k) = p$ and let $K^p = \{\, a^p \; ; \; a \in K \,\}$. Show that K if formally unramified over k if and only if $K = k(K^p)$.

Exercise 18.8. Let X be a scheme and let $g \colon Y \to X$ be a morphism of schemes. A *universal thickening of order at most 1 of g* is a thickening $Y \to Y'$ of order at most 1 of X-schemes such that for every thickening $T \to T'$ of order at most 1 of X-schemes and for every morphism $f \colon T \to Y$ of X-schemes there exists a unique morphism $f' \colon T' \to Y'$ of X-schemes such that

$$
\begin{array}{ccc}
T & \xrightarrow{\ f\ } & Y \\
\downarrow & & \downarrow \\
T' & \xrightarrow{\ f'\ } & Y'
\end{array}
$$

commutes.

(1) Define a category of thickenings of order at most 1 over X and show that a universal thickening of order at most 1 of a morphism $Y \to X$ is unique up to unique isomorphism in this category.

(2) Suppose that g is an immersion. Show that the first infinitesimal neighborhood $Y_g^{(1)}$ of Y on X is the universal thickening of order at most 1 of g.

(3) Show that the universal thickening $Y \to Y'$ of order at most 1 of g exists if g is formally unramified.

   Hint: Reduce to the case that $Y = \operatorname{Spec} B$ and $X = \operatorname{Spec} A$ are affine. Let $P = A[(T_i)_i]$ be a polynomial A-algebra such that there exists a surjection of A-algebras $P \to B$. Let I be its kernel. Use that g is formally unramified to see that $d \colon I/I^2 \to \Omega^1_{P/A} \otimes_P B$ has a B-linear section and deduce that there exists an ideal $I^2 \subseteq I' \subseteq I$ such that d induces an isomorphism $I'/I^2 \cong \Omega^1_{P/A} \otimes_P B$. Define $Y' := \operatorname{Spec} P/I'$.

(4) Suppose that g is formally unramified and that $Y \to Y'$ is the universal thickening of order at most 1 for g. Show that $Y' \to X$ is formally unramified.

Let g be formally unramified and let $Y \to Y'$ be the universal thickening of order at most 1 for g. Then the quasi-coherent ideal defining $Y \to Y'$, considered as a quasi-coherent $\mathcal{O}_Y$-module is called the *conormal sheaf of g* and denoted by $\mathscr{C}_g$.

Exercise 18.9. Let $R \to A \to B$ be ring homomorphism. Then the B-module

$$
\Gamma_{B/A/R} := \operatorname{Ker}(\Omega^1_{A/R} \otimes_A B \to \Omega^1_{B/R})
$$

is called the *imperfection module of the A-algebra B over R*.

(1) Let $B \to C$ be a flat homomorphism. Show that there exists a natural exact sequence of C-modules

$$
0 \to \Gamma_{B/A/R} \otimes_B C \to \Gamma_{C/A/R} \to \Gamma_{C/B/R} \to \Omega^1_{B/A} \otimes_B C \to \Omega^1_{C/A} \to \Omega^1_{C/B} \to 0.
$$

(2) Show that if B is a formally smooth A-algebra, then $\Gamma_{B/A/R} = 0$.

(3) Now let $k \to K \to L$ be field extensions. Assume that k is perfect. Show that L is a separable extension of K if and only if $\Gamma_{L/K/k} = 0$.

Exercise 18.10. Let k be a perfect field, let K be an extension of k and let L be a finitely generated extension of K. Show the "Cartier equality"

$$
\dim_L(\Omega^1_{L/K}) = \operatorname{trdeg}_K L + \dim_L \Gamma_{L/K/k}.
$$

Hint: Exercise 18.9.

Exercise 18.11. Let $f: X \to S$ be a morphism locally of finite type. Show that f is unramified if and only if for every S-scheme $S' \to S$ every section of $f_{S'} \times X \times_S S' \to S'$ is an open immersion.

Exercise 18.12. Let $(R_i)_{i \in I}$ be a filtered system of rings and let R be its colimit. Let A be an unramified R-algebra (resp. an unramified R-algebra of finite presentation). Show that there exists $i \in I$ and an unramified R_i-algebra A_i such that A is a quotient of $R \otimes_{R_i} A_i$ (resp. such that $A \cong R \otimes_{R_i} A_i$).

Exercise 18.13. Let $f: X \to S$ be a morphism of schemes, let $x \in X$, $s := f(x)$. Show that f is unramified at x if and only if there exists an open affine neighborhood $V = \operatorname{Spec} R$ of s and an open affine neighborhood $U = \operatorname{Spec} A \subseteq f^{-1}(V)$ of x such that A is isomorphic to a quotient of a standard étale algebra.
Hint: Use Exercise 18.12 to reduce to the noetherian case.

Exercise 18.14. Let S be a scheme, let $\mathscr{L}$ be an invertible $\mathscr{O}_S$-module, let $n \geq 1$ be an integer, and let $u: \mathscr{L}^{\otimes n} \xrightarrow{\sim} \mathscr{O}_S$ be a fixed isomorphism of $\mathscr{O}_S$-modules. Using u we may identify $\mathscr{L}^{\otimes(d+n)} \xrightarrow{\sim} \mathscr{L}^{\otimes d}$ for all $d \geq 0$. Via this identification the canonical morphism $\mathscr{L}^{\otimes d} \otimes \mathscr{L}^{\otimes e} \to \mathscr{L}^{\otimes(d+e)}$ defines the structure of a $\mathbb{Z}/n\mathbb{Z}$-graded $\mathscr{O}_S$-algebra on $\mathscr{A} := \bigoplus_{d=0}^{n-1} \mathscr{L}^{\otimes d}$.

Show that $X := \operatorname{Spec}(\mathscr{A})$ is a finite locally free S-scheme which is étale if and only if n is invertible in $\mathscr{O}_S$.

Exercise 18.15. A morphism $f: X \to S$ of schemes is called *weakly étale* if f and the diagonal Δ_f are flat. Show that a morphism f of schemes is étale if and only if f is weakly étale and locally of finite presentation.
Remark: See also Section (20.4).

Exercise 18.16. Let $f: X \to S$ be a morphism locally of finite presentation. Show that f is unramified (resp. smooth, resp. étale) if and only if for every square (18.0.1), where $T = \operatorname{Spec} C$ is a local ring and where i is a thickening of order at most 1, there exists at most one (resp. there exists a, resp. there exists a unique) morphism $b: T \to X$ commuting with (18.0.1).
Hint: Exercise 10.22.

Exercise 18.17. Let S be a scheme and let $\mathscr{E}$ be a quasi-coherent $\mathscr{O}_S$-module. Then the attached quasi-coherent bundle $\mathbb{V}(\mathscr{E})$ is smooth over S if and only if $\mathscr{E}$ is finite locally free.

Exercise 18.18. Let X be a scheme locally of finite type over a field k and let $x \in X$. Show that X is smooth at x if and only if $\Omega^1_{X/k,x}$ is a free $\mathscr{O}_{X,x}$-module and X is geometrically reduced at x.
Hint: Use Cartier's equality (Exercise 18.10).

Exercise 18.19. Let k be a non-perfect field of characteristic $p > 2$ and let $a \in k \setminus \{\alpha^p \; ; \; \alpha \in k\}$. Let $A := k[T, S]/(T^2 - S^p + a)$ and let $X = \operatorname{Spec} A$. Let $x \in X$ be the point corresponding to the maximal ideal $(T, S^p - a)$.
(1) Show that X is a regular, geometrically reduced, and integral scheme of dimension 1.
(2) Show that the smooth locus of X over k is $X \setminus \{x\}$.
(3) Show that $\Omega^1_{X/k,x} \cong \mathscr{O}_{X,x} \oplus \kappa(x)$.

Exercise 18.20. Let $f: X \to S$ be a morphism locally of finite presentation, $x \in X$, $s := f(x)$. Show that the following assertions are equivalent.

(i) f is smooth at x.

(ii) $\mathscr{O}_{X,x}$ is a formally smooth $\mathscr{O}_{S,s}$-algebra.

If S is locally noetherian, show that (i) and (ii) are equivalent to:

(iii) $\widehat{\mathscr{O}}_{X,x}$ is a formally smooth $\widehat{\mathscr{O}}_{S,s}$-algebra.

Exercise 18.21. Let $R \to A$ be a ring homomorphism such that $\operatorname{Spec} A \to \operatorname{Spec} R$ is smooth. Show that there exists an open covering of $\operatorname{Spec} A$ be principal open subsets $D(g)$ such that $A_g \cong R[T_1, \ldots, T_n]/(f_1, \ldots, f_r)$ where $f_i \in R[T_1, \ldots, T_n]$ such that the determinant of the Jacobian matrix $\det(J_{f_1, \ldots, f_r}) \in R[T_1, \ldots, T_n]$ maps to an invertible element in A_g.

Exercise 18.22. Let S be a scheme, let $\mathscr{E}$ be a finite locally free $\mathscr{O}_S$-module of rank $n \geq 1$, and let $0 \leq e \leq n$ be an integer. Show that the Grassmannian $\operatorname{Grass}^e(\mathscr{E})$ of quotients of $\mathscr{E}$ of rank e is smooth of relative dimension $e(n-e)$ over S. Deduce that the projective bundle $\mathbb{P}(\mathscr{E})$ is smooth of relative dimension $n-1$ over S.

Exercise 18.23. Let $f\colon X \to S$ be a separated unramified morphism of schemes. Show that every section of f is an open and closed immersion and that this defines a bijection between the set of sections of f and the set of open and closed subscheme $Y \subseteq X$ such that $f_{|Y}\colon Y \to S$ is an isomorphism.

Exercise 18.24. Let $f\colon X \to S$ be a morphism locally of finite presentation. Show that f is étale if and only if for every diagram (18.0.1), where i is a nilpotent thickening of order at most 1 of arbitrary (not necessarily affine) schemes, there exists a morphism $T \to X$ commuting with the diagram.

Hint: To show that the condition implies that f is étale, it is sufficient to show that all geometric fibers of f are 0-dimensional. Hence one can assume that $S = \operatorname{Spec} k$ for an algebraically closed field and that X is étale over $\mathbb{A}_k^n$ for some $n \geq 0$. Let Z be a proper integral k-scheme and let $\mathscr{F} \to \mathscr{G}$ be a surjection of coherent $\mathscr{O}_Z$-schemes that is not surjective on global sections. It induces a nilpotent thickening $T_0 := D_Z(\mathscr{G}) \to T := D_Z(\mathscr{F})$ of order at most 1 (Remark 17.26). Use that every morphism $Z \to \mathbb{A}_k^n$ of k-schemes is constant to show that n is necessarily 0.

Remark: The result is due to Björn Poonen, https://mathoverflow.net/questions/ 22015

Exercise 18.25. Let S be a scheme of characteristic $p > 0$, let $f\colon X \to S$ be a morphism of schemes, and let $F_{X/S}\colon X \to X^{(p)}$ be the relative Frobenius.

(1) Show that $F_{X/S}$ is a universal homeomorphism.

(2) Show that the homomorphism $w\colon \Omega^1_{X/S} \to \Omega^1_{X/X^{(p)}}$ (17.5.6) is an isomorphism. Deduce that f is formally unramified if and only if $F_{X/S}$ is formally unramified.

(3) From now on let f be locally of finite presentation. Show that $F_{X/S}$ is locally of finite presentation and quasi-finite.

(4) Show that f is étale if and only if $F_{X/S}$ is an isomorphism.

(5) Show that if f is smooth, then $F_{X/S}$ is flat.

(6) Show that if f is smooth of relative dimension r, then $F_{X/S}$ is finite locally free of degree p^r.

Exercise 18.26. Let k be a field of characteristic $p > 0$ and let X be an integral k-scheme of finite type. Show that the following assertions are equivalent.

(i) The k-scheme X is geometrically reduced.

(ii) The k-scheme $X^{(p)}$ is reduced.

(iii) The ring $K(X) \otimes_k k^{1/p}$ is a field.

(iv) The k-scheme X is generically smooth over k, i.e., there exists a non-empty open subset $U \subseteq X$ such that U is smooth over k.

Hint: Use Proposition G.33.

Exercise 18.27. Let S be a scheme and let

$$
\begin{array}{ccc}
Y' & \xrightarrow{\ g\ } & Y \\
{\scriptstyle i'}\downarrow & & \downarrow{\scriptstyle i} \\
X' & \xrightarrow{\ f\ } & X
\end{array}
$$

be a cartesian diagram of S-schemes, where i is an immersion. Suppose that X, Y, and X' are smooth over S. Show that the following assertions are equivalent.

(i) The scheme Y' is smooth over S.

(ii) The canonical map $g^*(\mathscr{C}_i) \to \mathscr{C}_{i'}$ is bijective.

If these equivalent conditions are satisfied, f is said to be *transversal for Y*. Show the following assertions.

(1) The condition to be transversal is stable under base change and under fpqc descent.

(2) The morphism f is transversal for Y if and only if for every $s \in S$ the morphism of the fibers $f_s \colon X'_s \to X_s$ is transversal for Y_s.

Remark: Combining (2) with the faithfully flat base change to some algebraic closure of $\kappa(s)$ for all s allows to check transversality in the case that the base scheme is the spectrum of an algebraically closed field. See Exercise 18.28.

Exercise 18.28. With the notation of Exercise 18.27 suppose that in addition $S = \operatorname{Spec} k$ for an algebraically closed field k. Show that f is transversal for Y if and only if for every $y' \in Y'(k)$ the sum of the tangent map $T_{y'}(X') \oplus T_{f(y')}(Y) \to T_{f(y')}(X)$ is surjective.

Exercise 18.29. Let $f \colon X \to Y$ be a morphism locally of finite type. Show that f is a monomorphism if and only if for all $y \in Y$ the fiber $f^{-1}(y)$ is empty or $f^{-1}(y) \to \operatorname{Spec} \kappa(y)$ is an isomorphism.

19   Local complete intersections

Content

- The Koszul complex and regular immersions
- Local complete intersection morphisms

If k is a field and $X = \operatorname{Spec} A$ is a closed subscheme of $\mathbb{A}_k^n$, then in general it is not possible to write X as the vanishing scheme of $r = \operatorname{codim}_{\mathbb{A}_k^n}(X)$ polynomials $f_i \in k[T_1, \ldots, T_n]$ (cf. Exercise 19.17). If this is possible at least locally on X, then we call X a local complete intersection over k. By Krull's principal ideal theorem (Corollary B.61) we know that for a local noetherian ring A and for $f \in A$ one has $\dim A/(f) = \dim A - 1$ if f is a regular element. Hence it is not surprising that those subschemes that are locally the vanishing set of a regular sequence (Definition B.58) of polynomials are a local complete intersection. In fact the converse is also true (Proposition 19.50).

More generally, a morphism $f \colon X \to S$ of locally noetherian schemes will be defined a local complete intersection morphism if f is locally on X the composition of an immersion $X \hookrightarrow P$ which is locally defined by a regular sequence (a so-called regular immersion) followed by a smooth morphism $P \to S$ (Definition 19.37). This property is independent of the factorization. But it turns out that the same definition for arbitrary schemes yields a notion which is not known to be independent of the factorization into an immersion followed by a smooth morphism. Hence we replace the notion of a regular sequence by the slightly weaker notion of a completely intersecting sequence (Definition 19.10) and call a morphism of general schemes a local complete intersection if locally on the source it can be factorized as a completely intersecting immersion (an immersion defined locally by a completely intersecting sequence) followed by a smooth morphism. For locally noetherian schemes both notions of local complete intersection morphism are equivalent (Remark 19.22).

The Koszul complex and completely intersecting immersions

We start by defining the notion of a completely intersecting sequence. It is defined via its Koszul complex whose definition we explain first. The Koszul complex will also play an important role later elsewhere, for instance for computing the cohomology of twisted line bundles on projective spectra (Section (22.6)), for the notion of derived completion (Section (24.3)) and in Chapter 25.

(19.1) Koszul complex.

Let A be a ring, let L be an A-module, and let $u \colon L \to A$ be an A-linear map. For $p \geq 0$ let $d_u \colon \bigwedge^p(L) \to \bigwedge^{p-1}(L)$ be the contraction with u, i.e.,

© Springer Fachmedien Wiesbaden GmbH, ein Teil von Springer Nature 2023
U. Görtz und T. Wedhorn, *Algebraic Geometry II: Cohomology of Schemes*,
Springer Studium Mathematik – Master, https://doi.org/10.1007/978-3-658-43031-3_4

$$d_u(e_1 \wedge \cdots \wedge e_p) = \sum_{i=1}^{p} (-1)^{i+1} u(e_i) e_1 \wedge \cdots \wedge \widehat{e_i} \wedge \cdots \wedge e_p$$

for $e_1, \ldots, e_p \in L$, where as usual $\widehat{e_i}$ means that e_i is omitted. Then it is easy to check that

$$\cdots \xrightarrow{d_u} \textstyle\bigwedge^p(L) \xrightarrow{d_u} \bigwedge^{p-1}(L) \xrightarrow{d_u} \cdots \xrightarrow{d_u} \bigwedge^1(L) \xrightarrow{d_u=u} \bigwedge^0(L) \to 0$$

is a complex and that

(19.1.1) $\qquad d_u(a \wedge b) = d_u(a) \wedge b + (-1)^p a d_u(b), \qquad a \in \textstyle\bigwedge^p(L), b \in \bigwedge(L).$

In other words, $(\bigwedge(L), d_u)$ is a graded commutative differential graded algebra with differential of degree -1 (Section (17.8)).

Definition 19.1. *The differential graded algebra* $(\bigwedge(L), d_u)$ *is called the* Koszul complex. *It is denoted by* $K_\bullet(u)$. *If* $\mathbf{f} = (f_1, \ldots, f_n)$ *is a tuple of elements* $f_i \in A$ *and* $u \colon A^n \to A$ *is the associated linear map* $(a_1, \ldots, a_n) \mapsto \sum_i f_i a_i$, *we also write* $K_\bullet(\mathbf{f})$ *instead of* $K_\bullet(u)$.

If M *is an* A-*module, considered as complex of* A-*modules concentrated in degree 0, we define* $K_\bullet(u, M) := K_\bullet(u) \otimes_A M$ *and* $K_\bullet(\mathbf{f}, M) := K_\bullet(\mathbf{f}) \otimes_A M$.

The i-*th homology of the Koszul complex is denoted by* $H_i(u)$ *(or* $H_i(\mathbf{f})$, $H_i(u, M)$, *or* $H_i(\mathbf{f}, M)$, *respectively)*.

Remark 19.2.
(1) For $f \in A$ and an A-module M, considered as the endomorphism $A \to A$ given by multiplication by f, or as the one-element tuple consisting of f, the Koszul complex $K_\bullet(f, M)$ is the chain complex

$$0 \to M \xrightarrow{f} M \to 0$$

concentrated in degrees 1 and 0.

(2) If $\mathbf{f} = (f_1, \ldots, f_n)$ is a sequence of elements of A of length n, then $H_i(\mathbf{f}, M) = 0$ for all $i > n$ because $\bigwedge^i(A^n) = 0$ for $i > n$.

In particular, the Koszul complex $K_\bullet(f)$ is simply the complex

$$0 \longrightarrow A \xrightarrow{f} A \longrightarrow 0.$$

For an arbitrary finite sequence $\mathbf{f} = (f_1, \ldots, f_r)$ of elements of A we will see that $K_\bullet(\mathbf{f}) \cong K_\bullet(f_1) \otimes^g \cdots \otimes^g K_\bullet(f_r)$, where the tensor product is the graded tensor product of differential graded algebras that we define now.

Definition and Remark 19.3. Let R be a (commutative) ring and let $A = \bigoplus_{n \in \mathbb{Z}} A^n$ and $B = \bigoplus_{n \in \mathbb{Z}} B^n$ be graded (not necessarily commutative) R-algebras. Define the *graded tensor product* $A \otimes^g_R B$ as the graded R-module $A \otimes_R B$ with $(A \otimes_R B)^n = \bigoplus_{i+j=n} A^i \otimes_R B^j$ endowed with the unique R-algebra structure such that

$$(a \otimes b)(a' \otimes b') = (-1)^{\deg(b) \deg(a')} a a' \otimes b b'$$

for homogeneous elements a, a' of A and b, b' of B.

It is easy to check that if A and B are both (strictly) graded commutative, then $A \otimes^g_R B$ is (strictly) graded commutative.

Now suppose that (A, d) and (B, d) are differential graded algebras with differential of degree -1. Then $A \otimes_R^g B$ becomes a differential graded algebra with differential of degree -1 by endowing it with the structure of the tensor product of chain complexes (see Section (F.19) for the definition of the tensor product of cochain complexes). In other words, we have for $a \in A$ homogeneous and for $b \in B$

$$d(a \otimes b) = d(a) \otimes b + (-1)^{\deg(a)} a \otimes d(b).$$

If we are just interested in the underlying tensor product of chain complexes, we also write $A \otimes B$ instead of $A \otimes^g B$.

Example 19.4. Let A be a ring and let L_1 and L_2 be A-modules. Then the isomorphism $\bigwedge^n (L_1 \oplus L_2) \cong \bigoplus_{p+q=n} (\bigwedge^p L_1 \otimes \bigwedge^q L_2)$ induces an isomorphism of graded algebras

$$(19.1.2) \qquad \bigwedge (L_1 \oplus L_2) \cong \bigwedge L_1 \otimes_A^g \bigwedge L_2.$$

Remark 19.5. Let A be a ring and let M be an A-module.
(1) The formation of the Koszul complex is functorial: Let L and L' be A-modules, let $u \colon L \to A$ be a linear form and let $w \colon L' \to L$ be an A-linear map. Then w induces a homomorphism of complexes $K_\bullet(u \circ w, M) \to K_\bullet(u, M)$, given by $\bigwedge^i(w) \otimes \mathrm{id}_M$ in degree i.
(2) Let L_1 and L_2 be A-modules, $u_i \colon L_i \to A$ linear forms, and $u := u_1 + u_2 \colon L_1 \oplus L_2 \to A$. Then it is easy to check that the isomorphism (19.1.2) is an isomorphism of differential graded algebras

$$(19.1.3) \qquad K_\bullet(u) \cong K_\bullet(u_1) \otimes_A^g K_\bullet(u_2).$$

(3) Let $\mathbf{f} = (f_1, \dots, f_r)$ with $f_i \in A$. Then (2) implies that

$$K_\bullet(\mathbf{f}) \cong K_\bullet(f_1) \otimes_A^g \cdots \otimes_A^g K_\bullet(f_r).$$

(4) The formation of the Koszul complex is compatible with arbitrary base change in the following sense. Let $A \to A'$ be a ring homomorphism, $u \colon L \to A$ a linear form and $u' := \mathrm{id}_{A'} \otimes u \colon A' \otimes_A L \to A'$. Then the isomorphism $A' \otimes_A \bigwedge L \xrightarrow{\sim} \bigwedge (A' \otimes_A L)$ induces an isomorphism

$$(19.1.4) \qquad A' \otimes_A K_\bullet(u, M) \xrightarrow{\sim} K_\bullet(u', A' \otimes_A M)$$

and hence homomorphisms of A'-modules

$$(19.1.5) \qquad A' \otimes_A H_i(u, M) \to H_i(u', A' \otimes_A M),$$

which are isomorphisms if A' is a flat A-algebra.

Remark 19.6. Let A be a ring and let $u \colon L \to A$ be a map of A-modules. Let $\ell \in L$ with image $f = u(\ell) \in A$. Denote by $h_\ell \colon \bigwedge(L) \to \bigwedge(L)$ the left multiplication by ℓ, i.e., $h_\ell(c) = \ell \wedge c$. Then by (19.1.1) the left multiplication by f satisfies

$$f = h_\ell \circ d_u + d_u \circ h_\ell.$$

In particular, it is homotopic to zero. Then left multiplication by f is also homotopic to zero on $K_\bullet(u, M)$ for every A-module M. As a consequence one sees that the ideal $u(L)$ annihilates the homology groups $H_i(K_\bullet(u, M))$ for all i.

If $u(L) = A$, then we can take $f = 1$ and see that $K_\bullet(u, M)$ homotopic to zero.

Remark 19.7. Sometimes it is useful that the Koszul complex is independent of the base ring in the following sense. Suppose that R is a ring and that A is an R-algebra. Let M be an A-module which we also can view as an R-module by restriction of scalars. Let $g_1, \ldots, g_r \in R$ and denote by $f_1, \ldots, f_r$ their images in A. Then the actions of g_i and of f_i on M are the same. Hence $K_\bullet(\mathbf{g}, M)$ is simply the complex obtained $K_\bullet(\mathbf{f}, M)$ by restriction of scalars. Conversely, one can obtain from the complex of R-modules $K_\bullet(\mathbf{g}, M)$ the complex of A-modules $K_\bullet(\mathbf{f}, M)$ as follows. Every $a \in A$ yields by multiplication an R-linear map on M, and since $K_\bullet(\mathbf{f}, M)$ is functorial in M, we recover the scalar multiplication of A on $K_\bullet(\mathbf{f}, M)$.

(19.2) Regular and completely intersecting sequences.

We continue to denote by A a ring. Let $\mathbf{f} = (f_1, \ldots, f_n) \in A^n$ and let M be an A-module. Recall the following definition (Definition B.60).

Definition 19.8. *The sequence* $\mathbf{f}$ *is called* weakly M-regular *if for all* $1 \le i \le n$ *the multiplication*

$$f_i \colon M/(\sum_{j<i} f_j M) \to M/(\sum_{j<i} f_j M)$$

by f_i *is injective. If in addition* $M/(f_1, \ldots, f_n)M \ne 0$, *then* $\mathbf{f}$ *is called* M-regular. *If* $M = A$ *we call the sequence simply* weakly regular, *resp.* regular.

Remark 19.9. Some texts (e.g., [BouA10]$^\circ$) call $\mathbf{f}$ already M-regular when it is only weakly regular in our sense. Here (as in Volume I) we follow Matsumura [Mat2].

This convention has the following drawback. A sequence consisting of a single element $f \in A$ is weakly A-regular if and only if f is not a zero-divisor. It is A-regular if it is in addition not a unit. Hence an element $f \in A$ is regular (i.e., not a zero divisor, as defined in Section (B.1)) if and only if the sequence consisting of f is weakly A-regular.

Very often $f_1, \ldots, f_n$ will be in the radical of A and M will be a non-zero finitely generated A-module. Then $M/(f_1, \ldots, f_n)M \ne 0$ is satisfied automatically because of Nakayama's Lemma.

The notion of a regular sequence depends on its order (see Exercise 19.1 for a ring A and a sequence (x, y) of elements in A such that (x, y) is not regular but such that (y, x) is regular).

Another closely related notion is that of a completely intersecting sequence which works better in a non-noetherian setting. Recall that by definition for every linear form $u \colon L \to A$ and for every A-module M we have $H_0(K_\bullet(u, M)) = M/u(L)M$. In particular for a finite sequence $\mathbf{f} = (f_1, \ldots, f_r)$ of elements in A we find that

$$H_0(K_\bullet(\mathbf{f}, M)) = M/(f_1, \ldots, f_n)M.$$

Definition 19.10. *The sequence* $\mathbf{f}$ *is called* M-completely intersecting *if* $K_\bullet(\mathbf{f}, M) \to M/(f_1, \ldots, f_n)M$ *is a quasi-isomorphism (i.e., if* $H_q K_\bullet(\mathbf{f}, M) = 0$ *for all* $q > 0$). *If* $M = A$ *we call the sequence simply* completely intersecting.

With this terminology we follow [BouA10]$^\circ$ Chap. X. In [Sta] completely intersecting sequences are called Koszul regular.

Remark 19.11.
(1) A sequence consisting of a single element $f \in A$ is weakly M-regular if and only if it is M-completely intersecting (if and only if multiplication $M \to M$ by f is injective).
(2) If a sequence $(f_1, \ldots, f_r)$ is completely intersecting, then every permutation of it is also completely intersecting (Remark 19.5 (3)).

To explain the relation between these notions for sequences consisting of more than one element we introduce some notation.

Let A be a ring, M an A-module, let $\mathbf{f} = (f_1, \ldots, f_n) \in A^n$ be a sequence, and let $I = (f_1, \ldots, f_n)$ be the ideal generated by $\mathbf{f}$. Consider the graded A-module $\bigoplus_{d \geq 0} I^d M$ and the homogeneous ideal $\mathfrak{a}$ of $A[T_1, \ldots, T_n]$ generated by $(f_i T_j - f_j T_i)$ for $1 \leq i < j \leq n$. Then the homomorphism of graded A-modules

$$\alpha_M^{\mathbf{f}} \colon (A[T_1, \ldots, T_n]/\mathfrak{a}) \otimes_A M \to \bigoplus_{d \geq 0} I^d M, \qquad P \otimes m \mapsto P(f_1, \ldots, f_n)m,$$

where $P \in A[T_1, \ldots, T_n]$ is a homogeneous polynomial of degree d (which implies that $P(f_1, \ldots, f_n) \in I^d$) and $m \in M$, is well defined and surjective.

The following two results now give the relation between various properties of such sequences.

Theorem 19.12. *Consider the following assertions.*
(i) *The sequence $\mathbf{f}$ is weakly M-regular.*
(ii) *The sequence $\mathbf{f}$ is M-completely intersecting.*
(iii) $H_1(\mathbf{f}, M) = 0.$
(iv) *The homomorphism $\alpha_M^{\mathbf{f}} \colon (A[T_1, \ldots, T_n]/\mathfrak{a}) \otimes_A M \to \bigoplus_{d \geq 0} I^d M$ is an isomorphism.*
(v) *The homomorphism $\alpha_M^{\mathbf{f}} \otimes \mathrm{id}_{A/I} \colon (A/I)[T_1, \ldots, T_n] \otimes_A M \to \bigoplus_{d \geq 0}(I^d M/I^{d+1} M)$ is an isomorphism.*
Then one has the implications "(i) $\Rightarrow$ (ii) $\Rightarrow$ (iii) $\Leftrightarrow$ (iv) $\Rightarrow$ (v)". If for all $1 \leq i \leq n$ the A-module $M/(f_1 M + \ldots f_{i-1} M)$ is separated for the I-adic topology, then all assertions are equivalent.

For the proof we refer to [BouA10] ⁰ Chap. X, §9, 7–9.

Corollary 19.13. *If A is noetherian, $M \neq 0$ is a finitely generated A-module, and if $f_1, \ldots, f_n$ are in the Jacobson radical of A, then the assertions (i) to (v) of Theorem 19.12 are each equivalent to the assertion that $\mathbf{f}$ is a M-regular sequence.*

This result particularly applies if A is a local noetherian ring and none of the f_i is a unit.

Proof. By Proposition B.42 every finitely generated A-module is separated for the $\mathfrak{a}$-adic topology if $\mathfrak{a}$ is an ideal contained in the Jacobson radical of the noetherian ring A. Therefore the corollary follows from Theorem 19.12 and Remark 19.9. □

Lemma 19.14. *Let $\varphi \colon A \to B$ be a flat local ring homomorphism between noetherian local rings (resp. between local rings). Let $f_1, \ldots, f_r$ be elements of the maximal ideal of A. Then $(f_1, \ldots, f_r)$ is a regular sequence (resp. a completely intersecting sequence) in A if and only if $(\varphi(f_1), \ldots, \varphi(f_r))$ is a regular sequence (resp. a completely intersecting sequence) in B.*

Proof. The assumption implies that φ is faithfully flat. By Corollary 19.13, it suffices to consider the case of completely intersecting sequences. Since the Koszul complex is compatible with base change (Remark 19.5), the lemma follows. □

Let A be a ring and let $I \subseteq A$ be an ideal. Let

$$\sigma \colon \mathrm{Sym}_A(I) \to \bigoplus_{d \geq 0} I^d,$$

$$\tau \colon \mathrm{Sym}_A(I/I^2) \to \bigoplus_{d \geq 0} I^d/I^{d+1}$$

be the homomorphisms induced by the inclusions $I \to \bigoplus_{d \geq 0} I^d$ and $I/I^2 \to \bigoplus_{d \geq 0} I^d/I^{d+1}$.

Corollary 19.15. *Let* $\mathbf{f} = (f_1, \ldots, f_r)$ *be a generating sequence of* I *such that* $H_1(\mathbf{f}, A) = 0$ *(e.g., if* $\mathbf{f}$ *is completely intersecting). Then the images of* $f_1, \ldots, f_r$ *in* I/I^2 *are an* A/I-*basis of the* A/I-*module* I/I^2. *Moreover* σ *and* τ *are isomorphisms. In particular there exists an isomorphism of graded* (A/I)-*algebras*

$$\bigoplus_{d \geq 0} I^d/I^{d+1} \cong (A/I)[T_1, \ldots, T_r].$$

Proof. Let $(e_i)_i$ be the standard A-basis of A^n. As $H_1(\mathbf{f}, A) = 0$, the kernel of the surjective homomorphism $u \colon A^n \to I$, $\sum a_i e_i \mapsto \sum a_i f_i$ is generated by the elements $(f_j e_i - f_i e_j)$ for $1 \leq i < j \leq n$. Hence $A[T_1, \ldots, T_n]/\mathfrak{a} \cong \mathrm{Sym}_A(I)$, where $\mathfrak{a}$ is the ideal generated by the elements $f_j T_i - f_i T_j$. Therefore Theorem 19.12 shows that τ (and hence σ) is an isomorphism. □

See Section (19.5) for an interpretation of this corollary in terms of blow-ups.

Example 19.16. Let R be a ring, $A = R[T_1, \ldots, T_n]$ and $s_1, \ldots, s_r \in R$ $(r \leq n)$. Then $(T_1 - s_1, \ldots, T_r - s_r)$ is a regular sequence in A. In particular, it is completely intersecting.

Much more generally, if $f_1, \ldots, f_r \in R[T_1, \ldots, T_n]$ are elements $(r \leq n)$ such that $\mathrm{rk}(\frac{\partial f_i}{\partial T_j}(x)) = r$ for all $x \in \mathbb{A}^n_R$ with $f_1(x) = \cdots = f_r(x) = 0$, then $(f_1, \ldots, f_r)$ is a regular sequence. This will follow from Theorem 19.30 below.

Proposition 19.17. *Let* A *be a ring and let* M *be an* A-*module. Let* $\mathbf{f} = (f_1, \ldots, f_r)$ *be a finite sequence of elements of* A *and let* $n_1, \ldots, n_r \geq 1$ *be integers. Then* $(f_1, \ldots, f_r)$ *is* M-*completely intersecting if and only if* $(f_1^{n_1}, \ldots, f_r^{n_r})$ *is* M-*completely intersecting.*

For the proof we refer to [BouA10] ° Chap. X, §9.6, Prop. 6.

Proposition 19.18. *Let* A *be a ring and let* $I \subset A$ *be an ideal that can be generated by a completely intersecting sequence* $\mathbf{f}$ *of length* r. *Let* $g_1, \ldots, g_r \in I$ *be elements that generate* I. *Then the sequence* $(g_1, \ldots, g_r)$ *is completely intersecting.*

Proof. As I/I^2 is a free A/I-module of rank r (Corollary 19.15), the image of $\mathbf{g}$ in I/I^2 is automatically a basis (Corollary B.4). Write $g_j = \sum a_{ij} f_i$ for $a_{ij} \in A$ and let $w \colon A^r \to A^r$ be the A-linear map given by the matrix (a_{ij}). Then $w \otimes \mathrm{id}_{A/I}$ is an isomorphism. By functoriality, w yields a homomorphism $K_\bullet(\mathbf{f}) \to K_\bullet(\mathbf{g})$ (Remark 19.5 (1)) which induces an isomorphism $H_\bullet(\mathbf{f}) \xrightarrow{\sim} H_\bullet(\mathbf{g})$ since the homology is annihilated by I (Remark 19.6). □

(19.3) Regular and completely intersecting immersions.

We now define the notion of a regular and of a completely intersecting immersion. In the development of the theory we then concentrate on completely intersecting immersions because this is the notion we use to define complete intersection morphisms. Nevertheless, sometimes instead of showing that an immersion is completely intersecting it is easier to check that an immersion is regular and deduce that it is completely intersecting. Therefore we consider both notions. For morphisms of locally noetherian schemes both notions are equivalent anyway.

Definition 19.19. *Let $i\colon Z \to X$ be an immersion of schemes, let X^0 be an open subscheme of X such that i identifies Z with a closed subscheme of X^0, and let $\mathscr{I} \subset \mathscr{O}_{X^0}$ be the corresponding quasi-coherent ideal.*

Let $z \in Z$. Then i is called regular *(resp. completely intersecting) at z if there exists an open neighborhood U of $i(z)$ in X^0 and a weakly regular (resp. completely intersecting) sequence $\mathbf{f}$ of $\Gamma(U, \mathscr{O}_X)$ which generates $\mathscr{I}_{|U}$.*

The immersion i is called regular *(resp. completely intersecting) if it is regular (resp. completely intersecting) at each $z \in Z$.*

Let $i\colon Z \to X$ be an immersion. Then the set of $z \in Z$ such that i is regular (resp. completely intersecting) at z is open in Z. Every open immersion is regular and completely intersecting.

The following proposition shows in particular that the definition of "regular immersion" and "completely intersecting immersion" is independent of the choice of X^0.

Proposition 19.20. *Let $i\colon Z \to X$ be an immersion of schemes defined by a quasi-coherent ideal $\mathscr{I} \subseteq \mathscr{O}_{X^0}$ of finite type (with $X^0 \subseteq X$ open such that $i(Z) \subseteq X^0$ is closed) and let $z \in Z$ and $x := i(z) \in X$. Then the immersion i is regular (resp. a completely intersecting) at z if and only if the ideal $\mathscr{I}_x$ of $\mathscr{O}_{X^0,x} = \mathscr{O}_{X,x}$ is generated by a weakly regular (resp. a completely intersecting) sequence.*

We prove the equivalence only for completely intersecting immersions. The equivalence for regular immersions is easy.

Proof. We may assume that $X = X^0$. The condition is necessary as the notion of completely intersecting immersion is stable under the flat base change $\operatorname{Spec}\mathscr{O}_{X,x} \to U$ by Remark 19.5 (4). To show the converse, we may assume that $X = \operatorname{Spec}A$, $Z = \operatorname{Spec}A/I$ for a finitely generated ideal I. Let $\mathfrak{p} \subset A$ be the prime ideal corresponding to x and let $\mathbf{f}_{\mathfrak{p}}$ be a completely intersecting sequence generating $I_{\mathfrak{p}}$. Replacing X be an open affine neighborhood of x we may assume that $\mathbf{f}_{\mathfrak{p}}$ is the image of a sequence $\mathbf{f}$ of elements of A which generates I (Corollary 7.32). By Remark 19.5 (4) we have $H_i(\mathbf{f}, A)_{\mathfrak{p}} = H_i(\mathbf{f}_{\mathfrak{p}}, A_{\mathfrak{p}}) = 0$ for all $i > 0$. As for any finite sequence $\mathbf{f}$ there are only finitely many i such that $H_i(\mathbf{f}, A) \neq 0$, there exists $g \in A$ with $g \notin \mathfrak{p}$ such that $H_i(\mathbf{f}, A)_g = 0$ for all i. Then the image of $\mathbf{f}$ in A_g is completely intersecting and hence $\mathscr{I}_{|D(g)}$ is generated by a completely intersecting sequence. $\qquad\square$

We will also use the following weaker notion of regularity for an immersion.

Definition 19.21. *Let $i\colon Z \to X$ be an immersion defined by a quasi-coherent ideal $\mathscr{I} \subseteq \mathscr{O}_{X^0}$ with $X^0 \subseteq X$ open such that $i(Z)$ is a closed subscheme of X^0. Then i is called* quasi-regular, *if the following three conditions are satisfied.*

(a) The ideal $\mathscr{I}$ is of finite type.

(b) *The conformal sheaf $\mathcal{C}_i = i^*(\mathcal{I}/\mathcal{I}^2)$ is a finite locally free $\mathcal{O}_Z$-module.*
(c) *The canonical homomorphism*

$$\mathrm{Sym}_{\mathcal{O}_Z}(\mathcal{I}/\mathcal{I}^2) \to \bigoplus_{d \geq 0} \mathcal{I}^d/\mathcal{I}^{d+1}$$

is an isomorphism.

Let $z \in Z$. Then i is called a quasi-regular at z *if there exists an open neighborhood U of $i(z)$ in X such that the restriction $i^{-1}(U) \to U$ of i is quasi-regular.*

For a quasi-regular immersion i we call the finite locally free conformal sheaf $\mathcal{C}_i$ also the conformal bundle. *Its dual $\mathcal{N} = \mathcal{N}_i := \mathcal{C}_i^\vee$ is called the* normal bundle.

Remark 19.22. By Theorem 19.12 we have for an immersion $i: Z \to X$ the implications

$$i \text{ regular} \Rightarrow i \text{ completely intersecting} \Rightarrow i \text{ quasi-regular.}$$

All properties are equivalent if X is locally noetherian (Corollary 19.13).

More precisely, if i is given by a quasi-coherent ideal $\mathcal{I} \subseteq \mathcal{O}_{X^0}$ with $X^0 \subseteq X$ open such that $i(Z)$ is a closed subscheme of X^0 and if $\mathbf{f} = (f_1, \ldots, f_n)$ is a completely intersecting sequence of $f_i \in \Gamma(X, \mathcal{I})$ which generates $\mathcal{I}$, then the images $\bar{f}_i$ of the f_i in $\mathcal{I}/\mathcal{I}^2 = \mathcal{C}_i$ form a basis of the $\mathcal{O}_Z$-module $\mathcal{C}_i$ and there is an isomorphism of graded $\mathcal{O}_Z$-algebras

$$\mathcal{O}_Z[T_1, \ldots, T_n] \xrightarrow{\sim} \bigoplus_{d \geq 0} \mathcal{I}^d/\mathcal{I}^{d+1}, \qquad T_i \mapsto \bar{f}_i \in \Gamma(Z, \mathcal{I}/\mathcal{I}^2).$$

Definition 19.23. *Let $i: Z \to X$ be a quasi-regular immersion of schemes. For $z \in Z$ we call $\mathrm{rk}_{\mathcal{O}_{Z,z}}(\mathcal{C}_i)_z$ the* codimension of i in z. *We also set $\dim_z(i) := -\mathrm{rk}_{\mathcal{O}_{Z,z}}(\mathcal{C}_i)_z$.*

As $\mathcal{C}_i$ is finite locally free if i is quasi-regular, the map $Z \to \mathbb{Z}$, $z \mapsto \dim_z(i)$ is locally constant.

Let us look at regular immersions of small codimension.

Remark 19.24. A completely intersecting immersion which is of codimension ≤ 1 in all points is a regular immersion, because completely intersecting sequences of length ≤ 1 are always weakly regular.

An immersion is quasi-regular of codimension 0 in all points if and only if it is an open immersion.

If the inclusion $i: D \to X$ of a closed subscheme is completely intersecting of codimension 1 in all points of D (Remark 11.27), then D is an effective Cartier divisor. Its conformal sheaf $\mathcal{C}_i$ is given by $i^*\mathcal{O}_X(-D)$.

Example 19.25. Let X be a scheme, let $\mathcal{E}$ be a finite locally free $\mathcal{O}_X$-module of rank r, let $\mathbb{V}(\mathcal{E}) = \mathrm{Spec}(\mathrm{Sym}(\mathcal{E}))$ be the attached geometric vector bundle (Proposition 11.7), and let $s_0: X \to \mathbb{V}(\mathcal{E})$ be its zero section. Then s_0 a regular immersion of codimension r with conformal sheaf $\mathcal{C}_{s_0} = \mathcal{E}$.

Indeed, s_0 is given by the ideal $\mathcal{I} := \bigoplus_{d \geq 1} \mathrm{Sym}^d(\mathcal{E})$ of $\mathrm{Sym}(\mathcal{E})$ and hence $\mathcal{C}_{s_0} = \mathcal{I}/\mathcal{I}^2 = \mathcal{E}$, where the second inequality is induced by the projection $\mathcal{I} \to \mathrm{Sym}^1(\mathcal{E}) = \mathcal{E}$. To see that s_0 is a regular immersion, we can work locally on X and hence assume that $\mathcal{E} = \mathcal{O}_X^r$. Then s_0 can be identified with the zero section of $\mathbb{A}_X^r$ which is clearly regular, see Example 19.16.

One has the following permanence properties for regular and completely intersecting immersions.

Proposition 19.26.

(1) *Let* $i\colon Z \to Y$ *and* $j\colon Y \to X$ *be regular immersions (resp. completely intersecting immersions). Then* $j \circ i$ *is a regular (resp. completely intersecting) immersion. For* $z \in Z$ *one has* $\dim_z(j \circ i) = \dim_z(i) + \dim_{i(z)}(j)$.

(2) *Let* $i\colon Z \to X$ *be an immersion and let* $g\colon X' \to X$ *be a flat morphism. If* i *is regular (resp. completely intersecting), then the base change* $i_{(X')}\colon Z \times_X X' \to X'$ *is a regular (resp. completely intersecting) immersion.*

(3) *Let* $i\colon Z \to X$ *be an immersion and let* $g\colon X' \to X$ *be faithfully flat and quasi-compact. If* $i_{(X')}$ *is a regular (resp. completely intersecting) immersion, then* i *is a regular (resp. completely intersecting) immersion.*

We give the proof only for completely intersecting immersions. For the case of regular immersions we refer to [EGAIV] ^{O} (19.1.5).

Proof. To show (1) we may assume that i and j are closed immersions and that $X = \operatorname{Spec} A$ is affine. Let $\mathbf{f}$ be a completely intersecting sequence in A generating an ideal I such that $Y \cong \operatorname{Spec} A/I$ and let $\mathbf{g}$ be sequence in A whose image $\bar{\mathbf{g}}$ in A/I is completely intersecting and such that the vanishing locus of $\bar{\mathbf{g}}$ is Z. Then there are quasi-isomorphisms of complexes of A-modules

$$K_\bullet(\mathbf{f}, \mathbf{g}) \overset{(19.1.3)}{=} K_\bullet(\mathbf{f}) \otimes_A K_\bullet(\mathbf{g}) \cong A/I \otimes K_\bullet(\mathbf{g}) \overset{(19.1.4)}{=} K_\bullet(\bar{\mathbf{g}}) \cong A/(\mathbf{f}, \mathbf{g}).$$

Therefore $(\mathbf{f}, \mathbf{g})$ is a completely intersecting sequence. This proves (1).

To show (2) and (3), we again can assume that $X = \operatorname{Spec} A$ and i is a closed immersion. To show (2) we may also assume that X' is also affine. For (3) we remark that X' is quasi-compact because g is quasi-compact. Thus we may replace X' by the affine scheme $\coprod_i U'_i$, where $(U'_i)_i$ is a finite affine cover of X'. Hence we may assume that $X' = \operatorname{Spec} A'$ in both cases. Then A' is a flat A-algebra in (2) and a faithfully flat A-algebra in (3). Hence (2) and (3) follow because the formation of $H_\bullet(\mathbf{f})$ is compatible with flat base change (Remark 19.5 (4)). $\qquad\square$

The composition of two quasi-regular immersions is not quasi-regular in general (Exercise 19.12), see however Exercise 19.11. The analogous statements of (2) and (3) for quasi-regular immersions hold (Exercise 19.13).

Proposition 19.27. *Let* X *be an* S-scheme. *For* $l = 1, \dots, r$, *let* $X_l \subseteq X$ *be a closed subscheme, and let* $Y := \bigcap_{l=1}^r X_l$ *be their schematic intersection. Assume that all the immersions* $X_l \hookrightarrow X$ *are completely intersecting of codimension* d_l, *say, and that the immersion* $i\colon Y \to X$ *is completely intersecting of codimension* $d := \sum_{l=1}^r d_l$. *Denote by* $\mathscr{C}_{X_l/X}$ *and* $\mathscr{C}_{Y/X}$ *the respective locally free conormal sheaves. Then*

$$\mathscr{C}_{Y/X} \cong \bigoplus_{l=1}^r i^* \mathscr{C}_{X_l/X}.$$

Proof. We have a natural map $i^* \mathscr{C}_{X_l/X} \to \mathscr{C}_{Y/X}$ for each l, because the ideal defining X_l inside X is contained in the ideal of Y, so we get a homomorphism $\bigoplus_{l=1}^r i^* \mathscr{C}_{X_l/X} \to \mathscr{C}_{Y/X}$. We can check that it is an isomorphism locally on X, and can hence assume that each X_l is defined by a completely intersecting sequence. Then our assumption on the codimension of Y implies that joining all these sequences gives a completely intersecting sequence (Proposition 19.18), which defines Y as a closed subscheme of X. Then all the conormal sheaves we have to consider are free, with bases induced by the respective defining equations, and the claim follows; cf. Remark 19.22. □

(19.4) Regular immersions of flat and of smooth schemes.

In the presence of flatness many of the notions introduced above fall together:

Proposition 19.28. *Let*

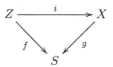

be a commutative diagram of schemes, where i is an immersion and let $z \in Z$, $s := f(z)$, and $x := i(z)$. Assume that S and X are locally noetherian or that f and g are locally of finite presentation. Then the following assertions are equivalent.

(i) *f is flat in a neighborhood of z and i is a regular immersion at z.*

(ii) *f is flat in a neighborhood of z and i is a completely intersecting at z.*

(iii) *f is flat in a neighborhood of z and i is a quasi-regular immersion at z.*

(iv) *There exists an open affine neighborhood $U = \operatorname{Spec} A$ of x in X and a regular sequence $(f_1, \ldots, f_n)$ in A such that i induces an isomorphism of $i^{-1}(U)$ with the closed subscheme $\operatorname{Spec} A/(f_1, \ldots, f_n)$ and such that $A/(f_1, \ldots, f_r)$ is flat over S for all $r = 0, \ldots, n$.*

(v) *f is flat in a neighborhood of z, g is flat in a neighborhood of x and for all morphisms $S' \to S$ the base change $i_{(S')} \colon Z \times_S S' \to X \times_S S'$ is regular in every point of $Z \times_S S'$ over z.*

(vi) *g is flat in a neighborhood of x and the induced immersion on the fiber $i_s \colon Z_s \to X_s$ is quasi-regular at z.*

The proof will show that if these equivalent conditions are satisfied, then $\dim_z(i) = \dim_z(i_s) = n$ with n as in (iv).

We will give the proof only in the locally noetherian case. See [EGAIV]^{O} (19.2.4) for the proof (using noetherian approximation) if f and g are locally of finite presentation.

Proof. Assertions (i), (ii), and (iii) are equivalent by our assumption that X is locally noetherian (Remark 19.22), and trivially (iv) implies (i) and (v) implies (vi).

To show the remaining assertions we may assume that $S = \operatorname{Spec} R$, $X = \operatorname{Spec} A$ and $Z = \operatorname{Spec} A/I$ for noetherian rings R and A and an ideal I of A. If A/I is flat over R, which we may assume under all hypotheses except (vi), then for every R-algebra R' the sequence

$$0 \to I \otimes_R R' \to A \otimes_R R' \to A/I \otimes_R R' \to 0$$

is exact (Proposition B.16). Hence for $S' = \operatorname{Spec} R'$ we see that $i_{(S')}$ is given by the ideal $I \otimes_R R'$. We will write $A_r := A/(f_1, \ldots, f_r)$.

(vi) $\Rightarrow$ *(iv).* By assumption $I_x/(\mathfrak{m}_s I_x + I_x^2)$ is a free $A_x/(\mathfrak{m}_s A_x + I_x)$-module of finite rank. Let $\mathbf{f} = (f_1, \ldots, f_n)$ be a sequence of elements $f_j \in I$ whose images in $I_x/(\mathfrak{m}_s I_x + I_x^2)$ form a basis. Let g_j be the image of f_j in $A/\mathfrak{m}_s A$. Then $\mathbf{g} = (g_1, \ldots, g_n)$ generates $I_x/\mathfrak{m}_s I_x$ by Nakayama's lemma. By Assumption and by Remark 19.22 we know that $I_x/\mathfrak{m}_s I_x$ is generated a regular sequence of length n. Hence $\mathbf{g}$ is a regular sequence in $A_x/\mathfrak{m}_s A_x$ (Proposition 19.18).

Replacing $X = \operatorname{Spec} A$ by a some open affine neighborhood of x, we may assume that
$\mathbf{g}$ is a regular sequence in $A/\mathfrak{m}_s A$. Again by Nakayama's lemma, $\mathbf{f}$ generates I after
passing to a smaller affine open neighborhood of x, if necessary. By hypothesis we may
also assume that A is flat over S.

By induction we may assume that for $0 \le r < n$ the $f_1, \ldots, f_r$ form a regular sequence
and that $A_r := A/(f_1, \ldots, f_r)$ is flat over R. To show that A_{r+1} is flat over R and that
the multiplication by f_{r+1} on A_r is injective we may assume that A_r is a local ring. But
then we can conclude by Proposition G.2 because the multiplication by g_{r+1} on $A_r/\mathfrak{m}_s A_r$
is injective.

(i) $\Rightarrow$ (iv). We may assume that I is generated by a regular sequence $(f_1, \ldots, f_n)$ in
A and that A/I is flat over R. By descending induction on r we may assume that A_{r+1}
is flat over R. Consider the sequence

$$(*) \qquad\qquad 0 \to A_r \xrightarrow{f_{r+1}} A_r \longrightarrow A_{r+1} \to 0$$

which is exact because $(f_1, \ldots, f_n)$ is regular. Then the local criterion for flatness Theorem B.51 shows that A_r is flat over R.

(iv) $\Rightarrow$ (v). We may assume that $S' = \operatorname{Spec} R'$ and I is generated by a regular sequence
$(f_1, \ldots, f_n)$ in A such that $A_r := A/(f_1, \ldots, f_r)$ is flat over S for all $r = 0, \ldots, n$. Let R'
be an R-algebra. For all $r = 0, \ldots, n-1$ we have then $A_{r+1} \otimes_R R' = (A \otimes_R R')/(f_1 \otimes
1, \ldots, f_{r+1} \otimes 1)$. As A_{r+1} is flat over R, tensoring the sequence $(*)$ with R' yields an exact
sequence

$$0 \to A_r \otimes_R R' \xrightarrow{f_{r+1} \otimes 1} A_r \otimes_R R' \longrightarrow A_{r+1} \otimes_R R' \to 0.$$

This proves the claim. $\square$

Note that for the proof of "(iv) $\Rightarrow$ (v)" we did not use any finiteness conditions except
that i is locally of finite presentation.

Our next goal is the study of regular immersions between smooth schemes. We start by
recalling (a geometric version of) a result from Commutative Algebra.

Proposition 19.29. *Let $i: Z \to X$ be an immersion of locally noetherian schemes. Let
$z \in Z$ such that $\mathscr{O}_{Z,z}$ is regular. Then $\mathscr{O}_{X,i(z)}$ is regular if and only if i is a regular
immersion at z.*

Proof. By replacing X by an open neighborhood V of $i(Z)$ such that $i(Z)$ is closed in
V we may assume that i is a closed immersion. By Proposition 19.20 we are reduced to
Proposition G.19. $\square$

Theorem 19.30. *Let*

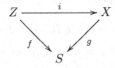

*be a commutative diagram of schemes, where i is an immersion and g is locally of finite
presentation. Let $z \in Z$, $x := i(z)$, and $s := f(z)$. Assume that f is smooth at z. Then
the following assertions are equivalent.*
(i) g is smooth at x.
(ii) i is a regular immersion at z.
(iii) i is a quasi-regular immersion at z.

(iv) *The immersion $i_s \colon Z_s \to X_s$ induced on fibers is a quasi-regular immersion at z.*
If these equivalent conditions are satisfied, one has $\dim_z(f) = \dim_z(i) + \dim_{i(z)}(g)$.

In particular, any immersion of smooth S-schemes is regular and in particular completely intersecting.

Proof. As f and g are locally of finite presentation in a neighborhood of z resp. x, they are both flat in a neighborhood of z resp. x if they are flat at z resp. x because the locus of flatness is open for morphisms locally of finite presentation (Theorem 14.44). Hence by Proposition 19.28 and as smoothness of flat morphisms can be checked on fibers (Corollary 18.57), we may assume that $S = \operatorname{Spec} k$ is a field. Then the equivalence of (iii) and (iv) is clear and the equivalence of (ii) and (iii) follows from Corollary 19.13. As the properties "regular immersion" and "smooth" can be checked after a faithfully flat quasi-compact base change, we may assume that k is algebraically closed. But then f is smooth at z (resp. g is smooth at x) if and only if $\mathcal{O}_{Z,z}$ (resp. $\mathcal{O}_{X,x}$) is regular (Theorem 6.28). Hence the equivalence of (i) and (ii) follows from Proposition 19.29. $\quad\square$

Corollary 19.31. *Let $f \colon X \to S$ be a smooth morphism of schemes of relative dimension d.*
(1) Let $i \colon S \to X$ be a section of f. Then i is a regular immersion of codimension d.
(2) Let $g \colon Y \to X$ be a morphism of S-schemes. Then the graph $\Gamma_g \colon Y \to Y \times_S X$ is a regular immersion of codimension d.
(3) The diagonal $\Delta \colon X \to X \times_S X$ is a regular immersion of codimension d.

Proof. Assertion (1) follows from Theorem 19.30 because every section is an immersion (Example 9.12). Assertion (2) follows from (1) because Γ_g is a section of the projection $Y \times_S X \to Y$ which is smooth because smoothness is stable under base change. Finally, (3) is a special case of (2). $\quad\square$

If $X \to S$ is smooth of relative dimension n, then $\Omega^1_{X/S}$ is locally free of rank n and $\Omega^n_{X/S} = \bigwedge^n \Omega^1_{X/S} = \det(\Omega^1_{X/S})$ it its top exterior power. Here we set $\det(\mathscr{E}) := \bigwedge^r \mathscr{E}$ for a locally free module $\mathscr{E}$ of rank r, which is a line bundle. To calculate it, sometimes the following result is useful which relies on the remark (Exercise 7.29) that if $0 \to \mathscr{E}' \to \mathscr{E} \to \mathscr{E}'' \to 0$ is a short exact sequence of finite locally free modules, then $\det(\mathscr{E}) \cong \det(\mathscr{E}') \otimes \det(\mathscr{E}'')$.

Proposition 19.32. *Let S be a scheme, let $X \to S$ and $Y \to S$ be smooth morphisms of constant relative dimension n and m, respectively. Let $i \colon Y \to X$ be a closed immersion of S-schemes. Then i is a regular immersion of codimension $n - m$ and*

$$\Omega^m_{Y/S} \cong \left(\bigwedge^{n-m} \mathscr{C}_i \right)^{\vee} \otimes_{\mathcal{O}_Y} i^* \Omega^n_{X/S}.$$

Proof. The immersion is regular of codimension $m - n$ by Theorem 19.30. We have a short exact sequence of finite locally free $\mathcal{O}_Y$-modules (Corollary 18.73)

$$0 \to \mathscr{C}_i \to i^* \Omega^1_{X/S} \to \Omega^1_{Y/S} \to 0$$

and hence we obtain

$$i^* \Omega^n_{X/S} = \det(i^* \Omega^1_{X/S}) \cong \det(\mathscr{C}_i) \otimes \det(\Omega^1_{Y/S})$$

which shows the claim. $\quad\square$

Corollary 19.33. *Let S be a scheme, let X be a smooth S-scheme, and for $j = 1, \ldots, r$ let D_j be an effective Cartier divisor on X. Assume that the embedding $i \colon Y := \bigcap_{j=1}^r D_j \to X$ is regular of codimension r and that Y is smooth over S. Then for the top exterior power of $\Omega^1_{Y/S}$ we obtain*

$$\det(\Omega^1_{Y/S}) \cong \left(\det(\Omega^1_{X/S}) \otimes \mathscr{O}_X \left(\sum_{i=1}^r D_i \right) \right)_{|Y}.$$

Proof. We have $\mathscr{C}_i \cong \bigoplus_{j=1}^r \mathscr{O}_X(-D_j)_{|Y}$ by Proposition 19.27 and therefore we obtain $\det(\mathscr{C}_i) \cong \bigotimes_{j=1}^r \mathscr{O}_X(-D_j)_{|Y}$. Hence we conclude by Proposition 19.32. $\square$

For instance, one often applies Proposition 19.32 and Corollary 19.33 to $X = \mathbb{P}_S^n$ in which case one has $\Omega^n_{X/S} = \det(\Omega^1_{X/S}) = \mathscr{O}_X(-n-1)$.

(19.5) Blow-up of regularly immersed smooth subschemes.

Remark 19.34. Let X be a scheme and let Z be a closed subscheme given by a quasi-coherent ideal $\mathscr{I}$ of $\mathscr{O}_X$ such that the inclusion $i \colon Z \to X$ is a completely intersecting immersion (e.g., if i is a regular immersion). Recall that $\operatorname{Proj}(\bigoplus_d \mathscr{I}^d)$ is the blow up $\operatorname{Bl}_Z(X)$ of X in the closed subscheme Z (Proposition 13.92) and that $E := \operatorname{Proj}(\bigoplus_d \mathscr{I}^d/\mathscr{I}^{d+1}) \subseteq \operatorname{Bl}_Z(X)$ is its exceptional divisor (Remark 13.94). Therefore Corollary 19.15 implies that we have isomorphisms of X-schemes

$$\operatorname{Bl}_Z(X) \cong \mathbb{P}(\mathscr{I}), \qquad E \cong \mathbb{P}(\mathscr{C}_i),$$

using that $\mathbb{P}(\mathscr{E}) \cong \operatorname{Proj}(\operatorname{Sym}(\mathscr{E}))$ for every quasi-coherent module $\mathscr{E}$ (Theorem 13.32). As $\mathscr{C}_i$ is a finite locally free $\mathscr{O}_Z$-module, the structure morphism $E \to Z$ is locally on Z isomorphic to $\mathbb{P}_Z^{n-1}$, where n is the codimension of i.

Proposition 19.35. *Let $f \colon X \to S$ be a smooth morphism of schemes and let Z be a closed subscheme of X which is smooth over S. Then the blow-up $\operatorname{Bl}_Z(X)$ of X along Z is smooth over S.*

Proof. Denote by $\pi \colon \operatorname{Bl}_Z(X) \to X$ the blow-up morphism and by $E := \pi^{-1}(Z)$ the exceptional divisor. In particular, the immersion $E \to \operatorname{Bl}_Z(X)$ is regular (of codimension 1). To show that $\operatorname{Bl}_Z(X) \to S$ is smooth in every point $\tilde{x}$ of $\operatorname{Bl}_Z(X)$ we distinguish two cases.

If $\tilde{x} \notin E$ then the isomorphism $\operatorname{Bl}_Z(X) \setminus E \cong X \setminus Z$ (Proposition 13.91 (3)) implies the smoothness at $\tilde{x}$ because X is smooth over S.

Now suppose that $\tilde{x} \in E$. By Theorem 19.30 we see that the inclusion $i \colon Z \to X$ is a regular immersion. Hence its conormal sheaf $\mathscr{C}_i$ is finite locally free and $E \cong \mathbb{P}(\mathscr{C}_i)$ as Z-schemes (Remark 19.34). In particular, $E \to Z$ is smooth and hence E is smooth over S. Now we can apply Theorem 19.30 to the regular immersion $E \to \operatorname{Bl}_Z(X)$ to see that $\operatorname{Bl}_Z(X) \to S$ is smooth at $\tilde{x} \in E$. $\square$

Remark 19.36. Using Proposition 19.29 instead of Theorem 19.30 the same argument as in the proof of Proposition 19.35 shows that if one blows up a regular scheme X in a closed regular subscheme Z, then $\operatorname{Bl}_Z(X)$ is regular.

Local complete intersection and syntomic morphisms

We now define the notion of a local complete intersection morphism as a morphism that can be locally factorized as a completely intersecting immersion followed by a smooth morphism. Flat local complete intersection morphisms will be called syntomic.

(19.6) Local complete intersection morphisms.

Definition 19.37. *Let $f: X \to S$ be a morphism of schemes.*
(1) *Let $x \in X$. Then f is called* locally completely intersecting *at x if there exists an open neighborhood U of x in X and a factorization of $f_{|U}$ of the form $U \xrightarrow{i} P \xrightarrow{\pi} S$, where i is a completely intersecting immersion and where π is a smooth morphism.*
(2) *The morphism f is called a* local complete intersection morphism *(or short: an* lci-morphism*) if it is a locally completely intersecting at x for all $x \in X$.*

By definition, the set $\{\, x \in X \;;\; f \text{ is locally completely intersecting at } x\,\}$ is open. An lci-morphism is locally of finite presentation. Every smooth morphism is an lci-morphism.

The next result shows that the property of being an lci-morphism does not depend on the chosen factorization.

Proposition 19.38. *Let $f: X \to S$ be a morphism locally of finite presentation and let*

$$X \xrightarrow{i} P \xrightarrow{\pi} S \qquad and \qquad X \xrightarrow{i'} P' \xrightarrow{\pi'} S$$

be factorizations of f, where i and i' are immersions and π and π' are smooth. Then i is a completely intersecting immersion if and only if i' is completely intersecting. In this case for all $x \in X$ one has

$$\dim_x(i) + \dim_{i(x)}(\pi) = \dim_x(i') + \dim_{i'(x)}(\pi').$$

The proposition shows in particular that the following notion is well defined.

Definition 19.39. *Let $f: X \to S$ be a morphism of schemes that is local complete intersection at $x \in X$. Let U be an open neighborhood of x such that there exists a factorization $f_{|U} = \pi \circ i$ with i a completely intersecting immersion and π smooth. Then*

$$\dim_x(f) := \dim_x(i) + \dim_{i(x)}(\pi) \in \mathbb{Z}$$

is called the relative dimension *of f at x.*

To prove Proposition 19.38 we follow [Sta] 069E. The essential argument is the following lemma.

Lemma 19.40. *Let*

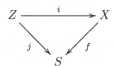

be a commutative diagram of schemes, where i and j are immersions and where $f: X \to S$ is smooth. Then i is completely intersecting if and only if j is completely intersecting. In this case one has $\dim_z(i) + \dim_{i(z)}(f) = \dim_z(j)$ for all $z \in Z$.

Proof. *(i)*. Assume that j is completely intersecting. The graph morphism Γ_i is a section of the smooth projection $X \times_S Z \to Z$ and hence it is a regular immersion (Corollary 19.31) and in particular completely intersecting (Theorem 19.12). As X is smooth over S, the projection $p\colon X \times_S Z \to X$ is a flat base change of the completely intersecting immersion j and hence it is again a completely intersecting immersion (Proposition 19.26 (2)). Therefore $i = p \circ \Gamma_i$ is completely intersecting (Proposition 19.26 (1)).

(ii). We proceed in several steps. *(I)*. Let us now assume that i is completely intersecting. The question whether j is completely intersecting is local on X and S. In particular we may assume that $S = \operatorname{Spec} R$ and $X = \operatorname{Spec} A$ is affine and that i and j are closed immersions. In particular j yields an isomorphism $Z \cong \operatorname{Spec} R/J$ for some ideal $J \subseteq R$. Writing A as a quotient of a polynomial ring, we obtain a closed immersion $i_1\colon X \to \mathbb{A}^n_S$ which is automatically regular by Theorem 19.30 and in particular completely intersecting. Hence $i_1 \circ i$ is completely intersecting (Proposition 19.26 (1)). Therefore we may assume that $X = \mathbb{A}^n_R$.

(II). Then i corresponds to a surjective R-algebra homomorphism

$$\varphi\colon A = R[T_1, \ldots, T_n] \to R/J.$$

Choose $s_\alpha \in R$ whose images in R/J are $\varphi(T_\alpha)$. Set $K := (T_1 - s_1, \ldots, T_n - s_n)$. Then the kernel of φ is the ideal $I := K + JA$. As $A/K \cong R$ it suffices to show that $j'\colon Z \to V(K)$ is completely intersecting.

The sequence $(T_1 - s_1, \ldots, T_n - s_n)$ is regular (Example 19.16). Therefore K/K^2 is a free A/K-module with basis $(T_\alpha - s_\alpha)_{1 \leq \alpha \leq n}$. Hence $K/K^2 \otimes_A A/I \cong K/KI$ is free over $A/I = R/J$ with the same basis. The composition of the canonical map $u\colon K/KI \to I/I^2$ with the A/I-linear map $I/I^2 \to \Omega^1_{A/R} \otimes_A A/I$ maps the basis $(T_\alpha - s_\alpha)_\alpha$ to the basis of $(dT_\alpha \otimes 1)_\alpha$. Hence this composition is an isomorphism. We deduce the existence of an A/I-linear map $v\colon I/I^2 \to K/KI$ with $v \circ u = \operatorname{id}$. Therefore u identifies K/KI with a direct summand of I/I^2.

(III). Thus we now have closed immersions $Z := V(I) \xrightarrow{j'} V(K) \to X = \operatorname{Spec} A$, where A is some ring, such that $V(I) \to X$ and $V(K) \to X$ are completely intersecting and $K/K^2 \otimes_A A/I$ is a direct summand of I/I^2. We claim that these hypotheses imply that j' is completely intersecting. Then the equality of relative dimensions follows because the relative dimension of $V(K) \to X$ is the negative of the relative dimension of X over S.

We may assume that A is a local ring. Moreover we may assume that I is contained in the maximal ideal of A, otherwise our claim is trivial as the empty subscheme is always regularly embedded. As $V(I) \to X$ (resp. $V(K) \to X$) is completely intersecting, I/I^2 (resp. K/K^2) is a finite free A/I-module (resp. A/K-module). As $K/K^2 \otimes_A A/I$ is a direct summand of I/I^2 we may choose $f_1, \ldots, f_n \in K$ such that their images in K/K^2 are an A/K-basis and we may choose $g_1, \ldots, g_s \in I$ such that the images of $f_1, \ldots, f_n$ and of $g_1, \ldots, g_s$ in I/I^2 are a basis of I/I^2. By Nakayama's lemma, $\mathbf{f} = (f_1, \ldots, f_n)$ generates K and $(\mathbf{f}, \mathbf{g}) = (f_1, \ldots, f_n, g_1, \ldots, g_s)$ generates I. Then $\mathbf{f}$ is a completely intersecting sequence generating K and $(\mathbf{f}, \mathbf{g})$ is a completely intersecting sequence generating I (Proposition 19.18).

It suffices to show that the image $\bar{\mathbf{g}}$ in $A/K = A/(\mathbf{f})$ is completely intersecting. This follows from the quasi-isomorphism of chain complexes

$$K_\bullet(\bar{\mathbf{g}}) \overset{(19.1.4)}{=} A/I \otimes K_\bullet(\mathbf{g}) \cong K_\bullet(\mathbf{f}) \otimes_A K_\bullet(\mathbf{g}) \overset{(19.1.3)}{=} K_\bullet(\mathbf{f}, \mathbf{g}) \cong A/(\mathbf{f}, \mathbf{g}). \qquad \square$$

We now come to the proof of Proposition 19.38.

Proof. Let i' be completely intersecting. By symmetry it suffices to show that i is completely intersecting. The morphism $i'' := (i, i') \colon X \to P'' := P \times_S P'$ is a closed immersion into a smooth S-scheme (because the properties "closed immersion" and "smooth" are both stable under composition and under base change). Applying Lemma 19.40 to $i' \colon X \xrightarrow{i''} P'' \xrightarrow{q} P'$, where q is the (smooth) second projection, shows that i'' is completely intersecting. Then applying Lemma 19.40 to $i \colon X \xrightarrow{i''} P'' \xrightarrow{p} P$, where p is the (smooth) first projection shows that i is completely intersecting. $\qquad\square$

Corollary 19.41. *An immersion is an lci-morphism if and only if it is completely intersecting.*

Proposition 19.42.
(1) *The property "lci-morphism" is local on the source and local on the target.*
(2) *If $f \colon X \to Y$ and $g \colon Y \to Z$ are lci-morphisms, then $g \circ f$ is an lci-morphism and $\dim_x(g \circ f) = \dim_x(f) + \dim_{f(x)}(g)$ for all $x \in X$.*
(3) *If $f \colon X \to Y$ is an lci-morphism and $Y' \to Y$ is a flat morphism, then the base change $f' \colon X \times_Y Y' \to Y'$ is an lci-morphism.*

The property "lci-morphism" is also stable under faithfully flat descent (Exercise 19.14).

Proof. Assertion (1) holds by definition and (3) holds because the property "smooth" is stable under base change and the property "completely intersecting immersion" is stable under flat base change.

Let us show (2): The question is local on X, Y, and Z. Hence we may assume that there are factorizations $X \xrightarrow{i} \mathbb{A}_Y^n \to Y$ of f and $Y \xrightarrow{j} \mathbb{A}_Z^m \to Z$ of g, where i and j are completely intersecting immersions. We obtain a factorization of $g \circ f$ as

$$X \xrightarrow{i} \mathbb{A}_Y^n \xrightarrow{j \otimes \mathrm{id}_{\mathbb{A}^m}} \mathbb{A}_Z^{n+m} \to Z.$$

Then the composition $(j \otimes \mathrm{id}_{\mathbb{A}^m}) \circ i$ is a completely intersecting immersion because "completely intersecting immersion" is stable under composition and flat base change. This shows that $g \circ f$ is an lci-morphism. The assertion about the relative dimension follows because the relative dimension for completely intersecting immersions and for smooth morphisms is additive under composition. $\qquad\square$

Remark 19.43. Let $f \colon X \to Y$ be an lci-morphism. Then $X \to \mathbb{Z}$, $x \mapsto \dim_x(f)$ is a locally constant function because the analogous assertion holds for completely intersecting immersions (Proposition 19.26 (1)) and for smooth morphisms (Proposition 18.59).

Corollary 19.44. *Let $f \colon X \to Y$ be a morphism of S-schemes. Assume that $Y \to S$ is smooth. Then f is an lci-morphism if and only if $X \to S$ is an lci-morphism.*

Proof. If f is an lci-morphism, then the composition $X \to Y \to S$ is an lci-morphism by Proposition 19.42 (2). Conversely, assume that $X \to S$ is an lci-morphism. Locally on X we may factorize f in $X \xrightarrow{i} \mathbb{A}_Y^n \to Y$, where i is an immersion. The composition $\mathbb{A}_Y^n \to Y \to S$ is smooth. Hence i is completely intersecting because $X \to S$ is an lci-morphism (Proposition 19.38). $\qquad\square$

(19.7) Complete intersection rings.

There is also an absolute notion of complete intersection for locally noetherian schemes. We will see in Proposition 19.50 that every local ring of a scheme X locally of finite type over a field k is complete intersection in this absolute sense if and only if the structure morphism $X \to \operatorname{Spec} k$ is a local complete intersection morphism. For the definition of a complete intersection ring we first introduce "the" Koszul complex of a local noetherian ring.

Remark 19.45. Let A be a local noetherian ring, $\mathfrak{m}$ its maximal ideal, k its residue field. Choose a sequence $\mathbf{f} = (f_1, \ldots, f_n)$ in $\mathfrak{m}$ whose image in $\mathfrak{m}/\mathfrak{m}^2$ is a k-basis. The corresponding Koszul complex $K_\bullet(\mathbf{f})$ is independent of the choice $\mathbf{f}$ up to isomorphism. Indeed, if $\mathbf{g} = (g_1, \ldots, g_n)$ is another such sequence, then any $n \times n$ matrix $L = (l_{ij})$ with $g_i = \sum_{j=1}^n l_{ji} f_j$ is invertible because modulo $\mathfrak{m}$ it maps a k-basis to a k-basis. The A-linear isomorphism $A^n \to A^n$ corresponding to L then yields an isomorphism $K_\bullet(\mathbf{g}) \xrightarrow{\sim} K_\bullet(\mathbf{f})$.

Moreover, $\mathfrak{m}H_i(\mathbf{f}) = 0$ for all i because $\mathbf{f}$ generates $\mathfrak{m}$. Hence we may define for $i \geq 0$ the numbers

$$(19.7.1) \qquad\qquad \varepsilon_i(A) := \dim_k H_i(\mathbf{f}),$$

which depend only on A.

One can show that $\varepsilon_1(A) \geq \dim_k \mathfrak{m}/\mathfrak{m}^2 - \dim(A)$ for every local noetherian ring ([Mat2] 21.1).

Definition 19.46. *A local noetherian ring A with maximal ideal $\mathfrak{m}$ and residue field k is called a* complete intersection ring *if $\varepsilon_1(A) = \dim_k \mathfrak{m}/\mathfrak{m}^2 - \dim(A)$.*

If X is a scheme locally of finite type over a field k and if $x \in X(k)$ is a k-rational point, then $\mathcal{O}_{X,x}$ is a complete intersection ring if and only if $\dim_k T_x(X) - \dim_x(X) = \varepsilon_1(\mathcal{O}_{X,x})$.

Let us collect some properties of complete intersection rings.

Remark 19.47. Let A be a local noetherian ring with maximal ideal $\mathfrak{m}$ and residue class field k.
(1) Definition B.73 and Corollary 19.13 show:

$$A \text{ is regular} \Leftrightarrow \varepsilon_1(A) = 0 \Leftrightarrow \dim_k \mathfrak{m}/\mathfrak{m}^2 = \dim(A).$$

In particular, every regular local ring is a complete intersection ring.
(2) If $\hat{A}$ is the completion of A with maximal ideal $\hat{\mathfrak{m}} = \mathfrak{m}\hat{A}$, then $\mathfrak{m}/\mathfrak{m}^2 = \hat{\mathfrak{m}}/\hat{\mathfrak{m}}^2$ (Proposition B.39) and $\dim A = \dim \hat{A}$ (Proposition B.64). Moreover, a sequence in $\mathfrak{m}$ that yields a basis of $\mathfrak{m}/\mathfrak{m}^2$ is also a sequence in $\mathfrak{m}\hat{A}$ generating $\hat{\mathfrak{m}}/\hat{\mathfrak{m}}^2$, hence if $K_\bullet$ is the Koszul complex for A, the Koszul complex $\hat{K}_\bullet$ for $\hat{A}$ is isomorphic to $\hat{A} \otimes_A K_\bullet$. As $\hat{A}$ is flat over A, we also have $H_i(\hat{K}_\bullet) \cong H_i(K_\bullet) \otimes_A \hat{A}$ for all i. Since $H_i(K_\bullet)$ is annihilated by $\mathfrak{m}$, this implies $H_i(\hat{K}_\bullet) \cong H_i(K_\bullet)$ and hence $\varepsilon_i(A) = \varepsilon_i(\hat{A})$ for all i.

In particular, A is a complete intersection ring if and only if $\hat{A}$ is a complete intersection ring.
(3) If A is a complete intersection ring, then A is Gorenstein (Definition G.26) by [Mat2] Theorem 21.3. In particular A is Cohen-Macaulay (Proposition G.26). Hence we have the implications

$$A \text{ regular} \Rightarrow A \text{ complete intersection} \Rightarrow A \text{ Gorenstein} \Rightarrow A \text{ Cohen-Macaulay.}$$

(4) Let $A \to B$ be a flat local homomorphism between local noetherian rings, and let $\mathfrak{m} \subset A$ denote the maximal ideal. Then B is a complete intersection ring if and only if A and $B/\mathfrak{m}B$ are complete intersection rings. ([Mat2], Remark on p. 182.)

(19.8) Local complete intersection morphisms over a field.

Remark 19.48. Let k be a field, A a finitely generated k-algebra, $x \in X := \operatorname{Spec} A$ and let $\mathbf{f} = (f_1, \ldots, f_r)$ be a sequence contained in $\mathfrak{p}_x$. Let $Z = \operatorname{Spec} A/\mathbf{f}A$ and $z \in Z$ the point corresponding to $\mathfrak{p}_x/\mathbf{f}$. Then Corollary B.61 implies that $\dim_z(Z) \geq \dim_x(X) - r$ with equality if $\mathbf{f}$ is a regular sequence.

We will use the following result from Commutative Algebra ([Mat2] Theorem 21.2).

Proposition 19.49. *Let A be a local noetherian ring.*
(1) *A is a complete intersection ring if and only if there exists a complete regular local ring R and an ideal I of R generated by an R-regular sequence such that $\hat{A} = R/I$, where $\hat{A}$ denotes the completion of A.*
(2) *Let R be a regular local ring such that $A \cong R/\mathfrak{a}$. Then A is a local intersection ring if and only if the ideal $\mathfrak{a}$ is generated by a regular sequence.*

Proposition 19.50. *Let k be a field, let $g \colon X \to \operatorname{Spec} k$ be a k-scheme locally of finite type and let $x \in X$. Then the following assertions are equivalent.*
(i) *g is locally completely intersecting at x.*
(ii) *There exists an open affine neighborhood $U = \operatorname{Spec} A$ of x and an isomorphism of k-algebras $A \cong k[T_1, \ldots, T_n]/(f_1, \ldots, f_c)$ with $\dim A = n - c$.*
(iii) *$\mathcal{O}_{X,x}$ is a complete intersection ring.*
In this case one has

$$(19.8.1) \qquad \dim_x(g) = \dim_x(X) = n - c$$

and U is equi-dimensional of dimension $n - c$.

Proof. Every irreducible component of $\operatorname{Spec} k[T_1, \ldots, T_n]/(f_1, \ldots, f_c)$ has dimension $\geq n - c$ (Corollary 5.33). Hence (ii) implies the last assertion. Then 19.8.1 holds by Remark 19.48 (and for all $x \in U$). It remains to show that (i), (ii), and (iii) are equivalent.

We may assume that $X = \operatorname{Spec} A$ is affine with $A = R/I$, where R is a smooth k-algebra of finite type, for instance $R = k[T_1, \ldots, T_m]$. Let $i \colon X \to \operatorname{Spec} R$ be the corresponding closed immersion. Let $\mathfrak{p} \subset A$ be the prime ideal corresponding to x and let $\mathfrak{q}$ be the prime ideal of R corresponding to $i(x)$. As R is regular (Theorem 6.28), $R_{\mathfrak{q}}$ is a local regular ring (Proposition B.74) and $A_{\mathfrak{p}} = R_{\mathfrak{q}}/I_{\mathfrak{q}}$.

(iii) $\Rightarrow$ (i). If $A_{\mathfrak{p}} = \mathcal{O}_{X,x}$ is a complete intersection ring, $I_{\mathfrak{q}}$ is generated by a regular sequence $\mathbf{f} = (f_1, \ldots, f_c)$ (Proposition 19.49). Replacing R by a localization R_g with $g \notin \mathfrak{q}$, we may assume that $f_1, \ldots, f_c \in R$ and that they form a regular sequence of R (Proposition 19.20). Then i is a regular immersion into the spectrum of a smooth k-algebra and hence g is a local complete intersection morphism.

(i) $\Rightarrow$ (ii). We may assume that $R = k[T_1, \ldots, T_l]$. Passing to a principal open subset, we may assume $A = R_h/I$, where $h \in R$ and I is generated by a regular sequence $(\bar{f}_1, \ldots, \bar{f}_d)$ (Proposition 19.49). We identify R_h with $k[T_1, \ldots, T_{l+1}]/(T_{l+1}h - 1)$ and we choose $f_i \in k[T_1, \ldots, T_{l+1}]$ whose images in R_h is $\bar{f}_i$. As $f := T_{l+1}h - 1$ is a regular element in $k[T_1, \ldots, T_{l+1}]$, we deduce from Proposition 19.26 (1) that $(f_1, \ldots f_d, f)$ is a regular sequence in $k[T_1, \ldots, T_{l+1}]$. Hence A is a equi-dimensional of dimension $(l+1) - (d+1)$.

(ii) $\Rightarrow$ *(iii).* After localization, (ii) implies that $\mathcal{O}_{X,x} \cong B/J$, where B is a regular local ring and where J is an ideal of B generated by $c := \dim B - \dim \mathcal{O}_{X,x}$ elements. As regular rings are Cohen-Macaulay, Proposition G.20 implies that J is generated by a regular sequence. $\square$

(19.9) Syntomic morphisms.

Definition 19.51. *A flat local complete intersection morphism is called* syntomic.

Every smooth morphism is syntomic and every syntomic morphism is flat and locally of finite presentation.

Proposition 19.52. *Let $f\colon X \to S$ be a morphism of schemes. Then the following assertions are equivalent.*
(i) *f is syntomic.*
(ii) *f is flat, locally of finite presentation and for all $s \in S$ the fiber $f^{-1}(s)$ is a local complete intersection over $\kappa(s)$.*
(iii) *For all $x \in X$ there exist open affine neighborhoods $\operatorname{Spec} A$ of x in X and $\operatorname{Spec} R$ of $f(x)$ in S such that $A \cong R[T_1, \ldots, T_n]/(f_1, \ldots, f_c)$ and such that fiber of $\operatorname{Spec} A \to \operatorname{Spec} R$ in $f(x)$ has dimension $n - c$.*

Proof. All assertions are local on source and target, so we may assume that $X = \operatorname{Spec} A$ and $S = \operatorname{Spec} R$ are affine and that there exists a closed immersion $X \to \mathbb{A}^n_R$. Then the implications (iii) $\Rightarrow$ (i) $\Leftrightarrow$ (ii) follow from Proposition 19.28 applied to $X \to \mathbb{A}^n_R \to S$ (for (iii) $\Rightarrow$ (ii) one also uses Proposition 19.50).

Let us show (ii) $\Rightarrow$ (iii). Let $x \in X$ and set $k := \kappa(f(x))$ and write $A = R[T_1, \ldots, T_n]/I$ for an ideal I, which is finitely generated because f is locally of finite presentation. By Proposition 19.50 we may assume (after possibly shrinking $\operatorname{Spec} A$) that $A \otimes_R k = k[T_1, \ldots, T_n]/(\bar{f}_1, \ldots, \bar{f}_c)$ is of dimension $n - c$, where $\bar{f}_i$ are elements in the image $\bar{I}$ of I in $A \otimes_R k$. Choose $f_i \in I$ whose image in $\bar{I}$ is $\bar{f}_i$ and set $A' := R[T_1, \ldots, T_n]/(f_1, \ldots, f_c)$. Let $J := \operatorname{Ker}(A' \to A)$. This is a finitely generated ideal because A and A' are both of finite presentation over R. As A is flat over R, $J \otimes_R k \to A' \otimes_R k$ is still injective (Proposition B.15), hence $J \otimes_R k = 0$ by construction. Let $\mathfrak{q} \subset B := R[T_1, \ldots, T_n]$ be the prime ideal corresponding to the image of x in $\mathbb{A}^n_R$. Then Nakayama's lemma (applied to the ring $B_{\mathfrak{q}}$, the finitely generated module $J_{\mathfrak{q}}$ and the ideal $\mathfrak{m}_{f(x)} B_{\mathfrak{q}}$) implies that $J_{\mathfrak{q}} = 0$. As J is finitely generated, there exists $g \in B$, $g \notin \mathfrak{q}$ such that $J_g = 0$. Replacing B by $R[T_1, \ldots, T_{n+1}]$ and $(f_1, \ldots, f_c)$ by $(f_1, \ldots, f_c, gT_{n+1} - 1)$ we are done. $\square$

Note that the proof shows, that if $f\colon X \to S$ is flat and locally of finite presentation and there exists $s \in S$ such that $f^{-1}(s)$ is a local complete intersection over $\kappa(s)$ in a neighborhood of some $z \in f^{-1}(s)$, then (iii) holds after replacing X by an open neighborhood of z.

Proposition 19.53. *The property "syntomic" is local on the source, local on the target, stable under composition, stable under base change, stable under fpqc descent and compatible with inductive limits.*

Proof. Except for the stability under base change all these permanence properties hold because "lci-morphism" and "flat" have the same permanence properties. The stability under base change follows from Proposition 19.52 and the fact that the property "lci-morphism" is stable under flat base change and hence in particular under change of base fields for the fibers. □

Corollary 19.54. *Let $f\colon X \to Y$ be a morphism of S-schemes. For $s \in S$ let $f_s\colon X_s \to Y_s$ be the morphism induced by f on the fiber over s. Let X and Y be locally of finite presentation over S and let X be flat over S. Then f is syntomic if and only if f_s is syntomic for all $s \in S$.*

Proof. If f is syntomic, then f_s is syntomic for $s \in S$ because "syntomic" is stable under base change (Proposition 19.53). Conversely, if all f_s are syntomic, then for all $s \in S$ and $y \in Y_s$ the fiber $f_s^{-1}(y) = f^{-1}(y)$ is syntomic. Moreover, as X is flat over S and f_s is flat for all $s \in S$, the fiber criterion for flatness (Theorem 14.25) shows that f is flat. Therefore f is syntomic by Proposition 19.52. □

Remark 19.55. Let $S = \operatorname{Spec} R$ be an affine scheme, $X = \operatorname{Spec} R[T_1, \dots, T_n]/(f_1, \dots, f_c)$. Then every irreducible component of every fiber of $X \to S$ has at least dimension $n - c$ (Corollary 5.33). By semicontinuity of the fiber dimension (Theorem 14.112), $U := \{ x \in X \; ; \; \dim_x f^{-1}(f(x)) = n - c \}$ is therefore open in X (but possibly empty), and $U \to S$ is syntomic by Proposition 19.52.

One important property of syntomic morphisms is that they can always be deformed locally:

Theorem 19.56. *Let $X \to S$ be a syntomic morphism of schemes and let $S \hookrightarrow \tilde{S}$ be a nil-immersion. Then for every point of X there exist an open neighborhood $U \subseteq X$ of x and a syntomic $\tilde{S}$-scheme $\tilde{U}$ such that $\tilde{U} \times_{\tilde{S}} S \cong U$.*

This theorem implies a similar result for smooth morphisms (see Exercise 19.22). Moreover, it can be generalized to arbitrary immersions $S \hookrightarrow \tilde{S}$ (Exercise 19.23).

Proof. We may assume that $\tilde{S} = \operatorname{Spec} \tilde{R}$, $S = \operatorname{Spec} R$ with $R = \tilde{R}/I$ for an ideal I of $\tilde{R}$, and $X = \operatorname{Spec} A$ are affine with $A = R[T_1, \dots, T_n]/(f_1, \dots, f_c)$ such that every non-empty fiber of $X \to S$ has dimension $n - c$ (Proposition 19.52). Choose any lifts $\tilde{f}_i \in \tilde{R}[T_1, \dots, T_n]$ of f_i and set $\tilde{A} := \tilde{R}[T_1, \dots, T_n]/(\tilde{f}_1, \dots, \tilde{f}_c)$. As the fibers of $\operatorname{Spec} \tilde{A} \to \tilde{S}$ and of $\operatorname{Spec} A \to S$ are the same, Proposition 19.52 implies that $\tilde{X} := \operatorname{Spec} \tilde{A}$ is syntomic over $\tilde{S}$. □

Exercises

Exercise 19.1. Let k be a field, $A = k[X, Y, Z]/(Z^2, ZX, Z(Y - 1))$ and let x and y be the image of X and Y in A. Show that (x, y) is a completely intersecting sequence in A but not a regular sequence. Show that (y, x) is a regular sequence in A.

Exercise 19.2. Let A be a local ring such that $A \cong R/I$ where R is a regular local ring and $I \subset R$ an ideal. Show that there exists a regular local ring S with maximal ideal $\mathfrak{m}_S$ and an ideal $J \subset S$ such that $A \cong S/J$ and $J \subseteq \mathfrak{m}_S^2$. Moreover, show that I is generated by a regular sequence if and only if J is generated by a regular sequence.

Exercise 19.3. Let R be a local regular ring with maximal ideal $\mathfrak{m}_R$ and residue field k_R, let $I \subsetneq R$ be an ideal, $A := R/I$. Let $\mathbf{x} = (x_1, \ldots, x_d)$ be a regular sequence of R generating $\mathfrak{m}_R$ (hence $d = \dim R$).

(1) Assume that $I \subseteq \mathfrak{m}_R^2$ (cf. Exercise 19.2). Show that $H_1(\mathbf{x}, A) \cong I/\mathfrak{m}_R I$.

(2) Show that the image of $\mathbf{x}$ in A is a minimal system of generators of the maximal ideal of A if and only if $I \subseteq \mathfrak{m}_R^2$.

(3) Let $\mathfrak{m}_A$ be the maximal ideal and k_A the residue field of A. Show that $\dim_{k_A} H_1(\mathbf{x}, A) = \dim_{k_R}(I/\mathfrak{m}_R I) - d + \dim_{k_A}(\mathfrak{m}_A/\mathfrak{m}_A^2)$.

Exercise 19.4. Let A be a ring and let $\mathbf{f} = (f_1, \ldots, f_n) \in A^n$, $n \geq 1$ an integer. For every A-module M we define a complex (with differential of degree $+1$)

$$K^\bullet(\mathbf{f}, M) := \mathrm{Hom}(K_\bullet(\mathbf{f}), M).$$

Denote its i-th cohomology by $H^i(\mathbf{f}, M)$.

Assume that for all $i = 1, \ldots, n$ the multiplication by f_i on the module

$$\ker(M \xrightarrow{f_1} M) \cap \cdots \cap \ker(M \xrightarrow{f_{i-1}} M)$$

is surjective (then $\mathbf{f}$ is called M-coregular). Show that $H_i(\mathbf{f}, M) = 0$ for $i < n$.

Exercise 19.5. Let A be a ring, let L be a flat A-module, and let $u \colon L \to A$ be an A-linear map. Show that $K_\bullet(u, -)$ is an exact functor from the category of A-modules to the category of chain complexes of A-modules. Deduce that $(H_i(u, -))_i$ form a homological δ-functor (here "homological" means that the boundary maps lower the degree).

Remark: Using the derived category of A-modules and the derived tensor product introduced in Chapter 21, this exercise can be also interpreted as follows: Let L be any A-module also consider $K_\bullet(u)$ as a cochain complex $K(u)$ concentrated in degrees ≤ 0 by setting $K^i(u) := K_{-i}(u)$ with the same differentials. Then $M \mapsto K(u) \otimes_A^L M$ defines a triangulated functor $D(A) \to D(A)$. Hence every distinguished triangle in $D(A)$, e.g., given by an exact sequence of complexes of A-modules, yields a long exact cohomology sequence. If L is flat, then $K(u)$ is a K-flat complex of A-modules and the derived tensor product $K(u) \otimes_A^L M$ is given by the usual tensor products $K(u) \otimes_A M$ of complexes.

Exercise 19.6. Let R be a ring, let M be an R-module, $I := \{1, \ldots, n\}$ and $p \geq 0$ an integer. A map $\omega \colon I^p \to M$ is called *alternating* if it satisfies the following conditions.

(a) For every permutation $\sigma \in S_p$ and $(a_1, \ldots, a_p) \in I^p$ one has $\omega(a_{\sigma(1)}, \ldots, a_{\sigma(p)}) = \mathrm{sgn}(\sigma)\omega(a_1, \ldots, a_p)$.

(b) For all $(a_1, \ldots, a_p) \in I^p$ such that two of the a_i are equal one has $\omega(a_1, \ldots, a_p) = 0$.

We denote by $C_n^p(M)$ the R-module of alternating maps $I^p \to M$.

Now let $A = R[X_1, \ldots, X_n]$ and assume M is an A-module (in other words, M is an R-module endowed with pairwise commuting R-linear endomorphisms f_i of M defined by X_i). Let $\mathbf{X} := (X_1, \ldots, X_n)$.

(1) Consider the map $\mathrm{Hom}_A(\bigwedge^p A^I, M) \to C_n^p(M)$ that sends f to the alternating map $(a_1, \ldots, a_p) \mapsto f(e_{a_1} \wedge \cdots \wedge e_{a_p})$, where $(e_i)_{1 \leq i \leq n}$ is the standard basis of A^I. Show that this map is an isomorphism of R-modules

$$K^i(\mathbf{X}, M) \xrightarrow{\sim} C_n^p(M).$$

By transport of structure one obtains a complex with differentials $\partial^p \colon C_n^p(M) \to C_n^{p+1}(M)$. It is denoted by $K^\bullet(\mathbf{f}, M)$. If f_i is given by multiplication by an element of R for all i, then $K^\bullet(\mathbf{f}, M)$ reduces to the complex defined in Exercise 19.4.

(2) Let $M = R[T_1, \ldots, T_n]$ and f_i the R-linear endomorphism of M given by $\partial/\partial T_i$. Show that attaching to $\omega \in C_n^p(M)$ the differential form $\sum_{a_1 < \cdots < a_p} \omega(a_1, \ldots, a_p) dT_{a_1} \wedge \cdots \wedge dT_{a_p}$ yields an isomorphism

$$K^\bullet((\partial/\partial T_1, \ldots, \partial/\partial T_n), M) \overset{\sim}{\to} \Omega_{A/R}^\bullet.$$

Exercise 19.7. Let R be a $\mathbb{Q}$-algebra. Show that the De Rham complex $\Omega_{R[T_1, \ldots, T_n]/R}^\bullet$ is acyclic in degrees > 0.
Hint: Exercises 19.6 and 19.4.

Exercise 19.8. Let S be a scheme, let $\mathscr{L}$ be a quasi-coherent $\mathscr{O}_S$-module, and let $u \colon \mathscr{F} \to \mathscr{O}_S$ be an $\mathscr{O}_S$-linear map.
(1) Define a complex of quasi-coherent $\mathscr{O}_S$-modules $K_\bullet(u)$, again called the *Koszul complex* by globalizing the construction in Section (19.1). Formulate and prove global variants of the assertions of Remark 19.5.
(2) Now suppose that $\mathscr{L}$ is finite locally free, set $\mathscr{I} := \mathrm{Im}(u)$, and let $i \colon Y := V(\mathscr{I}) \to S$ be the corresponding closed immersion. Show that i is completely intersecting if $K_\bullet(u) \to \mathscr{O}_S/\mathscr{J}$ is a quasi-isomorphism.
(3) Conversely, suppose that there exists an ample line bundle on S and let $i \colon Y \to S$ be a completely intersecting closed immersion, given by a quasi-coherent ideal $\mathscr{I}$. Show that there exists a finite locally free $\mathscr{O}_S$-module $\mathscr{L}$ and a surjective $\mathscr{O}_S$-linear map $u \colon \mathscr{L} \to \mathscr{I}$ such that $K_\bullet(u) \to \mathscr{O}_S/\mathscr{J}$ is a quasi-isomorphism.

Exercise 19.9. Let A be ring, let M be an A-module, and let $f_1, \ldots, f_r \in A$ be elements. Set $R := \mathbb{Z}[T_1, \ldots, T_r]$ be the polynomial ring and let $R \to A$ be the unique ring homomorphism sending T_i to f_i. Show that one has for all $i \geq 0$ an isomorphism of A-modules

$$H_i(\mathbf{f}, M) \cong \mathrm{Tor}_i^R(R/(T_1, \ldots, T_r), M),$$

where the right hand side becomes an A-module via multiplication by $a \in A$ on M and by functoriality of the Tor functor (for their definition see Section (21.20), in particular Remark 21.101).

Exercise 19.10. Let $i \colon Z \to Y$ and $j \colon Y \to X$ be immersions. Let j and $j \circ i$ be completely intersecting. Show that i is completely intersecting.

Exercise 19.11. Let $i \colon Z \to Y$ be a completely intersecting immersion and let $j \colon Y \to X$ be a quasi-regular immersion. Show that $j \circ i$ is quasi-regular.

Exercise 19.12. The following example is taken from [Sta] 065M. Let k be a field, $A = k[x, y, w, z_0, z_1, z_2, \ldots]/(y^2 z_0 - wx, z_0 - yz_1, z_1 - yz_2, \ldots)$. Let $X = \mathrm{Spec}\, A$, $Y = V(x)$ and $Z = V(x, y)$. Show that $Z \hookrightarrow Y$ is a quasi-regular immersion, that $Y \hookrightarrow X$ is a regular immersion, but that $Z \hookrightarrow X$ is not quasi-regular.

Exercise 19.13. Let $i \colon Z \to X$ be an immersion.
(1) Let $X' \to X$ be flat and let i be quasi-regular. Show that the base change $i_{(X')} \colon Z \times_X X' \to X'$ is quasi-regular.
(2) Let $X' \to X$ be faithfully flat and quasi-compact and let $i_{(X')}$ be a quasi-regular immersion. Show that i is a quasi-regular immersion.

Exercise 19.14. Let $f \colon X \to Y$ be a morphism of schemes, let $Y' \to Y$ be a faithfully flat quasi-compact morphism, and assume that the base change $f' \colon X \times_Y Y' \to Y'$ is an lci-morphism. Show that f is an lci-morphism.

Exercise 19.15. Let k be a field, $n \geq 2$ an integer. Show that $k[T]/(T^n)$ is a complete intersection ring which is not regular. What is its Koszul complex?

Exercise 19.16. Let A be a local noetherian ring that can be written as the quotient of a regular local noetherian ring. Show that if A is a complete intersection ring, then $A_{\mathfrak{p}}$ is a complete intersection ring for all prime ideals $\mathfrak{p}$ of A.
Remark: The assertion also holds for arbitrary local noetherian rings ([Avr]$^{\circ}$).

Exercise 19.17. Let k be a field, $A := k[\![T_1, \ldots, T_4]\!]/(T_1T_3 + T_2T_4, T_1T_2, T_2T_4)$. Show that depth $A = 0$ and dim $A = 2$. Determine $\varepsilon_i(A)$ for all i. Show that A is not a complete intersection ring.

Exercise 19.18. Let k be a field, $X = \operatorname{Spec} k[T, U, V]/(T^2 - U^2, U^2 - Z^2, TU, UV, TV)$ and let $x \in X$ be the origin in $\mathbb{A}^2_k$. Show that $\mathscr{O}_{X,x}$ is a Gorenstein ring but it is not a complete intersection ring.

Exercise 19.19. Let k be a field and let $X = V_+(f_1, \ldots, f_r) \subset \mathbb{P}^n_k$ be a projective k-scheme, where $f_i \in k[T_0, \ldots, T_n]$ are homogeneous polynomials. Let $x \in X$. Show that if $\dim_x(X) = n - r$, then $X \to \operatorname{Spec} k$ is a local complete intersection morphism at x.

Exercise 19.20. Let A be a ring and let $f \in A[T]$ be a monic polynomial. Show that $\operatorname{Spec} A[T]/(f) \to \operatorname{Spec} A$ is syntomic.

Exercise 19.21. Show that "splitting rings" of polynomials can be chosen syntomic: Let A be a ring and $f \in A[T]$ a monic polynomial. Show that there exists a ring $\tilde{A}$ and a syntomic, finite locally free, faithfully flat morphism $\operatorname{Spec} \tilde{A} \to \operatorname{Spec} A$ such that $f = (T - \tilde{a}_1) \cdots (T - \tilde{a}_n)$ for $\tilde{a}_i \in \tilde{A}$.

Exercise 19.22. Let $X \to S$ be a smooth morphism of schemes and let $S \hookrightarrow S'$ be a nil immersion. Show that for every point of X there exist an open neighborhood $U \subseteq X$ of x and smooth $\tilde{S}$-scheme $\tilde{U}$ such that $\tilde{U} \times_{\tilde{S}} S \cong U$.

Exercise 19.23. Generalize Theorem 19.56 to the case that $S \hookrightarrow \tilde{S}$ is an arbitrary immersion.
Hint: Define $\tilde{A}$ as in the proof of Theorem 19.56 and then localize $\tilde{A}$ in an element $h \in \tilde{A}$ whose image in A is a unit.

20   The étale topology

Content

- Henselian rings
- The étale topology
- The étale fundamental group of a scheme

Let $X = \operatorname{Spec} A$, where A is a finitely generated $\mathbb{C}$-algebra. Then X is the vanishing locus of finitely many polynomials $f_1, \dots, f_r \in \mathbb{C}[T_1, \dots, T_n]$. These polynomials also define a complex analytic space X^{an}, called the analytification of X, whose underlying topological space is $X(\mathbb{C}) = \{ z \in \mathbb{C}^n \; ; \; f_1(z) = \cdots = f_r(z) = 0 \}$ endowed with the subspace topology of $\mathbb{C}^n$. If X is any scheme locally of finite type over $\mathbb{C}$, then one can glue local analytifications to obtain a complex analytic space X^{an} (see Section (20.12) for a precise definition of X^{an}). If X is smooth over $\mathbb{C}$, then locally one can find $f_1, \dots, f_r$ such that the rank of the Jacobi matrix $((\partial f_i / \partial T_j)(x))_{i,j}$ is equal to r for all $x \in X(\mathbb{C})$. Hence X^{an} will be a complex manifold of (complex) dimension $n - r$.

The complex analytic topology on X^{an} is much finer than the Zariski topology on X, and a morphism $f \colon X \to Y$ of schemes locally of finite type over $\mathbb{C}$ induces a local isomorphism $X^{\mathrm{an}} \to Y^{\mathrm{an}}$ if and only if f is étale (as opposed to if and only if f is locally for the Zariski topology an isomorphism). To obtain an algebraic way of detecting this Grothendieck and his school introduced a finer "topology", the étale topology. To formalize this, note that to talk about sheaves on a topological space, it is enough to know what a covering is, and how to form finite intersections of open subsets. The key idea now is that one may replace the notion of open subset of a scheme X, in other words inclusions $U \hookrightarrow X$ of an open subscheme, by étale morphisms $U \to X$. Hence an "open" of X in the étale topology is simply an étale morphism $U \to X$. An "open covering" of X in the étale topology is a family of étale morphisms $(g_i \colon U_i \to X)_i$ such that $X = \bigcup_i g_i(U_i)$. The "intersection of two opens" $U \to X$ and $V \to X$ is the fiber product $U \times_X V \to X$ (which is again an étale morphism). This allows to develop a theory of sheaves for the étale topology and the notion of the stalk of a sheaf at a point.

There are several variants of such topologies, that is, instances of the very useful abstract notion of a *Grothendieck (pre-)topology* on a category with fiber products by specifying coverings (satisfying certain obvious axioms). We will not explain this formalism here but refer the reader to the literature (e.g. [SGA3] $^{\underline{0}}_{X}$ Exp. IV, [SGA4] ^{0}, [Art1], [Sta]).

For the étale topology the stalk of the structure sheaf at a point is a so-called strictly henselian ring. Hence (strictly) henselian rings are as ubiquitous when considering the étale topology as are local rings when considering the Zariski topology. Therefore we start this chapter by discussing henselian rings, i.e., those local rings that satisfy Hensel's lemma. Then we introduce the étale topology, sheaves in the étale topology, and stalks.

In the last part of this chapter we introduce the fundamental group of a connected scheme S and prove Grothendieck's far reaching generalization of Galois theory replacing (spectra of) fields by arbitrary connected schemes.

© Springer Fachmedien Wiesbaden GmbH, ein Teil von Springer Nature 2023
U. Görtz und T. Wedhorn, *Algebraic Geometry II: Cohomology of Schemes*,
Springer Studium Mathematik – Master, https://doi.org/10.1007/978-3-658-43031-3_5

Henselian rings

Hensel's lemma in number theory (e.g., [Neu]^O II (4.6)) says that if A is a complete discrete valuation ring with residue field k, then every monic polynomial $f \in A[T]$ whose image $\bar{f} \in k[T]$ has a factorization $\bar{f} = \bar{g}\bar{h}$ in $k[T]$, where $\bar{g}$ and $\bar{h}$ are prime to each other, has itself a factorization $f = gh$ in $A[T]$ such that the images of g and h in $k[T]$ are $\bar{g}$ and $\bar{h}$, respectively. An arbitrary local ring A is called henselian if it satisfies Hensel's lemma. In fact we will define "henselian" differently and then we will prove that our definition is equivalent to the above definition (Proposition 20.5). The main result is Theorem 20.12 which characterizes henselian rings in terms of sections of smooth schemes over R.

(20.1) Definition of henselian rings.

Let A be a local ring with maximal ideal $\mathfrak{m}$. If $A \to B$ is an integral injective homomorphism, all maximal ideals of B lie over the maximal ideal of A (Proposition 5.12). If B is a finite A-algebra, all fibers of $\operatorname{Spec} B \to \operatorname{Spec} A$ are finite. Hence B is semilocal, i.e., B has only finitely many maximal ideals, say $\mathfrak{n}_1, \ldots, \mathfrak{n}_r$. Moreover B is a product of local rings if and only if the canonical homomorphism

$$(20.1.1) \qquad\qquad B \longrightarrow \prod_{i=1}^{r} B_{\mathfrak{n}_i}$$

is an isomorphism.

Definition 20.1. *A local ring A is called* henselian *if every finite A-algebra is a product of local rings.*

Remark 20.2. Let A be a henselian local ring. If A' is a finite local A-algebra, then A' is also henselian. In particular, every non-zero quotient of A is again a henselian ring.

Every field k is henselian because any finite k-algebra is Artinian and hence a product of local Artinian rings. Much more generally we have the following important class of examples for henselian rings.

Example 20.3. Every local ring A that is complete with respect to the $\mathfrak{m}_A$-adic topology is henselian. Indeed, if A is noetherian, this follows from Proposition B.46. For the general case see Exercise 20.7 and the hint there.

In particular we see that for every complete local ring R the ring of formal power series $R[\![T_1, \ldots, T_n]\!]$ is henselian.

We will now prove some characterizations for henselian rings. In particular we will see that a local ring is henselian if and only if it satisfies Hensel's lemma. For further characterizations of henselian rings we refer to Theorem 20.12 below and to Exercises 20.5, 20.6, 20.8, and Exercise 20.11. We start with a lemma.

Lemma 20.4. *Let A be a ring, let $f \in B := A[T_1, \ldots, T_n]$ and call f* primitive *if the ideal in A generated by the coefficients of f is A.*
(1) If f is primitive, then f is a regular element of B and $B/(f)$ is flat over A.

(2) *Assume $n = 1$. Then f is primitive if and only if $\operatorname{Spec} A[T]/(f) \to \operatorname{Spec} A$ is quasi-finite.*

Proof. The polynomial f is primitive if and only if for all prime ideals $\mathfrak{p}$ of A the image $f_{\mathfrak{p}}$ of f in $\kappa(\mathfrak{p})[T_1, \ldots, T_n]$ is non-zero. Therefore the multiplication by f induced on fibers over $\operatorname{Spec} B \to \operatorname{Spec} A$ is injective. As B is a flat A-algebra of finite presentation, Proposition G.2 implies that the multiplication by f on B is injective and has A-flat cokernel. This shows (1).

Let us show (2). Clearly, $A[T]/(f)$ is an A-algebra of finite type. Hence $\operatorname{Spec} A[T]/(f) \to \operatorname{Spec} A$ is quasi-finite if and only if its fibers are finite, i.e., if and only if $\kappa(\mathfrak{p})[T]/(f_{\mathfrak{p}})$ is finite for all $\mathfrak{p} \in \operatorname{Spec} A$, where $f_{\mathfrak{p}}$ denotes the image of f in $\kappa(\mathfrak{p})[T]$. But this is equivalent to $f_{\mathfrak{p}} \neq 0$ for all $\mathfrak{p}$. $\square$

Proposition 20.5. *Let A be a local ring with maximal ideal $\mathfrak{m}$ and residue field k. Let s be the closed point of $S := \operatorname{Spec} A$. For a polynomial $f \in A[T]$ we denote by $\bar{f}$ its image in $k[T]$. Then the following assertions are equivalent.*
(i) *A is henselian.*
(ii) *For every monic polynomial $f \in A[T]$, the finite A-algebra $B = A[T]/(f)$ is a product of finite local A-algebras.*
(iii) *If $u \colon X \to S$ is a separated morphism of finite type, then X has a decomposition $X = Y \sqcup X_1 \sqcup \cdots \sqcup X_r$ into open and closed subschemes such that every irreducible component of the special fiber Y_s of Y has dimension ≥ 1 and such that for $i = 1, \ldots, r$ one has $X_i = \operatorname{Spec} B_i$ for a finite local A-algebras B_i.*
(iv) *For every $f \in A[T]$ and for every factorization $\bar{f} = g_0 h_0$ in $k[T]$ such that g_0 is monic and such that g_0 and h_0 are prime to each other there exists a factorization $f = gh$ in $A[T]$ such that g is monic, $\bar{g} = g_0$, and $\bar{h} = h_0$.*

The proof will show that in (iv) we obtain an equivalent assertion if we assume f to be monic.

Before giving the proof we make a further remark on property (iv) which in particular implies that the factorization $f = gh$ there is automatically unique.

Remark 20.6. Let A be a local ring with maximal ideal $\mathfrak{m}$ and residue field k.
(1) Let $g, h \in A[T]$ with g monic. If $\bar{g}$ and $\bar{h}$ are prime to each other in $k[T]$, then $(g) + (h) = A[T]$.

Indeed, as g is monic, $A[T]/(g)$ is a finite free A-module. Hence $M := A[T]/(g,h)$ is a finite A-module with $M \otimes_A k = 0$ because $\bar{g}$ and $\bar{h}$ are prime to each other. By Nakayama's lemma we have $M = 0$. See also Remark 18.47.
(2) Let $f = gh \in A[T]$ with g monic and $\bar{g}$ and $\bar{h}$ prime to each other. Then g and h are uniquely determined by f, $\bar{g}$, and $\bar{h}$.

Indeed, write $f = g_1 h_1 = g_2 h_2$ with g_i monic, $\bar{g}_1 = \bar{g}_2$ and $\bar{h}_1 = \bar{h}_2$. By (1), there exist $r, s \in A[T]$ such that $rg_1 + sh_2 = 1$ and hence $g_2 = rg_1 g_2 + sg_2 h_2 = g_1(rg_2 + sh_1)$. Hence g_1 divides g_2. As they are monic of the same degree, we have $g_1 = g_2$. As the multiplication by a monic polynomial in $A[T]$ is injective, this also implies $h_1 = h_2$.

Proof. Recall for the proof that any non-trivial decomposition of rings $B = B_1 \times B_2$ yields non-trivial idempotents in B (namely $(1,0)$ and $(0,1)$). Conversely if $e \neq 0, 1$ is an idempotent in a ring B, then one obtains a non-trivial decomposition of rings $B = Be \times B(1-e)$.

(i) ⇒ (iii). This is an application of Zariski's main theorem. Let $u\colon X \to S$ be separated and of finite type.

Let $x \in X_s = X \otimes_A k$ be isolated in X_s, i.e., $\{x\} \subseteq X_s$ is open and closed. We claim that we find an open and closed subscheme $Z_x \subseteq X$ containing x such that $Z_x = \operatorname{Spec} B_x$, where B_x is a finite local A-algebra.

Indeed, by Zariski's main theorem (Theorem 12.73) there exists an open quasi-compact neighborhood V of x and a factorization of $u_{|V}$ as an open immersion $j\colon V \to Z$ followed by a finite morphism $Z \to \operatorname{Spec} A$. As A is henselian, $Z = Z_1 \sqcup \cdots \sqcup Z_r$ where Z_i is the spectrum of a local finite A-algebra. Choose j such that $x \in Z_j$ and set $Z_x := Z_j$. As the unique closed point of Z_x is contained in the open subscheme V, the scheme Z_x is an open subscheme of V and hence an open subscheme of X. As $Z_x \to \operatorname{Spec} R$ is finite and in particular proper, the immersion $Z_x \to X$ is also closed (Proposition 12.58 (3)). This shows the claim.

Note that only the maximal ideal of B_x lies over the maximal ideal over A (Theorem B.56). Hence $Z_x \cap X_s = \{x\}$. As u is quasi-compact, X_s is quasi-compact and there exist only finitely many isolated points $x_1, \ldots, x_r$ in X_s. Moreover $Z_{x_i} \cap Z_{x_j} = \emptyset$ for $x_i \neq x_j$. Otherwise, if there existed $y \in Z_{x_i} \cap Z_{x_j}$, then $x_i, x_j \in \overline{\{y\}}$ because Z_{x_i} and Z_{x_j} are spectra of local rings. As Z_{x_i} is closed, this would imply $x_i, x_j \in Z_{x_i}$ which contradicts $Z_{x_i} \cap X_s = \{x_i\}$.

Therefore we can set $X_i := Z_{x_i}$ and $Y := X \setminus \bigcup_i X_i$.

(iii) ⇒ (iv). Let f, g_0 and h_0 as in (iv) and set $X := \operatorname{Spec} A[T]/(f)$. As $\bar{f} \neq 0$, at least one of the coefficients of f is a unit in A. Thus Lemma 20.4 shows that $X \to \operatorname{Spec} A$ is quasi-finite and flat. Let $X = Y \sqcup X_1 \sqcup \cdots \sqcup X_r$ be a composition as in (iii). Then X_i is finite flat over $\operatorname{Spec} A$ for $1 \leq i \leq r$ and $Y \otimes_A k = \emptyset$. As

$$X \otimes_A k = (\operatorname{Spec} k[T]/(g_0)) \sqcup (\operatorname{Spec} k[T]/(h_0)),$$

we can renumber the X_i for $1 \leq i \leq r$ such that

$$\operatorname{Spec} k[T]/(g_0) = (X_1 \otimes_A k) \sqcup \cdots \sqcup (X_m \otimes_A k),$$
$$\operatorname{Spec} k[T]/(h_0) = (X_{m+1} \otimes_A k) \sqcup \cdots \sqcup (X_r \otimes_A k)$$

for some $1 < m < r$. Then $X_1 \sqcup \cdots \sqcup X_m = \operatorname{Spec} C$ for a finite flat A-algebra C.

As A is local, C is a free A-module (Proposition G.1). Its rank is $\deg(g_0)$. Let $t \in C$ be the image of T in C. As C is a quotient of $A[T]/(f)$ ($\operatorname{Spec} C$ being a closed subscheme of X), C is generated by t as an A-algebra. Let $g \in A[T]$ be the characteristic polynomial of the A-linear multiplication $C \to C$ by t. Then g is monic of degree $\deg(g) = \deg(g_0)$ and $\bar{g} = g_0$. By the theorem of Cayley-Hamilton, $g(t) = 0$. Hence $T \mapsto t$ yields a surjective homomorphism $A[T]/(g) \to C$ of free A-modules of the same rank which is therefore an isomorphism (Corollary B.4). By construction, $A[T] \to C$ factors through $A[T]/(f)$, therefore $f \in (g)$ and we may write $f = gh$ for some $h \in A[T]$. As $\bar{f} = g_0\bar{h} = g_0 h_0$ one necessarily has $\bar{h} = h_0$ because $k[T]$ has no zero-divisors.

(iv) ⇒ (ii). Let $f \in A[T]$ be a monic polynomial and let $B = A[T]/(f)$. Write $\bar{f} = \bar{g}_1^{e_1} \bar{g}_2^{e_2} \ldots \bar{g}_r^{e_r}$ where the $\bar{g}_i \in k[T]$ are monic, irreducible and pairwise prime to each other and where $e_i \geq 1$ are integers. Then (iv) and induction show that there exists a decomposition $f = G_1 G_2 \ldots G_r$ into monic polynomials such that the image of G_i is $\bar{g}_i^{e_i}$. Then $(G_i) + (G_j) = A[T]$ for $i \neq j$ by Remark 20.6. Hence the Chinese remainder theorem yields $A[T]/(f) = \prod_i A[T]/(G_i)$ and $A[T]/(G_i)$ is local as its reduction modulo $\mathfrak{m}$ is local.

(ii) ⇒ (i). Let B be a finite A-algebra. By induction on the number of maximal ideals in B it suffices to show that there exists a non-trivial decomposition $B = B_1 \times B_2$ if B is not local. But if B is not local, then $\bar{B} := B/\mathfrak{m}B$ is not local and hence a non trivial product of finite local algebras (because k is henselian). Hence there exists a non-trivial idempotent $\bar{e} \in \bar{B}$. Let $e \in B$ be any lift of $\bar{e}$. As B is a finite A-algebra, e is integral over A and there exists a monic polynomial $f \in A[T]$ such that $f(e) = 0$. Let $C := A[T]/(f)$ and let $\varphi \colon C \to B$ be the homomorphism of A-algebras given by $T \mapsto e$. By (ii) we can decompose C into a product of finite local A-algebras. Hence we find an idempotent $c \in C$ such that $\overline{\varphi(c)} = \bar{e}$. Then $B = B\varphi(c) \times B(1 - \varphi(c))$ is a non-trivial decomposition. $\qquad\square$

Exercise 20.4 asserts that there is a bijection between decompositions $f = gh$ with $g, h \in A[T]$ monic and coprime, and decompositions of $A[T]/(f)$ as A-algebra.

(20.2) Sections of smooth morphisms.

Our next goal is to show that for a smooth scheme X over a henselian ring R with residue field k each k-valued point of X can be lifted to an R-valued point, we will use that "locally for the étale topology" smooth morphisms have always a section: We first start with a more precise local result and deduce the existence of sections as a corollary.

Proposition 20.7. *Let $f \colon X \to S$ be a smooth morphism. Let $s \in S$ and let x be a closed point of the fiber $X_s = f^{-1}(s)$ such that $\kappa(x)$ is a separable extension of $\kappa(s)$. Then there exist an open neighborhood U of s in S and a subscheme $Z \subseteq f^{-1}(U)$ with $x \in Z$ such that $f_{|Z} \colon Z \to U$ is étale.*

Proof. The assertions is local on X and on S. In particular we may assume that f is of finite presentation. As f is smooth, its fiber X_s is smooth over $\kappa(s)$. Hence the local ring $\mathscr{O}_{X_s,x}$ is regular (Proposition 18.67) and hence its maximal ideal is generated by a regular sequence $(\bar{g}_1, \dots \bar{g}_n)$. Replacing X by a sufficiently small open affine neighborhood of x we may assume that $X = \operatorname{Spec} A$ is affine and that there exists a sequence $(g_1, \dots, g_n)$ in A such that the image of $(g_i)_x \in \mathscr{O}_{X,x}$ in $\mathscr{O}_{X_s,x} = \mathscr{O}_{X,x}/\mathfrak{m}_s\mathscr{O}_{X,x}$ equals $\bar{g}_i$. Let $Z = V(g_1, \dots, g_n) \subseteq X$ and let $i \colon Z \to X$ be the inclusion. Then its fiber i_s is a regular immersion $\operatorname{Spec} \kappa(x) \to X_s$ and thus by shrinking X we may assume that $f_{|Z} \colon Z \to S$ is flat over S and such that i is a regular immersion (Proposition 19.28).

By construction we have $\mathscr{O}_{Z_s,x} = \kappa(x)$. As x is a closed point of Z_s, $\kappa(x)$ is a finite extension of $\kappa(s)$ which is separable by assumption. Hence $f_{|Z} \colon Z \to S$ is unramified at x (Proposition 18.27). Hence by shrinking X we may assume that $f_{|Z}$ is in addition unramified and hence étale by Theorem 18.44. $\qquad\square$

Corollary 20.8. *Let $f \colon X \to S$ be a smooth surjective morphism. Then there exists an étale surjective morphism $S' \to S$ and an S'-section of $X \times_S S'$.*

Proof. For all $s \in S$ the fiber X_s is smooth (and in particular geometrically reduced) over $\kappa(s)$. Hence Proposition 6.21 shows that the set of closed points x of X_s such that $\kappa(x)$ is a separable extension of $\kappa(s)$ is dense in X_s. In particular, there exists such an $x_s \in X_s$ because $X_s \neq \emptyset$. By Proposition 20.7 there exists a subscheme $Z(s)$ of X containing x_s such that $f_{|Z(s)}$ is étale. Then we can define S' as the disjoint union of the schemes $Z(s)$ for $s \in S$. $\qquad\square$

Remark 20.9. If in the situation of Corollary 20.8 the scheme S is quasi-compact, one can choose S' affine. Indeed (with the notation of the proof), we may assume that $Z(s)$ is affine by shrinking $Z(s)$. As $f_{|Z(s)}$ is flat and locally of finite presentation, its image $U(s)$ is open in S (Theorem 14.35). If S is quasi-compact, it is covered by finitely many $U(s_j)$ and we can define S' as the disjoint union of the $Z(s_j)$.

(20.3) Sections of étale and smooth schemes over henselian rings.

By definition, a local ring R with maximal ideal $\mathfrak{m}$ is henselian if and only if for every finite R-algebra B the map $B \to B/\mathfrak{m}B$ yields a bijection between idempotents of B and idempotents of $B/\mathfrak{m}B$. On the other hand, the set of idempotents of any ring A is in bijection to the set of open and closed subschemes of $\operatorname{Spec} A$.

For any scheme X we define

$$\operatorname{Clopen}(X) := \{ Z \subseteq X \; ; \; Z \text{ open and closed subscheme of } X \},$$

which is in bijection to the set of idempotent elements in $\Gamma(X, \mathcal{O}_X)$. Under this bijection an open and closed subscheme Z corresponds to the section $e_Z \in \Gamma(X, \mathcal{O}_X)$ such that $e_{Z|Z} = 1 \in \Gamma(Z, \mathcal{O}_X)$ and $e_{Z|X \setminus Z} = 0 \in \Gamma(X \setminus Z, \mathcal{O}_X)$.

As the inverse image of an open and closed subscheme under a scheme morphism is again open and closed, we obtain a contravariant functor Clopen on the category of schemes with values in the category of sets.

Lemma 20.10. *Let $f \colon X \to S$ be a finite locally free morphism of schemes. Then the functor on S-schemes*

$$\mathscr{C\!O}_{X/S} \colon S' \mapsto \operatorname{Clopen}(X \times_S S')$$

is representable by an affine étale S-scheme of finite presentation.

Proof. We can work locally on S and hence assume $S = \operatorname{Spec} R$ and $X = \operatorname{Spec} A$, where A is a finite free R-algebra. Choose an R-basis $(e_i)_{1 \le i \le r}$ of the R-module A. The fact that an element $a = \sum T_i e_i \in A$ is an idempotent can be expressed via the multiplication table of the e_i by equations $P_j(T_1, \dots, T_r) = 0$ $(1 \le j \le r)$, where $P_j \in R[T_1, \dots, T_r]$ are certain polynomials of degree ≤ 2. Let C be the quotient of $R[T_1, \dots, T_r]$ by the ideal generated by $P_1, \dots, P_r$. Then $Z := \operatorname{Spec} C$ represents the functor $\mathscr{C\!O}_{X/S}$. Clearly Z is affine and of finite presentation over S.

Hence it remains to show that Z is formally étale over S. But for any thickening $T_0 \to T$ of order at most 1 of affine S-schemes one clearly has $\mathscr{C\!O}_{X/S}(T) = \mathscr{C\!O}_{X/S}(T_0)$ because $T_0 \to T$ is a universal homeomorphism. $\qquad\square$

Lemma 20.11. *Let R be a ring and $I \subsetneq R$ an ideal contained in the Jacobson radical of R. Let X be a scheme unramified over $\operatorname{Spec} R$. Then the canonical map $X(R) \to X(R/I)$ is injective.*

Proof. Let $S = \operatorname{Spec} R$, $S_0 = \operatorname{Spec} R/I$, and let $t_1, t_2 \colon S \to X$ be two S-sections of X whose restrictions to S_0 are equal. The subscheme $\operatorname{Eq}(t_1, t_2)$, where t_1 and t_2 coincide (Definition 9.1), is open in S because X is unramified over S (Remark 18.30). As it contains S_0 one has $S = \operatorname{Eq}(t_1, t_2)$, i.e. $t_1 = t_2$. $\qquad\square$

Theorem 20.12. *Let R be a local ring with residue field k. The following assertions are equivalent.*

(i) *R is Henselian.*

(ii) *For every étale morphism $f\colon X \to \operatorname{Spec} R$ and every ideal $I \subsetneq R$ the canonical map $X(R) \to X(R/I)$ is bijective.*

(iii) *For every étale morphism $f\colon X \to \operatorname{Spec} R$ the canonical map $X(R) \to X(k)$ is surjective.*

(iv) *For every smooth morphism $f\colon X \to \operatorname{Spec} R$ the canonical map $X(R) \to X(k)$ is surjective.*

Proof. The implications "(iv) $\Rightarrow$ (iii)" and "(ii) $\Rightarrow$ (iii)" are clear. Now let $S = \operatorname{Spec} R$ and let $s \in S$ be the closed point of S. For every $f \in R[T]$ we denote by $\bar{f}$ its image in $k[T]$.

(i) $\Rightarrow$ (ii). It suffices to show (ii) for $I = \mathfrak{m}$ the maximal ideal of R. Indeed, if R is henselian and $I \subsetneq R$ is an arbitrary proper ideal, then R/I is henselian (Remark 20.2) and the bijectivity of $X(R) \to X(k)$ and of $X(R/I) \to X(k)$ implies the bijectivity of $X(R) \to X(R/I)$.

Moreover by Lemma 20.11 it then suffices to show that (i) implies (iii). Let $t_0\colon \operatorname{Spec} k \to X$ be an S-morphism. By Theorem 18.42 there exists an open affine neighborhood $U = \operatorname{Spec} A$ of $t_0(s)$ such that A is standard étale and we may replace X by $\operatorname{Spec} A$ with $A = (R[T]/(f))_g$ such that $f, g \in R[T]$, f monic and f' invertible in A.

The morphism t_0 corresponds to a root $\bar{a} \in k$ of $\bar{f}$ which is not a root of $\bar{g}$. As f' is invertible in A, we have $\bar{f}'(\bar{a}) \neq 0$. Therefore $\bar{a}$ is a simple root of $\bar{f}$ and hence $\bar{f} = (T - \bar{a})\bar{h}$ with $\bar{h}$ prime to $T - \bar{a}$. As R is henselian, we may write $f = (T - a)h$ for some $a \in A$ with image $\bar{a}$ in k (Proposition 20.5). This implies $g(a) \in R^\times$ and $(R[T]/(f))_g \to R$, $T \mapsto a$, defines a section $S \to X$ lifting t_0.

(iii) $\Rightarrow$ (i). By Proposition 20.5 (ii) it suffices to show that for every finite free R-algebra A the map, with $X = \operatorname{Spec} A$,

$$\mathscr{CO}_{X/S}(R) = \operatorname{Clopen}(X) \to \mathscr{CO}_{X/S}(k) = \operatorname{Clopen}(X \otimes_R k)$$

is surjective. This follows from (iii) by Lemma 20.10.

(iii) $\Rightarrow$ (iv). Let $f\colon X \to S$ be a smooth morphism, let $t_0\colon \operatorname{Spec} k \to X$ be an S-morphism and set $x := t_0(s)$. Then $\kappa(x) = \kappa(s)$ and there exists a subscheme Z of X containing x such that $f_{|Z}\colon Z \to S$ is étale (Proposition 20.7). By (iii) we can extend t_0 to an S-section of Z which yields an S-section of X by composing with the inclusion $Z \hookrightarrow X$. $\qquad\square$

Remark 20.13. The proof of Theorem 20.12 shows that in (iii) it suffices to check the surjectivity for $\operatorname{Spec} A \to \operatorname{Spec} R$, where A is a standard étale R-algebra.

One can also show that if $I \subsetneq R$ is an arbitrary proper ideal and X is a smooth R-scheme, then $X(R) \to X(R/I)$ is surjective, see also Theorem 20.15 below for a more general result.

By our definition of smoothness via the "Jacobian criterion" (Definition 6.14) we obtain the following application of Theorem 20.12.

Remark 20.14. Let R be a henselian ring with residue field k and let $f_1, \ldots, f_r \in R[T_1, \ldots, T_n]$. If there exists $\bar{a} = (\bar{a}_1, \ldots, \bar{a}_n) \in k^n$ such that $\bar{f}_i(\bar{a}) = 0$ for all $i = 1, \ldots, r$ and such that $\operatorname{rank}((\partial \bar{f}_i/\partial T_j)(a)) = r$, then there exists $a \in R^n$ with image $\bar{a}$ in k^n such that $f_i(a) = 0$ for all i.

Indeed, we apply Theorem 20.12 to $X = \operatorname{Spec} R[T_1, \ldots, T_n]/(f_1, \ldots, f_r)$. Then $\bar{a}$ corresponds to an R-morphism $t_0 \colon \operatorname{Spec} k \to X$ and the rank hypothesis on the Jacobi matrix implies that X is smooth at the image of t_0.

Moreover, if $r = n$, then X is étale at the image of t_0 and Theorem 20.12 implies that there exists a *unique* $a \in R^n$ as above.

(20.4) Henselian pairs.

Instead of considering the henselian property for a local ring and its maximal ideal, we can also consider the case of an arbitrary ring and an arbitrary ideal. Only recently, e.g. in [Čes] $\overset{\text{o}}{\text{x}}$ and [dJOl] x, there have been substantial new results and we collect here some of them – most of the time without proof.

Definition and Theorem 20.15. *Let A be a ring and let $I \subseteq A$ be an ideal. Then the pair (A, I) is called* henselian *if the following equivalent conditions hold.*
(i) *Every polynomial $f \in A[T]$ with $f(0) \in I$ and $\overline{f'(0)} \in (A/I)^{\times}$ has a root in I.*
(ii) *The ideal I is contained in the Jacobson radical of A and for every polynomial $f \in A[T]$ and every factorization $\bar{f} = \bar{g}\bar{h}$ with $\bar{g}, \bar{h} \in A/I[T]$ generating the unit ideal in $A/I[T]$ and $\bar{g}$ monic, there exists a factorization $f = gh$ in $A[T]$ with g, h lifts of $\bar{g}, \bar{h}$, respectively and g monic.*
(iii) *The ideal I is contained in the Jacobson radical of A and every monic polynomial $f \in A[T]$ of the form $f = T^n(T - 1) + a_n T^n + \cdots + a_1 T + a_0$ with $a_0, \ldots, a_n \in I$ and $n \geq 1$ has a root in $1 + I$.*
(iv) *For every smooth morphism $X \to \operatorname{Spec} A$ of schemes the canonical map $X(A) \to X(A/I)$ is surjective.*
(v) *For every affine étale morphism $X \to \operatorname{Spec} A$ of schemes the canonical map $X(A) \to X(A/I)$ is surjective.*
(vi) *For every étale morphism $X \to \operatorname{Spec} A$ of schemes the canonical map $X(A) \to X(A/I)$ is bijective.*
(vii) *For every integral A-algebra B the map $B \to B/IB$ induces a bijection on idempotents.*
(viii) *The ideal I is contained in the Jacobson radical of A and for every finite A-algebra B the map $B \to B/IB$ induces a surjection on idempotents.*
If these equivalent conditions are satisfied, then the root in (i), the polynomials g and h in (ii), and the root in (iii) are unique.

This is a deep and difficult theorem. Let us explain how it follows from existing results in the literature. By [Sta] 09XI, (ii), (v), (vii), and (iii) are equivalent, and the root in (iii) is unique. The uniqueness of g and h in (ii) follows as in the local case (Remark 20.6). The equivalence of (i), (ii), and (viii) is shown in [KPR] 2.6 together with [KPR] 3.4.3, the uniqueness of the root in (i) is shown in [KPR] 2.2.1.

Clearly, (iv) and (vi) both imply (v). In [Čes] $\overset{\text{o}}{\text{x}}$ 6.1.1 it is shown that (ii) implies (iv). In particular we see that (ii) implies that $X(A) \to X(A/I)$ is surjective for every étale A-scheme. But it is also injective by Lemma 20.11. Hence (ii) also implies (vi).

One has the following permanence properties for henselian pairs.

Proposition 20.16. *Let A be a ring and let $I \subseteq A$ be an ideal.*
(1) *Let $A \to B$ be an integral ring homomorphism. If (A, I) is a henselian pair, then (B, IB) is a henselian pair.*

(2) *Let $J \subseteq I$ be an ideal of A. Then (A, I) is henselian if and only if (A, J) and $(A/J, I/J)$ are henselian.*

(3) *Let $i \mapsto (A_i, I_i)$ be a diagram of pairs of a ring together with an ideal (maps $(A, I) \to (B, J)$ are ring maps $\rho \colon A \to B$ with $\rho(I) \subseteq J$). Suppose that (A_i, I_i) is henselian for all i. Then $\lim_i(A_i, I_i)$ is henselian.*

(4) *Let $i \mapsto (A_i, I_i)$ be a filtered diagram of henselian pairs. Then $(\operatorname{colim}_i A_i, \operatorname{colim} I_i)$ is a henselian pair.*

Proof. Assertion (1) follows from Theorem 20.15 (vii).

Let us show (2) again using Theorem 20.15 (vii). Let B be an integral A-algebra. Consider the ring homomorphisms

$$(*) \qquad\qquad B \longrightarrow B/IB \longrightarrow B/JB$$

If (A, J) and $(A/J, I/J)$ are henselian, then both maps induce bijections on idempotents, hence (A, J) is henselian. Conversely, suppose that (A, J) is henselian. Then $(A/J, I/J)$ is henselian by (1). Hence the ring homomorphisms in $(*)$ induce maps on idempotents such that their composition and the second induced map is bijective. Hence (A, I) is henselian.

The remaining assertions are easy applications of Theorem 20.15 (i) and the uniqueness statement at the end of Theorem 20.15. Let us show (3). Let $A := \lim A_i$ and $f \in A[T]$ with $f(0) \in I := \lim I_i$ and $\overline{f'(0)} \in (A/I)^\times$. Let f_i be the image of f in $A_i[T]$. Then $f_i(0) \in I_i$ and $\overline{f_i'(0)} \in (A_i/I_i)^\times$. Hence f_i has a unique root $a_i \in I_i$ and $(a_i)_i$ yields an element in I which is a root of f.

It remains to show (4). Let $A := \operatorname{colim} A_i$, $I := \operatorname{colim} I_i$ and $f \in A[T]$ with $f(0) \in I$ and $\overline{f'(0)} \in (A/I)^\times$. Then there exists an i such f is represented by $f_i \in A_i[T]$ with $f_i(0) \in I_i$ and $\overline{f_i'(0)} \in (A_i/I_i)^\times$. Hence f_i has a root in I_i and therefore f has a root in I. $\qquad\square$

Proposition 20.17. *Let A be a ring. Then there is a unique largest ideal I of A such that (A, I) is a henselian pair. Moreover, I is equal to its radical.*

In other words, there exists a unique closed subset Z of $\operatorname{Spec} A$ such that (A, I) is henselian for every ideal $I \subset A$ with $Z \subseteq V(I)$. The subset Z necessarily contains every closed point of $\operatorname{Spec} A$ since every such I is contained in the Jacobson radical of A.

This shows in particular that if all elements of I are nilpotent, then (A, I) is a henselian pair.

Proof. Let $\mathcal{H}$ be the set of ideals I of A such that (A, I) is a henselian pair. Clearly $(0) \in \mathcal{H}$ and every totally ordered subset $\mathcal{L} \subseteq \mathcal{H}$ has the upper bound $\bigcup_{I \in \mathcal{L}} I$ in $\mathcal{H}$ by Proposition 20.16 (4). Thus $\mathcal{H}$ has maximal elements by Zorn's lemma. If there existed two different maximal elements in $\mathcal{H}$, say I and J, then we claim that $I + J \in \mathcal{H}$ which would contradict the maximality of I and J. Hence $\mathcal{H}$ has a unique maximal element.

Let us show the claim. As (A, J) is henselian, so is $(A/I, (J + I)/I)$ by Proposition 20.16 (1). Hence $(A, I + J)$ is henselian by Proposition 20.16 (2) since (A, I) is henselian.

It remains to show that if $I \in \mathcal{H}$, then the radical of I is in $\mathcal{H}$. Again by Proposition 20.16 (2), we may replace A by A/I and hence it suffices to show that the nilradical of A is in $\mathcal{H}$. By Theorem 20.15 (i) we have to show the following claim. Let $n \geq 1$ and $f \in A[T]$ with $f(0)^n = 0$ and $f'(0) \in A^\times$. Then there exists a nilpotent root a of f. Here we use that $f'(0)$ is a unit if and only if its image in a quotient by some ideal consisting of nilpotent elements is a unit (Lemma 18.37).

We show the existence of a by induction on n. The case $n = 1$ is trivial. For $n > 1$ we write $f = f(0) + f'(0)T + gT^2$ for some $g \in A[T]$ and set $b := -f'(0)^{-1}f(0)$. Then $f(b) = g(b)b^2$ and hence $f(b)^{n-1} = 0$ since $b^n = 0$. By induction hypothesis applied to $f(b + T)$ we find a nilpotent element a such that $f(b + a) = 0$. As $b + a$ is nilpotent, this implies the claim.   □

Corollary 20.18. *Let A be a ring and let $I \subseteq A$ be an ideal such that $A = \lim_n A/I^n$. Then (A, I) is a henselian pair.*

Proof. By Proposition 20.17, (A, J) is henselian if J contains only nilpotent elements. In particular, the pair $(A/I^n, I/I^n)$ is henselian for all n and this shows the claim as the henselian property is stable under limits (Proposition 20.16 (3)).   □

Using henselian pairs one has the following variants of étale/smooth/unramified morphism.

Definition 20.19. *A morphism $f\colon X \to Y$ of schemes is said to be* weakly smooth *(resp.* weakly unramified, *resp.* weakly étale*) if for every henselian pair (A, I) and every commutative diagram*

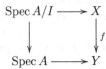

there exists at least (resp. at most, resp. exactly) one morphism $\operatorname{Spec} A \to X$ that makes the resulting diagram commutative.

We stress that the notions of being weakly smooth and weakly unramified are non-standard.

Remark 20.20. Let $f\colon X \to Y$ be a morphism of schemes.
(1) If f is smooth (resp. unramified, resp. étale), then f is weakly smooth (resp. weakly unramified, resp. weakly étale).

   Indeed, for "smooth" this follows from Theorem 20.15 (iv). For "unramified" this follows from Lemma 20.11. Formally, this implies the assertion for "étale".
(2) As any pair (A, I) with I nilpotent is henselian (Proposition 20.17), every weakly smooth (resp. weakly unramified, resp. weakly étale) morphism is formally smooth (resp. formally unramified, resp. formally étale).
(3) Gabber and Ramero defined in [GaRa]$^{\text{o}}_X$ a morphism f to be weakly unramified (resp. weakly étale) if $\Delta_f\colon X \to X \times_Y X$ is flat (resp. if f and Δ_f are flat). De Jong and Olander show in [dJOl]$_X$ Theorem 1, that a morphism is weakly étale in the sense of Gabber-Ramero if and only if it is weakly étale in the sense of Definition 20.19. They also show (loc. cit. Corollary 13) that every weakly unramified morphism in the sense of Gabber-Ramero is weakly unramified in the sense of Definition 20.19.

The étale topology

As explained above we will introduce the étale "topology" by replacing open subsets of a scheme X by étale morphisms $U \to X$. Of course this will not lead to a topology in the usual sense. For instance it does not make sense to speak of the union of opens in the étale topology. But by defining the notion of a covering for the étale topology it makes sense to say that properties are "local for the étale topology". Moreover we will consider for two étale morphisms $U \to X$ and $V \to X$ the fiber product $U \times_X V$ as the "intersection" of $U \to X$ and $V \to X$. Then having the notions of finite intersections and coverings at our disposal we may also speak of "sheaves for the étale topology". For instance the descent results on quasi-coherent modules in Section (14.16) will immediately imply that quasi-coherent modules define sheaves for the étale topology. We will also introduce the notion of a "point in the étale topology" and the stalk of a sheaf for the étale topology. It turns out that the stalk in the étale topology of the structure sheaf of a scheme X is always a henselian ring with separably closed residue field.

(20.5) Étale topology.

Definition 20.21. *Let X be a scheme.*
(1) *An* étale covering *of X is a family of étale morphisms $(g_i \colon U_i \to X)_{i \in I}$ such that $X = \bigcup_{i \in I} g(U_i)$.*
(2) *If $\mathcal{U} = (U_i \to X)_{i \in I}$ and $\mathcal{V} = (V_j \to X)_{j \in J}$ are étale coverings of X, then $\mathcal{V}$ is called a* refinement *of $\mathcal{U}$ if there exists a map $\alpha \colon J \to I$ and for each $j \in J$ an X-morphism $V_j \to U_{\alpha(j)}$.*

Note that $V_j \to U_{\alpha(j)}$ is automatically étale by Remark 18.36.

If one replaces "étale" by "inclusion of an open subscheme" in Definition 20.21, one obtains the usual topological notions.

Remark 20.22. Let X be a quasi-compact scheme. Then for every étale covering $\mathcal{U} = (g_i \colon U_i \to X)_{i \in I}$ of X there exists a finite refinement $(V_j \to X)_{j \in J}$ of $\mathcal{U}$ such that V_j is affine for all $j \in J$.

Indeed, choosing for each U_i an open affine covering, we can replace $\mathcal{U}$ by a refinement such that all U_i are affine. As $g_i(U_i)$ is open in X (g_i is flat and locally of finite presentation and hence open by Theorem 14.35) and as X is quasi-compact, there exists a finite subset J of I such that $X = \bigcup_{j \in J} g(U_j)$.

Remark 20.23. We say that a property **P** of a scheme is *local for the étale topology* if for any scheme X with an étale covering $(U_i \to X)_i$ the scheme X has property **P** if and only if all U_i have property **P**.

Examples for properties that are local for the étale topology on locally noetherian schemes are the properties "reduced", "normal", "regular", "Cohen-Macaulay" and, more generally, the properties "(R_k)" and "(S_k)" (for fixed k) by Corollary 14.60 and Remark 14.61.

Similarly one has the notion that a property **P** of a scheme morphism is local for the étale topology on the source or on the target.

Definition 20.24.
(1) **P** *is* local for the étale topology on the source *if for all morphisms of schemes*
    $f\colon X \to Y$ *and for every étale covering* $(g_i\colon U_i \to X)_i$ *of* X *the morphism* f *satisfies*
    P *if and only if* $f \circ g_i$ *satisfies* **P** *for all* i.
(2) **P** *is* local for the étale topology on the target *if for all morphisms of schemes*
    $f\colon X \to Y$ *and for every étale covering* $(V_j)_j$ *of* Y *the morphism* f *satisfies* **P** *if and*
    only if $f \times \mathrm{id}_{V_j}\colon X \times_Y V_j \to V_j$ *satisfies* **P** *for all* j.

Remark 20.25. Assume that the property "étale" implies **P** and that for all morphisms
$f\colon X \to Y$ and all étale surjective morphisms $g\colon U \to X$ the composition $f \circ g$ has
property **P** if and only if f has property **P**. Then **P** is local on the source for the étale
topology.

Indeed, if $(g_i\colon U_i \to X)_i$ is an étale covering of X, then one can pass to the induced
étale surjective morphism $g\colon \bigsqcup_i U_i \to X$ whose restriction to U_i is g_i.

In view of Appendix C the following proposition yields plenty of properties that are
local for the étale topology on the target.

Proposition 20.26. *Let* **P** *be a property of morphisms of schemes. Assume that* **P** *is*
local on the target for the Zariski topology, stable under étale base change and stable under
faithfully flat descent (Definition 14.54). Then **P** *is local for the étale topology on the*
target.

Proof. Let $f\colon X \to Y$ be a morphism of schemes and let $(V_j \to Y)_{j \in J}$ be an étale covering
of Y. If f satisfies **P**, then its étale base change $f_{(Y_j)}\colon X \times_Y V_j \to V_j$ satisfies **P** for all j
by hypothesis.

Conversely, assume that $f_{(Y_j)}$ satisfies **P** for all j. As **P** is local on the target for the
Zariski topology, we may assume that Y is affine. As **P** is stable under étale base change,
we may replace $(V_j \to Y)_{j \in J}$ by any refinement. Hence we may assume by Remark 20.22
that J is finite and that V_j is affine. Then the morphism $g\colon V := \bigsqcup_j V_j \to Y$, whose
restriction to each V_j is the given $V_j \to X$, is an étale surjective affine morphism. In
particular, it is faithfully flat and quasi-compact. As **P** is local on the target for the
Zariski topology, $f_{(V_j)}$ satisfies **P** if and only if $f_{(V)}$ satisfies **P**. As **P** is stable under
faithfully flat descent, this implies that f has the property **P**.   □

Example 20.27. The property "smooth" is local for the étale topology on the source
(we may apply Remark 20.25 by Theorem 18.74) and on the target (Proposition 20.26).

A morphism $f\colon X \to S$ is locally for the étale topology of the form $\mathbb{A}_S^n \to S$ if and only
if it is smooth (Corollary 18.57).

Remark 20.28. One can also work with other Grothendieck topologies on the category
of schemes by replacing "étale" in Definition 20.21 by other properties **P** of scheme
morphisms. To really get a Grothendieck topology one should assume that **P** is stable
under composition and base change. Usually one also assumes that **P** is local for the
Zariski topology on the source and on the target (to ensure that they define a Grothendieck
topology which is finer than the Zariski topology). As one would like to glue morphisms
of schemes given locally for such a topology one usually assumes that the topology is
coarser than the fpqc-topology, cf. Section (14.18). Moreover, to have an analogue of
Remark 20.22 it is useful to assume that **P** implies "open" (otherwise instead of a naive
analogue of Definition 20.21 a more refined version should be used).

Standard choices for **P** are "syntomic" and "flat, locally of finite presentation" (usually abbreviated as fppf from the French fidèlement plat de présentation finie). For these choices one can use the analogues of Definition 20.21 and Definition 20.24, and Remark 20.22, Remark 20.25, and Proposition 20.26 hold with the same proof. See also Section (27.6) below where fppf-sheaves are studied in more detail.

Another obvious choice would be **P** = "smooth" but Corollary 20.8 implies that every smooth covering can be refined by an étale covering.

For a careful discussion of these matters we refer to the Stacks project [Sta], in particular Chapter 020K.

(20.6) Lifting of étale schemes.

We consider the question what kind of morphisms should be considered as "homeomorphisms for the étale topology". It will turn out that certainly universal homeomorphisms do the job (Theorem 20.30). Hence let us first recall the following characterization of universal homeomorphisms, which is essentially already contained in Volume I.

Proposition 20.29. *Let* $f \colon X \to S$ *be a morphism of schemes. The following assertions are equivalent.*

(i) *f is a universal homeomorphism (i.e. for all morphisms of schemes $S' \to S$ the base change $X \times_S S' \to S'$ of f is a homeomorphism).*

(ii) *f is universally injective, integral, and surjective.*

If f is assumed to be of finite type, then this is Exercise 12.32 whose proof is sketched in Appendix D. Recall that according to Proposition 4.35, f is universally injective if and only if it is injective and all residue class extensions are purely inseparable.

Proof. *"(ii) $\Rightarrow$ (i)"*. As "surjective" is stable under base change, f is universally bijective. As f is integral and "integral" is stable under base change, f is universally closed by Proposition 12.12. Hence f is a universal homeomorphism.

"(i) $\Rightarrow$ (ii)". Let f be a universal homeomorphism. Then f is universally injective and surjective. As f is injective and closed, f is affine (Exercise 12.3). As f is affine and universally closed, f is integral (Exercise 12.19). $\qquad\square$

Examples for universal homeomorphisms are nil-immersions, the Frobenius morphism for a scheme in characteristic p (Exercise 4.17), and the projection $X \otimes_k K \to X$ if X is a scheme over a field k and K/k is a purely inseparable extension (Corollary 5.46).

We denote by $(\text{Ét}/X)$ the category of schemes étale over X. Morphisms in $(\text{Ét}/X)$ are morphisms of X-schemes, automatically étale by Remark 18.36.

Theorem 20.30. *Let* $i \colon X_0 \to X$ *be a universal homeomorphism of schemes. Then the functor $Y \mapsto Y_0 := Y \times_X X_0$ yields an equivalence of categories*

$$\Phi \colon (\text{Ét}/X) \xrightarrow{\sim} (\text{Ét}/X_0).$$

The functor Φ is fully faithful under the weaker hypothesis that i is universally submersive (Corollary 14.43), separated and has geometrically connected fibers (Exercise 20.2).

We will prove that Φ is essentially surjective only if i is a nil-immersion. For the general case see [SGA4]$^\text{O}$ Exp. VIII, Théorème 1.1.

Proof. Φ is fully faithful. Let Y, Y' be étale X-schemes and let $f_0 \colon Y_0 \to Y_0'$ be a morphism of X_0-schemes. Let $\Gamma_{f_0} \subseteq Y_0 \times_{X_0} Y_0' = (Y \times_X Y')_0$ be the graph of f_0. As $Y_0' \to X_0$ is unramified, Γ_{f_0} is an open subscheme of $(Y \times_X Y')_0$ (Remark 18.30). As $(Y \times_X Y')_0 \to Y \times_X Y'$ is a homeomorphism, there exists a unique open subscheme Γ of $Y \times_X Y'$ such that $\Gamma_0 = \Gamma_{f_0}$. Then the composition $\Gamma \to Y \times_X Y' \to Y$ is étale and a universal homeomorphism and hence an isomorphism by Proposition 18.76. In other words Γ is the graph of a necessarily unique morphism $f \colon Y \to Y'$ such that $\Phi(f) = f_0$.

Φ is essentially surjective if i is a nil-immersion. As we already know that Φ is fully faithful and as i is a universal homeomorphism, gluing data for étale schemes over X and for étale schemes over X_0 correspond to each other. Thus we may assume that $X = \operatorname{Spec} A$ and $X_0 = \operatorname{Spec} A_0$ are affine and that we are given an affine étale X_0-scheme $Y_0 = \operatorname{Spec} B_0$, where B_0 is a standard étale algebra (Definition 18.39).

We have $A_0 = A/I$, where I is an ideal consisting only of nilpotent elements. Write $B_0 = (A_0[T]/f_0)_{g_0}$, where $f_0, g_0 \in A_0[T]$ such that f_0 is monic and such that f_0' is invertible in B_0. Choose polynomials $f, g \in A[T]$ lifting f_0 and g_0, respectively, such that f is monic. Set $B := (A[T]/f)_g$. Then $B \otimes_A A_0 \cong B_0$ and f' is invertible in B by Lemma 18.34. Hence B is a standard étale A-algebra which lifts B_0. $\qquad\square$

(20.7) Sheaves in the étale topology.

We will now define what it means to be a sheaf for the étale topology on a scheme S. In Section (14.18) we sketched how to endow the category of all schemes over S with a "topology" and how to define sheaves for such topologies.

Here we consider only schemes that are "open in the étale topology of S", i.e., étale morphisms $U \to S$. Accordingly, we will think of a sheaf (as defined below) as a sheaf on S equipped with the étale topology. So let $(\text{Ét}/S)$ be the category of all étale S-schemes $U \to S$. A morphism $(U \to S) \to (V \to S)$ in $(\text{Ét}/S)$ is an S-morphism $U \to V$ which is automatically étale by Remark 18.36.

For morphisms $j_1 \colon U_1 \to V$ and $j_2 \colon U_2 \to V$ in $(\text{Ét}/S)$, their fiber product $U_1 \times_V U_2$ is again an étale S-scheme because "étale" is stable under base change and composition. This clearly is a fiber product in the category $(\text{Ét}/S)$. If j_1 and j_2 are inclusions of open subschemes, $U_1 \times_V U_2 = U_1 \cap U_2$. Hence we think in general of $U_1 \times_V U_2$ as the "intersection" of two opens in the étale topology.

With this in mind, the definition of a sheaf for the étale topology is verbatim as the definition of a sheaf on a usual topological space (Definition 2.18).

Definition 20.31. *Let S be a scheme.*

(1) *A* presheaf *(of sets) for the étale topology on S is a contravariant functor on $(\text{Ét}/S)$ with values in the category of sets. A morphism of presheaves is a morphism of functors. For a morphism $g \colon U \to V$ in $(\text{Ét}/S)$ we often simply write $s \mapsto s_{|U}$ for the map $\mathscr{F}(g) \colon \mathscr{F}(V) \to \mathscr{F}(U)$.*

(2) *A presheaf $\mathscr{F}$ on S for the étale topology is called a* sheaf *if for every étale S-scheme $U \to S$ and for every étale covering $(g_i \colon U_i \to U)_{i \in I}$ the diagram*

$$(20.7.1) \qquad \mathscr{F}(U) \xrightarrow{\ \rho\ } \prod_{i \in I} \mathscr{F}(U_i) \underset{\sigma'}{\overset{\sigma}{\rightrightarrows}} \prod_{(i,j) \in I \times I} \mathscr{F}(U_i \times_U U_j)$$

is exact, where ρ is the map $s \mapsto (s_{|U_i})_i$ and where σ and σ' are given by $(s_i)_i \mapsto (s_{i|U_i \times_U U_j})_{i,j \in I}$ and by $(s_i)_i \mapsto (s_{j|U_i \times_U U_j})_{i,j \in I}$, respectively. A morphism of sheaves is a morphism of presheaves.

In the same way one also defines (pre-)sheaves of groups or rings on S for the étale topology. Furthermore, given a sheaf of rings $\mathscr{A}_S$ on S for the étale topology one has the obvious notion of an $\mathscr{A}_S$-module for the étale topology.

Remark 20.32. Arguing similarly as in the proof of Proposition 20.26 one sees that a presheaf $\mathscr{F}$ for the étale topology on S is a sheaf if (20.7.1) is exact for all Zariski coverings (i.e., coverings $(j_i\colon U_i \to U)_{i\in I}$, where j_i is the inclusion of an open subscheme) and for all étale coverings of the form $(g\colon V \to U)$, where g is an étale surjective morphism of affine schemes in $(\text{Ét}/S)$.

Proposition 20.33. *Let S be a scheme and let $\mathscr{F}$ be a quasi-coherent $\mathscr{O}_S$-module. Consider $\mathscr{F}$ as a presheaf on S for the étale topology by defining $\mathscr{F}(U) := \Gamma(U, g^*\mathscr{F})$ for every étale morphism $g\colon U \to S$. Then $\mathscr{F}$ is a sheaf for the étale topology.*

Proof. As $\mathscr{F}$ satisfies the sheaf condition of Zariski coverings, it suffices to check that the sequence (20.7.1) is exact for an étale covering of the form $(g\colon U \to V)$ with U and V affine. As g is étale and surjective, g is faithfully flat. Hence Lemma 14.64 yields the exactness of (20.7.1) in this case. □

In fact the proof shows the (much stronger) assertion that $\mathscr{F}$ yields a sheaf for the fpqc-topology in the sense of Section (14.18).

Example 20.34. Let $f\colon X \to S$ be a morphism of schemes and $d \geq 0$ an integer. For every étale X-scheme $U \to X$ define $\mathscr{F}(U) := \Gamma(U, \Omega^d_{U/S})$. Then this is a sheaf for the étale topology on X.

Indeed, for every étale X-scheme $g\colon U \to X$ we have $g^*\Omega^d_{X/S} = \Omega^d_{U/S}$ (use Corollary 18.19 and that exterior powers are compatible with pullback g^*). Hence $\mathscr{F}$ is simply the sheaf for the étale topology attached to the quasi-coherent $\mathscr{O}_X$-module $\Omega^d_{X/S}$ (Proposition 20.33).

Remark 20.35. Once one has a notion of sheaf of abelian groups, one can define a cohomology theory by considering the right derived functors of the global section functor, and more generally the right derived functors of the direct image functor. For the étale topology, this gives rise to the theory of étale cohomology. While the starting point, that is the formalism of derived functors, is similar to the theory of cohomology of $\mathscr{O}_X$-modules and (quasi-)coherent sheaves (for the Zariski topology) that we will develop in the following chapters, it turns out that étale cohomology yields a theory that is "more topological" and closer to theories such as singular cohomology of topological spaces or de Rham cohomology of manifolds. We will not go into this any further, see for example [SGA4]$^\circ$, [SGA4$\frac{1}{2}$]$^\circ$, [Mil1]$^\circ$, [FrKi]$^\circ$, [Tam]$^\circ$ or [Fu]$^\circ$.

(20.8) Points and stalks in the étale topology.

We now come to the stalk of the structure sheaf at a point in the étale topology. We first have to clarify what we mean by a "point". There is the abstract notion of a point of a topos (i.e., of the category of sheaves on a category endowed with a Grothendieck topology), see [SGA4]$^\circ$ Exp. IV, §6. We will not need this abstract notion here. One can show ([Sta] 04HU) that for a scheme endowed with the étale topology the points in this abstract sense are given (up to isomorphism) by geometric points in the following sense.

Definition 20.36. *Let S be a scheme. A* geometric point *of S is a morphism $\operatorname{Spec} k \to S$, where k is an algebraically closed field.*

Such a geometric point $\operatorname{Spec} k \to S$ is usually denoted by $\bar{s}$ if $s \in S$ is the image of $\operatorname{Spec} k \to S$. One also says that $\bar{s}$ lies over s. The field k is then denoted by $\kappa(\bar{s})$. Of course there are different geometric points lying over the same (topological) point of S.

If $f\colon S \to T$ is a morphism of schemes, we often denote the geometric point $f \circ \bar{s}$ simply by $f(\bar{s})$.

More generally, every morphism $\operatorname{Spec} k \to S$, where k is separably closed, gives rise to a point for the étale topology on a scheme S in the abstract sense, and sometimes the notion "geometric point" is defined accordingly.

Definition 20.37. *Let S be a scheme, $s \in S$ and let $\bar{s}\colon \operatorname{Spec} \kappa(\bar{s}) \to S$ be a geometric point lying over s. An* étale neighborhood *of $\bar{s}$ is a pair $(U, \bar{u})$, where $U \to S$ is an étale S-scheme and where $\bar{u}\colon \operatorname{Spec} \kappa(\bar{s}) \to U$ is a geometric point such that $g(\bar{u}) = \bar{s}$.*

We will now define two notions of stalks of the structure sheaf, one where we consider only étale neighborhoods with trivial residue field extension and the second where we will consider étale neighborhoods whose residue field extension is a subextension of a given geometric point. We first define the corresponding categories of neighborhoods and then show that they are cofiltered to obtain a well behaved colimit.

Let S be a scheme, $s \in S$ and let $\bar{s}\colon \operatorname{Spec} \kappa(\bar{s}) \to S$ be a geometric point lying over s. Let $\mathcal{U}(\bar{s})$ be the category of étale neighborhoods $(U, \bar{u})$ of $(S, \bar{s})$. A morphism $(U, \bar{u}) \to (V, \bar{v})$ in this category is an S-morphism $h\colon U \to V$ such $h(\bar{u}) = \bar{v}$.

Let $\mathcal{U}_0(s)$ be the full subcategory of $\mathcal{U}(\bar{s})$ consisting of those étale neighborhoods $(U, \bar{u})$ such that if $u \in U$ is the image of $\bar{u}$, the étale morphism $U \to S$ induces an isomorphism $\kappa(s) \overset{\sim}{\to} \kappa(u)$. Then the choice of $u \in U$ already determines $\bar{u}$. Therefore $\mathcal{U}_0(s)$ can also be described as the category of pairs (U, u), where $g\colon U \to S$ is an étale S-scheme and where $u \in U$ is a point with $g(u) = s$ and $\kappa(s) \cong \kappa(u)$. In particular it depends only on s and not on $\bar{s}$.

Lemma 20.38. *The categories $\mathcal{U}(\bar{s})$ and $\mathcal{U}_0(s)$ are cofiltered (Definition F.12).*

Proof. If $(U_i, \bar{u}_i)$, $i = 1, 2$ are étale neighborhoods of $\bar{s}$, then there exists an étale neighborhood $(U, \bar{u})$ and morphisms $(U, \bar{u}) \to (U_i, \bar{u}_i)$, namely $(U, \bar{u}) = (U_1 \times_S U_2, (\bar{u}_1, \bar{u}_2)_S)$. Let $h_1, h_2\colon (U, \bar{u}) \to (U', \bar{u}')$ be two morphisms in $\mathcal{U}(\bar{s})$. Then the inclusion $j\colon U'' := \operatorname{Ker}(h_1, h_2) \to U$ satisfies $h_1 \circ j = h_2 \circ j$. As $h_1(\bar{u}) = h_2(\bar{u})$, $\bar{u}$ factors through a geometric point $\bar{u}''$ of U''. As $U' \to S$ is unramified, U'' is an open subscheme of U (Remark 18.30). Hence it is étale over S and $(U'', \bar{u}'')$ is an étale neighborhood of $\bar{s}$. This shows that $\mathcal{U}(\bar{s})$ is cofiltered.

The same argument (using Lemma 4.28 to see that under the above constructions extensions of residue fields stay trivial) shows that $\mathcal{U}_0(s)$ is cofiltered. $\square$

Definition 20.39. *Let S be a scheme, let $\mathscr{F}$ be a presheaf on S for the étale topology, and let $\bar{s}\colon \operatorname{Spec} \kappa(\bar{s}) \to S$ be a geometric point. The* stalk *of $\mathscr{F}$ in $\bar{s}$ is defined as*

$$\mathscr{F}_{\bar{s}} := \underset{\mathcal{U}(\bar{s})}{\operatorname{colim}} \mathscr{F}(U).$$

The colimit here "makes sense" because the category $\mathcal{U}(\bar{s})$ admits a cofinal small category.

(20.9) Stalks of the structure sheaf: (strict) henselization.

As every quasi-coherent module defines a sheaf for the étale topology (Proposition 20.33), this holds in particular for the structure sheaf of a scheme S. We will now study its stalk $\mathcal{O}_{S,\bar{s}}$ in a geometric point $\bar{s}$ lying over a point $s \in S$. We also define

(20.9.1)
$$\mathcal{O}^h_{S,s} := \operatorname*{colim}_{\mathcal{U}_0(s)} \mathcal{O}_S(U),$$

i.e., here we take the limit only over étale neighborhoods inducing a trivial residue field extension. As the category of Zariski open neighborhoods is a full subcategory of $\mathcal{U}_0(s)$, and as $\mathcal{U}_0(s)$ is a full subcategory of $\mathcal{U}(\bar{s})$ we obtain ring homomorphism

(20.9.2)
$$\mathcal{O}_{S,s} \longrightarrow \mathcal{O}^h_{S,s} \longrightarrow \mathcal{O}_{S,\bar{s}}.$$

Definition 20.40. *A local henselian ring with separably closed residue field is called strictly henselian.*

Proposition 20.41. *Let S be a scheme, let $s \in S$, let $\mathfrak{m}_s \subset \mathcal{O}_{S,s}$ be the maximal ideal, and let $\bar{s} \to S$ be a geometric point lying over s.*
(1) *Then $\mathcal{O}^h_{S,s}$ is a henselian ring with residue field $\kappa(s)$, and $\mathcal{O}_{S,\bar{s}}$ is a strictly henselian ring whose residue field is the separable closure of $\kappa(s)$ in $\kappa(\bar{s})$.*
(2) *The homomorphisms (20.9.2) are local and faithfully flat. Moreover, $\mathfrak{m}_s\mathcal{O}^h_{S,s}$ is the maximal ideal of $\mathcal{O}^h_{S,s}$, and $\mathfrak{m}_s\mathcal{O}_{S,\bar{s}}$ is the maximal ideal of $\mathcal{O}_{S,\bar{s}}$.*

Proof. As "étale" is compatible with filtered inductive limits, we may replace S by $\operatorname{Spec} R$, where $R = \mathcal{O}_{S,s}$. Let $\mathfrak{m}$ be the maximal ideal of R and k its residue field.

Let $a \in R^h \setminus \mathfrak{m}R^h$. Then a is represented by a triple (U, u, a_U), where $(U, u) \in \mathcal{U}_0(s)$ and $a_U \in \mathcal{O}_U(U)$. By shrinking U we may assume that $U = \operatorname{Spec} A$ is affine and that u is the only point of U lying over s. Then its fiber U_s is isomorphic to $\operatorname{Spec}\kappa(u)$ because $U \to S$ is étale (Corollary 18.45). In other words, $A/\mathfrak{m}A = \kappa(u)$ and hence $\mathfrak{p}_u = \mathfrak{m}A$, where $\mathfrak{p}_u \subset A$ is the prime ideal corresponding to u. As $a \notin \mathfrak{m}R^h$ we find $a_U \notin \mathfrak{m}A = \mathfrak{p}_u$. Replacing U by the open neighborhood $D(a_U)$ of u, we may assume that a_U is invertible. Thus a is invertible in R^h. This shows that R^h is a local ring with maximal ideal $\mathfrak{m}R^h$. Its residue field is the compositum of all residue field extensions $\kappa(u)$ of k within $\kappa(\bar{s})$ hence equal to k because $\kappa(u) = k$ for all $(U, u) \in \mathcal{U}_0(s)$. By restricting to the cofinal subcategory of $\mathcal{U}_0(s)$ of those (U, u) such that U is affine, one sees that R^h is a filtered colimit of flat R-algebras (because étale morphisms are flat). Therefore R^h is flat over R and even faithfully flat because $R \to R^h$ is local (Example B.18).

Let us show that R^h is henselian. Let A be a standard étale R^h-algebra and set $X = \operatorname{Spec} A$. By Remark 20.13 it suffices to show that $X(R^h) \to X(k)$ is surjective. By Remark 18.41 there exists an affine étale neighborhood $(\operatorname{Spec} C, u) \in \mathcal{U}_0(s)$ and a standard étale C-algebra $\tilde{A}$ such that $\tilde{A} \otimes_C R^h = A$. An element $x \in X(k)$ yields a point $\tilde{u} \in \tilde{U} := \operatorname{Spec} \tilde{A}$ such that $k = \kappa(\tilde{u})$. As C is étale over R, $(\tilde{U}, \tilde{u})$ is an object in $\mathcal{U}_0(s)$. Therefore by the definition of R^h there exists a C-algebra homomorphism $\tilde{A} \to R^h$ or, equivalently, an element of $X(R^h)$ lifting x.

The same arguments show the claims for $\mathcal{O}_{S,\bar{s}}$ using that for every $(U, \bar{u}) \in \mathcal{U}(\bar{s})$, $\bar{u}$ lying over $u \in U$, the residue extension $\kappa(s) \to \kappa(u)$ is finite and separable. $\qquad\square$

Definition and Remark 20.42. Let S be a scheme and let $s \in S$.
(1) The ring $\mathscr{O}_{S,s}^{\mathrm{h}}$ is called the *henselization of S in s*. In particular we have for any local ring R its *henselization* R^{h} (for $S = \operatorname{Spec} R$, s its closed point).
(2) Let $\bar{s}, \bar{s}' \to S$ be two geometric points lying over s. Then, after possibly switching $\bar{s}$ and $\bar{s}'$, we may assume that there exists an S-morphism $\bar{s} \to \bar{s}'$. Composition with this morphism induces an equivalence of categories $\mathcal{U}(\bar{s}') \to \mathcal{U}(\bar{s})$. In particular, $\mathscr{O}_{S,\bar{s}}$ and $\mathscr{O}_{S,\bar{s}'}$ are isomorphic.

Therefore $\mathscr{O}_{S,\bar{s}}$ up to isomorphism (but not up to unique isomorphism) depends only on S and s and we denote it by $\mathscr{O}_{S,s}^{\mathrm{hs}}$ and call it a *strict henselization of $\mathscr{O}_{S,s}$*. In particular we have defined for a local ring R a *strict henselization* R^{hs}.

As usual we base the abbreviation "hs" on the French expression "henselisé strict".

If R is a normal local ring, then Exercise 20.26 gives another description of R^{h} and of R^{hs}, see also Exercise 20.27.

Remark 20.43. Let R be a local ring. Then $R \to R^{\mathrm{h}}$ yields an isomorphism of the completions $\hat{R} \to \widehat{R^{\mathrm{h}}}$ (with respect to the adic topologies given by maximal ideals).

Indeed, if R is noetherian, this follows from Corollary 18.66. For the general case we refer to [EGAIV]^{O} (18.6.6).

Many properties that are local for the étale topology can therefore be checked after a base change to a strictly henselian ring:

Remark 20.44. Assume that $\mathbf{P}$ is a property of morphisms of schemes that is local for the étale topology on the target and that is compatible with filtered inductive limits. Let $f \colon X \to S$ be a scheme morphism such that for every $s \in S$ the base change $f_{(\mathscr{O}_{S,s}^{\mathrm{hs}})} \colon X \times_S \operatorname{Spec} \mathscr{O}_{S,s}^{\mathrm{hs}} \to \operatorname{Spec} \mathscr{O}_{S,s}^{\mathrm{hs}}$ has property $\mathbf{P}$. Then f has property $\mathbf{P}$.

As an example of this principle we prove the following variant of Zariski's main theorem.

Proposition 20.45. *Let $f \colon X \to S$ be a morphism locally of finite type, $x \in X$, $s = f(x)$. Assume that x is an isolated point of the fiber X_s. Then there exists (U, u) in $\mathcal{U}_0(s)$ such that every point $x' \in X_U := X \times_S U$ lying over x and over u has an open affine neighborhood V' in X_U such that $f_{(U)|V'} \colon V' \to U$ is a finite morphism.*

Moreover, if f is separated, then V' is open and closed in X_U.

We will give the proof only if f is locally of finite presentation. The general case then follows by a standard limit argument in the spirit of Chapter 10 (see [EGAIV]^{O} (18.12.1)).

Proof. If f is separated, $f_{(U)}$ is separated. If $j' \colon V' \to X_U$ denotes the inclusion, then if $f_{(U)} \circ j'$ is finite, j' is finite by the standard cancellation argument (Remark 9.11). Hence j' is closed. This proves the final assertion.

As the question is local on X and on S, we may assume that $S = \operatorname{Spec} R$ and $X = \operatorname{Spec} A$, where A is an R-algebra of finite presentation. By Zariski's main theorem (Corollary 12.79) the points of X that are isolated in their fiber form an open subset of X. Thus by restriction to an affine open neighborhood of x we can assume that f is quasi-finite.

After base changing f with $T := \operatorname{Spec} \mathscr{O}_{S,s}^{\mathrm{h}} \to S$ we find for every point $\tilde{x} \in \tilde{X} := X \times_S T$ lying over the closed point of T an open and closed affine neighborhood $\tilde{V}$ of $\tilde{x}$ in $\tilde{X}$ such that $f_{(T)|\tilde{V}}$ is finite (Proposition 20.5).

As $\tilde{V}$ is of finite presentation over T and as the property "finite" is compatible with filtered inductive limits, we obtain the desired result because of the definition of $\mathscr{O}_{S,s}^{\mathrm{h}}$. □

We conclude this section by giving examples of properties of rings that can be checked on the (strict) henselization. For later use we also consider the following definition.

Definition 20.46. *A local noetherian ring R is called a* G-ring *if all fibers of* $\operatorname{Spec} \hat{R} \to \operatorname{Spec} R$ *are geometrically regular.*

A quasi-excellent local noetherian ring is a G-ring by definition.

Proposition 20.47. *Let R be a local ring.*
(1) *The following assertions are equivalent.*
  (i) *R is noetherian (resp. normal, resp. reduced).*
  (ii) *R^{h} is noetherian (resp. normal, resp. reduced).*
  (iii) *R^{hs} is noetherian (resp. normal, resp. reduced).*
(2) *Assume that R is noetherian. Then the following assertions are equivalent.*
  (i) *R is regular (resp. Cohen-Macaulay, resp. satisfies (R_k), resp. satisfies (S_k), resp. G-ring).*
  (ii) *R^{h} is regular (resp. Cohen-Macaulay, resp. satisfies (R_k), resp. satisfies (S_k), resp. G-ring).*
  (iii) *R^{hs} is regular (resp. Cohen-Macaulay, resp. satisfies (R_k), resp. satisfies (S_k), resp. G-ring).*
(3) *If R is excellent, then R^{h} is excellent.*

For the proof we refer to [EGAIV] ⁰ 18.6 – 18.8.

There is also a global analogue of the henselization of a local ring. For this consider the category of pairs (A, I) with A a ring and $I \subseteq A$ an ideal. Morphisms $(A, I) \to (B, J)$ in this category are ring homomorphisms $\varphi \colon A \to B$ such that $\varphi(I) \subseteq J$. Then we have the following result (see [Sta] 0A02, [Sta] 0AGU, [Sta] 0F0L).

Theorem 20.48. *The inclusion functor from the full subcategory of henselian pairs (A, I) into the category of all pairs has a left adjoint $(A, I) \mapsto (A^{\mathrm{h}}, I^{\mathrm{h}})$. The unit map $A \to A^{\mathrm{h}}$ is flat, $I^{\mathrm{h}} = IA^{\mathrm{h}}$ and $A/I^n \cong A^{\mathrm{h}}/I^n A^{\mathrm{h}}$ for all $n \geq 1$. Moreover, $(A^{\mathrm{h}}, I^{\mathrm{h}})$ depends only on A and the closed subset $V(I)$ of $\operatorname{Spec} A$.*

The construction of A^{h} is the same as in the local case: One defines A^{h} as the filtered colimit over all étale A-algebras B such that $A/I \to B/IB$ is an isomorphism and set $I^{\mathrm{h}} = IA^{\mathrm{h}}$. In particular, for local rings A and I their maximal ideal we recover the notion of henselization from Definition 20.42.

(20.10) Unibranch schemes.

Irreducibility is *not* local for the étale topology. To make this more precise, recall that for a scheme X and a point $x \in X$ the irreducible components of X containing x are in bijection to the minimal prime ideals of $\mathscr{O}_{X,x}$. Hence x is contained only in a single irreducible component if and only if $\mathscr{O}_{X,x}$ contains a unique prime ideal (if and only if $\operatorname{Spec} \mathscr{O}_{X,x}$ is irreducible).

As the ring homomorphisms $\mathcal{O}_{X,x} \to \mathcal{O}_{X,x}^{\mathrm{h}} \to \mathcal{O}_{X,x}^{\mathrm{hs}}$ are faithfully flat and in particular injective, one has implications

$$\text{Spec } \mathcal{O}_{X,x}^{\mathrm{hs}} \text{ is irreducible} \Rightarrow \text{Spec } \mathcal{O}_{X,x}^{\mathrm{h}} \text{ is irreducible} \Rightarrow \text{Spec } \mathcal{O}_{X,x} \text{ is irreducible.}$$

But the example of a node of a curve (Exercise 20.22) shows that the converse does not hold in general (even if $\mathcal{O}_{X,x}$ is noetherian, reduced and of dimension 1).

We think of the irreducible components of Spec $\mathcal{O}_{X,x}^{\mathrm{h}}$ as "branches through x". This leads us to the following definition.

Definition 20.49. *A local ring A is called* unibranch *(resp. geometrically unibranch) if* Spec A^{h} *(resp. Spec A^{hs}) is irreducible. A scheme X is called* unibranch *(resp. geometrically unibranch) if $\mathcal{O}_{X,x}$ is a unibranch (resp. geometrically unibranch) local ring for all $x \in X$.*

The following result relates minimal prime ideals of the (strict) henselization with maximal ideals of the normalization. For this we denote the set of minimal (resp. of maximal) prime ideals of a ring R by $\mathrm{Min}(R)$ (resp. by $\mathrm{Max}(R)$).

Lemma 20.50. *Let A be a local integral domain with maximal ideal $\mathfrak{m}$ and let B be its normalization. Let $A \to \tilde{A}$ be a local ring homomorphism such that $\tilde{A}$ is henselian and such that $\tilde{A}$ is a filtered colimit of étale A-algebras (e.g., $\tilde{A} = A^{\mathrm{h}}$ or $\tilde{A} = A^{\mathrm{hs}}$). Let $\tilde{k}$ be the residue field of $\tilde{A}$ and set $\tilde{B} := B \otimes_A \tilde{A}$. Then there exist bijections*

$$\mathrm{Min}(\tilde{A}) \longleftrightarrow \mathrm{Min}(\tilde{B}) \longleftrightarrow \mathrm{Max}(\tilde{B}) \longleftrightarrow \mathrm{Spec}(B \otimes_A \tilde{k}).$$

Proof. As B is integral over A, $\tilde{B}$ is integral over $\tilde{A}$. Hence "going up" (Theorem B.56) shows that the maximal ideals of $\tilde{B}$ are the prime ideals lying over the maximal ideal of $\tilde{A}$. This proves the last bijection.

Let us construct the first bijection. As $\tilde{A}$ is a filtered inductive limit of étale A-algebras, it is flat over A. As flat homomorphisms are generizing (Lemma 14.9), every minimal prime ideal of $\tilde{A}$ lies over the unique minimal prime ideal of A. In other words $\mathrm{Min}(\tilde{A}) = \mathrm{Min}(\tilde{A} \otimes_A K)$, where K is the field of fractions of A. As $\tilde{A}$ is flat over A, $\tilde{B}$ is flat over B and hence we also have $\mathrm{Min}(\tilde{B}) = \mathrm{Min}(\tilde{B} \otimes_B K)$ (as B is the normalization of A, B has the same field of fractions as A). But by definition of $\tilde{B}$ we have $\tilde{B} \otimes_B K \cong \tilde{A} \otimes_A K$. This yields the first bijection.

It remains to construct a bijection between $\mathrm{Min}(\tilde{B})$ and $\mathrm{Max}(\tilde{B})$. If $\tilde{A}$ is the filtered colimit of étale A-algebras A_λ, then $B_\lambda := B \otimes_A A_\lambda$ is étale over B. As B is normal, B_λ is normal (Corollary 14.60), whence $\tilde{B} = \mathrm{colim} \, B_\lambda$ is normal. Hence for $\tilde{\mathfrak{n}} \in \mathrm{Max}(\tilde{B})$ the local ring $\tilde{B}_{\tilde{\mathfrak{n}}}$ is local and normal and hence an integral domain. Hence there exists a unique minimal prime ideal of $\tilde{B}$ contained in $\tilde{\mathfrak{n}}$, which corresponds to the zero ideal in $\tilde{B}_{\tilde{\mathfrak{n}}}$. This defines a map $\mu \colon \mathrm{Max}(\tilde{B}) \to \mathrm{Min}(\tilde{B})$. We claim that μ is bijective.

Clearly, it is surjective as every minimal prime ideal is contained in some maximal ideal. To show the injectivity of μ let $\tilde{\mathfrak{n}}$ and $\tilde{\mathfrak{n}}'$ be distinct maximal ideal of $\tilde{B}$. Now $\tilde{B}$ is integral over $\tilde{A}$ and hence a filtered colimit of finite $\tilde{A}$-algebras $\tilde{B}_\alpha$. As $\tilde{A}$ is henselian, each $\tilde{B}_\alpha$ is a product of local rings. As the set idempotents in $\tilde{B}$ is the filtered colimit of the sets of idempotents of the $\tilde{B}_\alpha$, there exists an idempotent $e \in \tilde{B}$ with value 0 in $\kappa(\tilde{\mathfrak{n}})$ and value 1 in $\kappa(\tilde{\mathfrak{n}}')$. Therefore Spec $\tilde{B}$ is the disjoint union of two open and closed subschemes Y and Y' with $\tilde{\mathfrak{n}} \in Y$ and $\tilde{\mathfrak{n}}' \in Y'$. But then $\mu(\tilde{\mathfrak{n}}) \in Y$ and $\mu(\tilde{\mathfrak{n}}') \in Y'$ and we obtain that $\mu(\tilde{\mathfrak{n}}) \neq \mu(\tilde{\mathfrak{n}}')$. $\square$

Proposition 20.51. *Let A be a local integral domain with residue field k.*
(1) *A is unibranch if and only if its normalization is a local ring.*
(2) *A is geometrically unibranch if and only if its normalization is a local ring whose residue field is a purely inseparable extension of k.*

It is not difficult to generalize this Proposition to arbitrary local rings A (Exercise 20.20).

Proof. Let B be the normalization of A. To see (1), we apply Lemma 20.50 to $\tilde{A} = A^{\mathrm{h}}$. As the residue field $\tilde{k}$ of $\tilde{A}$ is equal to k, we find a bijection between $\mathrm{Min}(A^{\mathrm{h}})$ and $\mathrm{Spec}(B \otimes_A k)$. As B is integral over A we deduce a bijection $\mathrm{Min}(A^{\mathrm{h}}) \leftrightarrow \mathrm{Max}(B)$. This shows (1).

Let us prove (2). Let A^{hs} be a strict henselization of A and let k^{sep} be its residue field. Applying Lemma 20.50 to $\tilde{A} = A^{\mathrm{hs}}$ we obtain a bijection $\mathrm{Min}(A^{\mathrm{hs}}) \leftrightarrow \mathrm{Spec}(B \otimes_A k^{\mathrm{sep}})$. Let $(\mathfrak{n}_i)_{i \in I}$ be the family of maximal ideals of B and let $k_i = \kappa(\mathfrak{n}_i)$ their residue fields. As B is integral over A, k_i is an algebraic extension of k and hence $\# \mathrm{Max}(k_i \otimes_k k^{\mathrm{sep}}) = [k_i : k]_{sep} \in \mathbb{Z}_{\geq 1} \cup \{\infty\}$. Therefore we obtain an equality in $\mathbb{Z}_{\geq 1} \cup \{\infty\}$

$$\# \mathrm{Min}(A^{\mathrm{hs}}) = \# \mathrm{Spec}(B \otimes_A k^{\mathrm{sep}}) = \sum_{i \in I} \# \mathrm{Max}(k_i \otimes_k k^{\mathrm{sep}}) = \sum_{i \in I} [k_i : k]_{sep}.$$

This shows in particular that A is geometrically unibranch if and only if $\#I = 1$ (i.e., B is local) and $[\kappa_B : k]_{sep} = 1$, where κ_B is the residue field of B. □

Corollary 20.52. *Let X be an integral scheme and let $\pi\colon X' \to X$ be its normalization.*
(1) *X is unibranch if and only if π is injective. In this case π is a homeomorphism.*
(2) *X is geometrically unibranch if and only if π is universally injective. In this case π is a universal homeomorphism.*

Proof. We may assume that $X = \mathrm{Spec}\, A$ is affine. Then $X' = \mathrm{Spec}\, A'$ is affine. For $\mathfrak{p} \in \mathrm{Spec}\, A$ the normalization of $A_{\mathfrak{p}}$ is $A'_{\mathfrak{p}}$ (Proposition B.55) and the prime ideals of $A'_{\mathfrak{p}}$ over the maximal ideal $\mathfrak{p} A_{\mathfrak{p}}$ are the maximal ideals of $A'_{\mathfrak{p}}$ by "going up". Hence π is injective (resp. universally injective) if and only if for all $\mathfrak{p} \in \mathrm{Spec}\, A$ the ring $A'_{\mathfrak{p}}$ is a local ring (resp. if and only if $A'_{\mathfrak{p}}$ is a local ring whose residue field is a purely inseparable extension of $\kappa(\mathfrak{p})$ (Proposition 4.35)), i.e., if and only if $A_{\mathfrak{p}}$ is unibranch (resp. if and only if $A_{\mathfrak{p}}$ is geometrically unibranch) for all $\mathfrak{p}$ (Proposition 20.51).

As π is surjective and integral, it is surjective and universally closed. Hence if π is (universally) injective, then π is a (universal) homeomorphism. □

Corollary 20.53. *Every normal scheme is geometrically unibranch.*

The converse does not hold as the example of a cusp shows (Exercise 20.22).

(20.11) Artin approximation.

Let A be a local noetherian ring. Very often it is possible to construct solutions for polynomial equations in the completion $\hat{A}$ of A and one would like to deduce the existence of solutions in A which are arbitrarily close to the given solution in $\hat{A}$. More precisely, we will consider the following property.

Definition 20.54. *A local noetherian ring A with maximal ideal $\mathfrak{m}$ is said to have the approximation property if given polynomials $f_1, \ldots, f_m \in A[T_1, \ldots, T_n]$, a solution $\hat{a} = (\hat{a}_1, \ldots, \hat{a}_n)$ in $\hat{A}$ of the system of equations $f_1 = \cdots = f_m = 0$, and an integer $c \geq 1$, there exists a solution $a = (a_1, \ldots, a_n) \in A^n$ such that $a_i \equiv \hat{a}_i \pmod{\mathfrak{m}^c}$ for all $i = 1, \ldots, n$.*

Clearly, not every local noetherian ring has the approximation theory: Let $2 \neq p \in \mathbb{Z}$ be a prime. Then $T^2 + 1 = 0$ has no solution in $\mathbb{Q}$ and in particular not in $\mathbb{Z}_{(p)}$. But it has a solution in the completion $\mathbb{Z}_p$ if (and only if) $p \equiv 1 \bmod 4$: As $\mathbb{Z}_p$ is complete, it is henselian and it suffices to find a solution in $\mathbb{F}_p$. But if $p = 1 + 4n$, then $(2n)!$ is a solution of $T^2 = -1$ in $\mathbb{F}_p$, as follows from Wilson's theorem.

In Remark 20.14 we have seen that A is henselian if and only if for certain polynomial equations any solution in the residue field of A lifts to a solution in A. As $\hat{A}$ is henselian (Example 20.3), and A and $\hat{A}$ have the same residue field, we see that every local noetherian ring with the approximation property is necessarily henselian. Moreover, in [Rot] ^{O} it is shown that any local noetherian ring with the approximation property is excellent (and henselian).

Recall Popescu's theorem (Theorem 10.76) that if $\operatorname{Spec} B \to \operatorname{Spec} A$ is a regular morphism of noetherian affine schemes, then B is a filtered colimit of smooth A-algebras. In particular Popescu's theorem implies that a local noetherian ring A is a G-ring (Definition 20.46) if and only if $\hat{A}$ is a filtered colimit of smooth A-algebras.

It is not difficult to see that if A is henselian and $\hat{A}$ is a filtered colimit of smooth A-algebras, then A has the approximation theory (cf. [Art5]). Finally, if A is excellent, then by definition A is a G-ring. Therefore Popescu's theorem implies the following result.

Theorem 20.55. *Let A be a local noetherian ring. Then the following assertions are equivalent.*
(i) *A has the approximation property.*
(ii) *A is henselian and excellent.*
(iii) *A is henselian and $\hat{A}$ is a filtered colimit of smooth A-algebras.*

Example 20.56. This result can in particular be applied to the case where A is the local ring of germs of analytic functions on a complex analytic space X in a point x. In fact, A is henselian by Exercise 20.15. Since quotients of excellent rings are excellent ([EGAIV] ^{O} (7.8.3) (ii)), to show that A is excellent it is enough to consider the case of a smooth point. But if x is a smooth point, then $\mathcal{O}_{X,x}$ is the ring of convergent complex power series in $\dim(X)$ variables, and $\widehat{\mathcal{O}}_{X,x}$ is the ring of complex formal power series. It follows from [Mat1] Theorem 102 that $\mathcal{O}_{X,x}$ is excellent.

Corollary 20.57. *Let $X \to S$ be a morphism locally of finite type of locally noetherian schemes, and let $s \in S$ such that $\mathcal{O}_{S,s}$ is a G-ring. Assume that the base change $X \times_S \operatorname{Spec} \widehat{\mathcal{O}}_{S,s} \to \operatorname{Spec} \widehat{\mathcal{O}}_{S,s}$ of f has a section t_0. Then there exist an étale morphism $g \colon U \to S$ and $u \in U$ with $g(u) = s$ and $\kappa(s) \cong \kappa(u)$ such that the base change $X \times_S U \to U$ of f has a section t.*

Moreover, given an integer $c \geq 1$ one can choose t such that the images of t and of t_0 in $X(\mathcal{O}_{S,s}^{\text{h}}/\mathfrak{m}_s^c \mathcal{O}_{S,s}^{\text{h}}) = X(\mathcal{O}_{S,s}/\mathfrak{m}_s^c) = X(\widehat{\mathcal{O}}_{S,s}/\mathfrak{m}_s^c \widehat{\mathcal{O}}_{S,s})$ coincide.

It is easy to generalize this assertion from sections of an S-scheme X, i.e., values of the functor h_X, to values of an arbitrary functor $F \colon (\text{Sch})S^{\text{opp}} \to (\text{Sets})$ which is locally of finite presentation in the sense of Remark 18.61 (see [Art2] ^{O} Corollary (1.8)).

Proof. We may assume that $S = \operatorname{Spec} A'$ and $X = \operatorname{Spec} A'[T_1, \ldots, T_n]/(f_1, \ldots, f_m)$ are affine. Let $A := \mathscr{O}_{S,s}$. Then a section of the base change $f_{(\operatorname{Spec} \hat{A})}$ corresponds to a solution $\hat{a} \in \hat{A}^n$ of $f_1 = \cdots = f_m = 0$. As A is a G-ring, its henselization A^{h} is a G-ring (Proposition 20.47) whose completion is $\hat{A}$ (Remark 20.43). Hence A^{h} has the approximation property and we find a section of the base change $f_{(\operatorname{Spec} A^{\mathrm{h}})}$. The corollary follows from the definition of A^{h} (20.9.1). $\qquad\square$

The hypothesis that $\mathscr{O}_{S,s}$ is a G-ring is in practice very often satisfied: Recall from Section (12.12) that if R is a complete local noetherian ring (e.g., if R is a field) or if R a Dedekind ring whose field of fractions has characteristic zero (e.g., if R is the ring of integers in a number field), then every scheme S locally of finite type over R is excellent. In particular, every point $s \in S$ satisfies the condition of the corollary.

(20.12) Analytification of schemes over $\mathbb{C}$.

Let X be a scheme locally of finite type over $\mathbb{C}$. It is possible to attach to X a complex analytic space X^{an} whose underlying set is $X(\mathbb{C})$. A similar construction is possible for other topological fields (such as $\mathbb{Q}_p$) with the correct replacement for "complex analytic space" (such as rigid analytic space or adic space). Here we concentrate on the complex case. Let us first recall the definition of a complex analytic space. For $n \geq 0$ we endow $\mathbb{C}^n$ with the analytic topology and for $U \subseteq \mathbb{C}^n$ open we denote by $\mathscr{O}_U$ the sheaf of holomorphic functions on U, i.e., for $V \subseteq U$ open, $\mathscr{O}_U(V)$ is the set of holomorphic maps $V \to \mathbb{C}$.

Let $U \subseteq \mathbb{C}^n$ be an open subset, let $f_1, \ldots, f_r \colon U \to \mathbb{C}$ be holomorphic functions, and let $\mathscr{J} \subseteq \mathscr{O}_U$ be the ideal generated by $f_1, \ldots, f_r$. Let $X = \{\, x \in U \;;\; f_1(x) = \cdots = f_r(x) = 0 \,\}$ and define $\mathscr{O}_X := (\mathscr{O}_U/\mathscr{J})_{|X}$. Then $(X, \mathscr{O}_X)$ is a space locally ringed in $\mathbb{C}$-algebras. Such a locally ringed space is called a *standard analytic space* and it is denoted by $\mathcal{V}(f_1, \ldots, f_r)$.

A *complex analytic space* is a space locally ringed in $\mathbb{C}$-algebras $(X, \mathscr{O}_X)$ with the property for all $x \in X$ there exists an open neighborhood U of x in X such that the locally ringed space $(U, \mathscr{O}_{X|U})$ is isomorphic (over $\mathbb{C}$) to a standard analytic space.

A *morphism of complex analytic spaces* is a morphism of spaces locally ringed in $\mathbb{C}$-algebras.

Theorem 20.58. *Let X be a scheme locally of finite type over $\mathbb{C}$. Then there exists a complex analytic space X^{an} and a morphism $\alpha \colon X^{\mathrm{an}} \to X$ of spaces locally ringed in $\mathbb{C}$-algebras such that for every complex analytic space $\mathcal{Y}$ and for every morphism $\varphi \colon \mathcal{Y} \to X$ of spaces locally ringed in $\mathbb{C}$-algebras there exists a unique morphism of analytic spaces $f \colon \mathcal{Y} \to X^{\mathrm{an}}$ such that $\alpha \circ f = \varphi$.*

Moreover, α induces a bijection between the underlying topological set of X^{an} and $X(\mathbb{C})$ and for all $x \in X^{\mathrm{an}}$ an isomorphism of the completed local rings $\widehat{\mathscr{O}}_{X,\alpha(x)} \xrightarrow{\sim} \widehat{\mathscr{O}}_{X^{\mathrm{an}},x}$. In particular α is flat.

The complex analytic space X^{an} is called the *analytification of X*. The idea of the proof is simple: Locally X is of the form $X = \operatorname{Spec} \mathbb{C}[T_1, \ldots, T_n]/(f_1, \ldots, f_m)$ and we set $X^{\mathrm{an}} = \mathcal{V}(f_1, \ldots, f_m)$, where we consider f_i as polynomial functions (and hence as holomorphic functions) on $\mathbb{C}^n$. In particular one has $(\mathbb{A}^n_{\mathbb{C}})^{\mathrm{an}} = \mathbb{C}^n$ with the usual structure of a complex manifold. If X is an arbitrary scheme locally of finite type over $\mathbb{C}$, we glue the analytifications. One obtains the description of the underlying set of X^{an} by applying the universal property of $(X^{\mathrm{an}}, \alpha)$ to the complex analytic space $\mathrm{Spec}\,\mathbb{C}$. For details we refer to [Ser2] ^{O} or [SGA1] $^{\mathrm{O}}_X$ Exp. XII, 1.1.

Due to its universal property $(X^{\mathrm{an}}, \alpha)$ is unique up to unique isomorphism. The universal property of X^{an} shows that for every morphism $f\colon X \to Y$ of schemes locally of finite type over $\mathbb{C}$ there exists a unique morphism of complex analytic spaces $f^{\mathrm{an}}\colon X^{\mathrm{an}} \to Y^{\mathrm{an}}$ making the following diagram commutative

$$\begin{array}{ccc} X & \xrightarrow{\ f\ } & Y \\ \downarrow & & \downarrow \\ X^{\mathrm{an}} & \xrightarrow{\ f^{\mathrm{an}}\ } & Y^{\mathrm{an}}. \end{array}$$

We obtain a functor $X \mapsto X^{\mathrm{an}}$ from the category of schemes locally of finite type over $\mathbb{C}$ to the category of complex analytic spaces.

Remark 20.59. Many properties of X can be seen from properties of the attached complex analytic space X^{an}. We refer to [SGA1] $^{\mathrm{O}}_X$ Exp. XII for details. For instance, a scheme locally of finite type over $\mathbb{C}$ is separated (resp. proper, resp. smooth) over $\mathbb{C}$ if and only if X^{an} is Hausdorff (resp. X^{an} is compact[1], resp. X^{an} is a complex manifold[2]).

Similarly, many properties are equivalent for a morphism f and its analytification f^{an} (loc. cit.). For instance f is an isomorphism (resp. a closed immersion) if and only if f^{an} is an isomorphism (resp. a closed immersion). Note that this does not imply that if X and Y are schemes locally of finite type over $\mathbb{C}$ such that X^{an} and Y^{an} are isomorphic, then X and Y are isomorphic (see [Har2] ^{O} Chap. VI, Example 3.2 for a counterexample).

If one restricts to proper or projective schemes, one obtains better results. The main result is the following result, which we do not prove here (see [SGA1] $^{\mathrm{O}}_X$ Exp. XII, Théorème 4.4 for a proof).

Theorem 20.60. *Let X be a scheme that is proper over $\mathbb{C}$ and let $\alpha\colon X^{\mathrm{an}} \to X$ be its analytification. Then $\mathscr{F} \mapsto \mathscr{F}^{\mathrm{an}} := \alpha^*(\mathscr{F})$ yields an equivalence of the category of coherent $\mathcal{O}_X$-modules with the category of coherent $\mathcal{O}_{X^{\mathrm{an}}}$-modules.*

Here we use that the notion of coherence is defined for $\mathcal{O}_X$-modules on an arbitrary ringed space $(X, \mathcal{O}_X)$ (Definition 7.45). This theorem is a special case of the GAGA theorems that will be discussed in Section (23.9) below. Properness is essential here (e.g., see Exercise 26.14).

Similarly as for (locally noetherian) schemes, for every complex analytic space $\mathcal{X}$ there exists a bijective correspondence between closed analytic subspaces of $\mathcal{X}$ and coherent ideals of $\mathcal{O}_{\mathcal{X}}$ ([Ser2] ^{O} Proposition 1). Hence Theorem 20.60 also implies that if X is a proper $\mathbb{C}$-scheme and $\mathcal{Y}$ is a closed analytic subspace of X^{an}, then there exists a unique closed subscheme Y of X such that $Y^{\mathrm{an}} = \mathcal{Y}$.

As morphisms $X \to Y$ are given by their graphs $\Gamma_f \subset X \times Y$, or equivalently, by the ideal sheaves of $\mathcal{O}_{X \times Y}$ defining Γ_f, Theorem 20.60 implies the following corollary (we refer to [SGA1] $^{\mathrm{O}}_X$ Exp. XII, Corollaire 4.5 for details).

Corollary 20.61. *The functor $X \mapsto X^{\mathrm{an}}$ from the category of schemes proper over $\mathbb{C}$ to the category of compact complex analytic spaces is fully faithful.*

---

[1] In this book, "compact" is defined to be quasi-compact and Hausdorff.
[2] Here, a manifold is not necessarily Hausdorff or second countable.

In particular two proper $\mathbb{C}$-schemes X and Y are isomorphic if and only if the complex analytic spaces X^{an} and Y^{an} are isomorphic.

The analytification of $X = \mathbb{P}^n_{\mathbb{C}}$ is the usual complex projective space $\mathbb{P}^n(\mathbb{C})$. Hence every closed analytic subspace of $\mathbb{P}^n(\mathbb{C})$ is the analytification of a projective $\mathbb{C}$-scheme:

Corollary 20.62. *(Chow's theorem) The functor $X \mapsto X^{\mathrm{an}}$ yields an equivalence of the category of projective $\mathbb{C}$-schemes and the category of complex analytic spaces $\mathcal{X}$ such that there exists a closed analytic embedding $\mathcal{X} \hookrightarrow \mathbb{P}^n(\mathbb{C})$.*

On the other hand, the functor in Corollary 20.61 is far from being essentially surjective. We finish by explaining that étale morphisms of schemes correspond to local isomorphisms of complex analytic spaces.

Remark 20.63. Let $f \colon X \to Y$ be a morphism of schemes locally of finite type over $\mathbb{C}$. Then $\{\, x \in X \; ; f \text{ is étale in } x \,\}$ is open in X by definition. As the set of closed points in X is very dense (Proposition 3.35), f is étale if and only if it is étale in every closed point. Hence by Corollary 18.66, f is étale if and only if f induces isomorphisms $\widehat{\mathscr{O}}_{Y,f(x)} \to \widehat{\mathscr{O}}_{X,x}$ for all $x \in X(\mathbb{C})$. This in turn is equivalent to asking that f^{an} induces isomorphisms $\widehat{\mathscr{O}}_{Y^{\mathrm{an}},f^{\mathrm{an}}(x)} \overset{\sim}{\to} \widehat{\mathscr{O}}_{X^{\mathrm{an}},x}$ for all $x \in X^{\mathrm{an}}$. But this last condition holds if and only if $f^{\mathrm{an}} \colon X^{\mathrm{an}} \to Y^{\mathrm{an}}$ is a local isomorphism ([Gro5] ^{O} Proposition 1.9).

The étale fundamental group of a scheme

Our goal is to define a notion of fundamental group and fundamental groupoid for schemes which rests on the étale topology. Even though the étale topology is much more suitable at this point than the Zariski topology, there is no way to define a notion of path between two points of a scheme in a naive way. Let us begin by recalling two possible definitions of the fundamental groupoid of a topological space X from algebraic topology (see for instance [tDi] ^{O} or [May] $^{\mathrm{O}}_{\mathrm{X}}$).

The first definition of the fundamental groupoid $\Pi(X)$ is as the category whose objects are the points of X and whose morphisms $x \to y$ are the homotopy classes $[\gamma]$ of paths $\gamma \colon [0,1] \to X$ such that $\gamma(0) = x$ and $\gamma(1) = y$ (homotopies of paths are always assumed to fix start and end point of the paths). For paths $\gamma \colon x \to y$ and $\delta \colon y \to z$ the composition $[\delta] \circ [\gamma]$ is defined as the homotopy class of the concatenation $\delta \cdot \gamma$ obtained by first traversing γ and then δ, each with doubled velocity[3]. Then the category $\Pi(X)$ is indeed a groupoid, i.e., a category in which every morphism has an inverse: $[\gamma]^{-1}$ is the homotopy class of the path $\gamma^-(t) = \gamma(1-t)$. For a point $x \in X$, the fundamental group

$$\pi_1(X, x) := \mathrm{Aut}_{\Pi(X)}(x)$$

is the group of homotopy classes of closed paths with start point x in X.

An equivalent description of $\Pi(X)$ can be obtained as follows. Let $(\mathrm{Cov}(X))$ be the category of covers of X (not to be confused with an open covering of X) in the topological sense. A trivial cover of X is a continuous map of the form of a projection $X \times E \to X$, where E is a non-empty discrete topological space. A cover is a continuous map $p \colon Y \to X$

---

[3] Classically, concatenation is often defined in the other order. Here we follow [May] $^{\mathrm{O}}_{\mathrm{X}}$ which fits better into the categorical frame work explained here.

which is locally on X isomorphic to a trivial cover. Then $(\mathrm{Cov}(X))$ is the full subcategory of the category (Top/X) of topological spaces over X consisting of covers of X. Taking fibers gives rise to a natural functor

$$F\colon \Pi(X) \times (\mathrm{Cov}(X)) \to (\mathrm{Sets}).$$

Indeed, let $x \in X$ be a point (i.e., an object in $\Pi(X)$) and let $p\colon X' \to X$ a cover. Then we define $F(x, p)$ as the fiber $p^{-1}(x)$. If $[\gamma]\colon x \to y$ is a morphism in $\Pi(X)$ and if $\sigma\colon X' \to X''$ is a morphism of covers $p\colon X' \to X$ and $q\colon X'' \to X$, we define $F([\gamma], \sigma)\colon p^{-1}(x) \to q^{-1}(y)$ as follows. For all $c \in p^{-1}(x)$ there exists a unique path $\tilde{\gamma}_c\colon [0,1] \to X'$ such that $\gamma_c(0) = c$ and $p \circ \gamma_c = \gamma$. We set

$$F([\gamma], \sigma)(c) := \sigma(\tilde{\gamma}_c(1)).$$

This is well defined. The bi-functor F yields functors

$$L\colon \Pi(X) \to \mathrm{Func}((\mathrm{Cov}(X)), (\mathrm{Sets})), \qquad T\colon (\mathrm{Cov}(X)) \to \mathrm{Func}(\Pi(X), (\mathrm{Sets})),$$

where $\mathrm{Func}(\cdot, \cdot)$ denotes the category of functors.

From now on we assume that X satisfies certain additional connectedness assumptions (namely that X is path connected, locally path connected, and semilocally simply connected). Then T is an equivalence of categories and L is fully faithful ([tDi]$^\circ$ Theorem 3.3.2 and Proposition 3.4.1). Hence L yields an equivalence between $\Pi(X)$ and the full subcategory of fiber functors, i.e., those functors that are of the form $F_x\colon (p\colon Y \to X) \mapsto p^{-1}(x)$ for some point $x \in X$. In particular L induces for all $x \in X$ an isomorphism

$$(*) \qquad\qquad \pi_1(X, x) = \mathrm{Aut}_{\Pi(X)}(x) \overset{\sim}{\to} \mathrm{Aut}(F_x).$$

For a group G let BG be the groupoid that has a single object $*$ and such that $\mathrm{Aut}_{BG}(*) = G$. Fix a point $x \in X$. Since X is path connected the inclusion $B\pi_1(X, x) \hookrightarrow \Pi(X)$ sending $*$ to x is an equivalence. We obtain an equivalence of categories

$$\mathrm{Func}(\Pi(X), (\mathrm{Sets})) \simeq \mathrm{Func}(B\pi_1(X, x), (\mathrm{Sets})) = (\pi_1(X, x)\text{-Sets}),$$

where the right hand side denotes the category of sets endowed with a left action of $\pi_1(X, x)$. Composing this equivalence with T we obtain an equivalence of categories

$$\varepsilon_x\colon (\mathrm{Cov}(X)) \simeq (\pi_1(X, x)\text{-Sets})$$

which attaches to a cover $p\colon X' \to X$ the $\pi_1(X, x)$-set $p^{-1}(x)$. Under this equivalence a cover $X' \to X$ is path connected if and only if $\pi_1(X, x)$ acts transitively on the associated $\pi_1(X, x)$-set.

Let $p\colon X' \to X$ be a path connected cover, and choose an element $c \in p^{-1}(x)$. We consider $\pi_1(X', c)$ as a subgroup of $\pi_1(X, x)$ via the injective map $p_*\colon \pi_1(X', c) \to \pi_1(X, x)$ given by functoriality of the fundamental group. Then the corresponding $\pi_1(X, x)$-set is isomorphic to $\pi_1(X, x)/\pi_1(X', c)$ with the $\pi(X, x)$-action by left multiplication. A different choice of c replaces $\pi_1(X', c)$ by a conjugate subgroup. If $\pi_1(X', c)$ is a normal subgroup, then $p\colon X' \to X$ is called a Galois cover. In this case, the functor ε_x induces an isomorphism of the group of automorphisms of the cover X' over X and the quotient group $\pi_1(X, x)/\pi_1(X', c)$.

There exists a universal cover $p\colon \tilde{X} \to X$ which is simply connected and the above construction yields in particular an isomorphism between the group of automorphism of a universal cover p and $\pi_1(X, x)$.

Our goal is to develop a similar theory for connected schemes S. As there is no good notion of "path" in algebraic geometry, we will start by defining the algebraic version of a topological cover: étale covers. More precisely, what we will define should be seen as the analogue of a finite topological cover and this will be the reason why the theory is not entirely parallel to the topological theory. For every geometric point $\bar{s}$ of S we obtain a fiber functor $\mathcal{F}_{\bar{s}}$ from the category of étale covers to the category of finite sets and the fundamental groupoid $\Pi(S)$ of S will be defined as the category of all fiber functors. In analogy to (*) above we will define $\pi_1(S, \bar{s})$ as the group of automorphisms of the fiber functor $\mathcal{F}_{\bar{s}}$. Then $\pi_1(S, \bar{s})$ carries a natural topology which is in general non-discrete. It is the unique topology making $\pi_1(S, \bar{s})$ into a topological group such that the stabilizers of $c \in \mathcal{F}_{\bar{s}}(X)$, where X runs through the étale covers and c through the elements of $\mathcal{F}_{\bar{s}}(X)$, form a basis of neighborhoods of the neutral element of $\pi_1(S, \bar{s})$. It will turn out that with this topology $\pi_1(S, \bar{s})$ is a profinite group. This is in contrast to the topological situation, where the existence of a universal cover implies that the topology on the topological fundamental group defined as above is the discrete topology.

The main result of our discussion is an equivalence of the category of étale covers with the category of finite sets endowed with a continuous action by the profinite group $\pi_1(S, \bar{s})$ (Theorem 20.101).

At the end we will give several examples. In particular we show that this theory yields a vast generalization of classical Galois theory.

We recall that according to our conventions, the empty topological space is not connected.

(20.13) Étale covers.

Definition 20.64. *Let S be a scheme. An* étale cover *of S is a finite étale morphism of schemes $\pi \colon X \to S$. A morphism of* étale covers *of S is a morphism of S-schemes.*

The category of étale covers of S is denoted by $(\mathrm{F\acute{E}t}/S)$. If $S = \operatorname{Spec} R$ is affine, we also write $(\mathrm{F\acute{E}t}/R)$ instead of $(\mathrm{F\acute{E}t}/S)$. An étale cover $X \to S$ is called *split* if it is isomorphic as an S-scheme to $\coprod_{i=1}^{r} S$ for some integer $r \geq 0$.

The notion of an "étale cover" should not be confused with the concept of a "covering for the étale topology". Moreover we do *not* assume that an étale cover is surjective (but see Remark 20.66 (1)).

Example 20.65.
(1) If K is a finite separable extension of a field k, then $\operatorname{Spec} K \to \operatorname{Spec} k$ is an étale cover.
(2) Let R be a ring. Let $f \in R[T]$ be a monic polynomial such that f and f' generate $R[T]$ as ideal. Then $\operatorname{Spec} R[T]/(f) \to \operatorname{Spec} R$ is an étale cover by Example 18.47.
   This is in fact a generalization of (1), since $K \cong k[T]/(f)$ for a separable polynomial f by the theorem of the primitive element (Proposition B.98).

Let us collect some properties of étale covers that we have already seen.

Remark 20.66. Let S be a scheme.
(1) An étale cover π is finite, flat (Theorem 18.44), and locally of finite presentation and hence a finite locally free morphism (Proposition 12.19).

In particular, π is open and closed. Therefore π is surjective if $X \neq \emptyset$ and S is connected.

(2) It follows from (1) that the property to be an étale cover is stable under base change, under composition, is stable under faithfully flat descent, and is compatible with forming cofiltered limits of schemes with affine transition maps, since these permanence properties hold for the properties "finite", "flat", and "locally of finite presentation" (see Appendix C).

(3) Any morphism between étale covers of S is finite (Proposition 12.11) and étale (Remark 18.36).

(4) Let Z be any S-scheme, let $\pi\colon X \to S$ be an étale cover, and let $f, g\colon Z \to X$ be morphisms of S-schemes. If Z is connected and there exists a geometric point $\bar{z}$ of Z such that $f(\bar{z}) = g(\bar{z})$, then $f = g$ (Proposition 18.31).

(5) Every section of an étale cover is an open and closed immersion (Remark 18.30 (3) and Example 9.12).

(6) Every étale proper map is an étale cover. Indeed, quasi-compact étale maps are quasi-finite (Corollary 18.45) and quasi-finite and proper maps are finite by Zariski's main theorem (Corollary 12.89).

As for every finite locally free morphism, we have the notion of the *degree of a finite étale cover* $\pi\colon X \to S$, which is a locally constant function

$$\deg(X/S) := \deg(\pi)\colon S \to \mathbb{Z}_{\geq 0}.$$

Let us collect some properties of the degree which we have essentially already seen in Section (12.6).

Remark 20.67. Let $\pi\colon X \to S$ be a finite locally free morphism (e.g., an étale cover). Then the $\mathscr{O}_S$-algebra $\pi_* \mathscr{O}_X$ is a locally free $\mathscr{O}_S$-module of finite type and $X = \operatorname{Spec}(\pi_* \mathscr{O}_X)$ since π is affine (Corollary 12.2).

(1) The morphism π is an isomorphism if (and only if) $\deg(\pi) = 1$. Indeed, if $\deg(\pi) = 1$, then $\mathscr{O}_S \to \pi_* \mathscr{O}_X$ is an isomorphism and hence π is an isomorphism.

(2) If $g\colon S' \to S$ is a morphism of schemes, then the base change $\pi'\colon X \times_S S' \to S'$ of π is finite locally free and $\deg(\pi') = \deg(\pi) \circ g$.

(3) If $\pi'\colon X' \to S$ is a second finite locally free morphism. Let $X \amalg X'$ be the disjoint union of X and X'. Then the morphism $\varpi\colon X \amalg X' \to S$ with $\varpi_{|X} = \pi$ and $\varpi_{|X'} = \pi'$ is finite locally free and $\deg(\varpi) = \deg(\pi) + \deg(\pi')$.

(4) Let $\varpi\colon X' \to X$ be finite locally free. Then $\pi \circ \varpi$ is finite locally free. If $\deg(\pi)$ and $\deg(\varpi)$ are constant, then $\deg(\pi \circ \varpi) = \deg(\pi)\deg(\varpi)$.

The above properties of the degree function easily imply the following lemma.

Lemma 20.68. *Let $\pi\colon X \to S$ be an étale cover. Suppose that S and X are connected. Then every S-endomorphism of X is an automorphism.*

Proof. Let f be an S-endomorphism of X. Then f is an étale cover (Remark 20.66 (3)) with $\pi \circ f = \pi$ and hence $\deg(\pi) = \deg(\pi \circ f) = \deg(\pi)\deg(f)$. This implies that f is an isomorphism by Remark 20.67 (1). $\qquad\qquad\qquad\qquad\qquad\qquad\qquad\qquad\square$

"Locally" every finite étale cover is split, more precisely:

Proposition 20.69. *Let S be a scheme, let $\pi\colon X \to S$ be a morphism, and let $n \geq 1$ be an integer. Then the following assertions are equivalent.*

(i) *The morphism π is an étale cover of degree n.*

(ii) *There exists a faithfully flat quasi-compact morphism $T \to S$ and an isomorphism $X \times_S T \cong \coprod_{i=1}^{n} T$ of T-schemes.*

(iii) *There exists a surjective étale cover $T \to S$ and an isomorphism $X \times_S T \cong \coprod_{i=1}^{n} T$ of T-schemes. Moreover, $T \to S$ can be chosen such that it is of constant degree $n!$.*

Proof. Clearly, (iii) implies (ii), and (ii) implies (i) because being an étale cover of a fixed degree can be checked after faithfully flat quasi-compact base change.

Let us show that (i) implies (iii) by induction on n. For $n = 1$, π is an isomorphism and we can take $T \to S$ to be id_S. Let $n > 1$. By base change, we know that the first projection $X \times_S X \to X$ is an étale cover of degree n. Remark 20.66 (5), applied to $X \times_S X \to X$, shows that the diagonal $X \to X \times_S X$ is an open and closed embedding. Hence we have $X \times_S X = X \amalg X'$ for some open and closed subscheme X' of $X \times_S X$. Its restriction $X' \to X$ is an étale cover of degree $n - 1$. Applying the induction hypothesis, we find an étale cover $T \to X$ of degree $(n-1)!$ such that $X' \times_X T \cong \coprod_{i=1}^{n-1} T$. Then the composition $T \to X \to S$ is an étale cover of degree $n!$ and

$$X \times_S T = X \times_S X \times_X T = (X \amalg X') \times_X T = T \amalg \coprod_{i=1}^{n-1} T. \qquad \square$$

To characterize the étale covers among the finite locally free morphisms we introduce the trace form.

Remark and Definition 20.70. Let $\pi \colon X \to S$ be a finite locally free morphism. Then $\mathscr{A} := \pi_* \mathscr{O}_X$ is a finite locally free $\mathscr{O}_S$-algebra such that $X = \operatorname{Spec} \mathscr{A}$ (Proposition 12.19). We define a trace homomorphism and a trace bilinear form as usual: For $U \subseteq S$ open affine, $\Gamma(U, \mathscr{A})$ is a finitely generated projective $\Gamma(U, \mathscr{O}_S)$-module. For $a \in \Gamma(U, \mathscr{A})$ let $m_a \colon \Gamma(U, \mathscr{A}) \to \Gamma(U, \mathscr{A})$ be the multiplication by a. As $\mathscr{A}$ is a finite locally free $\mathscr{O}_S$-module, we can define $\operatorname{tr}_\pi(a) := \operatorname{tr}(m_a) \in \Gamma(U, \mathscr{O}_S)$ and $\tau_\pi(a \otimes a') := \operatorname{tr}(m_{aa'}) \in \Gamma(U, \mathscr{O}_S)$ for $a, a' \in \Gamma(U, \mathscr{A})$, see (7.20.8). We obtain the *trace homomorphism of π*,

$$\operatorname{tr}_\pi := \operatorname{tr}_{X/S} \colon \mathscr{A} \to \mathscr{O}_S,$$

and the symmetric *trace form of π*

$$\tau_\pi := \tau_{X/S} \colon \mathscr{A} \otimes_{\mathscr{O}_S} \mathscr{A} \to \mathscr{O}_S.$$

The bilinear form τ_π induces an $\mathscr{O}_S$-linear map $\mathscr{A} \to \mathscr{A}^\vee := \mathscr{H}\!om_{\mathscr{O}_S}(\mathscr{A}, \mathscr{O}_S)$ and τ_π is called *perfect* if this $\mathscr{O}_S$-linear map is an isomorphism of $\mathscr{O}_S$-modules.

For the formulation of the following proposition recall the ramification indices $e_{x/s}$ and inertia indices $f'_{x/s}$ and $f''_{x/s}$ for quasi-finite morphisms that were introduced in Section (12.5).

Proposition 20.71. Let $\pi \colon X \to S$ be a quasi-finite flat morphism locally of finite presentation.

(1) Let $x \in X$ and $s := \pi(x)$. Then the following assertions are equivalent.

   (i) π is étale at x.

   (ii) $e_{x/s} = f'_{x/s} = 1$.

   (iii) $\Omega^1_{X/S,x} = 0$.

(2) *Let π be a finite locally free morphism. Then the following assertions are equivalent.*

(i) *π is an étale cover.*

(ii) *For all $x \in X$ one has $e_{x/\pi(x)} = f'_{x/\pi(x)} = 1$.*

(iii) *$\Omega^1_{X/S} = 0$.*

(iv) *The trace form τ_π is perfect.*

The last criterion generalizes the well known fact ([BouAII] ^{O} V, §8.2, Proposition 1) that a finite field extension is separable if and only if the attached trace form is non-degenerate and we will reduce to this case.

Proof. Assertion (1) follows from Corollary 18.28 and Proposition 18.29 because a morphism locally of finite presentation is étale if and only if it is flat and unramified (Theorem 18.44). Moreover, the implications "(i) $\Leftrightarrow$ (ii) $\Leftrightarrow$ (iii)" in (2) follow from (1).

It remains to show "(iii) $\Leftrightarrow$ (iv)" in (2). Proposition 18.6 shows that $\Omega^1_{X/S} = 0$ if and only if π is unramified. This can be checked on fibers (Proposition 18.29). On the other hand, the formation of the trace of an endomorphism of a finite locally free $\mathcal{O}_S$-module is compatible with arbitrary base change $S' \to S$. Moreover, a homomorphism $u \colon \mathscr{E} \to \mathscr{F}$ of finite locally free $\mathcal{O}_S$-modules is an isomorphism if and only if the homomorphisms $u \otimes \mathrm{id}_{\kappa(s)}$ induced on fibers are all bijective (Corollary 8.12). Hence (iv) can also be checked on fibers. We thus may assume that $S = \mathrm{Spec}\, k$ for a field k.

But then π is unramified if and only if X is isomorphic to the spectrum of a finite product of finite separable field extensions of k (by Proposition 18.24 and because π is quasi-compact). Therefore the equivalence of (iii) and (iv) follows from the characterization of separable extensions recalled before the proof. $\square$

To apply the proposition, recall that if X and S are schemes of finite type over a noetherian ring, then a morphism $\pi \colon X \to S$ is quasi-finite and of finite presentation if and only if it has finite fibers. If in addition S is regular and X is Cohen-Macaulay, then π is automatically flat (Theorem 14.128). Finally, Exercise 20.17 gives a fiber criterion for a quasi-finite flat morphism of finite presentation to be finite locally free.

Proposition 20.71 can be reformulated by introducing the discriminant and the different:

Remark and Definition 20.72. We first define the discriminant. Let $\pi \colon X \to S$ be a finite locally free morphism. Let $\mathscr{A} := \pi_* \mathcal{O}_X$, and let $u \colon \mathscr{A} \to \mathscr{A}^\vee$ be the $\mathcal{O}_S$-linear homomorphism corresponding to the trace form τ_π. Then π is an étale cover if and only if $\det(u) \colon \det(\mathscr{A}) \to \det(\mathscr{A}^\vee)$ is an isomorphism of line bundles (combine Proposition 20.71 and Corollary 8.12). As $\det(\mathscr{A}^\vee) = \det(\mathscr{A})^\vee$, $\det(u)$ corresponds to an $\mathcal{O}_S$-linear homomorphism

$$D_{X/S} \colon \det(\mathscr{A}) \otimes \det(\mathscr{A}) \to \mathcal{O}_S$$

which is surjective if and only if $\det(u)$ is an isomorphism (again by Corollary 8.12). We call the image of $D_{X/S}$ the *discriminant of* π and denote it by $\mathscr{D}_{X/S}$ or by $\mathscr{D}_\pi$. Then $\mathscr{D}_{X/S}$ is a quasi-coherent ideal of $\mathcal{O}_S$ of finite type. As π is affine, the formation of $\pi_* \mathcal{O}_X$ commutes with arbitrary base change $g \colon S' \to S$ (Proposition 12.6). As the trace form and the determinant also commute with base change, we find

$$g^*(\mathscr{D}_{X/S}) = \mathscr{D}_{X \times_S S'/S'}.$$

Hence Proposition 20.71 implies:

Corollary 20.73. *Let $\pi\colon X \to S$ be finite locally free. Then a morphism $g\colon S' \to S$ factors through the open subscheme $S \setminus V(\mathscr{D}_{X/S})$ if and only if the base change $X \times_S S' \to S'$ is an étale cover.*

For the relation between the discriminant defined here and the discriminant of a monic polynomial we refer to Exercise 20.30.

Remark and Definition 20.74. Let us now define the different and the branch locus. Let $\pi\colon X \to S$ be a quasi-finite morphism locally of finite presentation. The *different of π* is defined as the zero-th Fitting ideal of $\Omega^1_{X/S}$, see Section (16.9). It is therefore a quasi-coherent ideal of $\mathcal{O}_X$ of finite type. We denote it by $\mathfrak{d}_\pi$ or $\mathfrak{d}_{X/S}$. The closed subscheme $V(\mathfrak{d}_{X/S}) \subseteq X$ is called the *branch locus of π*.

By definition of the Fitting ideal one has $\Omega^1_{X/S,x} = 0$ if and only if $x \in X \setminus V(\mathfrak{d}_{X/S})$. If π is in addition flat, then Proposition 20.71 shows that

$$\{\, x \in X \;;\; \pi \text{ is étale in } x \,\} = X \setminus V(\mathfrak{d}_{X/S}).$$

Sometimes the different is also defined as the annihilator of $\Omega^1_{X/S}$. The resulting closed subscheme of X has the same underlying topological space as $V(\mathfrak{d}_{X/S})$, see Section (16.9). Taking the zero-th Fitting ideal has the advantage that the formation of the different is compatible with base change $S' \to S$ because this holds for the formation of $\Omega^1_{X/S}$ (Proposition 17.30) and for the formation of the Fitting ideals (Proposition 16.30).

There exist other versions of the different ([Kun]$^\mathrm{O}$ Chap. 8). All of them agree if π is also syntomic and étale over an open dense subscheme of S (loc. cit. Theorem 8.15).

(20.14) Lifting of étale covers.

Theorem 20.75. *Let R be a local henselian ring with residue field k. Then the functor*

$$\mathcal{F}\colon (\mathrm{F\acute{E}t}/R) \to (\mathrm{F\acute{E}t}/k), \quad X \mapsto X \otimes_R k,$$

is an equivalence of categories.

Moreover, an étale cover $\operatorname{Spec} A \to \operatorname{Spec} R$ is the spectrum of a local R-algebra A if and only if $\bar{A} := A \otimes_R k$ is a (necessarily finite separable) field extension of k. In this case A is a standard étale R-algebra.

We first show a lemma which in particular implies that $\mathcal{F}$ is fully faithful.

Lemma 20.76. *Let R be a local ring with maximal ideal $\mathfrak{m}$ and residue field k. Then R is henselian if and only if for every finite morphism $f\colon X \to \operatorname{Spec} R$ and for every étale morphism $g\colon Y \to \operatorname{Spec} R$ the map*

$$\operatorname{Hom}_R(X, Y) \to \operatorname{Hom}_k(X \otimes_R k, Y \otimes_R k), \quad h \mapsto h \otimes \mathrm{id}_k$$

is bijective.

Proof. Let $S = \operatorname{Spec} R$. If we apply the condition to $f = \mathrm{id}_S$, Theorem 20.12 shows that R is henselian.

Conversely, assume that R is henselian. We remark that Theorem 20.12 shows that the canonical map $Y(R) \to Y(R/\mathfrak{m})$ is bijective for every étale R-scheme Y.

Let $f\colon X \to S$ be a finite morphism. Then $X = \operatorname{Spec} A$, where A is a finite product of local henselian finite R-algebras. Hence we may assume that $X = \operatorname{Spec} A$ for a local henselian finite R-algebra. Denote by K its residue field. Now let $g\colon Y \to \operatorname{Spec} R$ be an étale morphism and set $Y' := X \times_S Y$ which is an étale A-scheme. Hence the remark above, applied to A in place of R, shows that

$$\operatorname{Hom}_S(X, Y) = \operatorname{Hom}_X(X, Y') = Y'(A) \to Y'(A \otimes_R k) = \operatorname{Hom}_k(X \otimes_R k, Y \otimes_R k)$$

is bijective.   □

Proof. (of Theorem 20.75) As R is henselian, a finite R-algebra A is local if and only if $\bar{A} = A \otimes_R k$ is local. But an étale k-algebra is a finite product of finite separable field extensions. This shows the last assertion.

It remains to show that for every (finite) étale k-algebra A_0 there exists a finite étale R-algebra A such that $A \otimes_R k \cong A_0$. We may assume that $A_0 = K$ is a finite separable extension field of k. By the theorem of the primitive element (Proposition B.98), $K = k[T]/(f_0)$ for some monic polynomial $f_0 \in k[T]$ which is separable, i.e. f_0 is coprime to its derivative. Let $f \in R[T]$ be any monic polynomial with image f_0 in $k[T]$ and set $A := R[T]/(f)$. Then A is a finite free R-algebra with $A \otimes_R k \cong K$. As $(f) + (f') = R[T]$ (Remark 20.6), the image of f' in A is invertible. Therefore $A = A_{f'}$ is a standard étale R-algebra.   □

Since every étale cover of $\operatorname{Spec} k$ is a finite disjoint union of spectra of finite separable field extensions of k, Theorem 20.75 implies the following corollary.

Corollary 20.77. *Let R be a local henselian ring and let $X \to \operatorname{Spec} R$ be an étale cover. Then $X \cong \coprod_{i=1}^n \operatorname{Spec} A_i$, where A_i is a finite standard étale algebra. If R is strictly henselian, then X is a finite disjoint union of copies of $\operatorname{Spec} R$.*

Theorem 20.75 can be generalized to proper schemes over henselian rings, see Theorem 20.118 below.

For an arbitrary Dedekind domain or for a connected normal curve the connected étale covers are simply the integral closures in a finite separable extension of the function field which is unramified in all points. More precisely:

Example 20.78. Let S be a Dedekind scheme (i.e., S is noetherian, regular, irreducible, of dimension ≤ 1). Examples are $S = \operatorname{Spec} R$ for a Dedekind domain R and regular curves over a field. Let $K := K(S)$ be its function field.

(1) Let $\pi\colon X \to S$ be a connected étale cover. As $X \neq \emptyset$ and S is connected, π is surjective (Remark 20.66 (1)). Since S is regular and π is étale, X is regular (Corollary 14.60) and $\dim X = \dim S$ (Corollary 14.97). Hence X is a Dedekind scheme because X is connected. Because π is integral and X is normal, X is the integral closure of S in $L := K(X)$.

Let $\eta = \operatorname{Spec} K$ be the generic point of S. Then $\pi^{-1}(\eta) = \operatorname{Spec} L$ (Proposition B.55). As π is finite, L is a finite extension of K. Further, π is unramified, so L is a separable extension of K (Proposition 18.24).

Let $s \in S$ be a closed point. As π is étale, the fiber $\pi^{-1}(s)$ is an étale $\kappa(s)$-scheme or, equivalently by Proposition 20.71, for all $x \in \pi^{-1}(s)$ the ramification index $e_{x/s}$ is 1 and $\kappa(x)$ is a separable field extension of $\kappa(s)$.

(2) Conversely, let L be a finite separable extension of K and let X be the integral closure of S in L. Then X is a finite S-scheme (Proposition 12.53). Since S is a Dedekind scheme, $\pi\colon X \to S$ is dominant, X is reduced and $X \to S$ is flat (Proposition 14.14), so π is finite locally free. Moreover, π is an étale cover if and only if $e_{x/\pi(x)} = 1$ and the extension $\kappa(x)/\kappa(\pi(x))$ is separable for all $x \in X$ (Proposition 20.71).

Example 20.79. Let K be a local field, i.e., a locally compact topological field with a non-discrete topology. Then the topology of K can be defined by an absolute value $|\cdot|\colon K \to \mathbb{R}_{\geq 0}$ ([BouAC] ^{O} VI, §9.3, Theorem 1). Assume that K is non-archimedean, i.e., $|\cdot|$ satisfies the strong triangle inequality $|a + b| \leq \max\{|a|, |b|\}$ for all $a, b \in K$. Then K is isomorphic to a finite extension of the field $\mathbb{Q}_p$ of p-adic numbers or to the field of Laurent series $\mathbb{F}_q((T))$ over a finite field (loc. cit.), and $O_K := \{\, a \in K \;;\; |a| \leq 1 \,\}$ is a subring of K which is a complete discrete valuation ring whose residue field k is a finite field. Let $\pi \in O_K$ be a uniformizing element.

As O_K is complete, it is henselian (Example 20.3). Hence $(-) \otimes_{O_K} k$ is an equivalence between the category of local finite étale O_K-algebras and the category of finite separable extensions of k (Theorem 20.75).

Because k is finite, every extension of k is separable and for every integer $n \geq 1$ there exists a unique finite extension of degree n (up to isomorphism). On the other hand, every local finite étale O_K-algebra is the integral closure O_L of O_K in a finite separable extension L of K such that $\ell := O_L/\pi O_L$ is a field (Example 20.78). In algebraic number theory, one expresses this situation by saying that L is an *unramified* extension of K (e.g., [Neu] ^{O} II, §7). So by definition this means that the morphism $\operatorname{Spec} O_L \to \operatorname{Spec} O_K$ being unramified, which is much stronger than the condition that $\operatorname{Spec} L \to \operatorname{Spec} K$ is unramified in the sense of Definition 18.22.

Hence Theorem 20.75 implies that for every integer $n \geq 1$ there exists an unramified (in the sense of number theory) extension L with $[L : K] = n$ which is unique up to isomorphism. By the "fundamental equality" (12.6.2) one has $n = [\ell : k]$ and hence

$$\operatorname{Aut}_K(L) \cong \operatorname{Aut}_{O_K}(O_L) \cong \operatorname{Gal}(\ell/k) \cong \mathbb{Z}/n\mathbb{Z},$$

because every K-automorphism of L preserves the ring O_L of elements which are integral over O_K. In particular L is a Galois extension of K.

Remark 20.80. Let S be a scheme locally of finite type over $\mathbb{C}$ and let S^{an} be its analytification (Section (20.12)). Then [SGA1] $^{\text{O}}_{\times}$ Exp. XII, Théorème 5.1 says that étale covers of S correspond to finite covers of S^{an}: Attaching to an étale cover $\pi\colon X \to S$ of schemes its analytification π^{an} yields an equivalence of the category (FÉt/S) with the category of finite morphisms of analytic spaces $\mathcal{X} \to S^{\text{an}}$ that are isomorphisms locally for the analytic topology (i.e., a finite (not necessarily surjective) cover in the sense of classical topology). This result is called the *Riemann existence theorem* because it partially generalizes the classical Riemann existence theorem which asserts that on a compact Riemann surface there exist "many" meromorphic functions, so that it admits a closed embedding into a projective space. Compare Section (26.7). To relate the two versions, one can consider étale covers of open subschemes of the projective line.

(20.15) Fibers of étale covers and the fundamental groupoid.

As recalled above, every finite locally free morphism $\pi\colon X \to S$ has a degree (12.6.1) which is a locally constant function $S \to \mathbb{Z}_{\geq 0}$. Moreover π is an isomorphism if and only if $\deg(\pi) = 1$ (Remark 20.67 (1)). This will be often used in the sequel, for instance to prove the following lemma which will often allow us to reduce to a connected cover.

Lemma 20.81. *Let S be a connected scheme, and let $\pi\colon X \to S$ be a finite locally free morphism. Then X is the finite disjoint union of open and closed connected subschemes.*

In particular, we see that for an étale cover X of a connected scheme S, the connected components of X agree with the open and closed subspaces of X.

Proof. As S is connected, the degree of π (12.6.1) is constant. We show the claim by induction on $\deg(\pi)$. If $\deg(\pi) = 0$ (resp. $\deg(\pi) = 1$), then $X = \emptyset$ (resp. then π is an isomorphism) and the claim is clear. Now assume that X is not connected. Then $X = X_1 \sqcup X_2$ for non-empty open and closed subschemes X_i of X. As $\deg(\pi) = \deg(\pi_{|X_1}) + \deg(\pi_{|X_2})$, the induction hypothesis implies that each X_i is the finite disjoint union of open and closed connected subschemes. This shows the lemma. $\quad\square$

Let S be a scheme and let $\bar{s}\colon \operatorname{Spec}\kappa(\bar{s}) \to S$ be a geometric point. By abuse of notation we also denote by $\bar{s}$ the scheme $\operatorname{Spec}\kappa(\bar{s})$. If $\pi\colon X \to S$ is an étale cover, then its fiber $X_{\bar{s}} := X \times_S \kappa(\bar{s})$ over $\bar{s}$ is isomorphic to a finite disjoint union of copies of $\operatorname{Spec}\kappa(\bar{s})$ and hence as a $\kappa(\bar{s})$-scheme is determined by its underlying set, which can also be described as $\operatorname{Hom}_S(\bar{s}, X)$. We obtain the *fiber functor*

$$(20.15.1) \qquad \mathcal{F}_{\bar{s}}\colon (\mathrm{F\acute{E}t}/S) \to (\mathrm{sets}), \qquad X \mapsto \operatorname{Hom}_S(\bar{s}, X),$$

where we denote by (sets) the category of finite sets (morphisms are arbitrary maps of finite sets).

Definition 20.82. *Let S be a connected scheme. The full subcategory of the category of all functors $(\mathrm{F\acute{E}t}/S) \to (\mathrm{Sets})$ consisting of functors that are isomorphic to a fiber functor $\mathcal{F}_{\bar{s}}$ for some geometric point $\bar{s}$ of S is called the* fundamental groupoid *of S and denoted by $\Pi(S)$.*

Below in Proposition 20.93 we will see that $\Pi(S)$ is a connected groupoid, i.e., any morphism in $\Pi(S)$ is an isomorphism and any two objects in $\Pi(S)$ are isomorphic. We view $\Pi(S)$ as the algebraic analogue of the topological fundamental groupoid of a path-connected space.

More generally, one can define the fundamental groupoid $\Pi(S)$ in the same way for all schemes in which all connected components are open (and hence open and closed). This is for instance the case if S is locally noetherian. Formally, the above definition also makes sense for arbitrary schemes S but it does not seem clear, whether this is conceptually the "right" definition.

Let S be a connected scheme, so that the degree of any finite étale cover $\pi\colon X \to S$ is constant and stable under base change $S' \to S$. In particular, for every geometric point $\bar{s}$ of S we have

$$(20.15.2) \qquad \deg(\pi) = \#\mathcal{F}_{\bar{s}}(X).$$

If S is the spectrum of a separably closed field, the functor (20.15.1) is an equivalence of categories. More generally, Corollary 20.77 then implies:

Proposition 20.83. *Let S be the spectrum of a strictly henselian ring and let $\bar{s}$ be a geometric point of S lying over the closed point of S. Then $\mathcal{F}_{\bar{s}}\colon (\mathrm{F\acute{E}t}/S) \to (\mathrm{sets})$ yields an equivalence of categories.*

Definition 20.84. *Let S be a connected scheme and let $\bar{s}$ be a geometric point of S. A pointed étale cover is a pair $(X, \bar{x})$, where X is an étale cover of S and $\bar{x} \in \mathcal{F}_{\bar{s}}(X)$. A morphism of pointed étale covers $(X, \bar{x}) \to (Y, \bar{y})$ is a morphism of étale covers $f\colon X \to Y$ such that $f \circ \bar{x} = \bar{y}$.*

Let $\mathcal{I}_{\bar{s}}$ be the category of pointed étale covers $(X, \bar{x}) \to (S, \bar{s})$, where X is connected.

Remark 20.85.
(1) Let $(X, \bar{x})$ be an object in $\mathcal{I}_{\bar{s}}$ and let $Z \to S$ be an étale cover. Then the map

$$(20.15.3) \qquad \iota_{\bar{x}}\colon \mathrm{Hom}_{(\mathrm{F\acute{E}t}/S)}(X, Z) \to \mathcal{F}_{\bar{s}}(Z), \qquad f \mapsto f \circ \bar{x}$$

is injective (Remark 20.66 (4)).
(2) In particular, we see that for objects $(X, \bar{x})$ and $(Y, \bar{y})$ there exists at most one morphism $(X, \bar{x}) \to (Y, \bar{y})$ in $\mathcal{I}_{\bar{s}}$.
(3) The category $\mathcal{I}_{\bar{s}}$ is cofiltered: If $(X, \bar{x})$ and $(Y, \bar{y})$ are objects in $\mathcal{I}_{\bar{s}}$, set $\bar{z} := (\bar{x}, \bar{y})\colon \bar{s} \to X \times_S Y$ and let Z be the connected component of $X \times_S Y$ such that $\bar{z}$ factors through Z. Then $(Z, \bar{z})$ is an object of $\mathcal{I}_{\bar{s}}$ such that there exist morphisms $(Z, \bar{z}) \to (X, \bar{x})$ and $(Z, \bar{z}) \to (Y, \bar{y})$. By (2) this implies also that we can always equalize morphisms.

For every étale cover $X \to S$ a morphism $(Z, \bar{z}) \to (Y, \bar{y})$ in $\mathcal{I}_{\bar{s}}$ induces a map $\mathrm{Hom}_{(\mathrm{F\acute{E}t}/S)}(Y, X) \to \mathrm{Hom}_{(\mathrm{F\acute{E}t}/S)}(Z, X)$ compatible with the maps (20.15.3). We obtain a map

$$(20.15.4) \qquad \operatorname*{colim}_{(Z, \bar{z}) \in \mathcal{I}_{\bar{s}}} \mathrm{Hom}_{(\mathrm{F\acute{E}t}/S)}(Z, X) \to \mathcal{F}_{\bar{s}}(X)$$

which is functorial in Z.

Theorem 20.86. *The morphism (20.15.4) is an isomorphism of functors in X.*

Proof. The map (20.15.4) is injective because it is the filtered colimit of injective maps. If $\bar{x} \in \mathcal{F}_{\bar{s}}(X)$ then let X^0 be the connected component such that $\bar{x}$ factors through X^0. Then $(X^0, \bar{x})$ is an object of $\mathcal{I}_{\bar{s}}$ and hence $\bar{x}$ lies in the image of (20.15.4). $\qquad\square$

The theorem can be expressed by saying that $\mathcal{F}_{\bar{s}}$ is pro-representable by the cofiltered diagram $\mathcal{I}_{\bar{s}} \to (\mathrm{F\acute{E}t}/S)$, $(X, \bar{x}) \mapsto X$ (see for instance [SGA4]$^\circ$ Exp. I, 8.10 for the notion of pro-representable functors).

(20.16) Galois covers.

Roughly speaking, a Galois cover is an étale cover $X \to S$ such that the induced map $X/\mathrm{Aut}_S(X) \to S$ is an isomorphism (cf. the analogy with topology). To make this more precise, we start with the following easy globalization of Proposition 12.27.

Proposition 20.87. *Let S be a scheme, let $f\colon X \to S$ be an affine morphism, and let G be a finite group of S-automorphisms of X. Then there exists a unique pair (Y, p), where Y is an S-scheme and $p\colon X \to Y$ is a morphism of S-schemes with $p \circ g = p$ for all $g \in G$, such that for every morphism of S-schemes $q\colon X \to Z$ with $q \circ g = q$ for all $g \in G$ there exists a unique morphism $\bar{q}\colon Y \to Z$ of S-schemes such that $\bar{q} \circ p = q$.*

Moreover, p is integral, surjective and has finite fibers. For $x, x' \in X$ one has $p(x) = p(x')$ if and only if there exists $g \in G$ with $g(x) = x'$.

We denote Y by X/G and call it the *quotient of X by G*.

Proof. As X is affine over S we have $X = \operatorname{Spec} f_* \mathscr{O}_X$ (Proposition 12.1). Set $Y :=$ $\operatorname{Spec}(f_* \mathscr{O}_X)^G$ and let $p\colon X \to Y$ be the morphism of S-schemes corresponding to the inclusion $(f_* \mathscr{O}_X)^G \to f_* \mathscr{O}_X$. As taking G-invariants is compatible with flat base change (Remark 12.28), in particular with passing to open subschemes, we may assume that S is affine. Then all claims follow from Proposition 12.27. $\square$

Remark 20.88. Globalizing Remark 12.28 we see that in the situation of Proposition 20.87 for every morphism $S' \to S$ of schemes there is a morphism

$$(20.16.1) \qquad\qquad (X \times_S S')/G \longrightarrow (X/G) \times_S S',$$

which is an isomorphism if $S' \to S$ is flat.

Proposition 20.89. *Let $f\colon X \to S$ be an étale cover and let G be a finite group of S-automorphism of X. Then X/G is an étale cover of S.*

Moreover, for every geometric point $\bar{s}$ of S one has $\mathcal{F}_{\bar{s}}(X/G) = \mathcal{F}_{\bar{s}}(X)/G$, where G acts by composition on $\mathcal{F}_{\bar{s}}(X) = \operatorname{Hom}_S(\bar{s}, X)$. In particular $\deg(X/G \to S) = \#(\mathcal{F}_{\bar{s}}(X)/G)$.

Proof. As the formation of the quotient commutes with flat base change (Remark 20.88) and in particular with base change of the form $\operatorname{Spec} \mathscr{O}_{S,s}^{\mathrm{hs}} \to S$ for $s \in S$, we may assume that $S = \operatorname{Spec} R$, where R is a strictly henselian ring (Remark 20.44). But then X is isomorphic to a sum of copies of $\operatorname{Spec} R$ indexed by $\mathcal{F}_{\bar{s}}(X)$, where $\bar{s}$ is some geometric point of S lying over the closed point of s (Proposition 20.83), and G necessarily acts by permuting these copies. Hence X/G is also isomorphic to a finite sum of copies of $\operatorname{Spec} R$, one for each G-orbit in $\mathcal{F}_{\bar{s}}(X)$. $\square$

Let S be a connected scheme, let $\bar{s}$ be a geometric point, and let $X \to S$ be a connected étale cover. Then (20.15.3) shows that the action of $\operatorname{Aut}_{(\mathrm{FÉt}/S)}(X)$ on $\mathcal{F}_{\bar{s}}(X)$ is free. Therefore one has

$$(20.16.2) \qquad\qquad \#\operatorname{Aut}_{(\mathrm{FÉt}/S)}(X) \le \#\mathcal{F}_{\bar{s}}(X)$$

and $\operatorname{Aut}_{(\mathrm{FÉt}/S)}(X)$ is a finite group.

Definition and Remark 20.90. Let S be a connected scheme. A connected étale cover $Z \to S$ is called a *Galois cover* if the following equivalent conditions are satisfied.
(i) There exists a geometric point $\bar{s}$ of S such that $\#\operatorname{Aut}_{(\mathrm{FÉt}/S)}(Z) \ge \#\mathcal{F}_{\bar{s}}(Z)$.
(ii) For all geometric points $\bar{s}$ of S the group $\operatorname{Aut}_{(\mathrm{FÉt}/S)}(Z)$ acts simply transitively on $\mathcal{F}_{\bar{s}}(Z)$.
(iii) $\#\operatorname{Aut}_{(\mathrm{FÉt}/S)}(Z) = \deg(Z/S)$.
(iv) The morphism $Z/\operatorname{Aut}_{(\mathrm{FÉt}/S)}(Z) \to S$ is an isomorphism.
Let us show the equivalence of these conditions. Set $G := \operatorname{Aut}_{(\mathrm{FÉt}/S)}(Z)$. By Proposition 20.89, (iv) is equivalent to the transitivity of the G-action on $\mathcal{F}_{\bar{s}}(Z)$ for one or for all $\bar{s}$. As $\#G \le \#\mathcal{F}_{\bar{s}}(Z) = \deg(Z/S)$ such a transitive action is automatically simply transitive. Hence (ii), (iii), and (iv) are equivalent. Clearly, (ii) implies (i). Conversely, by Remark 20.85 (2) no element $\bar{x} \in \mathcal{F}_{\bar{s}}(Z)$ has a nontrivial stabilizer in G, hence the action of G on $\mathcal{F}_{\bar{s}}(Z)$ must be simply transitive if $\#\operatorname{Aut}_{(\mathrm{FÉt}/S)}(Z) \ge \#\mathcal{F}_{\bar{s}}(Z)$.

Lemma 20.91. *Let S be a connected scheme and let $\pi\colon X \to S$ and $\varpi\colon Z \to X$ be connected étale covers. If the composition $Z \to S$ is a Galois cover, then $Z \to X$ is a Galois cover.*

Proof. Let $\bar{x}$ be a geometric point of X. We show that $H := \mathrm{Aut}_{(\mathrm{FÉt}/X)}(Z)$ acts transitively on $\mathcal{F}_{\bar{x}}(Z)$. Let $\bar{s}$ be the image of $\bar{x}$ in S. Let $\bar{z}, \bar{z}' \in \mathcal{F}_{\bar{x}}(Z) \subseteq \mathcal{F}_{\bar{s}}(Z)$. As $Z \to S$ is a Galois cover, there exists $\alpha \in \mathrm{Aut}_{(\mathrm{FÉt}/S)}(Z)$ sending $\bar{z}$ to $\bar{z}'$. We have to show that $\varpi \circ \alpha = \varpi$, i.e., $\mathrm{Ker}(\varpi \circ \alpha, \varpi) = Z$. But $\mathrm{Ker}(\varpi \circ \alpha, \varpi)$ is an open and closed subscheme of the connected scheme Z (Remark 20.66 (4)), and it is non-empty because $\varpi \circ \alpha$ and ϖ both send $\bar{z}$ to $\bar{x}$. $\qquad\square$

Let $\mathcal{I}_{\bar{s}}$ be the category of pointed connected étale covers of $(S, \bar{s})$ defined in Definition 20.84.

Lemma 20.92. *For every étale cover $X \to S$ there exists an object $(Z, \bar{z})$ in $\mathcal{I}_{\bar{s}}$ such that the injective map $\iota_{\bar{z}}\colon \mathrm{Hom}_{(\mathrm{FÉt}/S)}(Z, X) \to \mathcal{F}_{\bar{s}}(X)$ (20.15.3) is bijective and such that Z is a Galois cover of S.*

Proof. Let $Y := X^{\mathcal{F}_{\bar{s}}(X)}$ be the fiber product over S of copies of X indexed by $\mathcal{F}_{\bar{s}}(X)$. Let $\bar{z}$ be the element of $\mathcal{F}_{\bar{s}}(Y) = \mathcal{F}_{\bar{s}}(X)^{\mathcal{F}_{\bar{s}}(X)}$ whose $\bar{x}$-th component is $\bar{x}$ for $\bar{x} \in \mathcal{F}_{\bar{s}}(X)$. Let Z be the connected component of Y such that $\bar{z}$ factors through Z.

For $\bar{x} \in \mathcal{F}_{\bar{s}}(X)$ let $p_{\bar{x}}\colon Z \to X$ be the restriction to Z of the projection $Y \to X$ onto the $\bar{x}$-th coordinate. Then the map (20.15.3) sends $p_{\bar{x}}$ to $\bar{x}$. This shows that the map $\iota_{\bar{z}}$ in (20.15.3) is bijective.

It remains to show that Z is a Galois cover of S. We show that $\mathrm{Aut}_{(\mathrm{FÉt}/S)}(Z)$ acts transitively on $\mathcal{F}_{\bar{s}}(Z)$. Let $\bar{z}'$ be any element of $\mathcal{F}_{\bar{s}}(Z)$. As $\#\mathrm{Hom}_{(\mathrm{FÉt}/S)}(Z, X) = \#\mathcal{F}_{\bar{s}}(X)$, the injective map $\iota_{\bar{z}'}$ is also bijective. This means that the coordinates of $\bar{z}'$, viewed as an element of $\mathcal{F}_{\bar{s}}(Y) = \mathcal{F}_{\bar{s}}(X)^{\mathcal{F}_{\bar{s}}(X)}$ are precisely the elements of $\mathcal{F}_{\bar{s}}(X)$. Therefore there exists an automorphism σ of the étale cover Y, permuting the factors, that sends $\bar{z}$ to $\bar{z}'$. This automorphism maps the connected component Z of Y to some connected component Z'. From $\bar{z}' \in \mathcal{F}_{\bar{s}}(Z) \cap \mathcal{F}_{\bar{s}}(Z')$ we conclude $Z = Z'$. $\qquad\square$

Lemma 20.92 implies that the full subcategory $\mathcal{G}_{\bar{s}}$ of objects $(Z, \bar{z})$ of $\mathcal{I}_{\bar{s}}$ such that Z is a Galois cover of S is a cofinal subcategory in $\mathcal{I}_{\bar{s}}$. In particular, for every étale cover X of S the bijection (20.15.4) induces a bijection, functorial in X,

$$(20.16.3) \qquad \underset{(Z,\bar{z})\in\mathcal{G}_{\bar{s}}}{\mathrm{colim}} \mathrm{Hom}_{(\mathrm{FÉt}/S)}(Z, X) \xrightarrow{\sim} \mathcal{F}_{\bar{s}}(X).$$

Proposition 20.93. *Let S be a connected. Then $\Pi(S)$ is a connected groupoid, i.e., any morphism in $\Pi(S)$ is an isomorphism and any two objects in $\Pi(S)$ are isomorphic.*

Proof. Le $\bar{s}$ and $\bar{s}'$ be geometric points of S. Let $\varphi\colon \mathcal{F}_{\bar{s}} \to \mathcal{F}_{\bar{s}'}$ be a morphism of functors. Let $(Z, \bar{z})$ in $\mathcal{G}_{\bar{s}}$ and set $\bar{z}' := \varphi_Z(\bar{z})$. Then $(Z, \bar{z}') \in \mathcal{G}_{\bar{s}'}$ and for every étale cover $X \to S$ one has a diagram

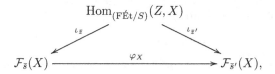

which is commutative since φ is a morphism of functors. Now given $(X, \bar{x})$, we choose $(Z, \bar{z})$ as in Lemma 20.92, so that we obtain a diagram as above with $\iota_{\bar{z}}$ bijective. As $\iota_{\bar{z}'}$ is injective (Remark 20.85 (1)), it follows that φ_X is injective. Since $\#\mathcal{F}_{\bar{s}}(X) = \deg(X/S) = \#\mathcal{F}_{\bar{s}'}(X)$ we see that φ_X is bijective. Hence φ is an isomorphism of functors.

It remains to show that $\operatorname{Hom}(\mathcal{F}_{\bar{s}}, \mathcal{F}_{\bar{s}'}) \neq \emptyset$, where $\operatorname{Hom}(-, -)$ denotes morphisms of functors $(\text{FÉt}/S) \to (\text{Sets})$. For an étale cover $Z \to S$ define the functor $h^Z \colon (\text{FÉt}/S) \to (\text{Sets})$, $h^Z(X) = \operatorname{Hom}_{(\text{FÉt}/S)}(Z, X)$. Then (20.16.3) yields an isomorphism of functors

$$\mathcal{F}_{\bar{s}} \cong \operatorname*{colim}_{(Z,\bar{z}) \in \mathcal{G}_{\bar{s}}} h^Z$$

and hence we have

$$\operatorname{Hom}(\mathcal{F}_{\bar{s}}, \mathcal{F}_{\bar{s}'}) \cong \lim_{(Z,\bar{z})} \operatorname{Hom}(h^Z, \mathcal{F}_{\bar{s}'}) = \lim_{(Z,\bar{z})} \mathcal{F}_{\bar{s}'}(Z),$$

where the second identity holds by the Yoneda lemma. The right hand side is a cofiltered limit of finite non-empty sets and therefore is non-empty. $\qquad\square$

The proof shows that for geometric points $\bar{s}$ and $\bar{s}'$ of a connected scheme S one has a bijection

(20.16.4) $\qquad \operatorname{Isom}(\mathcal{F}_{\bar{s}}, \mathcal{F}_{\bar{s}'}) \xrightarrow{\sim} \lim_{(Z,\bar{z}) \in \mathcal{G}_{\bar{s}}} \mathcal{F}_{\bar{s}'}(Z), \qquad \sigma \mapsto (\sigma_Z(\bar{z}))_{(Z,\bar{z}) \in \mathcal{G}_{\bar{s}}}.$

(20.17) Profinite groups and the topology on the automorphism group of a functor.

Let S be a connected scheme and let $\bar{s}$ be a geometric point of S. Below we will define the algebraic fundamental group $\pi_1(S, \bar{s})$ as the group of automorphisms of the fiber functor $\mathcal{F}_{\bar{s}}$. This group has a natural topology and with this topology it will be a profinite group. Let us now explain these notions briefly.

We first recall the definition of a profinite group (e.g., see [NSW] $^{\text{O}}_{\text{X}}$ (1.1.3)).

Definition 20.94. *A topological group G is called* profinite *if it satisfies the following equivalent conditions:*
  (i) *G is isomorphic (as a topological group) to cofiltered limit of finite discrete groups.*
  (ii) *G is compact[4] and the unit element has a basis of neighborhoods consisting of open and closed normal subgroups.*
  (iii) *G is compact and totally disconnected.*

Remark 20.95.
  (1) Every closed subspace of a compact (resp. totally disconnected) space is again compact (resp. totally disconnected). Hence every closed subgroup of a profinite group is profinite.
  (2) As limits of compact (resp. totally disconnected) spaces are again compact (resp. totally disconnected), every limit of profinite groups is again profinite. In particular, products of profinite groups are profinite.

Let us collect some properties of profinite groups that we will use later.

---

[4] We define a topological space to be compact if it is quasi-compact and Hausdorff.

Lemma 20.96. *Let G be a profinite group.*
(1) A subgroup of G is open if and only if it is closed and of finite index.
(2) One has

$$\bigcap_{\substack{U \subseteq G \text{ open} \\ \text{normal subgroup}}} U = 1.$$

(3) A subgroup $H \subseteq G$ is closed if and only if H is the intersection of all open subgroups of G containing H.

Proof. (1). This is true for any compact group G. If H is open, then G/H is discrete and compact. Hence it is finite. The complement of H in G is a union of H-cosets and therefore open. Hence H is closed in G. Conversely, if H is closed and of finite index, then the complement of H is the union of finitely many closed H-cosets. Hence $G \setminus H$ is closed and therefore H is open.

(2). A topological group is Hausdorff if and only if the intersection of all open neighborhoods of the unit element e is $\{e\}$. A profinite group is Hausdorff and the open normal subgroups form a neighborhood basis of e.

(3). By (1) the condition is sufficient. Conversely, let H be a closed subgroup of G. As every subgroup that contains an open subgroup is itself open, it suffices to show that

$$\bigcap_{\substack{U \subseteq G \text{ open} \\ \text{normal subgroup}}} HU = H\left(\bigcap_U U\right) = H,$$

where the second equality holds by (2). We have to show that $\bigcap_U (HU) \subseteq H(\bigcap_U U)$ since the converse inclusion is clear. Let $g \in \bigcap_U (HU)$. Assume $g \notin H(\bigcap_U U)$, i.e. $gH \cap \bigcap_U U = \emptyset$. By compactness we find finitely many open normal subgroups $U_1, \dots, U_n$ with $Hg \cap \bigcap_{i=1}^n U_i = \emptyset$. But $U' := \bigcap_i U_i$ is itself an open normal compact subgroup of G and hence $g \in HU'$ by assumption. This is a contradiction. $\qquad\square$

Definition and Remark 20.97. Let G be a topological group. An action of G on a set X is called *continuous* if the map defining the action $G \times X \to X$ is continuous, where we endow X with the discrete topology. This is the case if and only if for all $x \in X$ the stabilizer $G_x := \{\, g \in G \;;\; gx = x \,\}$ is an open subgroup of X.

If X is in addition finite, then the action is continuous if and only if there exists an open subgroup U of G such that U acts trivially on X. We denote by $(G\text{-sets})$ the category of finite sets with a continuous G-action, in which morphisms are G-equivariant maps.

If G is profinite and acts on a finite set X, then a G-action on X is continuous if and only if there exists an open normal subgroup of G that acts trivially on X. As G is compact, every open subgroup has finite index.

Let us also explain how to endow the group of automorphisms of a set-valued functor with a topology.

Remark 20.98. Let $\mathcal{C}$ be a small category and let $\mathcal{F} \colon \mathcal{C} \to (\text{sets})$ be a functor with values in the categories of finite sets. Then there is a canonical injective map

$$(20.17.1) \qquad \operatorname{Aut}(\mathcal{F}) \hookrightarrow \prod_{X \in \mathrm{Ob}(\mathcal{C})} \operatorname{Aut}(\mathcal{F}(X)).$$

We endow the finite set $\mathrm{Aut}(\mathcal{F}(X))$ with the discrete topology and the right hand side with the product topology. Then it is easy to see that (20.17.1) identifies $\mathrm{Aut}(\mathcal{F})$ with a closed subgroup of the right hand side. We endow $\mathrm{Aut}(\mathcal{F})$ with the induced topology. Then $\mathrm{Aut}(\mathcal{F})$ is a profinite group by Remark 20.95.

(20.18) The étale fundamental group.

Let S be a connected scheme, let $\bar{s}$ be a geometric point, and let $\mathcal{F}_{\bar{s}} \colon (\mathrm{F\acute{E}t}/S) \to (\mathrm{Sets})$ be the corresponding fiber functor. Then $\mathrm{Aut}(\mathcal{F}_{\bar{s}})$ is a profinite group by Remark 20.98.

Definition 20.99. *The profinite group*

$$\pi_1(S, \bar{s}) := \mathrm{Aut}(\mathcal{F}_{\bar{s}})$$

is called the algebraic fundamental group of S with respect to $\bar{s}$ *(or the* étale fundamental group*).*

The topology on $\pi_1(S, \bar{s})$ is by definition the unique topology making $\pi_1(S, \bar{s})$ into a topological group such that the stabilizers of $c \in \mathcal{F}_{\bar{s}}(X)$, where X runs through the étale covers and c through the elements of $\mathcal{F}_{\bar{s}}(X)$, form a basis of neighborhoods of the neutral element of $\pi_1(S, \bar{s})$. By (20.16.4) we have a homeomorphism

$$(20.18.1) \qquad \pi_1(S, \bar{s}) \cong \varprojlim_{(Z, \bar{z}) \in \mathcal{G}_{\bar{s}}} \mathcal{F}_{\bar{s}}(Z), \qquad \sigma \mapsto (\sigma_Z(\bar{z}))_{(Z, \bar{z})}.$$

For a pointed Galois cover $(Z, \bar{z})$ one has a bijection $\iota_{\bar{z}} \colon \mathrm{Aut}_{(\mathrm{F\acute{E}t}/S)}(Z) \xrightarrow{\sim} \mathcal{F}_{\bar{s}}(Z)$. A morphism $f \colon (Z, \bar{z}) \to (Z', \bar{z}')$ in $\mathcal{G}_{\bar{s}}$ induces a map $\mathcal{F}_{\bar{s}}(T) \to \mathcal{F}_{\bar{s}}(Z')$. Hence there is a unique map $\alpha_f \colon \mathrm{Aut}_{(\mathrm{F\acute{E}t}/S)}(Z) \to \mathrm{Aut}_{(\mathrm{F\acute{E}t}/S)}(Z')$ such that the following diagram with bijective vertical maps is commutative

$$(20.18.2) \qquad \begin{array}{ccc} \mathrm{Aut}_{(\mathrm{F\acute{E}t}/S)}(Z) & \xrightarrow{\ \alpha_f\ } & \mathrm{Aut}_{(\mathrm{F\acute{E}t}/S)}(Z') \\ {\scriptstyle \iota_{\bar{z}}}\Big\downarrow & & \Big\downarrow{\scriptstyle \iota_{\bar{z}'}} \\ \mathcal{F}_{\bar{s}}(Z) & \xrightarrow{\ z \mapsto f \circ z\ } & \mathcal{F}_{\bar{s}}(Z'). \end{array}$$

In other words, for $u \in \mathrm{Aut}_{(\mathrm{F\acute{E}t}/S)}(Z)$ its image $\alpha_f(u)$ is the unique automorphism of Z' such that $\alpha_f(u) \circ f = f \circ u$. As the lower horizontal map is surjective, α_f is surjective. Moreover, it is a group homomorphism: It suffices to check that for $u_1, u_2 \in \mathrm{Aut}_{(\mathrm{F\acute{E}t}/S)}(Z)$ one has $\alpha_f(u_1 \circ u_2) \circ f = \alpha_f(u_1) \circ \alpha_f(u_2) \circ f$ and both sides are equal to $f \circ u_1 \circ u_2$.

For $(Z, \bar{z}) \in \mathcal{G}_{\bar{s}}$, the composition $\pi_1(S, \bar{s}) \to \mathcal{F}_{\bar{s}}(Z) \xrightarrow{\sim} \mathrm{Aut}_{(\mathrm{F\acute{E}t}/S)}(Z)$ sends $\sigma \in \pi_1(S, \bar{s})$ to the unique automorphism u_σ of Z such that $u_\sigma \circ \bar{z} = \sigma_Z(\bar{z})$. Therefore one has $u_{\sigma \sigma'} = u_{\sigma'} \circ u_\sigma$ and by (20.18.1) one obtains an isomorphism of profinite groups

$$(20.18.3) \qquad \pi_1(S, \bar{s}) \xrightarrow{\sim} \varprojlim_{(Z, \bar{z}) \in \mathcal{G}_{\bar{s}}} \mathrm{Aut}_{(\mathrm{F\acute{E}t}/S)}(Z)^{\mathrm{opp}},$$

where the transition maps on the right hand side are all surjective. Here $-^{\mathrm{opp}}$ denotes the opposite group, i.e., where the order of multiplication is reversed.

Remark 20.100. For two geometric points $\bar{s}_1$ and $\bar{s}_2$ of a connected scheme S the corresponding fiber functors $\mathcal{F}_{\bar{s}_1}$ and $\mathcal{F}_{\bar{s}_2}$ are isomorphic (Proposition 20.93). An isomorphism $\mathcal{F}_{\bar{s}_1} \overset{\sim}{\to} \mathcal{F}_{\bar{s}_2}$ is called a *path from $\bar{s}_1$ to $\bar{s}_2$*. Such a path yields an isomorphism $\pi_1(S, \bar{s}_1) \overset{\sim}{\to} \pi_1(S, \bar{s}_2)$ and two isomorphisms $\pi_1(S, \bar{s}_1) \overset{\sim}{\to} \pi_1(S, \bar{s}_2)$ obtained from different paths differ by an inner automorphism of $\pi_1(S, \bar{s}_2)$ (or of $\pi_1(S, \bar{s}_1)$).

Hence $\pi_1(S, \bar{s})$ does not depend on $\bar{s}$ up to composition with inner automorphisms.

For every étale cover X of S and every element $\sigma \in \pi_1(S, \bar{s})$ there is an automorphism σ_X of $\mathcal{F}_{\bar{s}}(X)$. This defines a continuous action of $\pi_1(S, \bar{s})$ of $\mathcal{F}_{\bar{s}}(X)$, i.e., we can consider $\mathcal{F}_{\bar{s}}(X)$ as an object of $(\pi_1(S, \bar{s})$-sets), the category of finite sets with a continuous action by $\pi_1(S, \bar{s})$. In this way we obtain a functor

$$(20.18.4) \qquad\qquad T_{\bar{s}} \colon (\mathrm{F\acute{E}t}/S) \to (\pi_1(S, \bar{s})\text{-sets}).$$

By definition, the composition of $T_{\bar{s}}$ with the forgetful functor $(\pi_1(S, \bar{s})$-sets$) \to$ (Sets) is the fiber functor $\mathcal{F}_{\bar{s}}$.

Theorem 20.101. *Let S be a connected scheme and let $\bar{s}$ be a geometric point. Then the functor $T_{\bar{s}}$ (20.18.4) is an equivalence of categories. Moreover:*

(1) *The functor $T_{\bar{s}}$ induces an equivalence between the category of connected étale covers of S and the category of finite sets with transitive continuous $\pi_1(S, \bar{s})$-action.*

(2) *A connected étale cover $Z \to S$ is a Galois cover if and only if for one (or, equivalently, for all) $\bar{t} \in T_{\bar{s}}(Z)$ the stabilizer $\pi_{\bar{t}} := \{\, \sigma \in \pi_1(S, \bar{s}) \,;\, \sigma(\bar{t}) = \bar{t} \,\}$ is a normal subgroup of $\pi_1(S, \bar{s})$. In this case $\pi_{\bar{t}} = \pi_1(Z, \bar{t})$ and*

$$\mathcal{F}_{\bar{s}}(Z) = \pi_1(S, \bar{s})/\pi_{\bar{t}} \cong \mathrm{Aut}_{(\mathrm{F\acute{E}t}/S)}(Z).$$

Proof. Set $\pi := \pi_1(S, \bar{s})$ and $T := T_{\bar{s}}$. For an étale cover X of S we simply write $\mathrm{Aut}(X)$ instead of $\mathrm{Aut}_{(\mathrm{F\acute{E}t}/S)}(X)$ und $\mathcal{F}(X)$ instead of $\mathcal{F}_{\bar{s}}(X)$.

If X and Y are étale covers of S, then $T(X \sqcup Y) = T(X) \sqcup T(Y)$. Therefore it suffices to show (1) and (2), and we restrict to connected covers and π-sets with transitive action from now on.

(I) T is essentially surjective. Let Σ be a finite set with a continuous transitive π-action. As the action is continuous, the stabilizer of each point of Σ in π is an open subgroup. The intersection of these finitely many subgroups is again open and therefore contains an open normal subgroup H such that $\pi/H \cong \mathrm{Aut}(Z)$, where Z is a Galois cover. Hence Σ is of the form $\mathrm{Aut}(Z)/G$ for some finite subgroup G of $\mathrm{Aut}(Z)$. Then Z/G is an étale connected cover of S and $T(Z/G) = \Sigma$ by Proposition 20.89.

(II) T is fully faithful. We start with a remark. Let A be any group, G and H subgroups of A. Then every map $\varphi \colon A/G \to A/H$ that is equivariant with respect to the A-action by left multiplication is uniquely determined by $\varphi(1 \cdot G) \in A/H$. Conversely $\sigma H \in A/H$ yields a well defined A-equivariant map $A/G \to A/H$ with $1 \cdot G \mapsto \sigma H$ if and only if $G\sigma \subseteq \sigma H$.

Now let X and Y be connected étale covers of S. Combining Remark 20.85 and Lemma 20.92 we see that there exists an object $(Z, \bar{z}) \in \mathcal{G}_{\bar{s}}$ such that $X = Z/G$ and $Y = Z/H$ for finite subgroups G, H of $\mathrm{Aut}(Z)$ with $T(X) = \mathrm{Aut}(Z)/G$ and $T(Y) = \mathrm{Aut}(Z)/H$. By the above remark we may identify

$$(*) \qquad \mathrm{Hom}_{(\pi\text{-sets})}(T(X), T(Y)) = \{\, \sigma H \in \mathrm{Aut}(Z)/H \,;\, G\sigma \subseteq \sigma H \,\}.$$

Let $f\colon X = Z/G \to Y = Z/H$ be a morphism of étale covers. Choose an element $\bar{z}' \in \mathcal{F}(Z)$ whose image in $\mathcal{F}(Y)$ is the image of $\bar{z}$ under the composition

$$\mathcal{F}(Z) \longrightarrow \mathcal{F}(X) \xrightarrow{\mathcal{F}(f)} \mathcal{F}(Y).$$

By Remark 20.90 (ii) there exists $\sigma \in \operatorname{Aut}(Z)$ such that $\sigma(\bar{z}) = \bar{z}'$. Clearly f is uniquely determined by the coset σH. Conversely, a given element $\sigma \in \operatorname{Aut}(Z)$ induces a well defined morphism $X \to Y$ if and only if $G\sigma \subseteq \sigma H$. This proves the fully faithfulness by (*). $\square$

If $(X, \bar{x})$ is a pointed connected étale cover of $(S, \bar{s})$, then we set

$$\pi_{(X,\bar{x})} := \{\, \sigma \in \pi_1(S, \bar{s}) \;;\; \sigma(\bar{x}) = \bar{x} \,\}.$$

Then Theorem 20.101 and its proof show the following Galois correspondence.

Corollary 20.102. *Attaching* $\pi_{(X,\bar{x})}$ *to* $(X, \bar{x})$ *yields a bijective map between the set of isomorphism classes of pointed connected étale covers and the set of open subgroups of* $\pi_1(S, \bar{s})$*. Via this bijection, pointed Galois covers correspond to normal subgroups.*

Proof. The inverse map is given by sending an open subgroup U of $\pi_1(S, \bar{s})$ to the finite $\pi_1(S, \bar{s})$-set $\pi_1(S, \bar{s})/U$ and then applying the inverse of the functor $T_{\bar{s}}$. $\square$

(20.19) Functoriality of the fundamental group.

To formulate functoriality statements for the fundamental group it is convenient to introduce the following notion. A *pointed scheme* is a pair $(S, \bar{s})$ consisting of a scheme S and a geometric point $\bar{s}$ of S. A *morphism of pointed schemes* $f\colon (T, \bar{t}) \to (S, \bar{s})$ is a morphism of schemes $f\colon T \to S$ such that $f(\bar{t}) := f \circ \bar{t} = \bar{s}$. By definition, this implies $\kappa(\bar{t}) = \kappa(\bar{s})$.

The fundamental group is functorial in the following sense.

Definition and Remark 20.103. Let $f\colon (T, \bar{t}) \to (S, \bar{s})$ be a morphism of connected pointed schemes. To emphasize the base scheme, we write $\mathcal{F}_{\bar{s}}(X \to S)$ instead of $\mathcal{F}_{\bar{s}}(X)$ for an étale cover X of S, and likewise for the base change to T. One has

$$\mathcal{F}_{\bar{s}}(X \to S) = \operatorname{Hom}_S(\operatorname{Spec}\kappa(\bar{s}), X) = \operatorname{Hom}_T(\operatorname{Spec}\kappa(\bar{t}), X \times_S T) = \mathcal{F}_{\bar{t}}(X \times_S T \to T).$$

Hence every isomorphism of fiber functors $\mathcal{F}_{\bar{t}} \to \mathcal{F}_{\bar{t}'}$ for geometric points $\bar{t}$ and $\bar{t}'$ of T yields an isomorphism $\mathcal{F}_{f(\bar{t})} \to \mathcal{F}_{f(\bar{t}')}$ and one obtains a functor

$$f_*\colon \Pi(T) \longrightarrow \Pi(S)$$

For every geometric point $\bar{t}$ of T it induces a continuous group homomorphism

$$f_*\colon \pi_1(T, \bar{t}) \longrightarrow \pi_1(S, f(\bar{t})).$$

By composition the functor f_* yields a functor

$$f^*\colon (\pi_1(S, f(\bar{t}))\text{-sets}) \to (\pi_1(T, \bar{t})\text{-sets})$$

and by construction one has an isomorphism of functors $(\text{FÉt}/S) \to (\pi_1(T, \bar{t})\text{-sets})$

(20.19.1) $$f^* \circ T_{f(\bar{t})} \cong T_{\bar{t}} \circ f^*,$$

where the f^* on the right hand side denotes the base change functor $f^*\colon (\mathrm{F\acute{E}t}/S) \to (\mathrm{F\acute{E}t}/T)$.

If $g\colon U \to T$ is a second morphism of connected schemes, then

$$(f \circ g)_* = g_* \circ f_*, \qquad (f \circ g)^* \cong g^* \circ f^*.$$

One has the following dictionary.

Proposition 20.104. *Let $f\colon (T,\bar{t}) \to (S,\bar{s})$ be a morphism of connected pointed schemes. Let*

$$f^*\colon (\mathrm{F\acute{E}t}/S) \to (\mathrm{F\acute{E}t}/T)$$

be the base change functor that sends an étale cover $X \to S$ to $X \times_S T \to T$ and let

$$f_*\colon \pi_1(T,\bar{t}) \to \pi_1(S, f(\bar{t}))$$

be the morphism of topological groups induced by f.

Let $(S',\bar{s}')$ (resp. $(T',\bar{t}')$) be a pointed connected étale cover of $(S,\bar{s})$ (resp. of $(T,\bar{t})$) and let $\pi_{(S',\bar{s}')}$ (resp. $\pi_{(T',\bar{t}')}$) be the corresponding open subgroup of $\pi_1(S,\bar{s})$ (resp. of $\pi_1(T,\bar{t})$) (Corollary 20.102).

(1) *The homomorphism f_* is trivial if and only if f^* sends every connected étale cover to a split étale cover.*

(2) *The homomorphism f_* is an isomorphism if and only if f^* is an equivalence of categories.*

(3) *There exists a morphism of pointed schemes $f'\colon (T,\bar{t}) \to (S',\bar{s}')$ (necessarily unique by Remark 20.66 (4)) such that the diagram*

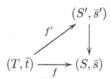

of pointed schemes commutes if and only if $f_*\pi_1(T,\bar{t}) \subseteq \pi_{(S',\bar{s}')}$.

In this case, $\pi_{(S',\bar{s}')}$ contains the normal subgroup generated by $f_*\pi_1(T,\bar{t})$ if and only if the étale cover $T \times_S S'$ of T is split.

(4) *The homomorphism f_* is surjective if and only if f^* is fully faithful.*

(5) *Suppose that f_* is surjective. Then the subgroup $\pi_{(T',\bar{t})}$ contains $\mathrm{Ker}(f_*)$ if and only if there exists an étale cover $X \to S$ and an isomorphism of étale T-covers $X \times_S T \cong T'$.*

Proof. We set $G := \pi_1(S,\bar{s})$, $H := \pi_1(T,\bar{t})$, $G' := \pi_{(S',\bar{s}')}$ and $H' := \pi_{(T',\bar{t}')}$. We call a G-set trivial (resp. homogeneous) if G acts trivially (resp. transitively) on it. Every homogeneous G-set is isomorphic to a G-set of the form G/U for an open subgroup $U \subseteq G$. By (20.19.1) and Theorem 20.101 the functor f^* has some given property that is stable under equivalences of categories if and only if the functor

$$C_f\colon (G\text{-sets}) \to (H\text{-sets})$$

induced by $f_*\colon H \to G$ has that same property.

Let us show (1). An étale cover of T is split if and only if the corresponding $\pi_1(T, \bar{t})$-set is trivial. Therefore f^* sends every connected étale cover to a split étale cover if and only if C_f sends every homogeneous G-set G/U, $U \subseteq G$ some open subgroup, to a trivial H-set. This means that $\operatorname{Im}(f_*) \subseteq U$ for every open subgroup U of G and hence that f_* is trivial by Lemma 20.96 (2).

For (2) one has to show that f_* is an isomorphism if and only if C_f is an equivalence of categories. The condition is clearly necessary. Conversely, if C_f is an equivalence of categories, then f_* induces a bijection between open normal subgroups V of H and open normal subgroups U of G given by $V \mapsto f_*(V)$ and f_* yields an isomorphism of groups $H/V \xrightarrow{\sim} G/f_*(V)$ all $V \subseteq H$ open normal subgroup. Passing to the limit, we see that f_* is an isomorphism of profinite groups.

Let us show (3). The finite G-set corresponding to $(S', \bar{s}')$ is G/G'. Then $\operatorname{Im}(f_*)$ (resp. the normal subgroup generated by $\operatorname{Im}(f_*)$) is contained in G' if and only if the coset G' in $C_f(G/G')$ has all of H as stabilizer (resp. if and only if $C_f(G/G')$ is the trivial H-set).

We show (4). If f_* is surjective, then C_f is fully faithful. Conversely, assume that f_* is not surjective. Then there exists an open subgroup $U \subsetneq G$ such that the image of f_* is contained in U (Lemma 20.96 (3)). Then G/U is a nontrivial finite G-set whose image under C_f is trivial. If C_f was fully faithful, we would find, denoting by $*$ the singleton with its trivial action

$$\emptyset = \operatorname{Hom}_{(G\text{-sets})}(*, G/U) \xrightarrow{\sim} \operatorname{Hom}_{(H\text{-sets})}(*, C_f(G/U)) \neq \emptyset,$$

a contradiction.

For (5) we have to show that the finite H-set H/H' is in the essential image of C_f if and only if H' contains $\operatorname{Ker}(f_*)$. The condition is clearly necessary. Conversely assume that H' contains $\operatorname{Ker}(f_*)$. Define an action of $g \in G$ on H/H' by $(g, hH') \mapsto \tilde{g}hH'$, where $\tilde{g} \in H$ with $f_*(\tilde{g}) = g$ which is possible because f_* is surjective. This is well defined because $\operatorname{Ker}(f_*) \subseteq H'$ and because $\operatorname{Ker}(f_*)$ is a normal subgroup of H. $\qquad\square$

There is the following invariance property for fundamental groups.

Proposition 20.105. *Let $g \colon X \to X'$ be a morphism of schemes that is a universal homeomorphism. Let $\bar{x}$ be a geometric point of X and let $\bar{x}'$ be its image in X'. Then the functor $Y' \mapsto Y' \times_{X'} X$ yields an equivalence $(\mathrm{F\acute{E}t}/X') \to (\mathrm{F\acute{E}t}/X)$. If X or equivalently X' is connected it induces an isomorphism of profinite groups*

$$\pi_1(X, \bar{x}) \xrightarrow{\sim} \pi_1(X', \bar{x}').$$

In particular, the fundamental group of a connected scheme depends only on the underlying reduced scheme.

Proof. It suffices to show the first assertion. By Theorem 20.30 it remains to show that an étale X'-scheme Y' is finite over X' if and only if $Y := Y' \times_{X'} X$ is finite over X.

As being finite is stable under base change, the condition is clearly necessary. Conversely, suppose that Y is finite over X. As $Y \to Y'$ is a universal homeomorphism, it is surjective and integral by Proposition 20.29. Therefore Y' is affine over X' by Lemma 20.106 (1) below. Moreover, because the composition $Y \to Y' \to X'$ is universally closed, $Y' \to X'$ is universally closed. Hence Y' is integral over X' by Lemma 20.106 (2) below. As $Y' \to X'$ is étale and in particular locally of finite type, $Y' \to X'$ is finite. $\qquad\square$

In the proof above we used the following facts about integral morphisms.

Lemma 20.106. *Let $f\colon X \to Y$ be a morphism of schemes.*
(1) Suppose that f is integral and surjective. Then X is affine if and only if Y is affine.
(2) The morphism f is integral if and only if it is affine and universally closed.

Proof. [Sketch] The first assertion is a generalization of Chevalley's theorem, that we proved when Y is noetherian and f is finite (Theorem 12.39). The general case is deduced by noetherian approximation and by approximating integral morphisms by finite morphisms. For details we refer to [Sta] 05YU. The second assertion is Exercise 12.19. □

If for a morphism of connected schemes $X \to S$ the base change functor $(\text{FÉt}/S) \to (\text{FÉt}/X)$ is fully faithful, then $\pi_1(X, \bar{x}) \to \pi_1(S, f(\bar{x}))$ is surjective by Proposition 20.104. This holds for instance in the following case, where for a scheme T we denote by $(\text{VB}(T))$ the category of locally free $\mathscr{O}_T$-modules of finite type.

Proposition 20.107. *Let $f\colon X \to S$ be morphism of schemes such that $\mathscr{O}_S \to f_*\mathscr{O}_X$ is an isomorphism.*
(1) The functor $(\text{VB}(S)) \to (\text{VB}(X))$, $\mathscr{E} \mapsto f^*\mathscr{E}$, is fully faithful.
(2) The functor $T \mapsto T \times_S X$ from the category of finite locally free S-schemes to the category of finite locally free X-schemes is fully faithful.
(3) The functor $(\text{FÉt}/S) \to (\text{FÉt}/X)$ is fully faithful.
(4) Suppose that S is connected. Then X is connected and $\pi_1(X, \bar{x}) \to \pi_1(S, f(\bar{x}))$ is surjective for every geometric point $\bar{x}$ of X.

See also Lemma 24.67 below for a more precise version of Part (1) of the proposition.

Proof. To show (1), we prove that for finite locally free $\mathscr{O}_S$-modules $\mathscr{E}$ and $\mathscr{F}$ the canonical map of $\mathscr{O}_S$-modules

$$(*) \qquad \mathscr{H}om_{\mathscr{O}_S}(\mathscr{E}, \mathscr{F}) \to f_*\mathscr{H}om_{\mathscr{O}_X}(f^*\mathscr{E}, f^*\mathscr{F})$$

is an isomorphism. Taking global sections of $(*)$ we obtain (1). To show that $(*)$ is an isomorphism, we may work locally on S. Hence we may assume $\mathscr{E} = \mathscr{O}_S^n$ and $\mathscr{F} = \mathscr{O}_S^m$. As $(*)$ is compatible with taking finite direct sums, we may assume $\mathscr{E} = \mathscr{F} = \mathscr{O}_S$, but then $(*)$ holds by assumption.

The functor $\mathscr{E} \mapsto f^*\mathscr{E}$ is also compatible with forming tensor products and hence induces a fully faithful functor from the category of finite locally free $\mathscr{O}_S$-algebras to the category of finite locally free $\mathscr{O}_X$-algebras. This shows (2) because the categories of finite locally free $\mathscr{O}_Y$-algebras and of finite locally free Y-schemes are equivalent for every scheme Y (Proposition 12.19). Now (3) is an immediate corollary.

If S is connected, i.e. 0 and 1 are the only idempotent elements in $\Gamma(S, \mathscr{O}_S)$, then X is connected as well because $\Gamma(X, \mathscr{O}_X) = \Gamma(S, \mathscr{O}_S)$ by hypothesis. Hence (4) follows from (3) using Proposition 20.104 (4). □

Using properties of the Stein factorization (proved in Section (24.11) below) one obtains the following corollary.

Corollary 20.108. *Let $f\colon X \to S$ be a proper surjective morphism with geometrically connected fibers. Then $(\text{FÉt}/S) \to (\text{FÉt}/X)$ is fully faithful and, for S connected, $\pi_1(X, \bar{x}) \to \pi_1(S, f(\bar{x}))$ is surjective.*

More generally, $(\text{FÉt}/S) \to (\text{FÉt}/X)$ is fully faithful if $X \to S$ is universally submersive with geometrically connected fibers (Exercise 20.2).

Proof. The Stein factorization of f is of the form $X \xrightarrow{f'} S' \longrightarrow S$ where $\mathscr{O}_{S'} \to f'_* \mathscr{O}_X$ is an isomorphism and $S' \to S$ is a universal homeomorphism (Proposition 24.56 below). We conclude by combining Proposition 20.107 and Proposition 20.105. $\qquad\qquad\square$

(20.20) Fundamental groups of fields.

Example 20.109. Let k be a field. Recall that for a k-algebra A the following assertions are equivalent (Remark 18.25 and Theorem 18.44).
(i) $\operatorname{Spec} A$ is an étale cover of k.
(ii) A is a finite separable k-algebra.
(iii) A is isomorphic to a finite product of finite separable field extensions of k.
In this case, $\operatorname{Spec} A$ is connected if and only if A is a separable field extension K of k. Such a connected étale cover $\operatorname{Spec} K \to \operatorname{Spec} k$ is a Galois cover (Definition 20.90) if and only if $\#\operatorname{Aut}_k(K) = [K : k]$, i.e., if and only if K is a Galois extension of k. In this case we have a group isomorphism

$$(20.20.1) \qquad \operatorname{Gal}(K/k) \xrightarrow{\sim} \operatorname{Aut}_{(\text{FÉt}/\operatorname{Spec} k)}(\operatorname{Spec} K)^{\mathrm{opp}}, \qquad \sigma \mapsto \operatorname{Spec} \sigma.$$

A geometric point $\bar{s}$ of $\operatorname{Spec} k$ corresponds to an algebraically closed extension field Ω of k. Let k^{sep} be the separable closure of k in Ω. A Galois cover in the category of connected pointed Galois covers $\mathcal{G}_{\bar{s}}$ as in Section (20.18) is a finite Galois extension of k together with a k-embedding $K \to \Omega$ or, equivalently, a k-embedding $K \to k^{\mathrm{sep}}$. Taking the limit over all finite Galois extension K of k contained in k^{sep} the isomorphism (20.20.1) induces an isomorphism of profinite groups

$$(20.20.2) \qquad \begin{aligned} \operatorname{Gal}(k^{\mathrm{sep}}/k) &= \varprojlim_K \operatorname{Gal}(K/k) \\ &= \varprojlim_K \operatorname{Aut}_{(\text{FÉt}/\operatorname{Spec} k)}(\operatorname{Spec} K)^{\mathrm{opp}} \xrightarrow{\sim} \pi_1(\operatorname{Spec} k, \bar{s}), \end{aligned}$$

where the last isomorphism is (20.18.3).

In this special case, Theorem 20.101 yields the following result.

Corollary 20.110. *There exists a contravariant equivalence T between the category of finite separable k-algebras A and the category of finite sets with continuous action by $\operatorname{Gal}(k^{\mathrm{sep}}/k)$, given by $A \mapsto T(A) := \operatorname{Hom}_k(A, k^{\mathrm{sep}})$, where $\operatorname{Gal}(k^{\mathrm{sep}}/k)$ acts on $T(A)$ by $(\sigma, t) \mapsto \sigma \circ t$.*

Moreover, A is a finite separable field extension of k if and only if the $\operatorname{Gal}(k^{\mathrm{sep}}/k)$-action on $T(A)$ is transitive.

Let A be a finite separable k-algebra. For $t \in T(A)$ denote by Γ_t the stabilizer $\{\sigma \in \operatorname{Gal}(k^{\mathrm{sep}}/k) \; ; \; \sigma(t) = t\}$. This is an open subgroup of $\operatorname{Gal}(k^{\mathrm{sep}}/k)$.

Assume that $A = K$ is a finite separable field extension. Then $T(K) = \operatorname{Hom}_k(K, k^{\mathrm{sep}})$ can be identified with $\operatorname{Gal}(k^{\mathrm{sep}}/k)/\Gamma_t$ (where $\operatorname{Gal}(k^{\mathrm{sep}}/k)$ acts on the quotient by left multiplication). This identification depends on the choice of t. For $t \in T(K)$ the field of elements in k^{sep} fixed by Γ_t is given by

$$(k^{\mathrm{sep}})^{\Gamma_t} = t(K).$$

Moreover, Γ_t is normal in $\mathrm{Gal}(k^{\mathrm{sep}}/k)$ for one $t \in T(K)$ if and only if one has $\Gamma_t = \Gamma_{t'}$ for all $t' \in T(K)$ if and only if K is a Galois extension of k. In this case $\mathrm{Gal}(K/k) = \mathrm{Aut}_{(\text{FÉt}/\mathrm{Spec}\,k)}(\mathrm{Spec}\,K)^{\mathrm{opp}} \cong \mathrm{Gal}(k^{\mathrm{sep}}/k)/\Gamma_t$.

(20.21) Examples from number theory.

Proposition 20.111. *Let R be a local henselian ring with residue field k and set $S = \mathrm{Spec}\,R$. Let $\bar{s}$ be a geometric point lying over the closed point s of S and let k^{sep} be the separable closure of k in $\kappa(\bar{s})$. Then by functoriality the closed immersion $s \to S$ yields an isomorphism*

$$\mathrm{Gal}(k^{\mathrm{sep}}/k) = \pi_1(\mathrm{Spec}\,k, \bar{s}) \xrightarrow{\sim} \pi_1(S, \bar{s}).$$

Proof. This follows from Example 20.109 and the fact that the functor $X \mapsto X \otimes_R k$ yields an equivalence $(\text{FÉt}/R) \to (\text{FÉt}/k)$, see Theorem 20.75. $\square$

Example 20.112. This proposition can in particular be applied if R is the ring of integers of a non-archimedean local field (Example 20.79). In this case, its residue field k is a finite field and hence $\mathrm{Gal}(k^{\mathrm{sep}}/k) \cong \widehat{\mathbb{Z}} := \lim_n \mathbb{Z}/n\mathbb{Z} \cong \prod_p \mathbb{Z}_p$, where p runs through all prime numbers in $\mathbb{Z}$ (e.g., [BouAII]$^\circ$ V, §12, No.3), and hence $\pi_1(\mathrm{Spec}\,R, \bar{s}) \cong \widehat{\mathbb{Z}}$.

Example 20.113. Let S be a Dedekind scheme with function field K. Fix an algebraic closure of K and let $\bar{s}$ be the corresponding geometric point of S. By Example 20.78 there is a bijective correspondence between étale covers of S and finite separable extensions of K that are unramified at all points of S. This correspondence is given by attaching to an étale cover $X \to S$ the function field L_X of X and by attaching to a finite separable extension L of K the normalization of S in L.

Every S-automorphism of X induces a K-automorphism of L_X. Conversely, every K-automorphism of L_X extends uniquely to an S-automorphism of X because of the functoriality of the normalization (Proposition 12.44). Therefore we see that $\pi_1(S, \bar{s})$ can be identified with the profinite Galois group of the maximal subextension L of K in $\bar{K}$ which is unramified at all points of S.

A typical example from number theory is the following. Let K be a number field, i.e., a finite extension of $\mathbb{Q}$, let O_K be its ring of integers and let $T \subset \mathrm{Spec}\,O_K$ be a finite closed subset. Set $U := (\mathrm{Spec}\,O_K) \setminus T$. Then $\pi_1(U, \bar{s})$ is the profinite Galois group of the maximal subextension K_T of K in $\bar{K}$ which is unramified outside T.

In the special case $K = \mathbb{Q}$, $O_K = \mathbb{Z}$ and $T = \emptyset$, it is known that $K_T = \mathbb{Q}$ by Minkowski's theorem ([Neu]$^\circ$ III, (2.18)). Hence $\pi_1(\mathrm{Spec}\,\mathbb{Z}, \bar{s}) = 1$. In this sense $\mathrm{Spec}\,\mathbb{Z}$ is simply connected.

(20.22) The fundamental group of $\mathbb{P}^1$.

We will now compute the fundamental group of $\mathbb{P}^1_k$ for a field k.

Theorem 20.114. *Let k^{sep} be a separable closure of k and let $\bar{x} \in \mathbb{P}^1(k^{\mathrm{sep}})$. Then the structure morphism $\mathbb{P}^1_k \to \mathrm{Spec}\,k$ induces an isomorphism of profinite groups*

$$\pi_1(\mathbb{P}^1_k, \bar{x}) \xrightarrow{\sim} \pi_1(\mathrm{Spec}\,k, k^{\mathrm{sep}}) = \mathrm{Gal}(k^{\mathrm{sep}}/k).$$

In particular the fundamental group of $\mathbb{P}^1_k$ is trivial if and only if k is separably closed.

Proof. We show that $T \mapsto T \times_k \mathbb{P}^1_k$ is an equivalence of categories (FÉt/Spec k) $\rightarrow$ (FÉt/$\mathbb{P}^1_k$). As $\Gamma(\mathbb{P}^1_k, \mathscr{O}_{\mathbb{P}^1_k}) = k$ (Example 11.45), by Proposition 20.107 we know already that the functor is fully faithful. Its essential image consists of those finite étale morphisms $g\colon \tilde{T} \rightarrow \mathbb{P}^1_k$ such that the finite locally free module $g_*\mathscr{O}_{\tilde{T}}$ is isomorphic to $\mathscr{O}^n_{\mathbb{P}^1_k}$ for some n.

Let $g\colon \tilde{T} \rightarrow \mathbb{P}^1_k$ be finite étale and set $\mathscr{A} := g_*\mathscr{O}_{\tilde{T}}$. Any finite locally free module $\mathscr{A}$ is a finite direct sum of line bundles $\mathscr{O}(d_i)$ for unique integers $d_1 \geq \cdots \geq d_n$ (Theorem 11.53, see also Section (26.22)). As the trace form is perfect (Proposition 20.71), we find $\mathscr{A} \cong \mathscr{A}^\vee$ as locally free modules. Because $\mathscr{O}(d)^\vee = \mathscr{O}(-d)$, this implies

$$(*) \qquad\qquad d_i + d_{n+1-i} = 0, \qquad\qquad i = 1, \dots, n.$$

In particular $d_1 \geq 0$ and $d_n = -d_1$.

Let $0 \neq s \in \Gamma(\mathbb{P}^1_k, \mathscr{O}(d_1)) \subseteq \Gamma(\mathbb{P}^1_k, \mathscr{A})$. Now $\Gamma(\mathbb{P}^1_k, \mathscr{A}) = \Gamma(\tilde{T}, \mathscr{O}_{\tilde{T}})$ is reduced because $\tilde{T}$ is reduced. Hence $s^2 \neq 0$. Assume that $d_1 > 0$. If we restrict the multiplication $m\colon \mathscr{A} \otimes \mathscr{A} \rightarrow \mathscr{A}$ to $\mathscr{O}(d_1) \otimes \mathscr{O}(d_1) = \mathscr{O}(2d_1)$, this corresponds to a global section of $\mathscr{A}(-2d_1)$ which has to be zero, because $\mathscr{A}(-2d_1)$ is the direct sum of line bundles $\mathscr{O}(e)$ with $e < 0$. But this contradicts $s^2 \neq 0$. Hence $d_1 = 0$ and therefore $d_i = 0$ for all i by $(*)$. $\qquad\square$

(20.23) Algebraic and analytic fundamental group.

Let S be a connected scheme locally of finite type over $\mathbb{C}$ and let $\alpha\colon S^{\mathrm{an}} \rightarrow S$ be its analytification (Section (20.12)). Let $s \in S^{\mathrm{an}}$ and let $\bar{s}$ be a geometric point of S lying over $\alpha(s)$. Let us compare $\pi_1(S^{\mathrm{an}}, s)$ and $\pi_1(S, \bar{s})$.

To do this we first recall that if G is any group, then the normal subgroups of G of finite index form a filtered inductive system $(H_i)_{i \in I}$ and the profinite group $\hat{G} := \lim_{i \in I} G/H_i$ is called the *profinite completion of* G, see also Exercise 20.40.

Now étale covers of S correspond bijectively to finite covers of S^{an} in a functorial way by the Riemann existence theorem (Remark 20.80). As every finite cover of S^{an} is a quotient of the universal cover of S^{an} by a subgroup of finite index of $\pi_1(S^{\mathrm{an}}, s)$, we hence obtain an isomorphism of profinite groups

$$\pi_1(S, \bar{s}) \xrightarrow{\sim} \widehat{\pi_1(S^{\mathrm{an}}, s)}.$$

(20.24) The fundamental exact sequence of fundamental groups.

Theorem 20.115. *Let k be a field, let $\bar{k}$ be an algebraic closure of k, and let k^{sep} be the separable closure of k in $\bar{k}$. Let X be a geometrically connected qcqs k-scheme and set $\bar{X} := X \otimes_k \bar{k}$. Let $\bar{x}$ be a geometric point of $\bar{X}$ and denote its image in X also by $\bar{x}$. Then the sequence*

$$(20.24.1) \qquad 1 \longrightarrow \pi_1(\bar{X}, \bar{x}) \longrightarrow \pi_1(X, \bar{x}) \longrightarrow \mathrm{Gal}(k^{\mathrm{sep}}/k) \longrightarrow 1$$

is exact.

Here we identify $\mathrm{Gal}(k^{\mathrm{sep}}/k)$ with $\pi_1(\mathrm{Spec}\, k, \bar{s})$ (Example 20.109), where $\bar{s}$ is the image of $\bar{x}$ in $\mathrm{Spec}\, k$, and the maps in the exact sequence arise by functoriality from the morphisms $\bar{X} \rightarrow X \rightarrow \mathrm{Spec}\, k$.

Proof. Let k^p be the perfect closure of k in $\bar{k}$. Then $\operatorname{Spec} k^p \to \operatorname{Spec} k$ and $X^p := X \otimes_k k^p \to X$ are universal homeomorphisms. Replacing k by k^p and X by X^p, by Proposition 20.105 we may assume that k is perfect.

Then $\bar{k}$ is the filtered colimit of finite Galois subextensions k_i of k. Set $X_i := X \otimes_k k_i$ and denote the image of $\bar{x}$ in X_i again by $\bar{x}$. If $Y \to \bar{X}$ is an étale covering, then there exists an i and a morphism of finite presentation $Y_i \to X_i$ such that $Y_i \times_{X_i} \bar{X} = Y$ and any two such schemes are isomorphic after possibly enlarging i (Theorem 10.66 and Remark 10.68). As the properties "finite" and "étale" are compatible with inductive limits, $Y_i \to X_i$ is an étale cover for i large enough. Using the notion of colimit of categories (which we do not define here), this can be expressed by saying that there is an equivalence of categories $\operatorname{colim}_i (\text{FÉt}/X_i) \cong (\text{FÉt}/X)$. It follows formally that the canonical homomorphism of profinite groups

$$(*) \qquad \pi_1(\bar{X}, \bar{x}) \longrightarrow \lim_i \pi_1(X_i, \bar{x})$$

is an isomorphism.

On the other hand $\operatorname{Spec} k_i \to \operatorname{Spec} k$ is a torsor (for the étale or for the fppf-topology) under the Galois group $\operatorname{Gal}(k_i/k)$ (cf. Definition 14.84). Hence $X_i \to X$ is a Galois cover with automorphism group $\operatorname{Gal}(k_i/k)$. By Theorem 20.101 (2) we have an exact sequence

$$1 \longrightarrow \pi_1(X_i, \bar{x}) \longrightarrow \pi_1(X, \bar{x}) \longrightarrow \operatorname{Gal}(k_i/k) \longrightarrow 1.$$

Passing to the limit over i and using $(*)$ we obtain the desired sequence. It is exact as a limit of exact sequences of compact topological groups (recall that a limit of non-empty compact spaces is non-empty and apply this to the fibers of $\pi_1(X, \bar{x}) \longrightarrow \operatorname{Gal}(k^{\text{sep}}/k)$). $\square$

Remark 20.116. Let X be as in Theorem 20.115. Every k-valued point of X defines a section of the exact sequence (20.24.1) by functoriality. The famous *section conjecture* by Grothendieck asserts that for $k/\mathbb{Q}$ finitely generated and X a geometrically connected, smooth projective curve of genus ≥ 2 (see Section (26.8)) every such section arises from a k-rational point.

There is the following variant of Theorem 20.115 for arbitrary base schemes and proper morphisms. It uses that the Stein factorization yields an étale cover for flat proper finitely presented morphisms with geometrically reduced fibers (see Theorem 24.61 below).

Proposition 20.117. *Let $f\colon X \to S$ be a flat proper morphism of finite presentation with geometrically connected and geometrically reduced fibers. Assume that S is connected. Let $\bar{x}$ be a geometric point of X and let $\bar{s}$ be its image in S. Then there is an exact sequence*

$$\pi_1(X_{\bar{s}}, \bar{x}) \longrightarrow \pi_1(X, \bar{x}) \longrightarrow \pi_1(S, \bar{s}) \longrightarrow 1$$

of profinite groups.

Note that the hypotheses imply that X and $X_{\bar{s}}$ are also connected.

Proof. We have already seen in Corollary 20.108 that $\pi_1(X, \bar{x}) \longrightarrow \pi_1(S, \bar{s})$ is surjective.

As the composition $(\text{FÉt}/S) \to (\text{FÉt}/X) \to (\text{FÉt}/X_{\bar{s}})$ sends any étale covering of S to a split covering, the composition $\pi_1(X_{\bar{s}}) \to \pi_1(X) \to \pi_1(S)$ is trivial.

It remains to show that if $Y \to X$ is an étale covering with Y connected and such that $Y \times_X X_{\bar{s}} = Y_{\bar{s}}$ has a section over $X_{\bar{s}}$, then there exists an étale covering $T \to S$ and an X-isomorphism $Y \xrightarrow{\sim} X \times_S T$ (Proposition 20.104). To see this let $Y \to T \to S$ be the Stein factorization of $Y \to S$. Then $T \to S$ is finite étale by Theorem 24.61 below. We claim that the X-morphism $u\colon Y \to X \times_S T$ is an isomorphism.

As u is a morphism of étale covers, it is finite étale (Remark 20.66 (3)). Because Y is connected and $Y \to T$ is surjective, T is connected. As $X \times_S T \to T$ is closed with connected fibers, $X \times_S T$ is connected as well by an elementary topological argument (see Lemma 24.52 below for a more general statement and its proof). Hence the degree of u is a constant function and it suffices to show that it equals 1 in a single point. This can be done after base change $\bar{s} \to S$, where it follows from the assumption that $Y_{\bar{s}}$ has a section over $X_{\bar{s}}$. $\square$

(20.25) Specialization of fundamental groups.

Let R be a local henselian ring with residue field k. In Theorem 20.75 we showed that the base change $U \mapsto U \otimes_R k$ is an equivalence of categories $(\text{FÉt}/R) \to (\text{FÉt}/k)$. This result can be generalized considerably.

Theorem 20.118. *Let (R, I) be a henselian pair (Definition 20.15). Let $X \to \operatorname{Spec} R$ be a proper morphism and set $X_0 := X \otimes_R R/I$. Then the functor*

$$(\text{FÉt}/X) \longrightarrow (\text{FÉt}/X_0), \qquad U \mapsto U \times_X X_0$$

is an equivalence of categories. In particular, if X and X_0 are connected, then for every geometric point $\bar{x}$ of X_0 the inclusion $X_0 \to X$ yields an isomorphism

$$\pi_1(X_0, \bar{x}) \xrightarrow{\sim} \pi_1(X, \bar{x}).$$

We will give a proof of this theorem below (Corollary 24.111) in the case that R is an I-adically complete noetherian ring, as an application of Grothendieck's existence theorem. The general case is reduced to this special case by Artin approximation (Section (20.11)) and by noetherian approximation. We refer to [Sta] 0GS2 for details.

We can use the theorem to show the following invariance of fundamental groups.

Proposition 20.119. *Let $k \to K$ be an extension of separably closed fields and let X be a proper scheme over k. Then the functor*

$$\mathcal{F}\colon (\text{FÉt}/X) \longrightarrow (\text{FÉt}/X_K), \qquad U \mapsto U_K := U \otimes_k K$$

is an equivalence of categories. In particular, if X is connected, for every geometric point $\bar{x}$ of X one has an isomorphism $\pi_1(X_K, \bar{x}) \xrightarrow{\sim} \pi_1(X, \bar{x})$.

Note that if X is connected, then it is geometrically connected because k is separably closed (Proposition 5.53). Hence X_K is connected as well.

Proof. By Proposition 20.105 we may assume that K and k are algebraically closed. The k-algebra K is the filtered union of its finitely generated k-subalgebras R.

We first show that $\mathcal{F}$ is essentially surjective. Let U be a finite étale scheme over X_K. Then we can find a finitely generated k-subalgebra R of K and an R-scheme U_R whose base change to X_K is U (Section (10.17)). As the properties "finite" and "étale" are compatible with inductive limits (Proposition 12.11, and Corollary 18.43) we may assume after enlarging R that U_R is finite étale over X_R.

Let $\mathfrak{m} \subset R$ be a maximal ideal, then $R/\mathfrak{m} = k$ since R is a finitely generated k-algebra and k is algebraically closed. Let $U_0 := U_R \otimes_R k$ be the special fiber of U_R. Let R^{h} be the henselization of $R_\mathfrak{m}$. We may choose a homomorphism of R-algebras $R^{\mathrm{h}} \to K$. Indeed, let L be the field of fractions of $R_\mathfrak{m}$ which is a subfield of K. As R^{h} is the colimit of localizations of étale $R_\mathfrak{m}$-algebras, we may view R^{h} as a subring of an algebraic closure $\bar{L}$ of L. As K is algebraically closed, we find an L-embedding $\bar{L} \to K$ and in particular a homomorphism of R-algebras $R^{\mathrm{h}} \to K$.

By Theorem 20.118 we have $U_0 \otimes_k R^{\mathrm{h}} \cong U_R \otimes_R R^{\mathrm{h}}$. Hence we see that U is isomorphic to the base change of U_0. This proves that $\mathcal{F}$ is essentially surjective.

To show that $\mathcal{F}$ is fully faithful we first remark that $\mathcal{F}$ is faithful because $k \to K$ is faithfully flat (Proposition 14.70). Now let U and U' be finite étale schemes over X and let $\pi_K \colon U_K \to U'_K$ be a morphism of finite étale schemes over X_K. As above we find a k-algebra R^{h} that is a local henselian ring with residue field k, a k-algebra homomorphism $R^{\mathrm{h}} \to K$, and a morphism $\pi_{R^{\mathrm{h}}} \colon U_{R^{\mathrm{h}}} \to U'_{R^{\mathrm{h}}}$ whose base change to X_K is π. We set $\pi := \pi_{R^{\mathrm{h}}} \otimes_{R^{\mathrm{h}}} \mathrm{id}_k$.

Again using Theorem 20.118 we see that $U_{R^{\mathrm{h}}} \otimes_{R^{\mathrm{h}}} k \cong U$ and $U'_{R^{\mathrm{h}}} \otimes_{R^{\mathrm{h}}} k \cong U'$ and that via these isomorphisms we have $\pi \otimes_k \mathrm{id}_{R^{\mathrm{h}}} = \pi_{R^{\mathrm{h}}}$ and therefore $\pi \otimes_k \mathrm{id}_K = \pi_K$. $\qquad \square$

Remark 20.120. The assertion in Proposition 20.119 does not hold for non-proper schemes over a separably closed field k in general (Exercise 20.44). Following essentially Kedlaya [Ked] §4.1 let us say that a k-scheme X is π_1-*proper* if for every separably closed extension K of k the pullback via the projection $X \otimes_k K \to X$ induces an equivalence of categories (FÉt$/X$) $\longrightarrow$ (FÉt$/X \otimes_k K$). Then Proposition 20.119 means that every proper k-scheme is π_1-proper. Moreover, in loc. cit. the following assertions are shown.
(1) If $\mathrm{char}(k) = 0$, then any k-scheme X is π_1-proper.
(2) A connected qcqs k-scheme X is π_1-proper if and only if for every connected k-scheme Y and every geometric point $\bar{z}$ of (the connected scheme) $X \times_k Y$ the map induced by functoriality
$$\pi_1(X \times_k Y, \bar{z}) \longrightarrow \pi_1(X, \bar{z}) \times \pi_1(Y, \bar{z})$$
is an isomorphism of topological groups.

The above results allow us to define specialization maps for fundamental groups as follows. Let S be a scheme, let $f \colon X \to S$ be a proper morphism with geometrically connected fibers. Let $s, \eta \in S$ be points such that s is a specialization of η, i.e., $s \in \overline{\{\eta\}}$. Hence η corresponds to a point of $\mathrm{Spec}\,\mathscr{O}_{S,s}$.

Let $\bar{s}$ and $\bar{\eta}$ be geometric points lying over s and η, respectively. Let $\mathscr{O}_{S,\bar{s}}$ be the strict henselization of $\mathscr{O}_{S,s}$ with respect to $\bar{s}$ (Def. 20.39). Its residue field is a separable closure $\kappa(s)^{\mathrm{sep}}$ of $\kappa(s)$ in $\kappa(\bar{s})$. Choose an S-morphism $\bar{\eta} \to \mathrm{Spec}(\mathscr{O}_{S,\bar{s}})$. Let $\bar{x}$ and $\bar{x}'$ be geometric points of $X_{\kappa(s)^{\mathrm{sep}}}$ and $X_{\bar{\eta}}$. Then one defines a homomorphism of pro-finite groups, the *specialization map*, as the composition
$$\pi_1(X_{\bar{\eta}}, \bar{x}') \longrightarrow \pi_1(X_{\mathscr{O}_{S,\bar{s}}}, \bar{x}) \xrightarrow{\sim} \pi_1(X_{\mathscr{O}_{S,\bar{s}}}, \bar{x}') \xrightarrow{\sim} \pi_1(X_{\kappa(s)^{\mathrm{sep}}}, \bar{x}') \xrightarrow{\sim} \pi_1(X_{\bar{s}}, \bar{x}'),$$
where the first map is given by functoriality, the second by the choice of a path from $\bar{x}$ to $\bar{x}'$ (Remark 20.100), the third is the isomorphism of Theorem 20.118, and the last isomorphism is given by Proposition 20.119.

For smooth proper morphisms the specialization map is surjective. More precisely, let p be a prime number or $p = 1$. We define for a profinite group π it *maximal prime-to-p-quotient*

$$\pi^{(p')} := \lim_U \pi/U,$$

where U runs through the open normal subgroups U of π such that π/U has order prime to p. Hence $\pi^{(1')} = \pi$. Then there is the following result, for whose proof we refer to [Sta] 0COQ and 0COR.

Theorem 20.121. *Let $f\colon X \to S$ be a smooth proper morphism of schemes with geometrically connected fibers. Let $s, \eta \in S$ with s a specialization of η. Let p be the characteristic exponent of $\kappa(s)$ (i.e., $p = 1$ if the characteristic of $\kappa(s)$ is zero and otherwise $p = \mathrm{char}(\kappa(s))$). Let $\bar{x}$ be a geometric point of X_s and let $\bar{x}'$ be a geometric point of X_η. Then the specialization map*

$$\pi_1(X_{\bar{\eta}}, \bar{x}') \longrightarrow \pi_1(X_{\bar{s}}, \bar{x})$$

is surjective and induces an isomorphism

$$\pi_1(X_{\bar{\eta}}, \bar{x}')^{(p')} \longrightarrow \pi_1(X_{\bar{s}}, \bar{x})^{(p')}.$$

Exercises

Exercise 20.1. Let A be a ring. For a polynomial $f \in A[T]$ let $c(f)$ be the ideal in A generated by the coefficients of f. It is called the *content of f*[5]. A polynomial f is called *primitive* if $c(f) = A$. Let $f, g \in A[T]$ be polynomials, $n := \deg(g)$. Show the following assertions.

(1) $c(fg) \subseteq c(f)c(g) \subseteq \sqrt{c(fg)}$. Deduce that f and g are primitive if and only if fg is primitive.

(2) $c(f)^{n+1}c(fg) = c(f)^n c(fg)$.
    Remark: This result is called the Dedekind-Mertens lemma.

(3) Let $A = k[x, y]$ for a field k and set $f = xT + y$ and $g = xT - y$. Show that $c(fg) \neq c(f)c(g)$.

Exercise 20.2. Let $h\colon X_0 \to X$ be a morphism of schemes. For every X-scheme Y let $Y_0 := X_0 \times_X Y$ be its base change with h. We also set $X_{00} := X_0 \times_X X_0$ and denote by Y_{00} the base change of Y to X_{00} (with respect to $X_{00} \to X_0 \to X$, which is independent of the choice which projection $X_{00} \to X_0$ is used). Let Y, Y' be X-schemes.

(1) Let h be surjective and assume that Y' is unramified over X. Show that the base change with h yields an injective map $\mathrm{Hom}_X(Y, Y') \to \mathrm{Hom}_{X_0}(Y_0, Y'_0)$.

(2) Let h be universally submersive and assume that Y' is étale over X. Show that the diagram

$$\mathrm{Hom}_X(Y, Y') \xrightarrow{\ h^*\ } \mathrm{Hom}_{X_0}(Y_0, Y'_0) \underset{\mathrm{pr}_2^*}{\overset{\mathrm{pr}_1^*}{\rightrightarrows}} \mathrm{Hom}_{X_{00}}(Y_{00}, Y'_{00})$$

is exact.

---

[5] This notion of content is not the same as the content of a polynomial over a factorial ring A as for instance defined in [BouAC] ^{0}, except if A is a principal domain.

(3) Assume that h is universally submersive, separated, and has geometrically connected fibers. Assume that Y and Y' are étale over X. Show that

$$\operatorname{Hom}_X(Y, Y') \xrightarrow{\ h^*\ } \operatorname{Hom}_{X_0}(Y_0, Y_0')$$

is bijective.

Exercise 20.3. Let $p \in \mathbb{Z}$ be a prime number and let $\mathbb{Z}_{(p)}$ the localization in (p). Show that $\mathbb{Z}_{(p)}$ is not henselian.

Exercise 20.4. Let A be a local ring, let $f \in A[T]$ be a monic polynomial, and let $B = A[T]/(f)$. Show that there exists a bijective correspondence between decompositions $f = gh$ as a product of two monic polynomials $g, h \in A[T]$ such that $(g) + (h) = A[T]$ and decompositions $B = C \times D$ as a product of A-algebras, which is given by sending (g, h) to the decomposition $B \xrightarrow{\sim} A[T]/(g) \times A[T]/(h)$.

Exercise 20.5. Let A be a local ring with maximal ideal $\mathfrak{m}$ and residue field k. For $f \in A[T]$ let $\bar{f}$ be the image of f in $k[T]$. Show that the following assertions are equivalent.
(i) A is henselian.
(ii) For every $f \in A[T]$ and every root $\bar{a}$ of the image $\bar{f}$ of f in $k[T]$ such that $\bar{f}'(\bar{a}) \neq 0$ there exists a unique $a \in A$ whose image in k is $\bar{a}$ and such that $f(a) = 0$.
(iii) For every monic $f \in A[T]$ such that $\bar{f}(0) = 0$ and $\bar{f}'(0) \neq 0$ there exists $a \in \mathfrak{m}$ such that $f(a) = 0$.
(iv) For every primitive $f \in A[T]$ and every factorization $\bar{f} = g_0 h_0$ in $k[T]$ such that g_0 and h_0 are prime to each other there exists a factorization $f = gh$ in $A[T]$ such that $\deg g = \deg g_0$, $\bar{g} = g_0$, and $\bar{h} = h_0$.

Exercise 20.6. Let K be a field, v a valuation on K (not necessarily discrete), and let A be the valuation ring of v. Show that A is henselian if and only if for every algebraic extension K' of K any two extensions of v to K' are equivalent.

Exercise 20.7. Let A be a local ring with maximal ideal $\mathfrak{m}$ and let $I \subset A$ be an ideal.
(1) Assume first that I consists of nilpotent elements. Show that A is henselian if and only if A/I is henselian.
   Hint: Use that the inclusion $\operatorname{Spec} B/IB \to \operatorname{Spec} B$ is a homeomorphism for every A-algebra B.
(2) More generally, assume that A is I-adically complete (Definition B.39). Show that A is henselian if and only if A/I is henselian.
   Hint: Use that every finite free A-algebra B is complete for the I-adic topology and apply (1) to $B/I^n B$. Alternatively, use Exercise 20.5 and construct (with the notations there) $a \bmod I^n$ successively.
(3) Deduce that if A is $\mathfrak{m}$-adically complete, then A is henselian.

Exercise 20.8. Let R be a local ring, $\mathfrak{m}$ its maximal ideal, k its residue field, $S = \operatorname{Spec} R$, $s \in S$ its closed point. Show that the following assertions are equivalent.
(i) R is henselian.
(ii) Every R-algebra A of finite type is isomorphic to a product $A = B \times A_1 \times \cdots \times A_n$, where the A_i are finite local R-algebras and all irreducible components of $\operatorname{Spec} B \otimes_R k$ have dimension ≥ 1.
(iii) For every finitely generated R-algebra C and every prime ideal $\mathfrak{p}$ lying over $\mathfrak{m}$ such that $C_{\mathfrak{p}}/\mathfrak{m} C_{\mathfrak{p}}$ is a finite k-algebra, $C_{\mathfrak{p}}$ is a finite R-algebra.

(iv) For every morphism $f\colon X \to S$ locally of finite type and for every x in the fiber X_s such that $\kappa(s) = \mathscr{O}_{X,x}/\mathfrak{m}_s\mathscr{O}_{X,x}$ there exists an open neighborhood U of x in X such that $f_{|U}\colon U \to S$ is an open immersion.

Exercise 20.9. Let $(R_\alpha, \varphi_{\beta\alpha})$ be a filtered inductive system of local rings, such that all homomorphisms $\varphi_{\beta\alpha}$ are local. Let R be its inductive limit.
(1) Show that R is local and that the ring homomorphisms $R_\alpha \to R$ are local.
(2) Show that if R_α is henselian for all α, then R is henselian.

Exercise 20.10. Let A be a local henselian ring with residue field k and let B be an A-algebra that is integral over A.
(1) Show that the canonical map $B \to B \otimes_A k$ induces a bijection

$$\{\text{idempotents of } B\} \leftrightarrow \{\text{idempotents of } B \otimes_A k\}.$$

(2) Show that for every maximal ideal $\mathfrak{n}$ of B the localization $B_\mathfrak{n}$ is integral over A and henselian.
Hint: Write B as the inductive limit of its finite A-subalgebras and use Exercise 20.9.

Exercise 20.11. Let A be a local ring. Show that A is henselian if and only if for every finite A-algebra B and for every maximal ideal $\mathfrak{n}$ of B the localization $B_\mathfrak{n}$ is integral over A.

Exercise 20.12. Let R be a local ring and let $\iota\colon R \to R^\mathrm{h}$ be its henselization. Show that for every local henselian ring A and every local homomorphism $\varphi\colon R \to A$ there exists a unique local homomorphism $\varphi^\mathrm{h}\colon R^\mathrm{h} \to A$ such that $\varphi^\mathrm{h} \circ \iota = \varphi$.

Exercise 20.13. Let $\varphi\colon A \to B$ be a local homomorphism of local rings.
(1) Show that there exists a unique (necessarily local) homomorphism $\varphi^\mathrm{h}\colon A^\mathrm{h} \to B^\mathrm{h}$ making the following diagram commutative

$$\begin{array}{ccc} A & \longrightarrow & A^\mathrm{h} \\ {\scriptstyle\varphi}\downarrow & & \downarrow{\scriptstyle\varphi^\mathrm{h}} \\ B & \longrightarrow & B^\mathrm{h} \end{array}$$

Hint: Exercise 20.12.
(2) Let k^sep be a separable closure of the residue field k of A, let ℓ^sep be a separable closure of the residue field ℓ of B, let A^hs and B^hs be the corresponding strict henselizations, and let $\iota\colon k^\mathrm{sep} \to \ell^\mathrm{sep}$ be a homomorphism which extends the homomorphism $k \to \ell$ induced by φ. Show that there exists a unique (necessarily local) homomorphism $\varphi^\mathrm{hs}\colon A^\mathrm{hs} \to B^\mathrm{hs}$ making the following diagram commutative

$$\begin{array}{ccccc} A & \longrightarrow & A^\mathrm{hs} & \longrightarrow & k^\mathrm{sep} \\ {\scriptstyle\varphi}\downarrow & & \downarrow{\scriptstyle\varphi^\mathrm{hs}} & & \downarrow{\scriptstyle\iota} \\ B & \longrightarrow & B^\mathrm{hs} & \longrightarrow & \ell^\mathrm{sep} \end{array}$$

Assertion (2) means that one should see strict henselization as a functor on the category of pairs consisting of a local ring R and a separable closure of the residue field of R.

Exercise 20.14. Let R be a local ring and fix a separable closure k^{sep} of the residue field of R.
(1) Assume that R is henselian (resp. strictly henselian). Show that $R \to R^{\mathrm{h}}$ (resp. $R \to R^{\mathrm{hs}}$) is an isomorphism.
(2) Show that $R^{\mathrm{hs}} \cong (R^{\mathrm{h}})^{\mathrm{hs}}$.
Here all strict henselizations are formed with respect to k^{sep}.

Exercise 20.15. Consider the following examples of a locally ringed space $(X, \mathscr{O}_X)$.
(1) X is a topological space and $\mathscr{O}_X$ is the sheaf of continuous $\mathbb{R}$-valued functions on X.
(2) X is a real C^α-manifold ($\alpha \in \mathbb{N} \cup \{\infty\}$) and $\mathscr{O}_X$ is the sheaf of $\mathbb{R}$-valued C^α-functions on X.
(3) X is a complex analytic space (e.g., a complex manifold) and $\mathscr{O}_X$ is the sheaf of complex analytic $\mathbb{C}$-valued functions on X.
Show that in all cases for all $x \in X$ the local ring $\mathscr{O}_{X,x}$ is henselian.

Exercise 20.16. Let K be a field endowed with an absolute value $|\cdot|: K \to \mathbb{R}^{\geq 0}$ (i.e., for all $a,b \in K$ one has: $|a| = 0 \Leftrightarrow a = 0$, $|ab| = |a||b|$, and $|a + b| \leq |a| + |b|$). A formal power series $f = \sum a_{i_1\ldots i_n} T_1^{i_1} \cdots T_n^{i_n} \in\in K[\![T_1,\ldots,T_n]\!]$ is called *convergent* if there exist $c_1,\ldots,c_n, M \in \mathbb{R}^{>0}$ such that $|a_{i_1\ldots i_n}| c_1^{i_1} \cdots c_n^{i_n} \leq M$ for all $(i_1,\ldots,i_n) \in \mathbb{N}^n$. Let $K\langle\!\langle T_1,\ldots,T_n \rangle\!\rangle$ be the subset of convergent power series in $K[\![T_1,\ldots,T_n]\!]$.
Show that $K\langle\!\langle T_1,\ldots,T_n \rangle\!\rangle$ is a regular noetherian henselian local ring whose completion is $K[\![T_1,\ldots,T_n]\!]$.
Hint: This is a difficult exercise; see [Nag] §45.

Exercise 20.17. Let $f: X \to Y$ be a separated quasi-finite flat morphism of finite presentation.
(1) Show that the degree function
$$\deg(f): Y \to \mathbb{Z}, \qquad y \mapsto \dim_{\kappa(y)} \Gamma(X_y, \mathscr{O}_{X_y})$$
is lower semicontinuous and constructible, i.e., $\{s \in S \; ; \; \deg(f)(s) \geq n\}$ is open and constructible in S for all $n \in \mathbb{Z}$.
(2) Show that f is finite locally free if and only if the function $y \mapsto \dim_{\kappa(y)} \Gamma(X_y, \mathscr{O}_{X_y})$ is locally constant.
Hint: For the constructibility reduce to Y noetherian. Then it remains to show that $\deg(f)$ jumps up under generization. For this reduce to the case that Y is the spectrum of a henselian local ring.

Exercise 20.18. Let $f: X \to S$ be a quasi-finite morphism. For every geometric point $\bar{s}$ lying over a point s of S we call $n(s) := \# \mathrm{Hom}_S(\bar{s}, X)$ the *number of geometric points of* $f^{-1}(s)$.
(1) Show that $n(s) = \sum_{x \in f^{-1}(s)} f''_{x/s}$, where $f''_{x/s}$ is the inertia index defined in Section (12.5).
(2) Let f in addition be flat and of finite presentation. Show that $n: S \to \mathbb{Z}$ is constructible and lower semicontinuous.
(3) Let f be quasi-compact, separated, and étale. Show that f is an étale cover if and only if the map $S \to \mathbb{Z}$, $s \mapsto n(s)$ is locally constant.
(4) Let f be finite locally free. Show that $s \mapsto n(s)$ is locally constant if and only if there exist a factorization of f
$$f: X \xrightarrow{f_r} X' \xrightarrow{f_e} S,$$

with f_r a finite locally free universal homeomorphism and f_e finite étale. In this case the factorization is unique up to unique isomorphism, is functorial in $X \to S$, and is compatible with base change $S' \to S$.
Hint: Exercise 20.17.

Exercise 20.19. Let A be a reduced local noetherian ring and let B be its normalization (i.e., $B = \{ a \in \operatorname{Frac} A \; ; \; a \text{ is integral over } A \}$, where $\operatorname{Frac} A$ denotes the total ring of fractions of A). Show that B is unramified over A if and only if the irreducible components of $\operatorname{Spec} A^{\mathrm{hs}}$ are normal.

Exercise 20.20. Let A be a local integral domain with residue field k and let A_{red} be the quotient of A by its nilradical.
(1) Show that A is unibranch if and only if A_{red} is an integral domain and its normalization (see Exercise 20.19) is local.
(2) Show that A is geometrically unibranch if and only if A_{red} is an integral domain and its normalization is a local ring whose residue field is a purely inseparable extension of k.
Hint: Show first that $(A^{\mathrm{h}})_{\mathrm{red}} = (A_{\mathrm{red}})^{\mathrm{h}}$ and $(A^{\mathrm{hs}})_{\mathrm{red}} = (A_{\mathrm{red}})^{\mathrm{hs}}$ (with respect to some chosen separable closure of k).

Exercise 20.21. Let k be a field and let $0 \neq f \in k[T, U]$ with $f(0, 0) = 0$. Set $X = \operatorname{Spec} k[T, U]/(f)$ and $x := (0, 0) \in X(k)$. Let f^* be the leading term of f and let n be its degree (Exercise 6.10). Assume $n \geq 1$.
(1) Show that $\operatorname{Spec} \mathscr{O}_{X,x}^{\mathrm{hs}}$ has at most n irreducible components (i.e., there are at most n "geometric branches" passing through x).
(2) Show that $\operatorname{Spec} \mathscr{O}_{X,x}^{\mathrm{hs}}$ has exactly n irreducible components if and only if every irreducible component of $\operatorname{Spec} A^{\mathrm{hs}}$ is normal (cf. Exercise 20.19).
      Assume $n \geq 2$. If the above condition is satisfied, then x is called a *non-cuspidal singularity of X*; otherwise x is called a *cuspidal singularity*.
(3) Write f^* over an algebraic closure $\bar{k}$ of k as a product $f^* = \ell_1 \cdots \ell_n$ of homogeneous polynomials of degree 1. Show that if the lines $V(\ell_i) \subset \mathbb{A}_{\bar{k}}^2$ are pairwise distinct, then $\mathscr{O}_{X,x}^{\mathrm{hs}}$ is isomorphic to the strict henselization of the tangent cone of X in x (i.e., X and its tangent cone are locally for the étale topology isomorphic).
This generalizes Exercise 6.11.

Exercise 20.22. Let k be a field of characteristic $\neq 2$.
(1) Let R be the local ring in $(0, 0)$ of $V(Y^2 - X^2(X + 1)) \subset \mathbb{A}_k^2$. Show that R is an integral domain but that $\operatorname{Spec} R^{\mathrm{h}}$ has two irreducible components. In particular R is not unibranch.
(2) Let R be the local ring in $(0, 0)$ of $V(Y^2 - X^3) \subset \mathbb{A}_k^2$. Show that R is geometrically unibranch but not normal.

Exercise 20.23. Let A be a henselian local noetherian ring and let $\hat{A}$ be its completion. Show that $B \mapsto B \otimes_A \hat{A}$ yields an equivalence between the category of finite étale A-algebras and the category of finite étale $\hat{A}$-algebras.

Exercise 20.24. Let X be a scheme, let G be a finite group of automorphisms of X, let $x \in X$. If $g(x) = x$ for some $g \in G$, then g induces an automorphism $g_{\kappa(x)}$ of $\kappa(x)$. The subgroup
$$I_x := \{ g \in G \; ; \; g(x) = x, \; g_{\kappa(x)} = \operatorname{id}_{\kappa(x)} \}$$
is called the *inertia subgroup of G at x*.

Now assume that there exists an affine morphism $X \to S$ such that G acts by S-automorphisms. Let H be a subgroup of G and set $Y := X/H$ and $Z := X/G$ (cf. Proposition 20.87). Let $p\colon X \to Y$ and $q\colon Y \to Z$ be the canonical morphisms. Consider the following assertions.

(i) $I_x \subseteq H$.

(ii) There exists an open neighborhood V of $p(x)$ in Y such that $q_{|V}\colon V \to Z$ is étale.

(iii) There exists an open neighborhood V of $p(x)$ in Y such that $q_{|V}\colon V \to Z$ is unramified.

Show the implications "(i) $\Rightarrow$ (ii) $\Rightarrow$ (iii)". Show that all assertions are equivalent if X is an integral scheme.

Hint: One may assume $S = Z$. To show "(i) $\Rightarrow$ (ii)" reduce to the case that $Z = \operatorname{Spec} R$ is the spectrum of a strictly henselian ring. Then prove that $R \to \mathscr{O}_{Y,p(x)}$ can be identified with $\mathscr{O}_{X,x}^{I_x} \to \mathscr{O}_{X,x}^{I_x \cap H}$. To show "(iii) $\Rightarrow$ (i)" consider $U_g := \operatorname{Ker}(p, p \circ g)$ for $g \in I_x$. Show that U_g is closed in X and contains an open neighborhood of x.

Remark: The existence of an affine morphism $X \to S$ such that G acts by S-automorphisms is only needed to ensure the existence of X/G and X/H. One can show that such a quotient, satisfying all the properties stated in Proposition 20.87, exists if every G-orbit is contained in an open affine subscheme of X ([SGA3] Exp. V, Théorème 4.1). This conditions is for instance satisfied if X is qcqs and there exists an ample line bundle on X (Proposition 13.49). The same remark applies to Exercise 20.25.

Exercise 20.25. Let X be a scheme and let G be a finite group of automorphisms of X. Assume that there exists an affine morphism $X \to S$ such that G acts by S-automorphisms.

(1) Assume that one has $I_x = 1$ for all $x \in X$, where I_x is the inertia group (Exercise 20.24). Show that $X \to X/G$ is finite étale.

(2) Assume that X is connected, that G acts faithfully on X, and that $X \to X/G$ is unramified. Show that $I_x = 1$ for all $x \in X$.

Exercise 20.26. Let R be a local normal ring with field of fractions K. Let K^{sep} be a separable closure of K. Then $\Gamma := \operatorname{Gal}(K^{\mathrm{sep}}/K)$ acts on the integral closure A of R in K^{sep}. Fix a maximal ideal $\mathfrak{m}$ of A, let $D := \{\, \sigma \in \Gamma \;;\; \sigma(\mathfrak{m}) = \mathfrak{m} \,\}$ be the decomposition group and let $I := \{\, \sigma \in D \;;\; \sigma \text{ induces on } \kappa(\mathfrak{m}) \text{ the identity} \}$ be the inertia group of $\mathfrak{m}$. Set $B := A^D$, $\mathfrak{n} := \mathfrak{m} \cap B$, and $B' := A^I$, $\mathfrak{n}' = \mathfrak{m} \cap B'$. Show

$$R^{\mathrm{h}} \cong B_{\mathfrak{n}}, \qquad R^{\mathrm{hs}} \cong B'_{\mathfrak{n}'}.$$

Hint: Exercise 20.24

Exercise 20.27. Let R be a local noetherian integral domain with field of fractions K. Let R^{h} be the henselization and $\hat{R}$ be the completion of R. Show that R^{h} is the subring of $\hat{R}$ consisting of elements $f \in \hat{R}$ that are algebraic over K, i.e., there exist $a_0, \ldots, a_n \in K$ with $n > 0$ and $a_n \neq 0$ such that $a_n f^n + a_{n-1} f^{n-1} + \cdots + a_1 f + a_0 = 0$.

Exercise 20.28. Let S be a $\mathbb{Q}$-scheme and let $f\colon X \to S$ be a smooth morphism. Show that the De Rham complex $\Omega^{\bullet}_{X/S}$ is acyclic in degrees > 0.
Hint: Exercise 19.7

Exercise 20.29. Let A be a ring and let B be a finite A-algebra. Show that $\operatorname{Spec} B \to \operatorname{Spec} A$ is an étale cover if and only if B is a projective A-module and a projective $B \otimes_A B$-module.

Exercise 20.30. Let R be a ring.
(1) Let A be a finite R-algebra which is free as an R-module with basis $(x_1, \dots, x_n)$. Show that the discriminant $\mathscr{D}_{A/R}$ is a principal ideal of R generated by $\det(\mathrm{tr}_{A/R}(x_i x_i))_{i,j}$.
(2) Let $f \in R[T]$ be a monic polynomial and set $A := R[T]/(f)$. Show that $\mathscr{D}_{A/R}$ is a principal ideal of R generated by the discriminant $\mathscr{D}_f$ of f (in the sense of Section (B.20), or [BouAII]$^\circ$ IV §6, No. 7).
(3) Let $f \in R[T]$ be a monic polynomial. Show that $R[T]/(f)$ is a finite étale R-algebra if and only if $\mathscr{D}_f$ is a unit in R.

Exercise 20.31. Let (A, I) be a henselian pair. Show that $P \mapsto P/IP$ induces a bijection between the set of isomorphism classes of finite projective A-modules and the set of isomorphism classes of finite projective A/I-modules.

Exercise 20.32. A pair (A, I) consisting of a ring A and an ideal $I \subseteq A$ is called a *Zariski pair* if I is contained in the Jacobson radical of A. Show that the inclusion functor from the full subcategory of Zariski pairs into the category of all pairs (A, I) consisting of a ring A and an ideal $I \subseteq A$ has a left adjoint functor which is given by $(A, I) \mapsto (S^{-1}A, S^{-1}I)$, where $S := 1 + I$.

Exercise 20.33. Show that the inclusion functor from the full subcategory of henselian pairs to the category of all pairs (A, I) consisting of a ring A and an ideal $I \subseteq A$ has a left adjoint functor $(A, I) \mapsto (A, I)^{\mathrm{h}}$, called *henselization of the pair* (A, I).
Hint: Consider the category $\mathcal{C}$ of étale ring homomorphism $A \to B$ that induce an isomorphism $A/IA \xrightarrow{\sim} B/IB$. Show that $\mathcal{C}$ is filtered. Set $A^{\mathrm{h}} := \mathrm{colim}_{B \in \mathcal{C}} B$ and $I^{\mathrm{h}} := IA^{\mathrm{h}}$ and show that $(A, I) \mapsto (A^{\mathrm{h}}, I^{\mathrm{h}})$ gives the desired left adjoint functor.
  Show the following properties of the henselization $(A^{\mathrm{h}}, I^{\mathrm{h}})$ of a pair (A, I).
(1) If A is a local ring with maximal ideal $\mathfrak{m}$, then $(A^{\mathrm{h}}, \mathfrak{m}^{\mathrm{h}})$ is the henselization defined in Definition 20.42.
(2) The ring homomorphism $A \to A^{\mathrm{h}}$ is flat. It is faithfully flat if and only if (A, I) is a Zariski pair (Exercise 20.32).
(3) If $I, J \subseteq A$ are ideals with $V(I) = V(J)$, then $(A, I)^{\mathrm{h}} \cong (A, J)^{\mathrm{h}}$.
(4) For all $n \geq 1$ the map $A \to A^{\mathrm{h}}$ induces isomorphisms $A/I^n \xrightarrow{\sim} A^{\mathrm{h}}/(I^{\mathrm{h}})^n$.

Exercise 20.34. Let X and S be integral schemes and let $f \colon X \to S$ be a finite dominant morphism of finite presentation. Let S be unibranch. Show that f is open.
Hint: Use the going down property (Theorem B.54 (3)).

Exercise 20.35. Let S be an integral geometrically unibranch scheme and let $f \colon X \to S$ be a dominant morphism. Assume that X is connected. Show that f is unramified if and only if f is étale.
Hint: Reduce to the case that $S = \mathrm{Spec}\, R$ and $X = \mathrm{Spec}\, A$ such that $R \hookrightarrow A$ is injective and such that A is the quotient of a standard étale R-algebra. Then reduce to the case that R is strictly henselian.

Exercise 20.36. Let X be a scheme, $U \subseteq X$ a non-empty open subscheme. Show that the functor $(\mathrm{FÉt}/X) \to (\mathrm{FÉt}/U)$ that sends X' to $X' \times_X U$ is fully faithful in the following two cases.
(1) The scheme X is the spectrum of a local geometrically unibranch ring and U is the complement of its closed point.
(2) The scheme X is geometrically unibranch and has a noetherian underlying topological space and U is dense in X.

Exercise 20.37. Let $\pi\colon X \to S$ be a finite locally free morphism of locally noetherian schemes and let Z be the branch locus. Show that one has "purity of the branch locus", i.e, each irreducible component of the branch locus of π has codimension 1 in X.

Remark: One also has purity of the branch locus if X and S are integral and locally noetherian, S is regular, X is normal, and f is quasi-finite and dominant ([SGA2] Exp. X, Thérème 3.4).

Exercise 20.38. Let k be an algebraically closed field of characteristic $p > 0$ and let $q = p^n$ for some integer $n \geq 1$. Show that the k-algebra homomorphism $k[T] \to k[T]$, $T \mapsto T^q - T$, defines a finite étale Galois cover $\mathbb{A}^1_k \to \mathbb{A}^1_k$, called *Artin-Schreier cover* with Galois group $(\mathbb{Z}/p\mathbb{Z})^n$.

Exercise 20.39. Let X be a scheme locally of finite type over a field k.
(1) Show that there exists an étale k-scheme $\pi_0(X)$ and a morphism $q_X \colon X \to \pi_0(X)$ with the following universal property. For every morphism $f\colon X \to Y$ of k schemes, where Y is an étale k-scheme, there exists a unique morphism $g\colon \pi_0(X) \to Y$ such that $f = g \circ q_X$.
(2) Show that q_X is faithfully flat and the fibers of q_X are the connected components of X.
(3) Show that for every morphism $f\colon X \to Y$ of k-schemes locally of finite type there exists a unique morphism $\pi_0(f)\colon \pi_0(X) \to \pi_0(Y)$ such that $q_y \circ f = \pi_0(f) \circ g$ and that we obtain a functor π_0 from the category of k-schemes locally of finite type to the category of étale k-schemes.
(4) Let K be a field extension of k. Show that there exists an isomorphism of K-schemes, functorial in X,
$$\pi_0(X \otimes_k K) \overset{\sim}{\to} \pi_0(X) \otimes_k K.$$
(5) Show that for two k-schemes X and Y locally of finite type over k, the canonical morphism $\pi_0(X \times_k Y) \to \pi_0(X) \times_k \pi_0(Y)$ is an isomorphism, functorial in X and Y.
(6) Show that X is geometrically connected over k if and only if $\pi_0(X) = \operatorname{Spec} k$.
The scheme $\pi_0(X)$ is called the *scheme of connected components of X*.

Exercise 20.40.
(1) Show that the inclusion functor from the category of profinite groups into the category of topological groups admits a left adjoint functor $G \mapsto \hat{G}$, where
$$\hat{G} = \lim_U G/U$$
where $U \subseteq G$ runs through the open normal subgroups of finite index. The profinite group $\hat{G}$ is called the *profinite completion of G*.
(2) Let G be a topological group and let $F\colon (G\text{-sets}) \to (\text{Sets})$ be the forgetful functor. Show that one has a functorial isomorphism $\hat{G} \overset{\sim}{\to} \operatorname{Aut}(F)$ of profinite groups.

Exercise 20.41. Let S be a connected scheme, let $\bar{s}$ be a geometric point of S, let $X \to S$ be an étale cover, and let F be the corresponding finite set with continuous $\pi_1(S, \bar{s})$-action.
(1) Show that the action of $\pi_1(S, \bar{s})$ on F is trivial if and only if $X \to S$ is a split étale cover.
(2) Show that there exists a Galois cover $T \to S$ such that $X \times_S T \to T$ is a split étale cover.

Exercise 20.42. Let Γ be a profinite group, let (t-Γ-sets) be the full subcategory of (Γ-sets) consisting of finite sets with transitive Γ-action. Let $\mathcal{V}^t\colon$ (t-Γ-sets) $\to$ (sets) and $\mathcal{V}\colon$ (Γ-sets) $\to$ (sets) be the forgetful functors.

(1) Show that the inclusion (t-Γ-sets) $\to$ (Γ-sets) yields an isomorphism $\mathrm{Aut}(\mathcal{V}) \overset{\sim}{\to} \mathrm{Aut}(\mathcal{V}^t)$.

(2) Let Δ be an open normal subgroup of Γ. Show that every map $\Gamma/\Delta \to \Gamma/\Delta$ in (sets) commuting with all automorphisms in (Γ-sets) of Γ/Δ is given by left multiplication by some element of Γ/Δ.

(3) Let Δ as in (2) and for $\sigma \in \mathrm{Aut}(\mathcal{V}^t)$ let $\gamma\Delta \in \Gamma/\Delta$ be the coset such that $\sigma_{\Gamma/\Delta}(\gamma'\Delta) = \gamma\gamma'\Delta$. Show that one obtains a surjective group homomorphism $\mathrm{Aut}(\mathcal{V}^t) \to \Gamma/\Delta$ which yields an isomorphism $\mathrm{Aut}(\mathcal{V}^t) \overset{\sim}{\to} \lim_\Delta \Gamma/\Delta$.

Exercise 20.43. Let k be a separably closed field and let X be a connected k-scheme. Let $k \to K$ be a separably closed extension of k and let $\bar{x}$ be a geometric point of $X \otimes_k K$. Show that the canonical map $\pi_1(X \otimes_k K, \bar{x}) \longrightarrow \pi_1(X, \bar{x})$ is surjective.

Exercise 20.44. Let k be an algebraically closed field of characteristic $p > 0$ and set $X = \mathbb{A}^1_k$. Show that the Artin-Schreier construction (Exercise 20.38) provides for any geometric point $\bar{x}$ of $\mathbb{A}^1_k$ an isomorphism between the group of continuous group homomorphisms $\pi_1(\mathbb{A}^1_k, \bar{x}) \to \mathbb{Z}/p\mathbb{Z}$ and the group $\{\sum_{i \geq 0} a_i T^i \in k[T] \;;\; a_i = 0$ if p divides $i\}$. Deduce that for a non-trivial algebraically closed extension $k \to K$ and for any geometric point $\bar{x}$ of $\mathbb{A}^1_K$ the surjective map $\pi_1(X \otimes_k K, \bar{x}) \longrightarrow \pi_1(X, \bar{x})$ (Exercise 20.43) is not injective.

21 Cohomology of $\mathcal{O}_X$-modules

Content

- Categories of abelian sheaves and of $\mathcal{O}_X$-modules
- Cohomology and derived direct image
- Čech cohomology
- Derived inverse image, Hom sheaves, and tensor products
- Relations between derived functors
- Perfect and pseudo-coherent complexes

In this chapter we develop a "four functor formalism" for the category of modules over a ringed space. For every ringed space $(X, \mathcal{O}_X)$ we will denote by $D(X)$ the (unbounded) derived category of the abelian category of $\mathcal{O}_X$-modules. More precisely, we will define for every morphism $f\colon (X, \mathcal{O}_X) \to (Y, \mathcal{O}_Y)$ of ringed spaces adjoint functors

$$Lf^* : D(Y) \rightleftarrows D(X) : Rf_*$$

where Rf_* is the right derived functor of the direct image functor and Lf^* is the left derived functor of the inverse image functor. The notation implies that the functor on the left Lf^* is left adjoint to Rf_* (and that Rf_* is right adjoint to Lf^*). Moreover, for every ringed space $(X, \mathcal{O}_X)$ we will define bi-functors "derived tensor product" and "derived inner Hom"

$$- \otimes^L_{\mathcal{O}_X} - : D(X) \times D(X) \longrightarrow D(X),$$
$$R\,\mathscr{H}om_{\mathcal{O}_X}(-,-)\colon D(X)^{\mathrm{opp}} \times D(X) \longrightarrow D(X),$$

such that $(-) \otimes^L_{\mathcal{O}_X} \mathscr{G}$ is left adjoint to $R\,\mathscr{H}om_{\mathcal{O}_X}(\mathscr{G}, -)$ for every $\mathcal{O}_X$-module $\mathscr{G}$. We will show that these functors satisfy several properties, e.g., functoriality of Rf_* and Lf^* (i.e., $R(g \circ f)_* = Rf_* \circ Rg_*$ and $L(g \circ f)^* = Lf^* \circ Lg^*$), symmetry and associativity of $- \otimes^L_{\mathcal{O}_X} -$, and compatibility of Lf^* with derived tensor product.

Ideally one would like to extend this "four functor formalism" to a six functor formalism in the sense of Grothendieck, see Section (21.30) below for further remarks what this means and why this is not possible without enlarging the category of schemes.

A special case of f_* (namely if Y consists of a single point with $\mathcal{O}_Y(Y) = \Gamma(X, \mathcal{O}_X)$) is the global section functor $\Gamma(X, -)$. In this case Rf_* identifies with the derived functor

$$R\Gamma(X, -)\colon D(X) \longrightarrow D(\Gamma(X, \mathcal{O}_X)),$$

where $D(A)$ denotes the derived category of the abelian category of A-modules for a ring A. Its cohomology modules define the sheaf cohomology of a (complex of) $\mathcal{O}_X$-module $\mathscr{F}$ by

$$H^i(X, \mathscr{F}) := H^i R\Gamma(X, \mathscr{F}), \qquad i \in \mathbb{Z}.$$

© Springer Fachmedien Wiesbaden GmbH, ein Teil von Springer Nature 2023
U. Görtz und T. Wedhorn, *Algebraic Geometry II: Cohomology of Schemes*,
Springer Studium Mathematik – Master, https://doi.org/10.1007/978-3-658-43031-3_6

The chapter develops the theory in reverse order as stated above. We start by proving that the abelian category of $\mathscr{O}_X$-modules is a Grothendieck abelian category which in particular ensures that all additive functors defined on it can be right derived (Corollary F.187). Then we study cohomology and, more generally, right derived image. Next we introduce the formalism of Čech cohomology, which is sometimes easier to compute than sheaf cohomology, and we explain the relation between Čech cohomology and sheaf cohomology. The next two parts of the chapter define the other three functors and study formal relations between the four functors. In other words, they develop the "four functor formalism". We conclude the chapter with defining and studying finiteness conditions for objects in $D(X)$.

Everything in this chapter is quite formal and will be done for arbitrary ringed spaces[1]. Only in the following chapters, we will prove deeper results that require the ringed spaces to be schemes such as the projection formula or compatibility of cohomology with base change. Hence we suggest to start with the definition of cohomology and derived direct image and then to come back to the other constructions, definitions, and results only later when they are needed.

Ringed spaces and rings

Let A be a ring. Every construction or property **P** defined for complexes of $\mathscr{O}_X$-modules for an arbitrary ringed space $(X, \mathscr{O}_X)$ yields the notion of property **P** for complexes of A-modules as follows. The category of complexes of A-modules is isomorphic to the category of complexes of $\mathscr{O}_X$-modules, where $(X, \mathscr{O}_X)$ is the ringed space consisting of a point with $\Gamma(X, \mathscr{O}_X) = A$. This ringed space is sometimes denoted by $(*, A)$. Then we can identify the category of A-modules and the category of $\mathscr{O}_X$-modules via taking global sections over X. We obtain in particular an identification of their derived categories, $D(A)$ and $D(X)$. Hence one defines a complex of A-modules E to have property **P** if E considered as a complex of $\mathscr{O}_X$-modules has the property **P**.

Similarly, a ring homomorphism $\varphi\colon A \to B$ corresponds to a morphism of ringed spaces $(*, B) \to (*, A)$, again denoted by φ. More precisely, attaching to a ring A the ringed space $(*, A)$ defines a contravariant fully faithful functor from the category of rings into the category of ringed spaces. Then pullback becomes the base change functor $M \mapsto \varphi^*(M) = B \otimes_A M$, and pushforward becomes the functor $N \mapsto \varphi_* N$ of restriction of scalars, i.e., of viewing a B-module as an A-module via φ.

In this way, all definitions and constructions for ringed spaces specialize to those for rings. When we introduce a notion for ringed spaces (or morphisms of them) we will always define the same notion for rings (or ring homomorphisms) in this manner.

Of course, there is a different definition for an A-module or a complex of A-module to have a property **P** defined for modules or complexes of modules over arbitrary ringed spaces: One could define that the associated complex $\tilde{E}$ of quasi-coherent $\mathscr{O}_S$-modules should have property **P**, where $S = \operatorname{Spec} A$. This will never be our definition if not explicitly stated otherwise. Nevertheless, we will see that this often (but not always) gives the same notion. Let us give some examples.

(1) For the properties "of finite type", "of finite presentation", and "flat" of an A-module E we have seen that E has this property if and only if the quasi-coherent $\mathscr{O}_S$-module $\tilde{E}$ has this property (Proposition 7.26 and Remark 7.39).

---

[1] In fact, almost all of the results generalize to modules over arbitrary ringed topoi, often with the same proofs.

(2) Let E be a complex of A-modules. Then E is K-flat (see Definition 21.91 below) if and only if $\tilde{E}$ is K-flat (Lemma 22.39).

(3) For the properties "perfect", "pseudo-coherent", and "of finite tor dimension" defined below, we will see in Lemma 22.44 that E has one of these property if and only if $\tilde{E}$ has the same property.

(4) But note that if E is K-injective (Definition F.179), even if E is concentrated in degree 0 and hence an injective A-module, then $\tilde{E}$ is not a K-injective object in $D(\operatorname{Spec} A)$ in general (see Caveat 22.77 (3) below).

Notation

If X is a topological space, we denote by $(\mathrm{PSh}(X))$ the category of presheaves (of sets) on X and by $(\mathrm{Sh}(X))$ the category of sheaves of sets on X.

For a ringed space $(X, \mathscr{O}_X)$ let $(\mathscr{O}_X\text{-Mod})$ be the abelian category of $\mathscr{O}_X$-modules. We work in this general setting because this allows us to treat the case $\mathscr{O}_X = \mathbb{Z}$ (the constant sheaf), the case that X consists of a single point and $\mathscr{O}_X(X) = A$ for a given ring A, and the case of a scheme simultaneously. In the first case, $(\mathscr{O}_X\text{-Mod})$ is the category of all abelian sheaves on X and in this case we also write $(\mathrm{Ab}(X))$ instead of $(\mathscr{O}_X\text{-Mod})$. In the second case, $(\mathscr{O}_X\text{-Mod})$ can be identified with the category of A-modules.

For more technical reasons, we will also consider the abelian category of *presheaves of $\mathscr{O}_X$-modules* which we denote by $(\text{P-}\mathscr{O}_X\text{-Mod})$. Objects are presheaves $\mathscr{F}$ of abelian groups together with a scalar multiplication $\mathscr{O}_X \times \mathscr{F} \to \mathscr{F}$ such that $\mathscr{F}(U)$ is an $\mathscr{O}_X(U)$-module for all $U \subseteq X$ open. As a special case we obtain the category $(\mathrm{PAb}(X))$ of abelian presheaves on X.

We denote by $C(X) := C(\mathscr{O}_X\text{-Mod})$ the category of complexes of $\mathscr{O}_X$-modules (Section (F.14)), by $K(X) := K(\mathscr{O}_X\text{-Mod})$ the category of complexes up to homotopy (Section (F.15)), and by $D(X) := D(\mathscr{O}_X\text{-Mod})$ the derived category of the category of $\mathscr{O}_X$-modules (Section (F.37)). For $* \in \{b, +, -\}$ we denote by $C^*(X)$, $K^*(X)$, and $D^*(X)$ the full subcategories of (left or right) bounded complexes. We also identify $D^*(X)$ with the full subcategory of $D(X)$ of (left or right) cohomologically bounded complexes (Proposition F.154).

Similarly, if R is a ring we write $C^*(R)$, $K^*(R)$, and $D^*(R)$ instead of $C^*(R\text{-Mod})$, $K^*(R\text{-Mod})$, and $D^*(R\text{-Mod})$ for $* \in \{\emptyset, +, -, b\}$.

Categories of abelian sheaves and of $\mathscr{O}_X$-modules

Our goal in the next sections is to prove that the category $(\mathscr{O}_X\text{-Mod})$ of modules on a ringed space $(X, \mathscr{O}_X)$ is a Grothendieck abelian category (Definition F.54). For this one has to show that direct sums exist (which we have already seen in Section (7.4)), that filtered colimits are exact (which is easily seen on stalks using that filtered colimits in the category of modules over a ring are exact), and that $(\mathscr{O}_X\text{-Mod})$ has a generator. We will see that a system of generators is given by the proper direct images of the structure sheaves of open subsets. Therefore we first study the functor of proper direct image for inclusions of locally closed subspaces.

(21.1) Sheaves of sections with proper support.

Let X be a topological space and let $\mathscr{F}$ be a presheaf of abelian groups on X. Recall that we define for $s \in \mathscr{F}(U)$, $U \subseteq X$ open, the support $\operatorname{Supp}(s) = \{\, x \in U \;;\; s_x \neq 0\,\}$. Then $\operatorname{Supp}(s)$ is always closed in U.

Definition 21.1. *Let X be a topological space and let $W \subseteq X$ be a locally closed subspace. Let $j \colon W \to X$ be the inclusion. For a sheaf $\mathscr{G}$ of abelian groups on W we define a sheaf $j_!\mathscr{G}$ on X by*

$$j_!\mathscr{G}(U) := \{\, s \in \mathscr{G}(W \cap U) \;;\; \operatorname{Supp}(s) \text{ is closed in } U \,\}$$

and call it the proper direct image *of $\mathscr{G}$.*

Clearly, this construction is functorial and we obtain a functor

$$j_! \colon (\mathrm{Ab}(W)) \longrightarrow (\mathrm{Ab}(X)).$$

Remark 21.2. Let $j \colon W \to X$ be the inclusion of a locally closed subspace W of a topological space X. Let $\mathscr{G}$ be a sheaf of abelian groups on W.
(1) By definition $j_!\mathscr{G}$ is a subsheaf of $j_*\mathscr{G}$. If W is closed in X, then $j_!\mathscr{G} = j_*\mathscr{G}$.
(2) For $x \in X$ one has

$$(21.1.1) \qquad\qquad (j_!\mathscr{G})_x = \begin{cases} \mathscr{G}_x, & \text{if } x \in W; \\ 0, & \text{if } x \in X \setminus W. \end{cases}$$

In particular, $j_!$ is an exact functor and fully faithful (Proposition 2.23).
(3) We have

$$(21.1.2) \qquad\qquad\qquad j^{-1}j_!\mathscr{G} = \mathscr{G}.$$

If $\mathscr{F}$ is an abelian sheaf on X with $\mathscr{F}_x = 0$ for $x \in X \setminus W$, then the adjunction morphism $\mathscr{F} \to j_*j^{-1}\mathscr{F}$ factors through an isomorphism

$$(21.1.3) \qquad\qquad\qquad \mathscr{F} \xrightarrow{\sim} j_!j^*\mathscr{F}.$$

This shows that the functor $j_!$ induces an equivalence between $(\mathrm{Ab}(W))$ and the full subcategory of $(\mathrm{Ab}(X))$ consisting of sheaves $\mathscr{F}$ with $\mathscr{F}_x = 0$ for all $x \in X \setminus W$. The inverse functor is induced by j^{-1}.

Remark 21.3. Let X be a topological space, let $Z \subseteq X$ be a closed subspace and let $U = X \setminus Z$ be its complement. Denote by $i \colon Z \to X$ and $j \colon U \to X$ the inclusions. Then one has for every abelian sheaf $\mathscr{F}$ on X a functorial exact sequence

$$0 \longrightarrow j_!j^{-1}\mathscr{F} \longrightarrow \mathscr{F} \longrightarrow i_*i^*\mathscr{F} \to 0,$$

where the morphisms are given by adjunction. The exactness is seen on stalks.

If $(X, \mathscr{O}_X)$ is a ringed space, W is open in X, and $\mathscr{G}$ is an $\mathscr{O}_W$-module, then $j_!\mathscr{G}$ is an $\mathscr{O}_X$-submodule of $j_*\mathscr{G}$ and we obtain a functor

$$j_! \colon (\mathscr{O}_W\text{-Mod}) \longrightarrow (\mathscr{O}_X\text{-Mod}).$$

There is a general construction of a proper direct image with respect to an arbitrary continuous map (Exercise 21.1) which we will not use.

We will mainly use the case that j is an open immersion. Then the functor $j_!$ can alternatively be described as the extension by zero as follows. For this we define the following functor of presheaves.

Let X be a topological space and let $j \colon U \to X$ be the inclusion of an open subspace. Let $\mathscr{G}$ be a presheaf of abelian groups on U and define a presheaf $j_!^p \mathscr{G}$ of abelian groups on X by

(21.1.4)
$$(j_!^p \mathscr{G})(V) := \begin{cases} \mathscr{G}(V), & \text{if } V \subseteq U; \\ 0, & \text{otherwise} \end{cases}, \qquad V \subseteq X \text{ open.}$$

We obtain a functor $j_!^p \colon (\mathrm{PAb}(U)) \to (\mathrm{PAb}(X))$ from the category of abelian presheaves on U to the category of abelian presheaves on X. If $(X, \mathscr{O}_X)$ is a ringed space, then $j_!^p$ induces a functor
$$j_!^p \colon (\text{P-}\mathscr{O}_U\text{-Mod}) \to (\text{P-}\mathscr{O}_X\text{-Mod}).$$

Proposition 21.4. *Let X be a topological space and let $j \colon U \to X$ be the inclusion of an open subspace.*
(1) *Let $\mathscr{G}$ be a sheaf of abelian group on U. Then there is a functorial isomorphism*
$$(j_!^p \mathscr{G})^{\#} \xrightarrow{\sim} j_!(\mathscr{G}),$$
where $(\ )^{\#}$ denotes the sheafification functor.
(2) *The functor $j_!$ is left adjoint to $j^{-1} \colon (\mathrm{Ab}(X)) \to (\mathrm{Ab}(U))$.*

Proof. By definition there is a functorial monomorphism of presheaves $j_!^p \mathscr{G} \to j_! \mathscr{G}$ that induces an isomorphism on all stalks. This shows (1).

For an abelian sheaf $\mathscr{G}$ on U and an abelian sheaf $\mathscr{F}$ on X one has functorial bijective maps
$$\mathrm{Hom}_{(\mathrm{Ab}(X))}(j_! \mathscr{G}, \mathscr{F}) \cong \mathrm{Hom}_{(\mathrm{PAb}(X))}(j_!^p \mathscr{G}, \mathscr{F})$$
$$\cong \mathrm{Hom}_{(\mathrm{PAb}(U))}(\mathscr{G}, \mathscr{F}_{|U}) = \mathrm{Hom}_{(\mathrm{Ab}(U))}(\mathscr{G}, j^{-1}\mathscr{F}),$$
where the first bijection is due to (1). This shows (2). $\qquad\square$

Remark 21.5. Let $(X, \mathscr{O}_X)$ be a ringed space, and let $j \colon U \to X$ be the inclusion of on open subspace. Then define the presheaf of $\mathscr{O}_X$-modules

(21.1.5)
$$\mathscr{O}_U^p := j_!^p (\mathscr{O}_{X|U})$$

and its sheafification

(21.1.6)
$$\mathscr{O}_{U \subseteq X} := j_! (\mathscr{O}_{X|U}).$$

Then for every presheaf of $\mathscr{O}_X$-modules $\mathscr{F}$, evaluation in $1 \in \mathscr{O}_U^p(U) = \mathscr{O}_X(U)$ yields a functorial isomorphism of $\mathscr{O}_X(U)$-modules

(21.1.7)
$$\mathrm{Hom}_{(\text{P-}\mathscr{O}_X\text{-Mod})}(\mathscr{O}_U^p, \mathscr{F}) \xrightarrow{\sim} \mathscr{F}(U).$$

If $\mathscr{F}$ is a sheaf, then by the universal property of the sheafification we also obtain an isomorphism

(21.1.8)
$$\mathrm{Hom}_{(\mathscr{O}_X\text{-Mod})}(\mathscr{O}_{U \subseteq X}, \mathscr{F}) \xrightarrow{\sim} \mathscr{F}(U)$$

which is functorial in $\mathscr{F}$. Both these isomorphisms are compatible with restriction to smaller open subsets of X.

(21.2) Categories of sheaves on a ringed space.

We aim to show that ($\mathscr{O}_X$-Mod) and (P-$\mathscr{O}_X$-Mod) are Grothendieck abelian categories (Definition F.54). By the following remark, it suffices to show that (P-$\mathscr{O}_X$-Mod) is a Grothendieck abelian category.

Remark 21.6. The abelian category ($\mathscr{O}_X$-Mod) is a Giraud subcategory (Definition F.60) of (P-$\mathscr{O}_X$-Mod). Indeed, the inclusion functor $\iota\colon (\mathscr{O}_X$-Mod) $\to$ (P-$\mathscr{O}_X$-Mod) has a left adjoint functor given by sheafification. It remains to show that sheafification is an exact functor. A sequence $0 \to \mathscr{F}' \to \mathscr{F} \to \mathscr{F}'' \to 0$ of presheaves of $\mathscr{O}_X$-modules is exact in (P-$\mathscr{O}_X$-Mod) if and only if for all $U \subseteq X$ open the sequence $0 \to \mathscr{F}'(U) \to \mathscr{F}(U) \to \mathscr{F}''(U) \to 0$ is exact. As filtered colimits are exact in the category of abelian groups, this implies that the sequence of stalks $0 \to \mathscr{F}'_x \to \mathscr{F}_x \to \mathscr{F}''_x \to 0$ is exact for all $x \in X$. Hence sheafification preserves stalks and one can check exactness of a sequence of $\mathscr{O}_X$-modules on stalks, sheafification is an exact functor.

Proposition 21.7. *The categories ($\mathscr{O}_X$-Mod) and (P-$\mathscr{O}_X$-Mod) are Grothendieck abelian categories.*

In particular (Theorem F.185), for every complex $\mathscr{F}$ of $\mathscr{O}_X$-modules there exists a quasi-isomorphism $\mathscr{F} \to \mathscr{I}$, where $\mathscr{I}$ is K-injective with injective components.

If there exists $N \in \mathbb{Z}$ such that $H^n(\mathscr{F}) = 0$ for all $n < N$, then $\mathscr{I}$ can be chosen such that $\mathscr{I}^n = 0$ for all $n < N$.

By our definition, the structure sheaf $\mathscr{O}_X$ of a ringed space is always a sheaf of commutative rings. But the proposition also holds for categories of (pre-)sheaves of left $\mathscr{O}_X$-modules if $\mathscr{O}_X$ is a sheaf of not necessarily commutative rings (with the same proof).

Proof. As ($\mathscr{O}_X$-Mod) is a Giraud subcategory of (P-$\mathscr{O}_X$-Mod), it suffices to show that (P-$\mathscr{O}_X$-Mod) is a Grothendieck abelian category (Proposition F.61).

Direct sums and filtered colimits in (P-$\mathscr{O}_X$-Mod) are formed section-wise, hence they exist and are exact. It remains to show that (P-$\mathscr{O}_X$-Mod) has a generator. We use Remark F.55.

For $U \subseteq X$ consider the presheaf of $\mathscr{O}_X$-modules $\mathscr{O}_U^p$ (21.1.5). If $\alpha\colon \mathscr{F} \to \mathscr{G}$ is a non-zero morphism and $s \in \mathscr{F}(U)$ is a section over some open set U such that $\alpha(s) \neq 0$, this section s corresponds to a morphism $\mathscr{O}_U^p \to \mathscr{F}$ (21.1.7) whose composition with α is non-zero. We conclude by Remark F.55. $\qquad\square$

Remark 21.8. The proof shows that the family of presheaves $\mathscr{O}_U^p$, for $U \subseteq X$ open, forms a system of generators of (P-$\mathscr{O}_X$-Mod). Moreover, Proposition F.61 shows that their sheafifications $\mathscr{O}_{U \subseteq X}$ form a system of generators of ($\mathscr{O}_X$-Mod). This is also easily seen directly as in the proof of Proposition 21.7. Hence the same proof allows to show directly that ($\mathscr{O}_X$-Mod) is a Grothendieck abelian category without invoking the formalism of Giraud subcategories. In fact, the argument in the proof shows that the sheaves $\mathscr{O}_{U \subseteq X}$ where U runs through a fixed basis of the topology already generate ($\mathscr{O}_X$-Mod).

Proposition 21.7 implies that all functors on ($\mathscr{O}_X$-Mod) are right derivable by Corollary F.187:

Corollary 21.9. *Let $\mathcal{B}$ be an abelian category, and let $F\colon (\mathscr{O}_X$-Mod) $\to \mathcal{B}$ be an additive functor. We also denote by F the induced functor $K(X) = K(\mathscr{O}_X$-Mod) $\to D(\mathcal{B})$.*

(1) *Then F admits a right derived functor $RF\colon D(X) \to D(\mathcal{B})$, and for every complex $\mathscr{F}$ of $\mathscr{O}_X$-modules one has $RF(\mathscr{F}) = F(I_{\mathscr{F}})$, where $\mathscr{F} \to I_{\mathscr{F}}$ is a quasi-isomorphism to a K-injective complex of $\mathscr{O}_X$-modules.*
(2) *The restriction of F to $K^+(X)$ also admits a right derived functor $R^+F\colon D^+(X) \to D^+(\mathcal{B})$, and R^+F is the restriction of RF to $D^+(X)$.*

The second assertion follows from Corollary F.178.

Again by Theorem F.185, Condition (Ac) (see F.198) is satisfied for every additive functor $F\colon (\mathscr{O}_X\text{-Mod}) \to \mathcal{B}$ of abelian categories (Example F.199). Hence we obtain the following version of Corollary 21.9 in the language of the higher derived functor $R^iF = H^i \circ RF\colon (\mathscr{O}_X\text{-Mod}) \to \mathcal{B}$.

Corollary 21.10. *Let $\mathcal{B}$ be an abelian category, and let $F\colon (\mathscr{O}_X\text{-Mod}) \to \mathcal{B}$ be a left exact functor.*
(1) *The higher right derived functors $(R^iF)_{i\geq 0}$ exist and form a universal δ-functor with $R^0F = F$. In particular, if $0 \to \mathscr{F}' \to \mathscr{F} \to \mathscr{F}'' \to 0$ is a short exact sequence of $\mathscr{O}_X$-modules, then one has a long exact sequence*

(21.2.1)
$$0 \to F(\mathscr{F}') \to F(\mathscr{F}) \to F(\mathscr{F}'')$$
$$\to R^1F(\mathscr{F}') \to R^1F(\mathscr{F}) \to R^1F(\mathscr{F}'') \to R^2F(\mathscr{F}') \to \ldots$$

(2) *An $\mathscr{O}_X$-module $\mathscr{A}$ is right F-acyclic if and only if $R^iF(\mathscr{A}) = 0$ for all $i > 0$. Every injective $\mathscr{O}_X$-module is right F-acyclic.*

Proof. By Remark F.200 one has $R^iF = 0$ for $i < 0$ and $R^0F = F$ because F is left exact. Injective $\mathscr{O}_X$-modules are right F-acyclic for every additive functor F (Example F.203). Hence the remaining assertions follow from Proposition F.204. $\qquad\square$

(21.3) Restriction to open subsets.

Let $(X, \mathscr{O}_X)$ be a ringed space and let $j\colon U \to X$ be the inclusion of an open subset. As the functor j^{-1} from the category of $\mathscr{O}_X$-modules to the category of $\mathscr{O}_U$-modules is exact, it induces a functor $j^{-1}\colon D(X) \to D(U)$ which we usually denote by $\mathscr{F} \mapsto \mathscr{F}_{|U}$.

Remark 21.11. *Let $(U_i)_{i\in I}$ be an open covering of X.*
(1) *Let $\mathscr{F}$ be a complex of $\mathscr{O}_X$-modules. If $\mathscr{F}_{|U_i} = 0$ in $D(U_i)$ for all i, then $\mathscr{F} = 0$ in $D(X)$.*

   Indeed, $\mathscr{F} = 0$ if and only if $H^p(\mathscr{F}) = 0$ for all $p \in \mathbb{Z}$. Now we use that the formation of $H^p(\mathscr{F})$ commutes with restriction to open subsets.
(2) *Let $v\colon \mathscr{F} \to \mathscr{G}$ be a morphism in $D(X)$. As being an isomorphism can be checked after applying $H^p(\ )$ (Remark F.151 (2)), v is an isomorphism if and only if $v_{|U_i}$ is an isomorphism in $D(U_i)$ for all i.*
(3) *In general, there exist non-zero morphisms in $D(X)$ whose restrictions to each U_i are zero (Exercise 22.1).*

Lemma 21.12. *Let $(X, \mathscr{O}_X)$ be a ringed space.*
(1) *Let $U \subseteq X$ be an open subset. If $\mathscr{I}$ is a K-injective complex of $\mathscr{O}_X$-modules (resp. an injective $\mathscr{O}_X$-module), then $\mathscr{I}_{|U}$ is a K-injective complex of $\mathscr{O}_{X|U}$-modules (resp. an injective $\mathscr{O}_{X|U}$-module).*
(2) *Conversely, let $\mathscr{I}$ be an $\mathscr{O}_X$-module and let $(U_i)_i$ be an open covering such that $\mathscr{I}_{|U_i}$ is an injective $\mathscr{O}_{U_i}$-module for all i. Then $\mathscr{I}$ is an injective $\mathscr{O}_X$-module.*

Proof. Let $j\colon U \to X$ be the inclusion such that $j^{-1}\mathscr{F} = \mathscr{F}_{|U}$. The functor j^{-1} has an exact left adjoint functor, namely $j_!$ (Proposition 21.4 (2)). This shows the first assertion by Corollary F.183.

Let us show the second assertion. The $\mathcal{O}_X$-modules $\mathcal{O}_{U \subseteq X}$ for U open and contained in some U_i form a system of generators of $(\mathcal{O}_X\text{-Mod})$ (Remark 21.8). Hence it is enough to show that for every monomorphism $\mathscr{G} \to \mathcal{O}_{U \subseteq X}$ for U contained in some U_i every map $u\colon \mathscr{G} \to \mathscr{I}$ can be extended to $\mathcal{O}_{U \subseteq X}$ (Proposition F.63). Denote by $j\colon U \to X$ the inclusion. As $\mathscr{G}_x = 0$ for $x \in X \setminus U$, we have $\mathscr{G} = j_!\mathscr{G}'$ for some $\mathcal{O}_U$-module $\mathscr{G}'$ and u corresponds to a map $\mathscr{G}' \to \mathscr{I}_{|U}$ that can be extended by hypothesis and by (1) to a map $\mathcal{O}_U \to \mathscr{I}_{|U}$ or, equivalently, to a map $\mathcal{O}_{U \subseteq X} \to \mathscr{I}$. $\qquad\square$

Cohomology and derived direct image

By now we have seen (Corollary 21.9) that we can right derive all additive functors defined on the category of $\mathcal{O}_X$-modules for an arbitrary ringed space $(X, \mathcal{O}_X)$. In particular we obtain the right derived functors of the direct image functor and the global section functor, i.e., higher direct images and cohomology, respectively.

The right derivations of both functors determine each other. In fact, the global section functor can be viewed as a very special case of the direct image functor (Remark 21.25 below). Conversely, higher direct images can be calculated via cohomology (Proposition 21.27 below).

By definition one can calculate the cohomology of an $\mathcal{O}_X$-module, or even of a whole complex of $\mathcal{O}_X$-modules, $\mathscr{F}$ by choosing a quasi-isomorphism to a $\Gamma(X, -)$-acyclic complex $\mathscr{A}$ and applying the global section functor to $\mathscr{A}$. Here one can take for $\mathscr{A}$ for instance a K-injective complex as K-injective complexes are acyclic for all additive functors. But usually it is difficult to have a good description of $\mathscr{A}$ in terms of $\mathscr{F}$. Hence we will define a different class of $\Gamma(X, -)$-acyclic complexes, namely bounded below complexes consisting of flasque sheaves. These complexes are also acyclic for the direct image functor and every $\mathcal{O}_X$-module has a rather concrete resolution by flasque sheaves (the Godement resolution, Remark 21.31). Moreover, we can use flasque sheaves to show a number of results (such as that H^1 defined via derived functors coincides with the definition of H^1 via torsors given in Volume I, the Leray spectral sequence, or the Mayer-Vietoris sequence). In the next part of this chapter, we will develop even better techniques to give concrete calculations of some cohomology groups using Čech cohomology.

In Section (21.10) we will apply the techniques developed so far to compute some cohomology groups of line bundles on the projective line. The awkwardness of the arguments there is also thought as a motivation to develop the theory further in the following chapters.

The remainder of this part of the chapter is devoted to some further general results about cohomology. We study the question under what hypotheses cohomology and higher direct images commute with filtered colimits. Although this might seem to be a rather technical question, it will be very useful in the sequel, for instance for the proof of the projection formula in Section (22.19) or in Chapter 25 on Grothendieck duality. We also state a general vanishing result of sheaf cohomology in degrees $> \dim(X)$ if X is a spectral space, i.e., the underlying topological space of a qcqs scheme, and prove this if the topological space X is noetherian (a result due to Grothendieck). We conclude by defining cohomology with support in a closed subset. For us this will be only a technical tool and we show only some very basic properties about it.

(21.4) Cohomology as derived functor.

Let $(X, \mathcal{O}_X)$ be a ringed space and set $A := \Gamma(X, \mathcal{O}_X)$. Consider the global section functor

$$\Gamma := \Gamma(X, -) \colon (\mathcal{O}_X\text{-Mod}) \to (A\text{-Mod}).$$

This is a left exact functor of abelian categories. In the special case that $\mathcal{O}_X$ is the constant sheaf $\mathbb{Z}_X$ the category $(\mathcal{O}_X\text{-Mod})$ is the category of abelian sheaves on X. In this case we also set $A := \mathbb{Z}$ and consider $\Gamma(X, -)$ as a functor from the category of abelian sheaves to the category of abelian groups. The right derived functor is denoted by

$$R\Gamma(X, -) \colon D(X) \to D(A).$$

It exists by Corollary 21.9.

Definition 21.13. *Let $\mathcal{F}$ be a complex of $\mathcal{O}_X$-modules. For $i \in \mathbb{Z}$, the i-th cohomology of X with coefficients in $\mathcal{F}$ or simply the i-th cohomology of $\mathcal{F}$ is the A-module $H^i(X, \mathcal{F}) := R^i\Gamma(X, \mathcal{F})$.*

This definition can in particular be applied to an $\mathcal{O}_X$-module $\mathcal{F}$ which we always consider as a complex of $\mathcal{O}_X$-modules concentrated in degree 0.

If one considers $H^i(X, \mathcal{F})$ for a complex $\mathcal{F}$ not concentrated in a single degree, then $H^i(X, \mathcal{F})$ is often also called the *i-th hypercohomology* of $\mathcal{F}$. We will not use this terminology.

Caveat 21.14. For a complex $\mathcal{F}$ of $\mathcal{O}_X$-modules one has to distinguish between

$$H^i(\mathcal{F}) = \operatorname{Ker}(\mathcal{F}^i \to \mathcal{F}^{i+1})/\operatorname{Im}(\mathcal{F}^{i-1} \to \mathcal{F}^i),$$

which is an $\mathcal{O}_X$-module, and $H^i(X, \mathcal{F})$, which is an A-module. Moreover, usually one has $\Gamma(X, H^i(\mathcal{F})) \neq H^i(X, \mathcal{F})$. For instance, if $\mathcal{F}$ is concentrated in degree 0, then $\Gamma(X, H^i(\mathcal{F})) = 0$ for all $i \neq 0$ but very often $H^i(X, \mathcal{F}) \neq 0$ for some $i > 0$.

The general results on derived functors collected in Appendix F specialize to the current situations as follows.

Remark 21.15. For any complex $\mathcal{F}$ of $\mathcal{O}_X$-modules, we can compute $H^i(X, \mathcal{F})$ as follows by Theorem F.173: Choose a quasi-isomorphism $\mathcal{F} \to A_{\mathcal{F}}$ to a right Γ-acyclic complex $A_{\mathcal{F}}$. For instance, $A_{\mathcal{F}}$ can be chosen to be a K-injective complex. Then

$$H^i(X, \mathcal{F}) = \operatorname{Ker}\big(\Gamma(X, A_{\mathcal{F}}^i) \to \Gamma(X, A_{\mathcal{F}}^{i+1})\big)/\operatorname{Im}\big(\Gamma(X, A_{\mathcal{F}}^{i-1}) \to \Gamma(X, A_{\mathcal{F}}^i)\big).$$

If $\mathcal{F}$ is a single $\mathcal{O}_X$-module, such a quasi-isomorphism can be chosen to be an exact sequence of $\mathcal{O}_X$-modules

$$0 \to \mathcal{F} \to A_{\mathcal{F}}^0 \to A_{\mathcal{F}}^1 \to \dots,$$

where for all $n \geq 0$ the $\mathcal{O}_X$-module $A_{\mathcal{F}}^n$ is right Γ-acyclic, i.e., $H^i(X, \mathcal{A}_{\mathcal{F}}^n) = 0$ for all $i > 0$.

In Section (21.7) below we will give an example of a useful class of right Γ-acyclic $\mathcal{O}_X$-modules.

Remark 21.16. Let $\mathscr{F}$ be a complex of $\mathcal{O}_X$-modules.

(1) If $\mathscr{F}$ is an $\mathcal{O}_X$-module, considered as a complex concentrated in degree 0, then one has $H^0(X, \mathscr{F}) = \Gamma(X, \mathscr{F})$ (Remark F.200 (3)).

(2) If there exists $n \in \mathbb{Z}$ with $H^i(\mathscr{F}) = 0$ for all $i < n$, then $H^i(X, \mathscr{F}) = 0$ for all $i < n$ (Remark F.200 (1)).

(3) Considered as functors $H^i(X, -) \colon (\mathcal{O}_X\text{-Mod}) \to (A\text{-Mod})$ the family $(H^i(X, -))_{i \geq 0}$ forms a universal δ-functor by Remark F.205.

(4) Every short exact sequence of complexes of $\mathcal{O}_X$-modules $0 \to \mathscr{F} \to \mathscr{G} \to \mathscr{H} \to 0$ yields in $D(X)$ a distinguished triangle (F.37.3) and hence by Remark F.167 there is a long exact cohomology sequence, functorial in the short exact sequence

$$(21.4.1) \quad \ldots \longrightarrow H^i(X, \mathscr{F}) \longrightarrow H^i(X, \mathscr{G}) \longrightarrow H^i(X, \mathscr{H}) \longrightarrow H^{i+1}(X, \mathscr{F}) \longrightarrow \ldots$$

(5) As the derived functor $R\Gamma$ is triangulated and in particular preserves shifts, one has for all $i, n \in \mathbb{Z}$

$$(21.4.2) \qquad\qquad H^i(X, \mathscr{F}[n]) = H^{i+n}(X, \mathscr{F}).$$

Every complex $\mathscr{F}$ of $\mathcal{O}_X$-modules can also be considered as a complex $\mathscr{F}^{\mathrm{ab}}$ of abelian sheaves. Below (Corollary 21.43 for bounded below complexes, and Corollary 21.116 in general) we will see that the underlying abelian group of the A-module $H^i(X, \mathscr{F})$ is the group $H^i(X, \mathscr{F}^{\mathrm{ab}})$ for all i.

There is a different way to express cohomology, namely via Ext groups (Section (F.52)) as we explain now.

Remark 21.17. Let $U \subseteq X$ be an open subset. Recall from Remark 21.5 that for every $\mathcal{O}_X$-module $\mathscr{F}$ there is a isomorphism of $\Gamma(U, \mathcal{O}_X)$-modules

$$\mathrm{Hom}_{(\mathcal{O}_X\text{-Mod})}(\mathcal{O}_{U \subseteq X}, \mathscr{F}) \xrightarrow{\sim} \Gamma(U, \mathscr{F})$$

which is functorial in $\mathscr{F}$ and compatible with restriction to smaller open subsets of X. Hence we obtain a triangulated isomorphism of derived functors

$$(21.4.3) \qquad\qquad R\,\mathrm{Hom}_{\mathcal{O}_X}(\mathcal{O}_{U \subseteq X}, -) \xrightarrow{\sim} R\Gamma(U, -).$$

Using $H^i \circ R\,\mathrm{Hom} = \mathrm{Ext}^i$ (F.52.5), we get isomorphisms

$$(21.4.4) \qquad\qquad \mathrm{Ext}^i_{\mathcal{O}_X}(\mathcal{O}_{U \subseteq X}, \mathscr{F}) \xrightarrow{\sim} H^i(U, \mathscr{F}).$$

which are functorial in the complex $\mathscr{F}$ of $\mathcal{O}_X$-modules.

(21.5) Cohomology and restriction to open subspaces.

Let $(X, \mathcal{O}_X)$ be a ringed space and let $U \subseteq X$ be open. If $\mathscr{F}$ is a complex of $\mathcal{O}_X$-modules, we can restrict each component of $\mathscr{F}$ to U and obtain a complex $\mathscr{F}_{|U}$ of $(\mathcal{O}_{X|U})$-modules.

Lemma 21.18. *Denote by $R\Gamma(U, -)$ the right derived functor of the functor $(\mathcal{O}_X\text{-Mod}) \to (\Gamma(U, \mathcal{O}_X)\text{-Mod})$, $\mathscr{F} \mapsto \mathscr{F}(U)$. Then we have for every complex of $\mathcal{O}_X$-modules $\mathscr{F}$ a functorial isomorphism*

$$(21.5.1) \qquad\qquad R\Gamma(U, \mathscr{F}_{|U}) \xrightarrow{\sim} R\Gamma(U, \mathscr{F}).$$

In particular $H^i(U, \mathscr{F}_{|U}) = H^i(U, \mathscr{F})$ for all $i \in \mathbb{Z}$.

In the sequel will usually identify $R\Gamma(U, \mathscr{F}_{|U})$ and $R\Gamma(U, \mathscr{F})$.

Proof. Let $\mathscr{F}$ be a complex of $\mathscr{O}_X$-modules and let $\mathscr{F} \to \mathscr{I}$ be a quasi-isomorphism to a K-injective complex. Then

$$R\Gamma(U, \mathscr{F}_{|U}) = \Gamma(U, \mathscr{I}_{|U}) = \Gamma(U, \mathscr{I}) = R\Gamma(U, \mathscr{F}).$$

where the first equality holds by Lemma 21.12. $\qquad\square$

Remark 21.19. Let $(X, \mathscr{O}_X)$ be a ringed space and let $v \colon \mathscr{F} \to \mathscr{G}$ be a morphism in $D(X)$. Let $\mathcal{B}$ be a basis of open subsets of X. Then v is an isomorphism (resp. $v = 0$) if and only the induced map $R\Gamma(U, \mathscr{F}) \to R\Gamma(U, \mathscr{G})$ is an isomorphism (resp. is zero) for all $U \in \mathcal{B}$.

Indeed, the condition is clearly necessary. To show that it is sufficient we may replace $\mathscr{F}$ and $\mathscr{G}$ by in $D(X)$ isomorphic objects and can assume that both are K-injective. But then $R\Gamma(U, \mathscr{F}) = \mathscr{F}(U)$ and $R\Gamma(U, \mathscr{G}) = \mathscr{G}(U)$ and v can be represented by a morphism in $K(X)$ (Prop. F.179 (7)). This is an isomorphism (resp. is zero) if $\mathscr{F}(U) \to \mathscr{G}(U)$ is an isomorphism (resp. is zero) for U in $\mathcal{B}$.

Remark 21.20. Let $(X, \mathscr{O}_X)$ be a ringed space. For $U \subseteq V \subseteq X$ open, restriction of sections defines a morphism of additive functors $\Gamma(V, -) \to \Gamma(U, -)$. Hence we obtain a morphism of triangulated functors $R\Gamma(V, -) \to R\Gamma(U, -)$ (Remark F.166) and in particular for all $i \in \mathbb{Z}$ homomorphisms of $\Gamma(V, \mathscr{O}_X)$-modules

$$(21.5.2) \qquad\qquad H^i(V, \mathscr{F}) \longrightarrow H^i(U, \mathscr{F})$$

functorial in complexes $\mathscr{F}$ of $\mathscr{O}_X$-modules. In particular we obtain a presheaf

$$\mathscr{H}^i(\mathscr{F}) \colon U \mapsto H^i(U, \mathscr{F})$$

of $\mathscr{O}_X$-modules.

These presheaves are in fact the higher derived functors of some functor:

Remark 21.21. Consider the inclusion ι from the category of $\mathscr{O}_X$-modules to the category of presheaves of $\mathscr{O}_X$-modules. This is a left exact functor. For all $i \in \mathbb{Z}$ and every complex $\mathscr{F}$ of $\mathscr{O}_X$-modules there is an isomorphism

$$(21.5.3) \qquad\qquad R^i\iota(\mathscr{F}) \xrightarrow{\sim} \mathscr{H}^i(\mathscr{F}).$$

of cohomological functors in $\mathscr{F}$.

Indeed, choose a K-injective resolution $\mathscr{F} \to \mathscr{I}$. Then one has for $U \subseteq X$ open

$$R^i\iota(\mathscr{F})(U) = H^i(\mathscr{I})(U) \overset{(*)}{=} H^i(\mathscr{I}(U)) = H^i(U, \mathscr{F}),$$

where for $(*)$ it is important to note that one takes cohomology as *presheaves*.

Lemma 21.22. *Let X be a ringed space, let $\mathscr{F}$ be a complex of $\mathscr{O}_X$-modules, and let $n \in \mathbb{Z}$. Then the sheafification of the presheaf $\mathscr{H}^n(\mathscr{F})$ is the n-th cohomology sheaf $H^n(\mathscr{F})$ of $\mathscr{F}$.*

Proof. Let $\mathscr{F} \to \mathscr{I}$ be a K-injective resolution. For all $U \subseteq X$ open one has by Lemma 21.18

$$H^n(U, \mathscr{F}) = H^n(\dots \to \mathscr{I}^{i-1}(U) \to \mathscr{I}^i(U) \to \mathscr{I}^{i+1}(U) \to \dots).$$

Hence the sheafification of $U \mapsto H^n(U, \mathscr{F})$ is given by $H^n(\mathscr{I}) = H^n(\mathscr{F})$. $\qquad\square$

Corollary 21.23. *Let X be a ringed space and let $\mathscr{F}$ be a complex of $\mathcal{O}_X$-modules. Let $n \in \mathbb{Z}$ such that $H^n(\mathscr{F}) = 0$ and let $\xi \in H^n(X, \mathscr{F})$. Then there exists an open covering $(U_i)_i$ of X such that $\xi_{|U_i} = 0$ for all i.*

(21.6) Higher direct images as derived functors.

The higher direct image functors provide relative versions of cohomology groups. Let $f \colon (X, \mathcal{O}_X) \to (Y, \mathcal{O}_Y)$ be a morphism of ringed spaces. Then the direct image is a left exact functor

$$f_* \colon (\mathcal{O}_X\text{-Mod}) \longrightarrow (\mathcal{O}_Y\text{-Mod}).$$

As all additive functors on $(\mathcal{O}_X\text{-Mod})$ it has a right derived functor (Corollary 21.9).

Definition 21.24. *The right derived functor $Rf_* \colon D(X) \to D(Y)$ of the direct image functor f_* is called* derived direct image *functor. If $\mathscr{F}$ is a complex of $\mathcal{O}_X$-modules, the $\mathcal{O}_Y$-module $R^i f_* \mathscr{F}$ ($i \geq 0$) is called the i-th direct image of $\mathscr{F}$ (under f).*

Remark 21.25. Taking cohomology is in fact a special case of forming higher derived images: Let $(X, \mathcal{O}_X)$ be a ringed space and consider the ringed space $(*, \Gamma(X, \mathcal{O}_X))$ consisting of one point $*$ with structure sheaf given by $\mathcal{O}_*(*) := \Gamma(X, \mathcal{O}_X)$. Modules over this ringed space are simply $\Gamma(X, \mathcal{O}_X)$-modules.

If we denote by $f \colon X \to *$ the obvious morphism of ringed spaces, then $f_* \mathscr{F} = \Gamma(X, \mathscr{F})$ for every $\mathcal{O}_X$-module $\mathscr{F}$. Therefore $Rf_* \mathscr{F} = R\Gamma(X, \mathscr{F})$ and $R^i f_* \mathscr{F} = H^i(X, \mathscr{F})$ for every complex $\mathscr{F}$ of $\mathcal{O}_X$-modules and for all $i \in \mathbb{Z}$.

Remark 21.26. Let $f \colon X \to Y$ be a morphism of ringed spaces and let $\mathscr{F}$ be a complex of $\mathcal{O}_X$-modules.
(1) Every exact sequence $0 \to \mathscr{F}' \to \mathscr{F} \to \mathscr{F}'' \to 0$ of complexes of $\mathcal{O}_X$-modules yields a long exact sequence of higher direct images

$$(21.6.1) \qquad \cdots \longrightarrow R^{i-1} f_* \mathscr{F}'' \longrightarrow R^i f_* \mathscr{F}' \longrightarrow R^i f_* \mathscr{F} \longrightarrow R^i f_* \mathscr{F}'' \longrightarrow \cdots$$

(2) Suppose there exists $n \in \mathbb{Z}$ such that $H^i(\mathscr{F}) = 0$ for all $i < n$. Then $R^i f_* \mathscr{F} = 0$ for all $i < n$ (Remark F.200 (1)).
(3) Let $\mathscr{F}$ be an $\mathcal{O}_X$-module (as always considered as a complex concentrated in degree 0). Then $R^i f_* \mathscr{F} = 0$ for all $i < 0$ by (2) and $R^0 f_* \mathscr{F} = f_* \mathscr{F}$ (Remark F.200 (3)).

Proposition 21.27. *Let $f \colon X \to Y$ be a morphism of ringed spaces, and let $\mathscr{F}$ be a complex of $\mathcal{O}_X$-modules. For all $i \in \mathbb{Z}$, the $\mathcal{O}_Y$-module $R^i f_* \mathscr{F}$ is the sheaf associated to the presheaf*

$$V \mapsto H^i(f^{-1}(V), \mathscr{F}), \qquad V \subseteq Y \text{ open}$$

(with restriction maps given by (21.5.2)).

Proof. Consider the commutative diagram of functors

$$
\begin{array}{ccc}
(\mathcal{O}_X\text{-Mod}) & \xrightarrow{\ \iota\ } & (\text{P-}\mathcal{O}_X\text{-Mod}) \\
{\scriptstyle f_*} \downarrow & & \downarrow {\scriptstyle f_*^{\mathrm{P}}} \\
(\mathcal{O}_Y\text{-Mod}) & \xleftarrow{\ (-)^\#\ } & (\text{P-}\mathcal{O}_Y\text{-Mod}),
\end{array}
$$

where ι denotes the inclusion, f_*^p the direct image of presheaves of $\mathcal{O}_X$-modules, and $(-)^\#$ the sheafification functor. Note that f_*^p and $(-)^\#$ are both exact functors. Hence we obtain by Remark F.175 isomorphisms of triangulated functors

$$Rf_* = R((-)^\# \circ f_*^p \circ \iota) \cong (-)^\# \circ f_*^p \circ R\iota.$$

Applying H^i one obtains the claim by Remark 21.21. $\qquad\square$

Derived direct image also commutes with restriction to open subsets:

Lemma 21.28. *let $f\colon X \to Y$ be a morphism of ringed spaces, let $V \subseteq Y$ be open and denote by $f_V\colon f^{-1}(V) \to V$ the restriction of f. Then there is for all complexes $\mathscr{F}$ of $\mathcal{O}_X$-modules a functorial isomorphism in $D(V)$*

$$Rf_*\mathscr{F}_{|V} \xrightarrow{\sim} R(f_V)_*(\mathscr{F}_{|f^{-1}(V)}).$$

Proof. Let $\mathscr{F} \to I_{\mathscr{F}}$ be a K-injective resolution. As $(I_{\mathscr{F}})_{|f^{-1}(V)}$ is again K-injective (Lemma 21.12) we find

$$Rf_*(\mathscr{F})_{|V} = f_*(I_{\mathscr{F}})_{|V} = (f_V)_*(I_{\mathscr{F}})_{|f^{-1}(V)} = R(f_V)_*(\mathscr{F}_{|f^{-1}(V)}). \qquad\square$$

(21.7) Flasque Sheaves.

In this section we will study "flasque sheaves" and we will show that they are right Γ-acyclic. This yields a – somewhat – more concrete way to calculate cohomology via flasque sheaves.

Definition 21.29. *Let X be a topological space. A sheaf $\mathscr{F}$ on X is called* flasque *or* flabby, *if all restriction maps $\Gamma(U, \mathscr{F}) \to \Gamma(V, \mathscr{F})$ for $V \subseteq U \subseteq X$ open, are surjective.*

Remark 21.30.
(1) Let $f\colon X \to Y$ be a continuous map of topological spaces and let $\mathscr{F}$ be a flasque sheaf on X. Then $f_*\mathscr{F}$ is a flasque sheaf on Y.
(2) Let X be an irreducible topological space, and let A be a set. Then the constant sheaf $\underline{A}$ attached to A is flasque.

Indeed, since X is irreducible, all non-empty open subsets are connected. Hence the presheaf $U \mapsto A$ (for $\emptyset \neq U \subseteq X$ open) is a sheaf, so this is the sheaf $\underline{A}$. The restriction maps between the sections on non-empty open subsets of X are simply the identity maps id_A.

Remark 21.31. (Godement resolution) Let $\mathscr{F}$ be a sheaf on a topological space X. Define a sheaf $\mathscr{F}^{[0]}$ on X by $\mathscr{F}^{[0]}(U) := \prod_{x \in U} \mathscr{F}_x$ for $U \subseteq X$ open, where the restriction maps are given by the projections. The sheaf $\mathscr{F}^{[0]}$ is flasque and there is an injective morphism of sheaves

$$\iota_{\mathscr{F}}\colon \mathscr{F} \hookrightarrow \mathscr{F}^{[0]}, \qquad \mathscr{F}(U) \ni s \mapsto (s_x)_{x \in U} \in \mathscr{F}^{[0]}(U), \quad U \subseteq X \text{ open}.$$

Every morphism $\varphi\colon \mathscr{F} \to \mathscr{G}$ of sheaves induces a morphism of flasque sheaves $\varphi^{[0]}\colon \mathscr{F}^{[0]} \to \mathscr{G}^{[0]}$ with $\varphi_U^{[0]} = \prod_{x \in U} \varphi_x$. We obtain a functor $(\ )^{[0]}$ from the category of sheaves to the full subcategory of flasque sheaves. The morphism $\iota_{\mathscr{F}}$ is functorial in $\mathscr{F}$.

Now suppose that $\mathscr{F}$ is an $\mathscr{O}_X$-module. Then $\mathscr{F}^{[0]}$ is an $\mathscr{O}_X$-module and we obtain a functor $(\ )^{[0]}$ from the category of $\mathscr{O}_X$-modules to the full subcategory of flasque $\mathscr{O}_X$-modules. The morphism $\iota_{\mathscr{F}}$ is a functorial homomorphism of $\mathscr{O}_X$-modules.

Applying Lemma F.194 we find for every bounded below complex $\mathscr{F}$ of $\mathscr{O}_X$-modules a quasi-isomorphism $\mathscr{F} \to \mathscr{G}_{\mathscr{F}}$, functorial in $\mathscr{F}$, where $\mathscr{G}_{\mathscr{F}}$ is a bounded below complex whose components are flasque sheaves. This resolution of $\mathscr{F}$ is called the *Godement resolution*.

Lemma 21.32. *Let $(X, \mathscr{O}_X)$ be a ringed space. Every injective $\mathscr{O}_X$-module is flasque.*

Proof. Let $\mathscr{I}$ be an injective $\mathscr{O}_X$-module. By Remark 21.31 there exists an injective homomorphism of $\mathscr{O}_X$-modules $i\colon \mathscr{I} \to \mathscr{G}$ where $\mathscr{G}$ is flasque. As $\mathscr{I}$ is injective, this makes $\mathscr{I}$ into a direct summand of $\mathscr{G}$. Hence $\mathscr{I}$ is flasque. $\qquad\square$

Recall that for any sheaf $\mathscr{G}$ of (not necessarily abelian) groups we defined the notion of a $\mathscr{G}$-torsor (Section (11.5)). We denoted by $H^1(X, \mathscr{G})$ the pointed set of isomorphism classes of $\mathscr{G}$-torsors. If $\mathscr{G}$ is an abelian sheaf, then $H^1(X, \mathscr{G})$ has the structure of an abelian group. Below (Proposition 21.40) we will show that this group is functorially isomorphic to the cohomology $H^1(X, \mathscr{G})$ defined via derived functors. Until then, we will denote the first cohomology defined via torsors by $H^1_{\mathrm{Tors}}(X, \mathscr{G})$.

Lemma 21.33. *Let X be a topological space and let $\mathscr{G}$ be a flasque sheaf of groups on X. Then every $\mathscr{G}$-torsor is trivial.*

Proof. Let T be a $\mathscr{G}$-torsor, let $(U_i)_{i \in I}$ be an open covering such that there exist $t_i \in T(U_i)$ for all i. For $J \subseteq I$ set $U_J := \bigcup_{i \in J} U_i$. Then $U_I = X$. Define $\mathcal{E} := \{\, (t, J) \;;\; J \subseteq I, t \in T(U_J) \,\}$. Then $\mathcal{E} \neq \emptyset$ because $(*, \emptyset) \in \mathcal{E}$. It is partially ordered by $(t, J) \leq (t', J')$ if $J \subseteq J'$ and $t'|_{U_J} = t$. As T is a sheaf, every totally ordered subset of $\mathcal{E}$ has an upper bound in $\mathcal{E}$. Hence there exists a maximal element (t, J) in $\mathcal{E}$ by Zorn's lemma. It suffices to show that $J = I$.

Assume there exists $i \in I \setminus J$ and let $g \in \mathscr{G}(U_J \cap U_i)$ with $t|_{U_J \cap U_i} = g t_i|_{U_J \cap U_i}$. As $\mathscr{G}$ is flasque, we can extend g to $\tilde{g} \in \mathscr{G}(X)$. Replacing t_i by $\tilde{g}|_{U_i} t_i$ we may assume $t|_{U_J \cap U_i} = t_i|_{U_J \cap U_i}$ and hence we can glue t and t_i to a section over $U_{J \cup \{i\}}$. This contradicts the maximality of (t, J). $\qquad\square$

Corollary 21.34. *Let X be a topological space, and let*

$$ 1 \longrightarrow \mathscr{F} \overset{\alpha}{\longrightarrow} \mathscr{G} \overset{\beta}{\longrightarrow} \mathscr{H} \longrightarrow 1 $$

be an exact sequence of sheaves of groups on X.
(1) *Suppose that $\mathscr{F}$ is flasque. Then for every open subset $U \subseteq X$ the sequence*

$$ 1 \to \mathscr{F}(U) \to \mathscr{G}(U) \to \mathscr{H}(U) \to 1 $$

is exact.
(2) *Suppose that $\mathscr{F}$ and $\mathscr{G}$ are flasque. Then $\mathscr{H}$ is flasque.*

Proof. To prove part (1), we may assume that $U = X$. By Proposition 11.14 one has the exact sequence of non-abelian cohomology $1 \to \mathscr{F}(X) \to \mathscr{G}(X) \to \mathscr{H}(X) \to H^1_{\mathrm{Tors}}(X, \mathscr{F})$ and $H^1_{\mathrm{Tors}}(X, \mathscr{F}) = 1$ by Lemma 21.33.

By an easy diagram chase, we obtain part (2) as a consequence of part (1). $\qquad\square$

Proposition 21.35. *Let $(X, \mathcal{O}_X)$ be a ringed space and let $U \subseteq X$ be open. Then flasque $\mathcal{O}_X$-modules are right $\Gamma(U, -)$-acyclic, i.e., for all flasque $\mathcal{O}_X$-modules $\mathcal{I}$ and $i > 0$ we have $H^i(U, \mathcal{I}) = 0$. Moreover, for every bounded below complex $\mathcal{F}$ of $\mathcal{O}_X$-modules there exists a quasi-isomorphism $\mathcal{F} \to A_{\mathcal{F}}$, where $A_{\mathcal{F}}$ is a bounded below complex whose components are flasque $\mathcal{O}_X$-modules.*

One can choose $A_{\mathcal{F}}$ to depend functorially on $\mathcal{F}$, for instance the Godement resolution (Remark 21.31).

Proof. As restrictions of flasque sheaves to open subsets are again flasque, we can assume that $U = X$. We can apply Proposition F.207: Condition (a) there is satisfied by Remark 21.31, Condition (b) is clear, and Condition (c) is satisfied by Corollary 21.34. $\square$

Below, we will use the following technical remark.

Remark 21.36. In some cases, we can enlarge the class of flasque sheaves slightly: Let X be a topological space such that the set $\mathcal{B}$ of open quasi-compact subsets of X is a basis of the topology of X that is stable under finite intersections (in particular X is quasi-compact).

Let us call a sheaf $\mathcal{F}$ in X *quasi-flasque* if $\mathcal{F}(X) \to \mathcal{F}(U)$ is surjective for all $U \in \mathcal{B}$. Then the same proof as in Lemma 21.33, using only finite coverings $(U_i)_i$ with $U_i \in \mathcal{B}$ for all i, shows that for a quasi-flasque sheaf of groups $\mathcal{G}$ every $\mathcal{G}$-torsor is trivial.

Then the same arguments as above show that bounded below complexes of quasi-flasque $\mathcal{O}_X$-modules are right Γ-acyclic.

Corollary 21.37. *Let $f \colon X \to Y$ be a morphism of ringed spaces. Then flasque $\mathcal{O}_X$-modules are right f_*-acyclic.*

Proof. By Proposition F.204 we have to show that $R^i f_* \mathcal{F} = 0$ for all flasque $\mathcal{O}_X$-modules $\mathcal{F}$ and all $i > 0$. This follows by Proposition 21.27 from Proposition 21.35. $\square$

Hence we can calculate cohomology and, more generally (Remark 21.25), higher direct images of bounded below complexes by flasque resolutions:

Upshot 21.38. Let X be a ringed space and let $\mathcal{F}$ be a bounded below complex of $\mathcal{O}_X$-modules. Let $\mathcal{F} \to A_{\mathcal{F}}$ be any quasi-isomorphism where $A_{\mathcal{F}}$ is a bounded below complex with flasque components (these exist, for instance the Godement resolution). Then for all $i \in \mathbb{Z}$ one has

$$H^i(X, \mathcal{F}) = \frac{\operatorname{Ker}\left(\Gamma(X, A_{\mathcal{F}}^i) \to \Gamma(X, A_{\mathcal{F}}^{i+1})\right)}{\operatorname{Im}\left(\Gamma(X, A_{\mathcal{F}}^{i-1}) \to \Gamma(X, A_{\mathcal{F}}^i)\right)}.$$

More generally, if $f \colon X \to Y$ is a morphism of ringed spaces, then for all $i \in \mathbb{Z}$ one has

$$R^i f_* \mathcal{F} = \frac{\operatorname{Ker}\left(f_* A_{\mathcal{F}}^i \to f_* A_{\mathcal{F}}^{i+1}\right)}{\operatorname{Im}\left(f_* A_{\mathcal{F}}^{i-1} \to f_* A_{\mathcal{F}}^i\right)}.$$

Example 21.39. Let k a field, let $X := \mathbb{A}_k^1$ be the affine line over k, and let $U := \mathbb{A}_k^1 \setminus \{0, 1\}$. Denote by $j \colon U \hookrightarrow X$ the inclusion. Let us compute the cohomology group $H^1(\mathbb{A}_k^1, j_! \mathbb{Z}_U)$.

Consider the exact sequence

$$0 \to j_! \mathbb{Z}_U \to \mathbb{Z}_X \to i_{0,*} \mathbb{Z} \oplus i_{1,*} \mathbb{Z}_X \to 0$$

of abelian sheaves on X, where i_x the inclusion $\operatorname{Spec} \kappa(x) \hookrightarrow X$, and the homomorphism $\mathbb{Z}_X \to i_{0,*}\mathbb{Z} \oplus i_{1,*}\mathbb{Z}$ is the direct sum of the natural homomorphisms. The exactness is easily checked on stalks.

The sheaves $\mathbb{Z}_X$ and $i_{0,*}\mathbb{Z} \oplus i_{1,*}\mathbb{Z}$ are flasque, hence this is an acyclic resolution of $j_!\mathbb{Z}_U$ which we can use to compute its cohomology. Passing to global sections of this resolution, we obtain a complex

$$0 \longrightarrow \mathbb{Z} \xrightarrow{m \mapsto (m,m)} \mathbb{Z}^2 \longrightarrow 0$$

with $\mathbb{Z}$ sitting in degree 0. This shows that $H^1(X, j_!\mathbb{Z}_U) \cong \mathbb{Z}$ and $H^n(X, j_!\mathbb{Z}_U) = 0$ for all $n \neq 1$.

(21.8) Applications of flasque resolutions.

Our main applications of the results of the previous section will be
(1) that the H^1 defined via derived functors and the H^1 defined via torsors in Section (11.5) coincide (Proposition 21.40),
(2) an isomorphism $R(g_* \circ f_*)\mathscr{F} \xrightarrow{\sim} (Rg_* \circ Rf_*)\mathscr{F}$ for bounded below complexes $\mathscr{F}$ (Proposition 21.41) and the corollary that the cohomology of a bounded below complex of $\mathscr{O}_X$-modules is the same as the cohomology of the underlying complex of abelian sheaves (Corollary 21.43),
(3) the Leray spectral sequence (Corollary 21.45) as an application of (2),
(4) and the Mayer-Vietoris sequence (Proposition 21.47) in the next section.
Later (Proposition 21.115) we will prove (2) also for arbitrary complexes and the proof there will be independent of the result obtained here.

Proposition 21.40. *Let X be a topological space and let $\mathscr{F}$ be a sheaf of abelian groups which we may consider a $\mathbb{Z}_X$-module. Then there is a functorial isomorphism $H^1(X, \mathscr{F}) \cong H^1_{\mathrm{Tors}}(X, \mathscr{F})$.*

Hence from now on we will not distinguish between $H^1(X, \mathscr{F})$ and $H^1_{\mathrm{Tors}}(X, \mathscr{F})$.

Proof. By Remark 21.31 there exists a functorial exact sequence of sheaves of abelian groups $0 \to \mathscr{F} \to \mathscr{A}^0 \to \mathscr{C} \to 0$ with $\mathscr{A}^0$ flasque. Then $H^1(X, \mathscr{A}^0) = 0$ by Proposition 21.35 and $H^1_{\mathrm{Tors}}(X, \mathscr{F}^0) = 0$ by Lemma 21.33. Hence the long exact cohomology sequence for $H^i(X, -)$ and the long exact sequence for $H^1_{\mathrm{Tors}}(X, -)$ imply that

$$H^1(X, \mathscr{F}) = H^1_{\mathrm{Tors}}(X, \mathscr{A}) = \operatorname{Coker}(\mathscr{A}^0(X) \to \mathscr{C}(X)). \qquad \square$$

Proposition 21.41. *Let $f: X \to Y$ and $g: Y \to Z$ be morphisms of ringed spaces. Then the canonical morphism $R(g_* \circ f_*) \to R(g_*) \circ R(f_*)$ of derived functors $D^+(X) \to D^+(Z)$ is an isomorphism.*

Proof. We apply Proposition F.211. As every module over a sheaf of rings can be embedded into a flasque module, it suffices to show that if $\mathscr{I}$ is a flasque $\mathscr{O}_X$-module, then $f_*\mathscr{I}$ is right g_*-acyclic. But $f_*\mathscr{I}$ is clearly a flasque $\mathscr{O}_Y$-module and hence right g_*-acyclic (Corollary 21.37). $\qquad \square$

Let $f: X \to Y$ be a morphism of ringed spaces, and set $A := \Gamma(X, \mathscr{O}_X)$ and $B := \Gamma(Y, \mathscr{O}_Y)$. The morphism f yields a ring homomorphism $\varphi: A \to B$ which allows to consider every B-module M as an A-module $\varphi_*(M)$. This yields an exact functor

$$\varphi_*: (B\text{-Mod}) \longrightarrow (A\text{-Mod}).$$

By the definition of the direct image we have

$$\Gamma(Y, -) \circ f_* = \varphi_* \circ \Gamma(X, -).$$

Let Z be the ringed space consisting of one point $*$ with $\mathscr{O}_Z(*) = A$ and $g\colon Y \to Z$ the canonical morphism. Then we can identify $g_* = \Gamma(Y, -)$. Hence we deduce from Proposition 21.41:

Corollary 21.42. *Let $f\colon X \to Y$ be a morphism of ringed spaces. There is an isomorphism of functors $D^+(X) \to D^+(\Gamma(Y, \mathscr{O}_Y))$*

$$\varphi_* \circ R\Gamma(X, -) \cong R\Gamma(Y, -) \circ Rf_*.$$

We use that $R(\varphi_* \circ \Gamma(X, -)) = \varphi_* \circ R\Gamma(X, -)$ because φ_* is exact (Remark F.175).

Applying Corollary 21.42 to the unique morphism of ringed spaces $(X, \mathscr{O}_X) \to (X, \mathbb{Z}_X)$ whose underlying topological map is the identity we deduce:

Corollary 21.43. *Let X be a ringed space and let $\mathscr{F}$ be a bounded below complex of $\mathscr{O}_X$-modules. Denote by $\mathscr{F}^{\mathrm{ab}}$ the underlying complex of abelian sheaves. Then the underlying abelian group of the $\Gamma(X, \mathscr{O}_X)$-module $H^i(X, \mathscr{F})$ is $H^i(X, \mathscr{F}^{\mathrm{ab}})$ for all $i \in \mathbb{Z}$.*

Below (Section (21.23)) we will prove by totally different techniques the assertions in Proposition 21.41 (and hence Corollaries 21.42 and 21.43) also for unbounded complexes. There is one important special case where we can do this right now.

Remark 21.44. Let $f\colon X \to Y$ and $g\colon Y \to Z$ be continuous maps of topological spaces. Then the direct image functor $f_*\colon (\mathrm{Ab}(X)) \to (\mathrm{Ab}(Y))$ of abelian sheaves has an exact left adjoint, namely f^{-1}. Hence f_* preserves K-injective complexes (Corollary F.183) and therefore the canonical morphism of functors $D(\mathrm{Ab}(X)) \to D(\mathrm{Ab}(Z))$

$$R(g_* \circ f_*) \longrightarrow Rg_* \circ Rf_*$$

is an isomorphism (Proposition F.176). Applying this to the case that Z consists of a single point we obtain an isomorphism of functors $D(\mathrm{Ab}(X)) \to D(\mathbb{Z})$

$$(21.8.1) \qquad\qquad R\Gamma(Y, -) \circ Rf_* \xrightarrow{\sim} R\Gamma(X, -).$$

If $f = i$ is the inclusion of a closed subspace, then i_* is exact and we obtain by Remark F.177 for every complex $\mathscr{F}$ of abelian sheaves on X a functorial isomorphism

$$(21.8.2) \qquad\qquad R(g_{|X})_* \mathscr{F} \xrightarrow{\sim} Rg_*(i_* \mathscr{F}).$$

We can also apply this in the case where g is the map to a point such that we can identify g_* and $\Gamma(Y, -)$. Then (21.8.2) becomes an isomorphism in $D(\mathbb{Z})$

$$(21.8.3) \qquad\qquad R\Gamma(X, \mathscr{F}) \xrightarrow{\sim} R\Gamma(Y, i_* \mathscr{F})$$

which is functorial for every complex $\mathscr{F}$ of abelian sheaves on X. Passing to cohomology we obtain for all $p \in \mathbb{Z}$ isomorphisms of abelian groups

$$(21.8.4) \qquad\qquad H^p(X, \mathscr{F}) \xrightarrow{\sim} H^p(Y, i_* \mathscr{F}).$$

Corollary 21.42 and Proposition 21.41 also imply the existence of a Grothendieck spectral sequence (Proposition F.212):

Corollary and Definition 21.45. *Let $f\colon X \to Y$ be a morphism of ringed spaces and let $\mathscr{F}$ be a bounded below complex of $\mathscr{O}_X$-modules.*

(1) *There is a convergent spectral sequence of $\Gamma(Y, \mathscr{O}_Y)$-modules, called the* Leray spectral sequence *for the morphism f,*

$$(21.8.5) \qquad E_2^{p,q} = H^p(Y, R^q f_* \mathscr{F}) \Longrightarrow H^{p+q}(X, \mathscr{F})$$

which is functorial in $\mathscr{F}$.

(2) *Let $g\colon Y \to Z$ be a morphism of ringed spaces. Then there is a convergent spectral sequence of $\mathscr{O}_Z$-modules*

$$(21.8.6) \qquad E_2^{p,q} = R^p g_*(R^q f_* \mathscr{F}) \Longrightarrow R^{p+q}(g \circ f)_* \mathscr{F}$$

which is functorial in $\mathscr{F}$.

If $N \in \mathbb{Z}$ is an integer such that $H^n(\mathscr{F}) = 0$ for all $n < N$, then both spectral sequences are concentrated in degrees $q \geq N$ and $p \geq 0$.

If $f = \mathrm{id}_X$, then $Rf = \mathrm{id}_{D(X)}$ and hence $R^q(\mathrm{id}_X)_* \mathscr{F} = H^q(\mathscr{F})$. Hence we obtain the following Corollary.

Corollary 21.46. *Let $(X, \mathscr{O}_X)$ be a ringed space and let $\mathscr{F}$ be a bounded below complex of $\mathscr{O}_X$-modules. Then there is a convergent spectral sequence of $\Gamma(X, \mathscr{O}_X)$-modules, sometimes called* hypercohomology spectral sequence,

$$(21.8.7) \qquad E_2^{p,q} = H^p(X, H^q(\mathscr{F})) \Longrightarrow H^{p+q}(X, \mathscr{F})$$

which is functorial in $\mathscr{F}$. If $f\colon X \to Y$ is a morphism of ringed spaces, then there is a convergent spectral sequence of $\mathscr{O}_Y$-modules

$$(21.8.8) \qquad E_2^{p,q} = R^p f_* H^q(\mathscr{F}) \Longrightarrow R^{p+q} f_* \mathscr{F}.$$

(21.9) Mayer-Vietoris sequences and gluing of complexes.

As an application of Lemma 21.32, we obtain the following useful exact sequence.

Proposition 21.47. *(Mayer-Vietoris sequence) Let X be a ringed space and suppose that $X = U \cup V$ for two open subsets $U, V \subseteq X$. Let $\mathscr{F}$ be a complex of $\mathscr{O}_X$-modules. Then there exists a long exact cohomology sequence*

$$(21.9.1) \qquad \begin{aligned} \cdots &\longrightarrow H^n(X, \mathscr{F}) \longrightarrow H^n(U, \mathscr{F}) \oplus H^n(V, \mathscr{F}) \longrightarrow H^n(U \cap V, \mathscr{F}) \\ &\longrightarrow H^{n+1}(X, \mathscr{F}) \longrightarrow \cdots, \end{aligned}$$

which is functorial in $\mathscr{F}$.

Proof. By Proposition 21.7 there exists a quasi-isomorphism $\mathscr{F} \to \mathscr{I}$, where $\mathscr{I}$ is a K-injective complex whose components are all injective and hence flasque $\mathscr{O}_X$-modules. Therefore the sequence of complexes

$$(*) \quad 0 \to \mathscr{I}(X) \xrightarrow{\; s \mapsto (s_{|U}, s_{|V}) \;} \mathscr{I}(U) \oplus \mathscr{I}(V) \xrightarrow{\; (s,t) \mapsto s_{|U \cap V} - t_{|U \cap V} \;} \mathscr{I}(U \cap V) \to 0$$

is exact. By Lemma 21.18 we have $H^n(W, \mathscr{F}) = H^n(\mathscr{I}(W))$ for all $W \subseteq X$ open. Hence taking cohomology of $(*)$ gives the result by the Snake Lemma F.48. $\qquad\square$

If $\mathscr{F}$ is a bounded below complex, one could have worked in the proof with $\mathscr{I}$ the Godement resolution of $\mathscr{F}$ (and hence avoid the usage of Theorem 21.7). In this case the Mayer-Vietoris sequence is also a special case of a hypercohomology spectral sequence attached to the Čech double complex (see Exercise 21.10).

Proposition 21.48. (Mayer-Vietoris triangle) *Let X be a ringed space, and let $U, V \subseteq X$ be open subspaces with $U \cup V = X$. Let j_U, j_V, $j_{U \cap V}$ denote the inclusions of these open subspaces into X.*

(1) *For every $\mathscr{O}_X$-module $\mathscr{F}$, there is a short exact sequence*

$$0 \to \mathscr{F} \to j_{U,*}(\mathscr{F}_{|U}) \oplus j_{V,*}(\mathscr{F}_{|V}) \to j_{U \cap V,*}(\mathscr{F}_{|U \cap V}) \to 0,$$

of $\mathscr{O}_X$-modules, and these sequences are functorial in $\mathscr{F}$.

(2) *There is an exact triangle*

(21.9.2) $$F \to Rj_{U,*}j_U^* F \oplus Rj_{V,*}j_V^* F \to Rj_{U \cap V,*}j_{U \cap V}^* F \to$$

for every $F \in D(X)$.

Proof. For (1), note that the sheaf property of $\mathscr{F}$ immediately implies the exactness of the sequence

$$0 \to \mathscr{F} \to j_{U,*}(\mathscr{F}_{|U}) \oplus j_{V,*}(\mathscr{F}_{|V}) \to j_{U \cap V,*}(\mathscr{F}_{|U \cap V}),$$

in fact that sequence is even exact after passing to sections on any open subscheme of X. It therefore only remains to prove the surjectivity of the final arrow.

This follows from Proposition 21.47 and the fact that for every element $\eta \in H^1(X, \mathscr{F})$, there exists an open cover $X = \bigcup_i U_i$ such that all restrictions $\eta_{|U_i}$ vanish (Corollary 21.23).

Using part (1), part (2) follows by choosing a K-injective resolution of F and computing the derived pushforwards using the restrictions of this resolution to U, V and $U \cap V$ respectively. In fact, by Lemma 21.12 these restrictions are K-injective resolutions of the corresponding restrictions of $\mathscr{F}$. □

Reverse engineering from (21.9.2) we obtain the following result about gluing of complexes.

Proposition 21.49. *We keep the notation of Proposition 21.48. Let $\mathscr{F}_U \in D(U)$ and $\mathscr{F}_V \in D(V)$ and let $\alpha \colon \mathscr{F}_{U|U \cap V} \xrightarrow{\sim} \mathscr{F}_{V|U \cap V}$ be an isomorphism. Then there exists an object $\mathscr{F} \in D(X)$ and isomorphisms $\mathscr{F}_{|U} \cong \mathscr{F}_U$ and $\mathscr{F}_{|V} \cong \mathscr{F}_V$.*

Proof. Set $\mathscr{F}_{U \cap V} := (\mathscr{F}_U)_{U \cap V} \cong (\mathscr{F}_V)_{|U \cap V}$. We have natural maps

$$Rj_{U*}\mathscr{F}_U \longrightarrow Rj_{U \cap V*}\mathscr{F}_{U \cap V},$$
$$Rj_{V*}\mathscr{F}_V \longrightarrow Rj_{U \cap V*}\mathscr{F}_{U \cap V},$$

and taking the difference of these maps, a map

$$Rj_{U*}\mathscr{F}_U \oplus Rj_{V*}\mathscr{F}_V \longrightarrow Rj_{U \cap V,*}\mathscr{F}_{U \cap V}$$

which we complete to an exact triangle

$$\mathscr{F} \longrightarrow Rj_{U*}\mathscr{F}_U \oplus Rj_{V*}\mathscr{F}_V \longrightarrow Rj_{U \cap V,*}\mathscr{F}_{U \cap V} \xrightarrow{+}$$

in $D(X)$.

We have $(Rj_{U,*}\mathscr{F}_U)_{|U} \cong \mathscr{F}_U$ and the restriction of the map $Rj_{V,*}\mathscr{F}_V \to Rj_{U \cap V*}(\mathscr{F}_{U \cap V})$ to U is an isomorphism. Hence the first map of the triangle induces $\mathscr{F}_{|U} \cong \mathscr{F}_U$. Likewise, we obtain $\mathscr{F}_{|V} \cong \mathscr{F}_V$. □

(21.10) Cohomology groups — a first example.

Let k be an algebraically closed field, let k be an integral proper smooth curve over k, and let $x \in C$ be a closed point. For any $d \in \mathbb{Z}$ we can consider the divisor $d[x]$ on C and the associated line bundle $\mathscr{O}(d[x])$. We want to investigate the cohomology groups $H^i(C, \mathscr{O}(d[x]))$.

Let $\mathscr{K}$ denote the constant sheaf associated with the field $K(C)$ of rational functions on C. Since this is a constant sheaf, it is flasque. We can identify $\mathscr{O}(d[x])$ as the subsheaf of $\mathscr{K}$ consisting of those sections which have a pole of order $\leq d$ at x and have no pole elsewhere. We obtain an embedding $\mathscr{O}(d[x]) \hookrightarrow \mathscr{K}$.

Let us analyze the quotient $\mathscr{K}/\mathscr{O}(d[x])$. Its stalk at a point $z \in C$ can be identified with the quotient $\mathscr{K}_z/\mathscr{O}(d[x])_z$. If z is the generic point of C, then this quotient vanishes. For a closed point $z \neq x$ the stalk $\mathscr{O}(d[x])_z$ can be identified, as a subring of $\mathscr{K}_z = K(C)$, with $\mathscr{O}_{C,z}$. For $z = x$, the stalk $\mathscr{O}(d[x])_x$ consists of all functions in $K(C)$ which have a pole of order $\leq d$ at x.

Since every rational function on C is regular at almost all closed points, we have a homomorphism from $\mathscr{K}/\mathscr{O}(d[x])$ into the direct sum

$$\bigoplus_{z \in C(k)} \iota_{z,*}\mathscr{K}_z/\mathscr{O}(d[x])_z,$$

the direct sum over all closed points of C, where we consider each stalk as a sheaf on the corresponding one-point space and denote by ι_z the inclusion $\{z\} \hookrightarrow C$.

In this way we obtain an exact sequence

$$0 \to \mathscr{O}(d[x]) \to \mathscr{K} \to \bigoplus_{z \in C(k)} \iota_{z,*}\mathscr{K}_z/\mathscr{O}(d[x])_z \to 0.$$

The exactness can be checked on stalks, where it is obvious.

Since the sheaves $\mathscr{K}$ and $\bigoplus_{z \in C(k)} \iota_{z,*}\mathscr{K}_z/\mathscr{O}(d[x])_z$ are flasque (use Example 21.30 and Lemma 21.52), we can use it to compute the cohomology groups $H^n(C, \mathscr{O}(d[x]))$. We see immediately that $H^n(C, \mathscr{O}(d[x])) = 0$ for $n > 2$ (compare Grothendieck's vanishing theorem, Theorem 21.57). Furthermore, we have an exact sequence

$$0 \to H^0(C, \mathscr{O}(d[x])) \to K(C) \to \bigoplus_{z \in C(k)} \mathscr{K}_z/\mathscr{O}(d[x])_z \to H^1(C, \mathscr{O}(d[x])) \to 0.$$

The global sections $H^0(C, \mathscr{O}(d[x]))$ are those rational functions on C which are regular outside x and have a pole of order $\leq d$ at x; of course, this is part of the definition of the line bundle $\mathscr{O}(d[x])$.

So the interesting cohomology group in this case is the group $H^1(C, \mathscr{O}(d[x]))$. The question whether $H^1(C, \mathscr{O}(d[x])) = 0$, or equivalently whether the map in the middle is surjective, is called the *Cousin problem* for C and $\mathscr{O}(d[x])$. In the case $C = \mathbb{P}^1(\mathbb{C})$, $d = 0$, in the context of complex geometry, one can interpret the middle map as attaching to a meromorphic function on $\mathbb{P}^1(\mathbb{C})$ its principal parts in the (finitely many) poles of f. Then the positive solution of the Cousin problem is the Theorem of Mittag-Leffler.

For an arbitrary algebraically closed field k and $C = \mathbb{P}^1_k$ (now again considered as a scheme), we can use the above sequence to compute all the cohomology groups $H^1(\mathbb{P}^1, \mathscr{O}(d))$. See also Theorem 22.22 for a generalization to $\mathbb{P}^n_k$.

Proposition 21.50. *With notation as above, for $C = \mathbb{P}_k^1$ we obtain*

$$\dim_k H^1(\mathbb{P}^1, \mathscr{O}(d)) = \begin{cases} 0 & \text{if } d \geq 0, \\ -d-1 & \text{if } d < 0. \end{cases}$$

Proof. In this case, $K(C) = k(T)$, the field of rational functions in one variable over k, and we need to compute the cokernel of the natural map

$$k(T) \longrightarrow \bigoplus_{a \in k} k(T)/k[T]_{(T-a)} \oplus k(T)/T^d k[T^{-1}]_{(T^{-1})}.$$

The right hand side is generated by the powers T^i, $i > d$, over k. Considering the images of the elements $(T-a)^{-i}$ and T^i for $a \in k$, $i \geq 0$, we see that this map is surjective for $d = -1$, and a fortiori for $d > -1$. Similarly, for $d < -1$, the elements $T^i \in k(T)/T^d k[T^{-1}]_{(T^{-1})}$ for $i = -1, \ldots, d+1$ induce a k-basis of the cokernel. $\qquad\square$

In particular, we see that $\mathscr{O}(-1)$ is the only line bundle on $\mathbb{P}^1$ which has vanishing H^0 and H^1.

Below in Corollary 25.129 we will prove Serre duality for a general proper smooth curve as an application of general Grothendieck duality. It implies that $\dim_k H^1(C, \mathscr{O}(d[x])) = \dim_k H^0(C, \mathscr{O}(-d[x] + K_C))$, where K_C is a canonical divisor (any divisor on C such that $\mathscr{O}(K_C) \cong \Omega^1_{C/k}$). If $C = \mathbb{P}_k^1$, then $\mathscr{O}(K_C) = \mathscr{O}(-2)$ and hence $\dim_k H^1(\mathbb{P}_k^1, \mathscr{O}(d)) = \dim_k H^0(\mathbb{P}_k^1, \mathscr{O}(-d-2))$ which again shows Proposition 21.50 since we already computed $\dim H^0(\mathbb{P}_k^1, \mathscr{O}(d)) = d+1$ in (11.14.5).

Coming back to the general case, we find the following criterion when the curve C is isomorphic to the projective line. Compare Section (26.16).

Proposition 21.51. *In the above situation, we have $H^1(C, \mathscr{O}_C) = 0$ if and only if $C \cong \mathbb{P}_k^1$.*

Proof. We have already seen that $H^1(\mathbb{P}_k^1, \mathscr{O}_{\mathbb{P}_k^1}) = 0$. Conversely, if $H^1(C, \mathscr{O}_C) = 0$ then for any two points $x_0 \neq x_1 \in C(k)$ we find $f \in K(C)$ such that $\operatorname{div}(f) = [x_1] - [x_0]$. We may consider f as a finite surjective morphism $C \to \mathbb{P}_k^1$, see Corollary 15.22.

Let us show that this morphism is an isomorphism. By Theorem 15.21 it is enough to show that the inclusion $K(\mathbb{P}_k^1) \subseteq K(C)$ of function fields given by f is an equality. It follows from the definition of f that $f^*[0] = [x_1]$, hence $[K(C) : K(\mathbb{P}_k^1)] = \deg(f) = 1$ by Proposition 15.30. $\qquad\square$

We will come back to the cohomology of curves in much greater detail in Chapter 26.

(21.11) Compatibility with colimits.

In this section we will give criteria when cohomology commutes with filtered colimits. This is not true in general, not even for global sections (see Exercise 7.1 considering a direct sum as a filtered colimit of finite direct sums). But it holds under the following hypothesis for a topological space X:

(COH) The space X is quasi-compact and has a basis $\mathcal{B}$ consisting of quasi-compact open subsets that is stable under finite intersections[2].

---

[2] Formally, the condition "X quasi-compact" is superfluous because we demanded that the empty intersection is part of the basis.

Such spaces are called *coherent* in [SGA4]^O Exp. VI, 2.2. We will not use this terminology. If X satisfies (COH), then every open quasi-compact subspace of X satisfies (COH).

Examples for spaces satisfying (COH) are underlying topological spaces of qcqs schemes (in particular, of affine schemes, and of noetherian schemes).

Lemma 21.52. *Let X be a topological space satisfying (COH). Let $\mathcal{I}$ be a filtered category and let $\mathscr{F}\colon \mathcal{I} \to (\mathrm{Sh}(X))$, $i \mapsto \mathscr{F}_i$, be a diagram of sheaves on X. Then the canonical map*

$$\varphi\colon \operatorname*{colim}_i \mathscr{F}_i(X) \longrightarrow (\operatorname*{colim}_i \mathscr{F}_i)(X)$$

is bijective.

Proof. (i). We first show that for every quasi-compact topological space the map φ is injective. Let $s \in \mathscr{F}_i(X)$ and $s' \in \mathscr{F}_{i'}(X')$ be sections that have the same image in the right hand side. As $\mathcal{I}$ is filtered, we find an open covering $(U_\alpha)_{\alpha \in A}$ of X, for all α an i_α in $\mathcal{I}$ and morphisms $\kappa_\alpha\colon i \to i_\alpha$ and $\kappa'_\alpha\colon i' \to i_\alpha$ such that $\mathscr{F}_{\kappa_\alpha}(s) = \mathscr{F}_{\kappa'_\alpha}(s')$. As X is quasi-compact, we can choose A to be finite. As $\mathcal{I}$ is filtered, we can find i_∞ in $\mathcal{I}$ and morphisms $i_\alpha \to i_\infty$ for all α such that the compositions

$$\nu\colon i \xrightarrow{\kappa_\alpha} i_\alpha \longrightarrow i_\infty \qquad \text{and} \qquad \nu'\colon i' \xrightarrow{\kappa'_\alpha} i_\alpha \longrightarrow i_\infty$$

are independent of α. Then $\mathscr{F}_\nu(s)_{|U_\alpha} = \mathscr{F}_{\nu'}(s')_{|U_\alpha}$ for all α and hence $\mathscr{F}_\nu(s) = \mathscr{F}_{\nu'}(s')$ which shows that the images of s and s' in $\operatorname{colim}_i \mathscr{F}_i(X)$ are equal.

(ii). It remains to show that φ is surjective if the Condition (COH) is satisfied. Let s be an element of the right hand side. By assumption there exists a finite covering $(U_\alpha)_\alpha$ of X such that $U_\alpha \cap U_{\alpha'}$ is quasi-compact for all α, α' and for each α an i_α in $\mathcal{I}$ and $s_\alpha \in \mathscr{F}_{i_\alpha}(U_\alpha)$ such that $s_{|U_\alpha}$ is the image of s_α. As $U_\alpha \cap U_{\alpha'}$ is quasi-compact, we can apply (i) and find $i_{\alpha\alpha'}$ in $\mathcal{I}$ and morphisms $\kappa_{\alpha\alpha'}\colon i_\alpha \to i_{\alpha\alpha'}$ and $\kappa'_{\alpha\alpha'}\colon i_{\alpha'} \to i_{\alpha\alpha'}$ such that the restrictions of $\mathscr{F}_{\kappa_{\alpha\alpha'}}(s_\alpha)$ and $\mathscr{F}_{\kappa'_{\alpha\alpha'}}(s_{\alpha'})$ to $U_\alpha \cap U_{\alpha'}$ are equal. As $\mathcal{I}$ is filtered, we can choose i in $\mathcal{I}$ and morphisms $i_{\alpha\alpha'} \to i$ such that the images of s_α in $\mathscr{F}_i(U_\alpha)$ glue to a section in $\mathscr{F}_i(X)$. This yields a preimage of s. $\square$

Corollary 21.53. *Let $(X, \mathscr{O}_X)$ be a ringed space. Suppose that X has a basis of open subspaces satisfying (COH) (e.g., if X is the underlying topological space of a scheme). Then the Grothendieck abelian category $(\mathscr{O}_X\text{-Mod})$ is locally finitely generated (Definition F.66).*

Proof. By Remark 21.8 it suffices to show that the $\mathscr{O}_X$-modules of the form $\mathscr{O}_{U \subseteq X}$ for $U \subseteq X$ open and satisfying (COH) are finitely generated. As $\mathrm{Hom}_{\mathscr{O}_X}(\mathscr{O}_{U \subseteq X}, \mathscr{F}) = \mathscr{F}(U)$ for every $\mathscr{O}_X$-module $\mathscr{F}$, this follows from Lemma 21.52. $\square$

Let X be a ringed space such that the underlying topological space of X satisfies (COH) and let $\mathcal{I}$ be a filtered category. Suppose that $\mathscr{F}$ is an $\mathcal{I}$-diagram of $\mathscr{O}_X$-modules. Then the underlying sheaf of the colimit of $\mathscr{F}$ in the category of $\mathscr{O}_X$-modules is the colimit of the underlying diagram of sheaves. Hence in this situation we have the bijectivity of φ also for filtered colimits of $\mathscr{O}_X$-modules. Hence we obtain an isomorphism of functors

$$(21.11.1) \qquad \operatorname*{colim}_{\mathcal{I}} \circ \Gamma(X, -) \cong \Gamma(X, -) \circ \operatorname*{colim}_{\mathcal{I}}\colon (\mathscr{O}_X\text{-Mod})^{\mathcal{I}} \longrightarrow (\Gamma(X, \mathscr{O}_X)\text{-Mod}),$$

where we extend $\Gamma(X, -)$ to a functor $(\mathscr{O}_X\text{-Mod})^{\mathcal{I}} \to (\Gamma(X, \mathscr{O}_X)\text{-Mod})^{\mathcal{I}}$ by composition and where $\operatorname{colim}_{\mathcal{I}}\colon (\mathscr{O}_X\text{-Mod})^{\mathcal{I}} \to (\mathscr{O}_X\text{-Mod})$ and $\operatorname{colim}_{\mathcal{I}}\colon (\Gamma(X, \mathscr{O}_X)\text{-Mod})^{\mathcal{I}} \to (\Gamma(X, \mathscr{O}_X)\text{-Mod})$ are the exact functors of functors of Grothendieck categories given by forming filtered colimits.

Right derivation of (21.11.1) now yields:

Proposition 21.54. *Let X be a ringed space such that the underlying topological space of X satisfies* (COH). *Let $\mathcal{I}$ be a filtered category and let $\mathscr{F} \colon \mathcal{I} \to C(\mathscr{O}_X\text{-Mod})$, $i \mapsto \mathscr{F}_i$ be an $\mathcal{I}$-diagram. Suppose that there exists an $n \in \mathbb{Z}$ (independent of i) such that $H^p(\mathscr{F}_i) = 0$ for all $p < n$ and for all $i \in \mathcal{I}$. Then there is an isomorphism in $D(\Gamma(X, \mathscr{O}_X))$*

$$\operatorname*{colim}_i R\Gamma(X, \mathscr{F}_i) \xrightarrow{\sim} R\Gamma(X, \operatorname*{colim}_i \mathscr{F}_i).$$

which is functorial in $\mathscr{F}$.

In particular, we obtain for $p \in \mathbb{Z}$ an isomorphism of $\Gamma(X, \mathscr{O}_X)$-modules

$$\operatorname*{colim}_i H^p(X, \mathscr{F}_i) \xrightarrow{\sim} H^p(X, \operatorname*{colim}_i \mathscr{F}_i),$$

which is functorial in $\mathscr{F}$.

The proposition applies if X is a qcqs scheme (for instance a noetherian scheme).

Proof. One has

$$\operatorname*{colim}_{\mathcal{I}} \circ R\Gamma(X, -) \cong R(\operatorname*{colim}_{\mathcal{I}} \circ \Gamma(X, -)) \cong R(\Gamma(X, -) \circ \operatorname*{colim}_{\mathcal{I}}),$$

where the first isomorphism holds by Remark F.175 because filtered colimits are exact and the second isomorphism is obtained by right derivation of (21.11.1). It remains to show that

$$R(\Gamma(X, -) \circ \operatorname*{colim}_{\mathcal{I}}) = R\Gamma(X, -) \circ \operatorname*{colim}_{\mathcal{I}}.$$

Hence we have to show that for every complex $\mathscr{F}$ in $(\mathscr{O}_X\text{-Mod})^{\mathcal{I}}$ satisfying the hypothesis above there exists a quasi-isomorphism $\mathscr{F} \to \mathscr{I}$ such that $\operatorname{colim}_i \mathscr{I}$ is right $\Gamma(X, -)$-acyclic (Remark F.177).

(I). We first remark that filtered colimits of flasque sheaves $\mathscr{G}_i$ are quasi-flasque[3] (Remark 21.36): If $U \subseteq X$ is open quasi-compact, then

$$(\operatorname*{colim}_i \mathscr{G}_i)(X) = \operatorname*{colim}_i \mathscr{G}_i(X) \longrightarrow \operatorname*{colim}_i \mathscr{G}_i(U) = (\operatorname*{colim}_i \mathscr{G}_i)(U)$$

is surjective, where the equalities hold by Lemma 21.52.

(II). As the Godement resolution is functorial, we obtain a quasi-isomorphism $\mathscr{F} \to \mathscr{I}$, with $\mathscr{I}$ in $(\mathscr{O}_X\text{-Mod})^{\mathcal{I}}$, where $\mathscr{I}_i$ is the Godement resolution of $\mathscr{F}_i$. In particular, for all $i \in \mathcal{I}$ the $\mathscr{O}_X$-modules $\mathscr{I}_i^m$ are flasque for all m and $\mathscr{I}_i^m = 0$ for $m < n$. By Step (I), $\operatorname{colim}_i \mathscr{I}$ is a bounded below complex with quasi-flasque components hence it is right $\Gamma(X, -)$-acyclic by Remark 21.36. □

Corollary 21.55. *Let $f \colon X \to Y$ be a morphism of ringed spaces such that $f^{-1}(V)$ satisfies Condition* (COH) *for all V in some basis $\mathcal{B}$ of the topology of Y. Let $\mathcal{I}$ and $\mathscr{F}$ be as in Proposition 21.54. Then one has for all $p \in \mathbb{Z}$ an isomorphism of $\mathscr{O}_Y$-modules*

$$\operatorname*{colim}_i R^p f_* \mathscr{F}_i \xrightarrow{\sim} R^p f_* \operatorname*{colim}_i \mathscr{F}_i,$$

which is functorial in $\mathscr{F}$.

The corollary applies if f is a qcqs morphism of schemes.

---

[3] If X is noetherian, one can simply work with flasque sheaves because on a noetherian space all open subspaces are quasi-compact.

Proof. By Proposition 21.27 both sides are the sheaves associated to the presheaf on $\mathcal{B}$

$$V \mapsto \operatorname*{colim}_i H^p(f^{-1}(V), \mathscr{F}_i) \overset{21.54}{=} H^p(f^{-1}(V), \operatorname*{colim}_i \mathscr{F}_i). \qquad \square$$

In fact, the same argument as in the proof of Proposition 21.54, using that quasi-flasque $\mathscr{O}_X$-modules are also right f_*-acyclic by Proposition 21.27, shows that there is an isomorphism in $D(Y)$

$$(21.11.2) \qquad\qquad \operatorname*{colim}_i(Rf_*\mathscr{F}_i) \overset{\sim}{\longrightarrow} Rf_*(\operatorname*{colim}_i \mathscr{F}_i).$$

Finally, as direct sums are filtered colimits of finite direct sums, we deduce:

Corollary 21.56. *Let X be a ringed space and let $(\mathscr{F}_i)_i$ be a family of complexes such that there exists an $n \in \mathbb{Z}$ with $H^m(\mathscr{F}_i) = 0$ for all $m < n$ and all i.*
(1) *Suppose that X satisfies Condition* (COH). *Then for all $p \in \mathbb{Z}$ the functorial homomorphism of $\Gamma(X, \mathscr{O}_X)$-modules*

$$\bigoplus_i H^p(X, \mathscr{F}_i) \longrightarrow H^p(X, \bigoplus_i \mathscr{F}_i)$$

is an isomorphism.
(2) *Let $f\colon Y \to Y$ be a morphism of ringed spaces such that $f^{-1}(V)$ satisfies Condition* (COH) *for all V in some basis $\mathcal{B}$ of the topology of Y. Then for all $p \in \mathbb{Z}$ the functorial homomorphism of $\mathscr{O}_Y$-modules*

$$\bigoplus_i (R^p f_* \mathscr{F}_i) \longrightarrow R^p f_*(\bigoplus_i \mathscr{F}_i)$$

is an isomorphism.

Below (Lemma 22.82), we will prove a similar result for qcqs schemes and families of unbounded complexes of *quasi-coherent* modules.

(21.12) The Grothendieck-Scheiderer vanishing theorem.

Recall that the underlying topological space of a qcqs scheme has the following properties. It is quasi-compact, it has a basis of quasi-compact open subsets stable under finite intersections, and every closed irreducible subspace has a unique generic point. Such spaces are called *spectral*. It can be shown ([Hoc]$^\circ$) that every spectral space is homeomorphic to the underlying topological space of $\operatorname{Spec} A$ for some ring A. Note that the underlying topological space of every qcqs (and in particular every noetherian) scheme is spectral.

Theorem 21.57. (Grothendieck-Scheiderer) *Let X be a spectral topological space (for instance a noetherian space) of Krull dimension d. Then for every abelian sheaf $\mathscr{F}$ on X and for all $n > d$, we have $H^n(X, \mathscr{F}) = 0$.*

We explain the proof only for noetherian spaces. For the general case see [Sche]$^\circ$.

Proof. The proof consists of a series of reduction steps, which will eventually show that the following lemma implies the desired result.

Lemma 21.58. *Suppose that X is an irreducible spectral space of dimension d. For every inclusion $j \colon U \hookrightarrow X$ of an open subset and every $n > d$ we have*

$$H^n(X, j_!(\mathbb{Z}_U)) = 0.$$

Proof. We may assume that $U \neq \emptyset$. We argue by induction on d. If $d = 0$, then X consists of a point and the assertion is trivial. Let Z denote the complement of U in X, and let $i \colon Z \to X$ denote the inclusion. We have a short sequence of abelian sheaves on X

$$0 \to j_!(\mathbb{Z}_U) \to \mathbb{Z}_X \to i_*(\mathbb{Z}_Z) \to 0$$

which is seen to be exact by looking at stalks.

Since Z is a proper closed subset of the irreducible space X, we have $\dim Z < \dim X$. Using our induction hypothesis, we obtain that

$$H^n(X, i_*\mathbb{Z}_Z) \cong H^n(Z, \mathbb{Z}_Z) = 0, \qquad \text{for all } n \geq d.$$

Here the first isomorphism is (21.8.4). Furthermore, every constant sheaf on an irreducible space is flasque (Example 21.30), hence $H^n(X, \mathbb{Z}_X) = 0$ for all $n > 0$. Now the statement of the lemma follows from the long exact cohomology sequence corresponding to the short exact sequence above. $\qquad\square$

It remains to reduce the general situation to the statement of the lemma. Again we use induction on the dimension of X. Hence we may assume that the result holds for all noetherian topological spaces of dimension $< d$.

We start by showing that it is enough to consider the case where X is irreducible. In general, X has finitely many irreducible components, and we proceed by induction on the number of irreducible components. Let $Z \subset X$ denote one of the irreducible components of X. This is a closed subset and we denote by $U \subset X$ its open complement. We have a short exact sequence

$$0 \to j_!(\mathscr{F}_{|U}) \to \mathscr{F} \to i_*(\mathscr{F}_{|Z}) \to 0,$$

where i and j denote the inclusions of Z and U into X, respectively. Furthermore, the closure $\overline{U}$ of U has one irreducible component less than X, so by the induction hypothesis and by (21.8.4) we have for all $n > d$

$$H^n(X, i_*(\mathscr{F}_{|Z})) = H^n(Z, \mathscr{F}_{|Z}) = 0$$

and

$$H^n(X, j_!(\mathscr{F}_{|U})) = H^n(\overline{U}, j_!(\mathscr{F}_{|U})) = 0,$$

where we denote the inclusion $U \hookrightarrow \overline{U}$ by j, as well. The long exact cohomology sequence for the short exact sequence above now implies that $H^n(X, \mathscr{F}) = 0$, as desired.

Now assume that X is irreducible. We write

$$\mathscr{F} = \operatorname*{colim}_I \mathscr{F}_I,$$

where I runs over the set, ordered by inclusion, of all finite sets $I = \{(U_1, s_1), \dots, (U_r, s_r)\}$ with $U_\mu \subseteq X$ open, $s_\mu \in \Gamma(U_\mu, \mathscr{F})$, and for such a set I, $\mathscr{F}_I$ is the abelian subsheaf of $\mathscr{F}$ generated by the sections s_μ. In other words, if s_μ corresponds to the map $\alpha_\mu \colon j_{\mu,!}\mathbb{Z}_{U_\mu} \to \mathscr{F}$ (21.1.8), then $\mathscr{F}_I$ is the image of the sum of the α_μ. By Proposition 21.54 it is enough to show that the cohomology of the sheaves $\mathscr{F}_I$ vanishes in degrees $> d$.

If $I' \subset I$, then we have an exact sequence

$$0 \to \mathscr{F}_{I'} \to \mathscr{F}_I \to \mathscr{F}_I/\mathscr{F}_{I'} \to 0,$$

and the quotient $\mathscr{F}_I/\mathscr{F}_{I'}$ is generated by the sections in $I \setminus I'$. If the cohomology of the terms on the left and on the right vanishes in degrees $> d$, then the same holds for the cohomology of $\mathscr{F}_I$ because of the long exact cohomology sequence. By induction, it is hence enough to consider the case where $\mathscr{F}$ is generated by a single section $s \in \Gamma(U, \mathscr{F})$ for some open $U \subseteq X$, i.e., we may assume that we have an exact sequence

$$0 \to \mathscr{G} \to j_!(\mathbb{Z}_U) \to \mathscr{F} \to 0$$

for some open $j \colon U \hookrightarrow X$ and some sheaf $\mathscr{G}$ on X. It is then enough to show the desired cohomology vanishing for $\mathscr{G}$ and for $j_!(\mathbb{Z}_U)$. For the latter one, we can simply invoke Lemma 21.58 above.

If $\mathscr{G} = 0$, then there is nothing to do. Otherwise, there is a non-zero stalk $\mathscr{G}_x$ for some $x \in U$, and for all $x \in U$ we have $\mathscr{G}_x \subseteq j_!(\mathbb{Z}_U)_x = \mathbb{Z}$, so we can identify $\mathscr{G}_x = m_x\mathbb{Z}$ for an integer $m_x \geq 0$. Let m be the minimum of all non-zero m_x. Picking $x \in U$ with $m_x = m$ and lifting the element $m_x \in \mathscr{G}_x$ to a suitable open neighborhood of x, we see that there exists a non-empty open $U' \subseteq U$ such that $\mathscr{G}_{|U'} = m j_!(\mathbb{Z}_U)_{|U'} \cong \mathbb{Z}_{U'}$. We obtain a short exact sequence

$$0 \to j'_!(\mathbb{Z}_{U'}) \to \mathscr{G} \to \mathscr{G}/j'_!(\mathbb{Z}_{U'}) \to 0,$$

where $j' \colon U' \hookrightarrow X$ denotes the inclusion. Now for the left term we have cohomology vanishing by the lemma, and for the right term we have it by induction, since it is supported on $X \setminus U'$. Hence the claim follows for $\mathscr{G}$ by invoking once again the long exact cohomology sequence. $\qquad\square$

(21.13) The local cohomology triangle.

In this section let X be a topological space and let $i \colon Z \hookrightarrow X$ be the inclusion of a locally closed subset.

Definition and Remark 21.59. Let $\mathscr{F}$ be an abelian sheaf. We define the *group of sections of $\mathscr{F}$ with support in Z* as follows. Suppose first that Z is closed in X. Then set

$$\Gamma_Z(X, \mathscr{F}) := \{\, s \in \Gamma(X, \mathscr{F}) \ ; \ \mathrm{Supp}(s) \subseteq Z \,\} = \mathrm{Ker}(\Gamma(X, \mathscr{F}) \to \Gamma(X \setminus Z, \mathscr{F})).$$

If Z is only locally closed, we choose an open subset $V \subseteq X$ such that Z is closed in V and define

$$\Gamma_Z(X, \mathscr{F}) := \Gamma_Z(V, \mathscr{F}_{|V}).$$

We claim that $\Gamma_Z(V, \mathscr{F}_{|V})$ does not depend on the choice of V. Indeed, let $V' \subseteq X$ be another open subset such that $Z \subseteq V'$ is closed. Replacing V' by $V \cap V'$ we may assume that $V' \subseteq V$. We show that the restriction $\mathscr{F}(V) \to \mathscr{F}(V')$ induces an isomorphism $\Gamma_Z(V, \mathscr{F}_{|V}) \overset{\sim}{\to} \Gamma_Z(V', \mathscr{F}_{|V'})$. Indeed let $s \in \Gamma_Z(V, \mathscr{F}_{|V})$ with $s_{|V'} = 0$, then $s = 0$ because $s_{|V\setminus Z} = 0$ by definition and $(V', V \setminus Z)$ is an open covering of V. Conversely, an element $s' \in \Gamma_Z(V', \mathscr{F}_{|V'})$ and $0 \in \mathscr{F}(V \setminus Z)$ glue to a section $s \in \mathscr{F}(V)$ such that $s_{|V'} = s'$ and $s \in \Gamma_Z(V, \mathscr{F}_{|V})$.

We obtain for every locally closed subset Z of X a left exact functor

$$(21.13.1) \qquad\qquad \Gamma_Z \colon \mathscr{F} \mapsto \Gamma_Z(X, \mathscr{F})$$

from the category of abelian sheaves to the category of abelian groups.

Remark 21.60. Let $\mathscr{F}$ be an abelian sheaf on X. We define an abelian sheaf on Z by

$$(i^!\mathscr{F})(W) := \Gamma_W(X, \mathscr{F}), \qquad W \subseteq Z \text{ open}$$

and obtain a functor

(21.13.2) $$i^! \colon (\mathrm{Ab}(X)) \to (\mathrm{Ab}(Z)).$$

For $U \subseteq X$ open, the restriction of the canonical homomorphism $\Gamma(U, \mathscr{F}) \to \Gamma(U \cap Z, i^{-1}\mathscr{F})$ to $\Gamma_{U\cap Z}(U, \mathscr{F}_{|U}) \subseteq \Gamma(U, \mathscr{F})$ is injective, we can view $i^!\mathscr{F}$ as a subsheaf of $i^{-1}(\mathscr{F})$. We have

(21.13.3) $$\Gamma(X, i_* i^!\mathscr{F}) = \Gamma_Z(\mathscr{F})$$

Proposition 21.61. *For every inclusion* $i \colon Z \to X$ *of a locally closed subset, the functor* $i^!$ *is right adjoint to the functor* $i_!$ *(Definition 21.1).*

Proof. For an abelian sheaf $\mathscr{F}$ on X let $\mathscr{F}^Z$ be the sheaf on X with

$$\mathscr{F}^Z(U) = \{\, s \in \Gamma(U, \mathscr{F}) \ ; \ \mathrm{Supp}(s) \subseteq Z \,\}$$

for $U \subseteq X$ open. Then $i^!\mathscr{F} = i^{-1}\mathscr{F}^Z$. One has $(\mathscr{F}^Z)_x = 0$ for $x \notin Z$. Hence Remark 21.2 (3) shows that $i_! i^!\mathscr{F} = \mathscr{F}^Z$. In particular we obtain a monomorphism $i_! i^!\mathscr{F} \hookrightarrow \mathscr{F}$ and any morphism of abelian sheaves $\mathscr{G} \to \mathscr{F}$, where $\mathscr{G}$ is an abelian sheaf on X with $\mathscr{G}_x = 0$ for $x \notin Z$ factors through this monomorphism. Hence we obtain for every sheaf $\mathscr{E}$ on Z functorial bijections

$$\mathrm{Hom}(i_!\mathscr{E}, \mathscr{F}) = \mathrm{Hom}(i_!\mathscr{E}, i_! i^!\mathscr{F}) = \mathrm{Hom}(\mathscr{E}, i^!\mathscr{F}),$$

where the second equality holds because $i_!$ is fully faithful (Remark 21.2 (2)). $\qquad\square$

In particular, $i^!$ is left exact. Moreover, it has an exact left adjoint functor and hence preserves K-injective complexes (Proposition F.183). For its derived functor, we have the following *local cohomology triangle*.

Proposition 21.62. *Let* X *be a topological space, let* $i \colon Z \to X$ *be the inclusion of a closed subset, and let* $j \colon U := X \setminus Z \hookrightarrow X$ *be the inclusion of the open complement of* Z. *Then for every complex* $\mathscr{F}$ *of abelian sheaves on* X *there is a distinguished triangle*

(21.13.4) $$i_* Ri^!\mathscr{F} \longrightarrow \mathscr{F} \longrightarrow Rj_* j^{-1}\mathscr{F} \longrightarrow i_* Ri^!\mathscr{F}[1]$$

in $D(\mathrm{Ab}(X))$, *which is functorial in* $\mathscr{F}$.

Proof. Let $\mathscr{F} \to \mathscr{I}$ be a quasi-isomorphism to a K-injective complex with injective components (Theorem 21.7). Then $i_* Ri^!\mathscr{F} = i_* i^!\mathscr{I}$. Moreover, $j^{-1}\mathscr{I}$ is again a K-injective complex with injective components (Lemma 21.12) and hence $Rj_* j^{-1}\mathscr{F} = j_*(\mathscr{I}_{|U})$. As injective abelian sheaves are flasque (Lemma 21.32) there is an exact sequence of complexes of abelian sheaves

$$0 \longrightarrow i_* i^!\mathscr{I} \longrightarrow \mathscr{I} \longrightarrow j_*(\mathscr{I}_{|U}) \longrightarrow 0$$

which yields the distinguished triangle (21.13.4). $\qquad\square$

The same argument also shows:

Corollary 21.63. *For every complex $\mathscr{F}$ of abelian sheaves on X one has a distinguished triangle in $D(\mathbb{Z})$*

$$(21.13.5) \qquad R\Gamma_Z(X, \mathscr{F}) \longrightarrow R\Gamma(X, \mathscr{F}) \longrightarrow R\Gamma(U, \mathscr{F}_{|U}) \longrightarrow R\Gamma_Z(X, \mathscr{F})[1]$$

which is functorial in $\mathscr{F}$. Writing $H^n_Z(X, \mathscr{F}) = H^n(R\Gamma_Z(X, \mathscr{F}))$, we obtain a long exact sequence

$$(21.13.6) \qquad \cdots \to H^n_Z(X, \mathscr{F}) \to H^n(X, \mathscr{F}) \to H^n(U, \mathscr{F}_{|U}) \to H^{n+1}_Z(X, \mathscr{F}) \to \cdots$$

which is functorial in $\mathscr{F}$.

We also could have obtained (21.13.5) by applying the triangulated functor $R\Gamma(X, -)$ to the distinguished triangle (21.13.4) using the isomorphism $R\Gamma(X, -) \circ Rj_* = R\Gamma(U, -)$, see (21.8.1). Indeed, as i_* and $i^!$ both have an exact left adjoint, namely i^{-1} and $i^!$ respectively, they preserve K-injective complexes (Proposition F.183) and hence $R\Gamma \circ i_* \circ Ri^! = R(\Gamma \circ i_* \circ i^!) = R\Gamma_Z$ by Proposition F.176.

Čech cohomology

Although sheaf cohomology has very nice formal properties that often can be used to relate various cohomology groups to each other, and thus to compute unknown cohomology groups from other, known ones, it is usually not easy to compute examples of cohomology groups using injective or flasque resolutions. For computing specific cohomology groups, often the point of view of Čech cohomology is useful. For instance, this is what we will use to compute the cohomology groups of line bundles on projective space, see Section (22.6) which is the starting point of one of the central finiteness theorems in Algebraic Geometry, namely that higher direct images of coherent modules under proper morphisms between locally noetherian schemes are again coherent.

Let X be a ringed space and let $\mathcal{U} = (U_i)_{i \in I}$ be an open covering of X. We will define the Čech cohomology groups $\check{H}^n(\mathcal{U}, \mathscr{F})$ of an $\mathscr{O}_X$-module $\mathscr{F}$ with respect to $\mathcal{U}$. These groups admit a rather concrete description as the cohomology of a complex in which only the module $\mathscr{F}$ and its sections over intersections of the open subsets U_i enter (and not some usually difficult to handle flasque or injective resolution). One can even ease the bookkeeping further by working with the alternating or the ordered Čech complex as explained in Section (21.15).

Then we relate Čech cohomology to usual cohomology. This happens in two steps. We first define a version $\check{H}^n(X, \mathscr{F})$ of Čech cohomology dependent only on X and $\mathscr{F}$ but not on $\mathcal{U}$ be forming the colimit over all open coverings, where morphisms of coverings are given by refinements of coverings. Then we relate $\check{H}^n(X, \mathscr{F})$ and sheaf cohomology $H^n(X, \mathscr{F})$ by a spectral sequence. This allows us to prove theorems of Leray and Cartan (Corollary 21.82 and Proposition 21.83) that give criteria on X, $\mathscr{F}$, and $\mathcal{U}$ that ensure that $\check{H}^n(\mathcal{U}, \mathscr{F}) \cong H^n(X, \mathscr{F})$.

In the next chapter we will use these criteria to prove that for any quasi-compact separated scheme X, any quasi-coherent $\mathscr{O}_X$-module $\mathscr{F}$, and any open affine covering $\mathcal{U}$ of X one has $\check{H}^n(\mathcal{U}, \mathscr{F}) \cong H^n(X, \mathscr{F})$ for all n. This allows us to calculate the cohomology of quasi-coherent modules in many "reasonable situations".

In general, even $\check{H}^n(X,\mathscr{F})$ does not coincide with sheaf cohomology. The problem is that in higher degrees of the Čech complex we allow only sections over intersections of the open subsets of our given open covering. It is possible to generalize this construction arriving at the notion of a hypercovering and the cohomology of $\mathscr{F}$ with respect to a hypercovering. The colimit over all hypercoverings is indeed isomorphic to sheaf cohomology. We will however not explain the formalism of hypercoverings since for quasi-coherent modules on most schemes it suffices to consider Čech cohomology.

(21.14) The Čech complex.

Let X be a ringed space, let $\mathscr{U} = (U_i)_{i\in I}$ be a covering of X by open subsets, and let $\mathscr{F}$ be a presheaf of $\mathscr{O}_X$-modules. We write

$$U_{\mathbf{i}} := U_{i_0\cdots i_n} := U_{i_0} \cap \cdots \cap U_{i_n} \qquad \text{for } \mathbf{i} = (i_0,\ldots,i_n) \in I^{n+1}$$

and define the $\Gamma(X,\mathscr{O}_X)$-module

$$\check{C}^n(\mathscr{U},\mathscr{F}) = \prod_{\mathbf{i}\in I^{n+1}} \Gamma(U_{\mathbf{i}},\mathscr{F}).$$

We have $\Gamma(X,\mathscr{O}_X)$-linear maps

$$d\colon \check{C}^n(\mathscr{U},\mathscr{F}) \to \check{C}^{n+1}(\mathscr{U},\mathscr{F}), \quad (s_{\mathbf{i}})_{\mathbf{i}\in I^{n+1}} \mapsto \left(\sum_{\nu=0}^{n+1}(-1)^\nu s_{\mathbf{j}(\hat{\nu})|U_{\mathbf{j}}}\right)_{\mathbf{j}\in I^{n+2}},$$

where we denote by $\mathbf{j}(\hat{\nu})$ the $(n+1)$-tuple obtained by omitting the ν-th entry from $\mathbf{j}$. It is easy to check that $d\circ d = 0$.

Definition 21.64. *The complex $\check{C}(\mathscr{U},\mathscr{F})$ is called the Čech complex of $\mathscr{F}$ with respect to the covering $\mathscr{U}$. Its cohomology groups are denoted by $\check{H}^n(\mathscr{U},\mathscr{F})$ and are called the Čech cohomology groups of $\mathscr{F}$ with respect to the covering $\mathscr{U}$.*

The Čech complex and its cohomology groups clearly behave functorially in $\mathscr{F}$. In particular we obtain a functor $\check{C}(\mathscr{U},-)$ from the category of presheaves of $\mathscr{O}_X$-modules to the category of complexes of $\Gamma(X,\mathscr{O}_X)$-modules.

As we will see later (Section (21.17)), in certain favorable situations, the cohomology of the Čech complex coincides with the cohomology in the sense of derived functors. This will allow us to explicitly compute cohomology groups much more easily than using acyclic resolutions.

Lemma 21.65. *Let X be a ringed space, and let $\mathscr{F}$ be a presheaf of $\mathscr{O}_X$-modules. The following are equivalent:*
(i) *The presheaf $\mathscr{F}$ is a sheaf.*
(ii) *For all open subsets $U \subseteq X$ and all open coverings $\mathscr{U}$ of U, the natural map $\Gamma(U,\mathscr{F}) \to \check{H}^0(\mathscr{U},\mathscr{F})$ is an isomorphism.*

Proof. This follows immediately from the definitions, because we can express the sheaf property by requiring that for all coverings $\mathscr{U} = (U_i)_i$ of open subsets $U \subseteq X$, the sequence

$$0 \to \Gamma(U,\mathscr{F}) \to \prod_i \Gamma(U_i,\mathscr{F}) \to \prod_{i,j} \Gamma(U_{ij},\mathscr{F})$$

is exact, where the second map is given by $(s_i)_i \mapsto (s_{j|U_{ij}} - s_{i|U_{ij}})_{i,j}$, i.e., by the Čech differential. $\square$

Remark 21.66. If $\mathscr{F}$ is a sheaf of abelian groups on a topological space X, then $\check{H}^1(\mathscr{U}, \mathscr{F})$ is the same abelian group as defined in (11.5). Hence it can be identified with the group of isomorphism classes of $\mathscr{F}$-torsors T such that $T_{|U_i}$ is trivial for all $i \in I$ (Proposition 11.13).

(21.15) The alternating and the ordered Čech complex.

There are two useful variants of the Čech complex which have the same cohomology groups: The alternating Čech complex and the ordered Čech complex. As before, let X be a ringed space, and let $\mathscr{F}$ be a presheaf of $\mathcal{O}_X$-modules. We also fix an open covering $\mathscr{U} = (U_i)_{i \in I}$ of X.

Definition 21.67. *The* alternating Čech complex *is given by*

$$\check{C}^n_{\mathrm{alt}}(\mathscr{U}, \mathscr{F}) := \{(s_{\mathbf{i}})_{\mathbf{i}} \in \check{C}^n(\mathscr{U}, \mathscr{F});$$

$$s_{\mathbf{i}} = 0 \text{ whenever two entries of } \mathbf{i} \text{ are equal,}$$

$$s_{\sigma(\mathbf{i})} = \mathrm{sgn}(\sigma)s_{\mathbf{i}} \text{ for every permutation } \sigma \in S_{n+1}\},$$

where the differential d is given by restricting the differential of the Čech complex to $\check{C}^n_{\mathrm{alt}}(\mathscr{U}, \mathscr{F})$.

Now let $\mathscr{U} = (U_i)_{i \in I}$ be an open covering, and suppose that a total order $<$ of the index set I is given.

Definition 21.68. *The* ordered Čech complex *is given by*

$$\check{C}^n_{\mathrm{ord}}(\mathscr{U}, \mathscr{F}) = \prod_{i_0 < i_1 < \cdots < i_n} \Gamma(U_{i_0 \cdots i_n}, \mathscr{F}),$$

where the differential is given by the same formula as the differential of the Čech complex.

The three versions of the Čech complex are related by the following result.
Consider the natural homomorphisms

$$\check{C}_{\mathrm{alt}}(\mathscr{U}, \mathscr{F}) \xrightarrow{\iota} \check{C}(\mathscr{U}, \mathscr{F}) \xrightarrow{\pi} \check{C}_{\mathrm{ord}}(\mathscr{U}, \mathscr{F}),$$

the first one being given by the inclusion, the second one being given by the projection.

Proposition 21.69. *In the above situation,*
(1) *The composition $\pi \circ \iota$ is an isomorphism of complexes.*
(2) *The maps ι and π are homotopy equivalences.*
(3) *The homomorphisms*

$$H^n(\check{C}_{\mathrm{alt}}(\mathscr{U}, \mathscr{F})) \to H^n(\check{C}(\mathscr{U}, \mathscr{F})) \text{ and } H^n(\check{C}(\mathscr{U}, \mathscr{F})) \to H^n(\check{C}_{\mathrm{ord}}(\mathscr{U}, \mathscr{F}))$$

induced by ι and π are isomorphisms.

Proof. An inverse of $\pi \circ \iota$ is given by $\gamma \colon \check{C}_{\mathrm{ord}}(\mathcal{U}, \mathcal{F}) \to \check{C}_{\mathrm{alt}}(\mathcal{U}, \mathcal{F})$ with

$$\gamma(s)_{i_0 \dots i_n} := \begin{cases} 0, & \text{if } i_p = i_q \text{ for some } p \neq q; \\ \mathrm{sgn}(\sigma) s_{i_{\sigma(0)} \dots i_{\sigma(n)}}, & \text{if } i_{\sigma(0)} < i_{\sigma(1)} < \dots < i_{\sigma(n)} \end{cases}$$

which shows Assertion (1). Assertion (3) follows from (2).

It remains to show that ι is a homotopy equivalence with inverse $\gamma \circ \pi$. We omit this quite tedious verification and refer to [Sta] 01FM for details. $\qquad\square$

In particular, all three complexes are isomorphic in the homotopy category $K(\Gamma(X, \mathcal{O}_X))$ (once a total ordering of the index set I is fixed, and as a corollary we see that the isomorphism class of the ordered Čech complex in $K(\Gamma(X, \mathcal{O}_X))$ is independent of the ordering).

Corollary 21.70. *Let $\mathcal{U} = (U_i)_{i \in I}$ be a finite open covering of X. Then $\check{H}^n(\mathcal{U}, \mathcal{F}) = 0$ whenever $n \geq \#I$.*

Proof. We can use the alternating (or the ordered) Čech complex to compute the Čech cohomology, and those complexes vanish in degrees $\geq \#I$. $\qquad\square$

(21.16) Passing to refinements for Čech cohomology.

We make the coverings of a topological space into a category as follows.

Definition 21.71. *Let X be a topological space and let $\mathcal{U} = (U_i)_{i \in I}$ and $\mathcal{V} = (V_j)_{j \in J}$ be open coverings of X. A map of open coverings is a map $\alpha \colon J \to I$ such that for all $j \in J$ we have $V_j \subseteq U_{\alpha(j)}$. If such a map of coverings exists, we call $\mathcal{V} = (V_j)_{j \in J}$ a refinement of $\mathcal{U}$.*

If $\alpha \colon \mathcal{V} \to \mathcal{U}$ is a map of open coverings, then for every abelian presheaf $\mathcal{F}$ on X, we obtain maps $\check{C}^n(\mathcal{U}, \mathcal{F}) \to \check{C}^n(\mathcal{V}, \mathcal{F})$ induced by the restriction of sections of $\mathcal{F}$. These maps clearly are compatible with the differentials of the Čech complexes and hence induce homomorphisms of the Čech complexes (and hence of the Čech cohomology groups with respect to $\mathcal{U}$ and to $\mathcal{V}$).

Lemma 21.72. *Let $\mathcal{U} = (U_i)_{i \in I}$ and $\mathcal{V} = (V_j)_{j \in J}$ be open coverings of X. Assume that $\mathcal{V}$ is a refinement of $\mathcal{U}$ and that α and α' are both maps of coverings $\mathcal{V} \to \mathcal{U}$. Denote by $\varphi, \varphi' \colon \check{C}(\mathcal{U}, \mathcal{F}) \to \check{C}(\mathcal{V}, \mathcal{F})$ the homomorphisms of the Čech complexes induced by α and α'. Then φ and φ' are homotopic.*

Proof. In fact, we have $\varphi - \varphi' = d \circ h + h \circ d$ for

$$h \colon \check{C}^n(\mathcal{U}, \mathcal{F}) \to \check{C}^{n-1}(\mathcal{V}, \mathcal{F})$$

given by

$$(h(s_{\mathbf{j}}))_{i_0, \dots, i_{n-1}} = \sum_{\nu=0}^{n-1} (-1)^\nu s_{\alpha(i_0) \dots \alpha(i_\nu) \alpha'(i_\nu) \dots \alpha'(i_{n-1})}|_{V_\mathbf{i}}.$$

We omit the computation which is required to check this. For a detailed proof, see [God], Ch. II, Théorème 5.7.1. $\qquad\square$

The lemma implies that the homotopy class of the homomorphism $\check{C}(\mathscr{U},\mathscr{F}) \to \check{C}(\mathscr{V},\mathscr{F})$ which we obtain for a refinement $\mathscr{V}$ of $\mathscr{U}$ is independent of the choice of the map α. In particular, the induced homomorphisms $\check{H}^n(\mathscr{U},\mathscr{F}) \to \check{H}^n(\mathscr{V},\mathscr{F})$ are independent of the choice of the map α. The notion of refinement provides a cofiltered partial order on the collection of all open coverings of X, and we can pass to the colimit:

Definition 21.73. *We call*

$$\check{C}(X,\mathscr{F}) := \operatorname*{colim}_{\mathscr{U}} \check{C}(\mathscr{U},\mathscr{F}) \in K^+(\Gamma(X,\mathscr{O}_X)),$$

$$\check{H}^n(X,\mathscr{F}) := \operatorname*{colim}_{\mathscr{U}} \check{H}^n(\mathscr{U},\mathscr{F}),$$

where the limit is taken over all open coverings of X, the Čech complex of $\mathscr{F}$ and the n-th Čech cohomology group of $\mathscr{F}$.

Note that the entirety of all open coverings $(U_i)_{i\in I}$ of X does not form a set because the index sets I may become arbitrary large. But by choosing for every $x \in X$ an $i_x \in I$ with $x \in U_{i_x}$ and replacing I by $\{ i_x \ ; \ x \in X \}$ we can always pass to a subcovering whose index set is (in bijection with) a subset of X. Such coverings do form a set. Hence the opposite category of the category of open coverings of X contains a cofinal small subcategory. Hence strictly speaking we form the colimit over the opposite of this small subcategory, see Proposition F.16 for a justification.

As filtered colimits are exact in the category of $\Gamma(X,\mathscr{O}_X)$-modules, one has

$$(21.16.1) \qquad\qquad H^n(\check{C}(X,\mathscr{F})) = \check{H}^n(X,\mathscr{F}).$$

(21.17) Čech cohomology versus cohomology.

We continue to denote by X a ringed space and by $\mathscr{U} = (U_i)_{i\in I}$ an open covering of X. We denote by

$$\iota\colon (\mathscr{O}_X\text{-Mod}) \longrightarrow (\text{P-}\mathscr{O}_X\text{-Mod})$$

the left exact inclusion functor.

There are several results to the effect that under certain conditions Čech cohomology coincides with derived functor cohomology. To obtain them, we proceed in three steps.
  (I) We extend $\check{C}(\mathscr{U},-)$ to a functor $D^+(\text{P-}\mathscr{O}_X\text{-Mod}) \longrightarrow D^+(\Gamma(X,\mathscr{O}_X))$.
  (II) We show that $\check{C}(\mathscr{U},-) = R\check{H}^0(\mathscr{U},-)$ and in this way obtain an isomorphism $\check{C}(\mathscr{U},-) \circ R\iota \xrightarrow{\sim} R\Gamma(X,-)$ of functors $D^+(X) \to D^+(\Gamma(X,\mathscr{O}_X))$.
  (III) We use the Grothendieck spectral sequence to obtain criteria when this isomorphism yields an isomorphism between (usual) cohomology and Čech cohomology.
  Let us start with the first step.

Definition and Remark 21.74. Let $\mathscr{F}$ be a bounded below complex of presheaves of $\mathscr{O}_X$-modules. By functoriality of the Čech complex we obtain a double complex $(\check{C}^i(\mathscr{U},\mathscr{F}^j))_{i,j}$ and we define the *Čech complex of $\mathscr{F}$ with respect to the covering $\mathscr{U}$* as its total complex (Definition F.87)

$$\check{C}(\mathscr{U},\mathscr{F}) := \text{Tot}((\check{C}^i(\mathscr{U},\mathscr{F}^j))_{i,j}).$$

Its cohomology groups $\check{H}^n(\mathscr{U},\mathscr{F})$ are called the *Čech cohomology groups of $\mathscr{F}$ with respect to $\mathscr{U}$.*

As forming the total complex preserves homotopy (Remark F.89) we obtain a functor

(21.17.1) $\check{C}(\mathscr{U},-)\colon K^+(\text{P-}\mathscr{O}_X\text{-Mod}) \longrightarrow K^+(\Gamma(X,\mathscr{O}_X)).$

Lemma 21.75. *The functor* (21.17.1) *sends exact complexes to exact complexes, i.e., it induces a functor*

(21.17.2) $\check{C}(\mathscr{U},-)\colon D^+(\text{P-}\mathscr{O}_X\text{-Mod}) \longrightarrow D^+(\Gamma(X,\mathscr{O}_X)).$

Note that the analogous assertion for the functor $\check{C}(\mathscr{U},-)\colon K^+(X) \to K^+(\Gamma(X,\mathscr{O}_X))$ would almost always be wrong!

Proof. A complex $\mathscr{F}$ of *presheaves* of $\mathscr{O}_X$-modules is exact if and only if the complex $\mathscr{F}(U)$ is exact for all $U \subseteq X$ open. Let $\mathscr{F}$ be bounded below. Hence the components of $\check{C}(\mathscr{U},\mathscr{F})$ are finite direct sums (by the definition of the total complex) of products (by the definition of the Čech complex) of exact functors. This shows the lemma. $\square$

The next step is the description of $\check{C}(\mathscr{U},-)$ as derived functor. We start with a lemma.

Lemma 21.76. *Let $\mathscr{I}$ be an injective presheaf of $\mathscr{O}_X$-modules. Then*

$$\check{H}^n(\mathscr{U},\mathscr{I}) = 0 \quad \textit{for all } n > 0.$$

Proof. Recall that we defined for $U \subseteq X$ open presheaves of $\mathscr{O}_X$-modules by $\mathscr{O}^p_U(V) = \mathscr{O}_X(V)$ if $V \subseteq U$ and $= 0$ otherwise and that (21.1.7)

(*) $\text{Hom}_{(\text{P-}\mathscr{O}_X\text{-Mod})}(\mathscr{O}^p_U,\mathscr{F}) = \mathscr{F}(U)$

for every presheaf $\mathscr{F}$ of $\mathscr{O}_X$-modules. For $U' \subseteq U$ there is a canonical morphism $\mathscr{O}^p_{U'} \to \mathscr{O}^p_U$ which yields the restriction $\mathscr{F}(U) \to \mathscr{F}(U')$ by (*) and functoriality. Define a complex $Z(\mathscr{U})$ with descending differentials by setting for $n \geq 0$

$$Z(\mathscr{U})_n := \bigoplus_{\mathbf{i} \in I^{n+1}} \mathscr{O}^p_{U_{\mathbf{i}}},$$

where for $\mathbf{j} \in I^{n+2}$ and $\nu \in \{0,\dots,n\}$ the differential $d\colon \mathscr{O}^p_{U_{\mathbf{j}}} \to \mathscr{O}^p_{U_{\mathbf{j}(\nu)}}$ is given by $(-1)^\nu$ times the canonical morphism. We also set $Z(\mathscr{U})_n := 0$ for $n < 0$. Then (*) yields

$$\text{Hom}(Z(\mathscr{U}),\mathscr{F}) = \check{C}(\mathscr{U},\mathscr{F}).$$

As $\mathscr{I}$ is injective, $\text{Hom}(-,\mathscr{I})$ is exact and it suffices to show that the complex $Z(\mathscr{U})$ of presheaves of $\mathscr{O}_X$-modules is exact in all degrees > 0.

Let $W \subseteq X$ be open. We want to show that the complex $Z(\mathscr{U})(W)$ is exact in degrees > 0. Fix $i^0 \in I$ with $W \subseteq U_{i^0}$ (if such an index does not exist, then $Z(\mathscr{U})(W) = 0$ and nothing needs to be done). Define a "homotopy between id and 0 in degrees > 0" by

$$h\colon Z(\mathscr{U})_n(W) \longrightarrow Z(\mathscr{U})_{n+1}(W),$$

$$h(s)_{i_0 \dots i_n} := \begin{cases} 0, & \text{if } i_0 \neq i^0; \\ s_{i_1 \dots i_{n+1}}, & \text{if } i_0 = i^0 \end{cases}$$

for $n \geq 0$. Then $d \circ h + h \circ d = \text{id}$ in degrees ≥ 0 which shows the exactness of $Z(\mathscr{U})(W)$ in degrees > 0. $\square$

Lemma 21.77. *For all bounded below complexes $\mathscr{F}$ of presheaves of $\mathscr{O}_X$-modules there is a functorial isomorphism in $D^+(\Gamma(X, \mathscr{O}_X))$*

$$(21.17.3) \qquad\qquad \check{C}(\mathscr{U}, \mathscr{F}) \xrightarrow{\sim} R\check{H}^0(\mathscr{U}, \mathscr{F}),$$

where the right hand side is the right derived functor $D^+(\text{P-}\mathscr{O}_X\text{-Mod}) \to D^+(\Gamma(X, \mathscr{O}_X))$ of the left exact functor $\check{H}^0(\mathscr{U}, -)\colon (\text{P-}\mathscr{O}_X\text{-Mod}) \to (\Gamma(X, \mathscr{O}_X)\text{-Mod})$.

Proof. By Corollary F.186 we have $D^+(\text{P-}\mathscr{O}_X\text{-Mod}) \cong K^+(\mathcal{I})$, where $\mathcal{I}$ is the full subcategory of injective objects in $(\text{P-}\mathscr{O}_X\text{-Mod})$. Therefore it suffices to construct the isomorphism (21.17.3) for $\mathscr{F} = \mathscr{I}$ a bounded below complex of injective objects.

By Lemma 21.76,

$$(*) \qquad\qquad \check{H}^0(\mathscr{U}, \mathscr{I}) \longrightarrow (\cdots \to \check{C}^i(\mathscr{U}, \mathscr{I}) \to \check{C}^{i+1}(\mathscr{U}, \mathscr{I}) \to \dots)$$

is a quasi-isomorphism of complexes with components in $C^+(\Gamma(X, \mathscr{O}_X))$, where we consider the left hand side as a complex with components in $C^+(\Gamma(X, \mathscr{O}_X))$ concentrated in degree 0. Hence we have a functorial quasi-isomorphism in $K^+(\Gamma(X, \mathscr{O}_X))$

$$R\check{H}^0(\mathscr{U}, \mathscr{I}) = \check{H}^0(\mathscr{U}, \mathscr{I}) \longrightarrow \check{C}(\mathscr{U}, \mathscr{I}),$$

which is induced by applying the total complex functor to the quasi-isomorphism $(*)$ (Lemma F.113). $\qquad\square$

For an $\mathscr{O}_X$-module $\mathscr{F}$ we have $\check{H}^0(\mathscr{U}, \mathscr{F}) = \Gamma(X, \mathscr{F})$ (Lemma 21.65) and hence an isomorphism of functors

$$(21.17.4) \qquad\qquad \check{H}^0(\mathscr{U}, -) \circ \iota \xrightarrow{\sim} \Gamma(X, -)\colon (\mathscr{O}_X\text{-Mod}) \to (\Gamma(X, \mathscr{O}_X)\text{-Mod}).$$

As the inclusion $\iota\colon (\mathscr{O}_X\text{-Mod}) \to (\text{P-}\mathscr{O}_X\text{-Mod})$ has an exact left adjoint functor, namely sheafification, its extension $K^+(\mathscr{O}_X\text{-Mod}) \to K^+(\text{P-}\mathscr{O}_X\text{-Mod})$ preserves bounded below K-injective complexes (Corollary F.183) and hence by Proposition F.176 we obtain an isomorphism of functors

$$(21.17.5) \qquad R\Gamma(X, -) \xrightarrow{\sim} R\check{H}^0(\mathscr{U}, -) \circ R\iota\colon D^+(\mathscr{O}_X\text{-Mod}) \to D^+(\Gamma(X, \mathscr{O}_X)).$$

Using Lemma 21.77 we easily derive the main result of this section.

Theorem 21.78. *Let X be a ringed space and let $\mathscr{U}$ be an open covering of X. Let $\mathscr{F}$ be a bounded below complex of $\mathscr{O}_X$-modules.*
(1) There is a natural functorial morphism (in $D^+(\Gamma(X, \mathscr{O}))$)

$$(21.17.6) \qquad\qquad \check{C}(\mathscr{U}, \mathscr{F}) \to R\Gamma(X, \mathscr{F}).$$

Passing to cohomology, we obtain homomorphisms

$$(21.17.7) \qquad\qquad \check{H}^n(\mathscr{U}, \mathscr{F}) \to H^n(X, \mathscr{F}) \qquad \text{for all } n \in \mathbb{Z}$$

that are functorial in $\mathscr{F}$.
(2) There is a converging spectral sequence

$$(21.17.8) \qquad\qquad E_2^{p,q} = \check{H}^p(\mathscr{U}, \mathscr{H}^q(\mathscr{F})) \Rightarrow H^{p+q}(X, \mathscr{F}),$$

which is functorial in $\mathscr{F}$. Here $\mathscr{H}^q(\mathscr{F})$ is the presheaf defined in Remark 21.20.

Proof. The spectral sequence in (2) is the Grothendieck spectral sequence (Proposition F.212) attached to (21.17.4): By (21.5.3) we have $R^q\iota(\mathscr{F}) = \mathscr{H}^q(\mathscr{F})$ and by Lemma 21.77 we have $R^p\check{H}^0(\mathscr{U}, -) = \check{C}^p(\mathscr{U}, -)$.

The morphism (21.17.6) is the composition of functors

$$\check{C}(\mathscr{U}, -) \xrightarrow{\sim} R\check{H}^0(\mathscr{U}, -) \circ \iota \longrightarrow R\check{H}^0(\mathscr{U}, -) \circ R\iota \cong R\Gamma(X, -),$$

where the first isomorphism holds by Lemma 21.77 and the last isomorphism is (21.17.4). $\square$

Note that the isomorphism (21.17.5) exists also for unbounded derived functors (with the same argument) but that we proved Lemma 21.77 and hence Theorem 21.78 only for bounded below complexes. This suffices for our purposes because we are interested in Čech cohomology only to compute cohomology in concrete situations, where all complexes are bounded below (and usually are even concentrated in degree 0).

Remark 21.79. Let $\mathscr{F}$ be an $\mathscr{O}_X$-module. Then $\mathscr{H}^0(\mathscr{F}) = \mathscr{F}$. The spectral sequence (21.17.8) is a first quadrant spectral sequence (i.e., $E_2^{p,q} = 0$ if $p < 0$ or $q < 0$). The proof of Theorem 21.78 shows that the morphism (21.17.6) is the edge morphism

$$E_2^{p,0} = \check{H}^p(\mathscr{U}, \mathscr{F}) \longrightarrow H^p(X, \mathscr{F}).$$

The formation of the spectral sequence (21.17.8) is compatible with passing to refinements and hence induces for all $\mathscr{O}_X$-modules $\mathscr{F}$ a functorial converging spectral sequence

(21.17.9) $$E_2^{p,q} = \check{H}^p(X, \mathscr{H}^q(\mathscr{F})) \Rightarrow H^{p+q}(X, \mathscr{F}).$$

In particular the edge morphisms are functorial homomorphism of $\Gamma(X, \mathscr{O}_X)$-modules

(21.17.10) $$\check{H}^n(X, \mathscr{F}) = \operatorname*{colim}_{\mathscr{U}} \check{H}^n(\mathscr{U}, \mathscr{F}) \longrightarrow H^n(X, \mathscr{F}), \qquad n \geq 0.$$

Lemma 21.80. *Let $\mathscr{G}$ be a presheaf of $\mathscr{O}_X$-modules whose sheafification is 0. Then $\check{H}^0(X, \mathscr{G}) = 0$.*

Proof. If $(s_i)_i \in \check{C}^0(\mathscr{U}, \mathscr{G})$, then locally all sections s_i are zero, hence there exists a refinement $\mathscr{V}$ of $\mathscr{U}$ such that the image of $(s_i)_i$ in $\check{C}^0(\mathscr{V}, \mathscr{G})$ is zero. Hence $\operatorname{colim}_{\mathscr{U}} \check{C}^0(\mathscr{U}, \mathscr{G}) = 0$ and in particular $\check{H}^0(X, \mathscr{G}) = 0$. $\square$

Corollary 21.81. *Let X be a ringed space, and let $\mathscr{F}$ be an $\mathscr{O}_X$-module. Then the functorial homomorphism (21.17.10) is an isomorphism for $n = 0, 1$, and is injective for $n = 2$.*

Proof. The case $n = 0$ easily follows from the definitions (see Lemma 21.65). For the other cases we use the exact sequence in low degrees (Proposition F.105) for the spectral sequence (21.17.9) and get for every open covering $\mathscr{U}$ of X an exact sequence

$$0 \longrightarrow \check{H}^1(\mathscr{U}, \mathscr{F}) \longrightarrow H^1(X, \mathscr{F}) \longrightarrow \check{H}^0(\mathscr{U}, \mathscr{H}^1(\mathscr{F})) \longrightarrow \check{H}^2(\mathscr{U}, \mathscr{F}) \longrightarrow H^2(X, \mathscr{F}).$$

Hence we have to show that $\check{H}^0(X, \mathscr{H}^1(\mathscr{F})) = 0$.

But $\mathscr{H}^1(\mathscr{F})$ is a presheaf whose sheafification is zero (Lemma 21.22). We conclude by Lemma 21.80. $\square$

The same proof shows that the corollary also holds for complexes $\mathscr{F}$ of $\mathscr{O}_X$-modules concentrated in degrees ≥ 0 (to obtain the exact sequence of lower terms) such that $H^1(\mathscr{F}) = 0$ (to ensure that $\mathscr{H}^1(\mathscr{F})$ is a presheaf whose sheafification is zero).

For an abelian sheaf $\mathscr{F}$ and for $n = 1$ one could also have shown that (21.17.10) is equal to the isomorphism $\check{H}^1(X, \mathscr{F}) \xrightarrow{\sim} H^1(X, \mathscr{F})$ of (11.5.2), where we identify $H^1(X, \mathscr{F})$ with the group of isomorphism classes of $\mathscr{F}$-torsors (Proposition 21.40).

Corollary 21.82. (Leray) *Let X be a ringed space and let $\mathscr{U} = (U_i)_{i \in I}$ be an open covering of X. Let $\mathscr{F}$ be an $\mathscr{O}_X$-module such that for all $q > 0$, $p \geq 0$ and $i_0, \dots, i_p \in I$,*

$$H^q(U_{i_0 \cdots i_p}, \mathscr{F}) = 0.$$

Then the natural homomorphism $\check{H}^n(\mathscr{U}, \mathscr{F}) \to H^n(X, \mathscr{F})$ is an isomorphism for all $n \geq 0$.

Proof. The hypothesis means that $\check{C}^p(\mathscr{U}, \mathscr{H}^q(\mathscr{F})) = 0$ if $q > 0$. In particular for the spectral sequence (21.17.8) we see that

$$(*) \qquad\qquad E_2^{p,q} = 0 \quad \text{for all } (p,q) \text{ with } q \neq 0.$$

Hence the spectral sequence degenerates at E_2 which implies that $H^n(X, \mathscr{F})$ has a filtration whose graded pieces are the $E_2^{p,q}$ with $p + q = n$. Hence $(*)$ implies that the edge morphism $\check{H}^n(\mathscr{U}, \mathscr{F}) \to H^n(X, \mathscr{F})$ is an isomorphism. $\square$

Proposition 21.83. (Cartan) *Let X be a ringed space and let $\mathcal{B}$ be a basis of the topology of X such that $U, U' \in \mathcal{B}$ implies $U \cap U' \in \mathcal{B}$. Let $\mathscr{F}$ be an $\mathscr{O}_X$-module. Suppose that*

$$\check{H}^p(U, \mathscr{F}) = 0 \qquad \text{for all } p > 0 \text{ and all } U \in \mathcal{B}.$$

Then
(1) for $U \in \mathcal{B}$ one has $H^n(U, \mathscr{F}) = 0$ for all $n > 0$,
(2) for every open covering $\mathcal{U}$ of X consisting of open sets in $\mathcal{B}$ the canonical morphism $\check{H}^n(\mathcal{U}, \mathscr{F}) \to H^n(X, \mathscr{F})$ is an isomorphism for all $n \geq 0$, and
(3) the canonical homomorphism $\check{H}^n(X, \mathscr{F}) \to H^n(X, \mathscr{F})$ is an isomorphism for all $n \geq 0$.

Proof. We prove (1) by induction on $n \geq 1$. Hence we may assume that $H^q(U, \mathscr{F}) = 0$ for all $0 < q < n$ and all $U \in \mathcal{B}$. As every open covering has a refinement by an open covering $(U_i)_i$ with $U_i \in \mathcal{B}$ (because $\mathcal{B}$ is a basis of the topology) and as $U_{i_0 \dots i_p} \in \mathcal{B}$ (because $\mathcal{B}$ is stable under finite nonempty intersections), the induction hypothesis implies

$$\check{C}(X, \mathscr{H}^q(\mathscr{F})) = 0$$

for all $0 < q < n$. Hence $E_2^{p,q} = 0$ for $0 < q < n$ and all p in the spectral sequence (21.17.9). Hence we obtain from (F.22.5) an exact sequence

$$0 \longrightarrow \check{H}^n(X, \mathscr{F}) \longrightarrow H^n(X, \mathscr{F}) \longrightarrow \check{H}^0(X, \mathscr{H}^n(\mathscr{F})),$$

and the last term is 0 by Lemma 21.80 because the sheafification of $\mathscr{H}^n(\mathscr{F})$ is zero by Lemma 21.22. Replacing X by $U \in \mathcal{B}$, $\mathscr{F}$ by $\mathscr{F}_{|U}$, and $\mathcal{B}$ by $\{ U' \in \mathcal{B} \; ; \; U' \subseteq U \}$ we deduce that $0 = \check{H}^n(U, \mathscr{F}) \longrightarrow H^n(U, \mathscr{F})$ is bijective for all $U \in \mathcal{B}$ which shows the claim.

Now (2) follows by applying Corollary 21.82 using again that $\mathcal{B}$ is stable under finite intersections (with non-empty index set). As open coverings consisting of elements in $\mathcal{B}$ are cofinal in the set of all open coverings, also (3) follows. $\square$

Derived inverse image, Hom sheaves, and tensor products

We will now define the other functors in our "formalism of four functors" explained in the introduction of this chapter. More precisely we are going to derive the Hom- and the $\mathscr{H}om$-functor, the tensor product, and the functor f^* for a morphism f of ringed spaces. There are two main difficulties to overcome.

The first one is that the $\mathscr{H}om$-functor and the tensor product are both bi-functors, i.e., functors in two variables and we have to explain in which variable we derive the functors, whether we could also derive these functors in the other variable, and whether we obtain in this case the same functor. To solve these problems we will check that we can apply the general results of Section (F.51).

The second difficulty is that the tensor product and the functor f^* are right exact functors. Hence we want to take their left derivation. If we wanted simply to dualize our general construction of right derived functors, we would need for every complex $\mathscr{F}$ of $\mathscr{O}_X$-modules a K-projective resolution $\mathscr{P} \to \mathscr{F}$. But in the category of $\mathscr{O}_X$-modules such K-projective resolutions usually do not exist, even for standard schemes such as the projective line $X = \mathbb{P}^1_k$ over a field k (see Exercise 21.5). Hence we need a larger class of complexes that are acyclic for the tensor product and the functor f^*. These will be the class of K-flat complexes that we will introduce in Section (21.19).

This problem also would occur if we wanted to derive the Hom-functor in the first variable. Here we cannot use K-flat resolutions as these are in general not acyclic for the Hom-functor in the first variable. In fact, the only complexes that are acyclic for the Hom-functor in the first variable are by definition the K-projective complexes (see the dual of Definition F.179).

If we consider the category of modules over a ring A (corresponding modules over a ringed spaces whose underlying topological space consists of a single point), then every complex of A-modules has a K-projective resolution (Theorem F.189) and we can also derive the Hom-functor in the first variable. Then the results of Section (F.51) will show that this derived functor coincides with the one obtained by deriving in the second variable.

(21.18) Derived functor of Hom and $\mathscr{H}om$.

In this section, we always denote by $(X, \mathscr{O}_X)$ a ringed space.

Recall from Section (F.52), that for any abelian category $\mathcal{A}$ with enough K-injective objects, we have a derived Hom-functor

$$R\operatorname{Hom}_{\mathcal{A}}(-,-) \colon D(\mathcal{A})^{\mathrm{opp}} \times D(\mathcal{A}) \to D(\mathrm{AbGrp}).$$

For $p \in \mathbb{Z}$ one has for all complexes F and G in $D(\mathcal{A})$

$$\operatorname{Ext}^p_{\mathcal{A}}(F, G) = H^p(R\operatorname{Hom}_{\mathcal{A}}(F, G)) = \operatorname{Hom}_{D(\mathcal{A})}(F, G[p]),$$

where the second identity holds by (F.52.5).

Remark 21.84. If $\mathcal{A}$ is the category of $\mathscr{O}_X$-modules, then for complexes $\mathscr{F}$ and $\mathscr{G}$ of $\mathscr{O}_X$-modules the Hom complex (Section (F.18)) is a complex of $\Gamma(X, \mathscr{O}_X)$-modules and we obtain a derived Hom-functor

(21.18.1) $$R\operatorname{Hom}_{\mathscr{O}_X}(-,-) \colon D(X)^{\mathrm{opp}} \times D(X) \to D(\Gamma(X, \mathscr{O}_X)).$$

The construction in Section (F.52) shows that $R\operatorname{Hom}_{\mathcal{O}_X}(\mathcal{F},\mathcal{G})$ is calculated as follows. One chooses a quasi-isomorphism $\mathcal{G} \to \mathcal{I}$, where $\mathcal{I}$ is a K-injective complex of $\mathcal{O}_X$-modules. Then $R\operatorname{Hom}_{\mathcal{O}_X}(\mathcal{F},\mathcal{G})$ is represented by the complex $\operatorname{Hom}_{\mathcal{O}_X}(\mathcal{F},\mathcal{I})$ of $\Gamma(X,\mathcal{O}_X)$-modules.

If there exists a quasi-isomorphism $\mathcal{P} \to \mathcal{F}$ for a K-projective complex of $\mathcal{O}_X$-modules (which is always the case if the category of $\mathcal{O}_X$-modules can be identified with the category of modules over a ring), then $R\operatorname{Hom}_{\mathcal{O}_X}(\mathcal{F},\mathcal{G})$ can also be represented by $\operatorname{Hom}_{\mathcal{O}_X}(\mathcal{P},\mathcal{G})$.

Often, K-projective resolutions are more manageable if they exist. Consider the following example.

Example 21.85. Let A be a ring and let I be an ideal generated by a completely intersected sequence $\mathbf{f} = (f_1,\ldots,f_r)$ in A. Then the Koszul complex $K(\mathbf{f})$, considered as cochain complex sitting in degrees $[-r,0]$, defines a projective resolution $K(\mathbf{f}) \to A/I$. Hence for every complex M of A-modules we can calculate $R\operatorname{Hom}_A(A/I,M)$ by the complex $\operatorname{Hom}_A(K(\mathbf{f}),M)$.

We want to define also a sheaf version of $R\operatorname{Hom}$, i.e., a derived functor attached to the $\mathcal{H}om$ functor (see Section (7.4)). For this we first define a sheaf version of the Hom-complex in the same way as in Section (F.18).

Definition and Remark 21.86. Let $\mathcal{F}$ and $\mathcal{G}$ be complexes of $\mathcal{O}_X$-modules. Setting

$$\mathcal{H}om^i_{\mathcal{O}_X}(\mathcal{F},\mathcal{G}) := \prod_{k\in\mathbb{Z}} \mathcal{H}om_{\mathcal{O}_X}(\mathcal{F}^k,\mathcal{G}^{k+i}),$$

$$d^i((u^k)_k) = (d^{k+i}_{\mathcal{G}} \circ u^k - (-1)^i u^{k+1} \circ d^k_{\mathcal{F}})_k,$$

defines a complex $\mathcal{H}om_{\mathcal{O}_X}(\mathcal{F},\mathcal{G})$ of $\mathcal{O}_X$-modules. The construction is obviously functorial in $\mathcal{F}$ and $\mathcal{G}$, and we obtain extensions of the bi-functor

(21.18.2) $\mathcal{H}om_{\mathcal{O}_X}(-,-)\colon (\mathcal{O}_X\text{-Mod})^{\mathrm{opp}} \times (\mathcal{O}_X\text{-Mod}) \longrightarrow (\mathcal{O}_X\text{-Mod})$

to a bi-functor $\mathcal{H}om_{\mathcal{O}_X}(-,-)\colon C(X)^{\mathrm{opp}} \times C(X) \longrightarrow C(X)$ that induces a triangulated bi-functor

(21.18.3) $\mathcal{H}om_{\mathcal{O}_X}(-,-)\colon K(X)^{\mathrm{opp}} \times K(X) \longrightarrow K(X).$

We have

(21.18.4) $\Gamma(X,\mathcal{H}om_{\mathcal{O}_X}(\mathcal{F},\mathcal{G})) = \operatorname{Hom}_{\mathcal{O}_X}(\mathcal{F},\mathcal{G}) \in C(\Gamma(X,\mathcal{O}_X))$

for all complexes $\mathcal{F}$ and $\mathcal{G}$ of $\mathcal{O}_X$-modules.

To derive this functor, we show that Condition (ACII) of Section (F.51) is satisfied. In other words we have to show:

Lemma 21.87. *Let $\mathcal{I}$ be a K-injective complex in $K(X)$. Then the functor $\mathcal{H}om_{\mathcal{O}_X}(-,\mathcal{I})$ sends quasi-isomorphisms to quasi-isomorphisms in $K(X)$ (equivalently, to isomorphisms in $D(X)$).*

Proof. We have to show that $\mathcal{H}om_{\mathcal{O}_X}(-,\mathcal{I})$ sends exact complexes to exact complexes (Example F.169). Let $\mathcal{F}$ be a complex of $\mathcal{O}_X$-modules. For $i \in \mathbb{Z}$ the cohomology sheaf $H^i(\mathcal{H}om_{\mathcal{O}_X}(\mathcal{F},\mathcal{I}))$ is the sheaf associated to the presheaf

$$U \mapsto H^i(\Gamma(U, \mathcal{H}om_{\mathcal{O}_X}(\mathcal{F}, \mathcal{I})))$$

(*)
$$= H^i(\mathrm{Hom}_{\mathcal{O}_U}(\mathcal{F}_{|U}, \mathcal{I}_{|U}))$$

$$= \mathrm{Hom}_{K(U)}(\mathcal{F}_{|U}, \mathcal{I}_{|U}[i]),$$

where the last equality is by (F.18.2). Now $\mathcal{I}_{|U}$ (and hence $\mathcal{I}_{|U}[i]$) is K-injective by Lemma 21.12. Therefore, if $\mathcal{F}$ is an exact complex, then the presheaf (*) is zero by Proposition F.179 (2) and $\mathcal{H}om_{\mathcal{O}_X}(\mathcal{F}, \mathcal{I})$ is exact. $\qquad\square$

Definition and Remark 21.88. Now Section (F.51) shows that the functor (21.18.3) induces a triangulated bi-functor

(21.18.5) $$R\,\mathcal{H}om_{\mathcal{O}_X} : D(X)^{\mathrm{opp}} \times D(X) \to D(X),$$

called the *derived inner Hom functor*. Hence the partial functors $R\,\mathcal{H}om_{\mathcal{O}_X}(-, \mathcal{G})$ and $R\,\mathcal{H}om_{\mathcal{O}_X}(\mathcal{F}, -)$ are triangulated functors ($D(X)^{\mathrm{opp}}$ viewed as triangulated category as in Section (F.30)), and in particular we are given for all $p \in \mathbb{Z}$ isomorphisms

(21.18.6) $$R\,\mathcal{H}om_{\mathcal{O}_X}(\mathcal{F}, \mathcal{G}[p]) \cong R\,\mathcal{H}om_{\mathcal{O}_X}(\mathcal{F}, \mathcal{G})[p] \cong R\,\mathcal{H}om_{\mathcal{O}_X}(\mathcal{F}[-p], \mathcal{G}),$$

functorial in both variables.

If we choose K-injective resolutions $\mathcal{G} \to \mathcal{I}_{\mathcal{G}}$ of every complex $\mathcal{G}$, then by definition

(21.18.7) $$R\,\mathcal{H}om_{\mathcal{O}_X}(\mathcal{F}, \mathcal{G}) = \mathcal{H}om_{\mathcal{O}_X}(\mathcal{F}, \mathcal{I}_{\mathcal{G}})$$

and we obtain a morphism of functors in $\mathcal{F}$ and $\mathcal{G}$

(21.18.8) $$\mathcal{H}om_{\mathcal{O}_X}(\mathcal{F}, \mathcal{G}) \longrightarrow \mathcal{H}om_{\mathcal{O}_X}(\mathcal{F}, \mathcal{I}_{\mathcal{G}}) = R\,\mathcal{H}om_{\mathcal{O}_X}(\mathcal{F}, \mathcal{G}).$$

As the restriction of a K-injective complex to an open subset is again K-injective (see Lemma 21.12), for all $U \subseteq X$ open we have

(21.18.9)
$$R\,\mathcal{H}om_{\mathcal{O}_X}(\mathcal{F}, \mathcal{G})_{|U} = \mathcal{H}om_{\mathcal{O}_X}(\mathcal{F}, \mathcal{I}_{\mathcal{G}})_{|U}$$

$$= \mathcal{H}om_{\mathcal{O}_U}(\mathcal{F}_{|U}, \mathcal{I}_{\mathcal{G}|U})$$

$$= R\,\mathcal{H}om_{\mathcal{O}_U}(\mathcal{F}_{|U}, \mathcal{G}_{|U}).$$

We have a functorial isomorphism in $D(X)$

(21.18.10) $$R\,\mathcal{H}om_{\mathcal{O}_X}(\mathcal{O}_X, \mathcal{G}) = \mathcal{G}$$

since both sides are represented by $\mathcal{H}om_{\mathcal{O}_X}(\mathcal{O}_X, \mathcal{I}_{\mathcal{G}}) = \mathcal{I}_{\mathcal{G}}$.

Finally, if $(X, \mathcal{O}_X) = (*, A)$ for a commutative ring A, we identify $D(X)$ and $D(A)$ and have $R\,\mathcal{H}om_{\mathcal{O}_X}(F, G) = R\,\mathrm{Hom}_A(F, G)$ for $F, G \in D(A)$.

Again, it is often useful to calculate $R\,\mathcal{H}om$ via the first variable. For this one has the following result.

Lemma 21.89. *Let $\mathcal{F}$ and $\mathcal{G}$ be complexes of $\mathcal{O}_X$-modules. Suppose that $\mathcal{G}$ is bounded below, that $\mathcal{F}$ is bounded above and that $\mathcal{F}^n$ is a direct summand of a finite free $\mathcal{O}_X$-module for all n. Then*

$$R\,\mathcal{H}om_{\mathcal{O}_X}(\mathcal{F}, \mathcal{G}) = \mathcal{H}om_{\mathcal{O}_X}(\mathcal{F}, \mathcal{G}).$$

For a similar assertion with $\mathcal{G}$ arbitrary and $\mathcal{F}$ bounded see Lemma 21.144 below.

Proof. Let $\mathscr{G} \to \mathscr{I}$ be a K-injective resolution. As $\mathscr{G}$ is bounded below, we may assume that $\mathscr{I}$ is bounded below (Theorem F.185). Then $R\mathscr{H}om(\mathscr{F}, \mathscr{G}) = \mathscr{H}om(\mathscr{F}, \mathscr{I})$ and it suffices to show that the canonical map $u\colon \mathscr{H}om(\mathscr{F}, \mathscr{G}) \to \mathscr{H}om(\mathscr{F}, \mathscr{I})$ is a quasi-isomorphism. Since $\mathscr{F}$ is bounded above and $\mathscr{I}$ is bounded below, we have

$$\mathscr{H}om(\mathscr{F}, \mathscr{I})^n = \prod_{p \in \mathbb{Z}} \mathscr{H}om(\mathscr{F}^p, \mathscr{I}^{p+n}) = \bigoplus_{p \in \mathbb{Z}} \mathscr{H}om(\mathscr{F}^p, \mathscr{I}^{p+n}).$$

Similarly, $\mathscr{H}om(\mathscr{F}, \mathscr{G})^n$ is in each degree the direct sum of $\mathscr{H}om(\mathscr{F}^p, \mathscr{G}^{p+n})$. Therefore both complexes are the total complex associated to a double complex and u comes from the canonical map $(\mathscr{H}om(\mathscr{F}^p, \mathscr{G}^q))_{p,q} \to (\mathscr{H}om(\mathscr{F}^p, \mathscr{I}^q))_{p,q}$ of double complexes. As the formation of the total complex preserves quasi-isomorphisms in each direction (Lemma F.113), it suffices to show that for a fixed $p \in \mathbb{Z}$ the map

$$\mathscr{H}om(\mathscr{F}^p, \mathscr{G}) \to \mathscr{H}om(\mathscr{F}^p, \mathscr{I})$$

is a quasi-isomorphism. This holds because $\mathscr{F}^p$ is the direct summand of a finite free $\mathcal{O}_X$-module. $\qquad\square$

(21.19) K-flat complexes.

Let $(X, \mathcal{O}_X)$ be a ringed space. Recall from Section (F.19) that for complexes $\mathscr{F}$ and $\mathscr{G}$ of $\mathcal{O}_X$-modules the tensor complex $\mathscr{F} \otimes_{\mathcal{O}_X} \mathscr{G}$ is defined, that we obtain functors

(21.19.1) $- \otimes - = - \otimes_{\mathcal{O}_X} -\colon C(X) \times C(X) \longrightarrow C(X),$
(21.19.2) $- \otimes - = - \otimes_{\mathcal{O}_X} -\colon K(X) \times K(X) \longrightarrow K(X),$

and there exist functorial isomorphism $\mathscr{F} \otimes \mathscr{G} \xrightarrow{\sim} \mathscr{G} \otimes \mathscr{F}$ and $(\mathscr{F} \otimes \mathscr{G}) \otimes \mathscr{H} \xrightarrow{\sim} \mathscr{F} \otimes (\mathscr{G} \otimes \mathscr{H})$. Moreover, (21.19.2) is a triangulated bi-functor (Example F.214).

As passing to stalks commutes with tensor products and with direct sums, there is for all $x \in X$ a functorial isomorphism of complexes of $\mathcal{O}_{X,x}$-modules

(21.19.3) $(\mathscr{F} \otimes_{\mathcal{O}_X} \mathscr{G})_x \cong \mathscr{F}_x \otimes_{\mathcal{O}_{X,x}} \mathscr{G}_x$

for all complexes $\mathscr{F}$ and $\mathscr{G}$ of $\mathcal{O}_X$-modules.

Remark 21.90. Let $\mathscr{F}, \mathscr{G}$, and $\mathscr{H}$ be complexes of $\mathcal{O}_X$-modules.
(1) A sheafified version of Proposition F.93 shows that there is a functorial isomorphism of complexes of $\mathcal{O}_X$-modules

(21.19.4) $\mathscr{H}om_{\mathcal{O}_X}(\mathscr{F} \otimes_{\mathcal{O}_X} \mathscr{G}, \mathscr{H}) \xrightarrow{\sim} \mathscr{H}om_{\mathcal{O}_X}(\mathscr{F}, \mathscr{H}om_{\mathcal{O}_X}(\mathscr{G}, \mathscr{H})).$

(2) By applying $Z^0(-) \circ \Gamma(X, -)$ to (21.19.4) one sees by (F.18.2) that for a fixed complex $\mathscr{G}$ the tensor functor $- \otimes \mathscr{G}\colon C(X) \to C(X)$ is left adjoint to the functor $\mathscr{H}om_{\mathcal{O}_X}(\mathscr{G}, -)\colon C(X) \to C(X)$.

In particular we see that $- \otimes \mathscr{F}$ commutes with arbitrary colimits. By symmetry we see that also the functor $\mathscr{F} \otimes -$ commutes with arbitrary colimits.

We obtain the same assertions for the induced functors $- \otimes \mathscr{F}\colon K(X) \to K(X)$ and $\mathscr{H}om_{\mathcal{O}_X}(\mathscr{G}, -)\colon K(X) \to K(X)$, now by applying $H^0(-) \circ \Gamma(X, -)$ to (21.19.4).

We now want to construct the left derivation of the tensor functor. The following class of complexes will be acyclic for the tensor functor, see Lemma 21.96 below.

Definition 21.91. *A complex $\mathscr{P}$ of $\mathscr{O}_X$-modules is called* K-flat, *if for every exact complex $\mathscr{F}$ of $\mathscr{O}_X$-modules, the tensor complex $\mathscr{F} \otimes_{\mathscr{O}_X} \mathscr{P}$ is exact (or equivalently: $\mathscr{P} \otimes_{\mathscr{O}_X} \mathscr{F}$ is exact).*

Applying this definition to the case where he underlying topological space of X consists of a single point, we obtain as a special case that for a ring A a complex P of A-modules is called *K-flat*, if for every exact complex F of A-modules, the tensor complex $P \otimes_A F$ is exact.

Remark 21.92.
(1) An $\mathscr{O}_X$-module $\mathscr{F}$ is flat if and only if $\mathscr{F}$, considered as a complex concentrated in degree 0, is K-flat.
(2) A complex $\mathscr{F}$ of $\mathscr{O}_X$-modules is K-flat if and only if for every quasi-isomorphism of complexes $\mathscr{G}_1 \to \mathscr{G}_2$ the induced map $\mathscr{F} \otimes \mathscr{G}_1 \to \mathscr{F} \otimes \mathscr{G}_2$ is again a quasi-isomorphism (Example F.169).
(3) The associativity of the tensor complex (Remark F.92) implies that if $\mathscr{P}$ and $\mathscr{Q}$ are K-flat complexes of $\mathscr{O}_X$-modules then the tensor complex $\mathscr{P} \otimes_{\mathscr{O}_X} \mathscr{Q}$ is a K-flat complex.
(4) A complex $\mathscr{P}$ of $\mathscr{O}_X$-modules is K-flat if and only if for every $x \in X$ the complex $\mathscr{P}_x$ of $\mathscr{O}_{X,x}$-modules is K-flat.
    Indeed, the condition is sufficient because exactness of complexes can be checked on stalks and because tensor products commute with forming stalks (21.19.3). Conversely, suppose that $\mathscr{P}$ is K-flat, let $x \in X$, and let F be an exact complex of $\mathscr{O}_{X,x}$-modules. Let $i_x \colon (\{x\}, \mathscr{O}_{X,x}) \to (X, \mathscr{O}_X)$ be the canonical morphism of ringed spaces. Then $i_{x,*}F$ is an exact complex of $\mathscr{O}_X$-modules and hence $\mathscr{P} \otimes_{\mathscr{O}_X} i_{x,*}F$ is exact. Therefore $(\mathscr{P} \otimes_{\mathscr{O}_X} i_{x,*}F)_x = \mathscr{P}_x \otimes_{\mathscr{O}_{X,x}} F$ is exact. This shows that $\mathscr{P}_x$ is K-flat.
(5) As tensoring with a fixed complex $\mathscr{F}$ is a triangulated functor, the long exact cohomology sequence (21.4.1) shows that the K-flat complexes form a triangulated subcategory of $K(X)$.
(6) Every filtered colimit of K-flat complexes in $C(X)$ is again K-flat. Indeed this follows because arbitrary colimits commute with tensor products (Remark 21.90) and filtered colimits commute with taking cohomology because filtered colimits are exact (Proposition 21.7).

Lemma 21.93. *Let $\mathscr{P}$ be a bounded above complex of flat $\mathscr{O}_X$-modules. Then $\mathscr{P}$ is a K-flat complex.*

Proof. We can write $\mathscr{P} = \operatorname{colim}_m \sigma^{\geq -m} \mathscr{P}$ and hence can assume by Remark 21.92 (6) that $\mathscr{P}$ is a bounded complex of flat modules, say of length $m - k + 1$:

$$\mathscr{P} = (\ldots \to 0 \to \mathscr{P}^k \to \ldots \to \mathscr{P}^m \to 0 \to \ldots).$$

Then the termwise split sequence $0 \to \sigma^{\geq m} \mathscr{P} \to \mathscr{P} \to \sigma^{\leq m-1} \mathscr{P} \to 0$ yields a distinguished triangle

$$\sigma^{\geq m} \mathscr{P} \to \mathscr{P} \to \sigma^{\leq m-1} \mathscr{P} \to \sigma^{\geq m} \mathscr{P}[1].$$

By induction on the length of the complex using Remark 21.92 (1) we can assume that $\sigma^{\geq m} \mathscr{P}$ and $\sigma^{\leq m-1} \mathscr{P}$ are K-flat. Hence $\mathscr{P}$ is K-flat by Remark 21.92 (5). $\qquad\square$

Proposition 21.94. *For each complex $\mathscr{F}$ of $\mathscr{O}_X$-modules there exists a K-flat complex $\mathscr{P}_{\mathscr{F}}$ together with a quasi-isomorphism $\alpha_{\mathscr{F}} \colon \mathscr{P}_{\mathscr{F}} \to \mathscr{F}$, such that $\mathscr{P}_{\mathscr{F}}$ and $\alpha_{\mathscr{F}}$ depend functorially on $\mathscr{F}$ and are compatible with shifts: $\mathscr{P}_{\mathscr{F}[1]} = \mathscr{P}_{\mathscr{F}}[1]$ and $\alpha_{\mathscr{F}[1]} = \alpha_{\mathscr{F}}[1]$.*

Proof. Let $\mathscr{F}$ be an $\mathcal{O}_X$-module. We will construct a functorial epimorphism $P^0(\mathscr{F}) \to \mathscr{F}$ with $P^0(\mathscr{F})$ a flat $\mathcal{O}_X$-module. For this recall from (21.1.6) the $\mathcal{O}_X$-module $\mathcal{O}_{U \subseteq X} = j_!(\mathcal{O}_{X|U})$ for an open subspace $j \colon U \hookrightarrow X$. Its stalk in $x \in X$ is $\mathcal{O}_{X,x}$ for $x \in U$ and zero otherwise. In particular, it is a flat $\mathcal{O}_X$-module. Define a flat $\mathcal{O}_X$-module

$$P^0(\mathscr{F}) := \bigoplus_{U \subseteq X \text{ open}} \bigoplus_{s \in \mathscr{F}(U)} \mathcal{O}_{U \subseteq X}$$

and a homomorphism $P^0(\mathscr{F}) \to \mathscr{F}$ whose restriction to the direct summand indexed by (U, s) is the homomorphism $\mathcal{O}_{U \subseteq X} \to \mathscr{F}$ corresponding to s via (21.1.8). Therefore every local section of $\mathscr{F}$ is in the image and in particular $P^0(\mathscr{F}) \to \mathscr{F}$ is an epimorphism.

Now we apply Corollary F.197 and see that every complex $\mathscr{F}$ of $\mathcal{O}_X$-modules has a functorial left resolution $\mathscr{P}_{\mathscr{F}} \to \mathscr{F}$, where $\mathscr{P}_{\mathscr{F}}$ is a filtered colimit of bounded above complexes with flat components. Hence $\mathscr{P}_{\mathscr{F}}$ is K-flat by Lemma 21.93 and Remark 21.92 (6). $\square$

Remark 21.95. The proof of Proposition 21.94 shows in fact the following more precise assertions.
(1) One can choose $\mathscr{P}_{\mathscr{F}}$ to be the colimit of an inductive system $(P_n)_{n \in \mathbb{N}}$ of bounded above complexes such that every component of each P_n is a direct sum of $\mathcal{O}_X$-modules of the form $\mathcal{O}_{U \subseteq X}$.
(2) If there exists $n \in \mathbb{Z}$ such that $H^i(\mathscr{F}) = 0$ for all $i > n$, one has a quasi-isomorphism $\tau^{\leq n}\mathscr{F} \to \mathscr{F}$ and hence there exists a quasi-isomorphism $\mathscr{P}_{\tau^{\leq n}\mathscr{F}} \to \mathscr{F}$ such that every component of $\mathscr{P}_{\tau^{\leq n}\mathscr{F}}$ is a direct sum of $\mathcal{O}_X$-modules of the form $\mathcal{O}_{U \subseteq X}$ and such that $\mathscr{P}^i_{\tau^{\leq n}\mathscr{F}} = 0$ for all $i > n$.

(21.20) Derived tensor product and Tor sheaves.

We now want to derive the bi-functor given by the tensor product. We check that (the dual versions of) Conditions (ACI) and (ACII) of Section (F.51) are satisfied using the class of K-flat complexes. We already know that there exist functorial quasi-isomorphisms $\mathscr{P}_{\mathscr{F}} \to \mathscr{F}$ for all complexes $\mathscr{F}$ with $\mathscr{P}_{\mathscr{F}}$ a K-flat complex (Proposition 21.94).

Lemma 21.96. *Let $\mathscr{F}$ be a complex of $\mathcal{O}_X$-modules. Denote by $T^1_{\mathscr{F}} \colon K(X) \to D(X)$ (resp. by $T^2_{\mathscr{F}} \colon K(X) \to D(X)$) the composition of the functor $(-) \otimes \mathscr{F} \colon K(X) \to K(X)$ (resp. of the functor $\mathscr{F} \otimes (-) \colon K(X) \to K(X)$) with the localization $K(X) \to D(X)$.*
(1) *For every K-flat complex $\mathscr{P}$ the functors $T^1_{\mathscr{P}}$ and $T^2_{\mathscr{P}}$ send quasi-isomorphisms to isomorphisms.*
(2) *For every exact K-flat complex $\mathscr{P}$ the complex $\mathscr{P} \otimes \mathscr{F}$ is exact for all complexes $\mathscr{F}$ of $\mathcal{O}_X$-modules.*
(3) *Every K-flat complex $\mathscr{P}$ of $\mathcal{O}_X$-modules is left acyclic for $T^1_{\mathscr{F}}$ and $T^2_{\mathscr{F}}$.*

Proof. *(1).* By symmetry, it suffices to consider T^1. The functor $T^1_{\mathscr{P}}$ sends quasi-isomorphisms to isomorphisms if and only if it maps exact complexes to the zero complex in $D(X)$ (Example F.169). This shows Assertion (1).

(2). Indeed, choose a quasi-isomorphism $\mathscr{Q} \to \mathscr{F}$ (Proposition 21.94). As $\mathscr{P}$ is K-flat, we may replace $\mathscr{F}$ by $\mathscr{Q}$ using (1). But then $\mathscr{P} \otimes \mathscr{Q}$ is exact because $\mathscr{P}$ is exact.

(3). Let $\mathscr{Q}' \to \mathscr{P}$ be a quasi-isomorphism. By Proposition 21.94 there exists a quasi-isomorphism $\mathscr{Q} \to \mathscr{Q}'$ with $\mathscr{Q}$ a K-flat complex. It suffices to show that by applying $T^1_{\mathscr{F}}$ to the quasi-isomorphism $\mathscr{Q} \to \mathscr{P}$ we obtain an isomorphism in $D(X)$.

Let $\mathscr{C}$ be the mapping cone of the quasi-isomorphism $\mathscr{Q} \to \mathscr{P}$. Then $\mathscr{C}$ is exact (Remark F.85) and K-flat (Remark 21.92 (5)). Therefore $\mathscr{C} \otimes \mathscr{F}$ is again exact by (2). As $(-) \otimes \mathscr{F}$ preserves distinguished triangles, we have a distinguished triangle

$$\mathscr{Q} \otimes \mathscr{F} \longrightarrow \mathscr{P} \otimes \mathscr{F} \longrightarrow \mathscr{C} \otimes \mathscr{F} \longrightarrow (\mathscr{Q} \otimes \mathscr{P})[1].$$

From its associated long cohomology sequence (Example F.138) we see that the homomorphism $\mathscr{Q} \otimes \mathscr{F} \to \mathscr{P} \otimes \mathscr{F}$ is a quasi-isomorphism, i.e., an isomorphism in $D(X)$. $\quad\square$

We can now apply the general formalism of Section (F.51) and obtain left derivations of the tensor product.

Proposition/Definition 21.97. *Let $(X, \mathscr{O}_X)$ be a ringed space. The* derived tensor product *is the triangulated bi-functor*

$$\otimes^L := \otimes^L_{\mathscr{O}_X} : D(X) \times D(X) \to D(X)$$

which maps a pair $(\mathscr{G}, \mathscr{H})$ of a complex $\mathscr{G}$ and a complex $\mathscr{H}$ with K-flat resolution $\mathscr{P}_{\mathscr{H}} \to \mathscr{H}$ to the tensor product $\mathscr{G} \otimes \mathscr{P}_{\mathscr{H}}$ (considered as an object of $D(X)$).

For fixed $\mathscr{G}$, the functors $\mathscr{G} \otimes^L -$ and $- \otimes^L \mathscr{G}$ are the left derived functors of the tensor product functors $\mathscr{G} \otimes -$ and $- \otimes \mathscr{G}$. There is a natural equivalence between these two functors.

The derived tensor product is associative in the obvious sense.

Definition 21.98. *Let $\mathscr{F}$ and $\mathscr{G}$ be complexes of $\mathscr{O}_X$-modules. For $n \in \mathbb{Z}$ we define the n-th* Tor sheaf *as the $\mathscr{O}_X$-module*

$$\mathscr{T}or_n(\mathscr{F}, \mathscr{G}) := \mathscr{T}or_n^{\mathscr{O}_X}(\mathscr{F}, \mathscr{G}) := H^{-n}(\mathscr{F} \otimes^L \mathscr{G}).$$

Remark 21.99. Let $\mathscr{F}$ and $\mathscr{G}$ be complexes of $\mathscr{O}_X$-modules.
(1) Given a short exact sequence $0 \to \mathscr{G}' \to \mathscr{G} \to \mathscr{G}'' \to 0$ of complexes of $\mathscr{O}_X$-modules or, more generally, given a distinguished triangle $\mathscr{G}' \to \mathscr{G} \to \mathscr{G}'' \to$ in $D(X)$, there is a long exact Tor sheaf sequence

$$
\begin{aligned}
\ldots \longrightarrow \mathscr{T}or_{n+1}(\mathscr{F}, \mathscr{G}') &\longrightarrow \mathscr{T}or_{n+1}(\mathscr{F}, \mathscr{G}) \longrightarrow \mathscr{T}or_{n+1}(\mathscr{F}, \mathscr{G}'') \\
\longrightarrow \mathscr{T}or_n(\mathscr{F}, \mathscr{G}') &\longrightarrow \mathscr{T}or_n(\mathscr{F}, \mathscr{G}) \longrightarrow \mathscr{T}or_n(\mathscr{F}, \mathscr{G}'') \\
\longrightarrow \mathscr{T}or_{n-1}(\mathscr{F}, \mathscr{G}') &\longrightarrow \ldots
\end{aligned}
$$

(21.20.1)

obtained as the long exact cohomology sequence of the distinguished triangle $\mathscr{F} \otimes^L \mathscr{G}' \to \mathscr{F} \otimes^L \mathscr{G} \to \mathscr{F} \otimes^L \mathscr{G}'' \to$.
(2) Suppose there are $m, n \in \mathbb{Z}$ such that $H^i(\mathscr{F}) = 0$ and $H^j(\mathscr{G}) = 0$ for all $i > n$ and all $j > m$. Then $\mathscr{T}or_k(\mathscr{F}, \mathscr{G}) = 0$ for all $k < -n - m$.

Indeed, we can choose a K-flat resolution $\mathscr{P}_{\mathscr{F}} \to \mathscr{F}$ with $\mathscr{P}_{\mathscr{F}}^i = 0$ for all all $i > n$ (Remark 21.95 (2)). Then $\mathscr{F} \otimes^L \mathscr{G} = \mathscr{P}_{\mathscr{F}} \otimes \mathscr{G}$. By Lemma 21.96 (1) we may also replace $\mathscr{G}$ by the complex $\tau^{\leq m}\mathscr{G}$ since $\tau^{\leq m}\mathscr{G} \to \mathscr{G}$ is a quasi-isomorphism. But then $(\mathscr{P}_{\mathscr{F}} \otimes \mathscr{G})^k = 0$ for all $k > n + m$ by the definition of the tensor product of complexes.
(3) Let $\mathscr{F}$ and $\mathscr{G}$ be $\mathscr{O}_X$-modules. Then $\mathscr{T}or_n(\mathscr{F}, \mathscr{G}) = 0$ for all $n < 0$ by (2) and

$$\mathscr{T}or_0(\mathscr{F}, \mathscr{G}) = \mathscr{F} \otimes_{\mathscr{O}_X} \mathscr{G}$$

by the dual version of Remark F.200 (3) because $\otimes$ is right exact in each variable as functor $(\mathscr{O}_X\text{-Mod}) \to (\mathscr{O}_X\text{-Mod})$.

Proposition 21.100. *Let $\mathscr{F}$ be an $\mathscr{O}_X$-module. Then the following are equivalent.*
(i) *$\mathscr{F}$ is a flat $\mathscr{O}_X$-module.*
(ii) *$\mathcal{T}or_1^{\mathscr{O}_X}(\mathscr{F},\mathscr{G}) = 0$ for every $\mathscr{O}_X$-module $\mathscr{G}$.*
(iii) *$\mathcal{T}or_i^{\mathscr{O}_X}(\mathscr{F},\mathscr{G}) = 0$ for all $i \geq 1$ and every complex of $\mathscr{O}_X$-modules $\mathscr{G}$ concentrated in degrees ≥ 0.*

Proof. If $\mathscr{F}$ is flat, then it is K-flat as a complex concentrated in degree 0. Hence for every complex of $\mathscr{O}_X$-modules we have $\mathscr{F} \otimes^L \mathscr{G} = \mathscr{F} \otimes \mathscr{G}$ which has only zero coefficients in degrees ≤ 0 if $\mathscr{G}$ is concentrated in degrees ≥ 0. This shows "(i) $\Rightarrow$ (iii)".

Clearly, (iii) implies (ii). Suppose that (ii) holds. If $0 \to \mathscr{G}' \to \mathscr{G} \to \mathscr{G}'' \to 0$ is an exact sequence of $\mathscr{O}_X$-modules, then by Remark 21.99 one obtains a long exact sequence

$$\cdots \to \mathcal{T}or_1(\mathscr{F},\mathscr{G}'') \to \mathscr{F} \otimes \mathscr{G}' \to \mathscr{F} \otimes \mathscr{G} \to \mathscr{F} \otimes \mathscr{G}'' \to 0$$

with $\mathcal{T}or_1(\mathscr{F},\mathscr{G}'') = 0$. In other words, $\mathscr{F} \otimes_{\mathscr{O}_X} (-)$ is an exact functor, i.e., $\mathscr{F}$ is flat. $\square$

Remark 21.101. If X consists of a single point and $A := \Gamma(X,\mathscr{O}_X)$, we identify the category of $\mathscr{O}_X$-modules with the category of A-modules and we obtain a triangulated bi-functor

$$\otimes_A^L \colon D(A) \times D(A) \longrightarrow D(A).$$

For complexes F and G of A-modules we define Tor A-modules

$$\mathrm{Tor}_i^A(F,G) := H^{-i}(F \otimes_A^L G).$$

Suppose that F and G are A-modules. Then the above general definitions show that $\mathrm{Tor}_i^A(F,G)$ can be calculated as follows. Choose a flat resolution $K^\bullet$ of one of the modules, say of F, in other words an exact sequence $\cdots \longrightarrow K^{-2} \xrightarrow{d^{-2}} K^{-1} \xrightarrow{d^{-1}} K^0 \longrightarrow F \longrightarrow 0$ with K^i flat. Then $\mathrm{Tor}_i^A(F,G)$ is the cohomology in degree $-i$ of the complex

$$\cdots \longrightarrow K^{-2} \otimes_A G \xrightarrow{d^{-2}\otimes\mathrm{id}} K^{-1} \otimes_A G \xrightarrow{d^{-1}\otimes\mathrm{id}} K^0 \otimes_A G \longrightarrow 0 \longrightarrow \cdots$$

By Proposition 21.100 an A-module M is flat if and only if $\mathrm{Tor}_1^A(M,N) = 0$ for every A-module N.

It is even enough to check that $\mathrm{Tor}_1^A(M,A/I) = 0$ for every finitely generated ideal I of A. Indeed, the exact sequence $0 \to I \to A \to A/I \to 0$ yields the exact sequence

$$0 = \mathrm{Tor}_1(M,A/I) \longrightarrow M \otimes I \longrightarrow M.$$

Hence $M \otimes I \to M$ is injective for every finitely generated ideal and we conclude by Proposition B.15.

Remark 21.102. Let $x \in X$. Then forming the stalk $(-)_x$ is an exact functor which commutes with forming tensor products and which sends K-flat complexes of $\mathscr{O}_X$-modules to K-flat complexes of $\mathscr{O}_{X,x}$-modules (Remark 21.92 (4)). Hence one has

(21.20.2) $(\mathscr{F} \otimes_{\mathscr{O}_X}^L \mathscr{G})_x = \mathscr{F}_x \otimes_{\mathscr{O}_{X,x}}^L \mathscr{G}_x$

for all complexes $\mathscr{F}$ and $\mathscr{G}$ of $\mathscr{O}_X$-modules. Again, since $(-)_x$ is exact, it commutes with passage to cohomology and we obtain a functorial isomorphism

(21.20.3) $\mathcal{T}or_p^{\mathscr{O}_X}(\mathscr{F},\mathscr{G})_x = \mathrm{Tor}_p^{\mathscr{O}_{X,x}}(\mathscr{F}_x,\mathscr{G}_x), \qquad p \in \mathbb{Z}.$

Remark 21.103. Let $\mathscr{F}$ and $\mathscr{G}$ be complexes in $D^-(X)$. Replacing $\mathscr{F}$ and $\mathscr{G}$ by K-flat bounded above complexes and applying Lemma F.112 yields two bounded spectral sequences, functorial in $\mathscr{F}$ and $\mathscr{G}$,

$$(21.20.4) \qquad \begin{aligned} {}_I E_2^{pq} &= H^p(\mathscr{F} \otimes_{\mathscr{O}_X}^L H^q(\mathscr{G})) \Rightarrow H^n(\mathscr{F} \otimes_{\mathscr{O}_X}^L \mathscr{G}), \\ {}_{II} E_2^{pq} &= H^q(H^p(\mathscr{F}) \otimes_{\mathscr{O}_X}^L \mathscr{G}) \Rightarrow H^n(\mathscr{F} \otimes_{\mathscr{O}_X}^L \mathscr{G}). \end{aligned}$$

Both induced filtrations on the limit term are finite.

Remark 21.104. Let $\mathscr{F}$ and $\mathscr{G}$ be complexes in $D(X)$. For all $p, q \in \mathbb{Z}$ there is an $\mathscr{O}_X$-linear map

$$(21.20.5) \qquad \cup \colon H^p(\mathscr{F}) \otimes_{\mathscr{O}_X} H^q(\mathscr{G}) \longrightarrow H^{p+q}(\mathscr{F} \otimes_{\mathscr{O}_X}^L \mathscr{G})$$

defined as follows. We replace $\mathscr{F}$ or $\mathscr{G}$ by a K-flat complex. Then $\mathscr{F} \otimes_{\mathscr{O}_X}^L \mathscr{G} = \mathscr{F} \otimes_{\mathscr{O}_X} \mathscr{G}$. Let α (resp. β) be a local section of $H^p(\mathscr{F})$ (resp. of $H^q(\mathscr{G})$) over an open subset $U \subseteq X$, represented locally by $a_i \in \mathrm{Ker}(\mathscr{F}^p(U_i) \to \mathscr{F}^{p+1}(U_i))$ and $b_i \in \mathrm{Ker}(\mathscr{G}^q(U_i) \to \mathscr{G}^{q+1}(U_i))$ for some open covering $(U_i)_i$ of U. Then

$$a_i \otimes b_i \in \mathrm{Ker}((\mathscr{F} \otimes \mathscr{G})^{p+q}(U_i) \longrightarrow (\mathscr{F} \otimes \mathscr{G})^{p+q+1}(U_i))$$

yields a class $a_i \cup b_i \in H^{p+q}(\mathscr{F} \otimes \mathscr{G})(U_i)$. Then it is straightforward to check that the $a_i \cup b_i$ agree on all intersections $U_i \cap U_j$ and are the restriction of a unique element $\alpha \cup \beta \in H^{p+q}(\mathscr{F} \otimes \mathscr{G})(U)$ which does not depend on the choice of $(U_i)_i$ or of the a_i and b_i.

(21.21) Ext sheaves.

Let X be a ringed space. Recall from Section (F.52) that we defined for complexes $\mathscr{F}$ and $\mathscr{G}$ of $\mathscr{O}_X$-modules and for all $i \in \mathbb{Z}$ the $\Gamma(X, \mathscr{O}_X)$-modules

$$(21.21.1) \qquad \mathrm{Ext}^i_{\mathscr{O}_X}(\mathscr{F}, \mathscr{G}) = \mathrm{Hom}_{D(X)}(\mathscr{F}, \mathscr{G}[i]) \cong H^i(R\,\mathrm{Hom}_{\mathscr{O}_X}(\mathscr{F}, \mathscr{G})).$$

Similarly to the second description, we define the $\mathscr{E}xt$ sheaves as the derived functors of the sheaf $\mathscr{H}om$ functor:

$$(21.21.2) \qquad \mathscr{E}xt^i(\mathscr{F}, \mathscr{G}) := H^i(R\,\mathscr{H}om_{\mathscr{O}_X}(\mathscr{F}, \mathscr{G})).$$

There the following easy properties of $\mathscr{E}xt$ sheaves.

Remark 21.105.
(1) Let $b, c \in \mathbb{Z}$. If $\mathscr{F}$ is in $D^{(-\infty, b]}(X)$ and $\mathscr{G}$ is in $D^{[c, \infty)}(X)$, then for $p < c - b$ one has $\mathrm{Ext}^p_{\mathscr{O}_X}(\mathscr{F}, \mathscr{G}) = 0$ and $\mathscr{E}xt^p_{\mathscr{O}_X}(\mathscr{F}, \mathscr{G}) = 0$ by Remark F.218.
(2) If $U \subseteq X$ is an open subspace, then

$$(21.21.3) \qquad \mathscr{E}xt^i_{\mathscr{O}_X}(\mathscr{F}, \mathscr{G})_{|U} = \mathscr{E}xt^i_{\mathscr{O}_U}(\mathscr{F}_{|U}, \mathscr{G}_{|U})$$

for all $i \in \mathbb{Z}$ by (21.18.9).

For every $\mathscr{O}_X$-module $\mathscr{F}$, one has an equality $\Gamma(X, -) \circ \mathscr{H}om(\mathscr{F}, -) = \mathrm{Hom}(\mathscr{F}, -)$ as functors $(\mathscr{O}_X\text{-Mod}) \to (\Gamma(X, \mathscr{O}_X)\text{-Mod})$. To obtain a derived version we will use the following lemma.

Lemma 21.106. *Let $(X, \mathscr{O}_X)$ be a ringed space, let $\mathscr{I}$ be a K-injective complexes of $\mathscr{O}_X$-modules, and let $\mathscr{P}$ be a K-flat complex of $\mathscr{O}_X$-modules. Then $\mathscr{H}om_{\mathscr{O}_X}(\mathscr{P}, \mathscr{I})$ is a K-injective complex of $\mathscr{O}_X$-modules.*

Proof. Let $\mathscr{F}$ be an exact complex of $\mathscr{O}_X$-modules. By Remark 21.90 we have a functorial isomorphism

$$\operatorname{Hom}_{K(X)}(\mathscr{F}, \mathscr{H}om_{\mathscr{O}_X}(\mathscr{P}, \mathscr{I})) \cong \operatorname{Hom}_{K(X)}(\mathscr{F} \otimes \mathscr{P}, \mathscr{I})$$

and $\mathscr{F} \otimes \mathscr{P}$ is exact because $\mathscr{P}$ is K-flat. We conclude by Proposition F.179 (2). □

Proposition 21.107. *There is an isomorphism of triangulated bi-functors*

$$(21.21.4) \qquad\qquad R\operatorname{Hom}_{\mathscr{O}_X}(-, -) \xrightarrow{\sim} R\Gamma\, R\mathscr{H}om_{\mathscr{O}_X}(-, -).$$

Proof. By Proposition F.176 it suffices to show that for a fixed complex $\mathscr{F}$ in $C(X)$ the functor $\mathscr{H}om_{\mathscr{O}_X}(\mathscr{F}, -)$ sends a K-injective complex $\mathscr{I}$ to a right Γ-acyclic complex. By Lemma 21.87 we may replace $\mathscr{F}$ by a K-flat resolution which exists by Proposition 21.94. Then Lemma 21.106 shows that $\operatorname{Hom}_{\mathscr{O}_X}(\mathscr{F}, \mathscr{I})$ is K-injective and in particular acyclic for every functor. □

In particular we find that for all complexes $\mathscr{F}$ and $\mathscr{G}$ of $\mathscr{O}_X$-modules we have

$$(21.21.5) \qquad\qquad H^p(X, R\mathscr{H}om_{\mathscr{O}_X}(\mathscr{F}, \mathscr{G})) = \operatorname{Ext}^p_{\mathscr{O}_X}(\mathscr{F}, \mathscr{G})$$

for all $p \in \mathbb{Z}$.

Restricting the isomorphism (21.21.4) to complexes $\mathscr{F}$ that are bounded below in the opposite category and to bounded below complexes $\mathscr{G}$ we obtain the following Grothendieck spectral sequence by Proposition F.212.

Corollary 21.108. *(Local-to-global Ext spectral sequence) Let $(X, \mathscr{O}_X)$ be a ringed space, let $\mathscr{F}$ and $\mathscr{G}$ be complexes of $\mathscr{O}_X$-modules. Suppose that $\mathscr{F}$ is bounded above and that $\mathscr{G}$ is bounded below. Then we have a convergent spectral sequence*

$$E_2^{pq} = H^p(X, \mathscr{E}xt^q_{\mathscr{O}_X}(\mathscr{F}, \mathscr{G})) \Rightarrow \operatorname{Ext}^{p+q}_{\mathscr{O}_X}(\mathscr{F}, \mathscr{G}),$$

which is functorial in $\mathscr{F}$ (contravariant) and $\mathscr{G}$ (covariant).

(21.22) Derived functor of the inverse image functor.

In this section we denote by $f\colon X \to Y$ a morphism of ringed spaces. We want to define the left derived functor of the pullback functor f^*. We first convince ourselves that K-flat complexes are acyclic for f^*.

Remark 21.109. Let $\mathscr{Q}$ be a K-flat complex of $\mathscr{O}_Y$-modules.
(1) As we can test K-flatness on stalks (Remark 21.92 (4)) we see that its pullback $f^*\mathscr{Q}$ is a K-flat complex of $\mathscr{O}_X$-modules: Let $x \in X$. If G is an exact complex of $\mathscr{O}_{X,x}$-modules, then

$$f^*(\mathscr{Q})_x \otimes_{\mathscr{O}_{X,x}} F = \mathscr{Q}_{f(x)} \otimes_{\mathscr{O}_{Y,f(x)}} F$$

is again an exact complex of $\mathscr{O}_{X,x}$-modules.

(2) If $\mathscr{Q}$ is in addition exact, then $f^*(\mathscr{Q})_x = \mathscr{Q}_{f(x)} \otimes_{\mathscr{O}_{Y,f(x)}} \mathscr{O}_{X,x}$ is exact for all $x \in X$ because of Lemma 21.96 (2). Hence $f^*(\mathscr{Q})$ is also exact. Now the same argument as in the proof of Lemma 21.96 (3) shows that K-flat complexes are left f^*-acyclic.

Therefore, using K-flat resolutions, we obtain the derived functor Lf^* of the inverse image functor f^*:

Proposition/Definition 21.110. *The pullback functor* $f^* \colon (\mathscr{O}_Y\text{-Mod}) \to (\mathscr{O}_X\text{-Mod})$ *has a left derived functor*

$$Lf^* \colon D(Y) \to D(X)$$

which maps a complex $\mathscr{G}$ *of* $\mathscr{O}_Y$-modules with K-flat resolution $\mathscr{P}_{\mathscr{G}} \to \mathscr{G}$ *to* $f^*\mathscr{P}_{\mathscr{G}}$. *We define for* $i \in \mathbb{Z}$

$$L_i f^*(\mathscr{G}) := H^{-i} Lf^*(\mathscr{G}).$$

Of course, if the morphism f is flat, then f^* is an exact functor and hence $Lf^* = f^*$ (Example F.169).

Remark 21.111. Let $\mathscr{G}$ be a complex of $\mathscr{O}_Y$-modules.
(1) For every exact sequence $0 \to \mathscr{G}' \to \mathscr{G} \to \mathscr{G}'' \to 0$ in $C(Y)$ or, more generally, for every distinguished triangle $\mathscr{G}' \to \mathscr{G} \to \mathscr{G}'' \to$ in $D(Y)$ we obtain a long exact sequence of $\mathscr{O}_X$-modules

$$(21.22.1) \qquad \cdots \longrightarrow L_{n+1}f^*\mathscr{G}'' \longrightarrow L_n f^*\mathscr{G}' \longrightarrow L_n f^*\mathscr{G} \longrightarrow L_n f^*\mathscr{G}'' \longrightarrow \cdots$$

(2) Suppose there exists $n \in \mathbb{Z}$ such that $H^i(\mathscr{G}) = 0$ for all $i > n$. Then $L_i f^*\mathscr{G} = 0$ for all $i < -n$.
(3) In particular, if $\mathscr{G}$ is an $\mathscr{O}_Y$-module, then $L_i f^*\mathscr{G} = 0$ for $i < 0$ and $L_0 f^*\mathscr{G} = f^*\mathscr{G}$.

Proposition 21.112. *Let* $f \colon X \to Y$ *and* $g \colon Y \to Z$ *be morphisms of ringed spaces. Then there is a natural isomorphism*

$$L(g \circ f)^* \cong Lf^* \circ Lg^*$$

of functors $D(Z) \to D(X)$.

Proof. By Remark 21.109 (2), K-flat complexes are left acyclic for the inverse image functor. Hence it suffices to show that g^* sends K-flat complexes to K-flat complexes by Proposition F.176. This we have seen in Remark 21.109 (1). □

(21.23) Derived direct image and composition.

Let $f \colon X \to Y$ and $g \colon Y \to Z$ be morphisms of ringed spaces. Recall that for bounded below complexes $\mathscr{F}$ of $\mathscr{O}_X$-modules we already have seen that the functorial morphism

$$R(g_* \circ f_*)(\mathscr{F}) \xrightarrow{\sim} (Rg_* \circ Rf_*)(\mathscr{F})$$

is an isomorphism (Proposition 21.41). We now want to extend this result to unbounded complexes.

As f^* is left adjoint to f_* we obtain by Proposition F.191 triangulated adjoint pairs of functors (Lf^*, Rf_*), (Lg^*, Rg_*), and $(L(f^*) \circ L(g_*), R(g_* \circ f_*))$, where we identify $L(f^*) \circ L(g_*) = L(f^* \circ g^*)$ by Proposition 21.112. One can show that this yields purely formally using Proposition F.191 an isomorphism of triangulated functors

$$R(g_* \circ f_*) \xrightarrow{\sim} Rg_* \circ Rf_*.$$

But for concrete calculations it is sometimes helpful to know that this isomorphism is obtained by right G-acyclicity of $f_*\mathscr{F}$ for K-injective complexes $\mathscr{F}$ (see Proposition F.176). To show this we start with a lemma that will also be used later.

Lemma 21.113. *Let $f\colon X \to Y$ be a morphism of ringed spaces. Let $\mathscr{G}$ be a K-flat complex of $\mathscr{O}_Y$-modules and let $\mathscr{F}$ be a K-injective complex of $\mathscr{O}_X$-modules. Then the natural morphism in $D(\Gamma(Y, \mathscr{O}_Y))$*

$$\gamma\colon \operatorname{Hom}_{\mathscr{O}_Y}(\mathscr{G}, f_*\mathscr{F}) \longrightarrow R\operatorname{Hom}_{\mathscr{O}_Y}(\mathscr{G}, f_*\mathscr{F})$$

is an isomorphism.

Proof. By the last assertion of Proposition F.191, the morphisms $H^i(\gamma)$ are isomorphisms

$$H^i\big(\operatorname{Hom}_{\mathscr{O}_Y}(\mathscr{G}, f_*\mathscr{F})\big) \stackrel{\text{(F.18.2)}}{=} \operatorname{Hom}_{K(Y)}(\mathscr{G}, f_*\mathscr{F}[i]) \longrightarrow \operatorname{Hom}_{D(Y)}(\mathscr{G}, f_*\mathscr{F}[i])$$

$$\stackrel{\text{(F.52.5)}}{=} \operatorname{Ext}^i_{\mathscr{O}_Y}(\mathscr{G}, f_*\mathscr{F}[i]) = H^i\big(R\operatorname{Hom}_{\mathscr{O}_Y}(\mathscr{G}, f_*\mathscr{F}[i])\big)$$

for all $i \in \mathbb{Z}$. Hence γ is an isomorphism in $D(\Gamma(Y, \mathscr{O}_Y))$. □

Proposition 21.114. *Let $f\colon X \to Y$ be a morphism of ringed spaces. Let $V \subseteq Y$ be open.*
(1) *For every K-injective complex $\mathscr{F}$ of $\mathscr{O}_X$-modules $f_*\mathscr{F}$ is right $\Gamma(V, -)$-acyclic.*
(2) *The canonical morphism of functors $D(X) \to D(\Gamma(V, \mathscr{O}_Y))$*

$$R\Gamma(f^{-1}(V), -) \longrightarrow R\Gamma(V, -) \circ Rf_*$$

is an isomorphism.

Proof. Assertion (2) follows from (1) by Proposition F.176.

To show (1) recall that we have

$$\Gamma(V, -) = \operatorname{Hom}_{\mathscr{O}_Y}(\mathscr{O}_{V \subseteq Y}, -), \qquad R\Gamma(V, -) = R\operatorname{Hom}_{\mathscr{O}_Y}(\mathscr{O}_{V \subseteq Y}, -)$$

by Remark 21.17. As $\mathscr{O}_{V \subseteq Y}$ is a flat $\mathscr{O}_Y$-module, i.e., a K-flat complex concentrated in degree 0, the map $\Gamma(V, f_*\mathscr{F}) \to R\Gamma(V, f_*\mathscr{F})$ is an isomorphism by Lemma 21.113. This shows that $f_*\mathscr{F}$ is right $\Gamma(V, -)$-acyclic by Corollary F.174 (1). □

Proposition 21.115. *Let $f\colon X \to Y$ and $g\colon Y \to Z$ be morphisms of ringed spaces.*
(1) *For every K-injective complex $\mathscr{F}$ of $\mathscr{O}_X$-modules, $f_*\mathscr{F}$ is right g_*-acyclic.*
(2) *The canonical morphism of functors $D(X) \to D(Z)$*

$$R(g_* \circ f_*) \longrightarrow Rg_* \circ Rf_*$$

is an isomorphism.

Proof. Let $s\colon f_*\mathscr{F} \to \mathscr{G}$ be a quasi-isomorphism in $K(Y)$ and let $t\colon \mathscr{G} \to \mathscr{I}$ be a quasi-isomorphism to a K-injective complex in $K(Y)$. For (1) it suffices to show that

(*) $g_*(t \circ s)\colon g_*f_*\mathscr{F} \to g_*\mathscr{I}$

is a quasi-isomorphism in $K(Z)$. Let $W \subseteq Z$ be open. Then $f_* \mathscr{F}$ is right $\Gamma(g^{-1}(W), -)$-acyclic by Proposition 21.114 and $\mathscr{I}$ is right $\Gamma(g^{-1}(W), -)$-acyclic because it is K-injective. Hence $\Gamma(g^{-1}(W), -)$ sends $t \circ s$ to a quasi-isomorphism

$$(g_* f_* \mathscr{F})(W) = (f_* \mathscr{F})(g^{-1}(W)) \longrightarrow \mathscr{I}(g^{-1}(W) = (g_* \mathscr{I})(W)$$

by Lemma F.172. This shows that $(*)$ is a quasi-isomorphism.

Again, Assertion (2) follows from (1) by Proposition F.176. $\qquad \square$

By applying Proposition 21.114 and Proposition 21.115 to $f : (X, \mathscr{O}_X) \to (X, \mathbb{Z}_X)$ we obtain an unbounded version of Corollary 21.43:

Corollary 21.116. *Let X be a ringed space and let $\mathscr{F}$ be a complex of $\mathscr{O}_X$-modules. Denote by $\mathscr{F}^{\mathrm{ab}}$ the underlying complex of abelian sheaves. Then the underlying abelian group of the $\Gamma(X, \mathscr{O}_X)$-module $H^i(X, \mathscr{F})$ is $H^i(X, \mathscr{F}^{\mathrm{ab}})$ for all $i \in \mathbb{Z}$.*

More generally, let $g : X \to Z$ be a morphism of ringed spaces. Then the higher direct images $R^i f_* \mathscr{F}$ of the complex $\mathscr{F}$ of $\mathscr{O}_X$-modules and of abelian sheaves are identical as abelian sheaves.

Relations between derived functors

For any ringed space X and any morphism $f : X \to Y$ of ringed spaces we have now defined functors Rf_*, Lf^*, $R\mathscr{H}om_{\mathscr{O}_X}(-, -)$, and $- \otimes^L_{\mathscr{O}_X} -$. Next we want to show relations between this functors.

For instance, it follows formally from the fact that f^* is left adjoint to f_* that Lf^* is left adjoint to Rf_* (Proposition F.191), in other words, we have a functorial isomorphism of $\Gamma(Y, \mathscr{O}_Y)$-modules

$$(*) \qquad \operatorname{Hom}_{D(X)}(Lf^* \mathscr{G}, \mathscr{F}) \cong \operatorname{Hom}_{D(Y)}(\mathscr{G}, Rf_* \mathscr{F})$$

for $\mathscr{F} \in D(X)$ and $\mathscr{G} \in D(Y)$. We want to "upgrade" this isomorphism to a functorial isomorphism in $D(Y)$

$$(**) \qquad Rf_* \, \mathscr{H}om_{\mathscr{O}_X}(Lf^* \mathscr{G}, \mathscr{F}) \cong \mathscr{H}om_{\mathscr{O}_Y}(\mathscr{G}, Rf_* \mathscr{F}).$$

Here "upgrade" should mean the following. If we apply $R\Gamma(Y, -)$ to $(**)$, we obtain a functorial isomorphism in $D(\Gamma(Y, \mathscr{O}_Y))$

$$R\operatorname{Hom}_{\mathscr{O}_X}(Lf^* \mathscr{G}, \mathscr{F}) \cong R\operatorname{Hom}_{\mathscr{O}_Y}(\mathscr{G}, Rf_* \mathscr{F})$$

since we have $R\Gamma(Y, -) \circ Rf_* = R\Gamma(X, -)$ by Proposition 21.114 (2) and $R\Gamma \circ R\mathscr{H}om = R\operatorname{Hom}$ by Proposition 21.107. If we further apply $H^0(-)$ we then should obtain $(*)$ by (F.52.6). We also want a similar upgrade to the $\otimes$-$\mathscr{H}om$-adjunction.

This part starts by showing that Lf^* respects derived tensor products which is easy to see since both are defined via K-flat resolutions. Then we give a general recipe how to upgrade an adjunction to a sheaf theoretic version. This is then applied to the both cases mentioned above.

After this, we add some further formal constructions. We define for every commutative square of ringed spaces a derived base change morphism. The deep question when this base change morphism is an isomorphism will be only studied for schemes in the last part of the next chapter. The same holds for the projection formula which should be part of our "formalism of four functors". We also define the cup product, again postponing any non-formal questions to the next two chapters.

This part is concluded with a remark what it would mean to extend our "formalism of four functors" to a quasi-coherent six-functor formalism for schemes in the sense of Grothendieck – and why for schemes the best one can hope for is a "five-functor formalism". This fifth functor, the exceptional inverse image, will be defined and studied in Chapter 25 on Grothendieck duality.

(21.24) Derived inverse images and $\otimes^L$.

Let $f: X \to Y$ be a morphism of ringed spaces. Recall that for $\mathscr{O}_Y$-modules $\mathscr{F}$ and $\mathscr{G}$ one has a functorial isomorphism $f^*\mathscr{F} \otimes_{\mathscr{O}_X} f^*\mathscr{G} \xrightarrow{\sim} f^*(\mathscr{F} \otimes_{\mathscr{O}_Y} \mathscr{G})$. As f^* commutes with direct sums we also obtain isomorphisms

(21.24.1) $$ f^*\mathscr{F} \otimes_{\mathscr{O}_X} f^*\mathscr{G} \xrightarrow{\sim} f^*(\mathscr{F} \otimes_{\mathscr{O}_Y} \mathscr{G}). $$

of complexes of $\mathscr{O}_X$-modules, for all complexes $\mathscr{F}$ and $\mathscr{G}$ of $\mathscr{O}_Y$-modules.

Proposition 21.117. *Let $f: X \to Y$ be a morphism of ringed spaces. Then there is a unique isomorphism of triangulated bi-functors in $\mathscr{F}, \mathscr{G} \in D(Y)$*

$$ Lf^*(\mathscr{F} \otimes_{\mathscr{O}_Y}^{L} \mathscr{G}) \xrightarrow{\sim} Lf^*\mathscr{F} \otimes_{\mathscr{O}_X}^{L} Lf^*\mathscr{G} $$

which is compatible with the corresponding non-derived isomorphism, i.e., such that the following diagram commutes

(21.24.2)

$$
\begin{array}{ccc}
f^*(\mathscr{F} \otimes_{\mathscr{O}_Y} \mathscr{G}) & \longleftarrow & Lf^*(\mathscr{F} \otimes_{\mathscr{O}_Y}^{L} \mathscr{G}) \\
{\scriptstyle (21.24.1)} \downarrow {\scriptstyle \cong} & & \downarrow {\scriptstyle \cong} \\
f^*\mathscr{F} \otimes_{\mathscr{O}_X} f^*\mathscr{G} & \longleftarrow & Lf^*\mathscr{F} \otimes_{\mathscr{O}_X}^{L} Lf^*\mathscr{G}
\end{array}
$$

Proof. We may assume that $\mathscr{F}$ and $\mathscr{G}$ are K-flat. Then $f^*\mathscr{F}$ and $f^*\mathscr{G}$ are K-flat (Remark 21.109 (1)) and $f^*\mathscr{F} \otimes f^*\mathscr{G}$ is K-flat (Remark 21.92 (3)). Therefore the horizontal arrows in (21.24.2) are isomorphisms. $\qquad\square$

(21.25) Adjointness and derived functors.

Let us consider the following situation. Let $f: X \to Y$ be a morphism of ringed spaces and let $\Phi: (\mathscr{O}_X\text{-Mod}) \to (\mathscr{O}_Y\text{-Mod})$ and $\Psi: (\mathscr{O}_Y\text{-Mod}) \to (\mathscr{O}_X\text{-Mod})$ be additive functors. Suppose that for all $\mathscr{O}_X$-modules $\mathscr{F}$ and $\mathscr{O}_Y$-modules $\mathscr{G}$ there is a functorial isomorphism of $\mathscr{O}_Y$-modules

$$ f_* \mathscr{H}om_{\mathscr{O}_X}(\Psi(\mathscr{G}), \mathscr{F}) \cong \mathscr{H}om_{\mathscr{O}_Y}(\mathscr{G}, \Phi(\mathscr{F})). $$

Applying $\Gamma(Y, -)$ one sees that in particular Ψ is a left adjoint functor of Φ. By Remark F.190 we obtain for $\mathscr{F}$ in $C(X)$ and $\mathscr{G}$ in $C(Y)$ a functorial isomorphism of complexes of $\mathscr{O}_Y$-modules

(21.25.1) $\qquad f_* \mathcal{H}om_{\mathcal{O}_X}(\Psi(\mathcal{G}), \mathcal{F}) \cong \mathcal{H}om_{\mathcal{O}_Y}(\mathcal{G}, \Phi(\mathcal{F})).$

Now let us also suppose that K-flat complexes $\mathcal{O}_Y$-modules are left Ψ-acyclic. Then we can apply Proposition F.191 and see that $L\Psi\colon D(Y) \to D(X)$ is left adjoint to $R\Phi\colon D(X) \to D(Y)$, i.e., we obtain a functorial isomorphism

(21.25.2) $\qquad \mathrm{Hom}_{D(X)}(L\Psi(\mathcal{G}), \mathcal{F}) \cong \mathrm{Hom}_{D(Y)}(\mathcal{G}, R\Phi(\mathcal{F}))$

of $\Gamma(Y, \mathcal{O}_Y)$-modules. One would like to "sheafify and derive" this isomorphism as follows.

Remark 21.118. In the situation above suppose that the canonical morphisms

$$f_* \mathcal{H}om_{\mathcal{O}_X}(\Psi(\mathcal{G}), \mathcal{F}) \longrightarrow Rf_* \mathcal{H}om_{\mathcal{O}_X}(\Psi(\mathcal{G}), \mathcal{F}),$$
$$\mathcal{H}om_{\mathcal{O}_Y}(\mathcal{G}, \Phi(\mathcal{F})) \longrightarrow R\mathcal{H}om_{\mathcal{O}_Y}(\mathcal{G}, \Phi(\mathcal{F}))$$

are both isomorphisms if $\mathcal{F}$ is a K-injective complex and $\mathcal{G}$ is a K-flat complex.
(1) Then there is for all $\mathcal{F}$ in $D(X)$ and $\mathcal{G}$ in $D(Y)$ an isomorphism in $D(Y)$

(21.25.3) $\qquad \rho\colon Rf_* R\mathcal{H}om_{\mathcal{O}_X}(L\Psi(\mathcal{G}), \mathcal{F}) \cong R\mathcal{H}om_{\mathcal{O}_Y}(\mathcal{G}, R\Phi(\mathcal{F}))$

which is triangulated functorial in $\mathcal{F}$ and in $\mathcal{G}$ and which is the unique such isomorphism such that for all $\mathcal{F}$ and $\mathcal{G}$ the following diagram commutes

(21.25.4)

$$
\begin{array}{ccc}
f_* \mathcal{H}om_{\mathcal{O}_X}(\Psi(\mathcal{G}), \mathcal{F}) & \longrightarrow & Rf_* R\mathcal{H}om_{\mathcal{O}_X}(L\Psi(\mathcal{G}), \mathcal{F}) \\
{\scriptstyle (21.25.1)} \downarrow \cong & & \downarrow \rho \\
\mathcal{H}om_{\mathcal{O}_Y}(\mathcal{G}, \Phi(\mathcal{F})) & \longrightarrow & R\mathcal{H}om_{\mathcal{O}_Y}(\mathcal{G}, R\Phi(\mathcal{F})).
\end{array}
$$

Indeed, replacing $\mathcal{F}$ by a K-injective resolution and $\mathcal{G}$ by a K-flat resolution we may assume that $\mathcal{F}$ is K-injective and $\mathcal{G}$ is K-flat. Then the upper horizontal line in (21.25.4) is the composition of isomorphisms

$$f_* \mathcal{H}om_{\mathcal{O}_X}(\Psi(\mathcal{G}), \mathcal{F}) \xrightarrow{(a)} Rf_* \mathcal{H}om_{\mathcal{O}_X}(\Psi(\mathcal{G}), \mathcal{F}) \xrightarrow{(b)} Rf_* R\mathcal{H}om_{\mathcal{O}_X}(\Psi(\mathcal{G}), \mathcal{F})$$
$$\xrightarrow{(c)} Rf_* R\mathcal{H}om_{\mathcal{O}_X}(L\Psi(\mathcal{G}), \mathcal{F}),$$

where (a) is an isomorphism by hypothesis, (b) is an isomorphism because $\mathcal{F}$ is K-injective, and (c) is an isomorphism because $\mathcal{G}$ is K-flat. Similarly, the lower horizontal line in (21.25.4) is the composition

$$\mathcal{H}om_{\mathcal{O}_Y}(\mathcal{G}, \Phi(\mathcal{F})) \xrightarrow{\sim} R\mathcal{H}om_{\mathcal{O}_Y}(\mathcal{G}, \Phi(\mathcal{F})) \cong \mathcal{H}om_{\mathcal{O}_Y}(\mathcal{G}, R\Phi(\mathcal{F})),$$

the first morphism is an isomorphism by hypothesis and the second because $\mathcal{F}$ is K-injective.
(2) Next, we can apply $R\Gamma(Y, -)$ to (21.25.3). As $R\Gamma(Y, -) \circ Rf_* = R\Gamma(X, -)$ (Proposition 21.114) and $R\Gamma \circ R\mathcal{H}om = R\mathrm{Hom}$ (Proposition 21.107) we obtain an isomorphism in $D(\Gamma(Y, \mathcal{O}_Y))$

(21.25.5) $\qquad \sigma\colon R\mathrm{Hom}_{\mathcal{O}_X}(L\Psi(\mathcal{G}), \mathcal{F}) \cong R\mathrm{Hom}_{\mathcal{O}_Y}(\mathcal{G}, R\Phi(\mathcal{F}))$

which is triangulated functorial in $\mathcal{F}$ in $D(X)$ and $\mathcal{G}$ in $D(Y)$ and which is the unique such isomorphism such that for all $\mathcal{F}$ and $\mathcal{G}$ the following diagram commutes

$$(21.25.6) \quad \begin{array}{ccc} \mathrm{Hom}_{\mathscr{O}_X}(\Psi(\mathscr{G}),\mathscr{F}) & \longrightarrow & R\,\mathrm{Hom}_{\mathscr{O}_X}(L\Psi(\mathscr{G}),\mathscr{F}) \\ \cong \downarrow & & \downarrow \rho \\ \mathrm{Hom}_{\mathscr{O}_Y}(\mathscr{G},\Phi(\mathscr{F})) & \longrightarrow & R\,\mathrm{Hom}_{\mathscr{O}_Y}(\mathscr{G},R\Phi(\mathscr{F})). \end{array}$$

We will now specialize the previous remark to adjointness of $R\mathscr{H}om$ and $\otimes^L$ and to Rf_* and Lf^*.

(21.26) Relations between $R\mathscr{H}om$ and $\otimes^L$.

Let X be a ringed space. By Section (7.5) one has for arbitrary $\mathscr{O}_X$-modules $\mathscr{F}$, $\mathscr{G}$, and $\mathscr{H}$ functorial isomorphisms of $\mathscr{O}_X$-modules

$$\mathscr{H}om_{\mathscr{O}_X}(\mathscr{F}\otimes_{\mathscr{O}_X}\mathscr{G},\mathscr{H}) \xrightarrow{\sim} \mathscr{H}om_{\mathscr{O}_X}(\mathscr{F},\mathscr{H}om_{\mathscr{O}_X}(\mathscr{G},\mathscr{H}))$$

and functorial homomorphisms

$$\mathscr{H}om_{\mathscr{O}_X}(\mathscr{F},\mathscr{G})\otimes_{\mathscr{O}_X}\mathscr{H} \longrightarrow \mathscr{H}om_{\mathscr{O}_X}(\mathscr{F},\mathscr{G}\otimes_{\mathscr{O}_X}\mathscr{H}).$$

We will now construct derived versions of these isomorphisms and homomorphisms. Recall that the second map is an isomorphisms if $\mathscr{F}$ is finite locally free (Proposition 7.7).

Theorem 21.119. *(Adjointness of $R\mathscr{H}om$ and $\otimes^L$) Let X be a ringed space, and let $\mathscr{F},\mathscr{G},\mathscr{H}\in D(\mathscr{O}_X\text{-Mod})$.*
(1) *There is an isomorphism*

$$(21.26.1) \qquad R\mathscr{H}om(\mathscr{F}\otimes^L\mathscr{G},\mathscr{H}) \cong R\mathscr{H}om(\mathscr{F},R\mathscr{H}om(\mathscr{G},\mathscr{H})),$$

which is triangulated-functorial in $\mathscr{F}$, $\mathscr{G}$ and $\mathscr{H}$.
(2) *There is a natural isomorphism*

$$(21.26.2) \qquad R\,\mathrm{Hom}(\mathscr{F}\otimes^L\mathscr{G},\mathscr{H}) \cong R\,\mathrm{Hom}(\mathscr{F},R\mathscr{H}om(\mathscr{G},\mathscr{H})),$$

which is triangulated-functorial in $\mathscr{F}$, $\mathscr{G}$ and $\mathscr{H}$.

Proof. Replacing $\mathscr{G}$ by a K-flat complex isomorphic to $\mathscr{G}$ in $D(X)$, we may assume that $\mathscr{G}$ is K-flat. We apply Remark 21.118 to $f=\mathrm{id}_X$, $\Phi=\mathscr{H}om(\mathscr{G},-)$ and $\Psi=-\otimes\mathscr{G}$. Hence it suffices to show that

$$\mathscr{H}om(\mathscr{F},\mathscr{H}om(\mathscr{G},\mathscr{I})) \longrightarrow R\mathscr{H}om(\mathscr{F},\mathscr{H}om(\mathscr{G},\mathscr{I}))$$

is an isomorphism if $\mathscr{I}$ is K-injective. This holds because $\mathscr{H}om(\mathscr{G},\mathscr{I})$ is K-injective (Lemma 21.106). □

By Remark 21.118, the isomorphism (21.26.1) is compatible with the non-derived adjointness, i.e., it is the unique such isomorphism making the following diagram commutative

$$\begin{array}{ccccc} \mathscr{H}om(\mathscr{F}\otimes\mathscr{G},\mathscr{H}) & \longrightarrow & R\mathscr{H}om(\mathscr{F}\otimes\mathscr{G},\mathscr{H}) & \longrightarrow & R\mathscr{H}om(\mathscr{F}\otimes^L\mathscr{G},\mathscr{H}) \\ \cong\downarrow & & \downarrow & & \downarrow\alpha \\ \mathscr{H}om(\mathscr{F},\mathscr{H}om(\mathscr{G},\mathscr{H})) & \longrightarrow & R\mathscr{H}om(\mathscr{F},\mathscr{H}om(\mathscr{G},\mathscr{H})) & \longrightarrow & R\mathscr{H}om(\mathscr{F},R\mathscr{H}om(\mathscr{G},\mathscr{H})), \end{array}$$

where the left vertical arrow is the isomorphism (21.19.4) and the horizontal arrows are induced by the canonical morphism $F \to RF$ and $LF \to F$ for functors F that are right derivable and left derivable, respectively.

Similarly, the isomorphism (21.26.2) is compatible with non-derived adjointness.

Remark 21.120. Let X be a ringed space, and let $\mathscr{F}, \mathscr{G}, \mathscr{H} \in D(X)$.
(1) Applying the functor $H^0(-)$ to (21.26.2) we obtain by (F.52.6) a functorial isomorphism of $\Gamma(X, \mathscr{O}_X)$-modules

$$(21.26.3) \qquad \mathrm{Hom}_{D(X)}(\mathscr{F} \otimes^L \mathscr{G}, \mathscr{H}) \cong \mathrm{Hom}_{D(X)}(\mathscr{F}, R\mathscr{H}om(\mathscr{G}, \mathscr{H})).$$

This shows in particular that for every complex $\mathscr{G}$ in $D(X)$ the functor $(-) \otimes^L \mathscr{G}$ is left adjoint to the functor $R\mathscr{H}om(\mathscr{G}, -)$.
(2) In particular, $(-) \otimes^L \mathscr{G}$ commutes with arbitrary colimits and $R\mathscr{H}om(\mathscr{G}, -)$ commutes with arbitrary limits if they exist. For instance arbitrary products and direct sums exist by Lemma F.188.

The functor $R\mathscr{H}om_{\mathscr{O}_X}$ also sends direct sums in $D(X)$ in the first component to products.

Corollary 21.121. *Let X be a ringed space, $\mathscr{H}$ be an object of $D(X)$, and let $(\mathscr{G}_i)_i$ be a family of objects in $D(X)$. Then there is a functorial isomorphism*

$$(21.26.4) \qquad R\mathscr{H}om_{\mathscr{O}_X}\left(\bigoplus_i \mathscr{G}_i, \mathscr{H}\right) \cong \prod_i R\mathscr{H}om_{\mathscr{O}_X}(\mathscr{G}_i, \mathscr{H}).$$

of objects in $D(X)$.

Proof. Let $\mathscr{F}$ be an object in $D(X)$. Then we have functorial isomorphisms

$$\mathrm{Hom}_{D(X)}\left(\mathscr{F}, \prod_i R\mathscr{H}om(\mathscr{G}_i, \mathscr{H})\right) = \prod_i \mathrm{Hom}_{D(X)}(\mathscr{F}, R\mathscr{H}om(\mathscr{G}_i, \mathscr{H}))$$

$$= \prod_i \mathrm{Hom}_{D(X)}(\mathscr{F} \otimes^L \mathscr{G}_i, \mathscr{H})$$

$$= \mathrm{Hom}_{D(X)}\left(\mathscr{F} \otimes^L \bigoplus_i \mathscr{G}_i, \mathscr{H}\right)$$

$$= \mathrm{Hom}_{D(X)}\left(\mathscr{F}, R\mathscr{H}om\left(\bigoplus_i \mathscr{G}_i, \mathscr{H}\right)\right)$$

and the claim follows from the Yoneda Lemma. $\qquad\square$

Remark 21.122. Let X be a ringed space, and let $\mathscr{E}, \mathscr{F}$, and $\mathscr{G}$ be complexes in $D(X)$. We will construct a functorial map

$$(21.26.5) \qquad R\mathscr{H}om_{\mathscr{O}_X}(\mathscr{E}, \mathscr{F}) \otimes^L_{\mathscr{O}_X} \mathscr{G} \longrightarrow R\mathscr{H}om_{\mathscr{O}_X}(\mathscr{E}, \mathscr{F} \otimes^L \mathscr{G})$$

in $D(X)$ functorial in $\mathscr{E}, \mathscr{F}$, and $\mathscr{G}$.

We may replace $\mathscr{F}$ by a K-injective complex $\mathscr{I}$ and $\mathscr{G}$ by a K-flat complex $\mathscr{K}$. Also choose a quasi-isomorphism of the tensor complex $\mathscr{I} \otimes \mathscr{K}$ to a K-injective complex $\mathscr{J}$. It is easy to see that there is a functorial map of complexes

$$\mathscr{H}om_{\mathscr{O}_X}(\mathscr{E}, \mathscr{I}) \otimes_{\mathscr{O}_X} \mathscr{K} \longrightarrow \mathscr{H}om_{\mathscr{O}_X}(\mathscr{E}, \mathscr{I} \otimes \mathscr{K}).$$

The left hand side computes $R\,\mathscr{H}om_{\mathscr{O}_X}(\mathscr{E},\mathscr{F})\otimes^L_{\mathscr{O}_X}\mathscr{G}$. Now by composing with the map $\mathscr{H}om_{\mathscr{O}_X}(\mathscr{E},\mathscr{I}\otimes\mathscr{K})\to\mathscr{H}om_{\mathscr{O}_X}(\mathscr{E},\mathscr{J})$ we get the desired map because $\mathscr{H}om_{\mathscr{O}_X}(\mathscr{E},\mathscr{J})$ computes $R\,\mathscr{H}om_{\mathscr{O}_X}(\mathscr{E},\mathscr{F}\otimes^L\mathscr{G})$.

Remark 21.123. Let X be a ringed space and let $\mathscr{E}$, $\mathscr{F}$, and $\mathscr{G}$ be complexes in $D(X)$. We will construct a functorial "derived composition" map in $D(X)$

$$(21.26.6)\qquad R\,\mathscr{H}om(\mathscr{F},\mathscr{G})\otimes^L R\,\mathscr{H}om(\mathscr{E},\mathscr{F})\longrightarrow R\,\mathscr{H}om(\mathscr{E},\mathscr{G}).$$

For this replace $\mathscr{F}$ and $\mathscr{G}$ be a K-injective complexes $\mathscr{I}$ and $\mathscr{J}$, respectively. Choose also a quasi-isomorphism $\mathscr{K}\to\mathscr{H}om(\mathscr{E},\mathscr{I})$ with $\mathscr{K}$ a K-flat complex. Consider the composition

$$\mathscr{H}om(\mathscr{I},\mathscr{J})\otimes\mathscr{K}\longrightarrow\mathscr{H}om(\mathscr{I},\mathscr{J})\otimes\mathscr{H}om(\mathscr{E},\mathscr{I})\longrightarrow\mathscr{H}om(\mathscr{E},\mathscr{J}),$$

where the second map is given by composition. The source calculates $R\,\mathscr{H}om(\mathscr{F},\mathscr{G})\otimes^L R\,\mathscr{H}om(\mathscr{E},\mathscr{F})$ and the target calculates $R\,\mathscr{H}om(\mathscr{E},\mathscr{G})$.

Setting $\mathscr{E}=\mathscr{O}_X$ and using that $R\,\mathscr{H}om_{\mathscr{O}_X}(\mathscr{O}_X,\mathscr{F})=\mathscr{F}$ (21.18.10) we deduce from (21.26.6) in particular a functorial derived evaluation map

$$(21.26.7)\qquad R\,\mathscr{H}om(\mathscr{F},\mathscr{G})\otimes^L\mathscr{F}\longrightarrow\mathscr{G}.$$

Remark 21.124. Let X be a ringed space and $\mathscr{E},\mathscr{G}\in D(X)$ be complexes. Then there is a functorial map

$$(21.26.8)\qquad \mathscr{G}\longrightarrow R\,\mathscr{H}om_{\mathscr{O}_X}(R\,\mathscr{H}om_{\mathscr{O}_X}(\mathscr{G},\mathscr{E}),\mathscr{E}).$$

Indeed, the evaluation map $R\,\mathscr{H}om(\mathscr{G},\mathscr{E})\otimes^L\mathscr{G}\to\mathscr{E}$ (21.26.7) corresponds to (21.26.8) by applying H^0 to (21.26.2).

If $\mathscr{F}$ is a further complex in $D(X)$, then we obtain by composing (21.26.8) with the derived composition map (21.26.6) a functorial map

$$
\begin{aligned}
(21.26.9)\qquad & R\,\mathscr{H}om_{\mathscr{O}_X}(\mathscr{E},\mathscr{F})\otimes^L_{\mathscr{O}_X}\mathscr{G}\\
&\longrightarrow R\,\mathscr{H}om_{\mathscr{O}_X}(\mathscr{E},\mathscr{F})\otimes^L_{\mathscr{O}_X}R\,\mathscr{H}om_{\mathscr{O}_X}(R\,\mathscr{H}om_{\mathscr{O}_X}(\mathscr{G},\mathscr{E}),\mathscr{E})\\
&\longrightarrow R\,\mathscr{H}om_{\mathscr{O}_X}(R\,\mathscr{H}om_{\mathscr{O}_X}(\mathscr{G},\mathscr{E}),\mathscr{F}).
\end{aligned}
$$

Remark 21.125. Let $f\colon X\to Y$ be a morphism of ringed spaces, let $\mathscr{F}$ and $\mathscr{G}$ be complexes in $D(Y)$. We will construct a functorial map

$$(21.26.10)\qquad Lf^*R\,\mathscr{H}om_{\mathscr{O}_Y}(\mathscr{F},\mathscr{G})\longrightarrow R\,\mathscr{H}om_{\mathscr{O}_X}(Lf^*\mathscr{F},Lf^*\mathscr{G})$$

in $D(X)$. By adjointness of $R\,\mathscr{H}om$ and derived tensor product (21.26.3) it suffices to construct a map

$$Lf^*(R\,\mathscr{H}om_{\mathscr{O}_Y}(\mathscr{F},\mathscr{G})\otimes^L_{\mathscr{O}_Y}\mathscr{F})=Lf^*R\,\mathscr{H}om_{\mathscr{O}_Y}(\mathscr{F},\mathscr{G})\otimes^L_{\mathscr{O}_X}Lf^*\mathscr{F}\longrightarrow Lf^*\mathscr{G},$$

where the first equality holds by Proposition 21.117. This is done by applying Lf^* to (21.26.7).

(21.27) Relations between Rf_* and Lf^*.

Next we look at the adjoint pair (f^*, f_*). Proposition F.191 immediately implies that Lf^* is left adjoint to Rf_*:

Proposition 21.126. *Let $f\colon X \to Y$ be a morphism of ringed spaces. Let $\mathscr{F} \in D(X)$ and $\mathscr{G} \in D(Y)$. There is a unique isomorphism*

$$\mathrm{Hom}_{D(X)}(Lf^*\mathscr{G}, \mathscr{F}) \cong \mathrm{Hom}_{D(Y)}(\mathscr{G}, Rf_*\mathscr{F}),$$

which is triangulated functorial in $\mathscr{F}$ and $\mathscr{G}$ and compatible with non-derived adjointness.

Remark 21.127. The unit and counit of the adjunction are functorial morphisms

$$\mathscr{G} \to Rf_*Lf^*\mathscr{G}, \qquad Lf^*Rf_*\mathscr{F} \to \mathscr{F}$$

which are usually not isomorphisms.

However, if $f = j\colon U \to Y$ is the embedding of an open subspace, then the counit $Lj^*Rj_*\mathscr{F} \to \mathscr{F}$ is an isomorphism for all $\mathscr{F}$ in $D(U)$. Indeed, if $\mathscr{F} \to \mathscr{I}$ is a K-injective resolution, then

$$Lj^*Rj_*\mathscr{F} \cong Lj^*j_*\mathscr{I} = j^*j_*\mathscr{I} \cong \mathscr{I} \cong \mathscr{F},$$

where we have $Lj^* = j^*$ because j is flat.

We also apply Remark 21.118 to (f^*, f_*).

Theorem 21.128. *Let $f\colon X \to Y$ be a morphism of ringed spaces. Let $\mathscr{F} \in D(X)$, $\mathscr{G} \in D(Y)$.*
(1) There is a unique isomorphism

$$Rf_*R\mathscr{H}om_X(Lf^*\mathscr{G}, \mathscr{F}) \cong R\mathscr{H}om_Y(\mathscr{G}, Rf_*\mathscr{F})$$

which is compatible with the non-derived adjointness of f_* and f^* and which is triangulated functorial in $\mathscr{F}$ and $\mathscr{G}$.
(2) There is a unique isomorphism

$$R\mathrm{Hom}_{\mathcal{O}_X}(Lf^*\mathscr{G}, \mathscr{F}) \cong R\mathrm{Hom}_{\mathcal{O}_Y}(\mathscr{G}, Rf_*\mathscr{F})$$

which is compatible with the non-derived adjointness of f_* and f^* and which is triangulated functorial in $\mathscr{F}$ and $\mathscr{G}$.

Proof. We apply Remark 21.118 to the functors $\Phi = f_*$ and $\Psi = f^*$. Hence we have to show that for $\mathscr{F}$ a K-injective complex and $\mathscr{G}$ a K-flat complex the morphisms

(*) $\qquad f_*\mathscr{H}om_{\mathcal{O}_X}(f^*(\mathscr{G}), \mathscr{F}) \longrightarrow Rf_*\mathscr{H}om_{\mathcal{O}_X}(f^*(\mathscr{G}), \mathscr{F}),$

(**) $\qquad \mathscr{H}om_{\mathcal{O}_Y}(\mathscr{G}, f_*(\mathscr{F})) \longrightarrow R\mathscr{H}om_{\mathcal{O}_Y}(\mathscr{G}, f_*(\mathscr{F}))$

are isomorphisms.

As $f^*\mathscr{G}$ is again K-flat (Remark 21.109), Lemma 21.106 shows that $\mathscr{H}om_{\mathcal{O}_X}(f^*(\mathscr{G}), \mathscr{F})$ is K-injective. Hence (*) is an isomorphism.

Let $f_*\mathscr{F} \to \mathscr{I}$ be a K-injective resolution, so $R\mathscr{H}om_{\mathcal{O}_Y}(\mathscr{G}, f_*(\mathscr{F})) = \mathscr{H}om_{\mathcal{O}_Y}(\mathscr{G}, \mathscr{I})$ and to see that (**) is an isomorphism it suffices to see that for all open subsets $V \subseteq Y$ the morphism (**) yields an isomorphism

$$\alpha\colon \operatorname{Hom}_{\mathscr{O}_V}(\mathscr{G}_{|V}, f_*\mathscr{F}_{|V}) = \Gamma(V, \mathscr{H}om_{\mathscr{O}_Y}(\mathscr{G}, f_*\mathscr{F}))$$
$$\longrightarrow \Gamma(V, \mathscr{H}om_{\mathscr{O}_Y}(\mathscr{G}, \mathscr{I})) = \operatorname{Hom}_{\mathscr{O}_V}(\mathscr{G}_{|V}, \mathscr{I}_{|V}).$$

Setting $g := f_{|f^{-1}(V)}\colon f^{-1}(V) \to V$ this follows from the commutative diagram

$$
\begin{array}{ccc}
\operatorname{Hom}_{\mathscr{O}_V}(\mathscr{G}_{|V}, f_*\mathscr{F}_{|V}) & \xrightarrow{\ \ \alpha\ \ } & \operatorname{Hom}_{\mathscr{O}_V}(\mathscr{G}_{|V}, \mathscr{I}_{|V}) \\
\| & & \cong \downarrow \beta \\
\operatorname{Hom}_{\mathscr{O}_V}(\mathscr{G}_{|V}, g_*(\mathscr{F}_{|f^{-1}(V)})) & \xrightarrow[\cong]{\ \ \gamma\ \ } & R\operatorname{Hom}_{\mathscr{O}_V}(\mathscr{G}_{|V}, g_*(\mathscr{F}_{|f^{-1}(V)}))
\end{array}
$$

where β is an isomorphism because $g_*(\mathscr{F}_{|f^{-1}(V)}) = (f_*\mathscr{F})_{|V} \to \mathscr{I}_{|V}$ is a K-injective resolution (Lemma 21.12) and γ is an isomorphism by Lemma 21.113, again using that restrictions of K-injective complexes to open subspaces are K-injective. $\square$

(21.28) The base change morphism.

We consider a commutative diagram

(21.28.1)
$$
\begin{array}{ccc}
X' & \xrightarrow{\ u'\ } & X \\
f' \downarrow & & \downarrow f \\
Y' & \xrightarrow{\ u\ } & Y
\end{array}
$$

of morphisms of ringed spaces.

Definition and Remark 21.129. Given a diagram as above, there exists a canonical *base change morphism* of functors $D(X) \to D(Y')$

$$Lu^*Rf_* \longrightarrow Rf'_*L(u')^*.$$

defined in the following equivalent ways.

(i) By adjunction between $L(f')^*$ and Rf'_* (Proposition 21.126) and because forming the derived pullback is compatible with composition (Proposition 21.112), specifying the base change morphism is the same as specifying a morphism $L(u')^*Lf^*Rf_* = L(f')^*Lu^*Rf_* \to L(u')^*$, and to get this morphism, we can simply apply the functor $L(u')^*$ to the adjunction counit map $Lf^*Rf_* \to \mathrm{id}$.

(ii) By adjunction between Lu^* and Ru_*, defining the base change morphism is equivalent to specifying a morphism $Rf_* \to Ru_*Rf'_*Lu'^* = Rf_*Ru'_*Lu'^*$ which we can get by applying Rf_* to the adjunction unit map $\mathrm{id} \to Ru'_*Lu'^*$.

It is a non-trivial fact that these definitions are equivalent. For the proof we refer to [Lip2] $\overset{\mathrm{O}}{\mathrm{X}}$ (3.7.2) or, for a much more general statement, to [SGA4] $^\mathrm{O}$ Exp. XVII, 2.1.3.

It is an interesting question, under which conditions the canonical base change morphism is an isomorphism. For morphisms of schemes we will study this question in Theorem 22.99.

(21.29) Cup Product.

Let $f\colon X \to Y$ be a morphism of ringed spaces. For complexes $\mathscr{F}$ and $\mathscr{G}$ of $\mathscr{O}_X$-modules we will construct a morphism in $D(Y)$, called the *relative cup product*

(21.29.1) $$Rf_*\mathscr{F} \otimes^L_{\mathscr{O}_Y} Rf_*\mathscr{G} \longrightarrow Rf_*(\mathscr{F} \otimes^L_{\mathscr{O}_X} \mathscr{G})$$

as follows. By definition it is the map adjoint to the composition

$$Lf^*(Rf_*\mathscr{F} \otimes^L_{\mathscr{O}_Y} Rf_*\mathscr{G}) \xrightarrow{\sim} Lf^*Rf_*\mathscr{F} \otimes^L_{\mathscr{O}_X} Lf^*Rf_*\mathscr{G} \longrightarrow \mathscr{F} \otimes^L_{\mathscr{O}_X} \mathscr{G},$$

where the first isomorphism is given by compatibility of derived pullback with tensor product (Proposition 21.117) and the second morphism is induced by the counit $Lf^* \circ Rf_* \longrightarrow \mathrm{id}$.

If $Y = (*, A)$, i.e., the underlying topological space of Y consists of a single point and A is the ring of global sections of Y, then (21.29.1) is a morphism in $D(A)$, simply called *cup product*

(21.29.2) $$R\Gamma(X, \mathscr{F}) \otimes^L_A R\Gamma(X, \mathscr{G}) \longrightarrow R\Gamma(X, \mathscr{F} \otimes^L_{\mathscr{O}_X} \mathscr{G}).$$

It induces for all $p, q \in \mathbb{Z}$ an A-bilinear map, also called *cup product*

(21.29.3) $$\cup\colon H^p(X, \mathscr{F}) \times H^q(X, \mathscr{G}) \longrightarrow H^{p+q}(X, \mathscr{F} \otimes^L_{\mathscr{O}_X} \mathscr{G})$$

as follows. By Remark 21.104 we have an A-linear map $H^p(X, \mathscr{F}) \otimes_A H^q(X, \mathscr{G}) \longrightarrow H^{p+q}(R\Gamma(X, \mathscr{F}) \otimes^L_A R\Gamma(X, \mathscr{G}))$ which we compose with (21.29.2) to obtain (21.29.3).

It can also be described as follows (we leave it as Exercise 21.27 to check this). Fix cohomology classes

$$\alpha \in H^p(X, \mathscr{F}) = \mathrm{Ext}^p_{\mathscr{O}_X}(\mathscr{O}_X, \mathscr{F}) = \mathrm{Hom}_{D(X)}(\mathscr{O}_X[-p], \mathscr{F}),$$
$$\beta \in H^q(X, \mathscr{G}) = \mathrm{Hom}_{D(X)}(\mathscr{O}_X[-q], \mathscr{G}),$$

where the equalities are given by (21.4.4) and by our definition of the Ext modules. Using $\mathscr{O}_X[-p] \otimes^L \mathscr{O}_X[-q] = \mathscr{O}_X[-p-q]$, the functoriality of the derived tensor product yields an element

$$\alpha \cup \beta \in \mathrm{Hom}_{D(X)}(\mathscr{O}_X[-p-q], \mathscr{F} \otimes^L_{\mathscr{O}_X} \mathscr{G}) = H^{p+q}(X, \mathscr{F} \otimes^L_{\mathscr{O}_X} \mathscr{G}).$$

Remark 21.130. Let $f\colon X \to Y$ be a morphism of ringed spaces and let $\mathscr{B}$ be a differential graded $\mathscr{O}_Y$-algebra over X (Section (17.9)). Considering them as complexes of $f^{-1}(\mathscr{O}_Y)$-modules we obtain a homomorphism in $D(X, f^{-1}(\mathscr{O}_Y))$

$$\mathscr{B} \otimes^L_{f^{-1}(\mathscr{O}_Y)} \mathscr{B} \longrightarrow \mathscr{B} \otimes_{f^{-1}(\mathscr{O}_Y)} \mathscr{B} \to \mathscr{B},$$

where the first map is the canonical map from the derived tensor product to the usual tensor product of complexes of $\mathscr{O}_X$-modules and where the second map is the multiplication on $\mathscr{B}$. Viewing f as a morphism of ringed spaces $(X, f^{-1}(\mathscr{O}_Y)) \to (Y, \mathscr{O}_Y)$, we can apply Rf_* to this map and precompose with the relative cup product (21.29.1). We obtain a multiplication

$$m\colon Rf_*\mathscr{B} \otimes^L_{\mathscr{O}_Y} Rf_*\mathscr{B} \longrightarrow Rf_*\mathscr{B}.$$

By Remark 21.104 we also deduce for all $p, q \in \mathbb{Z}$ maps of $\mathscr{O}_Y$-modules

$$R^p f_*\mathscr{B} \otimes R^q f_*\mathscr{B} \longrightarrow R^{p+q} f_*\mathscr{B}.$$

The associativity of the tensor product of complexes (F.19.3) implies that this defines on $\bigoplus_{p \in \mathbb{Z}} R^p f_*\mathscr{B}$ the structure of a $\mathbb{Z}$-graded $\mathscr{O}_Y$-algebra. If the differential graded algebra $\mathscr{B}$ is (strictly) graded commutative, then (F.19.2) shows that $\bigoplus_{p \in \mathbb{Z}} R^p f_*\mathscr{B}$ is (strictly) graded commutative.

If $\mathscr{B}$ is strictly graded commutative, then the identity of $R^1 f_* \mathscr{B}$ induces a homomorphism of strictly graded commutative graded $\mathscr{O}_Y$-algebras

$$(21.29.4) \qquad \bigwedge_{\mathscr{O}_Y}^{\bullet} R^1 f_* \mathscr{B} \longrightarrow \bigoplus_{p \in \mathbb{Z}} R^p f_* \mathscr{B}$$

by Proposition 17.50.

If $Y = (*, A)$ for a ring A, then the above maps correspond to a multiplication map

$$R\Gamma(X, \mathscr{B}) \otimes_R^L R\Gamma(X, \mathscr{B}) \longrightarrow R\Gamma(X, \mathscr{B})$$

in $D(A)$ and induce the structure of a graded A-algebra on $\bigoplus_{p \in \mathbb{Z}} H^p(X, \mathscr{B})$. It is (strictly) graded commutative if $\mathscr{B}$ is it.

Example 21.131. Let X be a ringed space, let A be a ring, and let $A \to \Gamma(X, \mathscr{O}_X)$ be a ring homomorphism, i.e., we are given a morphism of ringed spaces $X \to (*, A)$. We can consider $\mathscr{O}_X$ as a strictly graded commutative differential graded A-algebra over X concentrated in degree 0. By Remark 21.130 we obtain the structure of a strictly graded commutative graded A-algebra on

$$H^{\bullet}(X, \mathscr{O}_X) := \bigoplus_{p \geq 0} H^p(X, \mathscr{O}_X)$$

and a canonical map of graded A-algebras

$$(21.29.5) \qquad \bigwedge^{\bullet} H^1(X, \mathscr{O}_X) \longrightarrow H^{\bullet}(X, \mathscr{O}_X).$$

(21.30) Formalism of six functors.

Until now we have studied four functors on derived categories: For a morphism $f \colon X \to Y$ we have defined an adjoint pair

$$Lf^* \colon D(Y) \rightleftarrows D(X) \colon Rf_*$$

and for every ringed space X bi-functors

$$\otimes^L \colon D(X) \times D(X) \longrightarrow D(X), \qquad R\mathscr{H}om \colon D(X)^{\mathrm{opp}} \times D(X) \to D(X)$$

such that $\otimes^L$ is associative and symmetric and with unit ($\mathscr{O}_X$ in this case) up to natural isomorphism which are subject to certain coherence conditions and such that $(-) \otimes^L \mathscr{G}$ is left adjoint to $R\mathscr{H}om(\mathscr{G}, -)$. Moreover Lf^* is monoidal, i.e., compatible with derived tensor product and preserving the unit object (again up to natural isomorphisms satisfying certain coherence conditions).

Functors analogous to these exist for many cohomology theories, and their natural properties and compatibilities are codified by *Grothendieck's Six Functor Formalism*. For instance étale cohomology and the theory of $\mathscr{D}$-modules admit all six functors, for other theories such as cohomology for quasi-coherent sheaves as discussed in this book, only part of the formalism can be realized. The six functor formalism predicts, in addition to the four functors listed above, the existence of another adjoint pair

$$f_! \colon D(X) \rightleftarrows D(Y) \colon f^!$$

for "some" morphisms $f: X \to Y$ of ringed spaces. In our context, that of Algebraic Geometry, "some" should mean morphisms of *schemes* that are "sufficiently finite", e.g., separated morphisms of finite type between noetherian schemes. Note that it is not required that $f^!$ or $f_!$ are derived functors of functors between categories of modules over the structure sheaves. The functor $f_!$ is called *direct image with proper support*. The functor $f^!$ is called *exceptional inverse image* and is usually *not* the (left) derived functor of any functor $(\mathscr{O}_Y\text{-Mod}) \to (\mathscr{O}_X\text{-Mod})$.

These six functors should satisfy the following properties, which we deliberately formulate somewhat vaguely.

(1) The formation of Lf^*, Rf_*, $f^!$ and of $f_!$ is compatible with compositions of morphisms.

(2) There is a morphism of functors $\alpha_f: f_! \to Rf_*$, compatible with compositions, which is an isomorphism if f is proper.

(3) For any étale "sufficiently finite" morphism $f: X \to Y$ there exists an isomorphism, $Lf^* \xrightarrow{\sim} f^!$, compatible with compositions.

(4) For any "sufficiently finite" cartesian diagram of morphisms of schemes

$$
\begin{array}{ccc}
X' & \xrightarrow{\ f'\ } & Y' \\
{\scriptstyle g'}\downarrow & & \downarrow{\scriptstyle g} \\
X & \xrightarrow{\ f\ } & Y
\end{array}
$$

there exist natural base change isomorphisms

(BC1) $\qquad\qquad\qquad Lg^* \circ f_! \xrightarrow{\sim} f'_! \circ Lg'^*,$

(BC2) $\qquad\qquad\qquad Rg'_* \circ f'^! \xrightarrow{\sim} f^! \circ Rg_*.$

(5) For any "sufficiently finite" morphism $f: X \to Y$ there exist natural isomorphisms

(Projection formula) $\qquad (f_!\mathscr{F}) \otimes^L \mathscr{G} \xrightarrow{\sim} f_!(\mathscr{F} \otimes^L Lf^*\mathscr{G}),$

$$\mathscr{H}om_{\mathscr{O}_Y}(f_!\mathscr{F}, \mathscr{G}) \xrightarrow{\sim} Rf_*\,\mathscr{H}om_{\mathscr{O}_X}(\mathscr{F}, f^!\mathscr{G}),$$

$$f^!\,\mathscr{H}om_{\mathscr{O}_Y}(\mathscr{G}, \mathscr{G}') \xrightarrow{\sim} \mathscr{H}om_{\mathscr{O}_X}(Lf^*\mathscr{G}, f^!\mathscr{G}').$$

(6) For any closed immersion $i: Z \to X$ with complementary open immersion $j: U := X \setminus Z \to X$ the units and counits, respectively, of the relevant adjunctions yield a distinguished triangle

$$j_!j^! \longrightarrow \mathrm{id} \longrightarrow i_*Li^* \longrightarrow j_!j^![1]$$

For noetherian schemes X and for (bounded below) complexes of $\mathscr{O}_X$-modules with quasi-coherent cohomology we will establish a "formalism of five functors", i.e., we construct all functors as above except $f_!$ and show that they satisfy these properties. The construction of $f^!$ will be given in Chapter 25. By construction (Definition 25.61), $f^!$ will also be right adjoint to Rf_* for proper morphisms f of noetherian schemes.

Let us explain why we cannot expect a functor $f_!$ in this setting. In fact the above properties would determine $f_!$ for every separated finite type morphism $f: X \to Y$ of noetherian scheme because by Nagata compactification (Theorem 12.70) we can factorize f as an open immersion j followed by a proper morphism $\bar{f}$. Then we necessarily have $f_! = \bar{f}_! \circ j_! = R\bar{f}_* \circ j_!$ by (2), where $j_!$ is left adjoint to $j^! = Lj^* = j^*$ by (3). So $f_!$ would be determined by the functors that we already constructed if it existed.

But this cannot be the case since j^* cannot be right adjoint to another functor. Otherwise j^* would commute with arbitrary limits. In particular, for any index set I, for any ring A and for every $f \in A$ we would have

$$\left(\prod_I A \right) \otimes_A A_f = \prod_I A_f .$$

But on the left side powers of f in the denominator are bounded independent of $i \in I$ and on the right side these powers can get arbitrarily large. Hence we cannot expect such an equality and a full 6-functor formalism for quasi-coherent sheaves[4].

Note that we have already shown (1) for Rf_* and Lf^* (Proposition 21.115 and Proposition 21.112) and that we already constructed a morphism as in (BC1) (Proposition 21.129) with $f_!$ replaced by Rf_* (as we would expect at least for proper morphisms for which $f_! = Rf_*$ should hold). Finally, we already established a variant of a distinguished triangle as in (6) for abelian sheaves on topological spaces (Proposition 21.62).

Perfect and pseudo-coherent complexes

We now define finiteness properties in the derived category of $\mathscr{O}_X$-modules: Perfect and pseudo-coherent complexes. Here one can think of a perfect complex as being the derived analogue of an $\mathscr{O}_X$-module that is finite locally free (see Remark 21.149 below for a justification of this slogan) and of a pseudo-coherent complex as the derived analogue of being "sufficiently finitely generated". We will also introduce the notion to be of finite tor-amplitude. One of the main results in this part will be that a complex is perfect if and only if it is pseudo-coherent and locally of finite tor-amplitude (Theorem 21.174).

For instance, if A is a ring, then an A-module M considered as an object in $D(A)$ is perfect (resp. pseudo-coherent) if and only if M admits a finite left resolution (resp. a left resolution) by finite projective modules. It is of finite tor-amplitude if and only if it admits a finite left resolution by flat modules.

(21.31) Perfect complexes.

We fix a ringed space $(X, \mathscr{O}_X)$. Our basic finiteness condition for a complex will be to consist only of direct summands of finite free modules. Let us recall some properties of such modules.

Remark 21.132. Let $\mathscr{F}$ be a direct summand of a finite free $\mathscr{O}_X$-module.
(1) The $\mathscr{O}_X$-module $\mathscr{F}$ is flat and of finite presentation.
      Conversely, one can show that every flat $\mathscr{O}_X$-module of finite presentation is locally on X a direct summand of a finite free $\mathscr{O}_X$-module (Exercise 21.28).
(2) If $(X, \mathscr{O}_X)$ is a locally ringed space, $\mathscr{F}$ is finite locally free (Proposition 7.41).
(3) Let A be a ring. Then a direct summand of a finite free A-module is the same as a finite projective A-module.

---

[4] By enlarging the category of schemes it is possible to define a functor $f_!$ using the formalism of condensed mathematics, see [Scho]$_X$ and [Man]$_X$ 2.9.

Definition 21.133. *Let $\mathscr{F}$ be a complex of $\mathcal{O}_X$-modules.*
(1) *The complex $\mathscr{F}$ is called* strictly perfect *if $\mathscr{F}^p$ is zero for all but finitely many p and $\mathscr{F}^p$ is a direct summand of a finite free $\mathcal{O}_X$-module for all p.*
(2) *The complex $\mathscr{F}$ is called* perfect *if there exists an open covering $(U_i)_i$ of X such that for all i there exists a strictly perfect complex $\mathscr{E}_i$ of $\mathcal{O}_{U_i}$-modules and a quasi-isomorphism $\mathscr{E}_i \to \mathscr{F}_{|U_i}$.*

The restriction of a (strictly) perfect complex to an open subspace is again (strictly) perfect.

We want to define an object in the derived category to be perfect if it is perfect as a complex of $\mathcal{O}_X$-modules. But a priori this definition is problematic because it is not clear that being perfect is preserved under isomorphisms in $D(X)$. To see that this is indeed the case we need the following preparative lemma.

Lemma 21.134. *Let $\mathscr{E}$, $\mathscr{F}$, and $\mathscr{G}$ be complexes of $\mathcal{O}_X$-modules, where $\mathscr{E}$ is strictly perfect. Let $m \in \mathbb{Z}$.*
(1) *Let $\alpha\colon \mathscr{E} \to \mathscr{F}$ be a morphism of complexes. Suppose that $H^p(\mathscr{F}) = 0$ for $p \geq m$ and that $\mathscr{E}^p = 0$ for $p < m$. Then there exists an open covering $(U_i)_i$ of X such that $\alpha_{|U_i}$ is homotopic to 0.*
(2) *Let $\alpha\colon \mathscr{E} \to \mathscr{G}$ and $u\colon \mathscr{F} \to \mathscr{G}$ be maps in $C(X)$. Suppose that $\mathscr{E}^p = 0$ for all $p < m$ and that u is an m-isomorphism (Definition F.159). Then locally on X there exists a map $\beta\colon \mathscr{E} \to \mathscr{F}$ such that α and $u \circ \beta$ are homotopic.*
(3) *For any morphism $\alpha\colon \mathscr{E} \to \mathscr{F}$ in $D(X)$ there exists an open covering $(U_i)_i$ of X such that $\alpha_{|U_i}$ is given by a morphism of complexes $\mathscr{E}_{|U_i} \to \mathscr{F}_{|U_i}$ in $C(U_i)$.*

Proof. Let us show (1). Let $a \leq b$ be integers such that $\mathscr{E}^p = 0$ for all $p \notin [a,b]$. We show (1) by induction on $b - a$.

Suppose that $a = b =: n$. Then $\mathscr{E} = \mathscr{E}^n[-n]$ for a direct summand $\mathscr{E}^n$ of a finite free $\mathcal{O}_X$-module and an integer $n \geq m$. We can consider α as a morphism $\mathscr{E}^n \to \mathrm{Ker}(\mathscr{F}^n \to \mathscr{F}^{n+1})$. We have to find locally on X a map $h\colon \mathscr{E}^n \to \mathscr{F}^{n-1}$ that lifts α. For this we may assume that $\mathscr{E}^n = \mathcal{O}_X^r$ is a free module. Then α corresponds to r global sections of $\mathrm{Ker}(\mathscr{F}^n \to \mathscr{F}^{n+1})$. As $H^n(\mathscr{F}) = 0$, the map $\mathscr{F}^{n-1} \to \mathrm{Ker}(\mathscr{F}^n \to \mathscr{F}^{n+1})$ is surjective. Hence the r global sections can be locally lifted to $\mathscr{F}^{n-1}$ defining h locally.

In general there is a split exact sequence of complexes

$$0 \longrightarrow \mathscr{E}^b[-b] \longrightarrow \mathscr{E} \longrightarrow \sigma^{\leq b-1}\mathscr{E} \longrightarrow 0$$

yielding a distinguished triangle in $K(X)$. As $\mathrm{Hom}_{K(X)}(-, \mathscr{F})$ is a cohomological functor (Example F.137), we have an exact sequence

$$\mathrm{Hom}_{K(X)}(\sigma^{\leq b-1}\mathscr{E}, \mathscr{F}) \longrightarrow \mathrm{Hom}_{K(X)}(\mathscr{E}, \mathscr{F}) \longrightarrow \mathrm{Hom}_{K(X)}(\mathscr{E}^b[-b], \mathscr{F}).$$

As we have seen above, the composition $\mathscr{E}^b[-b] \to \mathscr{F}$ is locally zero in $K(X)$. Hence we may assume that α factors through a map $\sigma^{\leq b-1}\mathscr{E} \to \mathscr{F}$ which is locally zero by induction hypothesis. This shows (1).

To prove (2) let C_u be the mapping cone of u. Then $H^p(C_u) = 0$ for all $p \geq m$ (Remark F.159). As we have for every open subset $V \subseteq X$ a distinguished triangle

$$\mathscr{G}_{|V} \xrightarrow{u} \mathscr{F}_{|V} \longrightarrow C_{u|V} \xrightarrow{+1},$$

it suffices to show that locally on X the composition $\mathscr{E} \to \mathscr{F} \to C_u$ is homotopic to 0. This follows from (1).

To show (3) it suffices to show that we can choose $(U_i)_i$ such that $\alpha_{|U_i}$ is given by a morphism in $K(U_i)$. By the construction of the derived category, α is given by $s^{-1}\beta$, where $s\colon \mathcal{F} \to \mathcal{G}$ is a quasi-isomorphism and $\beta\colon \mathcal{E} \to \mathcal{G}$ is a map of complexes. Hence we are done by (2).   □

Remark 21.135. If u in Lemma 21.134 (2) is in addition surjective, then one can show that one can find β such that $u \circ \beta = \alpha$, see [Sta] 0649.

Now we can see that all plausible definitions for an object in the derived category $D(X)$ to be perfect are in fact equivalent.

Definition and Lemma 21.136. *Let X be a ringed space. An object $\mathcal{E}$ in $D(X)$ is called* perfect, *if it satisfies the following equivalent conditions.*
(i) *It is perfect as a complex of $\mathcal{O}_X$-modules.*
(ii) *Every complex of $\mathcal{O}_X$-modules $\mathcal{F}$ such that $\mathcal{F} \cong \mathcal{E}$ in $D(X)$ is perfect.*
(iii) *There exists an open covering $(U_i)_i$ of X and for all i strictly perfect complexes $\mathcal{E}_i$ of $\mathcal{O}_{U_i}$-modules such that $\mathcal{E}_{|U_i} \cong \mathcal{E}_i$ in $D(U_i)$.*

Proof. Clearly one has "(ii) $\Rightarrow$ (i) $\Rightarrow$ (iii)". So suppose that (iii) holds and let $\mathcal{F}$ be a complex of $\mathcal{O}_X$-modules which is isomorphic to $\mathcal{E}$ in $D(X)$. By Lemma 21.134 the isomorphisms $\mathcal{E}_i \cong \mathcal{E}_{|U_i} \cong \mathcal{F}_{|U_i}$ can be represented, after refining the open covering if necessary, by maps of complexes $\alpha_i\colon \mathcal{E}_i \to \mathcal{F}_{|U_i}$ that are necessarily quasi-isomorphisms.   □

Although one may think of a perfect complex as the derived version of a finite locally free module, note that even if a perfect complex $\mathcal{F}$ is concentrated in degree 0, then $\mathcal{F}^0$ is not necessarily finite locally free (see Example 21.138 (1) below).

Remark 21.137. Let X be a ringed space.
(1) Let $(U_i)_i$ be an open covering of X. A complex $\mathcal{E}$ in $D(X)$ is perfect if and only if $\mathcal{E}_{|U_i}$ is perfect for all $i \in I$.
      Indeed, this follows from Definition 21.136 (iii) since restrictions of strictly perfect complexes to open subspaces are again strictly perfect.
(2) If $\mathcal{E}$ is a perfect complex in $D(X)$, then $\mathcal{E}[n]$ is perfect for all $n \in \mathbb{Z}$.

Example 21.138.
(1) Let A be a ring made into a ringed space with underlying topological space consisting of a single point and A being the global section of the sheaf of rings. Then the category of A-modules is isomorphic to the category of modules over this ringed space.
      A complex of A-modules F is then perfect if and only if it is isomorphic in $D(A)$ to a finite complex E consisting of finitely generated projective A-modules. In fact Lemma 21.134 (3) shows that every isomorphism $E \xrightarrow{\sim} F$ in $D(A)$ is induced by a quasi-isomorphism of complexes $E \to F$.
      For instance, every A-module considered as a complex concentrated in degree 0 that has a finite resolution by finite projective modules is perfect.
(2) Let $(X, \mathcal{O}_X)$ be a locally ringed space (e.g. a scheme). Then an object $\mathcal{E}$ in $D(X)$ is perfect if and only if there exists an open covering $(U_i)_i$ of X such that $\mathcal{E}_{|U_i}$ can be represented by a finite complex of finite free $\mathcal{O}_{U_i}$-modules (Remark 21.132 (2)).
(3) If $X = \mathrm{Spec}(A)$, then we will see in Lemma 22.44 below that a complex of A-modules is perfect if and only if the associated complex of quasi-coherent $\mathcal{O}_X$-modules is perfect.

If one wants to show some assertion for perfect complexes, one can often work locally and then can reduce to show such an assertion for strictly perfect complexes. In this case the following result is often useful.

Proposition 21.139. *Let X be a ringed space and let $\mathcal{D}$ be a triangulated subcategory of $D(X)$ that contains $\mathscr{O}_X$ and is stable under direct summands in $D(X)$. Then $\mathcal{D}$ contains every strictly perfect complex.*

Proof. The hypotheses imply immediately that $\mathscr{M}[n]$ is in $\mathcal{D}$ for every direct summand $\mathscr{M}$ of a finite free $\mathscr{O}_X$-module and for all $n \in \mathbb{Z}$. Now let $\mathscr{F}$ be a strictly perfect complex, say concentrated in degrees $[a, b]$. We show by induction on $b - a$ that $\mathscr{F}$ is contained in $\mathcal{D}$. For $b = a$ we have $\mathscr{F} = \mathscr{M}[-a]$ for some direct summand $\mathscr{M}$ of a finite free module. In general, the exact sequence of complexes $0 \to \sigma^{>a}\mathscr{F} \to \mathscr{F} \to \mathscr{F}^a[-a] \to 0$ yields a distinguished triangle $\sigma^{>a}\mathscr{F} \to \mathscr{F} \to \mathscr{F}^a[-a] \to$. Then $\sigma^{>a}\mathscr{F}$ is contained in $\mathcal{D}$ by induction hypothesis and $\mathscr{F}^a[-a]$ is contained in $\mathcal{D}$ by the initial remark. Hence $\mathscr{F}$ is in $\mathcal{D}$ since $\mathcal{D}$ is triangulated. $\qquad\square$

If A is a ring, then every perfect complex in $D(A)$ is isomorphic to a strictly perfect complex. Hence we obtain the following useful corollary.

Corollary 21.140. *Let A be a ring and let $\mathcal{D}$ be a triangulated subcategory of $D(A)$ that is closed under isomorphisms, stable under direct summands and that contains A. Then $\mathcal{D}$ contains every perfect complex in $D(A)$.*

Often one can use Proposition 21.139 in the following form.

Corollary 21.141. *Let X be a ringed space, let $\mathcal{T}$ be any triangulated category, let $\Phi, \Psi \colon D(X) \to \mathcal{T}$ be triangulated functors and let $\Phi \to \Psi$ be a morphism of triangulated functors. Suppose $\mathscr{F}(\mathscr{O}_X) \to \mathscr{G}(\mathscr{O}_X)$ is an isomorphism. Then $\Phi(\mathscr{F}) \to \Psi(\mathscr{F})$ is an isomorphism for all strictly perfect complexes $\mathscr{F}$.*

Proof. Let $\mathcal{D}$ be the full subcategory of $D(X)$ of complexes $\mathscr{F}$ such that $\Phi(\mathscr{F}) \to \Psi(\mathscr{F})$ is an isomorphism. As Φ and Ψ are triangulated, $\mathcal{D}$ is a triangulated subcategory. Indeed, it is additive and stable under shifts and the five lemma in triangulated categories (Lemma F.119) shows that if two vertices of a distinguished triangle are in $\mathcal{D}$ then so is the third. Moreover, it is stable under direct summands and it contains $\mathscr{O}_X$ by hypothesis. One concludes by Lemma 21.139. $\qquad\square$

Remark 21.142. Let $f \colon X \to Y$ be a morphism of ringed spaces. If $\mathscr{E}$ is a perfect complex on Y, then $Lf^*\mathscr{E}$ is a perfect complex on X.

Indeed, the question is local on Y and hence we can assume that $\mathscr{E}$ is a strictly perfect complex. Such a complex is K-flat by Lemma 21.93. Hence one has $Lf^*\mathscr{E} = f^*\mathscr{E}$. As the pullback of a strictly perfect complex is strictly perfect, $Lf^*\mathscr{E}$ is again perfect.

Remark 21.143. Let $\mathscr{E}$ and $\mathscr{F}$ be perfect complexes of $\mathscr{O}_X$-modules. Then $\mathscr{E} \otimes^L_{\mathscr{O}_X} \mathscr{F}$ is a perfect complex in $D(X)$.

Indeed, we may assume that $\mathscr{E}$ and $\mathscr{F}$ are strictly perfect and in particular K-flat. Then $\mathscr{E} \otimes^L \mathscr{F} = \mathscr{E} \otimes \mathscr{F}$ is also strictly perfect.

(21.32) The dual of a perfect complex.

As for finite locally free modules we will define the notion of the dual of a perfect complex, and a canonical isomorphism with its double dual. As a preparation we need the following lemma.

Lemma 21.144. *Let X be a ringed space and let $\mathscr{E}$ and $\mathscr{F}$ be complexes of $\mathscr{O}_X$-modules. Suppose that $\mathscr{E}$ is strictly perfect. Then $R\mathscr{H}om_{\mathscr{O}_X}(\mathscr{E},\mathscr{F})$ is represented by the complex $\mathscr{H}om_{\mathscr{O}_X}(\mathscr{E},\mathscr{F})$ (Definition 21.86).*

For complexes of modules over a ring A, which is a special case (Example 21.138), there is nothing to prove because in this case strictly perfect complexes are K-projective (by the dual of Remark F.181 (2)) and we can calculate $R\operatorname{Hom}$ with K-projective resolutions in the first argument. But in general even a finite free $\mathscr{O}_X$-module is not a projective object in the category of $\mathscr{O}_X$-modules (Exercise 21.5) and hence a strictly perfect complex is typically not a K-projective complex.

Proof. Let $\mathscr{F} \to \mathscr{I}$ be a quasi-isomorphism of $\mathscr{F}$ with a K-injective complex $\mathscr{I}$. Then the complex $\mathscr{H}om_{\mathscr{O}_X}(\mathscr{E},\mathscr{I})$ represents $R\mathscr{H}om_{\mathscr{O}_X}(\mathscr{E},\mathscr{F})$. Hence it suffices to show that

$$\iota: \mathscr{H} := \mathscr{H}om_{\mathscr{O}_X}(\mathscr{E},\mathscr{F}) \longrightarrow \mathscr{H}' := \mathscr{H}om_{\mathscr{O}_X}(\mathscr{E},\mathscr{I})$$

is a quasi-isomorphism. Replacing $\mathscr{E}$ by shifts it suffices to show that ι induces an isomorphism after applying H^0. For $U \subseteq X$ open we have $H^0(\mathscr{H}(U)) = \operatorname{Hom}_{K(X)}(\mathscr{E}_{|U}, \mathscr{F}_{|U})$ by (F.18.2) and $H^0(\mathscr{H}'(U)) = \operatorname{Hom}_{D(X)}(\mathscr{E}_{|U}, \mathscr{F}_{|U})$ by (F.52.5). Now Lemma 21.134 (3) shows that the sheafifications of $U \mapsto H^0(\mathscr{H}(U))$ and of $U \mapsto H^0(\mathscr{H}'(U))$ are equal. $\square$

Lemma 21.145. *Let X be a ringed space, and let $\mathscr{E}$, $\mathscr{F}$, and $\mathscr{G}$ be complexes in $D(X)$ and suppose that $\mathscr{E}$ or $\mathscr{G}$ is perfect. Then the functorial map (21.26.5) is an isomorphism*

$$R\mathscr{H}om_{\mathscr{O}_X}(\mathscr{E},\mathscr{F}) \otimes^L_{\mathscr{O}_X} \mathscr{G} \xrightarrow{\sim} R\mathscr{H}om_{\mathscr{O}_X}(\mathscr{E}, \mathscr{F} \otimes^L \mathscr{G})$$

Proof. We can work locally on X and hence can assume that $\mathscr{E}$ or $\mathscr{G}$ is strictly perfect. As both sides are triangulated functors in $\mathscr{E}$ and $\mathscr{G}$, we can even assume that $\mathscr{E} = \mathscr{O}_X$ or $\mathscr{G} = \mathscr{O}_X$ by Corollary 21.141. In this case, the assertion is clear. $\square$

Definition 21.146. *Let X be a ringed space. For a complex $\mathscr{E}$ in $D(X)$ we call*

$$\mathscr{E}^\vee := R\mathscr{H}om_{\mathscr{O}_X}(\mathscr{E},\mathscr{O}_X)$$

its dual.

Remark 21.147. Let X be a ringed space. For every complex $\mathscr{E}$ in $D(X)$ there is functorial map

(21.32.1) $\mathscr{E} \longrightarrow (\mathscr{E}^\vee)^\vee.$

Indeed, by applying H^0 to (21.26.2) one sees that it suffices to construct a functorial map $\mathscr{E}^\vee \otimes^L \mathscr{E} \to \mathscr{O}_X$. But this is a special case of (21.26.7).

Proposition 21.148. *Let X be a ringed space and let $\mathscr{E}$ be a perfect complex in $D(X)$.*
(1) *The dual complex $\mathscr{E}^\vee$ is perfect and the functorial map $\mathscr{E} \to (\mathscr{E}^\vee)^\vee$ is an isomorphism.*

(2) *For all complexes $\mathscr{G}$ there is an isomorphism*

$$(21.32.2) \qquad \mathscr{E}^\vee \otimes^L_{\mathscr{O}_X} \mathscr{G} \xrightarrow{\sim} R\,\mathscr{H}om_{\mathscr{O}_X}(\mathscr{E},\mathscr{G})$$

in $D(X)$, functorial in $\mathscr{G}$ and $\mathscr{E}$.

(3) *For all complexes $\mathscr{F}$ and $\mathscr{G}$ there is an isomorphism*

$$(21.32.3) \qquad R\,\mathscr{H}om_{\mathscr{O}_X}(\mathscr{E}^\vee \otimes^L_{\mathscr{O}_X} \mathscr{F},\mathscr{G}) \xrightarrow{\sim} R\,\mathscr{H}om_{\mathscr{O}_X}(\mathscr{F},\mathscr{E} \otimes^L_{\mathscr{O}_X} \mathscr{G})$$

in $D(X)$, functorial in $\mathscr{E}$, $\mathscr{F}$, and $\mathscr{G}$.

Proof. The morphism (21.32.2) is given by the functorial map (21.26.5). Hence all assertions can be checked locally on X. Hence we can assume that $\mathscr{E}$ is given by a strictly perfect complex.

By Lemma 21.144 the dual $\mathscr{E}^\vee$ is then given by the complex $\mathscr{H}om(\mathscr{E},\mathscr{O}_X)$ which is again a strictly perfect complex. Hence $\mathscr{E}^\vee$ is perfect. Moreover $\mathscr{E} \to (\mathscr{E}^\vee)^\vee$ is an isomorphism as this is the case for strictly perfect complexes. This shows (1).

Assertion (2) is a special case of Lemma 21.145.

Now (3) is obtained from the following functorial isomorphisms

$$R\,\mathscr{H}om_{\mathscr{O}_X}(\mathscr{E}^\vee \otimes^L_{\mathscr{O}_X} \mathscr{F},\mathscr{G}) \cong R\,\mathscr{H}om_{\mathscr{O}_X}(\mathscr{F}, R\,\mathscr{H}om_{\mathscr{O}_X}(\mathscr{E}^\vee,\mathscr{G}))$$
$$\cong R\,\mathscr{H}om_{\mathscr{O}_X}(\mathscr{F},\mathscr{E} \otimes^L_{\mathscr{O}_X} \mathscr{G}),$$

where the first isomorphism is (21.26.1) and the second holds by (2) and (1). □

Definition and Remark 21.149. Let X be a ringed space and let $\mathscr{E}$ be in $D(X)$. One says that $\mathscr{E}$ is *dualizable* if there exists a dual, i.e., an object $\mathscr{F}$ in $D(X)$ and morphisms $\eta\colon \mathscr{O}_X \to \mathscr{E} \otimes^L \mathscr{F}$ and $\epsilon\colon \mathscr{F} \otimes^L \mathscr{E} \to \mathscr{O}_X$ such that the following diagrams commute

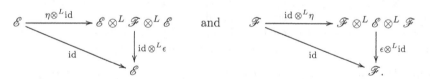

Then Proposition 21.148 shows that every perfect complex $\mathscr{E}$ is dualizable and that $\mathscr{E}^\vee = R\,\mathscr{H}om_{\mathscr{O}_X}(\mathscr{E},\mathscr{O}_X)$ is a dual. Conversely, one can show that every dualizable object in $D(X)$ is perfect ([Sta] OFPV, see also Exercise 25.1 if X is a scheme). Hence in $D(X)$ a complex is perfect if and only if it is dualizable. This gives a more concrete meaning to the slogan that being perfect is the "derived analogue" of being finite locally free since dualizable objects in the category of $\mathscr{O}_X$-modules are precisely those that are locally direct summands of a finite free $\mathscr{O}_X$-module (Exercise 21.28).

In Proposition 25.16 below, we will also see a further categorical characterization of perfect complexes: If X is qcqs scheme, then a complex $\mathscr{E}$ is perfect if and only if $\mathscr{E}$ is a compact object (Definition 25.1 below) of $D_{\mathrm{qcoh}}(X)$, the full subcategory of $D(X)$ of complexes all of whose cohomology objects are quasi-coherent. This characterization is rather particular to qcqs schemes and does not generalize to more general situations such as ringed spaces or qcqs algebraic stacks.

(21.33) Pseudo-coherent complexes.

Let $(X, \mathscr{O}_X)$ be a ringed space. Let $m \in \mathbb{Z}$. Recall from Definition F.159 that an m-isomorphism $\mathscr{E} \to \mathscr{F}$ is a morphism in $D(X)$ sitting in a distinguished triangle $\mathscr{E} \longrightarrow \mathscr{F} \longrightarrow \mathscr{C} \overset{+}{\longrightarrow}$ with $H^p(\mathscr{C}) = 0$ for all $p \geq m$, i.e., $H^m(\mathscr{E}) \to H^m(\mathscr{F})$ is surjective and $\tau^{\geq m+1}\mathscr{E} \to \tau^{\geq m+1}\mathscr{F}$ is an isomorphism in $D(X)$.

Definition and Lemma 21.150. *Let $(X, \mathscr{O}_X)$ be a ringed space. Let $m \in \mathbb{Z}$.*

(1) *A complex $\mathscr{F}$ of $\mathscr{O}_X$-modules is called m-pseudo-coherent if there exists an open covering $(U_i)_i$ of X and for each i an m-isomorphism $\mathscr{E}_i \to \mathscr{F}_{|U_i}$, where $\mathscr{E}_i$ is a strictly perfect complex on U_i.*

   The complex $\mathscr{F}$ is called pseudo-coherent if it is m-pseudo-coherent for all $m \in \mathbb{Z}$.

(2) *An object $\mathscr{E}$ in $D(X)$ is called m-pseudo-coherent if it satisfies the following equivalent conditions.*

   (i) *The complex $\mathscr{E}$ of $\mathscr{O}_X$-modules is m-pseudo-coherent.*

   (ii) *Every complex $\mathscr{F}$ of $\mathscr{O}_X$-modules such that $\mathscr{F} \cong \mathscr{E}$ in $D(X)$ is m-pseudo-coherent.*

   (iii) *There exist an open covering $(U_i)_i$ of X, strictly perfect complexes $\mathscr{E}_i$ on U_i, and m-isomorphisms $\mathscr{E}_i \to \mathscr{E}_{|U_i}$ in $D(U_i)$.*

   The object $\mathscr{E}$ in $D(X)$ is called pseudo-coherent if it is m-pseudo-coherent for all $m \in \mathbb{Z}$.

(3) *An $\mathscr{O}_X$-module is called m-pseudo-coherent (resp. pseudo-coherent) if it is m-pseudo-coherent (resp. pseudo-coherent) considered as a complex concentrated in degree 0.*

Proof. The equivalence of the conditions in (2) is proved as in Lemma 21.136.   □

Any perfect complex is pseudo-coherent. In general, the converse does not hold (Exercise 22.36). In Theorem 21.174 below we will give a criterion for a pseudo-coherent complex to be perfect.

Remark 21.151. Let $m \in \mathbb{Z}$. Let $\mathscr{F}$ be an m-pseudo-coherent object in $D(X)$.
(1) The object $\mathscr{F}$ is n-pseudo-coherent for all $n \geq m$.
(2) For all $i \in \mathbb{Z}$ the shifted complex $\mathscr{F}[i]$ is $(m - i)$-pseudo-coherent.
(3) Let X be quasi-compact. Then every m-pseudo-coherent complex on X lies in $D^-(X)$.

Lemma 21.152. *Let $\mathscr{E}$ be a bounded above complex of direct summands of finite free $\mathscr{O}_X$-modules. Then there exists a homotopy equivalence to a bounded above complex of finite free $\mathscr{O}_X$-modules.*

Proof. Let $n \in \mathbb{Z}$ such that $\mathscr{E}^p = 0$ for all $p > n$. Choose $\mathscr{F}^n = \mathscr{E}^n \oplus \mathscr{G}^n$ with $\mathscr{F}^n$ a finite free $\mathscr{O}_X$-module and for $p < n$ inductively $\mathscr{F}^p = \mathscr{E}^p \oplus \mathscr{G}^{p+1} \oplus \mathscr{G}^p$ with $\mathscr{F}^p$ finite free for all p. Set also $\mathscr{F}^p = 0$ for $p > n$. For $p < n$ define maps $\mathscr{F}^p \to \mathscr{F}^{p+1}$ as the sum of the given map $\mathscr{E}^p \to \mathscr{E}^{p+1}$, of the identity $\mathscr{G}^{p+1} \to \mathscr{G}^{p+1}$, and of the zero map on $\mathscr{G}^p$. This makes $\mathscr{F}$ into a complex and the projection $\mathscr{F} \to \mathscr{E}$ is a homotopy equivalence.   □

Proposition 21.153. *Let $\mathscr{F}$ be an $\mathscr{O}_X$-module viewed as an object in $D(X)$. Let $r \geq 0$ be an integer. Then $\mathscr{F}$ is $(-r)$-pseudo-coherent if and only if locally on X there exists an exact sequence of $\mathscr{O}_X$-modules*

$$(*) \qquad\qquad \mathscr{E}^{-r} \longrightarrow \mathscr{E}^{-r+1} \longrightarrow \cdots \longrightarrow \mathscr{E}^0 \longrightarrow \mathscr{F} \longrightarrow 0,$$

where $\mathscr{E}^i$ is a finite free $\mathscr{O}_X$-module.

For $r = 0$ (resp. $r = 1$) this means that $\mathscr{F}$ is of finite type (resp. of finite presentation).

Proof. Suppose that $\mathscr{F}$ is $(-r)$-pseudo-coherent. We can work locally on X and hence can assume that there exists a strictly perfect complex $\mathscr{E}$ and a $(-r)$-isomorphism $\mathscr{E} \to \mathscr{F}$.

We claim that we may assume that $\mathscr{E}^p = 0$ for all $p > 0$. Indeed, let n be the largest integer such that $\mathscr{E}^n \neq 0$ and assume that $n > 0$. As $H^n(\mathscr{E}) = 0$, we see that $\mathscr{E}^{n-1} \to \mathscr{E}^n$ is surjective. Hence locally on X we can write $\mathscr{E}^{n-1} = \mathscr{E}' \oplus \mathscr{E}^n$ and replace $\mathscr{E}$ by $\cdots \to \mathscr{E}^{n-3} \to \mathscr{E}^{n-2} \to \mathscr{E}' \to 0$. Hence induction on n shows our claim.

Now the $-r$-isomorphism $\mathscr{E} \to \mathscr{F}$ yields a resolution as in (*) with $\mathscr{E}^p$ a direct summand of a finite free $\mathscr{O}_X$-module for $p = -r, \ldots, 0$. By Lemma 21.152 we may assume that $\mathscr{E}^p$ is a finite free $\mathscr{O}_X$-module for $p = -r, \ldots, 0$.

Conversely, suppose that a resolution as in (*) exists, then $\mathscr{E} \to \mathscr{F}$ is a $(-r)$-isomorphism. $\qquad\square$

Proposition 21.154. *Let $m \in \mathbb{Z}$ and let*

$$\mathscr{E} \xrightarrow{u} \mathscr{F} \xrightarrow{v} \mathscr{G} \xrightarrow{w} \mathscr{E}[1]$$

be a distinguished triangle in $D(X)$. If $\mathscr{E}$ and $\mathscr{G}$ are m-pseudo-coherent, then $\mathscr{F}$ is m-pseudo-coherent.

By rotating the triangle one also sees:
(1) If $\mathscr{E}$ is $(m+1)$-pseudo-coherent and $\mathscr{F}$ is m-pseudo-coherent, then $\mathscr{G}$ is m-pseudo-coherent.
(2) If $\mathscr{F}$ is m-pseudo-coherent and $\mathscr{G}$ is $(m-1)$-pseudo-coherent, then $\mathscr{E}$ is m-pseudo-coherent.

Proof. By rotating the triangle it suffices to show that if $\mathscr{E}$ is $(m+1)$-pseudo-coherent and $\mathscr{F}$ is m-pseudo-coherent, then $\mathscr{G}$ is m-pseudo-coherent. This can be shown locally on X. Hence we may assume that there exist strictly perfect complexes $\mathscr{K}$ and $\mathscr{L}$, an $(m+1)$-isomorphism $\alpha \colon \mathscr{K} \to \mathscr{E}$ and an m-isomorphism $\beta \colon \mathscr{L} \to \mathscr{F}$. Replacing $\mathscr{K}$ by $\sigma^{\geq m+1}K$ we may assume that $\mathscr{K}^p = 0$ for $p < m+1$.

Working locally and using Lemma 21.134, we can first assume that the composition $\mathscr{K} \to \mathscr{E} \to \mathscr{F}$ and the map β are given by maps of complexes and then that there exists a map of complexes $\tilde{u} \colon \mathscr{K} \to \mathscr{L}$ such that $\beta \circ \tilde{u} = u \circ \alpha$ in $D(X)$. The cone $C_{\tilde{u}}$ of $\tilde{u}$ is strictly perfect by definition of a cone. By the axioms of a triangulated category there exists a morphism

$$(\mathscr{K} \to \mathscr{L} \to C_{\tilde{u}} \to) \longrightarrow (\mathscr{E} \to \mathscr{F} \to \mathscr{G} \to)$$

of distinguished triangles. Looking at the induced map on long exact cohomology sequences the Four Lemma F.46 shows that $C_{\tilde{u}} \to \mathscr{G}$ is an m-isomorphism. $\qquad\square$

Corollary 21.155. *Let $\mathscr{E} \to \mathscr{F} \to \mathscr{G} \to$ be a distinguished triangle in $D(X)$. If two of the three complexes are pseudo-coherent, then the third is pseudo-coherent.*

Hence the full subcategory of pseudo-coherent complexes is a triangulated subcategory of $D(X)$.

Proposition 21.156. *Let $\mathscr{K}$ and $\mathscr{L}$ be complexes in $D(X)$ and let $m \in \mathbb{Z}$. Then $\mathscr{K}$ and $\mathscr{L}$ are m-pseudo-coherent (resp. pseudo-coherent) if and only if $\mathscr{K} \oplus \mathscr{L}$ is m-pseudo-coherent (resp. pseudo-coherent).*

Proof. The distinguished triangle $\mathscr{K} \to \mathscr{K} \oplus \mathscr{L} \to \mathscr{L} \to$ shows the necessity of the condition by Proposition 21.154. Suppose that $\mathscr{K} \oplus \mathscr{L}$ is m-pseudo-coherent. It suffices to show that $\mathscr{L}$ is m-pseudo-coherent. Working locally on X, we may assume that $\mathscr{K} \oplus \mathscr{L}$ is bounded above. Then $\mathscr{K}$ and $\mathscr{L}$ are bounded above. Hence $\mathscr{L}[N]$ is m-pseudo-coherent for large N.

Using the direct sum of the distinguished triangles $\mathscr{K} \to \mathscr{K} \to 0 \to$ and $\mathscr{L} \to \mathscr{L} \to \mathscr{L} \oplus \mathscr{L}[1] \to$ we see by Proposition 21.154 that $\mathscr{L} \oplus \mathscr{L}[1]$ is m-pseudo-coherent. Then for every $n \geq 0$ the shift $\mathscr{L}[n] \oplus \mathscr{L}[n+1]$ is $(m-n)$-pseudo-coherent and in particular m-pseudo-coherent by Remark 21.151. As $\mathscr{L}[N]$ is m-pseudo-coherent for large N the distinguished triangle $\mathscr{L}[n+1] \to \mathscr{L}[n] \oplus \mathscr{L}[n+1] \to \mathscr{L}[n]$ shows that $\mathscr{L}[n]$ is m-pseudo-coherent for all $n \geq 0$. In particular, $\mathscr{L}$ is m-pseudo-coherent. $\qquad\square$

Proposition 21.157. *Let $f\colon X \to Y$ be a morphism of ringed spaces, let $m \in \mathbb{Z}$, and let $\mathscr{E}$ be a complex in $D(Y)$. If $\mathscr{E}$ is m-pseudo-coherent, then $Lf^*\mathscr{E}$ is m-pseudo-coherent.*

In particular, if $\mathscr{E}$ is pseudo-coherent, then $Lf^*\mathscr{E}$ is pseudo-coherent.

Proof. Working locally on Y we may assume there exists a strictly perfect complex $\mathscr{K}$ of $\mathscr{O}_Y$-modules and an m-isomorphism $\mathscr{K} \to \mathscr{E}$ in $D(Y)$. As $Lf^*\mathscr{K}$ is strictly perfect (Remark 21.142), it suffices to show that for every m-isomorphism $u\colon \mathscr{L} \to \mathscr{M}$ in $D(Y)$ its derived pullback Lf^*u is again an m-isomorphism. Indeed, we complement to a distinguished triangle
$$\mathscr{L} \xrightarrow{u} \mathscr{M} \longrightarrow \mathscr{C} \xrightarrow{+} .$$
Then $H^p(\mathscr{C}) = 0$ for all $p \geq m$ and hence $H^p(Lf^*\mathscr{C}) = 0$ for all $p \geq m$ (Remark 21.111). As Lf^* preserves distinguished triangles, this shows that Lf^*u is an m-isomorphism. $\qquad\square$

The question whether the derived image of pseudo-coherent complexes is again pseudo-coherent is one of the fundamental finiteness questions in Algebraic Geometry. We will for instance see that this is the case for proper morphisms between locally noetherian schemes (Section (23.6)).

Lemma 21.158. *Let X be a ringed space and let $\mathscr{E}$ be a locally bounded above complex of $\mathscr{O}_X$-modules such that $\mathscr{E}^i$ is a pseudo-coherent $\mathscr{O}_X$-module for all $i \in \mathbb{Z}$. Then $\mathscr{E}$ is pseudo-coherent.*

Proof. Fix $m \in \mathbb{Z}$. We will show that if $\mathscr{E}^i$ is $(m-i)$-pseudo-coherent for all i, then $\mathscr{E}$ is m-pseudo-coherent. We may work locally on X and hence can assume that $\mathscr{E}$ is bounded above. To check that $\mathscr{E}$ is m-pseudo-coherent, we may replace $\mathscr{E}$ by $\sigma^{\geq m-1}\mathscr{E}$ and hence can assume that $\mathscr{E}$ is bounded. By hypothesis, $\mathscr{E}^i[-i]$ is m-pseudo-coherent for all i. We conclude by induction on the length of the complex using the distinguished triangle $\sigma^{\geq i+1}\mathscr{E} \to \sigma^{\geq i}\mathscr{E} \to \mathscr{E}^i[-i] \to$ and Proposition 21.154. $\qquad\square$

By definition, for any pseudo-coherent complex $\mathscr{F}$ for all $m \in \mathbb{Z}$ there locally exists an m-isomorphism $\mathscr{E} \to \mathscr{F}$ for some strictly perfect complex $\mathscr{E}$, depending on m. Hence the following result is sometimes useful to prove that certain functors are isomorphic on pseudo-coherent complexes.

Proposition 21.159. *Let X be a ringed space and let $\mathcal{A}$ be an abelian category. Let $\Phi, \Psi\colon D(X) \to D(\mathcal{A})$ be triangulated functors and let $\eta\colon \Phi \to \Psi$ be a morphism of triangulated functors. Suppose that the following conditions hold.*
(a) *$\eta(\mathscr{O}_X)\colon \Phi(\mathscr{O}_X) \to \Psi(\mathscr{O}_X)$ is an isomorphism.*
(b) *There exists an integer N such that Φ and Ψ map $D^{<i}(X)$ to $D^{<N+i}(\mathcal{A})$ for all $i \in \mathbb{Z}$.*

Then $\eta(\mathscr{F})\colon \Phi(\mathscr{F}) \to \Psi(\mathscr{F})$ is an isomorphism for all complexes $\mathscr{F}$ in $D(X)$ such that for all $m \in \mathbb{Z}$ there exists an m-isomorphism $\mathscr{E} \to \mathscr{F}$ for some strictly perfect complex $\mathscr{E}$ (which may depend on m).

By passing to the opposite category of $D(\mathcal{A})$ we have the same assertion for contravariant functors Φ, Ψ such that there exists an integer N such that Φ and Ψ both map $D^{<i}(X)$ into $D^{>N-i}(\mathcal{A})$ for all $i \in \mathbb{Z}$.

Proof. Let $\mathcal{D}$ be the full subcategory of $D(X)$ of complexes $\mathscr{F}$ such that $\Phi(\mathscr{F}) \to \Psi(\mathscr{F})$ is an isomorphism. By Corollary 21.141 and its proof we know that $\mathcal{D}$ is a triangulated subcategory of $D(X)$ and that every strictly perfect complex is in $\mathcal{D}$. Let $\mathscr{F} \in D(X)$ such that for all $m \in \mathbb{Z}$ there exists an m-isomorphism $\mathscr{E}_m \to \mathscr{F}$ with $\mathscr{E}_m$ strictly perfect. We complete to a distinguished triangle $\mathscr{E}_m \longrightarrow \mathscr{F} \longrightarrow \mathscr{C}_m \overset{+1}{\longrightarrow}$ with $\mathscr{C}_m \in D^{<m}(X)$. As Φ and Ψ are triangulated, we obtain a map of distinguished triangles

$$
(*) \quad
\begin{array}{ccccccc}
\Phi(\mathscr{E}_m) & \longrightarrow & \Phi(\mathscr{F}) & \longrightarrow & \Phi(\mathscr{C}_m) & \longrightarrow & \Phi(\mathscr{E}_m)[1] \\
\downarrow{\scriptstyle\cong} & & \downarrow & & \downarrow & & \downarrow{\scriptstyle\cong} \\
\Psi(\mathscr{E}_m) & \longrightarrow & \Psi(\mathscr{F}) & \longrightarrow & \Psi(\mathscr{C}_m) & \longrightarrow & \Psi(\mathscr{E}_m)[1]
\end{array}
$$

As $\Phi(\mathscr{C}_m), \Psi(\mathscr{C}_m) \in D^{<N+m}(\mathcal{A})$ we see from the morphism of long exact sequences induced by $(*)$ that $H^p(\eta(\mathscr{F}))$ are isomorphisms $H^p\Phi(\mathscr{F}) \overset{\sim}{\to} H^p\Psi(\mathscr{F})$ for all $p > N+m$. Since we may choose m arbitrarily negative, this concludes the proof. $\qquad\square$

As similar argument shows the following variant of Proposition 21.159.

Remark 21.160. Let X be a ringed space and let $\mathcal{A}$ be an abelian category. Let $\Phi\colon D(X) \to D(\mathcal{A})$ be a triangulated functor such that there exists an integer N such that $\Phi(D^{<m}(X)) \subseteq D^{<N+m}(\mathcal{A})$ for all $m \in \mathbb{Z}$. Let $\mathcal{A}' \subseteq \mathcal{A}$ be a plump subcategory (Definition F.43). Suppose that $H^p\Phi(\mathscr{O}_X) \in \mathcal{A}'$ for all $p \in \mathbb{Z}$. Then $H^p\Phi(\mathscr{F}) \in \mathcal{A}'$ for all complexes $\mathscr{F}$ in $D(X)$ such that for all $m \in \mathbb{Z}$ there exists an m-isomorphism $\mathscr{E} \to \mathscr{F}$ for some strictly perfect complex $\mathscr{E}$.

Indeed, let $\mathcal{D}$ be the full subcategory of $D(X)$ of complexes $\mathscr{F}$ such that $H^p\Phi(\mathscr{F}) \in \mathcal{A}'$ for all $p \in \mathbb{Z}$. As $\mathcal{A}'$ is plump, this is a triangulated subcategory of $D(X)$ stable under direct summands. Hence Proposition 21.139 shows that $\mathcal{D}$ contains every strictly perfect complex. Now the same argument as in the proof of Proposition 21.159 finishes the claim.

(21.34) Pseudo-coherent complexes over a ring.

As before, let $(X, \mathscr{O}_X)$ be a ringed space. Let $\mathscr{E}$ be a complex in $D(X)$, let $m \in \mathbb{Z}$. Then one can show that $\mathscr{E}$ is m-pseudo-coherent if and only if locally on X there exists a bounded above complex $\mathscr{K}$ with $\mathscr{K}^p$ a direct summand of a finite free module for all $p \geq m$ and an isomorphism $\mathscr{K} \overset{\sim}{\to} \mathscr{E}$ in $D(X)$ ([SGA6]^0 I, 2.2).

But in general a pseudo-coherent complex might not locally on X be isomorphic in $D(X)$ to a bounded above complex $\mathscr{K}$ with $\mathscr{K}^p$ a direct summand of a finite free module for *all* p, since in general one cannot find, even locally, a complex $\mathscr{K}$ that works for all m. But this is true if X consists of a point, i.e., if we consider the complexes of A-modules for a ring A. From this we will deduce that for schemes X an object in $D(X)$ is pseudo-coherent if and only if it is locally on X isomorphic to a bounded above complex of finite free modules (Corollary 22.47).

To show this we will need the following lemma.

Lemma 21.161. *Let $m \in \mathbb{Z}$ and let $\mathscr{F}$ be an m-pseudo-coherent complex in $D(X)$. Suppose that $H^p(\mathscr{F}) = 0$ for all $p > m$. Then $H^m(\mathscr{F})$ is an $\mathscr{O}_X$-module of finite type.*

Proof. We may work locally on X. By hypothesis we have $\tau^{\leq m}\mathscr{F} \cong \mathscr{F}$ in $D(X)$ and hence we may assume that $\mathscr{F}$ is given by a complex with $\mathscr{F}^p = 0$ for $p > m$. Then $H^m(\mathscr{F})$ is a quotient of $\mathscr{F}^m$. We can work locally on X and hence can assume that there exists a strictly perfect complex $\mathscr{E}$ and an m-isomorphism $\mathscr{E} \to \mathscr{F}$. As in the proof or Proposition 21.153 we may assume that $\mathscr{E}^p = 0$ for all $p > m$.

But then $H^m(\mathscr{E})$ is a quotient of $\mathscr{E}^m$ and hence of finite type. And $H^m(\mathscr{F})$ is a quotient of $H^m(\mathscr{E})$ and hence also of finite type. $\qquad\square$

Proposition 21.162. *Let A be a ring. Then a complex E of A-modules is pseudo-coherent if and only if there exists a bounded above complex F of finite free A-modules and a quasi-isomorphism $F \to E$. If $m \in \mathbb{Z}$ is an integer such that $H^p(E) = 0$ for all $p \geq m$, then we may choose F such that $F^p = 0$ for all $p \geq m$.*

Proof. We apply Lemma F.160, where $\mathcal{A}$ is the category of A-modules, $\mathcal{D}$ is the full subcategory of finite free A-modules, and $\mathcal{C}$ is the category of pseudo-coherent complexes of A-modules. Let us check that all hypotheses are satisfied in this case.

This is clear for Hypothesis (a). As strictly perfect complexes are pseudo-coherent, one has $C^b(\mathcal{D}) \subseteq \mathcal{C}$. Moreover the cone of every morphism of pseudo-coherent complexes is again pseudo-coherent by a rotated version of Proposition 21.154. This shows that Hypothesis (b) holds. Finally, Hypothesis (c) holds because of Lemma 21.161.

Therefore Lemma F.160 applied to $Z = 0$ yields the proposition. $\qquad\square$

Lemma 21.163. *Let $A \to B$ be a faithfully flat ring homomorphism and let E be a complex of A-modules. Let $m \in \mathbb{Z}$. Then E is m-pseudo-coherent (resp. pseudo-coherent) if and only if $E \otimes_A^L B$ is an m-pseudo-coherent (resp. pseudo-coherent) complex of B-modules.*

Proof. It suffices to show the equivalence for "m-pseudo-coherent". As B is flat over A, one has $E \otimes_A^L B = E \otimes_A B$. Therefore, if E is m-pseudo-coherent, then so is $E \otimes_A B$ by Proposition 21.157.

Conversely, assume that $E \otimes_A B$ is m-pseudo-coherent and let $n \in \mathbb{Z}$ be the largest integer such that $H^n(E) \neq 0$. Then n is also the largest integer such that $H^n(E \otimes_A B) = H^n(E) \otimes_A B \neq 0$ because B is faithfully flat over A. We argue by induction on $n - m$.

If $n < m$, then $H^p(E) = 0$ for all $p \geq m$ and $0 \to E$ is an m-quasi-isomorphism. If $n \geq m$, then $H^n(E) \otimes_A B$ is a finitely generated B-module by Lemma 21.161. Therefore $H^n(E)$ is a finitely generated A-module by faithfully flat descent (Proposition 14.48). Choose a finite free A-module F and a surjection $F \to H^n(E)$. As F is projective, we can lift this to a map of complexes $u\colon F[-n] \to E$. Then for its cone C_u one has $H^p(C_u) = 0$ for all $p > n - 1$ by the long exact cohomology sequence. As $F[-n] \otimes_A B \to E \otimes_A B$ is a map from a pseudo-coherent complex to an m-pseudo-coherent complex, its cone $C_u \otimes_A B$ is m-pseudo-coherent by Proposition 21.154. Hence the induction hypothesis implies that C_u is m-pseudo-coherent. Applying now Proposition 21.154 to the distinguished triangle $F[-n] \to E \to C_u \to$ one sees that E is m-pseudo-coherent. $\qquad\square$

(21.35) Tor-amplitude and tor-dimension.

In this section $f\colon X \to Y$ denotes a morphism of ringed spaces.

Definition 21.164. *Let $\mathscr{K}$ be a complex in $D(X)$. Let $I \subseteq \mathbb{Z}$ be an interval.*
(1) *The complex $\mathscr{K}$ is said to have* tor-amplitude in I relative to f *if for every $\mathscr{O}_Y$-module $\mathscr{F}$*

$$H^p(\mathscr{K} \otimes_{\mathscr{O}_X}^L Lf^* \mathscr{F}) = 0 \qquad \text{for all } p \notin I.$$

We write tor-amp$_f$ $\mathscr{K} \subseteq I$ *in this case.*

One says that $\mathscr{K}$ has finite tor-dimension relative to f *if it has tor-amplitude in some bounded interval relative to f.*

One says that $\mathscr{K}$ has locally finite tor-dimension relative to f *if there exists an open covering $(V_i)_i$ of Y such that for all i the restriction $\mathscr{K}|_{f^{-1}(V_i)}$ has finite tor-dimension relative to f.*

(2) *If $f = \mathrm{id}_X$, then we simply say that $\mathscr{K}$ has tor-amplitude in I and write tor-amp $\mathscr{K} \subseteq I$ or that $\mathscr{K}$ is (locally) of finite tor-dimension if tor-amp $\mathscr{K}$ is (locally) contained in some bounded interval.*

(3) *If $\mathscr{K} = \mathscr{O}_X$ and $t \geq 0$ an integer, we say that f has tor-dimension $\leq t$ and write tor-dim $f \leq t$ if tor-amp$_f$ $\mathscr{O}_X \subseteq [-t, 0]$.*

The morphism f is called (locally) of finite tor-dimension if (locally on Y) there exists $t \geq 0$ such that f has tor-dimension $\leq t$.

Remark 21.165. Let $\mathscr{K}$ be a complex of $\mathscr{O}_X$-modules and let $a \leq b$ be integers.
(1) If $\mathscr{K}$ has tor-amplitude in $[a, b]$ relative to f, then the shifted complex $\mathscr{K}[i]$ has tor-amplitude in $[a - i, b - i]$ relative to f for all $i \in \mathbb{Z}$.
(2) If $H^p(\mathscr{K}) = 0$ for all $p > b$, then $\mathscr{K}$ has tor-amplitude in $(-\infty, b]$ relative to f. In particular, every morphism has tor-amplitude in $(-\infty, 0]$.

Indeed, there exists a quasi-isomorphism from a K-flat complex $\mathscr{K}'$ concentrated in degrees $\leq b$ to $\mathscr{K}$. Hence for $p > m$ one has $H^p(\mathscr{K} \otimes^L Lf^* \mathscr{F}) = H^p(\mathscr{K}' \otimes Lf^* \mathscr{F}) = 0$ for all $\mathscr{O}_X$-modules $\mathscr{F}$ because $Lf^* \mathscr{F}$ has only cohomology in degrees ≤ 0.
(3) If $\mathscr{K}$ has tor-amplitude in $[a, b]$ relative to f, then $\mathscr{K}$ is in $D^{[a,b]}(X)$ because $H^p(\mathscr{K}) = H^p(\mathscr{K} \otimes_{\mathscr{O}_X}^L Lf^* \mathscr{O}_Y) = 0$ for $p \notin [a, b]$.
(4) If $\mathscr{K}$ is a complex of flat $\mathscr{O}_X$-modules concentrated in degrees $[a, b]$, then $\mathscr{K}$ is K-flat by Lemma 21.93 and hence $\mathscr{K} \otimes_{\mathscr{O}_X}^L \mathscr{F} = \mathscr{K} \otimes_{\mathscr{O}_X} \mathscr{F}$ is concentrated in degrees $[a, b]$ for every $\mathscr{O}_X$-module $\mathscr{F}$. Therefore $\mathscr{K}$ has tor-amplitude in $[a, b]$.

There is also a converse, see Proposition 21.169 below.
(5) An $\mathscr{O}_X$-module has tor-amplitude in $[0, 0]$ if and only if it is flat by Proposition 21.100. A morphism f of ringed spaces is flat if and only if it has tor-dimension ≤ 0.
(6) Every strictly perfect complex concentrated in degrees $[a, b]$ has tor-amplitude in $[a, b]$ by (4). Hence every perfect complex is locally of finite tor-dimension.

Proposition 21.166. *Let $\mathscr{K} \to \mathscr{L} \to \mathscr{M} \to$ be a distinguished triangle in $D(X)$ and suppose that $\mathscr{K}$ and $\mathscr{M}$ have tor-amplitude in $[a, b]$. Then $\mathscr{L}$ has tor-amplitude in $[a, b]$.*

By rotation of distinguished triangles, one obtains also the implications

$$
(21.35.1) \quad
\begin{aligned}
\text{tor-amp } \mathscr{K} \subseteq [a+1, b+1], \quad \text{tor-amp } \mathscr{L} \subseteq [a, b] &\Rightarrow \text{tor-amp } \mathscr{M} \subseteq [a, b], \\
\text{tor-amp } \mathscr{L} \subseteq [a, b], \quad \text{tor-amp } \mathscr{M} \subseteq [a-1, b-1] &\Rightarrow \text{tor-amp } \mathscr{K} \subseteq [a, b]
\end{aligned}
$$

Proof. This follows from the long exact cohomology sequence associated to the distinguished triangle

$$\mathscr{K} \otimes^L \mathscr{F} \longrightarrow \mathscr{L} \otimes^L \mathscr{F} \longrightarrow \mathscr{M} \otimes^L \mathscr{F} \xrightarrow{+1}$$

for every $\mathscr{O}_X$-module $\mathscr{F}$. $\qquad\square$

Lemma 21.167. *Let $[a,b]$ and $[c,d]$ be bounded intervals in $\mathbb{Z}$ and let $t \geq 0$ be an integer. Let $f\colon X \to Y$ be a morphism of ringed spaces of tor-dimension $\leq t$ and let $\mathcal{K}$ be a complex of $\mathcal{O}_X$-modules with tor-amplitude in $[a,b]$. Let $\mathcal{G}$ be a complex of $\mathcal{O}_Y$-modules with $H^i(\mathcal{G}) = 0$ for all $i \notin [c,d]$. Then $H^j(\mathcal{K} \otimes^L_{\mathcal{O}_X} Lf^*\mathcal{G}) = 0$ for all $j \notin [a+c-t, b+d]$.*

Proof. We claim that $H^j(Lf^*\mathcal{G}) = 0$ for all $j \notin [c-t, d]$. Indeed, by induction on $d - c$ using the distinguished triangle $H^c(\mathcal{G})[-c] \longrightarrow \mathcal{G} \longrightarrow \tau^{\geq c+1}\mathcal{G} \longrightarrow$ and shifting, one can assume that $\mathcal{G}$ is an $\mathcal{O}_Y$-module, i.e., $c = d = 0$. Then the claim holds by definition.

The claim implies that it suffices to show the lemma for $f = \mathrm{id}$. This can again be done by induction on $d - c$ as above.   □

Proposition 21.168. *Let $f\colon X \to Y$ and $g\colon Y \to Z$ be morphisms of ringed spaces. Suppose that f has tor-dimension $\leq t$. Let $\mathcal{K}$ be a complex of $\mathcal{O}_X$-modules and let $\mathcal{L}$ be a complex of $\mathcal{O}_Y$-modules with tor-$\mathrm{amp}_f \mathcal{K} \subseteq [a,b]$ and tor-$\mathrm{amp}_g \mathcal{L} \subseteq [c,d]$ for bounded intervals $[a,b]$ and $[c,d]$ of $\mathbb{Z}$. Then tor-$\mathrm{amp}_{g\circ f}(\mathcal{K} \otimes^L Lf^*\mathcal{L}) \subseteq [a+c-t, b+d]$.*

We have the following special cases.
(1) Taking $f = \mathrm{id}_X$, we see that for two complexes $\mathcal{K}$ and $\mathcal{L}$ in $D(X)$ one has

$$
(21.35.2) \qquad
\begin{aligned}
&\text{tor-}\mathrm{amp}_g\,\mathcal{K} \subseteq [a,b], \quad \text{tor-}\mathrm{amp}_g\,\mathcal{L} \subseteq [c,d] \\
&\Rightarrow \text{tor-}\mathrm{amp}_g(\mathcal{K} \otimes^L_{\mathcal{O}_X} \mathcal{L}) \subseteq [a+c, b+d].
\end{aligned}
$$

(2) Taking $\mathcal{K} = \mathcal{O}_X$ and $[a,b] = [0,0]$, we see that for every complex $\mathcal{L}$ in $D(Y)$ one has

$$
(21.35.3) \qquad \text{tor-}\mathrm{amp}_g\,\mathcal{L} \subseteq [c,d] \Rightarrow \text{tor-}\mathrm{amp}_{g\circ f}(Lf^*\mathcal{L}) \subseteq [c-t, d]
$$

If in addition $\mathcal{L} = \mathcal{O}_Y$ one deduces

$$
(21.35.4) \qquad \text{tor-dim}\, f \leq t, \text{tor-dim}\, g \leq s \Rightarrow \text{tor-dim}\, g \circ f \leq s + t.
$$

Proof. Let $\mathcal{F}$ be an $\mathcal{O}_Z$-module. As the cohomology of $\mathcal{L} \otimes^L Lg^*\mathcal{F}$ is concentrated in degrees $[c,d]$, the cohomology of

$$
\mathcal{K} \otimes^L Lf^*\mathcal{L} \otimes^L L(g \circ f)^*\mathcal{F} = \mathcal{K} \otimes^L Lf^*(\mathcal{L} \otimes^L Lg^*\mathcal{F})
$$

is concentrated in degrees $[a + c - t, b + d]$ by Lemma 21.167.   □

Proposition 21.169. *Let $\mathcal{K}$ be a complex in $D(X)$ and let $a \leq b$ be integers. The following assertions are equivalent.*
(i) *The complex $\mathcal{K}$ has tor-amplitude in $[a,b]$.*
(ii) *There exists a complex $\mathcal{E}$ of flat $\mathcal{O}_X$-modules with $\mathcal{E}^p = 0$ for $p \notin [a,b]$ and an isomorphism $\mathcal{E} \cong \mathcal{K}$ in $D(X)$.*

Proof. We have already seen in Remark 21.165 (4) that (ii) implies (i).

Conversely, suppose that (i) holds. As $H^p(\mathcal{K}) = 0$ for all $p > b$ (Remark 21.165 (3)) there exists a complex $\mathcal{G}$ consisting of flat $\mathcal{O}_X$-modules with $\mathcal{G}^p = 0$ for $p > b$ and an isomorphism $\mathcal{G} \xrightarrow{\sim} \mathcal{K}$ in $D(X)$ by Remark 21.95 (2). As $H^p(\mathcal{G}) = H^p(\mathcal{K}) = 0$ for all $p < a$ we have $\mathcal{G} \cong \tau^{\geq a}\mathcal{G} =: \mathcal{E}$ in $D(X)$.

It remains to show that $\mathcal{E}^a$ is flat. There is a distinguished triangle $\mathcal{E}' \to \mathcal{E} \to \mathcal{E}^a[-a] \to$, where $\mathcal{E}'$ has flat components that are zero outside degrees $[a+1, b]$. Now apply the triangulated functor $\mathcal{F} \otimes^L (-)$ for an $\mathcal{O}_X$-module $\mathcal{F}$. We obtain an exact sequence

$$H^{a-1}(\mathscr{F} \otimes^L \mathscr{E}) \to H^{a-1}(\mathscr{F} \otimes^L \mathscr{E}^a[-a]) \to H^a(\mathscr{F} \otimes^L \mathscr{E}')$$

As $\mathscr{E}$ has tor-amplitude in $[a, b]$ the left module is zero. As $\mathscr{E}'$ is K-flat, we have $\mathscr{F} \otimes^L \mathscr{E}' = \mathscr{F} \otimes \mathscr{E}'$ and hence the right module is zero because $\mathscr{E}'$ is concentrated in degrees $> a$. Therefore $H^{a-1}(\mathscr{F} \otimes^L \mathscr{E}^a[-a]) = H^{-1}(\mathscr{F} \otimes^L \mathscr{E}^a) = \mathscr{T}or_1(\mathscr{F}, \mathscr{E}^a) = 0$ for all $\mathscr{F}$. Hence $\mathscr{E}^a$ is flat (Proposition 21.100). $\qquad\square$

Corollary 21.170. *Let $f : X \to Y$ be a morphism of ringed spaces. Let $\mathscr{K}$ be a complex in $D(Y)$. If $\mathscr{K}$ has tor-amplitude in $[a, b]$, then $Lf^* \mathscr{K}$ has tor-amplitude in $[a, b]$.*

Proof. By Proposition 21.169 we may assume that $\mathscr{K}$ is a complex of flat modules concentrated in degrees $[a, b]$. Then $Lf^* \mathscr{K} = f^* \mathscr{K}$ because $\mathscr{K}$ is K-flat, and $f^* \mathscr{K}$ is a complex of flat $\mathscr{O}_X$-modules concentrated in degrees $[a, b]$. Hence $Lf^* \mathscr{K}$ has tor-amplitude in $[a, b]$. $\qquad\square$

Tor-amplitude can be checked locally in the following sense.

Lemma 21.171. *Let $f : X \to Y$ be a morphism of ringed spaces, let $\mathscr{K}$ be a complex of $\mathscr{O}_X$-modules, and let $a \le b$ be integers. Then $\mathscr{K}$ has tor-amplitude in $[a, b]$ relative to f if and only if for every $x \in X$ the complex $\mathscr{K}_x$ of $\mathscr{O}_{X,x}$-modules has tor-amplitude in $[a, b]$ relative to $f_x^\# : \mathscr{O}_{Y,f(x)} \to \mathscr{O}_{X,x}$.*

Proof. Consider the commutative diagram of ringed spaces

$$(*) \qquad \begin{array}{ccc} (\{x\}, \mathscr{O}_{X,x}) & \xrightarrow{\;\; i_x \;\;} & X \\ {\scriptstyle f_x^\#} \downarrow & & \downarrow {\scriptstyle f} \\ (\{f(x)\}, \mathscr{O}_{Y,f(x)}) & \xrightarrow{\;\; i_{f(x)} \;\;} & Y. \end{array}$$

Then i_x and $i_{f(x)}$ are flat and we have $\mathscr{K}_x = i_x^* \mathscr{K} = L i_x^* \mathscr{K}$.

Suppose that $\mathscr{K}$ has tor-amplitude in $[a, b]$ relative to f. By (21.35.3) we have

$$\text{tor-amp}_{i_{f(x)} \circ f_x^\#}(\mathscr{K}_x) = \text{tor-amp}_{f \circ i_x}(\mathscr{K}_x) \subseteq [a, b].$$

Every $\mathscr{O}_{Y,f(x)}$-module G is of the form $i_{f(x)}^* \mathscr{G}$ for some $\mathscr{O}_Y$-module $\mathscr{G}$, for instance $\mathscr{G} = i_{f(x),*} G$. Therefore $\text{tor-amp}_{f_x^\#} \mathscr{K}_x \subseteq [a, b]$.

Conversely, suppose that $\text{tor-amp}_{f_x^\#} \mathscr{K}_x \subseteq [a, b]$ for all $x \in X$. Let $\mathscr{F}$ be an $\mathscr{O}_Y$-module. The commutative diagram $(*)$ shows $(Lf^* \mathscr{F})_x = Lf_x^{\#*} \mathscr{F}_{f(x)}$. As forming stalks is exact and commutes with derived tensor products (21.20.2), we have for $i \notin [a, b]$ that

$$H^i(\mathscr{K} \otimes^L_{\mathscr{O}_X} Lf^* \mathscr{F})_x = H^i(\mathscr{K}_x \otimes^L_{\mathscr{O}_{X,x}} (Lf^* \mathscr{F})_x) = H^i(\mathscr{K}_x \otimes^L_{\mathscr{O}_{X,x}} Lf_x^{\#*} \mathscr{F}_{f(x)}) = 0,$$

hence $\mathscr{K}$ has tor-amplitude in $[a, b]$. $\qquad\square$

Lemma 21.172. *Let $A \to B$ be a faithfully flat ring homomorphism and let E be a complex of A-modules. Then E has tor-amplitude in $[a, b]$ if and only if the complex of B-modules $E \otimes_A B$ has tor-amplitude in $[a, b]$.*

Proof. The ring homomorphism $A \to B$ corresponds to a morphism of ringed spaces $f : (*, B) \to (*, A)$, where $*$ denotes a singleton. As $A \to B$ is flat, the functor $f^* = B \otimes_A (-)$ is exact and hence $Lf^* = f^*$. Hence the condition is necessary by Corollary 21.170.

Conversely, suppose that $E \otimes_A B$ has tor-amplitude in $[a, b]$ and let M be an A-module. As $A \to B$ is flat, taking cohomology commutes with $(-)_B := (-) \otimes_A B$ and we have $(M \otimes_A^L E)_B = M_B \otimes_B^L E_B$. Therefore $\mathrm{Tor}_p^A(M, E)_B = \mathrm{Tor}_p^B(M_B, E_B)$. By hypothesis $\mathrm{Tor}_p^B(M_B, E_B) = 0$ for $p \notin [a, b]$ and hence $\mathrm{Tor}_p^A(M, E) = 0$ for $p \notin [a, b]$ because $A \to B$ is faithfully flat. $\qquad\square$

(21.36) Pseudo-coherent complexes of finite tor-dimension are perfect.

Let $(X, \mathscr{O}_X)$ be a ringed space. The goal of this section is to show that pseudo-coherent complexes of finite tor-dimension are perfect. This follows from the following more precise lemma.

Lemma 21.173. *Let $\mathscr{E}$ be an object in $D(X)$ and let $a \leq b$ be integers. If $\mathscr{E}$ has tor-amplitude in $[a, b]$ and is $(a-1)$-pseudo-coherent, then $\mathscr{E}$ is perfect.*

Proof. Working locally we can assume that there exists a strictly perfect complex $\mathscr{K}$ and an $(a-1)$-isomorphism $\alpha \colon \mathscr{K} \to \mathscr{E}$. Replacing $\mathscr{K}$ by $\sigma^{\geq a-1} \mathscr{K}$ we may assume that $\mathscr{K}^p = 0$ for $p < a - 1$. By assumption we have $H^p(\mathscr{E}) = 0$ for $p < a$. Hence we may replace $\mathscr{E}$ by $\tau^{\geq a} \mathscr{E}$ and can assume that $\mathscr{E}^p = 0$ for all $p < a$.

Complete α to a distinguished triangle $\mathscr{C} \to \mathscr{K} \to \mathscr{E} \to$. Looking at the long exact cohomology sequence and using that $H^p(\alpha)$ is an isomorphism for $p \geq a$ one sees that $H^{a-1}(\mathscr{C}) = \mathrm{Ker}(\mathscr{K}^{a-1} \to \mathscr{K}^a) =: C$ and that $H^p(\mathscr{C}) = 0$ for all $p \neq a - 1$. Hence $\mathscr{C} = C[1-a] = \tau^{\leq a-1} \mathscr{K}$. Thus the distinguished triangle $\tau^{\leq a-1} \mathscr{K} \to \mathscr{K} \to \tau^{\geq a} \mathscr{K} \to$ (Proposition F.157) shows that we have an isomorphism $\tau^{\geq a} \mathscr{K} \cong \mathscr{E}$ in $D(X)$. It therefore suffices to show that $\mathscr{L} := \tau^{\geq a} \mathscr{K}$ is strictly perfect.

Now $\mathscr{L}$ is a finite complex concentrated in degrees $\geq a$ and $\mathscr{L}^p = \mathscr{K}^p$ is a direct summand of a finite free $\mathscr{O}_X$-module for $p > a$. One has $\mathscr{L}^a = \mathrm{Coker}(\mathscr{K}^{a-1} \to \mathscr{K}^a)$. In particular $\mathscr{L}^a$ is an $\mathscr{O}_X$-module of finite presentation.

There is a distinguished triangle $\mathscr{L}' \to \mathscr{L} \to \mathscr{L}^a[-a] \to$, where $\mathscr{L}'$ is a finite complex concentrated in degrees $> a$ with flat components. In particular it is K-flat. Applying the functor $\mathscr{F} \otimes^L (-)$ for an arbitrary $\mathscr{O}_X$-module $\mathscr{F}$, the long exact cohomology sequence yields an exact sequence

$$H^{a-1}(\mathscr{L} \otimes^L \mathscr{F}) \longrightarrow H^{-1}(\mathscr{L}^a \otimes^L \mathscr{F}) \longrightarrow H^a(\mathscr{L}' \otimes \mathscr{F}).$$

Now the left cohomology vanishes because $\mathscr{L} \cong \mathscr{E}$ has tor-amplitude in $[a, b]$ and the right cohomology vanishes because the a-th component of $\mathscr{L}' \otimes \mathscr{F}$ is zero. This shows that $\mathrm{Tor}_1(\mathscr{L}^a, \mathscr{F}) = H^{-1}(\mathscr{L}^a \otimes^L \mathscr{F}) = 0$ for all $\mathscr{F}$. Hence $\mathscr{L}^a$ is a flat $\mathscr{O}_X$-module.

Being flat and of finite presentation, $\mathscr{L}^a$ is locally a direct summand of a finite free $\mathscr{O}_X$-module (see Proposition 7.41 if $(X, \mathscr{O}_X)$ is a locally ringed space, Proposition B.29 if X consists of a single point, and Exercise 21.28 in general). $\qquad\square$

As perfect complexes are pseudo-coherent and locally of finite tor-dimension (Remark 21.165 (6)) we obtain the following result.

Theorem 21.174. *Let $\mathscr{E}$ be an object in $D(X)$. Then $\mathscr{E}$ is perfect if and only if $\mathscr{E}$ is pseudo-coherent and locally on X has finite tor-dimension.*

Proposition 21.175. *Let $\mathscr{K} \to \mathscr{L} \to \mathscr{M} \to$ be a distinguished triangle in $D(X)$ and suppose that two of the three complexes are perfect. Then the third is perfect.*

Proof. By Theorem 21.174 this follows from the analogue assertions for pseudo-coherent complexes (Proposition 21.154) and complexes of finite tor-amplitude (Proposition 21.166). □

Proposition 21.176. *Let $\mathscr{K}$ and $\mathscr{L}$ be complexes in $D(X)$. Then $\mathscr{K}$ and $\mathscr{L}$ are perfect if and only if $\mathscr{K} \oplus \mathscr{L}$ is perfect.*

Proof. It is clear that $\mathscr{K}$ and $\mathscr{L}$ are locally of finite tor-dimension if and only if $\mathscr{K} \oplus \mathscr{L}$ is locally of finite tor-dimension. The analogous assertion holds for pseudo-coherence (Proposition 21.156). Hence we conclude by Theorem 21.174. □

We finish the section by recording two lemmas for later use.

Lemma 21.177. *Let X be a ringed space.*
(1) *If X is quasi-compact and $\mathscr{E}$ in $D(X)$ is perfect, then $\mathscr{E} \in D^b(X)$.*
(2) *Let $\mathscr{E}$ be in $D^b(X)$. If $H^i(\mathscr{E})$ is perfect (as a complex concentrated in degree 0) for all $i \in \mathbb{Z}$, then $\mathscr{E}$ is perfect.*

The converse to the second assertion does not hold in general, see Exercise 22.36.

Proof. The first assertion follows from the Definition 21.136 (iii). Let us show the second assertion. We may assume that $\mathscr{E}$ is cohomologically bounded, say $H^i(\mathscr{E}) = 0$ for $i \notin [a, b]$ for some integers $a \leq b$. We proceed by induction an $b - a$. Consider the distinguished triangle

$$H^a(X)[-a] \longrightarrow \mathscr{E} \longrightarrow \tau^{\geq a+1}\mathscr{E} \longrightarrow .$$

Then $\tau^{\geq a+1}\mathscr{E}$ has cohomology concentrated in degrees $[a+1, b]$ and $H^i(\tau^{\geq a+1}\mathscr{E}) = H^i(\mathscr{E})$ for all $i \geq a + 1$. Hence $\tau^{\geq a+1}\mathscr{E}$ is perfect by induction hypothesis. Moreover, $H^a(X)[-a]$ is perfect by assumption. Therefore we conclude by Proposition 21.175. □

Lemma 21.178. *Let $A \to B$ be a faithfully flat ring homomorphism, and let E be a complex of A-modules. Then E is perfect if and only if the complex of B-modules $E \otimes_A B$ is perfect.*

Proof. Again by Theorem 21.174 this follows from the analogues statements for pseudo-coherent complexes (Proposition 21.163) and complexes of finite tor-dimension (Proposition 21.163). □

Exercises

Exercise 21.1. Recall (for instance from [Wed]^0 Chap. 1) the following properties of continuous maps. A continuous map of topological spaces $f\colon X \to Y$ is called *proper* if it satisfies the following equivalent conditions.
(i) For every topological space Z the map $f \times \mathrm{id}_Z\colon X \times Z \to Y \times Z$ is closed.
(ii) The map f is closed and for every quasi-compact subspace V of Y the preimage $f^{-1}(V)$ is a quasi-compact subspace of X.
The map f is called *separated* if it satisfies the following equivalent conditions.
(i) The diagonal $\Delta_f\colon X \to X \times_Y X$ is a closed topological embedding.

(ii) For any relatively Hausdorff subspace B of Y (i.e., any two distinct points in B have disjoint neighborhoods in Y) the preimage $f^{-1}(B)$ is a relatively Hausdorff space in X.

Let $f\colon X \to Y$ be a continuous map of topological spaces. For an abelian sheaf $\mathscr{F}$ on X define a presheaf $f_! \mathscr{F}$ on Y by

$$f_! \mathscr{F}(V) := \{\, s \in \mathscr{F}(f^{-1}(V)) \;;\; f\colon \mathrm{Supp}(s) \to U \text{ is proper}\,\}.$$

(1) Show that $f_! \mathscr{F}$ is an abelian sheaf on Y and that one obtains a functor

$$f_! \colon (\mathrm{Ab}(X)) \to (\mathrm{Ab}(Y)).$$

Show that if f is proper, then $f_! = f_*$. Show that if $f = j$ is the inclusion of a locally closed subspace, then $j_!$ is the functor defined in Definition 21.1.
(2) Let $f\colon X \to Y$ and $g\colon Y \to Z$ continuous maps of topological spaces and suppose that g is separated. Show that the identity of functors $(g \circ f)_* = g_* \circ f_*$ restricts to an identity $(g \circ f)_! = g_! \circ f_!$.

Exercise 21.2. We continue to use the notions from Exercise 21.1. Consider a cartesian diagram in the category of topological spaces

$$
\begin{array}{ccc}
W & \xrightarrow{\ q\ } & Y \\
{\scriptstyle g}\big\downarrow & & \big\downarrow{\scriptstyle f} \\
Z & \xrightarrow[\ p\ ]{} & X.
\end{array}
$$

(1) Show that the identity $f_* \circ q_* = p_* \circ g_*$ induces a morphism of functors $f_! \circ q_* \to p_* \circ g_!$.
(2) Consider the morphisms of functors $f_! \to f_! q_* q^{-1} \to p_* g_! q^{-1}$, where the first morphism is given by the adjunction (q^{-1}, q_*) and the second is given by (1). By the adjunction (p^{-1}, p_*) we obtain a morphism

$$\beta \colon p^{-1} f_! \longrightarrow g_! q^{-1}$$

of functors $(\mathrm{Ab}(Y)) \to (\mathrm{Ab}(Z))$. Show that if f is proper and separated, then g is also proper and separated, and β is an isomorphism $p^{-1} f_* \xrightarrow{\sim} g_* q^{-1}$.
(3) Let f again be proper and separated. Show that β induces an isomorphism

$$p^{-1} \circ Rf_* \xrightarrow{\sim} Rg_* \circ q^{-1}$$

of functors $D^+(\mathrm{Ab}(Y)) \to D^+(\mathrm{Ab}(Z))$.
The conclusions in (2) and (3) are called the *proper base change theorems*.

Exercise 21.3. Let X be an irreducible topological space and let G be a constant abelian sheaf. Show that $H^p(X, G) = 0$ for all $p > 0$.

Exercise 21.4. Let S^1 be the unit circle with its usual topology and let $\mathbb{Z}$ be the constant sheaf of the abelian group of integers. Show that $H^1(S^1, \mathbb{Z}) \cong \mathbb{Z}$.

Exercise 21.5. Let k be a field and let X be an integral scheme of finite type over k of positive dimension. Show that there exists no surjective homomorphism $\mathscr{P} \to \mathscr{O}_X$ of $\mathscr{O}_X$-modules where $\mathscr{P}$ is a projective objective in the category of $\mathscr{O}_X$-modules.
Hint: Let $\mathscr{P}$ be a projective object in the category of $\mathscr{O}_X$-modules and let $u\colon \mathscr{P} \to \mathscr{O}_X$ be a map of $\mathscr{O}_X$-modules. Then necessarily $u = 0$. To see this, consider for $\emptyset \neq V \subseteq X$ open the $\mathscr{O}_X$-module $\mathscr{O}_{V \subseteq X}$ (21.1.6). Then $\mathscr{O}_{V \subseteq X}(U) = 0$ for all $U \subseteq X$ open with $U \not\subseteq V$. Now choose a closed point $x \in X$ and set $W := X \setminus \{x\}$. Let U be any open neighborhood of x. Choose a strictly smaller open neighborhood V of x. Then u factors through the surjective map $\mathscr{O}_{V \subseteq X} \oplus \mathscr{O}_{W \subseteq X} \to \mathscr{O}_X$ and hence is zero on sections over U. This shows $u_x = 0$ which implies $u = 0$.
Remark: If $X = \operatorname{Spec} A$ is affine, every quasi-coherent $\mathscr{O}_X$-module admits a surjection from a projective object in the category of *quasi-coherent* $\mathscr{O}_X$-modules since in this case $\operatorname{QCoh}(X)$ is equivalent to the category of A-modules. But one can show that for $X = \mathbb{P}_k^1$ there are no non-zero projective objectives in the abelian category of quasi-coherent $\mathscr{O}_X$-modules (e.g., [EEGRO]$^\lozenge$ 2.3).

Exercise 21.6. Let X be a topological space, $x \in X$ a point, and denote by $i_x\colon \{x\} \to X$ the inclusion. Let A be an abelian group considered as abelian sheaf on $\{x\}$. Show that $H^p(X, i_{x,*}A) = 0$ for all $p > 0$.

Exercise 21.7. Let $(X, \mathscr{O}_X)$ be a ringed space and let $Z \subseteq X$ be a closed subset. Consider the functor

$$\Gamma_Z(X, -)\colon (\mathscr{O}_X\text{-Mod}) \to (\Gamma(X, \mathscr{O}_X)\text{-Mod}),$$
$$\mathscr{F} \mapsto \Gamma_Z(X, \mathscr{F}) := \{\, s \in \mathscr{F}(X) \,;\, \operatorname{Supp}(s) \subset Z \,\}.$$

Show that this functor is left exact and has a right derived functor

$$R\Gamma_Z(X, -)\colon D(X) \to D(\Gamma(X, \mathscr{O}_X)).$$

For $\mathscr{F}$ in $D(X)$ and $i \in \mathbb{Z}$ one defines $H_Z^i(X, \mathscr{F}) := H^i R\Gamma_Z(X, -)$ the *cohomology of $\mathscr{F}$ with support in Z*. Moreover, show the following assertions.
(1) Set $U := X \setminus Z$. Show that there is a functorial distinguished triangle in $D(\Gamma(X, \mathscr{O}_X))$

$$R\Gamma_Z(X, \mathscr{F}) \longrightarrow R\Gamma(X, \mathscr{F}) \longrightarrow R\Gamma(U, \mathscr{F}) \xrightarrow{+1} .$$

(2) For $\mathscr{F}$ in $D(X)$ let $\mathscr{F}^{\mathrm{ab}}$ be the underlying complex of abelian sheaves. Show that there is a functorial isomorphism $R\Gamma_Z(X, \mathscr{F}) \xrightarrow{\sim} R\Gamma_Z(X, \mathscr{F}^{\mathrm{ab}})$ in $D(\mathbb{Z})$.
Hint: Use (1)

Exercise 21.8. Let X be ringed space, let $\mathscr{F}$ be a complex of $\mathscr{O}_X$-modules. Suppose that there exists a basis $\mathcal{B}$ of the topology of X and an integer $d \geq 0$ such that $H^p(U, H^q(\mathscr{F})) = 0$ for all $U \in \mathcal{B}$, $p > d$ and $q < 0$.
(1) Show that the canonical map $\mathscr{F} \longrightarrow R\lim_n \tau^{\geq n}\mathscr{F}$ is an isomorphism.
(2) Let $(\mathscr{I}_n)_n$ be a system of bounded below complexes of injective $\mathscr{O}_X$-modules that form a resolution of $(\tau^{\geq -n}\mathscr{F})_n$ as in Lemma F.195. Show that the limit map $\mathscr{F} = \lim_n \tau^{\geq -n}\mathscr{F} \to \lim I_n$ (F.47.2) is a quasi-isomorphism.
Hint: Exercise F.39

Exercise 21.9. Let X be a ringed space, let $\mathcal{U} = (U_i)_{i \in I}$ be an open covering and let $\mathscr{F}$ be an $\mathscr{O}_X$-module. For every $\mathbf{i} \in I^{n+1}$ we denote by $j_{\mathbf{i}}\colon U_{\mathbf{i}} \to X$ the inclusion. For $n \geq 0$ define an $\mathscr{O}_X$-module

$$\check{\mathscr{C}}^n(\mathscr{U},\mathscr{F}) := \prod_{i \in I^{n+1}} (j_i)_* j_i^{-1} \mathscr{F}$$

and define differentials as in Section (21.14) to obtain a complex $\check{\mathscr{C}}^\bullet(\mathscr{U},\mathscr{F})$ of $\mathcal{O}_X$-modules.

More generally, let $\mathscr{F}$ be a bounded below complex of $\mathcal{O}_X$-module. Then $\check{\mathscr{C}}^\bullet(\mathscr{U},\mathscr{F})$ is defined as the total complex of the double complex $(\check{\mathscr{C}}^p(\mathscr{U},\mathscr{F}^q))_{p,q}$.

For $\mathscr{F}$ a bounded below complex let $\mathscr{F} \to \check{\mathscr{C}}^0(\mathscr{U},\mathscr{F})$ be the map of complexes of $\mathcal{O}_X$-modules which is in each component the canonical map $\mathscr{F}^q \to (j_i)_* j_i^{-1} \mathscr{F}^q$. Show that this maps yields a quasi-isomorphism $\mathscr{F} \to \check{\mathscr{C}}^\bullet(\mathscr{U},\mathscr{F})$ in $C(X)$.

Exercise 21.10. Let X be a ringed space, let $\mathscr{U}$ be an open covering of X, and let $\mathscr{F}$ be a bounded below complex of $\mathcal{O}_X$-modules.
(1) Show that if one applies the first hypercohomology spectral sequence (F.49.5) to the double complex $(\check{\mathscr{C}}^p(\mathscr{U},\mathscr{F}^q))_{p,q}$ (Exercise 21.9) one obtains a convergent spectral sequence

$$E_1^{pq} = H^q(X, \check{\mathscr{C}}^p(\mathscr{U},\mathscr{F})) \Rightarrow H^n(X,\mathscr{F})$$

   Hint: Use Exercise 21.9.
(2) Show that if $\mathscr{U} = (U,V)$ consists of only two open subsets, then the spectral sequence degenerates at E_2 and yields the Mayer-Vietoris sequence for $\mathscr{F}$ and (U,V).

Exercise 21.11. Let X be a ringed space and let $\mathcal{U}$ be a finite open covering of X. Show that for every complex $\mathscr{F}$ of $\mathcal{O}_X$-modules there is a functorial map in $D(\Gamma(X,\mathcal{O}_X))$

$$\mathrm{Tot}(\check{C}_{\mathrm{alt}}(\mathcal{U},\mathscr{F})) \longrightarrow R\Gamma(X,\mathscr{F}).$$

Show that this map is an isomorphism if there exists a basis $\mathcal{B}$ of the topology of X that contains all finite intersections of the members of $\mathcal{U}$ and such that for all $U \in \mathcal{B}$, $q \in \mathbb{Z}$, and $p > 0$ one has

$$H^p(U,\mathscr{F}^q) = H^p(U, \mathrm{Coker}(\mathscr{F}^{q-1} \to \mathscr{F}^q)) = H^p(U, H^q(\mathscr{F})) = 0.$$

Hint: Prove the assertions first for $\mathscr{F}$ bounded below, then use Exercise 21.8.

Exercise 21.12. Let X be a ringed space and let $\mathscr{F}$ and $\mathscr{G}$ be $\mathcal{O}_X$-modules. Show that one has $\mathrm{Supp}\,\mathcal{T}or_p^{\mathcal{O}_X}(\mathscr{F},\mathscr{G}) \subseteq \mathrm{Supp}(\mathscr{F}) \cap \mathrm{Supp}(\mathscr{G})$ for all $p \geq 0$.

Exercise 21.13. Let $f\colon X \to Y$ and $g\colon Y \to Z$ be morphisms of ringed spaces, and let $\mathscr{G}$ be a complex of $\mathcal{O}_Y$-modules. Construct by adjunction a functorial morphism $Rg_*\mathscr{G} \to R(g \circ f)_* Lf^*\mathscr{G}$ in $D(Z)$.

Exercise 21.14. Let X be a ringed spaces. Show that every direct sum and every filtered colimit of K-flat complexes of $\mathcal{O}_X$-modules is again K-flat.

Are arbitrary colimits of K-flat complexes again K-flat?

Exercise 21.15. Let X be a ringed space. For every complex of $\mathcal{O}_X$-modules $\mathscr{G}$ consider the functor $T_{\mathscr{G}}\colon K(X) \to D(X)$, $\mathscr{E} \mapsto \mathscr{G} \otimes_{\mathcal{O}_X} \mathscr{E}$. Let $\mathscr{F}$ be a complex of $\mathcal{O}_X$-modules. Show that the following assertions are equivalent.
(i) $\mathscr{F}$ is K-flat.
(ii) $\mathscr{F}$ is left-$T_{\mathscr{G}}$-acyclic for all complexes $\mathscr{G}$ of $\mathcal{O}_X$-modules.
(iii) $\mathscr{F}$ is left-$T_{\mathscr{G}}$-acyclic for all exact complexes $\mathscr{G}$ of $\mathcal{O}_X$-modules.

Exercise 21.16. Let X be a ringed space and let $\mathscr{F}$ and $\mathscr{G}$ be complexes of $\mathscr{O}_X$-modules. Suppose that for all $a, b, c, d \in \mathbb{Z}$ the complex $\tau^{\leq a}\tau^{\geq b}\mathscr{F} \otimes \tau^{\leq c}\tau^{\geq d}\mathscr{G}$ is exact. Show that $\mathscr{F} \otimes \mathscr{G}$ is exact.
Hint: Use that $\mathscr{F} = \operatorname{colim}_a \tau^{\leq a}\mathscr{F}$.

Exercise 21.17. Let X be a ringed space and let $\mathscr{F}$ be a complex of $\mathscr{O}_X$-modules such that $H^n(\mathscr{F})$ is flat for all $n \in \mathbb{Z}$ and such that the image of $\mathscr{F}^n \to \mathscr{F}^{n+1}$ is a direct summand of $\mathscr{F}^{n+1}$ for all n. Show that $\mathscr{F}$ is K-flat.
Hint: Use Exercise 21.16 to reduce to the case that $\mathscr{F}$ is a bounded complex.

Exercise 21.18. Let A be a ring and let F be a complex of R-modules. Show that the following assertions are equivalent.
(i) The complex F is K-flat and exact.
(ii) The tensor complex $F \otimes_A E$ is exact for every complex E of A-modules.
(iii) The tensor complex $F \otimes_A M$ is exact for every A-module M of finite presentation.

Exercise 21.19. Let A be a ring, let $\mathfrak{a}$ and $\mathfrak{b}$ be two ideals of A. Show that
$$\operatorname{Tor}_1^A(A/\mathfrak{a}, A/\mathfrak{b}) \cong (\mathfrak{a} \cap \mathfrak{b})/\mathfrak{ab}.$$

Exercise 21.20. Let A be a ring, let $\mathbf{f} = (f_1, \ldots, f_r)$ be a sequence of elements in A, and let $K(\mathbf{f})$ be the corresponding Koszul complex considered as a cochain complex sitting in degrees $[-r, 0]$. Let $\mathbb{Z}[T_1, \ldots, T_r]$ be the unique ring homomorphism sending T_i to f_i. Show that
$$K(\mathbf{f}) \cong A \otimes_{\mathbb{Z}}^L \mathbb{Z}[T_1, \ldots, T_r]/(T_1, \ldots, T_r).$$

Exercise 21.21. Let A be a ring, let $\mathbf{f} = (f_1, \ldots, f_r)$ be a completely intersecting sequence, and let $I \subseteq A$ be the ideal generated by $\mathbf{f}$. Show that one has for every complex E in $D(A)$ and for all $p \in \mathbb{Z}$
$$\operatorname{Tor}_A^{-p}(A/I, E) = H^p(K(\mathbf{f}) \otimes_R E),$$
where $K(\mathbf{f})$ is the Koszul complex associated to $\mathbf{f}$ considered as a cochain complex sitting in degrees $[-r, 0]$.

Exercise 21.22. Let X be a ringed space and let $(\mathscr{F}_i)_i$ be a filtered diagram of complexes of $\mathscr{O}_X$-modules.
(1) Show that for every complex $\mathscr{G}$ of $\mathscr{O}_X$-modules there is a functorial isomorphism $\operatorname{colim}_i \mathscr{T}or_n(\mathscr{F}_i, \mathscr{G}) \xrightarrow{\sim} \mathscr{T}or_n(\operatorname{colim}_i \mathscr{F}_i, \mathscr{G})$ for all $n \in \mathbb{Z}$.
(2) Let $f \colon Y \to X$ be a morphism of ringed spaces. Show that for all $n \in \mathbb{Z}$ there exists a functorial isomorphism $\operatorname{colim}_i L_n f^* \mathscr{F}_i \xrightarrow{\sim} L_n f^* \operatorname{colim}_i \mathscr{F}_i$.

Exercise 21.23. Let A be a ring and let E and F be complexes of R-modules.
(1) Let $\alpha \in H^p(E)$ and $\beta \in H^q(F)$ be represented by elements $a \in \operatorname{Ker}(E^p \to E^{p+1})$ and $b \in \operatorname{Ker}(F^q \to F^{q+1})$. Let $\alpha \cdot \beta$ be the class of $a \otimes b$ in $H^{p+q}(E \otimes_A F)$. Show that this yields a well defined A-linear map
$$(*) \qquad H^p(E) \otimes_R H^q(F) \longrightarrow H^{p+q}(E \otimes_A F).$$
(2) Show that there is a unique A-linear functorial map
$$(**) \qquad H^p(E) \otimes_R H^q(F) \longrightarrow H^{p+q}(E \otimes_A^L F).$$
whose composition with the map induced by $E \otimes_A^L F \to E \otimes_A F$ is $(*)$.

Exercise 21.24. Let $\varphi \colon R \to A$ be a ring homomorphism and let E and F be in $D(R)$.
(1) Show that there is a functorial isomorphism in $D(A)$

$$(E \otimes_R^L A) \otimes_A^L (F \otimes_R^L A) \cong (E \otimes_R^L F) \otimes_R^L A$$

and use Exercise 21.23 to obtain A-linear maps

(*) $\operatorname{Tor}_p^R(E, A) \otimes_A \operatorname{Tor}_q^R(F, A) \longrightarrow \operatorname{Tor}_{p+q}^R(E \otimes^L F, A).$

(2) Let $R \to B$ be a second ring homomorphism. Show that (*) for $E = F = B$ yields a map

$$\operatorname{Tor}_p^R(B, A) \otimes_A \operatorname{Tor}_q^R(B, A) \longrightarrow \operatorname{Tor}_{p+q}^R(B \otimes_R B, A) \longrightarrow \operatorname{Tor}_{p+q}^R(B, A),$$

where the second map is induced by the multiplication $B \otimes_R B \to B$. Show that this makes $\operatorname{Tor}_\bullet^R(B, A)$ into a strictly graded commutative graded A-algebra.

Exercise 21.25. Let A be a ring and let I be an ideal.
(1) Show that the isomorphism $I/I^2 \overset{\sim}{\to} \operatorname{Tor}_1^A(A/I, A/I)$ (Exercise 21.19) induces a map of graded A/I-algebras (Exercise 21.24)

(*) $\bigwedge_A I/I^2 \longrightarrow \operatorname{Tor}_\bullet^A(A/I, A/I).$

(2) Show that (*) is an isomorphism if I is generated by a completely intersecting sequence.
    Hint: Reduce to the case that I is generated by one element.

Exercise 21.26. Let X be a locally ringed space and let $\mathscr{L}$ be an object of $D(X)$. Show that the following assertions are equivalent.
(i) The object $\mathscr{L}$ is invertible in $D(X)$, i.e., the functor $D(X) \to D(X)$, $\mathscr{E} \mapsto \mathscr{L} \otimes^L \mathscr{E}$ is an equivalence of categories.
(ii) There exists an open covering $(U_i)_i$ of X and for each i an integer n_i such that $\mathscr{L}_{|U_i} \cong \mathscr{L}_i[n_i]$ for some invertible $\mathcal{O}_{U_i}$-module $\mathscr{L}_i$.
Show that every invertible object in $D(X)$ is a perfect complex.

Exercise 21.27. Show that the definitions of the cup product in Section (21.29) agree.

Exercise 21.28. Let $(X, \mathcal{O}_X)$ be a ringed space. We define an $\mathcal{O}_X$-module $\mathscr{E}$ to be *dualizable* if it has a dual $\mathscr{F}$, where we use the same definition as in Definition 21.149 replacing derived tensor products by usual tensor products. Let $\mathscr{E}$ be an $\mathcal{O}_X$-module. Show that the following assertions are equivalent.
(i) $\mathscr{E}$ is dualizable.
(ii) $\mathscr{E}$ is locally the direct summand of a finite free $\mathcal{O}_X$-module.
(iii) $\mathscr{E}$ is of finite presentation and flat.

Exercise 21.29. Let X be a ringed space. Let $m \in \mathbb{Z}$. A complex $\mathscr{F}$ of $\mathcal{O}_X$-modules is called *strictly m-pseudo-coherent* if $\mathscr{F}^i$ is a direct summand of a finite free $\mathcal{O}_X$-module for $i \geq m$ and $\mathscr{F}^i = 0$ for i sufficiently large. A complex $\mathscr{F}$ is called *strictly pseudo-coherent* if it is a bounded above complex of direct summands of finite free $\mathcal{O}_X$-modules.
(1) Let $\mathscr{F}$ be a strictly m-pseudo-coherent complex of $\mathcal{O}_X$-modules with $H^i(\mathscr{F}) = 0$ for $i > m$. Show that $H^m(\mathscr{F})$ is an $\mathcal{O}_X$-module of finite type.
(2) Show that the following conditions for a complex $\mathscr{F}$ of $\mathcal{O}_X$-modules are equivalent.

(i) The complex $\mathscr{F}$ is m-pseudo-coherent.

(ii) Locally on X there exists a quasi-isomorphism $\mathscr{E} \to \mathscr{F}$ of complexes with $\mathscr{E}$ strictly m-pseudo-coherent.

(iii) Locally on X there exists an isomorphism $\mathscr{E} \to \mathscr{F}$ in $D(X)$ with $\mathscr{E}$ strictly m-pseudo-coherent.

Exercise 21.30. Let X be a ringed space and let $\mathscr{F}$ and $\mathscr{G}$ be in $D(X)$.

(1) Suppose that there exist $a, b \in \mathbb{Z}$ such that $H^i(\mathscr{F}) = 0$ for $i > a$ and $H^j(\mathscr{G}) = 0$ for $j > b$. Let $\mathscr{K}$ and $\mathscr{L}$ be a strictly perfect complexes and let $\mathscr{K} \to \mathscr{F}$ be an m-isomorphism and $\mathscr{L} \to \mathscr{G}$ be an n-isomorphism, $m, n \in \mathbb{Z}$. Show that the canonical map $\mathscr{K} \otimes \mathscr{L} \to \mathscr{F} \otimes_{\mathscr{O}_X}^L \mathscr{G}$ is a t-isomorphism for $t = \max(m + a, n + b)$.

(2) Show that if $\mathscr{F}$ and $\mathscr{G}$ are pseudo-coherent, then $\mathscr{F} \otimes_{\mathscr{O}_X}^L \mathscr{G}$ is pseudo-coherent.

Exercise 21.31. Let R be a ring. Show that a complex $E \in D^-(R)$ is pseudo-coherent if and only if the functors $\mathrm{Ext}_R^i(E, -) \colon (R\text{-Mod}) \to (R\text{-Mod})$ commute with filtered colimits for all $i \in \mathbb{Z}$.

Hint: Use that an R-module M is of finite presentation if and only if $\mathrm{Hom}_R(M, -)$ commutes with filtered colimits.

Exercise 21.32. Let $f \colon X \to Y$ be a morphism of ringed spaces and let $t \geq 0$ be an integer. Show that the following assertions are equivalent.

(i) The morphism f has tor-dimension $\leq t$.

(ii) For every $\mathscr{O}_Y$-module $\mathscr{F}$ one has $L_{t+1} f^* \mathscr{F} = 0$.

(iii) For all $x \in X$ there exists an exact sequence of $\mathscr{O}_{Y, f(x)}$-modules

$$0 \longrightarrow P_t \longrightarrow P_{t-1} \longrightarrow \cdots \longrightarrow P_1 \longrightarrow P_0 \longrightarrow \mathscr{O}_{X,x} \longrightarrow 0$$

with P_i a flat $\mathscr{O}_{Y, f(x)}$-module for all $0 \leq i \leq t$.

Exercise 21.33. Let X be a ringed space and let $\mathscr{F}$ be an $\mathscr{O}_X$-module. The *tor-dimension* or *flat dimension* of $\mathscr{F}$ is defined as

$$\mathrm{tor\text{-}dim}(\mathscr{F}) := \inf\{\, n \in \mathbb{Z} \; ; \; \mathrm{tor\text{-}amp}\,\mathscr{F} \subseteq [-n, 0] \,\}.$$

In particular, we have $\mathrm{tor\text{-}dim}(\mathscr{F}) = -\infty$ if and only if $\mathscr{F} = 0$. Otherwise $\mathrm{tor\text{-}dim}(\mathscr{F}) \geq 0$. As a special case (where X consists of a single point) we obtain the notion of the tor-dimension of an A-module M for a ring A.

(1) Let $n \geq 0$. Show that the following assertions are equivalent.

    (i) $\mathrm{tor\text{-}dim}\,\mathscr{F} \leq n$.

    (ii) $\mathscr{T}or_{n+1}^{\mathscr{O}_X}(\mathscr{F}, -)$ is the zero functor.

    (iii) $\mathscr{T}or_n^{\mathscr{O}_X}(\mathscr{F}, -)$ is left exact.

    (iv) For any exact sequence of $\mathscr{O}_X$-modules $0 \to \mathscr{F}' \to \mathscr{E}_{n-1} \to \cdots \to \mathscr{E}_0 \to \mathscr{F} \to 0$ with each $\mathscr{E}_i$ flat, $\mathscr{F}'$ is flat.

    (v) There exists an exact sequence of $\mathscr{O}_X$-modules $0 \to \mathscr{E}_n \to \cdots \to \mathscr{E}_0 \to \mathscr{F} \to 0$ with all $\mathscr{E}_i$ flat $\mathscr{O}_X$-modules.

(2) Let $(U_i)_{i \in I}$ be an open covering of X. Show that

$$\mathrm{tor\text{-}dim}\,\mathscr{F} = \sup\{\, \mathrm{tor\text{-}dim}_{\mathscr{O}_{X,x}} \mathscr{F}_x \; ; \; x \in X \,\} = \sup\{\, \mathrm{tor\text{-}dim}\,\mathscr{F}_{|U_i} \; ; \; i \in I \,\}.$$

22 Cohomology of quasi-coherent modules

Content

- – Cohomology of quasi-coherent modules and Čech cohomology
- – Derived categories of quasi-coherent modules
- – Finiteness properties of complexes on schemes
- – Projection formula, base change and the Künneth formula

We now study the cohomology of complexes of quasi-coherent modules on schemes. The first main result is the vanishing of higher cohomology of quasi-coherent modules on affine schemes (Theorem 22.2) and a relative version for affine morphisms (Corollary 22.5). This follows quite formally from the vanishing of Čech cohomology for open coverings of affine schemes by principal open subschemes (Lemma 22.1).

This lemma also implies that cohomology and Čech cohomology for open affine coverings are the same for quasi-coherent modules on a separated scheme (Theorem 22.9). This allows us to calculate several important examples of cohomology for quasi-coherent modules. We explain a general procedure how to do this on quasi-affine schemes and on projective spectra using the Koszul complex (Sections (22.4)–(22.6)) and apply this to calculate the cohomology of twisted line bundles on projective space (Theorem 22.22).

Next we study complexes of quasi-coherent modules on a scheme X. We show that the quasi-coherent modules form a Grothendieck abelian category (Theorem 22.32). If X is quasi-compact and separated, then its derived category can be identified with the full subcategory $D_{\mathrm{qcoh}}(X)$ of all $\mathscr{F}$ in $D(X)$ such that all $H^p(\mathscr{F})$ are quasi-coherent (Theorem 22.35). In particular, for a ring R we can identify $D(R)$ and $D_{\mathrm{qcoh}}(\operatorname{Spec} R)$. We show similar results for locally noetherian schemes and coherent modules (Theorem 22.42).

For any scheme X, every object in $D_{\mathrm{qcoh}}(X)$ is the homotopy limit of bounded below complexes (Lemma 22.24). This allows to prove many results on unbounded complexes by reducing to bounded below complexes. For instance, we use this result to show that for a qcqs morphism $f: X \to Y$ of schemes the functor Rf_* maps $D_{\mathrm{qcoh}}(X)$ to $D_{\mathrm{qcoh}}(Y)$ (Theorem 22.31). In particular $R^p f_* \mathscr{F}$ is quasi-coherent for every quasi-coherent $\mathscr{O}_X$-module $\mathscr{F}$ and for all p (Theorem 22.27).

In the next part of the chapter we more closely study how certain properties of complexes which we have introduced in the setting of ringed spaces behave on schemes. For instance the properties to be pseudo-coherent, to be perfect, to be of bounded tor-amplitude, and to be of bounded injective amplitude.

Finally, we prove two central results, the projection formula (Section (22.19)) and the derived base change theorem (Theorem 22.99).

Cohomology of quasi-coherent modules and Čech cohomology

(22.1) Quasi-coherent cohomology of affine schemes.

Let us start by rephrasing a result obtained in Volume I:

Lemma 22.1. *Let X be an affine scheme, and let $\mathscr{F}$ be a quasi-coherent $\mathscr{O}_X$-module. Then $\check{H}^0(X,\mathscr{F}) = \Gamma(X,\mathscr{F})$ and all higher Čech cohomology groups vanish:*

$$\check{H}^i(X,\mathscr{F}) = 0 \qquad \text{for all } i > 0.$$

Proof. It is clearly enough to check that $\check{H}^0(\mathcal{U},\mathscr{F}) = \Gamma(X,\mathscr{F})$, and $\check{H}^i(\mathcal{U},\mathscr{F}) = 0$ for $i > 0$ for all finite coverings $\mathcal{U} = (D(f_i))_i$ of X by principal open affine subschemes. This statement can be checked directly, and we have actually done so in Lemma 12.33. Another way to obtain this result is the following: Evaluating the definition of the Čech complex in this situation shows that we have to prove that the complex

$$0 \to M \to \bigoplus_i M_{f_i} \to \bigoplus_{i,j} M_{f_i f_j} \to \cdots$$

is exact, where $M := \Gamma(X,\mathscr{F})$.

But the exactness of this complex is a standard result in the theory of faithfully flat descent, applied to the faithfully flat ring homomorphism $\Gamma(X,\mathscr{O}_X) \to \prod_i \Gamma(D(f_i),\mathscr{O}_X)$. See Lemma 14.64. $\qquad\square$

Note that this result, as explained in Corollary 12.34, implies that for every exact sequence

$$0 \to \mathscr{F}' \to \mathscr{F} \to \mathscr{F}'' \to 0$$

of $\mathscr{O}_X$-modules with $\mathscr{F}'$ quasi-coherent, the sequence

$$0 \to \Gamma(X,\mathscr{F}') \to \Gamma(X,\mathscr{F}) \to \Gamma(X,\mathscr{F}'') \to 0$$

is exact. Since every (quasi-coherent) $\mathscr{O}_X$-module $\mathscr{F}'$ can be embedded into an injective $\mathscr{O}_X$-module $\mathscr{F}$, applying this reasoning to the short exact sequence thus obtained (with $\mathscr{F}'' = \mathscr{F}/\mathscr{F}'$), we conclude that $H^1(X,\mathscr{F}') = 0$. (We can also get this result from the equality of $\check{H}^1$ and H^1, see Corollary 21.81). But we can get more:

Theorem 22.2. *Let X be an affine scheme, and let $\mathscr{F}$ be a quasi-coherent $\mathscr{O}_X$-module. Then the higher cohomology groups vanish:*

$$H^i(X,\mathscr{F}) = 0 \qquad \text{for all } i > 0.$$

Proof. We apply Cartan's theorem, Proposition 21.83, to the basis of the topology of X consisting of affine open subsets of X. By Lemma 22.1, the hypotheses of that theorem are satisfied, and we obtain the desired result. $\qquad\square$

Recall that the vanishing of higher cohomology groups of quasi-coherent sheaves (or just the vanishing of H^1 for all ideal sheaves) characterizes affine schemes, Theorem 12.35.

Remark 22.3. The above theorem is not true without the condition that $\mathscr{F}$ be quasi-coherent. For an example, see Exercise 22.3

Of course, it is also easy to find non-vanishing cohomology groups $H^1(X, \mathscr{F})$ with X affine where $\mathscr{F}$ is not even an $\mathscr{O}_X$-module, but just an abelian sheaf, e. g. for $\mathscr{F} = \mathscr{O}_X^\times$ we have an identification $H^1(X, \mathscr{O}_X^\times) = \mathrm{Pic}(X)$, and this group is usually non-trivial.

Corollary 22.4. *Let X be a scheme, and let*

$$0 \to \mathscr{F} \to \mathscr{G} \to \mathscr{H} \to 0$$

be a short exact sequence of $\mathscr{O}_X$-modules. If two of the three modules are quasi-coherent, then so is the third.

Proof. We know already that kernels and cokernels of morphisms of quasi-coherent modules are again quasi-coherent (Corollary 7.19). Hence it suffices to show that if $\mathscr{F}$ and $\mathscr{H}$ are quasi-coherent, then so is $\mathscr{G}$. Since quasi-coherence is a local property, we may assume that X is affine. We have a commutative diagram

$$
\begin{array}{ccccccccc}
0 & \longrightarrow & \Gamma(X, \mathscr{F})^{\sim} & \longrightarrow & \Gamma(X, \mathscr{G})^{\sim} & \longrightarrow & \Gamma(X, \mathscr{H})^{\sim} & \longrightarrow & 0 \\
 & & \downarrow & & \downarrow & & \downarrow & & \\
0 & \longrightarrow & \mathscr{F} & \longrightarrow & \mathscr{G} & \longrightarrow & \mathscr{H} & \longrightarrow & 0
\end{array}
$$

with exact rows; for the top row, we use Theorem 22.2 (or Corollary 12.34) and that $\sim$ is exact. The outer two vertical maps are isomorphisms, hence so is the middle one. $\square$

Corollary 22.5. *Let $f\colon X \to Y$ be an affine morphism of schemes, and let $\mathscr{F}$ be a quasi-coherent $\mathscr{O}_X$-module. Then $R^i f_* \mathscr{F} = 0$ for all $i > 0$.*

Proof. In view of Theorem 22.2, this is an immediate consequence of the description of $R^i f_* \mathscr{F}$ as the sheaf associated to the presheaf

$$V \mapsto H^i(f^{-1}(V), \mathscr{F}),$$

see Proposition 21.27. $\square$

Corollary 22.6. *Let $f\colon X \to Y$ be an affine morphism of schemes, and let $\mathscr{F}$ be a quasi-coherent $\mathscr{O}_X$-module.*
(1) We have functorial isomorphisms

(22.1.1) $H^p(X, \mathscr{F}) \cong H^p(Y, f_* \mathscr{F})$ *for all $p \geq 0$.*

(2) Let $g\colon Y \to Z$ be a morphism of schemes, then we have functorial isomorphisms

(22.1.2) $R^p(g \circ f)_* \mathscr{F} \cong R^p g_* f_* \mathscr{F}$ *for all $p \geq 0$.*

Proof. For (1) consider the Leray spectral sequence (21.8.5):

$$E_2^{p,q} = H^p(Y, R^q f_* \mathscr{F}) \Rightarrow H^{p+q}(X, \mathscr{F}).$$

By Corollary 22.5 we have $E_2^{pq} = 0$ for $q \neq 0$. Hence the spectral sequence degenerates at E_2 and the edge morphism $E_2^{p0} \to H^p(X, \mathscr{F})$ is an isomorphism for all $p \geq 0$.

For (2) use the same argument with the relative Leray spectral sequence (21.8.6). $\square$

For a corresponding statement in terms of derived categories, see Proposition 22.33.

Remark 22.7. The argument in the proof of Corollary 22.6 shows that for every morphism of ringed spaces $f\colon X \to Y$ and every $\mathscr{O}_X$-module $\mathscr{F}$ one has isomorphisms as in (22.1.1) and in (22.1.2) (for $g\colon Y \to Z$ any morphism of ringed spaces) if $R^q f_* \mathscr{F} = 0$ for $q > 0$.

Remark 22.8. Theorem 22.2 and Corollary 22.5 are central results in the theory of the cohomology of quasi-coherent modules on schemes. To prove these results we first showed the vanishing of the higher Čech cohomology (Lemma 22.1, in fact we gave two proofs for this lemma, see also Remark 22.19 below for a third proof) and then used the spectral sequence linking Čech cohomology and cohomology in form of Cartan's theorem (Proposition 21.83). Let us now sketch an argument that avoids Čech cohomology beyond degree 1 and spectral sequences which we learned from Johannes Anschütz.

In fact we show the following two assertions by induction on $p \geq 1$ simultaneously.

(A_p) One has $H^i(X, \mathscr{F}) = 0$ for $1 \leq i \leq p$ for all affine schemes X and all quasi-coherent $\mathscr{O}_X$-modules $\mathscr{F}$.

(B_p) One has $R^i f_* \mathscr{F} = 0$ for $1 \leq i \leq p$ for all affine morphisms $f\colon X \to Y$ of schemes and for all quasi-coherent $\mathscr{O}_X$-modules $\mathscr{F}$.

Assertion (A_p) implies (B_p) for all p by the same argument as in in the proof of Corollary 22.5. Assertion (A_1) follows from Lemma 12.33 for $p = 1$ (here we use that cohomology agrees with Čech cohomology in degree 1 by Proposition 11.13 and Proposition 21.40). It remains to show (A_p) for $p \geq 2$ assuming that we already have proved (A_{p-1}) and (B_{p-1}).

Let $\xi \in H^p(X, \mathscr{F})$. By Corollary 21.23 there exists an affine open covering $(U_k)_{1 \leq k \leq n}$ of X such that $\xi_{|U_k} = 0$ for all k. Let $j_k\colon U_k \to X$ be the inclusion which is an open affine immersion. Complete $\mathscr{F} \to \bigoplus_k Rj_{k,*}(\mathscr{F}_{|U_k})$ to a distinguished triangle

$$\mathscr{F} \longrightarrow \bigoplus_k Rj_{k,*}(\mathscr{F}_{|U_k}) \longrightarrow \mathscr{G} \xrightarrow{+1}$$

and consider the associated long exact cohomology sequence

$$\cdots \to \bigoplus_{k=1}^n H^{p-1}(U_k, \mathscr{F}_{|U_k}) \to H^{p-1}(X, \mathscr{G}) \to H^p(X, \mathscr{F}) \xrightarrow{a} \bigoplus_{k=1}^n H^p(U_k, \mathscr{F}_{|U_k}) \to \cdots.$$

By induction hypothesis, we have $H^{p-1}(U_k, \mathscr{F}_{|U_k}) = 0$. By assumption, we have $a(\xi) = 0$. Hence it suffices to show that $H^{p-1}(X, \mathscr{G}) = 0$. For this consider the distinguished triangle

$$(*) \qquad\qquad \tau^{\leq 0}\mathscr{G} \longrightarrow \mathscr{G} \longrightarrow \tau^{\geq 1}\mathscr{G} \xrightarrow{+1}.$$

Then $\tau^{\leq 0}\mathscr{G}$ is the quasi-coherent module $\mathrm{Coker}(\mathscr{F} \to \bigoplus_k j_{k,*}(\mathscr{F}_{|U_k}))$ considered as a complex concentrated in degree 0. Moreover, since by induction hypothesis we have $R^i j_{k,*}(\mathscr{F}_{|U_k}) = 0$ for $1 \leq i < p$, we have $\tau^{\geq 1}\mathscr{G} = \tau^{\geq p}\mathscr{G}$. Therefore the long exact cohomology sequence attached to $(*)$ shows that

$$H^{p-1}(X, \mathscr{G}) \cong H^{p-1}(X, \tau^{\leq 0}\mathscr{G}) = 0,$$

where the second equality holds by induction hypothesis.

(22.2) Cohomology versus Čech Cohomology.

Theorem 22.9. *Let X be a separated scheme, and let $\mathscr{F}$ be a quasi-coherent $\mathscr{O}_X$-module. Then for every open covering $\mathcal{U}$ of X by open affine subschemes, the canonical homomorphisms*

$$H^i(\mathcal{U}, \mathscr{F}) \to \check{H}^i(X, \mathscr{F}) \to H^i(X, \mathscr{F})$$

are isomorphisms for all $i \geq 0$.

Proof. In view of Lemma 22.1, this follows immediately from Cartan's Theorem, Proposition 21.83, using that in a separated scheme the intersection of two open affine subschemes is again affine (Proposition 9.15). □

Corollary 22.10. *Let X be a separated scheme which can be covered by n affine open subschemes. Then for all $i \geq n$ and all quasi-coherent $\mathscr{O}_X$-modules $\mathscr{F}$, $H^i(X, \mathscr{F}) = 0$.*

Proof. Use the above identification of cohomology with Čech cohomology and the corresponding fact for Čech cohomology, which is a direct consequence of the fact that we can compute Čech cohomology using alternating cocycles, see Corollary 21.70. □

Remark 22.11. Let X be a separated scheme, let $\mathcal{U}$ be an affine open cover of X and let

$$0 \to \mathscr{F}' \to \mathscr{F} \to \mathscr{F}'' \to 0$$

be a short exact sequence of quasi-coherent $\mathscr{O}_X$-modules. The vanishing of higher degree cohomology of quasi-coherent modules on affine schemes implies that the sequence

$$0 \to C^\bullet(\mathcal{U}, \mathscr{F}') \to C^\bullet(\mathcal{U}, \mathscr{F}) \to C^\bullet(\mathcal{U}, \mathscr{F}'') \to 0$$

of Čech complexes is exact. It hence gives rise to a long exact cohomology sequence. It follows from the comparison between Čech cohomology and usual cohomology (Theorem 21.78) that under the identification of Theorem 22.9 this long exact sequence coincides with the long exact sequence for usual cohomology. This gives a handle on actually computing the maps in the long exact cohomology sequence.

Remark and Definition 22.12. The proof of Theorem 22.9 shows that the conclusions of Theorem 22.9 and Corollary 22.10 also hold if X has an affine diagonal (i.e. the intersection of any two open affine subschemes is again affine) instead of assuming that X is separated. Such schemes are called *semiseparated*. See Exercise 22.4 for more details.

For instance, let k be a field and let X be the k-scheme obtained from gluing two copies of $\mathbb{A}_k^n$ along $\mathbb{A}_k^n \setminus \{0\}$ via the identity. Then X is not separated for $n \geq 1$. It is semiseparated if and only if $n = 1$. Indeed, one can show that Čech cohomology does not agree with cohomology for $n \geq 2$ (cf. Exercise 22.5).

(22.3) Elementary examples: $\mathbb{P}^1$ and $\mathbb{A}^2 \setminus \{0\}$.

Let us give some examples of (and references to) explicit computations of cohomology groups using Čech cohomology.

Example 22.13. Let $R \neq 0$ be a ring, and consider the standard covering

$$\mathbb{P}^1_R = \text{Proj}(R[T_0, T_1]) = U_0 \cup U_1, \qquad U_i = D_+(T_i).$$

By Theorem 22.9 we can compute the cohomology groups of quasi-coherent sheaves using the alternating Čech complex for this covering, in particular for $d \in \mathbb{Z}$ the R-modules $H^i(\mathbb{P}^1_R, \mathcal{O}(d))$ are the cohomology modules of the sequence

$$0 \to H^0(U_0, \mathcal{O}(d)) \times H^0(U_1, \mathcal{O}(d)) \to H^0(U_0 \cap U_1, \mathcal{O}(d)) \to 0.$$

As in Example 11.45 we can consider the components of the sequence as R-submodules of $R(T_0, T_1)$ and can by (11.14.4) identify the above sequence with

$$0 \to T_0^d R[T_1/T_0] \times T_1^d R[T_0/T_1] \to T_0^d R[T_0/T_1, T_1/T_0] = T_1^d R[T_0/T_1, T_1/T_0] \to 0,$$
$$(f, g) \mapsto f - g.$$

We obtain that $H^0(\mathbb{P}^1_R, \mathcal{O}(d))$ and $H^1(\mathbb{P}^1_R, \mathcal{O}(d))$ are free R-modules of rank

$$\text{rk } H^0(\mathbb{P}^1_R, \mathcal{O}(d)) = \begin{cases} d+1 & \text{if } d \geq 0, \\ 0 & \text{if } d < 0, \end{cases}$$

and

$$\text{rk } H^1(\mathbb{P}^1_R, \mathcal{O}(d)) = \begin{cases} 0 & \text{if } d \geq -1, \\ -d-1 & \text{if } d < -1. \end{cases}$$

For $H^0(\mathbb{P}^1_R, \mathcal{O}(d))$ this reproduces the results of Example 11.45. Together with some more careful bookkeeping, the same method can be used to compute all cohomology groups $H^i(\mathbb{P}^r_R, \mathcal{O}(d))$. See Theorem 22.22 below.

Example 22.14. Let $R \neq 0$ be a ring, $A = R[T, S]$, and let $X = \mathbb{A}^2_R \setminus \{0\} \subset \mathbb{A}^2_R = \text{Spec } A$. We have the affine open cover $\mathcal{U} = (D(T), D(S))$ of X, and the corresponding alternating Čech complex for $\mathcal{O}_X$ is

$$0 \to A_T \times A_S \xrightarrow{d} A_{TS} \to 0, \qquad d(f, g) = g - f.$$

Hence

$$H^0(X, \mathcal{O}_X) = \check{H}^0(\mathcal{U}, \mathcal{O}_X) = A,$$
$$H^1(X, \mathcal{O}_X) = \check{H}^1(\mathcal{U}, \mathcal{O}_X) = \bigoplus_{i,j>0} RT^{-i}S^{-j}.$$

In particular, both are of infinite rank over R and we see (again, at least for $R = k$ an algebraically closed field, cf. Exercise 1.13) that X is not affine because otherwise we would have $H^1(X, \mathcal{O}_X) = 0$ by Theorem 22.2. We also see that the cohomology, even for "nice" sheaves such as the structure sheaf, may be not finitely generated. In fact, $H^1(X, \mathcal{O}_X)$ is not even finitely generated as a $\Gamma(X, \mathcal{O}_X)$-module.

We will see later (Corollary 23.18) that for R a noetherian ring, a *proper* R-scheme Z, and a coherent $\mathcal{O}_Z$-module $\mathscr{F}$, all cohomology groups $H^i(Z, \mathscr{F})$ are finitely generated R-modules.

(22.4) The extended ordered Čech complex and Koszul complexes.

The examples in Section (22.3) are a special case of a general procedure to calculate the Čech cohomology of quasi-affine schemes U and of projective spectra. This is based on identifying the extended ordered Čech complex for an open covering by principal open affine subsets with a colimit of Koszul complexes (Section (19.1)), as we explain now.

We put ourselves in the following situation. Let A be a ring, $X = \operatorname{Spec} A$, and let $f_1, \ldots, f_r$ be elements of A. The ordered Čech complex of the structure sheaf associated to the open covering $\mathcal{U} := (D(f_i))_i$ of $U := X \setminus V(f_1, \ldots, f_r)$, augmented by A, then yields the *extended ordered Čech complex*

$$(22.4.1) \quad C_A(f_1, \ldots, f_r): \qquad A \longrightarrow \prod_{i_1} A_{f_{i_1}} \longrightarrow \prod_{i_1 < i_2} A_{f_{i_1} f_{i_2}} \longrightarrow \cdots \longrightarrow A_{f_1 f_2 \cdots f_r},$$

which we consider as a cochain complex concentrated in degrees $[0, r]$. For every quasi-coherent $\mathcal{O}_X$-module $\mathscr{F} = \tilde{M}$, M an A-module, the extended and the usual ordered Čech complex are related by

$$(22.4.2) \qquad \check{C}_{\mathrm{ord}}(\mathcal{U}, \mathscr{F}) = \sigma^{\geq 0}(C_A(f_1, \ldots, f_r) \otimes_A M[1])$$

because all products in (22.4.1) are finite and hence commute with tensor products. In particular we find that for $p > 0$ one has

$$(22.4.3) \qquad H^p(U, \mathscr{F}) = \check{H}^p(\mathcal{U}, \mathscr{F}) = H^{p+1}(C_A(\mathbf{f}) \otimes_A M),$$

where the first equality holds by Theorem 22.9.

Note that $C_A(f_i)$ is the complex $A \to A_{f_i}$, concentrated in degrees 0 and 1, and that

$$(22.4.4) \qquad C_A(f_1, \ldots, f_r) = C_A(f_1) \otimes_A \cdots \otimes_A C_A(f_r).$$

We now relate the complex $C_A(f_1, \ldots, f_r)$ of (22.4.1) to the Koszul complex. For a sequence of elements $\mathbf{f} = (f_1, \ldots, f_r)$ in A we view the attached Koszul complex as a cochain complex $K(\mathbf{f}) = K^\bullet(\mathbf{f})$ in degrees $[-r, 0]$. Hence the entry of $K(\mathbf{f})$ in degree $-p$ is the A-module $\bigwedge^p(A^r)$ and the differential $K^{-p}(\mathbf{f}) \to K^{-p+1}(\mathbf{f})$ is given by

$$x_1 \wedge \cdots \wedge x_p \mapsto \sum_{i=1}^{p} (-1)^{i+1} \Big(\sum_{j=1}^{r} f_j x_{i,j} \Big) x_1 \wedge \cdots \wedge \widehat{x_i} \wedge \cdots \wedge x_p.$$

For all i we have that $K(f_i)$ is the complex $A \xrightarrow{f_i} A$ concentrated in degrees -1 and 0 and

$$(22.4.5) \quad K(\mathbf{f}) = K(f_1) \otimes_A K(f_2) \otimes_A \cdots \otimes_A K(f_r) = K(f_1) \otimes_A^L K(f_2) \otimes_A^L \cdots \otimes_A^L K(f_r),$$

where the second equality holds because $K(f_i)$ is K-flat.

We also consider the *dual Koszul complex*

$$K(\mathbf{f})^\vee := R\operatorname{Hom}_A(K(\mathbf{f}), A) = \operatorname{Hom}_A(K(\mathbf{f}), A),$$

where the equality holds because $K(\mathbf{f})$ is a K-projective complex of A-modules. For all i one has $K(f_i)^\vee = A \xrightarrow{-f_i} A$, now concentrated in degrees 0 and 1, i.e., $K(f_i)^\vee \cong K(f_i)[-1]$ (recall that by definition, also the shift by i introduces a sign $(-1)^i$). By (22.4.5) one deduces

$$(22.4.6) \qquad K(\mathbf{f})^\vee = K(f_1)^\vee \otimes_A \cdots \otimes_A K(f_r)^\vee \cong K(\mathbf{f})[-r].$$

Remark 22.15. Set $\mathbf{f}^n := (f_1^n, \ldots, f_r^n)$ for $n \geq 1$.

(1) The functoriality of the Koszul complex (Remark 19.5) then yields an $\mathbb{Z}_{\geq 1}^{\mathrm{opp}}$-diagram of complexes $\cdots \to K(\mathbf{f}^3) \to K(\mathbf{f}^2) \to K(\mathbf{f})$. If $(e_1, \ldots, e_r)$ denotes the standard basis of A^r, then in degree $-p$ the transition maps are given by

$$K^{-p}(\mathbf{f}^{n+1}) = \bigwedge^p A^r \longrightarrow K^{-p}(\mathbf{f}^n) = \bigwedge^p A^r,$$
$$e_{i_1} \wedge \cdots \wedge e_{i_p} \longmapsto ((-1)^p f_{i_1} \cdots f_{i_p}) e_{i_1} \wedge \cdots \wedge e_{i_p}.$$

(2) Passing to dual Koszul complexes and using (22.4.6), we obtain an $\mathbb{Z}_{\geq 1}$-diagram of complexes $K(\mathbf{f}^n)[-r] \to K(\mathbf{f}^{n+1})[-r]$ of complexes sitting in degrees $[0, r]$ by functoriality. The transition maps are given by

$$\varepsilon_{i_1} \wedge \cdots \wedge \varepsilon_{i_p} \longmapsto \left(\prod_{i \notin \{i_1, \ldots, i_p\}} f_i \right) \varepsilon_{i_1} \wedge \cdots \wedge \varepsilon_{i_p},$$

where $(\varepsilon_1, \ldots, \varepsilon_r)$ now denotes the dual basis of the standard basis. In particular, the transition maps are the identity in degree 0 and equal to $f_1 \cdots f_r$ in degree r. Using that $A_f = \mathrm{colim}(A \xrightarrow{f} A \xrightarrow{f} A \xrightarrow{f} \cdots)$ one sees that

(22.4.7)
$$\mathrm{colim}\, K(\mathbf{f}^n)^\vee = \mathrm{colim}\, K(f_1^n)[-1] \otimes \cdots \otimes \mathrm{colim}\, K(f_r^n)[-1]$$
$$= (A \to A_{f_1}) \otimes \cdots \otimes (A \to A_{f_r}) = C_A(f_1, \ldots, f_r)$$

is the extended ordered Čech complex defined in (22.4.1).

By (22.4.6) we obtain from (22.4.7) the following result.

Proposition 22.16. *There exists an isomorphism of complexes of A-modules*

(22.4.8)
$$\mathrm{colim}\, K(\mathbf{f}^n)[-r] \cong \mathrm{colim}\, K(\mathbf{f}^n)^\vee = C_A(f_1, \ldots, f_r).$$

(22.5) Example: Cohomology of quasi-affine schemes.

We can now use (22.4.8) to compute some cohomology groups for quasi-affine schemes.

Example 22.17. Let U be a quasi-affine scheme, let $A := \Gamma(U, \mathscr{O}_U)$ and let $j \colon U \to \mathrm{Spec}\, A$ be the canonical open quasi-compact schematically dominant immersion (Proposition 13.80). As U is quasi-compact, there exists a finite family $\mathbf{f} = (f_1, \ldots, f_r)$ of elements of A such that $\mathcal{U} := (D(f_i))_i$ is an affine open covering of U. Let $\mathscr{F}$ be a quasi-coherent $\mathscr{O}_U$-module. Then $j_* \mathscr{F}$ is quasi-coherent (Corollary 10.27) and hence $j_* \mathscr{F} \cong \tilde{M}$ for an A-module M. We have

(22.5.1)
$$H^0(U, \mathscr{F}) = H^0(\mathrm{Spec}\, A, j_* \mathscr{F}) = M.$$

Now assume that $p > 0$. If $r = 1$, then U is affine and hence $A = A_{f_1}$, i.e., f_1 is a unit and $H^p(U, \mathscr{F}) = 0$. Hence let us assume from now on that $r \geq 2$.

We can calculate Čech cohomology of $\mathscr{F}$ by (22.4.3) and hence get by (22.4.8)

(22.5.2)
$$\begin{aligned} H^p(U, \mathscr{F}) &= \check{H}^p(\mathcal{U}, \mathscr{F}) \\ &= H^{p+1}(C_A(\mathbf{f}) \otimes_A M) \\ &= H^{p+1}(\mathrm{colim}_n K(\mathbf{f}^n)[-r] \otimes_A M) \qquad \text{for } p > 0. \\ &= \mathrm{colim}_n H^{p-r+1}(K(\mathbf{f}^n, M)) \end{aligned}$$

Now suppose that $\mathbf{f}$ is M-completely intersecting. Then $\mathbf{f}^n$ is M-completely intersecting (Proposition 19.17) and hence $H^q(K(\mathbf{f}^n, M)) = 0$ for $q < 0$ and $H^0(K(\mathbf{f}^n, M)) = M/\mathbf{f}^n M := M/(f_1^n, \ldots, f_r^n)M$. Hence in this case by (22.5.2) and (22.5.1) we obtain

(22.5.3) $$H^p(U, \mathscr{F}) = \begin{cases} M, & \text{if } p = 0; \\ 0, & \text{if } p \neq 0, r - 1; \\ \operatorname{colim}_n M/\mathbf{f}^n M, & \text{if } p = r - 1, \end{cases}$$

where the transition maps in the colimit are given by multiplication by $f_1 f_2 \cdots f_r$.

Example 22.18. Let R be a ring and let $A = R[T_1, \ldots, T_r]$ with $r \geq 2$. We apply (22.5.3) to the family $(T_1, \ldots, T_r)$ which is regular and in particular completely intersecting (Theorem 19.12). One has $U = \bigcup_i D(T_i) = \mathbb{A}_R^r \setminus \{0\}$ and $\Gamma(U, \mathscr{O}_U) = A$. Hence we obtain

$$H^p(U, \mathscr{O}_U) = \begin{cases} A, & \text{if } p = 0; \\ 0, & \text{if } p \neq 0, r - 1; \\ \operatorname{colim}_n A/(T_1^n, \ldots, T_r^n)A, & \text{if } p = r - 1, \end{cases}$$

Let us make the term $\operatorname{colim}_n A/(T_1^n, \ldots, T_r^n)A$ more explicit. For any subset $I \subseteq \mathbb{Z}^r$, we denote the free R-submodule of $R[T_1^{\pm 1}, \ldots, T_r^{\pm 1}]$ generated by $T_1^{i_1} T_2^{i_2} \cdots T_r^{i_r}$ for $(i_1, \ldots, i_r) \in I$ by M^I. Then M^I is an A-submodule if and only if I is stable under addition with elements in $\mathbb{N}^r$. Set

$$\mathbb{Z}_+^r = \{ (i_1, \ldots, i_r) \in \mathbb{Z}^r \; ; \; \exists j \colon i_j \geq 0 \}.$$

As an A-module, we can identify $A/(T_1^n, \ldots, T_r^n)A$ with $M^{[-n,\infty)^r}/(M^{[-n,\infty)^r} \cap M^{\mathbb{Z}_+^r})$ such that the transition map $A/(T_1^n, \ldots, T_r^n)A \to A/(T_1^{n+1}, \ldots, T_r^{n+1})A$ given by multiplication by $T_1 \cdots T_r$ is induced by the inclusion $M^{[-n,\infty)^r} \hookrightarrow M^{[-n-1,\infty)^r}$. Then $\operatorname{colim}_n A/(T_1^n, \ldots, T_r^n)A$ becomes the A-module

$$H^{r-1}(U, \mathscr{O}_U) = M^{\mathbb{Z}^r}/M^{\mathbb{Z}_+^r}$$

$$= R[T_1^{\pm 1}, \ldots, T_r^{\pm 1}] \Big/ \sum_i R[T_1^{\pm 1}, \ldots, T_{i-1}^{\pm 1}, T_i, T_{i+1}^{\pm 1}, \ldots, T_r^{\pm 1}]$$

which we can identify as an R-module with $M^{(-\infty,-1]^r}$, i.e., the free R-module generated by the monomials $T_1^{i_1} \cdots T_r^{i_r}$ with $i_j < 0$ for all $j = 1, \ldots, r$.

Remark 22.19. Example 22.17 gives also a new proof that $\check{H}^p(\operatorname{Spec} A, \mathscr{F}) = 0$ for all quasi-coherent sheaves $\mathscr{F} = \tilde{M}$ on $\operatorname{Spec} A$ and all $p > 0$: If $(D(f_i))_i$ is an open covering of $\operatorname{Spec} A$, i.e., $f_1, \ldots, f_r$ generate the unit ideal, then $K(\mathbf{f}^n)$ is homotopy equivalent to 0 by Remark 19.6. Hence $K(\mathbf{f}^n, M)$ is homotopy equivalent to 0 and we conclude by (22.5.2).

(22.6) Example: Cohomology of twisted line bundles on projective spectra and on projective space.

Example 22.13 can also be considered as a special case of the technique to use the Koszul complex to calculate (Čech) cohomology.

Example 22.20. Let R be a ring and let $A = \bigoplus_{d \geq 0} A_d$ be a graded R-algebra with $A_0 = R$ which is generated by a finite family $\mathbf{f} = (f_1, \ldots, f_r)$ of elements $f_i \in A_1$ of degree 1. Set $X := \operatorname{Proj} A$. This is a projective scheme over R with an open affine covering $\mathcal{U} := (D_+(f_i))_i$ (Proposition 13.12). Hence we can calculate the cohomology of quasi-coherent $\mathcal{O}_X$-modules as Čech cohomology for the covering $\mathcal{U}$.

We would like to compute $H^p(X, \mathcal{O}_X(d))$ for all $d \in \mathbb{Z}$. As X is separated, we have $H^p(X, \mathcal{F}) = \check{H}^p(\mathcal{U}, \mathcal{F})$ for every quasi-coherent $\mathcal{O}_X$-module $\mathcal{F}$ by Theorem 22.9. Moreover, cohomology commutes with direct sums (Corollary 21.56) because X is qcqs. Hence we can compute the cohomology of $\mathcal{F} := \bigoplus_{d \in \mathbb{Z}} \mathcal{O}_X(d)$, and split up the result according to d afterwards.

Recall that we denote by A_{f_i} the graded localization of A by f_i and by $A_{(f_i)}$ the subring of degree 0 elements (Section (13.1)). By definition of $\mathcal{O}_X(d)$ (Section (13.4)) we have an isomorphism $\Gamma(D_+(f_i), \mathcal{F}) = A_{f_i}$ of graded $A_{(f_i)}$-modules. Therefore the ordered Čech complex of $\mathcal{F}$ with respect to $\mathcal{U}$ is given by the complex

(*)
$$0 \longrightarrow \prod_{i_1} A_{f_{i_1}} \longrightarrow \prod_{i_1 < i_2} A_{f_{i_1} f_{i_2}} \longrightarrow \cdots \longrightarrow A_{f_1 f_2 \cdots f_r} \longrightarrow 0$$

of graded A-modules which is equal to the complex $(\sigma^{\geq 1} C_A(f_1, \ldots, f_r))[1]$ of graded A-modules.

The Koszul complex $K(\mathbf{f}^n)$ is also a complex of graded A-modules and the isomorphism $\operatorname{colim} K(\mathbf{f}^n)[-r] \cong C_A(f_1, \ldots, f_r)$ (22.4.8) is an isomorphism of complexes of graded A-modules. As in Remark 22.17 we see that for $p > 0$,

(22.6.1)
$$H^p(X, \mathcal{O}_X(d)) = \operatorname*{colim}_n H^{p-r+1}(K(\mathbf{f}^n))_d,$$

where $(-)_d$ denotes the R-submodule of homogeneous elements of degree d.

Now suppose that $\mathbf{f}$ is completely intersecting and that $r \geq 2$, then we obtain, again as in Remark 22.17, that

(22.6.2)
$$H^p(X, \mathcal{O}_X(d)) = \begin{cases} 0, & \text{if } p \neq 0, r-1; \\ \operatorname{colim}_n (A/\mathbf{f}^n A)_d, & \text{if } p = r-1, \end{cases}$$

Moreover, we have $H^0(C_A(f_1, \ldots, f_r)) = H^1(C_A(f_1, \ldots, f_r)) = 0$ in this case because $K(\mathbf{f}^n)$ is acyclic in degrees $-r$ and $-r+1$. Hence the cohomology of the ordered Čech complex (*) in degree 0 is A. Therefore $H^0(X, \mathcal{F}) = H^0(\sigma^{\geq 0}(C_A(f_1, \ldots, f_r)[1]) = A$ and

(22.6.3)
$$H^0(X, \mathcal{O}_X(d)) = A_d.$$

Similarly as in Example 22.20 one can also compute $H^p(X, \mathcal{F}(d))$ for every quasi-coherent module $\mathcal{F}$ on $\operatorname{Proj} A$, see Exercise 22.6. All of these ideas can be further generalized, see Exercise 22.7.

Remark 22.21. The similarity of the results in Example 22.20 and in Example 22.17 is no happenstance. Let A be a graded R-algebra as in Example 22.20, let $f_1, \ldots, f_r \in A_1$ be generators of the R-algebra A, and set $X = \operatorname{Proj}(A)$. Suppose that $(f_1, \ldots, f_r)$ is completely intersecting with $r \geq 2$. Then

(*)
$$A = \Gamma_*(\mathcal{O}_X) = \bigoplus_d \Gamma(X, \mathcal{O}_X(d))$$

by (22.6.3). Let $C^0 = \operatorname{Spec} A \setminus V(f_1, \dots, f_r)$ be the pointed cone of A (Section (13.9)). The canonical morphism

$$\pi\colon C^0 \to \operatorname{Proj}(A)$$

is a surjective morphism such that $\pi^{-1}(D_+(f_i)) = D(f_i) \cong D_+(f_i) \times_R (\mathbb{A}_R^1 \setminus \{0\})$ (Proposition 13.37). In particular, π is affine. Moreover $\pi_* \mathscr{O}_{C^0} = \bigoplus_{d \in \mathbb{Z}} \mathscr{O}_X(d)$ by (*) and hence we find by Corollary 22.6 for all $p \geq 0$ an isomorphism

$$H^p(C^0, \mathscr{O}_{C^0}) = \bigoplus_{d \in \mathbb{Z}} H^p(X, \mathscr{O}_X(d)).$$

As an application we can calculate the cohomology of the twisted line bundles on projective space.

Theorem 22.22. *Let R be a ring, let $r > 0$ and consider $\mathbb{P}_R^r = \operatorname{Proj} R[T_0, \dots, T_r]$.*
(1) *For all $d \in \mathbb{Z}$ and all $i \neq 0, r$, we have $H^i(\mathbb{P}_R^r, \mathscr{O}_{\mathbb{P}_R^r}(d)) = 0$.*
(2) *The canonical homomorphism of graded R-algebras*

$$R[T_0, \dots, T_r] \to \bigoplus_{d \in \mathbb{Z}} H^0(\mathbb{P}_R^r, \mathscr{O}_{\mathbb{P}_R^r}(d))$$

is an isomorphism. In particular, $H^0(\mathbb{P}_R^r, \mathscr{O}_{\mathbb{P}_R^r}(d)) = 0$ for $d < 0$.
(3) *The A-module $H^r(\mathbb{P}_R^r, \mathscr{O}_{\mathbb{P}_R^r}(d))$ is free, and is isomorphic to the A-submodule of $R[T_0^{-1}, \dots, T_r^{-1}]$ generated by all monomials $T_0^{p_0} \cdots T_r^{p_r}$ with $p_i < 0$ for all i and $\sum_i p_i = d$. In particular, $H^r(\mathbb{P}_R^r, \mathscr{O}_{\mathbb{P}_R^r}(d)) = 0$ for $d > -r - 1$.*

The theorem can be generalized to arbitrary base schemes and projective bundles, see Theorem 22.86 below.

Proof. As $(T_0, \dots, T_r)$ is a regular sequence (of length $r + 1$) in $R[T_0, \dots, T_r]$, it is completely intersecting (Theorem 19.12). Hence (2) (which we also had already seen in Example 13.16) follows from (22.6.3). Assertions (1) and (3) are implied by (22.6.2), where for the description of $H^r(\mathbb{P}_R^r, \mathscr{O}_{\mathbb{P}_R^r}(d))$ we argue as in Example 22.18.

Instead of referring to (22.6.2) we could also have used the pointed affine cone $\pi\colon \mathbb{A}_R^{r+1} \setminus \{0\} \to \mathbb{P}_R^r$ and Example 22.18. $\square$

Corollary 22.23. *Let R be a ring, and let $r > 0$. The A-module $H^r(\mathbb{P}_R^r, \mathscr{O}_{\mathbb{P}_R^r}(-r-1))$ is free of rank 1, and for every $d \in \mathbb{Z}$ there is a perfect pairing*

$$(*) \quad H^0(\mathbb{P}_R^r, \mathscr{O}_{\mathbb{P}_R^r}(-r-1-d)) \times H^r(\mathbb{P}_R^r, \mathscr{O}_{\mathbb{P}_R^r}(d)) \to H^r(\mathbb{P}_R^r, \mathscr{O}_{\mathbb{P}_R^r}(-r-1)) \cong R.$$

Note that $\omega := \mathscr{O}_{\mathbb{P}_R^r}(-r-1) = \Omega_{\mathbb{P}_R^r/R}^r$ as one sees by applying the determinant to the short exact sequence (17.7.3) (Remark 17.58). This line bundle is in fact the relative dualizing sheaf for $\mathbb{P}_R^r$, see Section (25.9) below. The isomorphism $R \xrightarrow{\sim} H^r(\mathbb{P}_R^r, \mathscr{O}_{\mathbb{P}_R^r}(-r-1))$ is canonical (Theorem 22.86 below) and the pairing (*) is a special case of Grothendieck duality (Proposition 25.55).

Derived categories of quasi-coherent modules

Similarly as in (F.41), we denote for every scheme X by $D_{\mathrm{qcoh}}(X)$ the full additive subcategory of $D(X)$ consisting of objects E such that $H^i(E)$ is a *quasi-coherent* $\mathscr{O}_X$-module for all i.

We now will show two central results. The first (Theorem 22.31) shows that if $f\colon X \to Y$ is a qcqs morphism of schemes, then Rf_* maps $D_{\mathrm{qcoh}}(X)$ to $D_{\mathrm{qcoh}}(Y)$. A special case is the assertion that for every quasi-coherent module $\mathscr{F}$ on X all higher direct images $R^i f_* \mathscr{F}$ are again quasi-coherent. In fact, we will prove the special case first (Theorem 22.27). Deducing Theorem 22.31 from this special case is not difficult for bounded below complexes. In the general case one has to approximate complexes by bounded below complexes. This technique is explained in Section (22.7).

The second central result (Theorem 22.35) is that for quasi-compact separated[1] schemes X the natural functor from the derived category of the abelian category of quasi-coherent $\mathscr{O}_X$-modules to $D_{\mathrm{qcoh}}(X)$ is an equivalence of categories. In particular one obtains for every ring A an equivalence $D(A) \cong D_{\mathrm{qcoh}}(\mathrm{Spec}\,A)$.

In the end we will show that this equivalence is compatible with derived tensor products and derived pullback.

(22.7) Homotopy Limits.

For the notion of homotopy limit, indexed by $\mathbb{N}$, see Definition F.228. Since products exist in $D(X)$ (Lemma F.188), all homotopy limits exist.

Definition and Lemma 22.24. *Let X be a scheme. Then $D_{\mathrm{qcoh}}(X)$ is left complete, i.e., for every $E \in D_{\mathrm{qcoh}}(X)$ the natural morphism*

$$E \xrightarrow{\sim} \mathrm{holim}\,\tau^{\geq -n} E$$

is an isomorphism in $D(X)$.

Proof. We write $K_n := \tau^{\geq -n} E$ and $K := \mathrm{holim}\,K_n$. The natural maps $E \to K_n$ induce a morphism $E \to K$ via the exact triangle defining $\mathrm{holim}\,K_n$, and we want to show that this is an isomorphism, i.e., that $H^m(E) \xrightarrow{\sim} H^m(K)$ for all m.

(I). We claim that for all m, there exists an integer $n(m) \geq -m$ such that $H^m(U, K) \xrightarrow{\sim} H^m(U, K_{n(m)})$ for all affine open subschemes $U \subseteq X$. In fact, we have distinguished triangles

$$H^{-n-1}(E)[n+1] \to K_{n+1} \to K_n \to .$$

Since E has quasi-coherent cohomology sheaves and U is affine, by Theorem 22.2 we have $H^i(U, H^n(E)) = 0$ for all n and all $i > 0$, so using the long exact cohomology sequence for this triangle, we find $n(m)$ such that for all $n \geq n(m)$, $H^m(U, K_{n+1}) \to H^m(U, K_n)$ and $H^{m-1}(U, K_{n+1}) \to H^{m-1}(U, K_n)$ are isomorphisms. Then $R^1 \lim_n H^{m-1}(R\Gamma(U, K_n)) = 0$ by Lemma F.227, and Lemma F.233 together with the isomorphisms $H^m(U, K_{n+1}) \to H^m(U, K_n)$ gives

$$H^m(\mathrm{holim}\,R\Gamma(U, K_n)) \xrightarrow{\sim} \lim H^m(R\Gamma(U, K_n)) = \lim H^m(U, K_n) \cong H^m(U, K_{n(m)}).$$

Since $H^m(\mathrm{holim}\,R\Gamma(U, K_n)) = H^m(R\Gamma(U, \mathrm{holim}\,K_n)) = H^m(U, K)$ by Lemma F.230, we obtain the desired isomorphisms. This shows our claim.

---

[1] We will see that it is possible to relax the condition to be separated somewhat.

(II). Now let $x \in X$. We claim that the map $H^m(K)_x \to H^m(K_{n(m)})_x$ on stalks is injective. To see this, let $\gamma \in \mathrm{Ker}(H^m(K)_x \to H^m(K_{n(m)})_x)$. Since $H^m(K)$ is the sheafification of the presheaf $U \mapsto H^m(U, K)$ (Lemma 21.22), we can lift γ to an element $\tilde{\gamma} \in H^m(U, K)$ for some sufficiently small affine open $U \subseteq X$. Since γ maps to 0 in $H^m(K_{n(m)})_x$, it follows that after shrinking U, if necessary, $\tilde{\gamma} \in \mathrm{Ker}(H^m(U, K) \to H^m(U, K_{n(m)}))$ which is 0 by Step (I), whence $\gamma = 0$.

(III). Now consider the maps

$$H^m(E)_x \to H^m(K)_x \to H^m(K_{n(m)})_x.$$

Since $n(m) \geq -m$, $H^m(E) \to H^m(K_{n(m)})$ is an isomorphism, so the composition of the two maps above is an isomorphism. By Step (ii), we also know that the second map is injective. It follows that $H^m(E)_x \to H^m(K)_x$ is an isomorphism for all x, whence $H^m(E) \to H^m(K)$ is an isomorphism, as desired. □

Lemma 22.25. *Let X be a scheme, let $F \colon (\mathscr{O}_X\text{-Mod}) \to (\mathrm{AbGrp})$ be an additive functor which commutes with countable products, and let $N \geq 0$ such that $R^i F(\mathscr{F}) = 0$ for all $i \geq N$ and all quasi-coherent $\mathscr{O}_X$-modules $\mathscr{F}$.*

Then for all $E \in D_{\mathrm{qcoh}}(X)$ and all $i \in \mathbb{Z}$, the natural morphism

$$R^i F(E) \to R^i F(\tau^{\geq i-N+1} E)$$

is an isomorphism.

Proof. Since the claimed statement is compatible with shifts, we may assume that $i = 0$. By Lemma 22.24, we have $E \cong \mathrm{holim}(\tau^{\geq -n} E)$. By Lemma F.230 and the following remark, the derived functor RF commutes with derived limits, so

$$RF(E) = \mathrm{holim}\, RF(\tau^{\geq -n} E).$$

As a consequence, Lemma F.233 gives us a short exact sequence

$$0 \to R^1 \lim R^{-1} F(\tau^{\geq -n} E) \to R^0 F(E) \to \lim R^0 F(\tau^{\geq -n} E) \to 0.$$

Claim. The morphism

$$R^j F(\tau^{\geq -n} E) \to R^j F(\tau^{\geq -n+1} E)$$

is an isomorphism whenever $j + n \geq N$.

Note that the claim implies the lemma. Namely, using the claim for $j = -1$ and n sufficiently large, we see that the inverse system $(R^{-1} F(\tau^{\geq -n} E))_n$ satisfies the Mittag-Leffler property, and as a consequence the $R^1 \lim$ term on the left of the above short exact sequence vanishes (Lemma F.227). Furthermore, for $j = 0$ we see that the inverse system $(R^0 F(\tau^{\geq -n} E))_n$ becomes constant at $n = N - 1$, so the term on the right of the short exact sequence can be identified with $R^0 F(\tau^{\geq -N+1} E)$, as desired.

It remains to prove the claim. To this end, we consider the exact triangle (Proposition F.157)

$$H^{-n}(E)[n] \to \tau^{\geq -n} E \to \tau^{\geq -n+1} E \to$$

relating E with its truncations. By the assumptions on F, we have

$$R^j F(H^{-n}(E)[n]) = R^{j+n} F H^{-n}(E) = 0 \text{ for } j + n \geq N,$$

so the long exact cohomology sequence for RF proves the claim. □

(22.8) Quasi-coherence of higher direct images.

We have already seen (Corollary 10.27) that the direct image $f_*\mathscr{F}$ of a quasi-coherent $\mathscr{O}_X$-module under a qcqs morphism $f\colon X \to Y$ is quasi-coherent. The same is true for the higher direct images. To prove this, we start with a lemma.

Lemma 22.26. *Let X be a qcqs scheme, and let $\mathscr{F}$ be a quasi-coherent $\mathscr{O}_X$-module. Let $t \in \Gamma(X, \mathscr{O}_X)$, $X_t = \{x \in X;\ t(x) \neq 0\}$, and $i \geq 0$. Then we have a functorial isomorphism $H^i(X, \mathscr{F})_t \to H^i(X_t, \mathscr{F})$.*

Proof. Here, as usual, we write $H^i(X_t, \mathscr{F})$ as short-hand notation for $H^i(X_t, \mathscr{F}_{|X_t})$. For $i = 0$, this lemma is a special case ($\mathscr{L} = \mathscr{O}_X$) of Theorem 7.22 (and the above lemma also admits a corresponding generalization, see [EGAIII] $^\circ$ (1.4.5) or Exercise 22.7).

We start with the following remark: Let R be a ring, $t \in R$, and M an R-module. Then the localization M_t is naturally isomorphic to the colimit $\operatorname{colim} M$ of the system $M \to M \to M \to \cdots$ whose transition maps are all given by multiplication by t. For example, we can view $\Gamma(X_t, \mathscr{O}_X) = \Gamma(X, \mathscr{O}_X)_t$ as $\operatorname{colim}_{t \cdot} \Gamma(X, \mathscr{O}_X)$.

Passing to the context of sheaves, let $j\colon X_t \to X$ be the inclusion, and let $\mathscr{F}$ be a quasi-coherent sheaf on X. Then j is an open affine immersion. We claim that $j_*j^*\mathscr{F} = \operatorname{colim}_{t \cdot} \mathscr{F}$. In fact, both sides are quasi-coherent $\mathscr{O}_X$-modules, and the sections over some open affine subset $U \subseteq X$ are $H^0(U_{t_{|U}}, \mathscr{F})$ and $\operatorname{colim}_{t \cdot} H^0(U, \mathscr{F})$ (in the second case, use Lemma 21.52), and the claim follows from the above remark and the case $i = 0$ of the lemma, which we know is true.

Combining this with Corollary 22.6, applied to the affine morphism j, we see that

$$H^i(X_t, \mathscr{F}) = H^i(X, j_*j^*\mathscr{F}) = H^i(X, \operatorname{colim}_{t \cdot} \mathscr{F}).$$

Since cohomology commutes with inductive limits (Proposition 21.54), we also have

$$H^i(X, \operatorname{colim}_{t \cdot} \mathscr{F}) = \operatorname{colim}_{t \cdot} H^i(X, \mathscr{F}) = H^i(X, \mathscr{F})_t.$$

Note that all maps we considered are $\Gamma(X, \mathscr{O}_X)$-module homomorphisms. Because both $H^i(X_t, \mathscr{F})$ and $H^i(X, \mathscr{F})_t$ are modules over $\Gamma(X, \mathscr{O}_X)_t = \Gamma(X_t, \mathscr{O}_X)$, the resulting identification is actually an isomorphism of $\Gamma(X, \mathscr{O}_X)_t$-modules. $\square$

Theorem 22.27. *Let $f\colon X \to Y$ be a qcqs morphism of schemes, and let $\mathscr{F}$ be a quasi-coherent $\mathscr{O}_X$-module. Then for all $i \geq 0$, the $\mathscr{O}_Y$-module $R^if_*\mathscr{F}$ is quasi-coherent. If Y is affine, then*

$$R^if_*\mathscr{F} = H^i(X, \mathscr{F})^\sim.$$

Proof. Since taking higher direct images is local on Y (Lemma 21.28), it is enough to prove that $R^if_*\mathscr{F} = H^i(X, \mathscr{F})^\sim$ provided that $Y = \operatorname{Spec} A$ is affine. As in Remark 21.20, we denote by $\mathscr{H}^i(\mathscr{F})$ the presheaf $U \mapsto H^i(U, \mathscr{F}_{|U})$ on X. Recall (Proposition 21.27) that $R^if_*\mathscr{F}$ equals the sheafification of $f_*\mathscr{H}^i(\mathscr{F})$.

Now let $s \in A$. We obtain

$$\Gamma(D(s), f_*\mathscr{H}^i(\mathscr{F})) = H^i(X_{\varphi(s)}, \mathscr{F}) = H^i(X, \mathscr{F})_s = \Gamma(D(s), H^i(X, \mathscr{F})^\sim),$$

where $\varphi\colon A \to \Gamma(X, \mathscr{O}_X)$ denotes the ring homomorphism given by f, where one defines $X_{\varphi(s)} = \{x \in X;\ \varphi(s)(x) \neq 0\} = f^{-1}(D(s))$, and the second equality follows from Lemma 22.26. These equalities are compatible with restriction maps with respect to inclusions $D(t) \subseteq D(s)$. We see that the presheaf $f_* \mathscr{H}^i(\mathscr{F})$ equals the sheaf $H^i(X, \mathscr{F})^\sim$ on the basis $(D(s))_{s \in A}$ of open subsets of Y, and hence its sheafification $R^i f_* \mathscr{F}$ is $H^i(X, \mathscr{F})^\sim$. $\square$

For a different proof for Y affine and X noetherian, see Exercise 22.18.

Lemma 22.28. *Let Y be a quasi-compact scheme, and let $f\colon X \to Y$ be a qcqs morphism. There exists N such that $R^i f_* \mathscr{F} = 0$ for all $i \geq N$ and all quasi-coherent $\mathscr{O}_X$-modules $\mathscr{F}$.*

If Y is affine and f is separated, one can take N to be the minimal number r such that X can be covered by r open affine subschemes.

Proof. We may assume that Y is affine. By Theorem 22.27 it is enough to show that there exists N with $H^i(X, \mathscr{F}) = 0$ for all $i \geq N$ and all quasi-coherent $\mathscr{F}$. If X is separated, then this follows from Corollary 22.10.

To deal with the general case, we can use the spectral sequence from Čech to usual cohomology, Theorem 21.78, for a covering $\mathscr{U}$ of X by finitely many affine open subschemes,

$$E_2^{p,q} = \check{H}^p(\mathscr{U}, \mathscr{H}^q(\mathscr{F})) \Rightarrow H^{p+q}(X, \mathscr{F}).$$

By Proposition 21.69 we can compute the $E_2^{p,q}$-terms using the alternating Čech complex, hence $E_2^{p,q} = 0$ whenever p is sufficiently large, independent of $\mathscr{F}$. On the other hand, all intersections of subsets in $\mathscr{U}$ are separated, so that we know, by the remark in the beginning of the proof, that the lemma applies to them. Therefore the (alternating) Čech complex for $\mathscr{U}$ and $\mathscr{H}^q(\mathscr{F})$ vanishes whenever q is sufficiently large. Altogether we see that $E_2^{p,q} = 0$ for $p + q$ sufficiently large and the desired result follows. $\square$

Proposition 22.29. *Let $f\colon X \to S$ be a qcqs morphism between quasi-compact schemes. There exists an integer $N \geq 0$ (depending on f) such that for all integers $b \in \mathbb{Z}$ and for all $E \in D_{\mathrm{qcoh}}(X)$ with $H^j(E) = 0$ for $j > b$ we have $R^i f_* E = 0$ for all $i \geq N + b$.*

The proof will show that one can take the same N as in Lemma 22.28.

Proof. By shifting the degree of E it suffices to show the existence of N for $b = 0$. Let N be as in Lemma 22.28.

First assume that E is bounded below. We then have a spectral sequence (Corollary 21.46)

$$E_2^{ij} = R^i f_* H^j(E) \Longrightarrow R^{i+j} f_* E.$$

Since the $H^j(E)$ are quasi-coherent, $R^i f_* H^j(E) = 0$ whenever $i \geq N$ by Lemma 22.28. By assumption we have $H^j(E) = 0$ whenever $j > 0$. Therefore the spectral sequence can yield a non-zero term only for $i + j < N$. This proves the proposition for bounded below complexes.

To prove the general case, it is enough to show that $H^i(U, Rf_* E) = 0$ for all $i \geq N$ and $U \subseteq S$ affine: Indeed, by Lemma 21.22, this implies that the stalks $H^i(Rf_* E)_x$ all vanish, whence $R^i f_* E = 0$, as desired. We want to put ourselves into a situation where we can apply Lemma 22.25 to the functor $F = \Gamma(f^{-1}(U), -)$. First note that F commutes with arbitrary products.

For $i \geq N$ and $\mathscr{F}$ a quasi-coherent $\mathscr{O}_X$-module, we have (using Theorem 22.27 and Lemma 21.28)

$$H^i(f^{-1}(U), \mathscr{F}) = H^0(U, (R^i f_* \mathscr{F})_{|U}) = 0 \quad \text{for } i \geq N.$$

We can identify $R\Gamma(U, Rf_* E) = R\Gamma(f^{-1}(U), E)$ by Proposition 21.114 and in particular $H^j(U, Rf_* E) = H^j(f^{-1}(U), E)$ for all j. We thus get from Lemma 22.25, applied to the functor $F = \Gamma(f^{-1}(U), -)$, for all $i \geq N$ that

$$\begin{aligned} H^i(U, Rf_* E) = H^i(f^{-1}(U), E) &= H^i(f^{-1}(U), \tau^{\geq i - N + 1} E) \\ &= H^i(U, Rf_*(\tau^{\geq i - N + 1} E)) = 0 \end{aligned}$$

as we have already proved the proposition for bounded below complexes. As this holds for all open affine subschemes $U \subseteq S$, this show that $H^i(Rf_* E) = R^i f_* E = 0$ for $i \geq N$ by Lemma 21.22. □

Corollary 22.30. *Let $f \colon X \to S$ be a qcqs morphism between quasi-compact schemes. Then there exists $N \geq 0$ such that for all $E \in D_{\mathrm{qcoh}}(X)$ and all $i \in \mathbb{Z}$, the natural morphism*

$$R^i f_* E \to R^i f_*(\tau^{\geq i - N + 1} E)$$

is an isomorphism.

Proof. Let N be as in Proposition 22.29, and consider the exact triangle (see Proposition F.157)

$$\tau^{\leq -N} E \to E \to \tau^{\geq -N + 1} E \to$$

By Proposition 22.29, the cohomology of $Rf_* \tau^{\leq -N} E$ vanishes in degrees $\geq -N + N = 0$, hence the long exact cohomology sequence gives that $R^i f_* E$ coincides with $R^i f_* \tau^{\geq -N + 1} E$ for all $i \geq 0$. Applying this to all shifts of E gives the result. □

Theorem 22.31. *Let $f \colon X \to Y$ be a qcqs morphism of schemes. Then Rf_* induces functors*

$$\begin{aligned} Rf_* &\colon D_{\mathrm{qcoh}}(X) \to D_{\mathrm{qcoh}}(Y), \\ Rf_* &\colon D_{\mathrm{qcoh}}^{\geq n}(X) \to D_{\mathrm{qcoh}}^{\geq n}(Y) \qquad \text{for each } n \in \mathbb{Z}. \end{aligned}$$

If Y is quasi-compact, then there exists an $N \geq 0$ such that Rf_* induces a functor

$$Rf_* \colon D_{\mathrm{qcoh}}^{[a,b]}(X) \to D_{\mathrm{qcoh}}^{[a,b+N]}(Y) \qquad \text{for all integers } a \leq b.$$

Proof. For E bounded below, we use the spectral sequence

$$R^i f_* H^j(E) \Longrightarrow R^{i+j} f_* E$$

as in the proof of Proposition 22.29, and apply Theorem 22.27 together with the fact that extensions of quasi-coherent $\mathscr{O}_Y$-modules are quasi-coherent (Corollary 22.4).

For the general case, we use Corollary 22.30 for the functor f_* and see that $R^i f_* E$ is isomorphic to a higher direct image of some bounded below complex, hence quasi-coherent by the above. □

(22.9) The category of quasi-coherent $\mathscr{O}_X$-modules and its derived category.

Let X be a scheme. Denote by $(\mathrm{QCoh}(X))$ the abelian category of quasi-coherent $\mathscr{O}_X$-modules. It is a plump (Definition F.43) abelian subcategory of $(\mathscr{O}_X\text{-Mod})$ (Corollary 22.4) which is stable under tensor products.

If $X = \mathrm{Spec}\, A$ is affine, then $(\mathrm{QCoh}(X))$ is equivalent to the category of A-modules. By reduction to the affine case, it follows that the category $(\mathrm{QCoh}(X))$ has finite limits, and all colimits, and the inclusion $\iota\colon \mathrm{QCoh}(X) \to (\mathscr{O}_X\text{-Mod})$ preserves finite limits and all colimits, for an arbitrary scheme X.

Returning to the affine case $X = \mathrm{Spec}\, A$ for a moment, the equivalence between $(\mathrm{QCoh}(X))$ and the category of A-modules also gives us that this category has a generator (namely, $\mathscr{O}_X$), enough injectives, and all limits. In order to generalize this, we look at this from the following point of view: Every quasi-coherent $\mathscr{O}_X$-module on X has the form $M^\sim$ for some A-module M. Let $\mathscr{G}$ be an arbitrary $\mathscr{O}_X$-module. It is easy to check that

$$\mathrm{Hom}_{(\mathscr{O}_X\text{-Mod})}(M^\sim, \mathscr{G}) \cong \mathrm{Hom}_A(M, \Gamma(X, \mathscr{G})) = \mathrm{Hom}_{\mathrm{QCoh}(X)}(M^\sim, \Gamma(X, \mathscr{G})^\sim),$$

functorially in M and $\mathscr{G}$, i.e., the functor $Q\colon \mathscr{G} \mapsto \Gamma(X, \mathscr{G})^\sim$ is the right adjoint functor of the inclusion ι.

Since Q is a right adjoint functor, it preserves all limits. This implies that the category $\mathrm{QCoh}(X)$ has all limits — just take the limit in $(\mathscr{O}_X\text{-Mod})$ and apply Q. As ι is exact, Q sends injectives in $(\mathscr{O}_X\text{-Mod})$ to injectives in $\mathrm{QCoh}(X)$. We obtain that $\mathrm{QCoh}(X)$ has enough injectives.

As the following theorem states, the existence of a right adjoint to ι generalizes to arbitrary schemes.

Theorem and Definition 22.32. *Let X be a scheme.*
(1) *The category $\mathrm{QCoh}(X)$ has all limits. It is a Grothendieck abelian category (Def. F.54) and in particular has K-injective resolutions (Theorem F.185).*
(2) *The inclusion $\iota\colon \mathrm{QCoh}(X) \to (\mathscr{O}_X\text{-Mod})$ admits a right adjoint functor Q, called the coherator. The adjunction morphism $\mathrm{id} \to Q \circ \iota$ is an isomorphism.*

Proof. We give the proof of (1) only in the case that X is qcqs. (See below for remarks and references concerning the general case.) The key point is to show the existence of a generator. This can be seen as a set-theoretic problem: In fact, if we could form the direct sum of all objects in the category, it would obviously have the property of a generator. Now if X is qcqs, every quasi-coherent sheaf on X is a filtered colimit of finitely presented $\mathscr{O}_X$-modules by Corollary 10.50. The family of isomorphism classes of all finitely presented $\mathscr{O}_X$-modules is a set, and hence the direct sum of a set of representatives of these isomorphism classes exists and is a generator. We have already noted that all colimits exist in $\mathrm{QCoh}(X)$. Furthermore, filtered colimits are exact, since the stalk of a colimit is equal to the colimit of the stalks at the given point. Every Grothendieck abelian category has enough K-injective complexes, see Theorem F.185.

For the general case, one needs to prove a replacement for the above result that quasi-coherent $\mathscr{O}_X$-modules for X qcqs are filtered colimits of finitely presented modules, where roughly speaking "finite" is replaced by "less than some suitable cardinal κ". See [EnEs] Corollary 3.5 (Gabber's Lemma).

For (2), the existence of an adjoint functor follows from an appropriate version of the adjoint functor theorem (Corollary F.57). $\square$

In terms of derived categories, we can now restate the above result on vanishing of higher direct images of quasi-coherent modules under affine morphisms as follows.

Proposition 22.33. *Let* $f\colon X \to Y$ *be an affine morphism. The push-forward functor* $f_*\colon (\mathrm{QCoh}(X)) \to (\mathrm{QCoh}(Y))$ *is exact and hence it "is" its own derived functor* $f_*\colon D(\mathrm{QCoh}(X)) \to D(\mathrm{QCoh}(Y))$. *We obtain a commutative diagram*

$$
\begin{array}{ccc}
D(\mathrm{QCoh}(X)) & \xrightarrow{\ \iota_X\ } & D(X) \\
\downarrow{\scriptstyle f_*} & & \downarrow{\scriptstyle Rf_*} \\
D(\mathrm{QCoh}(Y)) & \xrightarrow{\ \iota_Y\ } & D(Y),
\end{array}
$$

where the horizontal arrows are the natural inclusion functors.

This property in fact characterizes affine morphisms among qcqs morphisms, see Exercise 22.11.

Proof. We know by Corollary 22.5 that f_* is exact on the category of quasi-coherent $\mathscr{O}_X$-modules. This means that we can compute its derived functor by applying f_* to all terms of any complex representing the object of the derived category at hand. (This is what we mean when we say that f_* is its own derived functor.)

To prove the commutativity of the diagram, we will show that for all complexes $\mathscr{F}$ of quasi-coherent $\mathscr{O}_X$-modules the natural morphism $f_*\mathscr{F} \to Rf_*\mathscr{F}$ is a quasi-isomorphism, hence an isomorphism in $D(Y)$. (Note that this is a direct consequence of Corollary 22.5 in case $\mathscr{F}$ is concentrated in degree 0.) Since this morphism is compatible with shifts, it is enough to show that for all complexes $\mathscr{F}$ of quasi-coherent $\mathscr{O}_X$-modules, the induced morphism

$$
H^0(f_*\mathscr{F}) \to R^0 f_*\mathscr{F}
$$

is an isomorphism.

By Corollary 22.30 we can reduce to the case that $\mathscr{F}$ is bounded below. Then we can apply Proposition F.204 which says that a bounded below complex consisting of objects which are all acyclic for some functor is itself acyclic for that functor. Since higher direct images of quasi-coherent sheaves under f vanish, we conclude that $\mathscr{F}$ is f_*-acyclic, or equivalently, that the natural morphism $f_*\mathscr{F} \to Rf_*\mathscr{F}$ is a quasi-isomorphism (see Corollary F.174). In particular, $H^0(f_*\mathscr{F}) \to R^0 f_*\mathscr{F}$ is an isomorphism, as desired. $\square$

Proposition 22.34. *Let* $f\colon X \to Y$ *be a flat qcqs morphism of schemes. Let* RQ_X *be the right derived functor of the coherator* $\mathrm{QCoh}(X) \to (\mathscr{O}_X\text{-}\mathrm{Mod})$ *(see Theorem 22.32), and correspondingly for* Y. *Then we have a commutative diagram (up to natural isomorphism)*

$$
\begin{array}{ccc}
D(\mathrm{QCoh}(X)) & \xleftarrow{\ \ RQ_X\ \ } & D(X) \\
\downarrow{\scriptstyle R_{\mathrm{QCoh}(X)}f_*} & & \downarrow{\scriptstyle Rf_*} \\
D(\mathrm{QCoh}(Y)) & \xleftarrow[\ RQ_Y\ ]{} & D(Y),
\end{array}
$$

where $R_{\mathrm{QCoh}(X)}f_*$ *denotes the right derived functor of the direct image functor* $\mathrm{QCoh}(X) \to \mathrm{QCoh}(Y)$.

Proof. Since f is qcqs, it gives rise to a functor $f_*\colon \mathrm{QCoh}(X) \to \mathrm{QCoh}(Y)$. Since $\mathrm{QCoh}(X)$ is a Grothendieck abelian category (Theorem 22.32), the right derived functor exists (Corollary F.187).

Now consider the inverse image functor f^*. Since f is flat by assumption, f^* is exact and is hence equal to its own derived functor $f^* = Lf^*$, regardless of whether we consider f on all of $(\mathscr{O}_X\text{-Mod})$, or as a functor $\mathrm{QCoh}(Y) \to \mathrm{QCoh}(X)$. Denoting by $\iota_X\colon D\mathrm{QCoh}(X) \to D(X)$ the inclusion functor, and likewise for Y, we have $f^*\iota_Y(F) = \iota_X(f^*F)$.

Using this, and adjunction between Rf_* and Lf^* (Theorem 21.126), one easily checks that both compositions $D(X) \to D\mathrm{QCoh}(Y)$ in the above diagram are right adjoint to the functor $D\mathrm{QCoh}(Y) \to D(X)$, $F \mapsto f^*\iota_Y(F) = \iota_X(f^*F)$. $\square$

(22.10) Comparison of $D_{\mathrm{qcoh}}(X)$ and $D(\mathbf{QCoh}(X))$.

Recall that we denote by $D_{\mathrm{qcoh}}(X)$ the full additive subcategory of $D(X)$ consisting of objects E such that $H^i(E)$ is a *quasi-coherent* $\mathscr{O}_X$-module for all i. The goal of this section is the proof of the following theorem:

Theorem 22.35. *Let X be a quasi-compact and semiseparated (Definition 22.12) scheme. The natural functor $\iota_X\colon D(\mathrm{QCoh}(X)) \to D_{\mathrm{qcoh}}(X)$ is an equivalence of categories.*

The proof will proceed in two steps. First we will consider the case where $X = \mathrm{Spec}\,R$ is affine. For most applications of Theorem 22.35 in the sequel, this special case will be sufficient. In the second step we will then globalize from affine schemes to quasi-compact semiseparated schemes using the right derived functor of the coherator (Definition 22.32).

The first step is the following lemma. For $X = \mathrm{Spec}\,R$, we identify the categories $(R\text{-Mod})$ and $(\mathrm{QCoh}(X))$ via the usual construction $M \mapsto \tilde{M}$, and as a consequence, we get an identification of derived categories $D(R) = D(\mathrm{QCoh}(X))$.

Lemma 22.36. *Let $X = \mathrm{Spec}\,R$ be an affine scheme. The functor*

$$\iota_X\colon D(R) = D(\mathrm{QCoh}(X)) \to D_{\mathrm{qcoh}}(X)$$

is an equivalence of categories with quasi-inverse functor $R\Gamma(X, -)$.

Proof. For every $\mathscr{O}_X$-module E, $\Gamma(X, E)$ is an R-module, and we denote by Ψ the restriction of the derived functor of $\Gamma(X, -)$ to $D_{\mathrm{qcoh}}(X)$:

$$\Psi = R\Gamma(X, -)_{|D_{\mathrm{qcoh}}(X)}\colon D_{\mathrm{qcoh}}(X) \longrightarrow D(R) = D(\mathrm{QCoh}(X)).$$

We have to show that ι_X and Ψ are quasi-inverse to each other.

Let us first show that the functor ι_X is left adjoint to Ψ. Write Y for the one-point ringed space $(\{\mathrm{pt}\}, R)$. Then $(\mathscr{O}_Y\text{-Mod}) = (R\text{-Mod})$ in an obvious way, and $D(Y) = D(R\text{-Mod})$. We have an obvious morphism $f\colon X \to Y$ of ringed spaces. Since $f_* = \Gamma(X, -)$, we get $\Psi(E) = Rf_*(E)$ for all $E \in D_{\mathrm{qcoh}}(X)$. Furthermore, for an R-module M, we have $\tilde{M} = f^*M$, so we have $\iota_X = f^* = Lf^*$, the latter equality following from the fact that the functor $\tilde{\ }$ is exact. Now Proposition 21.126 implies that ι_X is left adjoint to Ψ.

This adjunction gives rise to natural maps

$$\widetilde{R\Gamma(X, E)} = \iota_X(\Psi(E)) \to E, \quad E \in D_{\mathrm{qcoh}}(X),$$

and
$$M \to \Psi(\iota_X(M)) = R\Gamma(X, \widetilde{M}), \quad M \in D(R\text{-Mod}).$$

Here we denote the derived functor of $\widetilde{\ }$ by the same symbol. Since the original functor is exact, it just means applying $\widetilde{\ }$ to each term of any representative.

To conclude the proof, we will show that both these adjunction morphisms are isomorphisms (in the respective derived categories), for all E and M. In other words, choosing complexes representing E and M, respectively, and considering the above morphisms as morphisms of complexes, we have to show that they are quasi-isomorphisms. Since they are compatible with shifts, it is enough to prove that applying $H^0(-)$ yields isomorphisms.

By Lemma 22.25 applied to the functor $\Gamma(X, -)$ and $N = 1$ (Theorem 22.2), we have $R^0\Gamma(X, E) = R^0\Gamma(X, \tau^{\geq 0}E)$ (note that this is immediate if E is concentrated in a single degree; the lemma requires quite some work in order to obtain the result for unbounded complexes). Since $\tau^{\geq 0}E$ is bounded below, the spectral sequence in Corollary 21.46 then shows that $R^0\Gamma(X, E) = \Gamma(X, H^0(E))$.

Using this together with the exactness of $\widetilde{\ }$ and the assumption that $H^0(E)$ is quasi-coherent, we can compute
$$H^0(\widetilde{R\Gamma(X, E)}) = \widetilde{R^0\Gamma(X, E)} = \widetilde{\Gamma(X, H^0(E))} = H^0(E).$$

Similarly, we get
$$H^0(M) = \Gamma(X, \widetilde{H^0(M)}) = \Gamma(X, H^0(\widetilde{M})) = R^0\Gamma(X, \widetilde{M}) = H^0(R\Gamma(X, \widetilde{M})).$$

One checks that these identifications coincide with the adjunction maps. $\qquad\square$

Recall the coherator $Q\colon (\mathscr{O}_X\text{-Mod}) \to \mathrm{QCoh}(X)$, the right adjoint of the inclusion functor $\mathrm{QCoh}(X) \to (\mathscr{O}_X\text{-Mod})$. As a right adjoint, Q is left exact, and passing to its derived functor, we obtain a functor $RQ\colon D(X) \to D\mathrm{QCoh}(X)$, which is right adjoint to the inclusion $D\mathrm{QCoh}(X) \to D(X)$ (Proposition F.191), and a fortiori its restriction to $D_{\mathrm{qcoh}}(X)$ is right adjoint to the functor $\iota_X\colon D\mathrm{QCoh}(X) \to D_{\mathrm{qcoh}}(X)$. The previous lemma gives us an explicit description of Q (on $D_{\mathrm{qcoh}}(X)$) in case X is affine.

Proof. [Proof of Theorem 22.35] It remains to extend the result to the general case. We again consider the derived version $RQ = RQ_X\colon D_{\mathrm{qcoh}}(X) \to D\mathrm{QCoh}(X)$ of the coherator, and will show that the natural morphisms obtained from the adjunction,
$$\iota_X(RQ_X(E)) \to E \qquad \text{and} \qquad F \to RQ_X(\iota_X(F)),$$
are isomorphisms for all $E \in D_{\mathrm{qcoh}}(X)$, $F \in D\mathrm{QCoh}(X)$. This implies that ι_X is an equivalence with inverse RQ_X.

The proofs for the two maps proceed in a similar way: In both cases, we do induction on the minimum number n such that E (or F, resp.) is supported on a closed subscheme of X which can be covered by n affine open subschemes. Here we say that an object of a derived category is supported on a closed subscheme if this is true for all its cohomology objects.

Let us first consider the map $\iota_X(RQ_X(E)) \to E$. In case $n = 0$, the object E is supported on the empty subscheme, and is hence $= 0$, so that the claim becomes trivial. To proceed, assume that we can write $X = U \cup U'$ where $U \subseteq X$ is open and affine, and $U' \subseteq X$ is open and can be written as the union of $n-1$ open affines, so that the induction hypothesis applies to U'. Denote by j the inclusion of U into X. Let $i\colon Y := X \setminus U \hookrightarrow X$ be the closed embedding of the complement of U, endowed with the reduced scheme structure, into X. Consider the local cohomology exact triangle (Proposition 21.62)

$$i_*Ri^!E \to E \to Rj_*j^*E \to i_*Ri^!E[1].$$

The morphism j is affine, since X is semiseparated, therefore $Rj_*j^*E \in D_{\mathrm{qcoh}}(X)$. Thus two of the three terms are in $D_{\mathrm{qcoh}}(X)$, and it follows that the term $E' := i_*Ri^!E$ on the left is too. Moreover, E' is supported on $Y \subseteq U'$, so the induction hypothesis ensures that the map $\iota_X(RQ_X(E')) \to E'$ is an isomorphism. (In fact, we do not need the exact description of E' in terms of $Ri^!$ — it is enough to see that the morphism $E \to Rj_*j^*E$ becomes an isomorphism when restricted to U, whence E' is supported on Y.) Let us show that for $E'' := Rj_*j^*E$, the morphism $\iota_X(RQ_X(E'')) \to E''$ is an isomorphism, as well. This implies that for E we also get an isomorphism, as desired.

In fact, we can write this map as the composition of a chain of isomorphisms,

$$\iota_X(RQ_X(E'')) = \iota_X RQ_X(Rj_*j^*E) \cong \iota_X j_* RQ_U j^*E \cong Rj_* \iota_U RQ_U j^*E \cong Rj_*j^*E = E'',$$

where we have used Proposition 22.34 for the first, Proposition 22.33 for the second, and the affine case of the theorem for the third isomorphisms (not counting the equalities in the beginning and the end).

Now let us consider the morphisms $F \to RQ_X(\iota_X(F))$ for $F \in DQ\mathrm{Coh}(X)$. As before, we define n as the minimum number such that there exists a closed subscheme in X which can be covered by n affine open subschemes and such that all cohomology objects of F are supported on this closed subscheme. If $n = 0$, then $F = 0$, and the above morphism is an isomorphism for trivial reasons.

For $n > 0$, writing U, Y, U', etc., as before, we consider a similar exact triangle as in the first case, now in $DQ\mathrm{Coh}(()X)$:

$$F' \to F \to j_*j^*F \to F'[1].$$

Since j is affine, we are in the situation of Proposition 22.33. When we restrict the morphism $F \to j_*j^*F$ to U, we obtain an isomorphism. This shows that F' is supported on $Y \subseteq U'$, so the induction hypothesis applies to F'. It remains to prove that for $F'' := j_*j^*F$ we also obtain an isomorphism $F'' \to RQ_X(\iota_X(F''))$. We have

$$F'' = j_*j^*F \cong j_* RQ_U \iota_U j^*F \cong RQ_X Rj_* \iota_U j^*F \cong RQ_X \iota_X j_*j^*F = RQ_X(\iota_X(F''))$$

using the affine case of the theorem for the first, Proposition 22.34 for the second, and Proposition 22.33 for the third isomorphism.   □

Corollary 22.37. *Let X, Y be quasi-compact semiseparated schemes, and let $f\colon X \to Y$ be a qcqs morphism. Let $R_{Q\mathrm{Coh}(X)}f_*\colon D(Q\mathrm{Coh}(X)) \to D(Q\mathrm{Coh}(Y))$ denote the derived functor of $f_*\colon Q\mathrm{Coh}(X) \to Q\mathrm{Coh}(Y)$. Then the diagram*

$$
\begin{array}{ccc}
D(Q\mathrm{Coh}(X)) & \xrightarrow{\ \iota_X\ } & D(X) \\
{\scriptstyle R_{Q\mathrm{Coh}(X)}f_*}\Big\downarrow & & \Big\downarrow{\scriptstyle Rf_*} \\
D(Q\mathrm{Coh}(Y)) & \xrightarrow{\ \iota_Y\ } & D(Y)
\end{array}
$$

is commutative (up to natural isomorphism).

Proof. We have a canonical map $R_{Q\mathrm{Coh}(X)}f_*E \to Rf_*E$ for $E \in D(Q\mathrm{Coh}(X))$, and we need to check that this is an isomorphism.

We first consider the case that X is affine. Since Y has affine diagonal, it follows that the morphism f is affine, and Proposition 22.33 yields the desired diagram. Now let X be general and suppose $j\colon U \to X$ is the inclusion of an affine open subscheme. We obtain a diagram

$$
\begin{array}{ccc}
D(\mathrm{QCoh}(U)) & \xrightarrow{\ \iota_U\ } & D(U) \\
{\scriptstyle j_*}\downarrow & & \downarrow{\scriptstyle Rj_*} \\
D(\mathrm{QCoh}(X)) & \xrightarrow{\ \iota_X\ } & D(X) \\
{\scriptstyle R_{\mathrm{QCoh}(X)}f_*}\downarrow & & \downarrow{\scriptstyle Rf_*} \\
D(\mathrm{QCoh}(Y)) & \xrightarrow{\ \iota_Y\ } & D(Y)
\end{array}
$$

where the upper square and the full rectangle are commutative by the affine case. By Proposition 21.115, the composition $Rf_* \circ Rj_*$ can be identified with $R(f \circ j)_*$. Furthermore, the composition $R_{\mathrm{QCoh}(X)}f_* \circ j_*$ can be identified with $(f \circ j)_*$ (again, in the sense of Proposition 22.33). In fact, to see this, it is enough to see that j_* preserves the property of being K-injective (Proposition F.176). This follows from the fact that it admits an exact left adjoint, the inverse image j^*, see Corollary F.183.

From this diagram we obtain the claim for all objects in $D\mathrm{QCoh}(X)$ of the form j_*F for some j and $F \in D\mathrm{QCoh}(U)$.

Next, let $E \in D\mathrm{QCoh}(X) \cong D_{\mathrm{qcoh}}(X)$ be supported on a closed subscheme of X which can be covered by n affine open subschemes $U_1, \dots U_n$. We may assume $n \geq 1$ and write $U := U_1$, and $j\colon U \to X$, $i\colon X \setminus U \to X$ for the inclusions. By induction, we may assume that the result holds for all E which are supported on a closed subscheme of X which can be covered by $n-1$ open subschemes, such as $X \setminus U$. Consider the exact triangle (Proposition 21.62)

$$
i_* Ri^! E \to E \to j_* j^* E \to
$$

By the above, and by induction, the claim holds for the left and right hand terms of the triangle, and it follows that it holds for the middle term E as well. $\qquad\square$

Remark 22.38.
(1) If X is a noetherian scheme, then X is qcqs but not necessarily semiseparated (Remark 22.12). Hence Theorem 22.35 does not apply to arbitrary noetherian schemes. Nevertheless, one can show that for X noetherian the natural functor $D(\mathrm{QCoh}(X)) \to D_{\mathrm{qcoh}}(X)$ is an equivalence of categories whose quasi-inverse is given by the derived coherator functor DQ_X (Exercise 22.20).
(2) On the other hand, an example by Verdier (see [SGA6][0] II, App. I) shows that for an arbitrary qcqs scheme X, the functor $D(\mathrm{QCoh}(X)) \to D_{\mathrm{qcoh}}(X)$ need not be fully faithful.

(22.11) Derived tensor product and pullback of quasi-coherent complexes.

Let $X = \mathrm{Spec}\, A$ be an affine scheme. The functor $M \mapsto \tilde{M}$ from A-modules to $\mathscr{O}_X$-modules is exact, commutes with tensor products, and with direct sums (Corollary 7.19). Therefore for all complexes E and F of A-modules for the tensor complexes we find that

(22.11.1) $$(E \otimes_A F)^{\sim} = \tilde{E} \otimes_{\mathscr{O}_X} \tilde{F}.$$

Lemma 22.39. *Let $X = \operatorname{Spec} A$ be an affine scheme and let E be a complex of A-modules. Then E is K-flat if and only if $\tilde{E}$ is a K-flat complex of $\mathscr{O}_X$-modules.*

Proof. If $\tilde{E}$ is a K-flat complex of $\mathscr{O}_X$-modules, then $\tilde{E} \otimes_{\mathscr{O}_X} \mathscr{F}$ is exact for every exact complex $\mathscr{F}$ of $\mathscr{O}_X$-modules, in particular for those that are of the form $\tilde{F}$ for an exact complex F of A-modules. Therefore E is K-flat by (22.11.1).

Conversely, if E is K-flat, then $\tilde{E}_x = E_{\mathfrak{p}_x}$ is a K-flat complex of $\mathscr{O}_{X,x}$-modules for all $x \in X$ (Remark 21.109 (1)). Hence $\tilde{E}$ is K-flat by Remark 21.92 (4). $\qquad\square$

Proposition 22.40. *Let X be a scheme.*
(1) *Let $\mathscr{E}$ and $\mathscr{F}$ be complexes in $D_{\mathrm{qcoh}}(X)$. Then $\mathscr{E} \otimes_{\mathscr{O}_X}^L \mathscr{F}$ is in $D_{\mathrm{qcoh}}(X)$.*
(2) *Let $f \colon X' \to X$ be a morphism of schemes and let $\mathscr{E}$ be a complex in $D_{\mathrm{qcoh}}(X)$. Then $Lf^*\mathscr{E}$ is in $D_{\mathrm{qcoh}}(X')$.*

Proof. The assertions are local on X and the formation of $\otimes^L$ and of Lf^* is compatible with restriction to open subsets. Hence we may assume that $X = \operatorname{Spec} A$ and X' are affine. Then $\mathscr{E} = \tilde{E}$ and $\mathscr{F} = \tilde{F}$ for complexes E and F of A-modules (Lemma 22.36). Hence it suffices to show the following lemma. $\qquad\square$

Lemma 22.41. *Let A be a ring, $X = \operatorname{Spec} A$.*
(1) *Let E und F be complexes of A-modules. Then*

$$(E \otimes_A^L F)^{\sim} = \tilde{E} \otimes_{\mathscr{O}_X}^L \tilde{F}.$$

(2) *Let E be a complex of A-modules and let $A \to B$ be a ring homomorphism corresponding to a morphism of schemes $f \colon \operatorname{Spec} B \to \operatorname{Spec} A$. Then*

$$(E \otimes_A^L B)^{\sim} = Lf^*\tilde{E},$$

where $(E \otimes_A^L B)^{\sim}$ denotes the complex of quasi-coherent modules on $\operatorname{Spec} B$ corresponding to the complex $E \otimes_A^L B$ in $D(B)$.

Proof. Let $P \to E$ be a K-flat resolution of E. Then $\tilde{P} \to \tilde{E}$ is a quasi-isomorphism with $\tilde{P}$ K-flat by Lemma 22.39. Therefore

$$(E \otimes_A^L F)^{\sim} = (P \otimes_A F)^{\sim} = \tilde{P} \otimes_{\mathscr{O}_X} \tilde{F} = \tilde{E} \otimes_{\mathscr{O}_X}^L \tilde{F}.$$

The second assertion is proved in the same way. $\qquad\square$

(22.12) Derived categories of coherent modules on noetherian schemes.

Let X be a locally noetherian scheme and let $(\mathrm{Coh}(X))$ be the category of coherent $\mathscr{O}_X$-modules, which is a plumb subcategory (Definition F.43) of the category of $\mathscr{O}_X$-modules (Proposition 7.47), in particular, it is an abelian category. We denote by $D_{\mathrm{coh}}(X)$ the full subcategory of complexes $\mathscr{F}$ in $D(X)$ such that $H^p(\mathscr{F})$ is coherent for all $p \in \mathbb{Z}$. It is a triangulated subcategory (Section (F.41)). As usual, we set $D_{\mathrm{coh}}^?(X) := D^?(X) \cap D_{\mathrm{coh}}(X)$ for $? \in \{+, -, b, I\}$, where $I \subseteq \mathbb{Z}$ is some interval.

We introduce similar notation in the affine case. If A is a noetherian ring, then we denote by $(\mathrm{Coh}(A))$ the abelian category of finitely generated A-modules, which is also a plumb subcategory of the category of A-modules, and $D_{\mathrm{coh}}(A)$ denotes the triangulated subcategory of $D(A)$ of complexes M such that $H^p(M)$ is a finitely generated A-module for all $p \in \mathbb{Z}$. We define $D_{\mathrm{coh}}^?(A)$ for $? \in \{+, -, b, I\}$ as above.

Theorem 22.42. *Let X be a noetherian scheme. Then the natural functors*

$$D^-(\mathrm{Coh}(X)) \longrightarrow D^-_{\mathrm{coh}}(X) \qquad and \qquad D^b(\mathrm{Coh}(X)) \longrightarrow D^b_{\mathrm{coh}}(X)$$

are equivalences of categories.

Proof. All of the above categories are by definition full subcategories of $D(X)$. Hence the two functors are fully faithful. Moreover, if the first functor is essentially surjective, so is the second. Hence it suffices to show that $D^-(\mathrm{Coh}(X)) \longrightarrow D^-_{\mathrm{coh}}(X)$ is essentially surjective.

Let $\mathscr{F}$ be in $D^-_{\mathrm{coh}}(X)$. By Theorem 22.35 (if X is semiseparated) or Remark 22.38 (in general) we may assume that $\mathscr{F}$ has quasi-coherent components. We use Proposition F.160, applied to $\mathcal{A}$ the abelian category of quasi-coherent $\mathscr{O}_X$-modules, $\mathcal{C}$ the category of complexes of $\mathscr{O}_X$-modules with coherent cohomology, and to $\mathcal{D}$ the category of coherent $\mathscr{O}_X$-modules. Hence it suffices to show that if $\mathscr{F}$ is a coherent quotient of a quasi-coherent $\mathscr{O}_X$-module $\mathscr{G}$, then there exists a coherent $\mathscr{O}_X$-submodule $\mathscr{H}$ of $\mathscr{G}$ such that the composition $\mathscr{H} \hookrightarrow \mathscr{G} \to \mathscr{F}$ is surjective.

By Corollary 10.50, $\mathscr{G}$ is the filtered colimit of its coherent submodules. We can choose for $\mathscr{H}$ one of these by Lemma 10.47. $\qquad\qquad\square$

Corollary 22.43. *Let A be a noetherian ring and write $X = \mathrm{Spec}\, A$. Then one has for $* \in \{b, -\}$ a commutative diagram of triangulated equivalences*

$$
\begin{array}{ccc}
D^*(\mathrm{Coh}(A)) & \xrightarrow{\;\cong\;} & D^*_{\mathrm{coh}}(A) \\
\cong \Big\downarrow & & \Big\downarrow \cong \\
D^*(\mathrm{Coh}(X)) & \xrightarrow{\;\cong\;} & D^*_{\mathrm{coh}}(X),
\end{array}
$$

where the horizontal maps are given by the inclusion functors and the vertical maps by $M \mapsto \tilde{M}$.

Finiteness properties of complexes on schemes

In the next sections, we study the finiteness properties for complexes on a ringed space which we have introduced in Chapter 21, specifically pseudo-coherence and perfectness, in the case of schemes. One of the important results is Corollary 22.47 which says that a complex of $\mathscr{O}_X$-modules on a scheme X is pseudo-coherent if and only if locally on X it is isomorphic in $D(X)$ to a bounded above complex of finite free modules. For noetherian schemes, the situation is particularly simple, see Sections (22.15) and (22.18).

(22.13) Perfect and pseudo-coherent complexes on schemes.

We revisit different finiteness conditions for complexes on ringed spaces given in Chapter 21 and study them for complexes on schemes.

Recall, for a complex of modules on a ringed space or over a ring, the notion of being perfect (Definition 21.136), to be pseudo-coherent (Definition 21.150) or of having tor-amplitude in an interval $[a, b]$ (Definition 21.164).

By definition, given a ringed space X and an open covering $(U_i)_i$ of X, a complex $\mathscr{F}$ of modules on X is perfect (resp. pseudo-coherent, resp. of tor-amplitude in $[a,b]$) if and only if $\mathscr{F}_{|U_i}$ has this property for all i. Moreover, recall (Theorem 21.174) that a complex is perfect if and only if it is pseudo-coherent and locally of finite tor-dimension (i.e., locally of tor-amplitude in some bounded interval $[a,b]$).

Using the stability of these notions under faithfully flat descent (Lemma 21.163 and Lemma 21.172) one shows that for affine schemes the notions defined for a ring A and the ringed space $\operatorname{Spec} A$ are equivalent:

Lemma 22.44. *Let A be a ring, $X = \operatorname{Spec} A$, and let E be a complex of A-modules.*
(1) *Let $m \in \mathbb{Z}$. The complex E is m-pseudo-coherent if and only if the associated complex $\tilde{E}$ of quasi-coherent $\mathscr{O}_X$-modules is m-pseudo-coherent. In particular, E is pseudo-coherent if and only if $\tilde{E}$ is pseudo-coherent.*
(2) *Let $a, b \in \mathbb{Z}$ with $a \le b$. The complex E has tor-amplitude in $[a,b]$ if and only if the associated complex $\tilde{E}$ of quasi-coherent $\mathscr{O}_X$-modules has tor-amplitude in $[a,b]$.*
(3) *The complex E is perfect if and only if the associated complex $\tilde{E}$ of quasi-coherent $\mathscr{O}_X$-modules is perfect.*

Proof. If E is m-pseudo-coherent, there exists an m-isomorphism of a finite complex of finite projective modules to E. Hence there exists an m-isomorphism of a finite complex of finite locally free $\mathscr{O}_X$-modules $\mathscr{E}_m$ to $\tilde{E}$. Then $\mathscr{E}_m$ is perfect which shows that $\tilde{E}$ is m-pseudo-coherent.

Conversely, assume that $\tilde{E}$ is m-pseudo-coherent. Then there exists a finite principal open covering $(D(f_i))_i$ of X and an m-isomorphism of a finite complex of finite free $\mathscr{O}_{D(f_i)}$-modules to $\tilde{E}_{|D(f_i)} = (E \otimes_A A_{f_i})^{\sim}$. Hence $E \otimes_A A_{f_i}$ is m-pseudo-coherent for all i. As $A \to \prod_i A_{f_i}$ is faithfully flat, this implies that E is m-pseudo-coherent by Lemma 21.163.

The proof for the other properties is similar using the stability of both notions under faithfully flat descent (Lemma 21.172 and Lemma 21.178). $\square$

Remark 22.45. Every strictly perfect complex on a scheme X has quasi-coherent cohomology sheaves. As quasi-coherence is a local property, it follows that every perfect complex is in $D_{\mathrm{qcoh}}(X)$. Hence also every pseudo-coherent complex on X lies in $D_{\mathrm{qcoh}}(X)$.

Therefore Lemma 22.44 shows that if $X = \operatorname{Spec} A$ is an affine scheme, then the equivalence of $D(A)$ and $D_{\mathrm{qcoh}}(X)$ (Lemma 22.36) induces an equivalence between the full subcategory of $D(A)$ of perfect (resp. pseudo-coherent) complexes of A-modules and the full subcategory of $D(X)$ of perfect (resp. pseudo-coherent) complexes of $\mathscr{O}_X$-modules. This implies that we have the following description of perfect and pseudo-coherent complexes on affine schemes.

Remark 22.46. Let $X = \operatorname{Spec} A$ be an affine scheme.
(1) Every perfect complex is isomorphic in $D(X)$ to a strictly perfect complex. More precisely, for every perfect complex $\mathscr{F}$ on X there exists a finite complex $\mathscr{E}$ of finite locally free $\mathscr{O}_X$-modules and a quasi-isomorphism of complexes $\mathscr{E} \to \mathscr{F}$ (Example 21.138 (1)). See also Proposition 22.53 below how to bound $\mathscr{E}$ in terms of the tor-amplitude of $\mathscr{F}$.
(2) Every pseudo-coherent complex $\mathscr{F}$ is isomorphic in $D(X)$ to a bounded above complex $\mathscr{E}$ of finite free modules. If m is an integer such that $H^i(\mathscr{F}) = 0$ for all $i > m$, then we may choose $\mathscr{E}$ such that $\mathscr{E}^i = 0$ for all $i > m$ (Proposition 21.162).

As perfectness and pseudo-coherence can be checked locally, we obtain the following corollary.

Corollary 22.47. *Let X be a scheme.*

(1) *A complex of $\mathscr{O}_X$-modules is perfect if and only if it is locally on X isomorphic in $D(X)$ to a bounded complex of finite locally free $\mathscr{O}_X$-modules.*

(2) *A complex of $\mathscr{O}_X$-modules is pseudo-coherent if and only if it is locally on X isomorphic in $D(X)$ to a bounded above complex of finite free $\mathscr{O}_X$-modules.*

If X is quasi-compact and E is a pseudo-coherent complex of $\mathscr{O}_X$-modules, then there exists $b \in \mathbb{Z}$ such that $H^i(E) = 0$ for all $i > b$. Furthermore, E is then of tor-amplitude in $(-\infty, b]$.

Example 22.48. Let k be a field. A perfect complex E over k is isomorphic in $D(k)$ to a finite complex of finite-dimensional vector spaces whose differentials are zero.

Indeed, we may assume that E is a finite complex of finite-dimensional vector spaces, say concentrated in degrees $[a, b]$. As all exact sequences of k-vector spaces split, we can find an isomorphism of complexes $E \cong \tau^{\geq b} E \oplus F$, where F is the cokernel of the inclusion $E \to \tau^{\geq b} E = H^b(E)[-b]$. We get $E \cong F \oplus H^b(E)[-b]$ in $D(k)$. Proceeding inductively, we see that $E \cong \bigoplus_p H^p(E)[-p]$.

This argument can be vastly generalized, see Exercise F.31.

Example 22.49. Let X be a scheme and let $\mathscr{E}$ be a perfect complex. Then $\mathscr{E}$ is isomorphic to a finite locally free $\mathscr{O}_X$-module (considered as a complex concentrated in degree 0) if and only if $\mathscr{E}$ has tor-amplitude in $[0, 0]$.

Indeed, the condition is clearly necessary. Conversely, if $\mathscr{E}$ has tor-amplitude in $[0, 0]$, then $H^p(\mathscr{E}) = 0$ for $p \neq 0$ and hence $\mathscr{E} \cong H^0(\mathscr{E})[0]$. As $\mathscr{E}$ is pseudo-coherent, $H^0(\mathscr{E})$ is of finite presentation. As $H^0(\mathscr{E})$ has tor-dimension ≤ 0, it is also flat. Hence $H^0(\mathscr{E})$ is finite locally free.

Example 22.50. Let X be a scheme, and let $i\colon Z \to X$ be a completely intersecting closed immersion (e.g., a regular closed immersion). Then $i_*\mathscr{O}_Z$ is a perfect object in $D(X)$. Indeed, by definition, locally on X there exists a completely intersecting family of sections $f_1, \ldots, f_r$ of $\mathscr{O}_X$ such that $Z = V(f_1, \ldots, f_r)$. Then there exists a quasi-isomorphism from the Koszul complex attached to $(f_1, \ldots, f_r)$, which is a finite complex of finite free $\mathscr{O}_X$-modules, to $i_*\mathscr{O}_Z$.

By definition a complex in $D(X)$ is of tor-amplitude in $[a, b]$ relative to some morphism $f\colon X \to Y$ of schemes if $H^p(E \otimes_{\mathscr{O}_X}^L Lf^*\mathscr{G}) = 0$ for all $p \notin [a, b]$ and for every $\mathscr{O}_Y$-module $\mathscr{G}$. The next lemma shows that if $E \in D_{\mathrm{qcoh}}(X)$ and if Y is quasi-separated, then it suffices to check the condition only for quasi-coherent $\mathscr{O}_Y$-modules $\mathscr{G}$.

Lemma 22.51. *Let $f\colon X \to Y$ be a morphism of schemes and let $a \leq b$ be integers. Suppose that Y is quasi-separated (e.g., if Y is locally noetherian). Then a complex E in $D_{\mathrm{qcoh}}(X)$ is of tor-amplitude in $[a, b]$ relative to f if and only if for all quasi-coherent $\mathscr{O}_Y$-modules $\mathscr{G}$ and for all $p \notin [a, b]$ one has $H^p(E \otimes_{\mathscr{O}_X}^L Lf^*\mathscr{G}) = 0$.*

Proof. The condition is clearly necessary. For the converse implication, it suffices to show that for all open affines $U \subseteq X$ and $V \subseteq Y$ with $f(U) \subseteq V$, $E_{|U}$ is of tor-amplitude in $[a, b]$ relative to the restriction $g\colon U \to V$ of f (Lemma 21.171). The functor $(-)^\sim$ commutes with derived tensor products (Lemma 22.41). By Lemma 22.44 it suffices to show that $H^p(E_{|U} \otimes_{\mathscr{O}_U}^L Lg^*\mathscr{G}) = 0$ for all $p \notin [a, b]$ and all quasi-coherent $\mathscr{O}_V$-modules $\mathscr{G}$.

For every quasi-coherent $\mathscr{O}_Y$-modules $\mathscr{G}'$ one has $H^p(E \otimes^L_{\mathscr{O}_X} Lf^*\mathscr{G}')_{|U} = H^p(E_{|U} \otimes^L_{\mathscr{O}_U} Lg^*\mathscr{G}'_{|V})$. Therefore it suffices to show that every quasi-coherent $\mathscr{O}_V$-module $\mathscr{G}$ is isomorphic to the restriction of a quasi-coherent $\mathscr{O}_Y$-module $\mathscr{G}'$. As Y is quasi-separated, the inclusion $j \colon V \to Y$ is quasi-compact (and quasi-separated). Hence one can take $\mathscr{G}' := j_*\mathscr{G}$ which is quasi-coherent by Corollary 10.27. $\qquad\square$

Proposition 22.52. *Let $f \colon X \to Y$ be a morphism of schemes. Let $\mathscr{E}$ be a complex of $\mathscr{O}_Y$-modules. If $\mathscr{E}$ is perfect (resp. pseudo-coherent), then $Lf^*\mathscr{E}$ is perfect (resp. pseudo-coherent). The converse holds if f is faithfully flat and quasi-compact.*

Proof. The first assertion is a special case of Remark 21.142 and Proposition 21.157.

Now suppose that f is faithfully flat and quasi-compact and that $Lf^*\mathscr{E}$ is perfect (resp. pseudo-coherent). To see whether $\mathscr{E}$ has the same property, we may assume that $Y = \operatorname{Spec} A$ is affine. As f is quasi-compact, we find a finite open affine covering $(U_i)_i$ of X. Then $X' := \coprod_i U_i$ is affine, $f' \colon X' \to Y$ is faithfully flat, and $Lf'^*\mathscr{E}$ is perfect (resp. pseudo-coherent). Hence we can assume that $X = \operatorname{Spec} B$ is also affine. Then we can apply Lemma 21.178 (resp. Lemma 21.163) by Remark 22.45. $\qquad\square$

Perfect complexes (resp. pseudo-coherent complexes) on affine schemes can always be represented by finite (resp. bounded above) complexes of finite locally free modules. This is a key point for the theory of cohomology and base change. See Section (23.28) below. For perfect complexes the degrees in which these complexes of finite locally free modules are non-zero is given by the tor-amplitude as follows.

Proposition 22.53. *Let $X = \operatorname{Spec} A$ be an affine scheme, let $\mathscr{F}$ be a complex of $\mathscr{O}_X$-modules, and let $a \leq b$ be integers. Then $\mathscr{F}$ is pseudo-coherent and of tor-amplitude in $[a, b]$ if and only if $\mathscr{F}$ is isomorphic in $D(X)$ to a complex $\mathscr{E}$ of finite locally free $\mathscr{O}_X$-modules concentrated in degrees $[a, b]$.*

A complex $\mathscr{F}$ satisfying these equivalent conditions is perfect. It follows from Example 21.138 (1) that the isomorphism $\mathscr{E} \xrightarrow{\sim} \mathscr{F}$ in $D(X)$ is induced by a quasi-isomorphism $\mathscr{E} \to \mathscr{F}$ of complexes.

Proof. The condition is clearly sufficient. Suppose that $\mathscr{F}$ is pseudo-coherent and of tor-amplitude in $[a, b]$. We may pass to the corresponding pseudo-coherent complex E of A-modules which is of tor-amplitude in $[a, b]$. In particular, $H^p(E) = 0$ for all $p \notin [a, b]$. By Proposition 21.162 we may assume that E is a complex of finite free A-modules with $E^i = 0$ for $i > b$. As $E \to \tau^{\geq a}E$ is an isomorphism in $D(X)$, it suffices to show that

$$C := \operatorname{Coker}(d^{a-1} \colon E^{a-1} \to E^a)$$

is a finite projective A-module. Since E^{a-1} and E^a are of finite type, it is of finite presentation. Therefore it suffices to show that C is flat. As $H^p(E) = 0$ for $p < a$, the map $\sigma^{\leq a}E \to C[-a]$ is a quasi-isomorphism. As $\sigma^{\leq a}E$ and E are K-flat, we find for all A-modules M that

$$\operatorname{Tor}_1^A(C, M) = H^{a-1}(C[a] \otimes^L_A M) = H^{a-1}(\sigma^{\leq a}E \otimes_A M) = H^{a-1}(E \otimes^L_A M) = 0$$

because E is of tor-amplitude $\geq a$. Therefore C is a flat A-module. $\qquad\square$

Locally, one can say even more.

Proposition 22.54. *Let X be a scheme and let $\mathscr{E}$ be a perfect complex of $\mathcal{O}_X$-modules. Let $x \in X$ be a point, let $i_x \colon \operatorname{Spec} \kappa(x) \to X$ be the canonical map, and set $d_p := \dim_{\kappa(x)} H^p(\mathscr{E} \otimes^L \kappa(x))$, where $\mathscr{E} \otimes^L \kappa(x) := Li_x^* \mathscr{E}$ is the fiber of $\mathscr{E}$ in x.*

Then there exists an open neighborhood U of x and a complex $\mathscr{F}$ of finite free $\mathcal{O}_U$-modules of the form

$$ \cdots \longrightarrow \mathcal{O}_U^{d_{p-1}} \longrightarrow \mathcal{O}_U^{d_p} \longrightarrow \mathcal{O}_U^{d_{p+1}} \longrightarrow \cdots $$

such that $\mathscr{E}_{|U}$ is isomorphic to $\mathscr{F}$ in $D(U)$.

Moreover, if $(U_1, \mathscr{F}_1)$ and $(U_2, \mathscr{F}_2)$ are two such pairs, then there exist an open neighborhood $V \subseteq U_1 \cap U_2$ of x and an isomorphism of complexes of $\mathcal{O}_V$-modules $\mathscr{F}_{1|V} \cong \mathscr{F}_{2|V}$.

Only finitely many of d_p are non-zero, thus $\mathscr{F}$ is a finite complex. We have an isomorphism $\mathscr{E} \otimes^L \kappa(x) \cong \mathscr{F} \otimes \kappa(x)$ in $D(\kappa(x))$ and $\mathscr{F} \otimes \kappa(x)$ is a finite acyclic complex since the dimension of its components equals the dimension of its cohomology vector spaces.

Proof. We can pass to stalks by Remark 22.55 below. Then we are in the situation that we are given a perfect complex E over a local ring A. Let k be its residue field. Then $d_p = \dim_k H^p(E \otimes_A^L k)$. We may assume that E is a finite complex of finite free A-modules.

Now $E \otimes_A^L k = E \otimes_A k$ is a perfect complex over a field and therefore it is isomorphic to a complex $\bar{E}'$ with zero differentials of the form $\cdots \to k^{d_p} \to k^{d_{p+1}} \to \cdots$ (Example 22.48). This isomorphism in $D(k)$ is given by a quasi-isomorphism $\bar{u} \colon \bar{E}' \to E \otimes_A k$ of complexes by Lemma 21.134. Let $\bar{C}$ be the cone of $\bar{u}$ which is an acyclic complex because $\bar{u}$ is a quasi-isomorphism. Let C be an acyclic complex of finite free A-modules lifting $\bar{C}$ with $C \otimes_A k \cong \bar{C}$. Then the natural map $C \to \bar{C}$ is a surjective quasi-isomorphism of complexes of A-modules and using Remark 21.135 we obtain a map of complexes $v \colon E \to C$ making

commutative. Then $v^i \colon E^i \to C^i$ is injective with free cokernel as this holds after base change to k (Proposition 8.10).

Now we set $F := \operatorname{Coker}(v)[-1]$. We showed that F is a complex of finite free A-modules. As v is injective, we have a distinguished triangle $E \xrightarrow{v} C \longrightarrow F[1] \longrightarrow$. Rotating the triangle, we obtain a morphism $F \to E$ in $D(A)$ which is an isomorphism because C is acyclic. Finally, $F \otimes_A k$ is $\operatorname{Coker}(E \otimes_A k \to \bar{C})[-1]$ which is equal to $\bar{E}'$ by definition of the cone. In particular F^p has rank d_p for all p.

It remains to show that F is unique up to isomorphism. If there is another complex G with G^p free of rank d_p that is isomorphic to E in $D(A)$, we find again by Lemma 21.134 a quasi-isomorphism of complexes $w \colon F \to G$ representing the isomorphism $F \cong E \cong G$. The induced map $w \otimes 1 \colon F \otimes_A k \to G \otimes_A k$ is a quasi-isomorphism of acyclic complexes and hence an isomorphism. Therefore w is an isomorphism by Corollary 8.12. $\square$

Remark 22.55. Let $A = \operatorname{colim}_\lambda A_\lambda$ be a filtered colimit of rings. Then the triangulated category of perfect complexes over A is the "colimit of the triangulated categories of perfect complexes over A_λ". If E is a perfect complex in $D(A)$ of tor-amplitude in some finite interval $[a, b]$, then there exists a λ and a perfect complex E_λ in $D(A_\lambda)$ of tor-amplitude in $[a, b]$ such that $E_\lambda \otimes_{A_\lambda}^L A \cong E$.

Indeed, we have a similar assertion for the category of finitely generated projective modules by Section (10.16). As perfect complexes can be represented by finite complexes of finitely generated projective modules, which are concentrated in degrees $[a, b]$ if E is of tor-amplitude in $[a, b]$, and morphisms in the derived category between such complexes can be represented by morphisms of complexes (Lemma 21.134), this implies the claim.

(22.14) The resolution property and representing perfect complexes.

Definition 22.56. *We say that a scheme X satisfies the* resolution property *if every quasi-coherent $\mathscr{O}_X$-module of finite type is the quotient of a finite locally free $\mathscr{O}_X$-module.*

Any affine scheme has the resolution property since every finitely generated module over a ring is the quotient of a finitely generated free module. More generally, one has:

Proposition 22.57. *Let X be a qcqs scheme such that there exists an ample line bundle $\mathscr{L}$ on X. Then X has the resolution property.*

Proof. If $\mathscr{F}$ is a quasi-coherent $\mathscr{O}_X$-module of finite type, then there exists an $n \geq 0$ such that $\mathscr{F} \otimes \mathscr{L}^{\otimes n}$ is generated by its global sections. Hence we find a surjection $\mathscr{O}_X^N \to \mathscr{F} \otimes \mathscr{L}^{\otimes n}$ for some N (Lemma 10.47). In other words, $\mathscr{F}$ is a quotient of the finite locally free $\mathscr{O}_X$-module $(\mathscr{L}^{\otimes -n})^N$. $\square$

Every perfect complex on an affine scheme X is (globally) isomorphic in $D(X)$ to a finite complex of finite locally free $\mathscr{O}_X$-modules, i.e., to a strictly perfect complex. This can be generalized as follows.

Proposition 22.58. *Let X be a quasi-compact and semiseparated[2] scheme that has the resolution property. Then every perfect complex on X is isomorphic in $D(X)$ to a strictly perfect complex.*

The proof will show that if $\mathscr{E}$ is a perfect complex of tor-amplitude in $[a, b]$, then there exists a strictly perfect complex $\mathscr{F}$ concentrated in degrees $[a, b]$ and an isomorphism $\mathscr{F} \xrightarrow{\sim} \mathscr{E}$ in $D(X)$.

Proof. Let E be a perfect complex in $D(X)$. Then E lies in $D_{\mathrm{qcoh}}(X)$, so we can represent E by a complex $\mathscr{E}$ of quasi-coherent $\mathscr{O}_X$-modules (Theorem 22.35). As X is quasi-compact, it has bounded tor-amplitude, say in $[a, b]$ for integers $a \leq b$. In particular $H^i(E) = 0$ for $i \notin [a, b]$. Replacing $\mathscr{E}$ by $\tau^{\leq b}\tau^{\geq a}\mathscr{E}$ we may assume that $\mathscr{E}$ is concentrated in degrees $[a, b]$. We claim that there exists a quasi-isomorphism $\mathscr{F} \to \mathscr{E}$ with $\mathscr{F}$ a complex of finite locally free modules concentrated in degrees $[a, b]$.

We now prove the claim by induction on $b - a$. If $a = b$, then $\mathscr{E} \cong H^a(\mathscr{E})[-a]$ and $H^a(\mathscr{E})$ is finite locally free (Example 22.49).

Now let $b > a$. Next we show that $H^b(\mathscr{E})$ is a quasi-coherent $\mathscr{O}_X$-module of finite type. Indeed, this can be checked locally on X, so that we may assume that X is affine and that $\mathscr{E}$ is a complex of finite locally free $\mathscr{O}_X$-modules concentrated in degrees $[a, b]$ (Proposition 22.53). In this case $H^b(\mathscr{E})$ is a quotient of $\mathscr{E}^b$ and hence of finite type.

By the resolution property, we may choose a finite locally free $\mathscr{O}_X$-module $\mathscr{F}^b$ and a surjection $\mathscr{F}^b \to \mathscr{E}^b$. This surjection defines a map of complexes $u\colon \mathscr{F}^b[-b] \to \mathscr{E}$. Let C_u be its cone, so that we have a distinguished triangle

---

[2] A result by Totaro ([To]$_X^0$ 1.3) shows that the condition to be semiseparated automatically follows from the resolution property.

$$\mathscr{F}^b[-b] \longrightarrow \mathscr{E} \longrightarrow C_u \xrightarrow{+1}.$$

Then C_u is concentrated in degrees $[a, b]$ by definition of the cone and since $b > a$, it has tor-amplitude in $[a, b]$ by Proposition 21.166 and has $H^b(C_u) = 0$ as u induces a surjection $H^b(\mathscr{F}^b[-b]) = \mathscr{F}^b \to H^b(\mathscr{E})$ by construction. Hence C_u has tor-amplitude in $[a, b-1]$ (Remark 21.165 (2)). Therefore, we can apply the induction hypothesis to C_u and find a complex $\mathscr{H}$ of finite locally free $\mathscr{O}_X$-modules concentrated in degrees $[a, b-1]$ and, by the definition of C_u, a quasi-isomorphism

$$
\begin{array}{ccccccc}
\cdots \longrightarrow & \mathscr{H}^{b-2} & \longrightarrow & \mathscr{H}^{b-1} & \longrightarrow & 0 & \longrightarrow \cdots \\
& \downarrow & & \downarrow{\scriptstyle (v,w)} & & \downarrow & \\
\cdots \longrightarrow & \mathscr{E}^{b-2} & \longrightarrow & \mathscr{E}^{b-1} \oplus \mathscr{F}^b & \longrightarrow & \mathscr{E}^b & \longrightarrow \cdots.
\end{array}
$$

It induces a homomorphism of complexes $(\mathscr{H}^a \longrightarrow \cdots \longrightarrow \mathscr{H}^{b-1} \xrightarrow{w} \mathscr{F}^b) \longrightarrow \mathscr{E}$ and an easy diagram chase shows that this is a quasi-isomorphism. □

Corollary 22.59. *Let X be a qcqs scheme such that there exists an ample line bundle $\mathscr{L}$ on X. Then every perfect complex on X is isomorphic in $D(X)$ to a strictly perfect complex.*

Proof. The existence of an ample line bundle implies that X is separated (Proposition 13.48). Hence we can apply Proposition 22.58 by Proposition 22.57. □

Remark 22.60. Proposition 22.57 and Corollary 22.59 hold more generally if X has an ample family of line bundles (Exercise 23.19 and Exercise 23.20), e.g., if X is noetherian, semiseparated and locally factorial (Exercise 23.17).

(22.15) Pseudo-coherent complexes on noetherian schemes.

Let X be a scheme. Recall (Proposition 21.153) that an $\mathscr{O}_X$-module $\mathscr{F}$, considered as a complex concentrated in degree 0, is pseudo-coherent if and only if locally on X there exists an exact sequence of $\mathscr{O}_X$-modules

$$\cdots \longrightarrow \mathscr{E}^{-2} \longrightarrow \mathscr{E}^{-1} \longrightarrow \mathscr{E}^0 \longrightarrow \mathscr{F} \longrightarrow 0,$$

where $\mathscr{E}^i$ is a finite free $\mathscr{O}_X$-module for all $i \leq 0$. From this one easily deduces that on a locally noetherian scheme X an $\mathscr{O}_X$-module is pseudo-coherent if and only if it is coherent.

More generally, for noetherian schemes we have the following criterion when a complex is pseudo-coherent.

Proposition 22.61. *Let X be a noetherian scheme. Then a complex $\mathscr{E}$ in $D(X)$ is pseudo-coherent if and only if $\mathscr{E}$ is in $D^-_{\mathrm{coh}}(X)$.*

Proof. Let $\mathscr{E}$ be pseudo-coherent and $m \in \mathbb{Z}$. As X is quasi-compact, we find a finite open covering $(U_i)_i$ and for all i a bounded complex $\mathscr{K}_i$ of finite locally free $\mathscr{O}_{U_i}$-modules such that there exists an $(m-1)$-isomorphism $\mathscr{K}_i \to \mathscr{E}_{|U_i}$. In particular $H^p(\mathscr{E})_{|U_i} \cong H^p(\mathscr{K}_i)$ for all $p \geq m$ and all i. This shows that $H^m(\mathscr{E})$ is coherent and that $\mathscr{E}$ is in $D^-(X)$.

Conversely, suppose that $\mathscr{E}$ is in $D^-(X)$ with coherent cohomology. Let $m \in \mathbb{Z}$ and $x \in X$. We now apply Proposition F.160 to $\mathcal{A}$ the category of $\mathscr{O}_{X,x}$-modules, $\mathcal{D}$ the full subcategory of finite free $\mathscr{O}_{X,x}$-modules, and $\mathcal{C}$ the category of complexes of $\mathscr{O}_{X,x}$-modules with bounded above cohomology and finitely generated cohomology modules. Hence we obtain a bounded above complex $\mathscr{F}_x$ of finite free $\mathscr{O}_{X,x}$-modules and a quasi-isomorphism $\mathscr{F}_x \to \mathscr{E}_x$. Then $\sigma^{\geq m}\mathscr{F}_x$ is strictly perfect and $\sigma^{\geq m}\mathscr{F}_x \to \mathscr{E}_x$ is an m-isomorphism.

To extend this map to an open neighborhood of x we now repeatedly use Proposition 7.27 and that coherent modules are the same as modules of finite presentation over noetherian schemes (Proposition 7.46) as follows. Replacing X by an open neighborhood of x we may assume that there exists a strictly perfect complex $\mathscr{F}$ whose stalk in x is $\sigma^{\geq m}\mathscr{F}_x$. After shrinking X further we may assume that $\sigma^{\geq m}\mathscr{F}_x \to \mathscr{E}_x$ is induced by a map of complexes $\mathscr{F} \to \mathscr{E}$ which we may assume after shrinking X once more to be an m-isomorphism because $H^p(\mathscr{F})$ and $H^p(\mathscr{E})$ are coherent.   □

Corollary 22.62. *Let A be a noetherian ring. Then a complex E in $D(A)$ is pseudo-coherent if and only if $E \in D_{\mathrm{coh}}^-(A)$.*

(22.16) Derived tensor products of pseudo-coherent and perfect complexes.

Let X be a scheme.

Proposition 22.63. *Let $\mathscr{E}$ and $\mathscr{F}$ be complexes in $D(X)$. If $\mathscr{E}$ and $\mathscr{F}$ are pseudo-coherent (resp. perfect), then $\mathscr{E} \otimes_{\mathscr{O}_X}^L \mathscr{F}$ is pseudo-coherent (resp. perfect).*

Proof. For perfect complexes, we have already seen the result for general ringed spaces in Remark 21.143.

Let $\mathscr{E}$ and $\mathscr{F}$ be pseudo-coherent. To show that $\mathscr{E} \otimes^L \mathscr{F}$ is pseudo-coherent, we may assume that X is affine. Then we may assume that $\mathscr{E}$ and $\mathscr{F}$ are bounded above complexes of finite free $\mathscr{O}_X$-modules (Remark 22.46). Then $\mathscr{E} \otimes^L \mathscr{F} = \mathscr{E} \otimes \mathscr{F}$ is represented by a bounded above complex of finite free modules and hence it is pseudo-coherent.   □

Proposition 22.63 also holds if X is a ringed space but the proof for pseudo-coherent complexes gets more complicated since we cannot use Remark 22.46. Instead one can argue as sketched in Exercise 21.30.

Lemma 22.64. *Let X be quasi-compact and let $\mathscr{E}$ be a perfect complex in $D(X)$.*
(1) Let $\mathscr{F} \in D^b(X)$. Then $\mathscr{E} \otimes_{\mathscr{O}_X}^L \mathscr{F} \in D^b(X)$.
(2) Let X be noetherian and let $\mathscr{F}$ be in $D_{\mathrm{coh}}^b(X)$. Then $\mathscr{E} \otimes_{\mathscr{O}_X}^L \mathscr{F} \in D_{\mathrm{coh}}^b(X)$.

Proof. As X is quasi-compact, both assertions are local on X and hence we may assume that $\mathscr{E}$ is finite complex of free $\mathscr{O}_X$-modules. Then $\mathscr{E}$ is K-flat and hence $\mathscr{E} \otimes_{\mathscr{O}_X}^L \mathscr{F} = \mathscr{E} \otimes_{\mathscr{O}_X} \mathscr{F}$ is bounded. This implies (1).

To see (2) we use that $\mathscr{F}$ can be represented by a bounded complex of coherent $\mathscr{O}_X$-modules (Theorem 22.42). Then $\mathscr{E} \otimes_{\mathscr{O}_X} \mathscr{F}$ is also a bounded complex of coherent $\mathscr{O}_X$-modules.   □

(22.17) Derived Hom and Ext on schemes.

Recall the notion of Ext groups and of $\mathscr{E}xt$ sheaves on ringed spaces (Sections (F.52), (21.21)).

Proposition 22.65. *Let X be a scheme, and let $\mathscr{F}$, $\mathscr{G}$ be complexes of $\mathscr{O}_X$-modules. Assume that $\mathscr{F}$ is pseudo-coherent and that $\mathscr{G}$ is bounded below. Then we have natural isomorphisms*

$$\mathscr{E}xt^i(\mathscr{F},\mathscr{G})_x \cong \operatorname{Ext}^i_{\mathscr{O}_{X,x}}(\mathscr{F}_x,\mathscr{G}_x)$$

for all $i \in \mathbb{Z}$.

The same assertion also holds if $\mathscr{F}$ is perfect and $\mathscr{G}$ is arbitrary (Exercise 22.38).

Proof. Since the $\mathscr{E}xt$ functor is compatible with restriction to open subsets of X (21.21.3), we may assume that $X = \operatorname{Spec} A$ is affine. By Remark 22.46 we may assume that $\mathscr{F}$ is a bounded above complex with $\mathscr{F}^i$ free of finite rank for all i. As $\mathscr{G}$ is bounded below we have

$$R\,\mathscr{H}om(\mathscr{F},\mathscr{G}) = \mathscr{H}om(\mathscr{F},\mathscr{G})$$

by Lemma 21.89. We also have

$$R\operatorname{Hom}_{\mathscr{O}_{X,x}}(\mathscr{F}_x,\mathscr{G}_x) = \operatorname{Hom}_{\mathscr{O}_{X,x}}(\mathscr{F}_x,\mathscr{G}_x)$$

since $\mathscr{F}_x$, being a bounded above complex of free $\mathscr{O}_{X,x}$-modules, is K-projective. As the exact localizing functor $(-)_x$ commutes with taking cohomology, it suffices to show that the canonical map

$$\mathscr{H}om(\mathscr{F},\mathscr{G})_x \cong \operatorname{Hom}_{\mathscr{O}_{X,x}}(\mathscr{F}_x,\mathscr{G}_x)$$

of complexes of $\mathscr{O}_{X,x}$-modules is an isomorphism. As $\mathscr{F}$ is bounded above and $\mathscr{G}$ is bounded below, $\mathscr{H}om(\mathscr{F},\mathscr{G})^n \cong \bigoplus_p \mathscr{H}om(\mathscr{F}^p,\mathscr{G}^{p+n})$. As localization commutes with direct sums, it suffices to show that $\mathscr{H}om(\mathscr{F}^p,\mathscr{G}^q)_x \to \operatorname{Hom}_{\mathscr{O}_{X,x}}(\mathscr{F}^p_x,\mathscr{G}^q_x)$ is an isomorphism for all p and q. But this holds because $\mathscr{F}^p$ is finite locally free and thus of finite presentation (Proposition 7.27). $\qquad\square$

Proposition 22.66. *Let $\mathbf{P}$ be a property of $\mathscr{O}_X$-modules for all schemes X satisfying the following conditions.*
(a) *For every scheme X the subcategory $\mathcal{C}_X$ of all $\mathscr{O}_X$-modules that have $\mathbf{P}$ is plump (Definition F.43).*
(b) *For every scheme X and every open covering $(U_i)_i$ of X an $\mathscr{O}_X$-module $\mathscr{F}$ has $\mathbf{P}$ if and only if $\mathscr{F}_{|U_i}$ has $\mathbf{P}$ for all i.*
(c) *For every scheme X all $\mathscr{O}_X$-modules that have $\mathbf{P}$ are quasi-coherent.*
Let X be a scheme and set $\mathcal{C} := \mathcal{C}_X$. Let $D_{\mathcal{C}}(X) \subseteq D_{\mathrm{qcoh}}(X)$ be the triangulated subcategory of all complexes E in $D(X)$ such that one has $H^p(E) \in \mathcal{C}$ for all $p \in \mathbb{Z}$. Let $\mathscr{F} \in D(X)$ and $\mathscr{G} \in D_{\mathcal{C}}(X)$. Suppose that one of the following hypotheses is satisfied.
(1) *$\mathscr{F}$ is perfect.*
(2) *$\mathscr{F}$ is pseudo-coherent and $\mathscr{G}$ is locally bounded below.*
Then $R\,\mathscr{H}om_{\mathscr{O}_X}(\mathscr{F},\mathscr{G}) \in D_{\mathcal{C}}(X)$, i.e. $\mathscr{E}xt^n_{\mathscr{O}_X}(\mathscr{F},\mathscr{G}) \in \mathcal{C}$ for all $n \in \mathbb{Z}$.

Proof. As the formation of $\mathscr{E}xt$ commutes with restriction to open subsets (21.21.3), by (b) we may work locally on X. In particular, we may assume that $X = \operatorname{Spec} A$ for a ring A. By (c) we have $D_{\mathcal{C}}(X) \subseteq D_{\mathrm{qcoh}}(X) \cong D(A)$. Let $\mathcal{D} = \mathcal{D}(X,\mathscr{G})$ be the full subcategory of $D(X)$ of complexes $\mathscr{F}$ such that $\mathscr{E}xt^n_{\mathscr{O}_X}(\mathscr{F},\mathscr{G})$ is in $\mathcal{C}$ for all n. As $\mathcal{C}$ is a plump subcategory of the category of $\mathscr{O}_X$-modules, $\mathcal{D}$ is a triangulated subcategory of $D(X)$. Clearly, $\mathscr{O}_X$ is contained in $\mathcal{D}$ since $R\,\mathscr{H}om_{\mathscr{O}_X}(\mathscr{O}_X,\mathscr{G}) = \mathscr{G}$.

Let us first show that every perfect complex $\mathscr{F}$ is in $\mathcal{D}$. As X is affine, $\mathscr{F}$ is represented by a strictly perfect complex. Hence we conclude by Proposition 21.139.

Now suppose that $\mathscr{G}$ is locally bounded below. As X now is quasi-compact, one has $\mathscr{G} \in D^+(X)$. We have to show that every pseudo-coherent complex $\mathscr{F}$ is in $\mathcal{D}$. As X is affine, there exists for all $m \in \mathbb{Z}$ an m-isomorphism $\mathscr{E} \to \mathscr{F}$ with $\mathscr{E}$ strictly perfect (Remark 22.45).

Let $a \in \mathbb{Z}$ be such that $\mathscr{G} \in D^{\geq a}(X)$ (i.e. $H^p(\mathscr{G}) = 0$ for all $p < a$). If $\mathscr{F}$ is a complex with $H^p(\mathscr{F}) = 0$ for all $p \geq n$, then $R\mathscr{H}om(\mathscr{F}, \mathscr{G})$ is acyclic in degrees $\leq a - n$ (use the final statement of Theorem F.185). Hence we can apply Remark 21.160 to conclude. $\square$

The properties $\mathbf{P}$ of being quasi-coherent on a scheme or of being coherent on a locally noetherian scheme both satisfy the hypotheses in Proposition 22.66. Hence we obtain the following corollary.

Corollary 22.67. *Let X be a scheme and let $\mathscr{F}$ and $\mathscr{G}$ be in $D(X)$. Suppose that one of the following hypotheses is satisfied.*
(a) *$\mathscr{F}$ is perfect.*
(b) *$\mathscr{F}$ is a pseudo-coherent complex and $\mathscr{G}$ is locally bounded below.*
Then one has the following assertions.
(1) *Let $\mathscr{G}$ be in $D_{\mathrm{qcoh}}(X)$. Then $R\mathscr{H}om_{\mathscr{O}_X}(\mathscr{F}, \mathscr{G})$ is in $D_{\mathrm{qcoh}}(X)$.*
(2) *Let X be locally noetherian and let $\mathscr{G}$ be in $D_{\mathrm{coh}}(X)$. Then $R\mathscr{H}om_{\mathscr{O}_X}(\mathscr{F}, \mathscr{G})$ is in $D_{\mathrm{coh}}(X)$, i.e., $\mathscr{E}xt^p_{\mathscr{O}_X}(\mathscr{F}, \mathscr{G})$ is coherent for all $p \in \mathbb{Z}$.*

Assertion (2) can be generalized to certain non-noetherian schemes, see Exercise 22.33 for details.

Using the local-to-global spectral sequence for Ext (Corollary 21.108) and the vanishing of cohomology of quasi-coherent sheaves on affine schemes, as a further corollary to Corollary 22.67 (1) we obtain that if hypothesis (a) or (b) is satisfied and $\mathscr{G}$ is in $D_{\mathrm{qcoh}}(X)$, then for every affine open subscheme $U \subseteq X$ and every $i \in \mathbb{Z}$ one has

$$(22.17.1) \qquad \Gamma(U, \mathscr{E}xt^i(\mathscr{F}, \mathscr{G})) = \mathrm{Ext}^i_A(\Gamma(U, \mathscr{F}), \Gamma(U, \mathscr{G})).$$

Proposition 22.68. *Let A be a ring, $X = \mathrm{Spec}\, A$, and let $M, N \in D(A)$. Suppose that M is pseudo-coherent and that N is bounded below. Then there is a functorial isomorphism*

$$R\,\mathrm{Hom}_A(M, N)^{\sim} \xrightarrow{\sim} R\,\mathscr{H}om_{\mathscr{O}_X}(\tilde{M}, \tilde{N}).$$

In particular one sees that $\mathrm{Ext}^i_A(M, N)^{\sim} \cong \mathscr{E}xt^i_{\mathscr{O}_X}(\tilde{M}, \tilde{N})$ for all $i \in \mathbb{Z}$.
One has a similar result if M is perfect and N is an arbitrary complex (Exercise 22.37).

Proof. We may replace M and N, respectively, by complexes isomorphic to them in $D(A)$, and hence may assume that they are bounded above, and bounded below, respectively, and that every term of M is a finite free A-module (Proposition 21.162). It follows from Lemma 21.89, applied to the ringed spaces X and the one-point space with structure sheaf A, that it is enough to show that $\mathscr{H}om_{\mathscr{O}_X}(\tilde{M}^{-q}, \tilde{N}^p) \cong \mathrm{Hom}_A(M^{-q}, N^p)^{\sim}$ for all p, q. But this is clear since M^{-q} is a finitely generated free A-module. $\square$

Corollary 22.69. *Let A be a noetherian ring, let $M \in D^-_{\mathrm{coh}}(A)$ and $N \in D^+_{\mathrm{coh}}(A)$. Then $\mathrm{Ext}^p_A(M, N)$ is a finitely generated A-module for all $p \in \mathbb{Z}$.*

Proof. The complex M is pseudo-coherent by Corollary 22.62 and we conclude by Corollary 22.67 (2) and Proposition 22.68. $\square$

Proposition 22.70. *Let* $f \colon X' \to X$ *be a morphism of schemes. Let* $\mathscr{F}, \mathscr{G} \in D(X)$. *Assume that we are in one of the following situations:*

(1) *The complex* $\mathscr{F}$ *is perfect.*

(2) *The morphism* $f \colon X' \to X$ *has locally finite tor-dimension (Definition 21.164),* $\mathscr{F}$ *is pseudo-coherent, and* $\mathscr{G}$ *is locally bounded below.*

Then the natural morphism (21.26.10)

$$Lf^* R\mathscr{H}om_X(\mathscr{F}, \mathscr{G}) \to R\mathscr{H}om_{X'}(Lf^*\mathscr{F}, Lf^*\mathscr{G})$$

is an isomorphism.

Proof. As the formation of $R\mathscr{H}om$ commutes with restriction to open subsets (21.18.9) we can work locally on X. In particular, we can assume that $X = \operatorname{Spec} A$ is affine. Fix $\mathscr{G} \in D(X)$. We consider the contravariant triangulated functors $\Phi, \Psi \colon D(X) \to D(X')$ given by $\Phi(\mathscr{F}) = Lf^* R\mathscr{H}om_X(\mathscr{F}, \mathscr{G})$ and $\Psi(\mathscr{F}) = R\mathscr{H}om_{X'}(Lf^*\mathscr{F}, Lf^*\mathscr{G})$. Then (21.26.10) yields a morphism $\eta \colon \Phi \to \Psi$ of triangulated functors. Clearly, $\eta(\mathscr{O}_X)$ is an isomorphism.

To show (1) it suffices to show that $\eta(\mathscr{F})$ is an isomorphism for every strictly perfect complex. This follows from Corollary 21.141.

Let us show (2). By working locally on X it suffices to show that $\eta(\mathscr{F})$ is an isomorphism for all $\mathscr{F} \in D(X)$ such that for all $m \in \mathbb{Z}$ there exists an m-isomorphism $\mathscr{E} \to \mathscr{F}$ for some strictly perfect complex $\mathscr{E}$. As X is now quasi-compact, f is of finite tor-dimension, say $\leq t$, and $\mathscr{G}$ is bounded below, say $\mathscr{G} \in D^{\geq a}(X)$. Then $Lf^*\mathscr{G} \in D^{\geq a-t}(X')$. The functors Φ and Ψ are contravariant and both map $D^{\leq m}(X)$ to $D^{\geq a-m-t}(X')$. Hence we can conclude by Proposition 21.159 and the remark following it. $\square$

We can apply Proposition 22.70 in particular to morphisms f that are flat, i.e., of tor-dimension ≤ 0. For instance we obtain the following affine variant of Proposition 22.70 using that under the given hypothesis the equivalence $D(A) \cong D_{\operatorname{qcoh}}(\operatorname{Spec} A)$ is compatible with derived inverse images by Lemma 22.41 and derived Hom by Proposition 22.68.

Corollary 22.71. *Let* A *be a ring, let* A' *be a flat* A-*algebra, let* $F \in D(A)$ *be pseudo-coherent, and let* $G \in D^+(A)$. *Then the natural morphism*

$$R\operatorname{Hom}_A(F, G) \otimes_A^L A' \to R\operatorname{Hom}_{A'}(F \otimes_A^L A', G \otimes_A^L A')$$

is an isomorphism.

Lemma 22.72. *Let* X *be a scheme and let* $\mathscr{E}$, $\mathscr{F}$ *and* $\mathscr{G}$ *be complexes in* $D(X)$. *Then the functorial map* (21.26.9) *is an isomorphism*

$$(22.17.2) \qquad R\mathscr{H}om_{\mathscr{O}_X}(\mathscr{E}, \mathscr{F}) \otimes_{\mathscr{O}_X}^L \mathscr{G} \xrightarrow{\sim} R\mathscr{H}om_{\mathscr{O}_X}(R\mathscr{H}om_{\mathscr{O}_X}(\mathscr{G}, \mathscr{E}), \mathscr{F})$$

in the following two cases.

(1) $\mathscr{G}$ *is perfect.*

(2) $\mathscr{G}$ *is pseudo-coherent,* $\mathscr{E}$ *is locally bounded below and* $\mathscr{F}$ *is locally of finite injective dimension, i.e., there exists an open covering* $(U_i)_i$ *of* X *such that* $\mathscr{F}_{|U_i}$ *is of finite injective dimension (Definition F.221).*

Proof. The map (22.17.2) is an isomorphism if $\mathscr{G} = \mathscr{O}_X$. The formation of $R\mathscr{H}om$ and of the derived tensor product are compatible with restriction to open subsets, thus this is a local question. Hence we can assume for (1) that $\mathscr{G}$ is strictly perfect. Then (1) follows from Corollary 21.141.

To show (2) it suffices by Proposition 21.159 that both sides of (22.17.2) as functors in $\mathscr{G}$ are cohomologically bounded above if $\mathscr{E}$ is bounded below and $\mathscr{F}$ is of finite injective dimension. But then $\mathscr{F}$ can be represented by a finite complex $\mathscr{I}$ of injective $\mathscr{O}_X$-modules and $\mathscr{H} := R\mathscr{H}om_{\mathscr{O}_X}(\mathscr{E},\mathscr{F})$ is represented by $\mathscr{H}om_{\mathscr{O}_X}(\mathscr{E},\mathscr{I})$ which is zero in degrees $\gg 0$. Hence $H^i(\mathscr{H}) = 0$ for $i \gg 0$ and hence the left hand side $\mathscr{H} \otimes^L (-)$ of (22.17.2) is cohomologically bounded above. A similar argument shows that the right hand side $\mathscr{H} \otimes^L (-)$ of (22.17.2) is cohomologically bounded above. $\square$

(22.18) Injective modules on locally noetherian schemes.

We study injective modules on schemes. Recall that we can show injectivity locally (Lemma 21.12). Hence if X is a scheme, then an $\mathscr{O}_X$-module $\mathscr{I}$ is injective if and only if $\mathscr{I}_{|U_i}$ is injective for all i for some affine open covering $(U_i)_i$ of X.

Even if $X = \operatorname{Spec} A$ is affine, then an injective $\mathscr{O}_X$-module is in general not quasi-coherent, so that it does not correspond to an A-module. Even worse, if M is an injective A-module, then $\tilde{M}$ will be an $\mathscr{O}_X$-module which is injective in $\operatorname{QCoh}(X)$ but it will in general not be injective in the category of all $\mathscr{O}_X$-modules (see Caveat 22.77 (3) below). The situation becomes better in the noetherian case as we explain in the following. Some proofs will be only sketched, for details we refer to [Har1] ^{O} II §7.

Proposition 22.73. *Let X be a locally noetherian scheme. Then $(\mathscr{O}_X\text{-Mod})$ is a locally noetherian Grothendieck category (see Section (F.13)).*

By Remark 21.8 it suffices to show that the extension by zero $\mathscr{O}_{U\subseteq X}$ (21.1.6) is noetherian for every affine open $U \subseteq X$. Replacing X by U we may assume that $X = \operatorname{Spec} A$ is affine and it suffices to show that $\mathscr{O}_X$ is a noetherian object in the category of all $\mathscr{O}_X$-modules, i.e., that every ascending sequence $\mathscr{I}_1 \subseteq \mathscr{I}_2 \subseteq \ldots$ of ideals of $\mathscr{O}_X$ becomes stationary. If all these ideal were quasi-coherent, they would correspond to ideals of A and the claim would be clear since A is noetherian. In general one has to argue differently, see [Har1] ^{O} II 7.8.

Remark 22.74. Proposition 22.73 implies by Proposition F.67 and Corollary 21.53 that on a locally noetherian scheme every direct sum of injective $\mathscr{O}_X$-modules is again injective and that every injective $\mathscr{O}_X$-module is a direct sum of indecomposable injective $\mathscr{O}_X$-modules in a unique way (up to order).

In [Har1] ^{O} II 7.11 it is shown that the indecomposable injective $\mathscr{O}_X$-modules are precisely the $\mathscr{O}_X$-modules $\mathscr{J}(x,x')$ that are defined as follows. Let $x \in X$ and let $x' \in \overline{\{x\}}$ be a specialization of x. Let $j_x \colon \operatorname{Spec}(\mathscr{O}_{X,x}) \to X$ and $i_{x'} \colon \overline{\{x'\}} \to X$ be the inclusions. Let I_x be an injective hull of $\kappa(x)$ over $\mathscr{O}_{X,x}$ (Definition F.64) and set

$$\mathscr{J}(x,x') := i_{x',*}i_{x'}^{-1}j_{x,*}(\tilde{I}_x), \qquad \mathscr{J}(x) := \mathscr{J}(x,x).$$

Moreover, $\mathscr{J}(x,x')$ is quasi-coherent if and only if $x = x'$.

To see that on a locally noetherian scheme X, the embedding $\operatorname{QCoh}(X) \to (\mathscr{O}_X\text{-Mod})$ preserves injective objects, we will use the following lemma.

Lemma 22.75. *Let X be a locally noetherian scheme, and let $\mathscr{F}$ be a quasi-coherent $\mathscr{O}_X$-module. Then there exists a quasi-coherent $\mathscr{O}_X$-module $\mathscr{G}$ which is injective in $(\mathscr{O}_X\text{-Mod})$ together with a monomorphism $\mathscr{F} \hookrightarrow \mathscr{G}$.*

In fact, one can take for $\mathscr{G}$ the injective hull of $\mathscr{F}$ in $(\mathscr{O}_X\text{-Mod})$ (Definition F.64). By Remark 22.74 it is a direct sum of $\mathscr{O}_X$-modules of the form $\mathscr{J}(x, x')$ and one shows that one necessarily has $x = x'$ since $\mathscr{F}$ is quasi-coherent. But this implies that the injective hull is a direct sum of quasi-coherent $\mathscr{O}_X$-modules and hence quasi-coherent (see [Har1] $^\circ$ II 7.18 for details).

Proposition 22.76. *Let X be a locally noetherian scheme. Then the inclusion functor* $\mathrm{QCoh}(X) \to (\mathscr{O}_X\text{-Mod})$ *sends injectives to injectives.*

Hence we see that a quasi-coherent $\mathscr{O}_X$-module is injective in the category of all $\mathscr{O}_X$-modules if and only if it is injective in $\mathrm{QCoh}(X)$.

Proof. Let $\mathscr{F} \in \mathrm{QCoh}(X)$ be an injective object. As in the lemma, let $\iota\colon \mathscr{F} \hookrightarrow \mathscr{G}$ be a monomorphism from $\mathscr{F}$ into a quasi-coherent $\mathscr{O}_X$-module $\mathscr{G}$ which is injective as an $\mathscr{O}_X$-module. Since $\mathscr{F}$ is injective, there is a retraction $\rho\colon \mathscr{G} \to \mathscr{F}$ with $\rho \circ \iota = \mathrm{id}$. Hence $\mathscr{F}$ is a direct summand of $\mathscr{G}$, in $(\mathrm{QCoh}(X))$ as well as in $(\mathscr{O}_X\text{-Mod})$, and as a direct summand of an injective (in $(\mathscr{O}_X\text{-Mod})$) is injective, as well. $\qquad\square$

Caveat 22.77.
(1) An infinite product in $(\mathscr{O}_X\text{-Mod})$ of quasi-coherent $\mathscr{O}_X$-modules is not necessarily quasi-coherent.
(2) The coherator Q is not exact in general.
(3) There are examples of affine schemes with underlying noetherian topological space of finite dimension, for which the inclusion $(\mathrm{QCoh}(X)) \to (\mathscr{O}_X\text{-Mod})$ does not send injectives to injectives (not even to flasque sheaves). Cf. Verdier's example in [SGA6] $^\circ$ II App. I.

So for a locally noetherian scheme X, we can compute the value of right derived functors on a quasi-coherent $\mathscr{O}_X$-module using an injective resolution in $(\mathrm{QCoh}(X))$, while for general X this will not necessarily give the right result.

For quasi-coherent $\mathscr{O}_X$-modules on a locally noetherian scheme X, injectivity behaves very well.

Proposition 22.78. *Let X be a locally noetherian scheme and let $\mathscr{F}$ be a quasi-coherent* $\mathscr{O}_X$-*module. Then the following assertions are equivalent.*
(i) *$\mathscr{F}$ is an injective $\mathscr{O}_X$-module.*
(ii) *For every affine open covering $(U_i)_i$ of X and for all i one has that $\mathscr{F}_{|U_i} \cong \tilde{I}_i$ for some injective $\Gamma(U_i, \mathscr{O}_X)$-module I_i.*
(iii) *There exists an affine open covering $(U_i)_i$ of X such that for all i, $\mathscr{F}_{|U_i} \cong \tilde{I}_i$ for some injective $\Gamma(U_i, \mathscr{O}_X)$-module I_i.*
(iv) *For every $x \in X$ the stalk $\mathscr{F}_x$ is an injective $\mathscr{O}_{X,x}$-module.*
(v) *For every $\mathscr{O}_X$-module $\mathscr{G}$ and for all $i > 0$ one has $\mathscr{E}xt^i_{\mathscr{O}_X}(\mathscr{G}, \mathscr{F}) = 0$.*
(vi) *For every coherent ideal $\mathscr{J} \subseteq \mathscr{O}_X$ one has $\mathscr{E}xt^1_{\mathscr{O}_X}(\mathscr{O}_X/\mathscr{J}, \mathscr{F}) = 0$.*

Proof. We will proceed as follows:

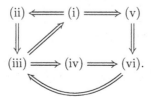

(i) $\Rightarrow$ (ii). Let $(U_i)_i$ be any open affine covering, $U_i = \operatorname{Spec} A_i$. Then $\mathscr{F}_{|U_i}$ is a quasi-coherent $\mathscr{O}_{U_i}$-module that is injective in the category of $\mathscr{O}_{U_i}$-modules (Lemma 21.12) and in particular it is injective in the category of quasi-coherent $\mathscr{O}_{U_i}$-modules. Hence it corresponds to an injective A_i-module.

(ii) $\Rightarrow$ (iii). This is clear.

(iii) $\Rightarrow$ (i). The hypothesis implies that $\mathscr{F}$ is locally on X an injective module in the category of quasi-coherent $\mathscr{O}_X$-modules. Hence it is locally injective in the category of all $\mathscr{O}_X$-modules by Proposition 22.76. Then Lemma 21.12 implies that $\mathscr{F}$ is an injective $\mathscr{O}_X$-module.

(iii) $\Rightarrow$ (iv). This follows from Proposition G.23.

(iv) $\Rightarrow$ (vi). As $\mathscr{O}/\mathscr{J}$ is coherent, the formation of $\mathscr{E}xt$ commutes with passage to stalks (Proposition 22.65) and $\operatorname{Ext}^1_{\mathscr{O}_{X,x}}(\mathscr{O}_x/\mathscr{J}_X,\mathscr{F}_x) = 0$ since $\mathscr{F}_x$ is injective.

(i) $\Rightarrow$ (v). As $\mathscr{F}$ is injective, we find $R\,\mathscr{H}om_{\mathscr{O}_X}(\mathscr{G},\mathscr{F}) = \mathscr{H}om_{\mathscr{O}_X}(\mathscr{G},\mathscr{F})$ which is a complex concentrated in degree 0.

(v) $\Rightarrow$ (vi). This is clear.

(vi) $\Rightarrow$ (iii). Assume first that $X = \operatorname{Spec} A$ is affine. Let $\tilde{I} = \mathscr{F}$. By Proposition G.21 it suffices to show that $\operatorname{Ext}^1_A(A/\mathfrak{a}, I) = 0$ for every ideal $\mathfrak{a}$ of A. As $\operatorname{Ext}^1_A(A/\mathfrak{a}, I)^\sim = \mathscr{E}xt^1_A(\mathscr{O}_X/\tilde{\mathfrak{a}}, \mathscr{F})$ one concludes by Proposition 22.68.

The general case follows since the formation of $\mathscr{E}xt$ commutes with restriction to open subsets (21.21.3) and since coherent ideals on an open subset of a locally noetherian scheme can be extended to a coherent ideal on all of X (see Proposition 10.48 if X is noetherian and [EGAInew] (6.9.7) in general).   □

Now we can use similar arguments to show that also the notion of injective dimension (Definition F.221) behaves well on locally noetherian schemes.

Proposition 22.79. *Let X be a locally noetherian scheme and let $\mathscr{F}$ be a complex in $D_{\mathrm{qcoh}}(X)$. Let $a \leq b$ be integers. Then the following assertions are equivalent.*
(i) *$\mathscr{F}$ has injective amplitude contained in $[a, b]$ in $D(X)$.*
(ii) *$\mathscr{F}$ is isomorphic in $D_{\mathrm{qcoh}}(X)$ to a finite complex of injective quasi-coherent $\mathscr{O}_X$-modules concentrated in degrees $[a, b]$.*
(iii) *For every open affine covering $(U_i)_i$ of X one has for all i that $\mathscr{F}_{|U_i} \cong \tilde{I}_i$ for a complex I_i of injective $\Gamma(U_i, \mathscr{O}_X)$-modules concentrated in degrees $[a, b]$.*
(iv) *There exists an open affine covering $(U_i)_i$ of X such that for all i, $\mathscr{F}_{|U_i} \cong \tilde{I}_i$ for some complex I_i of injective $\Gamma(U_i, \mathscr{O}_X)$-modules concentrated in degrees $[a, b]$.*
(v) *For every $x \in X$ the stalk $\mathscr{F}_x$ is a complex of $\mathscr{O}_{X,x}$-modules of injective amplitude contained in $[a, b]$.*
(vi) *For every $\mathscr{O}_X$-module $\mathscr{G}$ one has $\mathscr{E}xt^i_{\mathscr{O}_X}(\mathscr{G}, \mathscr{F}) = 0$ for all $i \notin [a, b]$.*
(vii) *For every coherent ideal $\mathscr{J} \subseteq \mathscr{O}_X$-module one has $\mathscr{E}xt^i_{\mathscr{O}_X}(\mathscr{O}_X/\mathscr{J}, \mathscr{F}) = 0$ for all $i \notin [a, b]$.*

Proof. Here we proceed as follows:

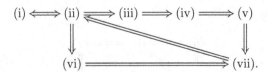

(ii) $\Rightarrow$ (i). By Proposition 22.76, $\mathscr{F}$ is isomorphic to a complex of injective $\mathscr{O}_X$-modules concentrated in degrees $[a, b]$. Hence one concludes by Proposition F.222.

(i) $\Rightarrow$ (ii). The proof of Proposition F.222 shows that $\mathscr{F} \in D^{[a,b]}(X)$. By Lemma 22.75, we find then an isomorphism $\mathscr{F} \xrightarrow{\sim} \mathscr{J}^\bullet$ such that $\mathscr{J}^i = 0$ for $i < a$ and such that $\mathscr{J}^i$ is a quasi-coherent injective $\mathscr{O}_X$-module. Then $\mathscr{F} \cong \tau^{\leq b}\mathscr{F} \cong \tau^{\leq b}\mathscr{J}^\bullet$ and all components of $\tau^{\leq b}\mathscr{J}^\bullet$ are quasi-coherent. The proof of Proposition F.222 shows that all components of $\tau^{\leq b}\mathscr{J}^\bullet$ are injective.

(ii) $\Rightarrow$ (iii) $\Rightarrow$ (iv) $\Rightarrow$ (v) *and* (ii) $\Rightarrow$ (vi). This follows as in Proposition 22.78.

(v) $\Rightarrow$ (vii). The hypothesis implies that $\mathscr{F}_x$ is in $D^+(\mathscr{O}_{X,x})$ for all x and hence $\mathscr{F} \in D^+(X)$. Hence one can again apply Proposition 22.65 and argue as in the proof of Proposition 22.78.

(vi) $\Rightarrow$ (vii). This is clear.

(vii) $\Rightarrow$ (ii). As $R\mathscr{H}om_{\mathscr{O}_X}(\mathscr{O}_X, \mathscr{F}) = \mathscr{F}$ (21.18.10), we have $\mathscr{E}xt^i_{\mathscr{O}_X}(\mathscr{O}_X, \mathscr{F}) = H^i(\mathscr{F})$. Hence (vii) shows that $\mathscr{F} \in D^{[a,b]}_{\text{qcoh}}(X)$. As in the proof of the implication "(i) $\Rightarrow$ (ii)", we see that $\mathscr{F}$ is isomorphic in $D_{\text{qcoh}}(X)$ to a complex $\mathscr{I}$ of quasi-coherent $\mathscr{O}_X$-modules concentrated in degrees $[a,b]$ and with $\mathscr{I}^i$ injective for all $i = a, \ldots, b-1$. To show that $\mathscr{I}^b$ is injective one can argue as in the proof of Proposition F.222, using $\mathscr{E}xt$ instead of Ext and that we have already seen that (ii) implies (vii). Then the argument shows that $\mathscr{E}xt^1_{\mathscr{O}_X}(\mathscr{O}_X/\mathscr{J}, \mathscr{I}^b) = 0$ for every coherent ideal $\mathscr{J}$ of $\mathscr{O}_X$. This implies that $\mathscr{I}^b$ is injective by Proposition 22.78. $\qquad\square$

Corollary 22.80. *Let A be a noetherian ring, let $X = \operatorname{Spec} A$, and let $a \leq b$ be integers. Then a complex M in $D(A)$ has injective amplitude in $[a,b]$ if and only if $\tilde{M}$ has injective amplitude contained in $[a,b]$ in $D(X)$.*

Projection formula, base change and the Künneth formula

The projection formula is a compatibility between pullbacks and pushforwards which holds in many contexts. The simplest prototype is the statement $f(F \cap f^{-1}(G)) = f(F) \cap G$ for a map $f: X \to Y$ of sets and subsets $F \subseteq X$, $G \subseteq Y$. Another particularly simple instance is Proposition 22.81 (3). Below we will prove a derived version of the projection formula.

Afterwards we will come back to the base change morphism (Definition 21.129) for a cartesian diagram of schemes and investigate when it is an isomorphism for all objects in $D_{\text{qcoh}}(X)$. As we will see, this is closely related to the Künneth morphism, which relates, roughly speaking, cohomology on a fiber product and cohomology of the factors of the product.

(22.19) The projection formula.

We will show two versions of projection formulas. The first is a non-derived version which for general quasi-coherent modules requires rather strong hypotheses, see (3) and (4) in the following proposition.

The second version will be a derived version that will hold in (almost) complete generality, see Proposition 22.84 (3) below.

Proposition 22.81. *(Projection formula) Let $f: X \to Y$ be a morphism of ringed spaces, let $\mathscr{F}$ be an $\mathscr{O}_X$-module, and let $\mathscr{G}$ be an $\mathscr{O}_Y$-module. There is a natural homomorphism*

$$(f_*\mathscr{F}) \otimes_{\mathscr{O}_Y} \mathscr{G} \to f_*(\mathscr{F} \otimes_{\mathscr{O}_X} f^*\mathscr{G})$$

which is functorial in $\mathscr{F}$ and $\mathscr{G}$. This homomorphism is an isomorphism in the following cases:

(1) If $\mathscr{G}$ is locally free of finite rank.
(2) If f is a homeomorphism onto a closed subspace of Y.
(3) If f is an affine morphism of schemes and $\mathscr{F}$, $\mathscr{G}$ are quasi-coherent.
(4) If f is a qcqs morphism of schemes, $\mathscr{F}$ and $\mathscr{G}$ are quasi-coherent, and $\mathscr{G}$ is flat over $\mathscr{O}_Y$.

Often, Case (4) is the most interesting case.

Proof. The homomorphism is obtained by adjunction from the homomorphism

$$f^*(f_*\mathscr{F} \otimes \mathscr{G}) \cong (f^*f_*\mathscr{F}) \otimes f^*\mathscr{G} \to \mathscr{F} \otimes f^*\mathscr{G}.$$

To check that this is an isomorphism, we may work locally on Y.

Case (1). We may assume that $\mathscr{G}$ is free, or even that it is isomorphic to $\mathscr{O}_Y$, by passing to a direct sum afterwards. In this case it is easy to check that the homomorphism in question is an isomorphism.

Case (2). We check this on stalks. Outside the image of f, the stalks of direct images under f vanish, hence the stalks of both the left and the right hand side vanish. For $y = f(x)$, the stalks are

$$\mathscr{F}_x \otimes_{\mathscr{O}_{Y,y}} \mathscr{G}_y \quad \text{and} \quad \mathscr{F}_x \otimes_{\mathscr{O}_{X,x}} \left(\mathscr{O}_{X,x} \otimes_{\mathscr{O}_{Y,y}} \mathscr{G}_y \right).$$

The natural map clearly induces an isomorphism between those, so we are done.

Case (3). We may assume that Y is affine. In this case, X is affine, as well, and the assertion is easily checked using the description of the functors f_*, f^* for a morphism between affine schemes (Proposition 7.24).

Case (4). We may assume that Y is affine. We want to use Lemma 10.26 with the property $\mathbf{P}(U)$, for $U \subseteq X$ open quasi-compact, to hold if for every quasi-coherent $\mathscr{O}_U$-module $\mathscr{F}'$ the homomorphism

$$(f_{|U})_*\mathscr{F}' \otimes_{\mathscr{O}_Y} \mathscr{G} \to (f_{|U})_*(\mathscr{F}' \otimes_{\mathscr{O}_U} (f_{|U})^*\mathscr{G})$$

is an isomorphism. We want to show that $\mathbf{P}(X)$ holds.

If $U \subseteq X$ is open affine, then $f_{|U}$ is affine and $\mathbf{P}(U)$ holds by (3). Now let $U \subseteq X$ be quasi-compact open and let $(U_i)_i$ be a finite open affine covering of U. Consider for a quasi-coherent $\mathscr{O}_U$-module $\mathscr{F}'$ the following commutative diagram

$$
(*) \quad
\begin{array}{ccccccc}
0 & \longrightarrow & (f'_*\mathscr{F}') \otimes \mathscr{G} & \longrightarrow & \prod_i f_{i,*}(\mathscr{F}'_{|U_i}) \otimes \mathscr{G} & \longrightarrow & \prod_{i,j} f_{ij,*}(\mathscr{F}'_{|U_{ij}}) \otimes \mathscr{G} \\
 & & \downarrow & & \downarrow & & \downarrow \\
0 & \longrightarrow & f'_*(\mathscr{F}' \otimes f^*\mathscr{G}) & \longrightarrow & \prod_i f_{i,*}(\mathscr{F}'_{|U_i} \otimes f_i^*\mathscr{G}) & \longrightarrow & \prod_{i,j} f_{ij,*}(\mathscr{F}'_{|U_{ij}} \otimes f_{ij}^*\mathscr{G}),
\end{array}
$$

where we write $f' := f_{|U}$, $U_{ij} := U_i \cap U_j$, $f_i := f_{|U_i}$, $f_{ij} := f_{|U_{ij}}$, as usual. Both rows are exact (see (10.3.2) and use the flatness assumption on $\mathscr{G}$). We have already seen that $\mathbf{P}(U_i)$ holds for all i and hence if $\mathbf{P}(U_{ij})$ holds for all i,j, then $\mathbf{P}(U)$ holds. Now Lemma 10.26 implies that $\mathbf{P}(X)$ holds. $\qquad\square$

We now come to the derived version of the projection formula. For its proof we will use the following two lemmas.

Lemma 22.82. *Let $f: X \to Y$ be a qcqs morphism. The functor $Rf_*: D_{\mathrm{qcoh}}(X) \to D_{\mathrm{qcoh}}(Y)$ commutes with arbitrary direct sums.*

Note that this property is not true when the assumptions on f are dropped; in fact, even the global sections functor $\Gamma(X, -)$ does not commute with coproducts in general. Also, a coproduct of K-injective objects is not K-injective in general.

Proof. Given a direct sum $E = \bigoplus_i E_i$ in $D_{\mathrm{qcoh}}(X)$, we have a natural morphism

$$\bigoplus_i Rf_*E_i \to Rf_*E.$$

We need to show that it induces isomorphisms after passing to cohomology objects. This can be done locally on Y. Hence we may assume Y is affine. Since the morphism is compatible with shifts, it is enough to prove that we obtain an isomorphism after applying H^0, i.e., we have to show that $\bigoplus_i R^0 f_* E_i \to R^0 f_* E$ is an isomorphism (using that applying H^0 commutes with direct sums).

Using Corollary 22.30, we see that we can compute these $R^0 f_*$ terms as $R^0 f_*$ of *bounded below* complexes in $D_{\mathrm{qcoh}}(X)$, allowing us to reduce to the case of bounded below complexes. For those, we can use the spectral sequence $R^p f_* H^q(-) \Rightarrow R^{p+q} f_* -$ (see Corollary 21.46). Again using that taking cohomology objects of a complex commutes with direct sums, this leaves us with showing that higher direct images of quasi-coherent sheaves commute with direct sums, which we have already seen in Corollary 21.56 (2). $\square$

Lemma 22.83. *Let A be a ring, and let $\mathbf{P}$ be a property of objects in $D(A)$ such that the following assertions hold.*
(a) *The ring A, considered as a complex concentrated in degree 0, has property $\mathbf{P}$.*
(b) *If $\mathbf{P}$ holds for objects $E_i \in D(A)$, $i \in I$, then it holds for the direct sum $\bigoplus_i E_i$.*
(c) *If $E \to F \to G \to$ is an exact triangle in $D(A)$ and $\mathbf{P}$ holds for two of its terms, then it holds for all of them.*
Then all objects of $D(A)$ have the property $\mathbf{P}$.

In particular, if $\mathcal{T}$ is a triangulated subcategory of $D(A)$ containing A that is stable under direct sums, then $\mathcal{T} = D(A)$.

Proof. Considering the empty direct sum, we see that 0 has property $\mathbf{P}$. If $E \xrightarrow{\sim} F$ is an isomorphism in $D(R)$ and E has $\mathbf{P}$, then F has $\mathbf{P}$ using the distinguished triangle $E \to F \to 0 \to$. For $E \in D(A)$ we have an exact triangle $E \to 0 \to E[1] \to$ (obtained by shifting the exact triangle $E \to E \to 0 \to$), so (c) implies that $\mathbf{P}$ is invariant under shifts.

Let M be a complex of A-modules. To finish the proof, we will construct a quasi-isomorphism $P \to M$ of complexes of A-modules such that there exists a filtration

$$0 = F_0 \subseteq F_1 \subseteq F_2 \subseteq \cdots \subseteq P$$

by subcomplexes with $P = \bigcup_i F_i$ and such that each inclusion $F_i \subseteq F_{i+1}$ is termwise split, and each subquotient F_{i+1}/F_i is isomorphic, as a complex, to a direct sum of shifts of A. The lemma follows from this: In fact, since $P \to M$ is a quasi-isomorphism, it is enough to check property $\mathbf{P}$ for P. We have a short exact sequence

$$0 \to \bigoplus_{i \geq 0} F_i \to \bigoplus_{i \geq 0} F_i \to P \to 0$$

of complexes, where the first map is given by the direct sum of the maps $F_i \to F_i \oplus F_{i+1} \subset \bigoplus_i F_i$, $f \mapsto (f, -f)$. It is termwise split, as is easily checked using that the inclusions $F_i \to F_{i+1}$ are termwise split, and hence we obtain an exact triangle (even in $K(R)$, see F.123)

$$\bigoplus_i F_i \to \bigoplus_i F_i \to P \to$$

Now it follows easily from the properties of the F_i and the assumptions in the lemma that P has property **P**.

To construct P, first note that we can find a (degree-wise) surjective morphism of complexes $Q \to M$ such that there is a short exact sequence $0 \to Q' \to Q \to Q'' \to 0$ of complexes where Q' and Q'' are direct sums of shifts of A (as complexes). In fact, for $m \in M^i$, the complex

(*) $$\cdots \to 0 \to A \to A \to 0 \to \cdots$$

with A in degrees i and $i + 1$ and differential the identity maps to M by mapping 1 to m in degree i, and mapping 1 to $d(m)$ in degree $i + 1$. Each complex as in (*) contains $A[-i-1]$ as subcomplex and $A[-i]$ as quotient complex. Taking a direct sum of complexes as in (*), we can construct the desired surjection $Q \to M$. Moreover, by adding further copies of shifts of A, we can ensure that in addition to the above properties, the map $\mathrm{Ker}(d_Q) \to \mathrm{Ker}(d_M)$ is (degree-wise) surjective, and we will always assume that this is true when we apply this construction below.

Now define families of complexes M_i and P_i by induction as follows: Let $M_0 = M$. For any i where M_i is defined, choose a surjection $P_i \to M_i$ as above, and let M_{i+1} be its kernel. We obtain an exact complex of complexes

$$\cdots \to P_2 \to P_1 \to P_0 \to M \to 0.$$

We define P as the total complex $P := \mathrm{Tot}^{\oplus}(P_{-\bullet}^\bullet)$ (where the lower index is the row index, i.e., each P_i is a horizontal complex).

We now define the filtration. By construction, for each P_i we have a short exact sequence

$$0 \to P_i' \to P_i \to P_i'' \to 0,$$

where P_i' and P_i'' are direct sums of shifts of A. We let

$$F_{2i}P := \bigoplus_{0 \le j \le i} P_i, \quad F_{2i+1}P := F_{2i}P \oplus P_{i+1}'.$$

One sees that the inclusions are termwise split and that the subquotients of this filtration are direct sums of shifts of A.

It remains to show that $P \to M$ is a quasi-isomorphism. This follows from Lemma F.114. $\square$

Proposition 22.84. *(Derived projection formula) Let $f \colon X \to Y$ be a morphism of ringed spaces. Let $\mathscr{F} \in D(X)$, $\mathscr{G} \in D(Y)$. There is a natural morphism in $D(Y)$,*

$$(Rf_*\mathscr{F}) \otimes_{\mathscr{O}_Y}^L \mathscr{G} \to Rf_*(\mathscr{F} \otimes_{\mathscr{O}_X}^L Lf^*\mathscr{G}),$$

which is functorial in $\mathscr{F}$ and $\mathscr{G}$.

This morphism is an isomorphism in any of the following situations:

(1) *If $\mathcal{G}$ is perfect (Definition 21.136).*
(2) *If f is a homeomorphism from X onto a closed subspace of Y.*
(3) *If $f\colon X \to Y$ is a qcqs morphism of schemes and $\mathcal{F} \in D_{\mathrm{qcoh}}(X)$, $\mathcal{G} \in D_{\mathrm{qcoh}}(Y)$.*

Proof. As in the non-derived case, the existence of the map in question follows from the adjunction of Lf^* and Rf_*, Proposition 21.126. Namely, we have the counit of the adjunction $Lf^*Rf_*\mathcal{F} \to \mathcal{F}$. It gives rise to

$$Lf^*(Rf_*\mathcal{F} \otimes^L \mathcal{G}) \cong (Lf^*Rf_*\mathcal{F}) \otimes^L Lf^*\mathcal{G} \to \mathcal{F} \otimes^L Rf_*\mathcal{G},$$

where the first isomorphism is given by Proposition 21.117. Using adjunction again, we obtain the desired morphism.

To check whether this is an isomorphism, we may work locally on Y, see Lemma 21.28.

Note that the statement holds for $\mathcal{G} = \bigoplus \mathcal{G}_i$ if and only if it holds for all $\mathcal{G}_i$ To show this, observe that Lf^*, Rf_* and $\otimes^L$ are compatible with direct sums. For Lf^*, this is a formal consequence of the fact that it is left adjoint to some functor, namely Rf_*. For the other two see Lemma 22.82 and Remark 21.120 (2). Similarly, if

$$\mathcal{G}_1 \to \mathcal{G}_2 \to \mathcal{G}_3 \to$$

is an exact triangle in $D(Y)$, and the statement holds for two of the terms of this triangle (in the role of $\mathcal{G}$), then it holds for the third one. This follows from the general fact that exactness of triangles is preserved by derived functors.

Case (1). After shrinking Y, if necessary, we may assume that $\mathcal{G}$ is represented by a strictly perfect complex. Using the "stupid truncation" functor (see (F.14.3)) to cut off the first non-zero term, and the compatibility with exact triangles discussed above, we inductively reduce to the case that $\mathcal{G}$ is a single direct summand of a finite free $\mathscr{O}_Y$-module. Then we may even assume that $\mathcal{G}$ is a finite free module and finally that $\mathcal{G} = \mathscr{O}_Y$ because of the compatibility of the statement with direct sums. In this case, the assertion is obviously true.

Case (2). In this case, the functor f_* is exact, as is easy to check by looking at stalks, and thus we can compute the functor Rf_* by applying f_* to any representative. To make explicit the computation of Lf^* and $\otimes^L$, choose a K-flat representative $\mathcal{G}$ of $\mathcal{G}$. Then $Lf^*\mathcal{G}$ is represented by $f^*\mathcal{G}$. We also fix an arbitrary representative $\mathcal{F}$ of $\mathcal{F}$, so the left hand side of the morphism above is represented by $\mathrm{Tot}((f_*\mathcal{F}) \otimes \mathcal{G})$. Similarly the right hand side is represented by $f_*(\mathrm{Tot}(\mathcal{F} \otimes f^*\mathcal{G}))$. Since f_* commutes with direct sums, we are reduced to showing that the natural map

$$(f_*\mathcal{F}) \otimes \mathcal{G} \to f_*(\mathcal{F} \otimes f^*\mathcal{G})$$

is an isomorphism for an $\mathscr{O}_X$-module $\mathcal{F}$ and an $\mathscr{O}_Y$-module $\mathcal{G}$. This is just the usual projection formula Proposition 22.81 (2).

Case (3). We may assume that Y is affine, say $S = \mathrm{Spec}\, A$. Because of Theorem 22.35, we may assume that $\mathcal{G}$ is a complex of quasi-coherent $\mathscr{O}_Y$-modules. Because of the compatibilities discussed above, we can apply Lemma 22.83 and reduce to the case $\mathcal{G} = \widetilde{A}$. But in this case, the above morphism is just the identity morphism of $Rf_*\mathcal{F}$. $\qquad\square$

Example 22.85. *See Exercise 22.24 for an example illustrating that in Case (3) the condition that $\mathcal{G} \in D_{\mathrm{qcoh}}(Y)$ cannot be dropped.*

(22.20) Example: Cohomology of projective bundles.

Let us globalize our results on higher direct images of Serre's twisting line bundles on projective spaces (Theorem 22.22).

Theorem 22.86. *Let S be a scheme, let $\mathcal{E}$ be a finite locally free $\mathcal{O}_S$-module of constant rank $r + 1$, let $X := \mathbb{P}(\mathcal{E})$ be the corresponding projective bundle, and let $\pi\colon X \to S$ be the structure map.*
(1) *Then one has $R^i\pi_*\mathcal{O}_X(d) = 0$ for $i \neq 0, r$ and for all $d \in \mathbb{Z}$.*
(2) *The canonical map (13.7.2)*

$$\mathrm{Sym}(\mathcal{E}) \longrightarrow \bigoplus_{d \in \mathbb{Z}} \pi_*\mathcal{O}_X(d)$$

is an isomorphism of graded $\mathcal{O}_S$-algebras.
(3) *One has*

(22.20.1) $$R^r\pi_*\mathcal{O}_X(d) \cong (\mathrm{Sym}^{-r-1-d}\,\mathcal{E} \otimes_{\mathcal{O}_S} \det(\mathcal{E}))^\vee.$$

(4) *Set $\omega := \pi^*(\det\mathcal{E})(-r-1)$. There exists a canonical isomorphism $\mathcal{O}_S \xrightarrow{\sim} R^r\pi_*(\omega)$ and for all $d \in \mathbb{Z}$ a perfect pairing*

(22.20.2) $$R^r\pi_*\mathcal{O}_X(d) \otimes_{\mathcal{O}_S} \pi_*\omega(-d) \longrightarrow R^r\pi_*(\omega) = \mathcal{O}_S.$$

Here $\det(\mathcal{E}) = \bigwedge^{r+1}\mathcal{E}$ is the top non-zero exterior power of $\mathcal{E}$, a line bundle. Note that $\omega \cong \Omega^r_{\mathbb{P}(\mathcal{E})/S}$ as one sees by applying the determinant to the short exact sequence (17.7.3) (Exercise 7.29). We refer to Section (25.9) for the relation of Assertion (4) with Grothendieck duality.

If $S = \mathrm{Spec}\, R$ is affine and $\mathcal{E} = \mathcal{O}_S^{r+1}$, then $\mathbb{P}(\mathcal{E}) = \mathbb{P}_R^r$ and we have already seen all assertions in Theorem 22.22 and Corollary 22.23. The proof will reduce to this case. To achieve this, we will use the following global version of the Koszul complex, see also Exercise 19.8.

Remark 22.87. Let X be a scheme, let $\mathcal{L}$ be an $\mathcal{O}_X$-module, and let $u\colon \mathcal{L} \to \mathcal{O}_X$ be an $\mathcal{O}_X$-linear map. As in Section (19.1) one constructs an attached Koszul complex $K(u)$, viewed as a cochain complex of $\mathcal{O}_X$-modules concentrated in degrees ≤ 0 with $K(u)^{-p} = \bigwedge^p \mathcal{L}$ for $p \geq 0$.

From now on we suppose that $\mathcal{L}$ is locally free of finite rank r. Then $K(u)^p = 0$ for $p \notin [-r, 0]$ and $K(u)^p$ is locally free of rank $\binom{r}{-p}$ for $-r \leq p \leq 0$. For every quasi-coherent $\mathcal{O}_X$-module $\mathcal{F}$ we also set $K(u, \mathcal{F}) = K(u) \otimes_{\mathcal{O}_X} \mathcal{F}$ in this case.

Let $\mathcal{I} \subseteq \mathcal{O}_X$ be the image of u. If $V \subseteq X$ is an open affine subscheme and $f \in \mathcal{I}(V)$ a section, we can choose $\ell \in \mathcal{L}(V)$ with $u(\ell) = f$ since V is affine. As in Remark 19.6, one sees that the multiplication by ℓ on $\bigwedge \mathcal{L}$ induces a homotopy between the multiplication by f and 0. In particular, for all p one has

(22.20.3) $$\mathcal{I}\,H^p(K(u, \mathcal{F})) = 0.$$

We have $\tau^{\geq 0}K(u, \mathcal{F}) = H^0(K(u, \mathcal{F})) = \mathcal{F}/\mathcal{I}\mathcal{F}$. We say that u is *completely intersecting* for $\mathcal{F}$ if the functorial map $K(u, \mathcal{F}) \to \mathcal{F}/\mathcal{I}\mathcal{F}$ is a quasi-isomorphism.

A trivial example is the case that u is surjective. Then (22.20.3) shows that $K(u, \mathcal{F})$ is homotopy equivalent to 0.

Proof. [of Theorem 22.86] To see (1) we may work locally on S and hence can assume that $S = \operatorname{Spec} R$ and that $\mathscr{E}$ is quasi-coherent module corresponding to the free R-module R^{r+1}. Then $\mathbb{P}(\mathscr{E}) = \mathbb{P}^r_R$ and (1) follows from Theorem 22.22.

For (2) note that $\mathbb{P}(\mathscr{E}) = \operatorname{Proj}(\operatorname{Sym}(\mathscr{E}))$ by Theorem 13.32. Hence by (13.7.2) there is a functorial homomorphism of graded $\mathscr{O}_S$-algebras $\operatorname{Sym}(\mathscr{E}) \longrightarrow \bigoplus_{d \in \mathbb{Z}} \pi_*(\mathscr{O}_X(d))$ which is easily seen to be a homomorphism of graded algebras. To see that it is an isomorphism, we can again work locally on S and hence use Theorem 22.22.

Similarly, we can check that $R^r \pi_* \mathscr{O}_X(d) = 0$ for $d > -(r+1)$ locally on S and again get the result from Theorem 22.22.

To show (3) for $d \leq -r-1$ and (4) we twist the canonical surjective map $\pi^*(\mathscr{E}) \to \mathscr{O}_X(1)$ by -1 and obtain a surjective map $u\colon \pi^*(\mathscr{E})(-1) \to \mathscr{O}_X$. The corresponding Koszul complex $K(u)$ is the complex of locally free $\mathscr{O}_X$-modules

$$ 0 \to \pi^*(\det \mathscr{E})(-r-1) \longrightarrow \dots \longrightarrow \pi^*(\textstyle\bigwedge^p \mathscr{E})(-p) \longrightarrow \dots \longrightarrow \pi^*\mathscr{E}(-1) \longrightarrow \mathscr{O}_X \longrightarrow 0, $$

where $\pi^*(\bigwedge^p \mathscr{E})(-p)$ sits in degree $-p$. As u is surjective, it is $= 0$ in the derived category $D(X)$ (Remark 22.87). Hence $R\pi_* K(u) = 0$ and the first hypercohomology spectral sequence for π_* (F.49.3) takes the form

$$ E_1^{pq} = R^q \pi_* (\pi^*(\textstyle\bigwedge^{-p} \mathscr{E})(p)) = \textstyle\bigwedge^{-p} \mathscr{E} \otimes_{\mathscr{O}_S} R^q \pi_*(\mathscr{O}_X(p)) \Rightarrow 0, $$

where the second equality holds by the projection formula. As $\bigwedge^{-p} \mathscr{E} = 0$ for $p < -(r+1)$ or $p > 0$, it follows from what we have already seen that $E_1^{pq} = 0$ unless (p,q) is $(0,0)$ or $(-r-1, r)$. Therefore there exists a unique nonzero differential in the spectral sequence, namely $d_{r+1}^{-r-1,r}\colon E_{r+1}^{-r-1,r} \to E_{r+1}^{0,0}$ which is necessarily an isomorphism since the spectral sequence converges to 0. As all differentials on level $< r+1$ are zero, we have $E_1^{pq} = E_{r+1}^{pq}$. Hence $d_{r+1}^{-r-1,r}$ yields an isomorphism

$$ E_1^{-r-1,r} = R^r \pi_*(\omega) \xrightarrow{\sim} E_1^{0,0} = \mathscr{O}_S. $$

This yields (22.20.1) for $d = -r-1$. To show (22.20.1) for $d < -r-1$, note that $\pi_* \omega(-d) = \det(\mathscr{E}) \otimes \operatorname{Sym}^{-r-1-d}(\mathscr{E})$ by the projection formula. Therefore it suffices to show the existence of the perfect pairing (22.20.2).

For this consider the map

$$ \pi^* \operatorname{Sym}^{-r-1-d} \mathscr{E} \otimes_{\mathscr{O}_X} \mathscr{O}_X(d) \longrightarrow \mathscr{O}_X(-r-1). $$

If we apply $R^r \pi_*$ and use the projection formula, we obtain a map

$$ \operatorname{Sym}^{-r-1-d} \mathscr{E} \otimes_{\mathscr{O}_S} R^r \pi_* \mathscr{O}_X(d) \longrightarrow R^r \pi_* \mathscr{O}_X(-r-1) $$

which yields by tensoring with $\det(\mathscr{E})$ the pairing (22.20.2). To show that this pairing is perfect, we may again work locally and can conclude by Corollary 22.23. $\square$

Theorem 22.86 will enable us (Remark 24.71 below) to calculate the higher direct images of an arbitrary line bundle $\mathscr{M}$ on $\mathbb{P}(\mathscr{E})$ using that $\mathscr{M} \cong \pi^*(\mathscr{L})(d)$ for some line bundle $\mathscr{L}$ on S and some integer $d \in \mathbb{Z}$, see Proposition 24.69 below.

(22.21) Special case of base change: Non-derived flat base change.

Consider a commutative diagram of schemes

(22.21.1)
$$
\begin{array}{ccc}
X' & \xrightarrow{u'} & X \\
\downarrow{f'} & & \downarrow{f} \\
S' & \xrightarrow{u} & S.
\end{array}
$$

If the diagram is cartesian and $X = \operatorname{Spec} B$, $S = \operatorname{Spec} A$ and $S' = \operatorname{Spec} A'$ are all affine, we have isomorphisms

$$ A' \otimes_A M \xrightarrow{\sim} (A' \otimes_A B) \otimes_B M $$

for every B-module M. In terms of quasi-coherent sheaves, we can rephrase this by saying that we have isomorphisms

$$ u^* f_* \mathscr{F} \xrightarrow{\sim} (f')_* (u')^* \mathscr{F} $$

for every quasi-coherent $\mathscr{O}_X$-module $\mathscr{F}$. Recall (Proposition 12.6), that more generally we have

Lemma 22.88. *Consider a diagram (22.21.1) of schemes where f is affine or f is qcqs and u is flat. Assume that the diagram is cartesian. For every quasi-coherent $\mathscr{O}_X$-module $\mathscr{F}$, there is a natural isomorphism*

$$ u^* f_* \mathscr{F} \xrightarrow{\sim} (f')_* (u')^* \mathscr{F}. $$

Let us now pass to the derived version of the base change morphism for the commutative diagram (22.21.1) which we constructed in Definition 21.129. The base change morphism is a natural morphism

(22.21.2) $$ \theta \colon Lu^* Rf_* \to R(f')_* L(u')^* $$

of functors $D(X) \to D(S')$. We will give in Theorem 22.99 below a criterion when θ is an isomorphism if the functors are restricted to $D_{\mathrm{qcoh}}(X)$.

Remark 22.89. It follows from the compatibility of derived direct and inverse images with restriction to open subsets, that the base change θ commutes with restriction to open subschemes first on S and then on S' and X. In other words, let $U \subseteq S$, $U' \subseteq u^{-1}(U)$ and $V \subseteq f^{-1}(U)$ be open affine subsets, and set $V' := u'^{-1}(V) \cap f'^{-1}(U')$ (we do not assume the diagram to be cartesian). Then the base change morphism ϑ for the commutative diagram

$$
\begin{array}{ccc}
V' & \longrightarrow & V \\
\downarrow & & \downarrow \\
U' & \longrightarrow & U
\end{array}
$$

satisfies $\vartheta \circ (-_{|V}) = (-_{|U'}) \circ \theta$ as morphisms of functors $D(X) \to D(U')$.

Let us describe $\theta_{\mathscr{F}} \colon Lu^* Rf_* \mathscr{F} \to R(f')_* L(u')^* \mathscr{F}$ for $\mathscr{F} \in D_{\mathrm{qcoh}}(X)$ if $S = \operatorname{Spec} R$, $S' = \operatorname{Spec} R'$ and $X = \operatorname{Spec} A$ are affine. Then $\mathscr{F} \cong \tilde{F}$ for a complex F of A-modules (Lemma 22.36). Choose a quasi-isomorphism $P \to F$, where P is a colimit of bounded above complexes of free A-modules (which is possible by Remark 21.95 (1) applied to the ringed space $(*, A)$). Then $Lu'^* \mathscr{F}$ is represented by $u'^* \tilde{P}$ (Lemma 22.39). Moreover, as f is affine and hence f_* is exact on quasi-coherent modules, we have $Rf_* \mathscr{F} \cong f_* \tilde{P}$. Now $\theta_{\mathscr{F}}$ is the composition

$$Lu^* Rf_* \mathscr{F} \cong Lu^* f_* \tilde{P} \to u^* f_* \tilde{P} \longrightarrow f'_* u'^* \tilde{P} \to Rf'_* u'^* \tilde{P} \cong Rf'_* Lu'^* \mathscr{F},$$

where the longer arrow in the middle is given by the non-derived base change map (12.2.2) extended to complexes of $\mathscr{O}_X$-modules.

Suppose that u is flat, so that $Lu^* = u^*$ (Example F.169). Passing to cohomology objects, in this case we obtain homomorphisms

$$(22.21.3) \qquad\qquad u^* R^i f_* \mathscr{F} \to R^i (f')_* (u')^* \mathscr{F}$$

for every $\mathscr{F} \in D(X)$ and $i \geq 0$. For their construction, as before, it is not necessary to assume that the diagram above is cartesian.

One can construct a homomorphism like this also without the flatness hypothesis, see Section (23.28) below.

Although the base change homomorphism is not in general an isomorphism (Exercise 22.42), it is so under certain circumstances. Interestingly, there are several quite different kinds of conditions one can impose to ensure this. At this point we prove a relatively simple case directly. See Theorem 23.140 for further important cases which are particularly significant because there u is not required to be flat.

Proposition 22.90. *Consider a cartesian diagram* (22.21.1) *of morphisms of schemes. If the morphism u is flat and the morphism f is qcqs, then for every complex of quasi-coherent $\mathscr{O}_X$-modules and every $i \in \mathbb{Z}$ the base change morphism induces an isomorphism*

$$u^* R^i f_* \mathscr{F} \cong R^i (f')_* (u')^* \mathscr{F}.$$

This generalizes Lemma 22.88 to higher direct images. Note that the base change morphism is also an isomorphism if f is affine because then all higher direct images vanish (Corollary 22.5).

We give here a direct proof in the case that f is separated and that $\mathscr{F}$ is a quasi-coherent $\mathscr{O}_X$-module and refer to Remark 22.108 below, where we will deduce the general result from the much more general base change Theorem 22.99.

Proof. Under our assumption that f is separated, we can compute cohomology as Čech cohomology, allowing for a simpler proof than in the general case. We can work locally on S' and on S, and hence assume that $S = \operatorname{Spec} A$ and $S' = \operatorname{Spec} A'$ are affine. Then u and hence u' are affine. Moreover, X and X' are quasi-compact and separated since we assumed f and hence f' to be so. In this situation, the base change morphism (22.21.3) is given by a homomorphism of A'-modules

$$(*) \qquad\qquad H^i(X, \mathscr{F}) \otimes_A A' \xrightarrow{\sim} H^i(X', \mathscr{F}'),$$

where we set $\mathscr{F}' := (u')^* \mathscr{F}$. Choose a finite affine open cover $X = \bigcup_i U_i$. We obtain an affine open cover $X' = \bigcup_i (u')^{-1}(U_i)$ and can compute the two cohomology groups above using Čech cohomology for this covering (Theorem 22.9).

As we have $\mathscr{F}(V) \otimes_A A' = \mathscr{F}(u'^{-1}(V), \mathscr{F}')$ for every open affine subscheme $V \subseteq X$ and as tensor products commute with finite products, we find for the Čech complexes

$$\check{C}^\bullet((U_i)_i, \mathscr{F}) \otimes_A A' \cong \check{C}^\bullet(((u')^{-1}(U_i)_i, \mathscr{F}').$$

As A' is a flat A-algebra, applying cohomology $H^i(-)$ commutes with the exact functor $- \otimes_A A'$. This yields an isomorphism (*).

We omit the check that this isomorphism is given by (22.21.3). $\qquad\square$

Corollary 22.91. *Let* $f \colon X \to \operatorname{Spec} R$ *be a qcqs morphism and let* $R \to R'$ *be a flat map of rings. Then for every quasi-coherent* $\mathscr{O}_X$*-module* $\mathscr{F}$ *the base change morphism induces for all* $i \geq 0$ *isomorphisms*

$$H^i(X, \mathscr{F}) \otimes_R R' \cong H^i(X_{R'}, \mathscr{F}_{R'}),$$

where $X_{R'} := X \otimes_R R'$ *and* $\mathscr{F}_{R'}$ *is the pullback of* $\mathscr{F}$ *to* $X_{R'}$.

Corollary 22.92. *Let* A *be a ring, let* X *an* A*-scheme, and* B *a faithfully flat* A*-algebra. Then* X *is affine if and only* $X \times_{\operatorname{Spec} A} \operatorname{Spec} B$ *is affine.*

We give a cohomological proof here. Note that X is affine if and only if the morphism $X \to \operatorname{Spec} A$ is affine. Hence the corollary is also a special case of the fact that being "affine" is a stable under fpqc-descent (Proposition 14.53).

Proof. Assume that $X \times_{\operatorname{Spec} A} \operatorname{Spec} B$ is affine. We want to show that X is affine. Note that X is qcqs by Proposition 14.51. Using Serre's criterion for affineness, Theorem 12.35, we see that it is enough to show that $H^1(X, \mathscr{F}) = 0$ for all quasi-coherent $\mathscr{O}_X$-modules $\mathscr{F}$. By Corollary 22.91 and since $X \otimes_A B$ is affine by assumption, we have

$$H^1(X, \mathscr{F}) \otimes_A B \cong H^1(X \otimes_A B, \mathscr{F}_B) = 0,$$

where $\mathscr{F}_B$ denotes the pullback of $\mathscr{F}$. Since B is faithfully flat over A, the result follows. $\square$

(22.22) Tor-independence.

Definition and Lemma 22.93. *Two morphisms of schemes*

$$
\begin{array}{c}
X \\
\downarrow f \\
S' \xrightarrow{\ u\ } S
\end{array}
$$

(22.22.1)

are said to be tor-independent, *if the following equivalent assertions are satisfied.*
  (i) *For all* $s' \in S'$ *and* $x \in X$ *with same image* $s \in S$ *one has*

$$\operatorname{Tor}_i^{\mathscr{O}_{S,s}}(\mathscr{O}_{S',s'}, \mathscr{O}_{X,x}) = 0 \qquad \text{for all } i > 0.$$

  (ii) *For all* $s' \in S'$ *and* $x \in X$ *with same image* $s \in S$ *there exist open affine neighborhoods* $\operatorname{Spec} R \subseteq S$ *of* s, $\operatorname{Spec} A \subseteq f^{-1}(\operatorname{Spec} R)$ *of* x, *and* $\operatorname{Spec} R' \subseteq u^{-1}(\operatorname{Spec} R)$ *of* s' *such that*

$$\operatorname{Tor}_i^R(R', A) = 0 \qquad \text{for all } i > 0.$$

  (iii) *For all open affine subschemes* $\operatorname{Spec} R \subseteq S$, $\operatorname{Spec} A \subseteq f^{-1}(\operatorname{Spec} R)$, *and* $\operatorname{Spec} R' \subseteq u^{-1}(\operatorname{Spec} R)$ *one has*

$$\operatorname{Tor}_i^R(R', A) = 0 \qquad \text{for all } i > 0.$$

Below (Definition 22.96) we will define tor-independence for complexes of quasi-coherent modules such that the complexes $\mathscr{O}_X$ and $\mathscr{O}_{S'}$, concentrated in degree 0, are tor-independent if and only if f and u are tor-independent (Remark 22.97 (1)). Then the equivalences of the conditions above will be a special case of Lemma 22.96. Hence, we do not give a proof here.

It is clear from the definition that tor-independence can be checked locally on S, S' and X. Furthermore, interchanging the roles of f and u does not affect the property of tor-independence.

There are many examples of morphisms that are not tor-independent. For instance every pair of closed immersions $Y \to X$ and $Z \to X$ given by quasi-coherent ideals $\mathscr{I}, \mathscr{J} \subseteq \mathscr{O}_X$ with $\mathscr{I}\mathscr{J} \neq \mathscr{I} \cap \mathscr{J}$ is not tor-independent (Exercise 21.19). A trivial but important example of tor-independent morphisms is the following.

Remark 22.94. If in the situation of Definition 22.93 one of the morphisms u or f is flat, then u and f are tor-independent.

An example of tor-independent closed immersions is the following.

Example 22.95. Let R be a ring and let $\mathbf{f} = (f_1, \ldots, f_r)$ be a completely intersecting sequence in R (Definition 19.10), choose some $1 \leq s \leq r$ and set $R' = R/(f_1, \ldots, f_{s-1})$ and $A = R/(f_s, \ldots, f_r)$. Then $\operatorname{Spec} A$ and $\operatorname{Spec} R'$ are tor-independent over $\operatorname{Spec} R$.

Indeed, in $D(R)$ one has isomorphisms

$$R' \otimes_R^L A \cong K_\bullet(f_1, \ldots, f_{s-1}) \otimes_R K_\bullet(f_s, \ldots, f_r) \cong K_\bullet(f_1, \ldots, f_r) \cong R/(f_1, \ldots, f_r)$$

for the first isomorphism using that Koszul complexes are K-flat, and we obtain

$$\operatorname{Tor}_i^R(R', A) = \begin{cases} R/(f_1, \ldots, f_r), & \text{if } i = 0, \\ 0, & \text{if } i > 0. \end{cases}$$

One can also define tor-independence more generally for complexes. The condition that $\operatorname{Tor}_i^R(R', A) = 0$ for $i > 0$ can be rephrased as the condition that the canonical map $R' \otimes_R^L A \to R' \otimes_R A$ is an isomorphism. This makes the following definition plausible.

Definition and Lemma 22.96. *Let $f \colon X \to S$ and $u \colon S' \to S$ be morphisms of schemes and let $\mathscr{F}$ (resp. $\mathscr{G}$) be a complex of quasi-coherent $\mathscr{O}_X$-modules (resp. of quasi-coherent $\mathscr{O}_{S'}$-modules). Then $\mathscr{F}$ and $\mathscr{G}$ are said to be* tor-independent over S *if the following equivalent conditions are satisfied.*

(i) *For all $s' \in S'$ and $x \in X$ with same image $s \in S$ the canonical homomorphism*

$$\mathscr{F}_x \otimes_{\mathscr{O}_{S,s}}^L \mathscr{G}_{s'} \longrightarrow \mathscr{F}_x \otimes_{\mathscr{O}_{S,s}} \mathscr{G}_{s'}$$

is an isomorphism.

(ii) *For all $s' \in S'$ and $x \in X$ with same image $s \in S$ there exist open affine neighborhoods $\operatorname{Spec} R \subseteq S$ of s, $\operatorname{Spec} A \subseteq f^{-1}(\operatorname{Spec} R)$ of x, and $\operatorname{Spec} R' \subseteq u^{-1}(\operatorname{Spec} R)$ of s' such that the canonical homomorphism*

$$F \otimes_R^L G \longrightarrow F \otimes_R G$$

is an isomorphism, where F (resp. G) is the complex of A-modules (resp. of R'-modules) corresponding to $\mathscr{F}_{|\operatorname{Spec} A}$ (resp. to $\mathscr{G}_{|\operatorname{Spec} R'}$).

(iii) *For all open affine subschemes $\operatorname{Spec} R \subseteq S$, $\operatorname{Spec} A \subseteq f^{-1}(\operatorname{Spec} R)$, and $\operatorname{Spec} R' \subseteq u^{-1}(\operatorname{Spec} R)$ the canonical homomorphism*

$$F \otimes_R^L G \longrightarrow F \otimes_R G$$

is an isomorphism, where F and G are as in (ii).

Proof. Clearly, (iii) implies (ii). To see that (ii) implies (i) let $\mathfrak{p} \subset R$, $\mathfrak{p}' \subset R'$ and $\mathfrak{q} \subset A$ be the prime ideals corresponding to s, s', and x, respectively. One has isomorphisms

(*)
$$F_x \otimes^L_{R_\mathfrak{p}} G_{s'} = F_x \otimes^L_R G_s = A_\mathfrak{q} \otimes F \otimes^L_R G \otimes R'_{\mathfrak{p}'},$$
$$F_x \otimes_{R_\mathfrak{p}} G_{s'} = F_x \otimes_R G_s = A_\mathfrak{q} \otimes F \otimes_R G \otimes R'_{\mathfrak{p}'}$$

where the second equality in the first line holds since localizations are flat modules. Hence if (ii) holds, then $F_x \otimes^L_{R_\mathfrak{p}} G_{s'} = F_x \otimes_{R_\mathfrak{p}} G_{s'}$.

It remains to show that (i) implies (iii). The map $F \otimes^L_R G \to F \otimes_R G$ is an isomorphism in $D(A \otimes_R R')$ if and only if it is an isomorphism on stalks, i.e., after localization in all prime ideals $\mathfrak{q}' \subset R' \otimes_R A$. Let $\mathfrak{p}$, $\mathfrak{p}'$, and $\mathfrak{q}$ be the inverse images of $\mathfrak{q}'$ in R, R', and A. Then (*) shows that $(F \otimes^L_R G)_{\mathfrak{q}'} \to (F \otimes_R G)_{\mathfrak{q}'}$ is a localization of $F_x \otimes^L_{R_\mathfrak{p}} G_{s'} \to F_x \otimes_{R_\mathfrak{p}} G_{s'}$, which is an isomorphism by hypothesis. $\square$

Remark 22.97. Let $f: X \to S$, $u: S' \to S$, and $\mathscr{F}$ and $\mathscr{G}$ be as in Definition 22.96.
(1) If $\mathscr{F}$ and $\mathscr{G}$ are quasi-coherent modules, considered as complexes concentrated in degree 0, then $\mathscr{F}$ and $\mathscr{G}$ are tor-independent if and only if $\mathrm{Tor}_i^{\mathscr{O}_{S,s}}(\mathscr{G}_{s'}, \mathscr{F}_x) = 0$ for all $i > 0$ and for all $s' \in S'$ and $x \in X$ with the same image s in S.

In particular, the morphisms f and u are tor-independent if and only if $\mathscr{O}_X$ and $\mathscr{O}_{S'}$ are tor-independent over S.
(2) Let $\mathscr{F}$ be K-flat as complex of $f^{-1}\mathscr{O}_S$-modules, e.g., if $\mathscr{F}$ is a bounded above complex of quasi-coherent modules that are flat over S (Lemma 21.93). Then $\mathscr{F}$ and $\mathscr{G}$ are tor-independent over S for all $\mathscr{G}$.

Indeed, we may assume that $S = \mathrm{Spec}\, R$, $S' = \mathrm{Spec}\, R'$, and $X = \mathrm{Spec}\, A$ are affine. Then $\mathscr{F}$ corresponds to a complex F of A-modules that is K-flat as complex of R-modules (Lemma 22.39). Hence $F \otimes^L_R G = F \otimes_R G$ for every complex G in $D(R')$.

(22.23) Base change and Künneth formula: Main Theorem.

In this section, we will more or less follow Lipman [Lip2] 3.10 with some generalizations (e.g. Proposition 22.102) needed for the general theory of base change of higher direct images in Section (23.28).

Let

(22.23.1)
$$\begin{array}{ccc} X' & \xrightarrow{u'} & X \\ \downarrow{f'} & & \downarrow{f} \\ S' & \xrightarrow{u} & S \end{array}$$

be a commutative diagram of schemes.

We write $g = f \circ u' = u \circ f'$ for the composition and define the *Künneth map* for $E \in D_{\mathrm{qcoh}}(S')$, $F \in D_{\mathrm{qcoh}}(X)$ as the composition

(22.23.2)
$$\eta: Ru_*E \otimes^L Rf_*F \to Rg_*Lg^*(Ru_*E \otimes^L Rf_*F)$$
$$\to Rg_*(L(f')^*Lu^*Ru_*E \otimes^L L(u')^*Lf^*Rf_*F)$$
$$\to Rg_*(L(f')^*E \otimes^L L(u')^*F).$$

Here we use the adjunction morphisms in the first and in the last step, and the compatibility between derived tensor product and derived pullback in the middle step. This construction is functorial in E and in F.

Example 22.98. Assume that in the above situation $S = \operatorname{Spec} k$ is the spectrum of a field. Then in particular u and f, and hence u' and f' are flat. In this case, we can compute derived pullback as usual pullback and derived tensor product over $\mathscr{O}_S$ as usual tensor product of complexes, using representatives of E and F. Then the Künneth morphism simplifies to give us a morphism

$$\eta\colon Ru_*E \otimes Rf_*F \to Rg_*((f')^*E \otimes (u')^*F).$$

Let $E \to I$ and $F \to J$ be K-injective resolutions. Then $Ru_*E \otimes Rf_*F = u_*I \otimes v_*J$. Passing to cohomology objects we obtain by Proposition F.95 for all $n \in \mathbb{Z}$ a homomorphism

$$\bigoplus_{i+j=n} H^i(S', E) \otimes_k H^j(X, F) \cong H^n(u_*I \otimes f_*J) \longrightarrow H^n(X', (f')^*E \otimes (u')^*F),$$

where we identify the categories of $\mathscr{O}_{\operatorname{Spec} k}$-modules and of k-vector spaces. We will see later (Corollary 22.110) that these homomorphisms always are isomorphisms.

The main result in this section is the following theorem.

Theorem 22.99. *Assume that the diagram* (22.23.1) *is cartesian, that all the schemes in the diagram are quasi-separated and that f and u are qcqs. Write $g := f \circ u' = u \circ f'$. The following are equivalent:*
(i) *The base change morphism $\theta\colon Lu^*Rf_* \to R(f')_*L(u')^*$* (22.21.2) *is an isomorphism of functors $D_{\mathrm{qcoh}}(X) \to D_{\mathrm{qcoh}}(S')$.*
(ii) *The Künneth morphism* (22.23.2) *for this diagram*

$$\eta\colon Ru_*E \otimes^L_{\mathscr{O}_S} Rf_*F \longrightarrow Rg_*(Lf'^*E \otimes^L_{\mathscr{O}_{X'}} Lu'^*F)$$

is an isomorphism for all $E \in D_{\mathrm{qcoh}}(S')$, $F \in D_{\mathrm{qcoh}}(X)$.
(iii) *The schemes X and S' are tor-independent over S (Definition 22.93).*

We will prove the theorem in the next sections. The proof will show that for some of the implications, the hypotheses can be further weakened, see Remark 22.106 below.
The following remark will show that (i) implies (ii).

Remark 22.100. With the notation of Theorem 22.99 (ii), we have a commutative diagram

(22.23.3)
$$\begin{CD} Ru_*E \otimes^L Rf_*F @>\eta>> Rg_*(L(f')^*E \otimes^L L(u')^*F) \\ @V\cong VV @AA\cong A \\ Ru_*(E \otimes^L Lu^*Rf_*F) @>>> Ru_*(E \otimes^L R(f')_*L(u')^*F), \end{CD}$$

where the lower horizontal arrow is obtained from the base change morphism by applying $E \otimes^L -$ and Ru_*, and the vertical isomorphisms come from the derived projection formula Proposition 22.84. The commutativity of the diagram follows formally from the properties of derived pullback and derived direct image (see [Lip2] $\overset{\mathrm{o}}{\times}$ (3.10.2.3) for details).

From the diagram, we obtain immediately that (i) implies (ii) in Theorem 22.99. Furthermore, whenever Ru_* is conservative (e.g., if u is affine, see Lemma 22.101 below), we even get that (ii) and (i) are equivalent.

If u is an open immersion, the base change morphism is an isomorphism (Lemma 21.28). Hence we conclude that the Künneth morphism is an isomorphism in this case.

(22.24) Tor-independence implies base change.

To show the other implications of Theorem 22.99 we will use the following lemmas.

Lemma 22.101. *Let $f\colon X \to S$ be a qcqs morphism of schemes. If f is an open immersion, or if f is affine, then the functor $Rf_*\colon D_{\mathrm{qcoh}}(X) \to D_{\mathrm{qcoh}}(S)$ is conservative (Definition F.3).*

The lemma in particular shows that $Rf_*\colon D_{\mathrm{qcoh}}(X) \to D_{\mathrm{qcoh}}(S)$ is conservative for every quasi-affine morphism $f\colon X \to S$ (Exercise 22.23).

Proof. If f is an open immersion, then the adjunction morphism $Lf^*Rf_*E \to E$ is an isomorphism for all E (Remark 21.127) and we can identify $\alpha = Lf^*Rf_*\alpha$, hence α is an isomorphism if $Rf_*\alpha$ is an isomorphism.

Now assume that f is affine, and let $\alpha\colon F \to G$ be a morphism in $D_{\mathrm{qcoh}}(X)$. Since the statement if local on S, we may assume that S and hence X are affine. In fact, by Proposition 22.33, we can equivalently consider complexes F, G of quasi-coherent $\mathscr{O}_X$-modules, and compute Rf_* by applying f_* to all terms (of a fixed representative). The fact that α, or $Rf_*\alpha$, is an isomorphism can equivalently be expressed by saying that the respective morphism fits into an exact triangle with third term an exact complex.

Thus it is enough to show that a complex E of quasi-coherent $\mathscr{O}_X$-modules such that f_*E is exact, is itself exact. But the functor f_* is exact on the category of quasi-coherent $\mathscr{O}_X$-modules, so $H^i(f_*E) = f_*(H^i(E))$, and if $\mathscr{F}$ is any quasi-coherent $\mathscr{O}_X$-module with $f_*\mathscr{F} = 0$, then $\mathscr{F} = 0$. It follows that Rf_* is indeed conservative. $\qquad\square$

We now come to the proof that (iii), i.e., tor-independence, implies (i), i.e., the base change formula. This will be the special case $\mathscr{F} = \mathscr{O}_X$ and $\mathscr{G} = \mathscr{O}_{S'}$ of Proposition 22.102 below.

To formulate this more general result we introduce the following notation. Consider a commutative diagram of morphism of schemes as in (22.23.1). Let $\mathscr{F}$ (resp. $\mathscr{G}$) be a complex of quasi-coherent $\mathscr{O}_X$-modules (resp. of quasi-coherent $\mathscr{O}_{S'}$-modules), and let E be in $D_{\mathrm{qcoh}}(X)$. We construct a general base change map in $D(S')$

$$(22.24.1) \qquad \mathscr{G} \otimes^L_{\mathscr{O}_{S'}} Lu^*Rf_*(\mathscr{F} \otimes^L_{\mathscr{O}_X} E) \longrightarrow Rf'_*(f'^*\mathscr{G} \otimes_{\mathscr{O}_{X'}} u'^*\mathscr{F} \otimes^L_{\mathscr{O}_{X'}} Lu^*E),$$

which is functorial in $\mathscr{G}$, $\mathscr{F}$ and E as follows. It is the composition in $D(S')$

$$\eta\colon \mathscr{G} \otimes^L Lu^*Rf_*(\mathscr{F} \otimes^L E) \longrightarrow \mathscr{G} \otimes^L Rf'_*Lu'^*(\mathscr{F} \otimes^L E)$$
$$\xrightarrow{\ \sim\ } Rf'_*\big(Lf'^*\mathscr{G} \otimes^L Lu'^*(\mathscr{F} \otimes^L E)\big)$$
$$\xrightarrow{\ \sim\ } Rf'_*\big(Lf'^*\mathscr{G} \otimes^L Lu'^*\mathscr{F} \otimes^L Lu'^*E)\big)$$
$$\longrightarrow Rf'_*\big(f'^*\mathscr{G} \otimes_{\mathscr{O}_{X'}} u'^*\mathscr{F} \otimes^L Lu'^*E)\big),$$

where the first map is given by the usual base change map, the second by the projection formula, the third by compatibility of derived pullback and derived tensor product (Proposition 21.117), and the last by the canonical map $Lf'^*\mathscr{G} \otimes^L Lu'^*\mathscr{F} \to f'^*\mathscr{G} \otimes u'^*\mathscr{F}$.

Proposition 22.102. *In the situation above suppose that Diagram (22.23.1) is cartesian, that f is qcqs, and that $\mathscr{F}$ and $\mathscr{G}$ are tor-independent over S (Definition 22.96). Then the base change map (22.24.1) is an isomorphism in $D_{\mathrm{qcoh}}(S')$.*

Proof. To show that the base change homomorphism is an isomorphism, we can work locally S and on S', and therefore we can assume that S and S' are affine. Then u, and hence u', are affine morphisms. Let

$$\theta_0 \colon Lf^*(u_*\mathscr{G}) \otimes^L_{\mathscr{O}_X} \mathscr{F} \longrightarrow f^*(u_*\mathscr{G}) \otimes_{\mathscr{O}_X} \mathscr{F}$$

be the canonical morphism in $D_{\mathrm{qcoh}}(X)$. We will show the following assertions (using the assumption that S and S' are affine). These assertions prove the proposition.
(1) The base change map (22.24.1) is an isomorphism if and only if the map

$$(*) \qquad Rf_*\big(Lf^*(u_*\mathscr{G}) \otimes^L \mathscr{F} \otimes^L E\big) \longrightarrow Rf_*\big(f^*(u_*\mathscr{G}) \otimes_{\mathscr{O}_X} \mathscr{F} \otimes^L E\big)$$

induced from θ_0 by functoriality is an isomorphism.
(2) The morphism θ_0 is an isomorphism if and only if $\mathscr{F}$ and $\mathscr{G}$ are tor-independent.
The proof of the first assertion will not need $\mathscr{F}$ and $\mathscr{G}$ to be tor-independent.

To show Assertion (1), note that by Lemma 22.101 (and noting that the terms involved have quasi-coherent cohomology) the base change map is an isomorphism if and only if it is an isomorphism after applying Ru_*. If we apply Ru_* to the left side of (22.24.1) we obtain

$$
\begin{aligned}
Ru_*(\mathscr{G} \otimes^L Lu^* Rf_*(\mathscr{F} \otimes^L E)) &= Ru_*\mathscr{G} \otimes^L Rf_*(\mathscr{F} \otimes^L E) \\
&= u_*\mathscr{G} \otimes^L Rf_*(\mathscr{F} \otimes^L E) \\
&= Rf_*(Lf^*(u_*\mathscr{G}) \otimes^L \mathscr{F} \otimes^L E),
\end{aligned}
$$

where the first and the last equality hold by the projection formula and where the second equality holds since u is affine and $\mathscr{G}$ is a complex of quasi-coherent modules (Proposition 22.33).

If we apply Ru_* to the right side of (22.24.1) we obtain

$$
\begin{aligned}
Ru_*\big(Rf'_*(f'^*\mathscr{G} \otimes u'^*\mathscr{F} \otimes^L Lu'^*E)\big) &= Rf_*\big(Ru'_*(f'^*\mathscr{G} \otimes u'^*\mathscr{F} \otimes^L Lu'^*E)\big) \\
&= Rf_*\big(Ru'_*(f'^*\mathscr{G} \otimes u'^*\mathscr{F}) \otimes^L E)\big) \\
&= Rf_*\big(u'_*(f'^*\mathscr{G} \otimes u'^*\mathscr{F}) \otimes^L E\big) \\
&= Rf_*\big(u'_*(f'^*\mathscr{G}) \otimes \mathscr{F} \otimes^L E\big) \\
&= Rf_*\big(f^*(u_*\mathscr{G}) \otimes \mathscr{F} \otimes^L E\big).
\end{aligned}
$$

Here the first equality holds since $u \circ f' = f \circ u$ using Proposition 21.115, the second by the projection formula. The remaining equalities holds since u' and u are affine, where for the last equality one uses also Lemma 22.88.

One checks, using Remark 21.129, that via these identifications the base change map is identified with the map in Assertion (1). This shows (1).

To see that Assertion (2) holds, we may work locally on X and hence assume that $X = \operatorname{Spec} A$ is affine. Let $S = \operatorname{Spec} R$ and $S' = \operatorname{Spec} R'$. We consider $\mathscr{F}$ as a complex F of A-modules and $\mathscr{G}$ as complex G of R'-modules. By Lemma 22.41, the source of θ_0 becomes (the complex of quasi-coherent $\mathscr{O}_X$-modules associated with) the complex of A-modules

$$G \otimes^L_R A \otimes^L_A F = G \otimes^L_R F,$$

and the target of θ_0 becomes

$$G \otimes_R A \otimes_A F = G \otimes_R F.$$

This shows Assertion (2). $\qquad\square$

In the next chapter we will use the proposition in the following form.

Corollary 22.103. *In the situation above suppose that Diagram (22.23.1) is cartesian and that f is qcqs. Let $\mathscr{F}$ be a bounded above complex of quasi-coherent modules that are flat over S and let E be in $D_{\mathrm{qcoh}}(X)$. Then there is a functorial isomorphism in $D_{\mathrm{qcoh}}(S')$*

$$(22.24.2) \qquad Lu^* Rf_* (\mathscr{F} \otimes^L_{\mathscr{O}_X} E) \xrightarrow{\sim} Rf'_* (u'^* \mathscr{F} \otimes^L_{\mathscr{O}_{X'}} Lu^* E).$$

Proof. This is the special case $\mathscr{G} = \mathscr{O}_{S'}$ of Proposition 22.102 using that $\mathscr{O}_S$ and $\mathscr{F}$ are tor-independent by Remark 22.97 (2). $\qquad\square$

The proof shows that instead of assuming that $\mathscr{F}$ is bounded above with S-flat components, it would be sufficient that $\mathscr{F}$ and $\mathscr{O}_{S'}$ are tor-independent over S.

(22.25) Künneth isomorphism implies tor-independence.

To finish the proof of Theorem 22.99, it remains to show that if the Künneth map is an isomorphism, then f und u are tor-independent. For this we show first that if the Künneth map is an isomorphism, then it is also locally an isomorphism.

Consider the commutative diagram (22.23.1). Let $S = \bigcup_{i \in I} S_i$, $u^{-1}(S_i) = \bigcup_{j \in J_i} S'_{ij}$, $f^{-1}(S_i) = \bigcup_{\ell \in L_i} X_{i\ell}$ be open covers, such that all the corresponding open immersions are quasi-compact morphisms (e.g., if all S_i, S'_{ij}, and $X_{i\ell}$ are affine). We also define $X'_{ij\ell} := f'^{-1}(S'_{ij}) \cap u'^{-1}(X_{i\ell})$. Then the $X'_{ij\ell}$ form an open covering of X' and the open immersions $X'_{ij\ell} \to X'$ are quasi-compact and in particular qcqs because all immersions are separated. We obtain for every $i \in I$, $j \in J_i$, $\ell \in L_i$, a commutative diagram

$$(22.25.1) \qquad
\begin{array}{ccc}
X_{ij\ell} & \xrightarrow{\;u'\;} & X_{i\ell} \\
{\scriptstyle f'}\downarrow & & \downarrow{\scriptstyle f} \\
S'_{ij} & \xrightarrow{\;u\;} & S_i.
\end{array}$$

If (22.23.1) is cartesian, then (22.25.1) is cartesian (Corollary 4.19).

Lemma 22.104. *Assume that the Künneth morphism for diagram (22.23.1) is an isomorphism for all $E \in D_{\mathrm{qcoh}}(S')$, $F \in D_{\mathrm{qcoh}}(X)$. Then the Künneth morphism for (22.25.1) is an isomorphism for all objects in $D_{\mathrm{qcoh}}(S'_{ij})$ and $D_{\mathrm{qcoh}}(X_{i\ell})$.*

As we can check (iii) of Theorem 22.99 locally, the theorem shows that a converse in the following sense is true if diagram (22.23.1) is cartesian: If the Künneth morphisms are isomorphisms for all objects in $D_{\mathrm{qcoh}}(S'_{ij})$ and $D_{\mathrm{qcoh}}(X_{i\ell})$ and the (then automatically cartesian) diagram (22.25.1), then the Künneth morphisms for all objects in $D_{\mathrm{qcoh}}(S')$ and $D_{\mathrm{qcoh}}(X)$ and the original cartesian diagram are isomorphisms.

Proof. [Sketch of proof] *(I).* First, we check that we may pass to open subsets $U \subseteq S$ (and correspondingly replace X by $X_U := X \times_S U$, S' by $U' := S' \times_S U$ and X' by $X'_U := X' \times_S U$) whenever the corresponding open immersion $i \colon U \to S$ is quasi-compact. In fact, denote by $i' \colon U' \to S'$ and $j \colon X_U \to X$ the morphisms obtained by base change from i. Let η be the Künneth morphism for the diagram

$$
\begin{array}{ccc}
X'_U & \longrightarrow & X_U \\
\downarrow & & \downarrow {\scriptstyle f} \\
U' & \longrightarrow & U
\end{array}
$$

and $E \in D_{\mathrm{qcoh}}(U')$, $F \in D_{\mathrm{qcoh}}(X_U)$. Then it is not difficult albeit quite tedious to check (see [Lip2]$\overset{0}{\underset{X}{}}$ Lemma (3.10.3.4) for the details, note that there the diagram (22.23.1) is assumed to be cartesian but this is not used in the proof) that $Ri_*\eta$ is isomorphic to the Künneth isomorphism for the original diagram and for Ri'_*E and Rj_*F (which lie in $D_{\mathrm{qcoh}}(S')$ and $D_{\mathrm{qcoh}}(X)$, resp., by Theorem 22.32). Hence by our assumption, $Ri_*\eta$ is an isomorphism. Since i is an open immersion, this implies that η is an isomorphism (Lemma 22.101), as desired.

(II). Second, given a diagram

$$
\begin{array}{ccccc}
U' \times_{S'} X' & \longrightarrow & X' & \overset{u'}{\longrightarrow} & X \\
\downarrow & & \downarrow {\scriptstyle f'} & & \downarrow {\scriptstyle f} \\
U' & \longrightarrow & S' & \overset{u}{\longrightarrow} & S
\end{array}
$$

where both squares have the property that the Künneth morphism always is an isomorphism, then the rectangle also has this property (cf. [Lip2]$\overset{0}{\underset{X}{}}$, Lemma 3.10.3.2).

(III). Finally, if u in (22.23.1) is an open immersion, then the Künneth morphism is an isomorphism (Remark 22.100). Using (II) this allows us to pass to open subsets of S'. Since the Künneth morphism is symmetric with respect to switching the roles of f and of u, by the same argument we obtain that we may pass to open subsets of X. $\qquad\square$

Proof. [Proof of Theorem 22.99] We have already seen that (ii) implies (i) (Remark 22.100) and that (iii) implies (i) (Proposition 22.102).

It remains to show that (ii) implies (iii). By Lemma 22.104 we know that (ii) for the original diagram implies that (ii) holds for all diagrams attached to suitable coverings of S, S', X. Since we can check (iii) locally on S, S' and X, this means that we may assume, without loss of generality, that S, S' and X are affine schemes, say $S = \operatorname{Spec} A$, $S' = \operatorname{Spec} A'$, $X = \operatorname{Spec} B$. Moreover, since the diagram is cartesian, $X' = X \times_S S' = \operatorname{Spec}(A \otimes_R R')$ is also affine.

Then u is affine and hence we have already seen that (ii) and (i) are equivalent (Remark 22.100). Hence it suffices to show that the diagram (22.23.1) is tor-independent if the base change morphism

$$
\theta_{\mathscr{O}_X} \colon Lu^* Rf_* \mathscr{O}_X \to Rf'_* Lu'^* \mathscr{O}_X
$$

is an isomorphism.

As f and f' are affine, we may identify by Proposition 22.33

$$
Lu^* Rf_* \mathscr{O}_X = Lu^* f_* \mathscr{O}_X = (R' \otimes_R^L A)^\sim
$$

and

$$
R(f')_* L(u')^* \mathscr{O}_X = (f')_* L(u')^* \mathscr{O}_X = (f')_* (u')^* \mathscr{O}_X = u^* f_* \mathscr{O}_X = (R' \otimes_R A)^\sim,
$$

where we have also used that $L(u')^* \mathscr{O}_X = (u')^* \mathscr{O}_X$ (since $\mathscr{O}_X$ is flat over itself), and in the final step the non-derived base change homomorphism $u^* f_* \mathscr{O}_X \to (f'_*)(u')^* \mathscr{O}_X$ which is an isomorphism by Lemma 22.88.

Since the non-derived and derived base change homomorphisms are compatible (Remark 22.89), one concludes that $\theta_{\mathscr{O}_X}$ is an isomorphism if and only if the natural map

$$A' \otimes_A^L B \to A' \otimes_A B$$

is an isomorphism, which means that f and u are tor-independent.   □

Remark 22.105. The condition that the Künneth morphism is an isomorphism (or the condition about the tor-independence) does not change when the roles of f and u in the base change diagram are exchanged; this actually was used in the proof. In particular it follows that the base change homomorphism for the cartesian diagram (22.23.1) is a functorial isomorphism if and only if the same holds for the diagram obtained by exchanging f and u.

Remark 22.106. The proof of Theorem 22.99 shows that for some of the implications some of the hypotheses are superfluous:
(1) The implications "(iii) $\Rightarrow$ (i) $\Rightarrow$ (ii)" hold without assuming that all schemes are quasi-separated.
(2) The implication "(iii) $\Rightarrow$ (i)" holds without assuming that u is qcqs.
(3) If Diagram (22.23.1) is merely commutative, but not necessarily cartesian, the implication "(i) $\Rightarrow$ (ii)" still holds, and (i) and (ii) are even equivalent if u is affine.
(4) Suppose again that the diagram (22.23.1) is merely commutative. Let f be affine and f' qcqs. Suppose that the base change morphism $\theta_{\mathscr{O}_X} : Lu^* Rf_* \mathscr{O}_X \to Rf'_* Lu'^* \mathscr{O}_X$ is an isomorphism.

　　As f_* is exact for quasi-coherent modules and $\mathscr{O}_X$ is K-flat as $\mathscr{O}_X$-module, we obtain an isomorphism $Lu^* f_* \mathscr{O}_X \cong Rf'_* u'^* \mathscr{O}_X = Rf'_* \mathscr{O}_{X'}$. The left hand side has cohomology in degrees ≤ 0, the right hand side has cohomology in degrees ≥ 0. This shows that one has isomorphisms of complexes in $D_{\mathrm{qcoh}}(S')$

$$Lu^* f_* \mathscr{O}_X \cong u^* f_* \mathscr{O}_X \cong f'_* \mathscr{O}_{X'} \cong Rf'_* \mathscr{O}_{X'},$$

in other words, one has
(a) $\mathrm{Tor}_i^{\mathscr{O}_S}(\mathscr{O}_X, \mathscr{O}_{S'}) = 0$ for all $i > 0$,
(b) $R^i f'_* \mathscr{O}_{X'} = 0$ for all $i > 0$,
(c) $u^* f_* \mathscr{O}_X \cong f'_* \mathscr{O}_{X'}$.
If f' is also affine, then (b) holds a priori (Corollary 22.5) and (c) shows that Diagram (22.23.1) is automatically cartesian by the equivalence of quasi-coherent $\mathscr{O}_{S'}$-algebras and affine morphisms with target S' (Corollary 12.2).

　　More generally, if f' is only quasi-affine, then (b) implies that f' is affine (Exercise 22.12) and the argument above again shows that (22.23.1) is cartesian.

Corollary 22.107. *Let*

$$(22.25.2) \qquad \begin{array}{ccc} X' & \xrightarrow{\;u'\;} & X \\ {\scriptstyle f'}\big\downarrow & & \big\downarrow{\scriptstyle f} \\ S' & \xrightarrow{\;u\;} & S \end{array}$$

be a cartesian diagram of schemes. Assume that f is qcqs, and that at least one of f and u is flat.

(1) *The base change morphism $Lu^* Rf_* \to R(f')_* L(u')^*$ is an isomorphism of functors $D_{\mathrm{qcoh}}(X) \to D_{\mathrm{qcoh}}(S')$.*

(2) *If u is addition qcqs, then the Künneth morphism (22.23.2) for this diagram is an isomorphism for all $E \in D_{\mathrm{qcoh}}(S')$, $F \in D_{\mathrm{qcoh}}(X)$.*

Proof. If f or u are flat, then they are tor-independent. Hence the corollary follows from Theorem 22.99 and Remark 22.106 (1). $\qquad \square$

Remark 22.108. If u is flat in the cartesian diagram (22.25.2), the functors u^* and u'^* are exact, and the base change morphism becomes a functorial isomorphism

$$(22.25.3) \qquad u^* Rf_* E \xrightarrow{\sim} R(f')_* (u')^* E$$

for every $E \in D_{\mathrm{qcoh}}(X)$. Applying H^i we obtain an isomorphism

$$u^* R^i f_* E \cong H^i(u^* Rf_* E) \xrightarrow{\sim} H^i(Rf'_* u'^{'}) = R^i f'_* u'^* E,$$

where the first isomorphism holds because u^* is exact and hence commutes with $H^i(-)$. This proves Proposition 22.90 in general.

Corollary 22.109. *Let $f \colon X \to S$ be a flat qcqs morphism, let $s \in S$ be a point, let $X_s = f^{-1}(s)$ be the fiber in s, and let $i \colon X_s \to X$ be the inclusion. Then for all E in $D_{\mathrm{qcoh}}(X)$ one has a functorial isomorphism in the derived category of $\kappa(s)$-vector spaces*

$$\kappa(s) \otimes^L_{\mathscr{O}_S} Rf_* E \xrightarrow{\sim} R\Gamma(X_s, Li^* E)$$

An important special case is the situation over a field k. For k-schemes X and Y and complexes of $\mathscr{O}_X$- and $\mathscr{O}_Y$-modules $\mathscr{F}, \mathscr{G}$, we denote by $\mathscr{F} \boxtimes \mathscr{G}$ the tensor product $p_1^* \mathscr{F} \otimes_{\mathscr{O}_{X \times_k Y}} p_2^* \mathscr{G}$ on $X \times_k Y$.

Corollary 22.110. *Let k be a field, and let X, Y be qcqs k-schemes. Then for all complexes $\mathscr{F}, \mathscr{G}$ of quasi-coherent modules over $\mathscr{O}_X$, and $\mathscr{O}_Y$ resp., the Künneth morphism induces an isomorphism*

$$(22.25.4) \qquad \bigoplus_{j=0}^{i} H^j(X, \mathscr{F}) \otimes_k H^{i-j}(Y, \mathscr{G}) \xrightarrow{\sim} H^i(X \times_k Y, \mathscr{F} \boxtimes \mathscr{G}).$$

Proof. As we have discussed in Example 22.98, in this case the Künneth morphism gives rise to homomorphisms as above. Since k is a field, the base change diagram is tor-independent, so by Theorem 22.99 the Künneth morphism is an isomorphism, which gives us the desired conclusion. $\qquad \square$

Remark 22.111. Let X and Y by qcqs schemes over a field k. Then the cup product makes

$$H^\bullet(X, \mathscr{O}_X) := \bigoplus_{i \geq 0} H^i(X, \mathscr{O}_X)$$

into a strictly graded commutative graded k-algebra (Example 21.131). The Künneth isomorphism (22.25.4) yields an isomorphism of graded k-vector spaces

$$\kappa \colon H^\bullet(X, \mathscr{O}_X) \otimes_k H^\bullet(Y, \mathscr{O}_Y) \xrightarrow{\sim} H^\bullet(X \times_k Y, \mathscr{O}_{X \times_k Y}).$$

It is not difficult to check (Exercise 22.45) that the map κ is an isomorphism of graded algebras if one endows the left hand side with the structure of the graded tensor product in the sense of Definition 19.3.

Exercises

Exercise 22.1. Let X be a ringed space and let $\mathscr{F}$ be a complex of $\mathscr{O}_X$-modules. Show that $H^p(X, \mathscr{F}) = \mathrm{Hom}_{D(X)}(\mathscr{O}_X[-p], \mathscr{F})$. Deduce that there are schemes X and non-zero morphisms u between objects in $D(X)$ such that the restriction of u to every open affine subscheme is zero.

Exercise 22.2. Give a different proof of the fact that for a quasi-coherent $\mathscr{O}_X$-module $\mathscr{F}$ on an affine scheme X we have $H^1(X, \mathscr{F}) = 0$, by writing the H^1 as $\mathrm{Ext}^1(\mathscr{O}_X, \mathscr{F})$ and using the Yoneda description of Ext^1 (see Remark F.219).

Exercise 22.3. Let X be an integral affine scheme with more than one closed point, let $z \in X$ be a closed point, $Z = \{z\}$, $U = X \setminus Z$. Denote by $i\colon Z \hookrightarrow X$, $j\colon U \hookrightarrow X$ the inclusions. Recall that $i^{-1}\mathscr{O}_X$ is a sheaf on the one-point space Z which we can identify with the stalk $\mathscr{O}_{X,z}$. Show that we have a short exact sequence

$$0 \to j_!\mathscr{O}_U \to \mathscr{O}_X \to i_*i^{-1}\mathscr{O}_X \to 0.$$

Show that neither of the outer two terms is quasi-coherent, and that $H^1(X, j_!\mathscr{O}_U) \neq 0$.

Exercise 22.4. A morphism of schemes $f\colon X \to Y$ is called *semiseparated* if the diagonal $\Delta_f\colon X \to X \times_Y X$ is an affine morphism. A scheme X is called *semiseparated* if $X \to \mathrm{Spec}\,\mathbb{Z}$ is semiseparated.
(1) Show that for a scheme X the following assertions are equivalent.
   (i) X is semiseparated.
   (ii) For all open affine subschemes $U, V \subseteq X$ the intersection $U \cap V$ is affine.
   (iii) There exists an open affine covering $(U_i)_i$ such that $U_i \cap U_j$ is affine for all i, j.
(2) Show that every separated morphism is semiseparated. In particular, every monomorphism is semiseparated. Show that every semiseparated morphism is quasi-separated.
(3) Show that the composition of two semiseparated morphisms is again semiseparated.
(4) Show that if $f\colon X \to Y$ is a semiseparated morphism, then for every morphism of schemes $Y' \to Y$ the base change $X \times_Y Y' \to Y'$ is semiseparated.
(5) Show that if the composition $g \circ f$ of scheme morphisms is semiseparated, then f is semiseparated.
Remark: Totaro has shown ([To] 8.1) that for a smooth scheme X of finite type over a field the following assertions are equivalent.
(i) X is semiseparated.
(ii) Every coherent $\mathscr{O}_X$-module is the quotient of a finite locally free $\mathscr{O}_X$-module.
(iii) X has an open affine covering by sets of the form $\{x \in X \; ; \; s(x) \neq 0\}$ with s a section of a line bundle on X.

Exercise 22.5. Let R be a ring, and let X be the R-scheme obtained from gluing two copies of $\mathbb{A}_R^2$ along $\mathbb{A}_R^2 \setminus \{0\}$ via the identity. Thus X is the "affine plane with doubled origin". Let $\mathcal{U}$ be the open affine covering of X given by the two copies of $\mathbb{A}_R^2$. Show that $\check{H}^2(\mathcal{U}, \mathscr{O}_X) \to H^2(X, \mathscr{O}_X)$ is not an isomorphism.

Exercise 22.6. Let R be a ring, let A be a graded R-algebra, let $\mathbf{f} = (f_1, \dots, f_r)$ be a finite family of homogeneous elements of A_+, and let M be a graded A-module. Set $X := \mathrm{Proj}\,A$, $U := \bigcup_i D_+(f_i)$, and let $\tilde{M}$ be the quasi-coherent $\mathscr{O}_X$-module attached to M.

(1) Show that for all $p > 0$ there exists an isomorphism of graded A-modules, functorial in M,

$$\bigoplus_{d \in \mathbb{Z}} H^p(U, \tilde{M}(d)) \xrightarrow{\sim} \operatorname*{colim}_n H^{p-r+1}(K(\mathbf{f}^n), M).$$

(2) Show that there exists an exact sequence, functorial in M,

$$0 \longrightarrow \operatorname*{colim}_n H^{-r}(K(\mathbf{f}^n), M) \longrightarrow M \longrightarrow \bigoplus_{d \in \mathbb{Z}} H^0(U, \tilde{M}(d))$$
$$\longrightarrow \operatorname*{colim}_n H^{-r+1}(K(\mathbf{f}^n), M) \longrightarrow 0.$$

Remark: If A is generated by finitely many elements in degree 1, then every quasi-coherent $\mathscr{O}_X$-module is of the form $\tilde{M}$ for some graded A-module M (Theorem 13.20).

Exercise 22.7. Let X be a qcqs scheme, let $\mathscr{L}$ be a line bundle on X, and let $\mathscr{F}$ be a quasi-coherent $\mathscr{O}_X$-module. We set $\mathscr{F}(m) := \mathscr{F} \otimes \mathscr{L}^{\otimes m}$, $\mathscr{S} := \bigoplus_{m \in \mathbb{Z}} \mathscr{O}_X(m)$, and $\mathscr{M} := \bigoplus_{m \in \mathbb{Z}} \mathscr{F}(m)$. We consider $f \in \Gamma(X, \mathscr{L}^{\otimes m})$ as a homogeneous element in the graded ring $S := \Gamma(X, \mathscr{S}) = \bigoplus_{m \in \mathbb{Z}} \Gamma(X, \mathscr{O}_X(m))$. Let X_f be the open quasi-compact subscheme of X of points $x \in X$ such that $f(x) \neq 0$.
(1) Show that for all $k \geq 0$ the structure of a graded $\mathscr{S}$-module on $\mathscr{M}$ endows

$$H^k(X, \mathscr{M}) = \bigoplus_{m \in \mathbb{Z}} H^k(X, \mathscr{F}(m))$$

with the structure of a graded S-module.
(2) Show that for all $k \geq 0$ there exists an isomorphism of graded S-modules

$$H^k(X_f, \mathscr{M}) \cong H^k(X, \mathscr{M})_f.$$

(3) For $i = 1, \dots, r$ let $f_i \in \Gamma(X, \mathscr{L}^{\otimes m_i})$ and set $U := \bigcup_i X_{f_i}$. Show that there exists an isomorphism of graded complexes

$$\bigoplus_{m \in \mathbb{Z}} \check{C}_{\mathrm{ord}}((X_{f_i})_i, \mathscr{F}(m)) \cong \sigma^{\geq 0}\left(\operatorname*{colim}_n K(\mathbf{f}^n, \Gamma(U, \bigoplus_{m \in \mathbb{Z}} \mathscr{F}(m)))\right)[1-r])$$

(4) Suppose that X is semiseparated (Definition 22.12) and that X_{f_i} is affine for all i. Show that for $p > 0$ one has an isomorphism of graded S-modules

$$\bigoplus_{m \in \mathbb{Z}} H^p(U, \mathscr{F}(m)) = \operatorname*{colim}_n H^{p-r+1}\left(K(\mathbf{f}^n, \Gamma(U, \bigoplus_{m \in \mathbb{Z}} \mathscr{F}(m)))\right).$$

Exercise 22.8. Let X be a separated scheme that can be covered by $n + 1$ open affine schemes. Let $b \in \mathbb{Z}$ and let E be a complex in $D_{\mathrm{qcoh}}(X)$, that is acyclic in degrees $> b$. Show that $H^p(X, E) = 0$ for all $p > n + b$.

Exercise 22.9. Let $R \neq 0$ be a ring and $n \geq 1$. Show that $\mathbb{P}_R^n$ cannot be covered by n open affine subschemes.

Exercise 22.10. Let A be a ring and let $I = (f_1, \dots, f_r)$ and $J = (g_1, \dots, g_s)$ be finitely generated ideals of A with the same radical. Show that the natural morphisms of extended ordered Čech complexes

$$C_A(f_1, \dots, f_r) \longleftarrow C_A(f_1, \dots, f_r, g_1, \dots, g_s) \longrightarrow C_A(g_1, \dots, g_s)$$

are quasi-isomorphisms.

Exercise 22.11. Let $f\colon X \to S$ be a qcqs morphism of schemes. Show that the following assertions are equivalent.

(i) f is affine.

(ii) The functor $f_*\colon \mathrm{QCoh}(X) \to \mathrm{QCoh}(S)$ is exact.

(iii) $f_*\colon \mathrm{QCoh}(X) \longrightarrow \mathrm{QCoh}(S)$ commutes with arbitrary colimits.

(iv) $f_*\colon \mathrm{QCoh}(X) \longrightarrow \mathrm{QCoh}(S)$ has a right adjoint functor.

(v) One has $R^i f_* \mathscr{F} = 0$ for every quasi-coherent $\mathscr{O}_X$-module $\mathscr{F}$ and for all $i > 0$.

(vi) One has $R^1 f_* \mathscr{I} = 0$ for every quasi-coherent ideal $\mathscr{I} \subseteq \mathscr{O}_X$ of finite type.

Exercise 22.12. Let $f\colon X \to S$ be a quasi-affine morphism. Show that f is affine if and only if $R^i f_* \mathscr{O}_X = 0$ for all $i > 0$.

Hint: It suffices to show that a quasi-affine scheme X is affine if $H^i(X, \mathscr{O}_X) = 0$ for all $i > 0$. Set $A := \Gamma(X, \mathscr{O}_X)$ and let $j\colon X \to \mathrm{Spec}\, A$ be the canonical quasi-compact open immersion. Cover X by finitely many principal open subsets $D(f_i)$ of $\mathrm{Spec}\, A$. Use the Čech complex of the covering and the hypothesis to conclude that $\Gamma(X, \mathscr{O}_X)$ is a flat A-module and that for every morphism $f\colon \mathrm{Spec}\, B \to \mathrm{Spec}\, A$ of affine schemes one has $\Gamma(f^{-1}(X), \mathscr{O}_{f^{-1}(X)}) = \Gamma(X, \mathscr{O}_X) \otimes_A B$. Deduce that the f_i generate the unit ideal of A and conclude by Serre's affineness criterion (Theorem 12.35).

Exercise 22.13. Let $f\colon X \to S$ be a quasi-separated and quasi-compact morphism of schemes. Let $\mathscr{F}$ be a complex of quasi-coherent $\mathscr{O}_X$-modules such that all components $\mathscr{F}^n$ are right acyclic for f_*. Show that $\mathscr{F}$ is right acyclic for f_*.

Exercise 22.14. Let A be a noetherian ring, and let M be an injective A-module.

(1) Prove that $\tilde{M}$ is flasque $\mathscr{O}_{\mathrm{Spec}\, A}$-module.

   Hint: Use noetherian induction on $\mathrm{Supp}(\tilde{M})$ and Exercise F.27.

(2) Prove the more difficult assertion that $\tilde{M}$ is an injective $\mathscr{O}_{\mathrm{Spec}\, A}$-module.

Remark: See [Har1]^{O} II 7.14.

Exercise 22.15. Let X be a noetherian scheme, and let $\mathscr{F}$ be a quasi-coherent $\mathscr{O}_X$-module. Show that there exists a quasi-coherent $\mathscr{O}_X$-module $\mathscr{G}$, which is injective in $(\mathscr{O}_X\text{-Mod})$, together with a monomorphism $\mathscr{F} \hookrightarrow \mathscr{G}$.

Hint: Exercise 22.14.

Exercise 22.16. Let X be a noetherian scheme. Show that the inclusion of abelian categories $\mathrm{QCoh}(X) \to (\mathscr{O}_X\text{-Mod})$ sends injective objects to injective objects.

Hint: Exercise 22.15.

Remark: See Caveat 22.77 (3).

Exercise 22.17. Let A be a discrete valuation ring, π a uniformizing element, $X = \mathrm{Spec}\, A$, and let $\mathscr{I}_n \subseteq \mathscr{O}_X$ be the quasi-coherent ideal corresponding to the ideal (π^n). Show that $\prod_n \mathscr{I}_n$ (product in the category of $\mathscr{O}_X$-modules, as usual) is not a quasi-coherent $\mathscr{O}_X$-module. What is the product of the $\mathscr{I}_n$ in the category of quasi-coherent $\mathscr{O}_X$-modules? What is its generic fiber?

Remark: Cf. Exercise 7.12.

Exercise 22.18. Give a different proof of Theorem 22.27 under the assumption that Y is affine and X is noetherian, by showing that $(H^i(X, \mathscr{F})^\sim)_i$ is an effaceable (hence universal) δ-functor $\mathrm{QCoh}(X) \to (\mathscr{O}_Y\text{-Mod})$.

Hint: Use Exercise 22.15. (Note that for the whole proof we need to work inside $\mathrm{QCoh}(X)$ since the statement fails for non-quasi-coherent $\mathscr{F}$ even for $i = 0$ and $f = \mathrm{id}$, so effaceability amounts to showing that every quasi-coherent $\mathscr{O}_X$-module $\mathscr{F}$ can be embedded into a *quasi-coherent* $\mathscr{O}_X$-module $\mathscr{G}$ with $H^i(X, \mathscr{G})^{\sim} = 0$ for all $i > 0$.)

Exercise 22.19. Give a direct proof of Lemma 22.82 in case f is semiseparated (Exercise 22.4), i.e., prove the following: Let $f\colon X \to Y$ be a semiseparated morphism. The functor $Rf_*\colon D_{\mathrm{qcoh}}(X) \to D_{\mathrm{qcoh}}(Y)$ commutes with arbitrary direct sums.
Hint: We may assume that S is affine. Do an induction over the number of affine open subschemes needed to cover X. Use the Mayer-Vietoris triangle, Proposition 21.48.

Exercise 22.20. Let X be a noetherian scheme. Show that the functor

$$DQ\mathrm{Coh}(X) \to D_{\mathrm{qcoh}}(X)$$

is a triangulated equivalence with quasi-inverse given by RQ_X, where Q_X is the coherator.
Hint: Exercise 22.13

Exercise 22.21. Let p be a prime number and let X be a scheme of characteristic p (i.e., $p\mathscr{O}_X = 0$). Let $\mathscr{I} \subseteq \mathscr{O}_X$ be a quasi-coherent ideal such that $\mathscr{I}^n = 0$ for some $n \geq 1$ and let $i\colon Y = V(\mathscr{I}) \to X$ be the corresponding closed immersion (e.g., if X is noetherian and $\mathscr{I}$ is the nilradical, then such an n exists; in this case $Y = X_{\mathrm{red}}$). Let $e \geq 1$ be an integer such that $p^e \geq n$.
(1) Show that the map $\mathscr{O}_Y \to \mathscr{O}_X$ given on local sections by $s \mapsto s^{p^e}$ is well defined and that it induces a group homomorphism $N_i\colon \mathrm{Pic}(Y) \to \mathrm{Pic}(X)$ such that $i^* N_i(\mathscr{L}) = \mathscr{L}^{\otimes p^e}$ for all $\mathscr{L} \in \mathrm{Pic}(Y)$.
(2) Suppose that X is qcqs. Show that $\mathscr{L} \in \mathrm{Pic}(Y)$ is ample if and only if $N_i(\mathscr{L})$ is ample.
Hint: Use arguments as in Remark 12.25 and Proposition 13.66.

Exercise 22.22. Let $f\colon X \to S$ be a quasi-affine morphism. Show that f can be written as a composition $f = g \circ j$, where g is affine and j is an open quasi-compact schematically dominant morphism.
Hint: Consider for j the canonical map $X \to \mathrm{Spec}\, f_* \mathscr{O}_X$.

Exercise 22.23. Let $f\colon X \to S$ be a quasi-affine morphism of schemes. Show that the functor $Rf_*\colon D_{\mathrm{qcoh}}(X) \to D_{\mathrm{qcoh}}(S)$ is conservative.
Hint: Exercise 22.22

Exercise 22.24. Let R be a local noetherian ring of dimension 2, let $S = \mathrm{Spec}\, R$, let $s \in S$ be the closed point, and let $j\colon U := S \setminus \{s\} \to S$ be the inclusion. Set $\mathscr{G} := j_! \mathscr{O}_U$ and $\mathscr{F} := \mathscr{O}_U$.
(1) Show that the stalk of $R^1 j_* \mathscr{F} \otimes \mathscr{G}$ at s in 0 and that the stalk of $R^1 j_*(\mathscr{F} \otimes j^* \mathscr{G}) = R^1 j_* \mathscr{O}_U$ is $H^1(U, \mathscr{O}_U)$.
(2) Show that $\mathscr{G}$ is not quasi-coherent and that $H^1(U, \mathscr{O}_U) \neq 0$.
Hint: One has $H^1(U, \mathscr{O}_U) = H^2_{\{s\}}(R)$, where $H_{\{s\}}$ denotes the local cohomology supported in $\{s\}$.

Exercise 22.25. Let A be a ring and let M be an A-module. Show the following assertions.
(1) M is finitely generated if and only if the map $M \otimes \prod_i N_i \to \prod_i (M \otimes N_i)$ is surjective for every family $(N_i)_i$ of A-modules.

(2) M is of finite presentation if and only if the map $M \otimes \prod_i N_i \to \prod_i (M \otimes N_i)$ is bijective for every family $(N_i)_i$ of A-modules.

Remark: One can also characterize the A-modules M for which the map $M \otimes \prod_i N_i \to \prod_i (M \otimes N_i)$ is injective for every family $(N_i)_i$ of A-modules. These are the so-called Mittag-Leffler modules, see [Sta] 059M.

Exercise 22.26. Let A be ring and let E be in $D^-(A)$. Show that the following assertions are equivalent.

(i) E is pseudo-coherent.
(ii) The canonical map $E \otimes_A^L (\prod_i M_i) \to \prod_i (E \otimes_A^L M_i)$ is an isomorphism in $D(A)$ for every family $(M_i)_i$ of A-modules.
(iii) For every set I, the canonical map $E \otimes_A^L A^I \to E^I$ is an isomorphism in $D(A)$.

Exercise 22.27. Let A be a ring. An A-module M is called *coherent* if M is finitely generated and every finitely generated A-submodule is of finite presentation (see also Section (7.19)). Let

$$M_1 \to M_2 \to M_3 \to M_4 \to M_5$$

be an exact sequence of A-modules. Show that if M_1, M_2, M_4, and M_5 are coherent, then M_3 is coherent. Deduce that the full subcategory of $(A\text{-Mod})$ of coherent A-modules is plump, in particular it is an abelian category. Show that it is a Serre subcategory if and only if A is noetherian.

Exercise 22.28. Let A be a ring. A ring A is called *coherent* if it is coherent as an A-module (Exercise 22.27), i.e., if every finitely generated ideal of A is an A-module of finite presentation.

(1) Show that the following properties are equivalent.

  (i) A is coherent.
  (ii) Every A-module of finite presentation is coherent.
  (iii) For every short exact sequence $0 \to M' \to M \to M'' \to 0$ of A-modules such that two of the modules M, M', M'' are of finite presentation, the third is of finite presentation.
  (iv) The full subcategory of $(A\text{-Mod})$ of finitely presented A-modules is a plump subcategory (in particular, it is an abelian category).
  (v) Every finitely generated submodule of an A-module of finite presentation is itself of finite presentation.
  (vi) For every complex M of A-modules such that M^i is of finite presentation for all $i \in \mathbb{Z}$, $H^i(M)$ is of finite presentation for all $i \in \mathbb{Z}$.
  (vii) Arbitrary products of flat A-modules are flat.

  Hint: To show that (i) and (vii) are equivalent one can use Exercise 22.25.

(2) Show that every noetherian ring is coherent.
(3) Let A be a coherent ring and let $S \subseteq A$ be a multiplicative subset. Show that $S^{-1}A$ is a coherent.
(4) Let A be a coherent ring and let B be an A-algebra that is of finite presentation as an A-module (e.g., $B = A/I$ for I a finitely generated ideal in A). Show that B is coherent.
(5) Let $A \to B$ be a faithfully flat ring homomorphism and let B be coherent. Show that A is coherent.
(6) For every filtered diagram of coherent rings with flat transition maps its colimit is coherent.

(7) Let R be a noetherian ring. Show that any polynomial algebra over R (in possibly infinitely many variables) is coherent.
Remark: An analogous assertion for coherent rings, even for one variable, does not hold (Exercise 22.32).

(8) Let R be a Dedekind domain, let K be an algebraic extension of the field of fractions of R and let A the integral closure of R in K. Show that A is coherent.

(9) Let k be a field, let $A = k[T_1, T_2, \dots]$ be the polynomial ring in countably many variables and let I be the ideal in A generated by $x_1 x_i$ for $i = 2, 3, \dots$. Show that A is coherent but A/I is not coherent.

Exercise 22.29. Let A be a ring and let M be an A-module. Consider the following finiteness properties for M.

(i) M is coherent (Exercise 22.27).
(ii) M is pseudo-coherent.
(iii) M is of finite presentation.
(iv) M is finitely generated.

Show the following assertions.

(1) In general, one has the implications (i) $\Rightarrow$ (ii) $\Rightarrow$ (iii) $\Rightarrow$ (iv).
(2) If A is noetherian, then all properties are equivalent.
(3) If A is coherent (Exercise 22.28), then (i) $\Leftrightarrow$ (ii) $\Leftrightarrow$ (iii).
(4) For each of the implications (iv) $\Rightarrow$ (iii) $\Rightarrow$ (ii) $\Rightarrow$ (i) give an example of a ring A and an ideal $I \subseteq A$ such that this implication does not hold for the A-module A/I.

Exercise 22.30. A scheme X is called *coherent*[3] if its structure sheaf $\mathscr{O}_X$ is a coherent $\mathscr{O}_X$-module (Definition 7.45).

(1) Show that an affine scheme $X = \operatorname{Spec} A$ is coherent if and only if A is a coherent ring (Exercise 22.28). More generally, show that for every ring A the functor $M \mapsto \tilde{M}$ yields an equivalence from the category of coherent A-modules to the category of coherent $\mathscr{O}_X$-modules.

(2) Show that any locally noetherian scheme is coherent.

(3) Let X be a coherent scheme. Show that an $\mathscr{O}_X$-module is coherent if and only if it is of finite presentation.

(4) Show that a scheme X is coherent if and only if the category of $\mathscr{O}_X$-modules of finite presentation, considered as a full subcategory of the category of all $\mathscr{O}_X$-modules, is abelian. Show that in this case it is a plump subcategory.

(5) Let $f \colon X \to Y$ be a faithfully flat quasi-compact morphism of schemes and suppose that X is coherent. Show that Y is coherent.
Hint: Exercise 22.28

Exercise 22.31. Let X be a coherent scheme (Exercise 22.30). Let $D_{\mathrm{coh}}^-(X)$ be the full subcategory of $D(X)$ consisting of bounded above complexes E such that $H^i(E)$ is a coherent $\mathscr{O}_X$-module (equivalently, an $\mathscr{O}_X$-module of finite presentation) for all i.

(1) Show that an $\mathscr{O}_X$-module $\mathscr{F}$ is of finite presentation if and only if it is pseudo-coherent (considered as complex concentrated in degree 0).

(2) Suppose that X is quasi-compact. Show that a complex E in $D(X)$ is pseudo-coherent if and only if $E \in D_{\mathrm{coh}}^-(X)$.

---

[3] Some authors call a scheme coherent if it is qcqs. This yields a totally different notion. For instance, not every affine scheme, which is always qcqs, is coherent in the sense defined here.

Exercise 22.32. Show that the following assertions for a ring A are equivalent.
(i) The set of integers $n \geq 0$ such that $A[T_1, \ldots, T_n]$ is coherent (Exercise 22.28) is unbounded.
(ii) Every A-algebra of finite presentation is coherent.
(iii) Every polynomial algebra over A (in possibly infinitely many variables) is coherent.
If A satisfies these equivalent conditions, A is said to be *universally coherent*. Moreover, a scheme X is called *universally coherent* if every X-scheme locally of finite presentation is coherent (Exercise 22.30). Show the following assertions.
(1) Let A be a ring. Show that A is universally coherent if and only if $\operatorname{Spec} A$ is a universally coherent scheme.
(2) Every absolutely flat ring A (i.e., every A-module is flat) is universally coherent. Deduce that reduced rings of dimension 0 (e.g., arbitrary products of fields) are coherent.
(3) Let $A = \mathbb{Q}^{\mathbb{N}}[T_1, T_2]$. Show that A is coherent but not universally coherent.

Exercise 22.33. Let X be a coherent scheme (Exercise 22.30), let $\mathscr{F} \in D(X)$ be pseudo-coherent and let $\mathscr{G} \in D(X)$ be a locally bounded below complex such that $H^p(\mathscr{G})$ is coherent for all $p \in \mathbb{Z}$. Show that $\mathscr{E}xt^n_{\mathscr{O}_X}(\mathscr{F}, \mathscr{G})$ is coherent for all $n \in \mathbb{Z}$.

Exercise 22.34. Let X be a quasi-compact semiseparated (Definition 22.12) scheme. Show that for every complex $\mathscr{F}$ in $D_{\mathrm{qcoh}}(X)$ there exists a complex $\mathscr{I}$ that is K-injective in $D_{\mathrm{qcoh}}(X)$ and a quasi-isomorphism $\mathscr{F} \to \mathscr{I}$.

Exercise 22.35. Let X be a quasi-compact semiseparated (Exercise 22.4) scheme. Show that for every complex $\mathscr{F}$ in $D_{\mathrm{qcoh}}(X)$ there exists a K-flat complex $\mathscr{P}$ of quasi-coherent modules and a quasi-isomorphism $\mathscr{P} \to \mathscr{F}$.

Exercise 22.36.
(1) Let R be a principal ideal domain. Show that every finitely generated R-module, considered as a complex in degree 0, is perfect.
(2) Give an example of a noetherian ring R and an R-module M such that M, considered as a complex concentrated in degree 0, is pseudo-coherent (in particular M is a finitely generated R-module) but not perfect.
(3) Give an example of a ring R and of a perfect complex E in $D^b(R)$ such that for some i the cohomology $H^i(E)$ (considered as a complex concentrated in degree 0) is not perfect.

Hint: For (b), one could compute $\operatorname{Tor}_i^{\mathbb{Z}/p^2\mathbb{Z}}(\mathbb{Z}/p\mathbb{Z}, \mathbb{Z}/p\mathbb{Z})$ for all i.

Remark: Part (a) can be vastly generalized, see Proposition 23.55.

Exercise 22.37. Let $X = \operatorname{Spec} A$ be an affine scheme and let M and N be complexes of A-modules. Suppose that M is perfect. Show that there is a functorial isomorphism in $D(X)$

$$R\widetilde{\operatorname{Hom}_A(M, N)} = R\mathscr{H}om_{\mathscr{O}_X}(\tilde{M}, \tilde{N}).$$

Exercise 22.38. Let X be a scheme, let $\mathscr{F}$ and $\mathscr{G}$ be complexes of $\mathscr{O}_X$-modules. Suppose that $\mathscr{F}$ is perfect.
(1) Let $f \colon X' \to X$ be a morphism of schemes. Show that one has a natural isomorphism

$$Lf^* R\mathscr{H}om_{\mathscr{O}_X}(\mathscr{F}, \mathscr{G}) \xrightarrow{\sim} R\mathscr{H}om_{\mathscr{O}_{X'}}(Lf^*\mathscr{F}, Lf^*\mathscr{G})$$

(2) Show that for all $x \in X$ one has a natural isomorphism

$$R \mathcal{H}om_{\mathcal{O}_X}(\mathcal{F}, \mathcal{G})_x \xrightarrow{\sim} R \operatorname{Hom}_{\mathcal{O}_{X,x}}(\mathcal{F}_x, \mathcal{G}_x).$$

Deduce that one has for all $p \in \mathbb{Z}$ a functorial isomorphism

$$\mathcal{E}xt^p_{\mathcal{O}_X}(\mathcal{F}, \mathcal{G})_x \xrightarrow{\sim} \operatorname{Ext}^p_{\mathcal{O}_{X,x}}(\mathcal{F}_x, \mathcal{G}_x).$$

Exercise 22.39. Let $f: X \to Y$ be a qcqs morphism of schemes, let $\mathcal{F}$ be in $D^+(X)$ and let $\mathcal{G}$ be in $D_{\mathrm{qcoh}}(Y)$ be locally of finite tor-dimension. Show that the projection formula map

$$Rf_* \mathcal{F} \otimes^L_{\mathcal{O}_Y} \mathcal{G} \longrightarrow Rf_*(\mathcal{F} \otimes^L_{\mathcal{O}_X} Lf^* \mathcal{G})$$

is an isomorphism.
Hint: Assume that Y is affine, replace $\mathcal{G}$ by a bounded complex of flat quasi-coherent modules using Exercise 22.35 and reduce to the case that $\mathcal{G}$ is a single flat quasi-coherent module. Write $\mathcal{G}$ as colimit of finite free modules (Theorem G.3).

Exercise 22.40. Let $X \to S$, $Y \to S$ be morphisms of schemes, let $\mathcal{F}$ (resp. $\mathcal{G}$) be a complex of quasi-coherent $\mathcal{O}_X$-modules (resp. $\mathcal{O}_Y$-modules), and let $S' \to S$ be a flat morphism. Let $v: X \times_S S' \to X$ and $w: Y \times_S S' \to Y$ be the projections. Show that if $\mathcal{F}$ and $\mathcal{G}$ are tor-independent over S and $S' \to S$ is flat, then $v^* \mathcal{F}$ and $w^* \mathcal{G}$ are tor-independent over S'.

Exercise 22.41. Let A be a noetherian local ring and let k be its residue field. Show that the diagram

$$
\begin{array}{ccc}
\operatorname{Spec} k & \xrightarrow{\mathrm{id}} & \operatorname{Spec} k \\
{\scriptstyle \mathrm{id}} \downarrow & & \downarrow {\scriptstyle i} \\
\operatorname{Spec} k & \xrightarrow{i} & \operatorname{Spec} A
\end{array}
$$

is cartesian, where i denotes the canonical map. Show that this diagram is tor-independent if and only if $A = k$.
Hint: Exercise 21.19.

Exercise 22.42. Show by an example that the base change homomorphism in Definition 21.129 is not an isomorphism, in general.
Hint: Exercise 22.41

Exercise 22.43. Let $u: S' \to S$ be a morphism of affine schemes and let $f: X \to S$ be a quasi-affine morphism and form the cartesian diagram

$$
\begin{array}{ccc}
X' & \xrightarrow{u'} & X \\
{\scriptstyle f'} \downarrow & & \downarrow {\scriptstyle f} \\
S' & \xrightarrow{u} & S.
\end{array}
$$

Let $\mathcal{F}$ be a complex of quasi-coherent $\mathcal{O}_X$-modules and let $\mathcal{G}$ be a complex of quasi-coherent $\mathcal{O}_{S'}$-modules. Show that $\mathcal{F}$ and $\mathcal{G}$ are tor-independent over S if and only if the base change morphism

$$\mathcal{G} \otimes^L_{\mathcal{O}_{S'}} Lu^* Rf_* \mathcal{F} \longrightarrow Rf'_*(f'^* \mathcal{G} \otimes_{\mathcal{O}_{X'}} u'^* \mathcal{F})$$

is an isomorphism.
Hint: Exercise 22.23.

Exercise 22.44. Show the following variant of the projection formula. Let $f\colon X \to S$ be a qcqs morphism of schemes, let $\mathscr{F}$ (resp. $\mathscr{G}$) be a complex of quasi-coherent $\mathscr{O}_X$-modules (resp. $\mathscr{O}_S$-modules) such that $\mathscr{F}$ and $\mathscr{G}$ are tor-independent (e.g., if $\mathscr{F}$ or $\mathscr{G}$ is bounded above with components that are flat over S). Show that there is a functorial isomorphism

$$\mathscr{G} \otimes_{\mathscr{O}_S}^L Rf_*\mathscr{F} \xrightarrow{\sim} Rf_*(f^*\mathscr{G} \otimes_{\mathscr{O}_X} \mathscr{F}).$$

Exercise 22.45. Let $f\colon X \to S$ be a morphism of ringed spaces, let $g\colon X \times_S X \to S$ be the structure morphism, and let $\Delta\colon X \to X \times_S X$ be the diagonal. Consider for $\mathscr{F}$ and $\mathscr{G}$ in $D(X)$ the Künneth map (22.23.2)

$$\eta\colon Rf_*\mathscr{F} \otimes^L Rf_*\mathscr{G} \to Rg_*(\mathscr{F} \boxtimes_{\mathscr{O}_{X \times_S X}}^L \mathscr{G})$$

for $S' = X$ and $u = f$. Compose it with Rg_* applied to the adjunction unit $\mathrm{id} \to R\Delta_* \circ L\Delta^*$ and show that the resulting morphism coincides with the relative cup product (21.29.1).

Exercise 22.46. Let $f\colon X \to Y$ be a morphism of schemes and let $t \geq 0$ be an integer. Show that the following assertions are equivalent.
(i) The morphism f has tor-dimension $\leq t$.
(ii) For every quasi-coherent $\mathscr{O}_Y$-module $\mathscr{F}$ one has $L_{t+1}f^*\mathscr{F} = 0$.
If Y is qcqs, show that these assertions are equivalent to
(iii) For every quasi-coherent ideal $\mathscr{J} \subseteq \mathscr{O}_Y$ of finite type one has $L_{t+1}f^*(\mathscr{O}_Y/\mathscr{J}) = 0$.
Hint: Exercise 21.32

Exercise 22.47. Let A be a ring and let $I \subseteq A$ be a finitely generated ideal. An A-module M is called an *I-power torsion module* if for all $m \in M$ there exists an $n \in \mathbb{N}$ such that $I^n m = 0$. Denote by I^∞-Tors the full subcategory of I-power torsion modules of the category of A-modules.
(1) Show that an A-module is I-power torsion if and only if its support is contained in $V(I)$. Deduce that the subcategory I^∞-Tors depends only on $V(I)$.
(2) Show that I^∞-Tors is a Serre subcategory of the category of A-modules.
(3) Show that for an object K of $D(A)$ the following assertions are equivalent.
    (i) $K \otimes_A^L A/I = 0$.
    (ii) $K \otimes_A^L M = 0$ for every I-power torsion A-module M.
    (iii) $K \otimes_A^L N = 0$ for every object N in $D^b(A)$ such that $H^i(N)$ is I-power torsion for all $i \in \mathbb{Z}$.

23   Cohomology of projective and proper schemes

Content

- Cohomology of projective schemes
- Coherence of higher direct images
- Numerical intersection theory, Euler characteristic, and Hilbert polynomial
- The Grothendieck-Riemann-Roch theorem
- Cohomology and base change
- Hilbert polynomials and flattening stratification

This chapter studies the question when cohomology or, more generally, derived direct images satisfy certain finiteness conditions (in the first two parts) and several applications of it. The first central result is the coherence of higher direct images of coherent sheaves under proper morphisms $f\colon X \to Y$ of locally noetherian schemes (Theorem 23.17) and its derived version that Rf_* preserves pseudo-coherent complexes. From this we deduce that it maps $D_{\mathrm{coh}}^b(X)$ to $D_{\mathrm{coh}}^b(Y)$ if Y is noetherian (Remark 23.24) and $D_{\mathrm{coh}}(X)$ to $D_{\mathrm{coh}}(Y)$ (Corollary 23.25).

The strategy for the proof of the coherence of higher direct images is to reduce the question to a similar and more precise version for projective morphisms (Theorem 23.1) using Chow's lemma. The result for projective morphisms will follow easily from the calculation of the higher direct images of Serre's twisting line bundles along the projection to the base (Theorem 22.22). We will also prove Serre's criterion for a line bundle to be ample (Theorem 23.6).

In the remaining parts of this chapter, we use these finiteness results to discuss further topics. We introduce Grothendieck groups and K-groups of schemes and develop numerical intersection theory for proper schemes over a field. In the following part, we will discuss K-theory as a special case of a cohomology theory. Relating K-theory to other cohomology theories will result in the Grothendieck-Riemann-Roch theorem, a vast generalization of the Theorem of Riemann-Roch for curves figuring in Chapters 15 and 26. These two parts have more exemplary character and we do not strive for greatest possible generality. In the part on numerical intersection theory we give full proofs hoping to convince the reader that the language of derived categories simplifies the exposition considerably. In the part about the Grothendieck-Riemann-Roch theorem we will only explain (Section (23.23)) how the theorem follows formally from the fact that K-theory is the initial multiplicative cohomology theory (Theorem 23.108) but we will give only a reference for the proof of this universal property of K-theory.

The last two parts of the chapter then develop the theory of cohomology and base change and apply this to the behavior of intersection numbers and Hilbert polynomials in flat families. In the last section we prove the existence of a flattening stratification, indexed by the Hilbert polynomial, for arbitrary projective morphisms of finite presentation.

© Springer Fachmedien Wiesbaden GmbH, ein Teil von Springer Nature 2023
U. Görtz und T. Wedhorn, *Algebraic Geometry II: Cohomology of Schemes*,
Springer Studium Mathematik – Master, https://doi.org/10.1007/978-3-658-43031-3_8

Cohomology of projective schemes

In this part we start with showing that for projective morphisms of noetherian schemes higher direct images of coherent modules are again coherent. Moreover, we show that after tensoring with a sufficiently high power of an ample line bundle, projective morphisms "behave as affine morphisms", at least for noetherian schemes and for coherent modules. Examples are the vanishing of higher direct images (Theorem 23.1) and the exactness if the direct image functor (Corollary 23.3). In fact, these kind of behavior characterizes ample line bundles (Theorem 23.6). In the final section of this part we apply these results to classify coherent modules on projective spectra of graded algebras.

(23.1) Coherence of direct images under projective morphisms and Serre's vanishing criterion.

The fundamental theorem on cohomology of coherent modules for projective schemes is the following result.

Theorem 23.1. *Let $f\colon X \to S$ be a proper morphism between locally noetherian schemes and let $\mathscr{L}$ be a line bundle on X that is ample for f. Let $\mathscr{F}$ be a coherent $\mathscr{O}_X$-module and set $\mathscr{F}(n) = \mathscr{F} \otimes \mathscr{L}^{\otimes n}$ for $n \in \mathbb{Z}$.*
(1) *For all $i \geq 0$ the higher direct image $R^i f_* \mathscr{F}$ is a coherent $\mathscr{O}_S$-module.*
(2) *If S is noetherian, there exists $n_0 \geq 0$ such that*

$$R^i f_* \mathscr{F}(n) = 0 \qquad \text{for all } n \geq n_0 \text{ and } i > 0.$$

(3) *If S is noetherian, there exists $n_1 \geq 0$ such that the canonical homomorphism $f^*(f_* \mathscr{F}(n)) \to \mathscr{F}(n)$ is surjective for all $n \geq n_1$.*

Proof. All assertions can be checked locally on S (for (2) and (3) we use that S is quasi-compact). Hence we may assume $S = \operatorname{Spec} A$ for a noetherian ring. Then $X \to S$ is projective (Corollary 13.72) and using Theorem 22.27 it suffices to show the following corollary. $\qquad\square$

Corollary 23.2. *Let X be a projective scheme over a noetherian ring A, and let $\mathscr{L}$ be an ample invertible $\mathscr{O}_X$-module. Let $\mathscr{F}$ be a coherent $\mathscr{O}_X$-module and set $\mathscr{F}(n) := \mathscr{F} \otimes \mathscr{L}^{\otimes n}$ for $n \in \mathbb{Z}$.*
(1) *For all $i \geq 0$ the cohomology modules $H^i(X, \mathscr{F})$ are finitely generated A-modules.*
(2) *There exists $n_0 \in \mathbb{Z}$ such that*

$$H^i(X, \mathscr{F} \otimes \mathscr{L}^{\otimes n}) = 0 \qquad \text{for all } n \geq n_0 \text{ and } i > 0.$$

(3) *There exists $n_1 \in \mathbb{Z}$ such that $\mathscr{F}(n)$ is generated by its global sections for all $n \geq n_1$.*

Proof. To see that (3) holds, recall (Summary 13.71) that we have $X \cong \operatorname{Proj} S$ where S is a graded A-algebra generated by finitely many elements in S_1 and we conclude by Proposition 13.22.

We now prove (1) and (2) simultaneously. It is enough to show the statement after replacing $\mathscr{L}$ with some positive tensor power $\mathscr{L}^{\otimes d}$ (because as the bound n_0 for $\mathscr{L}$ and $\mathscr{F}$ we can then take d times the maximum of the bounds for $\mathscr{L}^{\otimes d}$ and $\mathscr{F} \otimes \mathscr{L}^{\otimes j}$, $j = 0, \ldots, d-1$). Therefore we may assume that $\mathscr{L}$ is very ample (Theorem 13.59). By Proposition 13.56, there exists a closed immersion $i\colon X \to \mathbb{P}_A^N$ for some N, with $i^* \mathscr{O}_{\mathbb{P}_A^N}(1) \cong \mathscr{L}$. Then

$$i_*(\mathscr{F}(n)) = i_*(\mathscr{F} \otimes (i^* \mathscr{O}_{\mathbb{P}_A^N}(1))^{\otimes n}) = i_*(\mathscr{F}) \otimes \mathscr{O}_{\mathbb{P}_A^N}(1)^{\otimes n} = i_*(\mathscr{F})(n)$$

by the projection formula. The isomorphism (21.8.4) (or Corollary 22.6) shows that $H^i(X, \mathscr{F}(n)) = H^i(\mathbb{P}_A^N, i_*(\mathscr{F})(n))$, whence it is enough to show (1) and (2) for $X = \mathbb{P}_A^N$ and $\mathscr{L} = \mathscr{O}_X(1)$.

We now proceed by descending induction on i. In degrees $i > N$, the cohomology groups of all quasi-coherent sheaves vanish (Corollary 22.10). Assume we have proved the claims for degrees $> i$ and for all $\mathscr{F}$.

By Theorem 22.22 we know that (1) and (2) hold if $\mathscr{F}$ is a direct sum of $\mathscr{O}_X$-modules of the form $\mathscr{O}_X(d_i)$. As we already proved (3), we know that there exists a surjection $\mathscr{O}_X^r \to \mathscr{F}(n_1)$ for some $r \geq 0$ and hence a surjection $\varphi \colon \mathscr{G} := \mathscr{O}_X(-n_1)^r \to \mathscr{F}$. Let $\mathscr{E} = \mathrm{Ker}(\varphi)$ which is coherent since X is noetherian. As tensoring with a line bundle is exact, we obtain for all $n \in \mathbb{Z}$ an exact sequence

$$0 \to \mathscr{E}(n) \to \mathscr{G}(n) \to \mathscr{F}(n) \to 0.$$

We know that (1) and (2) hold for $\mathscr{G}(n)$ in all cohomology degrees and hold for $\mathscr{E}(n)$ in cohomology degrees $> i$. From the long exact cohomology sequence we obtain an exact sequence of A-modules

$$H^i(\mathbb{P}_A^N, \mathscr{G}(n)) \longrightarrow H^i(\mathbb{P}_A^N, \mathscr{F}(n)) \longrightarrow H^{i+1}(\mathbb{P}_A^N, \mathscr{E}(n))$$

The outer terms are finitely generated A-modules, hence $H^i(\mathbb{P}_A^N, \mathscr{F}(n))$ is finitely generated for all $n \in \mathbb{Z}$ since A is noetherian. In particular, $H^i(\mathbb{P}_A^N, \mathscr{F})$ is finitely generated.

Moreover, $H^i(\mathbb{P}_A^N, \mathscr{G}(n)) = 0$ and $H^{i+1}(\mathbb{P}^N, \mathscr{E}(n)) = 0$ for large n. Hence we obtain $H^i(\mathbb{P}_A^N, \mathscr{F}(n)) = 0$ for large n. $\qquad\square$

Corollary 23.3. *Let S be a noetherian scheme. With the notation of Theorem 23.1 let $\mathscr{F} \to \mathscr{G} \to \mathscr{H}$ be an exact sequence of coherent $\mathscr{O}_X$-modules. Then there exists an integer N such that for all $n \geq N$ the sequence $f_*\mathscr{F}(n) \to f_*\mathscr{G}(n) \to f_*\mathscr{H}(n)$ is exact.*

Proof. This is a routine argument by splitting the exact sequence into short exact sequences as follows. Let $\mathscr{K}$, $\mathscr{I}$, $\mathscr{C}$ be the kernel, the image, the cokernel of $\mathscr{F} \to \mathscr{G}$, respectively. Then $\mathscr{I}$ is the kernel and $\mathscr{C}$ is the image of $\mathscr{G} \to \mathscr{H}$. Let $\mathscr{D}$ be the cokernel of $\mathscr{G} \to \mathscr{H}$. These are all coherent $\mathscr{O}_X$-modules since X is noetherian. As $\mathscr{F}(n)$ is an exact functor in $\mathscr{F}$ for all n it suffices to show that for large n the sequences

$$0 \to f_*(\mathscr{K}(n)) \to f_*(\mathscr{F}(n)) \to f_*(\mathscr{I}(n)) \to 0$$
$$0 \to f_*(\mathscr{I}(n)) \to f_*(\mathscr{G}(n)) \to f_*(\mathscr{C}(n)) \to 0$$
$$0 \to f_*(\mathscr{C}(n)) \to f_*(\mathscr{H}(n)) \to f_*(\mathscr{D}(n)) \to 0$$

are exact. In other words, we may assume that $0 \to \mathscr{F} \to \mathscr{G} \to \mathscr{H} \to 0$ is exact. The cohomology sequence is then of the form

$$0 \to f_*(\mathscr{F}(n)) \to f_*(\mathscr{G}(n)) \to f_*(\mathscr{H}(n)) \to R^1 f_* \mathscr{F}(n) \to \cdots,$$

and $R^1 f_* \mathscr{F}(n) = 0$ for large n by Theorem 23.1. $\qquad\square$

(23.2) Serre's ampleness criterion.

We can now extend the results of Theorem 12.35 and Proposition 13.47, which relate ampleness of a line bundle $\mathscr{L}$ to the vanishing of cohomology groups. We start with a purely topological lemma (Exercise 3.13) that will also be useful at other occasions.

Lemma 23.4. *Let S be a quasi-compact Kolmogorov space (e.g., the underlying topological space of a quasi-compact scheme). Then for every point $s \in S$ there exists a specialization s_0 of s that is a closed point in S. In particular, S is the only open subset of S containing all closed points.*

Proof. Consider $\mathcal{S} := \{ \overline{\{x\}} \; ; \; x \in S \text{ is a specialization of } s\}$. It is non-empty. Let $\mathcal{T} \subseteq \mathcal{S}$ be a subset that is totally ordered by inclusion. As S is quasi-compact, $Z_0 := \bigcap_{Z \in \mathcal{T}} Z \neq \emptyset$. Choose $x_0 \in Z_0$. Then $Z_0 = \overline{\{x_0\}}$ and we see that $Z_0 \in \mathcal{S}$ is a lower bound for $\mathcal{T}$. By Zorn's lemma, $\mathcal{S}$ has a minimal element Z. As S is Kolmogorov, $Z = \overline{\{x\}}$ for some specialization x of s and $x \in S$ is a closed point.   □

Now we first state and prove the local case of Serre's ampleness criterion.

Lemma 23.5. *Let A be a noetherian ring, and let X be a scheme which is proper over $\operatorname{Spec} A$. For an invertible $\mathcal{O}_X$-module $\mathscr{L}$, the following are equivalent:*
(i) *The sheaf $\mathscr{L}$ is ample.*
(ii) *For every coherent $\mathcal{O}_X$-module $\mathscr{F}$, there exists $n_0 \geq 0$ such that $H^i(X, \mathscr{F} \otimes \mathscr{L}^{\otimes n}) = 0$ for all $n \geq n_0$, $i > 0$.*
(iii) *For every coherent ideal sheaf $\mathscr{I} \subseteq \mathcal{O}_X$, there exists $n_0 \geq 0$ such that $H^1(X, \mathscr{I} \otimes \mathscr{L}^{\otimes n}) = 0$ for all $n \geq n_0$.*

Proof. We have shown that (i) implies (ii) in Corollary 23.2 above, and (ii) trivially implies (iii). Now let us show that (iii) implies (i). By Proposition 13.47 it is enough to show that for every point $x \in X$ and every affine open neighborhood $U \subseteq X$, there exist $N \geq 0$ and $h \in \Gamma(X, \mathscr{L}^{\otimes N})$ with $x \in X_h \subseteq U$ (note that then X_h is necessarily affine by Example 12.4 (4)). By Lemma 23.4 we may assume that x is a closed point.

So consider x and U as above. Let $Z = X \setminus U$, viewed as a reduced closed subscheme of X. Similarly, we view $Z \cup \{x\}$ as a reduced closed subscheme of X. Let $\mathscr{J}$ be the quasi-coherent ideal sheaf corresponding to Z, and let $\mathscr{I}$ be the quasi-coherent ideal sheaf given by $Z \cup \{x\}$. As X is noetherian, $\mathscr{I}$ and $\mathscr{J}$ are coherent. We obtain a short exact sequence

$$0 \to \mathscr{I} \to \mathscr{J} \to \kappa(x) \to 0.$$

By assumption, there exists $n > 0$ with $H^1(X, \mathscr{I} \otimes \mathscr{L}^{\otimes n}) = 0$, so from the long exact cohomology sequence for the above short exact sequence twisted by $\mathscr{L}^{\otimes n}$, we get that the map

$$H^0(X, \mathscr{J} \otimes \mathscr{L}^{\otimes n}) \to \kappa(x)$$

is surjective. Taking any global section of $\mathscr{J} \otimes \mathscr{L}^{\otimes n}$ which does not map to 0, we obtain a section h of $\mathscr{L}^{\otimes n}$ which vanishes on $X \setminus U$, but does not vanish at x.   □

Theorem 23.6. *Let $f \colon X \to Y$ be a proper morphism between noetherian schemes, and let $\mathscr{L}$ be an invertible $\mathcal{O}_X$-module. The following are equivalent:*
(i) *The sheaf $\mathscr{L}$ is relatively ample for f (see Definition 13.60).*
(ii) *For every coherent $\mathcal{O}_X$-module $\mathscr{F}$, there exists $n_0 \geq 0$ such that $R^i f_*(\mathscr{F} \otimes \mathscr{L}^{\otimes n}) = 0$ for all $n \geq n_0$, $i > 0$.*
(iii) *For every coherent ideal sheaf $\mathscr{I} \subseteq \mathcal{O}_X$, there exists $n_0 \geq 0$ such that $R^1 f_*(\mathscr{I} \otimes \mathscr{L}^{\otimes n}) = 0$ for all $n \geq n_0$.*

Proof. Working locally on Y and using Theorem 22.27, we reduce to Lemma 23.5.   □

We want to use the cohomological characterization of ampleness to prove a criterion for ampleness of a line bundle, provided that the pullback under a finite surjective morphism is ample (see Proposition 23.8 for the precise statement). The key ingredient that is still missing is the following lemma.

Lemma 23.7. *Let X be a quasi-compact scheme, and let $i \colon X' \to X$ be a closed immersion that is given by a nilpotent quasi-coherent ideal $\mathscr{I}$. Let $\mathscr{L}$ be an invertible $\mathscr{O}_X$-module such that $i^*\mathscr{L}$ is ample. Then $\mathscr{L}$ is ample.*

If X is noetherian, then X_{red} is a closed subscheme in X defined by a nilpotent ideal.

Proof. By induction, we may assume that $\mathscr{I}^2 = 0$. We will show that for f running through all sections of tensor powers $\mathscr{L}^{\otimes k}$ of $\mathscr{L}$, those open subsets X_f of X which are affine form a basis of the topology of X. By Proposition 13.47 this is equivalent to $\mathscr{L}$ being ample. By assumption the analogous statement is true when f runs through sections of $(\mathscr{L}_{|X'})^{\otimes k}$, $n \geq 1$; note that X and X' have the same topological space. Recall that the assumption implies that X' is separated (Proposition 13.48); it follows from Proposition 9.13 (4) that X is separated, too. Also recall that an open subset U of $X' = X$ defines an *affine* open subscheme of X' if and only if it defines an affine open subscheme of X (Lemma 12.38).

Therefore we have reduced to proving the following

Claim. Let $n \geq 1$, and let $g \in \Gamma(X', (\mathscr{L}_{|X'})^{\otimes n})$ such that $(X')_g$ is affine. Then there exists $m \geq 1$ such that $g^{\otimes(m+1)}$ is the image under pullback of a section of $\mathscr{L}^{\otimes(m+1)n}$.

Proof of claim. Consider the exact sequence

$$0 \to \mathscr{I} \otimes \mathscr{L}^{\otimes n} \to \mathscr{L}^{\otimes n} \to \mathscr{O}_{X'} \otimes_{\mathscr{O}_X} \mathscr{L}^{\otimes n} \to 0$$

of sheaves on $X = X'$, which gives us an exact sequence

$$\Gamma(X, \mathscr{L}^{\otimes n}) \longrightarrow \Gamma(X, (\mathscr{L}_{|X'})^{\otimes n}) \overset{\partial}{\longrightarrow} H^1(X, \mathscr{I} \otimes \mathscr{L}^{\otimes n}).$$

Since $\mathscr{I}^2 = 0$, we can consider $\mathscr{I}$ as an $\mathscr{O}_{X'}$-module. Tensoring by sections of $(\mathscr{L}_{|X'})^{\otimes k}$ induces maps on cohomology which we will again denote as tensor products. Let us first show that $g^{\otimes m} \otimes \partial(g) = 0$ in $H^1(X, \mathscr{I} \otimes \mathscr{L}^{\otimes(m+1)n})$ for m sufficiently large. Indeed, writing $g' = g_{|(X')_g}$ for the restriction of g to the affine scheme $(X')_g$, we have $\partial(g)_{|(X')_g} = \partial(g') = 0$ since $H^1((X')_g, \mathscr{I} \otimes \mathscr{L}^{\otimes n}) = 0$ by Theorem 22.2 and since $\mathscr{I}$ is quasi-coherent. Let us spell out explicitly what that means, viewing the H^1 cohomology groups as Čech cohomology groups for a finite affine open cover $(U_i)_i$ of X. Lifting each $g_{|U_i}$ to an element $g_i \in \Gamma(U_i, \mathscr{L}^{\otimes n})$, $\partial(g)$ is described by the cocycle $(g_{j|U_{ij}} - g_{i|U_{ij}})_{i,j}$ (these differences lying in $\Gamma(U_{ij}, \mathscr{I} \otimes \mathscr{L}^{\otimes n})$). Here we use the notation $U_{ij} = U_i \cap U_j$. By separatedness, all $U_i \cap (X')_g$ are affine, too. Thus $\partial(g') = 0$ means that there exist $h_i \in \Gamma(U_i', \mathscr{I} \otimes \mathscr{L}^{\otimes n})$ with

$$g_{j|U_{ij}'} - g_{i|U_{ij}'} = h_{j|U_{ij}'} - h_{i|U_{ij}'}$$

for all i, j, where we write $U_i' = U_i \cap (X')_g$, $U_{ij}' = U_{ij} \cap (X')_g$. Now we can apply Theorem 7.22 which says that $g^m \otimes h_i$ lifts to $\Gamma(X, \mathscr{I} \otimes \mathscr{L}^{\otimes(m+1)n})$. Choosing an m which works for all i simultaneously, we find that $g^{\otimes m} \otimes \partial(g) = 0$, as desired.

Next, we use that the boundary maps $\partial = \partial_k \colon \Gamma(X, (\mathscr{L}_{|X'})^{\otimes k}) \to H^1(X, \mathscr{I} \otimes \mathscr{L}^{\otimes k})$ satisfy a Leibniz rule

$$\partial(s \otimes t) = \partial(s) \otimes t + s \otimes \partial(t) \in H^1(X, \mathscr{I} \otimes \mathscr{L}^{\otimes(k+\ell)})$$

for $s \in \Gamma(X, (\mathscr{L}_{|X'})^{\otimes k})$, $t \in \Gamma(X, (\mathscr{L}_{|X'})^{\otimes \ell})$. This can again be checked by an explicit computation with Čech cohomology groups. With notation similar to the above, the crucial point is the identity

$$(s_i \otimes t_i) - (s_j \otimes t_j) = (s_i - s_j) \otimes t_i + s_j \otimes (t_i - t_j)$$

on U_{ij} where we have omitted the restriction to U_{ij} from the notation everywhere.

Using this, we compute

$$\partial(g^{\otimes(m+1)}) = (m+1)g^{\otimes m} \otimes \partial(g) = 0$$

which shows that $g^{\otimes(m+1)}$ is in the image of the pullback map $\Gamma(X, \mathscr{L}^{\otimes(m+1)n}) \rightarrow \Gamma(X, (\mathscr{L}_{|X'})^{\otimes(m+1)n})$. $\qquad\square$

Proposition 23.8. *Let $f\colon X \to Y$ be a proper morphism between noetherian schemes, and let $g\colon X' \to X$ be a surjective finite morphism. Let $\mathscr{L}$ be an invertible $\mathscr{O}_X$-module. The following are equivalent:*
(i) *The sheaf $\mathscr{L}$ is relatively ample for f.*
(ii) *The sheaf $g^*\mathscr{L}$ is relatively ample for $f \circ g$.*

We also proved that if g is finite locally free and surjective, then $\mathscr{L}$ is ample if and only if $g^*\mathscr{L}$ is ample (Proposition 13.66). The proof here is entirely different and uses that X and X' are proper over some noetherian base scheme Y.

Proof. It is easy to see that (i) implies (ii), and for this implication it suffices that g is quasi-affine; see Proposition 13.83.

Now let us assume that $g^*\mathscr{L}$ is relatively ample for $f \circ g$. We know then that

$$R^1 f_*(g_*(\mathscr{H}) \otimes \mathscr{L}^{\otimes n}) = R^1 f_*(g_*(\mathscr{H} \otimes g^*\mathscr{L}^{\otimes n})) = R^1(f \circ g)_*(\mathscr{H} \otimes g^*\mathscr{L}^{\otimes n}) = 0$$

for every coherent $\mathscr{O}_{X'}$-module $\mathscr{H}$ and for sufficiently large n (depending on $\mathscr{H}$). Here we use the projection formula Proposition 22.81 and Theorem 23.6.

To show that $\mathscr{L}$ is f-ample, we may work locally on Y and hence assume that Y is affine. Then the notion of relative ampleness coincides with the notion of ampleness.

We can furthermore assume that X is reduced: In fact, the morphism $X'_{\text{red}} \to X_{\text{red}}$ is still finite and surjective, the pullback of $g^*\mathscr{L}$ to X'_{red} is ample by the easy implication of the proposition, and if we know that the pullback of $\mathscr{L}$ to X_{red} is ample, then $\mathscr{L}$ is ample by the previous Lemma 23.7. Thus we may replace X' by X'_{red}, X by X_{red} and g by g_{red}.

We now show that $\mathscr{L}_{|Z}$ is ample by noetherian induction on the closed subscheme $Z \subseteq X$. This means that in order to show that $\mathscr{L}$ is ample on X, we may assume that for every proper closed subscheme $Z \subsetneq X$, the restriction $\mathscr{L}_{|Z}$ is ample.

If X is not irreducible, let $i\colon Z \hookrightarrow X$ be the inclusion of an irreducible component of X. Let $\mathscr{F}$ be a coherent $\mathscr{O}_X$-module and consider the natural homomorphism $\rho\colon \mathscr{F} \to i_*i^*\mathscr{F}$. Then we have a short exact sequence

$$0 \to \text{Ker}(\rho) \to \mathscr{F} \to \text{Im}(\rho) \to 0.$$

But $i_*i^*\mathscr{F}$, and a fortiori $\operatorname{Im}(\rho)$ has support in Z. On the other hand, the generic point of Z is not contained in the support of $\operatorname{Ker}(\rho)$, since i is an isomorphism on a non-empty open subscheme of Z (here we use that X and Z are reduced). Hence both $\operatorname{Ker}(\rho)$ and $\operatorname{Im}(\rho)$ have support properly contained in X, which means that $R^1 f_*(\operatorname{Ker}(\rho) \otimes \mathscr{L}^{\otimes n}) = 0$ (and similarly for $\operatorname{Im}(\rho)$), for sufficiently large n by our noetherian induction hypothesis. The long exact cohomology sequence then implies that the same holds for $\mathscr{F}$.

It remains to consider the case where X is irreducible (and hence integral). By replacing, if necessary, X' by one of its irreducible components which surject onto X, we may assume that X' is also integral. We now apply the reasoning used in step (iii) of the proof of Theorem 12.39: Using Proposition 7.27, we see that $\mathscr{O}_{X'|V}$ is free over $\mathscr{O}_V$ for a suitable non-empty open $V \subseteq X$. This allows us to define an $\mathscr{O}_X$-module homomorphism $u \colon \mathscr{O}_X^n \to g_*\mathscr{O}_{X'}$ which induces an isomorphism after restriction to a suitable open $W \subseteq X$.

Now let $\mathscr{I} \subseteq \mathscr{O}_X$ be a coherent sheaf of ideals. Composition with u yields a homomorphism

$$v \colon \mathscr{G} := \mathscr{H}\!om_{\mathscr{O}_X}(g_*\mathscr{O}_{X'}, \mathscr{I}) \to \mathscr{H}\!om_{\mathscr{O}_X}(\mathscr{O}_X^n, \mathscr{I}) = \mathscr{I}^n$$

of $\mathscr{O}_X$-modules whose restriction to W is an isomorphism. Since X is integral, all restriction maps for $\mathscr{I}$ are injective, and hence v is injective. Furthermore, the cokernel of v is supported on a proper closed subscheme of X, so by induction hypothesis we have $R^1 f_*(\operatorname{Coker}(v) \otimes \mathscr{L}^{\otimes n}) = 0$ for all sufficiently large n. In view of the long exact cohomology sequence, it is then enough to show that $R^1 f_*(\mathscr{G} \otimes \mathscr{L}^{\otimes n}) = 0$ for n sufficiently large.

But $\mathscr{G}$ is an $g_*\mathscr{O}_{X'}$-module, and hence of the form $g_*\mathscr{H}$ for a coherent $\mathscr{O}_{X'}$-module $\mathscr{H}$ (use Proposition 12.5 and note that $\mathscr{H}$ is coherent since this is true for $\mathscr{G}$). As pointed out in the beginning of the proof, this implies $R^1 f_*(\mathscr{G} \otimes \mathscr{L}^{\otimes n}) = 0$ for n sufficiently large, as desired. $\qquad\square$

Corollary 23.9. *Let $f \colon X \to Y$ be a proper morphism between noetherian schemes, and let $\mathscr{L}$ be an invertible $\mathscr{O}_X$-module. The following are equivalent:*
(i) *The sheaf $\mathscr{L}$ is relatively ample for f.*
(ii) *The restriction of $\mathscr{L}$ to each irreducible component Z of X (considered as a reduced closed subscheme) is relatively ample for $f_{|Z}$.*

Proof. If $Z_1, \ldots, Z_r$ are the irreducible components of X with the reduced scheme structure, then we can apply Proposition 23.8 to the surjective finite morphism $\coprod_i Z_i \to X$. $\quad\square$

(23.3) Coherent modules on projective spectra revisited.

In this section we fix the following notation. Let Y be a scheme, let $\mathscr{S} = \bigoplus_{d \geq 0} \mathscr{S}_d$ be a graded quasi-coherent $\mathscr{O}_Y$-algebra. Set $X := \operatorname{Proj}(\mathscr{S})$ and let $\pi \colon X \to Y$ be the structure morphism, which is separated. For every graded quasi-coherent $\mathscr{S}$-module $\mathscr{M}$ let $\tilde{\mathscr{M}}$ be the associated quasi-coherent $\mathscr{O}_X$-module (Sections (13.4) and (13.7)).

Lemma 23.10. *The covariant functor $\mathscr{M} \mapsto \tilde{\mathscr{M}}$ from the category of graded quasi-coherent $\mathscr{S}$-modules to the category of quasi-coherent $\mathscr{O}_X$-modules is exact and commutes with colimits.*

Proof. We may assume that $Y = \operatorname{Spec} A$ is affine. Let S be the graded A-algebra corresponding to $\mathscr{S}$. The assertions can also be checked locally on X. As X can be covered by open subschemes of the form $D_+(f)$ for $f \in S$ homogeneous of degree > 0, it suffices to show that the functor $M \mapsto \tilde{M}_{|D_+(f)} = (M_{(f)})^{\sim}$ from the category of graded A-modules to the category of quasi-coherent $\mathscr{O}_{D_+(f)}$-modules has the asserted properties. But this functor is the composition of the functors $M \mapsto M_f$, of passing to degree 0, and of $(-)^{\sim}$ and each of these functors is exact and commutes with colimits. $\square$

From now on we also assume that $\mathscr{S}$ is generated by $\mathscr{S}_1$ and that $\mathscr{S}_1$ is an $\mathscr{O}_Y$-module of finite type. Then $\pi \colon X \to Y$ is projective (Summary 13.71). One has Serre's twisting line bundles $\mathscr{O}_X(n)$ for all $n \in \mathbb{Z}$. If $\mathscr{F}$ is a quasi-coherent $\mathscr{O}_X$-module we set

$$\Gamma_*(\mathscr{F}) := \bigoplus_{n \in \mathbb{Z}} \pi_* \mathscr{F}(n),$$

where $\mathscr{F}(n) = \mathscr{F} \otimes_{\mathscr{O}_X} \mathscr{O}_X(n)$. Then $\Gamma_*(\mathscr{F})$ is a graded quasi-coherent $\mathscr{S}$-module and by Theorem 13.29 one has a functorial isomorphism of quasi-coherent $\mathscr{O}_X$-modules

$$\beta \colon \Gamma_*(\mathscr{F})^{\sim} \xrightarrow{\sim} \mathscr{F}.$$

Recall (cf. Exercise 13.2) also the following definitions.

Definition 23.11. *Let $\mathscr{M}$ and $\mathscr{N}$ be graded quasi-coherent $\mathscr{S}$-modules.*
(1) *The graded module $\mathscr{M}$ is said to satisfy (TF) (resp. (TN)) if there exists an integer N such that the $\mathscr{S}$-module $\bigoplus_{n \geq N} \mathscr{M}_n$ is of finite type (resp. is zero).*
(2) *A homomorphism $u \colon \mathscr{M} \to \mathscr{N}$ of graded $\mathscr{S}$-modules of degree 0 is called (TN)-injective (resp. (TN)-surjective) if $\operatorname{Ker}(u)$ (resp. $\operatorname{Coker}(u)$) satisfies (TN). One calls u a (TN)-isomorphism if it is (TN)-injective and (TN)-surjective.*

Proposition 23.12. *Let $\mathscr{M}$ and $\mathscr{N}$ be graded quasi-coherent $\mathscr{S}$-modules.*
(1) *If $\mathscr{M}$ satisfies (TN), then $\tilde{\mathscr{M}} = 0$. Conversely, if $\mathscr{M}$ satisfies (TF) and $\tilde{\mathscr{M}} = 0$, then $\mathscr{M}$ satisfies (TN).*
(2) *If $\mathscr{M}$ satisfies (TF), then $\tilde{\mathscr{M}}$ is an $\mathscr{O}_X$-module of finite type.*
(3) *If a homomorphism $u \colon \mathscr{M} \to \mathscr{N}$ is (TN)-injective (resp. (TN)-surjective, resp. a (TN)-isomorphism), then $\tilde{u} \colon \tilde{\mathscr{M}} \to \tilde{\mathscr{N}}$ is injective (resp. surjective, resp. an isomorphism).*

In the proof we use only the assumption that $\bigoplus_{d > 0} \mathscr{S}_d$ is an ideal of $\mathscr{S}$ of finite type (cf. Lemma 13.9).

Proof. All assertions are local on Y and hence we can assume that $Y = \operatorname{Spec} A$. Then $\mathscr{S}$ corresponds to a graded A-algebra S such that S_1 generates S and is an A-module of finite type. Let M and N be the graded S-modules corresponding to $\mathscr{M}$ and $\mathscr{N}$.

If M satisfies (TN), then $M_{(f)} = 0$ for every homogeneous element of S of degree > 0. Hence $\tilde{M} = 0$. As $M \mapsto \tilde{M}$ is exact (Lemma 23.10), (3) follows.

Let us show (2). Suppose that M satisfies (TF). To see that $\tilde{M}$ is of finite type, we choose an integer N such that $M' := \bigoplus_{n \geq N} M_n$ is a finitely generated S-module. Then the inclusion $M' \to M$ is a (TN)-isomorphism and hence we may by (3) assume that M is finitely generated. To see that $\tilde{M}$ is of finite type it suffices to see that for $f \in S$ homogeneous of degree $d > 0$ the module $M_{(f)} = M^{(d)}/(f-1)M^{(d)}$ is a finitely generated module over $S_{(f)} = S^{(d)}/(f-1)S^{(d)}$. This follows from Lemma 13.10 (2), so we get (2).

Now suppose that M satisfies (TF) and that $\tilde{M} = 0$. As in the proof of (2) we may assume that M as an S-module is generated by finitely many homogeneous elements $x_1, \ldots, x_t$. Let $f_1, \ldots, f_r \in S$ be homogeneous elements that generate the ideal $\bigoplus_{d>0} S_d$. By assumption one has $M_{(f_i)} = 0$ for all i. Hence we can find an integer n such that $f_i^n x_j = 0$ for all i and j. This implies the existence of an integer m such that $S_k x_j = 0$ for all $k > m$ and all j. If d is the maximal degree of the x_j, then we conclude that $M_k = 0$ for all $k > d + m$. $\qquad\square$

Proposition 23.13. *Let Y be a noetherian scheme and let $\mathscr{M}$ be a graded quasi-coherent $\mathscr{S}$-module satisfying (TF). Then the canonical functorial homomorphism (13.5.4)*

$$\alpha \colon \mathscr{M} \longrightarrow \Gamma_*(\tilde{\mathscr{M}})$$

is a (TN)-isomorphism.

Proof. As $\pi \colon X \to Y$ is projective, X is noetherian as well. We have to show that $\mathrm{Ker}(\alpha)$ and $\mathrm{Coker}(\alpha)$ is zero in large degrees. As Y is quasi-compact, we may assume that $Y = \mathrm{Spec}\, A$ is affine for a noetherian ring A. Then $\mathscr{S}$ corresponds to a graded A-algebra S and $\mathscr{M}$ to a graded S-module M. By our general assumption, S_1 is a finitely generated A-module that generates S. As in the proof of Proposition 23.12 we may assume that M is a finitely generated S-module by replacing M with $\bigoplus_{n \geq N} M_n$ for some large N. Note that $\tilde{M}$ is a coherent $\mathscr{O}_X$-module by Proposition 23.12 (2).

We claim that it suffices to show the proposition for $M = S$. Indeed, as M is finitely generated and A is noetherian, we find an exact sequence $L' \to L \to M \to 0$, where L and L' are direct sums of graded S-modules of the form $S(m)$ for some $m \in \mathbb{Z}$. The sequence $\tilde{L}' \to \tilde{L} \to \tilde{M} \to 0$ is exact by Lemma 23.10. If the result holds for S, it also holds for $S(m)$ and hence for L and L'. For each $n \in \mathbb{Z}$ we have a commutative diagram of A-modules

$$
\begin{array}{ccccccc}
L'_n & \longrightarrow & L_n & \longrightarrow & M_n & \longrightarrow & 0 \\
\downarrow{\scriptstyle\alpha_n} & & \downarrow{\scriptstyle\alpha_n} & & \downarrow{\scriptstyle\alpha_n} & & \\
\Gamma(X, \tilde{L}'(n)) & \longrightarrow & \Gamma(X, \tilde{L}(n)) & \longrightarrow & \Gamma(X, \tilde{M}(n)) & \longrightarrow & 0.
\end{array}
$$

The upper row is exact by definition and the lower row is exact for large n by Corollary 23.3. Hence our claim follows by the five lemma.

Note that α is an isomorphism for $S = A[T_0, \ldots, T_r]$ by Example 13.16. This shows the proposition for $X = \mathbb{P}_A^r$.

In general, S is a quotient of $S' := A[T_1, \ldots, T_r]$. Let $i \colon X \to \mathbb{P}_A^r$ be the corresponding closed immersion. Then $i_*(\tilde{S}(n)) = \tilde{M}(n)$, where M denotes S considered as a graded S'-module. As we have already seen the proposition for S', we obtain that

$$S_n = M_n \longrightarrow \Gamma(\mathbb{P}_A^r, \tilde{M}(n)) = \Gamma(X, \tilde{S}(n))$$

is bijective for large n. $\qquad\square$

Lemma 23.14. *Let Y be qcqs and let $\mathscr{F}$ be a quasi-coherent $\mathscr{O}_X$-module of finite type. Then there exists a graded quasi-coherent $\mathscr{S}$-module $\mathscr{M}$ of finite type such that $\tilde{\mathscr{M}} \cong \mathscr{F}$.*

Proof. Set $\mathcal{N} := \Gamma_*(\mathcal{F})$. Then $\mathcal{N}_n$ is a quasi-coherent $\mathscr{O}_Y$-module for every $n \in \mathbb{Z}$. As Y is qcqs, we can write $\mathcal{N}_n$ as the filtered union of its finitely generated quasi-coherent submodules $\mathscr{P}_{n,\lambda_n}$ (Corollary 10.50). Then $\mathscr{Q}_{\lambda_n} := \mathscr{S} \cdot \mathscr{P}_{n,\lambda_n} \subseteq \mathcal{N}$ is a graded quasi-coherent $\mathscr{S}$-module of finite type for all n and all λ_n. If Φ denotes the set of all finite sums of submodules of $\mathcal{N}$ of the form $\mathscr{Q}_{\lambda_n}$, then $\mathcal{N}$ is the filtered union of the $\mathscr{S}$-submodule in Φ which are all of finite type.

By Lemma 23.10 we see that $\mathcal{F} \cong \tilde{\mathcal{N}}$ is then the filtered union of the submodules $\tilde{\mathcal{N}}_\mu$ for $\mathcal{N}_\mu \in \Phi$. Hence $\mathcal{F} = \tilde{\mathcal{N}}_\mu$ for some $\mathcal{N}_\mu \in \Phi$ by Lemma 10.47. $\qquad\square$

Corollary 23.15. *Let Y be a noetherian scheme and let $\mathcal{F}$ be a quasi-coherent $\mathscr{O}_X$-module. Then $\mathcal{F}$ is coherent if and only if the graded quasi-coherent $\mathscr{S}$-module $\Gamma_*(\mathcal{F})$ satisfies (TF).*

Proof. If $\Gamma_*(\mathcal{F})$ satisfies (TF), then $\mathcal{F} \cong \Gamma_*(\mathcal{F})^\sim$ is a quasi-coherent $\mathscr{O}_X$-module of finite type (Proposition 23.12 (2)) and hence it is coherent since X is noetherian.

Conversely, if $\mathcal{F}$ is coherent, then there exists a graded quasi-coherent $\mathscr{S}$-module $\mathcal{M}$ of finite type such that $\tilde{\mathcal{M}} \cong \mathcal{F}$ by Lemma 23.14. By Proposition 23.13, $\Gamma_*(\mathcal{F})$ is (TN)-isomorphic to $\mathcal{M}$. Therefore $\Gamma_*(\mathcal{F})$ satisfies (TF). $\qquad\square$

Remark 23.16. Proposition 23.13 and Corollary 23.15 can be reformulated as follows. Suppose that Y is noetherian. Then the functors $\mathcal{F} \mapsto \Gamma_*(\mathcal{F})$ and $\mathcal{M} \mapsto \tilde{\mathcal{M}}$ define an equivalence between the category of coherent $\mathscr{O}_X$-modules and the quotient category (Exercise F.14)

$$\mathrm{Mod}_{\mathrm{TF}}(\mathscr{S})/\mathrm{Mod}_{\mathrm{TN}}(\mathscr{S}),$$

where $\mathrm{Mod}_{\mathrm{TF}}(\mathscr{S})$ (resp. $\mathrm{Mod}_{\mathrm{TN}}(\mathscr{S})$) denotes the full abelian subcategory of the category of graded quasi-coherent $\mathscr{S}$-modules consisting of modules satisfying (TF) (resp. (TN)).

Coherence of higher direct images for proper morphisms

In this part we prove that for proper morphisms $f\colon X \to Y$ of locally noetherian schemes the higher direct images of a coherent module is again coherent. This is deduced from the analogue statement for projective morphism using a dévissage argument and the Lemma of Chow. Thereafter, it is a rather formal argument to generalize this result by showing that Rf_* preserves pseudo-coherent complexes and to deduce that Rf_* maps $D_{\mathrm{coh}}(X)$ to $D_{\mathrm{coh}}(Y)$ and, if Y is noetherian, maps $D_{\mathrm{coh}}^b(X)$ to $D_{\mathrm{coh}}^b(Y)$.

Next we study the question under which hypotheses Rf_* preserves pseudo-coherent (resp. perfect) complexes and give some general criteria. We conclude the part by showing some finiteness result of Ext groups and by explaining the GAGA principle which links cohomology of coherent modules on proper schemes over $\mathbb{C}$ and their analytic counterparts on the attached compact complex analytic spaces.

(23.4) Finiteness of higher direct images under proper morphisms.

Using the technique of dévissage (Lemma 12.63), we can generalize the result that higher direct images under projective morphisms of coherent modules are again coherent (Theorem 23.1) to the case of proper morphisms.

Theorem 23.17. *Let S be a locally noetherian scheme and let $f \colon X \to S$ be a proper morphism of schemes. If $\mathscr{F}$ is a coherent $\mathscr{O}_X$-module, then for all $i \geq 0$, the higher direct image $R^i f_* \mathscr{F}$ is a coherent $\mathscr{O}_S$-module.*

Proof. First note that we know that $R^i f_* \mathscr{F}$ is quasi-coherent by Theorem 22.27. Recall that a quasi-coherent module on a locally noetherian scheme is coherent if and only if it is of finite type (Proposition 7.46).

To show that $R^i f_* \mathscr{F}$ is coherent, we can work locally on S, so we may assume that S is noetherian. Then X is also noetherian because f is proper and in particular of finite type. Let $\mathcal{C}$ be the full subcategory of the category of coherent $\mathscr{O}_X$-modules of all $\mathscr{F}$ which satisfy the conclusion of the theorem. We want to show that $\mathcal{C}$ is all of $(\mathrm{Coh}(X))$. By Lemma 12.63 it suffices to show the following assertions.

(1) Let $0 \to \mathscr{F}' \to \mathscr{F} \to \mathscr{F}'' \to 0$ be an exact sequence of coherent $\mathscr{O}_X$-modules. If two of the modules are in $\mathcal{C}$, then the third is contained in $\mathcal{C}$.
(2) $\mathcal{C}$ is closed under taking direct summands.
(3) For every integral closed subscheme $Y \subseteq X$ with generic point η, there exists a coherent $\mathscr{O}_Y$-module $\mathscr{G}$ with non-vanishing stalk $\mathscr{G}_\eta$ and such that all higher direct images $R^i f_* \mathscr{G}$ are coherent.

Assertion (1) follows from the long exact cohomology sequence and the fact that if $\mathscr{E}_1 \to \mathscr{E}_2 \to \mathscr{E}_3$ is an exact sequence of quasi-coherent $\mathscr{O}_S$-modules and $\mathscr{E}_1$ and $\mathscr{E}_3$ are of finite type, then $\mathscr{E}_2$ is of finite type. Indeed, we can assume that $S = \mathrm{Spec}\, R$ for a noetherian ring and hence that the exact sequence corresponds to an exact sequence $M_1 \to M_2 \to M_3$ for R-modules where M_1 and M_3 are finitely generated. Then M_2 is finitely generated R-module since R is noetherian. Hence its attached quasi-coherent module $\mathscr{E}_2$ is coherent.

Assertion (2) is clear since direct summands of coherent modules are coherent and $R^i f_*$ is additive and hence compatible with finite direct sums.

It remains to show (3). We may as well replace X by Y. In other words, we may assume that X is integral, with generic point η, and need to show that there exists a coherent $\mathscr{O}_X$-module $\mathscr{G}$ such that $\mathscr{G}_\eta \neq 0$ and all $R^i f_* \mathscr{G}$ are coherent $\mathscr{O}_S$-modules.

By Chow's Lemma (Theorem 13.100), there exists a surjective projective morphism $\pi \colon X' \to X$ such that the composition $g := f \circ \pi$ is also projective and such that π is an isomorphism over a non-empty open subscheme of X. We will find an $\mathscr{O}_{X'}$-module $\mathscr{G}'$ such that $\mathscr{G} := \pi_*(\mathscr{G}')$ satisfies $\mathscr{G}_\eta \neq 0$ and has coherent higher direct images. Note that $\mathscr{G}$ is coherent by Theorem 23.1.

Let $\mathscr{L}$ be a π-ample invertible $\mathscr{O}_{X'}$-module. For n sufficiently large, we have that $\mathscr{G}' := \mathscr{L}^{\otimes n}$ satisfies $R^i \pi_* \mathscr{G}' = 0$ for all $i > 0$ by Theorem 23.1. Since π is an isomorphism over a non-empty open subscheme of X, $(\pi_* \mathscr{G}')_\eta \neq 0$.

By Remark 22.7 and Remark 21.44 we find that

$$R^i f_* \mathscr{G} = R^i f_* (\pi_* \mathscr{G}') \cong R^i (f \circ \pi)_* (\mathscr{G}') = R^i g_* \mathscr{G}',$$

which is coherent for all $i \geq 0$ by Theorem 23.1. $\qquad\square$

Corollary 23.18. *Let X be proper over an affine scheme $S = \mathrm{Spec}\, R$ with R noetherian, and let $\mathscr{F}$ be a coherent $\mathscr{O}_X$-module. Then for all i, $H^i(X, \mathscr{F})$ is a finite generated R-module.*

Proof. This follows from Theorem 23.17 by Theorem 22.27. $\qquad\square$

Remark 23.19. The analogue of Theorem 23.17 in complex analysis is also true: It is a theorem by Grauert that the higher direct images of a coherent sheaf under a proper morphism of complex spaces are coherent. The proof is considerably more difficult in this setting because of the intricate analytic problems it poses.

(23.5) Finiteness of cohomology of modules with proper support.

In this section we explain how to weaken the hypothesis in the above finiteness results that the morphism under which direct images are taken is proper.

Remark and Definition 23.20. Let X be a scheme and let $\mathscr{F}$ be a quasi-coherent $\mathcal{O}_X$-module of finite type. Then its annihilator $\mathscr{I} := \mathrm{Ann}(\mathscr{F})$ is a quasi-coherent ideal and $Z := V(\mathscr{I})$ is a closed subscheme whose underlying topological space is $\mathrm{Supp}(\mathscr{F})$ (Proposition 7.35). In the sequel, we will endow $\mathrm{Supp}(\mathscr{F})$ with this structure of a closed subscheme.

One has $\mathscr{I}\mathscr{F} = 0$ and hence $\mathscr{F} \to i_*(i^*\mathscr{F})$ is an isomorphism, where $i\colon Z \to X$ is the inclusion.

Definition and Lemma 23.21. *Let $X \to S$ be a morphism locally of finite type and let $\mathscr{F}$ be a quasi-coherent $\mathcal{O}_X$-module of finite type. Then $\mathscr{F}$ is said to have proper support over S, if the following equivalent conditions hold.*

(i) *There exists a structure of closed subscheme of X on $\mathrm{Supp}(\mathscr{F})$ such that the composition $\mathrm{Supp}(\mathscr{F}) \to X \to S$ is proper.*

(ii) *For every structure of closed subscheme of X on $\mathrm{Supp}(\mathscr{F})$ the composition $\mathrm{Supp}(\mathscr{F}) \to X \to S$ is proper.*

(iii) *There exists a closed immersion $i\colon Z \to X$ such that the composition $Z \to X \to S$ is proper and such that $\mathscr{F} \cong i_*\mathscr{G}$ for some quasi-coherent $\mathcal{O}_Z$-module of finite type.*

Proof. Assertions (i) and (ii) are equivalent by Proposition 12.58 and clearly (iii) implies (i). Finally (ii) implies (iii) by Remark 23.20. □

Corollary 23.22. *Let S be a locally noetherian scheme, let $f\colon X \to S$ be a morphism locally of finite type, and let $\mathscr{F}$ be coherent $\mathcal{O}_X$-module with proper support over S. Then $R^p f_*\mathscr{F}$ is a coherent $\mathcal{O}_S$-module for all $p \geq 0$. If $S = \mathrm{Spec}\, R$ is affine, then $H^p(X, \mathscr{F})$ is a finitely generated R-module for all $p \geq 0$.*

Proof. Coherence can be checked locally hence it suffices to show the second assertion. By Lemma 23.21 there exists a closed immersion $i\colon Z \to X$ such that $g\colon Z \to S$ is proper and such that $\mathscr{F}$ is of the form $i_*\mathscr{G}$ for a coherent $\mathcal{O}_Z$-module $\mathscr{G}$. We have $H^p(X, \mathscr{F}) = H^p(Z, \mathscr{G})$ by Corollary 22.6 and hence conclude by Corollary 23.18. □

(23.6) Derived image of pseudo-coherent complexes.

Let X be a locally noetherian scheme. Recall that for $? \in \{+, -, b, \emptyset\}$ we denote by $D^?_{\mathrm{coh}}(X)$ the triangulated subcategory of complexes $\mathscr{F}$ in $D^?(X)$ such that $H^p(\mathscr{F})$ is coherent for all $p \in \mathbb{Z}$. For instance, if X is noetherian, then $D^-_{\mathrm{coh}}(X)$ is the category of pseudo-coherent complexes on X by Proposition 22.61.

Proposition 23.23. *Let $f\colon X \to Y$ be a proper morphism between locally noetherian schemes. Let $E \in D(X)$ be pseudo-coherent. Then Rf_*E is pseudo-coherent.*

Proof. Since the question is local on Y, we may assume that Y is noetherian. Then X is noetherian because f is of finite type. Let us first consider the case that the cohomology of E is bounded, i.e., that $H^i(E) = 0$ for all i with $|i|$ larger than some bound (depending on E). By assumption, all $H^i(E)$ are coherent, so Theorem 23.17 shows that all $R^j f_* H^i(E)$ are coherent. Furthermore, we know that $R^j f_* H^i(E) = 0$ whenever j is larger than some constant N (depending only on f) by Proposition 22.29.

Now the hypercohomology spectral sequence (Corollary 21.46),

$$R^j f_* H^i(E) \implies R^{i+j} f_* E,$$

implies that all cohomology sheaves of $Rf_* E$ are coherent, and that the cohomology of $Rf_* E$ is bounded above, whence $Rf_* E$ is pseudo-coherent (Proposition 22.61).

Now we prove the general case. For an integer n, consider an exact triangle

$$E \to \tau^{\geq n} E \to C \to$$

where $E \to \tau^{\geq n} E$ is the natural morphism to the truncation. Now $\tau^{\geq n} E$ is pseudo-coherent and bounded, and hence $Rf_* \tau^{\geq n} E$ is pseudo-coherent by what we have already proved. Furthermore, we have that $\tau^{\geq n-1} C$ is exact, so using Proposition 22.29 again, we have $R^i f_* C = 0$ for all $i \geq n + N - 1$ for some N depending only on the morphism f. This shows that $\tau^{\geq n+N} Rf_* E \cong \tau^{\geq n+N} Rf_* \tau^{\geq n} E$, so $\tau^{\geq n+N} Rf_* E$ is pseudo-coherent for all n. It follows that $Rf_* E$ is pseudo-coherent. $\square$

Remark 23.24. The proof of Proposition 23.23 shows that if $f \colon X \to Y$ is a proper morphism of noetherian schemes, then Rf_* sends objects in $D^b_{\mathrm{coh}}(X)$ to objects in $D^b_{\mathrm{coh}}(Y)$.

Corollary 23.25. *Let $f \colon X \to Y$ be a proper morphism between locally noetherian schemes. Let $E \in D_{\mathrm{coh}}(X)$ be a complex with coherent cohomology. Then $Rf_* E \in D_{\mathrm{coh}}(Y)$, i.e., $R^p f_* E$ is coherent for all $p \in \mathbb{Z}$.*

Proof. We may work locally on Y and hence can assume that Y noetherian. Then X is also noetherian. Fix $p \in \mathbb{Z}$. Then $\tau^{\leq p} E$ is pseudo-coherent by Proposition 22.61. Consider a distinguished triangle

$$\tau^{\leq p} E \longrightarrow E \longrightarrow C \xrightarrow{+} .$$

The long exact cohomology sequence shows that $C \in D^{\geq p+1}(X)$ and hence $Rf_* C \in D^{\geq p+1}(Y)$. Applying Rf_* to the distinguished triangle, the long exact cohomology sequence yields an isomorphism $R^p f_*(\tau^{\leq p} E) \xrightarrow{\sim} R^p f_*(E)$. As $Rf_*(\tau^{\leq p} E)$ is pseudo-coherent by Proposition 23.23, $R^p f_*(\tau^{\leq p} E)$ is coherent. Therefore $R^p f_*(E)$ is coherent. $\square$

Proposition 23.23 can be generalized to non-noetherian situations, see Corollary 23.136 below, Kiehl's paper [Kie] $\overset{\mathrm{O}}{}$, [Lip2] $\overset{\mathrm{O}}{\mathrm{X}}$ Corollary 4.3.3.2, or [FuKa] $\overset{\mathrm{O}}{\mathrm{X}}$ I, 8.1.3.

(23.7) Morphisms of finite tor-dimension.

After we gave in Proposition 23.23 a criterion when Rf_* preserves pseudo-coherent complexes (see also Definition 23.31 below), we now study the question when Rf_* preserves complexes of bounded tor-amplitude. In the next section we will combine these results and give criteria when Rf_* preserves perfect complexes using that a complex is perfect if and only if it is pseudo-coherent and locally of finite tor-dimension (Theorem 21.174).

Let $t \geq 0$ be an integer. Recall (Definition 21.164) that a morphism $f: X \to Y$ of schemes is said to be of tor-dimension $\leq t$ if $H^i(Lf^*\mathscr{G}) = 0$ for all $i < -t$ and for all $\mathscr{O}_Y$-modules $\mathscr{G}$. If Y is quasi-separated (e.g., if Y is locally noetherian), then it suffices to check this condition for quasi-coherent $\mathscr{O}_Y$-modules $\mathscr{G}$ (Lemma 22.51). The morphism f is called locally of finite tor-dimension if there exists an open covering $(V_i)_i$ of Y such that the restriction $f^{-1}(V_i) \to V_i$ has tor-dimension $\leq t_i$ for some $t_i \geq 0$. Moreover, one has the following properties.

(1) The condition that f has tor-dimension $\leq t$ can be checked Zariski locally on X and on Y (Lemma 21.171).

(2) If $f\colon \operatorname{Spec} B \to \operatorname{Spec} A$ is a morphism of affine schemes, then f is of tor-dimension $\leq t$ if and only if B is of finite tor-dimension $\leq t$ as an A-module (Lemma 22.44).

(3) If f is of tor-dimension $\leq t$ and $g\colon Y \to Z$ is of tor-dimension $\leq s$, then $g \circ f$ is of tor-dimension $\leq t + s$ (21.35.4).

Example 23.26.

(1) Remark 21.165 (5) shows that a morphism of schemes is flat if and only if it is of tor-dimension ≤ 0.

(2) If $i\colon Y \to X$ is a completely intersecting immersion of codimension c, then i is of tor-dimension $\leq c$.

  Indeed, as we can check the condition locally on X and Y, we may assume that $X = \operatorname{Spec} A$ and $Y = \operatorname{Spec} A/I$ for a ring A and an ideal $I \subseteq A$ that is generated by a completely intersecting sequence $\mathbf{f} = (f_1, \ldots, f_c)$. Then we have a quasi-isomorphism between A/I and the Koszul complex $K(\mathbf{f})$ defined by $\mathbf{f}$ which we view as a complex of free A-modules concentrated in degrees $[-c, 0]$. Hence for every A-module M, $Li^*\tilde{M} \in D_{\mathrm{qcoh}}(Y)$ corresponds to the complex $M \otimes_A^L A/I = M \otimes_A K(\mathbf{f})$ that is concentrated in degrees $[-c, 0]$.

(3) Every locally completely intersecting morphism of locally of finite tor-dimension.

  Indeed, as this can be checked locally, we can assume that the morphism can be factorized as a completely intersecting immersion followed by a smooth (and in particular flat) morphism. Hence we can use (2) and (1).

(4) Let A be a ring and let $a \in A$ be a regular element that is not a unit. Then the closed immersion $\operatorname{Spec} A/(a) \to \operatorname{Spec} A/(a^2)$ is not of finite tor-dimension.

  Indeed $A/(a)$ has a quasi-isomorphism from the K-flat complex of $A/(a^2)$-modules

  $$\cdots \xrightarrow{\cdot a} A/(a^2) \xrightarrow{\cdot a} A/(a^2) \xrightarrow{\cdot a} A/(a^2) \longrightarrow 0 \longrightarrow \cdots,$$

  concentrated in degrees ≤ 0. Hence $A/(a) \otimes_{A/(a^2)}^L A/(a)$ is represented by the complex

  $$\cdots \xrightarrow{0} A/(a) \xrightarrow{0} A/(a) \xrightarrow{0} A/(a) \longrightarrow 0 \longrightarrow \cdots,$$

  which has cohomology in every degree ≤ 0.

Lemma 23.27. *Let $f\colon X \to Y$ be a morphisms of finite tor-dimension and let*

$$
\begin{array}{ccc}
X' & \longrightarrow & X \\
{\scriptstyle f'}\downarrow & & \downarrow{\scriptstyle f} \\
Y' & \xrightarrow{u} & X
\end{array}
$$

be a cartesian diagram of schemes such that u is flat. Then f' is of finite tor-dimension.

Proof. As we can check finite tor-dimension Zariski locally on source and target we may assume that all schemes are affine, say $Y = \operatorname{Spec} A$, $X = \operatorname{Spec} B$, $Y' = \operatorname{Spec} A'$. By hypothesis, B has finite tor-dimension as an A-module and hence $B \otimes_A A'$ has finite tor-dimension as an A'-module by Corollary 21.170, applied to u and using that $Lu^* = u^*$ since u is flat. □

Proposition 23.28. *Let Y be a qcqs scheme and let $f\colon X \to Y$ be a qcqs morphism of tor-dimension $\leq t$. Let $N \geq 0$ an integer such that $R^i f_* \mathscr{F} = 0$ for all $i \geq N$ and for all quasi-coherent $\mathscr{O}_X$-modules $\mathscr{F}$. Let $\mathscr{E}$ in $D_{\mathrm{qcoh}}(X)$ be of tor-amplitude in $[a,b]$. Then $Rf_* \mathscr{E}$ is of tor-amplitude in $[a-t, b+N]$.*

Such an integer N always exists by Lemma 22.28.

Proof. By Lemma 22.51 it suffices to show that $Rf_* \mathscr{E} \otimes^L_{\mathscr{O}_Y} \mathscr{G} \in D^{[a-t,b+N]}(Y)$ for every quasi-coherent $\mathscr{O}_Y$-module $\mathscr{G}$. But $Rf_* \mathscr{E} \otimes^L_{\mathscr{O}_Y} \mathscr{G} \cong Rf_*(\mathscr{E} \otimes^L_{\mathscr{O}_X} Lf^* \mathscr{G})$ by the projection formula. One has $\mathscr{E} \otimes^L_{\mathscr{O}_X} Lf^* \mathscr{G} \in D^{[a-t,b]}_{\mathrm{qcoh}}(X)$ by Lemma 21.167 and Proposition 22.40. Therefore $Rf_*(\mathscr{E} \otimes^L_{\mathscr{O}_X} Lf^* \mathscr{G}) \in D^{[a-t,b+N]}_{\mathrm{qcoh}}(Y)$ by Proposition 22.29. □

Corollary 23.29. *Let $f\colon X \to Y$ be a qcqs morphism of schemes which is locally of finite tor-dimension, and let $\mathscr{E}$ in $D_{\mathrm{qcoh}}(X)$ be locally of finite tor-dimension. Then $Rf_* \mathscr{E} \in D_{\mathrm{qcoh}}(Y)$ is locally of finite tor-dimension.*

(23.8) Derived image of perfect complexes.

Proposition 23.30. *Let Y be a locally noetherian scheme and let $f\colon X \to Y$ be a proper morphism locally of finite tor-dimension. Let $\mathscr{E}$ be a perfect complex on X. Then $Rf_* \mathscr{E}$ is perfect.*

There are in fact several nice characterizations of morphisms f such that Rf_* preserves perfect objects (see Remark 25.39 below).

Proof. Recall that a complex is perfect if and only if it is pseudo-coherent and has locally finite tor-dimension (Theorem 21.174). The complex $Rf_* \mathscr{E}$ is pseudo-coherent by Proposition 23.23 and has finite tor-dimension locally on Y by Corollary 23.29. □

The proof uses Theorem 21.174, which holds for arbitrary ringed spaces, Corollary 23.29, which holds for every qcqs morphism, and it uses the criterion Proposition 23.23 for the derived direct image of pseudo-coherent complexes to be pseudo-coherent.

Let us coin the following (non-standard) notion.

Definition 23.31. *A morphism $f\colon X \to Y$ of schemes is called* cohomologically proper[1] *if f is qcqs and if $Rf_* \mathscr{E}$ is pseudo-coherent for every pseudo-coherent complex $\mathscr{E}$ in $D(X)$.*

Then the same proof as in Proposition 23.30 yields the following result.

---

[1] Lipman calls such morphisms "quasi-proper" which however could be confusing since in general a proper morphism need not be quasi-proper (Exercise 23.8). A notion of a cohomologically proper morphism has also been defined by Kubrak and Prikhodko for morphisms of locally noetherian algebraic stacks in [KuPr]$^\mathrm{O}_\mathrm{X}$ which, if specialized to morphisms of locally noetherian schemes, is equivalent to the property of being proper and hence to the notion given here (Remark 23.33).

Proposition 23.32. *Let* $f\colon X \to Y$ *be a cohomologically proper morphism locally of finite tor-dimension. If* $\mathscr{E}$ *is a perfect complex on* X, *then* $Rf_*\mathscr{E}$ *is a perfect complex on* Y.

Remark 23.33. Proposition 23.23 just says that any proper morphism between locally noetherian schemes is cohomologically proper. The converse also holds for quasi-compact separated morphisms between locally noetherian schemes ([Hal]$_X$ 3.10, see also Exercise 26.6 if the morphism is in addition of finite type).

Let us give a criterion for affine morphisms to be cohomologically proper.

Proposition 23.34. *Let* $f\colon X \to Y$ *be an affine morphism of schemes. Then* f *is cohomologically proper if and only if* $f_*\mathscr{O}_X$ *is pseudo-coherent (i.e.,* $f_*\mathscr{O}_X$ *considered as a complex of* $\mathscr{O}_Y$-*modules concentrated in degree* 0 *is pseudo-coherent).*

These equivalent hypotheses imply that f is finite and of finite presentation by Proposition 21.153, in particular that f is proper.

Proof. As f is affine, we have $Rf_*\mathscr{O}_X = f_*\mathscr{O}_X$ and as $\mathscr{O}_X$ is pseudo-coherent, the condition is clearly necessary.

Now assume that $f_*\mathscr{O}_X$ is pseudo-coherent and let $\mathscr{E}$ be a pseudo-coherent complex in $D(X)$. To show that $f_*\mathscr{E}$ is pseudo-coherent, we may work locally on Y and hence can assume that $Y = \operatorname{Spec} A$ is affine. Then X is also affine, say $X = \operatorname{Spec} B$ and B is a pseudo-coherent A-module. The complex $\mathscr{E}$ corresponds to a pseudo-coherent complex E of B-modules (Remark 22.45) and we may represent E by a bounded above complex of finitely generated free B-modules (Proposition 21.162). As finite direct sums of pseudo-coherent complexes are again pseudo-coherent (Proposition 21.156) we deduce that all components of E are pseudo-coherent A-modules. Hence E is pseudo-coherent in $D(A)$ by Lemma 21.158.   □

There are further important examples of cohomologically proper morphisms. Below in Corollary 23.136 we will see that flat proper morphisms of finite presentation between arbitrary schemes are cohomologically proper. Exercise 23.10 also gives a criterion for projective morphisms to be cohomologically proper.

More generally, one calls a morphism $f\colon X \to Y$ of schemes *pseudo-coherent* if there exists an open covering $(U_\lambda)_\lambda$ of X such that for all λ the restriction $f_{|U_\lambda}$ can be factorized into $U_\lambda \xrightarrow{i_\lambda} P_\lambda \to S$, where $P_\lambda \to S$ is smooth and i_λ is a closed immersion such that $i_{\lambda*}\mathscr{O}_{U_\lambda}$ is a pseudo-coherent $\mathscr{O}_{P_\lambda}$-module. Then one has the following result for whose proof we refer to [Lip2]$_X^O$ (4.3.3.2).

Proposition 23.35. *Let* $f\colon X \to Y$ *be a proper pseudo-coherent morphism of schemes. Then* f *is cohomologically proper.*

Corollary 23.36. *Let* $f\colon X \to Y$ *be a proper lci-morphism and let* $\mathscr{E}$ *be a perfect complex on* X, *then* $Rf_*\mathscr{E}$ *is perfect.*

Proof. Example 23.26 (2) and (3) show that f is of finite tor-dimension and pseudo-coherent. Hence we conclude by Proposition 23.35 and Proposition 23.32.   □

Finally, [FuKa]$_X^O$ I.8.1.4 implies the following result, which generalizes the result that proper morphisms between locally noetherian schemes are cohomologically proper.

Proposition 23.37. *Let Y be a universally coherent scheme (Exercise 22.32) and let $f\colon X \to Y$ be a proper morphism of finite presentation. Then f is cohomologically proper.*

Proof. This can be checked locally on Y and hence we may assume that Y is quasi-compact. Then X is quasi-compact and coherent and hence a complex in $D(X)$ is pseudo-coherent if and only if it is in $D^-(X)$ (Exercise 22.31). Now we can apply loc. cit. which ensures that Rf_* sends $D^-(X)$ to $D^-(Y)$. $\qquad\square$

Corollary 23.38. *Let A be a noetherian ring and let $X \to \operatorname{Spec} A$ be a proper morphism. Let $\mathscr{F}$ and $\mathscr{G}$ be complexes in $D(X)$. Suppose that one of the following hypotheses is satisfied.*
(1) $\mathscr{F}$ is perfect and $\mathscr{G}$ is in $D_{\mathrm{coh}}(X)$.
(2) One has $\mathscr{F} \in D^-_{\mathrm{coh}}(X)$ and $\mathscr{G} \in D^+_{\mathrm{coh}}(X)$.
Then $\operatorname{Ext}^n_{\mathscr{O}_X}(\mathscr{F}, \mathscr{G})$ is a finitely generated A-module for all $n \in \mathbb{Z}$.

Proof. By Corollary 22.67, $H^n(R\,\mathscr{H}om_{\mathscr{O}_X}(\mathscr{F}, \mathscr{G}))$ is coherent for all $n \in \mathbb{Z}$. Hence $\operatorname{Ext}^n(\mathscr{F}, \mathscr{G}) = H^n(R\Gamma(X, R\,\mathscr{H}om_{\mathscr{O}_X}(\mathscr{F}, \mathscr{G})))$ is a finitely generated A-module for all n because f is proper (Corollary 23.25). $\qquad\square$

(23.9) GAGA.

Recall the notion of a complex analytic space and of the analytification from Section (20.12). Let X be a scheme that is locally of finite type over $\mathbb{C}$, let X^{an} be its analytification, which is a complex analytic space, and let $\alpha\colon X^{\mathrm{an}} \to X$ be the canonical flat morphism of locally ringed spaces (Theorem 20.58). For $\mathscr{F} \in D_{\mathrm{coh}}(X)$ we call $\mathscr{F}^{\mathrm{an}} := L\alpha^*\mathscr{F}$ its analytification. Since α is flat, for every complex $\mathscr{F}$ of $\mathscr{O}_X$-modules, its derived inverse image $L\alpha^*\mathscr{F}$ is represented by the complex $\alpha^*\mathscr{F}$.

In Theorem 20.60 we have seen that pullback by α yields an equivalence of the category of coherent $\mathscr{O}_X$-modules and the coherent $\mathscr{O}_{X^{\mathrm{an}}}$-modules if X is proper over $\mathbb{C}$. This extends to derived categories (e.g., [Hal]$_X$ Theorem A).

Theorem 23.39. *Let X be a scheme that is proper over $\mathbb{C}$ with analytification $\alpha\colon X^{\mathrm{an}} \to X$. Then $L\alpha^*$ induces triangulated equivalences of triangulated categories*

$$D^-_{\mathrm{coh}}(X) \xrightarrow{\sim} D^-_{\mathrm{coh}}(X^{\mathrm{an}}), \qquad D^b_{\mathrm{coh}}(X) \xrightarrow{\sim} D^b_{\mathrm{coh}}(X^{\mathrm{an}}).$$

yielding for all $\mathscr{F}$ in $D^-_{\mathrm{coh}}(X)$ and for all $p \in \mathbb{Z}$ an isomorphism

$$(23.9.1) \qquad H^p(X, \mathscr{F}) \xrightarrow{\sim} H^p(X^{\mathrm{an}}, \mathscr{F}^{\mathrm{an}}).$$

There is also a relative version of (23.9.1). Let $f\colon X \to Y$ be a morphism of schemes locally of finite type over $\mathbb{C}$. Consider the commutative diagram of morphisms of locally ringed spaces

$$
\begin{array}{ccc}
X^{\mathrm{an}} & \xrightarrow{\alpha_X} & X \\
{\scriptstyle f^{\mathrm{an}}}\downarrow & & \downarrow{\scriptstyle f} \\
Y^{\mathrm{an}} & \xrightarrow{\alpha_Y} & Y.
\end{array}
$$

By Proposition 21.129 one obtains a base change morphism

$$b\colon L\alpha_Y \circ Rf_* \longrightarrow Rf^{\mathrm{an}}_* \circ L\alpha_X$$

of functors $D(X) \to D(Y^{\mathrm{an}})$.

Theorem 23.40. *Let f be proper. Then b induces an isomorphism of functors $D^b_{\mathrm{coh}}(X) \to D^b_{\mathrm{coh}}(Y^{\mathrm{an}})$ and for every $\mathscr{F}$ in $D^b_{\mathrm{coh}}(X)$ and for all $p \in \mathbb{Z}$ one has a functorial isomorphism*

$$(R^p f_* \mathscr{F})^{\mathrm{an}} \xrightarrow{\sim} R^p f_*^{\mathrm{an}} \mathscr{F}^{\mathrm{an}}.$$

Numerical intersection theory, Euler characteristic, and Hilbert polynomial

In the next few sections, we will introduce the zeroth K-group of schemes, discuss the Euler characteristic and Hilbert polynomial of coherent sheaves, and study the basics of numerical intersection theory, i.e., we will attach to a family of line bundles and a coherent sheaf an integer. If the line bundles are the line bundles attached to effective divisors, and the coherent sheaf is the structure sheaf of a closed subscheme of the ambient scheme, we think of this number as the number of points in the intersection of the divisors and the closed subscheme, counted "correctly", i.e., with multiplicities. The formalism of the Euler characteristic, which is based on the cohomology of coherent sheaves, provides a relatively easy way to give a general definition which can then also be applied to cases where the geometric intuition fails or is problematic, e.g., taking the d-fold self-intersection of a divisor on a d-dimensional scheme.

For a more sophisticated approach of intersection theory, one can set up a theory of algebraic cycles and Chow groups, generalizing the notion of Weil divisor to higher codimensions, and define intersection products in this setting. See for instance [Ful2] ^{O}.

(23.10) Grothendieck group of abelian categories and triangulated categories.

Recall that a category is called *essentially small* if it is equivalent to a small category.

Definition 23.41.
(1) *Let $\mathcal{A}$ be an essentially small abelian category. The* Grothendieck group $K_0(\mathcal{A})$ *is defined as the abelian group with generators $[X]$ for each isomorphism class X of objects in $\mathcal{A}$ and relations of the form*

$$[X] = [X'] + [X'']$$

for each exact sequence $0 \to X' \to X \to X'' \to 0$ in $\mathcal{A}$.
(2) *Let $\mathcal{T}$ be an essentially small triangulated category. The* Grothendieck group $K_0(\mathcal{T})$ *of $\mathcal{T}$ is the abelian group with generators $[X]$ for each isomorphism class X of objects in $\mathcal{T}$ and relations of the form*

$$[X] = [X'] + [X'']$$

for each distinguished triangle $X' \to X \to X'' \to$ in $\mathcal{T}$.

The definition of $K_0(\mathcal{A})$ can be generalized to exact categories (Exercise F.15), see Exercise 23.16.

Remark 23.42. Let $\mathcal{T}$ be an essentially small triangulated category. Then in $K_0(\mathcal{T})$ one has $[X[1]] = -[X]$ for all objects X in $\mathcal{T}$.

Indeed, the distinguished triangle $0 \longrightarrow 0 \longrightarrow 0 \overset{+}{\longrightarrow}$ shows that $[0] = 0$ in $K_0(\mathcal{T})$, and rotating $X \overset{\mathrm{id}_X}{\longrightarrow} X \longrightarrow 0 \overset{+}{\longrightarrow}$ yields the relation $0 = [0] = [X] + [X[1]]$.

Proposition 23.43. *Let $\mathcal{A}$ be an essentially small abelian category. Then the map*

$$\Phi \colon K_0(\mathcal{A}) \longrightarrow K_0(D^b(\mathcal{A})), \qquad [M] \mapsto [M[0]]$$

is an isomorphism of abelian groups. Its inverse is given by

$$\Psi \colon [X] \mapsto \sum_{i \in \mathbb{Z}} (-1)^i [H^i(X)].$$

Proof. It is clear that Φ is well defined, and Ψ is well defined by the long exact cohomology sequence attached to a distinguished triangle. Moreover, $\Psi \circ \Phi = \mathrm{id}$ by definition. It remains to show that Φ is surjective.

By Remark 23.42, the class of $M[i]$ is in the image of Φ for every object M in $\mathcal{A}$ and all $i \in \mathbb{Z}$. Let X be in $D_{\mathcal{A}}^b(\mathcal{A}')$ and $a \leq c$ such that $X \in D_{\mathcal{A}}^{[a,c]}(\mathcal{A}')$. To show that $[X]$ is in the image of Φ, we use the distinguished triangle

$$\tau^{\leq c-1} X \to X \to H^c(X)[-c] \to$$

to argue by induction on $c - a$ applying the induction hypothesis to $\tau^{\leq c-1} X \in D_{\mathcal{A}}^{[a,c-1]}(\mathcal{A}')$. $\qquad\square$

(23.11) Grothendieck group of noetherian schemes.

Definition and Lemma 23.44. *Let X be a noetherian scheme. Then*

(23.11.1) $$K_0'(X) := K_0(\mathrm{Coh}(X)) = K_0(D_{\mathrm{coh}}^b(X)),$$

is called the Grothendieck group *of X.*

For each $r \geq 0$ let $K_0'(X)_r \subseteq K_0'(X)$ be the subgroup generated by those coherent $\mathcal{O}_X$-modules $\mathscr{F}$ with $\dim \mathrm{Supp}(\mathscr{F}) \leq r$.

If $X = \mathrm{Spec}\, A$ is affine with a noetherian ring A, we write $K_0'(A)$ and $K_0'(A)_r$ instead of $K_0'(\mathrm{Spec}\, A)$ and $K_0'(\mathrm{Spec}\, A)_r$.

The group $K_0'(X)$ is also sometimes denoted as $G_0(X)$.

Proof. We have to show the equality in (23.11.1), but $K_0(\mathrm{Coh}(X)) = K_0(D^b(\mathrm{Coh}(X)))$ by Proposition 23.43, and $D^b(\mathrm{Coh}(X))$ is equivalent to $D_{\mathrm{coh}}^b(X)$ by Theorem 22.42. $\qquad\square$

We obtain an ascending filtration

$$K_0'(X)_0 \subseteq K_0'(X)_1 \subseteq \ldots$$

We also set $K_0'(X)_r := 0$ for $r < 0$ and define

$$\mathrm{gr}_r K_0'(X) := K_0'(X)_r / K_0'(X)_{r-1}, \qquad r \in \mathbb{Z}.$$

Remark 23.45. The class of a complex $\mathscr{F} \in D^b_{\mathrm{coh}}(X)$ is contained in $K'_0(X)_r$ if $\dim \mathrm{Supp}\, H^i(\mathscr{F}) \leq r$ for all $i \in \mathbb{Z}$. Indeed, by Proposition 23.43 we know that $[\mathscr{F}] = \sum_i (-1)^i [H^i(\mathscr{F})]$ in $K'_0(X)$.

The converse does not hold as the example of the complex $0 \longrightarrow \mathscr{O}_X \overset{0}{\longrightarrow} \mathscr{O}_X \longrightarrow 0$ shows, whose class in $K_0(X)$ is $[\mathscr{O}_X] - [\mathscr{O}_X] = 0$ but which has cohomology that has support on all of X.

Remark 23.46. Let $f \colon X \to Y$ be a proper morphism of noetherian schemes. Then the triangulated functor $Rf_* \colon D^b_{\mathrm{coh}}(X) \to D^b_{\mathrm{coh}}(Y)$ (Remark 23.24) induces a group homomorphism

(23.11.2) $f_* \colon K'_0(X) \to K'_0(Y).$

Let $\mathscr{F}$ be a coherent $\mathscr{O}_X$-module.
(1) By Proposition 23.43 one has

(23.11.3) $$f_*[\mathscr{F}] = \sum_{i \geq 0} (-1)^i [R^i f_* \mathscr{F}].$$

(2) One has $\mathrm{Supp}\, R^i f_* \mathscr{F} \subseteq f(\mathrm{Supp}\, \mathscr{F})$ and $\dim f(\mathrm{Supp}\, \mathscr{F}) \leq \dim \mathrm{Supp}\, \mathscr{F}$ because f is closed. Therefore we see that for all r one has

$$f_*(K'_0(X)_r) \subseteq K'_0(Y)_r.$$

In particular, f_* induces for all $r \geq 0$ a homomorphism

(23.11.4) $f_* \colon \mathrm{gr}_r K'_0(X) \longrightarrow \mathrm{gr}_r K'_0(Y).$

(3) If f is finite (e.g., if f is a closed immersion), then $R^i f_* \mathscr{F} = 0$ for $i > 0$ (Corollary 22.5) and we see that $f_*[\mathscr{F}] = [f_* \mathscr{F}]$ in this case by (23.11.3).

By dévissage we obtain the following result.

Proposition 23.47. *Let X be a noetherian scheme and let $r \geq 0$. Let $\varphi, \psi \colon K'_0(X)_r \to A$ be group homomorphisms with values in an abelian group A such that $\varphi(i_* \mathscr{O}_Z) = \psi(i_* \mathscr{O}_Z)$ for each integral closed subscheme $i \colon Z \to X$ of dimension $\leq r$. Then $\varphi = \psi$.*

Proof. Considering $\varphi - \psi$ we may assume that $\psi = 0$. We want to show that the class of any coherent $\mathscr{O}_X$-module $\mathscr{F}$ with support of dimension $\leq r$ is in the kernel of φ. We equip $\mathrm{Supp}(\mathscr{F})$ with the structure of a closed subscheme by setting $\mathrm{Supp}(\mathscr{F}) = V(\mathrm{Ann}(\mathscr{F}))$ (Remark 23.20) and let $\iota \colon \mathrm{Supp}(\mathscr{F}) \to X$ be the inclusion. The image of $\iota_* \colon K'_0(\mathrm{Supp}(\mathscr{F})) \to K'_0(X)$ is contained in $K'_0(X)_r$ and contains the class of $\mathscr{F} = \iota_*(\iota^* \mathscr{F})$. Hence it suffices to show that $\varphi \circ \iota_* = 0$. Let Σ be the kernel of $\varphi \circ \iota_*$.

By hypothesis, Σ contains $i_* \mathscr{O}_Z$ for every closed integral subscheme $i \colon Z \to \mathrm{Supp}(\mathscr{F})$. Moreover, if $0 \to \mathscr{G}' \to \mathscr{G} \to \mathscr{G}'' \to 0$ is an exact sequence of coherent modules on $\mathrm{Supp}(\mathscr{F})$ and two of the three modules are in Σ, then the third is in Σ. Hence we conclude by Lemma 12.63. $\square$

Corollary 23.48. *Let X be a noetherian scheme, let $\mathscr{F}$ be a coherent sheaf on X, and $r := \dim \mathrm{Supp}\, \mathscr{F}$. Let $Z_1, \ldots, Z_m$ be the irreducible components of $\mathrm{Supp}\, \mathscr{F}$ of dimension r, considered as integral subschemes of X, and let $\eta_i \in Z_i$ be the generic point. Then one has*

$$[\mathscr{F}] \equiv \sum_{i=1}^{m} \lg_{\mathscr{O}_{X,\eta_i}}(\mathscr{F}_{\eta_i}) \, [\mathscr{O}_{Z_i}] \bmod K_0'(X)_{r-1}$$

in $\mathrm{gr}_r K_0'(X) = K_0'(X)_r / K_0'(X)_{r-1}$.

Proof. Both sides of the equation can be viewed as group homomorphisms $K_0'(X)_r \to \mathrm{gr}_r K_0'(X)$. They agree if $\mathscr{F} = i_* \mathscr{O}_Z$ for a closed integral subscheme $i \colon Z \to X$. Hence we can conclude by Proposition 23.47. □

Corollary 23.49. *Let X be a noetherian scheme and let $\mathscr{E}$ be a locally free $\mathscr{O}_X$-module of constant rank $e \geq 1$. Then $\mathscr{F} \mapsto \mathscr{E} \otimes \mathscr{F}$ induces an endomorphism of abelian groups $K_0'(X) \to K_0'(X)$ which preserves $K_0'(X)_r$ and induces on $\mathrm{gr}_r K_0'(X)$ the multiplication by e for all $r \geq 0$.*

Proof. As tensoring with $\mathscr{E}$ is an exact functor, it induces an endomorphism of abelian groups $K_0'(X) \to K_0'(X)$. Because $\mathrm{Supp}(\mathscr{E} \otimes \mathscr{F}) = \mathrm{Supp}\,\mathscr{F}$ for every coherent $\mathscr{O}_X$-module, it preserves $K_0'(X)_r$. Let $\mathscr{F}$ be a coherent module with $\dim \mathrm{Supp}\,\mathscr{F} \leq r$ and let $Z_1, \ldots, Z_m$ be the irreducible components of $\mathrm{Supp}\,\mathscr{F}$ of dimension r with generic points η_i. Then $\lg_{\mathscr{O}_{Z_i,\eta_i}}(\mathscr{E} \otimes \mathscr{F})_{\eta_i} = e \lg_{\mathscr{O}_{Z_i,\eta_i}} \mathscr{F}_{\eta_i}$. By Corollary 23.48, then one has modulo $K_0'(X)_{r-1}$ that $[\mathscr{E} \otimes \mathscr{F}] \equiv e[\mathscr{F}]$. □

(23.12) K-groups of quasi-compact schemes.

Let X be a scheme, then the full subcategory $(\mathrm{Perf}(X))$ of $D(X)$ consisting of perfect complexes is a triangulated subcategory by Proposition 21.175.

Definition 23.50. *Let X be a quasi-compact scheme. Then we set*

$$K_0(X) := K_0(\mathrm{Perf}(X)).$$

and call it the (zero-th) K-group *of X. If $X = \mathrm{Spec}\,A$, then we define the K-group $K_0(A) := K_0(\mathrm{Spec}\,A)$ of the ring A.*

If X is not necessarily quasi-compact, one defines $K_0(X)$ to be the K-group of all perfect complexes of (globally) finite tor-amplitude [ThTr] [O]. We will not use this generality in the sequel.

Moreover, one can also define higher K-groups but here we use only the 0-th K-group. Nevertheless we keep the index 0 to distinguish the K-group from $K(X)$, the triangulated category of complexes of $\mathscr{O}_X$-modules up to homotopy, and to remind the reader that this is only the top layer of a much deeper story.

Under certain circumstances, one can define $K_0(X)$ also just in terms of finite locally free modules as the following remark shows.

Remark 23.51. Let X be a quasi-compact scheme and let $\mathscr{E}$ be a strictly perfect complex, i.e., a finite complex of finite locally free $\mathscr{O}_X$-modules, say concentrated in degrees $[a, b]$. Then

$$0 = \sigma^{\geq b+1}\mathscr{E} \subseteq \sigma^{\geq b}\mathscr{E} \subseteq \cdots \sigma^{\geq a+1}\mathscr{E} \subseteq \sigma^{\geq a}\mathscr{E} = \mathscr{E}$$

is a filtration by strictly perfect complexes such that $\sigma^{\geq i+1}\mathscr{E}/\sigma^{\geq i}\mathscr{E} \cong \mathscr{E}^i[-i]$. As short exact sequences of complexes define distinguished triangles, we see that in $K_0(X)$ we have

$$[\mathscr{E}] = \sum_{i=a}^{b}(-1)^i[\mathscr{E}^i].$$

Now suppose that X is semiseparated and has the resolution property (Definition 22.56), e.g., if X admits an ample line bundle (Proposition 22.57). Then every perfect complex on X can be represented by a strictly perfect complex (Proposition 22.58). Hence $K_0(X)$ is generated as abelian group by the classes $[\mathscr{F}]$, where $\mathscr{F}$ is a finite locally free $\mathcal{O}_X$-module. One can show that the only relations between these generators are of the form $[\mathscr{F}] = [\mathscr{F}'] + [\mathscr{F}'']$ for short exact sequences $0 \to \mathscr{F}' \to \mathscr{F} \to \mathscr{F}'' \to 0$, see Exercise 23.23. See also Exercises 23.19 and 23.17 for further examples when the resolution property is satisfied.

The K-group $K_0(X)$ has a commutative ring structure and the Grothendieck group $K'_0(X)$ has the structure of a module over it:

Remark and Definition 23.52. Let X be a quasi-compact scheme.
(1) We define on the abelian group $K_0(X)$ the structure of a commutative ring by

$$K_0(X) \otimes_{\mathbb{Z}} K_0(X) \longrightarrow K_0(X), \qquad [\mathscr{E}] \otimes [\mathscr{F}] := [\mathscr{E} \otimes_{\mathcal{O}_X}^{L} \mathscr{F}]$$

for perfect complexes $\mathscr{E}$ and $\mathscr{F}$ in X. Indeed, $\mathscr{E} \otimes_{\mathcal{O}_X}^{L} \mathscr{F}$ is perfect by Remark 21.143 and since $- \otimes^{L} -$ is triangulated in both components we obtain an induced $\mathbb{Z}$-bilinear map $K_0(X) \times K_0(X) \to K_0(X)$ which is clearly associative and commutative. The unit object is the class of the structure sheaf.
(2) Let X be a noetherian scheme. By Lemma 22.64,

$$K_0(X) \otimes_{\mathbb{Z}} K'_0(X) \longrightarrow K'_0(X), \qquad [\mathscr{E}] \otimes [\mathscr{F}] \mapsto [\mathscr{E} \otimes_{\mathcal{O}_X}^{L} \mathscr{F}]$$

defines on the abelian group $K'_0(X)$ the structure of a $K_0(X)$-module.

Remark and Definition 23.53. Let $f\colon X \to Y$ be a morphism of quasi-compact schemes. Then sending a perfect complex $\mathscr{E}$ on Y to the perfect complex $Lf^*\mathscr{E}$ defines a homomorphism of rings

$$(23.12.1) \qquad\qquad f^*\colon K_0(Y) \to K_0(X).$$

Indeed, since Lf^* is a triangulated functor, it induces a homomorphism of abelian groups on K-groups which is a ring homomorphism by Proposition 21.117.

Note that the class of finite locally free modules $\mathscr{E}$, or more generally of strictly perfect complexes $\mathscr{E}$, on Y is sent to the class of $f^*\mathscr{E}$ since $Lf^*\mathscr{E} = f^*\mathscr{E}$ as $\mathscr{E}$ is K-flat in this case.

Remark 23.54. Let $f\colon X \to Y$ be a proper morphism of noetherian schemes. By $f^*\colon K_0(Y) \to K_0(X)$ we can view every $K_0(X)$-module as $K_0(Y)$-module. Then the projection formula (Proposition 22.84) shows that $f_*\colon K'_0(X) \to K'_0(Y)$ is a homomorphism of $K_0(Y)$-modules.

More generally, let

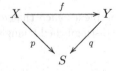

be a commutative diagram of proper morphisms of noetherian schemes. Then for $\alpha \in K_0'(X)$ and $\beta \in K_0(Y)$ one has

$$(23.12.2) \qquad q_*(\beta \cdot f_*\alpha) = q_*(f_*(f^*\beta \cdot \alpha)) = p_*(f^*\beta \cdot \alpha).$$

Let X be a noetherian scheme. Then every perfect complex is in $D^b_{\mathrm{coh}}(X)$ and the inclusion $(\mathrm{Perf}(X)) \to D^b_{\mathrm{coh}}(X)$ induces a group homomorphism

$$K_0(X) \to K_0'(X).$$

Proposition 23.55. *Let X be a regular noetherian scheme. Then every object in $D^b_{\mathrm{coh}}(X)$ is perfect. In particular, the map $K_0(X) \to K_0'(X)$ is an isomorphism.*

Proof. As the condition of being perfect can be checked locally, we may assume that $X = \mathrm{Spec}\, A$ is affine. Let $\mathscr{E}$ be in $D^b_{\mathrm{coh}}(X)$. By Lemma 21.177 it suffices to show that $H^i(\mathscr{E})$ is perfect. Hence it suffices to show that every finitely generated A-module M has a finite left resolution $0 \to F_{-n} \to \cdots F_0 \to M \to 0$ by finitely generated projective modules F_i. But this holds because A is regular (Proposition G.5). $\qquad\square$

Remark 23.56. If X is noetherian regular and semiseparated, then for every perfect complex $\mathscr{F}$ of tor-amplitude in $[a, b]$ there exists a complex $\mathscr{E}$ of finite locally free $\mathscr{O}_X$-modules concentrated in degrees $[a, b]$ and a quasi-isomorphism $\mathscr{E} \to \mathscr{F}$ (see [Sta] 0F8A, 0F8E (and its proof)). Hence if X is in addition of finite dimension n, then every coherent $\mathscr{O}_X$-module is isomorphic in $D(X)$ to a complex of finite locally free $\mathscr{O}_X$-modules concentrated in degrees $[-n, 0]$.

(23.13) Chern classes of line bundles on noetherian schemes.

In this section X denotes a noetherian scheme. We will now define *Chern classes*. These are invariants attached to line bundles, and more generally to vector bundles, which lie in the K-group $K_0(X)$ of the underlying scheme X, and are an important tool in the study of line bundles and vector bundles on X.

Definition 23.57. *Let $\mathscr{L}$ be a line bundle on X. The* first Chern class *of $\mathscr{L}$ is defined as*

$$c_1(\mathscr{L}) = [\mathscr{O}_X] - [\mathscr{L}^{\otimes -1}] \in K_0(X).$$

The Chern class induces an endomorphism of the abelian group $K_0'(X)$ given by

$$(23.13.1) \qquad [\mathscr{F}] \mapsto c_1(\mathscr{L}) \cdot [\mathscr{F}] = [\mathscr{F}] - [\mathscr{L}^{\otimes -1} \otimes \mathscr{F}],$$

and sometimes this endomorphism is called the first Chern class of $\mathscr{L}$, as well.

Proposition 23.58. *Let $\mathscr{L}$ be a line bundle on X.*
(1) Then $c_1(\mathscr{L}) \cdot K_0'(X)_r \subseteq K_0'(X)_{r-1}$ for all $r \geq 0$.
(2) For any two line bundles $\mathscr{L}_1$ and $\mathscr{L}_2$ the first Chern class endomorphisms $c_1(\mathscr{L}_1)$ and $c_1(\mathscr{L}_2)$ commute with each other.
(3) If $\mathscr{L} = \mathscr{O}_X(D)$ for an effective Cartier divisor $i \colon D \to X$, then $i_*\mathscr{O}_D$ is perfect and for all $\alpha \in K_0'(X)$ one has

$$c_1(\mathscr{L}) \cdot \alpha = i_*\mathscr{O}_D \otimes^L_{\mathscr{O}_X} \alpha,$$

where the right hand side is the multiplication defined in Definition 23.52 (2).

(4) *For each $r \geq 0$ and $\alpha \in K_0'(X)_r$ the map*

$$\mathrm{Pic}(X) \longrightarrow \mathrm{gr}_{r-1} K_0'(X), \qquad \mathscr{L} \mapsto c_1(\mathscr{L}) \cdot \alpha \mod K_0'(X)_{r-2}$$

is a group homomorphism.

Proof. By Corollary 23.49 one has $c_1(\mathscr{L}) \cdot \mathscr{F} \equiv 0$ modulo $K_0'(X)_{r-1}$ for all $[\mathscr{F}] \in K_0'(X)_r$. This shows (1).

Assertion (2) holds by commutativity of the tensor product, and Assertion (3) follows from the exact sequence $0 \to \mathscr{L}^{\otimes -1} \to \mathscr{O}_X \to i_*\mathscr{O}_D \to 0$.

Finally, we have for $\mathscr{L}_1, \mathscr{L}_2 \in \mathrm{Pic}(X)$ the formal equality

(23.13.2) $c_1(\mathscr{L}_1 \otimes \mathscr{L}_2) \cdot \alpha = c_1(\mathscr{L}_1) \cdot \alpha + c_1(\mathscr{L}_2) \cdot \alpha - c_1(\mathscr{L}_1) \cdot c_1(\mathscr{L}_2) \cdot \alpha.$

By (1) one has $c_1(\mathscr{L}_1) \cdot c_1(\mathscr{L}_2) \cdot \alpha \in K_0'(X)_{r-2}$ which shows (4). $\qquad\square$

(23.14) Euler characteristic of schemes over a field.

Let k be a field and let $X \to \mathrm{Spec}\, k$ be a proper morphism.

Definition and Remark 23.59. *For every complex $\mathscr{F}$ in $D^b_{\mathrm{coh}}(X)$ we define its Euler characteristic*

$$\chi(\mathscr{F}) := \chi(X, \mathscr{F}) := \sum_{i \in \mathbb{Z}} (-1)^i \dim_k H^i(X, \mathscr{F}).$$

The sum on the right side is finite by Remark 23.24.

The long exact cohomology sequence shows that χ induces a group homomorphism

$$\chi \colon K_0'(X) \longrightarrow \mathbb{Z}.$$

Note that $\chi(\mathscr{F})$ also depends on the ground field. Hence we sometimes write $\chi_k(\mathscr{F})$.

With the identification $K_0'(k) = \mathbb{Z}$ of Example 23.60, the map χ can be identified with the map induced by the derived direct image of the structure morphism $X \to \mathrm{Spec}\, k$ on K-groups (Exercise 23.14).

Example 23.60. Let k be a field. Then $K_0(k) = K_0'(k)$ and

$$\chi \colon K_0(k) \xrightarrow{\sim} \mathbb{Z}$$

is an isomorphism of rings. Indeed, clearly it is an isomorphism of abelian groups, as every vector space is free. To check that $\chi(E \otimes_k^L F) = \chi(E)\chi(F)$ one can assume that E and F are acyclic complexes of finite-dimensional k-vector spaces (Example 22.48), then $E \otimes^L F = E \otimes_k F$ is acyclic as well and

$$\chi(E)\chi(F) = \left(\sum_i (-1)^i \dim E^i\right)\left(\sum_j (-1)^j \dim F^j\right)$$

$$= \sum_n \sum_{i+j=n} (-1)^{i+j} \dim E^i \dim F^j$$

$$= \sum_n (-1)^n \sum_{i+j=n} \dim(E^i \otimes F^j)$$

$$= \chi(E \otimes_k^L F).$$

Example 23.61. Let $X = \mathbb{P}^r_k$ and $n \in \mathbb{Z}$. Then one has by Theorem 22.22

$$\chi(\mathscr{O}_X(n)) = \binom{n+r}{r}.$$

Here we define for every $\alpha \in \mathbb{R}$ (or, more generally, in any $\mathbb{Q}$-algebra)

(23.14.1)
$$\binom{\alpha}{r} := \frac{\alpha(\alpha-1)\cdots(\alpha-r+1)}{r!}.$$

Relations within $K'_0(X)$ yield identities for the Euler characteristic:

Remark 23.62. Let $X \to \operatorname{Spec} k$ be a proper morphism.
(1) Let $\mathscr{F} \in D^b_{\operatorname{coh}}(X)$. Then $\chi(\mathscr{F}[1]) = -\chi(\mathscr{F})$.
(2) Let $\mathscr{F}' \to \mathscr{F} \to \mathscr{F}'' \to$ be a distinguished triangle in $D^b_{\operatorname{coh}}(X)$ (e.g., given by a short exact sequence $0 \to \mathscr{F}' \to \mathscr{F} \to \mathscr{F}'' \to 0$ of coherent $\mathscr{O}_X$-modules), then

$$\chi(\mathscr{F}) = \chi(\mathscr{F}') + \chi(\mathscr{F}'').$$

(3) Let $\mathscr{F} \in D^b_{\operatorname{coh}}(X)$. Then

$$\chi(\mathscr{F}) = \sum_{i \in \mathbb{Z}} (-1)^i \chi(H^i(\mathscr{F})).$$

Corollary 23.63. *Let $f\colon X \to Y$ be a morphism of proper k-schemes. Then for every complex $\mathscr{F} \in D^b(X)$ one has*

$$\chi(X, \mathscr{F}) = \chi(Y, Rf_*\mathscr{F}) = \sum_{i \in \mathbb{Z}} (-1)^i \chi(Y, R^i f_* \mathscr{F}).$$

Proof. One has $R\Gamma(X, \mathscr{F}) = R\Gamma(Y, Rf_*\mathscr{F})$ (Proposition 21.114) and hence we conclude by Remark 23.62 (3). $\qquad\square$

Lemma 23.64. *Let $X \to \operatorname{Spec} k$ be a proper morphism and let $\mathscr{F}$ be in $D^b_{\operatorname{coh}}(X)$. Let K be a field extension, set $X_K := X \otimes_k K$ and let $\mathscr{F}_K$ be the (derived) pullback of $\mathscr{F}$ to X_K. Then*

$$\chi_K(\mathscr{F}_K) = \chi_k(\mathscr{F}).$$

If $\mathscr{F}$ a coherent $\mathscr{O}_X$-module, then $\mathscr{F}_K$ is the usual (non-derived) pullback of $\mathscr{F}$ to X_K since $X_K \to X$ is flat.

Proof. This follows from $H^i(X, \mathscr{F}) \otimes_k K = H^i(X_K, \mathscr{F}_K)$ by Corollary 22.91. $\qquad\square$

Proposition 23.65. (Projection formula) *Let $f\colon X \to Y$ be a morphism of proper k-schemes. For $\beta \in K_0(Y)$ and $\alpha \in K'_0(X)$ one has*

$$\chi(Y, \beta \cdot f_*\alpha) = \chi(X, f^*\beta \cdot \alpha).$$

Proof. This is the special case $S = \operatorname{Spec} k$ of (23.12.2). $\qquad\square$

Remark 23.66. All definitions and results in this section also hold for proper schemes over a local Artinian ring R by replacing "dimension of k-vector spaces" by "length of R-modules".

(23.15) Numerical intersection number for proper schemes over a field.

We now aim to define intersection numbers and study their properties. We will use the following formal lemma.

Lemma 23.67. *In $\mathbb{Z}[Y^{\pm 1}][\![1 - Y^{-1}]\!]$ one has for all $m \in \mathbb{Z}$ the identity*

$$Y^m = \sum_{i \geq 0} \binom{m + i - 1}{i} (1 - Y^{-1})^i.$$

Recall that by definition (23.14.1) we have

$$\binom{m + i - 1}{i} = \frac{(m + i - 1)(m + i - 2) \cdots (m + 1)m}{i!}.$$

Proof. In $\mathbb{Z}[(1 + X)^{\pm 1}][\![X]\!]$ we have for all $n \in \mathbb{Z}$ the formal identity

$$(1 + X)^n = \sum_{i \geq 0} \binom{n}{i} X^i.$$

Substituting $m := -n$ and $Y := (1 + X)^{-1}$ and using

$$\binom{-m}{i} = (-1)^i \binom{m + i - 1}{i}$$

we obtain the desired identity. $\square$

Let X be a proper scheme over a field k and let $\mathscr{L}$ be a line bundle. Then $\chi(\mathscr{L} \otimes \mathscr{F})$ is additive in exact sequences in coherent $\mathscr{O}_X$-modules $\mathscr{F}$ and therefore the map $\mathscr{F} \mapsto \chi(\mathscr{L} \otimes \mathscr{F})$ factors through $K_0'(X)$. We can describe the resulting group homomorphism as

$$K_0'(X) \longrightarrow \mathbb{Z}, \qquad \alpha \mapsto \chi(\mathscr{L} \cdot \alpha),$$

where $\mathscr{L} \cdot \alpha$ is the scalar multiplication of α by the class of $\mathscr{L}$ in $K_0(X)$, see Remark 23.52 (2).

Theorem 23.68. *Let X be a proper k-scheme, let $t \geq 0$ be integers, and let $\mathscr{L}_1, \ldots, \mathscr{L}_t$ be line bundles on X. Then for all $m_1, \ldots, m_t \in \mathbb{Z}$ and for all $\alpha \in K_0'(X)$ one has*

$$\chi(\mathscr{L}_1^{\otimes m_1} \otimes \cdots \otimes \mathscr{L}_t^{\otimes m_t} \cdot \alpha)$$

$$= \sum_{i_1, \ldots, i_t \geq 0} \chi(c_1(\mathscr{L}_1)^{i_1} \cdots c_1(\mathscr{L}_t)^{i_t} \cdot \alpha) \binom{m_1 + i_1 - 1}{i_1} \cdots \binom{m_t + i_t - 1}{i_t}$$

Proof. By Lemma 23.67 we have the identity

$$(*) \quad Y_1^{m_1} \cdots Y_t^{m_t} = \sum_{i_1, \ldots, i_t \geq 0} (1 - Y_1^{-1})^{i_1} \cdots (1 - Y_t^{-1})^{i_t} \binom{m_1 + i_1 - 1}{i_1} \cdots \binom{m_t + i_t - 1}{i_t}$$

of formal power series. We let Y_j act on $K_0'(X)$ by $Y_j \cdot \alpha := \mathscr{L}_j \cdot \alpha$. Then $c_1(\mathscr{L}_j) \cdot \alpha = (1 - Y_j^{-1}) \cdot \alpha$ and hence $1 - Y_j^{-1}$ acts nilpotently by Proposition 23.58 (1). Then applying $(*)$ and taking Euler characteristics yield the theorem. $\square$

If $\alpha \in K_0'(X)_r$ for some $r \geq 0$ and $i_1 + \cdots + i_t > r$, then $\chi(c_1(\mathscr{L}_1)^{i_1} \cdots c_1(\mathscr{L}_t)^{i_t} \cdot \alpha) = 0$ by Proposition 23.58 (1). In particular the sum in Theorem 23.68 is finite. We obtain the following corollary.

Corollary and Definition 23.69. *Let X be a proper scheme over a field k, let $t \geq 0$, let $\mathscr{L}_1, \ldots, \mathscr{L}_t$ be line bundles on X, and let $\mathscr{F} \in D_{\mathrm{coh}}^b(X)$. Then the function*

$$\sigma_{\mathscr{F}} \colon (n_1, \ldots, n_t) \mapsto \chi(\mathscr{L}_1^{\otimes n_1} \otimes \cdots \otimes \mathscr{L}_t^{\otimes n_t} \otimes_{\mathscr{O}_X}^L \mathscr{F})$$

is a numerical polynomial, i.e., a polynomial function given by a (necessarily unique) polynomial with rational coefficients, again denoted by $\sigma_{\mathscr{F}}$, that is integer-valued on integers. It is of total degree $\leq r$ if $\mathscr{F} \in K_0'(X)_r$.
The polynomial $\sigma_{\mathscr{F}}$ is called the Snapper polynomial *of $\mathscr{F}$ and $\mathscr{L}_1, \ldots, \mathscr{L}_t$.*

For more properties of numerical polynomials see Exercise 23.25.

Definition 23.70. *Let k be a field and let X be a proper k-scheme. Let $t \geq 0$ be an integer, let $\mathscr{L}_1, \ldots, \mathscr{L}_t$ be line bundles on X, and let $\mathscr{F}$ be an object in $D_{\mathrm{coh}}^b(X)$ such that $[\mathscr{F}] \in K_0'(X)_t$ (e.g., if $\dim \mathrm{Supp}(H^i(\mathscr{F})) \leq t$ for all i by Remark 23.45). Then*

$$(\mathscr{L}_1 \cdots \mathscr{L}_t \cdot \mathscr{F}) := \chi(c_1(\mathscr{L}_1) \cdots c_1(\mathscr{L}_t) \cdot \mathscr{F})$$

is called the intersection number *of $\mathscr{L}_1, \cdots, \mathscr{L}_t$ on $\mathscr{F}$.*
If $i \colon Z \to X$ is a closed subscheme of dimension $\leq t$, then we set

$$(\mathscr{L}_1 \cdots \mathscr{L}_t \cdot Z) := (\mathscr{L}_1 \cdots \mathscr{L}_t \cdot i_* \mathscr{O}_Z).$$

In the special case that $\mathscr{L}_1 = \cdots = \mathscr{L}_t = \mathscr{L}$ we write $(\mathscr{L}^t \cdot \mathscr{F})$ or $(\mathscr{L}^t \cdot Z)$. If $Z = X$ with $\dim(X) \leq t$, we also write $(\mathscr{L}_1 \cdots \mathscr{L}_t)$ (resp. $(\mathscr{L}^t)$) instead of $(\mathscr{L}_1 \cdots \mathscr{L}_t \cdot X)$ (resp. $(\mathscr{L}^t \cdot X)$).
If $\mathscr{L}_i = \mathscr{O}_X(D_i)$ for Cartier divisors D_i, one also writes $(D_1 \cdots D_t \cdot \mathscr{F})$ instead of $(\mathscr{L}_1 \cdots \mathscr{L}_t \cdot \mathscr{F})$.

Theorem 23.68 shows that

(23.15.1) $\qquad (\mathscr{L}_1 \cdots \mathscr{L}_t \cdot \mathscr{F}) = \text{coefficient of } n_1 n_2 \cdots n_t \text{ in } \sigma_{\mathscr{F}},$

where $\sigma_{\mathscr{F}}$ is the degree $\leq t$ Snapper polynomial of Definition 23.69.
The formation of the Snapper polynomial is invariant under changing the base field by Lemma 23.64. In particular:

Remark 23.71. In the situation of Definition 23.70, let K be a field extension of k, and denote $\mathscr{L}_{i,K}$ and $\mathscr{F}_K$ the pullback of $\mathscr{L}_i$ and $\mathscr{F}$ to $X \otimes_k K$. Then

$$(\mathscr{L}_1 \cdots \mathscr{L}_t \cdot \mathscr{F}) = (\mathscr{L}_{1,K} \cdots \mathscr{L}_{t,K} \cdot \mathscr{F}_K).$$

Corollary 23.72. *In the situation of Definition 23.70 the following assertions hold.*
(1) *If $\mathscr{F} \in K_0'(X)_{t-1}$, then $(\mathscr{L}_1 \cdots \mathscr{L}_t \cdot \mathscr{F}) = 0$ and hence there is a well defined group homomorphism*

$$\mathrm{gr}_t K_0'(X) \to \mathbb{Z}, \qquad [\mathscr{F}] \mapsto (\mathscr{L}_1 \cdots \mathscr{L}_t \cdot \mathscr{F}).$$

In particular $(\mathscr{L}_1 \cdots \mathscr{L}_t) = 0$ if $t > \dim(X)$.
(2) *The intersection number is symmetric and $\mathbb{Z}$-linear in each $\mathscr{L}_j$.*

(3) Let $\mathscr{F}$ be a coherent $\mathscr{O}_X$-module and let $Z_1, \ldots, Z_m$ be the irreducible components of $\operatorname{Supp} \mathscr{F}$ of dimension t, considered as integral subschemes of X, and let $\eta_i \in Z_i$ be the generic point. Then one has

$$(23.15.2) \qquad (\mathscr{L}_1 \cdots \mathscr{L}_t \cdot \mathscr{F}) = \sum_{i=1}^{m} \lg_{\mathscr{O}_{X, \eta_i}}(\mathscr{F}_{\eta_i}) (\mathscr{L}_1 \cdots \mathscr{L}_t \cdot Z_i).$$

Proof. Assertion (1) follows from Proposition 23.68 since multiplication by the first Chern class of line bundles and the Euler characteristic are additive. Assertion (2) follows from Proposition 23.58. Finally, (3) follows by (1) from Corollary 23.48. $\qquad\square$

Proposition 23.73. *Let $f \colon X \to Y$ be a morphism of proper k-schemes, let $\mathscr{L}_1, \ldots, \mathscr{L}_t$ be line bundles on Y and let $\mathscr{F}$ be an object in $D_{\mathrm{coh}}^b(X)$ with $[\mathscr{F}] \in K_0'(X)_t$. Then one has*

$$(f^* \mathscr{L}_1 \cdots f^* \mathscr{L}_t \cdot \mathscr{F}) = (\mathscr{L}_1 \cdots \mathscr{L}_t \cdot Rf_* \mathscr{F}).$$

Proof. For every line bundle $\mathscr{L}$ on Y we have

$$c_1(f^* \mathscr{L}) \cdot [\mathscr{F}] = ([\mathscr{O}_X] - [f^* \mathscr{L}^{\otimes -1}]) \cdot [\mathscr{F}] = f^*([\mathscr{O}_Y] - [\mathscr{L}^{\otimes -1}]) \cdot [\mathscr{F}]$$

and hence the equality follows from the projection formula (Proposition 23.65). $\qquad\square$

If f is a closed immersion, then $Rf_* \mathscr{F} = f_* \mathscr{F}$. In particular for every closed immersion $i \colon Z \to Y$ and for all line bundles $\mathscr{L}_1, \ldots, \mathscr{L}_t$ on Y with $t \geq \dim Z$ we obtain

$$(23.15.3) \qquad (i^* \mathscr{L}_1 \cdots i^* \mathscr{L}_t \cdot Z) = (\mathscr{L}_1 \cdots \mathscr{L}_t \cdot Z).$$

Example 23.74. Let $X = \mathbb{P}_k^r$. Then every line bundle on X is of the form $\mathscr{O}_X(d)$ for an integer $d \in \mathbb{Z}$ (Example 11.45). Let $\mathscr{L}_i = \mathscr{O}_X(d_i)$ with $d_i \in \mathbb{Z}$ for $i = 1, \ldots, r$. Then the Snapper polynomial is of the form

$$\sigma_{\mathscr{O}_X}(n_1, \ldots, n_r) = \chi(\mathscr{O}_X(d_1)^{\otimes n_1} \otimes \cdots \otimes \mathscr{O}_X(d_r)^{\otimes n_r}) = \chi(\mathscr{O}_X(d_1 n_1 + \cdots + d_r n_r))$$
$$= \binom{d_1 n_1 + \cdots + d_r n_r + r}{r},$$

where the last equality holds by Example 23.61. In particular

$$(\mathscr{O}_X(d_1) \cdots \mathscr{O}_X(d_r) \cdot \mathbb{P}_k^r) = d_1 d_2 \cdots d_r.$$

Example 23.75. Let k be a field.
(0) Let $X \to \operatorname{Spec} k$ be a proper scheme, and let $\mathscr{F}$ be a coherent $\mathscr{O}_X$-module with $\dim \operatorname{Supp} \mathscr{F} = 0$. Then one has $(\mathscr{F}) = \chi(X, \mathscr{F}) = \dim_k \Gamma(X, \mathscr{F})$ for the intersection of $\mathscr{F}$ with 0 line bundles.
(1) Let $C \to \operatorname{Spec} k$ be a proper curve (i.e. $C \to \operatorname{Spec} k$ is proper and all irreducible components of C have dimension 1). Let $\mathscr{L}$ be a line bundle on C. Then

$$(\mathscr{L} \cdot C) = \chi(\mathscr{O}_C) - \chi(\mathscr{L}^{\otimes -1}) = \deg(\mathscr{L}),$$

where the last equality is the definition of $\deg(\mathscr{L})$ given in Section (15.9). Using $\deg(\mathscr{L}^{-1}) = -\deg(\mathscr{L})$ it follows that for all $n \in \mathbb{Z}$ one has

$$(23.15.4) \qquad \chi(\mathscr{L}^{\otimes n}) = n \deg(\mathscr{L}) + \chi(\mathscr{O}_C).$$

(2) Now let $X \to \operatorname{Spec} k$ be a proper surface (i.e. $X \to \operatorname{Spec} k$ is proper and all irreducible components of X have dimension 2). Let $C, D \subseteq X$ be effective Cartier divisors with no common irreducible component. Then $Z := C \cap D$ is either empty or finite of dimension 0 and

$$(C \cdot D \cdot X) = \sum_{z \in Z} \dim_k(\mathscr{O}_{Z,z}) = \dim_k(\Gamma(Z, \mathscr{O}_Z)).$$

Indeed, as C and D are equi-dimensional of dimension 1, it is clear that Z has dimension ≤ 0. Hence it is finite and the second equality holds (Proposition 5.20). Consider the exact sequence

$$0 \longrightarrow \mathscr{O}_X(-D) \otimes \mathscr{O}_C \to \mathscr{O}_C \to \mathscr{O}_Z \to 0.$$

Then

$$\begin{aligned} \dim_k \Gamma(Z, \mathscr{O}_Z) &= \chi(\mathscr{O}_Z) = \chi(\mathscr{O}_C) - \chi(\mathscr{O}_X(D)^{\otimes -1}) \\ &= \chi(c_1(\mathscr{O}_X(D)) \cdot \mathscr{O}_C) \overset{(*)}{=} \chi(c_1(\mathscr{O}_X(D))c_1(\mathscr{O}_X(C))\mathscr{O}_X) \\ &= (C \cdot D \cdot X), \end{aligned}$$

where $(*)$ holds by Proposition 23.58 (3).

This shows that for $X = \mathbb{P}_k^2$ the intersection number $(C \cdot D \cdot X)$ equals the intersection number $i(C, D)$ defined in Section (5.14) in a rather ad hoc way. We obtain the following quick proof of Bézout's theorem 5.61 using that $(C \cdot D \cdot X)$ depends only on $\mathscr{O}_X(C)$ and $\mathscr{O}_X(D)$ and is bilinear in the line bundles: If $C = V_+(f)$ and $D = V_+(g)$ for non-constant homogeneous polynomials $f, g \in k[T_0, T_1, T_2]$ with $\deg(f) =: n$ and $\deg(g) =: m$. Then $\mathscr{O}_X(C) \cong \mathscr{O}_X(n)$ and $\mathscr{O}_X(D) \cong \mathscr{O}_X(m)$. If $m = n = 1$, then we can choose C and D to be different lines, and the intersection number is 1. In general one has

$$(C \cdot D \cdot \mathbb{P}_k^2) = (\mathscr{O}(n) \cdot \mathscr{O}(m) \cdot \mathbb{P}_k^2) = (\mathscr{O}(1)^{\otimes n} \cdot \mathscr{O}(1)^{\otimes m} \cdot \mathbb{P}_k^2) = nm.$$

We could also have used Example 23.74.

Remark and Definition 23.76. Let $f \colon X \to Y$ be a morphism between integral k-schemes that are proper over k. Then f is proper and hence

$$f \text{ surjective} \iff \dim Y = \dim f(X).$$

If f is surjective and $\dim X = \dim Y$, then the function field $K(X)$ is a finite extension of $K(Y)$.

The *degree of f* is defined to be the following non-negative integer:

$$\deg(f) := \begin{cases} [K(X) : K(Y)], & \text{if } \dim(X) = \dim(Y) = \dim(f(X)); \\ 0, & \text{otherwise.} \end{cases}$$

If f is finite locally free, then the degree defined in Definition 23.76 is the same as the degree defined for finite locally free morphisms in (12.6.1).

Proposition 23.77. *Let k be a field and let $f\colon X \to Y$ be a morphism of proper schemes over k. Let $Z \subseteq X$ be an integral closed subscheme of dimension d and let $\mathscr{L}_1, \dots, \mathscr{L}_d$ be line bundles on Y. Then*

$$(f^*\mathscr{L}_1 \cdots f^*\mathscr{L}_d \cdot Z) = \deg(f_{|Z}\colon Z \to f(Z))(\mathscr{L}_1 \cdots \mathscr{L}_d \cdot f(Z)),$$

where $f(Z)$ is considered as closed integral subscheme of Y.

Proof. By (23.15.3) we may replace X by Z and Y by $f(Z)$ and hence may assume that X is integral of dimension d, Y is integral, and that f is surjective. In particular, $\dim Y \leq d$. We have

$$(*) \qquad\qquad (f^*\mathscr{L}_1 \cdots f^*\mathscr{L}_d \cdot X) = (\mathscr{L}_1 \cdots \mathscr{L}_d \cdot Rf_*\mathscr{O}_X)$$

by Proposition 23.73.

If $\dim Y < d$, then $\deg(f) = 0$ and the right hand side of (*) is zero by Corollary 23.72 (1). Hence we can assume that $\dim Y = d$. Then f is finite over a non-empty open subscheme V of Y. Therefore one has for all $i > 0$ that $R^i f_*\mathscr{F}_{|V} = 0$ (Corollary 22.5) and hence $\dim \operatorname{Supp} R^i f_*\mathscr{F} < d$. This shows that, denoting by $\eta \in Y$ the generic point,

$$(\mathscr{L}_1 \cdots \mathscr{L}_d \cdot Rf_*\mathscr{O}_X) \overset{(1)}{=} (\mathscr{L}_1 \cdots \mathscr{L}_d \cdot f_*\mathscr{O}_X)$$
$$\overset{(2)}{=} \dim_{K(Y)}(f_*\mathscr{O}_X)_\eta(\mathscr{L}_1 \cdots \mathscr{L}_d \cdot Y).$$

where (1) and (2) hold by Corollary 23.72 (1) and (3). As $\deg(f) = \dim_{K(Y)}(f_*\mathscr{O}_X)_\eta$, this shows the claim. $\qquad\square$

(23.16) Asymptotic Riemann Roch theorem.

Let k be a field.

Proposition 23.78. *(Asymptotic Riemann-Roch theorem) Let X be a proper k-scheme of dimension d, let $\mathscr{F}$ be in $D^b_{\mathrm{coh}}(X)$, and let $\mathscr{L}$ be a line bundle on X. Then*

$$\chi(\mathscr{L}^{\otimes n} \otimes^L \mathscr{F}) = \frac{(\mathscr{L}^d \cdot \mathscr{F})}{d!}n^d + g(n),$$

for all $n \in \mathbb{Z}$, where $g(n)$ is a polynomial of degree $< d$ which depends only on X, $\mathscr{F}$, and the line bundle $\mathscr{L}$.

Proof. We use Theorem 23.68 with $t = 1$. By definition, $\chi(c_1(\mathscr{L})^i \cdot [\mathscr{F}]) = (\mathscr{L}^i \cdot \mathscr{F})$, so

$$\chi(\mathscr{L}^{\otimes n} \otimes^L \mathscr{F}) = \sum_{i=0}^d (\mathscr{L}^i \cdot \mathscr{F})\binom{n+i-1}{i}.$$

We see that the left hand side is a sum of polynomials in n of degree $\leq i$ for $i = 0, \dots, d$ and that the d-th summand is

$$(\mathscr{L}^d \cdot \mathscr{F})\binom{n+d-1}{d} = \frac{(\mathscr{L}^d \cdot \mathscr{F})}{d!}n^d + \text{ terms of degree} < d. \qquad\square$$

Proposition 23.79. *Let k be a field and let X be a proper k-scheme. Let $\mathscr{F} \neq 0$ be a coherent $\mathscr{O}_X$-module and set $t := \dim \operatorname{Supp}\mathscr{F}$. Let $\mathscr{L}_1, \dots, \mathscr{L}_t$ be ample line bundles on X. Then*

$$(\mathscr{L}_1 \cdots \mathscr{L}_t \cdot \mathscr{F}) > 0.$$

Proof. By Corollary 23.72 (3) it suffices to show that $(\mathscr{L}_1 \cdots \mathscr{L}_t \cdot Z) > 0$ for every closed integral subscheme Z of dimension t of X.

We use induction on t, the case $t = 0$ holding by Example 23.75 (0). Suppose $t > 0$. By (23.15.3), we may assume that $Z = X$. In particular, X is integral. By the multilinearity of the intersection number we may replace $\mathscr{L}_t$ by some positive power, hence we can assume that there exists $0 \neq s \in \Gamma(X, \mathscr{L}_t)$ such that its non-vanishing locus X_s is affine since $\mathscr{L}_t$ is ample. This section is automatically regular because X is integral and its vanishing locus $V(s)$ is non-empty because otherwise $X_s = X$ is affine and proper over k which would imply that X is finite (Corollary 12.89), contradicting that $t > 0$. Hence $i\colon V(s) \to X$ is an effective Cartier divisor with $\mathscr{O}_X(V(s)) \cong \mathscr{L}_t$. By Proposition 23.58 (3) and using the induction hypothesis we then find

$$(\mathscr{L}_1 \cdots \mathscr{L}_t \cdot X) = (\mathscr{L}_1 \cdots \mathscr{L}_{t-1} \cdot i_* \mathscr{O}_{V(s)}) = (i^* \mathscr{L}_1 \cdots i^* \mathscr{L}_{t-1} \cdot \mathscr{O}_{V(s)}) > 0,$$

where we use (23.15.3) for the second equality. $\qquad\square$

We will prove the converse to this statement, the *Nakai-Moishezon criterion*, in Section (23.90).

(23.17) The degree of a closed subscheme.

Let k be a field.

Definition 23.80. *Let X be a proper scheme over k. Let $\mathscr{L}$ be a line bundle on X. For a closed subscheme Z of X the integer*

$$(23.17.1) \qquad\qquad \deg_{\mathscr{L}}(Z) := (\mathscr{L}^{\dim Z} \cdot Z)$$

is called the degree of Z with respect to $\mathscr{L}$. *We also call*

$$(23.17.2) \qquad\qquad \deg(\mathscr{L}) := \deg_{\mathscr{L}}(X)$$

the degree of $\mathscr{L}$. *If $X = \mathbb{P}_k^r$ for some $r \geq 1$, then*

$$(23.17.3) \qquad\qquad \deg(Z) := \deg_{\mathscr{O}_X(1)}(Z)$$

is simply called the degree of Z.

Hence for every projective scheme Z over k and for every closed immersion $i\colon Z \to \mathbb{P}_k^r$ we have the notion of the degree of (Z, i) by identifying Z with a closed subscheme of $\mathbb{P}_k^r$ via i. Note that the degree of (Z, i) depends on i, see Exercise 23.36.

The degree is preserved by base change in the following sense.

Remark 23.81. Let k be a field, X be a proper k-scheme, $Z \subseteq X$ a closed subscheme, and let $\mathscr{L}$ be a line bundle on X. Let k' be a field extension of k, and set $X' := X \otimes_k k'$, $Z' := Z \otimes_k k'$, and let $\mathscr{L}'$ be the pullback of $\mathscr{L}$ to X'. Then Remark 23.71 shows that

$$\deg_{\mathscr{L}'}(Z') = \deg_{\mathscr{L}}(Z).$$

Remark 23.82. If $\mathscr{L}$ is an ample line bundle on a proper k-scheme and $Z \subseteq X$ is a non-empty closed subscheme, then $\deg_{\mathscr{L}}(Z) > 0$ by Proposition 23.79.

By Serre's vanishing theorem we obtain the following special case of the asymptotic Riemann-Roch theorem.

Proposition 23.83. *Let X be a proper k-scheme of dimension d and let $\mathscr{L}$ be an ample line bundle on X. Then*

$$\Gamma(X, \mathscr{L}^{\otimes n}) = \frac{\deg_{\mathscr{L}}(X)}{d!} n^d + O(n^{d-1}).$$

Proof. This follows from Proposition 23.78 and the fact that $H^i(X, \mathscr{L}^{\otimes n}) = 0$ for large n and $i > 0$ because $\mathscr{L}$ is ample (Theorem 23.2). $\square$

Proposition 23.84. *Let X and Y be integral proper k-schemes and let $f\colon X \to Y$ be a finite dominant morphism. Then for every line bundle $\mathscr{L}$ on Y one has the equality*

$$\deg(f^*\mathscr{L}) = \deg(f)\deg(\mathscr{L})$$

Proof. This follows from Proposition 23.77. $\square$

Remark 23.85. Let k be a field and let X be a proper scheme over k of dimension d. Let $\mathscr{L}$ be a line bundle on X. Then

$$\deg(\mathscr{L}^{\otimes k}) = k^d \deg(\mathscr{L}), \qquad \text{for all } k \in \mathbb{Z}$$

by Corollary 23.72 (2).

We have by now introduced at several occasions the notion of degree of a line bundle or the degree of a subvariety. Let us show in the remainder of this section that all these notions coincide.

Remark 23.86. If X is a proper curve over k and $\mathscr{L}$ is a line bundle on X, then $\deg(\mathscr{L})$ is the degree defined in Section (15.9) by Example 23.75 (1).

Example 23.87. Let $X = \mathbb{P}_k^r$ and let $f \in k[T_0, \dots, T_r]$ be homogeneous of degree $d \geq 1$. Then

$$\deg(V_+(f)) = (\mathscr{O}_X(1)^{r-1} \cdot \mathscr{O}_{V_+(f)}) = (\mathscr{O}_X(1)^{r-1} \cdot \mathscr{O}_X(d)) = d$$

by Example 23.74.

Remark 23.88. Let X be a proper k-scheme and let $Z \subseteq X$ be a closed subscheme of dimension 0. Then $f\colon Z \to \operatorname{Spec} k$ is finite and hence finite locally free and by Example 23.75 (0) for every line bundle $\mathscr{L}$ on X one has

$$\deg_{\mathscr{L}}(Z) = \dim_k \Gamma(Z, \mathscr{O}_Z) = \deg(f),$$

where the second equality holds by the definition of the degree in Definition (12.6.1).

Recall that in Section (14.31) we defined the degree of a closed subscheme X of $\mathbb{P}_k^n$ as follows. Let $d := \dim(X)$. We defined the incidence k-scheme $H^X = H_{n-d}^X$ such that for any field extension k' of k the k'-valued points $H^X(k')$ are the set of pairs (x, Λ) where $x \in X(k')$ and $\Lambda \subseteq \mathbb{P}_{k'}^n$ is a linear subspace of dimension $n - d$ with $x \in \Lambda$ (see Remark 13.86 for a description of H^X as a k-scheme). The scheme H^X is projective, and it is irreducible if and only if X is irreducible (Proposition 14.134). Let $L = L_{n-d}$ be the scheme of linear subspaces of $\mathbb{P}_k^n$ of dimension $n - d$ and let

$$q\colon H^X \to L, \qquad (x, \Lambda) \mapsto \Lambda$$

be the projection. Hence for every field extension k' and for every linear subspace $\Lambda \in L(k')$ of $\mathbb{P}^n_{k'}$ one has $q^{-1}(\Lambda) = X(k') \cap \Lambda(k')$. Then q is generically finite, i.e., the fiber of q over the generic point η of L is finite (Proposition 14.135), and we defined $\deg(X)$ as the degree of the finite $\kappa(\eta)$-scheme $X_d := q^{-1}(\eta)$, i.e., as $\dim_{\kappa(\eta)} H^0(X_d, \mathscr{O}_{X_d})$.

This can be expressed more geometrically as follows: As the property "finite" is a constructible property (Proposition 10.96), we find an open neighborhood $U \subseteq L$ of η such that the restriction $q^{-1}(U) \to U$ of q is finite. By generic flatness (Corollary 10.85), we can shrink U such that $q^{-1}(U) \to U$ is finite and flat and hence finite locally free (Proposition 12.19). As L is irreducible, U is irreducible, and $q^{-1}(U) \to U$ has constant rank equal to $\deg(X)$.

In other words, there exists an open dense subscheme U of L such that for every field extension k' and every linear subspace $\Lambda \subseteq \mathbb{P}^n_{k'}$ corresponding to a k'-valued point of U, we have that $\deg(X)$ is the k'-dimension of $H^0(X_{k'} \cap \Lambda, \mathscr{O}_{X_{k'} \cap \Lambda})$, i.e., the "number of points of $X_{k'} \cap \Lambda$ counted with multiplicities". Note that it is crucial to consider extension fields of k here since in general we could have $U(k) = \emptyset$. As L is isomorphic to a Grassmannian and hence has an open covering by affine spaces, this can only happen if k is finite.

More classically this is expressed by writing Λ as the intersection of d hyperplanes $H_1, \ldots, H_d$ and by saying that $\deg(X)$ is the "number of points of $X_{k'} \cap H_1 \cap \cdots \cap H_d$ counted with multiplicities for hyperplanes H_i in general position".

Proposition 23.89. *For a closed subscheme $X \subseteq \mathbb{P}^n_k$ of projective space over a field k, the two notions of degree defined in (23.17.3) and in Section (14.31) coincide.*

Proof. For the moment, we denote the degree defined in (23.17.3) by $\deg'(X)$. We keep the above notation, in particular we choose U as above. As $\deg'(X)$ is compatible with base change $k \to k'$ (Remark 23.81), we may assume that $k' = k$ is algebraically closed.

Then we are given hyperplanes $H_1, \cdots, H_d$ in $\mathbb{P}^n_k$ such that $X_d := X \cap H_1 \cap \cdots \cap H_d$ has dimension 0. This implies that $X_i := X \cap H_1 \cap \cdots \cap H_i$ has dimension $d - i$ for all $0 \le i \le d$. By definition we have $\deg(X) = \dim_k H^0(X_d, \mathscr{O}_{X_d})$.

Since every noetherian scheme has only finitely many associated points, we even can find hyperplanes such that in addition for all $i \ge 1$ the hyperplane H_i does not contain any associated point of $X_{i-1} = X \cap H_1 \cap \cdots \cap H_{i-1}$.

Then X_i is a Cartier divisor in X_{i-1} whose line bundle is the restriction of $\mathscr{O}_{\mathbb{P}^n_k}(1)$ to X_{i-1}. Thus we see by (23.15.3) and Proposition 23.58 (3) that

$$c_1(\mathscr{O}_{\mathbb{P}^n_k}(1)) \cdot \mathscr{O}_{X_{i-1}} = \mathscr{O}_{X_i} \otimes^L_{\mathscr{O}_{X_{i-1}}} \mathscr{O}_{X_{i-1}} = \mathscr{O}_{X_i}$$

and hence

$$\deg'(X) = \chi(c_1(\mathscr{O}_{\mathbb{P}^n_k}(1))^d \cdot \mathscr{O}_X) = \chi(\mathscr{O}_{X_d}) = \dim_k H^0(X_d, \mathscr{O}_{X_d}) = \deg(X). \qquad \square$$

(23.18) The Nakai-Moishezon criterion for ampleness.

The following theorem is a useful criterion to check ampleness of a line bundle in terms of intersection numbers. We will use it in Section (25.31) to prove that every smooth proper surface over a field, i.e., every smooth proper scheme over a field that is equi-dimensional of dimension 2, is projective.

Theorem 23.90. (Nakai-Moishezon criterion for ampleness) *Let k be a field and let X be a proper k-scheme. A line bundle $\mathscr{L}$ on X is ample if and only if for every integral closed subscheme $Z \subseteq X$ with $d := \dim Z > 0$, one has $(\mathscr{L}^d \cdot Z) > 0$.*

If X is a proper curve over k, the test objects Z are the irreducible components of X with the reduced scheme structure, and $(\mathscr{L} \cdot Z) = \deg(\mathscr{L}_{|Z})$ by Example 23.75 (1). Compare Proposition 26.57. See also the version for surfaces in Theorem 25.146 below.

For globally generated line bundles, it suffices to check the criterion for integral closed curves (Exercise 26.7).

Proof. If $\mathscr{L}$ is ample, then Proposition 23.79 shows that the intersection numbers in question are all positive. To show the converse, we proceed in several steps.

(I). We may reduce to the case that X is integral. In fact, $\mathscr{L}$ satisfies the numerical criterion of the theorem if and only if its restriction to each irreducible component (with the reduced scheme structure) of X satisfies the condition, because the test objects Z are the same, and $(\mathscr{L}^d \cdot Z)$ depends only on Z and the restriction $\mathscr{L}_{|Z}$ (see (23.15.3)). On the other hand, by Corollary 23.9 we may check ampleness of a line bundle by restricting it to each of the reduced irreducible components.

(II). Now assume that X is integral and that $(\mathscr{L}^d \cdot Z) > 0$ for all integral closed subschemes $Z \subseteq X$, $d = \dim Z$. By induction we may assume that the theorem holds for all proper k-schemes of smaller dimension. (Note that for $\dim X = 0$, the statement is clearly true.)

In particular, the induction hypothesis shows that $\mathscr{L}_{|W}$ is ample on W for every closed subscheme $W \subsetneq X$. Note that it is enough to show that some positive power of $\mathscr{L}$ is ample, so we may replace $\mathscr{L}$ by a positive power whenever that is convenient.

Let us show that some positive power of $\mathscr{L}$ has a non-trivial global section (in other words, some power of $\mathscr{L}$ is the line bundle attached to an effective divisor on X). The key point is to prove the following

Claim. For all $q \geq 2$ and all $n \gg 0$, we have $\dim_k H^q(X, \mathscr{L}^n) = \dim_k H^q(X, \mathscr{L}^{n-1})$.

Of course, we expect this to hold, and more precisely expect these cohomology groups to vanish if n is large by Theorem 23.6. Once the claim is proved, we find n_0 such that

$$\chi(\mathscr{L}^n) = \dim_k H^0(X, \mathscr{L}^n) - \dim_k H^1(X, \mathscr{L}^n) + \sum_{i \geq 2} (-1)^i \dim_k H^i(X, \mathscr{L}^{n_0})$$

for all $n \geq n_0$, with the sum on the very right independent of n. But the asymptotic Riemann-Roch theorem, Proposition 23.78, together with our assumption on $\mathscr{L}$ for $Z = X$, shows that $\chi(\mathscr{L}^n)$ tends to infinity for $n \to \infty$. This is only possible if $H^0(X, \mathscr{L}^n) \neq 0$ for n large.

Proof of claim. Let $\mathscr{I}, \mathscr{J} \subset \mathscr{O}_X$ be non-zero quasi-coherent ideal sheaves such that $\mathscr{I} \otimes \mathscr{L} \cong \mathscr{J}$. To see that such ideal sheaves exist, we can identify $\mathscr{L}$ with an $\mathscr{O}_X$-submodule of the constant sheaf $\mathscr{K}_X$ of rational functions on X (see Proposition 11.29). This also gives us an embedding $\mathscr{L}^{-1} \subset \mathscr{K}_X$, and we may then set $\mathscr{I} = \mathscr{L}^{-1} \cap \mathscr{O}_X$, and $\mathscr{J} := \mathscr{I}\mathscr{L} \subseteq \mathscr{O}_X$.

Now consider the exact sequences

$$0 \to \mathscr{I} \otimes \mathscr{L}^n \to \mathscr{L}^n \to \mathscr{O}_X / \mathscr{I} \otimes \mathscr{L}^n \to 0,$$
$$0 \to \mathscr{J} \otimes \mathscr{L}^{n-1} \to \mathscr{L}^{n-1} \to \mathscr{O}_X / \mathscr{J} \otimes \mathscr{L}^{n-1} \to 0.$$

By definition of $\mathscr{I}$ and $\mathscr{J}$, the terms on the left are isomorphic. We may view the terms on the right as powers of the restriction of $\mathscr{L}$ to the closed subscheme defined by $\mathscr{I}$ and $\mathscr{J}$, respectively. These restrictions are ample by induction hypothesis, so for n sufficiently large the cohomology of the terms on the right vanishes in all positive degrees. Comparing the two long exact cohomology sequences then proves the claim.

(III). Let us show that some positive power of $\mathscr{L}$ is generated by its global sections. By the previous step we may assume that $\mathscr{L} \cong \mathscr{O}_X(D)$ for an effective Cartier divisor D on X. Consider the short exact sequence

$$0 \to \mathscr{L}^{n-1} \to \mathscr{L}^n \to \mathscr{L}^n \otimes \mathscr{O}_D \to 0.$$

By induction hypothesis, $\mathscr{L}_{|D}$ is ample, and hence $H^1(X, \mathscr{L}^n \otimes \mathscr{O}_D)$ vanishes for sufficiently large n. In that case we obtain an exact sequence

$$H^0(X, \mathscr{L}^n) \to H^0(X, \mathscr{L}^n \otimes \mathscr{O}_D) \to H^1(X, \mathscr{L}^{n-1}) \to H^1(X, \mathscr{L}^n) \to 0.$$

In particular, $\dim_k H^1(X, \mathscr{L}^n) \leq \dim_k H^1(X, \mathscr{L}^{n-1})$, and since these k-vector spaces are finite-dimensional, we get equality for n sufficiently large. If equality holds, the map $H^0(X, \mathscr{L}^n) \to H^0(X, \mathscr{L}^n \otimes \mathscr{O}_D)$ in the above exact sequence is surjective. We know already that $\mathscr{L}_{|D}$ is ample on D, and increasing n further, if required, $\mathscr{L}^n \otimes \mathscr{O}_D$ is generated by global sections. Lifting a generating family to $H^0(X, \mathscr{L}^n)$ and adding a section in $H^0(X, \mathscr{L}^n)$ which has zero set D (and hence does not vanish anywhere on $X \setminus D$), we obtain a family of global sections of $\mathscr{L}^n$ generating this line bundle.

(IV). We can now conclude. As we have shown, we may assume that $\mathscr{L}$ is globally generated, and hence defines a morphism $f \colon X \to \mathbb{P}^N_k$ into some projective space with $f^* \mathscr{O}_{\mathbb{P}^N_k}(1) \cong \mathscr{L}$. Then f is proper, since X is proper. Furthermore, f is quasi-finite because otherwise there would exist an integral curve $C \subseteq X$ which is mapped under f to a point. However, then $(\mathscr{L} \cdot C) = (f^* \mathscr{O}_{\mathbb{P}^N_k}(1) \cdot C) = 0$ by Proposition 23.77, in contradiction to our assumption. Thus f is quasi-finite and proper, hence finite (Corollary 12.89). Since $\mathscr{L}$ is isomorphic to the pullback of an ample line bundle under a finite morphism, it is itself ample by Proposition 13.83. $\qquad\square$

(23.19) Hilbert polynomials of proper schemes over a field.

In this section we fix a field k, a proper k-scheme X, and an *ample* line bundle $\mathscr{L}$ on X, in particular, X is projective over k. For every complex $\mathscr{F}$ of $\mathscr{O}_X$-modules we set $\mathscr{F}(n) := \mathscr{F} \otimes_{\mathscr{O}_X} \mathscr{L}^{\otimes n}$ for all $n \in \mathbb{Z}$.

Proposition and Definition 23.91. *In the above situation, let $\mathscr{F}$ be in $D^b_{\mathrm{coh}}(X)$. Then*

$$\Phi_{X,\mathscr{L},\mathscr{F}}(n) := \Phi_{\mathscr{F}}(n) := \chi(X, \mathscr{F}(n))$$

is a numerical polynomial in n, called the Hilbert polynomial of $\mathscr{F}$ and of $(X, \mathscr{L})$.
(1) If $\mathscr{F}$ is a coherent $\mathscr{O}_X$-module, one has

$$d := \deg \Phi_{X,\mathscr{L},\mathscr{F}} = \dim \mathrm{Supp}(\mathscr{F})$$

and $\Phi_{\mathscr{F}}$ has leading term

$$\frac{(\mathscr{L}^d \cdot \mathscr{F})}{d!}.$$

Moreover, $(\mathscr{L}^d \cdot \mathscr{F})$ is a positive integer if $\mathscr{F} \neq 0$.
(2) The map $\mathscr{F} \mapsto \Phi_{\mathscr{F}}$ induces a homomorphism of abelian groups

(23.19.1) $$\Phi \colon K_0'(X) \longrightarrow \mathbb{Q}[T].$$

If $\mathscr{F} = \mathscr{O}_X$, we simple write Φ_X or $\Phi_{X,\mathscr{L}}$.

Proof. The Hilbert polynomial is additive in distinguished triangles of objects in $D^b_{\mathrm{coh}}(X)$ because the Euler characteristic has this property and tensoring with line bundles is a triangulated functor. Therefore we obtain (2).

Let us show (1). Let $\mathscr{F}$ be a coherent $\mathscr{O}_X$-module, endow $Z := \operatorname{Supp} \mathscr{F}$ with the scheme structure given by the annihilator ideal of $\mathscr{F}$ (Definition 23.20), and let $i \colon Z \to X$ be the inclusion. Then $\mathscr{F} = i_* i^* \mathscr{F}$ and $\chi(X, \mathscr{F}(n)) = \chi(Z, i^* \mathscr{F}(n))$ by Proposition 23.65. Hence we may assume that $X = \operatorname{Supp} \mathscr{F}$.

By Proposition 23.78 we already know that $\Phi_{\mathscr{F}}$ is a numerical polynomial of degree $\le d := \dim X$ and that its degree d coefficient is $(\mathscr{L}^d \cdot \mathscr{F})/d!$. If $\mathscr{F} \ne 0$ is a coherent $\mathscr{O}_X$-module, it is positive by Proposition 23.79. $\square$

Remark 23.92. If $\mathscr{F}$ is a coherent $\mathscr{O}_X$-module, then the vanishing of higher cohomology of $\mathscr{F}(n)$ for large n (Theorem 23.2) shows that one has

$$\Phi_{\mathscr{F}}(n) = \dim_k H^0(X, \mathscr{F}(n)), \qquad n \gg 0.$$

Example 23.93. Let $S = \bigoplus_{d \ge 0} S_d$ be a graded k-algebra of finite type that is generated by S_1, let $X := \operatorname{Proj}(S)$ and fix the very ample line bundle $\mathscr{L} = \mathscr{O}_X(1)$. Let M be a graded S-module of finite type and let $\mathscr{F} := \tilde{M}$ be the associated coherent $\mathscr{O}_X$-module. Then there exists an $N \ge 1$ such that one has

$$(23.19.2) \qquad\qquad \Phi_{\mathscr{F}}(n) = \dim_k M_n, \qquad \text{for all } n \ge N.$$

Indeed, for large n one has $\dim_k M_n = \dim H^0(X, \mathscr{F}(n))$ (Section (13.5)) and one can use Remark 23.92.

The map $n \mapsto \dim_k M_n$ is sometimes called the *Hilbert function* of M.

Now let $f \in S$ be a homogeneous regular element of degree d and let $H := V_+(f)$ be the corresponding hypersurface in X. Then from the closed subscheme exact sequence

$$0 \longrightarrow \mathscr{O}_X(-d) \longrightarrow \mathscr{O}_X \longrightarrow \mathscr{O}_H \longrightarrow 0$$

one obtains by (23.19.1)

$$(23.19.3) \qquad\qquad \Phi_H(n) = \Phi_X(n) - \Phi_X(n - d).$$

If we apply the example to $S = k[T_0, \ldots, T_r]$ so that $X = \mathbb{P}^r_k$, one gets:

Example 23.94. Let $r \ge 1$ be an integer. Then

$$\Phi_{\mathbb{P}^r_k}(n) = \binom{r + n}{r}.$$

This also follows from Example 23.61. If $H \subseteq \mathbb{P}^r_k$ is a hypersurface of degree $d \ge 1$, then (23.19.3) shows

$$\Phi_H(n) = \binom{r + n}{r} - \binom{r + n - d}{r}.$$

Example 23.95. Let C be a proper curve over k and let $\mathscr{L}$ be an ample line bundle on C (we will see in Section (26.5) below that such a line bundle always exists). By (23.15.4) the Hilbert polynomial $\Phi_{C,\mathscr{L}}$ is the numerical polynomial $\deg(\mathscr{L})T + \chi(\mathscr{O}_C) \in \mathbb{Q}[T]$.

The Grothendieck-Riemann-Roch theorem

In this part of the chapter we will briefly explain Grothendieck's variant of the Riemann-Roch theorem, mostly without proof. To keep the exposition as simple as possible, we do not strive for a general version but work only with smooth quasi-projective varieties over a field. More precisely, in this section we denote by k a field and by $\mathrm{Sm} = \mathrm{Sm}_k$ the full subcategory of the category of k-schemes consisting of smooth quasi-projective k-schemes. Every morphism in Sm_k is then quasi-projective ([EGAII] $^\mathrm{O}$ (5.3.4)) and locally of complete intersection (Corollary 19.44). In particular, every closed immersion is regular.

Often the Grothendieck-Riemann-Roch theorem is formulated as a compatibility statement between K-theory and Chow theory. We will not introduce Chow groups here but choose a more axiomatic approach following Grothendieck [Gro2] $^\mathrm{O}$, Panin [Pan] $^\mathrm{O}$ and Navarro [Nav] $^\mathrm{O}_X$. Following in particular the article [Nav] $^\mathrm{O}_X$ very closely, we make the following definition.

(23.20) Cohomology theories.

The following definition axiomatizes some nice properties that a "cohomology theory" should have. Note however that these properties are inspired from algebraic topology more than from algebraic geometry. Cohomology of coherent sheaves does not satisfy these conditions and thus is not a cohomology theory in the sense of this definition. We will give examples in Section (23.22).

Definition 23.96. *A* cohomology theory over k *is a functor*

$$A: \mathrm{Sm}_k^{\mathrm{opp}} \longrightarrow (\mathrm{Ring}), \qquad X \mapsto A(X), \quad f \mapsto f^*,$$

endowed for any proper morphism $f: X \to Y$ *in* Sm_k *with a functorial morphism* $f_*: A(X) \to A(Y)$ *of* $A(Y)$*-modules, called* direct image, *i.e., one has* $(\mathrm{id}_X)_* = \mathrm{id}_{A(X)}$, $(f \circ g)_* = f_* \circ g_*$ *and the* projection formula

$$(23.20.1) \qquad f_*(f^*(y)x) = yf_*(x), \qquad x \in A(X), y \in A(Y).$$

If $\mathscr{L}$ *is a line bundle on* X, *its zero section yields the zero section* $s_0: X \to \mathbb{V}(\mathscr{L}^\vee)$ *(Section (11.3)). Then* $c_1^A(\mathscr{L}) := s_0^*(s_{0*}(1)) \in A(X)$ *is called the* Chern class *of* $\mathscr{L}$.

These data are supposed to satisfy the following conditions.

(a) *For* X_1, X_2 *in* Sm_k *with natural immersions* $i_j: X_j \to X_1 \coprod X_2$, $j = 1, 2$, *the ring homomorphism* $(i_1^*, i_2^*): A(X_1 \coprod X_2) \to A(X_1) \times A(X_2)$ *is an isomorphism.*

(b) *Strong homotopy invariance: Let* $\pi: E \to X$ *be a vector bundle torsor (i.e., there exists a geometric vector bundle* $E_0 \to X$ *such that* E *is an* E_0*-torsor over* X *for the Zariski topology). Then* $\pi^*: A(X) \to A(E)$ *is an isomorphism.*

(c) *If* $i: Y \to X$ *is a smooth closed subscheme and* $j: X \setminus Y \to X$ *the open immersion of the complement, then the sequence*

$$A(Y) \xrightarrow{i_*} A(X) \xrightarrow{j^*} A(X \setminus Y)$$

is exact.

(d) *Let*

$$
\begin{array}{ccc}
Y' & \xrightarrow{\ i'\ } & X' \\
{\scriptstyle g}\downarrow & & \downarrow{\scriptstyle f} \\
Y & \xrightarrow{\ i\ } & X
\end{array}
$$

be a cartesian diagram of k-schemes that are in Sm_k such that $i\colon Y \to X$ is a closed immersion and such that the natural homomorphism of conormal sheaves $g^*\mathscr{C}_i \to \mathscr{C}_{i'}$ is an isomorphism, i.e., f is transversal for Y (see Exercises 18.27 and 18.28). Then one has

$$
f^* \circ i_* = i'_* \circ g^*.
$$

(e) *Let X be in Sm_k and let $\mathscr{E}$ be a finite locally free $\mathcal{O}_X$-module of constant rank r and let $\pi\colon \mathbb{P}(\mathscr{E}) \to X$ be the corresponding projective bundle. Then for every morphism $f\colon Y \to X$ in Sm_k one has a commutative diagram*

$$
\begin{array}{ccc}
A(\mathbb{P}(\mathscr{E})) & \xrightarrow{\ \varphi^*\ } & A(\mathbb{P}(f^*\mathscr{E})) \\
{\scriptstyle \pi_*}\downarrow & & \downarrow{\scriptstyle \varpi_*} \\
A(X) & \xrightarrow{\ f^*\ } & A(Y),
\end{array}
$$

where $\varphi\colon \mathbb{P}(f^*(\mathscr{E})) = \mathbb{P}(\mathscr{E}) \times_X Y \to \mathbb{P}(\mathscr{E})$ and $\varpi\colon \mathbb{P}(f^*(\mathscr{E})) \to Y$ are the projections. Moreover, let $x_{\mathscr{E}} := c_1^A(\mathcal{O}_{\mathbb{P}(\mathscr{E})}(1))$ be the first Chern class of the tautological bundle $\mathcal{O}_{\mathbb{P}(\mathscr{E})}(1)$ on $\mathbb{P}(\mathscr{E})$. Then $A(\mathbb{P}(E))$, made into an $A(X)$-algebra via $\pi^*\colon A(X) \to A(\mathbb{P}(E))$, is a free $A(X)$-module with basis $(1, x_{\mathscr{E}}, x_{\mathscr{E}}^2, \ldots, x_{\mathscr{E}}^{r-1})$.

If A and $\tilde{A}$ are cohomology theories over k, then a morphism of cohomology theories $\varphi\colon A \to \tilde{A}$ is a morphism of contravariant functors preserving direct images for proper morphisms.

A cohomology theory A is called multiplicative *(resp.* additive*)* if one has

$$
c_1^A(\mathscr{L}_1 \otimes \mathscr{L}_2) = c_1^A(\mathscr{L}_1) + c_1^A(\mathscr{L}_2) - c_1^A(\mathscr{L}_1)c_1^A(\mathscr{L}_2)
$$

(resp.

$$
c_1^A(\mathscr{L}_1 \otimes \mathscr{L}_2) = c_1^A(\mathscr{L}_1) + c_1^A(\mathscr{L}_2) \qquad)
$$

for all line bundles $\mathscr{L}_1$ and $\mathscr{L}_2$ on a scheme X in Sm_k.

A cohomology theory A is graded *if $A(X)$ is endowed with the structure of a graded ring $A(X) = \bigoplus_{d \geq 0} A^d(X)$ such that f^* is homogeneous of degree 0 for every morphism f in Sm_k and such that f_* is homogeneous of degree $\dim Y - \dim X$ for every proper morphism $f\colon X \to Y$ between irreducible schemes X and Y in Sm_k.*

Remark and Definition 23.97. Let A be a cohomology theory over k.

(1) Let $i\colon Y \to X$ be a closed immersion in Sm_k. Then i is a regular immersion (Theorem 19.30) and $[Y]^A := i_*(1) \in A(X)$ is called the *fundamental class* of Y in $A(X)$.

(2) Property (a) implies that $A(\emptyset) = 0$.

(3) The hypotheses in (c) are satisfied if $Y' = \emptyset$. In this case the condition simply means that $f^* \circ i_* = 0$.

(4) We can apply (c) to the situation of (a). As both i_1 and i_2 are transversal to i_1 and i_2, one has $i_2^* i_{1*} = 0$, $i_1^* i_{2*} = 0$, $i_1^* i_{1*} = \mathrm{id}$, $i_2^* i_{2*} = \mathrm{id}$. Therefore the inverse of the isomorphism (i_1^*, i_2^*) is given by

$$i_{1*} + i_{2*} \colon A(X_1) \times A(X_2) \xrightarrow{\sim} A(X_1 \coprod X_2).$$

Hence we find for any proper morphism $f = f_1 \coprod f_2 \colon X_1 \coprod X_2 \to Z$ that

$$f_* = f_{1*} + f_{2*} \colon A(X_1) \times A(X_2) \to A(Z).$$

(5) The formation of Chern classes is compatible with pullback: Let $f \colon Y \to X$ be a morphism in Sm_k and let $\mathscr{L}$ be a line bundle on X. Consider the cartesian diagram

$$
\begin{array}{ccc}
Y & \xrightarrow{\;s_0'\;} & Y \times_X \mathbb{V}(\mathscr{L}^\vee) = \mathbb{V}(f^* \mathscr{L}^\vee) \\
{\scriptstyle f} \downarrow & & \downarrow {\scriptstyle p_2} \\
X & \xrightarrow{\;s_0\;} & \mathbb{V}(\mathscr{L}^\vee),
\end{array}
$$

where s_0' denotes the zero section of $\mathbb{V}(f^* \mathscr{L}^\vee) \to Y$. It shows that p_2 is transversal for s_0. We therefore find

$$f^* c_1^A(\mathscr{L}) = f^* s_0^* s_{0*}(1) = s_0'^* \varphi^* s_{0*}(1) = s_0'^* s_{0*}'(1) = c_1^A(f^* \mathscr{L}).$$

Remark 23.98. Let $i \colon D \to X$ be a smooth Cartier divisor in X, i.e., i is a regular closed immersion of codimension 1 given by an invertible ideal $\mathscr{I} \subseteq \mathscr{O}_X$. The canonical section of $\mathscr{O}_X(D) = \mathscr{I}^{\otimes -1}$ (Remark 11.33) is a global regular section $s_D \in \Gamma(X, \mathscr{O}_X(D))$ whose vanishing locus is D. We view it as a section $s_D \colon X \to \mathbb{V}(\mathscr{I}) = \mathrm{Spec}(\mathrm{Sym}(\mathscr{I}))$ of the geometric line bundle $p \colon \mathbb{V}(\mathscr{I}) \to X$. It corresponds to the unique homomorphism of $\mathscr{O}_X$-algebras $s_D^* \colon \mathrm{Sym}\, \mathscr{I} \to \mathscr{O}_X$ whose restriction to $\mathscr{I} = \mathrm{Sym}^1(\mathscr{I})$ is the inclusion $\mathscr{I} \hookrightarrow \mathscr{O}_X$. Then s_D is transversal to the zero section $s_0 \colon X \to \mathbb{V}(\mathscr{I})$ as we have a cartesian diagram

$$
\begin{array}{ccc}
D & \xrightarrow{\;i\;} & X \\
{\scriptstyle i} \downarrow & & \downarrow {\scriptstyle s_D} \\
X & \xrightarrow{\;s_0\;} & \mathbb{V}(\mathscr{I}).
\end{array}
$$

Therefore we find

(23.20.2) $$c_1(\mathscr{O}_X(D)) = s_0^*(s_{0*}(1)) = s_D^*(s_{0*}(1)) = i_*(i^*(1)) = i_*(1) = [D],$$

where the second equality holds since s_0^* and s_D^* are both inverse maps to the isomorphism p^*, hence they are equal.

(23.21) Chern classes of vector bundles.

The following proposition, called the *splitting principle* shows that given a vector bundle $\mathscr{E}$ on X in Sm_k, we can find a smooth proper cover X' of X such that the class $[\mathscr{E}] \in K_0(X')$ splits as a sum of classes of line bundles. This is very useful to reduce statements about vector bundle to the case of line bundles which typically is much easier to handle. More precisely, we have the following result.

Proposition 23.99. *(Splitting principle) Given X in Sm_k and a vector bundle $\mathscr{E}$ on X of rank r, there exists a proper smooth morphism $\pi\colon X' \to X$ satisfying the following properties.*

(i) *There exists a filtration of $\mathscr{O}_{X'}$-modules*

$$\pi^*\mathscr{E} = \mathscr{E}^0 \supset \mathscr{E}^1 \supset \cdots \supset \mathscr{E}^r = 0$$

   such that $\mathscr{L}_l := \mathscr{E}^{r-l}/\mathscr{E}^{r-l+1}$ is a line bundle for $l = 1, \ldots, r$.

(ii) *The homomorphism $\pi^*\colon A(X) \to A(X')$ is injective for all cohomology theories A on Sm_k.*

Proof. We proceed by induction on r, the cases $r = 0, 1$ being trivial. Let $\varpi\colon \mathbb{P}(\mathscr{E}) \to X$ be the projective bundle of $\mathscr{E}$. The universal line bundle on $\mathbb{P}(\mathscr{E})$ sits in an exact sequence (see Section (13.8))

$$0 \longrightarrow \mathscr{E}' \longrightarrow \varpi^*\mathscr{E} \longrightarrow \mathscr{O}_{\mathbb{P}(\mathscr{E})}(1) \longrightarrow 0,$$

of vector bundles on $\mathbb{P}(\mathscr{E})$, where $\mathscr{E}'$ is a vector bundle of rank $r-1$. The map $\varpi^*\colon A(X) \to A(\mathbb{P}(\mathscr{E}))$ is injective for all cohomology theories A by Definition 23.96 (c). Now we conclude by the induction hypothesis. $\square$

Definition 23.100. *Let A be a cohomology theory over k. Let X be in Sm_k and let $\mathscr{E}$ be a vector bundle over X of rank r. Let $x_\mathscr{E} = c_1^A(\mathscr{O}_{\mathbb{P}(\mathscr{E})}(1)) \in A(\mathbb{P}(\mathscr{E}))$ be the first Chern class of the universal line bundle on $\mathbb{P}(\mathscr{E})$. Multiplication by $x_\mathscr{E}$ defines an endomorphism $m_\mathscr{E}$ of the free $A(X)$-module $A(\mathbb{P}(\mathscr{E}))$. The characteristic polynomial of $m_\mathscr{E}$ is denoted by*

$$\chi_\mathscr{E}^A = T^r - c_1^A(\mathscr{E})T^{r-1} + \cdots + (-1)^r c_r^A(\mathscr{E}) \in A(X)[T]$$

and $c_i^A(\mathscr{E}) \in A(X)$ is called the i-th Chern class of $\mathscr{E}$. One also sets $c_0^A(\mathscr{E}) := 1$ and $c_i^A(\mathscr{E}) := 0$ for $i > r$. The polynomial

$$c^A(\mathscr{E}) := (-T)^r \chi_\mathscr{E}^A(-T^{-1}) = 1 + c_1^A(\mathscr{E})T + c_2^A(\mathscr{E})T^2 + \cdots + c_r^A(\mathscr{E})T^r$$

is called the Chern polynomial *of $\mathscr{E}$.*

If $\mathscr{E} = \mathscr{L}$ is a line bundle, then $\mathbb{P}(\mathscr{L}) = X$ and $\mathscr{O}_{\mathbb{P}(\mathscr{L})}(1) = \mathscr{L}$ and hence $c_1^A(\mathscr{L})$ as defined in Definition 23.100 equals $c_1^A(\mathscr{L})$ as given by the cohomology theory. The Chern polynomial of $\mathscr{L}$ has the form

$$c^A(\mathscr{L}) = 1 + c_1^A(\mathscr{L})T.$$

Remark 23.101. Let A be a cohomology theory. For any morphism $f\colon X \to Y$ in Sm_k and for every vector bundle $\mathscr{E}$ on Y one has $c_i^A(f^*\mathscr{E}) = f^*c_i^A(\mathscr{E})$.

Indeed, this follows easily because $X \times_Y \mathbb{P}(\mathscr{E}) = \mathbb{P}(f^*\mathscr{E})$ and since first Chern classes of line bundles are compatible with pullback (Remark 23.97 (5)).

Proposition 23.102. *Let A be a cohomology theory. For every exact sequence $0 \to \mathscr{E}' \to \mathscr{E} \to \mathscr{E}'' \to 0$ of vector bundles on a scheme in Sm_k one has*

$$\chi^A(\mathscr{E}) = \chi^A(\mathscr{E}')\chi^A(\mathscr{E}''),$$
$$c^A(\mathscr{E}) = c^A(\mathscr{E}')c^A(\mathscr{E}''),$$
$$c_n^A(\mathscr{E}) = \sum_{i+j=n} c_i^A(\mathscr{E}')c_j^A(\mathscr{E}'').$$

One can prove $c^A(\mathscr{E}) = c^A(\mathscr{E}')c^A(\mathscr{E}'')$ by induction on the rank of $\mathscr{E}'$. The other two identities are then immediate corollaries. We refer to [Nav] $\overset{\mathrm{o}}{\times}$ 1.2 for details.

Example 23.103. Let X be in Sm_k and let $\mathscr{E} = \mathscr{O}_X^r$ be the trivial vector bundle of rank r. As $c_1^A(\mathscr{O}_X) = 0$, an induction on r using Proposition 23.102 shows that $c_i^A(\mathscr{O}_X^r) = 0$ for all $i > 0$.

If $0 \to \mathscr{E}' \to \mathscr{O}_X^r \to \mathscr{L} \to 0$ is an exact sequence, where $\mathscr{L}$ is a line bundle, then Proposition 23.102 shows that for all $n > 0$ one has $0 = c_n^A(\mathscr{E}') + c_{n-1}^A(\mathscr{E}')c_1(\mathscr{L})$ and hence

$$c_n^A(\mathscr{E}') = (-1)^i c_{n-i}(\mathscr{E}')c_1(\mathscr{L})^i \qquad 0 \le i \le n.$$

In particular, one obtains for $i = n = r$ that $c_1(\mathscr{L})^r = 0$. Such an exact sequence exists for every line bundle that is generated by finitely many global sections, so we see that the Chern classes of such line bundles are nilpotent. See Proposition 23.106 for a generalization.

Remark and Definition 23.104. Choose $\pi\colon X' \to X$ and line bundles $\mathscr{L}_1, \ldots, \mathscr{L}_r$ on X' as in the splitting principle (Proposition 23.99). Then Proposition 23.102 shows

$$\pi^* \chi^A(\mathscr{E}) = \prod_{l=1}^r (T - c_1^A(\mathscr{L}_l)), \qquad \pi^* c^A(\mathscr{E}) = \prod_{l=1}^r (1 + c_1^A(\mathscr{L}_l)T) \in A(X')[T].$$

The elements $\alpha_l := c_1^A(\mathscr{L}_l) \in A(X')$ are called *Chern roots* of $\mathscr{E}$. Then $\pi^* c_i^A(\mathscr{E})$ is the i-th elementary symmetric polynomial of the Chern roots

$$(23.21.1) \qquad \pi^* c_i^A(\mathscr{E}) = \sum_{1 \le l_1 < \cdots < l_i \le r} \alpha_{l_1} \cdots \alpha_{l_i}.$$

Example 23.105. Let A be a cohomology theory and set $R := A(\mathrm{Spec}\, k)$. Let $d \ge 1$. Then the cohomology ring of $\mathbb{P}_k^d$ is given by

$$A(\mathbb{P}_k^d) \cong R[T]/(T^{d+1}),$$

where the isomorphism is given by sending T to $x := c_1(\mathscr{O}_{\mathbb{P}_k^d}(1))$.

Indeed, by the axioms of a cohomology theory, $A(\mathbb{P}_k^d)$ is a free R-module with basis $1, x, \ldots, x^d$. Hence it suffices to show that $x^{d+1} = 0$ in $A(\mathbb{P}_k^d)$. But this follows from Example 23.103 because $\mathscr{O}_{\mathbb{P}_k^d}(1)$ is a quotient of a trivial vector bundle of rank $d + 1$.

Proposition 23.106. *Let A be a cohomology theory. For every X in Sm_k, Chern classes of vector bundles are nilpotent in $A(X)$.*

Proof. By the splitting principle and Remark 23.104 it suffices to show that for every scheme X in Sm_k and for every line bundle $\mathscr{L}$ on X its Chern class $c_1^A(\mathscr{L})$ is nilpotent. By Jouanolou's trick (see Exercise 23.29 or [Wei1]$\times$ 4.3) we can find a vector bundle torsor $p\colon E \to X$ such that E is affine. As $p^*\colon A(X) \to A(E)$ is an isomorphism, it suffices to show that $p^* c_1^A(\mathscr{L}) = c_1^A(p^*\mathscr{L})$ is nilpotent. Hence we may assume that X is affine.

Then $\mathscr{L}$ is generated by finitely many global sections and we conclude by Example 23.103. $\square$

(23.22) Examples of Cohomology theories.

An important example for a cohomology theory is given by K-theory as follows.

Example: K-theory 23.107. Recall that for X in Sm_k one has $K_0(X) = K_0'(X)$ by Proposition 23.55. Hence $X \mapsto K_0(X)$ is contravariant for all morphisms, covariant for all proper morphisms, and the projection formula holds. Moreover $K_0(X)$ is generated by the classes of finite locally free $\mathscr{O}_X$-modules (Remark 23.51).

The fundamental class of a closed smooth subvariety $i\colon Y \to X$ is given by $i_*\mathscr{O}_Y \in K_0(X)$. The Chern class of a line bundle $\mathscr{L}$ on X corresponding to a geometric line bundle $\mathbb{V}(\mathscr{L}^\vee) \to X$ with zero section s_0 is given by

$$c_1^{K_0}(L) = [Ls_0^*(s_{0*}\mathscr{O}_X)] = [\mathscr{O}_X] - [\mathscr{L}^{-1}] \in K_0(X),$$

so the Chern class in the sense of the cohomology theory K_0 coincides with the Chern class defined previously (Definition 23.57). One can show that $X \mapsto K_0(X)$ is a cohomology theory in the sense of Definition 23.96, see [SGA6] $^\mathrm{O}$, [Wei3] $^\mathrm{O}_\mathrm{X}$. Moreover, K-theory is a multiplicative cohomology theory by (23.13.2).

Let $\mathscr{E}$ be a locally free module of rank r over X with Chern roots $\alpha_i = c_1^{K_0}(\mathscr{L}_i) = 1 - [\mathscr{L}_i^\vee]$ for line bundles $\mathscr{L}_i$ on X', where $\pi\colon X' \to X$ is a smooth proper morphism such that π^* is injective. One has $[\pi^*\mathscr{E}] = [\mathscr{L}_1] + \cdots + [\mathscr{L}_r]$ in $K_0(X')$. Hence one has in $K_0(X) \subseteq K_0(X')$

$$c_1^{K_0}(\mathscr{E}) = \alpha_1 + \cdots + \alpha_r = \sum_{i=1}^{r}(1 - [\mathscr{L}_i^\vee]) = r - [\mathscr{E}^\vee],$$

$$c_r^{K_0}(\mathscr{E}) = \prod_{i=1}^{r}(1 - [\mathscr{L}_i^\vee]) = \sum_{j=0}^{r}(-1)^j \left[\bigwedge^j \mathscr{E}\right].$$

In fact, in [Nav] $^\mathrm{O}_\mathrm{X}$ it is shown that K-theory is the universal multiplicative K-theory in the following sense.

Theorem 23.108. *For any multiplicative cohomology theory A there exists a unique morphism of cohomology theories $\varphi\colon K_0 \to A$. If $\mathscr{E}$ is a vector bundle over a scheme X in Sm_k, then we have*

$$(23.22.1) \qquad\qquad \varphi([\mathscr{E}]) = \mathrm{rk}(\mathscr{E}) - c_1^A(\mathscr{E}^\vee).$$

Many important cohomology theories are additive. In fact one can show that every graded cohomology theory is additive ([Nav] $^\mathrm{O}_\mathrm{X}$ Lemma 3.1).

Example: Graded K-theory 23.109. Let X be in Sm_k. If X is connected and hence irreducible we denote by $K_0(X)^c \subseteq K_0(X) = K_0'(X)$ the subgroup generated by classes of coherent $\mathscr{O}_X$-modules $\mathscr{F}$ with support of codimension $\geq c$. Then $K_0(X)^c$ is an ideal in the ring $K_0(X)$. We extend this filtration to K-groups of not necessarily connected schemes in Sm_k in the unique way such that $K_0(X_1 \coprod X_2) = K_0(X_1) \times K_0(X_2)$ is an isomorphism of filtered rings. We obtain a decreasing filtration $K_0(X) = K_0(X)^0 \supseteq K_0(X)^1 \subseteq \cdots$. Then

$$X \mapsto \mathrm{gr}^\bullet K_0(X) = \bigoplus_{c \geq 0} \mathrm{gr}^c K_0(X), \qquad \mathrm{gr}^c K_0(X) := K_0(X)^c / K_0(X)^{c+1}$$

defines a graded cohomology theory. The fundamental class of a closed smooth subvariety $i\colon Y \to X$ of codimension c is

$$[i_* \mathscr{O}_Y] \in \operatorname{gr}^c K_0(X).$$

and the first Chern class of a line bundle bundle $\mathscr{L}$ on X is

$$c_1^{\operatorname{gr} K_0}(\mathscr{L}) = [\mathscr{O}_X] - [\mathscr{L}^{-1}] \in \operatorname{gr}^1 K_0(X).$$

By Proposition 23.58 (4) this cohomology theory is additive.

Navarro shows in [Nav]$_X^O$ that if k is a perfect field, then graded K-theory is the universal additive cohomology theory of Sm_k.

Example: Chow rings 23.110. Another important example of an additive graded cohomology theory is $A^\bullet(X) := \operatorname{CH}^\bullet(X)$ for $X \in \operatorname{Sm}_k$, where $\operatorname{CH}^c(X)$ is the Chow group of codimension c cycles on X up to rational equivalence (e.g., see [Ful2]^O).

For $c = 0$, $\operatorname{CH}^0(X)$ is the free abelian group on the connected components of X. For $c = 1$, $\operatorname{CH}^1(X)$ is the group of rational equivalence classes of Weil divisors. Since X is smooth over k and hence regular, we can identify $\operatorname{CH}^1(X)$ with $\operatorname{Pic}(X)$ by Theorem 11.40.

(23.23) The Grothendieck-Riemann-Roch theorem for additive cohomology theories.

While compatibility of the Chern character of a vector bundle with pullback is basically built into the definitions (cf. Remark 23.101), the Chern character is not invariant under push-forwards, as can be seen already in simple examples. However, it is possible to precisely describe the behavior of the Chern character under push-forwards, and this is the content of the Theorem of Grothendieck-Riemann-Roch (Theorem 23.112 below). As a crucial ingredient, we need to "twist" the given cohomology theory, thus changing its direct image functor, as follows.

Remark 23.111. For a scheme X in Sm_k we denote by $T_{X/k}$ the tangent bundle of X and for a proper morphism $f \colon X \to Y$ in Sm_k we write

$$T_f := [T_{X/k}] - [f^* T_{Y/k}] \in K_0(X)$$

for the *virtual relative tangent bundle*.

Let A be a cohomology theory and let $F \in A(\operatorname{Spec} k)[\![t]\!]^\times$ be an invertible formal power series with coefficients in $A(\operatorname{Spec} k)$. We define a homomorphism of groups

$$F_\times \colon K_0(X) \to A(X)^\times$$

as follows. Let $\mathscr{E}$ be a vector bundle on X with Chern roots $\alpha_1, \dots, \alpha_r \in A(X')$, where $\pi \colon X' \to X$ is as in the splitting principle. As Chern classes are nilpotent (Proposition 23.106), $F_\times(\mathscr{E}) := \prod_{i=1}^r F(\alpha_i)$ is a polynomial in the elementary symmetric functions of the α_i, which are just the Chern classes of $\mathscr{E}$. Hence $F_\times(\mathscr{E}) \in A(X)^\times$. By Proposition 23.102 this construction induces a well defined homomorphism $F_\times \colon K_0(X) \to A(X)^\times$.

Now one can define a twist A^F of A by F, where A^F has the same underlying contravariant functor from Sm_k to (Ring) as A but where the direct image functor for A^F is defined by

(23.23.1) $$f_*^F(\alpha) := f_*(F_\times(T_f)^{-1} \cdot \alpha).$$

One checks that indeed A^F is a cohomology theory (see [Nav]$_X^O$ 2.3) and that

(23.23.2) $$c_1^{A^F}(\mathscr{L}) = c_1^A(\mathscr{L})F(c_1^A(\mathscr{L}))$$

for every line bundle $\mathscr{L}$. Moreover loc. cit. shows that every cohomology theory whose underlying contravariant functor is equal to A is of this form.

Let A be an additive cohomology theory. By setting $A_{\mathbb{Q}}(X) := A(X) \otimes_{\mathbb{Z}} \mathbb{Q}$ we obtain an additive cohomology theory that takes values in $\mathbb{Q}$-algebras. We would like to turn $A_{\mathbb{Q}}$ into a multiplicative cohomology theory in order to use that K_0-theory is the initial multiplicative cohomology theory (Theorem 23.108). To do this we use the exponential series e^{at} (for some number a to be determined later) as follows. We would like to find a new cohomology theory where the first Chern class is given by

$$c_1^{\text{new}}(\mathscr{L}) := 1 - e^{ac_1^A(\mathscr{L})} = -ac_1^A(\mathscr{L}) - \frac{a^2}{2}c_1^A(\mathscr{L})^2 - \cdots \in A_{\mathbb{Q}}(X).$$

As Chern classes are nilpotent, this expression makes sense. Then

$$c_1^{\text{new}}(\mathscr{L}_1 \otimes \mathscr{L}_2) = c_1^{\text{new}}(\mathscr{L}_1) + c_1^{\text{new}}(\mathscr{L}_2) - c_1^{\text{new}}(\mathscr{L}_1)c_1^{\text{new}}(\mathscr{L}_2)$$

and we have obtained a multiplicative cohomology theory. As $1 - e^{at} = -at + \ldots$, it is convenient to choose $a = -1$.

By Remark 23.111 we can construct such a cohomology theory by choosing a formal power series $F \in \mathbb{Q}[\![t]\!]^{\times}$ such that $tF = 1 - e^{-t}$, i.e., we set

$$F(t) = \frac{1 - e^{-t}}{t} = 1 - \frac{t}{2} + \frac{t^2}{3!} - \frac{t^3}{4!} + \cdots.$$

Then $A_{\mathbb{Q}}^F$ is a multiplicative cohomology theory with direct image given by (23.23.1)

$$f_*^F(\alpha) = f_*(F_{\times}(T_f)^{-1} \cdot \alpha).$$

Hence there is a unique morphism

(23.23.3) $$\text{ch} := \text{ch}^A \colon K_0 \to A_{\mathbb{Q}}^F, \qquad [\mathscr{E}] \mapsto \text{rk}(\mathscr{E}) - c_1^{\text{new}}(\mathscr{E}^{\vee})$$

of cohomology theories by Theorem 23.108. This morphism is often called the *Chern character* of A.

The inverse of F is given by

$$F(t)^{-1} = \frac{t}{1 - e^{-t}} = 1 + \frac{1}{2}t + \sum_{k=1}^{\infty}(-1)^{k-1}\frac{B_k}{(2k)!}t^{2k},$$

where B_k is the k-th Bernoulli number, and we call

$$\text{td} := \text{td}^A \colon K_0(X) \longrightarrow A(X)^{\times}, \qquad \text{td}^A(x) := F_{\times}^{-1}(x)$$

the *Todd class*. If $\mathscr{E}$ is a vector bundle of constant rank r on X with Chern roots $\alpha_1, \ldots, \alpha_r$ we thus have

$$\text{td}([\mathscr{E}]) = \prod_{i=1}^{r} F(\alpha_i)^{-1}.$$

Using that Chern classes are elementary symmetric polynomials in the Chern roots (23.21.1) one can calculate the first few terms

(23.23.4) $\qquad \mathrm{td}([\mathscr{E}]) = 1 + \dfrac{1}{2}c_1(\mathscr{E}) + \dfrac{1}{12}(c_1(\mathscr{E})^2 + c_2(\mathscr{E})) + \dfrac{1}{24}c_1(\mathscr{E})c_2(\mathscr{E}) + \cdots .$

Then for every proper morphism $f\colon X \to Y$ in Sm_k and for $x \in K_0(X)$ we have the equalities

(23.23.5) $\qquad \mathrm{ch}(f_*(x)) = f_*^F(\mathrm{ch}(x)) = f_*(\mathrm{td}(T_f) \cdot \mathrm{ch}(x)).$

This can be rewritten as

$$
\begin{aligned}
\mathrm{ch}(f_*(x)) &= f_*(F_\times(T_f)^{-1} \cdot \mathrm{ch}(x)) \\
&= f_*(F_\times(f^* T_{Y/k} - T_{X/k}) \cdot \mathrm{ch}(x) \\
&= f_*(f^* F_\times(T_{Y/k}) F_\times(T_{X/k})^{-1} \mathrm{ch}(x)) \\
&= F_\times(T_{Y/k}) f_*(F_\times(T_{X/k})^{-1} \mathrm{ch}(x)) \\
&= \mathrm{td}(T_{Y/k})^{-1} f_*(\mathrm{td}(T_{X/k}) \mathrm{ch}(x)).
\end{aligned}
$$

We obtain the Grothendieck-Riemann-Roch theorem:

Theorem 23.112. *(Grothendieck-Riemann-Roch theorem) Let A be an additive cohomology theory on Sm_k (e.g. A could be graded K-theory or Chow theory). Then one has for every proper morphism $f\colon X \to Y$ in Sm_k a commutative diagram*

$$
\begin{CD}
K_0(X) @>{f_*}>> K_0(Y) \\
@V{\mathrm{td}^A(T_{X/k}) \cdot \mathrm{ch}^A}VV @VV{\mathrm{td}^A(T_{Y/k}) \cdot \mathrm{ch}^A}V \\
A_{\mathbb{Q}}(X) @>{f_*}>> A_{\mathbb{Q}}(Y).
\end{CD}
$$

Corollary 23.113. *Let $f\colon X \to \mathrm{Spec}\,k$ be projective and smooth. Then for every additive cohomology theory A and for every complex $\mathscr{F}$ in $D^b_{\mathrm{coh}}(X)$ we have*

$$
\chi(X, \mathscr{F}) = f_*(\mathrm{td}^A(T_{X/k}) \mathrm{ch}^A([\mathscr{F}]))
$$

in $A(\mathrm{Spec}\,k)$.

If $A(X) = \mathrm{CH}(X)$ is the Chow ring of X, then $f_*\colon \mathrm{CH}(X) \to \mathrm{CH}(\mathrm{Spec}\,k) = \mathbb{Z}$ is given by the degree for 0-cycles and by 0 on the group of r-cycles for $r \geq 1$. Because of the analogy with complex geometry where this corresponds to integration of cohomology classes, this map is often denoted by $\int_X$ and hence Corollary 23.113 yields the *Hirzebruch-Riemann-Roch formula*

(23.23.6) $\qquad \chi(X, \mathscr{F}) = \displaystyle\int_X \mathrm{td}(T_{X/k}) \mathrm{ch}(\mathscr{F}).$

Proof. We apply the Grothendieck-Riemann-Roch theorem to $Y = \mathrm{Spec}\,k$. We have $K_0(\mathrm{Spec}\,k) = \mathbb{Z}$ and $f_*\colon K_0(X) \to \mathbb{Z}$ is given by the Euler Poincaré characteristic $[\mathscr{F}] \mapsto \chi(X, \mathscr{F})$ for $\mathscr{F}$ in $D^b_{\mathrm{coh}}(X)$. The map $\mathrm{ch}\colon K_0(\mathrm{Spec}\,k) \to A(\mathrm{Spec}\,k)$ is necessarily the unique ring homomorphism $\mathbb{Z} \to A(\mathrm{Spec}\,k)$ and $\mathrm{td}(T_{Y/k}) = 1$. This yields the corollary. $\qquad\square$

Remark 23.114. Let A be an additive cohomology theory. By Theorem 23.108 the Chern character $\mathrm{ch}^A \colon K_0(X) \to A_{\mathbb{Q}}^F(X)$ of a vector bundle $\mathscr{E}$ on X is given by

$$\mathrm{ch}^A(\mathscr{E}) = \mathrm{rk}(\mathscr{E}) - c_1^{A_{\mathbb{Q}}^F}(\mathscr{E}^\vee).$$

If $\mathscr{E} = \mathscr{L}$ is a line bundle we obtain by (23.23.2) $c_1^{A_{\mathbb{Q}}^F}(\mathscr{L}^\vee) = P(c_1^A(\mathscr{L}^\vee))$ for $P(t) = 1 - e^{-t}$. As A is additive, we have $c_1^A(\mathscr{L}^\vee) = -c_1^A(\mathscr{L})$ and therefore

(23.23.7) $$\mathrm{ch}^A(\mathscr{L}) = \exp(c_1^A(\mathscr{L})).$$

Example 23.115. Let us look at the Hirzebruch-Riemann-Roch formula for a geometrically connected projective smooth curve C. Then $\mathrm{CH}^0(C) = \mathbb{Z}$, $\mathrm{CH}^1(C) = \mathrm{Pic}(C)$, and $\mathrm{CH}^i(C) = 0$ for $i > 1$. One has $T_{C/k} = (\Omega_{C/k}^1)^\vee$, which is a line bundle, and for every line bundle $\mathscr{L}$ one has by (23.23.4) and by (23.23.7)

$$\mathrm{td}(\mathscr{L}) = 1 + \frac{1}{2}c_1(\mathscr{L}),$$
$$\mathrm{ch}(\mathscr{L}) = \exp(c_1(\mathscr{L})) = 1 + c_1(\mathscr{L}),$$
$$\int_C (m + c_1(\mathscr{L})) = \deg(\mathscr{L}), \qquad m \in \mathbb{Z} = \mathrm{CH}^0(C)$$

since $\mathrm{CH}^i(C) = 0$ for $i > 1$. As C is proper, smooth, and geometrically connected, we have $\dim H^0(C, \mathscr{O}_C) = 1$ by Proposition 12.66. Setting $g := \dim H^1(C, \mathscr{O}_C)$ we therefore have $\chi(C, \mathscr{O}_C) = 1 - g$. Hence (23.23.6) yields

$$1 - g = \chi(C, \mathscr{O}_C) = \int_C \left(1 + \frac{1}{2}c_1(T_{C/k})\right) = \frac{1}{2}\deg(T_{C/k}) = -\frac{1}{2}\deg(\Omega_{C/k}^1)$$

and hence

$$\deg(T_{C/k}) = 2 - 2g.$$

Now we can again apply Hirzebruch-Riemann-Roch and get for every line bundle $\mathscr{L}$ on C

$$\chi(C, \mathscr{L}) = \int_C \left(1 + \frac{1}{2}c_1(T_{C/k})\right)(1 + c_1(\mathscr{L}))$$

$$= \int_C \left(1 + \frac{1}{2}c_1(T_{C/k}) + c_1(\mathscr{L})\right)$$

$$= \frac{1}{2}\deg(T_{C/k}) + \deg(\mathscr{L})$$

$$= 1 - g + \deg(\mathscr{L}).$$

This is a variant of the Riemann-Roch theorem in the special case that C is smooth, see Section (26.11) below.

Cohomology and base change

We now come to the question how cohomology is compatible with base change. Let

$$
\begin{array}{ccc}
X' & \xrightarrow{\;g'\;} & X \\
{\scriptstyle f'}\downarrow & & \downarrow{\scriptstyle f} \\
S' & \xrightarrow{\;g\;} & S
\end{array}
$$

be a cartesian diagram of schemes. Then for every complex E in $D_{\mathrm{qcoh}}(X)$ there is a functorial base change morphism (Proposition 21.129)

$$ Lg^* Rf_* E \longrightarrow Rf'_* L(g')^* E. $$

If f is qcqs, and f and g are tor-independent, e.g., if f is flat, then the base change morphism is an isomorphism by Theorem 22.99 and Remark 22.106. Unfortunately, this gets more complicated if one is interested in non-derived base change for higher direct images, i.e., in the question whether for a quasi-coherent module $\mathscr{F}$ and an integer $i \geq 0$ one has an isomorphism

$$ g^* R^i f_* \mathscr{F} \xrightarrow{\;\sim\;} R^i f'_* g'^* \mathscr{F}. $$

This is the case if g is flat by Proposition 22.90. But a case which is not covered by this result but which is geometrically very interesting and where something useful can be said is the case where S' is the spectrum of a residue class field of a point of S.

We approach this problem in two main steps. We gave already several criteria when $Rf_* E$ is pseudo-coherent or even perfect. Hence first we study the "Linear Algebra" problem how pseudo-coherent or perfect complexes on S and their cohomology are compatible with derived and non-derived pullback. It will turn out that passage to fibers, i.e., to residue fields of points of S, is the crucial case to understand. This will be done in Sections (23.24)–(23.26). Then we apply these "Linear Algebra" results to the special case that the perfect complex is the derived direct image $Rf_* E$ of a complex E on X in Sections (23.27) and (23.28). The main result will be Theorem 23.140 and its many applications.

(23.24) Semicontinuity of Betti numbers for pseudo-coherent complexes.

Recall (Remark 21.138) that a complex of $\mathscr{O}_S$-modules E on a locally ringed space S is called perfect if it is locally on S isomorphic in $D(S)$ to a finite complex of finite free $\mathscr{O}_S$-modules. A complex E is called pseudo-coherent if locally on S for all m there exists a distinguished triangle $P \longrightarrow E \longrightarrow C \xrightarrow{+1}$ in $D(S)$ with P perfect and $H^p(C) = 0$ for all $p \geq m$.

If S is a scheme, then a complex E is pseudo-coherent if and only if it is locally on S isomorphic in $D(S)$ to a bounded above complex of finite free $\mathscr{O}_S$-modules (Corollary 22.47). For more basic facts about perfect and pseudo-coherent complexes on schemes see Sections (22.13) and (22.15).

Let S be a locally ringed space, let $s \in S$ be a point, and let

(23.24.1) $$ i_s \colon \operatorname{Spec} \kappa(s) \to S $$

be the canonical morphism of locally ringed spaces with image $\{s\}$. Let E be a complex in $D(S)$. Then we set

(23.24.2) $E \otimes^L_{\mathscr{O}_S} \kappa(s) := E \otimes^L \kappa(s) := Li^*_s E.$

Definition and Remark 23.116. Let S be a locally ringed space.
(1) Fix $i \in \mathbb{Z}$ and let E be a pseudo-coherent complex. Then the function

(23.24.3) $b^i := b^i_E \colon S \to \mathbb{Z}, \qquad s \mapsto \dim_{\kappa(s)} H^i(E \otimes^L_{\mathscr{O}_S} \kappa(s))$

   is called the *Betti function for E*. As $E \otimes^L \kappa(s)$ is a pseudo-coherent complex in $D(\kappa(s))$, $b^i(s)$ is indeed a finite number ≥ 0.
(2) Let E be a perfect complex in $D(S)$. Then the function

(23.24.4) $\chi_E \colon S \to \mathbb{Z}, \quad s \mapsto \chi(E \otimes^L \kappa(s)) = \sum_i (-1)^i \dim_{\kappa(s)} H^i(E \otimes^L \kappa(s))$

   is called the *Euler characteristic (function) of E*. As the perfect complex $E \otimes^L \kappa(s)$ is isomorphic in $D(\kappa(s))$ to a finite complex of finite-dimensional vector spaces, the sum is finite.

Proposition 23.117. *Let S be a scheme, and let $E \in D(S)$ be pseudo-coherent.*
(1) *Let $i \in \mathbb{Z}$. The Betti function $b^i \colon S \to \mathbb{Z}$ is upper semicontinuous and constructible, i.e., for all n the subsets of the form*

(23.24.5) $\{s \in S;\ b^i(s) \geq n\}$

   are closed and constructible.
(2) *For all integers $i \in \mathbb{Z}$ and $n \geq 0$ the set*

(23.24.6) $\{s \in S;\ b^i(s) = n\}$

   is locally closed and constructible.
(3) *Suppose that E is perfect. Then the Euler characteristic $\chi_E \colon S \to \mathbb{Z}$ is locally constant on S.*

Recall from Section (10.10) that a closed subset of a qcqs scheme S is constructible if and only if its open complement is quasi-compact. A locally closed subset of S is constructible if and only if it is the intersection of a closed constructible and an open quasi-compact subset. In a general scheme S, a subset C is constructible if and only if there exists an open covering $(U_i)_i$ by qcqs schemes such that $C \cap U_i$ is constructible in U_i for all i.

Proof. All assertions can be checked locally on S and after passing to an isomorphic complex in the derived category. We may therefore assume that E is a bounded above complex of free $\mathscr{O}$-modules of finite rank,

$$\cdots \longrightarrow E^{i-1} \xrightarrow{d^{i-1}} E^i \xrightarrow{d^i} E^{i+1} \xrightarrow{d^{i+1}} \cdots .$$

Since the E^i are free, $E \otimes^L_{\mathscr{O}_S} \kappa(s) = E \otimes_{\mathscr{O}_S} \kappa(s)$ and

$$b^i(s) = \dim H^i(E \otimes_{\mathcal{O}_S} \kappa(s))$$

(23.24.7)
$$= \dim \operatorname{Ker}(d^i \otimes \kappa(s)) - \dim \operatorname{Im}(d^{i-1} \otimes \kappa(s))$$
$$= \dim \operatorname{Coker}(d^i \otimes \kappa(s)) + \dim \operatorname{Coker}(d^{i-1} \otimes \kappa(s)) - \dim E^{i-1} \otimes \kappa(s)$$

where $\dim E^{i-1} \otimes \kappa(s) = \operatorname{rk} E^{i-1}$ is independent of s. If E is perfect, then we can assume that E is bounded. Taking the alternating sum of these terms, the dimensions of the cokernels cancel, and (3) follows.

In general, Corollary 7.31 and the remark following it shows that the set

(*)
$$\{s \in S; \ \dim_{\kappa(s)} \operatorname{Coker}(d^i) \otimes \kappa(s) \geq r\}$$

is closed for every $r \in \mathbb{Z}$. Since $\operatorname{Coker}(d^i) \otimes \kappa(s) \cong \operatorname{Coker}(d^i \otimes \kappa(s))$, this implies (1) by (23.24.7). Assertion (2) follows from (1).

It remains to show that the sets in (23.24.5) and (23.24.6) are constructible. We fix $i \in \mathbb{Z}$. As being constructible can be checked locally, we may assume that $S = \operatorname{Spec} A$ is affine. It suffices to show that (23.24.5) is constructible. Again by (23.24.7) it suffices to show that (*) is constructible. But this is the locus where the s-th exterior power of d^i vanishes, where $s = \operatorname{rk}(E^{i+1}) - r$. This is a closed constructible subset of S as it is defined by a finitely generated ideal of A. $\qquad\square$

Remark 23.118. Let S be a scheme. From Example 23.60 one deduces the following assertions.
(1) If $E' \to E \to E'' \to$ is a distinguished triangle of perfect complexes on S, then
$$\chi_E = \chi_{E'} + \chi_{E''}.$$
(2) For perfect complexes E and F on S one has $\chi_{E \otimes^L F} = \chi_E \chi_F$.

The Betti function and the Euler characteristic are compatible with derived pull back in the following sense.

Proposition 23.119. *Let $f: S' \to S$ be a morphism of schemes and let E be a pseudocoherent complex in $D(S)$. Then one has for all $i \in \mathbb{Z}$ that*
$$b^i_{Lf^*E} = b^i_E \circ f.$$

If E is perfect, then one also has
$$\chi_{Lf^*E} = \chi_E \circ f.$$

Proof. It suffices to show the first equality. For $s' \in S'$ set $s := f(s')$. Let $g: \operatorname{Spec} \kappa(s') \to \operatorname{Spec} \kappa(s)$ be the morphism corresponding to the field extension $\kappa(s) \to \kappa(s')$ induced by f. As g is flat, we have $Lg^* = g^*$ and $H^i(g^*V) = H^i(V) \otimes_{\kappa(s)} \kappa(s')$ for every complex V in $D(\kappa(s))$. We have $f \circ i_{s'} = i_s \circ g$ with i_s and $i_{s'}$ defined in (23.24.1). Therefore
$$b^i_{Lf^*E}(s') = \dim_{\kappa(s')}(H^i(Lg^*Li_s^*E)) = \dim_{\kappa(s)}(H^i(i_s^*E)) = b^i_E(s). \qquad\square$$

(23.25) Base change for pseudo-coherent complexes.

For A a ring, and $E \in D_{\mathrm{qcoh}}(\operatorname{Spec} A) = D(A)$, we introduce the following notation:
$$T^i := T^i_E \colon (A\text{-Mod}) \to (A\text{-Mod}), \quad M \mapsto H^i(E \otimes^L M)$$

Considering this functor on the category of all A-modules (in contrast to only looking at A-algebras) is a crucial point in the theory of cohomology and base change developed below. More generally, whenever S is a not necessarily affine scheme and $E \in D_{\mathrm{qcoh}}(S)$, then we have by Proposition 22.40 a functor

$$T^i := T_E^i \colon \mathrm{QCoh}(S) \to \mathrm{QCoh}(S), \quad \mathscr{M} \mapsto H^i(E \otimes^L \mathscr{M})$$

which we denote by the same symbol. These functors are compatible with restriction to open subschemes, and if S is affine, then we recover the previous version (up to taking global sections and applying the $\tilde{\ }$-construction, respectively).

Whenever $0 \to \mathscr{M}_1 \to \mathscr{M}_2 \to \mathscr{M}_3 \to 0$ is a short exact sequence of quasi-coherent $\mathscr{O}_S$-modules, we obtain a long exact sequence

$$\cdots \to T^i(\mathscr{M}_1) \to T^i(\mathscr{M}_2) \to T^i(\mathscr{M}_3) \to T^{i+1}(\mathscr{M}_1) \to \cdots .$$

More precisely, the family $(T_i)_i$ is a δ-functor (Definition F.201).

Returning to the affine situation $S = \mathrm{Spec}\, A$, $E \in D_{\mathrm{qcoh}}(S)$, for any A-module M we have a natural morphism

$$\beta^i(M) \colon H^i(E) \otimes M \to H^i(E \otimes^L M).$$

In fact, the existence of β^i follows from the following general lemma, applied to $T = T^i$.

Lemma 23.120. *Let A be a ring, and let $T \colon (A\text{-Mod}) \to (A\text{-Mod})$ be an A-linear functor (Definition F.28).*
(1) *For all A-modules M we have natural homomorphisms, functorial in M,*

$$\beta(M) \colon T(A) \otimes_A M \to T(M).$$

(2) *The morphism of functors β is an isomorphism if and only if T is right exact and commutes with arbitrary direct sums.*

The functor T is right exact and commutes with arbitrary direct sums if and only if it commutes with arbitrary colimits (Proposition F.19).

Proof. The functor T induces an A-module homomorphism

$$M = \mathrm{Hom}_A(A, M) \to \mathrm{Hom}_A(T(A), T(M)),$$

which by the adjunction between Hom and $\otimes$ gives the desired homomorphism.

To prove part (2), first note that the functor $M \mapsto T(A) \otimes_A M$ clearly is right exact and commutes with direct sums. Conversely, assume that T has these properties. Clearly, $\beta(A)$ is an isomorphism, and since T commutes with direct sums, the same holds for every free A-module. For every A-module M one can choose an exact sequence $E' \to E \to M \to 0$ with E and E' free A-modules. Using the right exactness, we obtain that $T(M)$ is an isomorphism. $\qquad\square$

Similarly, we have a version for schemes S which are not necessarily affine. Let E be in $D_{\mathrm{qcoh}}(S)$. With notation as before, we obtain a morphism of functors $(\mathscr{O}_S\text{-Mod}) \to (\mathscr{O}_S\text{-Mod})$

(23.25.1) $\beta^i(\mathscr{M}) \colon H^i(E) \otimes_{\mathscr{O}_S} \mathscr{M} \to H^i(E \otimes^L \mathscr{M}).$

If $g\colon S' \to S$ is a morphism of schemes, we obtain a homomorphism of quasi-coherent $\mathscr{O}_{S'}$-modules

$$(23.25.2) \qquad\qquad g^* H^i(E) \longrightarrow H^i(Lg^* E).$$

If $S = \operatorname{Spec} R$ and $S' = \operatorname{Spec} R'$ and we consider E as a complex of R-modules (Lemma 22.36), then this corresponds to the canonical homomorphism of R-modules

$$H^i(E) \otimes_R R' \longrightarrow H^i(E \otimes_R^L R').$$

If $s \in S$ is a point, we obtain in particular the base change map

$$(23.25.3) \qquad\qquad \beta^i(\kappa(s))\colon H^i(E) \otimes_{\mathscr{O}_S} \kappa(s) \to H^i(E \otimes^L \kappa(s)).$$

We will use the following two linear algebra lemmas which we will prove simultaneously.

Lemma 23.121. *Let R be a ring and let $d\colon M \to N$ be an R-linear map of finitely generated projective R-modules. Then the following assertions are equivalent.*

(i) $\operatorname{Coker}(d)$ *is a finitely generated projective R-module.*

(ii) $\operatorname{Im}(d)$ *is a direct summand of N and for every R-module Q one has $\operatorname{Im}(d) \otimes_R Q = \operatorname{Im}(d \otimes \operatorname{id}_Q)$.*

(iii) $\operatorname{Im}(d)$ *is a finitely generated projective R-module and for every $s \in \operatorname{Spec} R$ one has $\operatorname{Im}(d) \otimes_R \kappa(s) = \operatorname{Im}(d \otimes \operatorname{id}_{\kappa(s)})$.*

(iv) $\operatorname{Ker}(d)$ *is a direct summand of M and for every R-module Q one has $\operatorname{Ker}(d) \otimes_R Q = \operatorname{Ker}(d \otimes \operatorname{id}_Q)$.*

(v) *For every $s \in \operatorname{Spec} R$ the map $\operatorname{Ker}(d) \otimes_R \kappa(s) \to \operatorname{Ker}(d \otimes \operatorname{id}_{\kappa(s)})$ is surjective.*

Lemma 23.122. *Let R be a ring, let $d\colon M \to N$ be an R-linear map of finitely generated projective R-modules. Let $s \in \operatorname{Spec} R$. Then $\operatorname{Ker}(d) \otimes \kappa(s) \to \operatorname{Ker}(d \otimes \kappa(s))$ is surjective if and only if there exists $f \in R$ with $s \in D(f)$ such that $\operatorname{Coker}(d)_f$ is finitely generated projective (and hence $d_f\colon M_f \to N_f$ satisfies the equivalent conditions of Lemma 23.121).*

Proof. We start with the proof of Lemma 23.121. If (i) holds, the exact sequence $0 \to \operatorname{Im}(d) \to N \to \operatorname{Coker}(d) \to 0$ is split and hence stays exact after tensoring with any module Q. As $\operatorname{Coker}(d \otimes \operatorname{id}_Q) = \operatorname{Coker}(d) \otimes Q$, this implies (ii). Clearly, one has "(ii) $\Rightarrow$ (iii)". If $\operatorname{Im}(d)$ is projective, then the exact sequence $0 \to \operatorname{Ker}(d) \to M \to \operatorname{Im}(d) \to 0$ splits and if the formation of $\operatorname{Im}(d)$ commutes with tensoring by Q (resp. by $\kappa(s)$), the same holds for the formation of $\operatorname{Ker}(d)$. Hence (ii) implies (iv), and (iii) implies (v). Clearly, (iv) implies (v). It remains to show that (v) implies (i). For this it suffices to show that the condition in Lemma 23.122 is necessary.

Hence suppose that $\operatorname{Ker}(d) \otimes \kappa(s) \to \operatorname{Ker}(d \otimes \kappa(s))$ is surjective. As $\operatorname{Coker}(d)$ is of finite presentation, it suffices to show that the localization of $\operatorname{Coker}(d)$ in $\mathfrak{p}_s$ is a free module (Proposition 7.27). Therefore we may assume that R is a local ring, and s is the closed point of $\operatorname{Spec} R$. Then $\operatorname{Ker}(d) \to \operatorname{Ker}(d \otimes \kappa(s))$ is surjective and M and N are free of ranks m and n, say. Set $\kappa := \kappa(s)$ and $r := \operatorname{rk}(d \otimes \kappa)$. We will show that $\operatorname{Im}(d)$ is a direct summand of N of rank r. Then $\operatorname{Coker}(d)$ is isomorphic to a direct summand of N and hence projective. Let $\bar{d}\colon M/\operatorname{Ker}(d) \to N$ be the induced injective map. By Proposition 8.10 it suffices to show that $\bar{d} \otimes \operatorname{id}_\kappa$ is injective.

As $\mathrm{Ker}(d) \otimes \kappa \to \mathrm{Ker}(d \otimes \kappa)$ is surjective, we may choose $x_1, \ldots, x_t \in \mathrm{Ker}(d)$ that map to a basis of $\mathrm{Ker}(d \otimes \kappa)$. We also choose $x_{t+1}, \ldots, x_m$ in M whose images in the quotient $(M \otimes \kappa)/\mathrm{Ker}(d \otimes \kappa)$ are a basis. Then $x_1, \ldots, x_m$ generate M by Nakayama's lemma and hence form a basis because M is free of rank $m = \dim_\kappa(M \otimes \kappa)$. As $x_1, \ldots, x_t \in \mathrm{Ker}(d)$, for $i > t$, the images $\bar{x}_i$ of x_i in $M/\mathrm{Ker}(d)$ generate this module. Now $\bar{d} \otimes \kappa$ maps $\bar{x}_i \otimes 1$ to $(d \otimes \kappa)(x_i \otimes 1)$ for $i > t$ and these elements are linearly independent. This shows that $\bar{d} \otimes \kappa$ is injective.

This concludes the proof of the necessity of the condition in Lemma 23.122 and of all equivalences in Lemma 23.121.

The sufficiency in Lemma 23.122 now follows from applying Lemma 23.121 to the map $d_f = d \otimes \mathrm{id}_{R_f} \colon M_f \to N_f$ of R_f-modules.   □

The central local result in this section is the following criterion for splitting up a pseudo-coherent complex as a direct sum of complexes concentrated in degrees below and above some degree, respectively.

Proposition 23.123. *Let R be a ring, let E be a pseudo-coherent complex of R-modules, and fix $i \in \mathbb{Z}$. Let $s \in \mathrm{Spec}\,R$ be a point such that the base change map*

$$(23.25.4) \qquad\qquad \beta^i(\kappa(s)) \colon H^i(E) \otimes \kappa(s) \to H^i(E \otimes^L \kappa(s))$$

is surjective.

Then there exists $f \in R$ such that $s \in U := D(f)$, and such that we have, setting $E_{|U} := E \otimes_R R_f$:
(1) We have a decomposition in $D(R_f)$

$$(23.25.5) \qquad\qquad E_{|U} \cong \tau^{\le i}(E_{|U}) \oplus \tau^{\ge i+1}(E_{|U}).$$

(2) The truncated complex $\tau^{\ge i+1}(E_{|U})$ is a perfect complex of R_f-modules. If E is of tor-amplitude in $(-\infty, b]$ where $b \in \mathbb{Z}$, then $\tau^{\ge i+1}(E_{|U})$ is of tor-amplitude in $[i+1, b]$. The complex $\tau^{\le i}(E_{|U})$ is pseudo-coherent. It is perfect if $E_{|U}$ is perfect.
(3) The R_f-linear map $\beta^i(M) \colon H^i(E_{|U}) \otimes M \to H^i(E_{|U} \otimes^L M)$ is an isomorphism for every R_f-module M. Therefore the functor

$$T^i \colon (R_f\text{-Mod}) \to (R_f\text{-Mod}), \qquad M \mapsto H^i(E_{|U} \otimes^L_{R_f} M)$$

commutes with arbitrary colimits and in particular is right exact.

Conversely, if the functor T^i commutes with base change for $M = \kappa(s)$, i.e., if the natural map yields an isomorphism $H^i(E_{|U}) \otimes M \cong H^i(E_{|U} \otimes^L M)$ for $M = \kappa(s)$, this says precisely that (23.25.4) is an isomorphism.

Proof. Replacing E by an isomorphic complex in $D(R)$, we may assume that E is a bounded above complex of finite free modules. Then E is K-flat and hence $E \otimes^L_R M = E \otimes_R M$ for every R-module M. Moreover, if b is an integer such that E is of tor-amplitude in $(-\infty, b]$ (e.g., an integer b such that $H^p(E) = 0$ for all $p > b$), we may assume that $E^i = 0$ for all $i > b$ (Remark 22.46).

By hypothesis, the map

$$\mathrm{Ker}(E^i \xrightarrow{d^i} E^{i+1}) \otimes \kappa(s) \to \mathrm{Ker}(d^i \otimes \kappa(s))/\mathrm{Im}(d^{i-1} \otimes \kappa(s))$$

is surjective. Hence we may choose $y_1, \ldots, y_t \in \mathrm{Ker}(d^i)$ that map to a basis of $\mathrm{Ker}(d^i \otimes \kappa)/\mathrm{Im}(d^{i-1} \otimes \kappa)$. Choose also $x'_1, \ldots, x'_r \in E^{i-1}$ such that the images of $d^{i-1}(x'_k)$ in $E^i \otimes \kappa$ form a basis of $\mathrm{Im}(d^{i-1} \otimes \kappa)$. Then the $d^{i-1}(x'_k)$ and the y_l are contained in $\mathrm{Ker}(d^i)$ and generate $\mathrm{Ker}(d^i \otimes \kappa(s))$ and hence $\mathrm{Ker}(d^i) \otimes \kappa(s) \to \mathrm{Ker}(d^i \otimes \kappa(s))$ is surjective.

Therefore we can use Lemma 23.122. It shows that, after replacing $\mathrm{Spec}\,R$ by an open affine neighborhood of s, if necessary, $\mathrm{Im}(d^i)$ and $\mathrm{Coker}(d^i)$ are free. In particular $\tau^{\geq i+1} E$ is perfect of tor-amplitude in $[i+1, b]$. The projection $E^{i+1} \to \mathrm{Coker}(d^i)$ admits a section, yielding a homomorphism of complexes $\tau^{\geq i+1}(E) \to E$. Hence there is a quasi-isomorphism of complexes $\tau^{\leq i}(E) \oplus \tau^{\geq i+1}(E) \to E$ and therefore E has a decomposition as in (1). Moreover, as direct summands of pseudo-coherent (resp. perfect) complexes are again pseudo-coherent (resp. perfect) by Proposition 21.156 (resp. by Proposition 21.176), $\tau^{\leq i} E$ is again pseudo-coherent (resp. perfect if E is perfect).

It remains to show (3). We have $H^i(E) = \mathrm{Coker}(d^{i-1} \colon E^{i-1} \to \mathrm{Ker}(d^i))$. As we have already seen that $\mathrm{Im}(d^i) = E^i/\mathrm{Ker}(d^i)$ is free, $\mathrm{Ker}(d^i)$ is a direct summand of E^i and hence its formation commutes with $\otimes M$. As this is also true for formations of cokernels, we see that $H^i(E \otimes M) = H^i(E) \otimes M$. The last assertion follows from Lemma 23.120. $\square$

Corollary 23.124. *Let R be a ring, and let E be a pseudo-coherent complex in $D(R)$. Let $i \in \mathbb{Z}$ be such that the base change map*

$$\beta^i(\kappa(s)) \colon H^i(E) \otimes \kappa(s) \to H^i(E \otimes^L \kappa(s))$$

is surjective for every closed point $s \in \mathrm{Spec}\,R$. Then the base change homomorphism $\beta^i(M) \colon H^i(E) \otimes_R M \to H^i(E \otimes_R^L M)$ is an isomorphism for every R-module M.

Corollary 23.125. *Let S be a scheme, let $E \in D(S)$ be pseudo-coherent, and fix $i \in \mathbb{Z}$. Let $s \in S$ such that $H^i(E \otimes^L \kappa(s)) = 0$. Then there exists an affine open neighborhood U of s such that $H^i(E)_{|U} = 0$ and such that*

$$(23.25.6) \qquad E_{|U} \cong \tau^{\leq i-1} E_{|U} \oplus \tau^{\geq i+1} E_{|U}$$

with $\tau^{\geq i+1} E$ perfect of tor-amplitude in $[i+1, \infty)$.

See also Lemma 27.227 for a more general statement.

Proof. By Proposition 23.123 we can write $E = \tau^{\leq i} E \oplus \tau^{\geq i+1} E$ with $\tau^{\geq i+1} E$ perfect of tor-amplitude in $[i+1, \infty)$, $\tau^{\leq i} E$ pseudo-coherent, and

$$(*) \qquad H^i(E \otimes^L \kappa(s)) = H^i(E) \otimes \kappa(s) = 0$$

after replacing S be an open affine neighborhood of s. Therefore $\tau^{\leq i} E$ can be represented by a complex L of finite free modules with $L^p = 0$ for $p > i$ (Remark 22.46). But then $H^i(E) = H^i(L) = \mathrm{Coker}(L^{i-1} \to L^i)$ is of finite presentation and hence zero in an open affine neighborhood U of s by $(*)$. Hence $\tau^{\leq i} E_{|U} \to \tau^{\leq i-1} E_{|U}$ is an isomorphism in $D(U)$ and (23.25.6) follows from Proposition 23.123 (1). $\square$

Proposition 23.126. *Let S be a scheme, let E be a pseudo-coherent complex on S and let $a \leq b$ be integers.*
(1) *Suppose that E is locally bounded below (e.g., if E is perfect) and let $s \in S$. Then E is perfect of tor-amplitude in $[a, b]$ in an open neighborhood of s if and only if $H^i(E \otimes^L \kappa(s)) = 0$ for all $i \notin [a, b]$.*

(2) *The complex E is perfect of tor-amplitude contained in $[a, b]$ if and only if $H^i(E \otimes^L \kappa(s)) = 0$ for all $i \notin [a, b]$ and for all $s \in S$.*

Proof. For both assertions, the condition are clearly necessary. Let us show that in (1) the condition is sufficient. Working locally on S, we may assume that $E \in D^b(S)$ since E is pseudo-coherent and locally bounded below. Since $H^i(E \otimes^L \kappa(s)) = 0$ for all $i \notin [a, b]$, by Corollary 23.125, we may replace S by an open neighborhood of s such that $H^i(E) = 0$ for all $i \notin [a, b]$ and such that $E = \tau^{\geq a} E = \tau^{\leq b} \tau^{\geq a} E$ is perfect of tor-amplitude in $[a, b]$.

Now we show the sufficiency in (2). Corollary 23.125 implies that $H^i(E)_s = 0$ for all $i \notin [a, b]$ and all $s \in S$. Hence $E \in D^{[a,b]}(S)$. Now we can apply (1) as we can check the condition on the tor-amplitude locally on S by Lemma 21.171. $\square$

Proposition 23.127. *Let S be a scheme and let $E \in D(S)$ be pseudo-coherent. Let $s_0 \in S$ be a point. Assume that the base change homomorphism $\beta^i(\kappa(s_0))$ is surjective. Then there exists an open affine neighborhood U of s_0 such that*

$$\beta^i(\mathscr{G}) \colon H^i(E_{|U}) \otimes_{\mathscr{O}_U} \mathscr{G} \to H^i(E_{|U} \otimes^L_{\mathscr{O}_U} \mathscr{G})$$

is an isomorphism for every quasi-coherent $\mathscr{O}_U$-module $\mathscr{G}$. Moreover, the following assertions are equivalent:
(i) *The homomorphisms $\beta^{i-1}(\kappa(s_0))$ is surjective.*
(ii) *There exists an open affine neighborhood V of s_0 such that the $\mathscr{O}_V$-module $H^i(E_{|V})$ is finite locally free.*
In this case, the Betti function $s \mapsto \dim_{\kappa(s)} H^i(E \otimes^L \kappa(s))$ is constant on an open neighborhood of s_0.

Proof. We may assume without loss of generality that $S = \operatorname{Spec} R$ is affine. We denote the pseudo-coherent complex of R-modules corresponding to E again by E. Then the first assertion follows from Proposition 23.123. Replacing S by U, we may assume

(*) $H^i(E) \otimes_R M = H^i(E \otimes^L_R M)$

for every R-module M. Hence $T^i = H^i(E \otimes^L_R -)$ is right exact and it is left exact if and only if T^{i-1} is right exact.

If $\beta^{i-1}(\kappa(s_0))$ is surjective, then passing to an open affine neighborhood of s_0, we may assume that T^{i-1} is in fact right exact, so T^i is exact. Therefore $H^i(E)$ is flat. By Proposition 23.123, $\tau^{\leq i} E$ is a direct summand of E and hence is again pseudo-coherent. The map $\beta^{i-1}(\kappa(s_0))$ is still surjective for $\tau^{\leq i} E$. Hence we can apply Proposition 23.123 again for $i - 1$ and $\tau^{\leq i} E$. Then we see that $\tau^{\geq i} \tau^{\leq i} E = H^i(E)[-i]$ is a perfect complex (concentrated in degree i). In particular, $H^i(E)$ is pseudo-coherent and therefore of finite presentation by Proposition 21.153. Hence $H^i(E)$ is finitely generated and projective (Corollary 7.42).

Conversely, if $H^i(E)$ is flat, then $T^i = H^i(E) \otimes -$ is exact, so T^{i-1} is right exact. Since it also commutes with direct sums (Remark 21.120), it follows from Lemma 23.120 that β^{i-1} is an isomorphism of functors.

Finally the i-th Betti function is locally constant since $b^i(s) = \dim_{\kappa(s)} H^i(E \otimes^L \kappa(s)) = \dim_{\kappa(s)} H^i(E) \otimes \kappa(s)$ for all $s \in S$ by (*). $\square$

Proposition 23.127 has the following converse if S is reduced.

Proposition 23.128. *Let S be a reduced scheme, let E be a pseudo-coherent complex in $D(S)$, and fix $i \in \mathbb{Z}$. Suppose that the Betti function*

$$b^i \colon s \mapsto \dim_{\kappa(s)} H^i(E \otimes^L \kappa(s))$$

is locally constant on S. Then the base change homomorphisms

$$\beta^j(\mathscr{G}) \colon H^j(E) \otimes_{\mathscr{O}_S} \mathscr{G} \longrightarrow H^j(E \otimes^L_{\mathscr{O}_S} \mathscr{G})$$

are isomorphisms for $j = i, i-1$ and for all quasi-coherent $\mathscr{O}_S$-modules $\mathscr{G}$, and $H^i(E)$ is finite locally free.

Proof. We may assume that $S = \operatorname{Spec} R$ is affine and that E is a bounded above complex of finite free $\mathscr{O}_S$-modules. Looking back to the proof of Proposition 23.117, in particular at (23.24.7), we see that the sum of the upper semicontinuous maps

$$s \mapsto \dim \operatorname{Coker}(d^i) \otimes \kappa(s) \qquad \text{and} \qquad s \mapsto \dim \operatorname{Coker}(d^{i-1}) \otimes \kappa(s).$$

is locally constant since b^i is locally constant. Therefore these two maps are also lower semicontinuous and hence locally constant on S. Since S is reduced by assumption, we obtain that $\operatorname{Coker}(d^i)$ and $\operatorname{Coker}(d^{i-1})$ are finite locally free (Corollary 11.19). As $\operatorname{Coker}(d^i)$ is finite locally free, for every quasi-coherent $\mathscr{O}_S$-module $\mathscr{G}$ the short exact sequence

$$0 \to \operatorname{Im}(d^i) \otimes \mathscr{G} \to E^{i+1} \otimes \mathscr{G} \to \operatorname{Coker}(d^i) \otimes \mathscr{G} \to 0$$

splits. Hence $\operatorname{Im}(d^i)$ is finite locally free. Since $\operatorname{Coker}(d^i \otimes \mathscr{G}) \cong \operatorname{Coker}(d^i) \otimes \mathscr{G}$, the sequence shows that $\operatorname{Im}(d^i) \otimes \mathscr{G} = \operatorname{Im}(d^i \otimes \mathscr{G})$. As $\operatorname{Im}(d^i)$ is finite locally free, the short exact sequence

$$0 \to H^i(E) \otimes \mathscr{G} \to \operatorname{Coker}(d^{i-1}) \otimes \mathscr{G} \to \operatorname{Im}(d^i) \otimes \mathscr{G} \to 0$$

is split. As the formation of $\operatorname{Im}(d^i)$ and $\operatorname{Coker}(d^{i-1})$ commute with $- \otimes \mathscr{G}$ this also holds for the formation of $H^i(E)$ which shows that $\beta^i(\mathscr{G})$ is an isomorphism for all $\mathscr{G}$.

Moreover, the second exact sequence of $\mathscr{G} = \mathscr{O}_S$ shows that $H^i(E)$ is finite locally free because $\operatorname{Coker}(d^{i-1})$ is finite locally free. Hence Proposition 23.127 implies that $\beta^{i-1}(\mathscr{G})$ is an isomorphism for all quasi-coherent modules $\mathscr{G}$. $\qquad\square$

Remark 23.129. Let S be a scheme, let E be a complex in $D(S)$, and let $i \in \mathbb{Z}$. Suppose that for every quasi-coherent $\mathscr{O}_S$-module $\mathscr{G}$ the base change homomorphism

$$\beta^i(\mathscr{G}) \colon H^i(E) \otimes_{\mathscr{O}_S} \mathscr{G} \longrightarrow H^i(E \otimes^L_{\mathscr{O}_S} \mathscr{G})$$

is an isomorphism. Then for every morphism of schemes $g \colon S' \to S$ one has functorial isomorphisms of $\mathscr{O}_{S'}$-modules

(23.25.7) $$g^* H^i(E) \cong H^i(Lg^*E).$$

Indeed, to check that the functorial homomorphism $g^* H^i(E) \to H^i(Lg^*E)$ is an isomorphism we can assume that $S = \operatorname{Spec} R$ and $S' = \operatorname{Spec} R'$ are affine. Then we can apply the hypothesis to the quasi-coherent $\mathscr{O}_S$-algebra $\mathscr{G} = g_* \mathscr{O}_{S'}$.

(23.26) Subschemes classifying properties of perfect complexes.

If E is a perfect complex of tor-amplitude in an interval $[a, b]$, then the condition that $H^a(E)$ or $H^b(E)$ are locally free of a fixed rank are represented by locally closed subschemes, more generally:

Proposition 23.130. *Let S be a scheme, let E be a perfect complex in $D(S)$ of tor-amplitude in an interval $[a, b]$, and let $I \subseteq [a, b]$ be an interval containing a or b. Fix a map $r \colon I \to \mathbb{Z}_{\geq 0}$, $i \mapsto r_i$. Then there exists a unique locally closed subscheme $j \colon Z = Z_r \to S$ such that a morphism $f \colon T \to S$ factors through Z if and only if for all morphisms $g \colon T' \to T$ and for all $i \in I$ the $\mathscr{O}_{T'}$-module $H^i(L(f \circ g)^* E)$ is locally free of rank r_i. Moreover,*

(1) *the immersion $j \colon Z \to X$ is of finite presentation,*
(2) *as a set one has*

$$(23.26.1) \qquad\qquad Z = \{\, s \in S \,;\, b^i(s) = r_i \text{ for all } i \in I \,\},$$

(3) *if $f \colon T \to S$ factors as $T \xrightarrow{\bar{f}} Z \xrightarrow{j} X$, then $H^i(Lf^* E \otimes_{\mathscr{O}_T} \mathscr{G}) = \bar{f}^* H^i(Lj^* E) \otimes_{\mathscr{O}_T} \mathscr{G}$ for all $i \in I$ and for all quasi-coherent $\mathscr{O}_T$-modules $\mathscr{G}$.*

Proof. Since Z is characterized by a universal property, the uniqueness assertion is clear. Hence, we may work locally on S and can assume that $S = \operatorname{Spec} R$ is affine. By Proposition 22.53, we may assume that E is given by a complex of finitely generated projective R-modules concentrated in degrees $[a, b]$, which we again denote by E. Again working locally on S, we may assume that E^i is a free R-module of rank n_i, say, for all $i \in \mathbb{Z}$. Note that then $Lf^* E = f^* E$ for every morphism $f \colon T \to S$.

By hypothesis, the interval I is of the form $[a, b']$ for some $b' \in [a, b]$ or of the form $[a', b]$ for some $a' \in [a, b]$. We first consider the case that $I = [a, b']$. The condition that

$$H^a(E) = \operatorname{Ker}(E^a \xrightarrow{d^a} E^{a+1})$$

is locally free and that its formation commutes with base change is equivalent to the condition that $\operatorname{Coker}(d^a)$ is locally free by Lemma 23.121. In this case, $H^a(E)$ has rank r_a if and only if $\operatorname{Coker}(d^a)$ has rank $t_a := n_{a+1} - n_a + r_a$. Furthermore, in this case $\operatorname{Im}(d^a)$ is a direct summand of E^{a+1}, it is locally free of rank $n_{a+1} - t_a = n_a - r_a$ and its formation commutes with base change (again by Lemma 23.121). Hence we now can apply Lemma 23.121 to $E^{a+1}/\operatorname{Im}(d^a) \to E^{a+2}$ and see that $H^{a+1}(E)$ is locally free of rank r_{a+1} and its formation commutes with base change, and hence it is a direct summand of $E^{a+1}/\operatorname{Im}(d^a)$, if and only if $\operatorname{Coker}(d^{a+1})$ is locally free of rank $t_{a+1} := (n_{a+2} - n_{a+1} + n_a) + (r_{a+1} - r_a)$. Proceeding by induction one sees that for $i \geq a$ the R-module $H^i(E)$ is locally free of rank r_i and that its formation commutes with base change if and only if $\operatorname{Coker}(d^i)$ is locally free of rank

$$t_i := \sum_{j=-1}^{i-a} (-1)^{j+1} n_{i-j} + \sum_{j=0}^{i-a} (-1)^j r_{i-j}.$$

In fact Lemma 23.121 shows that in this case, the formation of $\operatorname{Ker}(d^i)$ and of $\operatorname{Im}(d^{i-1})$ also commutes with tensoring by any R-module Q. Hence for every R-module Q we have

$$(*) \qquad\qquad H^i(E \otimes_R Q) = H^i(E) \otimes_R Q.$$

By Theorem 11.18 there is a subscheme $Z_i = F_{=t_i}(\mathrm{Coker}(d^i))$ such that a morphism $f\colon T \to S$ factors though Z_i if and only of $f^* \mathrm{Coker}(d^i)$ is locally free of rank t_i. As formation of $\mathrm{Coker}(d^i)$ always commutes with base change, we can take $Z = \bigcap_{i \in I} Z_i$, the scheme-theoretic intersection.

To construct Z in the case that $I = [a', b]$, one proceeds similarly by starting with

$$H^b(E) = \mathrm{Coker}(E^{b-1} \xrightarrow{d^{b-1}} E^b)$$

and shows inductively for $i = b, b-1, \ldots, a'$ that $H^i(E)$ is locally free of rank r_i and that its formation commutes with base change if and only if $\mathrm{Coker}(d^{i-1})$ is locally free of rank

$$s_i := \sum_{j=1}^{b-i} (-1)^{j-1} n_{i+j} + \sum_{j=0}^{b-i} (-1)^j r_{i+j}.$$

Hence one can set $Z := \bigcap_{a' \le i \le b} F_{=s_i}(\mathrm{Coker}(d^{i-1}))$.

As $\mathrm{Coker}(d^i)$ is of finite presentation, all immersions $F_{=r}(\mathrm{Coker}(d^i)) \to S$ are of finite presentation (see the remark after Theorem 11.18), hence $Z \to S$ is of finite presentation.

A point $s \in S$ is contained in Z if and only if $H^i(E \otimes_{\mathscr{O}_S} \kappa(s))$ is of rank r_i and its formation commutes with base change $T \to \mathrm{Spec}\,\kappa(s)$. But this base change is flat and hence the second conditions holds automatically. This shows (23.26.1).

The last assertion holds by construction of Z using (*). $\qquad\square$

Proposition 23.131. *Let S be a scheme and let E be a perfect complex in $D(S)$. Let $a \le b$ be integers. Then there exists a unique open constructible subscheme U of S such that a morphism $f\colon T \to S$ factors through U if and only if Lf^*E is of tor-amplitude in $[a, b]$.*

Proof. The uniqueness assertion is clear. Hence we can assume that $S = \mathrm{Spec}\,R$ and that E is of some tor-amplitude contained in a finite interval I (Theorem 21.174). Then Lf^*E also has tor-amplitude in I for every morphism $f\colon T \to S$. By using Proposition 23.126 and using that the formation of the Betti function is compatible with inverse image (Proposition 23.119) we see that the subscheme

$$U := \{\, s \in S \;;\; \mathrm{rk}\,H^i(E \otimes^L \kappa(s)) \le 0 \text{ for all } i \in I \setminus [a, b] \},$$

which is open and constructible by Proposition 23.117, has the desired properties. $\qquad\square$

Corollary 23.132. *Let S be a scheme, and let $E \in D(S)$ be perfect. Let $a, r \in \mathbb{Z}$ with $r \ge 0$. There exists a unique open subscheme $U \subseteq S$ such that a morphism $f\colon T \to S$ factors through U if and only if Lf^*E is isomorphic to a locally free $\mathscr{O}_T$-module of rank r placed in degree a.*

Proof. We apply Proposition 23.131 to the interval $[a, a]$ using that a perfect complex is of tor-amplitude in $[a, a]$ if and only if it is isomorphic to a finite locally free module placed in degree a (Proposition 22.53). $\qquad\square$

(23.27) Criteria when direct images are perfect and commute with base change.

Let $f\colon X \to S$ be a morphism of schemes, let $E \in D(X)$ and let $\mathscr{F}$ be a complex of $\mathscr{O}_X$-modules. Then we say that the formation of $Rf_*(E \otimes^L_{\mathscr{O}_X} \mathscr{F})$ *commutes with base change* if the following holds. For every morphism $S' \to S$ of schemes let $X' = X \times_S S'$, and consider the cartesian diagram

$$X' \xrightarrow{g'} X$$
$$\downarrow{f'} \qquad \downarrow{f}$$
$$S' \xrightarrow{g} S.$$

Then the following natural morphism is an isomorphism

$$\theta\colon Lg^*Rf_*(E \otimes^L_{\mathscr{O}_X} \mathscr{F}) \xrightarrow{\sim} Rf'_*(L(g')^*E \otimes^L_{\mathscr{O}_{X'}} (g')^*\mathscr{F})$$

in $D(S')$.

Note that we do not use $L(g')^*\mathscr{F}$ in the term on the right hand side of the isomorphism θ. See Exercise 23.42 for an example where the variant of Theorem 23.133 below with $L(g')^*\mathscr{F}$ in place of $(g')^*\mathscr{F}$ fails.

We have seen in Corollary 22.103 that $Rf_*(E \otimes^L_{\mathscr{O}_X} \mathscr{F})$ commutes with base change if $E \in D_{\mathrm{qcoh}}(X)$ and $\mathscr{F}$ is a bounded above complex of quasi-coherent $\mathscr{O}_X$-modules that are flat over S. We will now study criteria, when $Rf_*(E \otimes^L_{\mathscr{O}_X} \mathscr{F})$ is in addition perfect or at least pseudo-coherent.

Theorem 23.133. *Let $f\colon X \to S$ be a proper morphism of finite presentation (e.g., S locally noetherian and f proper). Let $E \in D(X)$ be perfect (resp. pseudo-coherent), and let $\mathscr{F}$ be a bounded complex of $\mathscr{O}_X$-modules of finite presentation that are flat over S.*

Then $Rf_*(E\otimes^L_{\mathscr{O}_X} \mathscr{F}) \in D(\mathscr{O}_S\text{-Mod})$ is perfect (resp. pseudo-coherent) and the formation of $Rf_*(E \otimes^L_{\mathscr{O}_X} \mathscr{F})$ commutes with base change.

We will prove the theorem only if S is locally noetherian and sketch the proof for general S if E is perfect. Then one can show the result for pseudo-coherent complexes by approximating pseudo-coherent complexes by perfect complexes. We refer to [Sta] OCSC for this.

Proof. The formation of $Rf_*(E \otimes^L_{\mathscr{O}_X} \mathscr{F})$ commutes with base change by Corollary 22.103. To show that $F := Rf_*(E \otimes^L_{\mathscr{O}_X} \mathscr{F})$ is perfect (resp. pseudo-coherent), we may assume that $S = \operatorname{Spec} R$ is affine.

(I). Let us first assume that R is noetherian and that E is pseudo-coherent. Then X is noetherian and $\mathscr{F}$ is pseudo-coherent (Proposition 22.61). Hence $E \otimes^L_{\mathscr{O}_X} \mathscr{F}$ is pseudo-coherent (Proposition 22.63). Therefore F is pseudo-coherent by Proposition 23.23.

(II). Now suppose that E is perfect and R arbitrary. We will show that F is of finite tor-amplitude. As X is quasi-compact, there exist integers $a \le b$ such that locally on X, the complex E can be represented by a complex of finite free modules $\mathscr{E}$ which is concentrated in degrees $[a, b]$ (Proposition 22.53). Suppose that $\mathscr{F}$ is concentrated in degrees $[c, d]$ for integers $c \le d$. Then $E' := \mathscr{E} \otimes^L_{\mathscr{O}_X} \mathscr{F} = \mathscr{E} \otimes_{\mathscr{O}_X} \mathscr{F}$ is locally on X represented by a complex concentrated in degrees $[a + c, b + d]$ such that each term is a direct sum of the $\mathscr{F}^i$. In particular it is represented by a complex whose terms are flat over S. As tor-amplitude can be checked locally (Lemma 21.171), this implies that E' has tor-amplitude in $[a + c, b + d]$ as an object of $D(X, f^{-1}\mathscr{O}_S)$ (Proposition 21.169).

Let us show that $F = Rf_*(E')$ is of tor-amplitude in $[a + c, b + d + N]$, where $N \ge 1$ is some number such that X can be covered $N + 1$ open affine subschemes. By Lemma 22.51, it is enough to show that $H^i(F \otimes^L_{\mathscr{O}_S} \mathscr{K}) = 0$ for all $i \notin [a + c, b + d + N]$ and all quasi-coherent $\mathscr{O}_S$-modules $\mathscr{K}$. For the upper bound $b + d + N$ it suffices to show that $H^i(F) = 0$ for all $i > b + d + N$ (Remark 21.165 (2)). This follows from Corollary 22.10.

To obtain the lower bound $a + c$, we apply the projection formula (Proposition 22.84) and obtain

$$F \otimes^L_{\mathscr{O}_S} \mathscr{K} = Rf_*(E') \otimes^L_{\mathscr{O}_S} \mathscr{K} \cong Rf_*(E' \otimes^L_{\mathscr{O}_X} Lf^*\mathscr{K}) \cong Rf_*(E' \otimes^L_{f^{-1}\mathscr{O}_S} f^{-1}\mathscr{K}).$$

As E' is represented by a complex concentrated in degrees $\geq a + c$ that is K-flat as a complex of $f^{-1}\mathscr{O}_S$-modules, the cohomology of $E' \otimes^L_{f^{-1}\mathscr{O}_S} f^{-1}\mathscr{K}$ is concentrated in degrees $\geq a + c$ for all quasi-coherent $\mathscr{O}_S$-modules $\mathscr{K}$, and Rf_* preserves this property.

(III). Combining Steps (I) and (II) proves that F is perfect if E is perfect and R noetherian by Theorem 21.174.

(IV). Now we show that F is perfect if E is perfect and if R is arbitrary. For this write R as a filtered colimit of noetherian rings R_λ. By the techniques explained in Chapter 10 (for details we refer to [Sta] 0A1H, see also Remark 22.55), we find for some λ a proper morphism $f_\lambda \colon X_\lambda \to S_\lambda := \operatorname{Spec} R_\lambda$ such that $X \cong S \times_{S_\lambda} X_\lambda$, a perfect complex E_λ on X_λ whose derived pullback to X is E, and a bounded complex $\mathscr{F}_\lambda$ of coherent $\mathscr{O}_{X_\lambda}$-modules that are flat over S_λ whose pullback to X is $\mathscr{F}$. If E is of tor-amplitude contained in $[a, b]$ (resp. $\mathscr{F}$ is concentrated in degrees $[c, d]$), we may assume that E_λ (resp. $\mathscr{F}_\lambda$) has the same property. If N is an integer such that X can be covered by $N + 1$ open affine subschemes, then we may assume that X_λ can be covered by $N + 1$ open affine subschemes.

We have already shown in Step (III) that $Rf_{\lambda,*}(E_\lambda \otimes^L \mathscr{F}_\lambda)$ is perfect of tor-amplitude in $[a + c, b + d + N]$. Moreover, its formation commutes with base change and hence F is its derived pullback under $S \to S_\lambda$. This show that F is perfect of tor-amplitude in $[a + c, b + d + N]$. $\qquad\square$

With only a little more work, the assumption that f be proper can be replaced by requiring that the support of the terms of the complex $\mathscr{F}$ are proper over S (Definition 23.21), see [Sta] 0A1H and 0CSC.

Remark 23.134. Suppose that E is perfect of tor-amplitude in an interval $[a, b]$ and that $\mathscr{F}$ is concentrated in degrees $[c, d]$ in the situation of Theorem 23.133. Then the proof shows that $Rf_*(E \otimes^L_{\mathscr{O}_X} \mathscr{F})$ is perfect of tor-amplitude $\geq a + c$. Moreover, if S is affine and X can be covered by $N + 1$ open affine subschemes, then the tor-amplitude of $Rf_*(E \otimes^L_{\mathscr{O}_X} \mathscr{F})$ is contained in $[a + c, b + d + N]$. For instance, if $\mathscr{F}$ is a module of finite presentation that is flat over S, then the tor-amplitude of $Rf_*\mathscr{F}$ is contained in $[0, N]$.

Let us record the following special cases of the previous result:

Corollary 23.135. *Let $f \colon X \to S$ be a proper morphism of finite presentation and let $\mathscr{F}$ be a bounded complex of $\mathscr{O}_X$-modules of finite presentation that are flat over S. Then $Rf_*\mathscr{F}$ is perfect and for all cartesian diagrams of schemes*

$$
\begin{array}{ccc}
X' & \xrightarrow{\;g'\;} & X \\
{\scriptstyle f'}\downarrow & & \downarrow{\scriptstyle f} \\
S' & \xrightarrow{\;g\;} & S.
\end{array}
$$

one has

$$Lg^* Rf_*\mathscr{F} = Rf'_*((g')^*\mathscr{F}).$$

Corollary 23.136. *Let $f\colon X \to S$ be a flat proper morphism of finite presentation and let E be a perfect (resp. pseudo-coherent) complex on X. Then Rf_*E is a perfect (resp. pseudo-coherent) complex in $D(S)$ and for all cartesian diagrams of schemes*

$$
\begin{array}{ccc}
X' & \xrightarrow{g'} & X \\
{\scriptstyle f'}\downarrow & & \downarrow{\scriptstyle f} \\
S' & \xrightarrow{g} & S.
\end{array}
$$

one has

$$Lg^*Rf_*(E) = Rf'_*(L(g')^*E).$$

Corollary 23.137. *Let $S = \operatorname{Spec} A$ be an affine noetherian scheme, and let $f\colon X \to S$ be a proper morphism. Let $\mathscr{F}$ be a coherent $\mathscr{O}_X$-module which is flat over S. Then there exists a finite complex $K^\bullet$ of finitely generated projective A-modules and isomorphisms, functorial in the A-algebra B,*

$$H^i(X \times_S \operatorname{Spec} B, \mathscr{F} \otimes_A B) \xrightarrow{\sim} H^i(K^\bullet \otimes_A B), \qquad \text{for all } i \geq 0.$$

For a direct proof of this version of the result see [Mum1] § 5. By Remark 23.134, we can even find $K^\bullet$ such that $H^i(K^\bullet) = 0$ for all $i \notin [0, N]$, where $N \geq 0$ is the minimal number such that X can be covered by $N + 1$ open affine subschemes.

(23.28) Semicontinuity theorems and base change of higher direct images.

We fix a cartesian diagram of schemes

(23.28.1)
$$
\begin{array}{ccc}
X' & \xrightarrow{g'} & X \\
{\scriptstyle f'}\downarrow & & \downarrow{\scriptstyle f} \\
S' & \xrightarrow{g} & S.
\end{array}
$$

Let f be qcqs and let $\mathscr{F}$ be a complex of quasi-coherent $\mathscr{O}_X$-modules. Then we have a functorial homomorphism of $\mathscr{O}_{S'}$-modules

(23.28.2)
$$g^*R^if_*\mathscr{F} \longrightarrow R^if'_*g'^*\mathscr{F}.$$

It is the composition of the following two morphisms.
(1) The base change morphism

(23.28.3)
$$\beta\colon g^*R^if_*\mathscr{F} \to H^i(Lg^*Rf_*\mathscr{F})$$

given by (23.25.2) applied to $Rf_*\mathscr{F}$ which is in $D_{\mathrm{qcoh}}(S)$ since f is qcqs (Theorem 22.31).
(2) The homomorphism

(23.28.4)
$$\gamma\colon H^i(Lg^*Rf_*\mathscr{F}) \longrightarrow H^i(Rf'_*g'^*\mathscr{F}) = R^if'_*g'^*\mathscr{F},$$

obtained by applying $H^i(-)$ to the composition

$$Lg^*Rf_*\mathscr{F} \longrightarrow Rf'_*Lg'^*\mathscr{F} \longrightarrow Rf'_*g'^*\mathscr{F},$$

where the first map is the derived base change morphism (Definition 21.129) and the second is induced by the functorial map $Lg'^*\mathscr{F} \to g'^*\mathscr{F}$.

If g is flat, the construction of (23.28.2) given here specializes to the construction of (22.21.3) and we already showed in Proposition 22.90 and Remark 22.108 that (23.28.2) is an isomorphism in this case.

Definition 23.138. *Let $i \in \mathbb{Z}$. Then we say that the* formation of $R^i f_* \mathscr{F}$ commutes with base change *if (23.28.2) is an isomorphism for every morphism of schemes $g: S' \to S$.*

If $U \subseteq S$ is an open subscheme, then we say that the formation of $R^i f_* \mathscr{F}_{|U}$ commutes with base change *if the formation of $R^i f_{U,*} \mathscr{F}_{|f^{-1}(U)}$ commutes with every base change $U' \to U$. Here $f_U: f^{-1}(U) \to U$ is the restriction of f.*

In this section we will focus on the following special situation in which we know $Rf_* \mathscr{F}$ to be perfect by Theorem 23.133 and γ (23.28.4) to be an isomorphism by Corollary 22.103.

(S) Let $f: X \to S$ be a proper morphism of finite presentation. Let $\mathscr{F}$ be a bounded complex of $\mathscr{O}_X$-modules of finite presentation that are flat over S.

These hypotheses are for instance satisfied if S is locally noetherian, f is proper, and $\mathscr{F}$ is a coherent $\mathscr{O}_X$-module that is flat over S. We will also show in Theorem 23.159 below that for every $\mathscr{O}_X$-module $\mathscr{F}$ of finite presentation there exists a stratification on S by locally closed subschemes, indexed by the Hilbert polynomial with respect to some relatively ample line bundle, such that the restriction of $\mathscr{F}$ to the inverse image of each stratum is flat over S.

Instead of (S) one can also consider other hypotheses to apply our general results of the previous sections, see Remark 23.149 below.

By the construction of the base change map (23.28.2), the formation of $R^i f_* \mathscr{F}$ commutes with base change if and only if the base change map β (23.28.3), which was based on (23.25.2), is an isomorphism. Conditions when β is an isomorphism were studied in detail in Section (23.25) for the perfect complex $Rf_* \mathscr{F}$. The results there show that the question whether $R^i f_* \mathscr{F}$ commutes with base change is controlled by the base change to residue fields of S. Therefore let us look at this situation more closely.

Consider for $s \in S$ the cartesian diagram

$$
\begin{array}{ccc}
X_s & \xrightarrow{\;i_s\;} & X \\
{\scriptstyle f_s}\downarrow & & \downarrow{\scriptstyle f} \\
\operatorname{Spec} \kappa(s) & \xrightarrow{\;g\;} & S.
\end{array}
$$

For a bounded complex $\mathscr{F}$ of S-flat $\mathscr{O}_X$-modules of finite presentation we denote by $\mathscr{F}_s$ the restriction $\mathscr{F}_{|X_s} = i_s^* \mathscr{F}$ (no derived inverse image here). Then the functorial homomorphism (23.28.4), which is an isomorphism by Corollary 22.103, becomes the isomorphism

$$
Rf_* \mathscr{F} \otimes_{\mathscr{O}_S}^L \kappa(s) = Lg^* Rf_* \mathscr{F} \cong Rf_{s,*}(\mathscr{F}_s).
$$

Taking for $i \in \mathbb{Z}$ the i-th cohomology on both sides, we obtain a functorial isomorphism

$$(23.28.5) \qquad H^i(Rf_* \mathscr{F} \otimes_{\mathscr{O}_S}^L \kappa(s)) = H^i(X_s, \mathscr{F}_s).$$

Therefore the base change map (23.25.3) takes the form

$$(23.28.6) \qquad \beta^i(\kappa(s)): R^i f_*(\mathscr{F}) \otimes_{\mathscr{O}_S} \kappa(s) \longrightarrow H^i(X_s, \mathscr{F}_s).$$

We can now apply the results from Sections (23.24) and (23.25). By applying Proposition 23.117 we obtain:

Theorem 23.139. *In the Situation* (S) *one has the following assertions.*
(1) *The Euler characteristic*

$$\chi_{\mathscr{F}} : S \to \mathbb{Z}, \quad s \mapsto \sum_{i \geq 0} (-1)^i \dim_{\kappa(s)} H^i(X_s, \mathscr{F}_s)$$

   is locally constant on S.
(2) *For each* $i \in \mathbb{Z}$ *the function*

$$S \to \mathbb{Z}, \quad s \mapsto \dim_{\kappa(s)} H^i(X_s, \mathscr{F}_s)$$

   is upper semicontinuous and constructible, i.e., for all $n \geq 0$ *the subset*

$$\{ s \in S; \ \dim_{\kappa(s)} H^i(X_s, \mathscr{F}_s) \geq n \}$$

   is closed and constructible in S.

Theorem 23.140. (Cohomology and base change) *In Situation* (S) *fix* $i \in \mathbb{Z}$ *and a point* $s \in S$.
(1) *The following conditions are equivalent:*
   (i) *The map* $\beta^i(\kappa(s)) \colon R^i f_* \mathscr{F} \otimes \kappa(s) \to H^i(X_s, \mathscr{F}_s)$ *is surjective.*
   (ii) *There exists an open neighborhood* U *of* s *such that the formation of* $R^i f_* \mathscr{F}_{|U}$ *commutes with base change (Definition 23.138).*
(2) *Assume that* $\beta^i(\kappa(s))$ *is surjective. Then the following conditions are equivalent:*
   (i) *The map* $\beta^{i-1}(\kappa(s))$ *is surjective (and hence the formation of* $R^{i-1} f_* \mathscr{F}$ *commutes with base change in an open neighborhood of* s*).*
   (ii) *There exists an open neighborhood* V *of* s *such that the* $\mathscr{O}_V$*-module* $R^i f_* \mathscr{F}_{|V}$ *is finite locally free.*
   In this case, the function $s \mapsto \dim_{\kappa(s)} H^i(X_s, \mathscr{F}_s)$ *is locally constant on* V.
(3) (Grauert's Theorem) *Conversely: Suppose that* S *is reduced and that the function* $s \mapsto \dim_{\kappa(s)} H^i(X_s, \mathscr{F}_s)$ *is locally constant on* S. *Then* $R^i f_* \mathscr{F}$ *is a finite locally free* $\mathscr{O}_S$*-module, and the formation of* $R^j f_* \mathscr{F}$ *commutes with base change for* $j = i, i - 1$. *In particular one has for all* $s \in S$ *and* $j = i, i-1$ *functorial isomorphisms of* $\kappa(s)$*-vector spaces*

$$R^j f_* \mathscr{F} \otimes \kappa(s) \to H^j(X_s, \mathscr{F}_s).$$

Proof. We apply the general base change results for pseudo-coherent complexes of Section (23.25) to the complex $Rf_* \mathscr{F}$ which is perfect and whose formation commutes with base change by Corollary 23.135.

Let us show Assertion (1). Clearly, (ii) implies (i). Conversely, if $\beta^i(\kappa(s))$ is surjective, we apply Proposition 23.127 and Remark 23.129 to the perfect complex $Rf_* \mathscr{F}$. Therefore we find after replacing S be an open neighborhood of s for every morphism $g \colon S' \to S$ yielding a cartesian diagram (23.28.1) that

$$g^* R^i f_* \mathscr{F} = H^i(Lg^* Rf_* \mathscr{F}) = R^i f'_*(g'^* \mathscr{F}),$$

where the second equality holds by Corollary 23.135. Therefore the formation of $R^i f_* \mathscr{F}$ commutes with base change.

Part (2) follows from Proposition 23.127. Part (3) follows from Proposition 23.128 using Assertion (1).   □

Corollary 23.141. *In Situation* (S) *fix* $m \in \mathbb{Z}$. *Then the formation of* $R^i f_* \mathscr{F}$ *commutes with base change for all* $i \geq m$ *if and only if* $R^i f_* \mathscr{F}$ *is finite locally free for all* $i > m$.

Proof. We may assume that S is quasi-compact. Then $s \mapsto \dim(X_s)$ is bounded (Proposition 14.107) and hence $H^i(X_s, \mathscr{F}_s) = 0$ for i large enough. Therefore we may argue by descending induction on m and can assume that the assertion holds for $m + 1$.

If the formation of $R^i f_* \mathscr{F}$ commutes with base change for $i \geq m$, then by induction hypothesis $R^i f_* \mathscr{F}$ is finite locally free for $i > m + 1$. Moreover, by Theorem 23.140 (2), $R^{m+1} f_* \mathscr{F}$ is finite locally free if and only if the formation of $R^m f_* \mathscr{F}$ commutes with base change.

Conversely, if $R^i f_* \mathscr{F}$ is finite locally free for all $i > m$, then by induction hypothesis the formation of $R^i f_* \mathscr{F}$ commutes with base change for all $i > m$. Again applying Theorem 23.140 (2) shows that also the formation of $R^m f_* \mathscr{F}$ commutes with base change since $R^{m+1} f_* \mathscr{F}$ is locally free. $\qquad\square$

Proposition 23.142. *Let* S *be a scheme, let* $f \colon X \to S$ *be a proper morphism of finite presentation, and let* $\mathscr{F}$ *be a bounded complex of* $\mathscr{O}_X$-modules of finite presentation that are flat over S. Fix $i \in \mathbb{Z}$ and set*

$$U := \{\, s \in S \; ; \; H^i(X_s, \mathscr{F}_s) = 0 \,\}.$$

Then U *is open and constructible,* $R^i f_* \mathscr{F}_{|U} = 0$, *and the formation of* $R^j f_* \mathscr{F}_{|U}$ *commutes with base change for* $j = i - 1, i$.

Proof. The first assertion holds because $\{\, s \in S \; ; \; \dim_{\kappa(s)} H^i(X_s, \mathscr{F}_s) \leq 0 \,\}$ is open and constructible by Theorem 23.139. Moreover, for all $s \in U$ the base change map $\beta^i(\kappa(s))$ has target $H^i(X_s, \mathscr{F}_s) = 0$ and hence is surjective. Therefore the formation of $R^i f_* \mathscr{F}_{|U}$ commutes with base change by Theorem 23.140 (1). Moreover, $R^i f_* \mathscr{F}_{|U} = 0$ by Corollary 23.125 applied to the perfect complex $E = Rf_* \mathscr{F}$. We now can use Theorem 23.140 (2) and hence conclude that the formation of $R^{i-1} f_* \mathscr{F}_{|U}$ commutes with base change. $\qquad\square$

Corollary 23.143. *Let* $f \colon X \to S$ *be a proper morphism of finite presentation, and let* $\mathscr{F}$ *be a bounded complex of* $\mathscr{O}_X$-modules of finite presentation that are flat over S. Fix $i \in \mathbb{Z}$. Suppose that there exists an $s \in S$ such that*

$$H^{i+1}(X_s, \mathscr{F}_s) = H^{i-1}(X_s, \mathscr{F}_s) = 0.$$

Then there exists an open neighborhood U *of* s *such that the following assertions hold.*
(1) $R^{i+1} f_* \mathscr{F}_{|U} = R^{i-1} f_* \mathscr{F}_{|U} = 0$.
(2) *The formation of* $R^j f_* \mathscr{F}_{|U}$ *commutes with base change for* $j = i - 2, i - 1, i, i + 1$.
(3) $R^i f_* \mathscr{F}_{|U}$ *is finite locally free.*

Proof. Proposition 23.142 implies (1) and (2). Then Theorem 23.140 (2) shows that $R^i f_* \mathscr{F}$ is finite locally free in a neighborhood of s. $\qquad\square$

Corollary 23.144. *Let* $f \colon X \to S$ *be a proper morphism of finite presentation, and let* $\mathscr{F}$ *be an* $\mathscr{O}_X$-module of finite presentation that is flat over S. Suppose that there exists an $s \in S$ such that $H^1(X_s, \mathscr{F}_s) = 0$. Then there exists an open neighborhood U of s such that $R^1 f_* \mathscr{F}_{|U} = 0$, $f_* \mathscr{F}_{|U}$ is finite locally free and its formation commutes with base change.*

Proof. This follows from Corollary 23.143 since there is no cohomology in degree -1 if $\mathscr{F}$ is concentrated in degrees ≥ 0. $\square$

Remark 23.145. If one wants to use Theorem 23.140 and Corollaries 23.143 and 23.144 to prove local freeness of higher direct images and compatibility with base change on all of S (and not only locally in a neighborhood of a point) one sees that all the conditions on the base change maps $\beta^i(\kappa(s))$ and $\beta^{i-1}(\kappa(s))$ and on the cohomology of the fibers have only to be checked for s in a subset T of S such that the only open subset of S containing T is all of S.

For instance, if S is quasi-compact, one can choose T to be the set of closed points of S (Lemma 23.4).

Corollary 23.146. *Let* $f\colon X \to S$ *be a proper morphism of finite presentation and let* $\mathscr{F}$ *be an* $\mathscr{O}_X$*-module of finite presentation that is flat over* S. *Let* $d \geq 0$ *be an integer such that* $\dim(X_s) \leq d$ *for all* $s \in S$. *Then* $R^i f_* \mathscr{F} = 0$ *for all* $i > d$ *and the formation of* $R^i f_* \mathscr{F}$ *commutes with base change for all* $i \geq d$.

See also Corollary 24.44 and the remark following it for a similar stronger result.

Proof. The hypothesis implies that $H^i(X_s, \mathscr{F}_s) = 0$ for all $s \in S$ and $i > d$ by the vanishing theorem of Grothendieck-Scheiderer 21.57. Hence we conclude by Proposition 23.142. $\square$

Proposition 23.147. *Let* $f\colon X \to S$ *be a proper morphism of finite presentation and let* $\mathscr{F}$ *be an* $\mathscr{O}_X$*-module of finite presentation that is flat over* S. *Let* $b \geq 0$ *be an integer and let* $r\colon [0,b] \to \mathbb{Z}_{\geq 0}$ *be a map. Let* $u\colon T \to S$ *be a morphism of schemes yielding a cartesian diagram*

$$
\begin{array}{ccc}
X_T & \xrightarrow{\;v\;} & X \\
{\scriptstyle f_T}\downarrow & & \downarrow{\scriptstyle f} \\
T & \xrightarrow{\;u\;} & S.
\end{array}
$$

Then there exists a unique subscheme Z *of* S *such that a morphism of schemes* $u\colon T \to S$ *factors through* Z *if and only if* $R^i f_{T,*}(v^*\mathscr{F})$ *is locally free of rank* $r(i)$ *for all* $i \in [0,b]$ *and the formation of* $R^i f_{T,*}(v^*\mathscr{F})$ *commutes with base change.*

Moreover, the inclusion $Z \to S$ *is of finite presentation and as sets one has*

$$
Z = \{\, s \in S \,;\, \dim_{\kappa(s)} H^i(X_s, \mathscr{F}_s) = r(i) \text{ for all } i \in [0,b]\,\}.
$$

Proof. The subscheme Z is clearly unique if it exists. Therefore we may assume that $S = \operatorname{Spec} R$ is affine. By Theorem 23.133 and Remark 23.134, $E := Rf_* \mathscr{F}$ is a perfect complex of tor-amplitude in $[0, c]$ for some $c \geq 0$, and its formation commutes with base change. Hence, all assertions follow from Proposition 23.130 using that $Lu^* E = Rf_{T,*} v^* \mathscr{F}$ and $H^i(X_s, \mathscr{F}_s) = H^i(E \otimes^L_{\mathscr{O}_s} \kappa(s))$. $\square$

The same proof combined with Corollary 23.146 also shows the following variant of Proposition 23.147.

Remark 23.148. Under the hypotheses in Proposition 23.147, let d be an integer such that $\dim X_s \leq d$ for all $s \in S$, let $0 \leq a \leq d$ be an integer and let $r\colon [a,d] \to \mathbb{Z}_{\geq 0}$ be a map. Then there exists a unique subscheme Z of S parametrizing, similarly as in Proposition 23.147, the locus where $R^i f_* \mathscr{F}$ is locally free of rank $r(i)$ and its formation commutes with base change for all $i \in [a, d]$.

Indeed, to apply Proposition 23.130 we have to show that $Rf_*\mathscr{F}$ has tor-amplitude in $[a', d]$ for some $a' \le d$. But Corollary 23.146 shows that $Rf_*\mathscr{F}$ has no cohomology in degrees $> d$ and hence we conclude by Remark 21.165 (2).

Remark 23.149. There are the following variants of the assumptions made in (S) and of the results shown above.

(1) All the results above using Situation (S) also hold if one assumes that f is only of finite presentation and that $\mathscr{F}$ is a bounded complex of $\mathscr{O}_X$-modules of finite presentation that are flat and of proper support over S.

(2) Let f be flat, proper and of finite presentation and let $E \in D_{\mathrm{qcoh}}(X)$ be pseudo-coherent. Then Corollary 23.136 shows that Rf_*E is pseudo-coherent and commutes with derived base change. In this case, for fixed $i \in \mathbb{Z}$ we say that the formation of $R^i f_*E$ commutes with base change, if for every cartesian diagram (23.28.1) the functorial morphism

$$(23.28.7) \qquad g^* R^i f_*E \longrightarrow R^i f'_* Lg'^* E$$

is an isomorphism. Note the derived inverse image on the right side which makes this definition different from Definition 23.138. By the same argument as above, this is the case if and only if the base change morphism

$$\beta \colon g^* R^i f_*E \longrightarrow H^i(Lg^* Rf_*E)$$

is an isomorphism. In the special case that $g \colon \operatorname{Spec} \kappa(s) \to S$ is the canonical morphism for a point $s \in S$ and $i_s \colon X_s \to X$ denotes the inclusion of the fiber, the base change map $\beta^i(\kappa(s))$ takes the form

$$\beta^i(\kappa(s)) \colon H^i(Rf_*E \otimes^L_{\mathscr{O}_S} \kappa(s)) \longrightarrow H^i(X_s, Li_s^* E).$$

All results from Theorem 23.139 to Corollary 23.146 have an analogue in this setting. One simply has to replace in all assertions $R^i f_*\mathscr{F}$ by $R^i f_*E$ and $\mathscr{F}_s$ by $Li_s^* E$. In addition, one has to assume that E is perfect for Theorem 23.139 (1) as this result needs Rf_*E to be perfect (Corollary 23.136).

There is also a variant of Proposition 23.147. For this one has to assume in addition that E is perfect of tor-amplitude contained in $[a, \infty)$, that $b \ge a$ is an integer and that the map r is given as map $[a, b] \to \mathbb{Z}_{\ge 0}$. Then Rf_*E is also of tor-amplitude contained in $[a, \infty)$ (Remark 23.134). The subscheme Z in Proposition 23.147 then classifies the locus where $R^i f_{T,*}(Lv^* E)$ is locally free of rank $r(i)$ for $i \in [a, b]$.

Hilbert polynomials and flattening stratification

We now study how the Hilbert polynomial varies in proper families of finite presentation. It follows easily from previous results that it is locally constant in *flat* families (Section (23.29)). The converse is also true. For families over a reduced locally noetherian base the fact that the Hilbert polynomial is locally constant implies flatness (Theorem 23.155). In the last section we combine (and generalize) these results by showing that for every proper morphism of finite presentation $X \to S$ there exists a stratification of S into subschemes S_Φ that are characterized by the fact that over them the morphism is flat and has a Hilbert polynomial Φ universally (Theorem 23.159). This is the main theorem of this part.

(23.29) Local constancy of Hilbert polynomials and of intersection numbers.

Let $X \to S$ be a morphism of schemes. As usual, we denote for $s \in S$ its fiber $X \times_S \operatorname{Spec} \kappa(s)$ by X_s. For each complex $\mathscr{F}$ of $\mathscr{O}_X$-modules we denote by $\mathscr{F}_s$ its (non-derived) restriction to X_s.

Proposition 23.150. *Let $f\colon X \to S$ be a proper morphism of finite presentation, and let $\mathscr{F}$ be a bounded complex of $\mathscr{O}_X$-modules of finite presentation that are flat over S. Let $\mathscr{L}$ be an ample line bundle on X. Then the function $S \to \mathbb{Q}[T]$ that sends $s \in S$ to the Hilbert polynomial of X_s, $\mathscr{F}_s$, and $\mathscr{L}_s$ (Definition 23.91) is locally constant.*

Proof. As $\mathscr{F} \otimes \mathscr{L}^{\otimes n}$ is for all $n \in \mathbb{Z}$ again a bounded complex of $\mathscr{O}_X$-modules of finite presentation that are flat over S, the claim follows because the Euler characteristic of $Rf_*(\mathscr{F} \otimes \mathscr{L}^{\otimes n})$ is locally constant (Theorem 23.139 (1)). $\square$

Proposition 23.151. *Let $f\colon X \to S$ be a proper morphism of finite presentation, and let $\mathscr{F}$ be a bounded complex of $\mathscr{O}_X$-modules of finite presentation that are flat over S. Let $\mathscr{L}_1, \dots, \mathscr{L}_t$ be line bundles on X.*
(1) The map $S \to \mathbb{Q}[T_1, \dots, T_n]$ sending $s \in S$ to the Snapper polynomial of X_s, of $\mathscr{F}_s \in D^b_{\mathrm{coh}}(X_s)$, and of the line bundles $\mathscr{L}_{1,s}, \dots, \mathscr{L}_{t,s}$ (Definition 23.69) is locally constant.
(2) Suppose that $\dim \operatorname{Supp}(H^i(\mathscr{F}_s)) \le t$ for all $i \in \mathbb{Z}$ and all $s \in S$. Then the function $s \mapsto \mathbb{Z}$ that sends $s \in S$ to the intersection number $(\mathscr{L}_{1,s} \cdots \mathscr{L}_{t,s} \cdot \mathscr{F}_s)$ (Definition 23.70) is locally constant.

The conditions that $\dim \operatorname{Supp}(H^i(\mathscr{F}_s)) \le t$ for all $i \in \mathbb{Z}$ can be replaced by the weaker hypothesis that the class of $\mathscr{F}_s$ in the Grothendieck group $K_0'(X_s)$ is contained in $K_0'(X_s)_t$, see Remark 23.45.

Proof. The first assertion again follows from Theorem 23.139 (1). And the second assertion follows immediately from the first because the intersection number is a coefficient in the Snapper polynomial. $\square$

(23.30) Flatness on projective schemes.

In this section S denotes a scheme, $f\colon X \to S$ a proper morphism of finite presentation, and $\mathscr{L}$ an S-ample line bundle. For every quasi-coherent $\mathscr{O}_X$-module $\mathscr{F}$ we set $\mathscr{F}(n) := \mathscr{F} \otimes_{\mathscr{O}_X} \mathscr{L}^{\otimes n}$ for all $n \in \mathbb{Z}$. The following remark often allow us to reduce to the noetherian case.

Remark 23.152. Suppose that $\mathscr{F}$ is an $\mathscr{O}_X$-module of finite presentation and that $S = \operatorname{Spec} R$ is affine. By noetherian approximation, there exists a finitely generated $\mathbb{Z}$-subalgebra R_0 of R, a proper R_0-scheme X_0 with $X \cong X_0 \otimes_{R_0} R$, an ample line bundle $\mathscr{L}_0$ on X_0 whose pullback to X is isomorphic to $\mathscr{L}$, and a coherent $\mathscr{O}_{X_0}$-module $\mathscr{F}_0$ whose pullback to X is isomorphic to $\mathscr{F}$. If $\mathscr{F}$ is flat over S, we can find R_0, X_0, and $\mathscr{F}_0$ such that $\mathscr{F}_0$ is flat over R_0 ([Sta] 05LY).

Proposition 23.153. *Let $\mathscr{F}$ be an $\mathscr{O}_X$-module of finite presentation. Then $\mathscr{F}$ is flat over S if and only if there exists an integer $N \geq 1$ such that $f_*\mathscr{F}(n)$ is finite locally free for all $n \geq N$. In this case the formation of $f_*\mathscr{F}(n)$ commutes with arbitrary base change for all $n \geq N$.*

Proof. We may assume that $S = \operatorname{Spec} R$ is affine. By Remark 23.152 we may assume that R is noetherian. There exists $N \geq 1$ such that $H^p(X, \mathscr{F}(n)) = 0$ for all $p \geq 1$ and all $n \geq N$ (Theorem 23.6).

Assume that $\mathscr{F}$ is flat over S. Then the formation of $H^p(X, \mathscr{F}(n))$ commutes with base change for all $p \geq 0$ and all $n \geq N$ (Corollary 23.141). In particular $H^1(X_s, \mathscr{F}(n)_s) = 0$ for all $s \in S$ and $n \geq N$. Hence $f_*\mathscr{F}(n) = H^0(X, \mathscr{F})^{\sim}$ is finite locally free by Corollary 23.144 for $n \geq N$ and its formation commutes with base change.

Conversely, suppose that there exists $N \geq 1$ such that $f_*\mathscr{F}(n)$ is finite locally free for $n \geq N$. Then this also holds if we replace $\mathscr{L}$ by some positive tensor power. Hence we may assume that $\mathscr{L}$ is very ample. Then there exists a closed immersion $i\colon X \to \mathbb{P}_R^d$ for some d such that $\mathscr{L} \cong i^*\mathscr{O}_{\mathbb{P}_R^d}(1)$. Replacing $\mathscr{F}$ by $i_*\mathscr{F}$ and X by $\mathbb{P}_R^d$, we may assume that $X = \mathbb{P}_R^d = \operatorname{Proj} A$ with $A = R[T_0, \ldots, T_d]$.

By Theorem 13.20 and Proposition 23.12 (3), the $\mathscr{O}_X$-module $\mathscr{F}$ is the quasi-coherent module associated with the graded A-module $M := \bigoplus_{n \geq N} H^0(X, \mathscr{F}(n))$. By hypothesis, $H^0(X, \mathscr{F}(n))$ is a flat R-module for $n \geq N$. Hence M is a flat R-module. Therefore its localization M_{T_i} and its graded localization $M_{(T_i)}$, which is a direct summand of M_{T_i}, is flat over R. As $\mathscr{F}_{|D_+(T_i)}$ is the quasi-coherent module corresponding to $M_{(T_i)}$, we see that $\mathscr{F}$ is flat over R. $\qquad\square$

Lemma 23.154. *Let S be a noetherian scheme, let $\mathscr{A}$ be a quasi-coherent graded $\mathscr{O}_S$-algebra such that $\mathscr{A}_1$ is of finite type and generates $\mathscr{A}$. Set $X := \operatorname{Proj} \mathscr{A}$. Let $g\colon S' \to S$ be a morphism of noetherian schemes, and consider the following cartesian diagram (Remark 13.27).*

$$
\begin{array}{ccc}
X' := \operatorname{Proj} g^*\mathscr{A} & \xrightarrow{\ h\ } & X \\
f' \downarrow & & \downarrow f \\
S' & \xrightarrow{\ g\ } & S.
\end{array}
$$

Let $\mathscr{F}$ be a quasi-coherent $\mathscr{O}_X$-module of finite type. Then for large n the base change homomorphism is an isomorphism

$$
g^*f_*\mathscr{F}(n) \xrightarrow{\ \sim\ } f'_*h^*\mathscr{F}(n).
$$

Proof. As S and S' are quasi-compact, we may assume that $S = \operatorname{Spec} R$ and $S' = \operatorname{Spec} R'$ are affine. Let A be the graded R-algebra corresponding to $\mathscr{A}$. One has $\mathscr{F} = \tilde{M}$ for a finitely generated graded A-module M (Lemma 23.14) and $M_n = \Gamma(X, \mathscr{F}(n))$ for large n (Proposition 23.13). Moreover, $h^*\mathscr{F}$ is associated to the graded module $M \otimes_R R' = \bigoplus_n (M_n \otimes_R R')$ and hence $M_n \otimes_R R' = \Gamma(X', h^*\mathscr{F}(n))$ for large n, again by Proposition 23.13. This shows the claim. $\qquad\square$

Theorem 23.155. *Let S be a reduced locally noetherian scheme, let $f\colon X \to S$ be a proper morphism, and let $\mathscr{L}$ be an S-ample line bundle on X. Let $\mathscr{F}$ be a coherent $\mathscr{O}_X$-module. Then $\mathscr{F}$ is flat over S if and only if the map $S \to \mathbb{Q}[T]$ that sends $s \in S$ to the Hilbert polynomial $\Phi_{X_s, \mathscr{L}_s, \mathscr{F}_s}$ of the fiber X_s is locally constant.*

Proof. The condition is necessary (without any hypothesis on S) by Proposition 23.150. Hence suppose that $s \mapsto \Phi_s := \Phi_{X_s, \mathscr{L}_s, \mathscr{F}_s}$ is locally constant. We may assume that $S = \operatorname{Spec} R$ is affine. Then $X = \operatorname{Proj} A$ for the graded R-algebra $A := \bigoplus_{d \geq 0} \Gamma(X, \mathscr{L}^{\otimes d})$ (Corollary 13.75). To show that $\mathscr{F}$ is flat over S we will show that $H^0(X, \mathscr{F}(n))$ is a projective R-module for large n (Proposition 23.153).

As cohomology commutes with flat base change, we may assume that R is local. Then S is connected and hence $s \mapsto \Phi_s$ is constant. Let $s_0 \in \operatorname{Spec} R$ be the closed point. As R is noetherian, there are only finitely many irreducible components of $\operatorname{Spec} R$. Let $\eta_1, \ldots, \eta_m$ be their generic points. As R is reduced, it suffices to show that for large n the function

$$(*) \qquad\qquad s \mapsto \dim_{\kappa(s)} H^0(X, \mathscr{F}(n)) \otimes_R \kappa(s)$$

on $\operatorname{Spec} R$ is constant (Corollary 11.19). By semicontinuity of $(*)$ (Corollary 7.31 and the remark following it) it suffices to show that $(*)$ is constant on the finite set $\{s_0, \eta_1, \ldots, \eta_m\}$. But for each $s \in \operatorname{Spec} R$ we find by Remark 23.92 and Lemma 23.154 an integer $N_s \geq 1$ such that

$$\Phi_s(n) = \dim H^0(X_s, \mathscr{F}(n)_s) = \dim(H^0(X, \mathscr{F}(n)) \otimes_R \kappa(s))$$

for $n \geq N_s$. Therefore the constancy of the Hilbert polynomial shows that $(*)$ is constant on $\{s_0, \eta_1, \ldots, \eta_m\}$ for $n \geq \max\{N_{s_0}, N_{\eta_1}, \ldots, N_{\eta_m}\}$. $\qquad\qquad\square$

(23.31) Flattening stratification by Hilbert polynomials.

In this section, $f \colon X \to S$ is a morphism of schemes and $\mathscr{F}$ is a quasi-coherent $\mathscr{O}_X$-module. For every morphism $T \to S$ we set $X_T := X \times_S T$ and denote by $\mathscr{F}_T$ the pullback of $\mathscr{F}$ to X_T. If $T = \operatorname{Spec} B$ is affine, we write X_B and $\mathscr{F}_B$ instead of X_T and $\mathscr{F}_T$. If $T = \operatorname{Spec} \kappa(s) \to S$ is the canonical morphism for a point $s \in S$, then we write X_s and $\mathscr{F}_s$ instead of $X_{\kappa(s)}$ and $\mathscr{F}_{\kappa(s)}$.

Recall that a polynomial $\Phi \in \mathbb{Q}[T]$ is called *numerical* if $\Phi(n) \in \mathbb{Z}$ for all $n \in \mathbb{Z}$. We denote the subring of $\mathbb{Q}[T]$ of all numerical polynomials by $\mathbb{Q}[T]_{\mathrm{num}}$. We endow this subring with a total order by defining $\Phi \leq \Phi'$ if $\Phi(n) \leq \Phi'(n)$ for $n \gg 0$. In other words, one has

$$a_0 + a_1 T + a_2 T^2 + \ldots \qquad \leq \qquad b_0 + b_1 T + b_2 T^2 + \ldots$$

if and only if there exists $r \geq 0$ such that $a_i = b_i$ for $i \geq r$ and $a_{r-1} < b_{r-1}$ (where we set $a_{-1} := b_{-1} := 0$). In the proof of Theorem 23.159 below we will use the following lemma.

Lemma 23.156. *Let $d \geq 0$ and N be integers. Then the map*

$$\{\Phi \in \mathbb{Q}[T]_{\mathrm{num}} \; ; \; \deg \Phi \leq d\} \longrightarrow \mathbb{Z}^{d+1},$$
$$\Phi \longmapsto (\Phi(N), \Phi(N+1), \ldots, \Phi(N+d))$$

is an isomorphism of abelian groups.

Proof. It is clear that the map is the restriction of the homomorphism

$$\{\Phi \in \mathbb{Q}[T] \; ; \; \deg \Phi \leq d\} \to \mathbb{Q}^{d+1}, \qquad \Phi \mapsto (\Phi(N), \Phi(N+1), \ldots, \Phi(N+d))$$

which is an isomorphism because for every $(d+1)$-tuple of rational numbers there exists a unique polynomial of degree $\leq d$ that takes these values at a given set of $d+1$ rational numbers. It remains to show that if $\Phi(N+i) \in \mathbb{Z}$ for $i = 0, \ldots, d$, then Φ is numerical.

We argue by induction on d. The case $d = 0$ is clear. Let $d \geq 1$ and set $\Psi(n) = \Phi(n+1) - \Phi(n)$. Then $\deg \Psi \leq d - 1$ and $\Psi(n) \in \mathbb{Z}$ for $N \leq n \leq N + d - 1$. Hence $\Psi(n) \in \mathbb{Z}$ for all $n \in \mathbb{Z}$ by induction hypothesis. Therefore, if $\Phi(n) \in \mathbb{Z}$, then $\Phi(n \pm 1) \in \mathbb{Z}$. As $\Phi(N) \in \mathbb{Z}$, this finishes the proof. $\qquad\square$

Now let S be a scheme, let $f \colon X \to S$ be proper and of finite presentation, and let $\mathscr{L}$ be an S-ample line bundle on X. Let $\mathscr{F}$ be an $\mathscr{O}_X$-module of finite presentation.

For every numerical polynomial $\Phi \in \mathbb{Q}[T]$ define a subset of S by

$$(23.31.1) \qquad S_\Phi(X, \mathscr{L}, \mathscr{F}) := S_\Phi := \{\, s \in S \;;\; \Phi_{X_s, \mathscr{L}_s, \mathscr{F}_s} = \Phi \,\},$$

where $\Phi_{X_s, \mathscr{L}_s, \mathscr{F}_s}$ denotes the Hilbert polynomial of $\mathscr{F}_s$ on the fiber X_s with respect to $\mathscr{L}_s$.

Remark 23.157. As the Euler characteristic and hence the Hilbert polynomial are invariant under change of the base field (Lemma 23.64), for every morphism $g \colon T \to S$ of schemes one has an equality of subsets of T

$$(23.31.2) \qquad g^{-1}(S_\Phi(X, \mathscr{L}, \mathscr{F})) = S_\Phi(X_T, \mathscr{L}_T, \mathscr{F}_T).$$

Lemma 23.158. *Suppose that S is noetherian. Then S is the set-theoretic disjoint union of finitely many affine subschemes S_i such that $\mathscr{F}_{S_i}$ is flat over S_i and such that $s \mapsto \Phi_{X_s, \mathscr{L}_s, \mathscr{F}_s}$ is constant on S_i. In particular, there exist only finitely many numerical polynomials Φ such that S_Φ is non-empty.*

Proof. By Proposition 10.86, S is the set-theoretic disjoint union of finitely many affine subschemes S_j' such that $\mathscr{F}_{S_j'}$ is flat over S_j' and hence the Hilbert polynomial is locally constant on S_j' (Proposition 23.150). As the S_j' are noetherian (being locally closed in a noetherian topological space) and in particular quasi-compact, S_j' is the disjoint union of finitely many open and closed affine subschemes S_i on which the Hilbert polynomial is constant. $\qquad\square$

Theorem and Definition 23.159. *Let $f \colon X \to S$ be proper of finite presentation, let $\mathscr{L}$ be an S-ample line bundle, and let $\mathscr{F}$ be an $\mathscr{O}_X$-modules of finite presentation. Let $\Phi \in \mathbb{Q}[T]$ be a numerical polynomial.*

(1) *There exists on S_Φ (23.31.1) a unique structure of a locally closed subscheme of S such that a morphism $T \to S$ factors through S_Φ if and only if $\mathscr{F}_T$ is flat over T and has Hilbert polynomial Φ in each point of T. Moreover, the inclusion $S_\Phi \to S$ is of finite presentation.*

(2) *The subspace $\bigcup_{\Phi' \geq \Phi} S_{\Phi'}$ is closed, in particular every point in the closure of S_Φ is contained in some $S_{\Phi'}$ for some numerical polynomial Φ' with $\Phi' \geq \Phi$.*

The family of locally closed subschemes $(S_\Phi)_\Phi$ is called the flattening stratification *of $\mathscr{F}$ with respect to $(X, \mathscr{L})$. If S is quasi-compact, then there are only finitely many Φ such that $S_\Phi \neq \emptyset$.*

Moreover, the flattening stratification is compatible with base change $g \colon T \to S$, i.e., (23.31.2) is an identity of subschemes of T.

If $X = S$ and $f = \mathrm{id}_S$, then $S_\Phi = \emptyset$ whenever $\deg \Phi > 0$ or $\Phi(0) < 0$. Moreover, the Hilbert polynomial is constant of value $m \geq 0$ in a point $s \in S$ if and only if $\dim_{\kappa(s)}(\mathscr{F} \otimes \kappa(s)) = m$. Therefore the flattening stratification in this special case coincides with the flattening stratification defined in Section (11.8).

Proof. The uniqueness of the subscheme structure on S_Φ is clear by its characterization as a subfunctor of S. This also shows the compatibility with base change once we have shown the existence of the flattening stratification. Therefore we may assume that $S = \operatorname{Spec} R$ is affine. By Remark 23.152 and Remark 23.157 we also may assume that R is noetherian. In the noetherian case, any immersion is of finite presentation and hence the same holds for the inclusions $S_\Phi \to S$ is general once we have proved the theorem in the noetherian case.

(I). By Lemma 23.158, S is the set-theoretic disjoint union of finitely many affine subschemes S_i such that $\mathscr{F}_{S_i}$ is flat over S_i. Let $f_i \colon f^{-1}(S_i) \to S_i$ be the restriction of f. By Lemma 23.154 and since there are only finitely many S_i, there exists an integer $N' \geq 1$ such that

(*) $$ f_*\mathscr{F}(n)_{|S_i} = f_{i*}\mathscr{F}_{S_i}(n) $$

for all $n \geq N'$ and for all i. Let $N \geq N'$ such that $R^p f_{i*}(\mathscr{F}_{S_i}(n)) = 0$ for all $p \geq 1$, all $n \geq N$, and all i (Theorem 23.6 using again that there are only finitely many S_i). As $\mathscr{F}_{S_i}$ is flat over S_i, we may apply Corollary 23.141 to see that the formation of $R^p f_{i*}(\mathscr{F}_{S_i}(n))$ commutes for all i with base change for all $p \geq 0$ and for all $n \geq N$. Combining this with (*) we obtain that for $n \geq N$ we have the following properties.
(A) $H^p(X_s, \mathscr{F}(n)_s) = 0$ for $p \geq 1$ and for all $s \in S$,
(B) $f_*\mathscr{F}(n) \otimes \kappa(s) = H^0(X_s, \mathscr{F}_s(n))$ for all $s \in S$.
(II). Let $d := \sup_{s \in S} \dim X_s$, a finite integer by Proposition 14.107. We set

$$ \mathscr{E}_j := f_*\mathscr{F}(N+j), \qquad j = 0, \ldots, d, $$

which are coherent $\mathscr{O}_S$-modules by Theorem 23.17. Applying the flattening stratification of Section (11.8) to $\mathscr{E}_0, \ldots, \mathscr{E}_d$ and taking scheme-theoretic intersections of all strata, we obtain subschemes $W_{e_0,\ldots,e_d}$ for integers $e_0, \ldots, e_d \geq 0$ such that a morphism $g \colon T \to S$ of schemes factors through $W_{e_0,\ldots,e_d}$ if and only if $g^*\mathscr{E}_i$ is locally free of rank e_j for all j. Let $\Phi \in \mathbb{Q}[T]$ be the unique numerical polynomial of degree $\leq d$ with $\Phi(N+i) = e_j$ for $j = 0, \ldots, d$ (Lemma 23.156) and write W_Φ instead of $W_{e_0,\ldots,e_d}$.

We claim that the underlying topological space of W_Φ is the subspace S_Φ. Indeed, for $s \in S$ set $\Phi_s := \Phi_{X_s,\mathscr{L}_s,\mathscr{F}_s}$. It is a numerical polynomial of degree $\leq d$ and $\Phi_s(n) = \dim H^0(X_s, \mathscr{F}_s(n))$ for $n \geq N$ by (A). It is determined by its values $\Phi_s(N), \ldots, \Phi_s(N+d)$. Hence the claim follows from (B).

(III). Next we endow S_Φ with the correct scheme structure. This may differ from the scheme structure of the W_Φ which ensures only that $f_*\mathscr{F}(N+j)$ is locally free of rank $\Phi(N+j)$ for $j = 0, \ldots, d$. For arbitrary $j \geq 0$ the coherent module $f_*\mathscr{F}(N+j)$ has fibers of rank $\Phi(N+j)$ in all points $s \in W_\Phi$ by the claim at the end of Step (II). For $\ell \geq d$ define a sequence of closed subschemes $W_\Phi^{(\ell)}$ of W_Φ with the same topological space as W_Φ such that a morphism of schemes $g \colon T \to W_\Phi$ factors through $W_\Phi^{(\ell)}$ if and only if $g^*(f_*\mathscr{F}(N+j)_{|W_\Phi})$ is locally free of rank $\Phi(N+j)$ for all $d \leq j \leq \ell$. As W_Φ is a noetherian scheme, there exists an $\ell_0 \geq d$ such that $W_\Phi^{(\ell)} = W_\Phi^{(\ell_0)}$ for all $\ell \geq \ell_0$. Endow S_Φ with the scheme structure of $W_\Phi^{(\ell_0)}$.

(IV). We now show that the subschemes S_Φ have the desired universal property. By Lemma 23.154 there exists an integer $M \geq N$ such that $(f_*\mathscr{F}(n))_{|S_\Phi} = (f_{S_\Phi})_*\mathscr{F}_{S_\Phi}(n)$ for all $n \geq M$. Let $g \colon T \to S$ be a morphism of schemes. Then g factors through S_Φ if and only if the Hilbert polynomial of $\mathscr{F}_T$ is in each point of T equal to Φ and if $g^* f_*\mathscr{F}(n)$ is locally free of rank $\Phi(n)$ for all $n \geq M$. Hence we conclude by Proposition 23.153.

(V). It remains to prove Assertion (2). As only finitely many S_Φ are non-empty by Lemma 23.158, there exists an $m \geq N$ such that for any two polynomials Φ and Φ' with $S_\Phi \neq \emptyset \neq S_{\Phi'}$ one has $\Phi < \Phi'$ if and only if $\Phi(m) < \Phi'(m)$. Then the non-empty S_Φ are the flattening stratification of the coherent $\mathcal{O}_S$-module $f_* \mathscr{F}(m)$ and we conclude using that the fiber rank of a coherent $\mathcal{O}_S$-module is upper semicontinuous (Corollary 7.31 and the remark following it). $\qquad\square$

For the flattening stratification of a finitely presented module on S there is also a canonical structure of a closed finitely presented subscheme where the module is of rank $\geq r$ for every fixed r (Section (16.9)). Using similar ideas it is possible to define for every numerical polynomial Φ a natural structure of a closed finitely presented scheme on $\bigcup_{\Phi' \geq \Phi} S_{\Phi'}$ which is compatible with base change $T \to S$.

The morphism of schemes $S' := \coprod_\Phi S_\Phi \to S$ is a surjective monomorphism. A morphism $T \to S$ factors through S' if and only if $\mathscr{F}_T$ is flat over T. Such a universal flattening exists without assuming the existence of an ample line bundle:

Theorem 23.160. *Let $f \colon X \to S$ be a proper morphism of finite presentation between schemes, let $\mathscr{F}$ be an $\mathcal{O}_X$-module of finite presentation. Then there exists a surjective monomorphism of schemes $S' \to S$ such that a morphism $T \to S$ factors through S' if and only if $\mathscr{F}_T$ is flat over T.*

We refer to [Sta] 05UH for a proof and for further generalizations.

Exercises

Exercise 23.1. Let $i \colon Y \to X$ be a closed immersion of locally noetherian schemes, that is a homeomorphism on underlying topological spaces. Show that X is quasi-affine if and only if Y is quasi-affine.

Remark: By noetherian approximation one can show that the hypothesis "locally noetherian" is in fact superfluous.

Exercise 23.2. Let X be a locally noetherian scheme and let $Z \subseteq X$ be a closed subscheme such that there exists an irreducible component X_0 of X such that $\mathrm{codim}(Z \cap X_0, X_0) = 1$. Let $j \colon U := X \setminus Z \to X$ be the inclusion. Show that $j_* \mathcal{O}_U$ is not coherent.

Exercise 23.3. Let R be an Artinian ring and let X be a proper scheme over $\mathrm{Spec}\, R$. Show that the abelian category $(\mathrm{Coh}(X))$ of coherent $\mathcal{O}_X$-modules satisfies the bi-chain condition (Exercise F.7) and deduce that every coherent $\mathcal{O}_X$-module $\mathscr{F}$ has a unique (up to order and isomorphism) decomposition $\mathscr{F} \cong \mathscr{F}_1 \oplus \cdots \oplus \mathscr{F}_r$ into indecomposable coherent $\mathcal{O}_X$-modules $\mathscr{F}_i$.

Hint: Exercise F.8.

Exercise 23.4. Let k be a field, $n \geq 1$ and integer, and let $\mathscr{F}$ be a coherent $\mathcal{O}_{\mathbb{P}_k^n}$-module. Show that the following assertions are equivalent.

(i) For all $i = 1, \ldots, n$ one has $H^i(\mathbb{P}_k^n, \mathscr{F}(-i)) = 0$.

(ii) For all $i \geq 1$ and all $m \geq -i$ one has $H^i(\mathbb{P}_k^n, \mathscr{F}(m)) = 0$.

(iii) There exists an exact sequence of the form

$$0 \longrightarrow \mathcal{O}_{\mathbb{P}_k^n}(-n)^{r_n} \longrightarrow \mathcal{O}_{\mathbb{P}_k^n}(-n+1)^{r_{n-1}} \longrightarrow \cdots \longrightarrow \mathcal{O}_{\mathbb{P}_k^n}(-1)^{r_1} \longrightarrow \mathcal{O}_{\mathbb{P}_k^n}^{r_0} \longrightarrow \mathscr{F} \longrightarrow 0$$

for some integers $r_i \geq 0$.

If these conditions hold, show that the canonical map

$$H^0(\mathbb{P}_k^n, \mathscr{F}) \otimes H^0(\mathbb{P}_k^n, \mathcal{O}_{\mathbb{P}_k^n}(d)) \longrightarrow H^0(\mathbb{P}_k^n, \mathscr{F}(d))$$

is surjective for all $d \geq 0$.

Exercise 23.5. Let k be a field and let $X \to \operatorname{Spec} k$ be a proper smooth morphism. Then the numbers

$$h^{ij}(X) := \dim_k H^j(X, \Omega_{X/k}^i)$$

are called the *Hodge numbers* of X. Show that

$$h^{ij}(\mathbb{P}_k^n) = \begin{cases} 1, & \text{if } 0 \leq i = j \leq n; \\ 0, & \text{otherwise.} \end{cases}$$

Remark: See Exercise 23.33 for a generalization.

Exercise 23.6. Let X be a qcqs scheme and let $\mathscr{L}$ be an ample line bundle ob X. Show that every pseudo-coherent (resp. perfect) complex in $D(X)$ is isomorphic to a bounded above (resp. bounded) complex whose terms are finite direct sums of tensor powers of $\mathscr{L}$. *Hint*: Use Proposition F.160.

Exercise 23.7. Let X be a hypersurface of degree d over a field k, i.e., a closed subscheme of $\mathbb{P}_k^n$ defined by a homogeneous polynomial of degree d. Show that the dimension of the k-vector space $H^i(X, \mathcal{O}_X)$ depends only on d and n.

Hint: First compute the ideal sheaf of $\mathcal{O}_{\mathbb{P}_k^n}$ defining X (a line bundle on $\mathbb{P}_k^n$).

Exercise 23.8. Let A be a ring, let $I \subseteq A$ be an ideal, and consider the closed immersion $i\colon \operatorname{Spec} A/I \to \operatorname{Spec} A$ (which is in particular a proper morphism).
(1) Suppose that I is not finitely generated. Show that i is not cohomologically proper.
(2) Give an example of a ring A and a finitely generated ideal I (and hence i is proper of finite presentation) such that i is not cohomologically proper.
(3) Show that i is cohomologically proper if and only if I is a pseudo-coherent A-module.

Exercise 23.9. Let $Y = \operatorname{Spec} A$ be an affine scheme, let $r \geq 1$, and let $f\colon X := \mathbb{P}_Y^r \to Y$ be the projection. Let $\mathscr{E} \in D_{\operatorname{qcoh}}(X)$ be pseudo-coherent and $a \in \mathbb{Z}$.
(1) Show that $Rf_*\mathscr{E}$ is pseudo-coherent, i.e., f is cohomologically proper.
(2) Suppose that $\mathscr{E}$ is acyclic in degrees $> a$. Then $Rf_*\mathscr{E}$ is acyclic in degrees $> a + r$. Show that there exists an integer N such that $Rf_*(\mathscr{E} \otimes_{\mathcal{O}_X} \mathcal{O}_X(n))$ is acyclic in degrees $> a$ for all $n \geq N$.

Exercise 23.10. Show that every morphism of schemes $f\colon X \to Y$ that can be locally on Y factorized into a closed immersion $i\colon X \to \mathbb{P}_Y^r$ followed by the structure map $\mathbb{P}_Y^r \to Y$ with $i_*\mathcal{O}_X$ pseudo-coherent is cohomologically proper.
Hint: Exercise 23.9.

Exercise 23.11. Let R be a noetherian ring. Show that the following assertions are equivalent.
(i) R is regular.
(ii) For every perfect complex E of R-modules its truncation $\tau^{\geq 0}E$ is again perfect.
(iii) For every perfect complex E of R-modules its truncation $\tau^{\leq 0}E$ is again perfect.

Exercise 23.12. Let $f\colon X \to Y$ be a morphism of schemes, suppose that Y is qcqs, and let $t \geq 0$ be an integer. Show that f is of tor-dimension $\leq t$ if and only if $H^i(Lf^*\mathscr{G}) = 0$ for $i < -t$ and for every $\mathscr{O}_Y$-module $\mathscr{G}$ of finite presentation.

Exercise 23.13. Let k be a field and $S = \operatorname{Spec} k$. Show that $K_0(D_{\mathrm{coh}}^-(S))$, $K_0(D_{\mathrm{coh}}^+(S))$, and $K_0(D_{\mathrm{coh}}(S))$ are all 0.

Exercise 23.14. Let $f\colon X \to \operatorname{Spec} k$ be a proper morphism of noetherian schemes. Show that $f_*\colon K_0'(X) \to K_0'(Y) = \mathbb{Z}$ is given by the Euler characteristic, where the equality is Exercise 23.24.

Exercise 23.15. Let X be a scheme. Let $(\mathrm{VB}(X))$ be the category of locally free $\mathscr{O}_X$-modules of finite type. Define a short sequence $0 \to \mathscr{E}' \to \mathscr{E} \to \mathscr{E}'' \to 0$ of vector bundles to be *admissible exact* if it is exact as a sequence of $\mathscr{O}_X$-modules. Show that this defines on $(\mathrm{VB}(X))$ the structure of an exact category (Exercise F.15).
Hint: Exercise F.16

Exercise 23.16. Let $\mathcal{A}$ be an exact essentially small category (Exercise F.15). Define $K_0(\mathcal{A})$ as the abelian group with generators $[X]$ for each isomorphism class of objects X in $\mathcal{A}$ and relations of the form $[X] = [X'] + [X'']$ for each admissible exact sequence $0 \to X' \to X \to X'' \to 0$ in $\mathcal{A}$.

Let R be a ring and let $\mathcal{F}_R$ be the category of finitely generated free R-modules endowed with its natural exact structure. Show that $K_0(F_R) \cong \mathbb{Z}$.

Exercise 23.17. Let X be a quasi-compact scheme and let $(\mathscr{L}_i)_{i \in I}$ be a family of line bundles on X. Show that the following assertions are equivalent.
(i) The open subsets X_s for $s \in \Gamma(X, \mathscr{L}_i^{\otimes n})$, $i \in I$ and $n \geq 1$, form a basis of the topology of X.
(ii) The open subsets X_s that are in addition affine form a basis of the topology of X.
(iii) The open subsets X_s that are in addition affine cover X.
(iv) For every quasi-coherent $\mathscr{O}_X$-module $\mathscr{F}$ of finite type there exist families $(r_i)_{i \in I}$ and $(n_i)_{i \in I}$ of integers $r_i \geq 0$ and $n_i > 0$ and a surjective map of $\mathscr{O}_X$-modules

$$\bigoplus_{i \in I} \mathscr{O}_X^{r_i} \otimes \mathscr{L}_i^{\otimes -n_i} \longrightarrow \mathscr{F}.$$

(v) For every quasi-coherent $\mathscr{O}_X$-module $\mathscr{F}$ the evaluation map

$$\bigoplus_{i,n \geq 1} \Gamma(X, \mathscr{F} \otimes_{\mathscr{O}_X} \mathscr{L}^{\otimes n}) \otimes_{\Gamma(X, \mathscr{O}_X)} \mathscr{L}_i^{\otimes -n_i} \longrightarrow \mathscr{F}$$

is surjective.
If $(\mathscr{L}_i)_{i \in I}$ satisfies these conditions, it is called an *ample family*.
Remark: One can show that every noetherian, semiseparated (Exercise 22.4) scheme X such that $\mathscr{O}_{X,x}$ is factorial for all $x \in X$ (e.g., if X is regular) admits an ample family of line bundles ([SGA6]^0 II 2.2.7^2).

Exercise 23.18. Let X be a quasi-compact scheme that carries an ample family of line bundles (Exercise 23.17).
(1) Show that there exists a finite ample subfamily.

---

^{2} The result in [SGA6]^0 is formulated for separated schemes but the proof carries over to the case where X is only semiseparated.

(2) Show that X is semiseparated (Exercise 22.4).
(3) Let k be a field and let X be the affine line with zero doubled. Denote by $U_1, U_2 \subseteq X$ be the two copies isomorphic to $\mathbb{A}_k^1$. For $n \in \mathbb{Z}$ define the Cartier divisor D_n given by $(U_1, 1)$ and (U_2, z^n), where z is the coordinate of the second copy of the affine line. Let $\mathscr{L}_n = \mathscr{O}_X(D_n)$ be the attached line bundle. Show that $(\mathscr{L}_1, \mathscr{L}_{-1})$ is an ample family of line bundles on X.
   Remark: Note that there exists no ample line bundle on X since this would imply that X is separated.
(4) Let $f\colon Y \to X$ be a quasi-projective morphism of schemes. Show that Y carries an ample family of line bundles.

Exercise 23.19. Let X be a quasi-compact scheme with an ample family $(\mathscr{L}_i)_{i \in I}$ (Exercise 23.17). Show the following assertions.
(1) For every quasi-coherent $\mathscr{O}_X$-module $\mathscr{F}$, there exists a locally free $\mathscr{O}_X$-module $\mathscr{E}$ and a surjection $\mathscr{E} \to \mathscr{F}$.
(2) For every quasi-coherent $\mathscr{O}_X$-module $\mathscr{F}$ of finite type, there exists a finite locally free $\mathscr{O}_X$-module $\mathscr{E}$ and a surjection $\mathscr{E} \to \mathscr{F}$.
(3) For every surjection $\mathscr{G} \to \mathscr{F}$ of quasi-coherent $\mathscr{O}_X$-modules with $\mathscr{F}$ of finite type, there exists a finite locally free $\mathscr{O}_X$-module $\mathscr{E}$ and a map of $\mathscr{O}_X$-modules $\mathscr{E} \to \mathscr{G}$ such that the composition $\mathscr{E} \to \mathscr{G} \to \mathscr{F}$ is surjective
Show that in all three assertions one may take $\mathscr{E}$ as a direct sum of tensor powers of the $\mathscr{L}_i$.

Exercise 23.20. Let X be a quasi-compact scheme with an ample family of line bundles (Exercise 23.17). Show that for every perfect complex $\mathscr{E}$ on X there exists a strictly perfect complex $\mathscr{F}$ and an isomorphism $\mathscr{F} \cong \mathscr{E}$ in $D(X)$.
Hint: Use Exercise 23.18 to show that $\mathscr{E}$ can be represented by a complex of quasi-coherent $\mathscr{O}_X$-modules.

Exercise 23.21. Let C be a connected normal scheme of dimension 1. Show that C is semiseparated and deduce that C has the resolution property.
Hint: Lemma 15.17, the remark in Exercise 23.17, Exercise 23.19

Exercise 23.22. Let $f\colon X \to Y$ be a morphism of schemes and suppose that Y is regular noetherian of finite dimension d.
(1) Show that for every coherent $\mathscr{O}_Y$-module $\mathscr{G}$ there exists a strictly perfect complex $\mathscr{F}$ on Y concentrated in degrees $[-d, 0]$ and an isomorphism $\mathscr{F} \xrightarrow{\sim} \mathscr{G}$ in $D(Y)$.
   Hint: Use the Remark at the end of Exercise 23.17.
(2) Show that f is of tor-dimension $\leq d$.
   Hint: Exercise 23.12.

Exercise 23.23. Let X be a scheme and let $(\mathrm{VB}(X))$ be the exact category of finite locally free $\mathscr{O}_X$-modules (Exercise 23.15) and let $K_0^{VB}(X) := K_0(\mathrm{VB}(X))$ (Exercise 23.16).
(1) Show that the multiplication $([\mathscr{E}], [\mathscr{F}]) \mapsto [\mathscr{E} \otimes_{\mathscr{O}_X} \mathscr{F}]$ defines a commutative ring structure on $K_0^{VB}(X)$ and that for quasi-compact[3] schemes X there is a ring homomorphism $K_0^{VB}(X) \to K_0(X)$ given by $[\mathscr{E}] \mapsto [\mathscr{E}[0]]$ which is functorial with respect to pullback f^* for morphisms f of quasi-compact schemes.

---

[3] The quasi-compactness assumption is only needed because we defined $K_0(X)$ only for quasi-compact schemes.

(2) Let X be a quasi-compact scheme such that there exists an ample family of line bundles on X (Exercise 23.17). Show that the natural map $K_0^{VB}(X) \to K_0(X)$ is an isomorphism.
Hint: Exercise 23.20.
(3) Let k be a field and let X be $\mathbb{A}_k^n$ with a double origin, i.e. X is obtained by gluing two copies of $\mathbb{A}_k^n$ along $\mathbb{A}_k^n \setminus \{0\}$ (Examples 3.13, 9.10). Let $n \geq 2$. Show that $K_0(X) = K_0'(X) \cong \mathbb{Z} \oplus \mathbb{Z}$ and that $K_0^{VB}(X) \cong \mathbb{Z}$.

Exercise 23.24. Let R be a local Artinian ring. Show that $K_0'(R) \cong \mathbb{Z}$ given by sending a finitely generated R-module M to its length. Show that the composition $K_0(R) \to K_0'(R) \cong \mathbb{Z}$ is a ring isomorphism.

Exercise 23.25. Consider the polynomial ring $\mathbb{Q}[X_1, \ldots, X_t]$. For $i \geq 0$ set
$$\binom{X}{i} := \frac{X(X-1)\cdots(X-i+1)}{i!} \in \mathbb{Q}[X].$$
Show that the polynomials of the form $\binom{X_1}{i_1}, \ldots, \binom{X_t}{i_t}$ for $i_k \geq 0$ form a $\mathbb{Q}$-basis of $\mathbb{Q}[X_1, \ldots, X_t]$. Show that for a polynomial $p \in \mathbb{Q}[X_1, \ldots, X_t]$ the following conditions are equivalent.
(i) For every $(n_1, \ldots, n_t) \in \mathbb{Z}^t$ the value $p(n_1, \ldots, n_t)$ is an integer.
(ii) If we express $p = \sum_{i_k \geq 0} a_{i_1,\ldots,i_t} \binom{X_1}{i_1} \cdots \binom{X_t}{i_t}$, then $a_{i_1,\ldots,i_t} \in \mathbb{Z}$.
If p satisfies these conditions, then p is called *numerical*. Often a map $\mathbb{Z}^n \to \mathbb{Z}$, that is given by a (necessarily unique) numerical polynomial is also called *numerical polynomial*.

Exercise 23.26. Let $f \colon \mathbb{Z} \to \mathbb{Z}$ be a map such that $n \mapsto f(n) - f(n-1)$ is a numerical polynomial (Exercise 23.25), then f is a numerical polynomial.

Exercise 23.27. Let k be a field, let X be a proper k-scheme, and let $\mathscr{L}_1, \cdots, \mathscr{L}_t$ be line bundles on X and let $\mathscr{F} \in D_{\mathrm{coh}}^b(X)$ such that $[\mathscr{F}] \in K_0'(X)_t$. Show that
$$(\mathscr{L}_1 \cdots \mathscr{L}_t \cdot \mathscr{F}) = \sum_{I \subseteq \{1,\ldots,t\}} (-1)^{|I|} \chi(X, \mathscr{F} \otimes^L \bigotimes_{i \in I} \mathscr{L}_i^\vee).$$

Exercise 23.28. Let k by a field, $n \geq 1$ an integer and $X \subseteq \mathbb{P}_k^n$ a non-empty closed subscheme of dimension d. Show that for all $m_1, \ldots, m_d \in \mathbb{Z}$ one has
$$(\mathcal{O}_{\mathbb{P}_k^n}(m_1) \cdots \mathcal{O}_{\mathbb{P}_k^n}(m_d) \cdot X) = m_1 \cdots m_d \deg(X).$$

Exercise 23.29. Let X be a quasi-compact scheme with an ample line bundle. Show *Jouanolou's trick*, i.e., show that there exists a vector bundle torsor $E \to X$ such that E is an affine scheme.
Hint: Use the ample line bundle to construct an affine map $s \colon X \to \mathbb{P}_{\mathbb{Z}}^N$. Show that the scheme W of rank 1 idempotent matrices in $M_{N+1}(\mathbb{Z})$ is a vector bundle torsor over $\mathbb{P}_{\mathbb{Z}}^N$ and pull W back to X.

Exercise 23.30. Let X be an integral scheme and let $\mathscr{E}$ be a finite locally free $\mathcal{O}_X$-module. Show that there exists an integral scheme X' and a birational proper morphism $f \colon X' \to X$ such that $f^*\mathscr{E}$ has a filtration whose graded pieces are invertible $\mathcal{O}_{X'}$-modules.
Hint: It suffices to show that there exists an f as above such that $\mathbb{P}(f^*\mathscr{E})$ has a section over X'.

Exercise 23.31. This problem exhibits the jumping phenomenon in the upper semi-continuity theorem. Let C be a smooth projective connected curve of positive genus (see Chapter 26) over an algebraically closed field k, and x_0 a closed point on C. Let D be the divisor $\Delta_C - \{x_0\} \times_k C$ on $C \times_k C$, where Δ_C denotes the divisor on $C \times_k C$ given by the image of the diagonal embedding $(1_C, 1_C) : C \to C \times_k C$. Consider the projection $p : C \times_k C \to C$ to the second factor, then compute

$$\dim_k H^i(C \times \{x\}, \mathcal{O}_{C \times_k C}(D)|_{C \times \{x\}})$$

for $x \in C(k)$.

Hint: It might be useful to use the fact that, since C has genus > 0, we have $H^0(C, \mathcal{O}_C(x - x_0)) = 0$ for all $x \neq x_0$, see Corollary 26.20.

Exercise 23.32. Let $f : X \to S$ be a morphism of schemes and let $\Omega^{\bullet}_{X/S}$ be the de Rham complex. It is a complex of $f^{-1}\mathcal{O}_S$-modules (but the differentials are usually not $\mathcal{O}_X$-linear) and we can consider it as an object of $D(X, f^{-1}\mathcal{O}_S)$. Considering f as a morphism of ringed spaces $(X, f^{-1}\mathcal{O}_S) \to (S, \mathcal{O}_S)$ we can form $Rf_*\Omega^{\bullet}_{X/S} \in D(S)$. Forming global sections we also obtain $R\Gamma(X, \Omega^{\bullet}_{X/S}) \in D(\Gamma(S, \mathcal{O}_S))$ and in particular the *de Rham cohomology*

$$H^p_{\mathrm{dR}}(X/S) := H^p(R\Gamma(X, \Omega^{\bullet}_{X/S})) \in (\Gamma(S, \mathcal{O}_S)\text{-Mod}).$$

(1) Let $X \to S$ be a morphism of schemes given by a ring homomorphism $R \to A$. Show that in $D(R)$ one has an isomorphism $R\Gamma(X, \Omega^{\bullet}_{X/S}) = \Omega^{\bullet}_{A/R}$ and in particular $H^i_{\mathrm{dR}}(X/S) = H^i(\Omega^{\bullet}_{A/R})$.

(2) Show that the first hypercohomology sequence yields a converging spectral sequence

$$E_1^{pq} = R^q f_* \Omega^p_{X/S} \Rightarrow Rf_* \Omega^{\bullet}_{X/S},$$

called the *Hodge spectral sequence*.

(3) Let f be qcqs. Show that $Rf_*\Omega^{\bullet}_{X/S} \in D_{\mathrm{qcoh}}(S)$.

(4) Let S be locally noetherian and let f be proper. Show that $Rf_*\Omega^{\bullet}_{X/S} \in D_{\mathrm{coh}}(S)$.

(5) Let f be proper and smooth. Show that $Rf_*\Omega^p_{X/S}$ for $p \geq 0$ and $Rf_*\Omega^{\bullet}_{X/S}$ are perfect objects in $D(S)$ whose formation commutes with arbitrary base change.

(6) Show that one can apply Remark 21.130 to the de Rham complex $\Omega^{\bullet}_{X/S}$ and that one obtains on $\bigoplus_{p \in \mathbb{Z}} H^p_{\mathrm{dR}}(X/S)$ the structure of a strictly graded commutative graded $\Gamma(S, \mathcal{O}_S)$-algebra. The multiplication

$$\cup : H^p_{\mathrm{dR}}(X/S) \otimes H^q_{\mathrm{dR}}(X/S) \longrightarrow H^{p+q}_{\mathrm{dR}}(X/S)$$

is called the *cup product*.

Exercise 23.33. Let R be a ring, $n \geq 1$ an integer, and set $X := \mathbb{P}^n_R$.
(1) Show that $H^q(X, \Omega^p_{X/R})$ is a free R-module of rank

$$\mathrm{rk}_R H^q(X, \Omega^p_{X/R}) = \begin{cases} 1, & \text{if } 0 \leq p = q \leq n; \\ 0, & \text{otherwise.} \end{cases}$$

(2) Show that the de Rham cohomology (Exercise 23.32) $H^i_{\mathrm{dR}}(X/R)$ is a free R-module of rank 1 if $0 \leq i \leq 2n$ is even and of rank 0 otherwise.

Exercise 23.34. Let k be a field, let $n_1, n_2 \geq 1$ be integers, let $p_i \colon X := \mathbb{P}_k^{n_1} \times_k \mathbb{P}_k^{n_2} \to \mathbb{P}_k^{n_i}$ be the projections. Show that $\mathscr{O}_X(1,1) := p_1^* \mathscr{O}_{\mathbb{P}_k^{n_1}}(1) \otimes p_2^* \mathscr{O}_{\mathbb{P}_k^{n_2}}(1)$ is an ample line bundle on X. What is the Hilbert polynomial of $\mathbb{P}_k^n \times_k \mathbb{P}_k^m$ with respect to $\mathscr{O}_X(1,1)$?

Exercise 23.35. Let X be a finite scheme over a field. Show that the Hilbert polynomial Φ_X is a constant polynomial with value $\dim_k \Gamma(X, \mathscr{O}_X)$.

Exercise 23.36. Let k be a field and let $n, d \geq 1$ be integers and set $N := \binom{n+d}{d} - 1$. Let $Z = \mathbb{P}_k^n$ embedded into $\mathbb{P}_k^N$ via the d-fold Veronese embedding $v_d \colon \mathbb{P}_k^n \to \mathbb{P}_k^N$ (Exercise 13.8). Show that the degree of (Z, v_d) is d^n.

Exercise 23.37. Let S be a scheme and let $E \in D(S)$ be a pseudo-coherent complex.
(1) Show that E is perfect if and only if for every point $s \in S$ the complex $E \otimes^L \kappa(s)$ is perfect.
(2) Let $f \colon X \to S$ be a surjective morphism of schemes. Show that E is perfect if and only if Lf^*E is perfect.
(3) Let $S = \operatorname{Spec} R$ be affine and let $I \subseteq R$ be an ideal contained in the Jacobson radical of R. Show that if $E \otimes_R^L R/I$ is perfect, then E is perfect.

Exercise 23.38. Let $i \colon Z \to S$ be a thickening of schemes. Let $E \in D_{\mathrm{qcoh}}(S)$ be a complex. Show that E is pseudo-coherent (resp. perfect) if and only if Li^*E is pseudo-coherent (resp. perfect).

Exercise 23.39. Let A be a ring and $I \subseteq A$ be an ideal such that A is I-adically complete. Let $E \in D(A)$ be a complex. Show that E is pseudo-coherent (resp. perfect) if and only if $E \otimes_A^L A/I$ is pseudo-coherent (resp. perfect).
Hint: Use Exercise 23.38. Exercise 23.37 (3) might also be useful.

Exercise 23.40. Let S be a scheme, let E be a pseudo-coherent complex in $D(S)$, and let $b \in \mathbb{Z}$ such that $H^p(E) = 0$ for all $p > b$. Let $a \leq b$ and let $r \colon [a, b] \to \mathbb{Z}_{\geq 0}$, $i \mapsto r_i$, be a map. Show that there exists a unique locally closed subscheme $j \colon Z = Z_r \to S$ such that a morphism $f \colon T \to S$ factors through Z if and only if for all morphisms $g \colon T' \to T$ the $\mathscr{O}_{T'}$-modules $H^i(L(f \circ g)^*E)$ is locally free of rank r_i for all $i \in [a, b]$. Show that moreover the following assertions hold.
(1) The immersion $j \colon Z \to X$ is of finite presentation.
(2) As a set one has

(23.31.3) $Z = \{\, s \in S \;;\; b^i(s) = r_i \text{ for all } i \in [a, b] \,\}.$

(3) If $f \colon T \to S$ factors into $T \xrightarrow{\bar{f}} Z \xrightarrow{j} X$, then $H^i(Lf^*E \otimes_{\mathscr{O}_T} \mathscr{G}) = \bar{f}^*H^i(Lj^*E) \otimes_{\mathscr{O}_T} \mathscr{G}$ for all $i \in [a, b]$ and for all quasi-coherent $\mathscr{O}_T$-modules $\mathscr{G}$.

Exercise 23.41. Let $f \colon X \to S$ be a flat projective morphism of noetherian schemes. Assume that $H^0(X_s, \mathscr{O}_{X_s}) = \kappa(s)$ for all s, where X_s denotes the scheme-theoretic fiber of f over s. Show that $f_* \mathscr{O}_X = \mathscr{O}_S$.

Hint: As a lemma, prove the following: If $\varphi \colon A \to B$ is a ring homomorphism such that B is a locally free A-module of rank 1 via φ, then φ is an isomorphism.

Remark: The assumption $H^0(X_s, \mathscr{O}_{X_s}) = \kappa(s)$ holds if all fibers of f are geometrically reduced and geometrically connected (Proposition 12.66). See also Corollary 24.63 for a more general result.

Exercise 23.42. Let k be a field, $A := k[X,Y]/(XY,Y^2)$, and $M := A/(Y) = k[X]$ as an A-module.

(1) Find a partial flat resolution of the A-module M of length at least 4, i.e., an exact sequence $F_n \to \cdots \to F_2 \to F_1 \to F_0 \to M \to 0$ of A-modules with $n \geq 3$ and all F_i flat.

(2) Now consider the fiber product diagram

$$
\begin{array}{ccc}
\operatorname{Spec} k[Y]/(Y^2) & \xrightarrow{\ v\ } & \operatorname{Spec} A \\
{\scriptstyle g}\downarrow & & \downarrow{\scriptstyle f} \\
\operatorname{Spec} k & \xrightarrow{\ u\ } & \operatorname{Spec} k[X]
\end{array}
\quad,
$$

where u corresponds to the origin of the affine line $\operatorname{Spec} k[X]$, and f is given by the canonical inclusion $k[X] \to A$.

Using Part (1), show that the canonical map

$$
Lu^* Rf_* \tilde{M} \to Rg_* Lv^* \tilde{M}
$$

is not an isomorphism, but the composition

$$
Lu^* Rf_* \tilde{M} \to Rg_* Lv^* \tilde{M} \to Rg_* v^* \tilde{M}
$$

is an isomorphism.

Exercise 23.43. Let S be a locally noetherian scheme, and $f : X \to S$ a flat proper morphism. We say that f is *cohomologically flat (in dimension 0)*, if for any base change diagram

$$
\begin{array}{ccc}
X_T & \xrightarrow{\ v\ } & X \\
{\scriptstyle g}\downarrow & & \downarrow{\scriptstyle f} \\
T & \xrightarrow{\ u\ } & S
\end{array}
\quad,
$$

the canonical map $u^* f_* \mathscr{O}_X \to g_* v^* \mathscr{O}_X$ is an isomorphism. For example, a map f as in Exercise 23.41 is cohomologically flat.

(1) Let $T \to S$ be a faithfully flat morphism. Show that f is cohomological flat if and only if the base change $X \times_S T \to T$ is cohomologically flat.

(2) Let R be a discrete valuation ring, $\mathfrak{m}$ its maximal ideal, $S = \operatorname{Spec} R$, and let $f : X \to S$ be a proper flat morphism of schemes.

   Show that f is cohomologically flat if and only if $R^1 f_* \mathscr{O}_X$ is a free $\mathscr{O}_S$-module, in other words, if and only if $H^1(X, \mathscr{O}_X)$ has no torsion.

(3) In the setting of Part (2), now assume that $H^0(X, \mathscr{O}_X) = R$ and suppose that f is not cohomologically flat. For $n \geq 1$, we set $X_n = X \times_S \operatorname{Spec}(R/\mathfrak{m}^n)$. Show that for n sufficiently large, X_n is a flat $R/\mathfrak{m}^n$-scheme with the property that the module $H^0(X_n, \mathscr{O}_{X_n})$ of global sections is not flat over $R/\mathfrak{m}^n$.

Remark: For another, explicit example of a flat proper map which is not cohomologically flat, see [Liu]^{O}, Chapter 5, Exercise 3.15.

Exercise 23.44. Let k be a field and let X be a proper k-scheme of dimension d. Define the *arithmetic genus* of X by

$$p_a(X) := (-1)^d(\chi(\mathscr{O}_X) - 1).$$

Let $i\colon X \to \mathbb{P}^n_k$ be a closed embedding and let Φ be the Hilbert polynomial of X with respect to the ample line bundle $i^*\mathscr{O}_{\mathbb{P}^n_k}(1)$. Suppose that X is geometrically connected and geometrically reduced over k. Show that

$$p_a(X) = \sum_{i=1}^{d}(-1)^{d-i}\dim H^i(X, \mathscr{O}_X).$$

Show the following assertions.
(1) $p_a(\mathbb{P}^r_k) = 0$.
(2) Let $H \subset \mathbb{P}^r_k$ be a hypersurface of degree $d \geq 1$. Show that $p_a(H) = \binom{d-1}{r}$.
(3) Let $\mathscr{E}$ be a finite locally free $\mathscr{O}_X$-module of rank $r + 1$. Show that $p_a(\mathbb{P}(\mathscr{E})) = (-1)^r p_a(X)$.

Exercise 23.45. Let $f\colon X \to S$ be a morphism of schemes. We call an effective Cartier divisor $D \subset X$ a *relative effective Cartier divisor*, if D (considered as a closed subscheme of X) is flat over S.

(1) Let $f\colon X \to S$ be a morphism of schemes and let $D \subset X$ be a relative effective Cartier divisor. Let $S' \to S$ be a morphism of schemes. Show that the base change $D \otimes_S S'$ is a relative effective Cartier divisor in $X \times_S S'$ over S'.
(2) Let $f\colon X \to S$ be a flat morphism of finite type between noetherian schemes, and let $D \subset X$ be a closed subscheme of X that is flat over S. Show that D is a relative effective Cartier divisor (over S) if and only if for every $s \in S$, the fiber D_s of D over s (as a closed subscheme of the fiber $X_s = f^{-1}(s)$) is an effective Cartier divisor of X_s. Show that in this case the line bundle associated with D_s is $\mathscr{O}_X(D)_{|X_s}$.

Exercise 23.46. Let k be a field and let X be a proper k-scheme. Let $A \subseteq \mathrm{Div}(X)$ be the subgroup of the group of Cartier divisors on X generated by all differences $D_1 - D_0$ where T is a smooth integral k-scheme of finite type, $D \subset X \times T$ is a relative effective Cartier divisor over T (Exercise 23.45), $x_0, x_1 \in T(k)$, and D_i is the fiber of D over x_i, $i = 0, 1$.

We call divisors $D_0, D_1 \in \mathrm{Div}(X)$ *algebraically equivalent*, if $D_1 - D_0 \in A$. The quotient $\mathrm{Div}(X)/A$ is called the group of divisors on X up to algebraic equivalence.

(1) Let $D_0, D_1 \in \mathrm{Div}(X)$ be linearly equivalent. Prove that D_0 and D_1 are algebraically equivalent.
(2) Let $D_{i,0}, D_{i,1} \in \mathrm{Div}(X)$ be divisors, $i = 1, \ldots, d = \dim(X)$, such that for every i, $D_{0,i}$ and $D_{1,i}$ are algebraically equivalent. Prove that for the intersection numbers, we have $(D_{0,1} \cdots D_{0,d}) = (D_{1,1} \cdots D_{1,d})$.
Hint: : Use Proposition 23.151.
(3) Let X be a proper k-scheme of dimension 2, let $f\colon X \to C$ be a surjective flat morphism from X to a connected smooth proper curve C over k, let $x_0, x_1 \in C(k)$, and consider the divisors $D_i = f^*[x_i]$ as effective Cartier divisors on X. Show that for every Cartier divisor E on X, $(D_0 \cdot E) = (D_1 \cdot E)$.
Hint: Consider the graph of f.

24 Theorem on formal functions

Content

- Derived Completion
- The theorem of formal functions
- Stein factorization
- Algebraization

This chapter focuses on the theorem of formal functions and some of its applications. In the simplest form, this theorem says the following. Let S be a locally noetherian affine scheme and let $f \colon X \to S$ be proper. Let $\mathscr{J} \subseteq \mathscr{O}_S$ be a quasi-coherent ideal sheaf, so that we have the closed subschemes $X_n := f^{-1}(V(\mathscr{J}^n)) \subseteq X$, an ascending sequence of closed subschemes which all have the same topological space. Then the limit $\lim_n H^0(X_n, \mathscr{O}_{X_n})$ (these are the "formal functions") equals the completion of $H^0(X, \mathscr{O}_X)$ with respect to the topology induced by $\mathscr{J}$. More generally, the theorem covers non-affine base schemes, higher direct images rather than just global sections and arbitrary coherent modules. If one replaces the completion by derived completion, one obtains a version for arbitrary pseudo-coherent objects in $D(X)$ and even a version which does not require the assumption that the base be noetherian. See Theorems 24.28 and Proposition 24.35.

Derived Completion

Let A be a ring and let $I \subseteq A$ be an ideal. Then classically the I-adic completion of an A-module M is defined by $\hat{M} := \lim_n M/I^n M$. If A is noetherian, then this functor is exact on finitely generated A-modules M, and $\hat{M}$ is I-adically complete, i.e. the natural map $\hat{M} \to \lim_n \hat{M}/I^n \hat{M}$ is an isomorphism (Proposition B.41).

But beyond the case of finitely generated modules over noetherian rings this classical completion functor is often very badly behaved. It is the composition of the right exact functor that sends a module M to the projective system $\cdots \to M/I^2 M \to M/IM$ followed by the left exact limit functor. The composition is often not exact, not even exact in the middle even if A is noetherian (Exercise 24.5). Moreover, $\hat{M}$ is not necessarily I-adically complete (Exercise 24.6), although this can only happen if I is not finitely generated (Proposition 24.2).

Here we will introduce the more sophisticated approach of derived completion – at least if I is finitely generated. The idea is first to construct a "derived version" of the functor $M \mapsto M/I^n M$ and then to take the derived limit or, if one starts with an object M in $D(A)$, the homotopy limit. To explain this idea in more detail let us assume first that $I = (f)$ is generated by one element.

© Springer Fachmedien Wiesbaden GmbH, ein Teil von Springer Nature 2023
U. Görtz und T. Wedhorn, *Algebraic Geometry II: Cohomology of Schemes*,
Springer Studium Mathematik – Master, https://doi.org/10.1007/978-3-658-43031-3_9

Instead of $M/f^n M$ we consider the two-term complex $M \xrightarrow{f^n} M$, concentrated in degrees -1 and 0. There exists a quasi-isomorphism from this complex to $M/f^n M$ if f is M-regular. Then the derived completion of M is given by

$$(*) \qquad M^\wedge := R\lim_n (M \xrightarrow{f^n} M) = R\operatorname{Hom}_A(A \to A_f, M).$$

This is an object in $D(A)$ which is not necessarily concentrated in degree 0. Another possibility for a derived version of $M \mapsto M/f^n M$ would be to use the functor $M \mapsto M \otimes_A^L A/f^n$. In general, the homotopy limit of $M \otimes_A^L A/f^n$ might not be $M^\wedge$ although this is the case if the f-torsion is bounded in A (Exercise 24.7).

If $I = (f_1, \dots, f_r)$ is generated by finitely many elements, we first observe that the projective systems $(M/I^n M)_n$ and $(M/(f_1^n, \dots, f_r^n))_n$ are equivalent and then replace in the definition (*) of the completion the complex

$$(M \xrightarrow{f^n} M) = M \otimes_A (A \xrightarrow{f^n} A)$$

by $M \otimes_A K(f_1^n, \dots, f_r^n)[-r]$, where

$$K(f_1^n, \dots, f_r^n)[-r] = (A \xrightarrow{f_1^n} A) \otimes_A \cdots \otimes_A (A \xrightarrow{f_r^n} A)$$

is the Koszul complex of the sequence $(f_1^n, \dots, f_r^n)$ shifted into degrees $[0, r]$. This amounts to replacing $A \to A_f$ in (*) by the tensor products of the complexes $A \to A_{f_i}$.

If A is noetherian and M is finitely generated, we will see that $M^\wedge$ is the classical completion $\lim_n M/I^n M$.

All these constructions can be generalized to consider the derived completion with respect to an ideal sheaf of finite type on a ringed space. Here we will give the construction only if the ideal is globally generated by finitely many elements, which is the only case that we use in the sequel.

Let us finish this introduction with a note on derived limits. Let $\mathcal{A}$ be a Grothendieck abelian category. If

$$(**) \qquad \cdots \longrightarrow M_2 \longrightarrow M_1 \longrightarrow M_0$$

is a projective system of complexes in $C(\mathcal{A})$, then we can view (**) as a complex of projective systems, i.e., as an element of $C(\mathcal{A}^{\mathbb{N}^{\mathrm{opp}}})$, and form the termwise derived limit $R\lim_n M_n \in D(\mathcal{A})$. If (**) is a projective system of objects in $D(\mathcal{A})$, i.e., an object in $D(\mathcal{A})^{\mathbb{N}^{\mathrm{opp}}}$, then it is in general not possible to view it as an element of $D(\mathcal{A}^{\mathbb{N}^{\mathrm{opp}}})$. Instead one takes the homotopy limit, cf. Section (F.54). If (**) is a projective system of complexes in $C(\mathcal{A})$ one has

$$R\lim M_n = \operatorname{holim} M_n$$

in $D(\mathcal{A})$ by Proposition F.231, where on the right side we consider the image of $(M_n)_n$ in $D(\mathcal{A})^{\mathbb{N}^{\mathrm{opp}}}$.

(24.1) Reminder on completions.

Before defining derived completions, we collect some definitions and properties of the classical completion of modules.

Let A be a ring, let $I \subseteq A$ be an ideal, and let M be an A-module. The *I-adic completion* of M is defined as the limit

(24.1.1) $$\hat{M} := \lim_{n} M/I^n M.$$

In particular we have the I-adic completion

(24.1.2) $$\hat{A} = \lim A/I^n A$$

of A. We consider $\hat{A}$ as an A-algebra and $\hat{M}$ as an $\hat{A}$-module. An A-module M is called
I-adically complete if the map

$$M \longrightarrow \hat{M} = \lim_{n} M/I^n M$$

is an isomorphism. The ring A is called I-adically complete if it is I-adically complete as
an A-module.

Remark 24.1. Let us link these notions to topological algebra. Given a ring A and an
ideal I there exists on A a unique topology making the underlying abelian group $(A, +)$
into a topological group such that $(I^n)_n$ is a basis of neighborhoods of 0. Moreover, this
topology makes A into a topological ring ([BouGT] ^{O} III, §6.3, Example 3). Similarly, given
an A-module M, there exists on M a unique topology making $(M, +)$ into a topological
group such that $(I^n M)_{n \geq 1}$ is a basis of neighborhoods of 0. This topology makes M into
a topological module over the topological ring A. It is called the I-adic topology on A
and on M.

There is the general notion of the completion of an abelian topological group (with
respect to its canonical uniform structure), see [BouGT] ^{O} III, §3.4, §3.5. In this case, this
is easily described since the I-adic topology on M is by definition first countable: We
call a sequence $(x_n)_n$ of elements $x_n \in M$ a Cauchy sequence if for all integers $m \geq 0$
there exists an $N \geq 0$ such that $x_n - x_{n'} \in I^m M$ for all $n, n' \geq N$. Then the Hausdorff
completion of M is given as usual by equivalence classes of Cauchy sequences. In fact,
the I-adic topology on M is given by a non-archimedean pseudo-metric, i.e., by a map
$d \colon M \times M \to \mathbb{R}_{\geq 0}$ that satisfies the following conditions for all $m, m', m'' \in M$
(a) $m = m' \Rightarrow d(m, m') = 0$,
(b) $d(m, m') = d(m', m)$,
(c) $d(m, m'') \leq \max\{d(m, m'), d(m', m'')\}$.
For instance, d can be defined by $d(m, m') = c^{v(m,m')}$, where $0 < c < 1$ is a fixed real
number and

$$v(m, m') = \sup\{ n \in \mathbb{N} \; ; \; m - m' \in I^n M \}$$

with the convention $c^\infty := 0$. With this definition the completion becomes the usual
Hausdorff completion of a pseudo-metric space.

By [BouGT] ^{O} III, §7.3, Cor. 1 of Prop. 2, the Hausdorff completion of A is a metrizable
complete topological ring isomorphic to $\hat{A}$, as defined in (24.1.2), and the Hausdorff
completion of M is a metrizable complete topological A-module isomorphic to $\hat{M}$, as
defined in (24.1.1). Its topology is the unique topology making $(\hat{M}, +)$ into a topological
group such that the

$$\widehat{I^n M} = \lim_{m \geq n} I^n M/I^m M = \mathrm{Ker}(\hat{M} \to M/I^n M)$$

form a basis of neighborhoods of 0.

The topological module M is itself Hausdorff (resp. Hausdorff and complete) if and
only if $\bigcap_n I^n M = 0$ (resp. it it is I-adically complete in the sense defined above).

However, in general one does not have $\widehat{I^n M} = I^n \hat{M}$ and the topology on $\hat{M}$ will in general be neither the I-adic topology ($\hat{M}$ considered as an A-module) nor the $I\hat{A}$-adic topology ($\hat{M}$ considered as an $\hat{A}$-module), see Exercise 24.6. However, if I is finitely generated, then the situation is better, as the following result shows.

Proposition 24.2. *Let A be a ring, let $I \subseteq A$ be a finitely generated ideal, and let M be an A-module. Then $I^n \hat{M} = \widehat{I^n M}$ for all $n \geq 0$, in particular the topology on $\hat{M}$ is the I-adic topology and $\hat{M}$ is I-adically complete.*

Proof. As I is finitely generated, I^n is finitely generated, say $I^n = (f_1, \ldots, f_r)$. Then

$$u \colon M^r \longrightarrow I^n M, \qquad (m_1, \ldots, m_r) \mapsto \sum_i f_i m_i,$$

is surjective. Hence the induced map on I-adic completions

$$\hat{u} \colon (\hat{M})^r \to \widehat{I^n M} = \mathrm{Ker}(\hat{M} \to M/I^n M), \qquad (\hat{m}_1, \ldots, \hat{m}_r) \mapsto \sum_i f_i \hat{m}_i$$

is surjective (Proposition B.49). Since the image of $\hat{u}$ is $I^n \hat{M}$, we see that $I^n \hat{M} = \widehat{I^n M}$. $\quad\square$

(24.2) Derived complete complexes.

Let A be a ring. For a complex M of A-modules and $f \in A$ we set

$$T(M, f) := R\lim(M \xleftarrow{f} M \xleftarrow{f} M \longleftarrow \ldots) \in D(A).$$

As $A_f = \mathrm{colim}(A \xrightarrow{f} A \xrightarrow{f} A \xrightarrow{f} \ldots)$ we see that

$$\mathrm{Hom}_{C(A)}(A_f, M) = \lim(M \xleftarrow{f} M \xleftarrow{f} M \longleftarrow \ldots)$$

and hence obtain an isomorphism in $D(A)$, functorial in M,

(24.2.1) $$T(M, f) \cong R\mathrm{Hom}_A(A_f, M).$$

If M is an A-module, we define

$$\delta_f \colon \prod_{\mathbb{N}} M \longrightarrow \prod_{\mathbb{N}} M, \qquad (x_n) \mapsto (x_n - f x_{n+1}).$$

Then Proposition F.224 shows that

(24.2.2)
$$\begin{aligned}
H^0 T(M, f) &= \mathrm{Hom}_A(A_f, M) = \mathrm{Ker}(\delta_f), \\
H^1 T(M, f) &= \mathrm{Ext}_A^1(A_f, M) = \mathrm{Coker}(\delta_f), \\
H^p T(M, f) &= \mathrm{Ext}_A^p(A_f, M) = 0 \qquad \text{for all } p \neq 0, 1.
\end{aligned}$$

Lemma 24.3. *Let A be a ring, M a complex of A-modules and $f \in A$ such that $T(M, f) = 0$. Then $R\mathrm{Hom}_A(E, M) = 0$ for every complex E of A_f-modules.*

Proof. In view of (24.2.1), the claim follows from the following equality.

(24.2.3) $$R\mathrm{Hom}_A(E, M) = R\mathrm{Hom}_{A_f}(E, R\mathrm{Hom}_A(A_f, M)).$$

To show the equality, let $M \to I$ be a K-injective resolution. Then $\operatorname{Hom}_A(A_f, I)$ is a K-injective complex of A_f-modules. Indeed, let X be an exact complex of A_f-modules. Then using (*) and Proposition F.179 one sees that

$$\operatorname{Hom}_{K(A_f)}(X, \operatorname{Hom}_A(A_f, I)) = \operatorname{Hom}_{K(A_f)}(X \otimes_A A_f, \operatorname{Hom}_A(A_f, I))$$
$$= \operatorname{Hom}_{K(A)}(X, I) = 0,$$

which shows that $\operatorname{Hom}_A(A_f, I)$ is K-injective. As $E \otimes_A A_f = E$, we find

$$R\operatorname{Hom}_{A_f}(E, R\operatorname{Hom}_A(A_f, M)) = R\operatorname{Hom}_{A_f}(E, \operatorname{Hom}_A(A_f, I))$$
$$= \operatorname{Hom}_{A_f}(E, \operatorname{Hom}_A(A_f, I))$$
$$= \operatorname{Hom}_{A_f}(E \otimes_A A_f, I)$$
$$= R\operatorname{Hom}_A(E, M). \qquad \square$$

Lemma 24.4. *Let A be a ring and let M be a complex of A-modules. Then the set I of $f \in A$ such that $T(M, f) = 0$ is a radical ideal of A.*

Proof. Recall that $T(M, f) = R\operatorname{Hom}_A(A_f, M)$ by (24.2.1). Let $f \in I$ and $g \in A$. Then A_{fg} is an A_f-module. Hence $R\operatorname{Hom}_A(A_{fg}, M) = 0$ by Lemma 24.3 and thus $fg \in I$.

Now let $f, g \in I$. As f and g generate the unit ideal in A_{f+g}, there is an exact sequence

$$0 \longrightarrow A_{f+g} \longrightarrow A_{f(f+g)} \times A_{g(f+g)} \longrightarrow A_{gf(f+g)} \longrightarrow 0$$

It is the extended alternating Čech complex of the structure sheaf on $\operatorname{Spec}(A_{f+g})$ attached to the open covering $(D(f), D(g))$ which is exact because the higher cohomology groups vanish for the structure sheaf. All modules except possibly A_{f+g} in the sequence are modules over A_f or A_g and hence are annihilated by the functor $R\operatorname{Hom}_A(-, M)$. Hence $R\operatorname{Hom}_A(A_{f+g}, M) = 0$.

Finally, let $f \in A$ and $n \geq 1$ with $f^n \in I$. Then $f \in I$ because $A_f = A_{f^n}$. $\qquad \square$

Definition 24.5. *Let A be a ring and let $I \subseteq A$ be an ideal. Then a complex M in $D(A)$ is called* derived complete with respect to I *or* derived I-complete *if $T(M, f) = 0$ for all $f \in I$.*

The full subcategory of derived I-complete complexes in $D(A)$ is denoted by $D_{\operatorname{comp}}(A, I)$.

As usual we consider A-modules as complexes concentrated in degree 0 and hence have also the notion of derived I-complete A-module. This notion is closely linked to the notion of an I-adically complete module, as we will see in Proposition 24.7 below.

Remark 24.6. Let A be a ring, let $I \subseteq A$ be an ideal, and let $S \subseteq A$ be a subset such that the ideal generated by S and the ideal I have the same radical ideal.
(1) A complex M in $D(A)$ is derived I-complete if and only if $T(M, f) = 0$ for all $f \in S$. This follows from Lemma 24.4.

In particular, derived completeness with respect to I for a complex M depends only on the radical of I, i.e., only on the closed subspace $V(I)$ of $\operatorname{Spec} A$.

Moreover, for every complex M in $D(A)$ there exists a greatest ideal $I_M \subseteq A$ such that M is derived I_M-complete. Exercise 24.16 shows that I_A is always contained in the Jacobson radical of A.
(2) By (24.2.2) an A-module M is derived I-complete if and only if $\operatorname{Hom}_A(A_f, M) = \operatorname{Ext}^1_A(A_f, M) = 0$ for all $f \in S$.

(3) The category $D_{\text{comp}}(A, I)$ is a triangulated subcategory (Definition F.131) because $R \operatorname{Hom}_A(A_f, -)$ is a triangulated functor. Moreover, $D_{\text{comp}}(A, I)$ is stable under passing to direct summands (i.e., if $M \oplus N$ are in $D_{\text{comp}}(A, I)$, then M and N are in $D_{\text{comp}}(A, I)$).

(4) As the formation of $T(M, f)$ commutes with products in M, the category $D_{\text{comp}}(A, I)$ is stable under arbitrary products.

(5) By Lemma F.230, $R \operatorname{Hom}_A(A_f, -)$ commutes with homotopy limits, so $D_{\text{comp}}(A, I)$ is stable under homotopy limits in $D(A)$.

We now relate derived completeness and adic completeness.

Proposition 24.7. *Let A be a ring, $I \subseteq A$ an ideal, and let M be an A-module.*
(1) *Suppose that M is I-adically complete, i.e., $M \to \lim_n M/I^n M$ is an isomorphism. Then M is derived I-complete.*
(2) *Conversely, suppose that I is finitely generated and that M is derived I-complete. Then $M \to \lim_n M/I^n M$ is surjective.*

The proof will show that in (2) it suffices to assume that I is finitely generated and that $H^1 T(M, f) = 0$, where f runs through a generating system of I.

Proof. Part (1).. Assume that M is I-adically complete. By (24.2.2) it suffices to show that $\operatorname{Hom}_A(A_f, M) = 0$ and $\operatorname{Coker}(\delta_f) = 0$ for $f \in I$, where

$$\delta_f \colon \prod_{\mathbb{N}} M \to \prod_{\mathbb{N}} M, \qquad (x_n)_n \mapsto (x_n - f x_{n+1}).$$

But $\operatorname{Hom}_A(A_f, M) = \lim \operatorname{Hom}_A(A_f, M/I^n M)$ and $\operatorname{Hom}_A(A_f, M/I^n M) = 0$ since $f \in I$. To show that the map δ_f is surjective, we define for $(y_n)_n \in \prod_{\mathbb{N}} M$ elements

$$x_n := \sum_{k=0}^{\infty} f^k y_{n+k}.$$

The series converges because M is I-adically complete. Then $\delta_f((x_n)_n) = (y_n)_n$.

Part (2).. Conversely, assume that M is derived I-complete for an ideal generated by elements $f_1, \ldots, f_r$. We first show that $M \to \lim M/f_i^n M$ is surjective for all $i = 1, \ldots, r$. For this we may assume that $I = (f)$ and $T(M, f) = 0$. An element of $\lim M/f^n M$ is given by a family $(x_n)_n$ of elements $x_n \in M$ such that $x_{n+1} = x_n + f^n y_n$ for some $y_n \in M$. Setting $y_0 := x_1$ we hence have $x_{n+1} = \sum_{i=0}^n f^i y_i$. We look for an $x \in M$ such that $x \equiv \sum_{i=0}^{n-1} f^i y_i \bmod f^n M$, for all n.

As δ_f is surjective, we can find $z_n \in M$ such that $y_n = z_n + f z_{n+1}$ for all $n \geq 0$. We then have $\sum_{i=0}^{n-1} y_i f^i = z_0 - f^n z_n$, so that we can choose $x := z_0$.

To show that $M \to \lim M/I^n M$ is surjective, note that $I^{rn} \subseteq (f_1^n, \ldots, f_r^n) \subseteq I^n$ implies $\lim M/I^n M = \lim M/(f_1^n, \ldots, f_r^n)$. An element x in $\lim M/(f_1^n, \ldots, f_r^n)$ is given by a family of elements $x_n \in M$ such that $x_{n+1} = x_n + f_1^n y_{n,1} + \cdots + f_r^n y_{n,r}$ for elements $y_{n,i} \in M$. Setting $y_{0,1} := x_1$ and $y_{0,i} := 0$, we can represent it as a formal sum

$$x = \sum_{n \geq 0} \sum_{i=1}^{r} f_i^n y_{n,i}.$$

As $M \to \lim M/f_i^n M$ is surjective, there exists $y_i \in M$ mapping to $\sum_n f_i^n y_{n,i}$ in $\lim M/f_i^n M$. Then $y = \sum_i y_i$ maps to x in $\lim M/(f_1^n, \ldots, f_r^n)$. This shows the surjectivity of $M \to \lim M/I^n M$. $\qquad\square$

Corollary 24.8. *Let A be a ring, $I \subsetneq A$ be an ideal, and let M be an A-module.*
(1) *Let I be finitely generated by elements $f_1, \ldots, f_r$ and suppose $\bigcap_n I^n M = 0$. Then the following assertions are equivalent.*
  (i) *The A-module M is I-adically complete.*
  (ii) *The A-module is derived I-complete.*
  (iii) *One has $H^1 T(M, f_i) = 0$ for all $i = 1, \ldots, r$.*
(2) *Let A be noetherian and let M be a finitely generated A-module. Suppose that I is contained in the Jacobson radical of A. Then M is I-adically complete if and only if it is derived I-complete.*

Exercise 24.15 shows that if A is a reduced ring and I is a finitely generated ideal, then A is I-adically complete if and only if A is derived I-adically complete.

Proof. The first assertion follows immediately from Proposition 24.7. It implies the second because one has $\bigcap_n I^n M = 0$ in this situation by Proposition B.42. $\square$

Proposition 24.9. *Let A be a ring, let M in $D(A)$ be a complex and let $I \subseteq A$ be an ideal. Then M is derived I-complete if and only if $H^q(M)$ is derived I-complete for all $q \in \mathbb{Z}$.*

Proof. Let $f \in I$. We have to show that $\mathrm{Ext}_A^p(A_f, H^q(M)) = 0$ for all $p, q \in \mathbb{Z}$ if and only if $\mathrm{Ext}_A^p(A_f, M) = 0$ for all $p \in \mathbb{Z}$. The second hypercohomology spectral sequence for the functor $\mathrm{Hom}_A(A_f, -)$ (F.49.4) yields a spectral sequence

$$E_2^{pq} = \mathrm{Ext}_A^p(A_f, H^q(M)) \Rightarrow \mathrm{Ext}_A^{p+q}(A_f, M).$$

By (24.2.2) we have $\mathrm{Ext}_A^p(A_f, H^q(M)) = 0$ for $p \neq 0, 1$. Therefore the spectral sequence degenerates at E_2 and the filtration of the limit term has only two steps. Thus we get for all p exact sequences

$$0 \to \mathrm{Ext}_A^1(A_f, H^{q-1}(M)) \to \mathrm{Ext}_A^p(A_f, M) \to \mathrm{Hom}_A(A_f, H^q(M)) \to 0$$

which proves the claim. $\square$

(24.3) Derived completion.

Let A be a ring, and let $I \subseteq A$ be an ideal generated by finitely many elements $f_1, \ldots, f_r \in I$. In other words

$$\mathrm{Spec}(A) \setminus V(I) = \bigcup_{i=1}^r D(f_i)$$

is quasi-compact. Recall that we defined in Section (22.4) the extended ordered Čech complex

(24.3.1) $C_A(f_1, \ldots, f_r)$: $A \longrightarrow \prod_{i_1} A_{f_{i_1}} \longrightarrow \prod_{i_1 < i_2} A_{f_{i_1} f_{i_2}} \longrightarrow \cdots \longrightarrow A_{f_1 f_2 \cdots f_r}$,

where we put A in degree 0.

Proposition 24.10. *The functor*

$$M \mapsto M^\wedge := R\operatorname{Hom}_A(C_A(f_1,\dots,f_r), M)$$

defines a triangulated functor $D(A) \to D_{\text{comp}}(A, I)$ which is left adjoint to the inclusion functor $D_{\text{comp}}(A, I) \to D(A)$. We call $M^\wedge$ the derived completion *of M.*

As the inclusion $D_{\text{comp}}(A, I) \to D(A)$ is fully faithful, it follows formally that $(M^\wedge)^\wedge = M^\wedge$ for every complex M in $D(A)$ (Proposition F.22). We will also see this in the proof.

Proof. The canonical quotient map $C_A(f_1,\dots,f_r) \to A$, where A is considered as a complex concentrated in degree 0, yields a functorial morphism $M = R\operatorname{Hom}_A(A, M) \to M^\wedge$. It is then enough to show that $M^\wedge$ is derived I-complete and that $M \to M^\wedge$ is an isomorphism if M is derived I-complete.

Let $f \in I$. For all $g \in A$ one has $A_f \otimes_A^L A_g = A_f \otimes A_g = A_{fg}$ as localization is flat. Hence the adjointness of $R\operatorname{Hom}$ and $\otimes^L$ (Theorem 21.119) shows that

$$R\operatorname{Hom}(A_f, M^\wedge) = R\operatorname{Hom}(C_{A_f}(ff_1,\dots,ff_r), M).$$

But $C_{A_f}(ff_1,\dots,ff_r)$ is the extended ordered Čech complex of the open affine covering $\operatorname{Spec}(A_f) = \bigcup_i D(ff_i)$ hence it is exact by Theorem 22.1. Therefore it is zero in $D(A)$. This shows $R\operatorname{Hom}(A_f, M^\wedge) = 0$ for all $f \in I$ and hence $M^\wedge$ is derived I-complete.

If M is derived complete, then $R\operatorname{Hom}_A(A_f, M) = 0$ for all f of the form $f_{i_1}f_{i_2}\cdots f_{i_p}$ and hence $C_A(f_1,\dots,f_r) \to A$ induces an isomorphism after applying $R\operatorname{Hom}(-, M)$. $\square$

Remark 24.11. Let I be a finitely generated ideal of a ring A. As $D_{\text{comp}}(A, I)$ does not depend on the choice of a set of generators of I and even depends only on the radical of I (Remark 24.6 (1)), the same holds for the completion functor $M \mapsto M^\wedge$. In fact, Exercise 22.10 shows that this also true for the extended ordered Čech complex as an object in $D(A)$.

Note that if M is an A-module, considered as a complex concentrated in degree 0, then its derived completion $M^\wedge$ is in general not concentrated in degree 0 (Exercise 24.2). In particular, the derived completion of A itself is not necessarily a ring anymore[1].

We will now describe the derived completion differently using the Koszul complex (see Section (19.1)). We first need a general remark on the functorial property of homotopy limits (Definition F.228) for which we refer to the Stacks Project [Sta].

Remark 24.12. Let A be a ring, and let $(K_n)_{n\in\mathbb{N}}$ be an inverse system of objects in $D(A)$. By Remark F.232 we can lift $(K_n)_n$ to an object $(\tilde{K}_n)_n$ of $D((A\text{-Mod})^{\mathbb{N}})$. We then obtain by [Sta] 091J a triangulated functor

$$\operatorname{holim}(- \otimes_A^L K_n)\colon D(A) \longrightarrow D(A).$$

This functor depends on the choice of $(\tilde{K}_n)_n$ but for each X in $D(A)$ the isomorphism class of $\operatorname{holim}(X \otimes_A^L K_n)$ is independent of the choice. Moreover, we have by [Sta] 091K and 091L the following functoriality properties in the inverse system.

---

[1] However, it is still an animated ring in the sense of Higher Algebra.

(1) Let $\tilde{K} \to \tilde{L} \to \tilde{M} \to \tilde{K}[1]$ be a distinguished triangle in $D((A\text{-Mod})^{\mathbb{N}})$ and let $(K_n)_n$, $(L_n)_n$, and $(M_n)_n$ be the associated inverse systems in $D(A)$. Then for every X in $D(A)$ there is a distinguished triangle in $D(A)$, functorial in X,

$$\operatorname{holim}(X \otimes^L K_n) \longrightarrow \operatorname{holim}(X \otimes^L L_n) \longrightarrow \operatorname{holim}(X \otimes^L M_n) \xrightarrow{+1}$$

(2) Let $\tilde{K} \to \tilde{L}$ be a morphism in $D((A\text{-Mod})^{\mathbb{N}})$. Let $(K_n)_n$ and $(L_n)_n$ be the inverse systems associated to $\tilde{K}$ and $\tilde{L}$, respectively. Let $(K_n) \to (L_n)$ be an isomorphism of pro-objects in $D(A)$ (i.e., there is given an isomorphism $\operatorname{colim}_n \operatorname{Hom}_{D(A)}(K_n, -) \to \operatorname{colim}_n \operatorname{Hom}_{D(A)}(L_n, -)$ of functors $D(A) \to (\text{Sets})$). Then for all X in $D(A)$ the corresponding morphism

$$\operatorname{holim}(X \otimes_A^L K_n) \longrightarrow \operatorname{holim}(X \otimes_A^L L_n)$$

is an isomorphism.

Recall (Section (22.4)) that we related the extended ordered Čech complex $C_A(f_1, \dots, f_r)$ of (24.3.1) to the Koszul complex and showed that, writing $\mathbf{f}^n = (f_1^n, \dots, f_r^n)$,

$$(24.3.2) \qquad \operatorname*{colim}_n K(\mathbf{f}^n)[-r] \cong \operatorname*{colim}_n K(\mathbf{f}^n)^\vee = C_A(f_1, \dots, f_r) \qquad \text{in } C(A).$$

Proposition 24.13. *Let A be a ring, let $I \subseteq A$ be an ideal generated by a finite sequence $\mathbf{f} = (f_1, \dots, f_r)$, and let M be a complex in $D(A)$ with derived I-completion $M^\wedge$. Then there is an isomorphism in $D(A)$, functorial in M,*

$$M^\wedge \cong \operatorname{holim}(M \otimes_A^L K(\mathbf{f}^n)).$$

Proof. Set $K^n := K(\mathbf{f}^n)^\vee$. This is a strictly perfect complex of A-modules and hence $R\operatorname{Hom}_A(K^n, A) = K(\mathbf{f}^n)$ by double duality of perfect complexes (Proposition 21.148). We find $M \otimes_A^L K(\mathbf{f}^n) = R\operatorname{Hom}_A(K^n, M)$ by Proposition 21.148 and hence by Remark F.229 that

$$(*) \qquad\qquad \operatorname{holim}(M \otimes_A^L K(\mathbf{f}^n)) \cong \operatorname{holim} R\operatorname{Hom}_A(K^n, M).$$

By definition of the homotopy limit there is a distinguished triangle

$$\operatorname{holim} R\operatorname{Hom}_A(K^n, M) \longrightarrow \prod R\operatorname{Hom}_A(K^n, M) \longrightarrow \prod R\operatorname{Hom}_A(K^n, M) \xrightarrow{+1}.$$

On the other hand we have

$$M^\wedge = R\operatorname{Hom}_A(C_A(\mathbf{f}), M) = R\operatorname{Hom}_A(\operatorname{colim} K^n, M) \cong R\operatorname{Hom}_A(\operatorname{hocolim} K^n, M),$$

where the second equality holds by (24.3.2) and the isomorphism $\operatorname{colim} K^n \cong \operatorname{hocolim} K^n$ by Lemma F.235. By definition of $\operatorname{hocolim} K^n$ and because $R\operatorname{Hom}$ is a triangulated functor in both variables, there is a distinguished triangle

$$R\operatorname{Hom}_A(\operatorname{hocolim} K^n, M) \longrightarrow R\operatorname{Hom}_A(\bigoplus K^n, M) \longrightarrow R\operatorname{Hom}_A(\bigoplus K^n, M) \xrightarrow{+1}.$$

Because we have $\prod R\operatorname{Hom}_A(K^n, M) = R\operatorname{Hom}_A(\bigoplus K_n, M)$ by Corollary 21.121, we deduce that $R\operatorname{Hom}_A(\operatorname{hocolim} K^n, M) \cong \operatorname{holim} R\operatorname{Hom}_A(K^n, M)$. This shows the proposition by (*). $\qquad\square$

Corollary 24.14. *Let A be a ring and let $I \subseteq A$ be an ideal that can be generated by r elements. Let $M \in D^{[a,b]}(A)$ be a complex with $-\infty \leq a \leq b \leq \infty$. Then the derived completion $M^\wedge$ is in $D^{[a-r,b]}(A)$.*

Proof. By Proposition 24.13 it suffices to show that $\mathrm{holim}(M \otimes_A^L K(\mathbf{f}^n)) \in D^{[a-r,b]}$. As $K(\mathbf{f}^n)$ is a K-flat complex, one has

$$M \otimes_A^L K(\mathbf{f}^n) = M \otimes_A K(\mathbf{f}^n).$$

As $K(\mathbf{f}^n)$ is concentrated in degrees $[-r, 0]$ we find $M \otimes_A^L K(\mathbf{f}^n) \in D^{[a-r,b]}(A)$. Then Lemma F.233 shows that $H^p(M^\wedge) = 0$ for $p \notin [a-r, b+1]$ and for $b < \infty$ one has

$$H^{b+1}(M^\wedge) \cong R^1 \lim H^b(M \otimes_A K(\mathbf{f}^n)).$$

Representing M by a complex with $M^p = 0$ for $p > b$, we see that

$$H^b(M \otimes K(\mathbf{f}^n)) = H^b(M) \otimes_A H^0(K(\mathbf{f}^n)) = H^b(M) \otimes_A A/(f_1^n, \ldots, f_r^n).$$

Hence the transition maps of the inverse system $H^b(M \otimes_A K(\mathbf{f}^n))$ are all surjective. Therefore $R^1 \lim H^b(M \otimes_A K(\mathbf{f}^n)) = 0$ by Proposition F.227. $\square$

We now compare the derived I-completion of a complex $M \in D(A)$ to the "naive derived completion" $\mathrm{holim}(M \otimes^L A/I^n)$. There are several important cases in which this already yields the derived completion. One is that of noetherian rings, see Proposition 24.16 below. Another is the case of ideals generated by one element f provided the f-torsion is bounded. Both results rely on the following remark.

Remark 24.15. Let A be a ring and let $I \subseteq A$ be an ideal generated by a finite family $\mathbf{f} = (f_1, \ldots, f_r)$. By Proposition 24.13 we have $M^\wedge \cong \mathrm{holim}(M \otimes^L K(\mathbf{f}^n))$. So if the pro-objects $(K(\mathbf{f}^n))_n$ and $(A/I^n)_n$ are isomorphic (Remark 24.12 (2)), then it follows that $M^\wedge \cong \mathrm{holim}(M \otimes^L A/I^n)$. As $I^{rn} \subseteq (f_1^n, \ldots, f_r^n) \subseteq I^n$, the pro-objects $(A/I^n)_n$ and $(A/(f_1^n, \ldots, f_r^n))_n$ are isomorphic.

We see that if the pro-objects $(K(\mathbf{f}^n))_n$ and $(A/(f_1^n, \ldots, f_r^n))_n$ are isomorphic, in which the sequence $(f_1, \ldots, f_r)$ is said to be *weakly proregular*, then for every complex M in $D(A)$ one has an isomorphism

$$(24.3.3) \qquad\qquad M^\wedge \cong \mathrm{holim}(M \otimes^L A/I^n),$$

which is functorial in M. See also Exercise 24.8 for more results on weakly proregular sequences.

This is in particular the case if $\mathbf{f}$ is completely intersecting. Then $\mathbf{f}^n$ is also completely intersecting (Proposition 19.17) and hence $K(\mathbf{f}^n) \cong A/(f_1^n, \ldots, f_r^n)$ in $D(A)$ for all n.

Proposition 24.16. *Let A be a noetherian ring. Then for every complex M in $D(A)$ there is a functorial isomorphism $M^\wedge \cong \mathrm{holim}(M \otimes^L A/I^n)$.*

Proof. As explained in Remark 24.15 it suffices to show that for every n there exists $m \geq n$ such that $K(\mathbf{f}^m) \to K(\mathbf{f}^n)$ factors through $K(\mathbf{f}^m) \to \tau^{\geq 0}(K(\mathbf{f}^m)) = A/(f_1^m, \ldots, f_r^m)$.

(I). We first claim that for all $p < 0$ and for all $n \geq 1$ there exists an $m \geq n$ such that the homomorphism $H^p(K(\mathbf{f}^m)) \to H^p(K(\mathbf{f}^n))$ is zero. Indeed, as A is noetherian, $H^p(K(\mathbf{f}^n))$ is a finitely generated A-module for all p and $n \geq 1$. By Remark 22.15 (1) the transition maps $K(\mathbf{f}^m)^p \to K(\mathbf{f}^n)^p$ for $m > n$ and $p < 0$ are given by multiplying the vectors in the standard basis by elements of the ideal I_{m-n} with $I_s := (f_1^s, \ldots, f_r^s)$ for $s \geq 1$.

Now we have $I_{m-n} \subseteq I_n^t$ for $m = n + tn$. The Artin-Rees lemma (Proposition B.40) shows that for large t one has

$$I_n^t K(\mathbf{f}^n)^p \cap \mathrm{Ker}\big(K(\mathbf{f}^n)^p \to K(\mathbf{f}^n)^{p+1}\big) \subseteq I_n \,\mathrm{Ker}\big(K(\mathbf{f}^n)^p \to K(\mathbf{f}^n)^{p+1}\big).$$

Putting things together, we obtain

$$\mathrm{Im}\big(K(\mathbf{f}^m)^p \to K(\mathbf{f}^n)^p\big) \cap \mathrm{Ker}\big(K(\mathbf{f}^n)^p \to K(\mathbf{f}^n)^{p+1}\big)$$
$$\subseteq I_{m-n} K(\mathbf{f}^n)^p \cap \mathrm{Ker}\big(K(\mathbf{f}^n)^p \to K(\mathbf{f}^n)^{p+1}\big)$$
$$\subseteq I_n^t K(\mathbf{f}^n)^p \cap \mathrm{Ker}\big(K(\mathbf{f}^n)^p \to K(\mathbf{f}^n)^{p+1}\big)$$
$$\subseteq I_n \,\mathrm{Ker}\big(K(\mathbf{f}^n)^p \to K(\mathbf{f}^n)^{p+1}\big).$$

This shows the claim because I_n annihilates $H^p(K(\mathbf{f}^n))$ (Remark 19.6).

(II). Let $n \geq 1$. As $H^p(K(\mathbf{f}^m)) = 0$ for $p \notin [-r, 0]$ and for all m, we find $m = n_0 \geq n_1 \geq \cdots \geq n_r = n$ such that $H^p(K(\mathbf{f}^{n_{i-1}})) \to H^p(K(\mathbf{f}^{n_i}))$ is zero for all i and for all p by our claim. Set $K_i := K(\mathbf{f}^{n_i})$. We claim that $K_i \to K_r = K(\mathbf{f}^n)$ factors through $\tau^{\geq -i} K_i$. Then this claim for $i = 0$ shows the proposition.

This is clear for $i = r$ because $\tau^{\geq -r} K_i = K_i$ for all i. Consider now the case $i = r - 1$ assuming $r > 0$. There is a distinguished triangle (Proposition F.157) in $D(A)$

$$H^{-r}(K_{r-1})[r] \longrightarrow K_{r-1} \longrightarrow \tau^{\geq 1-r} K_{r-1} \xrightarrow{+1}$$

yielding an exact sequence

$$\mathrm{Hom}_{D(A)}(\tau^{\geq 1-r} K_{r-1}, K_r) \longrightarrow \mathrm{Hom}_{D(A)}(K_{r-1}, K_r) \longrightarrow \mathrm{Hom}_{D(A)}(H^{-r}(K_{r-1})[r], K_r).$$

The composition $H^{-r}(K_{r-1})[r] \to K_{r-1} \to K_r$ is zero because it is equal to the composition $H^{-r}(K_{r-1})[r] \to H^{-r}(K_r)[r] = \tau^{\leq -r} K_r \to K_r$, where the first map vanishes by construction of the K_i. Therefore $K_{r-1} \to K_r$ factors through $\tau^{\geq 1-r} K_{r-1}$.

Now the same argument applied to $\tau^{\geq 1-r} K_{r-2} \to \tau^{\geq 1-r} K_{r-1}$ shows that this morphism factors through $\tau^{\geq 2-r} K_{r-2}$. Hence $K_{r-2} \to K_r$ factors through $\tau^{\geq 2-r} K_{r-2}$. Proceeding by induction shows our claim.   □

Proposition 24.17. *Let A be a noetherian ring, let $I \subseteq A$ be an ideal, and let M be an object in $D(A)$ such that $H^p(M)$ is a finitely generated A-module for all $p \in \mathbb{Z}$. Then the cohomology modules $H^p(M^\wedge)$ of the derived I-completion are the I-adic completions of the cohomology modules $H^p(M)$.*

In particular we see that if M is as in the proposition and in $D^{[a,b]}(A)$, then $M^\wedge \in D^{[a,b]}(A)$.

Proof. (I). We first assume that M is in $D^-(A)$. Then M is pseudo-coherent by Proposition 22.61. Hence it is isomorphic in $D(A)$ to a bounded above complex F of finite free A-modules by Proposition 21.162. As F is K-flat by Lemma 21.93, we have $M \otimes_A^L A/I^n = F/I^n F$ and hence

$$M^\wedge = R \lim F/I^n F = \lim F/I^n F,$$

where the first equality holds by Proposition 24.16 and Proposition F.231 and the second because the transition maps are surjective (Proposition F.227). As the usual I-adic completion is exact on finite A-modules (Proposition B.39) this shows the assertion.

(II). Now let M in $D(A)$ be arbitrary with $H^p(M)$ finitely generated. Fix $p \in \mathbb{Z}$ and consider a distinguished triangle

$$\tau^{\leq p+1+r}M \longrightarrow M \longrightarrow C \xrightarrow{+1},$$

where r is the cardinality of some finite set of generators of I. Then C is in $D^{[p+r+1,\infty)}(A)$ and hence $C^\wedge$ is in $D^{[p+1,\infty)}(A)$ by Corollary 24.14. Hence the long exact cohomology sequence of the distinguished triangle $(\tau^{\leq p+1+r}M)^\wedge \to M^\wedge \to C^\wedge \to$ yields an isomorphism $H^p((\tau^{\leq p+1+r}M)^\wedge) \xrightarrow{\sim} H^p(M^\wedge)$ and we can conclude by Step (I). $\qquad\square$

Corollary 24.18. *Let A be a noetherian ring, let $I \subseteq A$ be an ideal, and let M be a finite A-module. Then the derived I-completion of M is the usual I-adic completion of M.*

(24.4) Globalization of derived completion.

Most of the above constructions about the derived completion of modules over a ring can be globalized to derived completions of modules over ringed spaces X (or even general "ringed topoi") with respect to an ideal of finite type. These generalizations are rather straightforward if the ideal is globally generated by finitely many elements, the only case that we will need in the sequel.

We fix a ringed space $(X, \mathscr{O}_X)$. For $U \subseteq X$ open, $f \in \mathscr{O}_X(U)$ and $\mathscr{F}$ a complex of $\mathscr{O}_U$-modules we set

$$T(\mathscr{F}, f) := R\lim(\mathscr{F} \xleftarrow{f} \mathscr{F} \xleftarrow{f} \mathscr{F} \xleftarrow{f} \ldots) \in D(U).$$

If $\mathscr{F}$ is a complex of $\mathscr{O}_X$-modules, we set $T(\mathscr{F}, f) := T(\mathscr{F}_{|U}, f)$.

We also denote by $\mathscr{O}_{U,f}$ the sheaf of rings on U attached to the presheaf of localizations $V \mapsto (\mathscr{O}_X(V))_f$. Then for every $\mathscr{F}$ in $D(U)$ one has functorial isomorphisms

(24.4.1) $$T(\mathscr{F}, f) \cong R\mathscr{H}om_{\mathscr{O}_U}(\mathscr{O}_{U,f}, \mathscr{F})$$

in $D(U)$. For $V \subseteq U$ open we therefore have by (21.18.9) that

(24.4.2) $$T(\mathscr{F}, f)_{|V} = T(\mathscr{F}, f_{|V}).$$

Lemma 24.19. *Let $\mathscr{I} \subseteq \mathscr{O}_X$ be an ideal and let $\mathscr{F}$ be a complex of $\mathscr{O}_X$-modules. Let $U \subseteq X$ be open and $f \in \mathscr{I}(U)$. Then the following assertions are equivalent.*
(i) One has $T(\mathscr{F}, f) = 0$.
(ii) For all $\mathscr{E}$ in $D(\mathscr{O}_{U,f})$ one has

$$R\mathscr{H}om_{\mathscr{O}_U}(\mathscr{E}, \mathscr{F}_{|U}) = 0.$$

Proof. By (24.4.1), (ii) implies (i). Let us assume that (i) holds and let $\mathscr{E}$ be in $D(\mathscr{O}_{U,f})$. Choose a K-injective resolution $\mathscr{F}_{|U} \to \mathscr{I}$ in $D(U)$. The proof of (24.2.3) shows that

$$R\mathscr{H}om_{\mathscr{O}_U}(\mathscr{E}, \mathscr{F}_{|U}) = R\mathscr{H}om_{\mathscr{O}_{U,f}}(\mathscr{E}, T(\mathscr{F}, f)) = 0.$$

Hence $R\mathscr{H}om_{\mathscr{O}_U}(\mathscr{E}, \mathscr{F}_{|U}) = 0.$ $\qquad\square$

Definition 24.20. *Let $\mathscr{I} \subseteq \mathscr{O}_X$ be an ideal and let $\mathscr{F}$ be a complex of $\mathscr{O}_X$-modules. Then $\mathscr{F}$ is called* derived $\mathscr{I}$-complete *if for all $U \subseteq X$ open and $f \in \mathscr{I}(U)$ one has $T(\mathscr{F}, f) = 0$. Denote by $D_{\mathrm{comp}}(X) := D_{\mathrm{comp}}(X, \mathscr{I})$ the full subcategory of $D(X)$ consisting of derived $\mathscr{I}$-complete complexes.*

Remark 24.21. Let $\mathscr{I} \subseteq \mathscr{O}_X$ be an ideal and let $\mathscr{F}$ be a complex of $\mathscr{O}_X$-modules.
(1) Let $(U_i)_i$ be an open covering of X. Then (24.4.2) implies that $\mathscr{F}$ is derived $\mathscr{I}$-complete if and only if $\mathscr{F}_{|U_i}$ is derived $\mathscr{I}_{|U_i}$-complete for all i.
(2) If $\mathscr{F}$ is derived $\mathscr{I}$-complete and $\mathscr{J} \subseteq \mathscr{I}$ is a subideal, then $\mathscr{F}$ is derived $\mathscr{J}$-complete.

Proposition and Definition 24.22. *Let $\mathscr{I} \subseteq \mathscr{O}_X$ be an ideal of finite type. Then the inclusion functor $D_{\mathrm{comp}}(X, \mathscr{I}) \to D(X)$ has a left adjoint called* derived $\mathscr{I}$-completion. *It is denoted by $\mathscr{F} \mapsto \mathscr{F}^\wedge$.*
One has $\mathscr{F}^\wedge = R\mathscr{H}om_{\mathscr{O}_X}(\mathscr{C}, \mathscr{F})$ for a complex $\mathscr{C}$ in $D(X)$.

For schemes and quasi-coherent ideals we will use the following terminology.

Definition 24.23. *Let X be a scheme, let $\mathscr{I}$ be a quasi-coherent ideal of $\mathscr{O}_X$, and let Z be the corresponding closed subscheme of X. Then a complex $\mathscr{F}$ in $D(X)$ is also called* derived complete along Z *if it is derived $\mathscr{I}$-complete and we set $D_{\mathrm{comp}}(X, Z) := D_{\mathrm{comp}}(X, \mathscr{I})$. Moreover, we also call $\mathscr{F}^\wedge$ the* derived completion of $\mathscr{F}$ along Z *and sometimes write $\mathscr{F}^\wedge_{/Z}$.*

We refer to [Sta] 099F for a proof of Proposition 24.22 in general. We will explain the proof in the special case that the ideal $\mathscr{I}$ is globally generated by finitely many sections, where we can use the same argument as before. This hypothesis is for instance satisfied if there exists a morphism $f\colon X \to S$ of schemes, where $S = \operatorname{Spec} R$ is affine, and if $\mathscr{I}$ is the quasi-coherent ideal of $\mathscr{O}_X$ defining a closed subscheme of the form $f^{-1}(Z)$ for some closed subscheme Z of S that is defined by a finitely generated ideal of R. In all applications of derived completion in this book we will be in this situation.

Proof. If $\mathscr{I}$ is globally generated by sections $f_1, \ldots, f_r$, then one can use the same argument as in the proof of Proposition 24.10 using as derived completion the functor

$$(24.4.3) \qquad \mathscr{F} \mapsto \mathscr{F}^\wedge := R\mathscr{H}om_{\mathscr{O}_X}(C_X(f_1, \ldots, f_r), \mathscr{F}),$$

where $C_X(f_1, \ldots, f_r)$ is the complex

$$(24.4.4) \qquad \mathscr{O}_X \longrightarrow \prod_{i_1} \mathscr{O}_{X, f_{i_1}} \longrightarrow \prod_{i_1 < i_2} \mathscr{O}_{X, f_{i_1} f_{i_2}} \longrightarrow \cdots \longrightarrow \mathscr{O}_{X, f_1 f_2 \cdots f_r}$$

sitting in degrees $[0, r]$.   □

To show the existence of derived completion in general one constructs a complex $\mathscr{C}$ of $\mathscr{O}_X$-modules depending only on X and $\mathscr{I}$ such that for every open subset $U \subseteq X$ such that $\mathscr{I}_{|U}$ is generated by finitely many global sections $f_1, \ldots, f_r$ the restriction $\mathscr{C}_{|U}$ is isomorphic to the complex $C_U(f_1, \ldots, f_r)$ (24.4.4), see [Sta] 099E for details. Then one has $\mathscr{F}^\wedge = R\mathscr{H}om_{\mathscr{O}_X}(\mathscr{C}, \mathscr{F})$.

Remark 24.24. If $\mathscr{I}$ is globally generated by a family $\mathbf{f} = (f_1, \ldots, f_r)$ of sections, then the arguments of Proposition 24.13 show that one has for every complex $\mathscr{F}$ in $D(X)$ a functorial isomorphism

(24.4.5)
$$\mathscr{F}^{\wedge} := \operatorname{holim}(\mathscr{F} \otimes^{L}_{\mathscr{O}_{X}} K(\mathbf{f}^{n})),$$

where $K(\mathbf{f}^{n})$ is the Koszul complex of the global sections $f^{n}_{1}, \ldots, f^{n}_{r}$, considered as a complex sitting in degrees $[-r, 0]$.

We will see in Remark 24.30 below that the notions of derived completions over rings and over affine schemes are compatible.

Proposition 24.25. *Let $f \colon X \to Y$ be a morphism of ringed spaces and let $\mathscr{J} \subseteq \mathscr{O}_{Y}$ and $\mathscr{I} \subseteq \mathscr{O}_{X}$ be ideals such that $f^{\sharp} \colon f^{-1}\mathscr{O}_{Y} \to \mathscr{O}_{X}$ sends $f^{-1}\mathscr{J}$ into $\mathscr{I}$.*
(1) Then Rf_{*} sends $D_{\operatorname{comp}}(X, \mathscr{I})$ into $D_{\operatorname{comp}}(Y, \mathscr{J})$.
(2) Suppose that $\mathscr{I}$ is an ideal of finite type. Then the functor $Rf_{*} \colon D_{\operatorname{comp}}(X, \mathscr{I}) \to D_{\operatorname{comp}}(Y, \mathscr{J})$ has a left adjoint functor given by Lf^{*} followed by $\mathscr{I}$-derived completion.

Proof. (1). Let $\mathscr{F}$ be a complex in $D(X)$ that is $\mathscr{I}$-derived complete. Fix $V \subseteq Y$ open and $s \in \mathscr{J}(V)$ and let $U := f^{-1}(V)$. By adjunction of Rf_{*} and Lf^{*} (Proposition 21.126) one has

(*)
$$\operatorname{Hom}_{D(V)}(\mathscr{O}_{V,s}, Rf_{*}\mathscr{F}_{|V}) = \operatorname{Hom}_{D(U)}(Lf^{*}\mathscr{O}_{V,s}, \mathscr{F}_{|U}).$$

As $\mathscr{O}_{V,s}$ is a flat $\mathscr{O}_{V}$-module, we have $Lf^{*}\mathscr{O}_{V,s} = f^{*}\mathscr{O}_{V,s} = \mathscr{O}_{U,f^{*}(s)}$ and $f^{*}(s) \in \mathscr{I}(U)$ because of our assumption on $\mathscr{I}$ and $\mathscr{J}$. Now $\mathscr{F}[p]_{|U'}$ is derived $\mathscr{I}_{|U'}$-complete for all $p \in \mathbb{Z}$ and all open subsets U' of U. Hence we see that for all $V' \subseteq V$ open with $U' := f^{-1}(V')$ one has, using (21.21.5) for the first equality,

$$H^{p}(V', R\mathscr{H}om(\mathscr{O}_{V,s}, Rf_{*}\mathscr{F}_{|V})) = \operatorname{Hom}_{D(V')}(\mathscr{O}_{V',s_{|V'}}, Rf_{*}\mathscr{F}[p]_{|V'})$$
$$= \operatorname{Hom}_{D(U')}(\mathscr{O}_{U',f^{*}(s)_{|U'}}, \mathscr{F}[p]_{|U'}) = 0,$$

In view of Lemma 21.22 this shows that $R\mathscr{H}om(\mathscr{O}_{V,s}, Rf_{*}\mathscr{F}_{|V}) = 0$.

(2). Let $\mathscr{F}$ be in $D_{\operatorname{comp}}(X, \mathscr{I})$ and let $\mathscr{G}$ be in $D_{\operatorname{comp}}(Y, \mathscr{J})$. By the universal property of derived completion, which exists because $\mathscr{I}$ is of finite type, one has functorial bijections

$$\operatorname{Hom}(\mathscr{G}, Rf_{*}\mathscr{F}) = \operatorname{Hom}(Lf^{*}\mathscr{G}, \mathscr{F}) = \operatorname{Hom}((Lf^{*}\mathscr{G})^{\wedge}, \mathscr{F}).$$

$\square$

Derived completion commutes with pullback to open subspaces:

Proposition 24.26. *Let X be a ringed space, let $\mathscr{I} \subseteq \mathscr{O}_{X}$ be an ideal of finite type, let $j \colon U \to X$ be the inclusion of an open subspace, and let $\mathscr{F}$ be in $D(X)$. Then $\mathscr{F}^{\wedge}_{|U}$ is the derived completion of $\mathscr{F}_{|U}$ with respect to $\mathscr{I}_{|U}$.*

Proof. The restriction $\mathscr{F}^{\wedge}_{|U}$ is derived $\mathscr{I}_{|U}$-complete and we obtain a map $v \colon (\mathscr{F}_{|U})^{\wedge} \longrightarrow \mathscr{F}^{\wedge}_{|U}$ which is functorial in $\mathscr{F}$. Conversely, the adjunction unit yields a map $\mathscr{F} \to Rj_{*}(\mathscr{F}_{|U}) \to Rj_{*}((\mathscr{F}_{|U})^{\wedge})$ that factors through $\mathscr{F}^{\wedge} \to Rj_{*}((\mathscr{F}_{|U})^{\wedge})$ since the right hand side is derived complete by Proposition 24.25. By adjunction this morphism corresponds to a map $w \colon \mathscr{F}^{\wedge}_{|U} \to (\mathscr{F}_{|U})^{\wedge}$ that is an inverse map to v. $\square$

Proposition 24.27. *Let X be a ringed space and let $\mathscr{I} \subseteq \mathscr{O}_{X}$ be an ideal of finite type. Then for all $\mathscr{F}$ and $\mathscr{G}$ in $D(X)$ one has functorial isomorphisms*

$$R\mathscr{H}om_{\mathscr{O}_{X}}(\mathscr{F}, \mathscr{G})^{\wedge} \cong R\mathscr{H}om_{\mathscr{O}_{X}}(\mathscr{F}, \mathscr{G}^{\wedge}) \cong R\mathscr{H}om_{\mathscr{O}_{X}}(\mathscr{F}^{\wedge}, \mathscr{G}^{\wedge}).$$

Proof. By Proposition 24.22 derived completion is of the form $R\,\mathcal{H}om_{\mathcal{O}_X}(\mathscr{C}, -)$ for some complex $\mathscr{C}$ in $D(X)$. By the adjointness of $R\,\mathcal{H}om$ and $\otimes^L$ (Theorem 21.119) we find

$$R\,\mathcal{H}om_{\mathcal{O}_X}(\mathscr{C}, R\,\mathcal{H}om_{\mathcal{O}_X}(\mathscr{F}, \mathscr{G})) = R\,\mathcal{H}om_{\mathcal{O}_X}(\mathscr{C} \otimes^L \mathscr{F}, \mathscr{G})$$
$$= R\,\mathcal{H}om_{\mathcal{O}_X}(\mathscr{F}, R\,\mathcal{H}om_{\mathcal{O}_X}(\mathscr{C}, \mathscr{G})),$$

which gives the first isomorphism. The map $\mathscr{F} \to \mathscr{F}^\wedge$ induces by functoriality a map $R\,\mathcal{H}om_{\mathcal{O}_X}(\mathscr{F}^\wedge, \mathscr{G}^\wedge) \to R\,\mathcal{H}om_{\mathcal{O}_X}(\mathscr{F}, \mathscr{G}^\wedge)$ in $D(X)$. To show that this is an isomorphism it suffices by Remark 21.19 to show that for all open subsets U of X the induced map

$$R\,\mathrm{Hom}_{\mathcal{O}_U}((\mathscr{F}_{|U})^\wedge, (\mathscr{G}_{|U})^\wedge) = R\Gamma(U, R\,\mathcal{H}om_{\mathcal{O}_X}(\mathscr{F}^\wedge, \mathscr{G}^\wedge))$$
$$\longrightarrow R\Gamma(U, R\,\mathcal{H}om_{\mathcal{O}_X}(\mathscr{F}, \mathscr{G}^\wedge)) = R\,\mathrm{Hom}_{\mathcal{O}_U}(\mathscr{F}_{|U}, (\mathscr{G}_{|U})^\wedge)$$

is an isomorphism in $D(\Gamma(U, \mathcal{O}_X))$, where the equalities hold by Proposition 24.26 and Proposition 21.107. Using (F.52.5) and the universal property of the derived completion we see that this map induces an isomorphism after applying $H^p(-)$. Hence it is an isomorphism. □

The theorem of formal functions

In this section we prove the theorem on formal functions. We start by the most general derived version, and afterwards discuss how the statement simplifies if the base scheme is locally noetherian. In Section (24.7) we give a direct proof of the noetherian version which does not require the machinery of derived completion.

(24.5) Theorem of formal functions, derived version.

We formulate a quite formal result that says that derived completion commutes with derived pushforward in complete generality.

Theorem 24.28. *Let $f\colon X \to Y$ be a morphism of ringed spaces, let $\mathscr{J} \subseteq \mathcal{O}_Y$ be an ideal of finite type and let $\mathscr{I}$ be the ideal of $\mathcal{O}_X$ generated by $f^\sharp(f^{-1}\mathscr{J})$. Then one has for every complex $\mathscr{F}$ in $D(X)$ a functorial isomorphism*

(24.5.1) $$(Rf_*\mathscr{F})^\wedge \xrightarrow{\sim} Rf_*(\mathscr{F}^\wedge).$$

Here $(Rf_*\mathscr{F})^\wedge$ denotes the $\mathscr{J}$-derived completion of $Rf_*\mathscr{F}$, and $\mathscr{F}^\wedge$ denotes the $\mathscr{I}$-derived completion of $\mathscr{F}$.

If X and Y are schemes and $\mathscr{J}$ is the quasi-coherent ideal sheaf corresponding to a closed subscheme Z of Y, then $\mathscr{I}$ is quasi-coherent and corresponds to the closed subscheme $f^{-1}(Z)$ of X.

Proof. By Proposition 24.25, $Rf_*(\mathscr{F}^\wedge)$ is $\mathscr{J}$-derived complete. Hence the universal property of derived completion yields a map $(Rf_*\mathscr{F})^\wedge \to Rf_*(\mathscr{F}^\wedge)$. To check that this map is an isomorphism we can work locally on Y and hence assume that $\mathscr{I}$ is generated by a family $\mathbf{s} = (s_1, \ldots, s_r)$ of global sections. Then $\mathscr{J}$ is generated by $\mathbf{t} := (t_1, \ldots, t_r)$ with $t_i := f^*(s_i)$. Hence by Remark 24.24 one has to prove that

$$\mathrm{holim}(Rf_*\mathscr{F} \otimes^L K(\mathbf{s}^n)) \cong Rf_*(\mathrm{holim}\,\mathscr{F} \otimes^L K(\mathbf{t}^n)).$$

As Rf_* has a left adjoint functor Lf^*, it commutes with products and hence with homotopy limits by Lemma F.230. Therefore it suffices to show that

$$Rf_*\mathscr{F} \otimes^L K(\mathbf{s}^n) \cong Rf_*(\mathscr{F} \otimes^L K(\mathbf{t}^n)).$$

This follows from the projection formula (Proposition 22.84) since $K(\mathbf{s}^n)$ is perfect and $Lf^*K(\mathbf{s}^n) = f^*K(\mathbf{s}^n) = K(\mathbf{t}^n)$. □

Corollary 24.29. *Let A be a ring and let $I \subseteq A$ be a finitely generated ideal. Let $(X, \mathscr{O}_X)$ be a ringed space such that $\Gamma(X, \mathscr{O}_X)$ is endowed with the structure of an A-algebra. Then for every complex $\mathscr{F}$ in $D(X)$ one has*

$$R\Gamma(X, \mathscr{F})^\wedge = R\Gamma(X, \mathscr{F}^\wedge),$$

where $\mathscr{F}^\wedge$ is the derived completion with respect to $I\mathscr{O}_X$ and where the left hand side is the I-derived completion.

Later, we will apply the corollary when $X \to \mathrm{Spec}\,A$ is a morphism of schemes.

Proof. Apply Theorem 24.28 to the morphism $(X, \mathscr{O}_X) \to (Y, \mathscr{O}_Y)$, where Y consists of single point and $\mathscr{O}_Y(Y) = A$. □

Remark 24.30. Let A be a ring, let $I \subseteq A$ be an ideal generated by elements $f_1, \ldots, f_r$, let $X = \mathrm{Spec}\,A$ and let $\mathscr{I} \subseteq \mathscr{O}_X$ be the quasi-coherent ideal corresponding to I. Let M be an object in $D(A)$ and let $\tilde{M}$ be the corresponding object in $D(\mathrm{QCoh}(X)) = D_{\mathrm{qcoh}}(X) \subseteq D(X)$, where the equality holds by Lemma 22.36. Then $M = R\Gamma(X, \tilde{M})$ by Theorem 22.2. It then follows from Corollary 24.29 that

$$(M^\wedge)^\sim = \tilde{M}^\wedge$$

and that M is derived I-complete if and only if $\tilde{M}$ is derived $\mathscr{I}$-complete.

Corollary 24.31. *Let A, I, and X as in Corollary 24.29. Let $\mathscr{F}$ and $\mathscr{G}$ be complexes of $\mathscr{O}_X$-modules. Then there is a functorial isomorphism*

$$R\Gamma(X, R\mathscr{H}om_{\mathscr{O}_X}(\mathscr{F}, \mathscr{G}))^\wedge \cong R\Gamma(X, R\mathscr{H}om_{\mathscr{O}_X}(\mathscr{F}^\wedge, \mathscr{G}^\wedge))$$

in $D(A)$ where $\mathscr{F}^\wedge$ and $\mathscr{G}^\wedge$ denotes the derived completion with respect to $I\mathscr{O}_X$ and where the left hand side is the derived completion with respect to I.

Proof. Applying Corollary 24.29 to $R\mathscr{H}om(\mathscr{F}, \mathscr{G})$ one obtains

$$R\Gamma(X, R\mathscr{H}om_{\mathscr{O}_X}(\mathscr{F}, \mathscr{G}))^\wedge = R\Gamma(X, R\mathscr{H}om_{\mathscr{O}_X}(\mathscr{F}, \mathscr{G})^\wedge)$$

and one has $R\mathscr{H}om_{\mathscr{O}_X}(\mathscr{F}, \mathscr{G})^\wedge = R\mathscr{H}om_{\mathscr{O}_X}(\mathscr{F}^\wedge, \mathscr{G}^\wedge)$ by Proposition 24.27. □

(24.6) Theorem of formal functions for locally noetherian schemes.

Let X be a ringed space, let $\mathcal{I} \subseteq \mathcal{O}_X$ be a finitely generated ideal, and let $\mathcal{F}$ be an object in $D(X)$. For all $n \geq 1$, any local section $f \in \mathcal{I}(U)$ acts nilpotently on $\mathcal{F} \otimes^L_{\mathcal{O}_X} \mathcal{O}_X/\mathcal{I}^n$ and thus $R\mathcal{H}om(\mathcal{O}_{U,f}, \mathcal{F}) = 0$ by (24.4.1). Therefore $\mathcal{F} \otimes^L_{\mathcal{O}_X} \mathcal{O}_X/\mathcal{I}^n$ is derived $\mathcal{I}$-complete and there is a natural map in $D(X)$

$$(24.6.1) \qquad \mathcal{F}^\wedge \longrightarrow \operatorname*{holim}_n(\mathcal{F} \otimes^L \mathcal{O}_X/\mathcal{I}^n).$$

Lemma 24.32. *Let X be a locally noetherian scheme, let $\mathcal{I} \subseteq \mathcal{O}_X$ be a quasi-coherent ideal, and let $\mathcal{F}$ be in $D(X)$. Then (24.6.1) is an isomorphism.*

Proof. Restriction to an open subscheme of X commutes with derived completion by Proposition 24.26 and with homotopy limits (by construction of holim). Hence we may assume that $X = \operatorname{Spec} A$ is affine with a noetherian ring A. Then the lemma follows by Remark 24.30 from the corresponding assertion for objects in $D(A)$ (Proposition 24.16). $\square$

Proposition 24.33. *Let X be a locally noetherian scheme, let $\mathcal{I}$ be a quasi-coherent ideal, let $\mathcal{F}$ be a pseudo-coherent complex in $D(X)$, and let $\mathcal{F}^\wedge$ be its derived completion. Then for all $p \in \mathbb{Z}$ one has*

$$(24.6.2) \qquad H^p(\mathcal{F}^\wedge) = \lim_n H^p(\mathcal{F})/\mathcal{I}^n H^p(\mathcal{F}).$$

Proof. Let $\operatorname{Spec} A = U \subseteq X$ be an open affine subscheme. As $\mathcal{F}$ is pseudo-coherent, $R\Gamma(U, \mathcal{F})$ is a pseudo-coherent object in $D(A)$ (Lemma 22.44). Hence

$$H^p(U, \mathcal{F}^\wedge) = H^p(R\Gamma(U, \mathcal{F})^\wedge) = \lim_n H^p(U, \mathcal{F})/\mathcal{I}(U)^n H^p(U, \mathcal{F}),$$

where the first equality holds by Corollary 24.29 and the second by Proposition 24.17 since A is noetherian. As this holds for all $U \subseteq X$ open affine, by sheafification we obtain (24.6.2). $\square$

Applying Proposition 24.33 to a complex concentrated in degree 0 we obtain:

Corollary 24.34. *Let X be a locally noetherian scheme, let $\mathcal{I}$ be a quasi-coherent ideal, and let $\mathcal{F}$ be a coherent $\mathcal{O}_X$-module. Then derived completion coincides with usual completion, i.e.,*

$$\mathcal{F}^\wedge = \lim_n \mathcal{F}/\mathcal{I}^n \mathcal{F}.$$

There is also the following noetherian variant of Theorem 24.28.

Proposition 24.35. *Let S be a locally noetherian scheme, let $\mathcal{J} \subseteq \mathcal{O}_S$ be a quasi-coherent ideal, let $f\colon X \to S$ be a proper morphism, and let $\mathcal{F}$ be a pseudo-coherent object in $D(X)$ (e.g., if $\mathcal{F}$ is a coherent $\mathcal{O}_X$-module). Then one has for all $p \geq 0$ a functorial isomorphism*

$$(R^p f_* \mathcal{F})^\wedge \xrightarrow{\sim} R^p f_*(\mathcal{F}^\wedge).$$

Here the left side denotes the $\mathcal{J}$-adic completion of $R^p f_*(\mathcal{F})$ and $\mathcal{F}^\wedge$ on the right side is the derived completion with respect to the ideal $\mathcal{I}$ generated by $f^\sharp(f^{-1}\mathcal{J})$ (equal to the $\mathcal{I}$-adic completion if $\mathcal{F}$ is a coherent module).

Proof. By Theorem 24.28 one has

$$H^p((Rf_*\mathscr{F})^\wedge) = R^p f_*(\mathscr{F}^\wedge).$$

As $\mathscr{F}$ is pseudo-coherent and f is proper, $Rf_*\mathscr{F}$ is pseudo-coherent by Proposition 23.23. Therefore one concludes by Proposition 24.33. □

From Theorem 24.28 and Proposition 24.33 one easily deduces Theorem 24.37 below, a weak form of the theorem of formal functions.

Lemma 24.36. *Let $S = \operatorname{Spec} A$ be an affine noetherian scheme, let $I \subseteq A$ be an ideal, let $f\colon X \to S$ be a proper morphism, and let $\mathscr{F}$ be a coherent $\mathscr{O}_X$-module. Then one has an isomorphism*

$$(24.6.3) \qquad H^p(X, \lim_n \mathscr{F}/I^n\mathscr{F}) \xrightarrow{\sim} \lim_n H^p(X, \mathscr{F})/I^n H^p(X, \mathscr{F}),$$

of A-modules, functorial in $\mathscr{F}$.

Here we denote by $\mathscr{F}/I^n\mathscr{F}$ the coherent $\mathscr{O}_X$-module $\mathscr{F}/\mathscr{I}^n\mathscr{F}$, where $\mathscr{I}$ is the quasi-coherent ideal of $\mathscr{O}_X$ corresponding to the closed subscheme $f^{-1}(V(I))$.

Proof. Applying to (24.5.1) the functor $R\Gamma(S, -)$ and using that $R\Gamma(S, -) \circ Rf_* = R\Gamma(X, -)$ (Proposition 21.114 (2)), we obtain

$$(*) \qquad R\Gamma(X, \mathscr{F}^\wedge) = R\Gamma(S, Rf_*(\mathscr{F}))^\wedge).$$

As $R\Gamma(S, -)$ commutes with derived completion (Corollary 24.29), the right hand side of (*) is equal to

$$R\Gamma(S, Rf_*(\mathscr{F})))^\wedge = R\Gamma(X, \mathscr{F})^\wedge.$$

Taking cohomology in degree p we obtain from (*) that

$$H^p(X, \mathscr{F}^\wedge) = H^p(R\Gamma(X, \mathscr{F})^\wedge) = H^p(X, \mathscr{F})^\wedge,$$

where the second equality holds by Proposition 24.17. Note that $R\Gamma(X, \mathscr{F})$ is pseudo-coherent in $D(A)$ because f is proper. As $\mathscr{F}^\wedge = \lim_n \mathscr{F}/I^n\mathscr{F}$ (Corollary 24.34) and $H^p(X, \mathscr{F})^\wedge = \lim_n H^p(X, \mathscr{F})/I^n H^p(X, \mathscr{F})$ (Corollary 24.18) we obtain (24.6.3). □

Theorem 24.37. *(Theorem of formal functions, noetherian case) Let A be a noetherian ring and let $I \subseteq A$ be an ideal. Let $X \to S := \operatorname{Spec} A$ be a proper morphism and let $\mathscr{F}$ be a coherent $\mathscr{O}_X$-module. Then for all $p \in \mathbb{Z}$ one has*

$$(24.6.4) \qquad \lim_n H^p(X, \mathscr{F}/I^n\mathscr{F}) = \lim_n H^p(X, \mathscr{F})/I^n H^p(X, \mathscr{F}).$$

Let $X_n := X \otimes_A A/I^n$ viewed as a closed subscheme of X and let $i_n\colon X_n \to X$ be the inclusion. Then $\mathscr{F}/I^n\mathscr{F} = (i_n)_* i_n^*\mathscr{F}$ and hence

$$H^p(X, \mathscr{F}/I^n\mathscr{F}) = H^p(X_n, i_n^*\mathscr{F})$$

by Corollary 22.6.

We will deduce the Theorem of formal functions from the derived version Theorem 24.28. The key lemma in the proof is the fact that the system $(H^p(X, \mathscr{F}/I^n\mathscr{F}))_n$ satisfies the Mittag-Leffler condition (Definition F.225). This we will show in Section (24.7) below, where we will also give a direct proof of Theorem 24.37 without using derived completion.

Proof. By Lemma 24.36 it suffices to show that the functorial map $H^p(X, \lim_n \mathscr{F}/I^n\mathscr{F}) \to \lim_n H^p(X, \mathscr{F}/I^n\mathscr{F})$ is an isomorphism. This map can be written as the composition

$$
\begin{aligned}
H^p(X, \lim_n \mathscr{F}/I^n\mathscr{F}) &= H^p(R\Gamma(X, \lim_n \mathscr{F}/I^n\mathscr{F})) \\
&\xrightarrow{\text{(A)}} H^p(R\Gamma(X, R\lim_n \mathscr{F}/I^n\mathscr{F})) \\
&= H^p(R\lim_n R\Gamma(X, \mathscr{F}/I^n\mathscr{F})) \\
&\xrightarrow{\text{(B)}} \lim_n H^p(R\Gamma(X, \mathscr{F}/I^n\mathscr{F})) \\
&= \lim_n H^p(X, \mathscr{F}/I^n\mathscr{F}).
\end{aligned}
$$

(24.6.5)

The equality in the middle holds because $R\lim$ and $R\Gamma(X, -)$ commute by Lemma F.230.

Now (A) is an isomorphism by Lemma 24.38 below. By Lemma F.233, (B) is surjective with kernel $R^1 \lim H^{p-1}(X, \mathscr{F}/I^n\mathscr{F})$. But by Lemma 24.40 below (see also Remark 24.41 below for a much simpler proof if I is a maximal ideal), the inverse system $(H^{p-1}(X, \mathscr{F}/I^n\mathscr{F}))_n$ satisfies the Mittag-Leffler condition and hence Proposition F.227 yields that $R^1 \lim H^{p-1}(X, \mathscr{F}/I^n\mathscr{F}) = 0$. $\qquad\square$

Lemma 24.38. *Let X be a scheme and let $(\mathscr{F}_n)_n$ be an inverse system of quasi-coherent $\mathcal{O}_X$-modules such that $\mathscr{F}_n \to \mathscr{F}_{n-1}$ is surjective for all n. Then $\lim_n \mathscr{F}_n = R\lim_n \mathscr{F}_n = \operatorname{holim} \mathscr{F}_n$.*

Proof. The second equality follows from Proposition F.231. It remains to show the following claim: The natural map $\lim \mathscr{F}_n = H^0(\lim \mathscr{F}_n) \to H^0(\operatorname{holim} \mathscr{F}_n)$ is an isomorphism and $H^p(\operatorname{holim} \mathscr{F}_n) = 0$ for all $p \neq 0$.

For $U \subseteq X$ open, $R\Gamma(U, -)$ and holim commute (Lemma F.230), so for all p there is by Lemma F.233 an exact sequence

$$
0 \longrightarrow R^1 \lim H^{p-1}(U, \mathscr{F}_n) \longrightarrow H^p(U, \operatorname{holim} \mathscr{F}_n) \longrightarrow \lim H^p(U, \mathscr{F}_n) \longrightarrow 0
$$

Let U be affine. For $p \neq 1$ one has $H^{p-1}(U, \mathscr{F}_n) = 0$. For $p = 1$ the inverse system of modules $(H^0(U, \mathscr{F}_n))_n$ has surjective transition maps since the $\mathscr{F}_n$ are quasi-coherent. In particular it satisfies the Mittag-Leffler condition. Therefore $R^1 \lim H^{p-1}(U, \mathscr{F}_n) = 0$ for all p. Hence the exact sequence yields that $H^p(U, \operatorname{holim} \mathscr{F}_n) = 0$ for $p \neq 0$ and that $H^0(U, \operatorname{holim} \mathscr{F}_n) = \lim H^0(U, \mathscr{F}_n) = H^0(U, \lim \mathscr{F}_n)$. As for every complex $\mathscr{G}$ in $D(X)$ the sheafification of $U \mapsto H^p(U, \mathscr{G})$ is $H^p(\mathscr{G})$ (Lemma 21.22), the claim is shown. $\qquad\square$

(24.7) Mittag-Leffler condition for cohomology and direct proof of the Theorem of formal functions.

Proposition 24.39. *Let A be a noetherian ring and let $f\colon X \to \operatorname{Spec} A$ be a proper morphism. Let B be a finitely generated A-algebra and set $\mathscr{B} := f^*\tilde{B}$, a sheaf of rings on X. Let $\mathscr{F}$ be a quasi-coherent $\mathscr{B}$-module of finite type.*
(1) *For every $p \geq 0$ the B-module $H^p(X, \mathscr{F})$ is a finite B-module.*
(2) *Let $\mathscr{L}$ be an ample line bundle on X. Then there exists an integer N such that*
$$H^p(X, \mathscr{F} \otimes \mathscr{L}^{\otimes n}) = 0 \text{ for all } p > 0 \text{ and all } n \geq N.$$

For the proof of the formal function theorem only Assertion (1) is applied. Assertion (2) will be used to prove openness of ampleness on proper schemes (Theorem 24.46 below).

Proof. To show (1) define a morphism $\pi\colon Y \to X$ by the cartesian diagram

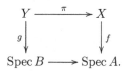

As the formation of the relative Spec commutes with base change (11.2.5), we have $Y = \operatorname{Spec}\mathscr{B}$. The $\mathscr{B}$-module $\mathscr{F}$ corresponds to a coherent $\mathscr{O}_Y$-module $\mathscr{F}'$ via the equivalence between the category of quasi-coherent $\mathscr{B}$-modules and quasi-coherent $\mathscr{O}_Y$-modules (Proposition 12.5). In particular one has $\pi_*\mathscr{F}' = \mathscr{F}$. As g is proper, $H^p(Y, \mathscr{F}')$ is a finite B-module (Corollary 23.18). As $\operatorname{Spec} B \to \operatorname{Spec} A$ is affine, π is affine and hence $H^p(Y, \mathscr{F}') = H^p(X, \mathscr{F})$ by Corollary 22.6. This proves (1).

To show (2) we set $\mathscr{M} := \pi^*\mathscr{L}$. As π is affine, $\mathscr{M}$ is an ample line bundle on Y by Proposition 13.83. By the projection formula (Proposition 22.81) we find that $\pi_*(\mathscr{F}' \otimes \mathscr{M}^{\otimes n}) = \mathscr{F} \otimes \mathscr{L}^{\otimes n}$ for all integers n and thus $H^p(Y, \mathscr{F}' \otimes \mathscr{M}^{\otimes n}) = H^p(X, \mathscr{F} \otimes \mathscr{L}^{\otimes n})$ by again using Corollary 22.6. Therefore (2) follows from Serre's vanishing criterion (Theorem 23.2). $\qquad\square$

Lemma 24.40. *Let A be a noetherian ring and let $I \subseteq A$ be an ideal. Let $f\colon X \to \operatorname{Spec} A$ be a proper morphism and let $\mathscr{F}$ be a coherent $\mathscr{O}_X$-module. Fix $p \geq 0$. Then the inverse system $(H^p(X, \mathscr{F}/I^n\mathscr{F}))_n$ satisfies the Mittag-Leffler condition.*

Proof. We set $H_k := H^p(X, \mathscr{F}/I^k\mathscr{F})$ and $H := H^p(X, \mathscr{F})$. We fix $n \geq 1$. We claim that for $k \geq n$ sufficiently large one has $\operatorname{Im}(H_k \to H_n) \subseteq \operatorname{Im}(H \to H_n)$ which in particular shows that $(H^p(X, \mathscr{F}/I^n\mathscr{F}))_n$ satisfies the Mittag-Leffler condition.

Let C_k be the cokernel of $H \to H_k$. It suffices to show that the map $C_k \to C_n$ is zero for $k \geq n$ sufficiently large. The exact sequence $0 \to I^k\mathscr{F} \to \mathscr{F} \to \mathscr{F}/I^k\mathscr{F} \to 0$ yields the long exact cohomology sequence

$$(24.7.1) \quad \cdots \to H^p(X, I^k\mathscr{F}) \to H^p(X, \mathscr{F}) \to H^p(X, \mathscr{F}/I^k\mathscr{F}) \to H^{p+1}(X, I^k\mathscr{F}) \to \cdots$$

One obtains the following two descriptions:

$$C_k = \operatorname{Im}(H^p(X, \mathscr{F}/I^k\mathscr{F}) \to H^{p+1}(X, I^k\mathscr{F})) = \operatorname{Ker}(H^{p+1}(X, I^k\mathscr{F}) \to H^{p+1}(X, \mathscr{F})).$$

Now $C := \bigoplus_{k \geq 0} C_k$ is a module over the A-algebra $B := \bigoplus_{k \geq 0} I^k$ via multiplication $I \otimes C_k \to C_{k+1}$. Since B is generated over A by a generating system of I, the ring B is noetherian. Now C is a B-submodule of $\bigoplus_k H^{p+1}(X, I^k\mathscr{F}) = H^{p+1}(X, \bigoplus_k I^k\mathscr{F})$ which is a finitely generated B-module by Proposition 24.39 (1) applied to the $f^*\tilde{B}$-module $\bigoplus_k I^k\mathscr{F}$. Therefore C is itself a finitely generated B-module.

The first description of C_k shows that C_k is annihilated by I^k. Since C is a finitely generated B-module, there exists an integer N such that the ideal $I^N \oplus I^{N+1} \oplus \cdots$ of B annihilates C. Consider the composition

$$I^r \otimes C_n \to C_{n+r} \to C_n.$$

As C is a finitely generated B-module, multiplication $I^r \otimes C_n \to C_{n+r}$ is surjective for large r. Moreover, the composition is zero for $r \geq N$. Hence $C_{n+r} \to C_n$ is zero for sufficiently large r. $\qquad\square$

Remark 24.41. Assume that in the situation of Lemma 24.40 the ideal I is a maximal ideal of A. This special case suffices for showing Theorem 24.42 below and its many applications (e.g., Corollary 24.44, Theorem 24.46, Theorem 24.49).

In this case one can simply show the Mittag-Leffler condition for $(H^p(X, \mathscr{F}/I^n\mathscr{F}))_n$ as follows. Let $X_n = X \otimes_A A/\mathfrak{m}^n$. Then $H^p(X, \mathscr{F}/I^n\mathscr{F}) = H^p(X_n, \mathscr{F}/I^n\mathscr{F})$ is a finite A/I^n-module by Corollary 23.18 and hence is an Artinian A-module since A/I^n is Artinian. But any inverse system of Artinian A-modules satisfies the Mittag-Leffler condition.

The above lemmas also allow to give a short direct proof of the noetherian version of the theorem of formal functions, Theorem 24.37, without referring to the formalism of derived completion. From this chapter it uses only Proposition 24.39 (1) and the proof of Lemma 24.40. Then it is completed by the following argument.

Proof. (Direct proof of the formal function theorem) We use the notation introduced in the proof of Lemma 24.40. Let E_k be the image of $H^p(X, I^k\mathscr{F}) \to H^p(X, \mathscr{F})$. Then (24.7.1) yields a projective system of short exact sequences

$$0 \to H/E_k \to H_k \to C_k \to 0.$$

We have already seen that for all $n \geq 1$ there exists a $k \geq n$ such that $C_k \to C_n$ is zero, hence $\lim_k C_k = 0$.

It remains to show that the family $(E_k)_k$ defines the I-adic topology on H since then taking limits we see that the I-adic completion of H is $\lim_k H_k$ which is the statement of the theorem of formal functions. We have $I^k H \subseteq E_k$, because $I_k H_k = 0$. Hence it suffices to show that there exists k_0 such that $IE_k = E_{k+1}$ for all $k \geq k_0$, since then $E_{r+k_0} = I^r E_{k_0} \subseteq I^r H$.

We have already seen that $M := \bigoplus_k H^p(X, I^k\mathscr{F})$ is a graded module over $B = \bigoplus I^k$ which is finitely generated as a B-module by Proposition 24.39 (1). In particular it is a noetherian B-module. Hence the ascending chain of submodules of the form

$$M_n := \left(\bigoplus_{k=0}^{n} H^p(X, I^k\mathscr{F})\right) \oplus \bigoplus_{r \geq 1} I^r H^p(X, I^n\mathscr{F})$$

has to stabilize. As $\bigcup_n M_n = M$ we find a k_0 such that $M_k = M$ for all $k \geq k_0$. This shows that $IH^p(X, I^k\mathscr{F}) = H^p(X, I^{k+1}\mathscr{F})$ for $k \geq k_0$. As the E_k are defined as images of $H^p(X, I^k\mathscr{F})$ we deduce our claim. $\square$

(24.8) Theorem of formal fibers.

Theorem 24.42. *Let S be a locally noetherian scheme, let $f: X \to S$ be a proper morphism, and let $\mathscr{F}$ be a coherent $\mathcal{O}_X$-module. Let $s \in S$ be a point and let $\mathfrak{m}_s \subset \mathcal{O}_{S,s}$ be the maximal ideal. Set $X_n := X \times_S \operatorname{Spec}(\mathcal{O}_{S,s}/\mathfrak{m}_s^n)$, let $i_n: X_n \to X$ be the projection, and set $\mathscr{F}_n := i_n^* \mathscr{F}$. Then for all $p \geq 0$ one has an isomorphism of $\widehat{\mathcal{O}}_{S,s}$-modules*

(24.8.1) $$(R^p f_* \mathscr{F})_s^\wedge \cong \lim_n H^p(X_n, \mathscr{F}_n).$$

Here the left hand side is the $\mathfrak{m}_s$-adic completion of the stalk $(R^p f_* \mathscr{F})_s$.

Proof. As the natural morphism $\operatorname{Spec} \mathcal{O}_{S,s} \to S$ is flat, we may replace S by $\operatorname{Spec} \mathcal{O}_{S,s}$ (Proposition 22.90). Then $(R^p f_* \mathscr{F})_s = \Gamma(S, R^p f_* \mathscr{F}) = H^p(X, \mathscr{F})$ (Theorem 22.27) and we conclude by the formal function theorem (Theorem 24.37). $\square$

The theorem is often applied in the following form.

Corollary 24.43. *In the situation of Theorem 24.42 suppose that* $\lim_n H^p(X_n, \mathscr{F}_n) = 0$. *Then there exists an open neighborhood* V *of* s *such that* $R^p f_*(\mathscr{F})_{|V} = 0$.

Proof. By Theorem 24.42 we have $(R^p f_* \mathscr{F})_s^{\wedge} = 0$. As f is proper, $R^p f_* \mathscr{F}$ is a coherent $\mathscr{O}_S$-module. Therefore the $\mathfrak{m}_s$-adic topology on the finite $\mathscr{O}_{S,s}$-module $(R^p f_* \mathscr{F})_s$ is Hausdorff (Proposition B.42). Together with the above, this implies $(R^p f_* \mathscr{F})_s = 0$. Since $R^p f_* \mathscr{F}$ is coherent, the subset $\{ s \in S \; ; \; (R^p f_* \mathscr{F})_s = 0 \}$ is open in S (Corollary 7.32). $\qquad\square$

Corollary 24.44. *Let* S *be a scheme, let* $f \colon X \to S$ *be a proper morphism, and let* $\mathscr{F}$ *be a quasi-coherent* $\mathscr{O}_X$-module. *Let* $s \in S$ *be a point and let* $d = \dim X_s$ *be the dimension of the fiber of* f *in* s. *Then for all* $p > d$ *there exists an open neighborhood* V *of* s *such that* $R^p f_* \mathscr{F}_{|V} = 0$.

We will prove the result only if S is locally noetherian and $\mathscr{F}$ is coherent. The general case is then obtained by approximation of proper morphisms by proper morphisms of finite presentation, by noetherian approximation and by writing every quasi-coherent module as a filtered colimit of modules of finite presentation. We refer to [Sta] 0E7D for the details. The reference also shows that the formation of $R^d f_* \mathscr{F}_{|V}$ commutes with arbitrary base change $T \to V$.

Proof. (if S is locally noetherian and $\mathscr{F}$ is coherent) The schemes X_n have the same underlying topological space as X_s for all $n \geq 1$. Hence by Grothendieck's vanishing theorem (Theorem 21.57) one has $H^p(X_n, \mathscr{F} \otimes (\mathscr{O}_{S,s}/\mathfrak{m}_s^n)) = 0$ for $p > d$. Hence we can apply Corollary 24.43. $\qquad\square$

(24.9) Ampleness is open on proper schemes.

We will use the following notation. For a morphism $f \colon X \to S$ of schemes, a point $s \in S$ and an $\mathscr{O}_X$-module $\mathscr{F}$, we denote by $\mathscr{F}_s$ the restriction (i.e., the pullback) of $\mathscr{F}$ to the schematic fiber $X_s = f^{-1}(s) = X \times_S \operatorname{Spec} \kappa(s)$.

Lemma 24.45. *Consider the standard set up for inductive limits of schemes (as given in Section (10.13) 1.–3.), and assume that* $X_0 \to S_0$ *is of finite presentation. Let* $\mathscr{L}_0$ *be a line bundle on* X_0 *and set* $\mathscr{L} := x_0^* \mathscr{L}_0$ *and* $\mathscr{L}_\lambda := x_{0,\lambda}^* \mathscr{L}_0$. *Then* $\mathscr{L}$ *is relatively ample over* S *(resp. very ample over* S*) if and only if there exists an index* λ_0 *such that* $\mathscr{L}_\lambda$ *is relatively ample over* S_λ *(resp. very ample over* S_λ*) for all* $\lambda \geq \lambda_0$.

Proof. As "ample over S" and "very ample over S" are compatible with base change $S' \to S$, it suffices to show that the conditions are necessary. As $X_\lambda \to S_\lambda$ is of finite type, we may replace $\mathscr{L}$ by some suitable power and can assume that $\mathscr{L}$ is very ample (Theorem 13.62). Then there exists an immersion $i \colon X \to \mathbb{P}_S^n =: P$ such that $i^* \mathscr{O}_P(1) = \mathscr{L}$. By Theorem 10.63, i is the base change of some S_λ-morphism $i_\lambda \colon X_\lambda \to \mathbb{P}_{S_\lambda}^n =: P_\lambda$. By Proposition 10.75, we can assume that i_λ is an immersion after enlarging λ. Finally by applying a gluing argument to Theorem 10.58 (see Exercise 10.32), we may assume that $i_\lambda^* \mathscr{O}_{P_\lambda}(1) \cong \mathscr{L}_\lambda$ which shows that $\mathscr{L}_\lambda$ is very ample. $\qquad\square$

Theorem 24.46. *Let* $f \colon X \to S$ *be a proper morphism of schemes and let* $\mathscr{L}$ *be an invertible* $\mathscr{O}_X$-module. *Let* $s \in S$ *be a point and assume that* $\mathscr{L}_s$ *is ample on the fiber* X_s. *Then there exists an open affine neighborhood* $U \subseteq S$ *of* s *such that* $\mathscr{L}_{|f^{-1}(U)}$ *is ample.*

We will prove the theorem only if S is locally noetherian. In general one first approximates X locally on S by proper S-schemes of finite presentation and uses noetherian approximation as explained in Chapter 10 and Lemma 24.45. For details see [Sta] 0D2S.

Proof. (if S is locally noetherian) We may assume that S is affine. Then by our additional assumption S is the spectrum of a noetherian ring. Since $\operatorname{Spec} \mathscr{O}_{S,s}$ is the filtered limit of its affine neighborhoods, we may replace S be $\operatorname{Spec} \mathscr{O}_{S,s}$ by Lemma 24.45. Hence we can assume that S is the spectrum of a local noetherian ring R and s is its closed point corresponding to the maximal ideal $\mathfrak{m}$ of R. We will show that for every coherent $\mathscr{O}_X$-module $\mathscr{F}$ there exists an integer N such that $H^1(X, \mathscr{F}(n)) = 0$ for $n \geq N$, where $\mathscr{F}(n) := \mathscr{F} \otimes \mathscr{L}^{\otimes n}$. This will prove that $\mathscr{L}$ is ample by Lemma 23.5.

As s is a closed point, X_s is a closed subscheme of X defined by the ideal sheaf $\mathscr{J} := \mathfrak{m}\mathscr{O}_X$. For every $\mathscr{O}_{X_s}$-module $\mathscr{G}$ we set $\mathscr{G}(n) := \mathscr{G} \otimes \mathscr{L}_s^{\otimes n}$. Let $i\colon X_s \to X$ be the inclusion. Set $C := \bigoplus_{j \geq 0} \mathfrak{m}^j/\mathfrak{m}^{j+1}$ which is a finitely generated R-algebra because R is noetherian. Hence $\mathscr{C} := f^*(\tilde{C})$ is a quasi-coherent $\mathscr{O}_X$-algebra of finite type. Since $\mathscr{J}\mathscr{C} = 0$ we have $\mathscr{C} = i_*(\mathscr{C}_s)$, where $\mathscr{C}_s := i^*\mathscr{C}$. We also set $\mathscr{M} := \bigoplus_{j \geq 0} \mathfrak{m}^j \mathscr{F}/\mathfrak{m}^{j+1}\mathscr{F}$. Again we have $\mathscr{M} = i_*\mathscr{M}_s$ with $\mathscr{M}_s := i^*\mathscr{M}$. As $\mathscr{F}$ is coherent, $\mathscr{M}$ is a quasi-coherent $\mathscr{C}$-module of finite type. Since $\mathscr{L}_s$ is ample, we can apply Proposition 24.39 (2) to the $\mathscr{C}_s$-module $\mathscr{M}_s$ and see that there exists an integer N such that $H^p(X_s, \mathscr{M}_s(n)) = 0$ for all $p > 0$ and all $n \geq N$. By the projection formula (Proposition 22.81) we have $i_*(\mathscr{M}_s(n)) = \mathscr{M}(n)$. As i is affine, we deduce that $H^p(X, \mathscr{M}(n)) = 0$ for all $p > 0$ and all $n \geq N$ (Corollary 22.6). Therefore we see that

$$(*) \qquad H^p(X, \mathfrak{m}^j \mathscr{F}/\mathfrak{m}^{j+1}\mathscr{F}(n)) = 0, \qquad \text{for all} \quad p > 0, j \geq 0, n \geq N.$$

In particular, $H^1(X, \mathscr{F}/\mathfrak{m}\mathscr{F}(n)) = 0$ for all $n \geq N$. Moreover, the long cohomology sequence attached to the short exact sequence

$$0 \longrightarrow \mathfrak{m}^j \mathscr{F}/\mathfrak{m}^{j+1}\mathscr{F}(n) \longrightarrow \mathscr{F}/\mathfrak{m}^{j+1}\mathscr{F}(n) \longrightarrow \mathscr{F}/\mathfrak{m}^j \mathscr{F}(n) \longrightarrow 0$$

shows that $(*)$ also implies that $H^1(X, \mathscr{F}/\mathfrak{m}^{j+1}\mathscr{F}(n)) \to H^1(X, \mathscr{F}/\mathfrak{m}^j \mathscr{F}(n))$ is an isomorphism for all j and $n \geq N$. In particular $\lim_j H^1(X, \mathscr{F}/\mathfrak{m}^j \mathscr{F}(n)) = 0$ for all $n \geq N$. Hence Corollary 24.43 implies that $(R^1 f_* \mathscr{F}(n))_s = H^1(X, \mathscr{F}(n))$ is zero for $n \geq N$. $\square$

Stein factorization

As an application of the theorem of formal functions, we come to the Stein factorization of a proper morphism, which is a factorization as a morphism with geometrically connected fibers followed by a finite morphism. Similarly as many other results on proper morphisms in this book, it has an analogue in complex geometry, that was proved several years earlier.

(24.10) Stein factorization.

Let $f\colon X \to S$ be a qcqs morphism of schemes. Then $f_*\mathscr{O}_X$ is a quasi-coherent $\mathscr{O}_S$-algebra (Corollary 10.27) and we obtain a factorization of f as

(24.10.1) $$X \xrightarrow{f'} S' = \mathrm{Spec}(f_* \mathscr{O}_X) \xrightarrow{\pi} S$$

with the following properties (Corollary 12.2)
(1) The homomorphism $\mathscr{O}_{S'} \to f'_* \mathscr{O}_X$ is an isomorphism.
(2) The morphism π is affine.
If S is affine, one has $S' = \mathrm{Spec}\, \Gamma(X, \mathscr{O}_X)$.

Remark 24.47. Let $f\colon X \to S$ be a qcqs morphism of schemes and let $T \to S$ be an affine morphism. Then every morphism of S-schemes $X \to T$ factorizes through $X \to \mathrm{Spec}(f_* \mathscr{O}_X)$.

Indeed, by Proposition 12.1 we find $T = \mathrm{Spec}\,\mathscr{A}$ for some quasi-coherent $\mathscr{O}_S$-algebra $\mathscr{A}$ and hence $\mathrm{Hom}_S(X, T) = \mathrm{Hom}_{\mathscr{O}_S\text{-Alg}}(\mathscr{A}, f_* \mathscr{O}_X) = \mathrm{Hom}_S(\mathrm{Spec}(f_* \mathscr{O}_X), T)$ by Proposition 11.1.

Remark 24.48. Let $f\colon X \to S$ be a qcqs morphism of schemes. The formation of $f_* \mathscr{O}_X$ commutes with flat base change (Proposition 12.6). Therefore the factorization (24.10.1) is compatible with base change by a flat morphism $S_1 \to S$, i.e., the factorization (24.10.1) of the base change $f_1\colon X_1 := X \times_S S_1 \to S_1$ of f is given by $X'_1 \to S'_1 \to S_1$ with $S'_1 = S' \times_S S_1$.

If f is proper, the factorization (24.10.1) is called the *Stein factorization*. It has the following properties.

Theorem 24.49. *Let $f\colon X \to S$ be a proper morphism of schemes. Then the factorization (24.10.1) of f has the following properties.*
(1) *The morphism f' is proper with geometrically connected fibers.*
(2) *The morphism π is integral with finite discrete fibers. For $s \in S$ and $s' \in \pi^{-1}(s)$ the field extension $\kappa(s) \to \kappa(s')$ is finite.*
(3) *If S is locally noetherian, then π is finite.*

Note that if π is finite, then all assertions of (2) are automatically satisfied. Recall that we assume connected spaces to be non-empty. In particular, f' is surjective.

We will prove the theorem only in the case that S is locally noetherian. The general case is deduced from this case by nontrivial noetherian approximation (see [Sta] 03H2, 0EOM for details).

Proof. (if S is locally noetherian) As π is affine, it is in particular separated. Hence f' is proper because f is proper (Proposition 12.58). By the coherence theorem (Theorem 23.17), $f_* \mathscr{O}_X$ is a coherent $\mathscr{O}_S$-module, hence π is finite. It remains to apply the following corollary to f'. $\qquad\square$

Corollary 24.50. *Let $f\colon X \to S$ be a proper morphism with $\mathscr{O}_S = f_* \mathscr{O}_X$. Then f has geometrically connected fibers.*

Proof. We again restrict to the case that S is locally noetherian. Fix $s \in S$. To show that the fiber $X_s = f^{-1}(s)$ is geometrically connected, we may replace S by $\mathrm{Spec}\,\mathscr{O}_{S,s}$ by Remark 24.48. Hence we may assume that $S = \mathrm{Spec}\, R$ for some local noetherian ring R with maximal ideal $\mathfrak{m}$.

To see that f is surjective it suffices to see that f is dominant because f is closed. But every morphism $f\colon X \to S$ with $\mathscr{O}_S = f_* \mathscr{O}_X$ is dominant because $R \to \Gamma(X, \mathscr{O}_X)$ is injective and thus f cannot factor through any closed subscheme $V(\mathfrak{a}) \subsetneq \mathrm{Spec}\, R$. To prove that X_s is geometrically connected we proceed in two steps.

(I). We first show that X_s is connected which proves that every proper morphism f with $\mathscr{O}_S = f_*\mathscr{O}_X$ has connected fibers. The fiber X_s is closed in X. The theorem of formal fibers 24.42 shows that

$$\lim_n \Gamma(X_n, \mathscr{O}_{X_n}) = R^\wedge,$$

where $X_n = V(\mathfrak{m}^{n+1}\mathscr{O}_X)$ is the n-th infinitesimal neighborhood of X_s. If X_s was not connected, then there existed a non-trivial idempotent element $(e_n)_n \in \lim_n \Gamma(X_n, \mathscr{O}_{X_n})$ as all X_n have the same underlying topological space as X_s. This contradicts the fact that $R^\wedge$ is local and hence cannot have a non-trivial idempotent element.

(II). To show that X_s is geometrically connected, it suffices to show that for every finite separable extension k of $\kappa(s)$ the base change $X_s \otimes_{\kappa(s)} k$ is connected (Proposition 5.53). By Lemma 18.65 there exists a finite flat local R-algebra R' with residue field k. Let $f'\colon X' := X \otimes_R R' \to S' := \operatorname{Spec} R'$ be the base change of f, and let $s' \in S'$ be the closed point. Again flatness and the corresponding fact for f imply that $\mathscr{O}_{S'} = f'_*\mathscr{O}_{X'}$. As we already proved Step (I), we know that the fiber $X'_{s'} = X_s \otimes_{\kappa(s)} k$ is connected. $\square$

Recall from Remark 5.55 that for a scheme X over a field k the geometric number of connected components is the number of connected components of $X \otimes_k K$, where K is some algebraically closed extension of k. This number is independent of the choice of K. It is finite if X is of finite type.

If $f\colon X \to S$ is a morphism of schemes, we denote for $s \in S$ by $n_{X/S}(s)$ the geometric number of connected components of the $\kappa(s)$-scheme X_s. This defines a map

(24.10.2) $n_{X/S}\colon S \to \mathbb{N} \cup \{\infty\}.$

Proposition 24.51. *Let $f\colon X \to S$ be a proper morphism of schemes and let*

$$X \xrightarrow{f'} S' \xrightarrow{\pi} S$$

be its Stein factorization. Let $s \in S$. Then $n_{X/S}(s)$ is finite and one has

$$n_{X/S}(s) = n_{S'/S}(s) = \sum_{s' \in \pi^{-1}(s)} [\kappa(s') : \kappa(s)]_{\mathrm{sep}},$$

where $[L : K]_{\mathrm{sep}}$ denotes the separability degree of a finite field extension $K \to L$.

Proof. Let us show the first equality. As in the proof of Corollary 24.50 we can reduce first to the case that $S = \operatorname{Spec} R$ for a local ring R and that $s \in S$ is the closed point. Then we may assume that $\kappa(s)$ is algebraically closed again using Lemma 18.65. The morphism $g\colon f^{-1}(s) \to \pi^{-1}(s)$ induced by f' is proper and surjective with connected fibers. Hence the first equality follows from the purely topological Lemma 24.52 below.

To show the second, essentially elementary, equality, we may assume that $S = \operatorname{Spec} k$. Then S' is a finite discrete k-scheme and hence we may assume the underlying topological space of S' consists of a single point s' with residue field k' such that $[k' : k]$ is finite. As the geometric connected components of S' and S'_{red} are the same, we may assume that $S' = \operatorname{Spec} k'$. Let K be an algebraically closed extension of k. Then the number of connected components of $\operatorname{Spec}(k' \otimes_k K)$ is equal to $\#\operatorname{Hom}_k(k', K) = [k' : k]_{\mathrm{sep}}$. $\square$

If X is a topological space, we define an equivalence relation on X by $x \sim x'$ if x and x' are in the same connected component. The quotient space is the space of connected components of X and denoted by $\pi_0(X)$. We obtain a functor π_0 from the category of topological spaces to the category of totally disconnected topological spaces. It is left adjoint to the inclusion functor.

If X is a scheme, then $\pi_0(X)$ denotes the space of connected components of the underlying topological space of X. If X has only finitely many connected components (e.g., if X is noetherian), then $\pi_0(X)$ is a finite discrete space, in other words, X has only finitely many connected components, and each of them is open and closed in X.

Lemma 24.52. *Let $g \colon Z \to Y$ be a continuous surjective map of topological spaces such that Y carries the quotient topology of Z (e.g., if g is open or closed). Suppose that all fibers of g are connected. Then the induced map $\pi_0(Z) \xrightarrow{\sim} \pi_0(Y)$ is a homeomorphism.*

The homeomorphism is given by sending a connected component Z' of Z to $g(Z')$ and its inverse is given by $Y' \mapsto g^{-1}(Y')$.

Proof. For topological spaces X and X' let $\mathcal{C}(X, X')$ be the set of continuous maps $X \to X'$. Let S be a totally disconnected space. We have functorial bijections

$$\mathcal{C}(\pi_0(Y), S) = \mathcal{C}(Y, S) = \{\, f \in \mathcal{C}(Z, S) \; ; \; f_{|g^{-1}(y)} \text{ is constant for all } y \in Y \,\}$$
$$= \mathcal{C}(Z, S) = \mathcal{C}(\pi_0(Z), S),$$

where the first and last equality hold by adjointness of π_0 and the inclusion functor, the second equality holds because Y carries the quotient topology of Z, and the third equality holds because all fibers of g are connected. Therefore $\pi_0(g) \colon \pi_0(Z) \xrightarrow{\sim} \pi_0(Y)$ is a homeomorphism by Yoneda's lemma. $\qquad\square$

Remark 24.53. Let $f \colon X \to S$ be a proper morphism and let $X \to T \to S$ be its Stein factorization.
(1) As $X \to T$ is closed and surjective with connected fibers, we can apply Lemma 24.52 to see that $\pi_0(X) \to \pi_0(T)$ is a homeomorphism.
(2) Let $s \in S$. Then we may apply Lemma 24.52 to the surjective proper morphism $X_s \to T_s$ induced by $X \to T$ to see that $\pi_0(X_s) = \pi_0(T_s)$ is a homeomorphism.

Corollary 24.54. *Let R be a local henselian ring, $S = \operatorname{Spec} R$, and let $s \in S$ be its closed point. Let $X \to S$ be a proper morphism. Then the inclusion $X_s \to X$ induces a bijection of finite discrete spaces $\pi_0(X_s) \xrightarrow{\sim} \pi_0(X)$.*

Proof. Let $X \to T \to S$ be the Stein factorization. The map $\pi_0(X_s) \xrightarrow{\sim} \pi_0(X)$ can be factorized as

$$\pi_0(X_s) \xrightarrow{\sim} \pi_0(T_s) \to \pi_0(T) \xrightarrow{\sim} \pi_0(X),$$

where one has the first and the third bijection by Remark 24.53. Moreover, $\pi_0(T_s)$ is a finite discrete space by Theorem 24.49 (2). It remains to show that the second map is bijective. If S is noetherian and hence $T \to S$ is finite, this follows from the definition of a local henselian ring (Definition 20.1). In general, it follows from Exercise 20.10. $\qquad\square$

Corollary 24.55. *Let k be a field and let $X \to \operatorname{Spec} k$ be a proper k-scheme. Then X is geometrically connected if and only if $\Gamma(X, \mathscr{O}_X)$ is a local ring of dimension 0 whose residue field is a purely inseparable extension of k.*

Proof. By Proposition 24.51, X is geometrically connected over k if and only if the affine scheme $\operatorname{Spec}\Gamma(X, \mathscr{O}_X)$ consists of only one point whose residue field is a purely inseparable extension of k. □

(24.11) Properties of Stein factorization.

Proposition 24.56. *Let $f\colon X \to S$ be a proper morphism with geometrically connected fibers and let $X \to S' \to S$ be its Stein factorization. Then $S' \to S$ is a universal homeomorphism.*

Proof. As f has non-empty fibers by hypothesis, it is surjective and hence the integral morphism $\pi\colon S' \to S$ is surjective. By Proposition 20.29 it suffices to show that $S' \to S$ is universally injective. By Proposition 4.35 we have to show that for all $s \in S$ the fiber $\pi^{-1}(s)$ consists of a single point s' and that $\kappa(s')$ is a purely inseparable extension of $\kappa(s)$. But $\pi^{-1}(s)$ is geometrically connected over $\kappa(s)$ by Proposition 24.51. As it is affine, we can conclude by Corollary 24.55. □

We can now prove Zariski's connectedness theorem, which also can be seen as a version of Zariski's main theorem.

Theorem 24.57. (Zariski's Connectedness Theorem) *Let $f\colon X \to S$ be a proper dominant morphism of integral schemes and let $\eta \in S$ be the generic point. Let $X \to S' \to S$ be its Stein factorization. Suppose that S is geometrically unibranch (Definition 20.49, e.g., if S is normal) and that the generic fiber X_η is geometrically connected. Then $S' \to S$ is a universal homeomorphism and all fibers of f are geometrically connected.*

Proof. We may assume that $S = \operatorname{Spec} R$ is affine. It suffices to show that the integral morphism $S' \to S$ is a universal homeomorphism by Proposition 24.51. Again it suffices to show that $S' \to S$ is universally injective by Proposition 20.29.

By construction of the Stein factorization we have $S' = \operatorname{Spec} R'$ with $R' = \Gamma(X, \mathscr{O}_X)$. As X is integral, S' is also integral. The properties of the Stein factorization imply that R' is integral over R (Theorem 24.49). The function field of S' is the field of fractions of $\Gamma(X, \mathscr{O}_X)$ or also the field of fractions of the localization $\Gamma(X_\eta, \mathscr{O}_{X_\eta})$. The latter is a purely inseparable extension K' of $\kappa(\eta)$ because X_η is integral and geometrically connected (Corollary 24.55).

Let $s \in S$ and let A be the localization of the R-algebra $\Gamma(X, \mathscr{O}_X)$ in s. It is integral over the geometrically unibranch ring $\mathscr{O}_{S,s}$. As $K' = \operatorname{Frac} A$ is a purely inseparable extension of $\kappa(\eta) = \operatorname{Frac} \mathscr{O}_{S,s}$, the normalization of $\mathscr{O}_{S,s}$ in K' is a local ring whose residue field is a purely inseparable extension of $\kappa(s)$ by Proposition 20.51 and Lemma G.32. Since A is integral over $\mathscr{O}_{S,s}$, its normalization equals the normalization of $\mathscr{O}_{S,s}$ in K'. In particular the fiber of $S' \to S$ over s consists of a single point and its residue field is purely inseparable over $\kappa(s)$. Therefore $S' \to S$ is universally injective (Proposition 4.35). □

Corollary 24.58. *Let $f\colon X \to S$ be a proper dominant morphism of integral schemes and let $\eta \in S$ be the generic point. Assume that S is normal and that the generic fiber of f is geometrically reduced. Then all fibers of f are geometrically connected.*

Proof. We have $\Gamma(X_\eta, \mathscr{O}_{X_\eta}) = \kappa(\eta)$ (Proposition 12.66). Therefore $S' \to S$ is birational and thus an isomorphism because $S' \to S$ is integral and S is normal. Therefore $f_*\mathscr{O}_X = \mathscr{O}_S$, and thus the morphism $S' \to S$ in the Stein factorization $X \to S' \to S$ of f is an isomorphism. □

Remark 24.59. In the situation of the corollary, if $\kappa(\eta)$ is a perfect field, e.g., if $\kappa(\eta)$ is a field of characteristic zero, then we may drop the assumption that the generic fiber of f be geometrically reduced. In fact, as X is assumed to be integral and hence reduced, the generic fiber X_η is certainly reduced. If $\kappa(\eta)$ is perfect, then X_η is necessarily geometrically reduced (Corollary 5.57).

The formation of $f_*\mathscr{O}_X$ does not commute with base change in general, but we gave several criteria for this to hold in Section (23.28). The results above show that this always holds "up to universal homeomorphism" in the following sense.

Corollary 24.60. *Let $f\colon X \to S$ be a morphism of proper schemes, let $g\colon T \to S$ be a morphism of schemes, and let $f_T\colon X_T := X \times_S T \to T$ be the base change of f. Let $S' = \operatorname{Spec} f_*\mathscr{O}_X$ and $S_T' := S' \times_S T = \operatorname{Spec} g^* f_*\mathscr{O}_X$ be its base change. Then the canonical morphism $\operatorname{Spec}(f_T)_*\mathscr{O}_{X_T} \to S_T'$ is a universal homeomorphism.*

Proof. Let $f'\colon X \to S'$ be the first morphism of the Stein factorization of f which is proper with geometrically connected fibers by Theorem 24.49. Therefore its base change $f_T'\colon X_T \to S_T'$ is proper with geometrically connected fibers. Hence the second morphism of its Stein factorization, which is $\operatorname{Spec}(f_T)_*\mathscr{O}_{X_T} \to S_T'$ is a universal homeomorphism by Proposition 24.56. $\qquad\square$

We have the following criterion when the Stein factorization is finite étale and commutes with arbitrary base change.

Theorem 24.61. *Let S be a scheme and let $f\colon X \to S$ be a proper flat morphism of finite presentation. Let $s \in S$ be a point such that the fiber X_s is geometrically reduced. Let $X \xrightarrow{f'} S' \xrightarrow{\pi} S$ be the Stein factorization of f. Then there exists an open neighborhood U of s such that $\pi^{-1}(U) \to U$ is finite étale and that the formation of the Stein factorization of $f^{-1}(U) \to U$ commutes with arbitrary base change.*

The theorem can in particular be applied if f is proper and smooth, and in that case we can take $U = S$.

Proof. We may assume that $S = \operatorname{Spec} R$ is affine.

(I). As X_s is geometrically reduced, $\Gamma(X_s, \mathscr{O}_{X_s})$ is a finite étale $\kappa(s)$-algebra by Example 18.26.

(II). We now claim that it suffices to show that $\beta^0(\kappa(s))\colon \Gamma(X, \mathscr{O}_X) \otimes_R \kappa(s) \to \Gamma(X_s, \mathscr{O}_{X_s})$ is surjective.

Indeed, if $\beta^0(\kappa(s))$ is surjective, then Theorem 23.140 shows that there exists an open neighborhood V of s such that the formation of $f_*\mathscr{O}_{X|V}$ commutes with base change. Hence Theorem 23.140 (2) shows that $f_*\mathscr{O}_{X|V}$ is a locally free $\mathscr{O}_S$-module. We conclude that the Stein factorization of $f^{-1}(V) \to V$ commutes with arbitrary base change. It follows that

$$(*) \qquad\qquad \pi^{-1}(s) = \operatorname{Spec}\Gamma(X_s, \mathscr{O}_{X_s}).$$

By Step (I), $\Gamma(X_s, \mathscr{O}_{X_s})$ is an étale $\kappa(s)$-algebra. Therefore π is étale over an open neighborhood of s because the locus where the fibers of a finite locally free morphism are étale is open (Corollary 20.73).

(III). To show that $\beta^0 := \beta^0(\kappa(s))$ is surjective, we may assume that R is local with closed point s. As cohomology commutes with flat base change, we may pass to a faithfully flat R-algebra and hence can assume that R is strictly henselian. Then $\pi^{-1}(s)$ is the scheme theoretic sum of finitely many copies of $\operatorname{Spec}\kappa(s)$ by Step (I) and (*). Passing to connected components of X, we may assume that $\pi^{-1}(s) = \operatorname{Spec}\kappa(s)$ by Corollary 24.54. But then β^0 is surjective because the composition

$$R \longrightarrow \Gamma(X, \mathscr{O}_X) \longrightarrow \Gamma(X, \mathscr{O}_X) \otimes \kappa(s) \xrightarrow{\beta^0} \Gamma(X_s, \mathscr{O}_{X_s}) = \kappa(s)$$

is surjective. □

Corollary 24.62. *Let $f\colon X \to S$ be a flat proper morphism of finite presentation with geometrically reduced fibers. Then the function $n_{X/S}$ of the geometric number of connected components is locally constant.*

Proof. Let $X \to S' \to S$ be the Stein factorization. By Proposition 24.51 it suffices to show $n_{S'/S}$ is locally constant. But this is clear because $S' \to S$ is finite étale by Theorem 24.61. □

Corollary 24.63. *Let $f\colon X \to S$ be a flat, proper morphism of finite presentation. If f has geometrically connected and geometrically reduced fibers, then $\mathscr{O}_S \to f_*\mathscr{O}_X$ is an isomorphism and the formation of $f_*\mathscr{O}_X$ is compatible with base change.*

Note that even if S is noetherian and one omits one of the properties "flat", "proper", "geometrically connected fibers" or "geometrically reduced fibers", then one does not have $\mathscr{O}_S = f_*\mathscr{O}_X$ in general: If one omits "flat", take a surjective closed immersion that is not an isomorphism. If one omits "proper", take $\mathbb{A}^1_S$. If one omits "geometrically connected fibers", consider $S \coprod S \to S$. If one omits "geometrically reduced fibers" take $\operatorname{Spec}\mathscr{O}_S[T]/(T^2) \to S$ or the Frobenius $\mathbb{A}^1_k \to \mathbb{A}^1_k$ for a field k of positive characteristic.

Proof. Let $X \to S' \to S$ be the Stein factorization. Then $S' \to S$ is finite étale and its formation commutes with arbitrary base change by Theorem 24.61, and it is of rank 1 by Proposition 24.51, hence $S' = \operatorname{Spec} f_*\mathscr{O}_X \to S$ is an isomorphism. □

Remark 24.64. We proved Theorem 24.49 about the main properties of the Stein factorization only in the locally noetherian case. To deduce the result for general base schemes requires a highly nontrivial noetherian approximation argument which is not covered by the techniques explained in Chapter 10. However, for Theorem 24.61 and its Corollaries 24.62 and 24.63 the noetherian approximation is easier and covered by the results and ideas described in Chapter 10. The main reason is that the formation of the Stein factorization commutes with arbitrary base change in this case.

Let us go into more detail. Suppose that $f\colon X \to S$ and $s \in S$ are as in Theorem 24.61 and we know the theorem for S locally noetherian. To prove it for an arbitrary scheme S, we may assume that $S = \operatorname{Spec} R$ is affine. We write $R = \operatorname{colim}_\lambda R_\lambda$ for a filtered inductive system $(R_\lambda)_\lambda$ of noetherian rings R_λ. Set $S_\lambda = \operatorname{Spec} R_\lambda$. As $X \to S$ is of finite presentation, there exists a scheme $f_\lambda\colon X_\lambda \to S_\lambda$ of finite type such that $X_\lambda \times_{S_\lambda} S \cong X$ (Theorem 10.66). As the properties "flat" and "proper" are compatible with filtered colimits of rings (Appendix C), we may assume that $X_\lambda \to S_\lambda$ is proper and flat. Moreover, let $s_\lambda \in S_\lambda$ be the image of s under $S \to S_\lambda$. Then $X_s = (X_\lambda)_{s_\lambda} \otimes_{\kappa(s_\lambda)} \kappa(s)$ and therefore the fiber $(X_\lambda)_{s_\lambda}$ is geometrically reduced because X_s is assumed to be geometrically reduced. Hence we can apply Theorem 24.61 in the noetherian case and we see that there exists an open neighborhood U_λ of s_λ such that $(f_\lambda)_*(\mathscr{O}_{X_\lambda})|_{U_\lambda}$ is finite locally free and that its formation commutes with arbitrary base change. Applying this to the base change $\rho_\lambda \colon S \to S_\lambda$ we deduce Theorem 24.61 for $X \to S$ and s by choosing $U = \rho_\lambda^{-1}(U_\lambda)$.

(24.12) The seesaw theorem.

In this section, we will prove the seesaw theorem, Theorem 24.66. We can motivate it by the following question. Assume that we have line bundles $\mathscr{L}$ and $\mathscr{L}'$ on an S-scheme $f \colon X \to S$ such that for all $s \in S$ the restrictions $\mathscr{L}_s$ and $\mathscr{L}_s'$ to the fiber $X \times_S \operatorname{Spec} \kappa(s)$ are isomorphic. Can we conclude that $\mathscr{L}$ and $\mathscr{L}'$ are isomorphic? It is easy to see that there are two reasons why this cannot be true: For one thing, the condition on the fibers will not change when we replace S by its underlying reduced subscheme, so we should certainly add the condition that S be reduced. Second, if we set $\mathscr{L}' := \mathscr{L} \otimes f^*\mathscr{M}$ for a line bundle $\mathscr{M}$ on S, then the restrictions of $\mathscr{L}$ and $\mathscr{L}'$ to the fibers will be isomorphic. So the best we can hope for is that for S reduced, if $\mathscr{L}_s \cong \mathscr{L}_s'$ for all $s \in S$, then $\mathscr{L}$ and $\mathscr{L}'$ are isomorphic up to tensoring one of them by the pullback of a line bundle from S. It is an immediate consequence of the seesaw theorem that this is true if X is a flat proper S-scheme of finite presentation with geometrically connected and geometrically reduced fibers. The term seesaw stems from the fact that for a line bundle on a product $X \times_S T$ one looks at restrictions to fibers $X \times \{t\}$ on the one hand, and to fibers $\{x\} \times T$ on the other hand, thus in a sense going back and forth between X and T.

Lemma 24.65. *Let k be a field, let $X \to \operatorname{Spec} k$ be proper and suppose that X is integral. Then a line bundle $\mathscr{L}$ on X is trivial if and only if $\Gamma(X, \mathscr{L}) \neq 0$ and $\Gamma(X, \mathscr{L}^{-1}) \neq 0$.*

Proof. The condition is clearly necessary. Conversely, let $0 \neq s \in \Gamma(X, \mathscr{L})$ and $0 \neq t \in \Gamma(X, \mathscr{L}^{-1})$ which we consider as non-zero linear maps $s \colon \mathscr{O}_X \to \mathscr{L}$ and $t \colon \mathscr{L} \to \mathscr{O}_X$. As X is integral, the composition $t \circ s \in \Gamma(X, \mathscr{O}_X)$ is non-zero. As X is proper over k and integral, $\Gamma(X, \mathscr{O}_X)$ is a finite field extension of k, hence $t \circ s$ is an isomorphism. Therefore t is an isomorphism as surjective homomorphisms of locally free $\mathscr{O}_X$-modules of the same rank are isomorphisms (Corollary 8.12). $\qquad\square$

Theorem 24.66. (Seesaw Theorem) *Let $f \colon X \to S$ be a flat, proper morphism of finite presentation with geometrically connected and geometrically reduced fibers, and let $\mathscr{E}$ be a finite locally free $\mathscr{O}_X$-module. Then there exists a unique subscheme Z of S with the following property.*

For $u \colon T \to S$ a morphism of schemes we consider the cartesian diagram

(C)
$$
\begin{array}{ccc}
X_T & \xrightarrow{\ v\ } & X \\
{\scriptstyle f_T}\big\downarrow & & \big\downarrow{\scriptstyle f} \\
T & \xrightarrow{\ u\ } & S.
\end{array}
$$

Then u factors through Z if and only if there exists a finite locally free $\mathscr{O}_T$-module $\mathscr{M}$ with $f_T^*\mathscr{M} \cong v^*\mathscr{E}$. Moreover one has:

(1) *The immersion $Z \to S$ is of finite presentation and as sets one has*

$$
Z = \{\, s \in S \,;\, \mathscr{E}_s \text{ is a free } \mathscr{O}_{X_s}\text{-module} \}.
$$

(2) *Suppose that there exists $s_0 \in Z$ such that $H^1(X_{s_0}, \mathscr{O}_{X_{s_0}}) = 0$. Then there exists an open subscheme U of S such that $s_0 \in U \subseteq Z$.*

(3) *Suppose that f has geometrically integral fibers and that $\mathscr{E}$ is a line bundle. Then Z is a closed subscheme of S.*

In Section (27.21) we will introduce the Picard space classifying line bundles on X up to pullback of line bundles from S and show that the condition $H^1(X_{s_0}, \mathscr{O}_{X_{s_0}}) = 0$ means that it has a 0-dimensional tangent space at s_0 (Proposition 27.122) and hence is étale (Exercise 27.3). This gives a more geometric interpretation of Assertion (2), at least if $\mathscr{E}$ is a line bundle.

For the proof of the seesaw theorem, we will use the following criterion when a vector bundle is the pullback of a vector bundle.

Lemma 24.67. *For a ringed space Y we denote by $(\mathrm{VB}(Y))$ the category of locally free $\mathscr{O}_Y$-modules of finite type. Let $f \colon X \to S$ be a morphism of ringed spaces. For $U \subseteq S$ open consider the functor*

$$f_U^* \colon (\mathrm{VB}(U)) \to (\mathrm{VB}(f^{-1}(U))), \qquad \mathscr{M} \mapsto f_U^* \mathscr{M},$$

where $f_U \colon f^{-1}(U) \to U$ denotes the restriction of f.

(1) *The functor f_U^* is fully faithful for all open subspaces $U \subseteq S$ if and only if $\mathscr{O}_S = f_* \mathscr{O}_X$.*

(2) *Suppose that $\mathscr{O}_S = f_* \mathscr{O}_X$. Let $\mathscr{E}$ be a finite locally free $\mathscr{O}_X$-module of rank r and let $\mathscr{E}^\vee$ be its dual. Then the following assertions are equivalent.*

(i) *The $\mathscr{O}_X$-module $\mathscr{E}$ is in the essential image of $f^* \colon (\mathrm{VB}(S)) \to (\mathrm{VB}(X))$.*

(ii) *The $\mathscr{O}_S$-module $f_* \mathscr{E}$ is locally free of finite type and the canonical homomorphism $f^* f_* \mathscr{E} \to \mathscr{E}$ is an isomorphism.*

(iii) *The $\mathscr{O}_S$-modules $f_* \mathscr{E}$ and $f_* \mathscr{E}^\vee$ are locally free of rank r and the canonical pairing*

$$(24.12.1) \qquad f_* \mathscr{E} \otimes_{\mathscr{O}_S} f_* \mathscr{E}^\vee \longrightarrow f_*(\mathscr{E} \otimes_{\mathscr{O}_X} \mathscr{E}^\vee) \longrightarrow f_* \mathscr{O}_X = \mathscr{O}_S$$

is perfect.

If these equivalent conditions are satisfied, then $\mathscr{E} \cong f^* f_* \mathscr{E}$. Moreover, the formation of $f_* \mathscr{E}$ commutes with base change if f is a morphism of schemes and $f_* \mathscr{O}_X = \mathscr{O}_S$ holds after arbitrary base change.

Proof. Let us show (1). The condition that f_U^* is fully faithful for all $U \subseteq S$ open holds if and only if for all open subspaces $U \subseteq S$ and for all $\mathscr{O}_U$-modules $\mathscr{M}_1$ and $\mathscr{M}_2$ the canonical map

$$(*) \qquad \mathscr{H}om_{\mathscr{O}_U}(\mathscr{M}_1, \mathscr{M}_2) \longrightarrow (f_U)_* f_U^* \mathscr{H}om_{\mathscr{O}_U}(\mathscr{M}_1, \mathscr{M}_2)$$

of $\mathscr{O}_U$-modules is an isomorphism. But if $(*)$ is an isomorphism for $U = S$ and $\mathscr{M}_1 = \mathscr{M}_2 = \mathscr{O}_S$, then $\mathscr{O}_S = f_* \mathscr{O}_X$.

Conversely, suppose that $\mathscr{O}_S = f_* \mathscr{O}_X$. Then for $U \subseteq S$ open one also has that $\mathscr{O}_U = (f_U)_* \mathscr{O}_{f^{-1}(U)}$ as the formation of f_* commutes with passing to open subspaces. Let $\mathscr{M}$ be any finite locally free $\mathscr{O}_U$-module, for instance $\mathscr{H}om_{\mathscr{O}_U}(\mathscr{M}_1, \mathscr{M}_2) = \mathscr{M}_1^\vee \otimes \mathscr{M}_2$ as above. We claim that the canonical map $(f_U)_* f_U^* \mathscr{M} \to \mathscr{M}$ is an isomorphism. Indeed, as this can be checked locally on U, we may assume that $\mathscr{M} = \mathscr{O}_U^n$ for some $n \geq 0$ and we conclude by $\mathscr{O}_U = (f_U)_* \mathscr{O}_{f^{-1}(U)}$. In particular, $(*)$ is an isomorphism.

Next we prove (2). Suppose that $\mathscr{E}$ is in the essential image of f^*, i.e., there exists a finite locally free $\mathscr{O}_S$-module $\mathscr{M}$ with $\mathscr{E} \cong f^*\mathscr{M}$, then necessarily $\mathscr{M} \cong f_*\mathscr{E}$ by the claim above, in particular $f_*\mathscr{E}$ is finite locally free, necessarily of rank r because $f^*f_*\mathscr{E} = \mathscr{E}$ is of rank r. The same argument shows that $f_*\mathscr{E}^\vee$ is finite locally free of rank r. Moreover, $f^*(\mathscr{M}^\vee) = \mathscr{E}^\vee$ and hence $f_*(\mathscr{E}^\vee) = \mathscr{M}^\vee$ and (24.12.1) is given by the canonical pairing $\mathscr{M} \otimes \mathscr{M}^\vee \to \mathscr{O}_S$ and in particular is perfect. To see that $f^*f_*\mathscr{E} \to \mathscr{E}$ is an isomorphism, we can work locally on X and hence assume that $\mathscr{E} = \mathscr{O}_X^n$ and we conclude by the hypothesis $f_*\mathscr{O}_X = \mathscr{O}_S$. Hence we have seen that (i) implies (ii) and (iii).

Suppose that (ii) holds. Then $\mathscr{E} \cong f^*\mathscr{M}$ with $\mathscr{M} := f_*\mathscr{E}$. Hence (ii) implies (i).

Let us show that (iii) implies (ii). Set $\mathscr{M} := f_*\mathscr{E}$ which is locally free of rank r by hypothesis. To see that the canonical homomorphism of locally free modules $\psi\colon f^*\mathscr{M} = f^*f_*\mathscr{E} \to \mathscr{E}$ is an isomorphism, we can work locally on S. Therefore, we may assume that $\mathscr{M} \cong \mathscr{O}_S^r \cong f_*(\mathscr{E}^\vee)$. Then ψ becomes a map $\mathscr{O}_X^r \to \mathscr{E}$. It corresponds to a homomorphism in $\mathrm{Hom}_{\mathscr{O}_S}(\mathscr{O}_S^r, f_*\mathscr{E})$. By hypothesis we find a homomorphism

$$\varphi \in \mathrm{Hom}_{\mathscr{O}_S}(\mathscr{O}_S^r, f_*(\mathscr{E}^\vee)) = \mathrm{Hom}_{\mathscr{O}_X}(\mathscr{O}_X^r, \mathscr{E}^\vee) = \mathrm{Hom}_{\mathscr{O}_X}(\mathscr{E}, \mathscr{O}_X^r)$$

such that $\varphi \circ \psi = \mathrm{id}_{\mathscr{O}_X^r}$. Hence φ and ψ are isomorphisms by Corollary 8.12.

It remains to show that the formation of $f_*\mathscr{E}$ is compatible with arbitrary base change for scheme morphisms. If $\mathscr{E} = f^*\mathscr{M}$, then for a cartesian diagram as in Theorem 24.66 (C) one has

$$(f_T)_*(v^*\mathscr{E}) = (f_T)_*f_T^*(u^*\mathscr{M}) = u^*\mathscr{M} = u^*f_*\mathscr{E}$$

which shows the claim. $\qquad\square$

Proof. (of Theorem 24.66) The hypotheses imply that $f_*\mathscr{O}_X = \mathscr{O}_S$, compatibly with base change (Corollary 24.63). All hypotheses and assertions are local on S. The uniqueness of Z is clear.

For an integer $r \geq 0$ let X_r be the open and closed subscheme where $\mathscr{E}$ has a fixed rank r. As all fibers of f are connected and as f is open and closed, $X_r = f^{-1}(S_r)$ for the open and closed subscheme $S_r := f(X_r)$ of S. Hence, we may assume that $\mathscr{E}$ is of constant rank r.

Now Lemma 24.67 shows that Z is the locus in S, where $f_*\mathscr{E}$ is locally free of rank of r with formations commuting with base change and where the pairing $f_*\mathscr{E} \otimes_{\mathscr{O}_S} f_*\mathscr{E}^\vee \to \mathscr{O}_S$ of locally free $\mathscr{O}_X$-modules (24.12.1) is perfect. The locus Z' defined by the first condition is given by an immersion $Z \to S$ of finite presentation by Proposition 23.147 applied with $b = 0$. The locus in Z' defined by the second condition is given by the non-vanishing of the r-th exterior power of the homomorphism of locally free rank r modules $f_*(\mathscr{E}^\vee)_{|Z'} \to f_*(\mathscr{E})^\vee_{|Z'}$ which defines an open subscheme $Z \to Z'$ of finite presentation.

Moreover $s \in S$ is in Z if and only if $\mathrm{Spec}\, \kappa(s) \to S$ factors through Z. This shows the set-theoretic description of Z in (1).

We now show (3). Lemma 24.65 shows that for $\mathscr{E} = \mathscr{L}$ a line bundle, Z as a set can also be described as

$$Z = \{\, s \in S \; ; \dim_{\kappa(s)} H^0(X_s, \mathscr{L}_s) > 0 \text{ and } \dim_{\kappa(s)} H^0(X_s, \mathscr{L}_s^{-1}) > 0\},$$

which is a closed subset by semicontinuity (Theorem 23.139).

It remains to show (2). As $s_0 \in Z$, $\mathscr{E}_{s_0} = \mathscr{E}_{|X_{s_0}}$ is a free $\mathscr{O}_{X_{s_0}}$-module, say isomorphic to $\mathscr{O}_{X_{s_0}}^r$. Hence by hypothesis we have $H^1(X_{s_0}, \mathscr{E}_{s_0}) = 0$. Thus Corollary 23.144 shows that, after possibly replacing S by an open neighborhood of s_0, $f_*\mathscr{E}$ is finite locally free and that its formation commutes with base change. Hence a basis $(e_1, \dots, e_r)$ of $H^0(X_{s_0}, \mathscr{E}_{s_0})$ lifts to a basis of $f_*\mathscr{E}$ again after possibly passing to a smaller neighborhood of s_0. This basis defines an isomorphism $\mathscr{O}_S^r \xrightarrow{\sim} f_*\mathscr{E}$ which corresponds to a homomorphism $u\colon f^*\mathscr{O}_S^r = \mathscr{O}_X^r \to \mathscr{E}$. As u comes from a basis of $H^0(X_{s_0}, \mathscr{E}_{s_0})$, its restriction to X_{s_0} is an isomorphism. Therefore u is surjective in an open neighborhood $W \subseteq X$ of X_{s_0} by Nakayama's lemma and hence an isomorphism on W, being a surjective map of finite free modules of the same rank. As f is closed, there exists an open neighborhood $U \subseteq S$ of s_0 such that $f^{-1}(U) \subseteq W$. Then $\mathscr{E}_{|f^{-1}(U)}$ is isomorphic to the pullback of $\mathscr{O}_U^r$ and hence the open immersion $U \to S$ factors through the subscheme Z by the universal property of Z. $\square$

Corollary 24.68. *Let $f\colon X \to S$ be a flat, proper morphism of finite presentation with geometrically connected and geometrically reduced fibers. Let $\mathscr{E}$ be a finite locally free $\mathscr{O}_X$-module and let $T \subseteq S$ be a subset such that $\mathscr{E}_{|X_s}$ is a free $\mathscr{O}_{X_s}$-module and such that $H^1(X_s, \mathscr{O}_{X_s}) = 0$ for all $s \in T$. Then there exists an open neighborhood U of T in S and a finite locally free $\mathscr{O}_U$-modules $\mathscr{M}$ such that $f_U^*\mathscr{M} \cong \mathscr{E}_{|f^{-1}(U)}$, where $f_U\colon f^{-1}(U) \to U$ is the restriction of f.*

(24.13) Application: Picard group of projective bundles and of products.

Proposition 24.69. *Let S be a non-empty scheme, let $\mathscr{E}$ be a finite locally free $\mathscr{O}_S$-module of rank $r + 1$ with $r \geq 1$, let $\pi\colon \mathbb{P}(\mathscr{E}) \to S$ be the structure morphism of the projective bundle defined by $\mathscr{E}$. Then the map*

$$\mu\colon \operatorname{Pic}(S) \times \mathbb{Z} \longrightarrow \operatorname{Pic}(\mathbb{P}(E)), \qquad (\mathscr{L}, d) \mapsto \pi^*\mathscr{L} \otimes \mathscr{O}_{\mathbb{P}(\mathscr{E})}(d)$$

is an isomorphism of groups.

Proof. Locally on S one has $\mathscr{E} \cong \mathscr{O}_S^{r+1}$ and hence $\mathbb{P}(\mathscr{E}) \cong \mathbb{P}_S^r$ locally on S. Therefore all fibers of π are projective spaces of dimension r over a field. So π is proper, flat, of finite presentation with geometrically integral fibers. In particular $\pi_*\mathscr{O}_{\mathbb{P}(\mathscr{E})} = \mathscr{O}_S$. Recall also that we have already seen that $\operatorname{Pic}(\mathbb{P}_k^r) = \mathbb{Z}$ for any field k (Example 11.45). Finally, we know by Proposition 22.22 that

(*) $H^1(\mathbb{P}_k^r, \mathscr{O}_{\mathbb{P}_k^r}) = 0.$

Set $X := \mathbb{P}(\mathscr{E})$ and write $\mathscr{F}(d) = \mathscr{F} \otimes_{\mathscr{O}_X} \mathscr{O}_X(d)$ for every $\mathscr{O}_X$-module $\mathscr{F}$, as usual. Let us show that μ is injective. Let $(\mathscr{L}, d) \in \operatorname{Ker}(\mu)$, i.e., $\pi^*\mathscr{L}(d)$ is trivial. Its restriction to some fiber $X_s \cong \mathbb{P}_{\kappa(s)}^r$ is $\mathscr{O}_{X_s}(d)$ hence $d = 0$. By Lemma 24.67 (1), $\pi^*\colon \operatorname{Pic}(S) \to \operatorname{Pic}(X)$ is injective and hence $\mathscr{L} \cong \mathscr{O}_S$.

If S is the disjoint union of open and closed subschemes S_i, then $\operatorname{Pic}(S) = \prod_i \operatorname{Pic}(S_i)$ and $\operatorname{Pic}(\mathbb{P}(\mathscr{E})) = \prod_i \operatorname{Pic}(\mathbb{P}(\mathscr{E}_{|S_i}))$. Hence to see that μ is surjective, we may pass to subschemes of S that are open and closed. Let $\mathscr{N}$ be a line bundle on X. Choose $s \in S$. Then $\mathscr{N}_{|X_s}$ is a line bundle on $\mathbb{P}_{\kappa(s)}^r$ and hence isomorphic to $\mathscr{O}_{X_s}(d)$ for some $d \in \mathbb{Z}$. Set $\mathscr{M} := \mathscr{N}(-d)$ such that $\mathscr{M}_{|X_s}$ is trivial. Let Z be the subscheme of S over which $\mathscr{M}$ is the pullback of a line bundle on the base, which is an open and closed subscheme of S because of (*) (Theorem 24.66). Hence after replacing S by Z, we see that $\mathscr{M} \cong \pi^*\mathscr{L}$ for some line bundle $\mathscr{L}$ on S. Hence $\mathscr{N} \cong \pi^*(\mathscr{L})(d)$. $\square$

A similar argument shows the following result.

Proposition 24.70. *Let k be a field, let T be a proper, geometrically connected, geometrically reduced scheme over k such that $H^1(T, \mathscr{O}_T) = 0$. Let S be an arbitrary k-scheme. Then there is an isomorphism of groups*

$$\mu \colon \operatorname{Pic}(S) \times \operatorname{Pic}(T) \xrightarrow{\sim} \operatorname{Pic}(S \times_k T), \qquad (\mathscr{L}, \mathscr{M}) \mapsto p_1^* \mathscr{L} \otimes p_2^* \mathscr{M}.$$

Examples of T satisfying the hypotheses are projective spaces (where the result is a special case of Proposition 24.69), K3-surfaces or, more generally, (strict) Calabi-Yau varieties (Exercise 25.18). On the other hand, in general $\operatorname{Pic}(S) \times \operatorname{Pic}(T) \not\cong \operatorname{Pic}(S \times T)$, even if S and T are smooth projective curves over a field.

Proof. The same argument as in the proof of Proposition 24.69 shows that μ is injective. The surjectivity of μ is also shown in the same way using that $p_2 \colon S \times_k T \to S$ is flat, proper, of finite presentation with geometrically connected and geometrically reduced fibers and that for all $s \in S$ one has

$$H^1(p_2^{-1}(s), \mathscr{O}_{p_2^{-1}(s)}) = H^1(T \otimes_k \kappa(s), \mathscr{O}_{T \otimes_k \kappa(s)}) = H^1(T, \mathscr{O}_T) \otimes_k \kappa(s) = 0. \qquad \square$$

Remark 24.71. In the situation of Proposition 24.69 any line bundle $\mathscr{M}$ on $\mathbb{P}(\mathscr{E})$ is of the form $\mathscr{M} \cong \pi^*(\mathscr{L})(d)$ for some line bundle $\mathscr{L}$ on S and some integer $d \in \mathbb{Z}$. By the projection formula one has for all $i \geq 0$

$$R^i \pi_*(\pi^* \mathscr{L}(d)) = R^i \pi_* \mathscr{O}_X(d) \otimes \mathscr{L}$$

and we calculated $R^i \pi_* \mathscr{O}_X(d)$ in Theorem 22.86.

(24.14) The Theorem of the Cube.

Lemma 24.72. *Let S be a scheme and let $f \colon X \to S$, $f_i \colon X_i \to S$, $i = 1, \dots, m$ be proper, flat S-schemes of finite presentation with geometrically reduced and geometrically connected fibers. Let $g_i \colon X_i \to X$ be morphisms of S-schemes, and let $\mathscr{E}$ be a finite locally free $\mathscr{O}_X$-module of rank r such that $g_i^* \mathscr{E}$ is trivial, i.e., $g_i^* \mathscr{E} \cong \mathscr{O}_{X_i}^r$ for all i. Let $Z = Z(\mathscr{E}, f) \subseteq S$ be the locus on which $\mathscr{E}$ is the pullback of a finite locally free $\mathscr{O}_S$-module (Theorem 24.66). Let $s \in Z$ be a point such that*

$$(24.14.1) \qquad H^1(X_s, \mathscr{O}_{X_s}) \to \bigoplus_i H^1(X_{i,s}, \mathscr{O}_{X_{i,s}})$$

is injective. Then there exists an open neighborhood U of s in S such that $f_* \mathscr{E}_{|U} \cong \mathscr{O}_U^r$ and such that $f^* f_* \mathscr{E} \to \mathscr{E}$ is an isomorphism over U. In particular, $U \to S$ factors through $Z \to S$.

We will give the proof only if S is locally noetherian. For a (different) proof for general schemes in case that $\mathscr{E}$ is a line bundle (the only case that we will use), see [Sta] 0BF2.

Proof. Suppose that S is locally noetherian. Let A be a local Artinian quotient ring of $\mathscr{O}_{S,s}$ and write X_A and $X_{i,A}$ instead of $X \times_S \operatorname{Spec} A$ and $X_i \times_S \operatorname{Spec} A$, respectively. By Corollary 24.63 one has

(*) $$H^0(X_A, \mathscr{O}_{X_A}) = H^0(X_{i,A}, \mathscr{O}_{X_{i,A}}) = A.$$

(I). We claim that $\mathscr{E}_{|X_A}$ is trivial. We prove this by induction on the length of A. If A has length 1, then $A = \kappa(s)$ and $\mathscr{E}_{|X_s}$ is trivial since $s \in Z$. If the length of A is $l > 1$, we can choose $0 \neq \varepsilon \in \mathfrak{m}_A$ such that $\varepsilon \mathfrak{m}_A = 0$. Then $A_0 := A/(\varepsilon)$ has length $l - 1$ and $\mathrm{Ker}(A \to A_0) \cong \kappa(s)$ as $\mathscr{O}_{S,s}$-modules. As $\mathscr{E}$ is flat over S we obtain an exact sequence of $\mathscr{O}_X$-modules

$$0 \longrightarrow \mathscr{E}_{|X_s} \longrightarrow \mathscr{E}_{|X_A} \longrightarrow \mathscr{E}_{|X_{A_0}} \longrightarrow 0.$$

By induction hypothesis, there exists an isomorphism $\mathscr{O}_{X_{A_0}}^r \xrightarrow{\sim} \mathscr{E}_{|X_{A_0}}$ which we view as a tuple of sections $\sigma_1, \dots, \sigma_r \in H^0(X_{A_0}, \mathscr{E}_{|X_{A_0}})$. It suffices to show that we can lift these sections to elements of $H^0(X_A, \mathscr{E}_{|X_A})$, i.e., that their images in $H^1(X_s, \mathscr{E}_{|X_s}) = H^1(X_s, \mathscr{O}_{X_s}^r)$ are zero. By hypothesis, it suffices to show that their images in $\bigoplus_i H^1(X_{i,s}, \mathscr{O}_{X_{i,s}})$ are zero.

As $g_i^* \mathscr{E}$ is trivial, by (*) the sections $g_i^*(\sigma_j) \in H^0(X_{i,A_0}, g_i^* \mathscr{E}_{|X_{i,A_0}})$ can be lifted to $H^0(X_{i,A}, g_i^* \mathscr{E}_{|X_{i,A}})$ and therefore their images in $H^1(X_{i,s}, \mathscr{O}_{X_{i,s}})$ are indeed zero. This shows the claim.

(II). Let $\mathscr{M} := f_* \mathscr{E}$, which is a coherent $\mathscr{O}_S$-module. Set $X_n := X_{\mathscr{O}_{S,s}/\mathfrak{m}_s^n}$. By the formal function theorem (Theorem 24.42) we find

$$\mathscr{M}_s^\wedge = \lim_n H^0(X_n, \mathscr{E}_{|X_n}) \cong (\mathscr{O}_{S,s}^\wedge)^r,$$

where the second equality holds by Step (I) using (*). Since $\mathscr{O}_{S,s}$ is noetherian, the homomorphism $\mathscr{O}_{S,s} \to \mathscr{O}_{S,s}^\wedge$ is faithfully flat, so we obtain that $\mathscr{M}_s$ is a free $\mathscr{O}_{S,s}$-module of rank r and the map $\mathscr{M}_s \to H^0(X_s, \mathscr{E}_{|X_s})$ is surjective. Therefore there exists an open neighborhood U of s such that $\mathscr{M}_{|U} \cong \mathscr{O}_U^r$ and such that $f^* \mathscr{M} \to \mathscr{E}$ is surjective over U. As both $f^* \mathscr{M}$ and $\mathscr{E}$ are finite locally free of the same rank over U, we see that $f^* \mathscr{M} \to \mathscr{E}$ is an isomorphism over U. □

The following result, usually called the theorem of the cube, will be important for example in Chapter 27 on abelian schemes.

Theorem 24.73. (Theorem of the Cube) *Let S be a scheme, let $X \to S$ and $Y \to S$ be flat proper morphisms of finite presentation with geometrically integral fibers, and let T be an S-scheme. Let $\mathscr{L}$ be a line bundle on $X \times_S Y \times_S T$. Suppose that there exist sections $x \in X(S)$ and $y \in Y(S)$ and a point $t_0 \in T$ such that the restrictions of $\mathscr{L}$ to $X \times_S Y \times_S \{t_0\}$, to $X \times_S \{y\} \times_S T$, and to $\{x\} \times_S Y \times_S T$ are trivial. Then there exists an open and closed neighborhood W of t_0 in T such that $\mathscr{L}_{|X \times_S Y \times_S W}$ is trivial.*

Proof. We apply Lemma 24.72 to the projection $f: X \times_S Y \times_S T \to T$ and to $f_1: \{x\} \times_S Y \times_S T \to T$ and $f_2: X \times_S \{y\} \times_S T \to T$ with g_i given by the inclusions induced by x and y, respectively. For a point $t \in T$ with image $s \in S$ the fiber $f^{-1}(t)$ is given by

$$X \times_S Y \times_S \mathrm{Spec}\, \kappa(t) = (X_s \times_{\kappa(s)} Y_s) \otimes_{\kappa(s)} \kappa(t).$$

Let $Z \to T$ be the closed subscheme associated to $(f, \mathscr{L})$ given by Theorem 24.66. It is non-empty since $t_0 \in Z$. By Lemma 24.72 it suffices to show that for every $t \in Z$ the map (24.14.1) is injective in our situation here. But for all $t \in T$ this map is of the form, omitting the respective structure sheaves from the notation,

(*) $$H^1(X_s \times_{\kappa(s)} Y_s) \otimes_{\kappa(s)} \kappa(t) \longrightarrow (H^1(Y_s) \otimes_{\kappa(s)} \kappa(t)) \oplus (H^1(X_s) \otimes_{\kappa(s)} \kappa(t))$$

and a right inverse is given by the base change to $\kappa(t)$ of the Künneth isomorphism

$$H^1(Y_s) \oplus H^1(X_s) = \bigoplus_{p+q=1} H^p(Y_s) \otimes_{\kappa(s)} H^q(X_s) \xrightarrow{\sim} H^1(X_s \times_{\kappa(s)} Y_s),$$

where the first equality holds since $X \to S$ and $Y \to S$ have geometrically integral fibers and hence $H^0(X_s) = H^0(Y_s) = \kappa(s)$. This shows that (*) is even an isomorphism. $\quad\square$

Algebraization

We will now consider the situation that for an I-adically complete ring A, where I is a finitely generated ideal of A, we are given a compatible family of schemes X_n over A/I^{n+1} (see Definition 24.82 below for a precise definition). This family is said to be algebraizable if there exists a scheme X over A such that $X \otimes_A A/I^{n+1} \cong X_n$ for all $n \geq 0$. Moreover if we are given compatible quasi-coherent $\mathscr{O}_{X_n}$-modules $\mathscr{F}_n$, we say that $(\mathscr{F}_n)_n$ is algebraizable if there exists a quasi-coherent $\mathscr{O}_X$-module $\mathscr{F}$ whose pullback to X_n is isomorphic to $\mathscr{F}_n$. The main results in this part are Grothendieck's algebraization theorem (Theorem 24.113) and Grothendieck's existence theorem (Theorem 24.94). For both theorems A has to be noetherian. Then the algebraization theorem states that a family $(X_n)_n$ of proper schemes can be algebraized (uniquely up to isomorphism) by a proper scheme over A if the X_n carry a compatible system of ample line bundles. The existence theorem states that if a proper family $(X_n)_n$ is algebraizable by a proper scheme X, then one has an equivalence between coherent modules over X and compatible systems $(\mathscr{F}_n)_n$ of coherent modules over $(X_n)_n$.

Notation

For any scheme X we will denote by $|X|$ its underlying topological space. Then $|\cdot|$ is a functor from the category of schemes to the category of topological spaces that sends every morphism f of schemes to the underlying continuous map $|f|$.

(24.15) Ideals of definition for constructible closed subsets.

Let Y be a qcqs scheme. Recall that a closed subset $Z \subseteq |Y|$ is constructible if and only if its complement $|Y| \setminus Z$ is quasi-compact (Proposition 10.44). In a noetherian scheme, any closed subset is constructible (Lemma 1.25). In general, the closed constructible subspaces are those that are locally defined by the vanishing of finitely many functions:

Lemma 24.74. *Let Y be a qcqs scheme, and let $Z \subseteq |Y|$ be a closed subspace. Then the following assertions are equivalent.*
(i) *The complement $U := |Y| \setminus Z$ is quasi-compact.*
(ii) *There exists a quasi-coherent ideal $\mathscr{I}$ of $\mathscr{O}_Y$ of finite type with $|V(\mathscr{I})| = Z$.*
(iii) *For every quasi-coherent ideal $\mathscr{J}$ of $\mathscr{O}_Y$ with $|V(\mathscr{J})| = Z$ there exists a finite type quasi-coherent ideal $\mathscr{I} \subseteq \mathscr{J}$ such that $|V(\mathscr{I})| = Z$.*

Proof. There exists a quasi-coherent ideal $\mathscr{J}$ of $\mathscr{O}_X$ such that $|V(\mathscr{J})| = Z$ by Proposition 3.52. Hence (iii) implies (ii).

Let us show that (ii) implies (i). As Y is quasi-compact, we can cover Y by finitely many open affine subschemes V. To show that U is quasi-compact it suffices to show that $V \cap U$ is quasi-compact for all V. Hence we may assume $Y = \operatorname{Spec} A$ is affine and that Z is the underlying topological space of $\operatorname{Spec} A/I$, where I is generated by finitely many elements $f_1, \ldots, f_r$. But then U is the union of the quasi-compact principal open subsets $D(f_i)$. Hence it is quasi-compact.

It remains to show that (i) implies (iii). If $|V(\mathscr{J})| = Z$, then $\mathscr{J}_{|U} = \mathscr{O}_U$ is an $\mathscr{O}_U$-module of finite type. As Y is qcqs, there exists a finite type quasi-coherent ideal $\mathscr{I} \subseteq \mathscr{J}$ with $\mathscr{I}_{|U} = \mathscr{O}_U$ (Proposition 10.48). Then $|V(\mathscr{J})| \subseteq |V(\mathscr{I})| \subseteq X \setminus U$ and hence $|V(\mathscr{I})| = Z$. $\square$

From now on Y will denote a qcqs scheme and Z will be a closed constructible subspace.

Definition 24.75. *An* ideal of definition *for Z is a quasi-coherent ideal $\mathscr{I} \subseteq \mathscr{O}_Y$ of finite type such that $|V(\mathscr{I})| = Z$.*

Let $\mathcal{I}(Z)$ be the set of ideals of definition for Z and let $\mathcal{S}(Z)$ be the set of closed subschemes of Y whose underlying topological space is Z and that are defined by a finite type quasi-coherent ideal.

Sending $\mathscr{I} \in \mathcal{I}(Z)$ to its vanishing scheme $V(\mathscr{I})$ yields an inclusion reversing bijection between $\mathcal{I}(Z)$ and $\mathcal{S}(Z)$. Both sets are non-empty by Lemma 24.74.

Example 24.76. Let $Y = \operatorname{Spec} A$ be affine. Then each ideal of definition for Z corresponds to a finitely generated ideal $I \subseteq A$ such that $\operatorname{rad}(I) = \bigcap_{z \in Z} \mathfrak{p}_z$. We call such an ideal also an ideal of definition for Z. In general $\operatorname{rad}(I)$ is not finitely generated and therefore it is not an ideal of definition.

Let $I \subseteq A$ be an ideal of definition for Z and $J \subseteq I$ be a finitely generated ideal. Then J is an ideal of definition for Z if and only if all elements of the image of I in A/J are nilpotent in A/J.

If $\mathscr{I}$ is an ideal of definition, then $\mathscr{I}^n$ is an ideal of definition for all $n \geq 1$. More generally, we have:

Lemma 24.77. *Let Y be a qcqs scheme, let $Z, Z' \subseteq |Y|$ be closed subsets and suppose that Z is constructible. Let $\mathscr{I}$ be an ideal of definition for Z and let $\mathscr{I}' \subseteq \mathscr{O}_Y$ be any quasi-coherent ideal with $|V(\mathscr{I}')| = Z'$.*
(1) Suppose $Z' \subseteq Z$. Then there exists an $n \geq 1$ such that $\mathscr{I}^n \subseteq \mathscr{I}'$.
(2) Suppose that $|Y| \setminus Z'$ is quasi-compact and that $\mathscr{I}'$ is an ideal of definition for Z'. Then $\mathscr{I}\mathscr{I}'$ is an ideal of definition for $Z \cup Z'$, and $\mathscr{I} + \mathscr{I}'$ is an ideal of definition for $Z \cap Z'$.

Proof. We may assume that $Y = \operatorname{Spec} A$ is affine, using that Y is quasi-compact for (1). Let $I, I' \subseteq A$ be the finitely generated ideals corresponding to $\mathscr{I}$ and $\mathscr{I}'$, respectively. Then II' and $I + I'$ are finitely generated. Hence (2) follows because for the radicals of ideals one has $\operatorname{rad}(II') = \operatorname{rad}(I) \cap \operatorname{rad}(I')$ and $\operatorname{rad}(I + I') = \operatorname{rad}(\operatorname{rad}(I) + \operatorname{rad}(I'))$.

Let us show (1). As $I \subseteq \operatorname{rad}(I) \subseteq \operatorname{rad}(I')$, we find generators $a_1, \ldots, a_r$ of I and integers $m_i \geq 1$ such that $a_i^{m_i} \in I'$. Then $I^{m_1 + \cdots + m_r} \subseteq I'$. $\square$

Proposition 24.78. *Let X be a qcqs scheme and let Z be a closed constructible subset. Let $\mathscr{F}$ be a quasi-coherent $\mathscr{O}_X$-module of finite type. Then $\operatorname{Supp}(\mathscr{F}) \subseteq Z$ if and only if there exists an ideal of definition $\mathscr{I}$ for Z with $\mathscr{I}\mathscr{F} = 0$.*

Proof. The condition is clearly sufficient. Hence suppose that $\text{Supp}(\mathscr{F}) \subseteq Z$. Let $\mathscr{I}'$ be the annihilator of $\mathscr{F}$ which is a quasi-coherent ideal with $|V(\mathscr{I}')| = \text{Supp}\,\mathscr{F}$ by Proposition 7.35. By Lemma 24.77 we find an ideal of definition $\mathscr{I}$ for Z with $\mathscr{I} \subseteq \mathscr{I}'$. Then $\mathscr{I}\mathscr{F} = 0$ since $\mathscr{F}$ is annihilated by its annihilator. $\qquad\square$

(24.16) Formal completion of qcqs schemes along closed subspaces.

Let S be a scheme. Recall from Section (4.2) that the Yoneda functor $Y \mapsto h_Y$ with $h_Y(T) = \text{Hom}_S(T, Y)$ yields a fully faithful embedding from the category (Sch/S) of S-schemes into the category $\widehat{(\text{Sch}/S)}$ of contravariant functors from (Sch/S) to the category of sets. We will usually identify a scheme with the functor it represents. In $\widehat{(\text{Sch}/S)}$ we have the full subcategory of Zariski sheaves $\widehat{(\text{Sch}/S)}_{Zar}$, i.e., of those functors $F\colon (\text{Sch}/S)^{\text{opp}} \to (\text{Sets})$ such that for every S-scheme T and for every open covering $(T_i)_i$ of T one has

$$F(T) = \text{Eq}\big(\, \textstyle\prod_i F(T_i) \rightrightarrows \prod_{i,j} F(T_i \times_T T_j)\,\big),$$

where $\text{Eq}(-)$ denotes the equalizer of a pair of parallel arrows. By Proposition 14.76 the Yoneda embedding factorizes through $\widehat{(\text{Sch}/S)}_{Zar}$.

If $\mathcal{I} \to (\text{Sch}/S)$, $i \mapsto Y_i$, is a diagram in (Sch/S) (in other words, simply a functor from a small category $\mathcal{I}$ to (Sch/S); the most important case for us will be $\mathcal{I} = \mathbb{N}$), then we write

$$\operatorname*{colim}_{\mathcal{I}} Y_i$$

for the colimit of the functors h_{Y_i} in the category $\widehat{(\text{Sch}/S)}_{Zar}$ of Zariski sheaves on (Sch/S). Note however that the embedding $(\text{Sch}/S) \to \widehat{(\text{Sch}/S)}_{Zar}$ does not commute with colimits (except if S is the empty scheme), therefore this is usually *not* the colimit in the category of schemes, and colimits do not exist in (Sch/S) in general. Hence $\operatorname*{colim}_{\mathcal{I}} Y_i$ is a Zariski sheaf on the category of S-schemes, but it is usually not represented by a scheme.

Proposition and Definition 24.79. *Let S be a scheme, let Y be a qcqs S-scheme and let $Z \subseteq |Y|$ be a closed constructible subset. Fix an ideal of definition $\mathscr{J} \in \mathcal{I}(Z)$ and set $Y_n := V(\mathscr{J}^{n+1})$ for $n \geq 0$.*
(1) *The partially ordered set $\mathcal{S}(Z)$ (Definition 24.75) is filtered. The system $(Y_n)_n$ is a final subset of $\mathcal{S}(Z)$.*
(2) *We call the Zariski sheaf*

(24.16.1) $$\hat{Y} := Y_{/Z} := \operatorname*{colim}_{Y' \in \mathcal{S}(Z)} Y' = \operatorname*{colim}_n Y_n$$

the formal completion of Y along Z. If T is a qcqs S-scheme, then

(24.16.2) $$Y_{/Z}(T) = \operatorname*{colim}_{Y' \in \mathcal{S}(Z)} Y'(T) = \operatorname*{colim}_n Y_n(T).$$

In particular $Y_{/Z}$ is the unique Zariski sheaf such that (24.16.2) holds for all S-schemes T that are affine.
(3) *For any S-scheme T one has*

(24.16.3) $$Y_{/Z}(T) = \{\, f\colon T \to Y \;;\; |f|(|T|) \subseteq Z \,\},$$

where $|f|\colon |T| \to |Y|$ denotes the underlying continuous map of topological spaces.

(4) The functor $Y_{/Z}$ is an fpqc-sheaf.

The equality (24.16.2) shows that after restriction to qcqs S-schemes, $Y_{/Z}$ is the colimit in the category of all contravariant functors from the category of qcqs S-schemes to the category of sets.

Proof. Assertion (1) follows from Lemma 24.77. It also shows the equality in (24.16.1) and the second equality in (24.16.2).

(I). The canonical map $\psi\colon \operatorname{colim}_n Y_n(T) \to Y_{/Z}(T)$ is bijective for qcqs schemes T by the same argument as in the proof of Lemma 21.52. This proves parts (1) and (2).

(II). Let F be the functor on S-schemes defined by the right side of (24.16.3). We show that F is an fpqc-sheaf. Indeed, let $(T_i \to T)_i$ be an fpqc cover and let $s_i \in F(T_i)$ such that the images of s_i and s_j in $F(T_i \times_T T_j)$ are equal for all i and j. As Y is an fpqc-sheaf by Proposition 14.76, there exists a unique $s \in Y(T)$ whose image in $F(T_i)$ is s_i for all i. Since $\coprod_i T_i \to T$ is surjective one necessarily has $s \in F(T)$.

(III). As the underlying topological space of Y_n is Z, one has a compatible system of monomorphisms of functor $f\colon Y_n \to F$. As F is a sheaf for the Zariski topology, one obtains an induced morphism α sitting in a commutative diagram

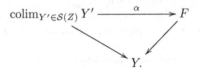

It remains to show that α is an isomorphism. As source and target of α are Zariski sheaves, we can do so locally on Y. Therefore we may assume that $Y = \operatorname{Spec} A$ is affine. Let $I \subseteq A$ be the radical ideal with $V(I) = Z$.

Clearly α is a monomorphism. To show that α is an epimorphism of Zariski sheaves, it suffices to show that $\alpha(T)$ is surjective for $T = \operatorname{Spec} B$ an affine scheme. Let $f\colon T \to Y$ be an element of $F(T)$. It is given by a ring homomorphism $\varphi\colon A \to B$ such that every prime ideal of B contains $\varphi(I)$, in other words, all elements in $\varphi(I)$ are nilpotent. Let $J := \operatorname{Ker}(\varphi)$. Then $J \subseteq I$ and the injective homomorphism $A/J \to B$ sends all elements of the ideal I/J in A/J to nilpotent elements. Hence all elements of I/J are nilpotent in A/J. Therefore $\operatorname{rad}(J) = I$ and the element $f \in F(T)$ is the image of a T-valued point of $\operatorname{Spec} A/J$. $\square$

For affine schemes we have the following situation.

Example 24.80. Let A be a ring, $X := \operatorname{Spec} A$, and let $Z \subseteq |X|$ be a closed subset. Then its complement is quasi-compact if and only if there exists a finitely generated ideal I of A such that $Z = |\operatorname{Spec} A/I|$. We assume this from now on. Every such ideal I is then an ideal of definition for Z. The set of ideals of definitions for Z is again denoted by $\mathcal{I}(Z)$. Hence

$$X_{/Z} = \operatorname*{colim}_{I \in \mathcal{I}(Z)} \operatorname{Spec} A/I = \operatorname*{colim}_{n \geq 0} \operatorname{Spec} A/J^{n+1},$$

where J is a fixed ideal of definition for Z. There is a unique topology on A making A into a topological ring such that the ideals $I \in \mathcal{I}(Z)$ form a basis of neighborhoods of 0 in A. For this topology $(J^{n+1})_{n \geq 0}$ is also a basis of neighborhoods of 0. Hence this topology is the J-adic topology. The *Z-adic completion* of this topological ring is given by

$$\hat{A} := \lim_{I \in \mathcal{I}(Z)} A/I = \lim_n A/J^{n+1},$$

where A/I and A/J^{n+1} are rings endowed with the discrete topology (Remark 24.1). We obtain morphisms of Zariski sheaves

$$\mathrm{Spec}(A)_{/Z} \longrightarrow \mathrm{Spec}\,\hat{A} \longrightarrow \mathrm{Spec}\,A.$$

As J is finitely generated, the topology on $\hat{A}$ is the $\hat{J}$-adic topology where $\hat{J} := J\hat{A}$ and $\hat{A} = \lim_n \hat{A}/\hat{J}^{n+1}$ (Proposition 24.2). Then the underlying closed subset of $V(\hat{J})$ is the preimage $\hat{Z} \subseteq \mathrm{Spec}\,\hat{A}$ of Z. We set

$$\mathrm{Spf}(\hat{A}) := (\mathrm{Spec}\,A)_{/Z} = (\mathrm{Spec}\,\hat{A})_{\hat{Z}}.$$

This Zariski sheaf is then even an fpqc-sheaf by Proposition 24.79 (4) and it is called the *formal spectrum* of the topological ring A.

The formation of the completion along a closed subspace is functorial in the following sense.

Remark 24.81. Let $f \colon X \to Y$ be a morphism of qcqs schemes. Let $Z \subseteq |Y|$ be closed constructible. The morphism f is itself quasi-compact (Remark 10.4) and quasi-separated (Proposition 10.25). If $\mathscr{I}$ is an ideal of definition for Z and hence $|V(\mathscr{I})| = Z$, then $\mathscr{J} := f^{-1}(\mathscr{I})\mathcal{O}_X$ is the quasi-coherent ideal of $\mathcal{O}_X$ defining the closed subscheme $f^{-1}(V(\mathscr{I}))$ whose underlying topological space is $f^{-1}(Z)$. Hence $\mathscr{J}$ is an ideal of definition for $f^{-1}(Z)$.

Let $T \subseteq f^{-1}(Z) \subseteq X$ be any closed subspace such that $|X| \setminus T$ is quasi-compact. Then by Lemma 24.77 there exists for every ideal of definition $\mathscr{K}$ for T an $n \geq 1$ such that $\mathscr{J}^n \subseteq \mathscr{K}$. In particular, for every X' in $\mathcal{S}(T)$ there exists an $Y' \in \mathcal{S}(Z)$ such that f restricts to a morphism of schemes $X' \to Y'$. Hence f induces a morphism of Zariski sheaves

$$f_{/Z,T} \colon X_{/T} \longrightarrow Y_{/Z}.$$

If $T = f^{-1}(Z)$, then we often write $X_{/Z}$ instead of $X_{/f^{-1}(Z)}$ and $f_{/Z}$ instead of $f_{/Z,f^{-1}(Z)}$.

One obtains a functor $(-)_{/Z}$ of completion along Z from the category of qcqs Y-schemes to the category of Zariski sheaves over $Y_{/Z}$.

(24.17) Adic formal schemes over complete rings.

In this section we denote by A a ring and by $I \subseteq A$ a finitely generated ideal such that $A = \lim_n A/I^n$. Set $S := \mathrm{Spec}\,A$, $Z := V(I)$, and $S_n := \mathrm{Spec}\,A/I^{n+1}$ such that Z is the underlying topological space of S_n for all $n \geq 0$.

Definition 24.82. *An adic formal scheme $\mathcal{X}$ over $S_{/Z}$ consists of a family $(X_n \to S_n)_n$ of scheme morphisms and for all $n \geq 0$ a morphism of S_{n+1}-schemes $X_n \to X_{n+1}$ such that the diagram*

(24.17.1)

$$\begin{array}{ccc} X_n & \longrightarrow & X_{n+1} \\ \downarrow & & \downarrow \\ S_n & \longrightarrow & S_{n+1} \end{array}$$

is cartesian for all $n \geq 0$. A morphism of adic formal schemes $\mathcal{X} \to \mathcal{X}'$ over $S_{/Z}$ is an inductive system of S_n-morphism $u_n \colon X_n \to X'_n$ such that the base change $(u_{n+1})_{X_n}$ is u_n for all $n \geq 0$. The set of these morphisms is denoted by $\operatorname{Hom}_{S_{/Z}}(\mathcal{X}, \mathcal{X}')$.

We obtain the category of adic formal schemes over $S_{/Z}$.

Geometrically, we think of formal schemes as a variants of usual schemes which are allowed to carry "infinitely thick infinitesimal neighborhoods". It is possible to develop the theory in a more geometric way by setting up a theory of topologically ringed spaces, see for instance [EGAInew] I.10.

By Proposition 24.79 (1) to give an adic formal scheme over $S_{/Z}$ is the same as to give a family $X_{S'} \to S'$ of scheme morphisms for all closed subschemes S' with Z as underlying topological space together with S-morphisms $\alpha_{S'_2, S'_1} \colon X_{S'_1} \to X_{S'_2}$ such that

$$
\begin{array}{ccc}
X_{S'_1} & \xrightarrow{\alpha_{S'_2, S'_1}} & X_{S'_1} \\
\downarrow & & \downarrow \\
S'_1 & \longrightarrow & S'_2
\end{array}
$$

is cartesian for all such closed subschemes S'_1 and S'_2 with $S'_1 \subseteq S'_2$ and such that for $S'_1 \subseteq S'_2 \subseteq S'_3$ one has $\alpha_{S'_3, S'_2} \circ \alpha_{S'_2, S'_1} = \alpha_{S'_3, S'_1}$. Hence the notion of an adic formal scheme over $S_{/Z}$ does not depend on the choice of the ideal I but only on the subspace Z.

Definition and Proposition 24.83. *Let $\mathcal{X} = (X_n)_n$ and $\mathcal{Y} = (Y_n)_n$ be adic formal schemes over $S_{/Z}$, and let $f = (f_n \colon X_n \to Y_n)_n$ be a morphism of adic formal schemes $\mathcal{X} \to \mathcal{Y}$.*

Let $\mathbf{P}$ be one of the properties "separated", "of finite type", "proper", "finite", "affine", "closed immersion", "quasi-compact", "quasi-separated". Then f is said to have $\mathbf{P}$ if the following equivalent conditions hold.

(i) *The morphisms $f_n \colon X_n \to Y_n$ have the property $\mathbf{P}$ for all $n \geq 0$.*

(ii) *The morphism $f_0 \colon X_0 \to Y_0$ has the property $\mathbf{P}$.*

We also say that f is finite locally free *if f_n is finite locally free for all n.*

Proof. As $S_n \to S_{n+1}$ is a universal homeomorphism and a closed immersion for all n, so are $X_n \to X_{n+1}$ and $Y_n \to Y_{n+1}$ for all $n \geq 0$ in view of the cartesian diagram (24.17.1). In particular, for the underlying reduced subschemes one has $(X_n)_{\mathrm{red}} = (X_m)_{\mathrm{red}}$ and $(Y_n)_{\mathrm{red}} = (Y_m)_{\mathrm{red}}$ as schemes over $(S_n)_{\mathrm{red}} = (S_m)_{\mathrm{red}}$ for all $n, m \geq 0$. Hence for all properties that depend only on the underlying reduced scheme (i) and (ii) are equivalent. This holds for the properties "quasi-compact" and "quasi-separated" which depend only on the underlying topological spaces. It also holds for the properties "separated" (Proposition 9.13) and "affine" (Lemma 12.38).

Now suppose that f_0 is of finite type. Then f is quasi-compact. To show that f_n is locally of finite type, we may assume that $Y_m = \operatorname{Spec} B_m$ and $X_m = \operatorname{Spec} C_m$ is affine for all m. By induction, we find a surjection of B_{n-1}-algebras $B_{n-1}[t_1, \ldots, t_r] \to C_{n-1}$ which we can lift to a map of B_n-algebras $B_n[t_1, \ldots, t_r] \to C_n$. This is surjective by Nakayama's lemma because $C_n \to C_{n-1}$ has nilpotent kernel. Hence C_n is a finitely generated B_n-algebra.

Now the equivalence of (i) and (ii) for "proper" follows from Proposition 12.58. We can also deduce the equivalence for "finite" because a morphism of schemes is finite if and only if it is affine and proper (Corollary 12.89).

If f_0 is a closed immersion, then f_n is finite for all n. To see that f_n is a closed immersion, we may assume that $Y_m = \operatorname{Spec} B_m$ is affine for all m. Then $X_m = \operatorname{Spec} C_m$ for a finite B_m-algebra for all m. We conclude by Nakayama's lemma which shows that $B_n \to C_n$ is surjective if and only if $B_0 \to C_0$ is surjective. □

Let $X \to S = \operatorname{Spec} A$ be a morphism of schemes. Then the family of schemes $X_n := X \times_S S_n$ forms an adic formal scheme over $S_{/Z}$.

Lemma and Definition 24.84. *If $X \to \operatorname{Spec} A$ is qcqs, then we have for all $n \geq 0$ a commutative diagram of Zariski sheaves in which all rectangles are cartesian*

(24.17.2)

$$
\begin{array}{ccc}
X_n \longrightarrow X_{/Z} \longrightarrow X \\
\downarrow \qquad \downarrow \qquad \downarrow \\
S_n \longrightarrow S_{/Z} \longrightarrow S.
\end{array}
$$

We call an qcqs adic formal scheme $(X_n)_n$ over $S_{/Z}$ algebraizable if it is of the form $(X \times_S S_n)_n$ for a qcqs scheme X over A.

Proof. It suffices to show that the right small rectangle is cartesian on T-valued points, where T is an affine scheme. But in this case one has $X_{/Z}(T) = \operatorname{colim}_n X_n(T)$ and $S_{/Z}(T) = \operatorname{colim}_n S_n(T)$ by (24.16.2). Then it is clear that the diagram is cartesian. □

This shows in particular that $X_{/Z} = \operatorname{colim}_n X_n$ and the qcqs adic formal scheme $(X_n = X_{/Z} \times_{S_{/Z}} S_n)_n$ determine each other in the algebraizable case. This holds indeed in general (Exercise 24.20), but we will not need this fact.

We will view the functor $(-)_{/Z}$ as a functor from qcqs schemes over A to qcqs adic formal schemes over $S_{/Z}$. If $f \colon X \to Y$ is a morphism of qcqs S-schemes, then $f_{/Z}$ is the family of morphisms $f_n := f \times \operatorname{id}_{S_n} \colon X_n \to Y_n$. If A is noetherian, we will study some of its properties in Section (24.23).

(24.18) Modules over formal schemes.

Consider the following situation. For $n \geq 0$ let $\iota_n \colon X_n \to X_{n+1}$ be closed immersions of schemes defined by a nilpotent ideal. In the sequel we will be only interested in one of the following examples.
(1) Let X be a qcqs scheme and let Z be a closed constructible subset. Let $\mathscr{I}$ be an ideal of definition for Z and set $X_n := V(\mathscr{I}^{n+1})$.
(2) The family $\mathcal{X} = (X_n)_n$ is a qcqs adic formal scheme over a $S_{/Z}$ as in Definition 24.82.

Definition 24.85.
(1) A module over $(X_n)_n$ consists of a family of $\mathscr{O}_{X_n}$-modules $\mathscr{F}_n$ together with isomorphisms of $\mathscr{O}_{X_n}$-modules $\alpha_n \colon \iota_n^* \mathscr{F}_{n+1} \xrightarrow{\sim} \mathscr{F}_n$ for all $n \geq 0$.
A morphism $\mathscr{F} \to \mathscr{G}$ of modules over $(X_n)_n$ is a family of homomorphisms $u_n \colon \mathscr{F}_n \to \mathscr{G}_n$ of $\mathscr{O}_{X_n}$-modules such that $\iota_n^*(u_{n+1}) = u_n$ for all $n \geq 0$. We obtain the category of modules over $(X_n)_n$ which we denote by $((X_n)_n\text{-Mod})$.
(2) Let $\mathbf{P}$ be one of the properties "quasi-coherent", "of finite type", "coherent", "locally free", "locally free of rank r" for a fixed $r \geq 0$. Then a module $(\mathscr{F}_n, \alpha_n)_n$ is said to have $\mathbf{P}$, if the $\mathscr{O}_{X_n}$-module $\mathscr{F}_n$ has property $\mathbf{P}$ for all n.

(3) *Let $\mathscr{F} = (\mathscr{F}_n)_n$ and $\mathscr{G} = (\mathscr{G}_n)_n$ be modules over $(X_n)_n$. Then we define their tensor product by $\mathscr{F} \otimes \mathscr{G} := (\mathscr{F}_n \otimes_{\mathscr{O}_{X_n}} \mathscr{G}_n)_n$.*

(4) *An algebra over $(X_n)_n$ is a module $\mathscr{A}$ over $(X_n)_n$ together with a map $\mathscr{A} \otimes \mathscr{A} \to \mathscr{A}$ of modules over $(X_n)_n$ that makes $\mathscr{A}_n$ into an $\mathscr{O}_{X_n}$-algebra for all n.*

In particular, the system $(\mathscr{O}_{X_n})_n$ is an algebra over $(X_n)_n$, which we call the *structure sheaf* of $(X_n)_n$.

Remark and Definition 24.86. Let X be a qcqs scheme, let Z be a closed constructible subset of $|X|$, let $\mathscr{J}$ be an ideal of definition for Z, and set $X_n := V(\mathscr{J}^{n+1})$. Then we call a module over $(X_n)_n$ also a *module over $X_{/Z}$*. If $(\mathscr{F}_n, \alpha_n)_n$ is such a module, then it is often notationally easier to view $\mathscr{F}_n$ as an $\mathscr{O}_X$-module that is annihilated by $\mathscr{J}^{n+1}$ and to view α_n as an isomorphism $\mathscr{F}_{n+1}/\mathscr{J}^{n+1}\mathscr{F}_{n+1} \xrightarrow{\sim} \mathscr{F}_n$ of $\mathscr{O}_X$-modules. Often one omits the α_n from the notation. By definition one has

$$\operatorname{Hom}_{(X_{/Z}\text{-Mod})}((\mathscr{F}_n)_n, (\mathscr{G}_n)_n) = \lim_n \operatorname{Hom}_{\mathscr{O}_X}(\mathscr{F}_n, \mathscr{G}_n).$$

If $\mathscr{F}$ is an $\mathscr{O}_X$-module, then $\mathscr{F}_{/Z} := (\mathscr{F}/\mathscr{J}^{n+1}\mathscr{F})_n$ is a module over $X_{/Z}$, called the *formal completion of $\mathscr{F}$ along Z*. We obtain an additive functor

$$(24.18.1) \qquad (\mathscr{O}_X\text{-Mod}) \longrightarrow (X_{/Z}\text{-Mod}), \qquad \mathscr{F} \to \mathscr{F}_{/Z}.$$

Moreover, as the pullback of modules via $X_n \to X$ commutes with tensor products we see that

$$(24.18.2) \qquad (\mathscr{F} \otimes_{\mathscr{O}_X} \mathscr{G})_{/Z} = \mathscr{F}_{/Z} \otimes \mathscr{G}_{/Z}.$$

Remark 24.87. By Proposition 24.79 (1) to give a module over $X_{/Z}$ is the same as to give for every ideal of definition $\mathscr{I}$ for Z an $\mathscr{O}_X$-module $\mathscr{F}_{\mathscr{I}}$ with $\mathscr{I}\mathscr{F}_{\mathscr{I}} = 0$ together with isomorphisms of $\mathscr{O}_X$-modules $\mathscr{F}_{\mathscr{I}}/\mathscr{I}'\mathscr{F}_{\mathscr{I}} \xrightarrow{\sim} \mathscr{F}_{\mathscr{I}'}$ for all ideals of definitions $\mathscr{I} \subseteq \mathscr{I}'$. Hence the notion of a module over $X_{/Z}$ given in Definition 24.86 does not depend on the choice of the ideal $\mathscr{J}$.

Similarly, the notion of a module over a qcqs adic formal scheme $\mathcal{X}$ over $S_{/Z}$ as in Definition 24.82 depends only on Z.

Next we study modules over $(X_n)_n$ in the affine situation. Let A be a ring and let $I \subseteq A$ be an ideal. Let $\lim_n (A/I^{n+1}\text{-Mod})$ be the category of modules over the projective system $(A/I^{n+1})_n$, i.e., of families of A/I^{n+1}-modules M_n together with isomorphisms $M_n/I^n M_n \xrightarrow{\sim} M_{n-1}$ for all n. The category $\lim_n (A/I^{n+1}\text{-Mod})$ can be identified with the category of quasi-coherent modules over $(\operatorname{Spec} A/I^{n+1})_n$ defined in Definition 24.85. Consider the functor

$$(24.18.3) \qquad (A\text{-Mod}) \longrightarrow \lim_n (A/I^{n+1}\text{-Mod}), \qquad M \mapsto (M/I^{n+1}M).$$

If I is finitely generated, it corresponds to the functor of formal completion (24.18.1) for quasi-coherent modules.

Proposition 24.88. *Let A be a ring and let $I \subseteq A$ be an ideal such that A is I-adically complete, i.e., $A \cong \lim_n A/I^{n+1}$.*

(1) *Let $(M_n)_n$ be in $\lim_n (A/I^{n+1}\text{-Mod})$ such that M_0 is a finitely generated A/I-module. Then M_n is a finitely generated A/I^{n+1}-module for all n and $M := \lim_n M_n$ is a finitely generated I-adically complete A-module. Moreover, $M/I^{n+1}M \cong M_n$.*

(2) *The functor* (24.18.3) *yields an equivalence between the category of finite projective A-modules M and the category of families* $(M_n)_n$ *of finite projective* A/I^{n+1}-*modules* M_n. *For a fixed integer* $r \geq 0$, M *it is of rank* r *if and only if* M_n *is of rank* r *for all* n.

(3) *The functor* (24.18.3) *yields an equivalence between the category of finitely generated I-adically complete A-modules and the category of families* $(M_n)_n$ *with* M_n *a finitely generated* A/I^{n+1}-*module*.

The proof will show that a quasi-inverse functor is given by $(M_n)_n \mapsto \lim_n M_n$ in (2) and (3).

Proof. Let us show (1). As M_0 is finitely generated, we can choose a surjection of A/I-modules $p_0 \colon (A/I)^N \to M_0$. Inductively one constructs a family of A/I^{n+1}-linear maps

$$p_n \colon (A/I^{n+1})^N \to M_n$$

such that $p_n \equiv p_{n-1} \pmod{I^n}$. As I/I^{n+1} is nilpotent in A/I^{n+1}, the maps p_n are surjective by Nakayama's lemma. In particular, M_n is a finitely generated A/I^{n+1}-module. Set $K_n := \operatorname{Ker}(p_n)$. As

$$\operatorname{Ker}((A/I^{n+1})^N \to (A/I^n)^N) = (I^n/I^{n+1})^N \longrightarrow I^n M_n = \operatorname{Ker}(M_n \to M_{n-1})$$

is surjective, the snake lemma implies that $K_n \to K_{n-1}$ is surjective. In particular, the system $(K_n)_n$ satisfies the Mittag-Leffler condition (Definition F.225). Hence applying the functor $\lim_n$ we obtain an exact sequence

$$0 \to K \longrightarrow A^N \longrightarrow \lim_n M_n \to 0,$$

where $K := \lim_n K_n$ by Proposition F.227. Hence $\lim_n M_n$ is a finitely generated A-module. Moreover $K \to K_n$ is surjective and therefore

$$M/I^{n+1}M = \operatorname{Coker}(K \to (A/I^{n+1})^N) = (A/I^{n+1})^N/K_n = M_n.$$

In particular $M = \lim_n M/I^{n+1}M$ is I-adically complete.

Let us now prove (2). As $A = \lim A/I^n$, the functor (24.18.3) is fully faithful on finite free modules and hence on finite projective modules by passage to direct summands. It remains to show that if $(M_n)_n$ is a system of finite projective A/I^{n+1}-modules, then $M := \lim_n M_n$ is a finite projective A-module. Construct p_n as above. As M_n is projective, we find sections $s_n \colon M_n \to (A/I^{n+1})^N$ of p_n for all n.

We claim that we can arrange the sections inductively such that $s_{n-1} \equiv s_n \pmod{I^n}$ for all $n \geq 1$. Indeed, write $\bar{s}_n := s_n \pmod{I^n}$ and consider the composition

$$M_n \longrightarrow M_{n-1} \xrightarrow{\bar{s}_n - s_{n-1}} K_{n-1}.$$

As M_n is projective and $K_n \to K_{n-1}$ is surjective, we can lift this homomorphism to a homomorphism $t_n \colon M_n \to K_n$ and we can replace s_n by $s_n - t_n$. This proves the claim.

Hence, we obtain a compatible system of idempotent endomorphisms $e_n := s_n \circ p_n \in M_N(A/I^{n+1})$ with $\operatorname{Im}(e_n) = M_n$. They define an idempotent endomorphism $e \in M_N(A)$ with $\operatorname{Im}(e) = M$. Hence M is finite projective. The assertion about the ranks is clear.

It remains to show (3). We have already seen in (1) that the functor (24.18.3) has $(M_n)_n \mapsto M := \lim_n M_n$ as a right inverse (up to isomorphism of functors). It is clear that this is also a left inverse. $\qquad\square$

Remark 24.89. The proof shows that all assertions of Proposition 24.88 also hold if one replaces I^n by I_n, where $(I_n)_n$ is a sequence of ideals with $I_{n+1} \subseteq I_n$ for all n, with $I_n/I_{n+1} \subseteq A/I_{n+1}$ nilpotent, and with $A = \lim_n A/I_n$.

If I is a finitely generated ideal in a ring A that is I-adically complete, then a finitely generated A-module M is I-adically complete if and only if $\bigcap_n I^n M = 0$ ([Sta] 031B). This condition is automatic if A is noetherian (Corollary B.44). Hence we obtain the following corollary.

Corollary 24.90. *Let $X = \operatorname{Spec} A$ be an affine noetherian scheme, let $J \subseteq A$ be an ideal, and let $Z := V(J)$ be its vanishing locus. Then the category of coherent modules over $X_{/Z}$ is equivalent to the category of finitely generated $\hat{A}$-modules, where $\hat{A}$ is the J-adic completion of A. In particular, it is an abelian category.*

The equivalence is given by sending a coherent module $(\mathscr{F}_n)_n$ over $X_{/Z}$ to $\lim_n \Gamma(X, \mathscr{F}_n)$ with quasi-inverse sending a $\hat{A}$-module M to $((M/J^{n+1}M)^\sim)_n$.

Proposition 24.91. *Let X be a noetherian scheme, let Z be a closed subspace of X.*
(1) *The category of coherent modules over $X_{/Z}$ is abelian.*
(2) *The functor $\mathscr{F} \mapsto \mathscr{F}_{/Z}$ from the category of coherent $\mathscr{O}_X$-modules to the category of coherent modules over $X_{/Z}$ is exact.*

Proof. Let $U = \operatorname{Spec} A \subseteq X$ be open affine. Then the category of modules over $U_{/(U \cap Z)}$ is abelian by Corollary 24.90. Hence kernels and cokernels of morphisms of coherent modules over $X_{/Z}$ exist locally on affine schemes. As these form a basis of the topology, they exist globally by gluing. Moreover, the question whether for a morphism u of coherent modules over $X_{/Z}$ the map $\operatorname{Coim}(u) \to \operatorname{Im}(u)$ is an isomorphism also immediately reduces to the affine situation, where it is answered affirmatively again by Corollary 24.90. This shows (1).

The exactness of $\mathscr{F} \mapsto \mathscr{F}_{/Z}$ can be checked locally. Hence we may assume that $X = \operatorname{Spec} A$ is affine. Let $J \subseteq A$ be an ideal of definition for Z, and set $\hat{A} := \lim_n A/J^{n+1}$. Then the functor $\mathscr{F} \mapsto \mathscr{F}_{/Z}$ is identified via the equivalence of categories of Corollary 24.90 with the functor $M \mapsto \hat{M} = \lim_n M/J^{n+1}M$ which is exact by Proposition B.41. $\qquad\square$

Remark 24.92. In the situation of Proposition 24.91 let $u = (u_n)_n \colon \mathscr{F} \to \mathscr{G}$ be a map of coherent modules over $X_{/Z}$. Then the cokernel of u in the abelian category of coherent modules over $X_{/Z}$ is the coherent module $(\operatorname{Coker}(u_n))_n$ over $X_{/Z}$.

Indeed, to check this we may assume that $X = \operatorname{Spec} A$ is affine. Let $\hat{A}$ be the I-adic completion for some ideal of definition I for Z. Then u corresponds to a projective system of A-linear maps $u_n \colon M_n \to P_n$. Let $\hat{u} \colon \hat{M} = \lim_n M_n \to \lim_n P_n$ the corresponding map of $\hat{A}$ modules. The proof of Proposition 24.91 shows that the cokernel of u is given by the system $(\operatorname{Coker}(\hat{u}) \otimes_{\hat{A}} \hat{A}/I^{n+1}\hat{A})_n = (\operatorname{Coker}(u_n))_n$.

Similarly, in the affine situation the system $(\operatorname{Ker}(\hat{u}) \otimes_{\hat{A}} \hat{A}/I^{n+1}\hat{A})_n$ gives the kernel of u but it is more difficult to describe it in terms of the kernels of the u_n.

Remark 24.93. Let X be a noetherian scheme, let Z be a closed subspace of X, and let $\mathscr{J}$ be an ideal of definition for Z. If $(\mathscr{F}_n)_n$ is a coherent module over $X_{/Z}$, define an $\mathscr{O}_X$-module $\mathscr{F} := \lim_n \mathscr{F}_n$. This module is usually not a coherent $\mathscr{O}_X$-module but it determines $(\mathscr{F}_n)_n$ in the following sense. The canonical map $\mathscr{F} \to \mathscr{F}_n$ factors through $\mathscr{F}/\mathscr{J}^{n+1}\mathscr{F} \to \mathscr{F}_n$. Working locally one deduces from Corollary 24.90 that this is an isomorphism.

(24.19) Grothendieck's existence theorem for coherent modules.

In this section let A be a noetherian ring, let $I \subseteq A$ be an ideal, and let $Z := V(I)$ be its vanishing space. We suppose that A is I-adically complete, i.e., the homomorphism $A \to \lim_n A/I^n$ is an isomorphism. Set $Y := \operatorname{Spec} A$ and $Y_n := \operatorname{Spec} A/I^{n+1}$.

Let $X \to \operatorname{Spec} A$ be a proper morphism of schemes. We set $X_n := X \times_Y Y_n$ and denote by $i_n \colon X_n \to X$ the canonical closed immersion. The following theorem has some similarity with the GAGA theorems of Section (20.12) and is sometimes called the "formal GAGA" principle: In usual GAGA we compare complex spaces (with holomorphic functions) and schemes (with "polynomial functions"), whereas here we have "formal schemes" (with complete coordinate rings, think of formal power series) and schemes.

Theorem 24.94. *(Grothendieck's existence theorem) Let* $X \to \operatorname{Spec} A$ *be a proper morphism. Then the functor* $\mathscr{F} \mapsto \mathscr{F}_{/Z}$ *is an equivalence between the category of coherent* $\mathscr{O}_X$*-modules and the category of coherent modules over* $X_{/Z}$*.*

We will prove Theorem 24.94 in the following sections. Full faithfulness of $\mathscr{F} \mapsto \mathscr{F}_{/Z}$ will be proved in Corollary 24.100. Essential surjectivity of $\mathscr{F} \mapsto \mathscr{F}_{/Z}$ will be proved in Lemma 24.103.

Below in Theorem 24.108 we will also see a generalization in which X is only assumed to be separated and of finite type over A and where one obtains an equivalence of coherent modules with proper support.

Via the equivalence of Theorem 24.94 finite locally free modules correspond to each other.

Proposition 24.95. *Let* $f \colon X \to \operatorname{Spec} A$ *be a proper morphism and let* $r \geq 0$ *be an integer. Then the functor* $\mathscr{F} \mapsto \mathscr{F}_{/Z}$ *yields an equivalence between the category of locally free* $\mathscr{O}_X$*-modules of rank* r *and the category of locally free modules over* $X_{/Z}$ *of rank* r*.*

Proof. If $\mathscr{F}$ is locally free of rank r, then $\mathscr{F}_{/Z} = (i_n^* \mathscr{F})_n$ is clearly locally free of rank r. Conversely, let $\mathscr{F}$ be a coherent $\mathscr{O}_X$-module such that $\mathscr{F}_{/Z}$ is locally free of rank r. We claim that it suffices to show that the stalk $\mathscr{F}_x$ is a flat $\mathscr{O}_{X,x}$-module for all $x \in f^{-1}(Z)$. Indeed, then $\mathscr{F}$ is locally free of rank r in an open neighborhood of $f^{-1}(Z)$ (Proposition 7.41), which is necessarily X by Lemma 24.96 below.

Now the condition $x \in f^{-1}(Z)$ means that the image of I is contained in the maximal ideal of $\mathscr{O}_{X,x}$. By hypothesis, $\mathscr{F}_x \otimes_{\mathscr{O}_{X,x}} \mathscr{O}_{X,x}/I^n\mathscr{O}_{X,x}$ is a projective module over $\mathscr{O}_{X,x}/I^n\mathscr{O}_{X,x}$ for all n. Hence $\mathscr{F}_x$ is a flat $\mathscr{O}_{X,x}$-module by the local criterion for flatness (Theorem B.51). $\qquad\square$

We used the following lemma.

Lemma 24.96. *Let* $f \colon X \to \operatorname{Spec} A$ *be a closed morphism and let* U *be an open neighborhood of* $f^{-1}(Z)$*. Then* $U = X$*.*

Proof. As f is closed, $f(X \setminus U)$ is a closed subset of $\operatorname{Spec} A$ that does not meet $Z = V(I)$. But I is contained in the Jacobson radical by Proposition B.42, i.e. Z contains all closed points of $\operatorname{Spec} A$. Therefore $f(X \setminus U) = \emptyset$ and hence $U = X$. $\qquad\square$

There are also derived versions of the existence theorem. It works also in the non-noetherian context if one imposes a stronger assumption than "proper" on the morphism. We state the following result as an example but refer to [Sta] 0DIG for the proof.

Theorem 24.97. *Let* $\cdots \to A_n \to A_{n-1} \to \cdots \to A_0$ *be a projective system of surjective ring homomorphisms whose kernels consist of nilpotent elements and set* $A := \lim_n A_n$. *Let* $X \to \operatorname{Spec} A$ *be a proper, flat morphism of finite presentation. We view* $X_n := X \otimes_A A_n$ *as a closed subscheme of* X *for all* n *and denote by* $i_n \colon X_n \to X_{n+1}$ *and* $I_n \colon X_n \to X$ *the inclusions.*

Let $(K_n, \varphi_n)_n$ *be a family consisting of pseudo-coherent objects in* $D(X_n)$ *and isomorphisms* $\varphi_n \colon Li_n^* K_{n+1} \to K_n$ *in* $D(X_{n-1})$. *Then there exists a pseudo-coherent object in* $D(X)$ *and isomorphisms* $LI_n^* K \overset{\sim}{\to} K_n$ *for all* n *that are compatible with the* φ_n.

(24.20) Full faithfulness of $\mathscr{F} \mapsto \mathscr{F}_{/Z}$.

If X is a locally noetherian scheme, we denote by $D_{\mathrm{coh}}(X)$ the full subcategory of $D(X)$ consisting of complexes $\mathscr{F}$ such that $H^p(\mathscr{F})$ is coherent for all $p \in \mathbb{Z}$.

Proposition 24.98. *Let* A *be a noetherian ring and* $I \subseteq A$ *an ideal with* $A = \lim_n A/I^{n+1}$. *Let* $f \colon X \to Y := \operatorname{Spec} A$ *be a proper morphism. Let* $\mathscr{F} \in D_{\mathrm{coh}}^-(X)$ *and* $\mathscr{G} \in D_{\mathrm{coh}}^+(X)$. *Then for all* $p \in \mathbb{Z}$ *one has functorial isomorphisms*

$$(24.20.1) \qquad \operatorname{Ext}_{\mathscr{O}_X}^p(\mathscr{F}, \mathscr{G}) \overset{\sim}{\to} \operatorname{Ext}_{\mathscr{O}_X}^p(\mathscr{F}^\wedge, \mathscr{G}^\wedge)$$

of finitely generated A-*modules, where* $(-)^\wedge$ *denotes the derived completion with respect to* $I\mathscr{O}_X$.

Proof. By Corollary 24.31 one has

$$(*) \qquad R\Gamma(X, R\mathscr{H}om_{\mathscr{O}_X}(\mathscr{F}, \mathscr{G}))^\wedge = R\Gamma(X, R\mathscr{H}om_{\mathscr{O}_X}(\mathscr{F}^\wedge, \mathscr{G}^\wedge)).$$

One has $R\Gamma(X, R\mathscr{H}om_{\mathscr{O}_X}(\mathscr{F}^\wedge, \mathscr{G}^\wedge)) = R\operatorname{Hom}_{\mathscr{O}_X}(\mathscr{F}^\wedge, \mathscr{G}^\wedge)$ by Proposition 21.107 and hence applying $H^p(-)$ to the right side of $(*)$ yields $\operatorname{Ext}^p(\mathscr{F}^\wedge, \mathscr{G}^\wedge)$.

As f is proper, $\operatorname{Ext}^p(\mathscr{F}, \mathscr{G})$ is a finite A-module (Corollary 23.38). Applying Proposition 24.17 one obtains

$$H^p(R\Gamma(X, R\mathscr{H}om_{\mathscr{O}_X}(\mathscr{F}, \mathscr{G}))^\wedge) = \operatorname{Ext}^p(\mathscr{F}, \mathscr{G})^\wedge = \operatorname{Ext}^p(\mathscr{F}, \mathscr{G}),$$

where the second equality holds because every finitely generated A-module is I-complete (Proposition B.44). $\square$

As the derived completion for a coherent module on a noetherian scheme coincides with the classical completion (Corollary 24.34) one obtains the following result.

Corollary 24.99. *Let* A *be a noetherian ring,* $I \subseteq A$ *an ideal with* $A = \lim_n A/I^{n+1}$. *Let* $f \colon X \to Y := \operatorname{Spec} A$ *be a proper morphism. Let* $\mathscr{F}$ *and* $\mathscr{G}$ *be coherent* $\mathscr{O}_X$-*modules. Then one has for all* $p \in \mathbb{Z}$ *a functorial isomorphism*

$$\operatorname{Ext}_{\mathscr{O}_X}^p(\mathscr{F}, \mathscr{G}) \overset{\sim}{\to} \operatorname{Ext}_{\mathscr{O}_X}^p(\mathscr{F}^\wedge, \mathscr{G}^\wedge),$$

where $(-)^\wedge$ *denotes the usual* I-*adic completion.*

The special case $p = 0$ yields the full faithfulness of the functor $\mathscr{F} \mapsto \mathscr{F}_{/Z}$.

Corollary 24.100. *Let* A *be a noetherian ring,* I *an ideal of* A, *and let* $Z = V(I)$ *be the closed subscheme of* $Y := \operatorname{Spec} A$ *defined by* I. *Let* $f \colon X \to Y$ *be a proper morphism. Then the functor* $\mathscr{F} \mapsto (\mathscr{F}/I^{n+1}\mathscr{F})_n$ *from the category of coherent* $\mathscr{O}_X$-*modules to the category of coherent modules over* $X_{/Z}$ *is fully faithful.*

Proof. By Corollary 24.99 one has

$$\operatorname{Hom}_{\mathscr{O}_X}(\mathscr{F},\mathscr{G}) = \operatorname{Hom}_{\mathscr{O}_X}(\mathscr{F}^\wedge, \lim_n \mathscr{G}/I^{n+1}\mathscr{G}) = \lim_n \operatorname{Hom}_{\mathscr{O}_X}(\mathscr{F}^\wedge, \mathscr{G}/I^{n+1}\mathscr{G})$$
$$= \lim_n \operatorname{Hom}_{\mathscr{O}_X}(\mathscr{F}^\wedge/I^{n+1}\mathscr{F}^\wedge, \mathscr{G}/I^{n+1}\mathscr{G})$$
$$= \lim_n \operatorname{Hom}_{\mathscr{O}_X}(\mathscr{F}/I^{n+1}\mathscr{F}, \mathscr{G}/I^{n+1}\mathscr{G}),$$

where the equality $\mathscr{F}^\wedge/I^{n+1}\mathscr{F}^\wedge = \mathscr{F}/I^{n+1}\mathscr{F}$ can be checked locally on X and hence holds by Proposition B.41. $\qquad\square$

(24.21) Essential surjectivity of $\mathscr{F} \mapsto \mathscr{F}_{/Z}$.

We keep the notation from the beginning of Section (24.19), i.e., A is a noetherian ring, $I \subseteq A$ an ideal, and $Z := V(I)$ its vanishing space. We suppose that A is I-adically complete and set $Y := \operatorname{Spec} A$, $Y_n := \operatorname{Spec} A/I^{n+1}$. Let $X \to \operatorname{Spec} A$ be a separated morphism of schemes of finite type. We set $X_n := X \times_Y Y_n$.

To prove the essential surjectivity if $X \to \operatorname{Spec} A$ is proper we proceed somewhat similarly as in the proof that higher direct images of coherent modules are again coherent. In other words, we prove the result first if X is projective and then generalize to the proper case using the lemma of Chow and a dévissage argument.

We say that a coherent module $(\mathscr{F}_n)_n$ over $X_{/Z}$ is *algebraizable* if it is isomorphic to a module of the form $\mathscr{F}_{/Z}$ for a coherent $\mathscr{O}_X$-module $\mathscr{F}$. We then also say that $(\mathscr{F}_n)_n$ is *algebraizable by* $\mathscr{F}$. We want to prove that every coherent module over $X_{/Z}$ is algebraizable if $X \to \operatorname{Spec} A$ is proper. Recall that the coherent modules over $X_{/Z}$ form an abelian category and that $\mathscr{F} \mapsto \mathscr{F}_{/Z}$ is exact (Proposition 24.91).

Lemma 24.101. *Let $X \to \operatorname{Spec} A$ be proper.*
(1) Let $u\colon (\mathscr{F}_n)_n \to (\mathscr{G}_n)_n$ be a morphism of coherent modules over $X_{/Z}$. If $(\mathscr{F}_n)_n$ and $(\mathscr{G}_n)_n$ are algebraizable, then $\operatorname{Ker}(u)$, $\operatorname{Coker}(u)$, and $\operatorname{Im}(u)$ are algebraizable.
(2) Let $0 \to (\mathscr{F}_n)_n \to (\mathscr{G}_n)_n \to (\mathscr{H}_n)_n \to 0$ be an exact sequence of modules over $X_{/Z}$. If $(\mathscr{F}_n)_n$ and $(\mathscr{H}_n)_n$ are algebraizable, then $(\mathscr{G}_n)_n$ is algebraizable.

Proof. (1). Let $\mathscr{F}$ and $\mathscr{G}$ be coherent $\mathscr{O}_X$-modules with $\mathscr{F}_{/Z} = (\mathscr{F}_n)_n$ and $\mathscr{G}_{/Z} = (\mathscr{G}_n)_n$. Then u corresponds to a map of $\mathscr{O}_X$-module $v\colon \mathscr{F} \to \mathscr{G}$ because we have already seen that the functor $(-)_{/Z}$ is fully faithful (Corollary 24.100). As $(-)_{/Z}$ is also exact (Proposition 24.91) we deduce that $\operatorname{Ker}(u)$, $\operatorname{Coker}(u)$ and $\operatorname{Im}(u)$ are algebraizable by $\operatorname{Ker}(v)$, $\operatorname{Coker}(v)$ and $\operatorname{Im}(v)$, respectively.

(2). Let $\mathscr{F}$ and $\mathscr{H}$ be coherent $\mathscr{O}_X$-modules with $\mathscr{F}_{/Z} = (\mathscr{F}_n)_n$ and $\mathscr{H}_{/Z} = (\mathscr{H}_n)_n$. Then $(\mathscr{G}_n)_n$ defines a class in $\operatorname{Ext}^1(\mathscr{F}^\wedge, \mathscr{H}^\wedge)$ by Proposition F.220. Hence we conclude by Corollary 24.99. $\qquad\square$

The following proposition will be the key step for the proof of the essential surjectivity of $\mathscr{F} \mapsto \mathscr{F}_{/Z}$ when X is projective over A. It will be also helpful for the proof of Theorem 24.113 below which gives a criterion when an adic formal scheme is algebraizable.

Let $\mathscr{X} = (X_n)_n$ be a proper adic formal scheme over $Y_{/Z}$ and let $\mathscr{L} = (\mathscr{L}_n)_n$ be a locally free module over $\mathscr{X}$ of rank 1 such that $\mathscr{L}_0$ is ample. For every coherent module $\mathscr{E} = (\mathscr{E}_n)_n$ over $\mathscr{X}$ and for every $m \in \mathbb{Z}$ we define the coherent module $\mathscr{E}(m) = (\mathscr{E}_n(m))_n$ over $\mathscr{X}$ by $\mathscr{E}_n(m) := \mathscr{E}_n \otimes_{\mathscr{O}_{X_n}} \mathscr{L}_n^{\otimes m}$. Let $\mathscr{O}_{\mathscr{X}} = (\mathscr{O}_{X_n})_n$ be the structure sheaf.

Proposition 24.102. *Let $\mathscr{E}$ be a coherent module over $\mathcal{X}$. Then there exists an integer $m_0 \geq 0$ such that*

(1) *The map $\Gamma(\mathcal{X}, \mathscr{E}(m)) := \lim_n \Gamma(X_n, \mathscr{E}_n(m)) \to \Gamma(X_0, \mathscr{E}_0(m))$ is surjective for all $m \geq m_0$.*

(2) *For all $m \geq m_0$ there exists an $r \geq 0$ and a map $u \colon \mathscr{O}_{\mathcal{X}}^r \to \mathscr{E}(m)$ such that $u_n \colon \mathscr{O}_{X_n}^r \to \mathscr{E}(m)_n$ is a surjective map of $\mathscr{O}_{X_n}$-modules for all n.*

Proof. Set $A_n := A/I^{n+1}$ and denote by $f = (f_n)_n$ the morphism $\mathcal{X} \to Y_{/Z}$ given by the proper morphisms $f_n \colon X_n \to \operatorname{Spec} A_n$. Let $B := \bigoplus_{n \geq 0} I^n/I^{n+1}$. As I/I^2 is a finitely generated A_0-module, this is a finitely generated A_0-algebra. Set $\mathscr{J} := I\mathscr{O}_X$ and $\mathscr{B} := f_0^*\tilde{B} = \bigoplus_n \mathscr{J}^n/\mathscr{J}^{n+1}$ which is a quasi-coherent $\mathscr{O}_{X_0}$-algebra of finite type.

We view the $\mathscr{O}_{X_n}$-modules $\mathscr{E}_n$ as $\mathscr{O}_X$-modules annihilated by $\mathscr{J}^{n+1}$ and set $\mathscr{F}_n := \operatorname{Ker}(\mathscr{E}_n \to \mathscr{E}_{n-1})$. As $\mathscr{J}\mathscr{F}_n = 0$ we may view $\mathscr{F}_n$ as an $\mathscr{O}_{X_0}$-module. Set $\mathscr{F} := \bigoplus_{n \geq 0} \mathscr{F}_n$ which is a quasi-coherent $\mathscr{B}$-module. We claim that it is a finitely generated $\mathscr{B}$-module. Indeed, this may be checked locally on X. Hence we can assume that $X = \operatorname{Spec} C$ is affine. If $J \subseteq C$ is the ideal corresponding to $\mathscr{J}$, then $\mathscr{B}$ corresponds to $B := \bigoplus_n J^n/J^{n+1}$. The module $(\mathscr{E}_n)_n$ over $X_{/Z}$ corresponds to a finite $\hat{C}$-module E by Corollary 24.90 and $\mathscr{F}$ corresponds to $\bigoplus_n J^n E/J^{n+1}E$ which is therefore a finite module over B. This proves the claim.

Hence we may apply Proposition 24.39 and see that there exists an integer m_0 such that $H^1(X_0, \mathscr{F}_n(m)) = 0$ for all n and for all $m \geq m_0$. Hence the transition maps $\Gamma(X, \mathscr{E}_n(m)) \to \Gamma(X, \mathscr{E}_{n-1}(m))$ are surjective for all n and for all $m \geq m_0$. This implies (1).

As $\mathscr{L}_0$ is ample, we may also assume after possibly enlarging m_0 that $\mathscr{E}_0(m)$ is generated by finitely many global sections for all $m \geq m_0$. By (1) we may lift these global sections to $\Gamma(\mathcal{X}, \mathscr{E}(m))$ and obtain a map $u \colon \mathscr{O}_{\mathcal{X}}^r \to \mathscr{E}(m)$ of modules over $\mathcal{X}$. As u_0 is surjective, u_n is surjective for all n by Nakayama's lemma. $\qquad\square$

We can now conclude the proof of Grothendieck's existence theorem, Theorem 24.94. As we already have seen that $\mathscr{F} \mapsto \mathscr{F}_{/Z}$ is fully faithful (Corollary 24.100), it suffices to show the following lemma.

Lemma 24.103. *Let $X \to \operatorname{Spec} A$ be a proper morphism. Then every coherent module over $X_{/Z}$ is algebraizable.*

Proof. Recall that the category of coherent modules of $X_{/Z}$ is abelian (Proposition 24.91) and that a map $u = (u_n)_n \colon \mathscr{E} \to \mathscr{F}$ of coherent modules over $X_{/Z}$ is surjective if and only if $u_n \colon \mathscr{E}_n \to \mathscr{F}_n$ is surjective for all n (Remark 24.92).

We first prove the lemma under the additional assumption that there exists an ample line bundle $\mathscr{L}$ on X, i.e., if $X \to \operatorname{Spec} A$ is projective. As usual we write $\mathscr{G}(m)$ for $\mathscr{G} \otimes \mathscr{L}^{\otimes m}$. Let $\mathscr{F}$ be a coherent module over $X_{/Z}$. Then Proposition 24.102 (2) shows that there exist integers $m, m', r, r' \geq 0$ and an exact sequence

$$\mathscr{O}_{X_{/Z}}(-m')^{r'} \xrightarrow{w} \mathscr{O}_{X_{/Z}}(-m)^r \longrightarrow \mathscr{F} \longrightarrow 0.$$

Hence $\mathscr{F}$ is algebraizable by Lemma 24.101 (1).

To prove the lemma for arbitrary proper morphisms $X \to \operatorname{Spec} A$ we proceed by noetherian induction on X. Hence we can assume that for every closed subscheme $T \subsetneq X$ all coherent modules over $T_{/Z}$ are algebraizable. We use Chow's lemma (Theorem 13.100) by which we can find a surjective projective morphism $\pi \colon X' \to X$ such that X' is projective over A and such that $\pi^{-1}(U) \to U$ is an isomorphism for some open dense subscheme U of X. Now we proceed as follows.

Let $\mathscr{F}$ be a coherent module over $X_{/Z}$.

(1) We first construct a coherent module $\pi^*\mathscr{F}$ on $X'_{/Z}$, which is algebraizable because we showed the lemma already in the projective case. Next we construct an algebraizable coherent module $\pi_*\pi^*(\mathscr{F})$ on $X_{/Z}$ and a canonical homomorphism $\pi_*\pi^*(\mathscr{F}) \to \mathscr{F}$.

(2) Let $\mathscr{K}$ (resp. $\mathscr{C}$) be its kernel (resp. its cokernel) such that we have an exact sequence

$$(24.21.1) \qquad 0 \longrightarrow \mathscr{K} \longrightarrow \pi_*\pi^*(\mathscr{F}) \longrightarrow \mathscr{F} \longrightarrow \mathscr{C} \longrightarrow 0$$

of coherent modules on $X_{/Z}$. Since $\pi_{|\pi^{-1}(U)}$ is an isomorphism, we will see that $\mathscr{K}$ and $\mathscr{C}$ can already be considered as coherent modules over $(X \setminus U)_{/Z}$ where $X \setminus U$ is endowed with a suitable structure of a closed subscheme of X. Hence they are both algebraizable by the induction hypothesis.

(3) Then the exact sequence (24.21.1) shows that $\mathscr{F}$ is algebraizable by Lemma 24.101. We will carry out these three steps below in Construction 24.104 for step (1), in Lemma 24.105 for step (2), and in Lemma 24.106 for step (3). □

To construct inverse and direct images for coherent modules over formal completions a conceptual approach would be to develop first a theory of formal schemes as topologically ringed spaces, see [EGAInew] Ch. X, such that formal completions are special cases of formal schemes. As we did not do this, our definitions will be rather ad hoc. They can be justified by showing that these definitions are equivalent to the more natural definitions found in loc. cit. – and because they allow us to complete the proof of Grothendieck's existence theorem which is the only reason why they are introduced here.

Construction 24.104. Let X be a noetherian scheme and let $Z \subseteq X$ be a closed subspace. We fix an ideal of definition $\mathscr{J}$ for Z. Let $\pi\colon X' \to X$ be a morphism of noetherian schemes, $Z' := \pi^{-1}(Z)$ and let $\mathscr{J}'$ be the ideal of $\mathscr{O}_{X'}$ generated by $\pi^\sharp(\pi^{-1}\mathscr{J})$. It is an ideal of definition for Z'. We set $X_n := V(\mathscr{J}^{n+1})$ and $X'_n := V(\mathscr{J}'^{n+1})$ for all $n \geq 0$. Then $\pi^{-1}(X_n) = X'_n$. We usually consider a module over $X_{/Z}$ as a family of $\mathscr{O}_X$-modules $\mathscr{F}_n$ with $\mathscr{J}^{n+1}\mathscr{F}_n = 0$ and $\mathscr{F}_n/\mathscr{J}^n\mathscr{F}_n \cong \mathscr{F}_{n-1}$, similarly for modules over $X'_{/Z'}$.

(1) Let $\mathscr{F} = (\mathscr{F}_n)_n$ be a coherent module over $X_{/Z}$. Then we define its pullback $\pi^*\mathscr{F} := (\pi^*\mathscr{F}_n)_n$. This is a coherent module over $X'_{/Z'}$.

If $\mathscr{F}$ is algebraizable by a coherent $\mathscr{O}_X$-module $\mathscr{G}$, then $\pi^*\mathscr{F}$ is algebraizable by $\pi^*\mathscr{G}$.

(2) Let $\mathscr{F}' = (\mathscr{F}'_n)_n$ be a coherent module over $X'_{/Z'}$ and suppose that π is proper. Set $\widehat{\mathscr{F}'} := \lim_n \mathscr{F}'_n$ which is an $\mathscr{O}_{X'}$-module and $\pi_*\mathscr{F}' := (\pi_*\widehat{\mathscr{F}'})_{/Z}$.

If $\mathscr{F}'$ is algebraizable by a coherent $\mathscr{O}_X$-module $\mathscr{G}$, then $\widehat{\mathscr{F}'}$ is the $\mathscr{J}'$-adic completion $\mathscr{G}^\wedge$ of $\mathscr{G}$ and $\pi_*\widehat{\mathscr{F}'} = \pi_*\mathscr{G}^\wedge = (\pi_*\mathscr{G})^\wedge$ by Proposition 24.35. And hence $\pi_*\mathscr{F}' = ((\pi_*\mathscr{G})^\wedge)_{/Z} = (\pi_*\mathscr{G})_{/Z}$ is a coherent module over $X_{/Z}$ which is algebraizable by $\pi_*\mathscr{G}$.

(3) We continue to assume that π is proper. Let $\mathscr{F} = (\mathscr{F}_n)_n$ be a coherent module over $X_{/Z}$ such that $\pi^*\mathscr{F}$ is algebraizable. There is a functorial homomorphism of modules over $X_{/Z}$

$$(24.21.2) \qquad \theta\colon \pi_*\pi^*\mathscr{F} \to \mathscr{F}.$$

Indeed, for each $n \geq 0$ we have by definition

$$(\pi_*\pi^*\mathscr{F})_n = \pi_*(\lim_k \pi^*\mathscr{F}_k)/\mathscr{J}^{n+1}\pi_*(\lim_k \pi^*\mathscr{F}_k)$$

and we define $\theta_n\colon (\pi_*\pi^*\mathscr{F})_n \to \mathscr{F}_n$ to be induced by the composition

$$\pi_*(\lim_k \pi^*\mathscr{F}_k) = \lim_k \pi_*\pi^*\mathscr{F}_k \to \pi_*\pi^*\mathscr{F}_n \to \mathscr{F}_n,$$

where the first equality holds because π_* has a left adjoint functor, namely π^*, and hence commutes with limits.

Lemma 24.105. *In the situation of Construction 24.104 (3) suppose that there exists an open subscheme $U \subseteq X$ such that π induces an isomorphism $p^{-1}(U) \xrightarrow{\sim} U$. Let $(\mathscr{K}_n)_n$ (resp. $(\mathscr{C}_n)_n$) be the kernel (resp. cokernel) of the map $\theta\colon \pi_*\pi^*\mathscr{F} \to \mathscr{F}$ (24.21.2). Then there exists an ideal of definition $\mathscr{A}$ for $X \setminus Z$ such that $\mathscr{A}$ annihilates $\mathscr{C}_n$ and $\mathscr{K}_n$ for all n.*

Proof. We may work locally on X and hence assume that $X = \operatorname{Spec} B$ is affine. Let $J \subseteq B$ be the ideal corresponding to $\mathscr{J}$.

(I). We first assume that $B = \lim_n B/J^n$. In this case we may apply Proposition 24.90 and see that $\mathscr{F}$ is algebraizable by the coherent $\mathscr{O}_X$-module $\widehat{\mathscr{F}} = \lim_n \mathscr{F}_n$. Then $\pi^*\mathscr{F}$ is algebraizable by $\pi^*\widehat{\mathscr{F}}$ and therefore $\pi_*\pi^*\mathscr{F}$ is algebraizable by $\pi_*\pi^*\widehat{\mathscr{F}}$ by Construction 24.104 (2). Hence $\operatorname{Ker}(\theta)$ and $\operatorname{Coker}(\theta)$ are algebraizable by the kernel $\mathscr{K}$ respective cokernel $\mathscr{C}$ of the canonical homomorphism $\pi_*\pi^*\widehat{\mathscr{F}} \to \widehat{\mathscr{F}}$ of coherent $\mathscr{O}_X$-modules. As π is an isomorphism over U, the support of $\mathscr{K}$ and $\mathscr{C}$ are contained in $X \setminus U$ hence there exist ideals of definition $\mathscr{I}$ and $\mathscr{I}'$ for $X \setminus U$ with $\mathscr{I}\mathscr{K} = 0$ and $\mathscr{I}'\mathscr{C} = 0$ (Proposition 24.78). Then $\mathscr{A} := \mathscr{I}\mathscr{I}'$ annihilates $\mathscr{K}$ and $\mathscr{C}$ and a fortiori all $\mathscr{K}_n$ and $\mathscr{C}_n$.

(II). We now show that we can replace B by $\hat{B} = \lim_n B/J^n$ and thus are reduced to the case that we already proved in Step (I). Set $\tilde{X} := \operatorname{Spec} \hat{B}$, let $g\colon \tilde{X} \to X$ be the canonical morphism and define $\tilde{X}'$, $\tilde{\pi}$, and g' by the following cartesian diagram

$$
\begin{array}{ccc}
\tilde{X}' & \xrightarrow{\;g'\;} & X' \\
{\scriptstyle \tilde{\pi}}\big\downarrow & & \big\downarrow{\scriptstyle \pi} \\
\tilde{X} & \xrightarrow{\;g\;} & X.
\end{array}
$$

Now g is flat because completions are flat for noetherian rings (Proposition B.41). By hypothesis, $\pi^*\mathscr{F}$ is algebraizable by a coherent $\mathscr{O}_{X'}$-module $\mathscr{G}$ and hence $g'^*\pi_*\pi^*\mathscr{F}$ is algebraizable by $g'^*\pi_*\mathscr{G} = \tilde{\pi}_*g'^*\mathscr{G}$, where the equality holds because push forward commutes with flat pullback (Proposition 12.6). Hence $g'^*\pi_*\pi^*\mathscr{F} = \tilde{\pi}_*\tilde{\pi}^*g^*\mathscr{F}$. By (i) we know therefore that the kernel and cokernel of $g^*\theta$ are annihilated by an ideal of definition $\tilde{\mathscr{A}}$ for $g^{-1}(X \setminus U)$.

We may assume $\tilde{\mathscr{A}}$ to be of the form $g^*(\mathscr{A})$ for some ideal of definition $\mathscr{A}$ for $X \setminus U$. Indeed, if $\mathscr{B}$ is any ideal of definition for $X \setminus U$, then $g^*\mathscr{B}$ is an ideal of $\mathscr{O}_{\tilde{X}}$ because g is flat and it is an ideal of definition for $g^{-1}(X \setminus U)$. Then for some $n \geq 1$ one has $g^*(\mathscr{B})^n = g^*(\mathscr{B}^n) \subseteq \tilde{\mathscr{A}}$ by Lemma 24.77 (1). Hence we may replace $\tilde{\mathscr{A}}$ by $g^*(\mathscr{A})$ for $\mathscr{A} := \mathscr{B}^{\otimes n}$.

Let $\mathfrak{a} \subseteq B$ be the ideal corresponding to $\mathscr{A}$ and let K_n (resp. C_n) be the finitely generated module over B/J^{n+1} corresponding to $\mathscr{K}_n$ and $\mathscr{C}_n$, respectively. Then $g^*\mathscr{K}_n$ corresponds to the $\hat{B}$-module $K_n \otimes \hat{B}/(J\hat{B})^{n+1}$ which can be identified with K_n since $\hat{B}/(J\hat{B})^{n+1} = B/J^{n+1}$. Hence K_n is annihilated by $\mathfrak{a}$ because it is annihilated by $\mathfrak{a}\hat{B}$. Hence $\mathscr{K}_n$ is annihilated by $\mathscr{A}$ for all n. The same argument shows that $\mathscr{C}_n$ is annihilated by $\mathscr{A}$ for all n. $\qquad\square$

Lemma 24.106. *Let* $f\colon X \to Y = \operatorname{Spec} A$ *be as in the beginning of this section and suppose that* f *is proper. Let*

$$0 \longrightarrow \mathscr{F}_1 \xrightarrow{u_1} \mathscr{F}_2 \xrightarrow{u_2} \mathscr{F}_3 \xrightarrow{u_3} \mathscr{F}_4 \longrightarrow 0$$

be an exact sequence of coherent modules over $X_{/Z}$. *If* $\mathscr{F}_1$, $\mathscr{F}_2$, *and* $\mathscr{F}_4$ *are algebraizable, then* $\mathscr{F}_3$ *is algebraizable.*

Proof. By Lemma 24.101 (1), $\operatorname{Im}(u_1)$ is algebraizable. By Lemma 24.101 (2) to the short exact sequence $0 \to \operatorname{Im}(u_1) \to \mathscr{F}_3 \to \mathscr{F}_4 \to 0$, $\mathscr{F}_3$ is algebraizable. □

(24.22) Grothendieck's existence theorem for coherent modules with proper support.

There is also a refined version of Theorem 24.94 where f is not necessarily proper but all modules have proper support. Compare Section (23.5). Let us explain what we mean by this.

Let $X \to \operatorname{Spec} A$ be a separated morphism of finite type. Recall (Definition 23.21) that a closed subspace T of X is said to be *proper over* Y if the composition $T \to X \to \operatorname{Spec} A$ is proper for some scheme structure on T making T into a closed subscheme of X.

Remark and Definition 24.107. Let $\mathscr{F} = (\mathscr{F}_n)_n$ be a quasi-coherent module of finite type over $X_{/Z}$. By Nakayama's lemma for nilpotent ideals (Proposition B.3), the homeomorphism $X_n \to X_{n+1}$ induces an identification between the support of $\mathscr{F}_n$ and the support of $\mathscr{F}_{n+1}$. This closed subspace of $|X_0| = |X_1| = \cdots$ is called the *support of* $\mathscr{F}$.

Let $\mathscr{F}$ be a coherent $\mathscr{O}_X$-module with proper support over Y. Then $\mathscr{F}_{/Z} := (i_n^*\mathscr{F})_n$ is a coherent module over the formal completion $X_{/Z}$ with proper support over Y.

Theorem 24.108. *(Grothendieck's existence theorem) Let* $f\colon X \to Y := \operatorname{Spec} A$ *be a separated morphism of finite type. Then the functor* $\mathscr{F} \mapsto \mathscr{F}_{/Z}$ *from the category of coherent* $\mathscr{O}_X$-*modules with proper support over* Y *to the category of coherent modules over* $X_{/Z}$ *with proper support over* Y_0 *is an equivalence of categories.*

A proof of this more general version of Grothendieck's existence theorem can be given along the same lines as the proof of Theorem 24.94 given above, taking into account the following additional remarks.

(1) As basic finiteness result one uses that under the hypotheses on X and Y the higher direct images $R^p f_* \mathscr{F}$ are coherent for all coherent $\mathscr{O}_X$-modules $\mathscr{F}$ whose support is proper over Y (Corollary 23.22).

(2) If $\mathscr{F}$ and $\mathscr{G}$ are coherent $\mathscr{O}_X$-modules, then the support of $\mathscr{E}\!xt^p_{\mathscr{O}_X}(\mathscr{F}, \mathscr{G})$ is contained in $\operatorname{Supp}(\mathscr{F}) \cap \operatorname{Supp}(\mathscr{G})$ (Proposition 22.65). Using (1), then one can argue as in the proof of Proposition 24.98 to obtain an isomorphism of finitely generated A-modules

$$\operatorname{Ext}^p_{\mathscr{O}_X}(\mathscr{F}, \mathscr{G}) \cong \operatorname{Ext}^p_{\mathscr{O}_X}(\mathscr{F}^\wedge, \mathscr{G}^\wedge).$$

(3) As in Corollary 24.100 one deduces an isomorphism of finitely generated A-modules

$$(24.22.1) \qquad \operatorname{Hom}_{\mathscr{O}_X}(\mathscr{F}, \mathscr{G}) \xrightarrow{\sim} \operatorname{Hom}_{(X_{/Z}\text{-Mod})}(\mathscr{F}_{/Z}, \mathscr{G}_{/Z})$$

if $\operatorname{Supp}(\mathscr{F}) \cap \operatorname{Supp}(\mathscr{G})$ is proper over A. This proves that the functor $\mathscr{F} \mapsto \mathscr{F}_{/Z}$ in Theorem 24.108 is fully faithful.

(4) Then one shows that all coherent modules over $X_{/Z}$ of proper support are algebraizable if X is quasi-projective over A. For this one chooses an open embedding $j\colon X \to P$ into a projective A-scheme P. Let $(\mathscr{F}_n)_n$ be a coherent module over $X_{/Z}$ whose support is proper over Y. Then $(j_*\mathscr{F}_n)_n$ is a coherent module over $P_{/Z}$. Indeed, each $\mathscr{F}_n$ is of the form $i_{n,*}\mathscr{G}_n$, where $i_n\colon T_n \to X$ is a closed immersion such that $T_n \to Y$ is proper and such that $\mathscr{G}_n$ is a coherent $\mathcal{O}_{T_n}$-module. Then the immersion $j \circ i_n$ is a proper morphism (Proposition 12.58) and hence a closed immersion. Therefore $j_*\mathscr{F}_n = (j \circ i_n)_*\mathscr{G}_n$ is coherent. Then $(j_*\mathscr{F}_n)_n$ is algebraizable by a coherent $\mathcal{O}_P$-module $\mathscr{H}$ with $\mathrm{Supp}(\mathscr{G}) = \mathrm{Supp}((j_*\mathscr{F}_n)_n) = \mathrm{Supp}(\mathscr{F})$ and hence $\mathscr{F} = (j^*j_*\mathscr{F}_n)_n$ is algebraizable by $j^*\mathscr{G}$ which has proper support over Y.

(5) Finally, one reduces the general case to the quasi-projective case with the same arguments as in the proper case using Construction 24.104, Lemma 24.105, and Lemma 24.106.

(24.23) Algebraization of proper schemes.

In this section let A be a noetherian ring and let $I \subseteq A$ be an ideal such that $A = \lim_n A/I^n$. Let $S := \mathrm{Spec}\, A$ and $Z := V(I) \subseteq S$. Set $S_n := \mathrm{Spec}\, A/I^{n+1}$. If $\mathcal{X}$ is an adic formal scheme over $S_{/Z}$, we set $X_n := \mathcal{X} \times_{S_{/Z}} S_n$. For instance, if $\mathcal{X} = X_{/Z}$ for a qcqs S-scheme X, then $X_n = X \times_S S_n$ (Lemma 24.84).

Proposition 24.109. *Let $f\colon Y \to S$ be a separated morphism of finite type. Then $X \mapsto X_{/Z}$ induces an equivalence between the category of finite Y-schemes X that are proper over S and the category of finite adic formal schemes over $Y_{/Z}$ that are proper over $S_{/Z}$. Moreover, a finite morphism $X \to Y$ is a closed immersion if and only if $X_{/Z} \to Y_{/Z}$ is a closed immersion.*

Proof. As formal completion of modules is compatible with tensor products (24.18.2), Grothendieck's existence theorem, Theorem 24.108 implies that $\mathscr{F} \mapsto \mathscr{F}_{/Z}$ induces an equivalence between the category of coherent $\mathcal{O}_X$-algebras that have proper support over S and the category of coherent algebras over $X_{/Z}$ that have proper support over $S_{/Z}$. This implies the first assertion by the equivalence of finite schemes and finite quasi-coherent algebras (Corollary 12.2 and Remark 12.10 (3)).

Now let $\iota\colon \mathcal{T} = (T_n)_n \to X_{/Z}$ be a closed immersion of adic formal schemes such that T_n is proper over Y for all n. By what we already showed there exists a finite morphism $i\colon T \to X$ of Y-schemes with $i_{/Z} = \iota$ such that T is proper over Y and such that $T_{/Z} = \mathcal{T}$. To see that i is a closed immersion we have to show that $\mathscr{C} := \mathrm{Coker}(i^\flat\colon \mathcal{O}_X \to i_*\mathcal{O}_T)$ is zero. It is a coherent $\mathcal{O}_X$-module whose support is proper because T is proper over Y. Now $i^\flat$ corresponds under (24.22.1) to the surjective morphism $(\mathcal{O}_{X_n} \to i_*\mathcal{O}_{T_n})_n$ of coherent modules over $X_{/Z}$. As formal completion of coherent modules is exact, $\mathscr{C}_{/T} = 0$. Hence $\mathscr{C} = 0$ because formal completion on coherent modules with proper support is an equivalence of categories (Theorem 24.108). $\qquad\square$

By Proposition 24.95 we deduce the following result.

Corollary 24.110. *Let $f\colon Y \to S$ be a proper morphism and let $r \geq 1$ be an integer. Then $X \mapsto X_{/Z}$ induces an equivalence between the category of finite locally free schemes X of rank r over Y and the category of finite locally free adic formal schemes of rank r over $Y_{/Z}$.*

Corollary 24.111. *Let* $f \colon Y \to S$ *be a proper morphism. Then* $X \mapsto X \times_Y Y_0$ *yields an equivalence from the category of étale covers of* Y *to the category of étale covers of* $Y_0 := Y \otimes_A A/I$.

In particular, the inclusion $Y_0 \to Y$ yields for every geometric point $\bar{y}$ of Y_0 an isomorphism of étale fundamental groups

$$\pi_1(Y_0, \bar{y}) \xrightarrow{\sim} \pi_1(Y, \bar{y})$$

if Y and Y_0 are connected (e.g., if S and S_0 are connected and f has connected fibers).

Proof. Set $Y_n := Y \otimes_A A/I^{n+1}$. We call an adic formal scheme $(X_n)_n$ étale (resp. finite étale) over $Y_{/Z}$ if $X_n \to Y_n$ is étale (resp. finite étale) for all n. By Theorem 20.30, there is an equivalence between the category of étale Y_0-schemes and the category of adic formal scheme that are étale over $Y_{/Z}$. By Proposition 24.83 this induces an equivalence between the category of étale covers of Y_0 and the category of adic formal scheme that are finite étale over $Y_{/Z}$.

Hence it remains to show that under the equivalence of Corollary 24.110 a Y-scheme X is an étale cover if and only if $X_{/Z} \to Y_{/Z}$ is finite étale. Clearly the condition is necessary. Conversely, if $X_{/Z} \to Y_{/Z}$ is finite étale, then $X \to Y$ is finite locally free. The locus V in Y where $X \to Y$ is étale is the non-vanishing locus of the discriminant of X over Y, hence it is open in Y (Corollary 20.73). As $X \to Y$ is flat, we have

$$V = \{ y \in Y \ ; \ X_y \text{ is an étale } \kappa(y)\text{-scheme}\}$$

by Corollary 18.45. It contains $|f^{-1}(Z)| = |Y_n|$ because for $y \in f^{-1}(Z)$ the fiber of X in y is the same as the fiber of X_n in y for all n. Hence $V = Y$ by Lemma 24.96. $\qquad \square$

We will now show that on proper schemes X over A the functor $X \mapsto X_{/Z}$ is fully faithful, more precisely:

Theorem 24.112. *Let* $g \colon X \to S$ *be a proper morphism and let* $h \colon Y \to S$ *be a separated morphism of finite type. Then* $f \mapsto f_{/Z}$ *is a bijection*

$$\mathrm{Hom}_S(X, Y) \xrightarrow{\sim} \mathrm{Hom}_{S_{/Z}}(X_{/Z}, Y_{/Z}).$$

To prove the theorem we will apply Proposition 24.109 to the graphs of morphisms.

Proof. Let $(f_n)_n \colon X_{/Z} \to Y_{/Z}$ be a morphism given by a family of morphisms $f_n \colon X_n \to Y_n$. Let $\Gamma_{f_n} \subseteq T_n := X_n \times_{S_n} Y_n$ be the graph of f_n. As the first projection induces an isomorphism $\Gamma_{f_n} \xrightarrow{\sim} X_n$, Γ_{f_n} is proper over S_n. Set $T := X \times_S Y$. Then $(\Gamma_{f_n})_n \to T_{/Z}$ is a closed immersion that corresponds by Proposition 24.109 to a closed subscheme $\Gamma \subseteq T$ that is proper over S. As $\Gamma_{f_n} \to X_n$ is finite locally free of rank 1, this is also true for $\Gamma \to X$ by Corollary 24.110. Hence $\Gamma \to X$ is an isomorphism and therefore the graph of a morphism $F \colon X \to Y$ such that $\Gamma_F \times_S S_n = \Gamma_{f_n}$, i.e., such that $F_{/Z} = (f_n)_n$.

Conversely, let $f \colon X \to Y$ be a morphism and let $F \colon X \to Y$ be the morphism attached to $(f_n)_n = f_{/Z}$ as above. Then Γ_f and Γ_F are both closed subschemes of T which are proper over S such that the corresponding closed adic formal subscheme of $T_{/Z}$ is $(\Gamma_{f_n})_n$. Hence $\Gamma_f = \Gamma_F$ by Proposition 24.109 and hence $f = F$. $\qquad \square$

Let $\mathcal{X}$ be a proper adic formal scheme over $S_{/Z}$. If $\mathcal{X}$ is algebraizable by a proper S-scheme X, then X is unique up to unique isomorphisms inducing the identity on $\mathcal{X}$ by Theorem 24.112. In general, there exist proper adic formal schemes over $S_{/Z}$ that are not algebraizable (see for instance [Har2] ^{O} Example 3.3). But if $\mathcal{X}$ carries an ample line bundle, then this cannot happen, as the next result shows.

Theorem 24.113. *(Grothendieck's algebraization theorem) Let $\mathcal{X} = (X_n)_n$ be a proper adic formal scheme over $S_{/Z}$. Let $\mathcal{L} = (\mathcal{L}_n)_n$ be a locally free module over $\mathcal{X}$ of rank 1 such that the line bundle $\mathcal{L}_0$ on X_0 is ample. Then $\mathcal{X}$ is algebraizable by a proper S-scheme X and there exists a unique (up to isomorphism) line bundle $\mathcal{M}$ on X such that $\mathcal{M}_{/Z} \cong \mathcal{L}$. Moreover, $\mathcal{M}$ is ample.*

Proof. We find an integer $m \geq 1$ such that $\mathcal{L}_0^{\otimes m}$ is very ample (Theorem 13.62) and that $\lim_n \Gamma(X_n, \mathcal{L}_n^{\otimes m}) \to \Gamma(X_0, \mathcal{L}_0^{\otimes m})$ is surjective (Proposition 24.102). Let $t_0, \ldots, t_r \in \Gamma(X_0, \mathcal{L}_0^{\otimes m})$ be sections, corresponding to a surjective map of $\mathcal{O}_{X_0}$-modules $u_0 \colon \mathcal{O}_{X_0}^{r+1} \to \mathcal{L}_0^{\otimes m}$, that define a closed embedding $i_0 \colon X_0 \to \mathbb{P}_{S_0}^r$ of S_0-schemes with $i_0^* \mathcal{O}_{\mathbb{P}_{S_0}^r}(1) = \mathcal{L}_0^{\otimes m}$.

As $\Gamma(X_n, \mathcal{L}_n) \to \Gamma(X_0, \mathcal{L}_0)$ is surjective for all n, we can lift the sections to $\Gamma(X_n, \mathcal{L}_n^{\otimes m})$ and obtain a map of $\mathcal{O}_{X_n}$-modules $u_n \colon \mathcal{O}_{X_n}^{r+1} \to \mathcal{L}_n^{\otimes m}$ that is equal to u_0 modulo the nilpotent ideal generated by I in $\mathcal{O}_{X_n}$. Because u_0 is surjective, by Nakayama's lemma u_n is surjective for all n. The maps u_n define morphisms of S_n-schemes $i_n \colon X_n \to \mathbb{P}_{S_n}^r$ such that $(i_n)_n$ is a morphism of adic formal schemes $\mathcal{X} \to (\mathbb{P}_{S_n}^r)_n = (\mathbb{P}_S^r)_{/Z}$. As i_0 is a closed immersion, i_n is a closed immersion for all n by Proposition 24.83. Hence $(i_n)_n$ is algebraizable by a closed immersion $i \colon X \to \mathbb{P}_S^r$ by Proposition 24.109. Moreover, there exists a unique line bundle $\mathcal{M}$ on X such that $\mathcal{M}_{/Z} \cong \mathcal{L}$ by Grothendieck's existence theorem, Theorem 24.94. As one has $\mathcal{M}^{\otimes m} = i^* \mathcal{O}_{\mathbb{P}_S^r}(1)$, $\mathcal{M}$ is ample. $\qquad\square$

(24.24) Remarks on the literature.

The notion of derived completion seems to be relatively new. Our main source is the Stacks project [Sta] 091N. Precursors and other sources for derived completions can be found in [GrMay] ^{O}, [Lu-DAGXII] $_X$ Chapter 4 and [BhSc] $_X^O$. There are several variants of derived completions, for instance an idealistic variant which is closer to the original approach of Greenlees and May in [GrMay] ^{O}. Nice overviews together with comparisons between the different notions of derived completions can be found in [Yek3] $_X^O$ and [Pos] $_X$.

The general compatibility of derived direct image and derived completion (Theorem 24.28) can also be found in the Stacks project [Sta] 0995. The classical theorem of formal functions for proper morphisms between noetherian schemes and for coherent modules (Theorem 24.37) and the Stein factorization theorem in Algebraic Geometry are of course much older. Still one of the best references is [EGAIII] ^{O}. This is also presented in a slightly more modern form, nicely explaining the main ideas, in [FGAex] $_X$, Chapter 8, by Illusie. Somewhat weaker statements are also shown in [Har3] ^{O} Chapter III.

The formal completion of a scheme along a closed subset can either be seen as a topologically ringed space or as a colimit of schemes in the category of Zariski (or, equivalently, fpqc-) sheaves. The first point of view is taken in [EGAInew] and also briefly explained in [FGAex] $_X$ Chapter 8. In the noetherian context they can also be seen as special cases of Huber's adic spaces [Hu] ^{O}. The point of view to consider them (maybe even only locally) as special cases of colimits of schemes, as we do here, can also be found in [Yas] $_X^O$ or in great generality in the Stacks project [Sta] 0AHW. Good references for

Grothendieck algebraization theorem are again [EGAIII] ^{O}, [FGAex] $_{\text{X}}$ Chapter 8, and the Stacks project [Sta] 0898.

Exercises

Exercise 24.1. Let A be a ring. What does it mean for a complex M in $D(A)$ to be derived complete with respect to the ideal $I = A$ or with respect to an ideal that contains only nilpotent elements?

Exercise 24.2. Let A be a principal ideal domain, $p \in A$ a prime element, $I = (p) \subseteq A$, and let K be the field of fractions of A. Show that the derived I-completion of the A-module K/A is $(\lim A/p^n A)[1]$.
Hint: Use that K/A is an injective A-module (Proposition G.22).

Exercise 24.3. (Derived Lemma of Nakayama) Let A be a ring and let $I \subseteq A$ be a finitely generated ideal. Show the following assertions.
(1) Let M be a derived I-complete A module such that $M/IM = 0$. Then $M = 0$.
(2) Let $K \in D(A)$ be a derived I-complete complex such that $K \otimes_A^L A/I = 0$. Then $K = 0$.
Hint: Exercise 22.47

Exercise 24.4. Let A be a ring and $I \subseteq A$ an ideal. Suppose that A is derived I-complete (e.g., if A is I-adically complete). Show that every pseudo-coherent complex in $D(A)$ is derived I-complete.

Exercise 24.5. Let k be a field, set $A := k[\![t]\!]$ and $I := (t) \subseteq k[\![t]\!]$. Let $M \mapsto \hat{M}$ be the functor of t-adic completion on the category of A-modules. This exercise gives an example which shows that this functor is not exact in the middle (and in particular neither left nor right exact).
If M is an A-module and $m \in M$ we set $\mathrm{ord}_t(m) := \sup\{ i \in \mathbb{N} \; ; \; m \in t^i M \}$. Let P be the A-module of maps $f \colon \mathbb{N} \to A$ such that $\{ n \in \mathbb{N} \; ; \; \mathrm{ord}_t(f(n)) \leq i \}$ is finite for all i. We endow it with the t-adic topology.
(1) Show that P is t-adically complete.
(2) Show that there exists a unique A-linear continuous map $u \colon P \to P$ with $u(\delta_n) = t^n \delta_n$, where $\delta_n \in P$ sends $m \in \mathbb{N}$ to 1 if $m = n$ and to 0 otherwise. Show that u is injective and that its image is not closed in P.
(3) Let Q be the cokernel of u such that one has an exact sequence

(*) $$0 \longrightarrow P \xrightarrow{u} P \xrightarrow{v} Q \longrightarrow 0$$

and let $\hat{Q}$ be the t-adic completion of Q. Show that $\hat{v} \colon P = \hat{P} \to \hat{Q}$ is surjective and that its kernel is the closure of the image of u.
(4) Deduce that the completion of (*) is exact at the left $P = \hat{P}$ and at $\hat{Q}$ but not exact in the middle.

Exercise 24.6. Let k be a field, let $A := k[t_1, t_2, \dots]$ be the polynomial ring in countably many variables, and let $I := (t_1, t_2, \dots)$ be the ideal generated by the variables. Let $\hat{A}$ be the I-adic completion of A.
(1) Show that $\hat{A}$ is isomorphic to the ring $k[\![t_1, t_2, \dots]\!]$ of formal power series (each of which has only a finite number of terms of given degree).

(2) Let $J := \mathrm{Ker}(\pi)$ where $\pi \colon \hat{A} \to k = A/I$ is the canonical map. Then J is the closure of I in $\hat{A}$. Show that $J \neq I\hat{A}$ by considering the formal power series $\sum_{n \geq 1} t_n^n$.

(3) Show that $\hat{A}$ is not I-adically complete.

(4) Show that $\hat{A}$ is not J-adically complete if k is a finite field.

Exercise 24.7. Let A be a ring and let $f \in A$ be an element. Let $A[f] := \{\, a \in A \,;\, fa = 0 \,\}$ be the ideal of f-torsion. Suppose that there exists $N \geq 1$ such that $A[f^n] = A[f^N]$ for all $n \geq N$. Show that for every object M in $D(A)$ one has

$$M^\wedge \cong \mathrm{holim}(M \otimes_A^L A/f^n).$$

Remark: See Exercise 24.9 for a generalization.

Exercise 24.8. Let A be a ring. A sequence $\mathbf{f} := (f_1, \ldots, f_r)$ of elements of A is called *weakly proregular* if for all $p < 0$ the inverse system of A-modules $(H^p(K(\mathbf{f}^n)))_{n \geq 1}$ is pro-zero, i.e., for every $n \geq 1$ there exists $m \geq n$ such that $H^p(K(\mathbf{f}^m)) \to H^p(K(\mathbf{f}^n))$ is zero. An ideal $I \subseteq A$ is called *weakly proregular* if it is generated by a weakly proregular sequence.

(1) Show that if I is weakly proregular, then one has for all $M \in D(A)$ a functorial isomorphism $M^\wedge \longrightarrow \mathrm{holim}(M \otimes^L A/I^n)$, where the left hand side denotes the derived I-completion of M.

(2) Show that every ideal of a noetherian ring is weakly proregular.

(3) Show that a sequence consisting of a single element $f \in A$ is weakly proregular if and only if there exists an $n \geq 1$ such that $A[f^m] = A[f^n] := \{\, a \in A \,;\, f^n a = 0 \,\}$ for all $m \geq n$, cf. Exercise 24.7.

(4) Let $A \to A'$ be a faithfully flat ring homomorphism and let $I \subseteq A$ be an ideal. Show that I is weakly proregular if and only if IA' is weakly proregular.

(5) Let $I \subseteq A$ be a weakly proregular ideal and let $J \subseteq A$ be a finitely generated ideal such that $\sqrt{I} = \sqrt{J}$. Show that J is weakly proregular.

(6) Let $I \subseteq A$ be a weakly pro-regular ideal and consider the A-linear I-adic completion functor $\Lambda_I \colon (A\text{-Mod}) \to (A\text{-Mod})$, $M \mapsto \lim_n M/I^n M$. Let $L\Lambda_I \colon D(A) \to D(A)$ be its left derived functor (note that since Λ_I is in general not right exact, we do not have necessarily $L^0 \Lambda_I = \Lambda_I$). Show that there exists a unique functorial morphism

$$\tau_M \colon M \to L\Lambda_I(M)$$

whose composition with the canonical morphism $L\Lambda_I(M) \to \Lambda_I(M)$ is the canonical map $M \to \Lambda_I(M)$. Show that a complex $M \in D(A)$ is derived I-complete if and only if τ_M is an isomorphism.

Remark: See [Yek3]$\overset{\mathrm{O}}{\underset{\mathrm{X}}{}}$.

Exercise 24.9. Let X be a scheme and let $\mathscr{I} \subseteq \mathscr{O}_X$ a quasi-coherent ideal. Show that the following assertions are equivalent.

(i) There exists an open affine covering $(U_i)_i$ of X such that $\Gamma(U_i, \mathscr{I})$ is a weakly proregular ideal (Exercise 24.8) of $\Gamma(U_i, \mathscr{O}_X)$ for all i.

(ii) For every $U \subseteq X$ open affine $\Gamma(U, \mathscr{I})$ is a weakly proregular ideal of $\Gamma(U, \mathscr{O}_X)$.

If these conditions are satisfied, we call the corresponding closed subscheme $V(\mathscr{I})$ a *weakly proregular subscheme* of X.

(1) Show that the underlying topological subspace of a weakly proregular subscheme is closed and constructible and that the property to be weakly proregular depends only on the underlying topological closed subspace.

(2) Show that any closed subscheme of a locally noetherian scheme is weakly proregular.

(3) Let $\mathscr{I} \subseteq \mathscr{O}_X$ be defining a weakly proregular subscheme. Show that for any $\mathscr{F}$ in $D(X)$ the natural map

$$\mathscr{F}^\wedge \longrightarrow \operatorname*{holim}_n (\mathscr{F} \otimes^L_{\mathscr{O}_X} \mathscr{O}_X / \mathscr{I}^n)$$

is an isomorphism, where $\mathscr{F}^\wedge$ denotes the $\mathscr{I}$-derived completion of $\mathscr{F}$.

Exercise 24.10. Let A be a ring and let $I \subseteq A$ be an ideal. An A-module M is called *I-completely flat* if $\operatorname{Tor}^A_i(M, N) = 0$ for all $i > 0$ and every I-torsion module N. Show that the following assertions are equivalent for an A-module M.

(i) M is I-completely flat.

(ii) For all $k \geq 1$ and all $i > 0$ one has $\operatorname{Tor}^A_i(M, A/I^k) = 0$ and $M \otimes_A A/I^k$ is a flat A/I^k-module.

(iii) For all $i > 0$ one $\operatorname{Tor}^A_i(M, A/I) = 0$ and $M \otimes_A A/I$ is a flat A/I-module.

Exercise 24.11. Let A be a ring, let $\mathbf{f} = (f_1, \ldots, f_r)$ be a finite family of elements of A, let I be the ideal generated by $\mathbf{f}$. Set $X := \operatorname{Spec} A$, $U := X \setminus Z$, and let $\mathcal{U}$ be the open covering $(D(f_i))_i$ of U. Show that there is a functorial isomorphism

$$\operatorname{Tot}(\check{C}^\bullet_{\mathrm{alt}}(\mathcal{U}, \mathscr{F}^\bullet)) \xrightarrow{\sim} R\Gamma(U, \mathscr{F})$$

for every $\mathscr{F}$ in $D_{\mathrm{qcoh}}(X)$.

Hint: Exercise 21.11

Exercise 24.12. Let A be a ring, let $I \subset A$ be an ideal generated by a finite family $\mathbf{f} = (f_1, \ldots, f_r)$, and set $Z = V(I)$. Let $C_A(\mathbf{f})$ be the associated extended Čech complex (24.3.1). Let $D_{I^\infty - \mathrm{Tors}}(A)$ be the full subcategory of $D(A)$ consisting of complexes whose cohomology modules are all I-power torsion (Exercise 22.47).

(1) Show that

$$R\Gamma_Z \colon D(A) \longrightarrow D_{I^\infty - \mathrm{Tors}}(A), \qquad E \mapsto C_A(\mathbf{f}) \otimes^L_A E$$

defines a triangulated functor and that $R\Gamma_Z$ is right adjoint to the inclusion functor $D_{I^\infty\text{-Tors}}(A) \to D(A)$.

(2) For E in $D(A)$ show that there is a functorial isomorphism $R\Gamma_Z(E) \cong R\Gamma_Z(X, \tilde{E})$, where the right hand side denotes the cohomology with support in Z (Exercise 21.7). *Hint*: Exercise 24.11

Exercise 24.13. Let $f \colon X \to S$ be a flat, proper morphism of finite presentation with geometrically integral fibers and let $\mathscr{L}$ be a line bundle on X. Suppose that S is reduced and that $\mathscr{L}_{|X_s}$ is trivial for every maximal point s of S (i.e., for generic points of all irreducible component of S). Show that there exists a line bundle $\mathscr{M}$ on S such that $f^* \mathscr{M} \cong \mathscr{L}$.

Remark: For a variant without the assumption that f is proper see Lemma 27.70.

Exercise 24.14. Let A be a ring and let $I \subseteq A$ be an ideal. Show that the full subcategory of derived I-complete A-modules is a plump subcategory of the category of all A-modules.

Exercise 24.15. Let A be a ring, and let $f \in A$. Let $J := \bigcap_n (f^n) \subseteq A$.

(1) Let M be a derived (f)-complete A-module. Show that $J \operatorname{Ker}(M \to \lim_n M/f^n M) = 0$.

(2) Deduce that if A is derived (f)-complete, then $J^2 = 0$.

(3) Let $I \subseteq A$ be an ideal that can be generated by r elements and suppose that A is derived I-complete. Show that $\left(\bigcap_n I^n\right)^{2^r} = 0$.

(4) Let I be a finitely generated ideal of A and assume that A is reduced. Show that A is I-adically complete if and only if A is derived I-adically complete.

Exercise 24.16. Let A be a ring and let I_A be the greatest ideal I of A such that A is derived I-adically complete. Show that I is contained in the Jacobson radical of A.
Hint: Exercise 24.15

Exercise 24.17. Let A be a ring, let $I \subseteq A$ be an ideal.
(1) Suppose that A is derived I-complete. Show that (A, I) is a henselian pair.
   Hint: Reduce to the case that I is generated by one element using Proposition 20.17.
(2) Give an example of a ring A and an ideal I such that (A, I) is henselian but A is not derived I-complete.

Exercise 24.18. Let $f\colon X \to S$ be a proper morphism and let $X \to X' \to S$ be its Stein factorization. Show that if X is normal, then X' is normal.

Exercise 24.19. Let S be a reduced scheme and let $f\colon X \to S$ be a flat proper morphism of finite presentation with geometrically integral fibers. Let $\mathscr{L}$ be a line bundle on X and let $U \subseteq S$ be an open dense subscheme such that the restriction of $\mathscr{L}$ to the fiber X_s is trivial for all $s \in U$. Show that $f_*\mathscr{L}$ is a line bundle and that $f^*f_*\mathscr{L} \cong \mathscr{L}$.

Exercise 24.20. With the notation used in Section (24.17). Show that the following categories are equivalent.
(a) The category of qcqs adic formal schemes over $S_{/Z}$.
(b) The category of morphisms of Zariski sheaves $\mathcal{X} \to S_{/Z}$ that are representable by schemes and qcqs.
Hint: If $\mathcal{X} \to S_{/Z}$ is a qcqs representable morphism, define $(X_n \to S_n)_n$ by $X_n := S_n \times_{S_{/Z}} \mathcal{X}$. Conversely, if $(X_n \to S_n)_n$ is a qcqs adic formal schemes over $S_{/Z}$, then define a morphism $X := \operatorname{colim}_n X_n \to S_{/Z}$. To show that these functors are quasi-inverse to each other use that every morphism $T \to S_{/Z}$ with T qcqs scheme factors through some S_n by Proposition 24.79.

Exercise 24.21. Let X be a noetherian integral scheme, and let Y be a separated scheme. Let f and g be morphisms $X \to Y$ such that for some non-empty closed subscheme $Z \subseteq X$, the restrictions $f_{|X_{/Z}}$ and $g_{|X_{/Z}}$ are equal. Show that then $f = g$.

25 Duality

Content

To motivate the theme of *coherent duality*, the topic of this chapter, let us start with the classical Riemann-Roch theorem for a geometrically connected smooth projective curve C over a field k, see Theorem 15.35, Theorem 26.48. One approach to proving it is to first express the Euler characteristic of a line bundle $\mathscr{L}$ in terms of the degree of the line bundle and the Euler characteristic of the structure sheaf (which can easily be expressed in terms of the genus of the curve). What remains to be done is to obtain a more manageable expression for the Euler characteristic

$$\chi(\mathscr{L}) = \dim_k H^0(C, \mathscr{L}) - \dim_k H^1(C, \mathscr{L})$$

of $\mathscr{L}$. It turns out that one can express the dimension of $H^1(C, \mathscr{L})$ as the dimension of a space of global sections of some other line bundle on C. More precisely,

$$\dim_k H^1(C, \mathscr{L}) = \dim_k H^0(C, \mathscr{L}^\vee \otimes_{\mathscr{O}_C} \Omega^1_{C/k}).$$

(For $C = \mathbb{P}^1_k$, we have already seen this in Theorem 22.22.)

In this chapter, we will prove and vastly generalize this result. To give an outline of how we will proceed, we first formulate a souped up version of the above statement. Rather than going for an equality of dimensions, it will turn out that there is even a perfect pairing

$$H^0(C, \mathscr{L}^\vee \otimes_{\mathscr{O}_C} \Omega^1_{C/k}) \times H^1(C, \mathscr{L}) \to H^1(C, \Omega^1_{C/k}) \cong k,$$

or in other words an isomorphism

$$\operatorname{Hom}_k(H^1(C, \mathscr{L}), k) \xrightarrow{\sim} H^0(C, \mathscr{L}^\vee \otimes_{\mathscr{O}_C} \Omega^1_{C/k}).$$

Even better, we can replace this formulation on the level of cohomology groups by a statement about objects of the derived categories of C and k. Denote by $f\colon C \to \operatorname{Spec} k$ the structure morphism. Writing $H^0(C, \mathscr{L}^\vee \otimes_{\mathscr{O}_C} \Omega^1_{C/k}) = \operatorname{Hom}_{\mathscr{O}_C}(\mathscr{L}, \Omega^1_{C/k}) = \operatorname{Hom}_{D(C)}(\mathscr{L}, \Omega^1_{C/k})$, the desired statement takes the following form:

$$\operatorname{Hom}_{D(k)}(Rf_*\mathscr{L}[1], k) \cong \operatorname{Hom}_{D(C)}(\mathscr{L}, \Omega^1_{C/k}).$$

We will prove below that for any morphism $f\colon X \to S$ between qcqs schemes,

© Springer Fachmedien Wiesbaden GmbH, ein Teil von Springer Nature 2023
U. Görtz und T. Wedhorn, *Algebraic Geometry II: Cohomology of Schemes*,
Springer Studium Mathematik – Master, https://doi.org/10.1007/978-3-658-43031-3_10

(1) the functor $Rf_* \colon D_{\mathrm{qcoh}}(X) \to D_{\mathrm{qcoh}}(S)$ admits a right adjoint $f^\times$ (Theorem 25.17),
(2) for suitable f (e.g., proper, flat, and of finite presentation), we can compute $f^\times K$ as
      $Lf^* K \otimes^L_{\mathscr{O}_X} f^\times \mathscr{O}_X$ (Theorem 25.37), and
(3) for f smooth of relative dimension n and proper we have $f^\times \mathscr{O}_S = \Omega^n_{X/S}[n]$ (Theo-
      rem 25.58).

The above isomorphism is an immediate corollary of this. We will obtain the existence of $f^\times$
from "abstract" category-theoretic results, namely a version of the Brown representability
theorem. See Section (25.33) for a short discussion of some other approaches.

While for proper morphisms f the functor $f^\times$ behaves well and is the "right" functor to
consider, it has the defect that it cannot be computed locally on the source. Moreover, even
if f is simply the structure morphism of the affine line over a field k, the functor $f^\times$ yields
extremely large and unwieldy modules (Remark 25.45). It is much nicer to work with a
functor which for open immersions coincides with the usual pullback functor. Interestingly,
one can combine the functors $f^\times$ for f proper and j^* for j an open immersion in the sense
that one can construct a *twisted inverse image functor* $f^! \colon D^+_{\mathrm{qcoh}}(S) \to D^+_{\mathrm{qcoh}}(X)$ for
every separated morphism of finite type between noetherian schemes, such that $f^! = f^\times$
for f proper, $f^! = f^*$ for f an open immersion, and even for every étale morphism f, and
such that $(f \circ g)^! \cong g^! \circ f^!$. In view of Nagata's compactification theorem, these conditions
determine the functors $f^!$ uniquely. Giving the construction and some of the properties of
$f^!$ (but at times without full proofs) is the second main goal in this chapter.

Coming back to the example of a smooth proper curve $f \colon C \to \operatorname{Spec} k$ over a field k,
we can consider the functor $\mathscr{L} \mapsto \mathscr{L}^\vee \otimes^L \Omega^1_{C/k}[1] = R\,\mathscr{H}\!om_{\mathscr{O}_C}(\mathscr{L}, f^\times \mathscr{O}_{\operatorname{Spec} k})$ (where we
can take any $\mathscr{L} \in D^b_{\mathrm{coh}}(C)$, not just line bundles) as a duality on the bounded derived
category $D^b_{\mathrm{coh}}(C)$ of coherent sheaves on C. More generally, an object $\omega^\bullet_X \in D^b_{\mathrm{coh}}(X)$, for a
noetherian scheme X, such that the functor $R\,\mathscr{H}\!om_{\mathscr{O}_X}(-, \omega^\bullet_X)$ induces an auto-equivalence
of $D^b_{\mathrm{coh}}(X)$ (and that is of finite injective dimension) is called a dualizing complex, a
notion we will study in the third part of the chapter, starting with Section (25.14).

At the end, in Section (25.26) we will summarize some of the results in the special
situation of proper schemes of finite type over a field.

The book [Lip2] $\overset{\mathrm{o}}{\underset{\mathrm{X}}{}}$ and the Stacks project [Sta] were very helpful when writing this
chapter, and their impact on the exposition will be very visible to the reader. For further
pointers to the literature see Section (25.33).

The right adjoint $f^\times$ of Rf_*

Let $f \colon X \to S$ be a morphism of qcqs schemes. We have the pair $Lf^* \colon D_{\mathrm{qcoh}}(S) \to$
$D_{\mathrm{qcoh}}(X)$ and $Rf_* \colon D_{\mathrm{qcoh}}(X) \to D_{\mathrm{qcoh}}(S)$ of adjoint functors, see Section (21.27). In this
chapter we will show the existence of a right adjoint $f^\times$ of Rf_* and study its properties.
In fact, the existence of $f^\times$ follows immediately from a suitable *Brown representability
theorem/adjoint functor theorem* for triangulated categories. But note that in general, $f^\times$
is not the derived functor of a functor $(\mathscr{O}_S\text{-Mod}) \to (\mathscr{O}_X\text{-Mod})$.

(25.1) Brown representability for triangulated categories.

We start out by some general considerations about representability of triangulated functors between triangulated categories that admit arbitrary coproducts, e.g., the derived category of a Grothendieck abelian category (Lemma F.188).

Definition 25.1. *Let $\mathcal{D}$ be a triangulated category which has arbitrary coproducts.*
(1) *A set $\mathcal{S}$ of objects in $\mathcal{D}$ is said to* generate $\mathcal{D}$ *if whenever A is an object in $\mathcal{D}$ with $\mathrm{Hom}(S, A) = 0$ for all $S \in \mathcal{S}$, then $A = 0$.*
(2) *We call an object $S \in \mathcal{D}$* compact *if for every index set I and every family $(A_i)_{i \in I}$ of objects in $\mathcal{D}$, the natural map*

$$\mathrm{Hom}(S, \bigoplus_{i \in I} A_i) \longrightarrow \bigoplus_{i \in I} \mathrm{Hom}(S, A_i)$$

is a bijection.
(3) *We say that the category $\mathcal{D}$ is* compactly generated, *if there exists a generating set $\mathcal{S}$ of compact objects in $\mathcal{D}$.*

Regarding Part (2) of this definition, note that the universal property of the coproduct talks about morphisms going out of a coproduct and is therefore not directly related to this condition. We will use the following general representability result.

Theorem 25.2. (Brown representability for triangulated categories) *Let $\mathcal{D}$ be a triangulated category which has arbitrary coproducts and which is compactly generated. Let $H \colon \mathcal{D}^{\mathrm{op}} \to (\mathrm{AbGrp})$ be an additive functor.*
Then H is representable if and only if H is cohomological (Definition F.136) and for every index set I and every family $(A_i)_{i \in I}$ of objects in $\mathcal{D}$, we have

$$H(\bigoplus_i A_i) = \prod_i H(A_i).$$

It is clear that every representable functor H has the properties stated in the theorem. For the converse we refer to [Nee3] ^{O}, Theorem 3.1.

Corollary 25.3. (Adjoint functor theorem for triangulated categories). *Let $\mathcal{D}$ be a triangulated category which has arbitrary coproducts and which is compactly generated, and let $\mathcal{E}$ be any triangulated category. Let $F \colon \mathcal{D} \to \mathcal{E}$ be a triangulated functor. Then F has a right adjoint $G \colon \mathcal{E} \to \mathcal{D}$ if and only if it preserves arbitrary coproducts, i.e., for every index set I and every family $(A_i)_{i \in I}$ of objects in $\mathcal{D}$, we have $F(\bigoplus_i A_i) = \bigoplus_i F(A_i)$.*

Proof. By general theory (Proposition F.23), the existence of a right adjoint ensures that F preserves arbitrary coproducts. Conversely, assume that F preserves all coproducts. Let $E \in \mathcal{E}$, and consider the functor

$$\mathcal{D}^{\mathrm{opp}} \to (\mathrm{AbGrp}), \quad A \mapsto \mathrm{Hom}_{\mathcal{E}}(F(A), E).$$

This is a cohomological functor and by our assumption on F we have

$$\mathrm{Hom}_{\mathcal{E}}(F(\bigoplus_i A_i), E) = \mathrm{Hom}(\bigoplus_i F(A_i), E) = \prod_i \mathrm{Hom}(F(A_i), E)$$

for every family $(A_i)_{i \in I}$.

Therefore Theorem 25.2 shows that this functor is representable, i.e., there exists $D \in \mathcal{D}$ such that

(*) $\mathrm{Hom}_{\mathcal{E}}(F(A), E) = \mathrm{Hom}_{\mathcal{D}}(A, D),$

functorially in A.

This means precisely that we can define the functor G that we are looking for by $G(E) := D$ on objects. By functoriality in E of the left hand side of (*) we can extend this definition to morphisms, and one obtains a functor G which is right adjoint to F.   □

We give two further category theoretic lemmas here that we will use later.

Lemma 25.4. *Let $\mathcal{D}$ be a triangulated category and let $\mathcal{S}$ be a set of objects generating $\mathcal{D}$. Then a morphism $X \to Y$ in $\mathcal{D}$ is an isomorphism if and only if for all S in $\mathcal{S}$ the induced map $\mathrm{Hom}_{\mathcal{D}}(S, X) \to \mathrm{Hom}_{\mathcal{D}}(S, Y)$ is bijective.*

Proof. The "only if" part is clear. For the other direction, extend the morphism $X \to Y$ to a distinguished triangle $X \to Y \to Z \to$. Applying the cohomological functors $\mathrm{Hom}_{\mathcal{D}}(S, -)$ to this triangle, we see that our assumption implies $\mathrm{Hom}_{\mathcal{D}}(S, Z) = 0$ for all S in $\mathcal{S}$. So $Z = 0$, which means that $X \to Y$ is an isomorphism (Remark F.122 (2)).   □

Lemma 25.5. *Let $\mathcal{D}$ be a triangulated category that is generated by a set $\mathcal{S}$ consisting of compact objects and let $\mathcal{E}$ be a triangulated category which has arbitrary coproducts. Let $F : \mathcal{D} \to \mathcal{E}$ be a triangulated functor which preserves coproducts, and let $G : \mathcal{E} \to \mathcal{D}$ be its right adjoint. Then the following are equivalent:*
(i) *The functor G respects all coproducts.*
(ii) *For all $S \in \mathcal{S}$, $F(S)$ is a compact object of $\mathcal{E}$.*

Recall that the existence of G follows from Corollary 25.3.

Proof. For S a compact object of $\mathcal{D}$, and any family $(A_i)_i$ of objects in $\mathcal{E}$, we have

$$\mathrm{Hom}_{\mathcal{D}}(S, G(\bigoplus_i A_i)) = \mathrm{Hom}_{\mathcal{E}}(F(S), \bigoplus_i A_i)$$

$$\leftarrow \bigoplus_i \mathrm{Hom}_{\mathcal{E}}(F(S), A_i)$$

$$= \bigoplus_i \mathrm{Hom}_{\mathcal{D}}(S, G(A_i))$$

$$= \mathrm{Hom}_{\mathcal{D}}(S, \bigoplus_i G(A_i)),$$

where we have used the adjunction between F and G (twice) and the compactness of S. If G respects coproducts, the composition is an isomorphism and hence the arrow is an isomorphism which shows that $F(S)$ is compact. Conversely, if $F(S)$ is compact, then the arrow is an isomorphism and by Lemma 25.4 this implies that $G(\bigoplus_i A_i) \cong \bigoplus_i G(A_i)$, as desired.   □

(25.2) $D_{\mathrm{qcoh}}(X)$ is compactly generated.

We will use the general considerations of the previous section to prove the existence of a right adjoint of the functor $Rf_* \colon D_{\mathrm{qcoh}}(X) \to D_{\mathrm{qcoh}}(S)$ for a morphism $f \colon X \to S$ between qcqs schemes (see Theorem 25.17 in the next section). For this we will study the triangulated category $D_{\mathrm{qcoh}}(X)$ and in particular show the following results for a qcqs scheme X.

(1) An object of $D_{\mathrm{qcoh}}(X)$ is compact if and only if it is perfect (Proposition 25.16).
(2) If $U \subseteq X$ is a quasi-compact open subscheme and $\mathscr{E}$ is a perfect complex on U, then $\mathscr{E} \oplus \mathscr{E}[1]$ can be extended to a perfect complex on X (Proposition 25.12, see also Remark 25.13 for a more precise version).
(3) $D_{\mathrm{qcoh}}(X)$ is compactly generated (Theorem 25.14).

Remark 25.6. Let X be a scheme. Then arbitrary coproducts of objects in $D_{\mathrm{qcoh}}(X)$ exist (in $D(X)$) and lie in $D_{\mathrm{qcoh}}(X)$. This follows by reduction to the affine case.

The following example gives a quick argument that $D_{\mathrm{qcoh}}(X)$ is compactly generated in the special case that X carries an ample line bundle.

Example 25.7. Let X be a qcqs scheme and let $\mathscr{L}$ be a line bundle on X.
(1) Let us show that $\mathscr{L}$ is compact as an object of $D_{\mathrm{qcoh}}(X)$, and so are all tensor powers of $\mathscr{L}$, and all shifts of those (this also follows from the more general Lemma 25.8 below).

In fact, we can identify

$$\operatorname{Hom}_{D(X)}(\mathscr{L}, A) = H^0(X, \mathscr{L}^{-1} \otimes A)$$

for any $A \in D_{\mathrm{qcoh}}(X)$, since tensoring by $\mathscr{L}^{-1}$ is exact and $\operatorname{Hom}_{D(X)}(\mathscr{O}_X, -)$ is just the global sections functor on X. Since tensor product and the global sections functor on a qcqs scheme commute with direct sums (Corollary 21.56 (2)), this shows the compactness of $\mathscr{L}$.

(2) Now suppose that $\mathscr{L}$ is an ample line bundle on X. We claim that the set $\mathcal{S} = \{\mathscr{L}^{\otimes m}[n];\ m, n \in \mathbb{Z}\}$ generates the category $D_{\mathrm{qcoh}}(X)$.

Let A be any non-zero object in $D_{\mathrm{qcoh}}(X)$. The existence of $\mathscr{L}$ implies that X is separated (Proposition 13.48). Hence by Theorem 22.35, we may represent A by a complex all of whose entries are quasi-coherent $\mathscr{O}_X$-modules. We need to show that there exist m and n and a non-zero morphism $\mathscr{L}^{\otimes -m}[-n] \to A$ in $D_{\mathrm{qcoh}}(X)$. Such a morphism corresponds to a non-zero morphism $\mathscr{O}_X \to A \otimes^L \mathscr{L}^{\otimes m}[n]$, i.e. to a non-zero element in $H^n(X, A \otimes^L \mathscr{L}^{\otimes m})$.

Since $A \neq 0$, there exists n such that the quasi-coherent $\mathscr{O}_X$-module $\mathscr{H}^n := H^n(A)$ does not vanish. Choose $x \in X$ such that $\mathscr{H}^n_x \neq 0$. As $\mathscr{L}$ is ample, we find an integer $p \geq 0$ and $s \in \Gamma(X, \mathscr{L}^{\otimes p})$ such that X_s is an open affine neighborhood of x. As $\mathscr{H}^n$ is quasi-coherent and X_s is affine, $\mathscr{H}^n_x$ is a localization of $\mathscr{H}^n(X_s)$. Hence we can find $h \in \mathscr{H}^n(X_s)$ which is send to a non-zero element in $\mathscr{H}^n_x$. Now $\mathscr{H}^n$ is a quotient of the quasi-coherent module $\operatorname{Ker}(A^n \to A^{n+1})$. As X_s is affine, we find

$$a \in \operatorname{Ker}(A^n \to A^{n+1})(X_s) = \operatorname{Ker}(A^n(X_s) \to A^{n+1}(X_s))$$

that maps to h. By Theorem 7.22 there exist $m', m'' \geq 1$ such that we can extend $h' \otimes s^{\otimes m'}$ to a section $a' \in \Gamma(X, A^n \otimes \mathscr{L}^{\otimes pm'})$ and such that $a'' := a' \otimes s^{m''}$ maps to zero in $\Gamma(X, A^{n+1} \otimes \mathscr{L}^{\otimes p(m'+m'')})$. Then the image of a'' gives a non-zero class in $H^n(X, A \otimes^L \mathscr{L}^{\otimes p(m'+m'')})$.

One has a similar assertion (with a similar proof) if X carries an ample family of line bundles (Exercise 25.3).

Next we will study the compact objects in $D_{\mathrm{qcoh}}(X)$. We will show that if X is a qcqs scheme, then an object E in $D_{\mathrm{qcoh}}(X)$ is compact if and only if it is perfect. We will now show one direction and see the other direction later (Proposition 25.16).

Lemma 25.8. *Let X be a qcqs scheme. Let $E \in D(X)$ be perfect. Then E is a compact object.*

Proof. The main point of the proof is to find a suitable stronger statement that we can check locally (and then eventually reduce to the case $E = \mathscr{O}_X$). Therefore, given a family $(A_i)_{i \in I}$ of objects in $D_{\mathrm{qcoh}}(X)$, we consider the derived Hom sheaf $R\mathscr{H}om(E, \bigoplus_i A_i)$. By the definition of perfect complexes, it is clear that E is an object of $D_{\mathrm{qcoh}}(X)$.

First we remark that the natural morphism

$$(*) \qquad \bigoplus_i R\mathscr{H}om_{\mathscr{O}_X}(E, A_i) \to R\mathscr{H}om_{\mathscr{O}_X}(E, \bigoplus_i A_i)$$

is an isomorphism. Indeed, we can check this locally and hence we may assume that E can be represented by a strictly perfect complex. Then by Corollary 21.141, it suffices to consider the case $E = \mathscr{O}_X$. Then $(*)$ is clearly an isomorphism by (21.18.10).

Moreover, if B is a complex in $D(X)$, then we have

$$(**) \qquad \begin{aligned} \mathrm{Hom}_{D(X)}(E, B) &= H^0(R\mathrm{Hom}(E, B)) \\ &= H^0(R\Gamma R\mathscr{H}om(E, B)) \\ &= H^0(X, R\mathscr{H}om(E, B)), \end{aligned}$$

where the first equality holds by (F.52.6) and for the second step we use Proposition 21.107. With these remarks, we conclude as follows.

$$\begin{aligned} \mathrm{Hom}_{D(X)}(E, \bigoplus_i A_i) &\overset{(**)}{=} H^0(X, R\mathscr{H}om(E, \bigoplus_i A_i)) \\ &\overset{(*)}{=} H^0(X, \bigoplus_i R\mathscr{H}om(E, A_i)) \\ &= \bigoplus_i H^0(X, R\mathscr{H}om(E, A_i)) \\ &\overset{(**)}{=} \bigoplus_i \mathrm{Hom}_{D(X)}(E, A_i), \end{aligned}$$

where we use for the unmarked equality that derived direct image, hence in particular $H^0(X, -)$, commutes with coproducts since X is qcqs (Lemma 22.82). $\qquad\square$

Our goal is to prove that for a qcqs scheme X the triangulated category $D_{\mathrm{qcoh}}(X)$ is generated by the shifts of a single perfect complex (see Theorem 25.11 below), and in particular is compactly generated by Lemma 25.8.

Lemma 25.9. *Let $V = \mathrm{Spec}\, R$ be an affine scheme, $f_1, \ldots, f_r \in R$, and let $j \colon U \hookrightarrow V$ be the inclusion of the open quasi-compact subscheme $U := \bigcup_{i=1}^r D(f_i)$. Denote by K the Koszul complex associated with the family $f_1, \ldots, f_r$. Let $E \in D_{\mathrm{qcoh}}(V)$. Then the natural morphism $E \to Rj_*(E_{|U})$ is an isomorphism if and only if $\mathrm{Hom}(K[n], E) = 0$ for all $n \in \mathbb{Z}$.*

Proof. First note that the final condition, $\operatorname{Hom}_{D(V)}(K[n], E) = \operatorname{Ext}_{\mathscr{O}_V}^{-n}(K, N) = 0$ for all $n \in \mathbb{Z}$, is equivalent to saying that $R\operatorname{Hom}_{\mathscr{O}_V}(K, E) = 0$, see Remark F.217. If $E \cong Rj_*(E_{|U})$, then in fact we have

$$R\operatorname{Hom}_{\mathscr{O}_V}(K, E) \cong R\operatorname{Hom}_{\mathscr{O}_V}(K, Rj_*(E_{|U})) = R\operatorname{Hom}_{\mathscr{O}_U}(K_{|U}, E_{|U}) = 0,$$

since $K_{|U} = 0$ (Remark 19.6).

Now suppose that $R\operatorname{Hom}_{\mathscr{O}_V}(K, E) = 0$. Write $I = (f_1, \ldots, f_r)$. Let us first consider the case $E_{|U} = 0$, so that the cohomology objects of E are I-torsion. By self-duality of the Koszul complex (22.4.6), we get

$$0 = R\operatorname{Hom}_{\mathscr{O}_V}(K, E) \cong R\operatorname{Hom}_{\mathscr{O}_V}(K, R) \otimes_R^L E \cong K[-r] \otimes_R^L E.$$

In the second step, we have used Lemma 21.145 and that K is perfect.

We can write the Koszul complex as $K = \bigotimes_{i=1}^r (R \xrightarrow{f_i} R)$ by Remark 19.5 (3). By induction, this reduces us to proving the following statement:

Let R be a ring, $f \in R$, and let $E \in D_{\mathrm{qcoh}}(\operatorname{Spec} R)$ such that $E_{|D(f)} = 0$ and that $(R \xrightarrow{f} R) \otimes E = 0$. Then $E = 0$.

But $E_{|D(f)} = 0$ means that the cohomology of E is killed by localizing with respect to f. The second condition implies that multiplication by f is an isomorphism on the cohomology of E. It follows that $E = 0$.

Now we come to the general case. Let $Z = V(I)$ be the closed subscheme defined by the ideal I, and denote by $i\colon Z \to V$ the inclusion. Recall the functor $i^!$ defined in Section (21.13), and the distinguished triangle (Proposition 21.62)

$$i_* Ri^! E \to E \to Rj_* j^* E \to$$

Since E is in $D(V)$ (rather than just a complex of abelian sheaves), this is in fact a distinguished triangle in $D(V)$, cf. Exercise 21.7.

Applying $R\operatorname{Hom}_{\mathscr{O}_V}(K, -)$ to this triangle, and applying the implication we have shown already to see that $R\operatorname{Hom}_{\mathscr{O}_V}(K, Rj_* j^* E) = 0$, we obtain

$$R\operatorname{Hom}_{\mathscr{O}_V}(K, i_* Ri^! E) \cong R\operatorname{Hom}_{\mathscr{O}_V}(K, E) = 0.$$

Since $(i_* Ri^! E)_{|U} = 0$, this gives us $i_* Ri^! E = 0$, and looking at the above exact triangle once more, that $E \cong Rj_* j^* E$. □

Corollary 25.10. *Let X be an affine scheme, and let $U \subseteq X$ be a quasi-compact open subscheme. Let $Z = X \setminus U$, and consider the full subcategory*

$$D_Z = \{F \in D_{\mathrm{qcoh}}(X);\ F_{|U} \cong 0\} \ \subseteq D_{\mathrm{qcoh}}(X).$$

There exists a complex $P \in D_Z$ which is perfect as an object of $D_{\mathrm{qcoh}}(X)$ and such that all its shifts generate D_Z, i.e., if F is in D_Z and $\operatorname{Hom}_{D_Z}(P[i], F) = 0$ for all i, then $F = 0$.

Proof. We identify $D_{\mathrm{qcoh}}(X) = D(\Gamma(X, \mathscr{O}_X))$, and correspondingly work with complexes of modules rather than complexes of sheaves. Since U is assumed to be quasi-compact, there is a finite cover $U = \bigcup_{i=1}^r D(f_i)$, $f_i \in R$. We define P as the Koszul complex of the sequence $(f_1, \ldots, f_r)$. It is perfect as an object of $D_{\mathrm{qcoh}}(X)$ and lies in D_Z. It follows from Lemma 25.9 that P together with all its shifts generates D_Z. □

We will also use the following general result about compactly generated triangulated categories.

Theorem 25.11. ([Nee3][O] *Theorem 2.1) Let $\mathcal{D}$ be a compactly generated triangulated category. Let P be a set of compact objects of $\mathcal{D}$ which is stable under shifts, and let $\mathcal{P}$ be the smallest triangulated subcategory of $\mathcal{D}$ that contains P and is closed under taking coproducts.*

Let $\mathcal{T} = \mathcal{D}/\mathcal{P}$ be the localization of $\mathcal{D}$ with respect to the null system $\mathcal{P}$, cf. Section (F.36).

(1) *The triangulated category $\mathcal{P}$ is compactly generated with P as a generating set. An object of $\mathcal{P}$ is compact in $\mathcal{P}$ if and only if it is compact as an object of $\mathcal{D}$.*

   If P is closed under formation of triangles and under passing to direct summands, then P equals the set of compact objects of $\mathcal{P}$.

(2) *If P is a generating set for $\mathcal{D}$, then $\mathcal{P} = \mathcal{D}$.*

(3) *If F is a compact object in $\mathcal{T}$, then there exists a compact object G in $\mathcal{T}$ and a compact object E in $\mathcal{D}$ such that $F \oplus G$ and E are isomorphic in $\mathcal{T}$.*

   Moreover, G may be chosen to be $F[1]$ or any other object such that $[F] + [G] = 0$ in $K_0(\mathcal{T})$ (Definition 23.41).

We omit the proof. Its main part are Lemmas 2.2 and 2.3 in [Nee1][O] which are stated and proved entirely in the context of triangulated categories. Fun fact: The octahedral axiom is used.

Proposition 25.12. *Let X be a qcqs scheme, let $U \subseteq X$ be a quasi-compact open subscheme, and let $Q \in D(U)$ be perfect. Then there exists a perfect object $P' \in D(X)$ such that $Q \oplus Q[1] \cong P'_{|U}$.*

Proof. First assume that $X = \mathrm{Spec}(R)$ is an affine scheme. In this case $D_{\mathrm{qcoh}}(X) = D(R)$ is obviously compactly generated, in fact, every non-zero complex of R-modules admits a non-zero morphism from some $R[d]$.

Let $Z = X \setminus U$ and consider the full subcategory

$$D_Z = \{F \in D_{\mathrm{qcoh}}(X);\ F_{|U} \cong 0\},$$

of $D_{\mathrm{qcoh}}(X)$. By Corollary 25.10, D_Z is generated by objects that are perfect and hence compact (Lemma 25.8) in $D_{\mathrm{qcoh}}(X)$.

Now we apply Theorem 25.11 with $\mathcal{D} = D_{\mathrm{qcoh}}(X)$, and with P a generating set for D_Z consisting of compact objects of $\mathcal{D}$. Then $\mathcal{P} = D_Z$ and D_Z is compactly generated.

Furthermore, in this situation, the category $\mathcal{T}$ of the theorem can be identified with the localization $D_{\mathrm{qcoh}}(X)/D_Z$. In fact, it is clear that the restriction functor $D_{\mathrm{qcoh}}(X) \to D_{\mathrm{qcoh}}(U)$ annihilates all objects of D_Z, so that we obtain a functor $D_{\mathrm{qcoh}}(X)/D_Z \to D_{\mathrm{qcoh}}(U)$ by the universal property of the localization. This functor is an equivalence of categories. This follows formally since j^* has a fully faithful right adjoint (namely Rj_*) by Lemma F.147.

Part (3) of the theorem then implies the statement of the lemma, for X affine.

Finally, we deduce the general case by a kind of gluing argument. Using the assumption that X is quasi-compact and an induction argument, we reduce to the case that $X = U \cup W$ with $U \subset X$ the open subscheme we are given in the statement of the lemma, and W an affine open subscheme. In fact, in general we can write $X = \bigcup_{i=1}^r U_i$ with affine open subschemes U_i, and then "extend" in the way specified in the lemma in steps: from U to $U \cup U_1$, then to $(U \cup U_1) \cup U_2$, and so on.

So assume that $X = U \cup W$ with $W \subseteq X$ affine open, and let $Q \in D(U)$ be perfect. By the case we have already discussed, we find $P_W \in D(W)$ such that $P_{W|U \cap W}$ is isomorphic to $Q_{|U \cap W} \oplus Q[1]_{|U \cap W}$. Write $P_U := Q \oplus Q[1]$. We obtain perfect complexes P_U on U and P_W on W together with an isomorphism $P_{U|U \cap W} \xrightarrow{\sim} P_{W|U \cap W}$. By Proposition 21.49 we can glue P_U and P_W to some object P in $D(X)$ which is then automatically perfect, since this can be checked locally on X. □

Remark 25.13. Let X be a qcqs scheme. Recall from Section (23.12) that to each perfect complex on X we can associate its class in $K_0(X)$. Let $U \subseteq X$ be an open quasi-compact subscheme. If a perfect complex on U can be extended to a perfect complex on X, then its class in $K_0(U)$ obviously must lie in the image of the map $K_0(X) \to K_0(U)$ given by restriction of perfect complex. A theorem by Thomason and Trobaugh ([ThTr]$^\circ$ 5.2.2) states that also the converse holds.

Granting the theorem, we obtain immediately that a complex of the form $Q \oplus Q[1]$ for a perfect $Q \in D(U)$ can always be extended to X, because the class of $Q \oplus Q[1]$ in $K_0(U)$ is trivial.

It is not difficult to give examples where the map $K_0(X) \to K_0(U)$ fails to be surjective. In that case, not every perfect complex on U can be extended to X. See Exercise 25.6. But if X (and hence every open subscheme $U \subseteq X$) is in addition assumed to be regular and noetherian, then $K_0(X) = K_0'(X)$ equals the Grothendieck group of the category of coherent $\mathscr{O}_X$-modules (Proposition 23.55), and since a coherent $\mathscr{O}_U$-module can always be extended to X (Corollary 10.49), in this situation the restriction map $K_0(X) \to K_0(U)$ is surjective.

An analogue of the theorem of Thomason and Trobaugh for vector bundles does not hold: Even if the map $K_0(X) \to K_0(U)$ is an isomorphism, there may exist vector bundles on U which do not extend to a *vector bundle* on X. See Exercise 25.7.

Theorem 25.14. *Let X be a qcqs scheme. Then the category $D_{\mathrm{qcoh}}(X)$ is compactly generated. In fact, there exists a perfect object $P \in D(X)$ such that for all $E \in D_{\mathrm{qcoh}}(X)$:*

$$E = 0 \quad \Longleftrightarrow \quad \forall m \in \mathbb{Z} : \mathrm{Hom}_{D(X)}(P[m], E) = 0.$$

Proof. To construct P as in the statement of the theorem, we proceed by induction on the number of open affine subschemes required to cover X. So fix an affine open cover $X = \bigcup_{i=1}^{n} U_i$. If $n = 1$, i.e., X is affine, then we can take $P = \mathscr{O}_X$.

Now assume that $n > 1$. Let $U = \bigcup_{i=1}^{n-1} U_i$, $V = U_n = \mathrm{Spec}\, A$, so that U and V are qcqs and such that the statement of the theorem holds for U (and for V) by induction hypothesis. Let $Q \in D(U)$ be a perfect object on U such that Q and all its shifts generate $D_{\mathrm{qcoh}}(U)$. Denote by j_U and j_V the inclusions of U and V, respectively, into X. Let P' be a perfect complex on X such that $P'_{|U} \cong Q \oplus Q[1]$ (Proposition 25.12).

Since $U \cap V$ is quasi-compact, X being quasi-separated, we can write it as a finite union $U \cap V = \bigcup_{i=1}^{r} D(f_i)$ of principal subsets of V, $f_i \in A$. Let $Z = V(f_1, \ldots, f_r)$, a closed subscheme of V with underlying topological space $X \setminus U$. Denote by $K \in D(V)$ the Koszul complex of $f_1, \ldots, f_r$, and by ι the closed immersion $Z \hookrightarrow X$.

The complex K has support contained in Z, so $(Rj_{V,*}K)_{|U} = 0$, and $(Rj_{V,*}K)_{|V} = K$. We conclude that $Rj_{V,*}K = R\iota_*(K_{|Z})$ is a *perfect* object of $D(X)$, because this can be checked locally on X. We define a perfect complex

$$P := P' \oplus Rj_{V,*}K.$$

Let us show that P has the desired property. Consider an object $E \in D_{\mathrm{qcoh}}(X)$ with $\mathrm{Hom}_{D(X)}(P[n], E) = 0$ for all n. We need to show that $E = 0$. First note that $Rj_{V,*}K = j_{V,!}K$ (recall that $j_{V,!}$ is exact) since K is supported on Z which is closed in X. By adjunction of the exact functors $j_{V,!}$ and j_V^* (Proposition 21.4), this gives us

$$\mathrm{Hom}(K[n], E_{|V}) = \mathrm{Hom}(Rj_{V,*}K[n], E) = 0.$$

By Lemma 25.9 this implies that $E_{|V} \cong Rj_* E_{|U \cap V}$, where j denotes the inclusion $U \cap V \hookrightarrow V$.

Now consider the natural map $E \to Rj_{U,*}E_{|U}$. Again this is an isomorphism: We can check this locally on X, and restricting to either U (clearly) or to V (by the above) we obtain an isomorphism.

Hence adjunction between $Rj_{U,*}$ and j_U^* yields

$$\mathrm{Hom}_{D(U)}(P'_{|U}[n], E_{|U}) = \mathrm{Hom}_{D(X)}(P'[n], E) = 0,$$

and hence $\mathrm{Hom}_{D(U)}(Q, E_{|U}) = 0$ for all n, since $P'_{|U}$ contains Q as a direct summand. The defining property of Q then gives us $E_{|U} = 0$, whence $E = Rj_{U,*}E_{|U} = 0$, as well. $\square$

Remark 25.15. Let $u \colon X' \to X$ be a quasi-affine morphism of qcqs schemes and let $\mathscr{E} \in D_{\mathrm{qcoh}}(X)$ be a perfect complex such that the $\mathscr{E}[n]$, $n \in \mathbb{Z}$, generate $D_{\mathrm{qcoh}}(X)$. Then the perfect complexes $Lu^*\mathscr{E}[n]$, $n \in \mathbb{Z}$, generate $D_{\mathrm{qcoh}}(X')$.

Indeed, let $\mathscr{F} \in D_{\mathrm{qcoh}}(X')$ be non-zero. Then $Ru_*\mathscr{F} \neq 0$ by Lemma 22.101 and hence there exists an $n \in \mathbb{Z}$ such that

$$\mathrm{Hom}_{D(X')}(Lu^*\mathscr{E}[n], \mathscr{F}) = \mathrm{Hom}_{D(X)}(\mathscr{E}[n], Ru_*\mathscr{F}) \neq 0.$$

Proposition 25.16. *Let X be a qcqs scheme. An object $E \in D_{\mathrm{qcoh}}(X)$ is compact if and only if it is perfect.*

Proof. We have already seen (Lemma 25.8) that every perfect complex is a compact object.

We can reduce the converse to Theorem 25.11 as follows. Let $\mathcal{D} = D_{\mathrm{qcoh}}(X)$ in the theorem, and take for P the set of perfect complexes. We know by Theorem 25.14 that P generates $\mathcal{D}$, and so $\mathcal{P} = \mathcal{D}$ with the notation of Theorem 25.11 by (2) of the theorem.

To conclude, it is now enough by Theorem 25.11 (1) to observe that P is closed under formation of triangles (Proposition 21.175) and under passing to direct summands (Proposition 21.176). $\square$

(25.3) Construction and first properties of $f^\times$.

Putting things together, we now obtain the existence of an adjoint of Rf_*.

Theorem 25.17. *Let $f \colon X \to S$ be a morphism of qcqs schemes. The derived push forward functor $Rf_* \colon D_{\mathrm{qcoh}}(X) \to D_{\mathrm{qcoh}}(S)$ admits a right adjoint triangulated functor $f^\times \colon D_{\mathrm{qcoh}}(S) \to D_{\mathrm{qcoh}}(X)$.*

Proof. It follows from Corollary 25.3, Lemma 22.82 and Theorem 25.14 that Rf_* has a right adjoint functor. It is triangulated since Rf_* is triangulated (Lemma F.129). $\square$

The adjointness between Rf_* and $f^\times$ means that we have isomorphisms

$$(25.3.1) \qquad \operatorname{Hom}_{D(X)}(F, f^\times G) \overset{\sim}{\to} \operatorname{Hom}_{D(S)}(Rf_* F, G),$$

functorially in $F \in D_{\mathrm{qcoh}}(X)$ and $G \in D_{\mathrm{qcoh}}(S)$. In particular, putting $F = f^\times G$, the identity of $f^\times G$ corresponds to a morphism $Rf_* f^\times G \to G$ by adjunction. We obtain the counit of the adjunction, a morphism of functors

$$(25.3.2) \qquad \operatorname{Tr}_f \colon Rf_* f^\times \longrightarrow \mathrm{id},$$

called the *trace map*. It yields the adjunction (25.3.1) by $u \mapsto \operatorname{Tr}_f(G) \circ Rf_* u$.

The first main goal we will set for ourselves is studying $f^\times G$ and the trace map $\operatorname{Tr}_f(G) \colon Rf_* f^\times G \to G$ for all G in $D_{\mathrm{qcoh}}(S)$. Below in Theorem 25.37 we will see that for many proper morphisms $f \colon X \to S$ one has $f^\times G \cong Lf^* G \otimes^L_{\mathscr{O}_X} f^\times \mathscr{O}_S$ and that one can describe Tr_f in terms of of $\operatorname{Tr}_f(\mathscr{O}_S) \colon Rf_* f^\times \mathscr{O}_S \to \mathscr{O}_S$. This reduces our goal to handling the case $G = \mathscr{O}_S$.

Remark 25.18. Consider a morphism $f \colon X \to S$ between qcqs schemes. To obtain a good theory, for instance to ensure that the functor $f^\times$ is compatible with base change one needs to impose further assumptions, for instance that f cohomologically proper (e.g., if S is noetherian and f is proper), see Theorem 25.31 below. Moreover, even if S is the spectrum of a field k and $X = \mathbb{A}^1_k$, $f^\times$ sends coherent modules not to objects in $D_{\mathrm{coh}}(X)$, see Remark 25.45 below. Finally, $f^\times$ cannot be computed locally on the source X. This fails even if $f \colon \mathbb{P}^1_k \to \operatorname{Spec} k$ is the structure map of the projective line over a field k, see Exercise 25.4.

This is one of the reasons why we will later modify it and construct a functor $f^!$ which combines the good properties of $f^\times$ for proper morphisms f and of f^* for étale morphisms f, at least for separated morphisms of finite type between noetherian schemes. See Section (25.11). We will see that this functor has good finiteness properties (Proposition 25.64 and Proposition 25.65) and can be computed locally on source and target (Proposition 25.62).

We start by recording some easy properties of $f^\times$ that follow from properties of Rf_* formally.

Proposition 25.19. *Let* $f \colon X \to Y$, $g \colon Y \to Z$ *be morphisms of qcqs schemes. Then there is a natural isomorphism*

$$f^\times g^\times \cong (gf)^\times$$

of functors $D_{\mathrm{qcoh}}(Z) \to D_{\mathrm{qcoh}}(X)$.

Proof. This follows immediately from the adjunction with derived direct image, and the corresponding property for that (Proposition 21.115). $\qquad\square$

Lemma 25.20. *Consider a cartesian diagram*

$$\begin{array}{ccc} X' & \overset{u'}{\longrightarrow} & X \\ {\scriptstyle f'}\downarrow & & \downarrow{\scriptstyle f} \\ S' & \underset{u}{\longrightarrow} & S \end{array}$$

of qcqs schemes.

(1) *The base change morphism* $Lu^* \circ Rf_* \to Rf'_* \circ L(u')^*$ *(Definition 21.129) induces, by passing to the right adjoints of all the four functors involved, a morphism*

$$(25.3.3) \qquad\qquad R(u')_* \circ (f')^\times \to f^\times \circ Ru_*$$

of functors $D_{\mathrm{qcoh}}(S') \to D_{\mathrm{qcoh}}(X)$.
(2) *If f and u are tor-independent, then (25.3.3) is an isomorphism of functors.*

Proof. Part (1) follows formally from the adjunction. For part (2), we only need to observe that under the assumption that f and u are tor-independent, the base change morphism $Lu^* \circ Rf_* \to Rf'_* \circ L(u')^*$ is an isomorphism of functors $D_{\mathrm{qcoh}}(X) \to D_{\mathrm{qcoh}}(S')$, see Theorem 22.99, so we can apply the previous reasoning to its inverse, as well. $\qquad\square$

(25.4) Variants of the adjunction of Rf_* and $f^\times$ for $R\operatorname{Hom}$ and $R\mathscr{H}om$.

Lemma 25.21. *Let $f\colon X \to S$ be a morphism between qcqs schemes. Consider $E \in D_{\mathrm{qcoh}}(X)$ and $F \in D_{\mathrm{qcoh}}(S)$. The composition*

$$R\operatorname{Hom}_{\mathscr{O}_X}(E, f^\times F) \to R\operatorname{Hom}_{\mathscr{O}_X}(Lf^* Rf_* E, f^\times F)$$
$$\to R\operatorname{Hom}_S(Rf_* E, Rf_* f^\times F) \to R\operatorname{Hom}_S(Rf_* E, F)$$

is an isomorphism.

The individual maps arise from the adjunctions between Lf^* and Rf_*, and between Rf_* and $f^\times$. Even though this method to obtain the resulting morphism may seem a bit roundabout, it is reasonable. Note that we cannot (without further justification) say that "applying a functor" (in this case: Rf_*) gives a map $R\operatorname{Hom}_{\mathscr{O}_X}(E, f^\times F) \to R\operatorname{Hom}_S(Rf_* E, Rf_* f^\times F)$.

Proof. To check that the given morphism is an isomorphism (in $D(S)$), it is enough to check that applying H^i gives an isomorphism for every i. Now

$$H^i(R\operatorname{Hom}_S(A, B)) = \operatorname{Hom}_{D(S)}(A, B[i]),$$

see Remark F.217. Since all of the above is compatible with shifts, we can replace all the $R\operatorname{Hom}$ terms above with usual Hom groups in the respective derived categories. In that case we get back the adjunction map relating Rf_* and $f^\times$. $\qquad\square$

Next we aim for a $R\mathscr{H}om$-version of $R\operatorname{Hom}_{\mathscr{O}_X}(E, f^\times F) \cong R\operatorname{Hom}_S(Rf_* E, F)$, i.e., an isomorphism

$$Rf_* R\mathscr{H}om_X(E, f^\times F) \overset{?}{\cong} R\mathscr{H}om_S(Rf_* E, F).$$

We first prove the following version.

Lemma 25.22. *Let $f\colon X \to S$ be a morphism between qcqs schemes. Let $E, G \in D_{\mathrm{qcoh}}(S)$, $F \in D_{\mathrm{qcoh}}(X)$. Then the composition*

$$\operatorname{Hom}_{D(S)}(G, Rf_* R\mathscr{H}om_X(E, f^\times F)) \to \operatorname{Hom}_{D(S)}(G, R\mathscr{H}om_S(Rf_* E, Rf_* f^\times F))$$
$$\to \operatorname{Hom}_{D(S)}(G, R\mathscr{H}om_S(Rf_* E, F))$$

is an isomorphism.

As in Lemma 25.21 the maps are obtained from the adjunctions between Lf^* and Rf_*, and between Rf_* and $f^\times$.

Proof. Using Remark 21.120 we rewrite the left hand side as

$$\operatorname{Hom}(G, Rf_* R\mathscr{H}om_{\mathscr{O}_X}(E, f^\times F)) = \operatorname{Hom}(Lf^*G, R\mathscr{H}om_{\mathscr{O}_X}(E, f^\times F))$$
$$= \operatorname{Hom}(Lf^*G \otimes^L E, f^\times F)$$
$$= \operatorname{Hom}(Rf_*(Lf^*G \otimes^L E), F),$$

and the right hand side as

$$\operatorname{Hom}(G, R\mathscr{H}om_{\mathscr{O}_S}(Rf_*E, F)) = \operatorname{Hom}(G \otimes^L Rf_*E, F).$$

We can identify these two terms by the derived projection formula Proposition 22.84. (We omit the check that this actually gives rise to the morphism between the two terms that is constructed in the way outlined above.) $\square$

Note that we cannot invoke the Yoneda lemma here to simply omit the $\operatorname{Hom}(G, -)$ in the statement of the lemma and conclude that the respective second entries are isomorphic. The problem is that the $R\mathscr{H}om$ complexes need not lie in $D_{\mathrm{qcoh}}(X)$, even if E and F do. But we have seen in Corollary 22.67 that this is the case if E is pseudo-coherent and F is bounded below. We start by proving a lemma.

Lemma 25.23. *Let $f\colon X \to S$ be a morphism between qcqs schemes. Then there exists an $N \geq 0$ such that $f^\times$ maps objects of $D_{\mathrm{qcoh}}^{\geq n}(S)$ to $D_{\mathrm{qcoh}}^{\geq n-N}(X)$. In particular, $f^\times$ maps $D_{\mathrm{qcoh}}^+(S)$ to $D_{\mathrm{qcoh}}^+(X)$.*

Proof. By Proposition 22.29 there exists $N \geq 0$ such that for all n and all $E \in D_{\mathrm{qcoh}}^{\leq n}(X)$, we have $Rf_*E \in D_{\mathrm{qcoh}}^{\leq n+N}(S)$.

Now let $F \in D_{\mathrm{qcoh}}^{\geq m}(S)$ for some $m \in \mathbb{Z}$. We obtain

$$\operatorname{Hom}_{D(X)}(E, f^\times F) = \operatorname{Hom}_{D(S)}(Rf_*E, F) = 0$$

whenever $Rf_*E \in D_{\mathrm{qcoh}}^{<m}(S)$ (see Proposition F.156), which by the above holds for all $E \in D_{\mathrm{qcoh}}^{<m-N}(X)$. This implies that $f^\times F \in D_{\mathrm{qcoh}}^{\geq m-N}(X)$. $\square$

Reminder 25.24. In the sequel, we will often use the hypothesis that a morphism f of schemes is cohomologically proper which means that Rf_* preserves pseudo-coherent complexes (Definition 23.31). Recall that a morphism f of qcqs schemes is cohomologically proper in each of the following cases (Proposition 23.23, Corollary 23.136).
(1) The morphism f is proper and S is noetherian.
(2) The morphism f is proper, flat, and of finite presentation.
For further examples of cohomologically proper morphisms see Section (23.8).

One has also the following nice criterion by Lipman and Neeman for morphisms to be cohomologically proper.

Proposition 25.25. *Let $f\colon X \to S$ be a morphism of qcqs schemes. Let $\mathscr{E} \in D_{\mathrm{qcoh}}(X)$ be a perfect complex such that the $\mathscr{E}[n]$, $n \in \mathbb{Z}$, generate $D_{\mathrm{qcoh}}(X)$. Then f is cohomologically proper if and only if $Rf_*\mathscr{E}$ is pseudo-coherent.*

We will use Proposition 25.25 only in the proof that cohomologically proper morphisms are stable under tor-independent base change (Lemma 25.30 below) and in the exercises. For our two main examples of cohomologically proper morphisms (Reminder 25.24) this is clear (in the first example, all schemes have to be noetherian) and does not even require the base change to be tor-independent.

For the proof of Proposition 25.25 we refer to [LN] $\overset{\circ}{\times}$ 4.3.2. Here we explain only very briefly its strategy. The condition is clearly necessary. If $Rf_*\mathscr{E}$ is pseudo-coherent, then one shows by general facts on compactly generated triangulated categories that $Rf_*\mathscr{P}$ is pseudo-coherent for every perfect complex $\mathscr{P}$ in $D(X)$. To extend this result to pseudo-coherent complexes one uses a deep result that allows to approximate pseudo-coherent complexes (globally) by perfect complexes. More precisely, for any qcqs scheme X there exists an integer N (only depending on X) such that for every $\mathscr{F} \in D_{\mathrm{qcoh}}(X)$ and for each integer m such that $\mathscr{F}$ is $(m - N)$-pseudo-coherent there exists a perfect complex $\mathscr{P}$ and an m-isomorphism $\mathscr{P} \to \mathscr{F}$.

Proposition 25.26. *Let $f: X \to S$ be a cohomologically proper morphism of qcqs schemes (Reminder 25.24). Let $E \in D_{\mathrm{qcoh}}(X)$ be pseudo-coherent, and let $F \in D_{\mathrm{qcoh}}^+(S)$. Then we have a natural isomorphism*

$$Rf_* R\mathscr{H}om_X(E, f^\times F) \cong R\mathscr{H}om_S(Rf_*E, F).$$

Compare Corollary 25.34 below for a more general statement.

Proof. Under these assumptions, both sides of the claimed isomorphism lie in $D_{\mathrm{qcoh}}(S)$. In fact, the assumption on f implies by definition that Rf_*E is again pseudo-coherent. Therefore the right hand side is in $D_{\mathrm{qcoh}}(S)$ by Corollary 22.67. By Lemma 25.23, $f^\times F$ lies in $D_{\mathrm{qcoh}}^+(X)$ and hence $R\mathscr{H}om_X(E, f^\times F) \in D_{\mathrm{qcoh}}(X)$ again by Corollary 22.67. Then the left hand side lies in $D_{\mathrm{qcoh}}(S)$ by Theorem 22.31.

Now the claim follows from Lemma 25.22 by the Yoneda lemma. $\square$

(25.5) $f^\times$ and base change.

In all of this section, we consider a cartesian diagram

(25.5.1)
$$\begin{array}{ccc} X' & \overset{u'}{\longrightarrow} & X \\ {\scriptstyle f'}\downarrow & & \downarrow{\scriptstyle f} \\ S' & \overset{u}{\longrightarrow} & S \end{array}$$

of qcqs schemes where f and u are tor-independent, e.g., if f or u is flat.

Proposition 25.27. *There is a natural transformation of functors*

(25.5.2)
$$L(u')^* f^\times \to (f')^\times Lu^*.$$

Proof. We construct the desired morphism of functors as the composition

$$L(u')^* \circ f^\times \to L(u')^* \circ f^\times \circ Ru_* \circ Lu^* \overset{\sim}{\to} L(u')^* \circ Ru'_* \circ (f')^\times Lu^* \to (f')^\times Lu^*,$$

where the first and third arrow come from the adjunction between Ru_* and Lu^*, and the second arrow is obtained from the inverse of the isomorphism of Lemma 25.20. $\square$

We want to give criteria when (25.5.2) is an isomorphism. We will prove these criteria only up to some (non-trivial) diagram chase for which we give only references. We need some lemmas.

Lemma 25.28. *Let $f\colon X \to S$ be a morphism between qcqs schemes. Let $\varphi\colon F \to G$ be a morphism in $D_{\mathrm{qcoh}}(X)$. Suppose that for every perfect $E \in D(X)$ the induced map*

$$Rf_* R\,\mathscr{H}om_{\mathscr{O}_X}(E, F) \to Rf_* R\,\mathscr{H}om_{\mathscr{O}_X}(E, G)$$

is an isomorphism. Then φ is an isomorphism.

Proof. Applying $H^0(R\Gamma(S, -))$ we obtain

$$
\begin{aligned}
H^0(R\Gamma(S, Rf_* R\,\mathscr{H}om_{\mathscr{O}_X}(E, F))) &\cong H^0(R\Gamma(X, R\,\mathscr{H}om_{\mathscr{O}_X}(E, F))) \\
&= H^0(R\operatorname{Hom}_{\mathscr{O}_X}(E, F)) \\
&= \operatorname{Hom}_{D(X)}(E, F)
\end{aligned}
$$

by Proposition 21.115, Proposition 21.107 and Remark F.217. Hence by the hypothesis we obtain isomorphisms

$$\operatorname{Hom}_{D(X)}(E, F) \to \operatorname{Hom}_{D(X)}(E, G)$$

for all perfect E, i.e., for all compact objects E of $D_{\mathrm{qcoh}}(X)$ (Proposition 25.16). As $D_{\mathrm{qcoh}}(X)$ is generated by the perfect complexes (Theorem 25.14), one concludes that $F \to G$ is an isomorphism (Lemma 25.4). $\qquad\square$

The second lemma is the following compatibility result.

Lemma 25.29. *Consider a commutative diagram*

$$
\begin{array}{ccccc}
X'' & \xrightarrow{\;v'\;} & X' & \xrightarrow{\;u'\;} & X \\
\downarrow{\scriptstyle f''} & & \downarrow{\scriptstyle f'} & & \downarrow{\scriptstyle f} \\
S'' & \xrightarrow{\;v\;} & S' & \xrightarrow{\;u\;} & S
\end{array}
$$

of morphisms of qcqs schemes, where both squares are cartesian and tor-independent. Then the morphism

$$L(u' \circ v')^* f^\times = L(v')^* L(u')^* f^\times \to L(v')^* (f')^\times Lu^* \to (f'')^\times Lv^* Lu^* = (f'')^\times L(u \circ v)^*$$

induced by the base change morphisms $L(v')^*(f')^\times \to (f'')^\times Lv^*$ and $L(u')^* f^\times \to (f')^\times Lu^*$ for the two squares is equal to the base change morphism for the outer rectangle.

Proof. We omit the proof, see for instance [Sta] OATR. $\qquad\square$

Finally, we need that being cohomologically proper is stable under tor-independent base change.

Lemma 25.30. *If in the tor-independent cartesian diagram (25.5.1) of qcqs schemes the morphism f is cohomologically proper (Reminder 25.24), then f' is cohomologically proper.*

Proof. We first assume that u (and hence u') is quasi-affine. Let $\mathscr{E} \in D_{\mathrm{qcoh}}(X)$ be a perfect complex such that all shifts of $\mathscr{E}$ generate $D_{\mathrm{qcoh}}(X)$ (Theorem 25.14). Then the perfect complexes $L(u')^*\mathscr{E}[n]$, $n \in \mathbb{Z}$, generate $D_{\mathrm{qcoh}}(X')$ (Remark 25.15). Hence it suffices by Proposition 25.25 to show that $Rf'_* L(u')^*\mathscr{E} \cong Lu^* Rf_*\mathscr{E}$ (Theorem 22.99) is pseudo-coherent. But by hypothesis $Rf_*\mathscr{E}$ is pseudo-coherent and hence $Lu^* Rf_*\mathscr{E}$ is pseudo-coherent.

For general u we can now argue as follows. Let $\mathscr{G}$ be a pseudo-coherent complex in $D_{\mathrm{qcoh}}(X')$. To see that $Rf'_*\mathscr{G}$ is pseudo-coherent we can assume that S is affine using that we already have seen that f stays cohomologically proper after base change to some open quasi-compact subscheme. We may also work locally on S' and hence can assume that S' is affine. But then is u is affine. $\qquad\square$

Theorem 25.31. *Consider the tor-independent cartesian diagram (25.5.1) of qcqs schemes and let $F \in D^+_{\mathrm{qcoh}}(S)$ be bounded below. Suppose that u has finite tor-dimension and that f is cohomologically proper (Reminder 25.24). Then the functorial morphism (25.5.2) is an isomorphism in $D_{\mathrm{qcoh}}(X')$*

$$L(u')^* f^\times F \xrightarrow{\sim} (f')^\times Lu^* F.$$

As a first step towards the proof of the theorem, we consider the special cases where u is quasi-affine, e.g., if u an open immersion or an affine morphism.

Lemma 25.32. *Theorem 25.31 holds if u is quasi-affine.*

Proof. Because of the compatibility with composition, Lemma 25.29, it is sufficient to consider the cases of an open immersion and of an affine morphism. To check that $L(u')^* f^\times \to (f')^\times Lu^* F$ is an isomorphism, by Lemma 25.28 and Lemma 22.101 in both cases it is enough to check that for all perfect $E \in D(X)$, the induced map

$$Rf_* R\mathscr{H}om(E, R(u')_*(L(u')^*(f^\times F))) \to Rf_* R\mathscr{H}om(E, R(u')_*(f')^\times(Lu^* F))$$

is an isomorphism. While at first sight things look more complicated now, this is a helpful reformulation because after some manipulations it will allow us to apply Proposition 25.26 and in that way to "get rid of" the occurrences of $f^\times$ and $(f')^\times$.

By Proposition 25.26, the morphism

$$(*) \qquad\qquad Rf_* R\mathscr{H}om_X(E, f^\times F) \cong R\mathscr{H}om_S(Rf_* E, F)$$

is an isomorphism. We apply $Ru_* Lu^*$ to both sides of $(*)$. For the left hand side, we obtain

$$Ru_* Lu^* Rf_* R\mathscr{H}om_X(E, f^\times F) \cong Ru_* Rf'_* L(u')^* R\mathscr{H}om_X(E, f^\times F)$$
$$\cong Rf_* Ru'_* R\mathscr{H}om_{X'}(L(u')^* E, L(u')^*(f^\times F))$$
$$\cong Rf_* R\mathscr{H}om_X(E, R(u')_* L(u')^*(f^\times F))),$$

where we have used Theorem 22.99 in the first step, Proposition 22.70 in the second, and Theorem 21.128 (1) in the final step.

For the right hand side of $(*)$, we get

$$Ru_*Lu^*R\,\mathscr{H}om_S(Rf_*E, F) \cong Ru_*R\,\mathscr{H}om_{S'}(Lu^*Rf_*E, Lu^*F)$$
$$\cong Ru_*R\,\mathscr{H}om_{S'}(Rf'_*L(u')^*E, Lu^*F)$$
$$\cong Ru_*Rf'_*R\,\mathscr{H}om_{X'}(L(u')^*E, (f')^\times L(u')^*F)$$
$$\cong Rf_*Ru'_*R\,\mathscr{H}om_{X'}(L(u')^*E, (f')^\times L(u')^*F)$$
$$\cong Rf_*R\,\mathscr{H}om_X(E, R(u')_*(f')^\times(F_{|S'})).$$

Here we use, in that order, Proposition 22.70 and the fact that Rf_*E is pseudo-coherent, Theorem 22.99, Proposition 25.26 for the pseudo-coherent $L(u')^*E$ on X' and for f' using that f' is cohomologically proper by Lemma 25.30, Proposition 21.115 and Theorem 21.128 (1).

One checks that the resulting map is precisely the map from above that we are interested in. See [Lip2] $\overset{\circ}{\underset{X}{}}$ Section 4.4 for further details. $\qquad\square$

Proof. (of Theorem 25.31) By the lemma, it is enough to reduce the general case to the case that both S and S' are affine schemes. Starting from the general situation, consider affine open subschemes $V \subseteq S$, $V' \subseteq S'$ such that $u(V') \subseteq V$, and the following cartesian diagrams

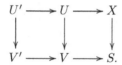

In both diagrams, the horizontal arrows are open immersions, a situation we have discussed in the previous lemma. It follows that the morphism (25.5.2) yields isomorphisms for $F \in D^+_{\mathrm{qcoh}}(S)$ and the left diagram, as well as for Lu^*F (which is in $D^+_{\mathrm{qcoh}}(S')$ since u has finite tor-dimension) and the right diagram.

Now we apply Lemma 25.29 – first to the diagram

$$
\begin{array}{ccccc}
U' & \longrightarrow & U & \longrightarrow & X \\
\downarrow & & \downarrow & & \downarrow \\
V' & \longrightarrow & V & \longrightarrow & S.
\end{array}
$$

where the case of open immersions and the case where S and S' are affine imply that for both squares and hence for the complete rectangle the morphism in question is an isomorphism. Next to the diagram

$$
\begin{array}{ccccc}
U' & \longrightarrow & X' & \longrightarrow & X \\
\downarrow & & \downarrow & & \downarrow \\
V' & \longrightarrow & S' & \longrightarrow & S
\end{array}
$$

where the previous case gives us that the base change morphisms give rise to isomorphisms for the complete rectangle, and where the case of open immersions takes care of the square on the left hand side. By the compatibility of the base change morphisms for the two squares and the rectangle, we see that after pullback to U' the base change morphism of Theorem 25.31 is an isomorphism. Letting V and V' run through affine open subschemes which cover S and S', respectively, the theorem follows, because the case where S' and S are affine is a special case of the above lemma. $\qquad\square$

Remark 25.33. In Theorem 25.31 the assumption that F is bounded below cannot be dropped even if u is an open immersion, see Exercise 25.5.

From the base change theorem, Theorem 25.31, we also get the following variant of Proposition 25.26.

Corollary 25.34. *Let $f\colon X \to S$ be a cohomologically proper morphism of qcqs schemes (Reminder 25.24). Let $E \in D_{\mathrm{qcoh}}(X)$ and let $F \in D_{\mathrm{qcoh}}^+(S)$. Then we have a natural isomorphism*

$$Rf_* R\,\mathcal{H}om_X(E, f^\times F) \cong R\,\mathcal{H}om_S(Rf_* E, F).$$

Proof. To prove the claim, because of Remark 21.19 it is enough to check that the natural morphism $Rf_* R\,\mathcal{H}om_X(E, f^\times F) \to R\,\mathcal{H}om_S(Rf_* E, F)$ induces an isomorphism after applying $R\Gamma_V u^*$ for $u\colon V \to S$ the inclusion of an affine open subscheme. Here we denote by Γ_V the functor $\Gamma(V, -)$.

We now consider the following cartesian diagram,

$$
\begin{array}{ccc}
U & \xrightarrow{\;u'\;} & X \\
{\scriptstyle f'}\big\downarrow & & \big\downarrow{\scriptstyle f} \\
V & \xrightarrow{\;u\;} & S
\end{array}
$$

which is tor-independent as u is flat.

Using, in that order, base change (Theorem 22.99), that $R\,\mathcal{H}om$ is compatible with restriction to open subsets (21.18.9), that $R\Gamma_V Rf'_* = R\Gamma_U$ by Proposition 21.114, Lemma 25.32, and Proposition 21.107 we then obtain

$$
\begin{aligned}
R\Gamma_V u^* Rf_* R\,\mathcal{H}om_X(E, f^\times F) &= R\Gamma_V Rf'_* (u')^* R\,\mathcal{H}om_X(E, f^\times F) \\
&= R\Gamma_V Rf'_* R\,\mathcal{H}om_U((u')^* E, (u')^* f^\times F) \\
&= R\Gamma_U R\,\mathcal{H}om_U((u')^* E, (u')^* f^\times F) \\
&= R\Gamma_U R\,\mathcal{H}om_U((u')^* E, (f')^\times (F_{|V})) \\
&= R\operatorname{Hom}_U(E_{|U}, (f')^\times (F_{|V}))
\end{aligned}
$$

for the left hand side. For the right hand side, we obtain using that $R\,\mathcal{H}om$ is compatible with restriction to open subsets (21.18.9), Theorem 22.99 and Proposition 21.107,

$$
\begin{aligned}
R\Gamma_V u^* R\,\mathcal{H}om_S(Rf_* E, F) &= R\Gamma_V R\,\mathcal{H}om_V(u^* Rf_* E, u^* F) \\
&= R\operatorname{Hom}_V(Rf'_*(E_{|U}), u^* F).
\end{aligned}
$$

One checks that the induced map is the isomorphism of Lemma 25.21 for the morphism $U \to V$ and for $E_{|U}$ and $F_{|V}$. See [Lip2]$^{\text{O}}_{\text{X}}$, Section 4.4, for further details. $\square$

(25.6) Calculation of $f^\times$ in terms of $f^\times \mathcal{O}_S$.

Let $f\colon X \to S$ be a morphism between qcqs schemes. Let $\mathrm{Tr}_f\colon Rf_* f^\times \to \mathrm{id}$ be the counit of the adjunction of Rf_* and $f^\times$. Let $F \in D_{\mathrm{qcoh}}(S)$. Then we claim that we have a natural morphism of triangulated functors in F

(25.6.1) $\chi\colon Lf^* F \otimes^L f^\times \mathcal{O}_S \to f^\times F,$

such that the diagram

$$
\begin{array}{ccc}
F \otimes^L_{\mathscr{O}_S} Rf_* f^\times \mathscr{O}_S & \xrightarrow{\mathrm{id} \otimes \mathrm{Tr}_f(\mathscr{O}_S)} & F \otimes^L_{\mathscr{O}_S} \mathscr{O}_S \\
\cong \downarrow & & \parallel \\
Rf_*(Lf^*F \otimes^L_{\mathscr{O}_X} f^\times \mathscr{O}_S) \xrightarrow{Rf_*\chi} Rf_* f^\times F & \xrightarrow{\mathrm{Tr}_f(F)} & F
\end{array}
$$

commutes. Here the left vertical arrow is given by the derived projection formula (22.84).

If χ is an isomorphism, it essentially reduces the calculation of $f^\times$ and of the trace map $\mathrm{Tr}_f \colon Rf_* f^\times \to \mathrm{id}$ to the calculation of $f^\times \mathscr{O}_S$ and the counit $\mathrm{Tr}_f(\mathscr{O}_S) \colon Rf_* f^\times \mathscr{O}_S \to \mathscr{O}_S$.

To construct the morphisms χ, or in fact more generally morphisms

(25.6.2) $$Lf^*F \otimes^L f^\times F' \longrightarrow f^\times(F \otimes^L F')$$

for $F, F' \in D_{\mathrm{qcoh}}(S)$, note that we have a morphism

$$Rf_*(Lf^*F \otimes^L f^\times F') = F \otimes^L Rf_* f^\times F' \longrightarrow F \otimes^L F',$$

where the first isomorphism is the inverse of the isomorphism given by the derived projection formula (22.84). The morphism $Rf_* f^\times F' \to F'$ comes from the adjunction between Rf_* and $f^\times$. Using the adjunction between Lf^* and Rf_* gives the above morphisms, and in the special case $F' = \mathscr{O}_S$ the morphisms χ. The commutativity of the diagram follows from the definition of χ by formal considerations.

We will see in Theorem 25.37 below that χ is an isomorphism for all F if and only if $f^\times$ commutes with coproducts. Hence let us first give the following criterion.

Proposition 25.35. *Let X and S be qcqs schemes and let $f \colon X \to S$ be a cohomologically proper morphism (Reminder 25.24) of finite tor-dimension. Then the functor $f^\times$ preserves all coproducts.*

Proof. By Lemma 25.5, it is enough to check that Rf_* sends compact objects to compact objects. According to Proposition 25.16 the compact objects in $D_{\mathrm{qcoh}}(X)$ are precisely the perfect complexes. Hence we can apply Proposition 23.32. $\square$

The following lemma shows that χ is an isomorphism in general for perfect complexes.

Lemma 25.36. *Let $f \colon X \to S$ be a morphism between qcqs schemes and let $F \in D_{\mathrm{qcoh}}(S)$ be a perfect complex. Then the morphism (25.6.1) is an isomorphism*

$$\chi \colon Lf^*F \otimes^L f^\times \mathscr{O}_S \xrightarrow{\sim} f^\times(F).$$

Proof. Let $F^\vee = R\mathscr{H}om_S(F, \mathscr{O}_S)$ be the dual of F.

It is enough to show that $\mathrm{Hom}(T, \chi)$ is an isomorphism for all $T \in D_{\mathrm{qcoh}}(X)$. We identify $\mathrm{Hom}(T, f^\times F) = \mathrm{Hom}(Rf_*T, F)$ by adjunction, so that we can view the morphism induced by χ as a morphism $\mathrm{Hom}(T, Lf^*F \otimes^L f^\times \mathscr{O}_S) \to \mathrm{Hom}(Rf_*T, F)$. One checks that this map can be identified with the following chain of isomorphisms:

$$
\begin{aligned}
\mathrm{Hom}(T, Lf^*F \otimes^L f^\times \mathscr{O}_S) &= \mathrm{Hom}(T \otimes^L Lf^*F^\vee, f^\times \mathscr{O}_S) \\
&= \mathrm{Hom}(Rf_*(T \otimes^L Lf^*F^\vee), \mathscr{O}_S) \\
&= \mathrm{Hom}(Rf_*T \otimes^L F^\vee, \mathscr{O}_S) \\
&= \mathrm{Hom}(Rf_*T, F),
\end{aligned}
$$

where we have used (in that order) $H^0 \circ R\Gamma$ of (21.32.3) and $Lf^*(F^\vee) = (Lf^*F)^\vee$ (Lemma 22.70), the adjunction of $f^\times$ and Rf_*, the derived projection formula, and again (21.32.3). $\square$

Theorem 25.37. *Let* $f\colon X \to S$ *be a morphism between qcqs schemes. The morphism* $\chi\colon Lf^*F \otimes^L f^\times \mathscr{O}_S \to f^\times F$ (25.6.1) *is an isomorphism for all* $F \in D_{\mathrm{qcoh}}(S)$ *if and only if* $f^\times$ *preserves all coproducts.*

Proof. Assume that $f^\times$ preserves coproducts. Consider the full subcategory of $D_{\mathrm{qcoh}}(S)$ with objects
$$\{ F \; ; \chi \text{ is an isomorphism for } F\}\}.$$

This is a triangulated subcategory, as one checks using the five lemma since all functors involved are triangulated. It contains all compact objects (which are the perfect complexes) by Lemma 25.36, and it is closed under coproducts since $f^\times$ preserves coproducts. Since $D_{\mathrm{qcoh}}(S)$ is compactly generated (Theorem 25.14), it must coincide with $D_{\mathrm{qcoh}}(S)$, as desired.

Finally, if χ is an isomorphism of functors, then it is easy to see that $f^\times$ preserves coproducts, because this is true for Lf^* and for $- \otimes^L f^\times \mathscr{O}_S$. $\square$

Corollary 25.38. *Let* $f\colon X \to S$ *be a morphism of qcqs schemes. Suppose that one of the following conditions is satisfied.*
(1) S *is noetherian and* f *is proper and of finite tor-dimension.*
(2) f *is proper, flat and of finite presentation.*
Then the morphism (25.6.1) *is an isomorphism*
$$\chi\colon Lf^*F \otimes^L f^\times \mathscr{O}_S \xrightarrow{\sim} f^\times F$$
for all $F \in D_{\mathrm{qcoh}}(S)$.

Proof. Combine Theorem 25.37, Proposition 25.35, and Reminder 25.24. $\square$

In particular, the theorem gives us that for f cohomologically proper and of finite tor-dimension knowing $f^\times \mathscr{O}_S$ gives us all the $f^\times F$, $F \in D_{\mathrm{qcoh}}(S)$. One of our main results will be that for a proper and smooth morphism f of relative dimension n, $f^\times \mathscr{O}_S$ is isomorphic to $\Omega_f^n[n]$, where Ω_f^n denotes the n-th exterior power of the relative sheaf of differentials of the morphism f. See Theorem 25.58.

Remark 25.39. In fact, in [LN]$_X^\circ$ it is shown that the following assertions are equivalent for a morphism $f\colon X \to S$ of qcqs schemes.
(i) $f^\times$ commutes with coproducts.
(ii) f is cohomologically proper and of finite tor-dimension.
(iii) f is cohomologically proper and $f^\times$ is bounded (Section (F.46)).
(iv) χ (25.6.1) is an isomorphism for all $F \in D_{\mathrm{qcoh}}(S)$.
(v) Rf_* takes perfect complexes to perfect complexes.
(vi) If $(\mathscr{E}_i)_{i \in I}$ be a family of perfect complexes in $D_{\mathrm{qcoh}}(X)$ such that the $\mathscr{E}_i[n]$, $i \in I$, $n \in \mathbb{Z}$, generate $D_{\mathrm{qcoh}}(X)$, then $Rf_*\mathscr{E}_i$ is perfect.
If S is noetherian and f is separated and of finite type, then these conditions are moreover equivalent by [Lip2]$_X^\circ$ (4.3.9) to
(vii) f is proper and of finite tor-dimension.
We collcect some examples of morphisms $f\colon X \to S$ satisfying these equivalent conditions in the following list.

(1) f is proper and of finite tor-dimension and S is noetherian (Proposition 23.30).

(2) f is proper and S is noetherian and regular.

Indeed, every perfect complex $\mathscr{E}$ on X is in $D^b_{\mathrm{coh}}(X)$ and hence $Rf_*\mathscr{E} \in D^b_{\mathrm{coh}}(S)$ by Remark 23.24. Hence it is perfect since S is regular (Proposition 23.55).

(3) f is proper, flat, and of finite presentation (Corollary 23.136).

(4) Criterion (vi) shows that if X carries an ample line bundle $\mathscr{L}$, then the equivalent conditions are satisfied if and only if $Rf_*\mathscr{L}^{\otimes i}$ is perfect for all $i \in \mathbb{Z}$ (Example 25.7).

(5) f is quasi-projective and for some f-ample line bundle $\mathscr{L}$, $Rf_*(\mathscr{L}^{\otimes i})$ is perfect for all $i \in \mathbb{Z}$.

Indeed, (v) shows that the above equivalent conditions can be shown locally on S. Hence one is reduced to (4).

(6) f is quasi-affine and $Rf_*\mathscr{O}_X$ is a perfect $\mathscr{O}_S$-module (this is a special case of (5)). If f is affine, then $Rf_*\mathscr{O}_X = f_*\mathscr{O}_X$ and the perfectness of $f_*\mathscr{O}_X$ implies that f is finite and of finite presentation (Proposition 21.153).

Let us give the following non-trivial example for an affine morphism satisfying (6).

Proposition 25.40. *Let S be a scheme and let $f\colon X \to Y$ be an affine morphism of S-schemes such that $f_*\mathscr{O}_X$ is a pseudo-coherent $\mathscr{O}_Y$-module. Suppose that $X \to S$ is flat and that $Y \to S$ is smooth. Then $f_*\mathscr{O}_X$ is a perfect $\mathscr{O}_Y$-module.*

If Y is locally noetherian, the hypothesis that $f_*\mathscr{O}_X$ is pseudo-coherent simply means that f is a finite morphism.

The proof will show that if $Y \to S$ is of relative dimension d, then $f_*\mathscr{O}_X$ is of tor-amplitude contained in $[-d, 0]$.

Proof. The question is local on S and on Y. Hence we may assume that $S = \operatorname{Spec} A$ and $Y = \operatorname{Spec} B$ and that $Y \to S$ is smooth of constant relative dimension d. Then $X = \operatorname{Spec} C$ and C is by hypothesis a pseudo-coherent B-module. As C is flat over A we conclude by the following lemma using that every smooth morphism has regular fibers (Theorem 18.56). $\qquad\square$

Lemma 25.41. *Let $f\colon \operatorname{Spec} B \to \operatorname{Spec} A$ be a flat morphism of schemes and $d \geq 0$ an integer. Suppose that all fibers of f are regular and of dimension $\leq d$. Let $E \in D(B)$ be pseudo-coherent, such that as a complex of A-modules E has tor-amplitude in $[a, b]$ for some $a \leq b \in \mathbb{Z}$.*

Then E is perfect as a complex of B-modules, with tor-amplitude in $[a - d, b]$.

Proof. By Proposition 23.126 (2) it suffices to show that $H^i(E \otimes^L_B \kappa(\mathfrak{q})) = 0$ for every $i \notin [a - d, b]$ and every prime ideal $\mathfrak{q}$ of B. By Proposition 21.162, we can assume that E is a bounded above complex of finite free B-modules. Then E is a bounded above complex of flat A-modules. Hence, if $\mathfrak{p}$ denotes the image of $\mathfrak{q}$ in $\operatorname{Spec} A$, we find

$$
\begin{aligned}
E \otimes^L_B \kappa(\mathfrak{q}) &= E \otimes_B \kappa(\mathfrak{q}) \\
&= E \otimes_A \kappa(\mathfrak{p}) \otimes_{B \otimes_A \kappa(\mathfrak{p})} \kappa(\mathfrak{q}) \\
&= E \otimes_A \kappa(\mathfrak{p}) \otimes^L_{B \otimes_A \kappa(\mathfrak{p})} \kappa(\mathfrak{q})
\end{aligned}
$$

using that $E \otimes_A \kappa(\mathfrak{p})$ is a bounded above complex of finite free $B \otimes_A \kappa(\mathfrak{p})$-modules. Hence it suffices to show that $E \otimes_A \kappa(\mathfrak{p})$ has tor-amplitude in $[a - d, b]$. Replacing A by $\kappa(\mathfrak{p})$, we may assume that A is a field and hence B is regular of dimension $\leq d$.

As $E \in D(B)$ has tor-amplitude in $[a, b]$ over A, we have $H^i(E) = 0$ for $i \notin [a, b]$. Hence it remains to show that if $E \in D^{[a,b]}_{\mathrm{coh}}(B)$ then E has tor-amplitude in $[a - d, b]$. We argue by induction on $b - a$.

If $a = b$, then $E = M[-b]$ for a B-module M of finite type and we conclude by Proposition G.5 and the easy direction of Proposition 21.169. For $b > 0$ we apply the induction hypothesis to the left and the right term of the distinguished triangle

$$\tau^{\leq b-1} E \longrightarrow E \longrightarrow H^b(E)[-b] \overset{+}{\longrightarrow}$$

using Proposition 21.166. $\square$

Computation of $f^\times$ in special cases

Next we describe the functor $f^\times$ first for morphisms between affine schemes and for closed immersions. Then the main goal of this part is a description of $f^\times$ for smooth proper morphisms (Theorem 25.58) and for completely intersecting closed immersions (Theorem 25.57). As a warm up we give a description for projective bundles first in Section (25.9).

(25.7) The functor $f^\times$ for morphisms f between affine schemes.

Let $\varphi \colon A \to B$ be a ring homomorphism. For an A-module M and a B-module N, we obtain a B-module structure on $\mathrm{Hom}_A(N, M)$ by setting $(bf)(n) = f(bn)$, $b \in B$, $n \in N$, $f \in \mathrm{Hom}_A(N, M)$. In this way, we obtain a functor

$$\mathrm{Hom}^B_A(B, -) \colon (A\text{-Mod}) \to (B\text{-Mod}).$$

One checks immediately that this functor is right adjoint to the "push-forward" functor $(B\text{-Mod}) \to (A\text{-Mod})$, i.e., considering a B-module as an A-module via φ.

Since the functor $\mathrm{Hom}^B_A(B, -)$ is left exact, it has a right derived functor.

Definition 25.42. *Let $\varphi \colon A \to B$ be a ring homomorphism. We denote by*

$$R\,\mathrm{Hom}^B_A(B, -) \colon D(A) \to D(B)$$

the right derived functor of the functor Hom^B_A introduced above.

In [Sta] this functor is denoted by $R\,\mathrm{Hom}(B, -)$; we prefer to add both A and B to the notation to distinguish it more visibly from $R\,\mathrm{Hom}_A(B, -)$. As the following lemma shows, the functors $R\,\mathrm{Hom}^B_A(B, -)$ and $R\,\mathrm{Hom}_A(B, -)$ are closely related.

Lemma 25.43.
(1) *The functor $R\,\mathrm{Hom}^B_A(B, -)$ is right adjoint to the restriction $D(B) \to D(A)$.*
(2) *Considered as an object of $D(A)$, $R\,\mathrm{Hom}^B_A(B, Y) = R\,\mathrm{Hom}_A(B, Y)$.*

Proof. The first part follows from the corresponding property of $\operatorname{Hom}^B_A(B,-)$ which passes to the derived functors by Proposition F.45.

Let us prove part (2). For an A-module M, clearly $\operatorname{Hom}^B_A(B,M)$ considered as an A-module is equal to $\operatorname{Hom}_A(B,M)$. For an object $K \in D(A)$, we can compute $R\operatorname{Hom}^B_A(B,K)$ as $\operatorname{Hom}^B_A(B,I^\bullet)$, where $I^\bullet$ is a K-injective complex isomorphic to K in $D(A)$. When we consider everything as A-modules, this computes $R\operatorname{Hom}_A(B,K)$. $\square$

Proposition 25.44. *Let $\varphi\colon A \to B$ be a ring homomorphism, and let $f\colon X = \operatorname{Spec}(B) \to Y = \operatorname{Spec}(A)$ be the corresponding morphism of affine schemes.*

Under the identifications $D_{\mathrm{qcoh}}(X) = D(B)$, $D_{\mathrm{qcoh}}(S) = D(A)$ (Lemma 22.36), the functor $f^\times$ is identified with the functor $R\operatorname{Hom}^B_A(B,-)$.

Proof. Both functors are right adjoint to the functor $Rf_* = f_*$ which is the natural functor $D(B) \to D(A)$ induced by considering B-modules as A-modules via φ, cf. Proposition 22.33. $\square$

Remark 25.45. For general morphisms f of affine schemes the functor $f^\times$ is not very useful. For instance, consider the case of an affine space over a field k, i.e., let $B := k[T_1,\ldots,T_n]$ for $n \geq 1$. Let K be the field of fractions of B. It can be shown ([Nee7] [6]) that $\operatorname{Hom}^B_k(B,k)$ is a huge injective B-module such that $\operatorname{Hom}^B_k(B,k) \otimes_B K$ has K-dimension $|k|^{\aleph_0}$.

The situation becomes better, if $f\colon \operatorname{Spec} B \to \operatorname{Spec} A$ is sufficiently finite. More precisely, Remark 25.39 (6) (see also Exercise 25.8) shows that the following assertions are equivalent (using that we can identify $D(R)$ and $D_{\mathrm{qcoh}}(\operatorname{Spec} R)$ for any ring R by Lemma 22.36).

(i) B is a perfect A-module (i.e., B considered as a complex of A-modules concentrated in degree 0 is perfect).

(ii) The functor $f^\times = R\operatorname{Hom}^B_A(B,-)$ of Proposition 25.44 is naturally isomorphic to the functor

$$D(A) \to D(B), \qquad K \mapsto K \otimes^L_A R\operatorname{Hom}^B_A(B,A)$$

If A is noetherian, then every finite A-algebra B is pseudo-coherent (Corollary 22.62). Hence in this case, B is perfect if and only if B is in addition of finite tor-dimension over A, e.g., if B is a finite flat A-algebra.

(25.8) The functor $i^\times$ for a closed immersion i.

For a morphism $f\colon Z \to X$ between qcqs schemes, the functor f_* in general does not have a right adjoint. In fact, one can show that $f_*\colon (Z\text{-QCoh}) \to (X\text{-QCoh})$ has a right adjoint if and only if f is affine (Exercise 22.11). For a closed immersion $i\colon Z \hookrightarrow X$ (of qcqs schemes), the situation is much better. In this case, even $i_*\colon (\mathscr{O}_Z\text{-Mod}) \to (\mathscr{O}_X\text{-Mod})$ is exact and has a right adjoint functor which can be described explicitly (see Lemma 25.47 below). Note that even if f is a finite morphism, then $f_*\colon (\mathscr{O}_Z\text{-Mod}) \to (\mathscr{O}_X\text{-Mod})$ is in general not exact (Exercise 25.10).

In an affine situation, we have defined the functor Hom^B_A (for a ring homomorphism $A \to B$) in Section (25.7), and we can globalize this construction as follows.

Let $i\colon Z \hookrightarrow X$ be a closed immersion of schemes with defining ideal sheaf $\mathscr{I}$. Consider an $\mathscr{O}_X$-module $\mathscr{F}$. Let $V \subseteq Z$ be open, and let $U \subseteq X$ be open with $U \cap Z = V$. We can identify

$$\mathscr{H}om_{\mathscr{O}_X}(i_*\mathscr{O}_Z,\mathscr{F})(U) = \{s \in \Gamma(U,\mathscr{F});\ \mathscr{I}(U)s = 0\}$$

as $\Gamma(U, \mathscr{O}_X)$-modules, by mapping a sheaf homomorphism $(i_*\mathscr{O}_Z)_{|U} \to \mathscr{F}_{|U}$ to the image of 1 in $\Gamma(U, \mathscr{F})$. Moreover, this set carries a $\Gamma(V, \mathscr{O}_Z)$-module structure and is independent of the choice of U with $U \cap Z = V$ (cf. the similar situation in Section (21.13)).

Thus $\mathscr{H}om_{\mathscr{O}_X}(i_*\mathscr{O}_Z, \mathscr{F})$ is of the form $i_*\mathscr{H}$ for a sheaf $\mathscr{H}(= i^*\,\mathscr{H}om_{\mathscr{O}_X}(i_*\mathscr{O}_Z, \mathscr{F}))$.

Definition 25.46. *Let $i\colon Z \hookrightarrow X$ be a closed immersion of schemes. For an $\mathscr{O}_X$-module $\mathscr{F}$, we denote the $\mathscr{O}_Z$-module $i^*\,\mathscr{H}om_{\mathscr{O}_X}(i_*\mathscr{O}_Z, \mathscr{F})$ by $\mathscr{H}om_{\mathscr{O}_X}^{\mathscr{O}_Z}(\mathscr{O}_Z, \mathscr{F})$. We obtain a left exact functor*

$$\mathscr{H}om_{\mathscr{O}_X}^{\mathscr{O}_Z}(\mathscr{O}_Z, -)\colon (\mathscr{O}_X\text{-Mod}) \to (\mathscr{O}_Z\text{-Mod}).$$

Lemma 25.47. *Let $i\colon Z \hookrightarrow X$ be a closed immersion of schemes. Then the functor $\mathscr{H}om_{\mathscr{O}_X}^{\mathscr{O}_Z}(\mathscr{O}_Z, -)$ is right adjoint to the functor $i_*\colon (\mathscr{O}_Z\text{-Mod}) \to (\mathscr{O}_X\text{-Mod})$.*

Proof. This is a direct consequence of the above description of $\mathscr{H}om_{\mathscr{O}_X}(i_*\mathscr{O}_Z, \mathscr{G})$ in terms of sections of $\mathscr{G}$ annihilated by the ideal sheaf $\mathscr{I}$ of Z. In fact, for an $\mathscr{O}_Z$-module $\mathscr{F}$ and an $\mathscr{O}_X$-module $\mathscr{G}$ a homomorphism $i_*\mathscr{F} \to \mathscr{G}$ is a family of homomorphisms $\mathscr{F}(U \cap Z) \to \mathscr{G}(U)$ for $U \subseteq X$ open, compatible with restrictions, and since all elements in $\mathscr{F}(U \cap Z)$ are annihilated by $\mathscr{I}$, the same automatically holds for their images in $\mathscr{G}(U)$. $\square$

Having a right adjoint of i_* at our disposal, we can pass to derived functors. This gives us a handle on the functor $i^\times$ in this case, however one must be careful because $\mathscr{H}om_{\mathscr{O}_X}^{\mathscr{O}_Z}(\mathscr{O}_Z, -)$ and its right derived functor are functors $(\mathscr{O}_X\text{-Mod}) \to (\mathscr{O}_Z\text{-Mod})$ and $D(X) \to D(Z)$, respectively, whereas the source of $i^\times$ is the full subcategory $D_{\mathrm{qcoh}}(Z)$ of $D(Z)$. Correspondingly, the property of being adjoint to i_* is different. But under further hypotheses, we will be able to identify the two constructions.

Proposition 25.48. *Let X be a qcqs scheme and let $i\colon Z \to X$ be a closed immersion. The functor*

$$R\,\mathscr{H}om_{\mathscr{O}_X}^{\mathscr{O}_Z}(\mathscr{O}_Z, -)\colon D(X) \to D(Z),$$

the right derived functor of the functor $\mathscr{H}om_{\mathscr{O}_X}^{\mathscr{O}_Z}(\mathscr{O}_Z, -)$, is a right adjoint of the functor $Ri_*\colon D(Z) \to D(X)$.

Proof. This follows from adjointness between i_* and $\mathscr{H}om_{\mathscr{O}_X}^{\mathscr{O}_Z}(\mathscr{O}_Z, -)$ by Proposition F.191. $\square$

In the following lemma we determine the push-forward of a complex of the form $R\,\mathscr{H}om_{\mathscr{O}_X}^{\mathscr{O}_Z}(\mathscr{O}_Z, K)$ to X.

Lemma 25.49. *Let X be a qcqs scheme and let $i\colon Z \to X$ be a closed immersion. Let $K \in D(X)$. We have*

$$i_*R\,\mathscr{H}om_{\mathscr{O}_X}^{\mathscr{O}_Z}(\mathscr{O}_Z, K) \cong R\,\mathscr{H}om_{\mathscr{O}_X}(i_*\mathscr{O}_Z, K)$$

Note that since i is a closed immersion, the functor i_* is exact, which allows us to just write i_* instead of Ri_*.

Proof. Let I be a K-injective complex such that there exists a quasi-isomorphism $K \to I$. Then we have $R\,\mathscr{H}om_{\mathscr{O}_X}^{\mathscr{O}_Z}(\mathscr{O}_Z, K) = \mathscr{H}om_{\mathscr{O}_X}^{\mathscr{O}_Z}(\mathscr{O}_Z, I)$. Thus we obtain

$$i_*R\,\mathscr{H}om_{\mathscr{O}_X}^{\mathscr{O}_Z}(\mathscr{O}_Z, I) \cong i_*\,\mathscr{H}om_{\mathscr{O}_X}^{\mathscr{O}_Z}(\mathscr{O}_Z, I) = \mathscr{H}om_{\mathscr{O}_X}(i_*\mathscr{O}_Z, I) = R\,\mathscr{H}om_{\mathscr{O}_X}(i_*\mathscr{O}_Z, K).$$

$\square$

Lemma 25.50. *Let X be a qcqs scheme and let $i\colon Z \to X$ be a closed immersion such that $i_* \mathcal{O}_Z$ is pseudo-coherent.*

Then $R\,\mathscr{H}om_{\mathcal{O}_X}^{\mathcal{O}_Z}(\mathcal{O}_Z, -)$ maps $D_{\mathrm{qcoh}}^+(X)$ to $D_{\mathrm{qcoh}}^+(Z)$. If X is noetherian, it maps $D_{\mathrm{coh}}^+(X)$ to $D_{\mathrm{coh}}^+(Z)$.

Proof. For objects in $D_{\mathrm{qcoh}}^+(X)$ we can find a bounded below injective resolution, whence $R\,\mathscr{H}om_{\mathcal{O}_X}^{\mathcal{O}_Z}(\mathcal{O}_Z, -)$ maps $D_{\mathrm{qcoh}}^+(X)$ to $D^+(Z)$. An $\mathcal{O}_Z$-module $\mathscr{F}$ for which $i_* \mathscr{F}$ is (quasi-)coherent, is (quasi-)coherent, since we can identify $\mathscr{F} = i^* i_* \mathscr{F}$. It is therefore sufficient to check that $i_* H^n(R\,\mathscr{H}om_{\mathcal{O}_X}^{\mathcal{O}_Z}(\mathcal{O}_Z, K))$ is a (quasi-)coherent $\mathcal{O}_X$-module for all n and all K in $D_{\mathrm{qcoh}}^+(X)$, or in $D_{\mathrm{coh}}^+(X)$, respectively.

In view of Lemma 25.49 we have to show that $\mathscr{E}xt^n(i_* \mathcal{O}_Z, K)$ is (quasi-)coherent for all $K \in D_{\mathrm{qcoh}}^+(X)$, or in $D_{\mathrm{coh}}^+(X)$. This follows from Corollary 22.67 since $i_* \mathcal{O}_Z$ is pseudo-coherent by assumption. $\qquad\square$

Remark 25.51. Let $i\colon Z \to X$ be a closed immersions. The hypothesis that $i_* \mathcal{O}_Z$ is pseudo-coherent is for instance satisfied if X is noetherian (Proposition 22.61), or if i is completely intersecting because locally $i_* \mathcal{O}_Z$ is isomorphic in $D(X)$ to a Koszul complex (Example 22.50).

Corollary 25.52. *Let X be a qcqs scheme and let $i\colon Z \to X$ be a closed immersion such that $i_* \mathcal{O}_Z$ is pseudo-coherent (cf. Lemma 25.50). For $K \in D_{\mathrm{qcoh}}^+(X)$, we have an identification*

$$i^\times K \cong R\,\mathscr{H}om_{\mathcal{O}_X}^{\mathcal{O}_Z}(\mathcal{O}_Z, K)$$

which is functorial in K.

Proof. Both $i^\times$ and $R\,\mathscr{H}om_{\mathcal{O}_X}^{\mathcal{O}_Z}(\mathcal{O}_Z, -)$ define functors $D_{\mathrm{qcoh}}^+(X) \to D_{\mathrm{qcoh}}^+(Z)$ which are right adjoint to i_*, therefore they are isomorphic. $\qquad\square$

In an affine situation, say $X = \operatorname{Spec} A$ and $Z = \operatorname{Spec} B$ with A a noetherian ring and $i\colon Z \to X$ given by a surjective ring homomorphism $A \to B$, we can therefore identify (in the sense of the identification $D_{\mathrm{qcoh}}^+(X) = D^+(A)$) $R\,\mathscr{H}om_{\mathcal{O}_X}^{\mathcal{O}_Z}(\mathcal{O}_Z, K)$ and $R\operatorname{Hom}_A^B(B, K)$ for objects $K \in D_{\mathrm{qcoh}}^+(X) = D^+(A)$.

Combining the corollary with Lemma 25.49, we obtain the following description which will later be useful.

Corollary 25.53. *Let X be a qcqs scheme and let $i\colon Z \to X$ be a closed immersion such that $i_* \mathcal{O}_Z$ is pseudo-coherent (cf. Lemma 25.50). Then*

$$i_* i^\times \mathcal{O}_X \cong R\,\mathscr{H}om_{\mathcal{O}_X}(i_* \mathcal{O}_Z, \mathcal{O}_X).$$

We can even say more for completely intersecting closed immersions.

Proposition 25.54. *Let X be a qcqs scheme and let $i\colon Z \to X$ be a completely intersecting closed immersion. Then we have for every K in $D_{\mathrm{qcoh}}(X)$ functorial isomorphisms*

$$i^\times K \cong Li^* K \otimes_{\mathcal{O}_Z}^L i^\times \mathcal{O}_X \cong Li^* K \otimes_{\mathcal{O}_Z}^L R\,\mathscr{H}om_{\mathcal{O}_X}^{\mathcal{O}_Z}(\mathcal{O}_Z, \mathcal{O}_X).$$

Proof. As i is completely intersecting, $i_* \mathscr{O}_Z$ is a perfect $\mathscr{O}_X$-module (Example 22.50), in particular, it is pseudo-coherent. Hence i is cohomologically proper by Proposition 23.34 and i is of finite tor-dimension (Example 23.26 (2)). Hence we can apply Theorem 25.37 by Proposition 25.35 to obtain the first isomorphism. The second isomorphism holds by Corollary 25.52. $\square$

(25.9) The example of projective space.

For $X \to S$ a smooth proper morphism of relative dimension n we set

$$\omega^\bullet := \omega_{X/S}^\bullet := \Omega_{X/S}^n[n]$$

and we will show that $f^\times \mathscr{O}_S \cong \omega^\bullet$ in the next section. Here, we give a direct proof for projective bundles using that we know their cohomology very well by Theorem 22.86.

Proposition 25.55. *Let A be a ring, let $n \geq 1$, and let $X = \mathbb{P}_A^n$. Denote by $f \colon X \to S := \operatorname{Spec} A$ the structure morphism. Then $f^\times \mathscr{O}_S \cong \omega^\bullet$.*

Proof. We have $\Omega_{X/S}^n \cong \mathscr{O}_X(-n-1)$ by Example 17.59. The starting point of the proof is the isomorphism

$$\varphi \colon R^n f_* \mathscr{O}_X(-n-1) = Rf_* \mathscr{O}_X(-n-1)[n] \xrightarrow{\sim} \mathscr{O}_S$$

of Theorem 22.86. By the adjunction between Rf_* and $f^\times$, it gives rise to a morphism

$$\psi \colon \mathscr{O}_X(-n-1)[n] \to f^\times \mathscr{O}_S,$$

and we will show that ψ is an isomorphism.

By Lemma 25.28, it is enough to prove that the induced map

$$Rf_* R\mathscr{H}om_{\mathscr{O}_X}(E, \mathscr{O}_X(-n-1)[n]) \to Rf_* R\mathscr{H}om_{\mathscr{O}_X}(E, f^\times \mathscr{O}_S) = R\mathscr{H}om_{\mathscr{O}_S}(Rf_* E, \mathscr{O}_S)$$

is an isomorphism for all perfect E, where we use Corollary 25.34 for the equality. As X carries an ample line bundle, we know that every perfect complex can be represented by a strictly perfect complex (Corollary 22.59). Hence by Corollary 21.141 it is even enough to prove this in case $E = \mathscr{O}_X$ which becomes the map

$$Rf_* \mathscr{O}_X(-n-1) \longrightarrow (Rf_* \mathscr{O}_X)^\vee,$$

which is an isomorphism since it corresponds to the perfect pairing (22.20.2). $\square$

Corollary 25.56. *Let S be a qcqs scheme, let $\mathscr{E}$ be a locally free $\mathscr{O}_S$-module of rank $n+1$, and let $X = \mathbb{P}(\mathscr{E})$ be the corresponding projective bundle over S. Denote by $f \colon X \to S$ the natural morphism.*

Then there is for every $F \in D_{\mathrm{qcoh}}(S)$ a functorial isomorphism

$$Lf^* F \otimes_{\mathscr{O}_X}^L f^*(\det \mathscr{E})(-n-1)[n] \xrightarrow{\sim} f^\times F,$$

where $\det(\mathscr{E}) := \bigwedge^{n+1} \mathscr{E}$.

Proof. Let $\mathscr{L} = \det(\mathscr{E})$. We then have $\Omega^n_{X/S} \cong f^*(\mathscr{L})(-n-1)$, see Example 17.59. We know (Theorem 22.86) that there is an isomorphism

$$R^n f_* \mathscr{L}(-n-1)[0] \cong Rf_*(f^*\mathscr{L}(-n-1))[n] \xrightarrow{\sim} \mathcal{O}_S.$$

From this morphism, we obtain a morphism $f^*\mathscr{L}(-n-1)[n] \to f^\times \mathcal{O}_S$ by adjunction. To check that it is an isomorphism, we can work locally on S using that the formation of $f^\times \mathcal{O}_S$ is local on S by Theorem 25.31. Then we can apply the previous proposition.

Moreover, f is smooth and proper and we conclude by Corollary 25.38. $\qquad\square$

(25.10) Computation of $f^\times \mathcal{O}_S$ for f smooth and proper.

Our next goal is to compute $f^\times \mathcal{O}_S$ for a smooth proper morphism $f: X \to S$ of qcqs schemes. We start by computing certain Ext sheaves; basically, this result can already be found in [Gro4] [o] (with even a slightly more precise statement) and is sometimes called the fundamental local isomorphism. It will be a crucial ingredient in the proof of Theorem 25.58, but it is also interesting in its own right.

Theorem 25.57. *Let X be a qcqs scheme, and let $i: Z \to X$ be a completely intersecting closed immersion of codimension n (Definition 19.19, Definition 19.23), with corresponding quasi-coherent ideal sheaf $\mathscr{I} \subseteq \mathcal{O}_X$.*
(1) There are natural isomorphisms

$$(25.10.1) \qquad \mathcal{E}xt^n_{\mathcal{O}_X}(i_*\mathcal{O}_Z, \mathscr{F}) \to \mathcal{H}om_{\mathcal{O}_Z}(\textstyle\bigwedge^n_{\mathcal{O}_Z} \mathscr{I}/\mathscr{I}^2, i^*\mathscr{F})$$

for every quasi-coherent $\mathcal{O}_X$-module $\mathscr{F}$, functorial in $\mathscr{F}$.
(2) Let $\mathscr{N} := \mathcal{H}om_Z(i^*\mathscr{I}/\mathscr{I}^2, \mathcal{O}_Z)$ be the normal bundle of i, i.e. $\mathscr{N} = \mathscr{C}_i^\vee$, where $\mathscr{C}_i$ is the conormal bundle of i. There is a natural isomorphism

$$(25.10.2) \qquad i^\times \mathcal{O}_X \cong \textstyle\bigwedge^n \mathscr{N}[-n].$$

More generally, there is for every $\mathscr{F}$ in $D_{\mathrm{qcoh}}(X)$ a natural isomorphism

$$(25.10.3) \qquad i^\times \mathscr{F} \cong Li^*\mathscr{F} \otimes^L_{\mathcal{O}_Z} \textstyle\bigwedge^n \mathscr{N}[-n].$$

Proof. (1). We first construct the isomorphism locally on X. Then we can assume that $X = \operatorname{Spec} A$ is affine and that $Z = \operatorname{Spec} A/I$, where I is an ideal generated by a completely intersecting sequence $\mathbf{f} = (f_1, \dots, f_n)$. The Koszul complex $K(\mathbf{f})$ is a projective resolution of A/I, which we can use to compute for every A-module M

$$\begin{aligned}
\operatorname{Ext}^n_A(A/I, M) &= H^n(R\operatorname{Hom}_A(K(\mathbf{f}), M)) \\
&\cong H^n(K(\mathbf{f})^\vee \otimes^L_A M) \\
&\cong H^n(K(\mathbf{f})[-n] \otimes^L_A M) \\
&= H^0(A/I \otimes^L_A M) \\
&= M/IM = \operatorname{Hom}_{A/I}(A/I, M/IM) \\
&\cong \operatorname{Hom}_{A/I}(\textstyle\bigwedge^n_{A/I} I/I^2, M/IM).
\end{aligned}$$

Here we use (22.4.6) to identify $K(\mathbf{f})^\vee \cong K(\mathbf{f})[-n]$ and in the last isomorphism we use that the images of f_i in I/I^2 form a basis as an A/I-module and that we therefore have an isomorphism $\bigwedge^n I/I^2 \xrightarrow{\sim} A/I$ given by $f_1 \wedge \cdots \wedge f_n \mapsto 1$. Passing to quasi-coherent sheaves on X, using Proposition 22.68, we obtain an isomorphism (25.10.1) in this local case.

To conclude, we need to check that these isomorphisms glue to give a global isomorphism, or in other words, that they are independent of the choice of regular sequence. But if $\mathbf{f} = (f_1, \ldots, f_n)$ and $\mathbf{g} = (g_1, \ldots, g_n)$ are regular sequences defining Z in X, as in the proof of Proposition 19.18, we can write $g_j = \sum_{i=1}^n a_{ij} f_i$ for suitable $a_{ij} \in A$ such that the map $A^n \to A^n$ given by the matrix $\alpha = (a_{ij})_{i,j}$ defines a quasi-isomorphism $\bigwedge \alpha \colon K(\mathbf{f}) \to K(\mathbf{g})$ which is given by $\det(\alpha)$ in degree $-n$. Hence the induced isomorphism $M/IM = H^n(K(\mathbf{f})^\vee \otimes_A^L M) \to H^n(K(\mathbf{g})^\vee \otimes_A^L M) = M/IM$ is multiplication by $\det({}^t\alpha) = \det(\alpha)$.

For the bases $f_1 \wedge \cdots \wedge f_n$ and $g_1 \wedge \cdots \wedge g_n$ of $\bigwedge^n \mathscr{I}/\mathscr{I}$ we likewise have $g_1 \wedge \cdots \wedge g_n = \det(\alpha) f_1 \wedge \cdots \wedge f_n$, and this implies the desired compatibility.

(2). The second isomorphism follows from the first one by Proposition 25.54. From Corollary 25.53, we know that $i_* i^\times \mathscr{O}_X \cong R\mathscr{H}om_{\mathscr{O}_X}(i_* \mathscr{O}_Z, \mathscr{O}_X)$, so it remains to show that

$$\mathscr{E}xt^d_{\mathscr{O}_X}(i_* \mathscr{O}_Z, \mathscr{O}_X) = 0 \text{ for } d \neq n$$

and that the morphism $\mathscr{O}_X \to i_* \mathscr{O}_Z$ induces an isomorphism

$$\mathscr{E}xt^n_{\mathscr{O}_X}(i_* \mathscr{O}_Z, \mathscr{O}_X) \cong \mathscr{E}xt^n_{\mathscr{O}_X}(i_* \mathscr{O}_Z, i_* \mathscr{O}_Z),$$

since we have

$$\mathscr{E}xt^n_{\mathscr{O}_X}(i_* \mathscr{O}_Z, i_* \mathscr{O}_Z) \cong \mathscr{H}om_{\mathscr{O}_Z}(\textstyle\bigwedge^n \mathscr{I}/\mathscr{I}^2, \mathscr{O}_Z) = \textstyle\bigwedge^n \mathscr{N}$$

by (1).

As in part (1), we can work locally on X and compute $R\mathscr{H}om_{\mathscr{O}_X}(\mathscr{O}_Z, \mathscr{O}_X)$ using the Koszul complex K for a suitable system of generators of the ideal sheaf of Z. The self-duality of the Koszul complex (22.4.6) gives us that $R\mathscr{H}om_{\mathscr{O}_X}(K, \mathscr{O}_X) = K[-n]$. Since K is a resolution of $\mathscr{O}_Z$, the only cohomology object of $K[-n]$ is in degree n, and is isomorphic to $\mathscr{O}_Z$, and more precisely the natural morphism $\mathscr{E}xt^n_{\mathscr{O}_X}(\mathscr{O}_Z, \mathscr{O}_X) \to \mathscr{E}xt^n_{\mathscr{O}_X}(\mathscr{O}_Z, \mathscr{O}_Z)$ is an isomorphism. $\qquad\square$

Theorem 25.58. *Let S be a qcqs scheme and let $f \colon X \to S$ be a proper morphism which is smooth of relative dimension $n \geq 0$. Then there is a canonical isomorphism $\vartheta \colon \Omega^n_{X/S}[n] \xrightarrow{\sim} f^\times \mathscr{O}_S$ and for every $\mathscr{F}$ in $D_{\mathrm{qcoh}}(S)$ one has a functorial isomorphism*

$$Lf^* \mathscr{F} \otimes_{\mathscr{O}_X}^L \Omega^n_{X/S}[n] \xrightarrow{\sim} f^\times \mathscr{F}.$$

If $X \to S$ is smooth of relative dimension n, then $\Omega^1_{X/S}$ is locally free of rank n and $\Omega^n_{X/S} = \bigwedge^n \Omega^1_{X/S} = \det(\Omega^1_{X/S})$ is its determinant (Remark 17.58).

For explicit computations of $\Omega^n_{X/S}$, Proposition 19.32 and Corollary 19.33 are then often useful.

The strategy of the following proof originates from Verdier's paper [Ver2][O].

Proof. By Corollary 25.38 it suffices to show the first assertion.

By Corollary 19.31 (3), the diagonal morphism $\Delta \colon X \to X \times_S X$ is completely intersecting of codimension n. By definition of the differentials as conormal sheaf of the diagonal, we have for its normal sheaf $\mathscr{N} \cong (\Omega^1_{X/S})^\vee$, the dual of the conormal sheaf $\mathscr{C}_\Delta$. As $X \to S$ is smooth of relative dimension n, $\Omega^1_{X/S}$ is locally free of rank n (Corollary 18.58). So Theorem 25.57 gives us an isomorphism

(*) $$ (\Omega^n_{X/S}[n])^\vee = \bigwedge\nolimits^n \mathscr{N}[-n] \xrightarrow{\sim} \Delta^\times \mathscr{O}_{X \times_S X}. $$

We denote by π_i the projections $X \times_S X \to X$. We have, using Proposition 25.19 for the first, and Proposition 25.54 for the second step,

$$ \mathscr{O}_X = \Delta^\times \pi_2^\times \mathscr{O}_X $$
$$ = L\Delta^* \pi_2^\times \mathscr{O}_X \otimes^L_{\mathscr{O}_X} \Delta^\times \mathscr{O}_{X \times_S X}. $$

Now we can identify

$$ \pi_2^\times \mathscr{O}_X = \pi_2^\times f^* \mathscr{O}_S = L\pi_1^* f^\times \mathscr{O}_S $$

by Theorem 25.31, and plugging in (*), we obtain

$$ \mathscr{O}_X = L\Delta^* L\pi_1^* f^\times \mathscr{O}_S \otimes^L_{\mathscr{O}_X} (\Omega^n_{X/S}[n])^\vee $$
$$ = f^\times \mathscr{O}_S \otimes^L_{\mathscr{O}_X} (\Omega^n_{X/S}[n])^\vee, $$

where we have used that derived pullback is compatible with composition of morphisms (Proposition 21.112) in the second step. Tensoring both sides with $\Omega^n_{X/S}[n]$ gives us the desired isomorphism. □

The functor $f^!$

The functor $f^\times$ which we constructed for a morphism $f \colon X \to S$ has the serious disadvantage that it is not local on the source X. It behaves well only for cohomologically proper morphisms, for instance for proper morphisms between noetherian schemes. For more general separated morphisms of finite type between noetherian schemes, there is a variant of $f^\times$, usually denoted by $f^!$, that coincides with $f^\times$ for proper f, but coincides with $f^* = Lf^*$ for open immersions f (and more generally for étale morphisms f). It turns out that sacrificing the property of being an adjoint of Rf_* for all f, and gaining the property of being local on the source, leads to an overall much more useful theory. Deligne in the Appendix to [Har1] ^{O} even says that the functor which we here call $f^\times$ does not deserve a symbol of its own.

Also from the point of view of dualizing complexes that we will discuss later in this chapter, it is natural to use f^* when f is an open immersion, and hence to prefer $f^!$ over $f^\times$ for general f.

(25.11) Construction of the functor $f^!$.

The method one uses to define the functor $f^!$ relies on the Nagata compactification theorem (cf. Theorem 12.70):

Theorem 25.59. *Let $f: X \to S$ be a separated morphism of finite type of qcqs schemes. Then there exists a proper S-scheme $\bar{X}$ and an open immersion $X \to \bar{X}$ of S-schemes.*

We call $(j, \bar{X})$ a *compactification* of the S-scheme X. Note that j is necessarily quasi-compact. Replacing $\bar{X}$ by its schematic closure, we even find always a compactification $j: X \to \bar{X}$ that is schematically dominant.

Obviously, for most X a compactification is not unique. To show that the construction of $f^!$ is independent of the choice of compactification, we will use the following lemma.

Lemma 25.60. *Let $f: X \to S$ be a separated morphism of finite type of qcqs schemes, and let $j_1: X \to \bar{X}_1$ and $j_2: X \to \bar{X}_2$ be compactifications of X as in Theorem 25.59, with structure morphisms $g_i: \bar{X}_i \to S$, $i = 1, 2$.*

Then there exists a schematically dominant compactification $j: X \to \bar{X}$ together with proper S-morphisms $\bar{X} \to \bar{X}_i$, $i = 1, 2$, such that the following diagram commutes

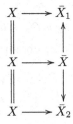

and both squares are cartesian.

Proof. We define $\bar{X}$ as the schematic image of X in $\bar{X}_1 \times_S \bar{X}_2$, and the maps $\bar{X} \to \bar{X}_i$ as the restrictions of the projections, $i = 1, 2$. It is then easy to check that all the properties are satisfied. $\qquad\square$

Using the results concerning compactifications, we can "combine" the functors $g^\times$ for proper g and j^* for open immersions j as follows:

Theorem and Definition 25.61. *Let $f: X \to S$ be a separated morphism of finite type of noetherian schemes. As in Theorem 25.59, let*

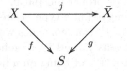

be a commutative diagram with g proper and j an open immersion with schematically dense image.

This diagram gives rise to a well defined functor

$$(25.11.1) \qquad f^!: D_{\mathrm{qcoh}}^+(S) \to D_{\mathrm{qcoh}}^+(X), \qquad f^!(F) := j^* g^\times(F).$$

The functor $f^!$ is independent of the choice of $\bar{X}$ up to canonical isomorphism. It is called the twisted inverse image functor *attached to f.*

The formation of $(\ )^!$ is compatible with composition: Let $f: X \to Y$ and $g: Y \to Z$ be separated morphisms of finite type of noetherian schemes. We then have a canonical isomorphism

(25.11.2) $$\alpha_{g,f}\colon (gf)^! \xrightarrow{\sim} f^! g^!$$

of functors $D^+_{\mathrm{qcoh}}(Z) \to D^+_{\mathrm{qcoh}}(X)$.

We will only sketch the proof and show what its essential ingredients are but refer to [Sta] 0A9Y for many of the (non-trivial) diagram chases. There it is also shown that the isomorphisms $\alpha_{g,f}$ satisfy a "cocycle condition" making $S \mapsto D^+_{\mathrm{qcoh}}(S)$, $f \mapsto f^!$, into a pseudo-functor from the category of noetherian schemes with finite type separated morphisms to the 2-category of triangulated categories in the sense of [Lip2] $\underset{X}{\overset{O}{}}$ (3.6.5) or [Sta] 003N.

To see this, one can alternatively choose a more axiomatized approach due to Deligne for the construction of the functors $f^!$ (see the Appendix to [Har1]^O, and [Lip2] $\underset{X}{\overset{O}{}}$ Section 4.8, in particular Theorem (4.8.1)) which relies on an abstract theorem about "pasting of pseudo-functors". This gives the functors $f^!$ as above, and one obtains even immediately that $f^! = f^*$ for étale morphisms f (which we will prove in Corollary 25.69 below) as part of the existence proof. It takes quite some effort, however, to make everything work.

Proof. [Sketch] First note that $f^!$ indeed induces a functor on bounded below complexes by Lemma 25.23. It remains to check the independence of $f^!$ of the choice of $\bar{X}$. More precisely, one defines for $f\colon X \to S$ the category of compactifications of f: If $X \xrightarrow{j_i} \bar{X}_i \xrightarrow{g_i} S$ for $i = 1, 2$ are compactifications of f, then a map of compactifications is a morphism $h\colon \bar{X}_1 \to \bar{X}_2$ of S-schemes such that $h \circ j_1 = j_2$ and $h^{-1}(j_2(X)) = j_1(X)$ (note that the second condition is automatic, if j_1 is dominant). Using Lemma 25.60 one shows that the category of compactifications is cofiltered and that the subcategory of compactification where $X \to \bar{X}$ is schematically dominant is initial. For each compactification $\bar{X}$, more precisely for each triple $(j, \bar{X}, g)$, one obtains a functor $f^!_{\bar{X}}$ by (25.11.1). Then one shows that for each morphism of compactifications $h\colon \bar{X}_1 \to \bar{X}_2$ one has an isomorphism $\alpha_h\colon f^!_{\bar{X}_1} \xrightarrow{\sim} f^!_{\bar{X}_2}$ of functors such that $\alpha_{k \circ h} = \alpha_k \circ \alpha_h$ for each composable pair (k, h) of morphisms of compactifications of f. Then one defines $f^!$ as the limit of the functors $f^!_{\bar{X}}$ in the category of triangulated functors $D^+_{\mathrm{qcoh}}(S) \to D^+_{\mathrm{qcoh}}(X)$. This ensures that for each compactification $\bar{X}$ one has a unique isomorphism $\alpha_{\bar{X}}\colon f^! \xrightarrow{\sim} f^!_{\bar{X}}$ of triangulated functors such that for any morphism $h\colon \bar{X}_1 \to \bar{X}_2$ of compactifications one has $\alpha_h \circ \alpha_{\bar{X}_1} = \alpha_{\bar{X}_2}$.

The essential idea of the above approach is contained in the following argument. Assume that two compactifications $j_i\colon X \to \bar{X}_i$, $i = 1, 2$, are given, so that we are in the situation of Lemma 25.60. Let $j\colon X \to \bar{X}$ be as in that lemma, together with the morphisms $h_i\colon \bar{X} \to \bar{X}_i$.

We will show that for each i, we have a canonical isomorphism

$$j_i^* \circ g_i^\times \cong j^* \circ g^\times$$

of functors $D^+_{\mathrm{qcoh}}(S) \to D^+_{\mathrm{qcoh}}(X)$. Together, we obtain the desired isomorphism $j_1^* \circ g_1^\times \cong j_2^* \circ g_2^\times$.

To simplify the notation, say $i = 1$ and consider the cartesian square

It is tor-independent since the horizontal morphisms are open immersions. Hence we can apply Theorem 25.31 to this square and obtain an isomorphism $j^* \circ h_1^\times \cong j_1^*$. (In order to use this theorem, we need to restrict to $D_{\text{qcoh}}^+(S)$.) By Proposition 25.19 we have an isomorphism $h_1^\times \circ g_1^\times \cong g^\times$. Combining these two, we obtain the desired isomorphism

$$j_1^* \circ g_1^\times \cong j^* \circ h_1^\times \circ g_1^\times \cong j^* \circ g^\times.$$

Now one has to show that the isomorphism we have produced is independent of the choice of $\bar{X}$ in Lemma 25.60 and that one has an isomorphism as in (25.11.2). We omit this; see [Sta] 0AAO and [Sta] 0ATX for the details. $\qquad\square$

In the proof we applied Theorem 25.31. This is the reason why we have to restrict to bounded below complexes and to noetherian schemes for the construction of $f^!$: We use that proper morphisms between noetherian schemes are cohomologically proper which does not hold for arbitrary proper morphisms of qcqs schemes. In fact, as pointed out by B. Zavyalov in [Zav]$_\times$ 5.5, the construction (and many of the properties of $f^!$ shown below) generalize to separated schemes of finite presentation over qcqs universally coherent schemes (Exercise 22.32) using Proposition 23.37 and that any separated morphism $X \to S$ of finite presentation can be factorized into an quasi-compact schematically dominant open immersion followed by a proper morphism of finite presentation ([CLO]$_\times^0$ 1.2.1).

(25.12) Properties of the functor $f^!$.

We first remark that the functor $f^!$ can be computed locally on the source and the target.

Proposition 25.62. *Let $f\colon X \to S$ be a morphism, and let $U \subseteq X$, $V \subseteq S$ be open subschemes with $f(U) \subseteq V$. Then for all $K \in D_{\text{qcoh}}^+(S)$ we have isomorphisms, functorial in K,*

$$(f^! K)_{|U} \cong (f_{|U})^!(K_{|V}).$$

Below in Corollary 25.67 we will see that the formation of $f^!$ commutes with arbitrary flat base change.

Proof. In fact, denoting by $j_U\colon U \hookrightarrow X$, $j_V\colon V \hookrightarrow S$ the open immersions attached to U and V, we have

$$(f^! K)_{|U} \cong j_U^* f^! K \cong j_U^! f^! K \cong (f_{|U})^! j_V^! K \cong (f_{|U})^!(K_{|V}). \qquad\square$$

Example 25.63. Let S be a noetherian scheme, $n \geq 0$, let $X = \mathbb{A}_S^n$ and let $f\colon X \to S$ be the projection. Then for all $K \in D_{\text{qcoh}}^+(S)$, there are isomorphisms

$$f^! K \cong Lf^*(K)[n] \cong f^*(K)[n],$$

functorial in K.

In fact, the second isomorphism holds since f is flat. For the first isomorphism consider the open embedding $j\colon X = \mathbb{A}_S^n \hookrightarrow \mathbb{P}_S^n$ into the projective line over S. This is a compactification of X. Denote by $p\colon \mathbb{P}_S^1 \to S$ the projection. We obtain

$$f^! K \cong (p^! K)_{|X} \cong (Lp^* K \otimes^L \mathscr{O}_{\mathbb{P}_S^n}(-n-1)[n])_{|X} \cong Lf^* K[1].$$

Here we have the first isomorphism by definition of $f^!$, the second one by Corollary 25.56 since $p^! = p^\times$, and the last one by identifying (non-canonically) $\mathscr{O}_{\mathbb{P}^n_S}(-n-1)_{|X}$ with $\mathscr{O}_X$.

Proposition 25.64. *Let* $f \colon X \to S$ *be a separated morphism of finite type between noetherian schemes. Then* $f^!$ *restricts to a functor*

$$f^! \colon D^+_{\mathrm{coh}}(S) \to D^+_{\mathrm{coh}}(X).$$

Proof. We can compute $f^!$ and check the relevant properties locally on X and S, so we may assume that they are both affine. In this case, we can factor the morphism f as a closed immersion followed by a projection $\mathbb{A}^N_S \to S$. As the formation of $(-)^!$ is compatible with composition, it is enough to consider these two cases separately. For closed immersions, the result follows from Corollary 25.53 and Lemma 25.50.

For the projection $\mathbb{A}^N_S \to S$, we use the computation in Example 25.63. $\qquad\square$

Recall that for every morphism $f \colon X \to S$ of noetherian schemes we have defined a morphism of functors $\chi_f \colon \colon Lf^*(-) \otimes^L f^\times \mathscr{O}_S \to f^\times(-)$ (25.6.1) which is an isomorphism if f is proper and of finite tor-dimension (Theorem 25.37 and Proposition 25.35). We will now prove an analogous result for the functor $f^!$. This reduces the calculation of $f^!$ to the calculation of $f^! \mathscr{O}_S$.

Let $f \colon X \to S$ be a separated morphism of finite type between noetherian schemes. Let $K \in D^+_{\mathrm{qcoh}}(S)$. Choose a compactification $X \xrightarrow{j} \bar{X} \xrightarrow{g} S$ of f. Then we define a morphism of functors $D^+_{\mathrm{qcoh}}(S) \to D^+_{\mathrm{qcoh}}(X)$

$$(25.12.1) \qquad\qquad \mu_f \colon Lf^*(-) \otimes^L_{\mathscr{O}_X} f^! \mathscr{O}_S \longrightarrow f^!(-)$$

by restricting χ_g to X. As before, one shows that μ_f is independent of the choice of the compactification and that the formation of μ_f is compatible with base change, i.e., one has for a morphism $g \colon S \to T$ and $K \in D^+_{\mathrm{qcoh}}(T)$

$$(25.12.2) \qquad f^! \mu_g(K) \circ \mu_f(Lg^*K \otimes^L g^! \mathscr{O}_T) = \mu_{g \circ f}(K) \circ (Lf^* Lg^* K \otimes^L \mu_f(g^! \mathscr{O}_T)).$$

Proposition 25.65. *Let* $f \colon X \to S$ *be a separated morphism of finite type between noetherian schemes. Suppose that* f *is of finite tor-dimension (e.g., if* f *is flat).*
(1) The map of functors μ_f *(25.12.1) is an isomorphism of functors.*
(2) The functor $f^!$ *maps* $D^b_{\mathrm{coh}}(S)$ *into* $D^b_{\mathrm{coh}}(X)$.

Proof. As $f^!$ and Lf^* can be computed locally on S and on X and also the condition to be of finite tor-dimension is local on S and on X (Lemma 21.171), we may assume that $S = \operatorname{Spec} R$ and $X = \operatorname{Spec} A$ is affine. Then we can factorize f into $X \xrightarrow{i} \mathbb{A}^n_R \to \operatorname{Spec} R$, where i is a closed immersion. As $\mathbb{A}^n_R \to R$ is flat, it is of finite tor-dimension. We claim that also i is of finite tor-dimension. For this we factorize i into its graph $\Gamma_i \colon X \to \mathbb{A}^n_X$, which is a regular immersion (Corollary 19.31) and hence of finite tor-dimension (Example 23.26 (2)), followed by $\mathbb{A}^n_X \to \mathbb{A}^n_R$, which is the flat base change of a map of finite tor-dimension and hence itself of finite tor-dimension (Lemma 23.27). Hence it suffices to prove both assertions in the case that f is a closed immersion or that f is the structure map of the affine space.

If $f \colon \mathbb{A}^n_R \to \operatorname{Spec} R$ is the structure map of affine space, both assertions are clear by Example 25.63 and the calculation. Hence it remains to show the case that $f = i$ is a closed immersion of finite tor-dimension of affine schemes. In this case i is proper and hence μ_i is an isomorphism.

To see the second assertion for $f = i$ write $X = \operatorname{Spec} R/I$ for an ideal $I \subseteq R$. Let $K \in D^b_{\mathrm{coh}}(R)$. We know already that $i^! K \in D^+_{\mathrm{coh}}(X)$ (Proposition 25.64). By Proposition 25.44 we have to show that $R\operatorname{Hom}_R(R/I, K)$ cohomologically bounded. But R/I is a pseudo-coherent R-module since R is noetherian and by hypothesis it has bounded tor-amplitude. Hence R/I is perfect and therefore isomorphic in $D(R)$ to a finite complex P of finitely generated projective modules. Hence $R\operatorname{Hom}_R(R/I, K) = \operatorname{Hom}_R(P, K)$ is cohomologically bounded. $\qquad\square$

From the base change theorem for $f^\times$ (Theorem 25.31) we also obtain the following result for $f^!$.

Proposition 25.66. *Consider a cartesian diagram*

$$
\begin{array}{ccc}
X' & \xrightarrow{\;u'\;} & X \\
\downarrow{\scriptstyle f'} & & \downarrow{\scriptstyle f} \\
S' & \xrightarrow{\;u\;} & S
\end{array}
$$

of schemes such that the following conditions are satisfied.
(a) *The schemes S and S' are noetherian.*
(b) *The morphism f is separated and of finite type.*
(c) *There exists a compactification $\bar f \colon \bar X \to S$ of f such that u and $\bar f$ are tor-independent.*
Then there is a natural isomorphism

$$(25.12.3) \qquad\qquad L(u')^* \circ f^! \xrightarrow{\sim} (f')^! \circ Lu^*$$

of functors on $D^+_{\mathrm{qcoh}}(S)$.

For instance, one has a base change isomorphism (25.12.3) for arbitrary morphisms u of noetherian schemes if f admits a flat compactification $\bar f$.

Proof. Let $\bar f \colon \bar X \to S$ be a compactification of X over S such that u and $\bar f$ are tor-independent, and denote by $j \colon X \hookrightarrow \bar X$ the open embedding of X into $\bar X$.

We define $\bar X' = \bar X \times_S S'$. We write $j' \colon X' \hookrightarrow \bar X'$ for the base change of j, $\bar u' \colon \bar X' \to \bar X$ for the base change of u, and $\bar f' \colon \bar X' \to S'$ for the base change of $\bar f$ to S'. For $K \in D^+_{\mathrm{qcoh}}(S)$, we obtain functorial isomorphisms

$$
\begin{aligned}
L(u')^* f^!(K) &\cong L(u')^* Lj^* \bar f^\times(K) \\
&\cong L(j')^* L(\bar u')^* \bar f^\times(K) \\
&\cong L(j')^* (\bar f')^\times Lu^*(K) \\
&\cong (f')^! Lu^*(K)
\end{aligned}
$$

For the third isomorphism, we use Theorem 25.31 for $\bar f$ using that by hypothesis u and $\bar f$ are tor-independent. $\qquad\square$

Corollary 25.67. *Consider a diagram of noetherian schemes as in Proposition 25.66, where f is separated and of finite type and u is flat. Then there is a natural isomorphism*

$$(u')^* \circ f^! \xrightarrow{\sim} (f')^! \circ u^*$$

of functors on $D^+_{\mathrm{qcoh}}(S)$.

(25.13) The functor $f^!$ for smooth morphisms f.

For smooth morphisms f, Theorem 25.58 carries over to $f^!$.

Theorem 25.68. *Let $f: X \to S$ be a separated morphism of noetherian schemes which is smooth of relative dimension n. Then there is a canonical isomorphism $\vartheta: \Omega^n_{X/S}[n] \xrightarrow{\sim} f^! \mathscr{O}_S$ and for every $\mathscr{F} \in D^+_{\mathrm{qcoh}}(S)$ (resp. $\mathscr{F} \in D^b_{\mathrm{coh}}(S)$) one has a functorial isomorphism in $D^+_{\mathrm{qcoh}}(X)$ (resp. in $D^b_{\mathrm{coh}}(X)$)*

$$Lf^* \mathscr{F} \otimes^L_{\mathscr{O}_X} \Omega^n_{X/S}[n] \xrightarrow{\sim} f^! \mathscr{F}.$$

Proof. By Proposition 25.65 it suffices to show the first assertion. Now the argument for Theorem 25.58 carries over to the situation at hand using the following modifications.
(1) The diagonal $\Delta: X \to X \times_S X$ is a closed immersion since $X \to S$ is separated and that therefore $\Delta^! = \Delta^\times$. This allows to replace $(-)^\times$ by $(-)^!$ everywhere in the proof.
(2) Instead of Theorem 25.31 we use Corollary 25.67. $\qquad\square$

Corollary 25.69. *Let $f: X \to S$ be an étale separated morphism between noetherian schemes. Then the functor $f^!$ on $D^+_{\mathrm{qcoh}}(S)$ is naturally isomorphic to the pullback functor $f^* = Lf^*$.*

Below in Section (25.24), we will study the more general case of Cohen-Macaulay morphisms.

Dualizing complexes

After studying the right adjoint $f^\times$ to Rf_* and the variant $f^!$, we will now take a slightly different point of view. Given a noetherian scheme X, an object $\omega^\bullet_X \in D^b_{\mathrm{coh}}(X)$ is called a dualizing complex if it is of finite injective dimension and the functor $R\mathscr{H}om_{\mathscr{O}_X}(-, \omega^\bullet_X)$ yields an equivalence of $D^b_{\mathrm{coh}}(X)$ with itself. For instance, if X is the spectrum of a field k, then $\omega^\bullet_X = k$ (in degree 0) is a dualizing complex.

We will make the connection to the previous results by showing (Proposition 25.17) that for a separated morphism $f: X \to S$ of finite type between noetherian schemes, $f^!$ preserves the property of being a dualizing complex. In particular, we see that every separated scheme of finite type over a field has a dualizing complex.

We will also define the relative version of the dualizing complex $\omega^\bullet_f := f^! \mathscr{O}_S$ for separated morphisms $f: X \to S$ of finite type of noetherian schemes. It "measures the difference" between a dualizing complex of S and of X if f is of finite tor-dimension (Remark 25.102).

(25.14) The dualizing complex.

Now we come to the problem of setting up a suitable duality. It turns out that the following is a reasonable definition that takes care of the requirements outlined above.

Definition 25.70. *Let X be a noetherian scheme. A dualizing complex on X is an object $\omega^\bullet_X \in D^b_{\mathrm{coh}}(X)$ of finite injective dimension, such that the functor $R\mathscr{H}om_{\mathscr{O}_X}(-, \omega^\bullet_X)$ induces an equivalence*

$$D^b_{\mathrm{coh}}(X)^{\mathrm{opp}} \to D^b_{\mathrm{coh}}(X)$$

of categories. We denote this functor by $\mathbb{D}_{\omega^\bullet_X}(-)$ *or simply by* $\mathbb{D}(-)$.

We will see later that a noetherian scheme X need not admit a dualizing complex (see Corollary 25.97), but that it does in many cases – for example if X is of finite type over a regular ring, in particular if X is of finite type over a field, see Section (25.19). Furthermore, while a dualizing complex is not uniquely determined, we will prove quite a strong uniqueness statement below (Proposition 25.84).

The condition of $\omega^\bullet_X$ being of finite injective dimension (see Proposition 22.79 for several characterizations) could be viewed as a technical condition that one would rather get rid of. While this is true to some extent (and see [Nee5] $\overset{\mathrm{o}}{\underset{\mathrm{X}}{}}$ for results without this assumption) it makes life easier in several places, and is a mild condition, especially because we require that $R\mathscr{H}om(-,\omega^\bullet_X)$ preserve the category $D^b_{\mathrm{coh}}(X)$. The *finite injective dimension* condition is crucial in our proof that an object $\omega^\bullet_X \in D^b_{\mathrm{coh}}(X)$ of finite injective dimension is a dualizing complex as soon as the natural morphism $\mathscr{O}_X \to R\mathscr{H}om(\omega^\bullet_X,\omega^\bullet_X)$ is an isomorphism (Proposition 25.76).

Lemma 25.71. *Let X be a noetherian scheme. If $K \in D^b_{\mathrm{coh}}(X)$ has finite injective dimension, then $R\mathscr{H}om_{\mathscr{O}_X}(-,K)$ preserves $D_{\mathrm{coh}}(X)$, maps $D^+_{\mathrm{coh}}(X)$ to $D^-_{\mathrm{coh}}(X)$ and conversely, and preserves $D^b_{\mathrm{coh}}(X)$.*

Proof. Choose a bounded complex I of injective objects, say supported in degrees $[a,b]$ that is isomorphic in $D(X)$ to K. Then I is K-injective and hence $R\mathscr{H}om(F,K) = \mathscr{H}om(F,I)$ for every $F \in D(X)$. Now consider $F \in D_{\mathrm{coh}}(X)$. To show that $\mathscr{E}xt^i_{\mathscr{O}_X}(F,K)$ is coherent, consider for $n \in \mathbb{Z}$ the distinguished triangle

$$\tau^{\leq n}F \to F \to \tau^{\geq n+1}F \to .$$

For n chosen sufficiently small, we get $H^j(R\mathscr{H}om(\tau^{\leq n}F,K)) = H^j(\mathscr{H}om(F,I^\bullet)) = 0$ for $j \geq i$, so that $\mathscr{E}xt^i(F,K) = \mathscr{E}xt^i(\tau^{\geq n+1}F,K)$. Now we can apply Corollary 22.67.

The definition of the $\mathscr{H}om$-complex shows that the functor $R\mathscr{H}om(-,K) = \mathscr{H}om(-,I)$ maps $D^+(X)$ to $D^-(X)$ and conversely. In particular, it preserves $D^b_{\mathrm{coh}}(X)$. □

Further, let us prove that a duality functor $\mathbb{D}$ is necessarily its own quasi-inverse.

Lemma 25.72. *Let X be a noetherian scheme, let $K \in D^b_{\mathrm{coh}}$ have finite injective dimension and write $\mathbb{D}(F) = R\mathscr{H}om_{\mathscr{O}_X}(F,K)$. Set $\mathcal{D} = D^b_{\mathrm{coh}}(X)$.*
(1) *The functor $\mathbb{D} \colon \mathcal{D}^{\mathrm{opp}} \to \mathcal{D}$ is its own right adjoint, now considered as a functor $\mathcal{D} \to \mathcal{D}^{\mathrm{opp}}$.*
(2) *The complex K is a dualizing complex if and only if the natural morphism $F \to \mathbb{D}(\mathbb{D}(F))$ is an isomorphism in $D^b_{\mathrm{coh}}(X)$ for every $F \in D^b_{\mathrm{coh}}(X)$.*

Proof. For (1), the adjunction between $R\mathscr{H}om$ and $\otimes^L$ (Theorem 21.119) implies that

$$\mathrm{Hom}_{\mathcal{D}^{\mathrm{opp}}}(\mathbb{D}G,F) = \mathrm{Hom}_{\mathcal{D}}(F,R\mathscr{H}om_X(G,K)) = \mathrm{Hom}_{\mathcal{D}}(F \otimes^L G,K) = \mathrm{Hom}_{\mathcal{D}}(G,\mathbb{D}F).$$

This says precisely that $R\mathscr{H}om_X(-,K)$ is its own right adjoint. The unit and counit of the adjunction are the same morphism $\mathrm{id} \to \mathbb{D}^2$ of functors.

Part (2) follows formally by Proposition F.22. □

(25.15) Local nature of dualizing complexes.

The notion of dualizing complex is well-adapted to being studied locally as we will show in Proposition 25.80 below. Over affine schemes $\operatorname{Spec} A$ we can relate the notion of dualizing complex on $\operatorname{Spec} A$ to the notion of dualizing complex in the category $D(A)$ which we define as above.

Definition 25.73. *Let A be a noetherian ring. A* dualizing complex *for A is an object $\omega_A^\bullet \in D_{\mathrm{coh}}^b(A)$ of finite injective dimension in $D(A)$, such that the functor*

$$R\operatorname{Hom}_A(-, \omega_A^\bullet) \colon D_{\mathrm{coh}}^b(A)^{\mathrm{opp}} \to D_{\mathrm{coh}}^b(A)$$

is an equivalence of categories.

Remark 25.74. Recall that for an affine noetherian scheme $X = \operatorname{Spec} A$ the functor $M \mapsto \tilde{M}$ yields an equivalence $D_{\mathrm{coh}}^b(A) \xrightarrow{\sim} D_{\mathrm{coh}}^b(X)$ (Corollary 22.43). Moreover one has
(1) $R\operatorname{Hom}_A(M, N)^\sim \cong R\mathcal{H}om_{\mathcal{O}_X}(\tilde{M}, \tilde{N})$ for $M, N \in D_{\mathrm{coh}}^b(A)$ (Proposition 22.68).
(2) $I \in D_{\mathrm{coh}}^b(A)$ is of finite injective dimension if and only if $\tilde{I} \in D_{\mathrm{coh}}^b(X)$ is of finite injective dimension (Proposition 22.78).

This remark immediately implies the following result by the results we already proved for noetherian schemes.

Proposition 25.75. *Let A be a noetherian ring, $X = \operatorname{Spec} A$, and let $K \in D_{\mathrm{coh}}^b(A)$ be in $D(A)$.*
(1) *The complex K is a dualizing complex if and only if $\tilde{K} \in D_{\mathrm{coh}}^b(X)$ is a dualizing complex.*
(2) *Let K be of finite injective dimension. Then the functor $R\operatorname{Hom}_A(-, K)$ maps objects of $D_{\mathrm{coh}}^b(A)$ to $D_{\mathrm{coh}}^b(A)$, the functor $R\operatorname{Hom}_A(-, K)$ is its own right adjoint functor. Furthermore, K is a dualizing complex if and only if for all $F \in D_{\mathrm{coh}}^b(A)$ the natural morphism $F \to R\operatorname{Hom}_A(R\operatorname{Hom}_A(F, K), K)$ is an isomorphism.*

Proposition 25.76. *Let X be a noetherian scheme. An object $\mathscr{D} \in D_{\mathrm{coh}}^b(X)$ of finite injective dimension is a dualizing complex if and only if the natural map $\mathcal{O}_X \to R\mathcal{H}om_{\mathcal{O}_X}(\mathscr{D}, \mathscr{D})$ is an isomorphism.*

Proof. By Lemma 25.72 it is clear that a dualizing complex has the desired property (we can identify $\mathscr{D} = R\mathcal{H}om_X(\mathcal{O}_X, \mathscr{D})$ and then the natural map of the proposition is identified with the morphism obtained by the adjunction between $R\mathcal{H}om_X(-, \mathscr{D})$ and itself).

On the other hand, assume that this map is an isomorphism. We need to show that for any $F \in D_{\mathrm{coh}}^b(X)$, the map

$$F \to R\mathcal{H}om_X(R\mathcal{H}om_X(F, \mathscr{D}), \mathscr{D})$$

is an isomorphism. This can be checked locally, so we may assume that $X = \operatorname{Spec} A$ is affine. Hence by Remark 25.74 it suffices to show the following lemma. $\square$

Lemma 25.77. *Let A be a noetherian ring and let $K \in D_{\mathrm{coh}}^b(A)$ be a complex with finite injective dimension such that the natural morphism $A \to R\operatorname{Hom}_A(K, K)$ is an isomorphism. Then K is a dualizing complex.*

Proof. We identify $R\operatorname{Hom}_A(A,K) = K$. Then the map $A \to R\operatorname{Hom}_A(K,K)$ in the statement of the proposition is the map $A \to R\operatorname{Hom}_A(R\operatorname{Hom}(A,K),K)$ given by the adjunction between $R\operatorname{Hom}_A(-,K)$ and itself.

By Lemma 22.72 (here we use that K has finite injective dimension), for $F \in D^b_{\mathrm{coh}}(A)$ we have a natural isomorphism

$$R\operatorname{Hom}_A(R\operatorname{Hom}(F,K),K) \cong R\operatorname{Hom}_A(K,K) \otimes^L_A F,$$

and therefore the assumption implies that the natural map

$$F \to R\operatorname{Hom}_A(R\operatorname{Hom}(F,K),K)$$

is an isomorphism. By Proposition 25.75, it follows that K is a dualizing complex. $\qquad\square$

Next we want to prove that the property of a complex to be dualizing can be checked locally. We start with the lemma that dualizing complexes are compatible with localization.

Lemma 25.78. *Let A be a noetherian ring, let $S \subset A$ be a multiplicative subset, and let $\omega_A^\bullet$ be a dualizing complex for A. Then $\omega_A^\bullet \otimes_A S^{-1}A$ is a dualizing complex for the localization $S^{-1}(A)$.*

Proof. Obviously $\omega_A^\bullet \otimes_A S^{-1}A$ lies in $D^b_{\mathrm{coh}}(S^{-1}A)$. Since over a noetherian ring, the localization functor maps injective modules to injective modules (Proposition G.23), it has finite injective dimension as an object of $D(S^{-1}A)$.

We can now apply the criterion of Lemma 25.77. It is enough to show that

$$S^{-1}A \to R\operatorname{Hom}_{S^{-1}A}(\omega_A^\bullet \otimes_A S^{-1}A, \omega_A^\bullet \otimes_A S^{-1}A)$$

is an isomorphism. This follows from the analogous fact for A in place of $S^{-1}A$ and Corollary 22.71. $\qquad\square$

Lemma 25.79. *Let A be a noetherian ring. Let $K \in D^b_{\mathrm{coh}}(A)$ a complex that has finite injective dimension. Then the following are equivalent.*
(i) K is a dualizing complex for A,
(ii) for all prime ideals $\mathfrak{p} \subset A$, $K_\mathfrak{p}$ is a dualizing complex for the local ring $A_\mathfrak{p}$,
(iii) for all maximal ideals $\mathfrak{m} \subset A$, $K_\mathfrak{m}$ is a dualizing complex for the local ring $A_\mathfrak{m}$.

Proof. The implication (i) $\Rightarrow$ (ii) follows directly from Lemma 25.78, and (ii) $\Rightarrow$ (iii) is trivial. (iii) $\Rightarrow$ (i) is a consequence of Lemma 25.77 and Corollary 22.71. $\qquad\square$

Proposition 25.80. *Let X be a noetherian scheme and let $\mathscr{D} \in D(X)$. The following are equivalent.*
(i) $\mathscr{D}$ is a dualizing complex on X.
(ii) For every open subscheme $U \subseteq X$, $\mathscr{D}_{|U}$ is a dualizing complex on U.
(iii) There exists an open covering $X = \bigcup_i U_i$ such that for every i, $\mathscr{D}_{|U_i}$ is a dualizing complex on U_i,
(iv) The complex $\mathscr{D}$ lies in $D^b_{\mathrm{coh}}(X)$ and for every $x \in X$, the stalk $\mathscr{D}_x \in D^b_{\mathrm{coh}}(\mathcal{O}_{X,x})$ is a dualizing complex for the local ring $\mathcal{O}_{X,x}$.

Proof. We use Proposition 25.76. The formation of $R\mathcal{H}om$ commutes with restriction to open subsets (21.18.9) and finite injective dimension can be checked locally on a quasi-compact scheme (Proposition 22.79). This shows that (i), (ii), and (iii) are equivalent. To see that (i) and (iv) are equivalent, we may therefore assume that $X = \operatorname{Spec} A$ is affine. By Proposition 25.75 this follows from Lemma 25.79. $\qquad\square$

(25.16) Uniqueness of dualizing complexes.

We next aim to show to what extent a dualizing complex is unique. We will see that a dualizing complex is unique up to tensoring with an invertible object. Let us first explain what we mean by this.

Definition 25.81. *Let X be a locally ringed space. An object $L \in D(X)$ is called invertible, if there exist an open covering $X = \bigcup_{i \in I} U_i$, invertible O_{U_i}-modules $\mathscr{L}_i$ and $n_i \in \mathbb{Z}$, $i \in I$, such that for each i the restriction $L_{|U_i}$ is isomorphic to $\mathscr{L}_i[n_i]$ in $D(U_i)$.*

If L is invertible, then L is a perfect object and $L \cong \bigoplus_{n \in \mathbb{Z}} H^n(L)[-n]$, where we consider the right hand side as a complex with zero differentials. We will use invertible object always in the above sense. The terminology is justified because of the following result which we will not use in the sequel and for which we will not give a full proof.

Proposition 25.82. *Let X be a locally ringed space, and let $L \in D(X)$. The following are equivalent.*
(i) *The object L is invertible in $D(X)$.*
(ii) *There exists an object $M \in D(X)$ such that $L \otimes^L M$ is isomorphic, in $D(X)$, to $\mathscr{O}_X$.*

Proof. (i) $\Rightarrow$ (ii). We set $M = R\mathscr{H}om_{\mathscr{O}_X}(L, \mathscr{O}_X)$. We have a natural evaluation map (21.26.7)
$$L \otimes^L_{\mathscr{O}_X} M \to \mathscr{O}_X$$
As the formation of $R\mathscr{H}om$ commutes with restriction to open subsets (21.18.9), we can check that this is an isomorphism locally on X, and it is clear that this is true on each open $U \subseteq X$ such that $L_{|U}$ is isomorphic in $D(U)$ to the shift of a line bundle on U.

(ii) $\Rightarrow$ (i). We omit the proof of this implication (which is the difficult direction of this proposition). For hints how to prove this if X is a scheme and $L \in D_{\mathrm{qcoh}}(X)$ see Exercise 25.2. For the general case see [Sta] OFPG. $\qquad\square$

Now to prove the desired uniqueness statement for dualizing complexes, we start with the following lemma, which is interesting by itself.

Lemma 25.83. *Let A be a noetherian ring, and let $F\colon D^b_{\mathrm{coh}}(A) \to D^b_{\mathrm{coh}}(A)$ be an A-linear equivalence of triangulated categories. Then $F(A)$ is an invertible object of $D^b_{\mathrm{coh}}(A)$.*

Proof. We first compute $F(\kappa)$, where κ is the residue class field of A at any fixed maximal ideal $\mathfrak{m} \subset A$. Since $\mathrm{Hom}_{D(A)}(\kappa, \kappa) = \mathrm{Hom}_A(\kappa, \kappa) = \kappa$, the A-module structure on this group factors through κ, so the same holds for $\mathrm{Hom}_{D(A)}(F(\kappa), F(\kappa))$ because F is A-linear. In particular, all cohomology groups of $F(\kappa)$ are κ-vector spaces. We also have

(*) $\qquad \mathrm{Hom}_{D(A)}(F(\kappa), F(\kappa)[i]) = \mathrm{Ext}^i_A(\kappa, \kappa) = 0 \qquad$ for $i < 0$.

Let $[a, b]$, $a \le b$, be the smallest interval with $H^i(F(\kappa)) = 0$ for all $i \notin [a, b]$. We then have natural maps $F(\kappa) \to H^b(F(\kappa))[-b]$ and $H^a(F(\kappa))[-a] \to F(\kappa)$ which induce isomorphisms on the cohomology in degrees b and a, respectively. Therefore any non-zero κ-vector space homomorphism $H^b(F(\kappa)) \to H^a(F(\kappa))$ gives rise to a non-zero map
$$F(\kappa) \to H^b(F(\kappa))[-b] \to H^a(F(\kappa))[-b] \to F(\kappa)[a - b].$$
Therefore, we have $a = b$ by (*). Using that $\mathrm{Hom}_{D(A)}(F(\kappa), F(\kappa)) = \kappa$, we see that
$$F(\kappa) = \kappa[-a].$$

We still need to get our hands on $F(A)$. Denote by G a quasi-inverse of F. By the same reasoning as above, and considering the composition of F and G, we find that $G(\kappa) = \kappa[a]$. Let E be a finite A-module which is annihilated by a power of $\mathfrak{m}$. As A is noetherian, this is an $A_{\mathfrak{m}}$-module of finite length. By induction on the length of E, we see that $G(E) = E'[a]$ for some finite A-module E' which is also annihilated by a power of $\mathfrak{m}$. Correspondingly $F(E') = E[-a]$.

Now consider

$$\mathrm{Hom}_{D(A)}(F(A), E[-a+i]) = \mathrm{Hom}_{D(A)}(F(A), F(E')[i]) = \mathrm{Ext}_A^i(A, E') = \begin{cases} E' & i = 0, \\ 0 & i \neq 0. \end{cases}$$

Plugging in $E = E' = \kappa$ and using that the complex $F(A) \otimes \kappa$ splits, this yields that $H^j(F(A)) \otimes_A \kappa$ vanishes for $j \neq a$ and is one-dimensional for $j = a$. Nakayama's lemma implies that $H^j(F(A))_{\mathfrak{m}} = 0$ for $j \neq a$ and that it is generated by one element for $j = a$. We conclude that the localization $F(A)_{\mathfrak{m}}$ is concentrated in degree a, and given there by an $A_{\mathfrak{m}}$-module M that is generated by one element. Now observe that

$$\mathrm{Hom}_{A_{\mathfrak{m}}}(M, M) = \mathrm{Hom}_{D(A)}(F(A), F(A))_{\mathfrak{m}} = \mathrm{Hom}_{D(A)}(A, A)_{\mathfrak{m}} = A_{\mathfrak{m}},$$

whence $M = A_{\mathfrak{m}}$. This shows that $F(A)$ is locally a free A-module of rank 1, up to shift. $\square$

Proposition 25.84. *Let X be a noetherian scheme. Let $\omega_X^{\bullet}$ be a dualizing complex on X, and let $\mathscr{D}$ be any object of $D_{\mathrm{coh}}^b(X)$. The following are equivalent:*
(i) *The complex $\mathscr{D}$ is a dualizing complex on X.*
(ii) *There exists an object $\mathscr{L}$ in $D_{\mathrm{coh}}^b(X)$ such that $\mathscr{D} \cong \omega_X^{\bullet} \otimes^L \mathscr{L}$, where $\mathscr{L} \in D(X)$ is an invertible object.*

Proof. It is clear that neither tensoring a dualizing complex by a line bundle, nor shifting it, destroy the defining properties. In fact, both operations can be "pulled out" of the $R\mathscr{H}om$ (Lemma 21.145). Using Lemma F.222, this also shows that the tensor product again has finite injective dimension. Therefore we have that (ii) implies (i).

In order to show that (i) implies (ii), we write $\mathscr{L} = R\mathscr{H}om(\omega_X^{\bullet}, \mathscr{D})$. First note that $\mathscr{L}$ is an invertible object of $D(X)$: This can be checked locally on X, and on an affine scheme $\mathrm{Spec}\,A$ it follows from Lemma 25.83 applied to the functor $F = R\,\mathrm{Hom}_A(R\,\mathrm{Hom}_A(-, \omega_X^{\bullet}), \mathscr{D})$ (identifying $\omega_X^{\bullet}$ and $\mathscr{D}$ with objects in $D_{\mathrm{coh}}^b(A)$).

We conclude the proof by noting that

$$\mathscr{D} \cong R\mathscr{H}om(\mathscr{O}_X, \mathscr{D})$$
$$\cong R\mathscr{H}om(R\mathscr{H}om(\omega_X^{\bullet}, \omega_X^{\bullet}), \mathscr{D})$$
$$\cong \omega_X^{\bullet} \otimes^L R\mathscr{H}om(\omega_X^{\bullet}, \mathscr{D}) = \omega_X^{\bullet} \otimes^L \mathscr{L},$$

where the second isomorphism holds by Proposition 25.76 and the third follows from Lemma 22.72. $\square$

Corollary 25.85. *Let A be a local noetherian ring which is of finite injective dimension as an A-module. Then the dualizing complexes on A are precisely the complexes of the form $A[n]$ for some $n \in \mathbb{Z}$.*

By definition, such rings A are called *Gorenstein rings* (see Definition G.26), a property that we will study more closely in Section (25.25) below.

Proof. It follows from Proposition 25.76 that $A[n]$ is indeed a dualizing complex and from Proposition 25.84 that then every dualizing complex is of this form. □

(25.17) Dualizing complexes and $f^!$.

For the next proposition, recall the functor $R\operatorname{Hom}_A^B$ which we introduced in Section (25.7).

Proposition 25.86. *Let $A \to B$ be a finite ring homomorphism between noetherian rings, and let $\omega_A^\bullet$ be a dualizing complex for A. Then $R\operatorname{Hom}_A^B(B, \omega_A^\bullet)$ is a dualizing complex for B.*

Proof. Since $R\operatorname{Hom}_A(B, \omega_A^\bullet)$ lies in $D_{\mathrm{coh}}^b(A)$, a fortiori $R\operatorname{Hom}_A^B(B, \omega_A^\bullet)$ lies in $D_{\mathrm{coh}}^b(B)$.

Since $\omega_A^\bullet$ has finite injective dimension, we can find a bounded complex $I^\bullet$ of injective A-modules representing it. Then the complex $\operatorname{Hom}_A(B, I^\bullet)$ is a bounded complex of injective B-modules (for an injective A-module I and any ring homomorphism $A \to A'$, $\operatorname{Hom}_A(A', I)$ is an injective A'-module, as follows directly from the definition) which represents $R\operatorname{Hom}_A^B(B, \omega_A^\bullet)$.

To finish the proof, we use Lemma 25.77. We have, writing $\mathscr{D} = R\operatorname{Hom}_A^B(B, \omega_A^\bullet)$ and using Lemma 25.43:

$$\operatorname{Hom}_{D(B)}(\mathscr{D}, \mathscr{D}) = \operatorname{Hom}_{D(A)}(R\operatorname{Hom}_A(B, \omega_A^\bullet), \omega_A^\bullet) = B$$

and similarly

$$\operatorname{Hom}_{D(B)}(\mathscr{D}, \mathscr{D}[n]) = \operatorname{Hom}_{D(A)}(R\operatorname{Hom}_A(B, \omega_A^\bullet), \omega_A^\bullet[n]) = 0$$

for $n \neq 0$. This gives the desired result. □

By Corollary 25.53, Proposition 25.86 means that if $f\colon \operatorname{Spec} B \to \operatorname{Spec} A$ is a finite morphism of affine noetherian schemes, then $f^!\omega_{\operatorname{Spec} A}^\bullet$ is a dualizing complex on $\operatorname{Spec} B$ for any dualizing complex $\omega_{\operatorname{Spec} A}^\bullet$ on $\operatorname{Spec} A$. From this, we can now easily deduce the stability of dualizing complexes under $f^!$ in full generality.

Proposition 25.87. *Let $f\colon X \to S$ be a separated morphism of finite type between noetherian schemes. Let $\omega_S^\bullet$ be a dualizing complex on S.*
Then $f^!\omega_S^\bullet$ is a dualizing complex on X.

Proof. By Proposition 25.62 and Proposition 25.80 we can work locally on X and S and hence assume that they both are affine. Then we can factor f as a closed immersion followed by the structure morphism of an affine space. As the formation of $(-)^!$ is compatible with composition, it is therefore sufficient to prove the claim for morphisms of the form $\mathbb{A}_S^n \to S$ for an affine scheme $S = \operatorname{Spec} A$ and for closed immersions of affine schemes.

The latter case follows from Proposition 25.86 as remarked above.

We have computed $f^!K$ for $f\colon X = \mathbb{A}_S^n \to S$ in Example 25.63, so in this case we need to show that $f^*\omega_S^\bullet[n] = Lf^*\omega_S^\bullet[n]$ is a dualizing complex on $\mathbb{A}_A^n$. We use the criterion of Proposition 25.76. The claim follows since we can identify

$$R\mathscr{H}om_X(f^*\omega_S^\bullet[n], f^*\omega_S^\bullet[n]) = f^*R\mathscr{H}om_S(\omega_S^\bullet, \omega_S^\bullet)$$

by Proposition 22.70 (2), compatibly with the natural maps $\mathscr{O}_S \to R\mathscr{H}om_S(\omega_S^\bullet, \omega_S^\bullet)$ and $\mathscr{O}_X \to R\mathscr{H}om_X(f^*\omega_S^\bullet[n], f^*\omega_S^\bullet[n])$. □

In view of Theorem 25.57, this proposition enables us to compute a dualizing complex on any regularly embedded closed subscheme of a smooth scheme, such as projective space, over a "nice" base scheme.

Corollary 25.88. *Let S be a noetherian scheme, let $N \in \mathbb{N}$, and let $P \to S$ be a separated, of finite type and smooth of relative dimension N. Let $i \colon X \to P$ be a regular closed immersion of codimension c, let $\mathscr{I}$ be its defining ideal sheaf and $\mathscr{N} := \mathscr{C}_i^\vee = \mathscr{H}om_X(i^*\mathscr{I}/\mathscr{I}^2, \mathscr{O}_X)$ be its normal bundle. Then*

$$f^!\mathscr{O}_S \cong i^\times \Omega_{P/S}^N[N] \cong \left(\bigwedge^c \mathscr{N} \otimes_{\mathscr{O}_X} i^*\Omega_{P/S}^N \right)[N-c]$$

(a line bundle concentrated in degree $c - N$).

If $\mathscr{O}_S$ is a dualizing complex on S (e.g., if S is the spectrum of a field, or more generally if S is finite-dimensional and Gorenstein, see Proposition 25.124 below), then $f^!\mathscr{O}_S$ is a dualizing complex for X.

If $P = \mathbb{P}_S^n$, we have $\Omega_{\mathbb{P}_S^N/S}^N \cong \mathscr{O}_{\mathbb{P}_S^N}(-N-1)$, see Example 17.59.

Proof. The last assertion holds by Proposition 25.87. Let $f \colon X \to S$ and $p \colon \mathbb{P}_S^N \to S$ be the structure morphisms. Then $f^!\mathscr{O}_S = i^!p^!\mathscr{O}_S = i^\times p^!\mathscr{O}_S$. We have $p^!\mathscr{O}_S = \Omega_{P/S}^N[N]$ by Theorem 25.68. Hence we find

$$f^!\mathscr{O}_S \cong i^\times \Omega_{P/S}^N[N]$$
$$\cong i^\times \mathscr{O}_P \otimes_{\mathscr{O}_X}^L i^*\Omega_{P/S}^N[N]$$
$$\cong \bigwedge^c \mathscr{N} \otimes_{\mathscr{O}_X} i^*\Omega_{P/S}^N[N-c],$$

where we use Theorem 25.37 for the second isomorphism and Theorem 25.57 for the third isomorphism. $\qquad\square$

Proposition 25.89. *Let $f \colon X \to S$ be a separated morphism of finite type between noetherian schemes. Let $\omega_S^\bullet$ be a dualizing complex on S. Write $\omega_X^\bullet = f^!\omega_S^\bullet$ (a dualizing complex on X by Proposition 25.87), and denote by $\mathbb{D}_S$, $\mathbb{D}_X$ the corresponding duality functors. Then there are canonical isomorphisms*

$$(25.17.1) \qquad\qquad f^!K \overset{\sim}{\to} \mathbb{D}_X(Lf^*\mathbb{D}_S(K))$$

for every $K \in D_{\mathrm{coh}}^b(S)$ that are functorial in K.

As $\mathbb{D}_X$ is its own inverse, (25.17.1) can also be viewed as an isomorphism

$$\mathbb{D}_X(f^!K) \overset{\sim}{\to} Lf^*\mathbb{D}_S(K).$$

Proof. According to its definition, we can express the functor $f^!$ as a composition $g^\times \circ j^*$ where $j \colon X \to \overline{X}$ is an open immersion and $g \colon \overline{X} \to S$ is proper. Then $g^!\omega_S^\bullet = g^\times \omega_S^\bullet$ is a dualizing complex on $\overline{X}$ and gives rise to a duality functor $\mathbb{D}_{\overline{X}}$ on $D_{\mathrm{coh}}^b(\overline{X})$.

We then have for all $F \in D_{\mathrm{coh}}^b(\overline{X})$ functorial isomorphisms

$$\mathrm{Hom}_{\overline{X}}(F, \mathbb{D}_{\overline{X}}(Lg^*\mathbb{D}_S(K)))$$
$$= \mathrm{Hom}_{\overline{X}}(F \otimes^L Lg^*\mathbb{D}_S(K), g^\times(\omega_S^\bullet))$$
$$= \mathrm{Hom}_{\overline{X}}(Rg_*(F \otimes^L Lg^*\mathbb{D}_S(K)), \omega_S^\bullet)$$
$$= \mathrm{Hom}_{\overline{X}}(Rg_*(F), \mathbb{D}_S\mathbb{D}_S(K))$$
$$= \mathrm{Hom}_{\overline{X}}(Rg_*(F), K)$$
$$= \mathrm{Hom}_{\overline{X}}(F, g^\times K),$$

where in the first and third step we use the adjunction between $\otimes^L$ and $R\mathscr{H}om$ (Remark 21.120), in the second and fifth step we use the adjunction between Rg_* and $g^\times$, and in the fourth step we use that $\mathbb{D}_S$ is its own quasi-inverse.

This gives us, by the Yoneda lemma, an isomorphism

$$\mathbb{D}_{\overline{X}}(Lg^*\mathbb{D}_S(K)) \cong g^\times K.$$

Applying j^*, we obtain an isomorphism

$$f^! K \cong j^* \mathbb{D}_{\overline{X}}(Lg^*\mathbb{D}_S(K)) \cong j^* R\mathscr{H}om(Lg^*\mathbb{D}_S(K), g^\times \omega_S^\bullet)$$
$$\cong R\mathscr{H}om(Lf^*\mathbb{D}_S(K), \omega_X^\bullet) \cong \mathbb{D}_X(Lf^*\mathbb{D}_S(K)),$$

where we have used that in the above situation $R\mathscr{H}om$ behaves well with respect to restriction along an open immersion. $\qquad\square$

(25.18) Dualizing complexes over local rings.

We now study dualizing complexes over a noetherian local ring A with maximal ideal $\mathfrak{m}$ and residue class field κ. We assume that A has a dualizing complex $\omega_A^\bullet$ (see Section (25.19) below for the existence of dualizing complexes). By the uniqueness result for dualizing complexes there is in fact a canonical choice. For this recall that for every dualizing complex $\omega_A^\bullet$ on A, $R\mathrm{Hom}_A^\kappa(\kappa, \omega_A^\bullet)$ is a dualizing complex for κ by Proposition 25.86, hence isomorphic to $\kappa[n]$ for some n by Proposition 25.85).

Definition 25.90. *A dualizing complex $\omega_A^\bullet$ is called* normalized *if $R\mathrm{Hom}_A^\kappa(\kappa, \omega_A^\bullet)$ is concentrated in degree 0.*

By Proposition 25.84, we see that this uniquely determines $\omega_A^\bullet$ (up to isomorphism).

We will use the following derived version of Matlis duality (Proposition G.24). To this end recall the notion of a Matlis module for A (Section (G.9)).

Proposition 25.91. *Let A be a noetherian local ring with maximal ideal $\mathfrak{m}$ and residue class field κ. Suppose that A has a dualizing complex, and denote by $\omega_A^\bullet$ the normalized dualizing complex. Then $R\mathrm{Hom}_A(-, \omega_A^\bullet)$ induces a contravariant equivalence from the category of finite length A-modules with itself which is its own quasi-inverse.*

More precisely, one has $R\mathrm{Hom}_A(A/\mathfrak{m}^n, \omega_A^\bullet) = E_n[0]$ for a finite length A-module E_n, $E := \mathrm{colim}_n E_n$ is a Matlis dual and there is for every A-module M of finite length an isomorphism, functorial in M,

$$R\mathrm{Hom}_A(M, \omega_A^\bullet) \cong \mathrm{Hom}_A(M, E)[0].$$

In other words, the functor $M \mapsto R\mathrm{Hom}_A(M, \omega_A^\bullet)$ restricts for finite length A-modules M to the Matlis duality of Proposition G.24.

Proof. By hypothesis we have $R\operatorname{Hom}_A(\kappa, \omega_A^\bullet) = \kappa[0]$. By induction on the length of a finite length A-module N we find that $R\operatorname{Hom}_A(N, \omega_A^\bullet)$ is a module of finite length sitting in degree 0. We conclude by Proposition G.25. $\qquad\square$

Proposition 25.91 gives a description of $R\operatorname{Hom}_A(M, \omega_A^\bullet)$ for modules of finite length, i.e., for A-modules M such that $\dim \operatorname{Supp} M = 0$. We now extend these results to arbitrary finitely generated A-modules and deduce properties of $\omega_A^\bullet$. For this recall the notion of the depth of a finitely generated A-module M (Definition B.61) and that one always has $\operatorname{depth}_A(M) \le \dim(M) < \infty$ (Corollary G.10), where $\dim(M) := \dim(\operatorname{Supp}(M))$.

Proposition 25.92. *Let A be a noetherian local ring with maximal ideal $\mathfrak{m}$, residue class field κ, and normalized dualizing complex $\omega_A^\bullet$.*
(1) *Let $M \ne 0$ be a finitely generated A-module. Then*
$$\operatorname{Ext}_A^i(M, \omega_A^\bullet) = 0, \qquad \text{for all } i \notin [-\dim(M), -\operatorname{depth}(M)].$$
Furthermore, $\operatorname{Ext}_A^{-\operatorname{depth}(M)}(M, \omega_A^\bullet) \ne 0$ and $\dim(\operatorname{Ext}_A^{-i}(M, \omega_A^\bullet)) \le i$ for all i.
(2) *If $\mathfrak{p} \subset A$ is a minimal prime, then $H^i(\omega_A^\bullet)_{\mathfrak{p}}$ is non-zero if and only if $i = -\dim(A/\mathfrak{p})$.*
(3) *We have*
$$H^i(\omega_A^\bullet) = 0 \qquad \text{for } i \notin [-\dim(A), -\operatorname{depth}(A)]$$
and $H^{-\dim(A)}(\omega_A^\bullet)$ and $H^{-\operatorname{depth}(A)}(\omega_A^\bullet)$ are non-zero.

Proof. (1). We proceed by induction on $d := \dim(M)$. The case $d = 0$ follows from Proposition 25.91.

Now suppose that $d > 0$, i.e., that M has positive-dimensional support.

If the depth of M is > 0, there exists $f \in \mathfrak{m}$ giving rise to a short exact sequence

$$0 \to M \xrightarrow{f \cdot} M \longrightarrow M/fM \to 0$$

and we can proceed by induction, as follows. Consider the long exact sequence of Ext groups attached to this short exact sequence:

$$\cdots \to \operatorname{Ext}_A^i(M/fM, \omega_A^\bullet) \to \operatorname{Ext}_A^i(M, \omega_A^\bullet) \to \operatorname{Ext}_A^i(M, \omega_A^\bullet) \to \operatorname{Ext}_A^{i+1}(M/fM, \omega_A^\bullet) \to \cdots.$$

The dimension of the support of M/fM is $d-1$ and $\operatorname{depth}(M/fM) = \operatorname{depth}(M) - 1$, so by induction we know that $\operatorname{Ext}^i(M/fM, \omega_A^\bullet) = 0$ for $i \notin [1-d, \ldots, 1-\operatorname{depth}(M)]$. It follows that for $i < -d$, and likewise for $i > -\operatorname{depth}(M)$, multiplication by f is surjective on $\operatorname{Ext}_A^i(M, \omega_A^\bullet)$, so by Nakayama's lemma we obtain $\operatorname{Ext}_A^i(M, \omega_A^\bullet) = 0$ for $i \notin [-d, -\operatorname{depth}(M)]$.

On the other hand, for $i = -\operatorname{depth}(M)$, we obtain that $\operatorname{Ext}_A^i(M, \omega_A^\bullet) \ne 0$, because $\operatorname{Ext}_A^{1-\operatorname{depth}(M)}(M/fM, \omega_A^\bullet) \ne 0$.

It remains to show the bound on the dimension of the support of $\operatorname{Ext}_A^{-i}(M, \omega_A^\bullet)$. But we have an injection

$$\operatorname{Ext}_A^{-i}(M, \omega_A^\bullet)/f \operatorname{Ext}_A^{-i}(M, \omega_A^\bullet) \hookrightarrow \operatorname{Ext}_A^{-i+1}(M/fM, \omega_A^\bullet),$$

whence

$$\dim \operatorname{Supp}(\operatorname{Ext}_A^{-i}(M, \omega_A^\bullet)/f \operatorname{Ext}_A^{-i}(M, \omega_A^\bullet)) \le \dim \operatorname{Supp} Ext_A^{-i+1}(M/fM, \omega_A^\bullet) \le i - 1.$$

The claim follows from this, because for any finitely generated A-module N and element $f \in \mathfrak{m}$ we have by Proposition G.8

479

$$\dim(N/fN) \le \dim(N) \le \dim(N/fN) + 1.$$

If, on the other hand, the depth of M is 0, let $N = M[\mathfrak{m}^\infty] \subseteq M$ be the union of all $\mathfrak{m}^r$-torsion submodules of M. Since M is finitely generated, N equals the $\mathfrak{m}^r$-torsion for some sufficiently large r, and hence is a finitely generated $A/\mathfrak{m}^r$-module, so in particular an A-module of finite length. Furthermore M/N has positive depth, and M and M/N have the same support. By the previous arguments, the result holds for N and for M/N. It then follows for M, as well, by using the long exact sequence of Ext groups attached to the short exact sequence $0 \to N \to M \to M/N \to 0$.

(2). Let $\mathfrak{p} \subset A$ be a minimal prime ideal with $\dim A/\mathfrak{p} = \dim A$. Since $A_\mathfrak{p}$ has dimension 0, the ideal $\mathfrak{p}A_\mathfrak{p}$ is nilpotent, say $\mathfrak{p}^n A_\mathfrak{p} = 0$. Let $B = A/\mathfrak{p}^n$. Then $B_\mathfrak{p} = A_\mathfrak{p}$, $\dim B = \dim A/\mathfrak{p}$ and by Proposition 25.86, $\omega_B^\bullet := R\operatorname{Hom}_A^B(B, \omega_A^\bullet)$ is a dualizing complex for the ring B. Moreover, we have

$$(\omega_B^\bullet)_\mathfrak{p} = R\operatorname{Hom}_A^B(B, \omega_A^\bullet) \otimes_A^L A_\mathfrak{p} = R\operatorname{Hom}_{A_\mathfrak{p}}^{B_\mathfrak{p}}(B_\mathfrak{p}, (\omega_A^\bullet)_\mathfrak{p}) = (\omega_A^\bullet)_\mathfrak{p}.$$

The identification in the middle follows from Corollary 22.71, the final one from the equality $B_\mathfrak{p} = A_\mathfrak{p}$.

Hence we may replace A by B and may therefore assume that $\mathfrak{p}$ is the unique minimal prime ideal of A and that $\dim A = \dim A/\mathfrak{p}$. Now part (1), applied to $M = A$, shows that for $-\dim(A) < i \le 0$, we have $\dim(\operatorname{Supp}(H^i(\omega_A^\bullet))) < \dim(A)$, hence $H^i(\omega_A^\bullet)_\mathfrak{p} = H^i((\omega_A^\bullet)_\mathfrak{p}) = 0$. Since $(\omega_A^\bullet)_\mathfrak{p}$ is a dualizing complex for $A_\mathfrak{p}$ by Lemma 25.78, it must have at least one non-zero cohomology object. In view of part (1), we conclude that $H^{-\dim(A)}(\omega_A^\bullet)_\mathfrak{p} = H^{-\dim(A)}((\omega_A^\bullet)_\mathfrak{p}) \ne 0$.

(3). We can apply part (1) for $M = A$ and obtain that $\omega_A^\bullet$ is supported in cohomological degrees contained in $[-\dim(A), -\operatorname{depth}(A)]$ and that $H^{-\operatorname{depth}(A)}(\omega_A^\bullet) \ne 0$. The other inequality follows from (2) as among the finitely many minimal prime ideals there exists a minimal prime ideal $\mathfrak{p}$ with $\dim(A/\mathfrak{p}) = \dim(A)$. $\square$

(25.19) Existence of dualizing complexes.

In general, a noetherian scheme X need not have a dualizing complex. One necessary condition is that X is catenary, as we will show below. We will summarize some of the known sufficient conditions for the existence of dualizing complexes in Remark 25.98, see also Theorem 25.99.

Definition 25.93. *Let X be a locally noetherian topological space. A dimension function on X is a function $\delta\colon X \to \mathbb{Z}$ such that $\delta(x) = \delta(y) + 1$ whenever y is a specialization of x (i.e., $y \in \overline{\{x\}}$) but there is no $z \in \overline{\{x\}} \setminus \{x\}$ that specializes to y.*

We will use this definition in the context of locally noetherian schemes X. Clearly, a dimension function can only exist if X is catenary (Definition 14.102).

Dimension functions are essentially unique on connected spaces, since we can "connect" every point to a fixed point by chains of specializations.

Lemma 25.94. *Let X be a locally noetherian sober topological space, e.g., the topological space of a locally noetherian scheme. If δ and δ' are dimension functions in X, then $\delta - \delta'$ is locally constant.*

Proof. Let us sketch the purely topological proof. Let $x_0 \in X$ be any point in X. As locally noetherian spaces are locally connected we may replace X by an open connected and noetherian neighborhood. Define $c = \delta'(x_0) - \delta(x_0)$.

Let $\sim$ be the equivalence relation generated by specialization, i.e., we define $x \sim y$ if either x is a specialization of y, or y is a specialization of x. By the properties of dimension functions, we have $\delta'(x) = \delta(x) + c$ for every x in the equivalence class of x_0. Hence it suffices to show that as X is connected, sober, and noetherian, there is only one equivalence class. This is easy using that X has only finitely many irreducible components. $\square$

Example 25.95. Let X be a noetherian scheme.
(1) If X is an irreducible catenary scheme, then $x \mapsto -\dim(\mathscr{O}_{X,x})$ is a dimension function on X.
(2) If X is catenary and local, i.e., has a unique closed point, then $x \mapsto \dim(\overline{\{x\}})$ is a dimension function on X.
(3) If X is a scheme of finite type over a field k, then

$$x \mapsto \mathrm{trdeg}_k(\kappa(x)) = \dim(\overline{\{x\}})$$

is a dimension function on X.

Note however that in general neither of the formulas of the example defines a dimension formula on a (non-irreducible, non-local) finite-dimensional catenary noetherian scheme. This is the reason why in the following proposition the notion of dimension function cannot be easily replaced by a more explicit expression, and is the reason why we introduced this notion here.

Proposition 25.96. *The function $\delta \colon X \to \mathbb{Z}$ that maps $x \in X$ to the unique integer $\delta(x)$ such that $\omega_{X,x}^\bullet[-\delta(x)]$ is a normalized dualizing complex for the local ring $\mathscr{O}_{X,x}$ is a dimension function on X.*

Proof. Let $x, y \in X$ be points where y is a specialization of x, and there is no intermediate specialization. We may assume that $X = \mathrm{Spec}\,A$ where A is a local noetherian ring with dualizing complex $\omega_A^\bullet$ and y is the unique closed point of X, because we can replace X by $\mathrm{Spec}\,\mathscr{O}_{X,y}$ with dualizing complex $\omega_{X,y}^\bullet$. Moreover, we can reduce to the case that x corresponds to the zero ideal in this ring. In fact, say x corresponds to $\mathfrak{p} \subset A$. Then $R\,\mathrm{Hom}_A^{A/\mathfrak{p}}(A/\mathfrak{p}, \omega_A^\bullet)$ is a dualizing complex for $A/\mathfrak{p}$. This is again a normalized dualizing complex since by Lemma 25.43 (1) the functor $R\,\mathrm{Hom}_A^B(B, -)$ is compatible with composition of ring homomorphisms, so

$$R\,\mathrm{Hom}_{A/\mathfrak{p}}^\kappa(\kappa, R\,\mathrm{Hom}_A^{A/\mathfrak{p}}(A/\mathfrak{p}, \omega_A^\bullet)) \cong R\,\mathrm{Hom}_A^\kappa(\kappa, \omega_A^\bullet),$$

where κ denotes the residue class field of A and $A/\mathfrak{p}$.

Hence, we may assume that A is a noetherian local integral domain of dimension 1 and $X = \mathrm{Spec}\,A$ has generic point x and special point y. Then $\omega_{X,x}^\bullet = \omega_{A,\mathfrak{p}}^\bullet$ is $\mathrm{Frac}(A)$ concentrated in a single degree (Corollary 25.85), namely in degree $-\delta(x)$. On the other hand, $\omega_A^\bullet[-\delta(y)]$ is a normalized dualizing complex for the ring $A = \mathscr{O}_{X,y}$, so Proposition 25.92 (2) shows that $H^i(\omega_A^\bullet[-\delta(y)])_\mathfrak{p} \neq 0$ if (and only if) $i = -\dim(A) = -1$. We obtain that $\delta(x) = \delta(y) + 1$, as desired. $\square$

Corollary 25.97. *Let X be a noetherian scheme which has a dualizing complex $\omega_X^\bullet$. Then X is universally catenary and has finite dimension.*

Proof. It follows directly from Proposition 25.96 that X is catenary. It is even universally catenary. To check this we need to check that for each $x \in X$ and each $\mathscr{O}_{X,x}$-algebra A of finite type, A is a catenary ring. But $\omega_{X,x}^{\bullet}$ then is a dualizing complex for $\mathscr{O}_{X,x}$, and we can apply Proposition 25.87 to the morphism $\operatorname{Spec} A \to \operatorname{Spec} \mathscr{O}_{X,x}$ to see that A has a dualizing complex.

Since $\omega_X^{\bullet}$ is bounded, by Proposition 25.80 and Proposition 25.92 we obtain a bound on the dimension of the local rings $\mathscr{O}_{X,x}$ which is independent of x. $\qquad\square$

Remark 25.98. While Corollary 25.97 shows that a dualizing complex need not exist in general, in many cases dualizing complexes do exist. It is clear that for k a field, the structure sheaf $\mathscr{O}_X = k$ is a dualizing complex on $X = \operatorname{Spec} k$. It then follows from Proposition 25.87 that every separated k-scheme of finite type admits a dualizing complex.

We will show later (Section (25.25)) that on every finite-dimensional noetherian scheme X which is Gorenstein (i.e., all local rings of X are Gorenstein local rings, Definition G.26) the structure sheaf $\mathscr{O}_X$ is a dualizing complex. Invoking Proposition 25.87 again, we see that every scheme which is separated and of finite type over a Gorenstein scheme has a dualizing complex.

For rings (equivalently, for affine schemes) this condition characterizes the class of noetherian rings which have a dualizing complex. This was conjectured by Sharp and proved by Kawasaki.

Theorem 25.99. (Sharp's conjecture [Sharp]^{O}, [Kaw]^{O}) *Let A be a noetherian ring. Then A admits a dualizing complex if and only if A is a quotient of a finite-dimensional Gorenstein ring.*

(25.20) Relative dualizing complexes.

Definition 25.100. *Let Y be a noetherian scheme and let $f\colon X \to Y$ be a separated morphism of finite type. Then*

$$\omega_{X/Y}^{\bullet} := f^{!}\mathscr{O}_Y$$

is called the relative dualizing complex *for f.*

Remark 25.101. Let $f\colon X \to Y$ be a separated morphism of finite type between noetherian schemes.
(1) One has $\omega_{X/Y}^{\bullet} \in D_{\text{coh}}^{+}(X)$ by Proposition 25.64.
(2) If f is of finite tor-dimension (e.g., if f is flat), then $\omega_{X/Y}^{\bullet} \in D_{\text{coh}}^{b}(X)$ and for all $K \in D_{\text{qcoh}}^{+}(Y)$ one has functorial isomorphisms

$$(25.20.1) \qquad\qquad f^{!}K \cong Lf^{*}K \otimes_{\mathscr{O}_X}^{L} \omega_{X/Y}^{\bullet}$$

by Proposition 25.65.

The following remark shows why $\omega_{X/Y}^{\bullet}$ is called relative dualizing complex.

Remark 25.102. Let S be a noetherian scheme that has a dualizing complex $\omega_S^{\bullet}$. Let $g\colon X \to S$ and $h\colon Y \to S$ be separated S-schemes of finite type and let $f\colon X \to Y$ be a morphism of S-schemes of finite tor-dimension. Then $\omega_X^{\bullet} := g^{!}\omega_S^{\bullet}$ and $\omega_Y^{\bullet} := h^{!}\omega_S^{\bullet}$ are dualizing complexes on X and Y, respectively, by Proposition 25.87. Moreover, (25.20.1) shows that

(25.20.2) $$\omega_X^\bullet \cong Lf^*\omega_Y^\bullet \otimes_{\mathscr{O}_X}^L \omega_{X/Y}^\bullet.$$

Next we show that for flat morphisms between affine schemes, the formation of the relative dualizing complex is compatible with arbitrary base change.

Remark 25.103. Let A be a noetherian ring and let B be an A-algebra of finite type. Let $X = \operatorname{Spec} B$, $Y = \operatorname{Spec} A$, and let $f\colon X \to Y$ denote the corresponding scheme morphism. We fix a closed embedding $i\colon X \to \mathbb{A}_Y^N$, i.e., a surjection $P := A[X_1, \dots, X_N] \twoheadrightarrow B$ of A-algebras, and denote by $g\colon \mathbb{A}_Y^N \to Y$ the projection. Then $f = g \circ i$.

Let $\omega_{B/A}^\bullet$ be the object in $D(B)$ corresponding to $f^!\mathscr{O}_Y = i^!g^!\mathscr{O}_Y$. By Example 25.63 and Proposition 25.44 we find

(25.20.3) $$\omega_{B/A}^\bullet = R\operatorname{Hom}_P^B(B, P[N]).$$

From now on suppose that B is flat over A. Then the formation of $\omega_{B/A}^\bullet$ is compatible with base change in the following sense. Let $A \to A'$ be any ring homomorphism and set $B' := B \otimes_A A'$. We claim that there is an isomorphism in $D(B')$

(25.20.4) $$\omega_{B/A}^\bullet \otimes_B^L B' \cong \omega_{B'/A'}^\bullet.$$

To show (25.20.4) note that B is perfect as a P-module by Proposition 25.40. Since B is flat over A, we have $B \otimes_A A' \cong B \otimes_A^L A'$. Together with Lemma 25.43 (2), it follows that

$$\omega_{B/A}^\bullet \otimes_B^L B' \cong R\operatorname{Hom}_P^B(B, P[N]) \otimes_B^L B' \cong R\operatorname{Hom}_P^B(B, P[N]) \otimes_A^L A'$$

in $D(A')$. The canonical map $R\operatorname{Hom}_P^B(B, P[N]) \to P[N]$ thus gives us, tensoring with $\operatorname{id}_{A'}$, a map

$$R\operatorname{Hom}_P^B(B, P[N]) \otimes_B^L B' \cong R\operatorname{Hom}_P^B(B, P[N]) \otimes_A^L A' \to P[N] \otimes_A^L A'$$

in $D(A')$, and hence by adjunction (Lemma 25.43 (1)) the top row in the following commutative diagram.

$$\begin{array}{ccc}
R\operatorname{Hom}_P^B(B, P[N]) \otimes_B^L B' & \longrightarrow & R\operatorname{Hom}_{P'}^{B'}(B', P[N] \otimes_A^L A') \\
\Big\downarrow {\scriptstyle =} & & \Big\downarrow {\scriptstyle =} \\
R\operatorname{Hom}_P(B, P[N]) \otimes_A^L A' & \longrightarrow & R\operatorname{Hom}_{P'}(B', P[N] \otimes_A^L A')
\end{array}$$

where the vertical identifications are identifications in $D(A')$ and the top arrow is a morphism in $D(B')$ and where $P' := P \otimes_A A' = P \otimes_A^L A'$. We can rewrite the bottom horizontal arrow, using Lemma 25.43 for the term on the right hand side, as

$$R\operatorname{Hom}_P(B, P[N]) \otimes_P^L P' \to R\operatorname{Hom}_P(B, P[N] \otimes_P^L P')$$

in $D(P)$. This is the canonical arrow as in (21.26.5) and it is an isomorphism since B is a perfect P-module (Lemma 21.145). Hence the top horizontal row in the above diagram is an isomorphism. This gives the isomorphism (25.20.4) by (25.20.3).

Next we collect some general properties of the relative dualizing complex.

Proposition 25.104. *Let $f\colon X \to Y$ be a separated morphism of noetherian schemes of finite type. Let $x \in X$, set $y := f(x)$, and let $X_y := f^{-1}(y)$ be the scheme-theoretic fiber.*

(1) *We have $H^i(\omega_{X/Y}^\bullet)_x = 0$ for all $i < -\dim_x X_y$.*

(2) *If f is flat, then we have $H^i(\omega_{X/Y}^\bullet)_x = 0$ for all $i > 0$.*

(3) *If f is flat and all fibers of f have dimension $\leq d$, then $\omega_{X/Y}^\bullet$, considered as an object of $D(X, f^{-1}(\mathscr{O}_Y))$, has tor-amplitude in $[-d, 0]$.*

Proof. It is enough to consider an affine situation, say $X = \operatorname{Spec} B$, $Y = \operatorname{Spec} A$, $\varphi\colon A \to B$ is the ring homomorphism defining f, and $\mathfrak{q}$ and $\mathfrak{p}$ are the prime ideals in B and A, respectively, corresponding to x and y. Now consider the setting of Remark 25.103, i.e., we choose a surjection $P := A[X_1, \ldots, X_N] \to B$ of A-algebras and get an isomorphism $\omega_{B/A}^\bullet \cong R\operatorname{Hom}_P^B(B, P[N])$, where $\omega_{B/A}^\bullet$ in $D(B)$ corresponds to $f^!\mathscr{O}_Y$. Let I be the kernel of $P \to B$ and let $\mathfrak{r} \subset P$ be the inverse image of $\mathfrak{q}$. Then $H^i(\omega_{X/Y}^\bullet)_x$ corresponds to

$$H^i(R\operatorname{Hom}_P^B(B, P[N])_{\mathfrak{q}}) \cong \operatorname{Ext}_P^i(B, P[N])_{\mathfrak{r}} \cong \operatorname{Ext}_{P_{\mathfrak{r}}}^{i+N}(P_{\mathfrak{r}}/IP_{\mathfrak{r}}, P_{\mathfrak{r}}),$$

where we use Corollary 22.71 for the second isomorphism.

(1). By the relation between vanishing of Ext and depth (Proposition G.9) the statement of (1) translates to

$$(*) \qquad \operatorname{depth}(IP_{\mathfrak{r}}, P_{\mathfrak{r}}) \geq N - d = \dim(P \otimes_A \kappa(\mathfrak{p})) - \dim(P/I \otimes_A \kappa(\mathfrak{p})),$$

where $d := \dim_x X_y$. As $P_{\mathfrak{r}}$ is a flat $A_{\mathfrak{p}}$-algebra, any sequence in $P_{\mathfrak{r}}$ that maps to a regular sequence in $P_{\mathfrak{r}} \otimes_A \kappa(\mathfrak{p})$ is regular (Proposition G.11). Hence to show $(*)$ we may assume that A is a field. Then $P_{\mathfrak{r}}$ is a regular local ring, and in particular Cohen-Macaulay. The above claim then follows from Proposition G.14.

(2). We need to show that $\operatorname{Ext}_P^i(B, P[N]) = \operatorname{Ext}_P^{i+N}(B, P) = 0$ for $i > 0$. By Lemma 25.41, B is a perfect P-module of tor-amplitude in $[-N, 0]$. Hence B is in $D(P)$ isomorphic to a complex K of finite projective P-modules concentrated in degrees $[-N, 0]$ (Proposition 22.53). In particular it is K-flat and we find $\operatorname{Ext}_P^{i+N}(B, P) = H^{i+N}(\operatorname{Hom}_P(K, P)) = 0$ for $i > 0$.

(3). We need to show that for every A-module M, $\omega_{B/A}^\bullet \otimes_A^L M$ has cohomology in degrees $[-d, 0]$ only. This statement is compatible with passing to colimits, so it is enough to consider finitely generated A-modules M. To deal with this case, we apply the base change (25.20.4) to the ring homomorphism $A \to A' := D_A(M) = A \oplus M$, where the multiplication on $A \oplus M$ is given by the scalar multiplication by A on M and the rule $mm' = 0$ for all $m, m' \in M$. As M is finitely generated, A' is noetherian.

Applying (1) and (2) to $A' \to B' := B \otimes_A A'$, we see that $\omega_{B'/A'}^\bullet$ has cohomology in degrees $[-d, 0]$ only. By (25.20.4), the same holds for $\omega_{B/A}^\bullet \otimes_A^L A'$, and hence in particular for its direct summand $\omega_{B/A}^\bullet \otimes_A^L M$. $\qquad\square$

Dualizing sheaves

Although the "correct" notion is the notion of dualizing *complex*, it turns out that the lowest cohomology sheaf of this dualizing complex already contains interesting information which makes it an object interesting in its own right, called a dualizing sheaf.

Moreover, we will see that for Cohen-Macaulay schemes, a dualizing complex is concentrated in a single degree and hence a dualizing sheaf determines a dualizing complex in this case. We will also consider a relative situation in which we introduce Cohen-Macaulay morphisms and study the relative dualizing complex.

(25.21) Definition and first properties of dualizing sheaves.

Let X be a noetherian scheme. We assume that X has a dualizing complex and fix one, denoted by $\omega_X^\bullet$.

Definition 25.105. *With notation as above, set* $n := \min\{i;\ H^i(\omega_X^\bullet) \neq 0\}$ *be the maximal value of the dimension function attached to* $\omega_X^\bullet$ *(Proposition 25.96). Then* $\omega_X := H^n(\omega_X^\bullet)$ *is called a* dualizing sheaf, *or* dualizing module, *on* X.

Analogously we can pass from a dualizing complex for a ring to the associated dualizing module. Similarly as for a dualizing complex, a dualizing sheaf (if it exists) is not unique, but only unique up to tensoring with a line bundle, cf. Proposition 25.84.

If $U \subseteq X$ is open, the integer n in the definition above may change when restricting to U such that one does not have $\omega_{X|U} = \omega_U$. This cannot happen if X is equi-dimensional, i.e., all irreducible components of X have the same dimension: Proposition 25.106 below shows that if X is equi-dimensional and ω_X is a dualizing sheaf on X, then $\mathrm{Supp}(\omega_X) = X$, and if $U \subseteq X$ is a non-empty open in X, then $\omega_{X|U}$ is a dualizing sheaf on U.

Proposition 25.106. *Let* X *be a connected noetherian scheme, let* $\omega_X^\bullet$ *be a dualizing complex and let* ω_X *be the induced dualizing sheaf on* X.
(1) *Let* δ *be a dimension function on* X *(which exists by Proposition 25.96). The support of* ω_X *is the union of those irreducible components of* X *such that* δ *takes its maximal value on the generic point of the irreducible component.*
(2) *If* X *is integral, then* ω_X *is torsion-free, i.e.,* $\Gamma(U, \omega_X)$ *is a torsion-free* $\Gamma(U, \mathscr{O}_X)$-*module for every open* $\emptyset \neq U \subseteq X$.
(3) *The induced coherent* $\mathscr{O}_X$-*module* ω_X *has the property* (S_2), *i.e., for all* $x \in X$ *the* $\mathscr{O}_{X,x}$-*module* $\omega_{X,x}$ *has depth at least* $\min(2, \dim(\mathrm{Supp}(\omega_{X,x})))$ *(Definition G.15).*

Proof. By Corollary 25.97 we know that $d := \dim(X)$ is finite. By shifting $\omega_X^\bullet$ appropriately, we may assume that d is the largest integer such that $H^{-d}(\omega_X^\bullet) \neq 0$, and hence that $\omega_X = H^{-d}(\omega_X^\bullet)$.

(1). Since any two dimension functions on X differ by a constant (Lemma 25.94), we may assume that δ is the dimension function given by $\omega_X^\bullet$ in the sense of Proposition 25.96.

The support of the coherent sheaf ω_X is closed (Corollary 7.32). If $x \in X$ is a point with $\delta(x) = d$ (hence maximal), then Proposition 25.92 applied to the local ring $\mathscr{O}_{X,x}$ shows that $\omega_{X,x} \neq 0$. Hence all irreducible components that are closures of points where δ is maximal are contained in the support of $\omega_{X,x}$.

If, on the other hand, $y \in X$ is a point with $\omega_{X,y} \neq 0$, then we apply Proposition 25.92 to the local ring $\mathscr{O}_{X,y}$ and the normalized dualizing complex $\omega_{X,y}[-\delta(y)]$ for this local ring. Part (3) of the proposition shows that $d - \delta(y) \leq \dim \mathscr{O}_{X,y}$. Let $x \in X$ correspond to a minimal prime ideal in $\mathscr{O}_{X,y}$ such that $\delta(x) - \delta(y) = \dim \mathscr{O}_{X,y}$, i.e., the closure of x in $\mathrm{Spec}\,\mathscr{O}_{X,y}$ is an irreducible component of (maximal) dimension $\dim \mathscr{O}_{X,y}$. We find that $d \leq \delta(x)$, hence $\delta(x) = d$. So y lies in one of the irreducible components named in the statement of the lemma.

(2). As X is irreducible, all dimension functions on X are of the form $d' - \dim(\mathscr{O}_{X,x})$ for some $d' \in \mathbb{Z}$ by Example 25.95 (1) and Lemma 25.94. By our normalization in the beginning of the proof, the dimension function defined by $\omega_X^\bullet$ is given by $x \mapsto d - \dim(\mathscr{O}_{X,x})$ (this can be checked at the generic point of X).

To see that ω_X is torsion-free can be checked on stalks (Proposition B.30). Let $x \in X$ and set $A := \mathscr{O}_{X,x}$, a noetherian local ring with dualizing complex $\omega_A^\bullet := \omega_{X,x}^\bullet$. As $\omega_{X,x} \neq 0$ by (1), we find $\omega_{X,x} = H^{-d}(\omega_A^\bullet) = \omega_A$ is the dualizing module for $\omega_A^\bullet$, i.e., its non-zero cohomology object in minimal degree. Hence we have to show that ω_A is a torsion-free A-module. Note that $\dim(A) \leq d$ and that $\omega_A^\bullet[\dim A - d]$ is normalized by Proposition 25.92 (3).

For $0 \neq s \in \mathfrak{m}$, the maximal ideal of A, the quotient $B = A/(s)$ has dimension $< \dim(A)$ and $\omega_B^\bullet := R\operatorname{Hom}_A^B(B, \omega_A^\bullet)$ is a dualizing complex on B (Proposition 25.86). We have $H^{-d}(\omega_B^\bullet) = 0$ by Proposition 25.92 (3). Applying $R\operatorname{Hom}_A(-, \omega_A^\bullet)$ to the short exact sequence $0 \to A \to A \to B \to 0$, where the map $A \to A$ is multiplication by s, gives us a distinguished triangle $\omega_B^\bullet \to \omega_A^\bullet \to \omega_A^\bullet \to \omega_B^\bullet[1]$ and the associated long exact cohomology sequence shows that multiplication by s defines an injective homomorphism $\omega_A \to \omega_A$.

(3). We can again reduce to the case of a local noetherian ring A. Let $\omega_A^\bullet$ be its normalized dualizing complex and let $\omega_A = H^{-\dim A}(\omega_A^\bullet)$ its dualizing module. Let $\mathfrak{p} \subset A$ be a prime ideal. We have to show that $\operatorname{depth}_{A_\mathfrak{p}}(\omega_A)_\mathfrak{p} \geq \inf\{2, \dim_{A_\mathfrak{p}}(\omega_A)_\mathfrak{p}\}$.

By (1) we know that the support of ω_A is the union of the irreducible components of $\operatorname{Spec} A$ of dimension $\dim A$. As $\operatorname{Spec} A$ is catenary (Corollary 25.97), this implies that for a prime ideal $\mathfrak{p}$ of A one has $(\omega_A)_\mathfrak{p} \neq 0$ if and only if $\dim(A/\mathfrak{p}) + \dim(A_\mathfrak{p}) = \dim A$. In this case, one has

$$(\omega_A)_\mathfrak{p} = H^{-\dim(A_\mathfrak{p})}((\omega_A^\bullet)_\mathfrak{p}[-\dim A/\mathfrak{p}]).$$

Hence it suffices to show that $\operatorname{depth}_A(\omega_A) \geq \inf\{2, \dim_A(\omega_A)\}$. We will show this by induction on $\dim A$. The case $\dim A = 0$ is clear. Hence assume that $\dim A > 0$.

If $\operatorname{depth}(A) > 0$, we can find a regular element $s \in \mathfrak{m}$, the maximal ideal of A. Set $B := A/(s)$. Then $\dim B < \dim A$. As in the end of the proof of (2) one sees that

(i) multiplication by s on ω_A is injective,

(ii) and one has an injective A-linear map $\omega_A/s\omega_A \hookrightarrow \omega_B$.

Then (i) shows that $\operatorname{depth}(\omega_A) \geq 1$. If $\dim A > 1$, then $\dim B > 0$ and by applying the induction hypothesis to B we see that ω_B has depth > 0. Hence ω_A has depth > 1 by (ii) which shows that ω_A satisfies (S_2) in this case.

If $\operatorname{depth}(A) = 0$, let $I \subset A$ be the ideal of elements annihilated by some power of the maximal ideal $\mathfrak{m}$ of A and set $B := A/I$. As A is noetherian, I is annihilated by some fixed power of $\mathfrak{m}$ and hence is of finite length. Applying $R\operatorname{Hom}_A(-, \omega_A^\bullet)$ to the short exact sequence $0 \to I \to A \to B \to 0$ we get by Proposition 25.86 and by derived Matlis duality (Proposition 25.91) a distinguished triangle

(*) $$\omega_B^\bullet \longrightarrow \omega_A^\bullet \longrightarrow \operatorname{Hom}_A(I, E)[0] \xrightarrow{+1},$$

where E is a Matlis module for A and $\omega_B^\bullet$ is a normalized dualizing complex for B. As $\dim A > 0$, we have depth $B \geq 1$ and (*) shows that $\omega_A \cong \omega_B$ (again, since $\dim A > 0$). As we proved above that ω_B satisfies (S_2), we are done. $\qquad\square$

Remark 25.107. Note that in Part (1) of the proposition it is not true in general that the support of ω_X equals the union of the irreducible components of X of dimension $\dim X$ (Exercise 25.11). However, if X is irreducible, or local, or of finite type over a field, then this is in fact true. See Example 25.95.

(25.22) Dualizing complexes and sheaves for proper schemes over local rings.

Proposition 25.108. *Let A be a local noetherian ring with a normalized dualizing complex $\omega_A^\bullet$. Let X be a proper A-scheme. Denote by $f\colon X \to \operatorname{Spec} A$ the structure morphism, and let $\omega_X^\bullet = f^! \omega_A^\bullet$. (This is a dualizing complex on X by Proposition 25.87.)*

(1) For all $i \notin [-\dim(X), \dots, 0]$, we have $H^i(\omega_X^\bullet) = 0$.

(2) We have $n := \min\{i;\ H^i(\omega_X^\bullet) \neq 0\} = -\dim X$. Therefore the $\mathcal{O}_X$-module $\omega_X = H^{-\dim(X)}(\omega_X^\bullet)$ is a dualizing module on X.

Proof. Let x be a closed point of X. As f is proper, $s := f(x)$ is the closed point of $\operatorname{Spec} A$. We first show that the dualizing complex $\omega_{X,x}^\bullet$ for the local ring $\mathcal{O}_{X,x}$ (Proposition 25.80) is normalized. Consider the commutative diagram

$$
\begin{array}{ccc}
\operatorname{Spec} \kappa(x) & \xrightarrow{\ i_x\ } & X \\[2pt]
{\scriptstyle g}\big\downarrow & & \big\downarrow{\scriptstyle f} \\[2pt]
\operatorname{Spec} \kappa(s) & \xrightarrow{\ i_s\ } & \operatorname{Spec} A.
\end{array}
$$

Using Proposition 25.44, we have

$$
R\operatorname{Hom}_{\mathcal{O}_{X,x}}^{\kappa(x)}(\kappa(x), \omega_{X,x}^\bullet) = i_x^! \omega_X^\bullet = i_x^! f^! \omega_A^\bullet = g^! i_s^! \omega_A^\bullet = g^! \kappa(s)
$$

and we need to show that this complex is concentrated in degree 0. But g is a finite morphism, so we can identify $g^! = R\operatorname{Hom}_{\kappa(s)}^{\kappa(x)}(\kappa(x), -)$ again by Proposition 25.44. Since $\kappa(x)$ is a free $\kappa(s)$-module, we obtain $R\operatorname{Hom}_{\mathcal{O}_{X,x}}^{\kappa(x)}(\kappa(x), \omega_{X,x}^\bullet) \cong \operatorname{Hom}_{\kappa(s)}(\kappa(x), \kappa(s))$. So we have proved that $\omega_{X,x}^\bullet$ is indeed normalized.

Hence for all closed points $x \in X$, we find that $H^i(\omega_X^\bullet)_x = H^i(\omega_{X,x}^\bullet) = 0$ for all $i \notin [-\dim(\mathcal{O}_{X,x}), -\operatorname{depth}(\mathcal{O}_{X,x})]$ by Proposition 25.92 (3). Since X is proper over A, it is quasi-compact, so every closed subset of X contains a closed point (Lemma 23.4). Part (1) follows, because the support of $H^i(\omega_X^\bullet)$ is closed.

For part (2), in view of part (1) we only need to prove that $H^{-\dim(X)}(\omega_X^\bullet) \neq 0$. This follows from Proposition 25.92 (3) and a similar reasoning as before, applied to a closed point $x \in X$ with $\dim \mathcal{O}_{X,x} = \dim X$. $\square$

Remark 25.109. The proof of Proposition 25.108 shows that one has more precisely $H^i(\omega_X^\bullet) = 0$ for $i \notin [-\dim(X), -\delta]$ with $\delta := \inf_x \operatorname{depth}(\mathcal{O}_{X,x})$ where x runs through the closed points of X.

Proposition 25.110. *Let A be a local noetherian ring with a normalized dualizing complex $\omega_A^\bullet$ and let ω_A be its dualizing sheaf. Let $f\colon X \to \operatorname{Spec} A$ be a proper A-scheme with dualizing sheaf $\omega_X = H^{-\dim X}(f^! \omega_A^\bullet)$ as above. Denote by $s \in S := \operatorname{Spec} A$ the closed point and by X_s the special fiber of X. Assume that $\dim X = \dim X_s + \dim S$ and write $d = \dim(X_s)$.*

Then there are isomorphisms, for every quasi-coherent $\mathcal{O}_X$-module $\mathscr{F}$,

$$
\operatorname{Hom}_{\mathcal{O}_X}(\mathscr{F}, \omega_X) \cong \operatorname{Hom}_A(H^d(X, \mathscr{F}), \omega_A)
$$

that are functorial in $\mathscr{F}$.

In other words, ω_X represents the functor $\mathscr{F} \mapsto \operatorname{Hom}_A(H^d(X, \mathscr{F}), \omega_A)$.

The condition that $\dim X = \dim X_s + \dim S$ holds for instance if S is irreducible, f is flat, and X is equi-dimensional (see Theorem 14.116 and the following remarks).

Proof. We use the adjunction between $f^!$ (which equals $f^\times$ since f is proper) and Rf_* and compute

$$
\begin{aligned}
\operatorname{Hom}_{\mathscr{O}_X}(\mathscr{F}, \omega_X) &= \operatorname{Hom}_{D(X)}(\tau^{\le -\dim X}\mathscr{F}[\dim X], \tau^{\ge -\dim(X)}f^!\omega_A^\bullet) \\
&= \operatorname{Hom}_{D(X)}(\mathscr{F}[\dim X], f^!\omega_A^\bullet) \\
&= \operatorname{Hom}_{D(A)}(Rf_*\mathscr{F}[\dim X], \omega_A^\bullet) \\
&= \operatorname{Hom}_{D(A)}(\tau^{\le -\dim(A)}Rf_*\mathscr{F}[\dim X], \tau^{\ge -\dim A}\omega_A^\bullet) \\
&= \operatorname{Hom}_A(H^d(X, \mathscr{F}), \omega_A).
\end{aligned}
$$

In the first and in the final step we use Proposition F.156, where for the final step we use the equality $-\dim A = d - \dim X$. For the second step we use Proposition 25.108. For the fourth step we use Proposition 25.92 (3) and Corollary 24.44 for Rf_*. The latter says that $R^i f_* \mathscr{F} = 0$ for $i > d$, so $H^i(Rf_*\mathscr{F}[\dim X]) = 0$ for $i > -\dim X + d = -\dim A$. $\square$

(25.23) Cohen-Macaulay schemes.

Again, let X be a Noetherian scheme with a dualizing complex $\omega_X^\bullet$. Recall from Sections (B.14) and (14.28) the notion of Cohen-Macaulay ring and Cohen-Macaulay scheme: A noetherian local ring is called Cohen-Macaulay if the depth, i.e., the maximal length of a regular sequence in its maximal ideal, equals its Krull dimension. A (locally) noetherian scheme is called Cohen-Macaulay, if all its local rings are Cohen-Macaulay. Assuming that X is connected and of finite type over a field, this implies that X is equi-dimensional by Proposition 14.126 (this does not hold for general connected Cohen-Macaulay schemes, see Exercise 25.12).

More generally, we call a coherent $\mathscr{O}_X$-module $\mathscr{F}$ *Cohen-Macaulay* if for all $x \in X$ its stalk $\mathscr{F}_x$ is a Cohen-Macaulay $\mathscr{O}_{X,x}$-module (Definition G.13).

Proposition 25.92 implies that the Cohen-Macaulay condition is characterized precisely by the condition that $\omega_X^\bullet$ has a single non-vanishing cohomology sheaf:

Proposition 25.111.
(1) *Let $(A, \mathfrak{m})$ be a local noetherian ring with a normalized dualizing complex $\omega_A^\bullet$. The following are equivalent.*
   (i) *The ring A is Cohen-Macaulay.*
   (ii) *The complex $\omega_A^\bullet$ is of the form $\omega_A[d]$ for some A-module $\omega_A \ne 0$ and $d \in \mathbb{Z}$.*
   In this case, $d = \dim A$ and ω_A is the dualizing module given by $\omega_A^\bullet$. Moreover ω_A is a Cohen-Macaulay module of depth d.
(2) *Let X be a connected noetherian scheme which has a dualizing complex $\omega_X^\bullet$. The following are equivalent:*
   (i) *X is Cohen-Macaulay,*
   (ii) *the complex $\omega_X^\bullet$ is of the form $\omega_X[d]$ for some $\mathscr{O}_X$-module ω_X and some $d \in \mathbb{Z}$.*
   In this case, ω_X is a dualizing sheaf on X in the sense of Definition 25.105, a coherent Cohen-Macaulay $\mathscr{O}_X$-module with $\operatorname{Supp} \omega_X = X$.

Proof. All assertions of (1) follow immediately from Proposition 25.92 (3) except the last one. To see that ω_A is Cohen-Macaulay it suffices to show that $\operatorname{depth}_A(\omega_A) = d = \dim A$ (Corollary G.10). By Proposition 25.76 we have $A = R\operatorname{Hom}_A(\omega_A^\bullet, \omega_A^\bullet) = R\operatorname{Hom}_A(\omega_A[d], \omega_A^\bullet) = R\operatorname{Hom}_A(\omega_A, \omega_A^\bullet)[-d]$. Hence we see that $\operatorname{Ext}^i(\omega_A, \omega_A^\bullet) \neq 0$ if and only if $i = -d$ which shows $\operatorname{depth}_A(\omega_A) = d$ by Proposition 25.92 (1).

Let us prove (2). First note that for $x \in X$ the stalk $\omega_{X,x}^\bullet$ is a dualizing complex for the local ring $\mathscr{O}_{X,x}$ by Proposition 25.80. Since the Cohen-Macaulay property of X is defined in terms of its local rings, this shows that the implication (ii) $\Rightarrow$ (i) in (2) follows directly from part (1).

To prove that (i) implies (ii), we need to show that the degree where the localizations $\omega_{X,x}^\bullet$ have their unique cohomology is constant on X. So let $x \in X$. By (i) and part (1), there exists a unique d_x such that $H^i(\omega_{X,x}^\bullet) = 0$ for all $i \neq d_x$. We show that $x \mapsto d_x$ is locally constant. Let U be an affine open neighborhood of x. Then only finitely many cohomology sheaves of $\omega_{X|U}^\bullet$ are non-zero.

For all $i \neq d_x$, the stalk of $H^i(\omega_{X|U}^\bullet)$ at x vanishes, so shrinking U further we may assume that $H^i(\omega_{X|U}^\bullet) = 0$. Since we have to do this for only finitely many i, the desired result follows.

Finally, $\operatorname{Supp}\omega_X = X$ follows from Proposition 25.106 (1). $\qquad\Box$

From the argument in the proof of part (2), we also obtain:

Corollary 25.112. *Let X be a noetherian scheme which has a dualizing complex $\omega_X^\bullet$. Then the subset*

$$U := \{x \in X;\ \mathscr{O}_{X,x} \text{ is Cohen-Macaulay}\}$$

is open and dense, and U is Cohen-Macaulay. Moreover, one has a decomposition $U = \coprod_d U_d$ into open and closed subschemes U_d of U where

$$U_d = \{\, x \in U \ ;\ H^i(\omega_{X,x}^\bullet) = 0 \text{ for all } i \neq d\}$$

Every scheme locally of finite type over a field k has a locally a dualizing complex.

Proof. The openness of U and the decomposition of U follow from the proof of Proposition 25.111 (2) and it is clear that U is Cohen-Macaulay. Moreover, since zero-dimensional rings are Cohen-Macaulay, U contains the generic point of every irreducible component. Hence U is dense in X. $\qquad\Box$

Remark 25.113. If X satisfies the equivalent conditions in Proposition 25.111 (2) and is of finite type over a field, then X is equi-dimensional and after shifting the degree of $\omega_X^\bullet$ one can assume that for all $x \in X$, $\omega_{X,x}^\bullet[-\dim\overline{\{x\}}]$ is a normalized dualizing complex for $\mathscr{O}_{X,x}$. In this case, the integer d of Proposition 25.111 (2) is $\dim X$.

(25.24) The relative dualizing complex for Cohen-Macaulay morphisms.

We will now transfer the results of Section (25.23) to a relative setting.

Definition 25.114. *Let $f \colon X \to Y$ be a morphism of schemes whose fibers are locally noetherian. Then f is called Cohen-Macaulay at $x \in X$, if f is flat at x and if the local ring $\mathscr{O}_{f^{-1}(f(x)),x}$ of x in the schematic fiber $f^{-1}(f(x))$ is Cohen-Macaulay.*

The morphism f is called a Cohen-Macaulay morphism *if it is Cohen-Macaulay at all points $x \in X$. This is equivalent to asking that f be flat and that all scheme-theoretic fibers of X be Cohen-Macaulay schemes.*

Note that we did not impose the condition that the fibers are "geometrically Cohen-Macaulay", i.e., stay Cohen-Macaulay after base change to any field extension K of $\kappa(f(x))$. The reason is that "Cohen-Macaulay" is already a "geometric property", see Exercise 25.13 for details.

Remark 25.115.
(1) Complete intersection rings are Cohen-Macaulay. This shows that any syntomic morphism is Cohen-Macaulay.
(2) Every quasi-finite flat morphism is Cohen-Macaulay since zero-dimensional rings are automatically Cohen-Macaulay.
(3) Let $f\colon X \to Y$ and $g\colon Y \to Z$ be morphisms such that f, g and $g \circ f$ have locally noetherian fibers and let $x \in X$.
  (a) If f is Cohen-Macaulay in x and g is Cohen-Macaulay in $f(x)$, then $g \circ f$ is Cohen-Macaulay in x.
  (b) If $g \circ f$ Cohen-Macaulay in x and f is flat in x, then f is Cohen-Macaulay in x and g is Cohen-Macaulay in $f(x)$.
Both assertions follow from Proposition B.82.

The main goal of this section is to show that if $f\colon X \to Y$ is a separated Cohen-Macaulay morphism of finite type of noetherian schemes, then the relative dualizing sheaf $\omega^{\bullet}_{X/Y} := f^!\mathscr{O}_Y$ is concentrated in one degree, see Proposition 25.120 below. If f is in addition proper, then we obtain a description for $f^!F$ for every F in $D_{\mathrm{qcoh}}(Y)$ by Corollary 25.38.

The next lemma will allow us to reduce the later discussion to a Cohen-Macaulay morphism with zero-dimensional fibers.

Lemma 25.116. *Let Y be a quasi-separated scheme, let $f\colon X \to Y$ be a morphism of finite presentation.*
(1) *Then for all $x \in X$ there exists an open affine neighborhood U of x and a quasi-finite morphism of Y-schemes $g\colon U \to \mathbb{A}^d_Y$ of finite presentation.*
(2) *If f is Cohen-Macaulay in $x \in X$, then the pair (U, g) in (1) can be chosen such that g is flat (and hence g is Cohen-Macaulay by Remark 25.115 (2)).*

If Y is locally noetherian, then Y is automatically quasi-separated and every finite type morphism $f\colon X \to Y$ is of finite presentation.

Proof. Let us first remark that any g as in (1) is of finite presentation. This follows from Remark 10.4 as follows. Since Y is quasi-separated and U is quasi-compact, the inclusion $U \to X$ is quasi-compact and hence of finite presentation. Therefore $f_{|U}$ is still of finite presentation. Hence g will be of finite presentation (Proposition 10.35). Therefore we can work locally on X and Y and may assume that $X = \operatorname{Spec} A$ and $Y = \operatorname{Spec} R$ are affine.

Let us show (1). For $x \in X$ and $y = f(x)$, using Noether normalization we can find a morphism $R[X_1, \dots, X_d] \to A$ from a polynomial ring over R which induces a finite injective ring homomorphism $\kappa(y)[X_1, \dots, X_d] \to A \otimes_R \kappa(y)$. By Corollary 12.79 (or Corollary 14.113) the locus of points that are isolated in their fiber is open. So replacing X by an open affine subscheme we may assume that the morphism $\operatorname{Spec} A \to \mathbb{A}^n_R$ is quasi-finite.

To show (2) we first remark that because of openness of flatness (Theorem 14.44) we may choose the open neighborhood U of x and g as in (1) such that f is flat. Replacing X by U it is therefore enough to prove the following. Given a flat morphism $f\colon X \to Y$ of affine schemes that is the composition of a quasi-finite morphism $g\colon X \to \mathbb{A}_Y^d$ and the projection $\mathbb{A}_Y^d \to Y$, we have that g is flat in x if f is Cohen-Macaulay in x (then again by openness of flatness we see that g is flat after possibly replacing X by a smaller open neighborhood of x).

In view of the flatness of f, to prove that g is flat, it is enough to show that the $\kappa(y)$-morphism g_y obtained by the base change $\operatorname{Spec}\kappa(y) \to Y$ is flat for all $y \in Y$ (by the fiber criterion for flatness, Corollary 14.27). On the other hand, whether f is Cohen-Macaulay can also be checked on the fibers over points $y \in Y$. Therefore, we may assume that Y is the spectrum of a field. Then X is Cohen-Macaulay in x and we can invoke Theorem 14.128 to see that g is flat in x. □

Remark 25.117. As compositions of Cohen-Macaulay morphisms are again Cohen-Macaulay (Remark 25.115 (3)), it follows conversely that every f that has locally a factorization as in (2) is automatically Cohen-Macaulay.

Corollary 25.118. *Let $f\colon X \to Y$ be a morphism locally of finite presentation. Then*

$$V := \{\, x \in X \; ; \; f \text{ is Cohen-Macaulay in } x \,\}$$

is open in X. If f is flat, then V is dense in every fiber of f.

Proof. We may assume that X and Y are affine and hence that f is of finite presentation. Let $x \in V$ and choose (U, g) as in Lemma 25.116 (2). Then Remark 25.117 shows that $U \subseteq V$, in particular V is open X. If f is flat, then V is dense in every fiber of f by Corollary 25.112. □

Recall that for morphisms $f\colon X \to Y$ locally of finite type the map

$$(25.24.1) \qquad\qquad X \longrightarrow \mathbb{Z}_{\geq 0}, \qquad x \mapsto \dim_x f^{-1}(f(x))$$

is upper semicontinuous (Theorem 14.112). For Cohen-Macaulay morphisms we have the following stronger result.

Proposition and Definition 25.119. *Let $f\colon X \to Y$ be a Cohen-Macaulay morphism locally of finite presentation. Then the map (25.24.1) is locally constant and it is called the* relative dimension *of the morphism f.*

Proof. We may assume that $Y = \operatorname{Spec} R$ and $X = \operatorname{Spec} A$. Then f is of finite presentation. By Lemma 25.116 we may assume that there exists a $d \geq 0$ such that f can be factorized in a flat quasi-finite morphism $g\colon X \to \mathbb{A}_Y^d$ followed by the structure map $\mathbb{A}_Y^d \to Y$. It suffices to show that $\dim_x f^{-1}(f(x)) = d$ for all $x \in X$.

For this we can make the base change to $\kappa(f(x))$ and hence may assume that $Y = \operatorname{Spec} k$ for a field k. Then it suffices to show that $\dim \mathscr{O}_{X,x} = \dim \mathscr{O}_{\mathbb{A}_Y^d, g(x)} = d$ for every closed point $x \in X$. But this holds since g is quasi-finite and flat (Corollary 14.97). □

Hence we see that in the situation of Proposition 25.119, X decomposes into open and closed subschemes X_d, $d \geq 0$, such that all non-empty fibers of $f_{|X_d}$ are equi-dimensional of dimension d.

If f is smooth, then the notion of relative dimension defined here coincides with the one in the definition of smooth morphism (Definition 6.14).

We can now prove a relative version of the fact (cf. Proposition 25.111) that the dualizing complex of a Cohen-Macaulay scheme is concentrated in a single degree.

Proposition 25.120. *Let Y be a noetherian scheme and let $f\colon X \to Y$ be a separated Cohen-Macaulay morphism of finite type and of relative dimension d. Then there exists a coherent $\mathscr{O}_X$-module $\omega_{X/Y}$, flat over Y, such that $f^!\mathscr{O}_Y \cong \omega_{X/Y}[d]$.*

Proof. First recall that the cohomology sheaves of $f^!\mathscr{O}_Y$ are all coherent, as shown by Proposition 25.64. To proceed, we can work locally on X and Y and hence assume that they are both affine and that d is constant.

If $d = 0$, then we are done by Proposition 25.104. If $d > 0$, we use Lemma 25.116 to factor f, locally on X and Y, as

$$X \xrightarrow{\ g\ } \mathbb{A}^d_Y \xrightarrow{\ p\ } Y,$$

where g is Cohen-Macaulay of relative dimension 0, and p is the projection. We then have

$$f^!\mathscr{O}_Y \cong g^!p^!\mathscr{O}_Y \cong (g^!\mathscr{O}_{\mathbb{A}^d_Y})[d]$$

by Example 25.63, and we can use the case $d = 0$ to conclude. □

In fact, the property that $f^!\mathscr{O}_Y$ is concentrated in a single degree characterizes Cohen-Macaulay morphisms.

Proposition 25.121. *Let Y be a noetherian scheme and let $f\colon X \to Y$ be a flat separated morphism of finite type. Let $x \in X$. The following are equivalent:*
(i) *The morphism f is Cohen-Macaulay at x,*
(ii) *In an open neighborhood of x, the complex $f^!\mathscr{O}_Y$ has a unique non-vanishing cohomology sheaf.*

Proof. If f is Cohen-Macaulay in x, then f is Cohen-Macaulay in an open neighborhood of x by Corollary 25.118. Then we have already proved the slightly more precise statement of Proposition 25.120 which shows that "(i) $\Rightarrow$ (ii)" holds.

Now let us show that (ii) implies (i). We write $y = f(x)$ and denote by X_y the fiber over y as before. We may work locally and hence assume that X and Y are both affine and that $f^!\mathscr{O}_Y = \omega[d]$ for some $d \in \mathbb{N}$ and some $\mathscr{O}_X$-module ω which is coherent by Proposition 25.64. Since f is flat by assumption, it is enough to show that the fiber X_y is Cohen-Macaulay at x. Let $i\colon X_y \to X$ be the inclusion. We will show that, after possibly passing to an open affine neighborhood of x, the pullback $i^*f^!\mathscr{O}_Y = i^*\omega[d]$ is a dualizing complex. As it is concentrated in one degree, we can then invoke Proposition 25.111 to finish the proof.

By Corollary 25.118 we see that f is Cohen-Macaulay on an open subset V of X such that $V \cap X_y$ is dense in X_y. Since $(f_{|V})^!\mathscr{O}_Y = \omega_{|V}[d]$, it follows from Proposition 25.120 that $f_{|V}$ is of constant relative dimension d and hence that all irreducible components of X_y have dimension d.

As $x \mapsto \dim_x f^{-1}(f(x))$ is upper semicontinuous (Theorem 14.112), we may assume that all fibers of f have dimension $\leq d$ after possibly shrinking X. By Proposition 25.104 above, ω (placed in degree 0) has tor-amplitude in $[0, d]$ over Y. But a module in degree 0 has tor-amplitude in $(-\infty, 0]$ and we conclude that ω is flat over Y.

The flatness of ω that we just proved and the flatness of f imply that the natural map $Li^*\omega \to i^*\omega$ is an isomorphism, which can be easily checked on stalks. Hence it remains to show that $Li^*f^!\mathscr{O}_Y$ is a dualizing complex on X_y. As f is flat, we have $Li^*f^!\mathscr{O}_Y \cong g^!\kappa(y)$ by (25.20.4) and this is a dualizing complex by Proposition 25.87. $\square$

Remark 25.122. Let S be a noetherian scheme. Let X, Y be separated S-schemes of finite type. Let $f\colon X \to Y$ be a Cohen-Macaulay morphism of S-schemes of constant relative dimension d.

We set $\omega_{X/Y} = H^{-d}(f^!\mathscr{O}_Y)$, the unique non-vanishing cohomology object of the relative dualizing complex $\omega_{X/Y}^\bullet = f^!\mathscr{O}_Y$ by Proposition 25.120. One calls $\omega_{X/Y}$ the *relative dualizing sheaf* for the morphism f.

If S admits a dualizing complex $\omega_S^\bullet$, we obtain dualizing complexes $\omega_X^\bullet = g^!\omega_S^\bullet$ and $\omega_Y^\bullet = h^!\omega_S^\bullet$ on X and Y (where g, h denote the structure morphisms of the S-schemes X, Y) by Proposition 25.87. We then have by (25.20.2)

$$\omega_X^\bullet \cong f^*\omega_Y^\bullet \otimes_{\mathscr{O}_X}^L \omega_{X/Y}[d].$$

In case X and Y are equi-dimensional, we can pass to the dualizing sheaves and obtain an isomorphism

$$\omega_X \cong f^*\omega_Y \otimes_{\mathscr{O}_X} \omega_{X/Y}$$

of coherent $\mathscr{O}_X$-modules.

(25.25) Gorenstein schemes.

Recall the notion of Gorenstein ring, see Definition G.26 and Definition G.29. Every complete intersection ring is Gorenstein (Remark 19.47 (3)). Every Gorenstein ring is Cohen-Macaulay, but not conversely. In fact the notion of being Gorenstein also admits a simple characterization in terms of the dualizing complex.

Definition 25.123. *Let X be a locally noetherian scheme. We call X a Gorenstein scheme if for every $x \in X$ the local ring $\mathscr{O}_{X,x}$ is a Gorenstein local noetherian ring.*

A noetherian ring A is Gorenstein if and only if $\operatorname{Spec} A$ is Gorenstein.

Proposition 25.124. *Let X be a finite-dimensional noetherian scheme. The following are equivalent.*
(i) *The scheme X is Gorenstein.*
(ii) *The structure sheaf $\mathscr{O}_X$ is a dualizing complex on X.*
(iii) *The scheme X has a dualizing complex $\omega_X^\bullet$ and $\omega_X^\bullet$ is an invertible object in $D(X)$.*

Proof. Because of Proposition 25.84, (ii) and (iii) are equivalent.

Let us prove (i) $\Rightarrow$ (ii). It follows from the Definition G.26 that $\mathscr{O}_X$ has injective amplitude in $[0, \dim X]$ as a module over itself, because we can check this on the local rings $\mathscr{O}_{X,x}$ (Proposition 22.79) and their dimensions are bounded by $\dim X$. By Proposition 25.80, it is then enough to show that for each x, $\mathscr{O}_{X,x}$ is a dualizing complex for the local ring $\mathscr{O}_{X,x}$. By Proposition G.26, $\mathscr{O}_{X,x}$ has finite injective dimension over itself. It is then clear that the criterion of Proposition 25.76 is satisfied.

Let us prove (ii) $\Rightarrow$ (i). For every $x \in X$, the complex $\mathscr{O}_{X,x}$ (placed in degree 0) is a dualizing complex for the local ring $\mathscr{O}_{X,x}$, and in particular has finite injective dimension. Therefore the ring $\mathscr{O}_{X,x}$ satisfies condition (i) of Proposition G.26, and therefore $\mathscr{O}_{X,x}$ is Gorenstein. $\square$

Example 25.125. Let S be a finite-dimensional and Gorenstein noetherian scheme. Then we can choose $\omega_S^\bullet = \mathscr{O}_S$ as a dualizing complex by Proposition 25.124. Let $g\colon P \to S$ be a separated morphism of finite type which is smooth of constant relative dimension N. Let $i\colon X \to P$ be a regular immersion of S-schemes of constant codimension c.

Then $f = g \circ i\colon X \to P$ is an lci-morphism of dimension $d := N - c$ (Proposition 19.42) and by Corollary 25.88 we find $\omega_X^\bullet := f^! \mathscr{O}_S = \omega_X[d]$ with

$$(25.25.1) \qquad \qquad \omega_X = i^* \Omega_{P/S}^N \otimes_{\mathscr{O}_X} \bigwedge^c \mathscr{C}_i^\vee.$$

In the important special case $P = \mathbb{P}_S^N$ we have $\Omega_{P/S}^N = \mathscr{O}_{\mathbb{P}_S^N}(-N-1)$ by Example 17.59.

Remark 25.126. There is also a relative version for the Gorenstein property, similarly as we detailed for Cohen-Macaulay morphisms in Section (25.24). One defines a morphism $f\colon X \to Y$ to be *Gorenstein* if f is flat and all fibers of f are locally noetherian Gorenstein schemes.

Every Gorenstein morphism is Cohen-Macaulay and every syntomic morphism is Gorenstein.

A relative version of Proposition 25.124 then says that a flat separated morphism $f\colon X \to Y$ of finite type between noetherian schemes is Gorenstein if and only if the relative dualizing complex $\omega_{X/Y}^\bullet = f^! \mathscr{O}_Y$ is an invertible object in $D(X)$ ([Sta] 0C08).

Hence if f is Gorenstein of constant relative dimension d, then $\omega_{X/Y}^\bullet = \omega_{X/Y}[d]$, where $\omega_{X/Y}$ is a line bundle.

Duality for schemes over fields

In this part, we consider the particular situation of schemes of finite type over a field, and make some of the above, fairly abstract results more explicit. In this context, duality for vector bundles or more generally for coherent sheaves is often called *Serre duality*, a reference to the work of Serre that was all-important as a starting point for the subject.

To apply Serre duality, it is useful to have a concrete description of the dualizing sheaf. For smooth schemes we have already seen (Theorem 25.68) that the dualizing sheaf is the determinant of the Kähler differentials. In Section (25.27) we will also give a description of the dualizing sheaf for normal schemes (Proposition 25.139). We conclude this part by an important application, the Lemma of Zariski-Enriques-Severi.

(25.26) Serre duality.

We start by collecting some immediate corollaries to what we have already shown.

Theorem 25.127. *Let X be a proper scheme over a field k. Denote by $f\colon X \to \operatorname{Spec} k$ the structure morphism.*

Then $\omega_X^\bullet := f^! \mathscr{O}_{\operatorname{Spec} k}(= f^\times \mathscr{O}_{\operatorname{Spec} k})$ is a dualizing complex on X. It has the following properties:

(1) $\omega_X^\bullet$ *is in $D_{\mathrm{coh}}^b(X)$ and $H^i(\omega_X^\bullet) = 0$ for all $i \notin [-\dim(X), 0]$,*

(2) $\omega_X := H^{-\dim X}(\omega_X^\bullet)$ *is a coherent (S_2)-module $\mathscr{O}_X$-module whose support is the union of the top-dimensional irreducible components of X, and is a dualizing sheaf on X. If X is integral, ω_X is torsion-free.*

(3) (Duality for complexes) *For $K \in D_{\mathrm{qcoh}}(X)$, there are isomorphisms for all $i \in \mathbb{Z}$*

(25.26.1) $$\mathrm{Ext}^i_X(K, \omega^\bullet_X) \xrightarrow{\sim} H^{-i}(X, K)^\vee$$

that are functorial in K and compatible with shifts and distinguished triangles, i.e., give rise to an isomorphism of the associated long exact cohomology sequences for $\mathrm{Ext}^i_X(-, \omega^\bullet_X)$ and $H^{-i}(X, -)$. (Here $-^\vee = \mathrm{Hom}_k(-, k)$ denotes the dual k-vector space.)

If $K \in D_{\mathrm{coh}}(X)$, then (25.26.1) is an isomorphism of finite-dimensional k-vector spaces.

(4) (Duality for perfect complexes or vector bundles) *If K is a perfect complex on X (e.g., a finite locally free $\mathscr{O}_X$-module), then there are functorial isomorphisms of finite-dimension k-vector spaces for all i*

(25.26.2) $$H^i(X, \omega^\bullet_X \otimes^L_{\mathscr{O}_X} K^\vee) \xrightarrow{\sim} H^{-i}(X, K)^\vee$$

compatible with shifts and distinguished triangles.

(5) (Duality for quasi-coherent $\mathscr{O}_X$-modules) *For a quasi-coherent $\mathscr{O}_X$-module $\mathscr{F}$, there are isomorphisms*

$$\mathrm{Hom}(\mathscr{F}, \omega_X) \xrightarrow{\sim} H^{\dim X}(X, \mathscr{F})^\vee$$

that are functorial in $\mathscr{F}$.

Proof. Since f is proper, by definition we have $f^! = f^\times$. It follows from Proposition 25.87 that $\omega^\bullet_X$ is a dualizing complex. Parts (1) and (2) follow from Proposition 25.106 with Remark 25.107 and Proposition 25.108.

For (3) consider the functorial isomorphisms

$$R\,\mathrm{Hom}(K, \omega^\bullet_X) = R\,\mathrm{Hom}(K, f^\times \mathscr{O}_{\mathrm{Spec}\,k}) \cong R\,\mathrm{Hom}(R\Gamma(X, K), k)$$

given by the adjunction between Rf_* and $f^\times$ (in the form of Lemma 25.21) and identifying Rf_*K with $R\Gamma(X, K)$. We have $H^i(R\,\mathrm{Hom}(K, \omega^\bullet_X)) = \mathrm{Ext}^i_X(K, \omega^\bullet_X)$.

Let us compute $H^i(-)$ of the right hand side. By Example F.153, we can represent $R\Gamma(X, K) \in D(k)$ by $\bigoplus_{i \in \mathbb{Z}} H^i(X, K)[-i]$. Since k is injective as a k-vector space, we find that $R\,\mathrm{Hom}(R\Gamma(X, K), k)$ is represented by the complex $\mathrm{Hom}_k(\bigoplus_{i \in \mathbb{Z}} H^i(X, K)[-i], k)$ whose cohomology in degree i is given by $H^{-i}(X, K)^\vee$.

If $K \in D_{\mathrm{coh}}(X)$, then $H^i(X, K)$ is a finite-dimensional k-vector space by Corollary 23.25.

Part (4) follows from (3) using that for perfect complexes one has, writing $R\Gamma(-)$ for the functor $R\Gamma(X, -)$,

$$\begin{aligned}
R\,\mathrm{Hom}(K, \omega^\bullet_X) &= R\Gamma R\,\mathscr{H}\!om(K, \omega^\bullet_X) \\
&= R\Gamma R\,\mathscr{H}\!om(\mathscr{O}_X, \omega^\bullet_X \otimes^L_{\mathscr{O}_X} K^\vee) \\
&= R\Gamma(X, \omega^\bullet_X \otimes^L_{\mathscr{O}_X} K^\vee),
\end{aligned}$$

where for the first identification we use Proposition 21.107, the second identification holds by Proposition 21.148 (3), and the last identification by (21.18.10).

Part (5) just restates Proposition 25.110.   □

The formation of dualizing complexes and dualizing sheaves is compatible with change of the base field.

Lemma 25.128. *Let k'/k be a field extension.*

(1) *Let X be a proper k-scheme and denote by X' the base change $X \otimes_k k'$. Then the dualizing complex $\omega_{X'}^\bullet$ and the $\mathscr{O}_{X'}$-module $\omega_{X'}$ (as in Theorem 25.127) coincide with the pullbacks of $\omega_X^\bullet$ and ω_X to X'.*

(2) *Let X be a proper k'-scheme and suppose that k'/k is finite (equivalently: X is a k'-scheme which is proper as a k-scheme). Then $\omega_X^\bullet$ and ω_X are independent of whether we consider X as a k'-scheme, or as a k-scheme.*

Proof. Part (1) follows directly from Theorem 25.31.

Part (2). The equivalence follows from Theorem 12.65, since $k' \subseteq H^0(X, \mathscr{O}_X)$. To prove the independence statement, let us denote by $f\colon X \to \operatorname{Spec} k$ and $g\colon X \to \operatorname{Spec} k'$ the structure morphisms, and by $h\colon \operatorname{Spec} k' \to \operatorname{Spec} k$ the morphism corresponding to the inclusion $k \subseteq k'$. Then $f^! = g^! \circ h^!$. The claim follows since $h^! k \cong R\operatorname{Hom}_k^{k'}(k', k) = \operatorname{Hom}_k^{k'}(k', k) \cong k'$, where we obtain the first isomorphism from Proposition 25.44. $\square$

In the remainder of this section, we always equip a k-scheme $f\colon X \to \operatorname{Spec} k$ with its dualizing complex $\omega_X^\bullet := f^! \mathscr{O}_{\operatorname{Spec} k}$ and the corresponding dualizing sheaf ω_X. For a (connected) Cohen-Macaulay scheme X with a dualizing complex, the dualizing complex is concentrated in one degree, and hence we can equivalently phrase the above results in terms of the dualizing sheaf by Proposition 25.111 and Proposition 25.124.

Corollary 25.129. *Let X be a connected proper scheme over a field k. Assume that X is Cohen-Macaulay (hence equi-dimensional, Proposition 14.126). Let $d := \dim X$. With notation as in Theorem 25.127, we have the following assertions.*

(1) *ω_X is a coherent Cohen-Macaulay module on X which is a dualizing module and $\operatorname{Supp}(\omega_X) = X$.*

(2) *X is Gorenstein (e.g., if X is lci over k) if and only if ω_X is a line bundle.*

(3) *(Duality on Cohen-Macaulay schemes for complexes) For $K \in D_{\mathrm{qcoh}}(X)$, there are isomorphisms for all i*

$$\operatorname{Ext}_X^i(K, \omega_X[d]) \xrightarrow{\sim} H^{-i}(X, K)^\vee$$

that are functorial in K and compatible with shifts and distinguished triangles.

(4) *(Duality on Cohen-Macaulay schemes for $\mathscr{O}_X$-modules) For $\mathscr{F} \in \operatorname{QCoh}(X)$, there are isomorphisms for all i*

$$\operatorname{Ext}_{\mathscr{O}_X}^{d-i}(\mathscr{F}, \omega_X) \xrightarrow{\sim} H^i(X, \mathscr{F})^\vee$$

that are functorial in $\mathscr{F}$.

(5) *(Duality on Cohen-Macaulay schemes for vector bundles) For every finite locally free $\mathscr{O}_X$-module $\mathscr{F}$ there are functorial isomorphisms of finite-dimensional k-vector spaces*

$$H^{d-i}(X, \omega_X \otimes_{\mathscr{O}_X} \mathscr{F}^\vee) \xrightarrow{\sim} H^i(X, \mathscr{F})^\vee.$$

Also recall the following explicit description of ω_X if the k-scheme X is smooth or, more generally, can be regularly embedded into a smooth scheme, obtained as a corollary to Theorem 25.68 and Example 25.125. Note that in these cases X is Gorenstein.

Corollary 25.130. *Let k be a field and let X be a connected separated k-scheme of dimension d.*

(1) *Suppose that X is smooth over k. Then*

$$\omega_X \cong \Omega^d_{X/k}.$$

(2) *Suppose that there exists a closed regular immersion $i\colon X \to P$, where P is a connected smooth and separated k-scheme of dimension N. Then X is equi-dimensional and*

$$\omega_X \cong i^*\Omega^N_{P/k} \otimes_{\mathscr{O}_X} \bigwedge^{N-d} \mathscr{C}_i^\vee,$$

where $\mathscr{C}_i$ denotes the conormal bundle of i, a finite locally free $\mathscr{O}_X$-module of rank $N - d$.

Note that under either of the assumptions X is Cohen-Macaulay, hence we can apply Corollary 25.129 if X is in addition proper over k. If $P = \mathbb{P}^N_k$, then $\Omega^N_{P/k} = \mathscr{O}_{\mathbb{P}^N_k}(-N-1)$ by Example 17.59.

Sometimes the following generalization of Serre duality is useful.

Corollary 25.131. *Let X be a proper scheme over a field k. Let $K \in D_{\mathrm{qcoh}}(X)$ and $\mathscr{E}$ be a perfect complex on X. Then there are functorial isomorphisms, compatible with shifts and distinguished triangles,*

$$\mathrm{Ext}^i_{\mathscr{O}_X}(K, \omega^\bullet_X \otimes^L_{\mathscr{O}_X} \mathscr{E}) \xrightarrow{\sim} \mathrm{Ext}^{-i}_{\mathscr{O}_X}(\mathscr{E}, K)^\vee. \qquad i \in \mathbb{Z}$$

For $K \in D_{\mathrm{coh}}(X)$ this is an isomorphism of finite-dimensional vector spaces.

Proof. Since $\mathscr{E}$ is perfect, we have for all $\mathscr{F}, \mathscr{G} \in D(X)$

$$(*) \qquad R\mathrm{Hom}_{\mathscr{O}_X}(\mathscr{F} \otimes^L \mathscr{E}^\vee, \mathscr{G}) \cong R\mathrm{Hom}_{\mathscr{O}_X}(\mathscr{F}, \mathscr{E} \otimes^L \mathscr{G}).$$

by applying $R\Gamma(X, -)$ to (21.32.3) and using $R\Gamma(X, R\mathscr{H}om(-,-)) = R\mathrm{Hom}(-,-)$ (Proposition 21.107). Therefore we find for all $i \in \mathbb{Z}$

$$\begin{aligned}
\mathrm{Ext}^i_{\mathscr{O}_X}(K, \omega^\bullet_X \otimes^L \mathscr{E}) &= \mathrm{Ext}^i_{\mathscr{O}_X}(K, \otimes^L \mathscr{E}^\vee, \omega^\bullet_X) \\
&= H^{-i}(X, K \otimes^L_{\mathscr{O}_X} \mathscr{E})^\vee \\
&= H^{-i}(X, R\mathscr{H}om_{\mathscr{O}_X}(\mathscr{E}, K))^\vee \\
&= (H^{-i}R\mathrm{Hom}_{\mathscr{O}_X}(\mathscr{E}, K))^\vee \\
&= \mathrm{Ext}^{-i}_{\mathscr{O}_X}(\mathscr{E}, K)^\vee.
\end{aligned}$$

Here the second equality holds by Serre duality in the form of Theorem 25.127 (3) and the third identity by (21.32.2). The last assertion follows from Corollary 23.38. □

Corollary 25.132. *Let X be a connected Cohen-Macaulay proper scheme over k of dimension d and let $\mathscr{E}$ be a finite locally free $\mathscr{O}_X$-module. Then we have functorial isomorphisms*

$$(25.26.3) \qquad \mathrm{Ext}^{d-i}(K, \omega_X \otimes_{\mathscr{O}_X} \mathscr{E}) \xrightarrow{\sim} \mathrm{Ext}^i(\mathscr{E}, K)^\vee$$

for all $i \in \mathbb{Z}$ and for all $K \in D_{\mathrm{qcoh}}(X)$. For $K \in D_{\mathrm{coh}}(X)$ this is an isomorphism of finite-dimensional vector spaces.

Proof. As X is Cohen-Macaulay and connected, it is equi-dimensional and $\omega^\bullet = \omega_X[d]$. We also have $\omega_X \otimes^L_{\mathscr{O}_X} \mathscr{E} = \omega_X \otimes_{\mathscr{O}_X} \mathscr{E}$ if $\mathscr{E}$ is finite locally free. Hence (25.26.3) is a special case of Corollary 25.131. □

(25.27) Dualizing sheaves on normal varieties.

Using the notion of reflexive $\mathscr{O}_X$-module introduced in the following definition, we can describe the dualizing sheaf on a normal variety (over a perfect field) in terms of the sheaf of differentials on its smooth locus.

Let X be a ringed space. For an $\mathscr{O}_X$-module $\mathscr{F}$, recall that we denote by $\mathscr{F}^\vee := \mathscr{H}om_{\mathscr{O}_X}(\mathscr{F}, \mathscr{O}_X)$ its dual.

Remark 25.133.
(1) If X is a scheme and $\mathscr{F}$ is of finite presentation, then $\mathscr{F}^\vee$ is again quasi-coherent (Proposition 7.29).
(2) If X is a locally noetherian scheme and $\mathscr{F}$ is a coherent $\mathscr{O}_X$-module, then $\mathscr{F}^\vee$ is again coherent. Indeed, working locally, by (1) it suffices to show that if M is a finitely generated module over a noetherian ring, then its dual $M^\vee = \operatorname{Hom}_A(M, A)$ is finitely generated. But a surjection $A^r \to M$ yields an injective map $M^\vee \hookrightarrow \operatorname{Hom}_A(A^r, A) = A^r$ which shows that $M^\vee$ is finitely generated since A is noetherian. Note that we also could have invoked Corollary 22.67.

Definition 25.134. *Let X be a locally noetherian scheme. A coherent $\mathscr{O}_X$-module $\mathscr{F}$ is called* reflexive, *if the natural homomorphism $\mathscr{F} \to \mathscr{F}^{\vee\vee}$ is an isomorphism.*

Every locally free $\mathscr{O}_X$-module of finite rank is reflexive.

Of course, this definition would make formally sense for an arbitrary $\mathscr{O}_X$-module over an arbitrary scheme X (or even an arbitrary ringed space), but if we want to have in the affine case the compatibility with the corresponding notion for modules, Remark 25.133 shows that additional conditions on X and on $\mathscr{F}$ are necessary.

Remark 25.135.
(1) Let A be a noetherian ring, $X = \operatorname{Spec} A$, and M a finitely generated A-module. Then M is reflexive (Definition G.16) if and only if $\tilde{M}$ is a reflexive $\mathscr{O}_X$-module.
(2) Let X be an integral scheme, $\mathscr{F}$ and $\mathscr{G}$ quasi-coherent $\mathscr{O}_X$-modules with $\mathscr{F}$ of finite presentation and $\mathscr{G}$ torsion-free. Then $\mathscr{H}om_{\mathscr{O}_X}(\mathscr{F}, \mathscr{G})$ is a torsion-free $\mathscr{O}_X$-module which can be checked easily by reduction to the affine case using Remark 25.133 (1).

In particular, we see that every reflexive $\mathscr{O}_X$-module is torsion-free if X is locally noetherian.

Here we will use this notion for normal integral schemes. Then we have the following simple properties.

Lemma 25.136. *Let X be a normal connected noetherian scheme.*
(1) *If $\mathscr{M}$ is a torsion-free coherent $\mathscr{O}_X$-module, then there exists an open subscheme $U \subseteq X$ such that $\operatorname{codim}_X(X \setminus U) \geq 2$ and such that $\mathscr{M}_{|U}$ is a locally free $\mathscr{O}_U$-module of finite rank.*
(2) *Let $\mathscr{M}$ be a coherent $\mathscr{O}_X$-module. For every open subscheme $j\colon U \hookrightarrow X$ such that $\operatorname{codim}_X(X \setminus U) \geq 2$ and such that $\mathscr{M}_{|U}$ is a locally free $\mathscr{O}_U$-module of finite rank, the natural homomorphism $\mathscr{M} \to \mathscr{M}^{\vee\vee}$ factors through $j_*(\mathscr{M}_{|U})$ and the resulting homomorphism $j_*(\mathscr{M}_{|U}) \to \mathscr{M}^{\vee\vee}$ is an isomorphism.*

Proof. For (1) note that for every $x \in X$ such that the local ring $\mathscr{O}_{X,x}$ has dimension 1, this ring is a discrete valuation ring, and hence every torsion-free finite module is free.

For (2), notice that our assumptions imply $j_*\mathscr{O}_U = \mathscr{O}_X$ (Theorem 6.45). Thus the sheaf-version of the adjunction between j^* and j_* gives us functorial identifications

$$\mathscr{M}^{\vee\vee} = \mathscr{H}om_{\mathscr{O}_X}(\mathscr{M}^\vee, j_*(\mathscr{O}_U)) = j_* \mathscr{H}om_{\mathscr{O}_U}(\mathscr{M}^\vee_{|U}, \mathscr{O}_U) = j_*\mathscr{M}_{|U},$$

and one checks that these are compatible with the natural maps from $\mathscr{M}$ to the left and right hand side, respectively. $\qquad\square$

As an immediate consequence we obtain the following characterization of reflexive $\mathscr{O}_X$-modules in this case.

Corollary 25.137. *Let X be a normal connected noetherian scheme. Let $\mathscr{M}$ be a coherent $\mathscr{O}_X$-module. Let $j\colon U \hookrightarrow X$ be an open subscheme such that $\operatorname{codim}_X(X \setminus U) \geq 2$ and such that $\mathscr{M}_{|U}$ is a locally free $\mathscr{O}_U$-module of finite rank. The following are equivalent.*
(i) *The $\mathscr{O}_X$-module $\mathscr{M}$ is reflexive.*
(ii) *The natural map*
$$\mathscr{M} \xrightarrow{\sim} j_*(\mathscr{M}_{|U})$$
is an isomorphism.

The relevance of this notion for the theory of dualizing sheaves comes from the following result.

Proposition 25.138. *Let X be a normal connected noetherian scheme, let $\omega_X^\bullet$ be a dualizing complex for X and let ω_X be its dualizing sheaf. Then ω_X is reflexive.*

Proof. By Proposition G.18 this follows from Proposition 25.106 (2) and (3). $\qquad\square$

Therefore we obtain the following description of the dualizing sheaf for normal varieties over perfect fields.

Proposition 25.139. *Let k be a perfect field, and let X be a normal connected k-scheme of finite type. Let $n = \dim X$ and let ω_X be the dualizing sheaf of X. Denote by $j\colon X_{\mathrm{sm}} \hookrightarrow X$ the open immersion of the smooth locus of X into X. Then*

$$\omega_X \cong j_*(\Omega^n_{X_{\mathrm{sm}}/k}) \cong (\Omega^n_{X/k})^{\vee\vee}.$$

Proof. Since k is perfect, the smooth locus X_{sm} of X is the same as the regular locus by Proposition 18.67. It is open and dense in X (Theorem 6.19). Its complement has at least codimension 2 since X is normal (Proposition 6.40). The restriction of ω_X to X_{sm} equals the dualizing sheaf $\Omega^n_{X_{\mathrm{sm}}/k}$ on X_{sm} (Corollary 25.130). The first isomorphism in the statement follows then from Corollary 25.137 since ω_X is reflexive by Proposition 25.138.

For the second one, note that the dual of any coherent $\mathscr{O}_X$-module is reflexive. In fact, this can be checked locally, and hence follows from Lemma G.17. Hence $(\Omega^n_{X/k})^{\vee\vee}$ is a reflexive $\mathscr{O}_X$-module whose restriction to X_{sm} is isomorphic to $\omega_{X|X_{\mathrm{sm}}}$. Hence $\omega_X \cong (\Omega^n_{X/k})^{\vee\vee}$ by Corollary 25.137. $\qquad\square$

(25.28) The Lemma of Enriques-Severi-Zariski.

Theorem 25.140. (Lemma of Enriques-Severi-Zariski) *Let k be a field and let X be a projective k-scheme of finite type. Let $\mathscr{L}$ be an ample line bundle on X. Let $\mathscr{F}$ be a coherent $\mathscr{O}_X$-module. Assume that $\operatorname{depth}_{\mathscr{O}_{X,x}}(\mathscr{F}_x) \geq 2$ for all closed points $x \in X$ (e.g., this holds, if X is normal of dimension ≥ 2 and $\mathscr{F}$ is locally free). Then there exists n_0 such that $H^1(X, \mathscr{F} \otimes \mathscr{L}^{-n}) = 0$ for all $n \geq n_0$.*

If X is Cohen-Macaulay of dimension ≥ 2 and $\mathscr{F}$ is locally free, then the theorem follows easily from Serre duality (Corollary 25.129 (5)) and Serre's criterion for ampleness (Lemma 23.5). In the general case, we first reduce to the case of projective space, and use Corollary 25.129 (4) in place of Corollary 25.129 (5), which allows us to weaken the assumption on $\mathscr{F}$.

Proof. Some power $\mathscr{L}^r$ of $\mathscr{L}$ is very ample and gives rise to a closed embedding $\iota\colon X \to \mathbb{P}_k^N$. Then $H^1(X, \mathscr{G}) \cong H^1(\mathbb{P}_k^N, \iota_*\mathscr{G})$ for every $\mathscr{O}_X$-module $\mathscr{G}$. Applying this to $\mathscr{F}, \mathscr{F} \otimes \mathscr{L}, \dots,$ $\mathscr{F} \otimes \mathscr{L}^{r-1}$, we reduce to the case that $X = \mathbb{P}_k^N$ and $\mathscr{L} = \mathscr{O}_{\mathbb{P}_k^N}(1)$. Note that the depth of $(\iota_*\mathscr{G})_{\iota(x)}$ (over $\mathscr{O}_{\mathbb{P}_k^N, \iota(x)}$) equals the depth of $\mathscr{G}_x$ (over $\mathscr{O}_{X,x}$) for $x \in X$, and $(\iota_*\mathscr{G})_y = 0$ has depth ∞ (by convention) for $y \notin \iota(X)$.

For the regular scheme $X = \mathbb{P}_k^N$, we may use Serre duality in the form of Corollary 25.129 (4) and obtain $H^1(X, \mathscr{F}(-n))^\vee \cong \mathrm{Ext}_{\mathscr{O}_{\mathbb{P}_k^N}}^{N-1}(\mathscr{F}(-n), \omega_{\mathbb{P}_k^N})$.

Now for n sufficiently large, we have

$$\mathrm{Ext}_{\mathscr{O}_{\mathbb{P}_k^N}}^{N-1}(\mathscr{F}(-n), \omega_{\mathbb{P}_k^N}) \cong \mathrm{Ext}_{\mathscr{O}_{\mathbb{P}_k^N}}^{N-1}(\mathscr{F}, \omega_{\mathbb{P}_k^N}(n)) \cong \Gamma(\mathbb{P}_k^N, \mathscr{E}xt_{\mathscr{O}_{\mathbb{P}_k^N}}^{N-1}(\mathscr{F}, \omega_{\mathbb{P}_k^N}(n))),$$

where the first equality follows from 21.148 (3) (and holds for all n), and the second equality follows from the local-to-global spectral sequence for Ext groups (Corollary 21.108) together with Serre's criterion for ampleness (in the form of Lemma 23.5) which implies the vanishing

$$H^p(X, \mathscr{E}xt^q(\mathscr{F}, \omega_{\mathbb{P}_k^N}(n))) \cong H^p(X, \mathscr{E}xt^q(\mathscr{F}, \omega_{\mathbb{P}_k^N})(n)) = 0$$

for all q, all $p > 0$ and all sufficiently large n. Note that by Corollary 22.67 (2), the sheaves $\mathscr{E}xt^q(\mathscr{F}, \omega_{\mathbb{P}_k^N})$ are coherent.

It is thus enough to show that $\mathscr{E}xt_{\mathscr{O}_{\mathbb{P}_k^N}}^{N-1}(\mathscr{F}, \omega_{\mathbb{P}_k^N}(n)) = 0$. We claim that in fact $\mathscr{E}xt_{\mathscr{O}_{\mathbb{P}_k^N}}^{N-1}(\mathscr{F}, \mathscr{G}) = 0$ for all $\mathscr{O}_X$-modules $\mathscr{G}$. To see this, it is enough to check that the stalks of these sheaves vanish, and those we can identify with the corresponding Ext groups over the regular local ring $\mathscr{O}_{\mathbb{P}_k^N, x}$ (see Proposition 22.65), which we can compute using a projective resolution of the $\mathscr{O}_{\mathbb{P}_k^N, x}$-module $\mathscr{F}_x$. By the Auslander-Buchsbaum formula, Theorem G.12, our assumption on the depth of $\mathscr{F}_x$ ensures that $\mathscr{F}$ has projective dimension $\leq N - 2$, and this yields the desired vanishing.

The condition on the depth is satisfied if X is normal of dimension ≥ 2 and $\mathscr{F}$ is locally free by Serre's criterion for normality, see Proposition B.81. $\qquad\square$

A beautiful geometric consequence of the Lemma of Enriques-Severi-Zariski is the following connectedness result.

Corollary 25.141. *Let k be a field and let X be a normal projective integral k-scheme of finite type, and such that $\dim X \geq 2$. Let D be an ample effective Cartier divisor on X. Then the support $\mathrm{Supp}(D)$ of D is connected.*

Proof. We may replace D by an arbitrary positive multiple, because that does not change the support. Thus by Theorem 25.140 we may assume that $H^1(X, \mathscr{O}_X(-D)) = 0$. We regard the effective divisor D as a closed subscheme of X. The short exact sequence

$$0 \to \mathscr{O}_X(-D) \to \mathscr{O}_X \to \mathscr{O}_D \to 0$$

gives rise to a surjection $H^0(X, \mathcal{O}_X) \to H^0(X, \mathcal{O}_D) = H^0(D, \mathcal{O}_D)$. Since X is integral and proper over k, $H^0(X, \mathcal{O}_X)$ is a domain and finite over k, hence a field, and it follows that $H^0(D, \mathcal{O}_D) \cong H^0(X, \mathcal{O}_X)$ is a field, too. In particular, D is connected. $\square$

Applications to algebraic surfaces

In the remaining sections of this chapter, we will discuss some applications to the theory of algebraic surfaces. For example we will show that every regular proper surface over a field is projective (Theorem 25.151). We also prove the Hodge index theorem (Theorem 25.156) which will be a crucial ingredient for the proof of the Weil conjectures for curves in Section (26.28).

(25.29) The theorem of Riemann-Roch for algebraic surfaces.

Let k be a field. We will study algebraic surfaces over k in the following sense.

Definition 25.142. *An* (algebraic) surface *is a separated k-scheme of finite type that is equi-dimensional of dimension 2.*

An important tool for the study of proper surfaces is the intersection pairing that we obtain as a special case of the intersection numbers introduced in Chapter 23. Let us recall the definition and the properties important for us in this special case. By a divisor on X we mean a Cartier divisor. If X is regular, then we can identify Cartier divisors and Weil divisors, see Section (11.13).

Definition and Proposition 25.143. *Let X be a proper algebraic surface over the field k. The intersection pairing $\mathrm{Pic}(X) \times \mathrm{Pic}(X) \to \mathbb{Z}$ given by*

$$(\mathcal{L}, \mathcal{M}) \mapsto (\mathcal{L} \cdot \mathcal{M} \cdot X) = \chi(\mathcal{O}_X) - \chi(\mathcal{L}^{-1}) - \chi(\mathcal{M}^{-1}) + \chi(\mathcal{L}^{-1} \otimes \mathcal{M}^{-1}),$$

is bilinear and has the following properties:
(1) *If $\mathcal{M} = \mathcal{O}_X(D)$ for an effective divisor X, then $(\mathcal{L} \cdot \mathcal{M} \cdot X) = (\mathcal{L} \cdot D) = \deg(\mathcal{L}_{|D})$, where we consider D as a closed subscheme of X.*
(2) *If C and D are effective divisors on X without a common irreducible component, and $Z = C \cap D$ is their schematic intersection, then*

$$(C \cdot D) = \sum_{z \in Z} \dim_k(\mathcal{O}_{Z,z}).$$

Proof. The formula for the pairing is a direct consequence of the definitions in Chapter 23, specifically Definition 23.57, Definition 23.70. The bilinearity was shown in Corollary 23.72 ((2)).

For part (1), see Proposition 23.58 (3) and Example 23.75 (1), for part (2) see part (2) of that example. $\square$

Given a proper surface, usually we denote the intersection product of line bundles $\mathscr{L}$ and $\mathscr{M}$, and of divisors C and D, respectively, by $(\mathscr{L}\cdot\mathscr{M})$ and $(C\cdot D)$. The self-intersection number $(\mathscr{L}^2) := (\mathscr{L}\cdot\mathscr{L})$, $(D^2) = (D\cdot D)$ is an important invariant. If D is a very ample divisor on a proper algebraic surface, then we have a simple interpretation of this number: (D^2) equals the degree of X with respect to the embedding $X \subseteq \mathbb{P}_k^N$ given by D, see Definition 23.80. On the other hand, in general the self-intersection number of a divisor may be negative, see Lemma 25.148 and Theorem 25.156 below.

The theorem of Riemann-Roch expresses the Euler characteristic of a line bundle in terms of the Euler characteristic of the structure sheaf and an intersection product involving the canonical divisor.

Theorem 25.144. (Theorem of Riemann-Roch for smooth proper algebraic surfaces) *Let k be a field and let X be a smooth proper algebraic surface over k. Let K be a canonical divisor on X, i.e., $\mathscr{O}_X(K) \cong \Omega^2_{X/k}$. Then for every divisor D on X we have*

$$\chi(\mathscr{O}_X(D)) = \frac{1}{2}(D\cdot(D-K)) + \chi(\mathscr{O}_X).$$

Proof. We compute the intersection product $(-D\cdot(D-K))$ using the definition:

$$(-D\cdot(D-K)) = \chi(\mathscr{O}_X) - \chi(\mathscr{O}_X(D)) - \chi(\mathscr{O}_X(K-D)) + \chi(\mathscr{O}_X(K)).$$

By Serre duality (Corollary 25.129) we have $\chi(\mathscr{O}_X) = \chi(\mathscr{O}_X(K))$ and $\chi(\mathscr{O}_X(D)) = \chi(\mathscr{O}_X(K-D))$. Hence we obtain

$$-(D\cdot(D-K)) = 2\chi(\mathscr{O}_X) - 2\chi(\mathscr{O}_X(D)),$$

as desired. $\qquad\square$

Proposition 25.145. (Adjunction formula) *Let k be a field, and let X be a connected smooth proper algebraic surface over k. Let $\omega_X = \Omega^2_{X/k}$ be its canonical bundle, and let K be a canonical divisor on X, i.e., $\mathscr{O}_X(K) \cong \omega_X$. Then for every effective Cartier divisor C on X (corresponding to a closed curve $C \subseteq X$) we have*

$$(C\cdot(C+K)) = 2g - 2,$$

where g is the genus of the curve C.

Proof. We have $(C\cdot(C+K)) = \deg((\mathscr{O}_X(C)\otimes_{\mathscr{O}_X}\omega_X)_{|C})$ by Proposition 25.143 (1). By Corollary 25.130 (2) and Remark 19.24, $(\mathscr{O}_X(C)\otimes_{\mathscr{O}_X}\omega_X)_{|C}$ is the dualizing sheaf of the curve C over k, which has degree $2g - 2$ by Corollary 26.52. If C is smooth, we could alternatively use Proposition 19.32. $\qquad\square$

For surfaces, the Nakai-Moishezon criterion for ampleness, Theorem 23.90, takes the following form.

Theorem 25.146. (Ampleness criterion of Nakai-Moishezon for algebraic surfaces) *Let k be a field, and let X be an integral proper algebraic surface over k. Let $\mathscr{L}$ be a line bundle on X.*

(1) *The line bundle $\mathscr{L}$ on X is ample if and only if $(\mathscr{L}^2) > 0$ and $(\mathscr{L}\cdot C) > 0$ for every integral curve $C \subset X$.*

(2) *If $H^0(X, \mathscr{L}) \neq 0$ and $(\mathscr{L} \cdot C) > 0$ for every integral curve $C \subset X$, then $\mathscr{L}$ is ample.*

Proof. The first formulation is a direct translation of the general version of the theorem. The second version follows by inspecting the proof of Theorem 23.90, and noting that the condition $(\mathscr{L}^2) > 0$ was used only to show that some power of $\mathscr{L}$ has non-trivial global sections. $\qquad\square$

(25.30) Resolution of indeterminacies for surfaces.

As an illustration of the usefulness of intersection numbers, let us prove the following theorem.

Theorem 25.147. *Let k be a field, let X be a regular proper algebraic surface over a field k, and let Y be a projective k-scheme. If $f \colon X \dashrightarrow Y$ is a rational map, then there exists a projective surjective morphism $\pi \colon \tilde{X} \to X$ which is an isomorphism over the maximal domain of definition of f and such that $f \circ \pi$ extends to a morphism $\tilde{X} \to Y$. Furthermore, π can be constructed as a composition of blow-ups of closed points.*

By saying that the rational map $f \circ \pi$ extends to a morphism, we mean that its (maximal) domain of definition (see Proposition 9.27) is equal to $\tilde{X}$. By abuse of terminology, we then also say that this rational map *is* a morphism.

Compare the much more general statement that we stated, without proof, as Theorem 13.98, and the special case of rational maps to projective schemes (Proposition 13.99) which almost shows the above theorem, except that it does not give that the morphism $\tilde{X} \to X$ can be constructed as a sequence of blow-ups of closed points, but on the other hand works in arbitrary dimension and does not require X to be regular. For the application in Section (25.31), that proposition is actually sufficient.

We start with a lemma which clarifies the intersection product on the Picard group of the surface obtained from X by blowing up a closed point in X. (See Sections (13.19) and (19.5).)

Lemma 25.148. *Let k be a field and let X be a regular proper surface over k. Let $x \in X$ be a closed point. Let $\pi \colon \tilde{X} \to X$ denote the blow-up of X in the point x, and let $E \subset \tilde{X}$ be the exceptional divisor.*
(1) *For divisors D, D' on X we have $(\pi^*D, \pi^*D') = (D, D')$.*
(2) *For every divisor D on X we have $(\pi^*D, E) = 0$.*
(3) *We have $(E^2) = -[\kappa(x) : k]$.*
(4) *The homomorphism*

$$\operatorname{Pic}(X) \oplus \mathbb{Z} \longrightarrow \operatorname{Pic}(\tilde{X}), \quad (\mathscr{L}, n) \mapsto \pi^*\mathscr{L} \otimes \mathscr{O}_X(nE),$$

is an isomorphism.

Proof. Part (1) follows from Proposition 23.77. We may compute the intersection number in Part (2) as $\deg(\pi^*\mathscr{O}_X(D)_{|E})$, which vanishes because $\mathscr{O}_X(D)$ is trivial on a neighborhood of x, whence $\pi^*\mathscr{O}_X(D)_{|E}$ is trivial.

Let us prove Part (3). Under the identification $E \cong \mathbb{P}^1_{\kappa(x)}$ (Example 13.95), the line bundle $\mathscr{O}_X(-E)_{|E}$ is identified with $\mathscr{O}_{\mathbb{P}^1_{\kappa(x)}}(1)$, cf. Remark 13.94. Therefore

$$(E^2) = \deg_k(\mathscr{O}_X(E)_{|E}) = [\kappa(x) : k] \deg_{\kappa(x)}(\mathscr{O}_{\mathbb{P}^1_{\kappa(x)}}(-1)) = -[\kappa(x) : k].$$

Finally, the surjectivity in Part (4) follows easily by interpreting the Picard groups as groups of Weil divisor classes (Proposition 11.42 using that $\tilde{X}$ is regular by Remark 19.36). But if $\pi^*\mathscr{L} \otimes \mathscr{O}_X(nE) \cong \mathscr{O}_X$, then it follows from Part (3) that $n = 0$, so $\pi^*\mathscr{L}$ is trivial. Writing $U = X \setminus \{x\}$, we find that $\mathscr{L}_{|U} = (\pi^*\mathscr{L})_{|\pi^{-1}(U)}$ is trivial, and using Proposition 11.42 once again, we conclude that $\mathscr{L}$ is trivial. □

We can now prove the theorem. The idea of the proof is rather simple. Consider a rational map $f\colon X \dashrightarrow Y$ from a regular proper surface X over k to a projective k-scheme Y. Since X is regular and Y is proper, the valuative criterion for properness applied to Y shows that f is defined on an open subset $U \subseteq X$ whose complement has codimension at least 2, i.e., consists of finitely many closed points. We expect that blowing up these points should improve the situation, even though the resulting rational map from the blow-up to Y might still be undefined at some of the points of the exceptional divisors. We repeat the process by blowing up further. To conclude the proof, we have to show that this process must eventually stop.

Proof. (of Theorem 25.147) We may embed Y in some projective space over k by assumption. Replacing Y we may then assume that $Y = \mathbb{P}_k^m$ for some m.

Then the (maximal) domain of definition of the rational map $f\colon X \dashrightarrow \mathbb{P}_k^m$ is an open subset U of X whose complement consists of (at most) finitely many closed points. In particular, we have $\mathrm{Pic}(X) \cong \mathrm{Pic}(U)$ by restriction of line bundles (Proposition 11.42), and the morphism $U \to \mathbb{P}_k^m$ given by f gives us a line bundle $\mathscr{L}$ on X. We have $\Gamma(X, \mathscr{L}) = \Gamma(U, \mathscr{L})$ because X is normal (locally, sections on U extend uniquely to sections on X by Hartogs's theorem, Theorem 6.45; since the extension is unique, the result follows by gluing). The points where f is not defined are precisely the points where all global sections of $\mathscr{L}$ vanish.

Claim 1. We have $(\mathscr{L}^2) \geq 0$.

Proof of claim. The property of $\mathscr{L}$ that we will use is that its "base locus", i.e., the common vanishing locus of all its global sections, has dimension 0, or in other words, does not contain a curve. This property is preserved by changing the base field, and by Remark 23.71 this does not change the intersection number, either. We may hence assume that k is infinite.

Let D be an effective divisor such that $\mathscr{L} \cong \mathscr{O}_X(D)$, and let $C_1, \ldots, C_r$ be the integral curves in the support of D. We will show that there exists a divisor D' which is linearly equivalent to D and such that none of the C_i lies in the support of D'. This implies $(\mathscr{L}^2) = (D \cdot D') \geq 0$ by Proposition 25.143 (2).

Finding such a D' amounts to finding a section $s \in \Gamma(X, \mathscr{L})$ which does not vanish along any of the C_i, i.e., s must not lie in the kernel of the evaluation map $\Gamma(X, \mathscr{L}) \to \mathscr{L}(\eta_i)$, for any i, where η_i denotes the generic point of C_i. For a fixed i, this kernel is a *proper* k-subvector space of $\Gamma(X, \mathscr{L})$, since by assumption C_i is not contained in the support of every divisor linearly equivalent to D. Since k is infinite, $\Gamma(X, \mathscr{L})$ does not equal the union of finitely many proper subvector spaces, and the existence of s satisfying the desired property for all i follows.

To continue, assume that $U \neq X$ and let $\pi\colon \tilde{X} \to X$ be the blow-up of X in a closed point X of the reduced closed subscheme $X \setminus U$. Denote by E the exceptional divisor. Similarly as before, the rational map $f \circ \pi\colon \tilde{X} \dashrightarrow \mathbb{P}_k^m$ gives rise to a line bundle $\tilde{\mathscr{L}}$ on $\tilde{X}$.

Claim 2. We have $(\tilde{\mathscr{L}}^2) < (\mathscr{L}^2)$.

Proof of claim. Using Proposition 11.42 for X and $\tilde{X}$, and their open subschemes $X \setminus \{x\}$ and $\tilde{X} \setminus E$, we see that $\tilde{\mathscr{L}} \otimes_{\mathscr{O}_{\tilde{X}}} (\pi^* \mathscr{L})^{-1} \cong \mathscr{O}_{\tilde{X}}(qE)$ for some $q \in \mathbb{Z}$. Therefore $(\tilde{\mathscr{L}}^2) = (\mathscr{L}^2) + q^2(E^2)$ by Lemma 25.148. The same lemma shows that $(E^2) < 0$, so it is enough to show that $q \neq 0$. But $q = 0$ would mean that $\tilde{\mathscr{L}} \cong \pi^* \mathscr{L}$. This is impossible because every global section of $\mathscr{L}$ vanishes at x, while all but finitely many points of E lie in the maximal domain of definition of $f \circ \pi$.

To finish the proof, note that we can apply Claim 1 to $\tilde{\mathscr{L}}$ just as well, so that we obtain $0 \leq (\tilde{\mathscr{L}}^2) < (\mathscr{L}^2)$ whenever we blow up a point where f is not defined. By induction, this finishes the proof. In fact, setting $X_0 := X$, letting X_i be the blow-up of X_{i-1} in one closed point where the rational map $X_{i-1} \to X_0 \to \mathbb{P}^m$ is not defined (if such a point exists), and denoting by $\mathscr{L}_i$ the line bundle on X_i constructed as above, the sequence $(\mathscr{L}_i^2)$ is a strictly decreasing sequence of non-negative numbers, so there exists an $n \geq 0$ such that the maximal domain of definition of the rational map $X_n \to \mathbb{P}_k^m$ equals X_n. $\quad\square$

(25.31) Projectivity of regular proper surfaces over a field.

In this section, we will prove that every regular proper surface over a field is projective. We start with the following lemma (which we will prove only for surfaces). It will be a key point of the proof of the theorem below and we will employ several of the difficult results that we have proved in the previous chapters.

Lemma 25.149. *Let k be a field, and let X be an irreducible proper k-scheme with $\dim X \geq 2$. Let $U \subseteq X$ be non-empty affine open. Then $X \setminus U$ is connected.*

Proof. We will prove the statement only for surfaces, similarly as in [Bad] ^{O} Thm. 1.28; this is the only case we will use below. For the general case, see [Har2] ^{O} Ch. II, Cor. 6.2 (there k is assumed to be algebraically closed, but it is easy to reduce to this situation).

First note that whenever $f \colon X' \to X$ is a surjective proper morphism of irreducible k-schemes such that $f^{-1}(U)$ is affine, we may replace X and U by X' and $f^{-1}(U)$.

Now choose a closed embedding $U \subseteq \mathbb{A}_k^n$ for some n, embed $\mathbb{A}_k^n \subset \mathbb{P}_k^n$ as usual, and denote by Y the schematic closure of U in $\mathbb{P}_k^n$. We regard the inclusion $U \hookrightarrow Y$ as a rational map $X \dashrightarrow Y$. Using Proposition 13.99 (or Theorem 25.147 in case X is regular), we find a proper surjective morphism $\pi \colon X' \to X$ which is an isomorphism over U and such that the composition $X' \to X \dashrightarrow Y$ is a morphism $X' \to Y$. Replacing X by X', we may hence assume that the birational map $X \dashrightarrow Y$ is a birational morphism $f \colon X \to Y$.

By construction, $U = f^{-1}(f(U))$ and $Y \setminus f(U)$ is the support of an ample effective divisor on Y. Replacing X by its normalization (which is finite over X, Corollary 12.52), and U by its inverse image in the normalization of X, we may assume that X is normal.

Let $X \xrightarrow{f'} Y' \xrightarrow{\pi} Y$ be the Stein factorization of f, see Section (24.10). Since X is normal, the same is true for Y' (Exercise 24.18). The morphism $Y' \to Y$ is finite, and thus pullbacks of ample line bundles are ample (Proposition 13.83). Hence $Y' \setminus f'(U) = \pi^{-1}(Y \setminus f(U))$ is again the support of an ample effective divisor, now on Y'. By Corollary 25.141, $Y' \setminus f'(U)$ is connected. Using that f' has connected fibers by Theorem 24.49, and that the map $(f')^{-1}(Y' \setminus f'(U)) \to Y' \setminus f'(U)$ is closed, we find that $X \setminus U = (f')^{-1}(Y' \setminus f'(U))$ is connected, as desired. $\quad\square$

Lemma 25.150. *Let X be a noetherian separated regular scheme and let $U \subseteq X$ be an open dense affine subscheme. Then every irreducible component of $X \setminus U$ has codimension*

1. *In particular, $X \setminus U$ endowed with its reduced scheme structure is an effective Cartier divisor.*

The proof will show that instead of supposing that X is regular it suffices to assume that the local rings $\mathscr{O}_{X,x}$ are factorial for all $x \in X \setminus U$.

Proof. Let $D := X \setminus U$ be endowed with the reduced scheme structure. If we have shown that D is a Weil divisor, then it is an effective Cartier divisor since X is regular and hence locally factorial. Hence it suffices to show the first assertion.

Let $\xi \in D$ be the generic point of an irreducible component. As U is affine and X is separated, the inclusion $U \to X$ is affine (Proposition 12.3 3). Hence

$$U_\xi := U \times_X \operatorname{Spec} \mathscr{O}_{X,\xi}$$

is affine. As ξ is in the closure of U, U_ξ is non-empty. As ξ is a maximal point of $X \setminus U$, topologically U_ξ is the complement of the special point ξ of $\operatorname{Spec} \mathscr{O}_{X,\xi}$. If ξ had codimension > 1 in $\operatorname{Spec} \mathscr{O}_{X,\xi}$, then $\Gamma(U_\xi, \operatorname{Spec} \mathscr{O}_{X,\xi}) = \mathscr{O}_{X,\xi}$ by Hartogs' theorem (Theorem 6.45) which is absurd since U_ξ is affine. Hence $\dim \mathscr{O}_{X,\xi} = 1$. $\qquad\square$

Theorem 25.151. *Let k be a field and let X be a regular proper surface over k. Then X is projective over k.*

More precisely, let $U \subset X$ be any non-empty affine open subscheme. Then there exists an effective ample divisor on X with support $Z = X \setminus U$.

Proof. It is clear that the more precise version implies the projectivity of X. Note that since X is regular, the notions of Cartier divisor and Weil divisor coincide. We now proceed in several steps. We may and will assume that X is connected. Clearly the support of an effective ample divisor is of pure codimension 1, and it is connected by Theorem 25.141. We will show first that these two properties do hold for Z as above.

(I). By Lemma 25.150, Z is of pure codimension 1, i.e., that every irreducible component of Z is a curve. It follows from Lemma 25.149 that Z is connected.

(II). We may replace k by a finite separable extension k' (we will use this in Step (III), if k is finite). Indeed, if k' is a separable extension, then $X \otimes_k k'$ is still regular (Proposition 18.52). If $U \subset X$ is non-empty open affine, then its inverse image U' in $X' := X \otimes_k k'$ is affine. If there exists a scheme structure on $D' := X' \setminus U'$ such that $\mathscr{O}_{X'}(D')$ is ample, then looking at local equations one sees that $N_{X'/X}(\mathscr{O}_{X'}(D'))$ is a line bundle on X of the form $\mathscr{O}_X(D)$ for some scheme structure on $D := X \setminus U$. It is ample by Proposition 13.66 (1).

(III). Let $Z_1, \ldots, Z_r$ be the irreducible components of Z, considered as prime divisors on X. Let us show that there exists $D' = \sum_{i=1}^r d_i Z_i$, $d_i \geq 0$, such that $(D' \cdot Z_i) \geq 0$ for all i, and $(D' \cdot Z_i) > 0$ for at least one i.

Let $s \in \Gamma(U, \mathscr{O}_X) \subset K(X)$ be non-constant and such that s does not vanish entirely along any of the Z_i. We can produce such an s by taking any non-constant section in $\Gamma(U, \mathscr{O}_X)$ and adding a suitable constant to it. If k is finite, then to find such a constant, we might have to enlarge k; this we can do by Step (II). Let D_1 be the divisor of zeros of s, and let D_2 the divisor of poles, i.e., D_1 and D_2 are effective divisors with $D_1 - D_2 = \operatorname{div}(s)$.

Now $s \in \Gamma(U, \mathscr{O}_X)$, so the locus of poles of s is disjoint from U, and hence as a Weil divisor, $D_2 = \sum_{i=1}^r d_i Z_i$, $d_i \geq 0$. We claim that we can set $D' := D_2$. The condition on the intersection numbers can be checked for D_1 in place of D_2, because these two divisors are linearly equivalent by construction.

Since s does not vanish along any of the Z_i, none of them is contained in the support of D_1, so $(D_1 \cdot Z_i) \geq 0$ for all i. Furthermore, the support of D_1 is closed in X and is thus a proper k-scheme, so it cannot be contained in U. Therefore we have $D_1 \cap Z_i \neq \emptyset$ for at least one i, and this implies $(D_1 \cdot Z_i) > 0$.

(IV). Let $F_0 := D'$ and $F_1 := Z_i$ as in Step (III), i.e., $(F_0 \cdot F_1) > 0$. Define $F_2, \ldots, F_s$ such that each F_j, $j \geq 2$, is one of the irreducible components of Z, and all of those occur among the F_i (possibly with repetitions), and such that $(F_{i-1} \cdot F_i) > 0$ for all $i = 2, \ldots, s$. To find such a sequence, we use that Z is connected.

It is then enough to find positive natural numbers e_i, $i = 1, \ldots, s$, such that for $D := \sum_{i=0}^{s} e_i F_i$ we have $(D \cdot Z_j) > 0$ for all $j = 1, \ldots, s$. In fact, then D is an effective divisor on X with support $X \setminus U$. To check that it is ample, we may use the Nakai-Moishezon criterion, Theorem 25.146 (2). If $C \subset X$ is an integral curve that is different from all the Z_i, then C, being proper, cannot be contained in U, and it follows that $(C \cdot D) > 0$. On the other hand, $(D \cdot Z_j) > 0$ holds by construction of D.

Let us spell out a sufficient condition on the coefficients e_i. Fix $j \geq 1$ for a moment and let i_0 be minimal with $F_{i_0} = Z_j$. Then

$$(D \cdot Z_j) = \sum_{i=0}^{i_0-2} e_i (F_i \cdot Z_j) + e_{i_0-1}(F_{i_0-1} \cdot F_{i_0}) + \sum_{i=i_0}^{s} e_i (F_i \cdot F_{i_0}),$$

where the first sum on the right hand side is non-negative. Since $(F_{i_0-1} \cdot F_{i_0}) > 0$ by construction of the F_i, we see that we can ensure that $(D \cdot Z_j) > 0$ by requiring that

$$e_{i_0-1} > -\frac{\sum_{i=i_0}^{s} e_i (F_i \cdot F_{i_0})}{(F_{i_0-1} \cdot F_{i_0})}.$$

Therefore we can set $e_s := 1$ and define the other coefficients by descending induction so that they are positive and satisfy the above inequality for all i_0.   □

Remark 25.152. The projectivity of regular surfaces was first proved by Zariski. The proof above is due to Goodman [Goo] $^\circ$; see also [Har2] $^\circ$. Slightly more generally, one can show that every proper surface X over a field k such that there exists an affine open $U \subset X$ containing all non-regular points of X is projective ([Goo] $^\circ$ Cor. to Thm. 2).

On the other hand, there exist non-projective proper surfaces and non-projective smooth proper varieties of dimension 3 (and hence of any higher dimension, as well).

(25.32) The Hodge index theorem.

Definition 25.153. *Let k be a field and let X be a proper surface over k.*
(1) *We call divisors D, D' on X numerically equivalent, if $(D \cdot E) = (D' \cdot E)$ for every divisor E on X. This also gives us the notion of numerical equivalence for classes of divisors up to linear equivalence.*
(2) *We denote by $\mathrm{Num}(X)$ the quotient of $\mathrm{Pic}(X)$ by the subgroup consisting of all divisor classes that are numerically equivalent to 0.*

By the definition of numerical equivalence, the intersection pairing on $\mathrm{Pic}(X)$ induces a non-degenerate pairing on $\mathrm{Num}(X)$, again called the intersection pairing.

It is known that $\mathrm{Num}(X)$ is a free finitely generated abelian group. In fact, it is clearly torsion-free. It follows from the Theorem of Néron-Severi (also called Theorem of the base) that it is finitely generated. This is a rather difficult result which we will not use below; see $[\mathrm{CJLO}]_X$ for a modern proof and further references. In particular, $\mathrm{Num}(X)_{\mathbb{R}}$ is a (finite-dimensional) $\mathbb{R}$-vector space, equipped with a non-degenerate symmetric bilinear form induced by the intersection pairing.

Lemma 25.154. *Let X be a smooth proper algebraic surface over a field k. Let H be an ample divisor on X. If D is a divisor on X such that $(D^2) > 0$, then the following are equivalent:*

(i) $(D \cdot H) > 0$,

(ii) *for all n sufficiently large, the divisor nD is linearly equivalent to an effective divisor, in other words, $H^0(X, \mathcal{O}_X(nD)) \neq 0$.*

Proof. (i) $\Rightarrow$ (ii). We have $\dim H^2(X, \mathcal{O}_X(nD)) = 0$ for n large. Indeed, Serre duality implies $\dim H^2(X, \mathcal{O}_X(nD)) = \dim H^0(X, \mathcal{O}_X(-nD) \otimes \omega_X)$. Here $\omega_X = \Omega^2_{X/k}$ is the dualizing sheaf of X. If $n(D \cdot H) > (\omega_X \cdot H)$, then we have $H^0(X, \mathcal{O}_X(-nD) \otimes \omega_X) = 0$ because the intersection number of an ample and an effective divisor is positive (Proposition 23.79).

It is thus enough to show that $\chi(\mathcal{O}_X(nD)) > 0$ for all n that are sufficiently large, and this follows from the Riemann-Roch theorem (Theorem 25.144) together with the assumption that $(D^2) > 0$.

(ii) $\Rightarrow$ (i). It is enough to show that $n(D \cdot H) = (nD \cdot H) > 0$ for n as in (ii). This follows from Proposition 23.79. $\square$

Corollary 25.155. *Let X be a smooth proper algebraic surface over a field k. Let D be a divisor on X such that $(D^2) > 0$. If H and H' are ample divisors on X, then one has $(D \cdot H) > 0$ if and only if $(D \cdot H') > 0$.*

Proof. This is an immediate consequence of Lemma 25.154, because the condition that all sufficiently high multiples of D are effective is independent of H and H'. $\square$

With these preparations we can prove the Hodge index theorem. In view of the theorem of Néron-Severi mentioned above, which ensures that $\mathrm{Num}(X)_{\mathbb{R}}$ is finite-dimensional, say of dimension n, it says that the "index" or signature of the non-degenerate quadratic space $\mathrm{Num}(X)_{\mathbb{R}}$ (with the intersection pairing) is $(1, n - 1)$, i.e., $\mathrm{Num}(X)_{\mathbb{R}}$ is the orthogonal direct sum of a positive definite line and a negative definite subspace of codimension 1.

Theorem 25.156. (Hodge index theorem) *Let k be a field and let X be a smooth proper surface over k. Let H be an ample divisor on X. If D is a divisor on X with $(D \cdot H) = 0$ that is not numerically equivalent to 0, then $(D^2) < 0$.*

In other words, the intersection pairing induces a negative definite pairing on the orthogonal complement

$$\{H\}^{\perp} = \{D \in \mathrm{Num}(X);\ (D \cdot H) = 0\}$$

of H in $\mathrm{Num}(X)$.

Proof. Let D be a divisor with $(D \cdot H) = 0$ and not numerically equivalent to 0.

We first exclude the case $(D^2) > 0$. Let $H' = D + nH$. For n sufficiently large, H' is ample by Proposition 13.50 (2). But then $(D \cdot H') = (D^2) > 0$ and $(D \cdot H) = 0$ in contradiction to Corollary 25.155.

Using this, we can exclude the case $(D^2) = 0$ by the following simple argument about non-degenerate bilinear forms. Let E be such that $(D \cdot E) \neq 0$, $(E \cdot H) = 0$. Then $((aD + E)^2) = 2a(D \cdot E) + (E^2) > 0$ for suitable a, and $((aD + E) \cdot H) = 0$, contradicting the case we have already handled.   □

(25.33) Further references.

Already in the fundamental paper [Ser1] ^{O} by Serre, at the very beginning of the study of the cohomology of (quasi-)coherent sheaves on algebraic varieties, instances of (Serre) duality were observed, in particular the duality for cohomology of line bundles on projective space (Corollary 22.23); see also Grothendieck's articles [Gro4] ^{O}, [Gro3] ^{O}. The theory was further developed by Grothendieck, Verdier, Hartshorne, Deligne, and others.

In Hartshorne's lecture notes [Har1] ^{O} of a seminar held by Grothendieck, the functor $f^!$ is constructed by starting from the cases of projective space and of closed immersions, and then showing the necessary compatibilities. See also Conrad's book [Con] ^{O} that provides a thorough discussion of several subtle points that were glossed over in [Har1] ^{O}.

Alternatively, one can construct a right adjoint $f^\times$ to Rf_* "directly", see [Ver1] ^{O}. More recently, the existence of $f^\times$ was proved more generally by Neeman [Nee3] ^{O}, making use of a Brown representability theorem for triangulated categories; this is the approach we have followed in this chapter. See also [Nee5] ^O_X, [LN] ^O_X.

Once one has the functors $f^\times$, one can construct the twisted inverse image functors $f^!$ "by hand" using Nagata's compactification theorem; this is the route taken, e. g., in [Sta] and also the one we have followed here. Another approach is to use a category-theoretic result about "pasting functors" as explained in the Appendix by Deligne in [Har1] ^{O}, cf. also [Lip2] ^O_X Section 4.8. See also [ILN] ^O_X for the relation between $f^\times$ and $f^!$ in general.

Instead of obtaining a dualizing complex on a (projective) scheme X by embedding it into some projective space, another approach, simpler in some respects, is to consider a finite surjective morphism $X \twoheadrightarrow \mathbb{P}^d$, and to "pullback" the dualizing complex on projective space to obtain one on X. See [Lip1] ^{O} for an exposition largely following this approach.

It is often (very...) difficult to align the different approaches, and in particular to make identifications obtained from abstract categorical considerations explicit in specific cases. See for instance [Yek2] ^{O}, [Sas1] ^{O}, [SaTo] ^{O}, [Sas2] ^O_X, [NS1] $_X$, [NS2] $_X$, [Lip3] $_X$ for further discussions and results in this direction.

The theory has been further extended, for instance to better cover non-noetherian schemes and formal schemes, and towards a non-commutative setting. Some pointers in this direction are [AJL] ^O_X, [LNS] ^{O}, [Yek1] ^{O}, [BDS] ^O_X.

If one is willing to accept further assumptions and/or to content oneself with less complete results, the technical machinery of derived categories can be avoided entirely. See for instance [AlKl] ^{O}, [Har3] ^{O} Chapter III.7, [Liu] ^{O} Section 6.4.

More recently, Clausen and Scholze have sketched a new approach to duality for coherent sheaves using their theory of condensed mathematics [Scho] $_X$. More details have been given by Mann in his thesis [Man] $_X$. In this setting, the category of schemes is embedded fully faithfully in the category of discrete adic spaces. In this category, every discrete adic space and every morphism of discrete adic spaces has a functorial "compactification" which simplifies the definition of $f^!$ if f is "sufficiently finite". One even obtains a full six-functor formalism (see Section (21.30)) if one restricts to "sufficiently finite" morphisms[1] for the

---

[1] i.e., separated and +-finite type morphisms as defined by Mann, which include all separated morphisms of finite type and all integral morphisms of arbitrary schemes

functors $f^!$ and $f_!$. The theory requires some mathematical machinery that is beyond the scope of this book, among other things the theory of ∞-categories. It is an exciting new perspective on the derived category of quasi-coherent $\mathscr{O}_X$-modules on a scheme X (and in fact also more general geometric objects such as derived schemes).

Regarding the applications to surfaces, the material above is covered in the books by Beauville [Beau] ^{O} and Bădescu [Bad] ^{O}, and partly also in [Har3] ^{O} Ch. V. Each of these sources provide a lot more material on algebraic surfaces, a beautiful topic which is more involved than the theory of algebraic curves, but still significantly simpler than the study of algebraic varieties of higher dimension. For instance, there are much stronger classification results, proved by Kodaira, Enriques, Mumford and Bombieri, than in higher dimension. Another example is resolution of singularities for surfaces, proved by Zariski and Abhyankar. See Artin's article in [CoSi] ^{O} for an account of the proof and for further references.

Exercises

Exercise 25.1. Let X be a scheme. Show that an object in $D_{\mathrm{qcoh}}(X)$ is perfect if and only if it is dualizable (Definition/Remark 21.149).
Hint: To show that the condition is sufficient reduce to the case that X is affine. Now show that every dualizable object is compact.

Exercise 25.2. Let X be a scheme, let $L \in D_{\mathrm{qcoh}}(X)$ be such that there exists $M \in D_{\mathrm{qcoh}}(X)$ with $M \otimes^L_{\mathscr{O}_X} L \cong \mathscr{O}_X$. Show that there exists an open covering $(U_i)_i$ of X such that $L_{|U_i} \cong \mathscr{L}_i[n_i]$, where $\mathscr{L}_i$ is a line bundle on U_i and $n_i \in \mathbb{Z}$.
Hint: One can assume that $X = \operatorname{Spec} A$ is affine. Show that L and M are dualizable and hence perfect by Exercise 25.1. Then use Proposition 22.54.

Exercise 25.3. Let X be a quasi-compact scheme that carries an ample family of line bundles $(\mathscr{L}_i)_{i \in I}$ (Exercise 23.17). Show that $\{ \mathscr{L}_i^{\otimes m}[n] \; ; \; i \in I, m, n \in \mathbb{Z} \}$ is a set of compact objects that generates $D_{\mathrm{qcoh}}(X)$.

Exercise 25.4. Let R be a ring, $n \geq 1$, and let $f \colon \mathbb{P}^n_R \to S := \operatorname{Spec} R$ be the structure morphism. Let $j \colon U := \mathbb{A}^n_R \to \mathbb{P}^n_R$ be the inclusion.
(1) Show that $(f^\times \mathscr{O}_S)_{|U} \cong \mathscr{O}_U[n]$.
(2) Let $R = k$ be a field. Show that $(f \circ j)^\times \mathscr{O}_S = \operatorname{Hom}_k(k[T_1, \ldots, T_n], k)^\sim$, considered as a complex of quasi-coherent $\mathscr{O}_U$-modules concentrated in degree 0.

Exercise 25.5. Let R be a noetherian ring, set $\tilde{R} := R[T]/(T^2)$ and let $\gamma \in R \subset \tilde{R}$ be neither a unit nor nilpotent. Form the cartesian diagram

$$
\begin{array}{ccc}
\operatorname{Spec} R[1/\gamma] & \xrightarrow{\;j'\;} & \operatorname{Spec} R \\
{\scriptstyle f'}\downarrow & & \downarrow{\scriptstyle f} \\
\operatorname{Spec} \tilde{R}[1/\gamma] & \xrightarrow{\;j\;} & \operatorname{Spec} \tilde{R},
\end{array}
$$

where f and f' correspond to the R-algebra homomorphisms sending T to 0 and where j and j' are the inclusions of principal open subschemes. Let $N \in D(\tilde{R})$ be the complex

$$
\cdots \xrightarrow{\;0\;} R \xrightarrow{\;0\;} R \xrightarrow{\;0\;} R \longrightarrow 0 \longrightarrow 0 \longrightarrow \cdots
$$

concentrated in degree ≤ 0, where we consider R as an $\tilde{R}$-module. We consider $f^\times$ as functor $D(\tilde{R}) \to D(R)$, and similarly for the functors j^*, $(f')^\times$ and $(j')^*$.

(1) Show that $f^! R = \prod_{i \geq 0} R[-i]$ is the complex with R in every non-negative degree and with zero differentials. Deduce that $f^! N = \prod_{k \geq 0} \prod_{i \geq 0} R[k - i]$.

   Hint: Use the projective resolution $\cdots \xrightarrow{T} \tilde{R} \xrightarrow{T} \tilde{R} \longrightarrow R \longrightarrow 0$ of the $\tilde{R}$-module R.

(2) Show that the base change map $(j')^* f^\times N \to (f')^\times j^* N$ is the map

$$\Big(\prod_{k \geq 0} \prod_{i \geq 0} R[k - i]\Big)[1/\gamma] \longrightarrow \prod_{k \geq 0} \prod_{i \geq 0} (R[k - i][1/\gamma])$$

and that it does not induce an isomorphism after applying H^0 since γ is not a unit and not nilpotent.

Exercise 25.6. Let X be a qcqs scheme and let $U \subseteq X$ be a quasi-compact open subset.

(1) Suppose that X is affine and let $\mathscr{L}$ be a line bundle on U. Suppose there exists a perfect complex $\mathscr{E}$ on X such that $\mathscr{E}_{|U} \cong \mathscr{L}$ in $D(U)$. Show that then there exists a line bundle $\mathscr{M}$ on X and an isomorphism $\mathscr{M}_{|U} \cong \mathscr{L}$ of line bundles on U.

   Hint: Represent $\mathscr{E}$ by a strictly perfect complex $\mathscr{E}^\bullet$, such that the isomorphism $\mathscr{E}_{|U} \cong \mathscr{L}$ is given by a quasi-isomorphism $\mathscr{E}^\bullet_{|U} \longrightarrow \mathscr{L}$. Then set $\mathscr{M} := \bigotimes_{i \in \mathbb{Z}} \det(\mathscr{E}^i)^{\otimes i}$.

   Remark: This construction is a special and imprecise case of the determinant of a perfect complex, see [KnMu] $^\circ$.

(2) Find an example of an affine scheme X of finite type over a field k, an open subscheme U and of a line bundle $\mathscr{L}$ on U that cannot be extended to a perfect complex on X.

   Hint: E.g., let X be the cone given by $z^2 = xy$ in $\mathbb{A}^3_k$ and U the complement of the origin. Show that $\mathrm{Pic}(X) = 0$, and that $\mathrm{Pic}(U) \neq 0$.

Exercise 25.7. Let k be a field, let $X = \mathbb{A}^n_k$ and $U = \mathbb{A}^n_k \setminus \{0\}$ for $n \geq 3$.

(1) Show that restriction of locally free sheaves induces an isomorphism $K_0(X) \xrightarrow{\sim} K_0(U)$.

(2) Show that a locally free $\mathscr{O}_U$-module is isomorphic to the restriction of a locally free $\mathscr{O}_X$-module if and only if it is isomorphic to a direct sum of line bundles, or equivalently, if and only if it is free.

   Hint: Use "Serre's conjecture" that every finite locally free $\mathscr{O}_X$-module is free (see [Lang] $^\circ$ XXI, 3.7).

(3) Show that there exists a locally free $\mathscr{O}_U$-module which is not isomorphic to the restriction of a locally free $\mathscr{O}_X$-module.

   Hint: Let $\mathscr{F}$ be the pullback of $\Omega^1_{\mathbb{P}^2_k}$ and show that $\mathscr{F}$ is not the direct sum of two line bundles.

Remark: See [Ser4] $^\circ$ §5, [Rou] $^\circ$, Remark 3.13.

Exercise 25.8. Let A be a ring and let B be an A-algebra. Let $f \colon \mathrm{Spec}\, B \to \mathrm{Spec}\, A$ be the corresponding morphism of affine schemes. Consider the following assertions.

(i) The A-module B is perfect.

(ii) Every perfect complex in $D(B)$ is perfect considered as an object in $D(A)$.

(iii) $\mathrm{Spec}\, B \to \mathrm{Spec}\, A$ is cohomologically proper and of finite tor-dimension.

(iv) The functor Rf_* sends perfect complexes in $D(B) = D_{\mathrm{qcoh}}(\mathrm{Spec}\, B)$ to perfect complexes in $D_{\mathrm{qcoh}}(\mathrm{Spec}\, A)$.

(v) The functor $f^\times \colon D_{\mathrm{qcoh}}(\mathrm{Spec}\, A) \to D_{\mathrm{qcoh}}(\mathrm{Spec}\, B)$ is naturally isomorphic to the functor

$$D(A) \to D(B), \qquad K \mapsto K \otimes^L_A R\,\mathrm{Hom}^B_A(B, A).$$

Show that (i) $\Leftrightarrow$ (ii) $\Leftrightarrow$ (iii) $\Leftrightarrow$ (iv) $\Rightarrow$ (v).

Remark: Remark 25.45 shows that in fact all these assertions are equivalent.

Exercise 25.9. A qcqs morphism $f\colon X \to Y$ of schemes is called *universally cohomologically proper*[2] if for every morphism $g\colon Y' \to Y$ of schemes the base change $X \times_Y Y' \to Y'$ of f is cohomologically proper. Show the following assertions.
(1) The property "universally cohomologically proper" is stable under composition and under base change.
(2) Let $f\colon X \to Y$ be a morphism of schemes and let $Y' \to Y$ be a faithfully flat surjective morphism of schemes. Suppose that the base change $X \times_Y Y' \to Y'$ of f is universally cohomologically proper. Then f is universally cohomologically proper.
(3) Every flat cohomologically proper morphism is universally cohomologically proper.

Exercise 25.10. Give an example of a finite morphism $f\colon X \to S$ of affine schemes such that $f_*\colon (\mathscr{O}_X\text{-Mod}) \to (\mathscr{O}_S\text{-Mod})$ is not exact.

Hint: Consider a Dedekind domain R with exactly two maximal ideals (e.g., $X = \operatorname{Spec} \mathbb{Z}[\sqrt{-1}]_{(p)}$ for a prime number $p \in \mathbb{Z}$ such that $p \equiv 1 \pmod 4$) and let $X = \{\eta, s, t\}$ be its spectrum, where s and t denote the closed points. Set $U := \{s, \eta\}$ and $V := \{t, \eta\}$. Let $K = \operatorname{Frac} R$ and define an $\mathscr{O}_X$-module $\mathscr{H}$ by $\mathscr{H}(X) = K \times K$, $\mathscr{H}(U) = \mathscr{H}(V) = K$ and $\mathscr{H}(\{\eta\}) = 0$. Let $\mathscr{F}$ be the constant sheaf with value K on X. Show that there is a unique surjective map $\mathscr{F} \to \mathscr{H}$ of $\mathscr{O}_X$-modules such that $\mathscr{F}(U) \to \mathscr{H}(U)$ and $\mathscr{F}(V) \to \mathscr{H}(V)$ are the identity. Let S be the spectrum of a discrete valuation ring and let $f\colon X \to S$ be a finite morphism, which necessarily sends s and t to the closed point of S (e.g., $S = \operatorname{Spec} \mathbb{Z}_{(p)}$ in the above example). Show that $f_*\mathscr{F} \to f_*\mathscr{H}$ is not surjective.

Exercise 25.11. Show that there exists a connected noetherian scheme X of dimension 2 which admits a dualizing complex and which has an irreducible component of dimension $< \dim(X)$ at whose generic point a dimension function takes its maximal value.

Hint: Exercise 14.24

Exercise 25.12. Let k be a field, $A := k[x, y]$, let $\mathfrak{p} := (x, y)$ and $\mathfrak{q} := (x - 1)$, and let $S^{-1}A$ with $S := A \setminus (\mathfrak{p} \cup \mathfrak{q})$ be the semilocalization of A in $\{\mathfrak{p}, \mathfrak{q}\}$. Show that $\operatorname{Spec} S^{-1}A$ is connected and regular (in particular Cohen-Macaulay) but not equi-dimensional.

Exercise 25.13. Let k be a field and let X be a locally noetherian scheme over k. Let $x \in X$ be a point. Show that the following assertions are equivalent.
(i) The local ring $\mathscr{O}_{X,x}$ is Cohen-Macaulay.
(ii) For every finitely generated extension $K \supseteq k$ and every point $x' \in X \otimes_k K$ mapping to x, the local ring $\mathscr{O}_{X \otimes_k K, x'}$ is Cohen-Macaulay.
(iii) There exists a field extension $K \supseteq k$ such that $X \otimes_k K$ is locally noetherian and a point $x' \in X \otimes_k K$ mapping to x such that $\mathscr{O}_{X \otimes_k K, x'}$ is Cohen-Macaulay.
If X is locally of finite type over k, show that these conditions are also equivalent to
(iv) For every field extension $K \supseteq k$ and every point $x' \in X \otimes_k K$ mapping to x, the local ring $\mathscr{O}_{X \otimes_k K, x'}$ is Cohen-Macaulay.

Hint: Proposition B.82

Exercise 25.14. Show the following properties of Cohen-Macaulay morphisms.
(1) The composition of Cohen-Macaulay morphisms is again Cohen-Macaulay.

---

[2] This is non-standard terminology.

(2) Let $f\colon X \to Y$ be a morphism of schemes such that all fibers are locally noetherian schemes, let $g\colon Y' \to Y$ be locally of finite type, and let $f'\colon X' \to Y'$ be the base change of f. Show that if f is Cohen-Macaulay, then f' is Cohen-Macaulay. Show that the converse holds if g is faithfully flat.
Hint: Exercise 25.13.

Exercise 25.15. Let A be a noetherian local ring with normalized dualizing complex $\omega_A^\bullet$, and let $d \geq 0$ be an integer. Show that $M \mapsto \mathrm{Ext}_A^{-d}(M, \omega_A^\bullet)$ is an anti-auto-equivalence of the category of finitely generated Cohen-Macaulay A-modules of depth d.

Exercise 25.16. Let $A \to B$ be a finite local homomorphism of noetherian local Cohen-Macaulay rings. Suppose that a dualizing module ω_A for A exists. Show that $\omega_B := \mathrm{Ext}_A^t(B, \omega_A)$ with $t := \dim A - \dim B$ is a dualizing module for B.

Exercise 25.17. Let k be a field and let X be a proper k-scheme. Use Proposition 25.110 to give another proof that a dualizing sheaf on X is torsion-free (cf. Proposition 25.106 (3)).

Exercise 25.18. Let k be a field. A *Calabi-Yau variety*[3] over k is a smooth proper geometrically connected k-scheme X of dimension $d > 0$ such that $\Omega_{X/k}^d \cong \mathcal{O}_X$ and $H^i(X, \mathcal{O}_X) = 0$ for all $0 < i < d$. A Calabi-Yau variety of dimension 2 is called a *K3-surface*.
(1) Let X be a Calabi-Yau variety of even dimension d. Prove $\dim H^0(X, \mathcal{O}_X) = \dim H^d(X, \mathcal{O}_X) = 1$ and deduce that $\chi(\mathcal{O}_X) = 1 + (-1)^d$.
(2) Show that any smooth geometrically connected hypersurface X of degree $d + 2$ in $\mathbb{P}_k^{d+1}$ is a Calabi-Yau variety of dimension d.
Hint: Consider the exact sequence $0 \to \mathcal{O}_{\mathbb{P}_k^{d+1}}(-d-2) \to \mathcal{O}_{\mathbb{P}_k^{d+1}} \to \mathcal{O}_X \to 0$ and use Exercise 17.7.

Exercise 25.19. Let X be a connected smooth proper scheme over a field k, let $d = \dim X$ and let $h^{ij}(X)$ be the Hodge numbers of X (Exercise 23.5). Show that $h^{ij}(X) = h^{d-i,d-j}(X)$.
Hint: Exercise 7.28.
Remark: If k is a field of characteristic 0, then one also has symmetry of the Hodge numbers, i.e. $h^{ij}(X) = h^{ji}(X)$.

---

[3] There are different (non-equivalent) definitions in the literature. For instance, sometimes the condition that $H^i(X, \mathcal{O}_X) = 0$ for all $0 < i < d$ is omitted. Then the Calabi-Yau varieties as defined here are called *strict Calabi-Yau varieties*.

26 Curves

Content

- Basic notions
- The Theorem of Riemann-Roch
- Special classes of curves
- Vector bundles on curves
- Further topics

In this chapter, we will study curves over fields. With the results of the preceding chapters, we can prove some of the highlights of the theory, such as the Theorem of Riemann-Roch, without much further effort. The intention of the chapter is not to give a systematic and comprehensive treatment of the theory, but rather to illustrate some important results and classes of particularly interesting curves, for instance elliptic and hyperelliptic curves, while at the same time providing examples of the usefulness of the general machinery of algebraic geometry we have built up so far. See Section (26.31) for further references and pointers to the literature.

Basic notions

We start by recalling (and in part, generalizing) a few facts that we have already proved in Volume I or in the preceding chapters. We then introduce the *genus* of a proper curve. At the end we study singularities of curves and single out some classes of particularly mild singularities.

(26.1) Recollections on curves.

Let k be a field. In this chapter we will mean by a *curve over k* the following.

Definition 26.1. *A curve over the field k is a separated k-scheme C of finite type that is equi-dimensional of dimension 1.*

In comparison to Definition 15.14 we add the requirement that C be separated.

Because a one-dimensional normal noetherian local ring is a discrete valuation ring, a curve C over a field k is regular, if and only if it is normal. If k is perfect, then this is equivalent to C being smooth over k (Proposition 18.67).

© Springer Fachmedien Wiesbaden GmbH, ein Teil von Springer Nature 2023
U. Görtz und T. Wedhorn, *Algebraic Geometry II: Cohomology of Schemes*,
Springer Studium Mathematik – Master, https://doi.org/10.1007/978-3-658-43031-3_11

Every reduced curve C is birational to a unique normal proper curve over k. In fact, replacing C by the disjoint union of its irreducible components, and handling each of them individually, we may assume that C is integral. Passing to the normalization, we may assume that C is normal, and then the unique normal compactification (Theorem 15.20) is the proper (and even projective, Theorem 26.16) curve that we are looking for. When C is integral, we can obtain it by using that C is quasi-projective (Theorem 15.18), and taking the normalization of the closure of the image of C under some embedding into a projective space. We extract the following definitions from this discussion:

Definition 26.2. *Let C be a curve over a field k. Let $C_1, \ldots, C_n$ be the irreducible components of C endowed with the reduced scheme structure, and for each i let $\tilde{C}_i$ be the normalization of C_i.*
(1) *The normalization of C is the disjoint union of the curves $\tilde{C}_i$, together with the natural morphism to C.*
(2) *The complete (or proper, or projective) normal model of C is the unique proper normal curve that is birational to the underlying reduced curve C_{red}.*

If C is integral, then this notion of normalization coincides with the one of Definition 12.42. As in Proposition 12.44 every morphism from a normal scheme X to C such that each irreducible component of C is dominated by an irreducible component of X factors through the normalization of C. In fact, this characterizes the normalization of C. On reduced affine open subschemes, it corresponds to passing to the integral closure of the coordinate ring in its ring of total fractions. The normalization morphism is the composition of the finite surjective morphisms

$$\coprod_i \tilde{C}_i \longrightarrow \coprod_i C_i \longrightarrow C$$

hence it is finite surjective.

Also recall the equivalence of categories between proper normal integral curves and their function fields, Theorem 15.21.

Theorem 26.3. *Let k be a field. Mapping an integral curve C to its function field $K(C)$, and a dominant (equivalently: non-constant) morphism between integral curves to the corresponding field extension of their function fields gives rise to a contravariant equivalence between*
(i) *the category of normal proper integral curves over k with non-constant morphisms, and*
(ii) *the category of extension fields K of k such that the extension K/k is finitely generated and has transcendence degree 1.*

(26.2) Reminder on geometrically connected schemes.

As before, let k be a field. We recall the notions of *geometrically* irreducible, reduced, and connected k-schemes (and particularly curves), see Section (5.13). We say that a k-scheme X has one of the properties irreducible, reduced, connected[1] geometrically, if X_K has the property for every field extension K/k, or equivalently if X_K has the property for some algebraically closed extension field K of k (Definition 5.48, Corollary 5.54).

---

[1] Recall that by definition connected spaces are non-empty.

Let X be an integral k-scheme with function field $K(X)$. Then X is geometrically integral if and only if k is algebraically closed in $K(X)$ and the extension $K(X)/k$ is separable (Proposition 5.51).

Every smooth k-scheme is reduced, and even geometrically reduced since smoothness is preserved by base change.

A scheme X is connected if and only if the ring $H^0(X, \mathscr{O}_X)$ has no non-trivial idempotent elements (equivalently, if this ring does not admit a non-trivial product decomposition, see Exercise 2.17). Now let X be a k-scheme and let Ω be an algebraically closed extension field of k. Then $H^0(X_\Omega, \mathscr{O}_{X_\Omega}) = H^0(X, \mathscr{O}_X) \otimes_k \Omega$, so X is geometrically connected if and only if $H^0(X, \mathscr{O}_X) \otimes_k \Omega$ has no non-trivial idempotents. For instance, this certainly holds whenever $H^0(X, \mathscr{O}_X) = k$. (Compare also Corollary 24.50 for a relative version of this statement.)

A connected k-scheme with a k-rational point is automatically geometrically connected. This follows from the following result (cf. Exercise 5.23) by applying it to $Y = \operatorname{Spec} k$.

Lemma 26.4. *Let k be a field and let X be a connected k-scheme. Assume that there exists a geometrically connected k-scheme Y and a k-morphism $f\colon Y \to X$. Then X is geometrically connected.*

Proof. Let Ω be an algebraic closure of k. We have to show that $X_\Omega = X \otimes_k \Omega$ is connected. Consider the surjective projection $p\colon X_\Omega \to X$. As Ω is an algebraic extension of k, p is integral and hence closed (Proposition 12.12). By Theorem 14.38 it is also open.

Assume that there existed an open and closed subset $\emptyset \neq Z \subsetneq X_\Omega$. Then $p(Z) = X$ since X is connected. Moreover $f_\Omega^{-1}(Z) = Y \times_X Z$ is open and closed in Y_Ω which is by hypothesis connected. As $Z \to X$ is surjective, $f^{-1}(Z) \to Y$ is surjective. In particular, $f^{-1}(Z)$ is non-empty and therefore $f^{-1}(Z) = Y_\Omega$. The same argument also shows that $f^{-1}(X_\Omega \setminus Z) = Y_\Omega$. This is absurd. $\qquad\square$

For X proper over k and reduced, we have the following simple criterion to be geometrically connected.

Lemma 26.5. *Let k be a field. Let X be a connected proper reduced k-scheme. Then $H^0(X, \mathscr{O}_X)$ is a finite extension field of k.*
(1) *The k-scheme X is geometrically connected if and only if the $H^0(X, \mathscr{O}_X)$ is a purely inseparable extension of k.*
(2) *If X is even geometrically reduced (e.g., if X is smooth over k), then X is geometrically connected if and only if $H^0(X, \mathscr{O}_X) = k$.*

Proof. Since X is proper, $H^0(X, \mathscr{O}_X)$ is a finite-dimensional k-vector space (Theorem 12.65, or Corollary 23.18) and hence a finite product of local Artin k-algebras. Now X is connected, so there can be only one factor, and since X is reduced, $H^0(X, \mathscr{O}_X)$ is a reduced ring, hence a field. Now (1) follows from Corollary 24.55. Under the assumptions of (2), we know that $H^0(X, \mathscr{O}_X)$ is a geometrically reduced k-algebra and hence a separable field extension of k (Proposition 18.24). Therefore (2) follows from (1). $\qquad\square$

(26.3) Singularities of curves.

Let us compare different forms of singularities for a curve C over a field k. In general, one has the implications

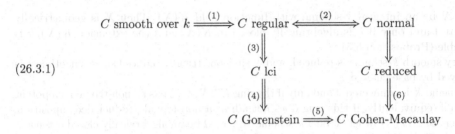

(26.3.1)

Implication (1) holds for any scheme locally of finite type over a field k and it is an equivalence if k is perfect (Proposition 18.67). The Equivalence (2) is particular to curves (Proposition 6.40). The Implications (3), (4), and (5) are general results for local noetherian rings (Remark 19.47). Finally, Implication (6) is again particular to curves and holds by the following lemma.

Lemma 26.6. *Let C be a noetherian scheme of dimension 1. Then C is Cohen-Macaulay if and only if C has no embedded components. This is true when $H^0(C, \mathscr{O}_C)$ is reduced.*

If C is reduced, then $H^0(C, \mathscr{O}_C)$ is reduced, but the converse does not hold in general, see Exercise 26.1.

Proof. Example B.80 (2) shows the characterization of being Cohen-Macaulay for curves. Now let us assume that $x \in C$ is an embedded associated point. As C is a curve, this is a closed point. We will construct a non-zero nilpotent element of $H^0(C, \mathscr{O}_C)$. To achieve this, let $U = \operatorname{Spec} A$ be an affine open neighborhood of x. The point x then corresponds to a maximal ideal $\mathfrak{m} \subset A$ of the form $\mathfrak{m} = \operatorname{Ann}(a)$ for some $0 \neq a \in A$. Let $\mathfrak{p}_1, \ldots, \mathfrak{p}_r$ be the minimal prime ideals of A. For each i, if $s \in \mathfrak{m} \setminus \mathfrak{p}_i$, then $sa = 0$ implies $a \in \mathfrak{p}_i$; thus a is nilpotent.

We show that one can extend a to a global section of C. In fact, by prime ideal avoidance (Proposition B.2 (2)), there exists $s \in \mathfrak{m} \setminus \bigcup_{i=1}^{r} \mathfrak{p}_i$. Then $a_{|D(s)} = 0$ and $V := D(s) \cup \{\mathfrak{m}\}$ is the complement in U of finitely many closed points, hence is open in U and therefore in C. We can then glue $a_{|V} \in H^0(V, \mathscr{O}_C)$ with $0 \in H^0(C \setminus \{\mathfrak{m}\}, \mathscr{O}_C)$ and obtain a non-zero nilpotent global section of $\mathscr{O}_C$, as desired. $\square$

All other possible implications in (26.3.1) do not hold, as the following examples show.

Example 26.7.
(1) Exercise 18.19 gives an example of regular, geometrically reduced curve which is not smooth (necessarily over a non-perfect field).
(2) Every reduced plane curve over a field k, i.e., every reduced curve which admits an embedding as a locally closed subscheme of $\mathbb{P}^2_k$, is locally a complete intersection (Proposition 5.31, Corollary 5.42). See also Proposition 26.78 below. This gives us plenty of examples of reduced lci curves which are not regular, e.g., $\operatorname{Spec} k[X, Y]/(Y^2 - X^3)$ or $\operatorname{Spec} k[X, Y]/(Y^2 - X^2(X - 1))$. It gives us also non-reduced curves which are lci, e.g., $\operatorname{Spec} k[X, Y]/(X^2)$.
(3) The subring $k[T^5, T^6, T^7, T^8]$ of $k[T]$ is Gorenstein but not lci (Exercise 26.3).
(4) Example 26.42 below shows that the union of the three coordinate axes in $\mathbb{A}^3_k$ gives an example of a reduced curve which is not Gorenstein. Compare this to (2) which shows that the union of finitely many lines in $\mathbb{A}^2_k$ is always lci, in particular Gorenstein.
(5) Finally, an example of an irreducible curve which is not Cohen Macaulay would be any curve with an embedded component, e.g. $\operatorname{Spec} k[X, Y]/(X^2, XY)$.

(26.4) Morphisms between curves.

Let us recall some results on morphisms between curves, and collect some new ones. First, we have the foundational fact, used for instance in the proof of Theorem 26.3, that every rational map from a normal curve C to a proper k-scheme X extends to a morphism $C \to X$. This is an immediate consequence of the valuative criterion of properness, Theorem 15.9.

Remark 26.8. (Morphisms to projective space) Let k be a field and X a k-scheme. Recall the description of morphisms $X \to \mathbb{P}^N_k$ (i.e., of X-valued points of $\mathbb{P}^N_k$) from Sections (8.5), (13.13). For a morphism $f\colon X \to \mathbb{P}^N_k$, the pullback $\mathscr{L} := f^* \mathcal{O}_{\mathbb{P}^N_k}(1)$ is a globally generated line bundle and the natural basis $T_0, \dots, T_N \in H^0(\mathbb{P}^N_k, \mathcal{O}(1))$ gives rise, by pullback, to sections $s_i = f^* T_i \in H^0(X, \mathscr{L})$ generating $\mathscr{L}$ everywhere.

Conversely, given a line bundle $\mathscr{L}$ on X with global sections $s_0, \dots, s_N \in H^0(X, \mathscr{L})$, not all $= 0$, we obtain a rational map $X \dashrightarrow \mathbb{P}^N_k$ which on k-valued points is given by $x \mapsto (s_0(x) : \dots : s_N(x))$ (where we choose an isomorphism $\mathscr{L}(x) \cong k$ and the resulting point in $\mathbb{P}^N_k$ is independent of the choice of isomorphism). This rational map is defined on the open subset $U \subseteq X$ which is the union of the non-vanishing loci of the s_i. Denoting by $f\colon U \to \mathbb{P}^N_k$ the corresponding morphism, we have $f^* \mathcal{O}_{\mathbb{P}^N_k}(1) \cong \mathscr{L}_{|U}$.

We call the subvector space of $H^0(X, \mathscr{L})$ generated by the s_i the corresponding *linear system*, and say that this linear system is *base-point free*, if the morphism described above is defined on all of X, i.e., if for every $x \in X$ there exists i with $s_i(x) \neq 0$. If $s_0, \dots, s_N$ generate $H^0(X, \mathscr{L})$, we speak about the *complete linear system* attached to $\mathscr{L}$. It is base-point free if and only if $\mathscr{L}$ is generated by its global sections.

As pointed out above, if $X = C$ is a *normal curve*, then every rational map $f\colon C \dashrightarrow \mathbb{P}^N_k$ can be extended to a morphism (but if the rational map comes from a linear system that is not base-point free, then the pullback $\tilde{f}^* \mathcal{O}_{\mathbb{P}^N_k}(1)$ under the morphism $\tilde{f}\colon C \to \mathbb{P}^N_k$ will be different from the line bundle giving rise to this linear system).

Lemma 26.9. *Let $f\colon C' \to C$ be a non-constant morphism of curves over a field k. Suppose that C' is irreducible.*
(1) *The morphism f is quasi-finite.*
(2) *Let C' be integral and let C be normal. Then f is flat.*
(3) *Let C' be proper over k. Then f is finite.*
(4) *Let C' be integral and proper over k and let C be normal, then f is finite locally free.*

Proof. As f is non-constant and C' is an irreducible scheme of dimension 1, the fibers of f must be 0-dimensional, hence f is quasi-finite.

Assertion (2) is a special case of Proposition 15.4 (3) which ultimately follows from the fact that a module over a discrete valuation ring is flat if and only if it is torsion-free.

Assertion (3) follows from Zariski's main theorem as follows. As C' is proper over k and C is separated over k, the morphism f is proper. It is quasi-finite by (1). Hence it is finite by Zariski's main theorem (Corollary 12.89).

Finally, (4) follows from (2) and (3) (Proposition 12.19). $\qquad\square$

Lemma 26.10. *Let k be a field, and let $f\colon X \to Y$ be a dominant morphism between integral k-schemes of finite type with $\dim(X) = \dim(Y)$. The following are equivalent*
(i) *The morphism f is generically étale, i.e., there exists a non-empty open subscheme $U \subseteq X$, such that $f_{|U}\colon U \to Y$ is étale.*

(ii) *The morphism f is generically unramified, i.e., there exists a non-empty open sub-scheme $U \subseteq X$, such that $f_{|U} : U \to Y$ is unramified.*
(iii) *The extension $K(X)/K(Y)$ of function fields induced by f is separable.*

Because of point (iii), in the literature morphisms with this property are often called *separable.*

Proof. The extension $K(X)/K(Y)$ is finite since $K(X)$ and $K(Y)$ are finitely generated field extensions of k with the same transcendence degree $\dim(X) = \dim(Y)$ over k. We remark that this shows that the morphism $\operatorname{Spec} K(X) \to \operatorname{Spec} K(Y)$ is étale if and only if it is smooth, if and only if it is unramified (Theorem 18.44), if and only if the extension is separable (e.g., use Proposition 18.24 and Theorem 18.44).

It is clear that (i) implies (ii). By the above remark and by Proposition 18.79 we have (iii) $\Rightarrow$ (i).

It remains to show that (ii) implies (iii). But since $\operatorname{Spec} K(X)$ is the scheme-theoretic fiber of f over the generic point $\operatorname{Spec} K(Y)$ of Y, this follows from the above remark and the fact that the property unramified is preserved by base change, see Remark 18.23. $\square$

An important invariant of a finite morphism between integral curves is its degree. (See also Definition 23.76, and (12.6.1) for the case of finite locally free morphisms).

Definition 26.11. *Let k be a field, and let $f : C' \to C$ be a non-constant morphism of integral curves over k. Then via f the function field $K(C)$ of C is contained in $K(C')$, and the degree $\deg(f)$ of f is defined as the degree of this finite field extension.*

It is clear that the field extension $K(C')/K(C)$ is indeed finite since both fields are finitely generated and have transcendence degree 1 over k.

Corollary 26.12. *Let k be a field. Let $f : C' \to C$ be a non-constant morphism between integral curves over k. If C' is proper over k, C is normal, and $\deg(f) = 1$ then f is an isomorphism.*

Proof. By Lemma 26.9 we know that f is finite locally free. Hence it is an isomorphism if and only if it is of degree 1. $\square$

The following example shows that the hypotheses in Corollary 26.12 are indeed necessary.

Example 26.13. Let C be an integral curve.
(1) Let $\pi : C' \to C$ be its normalization, which is a finite morphism by Corollary 12.52. Hence $K(C) = K(C')$ by definition and π is of degree 1. But π is an isomorphism if and only if C is normal.
(2) Every open immersion $j : C' \to C$ has degree 1. If j is not an isomorphism, then C' cannot be proper because otherwise j would be also a closed immersion and hence surjective.

Remark 26.14. Let k be a field, and let $\pi : C' \to C$ be a non-constant morphism of integral curves over k. Then π is quasi-finite by Lemma 26.9. Recall that we defined in Section (12.5) for $c' \in C'$ with image $c = \pi(c')$ positive integers

$$e_{c'/c} := \lg(\mathscr{O}_{\pi^{-1}(c),c'}), \qquad f'_{c'/c} := [\kappa(c') : \kappa(c)]_{\mathrm{insep}}, \qquad f''_{c'/c} := [\kappa(c') : \kappa(c)]_{\mathrm{sep}}.$$

Now suppose that π is flat (e.g., if C is normal, Lemma 26.9). Then π is étale in a point $c' \in C'$ if and only if $e_{c'/c} = f'_{c'/c} = 1$ by Corollary 18.28. In this case, π is étale in some open neighborhood of c' which is dense in C' since C' is integral, i.e., π is then generically étale and there exist only a finite set $S' \subseteq C'$ of closed points in which π is not étale. There exists a natural structure of a closed subscheme on S' given by the different of π (Definition 20.74). If C is normal, then S' will be a divisor which we will study in Section (26.13) below. The image $S := \pi(S')$ is a finite set of closed points of C. Let $U := C \setminus S$ be its open complement. Then the restriction of π to $\pi^{-1}(U)$ is étale.

Now suppose that π is finite locally free (e.g., if C' is proper over k and C is normal, see Lemma 26.9). Then we have by Proposition 12.21 for every $c \in C$

$$\deg(\pi) = \sum_{c' \in \pi^{-1}(c)} e_{c'/c} f'_{c'/c} f''_{c'/c}.$$

As all members of the right hand side sum are ≥ 1, we see in particular

$$\#\pi^{-1}(c) \leq \deg(f), \qquad c \in C$$

with equality if and only if $\mathcal{O}_{\pi^{-1}(c),c'} = \kappa(c)$ for all $c' \in \pi^{-1}(c)$. In this case we can endow S with a natural structure of a closed subscheme given by the discriminant of π (Definition 20.72).

(26.5) Quasi-projectivity of Curves.

In this section we complete the proof of the theorem, already stated in Volume I, that every separated curve over a field is quasi-projective. We will prove later, using the Theorem of Riemann-Roch, that a separated irreducible curve over a field is either affine or projective (Proposition 26.61).

The key ingredient that allows us to reduce the proof of quasi-projectivity in general to the case of reduced curves (which was handled in Chapter 15) is the cohomological characterization of ampleness. Furthermore, we need the following lemma.

Lemma 26.15. *Let* $i\colon X_0 \to X$ *be a closed immersion of schemes defined by a quasi-coherent ideal* $\mathscr{I} \subset \mathcal{O}_X$ *with* $\mathscr{I}^2 = 0$ *so that we can view* $\mathscr{I}$ *as* $\mathcal{O}_{X_0}$-*module. Then there exists an exact sequence of abelian groups*

(26.5.1) $$H^1(X_0, \mathscr{I}) \to \operatorname{Pic}(X) \to \operatorname{Pic}(X_0) \to H^2(X_0, \mathscr{I}).$$

In particular, $\operatorname{Pic}(X) \to \operatorname{Pic}(X_0)$ *is surjective if* $H^2(X_0, \mathscr{I}) = 0$.

Proof. Consider the short exact sequence of abelian sheaves

$$0 \longrightarrow \mathscr{I} \xrightarrow{f \mapsto 1+f} \mathcal{O}_X^\times \longrightarrow \mathcal{O}_{X_0}^\times \longrightarrow 1$$

on the underlying topological space of X_0 which is the same as the underlying topological space of X. Then the long exact cohomology sequence yields the exact sequence (26.5.1), because we can identify $H^1(X, \mathcal{O}_X^\times)$ with $\operatorname{Pic}(X)$, and likewise for X_0, see Section (11.7). $\square$

Theorem 26.16. *Let* k *be a field and let* $f\colon C \to \operatorname{Spec} k$ *be a curve. Then* f *is quasi-projective.*

Recall that in this chapters all curves are by definition separated.

Proof. Recall that we proved the theorem already if C is reduced (Theorem 15.18). We will reduce to this case. Since we know that C_{red} is quasi-projective, it admits an ample line bundle $\mathscr{L}_0$. By Lemma 26.15 and since the second cohomology group occurring there vanishes for dimension reasons (Theorem 21.57), we can lift $\mathscr{L}_0$ to a line bundle $\mathscr{L}$ on C (note that C_{red} is defined by a nilpotent ideal of $\mathscr{O}_C$ since C is noetherian). By Lemma 23.7, $\mathscr{L}$ is ample. It follows that C is quasi-projective, cf. Section (13.15). $\square$

See Proposition 26.168 for a result in this direction for relative curves. Also see Proposition 26.61.

(26.6) Divisors on curves.

In some of the discussions below we want to include curves C (always over a field) that are not necessarily normal. So the notions of Cartier divisors and of Weil divisors may not coincide. By a divisor, we always mean a Cartier divisor (Definition 11.26), i.e., an element of $\mathrm{Div}(C) = \Gamma(C, \mathscr{K}_C^\times / \mathscr{O}_C^\times)$, where $\mathscr{K}_C$ denotes the sheaf of meromorphic functions on C (Sections (11.10) and (11.11)).

For every divisor D on a curve C we have the associated line bundle $\mathscr{O}_C(D)$. If D is represented by $(U_i, f_i)_i$ with $U_i \subseteq C$ open and $f_i \in \mathscr{K}_C(U_i)$, then $\mathscr{O}_C(D)_{|U_i} = f_i^{-1}\mathscr{O}_{U_i}$. Since every curve C over a field is quasi-projective by Theorem 26.16, the map $D \mapsto \mathscr{O}_C(D)$ yields an isomorphism

$$(26.6.1) \qquad\qquad \mathrm{DivCl}(C) \xrightarrow{\sim} \mathrm{Pic}(C)$$

of the divisor class group $\mathrm{DivCl}(C)$ of Cartier divisors up to linear equivalence with the Picard group $\mathrm{Pic}(C)$, i.e., the group of isomorphism classes of line bundles on C, by Corollary 11.30.

Recall from Corollary 15.26 that every divisor on a curve can be expressed as the difference of two effective divisors, i.e., those divisors that are represented by $(U_i, f_i)_i$ with $f_i \in \mathscr{K}_C(U_i)^\times \cap \mathscr{O}_C(U_i)$ (Definition 11.26 (4)).

Given a Cartier divisor D on a curve C over a field k, we have the corresponding Weil divisor

$$\mathrm{cyc}(D) = \sum_{x \in C^1} n_x[x],$$

where the sum extends over the set C^1 of closed points of C, $n_x \in \mathbb{Z}$, and only finitely many summands are $\neq 0$. See Section (11.13).

The degree of D is then defined as

$$\deg(D) = \sum_{x \in C^1} n_x[\kappa(x) : k].$$

See Definition 15.28. If C is proper over k, then two linearly equivalent line bundles have the same degree (Theorem 15.31), i.e., by (26.6.1), deg induces a homomorphism of abelian groups

$$\deg \colon \mathrm{Pic}(C) \to \mathbb{Z}.$$

By Example 23.75 (1), the degree of a line bundle $\mathscr{L}$ is the same as the intersection number $(\mathscr{L} \cdot C)$.

If D is an effective divisor on an arbitrary curve C, then $n_x = \lg(\mathscr{O}_{C,x}/(f_x))$, where f_x is a local equation of D at x. In particular, $\deg(D) \geq 0$. In this case, D defines a closed subscheme, supported on a finite number of closed points of C. It is defined by the invertible quasi-coherent ideal $\mathscr{I}_C(D) := \mathscr{O}_C(D)^{\otimes -1}$. In particular, the closed immersion $D \to C$ is regular of codimension 1. Conversely, we can recover the divisor D from this closed subscheme, which we therefore usually also denote by D. We then see that $\deg(D) = \dim_k \Gamma(D, \mathscr{O}_D) = \dim_k \Gamma(C, \mathscr{O}_D)$.

If $x \in C$ is a closed point which lies in the normal locus of C, i.e., such that $\mathscr{O}_{C,x}$ is normal and hence a discrete valuation ring (Proposition B.71), then we can view $[x]$ as a Cartier divisor, i.e., the cycle $[x]$ is the Weil divisor attached to an (effective) Cartier divisor on C. The degree of this divisor is $[\kappa(x) : k]$. The corresponding closed subscheme of C is simply the scheme $\operatorname{Spec} \kappa(x)$ considered as a closed subscheme of C concentrated at x.

If C is normal, i.e., all local rings at closed points are discrete valuation rings, then we have a one-to-one correspondence between Cartier divisors and Weil divisors on C, i.e., the group of divisors is the free abelian group $\mathbb{Z}^{(C^1)}$.

If $D = (n_x)_{x \in C^1}$ is divisor on a connected normal curve C, Remark 11.41 shows that the corresponding line bundle $\mathscr{O}_C(D)$ has sections over some $U \subseteq C$ open given by

$$\Gamma(U, \mathscr{O}_C(D)) = \{ f \in K(C) \; ; \; v_x(f) \geq -n_x \text{ for all } x \in C^1 \cap U \}.$$

For suitable morphisms $f \colon C' \to C$ we can define a pullback operation on divisors, see Definition 11.49, Proposition 11.50. Here we record the following special cases:

Proposition 26.17. *Let k be a field and let $f \colon C' \to C$ be a morphism of curves over k. Assume that f is flat or that C' is reduced and f is non-constant on every irreducible component of C'. Then the homomorphism $\mathscr{O}_C \to f_* \mathscr{O}_{C'}$ induces a homomorphism $f^* \colon \operatorname{Div}(C) \to \operatorname{Div}(C')$, and we call $f^* D$ the pullback of D under f. For the associated line bundles we have $\mathscr{O}_{C'}(f^* D) \cong f^* \mathscr{O}_C(D)$.*

Remark 26.18. For an effective Cartier divisor, seen as a closed subscheme $D \subset C$ locally defined by a regular element t in some $\Gamma(U, \mathscr{O}_C)$, the conditions on f ensure that t is mapped to a regular element in $\Gamma(f^{-1}(U), \mathscr{O}_{C'})$ so that the scheme-theoretic inverse image $f^{-1}(D)$ is a closed subscheme in C' which corresponds to an effective Cartier divisor on C'.

The hypothesis on f is for instance satisfied if f is the normalization morphism of an arbitrary curve in the sense of Definition 26.2.

For a finite morphism f between integral curves, we can compute the degree of the pullback of D as the product of the degree of f and the degree of D. Compare Proposition 15.30, where we have proved this under the assumption that the target of the morphism is normal.

Proposition 26.19. *Let $f \colon C' \to C$ be a finite morphism of integral curves over k, and let D be a divisor on C. Then we have*

$$\deg(f^* D) = \deg(f) \deg(D).$$

If C' and C are proper over k, then Proposition 26.19 is a special case of Proposition 23.77.

Proof. We may assume that D is effective and can work on local rings, i.e., instead of C we consider the spectrum of a local domain A and assume that D corresponds to the closed subscheme A/tA for some regular element $t \in A$. Then f corresponds to a finite ring homomorphism $\varphi \colon A \to B$, and we have to show that $\dim_k B/tB = \deg(f)\dim_k A/tA$, where $\deg(f) = [\mathrm{Frac}(B) : \mathrm{Frac}(A)] =: d$.

Now let $b_1, \ldots, b_d \in B$ be a $\mathrm{Frac}(A)$-basis of $\mathrm{Frac}(B)$ and let $B' \subseteq B$ be the free A-module generated by $b_1, \ldots, b_n$. We then have isomorphisms

$$(B/tB')/(tB/tB') \cong B/tB, \quad tB/tB' \cong B/B', \quad (B/tB')/(B'/tB') \cong B/B',$$

where we use for the second isomorphism that t is regular in B by Remark 26.18. Since all occurring terms are finite-dimensional k-vector spaces, we obtain

$$\dim_k B/tB = \dim_k B'/tB' = d \dim_k A/tA,$$

where for the second equality we use that B' is free of rank d over A. $\qquad\square$

Corollary 26.20. *Let k be a field and let C be a connected normal proper curve. If there exist $x \neq y \in C(k)$ such that the divisors $[x]$ and $[y]$ are linearly equivalent, then $C \cong \mathbb{P}^1_k$.*

Proof. Say $[x] - [y] = \mathrm{div}(f)$ for $f \in K(C)$. As C is normal, f defines a non-constant morphism $f \colon C \to \mathbb{P}^1$ such that $f^*[0] = [x]$. By Proposition 26.19 we find $\deg(f) = 1$, so f is an isomorphism by Corollary 26.12. $\qquad\square$

Let X be an integral proper scheme over a field k. Then $\Gamma(X, \mathscr{O}_X)$ is a finite k-algebra without zero divisors, and hence a finite field extension of k. Given a divisor D on X, the line bundle $\mathscr{O}_X(D)$ is by definition contained in the sheaf $\mathscr{K}_X$ of rational functions on X, so $H^0(X, \mathscr{O}_X(D))$ is a $\Gamma(X, \mathscr{O}_X)$-sub-vector space of $K(X)$. The map

$$H^0(X, \mathscr{O}_X(D)) \to \mathrm{Div}(X), \qquad s \mapsto D + \mathrm{div}(s),$$

induces a bijection

(26.6.2)
$$\mathbb{P}(H^0(X, \mathscr{O}_X(D))^\vee) = (H^0(X, \mathscr{O}_X(D)) \setminus \{0\})/\Gamma(C, \mathscr{O}_C)^\times$$
$$\xrightarrow{\sim} \{D' \subseteq X \text{ effective divisor, linearly equivalent to } D\}$$

by Proposition 11.34. Here we use that X is integral to see that all non-zero sections of $H^0(X, \mathscr{O}_X(D))$ are regular.

Remark 26.21. Let C be a proper curve over a field k.
(1) As we have shown in Theorem 15.31, the degree of a principal divisor on X equals 0.
(2) Let C be integral and let D be a divisor of $\deg(D) < 0$. Then there cannot exist effective divisors that are linearly equivalent to D. Hence (26.6.2) shows

$$H^0(C, \mathscr{O}_C(D)) = 0.$$

(3) Let C be integral and D be a divisor of degree 0. Then $\mathscr{O}_C(D) \cong \mathscr{O}_C$ if and only if $H^0(C, \mathscr{O}_C(D)) \neq 0$. In fact, if $H^0(C, \mathscr{O}_C(D)) \neq 0$, then C is linearly equivalent to an effective divisor of degree 0, i.e., to the trivial divisor, thus $\mathscr{O}_C(D) \cong \mathscr{O}_C(0) = \mathscr{O}_C$.

A divisor D on a k-scheme X is called *base-point free*, if the invertible sheaf $\mathscr{O}_X(D)$ is generated by its global sections, or equivalently if it defines a morphism from X to the projective space $\mathbb{P}(H^0(X, \mathscr{O}_X(D)))$. The term *base-point free* refers to the latter condition – a base point is a point where all global sections of this line bundle vanish and correspondingly the rational map from X to projective space is not defined.

A divisor is called ample (resp. very ample) if the line bundle $\mathscr{O}_X(D)$ is ample (resp. very ample). A divisor is very ample if and only if it is base point free and it defines an immersion of X into projective space.

(26.7) Algebraic curves and compact Riemann surfaces.

In this section we work over the field $\mathbb{C}$ of complex numbers and compare the algebraic situation to the analytic setting, i.e., we have a look at complex manifolds X that have complex dimension 1, or equivalently, real dimension 2. So considered as differentiable manifolds, these objects are surfaces, which justifies the following terminology.

Definition 26.22. *A Riemann surface is a connected complex manifold of complex dimension 1.*

Simple examples are $\mathbb{C}$ and open subsets of $\mathbb{C}$, and the Riemann sphere $\mathbb{P}^1(\mathbb{C})$. More generally, every connected smooth algebraic curve C over $\mathbb{C}$ gives rise to a Riemann surface C^{an} by analytification (Section (20.12)). If the curve C is proper over $\mathbb{C}$, then the analytification C^{an} is a compact Riemann surface (cf. Remark 20.59). In the case of curves, we can also argue by using that a proper curve C admits a closed embedding into some projective space. The analytification of this embedding is then a closed embedding of C^{an} into a complex projective space $\mathbb{P}^n(\mathbb{C})$. Every smooth algebraic curve C over $\mathbb{C}$ can be compactified by adding finitely many closed points, to obtain the smooth projective model. Therefore C^{an} also admits a compactification by adding finitely many points. On the other hand, the open unit disk $D = \{z \in \mathbb{C};\ |z| < 1\}$ is a typical example of a Riemann surface that is not algebraizable, i.e., that is not of the form C^{an} for an algebraic curve C.

Surprisingly, at first, the result that every proper algebraic curve is projective, has an analog in the complex setting: Every compact Riemann surface is projective, i.e., it admits a closed embedding into some complex projective space. This implies that every compact Riemann surface is algebraizable. Therefore, once this fact is established, the theories of smooth proper algebraic curves over $\mathbb{C}$ and of compact Riemann surfaces are in a sense the same. However, different methods can of course be used on either side of this correspondence. We emphasize that the case of dimension ≤ 1 is completely exceptional in this respect.

Below we sketch one possible path to proving that every compact Riemann surface is algebraizable, based on the following analytic (and difficult) result which we will use as a black box.

Theorem 26.23. *Let X be a compact Riemann surface. Then the cohomology group $H^1(X, \mathscr{O}_X)$ (with coefficients in the structure sheaf of the complex manifold X, i.e., the sheaf of holomorphic functions on X) is a finite-dimensional $\mathbb{C}$-vector space.*

For a proof tailored to the setting of Riemann surfaces, see [For]^{O} §14. Of course, in the algebraic setting this finiteness is already known to us, it is a special case of the result that push-forward under proper morphisms preserves coherence (Theorem 23.1). Unlike what follows, this statement also holds for proper morphisms between higher-dimensional compact complex manifolds. See Remark 23.19.

With this result at hand, there are several routes to showing that X is algebraizable. The most direct one is to show that there exists a non-constant meromorphic function.

Definition 26.24. *Let X be a Riemann surface. A meromorphic function on X is a morphism $X \to \mathbb{P}^1(\mathbb{C})$ of Riemann surfaces.*

Proposition 26.25. *Let X be a compact Riemann surface. Then there exists a nonconstant meromorphic function on X.*

Proof. Let $x \in X$ be any point, and let (U, z) be a chart on X with $x \in U$ and $z(x) = 0$, i.e., $z \colon U \to \mathbb{C}$ is a holomorphic map identifying U with an open subset of $\mathbb{C}$ which maps x to 0. Let $V = X \setminus \{x\}$. Then $\mathscr{U} = (U, V)$ is an open cover of X, and the Čech cohomology group $H^1(\mathscr{U}, \mathscr{O}_X)$ embeds into $H^1(X, \mathscr{O}_X)$ (see the proof of Lemma 21.81). Thus by Theorem 26.23, $H^1(\mathscr{U}, \mathscr{O}_X)$ is finite-dimensional.

For $j \in \mathbb{N}_{>0}$, consider $z^{-j} \in \Gamma(U \setminus \{x\}, \mathscr{O}_X)$. Since $U \setminus \{x\} = U \cap V$, each z^{-j} defines a class in $H^1(\mathscr{U}, \mathscr{O}_X)$, and for sufficiently large r, the classes of $z^1, \ldots, z^{-r}$ must be linearly dependent. This means that there exist $a_1, \ldots, a_r \in \mathbb{C}$, not all zero, and $f \in \Gamma(U, \mathscr{O}_X)$, $g \in \Gamma(V, \mathscr{O}_X)$ with

$$\sum_{j=1}^{r} a_j z^{-j} = g - f.$$

Then $f + \sum_{j=1}^{r} a_j z^{-j} = g \in \Gamma(V, \mathscr{O}_X)$ extends to a non-constant meromorphic function $X \to \mathbb{P}^1(\mathbb{C})$ (by mapping x to $\infty \in \mathbb{P}^1(\mathbb{C})$). $\qquad\square$

Corollary 26.26. *Every compact Riemann surface X is algebraizable, i.e., is isomorphic as a complex manifold to the analytification C^{an} of a smooth proper curve C over $\mathbb{C}$.*

Proof. The previous proposition shows that there exists a non-constant morphism $f \colon X \to \mathbb{P}_1(\mathbb{C})$ of Riemann surfaces. One concludes by showing that this is necessarily a finite morphism of complex manifolds, which implies that X can be recovered from the coherent $\mathscr{O}_{\mathbb{P}^1(\mathbb{C})}$-algebra $f_* \mathscr{O}_X$ by an analogue of the relative Spec-construction, and using the GAGA principle for coherent sheaves over $\mathbb{P}^1_{\mathbb{C}}$ in the form of Theorem 20.60, cf. [SGA1]$_X^0$ Exp. XII, Corollaire 4.6. $\qquad\square$

With the corollary proved, we can apply the various instances of the GAGA principle to the curve C. For a connected proper smooth curve C over $\mathbb{C}$ with analytification C^{an}, the (derived) categories of coherent sheaves coincide (Theorem 20.60, Theorem 23.39), line bundles on C correspond to line bundles on C^{an}, and similarly for locally free sheaves of finite rank, divisors, etc. Moreover, the analytification functor on the category of proper $\mathbb{C}$-schemes is fully faithful (Corollary 20.61).

As a consequence, later results of this chapter, such as the Theorem of Riemann and Roch, carry over to the context of Riemann surfaces (of course, historically, all these results were first proved for Riemann surfaces). In particular, any line bundle $\mathscr{L}$ of sufficiently high degree on a compact Riemann surface X defines an embedding of X into projective space. This fact can also be shown more directly, using analytic methods. One way to do so is to develop Hodge theory (and in particular prove the *Hodge decomposition* $H^1(X, \mathbb{C}) = H^1(X, \mathscr{O}_X) \oplus H^0(X, \Omega^1_X)$, where Ω^1_X denotes the sheaf of holomorphic differentials on X). This gives a handle on proving the Kodaira vanishing theorem, a special case being that $H^1(X, \mathscr{L}) = 0$ for every (holomorphic) line bundle $\mathscr{L}$ on X which satisfies $\deg(\mathscr{L} \otimes \Omega_X^{1,\vee}) > 0$. As a consequence, one shows that $\mathscr{L}$ with $\deg(\mathscr{L}) > \deg(\Omega_X^1) + 2$ defines a closed embedding $X \to \mathbb{P}(H^0(X, \mathscr{L}))$ with arguments analogous to those in the proof of Proposition 26.59 below. See [GrHa] $^{\circ}$ Ch. 2.1.

In addition to complex-analytic methods, in the setting of Riemann surfaces it is natural and useful to study the underlying topological spaces. The following theorem states the classification of compact Riemann surfaces up to homeomorphism.

Theorem 26.27. *For every compact Riemann surface X there exists a unique natural number g, the genus of X such that X is homeomorphic to a "sphere with g handles". The topological fundamental group of X is isomorphic to the quotient of the free group on generators $a_1, \ldots, a_g, b_1, \ldots, b_g$ modulo the relation $\prod_{i=1}^g a_i b_i a_i^{-1} b_i^{-1} = 1$. The singular cohomology group $H^1(X, \mathbb{Z})$ is a free $\mathbb{Z}$-module of rank $2g$.*

More concretely, a "sphere with g handles" is the topological space that can be constructed by attaching g "handles" to a sphere. A relatively convenient way to write this down is to construct the space by starting with a regular $2g$-gon in the plane $\mathbb{R}^2$ and pasting, i.e., identifying some of its edges in a suitable way. See [FaKr] $^{\circ}$ I.2.5. Alternatively the genus g of a compact Riemann surface is often described as the "number of holes", in the sense that the sphere has no holes, a torus has one hole, etc.

Every Riemann surface of genus 0 is isomorphic to the Riemann sphere $\mathbb{P}_{\mathbb{C}}^{1,\mathrm{an}} = \mathbb{P}^1(\mathbb{C})$. For every $g > 0$ there exist infinitely many non-isomorphic Riemann surfaces of genus g. See Section (26.19) for the case of genus 1.

From the comparison theorem for singular cohomology and de Rham cohomology ([GrHa] $^{\circ}$ Ch. 0, Section 3), the Hodge decomposition and Serre duality, we obtain the following equivalent descriptions of the genus of a compact Riemann surface. The ones that use cohomology of coherent sheaves will allow us to define a useful notion of genus for proper algebraic curves over arbitrary fields.

Theorem 26.28. *Let X be a compact Riemann surface. Then*

$$\frac{1}{2} \operatorname{rk}_{\mathbb{Z}} H^1(X, \mathbb{Z}) = \frac{1}{2} \dim_{\mathbb{C}} H^1(X, \mathbb{C}) = \dim_{\mathbb{C}} H^1(X, \mathscr{O}_X) = \dim_{\mathbb{C}} H^0(X, \Omega_X^1).$$

We conclude this section by stating the uniformization theorem, another fundamental result in the theory of Riemann surfaces.

Theorem 26.29.
(1) *Let X be a Riemann surface. Then the universal cover $\tilde{X}$ of X as a topological space can be equipped, in a unique way, with the structure of Riemann surface such that the projection $\tilde{X} \to X$ is a morphism of Riemann surfaces.*
(2) *Every simply connected Riemann surface is isomorphic to precisely one of the following: $\mathbb{C}$, $\mathbb{P}^1(\mathbb{C})$, $D = \{z \in \mathbb{C}; \ |z| < 1\}$.*

In the theorem, the open unit disk D carries the structure of Riemann surface inherited from the open embedding $D \subset \mathbb{C}$. As Riemann surface, D is isomorphic to the complex upper half plane $\mathbb{H} = \{z \in \mathbb{C}; \ \operatorname{Im}(z) > 0\}$ (which we also view as an open submanifold of $\mathbb{C}$) via the Cayley transform. Part (1) is easy to prove, and in Part (2) it is clear that the three given Riemann surfaces are simply connected and pairwise non-isomorphic. The hard part is to show that there are no other simply connected Riemann surfaces, up to isomorphism. An equivalent formulation is that every simply connected Riemann surface X that is not isomorphic to either $\mathbb{C}$ or $\mathbb{P}^1(\mathbb{C})$ admits an isomorphism $X \xrightarrow{\sim} D$; this statement is also known as the *Riemann Mapping Theorem*. See [For] ^{O} §27.

It is then not very hard to show that the only Riemann surface with universal cover $\mathbb{P}^1(\mathbb{C})$ is $\mathbb{P}^1(\mathbb{C})$, up to isomorphism, and that only $\mathbb{C}$, $\mathbb{C}^\times$ and complex tori $\mathbb{C}/\Lambda$, where $\Lambda \subset \mathbb{C}$ is a lattice (Definition 26.109), have universal cover $\mathbb{C}$. See also Proposition 26.110.

For the compact case, one gets the following connection between genus and universal cover.

Theorem 26.30. *Let X be a compact Riemann surface of genus g, and let $\tilde{X}$ be its universal cover.*
(1) *We have $g = 0$ if and only if $\tilde{X} \cong \mathbb{P}^1(\mathbb{C})$.*
(2) *We have $g = 1$ if and only if $\tilde{X} \cong \mathbb{C}$.*
(3) *We have $g \geq 2$ if and only if $\tilde{X} \cong D$, the open unit disk.*

The division of compact Riemann surfaces into these three classes according to their universal cover has many other manifestations, for instance it reflects the curvature of the Riemann surface in the sense of differential geometry. Also in algebraic geometry, this trichotomy is visible, sometimes in situations that are quite far from complex geometry, compare for instance Section (26.30). Transferring it to algebraic varieties of higher dimension is a first step towards a theory of classification (up to birational maps, say) in that case.

Let us say some words about the proof of Theorem 26.30. From the discussion above we have Part (1), and also that a compact Riemann surface with universal cover $\mathbb{C}$ has genus 1. It then remains to prove that every compact Riemann surface of genus 1 has universal cover $\mathbb{C}$. One approach to this is the following: Let X have genus one and let $f \colon \tilde{X} \to X$ be its universal cover. We use that for a canonical divisor K on X, we have $\ell(K) = g(X) = 1$ and $\deg(K) = 2g(X) - 2 = 0$, so that K is (linearly equivalent to) the trivial divisor, compare Corollary 26.52 below. Therefore every non-zero holomorphic differential vanishes nowhere on X. Fixing such a differential form ω, integrating $f^*\omega$ along paths in $\tilde{X}$ from a fixed base point to any point on $\tilde{X}$ defines a map $\tilde{X} \to \mathbb{C}$. Because $\tilde{X}$ is simply connected, the value of the integral does not depend on the choice of path, but only on its end points, so that the map is well-defined. One can show that it is actually a covering map. Since $\mathbb{C}$ is simply connected, it follows that it is an isomorphism, so $\tilde{X} \cong \mathbb{C}$, as we wanted to show. Alternatively, one can consider the curvature of X and its universal cover. Yet another possibility is to show first that X carries a group structure (cf. Theorem 26.98), and then to use the exponential map for the compact complex Lie group X and to show that it induces an isomorphism between X and a quotient of $T_0 X \cong \mathbb{C}$ by a lattice. While it may look somewhat roundabout, this strategy has the advantage of working also in higher dimensions: Every connected compact complex Lie group (i.e., complex manifold which is a group object in the category of complex manifolds) is isomorphic as a complex manifold, to a complex torus $\mathbb{C}^{\dim(A)}/\Lambda$, for Λ a lattice in $\mathbb{C}^{\dim(A)}$, i.e., a subgroup generated by $2g$ elements that are $\mathbb{R}$-linearly independent. In particular this applies to the analytification A^{an} of an abelian variety A over $\mathbb{C}$ (but in dimension > 1 most complex tori are not of the form A^{an} for an algebraic A). See [Mum1] Ch. I.

(26.8) The arithmetic genus of a curve.

We start by discussing the *genus* of a proper curve. See the discussion in Section (26.7) for the notion of genus in the theory of compact Riemann surfaces, in particular Theorem 26.27

which states that the genus classifies compact Riemann surfaces up to homeomorphism, and Theorem 26.28 which gives expressions for the genus that can immediately be applied to proper algebraic curves over an arbitrary field.

For a smooth proper curve C over a field k, we have $\dim_k H^0(C, \Omega_{C/k}) = \dim_k H^1(C, \mathscr{O}_C)$ by duality (Proposition 25.110). In the non-smooth case, it turns out that working with cohomology of the structure sheaf has better technical properties. (But see Definition 26.44.) The following definition is even more flexible (if only slightly).

Definition 26.31. *Let k be a field, and let C be a proper curve over k. The (arithmetic) genus of C is defined as*

$$g(C) = 1 - \chi_k(\mathscr{O}_C) = 1 - \dim_k H^0(C, \mathscr{O}_C) + \dim_k H^1(C, \mathscr{O}_C).$$

Sometimes the arithmetic genus is denoted by $p_a(C)$ instead of $g(C)$. (Compare Definition 26.44.)

Note that the genus depends on the choice of base field and therefore should more precisely be denoted by $g(C/k)$, and we will do so, if necessary.

It seems, there is no generally accepted definition of the genus for an arbitrary proper curve. Our guiding principle is that the genus should have the following two properties.

(1) For smooth connected proper curves over algebraically closed fields (for instance, for those complex algebraic curves corresponding to compact Riemann surfaces, see Section (26.7)) there is a generally accepted definition of genus, and our definition should agree with this in that case.

(2) The genus should vary locally constantly in flat proper families of curves (see Section (26.26) below).

This leads us to the definition above since the Euler characteristic does vary locally constantly in flat proper families (Theorem 23.139, see also Proposition 26.164). Here, we follow essentially Serre who even advocates (for general projective varieties, see [Ser1] §80) to define the arithmetic genus simply as $\chi_k(\mathscr{O}_C)$. But this would violate the first of our guiding principles.

Note that our definition has the consequence that the genus can be negative (and so one has to be careful with arguments that conclude $g(C) = 0$ from $g(C) \le 0$). For instance, if $X = \mathbb{P}^1_{k'}$ for a finite extension k' of k, which we can consider as a proper curve of k, then

$$g(\mathbb{P}^1_{k'}/k) = 1 - [k' : k]$$

(using that $H^0(\mathbb{P}^1_{k'}, \mathscr{O}_{\mathbb{P}^1_{k'}}) = k'$ and $H^1(\mathbb{P}^1_{k'}, \mathscr{O}_{\mathbb{P}^1_{k'}}) = 0$).

Moreover, our definition behaves somewhat unexpected if the curve is non-connected: for two curves C and C' with arithmetic genus g and g', respectively, one has

$$g(C \amalg C') = 1 - \chi_k(\mathscr{O}_C) - \chi_k(\mathscr{O}_{C'}) = g + g' - 1.$$

In case that $\Gamma(C, \mathscr{O}_C) = k$ (which holds, e.g., whenever C is proper, geometrically reduced, and geometrically connected, see Lemma 26.5; in particular this is true if C is proper, smooth and geometrically connected over k), one has

$$g(C) = \dim_k H^1(C, \mathscr{O}_C) \ge 0.$$

The condition $\Gamma(C, \mathscr{O}_C) = k$ also implies that C is connected. Hence one could have defined the genus only for proper curves satisfying this additional property. But this has the unpleasant effect that there are curves for which the genus is defined but where the genus is not defined for its normalization (see Example 26.87 below).

528   26 Curves

Below in Section (26.10), we will also briefly consider the notion of the geometric genus
which depends only on a smooth proper model of the curve.

The genus is compatible with extension of the base field, as the following lemma shows.

Lemma 26.32. *Let k be a field, and let C be a proper curve over k. For every extension
field K of k, we have $g(C/k) = g(C \otimes_k K/K)$.*

Proof. This follows immediately from the invariance of the Euler characteristic under
extension of the base field, see Lemma 23.64.   □

For a non-smooth curve, we will replace the sheaf of differentials, which already played
a role above, by the dualizing sheaf of the curve in the following sense.

Definition 26.33. *Let C be a curve over a field k and denote by $f: C \to \operatorname{Spec} k$ the
structure morphism.*
(1) *Consider k as a coherent sheaf on $\operatorname{Spec} k$, or rather as a complex of coherent sheaves
    concentrated in degree 0. We then write $\omega_C^\bullet := f^! k$ and call $\omega_C^\bullet$ the dualizing complex
    of C.*
(2) *Furthermore, we call $\omega_C := H^{-1}(\omega_C^\bullet)$ the dualizing sheaf of C.*

Since k (in degree 0) is a dualizing complex for k, Proposition 25.87 shows that $\omega_C^\bullet$ is
indeed a dualizing complex, and hence ω_C is a dualizing sheaf on C.

If C is Cohen-Macaulay (which is true in particular, if $H^0(C, \mathcal{O}_C)$ is a reduced ring see
Lemma 26.6), then $\omega_C^\bullet$ is concentrated in degree -1 and hence is represented by $\omega_C[1]$.
See Proposition 25.111. If C is Gorenstein (and in particular Cohen-Macaulay), then ω_C
is a line bundle by Proposition 25.124. If C is smooth and proper over k, then we have
$\omega_C \cong \Omega^1_{C/k}$ and hence $\omega_C \cong \Omega^1_{C/k}$ by Theorem 25.58.

If a curve C admits a regular closed immersion $i: C \to X$ of codimension n into a
proper smooth k-scheme X, then Example 25.125 gives us $\omega_C^\bullet = \omega_C[1]$ with

$$(26.8.1) \qquad \omega_C = i^* \Omega^{n+1}_{X/k} \otimes_{\mathcal{O}_C} \bigwedge^n \mathscr{C}_i^\vee.$$

In this case C is a local complete intersection over k and in particular Gorenstein.

By Theorem 25.127 (5), we can express the genus in terms of global sections of the
dualizing sheaf as follows:

Corollary 26.34. *Let C be a proper curve over a field k, and denote by ω_C its dualizing
sheaf as above. If $H^0(C, \mathcal{O}_C) = k$, then $g(C) = \dim_k H^0(C, \omega_C)$.*

Let C be a reduced proper curve over a field k. We want to compare the (arithmetic)
genus of a curve and the genus of (the connected components of) its normalization. Let
$f: \tilde{C} \to C$ denote the normalization morphism. It is an isomorphism over the normal
locus C_{norm} of C, and the complement $C \setminus C_{\operatorname{norm}}$ consists of finitely many closed points
of C. We obtain a short exact sequence

$$0 \to \mathcal{O}_C \to f_* \mathcal{O}_{\tilde{C}} \to \mathscr{F} \to 0,$$

where we define $\mathscr{F}$ as the cokernel of the homomorphism on the left. This is an $\mathcal{O}_C$-module
supported on finitely many closed points, and in particular $H^1(C, \mathscr{F}) = 0$, so that we
obtain an exact sequence

$$0 \to H^0(C, \mathscr{O}_C) \to H^0(\tilde{C}, \mathscr{O}_{\tilde{C}}) \to \bigoplus_{x \in C \setminus C_{\mathrm{norm}}} \mathscr{F}_x$$

(26.8.2)

$$\to H^1(C, \mathscr{O}_C) \to H^1(\tilde{C}, \mathscr{O}_{\tilde{C}}) \to 0$$

which will be the basis of our comparison. In fact, we get immediately that

(26.8.3)
$$\chi(\mathscr{O}_{\tilde{C}}) = \chi(\mathscr{O}_C) + \chi(\mathscr{F}) = \chi(\mathscr{O}_C) + \dim_k \bigoplus_{x \in C \setminus C_{\mathrm{norm}}} \mathscr{F}_x.$$

Proposition 26.35. *Let k be a field and let C be a reduced proper curve with normalization $f \colon \tilde{C} \to C$. Then we have*

(26.8.4)
$$g(C) = g(\tilde{C}) + \sum_{x \in C^1} \delta_x,$$

where for a closed point x of C we define

$$\delta_x = \dim_k(\tilde{\mathscr{O}}_{C,x}/\mathscr{O}_{C,x}),$$

where $\tilde{\mathscr{O}}_{C,x}$ is the integral closure of $\mathscr{O}_{C,x}$ in its total ring of fractions.

In particular, for a closed point $x \in C$, we have $\delta_x = 0$ if and only x lies in the normal locus of C. Hence the sum in (26.8.4) is finite. In Section (26.9) below we will study δ_x more closely.

Proof. As remarked above, we have $g(C) = 1 - \chi(C) = 1 - \chi(\tilde{C}) + \sum \dim_k(\mathscr{F}_x)$. Furthermore, for each closed point x of C we have $\tilde{\mathscr{O}}_{C,x} = (f_*\mathscr{O}_{\tilde{C}})_x$ because of the compatibility between taking integral closure and localization, and thus $\delta_x = \dim_k(\mathscr{F}_x)$. From this we obtain the desired formula. $\qquad\square$

(26.9) Ordinary multiple points.

A model for a particularly simple curve singularity is the point of intersection of the coordinate axes in the plane, i.e., the singularity of $\operatorname{Spec} k[T_1, T_2]/(T_1 T_2)$ at the origin, or more generally, the point of intersection of the coordinate axes in m-dimensional affine space, that is the singularity of $\operatorname{Spec} k[T_1, \ldots, T_m]/(T_i T_j; i \neq j)$ at the origin. The normalization of this ring is $\operatorname{Spec} \prod_{i=1}^m k[T_i]$, the disjoint union of m copies of the affine line over k.

Passing to the completion, we get $k[\![T_1, \ldots, T_m]\!]/(T_i T_j; i \neq j)$ as the complete local ring at the origin, and $k[\![T_1]\!] \times \cdots \times k[\![T_m]\!]$ as its normalization (in its total ring of fractions). We call curve singularities where the complete local ring is isomorphic to $k[\![T_1, \ldots, T_m]\!]/(T_i T_j; i \neq j)$ *ordinary multiple points*, more precisely:

Definition 26.36.

(1) *Let k be an algebraically closed field and let C be a curve over k. A closed point $x \in C$ is called an* ordinary multiple point, *if there exists an isomorphism*

$$\mathscr{O}^{\wedge}_{C,x} \cong k[\![T_1, \ldots, T_m]\!]/(T_i T_j; i \neq j)$$

of k-algebras. In this case $m = \dim_k T_{C,x}$ is called the multiplicity *of x.*

(2) *Let k be a field, let $\overline{k}$ be an algebraic closure of k and let C be a curve over k. A closed point $x \in C$ is called an* ordinary multiple point of multiplicity m, *if every point of the base change $C \otimes_k \overline{k}$ lying over x is an ordinary multiple point of multiplicity m in the sense of (1).*

(3) *An ordinary multiple point of multiplicity 2 of a curve over a field is also called an* ordinary double point *(or a* node*).*

(4) *A curve over a field is called an* (at worst) nodal curve, *if every closed point is either normal or an ordinary double point.*

For a plane curve C, the multiplicity of an ordinary multiple point x on C is either 1 or 2, and it is easy to see that it coincides with the multiplicity in the sense of Exercise 6.11. See also [Full]^O Section 3.2 for a more extensive discussion, but note that there the notion of ordinary multiple point is different form ours.

Let C be a reduced curve over an algebraically closed field k. Let $f: \tilde{C} \to C$ be the normalization morphism. For $x \in C$ a closed point recall the number $\delta_x = \dim_k(f_* \mathscr{O}_{\tilde{C}})_x / \mathscr{O}_{C,x}$ we defined in Proposition 26.35. We denote by

(26.9.1) $m_x := \# f^{-1}(x)$

the number of points of the fiber $f^{-1}(x)$. We always have

(26.9.2) $m_x - 1 \le \delta_x.$

This follows from the following remark in which we study these numbers and their connection in a slightly more general situation.

Remark 26.37. Let $f: Y \to X$ be a finite birational morphism of noetherian schemes that are equi-dimensional of dimension 1. Then $f_* \mathscr{O}_Y$ is a finite $\mathscr{O}_X$-algebra. We assume that $\mathscr{O}_X \to f_* \mathscr{O}_Y$ is injective which is automatic if X is reduced since f is birational. The $\mathscr{O}_X$-module $f_* \mathscr{O}_Y / \mathscr{O}_X$ is zero over the open dense subset of X over which f is an isomorphism. Hence it has support in a finite closed subset T of X.

Let $x \in X$ be a closed point. Set

$$\delta_x(f) := \lg_{\mathscr{O}_{X,x}} \left((f_* \mathscr{O}_Y)_x / \mathscr{O}_{X,x} \right).$$

If X is a curve over the field k and $x \in X(k)$, then one has $\delta_x(f) = \dim_k(f_* \mathscr{O}_Y)_x / \mathscr{O}_{X,x}$.

Let us explain why $\delta_x(f)$ is a finite number. Set $M := (f_* \mathscr{O}_Y)_x / \mathscr{O}_{X,x}$. If $x \notin T$, then $M = 0$. For $x \in T$, the support of M is $\{x\}$. Since we have $\mathrm{Supp}(M) = V(\mathrm{Ann}(M))$ (as sets) for every finitely generated module M (Proposition 7.35) we see that $\mathrm{rad}(\mathrm{Ann}(M)) = \mathfrak{m}$, the maximal ideal of $\mathscr{O}_{X,x}$. Hence there exists $r \ge 1$ such that $\mathfrak{m}^r M = 0$. This shows in particular, that M is an $\mathscr{O}_{X,x}$-module of finite length.

Let $\mathfrak{m}$ be the maximal ideal of $\mathscr{O}_{X,x}$ and denote by $(\ )^{\wedge}$ the $\mathfrak{m}$-adic completion. As $(f_* \mathscr{O}_Y)_x$ is a finite $\mathscr{O}_{X,x}$-algebra, Proposition B.41 and Proposition B.46 show that

(26.9.3) $(f_* \mathscr{O}_Y)_x \otimes_{\mathscr{O}_{X,x}} \mathscr{O}_{X,x}^{\wedge} = (f_* \mathscr{O}_Y)_x^{\wedge} = \displaystyle\prod_{y \in f^{-1}(x)} \hat{\mathscr{O}}_{Y,y},$

where $\hat{\mathscr{O}}_{Y,y}$ denotes the $\mathfrak{m}_y$-adic completion of $\mathscr{O}_{Y,y}$.

As M is annihilated by some power of $\mathfrak{m}$, one has $M = M^{\wedge}$ and we obtain by (26.9.3), using that $(\ )^{\wedge}$ is an exact functor on finitely generated $\mathscr{O}_{X,x}$-modules,

(26.9.4)
$$M = (f_*\mathcal{O}_Y)_x/\mathcal{O}_{X,x} \cong \Big(\prod_{y \in f^{-1}(x)} \hat{\mathcal{O}}_{Y,y} \Big)/\mathcal{O}_{X,x}^\wedge,$$

where $\mathcal{O}_{X,x}^\wedge$ is embedded diagonally into the product.

The fiber $f^{-1}(x)$ is a finite $\kappa(x)$-scheme, i.e., of the form $\operatorname{Spec} B_x$ for a finite $\kappa(x)$-algebra B_x. As $(f_*\mathcal{O}_Y)_x \otimes_{\mathcal{O}_{X,x}} \kappa(x) = B_x$, we have $M \otimes_{\mathcal{O}_{X,x}} \kappa(x) = B_x/\kappa(x)$. This shows that

(26.9.5)
$$\#f^{-1}(x) - 1 \le \dim_{\kappa(x)} B_x - 1 \le \delta_x(f).$$

The first inequality is an equality if and only if B_x is reduced and $\kappa(y) = \kappa(x)$ for all $y \in f^{-1}(x)$. The second inequality is an equality if and only if $\mathfrak{m}M = 0$.

We return to the situation in which C is a curve over an algebraically closed field with normalization $f\colon \tilde{C} \to C$.

One extreme case in (26.9.2) is that $m_x = 1$, i.e., f is injective in a neighborhood of x. As f is surjective, f is then bijective in a neighborhood of x, i.e., C is unibranch in x (Corollary 20.52). Let $y \in \tilde{C}$ be the unique point with $f(y) = x$. Then C is geometrically unibranch if and only if $\kappa(y)$ is a purely inseparable extension of $\kappa(x)$, again by Corollary 20.52.

Example 26.38. Let k be a field of characteristic different from 2 and 3. The curve $C := V(Y^2 - X^3) \subseteq \mathbb{A}_k^2$ is smooth outside the origin $x = (0,0)$. Its normalization is given by $k[X,Y]/(Y^2 - X^3) \to k[T]$ with $X \mapsto T^2$, $Y \mapsto T^3$. This identifies $k[X,Y]/(Y^2 - X^3)$ with the subring $k[T^2, T^3] = \{\sum_n a_n T^n \in k[T] \; ; \; a_1 = 0\}$ of $k[T]$. The origin is a singularity and one computes that with the above notation $m_x = 1$ and $\delta_x = 1$. Moreover, the residue extension of the normalization map in the origin is trivial. This shows that C is geometrically unibranch.

The other extreme case is that $\delta_x = m_x - 1$. Over an algebraically closed field this equality holds if and only if x is an ordinary multiple point:

Proposition 26.39. *Let C be a reduced curve over an algebraically closed field k, let $f\colon \tilde{C} \to C$ be its normalization and let $x \in C$ be a closed point. Then $\delta_x \ge m_x - 1$ and the following are equivalent.*
(i) *The point x is an ordinary multiple point.*
(ii) *We have $\delta_x = m_x - 1$.*
(iii) *Writing $m = m_x$, $f^{-1}(x) = \{q_1, \ldots, q_m\}$ and denoting by $\Delta \subseteq k^m$ the diagonal and by $\eta\colon (f_*\mathcal{O}_{\tilde{C}})_x \to k^m$, $s \mapsto (s(q_1), \ldots, s(q_m))$, the evaluation map, we have*

$$\mathcal{O}_{C,x} = \{s \in (f_*\mathcal{O}_{\tilde{C}})_x; \; \eta(s) \in \Delta\} =: A.$$

There is a similar criterion (with more complicated notation) over non-algebraically closed fields taking into account the possibility that the normalization may have non-trivial residue field extensions in this case. We do not make this explicit.

Proof. Let $f\colon \tilde{C} \to C$ denote the normalization morphism.

(i) $\Rightarrow$ (ii). By assumption, we have an isomorphism $\mathcal{O}_{C,x}^\wedge \cong k[\![T_1, \ldots, T_n]\!]/(T_i T_j; i \ne j)$ and $f_*(\mathcal{O}_{\tilde{C}})_x^\wedge \cong \prod_{i=1}^n k[\![T_i]\!]$ by (26.9.3). The canonical map $\mathcal{O}_{C,x}^\wedge \to f_*(\mathcal{O}_{\tilde{C}})_x^\wedge$ identifies $k[\![T_1, \ldots, T_n]\!]/(T_i T_j; i \ne j)$ with the subring of $\prod_{i=1}^n k[\![T_i]\!]$ consisting of tuples of power series $(f_1, \ldots, f_m)$ with $f_1(0) = \cdots = f_m(0)$. Hence $m_x = n$ and by (26.9.4)

$$\delta_x = \dim_k(\prod_{i=1}^n k[\![T_i]\!]/(k[\![T_1,\ldots,T_n]\!]/(T_iT_j; i \neq j)) = n - 1.$$

(ii) $\Rightarrow$ (iii). As $\delta_x = m_x - 1$ we have by (26.9.5) and the following remark that $(f_*\mathscr{O}_{\tilde{C}})_x \otimes_{\mathscr{O}_{C,x}} \kappa(x) = k^m$ and we can view η as the map $(f_*\mathscr{O}_{\tilde{C}})_x \to (f_*\mathscr{O}_{\tilde{C}})_x \otimes_{\mathscr{O}_{C,x}} \kappa(x)$. Hence η induces an isomorphism $(f_*\mathscr{O}_{\tilde{C}})_x/B \cong k^m/\Delta \cong k^{m-1}$. By assumption, we obtain $\dim_k(f_*\mathscr{O}_{\tilde{C}})_x/A = \delta_x = \dim_k(f_*\mathscr{O}_{\tilde{C}})_x/\mathscr{O}_{C,x}$. Since $\mathscr{O}_{C,x}$ is contained in A, this implies the equality.

(iii) $\Rightarrow$ (i). As $\tilde{C}$ is regular over the perfect field k, it is smooth over k (Proposition 18.67) and we have $\hat{\mathscr{O}}_{\tilde{C},y} \cong k[\![T]\!]$ for every closed point $y \in \tilde{C}$ (Theorem 6.28). Denoting by $(\ )^\wedge$ the $\mathfrak{m}_x$-adic completion, (26.9.3) gives

$$(f_*\mathscr{O}_{\tilde{C}})_x^\wedge \cong \prod_{i=1}^m \mathscr{O}_{\tilde{C},q_i} \cong \prod_{i=1}^m k[\![T_i]\!].$$

Now assumption (iii) says that we have a short exact sequence

$$0 \to \mathscr{O}_{C,x} \to (f_*\mathscr{O}_{\tilde{C}})_x \to k^m/\Delta \to 0.$$

We see that the image of $\mathscr{O}_{C,x}^\wedge \subseteq (f_*\mathscr{O}_{\tilde{C}})_x^\wedge \cong \prod_{i=1}^m k[\![T_i]\!]$ is the subring of tuples $(f_1,\ldots,f_m)$ of power series with $f_1(0) = \cdots = f_m(0)$, i.e., of power series with identical absolute terms. As remarked in the beginning of the proof, this subring can be identified with $k[\![T_1,\ldots,T_m]\!]/(T_iT_j; i \neq j)$, giving us the desired isomorphism. $\qquad\square$

Example 26.40. The curve $V(Y^2 - X^2(X-1)) \subseteq \mathbb{A}_k^2$ over a field k of characteristic $\neq 2$ has a node at the origin and is smooth at all other points. In fact, the normalization of $k[X,Y]/(Y^2 - X^2(X-1))$ is given by

$$k[X,Y]/(Y^2 - X^2(X-1)) \to k[T], \quad X \mapsto T^2 + 1, \ Y \mapsto T(T^2 + 1),$$

and one has $\delta_x = 1$. The fiber over the origin (which corresponds to the prime ideal (X,Y)) is the spectrum of

$$k[T] \otimes_{k[X,Y]/(Y^2-X^2(X-1))} k \cong k[T]/(T^2 + 1, T(T^2 + 1)) \cong k^2,$$

so it consists of two points.

Lemma 26.41. *Let k be a field and let C be a curve over k that has at most nodal singularities. Then C is a locally complete intersection, and in particular Gorenstein (Definition G.26).*

Proof. In fact, we can check that the local rings of C are complete intersection rings (Definition 19.46) after base change to an algebraic closure of C (Remark 19.47 (4)), and after passing to complete local rings (Remark 19.47 (2)). But the ring $k[\![T_1,T_2]\!]/(T_1T_2)$ is a locally complete intersection. For the final part, use Remark 19.47 (3). $\qquad\square$

Example 26.42. Let k be a field. The curve $C = V(XY, YZ, XZ)$, that is the union of the coordinate axes in $\mathbb{A}_k^3$, is not Gorenstein.

In fact, using [Mat2] Theorem 18.11, it is enough to exhibit a reducible parameter ideal in the ring $A := k[X,Y,Z]/(XY,YZ,XZ)$, and $(X+Y+Z) = (X,X+Y+Z) \cap (Y,X+Y+Z)$ is such an ideal.

Alternatively, we can show that $\mathrm{Ext}^1_A(k, A) \neq k$ as follows. Using a Koszul complex argument, we have $\mathrm{Ext}^1_A(k, A) \cong \mathrm{Hom}_B(k, B)$ for $B = A/(X + Y + Z)$ (see the proof of the theorem in [Mat2] mentioned above), and $B \cong k[\![X, Z]\!]/(XY, X^2, Y^2)$, so that $\dim_k \mathrm{Hom}_B(k, B) = 2$. (The linear maps with $1 \mapsto X$ and $1 \mapsto Y$ are a basis of this Hom space.)

A similar argument shows that an ordinary multiple point of order > 2 is not Gorenstein.

(26.10) The geometric genus.

For a smooth proper curve C over a field k with $H^0(C, \mathscr{O}_C) = k$, we have $g(C) = 1 - \chi(\mathscr{O}_C) = \dim H^1(C, \mathscr{O}_C) = \dim_k H^0(C, \Omega^1_{C/k})$, as we have discussed above. Clearly for higher-dimensional varieties X over k, say $n = \dim X$, the quantities $(-1)^n(\chi(\mathscr{O}_X) - 1)$ and $\dim_k H^0(C, \Omega^n_{C/k})$ will in general be different. To distinguish these two numbers, both of which are interesting invariants of X, one calls $(-1)^n(\chi(\mathscr{O}_X) - 1)$ the *arithmetic genus* of X and $\dim_k H^0(X, \Omega^n_{X/k})$ the *geometric genus of X*. Returning to the case of curves, if we drop the smoothness assumption, then we can define the arithmetic genus as before, but for the geometric genus it is preferable to modify the definition because the sheaf of differentials on a singular curve is less well behaved. We thus define the *geometric genus* as the genus of a smooth proper curve birational to X (if necessary, after an extension of the base field).

The reason why we might have to extend the base field is the following. While we can always find a proper normal curve birational to the given one, over non-perfect base fields, a normal curve need not be smooth. In order to handle this case appropriately, we use the following lemma. Recall that every reduced curve C over a field k is birationally equivalent to a normal proper curve $\widetilde{C}$, which is unique up to isomorphism and which is called the normal proper model of the given curve (Definition 26.2).

Lemma 26.43. *Let k be a field and let C be a curve over k.*
(1) *There exists a finite, purely inseparable field extension k'/k such that the normalization of $C_{k'}$ and its normal proper model $(C_{k'})^\sim$ are smooth over k'.*
(2) *If k'/k is a field extension such that $(C_{k'})^\sim$ is smooth over k', and k''/k' is a further field extension, then $(C_{k''})^\sim = (C_{k'})^\sim \otimes_{k'} k''$, and in particular this is a smooth k''-scheme.*

Proof. (1) If k'' is a perfect field, then every regular k''-scheme is smooth (Proposition 18.67), and hence $(C_{k''})^\sim$ is smooth over k'', since every normal curve is regular. Thus the perfect closure of k in some algebraic closure gives a purely inseparable extension k'' of k with $(C_{k''})^\sim$ being smooth over k''. Now the claim follows from a standard limit argument: The algebraic extension k'' is the filtered union of its finite subextensions k' which are all purely inseparable extensions of k. By Theorem 10.66 and Theorem 10.63 we can find a finite purely inseparable extension k', a curve C_1, and a morphism of k'-schemes $\pi \colon C_1 \to C_{k'}$ whose base change to k'' is the normalization morphism $(C_{k''})^\sim \to C_{k''}$ which is by definition a birational map $(C_{k''})^\sim \to C_{k'',\mathrm{red}}$. After possibly enlarging k', one can assume that C_1 is smooth over k' (Proposition 18.59) and that there exists an open dense subset $V \subseteq C_{k'}$ such that $\pi^{-1}(V) \to V$ is a closed surjective immersion (Proposition 10.75). This implies that π is the normalization morphism and hence that the normalization of $C_{k'}$ is smooth over k'. Hence its base change to any field extension K is also smooth over K and in particular normal. Hence a similar approximation argument shows that, after possibly passing to some further finite purely inseparable extension of k', also its proper normal model is smooth.

(2) We have $((C^\sim) \otimes_k k')^\sim = (C \otimes_k k')^\sim$, both sides locally being given by the integral closure of rings $\Gamma(U, \mathcal{O}_C) \otimes_k k'$ in the same product of fields. If $(C_{k'})^\sim$ is smooth, then so is $(C_{k'})^\sim \otimes_{k'} k''$, and the claim follows. $\square$

Definition 26.44. *Let k be a field and let C be an irreducible proper curve over k. The geometric genus of C is defined as*

$$p_g(C) = \dim_{k'} H^0(\tilde{C}, \Omega^1_{\tilde{C}/k'}),$$

where $\tilde{C}$ is a smooth proper model of C (as in Lemma 26.43), defined over a suitable finite purely inseparable extension k' of k.

The definition of $p_g(C)$ is independent of the choice of k' by Lemma 26.43 (2). If k' is a purely inseparable extension k, then $\operatorname{Spec} k' \to \operatorname{Spec} k$ is a universal homeomorphism. Therefore, as C is irreducible, $C_{k'}$ and its smooth projective model are again irreducible.

Remark 26.45. Let k be a field, and let X be an n-dimensional smooth proper k-scheme with $H^0(X, \mathcal{O}_X) = k$. Both the geometric genus $p_g(X) := \dim_k H^0(X, \Omega^n_X)$ and the arithmetic genus $p_a(X) := (-1)^n(\chi(\mathcal{O}_X) - 1)$ are birational invariants in the following sense: If Y has the same properties as required above for X, and $f : X \dashrightarrow Y$ is a birational map, then $p_a(X) = p_a(Y)$ and $p_g(X) = p_g(Y)$. (It is essential that X and Y are both smooth, as already the case of curves shows.)

For the geometric genus, it is not too hard to show this, using that the domain of definition of the rational map $X \dashrightarrow Y$ has complement of codimension ≥ 2 and a version of Hartogs's principle which allows to extend sections (of $\Omega^n_{X/k}$) from an open whose complement has codimension ≥ 1 to all of X. Here it is preferable to work with the space of global sections of a sheaf rather than to consider $H^n(X, \mathcal{O}_X)$, even if by Serre duality both have the same dimension.

In fact, one can show that even $\dim_k H^i(X, \mathcal{O}_X)$ is a birational invariant for each i, which implies that both $p_a(X)$ and $p_g(X)$ are birational invariants. This result is more difficult. While in characteristic 0 there are several methods to prove this and the statement is a "classical fact", in positive characteristic it is a relatively recent result by Chatzistamatiou and Rülling [ChRü].

The Theorem of Riemann-Roch

We now come to the Theorem of Riemann-Roch, a cornerstone of the theory of curves (which also has a counterpart in complex geometry, specifically for Riemann surfaces) and at the same time the foundation of the far-reaching generalizations for higher-dimensional varieties developed by Hirzebruch and by Grothendieck, see Section (23.23), in particular Example 23.115. In the following sections, we see several applications of the theorem.

(26.11) The Theorem of Riemann-Roch.

We start with the following preliminary form of the Riemann-Roch theorem, cf. Proposition 15.33.

Proposition 26.46. *Let C be a proper curve over a field k and let $\mathscr{L}$ be a line bundle on C. Then we have*

(26.11.1) $\deg(\mathscr{L}) = \chi(\mathscr{L}) - \chi(\mathscr{O}_C) = \chi(\mathscr{L}) - 1 + g,$

where g is the genus of C.

Proof. The second equality holds by definition of the genus. Let us show the first equality. The assertion is clear for $\mathscr{L} = \mathscr{O}_C$. By Corollary 15.26, every Cartier divisor on C can be written as a difference of effective Cartier divisors. It is then enough to show that for any line bundle $\mathscr{L}$ on C and any effective Cartier divisor D, the statement holds for $\mathscr{L}$ if and only if it holds for $\mathscr{L} \otimes \mathscr{O}_C(-D)$.

To show this, consider the short exact sequence

$$0 \to \mathscr{O}_C(-D) \to \mathscr{O}_C \to \mathscr{O}_D \to 0.$$

Since the support of $\mathscr{O}_D$ consists of just finitely many closed points, we have $\mathscr{L} \otimes_{\mathscr{O}_C} \mathscr{O}_D \cong \mathscr{O}_D$, so by tensoring by $\mathscr{L}$ we obtain a short exact sequence

$$0 \to \mathscr{L} \otimes_{\mathscr{O}_C} \mathscr{O}_C(-D) \to \mathscr{L} \to \mathscr{O}_D \to 0.$$

Passing to the associated long exact cohomology sequence and using that $\deg(D) = \dim \Gamma(C, \mathscr{O}_D)$, we get that $\chi(\mathscr{L}) = \chi(\mathscr{L} \otimes_{\mathscr{O}_C} \mathscr{O}_C(-D)) + \deg(D)$, as desired. $\qquad\square$

Compare Example 23.75 (1). In view of Definition 23.80, we get the following result.

Proposition 26.47. *Consider a proper curve C over a field k and a very ample divisor D on C. Let $\iota\colon C \to \mathbb{P}_k^N$ be the closed embedding corresponding to D (and a choice of basis of $H^0(C, \mathscr{O}_C(D))$). Then the degree of C as a subvariety of $\mathbb{P}_k^N$ equals the degree of the divisor D.*

Combining Proposition 26.46 with Grothendieck-Serre duality (in the form of Proposition 25.110) we obtain the Riemann-Roch Theorem.

Theorem 26.48. (Theorem of Riemann-Roch) *Let k be a field and let C be a proper curve over k. Let ω_C denote its dualizing sheaf. For every line bundle $\mathscr{L}$ on C we have*

$$\dim_k H^0(C, \mathscr{L}) - \dim_k H^0(C, \mathscr{L}^\vee \otimes \omega_C) = \deg(\mathscr{L}) + 1 - g(C).$$

Proof. By definition we have $\chi(\mathscr{O}_C) = 1 - g(C)$, so it only remains to show that the left hand side of the equation equals $\chi(\mathscr{L})$, i.e., to showing

$$\dim_k H^0(C, \mathscr{L}^\vee \otimes \omega_C) = \dim_k H^1(C, \mathscr{L}).$$

But that is a direct consequence of Proposition 25.110 using that $\operatorname{Hom}_{\mathscr{O}_C}(\mathscr{L}, \omega_C) = \operatorname{Hom}_{\mathscr{O}_C}(\mathscr{O}_C, \omega_C \otimes \mathscr{L}^\vee) = H^0(C, \omega_C \otimes \mathscr{L}^\vee)$. $\qquad\square$

In the Riemann-Roch formula, the dimension of $H^0(C, \mathscr{L}^\vee \otimes \omega_C)$ is usually thought of as a correction term (which vanishes for instance if the degree of $\mathscr{L}$ is sufficiently large, see the corollaries below).

Replacing Proposition 26.46 by more general versions (e.g., cf. Section (23.23) for smooth curves) and using Serre duality to express the Euler characteristic, one obtains generalizations of Theorem 26.48. See Theorem 26.138 below for a Riemann-Roch version for vector bundles on curves.

Under further assumptions, the situation becomes easier to handle (see also the discussion in Section (26.8)):

- If the curve C is Gorenstein, then the dualizing sheaf is a line bundle, and hence corresponds to a linear equivalence class of Cartier divisors. Any divisor in this class is called a *canonical divisor*, and often one (implicitly) fixes one such divisor K_C and calls it *the* canonical divisor on C.

- If C is even normal (equivalently, regular and hence locally factorial), then the notions of Cartier divisor and of Weil divisor coincide.

- If C is smooth, then it is normal (and over a perfect field the two notions coincide). In this case, the dualizing sheaf is the sheaf $\Omega^1_{C/k}$ of Kähler differentials on C and hence is very explicit.

For Gorenstein curves, we can rewrite the Riemann-Roch theorem in terms of divisors. To do so, we make the following definition.

Definition 26.49. *Let C be a proper curve over a field k, and let D be a divisor on C. Then we write $\ell(D) := \dim_k H^0(C, \mathscr{O}_C(D))$, where $\mathscr{O}_C(D)$ is the line bundle on C attached to the divisor D.*

Corollary 26.50. *Let C be a proper Gorenstein curve over a field k of genus $g = g(C)$, and let K be a canonical divisor on C. Then for every divisor D on C we have*

$$\ell(D) - \ell(K - D) = \deg(D) + 1 - g.$$

Remark 26.51. Historically, as the name suggests, the theorem goes back to Riemann and his student Roch. Riemann proved what is sometimes called *Riemann's inequality*, i.e., using the notation of the previous corollary, $\ell(D) \geq \deg(D)+1-g$. Roch proved the formula with the appropriate correction term. They worked in the setting of compact Riemann surfaces and correspondingly used analytic methods. By now, there are far-reaching generalizations, in particular the Hirzebruch-Riemann-Roch and Grothendieck-Riemann-Roch theorems. Those theorems work for higher-dimensional varieties, as well, and express the Euler characteristic of coherent sheaves (or even objects in the derived category), or even more generally describe the behavior of the Chern character under push-forward along a proper map. So they may be seen as generalizations of Proposition 26.46. See Section (23.23).

Hirzebruch worked in the setting of complex geometry; he used cobordism to prove the theorem. A different approach is to prove it using the Atiyah-Singer index theorem. Grothendieck, on the other hand, worked in the setting of algebraic geometry.

For other proofs of Riemann-Roch in the case of curves, see [Ser6], [Kem] $^\circ$, [ACGH] $^\circ$ App. A.

Corollary 26.52. *Let C be a proper Gorenstein curve over a field k of genus $g = g(C)$, and let K be a canonical divisor on C. Then*

$$\deg(K) = 2g - 2 \quad and \quad \ell(K) = g - 1 + \ell(0).$$

In particular, if $H^0(C, \mathscr{O}_C) = k$, then $\ell(K) = g$.

Proof. Using the Riemann-Roch formula for the trivial divisor and for the divisor K, we have

$$\ell(0) - \ell(K) = 1 - g, \qquad \ell(K) - \ell(0) = \deg(K) + 1 - g$$

which implies $\deg(K) = 2g - 2$ and $\ell(K) = g - 1 + \ell(0)$. $\qquad\qquad\square$

Corollary 26.53. *Let C be a integral normal proper curve over a field k. If $g(C) = 0$ and $C(k) \neq \emptyset$, then $C \cong \mathbb{P}_k^1$.*

Proof. Since C is integral and proper over k, $k' := H^0(C, \mathcal{O}_C)$ is a finite field extension. Saying that $g(C) = 0$ is equivalent to $\chi(\mathcal{O}_C) = 1$. But the cohomology groups occurring here are k'-vector spaces, so $\chi(\mathcal{O}_C)$ is a multiple of $[k' : k]$. It follows that $H^0(C, \mathcal{O}_C) = k$. Let K be a canonical divisor on C.

Let $x \in C$ be a closed point with residue class field k (which exists by the assumption $C(k) \neq \emptyset$). Then $\deg(K - [x]) = -3$ (Corollary 26.52), so $\ell(K - [x]) = 0$ by Remark 26.21 and hence $\ell([x]) = 2$ by the Theorem of Riemann-Roch (Corollary 26.50). Thus the line bundle $\mathcal{O}_C([x])$ defines a non-constant rational function $f \colon C \dashrightarrow \mathbb{P}(H^0(C, \mathcal{O}_C([x]))) \cong \mathbb{P}_k^1$ which extends to f since C is normal (Proposition 15.5). Moreover, $\deg(f) \deg(\mathcal{O}_{\mathbb{P}_k^1}(1)) = \deg(f^* \mathcal{O}_{\mathbb{P}_k^1}(1)) = \deg([x]) = 1$ by Proposition 26.19, hence $\deg(f) = 1$. By Corollary 26.12, f is an isomorphism. $\qquad\square$

The Brauer-Severi curve attached to a quaternionic skew field (Section (8.11)) is an example of a smooth, proper, geometrically connected curve of genus 0 that has no k-rational point.

See Section (26.16) below for further, similar characterizations of the projective line.

(26.12) Divisors on curves, continued.

Let us collect some further applications of the Theorem of Riemann-Roch to divisors on curves. We start with the following immediate description of line bundles of degree 0.

Example 26.54. Let C be a connected proper normal curve over an algebraically closed field k. Then every divisor D of degree 0 is linearly equivalent to a divisor of the form $\sum_{i=1}^{n}([x_i] - [x_0])$ for some $n \geq 0$ and some points $x_0, x_1, \ldots, x_n \in C(k)$.

Indeed, as C is normal, Cartier divisors are simply formal linear combinations of closed points which are all k-rational since k is algebraically closed. We can choose any $x_0 \in C(k)$. By the Theorem of Riemann-Roch, we have $\dim H^0(C, \mathcal{O}_C(D + n[x_0])) \geq \deg(D) + n + 1 - g(C) = n + 1 - g(C)$ which is > 0 for $n \geq g(C)$. Since C is connected, $D + n[x_0]$ is linearly equivalent to an effective divisor, i.e., a divisor of the form $\sum_{i=1}^{n}[x_i]$. Note that we can take $n = g(C)$.

The Riemann-Roch theorem also allows us estimate how the global sections grow if one adds closed points. Let C be a curve over a field. Then we denote by C_{reg} the open subset of points $x \in C$ such that $\mathcal{O}_{C,x}$ is normal (or, equivalently, regular). If C is generically reduced, this is an open dense subset of C.

Proposition 26.55. *Let k be a field, let C be proper curve over k and let D be a divisor on C.*
(1) $\ell(D) \leq \ell(D + [x]) \leq \ell(D) + [\kappa(x) : k]$ for $x \in C_{\mathrm{reg}}$ closed.
(2) Suppose that C is integral Gorenstein with $H^0(C, \mathcal{O}_C) = 1$. Let g be the genus of C. If $\deg D \geq 2g - 2$, then $H^1(C, \mathcal{O}_C(D)) = 0$ and

$$\ell(D) = \deg(D) + 1 - g.$$

Proof. For Part (1) note that the Weil divisor $[x]$ is a (Cartier) divisor since x is regular. Then (1) follows by considering the short exact sequence

$$0 \to \mathscr{O}_C(D) \to \mathscr{O}_C(D + [x]) \to \kappa(x) \to 0$$

and the associated long exact cohomology sequence which shows that

$$\ell(D + [x]) - \ell(D) = \dim_k \operatorname{Ker}([\kappa(x)] \to H^1(C, \mathscr{O}_C(D))).$$

Let us show (2). If $\deg(D) > 2g - 2$, then $\deg(K - D) = 2g - 2 - \deg(D) < 0$ in view of Corollary 26.52, and hence $\ell(K - D) = 0$ (Remark 26.21). Thus the Riemann-Roch formula gives the statement in part (2). $\qquad\square$

Remark 26.56. If C is a proper reduced connected curve over an algebraically closed field k, then $H^0(C, \mathscr{O}_C) = k$ (Lemma 26.5). If C is Gorenstein (e.g., if C is normal), integral, and proper of genus g, then we can apply Proposition 26.55, where we have $\kappa(x) = k$ since k is algebraically closed.
(1) For $g > 0$, we see that if we have a chain of divisors

$$0 = D_0 < D_1 < D_2 < \cdots < D_{2g-2} < D_{2g-1} < \cdots$$

with $D_i = D_{i-1} + [x_i]$ for closed points $x_i \in C_{\mathrm{reg}}$, then $\ell(D_i) \in \{\ell(D_{i-1}), \ell(D_{i-1}) + 1\}$ for $i = 1, \ldots, 2g - 1$ and

$$\#\{\, 1 \le i \le 2g - 1 \;;\; \ell(D_i) = \ell(D_{i-1})\,\} = g - 1.$$

For $i \ge 2g - 1$ one has $\ell(D) = i + 1 - g$.
(2) If C is normal and $g = 0$, then $C \cong \mathbb{P}^1_k$ by Corollary 26.53 and $\ell(D) = \max(0, \deg D)$ for any divisor D on C.

For curves, the Nakai-Moishezon criterion (Theorem 23.90) takes the following form, and we give a direct proof here.

Proposition 26.57. *Let k be a field, let C be a proper curve over k, and let $\mathscr{L}$ be a line bundle on C. Then $\mathscr{L}$ is an ample line bundle if and only if $\deg(\mathscr{L}_{|C_i}) > 0$ for every irreducible component C_i of C.*

Proof. Since we can check ampleness of a line bundle on C after restricting it to each of its irreducible components (Corollary 23.9), we may assume that C is integral. Let g be the genus of C.

Suppose that $\mathscr{L}$ is ample. Then $\mathscr{L}^{\otimes n}$ is very ample for large n and in particular generated by its global sections. If $\deg(\mathscr{L}) < 0$, then $H^0(C, \mathscr{L}^{\otimes n}) = 0$ for all $n \ge 1$ by Remark 26.21 (2), a contradiction. If $\deg(\mathscr{L}) = 0$, then $\mathscr{L}^{\otimes n} \cong \mathscr{O}_C$ for large n by Remark 26.21 (3). Hence C would be quasi-affine and thus finite over k since C is proper over k (Corollary 13.82). So we must have $\deg(\mathscr{L}) > 0$.

Conversely, suppose that $\deg(\mathscr{L}) > 0$. To show that $\mathscr{L}$ is ample, we may always replace $\mathscr{L}$ by some positive power (Proposition 13.50). We will use Serre's ampleness criterion Lemma 23.5 to show that $\mathscr{L}$ is ample, i.e., we will show that for every coherent $\mathscr{O}_C$-module $\mathscr{F}$ we have $H^1(C, \mathscr{F} \otimes \mathscr{L}^{\otimes n}) = 0$ for large n. By Theorem 26.16 there exists an ample line bundle $\mathscr{M}$ on C. By Riemann-Roch (Theorem 26.48) we have $\dim H^0(C, \mathscr{L}^{\otimes m} \otimes \mathscr{M}^{\otimes -1}) \ge m \deg(\mathscr{L}) - \deg(\mathscr{M}) + 1 - g$ for all m and hence

$$H^0(C, \mathscr{L}^{\otimes m} \otimes \mathscr{M}^{\otimes -1}) = \operatorname{Hom}_{\mathscr{O}_C}(\mathscr{M}, \mathscr{L}^{\otimes m}) \ne 0, \qquad \text{for } m \gg 0.$$

Replacing $\mathscr{L}$ by $\mathscr{L}^{\otimes m}$ we have for all $n \ge 1$ an exact sequence

$$0 \longrightarrow \mathscr{M}^{\otimes n} \longrightarrow \mathscr{L}^{\otimes n} \longrightarrow \mathscr{G}_n \longrightarrow 0$$

for some skyscraper sheaf $\mathscr{G}_n$. Tensoring with a coherent $\mathscr{O}_C$-module $\mathscr{F}$ we obtain an exact sequence

$$\mathscr{F} \otimes \mathscr{M}^{\otimes n} \xrightarrow{u} \mathscr{F} \otimes \mathscr{L}^{\otimes n} \longrightarrow \mathscr{F} \otimes \mathscr{G}_n \longrightarrow 0.$$

As C is a curve, $H^2(C, \mathrm{Ker}(u)) = 0$ and hence $H^1(C, \mathscr{F} \otimes \mathscr{M}^{\otimes n}) \to H^1(C, \mathrm{Im}(u))$ is surjective. As $\mathscr{G}_n$ has support of dimension 0 so has $\mathscr{F} \otimes \mathscr{G}_n$, hence $H^1(C, \mathscr{F} \otimes \mathscr{G}_n) = 0$ and therefore $H^1(C, \mathrm{Im}(u)) \to H^1(C, \mathscr{F} \otimes \mathscr{L}^{\otimes n})$ is surjective. We deduce that $H^1(C, \mathscr{F} \otimes \mathscr{M}^{\otimes n}) \to H^1(C, \mathscr{F} \otimes \mathscr{L}^{\otimes n})$ is surjective. Now we can apply Serre's ampleness criterion to see that the ampleness of $\mathscr{M}$ implies the ampleness of $\mathscr{L}$. $\qquad\square$

We also have the following sufficient criteria for line bundles to be globally generated or very ample.

Proposition 26.58. *Let k be an algebraically closed field. Let C be a proper normal curve over k. Let D be a divisor on C, and let $\mathscr{L} = \mathscr{O}_C(D)$ be the corresponding line bundle.*
(1) *The following are equivalent:*
    (i) *The line bundle $\mathscr{L}$ is generated by global sections.*
    (ii) *For every closed point $x \in C$, the inclusion*

$$\Gamma(C, \mathscr{O}(D - [x])) \subseteq \Gamma(C, \mathscr{O}(D))$$

   is a proper inclusion.
    (iii) *For every closed point $x \in C$,*

$$\dim_k \Gamma(C, \mathscr{O}(D - [x])) = \dim_k \Gamma(C, \mathscr{O}(D)) - 1.$$

(2) *Then the following are equivalent:*
    (i) *The line bundle $\mathscr{L}$ is very ample.*
    (ii) *For all closed points $x, y \in C$ (not necessarily different) the inclusions*

$$\Gamma(C, \mathscr{O}(D - [x] - [y])) \subseteq \Gamma(C, \mathscr{O}(D - [x])) \subseteq \Gamma(C, \mathscr{O}(D))$$

   are proper inclusions.
    (iii) *For all closed points $x, y \in C$ (not necessarily different)*

$$\dim_k \Gamma(C, \mathscr{O}(D - [x] - [y])) = \dim_k \Gamma(C, \mathscr{O}(D)) - 2.$$

Proof. Part (1). Given a closed point $x \in C$, consider the short exact sequence

(*) $$0 \to \mathscr{L} \otimes \mathscr{O}(-[x]) \to \mathscr{L} \to \kappa(x) \to 0.$$

The line bundle $\mathscr{L}$ is generated by global sections if and only if $\Gamma(C, \mathscr{L}) \to \kappa(x) = k$ is surjective for all x. Passing to global sections in (*) we see that the surjectivity of the above map in turn is equivalent to the condition that the map $\Gamma(C, \mathscr{O}(D - [x])) \to \Gamma(C, \mathscr{O}(D))$ is not surjective, i.e., that (ii) holds. Furthermore, Proposition 26.55 (1) shows that the difference of the dimensions of the two spaces is at most 1, whence (ii) and (iii) are equivalent.

Part (2). Again Proposition 26.55 (1) shows that (ii) and (iii) are equivalent. It remains to show the equivalence of (i) and (ii). In view of part (1) either assumption implies that $\mathscr{L}$ is generated by its global sections and therefore defines a morphism $\iota\colon C \to \mathbb{P}_k^N$ for $N = \dim_k \Gamma(C, \mathscr{L}) - 1$.

We rephrase the condition that ι is a closed immersion by Theorem 12.94 which implies that this is equivalent to ι being injective on closed points, and inducing injective homomorphisms on the tangent spaces to closed points of C.

Now consider closed points $x \ne y$ of C. The condition $\iota(x) \ne \iota(y)$ means that for some $s \in \Gamma(C, \mathscr{L})$, $s(x) = 0$ (in the fiber $\mathscr{L}(x)$ of $\mathscr{L}$ at x) while $s(y) \ne 0$. But these conditions on s can be rephrased as $s \in \Gamma(C, \mathscr{L} \otimes \mathscr{O}(-[x])) \setminus \Gamma(C, \mathscr{L} \otimes \mathscr{O}(-[x] - [y]))$. (Here we view $\mathscr{L} = \mathscr{O}_C(D)$ as a subbundle of $\mathscr{K}_C$, compare the definition in Section (11.12), and similarly for the other divisors appearing here.) This gives the equivalence of injectivity on closed points and condition (ii) for all $x \ne y$.

Now fix a closed point x of C and consider the map $T_x C \to T_{\iota(x)} \mathbb{P}_k^N$ induced by ι on the tangent spaces. It is injective if and only if its dual homomorphism $\mathfrak{m}_{\mathbb{P}_k^N, \iota(x)}/\mathfrak{m}_{\mathbb{P}_k^N, \iota(x)}^2 \to \mathfrak{m}_{C,x}/\mathfrak{m}_{C,x}^2$ is surjective. Choosing an isomorphism $\mathscr{O}(1)_{\iota(x)} \cong \mathscr{O}_{\mathbb{P}^N, \iota(x)}$ we obtain an isomorphism $\mathscr{L}_x \cong \mathscr{O}_{C,x}$ and a commutative diagram

$$
\begin{array}{ccc}
\mathfrak{m}_{\mathbb{P}_k^N, \iota(x)}/\mathfrak{m}_{\mathbb{P}_k^N, \iota(x)}^2 & \longrightarrow & \mathfrak{m}_{C,x}/\mathfrak{m}_{C,x}^2 \\
\cong \Big\uparrow & & \Big\uparrow \\
\{s \in \Gamma(\mathbb{P}^N, \mathscr{O}(1));\ s(\iota(x)) = 0\} & \longrightarrow & \{s \in \Gamma(C, \mathscr{L});\ s(x) = 0\}.
\end{array}
$$

In this diagram, the left vertical map is an isomorphism and the lower horizontal map is onto. We see that the upper horizontal map is onto (which is what we are interested in) if and only if the right vertical map is onto.

Since $T_x C$ is one-dimensional, C being smooth over k (Section (26.3)), the map is onto if and only if it is non-zero, i.e., if there exists $s \in \Gamma(C, \mathscr{L})$ with $s \in \mathfrak{m}_{C,x}\mathscr{L}$ but $s \notin \mathfrak{m}_{C,x}^2 \mathscr{L}$. Again viewing the line bundles of divisors as subbundles of $\mathscr{K}_C$, this condition translates to the existence of an element $s \in \Gamma(C, \mathscr{L} \otimes \mathscr{O}(-[x])) \setminus \Gamma(C, \mathscr{L} \otimes \mathscr{O}(-2[x]))$. This gives the equivalence of (i) and (ii). $\qquad\square$

Proposition 26.59. *Let k be a field. Let $f\colon C \to \operatorname{Spec} k$ be a proper geometrically connected smooth curve of genus g. Let $\mathscr{L}$ be a line bundle on C.*
(1) *If $\deg(\mathscr{L}) \ge 2g$, then $\mathscr{L}$ is generated by its global sections.*
(2) *If $\deg(\mathscr{L}) \ge 2g + 1$, then $\mathscr{L}$ is very ample.*

Proof. The line bundle $\mathscr{L}$ is generated by its global sections if and only if the canonical homomorphism $f^* f_* \mathscr{L} \to \mathscr{L}$ is surjective (Proposition 13.30). This can be done after base change to some field extension of k. Similarly, $\mathscr{L}$ is very ample if and only if it is very ample after base change to some field extension (Proposition 14.58). Moreover, neither the genus of the curve nor the degree of $\mathscr{L}$ change by an extension of the base field. Hence we may assume that k is algebraically closed.

Let us show (1). Say $\mathscr{L} \cong \mathscr{O}(D)$ for a divisor D on C. By Proposition 26.58 (1) it is enough to show that $\ell(D - [x]) = \ell(D) - 1$ for every closed point $x \in C$. By assumption $\deg(D)$ and $\deg(D - [x]) = \deg(D) - 1$ are both larger than $2g - 2$, so that we have

$$
\ell(D - [x]) = \deg(D - [x]) + 1 - g(C) = \deg(D) - 1 + 1 - g(C) = \ell(D) - 1
$$

by Remark 26.55.

The same argument, now using Proposition 26.58 (2), shows (2) if $\mathscr{L}$ is a line bundle of degree at least $2g + 1$. □

Remark 26.60. Consider the situation of Proposition 26.59.

(1) For $g \geq 2$ the converse to (1) or (2) does not hold in general (Exercise 26.9).

(2) If $g = 0$, then $\mathscr{L}$ is ample if and only if it is very ample. Indeed by faithfully flat descent (Proposition 14.58) we may assume that k is algebraically closed. Then $C \cong \mathbb{P}^1_k$ by Corollary 26.53 and the claim follows since a line bundle on a projective space $\mathbb{P}^n_k$ is ample if and only if it is very ample (if and only if it is isomorphic to $\mathscr{O}_{\mathbb{P}^n_k}(d)$ for some $d > 0$), see Example 13.45 and Remark 13.58, also cf. Section (26.16).

(3) If $g = 1$, then $\mathscr{L}$ is very ample if and only if $\deg(\mathscr{L}) \geq 3$. Indeed, if there existed a very ample line bundle of degree 1 or 2, then this line bundle would yield a closed immersion into $\mathbb{P}^0_k$ or $\mathbb{P}^1_k$. The first case is obviously absurd and so is the second case because every closed immersion between integral schemes of the same finite dimension is necessarily an isomorphism (Lemma 5.7 (1)). Compare Theorem 27.279 which generalizes this result to abelian varieties.

Proposition 26.61. *Let C be a separated irreducible curve over a field. Then C is projective over k or C is affine.*

Proof. We start by some reduction steps based on Chevalley's criterion that for a finite surjective morphism $X \to Y$ between noetherian schemes, Y is affine, if (and only if) X is affine, Theorem 12.39, and the analogous fact for properness, Proposition 12.59. Moreover, it suffices to show the claim after base change to some field extension (Proposition 14.53 and Proposition 14.57).

After passing to a purely inseparable perfect field extension, which leaves C irreducible, passing to C_{red}, and passing to the normalization we may assume that C and its normal proper model $\tilde{C}$ are smooth over k, cf. Lemma 26.43. We know already that C is quasi-projective (Theorem 26.16).

Suppose that C is not projective (equivalently, not proper). Let $\bar{k}$ be an algebraic closure of k and let k' be a finite subextension of $\bar{k}$ such that every connected component of $C_{\bar{k}}$ is already defined over k'. If one of the connected components C_1 of $C_{k'}$ was proper over k', then C_1 would be proper over k. As $C_{k'} \to C$ is finite free and in particular open and closed, its restriction to C_1 yields a surjective morphism $C_1 \to C$ of k-schemes. Hence C would be proper. Therefore no connected component of $C_{k'}$ is proper and it suffices to show that every such connected component is affine. Hence we may assume that C is even geometrically connected.

If C is not projective, then $\tilde{C} \setminus C =: \{x_1, \ldots, x_r\}$ is a non-empty set of closed points of $\tilde{C}$. For n sufficiently large, then $n([x_1] + \cdots + [x_r])$ is a very ample divisor on $\tilde{C}$ by Proposition 26.59. We obtain an embedding of $\tilde{C}$ into projective space, such that $\tilde{C} \setminus C$ (topologically) equals the intersection of $\tilde{C}$ with a hyperplane. Its complement in $\tilde{C}$, namely C, is therefore closed in an affine space, hence affine. □

(26.13) The formula of Riemann and Hurwitz.

Let k be a field, let X and Y be connected smooth curves over k, and let $f \colon X \to Y$ be a generically étale morphism. Then f is non-constant and hence quasi-finite and flat by Lemma 26.9. If X is proper over k, then f is finite locally free.

We have a short exact sequence

(26.13.1) $$0 \to f^*\Omega_{Y/k} \to \Omega_{X/k} \to \Omega_{X/Y} \to 0,$$

see Proposition 17.31. For the injectivity on the left, we use that the domain and the target of the map are line bundles on X and that the given homomorphism is non-zero because f is generically étale. (We have already proved that the sequence is exact if f is assumed to be smooth, Corollary 18.72; however, this case is not interesting from the point of view of this section. In fact, if f is smooth, then it is necessarily étale and $\Omega_{X/Y} = 0$.)

We obtain a global section $s \in \Gamma(X, \Omega_{X/k} \otimes (f^*\Omega_{Y/k})^\vee)$ and call the associated effective divisor on X the *ramification divisor* of f. Denoting this divisor by R, we have the explicit description

$$R = \sum_{x \in X^1} \lg_{\mathscr{O}_{X,x}}(\Omega_{X/Y,x})[x]$$

of R as a Weil divisor, where X^1 denotes the set of closed points of the curve X. Using Theorem 18.74, we see that the support of R consists of the points of X where f is not étale. Since f is flat, being étale at x is equivalent to being unramified at x. This justifies the term *ramification divisor*. We arrive at the following definition.

Definition 26.62. *Let k be a field and let $f \colon X \to Y$ be a generically étale morphism of smooth curves over k. The* ramification divisor *of f is the divisor*

$$R = \sum_{x \in X^1} \lg_{\mathscr{O}_{X,x}}(\Omega_{X/Y,x}) \cdot [x]$$

on X.

From the above short exact sequence we see immediately how the ramification divisor is related to the canonical divisors on smooth proper curves X and Y.

Proposition 26.63. *Let k be a field and let $f \colon X \to Y$ be a generically étale morphism of smooth proper curves over k. Let K_X and K_Y denote canonical divisors on X and Y, respectively. Then we have the linear equivalence (of divisors on X)*

$$K_X \sim f^*K_Y + R.$$

The main result of this section is the Theorem of Riemann and Hurwitz which expresses the degree of the ramification divisor in terms of the genus of X and the genus of Y, and the degree of the morphism f. We will discuss below how to explicitly compute the degree of the ramification divisor in many cases.

Theorem 26.64. (Theorem of Riemann-Hurwitz) *Let $f \colon X \to Y$ be a generically étale morphism of geometrically connected smooth proper curves, and let R be the ramification divisor of f. Then*

$$2g(X) - 2 = \deg(f)(2g(Y) - 2) + \deg(R).$$

Proof. Again denoting by K_X and K_Y canonical divisors on X and Y, Proposition 26.63 shows that

$$\deg(R) = \deg(K_X) - \deg(f^*K_Y) = \deg(K_X) - \deg(f)\deg(K_Y).$$

Since $\deg(K_X) = 2g(X) - 2$ (Corollary 26.52), and similarly for Y, we obtain the formula stated in the theorem. $\qquad\square$

Recall that by definition of R one has

(26.13.2)
$$\deg(R) = \sum_{x \in X^1} \lg_{\mathscr{O}_{X,x}}(\Omega_{X/Y,x})[\kappa(x) : k].$$

Under some further assumptions, we can relate the degree of R to the ramification index we have introduced in Section (12.5); see below for a brief reminder. We will use the following lemma.

Lemma 26.65. *Let X be a smooth curve over the field k and let x be a closed point of X. If the extension $\kappa(x)/k$ is separable and $s \in \mathscr{O}_{X,x}$ is a uniformizer, then ds is a generator of the free $\mathscr{O}_{X,x}$-module $\Omega_{X/k,x}$ of rank 1.*

Proof. The k-algebra $A = \mathscr{O}_{X,x}$ is a discrete valuation ring, and $\Omega_{X/k,x} = \Omega_{A/k}$ is a free A-module of rank 1 because X is smooth over k and $x \in X$ is a closed point. By Nakayama's lemma it is enough to show that $ds \neq 0 \in \Omega_{A/k} \otimes_A \kappa(x)$.

But since $\Omega_{\kappa(x)/k} = 0$, the extension $\kappa(x)/k$ being finite separable, the differential d induces an isomorphism $(s)/(s^2) \to \Omega_{A/k} \otimes_k \kappa(x)$ (e.g., use Corollary 18.73). □

Let $f \colon X \to Y$ be a generically étale morphism of geometrically connected smooth proper curves, let x be a closed point of X and let $y = f(x)$. Let $X_y = X \times_Y \operatorname{Spec} \kappa(y)$ be the fiber of f over y, a finite k-scheme, and let A be the local ring of X_y at x. Recall from Section (12.5) the ramification index $e_{x/y} = \lg_A(A)$ and the inertia index $f_{x/y} = [\kappa(x) : \kappa(y)]$.

Proposition 26.66. *Let $f \colon X \to Y$ be a generically étale morphism of geometrically connected smooth proper curves over a field k, and let x be a closed point of X.*

(1) *If f is tamely ramified at x, i.e., the extension $\kappa(x)/\kappa(f(x))$ is separable and $\operatorname{char}(k) \nmid e$, then*
$$\lg(\Omega_{X/Y,x}) = e_{x/y} - 1.$$

(2) *If the assumptions in part (1) are not satisfied, then*
$$\lg(\Omega_{X/Y,x}) > e_{x/y} - 1.$$

Proof. We prove the proposition under the assumption that $\kappa(x)/k$ is separable. See [Liu] Theorem 7.4.16 or [Sta] 0C1F for the general case.

Choose uniformizers s and t of the discrete valuation rings $\mathscr{O}_{X,x}$ and $\mathscr{O}_{Y,y}$. We write $e = e_{x/y}$. Under the ring homomorphism $\mathscr{O}_{Y,y} \to \mathscr{O}_{X,x}$ given by f, $t \mapsto us^e$ for a unit $u \in \mathscr{O}_{X,x}^\times$, and under the natural homomorphism $\Omega_{Y/k,y} \otimes_{\mathscr{O}_{Y,y}} \mathscr{O}_{X,x} \to \Omega_{X/k,x}$,

$$dt \otimes 1 \mapsto d(us^e) = s^e du + es^{e-1}uds = s^e wds + es^{e-1}uds,$$

(where we write $du = wds$ for some $w \in \mathscr{O}_{X,x}$).

As $\kappa(x)/k$ is separable by our assumption, ds is a generator of $\Omega_{X/k,x}$ by Lemma 26.65. If in addition $e \neq 0$ in $\kappa(x)$, we obtain the statement in part (1) from this computation and the short exact sequence (26.13.1).

On the other hand, if $e = 0$ in $\kappa(x)$, then the length of the quotient in this short exact sequence is at least e. □

By this result and by (26.13.2), in the tamely ramified case and in particular in the case when the base field has characteristic 0, the following more explicit version of the Riemann-Hurwitz formula holds.

Corollary 26.67. (Theorem of Riemann-Hurwitz, tamely ramified case) *Let $f \colon X \to Y$ be a generically étale morphism of geometrically connected proper smooth curves over a field k. Assume that for all closed points x of X the extension $\kappa(x)/\kappa(f(x))$ is separable and the ramification index $e_{x/f(x)}$ is invertible in k (both assumptions are satisfied if k is of characteristic 0). Then*

$$2g(X) - 2 = \deg(f)(2g(Y) - 2) + \sum_{x \in X^1} (e_{x/f(x)} - 1)[\kappa(x) : k],$$

where the sum extends over all closed points of X (and only finitely many summands are non-zero).

As a useful consequence we spell out the bounds on the number of points where a generically étale morphism is ramified that we obtain in this way.

Corollary 26.68. *Let $f \colon X \to Y$ be a generically étale morphism of geometrically connected proper smooth curves. Let ρ_X be the number of points $x \in X^1$ in which f is ramified and let ρ_Y be the number of points $y \in Y^1$ such that there exists an $x \in f^{-1}(y)$ in which f is ramified.*

(1) *One has $\rho_Y \le \rho_X \le 2g(X) - 2 - \deg(f)(2g(Y) - 2)$.*

(2) *Suppose that k is algebraically closed, that all ramification indices $e_{x/f(x)}$ for f are invertible in k, and that f is not an isomorphism. Then*

$$\frac{2g(X) - 2 - \deg(f)(2g(Y) - 2)}{\deg(f) - 1} \le \rho_Y.$$

Proof. Clearly, one has $\rho_Y \le \rho_X$. Hence the first assertion follows from Theorem 26.64 since $\rho_X \le \deg(R)$. Let R_Y be the set of $y \in Y^1$ such that there exists an $x \in f^{-1}(y)$ in which f is ramified. Since k is algebraically closed, all residue extensions of closed points are trivial and one has by Corollary 26.67

$$2g(X) - 2 - \deg(f)(2g(Y) - 2) = \sum_{x \in X^1} (e_{x/f(x)} - 1)$$

$$= \sum_{y \in R_Y} \left(\sum_{x \in f^{-1}(y)} (e_{x/y} - 1) \right)$$

$$= \sum_{y \in R_Y} (\deg(f) - \#f^{-1}(y))$$

$$\le \rho_Y(\deg(f) - 1),$$

where for the third equality one uses $\deg(f) = \sum_{x \in f^{-1}(y)} e_{x/y}$ (12.6.2). $\square$

Remark 26.69. Let $f \colon X \to Y$ be a generically étale morphism of geometrically connected smooth proper curves over a field. Then Corollary 26.68 (1) implies the following assertions.

(1) One has $g(X) \ge g(Y)$. We will show below (Corollary 26.75) that this holds for any finite morphism between such curves.

(2) If $g(X) = g(Y) > 1$, then f is étale of degree 1 and hence is an isomorphism.

Finally, the Riemann-Hurwitz formula also allows to give a new proof that the projective line over an algebraically closed field is "simply connected" in the sense that every étale cover splits completely, i.e., is just a disjoint union of copies of the projective line (see Theorem 20.114). More precisely, we have the following result.

Corollary 26.70. *Let k be a field, let X be a geometrically connected curve and let $f\colon X \to \mathbb{P}^1_k$ be a finite étale morphism of k-schemes. Then f is an isomorphism.*

Proof. As f is finite and étale, X is proper and smooth. By assumption, the ramification divisor of f is trivial. Since $g(X) \geq 0$, the formula of Riemann-Hurwitz implies that $\deg(f) = 1$ and Corollary 26.12 shows that f is an isomorphism. $\qquad\square$

(26.14) Purely inseparable morphisms.

Let p be a prime number and let S be an $\mathbb{F}_p$-scheme. In Remark/Definition 4.24, we have defined the absolute Frobenius morphism $\mathrm{Frob}_S\colon S \to S$, which is the identity on topological spaces and the homomorphism $x \mapsto x^p$ on (sections of) the structure sheaves. For an S-scheme $f\colon X \to S$, we let $X^{(p)} := X \times_{S, \mathrm{Frob}_S} S$ (an S-scheme via projection to the second factor) and obtain the *relative Frobenius morphism* $F_{X/S} := (\mathrm{Frob}_X, f)\colon X \to X^{(p)}$, a morphism of S-schemes.

Note that the S-scheme $X^{(p)}$ arises from the S-scheme X by base change, so it shares with X all properties that are stable under base change, such as being smooth over S, proper over S, etc. If k is a perfect field of characteristic p, then $X^{(p)}$ is isomorphic to X as an abstract scheme (but not in general as a k-scheme).

Iterating the construction, we have the n-fold twist $X^{(p^n)} := X \times_{S, \mathrm{Frob}_S^n} S$ and the n-fold relative Frobenius $X \to X^{(p^n)}$.

Proposition 26.71. *Let k be a field of characteristic $p > 0$ and let X be a smooth irreducible k-scheme of finite type.*

The relative Frobenius morphism $F_{X/k}\colon X \to X^{(p)}$ is a finite homeomorphism and has degree $p^{\dim(X)}$. The k-scheme $X^{(p)}$ is integral and we have $K(X^{(p)}) = k \cdot K(X)^p$.

Proof. The absolute Frobenius morphism of any scheme is a universal homeomorphism. It is also integral (by definition of the relative Frobenius or by Proposition 20.29). As a morphism between schemes of finite type over a field, $F_{X/k}$ is also of finite type and hence finite. Now $X^{(p)}$ is the base change of a smooth scheme and in particular it is reduced and therefore integral. The spectrum of $K(X) \otimes_k k^{1/p}$ is the schematic fiber of the generic point of X under the projection $X \to X^{(p)}$. As $X^{(p)}$ is reduced and $X \to X^{(p)}$ is a homeomorphism, $K(X) \otimes_k k^{1/p}$ is a field. We conclude that $K(X) \otimes_k k^{1/p} = K(X^{(p)})$. Viewing $K(X^{(p)})$ as a subfield of $K(X)$ via the relative Frobenius $X \to X^{(p)}$, we obtain the identification

$$K(X^{(p)}) = \mathrm{Im}(K(X) \otimes_k k^{1/p} \to K(X)) = kK(X)^p.$$

Since X is smooth, $\Omega_{X/k}$ is locally free of rank $\dim X$, so $\Omega_{K(X)/k}$ is a $K(X)$-vector space of dimension $\dim X$. By Lemma G.34, we have $[K(X) : kK(X)^p] = p^{\dim X}$. $\qquad\square$

Remark 26.72. Exercise 18.25 shows further properties of the relative Frobenius in a more general situation. In particular, it shows that the relative Frobenius of a smooth scheme is automatically flat (and hence finite locally free). For morphisms of normal curves we also have the following argument.

Let k be a field and $f\colon X \to Y$ a surjective morphism of connected curves over k, where X is smooth over k and Y is normal. Then f is flat by Lemma 26.9 and Y is smooth over k by Proposition 18.60 (2).

Proposition 26.73. *Let k be a field of characteristic $p > 0$ and let $f\colon X \to Y$ be a morphism of connected proper normal curves over k such that the induced field extension $K(Y) \subseteq K(X)$ is purely inseparable, of degree p^n, say.*

If X is smooth over k, then there is a unique isomorphism $Y \cong X^{(p^n)}$ over k which identifies f with the n-fold relative Frobenius morphism $X \to X^{(p^n)}$.

Proof. There is a chain of intermediate extensions $K(Y) = K_0 \subset K_1 \subset \cdots \subset K_n = K(X)$ such that each extension K_{i+1}/K_i has degree p. Each K_i is the function field of a proper normal curve over k (Theorem 26.3) and the inclusions give rise to morphisms between these curves. In view of Remark 26.72, they are even smooth. By induction, it is then enough to consider the case $n = 1$.

In this case, we have an inclusion $kK(X)^p \subseteq K(Y)$ (clearly $k \subset K(Y)$, and since $K(X)/K(Y)$ is purely inseparable of degree p, we can write $K(X) = K(Y)[\alpha]$ for some $\alpha \in K(X)$ with $\alpha^p \in K(Y)$, so $K(X)^p \subseteq K(Y)$, as well). By Proposition 26.71 we have $kK(X)^p = K(X^{(p)})$ and $[K(X) : kK(X)^p] = p$. We thus obtain equalities $K(Y) = kK(X)^p = K(X^{(p)})$. Since a proper normal curve is determined by its function field, the proposition follows. $\qquad\square$

Corollary 26.74. *Let k be a field of characteristic $p > 0$. Let $f\colon X \to Y$ be a morphism between connected smooth proper curves over k such that the induced field extension $K(Y) \subseteq K(X)$ is purely inseparable. Then we have $g(X) = g(Y)$.*

Proof. By Proposition 26.73, it is enough to show that $g(X^{(p)}) = g(X)$. But $X^{(p)}$ is the base change of X/k with respect to the absolute Frobenius morphism of $\operatorname{Spec} k$, hence is still connected, and base change preserves the genus by Lemma 26.32. $\qquad\square$

Proposition 26.75. *Let $f\colon X \to Y$ be a finite morphism of geometrically connected smooth proper curves over a field. Then $g(X) \geq g(Y)$.*

Proof. Let L be the separable closure of $K(Y)$ in $K(X)$, so that the extension $K(X)/K(Y)$ is split into a purely inseparable extension $K(X)/L$ and a separable extension $L/K(Y)$. Accordingly, we can factor f as $X \to X' \to Y$ with $X \to X'$ purely inseparable and $X' \to Y$ separable, for some connected smooth proper curve X' (see Remark 26.72).

Then $g(X) = g(X')$ by Corollary 26.74 and $g(X') \geq g(Y)$ by the formula of Riemann-Hurwitz, see Remark 26.69. $\qquad\square$

Corollary 26.76. (Lüroth's theorem) *Let k be a field.*
(1) *Let X be a normal connected curve over k such that there exists a non-constant morphism $f\colon \mathbb{P}^1_k \to X$. Then $X \cong \mathbb{P}^1_k$.*
(2) *Let $K = k(T)$ be the field of rational functions in one variable over k. Let $L \neq k$ be an intermediate field of the extension K/k. Then there exists $x \in K$ with $L = k(x)$. (Here x is necessarily transcendental over k, so L is isomorphic to the field of rational functions over k in one variable.)*

Proof. We prove (1). By Lemma 26.9, f is flat. In particular, it is open. It is also closed since $\mathbb{P}^1_k$ is proper over k. Hence f is surjective because X is connected. Therefore X is proper over k because $\mathbb{P}^1_k$ is proper. By Remark 26.72, X is smooth over k. As $\mathbb{P}^1_k$ is geometrically connected, so is X. Then $H^0(X, \mathscr{O}_X) = k$ and hence $g(X) \geq 0$. Hence we can apply Proposition 26.75 to see that $g(X) = 0$. Moreover, X has a k-valued point since this is true for $\mathbb{P}^1_k$. The claim then follows from Corollary 26.53.

As L contains a non-constant rational function f of $k(T)$, L is not an algebraic extension and hence has transcendence degree 1 over k. As $k(T)$ is a finitely generated algebraic extension of $k(f)$ and hence a finite extension of $k(f)$, L is a finite extension of $k(f)$ and therefore a finitely generated extension of k. Hence we can use the dictionary between connected normal proper curves and their function fields (Theorem 26.3), to deduce (2) from (1). $\qquad\square$

Special classes of curves

At this point, we come to several important classes of examples. We start out by studying curves which can be embedded in the projective plane $\mathbb{P}^2_k$. Afterwards we take a look at curves of small genus and at hyperelliptic curves.

(26.15) Plane curves.

Definition 26.77. *A curve C over a field k is called a* plane curve *if there exists a locally closed immersion $C \to \mathbb{P}^2_k$.*

We start by collecting some basic properties of plane curves.

Proposition 26.78. *Let k be a field.*
(1) *If C is a reduced proper plane curve over k, then C is isomorphic to a closed subscheme of $\mathbb{P}^2_k$ of the form $V_+(f)$ for a homogeneous polynomial $f \in k[T_0, T_1, T_2]$.*
(2) *For a non-constant homogeneous polynomial $f \in k[T_0, T_1, T_2]$, $V_+(f)$ is a proper plane curve over k. The embedding $V_+(f) \hookrightarrow \mathbb{P}^2_k$ is a regular immersion of codimension 1 and in particular $V_+(f)$ is locally a complete intersection. The curve $V_+(f)$ is integral if and only if f is an irreducible polynomial.*

Proof. For C integral, Part (1) is Corollary 5.42. For a general reduced curve with irreducible components $C_1, \ldots, C_n$, each C_i is integral and hence of the form $V_+(f_i)$, and then $C = V_+(f_1 \cdots f_n)$. For part (2), we use Proposition 5.40 (with $n = 2$ and $X = \mathbb{P}^2_k$) to see that $V_+(f)$ is in fact a curve, i.e., equi-dimensional of dimension 1. The other assertions are easy to check. $\qquad\square$

Not surprisingly, curves defined by a single homogeneous equation in $\mathbb{P}^2_k$ are particularly explicit. Since they are complete intersections in $\mathbb{P}^2_k$, they are in particular Gorenstein. Hence the canonical sheaf is a line bundle.

Proposition 26.79. *Let k be a field and let $C = V_+(f) \subset \mathbb{P}^2_k$ be a plane curve, where $f \in k[T_0, T_1, T_2]$ is a non-constant homogeneous polynomial of degree d.*
(1) *We have an isomorphism $\omega_C \cong \mathcal{O}_{\mathbb{P}^2_k}(d-3)_{|C}$.*
(2) *The degree $\deg(C)$ of C as a subvariety of $\mathbb{P}^2_k$ equals d.*
(3) *We have $H^0(C, \mathcal{O}_C) = k$, C is connected, and the arithmetic genus of C is*

$$g(C) = \frac{(d-1)(d-2)}{2}.$$

Proof. Part (1). Let $i\colon C \to \mathbb{P}^2_k$ be the embedding. Then its conormal sheaf $\mathscr{C}_i$ is $i^*\mathscr{O}_{\mathbb{P}^2_k}(-d)$ and the top exterior power of $\Omega^1_{\mathbb{P}^2_k/k}$ is $\mathscr{O}_{\mathbb{P}^2_k}(-3)$ (Example 17.59). We conclude by (26.8.1).

Part (2). We have $d = \deg(C)$ by Example 23.87.

Part (3). We denote by $i\colon C \hookrightarrow \mathbb{P}^2_k$ the closed embedding. We have a short exact sequence (13.4.6)

$$0 \to \mathscr{O}_{\mathbb{P}^2_k}(-d) \to \mathscr{O}_{\mathbb{P}^2_k} \to \iota_*\mathscr{O}_C \to 0.$$

From the long exact cohomology sequence for this short exact sequence and our knowledge of the cohomology of line bundles on projective space (Theorem 22.22), we obtain isomorphisms

$$k = H^0(\mathbb{P}^2_k, \mathscr{O}_{\mathbb{P}^2_k}) \xrightarrow{\sim} H^0(C, \mathscr{O}_C) \quad \text{and} \quad H^1(C, \mathscr{O}_C) \xrightarrow{\sim} H^2(\mathbb{P}^2_k, \mathscr{O}_{\mathbb{P}^2_k}(-d)) \cong k^{(d-1)(d-2)/2}.$$

As $H^0(C, \mathscr{O}_C) = k$ is a field, C is connected and $g(C) = \dim H^1(C, \mathscr{O}_C) = (d-1)(d-2)/2$. $\qquad\square$

Remark 26.80. We could also have calculated the genus of a plane connected curve as in Proposition 26.79 as follows. We have $\omega_C \cong \mathscr{O}_{\mathbb{P}^2}(d-3)_{|C}$. Now $\deg \mathscr{O}_{\mathbb{P}^2_k}(1)_{|C} = \deg(C) = d$, by definition of $\deg(C)$. Hence ω_C has degree $d(d-3)$. But we know also that ω_C has degree $2g - 2$ (Corollary 26.52), so $g = (d(d-3)+2)/2 = (d-1)(d-2)/2$.

Remark 26.81. Since, for example, $2, 4, 5, 7$ are not of the form $\frac{(d-1)(d-2)}{2}$ for any d, we see in particular that a reduced proper plane curve cannot have genus 2 (or genus $4, 5, 7, \dots$). Correspondingly, reduced proper curves of genus 2 cannot be embedded in $\mathbb{P}^2_k$. On the other hand, we will see below that over an algebraically closed field k every smooth proper connected curve of genus 0 or 1 is a plane curve, see Proposition 26.89 and Proposition 26.95 for more precise assertions.

(26.16) Curves of genus 0.

In this section, we discuss curves of genus 0. The most prominent example, of course, is the projective line $\mathbb{P}^1_k$ over the base field k. We will give several characterizations of $\mathbb{P}^1_k$ as a genus 0 curve with some additional properties below. Recall that in Corollary 26.53 we have already seen that a connected normal proper curve of genus 0 over k with a rational point is isomorphic to $\mathbb{P}^1_k$.

On the other hand, Brauer-Severi varieties of dimension 1 (Section (8.11)) also have genus 0 by Lemma 26.32. They correspond to central quaternion algebras over k (Theorem 14.95), i.e., every Brauer-Severi curve corresponding to such an algebra that is not isomorphic to the algebra of 2×2 matrices $M_2(k)$ is an example of a smooth projective geometrically connected curve of genus 0 that is not isomorphic to $\mathbb{P}^1_k$.

Proposition 26.82. *Let k be a field and let X be an integral proper curve of genus $g(X) = 0$ over k such that $H^0(X, \mathscr{O}_X) = k$. Let $\mathscr{L}$ be a line bundle on X.*
(1) *If $\deg(\mathscr{L}) = 0$, then $\mathscr{L} \cong \mathscr{O}_X$.*
(2) *If $\deg(\mathscr{L}) > 0$, then $\dim_k H^0(X, \mathscr{L}) = \deg(\mathscr{L}) + 1$, $H^1(X, \mathscr{L}) = 0$ and $\mathscr{L}$ is very ample.*

Proof. Part (1). The Riemann-Roch theorem in the form of Proposition 26.46 shows that $\dim_k H^0(X, \mathscr{L}) = \deg(\mathscr{L}) + 1 - g(X) + \dim_k H^1(X, \mathscr{L}) > 0$, so we conclude by Remark 26.21 (3).

Part (2). As in the proof of part (1), we see that $H^0(X, \mathcal{L}) \neq 0$ and hence $\mathcal{L} \cong \mathcal{O}_X(D)$ for an effective Cartier divisor D on X. We have a short exact sequence

$$0 \to \mathcal{O}_X \to \mathcal{L} \to \mathcal{O}_D \to 0,$$

where the morphism $\mathcal{O}_X \to \mathcal{L}$ corresponds to a non-trivial global section of $\mathcal{L}$ which defines D. From the associated long exact cohomology sequence and since $H^1(X, \mathcal{O}_X) = 0$ by our assumption, we conclude that $H^1(X, \mathcal{L}) = 0$. Again using Proposition 26.46, this implies $\dim_k H^0(X, \mathcal{L}) = \deg(\mathcal{L}) + 1$.

The hypothesis that $H^0(X, \mathcal{O}_X) = k$ implies that X is geometrically connected (Section (26.2)). Hence we have already proved that in this situation $\mathcal{L}$ is very ample if X is known to be smooth, see Proposition 26.59. Moreover, Proposition 26.57 shows that $\mathcal{L}$ is ample in general.

We omit the proof that $\mathcal{L}$ is very ample in the general case, see Exercise 26.11 or [Sta] 0C6T. $\qquad\square$

If there exists a Cartier divisor (or equivalently, a line bundle) of degree 1 on a genus 0 curve as in the previous proposition, then the curve is necessarily isomorphic to the projective line.

Proposition 26.83. *Let k be a field and let X be an integral proper curve of genus $g(X) = 0$ over k such that $H^0(X, \mathcal{O}_X) = k$. Let $\mathcal{L}$ be a line bundle on X with $\deg(\mathcal{L}) = 1$. Then X is isomorphic to the projective line $\mathbb{P}^1_k$.*

Proof. By Proposition 26.82, $\mathcal{L}$ is very ample and $\dim_k H^0(X, \mathcal{L}) = 2$, so $\mathcal{L}$ induces a closed immersion $X \to \mathbb{P}^1_k$ which necessarily is an isomorphism. $\qquad\square$

Clearly, a k-rational point inside the normal locus of X gives rise to a divisor, and hence a line bundle, of degree 1. As a corollary we have the following result. Compare Theorem 14.93.

Corollary 26.84. *Let k be a field and let X be a geometrically integral proper curve of genus $g(X) = 0$ over k. For any separably closed extension field k' of k, $X \otimes_k k' \cong \mathbb{P}^1_{k'}$ as k'-schemes.*

In particular a curve is a Brauer-Severi curve if and only if it is geometrically integral, proper, and of genus 0.

Proof. The genus is preserved by base change of the field (Lemma 26.32). Hence we may assume that k is separably closed. We have $H^0(X, \mathcal{O}_X) = k$ by Lemma 26.5. It follows from Remark 6.20 and Proposition 6.21 that there exists a k-rational point $x \in X$ that lies in the smooth locus. The divisor $[x]$ then has degree 1. $\qquad\square$

Example 26.85.

(1) The curve $V_+(T_0^2 + T_1^2 + T_2^2) \subseteq \mathbb{P}^2_{\mathbb{R}}$ over $\mathbb{R}$ is a smooth proper geometrically connected curve of genus 0 without rational points. It is the Brauer-Severi curve corresponding to the Hamilton quaternions over $\mathbb{R}$.

(2) Let C be a reduced curve over the field k all of whose irreducible components are isomorphic to $\mathbb{P}^1_k$. Assume further that the intersections of the irreducible components are transversal, i.e., at most two irreducible components intersect in a point, and the intersection multiplicity is equal to 1. Let Γ denote the dual graph of this configuration, i.e., the graph with vertex set the set of irreducible components of C and where for each intersection point of irreducible components, there is an edge in Γ between the corresponding vertices. If Γ is a tree, then C has genus 0. This follows from Proposition 26.35 by induction on the number of irreducible components.

(3) Let k be a field of characteristic $\neq 2$ and let $d \in k$ be an element which is not a square. Consider $C = V_+(T_0^2 - dT_1^2) \subseteq \mathbb{P}^2_k$. This is an integral curve with $(0 : 0 : 1)$ as its only rational point. After base change to $k(\sqrt{d})$ it becomes isomorphic to the union of two projective lines intersecting transversally. Thus the base change has genus 0 over $k(\sqrt{d})$ by (2), and by Lemma 26.32, X has genus 0 over k. Compare Example 26.87 below.

We state some further variants where the condition on the existence of a degree 1 divisor is relaxed a little.

Corollary 26.86. *Let k be a field and let X be a geometrically integral proper curve of genus $g(X) = 0$.*

(1) *If X is Gorenstein (e.g., if X is normal) and there exists a line bundle $\mathscr{L}$ on X of odd degree, then $X \cong \mathbb{P}^1_k$.*

(2) *If the greatest common divisor of the degrees $[\kappa(x) : k]$, where x ranges over all closed point of the normal locus X_{norm} of X, is equal to 1 (e.g., if the normal locus contains a k-valued point), then $X \cong \mathbb{P}^1_k$.*

Proof. For Part (1) note that the Gorenstein condition implies that the canonical sheaf of X is a line bundle; it has degree -2 by our assumption on the genus (Corollary 26.52). Since there exists a line bundle of odd degree, there also exists a line bundle of degree 1, and we conclude by Proposition 26.83.

Part (2). By assumption, there exist natural numbers $a_1, \dots, a_r$ with greatest common divisor 1 and divisors $D_1, \dots, D_r$ on C with $\deg(D_i) = a_i$. Then 1 can be expressed as a $\mathbb{Z}$-linear combination of the a_i, and correspondingly, we get a divisor of degree 1 as a linear combination of the D_i, using the same coefficients. $\square$

Over finite fields, Condition (2) is satisfied for all geometrically connected smooth proper curves (see Corollary 26.176 below), so we see again that there are no Brauer-Severi curves over a finite field.

As we have seen in Corollary 26.84, a geometrically integral proper curve of genus 0 over an algebraically closed field is necessarily smooth. Over other fields, singular curves of genus 0 do exist, though, as the following example shows.

Example 26.87. If k is not algebraically closed, then there exist singular proper curves of genus 0, as the following construction shows. Let $\tilde{k}/k$ be a non-trivial finite field extension, and let

$$A_0 = \{f \in \tilde{k}[S];\ f(0) \in k\}, \qquad A_1 = \tilde{k}[T].$$

Mapping $S \mapsto T^{-1}$ yields an isomorphism $A_{0,S} \to A_{1,T}$ between the localizations, and using this for gluing the affine schemes $\operatorname{Spec} A_0$ and $\operatorname{Spec} A_1$, we obtain an integral proper curve C over k with $H^0(C, \mathscr{O}_C) = k$. Clearly, the normalization of A_0 is $\tilde{k}[S]$. In particular, since by assumption the extension $\tilde{k}/k$ is not trivial, A_0 is not normal. In fact, the point

x corresponding to the prime ideal $(S) \subset A_0$ is not normal. We have $\delta_x = 1$, since $\dim_k \tilde{k}[S]/A_0 = 1$.

The normalization of C is birational to $\mathrm{Spec}(A_1) \cong \mathbb{A}^1_{\tilde{k}}$ and is hence isomorphic to $\mathbb{P}^1_{\tilde{k}}$ which has genus $g(\mathbb{P}^1_{\tilde{k}}/k) = 1 - 2 + 0 = -1$. We thus have $g(C) = 0$ by Proposition 26.35.

Remark 26.88. Let k be a field and let C be a singular integral proper curve of genus 0 with $H^0(C, \mathscr{O}_C) = k$. Let $f\colon \tilde{C} \to C$ be the normalization morphism. From the exact sequence (26.8.2) we obtain that $H^1(\tilde{C}, \mathscr{O}_{\tilde{C}}) = 0$ and that

$$\dim_k H^0(\tilde{C}, \mathscr{O}_{\tilde{C}}) = 1 + \sum_x \delta_x,$$

where the sum is taken over all closed points of C and δ_x is defined as in Proposition 26.35. Since C is assumed to be singular, $\sum \delta_x \neq 0$, so writing $\tilde{k} = H^0(\tilde{C}, \mathscr{O}_{\tilde{C}})$, the extension $\tilde{k}/k$ is a non-trivial field extension.

We see that $g(\tilde{C}/\tilde{k}) = 0$. One can show that there is precisely one singular point in C, and that there is precisely one point in $\tilde{C}$ lying over it, and that that is a $\tilde{k}$-rational point. It then follows from Corollary 26.53 that $\tilde{C} \cong \mathbb{P}^1_{\tilde{k}}$. More precisely, one can show that C must be as in Example 26.87. See [Sta] ODJB.

Every smooth, and more generally every Gorenstein proper curve of genus 0 is isomorphic to a conic, i.e., a plane curve of degree 2:

Proposition 26.89. *Let X be a Gorenstein integral proper curve over a field k with $g(X) = 0$ and $H^0(X, \mathscr{O}_X) = k$. Then there exists a closed immersion $X \hookrightarrow \mathbb{P}^2_k$ of k-schemes, and the image is a plane curve of degree 2 (a conic).*

Proof. The canonical sheaf ω_X is a line bundle of degree -2, hence its dual $\omega_X^\vee$ is very ample and the first cohomology $H^1(X, \omega_X^\vee)$ vanishes by Proposition 26.82. We therefore have $\dim H^0(X, \omega_X^\vee) = \deg(\omega_X^\vee) + 1 - g(X) + \dim_k H^1(X, \omega_X^\vee) = 3$, so this very ample line bundle gives us a closed immersion $X \hookrightarrow \mathbb{P}^2_k$. Its image has degree $\deg(\omega_X^\vee) = 2$. $\square$

Example 26.90. Let C be a connected proper smooth curve over an algebraically closed field k.
(1) If $g(C) = 0$, and hence $C \cong \mathbb{P}^1_k$, then all divisors of degree 1 are linearly equivalent to each other.
(2) Suppose $g(C) \geq 1$. For $x, x' \in C$ the divisors $[x]$ and $[x']$ are linearly equivalent if and only if $x = x'$, by Corollary 26.20.

Remark 26.91. If C is a Gorenstein proper connected curve over a field with $H^0(C, \mathscr{O}_C) = k$ and $-K_C$ is ample, then $g(C) = 0$, because K_C must then have negative degree, but $\deg(K_C) = 2g(C) - 2$ can only be negative if $g(C) = 0$.

Remark 26.92. As a consequence of the Theorem of Hasse and Minkowski ([Ser5]) a conic over a finite extension field k of $\mathbb{Q}$ (more generally: over any global field) has a k-valued point if and only if it has a k_v valued point in the completion k_v of every place v of k. For $k = \mathbb{Q}$ this means that a conic has a $\mathbb{Q}$-valued point if and only if it has an $\mathbb{R}$-valued point and for every prime number p has a point with values in the field $\mathbb{Q}_p$ of p-adic numbers. This situation is often described by saying that genus 0 curves satisfy the *local-global principle* (or the *Hasse principle*).

(26.17) Curves of genus 1, elliptic curves.

After studying curves of genus 0, we now consider the case of genus 1. We have already briefly touched upon this topic in Section (16.35). While there is only a single geometrically connected smooth proper curve of genus 0 – the projective line, which we understand very well, in genus 1 we see a fascinating picture which is much richer and in parts still quite mysterious. Here we can only deal with the basics, and in particular mostly have to leave out the arithmetic questions and applications connected to this topic, which have been one of the main driving forces in arithmetic geometry and algebraic number theory over the last one or two centuries. See [Sil1] ^{O}, [Sil2] ^{O}, [CSS] ^{O} for further information and additional references.

The Riemann-Roch theorem gives us the following information in the case of genus 1 curves.

Corollary 26.93. *Let k be a field and let C be a geometrically connected smooth proper curve of genus $g(C) = 1$ over k.*
(1) *The canonical bundle ω_C is trivial.*
(2) *Let $\mathscr{L}$ be a line bundle on C. If $\deg(\mathscr{L}) > 0$, then $\ell(\mathscr{L}) = \deg(\mathscr{L})$.*

Proof. By the Riemann-Roch theorem (Corollary 26.52), ω_C has degree $2g(C) - 2 = 0$, and $\ell(\omega_C) = g(C) = 1$. This implies $\omega_C \cong \mathscr{O}_C$ by Remark 26.21. The second part then also follows from the Riemann-Roch formula, since $\deg(\mathscr{L}) > 0$ implies $\deg(\mathscr{L}^\vee \otimes \omega_C) = \deg(\mathscr{L}^\vee) = -\deg(\mathscr{L}) < 0$ and hence $\ell(\mathscr{L}^\vee \otimes \omega_C) = 0$. $\square$

It is an exciting problem to study the structure of the set of k-valued points of a smooth proper curve C of genus 1. Obviously, this problem only has content, if $C(k)$ is non-empty. As we will see, fixing an element of $C(k)$ gives rise to a structure of abelian group on the set $C(k)$, and we thus make the following definition.

Definition 26.94. *Let k be a field. An* elliptic curve *over k is a (geometrically) connected smooth proper curve E of genus $g(E) = 1$ over k together with a point $0 \in E(k)$.*

The term *elliptic curve* comes from the relation between elliptic curves over $\mathbb{C}$ (or equivalently, compact Riemann surfaces of genus 1, with a fixed base point, i.e., quotients $\mathbb{C}/\Lambda$ of $\mathbb{C}$ by a lattice, see Section (26.19)) and elliptic integrals, e.g., the integral to compute the arc-length of an ellipse. Cf. [Sil1] ^{O} Ch. VI.

We will study elliptic curves over general base schemes in Section (27.27) and reprove some of the results below in this generality. Here we focus on elliptic curves over a field.

Note that since an elliptic curve E has a k-valued point, the properties *connected* and *geometrically connected* are equivalent (Lemma 26.4).

Proposition 26.95. *Let k be a field, let E be a connected smooth proper curve over k and let $0 \in E(k)$.*
(1) *The following are equivalent:*
    (i) *The genus of E is equal to 1, i.e., E (with the fixed point 0) is an elliptic curve.*
    (ii) *The curve E is isomorphic to a smooth projective curve of degree 3 (a "cubic") in $\mathbb{P}^2_k$.*
(2) *If the equivalent conditions of part (1) are satisfied, then E is isomorphic to a cubic given by an equation in* Weierstraß form, *i.e., of the form*

$$V_+(Y^2Z + a_1XYZ + a_3YZ^2 - X^3 - a_2X^2Z - a_4XZ^2 - a_6Z^3), \quad a_i \in k,$$

where X, Y, Z denote homogeneous coordinates on $\mathbb{P}^2_k$. The isomorphism can be chosen so that $0 \in E(k)$ is mapped to the point $(0 : 1 : 0)$.

(3) *If the characteristic of the field k is different from 2 and 3, and the conditions of part (1) hold, then E is isomorphic to a cubic given by an equation of the form*

$$V_+(Y^2Z - X^3 - aXZ^2 + bZ^3), \quad a, b \in k.$$

Proof. First note that in part (1), the implication (ii) $\Rightarrow$ (i) follows immediately from the genus-degree-formula for plane curves, Proposition 26.79. (Cf. also Exercise 17.9.) Now assume (i). We find a Weierstraß equation as in (2) for E as follows. We have $\ell(\mathscr{O}_E(2[0])) = \deg(2[0]) = 2$, so there exists $x \in K(E)$ such that $1, x$ is a k-basis of $H^0(E, \mathscr{O}(2[0]))$. Similarly, $\ell(\mathscr{O}_E(3[0])) = 3$, so we can extend the family $1, x$ to a k-basis $1, x, y$ of $H^0(E, \mathscr{O}(3[0]))$.

Then the rational functions $1, x, y, xy, x^2, x^3, y^2$ are 7 elements in the 6-dimensional k-vector space $H^0(E, \mathscr{O}(6[0]))$, so they are linearly dependent over k. On the other hand, $1, x, y, xy, x^2$ are linearly independent, having pairwise different pole orders at 0. Therefore we find a non-trivial linear relation between the above-named elements where the coefficients u of x^3 and v of y^2 are both $\neq 0$. Replacing x by uvx and y by u^2vy and dividing by u^4v^3, we may assume that the coefficients of x and y are both $= 1$. We conclude that there exist $a_1, a_2, a_3, a_4, a_6 \in k$ with

$$y^2 + a_1xy + a_3y = x^3 + a_2x^2 + a_4x + a_6.$$

We set $f := Y^2Z + a_1XYZ + a_3YZ^2 - X^3 - a_2X^2Z - a_4XZ^2 - a_6Z^3 \in k[X, Y, Z]$. The image of the morphism $E \to \mathbb{P}^2$ given by the very ample sheaf $\mathscr{O}(3[0])$ and the basis $x, y, 1$ of $H^0(E, \mathscr{O}(3[0]))$ is contained in $V_+(f)$. Since the latter is integral (the polynomial f being irreducible as is easily checked), we obtain an isomorphism $E \cong V_+(f)$, as desired. In particular, (ii) holds.

By the choice of x and y, $0 \in E(k)$ is mapped to the point $[0 : 1 : 0]$.

To get part (3) from part (2), we first do the coordinate change replacing Y by $Y + \frac{1}{2}(a_1X + a_3)$ and then replace X by $X + \frac{1}{3}a_2$. $\qquad\square$

The numbering of the coefficients a_i in the Weierstraß equation reflects the behavior under a coordinate change of the form $X \mapsto u^2X$, $Y \mapsto u^3Y$ ($u \in k^\times$). Plugging u^2X and u^3Y into a Weierstraß equation as above and dividing by u^6, we obtain a Weierstraß equation with coefficients $u^{-i}a_i$.

See Proposition 27.150 for a more general version of the above proposition which handles the case of a general base scheme and yields a Weierstraß equation locally on the base. The following lemma gives an explicit criterion, when the curve defined by a Weierstraß equation is smooth.

Lemma 26.96. *Let k be a field.*
(1) *For $a_i \in k$, the projective plane curve*

$$V_+(Y^2Z + a_1XYZ + a_3YZ^2 - X^3 - a_2X^2Z - a_4XZ^2 - a_6Z^3)$$

is always smooth at the point $(0 : 1 : 0)$. It is smooth over k if and only if

$$\Delta := -b_2^2b_8 - 8b_4^3 - 27b_6^2 + 9b_2b_4b_6 \neq 0,$$

where

$$b_2 = a_1^2 + 4a_2, \ b_4 = 2a_4 + a_1a_3, \ b_6 = a_3^2 + 4a_6, \ b_8 = a_1^2a_6 + 4a_2a_6 - a_1a_3a_4 + a_2a_3^2 - a_4^2.$$

(2) *Assume that* char $k \neq 2$. *For* $a, b \in k$, *the projective plane curve*

$$V_+(Y^2 Z - X^3 - aXZ^2 + bZ^3)$$

is smooth over k *if and only if*

$$4a^3 + 27b^2 \neq 0,$$

i.e., if and only if the polynomial $x^3 + ax + b$ *has no multiple zeros in an algebraic closure of* k.

Proof. Part (1) can be shown by a direct, but somewhat tedious computation which we omit here. See [Sil1] $^{\circ}$ Section III.1 and Appendix A. In the situation of part (2), a point $(x : y : z) \in E(k)$ is smooth if and only if all partial derivatives of $f = Y^2 Z - X^3 - aXZ^2 + bZ^3$ vanish. We find that $y = 0$, and may assume $z = 1$ without loss of generality, since we are working with homogeneous coordinates, and the only point on E with $z = 0$ is $(0 : 1 : 0)$. Then the condition on x is that x is a zero of $X^3 + aX + b$ (since the chosen point lies on E) and that x is a zero of the derivative of this polynomial with respect to X (which equals $\partial f / \partial X$). From this, the statement of the lemma follows from the formula for the discriminant of a cubic polynomial of this form.   □

Remark 26.97. The quantity Δ in part (1) for an equation of the particular form of part (2) is $-16(4a^3 + 27b^2)$. In characteristic $\neq 2$, the vanishing of one of these expressions is of course equivalent to the vanishing of the other, but using $-16(4a^3 + 27b^2)$ is arguably a better normalization and reflects the fact that a curve of the form $V_+(Y^2 Z - X^3 - aXZ^2 + bZ^3)$ with $a, b \in k$ is necessarily singular, if char $k = 2$.

The following theorem gives us the above-mentioned group structure.

Theorem 26.98. *Let* k *be a field, let* E *be a connected smooth proper curve over* k *and let* $0 \in E(k)$. *The following are equivalent:*
(i) *The genus of* E *is equal to 1, i.e.,* E *(with the fixed point 0) is an elliptic curve.*
(ii) *The curve* E *can be equipped with the structure of a group variety over* k *with neutral element* $0 \in E(k)$, *i.e.,* E *is an abelian variety over* k *(Definition 16.53, Chapter 27).*

Proof. (i) $\Rightarrow$ (ii). Consider E of genus 1 together with a rational point 0. We then have a bijection

$$\iota^0 \colon E(k) \to \mathrm{Pic}^0(E) = \{ \mathscr{L} \in \mathrm{Pic}(E); \ \deg(\mathscr{L}) = 0 \}, \quad x \mapsto \mathscr{O}_E([x]) \otimes_{\mathscr{O}_E} \mathscr{O}_E([0])^{\vee}.$$

Indeed, if $\iota^0(x) = \iota^0(y)$, then $[x]$ and $[y]$ are linearly equivalent. Since $g(E) \neq 0$ this implies $x = y$ (Example 26.90). For the surjectivity, let $\mathscr{L}$ be a line bundle on E of degree 0. Then $\mathscr{L} \otimes_{\mathscr{O}_E} \mathscr{O}_E([0])$ has degree 1 and hence $\ell(\mathscr{L} \otimes_{\mathscr{O}_E} \mathscr{O}_E([0])) = 1$ by Corollary 26.93. Therefore, there exists an effective divisor of degree 1, say $[x]$, $x \in E(k)$, such that $\mathscr{L} \otimes_{\mathscr{O}_E} \mathscr{O}_E([0]) \cong \mathscr{O}_E([x])$.

Since $\mathrm{Pic}^0(E)$ is an abelian group (a subgroup of $\mathrm{Pic}(E)$, i.e., the group structure being given by tensor product), this gives us a structure of abelian group on $E(k)$. It is clear that 0 is the neutral element.

By applying the same construction to the base change of E to extension fields k' of k, we analogously obtain a group structure on each $E(k')$. It is clear that $E(k)$ is then a subgroup of $E(k')$. In particular, taking as k' an algebraic closure of k, we see that there is at most one morphism $E \times E \to E$ which on k'-valued points gives the multiplication and at most one morphism $E \to E$ which on k'-valued points gives the map mapping each element to its inverse, for all k'/k. Since the group structure is commutative, it is usually denoted as addition and we will follow this convention.

We give two ways of showing that indeed the group structure comes from morphisms $E \times E \to E$, $E \to E$ of k-schemes. The more elementary approach is to embed E into $\mathbb{P}_k^2$ via Proposition 26.95 and to compute explicit formulas describing the addition and the map mapping a point to its negative. See Remark 26.100. This shows directly and very explicitly that these maps are morphisms of projective k-varieties and gives the well-known geometric description of the group law. Cf. Section (16.35).

A different approach is to soup up the isomorphism $E(k) \cong \mathrm{Pic}^0(E)$ so as to obtain a group structure on $E(T)$ for every k-scheme T, functorially in T. This will give us a structure of group functor on the functor of T-valued points of E, and by the Yoneda lemma the desired morphisms $E \times E \to E$ and $E \to E$ defining the group variety structure. Compare Section (27.21) where the functor Pic^0 is considered in much greater generality.

Let us first set up some notation. For a k-scheme T, we abbreviate $E \times_k T$ as E_T. Similarly, for $t \in T$ we denote by $E_t = E \otimes_k \kappa(t)$ the fiber of E_T over t. Let $\mathrm{Pic}^0(E_T)$ be the subgroup of $\mathrm{Pic}(E_T)$ consisting of line bundles $\mathscr{L}$ such that for each $t \in T$, the pullback $\mathscr{L}_t$ of $\mathscr{L}$ to the fiber $E_t = E \times_k \mathrm{Spec}(\kappa(t))$ has degree $\deg(\mathscr{L}_t) = 0$. (Cf. Proposition 26.166.) Pullback along the projection $p \colon E \times_k T \to T$ gives us a map $\mathrm{Pic}(T) \to \mathrm{Pic}(E_T)$ which is injective since the point 0 gives rise to a section $T \to E \times_k T$ and whose image $p^* \mathrm{Pic}(T)$ lies in $\mathrm{Pic}^0(E \times_k T)$.

Given a morphism $x \colon T \to E$, we obtain a morphism $(x, \mathrm{id}_T) \colon T \to E \times_k T$ of T-schemes. This is a regular closed immersion, so its image is an effective Cartier divisor of E_T. We denote by $\mathscr{O}_{E_T/T}(x)$ the corresponding line bundle. Cf. Remark 26.163.

We denote by $0_T \colon T \to E$ the composition $T \to \mathrm{Spec}\, k \to E$, the second arrow being our fixed point $0 \in E(k)$.

Claim. The maps

$$E(T) \to \mathrm{Pic}^0(E \times_k T)/p^* \mathrm{Pic}(T), \quad x \mapsto \mathscr{O}_{E_T/T}(x) \otimes \mathscr{O}_{E_T/T}(0_T)^\vee,$$

for k-schemes T are bijective and functorial in T.

Proof of claim. It is easy to check that the given line bundles lie in $\mathrm{Pic}^0(E \times_k T)$ and that the construction is functorial in T. In particular, whenever T is the spectrum of a field, this bijection induces on $E(T)$ the same group structure as above.

To prove the bijectivity we construct a map in the other direction such that the two maps are inverse to each other. For a line bundle $\mathscr{L} \in \mathrm{Pic}^0(E \times_k T)$, we write $\mathscr{L}' := \mathscr{L} \otimes \mathscr{O}_{E_T/T}(0_T)$. Then $\mathscr{L}'$ is a line bundle on E_T which is (fiberwise) of degree 1. We denote its restriction to E_t by $\mathscr{L}'_t$.

Given $\mathscr{L} \in \mathrm{Pic}^0(E \times_k T)$, we want to show that there exists a unique morphism $x \colon T \to E$ such that for some $\mathscr{M} \in \mathrm{Pic}(T)$ we have

$$\mathscr{L} \cong \mathscr{O}_{E_T/T}(x) \otimes \mathscr{O}_{E_T/T}(0)^\vee \otimes p^* \mathscr{M},$$

or equivalently

$$\mathscr{L}' \cong \mathscr{O}_{E_T/T}(x) \otimes p^* \mathscr{M}.$$

For $t \in T$, we have $\dim_{\kappa(t)} H^0(E_t, \mathscr{L}'_t) = \deg(\mathscr{L}'_t) = 1$ and $\dim_{\kappa(t)} H^1(E_t, \mathscr{L}'_t) = 0$ by the Theorem of Riemann-Roch, Theorem 26.48. By cohomology and base change for the proper morphism $p\colon E_T \to T$, we conclude that $R^1 p_* \mathscr{L}' = 0$, and that $p_* \mathscr{L}'$ is locally free of rank 1. See Corollary 23.144, Remark 26.165.

Since we work up to tensoring with line bundles pulled back from T, we may replace $\mathscr{L}$ by $\mathscr{L} \otimes (p^* p_* \mathscr{L}')^\vee$, and therefore may assume without loss of generality that $p_* \mathscr{L}' \cong \mathscr{O}_T$.

In this situation, the section $1 \in \Gamma(T, \mathscr{O}_T) = \Gamma(E_T, \mathscr{L}')$ defines an effective Cartier divisor on E_T (with associated line bundle $\mathscr{L}'$) which we view as a closed subscheme $Z \subset E_T$. The construction is compatible with base change $T' \to T$. In particular, we obtain a Cartier divisor of degree 1 on each fiber E_t. Proposition 14.22 shows that Z is flat over T. Thus the morphism $Z \to T$ is finite locally free of rank 1, and hence an isomorphism. We obtain a morphism $T \xrightarrow{\sim} Z \to E_T$, and composing with the projection $E_T \to E$ we get a morphism $x\colon T \to E$. This concludes the definition of a map $\mathrm{Pic}^0(E \times_k)/p^* \mathrm{Pic}(T) \to E(T)$.

It is not hard to check that the two maps are inverse to each other.

(i) $\Rightarrow$ (ii). For the converse, we use the fact (Proposition 27.15 in the next chapter) that the existence of a group variety structure on E implies that $\omega_E \cong \mathscr{O}_E$. Hence $2g(E) - 2 = \deg(\omega_E) = 0$. $\qquad\square$

Remark 26.99. From the rigidity lemma, Lemma 16.55, we obtain the following consequences.

(1) The choice of the point $0 \in E(k)$ determines the group structure uniquely. In fact, suppose that $m_\nu\colon E \times E \to E$ are multiplication morphisms and $i_\nu\colon E \to E$ are morphisms giving the inverse of an element, so that an elliptic curve E is a group variety with neutral element $0\colon \mathrm{Spec}(k) \to E$ for $\nu = 1, 2$. Consider the morphism

$$E \times E \to E, \quad (x, y) \mapsto m_2(m_1(x, y), i_2(x)),$$

i.e., we add x to y with respect to the first group law, and then subtract x with respect to the second one. For $y = 0$, we obtain the constant morphism $E = E \times \{0\} \to E$, $x \mapsto 0$. By Lemma 16.55, the above morphism factors through the second projection, and that implies that $m_1 = m_2$.

(2) A scheme morphism between elliptic curves which maps the fixed rational point on the domain to the fixed rational point on the target, is a group morphism. More precisely, we have, as a special case of Corollary 16.56 (1): Let k be a field, let E, E' with rational points $0 \in E(k)$, $0' \in E'(k)$ be elliptic curves over k and let $f\colon E \to E$ be a morphism of k-schemes. If $f(0) = 0'$, then f is a morphism of group varieties for the group variety structure on E and E', respectively, given by Theorem 26.98.

Remark 26.100. As mentioned above, one can write down explicit formulas for the group law on an elliptic curve E which is given as the closed subscheme of $\mathbb{P}^2_k$ defined by a Weierstraß equation. In fact, we can reformulate the definition of the group law as follows. For points $P, Q, R \in E(k)$ we have

$$P + Q + R = 0 \quad \Leftrightarrow \quad [P] + [Q] + [R] \sim 3[0],$$

and since the embedding $E \subset \mathbb{P}^2_k$ is given by the line bundle $\mathscr{O}_E(3[0])$, the condition on the right hand side is equivalent to saying that P, Q and R are the three intersection points (counted with multiplicity) of E with some line in $\mathbb{P}^2_k$ (cf. Theorem 5.61). It follows from the above results, but can also easily be shown directly, that a line in $\mathbb{P}^2_k$ intersects

E in either 0, 1 or 3 points of $E(k)$, counted with multiplicity. If one starts with this, then one can define the addition on E by the rule stated above. However, it is then a non-trivial task to show that this structure is associative.

For writing out explicit formulas, let us for simplicity restrict to the case of an elliptic curve E of the form $V_+(f)$ for

$$f = Y^2 Z - X^3 - aXZ^2 - bZ^3, \quad a, b \in k.$$

(This covers the situation where k has characteristic $\neq 2, 3$. See [Sil1]^O Section III.2 for the general case.)

First note that $0 = (0 : 1 : 0)$ is the only point of E on the "line at infinity" $V_+(Z)$. For the point 0, being the neutral element for the group law, there is nothing to compute. Therefore we can pass to the affine curve

$$E' = V(Y^2 - X^3 - aX - b) \quad \subset \mathbb{A}_k^2,$$

where now X, Y are affine coordinates on $\mathbb{A}_k^2$. For a point $P = (x, y) \in E(k)$, the negative of P with respect to the group law is $-P = (x, -y)$. In fact, it is clear that the points $(x : y : 1)$, $(x : -y : 1)$ and $(0 : 1 : 0)$ are the intersection points of E with the line $V_+(X - xZ)$.

It remains to give formulas for the sum of points $P_i = (x_i, y_i) \in E'(k)$, $i = 1, 2$ where $x_1 \neq x_2$ or $P_1 = P_2$ but $2P_1 \neq 0$, i.e., $y_1 \neq 0$.

Case 1. $P_i = (x_i, y_i) \in E'(k)$, $i = 1, 2$, $x_1 \neq x_2$. We define

$$\lambda = \frac{y_2 - y_1}{x_2 - x_1}.$$

Case 2. $P_i = (x_i, y_i) \in E'(k)$, $i = 1, 2$, $P_1 = P_2$. We define

$$\lambda = \frac{3x_1^2 + a}{2y_1}.$$

With this case-by-case definition of λ, we can state the result in both cases in a uniform way. We have $P_1 + P_2 = (x_3, y_3)$ with

$$x_3 = \lambda^2 - x_1 - x_2, \quad y_3 = \lambda(x_1 - x_3) - y_1.$$

Remark 26.101. Let E be an elliptic curve over a field k. The curve E is a group variety, and the group of k-scheme-automorphisms of E contains all translation morphisms $x \mapsto x + a$, $a \in E(k)$. This implies that the group of k-scheme-automorphisms acts transitively on $E(k)$.

On the other hand, the group of group variety automorphisms of E is finite, as Proposition 26.103 shows.

Example 26.102. In contrast to genus 0 curves (Remark 26.92), curves of genus 1 do not satisfy a local-global principle. The following famous example is due to Selmer. Let C be the smooth projective plane cubic curve over $\mathbb{Q}$ given by the homogeneous equation $3X^3 + 4Y^3 + 5Z^3$. It is not hard to show that $C(\mathbb{R}) \neq \emptyset$ and $C(\mathbb{Q}_p) \neq \emptyset$ for every prime number p. Moreover, $C(\mathbb{Q}) = \emptyset$ (but proving this is more involved). See [Sil1]^O Chapter X for a more thorough discussion and further references.

(26.18) The Legendre family and the j-invariant.

Let k be a field of characteristic $\neq 2$. Let E be an elliptic curve over k, with affine Weierstraß equation $Y^2 = X^3 + a_2 X^2 + a_4 X + a_6$ for E (use Proposition 26.95 and that $\operatorname{char}(k) \neq 2$ to complete the square on the left hand side). The projection map $(x, y) \mapsto x$ yields a finite morphism $E \to \mathbb{P}_k^1$ of degree 2. In fact, it corresponds to the extension $K(\mathbb{P}_k^1) = k(T) \subseteq K(E)$ mapping T to a rational function x on E with a double pole at $0 = (0 : 1 : 0)$ and no other poles and zeros. Then $K(E) \cong k(x)[y]/(y^2 - x^3 - a_2 x^2 - a_4 x - a_6)$, since we can compute the function field on the standard affine chart of E, and hence the extension has degree 2. Since y, as a rational function on E, has a triple pole at 0, we also see that $0 \in E(k)$ is mapped to ∞, and is the only such point. For a k-valued point $(1 : x) \in \mathbb{P}_k^1(k)$, the fiber consists of 2 points, if and only if $x^3 + a_2 x^2 + a_4 x + a_6$ is $\neq 0$ and a square in k. Otherwise it consists of a single point.

Hence if k is algebraically closed, then the morphism $E \to \mathbb{P}_k^1$ is ramified precisely over ∞ and over the three zeros of the polynomial $X^3 + a_2 X^2 + a_4 X + a_6$, seen as elements of $k = \mathbb{A}_k^1(k) \subset \mathbb{P}_k^1(k)$. The ramification degree of each ramified point is 2. Since k has characteristic $\neq 2$, f is tamely ramified. Compare with the Riemann-Hurwitz formula (Corollary 26.67).

From a different point of view, the points P of the form $(x, 0)$ in (the affine part of) $E(k)$ are precisely the points $P \neq 0$ with $2P = 0$ with respect to the group law on E. For equations as above with $a_4 = 0$ this follows directly from the formulas in Remark 26.100, but the computation is basically the same if a_4 is arbitrary. In particular, the set of these points is preserved by every group variety automorphism of E. This gives us a handle on the automorphism group of the elliptic curve E.

Proposition 26.103. *Let k be a field of characteristic $\neq 2$ and let E be an elliptic curve over k. Let G be the group of automorphisms of E as a group variety, equivalently, the group of automorphisms of E fixing 0 (Remark 26.101). Then G is non-trivial and finite, of order dividing 12.*

Proof. The group G is non-trivial because multiplication by -1, i.e., mapping each point to its negative, is not the identity automorphism. For an affine point $P = (x, y) \in E(k)$, $P = -P$ is equivalent to $y = 0$, as remarked above, so there are at most 4 such points in $E(k)$ (including the point at infinity). Therefore multiplication by -1 cannot be the identity morphism, because otherwise it would give rise to the identity on $E(k')$ for every extension field k' of k. Compare Proposition 27.186 which shows that multiplication by 2 is a finite morphism, which in turn implies that multiplication by -1 is not the identity.

To show that G is finite, we may assume that k is algebraically closed, because passing to an extension field can only enlarge the automorphism group. We may thus assume that E is of the form E_λ for some $\lambda \in k$. Let $f: E \to \mathbb{P}_k^1$ be a morphism of degree 2 ramified over $0, 1, \lambda$ and ∞, and assume that $0 \in E(k)$ is mapped to ∞. The points in E where f is ramified are the 2-torsion points, i.e., the points in $E[2](k) := \{P \in E(k); \, 2P = 0\}$, hence every automorphism of E permutes these points. Since it fixes 0, it induces a permutation of $E[2](k) \setminus \{0\}$. In this way we obtain a homomorphism $\Phi: G \to S_3$ from G to the symmetric group S_3.

The kernel of Φ consists of those automorphisms of E that fix the four points where f is ramified. Hence every $g \in \operatorname{Ker}(\Phi)$ induces a Galois cover $E \setminus E[2](k) \to \mathbb{P}^1 \setminus \{0, 1, \lambda, \infty\}$ with Galois group $\mathbb{Z}/2$ (see Section (20.16)). There are precisely 2 such covers, the identity and a second one, namely multiplication by -1.

This implies the claim of the proposition. In general, the image of Φ will be a proper subgroup of S_3 (it is in bijection with the set of possible expressions for λ in Proposition 26.105 that equal λ). $\square$

For fields of characteristic $= 2$, the automorphism group of an elliptic curve is still necessarily finite, but may have order up to 24.

Now specifically, for every element $\lambda \in k \setminus \{0, 1\}$, we may consider the affine curve

$$E'_\lambda := V(Y^2 - X(X - 1)(X - \lambda)) \subset \mathbb{A}^2_k.$$

Its schematic closure $E_\lambda \subset \mathbb{P}^2_k$ is an elliptic curve over k (Lemma 26.96).

As before, we obtain a morphism $f\colon E \to \mathbb{P}^1_k$ extending the map $(x, y) \mapsto x$ on the affine part of E. Then f is ramified precisely over 0, 1, λ and ∞, and the ramification degree equals 2 for each of these points.

We can view the curves E_λ as the fibers of a smooth proper morphism

$$\mathscr{E} := V_+(Y^2Z - X(X - Z)(X - \lambda Z)) \to S := \mathbb{P}^1_k \setminus \{0, 1, \infty\},$$

where $\mathscr{E} \subset \mathbb{P}^2_k \times_k S$, the *Legendre family* of elliptic curves. With the terminology of Definition 27.144, $\mathscr{E}$ is a (relative) elliptic curve over $S = \mathbb{P}^1_k \setminus \{0, 1, \infty\}$.

Proposition 26.104. *For k algebraically closed of characteristic $\neq 2$, every elliptic curve over k is isomorphic to a curve of the form E_λ.*

Proof. Given an elliptic curve E with Weierstraß equation $Y^2 = X^3 + a_2X^2 + a_4X + a_6$, we can factor the polynomial $X^3 + a_2X^2 + a_4X + a_6$ as $(X - \alpha)(X - \beta)(X - \gamma)$, say, where $\alpha, \beta, \gamma \in k$ are pairwise distinct by Lemma 26.96, since k is algebraically closed. After a suitable change of coordinates, we may assume that $\alpha = 0$, $\beta = 1$. $\square$

Proposition 26.105. *Let k be a field of characteristic $\neq 2$ and let E, E' be elliptic curves over k. Suppose that there are morphisms $f\colon E \to \mathbb{P}^1_k$ and $f'\colon E' \to \mathbb{P}^1_k$ of degree 2 that are ramified over 0, 1, ∞ and, respectively, λ and λ'.*

Then the following are equivalent:

(i) The curves E and E' are isomorphic.

(ii) We have

$$\lambda' \in \left\{ \lambda, 1 - \lambda, \frac{1}{\lambda}, \frac{1}{1 - \lambda}, \frac{\lambda}{1 - \lambda}, \frac{1 - \lambda}{\lambda} \right\}.$$

Proof. First suppose that E and E' are isomorphic. We may then just as well assume that $E = E'$. The automorphism group of the curve E acts transitively on $E(k)$; this follows immediately from the existence of the group structure on E. Replacing f and f' by the composition with suitable automorphisms of E, we may therefore assume that there f and f' are ramified at $0 \in E(k)$ and that $f(0) = f'(0) = \infty$.

It follows that $(f')^*\mathscr{O}(1) \cong \mathscr{O}_E(2[\infty]) \cong g^*\mathscr{O}(1)$, i.e., f and f' are given by the same line bundle on E, but possibly different choices of basis of $H^0(E, \mathscr{O}_E(2[\infty]))$. A suitable change of basis corresponds to an automorphism of $\mathbb{P}^1_k$ fixing ∞ and mapping the set $\{0, 1, \lambda\}$ to $\{0, 1, \lambda'\}$. Using the explicit description $\mathrm{Aut}_k(\mathbb{P}^1_k) = PGL_2(k)$ of the automorphism group of the projective line, see Section (11.15), one arrives at the desired conclusion.

If, on the other hand, condition (ii) is satisfied, one sees similarly as in the first part of the proof that we may assume $\lambda = \lambda'$ after replacing f' by the composition with a suitable automorphism of $\mathbb{P}^1_k$. But then E and E' are both isomorphic to the elliptic curve with affine Weierstraß equation $y^2 = x(x - 1)(x - \lambda)$. $\square$

The proposition allows us to make the following definition:

Definition 26.106. *Let k be an algebraically closed field of characteristic $\neq 2$, and let E be an elliptic curve over k. The j-invariant $j(E) \in k$ is defined as*

$$j(E) = j(\lambda) := 2^8 \frac{(\lambda^2 - \lambda + 1)^3}{\lambda^2(\lambda - 1)^2},$$

where $\lambda \in k$ is chosen so that $E \cong E_\lambda$.

One has to check that $j(E)$ is independent of the choice of λ, i.e., that for all the 6 possibilities for λ' in the previous proposition one obtains the same value of the j-invariant. In fact, the map $\lambda \mapsto j(\lambda)$ describes a morphism $\mathbb{P}_k^1 \to \mathbb{P}_k^1$ of degree 6, ramified only over ∞.

A different approach is to express $j(E)$ in terms of a Weierstraß equation. For instance, if E is given by the affine equation $y^2 = x^3 + ax + b$, then $j(E) = \frac{(4a)^3}{(4a^3 + 27b^2)}$. In this way, one can define $j(E)$ for elliptic curves over arbitrary ground fields, not necessarily algebraically closed. The j-invariant classifies elliptic curves up to isomorphism over algebraically closed fields.

Theorem 26.107. *Let k be a field of characteristic $\neq 2$ and let $\bar{k}$ be an algebraic closure of k. Let E, E' be elliptic curves over k. Then $j(E) = j(E')$ if and only if $E \otimes_k \bar{k}$ and $E' \otimes_k \bar{k}$ are isomorphic as elliptic curves over $\bar{k}$, or equivalently as $\bar{k}$-schemes.*

Proof. The j-invariant does not change under base change of the ground field, so we have $j(E) = j(E')$ whenever E and E' become isomorphic after base change to any extension field of k.

Conversely, assume that $j(E) = j(E')$. We replace k by $\bar{k}$, i.e., we will show that assuming k is algebraically closed, it follows that E and E' are isomorphic. Choose ramified covers $E \to \mathbb{P}_k^1$ and $E' \to \mathbb{P}_k^1$, ramified over 0, 1, ∞ and λ, λ', respectively. (Here we need that k is algebraically closed, to be able to apply Proposition 26.104.) We conclude by Proposition 26.105. $\square$

This theorem also holds if k has characteristic 2 (with an appropriate definition of the j-invariant), and so does the following corollary.

Corollary 26.108. *Let k be an algebraically closed field. The map $E \mapsto j(E)$ induces a bijection between the set of isomorphism classes of elliptic curves over k and the set k.*

See [Sil1]$^\text{O}$ Chapter III for further details.

(26.19) Elliptic curves over the complex numbers.

Over the complex numbers, a smooth proper curve by analytification gives rise to a compact Riemann surface, see Section (26.7). In this section, we discuss the case of genus 1 in this setting in a bit more (but not always full) detail, because it gives some interesting insight into the general theory.

So let X be a compact Riemann surface of genus 1. By Theorem 26.30 the universal cover of X is $\mathbb{C} \to X$. This yields a very explicit description of X.

Definition 26.109. *A subgroup $\Lambda \subset \mathbb{C}$ is a* lattice *if it is generated as a $\mathbb{Z}$-module by two $\mathbb{R}$-linearly independent elements.*

Using the topological interpretation of the genus it is then easy to see that the quotient $\mathbb{C}/\Lambda$, with the natural structure of Riemann surface, has genus 1.

Proposition 26.110. *Let X be a compact Riemann surface of genus 1 and let $f \colon \mathbb{C} \to X$ be its universal cover. Then $\Lambda := f^{-1}(f(0))$ is a lattice in $\mathbb{C}$ and f induces an isomorphism $X \cong \mathbb{C}/\Lambda$ of Riemann surfaces.*

Proof. We sketch the key steps of the proof. The Riemann surface X is the quotient of $\mathbb{C}$ by the action of the fundamental group $\pi_1(X)$ of X on its universal covering. The automorphisms of the complex manifold $\mathbb{C}$ are the maps $z \mapsto \mu z + \lambda$, $\mu \in \mathbb{C}^\times$, $\lambda \in \mathbb{C}$. Covering transformations ($\neq \mathrm{id}$) do not have fix points, so for those we must have $\mu = 1$.

Therefore we can identify the fundamental group $\pi_1(X)$ with a subgroup of $\mathbb{C}$. One checks that it is a discrete subgroup, thus a subgroup generated by 0, or 1 or 2 elements of $\mathbb{C}$ that are linearly independent over $\mathbb{R}$. Since $\mathbb{C}$ and $\mathbb{C}/\mathbb{Z} \cong \mathbb{C}^\times$ are non-compact, the desired statement follows. $\square$

Corollary 26.111. *Let X be a compact Riemann surface of genus 1. Then X can be equipped with the structure of an abelian group so that it becomes a group object in the category of Riemann surfaces (i.e., multiplication and inverse are given by morphisms of complex manifolds).*

Proof. This is clear by the above, since the quotient of $\mathbb{C}$ by a lattice is an abelian group. $\square$

Corollary 26.112. *Let $\Lambda \subset \mathbb{C}$ be a lattice, and let $X = \mathbb{C}/\Lambda$, a compact Riemann surface of genus 1. As an abelian group, X is isomorphic to $\mathbb{R}/\mathbb{Z} \times \mathbb{R}/\mathbb{Z}$. In particular, for every $n \in \mathbb{N}_{>0}$, the kernel of multiplication by n on X (the n-torsion on X) is isomorphic, as an abelian group, to $(\mathbb{Z}/n\mathbb{Z})^2$.*

Morphisms between elliptic curves are easy to understand in terms of lattices.

Proposition 26.113. *Let $\Lambda_1, \Lambda_2 \subset \mathbb{C}$ be lattices. The map*

$$\{\alpha \in \mathbb{C};\ \alpha\Lambda_1 \subseteq \Lambda_2\} \to \{f \colon \mathbb{C}/\Lambda_1 \to \mathbb{C}/\Lambda_2;\ f\ \text{holomorphic},\ f(0) = 0\}$$

which maps α to the map $z \mapsto \alpha z$, is a bijection.

Proof. We omit the proof (that is not very difficult). See [Sil1]$^{\circ}$ Chapter VI, Theorem 4.1 (a). $\square$

From the proposition, we obtain the following corollary. This allows, for instance, to analyze the endomorphism ring of an elliptic curve over $\mathbb{C}$ in terms of the corresponding lattice. While in most cases, the endomorphism ring is just $\mathbb{Z}$ (i.e., every endomorphism has the form "multiplication by n" by some $n \in \mathbb{Z}$), some elliptic curves admit more endomorphisms. Those are called elliptic curves *with complex multiplication*. It turns out that this property is closely related to arithmetic questions. We do not go into this further here, however, and content ourselves with stating the following corollary about homomorphisms between elliptic curves.

Corollary 26.114. *Let $\Lambda_1, \Lambda_2 \subset \mathbb{C}$ be lattices.*

(1) *Every morphism $\mathbb{C}/\Lambda_1 \to \mathbb{C}/\Lambda_2$ which maps 0 to 0 is a group homomorphism.*

(2) *The Riemann surfaces $\mathbb{C}/\Lambda_1$ and $\mathbb{C}/\Lambda_2$ are isomorphic as Riemann surfaces if and only if there exists an isomorphism of Riemann surfaces preserving the group structures, if and only if the lattices Λ_1 and Λ_2 are homothetic (i.e., there exists $\alpha \in \mathbb{C}^\times$ with $\Lambda_2 = \alpha\Lambda_1$).*

We therefore obtain a bijection between the set of homothety classes of lattices in $\mathbb{C}$ and the set of isomorphism classes of elliptic curves over $\mathbb{C}$, which on the level of Riemann surfaces is given by mapping the homothety class of a lattice Λ to $\mathbb{C}/\Lambda$. It is easy to see that every homothety class can be represented by a lattice of the form $\mathbb{Z} \oplus \mathbb{Z}\tau$ for some $\tau \in \mathbb{H}$, the upper complex half-plane. Lattices $\mathbb{Z} \oplus \mathbb{Z}\tau$ and $\mathbb{Z} \oplus \mathbb{Z}\tau'$ are homothetic if and only if there exists a matrix $\begin{pmatrix} a & b \\ c & d \end{pmatrix} \in SL_2(\mathbb{Z})$ such that $\tau' = \frac{a\tau+b}{c\tau+d}$. Summarizing, we obtain the following corollary.

Corollary 26.115. *Let $SL_2(\mathbb{Z})$ act on the complex upper half-plane $\mathbb{H}$ by*

$$\begin{pmatrix} a & b \\ c & d \end{pmatrix} \cdot \tau = \frac{a\tau + b}{c\tau + d}.$$

Then the map $\tau \mapsto \mathbb{C}/(\mathbb{Z} \oplus \mathbb{Z}\tau)$ induces a bijection between the set $SL_2(\mathbb{Z})\backslash\mathbb{H}$ and the set of isomorphism classes of compact Riemann surfaces of genus 1 with a fixed base point, which in turn is in bijection with the set of isomorphism classes of elliptic curves over $\mathbb{C}$.

One can show that one can equip $SL_2(\mathbb{Z})\backslash\mathbb{H}$ with the structure of Riemann surface such that the projection $\mathbb{H} \to SL_2(\mathbb{Z})\backslash\mathbb{H}$ is a morphism of Riemann surfaces. The theory of the j-invariant yields an isomorphism $SL_2(\mathbb{Z})\backslash\mathbb{H} \cong \mathbb{C}$ of Riemann surfaces.

To conclude this section, let us discuss how in the case of genus 1, one can give a more direct proof of the algebraizability of compact Riemann surfaces (Corollary 26.26). From hindsight (say, the Theorem of Riemann-Roch) we know that a non-constant meromorphic function cannot have only a simple pole at a single point, but we expect that a meromorphic function regular outside a single point and with a double pole at that point should exist. Without loss of generality, we will look for such a function with the double pole located at the fixed point 0 of the given Riemann surface X. Via pullback along the universal cover, a function with this property corresponds to a Λ-invariant morphism $f \colon \mathbb{C} \to \mathbb{P}^1(\mathbb{C})$ (i.e., $f(\lambda z) = f(z)$ for all $\lambda \in \Lambda$) with a double pole at all points of Λ, and regular on $\mathbb{C} \setminus \Lambda$. The Weierstraß $\wp$-function

$$\wp(z) = \frac{1}{z^2} + \sum_{\lambda \in \Lambda \setminus \{0\}} \left(\frac{1}{(z-\lambda)^2} - \frac{1}{\lambda^2} \right)$$

is a function with these properties. We omit the verification that the series converges absolutely and uniformly, locally on $\mathbb{C} \setminus \Lambda$. Omitting all the squares in the definition would not give a convergent series; this gives a different perspective on the fact that a meromorphic function on X that is regular outside 0 and has a simple pole at 0 cannot exist.

The existence of a non-constant meromorphic function implies that X is algebraizable, as we have seen in the proof of Corollary 26.26. In fact, in the situation at hand, we can be more precise. Namely, the $\wp$-function and its derivative $\wp'$ satisfy the equation

$$(\wp')^2 = 4\wp^3 - g_2\wp - g_3,$$

where

$$g_2 = 60 \sum_{\lambda \in \Lambda \setminus \{0\}} \frac{1}{\lambda^4}, \qquad g_3 = 140 \sum_{\lambda \in \Lambda \setminus \{0\}} \frac{1}{\lambda^6}.$$

Again, we omit the necessary (and not very difficult) calculations. We obtain that the map

$$X \to \mathbb{P}^2(\mathbb{C}), \quad x \mapsto (\wp(x) : \wp'(x) : 1),$$

where $0 \mapsto (0 : 1 : 0)$, because $\wp$ has a pole of order 2, and $\wp'$ has a pole of order 3 at 0, is a morphism of Riemann surfaces. Its image is contained in the curve $V_+(Y^2Z - 4X^3 + g_2XZ^2 + g_3Z^3)$ by the differential equation for $\wp$ stated above. Since the algebraic curve $V_+(Y^2Z - 4X^3 + g_2XZ^2 + g_3Z^3)$ is a compact Riemann surface, equality follows.

Moreover, the curve $V_+(Y^2Z - 4X^3 + g_2XZ^2 + g_3Z^3)$ is smooth for every lattice Λ, and the above map is an isomorphism from X onto its image, i.e., a closed embedding of X into $\mathbb{P}^2(\mathbb{C})$. The equation we find here is almost a Weierstraß equation. The smooth cubic with "distinguished" point $(0 : 1 : 0)$ which it defines is an elliptic curve. This is even a group isomorphism, as follows from Corollary 26.114, or can be shown more directly by proving an "addition theorem" for the $\wp$-function. Since the group structure on the quotient $\mathbb{C}/\Lambda$ is obvious, this provides an approach to discovering the geometric description of the group law on smooth cubics in a natural way. Summarizing, we have the following result.

Theorem 26.116. *Let X be a compact Riemann surface of genus 1. There exists a lattice $\Lambda \subset \mathbb{C}$ such that $X \cong \mathbb{C}/\Lambda$, and with g_2, g_3 defined above (depending on Λ), X is isomorphic to the analytification of the algebraic elliptic curve $V_+(Y^2Z - 4X^3 + g_2XZ^2 + g_3Z^3) \subset \mathbb{P}^2_{\mathbb{C}}$, as Riemann surfaces and as groups.*

(26.20) Hyperelliptic curves.

In this section we study *hyperelliptic curves*, which share certain properties with elliptic curves, e.g., they admit a generically étale morphism to a smooth proper curve of genus 0, and can also be described by an explicit equation that has a form reminiscent of the Weierstraß equation of an elliptic curve. See Proposition 26.122 for a more precise statement.

Definition 26.117. *Let k be a field. A geometrically connected smooth proper curve C over k is called* hyperelliptic, *if $g(C) \geq 2$ and if there exists a generically étale morphism of degree 2 from C onto a smooth proper curve of genus 0.*

Curves of genus 0 and of genus 1 also admit a generically étale morphism of degree 2 to a genus 0 curve, but are of exceptional nature as far as the considerations of this section are concerned, and are therefore excluded in the above definition. In the literature, sometimes curves of genus 1 are also called hyperelliptic.

Over an algebraically closed field (or more generally, whenever C has a k-valued point) the morphism in the definition is a generically étale morphism of degree 2 from C to a genus 0 curve with a rational point, i.e., to the projective line $\mathbb{P}^1_k$ (Corollary 26.53). Sometimes in the literature a curve is called hyperelliptic, if it admits such a morphism to $\mathbb{P}^1_k$. For arbitrary fields our definition is more general, and for instance has the advantage that the property of being hyperelliptic descends along extensions of the base field, see Lemma 26.120.

Lemma 26.118. *Let C be a hyperelliptic curve over a field k of genus $g = g(C)$. Assume that there exists a generically étale morphism $f\colon C \to \mathbb{P}^1_k$ of degree 2, and denote by $\mathscr{L} := f^*\mathscr{O}(1)$ the corresponding line bundle on C. Then $\mathscr{L}^{\otimes(g-1)} \cong \omega_C$.*

In particular, the map $C \to \mathbb{P}^1_k$ in the definition is unique up to composition with an automorphism of $\mathbb{P}^1_k$ and of C.

Proof. Recall the Veronese embedding $\mathbb{P}^1_k \hookrightarrow \mathbb{P}^{g-1}_k$ given by the very ample line bundle $\mathscr{O}_{\mathbb{P}^1_k}(g - 1)$. On homogeneous coordinates, it is given by $(x_0 : x_1) \mapsto (x_0^{g-1} : x_0^{g-2}x_1 : \cdots : x_1^{g-1})$. Cf. Exercise 13.8. The composition $C \to \mathbb{P}^1_k \to \mathbb{P}^{g-1}_k$ of the morphism $C \to \mathbb{P}^1_k$ with the Veronese embedding corresponds to the line bundle $\mathscr{L}^{\otimes(g-1)}$. Since the map $C \to \mathbb{P}^1_k$ is surjective, and the image of $\mathbb{P}^1_k$ under the Veronese embedding is not contained in any hyperplane, the pullbacks of the standard basis vectors of $H^0(\mathbb{P}^{g-1}_k, \mathscr{O}_{\mathbb{P}^{g-1}}(1))$ are linearly independent in $H^0(C, \mathscr{L})$; it follows that $\ell(\mathscr{L}^{\otimes(g-1)}) \geq g$. Since $\deg(\mathscr{L}^{\otimes(g-1)}) = 2g - 2$, we see, using the Theorem of Riemann-Roch, that $\mathscr{L}^{\otimes(1-g)} \otimes \omega_C$ has degree 0 and has non-trivial global sections, and hence is trivial (Remark 26.21).

For the final part, note that by the above we can identify the given map $C \to \mathbb{P}^1_k$ with the morphism from C onto the image of "the" morphism $C \to \mathbb{P}^{g-1}_k$ induced by ω_C, up to isomorphisms of the source and the target. $\qquad\square$

Proposition 26.119. *Let C be a hyperelliptic curve over a field k of genus $g = g(C)$. The generically étale morphism $C \to P$ from C onto a smooth proper curve P of genus 0 is uniquely determined by C. In other words, there exists a unique subfield K of the function field $K(C)$ such that $K(C)/K$ is a quadratic Galois extension and K is the function field of a smooth proper curve of genus 0.*

The non-trivial Galois automorphism of $K(C)$ over K defines an automorphism $C \to C$ of order 2, called the hyperelliptic involution *of C.*

Proof. Let k'/k be a finite field extension such that $C(k') \neq \emptyset$ and hence $P(k') \neq \emptyset$. This implies that $P \otimes_k k' \cong \mathbb{P}^1_{k'}$. By Lemma 26.118, we have that $P \otimes_k k'$ is the image of C under the morphism $C \otimes_k k' \to \mathbb{P}^{g-1}_{k'}$ given by the canonical sheaf of $C \otimes_k k'$ and is hence independent of choices.

We claim that
$$K(P) = K(P) \otimes_k k' \cap K(C) \subseteq K(C) \otimes_k k'.$$

Clearly the left hand side is contained in the right hand side. The other inclusion holds because k is algebraically closed in $K(C)$, C being geometrically integral (Proposition 5.51). It follows that the subfield $K(P)$ of $K(C)$ depends only on C and not on the choice of morphism from C to a genus 0 curve.

The final part follows from the equivalence of categories between normal proper curves and their function fields. $\qquad\square$

On k-valued points, the hyperelliptic involution fixes the points of C that are ramified over $\mathbb{P}^1_k$. For all other points, the fiber of the image in $\mathbb{P}^1_k$ has two elements, and those are swapped by the involution. As a consequence, we see that the property of being hyperelliptic is preserved by base change and descends along change of the base field.

Lemma 26.120. *Let k'/k be a field extension and let C be a geometrically connected, smooth proper curve over k. Then C is hyperelliptic as a curve over k if and only if $C \otimes_k k'$ is hyperelliptic as a curve over k'.*

Proof. It is clear that $C \otimes_k k'$ is hyperelliptic, if C is. For the converse, we argue as follows. By passing to an extension of k' we may assume that $C(k') \neq \emptyset$. Let g be the common genus of C and C' (Lemma 26.32) and consider the morphism $C \to \mathbb{P}_k^{g-1}$ given by the canonical sheaf of C. Its base change to k' gives us a generically étale morphism of degree 2 from C' onto $\mathbb{P}_{k'}^1$.

This implies that the image of C on $\mathbb{P}_k^{g-1}$ is a smooth projective curve P of genus 0, and that the morphism $C \to P$ is generically étale of degree 2. $\qquad\square$

See [LoKl]$^\circ$ for a discussion of hyperelliptic relative curves. Next we will show that every curve of genus 2 is hyperelliptic.

Proposition 26.121. *Let k be a field and let C be a geometrically connected smooth proper curve of genus $g(C) = 2$. Then C admits a generically étale morphism of degree 2 to $\mathbb{P}_k^1$ and in particular is hyperelliptic.*

Proof. The canonical line bundle ω_C satisfies $\deg(\omega_C) = 2g(C) - 2 = 2$ and $\ell(\omega_C) = g = 2$, and hence defines a morphism $C \to \mathbb{P}_k^1$ of degree 2. Since the genus does not change under purely inseparable morphisms and the degree is a prime number, this morphism must be generically étale. $\qquad\square$

A hyperelliptic curve which admits a generically étale morphism of degree 2 to the projective line can be described by an equation of a particularly simple form (compare the description of an elliptic curve by a Weierstraß equation, Proposition 26.95). The situation is more delicate, however, since no hyperelliptic curve is a plane curve, see Remark 26.81 for the case of genus 2 curves and Exercise 26.15 for the general case.

Proposition 26.122. *Let C be a hyperelliptic curve over a field k of genus $g = g(C)$. Assume that there exists a generically étale morphism $f\colon C \to \mathbb{P}_k^1 = \mathrm{Proj}(k[T_0, T_1])$ of degree 2.*
(1) *There exist polynomials $p, q \in K[X]$ satisfying*

$$2g + 1 \leq \max(\deg(p), 2\deg(q)) \leq 2g + 2,$$

such that, for $X = \frac{T_1}{T_0}$,
(a) *the affine open subscheme $U := f^{-1}(D_+(T_0))$ of C is isomorphic to*

$$U' := V(Y^2 + q(X)Y - p(X)) \quad \subset \mathbb{A}_k^2 = \mathrm{Spec}\, k[X, Y],$$

(b) *the affine open subscheme $V := f^{-1}(D_+(T_1))$ of C is isomorphic to*

$$V' := V(Z^2 + X^{-(g+1)}q(X)Z - X^{-2(g+1)}p(X)) \quad \subset \mathbb{A}_k^2 = \mathrm{Spec}\, k[X^{-1}, Z],$$

and C is isomorphic to the curve obtained by gluing U' and V' via $Y = X^{g+1}Z$.
If $\mathrm{char}(k) \neq 2$, then we may arrange that $q = 0$.
(2) *In terms of this description the ramification divisor of f, seen as a closed subscheme of C, consists of the vanishing locus $V(4p + q^2) \subset U'$, and, if and only if $\deg(4p + q^2) < 2g + 2$, the one point in $C \setminus U$. (See the proof for information on the contribution of each of these components to the degree of the ramification divisor.)*

Proof. We follow [Liu]$^\circ$ Section 7.4.3. Part (1). Consider the short exact sequence

$$0 \to \mathscr{O}_{\mathbb{P}_k^1} \to f_* \mathscr{O}_C \to \mathscr{L} \to 0,$$

where $\mathscr{L}$ is defined as the cokernel of the map on the left hand side (which is injective since f is dominant).

Here $f_*\mathscr{O}_C$ is a locally free $\mathscr{O}_{\mathbb{P}^1_k}$-module of rank 2 (it is locally free because f is finite and flat by Lemma 26.9). We claim that $\mathscr{L}$ is locally free of rank 1. Because of Corollary 11.19 it is enough to show that all fibers of $\mathscr{L}$ are 1-dimensional vector spaces. But for $\xi \in \mathbb{P}^1_k$ the fiber of $\mathscr{O}_{\mathbb{P}^1_k} \to f_*\mathscr{O}_C$ is the ring homomorphism $\kappa(\xi) \to (f_*\mathscr{O}_C)(\xi)$ which is injective because its domain is a field.

So the restrictions $(f_*\mathscr{O}_C)_{|D_+(T_0)}$ and $(f_*\mathscr{O}_C)_{|D_+(T_1)}$ are free over $k[X]$ and $k[X^{-1}]$, respectively, and more precisely we can find bases $1, y \in \Gamma(U, \mathscr{O}_C)$ and $1, z \in \Gamma(V, \mathscr{O}_C)$ over $k[X]$ and $k[X^{-1}]$, respectively. Over the intersection $D_+(T_0) \cap D_+(T_1)$, we can express y in terms of 1 and z, say

$$y = a + bz, \quad a \in k[X, X^{-1}], \ b \in k[X, X^{-1}]^\times.$$

After a change of basis given by multiplying z by an element of $k^\times$, if necessary, we may assume that $b = X^r$ for some r. We write $a = a_+ + a_-$ with $a_+ \in k[X]$ and $a_- \in k[X^{-1}]$. Then $y - a_+ = a_- + T^r z$. If we had $r \leq 0$, then this expression would lie in $k[T] \cap k[T^{-1}] = k$ which is impossible, since $y \in K(C) \setminus k(T)$. Therefore $r \geq 1$. Replacing y by $y - a_+$ and z by $z + X^{-r}a_-$ we find bases $1, y$ of $\Gamma(U, \mathscr{O}_C)$ and $1, z$ of $\Gamma(V, \mathscr{O}_C)$ such that

$$y = X^r z, \quad r \geq 1,$$

and expressing y^2 with respect to this basis we find

$$y^2 + q(X)y = p(X) \quad \text{for polynomials } p, q \in k[X].$$

We define U' and V' as in the statement of the proposition. Dividing by X^{2r} we obtain

$$z^2 + \frac{q(X)}{X^r}z = \frac{p(X)}{X^{2r}}.$$

Now z (considered as an element of $K(C)$) is integral over $k[X^{-1}]$, hence its minimal polynomial has coefficients in $k[X^{-1}]$, i.e., $\frac{q(X)}{X^r}, \frac{p(X)}{X^{2r}} \in k[X^{-1}]$. This implies that $\deg(q) \leq r$, $\deg(p) \leq 2r$.

To get a lower bound on the degrees, we write $q_- = q/X^r, p_- = p/X^{2r} \in k[X^{-1}]$ and evaluate the condition that V' is smooth (at the point(s) of $V(X^{-1})$). If the characteristic of k is $\neq 2$, then this smoothness condition is equivalent to X^{-1} not being a multiple zero of $(4p + q^2)/X^{2r}$, which in turn is equivalent to $\deg(4p + q^2) \geq 2r - 1$. Altogether we obtain that $2r - 1 \leq \max(\deg(p), 2\deg(q)) \leq 2r$ in the case of characteristic $\neq 2$. If k has characteristic 2, then the smoothness of V' at the points of $V(X^{-1})$ is equivalent to the condition that $q_-(0) \neq 0$ or $p_-(0)q'_-(0) \neq p'_-(0)^2$. Therefore, 0 is not a zero of q_- (so $\deg(q) \geq r$) or 0 is not a multiple zero of p_- (so $\deg(p) \geq 2r - 1$). Thus we again obtain that $2r - 1 \leq \max(\deg(p), 2\deg(q)) \leq 2r$.

If k has characteristic $\neq 2$, we can use a coordinate change of $\mathbb{A}^2_k$ to complete the square, and thus may arrange $q = 0$.

At this point we have almost finished the proof of part (1). It only remains to show that $r = g + 1$. We will see this when we analyze the ramification divisor of f in the course of proving part (2).

Part (2). To compute the ramification divisor of $f_{|U}$, we compute the $\mathscr{O}_{U'}$-module $\Omega_{U'/\mathbb{A}^1_k}$. Write $f = Y^2 + q(X)Y - p(X)$. Then according to Remark 17.35, $\Omega_{U'/\mathbb{A}^1_k}$ is associated with the $\Gamma(U', \mathscr{O}_{U'})$-module

$$k[X,Y]dY/(fdY,df) \cong k[X,Y]dY/(fdY,(2Y+q)dY) \cong \Gamma(U',\mathscr{O}_{U'})/(2Y+q).$$

If char $k \neq 2$, then this is isomorphic to $k[X]/(4p+q^2)$. If char $k = 2$, then it is isomorphic to $\Gamma(U',\mathscr{O}_{U'})/(q) \cong \Gamma(U',\mathscr{O}_{U'}) \otimes_k k[X]/(q)$. In either case, the ramification divisor of $f_{|U}$ has degree $\deg(4p + q^2)$.

By the analogous computation for V' in place of U', we see that the component of the ramification divisor supported on $C \setminus U = V(X^{-1}) \subset V'$ is (as a closed subscheme of V) given by

$$\left(k[X^{-1},Z]/(2Z+q_-,Z^2+q_-Z-p_-)\right)_{(X^{-1})},$$

where we again write $q_- = q/X^r, p_- = p/X^{2r} \in k[X^{-1}]$.

If the characteristic of k is $\neq 2$, then this is isomorphic to $(k[X^{-1}]/(4p_-+q_-^2))_{(X^{-1})}$, and since the smoothness of V' implies that $4p_-+q_-^2$ is separable, it is zero- or one-dimensional over k, depending on whether $(4p_-+q_-^2)(0) \neq 0$, or equivalently, is zero if $\deg(4p+q^2) < 2r$ (and thus $\deg(4p+q^2) = 2r-1$) and is one-dimensional, if $\deg(4p+q^2) = 2r$. In either case, the degree of the ramification divisor for f is $2r$.

If the characteristic of k is 2, the situation is different in the sense that the ramification divisor of $f_{|U}$ may be trivial and the ramification of f may be concentrated at the single point $V(X^{-1}) \subset V$. See Exercise 26.18. To finish the computation in this case, write m_q and m_p for the multiplicity of $X^{-1} = 0$ as a zero of q_- and p_- (viewed as polynomials in $k[X^{-1}]$), respectively. In other words, $m_q = r - \deg(q)$, $m_p = 2r - \deg(p)$. Then the component of the ramification divisor supported at $V(X^{-1})$ is given by

$$\left(k[X^{-1},Z]/(2Z+q_-,Z^2+q_-Z-p_-)\right)_{(X^{-1})} \cong (k[X^{-1},Z]/(X^{-m_q},Z^2-uX^{-m_p}))_{(X^{-1})}$$

for some unit $u \in k[X^{-1}]^\times_{(X^{-1})}$. Since $\max(\deg(p), 2\deg(q)) \geq 2r-1$ (by the smoothness of V, see above), m_p and m_q cannot both be > 1. If $m_q = 0$, then the above ring is 0, so f is unramified at the point(s) of $V(X^{-1})$. Now let $m_q > 0$. If $m_p = 0$, then the k-dimension of the above local ring is $2m_q$ (and its residue class field is k or an inseparable quadratic extension field of k, depending on whether the constant term of u is a square in k, or not). If $m_q > 0$ and $m_p = 1$, then the above local ring is isomorphic to $k[Z]/(Z^{2m_q})$ and hence again has dimension $2m_q$ over k. So the outcome is the same in all cases, and putting things together, we see that the ramification divisor of f has degree $2\deg(q) + 2m_q = 2r$ in the characteristic 2 case, too.

This in particular proves the description of the support of the ramification divisor stated in part (2). Furthermore, by the Riemann-Hurwitz formula (Corollary 26.67), the ramification divisor R of f has degree $2g + 2$. Therefore we have also proved that $r = g + 1$. $\qquad\square$

The proof can be simplified, if there exists a point $0 \in C(k)$ with $\ell(2[0]) = 2$. If k is algebraically closed, then such a point always exists, but not in general (Example 26.126). See Exercise 26.17.

Remark 26.123.
(1) In the situation of the proposition, since $g(C) > 1$, the schematic closure in $\mathbb{P}^2_k$ of U' (as in (1)) is not smooth. In fact, smoothness would contradict the genus-degree formula Proposition 26.79.

(2) To remedy the situation and produce an embedding of a curve C as in the proposition into projective space, we can proceed as follows. We first recall the notion of weighted projective space (see also Exercise 13.1), or rather the special case that we are going to use. Consider the polynomial ring $A := k[T_0, T_1, Y]$ with the grading that attaches degree 1 to T_0 and to T_1, and degree $g + 1$ to Y. We set $\mathbb{P}_k(1, 1, g+1) :=$ $\mathrm{Proj}(k[T_0, T_1, Y])$.

According to the way the Proj construction works, $\mathbb{P}_k(1, 1, g+1)$ is covered by the open subschemes $D_+(T_0)$, $D_+(T_1)$ and $D_+(Y)$ which are the spectra of the homogeneous localizations $A_{(T_0)}$, $A_{(T_1)}$ and $A_{(Y)}$. Now $A_{(T_0)} = k\left[\frac{T_1}{T_0}, \frac{Y}{T_0^{g+1}}\right]$ is a polynomial ring in two variables, and the same holds for $A_{(T_1)}$. So the corresponding open charts of $\mathbb{P}_k(1, 1, g+1)$ are isomorphic to $\mathbb{A}_k^2$. They cover all of $\mathbb{P}_k(1, 1, g+1)$ except for one point. On the other hand, $A_{(Y)} = k\left[\frac{T_0^{g+1}}{Y}, \frac{T_0^g T_1}{Y}, \dots, \frac{T_1^{g+1}}{Y}\right]$, and the local ring at the "origin", i.e., at the prime ideal $\left(\frac{T_0^{g+1}}{Y}, \frac{T_0^g T_1}{Y}, \dots, \frac{T_1^{g+1}}{Y}\right)$ is not regular. (If the characteristic of k does not divide $g+1$ and k contains all $(g+1)$-th roots of unity, then the affine scheme $\mathrm{Spec}\, A_{(Y)}$ is the quotient of $\mathbb{A}_k^2$ under the action of the cyclic group $\mu_{g+1}(k)$ of $(g+1)$-th roots of unity given by $\zeta \cdot (t_0, t_1) = (\zeta t_0, \zeta t_1)$, in the sense of Proposition (12.27).)

There is a natural morphism $\mathbb{A}_k^3 \setminus \{0\} \to \mathbb{P}_k(1, 1, g+1)$ induced by the inclusions of the homogeneous localizations in the usual localizations. Correspondingly, we obtain a description of the k-valued points $\mathbb{P}_k(1, 1, g+1)(k)$ in terms of homogeneous coordinates, i.e., as equivalence classes $(t_0 : t_1 : y)$, not all entries $= 0$, where $(t_0 : t_1 : y) = (t_0' : t_1' : y')$ if and only if there exists $\lambda \in k^\times$ with $t_0' = \lambda t_0$, $t_1' = \lambda t_1$, $y' = \lambda^{g+1} y$. (And of course similarly for extension fields of k.)

We have a closed immersion $\mathbb{P}_k(1, 1, g+1) \hookrightarrow \mathbb{P}_k^{g+1}$ which in terms of homogeneous coordinates is given by $(t_0 : t_1 : y) \mapsto (t_0^{g+1} : t_0^g t_1 : \dots : t_1^{g+1} : y)$. In terms of the Proj construction we can describe this embedding as follows. As in Section (13.1), we write $A^{(g+1)} := \bigoplus_i A_{i(g+1)}$. Then $\mathbb{P}(1, 1, g+1) = \mathrm{Proj}\, A^{(g+1)}$ (Remark 13.7) and we have a surjective k-algebra homomorphism $k[T_0, \dots, T_{g+1}, Y] \twoheadrightarrow A^{(g+1)}$, $T_i \mapsto T_0^{g+1-i} T_1^i$, $Y \mapsto Y$, which induces the above morphism.

(3) The way in which the affine curves U' and V' have to be glued in order to obtain the curve C translates precisely to the statement, that we can embed C in the weighted projective plane $\mathbb{P}_k(1, 1, g+1)$. Composing this embedding with the embedding $\mathbb{P}(1, 1, g+1) \subset \mathbb{P}_k^{g+2}$, we in particular obtain an embedding $C \hookrightarrow \mathbb{P}_k^{g+2}$ whose restriction to $U' \cong U \subset C$ is given by $(x, y) \mapsto (1 : x : \dots : x^{g+1} : y)$.

The image of C in $\mathbb{P}(1, 1, g+1)$ does not contain the unique singular point $(0 : 0 : 1)$ of $\mathbb{P}_k(1, 1, g+1)$, in other words, it is contained in the union of the two charts $D_+(T_0)$, $D_+(T_1)$.

Let us compute the fiber of f over $(0 : 1) \in \mathbb{P}_k^1$; set-theoretically, this is the complement $C \setminus U = V \setminus (V \cap U)$ of U in C. This fiber is the closed subscheme of V' given by the vanishing of X^{-1}. Writing $p = \sum_{i=0}^{2r} a_i X^i$, $q = \sum_{i=0}^r b_i X^i$, we obtain

$$f^{-1}((0 : 1)) \cong k[X^{-1}, Z]/(Z^2 + X^{-r} q(X) Z - X^{-2r} p(X), X^{-1})$$
$$\cong k[Z]/(Z^2 + b_r Z - a_{2r}).$$

This is a finite k-scheme of degree 2 over k. Depending on whether the quadratic equation $Z^2 + b_r Z - a_{2r} = 0$ has 0, 1 or 2 solutions in k, it has 0, 1 or 2 points with values in k. If $C \setminus U$ has no k-valued point, then it consists of one point, and the residue class field of that point is a quadratic extension of k. If the characteristic of k is $\neq 2$ and $q = 0$, there is precisely one solution if and only if p has degree $2g + 1$, and in this case this is a point where f is ramified.

Remark 26.124. Let us discuss when an affine equation as in Proposition 26.122 defines a hyperelliptic curve.

(1) Consider an affine plane curve

$$U = V(Y^2 + q(X)Y - p(X)) \quad \subset \mathbb{A}_k^2 = \operatorname{Spec} k[X, Y] \quad \text{with } p, q \in k[X].$$

The curve U is smooth if and only if

- characteristic $\neq 2$: $4p + q^2$ is separable,

- characteristic $= 2$: q and $(q')^2 p + (p')^2$ are coprime.

Now assume that U is smooth. Let $g \in \mathbb{N}$ such that $\deg(p) \in \{2g+1, 2g+2\}$ and assume that $g > 1$ and $\deg(q) \leq g + 1$. Let $V = V(Z^2 + \frac{q}{X^{g+1}}Z - \frac{p}{X^{2(g+1)}}) \subset \operatorname{Spec} k[X^{-1}, Z]$. We can glue the affine curves U and V according to the identification $Y = X^{g+1}Z$ and obtain a projective curve C (which we can embed in $\mathbb{P}(1, 1, g+1)$). Similarly as for U, one obtains a criterion for the smoothness of V. In characteristic $\neq 2$, C is smooth if and only if $4p + q^2$ is separable of degree $\geq 2g + 1$.

If C is smooth, then it is a hyperelliptic curve of genus g. Indeed, the projection $(x, y) \mapsto x$ defines a morphism $U \to \mathbb{A}_k^1$ which extends to a morphism $f \colon C \to \mathbb{P}_k^1$. As in the proof of Proposition 26.122, one checks that the ramification divisor of f has degree $2g + 2$, so that C indeed has genus g. In particular, then C has genus > 0, thus f is generically étale of degree 2.

(2) If k is algebraically closed of characteristic $\neq 2$, then every hyperelliptic curve is isomorphic to the proper normal model of an affine curve of the form $V(y^2 - \prod_{i=1}^{2g+1}(X - a_i))$ for pairwise different $a_i \in k$, with $g = g(C)$, and conversely every such affine curve with $g > 1$ defines a hyperelliptic curve of genus g. In this case, there is a unique point "at infinity" which is ramified for the hyperelliptic cover. Alternatively, we can work with $2g + 2$ pairwise different elements $a_1, \ldots, a_{2g+2} \in k$ and consider the affine curve $V(y^2 - \prod_{i=1}^{2g+2}(X - a_i))$; then we obtain a hyperelliptic curve where the two points at infinity are unramified.

(3) In particular, we see that there exist hyperelliptic curves of any genus > 1.

Yet another characterization of hyperelliptic curves (over algebraically closed fields of characteristic $\neq 2$) is given by the following proposition.

Proposition 26.125. *Let k be an algebraically closed field of characteristic $\neq 2$, and let C be a connected smooth proper curve over k of genus $g > 1$. The following are equivalent.*

(a) *The curve C is hyperelliptic.*

(b) *If there exists an involution $\sigma \colon C \to C$ (i.e., an automorphism with $\sigma^2 = \operatorname{id}$) with precisely $2g + 2$ fix points.*

Proof. If C is hyperelliptic, let $f \colon C \to \mathbb{P}_k^1$ be the covering map and let σ be the hyperelliptic involution. The fix points of σ are precisely the points of C where f is ramified (cf. Proposition 12.21). Since $\deg(f) = 2$, the ramification index at each ramified points is equal to 2, so the Riemann-Hurwitz formula (in the tamely ramified case, Corollary 26.67) shows that there are $2g + 2$ such points.

Conversely, given an involution with $2g + 2$ fix points, the fix field of the corresponding field automorphism $K(C) \to K(C)$ is the function field of a connected smooth proper curve C', and the inclusion of function fields defines a morphism $C \to C'$ of degree 2, which is necessarily generically étale since we have excluded characteristic 2. Hence, as before, the fix points of σ are the points where this cover is ramified. The degree of the ramification divisor is $2g + 2$, and so $g(C') = 0$, once again using the formula of Riemann-Hurwitz. $\square$

Example 26.126. A hyperelliptic curve C over a field k which admits a generically étale morphism of degree 2 to the projective line need not have a k-valued point. For example, consider a smooth proper model of the affine curve given by the equation $y^2 = 2x^6 - 4$ over $\mathbb{Q}$.

Proposition 26.127. *Let k be a field. Let $f \in k[X, Y, Z]$ be a homogeneous polynomial of degree 4 such that $C := V_+(f) \subset \mathbb{P}_k^2$ is a smooth curve. Then C is not hyperelliptic. In particular, there exist curves over k of genus 3 that are not hyperelliptic.*

Proof. We have $\omega_C \cong \mathscr{O}_{\mathbb{P}_k^2}(1)_{|C}$ by Proposition 26.79, so ω_C is very ample. But this implies that C is not hyperelliptic by the discussion in the proof of Lemma 26.118. By the genus-degree formula, Proposition 26.79, C has genus 3.

For the final part, it remains to show that over every field there exists a degree 4 homogeneous polynomial f defining a smooth plane curve. If the characteristic of k is $\neq 2$, we can take $f = X^4 + Y^4 + Z^4$. In characteristic 2 we can take $f = X^3Y + Y^3Z + Z^4$. $\square$

Remark 26.128. There exist curves over k of any genus ≥ 3 that are not hyperelliptic. And in a sense (which we do not make precise here) for fixed $g \geq 3$ only a small part of all curves of genus g are hyperelliptic.

(26.21) Curves of genus > 2.

After discussing the classes of curves of genus 0, of genus 1 and of hyperelliptic curves, which include curves of genus 2, we now briefly look at non-hyperelliptic curves of genus > 2. In many respects, these behave differently.

Proposition 26.129. *Let k be a field and let C be a geometrically connected smooth proper curve over k of genus g.*
(1) *If $g \geq 1$, then the canonical bundle ω_C is generated by global sections and hence defines a morphism $C \to \mathbb{P}_k^{g-1}$.*
(2) *The following are equivalent:*
   (i) *The curve C has genus > 2 and is not hyperelliptic.*
   (ii) *The canonical bundle ω_C is very ample.*

Proof. All properties can be checked after base change to an extension field (see the proof of Proposition 26.59, Proposition 14.58, Lemma 26.120), so we may assume that k is algebraically closed. Let K be a fixed canonical divisor on C.

(1) By Proposition 26.58, it is enough to show that $\ell(K - [x]) = \ell(K) - 1 = g - 1$ for all closed points $x \in C$. From the Theorem of Riemann-Roch for the divisor $K - [x]$, we obtain

$$\ell(K - [x]) - \ell([x]) = 2g - 3 + 1 - g = g - 2,$$

and since $g \neq 0$, $\ell([x]) = 1$. In fact, there is no non-constant rational function on C with at most a single pole at x, because the existence of such a function would entail that $[x]$ is linear equivalent to some other divisor on C, necessarily also of degree 1. In view of Example 26.90, this would contradict that C has genus $\neq 0$.

(2), (i) $\Rightarrow$ (ii). By part (1) the sheaf ω_C is generated by its global sections, so by Proposition 26.58 it is enough to show that $\ell(K - [x] - [y]) = g - 2$ for all $x, y \in C(k)$. Applying the Theorem of Riemann-Roch to the divisor $K - [x] - [y]$, we find that

$$\ell(K - [x] - [y]) - \ell([x] + [y]) = 2g - 4 + 1 - g = g - 3.$$

Furthermore, we have $1 \leq \ell([x] + [y]) \leq 2$ (cf. Remark 26.55 (1)). If we had $\ell([x] + [y]) = 2$, then the line bundle $\mathcal{O}_C([x] + [y])$ would define a rational map which we could extend to a morphism $C \to \mathbb{P}^1_k$ of degree 2. Since $g(C) > 0$, this morphism cannot be purely inseparable, and thus C had to be hyperelliptic.

(ii) $\Rightarrow$ (i). If ω_C is ample, then it has positive degree, and hence $g \geq 2$. Since all curves of genus 2 are hyperelliptic by Proposition 26.121, it is enough to show that the canonical bundle of a hyperelliptic curve if not very ample. But we have seen this in the proof of Lemma 26.118 – in fact, the canonical bundle gives rise to a map $C \to \mathbb{P}^{g-1}_k$ which factors through the hyperelliptic cover and therefore is not injective. $\qquad \square$

Definition 26.130. *Let C be a geometrically connected smooth proper curve over a field k such that ω_C is very ample. We call the closed embedding $C \to \mathbb{P}^{g-1}_k$ that is defined by ω_C the* canonical embedding *of C.*

Vector bundles on curves

In this part of the chapter we study vector bundles on smooth proper geometrically connected curves. The main result (Theorem 26.156) is that every vector bundle $\mathcal{E}$ has a unique filtration, called the *Harder-Narasimhan filtration*,

$$\mathcal{E} = \mathcal{E}^0 \supsetneq \mathcal{E}^1 \supsetneq \cdots \supsetneq \mathcal{E}^m = 0$$

by subbundles such that $\mathcal{E}^{i-1}/\mathcal{E}^i$ is semistable of slope λ_i where $\lambda_1 < \lambda_2 < \cdots < \lambda_m$ are rational numbers, see Definition 26.148 for the definition of semistability and Definition 26.142 for the definition of the slope of a vector bundle. Moreover, as the second main result (Proposition 26.152) we show that the category of semistable vector bundles of a fixed slope is abelian and every object in that category has a Jordan-Hölder series whose successive quotients are so-called stable vector bundles (Definition 26.148).

As a warm-up we start by giving a new proof of the classification of vector bundles on $\mathbb{P}^1_k$ using cohomological methods. Then we state a version of the theorem of Riemann-Roch for vector bundles on arbitrary proper curves and prove it in the case that the curve is normal. We then introduce semistability and stability of vector bundles and investigate the category of semistable vector bundles of a fixed slope. Finally we prove the existence and uniqueness of the Harder-Narasimhan stratification.

(26.22) Vector bundles on $\mathbb{P}^1_k$.

Recall that a *vector bundle* $\mathscr{E}$ on a scheme X is a locally free $\mathscr{O}_X$-module of finite rank. A vector bundle of constant rank 1 is called a *line bundle*.

Definition 26.131. *An $\mathscr{O}_X$-submodule $\mathscr{F}$ of a vector bundle $\mathscr{E}$ is called a* subbundle *if $\mathscr{E}/\mathscr{F}$ is a vector bundle.*

Then $\mathscr{F}$ is a vector bundle and it is locally on X a direct summand of $\mathscr{E}$ (Proposition 8.10).

Now let k be a field. In Section (11.17) we have shown that every vector bundle on $\mathbb{P}^1_k$ is isomorphic to a sum of line bundles. There we used that every vector bundle can be glued from two copies of a free module on each of the two standard charts of $\mathbb{P}^1_k$ in order to reduce the problem to an explicit statement about matrices with entries in $k[T, T^{-1}]$. Here we will give another proof of this theorem. In fact, interpreting Ext^1 groups as Yoneda-Ext-groups (Proposition F.220), we will be able to solve the problem using cohomological methods.

We start by stating some facts, which we have essentially already seen, about coherent modules on Dedekind schemes.

Remark 26.132. Let C be a connected normal locally noetherian scheme of dimension 1. Such a scheme C is automatically integral and regular (Proposition 6.40) and we denote by $\eta \in C$ its generic point. As C is regular of dimension 1, all local rings $\mathscr{O}_{C,c}$ for $c \in C$ are principal ideal domains. We obtain the following assertions.

(1) A coherent $\mathscr{O}_C$-module is a vector bundle if and only if it is torsion free since finitely generated torsion free modules over a principal ideal domain are free.

(2) For a coherent $\mathscr{O}_C$-module $\mathscr{F}$ we denote by $\mathscr{F}_{\mathrm{tors}}$ its torsion submodule, i.e.,

$$\mathscr{F}_{\mathrm{tors}}(U) = \mathscr{F}(U)_{\mathrm{tors}} = \mathrm{Ker}(\mathscr{F}(U) \to \mathscr{F}_\eta)$$

for every $\emptyset \neq U \subseteq C$ open affine. Then $\mathscr{F}/\mathscr{F}_{\mathrm{tors}}$ is torsion free and hence a vector bundle.

(3) Assertion (1) implies that every coherent submodule $\mathscr{F}$ of a vector bundle $\mathscr{E}$ on C is again a vector bundle.

It is not necessarily a subbundle, but it is always contained in a subbundle $\overline{\mathscr{F}} \subseteq \mathscr{E}$ of the same rank, namely $\overline{\mathscr{F}} := \mathrm{Ker}(\mathscr{E} \to (\mathscr{E}/\mathscr{F})/(\mathscr{E}/\mathscr{F})_{\mathrm{tors}})$, which is called the *saturation of $\mathscr{F}$ in $\mathscr{E}$*.

(4) Let $\mathscr{E}$ be a vector bundle on C. Then $\mathscr{F} \mapsto \mathscr{F}_\eta$ yields a bijection between subbundles $\mathscr{F}$ of $\mathscr{E}$ and $K(C)$-subvector spaces of $\mathscr{E}_\eta$. The inverse map is given by attaching to $W \subseteq \mathscr{E}_\eta$ the subbundle $\mathscr{F}$ of $\mathscr{E}$ with $\mathscr{F}(U) = W \cap \mathscr{E}(U)$ for every open affine subset $U \subseteq C$.

(5) Let $\mathscr{E} \neq 0$ be a vector bundle on C. Then there exists an exact sequence of vector bundles

$$0 \longrightarrow \mathscr{E}' \longrightarrow \mathscr{E} \longrightarrow \mathscr{L} \longrightarrow 0,$$

where $\mathscr{L}$ has rank 1.

Indeed, line bundle quotients $\mathscr{E} \to \mathscr{L}$ are parametrized by the projective bundle $\mathbb{P}(\mathscr{E})$, i.e., we have to show that $\mathbb{P}(\mathscr{E})(C) \neq \emptyset$. Let $U \subseteq C$ be an open dense subset such that $\mathscr{E}_{|U} \cong \mathscr{O}_U^n$. Then there exists a section of $\mathbb{P}(\mathscr{E})$ defined over U. As $\mathbb{P}(\mathscr{E})$ is proper over C and C is normal and of dimension 1, we can extend this section to a section $C \to \mathbb{P}(\mathscr{E})$ by Proposition 15.5.

(6) Suppose that C admits an ample line bundle[2], e.g., if C is a curve over a field
(Theorem 26.16). For every coherent $\mathscr{O}_C$-module $\mathscr{F}$ there exists a short exact sequence

$$0 \longrightarrow \mathscr{E}^{-1} \longrightarrow \mathscr{E}^0 \longrightarrow \mathscr{F} \longrightarrow 0,$$

where $\mathscr{E}^{-1}$ and $\mathscr{E}^0$ are vector bundles. Indeed, by our assumption there exist a vector
bundle $\mathscr{E}^0$ and a surjection $u\colon \mathscr{E}^0 \to \mathscr{F}$ (Proposition 22.57). By (3), $\mathscr{E}^{-1} := \mathrm{Ker}(u)$
is a vector bundle.

Theorem 26.133. *Let k be a field. Let $\mathscr{E}$ be a locally free sheaf of finite rank r on $\mathbb{P}^1_k$.
Then there exist uniquely determined integers $d_1 \geq d_2 \geq \cdots \geq d_r$ such that*

$$\mathscr{E} \cong \bigoplus_{i=1}^{r} \mathscr{O}_{\mathbb{P}^1_k}(d_i).$$

Specializing to the case of locally free modules of rank 1 the theorem gives a new
proof of the isomorphism $\mathbb{Z} \cong \mathrm{Pic}(\mathbb{P}^1_k)$, $d \mapsto \mathscr{O}_{\mathbb{P}^1_k}(d)$, proved by elementary methods in
Example 11.45, see also Proposition 24.69 for a description of $\mathrm{Pic}(\mathbb{P}^n_S)$ for an arbitrary
base scheme S and for all $n \geq 1$. We cannot expect a generalization of Theorem 26.133 to
more general base schemes for vector bundles of rank > 1.

For the existence we start with the following lemma which gives us a way of finding a
candidate for the integer d_1.

Lemma 26.134. *Let k be a field, let C be a proper integral curve over k and let $\mathscr{L}$ be a
very ample line bundle on C. Let $\mathscr{E} \neq 0$ be a locally free sheaf of finite rank n on C. As
usual, we write $\mathscr{E}(d) := \mathscr{E} \otimes_{\mathscr{O}_C} \mathscr{L}^{\otimes d}$ for $d \in \mathbb{Z}$.*
*There exists an integer d_1 such that $H^0(C, \mathscr{E}(-d_1)) \neq 0$ and $H^0(C, \mathscr{E}(-d)) = 0$ for all
$d > d_1$.*

Proof. Since $\mathscr{L}$ is very ample, for d sufficiently negative, $\mathscr{E}(-d)$ is generated by its global
sections (Corollary 23.2). In particular we then have $H^0(C, \mathscr{E}(-d)) \neq 0$. It remains to
show that there exists an integer e such that $H^0(C, \mathscr{E}(-d)) = 0$ for all $d \geq e$.

For this we apply the remark at the beginning of the proof to the dual $\mathscr{E}^\vee$. Hence
we find an integer e such that for all $d \leq e + 1$ there exists an N and a surjection
$\mathscr{O}_C^N \twoheadrightarrow \mathscr{E}^\vee(-d)$ and hence an injection $\mathscr{E}(d) \hookrightarrow \mathscr{O}_C^N$. Thus for all $d \leq e$, there exists an N
such that $\mathscr{E}(d)$ injects into $\mathscr{O}_C^N(-1)$ which has no global sections (because $\deg(\mathscr{L}^\vee) < 0$
by Proposition 26.57). It follows that for $d \geq e$, $H^0(C, \mathscr{E}(-d)) = 0$.

On the other hand, since $H^0(C, \mathscr{L}) \neq 0$ ($\mathscr{L}$ being very ample, hence generated by its
global sections), $H^0(C, \mathscr{E}(-d)) = 0$ implies $H^0(C, \mathscr{E}(-d-1)) = 0$. Altogether we obtain
the desired result. $\qquad\square$

Proof. (of Theorem 26.133) Consider d_1 as in Lemma 26.134 (for $C = \mathbb{P}^1_k$ and $\mathscr{L} = \mathscr{O}_{\mathbb{P}^1}(1)$) and let $s \neq 0$ be an element of $H^0(\mathbb{P}^1_k, \mathscr{E}(-d_1))$. We can view s as a non-zero
homomorphism $\mathscr{O}_{\mathbb{P}^1_k}(d_1) \to \mathscr{E}$. The homomorphism $\mathscr{O}_{\mathbb{P}^1_k}(d_1) \to \mathscr{E}$ is injective (this is true
for the stalks at the generic point, and the sections over any non-empty open embed into
this stalk). Denoting by $\mathscr{F}$ the cokernel, we obtain a short exact sequence

(*) $$0 \to \mathscr{O}_{\mathbb{P}^1_k}(d_1) \to \mathscr{E} \to \mathscr{F} \to 0.$$

---

[2] This hypothesis is in fact superfluous, see Exercise 23.21.

We claim that $\mathscr{F}$ is locally free (necessarily of rank $\mathrm{rk}(\mathscr{E}) - 1$). To show this, by Proposition 8.10, it is enough to show that for every point $x \in \mathbb{P}^1_k$ the induced sequence of the fibers is also exact, or in other words that the global section of $\mathscr{E}(-d_1)$ giving rise to the sequence never vanishes. But if this section had a zero at a (closed) point x, then the image of the corresponding map $\mathscr{O}_{\mathbb{P}^1_k} \to \mathscr{E}(-d_1)$ would be contained in $\mathscr{E}(-d_1) \otimes \mathscr{J} \subset \mathscr{E}(-d_1) \otimes \mathscr{O}_{\mathbb{P}^1_k}$, where $\mathscr{J}$ denotes the ideal sheaf defining the closed subscheme $\{x\}$. Since $\mathscr{J} \cong \mathscr{O}_{\mathbb{P}^1_k}(-m)$ with $m = [\kappa(x) : k]$, we would obtain a non-zero global section of $\mathscr{E}(-d_1 - m)$, a contradiction to the definition of d_1.

By induction hypothesis, there is an isomorphism $\mathscr{F} \cong \bigoplus_{i=2}^{r} \mathscr{O}_{\mathbb{P}^1_k}(d_i)$ for integers $d_2 \geq \cdots \geq d_r$. Since $\mathscr{E}(-d_1 - 1)$ has no global sections, the same must hold for $\mathscr{F}(-d_1 - 1)$. In fact, tensoring the above short exact sequence (*) with $\mathscr{O}_{\mathbb{P}^1_k}(-d_1 - 1)$, the long exact cohomology sequence shows that

$$0 = H^0(\mathbb{P}^1_k, \mathscr{E}(-d_1 - 1)) \cong H^0(\mathbb{P}^1_k, \mathscr{F}(-d_1 - 1)) \cong \bigoplus_{i=2}^{r} H^0(\mathbb{P}^1_k, \mathscr{O}_{\mathbb{P}^1_k}(d_i - d_1 - 1))$$

Here we use that $H^0(\mathbb{P}^1_k, \mathscr{O}_{\mathbb{P}^1_k}(-1)) = H^1(\mathbb{P}^1_k, \mathscr{O}_{\mathbb{P}^1_k}(-1)) = 0$ by Theorem 22.22. We conclude that $d_1 \geq d_2 \geq \cdots \geq d_r$ since a line bundle $\mathscr{O}_{\mathbb{P}^1_k}(d_i - d_1 - 1)$ has no non-zero global sections if and only if $d_i - d_1 - 1 < 0$, i.e., $d_i \leq d_1$. But then

$$\mathrm{Ext}^1_{\mathbb{P}^1_k}(\mathscr{F}, \mathscr{O}_{\mathbb{P}^1_k}(d_1)) \cong \bigoplus_i \mathrm{Ext}^1_{\mathbb{P}^1}(\mathscr{O}_{\mathbb{P}^1}, \mathscr{O}_{\mathbb{P}^1_k}(d_1 - e_i)) \cong \bigoplus_i H^1(\mathbb{P}^1_k, \mathscr{O}_{\mathbb{P}^1_k}(d_1 - e_i)) = 0.$$

This proves that the short exact sequence (*) splits, and we obtain the desired isomorphism between $\mathscr{E}$ and a direct sum of line bundles.

Let us show the uniqueness of the decomposition. By the existence part of the theorem we have a decomposition

(*) $$\mathscr{E} \cong \bigoplus_{d \in \mathbb{Z}} V_d \otimes_k \mathscr{O}(d)$$

for finite-dimensional k-vector spaces V_d of which all but finitely many are zero. We have to show that $\mathscr{E}$ determines $\dim(V_d)$. For every $\alpha \in \mathbb{Z}$ let $\mathscr{E}^\alpha$ be the image of the evaluation map $H^0(\mathbb{P}^1_k, \mathscr{E}(-\alpha)) \otimes_k \mathscr{O}(\alpha) \to \mathscr{E}$. We obtain a decreasing filtration

(26.22.1) $$\cdots \supseteq \mathscr{E}^{-1} \supseteq \mathscr{E}^0 \supseteq \mathscr{E}^1 \supseteq \cdots$$

of $\mathscr{E}$ with $\mathscr{E}^\alpha = \mathscr{E}$ for $\alpha \ll 0$ since $\mathscr{E}(-\alpha)$ is globally generated for small α and with $\mathscr{E}^\alpha = 0$ for $\alpha \gg 0$ since $H^0(\mathbb{P}^1_k, \mathscr{E}(-\alpha)) = 0$ for large α by Lemma 26.134. For $\mathscr{E}$ as in (*) we have $\mathscr{E}^\alpha \cong \bigoplus_{d \geq \alpha} V_d \otimes_k \mathscr{O}(d)$ which shows $\dim(V_d) = \mathrm{rk}(\mathscr{E}^d / \mathscr{E}^{d+1})$. $\qquad \square$

The proof shows that every vector bundle $\mathscr{E}$ on $\mathbb{P}^1_k$ has a unique filtration (26.22.1) whose d-th graded step $\mathscr{E}^d / \mathscr{E}^{d+1}$ is (non-canonically) isomorphic to a direct sum of copies of the line bundle $\mathscr{O}(d)$. Moreover, the filtration is split, though the splitting homomorphisms are not unique. This filtration is called the Harder-Narasimhan filtration. When formulated in a suitable way, this statement can be generalized to arbitrary smooth proper geometrically connected curves as we will see in Section (26.25) below.

(26.23) The Riemann-Roch theorem for vector bundles on curves.

Let X be a scheme and let $\mathscr{E}$ be a finite locally free $\mathscr{O}_X$-module. Recall that $\det(\mathscr{E}) := \bigwedge^{\mathrm{rk}(\mathscr{E})}(\mathscr{E})$ denotes its determinant line bundle (Remark 17.58).

Definition 26.135. *Let C be a proper curve over a field k and let $\mathscr{E}$ be a vector bundle on C. Then* $\deg(\mathscr{E}) := \deg(\det(\mathscr{E}))$ *is called the* degree *of $\mathscr{E}$.*

We then have $\deg(\mathscr{E}) = \chi(C, \det(\mathscr{E})) + g - 1$ by Proposition 26.46, where g denotes the genus of C. As with the Euler characteristic, the degree depends on k and we also write $\deg_k(\mathscr{E})$ instead of $\deg(\mathscr{E})$.

Remark 26.136. Let C be a proper curve over a field k and let $\mathscr{E}$ be a vector bundle on C.

(1) Let k' be a field extension of k and let $\mathscr{E}'$ be the pullback of $\mathscr{E}$ to $C \otimes_k k'$. Then $\deg_{k'}(\mathscr{E}') = \deg_k \mathscr{E}$ since the degree of line bundles is preserved under passing to field extensions (Remark 15.29) and since forming exterior powers is compatible with pullback (Proposition 7.49).

(2) Let $\mathscr{L}$ be a line bundle and suppose that $\mathscr{E}$ has constant rank r. Then $\det(\mathscr{E} \otimes_{\mathscr{O}_C} \mathscr{L}) = \det(\mathscr{E}) \otimes \mathscr{L}^{\otimes r}$. Hence we have

(26.23.1) $$\deg(\mathscr{E} \otimes \mathscr{L}) = \deg(\mathscr{E}) + r \deg(\mathscr{L}).$$

See also Exercise 26.21 for a generalization.

(3) Since $\det(\mathscr{E}^\vee) = \det(\mathscr{E})^\vee$ and $\deg(\mathscr{L}^\vee) = -\deg(\mathscr{L})$ for every line bundle, we find that

(26.23.2) $$\deg(\mathscr{E}^\vee) = -\deg(\mathscr{E}).$$

Proposition 26.137. *Let k be a field, let C' and C be proper integral curves over k and let $f\colon C' \to C$ be a finite dominant morphism. Let $\mathscr{E}$ be a vector bundle on C. Then*

$$\deg(f^*\mathscr{E}) = \deg(f)\deg(\mathscr{E})$$

Proof. As forming exterior powers commutes with pullback f^* it suffices to show the equality for line bundles. This is a special case of Proposition 23.84. $\qquad\square$

Next, we prove a version of the Riemann-Roch theorem for vector bundles. It is in fact a generalization of the equality $\chi(\mathscr{L}) = \deg(\mathscr{L}) + 1 - g$ for line bundles $\mathscr{L}$ on a curve (Proposition 26.46).

Theorem 26.138. *Let C be a proper curve over a field k of genus g and let $\mathscr{E}$ be a vector bundle on C of constant rank. Then*

(26.23.3) $$\chi(C, \mathscr{E}) = \deg(\mathscr{E}) + \mathrm{rk}(\mathscr{E})(1 - g).$$

We will give the proof only if C is normal, which is the only case that we will use in the sequel. For the general case see [Sta] ODJ5.

Proof. (if C is normal) We proceed by induction on $n := \mathrm{rk}(\mathscr{E})$. As remarked above, we know the equality already for vector bundles of rank 1. As we assumed that C is normal, we have seen in Remark 26.132 (5) that there exists a short exact sequence $0 \to \mathscr{F} \to \mathscr{E} \to \mathscr{L} \to 0$ with $\mathscr{L}$ a line bundle and $\mathscr{F}$ a vector bundle of rank $\mathrm{rk}(\mathscr{E}) - 1$. By induction we know the description of the Euler characteristic already for $\mathscr{L}$ and $\mathscr{F}$, we obtain it for $\mathscr{E}$, since the rank, the degree, and the Euler characteristic of vector bundles on C are additive in short exact sequences of vector bundles (for the degree one uses (17.10.5)). $\qquad\square$

Now we can use Serre duality as in Theorem 26.48 to express $H^1(C,\mathscr{E})$ as global sections of another vector bundle.

Corollary 26.139. *Let k be a field and let C be a proper curve over k of genus g. Let ω_C be the dualizing sheaf. Then for every vector bundle $\mathscr{E}$ of constant rank one has*

$$\dim_k H^0(C,\mathscr{E}) - \dim_k H^0(C,\mathscr{E}^\vee \otimes \omega_C) = \deg(\mathscr{E}) + \mathrm{rk}(\mathscr{E})(1-g).$$

Remark 26.140. Let C be a proper curve of genus g over a field k. Suppose that C is connected to simplify the exposition (for non-connected curves one should regard the rank as a locally constant function on C as usual, and redefine the degree as a locally constant function, by considering each connected component individually).
(1) The rank and the degree of vector bundles on C are additive in short exact sequences of vector bundles (for the degree one uses (17.10.5)). Hence by Remark 23.51 we obtain induced maps
$$\mathrm{rk}, \deg, \chi(C,-)\colon K_0(C) \longrightarrow \mathbb{Z}.$$
and the Riemann-Roch formula (26.23.3) holds for every perfect complex $\mathscr{E}$ on C.
(2) Now suppose that C is in addition normal and hence regular. Then every complex in $D^b_{\mathrm{coh}}(C)$ is perfect and we have $K_0(C) = K_0(D^b_{\mathrm{coh}}(C))$ by Proposition 23.55. Hence we can define $\mathrm{rk}(\mathscr{E})$ and $\deg(\mathscr{E})$ for every $\mathscr{E}$ in $D^b_{\mathrm{coh}}(C)$ and obtain (26.23.3) for every $\mathscr{E} \in D^b_{\mathrm{coh}}(C)$, in particular for every coherent $\mathscr{O}_C$-module. These can be described as follows.

As C is normal and connected, it is integral. Denote by η the generic point of C. Let $\mathscr{F}$ be a coherent $\mathscr{O}_C$-module, then it follows from Remark 26.132 (6) that the rank of $\mathscr{F}$ is given by
$$\mathrm{rk}(\mathscr{F}) = \dim_{K(C)}(\mathscr{F}_\eta),$$
where $K(C) = \mathscr{O}_{C,\eta}$ is the function field of C. If $\mathscr{F}$ is an arbitrary object in $D^b_{\mathrm{coh}}(C)$, we have $\mathrm{rk}(\mathscr{F}) = \sum_{i\in\mathbb{Z}}(-1)^i\,\mathrm{rk}(H^i(\mathscr{F}))$ and therefore for every point $c \in C$

$$(26.23.4) \qquad \mathrm{rk}(\mathscr{F}) = \chi(\mathscr{F}\otimes_{\mathscr{O}_C} K(C)) = \chi(\mathscr{F}\otimes^L_{\mathscr{O}_C}\kappa(c)),$$

where the second equality holds since the Euler characteristic is constant by Proposition 23.117 (3). The degree is given by Riemann-Roch

$$(26.23.5) \qquad \deg(\mathscr{F}) = \chi(C,\mathscr{F}) - \mathrm{rk}(\mathscr{F})(1-g), \qquad \mathscr{F}\in D^b_{\mathrm{coh}}(C).$$

Example 26.141. Let $\mathscr{F}$ be a coherent $\mathscr{O}_C$-module with finite support or, equivalently, of rank 0. By Riemann-Roch we have

$$\deg(\mathscr{F}) = \dim H^0(C,\mathscr{F}) \geq 0.$$

Moreover, one has $\mathscr{F} \neq 0$ if and only if $\deg(\mathscr{F}) > 0$ since $\mathscr{F}$ is supported on an affine subscheme of C. See also Exercise 26.20 for a more concrete description of $\deg(\mathscr{F})$.
As $\deg$ is additive in exact sequences we obtain the following assertions.
(1) Let $\mathscr{E} \neq 0$ be a coherent $\mathscr{O}_C$-module and let $\mathscr{E}' \subsetneq \mathscr{E}$ be a coherent submodule of the same rank. Then $\mathscr{E}/\mathscr{E}'$ has rank 0 and therefore $\deg(\mathscr{E}') < \deg(\mathscr{E})$.
(2) Let $\mathscr{E} \neq 0$ be a coherent $\mathscr{O}_C$-module and let $\mathscr{E}''$ be a nontrivial coherent quotient module of the same rank. Then $\mathrm{Ker}(\mathscr{E}\to\mathscr{E}'')$ has rank 0 and it follows that $\deg(\mathscr{E}'') < \deg(\mathscr{E})$.

(26.24) Semistable vector bundles.

In this section C will denote a smooth proper geometrically connected curve over a field k. We denote by g its genus.

In Theorem 26.133 we have seen that every vector bundle on $\mathbb{P}_k^1$ is the direct sum of line bundles. This result does not generalize to curves of higher genus. But every vector bundle still has a canonical filtration whose graded pieces are of a "simple" form (semistable bundles in the sense of Definition 26.148). This filtration is called the Harder-Narasimhan stratification, see Theorem 26.156 below.

The following definition will be central for the construction of the Harder-Narasimhan stratification.

Definition 26.142. *Let $\mathscr{E} \neq 0$ be a vector bundle on C. The rational number*

$$\mu(\mathscr{E}) := \deg(\mathscr{E}) / \operatorname{rk}(\mathscr{E})$$

is called the slope *of $\mathscr{E}$. We also define the zero bundle to have every rational number as a slope, i.e., $\mu(0) = \lambda$ holds for every rational number λ.*

By Remark 26.140 (2), the slope $\mu(\mathscr{E})$ is defined for all $\mathscr{E}$ in $D_{\operatorname{coh}}^b(C)$ with $\operatorname{rk}(\mathscr{E}) \neq 0$.

Remark 26.143. Let $\mathscr{E} \neq 0$ be a vector bundle on C.
(1) Let $\mathscr{L}$ be a line bundle. Then (26.23.1) shows that

(26.24.1) $\mu(\mathscr{E} \otimes_{\mathscr{O}_C} \mathscr{L}) = \mu(\mathscr{E}) + \deg(\mathscr{L})$.

(2) By (26.23.2) we have $\mu(\mathscr{E}^\vee) = -\mu(\mathscr{E})$.

Example 26.144. Let $C = \mathbb{P}_k^1$. For $d \in \mathbb{Z}$ one has $\deg(\mathscr{O}_C(d)) = d$ and therefore $\mu(\mathscr{O}(d)) = d$. More generally, if $\mathscr{E} \cong \bigoplus_{i=1}^r \mathscr{O}_C(d_i)$ is a vector bundle of rank r over $\mathbb{P}_k^1$ (Theorem 26.133), then $\deg(\mathscr{E}) = d_1 + \ldots d_r$ and $\mu(\mathscr{E}) = (d_1 + \cdots + d_r)/r$.

Let us study how the slope behaves in short exact sequences.

Remark 26.145. Let $0 \longrightarrow \mathscr{F}' \longrightarrow \mathscr{F} \longrightarrow \mathscr{F}'' \longrightarrow 0$ be an exact sequence of coherent $\mathscr{O}_C$-modules.
(1) If $\mathscr{F}'' \neq 0$ and $\operatorname{rk}(\mathscr{F}'') = 0$, then $\operatorname{rk}(\mathscr{F}') = \operatorname{rk}(\mathscr{F})$ and Example 26.141 (1) shows that $\mu(\mathscr{F}') < \mu(\mathscr{F})$ if $\operatorname{rk}(\mathscr{F}) \neq 0$.
(2) If $\mathscr{F}' \neq 0$ and $\operatorname{rk}(\mathscr{F}') = 0$, then $\operatorname{rk}(\mathscr{F}'') = \operatorname{rk}(\mathscr{F})$ and Example 26.141 (2) shows that $\mu(\mathscr{F}'') < \mu(\mathscr{F})$ if $\operatorname{rk}(\mathscr{F}) \neq 0$.

For coherent $\mathscr{O}_C$-modules of rank > 0 one has the following result.

Lemma 26.146. *Let $0 \to \mathscr{E}' \to \mathscr{E} \to \mathscr{E}'' \to 0$ be an exact sequence of coherent $\mathscr{O}_C$-modules of rank > 0. Then always one of the following cases occurs.*

$$\mu(\mathscr{E}') = \mu(\mathscr{E}) = \mu(\mathscr{E}''), \qquad or$$
$$\mu(\mathscr{E}') < \mu(\mathscr{E}) < \mu(\mathscr{E}''), \qquad or$$
$$\mu(\mathscr{E}') > \mu(\mathscr{E}) > \mu(\mathscr{E}'').$$

Proof. This is completely elementary. Set $d' := \deg(\mathscr{E}')$, $r' := \operatorname{rk}(\mathscr{E}')$, $d'' := \deg(\mathscr{E}'')$, and $r'' := \operatorname{rk}(\mathscr{E}'')$. Then $\operatorname{rk}(\mathscr{E}) = r' + r''$ and $\deg(\mathscr{E}) = d' + d''$. If $\mu(\mathscr{E}') = \mu(\mathscr{E}'')$, then $d' = d''r'/r''$ and hence

(*) $$\mu(\mathscr{E}) = \frac{d''r'/r'' + d''}{r' + r''} = \frac{d''(r' + r'')/r''}{r' + r''} = \frac{d''}{r''} = \mu(\mathscr{E}'').$$

If $\mu(\mathscr{E}') < \mu(\mathscr{E}'')$, then $d' < d''r'/r''$ and we then in (*) instead of the first equality we have a $<$ which shows $\mu(\mathscr{E}) < \mu(\mathscr{E}'')$. Similarly one obtains $\mu(\mathscr{E}) > \mu(\mathscr{E}')$ in this case, using that $d'' > d'r''/r'$.

The argument that $\mu(\mathscr{E}') > \mu(\mathscr{E}'')$ implies $\mu(\mathscr{E}') > \mu(\mathscr{E}) > \mu(\mathscr{E}'')$ is then the same with opposite inequalities.   □

Remark 26.147. By the Riemann-Roch theorem (in the form of Theorem 26.138) we can describe the slope of a non-zero vector bundle $\mathscr{E}$ on C also by

(26.24.2) $$\mu(\mathscr{E}) = \frac{\chi(C, \mathscr{E})}{\mathrm{rk}(\mathscr{E})} + g - 1 \leq \dim_k H^0(C, \mathscr{E}) + g - 1.$$

Hence if $0 \neq \mathscr{F}$ is any coherent $\mathscr{O}_C$-submodule (and hence $\mathscr{F}$ is a vector bundle of rank > 0 by Remark 26.132 (3)), then we have

$$\mu(\mathscr{F}) \leq \dim_k H^0(C, \mathscr{F}) + g - 1 \leq \dim_k H^0(C, \mathscr{E}) + g - 1,$$

and we see that the slope of submodules of $\mathscr{E}$ is bounded.

If the slopes of all submodules are bounded by the slope of $\mathscr{E}$, then we call $\mathscr{E}$ semistable:

Definition and Proposition 26.148. *Let $\mathscr{E}$ be a vector bundle on C. Then $\mathscr{E}$ is called semistable (resp. stable) if the following equivalent conditions are satisfied (resp. if the following equivalent conditions are satisfied and $\mathscr{E} \neq 0$).*
(i) For all subbundles $0 \neq \mathscr{E}' \subsetneq \mathscr{E}$ one has $\mu(\mathscr{E}') \leq \mu(\mathscr{E})$ (resp. $\mu(\mathscr{E}') < \mu(\mathscr{E})$).
(ii) For all coherent submodules $0 \neq \mathscr{E}' \subsetneq \mathscr{E}$ one has $\mu(\mathscr{E}') \leq \mu(\mathscr{E})$ (resp. $\mu(\mathscr{E}') < \mu(\mathscr{E})$).
(iii) For any surjective homomorphism $\mathscr{E} \to \mathscr{E}''$ of vector bundles with $\mathscr{E}'' \neq 0$ one has
  $\mu(\mathscr{E}'') \geq \mu(\mathscr{E})$ (resp. $\mu(\mathscr{E}'') > \mu(\mathscr{E})$).
(iv) For any surjective homomorphism $\mathscr{E} \to \mathscr{E}''$ of coherent $\mathscr{O}_C$-modules with $\mathrm{rk}(\mathscr{E}'') > 0$
  one has $\mu(\mathscr{E}'') \geq \mu(\mathscr{E})$ (resp. $\mu(\mathscr{E}'') > \mu(\mathscr{E})$).

The zero bundle is semistable but not stable according to our definition.

Proof. We may assume that $\mathscr{E} \neq 0$. We show the assertion for semistability. The proof for stability is the same with strict inequalities instead. The implications of "(i) $\Leftrightarrow$ (iii)" and "(ii) $\Rightarrow$ (iv)" follow from Lemma 26.146 and it is clear that (ii) implies (i) and that (iv) implies (iii).

It remains to show that (i) implies (ii). Let $0 \neq \mathscr{E}' \subseteq \mathscr{E}$ be a coherent submodule that is not a subbundle and let $\overline{\mathscr{E}'}$ be its saturation (Remark 26.132 (3)). Then $\overline{\mathscr{E}'}$ is a subbundle of $\mathscr{E}$ containing $\mathscr{E}'$ with $\mathrm{rk}(\mathscr{E}') = \mathrm{rk}(\overline{\mathscr{E}'})$. Therefore $\mu(\mathscr{E}') < \mu(\overline{\mathscr{E}'}) \leq \mu(\mathscr{E})$ by Remark 26.145.   □

Remark 26.149. Let $\mathscr{L}$ be a line bundle on C.
(1) Then $\mathscr{L}$ is clearly stable since there exist no non-trivial subbundles of $\mathscr{L}$. Its slope is given by $\mu(\mathscr{L}) = \chi(C, \mathscr{L}) + g - 1$ by (26.24.2).
(2) Let $\mathscr{E}$ be a vector bundle on C. As tensoring by $\mathscr{L}$ induces a bijection between subbundles of $\mathscr{E}$ and of $\mathscr{E} \otimes \mathscr{L}$ and since $\mu(\mathscr{E}' \otimes \mathscr{L}) = \mu(\mathscr{E}') + \deg(\mathscr{L})$ for every such subbundle $\mathscr{E}'$ (26.24.1), we see that $\mathscr{E}$ is semistable (resp. stable) of slope λ if and only if $\mathscr{E} \otimes \mathscr{L}$ is semistable (resp. stable) of slope $\lambda + \deg(\mathscr{L})$.

(3) Let $\mathscr{E}$ be a semistable vector bundle. Then every direct summand $\mathscr{E}'$ of $\mathscr{E}$ can be viewed as a submodule or as quotient of $\mathscr{E}$. Therefore $\mathscr{E}'$ is again semistable of the same slope as $\mathscr{E}$.

Proposition 26.150. *Let $\mathscr{E}$ and $\mathscr{F}$ be semistable vector bundles on C with $\mu(\mathscr{E}) > \mu(\mathscr{F})$. Then $\mathrm{Hom}_{\mathscr{O}_C}(\mathscr{E}, \mathscr{F}) = 0$.*

Proof. Let $u\colon \mathscr{E} \to \mathscr{F}$ be a non-zero $\mathscr{O}_C$-linear map and let $\mathscr{G} \neq 0$ be its image. As a subsheaf of $\mathscr{F}$ this is a vector bundle. Then semistability of $\mathscr{E}$ and of $\mathscr{F}$ show

$$\mu(\mathscr{E}) \leq \mu(\mathscr{G}) \leq \mu(\mathscr{F}). \qquad \square$$

Lemma 26.151. *Let $\mathscr{E}$ be a semistable vector bundle of slope λ.*
(1) *Let $\mathscr{F} \neq 0$ be a coherent submodule of $\mathscr{E}$ of slope λ. Then $\mathscr{F}$ is semistable and a subbundle of $\mathscr{E}$.*
(2) *Let $\mathscr{G}$ be a coherent quotient of $\mathscr{E}$ of rank $\neq 0$ and of slope λ. Then $\mathscr{G}$ is a semistable vector bundle.*

Proof. We show (1). By Remark 26.132 (1), $\mathscr{F}$ is a vector bundle. By (ii) of the Definition 26.148 of semistability, it is clear that $\mathscr{F}$ is semistable. Let $\overline{\mathscr{F}}$ be the saturation of $\mathscr{F}$ in $\mathscr{E}$ (Remark 26.132 (3)). If $\mathscr{F} \neq \overline{\mathscr{F}}$, then $\mu(\mathscr{F}) < \mu(\overline{\mathscr{F}}) \leq \mu(\mathscr{E})$ by Remark 26.145 which contradicts that $\mu(\mathscr{F}) = \mu(\mathscr{E})$. Hence $\mathscr{F} = \overline{\mathscr{F}}$ is a subbundle.

Let us show (2). As $\mathscr{E}$ is semistable, the vector bundle quotient $\mathscr{G}/\mathscr{G}_{\mathrm{tors}}$ of $\mathscr{E}$ has slope $\geq \lambda$. On the other hand, Remark 26.145 shows that $\mu(\mathscr{G}/\mathscr{G}_{\mathrm{tors}}) \leq \mu(\mathscr{G}) = \lambda$ with equality if and only if $\mathscr{G}_{\mathrm{tors}} = 0$. Therefore $\mathscr{G}$ is a vector bundle. Moreover, every quotient vector bundle of $\mathscr{G}$ is also a quotient vector bundle of $\mathscr{E}$ and hence has slope $\geq \lambda$. Therefore $\mathscr{G}$ is semistable. $\qquad \square$

For the next statement, recall that we denote by $(\mathrm{Coh}(C))$ the abelian category of coherent $\mathscr{O}_C$-modules.

Proposition 26.152. *Fix $\lambda \in \mathbb{Q}$ and denote by $(\mathrm{Vect}_\lambda(C))$ the full subcategory of $(\mathrm{Coh}(C))$ whose objects are the semistable vector bundles of slope λ.*
(1) *Then $(\mathrm{Vect}_\lambda(C))$ is a plump subcategory (Definition F.43) of $(\mathrm{Coh}(C))$. In particular, it is an abelian category.*
(2) *Every object of $(\mathrm{Vect}_\lambda(C))$ has finite length (Definition F.41) and therefore admits a composition series whose graded pieces are simple objects of $(\mathrm{Vect}_\lambda(C))$. The simple objects of $(\mathrm{Vect}_\lambda(C))$ are the stable vector bundles of slope λ.*

Here, by a simple object in an abelian category we mean an object that has no nontrivial subobjects (see Section (F.8)). Note that in the literature on vector bundles on curves, often a vector bundle $\mathscr{E}$ is called simple, if $\mathrm{End}_k(\mathscr{E}) = k$; as pointed out above, this notion differs from ours.

Proof. Let us show (1). Let $u\colon \mathscr{E} \to \mathscr{F}$ be a map of semistable vector bundles of slopes λ. We want to show that $\mathrm{Ker}(u)$ and $\mathrm{Coker}(u)$, formed in the category of coherent $\mathscr{O}_C$-modules, are semistable vector bundles of slope λ. We may assume that $u \neq 0$. As $\mathscr{E}$ and $\mathscr{F}$ are both semistable of slope λ we have

$$\lambda = \mu(\mathscr{E}) \leq \mu(\mathrm{Im}(u)) \leq \mu(\mathscr{F}) = \lambda.$$

Hence $\mathrm{Im}(u)$ is a non-zero submodule of $\mathscr{F}$ with the same slope. Therefore $\mathrm{Im}(u)$ is a subbundle and semistable by Lemma 26.151. This shows that $\mathrm{Coker}(u)$ is a vector bundle.

To show that $\mathrm{Ker}(u)$ (resp. $\mathrm{Coker}(u)$) is semistable of slope λ, we may assume that $\mathrm{Ker}(u) \neq 0$ (resp. $\mathrm{Coker}(u) \neq 0$). Using Lemma 26.146 applied to the exact sequences $0 \to \mathrm{Im}(u) \to \mathscr{F} \to \mathrm{Coker}(u) \to 0$ and $0 \to \mathrm{Ker}(u) \to \mathscr{E} \to \mathrm{Im}(u) \to 0$ we see that $\mu(\mathrm{Ker}(u)) = \mu(\mathrm{Coker}(u)) = \lambda$. Therefore Lemma 26.151 shows that $\mathrm{Ker}(u)$ and $\mathrm{Coker}(u)$ are semistable.

It remains to show that if $0 \to \mathscr{E}' \to \mathscr{E} \to \mathscr{E}'' \to 0$ is an exact sequence of vector bundles with $\mathscr{E}'$ and $\mathscr{E}''$ semistable of slope λ, then $\mathscr{E}$ is semistable of slope λ. We may assume that $\mathscr{E}'$ and $\mathscr{E}''$ are non-zero. By Lemma 26.146 we know that $\mu(\mathscr{E}) = \lambda$. Let $\mathscr{F} \subseteq \mathscr{E}$ be any non-zero subbundle. Let $\mathscr{F}' = \mathscr{F} \cap \mathscr{E}'$ and let $\mathscr{F}''$ be the image of $\mathscr{F}$ in $\mathscr{E}''$. As $\mathscr{E}'$ and $\mathscr{E}''$ are semistable of slope λ, we see that $\mu(\mathscr{F}'), \mu(\mathscr{F}'') \leq \lambda$. Using the exact sequence $0 \to \mathscr{F}' \to \mathscr{F} \to \mathscr{F}'' \to 0$ we see by Lemma 26.146 that $\mu(\mathscr{F}) \leq \lambda$.

We now show (2). We have seen that every proper subobject in $(\mathrm{Vect}_\lambda(C))$ is a subbundle and therefore has strictly smaller rank. This shows that every object of $(\mathrm{Vect}_\lambda(C))$ has finite length. The vector bundles in $(\mathrm{Vect}_\lambda(C))$ that have no non-trivial subobject in $(\mathrm{Vect}_\lambda(C))$ are precisely those for which every subbundle has slope $< \lambda$, i.e., the stable vector bundles. $\square$

Corollary 26.153. *Let $\mathscr{E}$ be a stable vector bundle on C. Then $\mathrm{End}_{\mathscr{O}_C}(\mathscr{E})$ is a finite-dimensional skew field over k.*

If k is algebraically closed, it follows that $\mathrm{End}_{\mathscr{O}_C}(\mathscr{E}) = k$. In general it can happen that the skew field is not commutative (Exercise 26.26) or that the center of $\mathrm{End}_{\mathscr{O}_C}(\mathscr{E})$ is strictly larger than k (Remark 27.274).

Proof. As $\mathrm{End}_{\mathscr{O}_C}(\mathscr{E}) = \Gamma(C, \mathscr{E}^\vee \otimes_{\mathscr{O}_C} \mathscr{E})$, it is a finite-dimensional k-vector space since C is proper over k. By Proposition 26.152, every endomorphism u of $\mathscr{E}$ has semistable kernel and cokernel of the same slope as $\mathscr{E}$. As $\mathscr{E}$ is stable, the kernel and the cokernel have to be trivial, i.e., $u = 0$ or u is an automorphism. $\square$

Even if k is algebraically closed, there exist semistable, non-stable vector bundles $\mathscr{E}$ with $\mathrm{End}_{\mathscr{O}_C}(\mathscr{E}) = k$ (Exercise 26.23). This can only happen if the genus g of C is ≥ 2: If $g = 0$, then $C = \mathbb{P}^1_k$ and it by Theorem 26.133 a vector bundle $\mathscr{E}$ on $\mathbb{P}^1_k$ is stable if and only if it is a line bundle if and only if $\mathrm{End}(\mathscr{E}) = k$. If $g = 1$, then C is an elliptic curve and it follows from Atiyah's classification of vector bundles on elliptic curves (Theorem 27.271 below) that a vector bundle $\mathscr{E}$ is stable if and only if $\mathrm{End}(\mathscr{E}) = k$ (Corollary 27.275).

(26.25) Harder-Narasimhan filtration.

We continue to denote by k a field and by C a proper smooth geometrically connected curve over k. Let g be the genus of C. In Proposition 26.152 we have seen that the semistable vector bundles of a fixed slope form an abelian category. We will now show that an arbitrary vector bundle always has a filtration whose graded pieces are semistable. This filtration is canonically indexed by the rational numbers. Hence let us first introduce the following general notion.

Definition and Remark 26.154. Let $\mathscr{E}$ be a vector bundle on C. An $\mathbb{R}$-*filtration* of $\mathscr{E}$ is a decreasing map

$$\mathrm{Fil}\colon \mathbb{R} \longrightarrow \{\text{subbundles of } \mathscr{E}\}, \qquad \alpha \mapsto \mathrm{Fil}^\alpha(\mathscr{E}),$$

where the set of subbundles is partially ordered by inclusion, such that the following conditions are satisfied.

(1) There exist $\alpha, \beta \in \mathbb{R}$ such that $\mathrm{Fil}^\alpha(\mathscr{E}) = \mathscr{E}$ and $\mathrm{Fil}^\beta(\mathscr{E}) = 0$, i.e., the filtration is exhaustive and separating.

(2) For every $\alpha \in \mathbb{R}$ one has $\bigcap_{\beta<\alpha} \mathrm{Fil}^\beta(\mathscr{E}) = \mathrm{Fil}^\alpha(\mathscr{E})$.

We then write $\mathrm{Fil}^{\alpha+}(\mathscr{E}) := \sum_{\beta>\alpha} \mathrm{Fil}^\beta(\mathscr{E})$ and set $\mathrm{gr}_{\mathrm{Fil}}^\alpha(\mathscr{E}) := \mathrm{Fil}^\alpha(\mathscr{E})/\mathrm{Fil}^{\alpha+}(\mathscr{E})$. As each strict subbundle has a smaller rank, there are only finitely many α with $\mathrm{gr}_{\mathrm{Fil}}^\alpha(\mathscr{E}) \neq 0$. These are called the *jumps of the filtration*.

If $I \subseteq \mathbb{R}$, we call Fil an *I-filtration* if $\mathrm{gr}_{\mathrm{Fil}}^\alpha(\mathscr{E}) = 0$ for all $\alpha \in \mathbb{R} \setminus I$.

Definition 26.155. *Let $\mathscr{E}$ be a vector bundle on C. A Harder-Narasimhan filtration or shorter HN filtration of $\mathscr{E}$ is a $\mathbb{Q}$-filtration $\alpha \mapsto \mathrm{HN}^\alpha(\mathscr{E})$ of $\mathscr{E}$ such that $\mathrm{gr}_{\mathrm{HN}}^\alpha(\mathscr{E})$ is semistable of slope α for all $\alpha \in \mathbb{Q}$.*

The main result is the following.

Theorem 26.156. *Every vector bundle $\mathscr{E}$ on C has a unique Harder-Narasimhan filtration. Moreover, it is functorial in the following sense: For every $\mathcal{O}_C$-linear map $u\colon \mathscr{E} \to \mathscr{F}$ of vector bundles and for all $\alpha \in \mathbb{R}$ one has $u(\mathrm{HN}^\alpha(\mathscr{E})) \subseteq \mathrm{HN}^\alpha(\mathscr{F})$.*

Proof. Let us show the existence of a Harder-Narasimhan filtration such that the maximal jump equals the maximal slope λ of non-zero submodules of $\mathscr{E}$. Such a λ exists since the slope of all non-zero submodules is bounded by Remark 26.147 and the possible denominators are also bounded by $\mathrm{rk}(\mathscr{E})$. We proceed by induction on $\mathrm{rk}(\mathscr{E})$.

If $\mathscr{E}$ is a line bundle, and more generally if $\mathscr{E}$ is semistable of some slope μ, we can set $\mathrm{HN}^\alpha(\mathscr{E}) = \mathscr{E}$ for $\alpha \leq \mu$ and $\mathrm{HN}^\alpha(\mathscr{E}) = 0$ for $\alpha > \mu$.

If $\mathscr{E}$ is not semistable, choose a coherent $\mathcal{O}_C$-submodule $0 \neq \mathscr{F} \subset \mathscr{E}$ such that $\mathscr{F}$ has slope λ and has maximal rank among all submodules that have slope λ. Then $\mathscr{F}$ is semistable by construction and it is equal to its saturation $\overline{\mathscr{F}}$ in $\mathscr{E}$ (otherwise, $\mu(\overline{\mathscr{F}}) > \mu(\mathscr{F})$ by Remark 26.145 (1)). Hence $\mathscr{F}$ is a subbundle of $\mathscr{E}$. Every non-zero submodule of the vector bundle $\mathscr{E}/\mathscr{F}$ is of the form $\mathscr{G}/\mathscr{F}$ for some submodule $\mathscr{G}$ of $\mathscr{E}$ with $\mathrm{rk}(\mathscr{G}) > \mathrm{rk}(\mathscr{F})$. Hence $\mu(\mathscr{G}) < \mu(\mathscr{F})$ and therefore $\mu(\mathscr{G}/\mathscr{F}) < \mu(\mathscr{F})$ by Lemma 26.146.

By induction hypothesis, $\mathscr{E}/\mathscr{F}$ has a HN filtration such that

$$\max\{\alpha \; ; \; \mathrm{gr}_{\mathrm{HN}}^\alpha(\mathscr{E}/\mathscr{F}) \neq 0\} < \lambda.$$

Let $p\colon \mathscr{E} \to \mathscr{E}/\mathscr{F}$ be the canonical map. Setting $\mathrm{HN}^\alpha(\mathscr{E}) := p^{-1}(\mathrm{HN}^\alpha(\mathscr{E}/\mathscr{F}))$ for $\alpha \leq \lambda$ and $\mathrm{HN}^\alpha(\mathscr{E}) = 0$ for $\alpha > \lambda$ we obtain a HN filtration of $\mathscr{E}$.

It remains to show that HN filtrations are functorial in maps $u\colon \mathscr{E} \to \mathscr{F}$ of vector bundles (then their uniqueness follows by applying functoriality to $\mathrm{id}_\mathscr{E}$). Let $\alpha \in \mathbb{Q}$ with $u(\mathrm{HN}^\alpha(\mathscr{E})) \neq 0$. Let β be maximal such that $u(\mathrm{HN}^\alpha(\mathscr{E})) \subseteq \mathrm{HN}^\beta(\mathscr{F})$. Then $\mathrm{gr}_{\mathrm{HN}}^\beta(\mathscr{F}) \neq 0$ and the composition $\mathrm{HN}^\alpha(\mathscr{E}) \xrightarrow{u} \mathrm{HN}^\beta(\mathscr{F}) \xrightarrow{\pi} \mathrm{gr}_{\mathrm{HN}}^\beta(\mathscr{F})$ is non-zero since β was maximal.

Assume that $\mathrm{HN}^\beta(\mathscr{F}) \not\subseteq \mathrm{HN}^\alpha(\mathscr{F})$. Then $\alpha > \beta$. Now $\mathrm{gr}^\beta(\mathscr{F})$ is semistable of slope β and $\mathrm{HN}^\alpha(\mathscr{E})$ has a filtration whose graded pieces are semistable of slope $\geq \alpha > \beta$. Therefore $\pi \circ u$ induces the zero map on all these graded pieces by Proposition 26.150. Hence $\pi \circ u = 0$, a contradiction. $\square$

Remark and Definition 26.157. Let $r \geq 0$ be an integer. We set

$$\mathbb{Q}_{\mathrm{dom}}^r := \{(\alpha_1, \ldots, \alpha_r) \in \mathbb{Q}^r \; ; \; \alpha_1 \geq \cdots \geq \alpha_r\}.$$

Here dom stands for *dominant* and comes from the connection with the theory of linear algebraic groups and their root systems which we do not discuss further here. Let $\mathscr{E}$ be a vector bundle of rank r on C. If $\lambda_1 > \cdots > \lambda_m$ are the jumps in the HN filtration of $\mathscr{E}$ and $r_i := \mathrm{rk}(\mathrm{gr}_{\mathrm{HN}}^{\lambda_i}(\mathscr{E}))$, then we call

$$(\underbrace{\lambda_1, \ldots, \lambda_1}_{r_1 \text{ times}}, \underbrace{\lambda_2, \ldots, \lambda_2}_{r_2 \text{ times}}, \ldots, \underbrace{\lambda_m, \ldots, \lambda_m}_{r_m \text{ times}}) \in \mathbb{Q}_{\mathrm{dom}}^r$$

the *Harder-Narasimhan vector* of $\mathscr{E}$.

One can picture the Harder-Narasimhan vector by the *Harder-Narasimhan polygon*. This is the piecewise linear polygon connecting by line segments the points

$$(0,0), \quad (\mathrm{rk}\,\mathrm{HN}^{\lambda_1}(\mathscr{E}), \deg \mathrm{HN}^{\lambda_1}(\mathscr{E})), \quad \ldots, \quad (\mathrm{rk}\,\mathrm{HN}^{\lambda_m}(\mathscr{E}), \deg \mathrm{HN}^{\lambda_m}(\mathscr{E}))$$

in $\mathbb{R}^2$. It is a concave polygon with break points in $\mathbb{Z}^2$ that has slopes λ_i with multiplicities r_i.

Lemma 26.158. *Let $\mathscr{E}$ and $\mathscr{F}$ be vector bundles on C. Suppose that the lowest slope of the HN polygon of $\mathscr{E}$ is strictly bigger than the smallest slope of the HN polygon of $\mathscr{F}$. Then $\mathrm{Hom}_{\mathscr{O}_C}(\mathscr{E}, \mathscr{F}) = 0$.*

Proof. Indeed, assume that $u \colon \mathscr{E} \to \mathscr{F}$ is non-zero. Let $\lambda \in \mathbb{Q}$ be maximal with $u(\mathrm{HN}^\lambda(\mathscr{E})) \neq 0$ and let $\mu \in \mathbb{Q}$ be maximal such that $u(\mathrm{HN}^\lambda(\mathscr{E})) \subseteq \mathrm{HN}^\mu(\mathscr{F})$. Then u induces a non-zero map $\mathrm{gr}_{\mathrm{HN}}^\lambda(\mathscr{E}) \to \mathrm{gr}_{\mathrm{HN}}^\mu(\mathscr{F})$. This contradicts Proposition 26.150 because $\lambda > \mu$ by assumption. $\square$

Proposition 26.159. *Let k' be a field extension of k. Let $p \colon C \otimes_k k' \to C$ be the projection. Then for every vector bundle $\mathscr{E}$ on C and for all $\alpha \in \mathbb{Q}$ one has*

$$p^*(\mathrm{HN}^\alpha(\mathscr{E})) \cong \mathrm{HN}^\alpha(p^*(\mathscr{E})).$$

In particular, $\mathscr{E}$ is semistable of slope λ if and only if $p^*\mathscr{E}$ is semistable of slope λ.

Proof. Set $C' := C \otimes_k k'$. Since $p \colon C' \to C$ is faithfully flat, p^* defines an injective map from the set of submodules of $\mathscr{E}$ to the set of submodules of $p^*\mathscr{E}$. Moreover, degree and rank are preserved by p^*. Hence we see that if there exists a submodule $\mathscr{F} \neq 0$ of $\mathscr{E}$ with $\mu(\mathscr{F}) > \mu(\mathscr{E})$, then $p^*(\mathscr{F})$ is a submodule of $p^*(\mathscr{E})$ with $\mu(p^*\mathscr{F}) > \mu(p^*\mathscr{E})$. Hence if $p^*\mathscr{E}$ is semistable of slope λ, then $\mathscr{E}$ is semistable of slope λ.

It therefore suffices to show that there exists a $\mathbb{Q}$-filtration $\mathscr{E}^\bullet$ of $\mathscr{E}$ such that $p^*(\mathscr{E}^\alpha) = \mathrm{HN}^\alpha(p^*\mathscr{E})$. To do this we may enlarge k' since p^* is injective on the set of submodules of $\mathscr{E}$. There are only finitely many different $\mathrm{HN}^\alpha(p^*\mathscr{E})$ and each of them is of finite presentation. Hence all of them are defined over some finitely generated subfield $k \subseteq K \subseteq k'$. Filtering K by suitable subfields, we can assume that $k' = k(x)$ for some $x \in k'$ and that we are in one of the following cases.

(1) The extension k'/k is finite separable.
(2) The extension k'/k is purely inseparable in characteristic $p > 0$ and $x^p \in k$.
(3) The extension k'/k is purely transcendental.

Let $K = K(C)$ and $K' = K(C')$ be the function fields. Set $\mathscr{F}' := \mathrm{HN}^\alpha(p^*\mathscr{E})$ for some α. To see that $\mathscr{F}' = p^*(\mathscr{F})$ for some subbundle $\mathscr{F}$ of $\mathscr{E}$ it suffices to show that the K'-subspace $F' := \mathscr{F}'_\eta$ of $E' := p^*(\mathscr{E})_\eta$ descends to $E := \mathscr{E}_\eta$, i.e., that there exists a K-subspace $F \subseteq E$ such that $F \otimes_K K' = F'$ (Remark 26.132 (4)).

For (1) we may assume that k' is a Galois extension by passing to the normal hull. Let G be the Galois group of k' over k. Then $p\colon C' \to C$ is a Galois covering and K' is a Galois extension of K with Galois group G. For $\sigma \in G$ let $\sigma_{K'}$ be the automorphism of C' over C induced by σ. Then $\sigma_{C'}^* p^* \mathscr{E} = p^* \mathscr{E}$. As $\sigma_{C'}^*$ preserves degree and rank it follows that $\sigma_{C'}^* \mathrm{HN}^\bullet(p^* \mathscr{E})$ is again the HN filtration of $p^* \mathscr{E}$. In particular σ preserves F'. Hence by Galois descent (Theorem 14.85), F' descends.

Consider Case (2). Then $K' = Kk'$ is a purely inseparable extension of K with $(K')^p \subseteq K$. By Exercise 17.6, the subspace F' of $p^*(\mathscr{E})_\eta$ descends to $\mathscr{E}_\eta$ if and only if for every $D \in \mathrm{Der}_K(K')$ one has $D(F') \subseteq F'$. The restriction of D to k' is a k-derivation which we again denote by D.

Set $D_\mathscr{E} := D \otimes_k \mathrm{id}_\mathscr{E}$ and consider the composition

$$\psi\colon \mathscr{F}' \longrightarrow p^*(\mathscr{E}) \xrightarrow{D_\mathscr{E}} p^* \mathscr{E} \longrightarrow p^* \mathscr{E}/\mathscr{F}'.$$

Even though $D_\mathscr{E}$ is only k-linear but not k'-linear, ψ is k'-linear and hence $\mathscr{O}_{C'}$-linear because we have $\psi(af) = a\psi(f) + D(a)f = a\psi(f) \bmod \mathscr{F}'$ for local sections f of $\mathscr{F}'$ and $a \in k'$. Now Lemma 26.158 implies $\psi = 0$, which means $D_\mathscr{E}(\mathscr{F}') \subseteq \mathscr{F}'$. Hence we have $D(F') \subseteq F'$.

Case (3) is similar to (1). In this case $K' = K(x)$ is purely transcendental. The group $G := \mathrm{Aut}_K(K')$ of automorphisms of K' fixing K contains the group of K-automorphisms induced by translations $x \mapsto x + a$ for $a \in K$. Note that K is infinite since it is the function field of a curve. Now every rational function $f/g \in K(x)$, with $f, g \in K[x]$, such that $f(a)/g(a) = f(0)/g(0)$ for all $a \in K$ is constant. Otherwise $g(0)f(x) - f(0)g(x)$ is a non-zero polynomial and there exists $a \in K$ which is not a root of this polynomial which would imply that $f(x+a)/g(x+a) \neq f(x)/g(x)$. It follows that $K'^G = K$.

Every $\sigma \in G$ induces an automorphism on E' and F' descends if and only if $\sigma(F') = F'$ for all $\sigma \in G$ ([BouAI]$^\mathrm{O}$ II, §8.7, Theorem 1). Each $\sigma \in G$ induces an automorphism $\sigma_{C'}$ of C' with $\sigma_{C'}^* p^* \mathscr{E} = p^* \mathscr{E}$ such that $\sigma_{C'}^* \mathscr{F}' = \mathscr{F}'$ and hence $\sigma(F') = F'$. $\square$

Definition 26.160. *A non-zero vector bundle that is not isomorphic to the direct sum of two non-zero vector bundles is called* indecomposable.

Proposition 26.161. *Suppose that C has genus $g \leq 1$. Then the HN filtration of any vector bundle $\mathscr{E}$ on C is split. In particular, every indecomposable vector bundle is semistable.*

Proof. We proceed by induction on the rank of $\mathscr{E}$. Let λ be the smallest jump of the HN filtration of $\mathscr{E}$, i.e., the maximal rational number α such that $\mathscr{E} = \mathrm{HN}^\alpha(\mathscr{E})$. Then $0 \neq \mathrm{gr}_{\mathrm{HN}}^\lambda(\mathscr{E})$ is semistable of slope λ. By induction, the HN filtration for $\mathscr{E}' := \mathrm{HN}^{\lambda+}(\mathscr{E})$ is split, i.e., $\mathscr{E}' \cong \bigoplus \mathscr{E}_i$, where $\mathscr{E}_i$ is semistable of slope $\mu(\mathscr{E}_i) > \lambda$. We have to show that the exact sequence

$$0 \longrightarrow \mathscr{E}' \longrightarrow \mathscr{E} \longrightarrow \mathrm{gr}_{\mathrm{HN}}^\lambda(\mathscr{E}) \longrightarrow 0$$

is split. For this it suffices to show (Remark F.219) that

$$\mathrm{Ext}_{\mathscr{O}_C}^1(\mathrm{gr}_{\mathrm{HN}}^\lambda(\mathscr{E}), \mathscr{E}') = \bigoplus_i \mathrm{Ext}_{\mathscr{O}_C}^1(\mathrm{gr}_{\mathrm{HN}}^\lambda(\mathscr{E}), \mathscr{E}_i) = 0.$$

Now Serre duality in the form of (25.26.3) gives us

$$(*) \qquad \mathrm{Ext}_{\mathscr{O}_C}^1(\mathrm{gr}_{\mathrm{HN}}^\lambda(\mathscr{E}), \mathscr{E}_i)^\vee = \mathrm{Hom}_{\mathscr{O}_C}(\mathscr{E}_i, \mathrm{gr}_{\mathrm{HN}}^\lambda(\mathscr{E}) \otimes \omega_C),$$

where $\omega_C = \Omega^1_{C/k}$ is the canonical bundle. Then $\deg(\omega_C) = 2g - 2 \le 0$ by assumption and hence $\mu(\mathrm{gr}^\lambda_{\mathrm{HN}}(\mathscr{E}) \otimes \omega_C) \le \lambda$ by Remark 26.149 (2). Therefore the right hand side of (*) is 0 by Proposition 26.150.   □

Hence we see that if C is of genus ≤ 1, every vector bundle is the direct sum of semistable vector bundles. If $C = \mathbb{P}^1_k$, then the non-zero semistable vector bundles of slope λ necessarily have slope $\lambda \in \mathbb{Z}$ and are isomorphic to a direct sum of copies of $\mathscr{O}_{\mathbb{P}^1_k}(\lambda)$ by Theorem 26.133. If C has genus 1 and $C(k) \ne \emptyset$, then C is the underlying curve of an elliptic curve. In this case the category of semistable vector bundles of a fixed slope is not semisimple but can still be described quite explicitly by a theorem of Atiyah. We will explain this in Section (27.50) below.

It is also known how the Harder-Narasimhan filtration behaves in families, see Theorem 26.167 below.

Further topics

We conclude the chapter by mentioning, in part without proof, some further topics in the theory of curves that are particularly important and interesting.

(26.26) Relative Curves.

In this section we study which of the above notions behave well in families. To this end we introduce the notion of a relative curve.

Definition 26.162. *Let S be a scheme. An S-scheme $f\colon C \to S$ is called a (relative) curve over S if f is separated, flat and of finite presentation and all fibers of f are equi-dimensional of dimension 1.*

Remark 26.163. Let $C \to S$ be a smooth relative curve. Then every section $x\colon S \to C$ is a regular immersion of codimension 1 (Theorem 19.30). Since $C \to S$ is assumed to be separated, x is a closed immersion and hence defines an effective divisor in C, denoted by $[x]$ or just by x. We denote by $\mathscr{I}(x) \subseteq \mathscr{O}_C$ its defining quasi-coherent ideal sheaf, so that $x(S) = \mathrm{Spec}(\mathscr{O}_C/\mathscr{I}(x))$. It is a line bundle and we denote by $\mathscr{O}_{C/S}([x]) := \mathscr{I}(x)^\vee$ its dual.

Proposition 26.164. *Let $f\colon C \to S$ be a proper relative curve. Then the map*

$$S \to \mathbb{Z}, \qquad s \mapsto g(C_s)$$

is locally constant.

We call this locally constant function the *genus of the relative curve C.*

Proof. By definition of the genus it is enough to show that the Euler characteristic $\chi(\mathscr{O}_{C_s})$ is locally constant on S. But we have seen this in Theorem 23.139.   □

Let us specialize some results from cohomology and base change to the case of relative curves.

Remark 26.165. Let $f\colon C \to S$ be a proper relative curve.

(1) If $\mathscr{F}$ is a quasi-coherent $\mathscr{O}_C$-module, then $R^i f_* \mathscr{F} = 0$ for all $i \geq 2$ and the formation of $R^1 f_* \mathscr{F}$ commutes with base change $S' \to S$ (by Corollary 24.44 and the remark following it). If $\mathscr{F}$ is an $\mathscr{O}_C$-module of finite presentation that is flat over S, then $R^1 f_* \mathscr{F}$ is locally free if and only if the formation of $f_* \mathscr{F}$ commutes with base change. In this case, $f_* \mathscr{F}$ is locally free (Theorem 23.140).

(2) Assume that $f_* \mathscr{O}_C = \mathscr{O}_S$ and that this continues to hold after arbitrary base change $S' \to S$. By Corollary 24.63 this holds, if f has geometrically connected and geometrically reduced fibers. Applying part (1) to $\mathscr{F} = \mathscr{O}_C$, we obtain that $R^1 f_* \mathscr{O}_C$ is locally free of rank g, where g denotes the genus of the fibers of C. Note that we can consider g as a locally constant function on S by Proposition 26.164.

Proposition 26.166. *Let $f\colon C \to S$ be a proper relative curve and let $\mathscr{L}$ be a line bundle on C. For $s \in S$ let $\mathscr{L}_s$ be the restriction of $\mathscr{L}$ to the schematic fiber C_s of f over s. Then the map*

$$\deg\colon S \to \mathbb{Z}, \quad s \mapsto \deg(\mathscr{L}_s)$$

is locally constant.

Proof. This follows from Theorem 23.139 and Proposition 26.46. □

For vector bundles on relative smooth proper relative curves, the Harder-Narasimhan polygon is upper semicontinuous. More precisely one has the following result.

Let S be a scheme and let $f\colon C \to S$ be a smooth proper relative curve with geometrically connected fibers. Let $\mathscr{E}$ be a vector bundle on C. As usual, we denote for $s \in S$ the fiber $C_s = C \otimes_{\mathscr{O}_S} \kappa(s)$ and by $\mathscr{E}_s$ the pullback of $\mathscr{E}$ to C_s. Then C_s is a proper smooth geometrically connected curve over $\kappa(s)$ and $\mathscr{E}_s$ is a vector bundle on C_s. We want to explain how the Harder-Narasimhan filtration of $\mathscr{E}_s$ depends on s.

Since the rank of $\mathscr{E}_s$ is constant on C_s, the continuous rank function $C \to \mathbb{Z}$, where $\mathbb{Z}$ is endowed with the discrete topology, factors through a function

$$S \to \mathbb{Z}, \quad s \mapsto \mathrm{rk}(\mathscr{E}_s)$$

which is again continuous, i.e., locally constant, since f is closed and surjective and hence S carries the quotient topology of C. As also the map $s \mapsto \deg(\mathscr{E}_s)$ is locally constant (apply Proposition 26.166 to the determinant of $\mathscr{E}$), it is harmless to assume that $s \mapsto \mathrm{rk}(\mathscr{E}_s)$ and $s \mapsto \deg(\mathscr{E}_s)$ are constant and we denote their respective values by $\mathrm{rk}(\mathscr{E})$ and $\deg(\mathscr{E})$.

For every $s \in S$ let $\mathrm{HNP}(\mathscr{E}_s)$ be the HN polygon of $\mathscr{E}_s$ (Remark 26.157). This is an element of the set P of piecewise linear concave polygons in $\mathbb{R}^2$ with integral break points with start point $(0,0)$ and end point $(\mathrm{rk}(\mathscr{E}), \deg(\mathscr{E}))$. We endow P with a partial order by defining $p \leq p'$ for $p, p' \in P$ if p lies under p'. Then by [Shatz] Theorem 3 and Proposition 10 one has the following result[3].

Theorem 26.167. *The map*

$$S \to P, \quad s \mapsto \mathrm{HNP}(\mathscr{E}_s)$$

is constructible and upper semicontinuous, i.e., $\{\, s \in S \;;\; \mathrm{HNP}(\mathscr{E}_s) \geq p \,\}$ is closed and constructible for every polygon $p \in P$.

---

[3] In loc. cit., this is only shown if S is noetherian but it is easy to reduce to this situation by noetherian approximation using Proposition 26.159 and Theorem 10.57.

As a consequence of Proposition 26.57, let us show that smooth proper curves of genus ≥ 2 over any qcqs base scheme are projective. Recall that the genus of a relative curve is locally constant on the base, Proposition 26.164, and that the fiberwise degree of a line bundle on a relative curve is locally constant (Proposition 26.166).

Proposition 26.168. *Let S be a scheme, and let $f\colon C \to S$ be a proper curve over S with irreducible fibers.*

(1) *If $\mathscr{L}$ is a line bundle on C whose (fiberwise) degree is positive, then $\mathscr{L}$ is relatively ample over S.*

(2) *Suppose that f is in addition smooth and that the genus $g(C)$ is at least 2. Then the line bundle $\Omega^1_{C/S}$ is relatively ample for f. In particular, if S is qcqs, then f is projective.*

Proof. For part (1), we apply Proposition 26.57 and Theorem 24.46.

Part (2) follows from part (1). In fact, for $s \in S$, we have $(\Omega^1_{C/S})_s = \Omega^1_{C_s/\kappa(s)}$ (Proposition 17.30). Hence we can apply Corollary 26.52 and (1) to see that the canonical bundle $\Omega^1_{C/S}$ is relative ample over S. Then f is projective if S is qcqs by Proposition 13.68. $\square$

The result can be further generalized to certain nodal curves (so-called stable curves), see [DeMu] $^\mathrm{O}$ Theorem 1.2. One can also show that whenever A is a Dedekind domain and $f\colon C \to \operatorname{Spec} A$ is a proper curve where the scheme C is connected and regular, then f is projective ([Lic] $^\mathrm{O}$ Theorem 2.8).

For a study of relative curves of genus 0, see [LoKl] $^\mathrm{O}$ Section 3. Finally, let us mention the theorem, proved independently by Fontaine and Abrashkin, that there is no smooth proper relative curve over $\operatorname{Spec} \mathbb{Z}$ of genus > 0.

(26.27) Plane curves birational to a given curve.

Recall that every smooth proper curve over an infinite field k can be embedded in $\mathbb{P}^3_k$ (Theorem 14.132). Over finite fields this result is not true, in fact there exist finite fields k and smooth curves with more than $\#\mathbb{P}^3_k(k)$ rational points.

We also have seen that not every (smooth) curve can be embedded in the projective plane. But at least we can find, for every curve a "plane model", birational to the original curve, with only ordinary double points as singularities.

Proposition 26.169. *Let k be an infinite field. Let C be a curve over k. Then C is birationally equivalent to a projective plane curve which has at worst ordinary double points as singularities.*

The basic idea of the proof, similarly as for Theorem 14.132, is to study the "space" of all non-constant morphisms $C \to \mathbb{P}^2_k$, and to show that the locus of morphisms that do not satisfy the criterion above, has smaller dimension than the whole space. However, filling in all the details is quite involved. See [Sam] $^\mathrm{O}$, [Har3] $^\mathrm{O}$. The following weaker statement is easier to prove, and a map as in that proposition can be obtained quite explicitly in terms of blow-ups. See [Ful1] $^\mathrm{O}$ Chapter 7.

Proposition 26.170. *Let k be an algebraically closed field. Let C be a curve over k. Then C is birationally equivalent to a plane curve whose closed points are normal or ordinary multiple points.*

(26.28) Curves over finite fields and the Weil conjecture.

Let k be a finite field of characteristic p, with q elements. For any $r \geq 1$, we denote by k_r the unique extension field of k of degree r in a fixed algebraic closure $\bar{k}$ of k.

Let C be a geometrically connected smooth projective curve over k of genus g. Denote by $N_r = N_r(C)$ the cardinality of the finite set $C(k_r)$. It is interesting to study the numbers N_r, for instance, to give upper and lower bounds for them. A first naive guess might be that N_r should not be too far off from $\#\mathbb{P}^1_k(k_r) = q^r + 1$, and in fact, Corollary 26.175 below gives a bound of $|N_r - (q^r + 1)|$ in terms of q, r and the genus of C.

The right tool to study this question is the so-called zeta function of the curve C. Not only does it allow to prove this bound on the number of points, but it also exhibits an intriguing connection with number theory, specifically with the Riemann zeta function and the famous Riemann hypothesis.

This connection is made by the so-called *Weil conjectures*, formulated for curves by E. Artin in 1924, and for higher-dimensional varieties over finite fields by A. Weil in 1949. While the case of curves was proved by Weil, the Weil conjectures for higher-dimensional varieties have been one of the major open problems that led to the development of modern algebraic geometry by Grothendieck and his school, in particular to the construction and study of *étale cohomology*. In this case, too, the analogue of the Riemann hypothesis was the main stumbling block, until P. Deligne proved it and thus completed the proof of the Weil conjectures in [Del] ◦ (1974). By now, several approaches (all relying on a suitable cohomology theory in an essential way) are known. See the books [FrKi] ◦, [KiWe] ◦ for detailed expositions of two of them.

Definition 26.171. *Let k be a finite field, and let C be a geometrically connected smooth projective curve over k. The* zeta function *of C is the formal power series*

$$Z(C/k, T) = \exp\left(\sum_{r=1}^{\infty} N_r \frac{T^r}{r}\right) \quad \in \mathbb{Q}[\![T]\!].$$

Remark 26.172. In the sequel, we use some easy facts on formal power series. We give a brief summary here and refer to [BouAII] ◦ Ch. 4 §4 for details. For any ring R, we have the ring $R[\![T]\!] = \lim_i R[T]/(T^i) = \left\{\sum_{i \geq 0} a_i T^i; \ a_i \in R\right\}$ of formal power series over R which contains the polynomial ring $R[T]$ as a subring. Its units are those formal power series with absolute term $a_0 \in R^\times$. If $f, g \in R[\![T]\!]$ and g has constant term 0, the substitution $f(g(T)) \in R[\![T]\!]$ is defined.

We regard $R[\![T]\!] = \lim_i R[T]/(T^i)$ as a topological ring by equipping each $R[T]/(T^i)$ with the discrete topology and putting the limit topology on $R[\![T]\!]$. This gives us the notion of convergence, and in particular the notion of infinite series and infinite products.

If k is a field, then $k[\![T]\!]$ is a discrete valuation ring and T is a uniformizer. Its field of fractions is $k((T)) := \operatorname{Frac}(k[\![T]\!]) = k[\![T]\!]_T = \left\{\sum_{i \geq N} a_i T^i; \ N \in \mathbb{Z}, \ a_i \in k\right\}$ and is called the *field of Laurent series over k*. It contains the field of rational functions $k(T)$, and for $f, g \in k[T]$, $g \neq 0$, we call the image of f/g in $k((T))$ the *Laurent expansion* of f/g at $T = 0$. A key example is the geometric series

$$\frac{1}{1 - T} = \sum_{i \geq 0} T^i \in k[\![T]\!].$$

We say that a Laurent series which lies in (the image of) $k(T)$ has a pole of order n at $T = a \in k$, if this is true for the rational function with which it can be identified.

Now let k be a field of characteristic 0, or more generally a $\mathbb{Q}$-algebra. We have the exponential series

$$\exp(T) = \sum_{i \geq 0} \frac{T^i}{i!} \in k[\![T]\!]$$

and the map $f \mapsto \exp(f)$ is an isomorphism between the additive topological group of formal power series with absolute term 0 onto the multiplicative topological group of power series with constant term 1. Its inverse log is called the logarithm, and we have

$$\log(1 + T) = \sum_{i > 0} (-1)^{i-1} \frac{T^i}{i},$$

and more generally the same formula holds if T is replaced by any formal power series with constant term 0. Note that the fact that exp and hence log are group homomorphisms says precisely that the functional equations of the classical exp and log are true for them. Since exp and log are isomorphisms of topological groups, the functional equations can also be applied to convergent infinite series and infinite products, respectively.

For any $f \in k[\![T]\!]$ we have its formal derivative $f'(X)$. If f has constant term 1, then the derivative of $\log(f)$ equals f'/f and is called the *logarithmic derivative* of f. The logarithmic derivative of a product of such series is the sum of their logarithmic derivatives.

We can now state the Weil conjectures for the curve C.

Theorem 26.173. (Weil conjectures for curves) *Let k be a finite field, let $q := \#k$, let C be a geometrically connected smooth proper curve of genus g over k and let $Z(C/k, T)$ be its zeta function.*

(1) (rationality) *There exists a polynomial $P(T) \in \mathbb{Z}[T]$ of degree $2g$ such that*

$$Z(C/k, T) = \frac{P(T)}{(1 - T)(1 - qT)} \qquad \text{in } \mathbb{Q}(\!(T)\!).$$

(2) (functional equation) *The zeta function satisfies the following functional equation:*

$$Z(C/k, \frac{1}{qT}) = q^{1-g} T^{2-2g} Z(C/k, T) \qquad (\text{in } \mathbb{Q}(\!(T)\!)).$$

(3) ("Riemann hypothesis" for curves over finite fields) *The polynomial $P(T)$ in Part (1) is of the form*

$$P(T) = \prod_{i=1}^{2g} (1 - \alpha_i T), \qquad \alpha_i \in \mathbb{C},$$

with $|\alpha_i|^2 = q$ for all i.

We will give the proof in Section (26.29) below. From this theorem we easily obtain the following reformulation.

Theorem 26.174. *Under the assumptions of Theorem 26.173, there exist algebraic integers α_i, $i = 1, \ldots, 2g$, such that*

(1) *for all* r,

$$N_r = q^r + 1 - \sum_{i=1}^{2g} \alpha_i^r,$$

(2) *for all* $i = 1, \ldots, g$, $\quad \alpha_i \alpha_{2g-i+1} = q$,
(3) *for all* $i = 1, \ldots, 2g$, $\quad |\alpha_i|^2 = q$.

Proof. Being the zeros of the monic polynomial $T^{2g} P(T^{-1}) \in \mathbb{Z}[T]$, the α_i are algebraic over $\mathbb{Q}$ and even integral over $\mathbb{Z}$, i.e., are algebraic integers. Taking the logarithmic derivative of the equality

$$\exp\left(\sum_{r=1}^{\infty} N_r \frac{T^r}{r} \right) = Z(C/k, T) = \frac{\prod_{i=1}^{2g}(1 - \alpha_i T)}{(1-T)(1-qT)}$$

of Theorem 26.173 and comparing coefficients for T^{r-1}, we see that

$$N_r - (q^r + 1) = \sum_{i=1}^{2g} \alpha_i^r \quad \text{for all } r \geq 1.$$

The polynomial P has integer coefficients, hence for each zero α of P the complex conjugate $\overline{\alpha}$ is also a zero, and $\alpha \overline{\alpha} = |\alpha|^2 = q^{-1}$ by Part (3) of the Weil conjectures. This already shows Part (3) and also that we can group the non-real zeros of P in pairs as required by (2). Now it is enough to show that the multiplicities of $\sqrt{q}^{-1}$ and of $-\sqrt{q}^{-1}$ as zeros of P are even. Since P has even degree $2g$, there is an even number of real roots (counted with multiplicity). To conclude, we use that P has positive leading coefficient, as follows from Theorem 26.173 (2) and the observation that $P(0) = Z(C/k, 0) = 1$. $\square$

The information about the growth of the numbers N_r contained in the theorem is a key point. We formulate it as the following corollary which follows immediately from Parts (1) and (3) of Theorem 26.174 and the fact that $\mathbb{P}^1(k_r) = q^r + 1$.

Corollary 26.175. (Weil bound) *With notation as in Theorem 26.173,*

$$|N_r - \#\mathbb{P}^1(k_r)| \leq 2g q^{\frac{r}{2}} \quad \text{for all } r \geq 1.$$

Corollary 26.176. *Let e denote the greatest common divisor of the numbers $[\kappa(x) : k]$, where x ranges over the closed points of C. Then $e = 1$.*

In other words: There exists a divisor on C which has degree 1.

Proof. The bound in the corollary implies that for all r sufficiently large, we have $N_r > 0$. But for $x \in C(k_r)$, the degree $\deg([x]) = [\kappa(x) : k]$ is a divisor of r. Similarly, for $y \in C(k_{r+1})$, $[\kappa(y) : k]$ divides $r + 1$. Therefore these two degrees have greatest common divisor 1. $\square$

Remark 26.177.
(1) For C as above of genus $g = 0$, the corollary implies that $C \cong \mathbb{P}^1_k$ (and in particular $C(k) \neq \emptyset$) by Corollary 26.86. As such curves are Brauer-Severi curves and hence are classified by quaternion algebras over k, this follows also from Wedderburn's theorem, that over a finite field k every central simple algebra is isomorphic to $M_n(k)$ for some $n \geq 1$.

(2) If C has genus $g = 1$, then the Weil bound shows directly that $C(k) \neq \emptyset$. Hence C can be endowed with the structure of an elliptic curve.

(3) There exist curves C with $C(k) = \emptyset$ (see Exercise 26.19 for an example).

We can also write the zeta function of the curve C as an infinite product, a so-called *Euler product*. For a closed point $x \in C$, we denote by $\deg([x]) = [\kappa(x) : k]$ the degree of the divisor $[x]$.

Proposition 26.178. (Euler product expansion) *Let k be a finite field, let C be a geometrically connected smooth proper curve over k and let $Z(C/k, T)$ be its zeta function. Let C^1 be the set of closed points of C. Then*

$$Z(C/k, T) = \prod_{x \in C^1} (1 - T^{\deg([x])})^{-1}.$$

Proof. For any r, we can write $N_r = \sum_{d|r} d M_d$, where M_d denotes the number of closed points $x \in C$ with residue class field $\kappa(x) = k_d$ (in other words, with $\deg([x]) = d$). The sum runs over the positive divisors of r. In the ring of formal power series $\mathbb{Q}[\![T]\!]$ we then have

$$\log(Z(C/k, T)) = \sum_{r=1}^{\infty} N_r \frac{T^r}{r} = \sum_{r=1}^{\infty} \sum_{d|r} d M_d \frac{T^r}{r}$$

$$= \sum_{d=1}^{\infty} M_d \left(\sum_{n=1}^{\infty} \frac{T^{dn}}{n} \right) = -\sum_{x \in C^1} \log(1 - T^{\deg([x])}).$$

Applying exp on both sides, we obtain the desired result.   □

Remark 26.179. The classical Riemann zeta function, defined for complex numbers with real part > 1 by

$$\zeta(s) = \sum_{n \geq 1} \frac{1}{n^s}, \quad \mathrm{Re}(s) > 1,$$

can also be expressed as the Euler product

$$\zeta(s) = \prod_p \frac{1}{1 - p^{-s}}, \quad \mathrm{Re}(s) > 1,$$

where the product extends over all prime numbers. The zeta function admits an analytic continuation to a meromorphic function on $\mathbb{C}$, again denoted by ζ. Its only pole is $s = 1$ and it is a simple pole.

This meromorphic continuation satisfies a functional equation relating the values $\zeta(s)$ and $\zeta(1 - s)$. The Riemann hypothesis predicts that all zeros s of ζ with $\mathrm{Re}(s) > 0$ actually have $\mathrm{Re}(s) = \frac{1}{2}$. More generally, it is classical in algebraic number theory that one can define a zeta function for every finite extension field of $\mathbb{Q}$, i.e., for every number field (or, in other words, every *global field* of characteristic 0 in the sense of algebraic number theory). As is sketched below, the theory extends to global fields of positive characteristic, i.e., to the fields of rational functions of curves over finite fields. The zeta function of a curve that we have defined above is closely related to this construction. Even more generally, one can attach a zeta function to any scheme of finite type over Spec $\mathbb{Z}$; see [Kah1]$_\mathrm{X}$, [Kah2]$^\mathrm{O}$ and the references given there.

Let k be a finite field and let C/k be a curve as above. Replacing $\mathbb{Q}$ by the function field $K(C)$, we replace prime numbers (i.e., maximal ideals of $\operatorname{Spec} \mathbb{Z}$) by closed points of the curve C. In analogy to the Euler product expression above we are led to consider the series

$$\zeta(C/k, s) = \prod_{x \in C^1} \frac{1}{1 - q^{-s \deg([x])}},$$

where now the product extends over the closed points of C, and $\deg([x]) = [\kappa(x) : k]$, so that $q^{\deg([x])} = \#\kappa(x)$. One can show that the product converges for $\operatorname{Re}(s) > 1$. Replacing q^{-s} by a variable T and considering the result as a formal power series, we obtain by Proposition 26.178 the zeta function of C in the form $Z(C/k, T)$ as above. Denote by P the polynomial in Theorem 26.173. The expression for $Z(C/k, T)$ in Part (1) of the theorem gives us in particular a meromorphic continuation of $\zeta(C/k, s) = Z(C/k, q^{-s})$. Furthermore, we have $\zeta(C/k, s) = 0$ if and only if $P(q^{-s}) = 0$. In view of Part (3) of the theorem, this is equivalent to $q^s = \alpha_i$ for some i. Thus the final statement about the absolute values of the α_i can be expressed as a statement about the absolute values $|q^s|$, where s is a zero of $\zeta(C/k, -)$. Since $|q^s| = q^{\operatorname{Re}(s)}$, the statement of the theorem corresponds precisely to the condition that $\operatorname{Re}(s) = \frac{1}{2}$, perfectly reflecting the classical Riemann hypothesis.

(26.29) Proof of the Weil conjecture for curves.

For the proof of Theorem 26.173 we start with the following lemma that describes how the zeta function of the curve behaves under passing to a finite extension of k.

Lemma 26.180. *Let C be a geometrically connected, smooth proper curve over the finite field k, and let k_r/k be a field extension of degree r. Let $\overline{\mathbb{Q}}$ be an algebraic closure of $\mathbb{Q}$, and denote by $\mu_r \subset \overline{\mathbb{Q}}^\times$ the subgroup of r-th roots of unity. Then in $\overline{\mathbb{Q}}[T]$ we have*

$$Z(C \otimes_k k_r/k_r, T^r) = \prod_{\zeta \in \mu_r} Z(C/k, \zeta T).$$

Proof. We need to understand the fibers of the map $C \otimes_k k_r \to C$. For $x \in C$ a closed point, the fiber over x is isomorphic to the spectrum of $\kappa(x) \otimes_k k_r$. Writing $d = [\kappa(x) : k]$ and denoting by g the greatest common divisor of d and r, and by $\ell = dr/g$ their lowest common multiple, we have

$$\kappa(x) \otimes_k k_r \cong \kappa(x) \otimes_{k_g} k_g \otimes_k k_r \cong \kappa(x) \otimes_{k_g} (k_r)^g \cong k_l^g,$$

i.e., the fiber over x consists of g points, each of which has residue class field k_l, and hence has degree $l/r = d/g$ over k_r. Thus for the left hand side we obtain, using Proposition 26.178,

$$Z(C \otimes_k k_r/k_r, T^r) = \prod_{x \in (C \otimes_k k_r)^1} (1 - T^{r[\kappa(x):k_r]})^{-1} = \prod_{x \in C^1} (1 - T^{r[\kappa(x):k]/g})^{-g}.$$

Now writing $d = [\kappa(x) : k]$ as above, we have

$$(1 - T^{rd/g})^g = \prod_{\xi \in \mu_{r/g}} (1 - \xi T^d)^g = \prod_{\zeta \in \mu_r} (1 - \zeta^d T^d),$$

which gives the identification we are looking for. $\qquad\square$

Proof. (of Theorem 26.173) We divide the fairly long proof into several steps.

(I) Rationality. We fix the finite field k and the geometrically connected smooth proper curve C over k of genus g. Starting from the Euler product expression of Proposition 26.178, expanding each of the factors $(1 - T^{\deg([x])})^{-1}$ as a geometric series, and multiplying out the product, we may express the zeta function as

$$Z(C/k,T) = \prod_{x \in C^1} (1 - T^{\deg([x])})^{-1} = \sum_{D \geq 0} T^{\deg(D)},$$

the sum running over all effective divisors on C. Grouping effective divisors according to linear equivalence, we may rewrite this as a sum over isomorphism classes of line bundles on C. For a line bundle $\mathscr{L}$, write $\ell(\mathscr{L}) = \dim_k H^0(C,\mathscr{L})$, so that $(q^{\ell(\mathscr{L})} - 1)/(q-1)$ is the cardinality of the linear equivalence class of divisors D with $\mathscr{O}_C(D) \cong \mathscr{L}$. We then obtain

$$Z(C/k,T) = \sum_{\mathscr{L} \in \mathrm{Pic}(C)} \frac{q^{\ell(\mathscr{L})} - 1}{q - 1} T^{\deg(\mathscr{L})}.$$

Let g be the genus of the curve C. Let $e \geq 1$ be the generator of the image $\mathrm{Im}(\deg) \subseteq \mathbb{Z}$ of the degree map $\deg\colon \mathrm{Pic}(C) \to \mathbb{Z}$. We will see below that $e = 1$ (cf. Corollary 26.176).

Claim. The zeta function $Z(C/k,T)$ is a rational function, i.e., an element in $\mathbb{Q}(T) \subset \mathbb{Q}((T))$, and has a simple pole at $T = 1$.

Proof of claim. The case $g = 0$ can be handled easily starting from the above expression, since then $\ell(\mathscr{L}) = \deg(\mathscr{L}) + 1$ for all $\mathscr{L}$ of non-negative degree, by the Theorem of Riemann-Roch. From this one easily computes that for C of genus 0, we have $Z(C/k,T) = \frac{1}{(1-T^e)(1-(qT)^e)}$. This proves the claim, and once we have shown $e = 1$, the full Weil conjectures in this case.

Therefore we now assume $g \geq 1$. The key to expressing the zeta function of C as a rational function is the observation that for line bundles $\mathscr{L}$ of sufficiently high degree (namely for degree $> 2g - 2$), the dimension $\ell(\mathscr{L})$ of the space of global sections depends only on the degree d of $\mathscr{L}$, namely $\ell(\mathscr{L}) = d + 1 - g$, see Proposition 26.55 (2). We write $(q-1)Z(C/k,T)$ as the sum $(q-1)Z(C/k,T) = f(T) + g(T)$ with

$$f(T) := \sum_{\mathscr{L},\, 0 \leq \deg(\mathscr{L}) \leq 2g-2} q^{\ell(\mathscr{L})} T^{\deg(\mathscr{L})},$$

a polynomial in $\mathbb{Z}[T]$ of degree $\leq 2g - 2$, and

$$g(T) := \sum_{\mathscr{L},\, \deg(\mathscr{L}) \geq 2g-1} q^{\ell(\mathscr{L})} T^{\deg(\mathscr{L})} - \sum_{\mathscr{L},\, \deg(\mathscr{L}) \geq 0} T^{\deg(\mathscr{L})}$$

$$= \sum_{\mathscr{L},\, \deg(\mathscr{L}) \geq 2g-1} q^{\deg(\mathscr{L})+1-g} T^{\deg(\mathscr{L})} - \sum_{\mathscr{L},\, \deg(\mathscr{L}) \geq 0} T^{\deg(\mathscr{L})},$$

where for the final equality we use the Riemann-Roch theorem as explained before. By definition of e, for $d \in \mathbb{Z}$ there exists a line bundle of degree d if and only if $e \mid d$. In this case, the group $\mathrm{Pic}^0(C)$ of (isomorphism classes of) line bundles of degree 0 acts simply transitively on the set of line bundles of degree d. In particular, the above considerations show that $h := \#\mathrm{Pic}^0(C)$ is finite (since for given d, there are only finitely many effective divisors of degree d, and for d large, every line bundle of degree d is isomorphic to $\mathscr{O}_X(D)$ for some effective D). So when $e \mid d$, there exist h line bundles of degree d on C, up to isomorphism. Let $d_0 \geq 0$ be minimal such that $d_0 e \geq 2g - 1$. We obtain

$$g(T) = h \left(q^{1-g} \sum_{d \geq d_0} (qT)^{ed} - \sum_{d \geq 0} T^{ed} \right) = h \left(\frac{q^{1-g}(qT)^{ed_0}}{1 - (qT)^e} - \frac{1}{1 - T^e} \right).$$

The right hand side clearly lies in $\mathbb{Q}(T)$ and has a simple pole at $T = 1$. This proves the claim.

Let us show that $e = 1$. We apply Lemma 26.180 with $r = e$. The same reasoning as before, applied to the base change $C \otimes_k k_e$, shows that $Z(C \otimes_k k_e/k_e, T)$ has a simple pole at $T = 1$, whence $Z(C \otimes_k k_e, T^e)$ also has a pole of order 1 at $T = 1$. On the other hand, the expression we have found for $Z(C/k, T)$ is a rational function in T^e, so $Z(C/k, \zeta T) = Z(C/k, T)$ for all e-th roots of unity ζ. This shows that $\prod_{\zeta \in \mu_e} Z(C/k, \zeta T) = Z(C/k, T)^e$ has a pole of order e at $T = 1$, and thus that $e = 1$. This also gives us $d_0 = 2g - 1$.

Putting things together, we obtain
(26.29.1)
$$Z(C/k, T) = \frac{1}{q-1} \left(\sum_{\mathscr{L},\, 0 \leq \deg(\mathscr{L}) \leq 2g-2} q^{\ell(\mathscr{L})} T^{\deg(\mathscr{L})} + h \left(\frac{q^{1-g}(qT)^{2g-1}}{1-qT} - \frac{1}{1-T} \right) \right).$$

This gives us an expression of the form
$$Z(C/k, T) = \frac{P(T)}{(1-T)(1-qT)}, \quad P \in \mathbb{Q}[T],\ \deg(P) \leq 2g.$$

Since $Z(C/k, T)$ has coefficients in $\mathbb{Z}$, the same holds for $(1-T)(1-qT)Z(C/k, T) = P$. We will see that $\deg(P) = 2g$ in the next step of the proof.

(II) Functional equation. While for the rationality it was enough to use a weak form of the Riemann-Roch theorem, we now use its precise statement, including Serre duality on the curve C. We again may exclude the case $g = 0$. We can check the functional equation for both summands of the expression (26.29.1) separately. For the second summand, this is an easy computation which we leave to the reader. To deal with the first summand, note that the map $\mathscr{L} \mapsto \mathscr{L}^\vee \otimes_{\mathscr{O}_C} \Omega^1_{C/k}$ is an involution of the set of isomorphism classes of line bundles with degree in $\{0, \ldots, 2g-2\}$. Therefore

$$\sum_{\mathscr{L},\, 0 \leq \deg(\mathscr{L}) \leq 2g-2} q^{\ell(\mathscr{L})} T^{\deg(\mathscr{L})} = \sum_{\mathscr{L},\, 0 \leq \deg(\mathscr{L}) \leq 2g-2} q^{\ell(\mathscr{L}^\vee \otimes_{\mathscr{O}_C} \Omega^1_{C/k})} T^{\deg(\mathscr{L}^\vee \otimes_{\mathscr{O}_C} \Omega^1_{C/k})}$$

$$= \sum_{\mathscr{L},\, 0 \leq \deg(\mathscr{L}) \leq 2g-2} q^{\ell(\mathscr{L}) - \deg(\mathscr{L}) - 1 + g} T^{2g-2-\deg(\mathscr{L})}.$$

where the evaluation of $\ell(\mathscr{L}^\vee \otimes_{\mathscr{O}_C} \Omega^1_{C/k})$ follows from the Riemann-Roch theorem, Theorem 26.48, and we use that $\deg(\Omega^1_{C/k}) = 2g-2$, see Corollary 26.52. This immediately yields the functional equation for the first summand.

Let us summarize what we have seen so far. It is an easy consequence of the functional equation that $\deg(P) = 2g$. Furthermore, the polynomial P has absolute term $P(0) = Z(C/k, 0) = 1$, hence we can factorize it as $\prod_{i=1}^{2g} (1 - \alpha_i T)$ with $\alpha_i \in \mathbb{C}$. Similarly as in the proof of Theorem 26.174 we see that $\sqrt{q}$ and $-\sqrt{q}$ both occur with even multiplicity among the α_i. By the functional equation, $\alpha \mapsto q^{-1}\alpha^{-1}$ is an involution of the set of zeros of P. Renumbering the α_i if necessary, $Z(C/k, T)$ can therefore be written in the form

$$Z(C/k, T) = \frac{P(T)}{(1-T)(1-qT)} \quad \text{with } P(T) = \prod_{i=1}^{2g} (1 - \alpha_i T) \in \mathbb{Z}[T]$$

and $\alpha_i \in \mathbb{C}$, $\alpha_i \alpha_{2g+1-i} = q$ for $i = 1, \dots, g$.

(III) Riemann hypothesis. It remains to prove the Riemann hypothesis, the most difficult part of the Weil conjectures. The main idea of the proof below, i.e., using the intersection pairing on the surface $C_{\overline{k}} \times C_{\overline{k}}$, goes back to Weil. Recall that we have fixed an algebraic closure $\overline{k}$ of k. We start by showing that proving a suitable bound on the numbers $N_r = \#C(k_r)$ entails the Riemann hypothesis (compare Corollary 26.175).

Lemma 26.181. *With notation as above, the following are equivalent:*
(1) $|\alpha_i| = q^{\frac{1}{2}}$ *for all* $i = 1, \dots, 2g$,
(2) $|\alpha_i| \le q^{\frac{1}{2}}$ *for all* $i = 1, \dots, 2g$,
(3) $|N_r - (q^r + 1)| \le 2g q^{\frac{r}{2}}$ *for all* $r \ge 1$.

Proof. Since we have $\alpha_i \alpha_{2g+1-i} = q$ for $i = 1, \dots, g$, the equivalence (i) $\Leftrightarrow$ (ii) is clear. The implication (ii) $\Rightarrow$ (iii) follows by writing $N_r - (q^r + 1) = \sum_{i=0}^{2g} \alpha_i^r$, see the proof of Theorem 26.174.

Let us show (iii) $\Rightarrow$ (ii). In view of the above identity, (iii) says that $\left| \sum_{i=1}^{2g} \alpha_i^r \right| \le 2g q^{\frac{r}{2}}$ for all $r \ge 1$. The bound (ii) then follows from the following lemma, applied to $\beta_i = \alpha_i/q^{\frac{1}{2}}$, $i = 1, \dots, 2g$. Note that to apply that lemma, it would even be enough to know that there exists a constant $C > 0$ such that $|N_r - (q^r + 1)| \le C q^{\frac{r}{2}}$ for all $r \ge 1$. $\qquad\square$

Lemma 26.182. *Let* $\beta_1, \dots, \beta_n \in \mathbb{C}$ *such that* $\left| \sum_{i=1}^{n} \beta_i^r \right|$ *is bounded independently of* $r \ge 1$. *Then* $|\beta_i| \le 1$ *for all* i.

Proof. To simplify the notation, we assume β_1 is non-zero and has the largest absolute value among all β_i. We consider the power series $\sum_{r \ge 1} \left(\sum_{i=1}^{n} \beta_i^r \right) t^r$ over the complex numbers. For $t \in \mathbb{C}$ with $|\beta_1 t| < 1$ we have

$$\sum_{r \ge 1} \left(\sum_{i=1}^{n} \beta_i^r \right) t^r = \sum_{i=1}^{n} \frac{\beta_i t}{1 - \beta_i t},$$

whence the series on the left has radius of convergence $|\beta_1|^{-1}$. Using the Cauchy-Hadamard formula and the assumption, this yields

$$|\beta_1| = \limsup_{r \to \infty} \left| \sum_{i=1}^{n} \beta^r \right|^{\frac{1}{r}} \le 1,$$

as desired.

Alternatively, one can give a more direct proof applying the simultaneous version of Dirichlet's approximation theorem to the arguments of the β_i to show that there exist infinitely many r such that all β_i^r lie in some small sector of the complex plane containing the positive real axis. This implies (restricting from now on to elements r with the above property) that $\sum_i |\beta_i^r| \le c |\sum_i \beta_i^r|$ for some constant $c > 0$ that does not depend on r. But then $\sum_i |\beta_i^r|$ is bounded, and thus so are all $|\beta_i^r|$. $\qquad\square$

In the remainder of the proof, we will prove the bound $|N_r - (q^r + 1)| \le 2g q^{r/2}$. It is convenient to pass to $\overline{k}$ to carry out the argument. To recover the number of k_r-rational points of C after the base change from k to $\overline{k}$, we use the relative Frobenius morphism $F_{C_{\overline{k}}/\overline{k}} \colon C_{\overline{k}} \to (C_{\overline{k}})^{(q)}$, see Remark 4.24. By a slight abuse of notation, here and below we always use the q-th power Frobenius, where q is the number of elements of k. We collect the essential properties that we will use in the following general lemma.

Lemma 26.183. *Let k be a finite field with q elements, let $\overline{k}$ be an algebraic closure of k and let $\sigma\colon \overline{k} \to \overline{k}$, $x \mapsto x^q$, be the q-Frobenius homomorphism on $\overline{k}$. Consider a k-scheme X_0, and denote by $X := X_0 \otimes_k \overline{k}$ its base change to $\overline{k}$.*

(1) We have a canonical identification $X^{(q)} = X$ as $\overline{k}$-schemes. Making this identification, we always view the relative Frobenius $F_{X/\overline{k}}$ as a morphism $F\colon X \to X$. It agrees with the base change $F_{X_0/k} \otimes_k \overline{k}$.

(2) The map $X(\overline{k}) \to X(\overline{k})$ induced by F coincides with the map given by composition with the Frobenius morphism on $\operatorname{Spec}\overline{k}$ (which is induced by σ). In particular, the inclusion $X_0(k_r) \subset X(\overline{k})$ gives us an identification of $X_0(k_r)$ with the set $X(\overline{k})^{F^r}$ of fix points under the r-fold composition of F with itself.

(3) For every point $x \in X(\overline{k})$ fixed by F^r, the differential $(dF^r)_x\colon T_x X \to T_x X$ vanishes.

The third point of the lemma will be important below; note that for $X_0 = \mathbb{A}^1_k$ it is basically equivalent to the fact the polynomial T^q as derivative 0 in characteristic $p \mid q$.

Proof. For Part (1), note that the restriction of σ to k is the identity, so

$$X^{(q)} = X \otimes_{\overline{k},\sigma} \overline{k} = X_0 \otimes_k \overline{k} \otimes_{k,\sigma} k = X_0 \otimes_{k,\sigma} \overline{k} = X.$$

It is easy to check that via this identification, $F := F_{X/\overline{k}} = F_{X_0/k} \otimes_k \overline{k}$. It should be emphasized, however, that the identification $X^{(q)} = X$ crucially relies on the fact that X arises by base change from a k-scheme, and is the way in which the Frobenius allows us to keep track of the "k-structure on X".

The first statement of Part (2) follows from Part (1) since $X(\overline{k}) = X_0(\overline{k})$ and because the absolute Frobenius is compatible with all morphisms (cf. the outer part of diagram (4.6.2)). The second statement is an immediate consequence of this.

It remains to prove (3). We view the composition F^r as the relative q^r-Frobenius morphism. It is clear that the differential of the absolute Frobenius of X vanishes everywhere. By definition of the relative Frobenius $F_{X/\overline{k}}$, the absolute Frobenius of X equals the composition of $F_{X/\overline{k}}$ with an isomorphism (namely the base change of the absolute Frobenius of the perfect field $\overline{k}$). This implies the desired vanishing. (Cf. Exercise 17.13.) $\qquad\square$

Now let $X := C_{\overline{k}} \times C_{\overline{k}}$, a smooth projective surface over $\overline{k}$. Fix a point $x \in C(\overline{k})$ and consider the following divisors on X:

$$H = C_{\overline{k}} \times \operatorname{Spec}\kappa(x), \quad V = \operatorname{Spec}\kappa(x) \times C_{\overline{k}}, \quad \Delta \text{ the diagonal}, \quad \Gamma_r \text{ the graph of } F^r_{C_{\overline{k}}/\overline{k}}.$$

We will need to know the following intersection numbers.

Lemma 26.184. *With notation as above, we have*

(1) $(H \cdot H) = (V \cdot V) = 0$,

(2) $(H \cdot V) = (H \cdot \Delta) = (V \cdot \Delta) = 1$,

(3) $(\Gamma_r \cdot \Delta) = \#C(\overline{k})^{F^r} = \#C(k_r)$.

Proof. Part (1) follows from Proposition 23.77, applied to the projections from X to $C_{\overline{k}}$. Alternatively one could use that the divisor $[x]$ on $C_{\overline{k}}$ (where $H = C \times \operatorname{Spec}\kappa(x)$) is linearly equivalent to a divisor on $C_{\overline{k}}$ whose support does not contain x. Thus H is linearly equivalent to a divisor whose support is disjoint from H, and the result follows for H, and by an analogous argument for V. For Part (2), it is enough to observe that all the three intersections under consideration are transversal intersections in the single point (x, x).

In Part (3) the second equality follows from the above discussion, and for the first one, it is enough to check that the intersection is transversal, because $C(\overline{k})^{Fr}$ equals the set-theoretic intersection $\Gamma_r \cap \Delta$ (as subsets of $C_{\overline{k}}$). So let $z \in C(k_r)$, considered as a point of $C(\overline{k}) = C_{\overline{k}}(\overline{k})$, so that $(z, z) \in \Gamma_r \cap \Delta$. We identify the tangent space of X at (z, z) with $T_z(C_{\overline{k}})^2$. The tangent space to Δ at this point is simply the diagonal in $T_z(C_{\overline{k}})^2$. On the other hand, since passing to tangent spaces is functorial and compatible with products, we may identify $T_{(z,z)}\Gamma_r \subseteq T_{(z,z)}(C_{\overline{k}} \times C_{\overline{k}}) = T_z(C_{\overline{k}})^2$ with the graph of the differential $(dF_r)_z$. However $(dF_r)_z = 0$ by Lemma 26.183. Hence this graph is just the first coordinate axis, and together with the diagonal it spans this vector space. $\square$

Now consider the quotient $\mathrm{Num}(X)$ of $\mathrm{Pic}(X)$ by the subgroup of line bundles numerically equivalent to 0 (see Section (25.32)). Let $\mathrm{Num}(X)_{\mathbb{R}} = \mathrm{Num}(X) \otimes_{\mathbb{Z}} \mathbb{R}$, a real vector space equipped with the non-degenerate symmetric bilinear form induced by the intersection pairing, and decompose this space as $\mathrm{Num}(X)_{\mathbb{R}} = U \perp U'$, where $U := \langle H, V \rangle_{\mathbb{R}}$ is the two-dimensional subspace generated by H and V, and U' is its orthogonal complement. By Lemma 26.184, $U \cap U' = 0$, so this is indeed a direct sum decomposition, and moreover $H + V$ has self-intersection number $((H + V)^2) = 2 > 0$. It follows from the Hodge index theorem, Theorem 25.32, that the intersection pairing on U' is negative definite (either using Sylvester's theorem for suitable finite-dimensional subspaces, or observing that $H + V$ is ample).

Consider the pullback homomorphism $\Phi = (F \times \mathrm{id})^*\colon \mathrm{Pic}(X) \to \mathrm{Pic}(X)$, see Section (11.16). It induces an endomorphism of $\mathrm{Num}(X)_{\mathbb{R}}$, again denoted by Φ, which has the following properties.

Lemma 26.185. *With notation as above, we have the following equalities in* $\mathrm{Num}(X)$:
(1) $\Phi(H) = H$,
(2) $\Phi(V) = qV$,
(3) $\Phi^r(\Delta) = \Gamma_r$,
(4) $(\Phi(D) \cdot \Phi(E)) = q(D \cdot E)$ *for all* $D, E \in \mathrm{Num}(X)_{\mathbb{R}}$.

Proof. We start with Part (2). Part (1) can be shown in a similar, but simpler way (and holds already in $\mathrm{Pic}(X)$). We write V as the pullback $p_1^*[x]$, where $p_1 \colon X \to C_{\overline{k}}$ is the projection to the first factor. Then $(F \times \mathrm{id})^* p_1^*[x] = p_1^* F^*[x]$. Now $F^{-1}(x)$ set-theoretically consists of a single point, say y, and Proposition 26.71 shows that $\deg(F) = q$. It then follows from Corollary 11.51 and Proposition 26.19 that $F^*[x] = q[y]$. It is thus enough to show that $p_1^*[x]$ and $p_1^*[y]$ are numerically equivalent. But this follows from Proposition 23.151 (2).

For Part (3), which holds already in $\mathrm{Pic}(X)$, by Corollary 11.51 it is enough to observe the general fact that the schematic inverse image of Δ under the r-fold composition $(F \times \mathrm{id})^r$ equals the graph Γ_r, cf. the right hand square in diagram (9.1.4). Part (4) follows from Proposition 23.77. $\square$

Write $\Delta = u + u'$ with $u \in U$, $u' \in U'$. Then $u = H + V$, since $(\Delta \cdot H) = (\Delta \cdot V) = 1$, and using Lemma 26.184 and Lemma 26.185 we can compute

$$\#C(k_r) = (\Gamma_r \cdot \Delta) = (\Phi^r(\Delta) \cdot \Delta)$$
$$= ((H + q^r V + \Phi^r(u')) \cdot (H + V + u')) = q^r + 1 + (\Phi^r(u'), u').$$

Thus

$$|\#C(k_r) - (q^r + 1)| = |(\Phi^r(u'), u')|$$
$$\leq \sqrt{|(\Phi^r(u'), \Phi^r(u'))| \cdot |(u', u)|} = q^{\frac{r}{2}} |(u' \cdot u')|,$$

where the estimate in the middle follows from the Cauchy-Schwarz inequality on the negative definite space U'. In view of the remark at the end of the proof of Lemma 26.181, this would already be enough to finish the proof of the Weil conjectures. But it is also easy to compute $(u' \cdot u')$ explicitly, so that we can apply Lemma 26.181 in the form it was stated.

By definition and Lemma 26.184, we have $(u' \cdot u') = ((\Delta - H - V)^2) = (\Delta^2) - 2$. Let us show that $(\Delta^2) = 2 - 2g$; this will give us $(u' \cdot u') = -2g$ and hence the desired result. (Note that if $g = 0$, then $C \cong \mathbb{P}^1_k$ and $U' = 0$.) By the adjunction formula, Proposition 25.145, we have $(\Delta \cdot (\Delta + K)) = 2g - 2$, where K is a canonical divisor on X and g is the genus of C. By Corollary 17.32, we may write $K = p_1^* K_C + p_2^* K_C$ (in $\mathrm{Pic}(X)$) where K_C is a canonical divisor on $C_{\overline{k}}$ and $p_1, p_2 \colon X \to C_{\overline{k}}$ are the projections. We have $(p_i^* K_C \cdot \Delta) = \deg((p_i^* K)_{|\Delta}) = \deg(K_C) = 2g - 2$ for $i = 1, 2$ (see Example 23.75, Corollary 26.52), so it follows that $(\Delta^2) = 2 - 2g$. This finishes the proof. $\square$

Several other proofs of the Weil conjectures for curves are known. Weil gave two proofs, one using the Jacobian of the curve C (see Section (27.26) below for the notion of Jacobians), and another one similar to the proof given above, using intersection theory on the surface $C \times_k C$. See [Mil3] $_X$, [Ras] ^{O}. A more elementary proof, based on ideas of Stepanov, was given by Bombieri [Bom] ^{O}. See also Lorenzini's book [Lor] ^{O} which gives a detailed exposition of this proof, starting from the basic notions of algebraic geometry, and including many examples.

(26.30) Curves over number fields.

Let k be a number field, i.e., a finite extension field of the field $\mathbb{Q}$ of rational numbers. Similarly as for Riemann surfaces (Theorem 26.30) we see the following trichotomy for proper smooth geometrically connected curves over k.

Genus 0. A curve (always in the above sense) of genus 0 which has a k-valued point is isomorphic to $\mathbb{P}^1_k$, and thus has infinitely many rational points. On the other hand, there exist curves of genus 0 without k-valued points. As we have seen above (and is true over every field), a connected smooth proper curve of genus 0 is isomorphic to a smooth conic, i.e., a smooth degree 2 plane curve. For these curves, the Hasse principle holds (see Remark 26.92), i.e., the curve has a k-valued point if (and only if) it has a k_v-valued point for every completion of k with respect to an (archimedean or non-archimedean) absolute value v.

Genus 1. Curves of genus 1 may have no k-valued points, or finitely many k-valued points, or infinitely many k-rational points. As we have seen, if there is a k-valued point, then the set $C(k)$ can be equipped with the structure of an abelian group, arising from a group scheme structure on C. The Theorem of Mordell (sometimes called the Theorem of Mordell-Weil in view of its generalization to the case of abelian varieties) states that $C(k)$ is a finitely generated abelian group. The famous Conjecture of Birch and Swinnerton-Dyer gives a precise description of the rank of this group in terms of the L-function of the elliptic curve, a holomorphic function attached to C. (See [Sil1] ^{O}, [Sil2] ^{O} for further details.) The Hasse principle may fail in this situation, see Example 26.102.

Genus ≥ 2. Curves of genus ≥ 2 have at most finitely many k-valued points. This statement was conjectured by L. Mordell. It is still often called Mordell's conjecture, although it was proved by G. Faltings ([Fal]^{O}, see also [CoSi]^{O} for further background on the methods used in Faltings's proof).

(26.31) Literature on algebraic curves.

We give some further references on the theory of algebraic curves. Both Hartshorne [Har3]^{O}, and Liu [Liu]^{O} discuss curves in quite some detail. Fulton's book [Ful1]^{O} does not require as much background. In the book [HiSi]^{O} by Hindry and Silverman, there is a comprehensive survey on the theory of algebraic curves.

The two large volumes [ACGH]^{O} and [ACG]^{O} by Arbarello, Cornalba, Griffiths and Harris contain a lot of material. Another classic is Mumford's book [Mum2]^{O} on Curves and their Jacobians.

For elliptic curves, the books [Sil1]^{O}, [Sil2]^{O} by Silverman are standard references, with a particular focus on the arithmetic theory. Alternatives are the books [Kna]^{O} by Knapp and [Was]^{O} by Washington.

There seems to be no comprehensive account of hyperelliptic curves in a text book, but [Liu]^{O} and Vakil's book [Vak]$_{\text{X}}$ are good to get started in the theory. See also [Loc]^{O}, [Sto]$_{\text{X}}$, [Gross]$_{\text{X}}$. For relative hyperelliptic (and other) curves, also see [LoKl]^{O}.

Exercises

Exercise 26.1. Let k be a field, and let $C = V_+(T_0^2) \subset \mathbb{P}_k^2$. Show that C is not reduced and that $H^0(C, \mathscr{O}_C) = k$.
Hint: This can be checked by a direct computation. Alternatively use cohomology, cf. the proof of Corollary 25.141.

Exercise 26.2. Let k be a field. Show that $C = \mathbb{P}_k^1 \otimes_k k[\varepsilon]/(\varepsilon^2)$ cannot be embedded as a closed subscheme of $\mathbb{P}_k^2$. Specify a closed immersion $C \hookrightarrow \mathbb{P}_k^3$.

Exercise 26.3. Let k be a field. For a commutative monoid S we denote by $k[S]$ the corresponding monoid algebra. From now on let $S \subseteq \mathbb{N}$ be an additive submonoid[4] so that

$$k[S] = \left\{ \sum_n a_n T^n \in k[T] ;\ a_n = 0 \text{ for } n \notin S \right\}.$$

Set $C_S := \operatorname{Spec} k[S]$ which is called the *monomial curve of S*.
(1) Show that there exists a unique finite generating set $n_1 < n_2 < \cdots < n_p$ of S (i.e., $S = \sum_{i=1}^p \mathbb{N} n_i$) such that if $n_j = \sum_{i=1}^p m_i n_i$ with $m_i \in \mathbb{N}$, then $m_i = 0$ for $i \neq j$.
(2) Let $\rho \colon L := \mathbb{N}^p \to S$ be the homomorphism of monoids sending e_i to n_i with n_i defined as in (1) and $e_i = (0, \ldots, 0, 1, 0, \ldots, 0)$ with 1 at the i-th place. Then ρ yields a k-algebra map $k[\rho] \colon k[L] = k[T_1, \ldots, T_p] \to k[S]$, $T_i \mapsto T^{n_i}$. The latter is surjective and thus defines a closed immersion $C_S \to \mathbb{A}_k^p$. Show that its kernel is generated by $f_{\nu,\mu} := \underline{T}^\nu - \underline{T}^\mu$ for $(\nu, \mu) \in R := \{ (\nu, \mu) \in L \times L ;\ \rho(\nu) = \rho(\mu) \}$. Let r be the minimal number s such that $\ker(k[\rho])$ is generated by s polynomials of the form $f_{\nu,\mu}$ for $(\nu, \mu) \in R$.

---
[4] Recall that $\mathbb{N}$ denotes the natural numbers including 0.

(3) Show that $r \geq p - 1$.

(4) Show that C_S is lci if and only if $r = p - 1$.

(5) Let $d > 0$ be the greatest common divisor of $\{n_1, \ldots, n_p\}$. Show that $d\mathbb{Z}$ is the subgroup of $\mathbb{Z}$ generated by S and that the inclusion $k[S] \to k[T^d]$ is the normalization of $k[S]$.

(6) From now on we assume that $d = 1$. Show that there exists a greatest integer $F(S)$ that is not in S. Show that $c(S) := (T^{F(S)+1})$ is the conductor of $k[S]$, i.e., the ideal $\{ f \in k[S] \; ; \; fk[T] \subseteq k[S] \}$ of $k[S]$ and of $k[T]$.

(7) Show that C_S is Gorenstein if and only if there exists an integer F such that $S = \{ m \in \mathbb{N} \; ; \; F - m \notin S \}$. If such an F exists it is necessarily the integer $F(S)$. In the case that $p = 2$ show that $F = (n_1 - 1)(n_2 - 1) - 1$ is such a number, showing that C_S is Gorenstein in this case (which also follows from the fact that reduced plane curves are always lci).

(8) Suppose that $p \leq 3$. Show that $r \leq 3$ and that C_S is lci if and only if it is Gorenstein.

(9) Let $a \geq 1$. Show that $k[T^a, T^{a+1}, T^{a+2}]$ is Gorenstein (equivalently by (8), lci) if and only if $a = 1$, 2, or 4.

(10) Let S be the monoid generated by $\{5, 6, 7, 8\}$. Show that $F(S) = 9$ and that C_S is Gorenstein but not lci.

Remark: See [Her]$^{\circ}$, [HeKu]$^{\circ}$.

Exercise 26.4. Let X be a noetherian scheme, let $U \subseteq X$ be an open dense subspace, and set $Y := X \setminus U$. Show that for every closed point $y \in Y$ there exists an integral one-dimensional subscheme $C \subseteq X$ such that $\{y\}$ is open in $C \cap Y$.

Hint: Use that in any noetherian local ring A with $\dim A \geq 1$ the intersection of all prime ideals $\mathfrak{p} \subset A$ with $\dim A/\mathfrak{p} = 1$ is the nilradical of A.

Exercise 26.5. Let Y be a noetherian scheme, and let $f: X \to Y$ be a separated morphism of finite type.

(1) Let $Z \subseteq X$ be a closed subscheme, minimal among those for which the restriction of f is not proper. Show that Z is integral and that $\dim Z = 1$.

Hint: For the second assertion use a Nagata compactification $\bar{Z} \to Y$ of Z over Y and Exercise 26.4 for $Z \subseteq \bar{Z}$.

(2) Show that if $f_*(\mathscr{O}_X/\mathscr{I})$ is a coherent $\mathscr{O}_Y$-module for every coherent ideal $\mathscr{I} \subseteq \mathscr{O}_X$, then f is proper.

Hint: Argue by contradiction and assume that f is not proper. Use (1) to reduce to the case that X is integral of dimension 1 and that Y is integral of dimension ≤ 1. If $\dim Y = 0$, then one can assume that $Y = \operatorname{Spec} k$ for a field k and hence that X is affine. Then $f_*\mathscr{O}_X$ is not coherent. If $\dim Y = 1$, then f is quasi-finite and can be factorized in an open dominant immersion j followed by a finite morphism (Corollary 12.85). Show that $j_*\mathscr{O}_X$ has to be coherent and deduce that j is an isomorphism.

Remark: For this and the previous exercise, see [Lip2]$^{\circ}_X$ (4.3.9)

Exercise 26.6. Let Y be a locally noetherian scheme and let $f: X \to Y$ be a morphism of schemes. Show that f is proper if and only if f is cohomologically proper, separated, and of finite type.

Hint: Exercise 26.5

Exercise 26.7. Let k be a field, let X be a proper scheme over k, and let $\mathscr{L}$ be a line bundle that is generated by its global sections. Show that $\mathscr{L}$ is ample if and only if for every closed immersion $i\colon C \subseteq X$ of an integral curve C, the restriction $i^*\mathscr{L}$ has degree $\neq 0$.

Exercise 26.8. Let k be a field and let C be a reduced geometrically connected proper curve over k of genus g. Suppose that the characteristic of k is 0 or prime to $g-1$. Show that $H^0(C, \mathscr{O}_C) = k$.

Exercise 26.9. Show that in the situation of Proposition 26.59, for $g \geq 2$ the converse to (1) or (2) does not hold in general.

Exercise 26.10. Let X be a proper scheme over a field k and let $\mathscr{L}$ be an ample line bundle on X such that for all $n \geq 0$ the multiplication map $H^0(X, \mathscr{L}) \otimes_k H^0(X, \mathscr{L}^{\otimes n}) \to H^0(X, \mathscr{L}^{\otimes n+1})$ is surjective. Show that $\bigoplus_{n\geq 0} H^0(X, \mathscr{L}^{\otimes n})$ is a quotient of $\operatorname{Sym} H^0(X, \mathscr{L})$ and deduce that $\mathscr{L}$ is very ample.

Exercise 26.11. Let X be an integral proper curve over a field k of genus 0 such that $H^0(X, \mathscr{O}_X) = k$. Show that every line bundle on X of degree > 0 is very ample.
Hint: Exercise 26.10

Exercise 26.12. Let k be a field, and let $C = V_+(f) \subset \mathbb{P}^2_k$ be a projective curve over k, with $f \in k[X_0, X_1, X_2]$ homogeneous of degree d. A point $x \in C(k)$ is called an *inflexion point*, or a *flex*, for short, if x is a smooth point of C and the intersection multiplicity of C with the tangent line T_xC at x is ≥ 3. (Cf. Definition 16.60.)
(1) The *Hessian* of f is
$$H_f = \det\left(\frac{\partial^2 f}{\partial X_i \partial X_j}\right)_{i,j=0,1,2}$$

Show that H_f is a homogeneous polynomial which is invariant up to scalars in $k^\times$ under change of coordinates, i.e., under automorphisms of $\mathbb{P}^2_k$.
(2) Now suppose that $\operatorname{char}(k) \neq 2$. Show that the flexes of C are the smooth points of C that lie in $V_+(H_f)$.
(3) Suppose that $\operatorname{char}(k) \neq 2$. Determine all flexes of the *Fermat cubic* $V_+(X_0^3 + X_1^3 + X_2^3)$ (depending on k).

Exercise 26.13. Let k be a field.
(1) Let $E \subset \mathbb{P}^2_k$ be an elliptic curve defined by a Weierstraß equation. Prove that the point $(0:1:0)$ is a flex (Exercise 26.12) of E.
(2) Now let $E \subset \mathbb{P}^2_k$ be an elliptic curve whose fixed neutral element 0 is a flex. Denote by $E[3]$ the kernel of the multiplication $E \to E$ by 3, i.e., $x \mapsto x + x + x$ where addition is the group law on E. Show that $E[3](k)$ is precisely the set of inflexion points in $E(k)$.

Exercise 26.14. Let k be a field, and let E be an elliptic curve over k with neutral element $0 \in E(k)$. Let $C = E \setminus \{0\}$, an affine curve over k.
(1) Prove that the groups $\operatorname{Pic}(C)$ and $E(k)$ are isomorphic.
(2) Now let $k = \mathbb{C}$ be the field of complex numbers. Denote by $\operatorname{Pic}^{\mathrm{an}}(C^{\mathrm{an}})$ the complex analytic Picard group of the (non-compact) Riemann surface C^{an}, i.e., the group of isomorphism classes of locally free $\mathscr{O}_{C^{\mathrm{an}}}$-modules of rank 1. Prove that $\operatorname{Pic}^{\mathrm{an}}(C^{\mathrm{an}}) \cong H^2(C^{\mathrm{an}}, \mathbb{Z}) = 0$. In particular, the natural map $\operatorname{Pic}(C) \to \operatorname{Pic}^{\mathrm{an}}(C^{\mathrm{an}})$ is not injective (in contrast to Theorem 20.60).

Exercise 26.15. Show that a plane curve is not hyperelliptic.
Hint: Use Proposition 26.79 and Proposition 26.129.

Exercise 26.16. Let $g \in \mathbb{N}$, and let k be a field. Show that there exists a smooth proper curve C of genus g over k.

Exercise 26.17. Let k be a field and let $f \colon C \to \mathbb{P}^1$ be a generically étale morphism of connected smooth proper curves over k with $\deg(f) = 2$. In particular, C is hyperelliptic.

(1) Assume that there exists a point $0 \in C(k)$ with $\ell(2[0]) > 1$. Show that there exists a rational function $x \in K(C)$ which has a double pole at 0 and no other poles. Show that $\ell(2g[0]) = g + 1$, that $\ell((2g + 1)[0]) = g + 2$ and that there is an element $y \in H^0(C, \mathscr{O}_C((2g+1)[0]))$ such that $1, x, \ldots, x^g, y$ is a basis of this space. Show that, possibly after changing x and y by elements of $k^\times$, they satisfy an equation of the form
$$y^2 + q(x)y = p(x)$$
for a monic polynomial $p \in k[x]$ of degree $2g + 1$ and a polynomial $q \in k[x]$ of degree $\leq g$. Compare Proposition 26.95.
   Show that the polynomial $Y^2 + q(X)Y - p(X)$ is irreducible and conclude that $C \setminus \{0\}$ is isomorphic to the affine plane curve defined by the above equation.

(2) Assume that k is algebraically closed. Show that there exist at most $2g + 2$ points $p \in C(k)$ with $\ell(2[p]) > 1$, and precisely $2g + 2$ such points if $\operatorname{char} k \neq 2$. *Hint.* Show that the hyperelliptic cover of C is ramified at p. Apply the Theorem of Riemann-Hurwitz to f.

Exercise 26.18. Let k be a field of characteristic 2 and let C be the hyperelliptic curve over k defined by the affine equation $y^2 - y - x^5 = 0$. Compute the ramification divisor of the hyperelliptic cover $C \to \mathbb{P}^1_k$.

Exercise 26.19. Let $k = \mathbb{F}_3$ be the field with 3 elements, and let C be the hyperelliptic curve over k obtained as the smooth proper model of the affine curve given by the equation $y^2 = -x^6 + x^2 - 1$. Show that $C(k) = \emptyset$.

Exercise 26.20. Let k be a field and let C be a proper connected curve over k. Let $\mathscr{F}$ be a coherent $\mathscr{O}_C$-module of rank 0. Show that
$$\deg(\mathscr{F}) = \sum_c [\kappa(c) : k] \operatorname{lg}_{\mathscr{O}_{C,c}}(\mathscr{F}_c).$$
where the sum runs over all closed points c of C.

Exercise 26.21. Let C be a proper curve over a field and let $\mathscr{E}$ and $\mathscr{F}$ be vector bundles of constant rank over C. Show that
$$\deg(\mathscr{E} \otimes \mathscr{F}) = \operatorname{rk}(\mathscr{E}) \deg(\mathscr{F}) + \operatorname{rk}(\mathscr{F}) \deg(\mathscr{E})$$
Hint: Use Exercises 23.30 to reduce to the case that $\mathscr{E}$ is a line bundle.

Exercise 26.22. Let k be a field, let C be a smooth proper geometrically connected curve over k. Let $\mathscr{L}_0$ and $\mathscr{L}_1$ be line bundles on C of degree d_0 and d_1, respectively. Let $\mathscr{E}$ be a vector bundle that sits in an extension

(*)
$$0 \longrightarrow \mathscr{L}_0 \longrightarrow \mathscr{E} \longrightarrow \mathscr{L}_1 \longrightarrow 0.$$

(1) Show that $\mathscr{E}$ has slope $\frac{d_0+d_1}{2}$.

(2) Suppose that (*) is non-split and that $d_1 = d_0 + 1$. Show that $\mathscr{E}$ is a stable vector bundle.

(3) Suppose that $d_1 > d_0$. Show that $\dim_k \operatorname{Ext}^1_{\mathcal{O}_C}(\mathscr{L}_1, \mathscr{L}_0) = d_1 - d_0 - 1 + g$ and deduce that there exist non-split exact sequences of the form (*) if and only if $g \geq d_0 - d_1 + 2$.

Exercise 26.23. Let k be a field, let C be a smooth proper geometrically connected curve over k of genus $g \geq 2$. Le $\mathscr{E}_1$ and $\mathscr{E}_2$ be two non-isomorphic stable vector bundles of the same slope λ.

(1) Show $\operatorname{Hom}_{\mathcal{O}_C}(\mathscr{E}_2, \mathscr{E}_1) = 0$ and deduce that $\dim_k(\operatorname{Ext}^1_{\mathcal{O}_C}(\mathscr{E}_2, \mathscr{E}_1)) = \operatorname{rk}(\mathscr{E}_1)\operatorname{rk}(\mathscr{E}_2)(g - 1) > 0$.

(2) Let $0 \to \mathscr{E}_1 \to \mathscr{E} \to \mathscr{E}_2 \to 0$ be a non-trivial extension. Show that $\mathscr{E}$ is semistable of slope λ and that $\mathscr{E}$ is not stable. Show that $\operatorname{End}_{\mathcal{O}_C}(\mathscr{E}) = k$ if k is algebraically closed.

Exercise 26.24. Let C be a smooth proper geometrically connected curve over a field k. Let $\mathscr{E}$ be a vector bundle on C. Show that for every subbundle $\mathscr{E}'$ of $\mathscr{E}$ the point $(\deg(\mathscr{E}'), \operatorname{rk}(\mathscr{E}'))$ lies on or below the HN polygon of $\mathscr{E}$. Deduce that the HN polygon is the concave envelope of the points $(\deg(\mathscr{E}'), \operatorname{rk}(\mathscr{E}'))$, where $\mathscr{E}'$ runs through the set of subbundles of $\mathscr{E}$.

Exercise 26.25. Let C be a smooth proper geometrically connected curve and let $p\colon \tilde{C} \to C$ be an etale cover. Show that $p_*\mathcal{O}_{\tilde{C}}$ is a semistable vector bundle of slope 0.

Exercise 26.26. Let k be a field. Let C be a smooth proper geometrically connected curve over k of genus 0. Suppose that $C(k) = \emptyset$. Then C is a Brauer-Severi curve its corresponding central quaternion k-algebra D is a skew field. Let $\bar{k}$ be an algebraic closure, fix an isomorphism $\mathbb{P}^1_{\bar{k}} \cong C \otimes_k \bar{k}$ and denote by $p\colon \mathbb{P}^1_{\bar{k}} \to C$ the projection.

(1) Show that every non-zero vector bundle on C has an integral slope $d \in \mathbb{Z}$.

(2) Show that for every $d \in \mathbb{Z}$ there exists a unique (up to isomorphism) stable vector bundle $\mathscr{E}_d$ of slope d. If d is even, then $\mathscr{E}_d$ is a line bundle with $p^*\mathscr{E}_d \cong \mathcal{O}_{\mathbb{P}^1_{\bar{k}}}(d)$. If d is odd, then $\mathscr{E}_d$ is a non-split extension of $\mathscr{E}_{d-1}$ by $\mathscr{E}_{d+1}$ and $p^*\mathscr{E}_d \cong \mathcal{O}_{\mathbb{P}^1_{\bar{k}}}(d)^2$.

(3) Show that one has $\operatorname{End}_{\mathcal{O}_C}(\mathscr{E}_d) = D$ for d odd.

Exercise 26.27. Let k be a field and let C be a connected smooth projective curve of genus 3 over C. Assume that C is not hyperelliptic. Show that C is a plane quartic curve, i.e., $C \cong V_+(f) \subset \mathbb{P}^2_k$ for a polynomial $f \in k[X, Y, Z]$ of degree 4.

Exercise 26.28. Let k be a finite field. Prove the Weil conjectures (Theorem 26.173) for $C = \mathbb{P}^1_k$ by computing the zeta function explicitly in this case.

27 Abelian schemes

Content

- Preliminaries and general results about group schemes
- Definition and basic properties of abelian schemes
- The Picard functor
- Duality of abelian schemes
- Cohomology of line bundles on abelian schemes

In this chapter, we take up the topic of abelian varieties that we already touched in Chapter 16, see Sections (16.31) ff. Recall that an abelian variety over a field k is a connected proper smooth group scheme. While we will prove several further results on abelian varieties below, for instance that every abelian variety is projective (Proposition 27.174), the main focus in this chapter is putting this notion in families: An *abelian scheme A* over a scheme S is a proper smooth group scheme A over S such that all fibers are (geometrically) connected, see Definition 27.89. In particular, every fiber is an abelian variety. This generalization to families of abelian varieties will allow us to illustrate many of the results of the previous chapters in action.

An important feature of the theory is a duality on the category of abelian varieties, and more generally of abelian schemes over an arbitrary base scheme: To an abelian scheme A/S we can attach the *dual abelian scheme A^t* over S. The dual abelian scheme is defined in terms of its functor of points. Proving that it is in fact representable is quite subtle. See Sections (27.29), (27.39). Here we follow a proof by Deligne that is sketched in [FaCh]$^{\circ}$, see also [GrK].

With the dual abelian scheme at our disposal we can prove several results about abelian schemes. We show the Fourier-Mukai equivalence (Theorem 27.243) for abelian schemes over arbitrary base schemes. This result is based on a good understanding of the cohomology of the Poincaré bundle (Proposition 27.229). Then we study the cohomology of line bundles on abelian schemes and in particular prove the Riemann-Roch theorem for abelian varieties (Theorem 27.253) and characterize relative ample line bundles (Theorem 27.264). We also use the Fourier-Mukai equivalence to prove Atiyah's classification of vector bundles on elliptic curves (Theorem 27.271).

We define the notion of polarization (Definition 27.280). Altogether, we then have all the ingredients to write down the moduli functor giving rise to the *moduli space of principally polarized abelian varieties with full level n structure*, see Section (27.55), an object of great importance in classical algebraic geometry (when considered over $\mathbb{C}$) as much as in arithmetic algebraic geometry (when considered over $\mathbb{Q}$ or over $\mathbb{Z}$). Showing the representability of this functor and studying the moduli space in more detail requires methods beyond the scope of this book, however. Nevertheless we hope that this will open the door for the reader to a fascinating topic in this area.

In Section (27.54) at the end we will discuss abelian varieties over the complex numbers.

© Springer Fachmedien Wiesbaden GmbH, ein Teil von Springer Nature 2023
U. Görtz und T. Wedhorn, *Algebraic Geometry II: Cohomology of Schemes*,
Springer Studium Mathematik – Master, https://doi.org/10.1007/978-3-658-43031-3_12

The complex manifold associated with an abelian variety over $\mathbb{C}$ is a complex torus, i.e., a quotient of a complex vector space by a full lattice. This description allows for simpler proofs for several of the results in this chapter in this particular situation, and of course also historically was studied before the general theory.

Preliminaries and general results about group schemes

We start the chapter by collecting some results about general group schemes, including a digression on fppf-sheaves and algebraic spaces which is necessary since some of the functors that we define, e.g., quotients of group schemes or Picard schemes, are only representable by algebraic spaces.

(27.1) General facts on group schemes.

In this section we denote by S a scheme. Let $f\colon G \to S$ be an S-scheme. Recall the notion of a *group scheme over S* (a scheme G such that for every S-scheme T the T-valued points $G(T)$ are equipped with a group structure, functorially in T, see Section (4.15)) and the notion of a *homomorphism of group schemes over S*. A group scheme G is called *commutative*[1] if the group $G(T)$ of morphisms $T \to G$ of S-schemes is commutative for all S-schemes.

If $G \to S$ is a group scheme over S and $T \to S$ is a morphism of schemes, then the base change $G_T = G \times_S T \to T$ is a group scheme over T. In particular, if $T \to S$ is the canonical map $\operatorname{Spec} \kappa(s) \to S$ for a point $s \in S$, then the fiber $G_s = G \times_S \operatorname{Spec} \kappa(s)$ is a group scheme over the residue field $\kappa(s)$.

We have the following notion of translation.

Definition and Remark 27.1. Let $f\colon G \to S$ be a group scheme, let T be an S-scheme and let $g \in G(T) = \operatorname{Hom}_S(T, G)$. Then the *left translation* $t_g\colon G_T \to G_T$ is the composition

$$G_T = T \times_T G_T \xrightarrow{(g,\mathrm{id})} G_T \times_T G_T \longrightarrow G_T,$$

where the last map is the multiplication of the T-group scheme G_T. For any T-scheme T' the left translation t_g is given on T'-valued points by left multiplication by the image of g in $G(T')$.

The left translation by g is an isomorphism of T-schemes whose inverse is given by left translation by g^{-1}.

There is an analogous definition for the *right translation*[2] by a T-valued point of G.

We have the following criterion for a group scheme to be separated which is the analogue of the purely topological fact that a topological group is Hausdorff if and only if its unit point is closed.

---

[1] While for an abstract group the terms *commutative* and *abelian* are used interchangeably, in the situation at hand we reserve the term *abelian* for abelian schemes in the sense of Definition 27.89 below.

[2] We do not introduce a notation for the right translation since in the sequel we will mainly deal with commutative group schemes.

Proposition 27.2. *Let $G \to S$ be a group scheme. Then $G \to S$ is separated if and only if the unit section $e\colon S \to G$ is a closed immersion.*

Proof. The condition is necessary by Example 9.12. It is sufficient because we have the following cartesian diagram,

$$
\begin{array}{ccc}
G & \longrightarrow & S \\
\Delta_{G/S} \downarrow & & \downarrow e \\
G \times_S G & \xrightarrow{(g,h) \mapsto gh^{-1}} & G.
\end{array}
$$
$\square$

Definition 27.3. *Let G be an S-group scheme. A* subgroup scheme *is a subscheme G' such that the following equivalent assertions hold.*
(i) *The subset $G'(T)$ is a subgroup of $G(T)$ for all S-schemes T.*
(ii) *The multiplication on G induces by restriction a morphism $G' \times_S G' \to G'$, the inversion on G induces by restriction a morphism $G' \to G'$, and the unit section of G factors through G'.*
Such a subgroup scheme is called a normal subgroup scheme *if $G'(T)$ is a normal subgroup in $G(T)$ for all T.*

Example 27.4. For instance, the unit section $e\colon S \to G$ is always an immersion as is any section of a scheme morphism (Example 9.12). Hence it defines a subgroup scheme of G, where $S(T) \subseteq G(T)$ is the trivial group consisting of one element.

This notion of subgroup scheme differs from the notion defined in Definition 4.45 where we assumed that the immersion $H \to G$ is closed. However, we will show below (Proposition 27.11 and Exercise 27.2) that if S is the spectrum of a field, then these two notions coincide. Moreover, almost all subgroup schemes that we will consider in the sequel will be closed subschemes.

Let $f\colon G \to H$ be a homomorphism of S-group schemes, and let H' be a subgroup scheme of H (not necessarily closed). Then we define the preimage $f^{-1}(H') := G \times_H H'$ which is a subgroup scheme of G. In particular, the *kernel* $\mathrm{Ker}(f)$ is defined by the cartesian diagram

(27.1.1)
$$
\begin{array}{ccc}
\mathrm{Ker}(f) & \longrightarrow & S \\
\downarrow & & \downarrow e \\
G & \xrightarrow{f} & H,
\end{array}
$$

where $e\colon S \to H$ is the unit section of H. For every S-scheme T one has $\mathrm{Ker}(f)(T) = \mathrm{Ker}(f(T)\colon G(T) \to H(T))$. In particular $\mathrm{Ker}(f)$ is a normal subgroup scheme of G. If $H \to S$ is separated (which is automatic if $S = \mathrm{Spec}\, k$ for a field k by Corollary 27.9 below), then e is a closed immersion (Proposition 27.2) and hence $\mathrm{Ker}(f)$ will be a closed subgroup scheme of G.

Remark 27.5. Let $G \to S$ be a group scheme. Then there is a commutative diagram of S-schemes

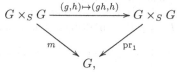

where m is the multiplication of the group scheme, pr_1 is the first projection, and where the horizontal arrow, given on T-valued points for any S-scheme T, is an isomorphism with inverse $(g, h) \mapsto (gh^{-1}, h)$. Hence if $\mathbf{P}$ is any property of scheme morphisms which is stable under base change and under composition with isomorphisms and if $G \to S$ has $\mathbf{P}$, then the multiplication m has $\mathbf{P}$.

(27.2) Affine Group schemes and Hopf algebras.

Let R be a ring and let $G = \operatorname{Spec} A$ be an affine group scheme over $\operatorname{Spec} R$. Then multiplication, inversion, and unit of G induce homomorphisms of R-algebras

$$(27.2.1) \qquad m^*: A \longrightarrow A \otimes_R A, \qquad i^*: A \to A, \qquad e^*: A \to R,$$

called *comultiplication*, *antipode*, and *counit*, respectively. The identities in the definition of a group scheme such as associativity of the group law correspond to the fact that (A, m^*, i^*, e^*) is a (commutative) Hopf algebra in the following sense.

Definition 27.6. *A commutative Hopf algebra over R is an R-algebra A together with maps as in (27.2.1) such that*

$$
\begin{array}{ccc}
A & \xrightarrow{\ m^*\ } & A \otimes A \\
{\scriptstyle m^*}\downarrow & & \downarrow{\scriptstyle m^* \otimes \mathrm{id}} \\
A \otimes A & \xrightarrow[\ \mathrm{id} \otimes m^*\ ]{} & A \otimes A \otimes A
\end{array}
\qquad\qquad
\begin{array}{ccc}
A & \xrightarrow{\ e^*\ } & R \\
{\scriptstyle m^*}\downarrow & & \downarrow \\
A \otimes A & \xrightarrow[\ a \otimes b \mapsto i^*(a)b\ ]{} & A
\end{array}
$$

commute and such that $\mathrm{id}_A = (e^* \otimes \mathrm{id}) \circ m^*$. A Hopf algebra A is called cocommutative if $m^* = s \circ m^*$, where s is the endomorphism of $A \otimes A$ with $s(a \otimes a') = a' \otimes a$. A homomorphism of Hopf algebras over R is a homomorphism of R-algebras that respects comultiplication, antipode, and counit.

The equivalence between rings and affine schemes then induces an equivalence between the category of affine group schemes over R and the category of commutative Hopf algebras over R. Via this equivalence, commutative affine group schemes correspond to cocommutative commutative Hopf algebras.

More generally, if S is a scheme, then one has the obvious notion of a Hopf $\mathscr{O}_S$-algebra, and an equivalence between the category of relatively affine (commutative) group schemes over S and the category of quasi-coherent (cocommutative) commutative Hopf $\mathscr{O}_S$-algebras.

Example 27.7. Let R be a ring, and let $\mathbb{G}_{a,R}$ be the additive group over R, i.e., for every R-algebra A we set $\mathbb{G}_{a,R}(A) = A$, with the group structure given by addition. The Hopf algebra corresponding to $\mathbb{G}_{a,R}$ is the R-algebra $R[T]$ with comultiplication, counit, and antipode given by

$$m^*(T) = T \otimes 1 + 1 \otimes T, \qquad e^*(T) = 0, \qquad i^*(T) = -T.$$

Example 27.8. Let R be a ring, and let $\mathbb{G}_{m,R}$ be the multiplicative group over R, i.e., for every R-algebra A one has $\mathbb{G}_{m,R}(A) = A^\times$. Then the Hopf algebra corresponding to $\mathbb{G}_{m,R}$ is the R-algebra $R[T, T^{-1}]$ with comultiplication, counit, and antipode given by

$$m^*(T) = T \otimes T, \qquad e^*(T) = 1, \qquad i^*(T) = T^{-1}.$$

(27.3) Generalities about group schemes over a field.

Corollary 27.9. *Let k be a field and let G be a group scheme over k. Then G is separated.*

Proof. Let $\pi\colon G \to \operatorname{Spec} k$ be the structure morphism. Let $x \in G$ be the image point of the unit section $e \in G(k)$. Let $U = \operatorname{Spec} A \subseteq G$ be any open affine neighborhood of x. As $\pi \circ e = \operatorname{id}_{\operatorname{Spec} k}$ the homomorphism $A \to k$ corresponding to $e\colon \operatorname{Spec} k \to U$ has a section and therefore is surjective. This shows that e is a closed immersion. Hence G is separated by Proposition 27.2. $\qquad\square$

The proof shows that, more generally, any group scheme $G \to S$ is separated over S if S consists of only one point. One deduces that every group scheme over a discrete base scheme is separated. But there exist group schemes over discrete valuation rings that are not separated (Exercise 27.1).

Recall that in Section (16.31) we proved already the following statements about group schemes over a field.

Proposition 27.10. *Let k be a field, and let G be a group scheme locally of finite type over k.*
(1) *Then G is smooth over k if and only if G is geometrically reduced over k.*
(2) *The group scheme G is geometrically irreducible if and only if it is connected.*
(3) *Let k be perfect. Then G_{red} is a subgroup scheme of G which is smooth over k.*

Proposition 27.11. *Let k be a field and let G be a group scheme locally of finite type over k.*
(1) *Let $U, V \subseteq G$ be open dense subsets. Then $U \cdot V$ (the image of $U \times_k V$ under multiplication) equals the underlying topological space of G.*
(2) *If G is irreducible, then G is quasi-compact.*
(3) *Let H be a k-subgroup scheme of G. Then H is closed in G.*

The hypothesis that G is locally of finite type over k is in fact superfluous (Exercise 27.2).

Proof. We claim that we may assume that k is algebraically closed. Indeed, let $\bar{k}$ be an algebraic closure of k. As the projection $G_{\bar{k}} \to G$ is an open morphism (Theorem 14.38), the inverse image of any dense subset of G is dense in $G_{\bar{k}}$. Moreover, $G_{\bar{k}} \to G$ is surjective, which shows the claim for (1). To see the claim for Assertion (2) we use that if G is irreducible, then $G_{\bar{k}}$ is irreducible by Proposition 27.10 (2). Moreover, quasi-compactness can be checked after surjective base change. The claim for (3) holds since being a closed immersion can be checked after faithfully flat base change (Proposition 14.53).

Hence we may assume that k is algebraically closed. Then it suffices to show that $U(k) \cdot V(k) = G(k)$. As the inversion is an isomorphism, $V(k)^{-1}$ is again open and dense in $G(k)$. Then for $g \in G(k)$, $g(V(k)^{-1})$ is still open and dense and hence meets $U(k)$. Therefore there exist $u \in U(k)$ and $v \in V(k)$ such that $gv^{-1} = u$. This shows (1).

To see (2) let $U \subseteq G$ be an open dense affine subscheme. Then $U \times_k U$ is quasi-compact and its image under multiplication map is G by (1). Therefore G is quasi-compact.

For part (3), it suffices to show that H is closed in G. Let $\bar{H}$ be the closure of H, endowed with the reduced scheme structure. As $H \to \operatorname{Spec} k$ and $\bar{H} \to \operatorname{Spec} k$ are universally open (Theorem 14.38) and since H is dense in $\bar{H}$, $H \times_k H$ is dense in $H \times_k \bar{H}$, and $H \times_k \bar{H}$ is dense in $\bar{H} \times_k \bar{H}$. Therefore $H \times_k H$ is dense in $\bar{H} \times_k \bar{H}$. Let m be the multiplication morphism. Then $m^{-1}(\bar{H})$ is closed in $G \times_k G$ and contains $H \times_k H$. Hence $m(\bar{H} \times_k \bar{H}) \subseteq \bar{H}$. As $\bar{H} \times_k \bar{H}$ is reduced (Corollary 5.57), m induces by restriction a multiplication on $\bar{H}$. As $\bar{H}$ is also stable under inversion, $\bar{H}$ is a subgroup scheme. Then (1) implies that $H = H \cdot H = \bar{H}$. $\square$

Definition and Proposition 27.12. *Let k be a field and let G be a group scheme locally of finite type over k. Then the image of the unit section is a point in G. Its connected component is denoted by G^0. As G is locally noetherian, G^0 is an open and closed subset and we endow it with the scheme structure making G^0 into an open subscheme. It is called the* identity component *of G. The following assertions hold.*
(1) *The subscheme G^0 is a quasi-compact normal open and closed subgroup scheme of G which is geometrically irreducible over k.*
(2) *Every homomorphism of group schemes $\varphi\colon G \to H$ locally of finite type over k induces a homomorphism $G^0 \to H^0$.*
(3) *For every field extension K of k one has $(G \otimes_k K)^0 = G^0 \otimes_k K$.*

Proof. As G^0 is connected and has a rational point, it is geometrically connected by Lemma 26.4. Hence $G^0 \times_k G^0$ is connected (Proposition 5.53). Therefore the restriction of the multiplication map to $G^0 \times_k G^0$ factors through G^0. Clearly, the restriction of the inversion map to G^0 factors also through G^0. Therefore G^0 is an open and closed connected subgroup scheme. It is geometrically irreducible by Proposition 27.10 and quasi-compact by Proposition 27.11 (2).

To show that G^0 is normal, consider the action of G on itself by conjugation. If G' is any connected component of G, then the restriction of the action to $G' \times_k G^0$ factors through G^0 since $G' \times_k G^0$ is connected. This shows that the conjugation action of all of G preserves G^0, i.e., G^0 is a normal subgroup.

Assertion (2) is clear, and (3) holds because G^0 is geometrically connected. $\square$

Lemma 27.13. *Let k be a field and let G be a k-group scheme locally of finite type. Then every connected component of G is quasi-compact and irreducible and G is equi-dimensional.*

Proof. Suppose first that k is algebraically closed. Let C be a connected component of G. As C is locally of finite type over k and k is algebraically closed, there exists $g \in C(k)$. Translation with g induces an isomorphism $G^0 \xrightarrow{\sim} C$. Therefore C is quasi-compact and irreducible by Proposition 27.12. Moreover, every irreducible component of G is isomorphic to G^0. In particular G is equi-dimensional of dimension $\dim G^0$.

Now let k be arbitrary and let $\bar{k}$ be an algebraic closure of k. Then $\operatorname{Spec} \bar{k} \to \operatorname{Spec} k$ is universally open (Theorem 14.38) and universally closed because $k \to \bar{k}$ is integral. Therefore the morphism $\pi\colon G_{\bar{k}} \to G$ obtained by base change is open, closed, integral, and surjective. Let C be a connected component of G and let C' be a connected component of $G_{\bar{k}}$ mapping to C. As π is open and closed, $\pi(C') = C$. As we have already seen that C' is quasi-compact and irreducible, so is C. As π is also integral, we have $\dim C' = \dim C$ (Proposition 12.12) which implies that G is equi-dimensional. $\square$

Over arbitrary fields in general there exist connected components of G that are not geometrically connected and in particular are not isomorphic to G^0 (Exercise 16.7).

Proposition 27.14. *Let k be a field and let $f\colon G \to H$ be a quasi-compact homomorphism of group schemes locally of finite type over k.*
(1) *The subspace $f(G)$ is closed in H.*
(2) *One has $\dim G = \dim f(G) + \dim \operatorname{Ker}(f)$.*
(3) *Suppose that H is smooth over k and that f is surjective. Then f is faithfully flat.*

Proof. Let $\bar{k}$ be an algebraic closure of k. Then $\pi\colon H_{\bar{k}} \to H$ is surjective and integral and in particular closed. Moreover for every closed subspace Z of $H_{\bar{k}}$ we have $\dim \pi(Z) = \dim Z$ by Proposition 12.12. Hence we may assume that k is algebraically closed.

We first prove (3). As H is reduced, its identity component H^0 is an integral scheme. Hence by generic flatness (Theorem 10.84) there exists a non-empty open subset V of H^0 such that the restriction $f^{-1}(V) \to V$ is flat. As f is a group homomorphism, it is flat over all translates $t_h(V)$ for $h \in H(k)$. As $H = \bigcup_{h \in H(k)} t_h(V)$, this shows that f is flat.

To show (1) and (2) we may replace G by G_{red} and assume that G is reduced and hence smooth over k (Proposition 27.10 (1)).

We show (1). Let C be the closed reduced subscheme of H whose underlying topological space is the closure of $f(G)$. Then C is stable under the inversion of H. The morphism $f\colon G \to C$ is quasi-compact and dominant, hence $f \times f\colon G \times G \to C \times C$ is also quasi-compact and dominant. Hence the multiplication of H maps $C \times C$ into C. Therefore C is a subgroup scheme of H. Replacing H by C we may assume that f is dominant. By Chevalley's theorem (Theorem 10.20), $f(G)$ is constructible. As it is dense in H it contains an open dense subset U of H. By Proposition 27.11 (1) we see that $H = U \cdot U \subseteq f(G)$. Therefore $f(G) = H$.

Let us show (2). We may replace H by $f(G)$ with the reduced scheme structure and can therefore assume that f is surjective and that H is smooth over k. Moreover f is then flat by (3). Then $f(G^0)$ is open (by openness of flat morphisms locally of finite presentation, Theorem 14.35) and closed (by (1)) in H^0. Hence the restriction $f^0\colon G^0 \to H^0$ is still surjective. We have $\dim(G^0) = \dim(G)$ and $\dim(H^0) = \dim(H)$. Moreover we have $\operatorname{Ker}(f)^0 \subseteq \operatorname{Ker}(f^0) \subseteq \operatorname{Ker}(f)$ and hence $\dim \operatorname{Ker}(f^0) = \dim \operatorname{Ker}(f)$. Therefore we may assume that G and H are integral schemes.

By Remark 14.119 there exists a non-empty open subscheme V of H such that $\dim(G) = \dim f^{-1}(v) + \dim H$ for all $v \in V(k)$. As all fibers have the same dimension by homogeneity this shows (2). $\square$

(27.4) Differentials of group schemes.

Let S be a scheme, let $f\colon G \to S$ be a group scheme, let $e\colon S \to G$ be its unit section which is an immersion (Remark 9.12). Let $\Omega^1_{G/S}$ be the sheaf of Kähler differentials (Section (17.1)). It is an $\mathscr{O}_G$-module which is of finite presentation if $G \to S$ is locally of finite presentation (Corollary 17.34).

Let $\mathscr{C}_e$ be the conormal sheaf of the immersion e (Section (17.3)). By (17.4.4) one has

$$(27.4.1) \qquad \mathscr{C}_e = e^*\Omega^1_{G/S}.$$

Proposition 27.15. *There are isomorphisms of $\mathscr{O}_G$-modules*

$$(27.4.2) \qquad \Omega^1_{G/S} \cong f^*e^*\Omega^1_{G/S} \cong f^*\mathscr{C}_e.$$

Thus $\Omega^1_{G/S}$ is a (locally) free $\mathscr{O}_G$-module whenever $\mathscr{C}_e$ is a (locally) free $\mathscr{O}_S$-module. In particular, if S is the spectrum of a field, then $\Omega^1_{G/S}$ is a free $\mathscr{O}_G$-module.

Proof. The second isomorphism follows from (27.4.1). To construct the first isomorphism consider the commutative diagram

$$
\begin{array}{ccc}
G \times_S G & \xrightarrow{\ m\ } & G \\
{\scriptstyle p_1}\downarrow & & \downarrow{\scriptstyle f} \\
G & \xrightarrow{\ f\ } & S,
\end{array}
$$

where m is multiplication of the group scheme and p_i is the i-th projection. It is easily checked on T-valued points, where T is some S-scheme, that this diagram is cartesian. Hence by Proposition 17.30 we get isomorphisms

$$
m^*\Omega^1_{G/S} \cong \Omega^1_{p_1} \cong p_2^*\Omega^1_{G/S}.
$$

Pulling back by $i := (\mathrm{id}_G, e)\colon G \to G \times_S G$ we obtain an isomorphism of $i^*m^*\Omega^1_{G/S} = \mathrm{id}_G^*\,\Omega^1_{G/S} = \Omega^1_{G/S}$ and $i^*p_2^*\Omega^1_{G/S} = f^*e^*\Omega^1_{G/S}$. $\qquad\square$

Applying exterior powers we obtain the following consequence.

Corollary 27.16. *For all $i \geq 0$ one has isomorphisms of $\mathscr{O}_G$-modules*

$$
(27.4.3) \qquad\qquad \Omega^i_{G/S} \cong f^*e^*\Omega^i_{G/S} \cong f^*\big(\textstyle\bigwedge^i_{\mathscr{O}_S}\mathscr{C}_e\big).
$$

Remark and Definition 27.17. Let $f\colon G \to S$ be a group scheme locally of finite presentation. Let $e\colon S \to G$ be its unit section with conormal sheaf $\mathscr{C}_e$. We set

$$
\mathrm{Lie}(G) := T_e(G/S) = \mathscr{C}_e^\vee = (e^*\Omega^1_{G/S})^\vee = e^*\mathscr{T}_{G/S},
$$

where $(-)^\vee$ denotes the $\mathscr{O}_S$-dual and $\mathscr{T}_{G/S}$ denotes the tangent sheaf. This is the underlying $\mathscr{O}_S$-module of the Lie algebra[3] of G. As $\Omega^1_{G/S}$ is of finite presentation (Corollary 17.34), $\mathrm{Lie}(G)$ is quasi-coherent by Proposition 7.29. The $\mathscr{O}_S$-module $\mathrm{Lie}(G)$ determines the tangent sheaf by $\mathscr{T}_{G/S} = f^*\mathrm{Lie}(G)$ (27.4.2).

One has the following functoriality properties for $\mathrm{Lie}(G)$.

Remark 27.18. Let G and H be group schemes locally of finite presentation over S.
(1) Let $f\colon G \to H$ be a morphism of S-schemes sending e_G to e_H. Then by functoriality (17.6.9) one obtains a homomorphism of $\mathscr{O}_S$-modules

$$
(27.4.4) \qquad\qquad T_e(f)\colon \mathrm{Lie}(G) \longrightarrow \mathrm{Lie}(H).
$$

(2) By (17.6.10) one has

$$
(27.4.5) \qquad\qquad \mathrm{Lie}(G \times_S H) = \mathrm{Lie}(G) \times \mathrm{Lie}(H).
$$

---

[3] As we are interested in the sequel mainly in commutative group schemes, we will not give the definition of the Lie bracket on $\mathrm{Lie}(G)$ here. It is obtained by pulling back the Lie bracket on $\mathscr{T}_{G/S}$ (Remark 17.40) with some caveat as the Lie bracket on $\mathscr{T}_{G/S}$ is only $\mathscr{O}_S$-linear. See [SGA3] Exp. II.

(3) The multiplication map $m\colon G \times_S G \to G$ induces the addition

$$T_e(m)\colon \mathrm{Lie}(G) \times \mathrm{Lie}(G) = \mathrm{Lie}(G \times_S G) \longrightarrow \mathrm{Lie}(G), \qquad (\xi, \zeta) \mapsto \xi + \zeta.$$

Indeed, write $m_* := T_e(m)$. As $m \circ (\mathrm{id}_G, e) = \mathrm{id}_G$, we find $m_*(\xi, 0) = \xi$. Similarly, $m_*(0, \zeta) = \zeta$. Hence

$$m_*(\xi, \zeta) = m_*((\xi, 0) + (0, \zeta)) = m_*(\xi, 0) + m_*(0, \zeta) = \xi + \zeta.$$

(4) For a scheme U we set $U[\varepsilon] := \mathrm{Spec}\, \mathscr{O}_U[T]/(T^2)$. Then the structure morphism $U[\varepsilon] \to U$ has a section that corresponds to $T \mapsto 0$. It is a thickening of order 1. Combining Proposition 17.43 with (3) one sees that one has for every open subscheme U of S an exact sequence of groups

(27.4.6) $$0 \to \Gamma(U, \mathrm{Lie}(G)) \to G(U[\varepsilon]) \to G(U) \to 1.$$

Every $a \in \Gamma(U, \mathscr{O}_S)$ yields an endomorphism of the U-scheme $U[\varepsilon]$ by multiplying ε, i.e., the residue class of T, by a. By functoriality we obtain an endomorphism of the functor $U \mapsto G(U[\varepsilon])$ which induces a functorial endomorphism of the abelian group $\Gamma(U, \mathrm{Lie}(G))$. It is not difficult to see that this is simply the scalar multiplication of the $\Gamma(U, \mathscr{O}_S)$-module $\mathrm{Lie}(G)$ (cf. Remark 6.8).

Remark 27.19. If $G\colon (\mathrm{Sch}/S)^{\mathrm{opp}} \to (\mathrm{AbGrp})$ is any functor, then one can define a sheaf of abelian groups $\mathrm{Lie}(G)$ on S by

$$\Gamma(U, \mathrm{Lie}(G)) := \mathrm{Ker}(G(U[\varepsilon]) \to G(U))$$

and endow $\mathrm{Lie}(G)$ with the structure of an $\mathscr{O}_S$-module by the construction of Remark 27.18 (4).

Proposition 27.20. *Let G be a group scheme locally of finite type over a field k. Then*

(27.4.7) $$\dim \mathrm{Lie}(G) \geq \dim G.$$

The following assertions are equivalent.
(i) The scheme G is smooth over k.
(ii) One has equality in (27.4.7).
(iii) The scheme G is smooth in $e \in G(k)$ (viewed as a closed point of G).

Proof. Inequality 27.4.7 holds since $\dim T_g(G) \geq \dim G$ for all $g \in G(k)$.

Clearly, (i) implies (iii). Moreover, (iii) and (ii) are equivalent by Corollary 18.68.

It remains to show that (iii) implies (i). We can check smoothness fpqc-locally and hence may assume that k is algebraically closed. Using that the translation with $g \in G(k)$ is an isomorphism we see that G is smooth in all $g \in G(k)$. As k is algebraically closed, $G(k)$ is very dense in G (Proposition 3.35). As the set of points in which G is smooth over k is open, this shows that G is smooth in all points. $\square$

As we can check smoothness of flat morphisms of locally finite presentation on fibers (Theorem 18.56), we obtain by (27.4.2) the following corollary.

Corollary 27.21. *Let S be a scheme. Let $G \to S$ be a group scheme locally of finite presentation and let $g \geq 0$ be an integer. Then the following assertions are equivalent.*
(i) *The morphism $G \to S$ is smooth of relative dimension g.*
(ii) *The morphism $G \to S$ is flat, all fibers are of dimension g, and $\mathrm{Lie}(G)$ is a locally free $\mathscr{O}_S$-module of rank g.*
In this case, $\mathscr{C}_e = \mathrm{Lie}(G)^\vee$ is locally free of rank g and hence $\Omega^i_{G/S}$ is a locally free $\mathscr{O}_G$-module of rank $\binom{g}{i}$. Moreover, one has for the sheaf of global i-forms

$$(27.4.8) \qquad\qquad R^p f_* \Omega^i_{G/S} = R^p f_* \mathscr{O}_G \otimes_{\mathscr{O}_S} \bigwedge\nolimits^i_{\mathscr{O}_S} \mathscr{C}_e$$

for all $p \geq 0$ and all $i \geq 0$.

The last assertion follows from the projection formula (Proposition 22.84).

The formation of the $\mathrm{Lie}(G)$ commutes with base change in the following cases.

Proposition 27.22. *Let $G \to S$ be a group scheme locally of finite presentation, let $h\colon T \to S$ be a scheme morphism, and set $G_T = G \times_S T$. Suppose that one of the following hypotheses is satisfied.*
(1) *The morphism $G \to S$ is smooth.*
(2) *The morphism $h\colon T \to S$ is flat.*
Then there is a functorial isomorphism $\mathrm{Lie}(G_T) \cong h^* \mathrm{Lie}(G)$.

Proof. Let $e\colon S \to G$ and $e_T\colon T \to G_T$ be the unit sections, then $h^*\mathscr{C}_e \cong \mathscr{C}_{e_T}$. Since $\mathrm{Lie}(G) = \mathscr{C}_e^\vee$, we obtain a functorial homomorphism $\alpha\colon h^*(\mathscr{C}_e^\vee) \to (h^*\mathscr{C}_e)^\vee$ and we have to show that α is an isomorphism. If $G \to S$ is smooth, then $\mathscr{C}_e$ is finite locally free and we can conclude by Proposition 7.7.

Now suppose that $T \to S$ is flat. To see that α is an isomorphism, one can work locally on S and T. Hence we may assume that $S = \mathrm{Spec}\,A$ and $T = \mathrm{Spec}\,B$ are affine. Let M be the A-module of finite presentation corresponding to $\mathscr{C}_e$. Then apply Lemma 27.23 with $P = B$ and $N = A$. $\qquad\square$

Lemma 27.23. *Let A be a ring, let M, N, and P be A-modules. Suppose that M is of finite presentation and that P is flat. Then the functorial A-linear map*

$$\mathrm{Hom}_A(M, N) \otimes_A P \to \mathrm{Hom}_A(M, N \otimes_A P)$$

is an isomorphism.

Proof. Consider both sides as functor in M. Choose a resolution $L_1 \to L_0 \to M \to 0$ with L_i finitely generated free A-module. As P is flat, both functors map this resolution to a short left exact sequence. Hence we may assume that M is finitely generated and free by the five lemma. As both sides commute with direct sums in M, we may even assume that $M = A$. Then the assertion is clear. $\qquad\square$

(27.5) Singularities of group schemes.

Over arbitrary base schemes we easily deduce the following corollary in the presence of flatness.

Corollary 27.24. *Let S be a scheme and let $G \to S$ be a flat group scheme locally of finite presentation with geometrically reduced fibers. Then $G \to S$ is smooth.*

Proof. By Proposition 27.10 (1) all fibers of $G \to S$ are smooth, hence G is smooth over S by Theorem 18.56. □

In characteristic zero flat group schemes locally of finite presentation are smooth.

Theorem 27.25. *Let S be a scheme of characteristic 0, i.e. S is a $\mathbb{Q}$-scheme. Let $f\colon G \to S$ be an S-group scheme that is flat and locally of finite presentation over S. Then G is smooth over S.*

Proof. By Corollary 27.24 we may assume that $S = \operatorname{Spec} k$ is a field of characteristic zero. By Proposition 27.15 we have $\Omega^1_{G/k} \cong f^* \mathscr{C}_e$, where $e \in G(k)$ is the unit section. As $\mathscr{C}_e$ is a k-vector space, $\Omega^1_{G/k}$ is a (globally) free $\mathscr{O}_G$-module. Hence G is a smooth k-scheme because $\operatorname{char}(k) = 0$ (Proposition 18.69). □

In positive characteristic there exist plenty of examples of non-smooth group schemes (see for instance Example 18.47). But the structure morphism of a flat group scheme locally of finite presentation is still syntomic (see (19.9)).

Proposition 27.26. *Let S be a scheme and let G be an S-group scheme that is flat and locally of finite presentation over S. Then $G \to S$ is syntomic.*

As we will not use this result, we will only reduce it to a local statement and then refer to $[\mathrm{SGA3}]\substack{0\\ \mathsf{X}}$ for the crucial point.

Proof. Indeed, by Proposition 19.52 we may assume that $S = \operatorname{Spec} k$ for a field k. If $\operatorname{char} k = 0$, then Theorem 27.25 shows that G is even smooth over k. Hence we may assume that $\operatorname{char}(k) = p > 0$. As "syntomic" is stable under faithfully flat descent, we may also assume that k is algebraically closed. By Proposition 19.50 it suffices to show that $\mathscr{O}_{G,g}$ is a complete intersection ring for every closed point $g \in G(k)$. But left multiplication by g yields an automorphism of the k-scheme G which induces an isomorphism $\mathscr{O}_{G,g} \xrightarrow{\sim} \mathscr{O}_{G,e}$, where $e \in G(k)$ is the unit section. Hence it suffices to show that $\mathscr{O}_{G,e}$ or, equivalently by Remark 19.47 (2), $\widehat{\mathscr{O}}_{G,e}$ is a complete intersection ring. But one can show (e.g., $[\mathrm{SGA3}]\substack{0\\ \mathsf{X}}$ Exp. VIIB, Corollaire 5.4) that there exist integers $s \geq r \geq 0$ and $n_1, \ldots, n_r \geq 1$ such that $\widehat{\mathscr{O}}_{G,e} \cong k[\![T_1, \ldots, T_r, \ldots, T_s]\!]/(T_1^{p^{n_1}}, \ldots, T_r^{p^{n_r}})$ which is clearly a complete intersection ring. □

Proposition 27.27. *Let G be a group scheme that is flat and locally of finite presentation over a scheme S. Then the function $s \mapsto d(s) := \dim G_s$ is locally constant.*

Proof. Let $\pi\colon G \to S$ be the structure morphism. We first claim that the function $\delta\colon g \mapsto \dim_g(G_{\pi(g)})$ from G to $\mathbb{Z}$ is locally constant, i.e., continuous if $\mathbb{Z}$ is given the discrete topology. If π is smooth, then this is clear since the map is given by the relative dimension of G over S by (18.10.3). In general, we can use Proposition 27.26 to see that all fibers of $G \to S$ are locally complete intersections and in particular Cohen-Macaulay. Therefore δ is locally constant by Proposition 25.119.

As all fibers of G are equi-dimensional (Lemma 27.13), δ factors into $d \circ \pi$ for some map $d\colon S \to \mathbb{Z}$. As π is open and surjective (Theorem 14.35), d is also continuous. □

(27.6) Digression: Sheaves for the fppf topology.

Next we would like to study quotients of group schemes. The notion of quotient is much more subtle than that of subgroup schemes. If G is a group scheme over a base scheme S and $H \subseteq G$ is a subgroup scheme one would like to define a homogeneous space G/H, i.e., (ideally) a scheme G/H with a "canonical projection" $G \to G/H$ with suitable properties reflecting the quotient property. As G and H can be both considered as functors $(\mathrm{Sch}/S)^{\mathrm{opp}} \to (\mathrm{Grp})$ (Section (4.15)) and $H(T)$ is a subgroup of $G(T)$ for every S-scheme T, a first idea would be to consider the functor $T \mapsto G(T)/H(T)$. But this will almost never be a scheme. In fact it will usually not even be a Zariski sheaf (Section (8.3)). Moreover, we know that S-schemes considered as functors $(\mathrm{Sch}/S)^{\mathrm{opp}} \to (\mathrm{Sets})$ (called presheaves in (Sch/S)) satisfy the sheaf criterion for finer "topologies on (Sch/S)" (more precisely, for suitable Grothendieck topologies on (Sch/S)) as explained in Section (14.18). In order to have a chance that the quotient G/H is a scheme one has to sheafify the presheaf $T \mapsto G(T)/H(T)$ for some Grothendieck topology.

It turns out that one possible choice of Grothendieck topology is the fppf topology that we have already briefly mentioned in Remark 14.77. Thus we will define G/H as the fppf-sheafification of the presheaf $T \mapsto G(T)/H(T)$ on (Sch/S), see below for precise definitions. Then it is still not at all clear that the fppf-sheaf G/H is representable by an S-scheme. In fact, in general this will not be the case. But most of the fppf-sheaves that we encounter are at least locally for the étale topology (Section (20.5)) representable. Such fppf-sheaves are called algebraic spaces.

Therefore we will digress in this and the following two sections from the theory of group schemes and explain fppf-sheaves, algebraic spaces and what it means to be a surjective map of fppf-sheaves.

The category (Sch/S) is a large category and hence one runs into certain set-theoretic issues if one considers categories of functors with domain (Sch/S) which we largely ignore. These problems can be resolved, see Section (F.1) for a brief discussion and references.

For a functor F on (Sch/S), a morphism $g \colon T' \to T$ of S-schemes and an element $f \in F(T)$ we think of $F(g)(f)$ as the pullback of f along g and hence usually denote this element of $F(T')$ by g^*f. We can also view $f \in F(T)$ as a morphism $T \to F$ of functors; then g^*f is just the composition $f \circ g$. Sometimes, specifically if we think of g^*f as the restriction of f to an "open" (in the sense of some Grothendieck topology), we simply write $f_{|T'} := g^*f$. Given a morphism $f \colon G \to H$ of functors on (Sch/S) and an element $g \in G(T)$ for some S-scheme T, we also write $f(g) = f(T)(g) = f \circ g$, where for the final expression we again adopt the point of view that g corresponds to a morphism $T \to G$ which can be composed with f.

Definition and Lemma 27.28. *Let S be a scheme.*

(1) *A presheaf on (Sch/S) is by definition a functor $F \colon (\mathrm{Sch}/S)^{\mathrm{opp}} \to (\mathrm{Sets})$.*

(2) *A family of scheme morphisms $g_i \colon T_i \to T$ is called an* fppf-covering *if g_i is flat and locally of finite presentation for all i and if $T = \bigcup g_i(T_i)$.*

(3) *Let F be a presheaf on (Sch/S). Then F is called a* sheaf for the fppf-topology *or an* fppf-sheaf *if it satisfies the following equivalent conditions.*

   (i) *For every fppf-covering of S-schemes $(g_i \colon T_i \to T)_i$ the functor F satisfies the sheaf property for this covering. This means, denoting by $p_1 \colon T_i \times_T T_j \to T_i$ and $p_2 \colon T_i \times_T T_j \to T_j$ the projections, that for every $(a_i)_i \in \prod_i F(T_i)$ such that $p_1^* a_i = p_2^* a_j \in F(T_i \times_T T_j)$ there exists a unique $a \in F(T)$ with $g_i^* a = a_i$.*

(ii) *The functor F is a sheaf for the Zariski topology (i.e., satisfies the sheaf property for all families $(g_i \colon T_i \to T)_i$, where g_i are open immersions with $\bigcup g_i(T_i) = T$) and F satisfies the sheaf property for each family consisting of a single faithfully flat finitely presented morphism $g \colon \operatorname{Spec} R' \to \operatorname{Spec} R$ of affine S-schemes.*

Proof. Clearly, (i) implies (ii). Conversely, suppose that F satisfies (ii) and let $(g_i \colon T_i \to T)_i$ be an arbitrary fppf-covering. Set $T' = \coprod_i T_i$. Then the g_i define a morphism $g \colon T' \to T$ such that $(T' \to T)$ is an fppf-covering and F satisfies the sheaf condition with respect to $(T_i \to T)_i$ if and only if it satisfies it with respect to the covering consisting of the single morphism $g \colon T' \to T$.

Now choose an open affine covering $(U_\alpha)_\alpha$ of T. As F is a Zariski sheaf, it suffices to show that F satisfies the sheaf condition for $g^{-1}(U_\alpha) \to U_\alpha$ for all α. Hence we may assume that T is affine. In this case we may g precompose with a morphism $T'' \to T'$ such that the composition $T'' \to T$ is an fppf covering and such that T'' is affine (Lemma 14.79). But F satisfies the sheaf condition for such fppf coverings by hypothesis. $\square$

Example 27.29. Let S be a scheme and let X be an S-scheme. Then the functor $X \colon (\operatorname{Sch}/S)^{\operatorname{opp}} \to (\operatorname{Sets})$ that sends an S-scheme T to $X(T) = \operatorname{Hom}_S(T, X)$ is an fppf-sheaf by faithfully flat descent for morphisms of schemes, Proposition 14.76.

As for usual sheaves, one can construct an fppf-sheafification for presheaves, i.e., functors on (Sch/S). We omit the proof, see [Art1] or [Sta] 00W1.

Proposition and Definition 27.30. *The inclusion functor from the category of fppf-sheaves to the category of presheaves on (Sch/S) admits a left adjoint functor $F \mapsto F^{\#}$. In other words, for every presheaf F on (Sch/S) there exists an fppf-sheaf $F^{\#}$ and a functorial morphism of presheaves $\iota \colon F \to F^{\#}$ such that for every fppf-sheaf G on (Sch/S) composition with ι yields a bijection*

$$\operatorname{Hom}(F^{\#}, G) \cong \operatorname{Hom}(F, G).$$

Moreover, we have
(1) *For all S-schemes T and all elements $a, b \in F(T)$ the images of a and b in $F^{\#}(T)$ are equal if and only if there exists an fppf-covering $(g_i \colon T_i \to T)_i$ such that $g_i^*(a) = g_i^*(b)$ for all i.*
(2) *For each S-scheme T and for any $\alpha \in F^{\#}(T)$ there exists an fppf-covering $(g_i \colon T_i \to T)_i$ and $a_i \in F(T_i)$ such that $\iota(a_i) = g_i^*(\alpha)$.*
The fppf-sheaf $F^{\#}$ is called the fppf-sheafification *of the presheaf F.*

Definition 27.31. *An* fppf-sheaf of groups *is a functor $(\operatorname{Sch}/S)^{\operatorname{opp}} \to (\operatorname{Grp})$ whose composition with the forgetful functor $(\operatorname{AbGrp}) \to (\operatorname{Sets})$ is an fppf-sheaf. A* morphism of fppf-sheaves of groups *is a morphism of functors $(\operatorname{Sch}/S)^{\operatorname{opp}} \to (\operatorname{Grp})$. An* abelian fppf-sheaf *is an fppf-sheaf of groups G such $G(T)$ is an abelian group for all S-schemes T.*

Remark 27.32. The category of abelian fppf-sheaves on (Sch/S) is an abelian category. For instance a sequence of abelian fppf-sheaves

$$0 \longrightarrow F \longrightarrow G \xrightarrow{\ f\ } H \longrightarrow 0$$

is exact if and only if $0 \to F(T) \to G(T) \to H(T)$ is exact for all S-schemes T and if $G \to H$ is an epimorphism in the category of fppf-sheaves, i.e., if for all S-schemes T and for all $h \in H(T)$ there exists an fppf-covering $(a_i \colon T_i \to T)_i$ and $g_i \in G(T_i)$ such that $f(g_i) = a_i^*(h)$. See Section (27.8) below for more details.

(27.7) Digression: Algebraic Spaces.

Let S be a scheme. We will encounter fppf-sheaves on (Sch/S) that are not representable by an S-scheme but are at least representable by an algebraic space in the following sense. Our most important example of such a functor is the relative Picard functor $\mathrm{Pic}_{X/S}$ introduced in Section (27.21) below.

For us algebraic spaces will form a technical tool and it is beyond the scope of this book to give a full introduction into this topic. Therefore we will omit some of the proofs and refer to the stacks project [Sta]. Roughly speaking, the notion of algebraic space will allow us to attach a "geometric meaning" to fppf sheaves Z which may not be representable by a scheme but which still are fairly close to a scheme in the sense that there exists a surjective étale representable morphism $U \to Z$ (and a further technical condition, see below). In the category of fppf sheaves we can then view Z as the quotient of U by some "equivalence relation", so by definition, the category of algebraic spaces is much more flexible regarding the construction of quotients. In fact, it turns out that many fppf sheaves defined as moduli functors, i.e., whose T-valued points parametrize algebro-geometric objects of some sort, are representable by an algebraic space. Artin gave a list of criteria that ensure representability by an algebraic space in terms of the "deformation theory" of the objects to be parametrized, i.e., their lifting behavior along nil-immersions of spectra of local Artin rings. See [Art3] ^{O}, [Art4] ^{O}, [BLR] ^{O} 8.3, [Sta] 07T0.

To introduce algebraic spaces we consider as usually the category of S-schemes as a full subcategory of the category of fppf-sheaves on (Sch/S) via the Yoneda embedding (Section (8.1) and Section (14.18)).

Recall from Section (8.2) the notion of representability of a morphism of functors $(\mathrm{Sch}/S)^{\mathrm{opp}} \to (\mathrm{Sets})$ and what it means that such a representable morphism has some property $\mathbf{P}$, e.g., being surjective or being étale. Clearly, the composition of two representable morphisms is again representable. If

$$\begin{array}{ccc} X' & \longrightarrow & X \\ f' \downarrow & & \downarrow f \\ Y' & \longrightarrow & Y \end{array}$$

is a cartesian diagram of functors $(\mathrm{Sch}/S)^{\mathrm{opp}} \to (\mathrm{Sets})$ and if f is representable, then f' is representable.

Lemma 27.33. *Let $Z \colon (\mathrm{Sch}/S)^{\mathrm{opp}} \to (\mathrm{Sets})$ be a functor. Then the diagonal $\Delta_Z \colon Z \to Z \times_S Z$ is representable if and only if for every S-scheme U every morphism $U \to Z$ is representable.*

Proof. We write $- \times -$ instead of $- \times_S -$. Suppose Δ_Z is representable. Let $U \to Z$ be a morphism of functors. We have to show that for every scheme V and every morphism $V \to Z$ the fiber product $U \times_Z V$ is representable by a scheme. This follows from the cartesian diagram

$$\begin{array}{ccc} U \times_Z V & \longrightarrow & U \times V \\ \downarrow & & \downarrow \\ Z & \xrightarrow{\;\Delta_Z\;} & Z \times Z. \end{array}$$

For the other direction let $U \to Z \times Z$ be a morphism, where U is an S-scheme. Then by hypothesis, $U \times_Z U$ is representable and we conclude by

$$Z \times_{\Delta_Z, Z \times Z} U = U \times_{\Delta_U, U \times U} (U \times_Z U). \qquad \square$$

Definition 27.34. *An* algebraic space *over S is an fppf-sheaf $Z \colon (\mathrm{Sch}/S)^{\mathrm{opp}} \to (\mathrm{Sets})$ such that*
(a) *the diagonal $Z \to Z \times_S Z$ is representable, and*
(b) *there exists an S-scheme U and a surjective étale morphism $U \to Z$ (automatically representable by Lemma 27.33).*
A morphism of algebraic spaces *is a morphism of functors $(\mathrm{Sch}/S)^{\mathrm{opp}} \to (\mathrm{Sets})$.*

Clearly, every scheme is an algebraic space. A surjective étale morphism $U \to Z$ for a scheme U is sometimes called an *atlas*.

Remark 27.35. Let S be a scheme and let $X \to Z$ and $Y \to Z$ be morphisms of algebraic spaces over S. Then the fiber product $X \times_S Y$ of fppf-sheaves on (Sch/S) is again an algebraic space.

Lemma 27.36. *Let Y be an algebraic space over S and let $g \colon X \to Y$ be a morphism of functors $(\mathrm{Sch}/S)^{\mathrm{opp}} \to (\mathrm{Sets})$. If g is representable, then X is an algebraic space.*

Proof. Let U be an S-scheme, and let $U \to X \times_Y X$ be a morphism given by $h_1, h_2 \colon U \to X$ with $g \circ h_1 = g \circ h_2$. Since the diagonal of Y is representable, so is $\mathrm{Eq}(h_1, h_2) \to U$ by (9.1.4). Therefore $\mathrm{Eq}(h_1, h_2)$ is representable by a scheme. Then the cartesian diagrams

$$\begin{array}{ccc} X \times_Y X & \longrightarrow & Y \\ \downarrow & & \downarrow{\scriptstyle \Delta} \\ X \times_S X & \longrightarrow & Y \times_S Y \end{array} \qquad\qquad \begin{array}{ccc} \mathrm{Eq}(h_1, h_2) & \longrightarrow & U \\ \downarrow & & \downarrow{\scriptstyle (h_1, h_2)} \\ X & \longrightarrow & X \times_Y X \end{array}$$

show first that $X \times_Y X \to X \times_S X$ is representable and then that $X \to X \times_Y X$ is representable. Therefore their composition, the diagonal of X, is representable.

If U is an S-scheme and $U \to Y$ is an étale surjective morphism, then $U \times_Y X$ is a scheme because $X \to Y$ is representable. As the properties "étale" and "surjective" are stable under base change, the projection $U \times_Y X \to X$ is étale and surjective. $\qquad \square$

We are now going to define the underlying topological space of an algebraic space. For a scheme Y we denote by $|Y|$ its underlying topological space. Let X be an algebraic space. Choose a scheme U and an étale surjective morphism $U \to X$. Set $R := U \times_X U$ which is a scheme by Lemma 27.33. The image of $|R|$ in $|U| \times |U|$ is an equivalence relation ([Sta] 03BW) and we define $|X|$ to be the quotient space of $|U|$ with respect to this equivalence relation. By loc. cit. this is well defined.

Lemma and Definition 27.37. *The above construction of the topological space $|X|$ does not depend on the choice of atlas $U \to X$ and $|X|$ is called the* underlying topological space *of X.*

Let $f\colon X \to Y$ be a morphism of algebraic spaces. Choose schemes U and V with étale surjective morphisms $V \to Y$ and $U \to V \times_Y X$. Then the composition $U \to X$ is surjective étale and $U \to V$ induces a continuous map $|X| \to |Y|$ independent of all choices ([Sta] 03BX). We obtain a functor from the category of algebraic spaces to the category of topological spaces.

Definition 27.38. *Let X be an algebraic space.*
(1) *Then X is called* quasi-compact *if there exists a quasi-compact scheme U and an étale surjective morphism $U \to X$. Then X is quasi-compact if and only if its underlying topological space $|X|$ is quasi-compact ([Sta] 03E4).*
(2) *Let $\mathbf{P}$ be a property of schemes that is local in the étale topology, i.e., if S is a scheme and $(S_i \to S)_{i \in I}$ is an étale covering (Definition 20.21), then S has the property $\mathbf{P}$ if and only if S_i has the property $\mathbf{P}$. Then an algebraic space X is said to have the property $\mathbf{P}$ if there exists a scheme U that has the property $\mathbf{P}$ and an étale surjective morphism $U \to X$.*

Properties of schemes that are local in the étale topology are for instance "locally noetherian", "reduced", "normal", "regular".

Definition 27.39. *Let $\mathbf{P}$ be a property of morphisms of schemes that is local on source and target for the étale topology. Then a morphism $f\colon X \to Y$ of algebraic spaces is said to have the property $\mathbf{P}$ if there exists a commutative diagram*

$$(*) \qquad\qquad \begin{array}{ccc} U & \overset{g}{\longrightarrow} & V \\ \downarrow & & \downarrow \\ X & \overset{f}{\longrightarrow} & Y \end{array}$$

with U and V schemes, $U \to X$ and $V \to Y$ étale, and $U \to X$ surjective such that g has property $\mathbf{P}$.

Examples for properties that are local on source and target for the étale topology are "locally of finite presentation", "locally of finite type", "smooth", "étale", "unramified".

Remark 27.40. Let $\mathbf{P}$ be a property of morphisms of schemes that is local on source and target for the étale topology. If the property $\mathbf{P}$ for scheme morphisms is stable under composition (resp. stable under base change), so is the property $\mathbf{P}$ for morphisms of algebraic spaces.

Working with an atlas one also easily deduces the following fact from the analogous property for morphisms of schemes (Proposition 10.35).

Remark 27.41. Let $X \to Y$ be a morphism of algebraic spaces over S. Suppose that $X \to S$ is locally of finite presentation and that $Y \to S$ is locally of finite type. Then $X \to Y$ is locally of finite presentation.

Remark 27.42. By definition, an algebraic space Z over S is smooth over S if there exist a smooth S-scheme U and an étale surjective S-morphism $U \to Z$. If Z is locally of finite presentation over S, then Z is smooth over S if and only if and for every thickening $T_0 \to T$ of affine S-schemes the canonical map $Z(T) \to Z(T_0)$ is surjective ([Sta] 04AM). If S is locally noetherian, then it suffices to check this condition for every thickening of local Artinian rings $\operatorname{Spec} R_0 \to \operatorname{Spec} R$ ([Sta] 0APN) defined by an ideal $I \subseteq R$ with $I\mathfrak{m}_R = 0$.

These are the analogues of Theorem 18.56 and Theorem 18.63 for schemes.

We will also need the following properties for morphisms of algebraic spaces which are not local on the source. Note that if $X \to Y$ is a morphism of algebraic spaces over a scheme S, then the diagonal $\Delta_{X/Y} \colon X \to X \times_Y X$ is representable, because the diagonal $X \to X \times_S X$ is representable, and $X \times_Y X \to X \times_S X$ is a monomorphism.

Definition 27.43. *Let* $f \colon X \to Y$ *be a morphism of algebraic spaces.*
(1) *The morphism* f *is called* separated *if the representable morphism* $\Delta_{X/Y}$ *is a closed immersion.*
(2) *The morphism* f *is called* quasi-compact *if for every affine scheme* T *and every morphism* $T \to Y$ *the fiber product* $X \times_Y T$ *is quasi-compact.*
(3) *The morphism* f *is called* proper *if* f *is separated, quasi-compact, locally of finite type, and satisfies the following valuative criterion: For every valuation ring* R *with fraction field* K *and for every commutative diagram*

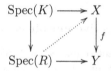

there exists a dotted arrow making the diagram commutative.
It is shown in [Sta] 03KU that the dotted arrow is necessarily unique since f *is separated.*

Remark 27.44.
(1) Similarly as for scheme morphisms one see that the properties "separated", "quasi-compact", and "proper" for morphisms of algebraic spaces are stable under composition and under base change.
(2) If S is an affine scheme, then a morphism of algebraic spaces $X \to S$ is quasi-compact if and only if X is quasi-compact (combine [Sta] 03KG with Remark 10.2).
(3) If Y is a locally noetherian algebraic space, then a morphism $X \to Y$ is proper as soon as the criterion in Definition 27.43 (3) holds for all *discrete* valuation rings R ([Sta] 0ARK).

Remark 27.45. Let X be an algebraic space over S. Then every section $e \colon S \to X$ of $X \to S$ is representable and an immersion since we have a cartesian diagram

$$
\begin{array}{ccc}
S & \xrightarrow{\ e\ } & X \\
{\scriptstyle e}\downarrow & & \downarrow{\scriptstyle \Delta_{X/S}} \\
X & \xrightarrow{(\mathrm{id},e)} & X \times_S X.
\end{array}
$$

If $X \to S$ is separated, then e is a closed immersion.

Conversely, the same argument as in Proposition 27.2 shows that if X is a group algebraic space (Definition 27.46) such that the unit section is a closed immersion, then X is separated over S.

In the sequel we will be mainly considering group algebraic spaces in the following sense.

Definition 27.46. *Let S be a scheme. An fppf-sheaf in groups on (Sch/S) that is an algebraic space is called a* group algebraic space

The following result allows to avoid the language of algebraic spaces if the base scheme is the spectrum of a field.

Theorem 27.47. *Let k be a field and let G be a separated group algebraic space over k. Then G is a scheme.*

We will use the theorem only if G is in addition locally of finite type over k. For the proof in this case one first reduces to the case that k is algebraically closed, using that every group scheme locally of finite type over a field is quasi-projective (see Corollary 27.73) and that for quasi-projective schemes Galois-descent is effective (Theorem 14.86). Then one uses that G contains an open dense subspace U that is represented by a scheme ([Sta] 06NH). By translating U one deduces that G is represented by a scheme. For details of the proof we refer to [Sta] 0B8G which also covers the general case.

(27.8) Digression: Fppf-surjective morphisms.

In order to have a notion of exact sequences of fppf-sheaves of groups we need to know what it means for a morphism of fppf-sheaves to be surjective. This is defined as usual for sheaves replacing open coverings by fppf-coverings.

We continue to denote by S a scheme.

Definition 27.48. *Let $f\colon X \to Y$ be a morphism of fppf-sheaves over S. Then f is called* fppf-surjective *if it is an epimorphism in the category of fppf-sheaves, i.e., if for all S-schemes T and for all $y \in Y(T)$ there exists an fppf-covering $(g_i\colon T_i \to T)_i$ and $x_i \in X(T_i)$ such that $f(x_i) = g_i^*(y)$. (See [MaMo] III.7, Cor. 5, for a proof of the equivalence.)*

Remark 27.49. Let $f\colon X \to Y$ be a morphism of fppf-sheaves over S.
(1) By passing to an open affine covering, which is a special case of an fppf-covering, it suffices to check the condition for fppf-surjectivity in Definition 27.48 only for affine schemes T. If f is fppf-surjective and T is affine, then Lemma 14.79 shows that we can find for all $y \in Y(T)$ a surjective flat morphism of finite presentation $T' \to T$ of *affine* schemes and $x \in X(T')$ with $f(g) = y_{|T'}$.

   Therefore, f is fppf-surjective if and only if for every affine S-scheme $T = \mathrm{Spec}\,R$ and for every $y \in Y(R)$ there exists a faithfully flat R-algebra R' of finite presentation such that the image y in $Y(R')$ lies in the image of $X(R') \to Y(R')$.
(2) If $(g_i\colon T_i \to T)_i$ is an fppf-covering, then the family consisting of the single morphism $T' := \coprod_i T_i \to T$ is also an fppf-covering. Moreover, $Z(T') = \prod_i Z(T_i)$ for every fppf-sheaf Z over S by the sheaf condition for the fppf-covering $(T_i \to T)_i$.
(3) If $f\colon X \to Y$ is fppf-surjective and $S' \to S$ is a morphism of schemes, then the base change $f_{S'}\colon X \times_S S' \to Y \times_S S'$ is fppf-surjective.

Lemma 27.50. *Suppose that X and Y are algebraic spaces over S and let $f\colon X \to Y$ be a morphism.*
(1) *If f is faithfully flat and locally of finite presentation, then f is fppf-surjective.*
(2) *If f is fppf-surjective, then f has a section fppf-locally and is surjective.*

Proof. To see (1) choose an atlas $V \to Y$ and an atlas $U \to X \times_Y V$. Then $U \to V \to Y$ is faithfully flat and locally of finite presentation. Let $y \in Y(T)$ for some S-scheme T. Then $g \colon T' := V \times_S T \to T$ is an fppf-covering and $T' \to U \to X$ is a point $x' \in X(T')$ with $f(x') = g^*(y)$.

To see (2) we apply Definition 27.48 to $T = Y$ and $y = \mathrm{id}_Y$. This shows that f has a section fppf-locally. In particular, it is surjective fppf-locally. Hence it is surjective by Proposition 14.50. □

Example 27.51. Let k be a field. Suppose that $f \colon X \to Y$ is an fppf-surjective morphism of fppf-sheaves over k. Then for $y \in Y(k)$ there exists a finite extension K of k such that the image of y in $Y(K)$ is in the image of $X(K) \to Y(K)$.

Indeed, by hypothesis there exists a k-algebra $R' \neq 0$ of finite presentation and $x' \in X(R')$ such that $f(x) = g^*(y)$ for the fppf-covering $\operatorname{Spec} R' \to \operatorname{Spec} k$. Let $\mathfrak{m} \subset R'$ be a maximal ideal, then $K := R'/\mathfrak{m}$ is a finite extension of k by the Nullstellensatz and $h \colon \operatorname{Spec} K \to \operatorname{Spec} k$ is still an fppf-covering. If $x \in X(K)$ is the image of x', then $f(x) = h^*(y)$.

Lemma 27.52. *Let $f \colon X \to Y$ and $g \colon Y' \to Y$ be morphisms of S-schemes. Let g be fppf-surjective. Let $\mathbf{P}$ be a property of scheme morphisms that is stable under base change and can be checked fppf-locally. Then f has property $\mathbf{P}$ if and only if its base change $f' \colon X \times_Y Y' \to Y'$ has property $\mathbf{P}$.*

Proof. The conditions is clearly necessary. To show that it is sufficient, we may work locally for the fppf-topology by hypothesis. Hence we may assume that g has a section i. Then f is the base change of f' along i, hence f has property $\mathbf{P}$ if f' has it. □

Lemma 27.53. *If $f \colon X \to Y$ is fppf-surjective and a monomorphism, then f is an isomorphism.*

Proof. By Lemma 27.52 we may check that f is an isomorphism after making the fppf-surjective base change with f itself. Hence it suffices to show that the second projection $X \times_Y X \to X$ is an isomorphism. But this morphism has as a section the diagonal $\Delta_f \colon X \to X \times_Y X$ which is an isomorphism because f is a monomorphism. Therefore the projection is an isomorphism. □

Proposition 27.54. *Let $f \colon G \to H$ be a homomorphism of S-group algebraic spaces that are separated and locally of finite presentation over S. Let G be flat over S, let H have geometrically reduced fibers over S, and let f be surjective. Then H is flat over S, f is faithfully flat locally of finite presentation, and in particular f is fppf-surjective.*

We will give the proof if G and H are S-group schemes and give references to the Stacks project for the necessary modifications in the general case.

Proof. First note that f is locally of finite presentation by Proposition 10.35 ([Sta] 06G4 for algebraic spaces). Hence it suffices to show that f is flat.

This follows from Proposition 27.14 (3) by the fiber criterion for flatness (Corollary 14.27, [Sta] 05X1 for algebraic spaces) and because a group scheme locally of finite type over a field is smooth if and only if it is geometrically reduced (Proposition 27.10). We can apply Proposition 27.14 (3) by Theorem 27.47 or by proving an analogue for Proposition 27.14 (3) for group algebraic spaces which is not difficult. □

(27.9) Quotient spaces and homogeneous spaces.

In this section, S always denotes a scheme.

Now having the notion of fppf-surjectivity at our disposal, we say that a (finite or infinite) sequence of homomorphisms of S-group schemes

$$\cdots \longrightarrow G^{i-1} \xrightarrow{f^{i-1}} G^i \xrightarrow{f^i} G^{i+1} \longrightarrow \cdots$$

is *exact* if for all i the homomorphism f^{i-1} factors through $\mathrm{Ker}(f^i)$ and the induced homomorphism $G^{i-1} \to \mathrm{Ker}(f^i)$ is fppf-surjective. In particular a short sequence of S-group schemes

$$1 \longrightarrow G' \xrightarrow{f} G \xrightarrow{g} G'' \longrightarrow 1$$

is exact if and only if $1 \to G'(T) \to G(T) \to G''(T)$ is exact for all S-schemes T (use Lemma 27.53 to see that $G' \to \mathrm{Ker}(g)$ is fppf-surjective if and only if it is an isomorphism) and g is fppf-surjective. If G', G, and G'' are commutative group schemes viewed as abelian fppf-sheaves this notion of exactness coincides with the exactness notion defined in Remark 27.32.

Example 27.55. Let S be a scheme. Let $\mathbb{G}_m$ be the multiplicative group over S, i.e., $\mathbb{G}_m(T) = \Gamma(T, \mathscr{O}_T)^\times$ for every S-scheme T. This functor is represented by the flat scheme $\mathrm{Spec}\, \mathscr{O}_S[X, X^{-1}]$. Let $n \neq 0$ be an integer. Then by Proposition 27.54 the n-th power homomorphism

$$\mathbb{G}_m \longrightarrow \mathbb{G}_m, \qquad \mathbb{G}_m(T) \ni t \mapsto t^n \in \mathbb{G}_m(T)$$

is fppf-surjective and hence we have an exact sequence of S-group schemes

$$1 \longrightarrow \mu_n \longrightarrow \mathbb{G}_m \xrightarrow{(-)^m} \mathbb{G}_m \longrightarrow 1,$$

where μ_n is the group scheme of n-th root of unity, i.e., for every S-scheme T one has $\mu_n(T) = \{\, \zeta \in \Gamma(T, \mathscr{O}_T)^\times \ ;\ \zeta^n = 1 \,\}$. It is represented by $\mathrm{Spec}\, \mathscr{O}_S[X]/(X^n - 1)$.

In the following definition, by a G-action $G \times_S X \to X$ we mean a morphism of fppf-sheaves such that for every S-scheme T the map $G(T) \times X(T) \to X(T)$ is a group action of the group $G(T)$ on the set $X(T)$.

Definition 27.56. *Let G be a group algebraic space over S that is flat and locally of finite presentation over S, and let X be an algebraic space over S with a G-action $G \times_S X \to X$, simply denoted by $(g, x) \mapsto gx$ on T-valued points, T an S-scheme. Then X is called a* homogeneous G-space *if the following conditions are satisfied.*
(a) *The structure morphism $X \to S$ is fppf-surjective.*
(b) *The morphism $\sigma_X \colon G \times_S X \to X \times_S X$, $(g, x) \mapsto (gx, x)$ is fppf-surjective.*

Definition 27.57. *Let G be an fppf-sheaf of groups on (Sch/S) and let $H \subseteq G$ be an fppf-sheaf of subgroups. The fppf-sheafification of the presheaf*

$$(G/H)' \colon (\mathrm{Sch}/S)^{\mathrm{opp}} \longrightarrow (\mathrm{Sets}), \qquad T \mapsto G(T)/H(T)$$

is called the quotient of G by H *and is denoted by G/H. The canonical map $f \colon G \to G/H$ is fppf-surjective.*

Even if G and H are group schemes, then G/H is in general not representable by an S-scheme. But it can be shown ([Sta] 06PH) that if G is a group algebraic space over S and H is a subgroup algebraic space of G that is flat and locally of finite presentation over S, then G/H is always an algebraic space and $G \to G/H$ is faithfully flat and locally of finite presentation.

Remark 27.58. Let G and H be as in Definition 27.57. Then the formation of the presheaf $(G/H)'$ is compatible with base change $S' \to S$. Hence this holds also for G/H, i.e., one has $(G/H) \times_S S' = (G \times_S S')/(H \times_S S')$.

Lemma 27.59. *Let G be a group algebraic space, let H be a subgroup fppf-sheaf such that G/H is a separated algebraic space. Then G/H with its natural left action is a homogeneous G-space. Moreover, H is a closed subgroup algebraic space of G and one has an isomorphism of algebraic spaces over S*

$$(27.9.1) \qquad G \times_S H \xrightarrow{\sim} G \times_{G/H} G, \qquad (g, h) \mapsto (gh, g).$$

Proof. As $G \to S$ has a section, it is fppf-surjective. As the composition $G \to G/H \to S$ is fppf-surjective, $G/H \to S$ is fppf-surjective. By definition, for all $x_1, x_2 \in (G/H)(T)$ there exists an fppf-covering $T' \to T$ and an element $g \in G(T')$ such that $g \cdot (x_1)_{T'} = (x_2)_{T'}$. But this shows that $\sigma_{G/H}$ is fppf-surjective. Therefore G/H is a homogeneous G-space. The image of the unit section $e \in G(S)$ is a section $x_0 \in (G/H)(S)$ which is a closed immersion because $G/H \to S$ is separated. Then $H = f^{-1}(x_0)$, i.e., H is defined by the cartesian diagram

$$(27.9.2)$$

$$
\begin{array}{ccc}
H & \longrightarrow & S \\
\downarrow & & \downarrow{\scriptstyle x_0} \\
G & \xrightarrow{f} & X.
\end{array}
$$

This shows that H is a closed subgroup algebraic space of G.

Let T be an S-scheme. If $g \in G(T)$ such that $g_{T'} \in H(T')$ for some fppf-covering $T' \to T$, then $g \in H(T)$ as fppf-coverings are fppf-surjective. Hence for $g_1, g_2 \in G(T)$ one has

$$(g_1, g_2) \in (G \times_{G/H} G)(T) = G(T) \times_{(G/H)(T)} G(T) \Leftrightarrow g_2^{-1} g_1 \in H(T).$$

This shows that $(g_1, g_2) \mapsto (g_2, g_2^{-1} g_1)$ is a well defined inverse map to (27.9.1). $\square$

Remark 27.60. Let G and X be as in Definition 27.56 with X a homogeneous G-space. Let $S' \to S$ be a morphism of schemes. Then $X \times_S S'$ is a homogeneous $(G \times_S S')$-space (Remark 27.49 (3)).

Lemma 27.61. *Let G and X be as in Definition 27.56 with X a separated homogeneous G-space. Then locally for the fppf-topology, X is of the form G/H for a closed subgroup algebraic space H of G.*

Proof. Working locally for the fppf-topology we may assume that there exists a section $x_0 \in X(S)$ which is a closed immersions because X is separated over S. Define a morphism of S-schemes $f: G \to X$ by $f(g) = g \cdot x_0$ (on T-valued points). Then f is G-equivariant, where we endow G with the action by itself via left multiplication. The cartesian diagram

$$
\begin{array}{ccc}
G \times_S S = G & \xrightarrow{f} & X = X \times_S S \\
{\scriptstyle \mathrm{id}_g \times x_0}\downarrow & & \downarrow{\scriptstyle \mathrm{id}_X \times s_0} \\
G \times_S X & \xrightarrow{\sigma_X} & X \times_S X
\end{array}
$$

shows that f is fppf-surjective since σ_X is fppf-surjective.

Let $H := f^{-1}(x_0)$. Then H is the stabilizer of x_0 in G, i.e., for every S-scheme T one has

$$H(T) = \{\, g \in G(T) \ ; \ gx_{0,T} = x_{0,T} \,\}.$$

Therefore, H is a subgroup algebraic space of G. As x_0 is a closed immersion, $H \to G$ is a closed immersion. As f is fppf-surjective, it induces an morphism of fppf-sheaves $G/H \to X$ which is a monomorphism and fppf-surjective. Hence it is an isomorphism by Lemma 27.53. $\qquad\square$

Proposition 27.62. *Let G be a group algebraic space that is flat and locally of finite presentation over S and let $H \subseteq G$ be a subgroup fppf-sheaf such that G/H is a separated algebraic space locally of finite presentation over S.*
(1) *Then H is an algebraic space which is flat and locally of finite presentation over S. The quotient G/H is faithfully flat over S, and $G \to G/H$ is locally of finite presentation and faithfully flat.*
(2) *The map $G \to G/H$ is an H-torsor for the fppf-topology.*
(3) *Let $\mathbf{P}$ be a property of morphisms of algebraic spaces that is stable under composition with isomorphisms, stable under base change, and that can be checked locally for the fppf-topology on the target. Then $G \to G/H$ has property $\mathbf{P}$ if and only if $H \to S$ has property $\mathbf{P}$.*
(4) *If G is smooth (resp. smooth with geometrically connected fibers) over S, then G/H is smooth (resp. smooth with geometrically connected fibers) over S.*

Proof. Let us show (1). The canonical fppf-surjective morphism $f \colon G \to G/H$ is surjective, and it is locally of finite presentation by Proposition 10.35 ([Sta] 06G4 for algebraic spaces). We have already seen that $G/H \to S$ is fppf-surjective and in particular surjective.

To see that f and $G/H \to S$ are flat it suffices to show that all fibers of f are flat by the fiber criterion for flatness (Corollary 14.27, [Sta] 05X1 for algebraic spaces). Hence we can assume that S is the spectrum of a field. Now we can base change by the fppf-surjective morphism f itself (Lemma 27.52) and it suffices to show that projection $G \times_S H \to G$ is flat by (27.9.1). But this is clear because the base is the spectrum of a field.

As f is flat and locally of finite presentation, the cartesian diagram (27.9.2) shows that $H \to S$ is flat and locally of finite presentation.

Now (2) follows from (27.9.1) since we have seen in (1) that $G \to G/H$ is an fppf-covering.

To see (3), also follows from (27.9.1): The morphism $G \to G/H$ has $\mathbf{P}$ if and only if $G \times_S H = G \times_{G/H} G \to G$ has $\mathbf{P}$ if and only if $H \to S$ has $\mathbf{P}$ since $G \to S$ is an fppf-covering.

It remains to show (4). As we already know that $G/H \to S$ is flat, we can show its smoothness on geometric fibers (Theorem 18.56). Therefore we can assume that $S = \operatorname{Spec} k$ for an algebraically closed field k. As $G \to G/H$ is surjective, G/H is connected if G is connected, regardless of smoothness.

Now suppose that G is smooth. As $G \to G/H$ is faithfully flat, G/H is regular (Corollary 14.60) and hence smooth over k (Theorem 6.28). $\qquad\square$

Corollary 27.63. *Let G and G' be group algebraic spaces locally of finite presentation over S. Suppose that G is flat over S and that G' is separated over S. Let $f \colon G \to G'$ be an fppf-surjective homomorphism of group algebraic spaces. Let $\mathbf{P}$ be a property of morphisms of algebraic spaces that is stable under composition with isomorphisms, stable* *under base change, and that can be checked locally for the fppf-topology on the target, i.e., after passing to an fppf-covering.*

Then G' is flat over S, and f is faithfully flat locally of finite presentation. Moreover, f has property **P** if and only if $H := \operatorname{Ker}(f) \to S$ has property **P**.

Proof. As f is fppf-surjective, it induces an isomorphism $G/\operatorname{Ker}(f) \xrightarrow{\sim} G'$. Now we conclude by Proposition 27.62. $\qquad\Box$

Corollary 27.64. *Let G be an S-group scheme which is flat and locally of finite presentation (resp. smooth) over S and let X be a separated homogeneous G-space locally of finite presentation over S. Then X is faithfully flat (resp. smooth and surjective) over S.*

Proof. The assertions may be checked fppf-locally and hence we may assume that $X = G/H$ for some subgroup algebraic space H of G (Lemma 27.61). We conclude by Proposition 27.62. $\qquad\Box$

Corollary 27.65. *Let G be an S-group scheme which is flat and locally of finite presentation over S and let X be a separated homogeneous G-space locally of finite presentation over S. Then the morphisms*

$$\sigma_X \colon G \times_S X \to X \times_S X, \qquad (g,x) \mapsto (gx, x),$$
$$a_X \colon G \times_S X \to X, \qquad (g,x) \mapsto gx$$

are faithfully flat and locally of finite presentation.

Proof. As X and G are both locally of finite presentation, both morphisms are locally of finite presentation. It suffices to show that σ_X is flat. Then it is also faithfully flat because it is fppf-surjective by hypothesis. And a_X is the composition of σ_X followed by the first projection $X \times_S X \to X$ which is faithfully flat by Corollary 27.64.

But σ_X is the G_X-equivariant morphism $G_X \to X_X$ of X-schemes given by the identity $x_0 := \operatorname{id}_X \in X(X)$ that is given on T-valued points by $g \mapsto g \cdot x_0$ for every X-scheme T. To see that σ_X is flat we can work fppf-locally and hence assume that σ_X is the projection $G_X \to G_X/H$, where H is a subgroup fppf-sheaf by Lemma 27.61. Now we conclude via Proposition 27.62. $\qquad\Box$

Corollary 27.66. *Let S be a scheme, let G be a smooth S-group scheme with connected fibers, and let $f \colon X \to S$ be a separated homogeneous G-space that is locally of finite presentation over S. Then $X \to S$ is smooth and has geometrically irreducible and quasi-compact fibers.*

Proof. By Corollary 27.64, X is smooth over S. Locally for the fppf topology on S there exists an fppf-surjective morphism $G \to X$ which is in particular surjective (Lemma 27.50). As G has geometrically irreducible and quasi-compact fibers (Proposition 27.10 (2) and Proposition 27.11 (2)), X also has geometrically irreducible and quasi-compact fibers. $\quad\Box$

Remark 27.67. The separatedness hypothesis on X in Lemma 27.61, on G/H in Proposition 27.62, on G' Corollary 27.63, and on X Corollary 27.65 and in Corollary 27.64 can be weakened. It was only used to see that any section is a closed immersion. If one assumes only that the diagonal of these algebraic spaces is an immersion (such an algebraic space is called *locally separated* in [Sta] 02X5), then any section is an immersion. This is for instance the case if X is a scheme. If the diagonal is an immersion, H will be a subgroup algebraic space of G but not necessarily be closed in G. Otherwise, all assertions and arguments remain the same.

Theorem 27.68. *Let S be a scheme, let G be a quasi-projective S-group scheme and let $H \subseteq G$ be a subgroup scheme that is finite locally free over S. Then G/H is representable by a scheme and the projection $G \to G/H$ is an H-torsor for the fppf-topology.*

We will not give a proof for this representability result here but refer to [SGA3]$\overset{\text{O}}{\text{X}}$ Exp. V, 4.1., or to [Sta] 07S7.

(27.10) Digression: Homotopy invariance of Picard group.

We digress briefly for the following result that we will use in the next section to prove quasi-projectivity of homogeneous spaces.

Proposition 27.69. *Let S be a noetherian scheme and let $U \subseteq \mathbb{A}_S^n$ be an open subscheme such that $U(S) \neq \emptyset$. Let Y be a normal noetherian integral scheme over S. Let $p \colon Y \times_S U \to Y$ be the projection. Then*

$$p^* \colon \mathrm{Pic}(Y) \to \mathrm{Pic}(Y \times_S U)$$

is an isomorphism.

The proposition in particular shows that the canonical map $\mathrm{Pic}(Y) \to \mathrm{Pic}(\mathbb{A}_Y^n)$ is bijective for all $n \geq 0$ if Y is a normal integral noetherian scheme. Without the normality assumption on Y, the assertion fails, see Exercise 27.6 for details.

Proof. Choose $u \in U(S)$. Then $u_Y \colon Y \to Y \times_S U$ is a section of p and hence $u^* \circ p^* = \mathrm{id}_{\mathrm{Pic}(Y)}$. In particular, p^* is injective. To show the surjectivity of p^* we will use Lemma 27.70 below. Clearly $U \to S$ is flat. As it has a section, it is faithfully flat. The generic fiber of p is an open non-empty subscheme U_η of $\mathbb{A}_{\kappa(\eta)}^n$. Now $\mathrm{Pic}(\mathbb{A}_{\kappa(\eta)}^n) \to \mathrm{Pic}(U_\eta)$ is surjective by Corollary 11.43 and $\mathrm{Pic}(\mathbb{A}_{\kappa(\eta)}^n) = 0$ because polynomial rings over fields are factorial. Hence we may apply Lemma 27.70. $\square$

Lemma 27.70. *Let Y be a normal integral locally noetherian scheme with generic point η, let $f \colon X \to Y$ be a faithfully flat morphism locally of finite type with integral fibers, and let $\mathscr{L}$ be a line bundle on X whose restriction to the general fiber X_η of f is trivial. Then there exists a line bundle $\mathscr{M}$ on Y such that $\mathscr{L} \cong f^*\mathscr{M}$.*
In particular, $f^* \colon \mathrm{Pic}(Y) \to \mathrm{Pic}(X)$ is surjective if $\mathrm{Pic}(X_\eta) = 0$.

We will not prove this result here but refer to [EGAIV]^{O} (21.4.11) or to [Sta] 0BD7.

(27.11) Quasi-projectivity of homogeneous spaces.

Theorem 27.71. *Let S be a normal noetherian scheme, let G be a smooth S-group scheme with connected fibers, and let $f \colon X \to S$ be a homogeneous G-space that is of finite type over S. Suppose that there exists an open subscheme U of X that is quasi-affine over S and that meets every fiber of $X \to S$. Then X is quasi-projective.*

We will not use the theorem in the sequel and give the proof only if $S = \mathrm{Spec}\, k$ for a field k. For a proof in the general case we refer to [Ray1]^{O} V 3.10. Raynaud shows the following more precise assertion. Let $D_1, \ldots, D_r$ be the irreducible components of $X \setminus U$ that are of codimension 1 in X and let $n_1, \ldots, n_r > 0$ be positive integers. Then the Weil divisor $\sum_{i=1}^r n_i D_i$ is a Cartier divisor D such that $\mathscr{O}_X(D)$ is ample.

Here we follow the Stacks Project [Sta], where the theorem is shown if $X = G$. Note that the existence of U is clear if S consists of only one point. In fact, it is shown in [Sta] 0BF7 that every group scheme of finite type over a field is quasi-projective. The method there can be generalized to prove that for every group scheme of finite type every homogeneous space is quasi-projective (see also Exercise 27.7 for a sketch how to deduce this result from Theorem 27.71).

Proof. (if $S = \operatorname{Spec} k$ for a field k) We will prove the theorem in several steps. Step (I) and (II) will not use any hypothesis on S (not even that S is normal and noetherian). Only from step (III) on we will assume that S is the spectrum of a field.

(I). Let $W \subseteq G$ be an open non-empty subscheme that meets every fiber of $G \to S$. Then we will show that $W \cdot U = X$, i.e., the restriction of the action morphism $a\colon G \times_S X \to X$ to $W \times_S U$ is surjective.

Indeed, the question is local for the fppf-topology. Hence we can assume that there exists a G-equivariant fppf-surjective morphism $\pi\colon G \to X$. Set $U' := \pi^{-1}(U)$. It suffices to show that $W \cdot U' = G$. This follows from Proposition 27.11 (1) because all fibers of G are irreducible by Proposition 27.10 (2).

(II). Next we show that X is separated over S. We will use the valuative criterion Theorem 15.9. As all hypotheses are stable under base change $S' \to S$ we may assume that $S = \operatorname{Spec} R$ for a complete discrete valuation ring R with algebraically closed residue field (Remark 15.11). Let $s \in S$ (resp. $\eta \in S$) be the special (resp. generic) point. Let $x_1, x_2 \in X(S)$ be sections that agree on η. We have to prove that $x_1 = x_2$.

For this we show that there exists $\tilde{g} \in G(S)$ such that $x_1, x_2 \in (\tilde{g}U)(S)$. As U is separated over S, $\tilde{g}U$ is also separated and therefore this implies $x_1 = x_2$. To find $\tilde{g}$ consider the morphisms of $\kappa(s)$-schemes

$$\pi_i\colon G_s \longrightarrow X_s, \qquad g \mapsto g \cdot (x_i)_s, \qquad i = 1, 2.$$

By step (I) we have $G \cdot U = X$ and therefore U meets the orbit of $(x_i)_s$. Hence $U_i := \pi_i^{-1}(U)$ is non-empty. As G_s is irreducible and $\kappa(s)$ is algebraically closed, there exists $\tilde{g}_s \in G(s)$ with $\tilde{g}_s^{-1} \in U_1 \cap U_2$. As R is complete and in particular henselian, we can lift $\tilde{g}_s$ to a section $\tilde{g} \in G(S)$ (Theorem 20.12). Then x_1 and x_2 both factor through $\tilde{g}U$.

(III). From now on we assume that $S = \operatorname{Spec} k$ for a field k. By Proposition 14.57 one can even assume that k is algebraically closed. We will now construct a family $(D_v)_v$ of effective divisors on X.

By Corollary 18.57 we find an open dense subscheme $W' \subseteq G$ and an étale morphism $\pi\colon W' \to \mathbb{A}_k^n$, where $n = \dim(G)$. As π is generically finite, we find a non-empty open subscheme $V \subseteq \mathbb{A}_k^n$ such that the restriction of π to a morphism $W := \pi^{-1}(V) \to V$ is finite étale (Proposition 12.11 (4)).

Let $U \subseteq X$ be any non-empty open affine subscheme. As X is smooth over S and irreducible (Corollary 27.66), X is regular (Corollary 14.60) and U is dense in X. By Lemma 25.150 there exists an effective Cartier divisor E whose underlying topological space is $X \setminus U$. For $v \in V(k)$ we set

$$D_v := \sum_{g \in \pi^{-1}(v)(k)} g^{-1}E,$$

where the sum is the sum of effective Cartier divisor, i.e., if D_1 and D_2 are effective Cartier divisors, locally of the form $V(f_1)$ and $V(f_2)$ for regular sections f_1 and f_2 of $\mathscr{O}_X$, then $D_1 + D_2$ is locally of the form $V(f_1 f_2)$.

(IV). We claim that the isomorphism class $\mathscr{L}$ of $\mathscr{O}_X(D_v)$ is independent of $v \in V(k)$.

As V is connected, the degree of π over V is constant. As π is also étale and k is algebraically closed we find $\pi^{-1}(v) = \operatorname{Spec} \prod_{g \in \pi^{-1}(v)(k)} k$ as k-schemes and the cardinality of $\pi^{-1}(v)$ does not depend on v.

Let $\mathcal{E} \subseteq G \times_S X$ be the preimage of E under the action morphism a. This is again an effective Cartier divisor because a is flat (Corollary 27.65). Then the fiber $\mathcal{E}_g$ of $\mathcal{E}$ over a k-valued point $g \in G(k)$ is the effective Cartier divisor $g^{-1}E$. Let $\mathcal{N} := \mathscr{O}_{G \times_S X}(\mathcal{E})$ be the line bundle corresponding to $\mathcal{E}$ and let $\mathcal{M} := N_{\pi^{-1}(V)/V}(\mathcal{N}_{|V \times_S X})$ be the norm (Remark 12.25) of its restriction to $V \times_S X$. For $v \in V(k)$ the restriction $\mathcal{M}_v$ of $\mathcal{M}$ to the fiber of $V \times_S X \to V$ over v is $\mathscr{O}_X(D_v)$.

In view of Corollary 11.43, V (and likewise the base change to any extension field of k), has trivial Picard group, thus Proposition 27.69 can be applied to $q \colon V \times_S X \to X$. This shows that there exists a line bundle $\mathscr{L}$ on X such that $\mathcal{M} \cong q^*(\mathscr{L})$ and hence $\mathcal{M}_v \cong \mathscr{L}$ for all $v \in V(k)$.

(V). Finally, we show that $\mathscr{L}$ is ample. By Proposition 13.47 it suffices to show that there exist finitely many sections $f_i \in \Gamma(X, \mathscr{L})$ such that $X_{f_i} := X_{f_i}(\mathscr{L})$, cf. (7.11.1), is affine and $X = \bigcup X_{f_i}$.

As $\mathscr{L} \cong \mathscr{O}_X(D_v)$ for all $v \in V(k)$, we find $f_v \in \Gamma(X, \mathscr{L})$ such that $X_{f_v} = X \setminus D_v$. Since X is separated, finite intersections of open affine subschemes are again affine (Proposition 9.15). Hence

$$X_{f_v} = \bigcap_{g \in \pi^{-1}(v)(k)} g^{-1}U$$

is affine. Moreover $\bigcup_{v \in V(k)} U_v = W \cdot U = X$ by step (I) of the proof. As X is quasi-compact, there exist $v_1, \ldots, v_n \in V(k)$ with $X = \bigcup_i X_{f_{v_i}}$. $\square$

Corollary 27.72. *Let S be a normal scheme, let G be a smooth S-group scheme with connected fibers, and let $f \colon X \to S$ be a homogeneous G-space that is locally of finite type over S. Then there exists an open covering $(S_i)_i$ of S such that X_{S_i} is quasi-projective over S_i.*

In particular, X is separated over S (Remark after Definition 13.60). We will give the proof only if S is in addition locally noetherian. For a proof in the general case by noetherian approximation we refer to [Ray1]$^\circ$ VI 2.5.

Proof. (if S is in addition locally noetherian) We may assume that S is affine. Let $s \in S$ and let $U \subseteq X$ be any open affine subscheme meeting the fiber X_s. As f is flat (Corollary 27.64), $V := f(U)$ is an open neighborhood of s in S by Theorem 14.35. Therefore it suffices to see that $X_V \to V$ is quasi-projective. But by definition, U meets every fiber of $X_V \to V$. Moreover, as S is separated, V is separated and therefore $U \to V$ is an affine morphism (Proposition 12.3). Hence we can apply Theorem 27.71. $\square$

Corollary 27.73. *Let R be a local normal ring and let $G \to \operatorname{Spec} R$ be a smooth group scheme with connected fibers. Then every homogeneous G-space of finite type is quasi-projective. In particular, G is quasi-projective.*

(27.12) The graded Hopf algebra structure on the cohomology ring of an algebraic group.

Let G be a quasi-compact group scheme over a field k. We know from Example 21.131 that the cup product induces on

$$H^\bullet(G, \mathscr{O}_G) = \bigoplus_{p \geq 0} H^p(G, \mathscr{O}_G)$$

the structure of a strictly graded commutative graded k-algebra. Moreover, the group law $m \colon G \times_k G \to G$ induces by functoriality a homomorphism of graded k-algebras

$$\mu \colon H^\bullet(G, \mathscr{O}_G) \longrightarrow H^\bullet(G \times_k G, \mathscr{O}_{G \times_k G}) \cong H^\bullet(G, \mathscr{O}_G) \otimes_k^g H^\bullet(G, \mathscr{O}_G),$$

where the isomorphism is given by the Künneth isomorphism (Corollary 22.110). Here $-\otimes^g-$ denotes the graded tensor product (Definition 19.3). The unit section $\operatorname{Spec} k \to G$ defines a counit

$$\varepsilon \colon H^\bullet(G, \mathscr{O}_G) \longrightarrow k$$

and the inverse map $G \to G$ defines an antipode

$$S \colon H^\bullet(G, \mathscr{O}_G) \to H^\bullet(G, \mathscr{O}_G).$$

Altogether we obtain on $H^\bullet(G, \mathscr{O}_G)$ the structure of a strictly graded commutative graded Hopf algebra over k in the following sense.

Definition 27.74. *Let R be a commutative ring.*
(1) *A graded bialgebra over R is an $\mathbb{N}$-graded R-algebra $H^\bullet = \bigoplus_{p \geq 0} H^p$ together with a comultiplication and a counit*

$$\mu \colon H^\bullet \longrightarrow H^\bullet \otimes_R^g H^\bullet, \qquad \varepsilon \colon H^\bullet \longrightarrow R,$$

that are both homomorphisms of graded R-algebras and such that the diagram

$$
\begin{array}{ccc}
H^\bullet & \xrightarrow{\ \mu\ } & H^\bullet \otimes_R^g H^\bullet \\
\mu \downarrow & & \downarrow \mu \otimes \mathrm{id} \\
H^\bullet \otimes_R^g H^\bullet & \xrightarrow{\ \mathrm{id} \otimes \mu\ } & H^\bullet \otimes_R^g H^\bullet \otimes_R^g H^\bullet
\end{array}
$$

commutes and such that

$$(\varepsilon \otimes \mathrm{id}) \circ \mu = (\mathrm{id} \otimes \varepsilon) \circ \mu = \mathrm{id} \colon H^\bullet \to H^\bullet,$$

using $H^\bullet \otimes_R^g R = H^\bullet$.
(2) *A map between graded bialgebras is called a homomorphism of graded bialgebras over R if it is a homomorphism of graded R-algebra that is compatible with comultiplication and counit.*
(3) *A graded bialgebra $H^\bullet$ over R is called (strictly) graded commutative if the underlying graded R-algebra is (strictly) graded commutative (Definition 17.49).*
(4) *A graded bialgebra $H^\bullet$ is called graded Hopf algebra if there exists an antipode $S \colon H^\bullet \to H^\bullet$, i.e., an isomorphism of graded R-modules such that*

$$m \circ (\mathrm{id} \otimes S) \circ \mu = i \circ \varepsilon = m \circ (S \otimes \mathrm{id}) \circ \mu,$$

where $m \colon H^\bullet \otimes^g H^\bullet \to H^\bullet$ is the multiplication and $i \colon R \to H^\bullet$ is the unit.

An antipode on a bialgebra is unique if it exists. Every homomorphism of graded bialgebras between graded Hopf algebras automatically preserves the antipode.

Example 27.75. Let R be a commutative ring and let M be an R-module. Then its exterior algebra $\bigwedge^\bullet(M)$ is endowed with the structure of a strictly graded commutative graded Hopf algebra: We have already seen that it carries the structure of a strictly graded commutative graded R-algebra (Section (17.8)). The diagonal map $M \to M \oplus M$ induces a comultiplication

$$\mu\colon \bigwedge{}^\bullet(M) \longrightarrow \bigwedge{}^\bullet(M \oplus M) = \bigwedge{}^\bullet(M) \otimes_R^g \bigwedge{}^\bullet(M),$$

the projection $\bigwedge^\bullet(M) \to \bigwedge^0(M) = R$ defines a counit, and the map $M \to M$, $m \mapsto -m$ induces the antipode.

Example 27.76. Let k be a field. We consider the finite-dimensional monogeneous case. Let $H^\bullet$ be a finite strictly graded commutative k-algebra generated by one homogeneous element $x \neq 0$ of degree $d = \deg(x) \geq 1$. Then $H^\bullet$ is the quotient of the (commutative) polynomial ring $k[x]$ by a homogeneous ideal, i.e. $H^\bullet \cong k[x]/(x^s)$ for some $s \geq 2$.

As $H^\bullet$ is strictly graded commutative, we necessarily have $s = 2$ if d is odd. If d is even and $H^\bullet$ can be endowed with the structure of a graded k-bialgebra, then one can show ([MiMo] $^\circ$ 7.8) that necessarily $p := \operatorname{char}(k) > 0$ and that s is a power of p.

For two graded bialgebras $H_1^\bullet$ and $H_2^\bullet$ we can form the tensor product of graded R-algebras $H_1^\bullet \otimes_R^g H_2^\bullet$ and endow it with the structure of a graded bialgebra by defining as comultiplication the composition

$$H_1^\bullet \otimes_R^g H_2^\bullet \xrightarrow{\mu_1 \otimes \mu_2} H_1^\bullet \otimes_R^g H_1^\bullet \otimes_R^g H_2^\bullet \otimes_R^g H_2^\bullet$$
$$\xrightarrow{\operatorname{id} \otimes T \otimes \operatorname{id}} (H_1^\bullet \otimes_R^g H_2^\bullet) \otimes_R^g (H_1^\bullet \otimes_R^g H_2^\bullet)$$

where $T\colon H_2^\bullet \otimes_R^g H_1^\bullet \longrightarrow H_1^\bullet \otimes_R^g H_2^\bullet$ is the map of graded R-modules with $T(x_2 \otimes x_1) = (-1)^{\deg(x_2)\deg(x_1)} x_1 \otimes x_2$ for homogeneous elements $x_i \in H_i^\bullet$.

Example 27.77. Let k be a field. If V is a k-vector space with basis $(x_1, \ldots, x_r)$, then there is an isomorphism of graded k-bialgebras

$$\bigwedge{}^\bullet(V) \cong k[x_1]/(x_1^2) \otimes_k^g \cdots \otimes_k^g k[x_r]/(x_r^2),$$

with $\deg(x_i) = 1$ for all i.

We will use the following structure theorem by Borel and Hopf, see [MiMo] $^\circ$ 7.11.

Theorem 27.78. *Let k be a perfect field and let $H^\bullet$ be a finite-dimensional strictly graded commutative graded k-bialgebra with $H^0 = k$. Then there exist monogeneous k-bialgebras $H_i^\bullet$, $i = 1, \ldots, r$, as in Example 27.76 and an isomorphism of k-bialgebras*

$$H^\bullet \cong H_1^\bullet \otimes_k^g \cdots \otimes_k^g H_r^\bullet.$$

Corollary 27.79. *Let k and $H^\bullet$ be as in Theorem 27.78 and assume that there exists an integer $g \geq 0$ such that $H^i = 0$ for all $i > g$. Then $\dim_k(H^1) \leq g$. If $\dim_k(H^1) = g$, then $H^\bullet \cong \bigwedge^\bullet(H^1)$ as graded Hopf algebras over k.*

Proof. Write $H^\bullet = k[x_1]/(x_1^{s_1}) \otimes_k^g \cdots \otimes_k^g k[x_r](x_r^{s_r})$ with $d_i = \deg(x_i)$ as in Theorem 27.78. Then $\dim_k(H^1)$ is the number of generators x_i such that $d_i = 1$. As $x_1 \otimes \cdots \otimes x_r \in H^{d_1 + \cdots + d_r}$ is nonzero we find $d_1 + \cdots + d_r \le g$. In particular $\dim_k(H^1) \le g$.

We have equality if and only if $d_i = 1$ for all i and $r = g$. If in this case there existed an i such that $x_i^2 \ne 0$, say $i = 1$, then $x_1^2 x_2 \cdots x_g$ is a nonzero element of degree $g + 1$, contradicting the hypothesis. Hence $H^\bullet \cong \bigwedge^\bullet H^1$ by Example 27.77. $\square$

Corollary 27.80. *Let k be a field and let G be a group scheme of finite type over k such that $H^\bullet(G, \mathscr{O}_G)$ is a finite-dimensional k-vector space and such that $H^0(G, \mathscr{O}_G) = k$. Then $\dim H^1(G, \mathscr{O}_G) \le \dim G$.*

If $\dim H^1(G, \mathscr{O}_G) = \dim G$, then the canonical map $\bigwedge^\bullet H^1(G, \mathscr{O}_G) \to H^\bullet(G, \mathscr{O}_G)$ is an isomorphism.

Proof. We can pass to a perfect extension of k because formation of cohomology and of exterior products commute with flat base change. Then we apply Corollary 27.79 with $g = \dim(G)$. $\square$

More generally, if G is any connected group scheme of finite type over a field k, then Brion [Bri]$_X^O$ has shown that there is an isomorphism of graded Hopf algebras

$$H^\bullet(G, \mathscr{O}_G) \cong \Gamma(G, \mathscr{O}_G) \otimes \bigwedge{}^\bullet(P^1),$$

where $P^i := \{ \gamma \in H^i(G, \mathscr{O}_G) \; ; \; \mu^*(\gamma) = \gamma \otimes 1 + 1 \otimes \gamma \}$ denotes the subspace of primitive elements of $H^i(G, \mathscr{O}_G)$. Moreover, one has $P^i = 0$ for $i \ge 2$.

(27.13) Cartier duality.

Let S be a scheme and let $\pi \colon G \to S$ be a group scheme over S. Then we can define its functor of characters that sends an S-scheme T to the abelian group

$$X^*(G)(T) := \mathrm{Hom}_{\mathrm{GrSch}/T}(G_T, \mathbb{G}_{m,T})$$

of homomorphisms $G \times_S T \to \mathbb{G}_{m,T}$ of group schemes over T.

Now suppose that $\pi \colon G \to S$ is a finite locally free commutative group scheme. Then $X^*(G)$ is representable by a finite locally free group scheme that can be described as follows. Let $\mathscr{A} := \pi_* \mathscr{O}_G$ be the finite locally free commutative cocommutative Hopf algebra corresponding to G (Section (27.2)). Then we can endow the linear dual $\mathscr{A}^\vee = \mathscr{H}om_{\mathscr{O}_S}(\mathscr{A}, \mathscr{O}_S)$ again with the structure of a finite locally free commutative cocommutative Hopf algebra. Its multiplication and unit is given by the dual of the comultiplication and of the counit, its comultiplication and counit is given by the dual of multiplication and of the unit, and its antipode is given by the dual of the antipode map. We obtain a finite locally free commutative group scheme $G^D := \mathrm{Spec}\,\mathscr{A}^\vee$ over S.

Proposition 27.81. *(Cartier duality) Let G be a commutative finite locally free group scheme over S. The functor $X^*(G)$ is representable by G^D.*

In the proof we will use the following notion. Let (B, m^*, i^*, e^*) be a commutative Hopf algebra over a ring R. An element $b \in B$ is called *group-like* if $m^*(b) = b \otimes b$ and $e^*(b) = 1$. Then one has $1 = e^*(b) = (i^*, id)m^*(b) = (i^*, \mathrm{id})(b \otimes b) = i^*(b)b$. Hence $b \in B^\times$ and $i^*(b) = b^{-1}$. Let B^{gl} be the set of group like elements of B. Then B^{gl} is a subgroup of $B^\times$.

Proof. The formation of G^D commutes with base change $T \to S$. Hence we may assume that $S = \operatorname{Spec} R$ is affine and it suffices that one has functorial isomorphism of abelian groups $G^D(S) \cong \operatorname{Hom}_{\operatorname{GrSch}/S}(G, \mathbb{G}_{m,S})$. Let $(A, m^*, i^*, e^*) = \Gamma(G, \mathscr{O}_G)$ be the Hopf algebra corresponding to G.

A character $G \to \mathbb{G}_{m,R}$ corresponds to a homomorphism of Hopf algebras $R[T, T^{-1}] \to A$ and hence by Example 27.8 to a unit $a \in A^\times$ such that $m^*(a) = a \otimes a$, $e^*(a) = 1$, and $i^*(a) = a^{-1}$. In other words we have a functorial bijection

$$(27.13.1) \qquad \operatorname{Hom}_{\operatorname{GrSch}/S}(G, \mathbb{G}_{m,S}) \cong \Gamma(G, \mathscr{O}_G)^{\mathrm{gl}}.$$

which is easily seen to be an isomorphism of abelian groups.

To identify $G^D(S) = \operatorname{Hom}_{R-\operatorname{Alg}}(A^\vee, R)$ with A^{gl} we remark that every R-linear map $A^\vee \to R$ is of the form $\operatorname{ev}_a \colon \lambda \mapsto \lambda(a)$ for some $a \in A$ because A is finite locally free and hence $A \cong (A^\vee)^\vee$. Moreover, for $a \in A$ one has $\operatorname{ev}_a(1) = 1$ if and only if $e^*(1) = 1$ and ev_a is a ring homomorphism if and only if $m^*(a) = a \otimes a$. This yields a functorial bijection

$$(27.13.2) \qquad\qquad G^D(S) \cong A^{\mathrm{gl}}$$

which is an isomorphism of groups because the multiplication of the group scheme G^D is given by the dual of the multiplication $A \otimes A \to A$ of the R-algebra A. $\qquad \square$

Definition 27.82. *Let G be a finite locally free commutative group scheme over a scheme S. The group scheme G^D is called the* Cartier dual *of G.*

Remark 27.83. Let G be a finite locally free commutative group scheme over a scheme S.
(1) The biduality isomorphism $\mathscr{A} \xrightarrow{\sim} (\mathscr{A}^\vee)^\vee$ is an isomorphism of Hopf algebras and hence defines an isomorphism of group schemes

$$(G^D)^D \xrightarrow{\sim} G.$$

(2) The functor $G \mapsto X^*(G)$ is contravariant in G. In particular $G \mapsto G^D$ yields an anti-equivalence of the category of finite locally free commutative group schemes over S with itself.
(3) One has $\operatorname{rk}(G^D) = \operatorname{rk}(\mathscr{A}^\vee) = \operatorname{rk}(\mathscr{A}) = \operatorname{rk}(G)$.
(4) Formation of Cartier dual is compatible with base change, i.e., for every morphism of schemes $T \to S$ one has $(G_T)^D \cong (G^D)_T$.

Remark 27.84. Let R be a ring. The proof of Proposition 27.81 shows the following assertions.
(1) For any affine group scheme G over R one has the functorial isomorphism (27.13.1) of abelian groups.
(2) Let $G = \operatorname{Spec} A$ be a commutative finite locally free group scheme for a finite locally free Hopf algebra A over R. Then (27.13.2), applied to G^D, shows that one has a functorial group isomorphism

$$(27.13.3) \qquad\qquad G(R) \to (A^\vee)^{\mathrm{gl}}.$$

Example 27.85. Let S be a scheme. Let $n \geq 1$ be an integer and let $\underline{\mathbb{Z}/n\mathbb{Z}}_S$ be the constant group scheme corresponding to the abelian group $\mathbb{Z}/n\mathbb{Z}$ (Example 4.43). Let $\mu_n(T)$ be the group scheme of n-th roots of unity, i.e., $\mu_n(T) = \{ a \in \Gamma(T, \mathscr{O}_T)^\times \; ; \; a^n = 1 \}$. In other words, μ_n is the kernel of the group scheme homomorphism $\mathbb{G}_m \to \mathbb{G}_m$ given (on T-valued points) by $z \mapsto z^n$. Then $X^*(\underline{\mathbb{Z}/n\mathbb{Z}}_S)(T) = \mu_n(T)$. Therefore we see that the group schemes $\underline{\mathbb{Z}/n\mathbb{Z}}_S$ and $\mu_{n,S}$ are Cartier dual to each other. If $\Gamma(S, \mathscr{O}_S)$ contains a primitive n-th root of unity (e.g., if S is a k-scheme for a field k which contains a primitive n-th root of unity), then a choice of one such yields an isomorphism $\mu_{n,S} \cong \underline{\mathbb{Z}/n\mathbb{Z}}_S$. If n is invertible in $\Gamma(S, \mathscr{O}_S)$, then such an isomorphism exists at least étale-locally on S. On the other hand, if S is a scheme over a field of positive characteristic p dividing n, then $\mu_{n,S}$ is not étale over S, and the structure morphism $\mu_{p,S} \to S$ is a homeomorphism.

(27.14) Annihilation of commutative finite locally free group schemes.

Let S be a scheme and let G be a commutative finite locally free group scheme over S. We want to show the following result.

Proposition 27.86. *Suppose that G is of constant rank $r \in \mathbb{N}_{>0}$. Then G is annihilated by r, i.e., $g^r = 1$ for all $g \in G(T)$ and any S-scheme T.*

To see this we use the following construction. Suppose that $S = \operatorname{Spec} R$ is affine. Then $G = \operatorname{Spec} A$ for a finite locally free Hopf algebra. Let $A^\vee$ be its dual (Section (27.13)). Let $\varphi \colon R \to R'$ be a finite locally free R-algebra. By (27.13.3) we have $G(R) = (A^\vee)^{\mathrm{gl}}$ and $G(R') = (A^\vee \otimes_R R')^{\mathrm{gl}}$. The norm map $N \colon (A^\vee \otimes_R R')^\times \to (A^\vee)^\times$ (Remark 12.25) then induces a homomorphism of abelian groups

$$N_{R'/R} \colon G(R') \longrightarrow G(R).$$

The properties of the norm map yield the following assertions.
(1) Let R' be of constant rank r' over R. If $\varphi^* \colon G(R) \to G(R')$ denotes the group homomorphism induced by φ by functoriality, then for all $g \in G(R)$ one has

$$(27.14.1) \qquad N_{R'/R}(\varphi^*(g)) = g^{r'}$$

(2) If $\psi \colon R' \to R'$ is an automorphism of R-algebras, then for all $g' \in G(R')$ one has

$$(27.14.2) \qquad N_{R'/R}(\psi^*(g')) = N_{R'/R}(g').$$

Proof. [of Proposition 27.86] Since $G(T) = \operatorname{Hom}_T(T, G \times_S T)$ it suffices to show that $g^r = 1$ for any $g \in G(S)$. If $(U_i)_i$ is an open covering of S, then $G(S) \to \prod_i G(U_i)$ is injective. Hence we may assume that $S = \operatorname{Spec} R$ is affine. Then $G = \operatorname{Spec} A$ for a finite locally free R-algebra $\varphi \colon R \to A$ of rank r.

Let $g \in G(R)$. Denote by $t_g \colon G \to G$ the translation by g corresponding to an R-algebra automorphism ψ_g of A. Consider $\mathrm{id}_G \in G(A)$. Then we have the following equality in the group $G(R)$ which shows that $g^r = 1$

$$N_{A/R}(\mathrm{id}_G) = N_{A/R}(\psi_g^*(\mathrm{id}_G)) = N_{A/R}(\mathrm{id}_G)N_{A/R}(\varphi^*(g)) = N_{A/R}(\mathrm{id}_G)g^r.$$

Here the first equality holds by (27.14.2), the second because $\psi_g^*(\mathrm{id}_G) = \mathrm{id}_G \cdot \varphi^*(g)$ (multiplication in $G(A)$), and the third by (27.14.1). $\qquad\square$

(27.15) Digression: Collection of some properties of schemes over inductive limits of rings.

Sometimes we will study schemes over inductive limits of rings, e.g., to reduce to the case of schemes over a noetherian ring or over a local ring. See Chapter 10. Here we collect some further results that we will use in order to give precise references. Very often we will not give proofs.

Let $(R_\lambda)_{\lambda \in \Lambda}$ be filtered inductive system of rings, let $R := \operatorname{colim} R_\lambda$ be their colimit, set $S_\lambda := \operatorname{Spec} R_\lambda$ and $S := \operatorname{Spec} R$. Most often we will apply this to one of the following cases.

(1) We write a ring R as the filtered union of its finitely generated $\mathbb{Z}$-subalgebras R_λ.
(2) We write the localization $R = T^{-1}A$ of a ring A with respect to a multiplicative subset $T \subseteq A$ as the filtered colimit of the localizations A_f for $f \in T$.

Remark 27.87. Let $G \to S$ be a group scheme of finite presentation. It follows from Corollary 10.67 and Theorem 10.63 (applied to multiplication, inverse and identity element) that there exists a λ and a group scheme of finite presentation $G_\lambda \to S_\lambda$ such that $G \cong G_\lambda \times_{S_\lambda} S$.

We now assume that Λ has a unique minimal element 0 and that we are given an S_0-scheme X_0. Set $X_\lambda := X_0 \times_{S_0} S_\lambda$ and $X := X_0 \times_{S_0} S$.

We now list a number of properties of X that descend to some X_λ which we will use in this chapter.

Lemma 27.88. *Suppose that $X_0 \to S_0$ is of finite presentation. Let* **P** *be one of the following properties of a morphism of schemes.*

(1) *flat,*
(2) *proper,*
(3) *separated,*
(4) *smooth,*
(5) *the morphism has geometrically reduced fibers,*
(6) *the morphism has geometrically connected fibers,*
(7) *the morphism has equi-dimensional fibers of dimension d, where $d \geq 0$ is a fixed integer.*

Then $X \to S$ has property **P** *if and only if there exists a λ such that $X_\lambda \to S_\lambda$ has property* **P**.

Proof. All of the above properties are stable under base change. Therefore if $X_\lambda \to S_\lambda$ has one of these properties for some λ, then $X \to S$ has the same property.

So the core of the lemma is the converse, for which we give references. For the property of being "flat" see [EGAIV]^{O} (11.2.6), for the properties "proper" and "separated" see [EGAIV]^{O} (8.10.5), for "smooth" see Proposition 18.59.

For all properties of fibers it suffices to show that for all λ the set of points s_λ in S_λ such that the fiber of $X_\lambda \to S_\lambda$ in s_λ has the stated property is constructible. Then one can conclude by Proposition 10.56.

For constructibility for the fiber properties "geometrically reduced" see [EGAIV]^{O} (9.7.7), for "geometrically connected" see [EGAIV]^{O} (9.7.9), for "equi-dimensional of dimension d" see [EGAIV]^{O} (9.8.5). $\qquad\square$

Definition and basic properties of abelian schemes

We now come to the definition of *abelian schemes* as a smooth proper group scheme with connected fibers, which we think of as families of abelian varieties (Definition 16.53). Using some rigidity results proved in Section (27.17) we then show that an abelian scheme is a commutative group scheme. As further applications of these rigidity results we show that all reduced connected fibers of morphisms to some scheme are translates of abelian varieties (Proposition 27.106) and that to define the group law on an abelian scheme it suffices to find a composition law that has an identity (Proposition 27.109).

(27.16) Definition of abelian schemes.

We continue to denote by S a scheme.

Recall the definition of an abelian variety over a field k. It is defined as a smooth (equivalently by Proposition 27.10 (1), geometrically reduced) proper connected group scheme over k. Moreover, a group scheme over k is an abelian variety if and only if its base change to an algebraic closure (or to any field extension) is an abelian variety since all defining properties are stable under base change and under faithfully flat descent (for "connected group scheme" use Proposition 27.10 (2)).

We would now like to define an abelian scheme over a general base scheme as a group scheme that is a "continuously varying family of abelian varieties". By this we mean a flat group scheme[4] $X \to S$ with certain addition properties. First of all we add as a finiteness condition "locally of finite presentation" as one would like that everything is generated locally by finitely many elements with finitely many relations. This is something not visible over a field (or a noetherian ring) as then the notions "locally of finite type" and "locally of finite presentation" coincide. For the other properties to add we distinguish between "global properties" and "fiber properties". For the defining properties of an abelian variety, the global property is "proper" and the fiber properties are "geometrically reduced" and "(geometrically) connected":

Definition 27.89. *A group scheme X over S is called an* abelian scheme *if it satisfies the following properties.*
(a) *The structure morphism $X \to S$ is proper, flat and locally of finite presentation.*
(b) *All fibers of $X \to S$ are geometrically reduced and connected.*

Proposition 27.92 below shows that an equivalent definition of an abelian scheme would be a group scheme X over S such that $X \to S$ is proper smooth and has connected fibers (which are then automatically geometrically integral).

As recalled before, an abelian scheme over a field k is the same as an *abelian variety* over k.

The next two remarks show that the property of being an abelian scheme satisfies the usual permanence properties.

Remark 27.90. All the defining properties of abelian schemes are stable under base change, composition and fpqc descent. Hence we see:

---

[4] In fact, usually it is better to work with algebraic spaces, e.g., since descent often behaves better for algebraic spaces than for schemes. Here it does not make a difference because an abelian algebraic space, defined exactly as an abelian scheme below with the word "scheme" replaced by "algebraic space" can be shown to be automatically a scheme (see Theorem 27.211 below).

(1) If $X \to S$ is an abelian scheme and $T \to S$ is a morphism of schemes, then the base change $X \times_S T \to T$ is an abelian scheme.

(2) Let $X \to S$ be a group scheme and let $S' \to S$ be a faithfully flat quasi-compact morphism such that the base change $X_{S'} \to S'$ is an abelian scheme, then $X \to S$ is an abelian scheme.

(3) Let X and Y be abelian schemes over S. Then $X \times_S Y$ is an abelian scheme over S.

Remark 27.91. Let $(R_\lambda)_{\lambda \in \Lambda}$ be a filtered inductive system of rings, let $R := \operatorname{colim} R_\lambda$ be their colimit, set $S_\lambda := \operatorname{Spec} R_\lambda$ and $S := \operatorname{Spec} R$. Let $X \to S$ be an abelian scheme. As $X \to S$ is of finite presentation, there exists a λ and an abelian scheme $X_\lambda \to S_\lambda$ such that $X \cong X_\lambda \times_{S_\lambda} S$ by Remark 27.87 and Lemma 27.88.

The following properties of the structure morphism of an abelian scheme now follow immediately from earlier results.

Proposition 27.92. *Let $f \colon X \to S$ be an abelian scheme. Then $X \to S$*
(1) *is faithfully flat and quasi-compact,*
(2) *is smooth,*
(3) *is of finite presentation,*
(4) *is universally open, and*
(5) *has geometrically integral fibers.*
(6) *One has $\mathscr{O}_S = f_* \mathscr{O}_X$ and the formation of $f_* \mathscr{O}_X$ commutes with base change.*

Proof. Assertion (1) holds since proper morphisms are quasi-compact and since $X \to S$ is surjective because it has a section. Assertion (2) follows from Corollary 27.24, (3) holds because $X \to S$ is proper and locally of finite presentation, (4) holds because every flat morphism locally of finite presentation is universally open by Theorem 14.35, (5) follows from Proposition 27.10 (2), and $\mathscr{O}_S = f_* \mathscr{O}_X$ holds universally by Corollary 24.63. $\square$

Proposition 27.93. *Let $f \colon X \to S$ be an abelian scheme of relative dimension g (g a locally constant function on S). Let $\mathscr{C}_e$ be the conormal sheaf of the unit section e. Then $\mathscr{C}_e$ is a locally free $\mathscr{O}_S$-module of rank g and one has isomorphisms*

$$f^* \mathscr{C}_e \cong \Omega^1_{X/S},$$
$$e^* \Omega^1_{X/S} \cong \mathscr{C}_e \cong f_* \Omega^1_{X/S},$$

that are functorial in X.

Proof. By Corollary 19.31, e is a regular immersion of codimension g. Hence $\mathscr{C}_e$ is locally free of rank g. The first (resp. second) isomorphism holds for arbitrary group schemes (resp. for arbitrary schemes) by Proposition 27.15 (resp. Equation (17.4.4)). To show the last isomorphism it therefore suffices to show that for every locally free $\mathscr{O}_S$-module $\mathscr{E}$ the canonical homomorphism $\mathscr{E} \to f_* f^* \mathscr{E}$ is an isomorphism. This can be shown locally, hence we may assume that $\mathscr{E} = \mathscr{O}_S^r$ for some $r \geq 0$. Then the claim follows since $f_* \mathscr{O}_X = \mathscr{O}_S$ (Proposition 27.92 (6)). $\square$

(27.17) The constancy locus of a morphism of schemes.

We continue to denote by S a scheme. Let $f \colon X \to S$ and $g \colon Y \to S$ be S-schemes. An S-morphism $h \colon X \to Y$ is called *constant* if it factorizes through f, i.e., there exists a section $t \colon S \to Y$ of g such that $h = t \circ f$.

For a map of schemes $T \to S$ we set $X_T := X \times_S T$ and write $h_T \colon X_T \to Y_T$ for $h \times_S \mathrm{id}_T$. If $T = \operatorname{Spec} R$ is affine, we write X_R, Y_R and h_R instead of X_T, Y_T, and h_T. If $T \to S$ is the canonical morphism $\operatorname{Spec} \kappa(s) \to S$ we obtain the fibers over s, denoted by X_s, Y_s, and h_s instead of $X_{\kappa(s)}$, $Y_{\kappa(s)}$, and $h_{\kappa(s)}$. Finally, we define the locus of points in S in which the fibers of h are constant,

(27.17.1) $\qquad \operatorname{Const}(h) := \{\, s \in S \,;\, X_s \neq \emptyset \text{ and } h_s \colon X_s \to Y_s \text{ is constant}\}.$

Remark 27.94. Suppose that h is constant.
(1) Then h_T is constant for every scheme map $T \to S$.
(2) If $f \colon X \to S$ is faithfully flat and quasi-compact (or, more generally, an epimorphism of schemes), then the section of Y through which h factors is unique.

Moreover, the locus, where morphisms induced on fibers are constant is stable under base change in the following sense.

Lemma 27.95. *In the situation above, let $\xi \colon S' \to S$ be a morphism of schemes. Then*

$$\xi^{-1}(\operatorname{Const}(h)) = \operatorname{Const}(h_{S'}).$$

Proof. Let $s' \in S$ and let $s := \xi(s')$. Then the morphism induced on the fiber in s' by $h' := h_{S'}$ is the base change of the morphism induced on the fiber in s by h via the field extension $\kappa(s) \to \kappa(s')$. One has $\xi^{-1}(\operatorname{Const}(h)) \subseteq \operatorname{Const}(h')$ since being constant is stable under base change.

For the other inclusion, we have to show that if $h'_{s'}$ is constant, then h_s is constant. Now if the fiber of $X_{S'}$ in s' is non-empty, then X_s is non-empty. Hence there exists for every field extension k' of $\kappa(s)$ at most one section $t' \colon \operatorname{Spec} k' \to Y_{k'}$ such that $t' \circ f_{k'} = h_{k'}$. Therefore the existence of such a section can be checked fpqc locally by Proposition 14.76. $\qquad \square$

Proposition 27.96. *Let $f \colon X \to S$ and $g \colon Y \to S$ be S-schemes and let $h \colon X \to Y$ be an S-morphism.*
(1) *Suppose that f is proper and that $\mathcal{O}_S = f_*\mathcal{O}_X$. Then $\operatorname{Const}(h)$ is open in S and $h_{\operatorname{Const}(h)} \colon X_{\operatorname{Const}(h)} \to Y_{\operatorname{Const}(h)}$ is constant.*
(2) *Suppose that $X \to S$ is proper, flat, of finite presentation, and has geometrically connected and geometrically reduced fibers and that $Y \to S$ is separated and of finite presentation. Then $\operatorname{Const}(h)$ is open and closed in S and $h_{\operatorname{Const}(h)} \colon X_{\operatorname{Const}(h)} \to Y_{\operatorname{Const}(h)}$ is constant.*

Under both hypotheses (1) or (2), the morphism f is surjective and hence

$$\operatorname{Const}(h) = \{\, s \in S \,;\, h_s \colon X_s \to Y_s \text{ is constant}\}$$

Proof. We may assume S is affine and that $\operatorname{Const}(h)$ is nonempty.

Let us prove (1). Let $s \in \operatorname{Const}(h)$. We have to show that there exists an open neighborhood W of s such that h_W is constant. Let $V \subseteq Y$ be an open affine neighborhood of the unique point in $f_s(X_s) \subseteq Y_s \subseteq Y$. Then $X_s \subseteq h^{-1}(V)$ and hence there exists an open neighborhood W of s such that $X_W \subseteq h^{-1}(V)$ because f is closed. Hence h restricts to a map of W-schemes $h'_W \colon X_W \to V_W$. Now $V \to S$ is affine as a morphism between affine schemes and hence V_W is affine over W. By Remark 24.47, h'_W factors through $f_{W,*}\mathcal{O}_{X_W} = (f_*\mathcal{O}_X)_{|W}$ which is equal to $\mathcal{O}_W$ by hypothesis. This shows that h_W is constant.

To prove ((2)), we now suppose that $X \to S$ is proper, flat, of finite presentation, and has geometrically connected and geometrically reduced fibers. Then $\mathscr{O}_S = f_* \mathscr{O}_X$ by Corollary 24.63. Therefore, we know already that $\mathrm{Const}(h)$ is open in S.

Let $S = \operatorname{Spec} R$. By Corollary 10.67 there exists a noetherian subring R_λ and a morphism $h_\lambda \colon X_\lambda \to Y_\lambda$ of finite type R_λ-schemes whose base change to S is h. By Lemma 27.88 we may assume that $X_\lambda \to S_\lambda := \operatorname{Spec} R_\lambda$ is proper, flat, and has geometrically connected and geometrically reduced fibers and $Y_\lambda \to S_\lambda$ is separated. Moreover, $\mathrm{Const}(h)$ is the inverse image under $S \to S_\lambda$ of $\mathrm{Const}(h_\lambda)$ by Lemma 27.95.

Hence it remains to show that $\mathrm{Const}(h)$ is closed if S is a noetherian scheme. As we already know that $\mathrm{Const}(h)$ is open in a noetherian scheme, it is constructible. Hence it suffices to prove that $\mathrm{Const}(h)$ is stable under specialization (Lemma 10.17). For this we may base chance to a discrete valuation ring by Proposition 15.7 using that the constancy locus is stable under base change (Lemma 27.95). Hence we can assume that $S = \operatorname{Spec} R$ for a discrete valuation ring R. Let $s \in S$ be the special point and let $\eta \in S$ be the generic point. We have to show that if $\eta \in \mathrm{Const}(h)$, then $s \in \mathrm{Const}(h)$.

Note that h is proper because $X \to S$ is proper and $Y \to S$ is separated (Proposition 12.58). Consider the schematic image $\mathrm{Im}(h)$ of h (Section (10.8)). As h is proper, the underlying topological space of the schematic image of $\mathrm{Im}(h)$ is $h(X)$. As images of proper schemes are proper (Proposition 12.59), $\mathrm{Im}(h)$ is proper. Therefore replacing Y by $\mathrm{Im}(h)$ we can assume that $Y \to S$ is proper (and that h is surjective).

Let $K = \kappa(\eta)$ be the field of fractions of R and let $t \colon \operatorname{Spec} K \to Y_K$ be the section of Y_K such that $h_K = t \circ f_K$. As Y is proper over R, there exists a unique extension $\tilde{t} \colon S \to Y$ of t by the valuative criterion (Theorem 15.9). It remains to show that $h = \tilde{t} \circ f$.

Let $E = \mathrm{Eq}(h, \tilde{t} \circ f) \subseteq X$ be the equalizer (Definition 9.1). Then E contains the generic fiber X_η. As $Y \to S$ is separated, E is a closed subscheme of X (Proposition 9.7). As X is flat over R, it is the smallest closed subscheme of X containing X_η (Proposition 14.14). Therefore $E = X$. This concludes the proof. $\square$

Remark 27.97. Under the assumptions in Proposition 27.96 (1) the proof shows that if there exists an $s \in S$ such that $f(X_s)$ is contained in an open affine subscheme of Y, then there exists an open neighborhood W of s such that the restriction $X_W \to Y_W$ is constant. If the hypotheses in (2) are satisfied, then W can be chosen open and closed.

Corollary 27.98. *Let S be a scheme, let*

be a diagram of S-schemes. Suppose that $g \colon X \to T$ is proper, flat, of finite presentation, and has geometrically connected and geometrically reduced fibers and that the structure morphism $Y \to S$ is separated and of finite presentation.

If there exists $t \in T$ such that the set-theoretic image $f(g^{-1}(t)) \subseteq Y$ consists of a single point, then there exists an open and closed neighborhood W of t in T such that $f_{|X \times_T W}$ factors through $g_{|X \times_T W}$.

Proof. We apply Proposition 27.96 (2) and Remark 27.97 to the morphism $(f, g)_S \colon X \to Y \times_S T$ of T-schemes and obtain that there exists an open and closed neighborhood W of t such that the restriction of $(f, g)_S$ to $X_W := X \times_T W$ factors through a section of the projection $Y \times_S W \to W$. This section corresponds to a morphism $h\colon W \to Y$ with $h \circ g_{|X_W} = f_{|X_W}$. □

By the following remark, this corollary generalizes the *rigidity lemma*, Proposition 16.55, and we sometimes refer to the results in this section as *rigidity*.

Remark 27.99. The condition that $Y \to S$ is of finite presentation in Proposition 27.96 (2) and in Corollary 27.98 is only used in the proof to reduce to the noetherian case. Hence it is superfluous if S is locally noetherian.

(27.18) Abelian schemes are commutative.

In this section S denotes a scheme.

Definition 27.100. *A pair (F, e) consisting of a presheaf $F\colon (\mathrm{Sch}/S)^{\mathrm{opp}} \to (\mathrm{Sets})$ and an element $e \in F(S)$ is called a* pointed presheaf. *A morphism of pointed presheaves $(F, e) \to (F', e')$ is a morphism $\alpha\colon F \to F'$ of functors such that $\alpha(S)(e) = e'$.*

As special cases we obtain the notions of pointed fppf-sheaves, of pointed algebraic spaces, and of pointed schemes and their morphisms. For instance, a *pointed S-scheme* is a pair (X, e) consisting of an S-scheme X and a section $e \in X(S)$ of the structure morphism $X \to S$. A *morphism of pointed S-schemes* $(X, e) \to (X', e')$ is then simply a morphism of S-schemes $f\colon X \to X'$ such that $f \circ e = e'$.

If F is a presheaf of groups, for instance a group scheme over S, we will consider F always as pointed via the unit section in $F(S)$.

Proposition 27.101. *Let X and Y be abelian schemes over S and let $f\colon X \to Y$ be a morphism of S-schemes. Then there exists $y \in Y(S)$ such that $t_y \circ f$ is a homomorphism of group schemes over S.*

Here $t_y\colon Y \to Y$ denotes the left translation by y (Definition 27.1).

Proof. Translating with the inverse of $f(e_X) \in Y(S)$, we may assume that f preserves unit sections. Hence it suffices to show the following corollary. □

Corollary 27.102. *Let X and Y be abelian schemes over S and let $f\colon X \to Y$ be a morphism of pointed S-schemes. Then f is a homomorphism of group schemes over S.*

In the proof we use Corollary 16.56 which gives the result if S is the spectrum of a field. The proof then relies only on Part (1) of Proposition 27.96. A direct proof that does not use the case over a field but that instead uses also Part (2) of Proposition 27.96 is sketched in Exercise 27.8.

Proof. Let $h\colon X \times_S X \to Y$ be the morphism of S-schemes that is given on T-valued points by $(x, x') \mapsto f(x)f(x')(f(xx'))^{-1}$, where we write the group law in $X(T)$ and in $Y(T)$ multiplicatively. We have to show that h factors through the unit section e_Y of Y. If h factors through some section t, then this section is necessarily unique because $X \times_S X \to S$ is surjective and faithfully flat (Remark 27.94). As the composition of h with $(e, \mathrm{id}_X)\colon X \to X \times_S X$ factors through e_Y, we have $t = e_Y$. Hence it suffices to show that h is constant. By Proposition 27.96 (1) we can check this on fibers. Hence we conclude by Corollary 16.56. □

Applying Corollary 27.102 to the inversion of the abelian scheme we deduce:

Corollary 27.103. *Abelian schemes are commutative group schemes.*

From now on we will use additive notions for abelian schemes, e.g., if $X \to S$ is an abelian scheme, we write the group law on $X(T)$ for T some S-scheme additively. The unit section in $X(S)$ will be denote by e or by 0 and will be called the zero section.

Corollary 27.104. *Let $X \to S$ be an S-scheme and let $0 \in X(S)$ be a section. Then there exists at most one structure of an abelian scheme over S on X such that 0 is the zero section.*

Proof. Let m_1 and m_2 be two group scheme structures on X with unit section 0 making X into an abelian scheme. Then id_X is a homomorphism of group schemes by Proposition 27.102. Hence $m_1 = m_2$. $\square$

(27.19) Further applications of rigidity.

Proposition 27.105. *Let $X \to S$ be an abelian scheme and let $Y \to S$ be a separated S-group scheme of finite presentation. Let $f, g \colon X \to Y$ be homomorphisms of group schemes and let $s \in S$ be a point such that the geometric fibers $f_{\bar{s}}$ and $g_{\bar{s}}$ agree. Then there exists an open and closed neighborhood W of s such that $f_W = g_W \colon X_W \to Y_W$.*

Proof. One has $f_{\bar{s}} = g_{\bar{s}}$ if and only if the fibers f_s and g_s are equal (Theorem 14.72). Consider $h := fg^{-1} \colon X \to Y$. Then h_s factors through the unit section of G_s. Hence there exists by Proposition 27.96 an open and closed neighborhood W of s such that h_W factors through a section $\sigma \colon W \to Y_W$. Let $\pi \colon X \to S$ be the structure morphism. Then $\sigma = \sigma \circ \pi_W \circ 0 = h_W \circ 0$. As h preserves unit sections, this shows that σ is the unit section of Y_W and hence that $f_W = g_W$. $\square$

The following amazing result, which we learned from [EGM] $\times$ 2.20, essentially states that all k-rational non-empty fibers of morphisms of an abelian variety to some other scheme are translates of abelian varieties after passing to reduced connected components.

Proposition 27.106. *Let k be a perfect field, let X be an abelian variety over k, let Y be a k-scheme, and let $f \colon X \to Y$ be a morphism of k-schemes. For $x \in X(k)$ let $F_x := (f^{-1}(f(x)))^0_{\mathrm{red}}$, where $(\ )^0$ denotes the connected component containing x.*
Then F_0 is an abelian subvariety of X and for $x \in X(k)$ one has $F_x = t_x(F_0)$.

Proof. As $f \circ x$ is a closed immersion $\operatorname{Spec} k \to Y$, the scheme F_x is a closed subscheme of X. It is by definition reduced and hence geometrically reduced because k is perfect (Corollary 5.57). It is connected and has x as a k-valued point. Hence it is geometrically connected (Lemma 26.4). In particular, the formation of F_x is compatible with passing to field extensions of k.
We will show that for all $x \in X(k)$ we have

(*) $$F_x = t_x(F_0)$$

Let us first show how (*) implies that F_0 is a subgroup scheme. As F_0 is a geometrically reduced k-scheme of finite type over a field, it suffices to show that $F_0(\bar{k})$ is a subgroup of $X(\bar{k})$ for some algebraic closure $\bar{k}$ of k. Hence it remains to show that for $y \in F_0(\bar{k})$ we have $y + F_0(\bar{k}) = F_0(\bar{k})$. After base change $k \to \bar{k}$ we may assume that $k = \bar{k}$. But then $t_y(F_0) = F_y = F_0$ by (*).

As F_0 is closed in X, it is a proper k-scheme. It is connected and smooth (Proposition 27.10 (3)) and hence an abelian variety.

It remains to prove (*). Showing the equality of two closed subschemes can be done fpqc-locally hence we may assume k is algebraically closed. We apply Corollary 27.98 to the projection $g\colon X \times_k F_x \to F_x$, to $\varphi\colon X \times_k F_x \to Y$ given by $(y,z) \mapsto f(y+z)$ and to $t = 0 \in X$ using that $f(0+z) = f(x)$ for all $z \in F_x(k)$. Hence φ factors through $\bar\varphi\colon X \to Y$, i.e., $f(y+z)$ is independent of $z \in F_x(k)$ for all $y \in X(k)$. In particular

$$f(y - x + F_x(k)) = \{f(y)\}, \qquad \text{for all } x, y \in X(k).$$

For $y = 0$ we obtain $-x + F_x(k) \subseteq F_0(k)$ and for $x = 0$ we obtain $y + F_0(k) \subseteq F_y(k)$. This proves (*). $\qquad\square$

Corollary 27.107. *Let X be an abelian variety over a field k. Let Y be a k-scheme and let $f\colon X \to Y$ be a morphism of k-schemes. Then all non-empty fibers of f are equi-dimensional of the same dimension.*

Proof. As the dimension of fibers does not change under field extensions (Proposition 5.38), we may assume that k is algebraically closed. We want to show that the function

$$d\colon X \to \mathbb{Z}, \qquad x \mapsto \dim_x f^{-1}(f(x))$$

is constant. As it is upper semicontinuous by Chevalley's theorem, Theorem 14.112, it suffices to show that it is constant on the very dense subspace of k-valued points. But Proposition 27.106 shows that for every $x \in X(k)$, the underlying reduced subscheme of every connected component of $f^{-1}(f(x))$ is isomorphic to the same abelian variety. In particular all irreducible components of $f^{-1}(f(x))$ have the same dimension for all $x \in X(k)$. $\qquad\square$

(27.20) Constructing abelian schemes.

The main result in this section is Proposition 27.109. For the proof we will use the following lemma.

Lemma 27.108. *Let k be a field, let X and Y be irreducible k-schemes of finite type. Suppose that X is proper over k and that Y is separated over k. Suppose that $\dim X \geq \dim Y$ and let $f\colon X \to Y$ be a morphism of k-schemes such that there exists a point $y \in Y$ such that $\dim f^{-1}(y) = 0$. Then f is surjective and $\dim X = \dim Y$.*

Proof. By passing to the underlying reduced subschemes we may assume that X and Y are integral. The morphism f is proper since X is proper and Y is separated. In particular $f(X) \subseteq Y$ is closed. If we show that $\dim X = \dim f(X)$, then necessarily $f(X) = Y$ since $\dim X \geq \dim Y$. Hence we may assume that f is surjective and it remains to show that $\dim X = \dim Y$. By Corollary 14.115 the subset $\{ y \in Y \ ; \ \dim f^{-1}(y) = 0 \}$ is open. By hypothesis it is non-empty and hence dense because Y is irreducible. Hence $\dim f^{-1}(\eta) = 0$ if η is the generic point and we conclude by Proposition 14.109 (2). $\qquad\square$

Proposition 27.109. *Let S be a scheme and let $X \to S$ be a proper flat morphism of finite presentation with geometrically integral fibers. Suppose that there is given a point $e \in X(S)$ and a morphism of S-schemes $m\colon X \times_S X \to X$ such that $m(x,e) = x = m(e,x)$ for all $x \in X(T)$, T any S-scheme. Then $X \to S$ is an abelian scheme with group law m and unit e.*

Lemma 27.110. *Let G be a set endowed with an associative composition law $G \times G \to G$, $(g, h) \mapsto gh$. For $g \in G$ we denote by*

$$l_g \colon G \to G, \quad h \mapsto hg \quad \text{and} \quad r_g \colon G \to G, \quad h \mapsto gh$$

the left and right translation. Suppose that r_g is surjective for all $g \in G$ and that there exists an $f \in G$ such that l_f is surjective.

Then the composition law defines a group structure on G.

Proof. As r_f is surjective, we find $e \in G$ with $ef = f$. Let $g \in G$. As l_f is surjective, we find $h \in G$ with $fh = g$. Then $eg = e(fh) = (ef)h = fh = g$, hence e is a left unit.

For every $g \in G$ we also find $h \in G$ such that $hg = e$ since r_g is surjective. Hence every g has a left inverse. To show that h is a also a right inverse choose $h' \in G$ with $h'h = e$ using the surjectivity of r_h. Then

$$gh = e(gh) = (h'h)(gh) = h'((hg)h) = h'(eh) = h'h = e.$$

It remains to show that e is also a right unit. But we find

$$ge = g(hg) = (gh)g = eg = g. \qquad \square$$

Proof. [of Proposition 27.109] If T is an S-scheme and $x, y \in X(T)$ we write $x \cdot y$ instead of $m(x, y)$. Consider the morphism

$$\sigma \colon X \times_S X \longrightarrow X \times_S X, \qquad \sigma(x, y) = (x \cdot y, y).$$

Consider the following two claims.

(1) σ is an isomorphism of S-schemes. This means that for all S-schemes T and for all $y \in X(T)$ the right translation $x \mapsto x \cdot y$ is bijective. In particular, there exists for every $y \in Y(T)$ a unique T-valued point y^{-1} such that $y^{-1} \cdot y = e$. As the right translation is functorial in T, so is $y \mapsto y^{-1}$ and we obtain a morphism $i \colon X \to X$ of S-schemes given on T-valued points by $y \mapsto y^{-1}$ with $i(e) = e$.

(2) The morphism of S-schemes

$$(*) \qquad X \times_S X \times_S X \longrightarrow X, \qquad (x, y, z) \mapsto i(x \cdot (y \cdot z)) \cdot ((x \cdot y) \cdot z)$$

is constant.

If we have shown both claims, then setting $x = y = z = e$ in (2) it follows that the morphism (*) necessarily factors through the unit section and hence that the composition law is associative. Now we apply Lemma 27.110 to $X(T)$ for every S-scheme T and see that (X, m, i, e) is an S-group scheme and hence an abelian scheme.

Therefore it suffices to show the claims (1) and (2). For claim (1) (resp. (2)) it suffices to do this on fibers by Corollary 18.77 (resp. by Proposition 27.96). Hence we may assume that $S = \operatorname{Spec} k$ is a field. By fpqc descent, see Proposition 14.53 (resp. Lemma 27.95), we may in addition assume that k is algebraically closed. Then it suffices to show that m defines a group structure on $X(k)$.

We have $\sigma^{-1}(e, e) = \{(e, e)\}$, therefore Lemma 27.108 shows that σ is surjective.

Consider the closed subscheme $\tilde{Z} \subseteq X \times X$ given by $\tilde{Z} = \{(x, y) \, ; \, x \cdot y = e\} = \sigma^{-1}(\{e\} \times X)$. The surjectivity of σ implies that the second projection $\tilde{p}_2 \colon \tilde{Z} \to X$ is surjective. Let $Z \subseteq \tilde{Z}$ be an irreducible component, considered as a closed integral subscheme, with $\tilde{p}_2(Z) = X$. Then Z is proper over k and $\dim Z \geq \dim X$. As $\tilde{p}_2^{-1}(e) = \{(e, e)\}$, we have $(e, e) \in Z$. Let $p_1 \colon Z \to X$ be the first projection. Then $p_1^{-1}(e) = (e, e)$ and Lemma 27.108 shows that p_1 is surjective.

Define $f\colon Z \times X \times X \to X$ by $f((x,y),z,w) = x \cdot ((y \cdot z) \cdot w)$. Then we have $f(Z \times \{e\} \times \{e\}) = \{e\}$ and hence we find by Corollary 27.98 that

$$(*) \qquad x \cdot ((y \cdot z) \cdot w) = z \cdot w \qquad \text{for } (x,y) \in Z \text{ and } w,z \in X.$$

Taking $w = e$ in (*) we obtain in particular

$$(**) \qquad x \cdot (y \cdot z) = z \qquad \text{for } (x,y) \in Z \text{ and } z \in X.$$

Now fix $y \in X(k)$. As we have shown that both projections $Z \to X$ are surjective, we find $x, x' \in X(k)$ with $(x,y) \in Z$ and $(y,x') \in Z$. Then (**) gives $x = x \cdot e = x \cdot (y \cdot x') = x'$. This shows that y has a unique left and right inverse in $X(k)$, which we call y^{-1}. Finally, we obtain for all $x,y,z \in X(k)$

$$y \cdot (z \cdot w) \overset{(*)}{=} y \cdot (x \cdot ((y \cdot z) \cdot w)) = y \cdot (x' \cdot ((y \cdot z) \cdot w)) \overset{(**)}{=} (y \cdot z) \cdot w,$$

which shows that the composition law is associative. $\qquad\square$

The Picard functor

As mentioned in the introduction to this chapter, one of our goals is to introduce the *dual abelian scheme* of an abelian scheme. The underlying functor will be defined in terms of line bundles on the abelian scheme, and as a preparation, we will study the *relative Picard functor* for general schemes in the next few sections. Its representability is a difficult result, useful also in many other places in algebraic geometry, which we cannot prove here.

We will also briefly digress to define the Jacobian of a family of curves. Finally we will generalize many results of Section (26.17) about elliptic curves to families of elliptic curves parametrized by a scheme. See below for further comments and references.

(27.21) The Picard functor.

In this section we collect some results on the Picard functor. Let $f\colon X \to S$ be a morphism of schemes and consider the functor

$$(27.21.1) \qquad P_{X/S}\colon (\mathrm{Sch}/S)^{\mathrm{opp}} \to (\mathrm{AbGrp}), \qquad T \mapsto \mathrm{Pic}(X_T).$$

Definition 27.111. *The* (relative) Picard functor $\mathrm{Pic}_{X/S}\colon (\mathrm{Sch}/S)^{\mathrm{opp}} \to (\mathrm{AbGrp})$ *is the fppf-sheafification of* $P_{X/S}$ *for the fppf-topology (Definition 27.30).*

Remark 27.112. There is a map of functors of abelian groups $\iota\colon P_{X/S} \to \mathrm{Pic}_{X/S}$, where $\mathrm{Pic}_{X/S}$ is an abelian fppf-sheaf, and the pair $(\mathrm{Pic}_{X/S}, \iota)$ is characterized up to unique isomorphism by the following two properties.
(1) For every element $L \in \mathrm{Pic}_{X/S}(T)$ there exists an fppf-covering $(g_i\colon T_i \to T)_i$ such that the image of L in $\mathrm{Pic}_{X/S}(T_i)$ lies in the image of $P_{X/S}(T_i) \to \mathrm{Pic}_{X/S}(T_i)$.
(2) A line bundle $\mathscr{L} \in \mathrm{Pic}(X_T)$ is in the kernel of $\mathrm{Pic}(X_T) \to \mathrm{Pic}_{X/S}(T)$ if and only if there exists an fppf-covering $(g_i\colon T_i \to T)_i$ such that the pullback of $\mathscr{L}$ to X_{T_i} is trivial.

Indeed, if $(\text{Pic}'_{X/S}, \iota')$ is a pair satisfying (1) and (2), then by the universal property of the fppf-sheafification there exists a unique homomorphism of abelian fppf-sheaves $\alpha\colon \text{Pic}_{X/S} \to \text{Pic}'_{X/S}$ because $\text{Pic}'_{X/S}$ is an abelian sheaf. Then (1) implies that α is surjective and (2) implies that α is injective.

Now consider the following two conditions on f.

(a) One has $\mathscr{O}_S = f_*\mathscr{O}_X$ compatibly with base change, i.e., for all S-schemes T the homomorphism $\mathscr{O}_T \to (f_T)_*\mathscr{O}_{X_T}$ is an isomorphism. This is for instance the case if f is flat, proper, of finite presentation and has geometrically reduced and geometrically connected fibers (Corollary 24.63).

(b) The morphism f has a section $\sigma\colon S \to X$.

Both conditions are satisfied if X is an abelian scheme over S.

If (b) is satisfied, then the base change $\sigma_T\colon T \to X_T$ is a section of $f_T\colon X_T \to T$ for every S-scheme T. In particular, the map $f_T^*\colon \text{Pic}(T) \to \text{Pic}(X_T)$ is injective because it has σ_T^* as a left inverse.

Lemma 27.113. *Suppose f satisfies* (a) *and* (b) *above. For every S-scheme T we consider* $\text{Pic}(T)$ *as a subgroup of* $\text{Pic}(X_T)$. *Then we have*

$$(27.21.2) \qquad\qquad \text{Pic}_{X/S}(T) = \text{Pic}(X_T)/\text{Pic}(T).$$

Proof. We will first show that the functor

$$T \mapsto P_{X/S}^{\#}(T) := \text{Pic}(X_T)/\text{Pic}(T) \cong K(T) := \text{Ker}(\sigma_T^*\colon \text{Pic}(X_T) \to \text{Pic}(T))$$

is an fppf-sheaf. We will show that K is an fppf-sheaf. Let $(T_i \to T)_i$ be an fppf-covering. As $K(\coprod_i T_i) = \prod_i K(T_i)$, it suffices to check the sheaf condition for an fppf-covering consisting of a single morphism $h\colon T' \to T$.

Set $T'' := T' \times_T T'$, $T''' := T' \times_T T' \times_T T'$, let $p_i\colon T'' \to T'$ and $p_{ij}\colon T''' \to T''$ be the projections, and let $p_{i,X}\colon X_{T''} \to X_{T'}$ and $p_{ij,X}\colon X_{T'''} \to X_{T''}$ be their base changes. Let $\mathscr{L}'$ be an element of $K(T')$ such that there exists an isomorphism

$$\varphi\colon p_{1,X}^*\mathscr{L}' \xrightarrow{\sim} p_{2,X}^*\mathscr{L}'.$$

Since $\mathscr{L}' \in K(T')$, there exists an isomorphism $\alpha\colon \mathscr{O}_{T'} \xrightarrow{\sim} \sigma_{T'}^*\mathscr{L}'$. Set

$$\gamma := p_2^*(\alpha^{-1}) \circ \sigma_{T''}^*(\varphi) \circ p_1^*(\alpha)\colon \mathscr{O}_{T''} \to \mathscr{O}_{T''}.$$

Scaling φ by γ^{-1} we may assume that $\gamma = 1$. Now $p_{X,23}^*\varphi \circ p_{X,12}^*\varphi$ differs from $p_{X,13}^*\varphi$ by some invertible section s of $\mathscr{O}_{X_{T'''}}$. By our scaling of φ we know that its pullback by $\sigma_{T'''}$ is 1. Hence $s = 1$ because $f_*\mathscr{O}_{X_{T'''}} = \mathscr{O}_{T'''}$ by assumption on f. Hence by faithfully flat descent there exists a line bundle $\mathscr{L}$ on X_T and an isomorphism $\mathscr{O}_T \xrightarrow{\sim} \sigma_T^*\mathscr{L}$ whose pullback to $X_{T'}$ is the pair $(\mathscr{L}', \alpha)$. This proves that K and hence $T \mapsto \text{Pic}(X_T)/\text{Pic}(T)$ is an fppf-sheaf.

It remains to show that the map of functors $P_{X/S} \to P_{X/S}^{\#}$ satisfies the properties (1) and (2) of Remark 27.112. This is clear for (1). It remains to show that for every S-scheme a line bundle $\mathscr{L}$ in $\text{Pic}(X_T)$ is fppf-locally on T trivial if and only if it is of the form $f_T^*\mathscr{M}$ for a line bundle $\mathscr{M}$ on T. Replacing S by T and X by X_T, we may assume that $S = T$.

The condition is clearly sufficient (take as fppf-covering of S a Zariski covering that trivializes $\mathcal{M}$). Conversely, let $(S_i \to S)_i$ be an fppf covering such that the pullback $\mathcal{L}_{S_i}$ of $\mathcal{L}$ to X_{S_i} is trivial for all i. Choose an isomorphism $\alpha_i \colon \mathcal{O}_{X_{S_i}} \xrightarrow{\sim} \mathcal{L}_{S_i}$. The pullbacks of α_i and α_j to $X_{S_i \times_S S_j}$ differ by an element

$$g_{ij} \in \Gamma(X_{S_i \times_S S_j}, \mathcal{O}^\times_{X_{S_i \times_S S_j}}) = \Gamma(S_i \times_S S_j, \mathcal{O}_{S_i \times_S S_j}),$$

where for the equality we used the Condition (a) on $X \to S$. The g_{ij} satisfy the cocycle condition and hence define a descent datum for an line bundle with respect to the fppf-covering $(S_i \to S)_i$. All such descent data are effective (by Theorem 14.68 combined with Proposition 14.48) and hence we obtain a line bundle $\mathcal{M}$ on S such that $f^*\mathcal{M}$ and $\mathcal{L}$ are given by the same descent datum with respect to the fppf-covering $(X_{S_i} \to X)_i$. Hence $\mathcal{L} \cong f^*\mathcal{M}$ again by Theorem 14.68. $\qquad\square$

Example 27.114. Let k be a field and let X be a k-scheme. If $X(k) \neq \emptyset$, then $\mathrm{Pic}_{X/k}(k) = \mathrm{Pic}(X)$.

More generally, this holds whenever k is a ring with $\mathrm{Pic}(k) = 0$, e.g., if k is a local ring or a factorial ring.

The formation of $\mathrm{Pic}_{X/S}$ is compatible with base change:

Remark 27.115. Let $X \to S$ be a morphism of schemes and let $S' \to S$ be a morphism of schemes. Then $P_{X/S} \times_S S' = P_{X \times_S S'/S'}$ and hence

$$\mathrm{Pic}_{X/S} \times_S S' = \mathrm{Pic}_{X \times_S S'/S'}.$$

To study formal smoothness properties of the Picard functor we will use the following lemma.

Lemma 27.116. *Let $f \colon X \to S$ be flat, proper, of finite presentation with geometrically reduced and geometrically connected fibers such that f has a section. Let T be an affine S-scheme and let $i \colon T_0 \to T$ be a nil-immersion defined by a quasi-coherent ideal $\mathscr{I}$ with $\mathscr{I}^2 = 0$, allowing to view $\mathscr{I}$ as quasi-coherent module on T_0. Then there exists an exact sequence*

$$(27.21.3) \qquad \mathrm{Pic}_{X/S}(T) \longrightarrow \mathrm{Pic}_{X/S}(T_0) \longrightarrow H^2(X_{T_0}, \mathscr{I}\mathcal{O}_{X_{T_0}}).$$

Here we set $X_T := X \times_S T$ and define X_{T_0} similarly.

Proof. By Lemma 27.113 we have $\mathrm{Pic}_{X/S}(T) = \mathrm{Pic}(X_T)/\mathrm{Pic}(T)$ and $\mathrm{Pic}_{X/S}(T_0) = \mathrm{Pic}(X_{T_0})/\mathrm{Pic}(T_0)$. By Lemma 26.15 we have an exact sequence

$$(*) \qquad \mathrm{Pic}(X_T) \longrightarrow \mathrm{Pic}(X_{T_0}) \longrightarrow H^2(X_{T_0}, \mathscr{I}\mathcal{O}_{X_{T_0}}).$$

The composition $\mathrm{Pic}(T_0) \to \mathrm{Pic}(X_{T_0}) \to H^2(X_{T_0}, \mathscr{I}\mathcal{O}_{X_{T_0}})$ factors through $H^2(T_0, \mathscr{I})$ which is zero because T_0 is affine and $\mathscr{I}$ is quasi-coherent. Hence the right map of $(*)$ induces a map $\mathrm{Pic}_{X/S}(T_0) \longrightarrow H^2(X_{T_0}, \mathscr{I}\mathcal{O}_{X_{T_0}})$. Now the surjectivity of $\mathrm{Pic}(X_T) \to \mathrm{Pic}_{X/S}(T)$ and $\mathrm{Pic}(X_{T_0}) \to \mathrm{Pic}_{X/S}(T_0)$ imply the exactness of $(27.21.3)$. $\qquad\square$

(27.22) Representability of the Picard functor.

We will use in the sequel the following representability theorem of $\mathrm{Pic}_{X/S}$.

Theorem 27.117. *Let S be a scheme, let $f\colon X \to S$ be a flat proper morphism of finite presentation with geometrically reduced and geometrically connected fibers. Then $\mathrm{Pic}_{X/S}$ is an algebraic space and $\mathrm{Pic}_{X/S} \to S$ is locally of finite presentation.*

The theorem can be proved by checking Artin's criteria for representability by an algebraic space, see [Art3] ^{O}, [Art4] ^{O}, [Sta] 07T0. We will not give the proof here but refer to [Sta] 0D2C and 0DNI.

The Seesaw theorem 24.66 just says (for line bundles) that the diagonal of $\mathrm{Pic}_{X/S}$ is indeed representable by an immersion. More precisely, we have:

Proposition 27.118. *Under the hypotheses on f in Theorem 27.117, the diagonal $\mathrm{Pic}_{X/S} \to \mathrm{Pic}_{X/S} \times_S \mathrm{Pic}_{X/S}$ is representable by an immersion of finite presentation. If f has in addition geometrically integral fibers, then the diagonal is represented by a closed immersion of finite presentation and hence $\mathrm{Pic}_{X/S}$ is separated over S.*

Proof. Suppose that f has a section. Let T be an S-scheme and let $[\mathscr{L}_1], [\mathscr{L}_2] \in \mathrm{Pic}_{X/S}(T)$ for $\mathscr{L}_1, \mathscr{L}_2 \in \mathrm{Pic}(X_T)$. We have to show that the locus on T, where $\mathscr{L}_1$ and $\mathscr{L}_2$ are isomorphic up to the pullback of a line bundle from T is a subscheme (resp. a closed subscheme if f has geometrically integral fibers). Set $\mathscr{M} := \mathscr{L}_1 \otimes \mathscr{L}_2^{\otimes -1}$. Then we are looking for a subscheme (resp. closed subscheme) Z of T such that a morphism $T' \to T$ factors through Z if and only if $\mathscr{M}_{X_{T'}}$ is the pullback of a line bundle on T'. This is given by Theorem 24.66.

If f does not necessarily have a section, then we choose a faithfully flat quasi-compact base change $S' \to S$ such that $X_{S'}$ has a section (e.g. $S' = X$ with the diagonal as a section) and use that the property of being a (closed) immersion of finite presentation is stable under fpqc-descent (see Proposition 14.53). We omit the details. $\qquad\square$

Many more representability results about $\mathrm{Pic}_{X/S}$ are known. We refer to [FGAex] $\times$ §9.4 for a thorough discussion. Here we mention only two further results that give criteria when $\mathrm{Pic}_{X/S}$ is represented by a scheme.

The first result shows that the Picard functor is representable for proper schemes over a field.

Theorem 27.119. *Let k be a field and let X be a proper k-scheme. Then $\mathrm{Pic}_{X/k}$ is representable by a separated scheme locally of finite type over k.*

We will consider in the proof only the case that X is geometrically integral over k (the only case that we will need in the sequel). For the general case we refer to [Mur] ^{O} (II.15) (note that once we know that $\mathrm{Pic}_{X/k}$ is a group scheme locally of finite type over k, then it is automatically separated by Corollary 27.9).

Proof. [if X is geometrically integral over k] By Theorem 27.117, $\mathrm{Pic}_{X/k}$ is an algebraic space locally of finite type over k. As X is geometrically integral over k, it is separated by Proposition 27.118. Therefore $\mathrm{Pic}_{X/k}$ is representable by Theorem 27.47. $\qquad\square$

The second representability result we state here, again without proof, is the following theorem by Grothendieck [FGA] $_X$. It was one of the first major theorems that Grothendieck proved after introducing the machinery of schemes and the functorial point of view. By relating line bundles to divisors, the representability of the Picard functor is reduced to the representability of the Hilbert functor, see Section (14.32), and thus ultimately to the representability of the Grassmannian functor.

Theorem 27.120. *Let $f \colon X \to S$ be flat and of finite presentation with geometrically integral fibers. Suppose that there exists an open covering $(U_i)_i$ and for all i a closed immersion $X_{|U_i} \to \mathbb{P}^n_{U_i}$ (with n depending on i) (e.g., if $X \to S$ is projective). Then $\mathrm{Pic}_{X/S}$ is represented by a separated S-scheme locally of finite presentation.*

We refer to [FGAex] $_X$ 9.4.8 for a proof, where this result is shown if S is locally noetherian. But by the following remark this implies the result in the general case.

Remark 27.121. Let $\mathbf{P}$ be a property of scheme morphisms and let $\mathbf{Q}$ be a property of morphisms from an fppf-sheaf to a scheme. Let $\mathbf{P}$ be Zariski local on the target and compatible with inductive limits of rings (Section (10.13)) and let $\mathbf{Q}$ be Zariski local on the target and stable under base change.

Suppose that one wants to show that for every morphism of schemes $X \to S$ of finite presentation that has $\mathbf{P}$ the morphism $\mathrm{Pic}_{X/S} \to S$ has property $\mathbf{Q}$. Then it suffices to show this in the case that $S = \mathrm{Spec}\, R_0$ where R_0 is a finitely generated $\mathbb{Z}$-algebra.

Indeed, since $\mathbf{P}$ and $\mathbf{Q}$ are both local on the target, we may assume that $S = \mathrm{Spec}\, R$ is affine. We write R as the filtered union of its finitely generated $\mathbb{Z}$-subalgebras R_λ. As $X \to S$ is of finite presentation and $\mathbf{P}$ is compatible with inductive limits of rings, we find an index λ and an R_λ-scheme X_λ of finite type such that $X_\lambda \to \mathrm{Spec}\, R_\lambda$ has property $\mathbf{P}$ and such that $X_\lambda \otimes_{R_\lambda} R \cong X$. Then $\mathrm{Pic}_{X_\lambda/\mathrm{Spec}\, R_\lambda} \to \mathrm{Spec}\, R_\lambda$ has property $\mathbf{Q}$ by hypothesis. Hence $\mathrm{Pic}_{X/S} \to S$ has property $\mathbf{Q}$ because $\mathbf{Q}$ is stable under base change and because the formation of $\mathrm{Pic}_{X/S}$ commutes with base change.

For a list of properties of scheme morphisms that are local on the target and compatible with inductive limits of rings we refer to Appendix C. An example for property $\mathbf{Q}$ as above is being representable by a scheme (Theorem 8.9 using that fppf-sheaves are in particular sheaves for the Zariski topology). If one already knows that under hypothesis $\mathbf{P}$ on $X \to S$ the Picard sheaf $\mathrm{Pic}_{X/S}$ is an algebraic space, then further examples for $\mathbf{Q}$ are "separated", "proper", or "smooth".

(27.23) The Lie algebra of the Picard functor.

Proposition 27.122. *Let $f \colon X \to S$ be a proper flat morphism of finite presentation with geometrically reduced and geometrically connected fibers. Then $\mathrm{Lie}(\mathrm{Pic}_{X/S}) \cong R^1 f_* \mathcal{O}_X$.*

Here we use the definition of the Lie algebra given in Remark 27.19. In the proof we will use that $\mathrm{Lie}(\mathrm{Pic}_{X/S})$ is a quasi-coherent $\mathcal{O}_S$-module. This holds by Remark 27.17 if $\mathrm{Pic}_{X/S}$ is a scheme locally of finite presentation and we will use the proposition only in this case. If $\mathrm{Pic}_{X/S}$ is only an algebraic space locally of finite presentation, then one can show that $\mathrm{Lie}(\mathrm{Pic}_{X/S})$ is still quasi-coherent (by the same argument, using that $\Omega^1_{\mathrm{Pic}_{X/S}/S}$ is of finite presentation by [Sta] 05ZF).

Proof. For a scheme U we set $U[\varepsilon] := \operatorname{Spec} \mathscr{O}_U[T]/(T^2)$. We denote by ε the image of the indeterminate T (in the sections over open subsets). Then the structure morphism $U[\varepsilon] \to U$ has a section that corresponds to $T \mapsto 0$. It is a thickening of order 1. In particular, U and $U[\varepsilon]$ have the same underlying topological space. We have $X[\varepsilon] = X \times_S S[\varepsilon]$ by (11.2.5).

(I). We first construct a homomorphism $\psi\colon R^1 f_* \mathscr{O}_X \to \operatorname{Lie}(\operatorname{Pic}_{X/S})$ of $\mathscr{O}_S$-modules. Consider the structure sheaf $\mathscr{O}_{X[\varepsilon]}$ as a sheaf of rings on X. Then for $U \subseteq X$ open, $\mathscr{O}_{X[\varepsilon]}(U) = \Gamma(U, \mathscr{O}_X)[T]/(T^2)$. We have a short exact sequence of abelian sheaves on X

$$(*) \qquad 0 \longrightarrow \mathscr{O}_X \longrightarrow \mathscr{O}_{X[\varepsilon]}^{\times} \longrightarrow \mathscr{O}_X^{\times} \longrightarrow 1,$$

where the first map is given on local sections by $a \mapsto 1 + a\varepsilon$. The sequence is split by the map $\mathscr{O}_X^{\times} \to \mathscr{O}_{X[\varepsilon]}^{\times}$ given by $a \mapsto a + 0\varepsilon$ on local sections. By Section (11.7) we have $\operatorname{Pic}(X) = H^1(X, \mathscr{O}_X^{\times})$ and similarly for $X[\varepsilon]$. Let $V \subseteq S$ be open. Taking cohomology over $f^{-1}(V) = X \times_S V$ we obtain from $(*)$ a split exact sequence

$$(**) \qquad 0 \longrightarrow H^1(f^{-1}(V), \mathscr{O}_X) \longrightarrow \operatorname{Pic}(X \times_S V[\varepsilon]) \longrightarrow \operatorname{Pic}(X \times_S V) \longrightarrow 1.$$

As $\operatorname{Pic}(X_T) \to \operatorname{Pic}_{X/S}(T)$ is functorial in S-schemes T and as we have

$$(***) \qquad \Gamma(V, \operatorname{Lie}(\operatorname{Pic}_{X/S})) = \operatorname{Ker}(\operatorname{Pic}_{X/S}(V[\varepsilon]) \to \operatorname{Pic}_{X/S}(V)),$$

we obtain a homomorphism of abelian groups

$$\psi_V\colon H^1(f^{-1}(V), \mathscr{O}_X) \to \Gamma(V, \operatorname{Lie}(\operatorname{Pic}_{X/S})),$$

compatible with restriction to open subsets. In view of Proposition 21.27 this construction yields a homomorphism of abelian sheaves

$$\psi\colon R^1 f_* \mathscr{O}_X \longrightarrow \operatorname{Lie}(\operatorname{Pic}_{X/S}).$$

Let us show that ψ is $\mathscr{O}_S$-linear. For $a \in \Gamma(V, \mathscr{O}_S)$ let $m_a\colon V[\varepsilon] \to V[\varepsilon]$ be the endomorphism of V-schemes given by $\varepsilon \mapsto a\varepsilon$. It induces on $\Gamma(V, \operatorname{Lie}(\operatorname{Pic}_{X/S}))$ a scalar multiplication which equips it with the structure of a $\Gamma(V, \mathscr{O}_S)$-module. On $\operatorname{Pic}_{X/S}(V[\varepsilon]) = H^1(X_V, \mathscr{O}_{X[\varepsilon]}^{\times})$ it is induced by the endomorphism of $\mathscr{O}_{X[\varepsilon]}^{\times}$ given by $\varepsilon \mapsto a\varepsilon$. Hence m_a induces the scalar multiplication by a on $H^1(f^{-1}(V), \mathscr{O}_X)$. This shows that ψ_V is $\Gamma(V, \mathscr{O}_S)$-linear and hence that ψ is $\mathscr{O}_S$-linear.

(II). It remains to show that ψ is an isomorphism. As the formation of $R^1 f_* \mathscr{O}_X$ and of $\operatorname{Lie}(\operatorname{Pic}_{X/S})$ commutes with flat base change (Proposition 22.90 and Proposition 27.22), we can do this after passing to an fppf-covering of S. Hence, we may assume that $X \to S$ has a section and hence that $\operatorname{Pic}_{X/S}(T) = \operatorname{Pic}(X_T)/\operatorname{Pic}(T)$ for every S-scheme T. We may assume that $S = \operatorname{Spec} R$ is affine. Because the source and target of ψ are quasi-coherent $\mathscr{O}_S$-modules, it suffices to show that the induced R-linear map on global sections

$$\psi_S\colon H^1(X, \mathscr{O}_X) \longrightarrow \Gamma(S, \operatorname{Lie}(\operatorname{Pic}_{X/S}))$$

is an isomorphism. By passing to stalks we may also assume that R is local. Then $R[\varepsilon]$ is also local and hence $\operatorname{Pic}(S) = \operatorname{Pic}(S[\varepsilon]) = 1$. Therefore $\operatorname{Pic}(X_S) \to \operatorname{Pic}_{X/S}(S)$ and $\operatorname{Pic}(X_{S[\varepsilon]}) \to \operatorname{Pic}_{X/S}(S[\varepsilon])$ are isomorphisms. Hence ψ_S is an isomorphism by $(**)$ and $(***)$. $\qquad\square$

(27.24) The identity component of the Picard functor.

We will be interested in the identity component of $\mathrm{Pic}_{X/S}$, so we want to extend this notion, defined for group schemes in Definition 27.12 for group schemes, to group functors. Let us explain what we mean by this. Let S be a scheme and let $G\colon (\mathrm{Sch}/S)^{\mathrm{opp}} \to (\mathrm{Grp})$ be a functor of groups. For an S-scheme T let $G^0(T)$ be the set of $g \in G(T)$ such that for every geometric point $\bar{t} \to T$ there exist a connected $\bar{t}$-scheme Z, points $z, z_0 \in Z(\bar{t})$, and an S-morphism $h\colon Z \to G$ such that $h(z) = g(\bar{t})$ and $h(z_0) = 1$ in $G(\bar{t})$.

This is a subfunctor of G. Clearly, the unit section is contained in $G^0(S)$ and G^0 is stable under the inversion ι of G by replacing h by $\iota \circ h$. If $g, g' \in G^0(T)$, let $\bar{t} \to T$ be a geometric point and let (Z, z, z_0, h) for g and (Z', z', z_0', h') for g' as above. Then $gg' \in G^0(T)$ because $Z \times_{\bar{t}} Z'$ is connected (Proposition 5.53) and one can take $(Z \times_{\bar{t}} Z', (z, z'), (z_0, z_0'), \tilde{h})$ for gg', where $\tilde{h}$ is the composition

$$Z \times_{\bar{t}} Z' \longrightarrow Z \times_S Z' \xrightarrow{h \times h'} G \times_S G \xrightarrow{m} G.$$

Definition 27.123. *The subgroup functor G^0 of G is called the* identity component *of G.*

In particular, for every morphism of schemes $f\colon X \to S$ we have the identity component $\mathrm{Pic}^0_{X/S}$ of the Picard functor.

Lemma 27.124. *Let k be a field and let G be a k-group scheme locally of finite type. Then the two notions of identity component defined in Definition 27.12 and in Definition 27.123 coincide.*

Proof. Let G^0 be the connected component of the unit section of G and let $\tilde{G}^0$ be the subfunctor defined in Definition 27.123. Let T be a k-scheme. Let $g \in G^0(T)$ and $\bar{t} \to T$ be a geometric point. As G^0 is geometrically connected, $G^0_{\bar{t}}$ is connected and we may take $Z = G^0_{\bar{t}}$, h the canonical map, $z := g(\bar{t})$ and z_0 the unit section of $G^0_{\bar{t}}$ to see that $g \in \tilde{G}^0(T)$. Conversely, let g be in $\tilde{G}^0(T)$ which we consider as a morphism $g\colon T \to G$. If g did not factor through G^0 we would find a geometric point $\bar{t} \to T$ such that $g(\bar{t}) \notin G^0$. But any morphism $h\colon Z \to G$ from a connected scheme Z such that the image of h contains the unit section has to factor through G^0. Therefore $g \in G^0(T)$. $\square$

Remark 27.125. Let S be a scheme and let $G\colon (\mathrm{Sch}/S)^{\mathrm{opp}} \to (\mathrm{Grp})$ be a functor of groups such that for all $s \in S$ the fiber $G_s\colon (\mathrm{Sch}/\kappa(s))^{\mathrm{opp}} \to (\mathrm{Grp})$ is representable by a group scheme over $\kappa(s)$ locally of finite type.
(1) Lemma 27.124 shows that for every S-scheme T and $g \in G(T)$ one has $g \in G^0(T)$ if and only if for every geometric point $\bar{t} \to T$ one has $g(\bar{t}) \in G^0_{\bar{t}}(\bar{t})$ with $G^0_{\bar{t}}$ the identity component of the group scheme $G_{\bar{t}}$ defined in Definition 27.12.
(2) By Proposition 27.12 (3) for every morphism $S' \to S$ of schemes one then has an isomorphism

$$(G \times_S S')^0 \cong G^0 \times_S S'.$$

The hypothesis on G is in particular satisfied if $G = \mathrm{Pic}_{X/S}$, where $X \to S$ is proper (Theorem 27.119).

Lemma 27.126. *Let $X \to S$ be a proper, smooth morphism of schemes with geometrically connected fibers such that the monomorphism $\mathrm{Pic}^0_{X/S} \to \mathrm{Pic}_{X/S}$ is representable by an open and closed immersion. Then the morphism $\mathrm{Pic}^0_{X/S} \to S$ of algebraic spaces is proper.*

We will see below that the assumptions of this lemma are satisfied for smooth proper relative curves with geometrically connected fibers (Proposition 27.136) and for abelian schemes (Remark 27.210).

Proof. In the proof we will use that the formation of $\mathrm{Pic}^0_{X/S}$ is compatible with base change which holds by Remark 27.125. As f is smooth and has geometrically connected fibers, it has geometrically integral fibers. Hence $\mathrm{Pic}_{X/S}$ is an algebraic space locally of finite type over S (Theorem 27.117) and $\mathrm{Pic}_{X/S} \to S$ is separated (Proposition 27.118). Hence $\mathrm{Pic}^0_{X/S} \to S$ is a group algebraic space which is separated and locally of finite presentation over S. The fibers $\mathrm{Pic}^0_{X_s/\kappa(s)}$ of $\mathrm{Pic}^0_{X/S} \to S$ are group schemes locally of finite type over $\kappa(s)$ by Theorem 27.119. By definition, they are connected, hence $\mathrm{Pic}^0_{X_s/\kappa(s)}$ is quasi-compact by Proposition 27.12 (1).

To see that $\mathrm{Pic}^0_{X/S} \to S$ is proper, we may assume that S is affine and that S is noetherian by noetherian approximation (Remark 27.121). As properness can be checked after passing to an fppf-covering, we may assume $X \to S$ has a section (e.g., by passing to the fppf-covering $X \to S$). As properness is compatible with filtered inductive limits of rings, we may even assume that S is local.

Let us show that $\mathrm{Pic}^0_{X/S} \to S$ satisfies the valuative criterion of Definition 27.43 (3). By Remark 27.44 (3) it suffices to show the criterion for discrete valuation rings. Hence let R be a discrete valuation ring with fraction field K and let $\operatorname{Spec} R \to S$ be a morphism. Let $L \in \mathrm{Pic}^0_{X/S}(K)$ be represented by a line bundle $\mathscr{L}$ on X_K. As X_R is smooth over the regular ring R, the scheme X_R is also regular. Note that $X_R \setminus X_K$ is the zero scheme of the uniformizer of R, pulled back to X_R. As $X_R \to \operatorname{Spec} R$ is flat, this is a Cartier divisor, and since X is proper, it is non-empty. So $X_R \setminus X_K$ is of codimension 1 in X_R. Hence $\mathrm{Pic}(X_R) \to \mathrm{Pic}(X_K)$ is surjective by Corollary 11.43 and we can extend $\mathscr{L}$ to a line bundle $\mathscr{L}'$ on X_R. As $\mathrm{Pic}^0_{X/S}$ is open and closed in $\mathrm{Pic}_{X/S}$ the induced morphism $\operatorname{Spec} R \to \mathrm{Pic}_{X/S}$ factors through $\mathrm{Pic}^0_{X/S}$.

It remains to show that the structure morphism $\pi \colon \mathrm{Pic}^0_{X/S} \to S$ is quasi-compact. As S is affine, it suffices to show that the underlying topological space of $\mathrm{Pic}^0_{X/S}$ is quasi-compact (Remark 27.44 (2)). Let $(U_i)_i$ be an open covering. Let $s \in S$ be the closed point. As the fiber $\mathrm{Pic}^0_{X_s/\kappa(s)}$ is quasi-compact, there exists a finite subset $J \subseteq I$ such that $V := \bigcup_{i \in J} U_i$ contains $\mathrm{Pic}^0_{X_s/\kappa(s)}$. It suffices to show that V equals the underlying topological space of $\mathrm{Pic}^0_{X/S}$. Let Z be its complement and assume that $Z \neq \emptyset$. As $\mathrm{Pic}^0_{X/S} \to S$ satisfies the valuative criterion above, $\pi(Z)$ is stable under specialization (Proposition 15.7) and in particular $s \in \pi(Z)$. This is a contradiction since Z does not meet the special fiber. $\square$

Remark 27.127. The proof shows that if S is locally noetherian and $X \to S$ is a proper, smooth morphism of schemes with geometrically connected fibers that has a section, then $\mathrm{Pic}_{X/S}$ satisfies the valuative criterion of Definition 27.43 (3) for discrete valuation rings R.

In fact, the hypothesis "locally noetherian" on S is superfluous and one can show that $\mathrm{Pic}_{X/S}$ satisfies the valuative criterion for properness for arbitrary valuation rings if one replaces the reference to Corollary 11.43 in the proof of Lemma 27.126 by the argument in the proof of [Sta] ODNG.

As a consequence of the lemma, $\mathrm{Pic}_{X/S}$ is formally proper in the following sense.

Definition 27.128. *Let* $f\colon Z \to S$ *be a morphism locally of finite type between algebraic spaces. Then* f *is called* formally proper *if for every valuation ring* R *with fraction field* K *and for every commutative diagram*

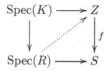

there exists a unique dotted arrow making the diagram commutative.

Note that this is not a standard notion.

Remark 27.129.
(1) Every proper morphism is formally proper.
(2) The composition of formally proper morphisms is formally proper.
(3) In particular, if $X \to S$ is formally proper and if Z is a closed subscheme of X, then $Z \to S$ is formally proper.

(27.25) The Picard functor of curves.

In this section S denotes a scheme. Let us first explain what we mean by a curve over S here and in the sequel.

Recall from Chapter 26 that a *(relative) curve over S* is a separated flat morphism $f\colon C \to S$ of finite presentation such that all fibers are equi-dimensional of dimension 1 (Definition 26.162).

If in addition $C \to S$ is proper and has geometrically reduced and geometrically connected fibers, then $\mathrm{Pic}_{C/S}$ is representable by an algebraic space (Theorem 27.117). In fact, Theorem 26.16 and Theorem 27.120 show that $\mathrm{Pic}_{C/S}$ is a scheme; see Theorem 27.137 below for a different approach to proving this. It is even smooth.

Proposition 27.130. *Let* $f\colon C \to S$ *be a relative proper curve with geometrically reduced and geometrically connected fibers. Then* $\mathrm{Pic}_{C/S}$ *is smooth.*

Proof. We can check the smoothness locally for the fppf topology. Working Zariski locally we may assume that S is affine. By noetherian approximation, we may assume that S is noetherian (Remark 27.121). By passing to an fppf-covering (e.g., to $C \to S$) we may assume that $C \to S$ has a section. By Remark 27.42 it suffices to show that $\mathrm{Pic}_{C/S}(T) \to \mathrm{Pic}_{C/S}(T_0)$ is surjective for every thickening $T_0 \to T$ of affine S-schemes $T_0 = \mathrm{Spec}\, R_0$ and $T = \mathrm{Spec}\, R$. We may assume that $R_0 = R/I$ with $I^2 = 0$ and that R is a local Artinian ring.

By Lemma 27.116 it suffices to show that $H^2(C_{T_0}, \mathscr{F}) = 0$, where $\mathscr{F}$ is the coherent module $\mathscr{I} \mathscr{O}_{C_{T_0}}$. As T_0 is affine, we have $H^2(X_{T_0}, \mathscr{F})^{\sim} = R^2 f_{T_0, *} \mathscr{F}$ which is zero because the fibers of $C_{T_0} \to T_0$ have dimension 1 (Corollary 24.44). $\qquad\square$

Remark 27.131. Let $f\colon C \to S$ be a relative proper curve and let $\mathscr{L}$ be a line bundle on C. Then the map $s \mapsto \deg(\mathscr{L}_{|C_s})$ is locally constant on S, because $\deg(\mathscr{L}_{|C_s}) = \chi(\mathscr{O}_{C_s}) - \chi(\mathscr{L}_s^{\otimes -1})$ (Section (15.9)) and the Euler characteristic is locally constant (Theorem 23.139).

Let $f\colon C \to S$ be a proper relative curve, let T be an S-scheme, and let $L \in \operatorname{Pic}_{C/S}(T)$. We are going to define a locally constant map

(27.25.1) $\deg_L \colon T \to \mathbb{Z}$.

For $t \in T$ let L_t be the image of L in $\operatorname{Pic}_{C/S}(\kappa(t))$. Fppf-locally, L_t comes from a line bundle on C_t, i.e., we can choose a field extension k of $\kappa(t)$ and $\mathscr{L} \in \operatorname{Pic}(C \otimes k)$ whose image in $\operatorname{Pic}_{C/S}(k)$ is the pullback of L_t. Set $\deg_L(t) := \deg_k(\mathscr{L})$. This is well defined, commutes with base change, and defines a locally constant function:

Lemma 27.132. *In the situation above we have the following assertions.*
(1) *The integer $\deg_L(t)$ does not depend on the choices made.*
(2) *If $g\colon T' \to T$ is a morphism of S-schemes, then*

$$\deg_{g^*L} = \deg_L \circ g \colon T' \to \mathbb{Z}.$$

(3) *The map $\deg_L$ is a locally constant.*

Proof. If k' and $\mathscr{L}' \in \operatorname{Pic}(C_{k'})$ is another pair as above, then we find by Remark 27.112 (2) an extension of K of both of k and k' such that the images of $\mathscr{L}$ and of $\mathscr{L}'$ in $\operatorname{Pic}(C_K)$ are isomorphic, say to $\mathscr{M}$. Therefore we have

$$\deg_k(\mathscr{L}) = \deg_K(\mathscr{M}) = \deg_{k'}(\mathscr{L}')$$

by Remark 15.29. The invariance of the degree under base change also shows the second assertion.

To show that $\deg_L$ is locally constant, i.e., continuous if we endow $\mathbb{Z}$ with the discrete topology, we choose an fppf-covering $g\colon T' \to T$ and $\mathscr{L} \in \operatorname{Pic}(C_{T'})$ whose image in $\operatorname{Pic}_{C/S}(T')$ is the inverse image of L. Then $\deg_{g^*L}$ is locally constant by Remark 27.131. As g is surjective and open (Theorem 14.35), it is a topological quotient map, therefore $\deg_L$ is locally constant by (2). $\square$

Definition 27.133. *Let $C \to S$ be a proper relative curve and let $d \in \mathbb{Z}$ be an integer. For any S-scheme T we define*

$$\operatorname{Pic}^{(d)}_{C/S}(T) = \{\, L \in \operatorname{Pic}_{C/S}(T) \,;\, \deg_L \ \text{constant with value } d\}.$$

By Lemma 27.132 (2) this defines a subfunctor of $\operatorname{Pic}_{C/S}$.

The inclusion $\operatorname{Pic}^{(d)}_{C/S} \to \operatorname{Pic}_{C/S}$ is representable by an open and closed immersion by Lemma 27.132 (3). In particular, $\operatorname{Pic}^{(d)}_{C/S}$ is an algebraic space if $\operatorname{Pic}_{C/S}$ is an algebraic space (Lemma 27.36), for instance, if $C \to S$ has geometrically reduced and geometrically connected fibers (Theorem 27.117).

Remark 27.134. Let $C \to S$ be a proper relative curve. The multiplication on $\operatorname{Pic}_{C/S}$ induces for all integers $d, d' \in \mathbb{Z}$ a multiplication morphism

$$\operatorname{Pic}^{(d)}_{C/S} \times_S \operatorname{Pic}^{(d')}_{C/S} \longrightarrow \operatorname{Pic}^{(d+d')}_{C/S},$$

and the inversion yields an isomorphism $\operatorname{Pic}^{(d)}_{C/S} \xrightarrow{\sim} \operatorname{Pic}^{(-d)}_{C/S}$.

In particular, $\operatorname{Pic}^{(0)}_{C/S}$ becomes an open and closed subgroup functor. Moreover, multiplication makes $\operatorname{Pic}^{(d)}_{C/S}$ into an $\operatorname{Pic}^{(0)}_{C/S}$-torsor for the fppf-topology.

(27.26) The Jacobian of a curve.

Definition and Remark 27.135. Let $f\colon C \to S$ be a smooth relative curve. Let $x \in C(S)$, i.e., $x\colon S \to C$ is a section of f. Then x is a regular immersion (Theorem 19.30) of codimension 1 (Proposition 19.42 (2)). It is a closed immersion because $C \to S$ is separated (Example 9.12). Therefore we may view x as an effective Cartier divisor on C (cf. Remark 26.163). Let $\mathscr{O}_C(x)$ be the corresponding line bundle, i.e., $\mathscr{O}_C(x)^{\otimes -1} \subseteq \mathscr{O}_C$ is the ideal defining the closed subscheme corresponding to x. On each fiber of $C \to S$ this line bundle is of degree 1. If $C \to S$ is proper, we denote its image in $\mathrm{Pic}^{(1)}_{C/S}(S)$ again by $\mathscr{O}_C(x)$.

Replacing S by an arbitrary S-scheme T and C by C_T we obtain a morphism of fppf-sheaves on (Sch/S)

$$(27.26.1) \qquad\qquad A_C\colon C \to \mathrm{Pic}^{(1)}_{C/S}$$

defined by $C_T(T) \ni x \mapsto \mathscr{O}_{C_T}(x)$ on T-valued points, called the *Abel morphism* or *Abel-Jacobi morphism*.

Now let $\mathscr{L}$ be a line bundle on C such that $\mathscr{L}_{|C_s}$ has degree 1 for all $s \in S$ (such a line bundle does not have to exist in general). Its image in $\mathrm{Pic}^{(1)}_{C/S}$ is again denoted by $\mathscr{L}$. Then we obtain the following variant of the Abel morphism

$$(27.26.2) \qquad\qquad \iota^{\mathscr{L}}_C\colon C \to \mathrm{Pic}^{(0)}_{C/S}, \qquad x \mapsto \mathscr{O}_C(x) \otimes \mathscr{L}^{\otimes -1}.$$

If $\mathscr{L} = \mathscr{O}_C(x_0)$ for some $x_0 \in C(S)$, then we also write $\iota^{x_0}_C$ instead of $\iota^{\mathscr{O}_C(x_0)}_C$.

Proposition 27.136. *Let $C \to S$ be a smooth proper relative curve with geometrically connected fibers. Then $\mathrm{Pic}^0_{C/S} = \mathrm{Pic}^{(0)}_{C/S}$.*

In particular, $\mathrm{Pic}^{(0)}_{C/S}$ has geometrically connected fibers over S.

Proof. By Remark 27.125 we may assume that $S = \operatorname{Spec} k$ for some algebraically closed field k. Then both are open subschemes of the scheme $\mathrm{Pic}_{C/k}$ which is locally of finite type. Hence it suffices to show that

$$\mathrm{Pic}^0_{C/k}(k) = \mathrm{Pic}^{(0)}_{C/k}(k)$$

We have $\mathrm{Pic}_{C/k}(k) = \mathrm{Pic}(C)$ by Example 27.114.

Let $\mathscr{L} \in \mathrm{Pic}^0_{C/k}(k)$ and let $\xi \in \mathrm{Pic}^0_{C/k}$ be the corresponding closed point. Then $\mathscr{L}$ is the fiber of the universal line bundle on $\mathrm{Pic}_{C/k} \times_k C$ at ξ. The fiber of the universal line bundle in $0 \in \mathrm{Pic}^0_{C/k}(k)$ is the trivial line bundle and hence has degree 0. As the degree is constant on connected schemes (Remark 27.131), $\mathscr{L}$ has degree 0.

Conversely, let $\mathscr{L} \in \mathrm{Pic}(C)$ with $\deg \mathscr{L} = 0$. Then there exists a divisor D on C of the form $D = \sum_{i=1}^{n}([x_i] - [x_0])$ for closed points $x_0, \ldots, x_n$ on C such that $\mathscr{O}_C(D) \cong \mathscr{L}$ (Example 26.54). Since C is connected, the morphism $\iota^{x_0}_C\colon C \to \mathrm{Pic}_{C/k}$ (27.26.2) factors through $\mathrm{Pic}^0_{C/k}$ which shows that $\mathscr{O}_C(x - x_0) \in \mathrm{Pic}^0_{C/k}(k)$ for all $x \in C(k)$. As $\mathrm{Pic}^0_{C/k}(k)$ is a subgroup of $\mathrm{Pic}(C)$, this shows that also $\mathscr{L} \in \mathrm{Pic}^0_{C/k}(k)$. $\qquad\square$

Theorem and Definition 27.137. *Let S be a scheme and let $C \to S$ be a smooth proper relative curve with geometrically connected fibers. Then $\mathrm{Pic}^0_{C/S}$ is an abelian scheme over S called the* Jacobian *of $C \to S$.*

In the proof we will use Theorem 27.211 below whose proof uses none of the results of this or the next section.

Proof. By Proposition 27.136, $\mathrm{Pic}^0_{C/S} \to \mathrm{Pic}_{C/S}$ is an open and closed immersion. Hence we can apply Lemma 27.126 and see that $\mathrm{Pic}^0_{C/S}$ is proper over S. Moreover, it is smooth over S by Proposition 27.130. By definition, $\mathrm{Pic}^0_{C/S} \to S$ has geometrically connected fibers. Hence we conclude by Theorem 27.211 below. $\qquad\square$

As $\mathrm{Pic}^{(d)}_{C/S}$ is a $\mathrm{Pic}^{(0)}_{C/S}$-torsor for the fppf-topology (Remark 27.134) we obtain the following corollary.

Corollary 27.138. *Let $C \to S$ be a smooth proper relative curve with geometrically connected fibers. Then $\mathrm{Pic}^{(d)}_{C/S} \to S$ is proper, smooth, and has geometrically connected fibers for all $d \in \mathbb{Z}$.*

One can show ([BLR] ^{0} 9.4, Prop. 4) that the Jacobian also has an S-ample line bundle which depends functorially on $C \to S$. In particular, $\mathrm{Pic}^0_{C/S}$ is projective over S if S is qcqs (Corollary 13.72). This implies that the algebraic spaces $\mathrm{Pic}^{(d)}_{C/S}$ are schemes for all $d \in \mathbb{Z}$ by Remark 14.74. Hence $\mathrm{Pic}_{C/S}$ is a smooth scheme over S. We will use this fact in the proofs of the results of the rest of this section and of the next section.

Let $C \to S$ be a smooth proper relative curve with geometrically connected fibers and let $\mathscr{L}$ be a line bundle on C. Recall that the functions

(27.26.3)
$$
\begin{aligned}
g_C \colon S \to \mathbb{Z}, &\quad s \mapsto g(C_s), \quad \text{the genus of } C, \\
\deg_{\mathscr{L}} \colon S \to \mathbb{Z}, &\quad s \mapsto \deg(\mathscr{L}_{|C_s}),
\end{aligned}
$$

are locally constant on S.

Corollary 27.139. *Let $C \to S$ be a smooth proper relative curve with geometrically connected fibers. Then the relative dimension of $\mathrm{Pic}^0_{C/S}$ over S is equal to g_C.*

Proof. This can be checked on geometric fibers, hence we may assume that $S = \mathrm{Spec}\, k$ for an algebraically closed field k. As $\mathrm{Pic}^0_{C/k}$ is smooth over k, we have

$$
\dim \mathrm{Pic}^0_{C/k} = \dim \mathrm{Lie}(\mathrm{Pic}^0_{C/k}) = \dim H^1(C, \mathcal{O}_C) = \mathrm{genus}(C),
$$

where the second equality holds because $\mathrm{Pic}^0_{C/k}$ is open in $\mathrm{Pic}_{C/k}$ and because of Proposition 27.122. $\qquad\square$

Lemma 27.140. *Let S be a scheme and let $f \colon C \to S$ be a smooth proper relative curve. Let $\mathscr{L}$ be a line bundle on C which is of degree 1 on each fiber.*
(1) *There exists $x \in C(S)$ such that $\mathscr{L} \cong \mathcal{O}_C(x)$ if and only if there exists a regular section $\ell \in \Gamma(C, \mathscr{L})$ such that $\mathrm{Coker}(\ell \colon \mathcal{O}_C \to \mathscr{L})$ is locally free of rank 1 over S.*
(2) *Suppose that $C \to S$ has geometrically connected fibers and that for all $s \in S$ the fiber C_s has genus ≥ 1. If $\mathscr{L}$ satisfies the equivalent assertions in (1), then $f_*\mathscr{L}$ is locally free of rank 1, its formation commutes with arbitrary base change $S' \to S$, and x is uniquely determined by $\mathscr{L}$.*

Proof. (1). Let $x\colon S \to C$ be such that $\mathscr{L} \cong \mathscr{O}_C(x)$. Let $\mathscr{I}(x) \subseteq \mathscr{O}_C$ be the ideal corresponding to the closed immersion x. Then $\mathscr{I}(x) \cong \mathscr{L}^{\otimes -1}$. Tensoring the exact sequence

$$(*) \qquad\qquad 0 \longrightarrow \mathscr{I}(x) \longrightarrow \mathscr{O}_C \longrightarrow x_*\mathscr{O}_S \longrightarrow 0$$

with $\mathscr{L}$ yields an exact sequence

$$(**) \qquad\qquad 0 \longrightarrow \mathscr{O}_C \longrightarrow \mathscr{L} \longrightarrow x_*(x^*\mathscr{L}) \longrightarrow 0,$$

where we use that $x_*\mathscr{O}_S \otimes_{\mathscr{O}_C} \mathscr{L} = x_*(x^*\mathscr{L})$ by the projection formula, Proposition 22.81. The injective homomorphism $\mathscr{O}_C \longrightarrow \mathscr{L}$ corresponds to a regular section $\ell \in \Gamma(C, \mathscr{L})$ and its cokernel is locally free of rank 1 over S.

Conversely, let $\ell\colon \mathscr{O}_C \to \mathscr{L}$ be an injection whose cokernel is locally free of rank 1 over S. Tensoring ℓ with $\mathscr{I} := \mathscr{L}^{\otimes -1}$ makes $\mathscr{I}$ into an invertible ideal of $\mathscr{O}_C$ that defines a closed subscheme D of C such that $D \to S$ is locally free of rank 1, i.e., an isomorphism. This yields the desired section $x\colon S = D \to C$.

(2). By hypothesis there exists $x \in C(S)$ and an exact sequence as in $(**)$. As $x_*(x^*\mathscr{L})$ is flat over S, $(**)$ stays exact after base change $S' \to S$. Using $f_*\mathscr{O}_C = \mathscr{O}_S$ we obtain by applying f_* to $(**)$ an exact sequence

$$0 \to \mathscr{O}_S \to f_*\mathscr{L} \to x^*\mathscr{L} \to R^1 f_*\mathscr{O}_C \to R^1 f_*\mathscr{L} \to 0.$$

As the formation of all terms, except possibly $f_*\mathscr{L}$, commute with base change $S' \to S$ (Corollary 24.63 and Corollary 23.146), the formation of $f_*\mathscr{L}$ also commutes with base change by the five lemma.

This implies that $f_*\mathscr{L}$ is finite locally free by Theorem 23.140 (2). To see that $f_*\mathscr{L}$ has rank 1, we may assume that $S = \operatorname{Spec} k$ for a field k. Then by Riemann-Roch, Theorem 26.48, one has

$$\dim_k H^0(C, \mathscr{L}) - \dim_k H^1(X, \mathscr{L}) = 2 - g(C) \le 1$$

which shows that $\dim_k H^0(C, \mathscr{L}) \le 1$. But by hypothesis, there exists a regular global section in $H^0(C, \mathscr{L})$. Therefore $\dim_k H^0(C, \mathscr{L}) = 1$.

To see that x is unique, we may work locally on S. Hence we may assume $S = \operatorname{Spec} R$ is affine and that $H^0(C, \mathscr{L})$ is a free R-module of rank 1. Choose ℓ as in (1) which we may consider as an R-linear map $R \to H^0(C, \mathscr{L})$. As ℓ stays injective after arbitrary base change $R \to R'$, ℓ is necessarily an isomorphism by Proposition 8.10. Hence any two choices of ℓ differ by some unit of R. Therefore the corresponding sections of C, as constructed in (1), are equal. $\qquad\square$

Lemma 27.141. *Let S be a scheme, let $f\colon C \to S$ be a smooth proper relative curve with geometrically connected fibers, and let $\mathscr{L}$ be a line bundle on C. Suppose that $\deg(\mathscr{L}) > 2g(C) - 2$ (27.26.3). Then $R^1 f_*\mathscr{L} = 0$ and $f_*\mathscr{L}$ is a locally free $\mathscr{O}_S$-module of rank $\deg(\mathscr{L}) + 1 - g(C)$ whose formation is compatible with base change $S' \to S$.*

Proof. By Remark 26.55 we have $H^1(C_s, \mathscr{L}_{|C_s}) = 0$ for all $s \in S$. Hence $R^1 f_*\mathscr{L} = 0$ and $f_*\mathscr{L}$ is finite locally free and its formation commutes with base change by Corollary 23.144. To calculate its rank, we can suppose that S is the spectrum of a field. Then $\dim H^0(C, \mathscr{L}) = \deg(\mathscr{L}) + 1 - g(C)$ by the Riemann-Roch theorem. $\qquad\square$

Proposition 27.142. *Let S be a scheme and let $C \to S$ be a smooth proper relative curve with geometrically connected fibers such that g_C is nowhere 0 on S. Then the Abel morphism $\iota\colon C \to \mathrm{Pic}^{(1)}_{C/S}$ is a closed immersion.*

Proof. One can check that a morphism between two proper schemes is a closed immersion fiberwise (Proposition 12.93). Hence we may assume that $S = \mathrm{Spec}\, k$ for a field k. As being a closed immersion can be checked after faithfully flat base change (Proposition 14.53), we may assume that k is algebraically closed. In particular $C(k) \neq \emptyset$. As C is proper over k and $\mathrm{Pic}^{(1)}_{C/k}$ is separated, ι is proper (Proposition 12.58). Hence it suffices to show that ι is injective on $k[\varepsilon]$-valued points (Remark 17.45). But then $\mathrm{Pic}_{C/k}(k[\varepsilon]) = \mathrm{Pic}(C_{k[\varepsilon]})$ and the injectivity of $C(k[\varepsilon]) \to \mathrm{Pic}(C_{k[\varepsilon]})$ follows from Lemma 27.140 (2). $\square$

Corollary 27.143. *Let $C \to S$ be as in Proposition 27.142 and let $\mathscr{L}$ be a line bundle on C which is on all fibers of degree 1. Then $\iota^{\mathscr{L}}\colon C \to \mathrm{Pic}^0_{C/S}$ is a closed immersion.*

(27.27) Elliptic curves.

We continue to denote by S an arbitrary scheme. In this section we will take a look at elliptic curves over S. See Section (26.17) for the case when S is the spectrum of a field.

Definition 27.144. *An elliptic curve over S is a pair consisting of a morphism of schemes $f\colon E \to S$ together with a section $0\colon S \to E$ of f such that*
(a) *the morphism $E \to S$ is flat, proper, and of finite presentation and*
(b) *all fibers of $E \to S$ are smooth, geometrically connected curves of genus 1.*

Remark 27.145. Let $(f\colon E \to S, 0)$ be an elliptic curve over S.
(1) If $T \to S$ is a morphism of schemes, then the base change $(E_T, 0_T)$ is an elliptic curve over T.
(2) The morphism $E \to S$ is smooth because flat morphisms locally of finite presentation with smooth fibers are smooth (Corollary 18.57).
(3) By Corollary 24.63 one has $f_* \mathscr{O}_E = \mathscr{O}_S$, compatibly with base change.

Example 27.146. Let $E \to S$ be an abelian scheme of dimension 1. Then $(E, 0)$ is an elliptic curve, where $0 \in E(S)$ is the zero section.

Indeed, we have to show that all fibers of E have genus 1. But this is clear because we have seen in Proposition 27.15 that the canonical bundle $\Omega^1_{E_s/\kappa(s)}$ of each fiber E_s is trivial. Hence the genus of E_s satisfies $2g(E_s) - 2 = 0$ by Corollary 26.52.

Theorem 27.147. *Let $(E, 0)$ be an elliptic curve over S. Then there exists a unique structure of a group scheme over S on E such that 0 is the unit section. This makes E into an abelian scheme of relative dimension 1.*

Hence we see that elliptic curves are precisely abelian schemes of relative dimension 1.

Proof. By definition, any structure of a group scheme on E makes E into an abelian scheme. Therefore the uniqueness assertion holds by Corollary 27.104. To show the existence it suffices to factorize the functor $T \mapsto E(T)$ from (Sch/S) to the category of sets over the category of abelian groups. Hence it suffices to show the following more precise proposition. $\square$

Proposition 27.148. *Let* $(f\colon E \to S, 0)$ *be an elliptic curve over* S. *Then the morphism*

$$\iota^0\colon E \to \mathrm{Pic}^0_{E/S}, \qquad x \mapsto \mathscr{O}_E(x) \otimes \mathscr{O}_E(0)^{\otimes -1},$$

defined in (27.26.2) *is an isomorphism.*

The proposition shows that every elliptic curve is self-dual, i.e., isomorphic to its dual abelian scheme (see Definition 27.159 and Corollary 27.212 below), and is also isomorphic to its own Jacobian.

Proof. By Proposition 14.28 we may assume $S = \mathrm{Spec}\, k$ for a field k. We know already that $\mathrm{Pic}^0_{E/S}$ is an abelian scheme, and by Proposition 27.122 it has dimension $\dim_k H^1(E, \mathscr{O}_E) = g(E) = 1$. So both sides are integral schemes of dimension 1. By Corollary 27.143, ι^0 is a closed immersion. Hence it must be an isomorphism. $\qquad\square$

Remark 27.149. The proposition shows that for points $P, Q, R \in E(T)$ one has $R = P + Q$ if and only if there exists a line bundle $\mathscr{M}$ on T such that

$$\mathscr{O}_{E_T}(P) \otimes \mathscr{O}_{E_T}(Q) \otimes \mathscr{O}_{E_T}(0_T)^{\otimes -1} \cong \mathscr{O}_{E_T}(R) \otimes f_T^*(\mathscr{M}).$$

If $\mathrm{Pic}(T) = 0$ (e.g., if T is the spectrum of a local ring or of a factorial ring), then the factor $f_T^*(\mathscr{M})$ can be omitted. Compare Theorem 26.98, Remark 26.100.

The following proposition is a relative version of Proposition 26.95.

Proposition 27.150. *Let* $(f\colon E \to S, 0)$ *be an elliptic curve. Then there exists an open affine covering* $(U_i)_i$ *of* S *with* $U_i = \mathrm{Spec}\, R_i$ *such that for all* i *one has* $E \times_S U_i \cong V_+(w_i) \subseteq \mathbb{P}^2_{R_i}$, *where* $w_i \in R_i[X, Y, Z]$ *is a cubic homogeneous polynomial of the form*

$$(27.27.1) \qquad w_i = Y^2 Z + a_1 XYZ + a_3 YZ^2 - X^3 - a_2 X^2 Z + a_4 XZ^2 + a_6 Z^3.$$

An equation as in (27.27.1) is called a *Weierstraß equation* for E.

The proof will show that if S is a scheme with $\mathrm{Pic}(S) = 0$ (e.g., if $S = \mathrm{Spec}\, R$ with R factorial), then such an embedding into $\mathbb{P}^2$ and the Weierstraß equation exist globally.

Proof. Let $\mathscr{L} := \mathscr{O}_E(0)$ and let $\ell\colon \mathscr{O}_E \to \mathscr{L}$ be the injective homomorphism given by the zero section (Lemma 27.140). Tensoring with $\mathscr{L}^{\otimes d}$ we obtain injective homomorphisms $\mathscr{L}^{\otimes d} \to \mathscr{L}^{\otimes d+1}$ for all d. By Lemma 27.141, $f_*\mathscr{L}^{\otimes d}$ is a locally free $\mathscr{O}_S$-module of rank d for all $d \geq 1$ and its formation commutes with base change of S. As the inclusions $f_*\mathscr{L}^{\otimes d} \to f_*\mathscr{L}^{\otimes(d+1)}$ stay injective after arbitrary base change, their cokernels are locally free (Proposition 8.10). We obtain a chain of injections

$$\mathscr{O}_S = f_*\mathscr{O}_E \cong f_*\mathscr{L} \to f_*\mathscr{L}^{\otimes 2} \to f_*\mathscr{L}^{\otimes 3} \to \cdots,$$

where all arrows have a locally free cokernel of rank 1. We view these injections as inclusions.

By working locally on S we can assume that $f_*\mathscr{L} \to f_*\mathscr{L}^{\otimes 2}$ and $f_*\mathscr{L}^{\otimes 2} \to f_*\mathscr{L}^{\otimes 3}$ have free cokernel. Then we can choose an $\mathscr{O}_S$-basis $(1, x, y)$ of $f_*\mathscr{L}^{\otimes 3}$ (i.e., a triple of elements $1, x, y \in \Gamma(E, \mathscr{L}^{\otimes 3}) = \Gamma(S, f_*\mathscr{L}^{\otimes 3})$ such that the corresponding morphism $\mathscr{O}_S^3 \to f_*\mathscr{L}^{\otimes 3}$ is an isomorphism) such that $(1, x)$ is an $\mathscr{O}_S$-basis of $f_*\mathscr{L}^{\otimes 2}$. As the formation of $f_*\mathscr{L}^{\otimes 3}$ commutes with base change, $1, x, y$ induce a basis of $H^0(E_s, \mathscr{L}_s^{\otimes 3})$ on each fiber E_s. As $\mathscr{L}_s$ is very ample (Proposition 26.59), the images of sections $1, x, y$ in $H^0(E_s, \mathscr{L}_s^{\otimes 3})$ generate $\mathscr{L}_s^{\otimes 3}$. Hence $1, x, y$ generate $\mathscr{L}^{\otimes 3}$ by Nakayama's lemma. The associated S-morphism $i\colon E \to \mathbb{P}_S^2$ is fiberwise a closed immersion. Hence it is a closed immersion by Proposition 12.93. It realizes E in $\mathbb{P}_S^2$ as the vanishing scheme of a homogeneous equation of degree 3.

To see that one can choose the equation of the special form (27.27.1), one argues as follows. Let x^2 be the image $x \otimes x$ in $\Gamma(E, \mathscr{L}^{\otimes 4})$. For each $s \in S$ the image of x^2 in $H^0(E_s, \mathscr{L}_s^{\otimes 4})$ is non-zero and not contained in $H^0(E_s, \mathscr{L}_s^{\otimes 3})$. Hence the images of 1, x, y, x^2 form a basis of $H^0(E_s, \mathscr{L}_s^{\otimes 4})$. Again Proposition 8.10 shows then that $(1, x, y, x^2)$ is an $\mathscr{O}_S$-basis of $f_*\mathscr{L}^{\otimes 4}$. Similarly, $(1, x, y, x^2, xy)$ is an $\mathscr{O}_S$-basis of $f_*\mathscr{L}^{\otimes 5}$ and that $(1, x, y, x^2, xy, y^2)$ and $(1, x, y, x^2, xy, x^3)$ are both $\mathscr{O}_S$-bases of $f_*\mathscr{L}^{\otimes 6}$. Moreover, $y^2 - x^3$ is a section of $f_*\mathscr{L}^{\otimes 6}$. This yields $E \xrightarrow{\sim} V_+(w)$ with w as in (27.27.1) such that $E \setminus 0(S)$ is given in $\mathbb{A}_S^2 \subseteq \mathbb{P}_S^2$ by the dehomogenization of w with respect to the variable Z. $\square$

Duality of abelian schemes

We now come to the definition of the dual X^t of an abelian scheme X as the identity component of the Picard scheme of X. A priori, this is only an algebraic space (but see below). Every line bundle $\mathscr{L}$ on X then defines a homomorphism of group schemes $\varphi_{\mathscr{L}}\colon X \to X^t$ (Theorem of the Square 27.168) which will play a central role in the sequel. The Theorem of the Square will also be instrumental in our proof that abelian varieties over a field are projective (Proposition 27.174), a result that we will generalize to normal base schemes in Theorem 27.291, and in the result that multiplication by non-zero integers is surjective (Proposition 27.186). In case that X is projective over its base scheme, we will show in Theorem 27.198 that X^t is representable by a scheme. A key step in the proof is to show that $\dim_k H^1(X, \mathscr{O}_X) = \dim(X)$ for an abelian variety over a field k. An abstract result about Hopf algebras then gives us that the whole cohomology of $\mathscr{O}_X$ is the exterior algebra of $H^1(X, \mathscr{O}_X)$ (Corollary 27.200). In Theorem 27.203 and its corollaries we will generalize these results to arbitrary base schemes.

In fact, for any base scheme S and any abelian scheme X over S, the dual X^t is representable by a scheme, see Theorem 27.211. We sketch the proof. The idea is to reduce to the case that S is noetherian and normal by noetherian approximation and by some gluing arguments. In this case the above mentioned results show that X is projective and hence representable by a scheme.

We conclude this part by studying the Poincaré bundle which gives us a proof that there is a functorial isomorphism $X \xrightarrow{\sim} (X^t)^t$ (Theorem 27.222).

(27.28) The space of correspondence classes.

In this section, S denotes a scheme. Given a morphism $X \to Y$ of S-schemes, its graph is a closed subscheme of the product $X \times_S Y$ (which projects isomorphically onto X, and this allows us to recover the given morphism from the graph subscheme). Therefore an arbitrary closed subscheme of $X \times_S Y$, sometimes called a *correspondence*, can be viewed as a generalization of a morphism between X and Y. For our purposes it will be useful to consider *divisorial correspondences* which we will define (Definition 27.151) as certain equivalence classes of divisors, or rather of line bundles, on the product $X \times_S Y$. It will turn out to be an important tool in our study of the Pic^0 functor of an abelian scheme, see Lemma 27.155, Proposition 27.163, Section (27.39). See Remark 27.288 for the case where X and Y are relative curves.

Definition 27.151. *Let $X \to S$ and $X' \to S$ be S-schemes and let $p\colon X \times_S X' \to X$, $p'\colon X \times_S X' \to X'$ be the projections. Consider the homomorphism of abelian fppf-sheaves on* (Sch/S)

$$(27.28.1) \qquad \alpha\colon \mathrm{Pic}_{X/S} \times \mathrm{Pic}_{X'/S} \longrightarrow \mathrm{Pic}_{X \times_S X'/S},$$

induced by $(\mathscr{L}, \mathscr{M}) \mapsto p^*\mathscr{L} \otimes p'^*\mathscr{M}$. *Then*

$$\mathrm{Corr}_S(X, X') := \mathrm{Coker}(\alpha)$$

is called the functor of divisorial correspondence classes between X and X'.

In other words, $\mathrm{Corr}_S(X, X')$ is the fppf-sheaf associated with the presheaf

$$\mathrm{Corr}'_S(X, X')\colon (\mathrm{Sch}/S) \longrightarrow (\mathrm{AbGrp}),$$
$$T \longmapsto \mathrm{Coker}(\mathrm{Pic}_{X/S}(T) \times \mathrm{Pic}_{X'/S}(T) \to \mathrm{Pic}_{X \times_S X'/S}(T)).$$

In fact, we will only be interested in the case where X and X' are abelian schemes over S. In this case, $\mathrm{Corr}'_S(X, X')$ is already an fppf-sheaf and hence $\mathrm{Corr}_S(X, X') = \mathrm{Corr}'_S(X, X')$. More precisely, we have:

Remark 27.152. Let $f\colon X \to S$ and $f'\colon X' \to S$ be morphisms of schemes such that $f_*\mathscr{O}_X = \mathscr{O}_S$ and $f'_*\mathscr{O}'_X = \mathscr{O}_S$ compatibly with base change $S' \to S$ and endowed with sections $e \in X(S)$ and $e' \in X'(S)$.

Let $p\colon X \times_S X' \to X$ and $p'\colon X \times_S X' \to X'$ be the projections. The sections $e' \in X(S')$ and $e \in X(S)$ yield by base change sections $s\colon X \to X \times_S X'$ of p and $s'\colon X' \to X \times_S X'$ of p'.

(1) We claim that there is a split exact sequence of fppf-sheaves

$$(27.28.2) \qquad 0 \longrightarrow \mathrm{Pic}_{X/S} \times \mathrm{Pic}_{X'/S} \xrightarrow{\alpha} \mathrm{Pic}_{X \times_S X'/S} \longrightarrow \mathrm{Corr}'_S(X, X') \longrightarrow 0.$$

A splitting is given by

$$(27.28.3) \qquad \beta := \beta_{e,e'}\colon \mathrm{Pic}_{X \times_S X'/S} \longrightarrow \mathrm{Pic}_{X/S} \times \mathrm{Pic}_{X'/S}$$

which is induced by $\mathscr{N} \mapsto (s^*\mathscr{N}, (s')^*\mathscr{N})$.

Indeed, we have to show that $\beta \circ \alpha = \mathrm{id}$ on T-valued points for an S-scheme T. By base change, we may assume that $T = S$. Let $\mathscr{L} \in \mathrm{Pic}(X)$ and $\mathscr{L}' \in \mathrm{Pic}(X')$. By symmetry it suffices to show that $\mathscr{L}$ and $s^*(p^*\mathscr{L}) \otimes s^*(p'^*\mathscr{L}')$ represent the same element in $\mathrm{Pic}_{X/S}(S) = \mathrm{Pic}(X)/\mathrm{Pic}(S)$. But $s^*p^*\mathscr{L} = \mathscr{L}$ and $s^*(p'^*\mathscr{L}')$ lies in the image of $\mathrm{Pic}(S) \to \mathrm{Pic}(X)$ because $p' \circ s$ factors through the chosen section $e'\colon S \to X'$.

(2) This shows that $\mathrm{Corr}'_S(X, X')$ is isomorphic to the kernel of the homomorphism β of abelian fppf-shaves and hence is already an fppf-sheaf. In particular $\mathrm{Corr}'_S(X, X') = \mathrm{Corr}_S(X, X')$.

Remark 27.153. Let $X \to S$ and $X' \to S$ be S-schemes. As the formation of $\mathrm{Pic}_{Z/S}$ commutes with base change $S' \to S$ for every S-scheme Z, so does the formation of $\mathrm{Corr}_S(X, X')$, i.e., for all scheme morphisms $T \to S$ one has

$$\mathrm{Corr}_S(X, X') \times_S T \cong \mathrm{Corr}_T(X \times_S T, X' \times_S T).$$

Notation and Remark 27.154. Let (X, e_X) and (Y, e_Y) be pointed presheaves on (Sch/S). Then we denote by $\underline{\mathrm{Hom}}_0(X, Y)$ the pointed presheaf on (Sch/S) such that $\underline{\mathrm{Hom}}_0(X, Y)(T)$ is the set of morphisms of pointed presheaves $X \times_S T \to Y \times_S T$ on (Sch/T) for every S-scheme T. The base point of $\underline{\mathrm{Hom}}_0(X, Y)$ is given by the morphism $X \to Y$ that sends all $x \in X(T)$ to the image of e_Y in $Y(T)$.

If Y is a presheaf of abelian groups on (Sch/S), then the sum of two pointed morphisms is again pointed and $\underline{\mathrm{Hom}}_0(X, Y)$ becomes a presheaf of abelian groups.

Lemma 27.155. *Let (X, e) and (X', e') be as in Remark 27.152. Then there is an isomorphism of abelian fppf-sheaves on (Sch/S), functorial in (X, e) and (X', e'),*

$$\mathrm{Corr}_S(X, X') \cong \underline{\mathrm{Hom}}_0(X, \mathrm{Pic}_{X'/S}).$$

Proof. Let $s' \colon X' \to X \times_S X'$ be the section of the second projection obtained by base change from e. Via the chosen section e' we identify

(*) $\mathrm{Pic}_{X'/S}(T) = \mathrm{Ker}(\mathrm{Pic}(X'_T) \xrightarrow{e'^*_T} \mathrm{Pic}(T)).$

To construct the above isomorphism we will work on T-valued points for an S-scheme T. As everything is compatible with base change $T \to S$, we may assume that $T = S$.

By the Yoneda lemma we have

$$\mathrm{Pic}_{X'/S}(X) \cong \mathrm{Hom}_S(X, \mathrm{Pic}_{X'/S})$$

and this isomorphism is given by sending a line bundle $\mathscr{L}$ on $X \times_S X'$ such that $e'^*_X \mathscr{L} \in \mathrm{Pic}(X)$ is trivial to the morphism $u \colon X \to \mathrm{Pic}_{X'/S}$ that maps $x \in X(T)$ to $x^*_{X'_T} \mathscr{L}_T$, where $x_{X'_T} \colon X'_T \to X_T \times_T X'_T$ is the section of the second projection obtained from x by base change. Hence $\mathscr{L}$ defines a pointed morphism u if and only if $s'^* \mathscr{L}$ is trivial in $\mathrm{Pic}(X')$. Via the identification (*) for $T = X$ we see that

$$\underline{\mathrm{Hom}}_0(X, \mathrm{Pic}_{X'/S})(S) = \mathrm{Ker}(\beta_{e,e'}) = \mathrm{Corr}_S(X, X')(S)$$

for the homomorphism $\beta_{e,e'}$ defined in (27.28.3).   □

Lemma 27.156. *Let S be a scheme, let $X \to S$ and $X' \to S$ be flat proper morphisms of finite presentation with geometrically integral fibers and suppose that one can choose $e \in X(S)$ and $e' \in X'(S)$.*

Then the morphism $\beta_{e,e'} \colon \mathrm{Corr}_S(X, X') \to \mathrm{Pic}_{X \times_S X'/S}$ (27.28.3) is representable by a closed immersion of finite presentation. In particular $\mathrm{Corr}_S(X, X')$ is a separated algebraic space locally of finite presentation over S.

Proof. By Remark 27.152, $\text{Corr}_S(X, X')$ is isomorphic to the kernel of a homomorphism of abelian fppf-sheaves $\text{Pic}_{X \times_S X'/S} \to \text{Pic}_{X/S} \times \text{Pic}_{X'/S}$ which are all separated group algebraic spaces locally of finite presentation by Theorem 27.117. In particular, the unit section of $\text{Pic}_{X/S} \times \text{Pic}_{X'/S}$ is representable by a closed immersion of finite presentation (Remark 27.45 and Remark 27.41). Therefore this holds for $\beta_{e,e'}$ by stability under base change (Remark 27.44 (1)). □

In fact, $\text{Corr}_S(X, X')$ is a separated algebraic space locally of finite presentation over S without the hypothesis that there exist sections of X or X': One has to show that all properties can be checked locally for the fppf topology because after passing to the fppf-covering $X \times_S X' \to S$, both X and X' have sections. For the property of being an algebraic space this is a difficult result ([Sta] `0ADV`), for the other properties this can be proved using the analogous result for schemes (Proposition 14.53).

Proposition 27.157. *Let $X \to S$ and $X' \to S$ be as in Lemma 27.156. Then $\text{Corr}_S(X, X')$ is a scheme which is separated, unramified, locally of finite presentation, and formally proper (Definition 27.128) over S.*

Proof. We already know that $\text{Corr}_S(X, X')$ is an algebraic space which is separated locally of finite presentation over S. It is formally proper because it is a closed algebraic subspace of $\text{Pic}_{X \times_S X'/S}$ (Lemma 27.156) which is formally proper by Remark 27.127.

Next we show that $\text{Corr}_S(X, X') \to S$ is unramified. It suffices to show that all fibers are unramified (for schemes locally of finite type over S this holds by Proposition 18.29, and for algebraic spaces this follows easily by choosing an étale atlas $U \to \text{Corr}_S(X, X')$ and to use that by definition, $\text{Corr}_S(X, X')$ is unramified over S if and only if U is unramified over S). As the formation of $\text{Corr}_S(X, X')$ is compatible with base change, we may assume $S = \text{Spec } k$ for a field k which we may even assume to be algebraically closed by Proposition 18.24.

To prove that $\text{Corr}_S(X, X') \to S$ is unramified we claim that it suffices to show that $\text{Lie}(\text{Corr}_S(X, X')) = 0$. For this we use that $\text{Corr}_S(X, X')$ is a scheme by Theorem 27.47[5]. By Proposition 18.27 and by Remark 17.44 it suffices to show that $T_z(\text{Corr}_S(X, X')) = 0$ for all k-valued points z of $\text{Corr}_S(X, X')$. By homogeneity it suffices to prove this for the unit section. This shows the claim.

To show that $\text{Lie}(\text{Corr}_S(X, X')) = 0$, consider the split exact sequence of abelian fppf-sheaves (27.28.2). It yields an exact sequence of k-vector spaces

$$0 \longrightarrow \text{Lie Pic}_{X/S} \times \text{Lie Pic}_{X'/S} \xrightarrow{\text{Lie } \alpha} \text{Lie Pic}_{X \times_S X'/S} \longrightarrow \text{Lie Corr}_S(X, X') \longrightarrow 0.$$

Now we use that $\text{Lie Pic}_{X/S} = H^1(X, \mathscr{O}_X)$ (Proposition 27.122). Hence using the Künneth isomorphism (Corollary 22.110) we can identify the term in the middle with

$$H^1(X \times X', \mathscr{O}_{X \times X'}) = \bigoplus_{i=0}^{1} H^i(X, \mathscr{O}_X) \otimes H^{1-i}(X', \mathscr{O}_{X'}).$$

Since $H^0(X, \mathscr{O}_X) = H^0(X', \mathscr{O}'_X) = k$ (Corollary 24.63), altogether we obtain that $\text{Lie}(\text{Corr}_S(X, X')) = 0$.

It remains to show that $\text{Corr}_S(X, X')$ is a scheme. As $\text{Corr}_S(X, X') \to S$ is unramified, it is locally quasi-finite. Moreover it is separated. Therefore $\text{Corr}_S(X, X')$ is a scheme by [Sta] `03XX`. □

---

[5] Instead of using the difficult Theorem 27.47 one could also argue as in the sequel by checking that all arguments are also valid for algebraic spaces, which is not too difficult.

(27.29) Definition of the dual functor to an abelian scheme.

Let S be a scheme and let X be an abelian scheme over S with zero section $0 \in X(S)$. Let us collect some notation and recall some results about $\mathrm{Pic}_{X/S}$ and $\mathrm{Corr}_S(X, X)$ in this special case.

We have the morphisms

$$m, p_1, p_2 \colon X \times_S X \longrightarrow X,$$

where m is the group law on X and p_1 and p_2 are the two projections. The zero section yields sections s_i of p_i given by

$$s_1 \colon X = X \times_S 0 \xrightarrow{\,\mathrm{id}_X \times 0\,} X \times X,$$
$$s_2 \colon X = 0 \times_S X \xrightarrow{\,0 \times \mathrm{id}_X\,} X \times_S X.$$

If T is an S-scheme, we will often also denote the base changes $X_T \times_T X_T \to X_T$ of m, p_1, p_2 (resp. the base changes $X_T \to X_T \times_T X_T$ of s_1 and s_2) again by m, p_1, p_2 (resp. by s_1 and s_2).

For every S-scheme T we have

$$\mathrm{Pic}_{X/S}(T) = \mathrm{Pic}(X_T)/\mathrm{Pic}(T)$$

and $\mathrm{Pic}_{X/S}$ is a separated group algebraic space locally of finite presentation.

Remark 27.158. As a special case of the results of Section (27.28) we have the S-group scheme $\mathrm{Corr}_S(X, X)$ of divisorial correspondences on X. It is separated, unramified, and locally of finite presentation. By definition it is equipped with an fppf-surjective homomorphism

$$(27.29.1) \qquad\qquad \mathrm{Pic}_{X \times_S X/S} \longrightarrow \mathrm{Corr}_S(X, X)$$

of group algebraic spaces over S. Its kernel consists by definition of those classes of line bundles $\mathcal{M}$ on $X \times_S X$ that are of the form $p_1^* \mathcal{L}_1 \otimes p_2^* \mathcal{L}_2$ for line bundles $\mathcal{L}_1$ and $\mathcal{L}_2$ on X.

(1) The zero section 0 of X defines a section of (27.29.1)

$$(27.29.2) \qquad\qquad \mathrm{Corr}_S(X, X) \longrightarrow \mathrm{Pic}_{X \times_S X/S}$$

that identifies $\mathrm{Corr}_S(X, X)(T)$ with those classes of line bundles $\mathcal{M}$ on $X_T \times_T X_T$ such that $s_1^* \mathcal{M} = s_2^* \mathcal{M} = \mathcal{O}_{X_T}$ in $\mathrm{Pic}_{X/S}(T)$. Via this description, (27.29.1) is induced by the map that sends a line bundle $\mathcal{M}$ on $X_T \times_T X_T$ to

$$(27.29.3) \qquad\qquad \mathcal{M} \otimes p_1^* s_1^* \mathcal{M}^{-1} \otimes p_2^* s_2^* \mathcal{M}^{-1}.$$

(2) We have seen in Lemma 27.155 that we can identify

$$(27.29.4) \quad \mathrm{Corr}_S(X, X)(T) = \{\, \psi \colon X_T \to \mathrm{Pic}_{X_T/T} \;;\; \psi \text{ pointed morphism over } T \,\}$$

for every S-scheme T. Via this description, the proof of Lemma 27.155 shows that (27.29.1) is induced by the map that sends a line bundle $\mathcal{M}$ on $X_T \times_T X_T$ to the following pointed morphism. A point $x \in X(T')$, T' a T-scheme is mapped to the pullback of the line bundle in (27.29.3) under $\tilde{x} := (x, \mathrm{id}_{X_{T''}}) \colon X_{T'} \to X_{T'} \times_{T'} X_{T'}$ which in $\mathrm{Pic}_{X/S}(T')$ equals the line bundle

$$(27.29.5) \qquad\qquad \tilde{x}^* \mathcal{M} \otimes s_2^* \mathcal{M}^{-1}.$$

The following definitions will be central for the rest of the chapter.

Definition 27.159. *Let X be an abelian scheme over S with group law $m\colon X \times_S X \to X$.*

(1) *Define the homomorphism $\varphi\colon \operatorname{Pic}_{X/S} \to \operatorname{Corr}_S(X,X)$ of group algebraic spaces over S as the composition*

$$\varphi\colon \operatorname{Pic}_{X/S} \xrightarrow{m^*} \operatorname{Pic}_{X\times_S X/S} \longrightarrow \operatorname{Corr}_S(X,X).$$

(2) *If $\mathscr{L}$ is a line bundle on X, we denote the image of the class of $\mathscr{L}$ in $\operatorname{Pic}_{X/S}(S)$ under the map φ by $\varphi_{\mathscr{L}}$. Via (27.29.4) we usually consider it as a morphism of pointed algebraic spaces*

(27.29.6) $$\varphi_{\mathscr{L}}\colon X \to \operatorname{Pic}_{X/S}.$$

(3) *The* dual abelian space *of X is defined as*

$$X^t := \operatorname{Ker}(\varphi).$$

From the above results we deduce easily:

Lemma 27.160. *The inclusion $X^t \to \operatorname{Pic}_{X/S}$ is representable by an open and closed immersion. Therefore X^t is a commutative separated group algebraic space over S locally of finite presentation.*

We will see below (Proposition 27.207) that $X^t = \operatorname{Pic}^0_{X/S}$, the identity component of $\operatorname{Pic}_{X/S}$ (Section (27.24)), and that X^t is always an abelian scheme over S (Corollary 27.212), which will be called the *dual abelian scheme of X*.

Proof. As $\operatorname{Corr}_S(X,X) \to S$ is unramified and separated, the unit section of $\operatorname{Corr}_S(X,X)$ is an open and closed immersion (Remark 18.30 and Remark 9.12). This shows that the inclusion $j\colon X^t \to \operatorname{Pic}_{X/S}$ is representable and an open and closed immersion. As $\operatorname{Pic}_{X/S}$ is a separated algebraic space locally of finite presentation over S, so is X^t. $\square$

We have the following more explicit description of φ.

Proposition 27.161. *Let $\mathscr{L}$ be a line bundle on the abelian scheme X.*
(1) *If one considers $\operatorname{Corr}_S(X,X)$ as a subgroup of $\operatorname{Pic}_{X\times_S X/S}$ via (27.29.2), then the image of the class of $\mathscr{L}$ under φ is given by the class of the line bundle*

$$\varphi(\mathscr{L}) = \Lambda(\mathscr{L}) := m^*\mathscr{L} \otimes p_1^*\mathscr{L}^{-1} \otimes p_2^*\mathscr{L}^{-1} \otimes [0]^*\mathscr{L}$$

on $X \times_S X$, where $[0]^*\mathscr{L}$ is obtained by pulling back $\mathscr{L}$ to S via the zero section of X and then by pulling back to $X \times_S X$ via the structure morphism.
(2) *The morphism $\varphi_{\mathscr{L}}\colon X \to \operatorname{Pic}_{X/S}$ is given by*

$$X(T) \ni x \mapsto \text{the class of } t_x^*\mathscr{L}_T \otimes \mathscr{L}_T^{-1}$$

for every S-scheme T. Here $X_T := X \times_S T$, $\mathscr{L}_T$ is the pullback of $\mathscr{L}$ to X_T, and t_x is the translation $X_T \to X_T$ by x.

The line bundle $\Lambda(\mathscr{L})$ is sometimes called the *Mumford bundle*. Its class in $\operatorname{Pic}_{X\times_S X/S}(S)$ is the same as the class of $m^*\mathscr{L} \otimes p_1^*\mathscr{L}^{-1} \otimes p_2^*\mathscr{L}^{-1}$ and often one can ignore the factor "$\otimes [0]^*\mathscr{L}$", e.g., if $\operatorname{Pic}(S) = 0$, for instance if $S = \operatorname{Spec} k$ for a field k. But the definition of $\Lambda(\mathscr{L})$ we have given has the advantage that there are canonical isomorphisms $\Lambda(\mathscr{L})_{|X\times 0} \cong \mathscr{O}_X \cong \Lambda(\mathscr{L})_{|0\times X}$.

Proof. This follows from Remark 27.158: One obtains the description in (1) (resp. in (2)) by applying (27.29.3) (resp. (27.29.5)) to $\mathscr{M} := m^*\mathscr{L}$. $\qquad\square$

Corollary 27.162. *Let $f\colon X \to S$ be an abelian scheme and let $\mathscr{L}$ be a line bundle on X.*
(1) *The following assertions are equivalent.*
  (i) *The class of $\mathscr{L}$ is in $X^t(S)$.*
  (ii) *For every geometric point $\bar{s}$ of S the class of $\mathscr{L}_{\bar{s}}$ is in $X^t(\bar{s})$.*
  (iii) *The Mumford bundle $\Lambda(\mathscr{L})$ is isomorphic to the pullback of a line bundle on S.*
  (iv) *For all S-schemes T and all $x \in X(T)$ the line bundle $t_x^*\mathscr{L}_T \otimes \mathscr{L}_T^{-1}$ is isomorphic to the pullback of a line bundle on T.*
(2) *There exists an open and closed immersion $S_0 \to S$ such that a morphism of schemes $f\colon T \to S$ factors through S_0 if and only if the class of $(\mathrm{id}_X \times f)^*\mathscr{L}$ is in $X^t(T)$.*

In particular, if S is connected and there exists a geometric point $\bar{s} \to S$ such that the class of $\mathscr{L}_{\bar{s}}$ is in $X^t(\bar{s})$, then the class of $\mathscr{L}$ is in $X^t(S)$.

Proof. The equivalence of (i), (iii), and (iv) holds by Proposition 27.161. The equivalence of (i) and (ii) and part (2) hold since $X^t \to \mathrm{Pic}_{X/S}$ is an open and closed immersion. $\qquad\square$

Proposition 27.163. *Let X and Y be abelian schemes over S.*
(1) *Every pointed morphism $f\colon X \to \mathrm{Pic}_{Y/S}$ factors through the open subgroup space Y^t. In particular*
$$\mathrm{Corr}_S(X, Y) \cong \underline{\mathrm{Hom}}_0(X, Y^t).$$
(2) *One has a functorial isomorphism*
$$\underline{\mathrm{Hom}}_0(X, Y^t) \cong \underline{\mathrm{Hom}}_0(Y, X^t).$$
(3) *Let $\mathscr{L}$ be a line bundle on X. The pointed morphism $\varphi_{\mathscr{L}}$ (27.29.6) factors uniquely through a pointed morphism*

$$(27.29.7) \qquad\qquad\qquad \varphi_{\mathscr{L}}\colon X \to X^t.$$

Proof. As $Y^t \to \mathrm{Pic}_{Y/S}$ is an open and closed immersion and f map the zero section into Y^t, (1) holds since X has connected fibers over S. Then the canonical isomorphism $\mathrm{Corr}_S(X, Y) \cong \mathrm{Corr}_S(Y, X)$ implies (2). Finally, (3) is an immediate corollary of (1). $\qquad\square$

Next we show that $(\ )^t$ commutes with base change and is a functor.

Remark 27.164. Let X be an abelian scheme over S. As the formation of $\mathrm{Pic}_{X/S}$ and $\mathrm{Corr}_S(X, X)$ is compatible with base change, so is the formation of X^t, i.e., for every scheme morphism $T \to S$ we have a functorial isomorphism

$$(X \times_S T)^t \cong X^t \times_S T.$$

Lemma and Definition 27.165. *Let $f\colon X \to Y$ be a homomorphism of abelian schemes over S.*
(1) *The homomorphism $f^*\colon \mathrm{Pic}_{Y/S} \to \mathrm{Pic}_{X/S}$ induces a homomorphism*

$$f^t\colon Y^t \to X^t$$

called the dual *of f.*

(2) *Let $\mathscr{L}$ be line bundle on Y. Then the diagram*

(27.29.8)

$$
\begin{array}{ccc}
X & \xrightarrow{\;f\;} & Y \\
{\scriptstyle \varphi_{f^*\mathscr{L}}}\Big\downarrow & & \Big\downarrow{\scriptstyle \varphi_{\mathscr{L}}} \\
X^t & \xleftarrow{\;f^t\;} & Y^t
\end{array}
$$

commutes.

Proof. We first show the commutativity of (27.29.8) if we replace the lower horizontal row of the diagram with $\mathrm{Pic}_{X/S} \xleftarrow{\;f^*\;} \mathrm{Pic}_{Y/S}$. We check this on T-valued points for an S-scheme T. As all constructions commute with base change, we may assume $T = S$. Let $x \in X(S)$. Using $t_{f(x)} \circ f = f \circ t_x$ and hence $t_x^* f^* \mathscr{L} = f^* t_{f(x)}^* \mathscr{L}$ we find

(27.29.9)
$$
\begin{aligned}
(f^t \circ \varphi_{\mathscr{L}} \circ f)(x) &= f^*(t_{f(x)}^* \mathscr{L} \otimes \mathscr{L}^{-1}) \\
&= t_x^* f^* \mathscr{L} \otimes (f^* \mathscr{L})^{-1} \\
&= \varphi_{f^* \mathscr{L}}(x).
\end{aligned}
$$

By Proposition 27.163 (3) it now is enough to show that (1) holds. As $\mathscr{L} \in X^t(S)$ if and only if $\varphi_{\mathscr{L}} = 0$, the commutative diagram shows that $f^* \mathscr{L} \in Y^t(S)$. $\qquad\square$

Remark 27.166. Let $\mathscr{L}$ be a line bundle on X whose class is contained in $X^t(S)$ (cf. Corollary 27.162). Then one has

$$
m^* \mathscr{L} = p_1^* \mathscr{L} \otimes p_2^* \mathscr{L}
$$

in $\mathrm{Pic}(X \times_S X)/\mathrm{Pic}(S)$. Pulling back along a pair (x, y) of T-valued points of X one obtains

$$
(x + y)^* \mathscr{L} = x^* \mathscr{L} \otimes y^* \mathscr{L}
$$

in $\mathrm{Pic}(X)$ up to tensoring with a line bundle that is pulled back from S. In particular, if $g_1, g_2 \colon Y \to X$ are homomorphisms of abelian schemes, then we have an equality of homomorphisms of commutative group algebraic spaces $X^t \to Y^t$

(27.29.10)
$$
(g_1 + g_2)^t = g_1^t + g_2^t.
$$

Hence, $X \mapsto X^t$ defines a contravariant additive functor $(\ )^t$ from the category of abelian schemes over S to the category of abelian algebraic group spaces over S. In the sequel, we will see that this functor defines a contravariant autoduality of the category of abelian schemes over S and that $(\ )^t \circ (\ )^t$ is isomorphic to the identity.

Note that the map

$$
\mathrm{Hom}(X, Y) \longrightarrow \mathrm{Hom}(\mathrm{Pic}_{Y/S}, \mathrm{Pic}_{X/S}), \qquad f \mapsto f^*
$$

is (except in trivial cases) not a group homomorphism. For instance, we will see in Proposition 27.234 below that the multiplication by an integer $[n]$ induces on the quotient $\mathrm{Pic}_{X/S}/X^t$ the multiplication by n^2.

(27.30) The Theorem of the Square.

In this section S denotes a scheme. The goal of this section is to show that $\varphi_{\mathscr{L}} \colon X \to X^t$ (27.29.6) is a homomorphism of group algebraic spaces. If we already knew that X^t is an abelian scheme (see Corollary 27.212 below), then we could simply apply Corollary 27.102 as $\varphi_{\mathscr{L}}$ preserves unit sections. But the homomorphism $\varphi_{\mathscr{L}}$ will be an important tool in the proof that X^t is an abelian scheme. Hence we have to argue differently.

Proposition 27.167. *(Cubical structure on line bundles) Let X be an abelian scheme over S, let $\mathscr{L}$ be a line bundle on X, let T be an S-scheme and let $x_1, x_2, x_3 \in X(T)$. Then the line bundle*

$$c_T(\mathscr{L}) := (x_1 + x_2 + x_3)^* \mathscr{L} \otimes (x_1 + x_2)^* \mathscr{L}^{-1} \otimes (x_1 + x_3)^* \mathscr{L}^{-1} \otimes (x_2 + x_3)^* \mathscr{L}^{-1}$$

$$\otimes \, x_1^* \mathscr{L} \otimes x_2^* \mathscr{L} \otimes x_3^* \mathscr{L} \otimes 0^* \mathscr{L}^{-1}$$

on T is trivial.

Proof. It suffices to treat the universal case $T = X \times_S X \times_S X$ and $x_i = p_i \colon T \to A$ the i-th projection. For $I \subseteq \{1, \dots, 3\}$ let $p_I \colon X \times_S X \times_S X$ be the morphism $\sum_{i \in I} p_i$. Then the line bundle $c_T(\mathscr{L})$ is given by

$$(27.30.1) \qquad\qquad c(\mathscr{L}) := \bigotimes_{I \subseteq \{1,2,3\}} p_I^*(\mathscr{L})^{(-1)^{|I|+1}}.$$

We claim that the restrictions of $c(\mathscr{L})$ to $\{0\} \times_S X \times_S X$, to $X \times_S \{0\} \times_S X$, and to $X \times_S X \times_S \{0\}$ are trivial. Indeed, by symmetry it suffices to treat the case $\{0\} \times_S X \times_S X$. Then

$$c(\mathscr{L})_{|\{0\} \times_S X \times_S X} = m^* \mathscr{L} \otimes q_1^* \mathscr{L}^{-1} \otimes q_2^* \mathscr{L}^{-1} \otimes m^* \mathscr{L}^{-1}$$

$$\otimes \, (0^* \mathscr{L})_{X \times_S X} \otimes q_1^* \mathscr{L} \otimes q_2^* \mathscr{L} \otimes (0^* \mathscr{L})^{-1}_{X \times_S X}$$

where m, q_1, q_2 are the multiplication and the projections $X \times_S X \to X$. Clearly this is trivial.

We now use the Theorem of the Cube 24.73. By the claim above, we find in every fiber of $X \to S$, considered as the third factor of $X \times_S X \times_S X$, a point $x \in X$ such that the restriction of $c(\mathscr{L})$ to $X \times_S X \times_S \{x\}$ is trivial. Namely, we can take x to be the image of the unit section of the fiber. Hence the Theorem of the Cube implies that we find for every $s \in S$ an open and closed subscheme U_s of X that meet the fiber X_s and such that the restriction of $c(\mathscr{L})$ to $X \times_S X \times_S U_s$ is trivial. As all fibers of $X \to S$ are connected one has $X_s \subseteq U_s$ and hence $(U_s)_{s \in S}$ is a covering of X. This shows that $c(\mathscr{L})$ is trivial. $\qquad\square$

Theorem of the Square 27.168. *Let X be an abelian scheme over S and let $\mathscr{L}$ be a line bundle on X. Then the pointed morphism $\varphi_{\mathscr{L}} \colon X \to X^t$ is a homomorphism of group algebraic spaces over S.*

Proof. By the description of $\varphi_{\mathscr{L}}$ in Proposition (27.161) (2) it suffices to show that for each S-scheme T and for all $x, y \in X(T)$ we have an equality of classes

$$(27.30.2) \qquad\qquad t_{x+y}^* \mathscr{L}_T \otimes \mathscr{L}_T = t_x^* \mathscr{L}_T \otimes t_y^* \mathscr{L}_T$$

in $\mathrm{Pic}_{X/S}(T)$. Let $p_X \colon X_T = X \times_S T \to X$ and $p_T \colon X_T \to T$ be the projections. Then $\mathscr{L}_T = p_X^* \mathscr{L}$. Define $(X \times_S T)$-valued points of X by $x_1 := p_X$, $x_2 := x \circ p_T$, and $x_3 := y \circ p_T$. Then

$$x_1 + x_2 = p_X \circ t_x, \quad x_1 + x_3 = p_X \circ t_y, \quad x_2 + x_3 = (x+y) \circ p_T, \quad x_1 + x_2 + x_3 = p_X \circ t_{x+y}.$$

Now we apply Proposition 27.167 and see that both sides of (27.30.2) are isomorphic line bundles on X_T up to tensoring with $p_T^*(\mathscr{M})$, where

$$\mathscr{M} := (x+y)^* \mathscr{L} \otimes x^* \mathscr{L}^{-1} \otimes y^* \mathscr{L}^{-1} \otimes [0]^* \mathscr{L} \in \operatorname{Pic}(T).$$

Hence we obtain the desired equality in $\operatorname{Pic}_{X/S}(T) = \operatorname{Pic}(X_T)/\operatorname{Pic}(T)$. $\qquad\square$

Remark 27.169. Let X be an abelian scheme over S and let $\mathscr{L}$ and $\mathscr{M}$ be line bundles on X.
(1) As φ is a group homomorphism, one has

(27.30.3)
$$\varphi_{\mathscr{L} \otimes \mathscr{M}} = \varphi_{\mathscr{L}} + \varphi_{\mathscr{M}},$$
$$\varphi_{\mathscr{L}^{-1}} = -\varphi_{\mathscr{L}}.$$

(2) Moreover for $x \in X(S)$ we have

(27.30.4)
$$\varphi_{t_x^* \mathscr{L}} = \varphi_{\mathscr{L}}.$$

Indeed, let T be an S-scheme and let $y \in X(T)$. We denote the image of x in $X(T)$ again by x. Then we have an equality in $\operatorname{Pic}_{X/S}(T)$

$$\varphi_{t_x^* \mathscr{L}}(y) = t_{x+y}^* \mathscr{L}_T \otimes t_x^* \mathscr{L}^{-1} = t_y^* \mathscr{L} \otimes \mathscr{L}^{-1} = \varphi_{\mathscr{L}}(y),$$

where the equality in the middle holds by (27.30.2).
(3) Let $x_1, \ldots, x_r \in X(S)$ be S-valued points such that $\sum_{i=1}^r x_i = 0$ in the abelian group $X(S)$. By induction, (27.30.2) implies the equality

$$\bigotimes_{i=1}^r t_{x_i}^* \mathscr{L} = \mathscr{L}^{\otimes r}$$

in $\operatorname{Pic}_{X/S}(S) = \operatorname{Pic}(X)/\operatorname{Pic}(S)$.

(27.31) The kernel of $\varphi_{\mathscr{L}}$.

In this section S denotes a scheme, $f \colon X \to S$ an abelian scheme over S, and $\mathscr{L}$ a line bundle on X. We set

(27.31.1)
$$K(\mathscr{L}) := \operatorname{Ker}(\varphi_{\mathscr{L}}).$$

This is a closed subgroup scheme of X.

By definition, $K(\mathscr{L})$ depends only on the class of $\mathscr{L}$ in $\operatorname{Pic}_{X/S}(S)$, i.e. $K(\mathscr{L} \otimes f^* \mathscr{N}) = K(\mathscr{L})$ for a line bundle $\mathscr{N}$ on $\operatorname{Pic}(S)$. By Proposition 27.161 there are the following descriptions of $K(\mathscr{L})$.

Remark 27.170. For any S-scheme T one has

$$K(\mathscr{L})(T) = \{ x \in X(T) \, ; \, t_x^* \mathscr{L} \otimes \mathscr{L}^{-1} \cong f_T^* \mathscr{N} \text{ for some } \mathscr{N} \in \operatorname{Pic}(T) \}.$$

Moreover, $K(\mathscr{L})$ is the maximal closed subscheme of X such that the restriction of the Mumford bundle $\Lambda(\mathscr{L})$ to $K(\mathscr{L}) \times_S X$ is isomorphic to the pullback of a line bundle on $K(\mathscr{L})$ (in the sense of the Seesaw Proposition 24.66).

Lemma 27.171. *Let S be a scheme.*
1. *The restriction $\Lambda(\mathscr{L})_{|K(\mathscr{L}) \times_S X}$ is trivial.*
2. *The restriction $\mathscr{L} \otimes [-1]^* \mathscr{L}$ to $K(\mathscr{L})$ is isomorphic to $[0]^* \mathscr{L}^{\otimes -2}$, in particular it is trivial if $\operatorname{Pic}(S) = 0$.*

Proof. By Remark 27.170 we know that $\Lambda(\mathscr{L})_{|K(\mathscr{L}) \times_S X}$ is the pullback of a line bundle $\mathscr{N}$ from $K(\mathscr{L})$. Let $s := (\mathrm{id}, e) \colon K(\mathscr{L}) \to K(\mathscr{L}) \times_S X$. Then

$$\mathscr{N} = s^*(\Lambda(\mathscr{L})_{|K(\mathscr{L}) \times_S X}) = (\mathscr{L} \otimes \mathscr{L}^{-1} \otimes [0]^*(\mathscr{L}^{-1}) \otimes [0]^*(\mathscr{L}))_{|K(\mathscr{L})} \cong \mathscr{O}_{K(\mathscr{L})}.$$

This shows (1).

By (1) we know that the restriction of $\Lambda(\mathscr{L})$ to $K(\mathscr{L}) \times_S K(\mathscr{L})$ is trivial. Let $\tau := (\mathrm{id}, -\mathrm{id}) \colon X \to X \times_S X$. Then

$$\tau^* \Lambda(\mathscr{L}) = [0]^* \mathscr{L} \otimes \mathscr{L}^{-1} \otimes [-1]^* \mathscr{L}^{-1} \otimes [0]^* \mathscr{L}$$

and the restriction of $\tau^* \Lambda(\mathscr{L})$ to $K(\mathscr{L})$ is trivial. This implies the claim. $\qquad\square$

Corollary 27.172. *Let $\mathscr{L}$ be an ample line bundle on X. Then $K(\mathscr{L})$ is a finite S-scheme.*

In Theorem 27.198 below we will see that $K(\mathscr{L}) \to S$ is finite locally free.

Proof. As $K(\mathscr{L})$ is a closed subscheme of X, it is proper over S. Hence we can check finiteness of $K(\mathscr{L}) \to S$ on fibers (Proposition 12.93). Therefore we may assume that $S = \operatorname{Spec} k$ for a field k.

As $\mathscr{L}$ is ample, $\mathscr{L} \otimes [-1]^* \mathscr{L}$ is also ample. By Lemma 27.171 we know that the ample line bundle $(\mathscr{L} \otimes [-1]^* \mathscr{L})_{|K(\mathscr{L})}$ is trivial. Hence $K(\mathscr{L}) \to S$ is quasi-affine (Definition 13.78) and therefore finite by Corollary 13.82. $\qquad\square$

(27.32) Projectivity of abelian varieties.

The results of Section (27.11) on general homogeneous spaces show the following general projectivity result.

Proposition 27.173. *Let S be a local normal noetherian scheme and let $X \to S$ be an abelian scheme. Let $U \subseteq X$ be an open affine neighborhood that meets the special fiber of $X \to S$. Let $D_1, \ldots, D_r$ be the irreducible components of $X \setminus U$ that are of codimension 1 in X, considered as reduced subschemes of X. Let $n_1, \ldots, n_r > 0$ be integers. Then the Weil divisor $D := \sum_{i=1}^r n_i D_i$ is a Cartier divisor and $\mathscr{O}_X(D)$ is ample. In particular, $X \to S$ is projective.*

Note that the image of U in S is open and contains the special point of S. Therefore U meets every fiber of $X \to S$.

We gave a proof of the general statement for homogeneous spaces only if $S = \operatorname{Spec} k$ is a field – and even in this case we omitted the proof of the difficult Lemma 27.70. Below in Theorem 27.291 we will prove, more generally, that every abelian scheme over a normal noetherian scheme is projective.

We will now give a self-contained proof for the fact that abelian varieties over a field are projective. In fact we will show the following more precise result.

Proposition 27.174. *Let X be an abelian variety over a field k and let $U \subseteq X$ be a nonempty open affine subscheme. Then every effective divisor D whose underlying topological space is $X \setminus U$ is ample. In particular, every abelian variety over a field is projective.*

For instance $(X \setminus U)_{\mathrm{red}}$ is an effective ample divisor by Lemma 25.150.

In Proposition 27.265 below we will see that conversely every ample line bundle is of the form $\mathscr{O}_X(D)$ with D an effective divisor such that $X \setminus D$ is affine.

To prove Proposition 27.174 we will use the following lemma.

Lemma 27.175. *Let X be an abelian variety over an algebraically closed field k, let $D \subseteq X$ be an effective Cartier divisor, and set $\mathscr{L} := \mathscr{O}_X(D)$. Then $\mathscr{L}^{\otimes 2}$ is globally generated.*

Below in Proposition 27.278 we will prove that the tensor product of any two ample line bundles on an abelian scheme is globally generated.

Proof. Fix $x \in X(k)$. We have to find an effective divisor E that is linearly equivalent to $2D$ such that $x \notin E$. Then the complete linear system of $\mathscr{L}^{\otimes 2}$ is base point free and hence $\mathscr{L}^{\otimes 2}$ is globally generated (Section (13.13)).

For $y \in X(k)$ we set $E_y := t_y(D) + t_{-y}(D)$ which is linearly equivalent to $2D$ by (27.30.2). Set $U := X \setminus D$. Then

$$x \notin t_y(D) \cup t_{-y}(D)$$
$$\Leftrightarrow x \in t_y(U) \cap t_{-y}(U)$$
$$\Leftrightarrow y \in t_{-x}(U) \cap t_x([-1](U)) =: V.$$

As X is irreducible, $V \neq \emptyset$, and for all $y \in V$ one has $x \notin E_y$. $\square$

Proof. [of Proposition 27.174] Set $\mathscr{L} := \mathscr{O}_X(D)$. Let $\bar{k}$ be an algebraic closure of k. Then the inverse image of D in $X_{\bar{k}}$ is still an effective Cartier divisor whose complement is a non-empty open affine subscheme of $X_{\bar{k}}$. To see that $\mathscr{L}$ is ample, we may work fpqc-locally (Proposition 14.58). Hence we can assume that k is algebraically closed.

By Lemma 27.175 there exists a morphism $f \colon X \to \mathbb{P}_k^N$ such that $f^* \mathscr{O}_{\mathbb{P}^N}(1) \cong \mathscr{L}^{\otimes 2}$ and there exists a hyperplane H of $\mathbb{P}_k^N$ such that $f^{-1}(H) = 2D$ and hence $U = f^{-1}(\mathbb{P}_k^N \setminus H)$. As X is proper over k and $\mathbb{P}_k^N$ is separated over k, the morphism f is proper. Hence for each $x \in U(k)$ the fiber $f^{-1}(f(x))$ is proper and contained in U. As U is affine, $f^{-1}(f(x))$ is finite (Corollary 12.89). As all non-empty fibers of f have the same dimension (Corollary 27.107), f is quasi-finite. Hence f is finite, again by Corollary 12.89.

Therefore, $\mathscr{L}^{\otimes 2}$ is the pullback of an ample line bundle by a finite morphism and hence is itself ample (Proposition 13.83). Hence $\mathscr{L}$ is ample (Proposition 13.50). $\square$

(27.33) Isogenies.

In this section, S denotes a scheme.

Definition and Proposition 27.176. *Let G and H be S-group schemes that are smooth, separated, of finite presentation[6], and with connected fibers over S. Then a homomorphism $f: G \to H$ of S-group scheme is called an* isogeny *if it satisfies the following equivalent conditions.*

(i) *The homomorphism f is quasi-finite, of finite presentation, and flat.*

(ii) *The homomorphism f is surjective and $\mathrm{Ker}(f) \to S$ is quasi-finite.*

(iii) *For all $s \in S$ one has $\dim(G_s) = \dim(H_s)$ and $\mathrm{Ker}(f_s)$ is finite over $\kappa(s)$.*

(iv) *For all $s \in S$, f_s is surjective and finite.*

Every isogeny $f: G \to H$ induces an isomorphism $G/\mathrm{Ker}(f) \xrightarrow{\sim} H$ of group schemes over S.

Proof. We recall the following facts that we have already seen.

(1) Every surjective homomorphism $f: G \to H$ is faithfully flat (Proposition 27.54). Then f is fppf-surjective and hence induces an isomorphism $G/\mathrm{Ker}(f) \xrightarrow{\sim} H$.

(2) Therefore Corollary 27.63 shows that if f is surjective, then f is quasi-finite if and only if $\mathrm{Ker}(f) \to S$ is quasi-finite.

(3) A scheme over a field is quasi-finite if and only if it is finite (Proposition 5.20).

Moreover, f is automatically of finite presentation since G and H are of finite presentation over S (Proposition 10.35).

Hence (ii) implies (i). Moreover, (ii) and (iv) are equivalent because of (3).

Let us show that (i) implies (iv). As f_s is flat, its image is open (Theorem 14.35) and dense because H_s is by hypothesis irreducible. Hence f_s is surjective (Proposition 27.14 (1)). As f is quasi-finite, each fiber f_s is quasi-finite. Then $\mathrm{Ker}(f_s) \to \mathrm{Spec}\,\kappa(s)$ is quasi-finite by (2) and hence finite by (3). Therefore f_s is finite by (2).

It remains to show that (iv) and (iii) are equivalent. For this we can assume that $S = \mathrm{Spec}\,k$ for a field k. Then (iv) implies (iii) by Proposition 27.14 (2). Conversely, if $\mathrm{Ker}(f)$ is finite over k, then f is finite by (2). Moreover $f(G)$ is a closed subspace of H of the same dimension, again by Proposition 27.14. As H is irreducible, we see $f(G) = H$. $\qquad\square$

Corollary 27.177. *Let X and Y be abelian schemes over S and let $f: X \to Y$ be a homomorphism of abelian schemes.*

(1) *If f is an isogeny, then f is surjective and finite locally free and the relative dimensions of X and of Y over S are equal.*

(2) *The homomorphism f is an isogeny if and only if any two of the following assertions hold.*

   (a) *f is finite (equivalently, f has finite fibers).*

   (b) *f is surjective.*

   (c) *The relative dimensions of X and of Y over S are equal.*

   Moreover all these assertions can be checked on fibers over S.

The *degree of f* is defined as the degree of the finite locally free morphism $\mathrm{Ker}(f) \to S$ which is a locally constant function on S.

---

[6] Actually *finite presentation* follows from the other hypotheses: As G and H are smooth and separated over S, to be in addition of finite presentation only means that $G \to S$ and $H \to S$ are also quasi-compact. But any group scheme G over a scheme S whose fibers are locally of finite type and connected is quasi-compact over S by [SGA3] Exp. VIB, 3.6.

Proof. As X and Y are proper and of finite presentation over S, so is f. Now any proper quasi-finite morphism is finite (Corollary 12.89) therefore f is finite if any only if it has finite fibers. Moreover, a finite flat morphism of finite presentation is finite locally free (Proposition 12.19). Hence Proposition 27.176 shows the first assertion and that any two of the assertions (a), (b), (c) imply that f is an isogeny. Clearly, all these assertions can be checked on fibers. $\square$

Proposition 27.178. *Let $f\colon X \to Y$ be an isogeny of abelian schemes.*
(1) *Let $g_1, g_2\colon Y \to Z$ be homomorphisms of abelian schemes such that $g_1 \circ f = g_2 \circ f$. Then $g_1 = g_2$.*
(2) *Let $h_1, h_2\colon Z \to X$ be homomorphisms of abelian schemes such that one has $f \circ h_1 = f \circ h_2$. Then $h_1 = h_2$.*

If S is connected, it suffices to assume that $(g_1)_{\bar{s}} \circ f_{\bar{s}} = (g_2)_{\bar{s}} \circ f_{\bar{s}}$ (resp. $f_{\bar{s}} \circ (h_1)_{\bar{s}} = f_{\bar{s}} \circ (h_2)_{\bar{s}}$) in a single geometric point $\bar{s} \to S$ to conclude that $g_1 = g_2$ (resp. that $h_1 = h_2$) by Proposition 27.105.

Proof. The first assertion holds since f is an epimorphism (Lemma 27.50). Now let $h := h_1 - h_2$. Then h factors through $\mathrm{Ker}(f)$ by assumption, and we have to show that $h = 0$. By Proposition 27.105 it suffices to show this on geometric fibers. Hence we may assume that $S = \mathrm{Spec}\, k$ for an algebraically closed field k.

As Z is connected and reduced, $h\colon Z \to X$ factors through a connected and reduced subscheme of $\mathrm{Ker}(f)$. Since $\mathrm{Ker}(f)$ is finite, it follows that h factors through its identity section. $\square$

Proposition 27.179. *Let X and Y be abelian varieties over a field k and let $f\colon X \to Y$ be an isogeny. Then for every $\mathscr{F}$ in $D^b_{\mathrm{coh}}(Y)$ one has*

$$(27.33.1) \qquad \chi(X, f^*\mathscr{F}) = \deg(f)\chi(Y, \mathscr{F}).$$

Proof. We denote by $\mathrm{gr}\, K_0(Z)$ the graded K-group of a smooth k-scheme Z, so that $Z \mapsto \mathrm{gr}\, K_0(Z)$ is endowed with the structure of an additive cohomology theory (Example 23.109): For any morphism $g\colon Z \to Z'$ of smooth k-schemes, $\mathscr{E} \mapsto Lf^*\mathscr{E}$ induces a ring homomorphism $f^*\colon \mathrm{gr}\, K_0(Z') \to \mathrm{gr}\, K_0(Z)$. For a proper morphism $g\colon Z \to Z'$ of smooth k-schemes we denote by $g_*\colon \mathrm{gr}\, K_0(Z) \to \mathrm{gr}\, K_0(Z')$ the push forward (23.11.4), which is a homomorphism of abelian groups. Finally, there is a Chern character (23.23.3)

$$\mathrm{ch}\colon K_0(-) \to \mathrm{gr}\, K_0(-)_{\mathbb{Q}},$$

which is compatible with pullback.

As tangent bundles of abelian varieties are trivial, the relative tangent bundle $T_f = [T_{X/k} - f^*T_{Y/k}] \in K_0(X)$ is trivial. Hence for every $\mathscr{E}$ in $D^b(X)$ the Grothendieck-Riemann-Roch theorem (23.23.5) yields that $\mathrm{ch}(Rf_*\mathscr{E}) = f_*(\mathrm{ch}(\mathscr{E}))$. Applying this to $\mathscr{E} = f^*\mathscr{F} = Lf^*\mathscr{F}$ we obtain

$$(27.33.2) \qquad \begin{aligned} \mathrm{ch}(Rf_*f^*\mathscr{F}) &\overset{(1)}{=} f_*\,\mathrm{ch}(f^*\mathscr{F}) = f_*f^*(\mathrm{ch}(\mathscr{F})) \\ &\overset{(2)}{=} \mathrm{ch}(\mathscr{F})f_*f^*(\mathscr{O}_Y) = \mathrm{ch}(\mathscr{F})f_*\mathscr{O}_X \\ &\overset{(3)}{=} \deg(f)\,\mathrm{ch}(\mathscr{F}). \end{aligned}$$

where (1) holds by the Grothendieck-Riemann-Roch theorem, Theorem 23.112, (2) by the projection formula (23.20.1), and (3) holds by Corollary 23.49 since $f_*\mathcal{O}_X$ is a finite locally free $\mathcal{O}_Y$-module of degree $\deg(f)$.

Let $h\colon Y \to \operatorname{Spec} k$ be the structure morphism. Multiplying (27.33.2) with $\operatorname{td}(T_{Y/k})^7$ and applying h_* we obtain by Corollary 23.113 the equality of

$$\deg(f)h_*(\operatorname{td}(T_{Y/k})\operatorname{ch}(\mathscr{F})) = \deg(f)\chi(Y,\mathscr{F})$$

and of

$$h_*(\operatorname{td}(T_{Y/k})\operatorname{ch}(Rf_*f^*\mathscr{F})) = \chi(Y, Rf_*f^*\mathscr{F}) = \chi(X, f^*\mathscr{F}),$$

where the last equality holds by Corollary 23.63. $\square$

The proof would have also worked verbatim by using the Chow ring (which we did not introduce) as an additive cohomology theory.

Remark 27.180. The proof shows that (27.33.1) holds more generally if $f\colon X \to Y$ is a finite locally free morphism of projective smooth schemes over a field k such that the classes of $f^*\Omega^1_{Y/k}$ and of $\Omega^1_{X/k}$ in K-theory $K_0(X)$ agree. These properties are for instance satisfied if f is finite étale because then $f^*\Omega^1_{Y/k} \cong \Omega^1_{X/k}$ (Corollary 18.19).

(27.34) The Frobenius isogeny.

Let S be a scheme of characteristic $p > 0$, i.e. $p\mathcal{O}_S = 0$. Recall from Definition 4.24 the absolute Frobenius and the relative Frobenius. The absolute Frobenius $\operatorname{Frob}_S\colon S \to S$ is the identity on underlying topological spaces and raises local sections to their p-th power. If $S = \operatorname{Spec} R$ for a ring R, then Frob_S corresponds to the usual Frobenius $\operatorname{Frob}_R\colon R \to R$ that sends a to a^p.

If $X \to S$ is a morphism of schemes, we set $X^{(p)} := X \times_{S,\operatorname{Frob}_S} S$. This defines a functor $X \mapsto X^{(p)}$ from the category of S-schemes to itself that commutes with fiber products over S. Hence it transforms S-group schemes over S into S-group schemes.

Let $F_{X/S}\colon X \to X^{(p)}$ be the relative Frobenius, i.e., the unique morphism of S-schemes that makes the diagram

(27.34.1)

commutative. The relative Frobenius $F_{X/S}$ is functorial in the S-scheme X. Hence if G is an S-group scheme, $F_{G/S}\colon G \mapsto G^{(p)}$ is a homomorphism of S-group schemes.

Example 27.181. Let S be a scheme of characteristic p and let G be an S-group scheme. Suppose that G is defined over $\mathbb{F}_p$, i.e., there exists a group scheme G_0 over $\mathbb{F}_p$ such that $G_0 \times_{\mathbb{F}_p} S \cong G$. Then

---

[7] This is not really necessary here because the tangent bundle of the abelian variety Y is trivial and hence $\operatorname{td}(T_{Y/k}) = 1$; it is only included in the proof for Remark 27.180 below.

$$G^{(p)} = G \times_{S,\mathrm{Frob}_S} S = G_0 \times_{\mathbb{F}_p} S \times_{S,\mathrm{Frob}_S} S$$
$$= G_0 \times_{\mathbb{F}_p,\mathrm{Frob}_{\mathbb{F}_p}} \mathrm{Spec}(\mathbb{F}_p) \times_{\mathbb{F}_p} S = G_0 \times_{\mathbb{F}_p} S = G$$

because the Frobenius on $\mathbb{F}_p$ is the identity. In this case, the relative Frobenius is an endomorphism. It is the base change of the absolute Frobenius on G_0, which is an endomorphism of group schemes over $\mathbb{F}_p$. Consider the following examples.

(1) $G = \mathbb{G}_{a,S}$ i.e. $\mathbb{G}_a(T) = \Gamma(T, \mathscr{O}_T)$ for every S-scheme T. In this case the relative Frobenius is given by

$$F_{\mathbb{G}_{a,S}/S}(T) \colon \mathbb{G}_{a,S}(T) \to \mathbb{G}_{a,S}(T), \qquad x \mapsto x^p.$$

(2) $G = \mathbb{G}_{m,S}$, i.e. $\mathbb{G}_m(T) = \Gamma(T, \mathscr{O}_T)^\times$ for every S-scheme T. In this case the relative Frobenius is again given by $x \mapsto x^p$.

(3) Let H be any abstract group and let $G = \underline{H}_S$ the corresponding constant S-group scheme, i.e., $G(T)$ consists of the group of locally constant maps $T \to H$. Then the relative Frobenius is the identity.

Proposition 27.182. *Let p be a prime number and let S be a scheme with $p\mathscr{O}_S = 0$. Let $G \to S$ be a separated smooth group scheme of finite presentation with connected fibers. Then the relative Frobenius $F_G \colon G \to G^{(p)}$ is an isogeny. Its kernel is finite locally free over S of degree $s \mapsto p^{\dim G_s}$.*

Proof. The relative Frobenius F_G is an isogeny by Exercise 18.25. In particular, the kernel of F_G is separated, quasi-finite, flat and of finite presentation over S. We will use Exercise 20.17 and show that its degree is equal to the function $s \mapsto p^{\dim G_s}$ which is locally constant by Proposition 27.27. But this follows again from Exercise 18.25 since $\dim(G_s)$ is the relative dimension of $G \to S$ in every point of G that is mapped to s. $\square$

(27.35) Torsion points.

In this section S denotes a scheme. Let X be an abelian fppf-sheaf over S. For all $n \in \mathbb{Z}$ we denote by

(27.35.1) $\qquad [n] = [n]_X \colon X \to X, \qquad X(T) \ni x \mapsto nx \in X(T), \qquad T$ an S-scheme

the multiplication by n. It is a homomorphism of abelian fppf-sheaves over S.

From now on let X and Y denote abelian schemes over S. Then $[n]$ is a homomorphism of abelian schemes. For $n = 0$ it is the composition $X \to S \xrightarrow{0} X$, where $X \to S$ is the structure morphism. For $n = 1, -1$ it is an isomorphism. For all $n \in \mathbb{Z}$ we denote by

(27.35.2) $\qquad\qquad\qquad X[n] := \mathrm{Ker}([n] \colon X \to X)$

its kernel, i.e. $X[n](T) = \{\, x \in X(T) \; ; \; nx = 0 \,\}$ for all S-schemes T. This is a closed subgroup scheme of X.

Remark 27.183. For $\mathscr{L} \in X^t(S) \subseteq \mathrm{Pic}(X)/\mathrm{Pic}(S)$ by (27.29.10) we have the equality $[n]^* \mathscr{L} = \mathscr{L}^{\otimes n}$ for all $n \in \mathbb{Z}$, i.e.

(27.35.3) $\qquad\qquad\qquad [n]^t_X = [n]_{X^t}$

Proposition 27.184. *Let X be an abelian scheme over S, let $\mathscr{L}$ be a line bundle on X, and let $n \in \mathbb{Z}$.*

(1) *One has an equality*

$$[n]^*\mathscr{L} = \mathscr{L}^{\otimes (n^2+n)/2} \otimes [-1]^*\mathscr{L}^{\otimes (n^2-n)/2}$$

in $\mathrm{Pic}_{X/S}(S)$, i.e., both sides are isomorphic line bundles on X up to multiplication by a line bundle that is obtained by pullback from a line bundle on S.

(2) *One has $[-1]^*\mathscr{L} \otimes \mathscr{L}^{-1} \in X^t(S)$.*

(3) *One has in $\mathrm{Pic}_{X/S}(S)$ that*

$$[n]^*\mathscr{L} = \mathscr{L}^{\otimes n^2} \otimes \mathscr{N}, \qquad \text{for some } \mathscr{N} \in X^t(S).$$

Proof. Let us show (1). The equality is clear for $n = 0, 1$. For general n we use Proposition 27.167 with $x_1 := [n+1]$, $x_2 := [1] = \mathrm{id}_X$ and with $x_3 := [-1]$ and get

$$\mathscr{O}_X \cong [n+1]^*\mathscr{L} \otimes [n+2]^*\mathscr{L}^{-1} \otimes [n]^*\mathscr{L}^{-1} \otimes [0]^*\mathscr{L}^{-1}$$
$$\otimes [n+1]^*\mathscr{L} \otimes \mathscr{L} \otimes [-1]^*\mathscr{L} \otimes [0]^*\mathscr{L}^{-1}.$$

Hence we get in $\mathrm{Pic}_{X/S}(S)$ the equality

$$[n+2]^*\mathscr{L} \otimes [n+1]^*\mathscr{L}^{-1} = ([n+1]^*\mathscr{L} \otimes [n]^*\mathscr{L}^{-1}) \otimes \mathscr{L} \otimes [-1]^*\mathscr{L}.$$

Arguing by increasing and by decreasing induction on n we obtain the claim.

Next we show (2). For every S-scheme T and every $x \in X(T)$ we have in $\mathrm{Pic}_{X/S}(T)$

$$\begin{aligned}
\varphi_{[-1]^*\mathscr{L}}(x) &= t_x^*[-1]^*\mathscr{L}_T \otimes [-1]^*\mathscr{L}_T^{-1} \\
&= [-1]^*(t_{-x}^*\mathscr{L}_T \otimes \mathscr{L}_T^{-1}) \\
&= t_{-x}^*\mathscr{L}_T^{-1} \otimes \mathscr{L}_T \\
&= \varphi_{\mathscr{L}^{-1}}(-x) = \varphi_{\mathscr{L}}(x),
\end{aligned}$$

where the third equality holds by Remark 27.183 since $t_{-x}^*\mathscr{L}_T \otimes \mathscr{L}_T^{-1} = \varphi_{\mathscr{L}}(-x)$ lies in $X^t(T)$ (Proposition 27.163 (3)). This shows $\varphi_{[-1]^*\mathscr{L} \otimes \mathscr{L}^{-1}} = 0$ by Remark 27.169 (1).

Now (3) follows because (1) shows that

$$[n]^*\mathscr{L} = \mathscr{L}^{\otimes n^2} \otimes \mathscr{N}, \qquad \text{with } \mathscr{N} := (\mathscr{L}^{-1} \otimes [-1]^*\mathscr{L})^{\otimes (n^2-n)/2}.$$

and $\mathscr{N} \in X^t(S)$ by (2). $\qquad \square$

Definition and Remark 27.185. Let $\mathscr{L}$ be a line bundle on an abelian scheme X over S. Then $\mathscr{L}$ is called *symmetric* if $\mathscr{L} = [-1]^*\mathscr{L}$ in $\mathrm{Pic}_{X/S}(S)$. For symmetric line bundles $\mathscr{L}$, Proposition 27.184 means that in $\mathrm{Pic}_{X/S}(S)$ one has

$$[n]^*\mathscr{L} = \mathscr{L}^{\otimes n^2}.$$

It is easy to produce symmetric line bundles: If $\mathscr{M}$ is any line bundle on X, then $\mathscr{M} \otimes [-1]^*\mathscr{M}$ is symmetric. If $\mathscr{M}$ is S-ample, then $[-1]^*\mathscr{M}$ is S-ample since $[-1]$ is an isomorphism. Therefore $\mathscr{M} \otimes [-1]^*\mathscr{M}$ is a symmetric S-ample line bundle.

Proposition 27.186. *Let X be an abelian scheme over S. For an integer $n \neq 0$, the multiplication $[n]: X \to X$ is an isogeny. In particular, $X[n]$ is a finite locally free S-group scheme. Let g be the relative dimension of X over S, considered as locally constant function on S. Then $\deg([n]) = n^{2g}$.*

Proof. By Proposition 27.176 we may assume that $S = \operatorname{Spec} k$ for a field k and it suffices to show that $X[n]$ is finite and $\deg([n]) = n^{2g}$. As X is projective over k (Proposition 27.174), we may choose a symmetric ample line bundle $\mathscr{L}$ on X (Remark 27.185).

Restricting $[n]^*\mathscr{L} \cong \mathscr{L}^{\otimes n^2}$ to $X[n]$ yields that the ample line bundle $\mathscr{L}^{\otimes n^2}{}_{|X[n]}$ is trivial. Hence $X[n]$ is quasi-affine over k and thus finite over k (Corollary 13.82).

It remains to show that $\deg([n]) = n^{2g}$. From Proposition 23.84 we obtain

$$\deg([n])\deg(\mathscr{L}) = \deg([n]^*\mathscr{L}) = \deg(\mathscr{L}^{n^2}) = n^{2g}\deg(\mathscr{L}),$$

where the last equality holds by Remark 23.85. As $\mathscr{L}$ is ample, one has $\deg(\mathscr{L}) > 0$ (Remark 23.82). This shows that $\deg([n]) = n^{2g}$. $\qquad\square$

If n is invertible on S (e.g., if S is a scheme over $\mathbb{Q}$), then a much more elementary argument shows a much stronger result:

Proposition 27.187. *Let $n \in \mathbb{Z}$ be invertible on S. Then $[n]\colon X \to X$ is finite étale. In particular $X[n]$ is finite étale over S.*

Proof. By Corollary 18.77 we may assume that $S = \operatorname{Spec} k$ for a field k that we may assume to be algebraically closed by flat descent. To see that $[n]$ is étale it suffices to show by Theorem 18.74 that $[n]$ induces an isomorphism on tangent spaces. By homogeneity, it suffices to show that $\operatorname{Lie}([n])\colon \operatorname{Lie}(X) \to \operatorname{Lie}(X)$ is an isomorphism. But $\operatorname{Lie}([n])$ is simply the multiplication by n (Remark 27.18 (3)) which is by hypothesis bijective. Since $[n]$ is proper as a morphism between proper schemes, it is finite étale. $\qquad\square$

Proposition 27.188. *Let X be an abelian scheme over S of relative dimension g.*
(1) *Let n be an integer that is invertible on S. Then there exists a finite étale surjective morphism $S' \to S$ such that $X[n]_{S'} \cong (\mathbb{Z}/n\mathbb{Z})^{2g}_{S'}$.*
(2) *Let $S = \operatorname{Spec} k$ be a field of characteristic $p > 0$. Then there exists an integer $f(X)$ with $0 \le f(X) \le g$ such that for all $m \ge 1$ and for every separably closed extension field K of k one has $X[p^m](K) = (\mathbb{Z}/p^m\mathbb{Z})^{f(X)}$.*

The structure of the finite group scheme $X[p^m]$ over an algebraically closed field of characteristic p can be quite complicated and is beyond the scope of this book. We will prove and use in the sequel only (1). For (2) we refer to [Mum1] §15.

Proof. [of (1)] By Proposition 27.186 and Proposition 27.187 we know that $X[n]$ is finite étale over S of degree n^{2g}. By Proposition 20.69 there exists a finite étale surjective morphism $S' \to S$ such that $X[n]_{S'} \cong G_{S'}$ for some finite abelian group G of order n^{2g} with $nG = 0$. Moreover, for every divisor d of n, the d-torsion $G[d]$ has the same properties, i.e., it is of order d^{2g} and $dG[d] = 0$. By the classification of finite abelian groups this implies hat $G \cong (\mathbb{Z}/n\mathbb{Z})^{2g}$. $\qquad\square$

In particular, if $S = \operatorname{Spec} R$ for a strictly henselian local ring R (e.g., if R is a separably closed field) and if n is prime to the characteristic of the residue field of R, then $X[n] \cong (\mathbb{Z}/n\mathbb{Z})^{2g}_S$.

Definition and Remark 27.189. Let p be a prime number and let X be an abelian variety over a field k of characteristic p. The integer $p\text{-rk}(X) := r$ in Proposition 27.188 (2) is called the *p-rank of X*.

One can show that for every algebraically closed field of characteristic p and every integer $0 \leq f \leq g$ there exist abelian varieties of dimension g over k whose p-rank is f. An abelian variety of dimension g over a field of characteristic p is called *ordinary* if its p-rank is g, i.e., maximal possible.

Let S be a scheme such that for all $s \in S$ one has $\mathrm{char}(\kappa(s)) = p$. Let X be an abelian scheme over S. Then one can show (Exercise 27.13) that the p-rank function

$$S \longrightarrow \mathbb{Z}, \qquad s \mapsto p\text{-}\mathrm{rk}(X_s)$$

is lower semicontinuous and constructible. In particular, if X is of relative dimension g, then the ordinary locus $\{\, s \in S \; ; \; p\text{-}\mathrm{rk}(X_s) = g \,\}$ is open and constructible in S.

An elliptic curve whose p-rank is 0 is called *supersingular*. There is also the notion of a supersingular abelian variety of higher dimension which is more complicated to define. All supersingular abelian varieties have p-rank 0 but the converse does not hold.

Proposition 27.190. *Let S be a scheme, let $n \geq 1$ be an integer, and let $f\colon X \to Y$ be an isogeny of degree n of abelian schemes over S. Then there exists a unique homomorphism $g\colon Y \to X$ such that $g \circ f = [n]_X$. Moreover, g is an isogeny and $f \circ g = [n]_Y$.*

Proof. As $\mathrm{Ker}(f)$ is a finite locally free group scheme of rank n, it is annihilated by n (Proposition 27.86). Hence $[n]_X$ factors as $[n]_X = g \circ f$ for some homomorphism g. By Proposition 27.178 (1), g is unique. As $[n]_X$ is surjective (Proposition 27.186), g is surjective. As f is an isogeny, X and Y have the same relative dimension over S, hence g is an isogeny (Corollary 27.177). Finally, we have

$$g \circ [n]_Y = [n]_X \circ g = g \circ (f \circ g),$$

where the first equality holds since g is a homomorphism of group schemes. Hence $f \circ g = [n]_Y$ by Proposition 27.178 (2). □

(27.36) Fundamental groups of abelian varieties.

In this section we will compute the étale fundamental group of an abelian variety. If A is an abelian variety over the complex numbers $\mathbb{C}$, then $A(\mathbb{C})$ is isomorphic as a complex manifold to a complex torus, i.e., a quotient $\mathbb{C}^g/\Lambda$, where $\Lambda \subset \mathbb{C}^g$ is a lattice (a free $\mathbb{Z}$-module of rank $2g$ with $\Lambda \otimes_{\mathbb{Z}} \mathbb{R} = \mathbb{C}^g$). Thus the universal cover of $A(\mathbb{C})$ in the sense of algebraic topology is $\mathbb{C}^g \to A(\mathbb{C})$ and the topological fundamental group is $\pi_1(A(\mathbb{C}), 0) = \Lambda$ (and in particular is abelian). We conclude that the multiplication by n morphisms $A(\mathbb{C}) \to A(\mathbb{C})$, which are unramified covers with deck transformation group $(\mathbb{Z}/n\mathbb{Z})^{2g}$ by Proposition 27.188, form a cofinal system of unramified connected covers. The latter result also makes sense algebraically, and we will show that the étale fundamental group of an abelian variety A over a field k is isomorphic to the limit $\lim A[n](k^s)$, where k^s is a separable closure of A (Proposition 27.194). The crucial point in the proof is Theorem 27.192 which shows that for any finite étale surjective k-morphism $f\colon B \to A$ together with a choice of k-valued point $e \in f^{-1}(0)(k)$, there exists a unique structure of abelian variety on B with neutral element e (and hence f is a homomorphism of abelian varieties).

Lemma 27.191. *Let $f\colon X \to S$ be a smooth proper morphism of connected schemes that has a section σ. Then all fibers of f are geometrically integral.*

The proof will show that if f is any flat, proper morphism of finite presentation with geometrically reduced fibers between connected schemes and if f has a section, then f has geometrically connected fibers.

Proof. Let $X \xrightarrow{f'} S' \xrightarrow{\pi} S$ be the Stein factorization of f. As X is connected and f' is surjective, S' is connected. Then π is finite étale by Theorem 24.61. Moreover, $\tau := f' \circ \sigma$ is a section of π and hence τ is an open and closed immersion (Remark 20.66 (5)). As S' is connected, τ is necessarily an isomorphism. Therefore $f = f'$ has geometrically connected fibers. As f is smooth, all fibers are hence geometrically integral. $\qquad\square$

Theorem 27.192. *Let S be a scheme, let $X \to S$ be an abelian scheme, let $Y \to S$ be an S-scheme with geometrically connected fibers, and let $f \colon Y \to X$ be a finite étale surjective S-morphism, and let $e_Y \in Y(S)$ be a section that is mapped by f to the zero section e_X of X.*

Then there exists on Y a unique structure of an abelian scheme over S such that e_Y is the zero section and such that f is an isogeny.

Proof. We use Proposition 27.109. Hence we have to construct a morphism $m_Y \colon Y \times_S Y \to Y$ of S-morphism for which e_Y is a left and a right unit.

As $Y \to S$ is the composition $Y \xrightarrow{f} X \to S$, the structure morphism $Y \to S$ is smooth and proper. Hence all fibers are geometrically integral. Moreover, $Y \to S$ is surjective because it has a section.

If the desired structure of an abelian scheme exists on Y, it is necessarily unique by Corollary 27.104. We may therefore work locally on S and hence can assume that $S = \operatorname{Spec} R$ is affine. By noetherian approximation we may assume that R is noetherian. Then $\operatorname{Spec} R$ is locally connected and hence equals the direct sum of its connected components. Therefore we can assume that S is connected. Then X^I and Y^I are connected as well for every finite set I, as the structure morphisms are proper surjective with geometrically connected fibers (Lemma 24.52).

Let $\Gamma_X \subseteq X \times_S X \times_S X$ be the graph of the group law on X and let $\Gamma_Y' \subseteq Y \times_S Y \times_S Y$ be the inverse image of Γ_X under $f \times f \times f$. As S is connected, there exists a connected component Γ_Y of Γ_Y' containing the image of (e_Y, e_Y, e_Y).

We claim that Γ_Y is the graph of a morphism $m_Y \colon Y \times_S Y \to Y$. For $I \subseteq \{1,2,3\}$ denote by $p_I \colon \Gamma_X \to X^I$ and by $q_I \colon \Gamma_Y \to Y^I$ the projections. To prove the claim we have to show that q_{12} is an isomorphism. Then we can set $m_Y := q_3 \circ q_{12}^{-1}$.

The morphism q_{12} has sections s_1 over $e_Y(S) \times Y$ and s_2 over $Y \times e_Y(S)$ given on points by $s_1(e_Y, y) = (e_Y, y, y)$ and $s_2(y, e_Y) = (y, e_Y, y)$. So once we have proved the claim, we have $m_Y(e_Y, y) = y = m_Y(y, e_Y)$ and therefore we can apply Proposition 27.109. Moreover, f respects these group laws as by construction $f \times f \times f$ maps the graph of m_Y to the graph of m_X.

To show that q_{12} is an isomorphism, consider the commutative diagram

$$
\begin{array}{ccc}
\Gamma_Y & \longrightarrow & \Gamma_X \\
\Big\downarrow{\scriptstyle q_{12}} & & \Big\downarrow{\scriptstyle p_{12}} \\
Y \times_S Y & \xrightarrow{\ f \times f\ } & X \times_S X.
\end{array}
$$

The horizontal maps are finite étale and p_{12} is an isomorphism. Hence q_{12} is finite étale (Remark 20.66 (3)). As $Y \times_S Y$ is connected and finite étale morphisms have open and closed images, q_{12} is surjective. It suffices to show that q_{12} has degree 1. As the degree is locally constant and as $Y \times_S Y$ is connected, it suffices to show that the restriction $r \colon Z := q_{12}^{-1}(Y \times e_Y(S)) \to Y \times_S e_Y(S)$ of q_{12} is an isomorphism.

The projection $q_2 \colon \Gamma_Y \to Y$ is the composition of q_{12} and of the second projection $Y \times_S Y$, hence it is proper and smooth. It has a section given on points by $y \mapsto (e_Y, y, y)$. Hence we can apply Lemma 27.191 to see that q_2 has geometrically integral fibers. In particular $Z = q_2^{-1}(e_Y(S))$ is connected (Lemma 24.52). But the étale covering r has a section given on points by $(y, e_Y) \mapsto (y, e_Y, y)$ which is an open and closed immersion (Remark 20.66 (5)) and hence an isomorphism. Therefore r is an isomorphism. $\qquad\square$

Remark and Definition 27.193. Let k be a field, let $\bar{k}$ be an algebraic closure of k, and let k^s be the separable closure of k in $\bar{k}$. Let X be an abelian variety of dimension g over k. Let $n \geq 1$ be an integer. Then $G = \mathrm{Gal}(k^s/k)$ acts on $X[n](k^s)$ continuously by functoriality. If n is invertible in k, we have $X[n](k^s) \cong (\mathbb{Z}/n\mathbb{Z})^{2g}$ by Proposition 27.188. If $\mathrm{char}(k) = p > 0$, then $X[p^m](k^s) \cong (\mathbb{Z}/p^m\mathbb{Z})^f$, where $0 \leq f \leq g$ is the p-rank of X (Definition 27.189). By forming the limit over all n (with respect to the divisibility order), we obtain a topological module

$$T(X) := \lim_n X[n](k^s)$$

over the ring $\hat{\mathbb{Z}} := \lim_n \mathbb{Z}/n\mathbb{Z}$ with a continuous G-action. It is called the *total Tate module of X*. By the Chinese remainder theorem one has

$$T(X) = \prod_\ell T_\ell(X),$$

where ℓ runs over all prime numbers. Here

$$T_\ell(X) = \lim_m X[\ell^m](k^s)$$

is the *ℓ-adic Tate module*. It is a $\mathbb{Z}_\ell$-module endowed with a continuous G-action, where $\mathbb{Z}_\ell = \lim_n \mathbb{Z}/\ell^n\mathbb{Z}$ denotes the ring of ℓ-adic integers.

If $\ell \neq \mathrm{char}(k)$, then $T_\ell(X)$ is a free $\mathbb{Z}_\ell$-module of rank $2g$. If $\mathrm{char}(k) = p > 0$, then $T_p(X)$ is a free $\mathbb{Z}_p$-module of rank f.

Proposition 27.194. *Let k be a separably closed field and let X be an abelian variety over k. Then there is an isomorphism of abelian pro-finite groups*

$$\pi_1(X, 0) \cong T(X).$$

Proof. By Proposition 20.105 we may assume that k is algebraically closed. If $f \colon Y \to X$ is an étale cover of degree n with Y connected, we can choose $0_Y \in Y(k)$ that is mapped to the zero section of X. By Theorem 27.192 we can equip Y with the structure of an abelian variety such that f becomes an isogeny of degree n. By Proposition 27.190 there exists an isogeny $g \colon X \to Y$ such that $f \circ g = [n]_X$.

By Exercise 27.12 we have $X[n] = X[n]^0 \times X[n]_{\mathrm{red}}$ with $X[n]_{\mathrm{red}}$ finite constant and $X[n](k) = X[n]_{\mathrm{red}}(k)$. Now $g(X[n]_0) \subseteq \mathrm{Ker}(f)$ is connected and $\mathrm{Ker}(f)$ is constant. Hence $X[n]_0 \subseteq \mathrm{Ker}(g)$. Therefore we find a factorization

$$[n]_0 \colon X/X[n]^0 \longrightarrow Y \xrightarrow{f} X,$$

where the composition $[n]_0$ is induced by the multiplication by n. Its kernel is the finite étale group scheme $X[n]_{\mathrm{red}}$, hence $[n]_0$ is finite étale (Corollary 27.63). This shows that the multiplication maps $[n]_0 \colon X/X[n]^0 \to X$ are cofinal in the category of all connected étale covers and hence

$$\pi_1(X,0) \cong \lim_n X[n]_{\mathrm{red}}(k) = \lim_n X[n](k) = T(X). \qquad \square$$

By Theorem 20.115 and Proposition 20.117 we obtain the following corollaries.

Corollary 27.195. *Let k be a field and let X be an abelian variety over k. Then via functoriality one obtains a split exact sequence of profinite groups*

$$0 \longrightarrow T(X) \longrightarrow \pi_1(X,0) \longrightarrow \mathrm{Gal}(k^s/k) \longrightarrow 0.$$

The splitting is given by the zero section (Remark 20.116) and one obtains an isomorphism of profinite groups $\pi_1(X,0) \cong T(X) \rtimes \mathrm{Gal}(k^s/k)$, where the action of $\mathrm{Gal}(k^s/k)$ on $T(X)$ is the action defined in Definition 27.193.

From Proposition 20.117 we now immediately obtain the following corollary.

Corollary 27.196. *Let S be a connected scheme, let $\bar{s}$ be a geometric point of S and let $\bar{x}$ be its image in X under the zero section of X. Then there is an exact sequence of profinite groups*

$$T(X_{\bar{s}}) \longrightarrow \pi_1(X,\bar{x}) \longrightarrow \pi_1(S,\bar{s}) \longrightarrow 1.$$

(27.37) The dual abelian scheme for projective abelian schemes.

Lemma 27.197. *Let k be a field and let X be an abelian scheme over k. Let $\mathscr{L} \in X^t(k)$. Then $\mathscr{L}$ is non-trivial if and only if $H^i(X,\mathscr{L}) = 0$ for all $i \geq 0$.*

Proof. Since $H^0(X,\mathscr{O}_X) = k$, the condition is clearly sufficient. Let $\mathscr{L}$ be non-trivial and let us show $H^i(X,\mathscr{L}) = 0$ by induction on i.

As $\mathscr{L} \in X^t(k)$ we have $[-1]^*\mathscr{L} \cong \mathscr{L}^{-1}$ (27.35.3). Hence if $H^0(X,\mathscr{L}) \neq 0$, then also $H^0(X,\mathscr{L}^{-1}) \neq 0$. This implies that $\mathscr{L}$ is trivial by Lemma 24.65. Now let $i \geq 1$ and assume by induction that

$$H^0(X,\mathscr{L}) = H^1(X,\mathscr{L}) = \cdots = H^{i-1}(X,\mathscr{L}) = 0.$$

Since id_X can be factorized as

$$X \xrightarrow{(\mathrm{id},0)} X \times X \xrightarrow{m} X$$

we see that $H^i(X,\mathscr{L})$ is embedded into $H^i(X \times X, m^*\mathscr{L})$. As $\mathscr{L} \in X^t(k)$, we have $m^*\mathscr{L} = p_1^*\mathscr{L} \otimes p_2^*\mathscr{L}$ and therefore the Künneth formula (Corollary 22.110) shows

$$H^i(X \times X, m^*\mathscr{L}) = \bigoplus_{j=0}^{i} H^j(X,\mathscr{L}) \otimes H^{i-j}(X,\mathscr{L}) = 0. \qquad \square$$

Theorem and Definition 27.198. *Let S be a scheme, let X be an abelian scheme over S and let $\mathscr{L}$ be an S-ample line bundle on X. Then the following assertions hold.*

(1) *The homomorphism $\varphi_{\mathscr{L}} \colon X \to X^t$ is finite locally free and surjective. In particular, $K(\mathscr{L})$ is a finite locally free S-group scheme.*

(2) *The dual abelian space X^t is representable by an abelian scheme with $\dim(X^t/S) = \dim(X/S)$ and we have $X^t = \mathrm{Pic}^0_{X/S}$.*

We call X^t the dual abelian scheme.

We have already seen that on an abelian scheme there exists an ample line bundle if $S = \mathrm{Spec}\,k$ for a field k (Proposition 27.174).

Proof. Let us first assume that $S = \mathrm{Spec}\,k$ for a field k. In this case we know that X^t is a scheme by Theorem 27.47.

(I). Suppose in addition that k is algebraically closed. We claim that the group homomorphism $\varphi_{\mathscr{L}}(k) \colon X(k) \to X^t(k)$ is surjective.

Assume there exists $\mathscr{N} \in X^t(k)$ that is not in the image of $\varphi_{\mathscr{L}}(k)$. Consider the line bundle

$$\mathscr{M} := \Lambda(\mathscr{L}) \otimes p_1^* \mathscr{N}^{-1}$$

on $X \times_k X$. For $x \in X(k)$ one has

$$\mathscr{M}_{|X \times \{x\}} = t_x^* \mathscr{L} \otimes \mathscr{L}^{-1} \otimes \mathscr{N}^{-1}, \qquad \mathscr{M}_{|\{x\} \times X} = t_x^* \mathscr{L} \otimes \mathscr{L}^{-1}.$$

In particular, $\mathscr{M}_{|X \times \{x\}} \in X^t(k)$ is non-trivial for all $x \in X(k)$.

By Lemma 27.197 we find

$$H^j(X, \mathscr{M}_{|X \times \{x\}}) = 0 \qquad \text{for all } x \in X(k) \text{ and all } j \geq 0.$$

Therefore $R^j p_{2,*} \mathscr{M} = 0$ for all $j \geq 0$ (Proposition 23.142). The Leray spectral sequence $H^i(X, R^j p_{2,*} \mathscr{M}) \Rightarrow H^{i+j}(X \times X, \mathscr{M})$ (Corollary 21.45) shows that $H^j(X \times X, \mathscr{M}) = 0$ for all $j \geq 0$.

If $x \in X(k)$, then Lemma 27.197 shows that

(*) $\qquad\qquad H^j(X, \mathscr{M}_{|\{x\} \times X}) = 0 \quad \text{for all } j \qquad \Leftrightarrow \qquad x \notin K(\mathscr{L})(k).$

Therefore $R^j p_{1,*} \mathscr{M}$ has support in $K(\mathscr{L})$, which is a finite k-scheme because $\mathscr{L}$ is ample (Corollary 27.172). Therefore $H^i(X, R^j p_{1,*} \mathscr{M}) = 0$ for $i > 0$. The Leray spectral sequence for p_1 then shows for all $j \geq 0$ that

$$H^0(X, R^j p_{1,*} \mathscr{M}) = H^j(X \times X, \mathscr{M}) = 0$$

which implies that $R^j p_{1,*} \mathscr{M} = 0$ for all j because $R^j p_{1,*} \mathscr{M}$ is coherent and has support on the affine scheme $K(\mathscr{L})$. Hence $H^j(X, \mathscr{M}_{|\{x\} \times X}) = 0$ for all j and for all $x \in X(k)$. This is a contradiction to (*), since $\mathscr{M}_{|\{0\} \times X} \cong \mathcal{O}_X$.

(II). Now let k be an arbitrary field. We show that X^t is smooth over k and that $\varphi_{\mathscr{L}}$ is finite locally free and surjective.

Indeed, by fpqc descent we may assume that k is algebraically closed. We know by Corollary 27.172 that the kernel $K(\mathscr{L})$ of $\varphi_{\mathscr{L}}$ is finite. Therefore $\dim X = \dim X^t$ because (I) shows that $\varphi_{\mathscr{L}}$ is surjective. Moreover, we have

(27.37.1) $\qquad\qquad H^1(X, \mathcal{O}_X) = \mathrm{Lie}(\mathrm{Pic}_{X/S}) = \mathrm{Lie}(X^t).$

Here the first identity holds by Proposition 27.122 and the second identity because X^t is open in $\mathrm{Pic}_{X/S}$ (Lemma 27.160). We deduce

$$(**) \qquad \dim X^t \leq \dim \mathrm{Lie}(X^t) = \dim H^1(X, \mathscr{O}_X) \leq \dim(X) = \dim X^t,$$

where the first inequality is (27.4.7) and the second inequality holds by Corollary 27.80. Hence we have equality everywhere in (**). This shows that X^t is smooth (Corollary 27.21) and that

$$(27.37.2) \qquad \dim H^1(X, \mathscr{O}_X) = \dim(X).$$

As X^t is smooth and $\varphi_{\mathscr{L}}$ is surjective, $\varphi_{\mathscr{L}}$ is faithfully flat (Proposition 27.54). As its kernel is finite, $\varphi_{\mathscr{L}}$ is finite and hence finite locally free.

From now on let S be an arbitrary scheme.

(III). The surjectivity of $\varphi_{\mathscr{L}}$ on fibers implies that $\varphi_{\mathscr{L}}$ is surjective. As X is flat over S and $\varphi_{\mathscr{L}}$ is flat on fibers, $\varphi_{\mathscr{L}}$ is flat and X^t is flat over S by the fiber criterion for flatness (Corollary 14.27, [Sta] 05X1 for algebraic spaces). As $K(\mathscr{L})$ is finite (Lemma 27.172), $\varphi_{\mathscr{L}}$ is finite (Corollary 27.63) and hence finite locally free (Proposition 12.19). Therefore $K(\mathscr{L})$ is finite locally free over S.

In particular $\varphi_{\mathscr{L}}$ is fppf-surjective and therefore identifies X^t with the fppf-quotient $X/K(\mathscr{L})$. Hence X^t is a scheme over S (Theorem 27.68) and $\varphi_{\mathscr{L}}$ makes X into a $K(\mathscr{L})$-torsor for the fppf-topology over X^t.

Moreover, since there exists a surjective map $X \to X^t$, the scheme X^t has connected fibers over S and is proper over S. Hence X^t is an abelian scheme with $X^t = \mathrm{Pic}^0_{X/S}$ by Lemma 27.160. □

Remark 27.199. Let X and $\mathscr{L}$ be as in Theorem 27.198. Then $\varphi_{\mathscr{L}} \colon X \to X^t$ is an isogeny with kernel $K(\mathscr{L})$. In particular, $\varphi_{\mathscr{L}}$ makes X into a $K(\mathscr{L})$-torsor for the fppf-topology over X^t. Hence Section (14.21) shows that $\varphi_{\mathscr{L}}^*$ induces an equivalence of categories

$$\mathrm{QCoh}(X^t) \xrightarrow{\sim} (K(\mathscr{L})\text{-equivariant quasi-coherent } \mathscr{O}_X\text{-modules}).$$

Via this equivalence, finite locally free $\mathscr{O}_{X^t}$-modules (resp. line bundles on X^t) correspond to finite locally locally free $\mathscr{O}_X$-modules (resp. line bundles on X) with a $K(\mathscr{L})$-equivariant structure.

(27.38) Cohomology of the structure sheaf and of the sheaves of differentials.

If X is an abelian variety over a field, we obtain as a direct corollary the cohomology of the structure sheaf on X.

Corollary 27.200. *Let k be a field and X be an abelian variety of dimension g over k. Then $\dim H^1(X, \mathscr{O}_X) = g$ and the canonical homomorphism (21.29.5) of graded k-bialgebras*

$$\bigwedge{}^{\bullet} H^1(X, \mathscr{O}_X) \longrightarrow H^{\bullet}(X, \mathscr{O}_X)$$

is an isomorphism. In particular, $\dim_k H^p(X, \mathscr{O}_X) = \binom{g}{p}$ for all $p \geq 0$.

Proof. We have seen that $\dim H^1(X, \mathscr{O}_X) = \dim X$ in (27.37.2). Then we can conclude by Corollary 27.80. □

Corollary 27.201. *Let X be an abelian variety of dimension g over a field k. For all p and q there are functorial isomorphisms*

$$H^q(X, \Omega^p_{X/k}) \cong \bigwedge^p \mathrm{Lie}(X)^\vee \otimes \bigwedge^q \mathrm{Lie}(X^t),$$

in particular $h^{p,q}(X) := \dim H^q(X, \Omega^p_{X/k}) = \binom{g}{p}\binom{g}{q}$.

In Corollary 27.205 below we will generalize this result to abelian schemes over arbitrary base schemes.

Proof. Let $e\colon \mathrm{Spec}\,k \to X$ be the zero section and let $\mathscr{C}_e = e^*\Omega^1_{X/S}$ be the space of invariant differential forms on X, considered as vector space over k. Then one has

$$H^q(X, \Omega^p_{X/k}) = H^q(X, \mathscr{O}_X) \otimes \bigwedge^p \mathscr{C}_e = \bigwedge^p \mathrm{Lie}(X)^\vee \otimes \bigwedge^q \mathrm{Lie}(X^t),$$

where the first identity holds by (27.4.8). The second identity holds because the conormal sheaf is the k-linear dual of the Lie algebra of X (Definition 27.17) and since we have $H^q(X, \mathscr{O}_X) = \bigwedge^q H^1(X, \mathscr{O}_X)$ by Corollary 27.204 and $H^1(X, \mathscr{O}_X) = \mathrm{Lie}(X^t)$ (27.37.1). $\qquad\square$

Corollary 27.202. *Let X be an abelian variety over a field. Then for $n \in \mathbb{Z}$ the multiplication $[n]$ on X induces the multiplication by n^{p+q} on $H^q(X, \Omega^p_X)$.*

Proof. The dual of $[n]$ is again the multiplication by n on X^t by (27.35.3). They induce multiplication by n on $\mathrm{Lie}(X)$ and $\mathrm{Lie}(X^t)$ by Remark 27.18 (3). Hence we conclude by Corollary 27.201. $\qquad\square$

We will now generalize Corollary 27.200 and Corollary 27.201 to arbitrary base schemes.

Theorem 27.203. *Let S be a scheme and let $f\colon X \to S$ be an abelian scheme of relative dimension g. Then $R^p f_* \mathscr{O}_X$ is a locally free $\mathscr{O}_S$-module of rank $\binom{g}{p}$ whose formation is compatible with arbitrary base change $T \to S$.*

We will give a proof of the theorem if S is reduced or if 2 is invertible in $\mathscr{O}_S$. For a proof in the general case we refer to [BBM] $^\mathrm{O}$ 2.5.2.

Proof. Once we have shown that $R^p f_* \mathscr{O}_X$ is locally free and that its formation commutes with base change, it follows from Corollary 27.200 that its rank is $\binom{g}{p}$.

(I). If S is a reduced scheme, then Grauert's theorem (Theorem 23.140 (3)) implies the result by Corollary 27.200.

(II). For non-reduced base schemes S, we assume that 2 is invertible on S. We will reduce below by a standard argument to the case that $S = \mathrm{Spec}\,R$ where R is a local Artinian ring.

Hence let us study this case first. Let k be the residue field of R and let $\mathfrak{m}$ be its maximal ideal. By Theorem 23.133, $E := R\Gamma(X, \mathscr{O}_X)$ is a perfect complex of R-modules whose formation commutes with base change. Hence we know that $H^p(E \otimes^L_R k) = H^p(X \otimes_R k, \mathscr{O}_{X \otimes_R k})$ and hence $\dim_k H^p(E \otimes^L_R k) = \binom{g}{p}$ by Corollary 27.200. By Proposition 22.54 we know that we can assume that E is a complex of finite free modules with $E^p = R^{\binom{g}{p}}$ and where the differentials $d^p\colon E^p \to E^{p+1}$ are given by matrices D_p with entries in $\mathfrak{m}$. We have to show that E is acyclic, i.e. $D_p = 0$ for all p. Then $H^p(X, \mathscr{O}_X) = E^p$ is free of rank $\binom{g}{p}$ and its formation commutes with base change to k and therefore with arbitrary base change by Theorem 23.140.

To show that $D_p = 0$ let n be an integer such that n and $n-1$ are invertible in R. Here our additional hypothesis $\mathrm{char}(k) \neq 2$ ensures that such an integer exists, e.g., $n = 2$. Multiplication by n on X induces by functoriality an endomorphism of E in $D(R)$. By Lemma 21.134 (3) we can represent this endomorphism by an endomorphism $u\colon E \to E$ of complexes. Moreover we know that $u^p \otimes 1$ on $E^p \otimes_R k = H^p(X \otimes_R k, \mathscr{O}_{X \otimes_R k})$ is given by multiplication by n^p (Corollary 27.202). Hence the matrix U_p of u^p is of the form $n^p I + X_p$, where I is the identity matrix and where X_p has entries in $\mathfrak{m}$. As $u^{p+1} \circ d^p = d^p \circ u^p$ we find

$$(*) \qquad D_p X_p - X_{p+1} D_p = (n^{p+1} - n^p) D_p = n^p(n-1) D_p.$$

If we had $D_p \neq 0$, we would find a minimal $r \geq 1$ such that all entries of D_p are in $\mathfrak{m}^r$, but for some entry d_{ij} we would have $d_{ij} \notin \mathfrak{m}^{r+1}$. As n and $n-1$ are invertible, the (i,j)-th entry of the right side of $(*)$ would again be in $\mathfrak{m}^r \setminus \mathfrak{m}^{r+1}$. But the (i,j)-th entry of the left side is a sum of elements of the form dx, where d is an entry of D_p and hence $d \in \mathfrak{m}^r$ and where x is an entry of X_p or X_{p+1} and hence $x \in \mathfrak{m}$. This is a contradiction.

(III). The reduction to the local Artinian case is a special case of Remark 14.56. It works without our additional hypothesis that 2 is invertible. Let us spell out the details. We first may assume that $S = \mathrm{Spec}\,R$ is affine. Next we write R as a filtered colimit of noetherian rings R_λ. For sufficiently large λ we find an abelian scheme $f_\lambda\colon X_\lambda \to \mathrm{Spec}\,R_\lambda$ such that its base change to $\mathrm{Spec}\,R$ is X (Remark 27.91). If we have shown that $R^p f_{\lambda *} \mathscr{O}_{X_\lambda}$ is locally free and its formation commutes with base change, we get the same assertion for $X \to S$. Hence we can assume that R is noetherian.

As f is proper, we know that $R^p f_* \mathscr{O}_X$ is a coherent $\mathscr{O}_S$-module (Theorem 23.17). To show that it is locally free, we may therefore pass to stalks of $\mathscr{O}_S$ (Proposition 7.27). Moreover, its formation commutes with arbitrary base change if it commutes with base change to residue fields (Theorem 23.140). Therefore we can in addition assume that R is a local ring. Let k be its residue field. It remains to show that $H^p(X, \mathscr{O}_X)$ is a free R-module and that its formation commutes with base change to k. As the formation of cohomology commutes with flat base change (Corollary 22.91) and because we can check (local) freeness after a faithfully flat base change (Proposition 14.48), we may even assume that R is a complete local noetherian ring by passing to the completion of R.

Let $\mathfrak{m}$ be the maximal ideal of R, set $R_n := R/\mathfrak{m}^n$ and $X_n := X \otimes_R R_n$. Then X_n is an abelian scheme over the local Artinian ring $R/\mathfrak{m}^n$. Hence we know the theorem already for X_n, i.e., $H^p(X_n, \mathscr{O}_{X_n})$ is a free module and $H^p(X_n, \mathscr{O}_{X_n}) \otimes_{R_n} R_{n-1} = H^p(X_{n-1}, \mathscr{O}_{X_{n-1}})$ for all n, in other words, the family $(H^p(X_n, \mathscr{O}_{X_n}))_n$ defines a finite free module over the formal scheme $(\mathrm{Spec}\,R_n)_n$ (Definition 24.85).

By the Theorem of Formal Functions, Theorem 24.37, we have

$$\lim_n H^p(X_n, \mathscr{O}_{X_n}) = \lim_n (H^p(X, \mathscr{O}_X) \otimes_R R_n)$$

Hence the family $(H^p(X_n, \mathscr{O}_{X_n}))_n$ corresponds to $H^p(X, \mathscr{O}_X)$ under the equivalence of finite free R-modules and finite free modules over the formal scheme $(\mathrm{Spec}\,R_n)_n$ (Proposition 24.88). This shows that $H^p(X, \mathscr{O}_X)$ is finite free and that $H^p(X, \mathscr{O}_X) \otimes_R R_n = H^p(X_n, \mathscr{O}_{X_n})$. In particular, the formation of $H^p(X, \mathscr{O}_X)$ commutes with base change to $k = R_1$. $\qquad\square$

Corollary 27.204. *Let* $f \colon X \to S$ *be an abelian scheme over a scheme* S. *The canonical homomorphism* (21.29.4) *of graded* $\mathscr{O}_S$-*algebras*

$$v \colon \bigwedge^{\bullet} R^1 f_* \mathscr{O}_X \cong \bigoplus_{p \geq 0} R^p f_* \mathscr{O}_X$$

is an isomorphism.

Proof. By Theorem 27.203 we know that v is a homomorphism of locally free modules of the same rank. Hence it suffices to show that v is an isomorphism on fibers, i.e., after base change $\operatorname{Spec} \kappa(s) \to S$ for $s \in S$ (Corollary 8.12). As the formation of $R^p f_* \mathscr{O}_X$ commutes with base change, we conclude by Corollary 27.200. $\qquad\square$

Now one shows as in Corollaries 27.201 and 27.202 the following result.

Corollary 27.205. *Let* $f \colon X \to S$ *be an abelian scheme over a scheme* S *of relative dimension* g. *Let* $\mathscr{C}_e = e^* \Omega^1_{X/S}$ *be the conormal sheaf of* e *(Remark 17.22). Then for all* $p, q \geq 0$ *there are functorial isomorphisms*

$$R^p f_* \Omega^q_{X/S} = R^p f_* \mathscr{O}_X \otimes \bigwedge^q \mathscr{C}_e = \bigwedge^p \operatorname{Lie}(X^t) \otimes \bigwedge^q \operatorname{Lie}(X)^{\vee}$$

of locally free $\mathscr{O}_S$-*modules of rank* $\binom{g}{p}\binom{g}{q}$ *and the formation of* $R^p f_* \Omega^q_{X/S}$ *commutes with base change.*

Multiplication by an integer n *on* X *induces multiplication by* n^{p+q} *on* $R^p f_* \Omega^q_{X/S}$.

(27.39) The dual abelian space is a scheme.

Let S be a scheme and let $f \colon X \to S$ be an abelian scheme over S. Recall that $\operatorname{Pic}_{X/S}$ is represented by a separated algebraic group space locally of finite presentation (Theorem 27.117, Proposition 27.118). In $\operatorname{Pic}_{X/S}$ we have two subgroup functors $\operatorname{Pic}^0_{X/S}$ and X^t. Moreover, we have seen that the inclusion $X^t \to \operatorname{Pic}_{X/S}$ is representable by an open and closed immersion (Lemma 27.160).

Lemma 27.206. *Let* k *be an algebraically closed field and let* X *be an abelian variety over* k. *Then* $\operatorname{Pic}_{X/k}$ *is a smooth group scheme over* k.

Proof. We know that $\operatorname{Pic}_{X/k}$ is a group scheme locally of finite type (Theorem 27.119) whose identity component X^t is smooth (Theorem 27.198). Hence it is smooth (Proposition 27.20). $\qquad\square$

In [Mum1], Sections 12, 13, the existence of the dual abelian variety of an abelian variety X over an algebraically closed field k is proved directly, i.e., without using general representability results for the relative Picard functor. The strategy is to start with an ample line bundle $\mathscr{L}$ and to prove that the scheme $K(\mathscr{L})$ is a finite group scheme. It is relatively simple to construct the quotient $X^{\vee} := X/K(\mathscr{L})$ by this finite group scheme, generalizing Proposition 12.27 where the case of a constant finite group scheme is dealt with. Then one has to show that $X^{\vee}$, defined in this way, satisfies the universal property of the dual abelian variety, i.e., of $\operatorname{Pic}^0_{X/k}$.

Proposition 27.207. *Let* X *be an abelian scheme over a scheme* S. *Then the two subgroup functors* $\operatorname{Pic}^0_{X/S}$ *and* X^t *are equal.*

Proof. By the definition of $\mathrm{Pic}^0_{X/S}$ (Definition 27.123) and by the description of X^t in Corollary 27.162 it suffices to show that $\mathrm{Pic}^0_{X/S}(\bar{s}) = X^t(\bar{s})$ for all geometric points $\bar{s}$ of S. As the formation of both functors is compatible with base change, we may assume that $S = \mathrm{Spec}\, k$ for an algebraically closed field. Then we can conclude by Theorem 27.198. $\square$

Proposition 27.208. *Let X be an abelian scheme over a scheme S. Then the morphism $\varphi\colon \mathrm{Pic}_{X/S} \to \mathrm{Corr}_S(X, X)$ (Definition 27.159) is smooth. In particular, X^t is smooth over S.*

Proof. Once we know that φ is smooth, it follows that X^t is smooth over S because X^t is the kernel of φ.

To show that φ is smooth, we may assume that S is affine and then by noetherian approximation that S is noetherian. By Remark 27.42 it suffices to show the infinitesimal lifting criterion for nil-immersions $i\colon T_0 = \mathrm{Spec}\, R_0 \to T = \mathrm{Spec}\, R$, where R is a locally Artinian ring and where i is defined by an ideal $I \subseteq R$ with $Im = 0$. Let k be the residue field of R. We may replace X by its base change to T and hence assume that $S = T$, and we set $X_0 := X \otimes_R R_0$ and $X_k := X \otimes_R k$. Since $Im = 0$, we can view I as a finite-dimensional k-vector space. Then $H^n(X_0, I\mathscr{O}_{X_0}) = H^n(X_k, I\mathscr{O}_{X_k})$ (Corollary 22.6) and we obtain from Lemma 27.116 a commutative diagram

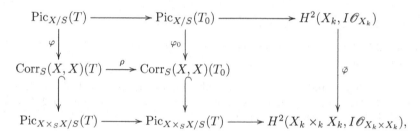

where the upper and the lower horizontal lines are exact and where $\bar{\varphi}$ is the map $m^* - p_1^* - p_2^*$ on cohomology (see Remark 27.158). Given $L_0 \in \mathrm{Pic}_{X/S}(T_0)$ and $\lambda \in \mathrm{Corr}_S(X, X)(T)$ with $\varphi_0(L_0) = \rho(\lambda)$ we have to show that there exists $L \in \mathrm{Pic}_{X/S}(T)$ that maps to L_0 and to λ. A diagram chase shows that it suffices to show that ρ and $\bar{\varphi}$ are injective.

The map ρ is injective since $\mathrm{Corr}_S(X, X)$ is unramified over S (Proposition 27.157). To see that $\bar{\varphi}$ is injective, we note that $H^2(X_k, I\mathscr{O}_{X_k}) = H^2(X_k, \mathscr{O}_{X_k}) \otimes_k I$ and similarly for $H^2(X_k \times_k X_k, I\mathscr{O}_{X_k \times X_k})$. Hence it suffices to show the following lemma. $\square$

Lemma 27.209. *Let X be an abelian variety over a field k and let $\alpha^p\colon H^p(X, \mathscr{O}_X) \to H^p(X \times_k X, \mathscr{O}_{X \times_k X})$ be the k-linear map induced by $m^* - p_1^* - p_2^*$. Then α^p is injective for $p \geq 2$.*

Proof. Let $\mu\colon H^\bullet(X, \mathscr{O}_X) \to H^\bullet(X, \mathscr{O}_X) \otimes_k H^\bullet(X, \mathscr{O}_X)$ be the comultiplication of the graded bialgebra $H^\bullet(X, \mathscr{O}_X)$ induced by m^* (Section (27.12)). Then the kernel of α^p are the primitive elements in $H^p(X, \mathscr{O}_X)$, i.e., those elements $x \in H^p(X, \mathscr{O}_X)$ such that $m^*(x) = x \otimes 1 + 1 \otimes x$. But we have $H^\bullet(X, \mathscr{O}_X) = \bigwedge^\bullet H^1(X, \mathscr{O}_X)$ (Corollary 27.200) and the primitive elements of an exterior algebra are homogeneous elements of degree 1. Therefore α^p is injective for $p \geq 2$. $\square$

Remark 27.210. Let X be an abelian scheme over a scheme S. We collect what we have seen about $\mathrm{Pic}_{X/S}$ and X^t by now.

(1) The functor $\mathrm{Pic}_{X/S}$ is an algebraic space locally of finite presentation over S (Theorem 27.117).

(2) The Picard space $\mathrm{Pic}_{X/S}$ is separated (Proposition 27.118).

(3) If $S = \mathrm{Spec}\,k$ for a field k, then $\mathrm{Pic}_{X/S}$ is representable by a group scheme locally of finite type (Theorem 27.47).

(4) If $S = \mathrm{Spec}\,k$ for a field k, then $\mathrm{Pic}_{X/S}$ is representable by a smooth group scheme.

(5) We have $X^t = \mathrm{Pic}^0_{X/S}$ (Proposition 27.207) and the inclusion $X^t \to \mathrm{Pic}_{X/S}$ is an open and closed immersion (Lemma 27.160).

(6) The algebraic space X^t is smooth (Proposition 27.208) and proper (Lemma 27.126) over S and has geometrically connected fibers by (5).

(7) If $X \to S$ is projective, then X^t is an abelian scheme (Theorem 27.198).

For the representability results (1) and (3) we only gave references. For all other properties we gave proofs using only some standard properties of morphisms of algebraic spaces.

We can now invoke the following result.

Theorem 27.211. *Let S be a scheme and let Y be a group algebraic space such that $Y \to S$ is proper smooth with geometrically connected fibers. Then Y is an abelian scheme over S.*

By Remark 27.210 we obtain the following corollary.

Corollary 27.212. *Let S be a scheme and let X be an abelian scheme. Then X^t is an abelian scheme over S.*

As we have already indicated above, the abelian scheme X^t is called the *dual abelian scheme* of X. We will only give a sketch of the proof of Theorem 27.211 and refer to [FaCh]^{O} I, 1.9 or to [GrK] for details.

Proof. [of Theorem 27.211, Sketch] *(I).* As Y is by definition an fppf-sheaf the question whether Y is a scheme is Zariski local (Theorem 8.9), in particular we may assume that S is affine. Next we may assume by noetherian approximation that S is of finite type over $\mathbb{Z}$. Moreover if $X \to Y$ is a surjective integral morphism of algebraic spaces and X is a scheme, then Y is a scheme ([Sta] 07VT). In particular, an algebraic space Y is a scheme if and only if its underlying reduced algebraic space is a scheme. Therefore we may assume that S is reduced.

(II). By noetherian induction one can now assume the base change of Y to any proper closed subscheme of S is a scheme. One uses gluing along closed subschemes (see [Fer]^{O} and [TeTy]$^{\mathrm{O}}_{\mathrm{X}}$) to reduce to the case that S is integral. Here the general idea is the following. Let $Z \to X$ be a closed immersion and let $Z \to T$ be a finite morphism of algebraic spaces. Then the pushout $T \amalg_Z X$ exists in the category of algebraic spaces ([TeTy]$^{\mathrm{O}}_{\mathrm{X}}$ 6.2.1) and it is a scheme if Z, T, and X are schemes. If $T = \mathrm{Spec}\,R$, $X = \mathrm{Spec}\,A$ and $Z = \mathrm{Spec}\,A/I$ are affine, then $T \amalg_Z X$ is affine and is the spectrum of the fiber product of rings $R \times_{A/I} A$. Hence gluing S from its irreducible components, viewed as closed integral subschemes, one can assume that S is integral.

Next let $\tilde{S} \to S$ be the normalization. Then the projection $Y_{\tilde{S}} \to Y$ is surjective and integral. Hence if $Y_{\tilde{S}}$ is a scheme, then Y will be a scheme, again by [Sta] 07VT. Therefore we may assume that S is noetherian and normal.

(III). Finally one shows that if S is noetherian and normal, then Y carries an ample line bundle and hence is a projective abelian scheme. If Y is an abelian scheme, this is proved below in Theorem 27.291 but using ampleness criteria whose proof assumed that one knows already that Y is a scheme. One has to check that the essential arguments in the proof of Theorem 27.291 also work without assuming that Y is a scheme. $\qquad\square$

(27.40) Dual homomorphisms.

Let S be a scheme, and let X and Y be abelian schemes over S. We denote by $\operatorname{Hom}(X, Y)$ the set of homomorphisms $X \to Y$ of abelian schemes over S. Let $f \in \operatorname{Hom}(X, Y)$. Then the pullback $f^* \colon \operatorname{Pic}_{Y/S} \longrightarrow \operatorname{Pic}_{X/S}$ induces the dual homomorphism (Definition 27.165)

$$f^t \colon X^t \longrightarrow Y^t.$$

We obtain a map

$$\operatorname{Hom}(X, Y) \longrightarrow \operatorname{Hom}(Y^t, X^t), \qquad f \mapsto f^t.$$

By Remark 27.166 this is a homomorphism of abelian groups.

Proposition 27.213. *Let $f \colon X \to Y$ be an isogeny of abelian schemes over S.*
(1) *Then the dual homomorphism $f^t \colon Y^t \to X^t$ is also an isogeny.*
(2) *Let $\mathscr{M}$ be a line bundle on Y. Then the class of $\mathscr{M}$ is in $Y^t(S)$ if and only if the class of $f^*\mathscr{M}$ is in $X^t(S)$.*
(3) *The kernel of f^t is isomorphic to the Cartier dual of $\operatorname{Ker}(f)$. In particular, one has $\deg(f) = \deg(f^t)$.*

Proof. The formation of f^t commutes with base change. To show (1) we may by Proposition 27.176 assume that S is the spectrum of a field. As f is an isogeny, we have $\dim(X^t) = \dim(X) = \dim(Y) = \dim(Y^t)$, hence it suffices to show that f^t is surjective. As the base is a field, we can choose an ample line bundle $\mathscr{L}$ on Y. As f is finite and in particular quasi-affine, $f^*\mathscr{L}$ is also ample. Hence $\varphi_{f^*\mathscr{L}}$ is an isogeny and therefore surjective. Hence f^t is surjective by Lemma 27.165.

Let us show (2). The condition is clearly necessary (Definition 27.165). For the converse we have to show that if $\varphi_{f^*\mathscr{L}} = 0$, then $\varphi_{\mathscr{L}} = 0$. By (27.29.9) we have $f^t \circ \varphi_{\mathscr{L}} \circ f = 0$.

As the locus on S, where a morphism of abelian schemes over S is constant, is open and closed in S (Proposition 27.96), we may assume that $S = \operatorname{Spec} k$ for a field k. As f is an epimorphism, we know that $f^t \circ \varphi_{\mathscr{L}} = 0$. As f^t has finite kernel by (1), $\varphi_{\mathscr{L}}$ factors through a finite k-scheme which is geometrically integral because Y is geometrically integral. Hence $\varphi_{\mathscr{L}} = 0$.

Now we show (3). Let T be an S-scheme. We have to show that $\operatorname{Ker}(f^t)(T) \cong \operatorname{Ker}(f)^D(T)$ functorially for all S-schemes T. Replacing $f \colon X \to Y$ by its base change to T we may assume that $T = S$ which simplifies the notation. We will construct functorial isomorphisms

$$
\begin{aligned}
\operatorname{Ker}(f^t)(S) &\overset{(a)}{=} \operatorname{Ker}(f^* \colon \operatorname{Pic}_{Y/S}(S) \to \operatorname{Pic}_{X/S}(S)) \\
&\overset{(b)}{=} \operatorname{Ker}(f^* \colon \operatorname{Pic}(Y) \to \operatorname{Pic}(X)) \\
&\overset{(c)}{=} \operatorname{Hom}_{S-\mathrm{Gr}}(K(\mathscr{L}), \mathbb{G}_{m,S}).
\end{aligned}
$$

We can identify $\operatorname{Pic}_{Y/S}(S)$ with the subgroup of those $\mathscr{L} \in \operatorname{Pic}(Y)$ such that $e^*\mathscr{L} \cong \mathscr{O}_S$, where e is the zero section of Y.

Let $\mathscr{L} \in \operatorname{Pic}_{Y/S}(S)$ be such that $f^* \mathscr{L}$ is trivial. Then $\mathscr{L} \in Y^t(S)$ by (2) which shows equality (a). Equality (b) follows from the above interpretation of $\operatorname{Pic}_{Y/S}(S)$.

To see (c) we use that f makes X into a $K(\mathscr{L})$-torsor over Y (Proposition 27.62) and that therefore to give a line bundle $\mathscr{L}$ on Y is the same as to give $f^* \mathscr{L}$ together with its canonical $K(\mathscr{L})$-equivariant structure (Section (14.21)). Hence line bundles in $\operatorname{Ker}(f^* \colon \operatorname{Pic}(Y) \to \operatorname{Pic}(X))$ correspond to $K(\mathscr{L})$-equivariant structures on $\mathscr{O}_X$, i.e., to actions of $K(\mathscr{L})$ on $\mathbb{A}^1_X$ lifting the action on X by translation. The trivial line bundle $\mathscr{O}_Y$ corresponds to the trivial lifting

$$K(\mathscr{L}) \times_S \mathbb{A}^1_X = K(\mathscr{L}) \times_S X \times_S \mathbb{A}^1_S \xrightarrow{(x',x,t) \mapsto (x'+x,t)} X \times_S \mathbb{A}^1_S = \mathbb{A}^1_X.$$

As $f_* \mathscr{O}_X = \mathscr{O}_S$ holds after any base change, it is easy to check that any other lifting differs by a homomorphism $K(\mathscr{L}) \to H$, where H is the group functor $T \mapsto \Gamma(X_T, \mathscr{O}^\times_{X_T}) = \Gamma(T, \mathscr{O}^\times_T)$, i.e. $H = \mathbb{G}_{m,S}$. This yields (c). $\qquad\square$

Corollary 27.214. *Let X be an abelian scheme over a scheme S and let $n \neq 0$ be an integer. Then $X[n]$ and $X^t[n]$ are Cartier dual to each other.*

Proof. This follows from Proposition 27.213 since $[n]^t_X = [n]_{X^t}$ (27.35.3). $\qquad\square$

Corollary 27.215. *Let k be a field and let $f \colon X \to Y$ be an isogeny of abelian varieties over k. Then for every line bundle $\mathscr{L}$ on X there exist only finitely many isomorphism classes of line bundles $\mathscr{M}$ on Y such that $f^* \mathscr{M} \cong \mathscr{L}$.*

Proof. As $f^* \colon \operatorname{Pic}(Y) \to \operatorname{Pic}(X)$ is a group homomorphism, it suffices to show that $\operatorname{Ker}(f^*)(k)$ is finite. By Proposition 27.213 we know that $\operatorname{Ker}(f^*)(k) \subseteq Y^t(k)$ is finite. $\qquad\square$

(27.41) The Poincaré bundle.

Let S be a scheme and let X be an abelian scheme over S. Let T be an S-scheme and let $\mathscr{F}$ be an $\mathscr{O}_{X \times_S T}$-module. Then we denote by $\mathscr{F}_{|0 \times T}$ the pullback of $\mathscr{F}$ under the zero section 0 of the abelian scheme $X \times_S T \to T$. By definition we have

$$\operatorname{Pic}_{X/S}(T) = \operatorname{Pic}(X \times_S T)/\operatorname{Pic}(T) \cong \{ \, \mathscr{L} \in \operatorname{Pic}(X \times_S T) \; ; \; \mathscr{L}_{|0 \times T} \cong \mathscr{O}_T \, \}.$$

We call an isomorphism $\iota \colon \mathscr{L}_{|0 \times T} \xrightarrow{\sim} \mathscr{O}_T$ a *rigidification* of $\mathscr{L}$. Given two line bundles $\mathscr{L}$ and $\mathscr{L}'$ with rigidifications ι and ι', respectively, any isomorphism $\alpha \colon \mathscr{L} \xrightarrow{\sim} \mathscr{L}'$ can be changed by precomposition with an automorphism of $\mathscr{L}$ such that the modified isomorphism α satisfies $\iota' \circ \alpha_{|0 \times T} = \iota$ because

$$\operatorname{Aut}(\mathscr{L}) = \Gamma(X \times_S T, \mathscr{O}_{X \times_S T})^\times = \Gamma(T, \mathscr{O}_T)^\times.$$

Therefore $\operatorname{Pic}_{X/S}(T)$ can also be identified with the group of isomorphism classes of rigidified line bundles $(\mathscr{L}, \iota)$ on $X \times_S T$.

By Corollary 27.162 one has for every S-scheme T

(27.41.1) $X^t(T) = \{ \, \mathscr{L} \in \operatorname{Pic}_{X/S}(T) \; ; \; m^* \mathscr{L} = p_1^* \mathscr{L} \otimes p_2^* \mathscr{L} \text{ in } \operatorname{Pic}_{X \times_S X}(T) \}$

and $\mathscr{L} \in \operatorname{Pic}_{X/S}(T)$ is in $X^t(T)$ if and only if for every geometric point $\bar{t} \colon \operatorname{Spec} \kappa \to T$ one has $(\operatorname{id}_X \times \bar{t})^* \mathscr{L} \in X^t(\bar{t})$.

Definition 27.216. *The rigidified line bundle* $(\mathscr{P}, \iota_{\mathscr{P}})$ *on* $X \times_S X^t$ *corresponding to* $\mathrm{id}_{X^t} \in X^t(X^t)$ *via* (27.41.1) *is called the* Poincaré bundle *of* X.

Hence $\mathscr{P}$ is a line bundle on $X \times_S X^t$ together with an isomorphism $\iota_{\mathscr{P}} \colon \mathscr{P}_{|0 \times X^t} \overset{\sim}{\to} \mathscr{O}_{X^t}$ such that for every S-scheme T and for every $(\mathscr{L}, \iota_{\mathscr{L}}) \in X^t(T)$ there exists a unique morphism $f \colon T \to X^t$ of S-schemes such that $(\mathrm{id}_X \times f)^*(\mathscr{P}, \iota_{\mathscr{P}}) = (\mathscr{L}, \iota_{\mathscr{L}})$.

Remark 27.217. If $\mathscr{L}$ is any line bundle on X and $\varphi_{\mathscr{L}} \colon X \to X^t$ is the corresponding homomorphism, then $(\mathrm{id}_X \times \varphi_{\mathscr{L}})^*(\mathscr{P}) = \Lambda(\mathscr{L})$ in $\mathrm{Pic}_{X/S}(X)$, where $\Lambda(\mathscr{L})$ is the Mumford bundle associated to $\mathscr{L}$. Indeed, by Proposition 27.161 (1) the composition

$$X \xrightarrow{\varphi_{\mathscr{L}}} X^t \to \mathrm{Pic}_{X/S}$$

corresponds to the class of $\Lambda(\mathscr{L})$.

Remark 27.218. Let $X \to S$ be an abelian scheme and let $\mathscr{P}$ be its Poincaré bundle.
(1) The formation of the Poincaré bundle is compatible with base change: If $T \to S$ is a morphism of schemes, then the Poincaré bundle of $X_T \to T$ is the pullback of $\mathscr{P}$ under $X_T \times_T (X^t)_T = X \times_S X^t \times_S T \to X \times_S X^t$.
(2) The condition that $\mathscr{P} \in \mathrm{Pic}_{X/S}(X^t)$ is an X^t-valued point of X^t means that $m^*\mathscr{P} \cong p_1^*\mathscr{P} \otimes p_2^*\mathscr{P}$, where

$$m, p_1, p_2 \colon (X \times_S X^t) \times_{X^t} (X \times_S X^t) \longrightarrow X \times_S X^t$$

are the multiplication and the projections of the abelian scheme $X \times_S X^t \to X^t$. Identifying the left hand side with $X \times_S X \times_S X^t$ we obtain

$$\mu^*\mathscr{P} \cong p_{13}^*\mathscr{P} \otimes p_{23}^*\mathscr{P},$$

where $\mu \colon X \times_S X \times_S X^t \to X \times_S X^t$ is given by $(x, y, \xi) \mapsto (x + y, \xi)$.

Remark 27.219. Let $f \colon X \to Y$ be a homomorphism of abelian schemes over a scheme. Its dual $f^t \colon Y^t \to X^t$ is an element of $X^t(Y^t)$ and hence corresponds to the element $L = (\mathrm{id}_X \times f^t)^*\mathscr{P}_X \in \mathrm{Pic}_{X/S}(Y^t)$. As f^t is defined by pullback of line bundles via f we find that there is an isomorphism

$$(27.41.2) \qquad (\mathrm{id}_X \times f^t)^*\mathscr{P}_X \cong (f \times \mathrm{id}_{Y^t})^*\mathscr{P}_Y$$

of rigidified line bundles on $X \times_S Y^t$ with a rigidification along $\{0\} \times Y^t$.

(27.42) Biduality.

In this section let S be a scheme and let X be an abelian S-scheme. For abelian schemes X and Y over S we denote by $\mathrm{Hom}(X, Y)$ the abelian group of homomorphisms $X \to Y$ of group schemes over S. Recall (Remark 27.152 and Proposition 27.163) that one has identifications

$$(27.42.1) \qquad \mathrm{Hom}(X, Y^t) = \mathrm{Corr}_S(X, Y)(S) = \mathrm{Ker}(\beta),$$

where β is given by

$$\beta \colon \mathrm{Pic}_{X \times_S Y/S}(S) \longrightarrow \mathrm{Pic}_{X/S}(S) \times \mathrm{Pic}_{Y/S}(S),$$
$$\mathscr{M} \mapsto (\mathscr{M}_{|X \times \{0\}}, \mathscr{M}_{|\{0\} \times Y}).$$

Observe that $\mathrm{Corr}_S(X, Y)$ is symmetric in X and Y, the identification of $\mathrm{Corr}_S(X, Y)$ with $\mathrm{Corr}_S(Y, X)$ being given by pullback along the morphism $\sigma_{X,Y} \colon X \times_S Y \to Y \times_S X$ which switches the two factors. We obtain an identification $\mathrm{Hom}(X, Y^t) = \mathrm{Hom}(Y, X^t)$.

Now we set $Y = X^t$. Then id_{X^t} corresponds to the Poincaré bundle $\mathscr{P}_X \in \mathrm{Ker}(\beta)$ and also corresponds to a homomorphism of abelian schemes

$$(27.42.2) \qquad\qquad \kappa_X \colon X \longrightarrow (X^t)^t.$$

By definition, we then have an isomorphism $(\mathrm{id}_{X^t} \times \kappa_X)^* \mathscr{P}_{X^t} \cong \sigma_{X,X^t}^* \mathscr{P}_X$ of line bundles on $X^t \times X$.

Proposition 27.220.
(1) *For abelian schemes X, Y over S, the isomorphism $\mathrm{Hom}(Y, X^t) \cong \mathrm{Hom}(X, Y^t)$ is given by $f \mapsto f^t \circ \kappa_X$.*
(2) *The formation of κ_X is functorial in X, i.e., for every homomorphism $f \colon X \to Y$ of abelian schemes over S, we have $\kappa_Y \circ f = f^{tt} \circ \kappa_X$.*

Proof. Any morphism $f \colon Y \to X^t$ is determined by the pullback $(\mathrm{id} \times f)^* \mathscr{P}_X$ of the Poincaré bundle, and by definition of the identification $\mathrm{Hom}(Y, X^t) \cong \mathrm{Hom}(X, Y^t)$ the morphism f corresponds to $\sigma_{X,Y}^* (\mathrm{id} \times f)^* \mathscr{P}_X$ under this bijection. In view of the definition of κ_X and of (27.41.2), we may compute

$$\begin{aligned}
\sigma_{X,Y}^* (\mathrm{id} \times f)^* \mathscr{P}_X &\cong (f \times \mathrm{id}_X)^* \sigma_{X,X^t}^* \mathscr{P}_X \\
&\cong (f \times \mathrm{id}_X)^* (\mathrm{id}_{X^t} \times \kappa_X)^* \mathscr{P}_{X^t} \\
&\cong (\mathrm{id}_Y \times \kappa_X)^* (f \times \mathrm{id}_{X^{tt}})^* \mathscr{P}_{X^t} \\
&\cong (\mathrm{id}_Y \times \kappa_X)^* (\mathrm{id}_Y \times f^t)^* \mathscr{P}_Y \\
&\cong (\mathrm{id}_Y \times (f^t \circ \kappa_X))^* \mathscr{P}_Y.
\end{aligned}$$

This proves part (1), and by symmetry also shows that the inverse of this identification is given by $g \mapsto g^t \circ \kappa_Y$. In particular, applying this to $Y = X^t$ and $f = \mathrm{id}$ we find that $\kappa_X^t \circ \kappa_{X^t} = \mathrm{id}_{X^t}$. (We will show in Theorem 27.222 that as a consequence κ_X is an isomorphism.)

Now to prove part (2), let $f \colon X \to Y$ be a homomorphism of abelian schemes, and consider $f^t \in \mathrm{Hom}(Y^t, X^t)$. In view of part (1) it corresponds to $f^{tt} \kappa_X \in \mathrm{Hom}(X, Y^{tt})$. On the other hand, rewriting it as $f^t = f^t \kappa_Y^t \kappa_{Y^t} = (\kappa_Y f)^t \kappa_{Y^t}$, we see that at the same time it corresponds to $\kappa_Y f$. $\qquad\square$

By Theorem 27.222 below, we can identify $X^{tt} = X$ (via κ_X) for every abelian scheme X. Then part (1) of the proposition says that $f^{tt} = f$ for every morphism f of abelian schemes.

Remark 27.221. Let X be an abelian scheme over a scheme S. As the identifications in (27.42.1) are compatible with base change, we find that $\kappa_{X \times_S T} = \kappa_X \times \mathrm{id}_T$ for every morphism $T \to S$.

Theorem 27.222. *Let X be an abelian scheme over a scheme S. Then the biduality morphism κ_X is an isomorphism.*

Proof. As the formation of κ_X is compatible with base change and since we can check whether a morphism of S-schemes is an isomorphism on fibers (Corollary 18.77), we may assume that S is the spectrum of a field which we can even assume to be algebraically closed by faithfully flat descent (Proposition 14.53). In the proof of Proposition 27.220 we have already seen that

$$(\kappa_X)^t \circ \kappa_{X^t} = \mathrm{id}_{X^t}.$$

Hence κ_{X^t} has trivial kernel. As a homomorphism between abelian varieties, it is therefore an isogeny of degree 1, i.e., an isomorphism. This shows that $(\kappa_X)^t$ is an isomorphism. Hence it is enough to show the following lemma. $\square$

Lemma 27.223. *Let $f\colon X \to Y$ be a homomorphism of abelian varieties over an algebraically closed field such that f^t is an isomorphism. Then f is an isomorphism.*

Proof. By Proposition 27.213 it suffices to show that f is an isogeny. One has $\dim(X) = \dim(X^t) = \dim(Y^t) = \dim(Y)$ because f^t is an isomorphism. Hence it suffices to show that f is surjective. As f is proper, $Y' := f(X)$ is closed in Y, and when endowed with the scheme structure that makes Y' into a reduced closed subscheme of Y, it is an abelian subvariety. By the functoriality of the formation of the dual homomorphism we see that f^t factorizes as $Y^t \to (Y')^t \to X^t$. As f^t is an isomorphism, the kernel of $Y^t \to (Y')^t$ is trivial which implies $\dim(Y')^t \geq \dim Y^t$ (Proposition 27.14) and hence $\dim Y' = \dim Y$. As Y is irreducible, we see that $Y' = Y$. $\square$

Remark 27.224. Let X be an abelian scheme over a scheme S. Let $\mathscr{P}_{X^t} \in \mathrm{Pic}(X^t \times_S (X^t)^t)$ be the Poincaré bundle of X^t. Let $\sigma\colon X \times_S X^t \to X^t \times_S X$ be the canonical morphism that switches the factors. Then by definition of κ_X we have an isomorphism of rigidified line bundles

$$(27.42.3) \qquad\qquad \sigma^*(\mathrm{id}_X \times \kappa_X)^* \mathscr{P}_{X^t} \cong \mathscr{P}_X.$$

Moreover, the homomorphism $\varphi_{\mathscr{P}_X}\colon X \times_S X^t \to X^t \times X^{tt}$ is given on T-valued points by

$$(27.42.4) \qquad\qquad \varphi_{\mathscr{P}}(x, \xi) = (\xi, \kappa_X(x)).$$

In the sequel we will usually identify X with $(X^t)^t$ using κ_X.

Remark 27.225. Let S be a scheme and let X be a pointed S-scheme. We say that an abelian scheme A/S together with a morphism $\alpha\colon X \to A$ of pointed S-schemes is an *Albanese scheme* for X (sometimes denoted by $\mathrm{Alb}_{X/S}$), if it satisfies the following universal property for morphisms from X into abelian schemes: For every abelian scheme B/S and every morphism $f\colon X \to B$ of pointed S-schemes, there exists a unique homomorphism $\varphi\colon A \to B$ of abelian schemes such that $f = \varphi \circ \alpha$. It is clear that an Albanese scheme for X is uniquely determined up to unique isomorphism, if it exists.

Now assume that $\mathrm{Pic}^0_{X/S}$ is representable by an abelian scheme. We claim that $(\mathrm{Pic}^0_{X/S})^\vee$ is an Albanese scheme for X. First note that we have a canonical morphism $X \to (\mathrm{Pic}^0_{X/S})^\vee$. In fact, consider the universal line bundle on $X \times \mathrm{Pic}^0_{X/S}$. We can view this as an actual line bundle on $X \times \mathrm{Pic}^0_{X/S}$, rather than just an equivalence class in $\mathrm{Pic}^0_{X/S}(\mathrm{Pic}^0_{X/S})$ by requiring that its restrictions to $\{x_0\} \times \mathrm{Pic}^0_{X/S}$ and to $X \times \{0\}$ be trivial. (Here $x_0 \in X(S)$ denotes the fixed point giving X the structure of a pointed scheme.) Since $X \times \mathrm{Pic}^0_{X/S} \cong \mathrm{Pic}^0_{X/S} \times X$, this line bundle defines a morphism $\alpha\colon X \to (\mathrm{Pic}^0_{X/S})^\vee$.

Now consider any morphism $f\colon X \to B$ of pointed S-schemes, where B is an abelian scheme. Pullback of line bundles gives us a morphism $B^{\vee} = \mathrm{Pic}^0_{B/S} \to \mathrm{Pic}^0_{X/S}$, and passing to duals we obtain a morphism $\varphi\colon \mathrm{Alb}_{X/S} \to B$. It is not hard to check that $f = \varphi \circ \alpha$ and that φ is the unique homomorphism with this property.

If S is the spectrum of a field k and X is a geometrically connected, smooth and proper pointed k-scheme, then $\mathrm{Pic}^0_{X/k}$ is representable by a proper k-group scheme (Theorem 27.119). In characteristic 0 it is necessarily reduced and therefore an abelian variety (Theorem 27.25) and the above discussion shows that its dual $(\mathrm{Pic}^0_{X/k})^{\vee}$ is an Albanese variety for X. In positive characteristic, $\mathrm{Pic}^0_{X/k}$ might be non-reduced. If k is perfect, then its underlying reduced scheme is an abelian variety (Proposition 27.10 (3)), and by similar arguments as above one can show that the dual of $(\mathrm{Pic}^0_{X/k})_{\mathrm{red}}$ is an Albanese variety for X. See [Moc]$_X$ Appendix for further details and a more comprehensive discussion.

Cohomology of line bundles on abelian schemes

Next we calculate the cohomology of the Poincaré bundle (Proposition 27.229) and use it to prove the Fourier-Mukai equivalence (Theorem 27.243) for abelian schemes. We then use the Fourier-Mukai equivalence to prove the Riemann-Roch theorem for abelian varieties (Theorem 27.253).

For line bundles with non-vanishing Euler characteristic, so-called non-degenerate line bundles, there exists a unique degree in which the cohomology does not vanish. This degree is called the *index* of the line bundle and it varies locally constant in families. Moreover, a line bundle $\mathscr{L}$ is relatively ample if and only if it is non-degenerate and of index 0 (Theorem 27.264). We then use the Fourier-Mukai equivalence to prove Atiyah's classification of vector bundles on elliptic curves (Theorem 27.271).

The characterization of ampleness is used to show that any tensor product of two (resp. three) relatively ample line bundles is globally generated locally on the base scheme (resp. very ample), see Proposition 27.278 and Theorem 27.279. We also introduce the notion of a polarization and show that the property to be a polarization can be checked on a single geometric fiber for a connected base scheme (Corollary 27.285).

We conclude the chapter by proving that abelian schemes over normal noetherian base schemes are projective (Theorem 27.291) and by briefly explaining many of the notions in this chapter in the special case of abelian varieties over the complex numbers, see Section (27.54).

(27.43) Cohomology of the Poincaré bundle.

Let $\pi\colon X \to S$ be an abelian scheme of relative dimension g, let $\pi'\colon X^t \to S$ be the structure morphism of the dual abelian scheme, let $p\colon X \times_S X^t \to X$ and $p'\colon X \times_S X^t \to X^t$ be the projections, and let $\varpi\colon X \times_S X^t \to S$ be the structure morphism. Let $e\colon S \to X$ and $e'\colon S \to X^t$ be the zero sections. Hence we have a commutative diagram

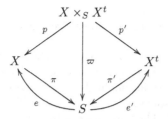

Lemma 27.226. *With the above notation the following assertions hold.*

(1) *The Leray spectral sequence for $\varpi = \pi' \circ p'$ induces for all $i \geq 0$ an isomorphism*

$$(*) \qquad \pi'_*(R^i p'_* \mathscr{P}) \xrightarrow{\sim} R^i \varpi_* \mathscr{P}.$$

(2) *One has $R^i \varpi_* \mathscr{P} = 0$ for all $i \neq g$, $R^g \varpi_* \mathscr{P}$ is finite locally free, and the formation of $R^i \varpi_* \mathscr{P}$ commutes with base change for all i.*

(3) *One has $R^i p'_* \mathscr{P} = 0$ for all $i \neq g$.*

(4) *The formation of $R^g p'_* \mathscr{P}$ commutes with arbitrary base change $T \to X^t$.*

We will prove (3) fully only if S is locally noetherian. For the general case we will give a proof using Lemma 27.227 below for whose proof we essentially only give a reference.

Proof. We show (1). Let $y \in X^t$ and $s' := \pi'(y)$. Then $p^{t,-1}(y) = X \otimes_{\kappa(s')} \kappa(y) =: X_y$. The line bundle $\mathscr{P}_y = \mathscr{P}_{|X_y}$ is by definition isomorphic to the line bundle in $X^t(\kappa(y)) \subseteq \mathrm{Pic}(X_y)$ corresponding to the morphism $\mathrm{Spec}\,\kappa(y) \to X^t$. In particular, $\mathscr{P}_y \not\cong \mathcal{O}_{X_y}$ if and only if $y \notin e'(S) \subset X^t$. Hence $H^i(X_y, \mathscr{P}_y) = 0$ for all i if $y \in X^t \setminus e'(S)$ by Lemma 27.197. So Proposition 23.142 shows that one has for all i

$$(27.43.1) \qquad R^i p'_* \mathscr{P}_{|(X^t \setminus e'(S))} = 0.$$

Thus for all $s \in S$ we have $\mathrm{Supp}(R^i p'_* \mathscr{P})_s \subseteq \{e'(s)\} \subseteq X^t_s$ and in particular we see that $H^n(X^t_s, (R^i p'_* \mathscr{P})_s) = 0$ for all $n > 0$. By Proposition 23.142 this shows that $R^n \pi'_*(R^i p'_* \mathscr{P}) = 0$ for all $n > 0$ and for all i. Now the Leray spectral sequence for $\varpi = \pi' \circ p_{X^t}$ yields the isomorphism $(*)$.

To show (2) we may assume that $S = \mathrm{Spec}\,k$ for a field k by Corollary 23.143 and because the formation of the Poincaré bundle is compatible with base change, and it suffices to show that $H^i(X \times_k X^t, \mathscr{P}) = 0$ for all $i \neq g$. Then the isomorphism $(*)$ becomes

$$H^0(X^t, R^i p'_* \mathscr{P}) \cong H^i(X \times_k X^t, \mathscr{P}).$$

As p' is proper with fibers of dimension g, we find that

$$(**) \qquad R^i p'_* \mathscr{P} = 0 \qquad \text{for all } i > g$$

by Corollary 24.44 and hence $H^i(X \times_k X^t, \mathscr{P}) = 0$ for $i > g$.

We now apply Serre duality to the Poincaré bundle. Applying (27.35.3) to the abelian scheme $X \times_k X^t \to X^t$, we have $([-1]^t_X, \mathrm{id}_{X^t})^* \mathscr{P} \cong \mathscr{P}^{-1} \otimes p^{t,*}(\mathscr{M})$ for some line bundle $\mathscr{M}$ on X^t. As $\mathscr{P}$ and $\mathscr{P}^{-1}$ are rigidified we find by restriction to $0 \times X^t$ that $\mathscr{M} \cong \mathcal{O}_{X^t}$. Therefore $\mathscr{P} \cong \mathscr{P}^{-1}$. As $X \times_k X^t$ is a smooth group scheme over k, $\Omega^1_{X \times_k X^t / k} \cong \mathcal{O}^{2g}_{X \times_k X^t}$ and hence $\Omega^{2g}_{X \times_k X^t / k} \cong \mathcal{O}_{X \times X^t}$. Therefore Serre duality (Corollary 25.129) yields

$$H^i(X \times X^t, \mathscr{P}) \cong H^{2g-i}(X \times X^t, \mathscr{P}^{-1})^\vee \cong H^{2g-i}(X \times X^t, \mathscr{P})^\vee.$$

This shows $H^i(X \times_k X^t, \mathscr{P}) = 0$ also for $i < g$.

Assertion (4) follows as top degree higher direct image always commutes with base change by Corollary 23.146.

It remains to show (3). By (**) it suffices to show that $R^i p'_* \mathscr{P} = 0$ for all $i < g$. If S is locally noetherian one can argue as follows. Then $R^i p'_* \mathscr{P}$ is a coherent $\mathscr{O}_{X^t}$-module (Theorem 23.17). Hence its annihilator is a quasi-coherent ideal (Proposition 7.35) which defines a closed subscheme $h \colon Z \hookrightarrow X^t$ and $R^i p'_* \mathscr{P} = h_* \mathscr{F}_i$ for the coherent $\mathscr{O}_Z$-module $\mathscr{F}_i := h^* R^i p'_* \mathscr{P}$. The underlying topological space of Z is contained in the closed subscheme defined by the closed immersion e' by (27.43.1). Hence the restriction of π' to Z is affine by Lemma 12.38. As $\pi'_*(R^i p'_* \mathscr{P}) = R^i \varpi_* \mathscr{P} = 0$ for $i \neq g$, this shows that $R^i p'_* \mathscr{P} = 0$ for $i \neq g$ by Proposition 12.5.

Let us now give a proof of (3) if S is not necessarily locally noetherian. Set $E := Rp'_* \mathscr{P}$, which is a perfect complex by Corollary 23.136. Let $s \in S$ and denote by Y_s the fiber in s of an S-scheme Y. By derived cohomology and base change (Theorem 23.133) applied to the cartesian diagram

$$
\begin{array}{ccc}
X_s \times_{\kappa(s)} X_s^t & \longrightarrow & X \times_S X^t \\
{\scriptstyle p'_s} \downarrow & & \downarrow {\scriptstyle p'} \\
X_s^t & \xrightarrow{\;\; i_s \;\;} & X^t
\end{array}
$$

we find $Li_s^* E = Rp'_{s*} \mathscr{P}_s$, where $\mathscr{P}_s$ is the Poincaré bundle for X_s. As we proved (3) already if S is the spectrum of a field, we have $R^i p'_{s*} \mathscr{P}_s = 0$ for all $i < g$. Hence Lemma 27.227 below shows that $H^i(E) = R^i p'_* \mathscr{P} = 0$ for all $i < g$. $\qquad\square$

Lemma 27.227. *Let $f \colon X \to S$ be a flat morphism locally of finite presentation and let E be a pseudo-coherent complex on X. For $s \in S$ denote by $i_s \colon X_s := X \times_S \kappa(s) \to X$ the canonical morphism. Then for all $n \in \mathbb{Z}$*

$$
U := \{\, x \in X \;;\; H^n(Li_{f(x)}^* E)_x = 0 \,\}
$$

is open and constructible in X and $H^n(E)_{|U} = 0$.

Proof. One can assume that S and X are affine. Then E can be represented by a bounded above complex of finite free $\mathscr{O}_X$-modules $\mathscr{E}$ (Remark 22.46). Then $\mathscr{E}$ is K-flat and $Li_{f(x)}^* E = i_{f(x)}^* \mathscr{E}$. Now we can apply [EGAIV]$^\circ$ (12.3.3) (the fact that U is constructible follows from the proof in loc. cit.). $\qquad\square$

Lemma 27.228. *With the notation above we have for every quasi-coherent $\mathscr{O}_S$-module $\mathscr{G}$ and every quasi-coherent $\mathscr{O}_{X^t}$-module $\mathscr{H}$ functorial isomorphisms*

$$
(27.43.2) \qquad \operatorname{Hom}_{\mathscr{O}_S}(R^g \pi_* \mathscr{O}_X, \mathscr{G}) \xrightarrow{\;\sim\;} \Gamma(X, \pi^* \mathscr{G} \otimes \Omega_{X/S}^g),
$$

$$
(27.43.3) \qquad \operatorname{Hom}_{\mathscr{O}_{X^t}}(R^g p'_* \mathscr{P}, \mathscr{H}) \xrightarrow{\;\sim\;} \operatorname{Hom}_{\mathscr{O}_{X \times X^t}}(\mathscr{P}, p^{t,*} \mathscr{H} \otimes p^* \Omega_{X/S}^g).
$$

As $\pi \colon X \to S$ is a smooth group scheme of relative dimension g, $\Omega_{X/S}^g$ is the pullback of a line bundle on S (Proposition 27.15, Corollary 27.21). Hence we find

$$
\operatorname{Hom}_{\mathscr{O}_S}(R^g \pi_* \mathscr{O}_X, \mathscr{G}) \xrightarrow{\;\sim\;} \Gamma(X, \pi^* \mathscr{G})
$$

if $\operatorname{Pic}(S) = 0$, e.g., if S is the spectrum of a field.

Proof. We show the first isomorphism. As π is proper and smooth, for every complex G in $D_{\mathrm{qcoh}}(S)$, one has $\pi^{\times}G = L\pi^*G \otimes^L_{\mathscr{O}_X} \Omega^g_{X/S}[g]$ (Theorem 25.58). Hence by Grothendieck duality (Corollary 25.17, Theorem 25.58) we have

$$\mathrm{Hom}_{D(S)}(R\pi_*\mathscr{O}_X, G) = \mathrm{Hom}_{D(X)}(\mathscr{O}_X, L\pi^*G \otimes^L_{\mathscr{O}_X} \Omega^g_{X/S}[g]).$$

Taking $G = \mathscr{G}[-g]$ we obtain

$$\begin{aligned} \mathrm{Hom}_{D(S)}(R\pi_*\mathscr{O}_X, \mathscr{G}[-g]) &= \mathrm{Hom}_{D(X)}(\mathscr{O}_X, L\pi^*\mathscr{G} \otimes^L_{\mathscr{O}_X} \Omega^g_{X/S}) \\ &= \mathrm{Hom}_{\mathscr{O}_X}(\mathscr{O}_X, \pi^*\mathscr{G} \otimes \Omega^g_{X/S}) \\ &= \Gamma(X, \pi^*\mathscr{G} \otimes \Omega^g_{X/S}) \end{aligned}$$

using that π and $\Omega^g_{X/S}$ are flat. Consider the natural map $R\pi_*\mathscr{O}_X \to \tau^{\geq g}R\pi_*\mathscr{O}_X = R^g\pi_*\mathscr{O}_X[-g]$, where the equality holds because $R\pi_*\mathscr{O}_X$ has cohomology concentrated in degrees $\leq g$. This map induces by functoriality an isomorphism

$$\mathrm{Hom}_{\mathscr{O}_S}(R^g\pi_*\mathscr{O}_X, \mathscr{G}) = \mathrm{Hom}_{D(S)}(R^g\pi_*\mathscr{O}_X[-g], \mathscr{G}[-g]) \xrightarrow{\sim} \mathrm{Hom}_{D(S)}(R\pi_*\mathscr{O}_X, \mathscr{G}[-g])$$

because $\tau^{\geq g}$ is left adjoint to the inclusion $D^{\geq g}(S) \to D(S)$. This gives the first isomorphism.

For the second isomorphism one argues similarly, using that $Rp'_*\mathscr{P} = R^gp'_*\mathscr{P}[-g]$ by Lemma 27.226 (3) and that $\Omega^g_{X\times_S X^t/X^t} = p^*\Omega^g_{X/S}$ by Proposition 17.30. $\qquad\square$

Proposition 27.229. *With the notation above let $\mathscr{C}_e = e^*\Omega^1_{X/S}$ be the conormal bundle of the zero section, which is a locally free $\mathscr{O}_S$-module of rank g, and let $\det(\mathscr{C}_e)$ be its determinant, i.e., its g-th exterior power. Then the following assertions hold.*
(1)

$$R^i\varpi_*\mathscr{P} = \begin{cases} 0, & \text{if } i \neq g, \\ \det(\mathscr{C}_e)^{-1}, & \text{if } i = g. \end{cases}$$

(2)

$$R^ip'_*\mathscr{P} = \begin{cases} 0, & \text{if } i \neq g, \\ e'_*\det(\mathscr{C}_e)^{-1}, & \text{if } i = g. \end{cases}$$

If $S = \mathrm{Spec}\, k$ for a field k (or, more generally, for a ring k such that $\mathrm{Pic}(k) = 0$), then $\det(\mathscr{C}_e)^{-1} \cong \mathscr{O}_S$.

Proof. We have already seen in Lemma 27.226 that $R^i\varpi_*\mathscr{P}$ and $R^ip'_*\mathscr{P}$ vanish for $i \neq g$. Hence it suffices by (1) of the lemma to show that $R^gp'_*\mathscr{P} \cong e'_*\det(\mathscr{C}_e)^{-1}$.

As the formation of $R^gp'_*\mathscr{P}$ commutes with base change (Lemma 27.226 (4)), we have $e^{t,*}R^gp'_*\mathscr{P} = R^g\pi_*((\mathrm{id}_X, e')^*\mathscr{P}) = R^g\pi_*\mathscr{O}_X$ by the cartesian diagram

$$\begin{array}{ccc} X & \xrightarrow{(\mathrm{id}_X, e')} & X \times_S X^t \\ \pi \downarrow & & \downarrow p' \\ S & \xrightarrow{e'} & X^t. \end{array}$$

Therefore we have

$$\operatorname{Hom}_{\mathscr{O}_{X^t}}(R^g p'_* \mathscr{P}, e'_* \det(\mathscr{C}_e)^{-1}) = \operatorname{Hom}_{\mathscr{O}_S}(e^{t,*} R^g p'_* \mathscr{P}, \det(\mathscr{C}_e)^{-1})$$
$$= \operatorname{Hom}_{\mathscr{O}_S}(R^g \pi_* \mathscr{O}_X, \det(\mathscr{C}_e)^{-1})$$
$$(27.43.4) \qquad\qquad\qquad = \Gamma(X, \pi^* \det(\mathscr{C}_e)^{-1} \otimes \Omega^g_{X/S})$$
$$= \Gamma(X, \mathscr{O}_X) = \Gamma(S, \mathscr{O}_S),$$

where the third equality holds by Lemma 27.228. Hence there is a natural homomorphism of $\mathscr{O}_{X^t}$-modules

$$\xi \colon R^g p'_* \mathscr{P} \to e'_* \det(\mathscr{C}_e)^{-1}$$

corresponding to $1 \in \Gamma(S, \mathscr{O}_S)$.

The identifications in (27.43.4) are compatible with base change $T \to S$, hence ξ is compatible with base change $T \to S$. To prove that ξ is an isomorphism, we may work Zariski locally on S. In particular we can assume that $S = \operatorname{Spec} R$ is affine. Moreover we may assume that R is noetherian by writing R as a filtered colimit of finitely generated $\mathbb{Z}$-algebras. As in the proof of Lemma 27.226 (3) we find a closed subscheme $h \colon Z \to X^t$, the vanishing locus of the annihilator of $R^g p'_* \mathscr{P}$, whose underlying topological space is contained in $e'(S)$ such that $h_* h^* R^g p'_* \mathscr{P} = R^g p'_* \mathscr{P}$. For $s \in S$ we have $\kappa(s) = \kappa(e'(s))$ because e' is an immersion. By Lemma 27.226 (4)

$$(*) \qquad R^g p'_* \mathscr{P} \otimes_{\mathscr{O}_{X^t}} \kappa(s) = H^g(X_{e'(s)}, \mathscr{P}_{e'(s)}) = H^g(X_s, \mathscr{O}_{X_s}) \cong \kappa(s).$$

Therefore, the underlying topological space of Z is equal to $e'(S)$ and $R^g p'_* \mathscr{P}$ is locally generated by one element.

We claim that $e' \colon S \to X^t$ factors through the subscheme Z and hence that Z is a nilpotent thickening of the closed subscheme $e'(S)$. For this let $Z' \subseteq Z$ be the maximal subscheme, where $R^g p'_* \mathscr{P}$ is locally free of rank 1 and its formation is compatible with base change (Remark 23.148). As $e^{t,*} R^g p'_* \mathscr{P} = R^g \pi_* \mathscr{O}_X$ is finite locally free of rank 1 (Theorem 27.203), e' factors through Z' and hence through Z. This shows the claim.

As S is affine, Z is affine as a nilpotent thickening by Corollary 12.40. Hence $Z = \operatorname{Spec} A$ and there exists a nilpotent ideal $I \subseteq A$ with $R = A/I$. Moreover, (*) shows that locally on Z and hence locally on S, which has the same underlying topological space, $h^* R^g p'_* \mathscr{P}$ corresponds to an A-module generated by one element. Therefore we see that $R^g p'_* \mathscr{P} = h_*(\widetilde{A/J})$ for some ideal $J \subseteq A$. Then J is contained in the ideal defining Z' in Z and in particular $J \subseteq I$. If we set $Z_J := \operatorname{Spec} A/J$ and $X_J := X \times_S Z_J$, then $X \times 0 = X_I \to X_J$ is a nilpotent thickening.

Further localizing on S we may assume that $\det(\mathscr{C}_e) \cong \mathscr{O}_S$ and therefore $\Omega^g_{X/S} = \pi^* \det(\mathscr{C}_e) \cong \mathscr{O}_X$. Then ξ is given by the canonical map $\mathscr{O}_{Z_J} \to \mathscr{O}_{Z_I}$ and it remains to see that $I = J$.

We set $\mathscr{P}_J := \mathscr{P}_{|X_J}$. Then $\mathscr{P}_I = \mathscr{P}_{|X \times 0} \cong \mathscr{O}_{X_I}$ and we have a commutative diagram

$$\operatorname{Hom}_{\mathscr{O}_{X_J}}(\mathscr{P}_J, \mathscr{O}_{X_J}) = \operatorname{Hom}_{\mathscr{O}_{X \times_S X^t}}(\mathscr{P}, \mathscr{O}_{X_J}) \xrightarrow{\sim} \operatorname{Hom}_{\mathscr{O}_{X^t}}(\mathscr{O}_{Z_J}, \mathscr{O}_{Z_J}) = A/J$$

$$\downarrow \qquad\qquad\qquad\qquad\qquad\qquad\qquad\qquad\qquad \downarrow$$

$$\operatorname{Hom}_{\mathscr{O}_{X_I}}(\mathscr{O}_{X_I}, \mathscr{O}_{X_I}) = \operatorname{Hom}_{\mathscr{O}_{X \times_S X^t}}(\mathscr{P}, \mathscr{O}_{X_I}) \xrightarrow{\sim} \operatorname{Hom}_{\mathscr{O}_{X^t}}(\mathscr{O}_{Z_J}, \mathscr{O}_{Z_I}) = A/I,$$

where the horizontal maps are the isomorphisms (27.43.3). As the right vertical map is surjective, the identity of $\mathscr{O}_{X_I}$ can be lifted to a map $u \colon \mathscr{P}_J \to \mathscr{O}_{X_J}$. As u is an isomorphism modulo the nilpotent ideal I, u is surjective by Nakayama's lemma and hence an isomorphism as a map between line bundles. This shows that the restriction of $\mathscr{P}$ to X_J is trivial. By the universal property of $\mathscr{P}$ this implies that $Z_J \to X^t$ factors through $X_I = X \times 0$. Hence $I = J$. □

Applying Proposition 27.229 to the Poincaré bundle of X^t and using Remark 27.224 we obtain also the following result.

Corollary 27.230. *With the notation above we have* $\det(\mathscr{C}_e) \cong \det(\mathscr{C}_{e'})$ *and*

$$R^i p_* \mathscr{P} = \begin{cases} 0, & \text{if } i \neq g, \\ e_* \det(\mathscr{C}_e)^{-1}, & \text{if } i = g, \end{cases}$$

(27.44) The Néron-Severi group of an abelian scheme.

Proposition and Definition 27.231. *Let k be an algebraically closed field and let X be an abelian variety over k. Let $\mathscr{L}_1$ and $\mathscr{L}_2$ be line bundles on X. Then the following assertions are equivalent.*
(i) *One has $\mathscr{L}_1 \otimes \mathscr{L}_2^{-1} \in X^t(k)$.*
(ii) *There exists a connected k-scheme Z, a line bundle $\mathscr{M}$ on $X \times_k Z$ and $z_1, z_2 \in Z(k)$ such that $\mathscr{L}_i = \mathscr{M}_{|X \times \{z_i\}}$ for $i = 1, 2$.*
Moreover, the k-scheme Z can be chosen to be smooth and of finite type.
If these conditions are satisfied, then $\mathscr{L}_1$ and $\mathscr{L}_2$ are called algebraically equivalent.

Proof. Condition (ii) means that $\mathscr{L}_1$ and $\mathscr{L}_2$ are in the same connected component Z' of $\mathrm{Pic}_{X/k}$, i.e., it is satisfied if and only if $\mathscr{L}_1 \otimes \mathscr{L}_2^{-1} \in \mathrm{Pic}^0_{X/k}(k) = X^t(k)$.
Moreover, one can then choose for Z this connected component Z', which is smooth and of finite type, and for $\mathscr{M}$ some representative of the universal line bundle in $\mathrm{Pic}_{X/S}(Z') = \mathrm{Pic}(X \times_k Z')/\mathrm{Pic}(Z')$ over it. □

Definition 27.232. *Let S be a scheme and let X be an abelian scheme over S. Then the quotient of abelian fppf-groups*

$$\mathrm{NS}(X) := \mathrm{Pic}_{X/S}/X^t$$

is called the Néron-Severi group *of X.*

As X^t is the identity component of $\mathrm{Pic}_{X/S}$, one can view $\mathrm{NS}(X)$ as the functor of connected components of $\mathrm{Pic}_{X/S}$.

Remark 27.233. Let S be a scheme and let X be an abelian scheme over S. Let $\underline{\mathrm{Hom}}(X, X^t)$ be the functor that sends an S-scheme T to the abelian group of homomorphisms of abelian group spaces $X_T \to X^t_T$. This is an fppf-sheaf by Theorem 14.72. It is isomorphic to $\mathrm{Corr}_S(X, X)$ (Proposition 27.163) and hence a scheme that is separated, unramified, and locally of finite presentation over S (Proposition 27.157).
(1) For an integer $n \in \mathbb{Z}$ and $f \in \underline{\mathrm{Hom}}(X, X^t)(T)$ we have $n \cdot f = f \circ [n]_X$. If $n \neq 0$, then $[n]_X$ is an epimorphism and hence $nf = 0$ implies $f = 0$. In other words, $\underline{\mathrm{Hom}}(X, X^t)$ is a sheaf of torsion free $\mathbb{Z}$-modules.

(2) The homomorphism of abelian fppf-sheaves $\varphi\colon \mathrm{Pic}_{X/S} \to \underline{\mathrm{Hom}}(X, X^t)$, $\mathscr{L} \mapsto \varphi_{\mathscr{L}}$, by definition has kernel X^t. Hence it induces a monomorphism

(27.44.1) $\bar\varphi\colon \mathrm{NS}(X) \hookrightarrow \underline{\mathrm{Hom}}(X, X^t)$.

In particular, $\mathrm{NS}(X)(T)$ is torsion free for all S-schemes T and $\mathrm{NS}(X)$ is formally unramified as a functor, i.e. $\mathrm{NS}(X)(T) \to \mathrm{NS}(X)(T_0)$ is injective for all closed immersions $T_0 \to T$ defined by a locally nilpotent quasi-coherent ideal of $\mathscr{O}_T$.

Let $f\colon X \to Y$ be a homomorphism of abelian schemes. Then $f^*\colon \mathrm{Pic}_{Y/S} \to \mathrm{Pic}_{X/S}$ induces a homomorphism of abelian fppf-sheaves $f^*\colon \mathrm{NS}(Y) \to \mathrm{NS}(X)$.

Proposition 27.234. *Let $n \in \mathbb{Z}$. The multiplication $[n]\colon X \to X$ induces on $\mathrm{NS}(X)$ the multiplication by n^2.*

Proof. This follows from Proposition 27.184 (3).   □

We can identify $\underline{\mathrm{Hom}}(X, X^t) = \mathrm{Corr}_S(X, X)$ with a subgroup functor of $\mathrm{Pic}_{X\times_S X/S}$ (Remark 27.152). Composing the inclusion $\underline{\mathrm{Hom}}(X, X^t) \hookrightarrow \mathrm{Pic}_{X\times_S X/S}$ with the map $\Delta^*\colon \mathrm{Pic}_{X\times_S X/S} \to \mathrm{Pic}_{X/S}$, where $\Delta\colon X \to X \times_S X$ is the diagonal, we obtain a map $\delta\colon \underline{\mathrm{Hom}}(X, X^t) \to \mathrm{Pic}_{X/S}$. Consider the composition

(27.44.2) $\mathrm{NS}(X) \xrightarrow{\ \bar\varphi\ } \underline{\mathrm{Hom}}(X, X^t) \xrightarrow{\ \delta\ } \mathrm{Pic}_{X/S} \longrightarrow \mathrm{NS}(X)$,

where the last map is the canonical projection.

Lemma 27.235. *The map (27.44.2) is the multiplication by 2 on $\mathrm{NS}(X)$.*

Proof. Indeed, the composition $\mathrm{Pic}_{X/S} \xrightarrow{\ \varphi\ } \mathrm{Corr}_S(X, X) \hookrightarrow \mathrm{Pic}_{X\times_S X/S}$ is given by

$$\mathscr{L} \mapsto m^*\mathscr{L} \otimes p_1^*\mathscr{L}^{-1} \otimes p_2^*\mathscr{L}^{-1}$$

by Proposition 27.161 (1). Applying Δ^* we obtain the class of $[2]^*\mathscr{L} \otimes \mathscr{L}^{\otimes -2}$ in $\mathrm{Pic}_{X/S}$. Since $[2]^*\mathscr{L}$ is algebraically equivalent to $\mathscr{L}^{\otimes 4}$ by Proposition 27.234, this shows the claim.   □

(27.45) Fourier-Mukai transforms.

In this section we discuss Fourier-Mukai transforms, a construction invented by S. Mukai which gives rise to autoequivalences of categories of the form $D_{\mathrm{qcoh}}(X)$ and in a sense is analogous to the classical Fourier transform (since push-forward can be thought of as an analogue of summation or integration, cf. formula (23.23.6)). In the next section we will show that the Fourier-Mukai transform given by the Poincaré bundle yields an equivalence $D_{\mathrm{qcoh}}(X) \cong D_{\mathrm{qcoh}}(X^t)$ for any abelian scheme X.

Let S be a scheme, let X and Y be S-schemes, and denote by $p\colon X \times_S Y \to X$ and $q\colon X \times_S Y \to Y$ the projections.

Definition 27.236. *Let K be a complex in $D(X \times_S Y)$. Then the triangulated functors*

$$\Phi_K\colon D(X) \longrightarrow D(Y), \qquad E \mapsto Rq_*(Lp^*E \otimes^L_{\mathscr{O}_{X\times_S Y}} K),$$

$$\Psi_K\colon D(Y) \longrightarrow D(X), \qquad F \mapsto Rp_*(Lq^*F \otimes^L_{\mathscr{O}_{X\times_S Y}} K)$$

are called Fourier-Mukai transforms with kernel K.

Remark 27.237. Usually, one restricts the Fourier-Mukai transforms to suitable subcategories of $D(X)$.
(1) Let $X \to S$ and $Y \to S$ be qcqs. Then the projections $p \colon X \times_S Y \to X$ and $q \colon X \times_S Y \to Y$ are also qcqs. Let $K \in D_{\mathrm{qcoh}}(X \times_S Y)$. Then the Fourier-Mukai transforms Φ_K and Ψ_K induce functors

$$\Phi_K \colon D_{\mathrm{qcoh}}(X) \longrightarrow D_{\mathrm{qcoh}}(Y), \qquad \Psi_K \colon D_{\mathrm{qcoh}}(Y) \longrightarrow D_{\mathrm{qcoh}}(X)$$

by Proposition 22.40 and Theorem 22.31.
(2) Let $X \to S$ and $Y \to S$ be proper flat morphisms of finite presentation. Then p and q have the same properties. Let $K \in D(X \times_S Y)$ be perfect. Then Φ_K and Ψ_K restrict to functors

$$\Phi_K \colon (\mathrm{Perf}(X)) \longrightarrow (\mathrm{Perf}(Y)), \qquad \Psi_K \colon (\mathrm{Perf}(Y)) \longrightarrow (\mathrm{Perf}(X))$$

by Remark 21.142, Remark 21.143 and Corollary 23.136.

Often one considers the case where $S = \operatorname{Spec} k$ for a field k and X and Y are smooth proper schemes over k. Then X, Y, and $X \times_k Y$ are regular of finite dimension and hence a complex over X, over Y, or over $X \times_k Y$ is perfect if and only if its cohomology modules are coherent and non-zero only in finitely many degrees (Proposition 23.55). Hence in this case we obtain functors

$$\Phi_K \colon D^b_{\mathrm{coh}}(X) \longrightarrow D^b_{\mathrm{coh}}(Y), \qquad \Psi_K \colon D^b_{\mathrm{coh}}(Y) \longrightarrow D^b_{\mathrm{coh}}(X).$$

The following example shows that derived pushforward, derived pullback and derived tensor product are all special cases of Fourier-Mukai transforms, at least for qcqs morphisms.

Example 27.238. Let $f \colon X \to S$ and $g \colon Y \to S$ be qcqs morphisms of schemes. Let $h \colon X \to Y$ be a morphism of S-schemes, let $\Gamma \colon X \to X \times_S Y$ be its graph morphism. Let $\mathscr{M}$ be in $D_{\mathrm{qcoh}}(X)$, and set $K := R\Gamma_* \mathscr{M}$. Then Γ is qcqs and hence $K \in D_{\mathrm{qcoh}}(X \times_S Y)$. The corresponding Fourier-Mukai transforms are given by

$$\Phi_K \colon D_{\mathrm{qcoh}}(X) \longrightarrow D_{\mathrm{qcoh}}(Y), \qquad \Phi_K(\mathscr{F}) = Rh_*(\mathscr{F} \otimes^L \mathscr{M}),$$
$$\Psi_K \colon D_{\mathrm{qcoh}}(Y) \longrightarrow D_{\mathrm{qcoh}}(Y), \qquad \Psi_K(\mathscr{G}) = Lh^* \mathscr{G} \otimes^L \mathscr{M}.$$

Indeed, for $\mathscr{F} \in D_{\mathrm{qcoh}}(X)$ and $\mathscr{G} \in D_{\mathrm{qcoh}}(Y)$ one has by the projection formula (Proposition 22.84)

$$\Phi_K(\mathscr{F}) = Rq_*(Lp^* \mathscr{F} \otimes^L R\Gamma_* \mathscr{M}) = Rq_*(R\Gamma_*(L\Gamma^* Lp^* \mathscr{F} \otimes^L \mathscr{M}))$$
$$= Rq_* R\Gamma_*(\mathscr{F} \otimes^L \mathscr{M}) = Rh_*(\mathscr{F} \otimes^L \mathscr{M}),$$
$$\Psi_K(\mathscr{G}) = Rp_*(Lq^* \mathscr{G} \otimes^L R\Gamma_* \mathscr{M})$$
$$= Rp_*(R\Gamma_*(L\Gamma^* Lq^* \mathscr{G} \otimes^L \mathscr{M})) = Lh^* \mathscr{G} \otimes^L \mathscr{M}.$$

As special cases we obtain
(1) If $\mathscr{M} = \mathscr{O}_X$, then $\Phi_K = Rh_*$ and $\Psi_K = Lh^*$.
(2) If $h = \mathrm{id}_X$, then $\Gamma \colon X \to X \times_S X$ is the diagonal, and $\Phi_K = \Psi_K = (- \otimes^L \mathscr{M})$.

Fourier-Mukai transforms for perfect complexes admit a left and a right adjoint by Grothendieck duality for smooth proper morphisms.

Proposition 27.239. *Let S be a scheme, let $f: X \to S$ and $g: Y \to S$ be smooth proper morphisms of relative dimension m and n, respectively, and denote by $p: X \times_S Y \to X$ and $q: X \times_S Y \to Y$ the projections. Let K be a perfect complex in $D(X \times_S Y)$ and set*

$$K_R := K^\vee \otimes^L Lp^* \Omega^m_{X/S}[m], \qquad K_L := K^\vee \otimes^L Lq^* \Omega^n_{Y/S}[n].$$

Then Ψ_{K_L} is a left adjoint and Ψ_{K_R} is a right adjoint to $\Phi_K: (\mathrm{Perf}(X)) \to (\mathrm{Perf}(Y))$.

Proof. As f and g are flat, so are p and q and therefore $p^* = Lp^*$ and $q^* = Lq^*$. For any E in $(\mathrm{Perf}(X))$ and F in $(\mathrm{Perf}(Y))$ one has functorial isomorphisms

$$
\begin{aligned}
\mathrm{Hom}_{D(X)}(\Psi_{K_L}(F), E) &= \mathrm{Hom}_{D(X)}(Rq_*(K_L \otimes^L q^*F), E) \\
&\cong \mathrm{Hom}_{D(X \times Y)}(K_L \otimes^L q^*F, p^! E) \\
&\cong \mathrm{Hom}_{D(X \times Y)}(K_L \otimes^L q^*F, p^*E \otimes^L \Omega^n_{X \times Y/X}[n]) \\
&\cong \mathrm{Hom}_{D(X \times Y)}(K^\vee \otimes^L q^*\Omega^n_{Y/S}[n] \otimes^L q^*F, p^*E \otimes^L q^*\Omega^n_{Y/S}[n]) \\
&\cong \mathrm{Hom}_{D(X \times Y)}(K^\vee \otimes^L q^*F, p^*E) \\
&\cong \mathrm{Hom}_{D(X \times Y)}(q^*F, K \otimes^L p^*E) \\
&\cong \mathrm{Hom}_{D(Y)}(F, Rq_*(K \otimes^L p^*E)) \\
&= \mathrm{Hom}_{D(Y)}(F, \Phi_K(E)).
\end{aligned}
$$

Here the first isomorphism holds by definition of the functor $p^! = p^\times$ as p is proper, the second by Theorem 25.58, the third by compatibility of Kähler differentials with base change (apply the n-th exterior power to Proposition 17.30), the fourth because $q^*\Omega^n_{Y/S}[n]$ is an invertible object in $D(X \times_S Y)$, the fifth by duality of perfect complexes (apply $H^0 \circ R\Gamma$ to (21.32.3)), and the sixth by adjointness of $q^* = Lq^*$ and Rq_* (Proposition 21.126). $\square$

Remark 27.240. A theorem of Orlov ([Orl]$_X^O$ 3.2.1) shows the following result. Let X and Y be smooth projective schemes over a field k (and hence $D^b_{\mathrm{coh}}(X) = (\mathrm{Perf}(X))$ and $D^b_{\mathrm{coh}}(Y) = (\mathrm{Perf}(Y))$ by Proposition 23.55). Let $F: D^b_{\mathrm{coh}}(X) \to D^b_{\mathrm{coh}}(Y)$ be a fully faithful triangulated functor. Then there exists $K \in D^b_{\mathrm{coh}}(X \times_k Y)$ such that $F \cong \Phi_K$, and K is unique up to isomorphism.

In loc. cit. this result is formulated with the additional hypothesis that F has a left or a right adjoint functor. This is in fact automatic by a result of Bondal and van den Bergh (e.g., [BBH]^O 1.20).

We now show how to compute the composition of Fourier-Mukai transforms. Let S be a scheme and let $X \to S$, $Y \to S$, and $Z \to S$ be S-schemes. We write $X \times Y$ instead of $X \times_S Y$ and similarly for other fiber products. Consider the commuting diagram of projections

(27.45.1)

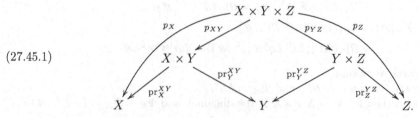

The middle diamond is cartesian. Moreover denote by $p_{XZ}: X \times Y \times Z \to X \times Z$ the projection.

Definition and Proposition 27.241. *For $K \in D(X \times Y)$ and $M \in D(Y \times Z)$ define their convolution by*

$$(27.45.2) \qquad K * M := Rp_{XZ,*}(Lp_{XY}^* K \otimes^L Lp_{YZ}^* M) \in D(X \times Z).$$

Suppose
(a) *that the morphisms $X \to S$, $Y \to S$, and $Z \to S$ are qcqs,*
(b) *that the two projections $\mathrm{pr}_Y^{XY} : X \times Y \to Y$ and $\mathrm{pr}_Y^{YZ} : Y \times Z \to Y$ are tor-independent (e.g., if $X \to S$ or $Z \to S$ is flat), and*
(c) *that $K \in D_{\mathrm{qcoh}}(X \times Y)$ and $M \in D_{\mathrm{qcoh}}(Y \times Z)$.*
Then one has

$$\Phi_M \circ \Phi_K = \Phi_{K*M} \colon D_{\mathrm{qcoh}}(X) \longrightarrow D_{\mathrm{qcoh}}(Z)$$

and similarly $\Psi_K \circ \Psi_M = \Psi_{K*M}$ as functors from $D_{\mathrm{qcoh}}(Z)$ to $D_{\mathrm{qcoh}}(X)$.

Proof. As $Y \to S$ is qcqs, p_{XZ} is qcqs, and hence $K * L \in D_{\mathrm{qcoh}}(X \times Z)$. Let $\mathrm{pr}_X^{XZ} \colon X \times Z \to X$ and $\mathrm{pr}_Z^{XZ} \colon X \times Z \to Z$ be the projections. By symmetry in X and Z it suffices to show $\Phi_M \circ \Phi_K = \Phi_{K*M}$. Let $\mathscr{F} \in D_{\mathrm{qcoh}}(X)$. Then

$$
\begin{aligned}
(\Phi_M \circ \Phi_K)(\mathscr{F}) &= \Phi_M(R\,\mathrm{pr}_{Y,*}^{XY}(L\,\mathrm{pr}_X^{XY,*}(\mathscr{F}) \otimes^L K)) \\
&= R\,\mathrm{pr}_{Z,*}^{YZ}\left(L\,\mathrm{pr}_Y^{YZ,*}\big(R\,\mathrm{pr}_{Y,*}^{XY}(L\,\mathrm{pr}_X^{XY,*}(\mathscr{F}) \otimes^L K)\big) \otimes^L M\right) \\
&\cong R\,\mathrm{pr}_{Z,*}^{YZ}\left(Rp_{YZ,*}\big(Lp_{XY}^*(L\,\mathrm{pr}_X^{XY,*}(\mathscr{F}) \otimes^L K)\big) \otimes^L M\right) \\
&= R\,\mathrm{pr}_{Z,*}^{YZ}\left(Rp_{YZ,*}\big(Lp_X^* \mathscr{F} \otimes^L Lp_{XY}^* K\big) \otimes^L M\right) \\
&\cong R\,\mathrm{pr}_{Z,*}^{YZ}\left(Rp_{YZ,*}\big(Lp_X^* \mathscr{F} \otimes^L Lp_{XY}^* K \otimes^L Lp_{YZ}^* M\big)\right) \\
&= Rp_{Z,*}\big(Lp_X^* \mathscr{F} \otimes^L Lp_{XY}^* K \otimes^L Lp_{YZ}^* M\big) \\
&= R\,\mathrm{pr}_{Z,*}^{XZ}\left(Rp_{XZ,*}\big(Lp_{XZ}^*(L\,\mathrm{pr}_X^{XZ,*} \mathscr{F}) \otimes^L Lp_{XY}^* K \otimes^L Lp_{YZ}^* M\big)\right) \\
&\cong R\,\mathrm{pr}_{Z,*}^{XZ}\left(L\,\mathrm{pr}_X^{XZ,*} \mathscr{F} \otimes^L Rp_{XZ,*}\big(Lp_{XY}^* K \otimes^L Lp_{YZ}^* M\big)\right) \\
&= R\,\mathrm{pr}_{Z,*}^{XZ}\left(L\,\mathrm{pr}_X^{XZ,*} \mathscr{F} \otimes^L (K * M)\right) \\
&= \Phi_{L*M}(\mathscr{F}),
\end{aligned}
$$

where the first isomorphism holds since the middle diamond in (27.45.1) is tor-independent (Theorem 22.99), the second isomorphism holds by the projection formula (Proposition 22.84) for p_{YZ}, and the third isomorphism holds by the projection formula for p_{XZ}. $\qquad\square$

Finally, we study when the Fourier-Mukai transform is compatible with base change.

Proposition 27.242. *Let $S' \to S$ be a morphism of schemes, set $X' = X \times_S S'$ and $Y' = Y \times_S S'$, and let*

$$\pi \colon X' \longrightarrow X, \qquad \varpi \colon Y' \to Y, \qquad \pi \times \varpi \colon X' \times_{S'} Y' \longrightarrow X \times_S Y$$

be the projections. Suppose that $X \to S$ and $Y \to S$ are qcqs and let K be an object in $D_{\mathrm{qcoh}}(X \times_S Y)$.

(1) *If $S' \to S$ or $X \to S$ are flat (or, more generally, if ϖ and $q\colon X \times_S Y \to Y$ are tor-independent), then we have for every $E \in D_{\mathrm{qcoh}}(X)$*

$$\Phi_{L(\pi \times \varpi)^* K}(L\pi^* E) = L\varpi^* \Phi_K(E).$$

(2) *Dually, if $S' \to S$ or $Y \to S$ are flat (or, more generally, if π and $p\colon X \times_S Y \to X$ are tor-independent), then we have for every $F \in D_{\mathrm{qcoh}}(X)$*

$$\Phi_{L(\pi \times \varpi)^* K}(L\varpi^* F) = L\pi^* \Phi_K(F).$$

Proof. We show the first assertion. The proof of (2) is similar. Let $p'\colon X' \times_{S'} Y' \to X'$ and $q'\colon X' \times_{S'} Y' \to Y'$ be the projections. We have

$$
\begin{aligned}
\Phi_{L(\pi \times \varpi)^* K}(L\pi^* E) &= Rq'_*(Lp'^* L\pi^* E \otimes^L L(\pi \times \varpi)^* K) \\
&= Rq'_*(L(\pi \otimes \varpi)^* Lp^* E \otimes^L L(\pi \times \varpi)^* K) \\
&= Rq'_* L(\pi \otimes \varpi)^* (Lp^* E \otimes^L K) \\
&= L\varpi^* Rq_*(Lp^* E \otimes^L K) \\
&= L\varpi^* \Phi_K(E).
\end{aligned}
$$

Here we used for the fourth equality that the cartesian diagram

$$
\begin{array}{ccc}
X' \times_{S'} Y' & \xrightarrow{\;\pi \times \varpi\;} & X \times_S Y \\
{\scriptstyle q'}\downarrow & & \downarrow{\scriptstyle q} \\
Y' & \xrightarrow{\;\varpi\;} & Y
\end{array}
$$

is tor-independent by hypothesis so that we can apply Theorem 22.99.   □

(27.46) Fourier-Mukai equivalence for abelian schemes.

Let S be a scheme, let $X \to S$ be an abelian scheme over S, and let X^t be its dual abelian scheme. Let $\mathscr{P}_X$ be the Poincaré bundle on $X \times_S X^t$. We obtain Fourier-Mukai transforms

$$S_X := \Phi_{\mathscr{P}_X}\colon D_{\mathrm{qcoh}}(X) \to D_{\mathrm{qcoh}}(X^t).$$

We consider $\mathscr{P}_X$ also as a line bundle on $X^t \times_S X$. This line bundle is identified with $\mathscr{P}_{X^t}$ if we identify $X^t \times_S X$ with $X^t \times_S X^{tt}$ via $\mathrm{id}_{X^t} \times \kappa_X$, where κ_X is the biduality isomorphism (Remark 27.224). Hence we obtain a Fourier-Mukai transform

$$S_{X^t} := \Phi_{\mathscr{P}_{X^t}}\colon D_{\mathrm{qcoh}}(X^t) \to D_{\mathrm{qcoh}}(X).$$

In the sequel we will use the following notation for various standard maps.
(a) We denote by $f\colon X \to S$ the structure morphism and by $e\colon S \to X$ its zero section.
(b) We write $\pi\colon X \times_S X^t \to X$ and $\pi'\colon X \times_S X^t \to X^t$ for the projections.
(c) We identify $X^t \times_S X$ with $X \times_S X^t$ and call this identification σ. This yields an identification $\mathscr{P}_X$ with $\mathscr{P}_{X^t}$.
(d) We denote by $p_1, p_2, m\colon X \times_S X \to X$ the first projection, the second projection, and the group law.

(e) We identify $X \times_S X^t \times_S X$ with $X \times_S X \times_S X^t$ and for $\emptyset \neq I \subset \{1,2,3\}$ write pr_I for the I-th projection on $X \times_S X \times_S X^t$, for instance the projections pr_{12} to $X \times_S X$ and pr_3 to X^t.

(f) We write $\mu \colon X \times_S X \times_S X^t \to X \times_S X^t$ for the morphism $(x, y, \xi) \mapsto (x + y, \xi)$

Theorem 27.243. *Let S be a scheme and let $X \to S$ be an abelian scheme of relative dimension g. One has*

$$S_{X^t} \circ S_X \cong [-1]^* \circ (- \otimes^L (\Omega^g_{X/S})^{-1}[-g]),$$

$$S_X \circ S_{X^t} \cong [-1]^* \circ (- \otimes^L (\Omega^g_{X^t/S})^{-1}[-g]).$$

In particular, S_X yields an equivalence of triangulated categories

$$S_X \colon D_{\mathrm{qcoh}}(X) \xrightarrow{\sim} D_{\mathrm{qcoh}}(X^t)$$

inducing an equivalence $(\mathrm{Perf}(X)) \xrightarrow{\sim} (\mathrm{Perf}(X^t))$. Moreover, these equivalences are compatible with base change $S' \to S$.

The relative canonical bundle $\Omega^g_{X/S}$ is a line bundle on X that is obtained via pullback from the line bundle $\det(\mathscr{C}_e)$ on S because $X \to S$ is a group scheme (Proposition 27.15). In particular, if $\mathrm{Pic}(S) = 0$, then

$$S_{X^t} \circ S_X \cong [-1]^*[-g].$$

Proof. The last assertion follows from Proposition 27.242 since $X \to S$ and $X^t \to S$ are flat. By Proposition 27.241 we may identify $S_{X^t} \circ S_X = \Phi_{\mathscr{P}_X * \mathscr{P}_{X^t}}$. As $[-1]$ is an automorphism of X of order 2, we have $[-1]^* = L[-1]^* = R[-1]_* = [-1]_*$ as functors $D(X) \to D(X)$. Hence by Example 27.238 it suffices to identify $\mathscr{P}_X * \mathscr{P}_{X^t}$ with $R\Gamma_*(\Omega^g_{X/S})[-g]$, where $\Gamma \colon X \to X \times_S X$ is the graph of $[-1]$.

We set $\mathscr{P} := \mathscr{P}_X$ and $\tilde{\mathscr{P}} := \mathscr{P}_{X^t}$. By Remark 27.218 (2) we have $\mathrm{pr}_{13}^* \mathscr{P} \otimes \mathrm{pr}_{23}^* \mathscr{P} = \mu^* \mathscr{P} = L\mu^* \mathscr{P}$. This shows the first of the following isomorphisms

$$\mathscr{P} * \tilde{\mathscr{P}} \cong R\mathrm{pr}_{12,*}(L\mu^* \mathscr{P})$$

$$\cong Lm^*(R\pi_* \mathscr{P})$$

$$\cong Lm^*(e_* \det(\mathscr{C}_e)^{-1}) = R\Gamma_*(\Omega^g_{X/S})[-g].$$

The second isomorphism follows by flat base change (m is flat by Remark 27.5) using the cartesian diagram

$$
\begin{array}{ccc}
X \times_S X \times_S X^t & \xrightarrow{\mu} & X \times_S X^t \\
{\scriptstyle \mathrm{pr}_{12}} \downarrow & & \downarrow {\scriptstyle \pi} \\
X \times_S X & \xrightarrow{m} & X.
\end{array}
$$

The isomorphism $R\pi_* \mathscr{P} = e_* \det(\mathscr{C}_e)^{-1}$ holds by Corollary 27.230. The last isomorphism follows from flat base change from the cartesian diagram

$$
\begin{array}{ccc}
X & \xrightarrow{f} & S \\
{\scriptstyle \Gamma} \downarrow & & \downarrow {\scriptstyle e} \\
X \times_S X & \xrightarrow{m} & X
\end{array}
$$

using that $\Omega^g_{X/S}$ is the pullback of $\det(\mathscr{C}_e)$ by Proposition 27.15. $\qquad\square$

Proposition 27.244. *Let S be a scheme and let $f\colon X \to Y$ be a homomorphism of abelian schemes over S. Then one has an isomorphism*

$$\mathcal{S}_Y \circ Rf_* \cong L(f^t)^* \circ \mathcal{S}_X.$$

of functors $D_{\mathrm{qcoh}}(X) \to D_{\mathrm{qcoh}}(Y^t)$.

Proof. Denote by $\rho\colon X \times_S Y^t \to X$ and $\sigma\colon X \times_S Y^t \to Y^t$ the projections. Then for $\mathscr{F}$ in $D_{\mathrm{qcoh}}(X)$ one has functorial isomorphisms

$$
\begin{aligned}
\mathcal{S}_Y(Rf_*\mathscr{F}) &= R\pi'_{Y,*}(L\pi_Y^* Rf_*\mathscr{F} \otimes^L \mathscr{P}_Y) \\
&\cong R\pi'_{Y,*}(R(f \times \mathrm{id}_{Y^t})_* L\rho^* \mathscr{F} \otimes^L \mathscr{P}_Y) \\
&\cong R\pi'_{Y,*}(R(f \times \mathrm{id}_{Y^t})_*(L\rho^* \mathscr{F} \otimes^L L(f \times \mathrm{id}_{Y^t})^* \mathscr{P}_Y)) \\
&\cong R\sigma_*(L\rho^* \mathscr{F} \otimes^L L(\mathrm{id}_X \times f^t)^* \mathscr{P}_X) \\
&= R\sigma_* L(\mathrm{id}_X \times f^t)^*(L\pi_X^* \mathscr{F} \otimes^L \mathscr{P}_X) \\
&\cong Lf^{t*}(R\pi'_{X,*}(L\pi_X^* \mathscr{F} \otimes^L \mathscr{P}_X)) \\
&= Lf^{t*}(\mathcal{S}_X(\mathscr{F})).
\end{aligned}
$$

Here the first and the last isomorphism hold by base change, the second isomorphism by the projection formula, and the third isomorphism by (27.41.2). $\square$

Remark 27.245. Let X be an abelian scheme over a scheme S of relative dimension g, let T be an S-scheme, let $\xi \in X^t(T)$ be a T-valued point and let $\mathscr{M}_\xi$ be the corresponding rigidified line bundle on $X \times_S T$.

(1) Let $p_T\colon X \times_S T \to T$ and $p_X\colon X \times_S T \to X$ be the projections. By derived base change one has for every $\mathscr{F}$ in $D_{\mathrm{qcoh}}(X)$ a functorial isomorphism

$$
\begin{aligned}
L\xi^* \mathcal{S}_X(\mathscr{F}) &= L\xi^*(R\pi'_*(L\pi^* \mathscr{F} \otimes^L \mathscr{P}_X)) \\
&\cong Rp_{T*} L(\mathrm{id}_X \times \xi)^*(L\pi^* \mathscr{F} \otimes^L \mathscr{P}_X) \\
&= Rp_{T*}(Lp_X^* \mathscr{F} \otimes^L \mathscr{M}_\xi).
\end{aligned}
$$

(2) Let $\mathscr{G}$ be in $D_{\mathrm{qcoh}}(X^t)$. Applying (1) to $\mathscr{F} = \mathcal{S}_{X^t}(\mathscr{G})$ one obtains by Theorem 27.243 a functorial isomorphism

$$Rp_{T*}(Lp_X^* \mathcal{S}_{X^t}(\mathscr{G}) \otimes^L \mathscr{M}_\xi) \cong L\xi^*(\mathscr{G}[-g] \otimes^L (\Omega^g_{X^t/S})^{-1}).$$

(3) If $T = S = \operatorname{Spec} R$ is affine with $\operatorname{Pic}(S) = 0$ (e.g., if R is a local ring), then $p_X = \mathrm{id}_X$, $p_T\colon X \to \operatorname{Spec} R$ is the structure morphism, and $\Omega^g_{X^t/S} \cong \mathcal{O}_{X^t}$. Hence (1) and (2) become

(27.46.1) $L\xi^* \mathcal{S}_X(\mathscr{F}) \cong R\Gamma(X, \mathscr{F} \otimes^L \mathscr{M}_\xi),$

(27.46.2) $L\xi^* \mathscr{G}[-g] \cong R\Gamma(X, \mathcal{S}_{X^t}(\mathscr{G}) \otimes^L \mathscr{M}_\xi).$

Let X be an abelian scheme over a scheme S. For $\mathscr{F}$ and $\mathscr{G}$ in $D_{\mathrm{qcoh}}(X)$ we define their *convolution*

$$\mathscr{F} \odot_X \mathscr{G} := Rm_*(p_1^* \mathscr{F} \otimes^L p_2^* \mathscr{G})$$

This is not the convolution defined in (27.45.2) (which is the reason why we use this slightly unconventional symbol to denote it) but rather a construction reminiscent of the convolution of functions on a locally compact group. Then the Fourier-Mukai equivalence $\mathcal{S}\colon D_{\mathrm{qcoh}}(X) \xrightarrow{\sim} D_{\mathrm{qcoh}}(X^t)$ interchanges tensor product and convolution, similarly as the usual Fourier transform of a convolution of two (suitable) functions is the pointwise product of their Fourier transforms.

Proposition 27.246. *For $\mathscr{F}$ and $\mathscr{G}$ in $D_{\mathrm{qcoh}}(X)$ we have a functorial isomorphism*

$$\mathcal{S}(\mathscr{F} \odot_X \mathscr{G}) \cong \mathcal{S}(\mathscr{F}) \otimes^L \mathcal{S}(\mathscr{G}).$$

Proof. Then we find

$$
\begin{aligned}
\mathcal{S}(\mathscr{F} \odot_X \mathscr{G}) &= R\pi'_*(\mathscr{P}_X \otimes^L L\pi^* Rm_*(Lp_1^*\mathscr{F} \otimes^L Lp_2^*\mathscr{G})) \\
&\cong R\pi'_*(\mathscr{P}_X \otimes^L R\mu_* L\,\mathrm{pr}_{12}^*(Lp_1^*\mathscr{F} \otimes^L Lp_2^*\mathscr{G})) \\
&\cong R\pi'_* R\mu_*(L\mu^*(\mathscr{P}_X) \otimes^L L\,\mathrm{pr}_1^*\,\mathscr{F} \otimes^L L\,\mathrm{pr}_2^*\,\mathscr{G}) \\
&\cong R\pi'_* R\mu_*(L\,\mathrm{pr}_{13}^*(\mathscr{P}_X) \otimes^L L\,\mathrm{pr}_{23}^*(\mathscr{P}_X) \otimes^L L\,\mathrm{pr}_1^*\,\mathscr{F} \otimes^L L\,\mathrm{pr}_2^*\,\mathscr{G}) \\
&= R\,\mathrm{pr}_{3,*}((L\,\mathrm{pr}_{13}^*(\mathscr{P}_X) \otimes^L L\,\mathrm{pr}_1^*\,\mathscr{F}) \otimes^L (L\,\mathrm{pr}_{23}^*(\mathscr{P}_X) \otimes^L L\,\mathrm{pr}_2^*\,\mathscr{G})) \\
&\cong R\pi'_*(\mathscr{P}_X \otimes^L L\pi^*\mathscr{F}) \otimes^L R\pi'_*(\mathscr{P}_X \otimes^L L\pi^*\mathscr{G}) \\
&= \mathcal{S}(\mathscr{F}) \otimes^L \mathcal{S}(\mathscr{G}).
\end{aligned}
$$

Here the first isomorphism holds by flat base change with the cartesian diagram

$$
\begin{array}{ccc}
X \times_S X \times_S X^t & \xrightarrow{\;\mu\;} & X \times_S X^t \\
{\scriptstyle \mathrm{pr}_{12}}\big\downarrow & & \big\downarrow{\scriptstyle \pi} \\
X \times_S X & \xrightarrow{\;m\;} & X.
\end{array}
$$

The second isomorphism holds by the projection formula, the third by Remark 27.218 (2), and the last isomorphism by the Künneth isomorphism applied to the cartesian diagram

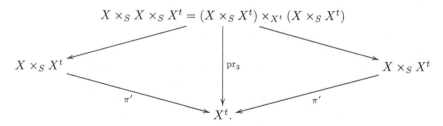

$\square$

(27.47) Riemann-Roch for abelian varieties.

Let $f\colon X \to S$ be an abelian scheme over a scheme S and denote by $\mathcal{S}_X\colon D_{\mathrm{qcoh}}(X) \xrightarrow{\sim} D_{\mathrm{qcoh}}(X^t)$ the Fourier-Mukai equivalence.

Definition 27.247. *A line bundle $\mathscr{L}$ on X is called* non-degenerate *if the following equivalent (Proposition 27.177) properties are satisfied.*
(i) $\varphi_{\mathscr{L}}$ *is an isogeny.*
(ii) $K(\mathscr{L})$ *is a finite group scheme over S.*
(iii) $K(\mathscr{L})$ *is a finite locally free group scheme over S.*

Example 27.248. We have seen in Corollary 27.172 that every ample line bundle on an abelian scheme is non-degenerate.

Remark 27.249. Let $\mathscr{L}$ be a line bundle on X. The formation of $K(\mathscr{L})$ commutes with base change. As $K(\mathscr{L})$ is proper over S, it is finite over S if and only if all fibers of $K(\mathscr{L}) \to S$ are finite (Corollary 12.89). Hence one obtains the following assertions.
(1) Let $g\colon T \to S$ be a morphism of schemes. If $\mathscr{L}$ is non-degenerate, then its pullback to $X \times_S T$ is non-degenerate. The converse holds if g is surjective.
(2) The line bundle $\mathscr{L}$ is non-degenerate if and only if the restriction of $\mathscr{L}$ to the fibers X_s for all $s \in S$ is a non-degenerate line bundle on the abelian variety $X_s \to \operatorname{Spec} \kappa(s)$.

Below in Proposition 27.258 we will see that if S is connected and $\mathscr{L}_{|X_s}$ is non-degenerate for one $s \in S$, then $\mathscr{L}$ is itself non-degenerate.

Proposition 27.250. *Let X be an abelian scheme and let $\mathscr{L}$ be a line bundle on X. Then*
$$L\varphi_{\mathscr{L}}^* S_X(\mathscr{L}) \cong f^*(Rf_*\mathscr{L}) \otimes \mathscr{L}^{-1} \otimes [0]^*\mathscr{L},$$
where $[0]^*\mathscr{L}$ denotes the line bundle $f^*e^*\mathscr{L}$.

If $\mathscr{L}$ is non-degenerate, then $\varphi_{\mathscr{L}}$ is faithfully flat and hence $L\varphi_{\mathscr{L}}^* = \varphi_{\mathscr{L}}^*$.

Proof. We use the notation set up in the beginning of Section (27.46). Then one has
$$\begin{aligned}
L\varphi_{\mathscr{L}}^*(S(\mathscr{L})) &= L\varphi_{\mathscr{L}}^*(R\pi'_*(\mathscr{P}_X \otimes \pi^*\mathscr{L}))\\
&\cong Rp_{2,*}L(\mathrm{id},\varphi_{\mathscr{L}})^*(\mathscr{P}_X \otimes \pi^*\mathscr{L})\\
&= Rp_{2,*}(\mathrm{id},\varphi_{\mathscr{L}})^*(\mathscr{P}_X \otimes \pi^*\mathscr{L})\\
&\cong Rp_{2,*}(\Lambda(\mathscr{L}) \otimes p_1^*\mathscr{L})\\
&= Rp_{2,*}(m^*\mathscr{L} \otimes p_2^*(\mathscr{L}^{-1} \otimes f^*e^*\mathscr{L}))\\
&\cong Rp_{2,*}(m^*\mathscr{L}) \otimes \mathscr{L}^{-1} \otimes [0]^*\mathscr{L}\\
&\cong f^*Rf_*\mathscr{L} \otimes \mathscr{L}^{-1} \otimes [0]^*\mathscr{L}.
\end{aligned}$$

Here the first isomorphism is base change via the cartesian diagram
$$\begin{array}{ccc}
X \times_S X & \xrightarrow{p_2} & X\\
{\scriptstyle(\mathrm{id},\varphi_{\mathscr{L}})}\downarrow & & \downarrow{\scriptstyle\varphi_{\mathscr{L}}}\\
X \times_S X^t & \xrightarrow{\pi'} & X^t.
\end{array}$$

The second isomorphism holds by Remark 27.217. The third isomorphism holds by the projection formula, and the last isomorphism by flat base change using the cartesian diagram

$$G \times_S G \xrightarrow{\ m\ } G$$

$$\begin{array}{ccc} G \times_S G & \xrightarrow{\ m\ } & G \\ {\scriptstyle p_2}\downarrow & & \downarrow{\scriptstyle f} \\ G & \xrightarrow{\ f\ } & S. \end{array}$$

$\square$

In the following we denote the Euler characteristic (Definition 23.59) of a sheaf (or complex) F on a scheme X by $\chi(X, F)$ rather than $\chi(F)$ to avoid confusion.

Corollary 27.251. *Let X be an abelian variety over a field k and $\mathscr{L}$ be a line bundle on X. Then*

$$\chi(X, L\varphi_{\mathscr{L}}^* \mathcal{S}_X(\mathscr{L})) = (-1)^g \chi(X, \mathscr{L})^2.$$

Proof. By Proposition 27.250 one has

$$L\varphi_{\mathscr{L}}^* \mathcal{S}_X(\mathscr{L}) \cong f^* Rf_* \mathscr{L} \otimes \mathscr{L}^{-1}.$$

Hence we get

$$\begin{aligned} \chi(X, L\varphi_{\mathscr{L}}^* \mathcal{S}_X(\mathscr{L})) &= \chi(X, f^* Rf_* \mathscr{L} \otimes \mathscr{L}^{-1}) \\ &= \chi(\operatorname{Spec} k, Rf_* \mathscr{L} \otimes Rf_* \mathscr{L}^{-1}) \\ &= \chi(\operatorname{Spec} k, Rf_* \mathscr{L})\chi(\operatorname{Spec} k, Rf_* \mathscr{L}^{-1}) \\ &= \chi(X, \mathscr{L})\chi(X, \mathscr{L}^{-1}). \end{aligned}$$

Here the second equality holds by the projection formula (Proposition 23.65) and the third by the multiplicativity of Euler-Poincaré characteristic (Example 23.60). As the dualizing sheaf on an abelian variety is trivial, we have $H^i(X, \mathscr{L}^{-1}) \cong H^{g-i}(X, \mathscr{L})^\vee$ by Serre duality (Corollary 25.129) and hence $\chi(X, \mathscr{L}^{-1}) = (-1)^g \chi(X, \mathscr{L})$. $\square$

Lemma 27.252. *Let X be an abelian variety over a field k and let $\mathscr{E}$ be a vector bundle on X of rank r. Then*

$$\chi(X^t, \mathcal{S}_X(\mathscr{E})) = (-1)^g r.$$

Proof. We apply (27.46.2) to the dual abelian variety of X whose dual we identify with X and to $\xi = e$ the zero section of X. Then we have

$$R\Gamma(X^t, \mathcal{S}_X(\mathscr{E})) \cong Le^* \mathscr{E}[-g] = e^* \mathscr{E}[-g] \cong k^r[-g]$$

because $\mathscr{E}$ is K-flat. Applying Euler characteristics, we obtain the desired equality. $\square$

Theorem 27.253. (Riemann-Roch Theorem) *Let k be a field, let $f \colon X \to \operatorname{Spec} k$ be an abelian variety of dimension g over k, and let $\mathscr{L}$ be a line bundle on X.*
(1) *One has $\chi(X, \mathscr{L}) = (\mathscr{L}^g)/g!$, where $(\mathscr{L}^g)$ denotes the g-fold self intersection of $\mathscr{L}$ defined in Definition 23.70.*
(2) *One has $\chi(X, \mathscr{L}) \neq 0$ if and only if $\mathscr{L}$ is non-degenerate. In this case*

$$(27.47.1) \qquad\qquad \chi(X, \mathscr{L})^2 = \deg(\varphi_{\mathscr{L}}).$$

Proof. Let us show (1). We use the Hirzebruch-Riemann-Roch Theorem (Corollary 23.113), where as additive cohomology theory we choose graded K-theory (Example 23.109). As the tangent bundle of X is trivial, its Todd class is 1 and we obtain

$$\chi(X, \mathscr{L}) = f_*(\mathrm{ch}(\mathscr{L})).$$

We have $\mathrm{ch}(\mathscr{L}) = \sum_{n=0}^{\infty} \frac{c_1([\mathscr{L}])^n}{n!}$ by Remark 23.114 where $\frac{c_1([\mathscr{L}])^n}{n!} \in \mathrm{gr}^n K_0(X)_{\mathbb{Q}}$. As the direct image f_* in graded K-theory is homogeneous of degree $-g$ we find that $f_* c_1([\mathscr{L}])^n = 0$ for $n \neq g$ and hence

$$f_*(\mathrm{ch}(\mathscr{L})) = f_* \frac{c_1([\mathscr{L}])^g}{g!} = \frac{\chi(c_1([\mathscr{L}])^g)}{g!} = \frac{(\mathscr{L}^g)}{g!}$$

by the definition of the intersection number.

Next we show (27.47.1) if $\mathscr{L}$ is non-degenerate. Then $L\varphi_{\mathscr{L}}^* = \varphi_{\mathscr{L}}^*$ and one has

$$\deg(\varphi_{\mathscr{L}})\chi(X^t, S_X(\mathscr{L})) = \chi(X, L\varphi_{\mathscr{L}}^* S_X(\mathscr{L}))$$
$$= (-1)^g \chi(X, \mathscr{L})^2.$$

Here the first equality holds by Proposition 27.179 and the second by Corollary 27.251. As $\chi(X^t, S_X(\mathscr{L})) = (-1)^g$ by Lemma 27.252, this shows (27.47.1).

It remains to show that $\chi(X, \mathscr{L}) = 0$ if $\mathscr{L}$ is not non-degenerate. For this we may assume that k is perfect. As $K(\mathscr{L})$ is a proper non-finite group scheme over k, $K(\mathscr{L})_{\mathrm{red}}^0$ is an abelian variety of positive dimension. In particular, there are finite subgroup schemes H of $K(\mathscr{L})$ of arbitrary large degree, e.g., the n-torsion points of $K(\mathscr{L})_{\mathrm{red}}^0$ for all n. Then $\varphi_{\mathscr{L}}$ factors into $X \xrightarrow{g} X/H \xrightarrow{\varphi} X^t$ and one has by Proposition 27.179 and Corollary 27.251

$$(-1)^g \chi(X, \mathscr{L})^2 = \chi(X, L\Phi_{\mathscr{L}}^* S_X(\mathscr{L})) = \deg(g)\chi(X/H, L\varphi^* S_X(\mathscr{L})).$$

Hence $\chi(X, \mathscr{L})$ is divisible by arbitrary large numbers and hence $\chi(X, \mathscr{L}) = 0$. □

(27.48) The index of non-degenerate line bundles.

Lemma 27.254. *Let X be an abelian variety over a field k of dimension g. Let $\mathscr{L}$ be a non-degenerate line bundle on X and let $\Lambda(\mathscr{L})$ be its Mumford bundle. Write $h^n(-)$ for $\dim_k H^n(-)$. Then for all $n \geq 0$ one has*

$$(27.48.1) \quad \sum_{p+q=n} h^p(X, \mathscr{L}) h^q(X, \mathscr{L}^{-1}) = h^n(X \times X, \Lambda(\mathscr{L})) = \begin{cases} 0, & \text{if } n \neq g; \\ \deg(\varphi_{\mathscr{L}}), & \text{if } n = g. \end{cases}$$

Proof. By Proposition 27.229 one has $R^n \pi'_* \mathscr{P} = 0$ for $n \neq g$ and $R^g \pi'_* \mathscr{P} = e'_* k$, where e' is the zero section of X^t. As $\varphi_{\mathscr{L}}$ is flat and $(\mathrm{id}_X \times \varphi_{\mathscr{L}})^*(\mathscr{P}) = \Lambda(\mathscr{L})$ we obtain by flat base change

$$(*) \qquad\qquad R^n p_{2*} \Lambda(\mathscr{L}) = \begin{cases} 0, & \text{if } n \neq g; \\ \iota_* \mathscr{O}_{K(\mathscr{L})}, & \text{if } n = g. \end{cases}$$

where $\iota\colon K(\mathscr{L}) \to X$ is the inclusion. As $K(\mathscr{L})$ is finite, the initial terms of the Leray spectral sequence $E_2^{ij} = H^i(X, R^j p_{2*}\Lambda(\mathscr{L})) \Rightarrow H^{i+j}(X \times X, \Lambda(\mathscr{L}))$ are all zero except for E_2^{0g} which is of dimension $\deg(\varphi_{\mathscr{L}})$. This shows the second equality of (27.48.1).

By definition of the Mumford bundle and by the projection formula we have

$$R^n p_{2*}\Lambda(\mathscr{L}) = R^n p_{2*}(m^*\mathscr{L} \otimes p_1^*\mathscr{L}^{-1}) \otimes \mathscr{L}^{-1} \cong R^n p_{2*}(m^*\mathscr{L} \otimes p_1^*\mathscr{L}^{-1}),$$

where the second isomorphism follows since $R^n p_{2*}\Lambda(\mathscr{L})$ is supported on the finite k-scheme $K(\mathscr{L})$ and any line bundle on a finite k-scheme is trivial. Again using the Leray spectral sequence for p_2 we see the first of the following isomorphisms

$$H^n(X \times X, \Lambda(\mathscr{L})) \cong H^n(X \times X, m^*\mathscr{L} \otimes p_1^*\mathscr{L}^{-1})$$
$$\cong H^n(X \times X, p_1^*\mathscr{L} \otimes p_2^*\mathscr{L}^{-1})$$
$$\cong \bigoplus_{p+q=n} H^p(X, \mathscr{L}) \otimes H^q(X, \mathscr{L}^{-1}).$$

Here one obtains the second isomorphism by observing that $\sigma\colon X \times X \to X \times X$, $(x, x') \mapsto (x + x', x)$ is an isomorphism with $\sigma^*(p_1^*\mathscr{L} \otimes p_2^*\mathscr{L}^{-1}) = m^*\mathscr{L} \otimes p_1^*\mathscr{L}^{-1}$. This shows the first equality of (27.48.1). $\qquad\square$

The lemma yields an alternative proof of (27.47.1) avoiding Fourier-Mukai transforms since it implies

$$(-1)^g \chi(X, \mathscr{L})^2 = \chi(X, \mathscr{L})\chi(X, \mathscr{L}^{-1}) = \chi(X \times X, \Lambda(\mathscr{L})) = (-1)^g \deg(\varphi_{\mathscr{L}}).$$

Proposition 27.255. *Let S be a scheme, let $f\colon X \to S$ be an abelian scheme, and let $\mathscr{L}$ be a non-degenerate line bundle on X. Then the formation of $R^p f_*\mathscr{L}$ commutes with arbitrary base change $T \to S$, and there exists a decomposition of S into open and closed subschemes*

$$S = \coprod_{i \geq 0} S_i$$

such that $R^p f_*\mathscr{L}_{|S_i} = 0$ for $p \neq i$ and such that $R^i f_*\mathscr{L}_{|S_i}$ is locally free of rank $\deg(\varphi_{\mathscr{L}})^{1/2} > 0$.

Proof. Let us first assume that $S = \operatorname{Spec} k$ for a field k. We have to show that there exists a unique integer i such that $h^i(\mathscr{L}) := \dim H^i(X, \mathscr{L}) \neq 0$. By (27.48.1) we have for all $n \geq 0$

$$\sum_{p+q=n} h^p(\mathscr{L})h^q(\mathscr{L}^{-1}) = \begin{cases} 0, & \text{if } n \neq g; \\ \deg(\varphi_{\mathscr{L}}), & \text{if } n = g. \end{cases}$$

This is only possible if there exists unique p and q such that $h^p(\mathscr{L}) \neq 0 \neq h^q(\mathscr{L}^{-1})$. Moreover, one has necessarily $p + q = g$. This shows in particular that $\chi(X, \mathscr{L}^{-1}) \neq 0$ and hence that $\mathscr{L}^{-1}$ is also non-degenerate by Theorem 27.253.

Now let S be arbitrary. We may assume that the relative dimension of X over S is a constant integer g. For $i = \{0, 1, \ldots, g\}$ we set

$$S_i := \{\, s \in S \;;\; H^i(X_s, \mathscr{L}_s) \neq 0 \,\}.$$

By semicontinuity (Theorem 23.139) we know that S_i is closed in S. By what we have already shown, we know that S is the disjoint union of the finitely many S_i. Hence every S_i is also open in S. Now we can use Corollary 23.143 and Proposition 23.142 to see that $R^i f_* \mathscr{L}_{|S_i}$ is locally free, that $R^p f_* \mathscr{L}_{|S_i} = 0$ for $p \neq i$, and that their formations commute with base change.

The compute the rank of $R^i f_* \mathscr{L}_{|S_i}$ we can look at the fiber in some $s \in S_i$. Then one has $(\dim H^i(X_s, \mathscr{L}_s))^2 = \chi(X_s, \mathscr{L}_s)^2 = \deg(\varphi_{\mathscr{L}_s})$ by Theorem 27.253.   □

Definition 27.256. *In the situation of Proposition 27.255 we call the locally constant function $i(\mathscr{L})\colon S \to \mathbb{N}$ that attaches to $s \in S$ the i with $s \in S_i$ the index of $\mathscr{L}$.*

If X has relative dimension g over S, then $i(\mathscr{L})$ takes values in $\{0, 1, \ldots, g\}$.

Remark 27.257. Let X be an abelian scheme of relative dimension g over a scheme S. If $\mathscr{L}$ is a non-degenerate line bundle, then the proof of Proposition 27.255 shows that $\mathscr{L}^{-1}$ is also non-degenerate and that $i(\mathscr{L}^{-1}) = g - i(\mathscr{L})$.

The property of being non-degenerate for a line bundle is also open and closed on the base scheme.

Proposition 27.258. *Let X be an abelian scheme over a scheme S and let $\mathscr{L}$ be a line bundle on X. Let $s_0 \in S$ such that $\mathscr{L}_{s_0} := \mathscr{L}_{|X_{s_0}}$ is a non-degenerate line bundle on the fiber X_{s_0}. Then there exists an open and closed neighborhood U of s_0 in S such that the restriction of $\mathscr{L}$ to $X \times_S U$ is non-degenerate of constant index equal to the index of $i(\mathscr{L}_{s_0})$.*

Proof. By Theorem 23.139 the map $s \mapsto \chi(X_s, \mathscr{L}_s)$ is locally constant. Hence by Theorem 27.253 there exists an open and closed neighborhood V of s_0 such that $\mathscr{L}_s$ is non-degenerate for all $s \in V$. Hence $\mathscr{L}_{|X \times_S V}$ is non-degenerate (Remark 27.249). By Proposition 27.255 we find in V an open and closed neighborhood U of s_0 such that the index of $\mathscr{L}$ is constant on U.   □

Corollary 27.259. *Let S be a scheme, let $f\colon X \to S$ be an abelian scheme, and let $\mathscr{L}$ be a non-degenerate line bundle on X of constant index $i(\mathscr{L})$. Then there exists a vector bundle $\mathscr{E}$ on X^t such that*

$$S_X(\mathscr{L}) \cong \mathscr{E}[-i(\mathscr{L})].$$

In particular, $i(\mathscr{L})$ is the unique integer i such that $H^i(S_X(\mathscr{L})) \neq 0$.

Proof. Set $i := i(\mathscr{L})$. As $\varphi_{\mathscr{L}}$ is faithfully flat, it suffices to show that $\varphi_{\mathscr{L}}^* S(\mathscr{L}) \cong \mathscr{E}'[-i]$ for some vector bundle $\mathscr{E}'$ on X. But by Proposition 27.259 and by Proposition 27.255 one has

$$\varphi_{\mathscr{L}}^* S_X(\mathscr{L}) \cong f^*(Rf_* \mathscr{L}) \otimes \mathscr{L}^{-1} \otimes [0]^* \mathscr{L}$$
$$= f^*(R^i f_* \mathscr{L}[-i]) \otimes \mathscr{L}^{-1} \otimes [0]^* \mathscr{L}$$
$$= (f^*(R^i f_* \mathscr{L}) \otimes \mathscr{L}^{-1} \otimes [0]^* \mathscr{L})[-i]$$

and $R^i f_* \mathscr{L}$ is finite locally free.   □

Proposition 27.260. *Let S be a scheme, let $f\colon X \to Y$ be an isogeny of abelian schemes, and let $\mathscr{L}$ be a non-degenerate line bundle on Y. Then $f^* \mathscr{L}$ is non-degenerate and $i(f^* \mathscr{L}) = i(\mathscr{L})$.*

Proof. One has $\varphi_{f_*\mathscr{L}} = f^t \circ \varphi_\mathscr{L} \circ f$ by Lemma 27.165 and hence $\varphi_{f_*\mathscr{L}}$ is a composition of three isogenies (Proposition 27.213) and therefore itself an isogeny. Hence $f^*\mathscr{L}$ is non-degenerate.

To compute the index of $f^*\mathscr{L}$ we may assume that $i(\mathscr{L})$ is constant. Let $\mathscr{E}$ be a vector bundle on X^t such that $\mathcal{S}_Y(\mathscr{L}) \cong \mathscr{E}[-i(\mathscr{L})]$ (Corollary 27.259). As f is flat, $Lf^* = f^*$, and as f^t is finite and in particular affine, $Rf^t_* = f_*$. Hence we have

$$\mathcal{S}_X(f^*\mathscr{L}) \cong f^t_* \mathcal{S}_Y(\mathscr{L}) \cong f^t_* \mathscr{E}[-i(\mathscr{L})]$$

by Proposition 27.244, and hence $i(\mathscr{L})$ is the unique integer i such that $H^i(\mathcal{S}_X(f^*\mathscr{L})) \neq 0$. This shows $i(f^*\mathscr{L}) = i(\mathscr{L})$. $\qquad\square$

Lemma 27.261. *Let S be a scheme, let X be an abelian scheme over S, and let $\mathscr{L}$ and $\mathscr{M}$ be line bundles on X such that the class of $\mathscr{L} \otimes \mathscr{M}^{-1}$ is in $X^t(S)$. If $\mathscr{L}$ is non-degenerate, then $\mathscr{M}$ is non-degenerate, and $i(\mathscr{L}) = i(\mathscr{M})$.*

Proof. As we have $\varphi_\mathscr{L} = \varphi_\mathscr{M}$, the first assertion is clear. To verify $i(\mathscr{L}) = i(\mathscr{M})$ we may base change to geometric points of S and hence can assume that $S = \operatorname{Spec} k$ for an algebraically closed field k. Then $\mathscr{L}$ and $\mathscr{M}$ are algebraically equivalent (Definition 27.231) and we find a connected k-scheme Z, k-valued points $z, w \in Z(k)$ and a line bundle $\mathscr{N}$ on $X \times_k Z$ with $\mathscr{N}_{|X\times\{z\}} \cong \mathscr{L}$ and $\mathscr{N}_{|X\times\{w\}} \cong \mathscr{M}$. By Proposition 27.258, $\mathscr{N}$ is a non-degenerate line bundle on the abelian scheme $X \times_k Z \to Z$ and $i(\mathscr{N})$ is constant on Z as a function on Z. In particular $i(\mathscr{M}) = i(\mathscr{N})(w) = i(\mathscr{N})(z) = i(\mathscr{L})$. $\qquad\square$

Proposition 27.262. *Let X be an abelian scheme over a scheme S and let $\mathscr{L}$ be a non-degenerate line bundle on X. For all $n > 0$, $\mathscr{L}^{\otimes n}$ is non-degenerate, and $i(\mathscr{L}^{\otimes n}) = i(\mathscr{L})$.*

We will use below only the fact that there exist arbitrary large integers n such that $i(\mathscr{L}^{\otimes n}) = i(\mathscr{L})$. Hence we will give the proof only in the case that $n = m^2$ is a square. In this case we easily find an isogeny $\alpha \colon X \to X$ such that one has an identity $\alpha^*(\mathscr{L}) = \mathscr{L}^{\otimes m^2}$ in $\operatorname{NS}(X)$, namely $\alpha = [m]$, see the proof below. In general one can reduce to this case by writing an arbitrary positive integer as a sum of four squares and by passing to X^4, an idea that is often called Zarhin's trick, see Exercise 27.16.

Proof. [if $n = m^2$] Consider the isogeny $[m] \colon X \to X$. Then $i([m]^*\mathscr{L}) = i(\mathscr{L})$ by Proposition 27.260. But by Proposition 27.184 (3) we have in $\operatorname{Pic}_{X/S}(S)$

$$[m]^*\mathscr{L} = \mathscr{L}^{\otimes m^2} \otimes \mathscr{N}, \qquad \text{with } \mathscr{N} \in X^t(S).$$

Hence $i([m]^*\mathscr{L}) = i(\mathscr{L}^{\otimes m^2})$ by Lemma 27.261. $\qquad\square$

Remark 27.263. In [Mum1] §16 it is shown that the index of a non-degenerate line bundle $\mathscr{L}$ on an abelian variety X can be computed by a Hilbert polynomial as follows. Fix an ample line bundle $\mathscr{M}$ on X and let $P_\mathscr{L} \in \mathbb{Q}[T]$ be the Hilbert polynomial of $\mathscr{L}$ with respect to $\mathscr{M}$, i.e. $P_\mathscr{L}(n) = \chi(X, \mathscr{L} \otimes \mathscr{M}^{\otimes n})$. This is a polynomial of degree $g = \dim(X)$.

Then all complex roots of $P_\mathscr{L}$ are real and $i(\mathscr{L})$ is the number of positive roots, counted with multiplicities.

(27.49) Characterization of ample line bundles on abelian schemes.

Theorem 27.264. *Let S be a scheme and let $f: X \to S$ be an abelian scheme. Then a line bundle $\mathscr{L}$ on X is relatively ample over S if and only if $\mathscr{L}$ is non-degenerate and $i(\mathscr{L}) = 0$.*

Proof. As ampleness is an open condition on proper schemes (Theorem 24.46) we can check it on fibers. The property of being non-degenerate can also be checked on fibers by Remark 27.249. Hence we can assume that S is the spectrum of a field and it suffices to show the following more precise result.   □

Proposition 27.265. *Let k be a field, let X be an abelian variety over k, and let $\mathscr{L}$ be an line bundle on X. Then the following assertions are equivalent.*
(i) *$\mathscr{L}$ is ample.*
(ii) *$\mathscr{L}$ is non-degenerate and $H^0(X, \mathscr{L}) \neq 0$.*
(iii) *There exists an effective divisor $D \subseteq X$ such that $\mathscr{L} \cong \mathcal{O}_X(D)$ and $X \setminus D$ is affine. In this case $X \setminus D$ is affine for all effective divisors D such that $\mathscr{L} \cong \mathcal{O}_X(D)$.*

Proof. The implication (iii) $\Rightarrow$ (i) has been shown in Proposition 27.174. Let us show that (i) implies (ii). Let $\mathscr{L}$ be ample. Then $\mathscr{L}$ is non-degenerate by Corollary 27.172. Moreover, any sufficiently high power $\mathscr{L}^{\otimes n}$ is very ample and in particular globally generated. Hence $i(\mathscr{L}^{\otimes n}) = 0$ and therefore $i(\mathscr{L}) = 0$ by Proposition 27.262.

Now suppose that (ii) holds. As $H^0(X, \mathscr{L}) \neq 0$ there exists an effective divisor D on X such that $\mathcal{O}_X(D) \cong \mathscr{L}$. Then $\mathscr{L}^{\otimes 2}$ is globally generated by Lemma 27.175 and hence defines a morphism $g: X \to \mathbb{P}^N_k$ such that $g^* \mathcal{O}_{\mathbb{P}^N_k}(1) \cong \mathscr{L}^{\otimes 2}$. If we show that g is finite, then $\mathscr{L}^{\otimes 2}$ is ample (Proposition 13.83) and hence $\mathscr{L}$ is ample (Proposition 13.50). Moreover for every D as above we find a hyperplane H in $\mathbb{P}^N_k$ such that $g^{-1}(H) = 2D$ and hence $X \setminus D$ is the inverse image of the open affine $\mathbb{P}^N_k \setminus H$ and therefore $X \setminus D$ is affine because g is affine.

Hence it remains to show that g is finite. As g is proper as a morphism between proper schemes, it suffices to show that g has finite fibers (Corollary 12.89). For this we may assume that k is algebraically closed. Let e be the zero section of X considered as a closed point of X. By Corollary 27.107 all non-empty fibers of g are equi-dimensional of the same dimension. Hence it suffices to show that the reduced connected component F_0 of $g^{-1}(g(e))$ containing e is contained in the finite scheme $K(\mathscr{L})$.

Let $x \in F_0(k)$. Let D be an effective divisor given by some non-zero section $s \in H^0(X, \mathscr{L})$. Now $g \circ t_x = g$ by Proposition 27.106 and hence $s^2 \in H^0(X, \mathscr{L}^{\otimes})$ and $t_x^* s^2$ have the same zero divisor. It follows that one has an equality of divisors $t_x^* D = D$ and in particular $x \in K(\mathscr{L})(k)$.   □

Corollary 27.266. *Let X be an abelian variety over a field k and let $\mathscr{L}$ be an ample line bundle on X. Then $H^p(X, \mathscr{L}) = 0$ for all $p > 0$ and in particular*

$$\dim H^0(X, \mathscr{L}) = \chi(X, \mathscr{L}) = (\deg \varphi_{\mathscr{L}})^{1/2}.$$

More generally, by Proposition 27.255 one sees the following result.

Corollary 27.267. *Let $\mathscr{L}$ be a relatively ample line bundle on X. Then $R^p f_* \mathscr{L} = 0$ for all $p \geq 1$ and $f_* \mathscr{L}$ is finite locally free of rank $\deg(\varphi_{\mathscr{L}})^{1/2}$ and its formation is compatible with base change.*

Using Proposition 27.258 one deduces from Theorem 27.264 that being ample is an open and closed condition on abelian schemes.

Corollary 27.268. *Let S be a scheme, let $f\colon X \to S$ be an abelian scheme, and let $\mathscr{L}$ be a line bundle on X. Let $s_0 \in S$ be a point, such that $\mathscr{L}_{|X_{s_0}}$ is ample. Then there exists an open and closed neighborhood U of s_0 such that $\mathscr{L}_{|X\times_S U}$ is relatively ample over U.*

A special case of Lemma 27.261 is the following assertion.

Corollary 27.269. *Let S be a scheme, let X be an abelian scheme over S, and let $\mathscr{L}$ and $\mathscr{M}$ be line bundles on X such that the class of $\mathscr{L} \otimes \mathscr{M}^{-1}$ is in $X^t(S)$. Then $\mathscr{M}$ is ample if and only if $\mathscr{L}$ is ample.*

(27.50) Vector bundles on elliptic curves.

Let us illustrate many of the previous notions for elliptic curves. As an application we will explain the classification of semistable vector bundles on elliptic curves. Here we follow [Pol][O] Chap. 14, see also Bhatt's lecture notes on abelian varieties [Bha][X]. Recall that every vector bundle on an elliptic curve is isomorphic to a direct sum of semistable vector bundles by Theorem 26.156 and Proposition 26.161.

We denote by k a field and by E an elliptic curve over k. Let us collect the information that we already know in this special case on line bundles.

Let us recall some facts about modules on elliptic curves.

Remark 27.270. As elliptic curves have genus 1, the Riemann-Roch theorem yields the following assertions.
(1) Let $\mathscr{L}$ be a line bundle on E. The line bundle $\mathscr{L}$ is non-degenerate if and only if $\deg(\mathscr{L}) \neq 0$. In this case one has

$$\deg(\mathscr{L}) = \chi(E, \mathscr{L}), \qquad \deg(\varphi_{\mathscr{L}}) = \deg(\mathscr{L})^2.$$

Indeed, the first equality holds by Riemann-Roch (Proposition 26.46) and implies the second equality by (27.47.1). This implies the first assertion by Proposition 27.253 (2).
Moreover, $\mathscr{L}$ is ample if and only if $\deg(\mathscr{L}) > 0$ (Proposition 26.57).
(2) More generally, by (26.23.5) one has for every $\mathscr{F}$ in $D^b_{\mathrm{coh}}(E)$

$$(27.50.1) \qquad\qquad \deg(\mathscr{F}) = \chi(E, \mathscr{F}).$$

The zero section of E determines a line bundle $\mathscr{O}_E([0])$ on E of degree 1 and by Part (1) of the previous remark, we obtain an isomorphism

$$\varphi_{\mathscr{O}_E([0])}\colon E \xrightarrow{\ \sim\ } E^t, \qquad x \mapsto \mathscr{O}_E([x]) \otimes \mathscr{O}_E([0])^{\otimes -1},$$

which we use to identify E and E^t, cf. Proposition 27.148. Via this identification, for the Fourier-Mukai transforms we have $\mathcal{S}_E = \mathcal{S}_{E^t}$ and we can view the Fourier-Mukai equivalence as an auto-equivalence

$$\mathcal{S}_E\colon D^b_{\mathrm{coh}}(E) \xrightarrow{\ \sim\ } D^b_{\mathrm{coh}}(E) \qquad \text{with} \quad \mathcal{S}_E(\mathcal{S}_E(\mathscr{F})) \cong [-1]^*\mathscr{F}[-1].$$

We want to classify vector bundles on E using the Harder-Narasimhan stratification introduced in Section (26.25). We have already understood line bundles quite well: If we denote by $\mathrm{Pic}^d(E)$ the set of degree $d \in \mathbb{Z}$ line bundles on E, then one has $E(k) \cong \mathrm{Pic}^1(E) \cong \mathrm{Pic}^d(E)$, where the first isomorphism is given by $x \mapsto \mathcal{O}_E([x])$ (Proposition 27.148) and the second is given by tensoring with $\mathcal{O}_E([0])^{\otimes d-1}$. In the language introduced in Section (26.24) this implies that the sets of isomorphism classes of rank 1 vector bundles of fixed slope (= degree) are in bijection to each other and also in bijection to the set of k-rational points of E. Attaching to $x \in E(k)$ the skyscraper sheaf $\kappa(x) = k$ with support in $\{x\}$ we also obtain for every $d \in \mathbb{Z}$ a bijection between isomorphism classes of slope d line bundles and torsion $\mathcal{O}_E$-modules $\mathscr{F}$ with $\dim_k H^0(E, \mathscr{F}) = 1$. This will be categorified and generalized to arbitrary vector bundles by the theorem of Atiyah below.

Recall (Theorem 26.156) that every vector bundle $\mathscr{E}$ has a unique and functorial $\mathbb{Q}$-filtration $(\mathrm{HN}^\lambda(\mathscr{E}))_\lambda$ such that $\mathrm{gr}_{\mathrm{HN}}^\lambda(\mathscr{E})$ is semistable of slope λ, which is called the Harder-Narasimhan filtration. Moreover, since elliptic curves have genus 1, the Harder-Narasimhan filtration is (non-canonically) split (Proposition 26.161).

Hence for every vector bundle $\mathscr{E}$ on E there exists a unique finite sequence of rational numbers $\lambda_1 > \lambda_2 > \cdots > \lambda_r$ and unique (up to isomorphism) semistable vector bundles $\mathscr{E}_i$ of slope λ_i such that

$$\mathscr{E} \cong \bigoplus_{i=1}^r \mathscr{E}_i,$$

where this isomorphism is in general not unique. Now for $\lambda \in \mathbb{Q}$ the abelian category $(\mathrm{Vect}_\lambda(E))$ of semistable vector bundles on E of slope λ can be described as follows. Let $(\mathrm{Coh}(E))_{\mathrm{tors}}$ be the category of coherent $\mathcal{O}_E$-modules with finite support.

Theorem 27.271. (Atiyah) *Let $\lambda \in \mathbb{Q}$. Then there is an equivalence of abelian categories*

$$T \colon (\mathrm{Vect}_\lambda(E)) \xrightarrow{\sim} (\mathrm{Coh}(E))_{\mathrm{tors}},$$

such that for every semistable vector bundle $\mathscr{E}$ of slope λ we have

(27.50.2) $$\dim_k H^0(E, T(\mathscr{E})) = \gcd(\deg(\mathscr{E}), \mathrm{rk}(\mathscr{E})),$$

the greatest common divisor of $\deg(\mathscr{E})$ and $\mathrm{rk}(\mathscr{E})$.

In fact, we will show that
(I) there exists an equivalence between the abelian categories $(\mathrm{Vect}_\lambda(E))$ and $(\mathrm{Vect}_0(E))$ preserving $\gcd(\deg(\mathscr{E}), \mathrm{rk}(\mathscr{E}))$ and that
(II) the shifted Fourier-Mukai equivalence $\mathcal{S}_E[1]$ induces an equivalence $T \colon (\mathrm{Vect}_0(E)) \xrightarrow{\sim} (\mathrm{Coh}(E))_{\mathrm{tors}}$ satisfying (27.50.2).

We start by recalling the following descriptions of the fibers of the Fourier-Mukai transform, which is a special case of Remark 27.245. Let $x \in E(k)$ be a k-rational point with corresponding degree zero line bundle

$$\mathscr{M}_x = \mathcal{O}_E([x]) \otimes \mathcal{O}_E([0])^{\otimes -1} \in \mathrm{Pic}^0(E).$$

Then for $\mathscr{F}$ in $D^b_{\mathrm{coh}}(E)$ we have

(27.50.3)
$$\mathcal{S}_E(\mathscr{F}) \otimes^L_{\mathcal{O}_E} \kappa(x) = R\Gamma(E, \mathscr{F} \otimes^L_{\mathcal{O}_E} \mathscr{M}_x),$$
$$\mathscr{F}[-1] \otimes^L_{\mathcal{O}_E} \kappa(x) = R\Gamma(E, \mathcal{S}_E(\mathscr{F}) \otimes^L_{\mathcal{O}_E} \mathscr{M}_x)$$

Recall from Remark 26.140 that we can define degree and rank for objects in $D^b_{\mathrm{coh}}(E)$. Let us show how these numbers behave under Fourier-Mukai equivalence.

Lemma 27.272. *For $\mathscr{F}$ in $D^b_{\mathrm{coh}}(E)$ we have*

$$\deg(\mathcal{S}_E(\mathscr{F})) = -\operatorname{rk}(\mathscr{F}), \qquad\qquad \operatorname{rk}(\mathcal{S}_E(\mathscr{F})) = \deg(\mathscr{F}).$$

Proof. Using (27.50.1), we obtain from (27.50.3) applied to $x = 0 \in E(k)$ (and viewing it as a morphism $\operatorname{Spec} k \to E$) that

$$\deg(\mathcal{S}_E(\mathscr{F})) = \chi(E, \mathcal{S}_E(\mathscr{F})) = \chi(L0^*\mathscr{F}[-1]) = -\chi(L0^*\mathscr{F}) \overset{(*)}{=} -\operatorname{rk}(\mathscr{F}),$$
$$\operatorname{rk}(\mathcal{S}_E(\mathscr{F})) \overset{(*)}{=} \chi(L0^*\mathcal{S}_E(\mathscr{F})) = \chi(E, \mathscr{F}) = \deg(\mathscr{F}),$$

where the equalities (*) hold by (26.23.4). □

The key step in the proof of Atiyah's theorem is the following.

Proposition 27.273. *Let $\lambda \in \mathbb{Q}$. Then the Fourier-Mukai transform induces equivalences*

$$\mathcal{S}_E[1]\colon (\mathrm{Vect}_\lambda(E)) \overset{\sim}{\longrightarrow} (\mathrm{Vect}_{-\lambda^{-1}}(E)), \qquad\qquad \text{if } \lambda < 0$$
$$\mathcal{S}_E\colon (\mathrm{Vect}_\lambda(E)) \overset{\sim}{\longrightarrow} (\mathrm{Vect}_{-\lambda^{-1}}(E)), \qquad\qquad \text{if } \lambda > 0$$
$$\mathcal{S}_E[1]\colon (\mathrm{Vect}_0(E)) \overset{\sim}{\longrightarrow} (\mathrm{Coh}(E))_{\mathrm{tors}}$$
$$\mathcal{S}_E\colon (\mathrm{Coh}(E))_{\mathrm{tors}} \overset{\sim}{\longrightarrow} (\mathrm{Vect}_0(E)).$$

Proof. Recall that $\mathcal{S}_E \circ \mathcal{S}_E[1] \cong [-1]^*$ by Fourier-Mukai equivalence. Since $[-1]^*$ preserves degrees of line bundles and one has $\det([-1]^*\mathscr{F}) = [-1]^* \det(\mathscr{F})$ for every vector bundle $\mathscr{F}$, we see that $[-1]^*$ preserves the degree of vector bundles. Clearly it also preserves the rank of vector bundles. Hence we have $\mu([-1]^*\mathscr{F}) = \mu(\mathscr{F})$ for every vector bundle $\mathscr{F} \neq 0$. Since $[-1]^*$ also induces a bijection between subbundles of $\mathscr{F}$ and of $[-1]^*\mathscr{F}$, we see that $[-1]^*$ is an auto-equivalence of $(\mathrm{Vect}_\lambda(E))$ for all $\lambda \in \mathbb{Q}$. Hence to show the equivalences it suffices to show the following assertions.
(1) If a vector bundle $\mathscr{F}$ on E is semistable of slope λ with $\lambda < 0$ (resp. with $\lambda > 0$), then $\mathcal{S}_E(\mathscr{F})[1]$ (resp. $\mathcal{S}_E(\mathscr{F})$) is a semistable vector bundle of slope $-\lambda^{-1}$.
(2) If $\mathscr{F}$ is a vector bundle of slope 0, then $\mathcal{S}_E(\mathscr{F})[1] \in (\mathrm{Coh}(E))_{\mathrm{tors}}$.
(3) If $\mathscr{F}$ is in $(\mathrm{Coh}(E))_{\mathrm{tors}}$, then $\mathcal{S}_E(\mathscr{F})$ is a semistable vector bundle of slope 0.
By Proposition 26.159 and since the Fourier-Mukai equivalence is compatible with base change, we may assume that k is algebraically closed.

Let us show (1). Let $\mathscr{F}$ be a semistable vector bundle of slope $\lambda \neq 0$. We write $\mathscr{F}$ as a finite direct sum of indecomposable vector bundles $\mathscr{F}_i$. Then each summand is again semistable of slope λ (Remark 26.149 (3)) and $\mathcal{S}_E(\mathscr{F})[1]$ (resp. $\mathcal{S}_E(\mathscr{F})$) is the direct sum of the $\mathcal{S}_E(\mathscr{F}_i)[1]$ (resp. $\mathcal{S}_E(\mathscr{F}_i)$). As finite direct sums of semistable vector bundles of the same slope $-\lambda^{-1}$ are again semistable of slope $-\lambda^{-1}$ by Proposition 26.152, we may assume that $\mathscr{F}$ is indecomposable. As $\mathcal{S}_E$ is an equivalence, it follows that $\mathcal{S}_E(\mathscr{F})$ is also indecomposable.

Now suppose that $\lambda < 0$. We find that for all $\mathscr{L} \in \mathrm{Pic}^0(E)$ one has

$$H^0(E, \mathscr{F} \otimes \mathscr{L}) = \mathrm{Hom}_{\mathscr{O}_E}(\mathscr{L}^{\otimes -1}, \mathscr{F}) = 0$$

by Proposition 26.150 because $\mathscr{L}^{\otimes -1}$ is semistable of slope 0. As $\dim E = 1$, we see that $H^i(E, \mathscr{F} \otimes \mathscr{L}) = 0$ for all $i \neq 1$. By (27.50.3) it follows that $H^i(\mathcal{S}_E(\mathscr{F}) \otimes^L_{\mathscr{O}_E} \kappa(x)) = 0$ for every $x \in E(k)$ and all $i \neq 1$. Therefore $\mathcal{S}_E(\mathscr{F})$ is of tor-amplitude in $[1, 1]$ by Proposition 23.126 (here we use that k is algebraically closed), i.e., $\mathcal{S}_E(\mathscr{F})[1]$ is a vector bundle. Its slope is $-\lambda^{-1}$ by Lemma 27.272. It is semistable because $\mathcal{S}_E(\mathscr{F})[1]$ is indecomposable by Proposition 26.161.

If $\lambda > 0$, then one can argue similarly. As the canonical bundle ω_E on E is trivial, Serre duality (Corollary 25.129) yields, for all $\mathscr{L} \in \mathrm{Pic}^0(E)$, that

$$H^1(E, \mathscr{F} \otimes \mathscr{L}) = H^0(E, \mathscr{F}^\vee \otimes_{\mathscr{O}_E} \mathscr{L}^\vee)^\vee = \mathrm{Hom}_{\mathscr{O}_E}(\mathscr{L}, \mathscr{E}^\vee)^\vee = 0,$$

since $\mu(\mathscr{E}^\vee) = -\lambda < 0$ by Remark 26.143 (2). As above it follows that $\mathcal{S}_E(\mathscr{F})$ is a semistable vector bundle of slope $-\lambda^{-1}$.

Let us show (2). Let $\mathscr{F} \neq 0$ be a semistable vector bundle of slope 0. We will argue by induction on $\mathrm{rk}(\mathscr{F})$. If $\mathscr{F}$ is a line bundle of degree 0, then $\mathscr{F} = \mathscr{M}_y$ for some $y \in E(k)$. By (27.50.3) we have for all $x \in E(k)$ with $x \neq -y$ that

$$\mathcal{S}_E(\mathscr{M}_y) \otimes^L \kappa(x) = R\Gamma(E, \mathscr{M}_{x+y}) = 0,$$

where the second equality follows from Lemma 27.197. Hence we have $\mathcal{S}_E(\mathscr{M}_y)|_{E \setminus \{-y\}} = 0$, in particular $\mathcal{S}_E(\mathscr{M}_y)$ is supported on an affine closed subscheme whose underlying topological space is $\{-y\}$. By (27.50.3) with $x = 0$ for its global sections we obtain

$$R\Gamma(E, \mathcal{S}_E(\mathscr{M}_y)) = \mathscr{M}_y[-1] \otimes_{\mathscr{O}_E} k \cong k[-1],$$

which shows that

(27.50.4) $$\mathcal{S}_E(\mathscr{M}_y) = \kappa(-y)[-1].$$

In particular $\mathcal{S}_E(\mathscr{M}_y) \in (\mathrm{Coh}(E))_{\mathrm{tors}}$.

Now consider $\mathscr{F}$ of rank > 1. We claim that there exists a line bundle $\mathscr{M}$ of degree 0 and a non-zero map $\mathscr{M} \to \mathscr{F}$. Otherwise we would have $H^0(E, \mathscr{E} \otimes \mathscr{L}) = \mathrm{Hom}_{\mathscr{O}_E}(\mathscr{L}^\vee, \mathscr{E}) = 0$ for all $\mathscr{L} \in \mathrm{Pic}^0(E)$ and hence as in the argument for $\lambda < 0$ one sees that $\mathcal{S}_E(\mathscr{F})[1]$ is a vector bundle. By Lemma 27.272 we have $\mathrm{rk}(\mathcal{S}_E(\mathscr{F})[1]) = -\mathrm{rk}(\mathcal{S}_E(\mathscr{F})) = -\deg(\mathscr{F}) = 0$. Hence $\mathcal{S}_E(\mathscr{F}) = 0$ and therefore $\mathscr{F} = 0$ since $\mathcal{S}_E$ is an equivalence. This contradiction shows the claim.

So let $u \colon \mathscr{M} \to \mathscr{F}$ be non-zero with $\mathscr{M}$ a line bundle of degree 0. Then u is injective since $\mathrm{rk}(\mathscr{M}) = 1$ and we can consider $\mathscr{M}$ as a submodule of $\mathscr{F}$. If $\mathscr{M}$ was not a subbundle, then its saturation would have slope > 0 (Remark 26.145) which is not possible, since $\mathscr{F}$ is semistable of slope 0. Therefore $\mathscr{F}/\mathscr{M}$ is a vector bundle which is semistable of slope 0 by Proposition 26.152. By induction, $\mathcal{S}_E(\mathscr{M})$ and $\mathcal{S}_E(\mathscr{F}/\mathscr{M})$ are rank 0 coherent sheaves, and hence the same holds for $\mathcal{S}_E(\mathscr{F})$.

It remains to show (3). Every torsion coherent sheaf $\mathscr{F}$ is an iterated extension of structure sheaves of closed points. Using that $\mathcal{S}_E \circ \mathcal{S}_E = [-1]_E^*[-1]$, (27.50.4) shows that for every closed point $y \in E(k)$ one has

$$\mathcal{S}(\kappa(y)) = [-1]^* \mathscr{M}_{-y} = \mathscr{M}_y.$$

Hence $\mathcal{S}_E(\mathscr{F})$ is an iterated extension of degree 0 line bundles and hence $\mathcal{S}_E(\mathscr{F}) \in (\mathrm{Vect}_0(E))$ by Proposition 26.152. $\square$

Proof. [of Theorem 27.271] Let $\lambda \in \mathbb{Q}$ be any rational number. We will show that there is an equivalence of $(\mathrm{Vect}_\lambda(E))$ and $(\mathrm{Vect}_0(E))$ preserving $\gcd(\deg(\mathscr{E}), \mathrm{rk}(\mathscr{E}))$ obtained by repeatedly applying $\mathcal{S}_E[1]$ and $(-) \otimes \mathscr{O}_E(0)^{\otimes d}$ for $d \in \mathbb{Z}$.

First note that the functors $\mathcal{S}_E[1]$ and $\mathcal{S}_E$ preserve $\gcd(\deg(\mathscr{E}), \mathrm{rk}(\mathscr{E}))$ by Lemma 27.272. The functor $(-) \otimes \mathscr{O}_E(0)$ preserves it since if $\frac{a}{b}$ is a rational number, then $\gcd(a, b) = \gcd(a + b, b)$.

Applying $\mathcal{S}_E[1]$ and $(-) \otimes \mathcal{O}_E(0)^{\otimes d}$ allows us to replace λ by $-\lambda^{-1}$ if $\lambda \neq 0$ and λ by $\lambda + d$.

Let us show that any rational number λ can be transformed into 0 by these operations. Let $\lambda \neq 0$. We argue by induction on b_λ, where $\lambda = \frac{a_\lambda}{b_\lambda}$ with a_λ and b_λ unique coprime integers with $b_\lambda \geq 1$. If $b_\lambda = 1$, λ is an integer and the claim is clear. If $b_\lambda > 1$ we may add same integer to λ so that $-1 < \lambda < 0$. Then $-b_\lambda < a_\lambda < 0$ and hence $b_{-\lambda^{-1}} = -a_\lambda < b_\lambda$ so that we know by induction hypothesis that $-\lambda^{-1}$ can be transformed into 0.

It remains to show (27.50.2) for the functor $\mathcal{S}_E[1]: (\mathrm{Vect}_0(E)) \xrightarrow{\sim} (\mathrm{Coh}(E))_{\mathrm{tors}}$, i.e., that $\dim H^0(E, \mathcal{S}_E(\mathscr{F})) = \gcd(\mathrm{rk}(\mathscr{F}), \deg(\mathscr{F})) = \mathrm{rk}(\mathscr{F})$ for a semistable vector bundle $\mathscr{F}$ of slope 0 and hence of degree 0. Since we have $\mathcal{S}_E(\mathcal{S}_E(\mathscr{F}[1])) = [-1]^*\mathscr{F}$ we find

$$\mathrm{rk}(\mathscr{F}) = \mathrm{rk}([-1]^*\mathscr{F}) = \deg \mathcal{S}_E(\mathscr{F})[-1] = \chi(E, \mathcal{S}_E(\mathscr{F})[-1]) = \dim H^0(\mathcal{S}_E(\mathscr{F})[-1]),$$

where the second equality holds by Lemma 27.272, the third by (27.50.1), and the last uses that $\mathcal{S}_E(\mathscr{F})[-1]$ is a coherent sheaf with finite support and hence $H^i(E, \mathcal{S}_E(\mathscr{F})[-1]) = 0$ for $i \neq 0$. $\square$

We obtain the following results about simple and indecomposable vector bundles.

Remark 27.274. Every torsion coherent sheaf on E is a finite direct sum of skyscraper sheaves $\mathcal{O}_{E,x}/(\pi_x)^e$ concentrated in $\{x\}$, where $x \in E$ is a closed point, π_x is a uniformizing element of the discrete valuation ring $\mathcal{O}_{E,x}$, and where $e \geq 1$ is an integer. These skyscraper sheaves are indecomposable. The simple objects of $(\mathrm{Coh}(E))_{\mathrm{tors}}$ are the skyscraper sheaves $\kappa(x)$ for $x \in E$ closed. Fix $\lambda \in \mathbb{Q}$.

(1) Via the equivalence of $(\mathrm{Vect}_\lambda(E))$ and $(\mathrm{Coh}(E))_{\mathrm{tors}}$, the skyscraper sheaves $\kappa(x)$ for $x \in E$ closed correspond to the simple objects of $(\mathrm{Vect}_\lambda(E))$, which are the stable vector bundles of slope λ by Proposition 26.152. This sets up a bijection

$$\{\text{closed points of } E\} \xrightarrow{\sim} \{\text{stable vector bundles of slope } \lambda\}, \qquad x \mapsto \mathscr{F}_\lambda(x).$$

The greatest common divisor of the rank and the degree of $\mathscr{F}_\lambda(x)$ is $[\kappa(x) : k]$.

(2) Let $\mathscr{F}$ be an indecomposable vector bundle and let λ be its slope. Then $\mathscr{F}$ is semistable by Proposition 26.161. Via the equivalence of $(\mathrm{Vect}_\lambda(E))$ and $(\mathrm{Coh}(E))_{\mathrm{tors}}$, it corresponds to the skyscraper sheaf $\mathcal{O}_{E,x}/(\pi_x)^e$ for a unique closed point x and a unique integer $e \geq 1$. Moreover, $\gcd(\mathrm{rk}(\mathscr{F}), \deg(\mathscr{F})) = e[\kappa(x) : k]$. Hence we obtain a bijection

$$\{\text{closed points of } E\} \xrightarrow{\sim} \{\text{indecomposable vector bundles of slope } \lambda\}.$$

Now $\mathcal{O}_{E,x}/(\pi_x)^e$ has decomposition series of length e whose graded pieces are all isomorphic to $\kappa(x)$. Therefore the indecomposable vector bundle $\widetilde{\mathscr{F}}_\lambda(x)$ of slope λ corresponding to the closed point x has a decomposition series whose graded pieces are all isomorphic to the stable vector bundle $\mathscr{F}_\lambda(x)$. The length e of this decomposition series is given by

$$e = \frac{\gcd(\mathrm{rk}(\widetilde{\mathscr{F}}_\lambda(x)), \deg(\widetilde{\mathscr{F}}_\lambda(x)))}{[\kappa(x) : k]}.$$

Corollary 27.275. *Let $\mathscr{F}$ be a vector bundle on E. Then $\mathscr{F}$ is stable if and only if $\mathrm{End}_{\mathcal{O}_E}(\mathscr{F})$ is a skew field. In this case $\mathrm{End}_{\mathcal{O}_E}(\mathscr{F})$ is a finite commutative field extension of k.*

Proof. If $\mathscr{F}$ is stable, say of slope λ, then $\mathscr{F}$ corresponds to the skyscraper sheaf $\kappa(x)$ for $x \in E$ closed via the equivalence $T \colon (\mathrm{Vect}_\lambda(E)) \cong (\mathrm{Coh}(E))_{\mathrm{tors}}$ (Remark 27.274) and $\mathrm{End}_{\mathscr{O}_E}(\mathscr{F}) = \mathrm{End}_{\mathscr{O}_E}(\kappa(x)) = \kappa(x)$.

Conversely, suppose that $\mathrm{End}_{\mathscr{O}_E}(\mathscr{F})$ is a skew field. Then $\mathscr{F}$ is necessarily indecomposable and therefore semistable (Proposition 26.161). Let λ be its slope. Hence $\mathscr{F}$ corresponds via T to an indecomposable coherent torsion sheaf on E, i.e., to a skyscraper sheaf $\mathscr{O}_{E,x}/\mathfrak{m}_x^e$ for some $x \in E$ closed and $e \geq 1$. Then $\mathrm{End}_{\mathscr{O}_E}(\mathscr{F}) = \mathrm{End}_{\mathscr{O}_E}(\mathscr{O}_{E,x}/\mathfrak{m}_x^e) = \mathscr{O}_{E,x}/\mathfrak{m}_x^e$ which is a skew field if and only if $e = 1$, i.e., if and only if $\mathscr{F}$ is stable (Remark 27.274). $\square$

(27.51) Very ample line bundles on abelian schemes.

In this section we show that the tensor product of any three relatively ample line bundles on an abelian scheme is already very ample.

We will need some properties of divisors on an abelian variety X over an algebraically closed field k. Recall that an abelian variety over a field is regular and hence the notions of (Cartier) divisor and of Weil divisor coincide.

For $x, y \in X(k)$ and a divisor D one has $\mathrm{Supp}(t_x^*D) = t_{-x}(\mathrm{Supp}(D))$; sometimes we denote this set by $\mathrm{Supp}(D) - x$. We have

$$y \in \mathrm{Supp}(t_x^*D) \Leftrightarrow x \in \mathrm{Supp}(t_y^*D).$$

Moreover, if $\mathscr{L}$ is a line bundle on X we denote by

$$|\mathscr{L}| := \mathbb{P}(\Gamma(X, \mathscr{L})^\vee)(k)$$

the corresponding complete linear system, i.e., the projective space of effective Cartier divisors D with $\mathscr{O}_X(D) \cong \mathscr{L}$ (Proposition 11.34 and Section (13.13)). Recall that $\mathscr{L}$ is globally generated if for all $x \in X(k)$ there exists a $D \in |\mathscr{L}|$ such that $x \notin \mathrm{Supp}(D)$.

Lemma 27.276. *Let k be an algebraically closed field, let X be an integral k-scheme of finite type, let D be a Cartier divisor on X and let $\mathscr{L} = \mathscr{O}_X(D)$ be the corresponding line bundle. Let $|\mathscr{L}| := \mathbb{P}(\Gamma(X, \mathscr{L})^\vee)(k)$ be the linear system given by D. Then there exists an open subscheme $U \subseteq \mathbb{P}(\Gamma(X, \mathscr{L})^\vee)$ such that $U(k)$ is the subset of points of $|\mathscr{L}|$ corresponding to reduced divisors.*

Proof. We will construct a divisor that is "universal" for the complete linear system $|\mathscr{L}|$, i.e., an effective divisor $\tilde{D} \subset X \times_k \mathbb{P}(\Gamma(X, \mathscr{L})^\vee)$ such that for every $p \in \mathbb{P}(\Gamma(X, \mathscr{L})^\vee)(k)$, the schematic fiber of $\tilde{D}$ over p is the effective divisor in $X = X \times \{p\}$ (considered as a closed subscheme) corresponding to p. The lemma follows from this because $\tilde{D}$ is proper and flat over $\mathbb{P}(\Gamma(X, \mathscr{L})^\vee)$ and we can apply (11) in Section (E.1), [EGAIV]^0 (12.2.1).

To construct $\tilde{D}$, write $P = \mathbb{P}(\Gamma(X, \mathscr{L})^\vee)$ and let $\mathscr{M} = \mathscr{L} \boxtimes \mathscr{O}_P(1)$ denote the exterior tensor product, i.e., the tensor product of the two pullbacks along the projections. Then $\Gamma(X \times P, \mathscr{M}) = \Gamma(X, \mathscr{L}) \otimes_k \Gamma(X, \mathscr{L})^\vee$ which we can canonically identify with $\mathrm{End}_k(\Gamma(X, \mathscr{L}))$. Denote by $s \in \Gamma(X \times P, \mathscr{M})$ the element corresponding to $\mathrm{id} \in \mathrm{End}_k(\Gamma(X, \mathscr{L}))$. Then s defines a Cartier divisor in $X \times P$. Fix $\dot{p} \in \Gamma(X, \mathscr{L}) \setminus \{0\}$ and let $p \in P$ be the point determined by $\dot{p}$. Then the fiber of $\tilde{D}$ over p is the Cartier divisor given by the line bundle $\mathscr{L}$ and the section obtained as the image of id under the map

$$\mathrm{End}_k(\Gamma(X, \mathscr{L})) = \Gamma(X, \mathscr{L}) \otimes_k \Gamma(X, \mathscr{L})^\vee \longrightarrow \Gamma(X, \mathscr{L}), \quad t \otimes \lambda \mapsto \lambda(\dot{p})t,$$

and this is just $\dot{p}$. The point in $|\mathscr{L}|$ corresponding to this divisor is p, as desired. $\square$

Lemma 27.277. *Let k be an algebraically closed field and let X be an abelian variety over k. Let D be an effective divisor on X and let $\mathscr{L} = \mathscr{O}_X(D)$ be the line bundle attached to D.*

(1) *The subset of points of $|\mathscr{L}|$ corresponding to reduced effective divisors is a dense open subset.*

(2) *Assume in addition that D is ample. Then for all D' in an open dense subset of $|\mathscr{L}|$ and all $0 \neq x \in X(k)$, one has $t_x^* D' \neq D'$.*

Note that in part (2) of the lemma we speak about inequality of divisors (as opposed to the two sides not being linearly equivalent).

Proof. Let us show (1). Suppose that $D = rE + F$ with E irreducible, $r > 1$, and $F \geq 0$ some divisor. Then rE is linearly equivalent to $\sum_{i=1}^r t_{x_i}^* E$ for all families of sections $x_i \in X(k)$ with $\sum_i x_i = 0$ (Remark 27.169 (3)). For a suitable choice of the x_i the $t_{x_i}^* E$ are all distinct and distinct from the other components of D. This shows that the subset in question is non-empty. In view of Lemma 27.276 it is thus open and dense.

Next we prove (2). Since $t_x^* D' = D'$ implies $t_x^* \mathscr{L} \cong \mathscr{L}$ and hence $x \in K(\mathscr{L})(k)$, we can fix one of these finitely many x. For the finiteness assertion we use Corollary 27.172 and the assumption that $\mathscr{L}$ is ample. We will now show that every reduced effective divisor D' in $|\mathscr{L}|$ with $t_x^* D' = D'$ lies in a union of linear subspaces of strictly smaller dimension. To avoid introducing further notation, let us assume that D is itself reduced and show this claim for D. The union of linear subspaces that we will find will be independent of D, so this finishes the proof. Let $G \subseteq K(\mathscr{L})$ be the finite group generated by x, considered as a finite étale group scheme over k, and consider the finite étale isogeny of abelian varieties $\pi \colon X \to X/G$ of degree $d := \#G > 1$. Then $\bar{D} := \pi(\operatorname{Supp} D)$ is a closed subset of pure codimension one in X/G which we consider as a reduced effective divisor. As π is étale, $\pi^{-1}(\bar{D})$ is again reduced with the same support as D and hence $D = \pi^{-1}(\bar{D})$ and $\mathscr{L} \cong \pi^* \mathscr{O}_{X/G}(\bar{D})$. As D is ample, $\bar{D}$ is ample as well (Proposition 13.66) and hence one has

$$
\begin{aligned}
\dim H^0(X, \mathscr{L}) &= \chi(X, \mathscr{L}) \\
&= d\chi(X/G, \mathscr{O}_{X/G}(\bar{D})) \\
&= d \dim H^0(X/G, \mathscr{O}_{X/G}(\bar{D})) \\
&> \dim H^0(X/G, \mathscr{O}_{X/G}(\bar{D})).
\end{aligned}
$$

Here the first and the last equality hold by Corollary 27.266, and the second equality by Proposition 27.179. Since $D = \pi^{-1}(\bar{D})$, considered as a point in $|\mathscr{L}|$, D lies in $\pi^* H^0(X/G, \mathscr{O}_{X/G}(\bar{D}))$. By Corollary 27.215 there are only finitely many isomorphism classes of line bundles $\mathscr{M}$ on X/G such that $\pi^* \mathscr{M} \cong \mathscr{L}$. The union in $H^0(X, \mathscr{L})$ of their global sections is the finite union of lower-dimensional subspaces we wanted to construct. $\qquad\square$

Proposition 27.278. *Let $S = \operatorname{Spec} R$ be an affine scheme and let $f \colon X \to S$ be an abelian scheme. Let $\mathscr{L}_1$ and $\mathscr{L}_2$ be ample line bundles on X. Then $\mathscr{L}_1 \otimes \mathscr{L}_2$ is globally generated.*

Proof. Let $\mathscr{L} := \mathscr{L}_1 \otimes \mathscr{L}_2$. In view of Proposition 13.30 we have to show that $f^*(f_* \mathscr{L}) \to \mathscr{L}$ is surjective. By Nakayama's lemma we can do this on fibers and hence can assume that $S = \operatorname{Spec} k$ for a field k. By faithfully flat descent we can in addition assume that k is algebraically closed.

Let $x \in X(k)$. We have to show that there exists $D \in |\mathscr{L}|$ such that $x \notin D$. Consider the homomorphism

$$\Phi := m_{X^t} \circ (\varphi_{\mathscr{L}_1} \times \varphi_{\mathscr{L}_2}) \colon X \times_k X \longrightarrow X^t,$$
$$(x_1, x_2) \longmapsto t_{x_1}^*(\mathscr{L}_1) \otimes \mathscr{L}_1^{-1} \otimes t_{x_2}^*(\mathscr{L}_2) \otimes \mathscr{L}_2^{-1}.$$

As $\mathscr{L}_1$ and $\mathscr{L}_2$ are ample, Φ is surjective and therefore $\dim \mathrm{Ker}(\Phi) = g := \dim(X)$. For $x_0 \in X(k)$ we have

$$\mathrm{Ker}(\Phi) \cap (\{x_0\} \times X) \cong \varphi_{\mathscr{L}_2}^{-1}(-\varphi_{\mathscr{L}_1}(x_0))$$

and similarly for $\mathrm{Ker}(\Phi) \cap (X \times \{x_0\})$. Therefore

(*) $\dim(\mathrm{Ker}(\Phi) \cap (\{x_0\} \times X)) = \dim(\mathrm{Ker}(\Phi) \cap (X \times \{x_0\})) = 0$

because $\varphi_{\mathscr{L}_i}$ is finite for $i = 1, 2$.

As $\mathscr{L}_1$ and $\mathscr{L}_2$ are ample, we can choose effective divisors $D_i \in |\mathscr{L}_i|$ (Proposition 27.265). By (*) we have

$$\dim(\mathrm{Ker}(\Phi) \cap (t_x^* D_1 \times X)) = \dim(\mathrm{Ker}(\Phi) \cap (X \times t_x^* D_2)) = g - 1.$$

Therefore we find $(x_1, x_2) \in \mathrm{Ker}(\Phi)(k)$ such that

$$(x_1, x_2) \notin (t_x^* D_1 \times X) + (X \times t_x^* D_2)$$

and hence $x_1 \notin t_x^* D_1$ and $x_2 \notin t_x^* D_2$. This shows

$$x \notin D := t_{x_1}^* D_1 + t_{x_2}^*(D_2).$$

As (x_1, x_2) are in the kernel of Φ we have

$$t_{x_1}^* \mathscr{L}_1 \otimes t_{x_2}^* \mathscr{L}_2 \cong \mathscr{L}_1 \otimes \mathscr{L}_2 = \mathscr{L}$$

and hence $D \in |\mathscr{L}|$. $\square$

Theorem 27.279. *Let S be a scheme, and let $X \to S$ be an abelian scheme. Let $\mathscr{L}_1, \mathscr{L}_2, \mathscr{L}_3$ be relatively ample line bundles on X. Then $\mathscr{L}_1 \otimes \mathscr{L}_2 \otimes \mathscr{L}_3$ is very ample over S.*

Proof. (I). We may assume that S is affine. Set $\mathscr{L} := \mathscr{L}_1 \otimes \mathscr{L}_2 \otimes \mathscr{L}_3$. We know by Proposition 27.278 that $\mathscr{L}$ is globally generated. In particular the corresponding morphism of S-schemes $r \colon X \to \mathbb{P}(f_* \mathscr{L})$ is defined on all of X. Now $f_* \mathscr{L}$ is a vector bundle on S (Corollary 27.267) and hence r is a morphism of proper S-schemes. We have to show that r is a closed immersion (Proposition 13.56). This can be done on fibers (Proposition 12.93) and hence we may assume that $S = \mathrm{Spec}\, k$ for a field k. By faithfully flat descent we can in addition assume that k is algebraically closed. By Proposition 12.94 it suffices to show that r is injective on k-valued points and on tangent spaces in k-valued points.

(II). Let us now show that r is injective on k-valued points. Let $y_1, y_2 \in X(k)$ such that $r(y_1) = r(y_2)$, so that for all $D \in |\mathscr{L}|$ one has

(*) $y_1 \in \mathrm{Supp}(D) \Leftrightarrow y_2 \in \mathrm{Supp}(D).$

As $H^0(X, \mathscr{L}_1) \neq 0$ (Proposition 27.265), there exists an effective divisor $D_1 \in |\mathscr{L}_1|$. We choose D_1 reduced and as in Lemma 27.277 (2). To prove that $y_1 = y_2$, it then suffices to show that

(**) $\qquad\qquad (\operatorname{Supp} D_1 - y_1) \subseteq (\operatorname{Supp} D_1 - y_2).$

In fact, (**) implies $t_{y_1}^* D_1 = t_{y_2}^* D_1$ because D_1 is reduced and Lemma 27.277 (2) shows that $y_1 = y_2$.

Let x_1 be any k-valued point of $\operatorname{Supp} D_1 - y_1$. Then

$$\mathscr{L} \otimes t_{x_1}^* \mathscr{L}_1^{-1} \cong \mathscr{L}_2 \otimes \mathscr{L}_3 \otimes (\mathscr{L}_1 \otimes t_{x_1}^* \mathscr{L}_1^{-1})$$

is the tensor product of the ample line bundles $\mathscr{L}_2$ and $\mathscr{L}_3 \otimes (\mathscr{L}_1 \otimes t_{x_1}^* \mathscr{L}_1^{-1})$ (Corollary 27.269) and hence is globally generated by Proposition 27.278. Therefore we find an effective divisor $D' \in |\mathscr{L} \otimes t_{x_1}^* \mathscr{L}_1^{-1}|$ such that $y_2 \notin D'$. The divisor $t_{x_1^*} D_1 + D' \in |\mathscr{L}|$ contains y_1 and hence y_2 by (*). By our choice of D' we must have $y_2 \in \operatorname{Supp}(t_{x_1^*} D_1)$, i.e. $x_1 \in \operatorname{Supp}(t_{y_2}^* D_1)$. This shows (**).

(III). Assume that r is not injective on tangent spaces. This means that there exist a point $x_0 \in X(k)$ and a tangent vector $0 \neq \xi \in T_{x_0}(X) = (\mathfrak{m}_{x_0}/\mathfrak{m}_{x_0}^2)^\vee$ such that for all effective divisors $D \in |\mathscr{L}|$ with $x_0 \in \operatorname{Supp}(D)$ the tangent vector ξ is tangential to D at x_0, i.e., if $s = 0$ is a local equation of D at x_0 with differential $ds \in x_0^* \Omega_{X/k} = \mathfrak{m}_{x_0}/\mathfrak{m}_{x_0}^2$ at x_0, then $\langle \xi, ds \rangle = 0$. Recall that the tangent bundle of X is free. Therefore the tangent vector ξ corresponds to a translation invariant section Σ of the tangent bundle (Section (17.6)) such that $\Sigma_{x_0} = \xi$.

As in Step (II) choose $D_1 \in |\mathscr{L}_1|$ reduced, and let $x_1 \in \operatorname{Supp}(t_{x_0}^* D_1)$. As before there exists $D' \in |\mathscr{L} \otimes t_{x_1}^* \mathscr{L}_1^{-1}|$ such that $x_0 \notin D'$. Then $D := D' + t_{x_1}^* D_1 \in |\mathscr{L}|$ and hence ξ is tangent to D at x_0 by assumption. As $x_0 \notin D'$, $\xi = \Sigma_{x_0}$ must be tangent to $t_{x_1}^* D_1$ at x_0, i.e., $\Sigma_{x_0+x_1}$ is tangent to D_1 in $x_0 + x_1$. Since $x_1 \in \operatorname{Supp}(D_1) - x_0$ was arbitrary, we conclude that Σ_x is tangent to D_1 at every point $x \in \operatorname{Supp}(D_1)$. We may consider Σ as a global derivation $\mathscr{O}_X \to \mathscr{O}_X$. We claim that it preserves the ideal sheaf defining D_1, in other words, that the following property holds.

(T) For $U \subseteq X$ affine open and for all $s \in \mathscr{O}_X(U)$ such that $D_1 \cap U = \operatorname{div}(s)$ one has $\Sigma(s) \in (s)$.

Since D_1 is reduced, the ideal (s) is a radical ideal, so to show that it contains $\Sigma(s)$, it is enough to show that $\Sigma(s)$ lies in every maximal ideal $\mathfrak{m} \subseteq \mathscr{O}_X(U)$ in the support of $D_1 \cap U$. Since the restriction of Σ to a map $\mathfrak{m} \to \mathscr{O}_X(U)$ induces, by tensoring with the residue class field, the map $\Sigma_x \colon \mathfrak{m}/\mathfrak{m}^2 \to \kappa(\mathfrak{m})$, the vanishing of $\Sigma_x(s)$ implies that $\Sigma(s) \in \mathfrak{m}$, as desired.

So when we consider Σ_0 as a $k[\varepsilon]/(\varepsilon^2)$-valued point of X whose underlying topological point is 0, the translation $t_\Sigma \colon X_{k[\varepsilon]/(\varepsilon^2)} \to X_{k[\varepsilon]/(\varepsilon^2)}$ preserves the divisor $D_{1,k[\varepsilon]/(\varepsilon^2)}$ and a fortiori preserves the associated line bundle. This implies $\Sigma_0 \in T_0(K(\mathscr{L}_1)) = \operatorname{Lie}(K(\mathscr{L}_1))$.

If the characteristic of k is zero, then all algebraic groups over k are smooth (Theorem 27.25) and the finite group scheme $K(\mathscr{L}_1)$ is étale and therefore $\operatorname{Lie}(K(\mathscr{L}_1)) = 0$, a contradiction.

Hence we can from now on assume that $\operatorname{char}(k) = p > 0$. Let $H \subseteq K(\mathscr{L})^0$ be the smallest group scheme with $\Sigma_0 \in \operatorname{Lie}(H)$. Consider the isogeny $\pi \colon X \to X/H$ which is an H-torsor. The scheme-theoretic action of H on X by translation preserves D_1 by (T), i.e., the locally free ideal sheaf $\mathscr{I}_{D_1}$ of rank 1 defining D_1 as a closed subscheme is H-equivariant. By descent along torsors (Section (14.21)) we obtain a locally free ideal sheaf of $\mathscr{O}_{X/H}$ that defines a divisor D_1' on X/H such that $\pi^* D_1' = D_1$. As in the proof of Lemma 27.277 we compute

$$\dim H^0(X, \mathscr{L}_1) = \deg(\pi) \dim H^0(X/H, \mathscr{O}_{X/H}(D_1')) > \dim H^0(X/H, \mathscr{O}_{X/H}(D_1'))$$

and conclude that all global sections of $\mathscr{L}_1$ which define reduced divisors lie in a finite set of proper subspaces. This contradicts Lemma 27.277 (1). $\qquad\square$

(27.52) Symmetric homomorphisms and polarizations.

Let S be a scheme and let X be an abelian scheme over S. We want to introduce the notion of *polarization* of an abelian scheme. Over an algebraically closed field, a polarization is an equivalence class of ample line bundles. For an abelian variety X over the complex numbers, we can view a polarization as a positive definite hermitian form on its Lie algebra, i.e., the tangent space at the zero element, with a certain integrality property; see Section (27.54). In general, polarizations can be seen, in some sense, as analogues of symmetric bilinear forms satisfying a positivity condition. More concretely, polarizations are indispensable for the construction of well-behaved moduli spaces of abelian varieties, see Section (27.55).

Definition 27.280. *A homomorphism* $\lambda\colon X \to X^t$ *is called*
(1) symmetric *if* $\lambda = \lambda^t$, *identifying* X *with* X^{tt},
(2) *a* polarization *if for every geometric point* $\bar{s} \to S$ *there exists an ample line bundle* $\mathscr{L}$ *on* $X_{\bar{s}}$ *such that* $\lambda_{\bar{s}} = \varphi_{\mathscr{L}}$.
    A polarization λ *is called a* principal polarization *if* λ *is an isomorphism.*

Given a polarization λ, for every geometric point $\bar{s} \to S$ the morphism $\lambda_{\bar{s}}$ is an isogeny, hence λ is an isogeny.

Lemma 27.281. *Let* $\lambda\colon X \to X^t$ *be a homomorphism of abelian schemes.*
(1) *Let* $s \in S$ *be a point such that* $\lambda_{\bar{s}}$ *is symmetric for some geometric point* $\bar{s} \to S$ *with image* s. *Then there exists an open and closed neighborhood* U *of* s *such that* $\lambda_U\colon X \times_S U \to X^t \times_S U$ *is symmetric.*
(2) *If* λ *is of the form* $\varphi_{\mathscr{L}}$ *for some line bundle* $\mathscr{L}$ *on* X, *then* λ *is symmetric.*
(3) *The homomorphism* λ *is symmetric if and only if there exists a line bundle* $\mathscr{L}$ *on* X *such that* $\varphi_{\mathscr{L}} = 2\lambda$.

Proof. Let us show (1). By hypothesis, $\lambda_{\bar{s}} - \lambda_{\bar{s}}^t$ factors through the zero section of $X_{\bar{s}}^t$. Hence we also have on fibers $\lambda_s - \lambda_s^t = 0$ since equality of morphisms can be checked after faithfully flat base change. As the constancy locus of $\lambda - \lambda^t$ is open and closed (Proposition 27.96), there exists an open and closed neighborhood U of s such that $\lambda_U - \lambda_U^t$ factors through a section. As $\lambda - \lambda^t$ is a homomorphism of group schemes, that section is necessarily the zero section. Hence λ_U is symmetric.

Assertion (2) holds because the automorphism σ of $X \times_S X$ that switches the factors satisfies $\sigma^*\Lambda(\mathscr{L}) \cong \Lambda(\mathscr{L})$, where $\Lambda(\mathscr{L})$ denotes the Mumford bundle.

It remains to show (3). The condition is sufficient by (2). Conversely, for every homomorphism $\lambda\colon X \to X^t$ set $\mathscr{L} := (\mathrm{id}_X, \lambda)^* \mathscr{P}_X$. If λ is symmetric, we have $\varphi_{\mathscr{L}} = \lambda + \lambda^t = 2\lambda$, where the first equality follows from Proposition 27.220 and (27.42.4). $\qquad\square$

Corollary 27.282. *Polarizations are symmetric.*

Remark 27.283. One can show that for every line bundle $\mathscr{L}$ on X and for every integer $n \geq 2$ there exists an fppf-covering $(U_i \to S)_i$ such that the pullback of $\mathscr{L}$ to $X \times_S U_i$ is isomorphic to $\mathscr{M}_i^{\otimes n}$ for line bundles $\mathscr{M}_i$ on $X \times_S U_i$ ([DePa] $^\circ$ 1.2 and its proof).

Hence Lemma 27.281 (3) implies that every symmetric homomorphism φ is fppf-locally of the form $\varphi_{\mathscr{M}}$ for some line bundle $\mathscr{M}$. If φ is a polarization, then $\mathscr{M}$ is necessarily ample. In fact this holds even locally for the étale topology (loc. cit. 1.3).

One can use this remark to show the following two results.

Proposition 27.284. *A homomorphism of abelian schemes* $\lambda\colon X \to X^t$ *is a polarization if and only if λ is symmetric and* $(\mathrm{id}_X, \lambda)^* \mathscr{P}_X$ *is a relative ample line bundle on X over S.*

Proof. By Lemma 27.281 (1) and since we can check relative ampleness over proper schemes on fibers (Theorem 24.46) and hence on geometric fibers (Proposition 14.58), we may assume that $S = \operatorname{Spec} k$ for an algebraically closed field k. We set $\mathscr{L} := (\mathrm{id}_X, \lambda)^* \mathscr{P}_X$.

Let λ be a polarization, i.e., of the form $\varphi_{\mathscr{M}}$ for an ample line bundle $\mathscr{M}$ on X. Then $\varphi_{\mathscr{M}^{\otimes 2}} = 2\lambda = \varphi_{\mathscr{L}}$. Therefore $\mathscr{M}^{\otimes 2}$ and $\mathscr{L}$ are algebraically equivalent and hence $\mathscr{L}$ is ample.

Conversely suppose that $\mathscr{L}$ is ample and that λ is symmetric. Then $\lambda = \varphi_{\mathscr{M}}$ for some line bundle $\mathscr{M}$ on X by Remark 27.283 and as above $\mathscr{M}^{\otimes 2}$ is algebraically equivalent to $\mathscr{L}$. Hence $\mathscr{M}$ is ample and λ is a polarization. $\square$

Corollary 27.285. *Let* $\lambda\colon X \to X^t$ *be a homomorphism of abelian schemes and let $s \in S$ be a point such that $\lambda_{\bar{s}}$ is polarization for some geometric point $\bar{s} \to S$ with image s. Then there exists an open and closed neighborhood U of s such that* $\lambda_U\colon X \times_S U \to X^t \times_S U$ *is a polarization.*

Proof. By Proposition 27.284 this follows from Lemma 27.281 (1) and from Corollary 27.268. $\square$

In general, given an abelian scheme there might not exist any polarization. For an abelian variety over a field, a polarization always exists, because every abelian variety is projective. But even if the base field is algebraically closed, a principal polarization need not exist.

Remark 27.286. It follows from Proposition 27.148 that every elliptic curve carries a canonical principal polarization.

Remark 27.287. Let S be a scheme and let $f\colon C \to S$ be a proper smooth curve with geometrically connected fibers. Then we have the Jacobian $\operatorname{Jac}_C = \operatorname{Pic}^0_{C/S}$ of C as in Section (27.26). One can show that the abelian scheme Jac_C carries a principal polarization. See [MFK]^{O} Prop. 6.9 and [Mil2]^{O} for a proof of this fact. If f has a section, then we have the Abel morphism $C \to \operatorname{Jac}_C$, (27.26.2). Pullback of line bundles along this morphism induces a morphism $\operatorname{Jac}_C^\vee \to \operatorname{Jac}_C$, and one can show that this morphism is an isomorphism.

The famous Theorem of Torelli states that given curves C, C' over a field k such that there exists an isomorphism between their Jacobians which is compatible with their canonical principal polarizations, then C and C' are isomorphic. For the case where k is algebraically closed, see [Mil2]^{O} Sections 12, 13 and the references given in the section at the end of that paper. See the appendix by Serre in [Lau]$_{\mathrm{X}}$ for a discussion of the result over an arbitrary field.

Remark 27.288. Let S be a scheme and let $f\colon X \to S$, $g\colon Y \to S$ be smooth proper relative curves over S with geometrically connected fibers. Let us also assume that X and Y are pointed S-schemes, i.e., are equipped with a fixed section in $X(S)$ and $Y(S)$, respectively. As in Lemma 27.155, we can identify divisorial correspondences in $\operatorname{Corr}_S(X,Y)(S)$ with morphisms $X \to \operatorname{Pic}^0_{Y/S}$ of pointed schemes. Using the Albanese property of $(\operatorname{Pic}^0_{X/S})^\vee$ (Remark 27.225) and the fact that $\operatorname{Pic}^0_{X/S} \cong (\operatorname{Pic}^0_{X/S})^\vee$ canonically (Remark 27.287), we obtain a morphism $\operatorname{Pic}^0_{X/S} \to \operatorname{Pic}^0_{Y/S}$ between the Jacobian of X and the Jacobian of Y. Using the same reasoning after base change to any S-scheme T, in this way we obtain isomorphisms

$$\operatorname{Corr}(X,Y) \xrightarrow{\sim} \underline{\operatorname{Hom}}_0(X, \operatorname{Pic}^0_{Y/S}) \xrightarrow{\sim} \underline{\operatorname{Hom}}(\operatorname{Pic}^0_{X/S}, \operatorname{Pic}^0_{Y/S}).$$

In fact, Lemma 27.155 implies that the first map is an isomorphism. The natural map $X \to (\operatorname{Pic}^0_{X/S})^\vee \cong \operatorname{Pic}^0_{X/S}$ induces an inverse of the second map.

(27.53) Projectivity of abelian schemes over normal base schemes.

Recall that in Proposition 27.174 we proved that every abelian variety (i.e., abelian scheme over a field) is projective. Equivalently, in view of the properness, there exists an ample line bundle. In this section we will generalize this result to abelian schemes over any normal noetherian scheme S.

Lemma 27.289. Let $f\colon Z \to S$ be a formally proper (Definition 27.128) morphism of schemes. Then the image $f(Z)$ is closed under specialization.

Proof. Let s be a point of $f(Z)$ and let $t \in S$ be a specialization of s. We have to show that t is in the image of f. We may assume that $t \neq s$. Let $z \in Z$ with $f(z) = s$. By Proposition 15.7 there exists a valuation ring R with field of fraction $K = \kappa(z)$ and a morphism $g\colon \operatorname{Spec} R \to S$ such that $g(\eta) = s$ and $g(x) = t$, where η (resp. x) is the generic point (resp. special point) of $\operatorname{Spec} R$. As f is formally proper, there exists a morphism $\tilde{g}\colon \operatorname{Spec} R \to Z$ with $f \circ \tilde{g} = g$. In particular, t lies in the image of f. $\qquad\square$

Lemma 27.290. Let $f\colon Z \to S$ be a separated, unramified, and formally proper (Definition 27.128) morphism of schemes. Suppose that S is locally noetherian and normal. Let $v\colon S \dashrightarrow Z$ be a rational section of f. Then v is defined on all of S.

Proof. We may assume that S is connected and hence integral. Let $U := \operatorname{dom}(v)$ be the domain of definition of v. As $Z \to S$ is separated there exists a representative $U \to Z$ of v which we again call v (Proposition 9.27). We have to show that $U = S$.

Assume that there exists $s \in S \setminus U$. As v is a section of the unramified morphism $f_{|f^{-1}(U)}$, it is an open immersion $v\colon U \to f^{-1}(U)$. Let Y be the closure of $v(U)$ in X, endowed with the reduced subscheme structure. Then Y is an integral scheme. Let $f' := f_{|Y}\colon Y \to S$. Then f' is unramified, separated, formally proper, and birational. The image of f' contains the generic point of S and is stable under specialization by Lemma 27.289. Hence f' is surjective and its fiber $Y_s := f^{-1}(s) \cap Y$ in s is non-empty. Let $y \in Y$ with $f'(y) = s$. Let $V \subseteq Y$ be an open affine neighborhood of y. Then $f'_{|V}\colon V \to S$ is quasi-compact, because S is locally noetherian and in particular quasi-separated. Therefore $f'_{|V}$ is quasi-finite (Corollary 18.28), separated, and birational. Hence it is an open immersion by Zariski's main theorem (Corollary 12.88). But this implies that the section v can be extended to an open neighborhood of s. Contradiction. $\qquad\square$

Theorem 27.291. *Let S be a normal noetherian scheme and let X be an abelian scheme over S. Then $X \to S$ is projective.*

Proof. We may assume that S is connected and hence integral. Let $\eta \in S$ be its generic point. By Proposition 27.174 there exists an ample line bundle $\mathscr{M}$ on the generic fiber X_η. By Lemma 24.45 there exists then also an open non-empty subscheme U of S and a relatively ample line bundle $\mathscr{M}'$ on $X \times_S U$ whose restriction to X_η is $\mathscr{M}$. The associated homomorphism $\varphi_{\mathscr{M}'}$ can be considered as a section of $\mathrm{Corr}_S(X, X)$ over U.

By Remark 27.127 we know that $\mathrm{Pic}_{X \times_S X/S}$ is formally proper. Since $\mathrm{Corr}_S(X, X) \to \mathrm{Pic}_{X \times_S X/S}$ is a closed immersion (Lemma 27.156), the same holds for $\mathrm{Corr}_S(X, X)$. Hence by Proposition 27.157 we can apply Lemma 27.290 and obtain a section of $\mathrm{Corr}_S(X, X)$, i.e., a homomorphism $\lambda \colon X \to X^t$. The morphism λ is generically on S symmetric (Lemma 27.281 (2)), hence it is symmetric (Lemma 27.281 (1)). Therefore there exists a line bundle $\mathscr{L}$ on X such that $\varphi_{\mathscr{L}} = 2\lambda$ (Lemma 27.281 (3)). Then the restriction $\mathscr{L}_\eta$ to X_η is algebraically equivalent to $\mathscr{M}^{\otimes 2}$. Therefore $\mathscr{L}_\eta$ is ample. As ampleness is open and closed on abelian schemes (Corollary 27.268), we see that $\mathscr{L}$ is relatively ample. Hence X is quasi-projective over S and hence projective because S is qcqs (Corollary 13.72). $\square$

The normality hypothesis in the theorem cannot be dropped. In [Ray1] ^{O} Ch. XII, Raynaud gives an example of an abelian scheme over a local noetherian ring (which is not normal) which does not have an ample line bundle.

(27.54) Abelian varieties over the complex numbers.

The theory of abelian varieties over the complex numbers forms a vast topic, and our treatment here will necessarily be very sketchy and incomplete. For further details, see the first chapter (and Sections 9 and 24) of Mumford's book [Mum1], and the comprehensive volume [BiLa] ^{O} by Birkenhake and Lange, which has a lot of material on abelian varieties over the complex numbers and also contains very informative introductions to the book and the individual chapters, including further references.

Let A be an abelian variety over $\mathbb{C}$. Its analytification A^{an} (Section (20.12), Section (23.9)) is a complex Lie group, i.e., a group object in the category of complex manifolds. For every complex Lie group G with neutral element e, we have the *exponential map* $\exp \colon T_eG \to G$, a homomorphism of complex Lie groups, with the property that $0 \mapsto e$ and that $d\exp_e = \mathrm{id}$. See [BouLie23] ^{O} Ch. III §6 no. 4.

Since A is proper over $\mathbb{C}$, A^{an} is compact (Remark 20.59), and we can apply the following proposition. Here, for any finite-dimensional complex vector space V, of dimension g, say, a *lattice* $\Lambda \subset V$ is a subgroup that is a free $\mathbb{Z}$-module of rank $2g$ which spans V as an $\mathbb{R}$-vector space; in other words, there exists a $\mathbb{Z}$-basis of Λ which is an $\mathbb{R}$-basis of V. It is then easy to see that the quotient V/Λ carries a unique structure of complex manifold so that the projection $V \to V/\Lambda$ is a morphism of complex manifolds which is locally on V an isomorphism.

Proposition 27.292. *Let G be a compact complex Lie group with neutral element e. Then the exponential map $T_eG \to G$ induces an isomorphism $T_eG/\Lambda \cong G$ of complex Lie groups, and $\Lambda \subset T_eG$, the kernel of the exponential map, is a lattice in T_eG.*

In particular, as an abstract group G is commutative.

Proof. See [Mum1] Section 1. $\square$

We call a compact complex Lie group of the form V/Λ a *complex torus*. It is useful to have a closer look at complex tori in general before we come back to the special case of abelian varieties.

Let $X = V/\Lambda$ be a complex torus of dimension g. (When we write this, it is always understood that V is a complex vector space and Λ is a lattice in V.) Viewing X as a real manifold, a choice of basis of Λ yields an identification $X \cong (\mathbb{R}/\mathbb{Z})^{2g}$ of real Lie groups. In particular, for the underlying topological space we conclude that the canonical projection $V \to X$ is the universal cover of X. For the fundamental group of X (with base point 0) we obtain the identification $\pi_1(X) = \Lambda$. Compare Section (27.36). In particular, the fundamental group is abelian and can therefore also be identified with the first singular homology $H_1(X, \mathbb{Z})$ of X. As an abstract group, the group $(\mathbb{R}/\mathbb{Z})^g$ is divisible, i.e., multiplication by any non-zero integer n is surjective, and the kernel of multiplication by n is isomorphic to $(\frac{1}{n}\mathbb{Z}/\mathbb{Z})^{2g} \cong (\mathbb{Z}/n\mathbb{Z})^{2g}$. Compare Proposition 27.186, Proposition 27.187. We also see that the set of all n-torsion points for $n \in \mathbb{N}_{>0}$ is dense in X.

Let $X = V/\Lambda$, $X' = V'/\Lambda'$ be complex tori, and let $f \colon X \to X'$ be a morphism of complex manifolds that maps 0 to 0. Then there is a unique lift $\tilde{f} \colon V \to V'$ of f to the universal covers with $\tilde{f}(0) = 0$. It is clear that $\tilde{f}$ is a morphism of complex manifolds. Then for $\lambda \in \Lambda$, the map $v \mapsto \tilde{f}(v + \lambda) - \tilde{f}(v)$ is continuous on V and takes images in Λ', and is thus constant. It follows that $\tilde{f}(v + \lambda) = \tilde{f}(v) + \tilde{f}(\lambda)$ for all $v \in V$, $\lambda \in \Lambda$. Fixing λ and taking partial derivatives, the constant term $\tilde{f}(\lambda)$ disappears, so all partial derivatives of $\tilde{f}$ are Λ-periodic, and are hence constant by Liouville's Theorem. This implies that $\tilde{f}$ is a homomorphism of $\mathbb{C}$-vector spaces. Clearly, $\tilde{f}(\Lambda) \subseteq \Lambda'$ and the restriction of $\tilde{f}$ to Λ determines $\tilde{f}$ and hence f. In particular, we have proved the following result. (Compare Corollary 16.56 (2) and Remark 27.233.)

Proposition 27.293. *Let $X = V/\Lambda$, $X' = V'/\Lambda'$ be complex tori, where V and V' are complex vector spaces of dimension g and g', respectively. Every morphism $X \to X'$ of complex manifolds which maps 0 to 0 is a homomorphism of complex Lie groups. The above construction defines an embedding $\operatorname{Hom}(X, X') \to \operatorname{Hom}(\Lambda, \Lambda')$ of the group of homomorphisms $X \to X'$ of complex Lie groups into the group of group homomorphisms between the corresponding lattices. In particular, $\operatorname{Hom}(X, X')$ is a free $\mathbb{Z}$-module of rank $\leq 4gg'$.*

Note that the GAGA principle (Corollary 20.61) implies that for abelian varieties A, A' over $\mathbb{C}$, the group $\operatorname{Hom}(A, A')$ of homomorphisms between them coincides with the group of homomorphisms between their analytifications, and hence is a free $\mathbb{Z}$-module of rank $\leq 4\dim(A)\dim(A')$ by the proposition. For a more precise analysis, covering also base fields of positive characteristic, see [Mum1] Section 19.

Let us describe the line bundles on a complex torus. First observe that, as in the algebraic situation, there are no non-trivial line bundles on a complex vector space.

Lemma 27.294. *Let V be a finite-dimensional complex vector space, considered as a complex manifold. Then every line bundle on V is trivial, i.e., every locally free module of rank 1 over the structure sheaf of the complex manifold V is free.*

Proof. We use the exponential sequence

$$0 \to \mathbb{Z} \to \mathscr{O}_V \to \mathscr{O}_V^\times \to 1.$$

Since $H^1(V, \mathscr{O}_V) = 0$ by the $\bar{\partial}$-Poincaré lemma and $H^2(V, \mathbb{Z}) = 0$ since V is a contractible topological space, it follows from the corresponding long exact cohomology sequence that $\mathrm{Pic}(V) = H^1(V, \mathscr{O}_V^\times)$ is trivial. $\qquad\square$

Thus if $X = V/\Lambda$ is a complex torus, then, similarly as in Section (14.21), we can identify line bundles on X and line bundles on V with a Λ-action that is compatible with the action of Λ on V by translations. Since for every line bundle $\mathscr{L}$ on the complex manifold X the pullback of $\mathscr{L}$ under the projection $V \to V/\Lambda$ is trivial, up to isomorphism a Λ-equivariant line bundle on V is the same thing as a Λ-action on the trivial line bundle $\mathscr{O}_V$ that lies over the natural action of Λ on V. Such an action is given by a map $\Lambda \to \mathrm{Aut}(\mathscr{O}_V) = \Gamma(V, \mathscr{O}_V)^\times = \Gamma(V, \mathscr{O}_V^\times)$ with certain properties (satisfying a cocycle condition), and two such maps give rise to isomorphic line bundles if and only if they differ by a coboundary. One obtains an identification $\mathrm{Pic}(X) = H^1(\Lambda, \Gamma(V, \mathscr{O}_V^\times))$.

Let us make this more explicit. The exponential sequence for X,

$$0 \to \underline{\mathbb{Z}} \to \mathscr{O}_X \to \mathscr{O}_X^\times \to 1,$$

gives rise to a map $\mathrm{Pic}(X) = H^1(X, \mathscr{O}_X^\times) \to H^2(X, \mathbb{Z})$. The kernel of this map is in bijection with the group of group homomorphisms $\mathrm{Hom}(\Lambda, \mathbb{C}_1^\times)$, where $\mathbb{C}_1^\times = \{z \in \mathbb{C}; |z| = 1\}$ is the unit circle, which naturally embeds into $\Gamma(V, \mathscr{O}_V^\times)$. On the other hand, we have an identification $H^2(X, \mathbb{Z}) = \bigwedge^2 H^1(X, \mathbb{Z})$, and we can view $\bigwedge^2 H^1(X, \mathbb{Z})$ as the space of $\mathbb{Z}$-valued alternating forms on $H_1(X, \mathbb{Z}) = \Lambda$. One can describe the image of $H^1(X, \mathscr{O}_X^\times)$ in this space as follows.

Lemma 27.295. *There are natural group isomorphisms between*
(i) *the image of the map $\mathrm{Pic}(X) = H^1(X, \mathscr{O}_X^\times) \to H^2(X, \mathbb{Z})$,*
(ii) *the set of $\mathbb{R}$-bilinear alternating forms $E\colon V \times V \to \mathbb{R}$ such that $E(iz, iz') = E(z, z')$ for all $z, z' \in V$ and $E(z, z') \in \mathbb{Z}$ for all $z, z' \in \Lambda$,*
(iii) *the set of hermitian forms H such that $\mathrm{Im}(H(z, z')) \in \mathbb{Z}$ for all $z, z' \in \Lambda$. We denote the set of hermitian forms with this property by $\mathscr{H}$.*

Proof. The map from the set (i) to the set (ii) is given by the restriction of the identification of $H^2(X, \mathbb{Z})$ with the set of $\mathbb{Z}$-valued alternating forms on Λ discussed above. Given a form E as in (ii), we define H by $H(z, z') = E(iz, z') + iE(z, z')$, and conversely given H, we set $E = \mathrm{Im}(H)$. We omit the further computations, see [Mum1] Chapter I or [BiLa] Chapter 2. $\qquad\square$

The image of $\mathscr{L} \in \mathrm{Pic}(X)$ in $H^2(X, \mathbb{Z})$ is the singular cohomology version of the first Chern class of $\mathscr{L}$; one can show that the first Chern class of $\mathscr{L}$ vanishes if and only if $\mathscr{L}$ is trivial as a *topological* line bundle.

Putting things together, one obtains the following description of the line bundles on a complex torus $X = V/\Lambda$. We define

$$\mathscr{P} = \{(H, \alpha); H \in \mathscr{H}, \alpha\colon \Lambda \to \mathbb{C}_1^\times,$$
$$\alpha(\lambda + \lambda') = \exp(i\pi \, \mathrm{Im}(H(\lambda, \lambda')))\alpha(\lambda)\alpha(\lambda') \text{ for all } \lambda, \lambda' \in \Lambda\}.$$

For $H = 0$, the condition for α simply says that α is a group homomorphism. We equip this set with the structure of an abelian group by defining

$$(H, \alpha)(H', \alpha') := (H + H', \alpha\alpha').$$

For $(H, \alpha) \in \mathscr{P}$, we denote by $\mathscr{L}(H, \alpha)$ the line bundle on X that is given by the Λ-action on the trivial line bundle $V \times \mathbb{C}$ on V corresponding to

$$\Lambda \to H^0(V, \mathscr{O}_V^\times), \quad \lambda \mapsto \left(z \mapsto \alpha(\lambda) \exp\left(\pi H(z, \lambda) + \frac{\pi}{2} H(\lambda, \lambda)\right)\right).$$

Theorem 27.296. (Theorem of Appell-Humbert) *Let $X = V/\Lambda$ be a complex torus and let $\mathscr{P}$ be as defined above. The map*

$$\mathscr{P} \xrightarrow{\sim} \mathrm{Pic}(X), \quad (H, \alpha) \mapsto \mathscr{L}(H, \alpha),$$

is a group isomorphism.

Proof. See [Mum1] Section 2 or [BiLa]$^\circ$ 2.2. $\qquad\square$

A choice of basis of Λ gives us an identification of the set $\mathrm{Hom}(\Lambda, \mathbb{C}_1^\times)$ with the real torus $(\mathbb{C}_1^\times)^{2g}$. It is not hard to show that $\mathrm{Hom}(\Lambda, \mathbb{C}_1^\times)$ even carries a natural structure of complex torus (the "dual complex torus" of X). Furthermore, this subgroup of $\mathrm{Pic}(X)$ "is" the Pic^0 of the complex torus in the sense that it consists precisely of those line bundles $\mathscr{L}$ on X such that the associated morphism $\varphi_{\mathscr{L}}$ as in 27.159 is trivial. See [Mum1] Section 9, [BiLa]$^\circ$ 2.4.

Let us next discuss the question under which conditions a complex torus $X = V/\Lambda$ is algebraizable. The crucial point is to identify a criterion for a line bundle on X being ample. As a first step, one computes the dimension of the space of global sections of line bundles $\mathscr{L}(H, \alpha)$ with H positive definite. We denote by $\det(E)$ the determinant of any structure matrix of the alternating form $E = \mathrm{Im}(H)$ for a basis of the lattice Λ. Since every unit in $\mathbb{Z}$ has square 1, it is independent of the choice of basis.

Lemma 27.297. *Let X be a complex torus, $(H, \alpha) \in \mathscr{P}_X$, and let $E = \mathrm{Im}(H)$ the alternating form attached to H. Assume that H is positive definite. Then $\dim H^0(X, \mathscr{L}(H, \alpha)) = \sqrt{\det(E)}$.*

The sections of $\mathscr{L}(H, \alpha)$ are called *theta functions* (with respect to $\mathscr{L}(H, \alpha)$).

Proof. For a proof of this by "complex analytic computations", see [Mum1] Section 3 or [BiLa]$^\circ$ Corollary 3.2.8. For abelian varieties we have given an algebraic proof above, see Corollary 27.266. $\qquad\square$

The understanding of these spaces of global sections is the key point to the proof of the following theorem which gives a characterization of ample line bundles on a complex torus.

Theorem 27.298. (Theorem of Lefschetz) *Let $X = V/\Lambda$ be a complex torus, and let $\mathscr{L}(H, \alpha)$ be a line bundle on X, where $H \in \mathscr{H}$ and $\alpha \in \mathrm{Hom}(\Lambda, \mathbb{C}_1^\times)$. The following are equivalent.*
(i) *The line bundle $\mathscr{L}(H, \alpha)$ is ample.*
(ii) *The hermitian form H is positive definite.*

Proof. The implication (i) $\Rightarrow$ (ii) is relatively easy, but the converse is more difficult. See [Mum1] Section 3 or [BiLa]$^\circ$ Proposition 4.5.2. $\qquad\square$

One then has the following characterization when a complex torus is algebraizable.

Theorem 27.299. *Let $X = V/\Lambda$ be a complex torus. The following are equivalent.*

(i) *The complex torus X is algebraizable, i.e. $X = A^{\mathrm{an}}$ for some variety A over $\mathbb{C}$.*

(ii) *The complex torus X is of the form $X = A^{\mathrm{an}}$ for an abelian variety A over $\mathbb{C}$, compatibly with the group laws on both sides.*

(iii) *The complex torus X is projective, i.e., there is a closed embedding $X \hookrightarrow (\mathbb{P}^n_{\mathbb{C}})^{\mathrm{an}}$ of complex spaces, for some n.*

(iv) *There exists a positive definite hermitian form H on V such that the imaginary part $\mathrm{Im}(H)$ of H takes values in $\mathbb{Z}$ on $\Lambda \times \Lambda$.*

Proof. The implication (i) $\Rightarrow$ (ii) follows from Corollary 20.61, since X is compact. Since abelian varieties are projective (Proposition 27.174), (ii) $\Rightarrow$ (iii) is clear. Obviously (iii) implies (i). The equivalence of (iii) and (iv) follows from Theorem 27.298. Also see [Mum1] Section 3 or [BiLa] $^\circ$ Section 4.5. $\qquad\square$

It is an easy consequence that every 1-dimensional complex torus X is algebraizable. In fact, up to isomorphism X can be written in the form $\mathbb{C}/\mathbb{Z} \oplus \tau\mathbb{Z}$ with τ in the upper complex half plane, and then $H(z,z') = \frac{1}{\mathrm{Im}(\tau)}z\overline{z'}$ is a positive definite hermitian form as in (iv). Compare Sections (26.7) and (26.19). On the other hand, for $g > 1$, "most" complex tori of dimension g are not algebraizable.

By the theorem of Appell-Humbert, the kernel of the map $\mathrm{Pic}(X) \to \mathrm{Hom}(X, X^\vee)$, $\mathscr{L} \mapsto \varphi_{\mathscr{L}}$, is $\mathrm{Pic}^0(X)$. Accordingly, we call a positive definite hermitian form H as in (iv) of the theorem a *polarization* of the complex torus X. Then X is algebraizable if and only if it admits a polarization, and in this case a polarization corresponds to a polarization in the sense of Section (27.52) of the corresponding abelian variety.

Let $X = V/\Lambda$ be an algebraizable complex torus and let H be a positive definite hermitian form as in (iv). For a suitable basis $e_1, \dots, e_g, f_1, \dots, f_g$ of the lattice Λ, the alternating form $\mathrm{Im}(H)$ has structure matrix $\begin{pmatrix} 0 & D \\ -D & 0 \end{pmatrix}$ for a uniquely determined diagonal matrix D with entries $d_1 \mid d_2 \mid \cdots \mid d_g$. The tuple $(d_1, \dots, d_g)$ is called the *type* of the polarization H. If $\varphi = \varphi_{\mathscr{L}(H,\alpha)}$ denotes the morphism $X \to X^\vee$ of abelian varieties corresponding to this polarization, the d_i describe the finite étale group scheme $\mathrm{Ker}(\varphi)$. We have $\mathrm{Ker}(\varphi) \cong \prod_{i=1}^g \mathbb{Z}/d_i\mathbb{Z}$, a product of constant group schemes. In particular, polarizations of type $(1, \dots, 1)$ are precisely the principal polarizations.

The elements $\frac{1}{d_i}f_i \in \Lambda \subset V$ form a $\mathbb{C}$-basis $\mathscr{B}$ of V. We can now express the $\mathbb{Z}$-basis $e_1, \dots, e_g, f_1, \dots, f_g$ of Λ in terms of $\mathscr{B}$ and we obtain a $(g \times 2g)$-matrix $(\lambda_{ij})_{ij} = \begin{pmatrix} Z & D \end{pmatrix}$ with $Z \in M_g(\mathbb{C})$. Clearly this matrix determines the complex vector space V, the lattice Λ and the form H (up to isomorphism). The condition that H be positive definite hermitian translates to the conditions

$$Z = Z^t, \text{ and } \mathrm{Im}(Z) \text{ positive definite,}$$

i.e., the matrix Z is symmetric, and its imaginary part $\mathrm{Im}(Z) \in M_g(\mathbb{R})$ is (symmetric and) positive definite. To abbreviate the latter condition one often writes $\mathrm{Im}(Z) > 0$.

Conversely, when we fix D, then every matrix Z with these properties defines an abelian variety of dimension g with a polarization of the fixed type. We denote by $\mathfrak{H}_g$ the set

$$\mathfrak{H}_g = \{\, Z \in M_g(\mathbb{C}) \,;\, Z = Z^t \text{ and } \mathrm{Im}(Z) \text{ positive definite}\,\}$$

of all these matrices, the so-called *Siegel upper half space*. Note that $\mathfrak{H}_1$ is just the complex upper half plane. We have constructed a map from $\mathfrak{H}_g$ to the set of isomorphism classes of abelian varieties over $\mathbb{C}$ with a polarization of type D.

For simplicity, let us now restrict to the principally polarized case, i.e., D is the unit matrix. We use the term *principally polarized abelian variety* for a pair of an abelian variety with a principal polarization. One checks that principally polarized abelian varieties corresponding to matrices Z, Z' are isomorphic if and only if there exists a matrix $T = \begin{pmatrix} a & b \\ c & d \end{pmatrix} \in \mathrm{Sp}_{2g}(\mathbb{Z})$, the symplectic group of matrices fixing the alternating form $\begin{pmatrix} 0 & 1_g \\ -1_g & 0 \end{pmatrix}$, such that

$$Z' = (aZ + b)(cZ + d)^{-1}.$$

Here $a, b, c, d \in M_g(\mathbb{Z})$ are the "block components" of T. For $g = 1$ we recover the usual action of the modular group $\mathrm{SL}_2(\mathbb{Z})$ on the complex upper half plane, see Corollary 26.115.
    We obtain a bijection

$$\mathrm{Sp}_{2g}(\mathbb{Z})\backslash \mathfrak{H}_g \xrightarrow{\sim} \{\text{princ. pol. abelian varieties over } \mathbb{C}\}/ \cong .$$

This hints at the question whether the set on the right hand side can be equipped with the structure of a complex space (or even an algebraic variety), since $\mathfrak{H}_g$ obviously is a complex manifold. In this sense, the above bijection is the starting point for the theory of moduli spaces of abelian varieties, see Section (27.55) for some further remarks in this direction. Note that $\mathfrak{H}_g$ has dimension $g(g+1)/2$, as one easily checks. Since $\mathrm{Sp}_{2g}(\mathbb{Z})$ is a discrete group, the moduli space should also have dimension $g(g+1)/2$. See Theorem 27.301.
    Finally, let us briefly discuss the construction of the Jacobian of a smooth projective curve C over $\mathbb{C}$. Let C be a connected smooth projective curve over $\mathbb{C}$. We implicitly identify C with its analytification C^{an}, a compact Riemann surface. See Section (26.7). Let g be the genus of C.
    Integration of differential 1-forms along closed paths on C induces a pairing

$$H_1(C, \mathbb{Z}) \times H^0(C, \Omega^1_{C/\mathbb{C}}) \to \mathbb{C}, \qquad (\gamma, \omega) \mapsto \int_\gamma \omega.$$

The resulting homomorphism $H_1(C, \mathbb{Z}) \to H^0(C, \Omega^1_{C/\mathbb{C}})^\vee$ is injective, so that we can view $H_1(C, \mathbb{Z})$ as a lattice in the g-dimensional complex vector space $H^0(C, \Omega^1_{C/\mathbb{C}})^\vee$. Thus the quotient $J := H^0(C, \Omega^1_{C/\mathbb{C}})^\vee / H_1(C, \mathbb{Z})$ is a complex torus. This gives the complex analytic construction of the Jacobian of C.
    This complex torus is algebraizable, and in fact carries a canonical principal polarization which can be described easily in terms of a suitable basis of $H_1(C, \mathbb{Z}) = \pi_1(C)^{\mathrm{ab}}$ consisting of classes of closed paths in C with simple intersection behavior. We do not go into further details at this point but refer to [BiLa]^{O} Chapter 11.
    Any divisor D of degree 0 on C can be written as a finite sum $\sum_i([p_i] - [q_i])$ for points p_i, q_i on C. For a differential form $\omega \in H^0(C, \Omega_{C/\mathbb{C}})$ a choice of path from p_i to q_i gives us the path integral $\int_{p_i}^{q_i} \omega$. We obtain a linear form in $H^0(C, \Omega_{C/\mathbb{C}})^\vee$ whose class in J is independent of the choice of path. This gives us a map $\mathrm{Div}^0(C) \to J$. It follows from classical theorems by Jacobi and Abel that this map induces a group isomorphism $\mathrm{Pic}^0(C) \to J$. One can then go on and construct a Poincaré bundle for J and show that J satisfies the universal property of the identity component of the Picard functor of C, i.e., it can be identified with the analytification of the Jacobian of the algebraic curve C.

(27.55) Outlook: The moduli space of principally polarized abelian varieties.

Constructing a moduli space (or parameter space) of abelian varieties means that we would like to equip the set of (isomorphism classes of) all abelian schemes of relative dimension g with the structure of a scheme. Compare also Section (16.33). The appropriate way to make this precise is to define a functor on the category of schemes and to study whether this functor is representable. Ideally (or: naively), we would like $\mathscr{A}_g^{\mathrm{naive}}(S)$ to be the set of isomorphism classes of abelian schemes of relative dimension g over S. However, it turns out that defining $\mathscr{A}_g^{\mathrm{naive}}(S)$ as the set of isomorphism classes of abelian schemes does not lead to a representable functor. The situation over the complex numbers hints at the fact that a better moduli problem (i.e., a better behaved functor) is obtained when we include a polarization.

Let us explain the necessity of considering polarized varieties in more detail since this is also tied nicely to other central topics considered in this book. M. Artin gave criteria when a functor is representable at least by an algebraic space (see [Art3] $^\mathrm{O}$ and [Art4] $^\mathrm{O}$ or [Sta] 07SZ). More precisely, let Z be a locally noetherian scheme such that $\mathscr{O}_{Z,z}$ is a G-ring (Definition 20.46) for every[8] $z \in Z$, which holds for instance, if Z is locally of finite type over a field or over a Dedekind ring whose field of fraction has characteristic 0. Let $\mathscr{F}\colon (\mathrm{Sch}/Z)^{\mathrm{opp}} \to (\mathrm{Sets})$ be a functor. Then Artin gave a list of criteria for $\mathscr{F}$ that are necessary and sufficient for $\mathscr{F}$ to be representable by an algebraic space locally of finite type over Z, see [Sta] 07Y1[9]. Let us single out one of these criteria. It says that for every complete local noetherian Z-algebra R with maximal ideal $\mathfrak{m}$ the canonical map $\mathscr{F}(R) \to \lim_n \mathscr{F}(R/\mathfrak{m}^n)$ is bijective.

If for any scheme (= Z-scheme) S we define $\mathscr{A}_g^{\mathrm{naive}}(S)$ to be the set of isomorphism classes of abelian schemes over S of relative dimension g, then this criterion means that given a formal abelian scheme over $\mathrm{Spf}(R)$, i.e., a compatible family $\mathcal{X} = (X_n)_n$ of abelian schemes X_n over $R/\mathfrak{m}^n$ (see Section (24.17)), we would always find an abelian scheme X over R (necessarily unique by Theorem 24.112) such that $X_n \cong X \otimes_R R/\mathfrak{m}^n$ for all n. In other words, we need $\mathcal{X}$ to be algebraizable. But this does not hold in general. To ensure that $\mathcal{X}$ is algebraizable we would like to argue with Grothendieck's algebraization Theorem 24.113. Hence we need to have a compatible system of ample line bundles $\mathscr{L}_n$ on X_n and should add this as a datum to our functor. Instead of specifying an ample bundle we may also specify a polarization since attaching to an ample line bundle $\mathscr{L}$ the polarization $\varphi_{\mathscr{L}}$ is fppf-surjective (Remark 27.283) and smooth (Proposition 27.208).

Therefore it is better for a scheme S to consider $\mathscr{A}_g^{\mathrm{coarse}}(S)$, the set of isomorphism classes of pairs (X, λ), where X is an abelian scheme of relative dimension g over S and where λ is a polarization of X. Usually one also fixes the degree d of the polarization, i.e., one considers the functor $\mathscr{A}_{g,d}^{\mathrm{coarse}}$, where $\mathscr{A}_{g,d}^{\mathrm{coarse}}(S)$ denotes isomorphism classes of pairs (X, λ) as above such that λ has fixed degree d. As the degree is locally constant in families, this defines an open and closed subfunctor of $\mathscr{A}_g^{\mathrm{coarse}}$.

Alas, $\mathscr{A}_{g,d}^{\mathrm{coarse}}$ is still not quite representable. It admits a coarse moduli scheme in the sense of [MFK] $^\mathrm{O}$ Definition 7.4, see loc. cit. Theorem 7.10. The problem with the representability of $\mathscr{A}_{g,d}^{\mathrm{coarse}}$ is the existence of non-trivial automorphisms of polarized abelian schemes, and even of polarized abelian schemes. Let us give a hint why this is problematic. If $\mathscr{A}_{g,d}^{\mathrm{coarse}}$ was representable it would be an fppf-sheaf, in fact even a fpqc-sheaf (Proposition 14.76), and in particular for every finite Galois extension k'/k we

---

[8] It suffices to make the assumption only for those $z \in Z$ that are closed in some open affine neighborhood.
[9] Artin proved this under somewhat stronger assumptions on Z.

would have that base change from k to k' yields an *injective* map $\mathscr{A}_{g,d}^{\text{coarse}}(k) \to \mathscr{A}_{g,d}^{\text{coarse}}(k')$ since $\operatorname{Spec} k' \to \operatorname{Spec} k$ is finite and faithfully flat. Hence given a polarized abelian variety (X', λ') over k' there should by at most one (up to isomorphism) k-forms of (X', λ'), i.e., at most one isomorphism class of a polarized abelian variety (X, λ) over k whose base change to k' is (X', λ'). But isomorphism classes of k-forms of (X', λ') are in bijection to $H^1(\operatorname{Gal}(k'/k), \operatorname{Aut}_{k'}(X', \lambda'))$, cf. Theorem 14.90, and we cannot expect this set to be trivial if $\operatorname{Aut}_{k'}(X', \lambda')$ is nontrivial.

There are (at least) three ways to deal with this problem. The first one would be to be satisfied with the existence of a coarse moduli space which at least over algebraically closed fields gives a satisfying answer. But this way falls short for many arithmetic applications of the theory.

The second and most conceptual way would be to take these automorphisms into account as they are an important feature of the moduli problem. This means, instead of a set-valued functor one considers the functor on the category of schemes that attaches to S the *groupoid* $\mathscr{A}_{g,d}(S)$, i.e., the category of abelian schemes of relative dimension g over S endowed with a polarization of degree d in which the only morphisms are the isomorphisms of such pairs. As this is not a set-valued functor any more, it cannot be representable by a scheme or an algebraic space. But it is an algebraic stack (see [Sta] 0ELS for the theory of algebraic stacks and [Lan] Theorem 1.4.1.11 for the proof that $\mathscr{A}_{g,d}$ is an algebraic stack), which allows to work with it quite similar as one works with schemes (although there are also some significant differences in the theories). Unfortunately, it is beyond the scope of this book to go into more detail here.

The third way to solve this problem is to modify the functor by including an auxiliary additional datum which serves as a rigidification. Below, we will add to the datum (X, λ) of a polarized abelian scheme an isomorphism $\eta\colon X[N] \xrightarrow{\sim} (\mathbb{Z}/N\mathbb{Z})^{2g}$ of group schemes for some fixed $N \geq 3$, a so-called full level N structure. Other rigidifications are possible. Serre's lemma tells us that there are no nontrivial automorphisms of the triple (X, λ, η) (by Proposition 27.105 it suffices to show this if X is defined over an algebraically closed field, then the result can be found in [Ser3]).

We arrive at the following definition.

Definition 27.300. *Let $g \geq 1$, $d \geq 1$ and $N \geq 3$ be natural numbers. The* moduli functor of abelian varieties of dimension g with a polarization of degree d and a full level N structure *is the functor*

$$\mathscr{A}_{g,d,N}\colon (\operatorname{Sch}/\mathbb{Z}[\tfrac{1}{N}])^{\text{opp}} \to (\text{Sets}),$$

$$T \mapsto \{(X, \lambda, \eta);$$

$$X \text{ an abelian scheme of relative dimension } g \text{ over } S,$$

$$\lambda \text{ a polarization of degree } d \text{ of } X,$$

$$\eta\colon X[N] \xrightarrow{\sim} (\mathbb{Z}/N\mathbb{Z})^{2g} \text{ an isomorphism}\}/\cong .$$

In particular, one has the functor $\mathscr{A}_{g,1,N}$ of principally polarized abelian varieties with a full level N structure.

One can show that $\mathscr{A}_{g,d,N}$ is representable by a scheme. More precisely, one has the following theorem.

Theorem 27.301. *The functor $\mathscr{A}_{g,d,N}$ is representable by a smooth quasi-projective* $\mathbb{Z}[\frac{1}{N}]$-*scheme, the* moduli space of abelian varieties of dimension g with polarization of degree d and with full level N structure. *Its relative dimension is* $g(g+1)/2$.

The proof of the theorem can be approached in several ways, but all of them require methods that we did not cover. Therefore we just give some references. In [MFK]$^\mathrm{O}$, a construction using geometric invariant theory is given. In [Lan]$^\mathrm{O}_\mathrm{X}$, the moduli space for $\mathscr{A}_{g,d}$ is first constructed as an algebraic stack using Artin's criteria from which the existence of the moduli space for $\mathscr{A}_{g,d,N}$ as an algebraic space is easily deduced using Serre's lemma (Theorem 1.4.1.11 and Corollary 1.4,1.12 of loc. cit.); representability of $\mathscr{A}_{g,d,N}$ as a (quasi-projective) scheme is obtained as a consequence of the theory of the Baily-Borel compactification (Corollary 7.2.3.10). For the theory over the complex numbers, see also Chai's article in [CoSi]$^\mathrm{O}$ or [BiLa]$^\mathrm{O}$ Chapter 8.

Note that one could write down the same moduli functor on the category of all schemes, but by Proposition 27.188 (2) an isomorphism η can only exist for abelian schemes over schemes S on which N is invertible, and therefore the fibers over $\operatorname{Spec}\mathbb{F}_p$ for $p \mid N$ would be empty.

These moduli spaces are important tools in the study of abelian varieties. They are also extremely interesting varieties which have been and are studied extensively. In fact, moduli of abelian varieties are a classical object of study in "classical" algebraic geometry. They are also useful for studying arithmetic questions.

(27.56) Literature on abelian varieties and abelian schemes.

The "modern classic" on abelian varieties, modern referring to the fact that is is written in the language of schemes, is Mumford's beautiful book [Mum1], which also contains a chapter on the analytic theory of abelian varieties over $\mathbb{C}$. In [MFK]$^\mathrm{O}$, some foundational material on abelian schemes over general base schemes can be found. The manuscript [EGM]$_\mathrm{X}$ written by Edixhoven, Moonen and van der Geer, with many finished or nearly finished chapters already and a large bibliography, is a comprehensive modern text on abelian varieties and abelian schemes. Other references are Milne's articles on abelian varieties and on Jacobians of curves in [CoSi]$^\mathrm{O}$, with extended bibliographical notes at the end of the article on Jacobians, and his lecture notes [Mil4]$_\mathrm{X}$.

For the theory over the complex numbers, there is Debarre's volume [Deb]$^\mathrm{O}$, Rosen's article in [CoSi]$^\mathrm{O}$ and the comprehensive volume [BiLa]$^\mathrm{O}$ by Birkenhake and Lange, which is a standard reference.

For more specialized topics, we mention Raynaud's Lecture Notes in Mathematics volume [Ray1]$^\mathrm{O}$, in particular for the question when homogeneous spaces are projective, and the book [FaCh]$^\mathrm{O}$, where a proof of the representability of the dual abelian scheme over a general base scheme is sketched, which has been worked out in detail by Große-Klönne in [GrK]. For the interplay of the theory of dual abelian varieties and Fourier-Mukai correspondence, good sources are, in addition to Mukai's original articles [Muk1]$^\mathrm{O}$ and [Muk2]$^\mathrm{O}$, the books by Huybrechts [Huy]$^\mathrm{O}$, Polishchuk [Pol]$^\mathrm{O}$, and the lecture notes of Bhatt [Bha]$_\mathrm{X}$.

Exercises

Exercise 27.1. Let S be a scheme that has a closed point $s \in S$. Let G be obtained by gluing two copies of S along $U := S \setminus \{s\}$.
(1) Show that there exists a unique structure of an S-group scheme on G such that $G \times_S U$ is the trivial group scheme and such that the fiber G_s is isomorphic to the constant group scheme attached to $\mathbb{Z}/2\mathbb{Z}$ over $\kappa(s)$.
(2) Assume that $\{s\}$ is not open in S. Show that $G \to S$ is not separated.

Exercise 27.2. Show that all assertions of Proposition 27.11 also hold for an arbitrary group scheme over a base scheme S consisting of a single point s.
Hint: Argue similarly as in Proposition 27.11 using that $\operatorname{Spec} \kappa(s) \to S$ is a universal homeomorphism. To show that $g \in G$ lies in the image in (1) make a base change to $\kappa(g)$.

Exercise 27.3. Let G be a group scheme locally of finite type over a field k. Suppose that $\operatorname{Lie}(G) = 0$. Show that G is étale over k.

Exercise 27.4. Let R be a ring. Show that $\operatorname{Pic}(R) \to \operatorname{Pic}(R_{\mathrm{red}})$ is an isomorphism.

Exercise 27.5. A ring R is called *seminormal* if for all $b, c \in R$ with $b^2 = c^3$ there exists $a \in R$ such that $a^3 = b$ and $a^2 = c$.
(1) Show that every normal domain is seminormal.
(2) Show that every seminormal ring is reduced.
(3) Let k be a field. Show that $k[x,y]/(xy)$ is seminormal but not normal. Show that $k[x,y]/(y^2 - x^3)$ is not seminormal.

Exercise 27.6. Let R be a ring. Show that the following assertions are equivalent.
(i) R_{red} is seminormal (Exercise 27.5).
(ii) $\operatorname{Pic}(R) \to \operatorname{Pic}(\mathbb{A}^1_R)$ is an isomorphism.
(iii) $\operatorname{Pic}(R) \to \operatorname{Pic}(\mathbb{A}^n_R)$ is an isomorphism for all $n \geq 0$.
Hint: This is difficult. See [Coq]$_X$ for an elementary proof. Note that one can assume that R is reduced by Exercise 27.4.

Exercise 27.7. Let k be a field. Let G be a group scheme of finite type. Show that every homogeneous G-space X is quasi-projective.
Hint: You might proceed as follows.
(1) Show that one can assume that k is algebraically closed and that G and X are connected.
(2) Show that G and X are smooth if $\operatorname{char}(k) = 0$ and that one can therefore assume that $\operatorname{char}(k) > 0$.
(3) Show that X_{red} is a homogeneous space for the smooth group scheme G_{red} and deduce that X_{red} is quasi-projective.
(4) Now use Exercise 22.21.

Exercise 27.8. Let S be a scheme, let X and Y be abelian schemes over S, and let $f \colon X \to Y$ be a morphism of S-schemes that preserves the unit section. Consider $g \colon X \times_S X \to Y$, $g(x, x') = f(x+x') - f(x) - f(x')$ on T-valued points $x, x' \in X(T)$. Show that (without assuming that f is a homomorphism of group schemes) that the restriction of g to $\{0\} \times X$ and to $X \times \{0\}$ is constant and deduce that f is a homomorphism of group schemes.

Exercise 27.9. Let S be a scheme and let $n \geq 1$. Show that $\mathrm{Pic}_{\mathbb{P}_S^n/S}$ is represented by the constant group scheme attached to the abelian group $\mathbb{Z}$.

Exercise 27.10. Let X be an abelian variety over a field k and let $f\colon X \to Y$ be a morphism of k-schemes. Show that every connected component of a fiber of f is geometrically irreducible.

Exercise 27.11. Let S be a scheme, and let G and H be smooth separated group schemes of finite presentation with connected fibers over S. Let $f\colon G \to H$ be an isogeny. Then the function

$$\deg(f)\colon S \longrightarrow \mathbb{Z}, \qquad s \mapsto \dim_{\kappa(s)} H^0(\mathrm{Ker}(f_s), \mathcal{O}_{\mathrm{Ker}(f_s)})$$

is called the *degree of f*.
(1) Show that $\deg(f)$ is a lower semicontinuous constructible function, i.e., for all $n \in \mathbb{Z}$ the set $\{\, s \in S \;;\; \deg(f)(s) \geq n \,\}$ is open and constructible in S.
(2) Show that f is finite locally free if and only if $\deg(f)$ is locally constant.
Hint: Exercise 20.17.

Exercise 27.12. Let k be a field and let G be a finite group scheme over k.
(1) Show that one has a functorial exact sequence of finite group schemes

$$(*) \qquad\qquad 1 \to G^0 \to G \to G^{\text{ét}} \to 1,$$

where G^0 is the identity component of G and $G^{\text{ét}}$ is étale over k. Show that $G^0 \to \mathrm{Spec}\, k$ is a universal homeomorphism.
(2) Now suppose that k is perfect. Show that $G^{\text{ét}} \cong G_{\mathrm{red}}$ and that the inclusion $G_{\mathrm{red}} \to G$ defines a splitting of $(*)$.
(3) Call $\mathrm{rk}_{\text{ét}}(G) := \dim_k \Gamma(G^{\text{ét}}, \mathcal{O}_{G^{\text{ét}}})$ the étale rank of G and $\mathrm{rk}_0(G) := \dim_k \Gamma(G^0, \mathcal{O}_{G^0})$ the local rank of G. Show that $\mathrm{rk}(G) = \mathrm{rk}_0(G) + \mathrm{rk}_{\text{ét}}(G)$.

Exercise 27.13. Let S be a scheme and let G be an S-group scheme such that $G \to S$ is separated, flat, quasi-finite, and of finite presentation. Show that the map $S \to \mathbb{Z}$, $s \mapsto \mathrm{rk}_{\text{ét}}(G_s)$ (Exercise 27.12) is constructible and lower semicontinuous.
Hint: Exercise 20.18

The following four exercises are taken from [EGM]x.

Exercise 27.14. Let k be a field, let X be an abelian variety over k, and let $\mathcal{L}$ be a line bundle on X.
(1) Show that for $n \in \mathbb{Z}$ one has $[n]^*\mathcal{L} \cong \mathcal{O}_X$ if and only if $\mathcal{L}^{\otimes n} \cong \mathcal{O}_X$.
(2) Show that for $n \in \mathbb{Z}$ with $n \neq -1, 0, 1$ one has $[n]^*\mathcal{L} \cong \mathcal{L}$ if and only if $\mathcal{L}^{\otimes n-1} \cong \mathcal{O}_X$.
(3) Let k be algebraically closed. Show that $\mathcal{L} \cong \mathcal{L}_1 \otimes \mathcal{L}_2$, where $\mathcal{L}_1$ is symmetric and $\mathcal{L}_2 \in \mathrm{Pic}^0_{X/k}(k)$.

Exercise 27.15. Let S be a scheme and let X be an abelian scheme over S. Let $m_X\colon X \times_S X \to X$ be the group law and $\Delta_X\colon X \to X \times_S X$ the diagonal. Show that $(m_X)^t = \Delta_{X^t}$ and $(\Delta_X)^t = m_{X^t}$.

Exercise 27.16. Let S be a scheme, let X be an abelian scheme over S, and let $\mathcal{L}$ be a line bundle on X. Let $Y := X^4$ and $\mathcal{M} := p_1^*\mathcal{L} \otimes p_2^*\mathcal{L} \otimes p_3^*\mathcal{L} \otimes p_4^*\mathcal{L}$, where $p_i\colon Y \to X$ is the i-th projection. Let $n \geq 1$ be an integer and write $n = a^2 + b^2 + c^2 + d^2$ for integers a, b, c, d (this is always possible by a theorem of Lagrange, see for instance [Ser5] Cor. 1 in Appendix to Chapter IV). Consider

$$\alpha := \begin{pmatrix} a & -b & -c & -d \\ b & a & -d & c \\ c & d & a & -b \\ d & -c & b & a \end{pmatrix}$$

which we consider as an isogeny $\alpha\colon Y \to Y$. Show that one has $\alpha^* \mathcal{M} = \mathcal{M}^{\otimes n}$ in $\mathrm{NS}(Y)$.

Exercise 27.17. Let k be a field and let X be an abelian variety over k. Let $\mathcal{L}_1$ and $\mathcal{L}_2$ be non-degenerate line bundles on X such that $\mathcal{L}_1 \otimes \mathcal{L}_2$ is also non-degenerate. Show $i(\mathcal{L}_1 \otimes \mathcal{L}_2) \leq i(\mathcal{L}_1) + i(\mathcal{L}_2)$.

Exercise 27.18. Let X be an abelian scheme over a scheme S and let $\underline{\mathrm{Hom}}(X, X^t)^{\mathrm{sym}}$ be the subgroup functor of $\underline{\mathrm{Hom}}(X, X^t)$ consisting of symmetric homomorphisms.
(1) Show that the inclusion $\underline{\mathrm{Hom}}(X, X^t)^{\mathrm{sym}} \hookrightarrow \underline{\mathrm{Hom}}(X, X^t)$ is representable by an open and closed immersion.
(2) Show that $\mathrm{NS}(X) \to \underline{\mathrm{Hom}}(X, X^t)^{\mathrm{sym}}$ is an isomorphism.

F Homological Algebra

Content

- Addenda to the language of categories
- Additive and abelian categories
- Complexes in additive and abelian categories
- Spectral sequences
- Triangulated categories
- Sign conventions
- Derived categories
- Derived functors

In this appendix we collect some notions and results from homological algebra that are needed in our exposition. We will not give any proofs but only references except if we were not able to find a suitable textbook reference. Sometimes results immediately follow from the definitions and in this case there are no references.

Addenda to the language of categories

In this part we introduce some more general categorical concepts. We begin with a remark on set-theoretic issues that become more serious when working with derived categories. We also introduce general limits and colimits: In [GWI]^{O}, Appendix A, we explained the notion of an inductive and a projective limit of a family of objects indexed by a preordered set. We now generalize this to "families indexed by a category", more precisely to functors from some small category into our given category. Finally, we recall the notion of adjoint functors and give some some general results about them.

(F.1) Set-theoretical remarks.

As in the first volume, we will largely ignore set-theoretic questions. However, in the context of homological algebra as we will use it, this is more problematic: The derived category of an abelian category is constructed as a localization of another category (see Section (F.37)), and the question of existence of localization of categories involves more subtle set-theoretic issues than most constructions within algebraic geometry.

© Springer Fachmedien Wiesbaden GmbH, ein Teil von Springer Nature 2023
U. Görtz und T. Wedhorn, *Algebraic Geometry II: Cohomology of Schemes*,
Springer Studium Mathematik – Master, https://doi.org/10.1007/978-3-658-43031-3

Because this is an important foundational point, we will give some pointers to the literature where possible approaches are discussed. We work within an axiomatic framework for set theory such as ZFC (the Zermelo-Fraenkel axioms plus the axiom of choice). In particular, we have at our disposal a notion of *set* (which we think of a collection of objects which is "not too large") and a notion of *class* (an arbitrary collection of objects). The ZFC axioms specify which set-theoretic constructions can be carried out inside the world of sets (e.g., we can take the power set of a set; but the collection of all sets is not a set itself).

However, for our purposes we need more fine-grained control. Defining the collection of morphisms of the localization of a category, we need to express that the size of this collection —while it might excess the size of "small" sets— is not arbitrarily large, i.e., is not an arbitrary class.

One way to do so is using the formalism of *universes*, and then working within a given universe $\mathcal{U}$, which for us will be a "large" infinite set (see [SGA4] ^{O} Exp. I for details). Then the $\mathcal{U}$-*small sets* are those sets which are elements of $\mathcal{U}$. The axioms defining a universe guarantee that the standard set-theoretic constructions can be carried out inside the fixed universe. The category of sets is defined as the category of all $\mathcal{U}$-small sets; in particular its collection of objects is a set (in ZFC), namely the set $\mathcal{U}$. A $\mathcal{U}$-*category* is a category $\mathcal{C}$ whose set of objects is a subset of $\mathcal{U}$ and such that for each pair of objects (X,Y) the set of morphisms $\mathrm{Hom}_\mathcal{C}(X,Y)$ is $\mathcal{U}$-small.

Constructions such as the derived category will not in general be possible within the fixed universe: I.e., the Hom sets will not necessarily be $\mathcal{U}$-small. Since they are ZFC sets nevertheless, the construction can be carried out inside a larger universe. So at certain points one may have to pass to a larger universe (but we will allow ourselves to do so without explicitly keeping track of it). For further details, see e.g. [SGA4] ^{O} Exp. I, [KaSh] ^{O} 1.1.

Another way is to restrict to objects (sets, groups, schemes, ...) whose size is bounded explicitly in terms of a cardinal number; cf. [Sta] 000H. Comparing this with fixing a universe, we see that given any universe $\mathcal{U}$, the cardinality of sets in $\mathcal{U}$ is bounded since $\mathcal{U}$ is a set. Conversely, bounding the size of sets is not enough to obtain a set (meaning that the collection of all sets of some bounded cardinality is still a proper class, not a set), so in order to reduce to a set one has to do something more, e.g., to restrict to a set which contains a representative of every isomorphism class. Nevertheless, this approach allows one to avoid the somewhat arbitrary choice of a universe.

Finally, we will use only derived categories of Grothendieck abelian categories (see Section (F.12)), or subcategories thereof. We will see, that in this case the derived category is equivalent to a category which lies in the same universe, so that, from hindsight, no set-theoretic issues arise.

(F.2) Categories and functors.

We will use the standard terminology of elementary category theory (categories, functors, morphisms of functors, isomorphisms, ...) as given in [GWI] ^{O}, Appendix A. See also [KaSh] ^{O} 1.2, 1.3, 1.4. If $\mathcal{C}$ is a category, we sometimes write $X \in \mathcal{C}$ instead of $X \in \mathrm{Ob}(\mathcal{C})$.

All functors are covariant. If we speak of a contravariant functor F from $\mathcal{C}$ to a category $\mathcal{D}$, this means that F is a (covariant) functor $F \colon \mathcal{C}^{\mathrm{opp}} \to \mathcal{D}$.

Occasionally, we will use the following notion.

Definition F.1. *Let C be a category. A subcategory $\mathcal{D}$ of C is called* strictly full *if it is a full subcategory (i.e., $\mathrm{Hom}_{\mathcal{D}}(x,y) = \mathrm{Hom}_{C}(x,y)$ for all objects x and y in $\mathcal{D}$) and for every object x in $\mathcal{D}$ all objects in C that are isomorphic to x are also in $\mathcal{D}$.*

Definition F.2. *Let $F\colon C \to \mathcal{D}$ be a functor of categories and let $d \in \mathcal{D}$ be an object. Then the* slice category *of F over d, denoted by $F_{/d}$ or $C_{/d}$, is the category such that*
(a) *an object is a pair (c, σ), where c is an object of C and $\sigma\colon F(c) \to d$ is a morphism in $\mathcal{D}$,*
(b) *a morphism $(c, \sigma) \to (c', \sigma')$ is a morphism $u\colon c \to c'$ such that $\sigma' \circ F(u) = \sigma$.*
Dually, define the slice category *of F under d, denoted by $F^{\backslash d}$ or $C^{\backslash d}$, as the category with*
(1) *objects being pairs $(c, \tau\colon d \to F(c))$,*
(2) *morphisms $(c, \tau) \to (c', \tau')$ being morphisms $v\colon c \to c'$ such that $F(v) \circ \tau = \tau'$.*

We will also use occasionally the following notion.

Definition F.3. *A functor $F\colon C \to \mathcal{D}$ is called* conservative *if it reflects isomorphisms, i.e., if whenever α is a morphism in C such that $F(\alpha)$ is an isomorphism in $\mathcal{D}$, then α is an isomorphism.*

Definition F.4. *A category C is called a* groupoid *if every morphism of C is an isomorphism.*

(F.3) Limits and Colimits.

In Appendix A we already explained the notion of a projective limit and an inductive limit. Here we generalize these concepts by replacing partially ordered index sets by index categories.

Let $\mathcal{I}$ always denote a small category (i.e., a category where the objects form a set, and for any two objects the collection of morphisms between them is a set). Let C be a category and let $C^{\mathcal{I}}$ be the category of functors $\mathcal{I} \to C$. For every object A of C let $c_A\colon \mathcal{I} \to C$ be the constant functor with value A, i.e., c_A sends every object of $\mathcal{I}$ to A and every morphism in $\mathcal{I}$ to id_A. Every morphism $A \to B$ in C induces a morphism of functors $c_A \to c_B$. We obtain a functor

(F.3.1) $$ C \longrightarrow C^{\mathcal{I}}, \qquad A \mapsto c_A. $$

Definition F.5. *Let $X\colon \mathcal{I} \to C$ be a functor. We also call such a functor an $\mathcal{I}$-diagram in C. We write X_i instead of $X(i)$ for an object $i \in \mathcal{I}$.*
(1) *Consider the covariant functor*

$$ \mathrm{colim}_{\mathcal{I}} X\colon C \to (\mathrm{Sets}) $$

sending every object A in C to the set $\mathrm{Hom}(X, c_A)$ of morphisms of functors $X \to c_A$. If this functor is representable, the representing object of C is called colimit *of X and it is denoted by*

$$ \mathrm{colim}_{\mathcal{I}} X \quad or \quad \mathrm{colim}_{i \in \mathcal{I}} X_i \quad or \quad \mathrm{colim}_{i \in \mathcal{I}} X_i. $$

In other words, $\mathrm{colim}_{\mathcal{I}} X$ is an object in C together with morphisms $s_i\colon X_i \to \mathrm{colim}_{\mathcal{I}} X$ in C for all objects i in $\mathcal{I}$ such that
(a) *for every morphism $\varphi\colon i \to j$ in $\mathcal{I}$ one has $s_i = s_j \circ X(\varphi)$,*

(b) *for every object Z in C and for all morphisms $t_i \colon X_i \to Z$ such that for all morphism $\psi \colon i \to j$ in $\mathcal{I}$ one has $t_i = t_j \circ X(\psi)$, there exists a unique morphism $t \colon \operatorname{colim}_{\mathcal{I}} X \to Z$ such that $t_i = t \circ s_i$.*

(2) *Dually, a limit of the functor X is an object*

$$\lim_{\mathcal{I}} X \quad or \quad \lim_{i \in \mathcal{I}} X_i \quad or \quad \varprojlim_{i \in \mathcal{I}} X_i$$

in C that represents the contravariant functor $\varprojlim_{\mathcal{I}} X \colon \mathcal{I} \to C,\ A \mapsto \operatorname{Hom}(c_A, X)$.

If $X, X' \colon \mathcal{I} \to C$ are functors, any morphism of functors $u \colon X \to X'$ induces a morphism

$$\operatorname{colim}_{\mathcal{I}}(u) \colon \operatorname{colim}_{\mathcal{I}} X' \to \operatorname{colim}_{\mathcal{I}} X, \quad \text{resp. } \varprojlim_{\mathcal{I}}(u) \colon \varprojlim_{\mathcal{I}} X \to \varprojlim_{\mathcal{I}} X'.$$

If $\operatorname{colim}_{\mathcal{I}} X$ and $\operatorname{colim}_{\mathcal{I}} X'$ (resp. $\varprojlim_{\mathcal{I}} X$ and $\varprojlim_{\mathcal{I}} X'$) are both representable, then this morphism corresponds to a unique morphism

$$\operatorname{colim}_{\mathcal{I}}(u) \colon \operatorname{colim}_{\mathcal{I}} X \to \operatorname{colim}_{\mathcal{I}} X', \quad \text{resp. } \lim_{\mathcal{I}}(u) \colon \lim_{\mathcal{I}} X \to \lim_{\mathcal{I}} X'.$$

Remark F.6. Let C be a category and let $\mathcal{I}$ be a small category such that for all functors $X \colon \mathcal{I} \to C$ the colimit (resp. the limit) exists. Then we obtain a functor

$$\operatorname{colim} \colon C^{\mathcal{I}} \longrightarrow C, \qquad (\text{resp.} \quad \lim \colon C^{\mathcal{I}} \longrightarrow C),$$

and this functor is left adjoint (resp. right adjoint, Definition F.21) to the functor $A \mapsto c_A$ (F.3.1).

Remark: Projective and inductive limits as limits and colimits F.7. Recall that every preordered set I can be considered as a category, still denoted by I. The objects are the elements of I and for any two elements $i, j \in I$ the set of morphisms $i \to j$ consists of one element if $i \le j$ and is empty otherwise.

If $X \colon I^{\mathrm{opp}} \to C$ is a projective system in a category C indexed by I, then $\lim_{I^{\mathrm{opp}}} X$ is the projective limit in the sense of Section (A.3). Similarly, if $X \colon I \to C$ is an inductive system in a category C indexed by I, then $\operatorname{colim}_I X$ is the inductive limit in the sense of Section (A.3).

In a given category C only some limits or colimits may exist.

Definition F.8. *A category in which arbitrary limits (resp. colimits) exist is called* complete *(resp.* cocomplete*). A category in which limits (resp. colimits) of all $\mathcal{I}$-diagrams exist for arbitrary finite categories $\mathcal{I}$ is called* finitely complete *(resp.* finitely cocomplete*).*

The category of sets is complete and cocomplete:

Example: Limits and colimits of sets F.9. Let $\mathcal{I}$ be a small category, $I := Ob(\mathcal{I})$, and let $X \colon \mathcal{I} \to (\text{Sets})$ be an $\mathcal{I}$-diagram in the category of sets.

(1) The limit $\lim_{\mathcal{I}} X$ exists in (Sets) and can be described by

$$(\text{F.3.2}) \qquad \lim_{\mathcal{I}} X = \{\, (x_i)_{i \in I} \in \prod_{i \in I} X_i \;;\; \forall \varphi \colon i \to j \text{ in } \mathcal{I} \colon X(\varphi)(x_i) = x_j \,\}.$$

For $j \in I$ the map $p_j \colon \lim_{\mathcal{I}} X \to X_j$ is given by the projection $(x_i)_{i \in I} \mapsto x_j$.

(2) The colimit $\operatorname{colim}_{\mathcal{I}} X$ exists in (Sets) and can be described by

$$\text{(F.3.3)} \qquad\qquad \operatorname*{colim}_{\mathcal{I}} X = (\coprod_{i \in I} X_i)/\sim,$$

where $(\coprod_{i \in I} X_i)$ is the disjoint union of the sets X_i and where $\sim$ is the equivalence relation generated by the relation $x_i \sim x_j$ if $x_i \in X_i$, $x_j \in X_j$ and $X(\varphi)(x_i) = x_j$ for some $\varphi: i \to j$. For $j \in I$ the map $s_j: X_j \to \operatorname{colim}_{\mathcal{I}} X$ is given by attaching to $x_j \in X_j$ the equivalence class of $x_j \in \coprod_{i \in I} X_i$.

Further examples for categories which are complete and cocomplete are
(1) the category of topological spaces,
(2) the category of groups,
(3) the category of left R-modules (R a fixed not necessarily commutative ring), more generally the category of $\mathcal{O}_X$-modules ($(X, \mathcal{O}_X)$ a ringed space).

Remark F.10. Let $X: \mathcal{I} \to \mathcal{C}$, $i \mapsto X_i$ be a diagram in a category $\mathcal{C}$. Then we can rephrase the universal property in the definition of $\lim X_i$ and $\operatorname{colim} X_i$ as follows.

An object $\lim_{\mathcal{I}} X$ in $\mathcal{C}$ together with morphisms $p_i: \lim_{\mathcal{I}} X \to X_i$ for all objects i of $\mathcal{I}$ is a limit of X in $\mathcal{C}$ if and only if for all objects Y in $\mathcal{C}$ the map

$$\operatorname{Hom}_{\mathcal{C}}(Y, \lim_{\mathcal{I}} X) \xrightarrow{u \mapsto (p_i \circ u)_i} \lim_{\mathcal{I}} \operatorname{Hom}_{\mathcal{C}}(Y, X_i)$$

is bijective, where the right hand side denotes the limit in the category of sets.

Similarly, an object $\operatorname{colim}_{\mathcal{I}} X$ in $\mathcal{C}$ together with morphisms $s_i: X_i \to \operatorname{colim}_{\mathcal{I}} X$ for all objects i of $\mathcal{I}$ is a colimit of X in $\mathcal{C}$ if and only if for all objects Y in $\mathcal{C}$ the map

$$\operatorname{Hom}_{\mathcal{C}}(\operatorname*{colim}_{\mathcal{I}} X, Y) \xrightarrow{u \mapsto (u \circ s_i)_i} \lim_{\mathcal{I}} \operatorname{Hom}_{\mathcal{C}}(X_i, Y)$$

is bijective.

Remark: Double limits F.11. Let $\mathcal{I}$ and $\mathcal{J}$ be (small) categories and let $\mathcal{C}$ be a category such that limits (resp. colimits) of all $\mathcal{I}$-diagrams and all $\mathcal{J}$-diagrams in $\mathcal{C}$ exist. Let $X: \mathcal{I} \times \mathcal{J} \to \mathcal{C}$, $(i,j) \mapsto X_{ij}$ be a diagram in $\mathcal{C}$.

Then because of the definition of limits (resp. colimits) via a universal property one obtains that $\lim_{\mathcal{I} \times \mathcal{J}} X$ (resp. $\operatorname{colim}_{\mathcal{I} \times \mathcal{J}} X$) exists and one has isomorphisms

$$\lim_{i,j} X_{ij} \cong \lim_i \lim_j X_{ij} \cong \lim_j \lim_i X_{ij}$$

$$\text{(resp.} \qquad \operatorname*{colim}_{i,j} X_{ij} \cong \operatorname*{colim}_i \operatorname*{colim}_j X_{ij} \cong \operatorname*{colim}_j \operatorname*{colim}_i X_{ij}).$$

Definition F.12. *A category $\mathcal{I}$ is called* filtered *if $\operatorname{Ob}(\mathcal{I})$ is non-empty and if the following two conditions are satisfied.*
(a) *For all objects i and j in $\mathcal{I}$ there exists an object k and morphisms $i \to k$ and $j \to k$.*
(b) *For all objects i and j and all morphisms $f, g: i \to j$ there exists a morphism $h: j \to k$ such that $h \circ f = h \circ g$.*
We also say that $\mathcal{I}$ is cofiltered *if the opposite category $\mathcal{I}^{\mathrm{opp}}$ is filtered.*

For instance, a partially ordered set I is filtered if and only if the attached category is a filtered category.

Example: Filtered colimits of sets F.13. Let $\mathcal{I}$ be a filtered category and let $X\colon \mathcal{I} \to (\text{Sets})$ be an $\mathcal{I}$-diagram in the category of sets (we speak of a *filtered diagram*). In this case one has $\operatorname{colim}_{\mathcal{I}} X = (\coprod_{i\in I} X_i)/\!\sim$ where for $x_i \in X_i$ and $x_j \in X_j$ one defines $x_i \sim x_j$ if there exist morphisms $\varphi\colon i \to k$ and $\psi\colon j \to k$ such that $X(\varphi)(x_i) = X(\psi)(x_j)$ (the properties of a filtered category imply that $\sim$ is an equivalence relation).

Definition F.14. *A category $\mathcal{I}$ is called* connected *if it is non-empty and for any pair of objects $X, Y \in \mathcal{I}$ there is a finite sequence of objects $(X_0 = X, X_1, \ldots, X_{n-1}, X_n = Y)$ such that $\operatorname{Hom}_{\mathcal{I}}(X_{j-1}, X_j)$ or $\operatorname{Hom}_{\mathcal{I}}(X_j, X_{j-1})$ is non-empty for all $j = 1, \ldots, n$.*

Definition F.15. *A functor $\alpha\colon \mathcal{J} \to \mathcal{I}$ is called* cofinal *if the slice category $\mathcal{J}^{\backslash i}$ of objects $i \to \alpha(j)$ under i is connected for all objects $i \in \mathcal{I}$. If α is the inclusion of a subcategory, we also say that $\mathcal{J}$ is cofinal in $\mathcal{I}$.*

Proposition F.16. *([KaSh]^{O} Proposition 2.5.2) Let $\alpha\colon \mathcal{J} \to \mathcal{I}$ be a cofinal functor of small categories. Then for every diagram $X\colon \mathcal{I} \to \mathcal{C}$ in a category $\mathcal{C}$ there is a functorial isomorphism*

$$\operatorname*{colim}_{j\in\mathcal{J}} X_{\alpha(j)} \xrightarrow{\sim} \operatorname*{colim}_{i\in\mathcal{I}} X_i$$

if either side exists.

Remark F.17. Let $\mathcal{J}$ be a filtered category.
(1) Then $\mathcal{J}$ is connected since for all objects $X, Y \in \mathcal{J}$ there exists an object Z in $\mathcal{J}$ and morphisms $X \to Z$ and $Y \to Z$.
(2) A functor $\alpha\colon \mathcal{J} \to \mathcal{I}$ is cofinal if and only if the following two conditions hold.
   (a) For every object $i \in \mathcal{I}$ there exists $j \in \mathcal{J}$ and a morphism $i \to \alpha(j)$.
   (b) For all objects $i \in \mathcal{I}$ and $j, j' \in \mathcal{J}$ there exists $\tilde{j} \in \mathcal{J}$ and morphisms $j \to \tilde{j}$, $j' \to \tilde{j}$ in $\mathcal{J}$ and morphisms $i \to \alpha(j)$ and $i \to \alpha(j')$ such that

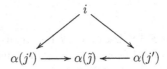

commutes.

(F.4) Special cases of limits and colimits.

Remark and Definition F.18. As already explained, one has several special cases of limits and colimits. Let $X\colon \mathcal{I} \to \mathcal{C}$ be an $\mathcal{I}$-diagram in a category $\mathcal{C}$.
(1) If $\mathcal{I}$ is a category with no morphisms except the identities, $\lim_{\mathcal{I}} X$ is the product $\prod_{i\in \mathrm{Ob}(\mathcal{I})} X_i$ and $\operatorname{colim}_{\mathcal{I}} X$ is the coproduct $\coprod_{i\in \mathrm{Ob}(\mathcal{I})} X_i$.
   A special case is if $\mathcal{I}$ is the empty category. Then there is a unique $\mathcal{I}$-diagram in every category $\mathcal{C}$. Its limit (resp. its colimit), if it exists, is a *final object* (resp. an *initial object*) of $\mathcal{C}$.
(2) Let $\mathcal{I}$ be the category with three objects j, i_1, and i_2 and whose only morphisms except the identities are two morphisms $i_1 \to j$ and $i_2 \to j$. We represent $\mathcal{I}$ schematically by

$$i_1 \longrightarrow j \longleftarrow i_2$$

Then an $\mathcal{I}$-diagram X in a category $\mathcal{C}$ is a diagram of morphisms in $\mathcal{C}$ of the form

$$X_1 \xrightarrow{f_1} Y \xleftarrow{f_2} X_2.$$

The limit of X, if it exists, is the *fiber product of X_1 and X_2 over Y*, denoted by $X_1 \times_Y X_2$. Sometimes a fiber product is called a *pullback*; the diagram formed by X_1, X_2, Y and their fiber product is called a *cartesian* diagram.

(3) Dually, there is the notion of a *pushout* in a category $\mathcal{C}$ which is the colimit of a diagram $X: \mathcal{J} \to \mathcal{C}$, where $\mathcal{J}$ is represented schematically by

The pushout of a $\mathcal{J}$-diagram $X_1 \longleftarrow Y \longrightarrow X_2$ in $\mathcal{C}$ is denoted by $X_1 \amalg_Y X_2$. Sometimes a pushout is called an *amalgamated sum*.

(4) Now consider the case $i_1 = i_2$, i.e., $\mathcal{I}$ is the category with two objects j and i whose only morphisms except the identities are two morphisms $i \to j$. Hence an $\mathcal{I}$-diagram X is given by a diagram

(F.4.1) $$X_i \underset{v}{\overset{u}{\rightrightarrows}} X_j.$$

In this case its limit is called the *kernel of u and v* or the *equalizer of u and v* and denoted by $\mathrm{Eq}(u, v)$.

The colimit of the diagram (F.4.1) is called the *cokernel of u and v* or the *coequalizer of u and v* and denoted by $\mathrm{Coeq}(u, v)$.

Let $F: \mathcal{C} \to \mathcal{D}$ be a functor between categories. Let $X: \mathcal{I} \to \mathcal{C}$ be a diagram in $\mathcal{C}$ such that the limits $\lim_{\mathcal{I}} X$ and $\lim_{\mathcal{I}}(F \circ X)$ exist in $\mathcal{C}$ and $\mathcal{D}$, respectively. For every object i in $\mathcal{I}$ the morphism $\lim_{\mathcal{I}} X \to X_i$ induces by application of F a morphism $F(\lim_{\mathcal{I}} X) \to F(X_i)$. The family of these morphisms corresponds by the universal property of $\lim_{\mathcal{I}}(F \circ X)$ to a morphism

(F.4.2) $$F(\lim_{\mathcal{I}} X) \longrightarrow \lim_{\mathcal{I}}(F \circ X).$$

We say that F *commutes with limits* if for every diagram $X: \mathcal{I} \to \mathcal{C}$ such that its limit $\lim_{\mathcal{I}} X$ exists in $\mathcal{C}$, the limit of $F \circ X$ exists in $\mathcal{D}$ and the morphism (F.4.2) is an isomorphism.

Dually, there is the notion of a functor that *commutes with colimits*.

Proposition F.19.

(1) *A category $\mathcal{C}$ is complete if and only if arbitrary products exist in $\mathcal{C}$ and for each pair of parallel arrows $u, v: X \to Y$ its equalizer exists.*

A functor $F: \mathcal{C} \to \mathcal{D}$ commutes with arbitrary limits if and only if it commutes with products and equalizers.

(2) *A category $\mathcal{C}$ is finitely complete if and only if final objects, products of two objects, and for each pair of parallel arrows $u, v: X \to Y$ its equalizer exist in $\mathcal{C}$.*

A functor $F: \mathcal{C} \to \mathcal{D}$ commutes with finite limits if and only if it commutes with final objects, products of two objects, and equalizers.

One has dual criteria for a category to be (finitely) cocomplete and for a functor to commute with (finite) colimits.

Proof. For a small category $\mathcal{I}$ let $\mathrm{Mor}(\mathcal{I})$ be the set of morphisms in $\mathcal{I}$ and for $\alpha \in \mathrm{Mor}(\mathcal{I})$ let $s(\alpha)$ and $t(\alpha)$ be its source respectively its target. The limit of a diagram $X \colon \mathcal{I} \to \mathcal{C}$ can be constructed as the kernel of the two morphisms

$$\sigma, \tau \colon \prod_{i \in \mathcal{I}} X_i \rightrightarrows \prod_{\alpha \in \mathrm{Mor}(\mathcal{I})} X_{t(\alpha)},$$

such that for all $\alpha \in \mathrm{Mor}(\mathcal{I})$ the composition with the projection is given by

$$\mathrm{pr}_\alpha \circ \sigma = X(\alpha) \circ \mathrm{pr}_{X_{s(\alpha)}},$$

$$\mathrm{pr}_\alpha \circ \tau = \mathrm{pr}_{X_{t(\alpha)}}.$$

The other assertions are shown similarly. For details see [KaSh]$^{\circ}$ 2.2.10, 2.2.11. $\square$

For instance the category of schemes is finitely complete because it has a final object (namely, $\mathrm{Spec}\,\mathbb{Z}$) and fiber products.

Remark F.20. Let $\mathcal{C}$ be a category.
(1) Suppose that finite coproducts and filtered colimits exist in $\mathcal{C}$. Then arbitrary coproducts exist in $\mathcal{C}$ because

$$\coprod_{i \in I} X_i = \operatorname*{colim}_{J \subseteq I \text{ finite}} \coprod_{j \in J} X_j$$

for any family $(X_i)_{i \in I}$ of objects in $\mathcal{C}$.
(2) Similarly, a functor $F \colon \mathcal{C} \to \mathcal{D}$ commutes with arbitrary coproducts if it commutes with finite coproducts and filtered colimits.
There are dual assertions for products and cofiltered limits.

(F.5) Adjoint functors.

Definition F.21. *Let $F \colon \mathcal{C} \to \mathcal{D}$ and $G \colon \mathcal{D} \to \mathcal{F}$ be functors. Recall that F is called* left adjoint *to G or, equivalently, that G is called* right adjoint *to F if there exist isomorphisms, functorial in $A \in \mathcal{C}$ and $B \in \mathcal{D}$,*

(F.5.1) $\mathrm{Hom}_{\mathcal{C}}(A, G(B)) \cong \mathrm{Hom}_{\mathcal{D}}(F(A), B).$

We also say that (F, G) is an adjoint pair. *Choosing $A = G(B)$ in (F.5.1), the identity of G corresponds to a morphism of functors*

$$\epsilon \colon F \circ G \longrightarrow \mathrm{id}_{\mathcal{D}}.$$

Choosing $B = F(A)$ one obtains in the same way a morphism of functors

$$\eta \colon \mathrm{id}_{\mathcal{C}} \longrightarrow G \circ F.$$

These morphism ϵ and η are called the adjunction morphisms. *Sometimes ϵ is called the* counit *and η is called the* unit *of the adjunction.*

If a functor admits a right (resp. left) adjoint functor, this adjoint functor is unique up to unique isomorphism.

Proposition F.22. ([KaSh] ^{O} 1.5.6) *Let (F, G) be a pair of adjoint functors with unit η and counit ϵ.*
(1) *The functor G is fully faithful if and only if ϵ is an isomorphism.*
(2) *The functor F is fully faithful if and only if η is an isomorphism.*
(3) *The following conditions are equivalent.*
  (i) *F is an equivalence of categories.*
  (ii) *G is an equivalence of categories.*
  (iii) *F and G are fully faithful.*
  In this case, F and G are quasi-inverse to each other and η and ϵ are isomorphisms.

Proposition F.23. ([KaSh] ^{O} 2.1.10) *Let $F\colon \mathcal{C} \to \mathcal{D}$ be a functor.*
(1) *Suppose that F is right adjoint to some functor $G\colon \mathcal{D} \to \mathcal{C}$. Then F commutes with limits.*
(2) *Dually, suppose that F is left adjoint to some functor. Then F commutes with colimits.*

Definition F.24. *Let $\mathcal{C}$ be a category that is finitely cocomplete. Then a functor $F\colon \mathcal{C} \to \mathcal{C}'$ is called* right exact *if it commutes with finite colimits.*
Similarly we define for a finitely complete category $\mathcal{C}$ a functor $F\colon \mathcal{C} \to \mathcal{C}'$ to be left exact *if it commutes with finite limits.*

Compare Proposition F.38 below which states that for additive functors between abelian categories this definition gives the usual notion of exactness.

Additive and abelian categories

We now reintroduce additive categories and abelian categories and give some more notions and results about them. A central notion for us will be the notion of a Grothendieck abelian category since all abelian categories that we will derive are Grothendieck categories.

(F.6) Additive categories.

See also Section (A.4). Proofs of the results below can be found in [KaSh] ^{O} Ch. 8.

Definition and Proposition F.25. *An* additive category *is a category $\mathcal{C}$ such that there exists the structure of an abelian group on the set $\mathrm{Hom}_{\mathcal{C}}(X, Y)$ for all objects X, Y of $\mathcal{C}$, such that the composition $\circ$ of morphisms is bilinear, and such that all finite products exist in $\mathcal{C}$.*
Such a structure is necessarily unique.

Note that being additive is really a property of the underlying category not the datum of an additional structure on the category. If $\mathcal{C}$ is additive, then there is a natural structure of additive category on the opposite category $\mathcal{C}^{\mathrm{opp}}$, as well.

Proposition F.26. *Let $\mathcal{C}$ be an additive category.*
(1) *The category $\mathcal{C}$ has a zero object 0 (i.e., an object which is both initial and terminal).*
(2) *For any objects X, Y in $\mathcal{C}$ the product $X \times Y$ and the coproduct $X \coprod Y$ exist, and the natural morphism $X \coprod Y \to X \times Y$ is an isomorphism. We write $X \oplus Y := X \times Y$.*

Definition F.27. *Let C and $\mathcal{D}$ be additive categories. A functor $F\colon C \to \mathcal{D}$ is called* additive, *if it satisfies the following equivalent conditions:*
(i) *For all objects X, Y of C, the map*

$$\mathrm{Hom}_C(X, Y) \to \mathrm{Hom}_{\mathcal{D}}(F(X), F((Y))$$

is a group homomorphism.
(ii) *The functor F commutes with finite products.*

Similarly, we will often consider a situation where all sets of morphisms carry the structure of an R-module for some given ring R, and thus make the following definition.

Definition F.28. *Let R be a (not necessarily commutative) ring. An additive category C is called* R-linear *if the abelian group $\mathrm{Hom}_C(X, Y)$ is endowed with the structure of a left R-module for all objects X and Y such that the composition is R-bilinear.*
A functor $f\colon C \to \mathcal{D}$ between R-linear categories is called R-linear *if for all objects the map $F\colon \mathrm{Hom}_C(X, Y) \to \mathrm{Hom}_{\mathcal{D}}(F(X), F(Y))$ is R-linear.*

For instance, the category of left modules over R is $\mathrm{Cent}(R)$-linear, where $\mathrm{Cent}(R)$ denotes the center of R.

Almost all functors between additive categories that we will consider are additive functors, and sometimes we will assume without further mention that functors between additive categories are additive.

Definition F.29. *Let C be an additive category. A subcategory C' of C is called an* additive subcategory *if there exists the structure of an additive category on C' (necessarily unique by Proposition F.25) such that the inclusion functor $C' \to C$ is additive.*

We can rephrase the definition as requiring that $\mathrm{Hom}_{C'}(X, Y)$ is a subgroup of $\mathrm{Hom}_C(X, Y)$ for all objects X, Y of C' and that C' is closed under taking finite products in C (up to isomorphism).

Definition: Kernel and Cokernel F.30. Let $u\colon X \to Y$ be a morphism in an additive category. If the equalizer $\mathrm{Eq}(u, 0)$ of u and the zero morphism $X \to Y$ exists (Definition F.18), it is called *kernel of u* and denoted by $\mathrm{Ker}(u)$. It is endowed with a morphism $i\colon \mathrm{Ker}(u) \to X$ such that $u \circ i = 0$. A morphism $w\colon Z \to X$ factors through $\mathrm{Ker}(u) \to X$ if and only if $u \circ w = 0$, and in this case the factorization is unique. The morphism u is a monomorphism if and only if its kernel exists and $\mathrm{Ker}(u) = 0$. For any two morphisms $u, v\colon X \to Y$ one has $\mathrm{Eq}(u, v) = \mathrm{Ker}(u - v)$.

Dually, one defines the *cokernel of u* by $\mathrm{Coker}(u) := \mathrm{Coeq}(u, 0)$ it it exists. Then u is an epimorphism if and only if its cokernel exists and $\mathrm{Coker}(u) = 0$.

Finally we also set $\mathrm{Im}(u) = \mathrm{Ker}(Y \to \mathrm{Coker}(u))$ and $\mathrm{Coim}(u) := \mathrm{Coker}(\mathrm{Ker}(u) \to X)$ if they exist.

Definition F.31. *A sequence*

$$0 \longrightarrow X \xrightarrow{\ f\ } Y \xrightarrow{\ g\ } Z \longrightarrow 0$$

of morphisms in an additive category C is called split *if there exists an isomorphism $Y \cong X \oplus Z$ such that via this isomorphism, f is identified with the natural embedding and g is identified with the natural projection.*

Definition F.32. *Let C be an additive category. A non-zero object X is called* indecomposable *if it is not isomorphic to the direct sum of two non-zero objects in C.*

(F.7) Abelian categories.

Definition F.33. *An additive category $\mathcal{C}$ is called an* abelian category, *if*
(a) *every morphism has a kernel and a cokernel, and*
(b) *every monomorphism is the kernel of some morphism, and every epimorphism is the cokernel of some morphism.*

Remark F.34. Let $\mathcal{C}$ be an additive category. By Remark F.30, Condition (a) of Definition F.33 is satisfied if and only if all equalizer and coequalizers exist. As in any additive category finite coproducts and finite products exist (Proposition F.26), we see that Condition (a) means that $\mathcal{C}$ is finitely complete and finitely cocomplete (Proposition F.19).

Let us assume that this is the case. For every morphism $f\colon X \to Y$ the universal properties of $\mathrm{Ker}(f)$ and $\mathrm{Coker}(f)$ show that there exists a unique morphism $\bar{f}\colon \mathrm{Coim}(f) \to \mathrm{Im}(f)$ making the diagram

$$
\begin{array}{ccc}
X & \xrightarrow{\ f\ } & Y \\
\downarrow & & \uparrow \\
\mathrm{Coim}(f) & \xrightarrow{\ \bar{f}\ } & \mathrm{Im}(f)
\end{array}
$$

commutative. Then Condition (b) of Definition F.33 is satisfied if and only if $\bar{u}$ is an isomorphism for every morphism u in $\mathcal{C}$ ([KaSh]^{O} 8.3.4).

Altogether we see that an additive category is abelian if and only if it is finitely complete and finitely cocomplete and for every morphism f the induced morphism $\mathrm{Coim}(f) \to \mathrm{Im}(f)$ is an isomorphism.

In particular a morphism f in an abelian category is an isomorphism if and only if $\mathrm{Ker}(f) = 0$ and $\mathrm{Coker}(f) = 0$, i.e., if and only if f is a monomorphism and an epimorphism.

Example F.35. The category (AbGrp) of abelian groups is the prototypical example of an abelian category. More generally, for any (possibly non-commutative) ring R, the category of left R-modules is an abelian category, as is the category of right R-modules.

The category of finitely generated *free* $\mathbb{Z}$-modules is an additive category which has all kernels and cokernels (kernels are defined as usual using that every submodule of a free $\mathbb{Z}$-module is again free, cokernels are formed by dividing the cokernel within the category of abelian groups by its torsion). But this category is not abelian: the homomorphism $\mathbb{Z} \to \mathbb{Z}$ given by multiplication by 2 is a monomorphism and an epimorphism in this category, but not an isomorphism.

Definition F.36. *Let $\mathcal{C}$ be an abelian category. A (finite or infinite) sequence*

$$
\cdots \longrightarrow X^{i-1} \xrightarrow{d^{i-1}} X^i \xrightarrow{d^i} X^{i+1} \longrightarrow \cdots
$$

in $\mathcal{C}$ is called exact at X^i *if $\mathrm{Ker}\, d^i = \mathrm{Im}\, d^{i-1}$. It is called* exact *if it is exact at every i. A* short exact sequence *is an exact sequence of the form*

$$
0 \to X \to Y \to Z \to 0.
$$

Proposition F.37. *Let C be an abelian category. For a short exact sequence*

$$0 \longrightarrow X \overset{f}{\longrightarrow} Y \overset{g}{\longrightarrow} Z \longrightarrow 0$$

in C the following conditions are equivalent.
(i) *The sequence is split (Definition F.31).*
(ii) *There exists a section of g, i.e., a morphism $s\colon Z \to Y$ with $g \circ s = \mathrm{id}_Z$.*
(iii) *There exists a retraction of f, i.e., a morphism $r\colon Y \to X$ with $r \circ f = \mathrm{id}_X$.*

Proof. Clearly, (i) implies (ii) and (iii). Conversely, if s is a section of g (resp. r is a retraction of f), then $s \circ g$ (resp. $f \circ r$) is an idempotent endomorphism u of Y (i.e., $u^2 = u$) and hence $Y = \mathrm{Ker}(u) \oplus \mathrm{Im}(u)$. $\square$

A functor commutes with finite limits if and only if it commutes with products and with kernels (Proposition F.19), hence:

Proposition F.38. *An additive functor $F\colon C \to D$ between abelian categories is*
(1) *left exact if and only if for any exact sequence*

$$0 \to X \to Y \to Z$$

  in C, the sequence

$$0 \to F(X) \to F(Y) \to F(Z)$$

  is exact.
(2) *right exact if and only if for any exact sequence*

$$X \to Y \to Z \to 0$$

  in C, the sequence

$$F(X) \to F(Y) \to F(Z) \to 0$$

  is exact.
It is enough to test left exactness or right exactness on short exact sequences.

Recall that we consider contravariant functors $C \to D$ as covariant functors $C^{\mathrm{opp}} \to D$. Hence for a contravariant functor to be *left* exact means that the 0 is on the *left* in the sequence obtained by applying the functor F.

For instance the Hom-functor is left exact in each variable (as it preserves limits in the second variable and sends colimits in the first variable to limits).

Definition: Exact Functors F.39. A functor $F\colon C \to D$ of abelian categories is called *exact* if it is left exact and right exact. Equivalently, for every exact sequence

$$\cdots \longrightarrow X^{i-1} \longrightarrow X^i \longrightarrow X^{i+1} \longrightarrow \cdots$$

the sequence

$$\cdots \longrightarrow F(X^{i-1}) \longrightarrow F(X^i) \longrightarrow F(X^{i+1}) \longrightarrow \cdots$$

is exact.

(F.8) Length and Jordan-Hölder series in abelian categories.

Definition F.40. *Let $\mathcal{C}$ be an abelian category and let X be an object of $\mathcal{C}$.*
(1) *X is called* simple *if $X \neq 0$ and 0 and X are the only quotient objects of X up to isomorphism.*
(2) *X is called* semisimple *if it is isomorphic to a finite direct sum of simple objects.*
(3) *X is said to be* of finite length *if there exists a finite sequence of subobjects*

$$0 = X_0 \hookrightarrow X_1 \hookrightarrow \cdots \hookrightarrow X_{n-1} \hookrightarrow X_n = X$$

such that X_i/X_{i-1} is a simple object. If such a sequence exists, it is called a Jordan-Hölder sequence *or* composition series.

An object X in an abelian category is of finite length if and only if it is artinian (i.e., it satisfies the descending chain condition on subobjects) and noetherian (i.e., it satisfies the ascending chain condition on subobjects).

Proposition and Definition F.41. *(Jordan-Hölder theorem, [Ste] IV, §5) Let X be an object of finite length in an abelian category $\mathcal{C}$. Let*

$$0 = X_0 \hookrightarrow X_1 \hookrightarrow \cdots \hookrightarrow X_n = X \quad and \quad 0 = X_0' \hookrightarrow X_1' \hookrightarrow \cdots \hookrightarrow X_m' = X$$

be two Jordan-Hölder sequences. Then $m = n$ and there exists a permutation π of $\{1, \ldots, n\}$ such that $X_{\pi(i)}'/X_{\pi(i)-1}' \cong X_i/X_{i-1}$. The common length n of all Jordan-Hölder sequences of X is called the length *of X. It is denoted by $\lg_{\mathcal{C}}(X)$ or simply by $\lg(X)$.*

The length is additive in exact sequences, more precisely:

Proposition F.42. *Let $\mathcal{C}$ be an abelian category and let $0 \to X' \to X \to X'' \to 0$ be a short exact sequence in $\mathcal{C}$. Then X is of finite length if and only if X' and X'' are of finite length. In this case one has $\lg(X) = \lg(X') + \lg(X'')$.*

(F.9) Subcategories of abelian categories.

Let $\mathcal{C}$ be an abelian category. A full subcategory $\mathcal{D}$ of $\mathcal{C}$ is additive if and only if $\mathcal{D}$ is closed under finite products in $\mathcal{C}$ (by Definition F.27). A full additive subcategory $\mathcal{D}$ of $\mathcal{C}$ is abelian and the inclusion functor is exact if and only if $\mathcal{D}$ is closed under kernels and cokernels in $\mathcal{C}$.

Definition and Proposition F.43. *We call a full subcategory $\mathcal{D}$ of an abelian category $\mathcal{C}$* plump[1] *if the following equivalent conditions are satisfied.*
(i) *$\mathcal{D}$ is closed under kernels, cokernels, and extensions in $\mathcal{C}$.*
(ii) *$\mathcal{D}$ is a full additive subcategory and for every exact sequence $X_1 \to X_2 \to X_3 \to X_4 \to X_5$ in $\mathcal{C}$ with X_1, X_2, X_4, X_5 in $\mathcal{D}$ also X_3 is in $\mathcal{D}$.*

The plump subcategory $\mathcal{D}$ is then itself abelian and the inclusion functor $\mathcal{D} \to \mathcal{C}$ is exact.

---

[1] Here we follow Lipman [Lip2] $\overset{\text{O}}{\underset{\text{X}}{}}$. Kashiwara and Schapira [KaSh] ^{O} use the notion of a thick subcategory which is very often defined differently. The Stacks project calls plump subcategories weakly Serre subcategories.

Definition and Proposition F.44. *Let $\mathcal{A}$ be an abelian category. A non-empty full subcategory $\mathcal{B}$ of $\mathcal{A}$ is called* Serre subcategory *if the following two equivalent conditions hold.*

(i) *For every short exact sequence $0 \to X' \to X \to X'' \to 0$ in $\mathcal{A}$ one has $X', X'' \in \mathcal{B}$ if and only if $X \in \mathcal{B}$.*

(ii) *For every exact sequence $Y' \to Y \to Y''$ in $\mathcal{A}$ with $Y', Y'' \in \mathcal{B}$ one has $Y \in \mathcal{B}$.*

Every Serre subcategory $\mathcal{B}$ is abelian and the inclusion $\mathcal{B} \to \mathcal{A}$ is an exact functor.

Every Serre subcategory is a plump subcategory. The converse does not necessarily hold (Exercise 22.27).

Example F.45. Let $\mathcal{C}$ be an abelian category. Then the full subcategory of objects of $\mathcal{C}$ of finite length (Definition F.40) is a Serre subcategory.

(F.10) Five Lemma and Snake Lemma.

Lemma F.46. *(Four Lemma) Let $\mathcal{C}$ be an abelian category. Let*

$$
\begin{array}{ccccccc}
X^0 & \longrightarrow & X^1 & \longrightarrow & X^2 & \longrightarrow & X^3 \\
\downarrow {\scriptstyle f^0} & & \downarrow {\scriptstyle f^1} & & \downarrow {\scriptstyle f^2} & & \downarrow {\scriptstyle f^3} \\
Y^0 & \longrightarrow & Y^1 & \longrightarrow & Y^2 & \longrightarrow & Y^3
\end{array}
$$

be a commutative diagram in $\mathcal{C}$ whose rows are exact.

(1) *If f^0 is an epimorphism and f^1 and f^3 are monomorphisms, then f^2 is a monomorphism.*

(2) *Dually, if f^3 is a monomorphism and f^0 and f^2 are epimorphisms, then f^1 is an epimorphism.*

In fact, it suffices that the rows are complexes (see Definition F.68 below) and that the sequences $X^1 \to X^2 \to X^3$ and $Y^0 \to Y^1 \to Y^2$ are exact.

In particular, we obtain the statement of the classical "five lemma":

Corollary F.47. *(Five lemma) Let $\mathcal{C}$ be an abelian category. Let*

$$
\begin{array}{ccccccccc}
X^0 & \longrightarrow & X^1 & \longrightarrow & X^2 & \longrightarrow & X^3 & \longrightarrow & X^4 \\
\downarrow {\scriptstyle f^0} & & \downarrow {\scriptstyle f^1} & & \downarrow {\scriptstyle f^2} & & \downarrow {\scriptstyle f^3} & & \downarrow {\scriptstyle f^4} \\
Y^0 & \longrightarrow & Y^1 & \longrightarrow & Y^2 & \longrightarrow & Y^3 & \longrightarrow & Y^4
\end{array}
$$

be a commutative diagram in $\mathcal{C}$ whose rows are exact sequences. If f^0, f^1, f^3, f^4 are isomorphisms, then f^2 is an isomorphism.

Lemma F.48. *(Snake lemma) Let $\mathcal{C}$ be an abelian category. Let*

$$
\begin{array}{ccccccc}
A & \overset{f}{\longrightarrow} & B & \overset{g}{\longrightarrow} & C & \longrightarrow & 0 \\
\downarrow {\scriptstyle \alpha} & & \downarrow {\scriptstyle \beta} & & \downarrow {\scriptstyle \gamma} & & \\
0 & \longrightarrow & X & \underset{u}{\longrightarrow} & Y & \underset{v}{\longrightarrow} & Z
\end{array}
$$

be a commutative diagram in $\mathcal{C}$ whose rows are exact sequences. Then there is an exact sequence

$$\operatorname{Ker}\alpha \xrightarrow{f'} \operatorname{Ker}\beta \xrightarrow{g'} \operatorname{Ker}\gamma \xrightarrow{\delta} \operatorname{Coker}\alpha \xrightarrow{u'} \operatorname{Coker}\beta \xrightarrow{v'} \operatorname{Coker}\gamma$$

where f', g', u', v' are the maps induced by f, g, u, and v, and where the "boundary map" δ is defined as follows: Set $P := B \times_C \operatorname{Ker}\gamma$ and $S := Y \amalg_X \operatorname{Coker}\alpha$. Then the projection $P \to \operatorname{Ker}\gamma$ is an epimorphism and $\operatorname{Coker}\alpha \to S$ is injective. The composition $P \to B \to Y \to S$ factors through a morphism

$$P \to \operatorname{Ker}\gamma \xrightarrow{\delta} \operatorname{Coker}\alpha \to S.$$

If there exists an exact fully faithful functor $\mathcal{C} \to (R\text{-Mod})$ for some not necessarily commutative ring (which is always the case if $\mathcal{C}$ is small by a Theorem of Mitchell), then δ can also be described on elements as follows. Let $c \in \operatorname{Ker}\gamma$, let $b \in B$ with $g(b) = c$, let $x \in X$ with $u(x) = \beta(b)$, and define $\delta(c)$ as the image of x in $\operatorname{Coker}\alpha$.

(F.11) Injective and projective Objects.

Definition F.49. *(Injective and projective objects) Let $\mathcal{C}$ be an abelian category.*
(1) *An object I in $\mathcal{C}$ is called* injective, *if the following equivalent conditions are satisfied.*
  (i) *The functor $\mathcal{C}^{\mathrm{opp}} \to (\mathrm{AbGrp})$, $X \mapsto \operatorname{Hom}_{\mathcal{C}}(X, I)$, is exact.*
  (ii) *Every exact sequence in $\mathcal{C}$ of the form $0 \to I \to X \to X'' \to 0$ is split.*
(2) *Dually, an object P in $\mathcal{C}$ is called* projective, *if the following equivalent conditions are satisfied.*
  (i) *The functor $\mathcal{C} \to (\mathrm{AbGrp})$, $X \mapsto \operatorname{Hom}_{\mathcal{C}}(P, X)$, is exact.*
  (ii) *Every exact sequence in $\mathcal{C}$ of the form $0 \to X' \to X \to P \to 0$ is split.*

The full subcategory of injective (resp. projective) objects of $\mathcal{C}$ is an additive subcategory of $\mathcal{C}$.

Definition F.50. *An abelian category $\mathcal{C}$ is said to*
(1) have enough injectives, *if for every object X in $\mathcal{C}$, there exists a monomorphism $X \to I$ from X to an injective object I.*
(2) have enough projectives, *if for every object X in $\mathcal{C}$, there exists an epimorphism $P \to X$ from a projective object P to X.*

Remark F.51. Let R be a not necessarily commutative ring. Then every free R-module is projective, and in particular the category $(R\text{-Mod})$ of left R-modules has enough projectives. Categories of abelian sheaves on some space typically do not have enough projectives.

The category $(R\text{-Mod})$ also has enough injectives (see Proposition F.62 below). More generally, for every ringed space $(X, \mathcal{O}_X)$ the category of $\mathcal{O}_X$-modules has enough injectives (combine Proposition 21.7 and Proposition F.62 below).

(F.12) Grothendieck abelian categories.

One criterion for an abelian category to have enough injectives is to be a Grothendieck category. We first introduce the notion of generators of a category.

Definition F.52. *Let C be a category. A family of objects $(G_i)_{i\in I}$ for a (small) index set I is called a* system of generators *if the functor $\prod_{i\in I} \mathrm{Hom}_C(G_i, -): C \to$ (Sets) is conservative (Definition F.3). If the family $(G_i)_{i\in I}$ consists of a single object G, then G is called a* generator.

If C admits coproducts and a system of generators $(G_i)_i$, then $\coprod_{i\in I} G_i$ is a generator of C.

Proposition F.53. ([KaSh]° 5.2.4) *Let $\mathcal{A}$ be an abelian category which admits all coproducts. For an object G of C, the following conditions are equivalent.*
(i) *G is a generator.*
(ii) *The functor $\mathrm{Hom}_C(G, -): C \to$ (Sets) is faithful.*
(iii) *For any object X in $\mathcal{A}$, there exist a (small) set I and an epimorphism $G^{(I)} \to X$.*

Definition F.54. *An abelian category C is called a* Grothendieck category, *if*
(a) *all (infinite) coproducts exist in C (and hence C is cocomplete by Proposition F.19),*
(b) *filtered colimits are exact,*
(c) *the category C admits a generator.*

For any ring R, the category of left (or: right) R-modules is a Grothendieck abelian category. The ring R, considered as an R-module, is a generator. Conversely, the Gabriel-Popescu theorem implies that every Grothendieck abelian category admits a fully faithful functor to the category of R-modules for a ring R (one takes as R the endomorphism ring of a generator G and the functor is given by $X \mapsto \mathrm{Hom}(G, X)$). This functor has an exact left adjoint functor.

The category of finite-dimensional vector spaces over a field is an example of an abelian category which does not have all coproducts, and hence is not a Grothendieck abelian category.

Remark F.55. Let C be a cocomplete abelian category. Let $(U_i)_{i\in I}$ be a family of objects in C such that for every epimorphism $A \to B$ in C with $B \neq 0$ there exists $i \in I$ and a morphism $U_i \to A$ such that composition $U_i \to B$ is non-zero. Then $\bigoplus U_i$ is a generator of C.

For contravariant functors on Grothendieck categories there is the following easy criterion to be representable.

Theorem F.56. ([Sta] 07D7) *Let C be a Grothendieck abelian category, and let $F: C^{\mathrm{opp}} \to$ (Sets) be a functor. The functor F is representable if and only if it commutes with all small limits, i.e., $F(\mathrm{colim}_i X_i) = \lim_i F(X_i)$ for any diagram $X: \mathcal{I} \to C$ and $\mathrm{colim}_i X_i$ the colimit in C.*

Corollary F.57. *Let C be a Grothendieck category and let $\mathcal{D}$ be a cocomplete category. Then a functor $F: C \to \mathcal{D}$ has a right adjoint functor if and only if F commutes with small colimits.*

Proof. The condition is clearly necessary. Now suppose that F commutes with colimits. Let X be an object in $\mathcal{D}$ and consider the functor $\tilde{G}_X \colon \mathcal{C}^{\mathrm{opp}} \to (\mathrm{Sets})$, $\tilde{G}_X(Y) := \mathrm{Hom}_\mathcal{D}(F(Y), X)$. Then $\tilde{G}_X$ sends colimits in $\mathcal{C}$ to limits. Therefore it is representable by Theorem F.56, i.e., there exists an object G_X in $\mathcal{C}$ and an isomorphism of functors $\mathrm{Hom}_\mathcal{D}(F(-), X) \cong \mathrm{Hom}_\mathcal{C}(-, G_X)$. By the Yoneda lemma, the construction $X \mapsto G_X$ is functorial in X and therefore defines a right adjoint functor to F. $\qquad\square$

Remark F.58.
(1) Every Grothendieck category is cocomplete, as it is finitely cocomplete, as every abelian category, and as arbitrary direct sums exist, because these are filtered colimits of finite direct sums. Now use Proposition F.19.
(2) One can also show that all Grothendieck categories are complete ([Bor] 5.2). This can also be easily deduced from Theorem F.56.

Remark F.59. Let $\mathcal{C}$ be a Grothendieck category and let $\mathcal{I}$ be a small category. Then the category $\mathcal{C}^\mathcal{I}$ of functors $\mathcal{I} \to \mathcal{C}$ is a Grothendieck category ([Gro1]^0 1.9.2). As Grothendieck categories are complete and cocomplete (Remark F.58), we obtain functors

$$\lim \colon \mathcal{C}^\mathcal{I} \to \mathcal{C}, \qquad \mathrm{colim} \colon \mathcal{C}^\mathcal{I} \to \mathcal{C}$$

As (co)limits commute with each other (see F.11), the functor $\lim$ is left exact and the functor colim is right exact.

The property to be a Grothendieck category is inherited by so-called Giraud subcategories:

Definition F.60. *A subcategory $\mathcal{D}$ of a Grothendieck abelian category $\mathcal{C}$ is called a* Giraud subcategory, *if the inclusion functor admits a left adjoint and if this left adjoint functor is left exact (and hence exact).*

For example, the category of sheaves of $\mathcal{O}_X$-modules on a ringed space $(X, \mathcal{O}_X)$ is a Giraud subcategory of the category of presheaves of $\mathcal{O}_X$-modules because sheafification is an exact functor.

Proposition F.61. ([Ste] X, §1) *Let $\mathcal{D}$ be a Giraud subcategory of a Grothendieck abelian category $\mathcal{C}$ and let $a \colon \mathcal{C} \to \mathcal{D}$ be the exact left adjoint to the inclusion. Then $\mathcal{D}$ is itself a Grothendieck abelian category. Moreover, an object of $\mathcal{D}$ is injective in $\mathcal{D}$ if and only if it is injective in $\mathcal{C}$. If G is a generator of $\mathcal{C}$, then $a(G)$ is a generator of $\mathcal{D}$.*

(F.13) Injective objects in Grothendieck abelian categories and locally noetherian categories.

In this section, $\mathcal{A}$ denotes a Grothendieck category. First note that $\mathcal{A}$ has enough injective objects.

Proposition F.62. *Let $\mathcal{C}$ be a Grothendieck abelian category. Then $\mathcal{C}$ has enough injectives.*

Below, we will state a stronger result (Theorem F.185).
We will use the following notions and results to study the category of $\mathcal{O}_X$-modules for a locally noetherian scheme X. A reference is [Ste] X, §2–5.

Proposition F.63. *Let $\mathcal{A}$ be a Grothendieck abelian category and let $(G_i)_{i \in I}$ be a family of generators. An object X of $\mathcal{A}$ is injective if and only if for all i, every monomorphism $\iota\colon U \to G_i$, and every morphism $\varphi\colon U \to X$, there exists $\tilde{\varphi}\colon G_i \to X$ such that $\tilde{\varphi} \circ \iota = \varphi$.*

For instance, if A is a ring, the category of A-modules is generated by A. Hence an A-module X is injective if and only if every homomorphism $\mathfrak{a} \to X$, where $\mathfrak{a}$ is an ideal of A, can be extended to A.

Definition F.64. *Let X be an object of $\mathcal{A}$. An* injective hull *of X is a monomorphism $\iota\colon X \to I$ such that I is injective and such that $\iota^{-1}(J) := X \times_I J \neq 0$ for every non-zero subobject J of I.*

Proposition F.65. *Every object X of $\mathcal{A}$ has an injective hull $\iota\colon X \to I$. If $\iota'\colon X \to I'$ is another injective hull, then there exists an isomorphism $\gamma\colon I \xrightarrow{\sim} I'$ (in general not unique) such that $\iota' = \gamma \circ \iota$.*

Definition and Proposition F.66. *Let $\mathcal{A}$ be a Grothendieck abelian category.*
(1) *An object X of $\mathcal{A}$ is called* finitely generated *if the following equivalent assertions hold.*
   (i) *Whenever X is the filtered colimit of subobjects X_i, then there exists an index i_0 such that $X = X_{i_0}$.*
   (ii) *The functor $\mathrm{Hom}_{\mathcal{A}}(X, -)$ preserves colimits of filtered diagrams $\mathcal{I} \to \mathcal{A}$ where the transition maps are monomorphisms.*
(2) *An object X of $\mathcal{A}$ is called* noetherian *if the following equivalent conditions hold.*
   (i) *Every ascending chain of subobjects of X becomes stationary.*
   (ii) *Every subobject of X is finitely generated.*
(3) *The category $\mathcal{A}$ is called* locally finitely generated *(resp. locally noetherian) it it has a family of finitely generated (resp. noetherian) generators.*

Let A be a ring. Then for the category of A-modules the notions of finitely generated and noetherian for A-modules defined above coincide with the usual notions. The category of A-modules is locally finitely generated (A is a finitely generated generator) and it is locally noetherian if and only if A is a noetherian ring.

Proposition F.67. *Let $\mathcal{A}$ be a locally finitely generated Grothendieck abelian category. Then $\mathcal{A}$ is locally noetherian if and only if every coproduct of injective objects is injective.*

In this case, every injective object Z is a coproduct of indecomposable injective objects. If $Z = \bigoplus_{i \in I} X_i = \bigoplus_{j \in J} Y_j$ are two such decompositions, then there exists a bijection $\alpha\colon I \to J$ such that $X_i \cong Y_{\alpha(i)}$ for all $i \in I$.

In particular, we see that a ring A is noetherian if and only if every direct sum of injective A-modules is again injective.

Complexes in additive and abelian categories

As a first step towards the definition of the derived category of an abelian category we recall the notion of a complex, the notion of homotopy between two morphisms of complexes, and the notion of a quasi-isomorphism. Then we give various constructions of complexes.

(F.14) Categories of Complexes.

In this section $\mathcal{A}$ always denotes an additive category.

Definition F.68.
(1) *A complex in $\mathcal{A}$ is a family $(X^j)_{j\in\mathbb{Z}}$ of objects and morphisms $(d^j\colon X^j \to X^{j+1})_{j\in\mathbb{Z}}$ of $\mathcal{A}$ such that $d^j \circ d^{j-1} = 0$ for all j.*
(2) *A morphism of complexes $f\colon X \to Y$ is a family of morphisms $f^j\colon X^j \to Y^j$ for $j \in \mathbb{Z}$ such that the diagram*

$$
\begin{array}{ccccccc}
\cdots \xrightarrow{d} & X^{j-1} & \xrightarrow{d} & X^j & \xrightarrow{d} & X^{j+1} & \xrightarrow{d} \cdots \\
& \downarrow{\scriptstyle f^{j-1}} & & \downarrow{\scriptstyle f^j} & & \downarrow{\scriptstyle f^{j+1}} & \\
\cdots \xrightarrow{d} & Y^{j-1} & \xrightarrow{d} & Y^j & \xrightarrow{d} & Y^{j+1} & \xrightarrow{d} \cdots
\end{array}
$$

commutes. We denote by $\mathrm{C}(\mathcal{A})$ the category of complexes in $\mathcal{A}$.
(3) *The translation functor $T\colon \mathrm{C}(\mathcal{A}) \to \mathrm{C}(\mathcal{A})$ is defined on objects by $(X^j)_j \mapsto (X^{j+1})_j$, $d^j_{TX} = -d^{j+1}_X$ and on morphisms by $(f^j)_j \mapsto (f^{j+1})_j$.*

As in the above diagram, we often write d instead of d^j. The translation functor is sometimes called the *shift functor*. It is an automorphism of the category of complexes. For $i \in \mathbb{Z}$ we set $(\ )[i] := T^i$. In other words, given a complex $X = (X^j)_j$, we have

$$
(X[i])^j := X^{j+i}, \qquad d^j_{X[i]} = (-1)^i d^{i+j}_X.
$$

If $\mathcal{A}$ admits countable direct sums, then we can also view a complex $(X_i)_i$ as a graded object $\bigoplus_i X_i$ together with an endomorphism $d\colon \bigoplus_i X_i \to \bigoplus_i X_i$ which shifts degrees by 1 and such that $d \circ d = 0$.

A complex $\cdots \to X^{i-1} \to X^i \to X^{i+1} \to \cdots$ is called *acyclic at X^i*, if $\operatorname{Ker} d^i = \operatorname{Im} d^{i-1}$. It is called *acyclic* if it is exact at every i. Clearly, an acyclic complex is the same as an exact sequence indexed by the integers.

The category $C(\mathcal{A})$ of complexes in $\mathcal{A}$ is additive. If $\mathcal{A}$ is abelian, then $C(\mathcal{A})$ is abelian, too. In particular, we have the following notions:

(1) Let X be a complex in $\mathcal{A}$. A *subcomplex* of X is a subobject of X in $C(\mathcal{A})$. Explicitly, a subcomplex of X is a complex Y such that $Y^i \subseteq X^i$ for all i, compatibly with the differentials.
(2) For every morphism $f\colon X \to Y$ of complexes there exists the notion of a *kernel* and a *cokernel*. If $\mathcal{A}$ is abelian, these always exist and they are formed component-wise: $\operatorname{Ker}(f)^i = \operatorname{Ker}(f^i)$, $\operatorname{Coker}(f)^i = \operatorname{Coker}(f^i)$, with differentials induced from the differentials of X. Similarly, we have the *image* $\operatorname{Im}(f)$ of f, $\operatorname{Im}(f)^i = \operatorname{Im}(f^i)$.

  If $Y \subset X$ is a subcomplex, we set $X/Y := \operatorname{Coker}(Y \to X)$ if this cokernel exists.
(3) Suppose that $\mathcal{A}$ is an abelian category. A sequence

$$
0 \to X \to Y \to Z \to 0
$$

of complexes is *exact* in $C(\mathcal{A})$ if and only if it is termwise exact, i.e. $0 \to X^i \to Y^i \to Z^i \to 0$ is exact for all $i \in \mathbb{Z}$.
(4) We can form finite direct sums and finite products in $C(\mathcal{A})$ componentwise.
(5) More generally, if $\mathcal{A}$ admits limits (resp. colimits) indexed by some category, then the same is true for $C(\mathcal{A})$ and these limits (resp. colimits) are formed componen-twise. In particular, if $\mathcal{A}$ is complete (resp. cocomplete), then $C(\mathcal{A})$ is complete (resp. cocomplete).

(6) If $\mathcal{A}$ is a Grothendieck abelian category, the same is true for $C(\mathcal{A})$.

Definition F.69. *Let $\mathcal{A}$ be an additive category. A sequence $0 \to X \to Y \to Z \to 0$ of complexes in $C(\mathcal{A})$ is called* termwise split *if $0 \to X^i \to Y^i \to Z^i \to 0$ is a split sequence (Definition F.31) for all $i \in \mathbb{Z}$.*

Definition F.70. *A complex $X = (X^j)$ is called* bounded *(bounded below, bounded above, resp.), if $X^j = 0$ for $|j| \gg 0$ (for $j \ll 0$, for $j \gg 0$, resp.). We denote by $C^b(\mathcal{A})$, $C^+(\mathcal{A})$, $C^-(\mathcal{A})$ the full subcategories of $C(\mathcal{A})$ consisting of all complexes which are bounded, bounded below, bounded above.*

Let $M \subseteq \mathbb{Z}$ be a subset. A complex X is said to be *concentrated in degree M* if $X^i = 0$ for all $i \notin M$. We have the full subcategory $C^M(\mathcal{A})$ of $C(\mathcal{A})$ of complexes concentrated in degrees M. We write $C^{\leq a}(\mathcal{A}) := C^{(-\infty, a]}(\mathcal{A})$, etc. All these categories are abelian.

Given a family of objects and morphisms as above with an "interval" in $\mathbb{Z}$ as index set, we can extend it to a complex by adding zeros. This gives obvious fully faithful exact functors between the categories above. In particular, we consider $\mathcal{A}$ as a full subcategory of $C(\mathcal{A})$ (and of $C^+(\mathcal{A})$, $\dots$) by sending an object $X \in \mathcal{A}$ to the complex $\cdots \to 0 \to X \to 0 \to \cdots$, where X sits in degree 0.

Let $\mathcal{A}$ be an abelian category. We frequently consider the following functors $C(\mathcal{A}) \to \mathcal{A}$:

$$\pi_n \colon (X^i)_i \mapsto X^n,$$
$$Z^n \colon (X^i)_i \mapsto \mathrm{Ker}(d \colon X^n \to X^{n+1}),$$
$$B^n \colon (X^i)_i \mapsto \mathrm{Im}(d \colon X^{n-1} \to X^n),$$
$$H^n \colon X \mapsto Z^n(X)/B^n(X).$$

We call $H^n(X)$ the *n-th cohomology object* of X. Note that $H^n(X) = H^0(X[n])$. As a reminiscence of the origin of these definitions in algebraic topology, sometimes Z^n is called the *object of cocycles* of degree n, and B^n the *object of coboundaries* of degree n.

If $M \subseteq \mathbb{Z}$ is a subset, then we say that a complex is *acyclic in degrees M* if $H^n(X) = 0$ for all $n \in M$.

Proposition F.71. *Let $\mathcal{A}$ be an abelian category. Let $0 \to X \to Y \to Z \to 0$ be a short exact sequence in $C(\mathcal{A})$. For $i \in \mathbb{Z}$ define a morphism $H^i(Z) \to H^{i+1}(X)$ by applying the Snake lemma F.48 to the diagram*

$$
\begin{array}{ccccccc}
X^i/\mathrm{Im}(d_X^{i-1}) & \longrightarrow & Y^i/\mathrm{Im}(d_Y^{i-1}) & \longrightarrow & Z^i/\mathrm{Im}(d_Z^{i-1}) & \longrightarrow & 0 \\
\downarrow{\scriptstyle d_X^i} & & \downarrow{\scriptstyle d_Y^i} & & \downarrow{\scriptstyle d_Z^i} & & \\
0 \longrightarrow \mathrm{Ker}(d_X^{i+1}) & \longrightarrow & \mathrm{Ker}(d_Y^{i+1}) & \longrightarrow & \mathrm{Ker}(d_Z^{i+1}). & &
\end{array}
$$

Then the long cohomology sequence

$$\cdots \longrightarrow H^{i-1}(Z) \longrightarrow H^i(X) \longrightarrow H^i(Y) \longrightarrow H^i(Z) \longrightarrow H^{i+1}(X) \longrightarrow \cdots$$

is exact.

Furthermore, we have the following *truncation functors*: For a complex X with components in an abelian category, we set

$$(\text{F.14.1}) \qquad \tau^{\leq n}(X) \colon \quad \cdots \longrightarrow X^{n-2} \longrightarrow X^{n-1} \longrightarrow \mathrm{Ker}\, d^n \longrightarrow 0 \longrightarrow \cdots,$$

and dually

(F.14.2) $\tau^{\geq n}(X)$: $\cdots \longrightarrow 0 \longrightarrow \operatorname{Coker} d^{n-1} \longrightarrow X^{n+1} \longrightarrow X^{n+2} \longrightarrow \cdots$.

Then $\tau^{\leq n}(X)$ is a subcomplex of X and $\tau^{\geq n}X$ is a quotient complex of X. For the cohomology objects, we have

$$H^i(\tau^{\leq n}X) = \begin{cases} H^i(X) & \text{if } i \leq n, \\ 0 & \text{if } i > n, \end{cases} \qquad H^i(\tau^{\geq n}X) = \begin{cases} H^i(X) & \text{if } i \geq n, \\ 0 & \text{if } i < n. \end{cases}$$

We also have the *stupid truncation functors* defined by

(F.14.3)
$$\sigma^{\geq n}(X) := (\ldots \to 0 \to 0 \to X^n \to X^{n+1} \to \ldots),$$
$$\sigma^{\leq n}(X) := (\ldots \to X^{n-1} \to X^n \to 0 \to 0 \to \ldots).$$

Then $\sigma^{\geq n}(X)$ is a subcomplex of X and $\sigma^{\leq n}(X)$ is a quotient of X.

Definition F.72. *(Mapping cone) For a morphism $f\colon X \to Y$ of complexes with components in an additive category $\mathcal{A}$, we define the* mapping cone C_f *of f by setting*

$$C_f^n = X^{n+1} \oplus Y^n, \qquad d_{C_f}^n = \begin{pmatrix} -d_X & 0 \\ f & d_Y \end{pmatrix}.$$

Below (Remark F.79) it is explained how to characterize the mapping cone by a universal property. The mapping cone encodes kernel and cokernel simultaneously in the following sense.

Example F.73. Let $\mathcal{A}$ be an abelian category and let $f\colon A \to B$ be a morphism in $\mathcal{A}$ which we consider as a morphism of complexes concentrated in degree 0. Then

$$C_f = (\ldots \longrightarrow 0 \longrightarrow A \xrightarrow{f} B \longrightarrow 0 \longrightarrow \ldots),$$

with B in degree 0. Hence

$$H^0(C_f) = \operatorname{Coker}(f), \qquad H^{-1}(C_f) = \operatorname{Ker}(f).$$

Remark F.74. Let $\mathcal{A}$ and $\mathcal{A}'$ be additive categories.

Let $F\colon \mathcal{A} \to \mathcal{A}'$ be an additive functor. Then on complexes F induces an additive functor $C(F)\colon C(\mathcal{A}) \to C(\mathcal{A}')$ by sending a complex $((X^i),(d^i))$ to $(F(X^i),F(d^i))$ and a morphism (u^i) to $(F(u^i))$. This functor preserves termwise split sequences and mapping cones (i.e., $F_{C_u} = C_{F(u)}$ for every morphism u of complexes).

(F.15) Homotopy of complexes.

Definition F.75. *Let $\mathcal{A}$ be an additive category. A* homotopy *between two morphisms $f,g\colon X \to Y$ of complexes of $\mathcal{A}$ is a family of morphisms $h^i\colon X^i \to Y^{i-1}$ of complexes such that*

$$f^i - g^i = h^{i+1} \circ d_X^i + d_Y^{i-1} \circ h^i.$$

If such a family $(h^i)_i$ exists, we call the morphisms $f,g\colon X \to Y$ homotopic.

Note that h is *not* assumed to be a morphism of complexes $X \to Y[-1]$. But if $u\colon X \to Y[-1]$ is a morphism of complexes, then $h + u$ is again a homotopy between f and g. More precisely the set of homotopies between f and g is a principal homogeneous space (possibly empty) under the abelian group $\mathrm{Hom}_{C(\mathcal{A})}(X, Y[-1])$.

Being homotopic is an equivalence relation on the set $\mathrm{Hom}_{C(\mathcal{A})}(X, Y)$. As is easily checked, composition of morphisms induces a pairing on the sets of equivalence classes. Hence we can define:

Definition F.76. *Let $\mathcal{A}$ be an additive category, and let $C(\mathcal{A})$ be the category of complexes in $\mathcal{A}$. The* homotopy category $K(\mathcal{A})$ *is defined by*

$$\mathrm{Ob}(K(\mathcal{A})) = \mathrm{Ob}(C(\mathcal{A})), \qquad \mathrm{Hom}_{K(\mathcal{A})}(X, Y) = \mathrm{Hom}_{C(\mathcal{A})}(X, Y)/ \sim,$$

where $f \sim g$ if and only if f and g are homotopic.

For $* \in \{b, +, -\}$ or $* \in \{\,[a, b]\; ;\; a, b \in \mathbb{Z} \cup \{\pm\infty\}, a \le b\,\}$ we denote by $K^*(\mathcal{A})$ the full subcategory of $K(\mathcal{A})$ consisting of complexes in $C^*(\mathcal{A})$.

Definition F.77. *Let $\mathcal{A}$ be an additive category. A morphism of complexes $f\colon X \to Y$ is called a* homotopy equivalence *if it is an isomorphism in $K(\mathcal{A})$, i.e., there exists a morphism $g\colon Y \to X$ of complexes, such that $g \circ f$ is homotopic to id_X, and $f \circ g$ is homotopic to id_Y. If two complexes are isomorphic in $K(\mathcal{A})$, they are also called* homotopy equivalent.

Remark F.78. The homotopy category $K(\mathcal{A})$ is an additive category. Finite direct sums are formed componentwise.

Note that even if $\mathcal{A}$ is an abelian category, $K(\mathcal{A})$ is not abelian in general. For instance if $\mathcal{A}$ is the abelian category of abelian groups, then the canonical morphism $\mathbb{Z} \to \mathbb{Z}/2\mathbb{Z}$, considered as morphisms of complexes concentrated in degree 0, cannot be factorized as an epimorphism followed by a monomorphism. This can be shown by a direct argument. Alternatively, one can also use that $K(\mathcal{A})$ can be endowed with the structure of a triangulated category (see Section (F.27) below) and that in a triangulated category every epimorphism has a right inverse (see Remark F.122 below).

Remark F.79. Let $u\colon X \to Y$ be a morphism in $C(\mathcal{A})$. Then the inclusions $Y^i \to X^{i+1} \oplus Y^i = C_u^i$ define a morphism of complexes $\iota\colon Y \to C_u$. Then the composition $X \to Y \to C_u$ is homotopic to zero via the homotopy h defined by the inclusion $h^i\colon X^i \to C_u^{i-1} = X^i \oplus Y^{i-1}$.

Moreover, the mapping cone C_u has the following universal property of a "homotopy cokernel". If $v\colon Y \to Z$ is a morphism of complexes and $\tilde{h}\colon v \circ u \simeq 0$ a homotopy, then there exists a unique morphism of complexes $w\colon C_u \to Z$ such that $v = w \circ \iota$ and $\tilde{h}^i = w^{i-1} \circ h^i$. Indeed, we define

$$w^i := \tilde{h}^{i+1} + v^i\colon X^{i+1} \oplus Y^i \to Z^i.$$

A similar argument shows that $C_u[-1]$ has "the universal property of a homotopy kernel", i.e., by composition one obtains a bijection between $\mathrm{Hom}_{C(\mathcal{A})}(Z, C_u[-1])$ and the set of pairs consisting of a morphism of complexes $v\colon Z \to X$ and a homotopy of $u \circ v$ to 0.

Remark F.80. Let $F\colon \mathcal{A} \to \mathcal{A}'$ be a (covariant or contravariant) additive functor of additive categories. Then the induced functor $C(F)\colon C(\mathcal{A}) \to C(\mathcal{A}')$ further induces a functor $K(F)\colon K(\mathcal{A}) \to K(\mathcal{A}')$.

Remark F.81. The truncation functors $\tau^{\leq n}$ and $\tau^{\geq n}$ induce functors $K(\mathcal{A}) \to K(\mathcal{A})$. Note that this is in general not true for the stupid truncation functors $\sigma^{\leq n}$ and $\sigma^{\geq n}$.

(F.16) Quasi-isomorphisms.

Definition F.82. *Let $\mathcal{A}$ be an abelian category. A morphism $f \colon X \to Y$ of complexes is called a* quasi-isomorphism *(sometimes abbreviated as* qis*), if the induced morphisms $H^i(f) \colon H^i(X) \to H^i(Y)$ on the cohomology are isomorphisms.*

A complex is exact if and only if there exists a quasi-isomorphism to the zero complex.

Note that if $f \colon X \to Y$ is a quasi-isomorphism, there does not necessarily exist a quasi-isomorphism $Y \to X$! For example, let X be given by $\mathbb{Z} \to \mathbb{Z}$, $x \to 2x$, (in degrees 0, 1, extended by 0), and let Y be given by $\mathbb{Z}/2$ in degree 1. While the natural map $X \to Y$ is a qis, there is no non-zero morphism of complexes $Y \to X$ at all.

The following example will be used very often.

Example F.83. Let $\cdots \to X^{i-1} \to X^i \to X^{i+1} \to \ldots$ be a sequence in an abelian category $\mathcal{A}$. Then this sequence is exact if and only if for every $i \in \mathbb{Z}$ the morphism of complexes

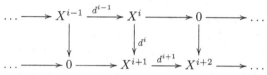

is a quasi-isomorphism.

The following lemma, though easy to prove, is a cornerstone of the theory of derived categories:

Lemma F.84. *Let $\mathcal{A}$ be an abelian category. If $f, g \colon X \to Y$ are homotopic morphisms in $C(\mathcal{A})$, then the induced maps on cohomology $H^i(f)$ and $H^i(g)$ are equal.*

In particular, every homotopy equivalence is a quasi-isomorphism.

The converse is not true: The above defined quasi-isomorphism

$$(\mathbb{Z} \xrightarrow{2} \mathbb{Z}) \to (0 \to \mathbb{Z}/2\mathbb{Z})$$

is not a homotopy equivalence.

Remark F.85. Let $\mathcal{A}$ be an abelian category. A morphism $f \colon X \to Y$ in $C(\mathcal{A})$ is a quasi-isomorphism if and only if C_f is an exact complex.

(F.17) Double complexes.

Let $\mathcal{A}$ be an additive category.

Definition F.86. *A* double complex *in $\mathcal{A}$ is given by a family $(X^{i,j})_{i,j \in \mathbb{Z}}$ of objects in $\mathcal{A}$ with morphisms $d_1^{i,j} \colon X^{i,j} \to X^{i+1,j}$ (horizontal differential) and $d_2^{i,j} \colon X^{i,j} \to X^{i,j+1}$ (vertical differential) for all $i, j \in \mathbb{Z}$, such that for all $i, j \in \mathbb{Z}$ we have*

$$d_1^{i+1,j} \circ d_1^{i,j} = 0, \quad d_2^{i,j+1} \circ d_2^{i,j} = 0, \quad d_2^{i+1,j} \circ d_1^{i,j} = d_1^{i,j+1} \circ d_2^{i,j}.$$

Together with the obvious notion of morphisms of double complexes, *we obtain the category* $C^2(\mathcal{A})$ *of double complexes in* $\mathcal{A}$.

We visualize a double complex as follows.

(F.17.1)

$$
\begin{array}{ccccccc}
 & \cdots & \vdots & & \vdots & & \cdots \\
 & & \uparrow & & \uparrow & & \\
\cdots \longrightarrow & X^{i,j+1} & \xrightarrow{\ d_1^{i,j+1}\ } & X^{i+1,j+1} & \longrightarrow & \cdots \\
 & d_2^{i,j}\uparrow & & d_2^{i+1,j}\uparrow & & \\
\cdots \longrightarrow & X^{i,j} & \xrightarrow{\ d_1^{i,j}\ } & X^{i+1,j} & \longrightarrow & \cdots \\
 & \uparrow & & \uparrow & & \\
 & \cdots & \vdots & & \vdots & & \cdots
\end{array}
$$

Each column and each row of a double complex is a complex and we can view every double complex as complex of vertical or of horizontal complexes. Both point of views yield an equivalence of categories $C^2(\mathcal{A}) = C(C(\mathcal{A}))$. In particular $C^2(\mathcal{A})$ is an additive category. It is abelian if $\mathcal{A}$ is an abelian category.

Definition F.87. *Let X be a double complex in $\mathcal{A}$. Suppose that $\mathcal{A}$ admits countable direct sums or that for each k there only finitely many non-zero components $X^{i,j}$ with $i + j = k$. The* total complex *attached to X is the complex* $\mathrm{Tot}(X)$ *given by*

$$
\mathrm{Tot}(X)^k = \bigoplus_{i+j=k} X^{i,j}, \quad d_{|X^{i,j}} = d_1^{i,j} \oplus (-1)^i d_2^{i,j} \colon X^{i,j} \to X^{i+1,j} \oplus X^{i,j+1}.
$$

This definition is *not* symmetric in (i,j). We could have also chosen to introduce a sign in the horizontal direction.

Remark F.88. Clearly the construction of the total complex is functorial yielding a functor Tot from the full subcategory of double complexes X, such that for each k there only finitely many non-zero components $X^{i,j}$ with $i + j = k$, to the category of complexes $C(\mathcal{A})$.

For the rest of the section let us agree to identify $C^2(\mathcal{A}) = C_{\mathrm{vert}}(C_{\mathrm{hor}}(\mathcal{A}))$, i.e., we view a double complex X as a complex in vertical direction of horizontal complexes

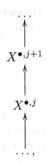

We obtain a fully faithful functor $\iota[j]\colon C(\mathcal{A}) \to C^2(\mathcal{A})$ by considering every (horizontal) complex as a double complex concentrated in vertical degree $-j$. By our choice of sign we find $\mathrm{Tot} \circ \iota[j] = \mathrm{id}_{C(\mathcal{A})}{}^2$.

Remark F.89. Suppose that $u\colon X \to Y$ is a morphism of double complexes which we view as a morphism of vertical complexes of horizontal complexes. Suppose that $(h^{\bullet,j}\colon X^{\bullet,j} \to Y^{\bullet,j-1})_j$ is a homotopy of u with 0, then one checks that

$$\bigoplus_{i+j=k} (-1)^i h^{i,j}\colon \mathrm{Tot}(X)^k \longrightarrow \mathrm{Tot}(Y)^{k-1}$$

yields a homotopy of $\mathrm{Tot}(u)$ with 0. In particular, Tot induces a functor

(F.17.2) $$\mathrm{Tot}\colon K^+(C^+(\mathcal{A})) \longrightarrow K^+(\mathcal{A}).$$

(F.18) The Hom complex.

Let $\mathcal{A}$ be an additive category. Let X and Y be complexes in $C(\mathcal{A})$. We have the complex of morphisms $\mathrm{Hom}_{\mathcal{A}}(X, Y)$,

$$\mathrm{Hom}_{\mathcal{A}}(X, Y)^i = \prod_{k\in\mathbb{Z}} \mathrm{Hom}_{\mathcal{A}}(X^k, Y^{k+i}),$$

$$d^i(f) = d_Y \circ f - (-1)^i f \circ d_X \qquad \text{for } f \in \mathrm{Hom}(X, Y)^i,$$

i.e., if $f = (f^k)_k \in \prod_{k\in\mathbb{Z}} \mathrm{Hom}_{\mathcal{A}}(X^k, Y^{k+i})$, then $d^i f \in \prod_{k\in\mathbb{Z}} \mathrm{Hom}_{\mathcal{A}}(X^k, Y^{k+i+1})$ is given by

$$(d^i f)^k = d_Y^{k+i} \circ f^k - (-1)^i f^{k+1} \circ d_X^k \in \mathrm{Hom}_{\mathcal{A}}(X^k, Y^{k+i+1}).$$

Then $\mathrm{Hom}_{\mathcal{A}}(X, Y)$ is a complex of abelian groups. This construction is functorial in X and in Y and we obtain a functor

(F.18.1) $$\mathrm{Hom}_{\mathcal{A}}(-, -)\colon C(\mathcal{A})^{\mathrm{opp}} \times C(\mathcal{A}) \longrightarrow C(\mathrm{AbGrp})$$

One easily checks that for all $i \in \mathbb{Z}$ one has

(F.18.2)
$$Z^i(\mathrm{Hom}_{\mathcal{A}}(X, Y)) = \mathrm{Hom}_{C(\mathcal{A})}(X, Y[i]),$$
$$H^i(\mathrm{Hom}_{\mathcal{A}}(X, Y)) = \mathrm{Hom}_{K(\mathcal{A})}(X, Y[i]).$$

Remark F.90. The functor (F.18.1) preserves homotopy in both variables ([BouA10]^{O} §5, Prop. 3) and hence induces a functor

(F.18.3) $$\mathrm{Hom}_{\mathcal{A}}(-, -)\colon K(\mathcal{A})^{\mathrm{opp}} \times K(\mathcal{A}) \longrightarrow K(\mathrm{AbGrp}).$$

---

[2] If we had viewed X as a complex in horizontal direction of vertical complexes, we would obtain inclusions $\iota'[i]\colon C(\mathcal{A}) \to C^2(\mathcal{A})$ with $\mathrm{Tot} \circ \iota'[i] = (-1)^i \mathrm{id}_{C(\mathcal{A})}$.

(F.19) Tensor product of complexes.

Assume that the additive category $\mathcal{A}$ has an associative and commutative tensor product. We will not make this precise (which can be done, see for instance [KaSh]^O, Chap. 4) because we will need the construction of tensor products of complexes only if $\mathcal{A}$ is the category of R-modules for some (commutative) ring R or the category of $\mathscr{O}_X$-modules for some ringed space $(X, \mathscr{O}_X)$, or some subcategory of these that is stable under the usual tensor product. In this case a tensor product is defined.

The *tensor product* $X \otimes Y$ *of complexes* X and Y in $C(\mathcal{A})$ is defined as the total complex of the double complex $X^i \otimes Y^j$ with differentials induced by those of X and Y. Explicitly, this means

$$(X \otimes Y)^n := \bigoplus_{i+j=n} X^i \otimes Y^j,$$

$$d^n_{|X^i \otimes Y^j} := d^i_X \otimes \mathrm{id}_{Y^j} + (-1)^i \, \mathrm{id}_{X^i} \otimes d^j_Y.$$

The complex $X \otimes Y$ exists if $\mathcal{A}$ admits countable direct sums or if for each n there are only finitely many (i,j) with $i + j = n$, $X^i \neq 0$, and $Y^j \neq 0$ (for instance if X and Y are both bounded below or are both bounded above). For the rest of the section we will always assume that all tensor products of complexes exist.

Remark F.91. The tensor product of complexes is functorial in both variables and we obtain a functor

$$- \otimes - : C(\mathcal{A}) \times C(\mathcal{A}) \longrightarrow C(\mathcal{A})$$

which is additive in each component. This functor preserves homotopy in both components ([BouA10]^O §4, Prop. 3) and hence induces a functor

(F.19.1) $$- \otimes - : K(\mathcal{A}) \times K(\mathcal{A}) \longrightarrow K(\mathcal{A})$$

Remark F.92. Let X, Y, and Z be complexes in $C(\mathcal{A})$.
(1) There is an isomorphism, functorial in X and Y

(F.19.2) $$X \otimes Y \xrightarrow{\sim} Y \otimes X$$

in $C(\mathcal{A})$ which is given on $X^i \otimes Y^j$ by $(-1)^{ij}\tau$, where $\tau : X^i \otimes Y^j \xrightarrow{\sim} Y^j \otimes X^i$ is given by the commutativity of the tensor product.
(2) There is an isomorphism, functorial in X, Y, and Z

(F.19.3) $$(X \otimes Y) \otimes Z \xrightarrow{\sim} X \otimes (Y \otimes Z)$$

in $C(\mathcal{A})$ whose restriction to $(X^i \otimes Y^j) \otimes Z^k$ is given by the associativity isomorphism $(X^i \otimes Y^j) \otimes Z^k \xrightarrow{\sim} X^i \otimes (Y^j \otimes Z^k)$ of the tensor product.

Let A be a commutative ring and let $C(A) := C(A\text{-Mod})$ be the category of complexes of A-modules. We denote by $K(A)$ its homotopy category. For complexes X and Y of A-modules the Hom complex $\mathrm{Hom}_A(X, Y)$ is a complex of A-modules and we obtain functors

(F.19.4)
$$\mathrm{Hom}_A(-, -) : C(A)^{\mathrm{opp}} \times C(A) \longrightarrow C(A)$$
$$\mathrm{Hom}_A(-, -) : K(A)^{\mathrm{opp}} \times K(A) \longrightarrow K(A).$$

Proposition F.93. ([BouA10]^{O} §5, 7) *Let A be a commutative ring. There is an isomorphism in $C(A)$*

(F.19.5) $$\operatorname{Hom}_A(X \otimes Y, Z) \xrightarrow{\sim} \operatorname{Hom}_A(X, \operatorname{Hom}_A(Y, Z)).$$

which is functorial in the complexes X, Y and Z of A-modules.

Remark F.94. Applying the functor Z^0 and H^0 to the isomorphism (F.19.5) we obtain by (F.18.2) functorial isomorphisms

(F.19.6)
$$\operatorname{Hom}_{C(A)}(X \otimes Y, Z) \xrightarrow{\sim} \operatorname{Hom}_{C(A)}(X, \operatorname{Hom}_A(Y, Z)),$$
$$\operatorname{Hom}_{K(A)}(X \otimes Y, Z) \xrightarrow{\sim} \operatorname{Hom}_{K(A)}(X, \operatorname{Hom}_A(Y, Z)).$$

Therefore the functor $X \mapsto X \otimes Y$ from $C(A)$ to $C(A)$ (resp. from $K(A)$ to $K(A)$) is left adjoint to the functor $Z \mapsto \operatorname{Hom}_A(Y, Z)$.

One has the following Künneth formula for the tensor product of complexes of modules.

Proposition F.95. *Let A be a ring, and let $E^\bullet$, $F^\bullet$ be complexes of A-modules and suppose that $Z^n(E)$ and $B^n(E)$ are flat A-modules for all $n \in \mathbb{Z}$.*
(1) For every $n \in \mathbb{Z}$ there is a functorial exact sequence

$$0 \longrightarrow \bigoplus_{i+j=n} H^i(E^\bullet) \otimes_A H^j(F^\bullet) \longrightarrow H^n(E^\bullet \otimes_A F^\bullet)$$

$$\longrightarrow \bigoplus_{i+j=n+1} \operatorname{Tor}_1^A(H^i(E), H^j(F)) \longrightarrow 0.$$

If in addition $B^n(E)$ and $B^n(F)$ are projective for all $n \in \mathbb{Z}$, then the exact sequence is (non-functorially) split.
(2) Suppose that $Z^n(E)$, $B^n(E)$ and $H^n(E)$ are flat A-modules for all $n \in \mathbb{Z}$. Then for all $n \in \mathbb{Z}$ there is a functorial isomorphism

$$\bigoplus_{i+j=n} H^i(E^\bullet) \otimes_A H^j(F^\bullet) \xrightarrow{\sim} H^n(E^\bullet \otimes_A F^\bullet).$$

For the proof we refer to [BouA10]^{O} §4.7.

Spectral sequences

We now give a very short introduction to spectral sequences tailored to our purposes. Definitions, constructions, and levels of generality that we will not need are usually not mentioned, even if they are very important in other areas of mathematics (such as topology). In the first three sections we start by defining spectral sequences and explain some information one gets from them such as certain exact sequences.

In the next three sections we will explain how one obtains spectral sequences, first from exact couples, then from filtered complexes, and finally from double complexes. Essentially each of these constructions is a special case of the previous construction.

(F.20) Graded and Filtered objects.

Definition F.96. *Let $\mathcal{A}$ be an additive category.*
(1) *A graded object in $\mathcal{A}$ is a family $(X^i)_{i \in \mathbb{Z}}$ of objects $X^i \in \mathcal{A}$.*
(2) *Let $d \in \mathbb{Z}$. A morphism $f \colon X = (X^i) \to Y = (Y^i)$ of graded objects of degree d is a family $(f^i)_{i \in \mathbb{Z}}$ of morphism $f^i \colon X^i \to Y^{i+d}$.*

A *bi-graded object* in $\mathcal{A}$ is a graded object in the category of graded objects of $\mathcal{A}$, i.e., it is a family $(X^{p,q})_{p,q \in \mathbb{Z}}$ of objects $X^{p,q} \in \mathcal{A}$. For $d, e \in \mathbb{Z}$ a family of morphisms $f^{p,q} \colon X^{p,q} \to Y^{p+d,q+e}$ is called a *morphism of bi-degree (d, e)*.

(Bi-)Graded objects and morphisms of (bi-)graded objects of degree 0 (resp. bi-degree $(0,0)$) form an additive category which is abelian if $\mathcal{A}$ is abelian (see also Exercise F.10).

Definition F.97. *Let $\mathcal{A}$ be an additive category and let X be an object of $\mathcal{A}$. A descending filtration on X is a family of subobjects*

$$X \supseteq \cdots \supseteq F^{p-1}X \supseteq F^p X \supseteq F^{p+1}X \supseteq \cdots$$

We call such a filtration finite *if there exist $p \leq p'$ with $F^p X = X$ and $F^{p'}X = 0$. A filtered object of $\mathcal{A}$ is an object of $\mathcal{A}$ together with a descending filtration.*

A morphism $f \colon (X, (F^p X)) \to (Y, (F^p Y))$ of filtered objects is a morphism $f \colon X \to Y$ such that $f_{|F^p X}$ factors through $F^p Y$ for all $p \in \mathbb{Z}$.

Suppose that $\mathcal{A}$ is an abelian category. Then we define for a filtered object $(F^p X)_p$ the associated graded object $(\mathrm{gr}_F^p(X))_{p \in \mathbb{Z}}$ by

$$\mathrm{gr}^p X := \mathrm{gr}_F^p(X) := F^p X / F^{p+1} X.$$

One obtains the category of filtered objects of $\mathcal{A}$. This category is usually not abelian even if $\mathcal{A}$ is abelian (Exercise F.11). If $\mathcal{A}$ is abelian, $(F^p X)_p \mapsto (\mathrm{gr}_F^p(X))_{p \in \mathbb{Z}}$ defines a functor from the category of filtered objects in $\mathcal{A}$ to the category of graded objects in $\mathcal{A}$.

Of course, one has analogous definitions for ascending filtrations.

(F.21) Definition of spectral sequences.

In this section, $\mathcal{A}$ always denotes an abelian category.

Definition F.98. *Let $r_0 \in \mathbb{Z}$.*
(1) *A spectral sequence (of cohomological type, starting at level r_0) consists of a family $(E_r, d_r)_{r \geq r_0}$ of bi-graded objects $E_r = (E_r^{p,q})_{p,q \in \mathbb{Z}}$ in $\mathcal{A}$ and morphisms $d_r \colon E_r \to E_r$ of bi-degree $(r, 1-r)$ such that $d_r \circ d_r = 0$ and of isomorphisms of bi-graded objects of bi-degree $(0, 0)$*

$$\mathrm{Ker}(d_r)/\mathrm{Im}(d_r) \overset{\sim}{\longrightarrow} E_{r+1}$$

for all $r \geq r_0$.
(2) *A morphism of spectral sequences $f \colon (E_r, d_r)_{r \geq r_0} \to (E_r', d_r')_{r \geq r_0}$ is given by a family of morphisms $f_r \colon E_r \to E_r'$ of bi-degree $(0, 0)$ such that $f_r \circ d_r = d_r' \circ f_r$ and such that f_{r+1} is the morphism induced by f_r via the given isomorphisms $\mathrm{Ker}(d_r)/\mathrm{Im}(d_r) \cong E_{r+1}$ and $\mathrm{Ker}(d_r')/\mathrm{Im}(d_r') \cong E_{r+1}'$.*

Remark F.99. Let $(E_r, d_r)_{r \geq r_0}$ be a spectral sequence.

(1) Each level E_r and d_r of a spectral sequence determines E_{r+1} up to isomorphism but it does not determine d_{r+1}.

(2) If $(E_r, d_r)_{r \geq r_0}$ is a spectral sequence starting at level r_0 and if $r_0' \geq r_0$ is an integer, then $(E_r, d_r)_{r \geq r_0'}$ is a spectral sequence starting at level r_0'.

Let $(E_r, d_r)_{r \geq r_0}$ be a spectral sequence. We define inductively bi-graded subobjects

$$0 = B_{r_0} \subseteq B_{r_0+1} \subseteq \ldots \subseteq Z_{r_0+1} \subseteq Z_{r_0} = E_{r_0}$$

as follows:

$$
\begin{aligned}
& B_{r_0} := 0, \qquad\qquad Z_{r_0} := E_{r_0}, \\
\text{(F.21.1)} \qquad & B_{r+1}/B_r := \mathrm{Im}(d_r \colon Z_r/B_r \to Z_r/B_r), \\
& Z_{r+1}/B_r := \mathrm{Ker}(d_r \colon Z_r/B_r \to Z_r/B_r),
\end{aligned}
$$

i.e., B_{r+1} is the unique subobject of Z_r containing B_r such that B_{r+1}/B_r is the image of d_r, and Z_{r+1} is the unique subobject of Z_r containing B_r such that Z_{r+1}/B_r is the kernel of d_r. Hence for all $r \geq r_0$ one has an isomorphism of bi-graded objects

$$Z_r/B_r \cong E_r$$

and a short exact sequence

$$0 \longrightarrow Z_{r+1}/B_r \longrightarrow Z_r/B_r \xrightarrow{d_r} B_{r+1}/B_r \longrightarrow 0.$$

If there exists a smallest bi-graded subobject $\bigcup_{r \geq r_0} B_r$ of E_{r_0} containing all B_r, we call this subobject B_∞. Similarly, if there exists a largest bi-graded subobject $\bigcap_{r \geq r_0} Z_r$ contained in all Z_r, we call this subobject Z_∞. We obtain a bi-graded object

$$E_\infty := Z_\infty/B_\infty.$$

From now on we will always assume that B_∞ and Z_∞ exist. This is for instance the case, if the abelian category $\mathcal{A}$ has countable direct sums and countable products. Then $B_\infty = \mathrm{Im}(\bigoplus_r B_r \to E_{r_0})$ and $Z_\infty = \mathrm{Ker}(E_{r_0} \to \prod_r E_{r_0}/Z_r)$. If filtered colimits are exact, we also have $B_\infty = \mathrm{colim}_r B_r$.

Remark/Definition F.100. A spectral sequence $(E_r, d_r)_{r \geq r_0}$ is said to *degenerate at* E_r if $d_k = 0$ for all $k \geq r$. In this case all ∞-terms exist and one has

$$B_r = B_{r+1} = \cdots = B_\infty, \qquad Z_r = Z_{r+1} = \cdots = Z_\infty, \qquad E_r \cong E_{r+1} \cong \ldots \cong E_\infty.$$

Definition/Remark F.101. Let $(E_r, d_r)_{r \geq r_0}$ be a spectral sequence in $\mathcal{A}$. We say that the spectral sequence is *bounded* if for each $n \in \mathbb{Z}$ there exists only finitely many $E_{r_0}^{pq} \neq 0$ with $p + q = n$.

Suppose that the spectral sequence is bounded. As E_r^{pq} is isomorphic to a subquotient of $E_{r_0}^{pq}$ for $r \geq r_0$, there exist for all $p, q \in \mathbb{Z}$ an integer $r(p, q) \geq 0$ such that both maps

$$d_r^{p-r,q-1+r} \colon E_r^{p-r,q-1+r} \longrightarrow E_r^{pq}, \qquad d_r^{pq} \colon E_r^{pq} \to E_r^{p+r,q+1-r}$$

are zero for all $r \geq r(p, q)$. Hence for all $r \geq r(p, q)$ one has

$$B_r^{pq} = B_{r+1}^{pq} = \cdots = B_\infty^{pq}, \qquad Z_r^{pq} = Z_{r+1}^{pq} = \cdots = Z_\infty^{pq}, \qquad E_r^{pq} = E_{r+1}^{pq} = \cdots = E_\infty^{pq}.$$

We say that the spectral sequence is a *first* (resp. *third*) quadrant spectral sequence if $E_{r_0}^{pq} = 0$ if $p < 0$ or $q < 0$ (resp. if $p > 0$ or $q > 0$). Such spectral sequences are bounded.

Definition F.102. Let $(E_r, d_r)_{r \geq r_0}$ be a bounded spectral sequence in $\mathcal{A}$. Let $(H^n)_{n \in \mathbb{Z}}$ be a graded object in $\mathcal{A}$ endowed with a descending filtration

$$\cdots \supseteq F^p H^n \supseteq F^{p+1} H^n \supseteq \cdots$$

We assume that for each $n \in \mathbb{Z}$ the filtration $(F^p H^n)_p$ is finite (i.e., there exist $p \leq p'$ (depending on n) such that $F^p H^n = H^n$ and $F^{p'} H^n = 0$).

Then we say that the *spectral sequence* $(E_r, d_r)_{r \geq r_0}$ *converges to* $(H^n, F^p H^n)_{n,p \in \mathbb{Z}}$ and write

$$E_{r_0}^{pq} \Rightarrow H^{p+q}$$

if we are given for all $p, q \in \mathbb{Z}$ an isomorphism

(F.21.2) $$E_\infty^{pq} \xrightarrow{\sim} \mathrm{gr}_F^p H^{p+q} = F^p H^{p+q} / F^{p+1} H^{p+q}.$$

The filtered graded object $(H^n, F^p H^n)_{n,p}$ is then also called the *limit of the spectral sequence*.

A *morphism* $(E_r, d_r, F^\bullet H^n) \to (\tilde{E}_r, \tilde{d}_r, F^\bullet \tilde{H}^n)$ *of spectral sequences with limit term* is a pair (f, g), where $f \colon (E_r, d_r) \to (\tilde{E}_r, \tilde{d}_r)$ is a morphism of bounded spectral sequences, $g \colon (H^n, F^\bullet H^n)_n \to (\tilde{H}^n, F^\bullet \tilde{H}^n)_n$ is a morphism of filtered graded objects such that for all $p, q \in \mathbb{Z}$ the following diagram commutes

$$\begin{array}{ccc} E_\infty^{pq} & \xrightarrow{\cong} & \mathrm{gr}_F^p H^{p+q} \\ \downarrow{\scriptstyle f_\infty^{pq}} & & \downarrow{\scriptstyle \mathrm{gr}_F^p(g^{p+q})} \\ \tilde{E}_\infty^{pq} & \xrightarrow{\cong} & \mathrm{gr}_F^p \tilde{H}^{p+q}, \end{array}$$

where the horizontal morphisms are the isomorphisms (F.21.2).

Remark F.103. Let $f \colon (E_r^{p,q}) \to (\tilde{E}_r^{p,q})$ be a morphism of spectral sequences. Suppose there exists r_0 such that f induces an isomorphism $E_{r_0}^{p,q} \xrightarrow{\sim} \tilde{E}_{r_0}^{p,q}$ for all p and q. Then the Five Lemma implies that f induces isomorphisms $E_s^{p,q} \xrightarrow{\sim} \tilde{E}_s^{p,q}$ for all $s \geq r_0$ and all p and q.

Suppose that both spectral sequences converge to (H^n) and $(\tilde{H}^n)$ respectively and that (f, g) is a morphism of spectral sequences with limit term, then g also induces an isomorphism of graded objects $(H^n)_n \xrightarrow{\sim} (\tilde{H}_n)_n$, again by the Five Lemma.

(F.22) Exact sequences attached to spectral sequences.

Let $\mathcal{A}$ be an abelian category. For each spectral sequence $(E_r, d_r)_{r \geq r_0}$ we continue to assume that the ∞-terms B_∞, Z_∞, and E_∞ exist (automatically satisfied if $\mathcal{A}$ has countable products and countable direct sums).

Remark/Definition F.104. Let $(E_r, d_r)_{r \geq r_0}$ be a spectral sequence.

(1) Let $(p, q) \in \mathbb{Z}$ and suppose that $d_r^{pq} = 0$ for all $r \geq r_0$. Then $Z_r^{pq} = E_{r_0}^{pq}$ for all $r \geq r_0$ and one obtains a sequence of quotient maps

$$E_{r_0}^{pq} = Z_{r_0}^{pq}/B_{r_0}^{pq} \twoheadrightarrow Z_{r_0}^{pq}/B_{r_0+1}^{pq} \twoheadrightarrow Z_{r_0}^{pq}/B_{r_0+2}^{pq} \twoheadrightarrow \ldots \twoheadrightarrow Z_\infty^{pq}/B_\infty^{pq} = E_\infty^{pq}.$$

(2) Dually, if the differentials $d_r^{p-r,q-1+r}$ with target E_r^{pq} are all zero, then $B_r^{pq} = 0$ for all $r \geq r_0$ and one obtains a sequence of monomorphisms

$$E_\infty^{pq} = Z_\infty^{pq} \hookrightarrow \ldots \hookrightarrow Z_{r_0+1}^{pq} \hookrightarrow Z_{r_0}^{pq} = E_{r_0}^{pq}$$

(3) Suppose that $r_0 \geq 2$ (i.e., all differentials go right and down) and that $E_{r_0}^{pq} = 0$ for all $q < 0$ (i.e., the spectral sequence is concentrated in the first and second quadrant). Then $d_r^{p,0} = 0$ for all $r \geq r_0$ and $p \in \mathbb{Z}$ and we obtain for all $p \in \mathbb{Z}$ quotient maps

(F.22.1) $$E_{r_0}^{p,0} \twoheadrightarrow E_\infty^{p,0}.$$

If the spectral sequence converges to $(H^n)_{n \in \mathbb{Z}}$, then H^p has a filtration with $E_\infty^{p,0}$ as last graded component (because $E_\infty^{-t,t} = 0$ for $t > 0$). Thus $E_\infty^{p,0}$ is a subobject of H^p and the composition with (F.22.1) yields a morphism

(F.22.2) $$E_{r_0}^{p,0} \twoheadrightarrow E_\infty^{p,0} \hookrightarrow H^p,$$

which is called *edge morphism.*

(4) Dually, suppose that $r_0 \geq 1$ (i.e., all differentials go right and not up), that $E_{r_0}^{pq} = 0$ for all $p < 0$, and that the spectral sequence converges to $(H^n)_{n \in \mathbb{Z}}$. Then one obtains for all $q \in \mathbb{Z}$ an *edge morphism* as composition

(F.22.3) $$H^q \twoheadrightarrow E_\infty^{0,q} \hookrightarrow E_{r_0}^{0,q}.$$

The hypotheses in (3) and (4) are both satisfied if $r_0 \geq 2$ and $(E_r, d_r)_{r \geq r_0}$ is a first quadrant spectral sequence.

Note that in (3) the same arguments show similar, notationally slightly more complicated results if one supposes that there exists $q_0 \in \mathbb{Z}$ with $E_{r_0}^{pq} = 0$ for all $q < q_0$. A similar remark holds for (4).

Proposition F.105. (Exact sequence of low degrees, [CaEi]$^\circ$ XV, 5.12) *Let $(E_r, d_r)_{r \geq 2}$ be a first quadrant spectral sequence converging against $(H^n)_n$.*

(1) *Then the sequence*

(F.22.4) $$0 \longrightarrow E_2^{1,0} \xrightarrow{e} H^1 \xrightarrow{e} E_2^{0,1} \xrightarrow{d_2^{0,1}} E_2^{2,0} \xrightarrow{e} H^2$$

is exact. Here all arrows labeled with e are edge morphisms.

(2) *More generally, let $n > 0$ and suppose that $E_2^{p,q} = 0$ for all $0 < q < n$ and for all p. Then for all $i < n$ the edge morphisms are isomorphisms*

$$E_2^{i,0} \xrightarrow{\sim} H^i$$

and the sequence

(F.22.5) $$0 \longrightarrow E_2^{n,0} \xrightarrow{e} H^n \xrightarrow{e} E_2^{0,n} \xrightarrow{\delta_2^{0,n}} E_2^{n+1,0} \xrightarrow{e} H^{n+1}$$

is exact, where the arrows labeled with e are edge morphisms and where $\delta_2^{0,n}$ is the composition

$$E_2^{0,n} \xrightarrow{\sim} E_3^{0,n} \xrightarrow{\sim} \cdots \xrightarrow{\sim} E_{n+1}^{0,n} \xrightarrow{d_{n+1}^{0,n}} E_{n+1}^{n+1,0} \xrightarrow{\sim} \cdots \xrightarrow{\sim} E_2^{n+1,0}.$$

(F.23) Spectral sequences associated to exact couples.

Let $\mathcal{A}$ be an abelian category. In the constructions below we will form certain limits and colimits in $\mathcal{A}$. We always assume that these exist. As we usually work with complete and cocomplete abelian categories, for instance if $\mathcal{A}$ is a Grothendieck abelian category (Remark F.58), this will be harmless for us.

Definition F.106. *An* exact couple *in $\mathcal{A}$ is a diagram in $\mathcal{A}$ of the form*

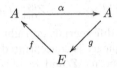

such that $\mathrm{Ker}(\alpha) = \mathrm{Im}(f)$, $\mathrm{Ker}(f) = \mathrm{Im}(g)$ *and* $\mathrm{Ker}(g) = \mathrm{Im}(\alpha)$. *A morphism of exact couples* $s\colon (A, E, \alpha, f, g) \to (\tilde{A}, \tilde{E}, \tilde{\alpha}, \tilde{f}, \tilde{g})$ *is a pair of morphisms* $t_A\colon A \to \tilde{A}$ *and* $t_E\colon E \to \tilde{E}$ *such that* $\tilde{\alpha} \circ t_A = t_A \circ \alpha$, $\tilde{f} \circ t_E = t_A \circ f$, *and* $\tilde{g} \circ t_A = t_E \circ g$.

Composition of morphisms of exact couples is defined in the obvious way and we obtain the category of exact couples in $\mathcal{A}$.

To every exact couple we can attach its derived exact couple as follows.

Remark F.107. Let (A, E, α, f, g) be an exact couple in $\mathcal{A}$. Set

$$d := g \circ f\colon E \to E.$$

As $f \circ g = 0$, one has $d \circ d = 0$. Set

$$E' := \mathrm{Ker}(d)/\mathrm{Im}(d), \qquad \text{and} \qquad A' := \mathrm{Im}(\alpha) = \mathrm{Ker}(g).$$

Let $\alpha'\colon A' \to A'$ be the map induced by α. As f is zero on $\mathrm{Im}(d)$ and sends $\mathrm{Ker}(d)$ to $\mathrm{Ker}(g)$, it induces a map $f'\colon E' \to A'$. Finally, let $g'\colon A' \to E'$ be the map induced by "$g \circ \alpha^{-1}$" using that $g(\mathrm{Ker}(\alpha)) = \mathrm{Im}(d)$. Then it is easily checked that $(A', E', \alpha', f', g')$ is again an exact couple which is called the *derived exact couple* of (A, E, α, f, g).

We now explain how to attach a spectral sequence to a bigraded exact couple.

Remark F.108. Let $(A, E, \alpha, f, g) = (A_1, E_1, \alpha_1, f_1, g_1)$ be an exact couple and define inductively for $r > 1$ an exact couple $(A_r, E_r, \alpha_r, f_r, g_r)$ as the derived exact couple of $(A_{r-1}, E_{r-1}, \alpha_{r-1}, f_{r-1}, g_{r-1})$. Then we have $A_r = \mathrm{Im}(\alpha^{r-1})$, where $\alpha^s := \alpha \circ \cdots \circ \alpha$ (s times) with $\alpha^0 := \mathrm{id}_A$.

Suppose also that (A, E, α, f, g) is equipped with a bigrading such that

$$\deg(\alpha) = (-1, 1), \qquad \deg(f) = (1, 0), \qquad \deg(g) = (0, 0).$$

Hence $\deg(d) = (1, 0)$, where $d := g \circ f$. All the derived exact couples $(A_r, E_r, \alpha_r, f_r, g_r)$ are endowed with an induced bigrading with $\deg(\alpha_r) = (-1, 1)$, $\deg(f_r) = (1, 0)$, $\deg(g_r) = (r - 1, 1 - r)$, and $\deg(d_r) = (r, 1 - r)$, where $d_r := g_r \circ f_r$.

Hence we obtain a spectral sequence $(E_r, d_r)_{r \geq 1}$ with

(F.23.1) $B_r = g(\mathrm{Ker}(\alpha^{r-1}))$, $Z_r = f^{-1}(\mathrm{Im}(\alpha^{r-1}))$, $d_r = \text{``}g \circ \alpha^{-(r-1)} \circ f\text{''}$.

The construction of a spectral sequence from an exact couple is functorial.

Let us consider the convergence of a spectral sequence attached to an exact couple. We give two examples (dual to each other) of possible criteria.

Proposition F.109. *Let* (A, E, α, f, g) *be an exact couple equipped with a bigrading as above and let* $(E_r, d_r)_{r \geq 1}$ *be the attached spectral sequence.*

(1) *Assume that for all* $n \in \mathbb{Z}$ *the map* $\alpha \colon A^{s,n-s} \to A^{s-1,n-s+1}$
  (a) *is zero for* $s \gg 0$ *(dependent on* n*) and*
  (b) *is an isomorphism for* $s \ll 0$ *(dependent on* n*).*
  Then the spectral sequence attached to the exact couple converges to

$$H^n := \operatorname*{colim}_s(\cdots \xrightarrow{\alpha} A^{s,n-s} \xrightarrow{\alpha} A^{s-1,n-s+1} \xrightarrow{\alpha} \cdots),$$

  i.e., if one identifies $A^{s,n-s} = A^{s-1,n-s+1} = \cdots$ *for large* s *via the isomorphism* α*, then* H^n *is this object. Its filtration is given by*

$$F^p H^n := \operatorname{Im}(A^{p,n-p} \to H^n).$$

(2) *Assume that for all* $n \in \mathbb{Z}$ *the map* $\alpha \colon A^{s,n-s} \to A^{s-1,n-s+1}$
  (a) *is an isomorphism for* $s \gg 0$ *(dependent on* n*) and*
  (b) *is zero for* $s \ll 0$ *(dependent on* n*).*
  Then the spectral sequence attached to the exact couple converges to

$$H^n := \lim_s(\cdots \xrightarrow{\alpha} A^{s,n-s} \xrightarrow{\alpha} A^{s-1,n-s+1} \xrightarrow{\alpha} \cdots).$$

Its filtration is given by

$$F^p H^n := \operatorname{Ker}(H^n \to A^{p,n-p}).$$

Proof. We only show (1), as the proof of (2) is dual. Using (F.23.1), Condition (a) implies that for all $p, q \in \mathbb{Z}$ there exists a ρ such that

$$(F.23.2) \qquad Z_r^{pq} = \operatorname{Ker}(f \colon E^{pq} \to A^{p+1,q}) \qquad \text{for all } r \geq \rho.$$

In particular $Z_\rho^{pq} = Z_{\rho+1}^{pq} = \cdots$. Similarly, Condition (b) implies that for all $p, q \in \mathbb{Z}$ there exists ρ such that

$$(F.23.3) \qquad g(\operatorname{Ker}(\alpha^{\rho-1})) \cap E^{pq} = B_\rho^{pq} = B_{\rho+1}^{pq} = \cdots.$$

We fix $\rho = \rho_{p,q}$ such that (F.23.2) and (F.23.3) both hold. Then with $\rho = \rho_{p,n-p}$ we have

$$\begin{aligned}
E_\infty^{p,n-p} &= Z_\infty^{p,n-p}/B_\infty^{p,n-p} = \left(\operatorname{Ker}(f)/g(\operatorname{Ker}(\alpha^{\rho-1}))\right)^{p,n-p} \\
&= \left(\operatorname{Im}(g)/g(\operatorname{Ker}(\alpha^{\rho-1}))\right)^{p,n-p} \\
&= A^{p,n-p}/\left(\operatorname{Ker}(\alpha^{\rho-1}) + \operatorname{Ker}(g)\right)^{p,n-p} \\
&= A^{p,n-p}/\left(\operatorname{Ker}(\alpha^{\rho-1}) + \operatorname{Im}(\alpha)\right)^{p,n-p}.
\end{aligned}$$

By enlarging ρ we may assume that $\alpha \colon A^{p-t,n-p+t} \to A^{p-t-1,n-p+t+1}$ is an isomorphism for $t \geq \rho$ using Condition (b). Then $A^{p,n-p}/\left(\operatorname{Ker}(\alpha^{\rho-1})\right)^{p,n-p}$ is the image of $A^{p,n-p}$ in H^n and hence

$$E_\infty^{p,n-p} \cong F^p H^n/F^{p+1} H^n. \qquad \square$$

(F.24) Spectral sequences associated to filtered complexes.

We continue to denote by $\mathcal{A}$ a complete and cocomplete abelian category. In addition we assume (for simplicity) that filtered colimits are exact in $\mathcal{A}$. All these hypotheses are satisfied is $\mathcal{A}$ is a Grothendieck abelian category.

Definition F.110. *A filtered complex X in $\mathcal{A}$ is a complex in $\mathcal{A}$ endowed with a descending filtration in the category of complexes of $\mathcal{A}$.*

If X is a filtered complex, each filtration step is a subcomplex $F^p X$ of X. The quotient $gr^p X := F^p X / F^{p+1} X$ is a complex of $\mathcal{A}$.

Remark F.111. Let X be a filtered complex. We attach an exact couple to X as follows. The exact sequence of complexes

$$0 \longrightarrow F^{p+1} X \longrightarrow F^p X \longrightarrow gr^p X \longrightarrow 0$$

yields a long exact cohomology sequence

$$\cdots \longrightarrow H^{p+q}(F^{p+1} X) \longrightarrow H^{p+q}(F^p X) \longrightarrow H^{p+q}(gr^p X) \longrightarrow H^{p+q+1}(F^{p+1} X) \longrightarrow \cdots .$$

Combining the morphisms of this sequence gives us a bigraded exact couple

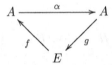

where

$$E^{pq} := H^{p+q}(gr^p X) \qquad\qquad A^{pq} := H^{p+q}(F^p X).$$

Then α has degree $(-1, 1)$, f has degree $(1, 0)$, and g has degree $(0, 0)$.

Hence the construction of Remark F.108 yields a spectral sequence $(E_r, d_r)_{r \geq 1}$ with

$$E_1^{pq} = H^{p+q}(gr^p X).$$

The convergence assumptions for Proposition F.109 (1) are in this case satisfied if for all $n \in \mathbb{Z}$

(a) there exists $s_1 \in \mathbb{Z}$ such that $H^n(F^s X) = 0$ for all $s \geq s_1$ and

(b) there exists $s_0 \in \mathbb{Z}$ such that $H^n(F^s X) \to H^n(F^{s-1} X)$ is an isomorphism for all $s \leq s_0$.

This is for instance the case (even uniformly for all n) if there exist $s_0 \leq s_1$ such that $F^{s_0} X = X$ and $F^{s_1} X = 0$. Then the spectral sequence converges to

$$\operatorname*{colim}_{s \in \mathbb{Z}^{\mathrm{opp}}} H^n(F^s X) = H^n\Big(\bigcup_s F^s X\Big)$$

because filtered colimits are assumed to be exact. Its filtration is given by

$$F^p H^n = \operatorname{Im}\Big(H^n(F^p X) \to H^n\big(\bigcup_s F^s X\big)\Big).$$

(F.25) Spectral sequences associated to double complexes.

We continue to denote by $\mathcal{A}$ a complete and cocomplete abelian category in which filtered colimits are exact, e.g., a Grothendieck abelian category.

Let X be a double complex with components in $\mathcal{A}$ which we visualize as in (F.17.1). Let $\text{Tot}(X)$ be the associated total complex. There are two natural filtrations on $\text{Tot}(X)$ given by

$$(F.25.1) \qquad {}_I F^p(\text{Tot}(X)^k) := \bigoplus_{\substack{i+j=k \\ i \geq p}} X^{i,j}, \qquad {}_{II} F^p(\text{Tot}(X)) := \bigoplus_{\substack{i+j=k \\ j \geq p}} X^{i,j}$$

By (F.24) we obtain two spectral sequences $({}_I E_r^{p,q}(X), {}_I d_r)_{r \geq 1}$ and $({}_{II} E_r^{p,q}(X), {}_{II} d_r)_{r \geq 1}$ attached to the double complex. These spectral sequences define for all $n \in \mathbb{Z}$ two filtrations on $H^n(\text{Tot}(X))$ which are denoted by ${}_I F$ and ${}_{II} F$. These spectral sequences are functorial in X.

Viewing X as a complex in horizontal direction of vertical complexes we can take its cohomology

$$H_I^p(X) := H^p(\cdots \to X^{i,\bullet} \to X^{i+1,\bullet} \to \dots),$$

which is a (vertical) complex. Similarly we define the cohomology $H_{II}^q(X)$ of X considered as complex in vertical direction of horizontal complexes. Then one checks that

$$(F.25.2) \qquad \begin{aligned} {}_I E_1^{p,q}(X) &= H^q(X^{p,\bullet}), & {}_{II} E_1^{p,q}(X) &= H^q(X^{\bullet,p}), \\ {}_I d_1^{p,q} &= H^q(d_1^{p,\bullet}), & {}_{II} d_1^{p,q} &= (-1)^q H^q(d_2^{\bullet,p}), \\ {}_I E_2^{p,q}(X) &= H^p(H_{II}^q(X)), & {}_{II} E_2^{p,q}(X) &= H^p(H_I^q(X)). \end{aligned}$$

Then (F.24) shows:

Lemma F.112. *Suppose that X is a double complex such that for all $n \in \mathbb{Z}$ there are only finitely many nonzero $X^{p,q}$ with $p + q = n$. Then both spectral sequences associated to X are bounded, converge to $H^*(\text{Tot}(X))$, and the two filtrations of $H^n(\text{Tot}(X))$ given by the spectral sequences are finite.*

As a trivial application (which one could also see directly) we will see that forming the total complex "preserves quasi-isomorphisms in each direction":

Lemma F.113. *Let $\mathcal{A}$ be an abelian category. Let X and Y be double complexes such that for all $k \in \mathbb{Z}$ there are only finitely many nonzero $X^{i,j}$ and $Y^{i,j}$ with $i + j = k$. Let $f \colon X \to Y$ be a morphism of double complexes which is a quasi-isomorphism of complexes of horizontal complexes, i.e.*

$$(\dots \to X^{\bullet,j} \to X^{\bullet,j+1} \to \dots) \xrightarrow{f} (\dots \to Y^{\bullet,j} \to Y^{\bullet,j+1} \to \dots)$$

is a quasi-isomorphism of complexes of complexes. Then $\text{Tot}(f) \colon \text{Tot}(X) \to \text{Tot}(Y)$ is a quasi-isomorphism.

Proof. By (F.25.2), f induces an isomorphism ${}_I E_1^{p,q}(X) \to {}_I E_1^{p,q}(Y)$. Hence we conclude by Remark F.103. $\qquad\qquad\square$

In the abelian category of modules over a ring there is also the following variant, see [Sta] 091Z.

Lemma F.114. *Let R be a (not necessarily commutative) ring, let M be a complex of left R-modules and let*

$$\cdots \longrightarrow P_2 \longrightarrow P_1 \longrightarrow P_0 \longrightarrow M \longrightarrow 0,$$

be an exact sequence of complexes of left R-modules such that for all $i \in \mathbb{Z}$ the sequence of left R-modules

$$\cdots \longrightarrow \operatorname{Ker}(d^i_{P_2}) \longrightarrow \operatorname{Ker}(d^i_{P_1}) \longrightarrow \operatorname{Ker}(d^i_{P_0}) \longrightarrow \operatorname{Ker}(d^i_M) \longrightarrow 0$$

is exact. Set $P^{i,j} := P^j_{-i}$ to obtain a double complex P. Then the morphism $\operatorname{Tot}(P) \to M$ induced by $P_0 \to M$ is a quasi-isomorphism.

Triangulated categories

Our final goal in this chapter is to introduce derived categories of an abelian category and derived functors. Derived categories will be additive categories endowed with the structure of a triangulated category and derived functors will be triangulated functors between triangulated categories. In this part we these notions. Our main references are [KaSh] $^\circ$ Ch. 10, [Lip2] $^{\text{O}}_{\text{X}}$ Ch. 1, and [Nee4] $^\circ$.

(F.26) Definition of triangulated categories.

Definition F.115.
(1) *An* additive category with translation *is a pair $(\mathcal{C}, T)$ consisting of an additive category $\mathcal{C}$ and an automorphism $T : \mathcal{C} \to \mathcal{C}$ (which is called the* translation functor*).*
(2) *A* functor of additive categories with translation $F : (\mathcal{C}, T) \to (\mathcal{C}', T')$ *is an additive functor $F : \mathcal{C} \to \mathcal{C}'$ together with an isomorphism $F \circ T \xrightarrow{\sim} T' \circ F$ of functors.*

As an automorphism a translation functor commutes with finite products, hence T is automatically additive.

Although the isomorphism $F \circ T \cong T' \circ F$ is not in general the identity, we will usually neglect it in the discussion, implicitly asserting that it can be chosen suitably.

For $i \in \mathbb{Z}$ the functor T^i is defined. We usually write $X[i] := T^i(X)$, and similarly for morphisms.

Our principal examples of an additive category with translations will be the category $C(\mathcal{A})$ of complexes with components in an additive category $\mathcal{A}$ (Definition F.68), its homotopy category $K(\mathcal{A})$ (Definition F.76), and for an abelian category $\mathcal{A}$ the derived category $D(\mathcal{A})$ (Definition F.148 below).

Definition F.116.
(1) *Let $(\mathcal{C}, T)$ be an additive category with translation. A* triangle *in $\mathcal{C}$ is a sequence of morphisms*

$$X \longrightarrow Y \longrightarrow Z \longrightarrow X[1].$$

This is sometimes written as

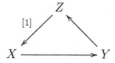

(hence the name triangle). We also often write

$$X \longrightarrow Y \longrightarrow Z \xrightarrow{+1}$$

or even just $X \to Y \to Z$.

(2) *A morphism of triangles is a commutative diagram*

$$
\begin{array}{ccccccc}
X & \longrightarrow & Y & \longrightarrow & Z & \longrightarrow & X[1] \\
\downarrow{\scriptstyle \alpha} & & \downarrow & & \downarrow & & \downarrow{\scriptstyle \alpha[1]} \\
X' & \longrightarrow & Y' & \longrightarrow & Z' & \longrightarrow & X[1].
\end{array}
$$

We obtain the category of triangles.

Definition F.117. *A* triangulated category *is an additive category* $(\mathcal{C}, T)$ *with translation together with a family of triangles (called* distinguished *triangles or* exact *triangles) such that the following conditions are satisfied:*

(TR0) *A triangle which is isomorphic to a distinguished triangle is itself distinguished.*
For all $X \in \mathcal{C}$, the triangle $X \xrightarrow{\mathrm{id}_X} X \longrightarrow 0 \longrightarrow X[1]$ is distinguished.

(TR1) *For every morphism* $f \colon X \to Y$ *in* $\mathcal{C}$ *there exists a distinguished triangle* $X \xrightarrow{f} Y \longrightarrow Z \longrightarrow X[1]$.

(TR2) *(Rotation of triangles) A triangle* $X \xrightarrow{f} Y \xrightarrow{g} Z \xrightarrow{h} X[1]$ *is distinguished if and only if the triangle* $Y \xrightarrow{g} Z \xrightarrow{h} X[1] \xrightarrow{-f[1]} Y[1]$ *is distinguished.*

(TR3) *Let* $X \xrightarrow{f} Y \xrightarrow{g} Z \xrightarrow{h} X[1]$ *and* $X' \xrightarrow{f'} Y' \xrightarrow{g'} Z' \xrightarrow{h'} X'[1]$ *be distinguished triangles, and let* $\alpha \colon X \to X'$ *and* $\beta \colon Y \to Y'$ *be morphisms with* $f' \circ \alpha = \beta \circ f$. *Then there exists a morphism* $\gamma \colon Z \to Z'$ *such that* (α, β, γ) *is a morphism of distinguished triangles.*

(TR4) *(Octahedral axiom) Given distinguished triangles*

$$X \xrightarrow{f} Y \xrightarrow{h} Z' \longrightarrow X[1]$$

$$Y \xrightarrow{g} Z \xrightarrow{k} X' \longrightarrow Y[1]$$

$$X \xrightarrow{g \circ f} Z \xrightarrow{\ell} Y' \longrightarrow X[1],$$

there exists a distinguished triangle

$$Z' \xrightarrow{u} Y' \xrightarrow{v} X' \xrightarrow{w} Z'[1]$$

such that the following diagram commutes:

$$
\begin{array}{ccccccc}
X & \xrightarrow{f} & Y & \xrightarrow{h} & Z' & \longrightarrow & X[1] \\
\downarrow{\scriptstyle id} & & \downarrow{\scriptstyle g} & & \downarrow{\scriptstyle u} & & \downarrow{\scriptstyle id} \\
X & \xrightarrow{g\circ f} & Z & \xrightarrow{\ell} & Y' & \longrightarrow & X[1] \\
\downarrow{\scriptstyle f} & & \downarrow{\scriptstyle id} & & \downarrow{\scriptstyle v} & & \downarrow{\scriptstyle f[1]} \\
Y & \xrightarrow{g} & Z & \xrightarrow{k} & X' & \longrightarrow & Y[1] \\
\downarrow{\scriptstyle h} & & \downarrow{\scriptstyle \ell} & & \downarrow{\scriptstyle id} & & \downarrow{\scriptstyle h[1]} \\
Z' & \xrightarrow{u} & Y' & \xrightarrow{v} & X' & \xrightarrow{w} & Z'[1].
\end{array}
$$

See for instance [KaSh]^{O} 10.1 for a visualization of the octahedron underlying the octahedral axiom.

Remark F.118. Using (TR1) – (TR3) one can show ([Hub]$_{\mathrm{X}}$ Appendix B) that (TR4) is also equivalent to the following variant of (TR3).
(TR4') Every commutative solid diagram with distinguished triangles as rows

$$
\begin{array}{ccccccc}
X & \xrightarrow{f} & Y & \xrightarrow{g} & Z & \xrightarrow{h} & X[1] \\
\| & & \downarrow{\scriptstyle u} & & \vdots{\scriptstyle v} & & \| \\
X & \xrightarrow{f'} & Y' & \xrightarrow{g'} & Z' & \xrightarrow{h'} & X[1]
\end{array}
$$

can be completed to a commutative diagram with dotted arrow such that the triangle

$$
Y \xrightarrow{\binom{-g}{u}} Z \oplus Y' \xrightarrow{(v,g')} Z' \xrightarrow{f[1]\circ h'} Y[1]
$$

is distinguished.

Lemma F.119. (Five lemma, [KaSh]^{O} Prop. 10.1.15) *Let $\mathcal{C}$ be a triangulated category, and let*

$$
\begin{array}{ccccccc}
X & \longrightarrow & Y & \longrightarrow & Z & \longrightarrow & X[1] \\
\downarrow{\scriptstyle \alpha} & & \downarrow{\scriptstyle \beta} & & \downarrow{\scriptstyle \gamma} & & \downarrow{\scriptstyle \alpha[1]} \\
X' & \longrightarrow & Y' & \longrightarrow & Z' & \longrightarrow & X[1].
\end{array}
$$

be a morphism of distinguished triangles. If two of the morphisms α, β, and γ are isomorphisms, then so is the third.

Remark F.120. Let $\mathcal{C}$ be a triangulated category.

(1) If $X \xrightarrow{f} Y \xrightarrow{g} Z \xrightarrow{h} X[1]$ is a distinguished triangle, then $g \circ f = 0$, as follows from axioms (TR0) and (TR3). Then using the rotation axiom (TR2) we also see that $h \circ g$ and $f[1] \circ h$ are zero.

(2) By the five lemma, for every morphism $f\colon X \to Y$ in $\mathcal{C}$ the distinguished triangle $X \xrightarrow{f} Y \longrightarrow Z \longrightarrow X[1]$ in (TR1) is unique up to isomorphism (but not up to unique isomorphism).

(3) In fact, the morphism γ in (TR3) is not required to be unique. However, we have: In the situation of condition (TR3) in the above definition, assume that $\operatorname{Hom}_{\mathcal{C}}(Y, X') = 0$ and that $\operatorname{Hom}_{\mathcal{C}}(X[1], Y') = 0$. Then γ is unique ([KaSh]$^{\circ}$, Prop. 10.1.17).

We remark that there are several problems built into the theory of triangulated categories. A triangulated category consists of an additive category together with the datum of an additional structure (a shift functor and the datum of distinguished triangles). Hence to be triangulated is not a property of an additive category. Moreover, by (2) and (3) of the previous remark it is possible to complete a morphism to a distinguished triangle but this usually cannot be done functorially. It is possible to remove these problems by passing to ∞-categories in the sense of Lurie [Lu-HTT]$^{\circ}_{\times}$. In that theory one can define what it means that an ∞-category is *stable* and one can view stable ∞-categories as "upgrades" of triangulated categories that result in a more satisfying theory. Unfortunately, this is beyond the scope of this book.

Proposition F.121. ([KaSh]$^{\circ}$ 10.1.19) *Let I be a set and let $\mathcal{C}$ be a triangulated category which admits direct sums indexed by I. Then direct sums indexed by I commute with the translation functor, and a direct sum of distinguished triangles indexed by I is distinguished. In particular for all $X, Y \in \mathcal{C}$, the obvious triangle*

$$(\text{F.26.1}) \qquad X \longrightarrow X \oplus Y \longrightarrow Y \xrightarrow{0} X[1]$$

is a distinguished triangle.

Remark F.122. Let $\mathcal{C}$ be a triangulated category.
(1) Any distinguished triangle of the form $X \longrightarrow Z \longrightarrow Y \xrightarrow{0} X[1]$ is isomorphic to the split triangle (F.26.1): Indeed by a rotated version of (TR3) there exist a morphism of triangles extending id_X and id_Y which is necessarily an isomorphism because of the five lemma Lemma F.119).
(2) Let $X \to Z$ be a morphism in $\mathcal{C}$. Then (1) shows that $X \longrightarrow Z \longrightarrow 0 \longrightarrow X[1]$ is a distinguished triangle if and only if $X \to Z$ is an isomorphism.
(3) If $u\colon Z \to Y$ is an epimorphism in $\mathcal{C}$, then u has a right inverse.
    Indeed, complete u to distinguished triangle $Z \xrightarrow{u} Y \xrightarrow{v} X \longrightarrow Z[1]$. As $v \circ u = 0$ (Remark F.120 (1)) and u is an epimorphism, we find $v = 0$. Now a rotated version of (1) shows the claim.
(4) Dually, any monomorphism has a left inverse.

(F.27) Triangulated Structures on categories of complexes.

Before continuing with the general theory of triangulated categories we explain the most important source of triangulated categories in this book: categories of complexes of an additive (often abelian) category. Throughout this section, let $\mathcal{A}$ denote an additive category.

We give the homotopy category $K(\mathcal{A})$ the structure of a triangulated category in the following way: The translation functor is given by translation of complexes, $X \mapsto X[1]$ (Definition F.68 (3)).

Given a morphism $f\colon X \to Y$ of complexes, there is an obvious "inclusion" $i\colon Y \to C_f$ and an obvious "projection" $p\colon C_f \to X[1]$. In this way, f gives rise to a triangle

$$\text{(F.27.1)} \qquad X \xrightarrow{\ f\ } Y \xrightarrow{\ i\ } C_f \xrightarrow{\ -p\ } X[1].$$

Note the sign! By definition, an exact (or distinguished) triangle in $K(\mathcal{A})$ is one which is isomorphic to a triangle of this form. One shows that with these definitions the axioms of a triangulated category are satisfied ([Lip2] $\overset{\mathrm{O}}{\times}$ 1.4).

The following construction gives the same set of distinguished triangles.

Remark F.123. Let $\mathcal{A}$ be an additive category. Let

$$0 \longrightarrow X \xrightarrow{\ f\ } Y \xrightarrow{\ g\ } Z \longrightarrow 0$$

be a termwise split sequence (Definition F.69) of complexes of $\mathcal{A}$. Choose morphisms $s^i\colon Z^i \to Y^i$ and $r^i\colon Y^i \to X^i$ such that $g^i \circ s^i = \mathrm{id}$, $r^i \circ f^i = \mathrm{id}$ and $r^i \circ s^i = 0$ for all $i \in \mathbb{Z}$ and define a triangle

$$X \xrightarrow{\ f\ } Y \xrightarrow{\ g\ } Z \xrightarrow{\ \delta\ } X[1],$$

where $\delta^i := r^{i+1} \circ d_Y^i \circ s^i$. Its isomorphism class as a triangle in $K(\mathcal{A})$ does not depend on the choice of the s^i and the r^i because every other choice of splittings defines a morphism homotopic to δ. Then the set of triangles in $K(\mathcal{A})$ that are isomorphic to a triangle of this form is the set of distinguished triangles in $K(\mathcal{A})$ defined above. Using that $K(\mathcal{A})$ is triangulated with the original definition of distinguished triangles, we see that indeed every triangle is isomorphic to one of the form $Y \to C_f \to X \to Y[1]$ for a morphism $f\colon X \to Y$ of complexes; one checks that the morphisms are those attached to the termwise split short exact sequence $0 \to Y \to C_f \to X[1] \to 0$ (for suitable choices of the splittings). To go in the other direction, the key point is to show that in $K(\mathcal{A})$ every morphism is isomorphic to one coming from a termwise split injection of complexes. Cf. [Sta] 014L.

The choice of sign in the definition of exact triangles (i.e., defining the final map as $-p$ in (F.27.1)) implies that the boundary maps in the long exact cohomology sequence for the exact triangle attached to a degree-wise split short exact sequence $0 \to X^\bullet \to Y^\bullet \to Z^\bullet \to 0$ of complexes coincide with those obtained via the snake lemma.

(F.28) Triangulated Functors.

Definition F.124.
(1) A *triangulated functor* or *exact functor* $F\colon \mathcal{C} \to \mathcal{C}'$ *between triangulated categories* $\mathcal{C}, \mathcal{C}'$ *is a functor of additive categories with translation which maps distinguished triangles to distinguished triangles.*
(2) *A triangulated functor which is an equivalence of categories is called an* equivalence *of triangulated categories.*
(3) *A morphism* $\alpha\colon F \to F'$ *of triangulated functors* $\mathcal{C} \to \mathcal{C}'$ *is a morphism of functors such that the following diagram commutes:*

$$
\begin{array}{ccc}
F \circ T & \xrightarrow{\ \alpha \circ T\ } & F' \circ T \\
\Big\downarrow{\scriptstyle \cong} & & \Big\downarrow{\scriptstyle \cong} \\
T' \circ F & \xrightarrow{\ T' \circ \alpha\ } & T' \circ F'.
\end{array}
$$

Here T and T' denote the translation functors of C and C', resp., and the vertical isomorphisms are the ones given by the structure of a functor of categories with translation.

A functor between triangulated categories compatible with translation and mapping distinguished triangles to distinguished triangles is automatically additive (use that F preserves the distinguished triangles $0 \to 0 \to 0 \to 0[1]$ and (F.26.1)) and hence a triangulated functor.

Clearly the composition of triangulated functors is again triangulated.

If F is an equivalence of triangulated categories in the sense of the above definition, then every quasi-inverse of F is also a triangulated functor. More generally:

Proposition F.125. ([KaSh] $^\mathrm{O}$ Cor. 10.1.16) *Let $F\colon C' \to C$ be a fully faithful triangulated functor of triangulated categories. Then a triangle of C' is distinguished if and only if the image triangle of C is distinguished.*

Remark F.126. Let $\mathcal{A}$ be an additive category and let C' be a triangulated category. By Remark F.123 a functor $F\colon K(\mathcal{A}) \to C'$ of additive categories with translation is triangulated if and only if it satisfies the following equivalent conditions.
(i) The functor F preserves mapping cones, i.e., for every morphism $f\colon X \to Y$ of complexes one has $F(C_f) = C_{F(f)}$.
(ii) The functor F sends the distinguished triangle given by a termwise split sequence of complexes to a distinguished triangle in C'.

Example F.127. Let $\mathcal{A}$ be an additive category. Fix Z in $K(\mathcal{A})$. The additive functor of complexes $X \mapsto \mathrm{Hom}_{\mathcal{A}}(Z, X)$ (the Hom complex defined in Section (F.18)) preserves homotopy and hence induces an additive functor

$$F\colon K(\mathcal{A}) \to K(\mathrm{AbGrp}), \qquad X \mapsto \mathrm{Hom}_{\mathcal{A}}(Z, X)$$

Then F commutes with the translation functors and preserves termwise split sequences of complexes. Hence it is a triangulated functor.

Definition F.128. Let $F\colon C \to C'$ and $G\colon C' \to C$ be triangulated functors of triangulated categories. Suppose that F is left adjoint to G and let $\eta\colon \mathrm{id}_C \to G \circ F$ and $\epsilon\colon F \circ G \to \mathrm{id}_{C'}$ be the adjunction morphisms. Then we say that (F, G) is a triangulated adjoint pair *if η (or, equivalently by [Lip2]$^\mathrm{O}_X$ (3.3.1), ϵ) is a morphism of triangulated functors.*

Lemma F.129. ([Nee2] $^\mathrm{O}$ 3.9) *Let C and $\mathcal{D}$ be triangulated categories and let $F\colon C \to \mathcal{D}$ be a triangulated functor. Suppose that F has a right (resp. left) adjoint $G\colon \mathcal{D} \to C$. Then G has the structure of a triangulated functor such that (F, G) (resp. (G, F)) is a triangulated adjoint pair.*

Example F.130. Let $F\colon \mathcal{A} \to \mathcal{B}$ and $G\colon \mathcal{B} \to \mathcal{A}$ be additive functors of additive categories. Suppose that F is left adjoint to G. Then for all objects A in $\mathcal{A}$ and B in $\mathcal{B}$ the adjunction isomorphism (F.5.1) is the composition

$$\mathrm{Hom}_{\mathcal{A}}(A, G(B)) \longrightarrow \mathrm{Hom}_{\mathcal{B}}(F(A), F(G(B))) \longrightarrow \mathrm{Hom}_{\mathcal{B}}(F(A), B),$$

where the first map is induced by F and the second from the adjunction morphism $F \circ G \to \mathrm{id}_{\mathcal{B}}$. In particular it is an isomorphism of abelian groups.

The functors F and G induce triangulated functors $F\colon K(\mathcal{A}) \to K(\mathcal{B})$ and $G\colon K(\mathcal{B}) \to K(\mathcal{A})$ and these two functors form a triangulated adjoint pair (F, G).

(F.29) Triangulated Subcategories.

Definition F.131. *Let C be a triangulated category. A triangulated subcategory C' of a triangulated category C is a full additive subcategory endowed with the structure of a triangulated category such that the restriction of the translation functor T of C induces the translation functor on C' and such that the inclusion functor $C' \to C$ is triangulated.*

Let C' be a full additive subcategory of C that is stable under isomorphism and stable under the translation automorphism and its inverse. Then C' carries at most one structure of triangulated category for which the translation is the restriction of that on C and such that the inclusion $C' \to C$ is a triangulated functor (Proposition F.125). Such a structure of triangulated category exists if and only if for every distinguished triangle $X \to Y \to Z \to X[1]$ of C with X and Y in C' one has that Z lies in C' (use Remark F.120 (2)).

Remark F.132. If C is the homotopy category $K(\mathcal{A})$ for an additive category $\mathcal{A}$, then a full additive isomorphism-stable subcategory C' of $K(\mathcal{A})$ is a triangulated subcategory if and only if:
(1) For every X in $K(\mathcal{A})$ the complex X is in C' if and only if $X[1]$ is in C'.
(2) The mapping cone of any morphism $X \to Y$ of complexes X and Y in C' is homotopy equivalent (i.e., isomorphic in $K(\mathcal{A})$) to a complex in C'.

The remark shows the full subcategories $K^b(\mathcal{A})$, $K^+(\mathcal{A})$, and $K^-(\mathcal{A})$ of (left/right) bounded complexes are examples of full triangulated subcategories of $K(\mathcal{A})$.

(F.30) The opposite triangulated category.

Remark F.133. Let C be a triangulated category. We obtain the *opposite triangulated category* by endowing its opposite category C^{opp} with the structure of a triangulated category as follows. Its translation automorphism is defined by $T^{\mathrm{opp}} \colon X \mapsto X[-1]$. If

$$(\mathrm{F.30.1}) \qquad\qquad X \xrightarrow{u} Y \xrightarrow{v} Z \xrightarrow{w} X[1]$$

is a distinguished triangle in C, we obtain a triangle

$$(\mathrm{F.30.2}) \qquad\qquad Z^{\mathrm{opp}} \xrightarrow{v^{\mathrm{opp}}} Y^{\mathrm{opp}} \xrightarrow{u^{\mathrm{opp}}} X^{\mathrm{opp}} \xrightarrow{w[-1]^{\mathrm{opp}}} T^{\mathrm{opp}}(Z^{\mathrm{opp}})$$

in C^{opp} and we define the distinguished triangles of C^{opp} to be all triangles of this form. Note that the distinguished triangle (F.30.2) differs by a sign from the triangle we get by applying the canonical contravariant functor $(-)^{\mathrm{opp}}$ to the distinguished triangle $Z[-1] \xrightarrow{-w[-1]} X \xrightarrow{u} Y \xrightarrow{v} Z$ obtained from (F.30.1).

If $\mathcal{D}$ is a triangulated category, then a *triangulated contravariant functor* $C \to \mathcal{D}$ is by definition a triangulated (covariant) functor $C^{\mathrm{opp}} \to \mathcal{D}$.

Example F.134. Let $\mathcal{A}$ be an additive category and fix Z in $K(\mathcal{A})$. Consider the additive functor

$$\mathrm{Hom}_{\mathcal{A}}(-, Z) \colon K(\mathcal{A})^{\mathrm{opp}} \longrightarrow K(\mathrm{AbGrp}), \qquad Y \mapsto \mathrm{Hom}_{\mathcal{A}}(Y, Z),$$

where the left hand side denotes the Hom complex defined in (F.18). To make $\mathrm{Hom}_{\mathcal{A}}(-, Z)$ into a triangulated functor, we define a functorial isomorphism

$$\theta_Y \colon \operatorname{Hom}_{\mathcal{A}}(Y[-1], Z) \xrightarrow{\sim} \operatorname{Hom}_{\mathcal{A}}(Y, Z)[1]$$

by multiplication by $(-1)^{i-1}$ in degree i. Then $(\operatorname{Hom}_{\mathcal{A}}(-, Z), \theta)$ is a functor of additive categories with translations. As $(\operatorname{Hom}_{\mathcal{A}}(-, Z)$ preserves termwise split sequences it is a triangulated functor.

Remark F.135. Let $\mathcal{A}$ be an additive category. Then the opposite category $\mathcal{A}^{\mathrm{opp}}$ is again additive. We will identify $K(\mathcal{A}^{\mathrm{opp}})$ and $K(\mathcal{A})^{\mathrm{opp}}$ as *triangulated* categories as follows. We define an isomorphism of categories

$$\Lambda \colon C(\mathcal{A})^{\mathrm{op}} \xrightarrow{\sim} C(\mathcal{A}^{\mathrm{op}})$$

by setting

$$\Lambda(X)^i := X^{-i}, \qquad d^i_{\Lambda(X)} := (-1)^i d_X^{-i-1}.$$

If a morphism f in $C(\mathcal{A})$ is homotopic to zero via a homotopy $(h^i)_{i \in \mathbb{Z}}$, then $\Lambda(f^{\mathrm{opp}})$ is homotopic to zero in $C(\mathcal{A}^{\mathrm{op}})$ via the homotopy $((-1)^i h^{i,\mathrm{opp}})_i$. Hence Λ induces an isomorphism of categories

(F.30.3) $$\Lambda \colon K(\mathcal{A})^{\mathrm{op}} \xrightarrow{\sim} K(\mathcal{A}^{\mathrm{op}}).$$

We have an isomorphism $\theta \colon \Lambda \circ [1] \cong [-1] \circ \Lambda$ of functors given by multiplication by $(-1)^{i-1}$ in degree $i \in \mathbb{Z}$, i.e., this isomorphism is given by

$$\theta^i_X \colon \Lambda(X[1])^i = X^{1-i} \xrightarrow{(-1)^{i-1}} X^{1-i} = (\Lambda(X)[-1])^i.$$

Then (Λ, θ) is an isomorphism of triangulated categories $K(\mathcal{A})^{\mathrm{op}} \xrightarrow{\sim} K(\mathcal{A}^{\mathrm{op}})$, where we endow $K(\mathcal{A})^{\mathrm{op}}$ with the structure of a triangulated category as in Remark F.133.

If F is a contravariant additive functor of additive categories from $\mathcal{A}$ to $\mathcal{A}'$, we consider F as usual as a (covariant) functor $F \colon \mathcal{A}^{\mathrm{op}} \to \mathcal{A}'$. By identifying $K(\mathcal{A}^{\mathrm{op}})$ with $K(\mathcal{A})^{\mathrm{op}}$ as above, we obtain a functor $K(F) \colon C(\mathcal{A})^{\mathrm{op}} \to C(\mathcal{A}')$, i.e.

$$K(F)(X)^i = F(X^{-i}), \qquad d^i_{K(F)(X)} = (-1)^i F(d_X^{-i-1}).$$

(F.31) Cohomological Functors.

Definition F.136. *Let $\mathcal{C}$ be a triangulated category and $\mathcal{A}$ an abelian category. An additive functor $\mathcal{C} \to \mathcal{A}$ is called* cohomological, *if for every distinguished triangle $X \to Y \to Z \to X[1]$ in $\mathcal{C}$ the sequence $F(X) \to F(Y) \to F(Z)$ in $\mathcal{A}$ is exact.*

The rotation axiom (TR2) shows that given a cohomological functor $F \colon \mathcal{C} \to \mathcal{A}$, every distinguished triangle $X \to Y \to Z \to X[1]$ gives rise to a long exact sequence

$$\cdots \to F(Z[-1]) \to F(X) \to F(Y) \to F(Z) \to F(X[1]) \to \cdots.$$

Example F.137. With notation as in the definition, for every $W \in \mathcal{C}$, the functors

$$\operatorname{Hom}_{\mathcal{C}}(W, -) \colon \mathcal{C} \to (\mathrm{AbGrp}), \qquad \operatorname{Hom}_{\mathcal{C}}(-, W) \colon \mathcal{C}^{\mathrm{opp}} \to (\mathrm{AbGrp})$$

are cohomological ([KaSh]^{O} 10.1.13).

Example F.138. Let $\mathcal{A}$ be an abelian category. Then the functor

$$H^0\colon K(\mathcal{A}) \to \mathcal{A}, \qquad X \mapsto H^0(X) = \mathrm{Ker}(d_X^0)/\mathrm{Im}(d_X^{-1})$$

is a cohomological functor. Indeed, by Remark F.123 we only have to show that if $0 \to X \to Y \to Z \to 0$ is a termwise split sequence of complexes, then $H^0(X) \to H^0(Y) \to H^0(Z)$ is exact and this follows from Proposition F.71. Hence using that $H^0(X[n]) = H^n(X)$, an exact triangle $0 \to X \to Y \to Z \to X[1]$ gives rise to a long exact cohomology sequence

$$\cdots H^i(X) \to H^i(Y) \to H^i(Z) \to H^{i+1}(X) \to \cdots.$$

Sign conventions

Unfortunately, in homological algebra, it is unavoidable to have signs come in in certain places (see the definitions of the Hom complex and the total complex of a double complex above, for instance), and some choices have to be made as to where to put which signs. There is (unfortunately, one could say) some freedom here, and almost every consistent choice of sign conventions occurs somewhere in the literature. We will survey our decision about sign conventions here, and give some references pointing to differing normalizations in other places.

Below, unless otherwise stated, all complexes are complexes in an abelian category $\mathcal{A}$. Sometimes we implicitly assume that $\mathcal{A}$ has infinite products (or that only finitely many terms in the products arising below are non-zero).

(F.32) Cones and Exact Triangles.

The shift $X[1]$ of a complex X is defined by $X[1]^i = X^{i+1}$, and the differential is changed by a sign: $d_{X[1]}^n = -d_X^n$. Similarly, we define the shift $X[j]$ for any integer j.

While this is the most common way to set up the shift, some authors do it differently: In [Wei2] [O] and [BouA10] [O], our $X[1]$ is called $X[-1]$. In [Law] [x], the shift defined above is denoted ΣX, while the notation $X[1]$ there is defined without changing the differential; while this convention is more in line with the slogan that a sign arises because certain operators (like d and Σ) anti-commute, it is so rare that we do not follow it.

Given a morphism $f\colon X \to Y$ of complexes, we defined the mapping cone $C = C_f$ by setting $C^n = X^{n+1} \oplus Y^n$, with differential given by the matrix

$$\begin{pmatrix} -d_X & 0 \\ f & d_Y \end{pmatrix}.$$

The sign conventions regarding the mapping cone are different in [BouA10] [O], [Har1] [O], [Wei2] [O].

Given a morphism $f\colon X \to Y$ of complexes, there is an obvious "inclusion" $i\colon Y \to C_f$ and an obvious "projection" $p\colon C_f \to X[1]$. We define a structure of triangulated category on the homotopy category $K(\mathcal{A})$ by saying that distinguished triangles are those isomorphic to a triangle of the form

$$X \xrightarrow{\ f\ } Y \xrightarrow{\ i\ } C_f \xrightarrow{\ -p\ } X[1].$$

Note the sign: The final map is given by $-p$.

With this convention, we follow [Con] ^{O}, [SGA4] ^{O}, [Sta]. In [Har1] ^{O}, [Law] $_X$, [Lip2] ^O_X, [KaSh] ^{O} (see Prop. 12.3.6), [Wei2] ^{O} the sign convention differs from ours. Our choice has the advantage that the boundary morphisms in the long exact cohomology sequence agree with the morphisms "obtained from the snake lemma", see Remark F.151 below for details.

(F.33) Double complexes and the total complex.

A double complex is a complex of complexes, i.e., a collection $X^{i,j}$ of objects, $i, j \in \mathbb{Z}$ with row differentials $d_1 \colon X^{i,j} \to X^{i+1,j}$ and column differentials $d_2 \colon X^{i,j} \to X^{i,j+1}$ such that all squares *commute*.

The total complex $\mathrm{Tot}(X^{\bullet,\bullet})$ of a double complex $X^{\bullet,\bullet}$ is given by

$$\mathrm{Tot}(X^{\bullet,\bullet})^n = \bigoplus_{i+j=n} X^{i,j},$$

$$d^n_{|X^{i,j}} = d_1 + (-1)^i d_2.$$

Often, in the literature a double complex (or bicomplex) is defined by requiring that the squares anti-commute, rather than commute (e.g., in [Con] ^{O}, [Wei2] ^{O}, [BouA10] ^{O}). In that case, the definition of the differential of the total complex does not involve a sign.

(F.34) Homomorphisms and tensor products.

Let X and Y be complexes. We have the complex of morphisms $\mathrm{Hom}(X,Y)$,

$$\mathrm{Hom}(X,Y)^n = \prod_i \mathrm{Hom}_{\mathcal{A}}(X^i, Y^{i+n}),$$

$$d^n(f) = d_X \circ f - (-1)^n f \circ d_X \qquad \text{for } f \in \mathrm{Hom}(X,Y)^n.$$

Usually the differential on the Hom complex is defined in this way, for instance in [BouA10] ^{O} taking into account that complexes there are defined with differentials lowering the degree and one has to pass from X_n there to X^{-n} here. But as always, there are exceptions: In [Har1] ^{O} and [Wei2] ^{O}, the sign differs by $(-1)^{n+1}$.

Assume that $\mathcal{A}$ has a tensor product. As defined above, the tensor product $X \otimes Y$ of complexes X and Y is the total complex of the double complex $X^i \otimes Y^j$ with differentials induced by those of X and Y. Explicitly, this means

$$(X \otimes Y)^n = \bigoplus_{i+j=n} X^i \otimes Y^j,$$

$$d^n_{|X^i \otimes Y^j}(x \otimes y) = d_X(x) \otimes y + (-1)^i x \otimes d_Y(y).$$

While this is definitively the most common way of defining the tensor product of complexes, in [Har1] ^{O} it is done differently. Our convention has the advantages that it fits well with the definition of the total complex of a double complex, and, more importantly, with adjunction between Hom and $\otimes$ (see Proposition F.93).

Derived categories

The derived category of an abelian category $\mathcal{A}$ will be defined as the homotopy category $K(\mathcal{A})$ "localized by the set of quasi-isomorphisms", i.e., where we invert all quasi-isomorphisms. This process of localizing a category will be explained first.

(F.35) Localization of Categories.

We follow [KaSh]^{O} Ch. 7. The localization of a category $\mathcal{C}$ with respect to a set S of morphisms in $\mathcal{C}$ is the "minimal" category together with a functor from $\mathcal{C}$ such that all elements of S are mapped to isomorphisms. More precisely, we define

Definition F.139. *Let $\mathcal{C}$ be a category, and let S be a set of morphisms in $\mathcal{C}$. A category $\mathcal{C}_S$ together with a functor $Q\colon \mathcal{C} \to \mathcal{C}_S$ is called a* localization *of $\mathcal{C}$ with respect to S, if for all $s \in S$, the morphism $Q(s)$ is an isomorphism in $\mathcal{C}_S$, and if the following universal property holds:*

For any category $\mathcal{D}$ and any functor $F\colon \mathcal{C} \to \mathcal{D}$ such that for all $s \in S$, $F(s)$ is an isomorphism, there exists a functor $G\colon \mathcal{C}_S \to \mathcal{D}$ and an isomorphism $F \cong G \circ Q$ of functors. Moreover, G is unique in the following sense: Given any two functors $G, G'\colon \mathcal{C}_S \to \mathcal{D}$, composition with Q induces a bijection

$$\mathrm{Hom}(G, G') \cong \mathrm{Hom}(G \circ Q, G' \circ Q)$$

between the respective sets of morphisms of functors. In particular, G as above is uniquely determined up to unique isomorphism, once the isomorphism $F \cong G \circ Q$ is fixed.

From the definition, we see immediately that the localization, if it exists, is unique up to equivalence of categories. If one ignores set-theoretic issues, which for instance can be safely done if $\mathcal{C}$ is a small category, then the localization $\mathcal{C}_S$ always exists ([GaZi]^{O} I.1). But the description of morphisms in $\mathcal{C}_S$ is somewhat awkward making it difficult to work with $\mathcal{C}_S$. The localization category becomes much more concrete if S is a (right or left) multiplicative system in the following sense.

Definition F.140. *Let $\mathcal{C}$ be a category, and let S be a set of morphisms in $\mathcal{C}$. We call S a* right multiplicative system, *if*
(MS1) *every isomorphism in $\mathcal{C}$ belongs to S, and for any two morphisms $s\colon X \to Y$ and $t\colon Y \to Z$ in S, the composite $t \circ s$ belongs to S,*
(MS2) *for every $s\colon X \to X'$ in S, and every morphism $f\colon X \to Y$, there exist morphisms $t\colon Y \to Y'$ in S and $g\colon X' \to Y'$ such that $g \circ s = t \circ f$, and*
(MS3) *Let $f, g \in \mathrm{Hom}_{\mathcal{C}}(X, Y)$ such that there exists $s\colon W \to X$ in S with $f \circ s = g \circ s$. Then there exists a morphism $t\colon Y \to Z$ in S such that $t \circ f = t \circ g$.*
Analogously, we have the notion of left multiplicative system, *where in (MS2) and (MS3) one reverses the arrows. A set of morphisms in $\mathcal{C}$ is called* multiplicative system *if it is a right multiplicative and a left multiplicative system.*

Here we follow [KaSh]^{O}. Some authors (e.g., [Sta]) call "right" what we call "left". For the application to the construction of the derived category this difference does not matter because there we will localize by a system that is (right and left) multiplicative. If S is a right (or left) multiplicative system, there is the following explicit construction of a localization of $\mathcal{C}$ by S.

Remark: Construction of a localization F.141. ([KaSh]^{O}, §7.1) Let S be a right multiplicative system in a category $\mathcal{C}$. Define a category $\mathcal{C}_S^r$ as follows.

(a) The objects of $\mathcal{C}_S^r$ are the objects of $\mathcal{C}$.

(b) For $X, Y \in \mathrm{Ob}(\mathcal{C}_S^r) = \mathrm{Ob}(\mathcal{C})$ we set

$$\mathrm{Hom}_{\mathcal{C}_S^r}(X, Y) := \operatornamewithlimits{colim}_{(Y \to Y') \in S} \mathrm{Hom}_{\mathcal{C}}(X, Y'),$$

where the colimit is taken over the category S^Y whose objects are morphisms $Y \to Y'$ in S and where a morphism $(s \colon Y \to Y') \longrightarrow (s' \colon Y \to Y'')$ is defined to be a morphism $h \colon Y' \to Y''$ in $\mathcal{C}$ (not assumed to be in S!) such that $h \circ s = s'$.

The axioms (MS1) – (MS3) show that S^Y is a filtered category. Hence a morphism $X \to Y$ can be also described as an equivalence class $s^{-1}f$ of a pair (f, s) of morphisms

$$X \xrightarrow{\ f\ } Y' \xleftarrow{\ s\ } Y$$

in $\mathcal{C}$ with $s \in S$. Here we call two such pairs $X \xrightarrow{f_1} Y_1 \xleftarrow{s_1} Y$ and $X \xrightarrow{f_2} Y_2 \xleftarrow{s_2} Y$ equivalent if there exists a third such pair $X \xrightarrow{\ f\ } Y' \xleftarrow{\ s\ } Y$ and a commutative diagram in $\mathcal{C}$,

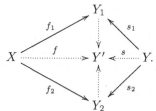

(c) The composition of two morphisms $s^{-1}f$ and $t^{-1}g$ represented by $X \xrightarrow{\ f\ } Y' \xleftarrow{\ s\ } Y$ and $Y \xrightarrow{\ g\ } Z' \xleftarrow{\ t\ } Z$ is defined by the diagram below with $u \in S$ (which exists by (MS2)):

$$X \xrightarrow{\ f\ } Y' \xleftarrow{\ s\ } Y \xrightarrow{\ g\ } Z' \xleftarrow{\ t\ } Z.$$

$$W$$

This makes $\mathcal{C}_S^r$ into a category. Denote by $Q \colon \mathcal{C} \to \mathcal{C}_S^r$ the functor that is the identity on objects and sends a morphism $f \colon X \to Y$ in $\mathcal{C}$ to $\mathrm{id}_Y^{-1} f$.

Theorem F.142. (Existence of localization, [KaSh]^{O} Thm. 7.1.16) *Let $\mathcal{C}$ be a category, and let S be a right multiplicative system of morphisms in $\mathcal{C}$. Then $(\mathcal{C}_S^r, Q)$ is a localization of $\mathcal{C}$ by S.*

From now on we write $\mathcal{C}_S$ instead of $\mathcal{C}_S^r$ and if we speak of the localization of $\mathcal{C}$ by S we mean the category $\mathcal{C}_S^r$ as above.

If T is a left multiplicative system in $\mathcal{C}$ we can define analogously a category $\mathcal{C}_T^l$. For instance morphisms in $\mathcal{C}_T^l$ are equivalence classes gt^{-1} of diagrams $X \xleftarrow{\ t\ } X' \xrightarrow{\ g\ } Y$ with $t \in T$, i.e.,

$$\mathrm{Hom}_{\mathcal{C}_T^l}(X, Y) = \operatornamewithlimits{colim}_{(X' \to X) \in T} \mathrm{Hom}_{\mathcal{C}}(X', Y).$$

If S is a (left and right) multiplicative system, then $\mathcal{C}_S^r$ and $\mathcal{C}_S^l$ are equivalent categories and we can write every morphism in $\mathcal{C}_S$ either as $s^{-1}f$ or as gt^{-1} with $s, t \in S$.

Proposition F.143. ([KaSh]^O, 7.1.20, 7.1.22) *Let C be a category, let S be a multiplicative system in C, and let $Q\colon C \to C_S$ be the localization functor.*

(1) *The following assertions are equivalent.*

    (i) *For all morphisms $X \xrightarrow{f} Y \xrightarrow{g} Z \xrightarrow{h} W$ in C with $h \circ g, g \circ f \in S$ one has $g \in S$.*[3]

    (ii) *A morphism f of C lies in S if and only if $Q(f)$ is an isomorphism in C_S.*

    If S satisfies these conditions, we call S saturated.

(2) *If C is finitely complete (resp. finitely cocomplete), then C_S is finitely complete (resp. finitely cocomplete) and Q is left exact (resp. right exact).*

(3) *If C is additive, then C_S is additive and Q is an additive functor.*

Definition F.144. (Localization of functors, [KaSh]^O 7.3)) *Let C be a category and let S be a set of morphisms in C such that the localization $Q\colon C \to C_S$ exists. Let $F\colon C \to \mathcal{E}$ be a functor to some category $\mathcal{E}$.*

(1) *The functor F is called* right localizable, *if there exists a functor $R_S F\colon C_S \to \mathcal{E}$ together with a morphism $F \to R_S F \circ Q$ such that for every functor $G\colon C_S \to \mathcal{E}$, the composition*

$$\mathrm{Hom}(R_S F, G) \to \mathrm{Hom}(R_S F \circ Q, G \circ Q) \to \mathrm{Hom}(F, G \circ Q)$$

    is bijective. We call the functor $R_S F$ the right localization *of F.*

(2) *The functor F is called* universally right localizable, *if for every functor $H\colon \mathcal{E} \to \mathcal{E}'$, the functor $H \circ F$ is right localizable, and $R_S(H \circ F) \cong H \circ R_S F$.*

Analogous notions and results are available for left localization.

(F.36) Localization of triangulated categories.

Throughout this section, let C be a triangulated category. We want to localize C by a multiplicative systems in C such that the localization can be endowed with the structure of a triangulated category. We follow [KaSh]^O 10.2. A source for such multiplicative systems are null systems in the following sense.

Definition F.145. *A* null system *in C is a strictly full (Definition F.1) triangulated subcategory $\mathcal{N}$ of C. Equivalently, $\mathcal{N}$ is a full subcategory of C such that*

(a) *$\mathcal{N}$ is a strictly full subcategory,*

(b) *the zero object 0 lies in $\mathcal{N}$,*

(c) *$\mathcal{N}$ is closed under the shift operator T and under its inverse,*

(d) *whenever $X \to Y \to Z \to X[1]$ is a distinguished triangle with $X, Z \in \mathcal{N}$, then $Y \in \mathcal{N}$.*

Given a null system $\mathcal{N}$ in C, we write

$$\mathcal{N}Q := \{f\colon X \to Y \text{ a morphism in } C\,;$$
$$\text{there ex. a dist. triangle } X \to Y \to Z \to X[1] \text{ with } Z \in \mathcal{N}\}.$$

---

[3] We can formulate this condition for S to be saturated more symmetrically than in [KaSh]^O 7.1.19 because we assumed that S is right and left multiplicative.

Then one can show that $\mathcal{N}\mathcal{Q}$ is a left and right multiplicative system. Denote by $Q\colon \mathcal{C} \to \mathcal{C}_{\mathcal{N}\mathcal{Q}}$ the localization functor from $\mathcal{C}$ to the localization with respect to this multiplicative system. The localization is additive, and the image of T provides it with the structure of a category with translation. We call a triangle in $\mathcal{C}_{\mathcal{N}\mathcal{Q}}$ distinguished if it is isomorphic to the image under Q of a distinguished triangle in $\mathcal{C}$.

Note that in general the localization is a "big" category, i.e., a category in which the set of morphisms between two objects does not belong to our chosen universe.

Theorem F.146. ([KaSh]^{o} 10.2.3)

(1) *The construction above endows $\mathcal{C}_{\mathcal{N}\mathcal{Q}}$ with the structure of a triangulated category and Q is a triangulated functor.*

(2) *For $X \in \mathcal{N}$ one has $Q(X) \cong 0$, and the localization is universal with respect to this property in the following sense: If $F\colon \mathcal{C} \to \mathcal{C}'$ is a triangulated functor of triangulated categories such that $F(X) \cong 0$ for all $X \in \mathcal{N}$, then F factors uniquely through Q.*

One often writes $\mathcal{C}/\mathcal{N}$ instead of $\mathcal{C}_{\mathcal{N}\mathcal{Q}}$.

By F.141 and using that $\mathcal{N}\mathcal{Q}$ is (right and left) multiplicative system, there is the following description of Hom-sets in the localization: With notation as above, let $X, Y \in \mathcal{C}/\mathcal{N}$. Then

$$\mathrm{Hom}_{\mathcal{C}/\mathcal{N}}(X,Y) \cong \operatorname*{colim}_{(Y\to Y')\in \mathcal{N}\mathcal{Q}} \mathrm{Hom}_{\mathcal{C}}(X,Y')$$

(F.36.1)
$$\cong \operatorname*{colim}_{(X'\to X)\in \mathcal{N}\mathcal{Q}} \mathrm{Hom}_{\mathcal{C}}(X',Y)$$

$$\cong \operatorname*{colim}_{(X'\to X)\in \mathcal{N}\mathcal{Q},(Y\to Y')\in \mathcal{N}\mathcal{Q}} \mathrm{Hom}_{\mathcal{C}}(X',Y')$$

Lemma F.147. ([Rou]^{o} Lemma 3.4) *Let $Q\colon \mathcal{C} \to \mathcal{C}'$ be a triangulated functor of triangulated categories. Suppose that Q admits a fully faithful right adjoint functor. Then the full subcategory $\mathrm{Ker}(Q)$ of objects X in $\mathcal{C}$ such that $Q(X) = 0$ is a null system stable under direct summands in $\mathcal{C}$ and Q induces an equivalence of triangulated categories*

$$\mathcal{C}/\mathrm{Ker}(Q) \cong \mathcal{C}'.$$

(F.37) The derived category of an abelian category.

In this section, and also in the remainder of this appendix, $\mathcal{A}$ will always denote an abelian category, unless something else is stated explicitly.

Recall the category $K(\mathcal{A})$ of complexes in $\mathcal{A}$ up to homotopy (Definition F.76). Denote by $\mathcal{N}$ the full subcategory of $K(\mathcal{A})$ consisting of objects X such that $H^n(X) \cong 0$ for all n (i.e., the unique map of complexes from X to the zero complex is a quasi-isomorphism). Then $\mathcal{N}$ is a null system in the triangulated category $K(\mathcal{A})$ (Definition F.145). The attached multiplicative system is the set of quasi-isomorphisms. It is saturated (Proposition F.143).

Definition F.148. *The localization $D(\mathcal{A}) := K(\mathcal{A})/\mathcal{N}$ is called the derived category of the abelian category $\mathcal{A}$.*

The objects in the derived category $D(\mathcal{A})$ are complexes of objects in $\mathcal{A}$. Hence one can define properties of objects in $D(\mathcal{A})$ by specifying properties of complexes. But any "good property" of objects in a category should be stable under passing to isomorphic objects and a "good property" for a complex as object in $C(\mathcal{A})$ might not be a "good property" for a complex as object in $D(\mathcal{A})$. For instance, the property of a complex X to be bounded below (i.e., $X^n = 0$ for sufficiently small n) is not a "good property" of X considered as an object in $D(\mathcal{A})$ since there are complexes that are isomorphic to to a bounded below complex in $D(\mathcal{A})$ and that are themselves not bounded below[4]. To stress this point, we will occasionally speak of a *representative* of an object X in $D(\mathcal{A})$ by which we mean a complex, now considered as object in $C(\mathcal{A})$, that is isomorphic to X in $D(\mathcal{A})$.

As noted in Section (F.1), the question of existence of the derived category entails some set-theoretic issues: In the framework of universes, the sets of morphisms in $D(\mathcal{A})$ are not necessarily members of the universe to which $\mathcal{A}$ belongs. One sometimes expresses this by saying that $D(\mathcal{A})$ is a *big* category. We will only consider (variants of) derived categories that are subcategories of $D(\mathcal{A})$ for some Grothendieck abelian category $\mathcal{A}$. In this case we will see (Corollary F.186 below) that the sets of morphisms between any two objects in $D(\mathcal{A})$ still form a set in the same universe. This allows us to ignore these set-theoretic issues in this book.

We could equivalently define $D(\mathcal{A})$ as the localization of the category $C(\mathcal{A})$ of complexes in $\mathcal{A}$ with respect to quasi-isomorphisms. However to obtain the structure of triangulated category on $D(\mathcal{A})$, it is convenient to work with $K(\mathcal{A})$. Moreover, the quasi-isomorphisms do not form a multiplicative system in $C(\mathcal{A})$ which would make the description of morphisms in $D(\mathcal{A})$ using only morphisms of complexes (not up to homotopy) very awkward.

By definition there is a *localization functor*

$$Q\colon K(\mathcal{A}) \to D(\mathcal{A}),$$

which maps quasi-isomorphisms to isomorphisms. The "universal property" of the triangulated localization yields:

Remark F.149. Composition with the localization functor $Q\colon K(\mathcal{A}) \to D(\mathcal{A})$ yields for any triangulated category $\mathcal{E}$ an isomorphism (not only an equivalence) of the category of triangulated functors $D(\mathcal{A}) \to \mathcal{E}$ to the full subcategory of the category of triangulated functors $K(\mathcal{A}) \to \mathcal{E}$ whose objects are triangulated functors F that transform quasi-isomorphisms in $K(\mathcal{A})$ to isomorphisms in $\mathcal{E}$ (equivalently $F(X) = 0$ for every complex X in $K(\mathcal{A})$ with $H^i(X) = 0$ for all $i \in \mathbb{Z}$).

Let us describe $D(\mathcal{A})$ more explicitly.

Remark F.150. As explained in F.141, objects in $D(\mathcal{A})$ are complexes of $\mathcal{A}$. A morphism $X \to Y$ in $D(\mathcal{A})$ can be described as an equivalence class $s^{-1}f$ of morphisms of complexes

$$X \xrightarrow{\ f\ } Y' \xleftarrow{\ s\ } Y,$$

where s is a quasi-isomorphism. As the quasi-isomorphisms in $K(\mathcal{A})$ form a saturated multiplicative system, a morphism $f\colon X \to Y$ in $K(\mathcal{A})$ is a quasi-isomorphism if and only if $Q(f)$ is an isomorphism in $D(\mathcal{A})$. Note that usually there exist isomorphisms in $D(\mathcal{A})$ that are not in the image of Q and hence are not given by a quasi-isomorphism of complexes.

---

[4] In this special example it is in fact possible to find a good replacement of the property to bounded below for objects in $D(\mathcal{A})$, see Section (F.38).

The structure of a triangulated category on $K(\mathcal{A})$ makes $D(\mathcal{A})$ into a triangulated category in the following way (Theorem F.146): As for $K(\mathcal{A})$ the translation functor is given by translation of complexes: $X \mapsto X[1]$, which obviously induces an automorphism of $D(\mathcal{A})$. Furthermore, we define a triangle in $D(\mathcal{A})$ to be distinguished (or exact) if it is isomorphic to the image of a distinguished triangle in $K(\mathcal{A})$ under the localization functor. This makes $D(\mathcal{A})$ into a triangulated category (Theorem F.146) and the localization functor $K(\mathcal{A}) \to D(\mathcal{A})$ is a triangulated functor.

The distinguished triangles in $D(\mathcal{A})$ can be given as follows. By definition, distinguished triangles in $D(\mathcal{A})$ are those triangles that are isomorphic to triangles of the form

(F.37.1) $$X \xrightarrow{f} Y \xrightarrow{i} C_f \xrightarrow{-p} X[1]$$

where $f \colon X \to Y$ is a morphism of complexes, C_f is its mapping cone (Definition F.72), $i \colon Y \to C_f$ is the canonical inclusion and $p \colon C_f \to X[1]$ is the canonical projection.

By Remark F.123 the distinguished triangles can equivalently be described as those that are isomorphic to a triangle given by a short exact sequence of complexes that is termwise split.

In $D(\mathcal{A})$ an arbitrary short exact sequence

(F.37.2) $$0 \longrightarrow X \xrightarrow{f} Y \xrightarrow{g} Z \longrightarrow 0$$

of complexes (not necessarily termwise split) also gives a distinguished triangle as follows. The maps

$$\chi^n \colon C_f^n \to Y^n \to Z^n$$

obtained by composing the projection and the map g give a surjective morphism

$$\chi \colon C_f \to Z$$

of complexes. We claim that this is a quasi-isomorphism. Indeed, the kernel of χ in $C(\mathcal{A})$ is the mapping cone C_{f_0} of the isomorphism $f_0 \colon X \to \mathrm{Im}(f)$ induced by f, hence $H^i(C_{f_0}) = 0$ for all i and the long exact cohomology sequence of $0 \to C_{f_0} \to C_f \xrightarrow{\chi} Z \to 0$ shows that χ is a quasi-isomorphism. Hence it is an isomorphism in $D(\mathcal{A})$ and we obtain in $D(\mathcal{A})$ a triangle

(F.37.3) $$X \xrightarrow{f} Y \xrightarrow{g} Z \xrightarrow{-p \circ \chi^{-1}} X[1]$$

which is called *the triangle associated to the exact sequence* (F.37.2). This triangle is isomorphic to the distinguished triangle (F.37.1). Hence it is distinguished. If the sequence (F.37.2) is termwise split, then χ is a homotopy equivalence (i.e., an isomorphism in $K(\mathcal{A})$) and the triangle (F.37.3) is the image of the distinguished triangle in $K(\mathcal{A})$ defined in Remark F.123.

Remark F.151. As the cohomology functors $H^i \colon K(\mathcal{A}) \to \mathcal{A}$ send quasi-isomorphisms to isomorphisms, they factor through functors $H^i \colon D(\mathcal{A}) \to \mathcal{A}$. By Example F.138, H^0 (in fact any H^i) is a cohomological functor. Hence for any distinguished triangle $X \to Y \to Z \to X[1]$ in $D(\mathcal{A})$ we have the long exact cohomology sequence

(F.37.4) $$\cdots \to H^i(X) \to H^i(Y) \to H^i(Z) \to H^{i+1}(X) \to \cdots .$$

If that distinguished triangle is the one associated to an exact sequence of complexes, the long exact cohomology sequence (F.37.4) is the same as long exact cohomology sequence given by the Snake Lemma in Proposition F.71.

(1) As a morphism in $K(\mathcal{A})$ is a quasi-isomorphism if and only if its image under Q is an isomorphism, for $X \in K(\mathcal{A})$ we have $Q(X) \cong 0$ if and only if $H^i(X) \cong 0$ for all i.

(2) A morphism f in $D(\mathcal{A})$ is an isomorphism if and only if $H^i(f)$ is an isomorphism for all $i \in \mathbb{Z}$.

Indeed, represent by f by $s^{-1}g$, where g is a morphism of complexes and s is a quasi-isomorphism. If $H^i(f)$ is an isomorphism, then $H^i(g)$ is an isomorphism because $H^i(s)$ is an isomorphism. Therefore if $H^i(f)$ is an isomorphism, then g is a quasi-isomorphism and hence an isomorphism in $D(\mathcal{A})$. This shows that f is an isomorphism.

(3) Note that usually there are non-zero morphisms f in $D(\mathcal{A})$ such that $H^i(f) = 0$ for all $i \in \mathbb{Z}$. In fact, for objects $A, B \in \mathcal{A}$, we can view the group $\operatorname{Hom}_{D(\mathcal{A})}(B, A[1]) = \operatorname{Ext}^1(B, A)$ as a Yoneda Ext group (see (F.52) below), and every non-split extension gives rise to a non-trivial morphism $f \colon B \to A[1]$ with $H^i(f) = 0$ for all i (since for a given i, at most one of the two sides has non-trivial cohomology in degree i).

Remark F.152. We can consider every object X of $\mathcal{A}$ as a complex $X[0]$ with this object in degree 0, and 0 elsewhere. This defines a functor

$$\mathcal{A} \to D(\mathcal{A})$$

and it is not difficult to see that this functor is fully faithful (see also Proposition F.156 below for a more general statement). Its essential image consists of objects X in $D(\mathcal{A})$ such that $H^i(X) = 0$ for all $i \neq 0$ (see Proposition F.154 (2) below for a stronger statement). Hence we can (and usually will) consider $\mathcal{A}$ as a full subcategory of $D(\mathcal{A})$. If we want to stress the difference between an object X in $\mathcal{A}$ and its image in $D(\mathcal{A})$ we will write $X[0]$ for this image. More generally, for $i \in \mathbb{Z}$ we will write $X[-i]$ for the complex with X in degree i and 0 elsewhere.

Example F.153. Let $\mathcal{A}$ be an abelian category such that all short exact sequences in $\mathcal{A}$ are split (see also Exercise F.29). Then for any complex $X \in C(\mathcal{A})$ there exists a functorial isomorphism $\bigoplus_{i \in \mathbb{Z}} H^i(X)[-i] \xrightarrow{\sim} X$ in $D(\mathcal{A})$ (see also Exercise F.31 for weaker hypotheses that ensure the existence of such an isomorphism which is then not necessarily functorial in X).

Indeed, note that $\bigoplus_{i \in \mathbb{Z}} H^i(X)[-i]$ denotes the complex

$$\cdots \xrightarrow{0} H^{i-1}(X) \xrightarrow{0} H^i(X) \xrightarrow{0} H^{i+1}(X) \xrightarrow{0} \cdots,$$

which exists even if $\mathcal{A}$ does not admit infinite coproducts. By assumption on $\mathcal{A}$ we find a section $u^i \colon H^i(X) \to \operatorname{Ker}(d_X^i)$ of the canonical map $\operatorname{Ker}(d_X^i) \to H^i(X)$. The u^i define a homomorphism of complexes $u \colon \bigoplus_{i \in \mathbb{Z}} H^i(X)[-i] \to \bigoplus_{i \in \mathbb{Z}} \operatorname{Ker}(d_X^i)[-i]$ and the inclusions $\operatorname{Ker}(d_X^i) \to X^i$ define a morphism of complexes $v \colon \bigoplus_{i \in \mathbb{Z}} \operatorname{Ker}(d_X^i)[-i] \to X$. The composition

$$v \circ u \colon \bigoplus_{i \in \mathbb{Z}} H^i(X)[-i] \to X$$

is a quasi-isomorphism and hence an isomorphism in $D(\mathcal{A})$. If one chooses other splittings $\tilde{u}^i$ defining a morphism $\tilde{u} \colon \bigoplus_{i \in \mathbb{Z}} H^i(X)[-i] \to \bigoplus_{i \in \mathbb{Z}} \operatorname{Ker}(d_X^i)[-i]$, then $u^i - \tilde{u}^i$ factors through $\operatorname{Im}(d_X^{i-1})$. Composing it with a section of $X^{i-1} \to \operatorname{Im}(d_X^{i-1})$, we obtain morphisms $h^i \colon H^i(X) \to X^{i-1}$ which defines a homotopy between $v \circ u$ and $v \circ \tilde{u}$. Hence the isomorphism $v \circ u$ in $D(\mathcal{A})$ is independent of all choices and one checks that it is functorial in X.

(F.38) Bounded derived categories.

Let $\mathcal{K} \subseteq K(\mathcal{A})$ be a triangulated subcategory. Localizing with respect to all quasi-isomorphisms in $\mathcal{K}$, we obtain a category $D^{\mathcal{K}}$ together with a triangulated localization functor $Q = Q_{\mathcal{K}} \colon \mathcal{K} \to D^{\mathcal{K}}$ which is universal for functors from $\mathcal{K}$ which map every quasi-isomorphism to an isomorphism. If $\mathcal{J} \subseteq \mathcal{K}$ is a second triangulated subcategory the universal property of the localization yields a unique triangulated functor

(F.38.1) $$D^{\mathcal{J}} \to D^{\mathcal{K}}$$

making the following diagram commutative

$$\begin{array}{ccc} \mathcal{J} & \longrightarrow & \mathcal{K} \\ {\scriptstyle Q_{\mathcal{I}}} \downarrow & & \downarrow {\scriptstyle Q_{\mathcal{K}}} \\ D^{\mathcal{J}} & \longrightarrow & D^{\mathcal{K}}. \end{array}$$

Typical choices for $\mathcal{K}$ are the homotopy categories obtained from the categories $C^b(\mathcal{A})$, $C^+(\mathcal{A})$, $C^-(\mathcal{A})$ of (left/right) bounded complexes. We denote their respective homotopy categories by $K^b(\mathcal{A})$, $K^+(\mathcal{A})$, $K^-(\mathcal{A})$, etc. By localization we obtain the bounded derived categories $D^{K^b(\mathcal{A})}$, etc. For these, we use the more common notation $D^b(\mathcal{A}) := D^{K^b(\mathcal{A})}$, $D^+(\mathcal{A})$, $D^-(\mathcal{A})$, etc. See Proposition F.154 below for a description in terms of the vanishing of cohomology objects and for the definition of further variants.

Let $I \subseteq \mathbb{Z}$ be an interval. The category $C^I(\mathcal{A})$ of complexes concentrated in degrees $\in I$ is an abelian subcategory of $C(\mathcal{A})$ and we denote by $K^I(\mathcal{A})$ the full subcategory of $K(\mathcal{A})$ of complexes concentrated in degrees $\in I$. Note that this is in general not a triangulated subcategory because the mapping cone of a complexes in $K^I(\mathcal{A})$ is not necessarily contained in $K^I(\mathcal{A})$. Hence it does not make sense to localize these categories as triangulated categories. Instead we denote by $D^I(\mathcal{A})$ the full additive subcategory of $D(\mathcal{A})$ consisting of all X with $H^p(X) = 0$ for all $p \notin I$. We write $D^{\leq n}(\mathcal{A}) := D^{(-\infty, n]}(\mathcal{A})$, $D^{\geq n}(\mathcal{A}) := D^{[n, \infty)}(\mathcal{A})$. This notation is justified by the following result.

Proposition F.154. ([KaSh]^{O} Prop. 13.1.12)
(1) *The triangulated functors $D^b(\mathcal{A}) \to D(\mathcal{A})$, $D^+(\mathcal{A}) \to D(\mathcal{A})$, $D^-(\mathcal{A}) \to D(\mathcal{A})$ (F.38.1) induce an equivalence of triangulated categories of $D^b(\mathcal{A})$ ($D^+(\mathcal{A})$, $D^-(\mathcal{A})$, resp.) to the full triangulated subcategories of $D(\mathcal{A})$ of objects X such that $H^i(X) = 0$ for $|i| \gg 0$ ($i \ll 0$, $i \gg 0$, resp.).*
(2) *Let $I \subseteq \mathbb{Z}$ be an interval. Then the restriction $K^I(\mathcal{A}) \to D^I(\mathcal{A})$ of the localization functor is essentially surjective.*

Hence we can consider the various categories $D^*(\mathcal{A})$ with $* \in \{b, +, -, I\}$ as full subcategories of $D(\mathcal{A})$ with obvious inclusions, e.g., $D^{\geq a}(\mathcal{A}) \subseteq D^+(\mathcal{A})$ for every integer $a \in \mathbb{Z}$.

Remark F.155. If $\mathcal{A}$ is an R-linear category for some ring R (Definition F.28), then the additive category $D^?(\mathcal{A})$ is R-linear for $? \in \{\emptyset, +, -, b, I\}$.

(F.39) Truncation in derived categories.

As the truncation functors $\tau^{\leq n}$, $\tau^{\geq n}$ preserve quasi-isomorphisms of complexes, they induce truncation functors on the derived categories

$$\tau^{\leq n} \colon D(\mathcal{A}) \to D^{\leq n}(\mathcal{A}), \qquad \tau^{\geq n} \colon D(\mathcal{A}) \to D^{\geq n}(\mathcal{A}).$$

Proposition F.156. ([KaSh] ^{O} Prop. 13.1.12) *Let* $a \in \mathbb{Z}$ *and* $X, Y \in D(\mathcal{A})$. *Then*

$$\operatorname{Hom}_{D(\mathcal{A})}(\tau^{\leq a} X, \tau^{\geq a} Y) \cong \operatorname{Hom}_{\mathcal{A}}(H^a(X), H^a(Y)).$$

In particular $\operatorname{Hom}_{D(\mathcal{A})}(\tau^{\leq a} X, \tau^{\geq a+1} Y) = 0$.

Note that the situation is not symmetric. In general $\operatorname{Hom}_{D(\mathcal{A})}(\tau^{\geq a+1} X, \tau^{\leq a} Y) \neq 0$.

Proposition F.157. (cf. [KaSh] ^{O} Prop. 13.1.15) *Let* X *be an object in* $D(\mathcal{A})$. *There are distinguished triangles*

$$\tau^{\leq n} X \to X \to \tau^{\geq n+1} X \to$$
$$\tau^{\leq n-1} X \to \tau^{\leq n} X \to H^n(X)[-n] \to$$
$$H^n(X)[-n] \to \tau^{\geq n} X \to \tau^{\geq n+1} X \to$$

in $D(\mathcal{A})$. *Furthermore, we have*

$$H^n(X)[-n] \cong \tau^{\leq n} \tau^{\geq n} X \cong \tau^{\geq n} \tau^{\leq n} X.$$

Proposition F.158. (cf. [KaSh] ^{O} Prop. 13.1.16) *The functor* $\tau^{\leq n} \colon D(\mathcal{A}) \to D^{\leq n}(\mathcal{A})$ *is right adjoint to the inclusion functor* $D^{\leq n}(\mathcal{A}) \to D(\mathcal{A})$. *Dually, the functor* $\tau^{\geq n} \colon D(\mathcal{A}) \to D^{\geq n}(\mathcal{A})$ *is left adjoint to the inclusion functor* $D^{\geq n}(\mathcal{A}) \to D(\mathcal{A})$.

(F.40) Construction of complexes.

Definition and Remark F.159. *Let* $\mathcal{A}$ *be an abelian category. Let* $m \in \mathbb{Z}$. *Let* $u \colon X \to Y$ *be a morphism in* $D(\mathcal{A})$, *and complete to a distinguished triangle* $X \to Y \to C \to$. *Then* u *is called an* m-*isomorphism if the following equivalent conditions are satisfied.*
(i) *One has* $H^p(C) = 0$ *for all* $p \geq m$.
(ii) *The induced maps* $H^p(u) \colon H^p(X) \to H^p(Y)$ *are isomorphisms for* $p > m$ *and surjective for* $p = m$.
A morphism of complexes in $C(\mathcal{A})$ *is called* m-*isomorphism if it is an* m-*isomorphism considered as a morphism in* $D(\mathcal{A})$.

The equivalence of the two conditions follows from the long exact cohomology sequence attached to the distinguished triangle.

If u is an m-isomorphism, then it is an n-isomorphism for all $n \geq m$. A morphism in $D(\mathcal{A})$ (resp. in $C(\mathcal{A})$) is an isomorphism (resp. a quasi-isomorphism) if it is an m-isomorphism for all m. The composition of m-isomorphisms is again an m-isomorphism. For every complex X in $C(\mathcal{A})$ the inclusion $\sigma^{\geq m} X \to X$ is an m-isomorphism.

Proposition F.160. ([ThTr] ^{O} 1.9.5) *Let* $\mathcal{A}$ *be an abelian category, let* $\mathcal{D}$ *be a full additive subcategory of* $\mathcal{A}$, *and let* $\mathcal{C}$ *be a full subcategory of* $C(\mathcal{A})$ *such that the following hypotheses are satisfied*
(a) *For any complex* X *in* $\mathcal{C}$ *one has* $H^p(X) = 0$ *for* p *large enough and if* $s \colon X \to Y$ *is a quasi-isomorphism in* $C(\mathcal{A})$, *then* X *is in* $\mathcal{C}$ *if and only if* Y *is in* $\mathcal{C}$.
(b) *One has* $C^b(\mathcal{D}) \subseteq \mathcal{C}$. *For every map of complexes* $u \colon X \to Y$ *with* X *in* $C^b(\mathcal{D})$ *and* Y *in* $\mathcal{C}$ *the mapping cone* C_u *lies in* $\mathcal{C}$.

(c) *Suppose given $m \in \mathbb{Z}$, a complex X in $\mathcal{C}$ with $H^p(X) = 0$ for $p \geq m$, and an epimorphism $M \to H^{m-1}(X)$ in $\mathcal{A}$. Then there exists an object P in $\mathcal{D}$ and a map $P \to M$ such that the composition $P \to M \to H^{m-1}(X)$ is an epimorphism.*

Then every map $u \colon Z \to X$ in $\mathcal{C}$ with Z in $C^-(\mathcal{D})$ can be factorized within $\mathcal{C}$ as

$$u \colon Z \xrightarrow{\ i\ } Z' \xrightarrow{\ u'\ } X,$$

where Z' is in $C^-(\mathcal{D})$, u' is a quasi-isomorphism, and i is a degree-wise split monomorphism.

Moreover, if $m \in \mathbb{Z}$ and u is an m-isomorphism, then one may choose Z' as above such that $i^p \colon Z^p \to Z'^p$ is an isomorphism for all $p \geq m$.

Often the proposition is applied to $Z = 0$. Then the assertion is that for every X in $\mathcal{C}$ there exists a complex Z' in $C^-(\mathcal{D})$ and a quasi-isomorphism $Z' \to X$.

(F.41) Variants of the derived category.

Let $\mathcal{A}$ be an abelian category, and let $\mathcal{A}' \subseteq \mathcal{A}$ be a plump abelian subcategory (Definition F.43). For $* \in \{\emptyset, b, +, -\}$, we denote by $K^*_{\mathcal{A}'}(\mathcal{A})$ the full subcategory of $K^*(\mathcal{A})$ of complexes X such that $H^i(X) \in \mathcal{A}'$ for all $i \in \mathbb{Z}$. These are triangulated subcategories of $K(\mathcal{A})$, and localizing with respect to all quasi-isomorphisms we obtain the category $D^{K_{\mathcal{A}'}(\mathcal{A})}$ as in Section (F.38).

Definition F.161. *Let $\mathcal{A}$ be an abelian category, and let $\mathcal{A}' \subseteq \mathcal{A}$ be a plump abelian subcategory (Definition F.43). For $* \in \{\emptyset, b, +, -\}$, we denote by $D^*_{\mathcal{A}'}(\mathcal{A})$ the full additive subcategory of $D^*(\mathcal{A})$ consisting of all X such that $H^i(X) \in \mathcal{A}'$ for all i.*

Lemma F.162. ([Lip2] $\overset{\text{O}}{\underset{\text{X}}{}}$ (1.9.1)) *The canonical functor $D^{K_{\mathcal{A}'}(\mathcal{A})} \to D(\mathcal{A})$ is fully faithful and yields an equivalence of $D^{K_{\mathcal{A}'}(\mathcal{A})}$ with the full subcategory $D^*_{\mathcal{A}'}(\mathcal{A})$ of $D^*(\mathcal{A})$.*

This defines triangulated subcategories, and we have natural functors

$$D^*(\mathcal{A}') \longrightarrow D^*_{\mathcal{A}'}(\mathcal{A}).$$

Note that even if $\mathcal{A}$ and $\mathcal{A}'$ are both Grothendieck abelian categories, these functors $D(\mathcal{A}') \to D_{\mathcal{A}'}(\mathcal{A})$ are not in general equivalences. We have the following general criterion:

Theorem F.163. (cf. [KaSh] ^{O} Theorem 13.2.8) *Let $\mathcal{A}'$ be a plump abelian subcategory of the abelian category $\mathcal{A}$. Assume that for every monomorphism $Y' \hookrightarrow X$ in $\mathcal{A}$ with $Y' \in \mathcal{A}'$ there exists a morphism $X \to Y$ with $Y \in \mathcal{A}'$ such that the composition $Y' \to Y$ is a monomorphism. Then the natural functors*

$$D^+(\mathcal{A}') \longrightarrow D^+_{\mathcal{A}'}(\mathcal{A}), \quad D^b(\mathcal{A}') \longrightarrow D^b_{\mathcal{A}'}(\mathcal{A})$$

are equivalences of categories.

Derived functors

We come now to the main topic in this chapter, the definition and construction of the right (resp. left) derived functor $RF\colon D(\mathcal{A}) \to D(\mathcal{B})$ (resp. $LF\colon D(\mathcal{A}) \to D(\mathcal{B})$) of an additive functor $F\colon \mathcal{A} \to \mathcal{B}$ between abelian categories. After stating the abstract definition we have to give criteria for their existence and how to compute them. For this we will define the notion of a right (resp. left) F-acyclic complex in $K(\mathcal{A})$ and see that RF (resp. LF) can be computed by evaluating the natural extension of F to a functor $K(\mathcal{A}) \to K(\mathcal{B})$ on these F-acyclic complexes. The full subcategory of $D(\mathcal{A})$ of objects that are isomorphic to right (resp. left) F-acyclic complexes is a triangulated subcategory and one can define RF (resp. LF) on that subcategory. Hence to have the derived functor RF (resp. LF) on all of $D(\mathcal{A})$ one has to study the question if each object in $D(\mathcal{A})$ is isomorphic in $D(\mathcal{A})$ to a right (resp. left) F-acyclic complex.

In fact, one can define the notion of a K-injective (resp. K-projective) complex (Definition F.179) and these are complexes that are right (resp. left) F-acyclic for *every* functor F and hence we can define RF (resp. LF) for all functors F if we know that every object in $D(\mathcal{A})$ is isomorphic to a K-injective (resp. K-projective) complex.

For right derived functors this idea just works fine: In this book we are almost always interested in the case that $\mathcal{A}$ is a Grothendieck abelian category since we show that the category of $\mathscr{O}_X$-modules for an arbitrary ringed space X is a Grothendieck abelian category (Proposition 21.7). In this case Theorem F.185 shows that indeed every object of $D(\mathcal{A})$ is isomorphic to a K-injective complex. Hence the right derived functor RF exists for every functor F.

For left derived functors the situation is more complicated in general. If $\mathcal{A}$ is the category of A-modules for some ring A, then one can show, using that every A-module has a left resolution by free modules, that every object in $D(\mathcal{A})$ is isomorphic to a K-projective complex (Theorem F.189). Hence in this case there exists the left derivation for every additive functor $(A\text{-Mod}) \to \mathcal{B}$. But for arbitrary ringed spaces X this does not hold any more, even for $X = \mathbb{P}^1_k$, k a field (see Exercise 21.5, which also shows that this problem cannot be removed by working only with quasi-coherent modules). Hence in this case one has to show for a given functor F of interest there exist sufficiently many left F-acyclic complexes. This is what we do in Section (21.19) and the following sections to define the left derived functor of the tensor product and of the pullback functor.

(F.42) Definition of derived functors.

Let $\mathcal{A}$ be an abelian category, and let $\mathcal{K} \subset K(\mathcal{A})$ be a triangulated subcategory (typical examples of $\mathcal{K}$ would be $K^*(\mathcal{A})$ for $* \in \{\emptyset, +, -, b\}$). We denote by $Q\colon \mathcal{K} \to D^{\mathcal{K}}$ the localization functor. Let $F\colon \mathcal{K} \to \mathcal{E}$ be a triangulated functor from $\mathcal{K}$ to some triangulated category $\mathcal{E}$.

Definition F.164. *We call the functor $F\colon \mathcal{K} \to \mathcal{E}$ right derivable on $\mathcal{K}$, if there exists a triangulated functor $RF\colon D^{\mathcal{K}} \to \mathcal{E}$ together with a morphism $\zeta\colon F \to RF \circ Q$ of triangulated functors such that for every triangulated functor $G\colon D^{\mathcal{K}} \to \mathcal{E}$ the composition*

$$\mathrm{Hom}(RF, G) \to \mathrm{Hom}(RF \circ Q, G \circ Q) \to \mathrm{Hom}(F, G \circ Q),$$

the final map being induced by ζ, is an isomorphism. Such a pair (RF, ζ) is called the right derived functor of F.

Note that we could express the existence of RF by saying that F is right localizable as a triangulated functor (i.e., in Definition F.144 we use only *triangulated* functors G as test functors). See also Remark F.168 below. If the dependence on $\mathcal{K}$ needs to be emphasized, we will denote RF by $R^{\mathcal{K}}F$ (or by $R^{+}F$ for $\mathcal{K} = K^{+}(\mathcal{A})$, etc.). Dually, we define

Definition F.165. *We call the functor $F\colon \mathcal{K} \to \mathcal{E}$ left derivable on $\mathcal{K}$, if there exists a triangulated functor $LF\colon D^{\mathcal{K}} \to \mathcal{E}$ together with a morphism $\xi\colon LF \circ Q \to F$ of triangulated functors such that for every triangulated functor $G\colon D^{\mathcal{K}} \to \mathcal{E}$ the composition*

$$\mathrm{Hom}(G, LF) \to \mathrm{Hom}(G \circ Q, LF \circ Q) \to \mathrm{Hom}(G \circ Q, F)$$

is an isomorphism. Such a pair (LF, ξ) is called the left derived functor of F.

Below we mostly omit the left derived case from the discussion, but of course, all the notions and results below can be dualized.

Remark F.166. Let $F, G\colon \mathcal{K} \to \mathcal{E}$ be triangulated functors and let $\alpha\colon F \to G$ be a morphism of triangulated functors. Suppose that F and G both admit right derived functors (RF, ζ) and (RG, ξ). Then the universal property of (RF, ζ) implies that there exists a unique morphism $R\alpha\colon RF \to RG$ of triangulated functors such that the diagram

$$\begin{array}{ccc} F & \xrightarrow{\ \ \alpha\ \ } & G \\ {\scriptstyle \zeta}\downarrow & & \downarrow{\scriptstyle \xi} \\ RF \circ Q & \xrightarrow{X \mapsto R\alpha(QX)} & RG \circ Q, \end{array}$$

where $Q\colon \mathcal{K} \to D^{\mathcal{K}}$ is the localization functor, commutes.

Instead of saying that F is right derivable, we often say that RF exists.

The most common case how the above functor F arises is the following

Remark F.167. Let $F\colon \mathcal{A} \to \mathcal{A}'$ be an additive functor between abelian categories $\mathcal{A}$, $\mathcal{A}'$. Then F induces a triangulated functor

$$K(F)\colon K(\mathcal{A}) \to K(\mathcal{A}')$$

by applying F to each degree separately, which we sometimes simply denote by F again. Similarly, we can restrict $K(F)$ to $K^{+}(\mathcal{A})$ or any other triangulated subcategory $\mathcal{K}$ of $K(\mathcal{A})$. Denote by $Q\colon K(\mathcal{A}) \to D(\mathcal{A})$ and $Q'\colon K(\mathcal{A}') \to D(\mathcal{A}')$ the localization functors. We now consider the functor

$$Q' \circ K(F)\colon K(\mathcal{A}) \to D(\mathcal{A}')$$

(or its restriction $\mathcal{K} \to \mathcal{E}$, where $\mathcal{E}$ is a triangulated subcategory of $D(\mathcal{A}')$). If this functor admits a right derived functor in the sense of the above definition, then we say that $F\colon \mathcal{A} \to \mathcal{A}'$ *has a right derived functor (on $\mathcal{K}$ with values in $\mathcal{E}$)*, and we denote that derived functor by RF (or by $R^{\mathcal{K}}F$).

In this case we set

$$R^{i}F := H^{i} \circ R^{\mathcal{K}}F\colon D^{\mathcal{K}} \to \mathcal{A}', \qquad i \in \mathbb{Z}.$$

As RF is a triangulated functor, the composition with the cohomological functor H^{0} is again a cohomological functor $R^{0}F\colon D^{\mathcal{K}} \to \mathcal{A}'$ and for every distinguished triangle $X \to Y \to Z \to X[1]$ in $\mathcal{K}$ we obtain a long exact sequence

$$\cdots \longrightarrow R^{i-1}F(Z) \longrightarrow R^i F(X) \longrightarrow R^i F(Y) \longrightarrow R^i F(Z) \longrightarrow R^{i+1}F(X) \longrightarrow \cdots$$

This applies in particular to a short exact sequence $0 \to X \to Y \to Z \to 0$ of complexes of $\mathcal{A}$ with X, Y, Z in $\mathcal{K}$.

Remark F.168. Note that in the literature there are different definitions of derived functors. Here we use Verdier's original definition following [Lip2] $\underset{X}{\overset{O}{}}$.

A stronger version of derived functors has been defined by Deligne in [SGA4]^{O} XVII, 1.7. This is also the approach followed in [KaSh]^{O}. It is based on the fact that the localization $\mathcal{K}_S$ of a category $\mathcal{K}$ by a right multiplicative system S can be embedded fully faithfully into the category $\mathrm{Ind}(\mathcal{K})$ of filtered inductive systems of $\mathcal{K}$. If $\iota_{\mathcal{K}} \colon \mathcal{K} \to \mathrm{Ind}(\mathcal{K})$ is the fully faithful canonical functor and $Q \colon \mathcal{K} \to \mathcal{K}_S$ is the localization functor, one obtains a diagram of functors

$$\mathcal{K} \xrightarrow{\quad Q \quad} \mathcal{K}_S$$
$$\iota_{\mathcal{K}} \searrow \qquad \downarrow \alpha$$
$$\mathrm{Ind}(\mathcal{K}),$$

which is *not* commutative. But there exists a natural morphism

(F.42.1) $\iota_{\mathcal{K}} \longrightarrow \alpha \circ Q.$

Let us suppose that $\mathcal{K}$ is a triangulated subcategory of $K(\mathcal{A})$ for an abelian category $\mathcal{A}$ and that $\mathcal{K}_S = D^{\mathcal{K}}$ is the localization by the system S of quasi-isomorphisms in $\mathcal{K}$. Then $\mathrm{Ind}(\mathcal{K})$ is also triangulated, all functors above are triangulated and the morphism (F.42.1) is a morphism of triangulated functors.

Let $F \colon \mathcal{K} \to \mathcal{E}$ be a triangulated functor of triangulated categories. Rather than requiring the existence of a right localization as a triangulated functor, in [KaSh]^{O} the existence of a universal right localization is required (Def. F.144). In particular, in the universal property, non-triangulated test functors are also allowed. Moreover, the derived functor $D^{\mathcal{K}} \to \mathcal{E}$ in the sense of [KaSh]^{O} exists if and only if the composition

$$D^{\mathcal{K}} \xrightarrow{\ \alpha\ } \mathrm{Ind}(\mathcal{K}) \xrightarrow{\ \mathrm{Ind}(F)\ } \mathrm{Ind}(\mathcal{E})$$

factors over the fully faithful functor $\iota_{\mathcal{E}} \colon \mathcal{E} \to \mathrm{Ind}(\mathcal{E})$. This factorization yields a triangulated functor $R \colon D^{\mathcal{K}} \to \mathcal{E}$ and a morphism of triangulated functors $\zeta \colon F \to R \circ Q$ induced by (F.42.1). The pair (R, ζ) is a right derived functor in the sense of [KaSh]^{O}. In particular, $(R = RF, \zeta)$ is a right derived functor in the sense of Definition F.164 above.

It is not clear (to us), whether a derived functor as in Definition F.164 is necessarily a derived functor in the sense of [KaSh]^{O}. In practice this distinction is not important because the results for the existence of derived functors in [KaSh]^{O} show that all techniques we use to construct derived functors (which all rely on Theorem F.173 below) also ensure the existence of the right derived functor in their stronger sense.

Compare also Corollary F.174 (2) below which shows that in the situation of Theorem F.173, F is "universally right localizable as a triangulated functor".

(F.43) Construction of derived functors.

We now give some existence results for derived functors on the unbounded derived category. We follow [Lip2] $\underset{X}{\overset{O}{}}$ 2.2. For comparisons with more classical situations see Section (F.48).

Classically derived functors are constructed via acyclic resolutions (or injective resolutions, which are acyclic resolutions for all left exact functors). Since unbounded complexes of acyclic (or injective) objects do not have the same good properties as bounded below such complexes, one cannot work with resolutions by complexes of acyclic objects (or with injective resolutions). But it turns out that there is a good generalization, namely the notion of right F-acyclic complexes (or of K-*injective* complexes).

Throughout, let $\mathcal{K}$ be a triangulated subcategory of $K(\mathcal{A})$, let $\mathcal{E}$ be a triangulated category, and let

$$F\colon \mathcal{K} \to \mathcal{E}$$

be a triangulated functor. Let $Q\colon \mathcal{K} \to D^{\mathcal{K}}$ be the localization functor. Usually we will apply the following results in the following "classical" situation:

(CL) F is the functor $K(\mathcal{A}) \to D(\mathcal{B})$ induced by an additive functor $\mathcal{A} \to \mathcal{B}$ of abelian categories, again denoted by F (Remark F.167).

There is one easy example where the existence of right derived functors follows immediately from the definition.

Example F.169. Recall (Remark F.149) that the following assertions for a triangulated functor $F\colon \mathcal{K} \to \mathcal{E}$ are equivalent.
 (i) There exists a (necessarily unique) triangulated functor $\bar{F}\colon D^{\mathcal{K}} \to \mathcal{E}$ such that $F = \bar{F} \circ Q$.
 (ii) The functor F maps quasi-isomorphisms in $\mathcal{K}$ to isomorphisms in $\mathcal{E}$.
 (iii) The functor F maps exact complexes in $\mathcal{K}$ to 0 in $\mathcal{E}$.
In this case, the functor $\bar{F}$ is the right derived functor of F. It is also the left derived functor of F and we usually simply write F instead of RF or LF in this case.

In Situation (CL), the functor $F\colon K(\mathcal{A}) \to D(\mathcal{B})$ maps quasi-isomorphisms to isomorphisms if and only if it is induced by an exact functor $\mathcal{A} \to \mathcal{B}$. Indeed, if $\mathcal{A} \to \mathcal{B}$ is exact, then the induced functor sends quasi-isomorphisms in $K(\mathcal{A})$ to quasi-isomorphisms in $K(\mathcal{B})$ and hence to isomorphisms in $D(\mathcal{B})$. Conversely, F sends quasi-isomorphisms in $K(\mathcal{A})$ to isomorphisms in $D(\mathcal{B})$ if and only if the functor $K(\mathcal{A}) \to K(\mathcal{B})$ preserves quasi-isomorphisms. As a complex is exact if and only if there exists a quasi-isomorphism to 0, this shows that the functor $F\colon \mathcal{A} \to \mathcal{B}$ is exact.

The idea to construct derived functors more generally is to reduce to the situation in Example F.169. For this we will use two ingredients:
 (I) To construct a (preferably large) triangulated subcategory $\mathcal{J}$ of $\mathcal{K}$ such that $F_{|\mathcal{J}}$ maps quasi-isomorphisms to isomorphisms.
 (II) A recipe to extend derived functors on a triangulated subcategory (such as $\mathcal{J}$ as in (I)) to larger triangulated subcategories.

We start with Step (II), i.e., we consider the following situation. Let $\mathcal{J} \subseteq \mathcal{K} \subseteq K(\mathcal{A})$ be triangulated subcategories yielding a commutative diagram

$$
\begin{array}{ccc}
\mathcal{J} & \xrightarrow{\;j\;} & \mathcal{K} \\
{\scriptstyle Q_{\mathcal{I}}}\downarrow & & \downarrow{\scriptstyle Q_{\mathcal{K}}} \\
D^{\mathcal{J}} & \xrightarrow{\;\bar{j}\;} & D^{\mathcal{K}},
\end{array}
$$

where j is the inclusion functor and $\bar{j}$ is the induced functor.

Proposition F.170. ([Lip2] $\overset{\text{O}}{\text{X}}$ (1.7.2) and (2.2.3)) *Suppose that for every object X in $\mathcal{K}$ there exists a quasi-isomorphism $\varphi_X\colon X \to A_X$, where A_X is an object of $\mathcal{J}$.*

(1) *Then $\bar{\jmath}$ is an equivalence of triangulated categories $D^{\mathcal{J}} \overset{\sim}{\longrightarrow} D^{\mathcal{K}}$ and there exists a quasi-inverse functor $\rho\colon D^{\mathcal{K}} \to D^{\mathcal{J}}$ with $\rho(X) = A_X$ and such that the φ_X yield isomorphisms of triangulated functors $\mathrm{id}_{D^{\mathcal{J}}} \overset{\sim}{\to} \rho \circ \bar{\jmath}$ and $\mathrm{id}_{D^{\mathcal{K}}} \overset{\sim}{\to} \bar{\jmath} \circ \rho$.*

(2) *Suppose that the restriction of $F\colon \mathcal{K} \to \mathcal{E}$ to $\mathcal{J}$ has a right derived functor*

$$R^{\mathcal{J}}F\colon D^{\mathcal{J}} \to \mathcal{E}, \qquad \zeta_{\mathcal{J}}\colon F_{|\mathcal{J}} \to R^{\mathcal{J}}F \circ Q_{\mathcal{J}}.$$

Then F has a right derived functor (RF, ζ) with

$$RF = R^{\mathcal{J}}F \circ \rho$$

and $\zeta(X)$ for X in $\mathcal{K}$ the composition

$$F(X) \xrightarrow{\;F(\varphi_X)\;} F(A_X) \xrightarrow{\;\zeta_{\mathcal{J}}(A_X)\;} RF^{\mathcal{J}}(A_X) = RF(X).$$

Now we come to step (I). The key definition is the following.

Definition F.171. *We call an object $X \in \mathcal{K}$ right F-acyclic, if for every quasi-isomorphism $X \to Y$ in $\mathcal{K}$, there exists a quasi-isomorphism $Y \to Z$ such that the map $F(X) \to F(Z)$ obtained by applying F to the composition is an isomorphism. (Reversing the directions of the arrows, we obtain the notion of* left F-acyclic *object.)*

Then F transforms quasi-isomorphisms between F-acyclic complexes to isomorphisms, more precisely:

Lemma F.172. ([Lip2] $\overset{\text{O}}{\text{X}}$ (2.2.5)) *The full subcategory $(F\text{-acycl})$ of $\mathcal{K}$ consisting of right F-acyclic objects is a triangulated subcategory, and the functor $D^{F\text{-acycl}} \to D^{\mathcal{K}}$ induced by the inclusion is fully faithful. Moreover the restriction of F to $(F\text{-acycl})$ transforms quasi-isomorphisms into isomorphisms.*

Combining Example F.169, Proposition F.170, and Lemma F.172 we see that if there are enough right F-acyclic objects in the sense of the following proposition, then the right derived functor of F exists:

Theorem F.173. *Let $\mathcal{K} \subseteq K(\mathcal{A})$ be a triangulated subcategory and let $F\colon \mathcal{K} \to \mathcal{E}$ be a triangulated functor of triangulated categories. Assume that for every object X of $\mathcal{K}$ there exists a quasi-isomorphism $\varphi_X\colon X \to A_X$, where A_X is right F-acyclic.*

Then F has a right derived functor $(RF\colon D^{\mathcal{K}} \to \mathcal{E}, \zeta\colon F \to RF\circ Q)$ with $RF(A) = F(A)$ for $A \in \mathrm{Ob}(D^{F\text{-acycl}}) = \mathrm{Ob}(F\text{-acycl})$. In particular,

$$RF(X) = F(A_X), \qquad for\ X \in \mathrm{Ob}(D^{\mathcal{K}}) = \mathrm{Ob}(\mathcal{K}).$$

The morphism $\zeta\colon F \to RF \circ Q$ is given by

$$F(\varphi_X)\colon F(X) \to F(A_X) = RF(X).$$

We have the following complement to the theorem.

Corollary F.174. ([Lip2] $\frac{0}{X}$ (2.2.6)) *Suppose the hypotheses of Theorem F.173 are satisfied.*
(1) *A complex X in $\mathcal{K}$ is right F-acyclic if and only if $\zeta(X)\colon F(X) \to RF(X)$ is an isomorphism.*
(2) *Let $G\colon \mathcal{E} \to \mathcal{E}'$ be any triangulated functor. Then $(G \circ RF, G(\zeta))$ is a right derived functor of $G \circ F$.*

Remark F.175. In practice, we will use Corollary F.174 (2) as follows. The triangulated categories $\mathcal{E}$ and $\mathcal{E}'$ will be a triangulated subcategories of $D(\mathcal{B})$ and $D(\mathcal{B}')$, respectively, for some abelian categories $\mathcal{B}$ and $\mathcal{B}'$. The functor $F\colon \mathcal{K} \to \mathcal{E}$ will be induced by some additive functor $F\colon \mathcal{A} \to \mathcal{B}$. And the functor G will be the functor induced by an exact functor $G\colon \mathcal{B} \to \mathcal{B}'$ (Example F.169). Then

(F.43.1) $$R(G \circ F) = G \circ RF.$$

In addition, we have the following compatibility with composition of functors:

Proposition F.176. (cf. [Lip2] $\frac{0}{X}$ Cor. 2.2.7) *Let $\mathcal{A}$, $\mathcal{A}'$ be abelian categories, $\mathcal{K} \subset K(\mathcal{A})$ and $\mathcal{K}' \subset K(\mathcal{A}')$ triangulated subcategories, and $Q\colon \mathcal{K} \to D^{\mathcal{K}}$, $Q'\colon \mathcal{K}' \to D^{\mathcal{K}'}$ the localization functors. Let $\mathcal{E}$ be a triangulated category.*
Let $G\colon \mathcal{K}' \to \mathcal{E}$ be a triangulated functor which has a right derived functor RG.
Let $F\colon \mathcal{K} \to \mathcal{K}'$ be a triangulated functor. Assume that there exists a family of quasi-isomorphisms $X \to A_X$, $X \in \mathcal{K}$, such that A_X is right $(Q' \circ F)$-acyclic, and $F(A_X)$ is right G-acyclic.
Then $Q' \circ F$ has a right derived functor RF and $G \circ F$ has a right derived functor $R(GF)$. There is a unique isomorphism $\alpha\colon R(GF) \overset{\sim}{\to} RG \circ RF$ of triangulated functors $D^{\mathcal{K}} \to \mathcal{E}$ such that the following diagram of morphisms of functors commutes:

$$
\begin{array}{ccc}
GF & \longrightarrow & R(GF) \circ Q \\
\downarrow & & \downarrow{\scriptstyle \alpha \circ Q} \\
RG \circ Q' \circ F & \longrightarrow & RG \circ RF \circ Q.
\end{array}
$$

Remark F.177. Let $\mathcal{A}$, $\mathcal{A}'$, $\mathcal{K} \subset K(\mathcal{A})$, $\mathcal{K}' \subset K(\mathcal{A}')$, and $\mathcal{E}$ be as in Proposition F.176. Let $F\colon \mathcal{K} \to \mathcal{K}'$ and $G\colon \mathcal{K}' \to \mathcal{E}$ be triangulated functors such that G has a right derived functor RG.
Suppose that F preserves quasi-isomorphisms (for instance, if F is induced by an exact functor $\mathcal{A} \to \mathcal{A}'$). Then $RF = F$ (omitting localization functors) and every X in $\mathcal{K}$ is right F-acyclic. Hence if there exists a family of quasi-isomorphisms $X \to A_X$, $X \in \mathcal{K}$, such that $F(A_X)$ is right G-acyclic, then

$$R(G \circ F) = RG \circ F.$$

We can apply Remark F.177 to inclusions of triangulated subcategories:

Corollary F.178. *Let $\mathcal{K}' \subseteq \mathcal{K}$ be triangulated subcategories of $K(\mathcal{A})$ and let $j\colon \mathcal{K}' \to \mathcal{K}$ be the inclusion. Let $F\colon \mathcal{K} \to \mathcal{E}$ be a triangulated functor such that $(RF\colon D^{\mathcal{K}} \to \mathcal{E}, \zeta)$ exists. Suppose that for every complex X in $\mathcal{K}'$ there exists a quasi-isomorphism $X \to A_X$ in $\mathcal{K}'$ such that A_X is right F-acyclic (as object of $\mathcal{K}$).*

Then the right derived functor $(R^{\mathcal{K}'}F, \zeta')$ of $F_{|\mathcal{K}'}$ exists and there exists a unique isomorphism

$$\alpha \colon R^{\mathcal{K}'}F \xrightarrow{\sim} RF \circ j$$

such that for all X in $\mathcal{K}'$ one has $\alpha(QX) \circ \zeta'(X) = \zeta(j(X))$.

(F.44) K-injective resolutions.

In many cases, we can find enough objects which are right acyclic for all triangulated functors on $\mathcal{K}$ simultaneously. The key notion is the notion of K-injective objects introduced by Spaltenstein. Dually we have the notion of K-projective objects. In the realm of algebraic geometry, such objects are however less abundant, so that in practice this notion turns out to be less relevant, and finding left derived functors often has to be done in different ways. In algebraic geometry, the notion of K-flat complexes, which are left acyclic for the tensor product, is the more important one.

Definition and Proposition F.179. ([Lip2] $\overset{\text{O}}{\text{X}}$ *Prop. (2.3.8)) Let $\mathcal{A}$ be an abelian category, and let $\mathcal{K}$ be a triangulated subcategory of $K(\mathcal{A})$. An object $I \in \mathcal{K}$ is called K-injective (or homotopically injective, or q-injective) in $\mathcal{K}$, if the following equivalent conditions are satisfied.*
(1) *I is right F-acyclic for every triangulated functor $F \colon \mathcal{K} \to \mathcal{E}$.*
(2) *One has $\operatorname{Hom}_{\mathcal{K}}(X, I) = 0$ for every exact complex X in $\mathcal{K}$.*
(3) *For every morphism $f \colon X \to I$ and quasi-isomorphism $s \colon X \to Y$ (both in $\mathcal{K}$), there exists $g \colon Y \to I$ in $\mathcal{K}$ with $gs = f$. (This g is necessarily unique.)*
(4) *Every quasi-isomorphism $s \colon I \to Y$ in $\mathcal{K}$ has a left inverse.*
(5) *Every quasi-isomorphism $s \colon I \to Y$ in $\mathcal{K}$ is a monomorphism.*
(6) *The triangulated functor $\operatorname{Hom}_{\mathcal{A}}(-, I) \colon \mathcal{K}^{\mathrm{op}} \to K(\mathrm{AbGrp})$ (Example F.127) preserves the property quasi-isomorphism.*
(7) *For every $X \in \mathcal{K}$, the natural map*

(F.44.1) $\operatorname{Hom}_{\mathcal{K}}(X, I) \cong \operatorname{Hom}_{D\mathcal{K}}(X, I)$

is a bijection.
One also defines dually the notion of an object $P \in \mathcal{K}$ that is called K-projective (or homotopically projective, or q-projective).

If an object I in $\mathcal{K}$ is K-injective (resp. K-projective) in $K(\mathcal{A})$, then it is also K-injective (resp. K-projective) in $\mathcal{K}$.

Remark F.180. Property (2) of Proposition F.179 shows that if I is K-injective and exact, then $\operatorname{id}_I = 0$ in $K(\mathcal{A})$, i.e., I is homotopy equivalent to 0.

Remark F.181. Let $\mathcal{A}$ be an abelian category. The connection between K-injective complexes and complexes of injective objects is the following ([Lip2] $\overset{\text{O}}{\text{X}}$ (2.3.3) and (2.3.4), see also Exercise F.20).
(1) An object A of $\mathcal{A}$ is injective if and only if it is K-injective in $K(\mathcal{A})$ (or, equivalently, in $K^b(\mathcal{A})$) if considered as a complex concentrated in degree 0.
(2) Every bounded below complex of injective objects of $\mathcal{A}$ is K-injective in $K(\mathcal{A})$. Conversely, if $\mathcal{A}$ has enough injective objects, then any complex I in $K^+(\mathcal{A})$ which is $K^+(\mathcal{A})$-injective is isomorphic in $K^+(\mathcal{A})$ to a bounded below complex of injective objects in $\mathcal{A}$.

Example F.182. It is not true that every complex consisting of injective objects is K-injective. Consider the category of modules over the ring $\mathbb{Z}/4\mathbb{Z}$, and the complex which in each degree has a free $\mathbb{Z}/4\mathbb{Z}$-module of rank 1, and with all differentials given by multiplication by 2. This is an exact complex of injective objects, but it is not K-injective. Because in that case it would necessarily be homotopic to 0 by Remark F.180, however tensoring by $- \otimes_{\mathbb{Z}/4} \mathbb{Z}/2\mathbb{Z}$ we obtain a complex which is not exact. Note that the same complex also is an example of a complex consisting of projective objects which is not K-projective.

Corollary F.183. *Let $F\colon \mathcal{A} \to \mathcal{B}$ be an additive functor of abelian categories. Suppose that F has an exact left adjoint functor G. Then F sends injective objects to injective objects and the induced functor $F\colon K(\mathcal{A}) \to K(\mathcal{B})$ sends K-injective complexes to K-injective complexes.*

Proof. We use (2) of Proposition F.179. Let I in $K(\mathcal{A})$ be a K-injective complex and let Y be an exact complex in $K(\mathcal{B})$. Then by Example F.130 we have

$$\mathrm{Hom}_{K(\mathcal{B})}(Y, F(I)) = \mathrm{Hom}_{K(\mathcal{A})}(G(Y), I) = 0$$

because $G(Y)$ is still exact. As an object is injective if and only if it is K-injective as a complex concentrated in degree 0, the first assertion also follows. $\qquad\square$

Definition F.184. *Let $\mathcal{A}$ be an abelian category, and let $\mathcal{K} \subset K(\mathcal{A})$ be a triangulated subcategory. We say that $\mathcal{K}$ has enough K-injective objects, if for every X in $\mathcal{K}$ there exists a quasi-isomorphism $X \to I_X$ where I_X is a K-injective object of $\mathcal{K}$. We then call such a quasi-isomorphism $X \to I_X$ a K-injective resolution of X.*

By Proposition F.179 (3), a K-injective resolution is unique up to unique isomorphism in $K(\mathcal{A})$.

Grothendieck abelian categories (Definition F.54) have enough K-injective objects. More precisely:

Theorem F.185. ([Sta] 079P) *Let $\mathcal{A}$ be a Grothendieck abelian category. Then for any complex X in $K(\mathcal{A})$ there exists a quasi-isomorphism and monomorphism of complexes $u\colon X \to I_X$ such that I_X is a K-injective complex in $K(\mathcal{A})$ and such that each component of I is an injective object of $\mathcal{A}$. Moreover, this construction is functorial in X.*

If there exists $n \in \mathbb{Z}$ with $H^i(X) = 0$ for $i < n$, then one can choose I such that $I^i = 0$ for all $i < n$.

Corollary F.186. *Let $\mathcal{A}$ be a Grothendieck abelian category, let $\mathcal{I}$ be the full subcategory of $\mathcal{A}$ of injective objects, and let $K_{\mathrm{K-inj}}(\mathcal{A}) \subseteq K(\mathcal{A})$ be the homotopy category of K-injective complexes. Then $K^+(\mathcal{I}) = K^+_{\mathrm{K-inj}}(\mathcal{A})$ and the natural functors*

(F.44.2) $$K_{\mathrm{K-inj}}(\mathcal{A}) \to D(\mathcal{A}), \qquad K^+(\mathcal{I}) \to D^+(\mathcal{A})$$

are triangulated equivalences.

In particular we see that for Grothendieck abelian categories $\mathcal{A}$, the derived category $D(\mathcal{A})$ is actually a category with small Hom sets.

For us the main application of the results above is the following corollary.

Corollary F.187. *Let $\mathcal{A}$ be a Grothendieck abelian category. Let $F\colon K(\mathcal{A}) \to \mathcal{E}$ be a triangulated functor to a triangulated category $\mathcal{E}$.*

(1) *Then F admits a right derived functor $RF\colon D(\mathcal{A}) \to \mathcal{E}$, and $RF(X) = F(I_X)$, where $X \to I_X$ is a quasi-isomorphism to a K-injective complex.*

(2) *The restriction of F to $K^+(\mathcal{A})$ also admits a right derived functor $R^+F\colon D^+(\mathcal{A}) \to \mathcal{E}$, and R^+F is the restriction of RF to $D^+(\mathcal{A})$.*

(3) *If $G\colon \mathcal{E} \to \mathcal{E}'$ is any triangulated functor, then $G \circ RF$ is the right derived functor of $G \circ F$.*

Furthermore, Proposition F.176 on the derived functor of a composition of functors specializes to the situation at hand (using K-injective resolutions) in the obvious way.

Lemma F.188. ([Sta] 07D9) *Let $\mathcal{A}$ be a Grothendieck abelian category. Then the derived category $D(\mathcal{A})$ has arbitrary products and coproducts.*

Coproducts are obtained by taking termwise direct sums of any representing complexes. Products are obtained by taking termwise products of K-injective representing complexes.

A Grothendieck category usually has not K-projective resolutions. But there is one important special case in which this is the case.

Theorem F.189. ([Spa]^0 Theorem C) *Let A be a ring. Then for every complex X of A-modules there exists a K-projective complex of A-modules P and a quasi-isomorphism $P \to X$.*

In fact, the localization functor $Q\colon K(A) \to D(A)$ has a fully faithful left adjoint functor P and for X in $K(A)$ the counit of the adjunction $P(Q(X)) \to X$ defines a functorial K-projective resolution of X ([Gil]$_X^0$).

(F.45) Adjointness of derived functors.

In this section, let $F\colon \mathcal{A} \to \mathcal{B}$ and $G\colon \mathcal{B} \to \mathcal{A}$ be additive functors of abelian categories such that G is left adjoint to F.

Remark F.190. Let X in $C(\mathcal{A})$ and Y in $C(\mathcal{B})$ be complexes. Then the adjunction isomorphisms $\operatorname{Hom}_{\mathcal{B}}(Y^i, F(X^j)) \xrightarrow{\sim} \operatorname{Hom}_{\mathcal{A}}(G(Y^i), X^j)$ for all $i, j \in \mathbb{Z}$ yield an isomorphism of complexes of abelian groups

(F.45.1) $$\operatorname{Hom}_{\mathcal{B}}(Y, F(X)) \xrightarrow{\sim} \operatorname{Hom}_{\mathcal{A}}(G(Y), X)$$

which is functorial in X and Y. Applying $Z^0(-)$ and $H^0(-)$ to this isomorphism one obtains functorial isomorphisms of abelian groups

(F.45.2)
$$\operatorname{Hom}_{C(\mathcal{B})}(Y, F(X)) \xrightarrow{\sim} \operatorname{Hom}_{C(\mathcal{A})}(G(Y), X),$$
$$\operatorname{Hom}_{K(\mathcal{B})}(Y, F(X)) \xrightarrow{\sim} \operatorname{Hom}_{K(\mathcal{A})}(G(Y), X).$$

Now suppose that for every complex X in $K(\mathcal{A})$ there exists a quasi-isomorphism $X \to I_X$ with I_X right F-acyclic and that for every complex Y in $K(\mathcal{B})$ there exists a quasi-isomorphism $P_Y \to Y$ with P_Y left G-acyclic. Then the right derived functor $RF\colon D(\mathcal{A}) \to D(\mathcal{B})$ and the left derived functor $LG\colon D(\mathcal{B}) \to D(\mathcal{A})$ exist.

Proposition F.191. ([Lip2]$_X^O$ (3.2.1), (3.2.2)[5]) *The pair* (LG, RF) *is a triangulated adjoint pair (Definition F.128). More precisely, there exists for* $X \in D(\mathcal{A})$ *and* $Y \in D(\mathcal{B})$ *a unique functorial isomorphism*

$$\rho \colon \mathrm{Hom}_{D(\mathcal{B})}(Y, RF(X)) \xrightarrow{\sim} \mathrm{Hom}_{D(\mathcal{A})}(LG(Y), X)$$

such that the following diagram is commutative

$$
\begin{array}{ccccc}
\mathrm{Hom}_{K(\mathcal{B})}(Y, F(X)) & \xrightarrow{\nu} & \mathrm{Hom}_{D(\mathcal{B})}(Y, F(X)) & \longrightarrow & \mathrm{Hom}_{D(\mathcal{B})}(Y, RF(X)) \\
{\scriptstyle (\mathrm{F.45.2})} \downarrow \cong & & & & \cong \downarrow \rho \\
\mathrm{Hom}_{K(\mathcal{A})}(G(Y), X) & \xrightarrow{\mu} & \mathrm{Hom}_{D(\mathcal{A})}(G(Y), X) & \longrightarrow & \mathrm{Hom}_{D(\mathcal{A})}(LG(Y), X).
\end{array}
$$

Moreover, if X is K-injective and Y is left G-acyclic (resp. if X is right F-acyclic and Y is K-projective), then ν (resp. μ) is an isomorphism.

(F.46) Bounded Functors.

Let $\mathcal{A}$ and $\mathcal{B}$ be abelian categories and let $\mathcal{A}'$ be a plump subcategory of $\mathcal{A}$. Recall that we defined for $* \in \{b, +, -, \emptyset\}$ the triangulated subcategory $D^*_{\mathcal{A}'}(\mathcal{A})$ of $D(\mathcal{A})$ consisting of complexes X such that $H^i(X) \in \mathcal{A}'$ for all $i \in \mathbb{Z}$ and that $H^i(X) = 0$ for $|i| \gg 0$ (resp. $i \ll 0$, resp. $i \gg 0$) if $* = b$ (resp. $* = +$, resp. $* = -$). Let $F \colon D^*_{\mathcal{A}'}(\mathcal{A}) \to D(\mathcal{B})$ be a triangulated functor. Typically, F could be the right or left derived functor of some functor $K^*_{\mathcal{A}'}(\mathcal{A}) \to D(\mathcal{B})$.

As $D^*_{\mathcal{A}'}(\mathcal{A})$ is a triangulated subcategory stable under the truncation functors $\tau^{\leq n}$ and $\tau^{\geq n}$ for all $n \in \mathbb{Z}$, we find (cf. [Lip2]$_X^O$ (1.11.2))

(F.46.1)
$$\{\, d \;;\; F(D^{\leq n}_{\mathcal{A}'}(\mathcal{A})) \subseteq D^{\leq n+d}(\mathcal{B}) \text{ for all (or one) } n \in \mathbb{Z} \,\}$$
$$= \{\, d \;;\; H^i F(X) \xrightarrow{\sim} H^i F(\tau^{\geq n} X) \text{ for all } X, \text{ all (or one) } n \in \mathbb{Z} \text{ and all } i \geq n+d \,\},$$

(F.46.2)
$$\{\, d \;;\; F(D^{\geq n}_{\mathcal{A}'}(\mathcal{A})) \subseteq D^{\geq n-d}(\mathcal{B}) \text{ for all (or one) } n \in \mathbb{Z} \,\}$$
$$= \{\, d \;;\; H^i F(\tau^{\leq n} X) \xrightarrow{\sim} H^i F(X) \text{ for all } X, \text{ all (or one) } n \in \mathbb{Z} \text{ and all } i \leq n-d \,\}.$$

More precisely, in the descriptions given second, the condition is to be understood as saying that the map induced by the exact triangle $\tau^{\leq n} X \to X \to \tau^{\geq n+1} X \to$ (Proposition F.157) is an isomorphism.

Let $\dim^+(F)$ (resp. $\dim^-(F)$) be the infimum of the set (F.46.1) (resp. of (F.46.2)). We say that F is *cohomologically bounded above* (resp. *cohomologically bounded below*) if $\dim^+ F < \infty$ (resp. if $\dim^- F < \infty$). The functor F is called *cohomologically bounded* if it is cohomological bounded above and below. If F is the (right or left) derived functor of some functor F', we also write $\dim^+(F')$ instead of $\dim^+(F)$, similarly for the other definitions.

---

[5] The proof in loc. cit. is given only for special functors. In the general case used here the proof is verbatim the same.

Usually, if F is the right (resp. left) derived functor on $D^+(\mathcal{A})$ (resp. on $D^-(\mathcal{A})$) of some additive non-zero functor $\mathcal{A} \to \mathcal{B}$, then we will have $\dim^-(F) = 0$ (resp. $\dim^+(F) = 0$), see Remark F.200 (2) below for details. Hence for right (resp. left) derived functors $\dim^+ F$ (resp. $\dim^- F$) is the interesting invariant. The cohomological dimension is usually determined as follows.

Lemma F.192. ([Lip2]$\substack{O\\X}$ (1.11.2)) *Let* $F \colon D^+_{\mathcal{A}'}(\mathcal{A}) \to D(\mathcal{B})$ *be a triangulated functor. Then*

$$\dim^+ F = \inf\{\, d \in \mathbb{Z} \;;\; H^i F(A) = 0 \text{ for all } i > d \text{ and } A \text{ in } \mathcal{A}'\,\}.$$

Similarly, if $L \colon D^-_{\mathcal{A}'}(\mathcal{A}) \to D(\mathcal{B})$ *is a triangulated functor, then*

$$\dim^- L = \inf\{\, d \in \mathbb{Z} \;;\; H^i F(A) = 0 \text{ for all } i < -d \text{ and } A \text{ in } \mathcal{A}'\,\}.$$

Proposition F.193. ([Lip2]$\substack{O\\X}$ (1.11.3)) *Let* $F, G \colon D^*_{\mathcal{A}'}(\mathcal{A}) \to D(\mathcal{B})$ *be triangulated functors and suppose that one of the following conditions hold.*
(a) $* = b$.
(b) $* = +$ *and both* F *and* G *are cohomologically bounded below.*
(c) $* = -$ *and both* F *and* G *are cohomologically bounded above.*
(d) $* = \emptyset$ *and both* F *and* G *are cohomologically bounded.*
Let $\eta \colon F \to G$ *a morphism of triangulated functors. Then* η *is an isomorphism if and only if* $\eta(A)$ *is an isomorphism for every object* A *of* $\mathcal{A}'$.

Moreover, suppose that Condition (b) (resp. Condition (c)) holds, and let I *(resp.* P*) be a set of objects of* $\mathcal{A}'$ *such that every object of* $\mathcal{A}'$ *admits a monomorphism into an object of* I *(resp. is the target of an epimorphism out of an object of* P*). Then* η *is an isomorphism if* $\eta(A)$ *is an isomorphism for every object* A *of* I *(resp. of* P*).*

(F.47) Construction of resolutions.

In this section we collect lemmas which allow us to construct resolutions. We will only consider the case of right resolutions of a complex X, i.e., of quasi-isomorphisms $X \to I_X$, where I_X lies in a special class of complexes (except for Corollary F.197). The case of left resolutions is entirely dual. We will always denote by $\mathcal{A}$ an abelian category.

The main technical tool to construct right resolutions of bounded below complexes that have components of a prescribed type is the following lemma.

Lemma F.194. *Let* $\mathcal{I}$ *be a class of objects of* $\mathcal{A}$ *containing* 0 *such that every object* A *of* $\mathcal{A}$ *admits a monomorphism into an object* $I^0(A)$ *in* $\mathcal{I}$*. We consider* $\mathcal{I}$ *as a full subcategory.*
(1) *Let* $n \in \mathbb{Z}$*. Every complex* $X \in C(\mathcal{A})$ *with* $H^i(X) = 0$ *for all* $i < n$ *admits a quasi-isomorphism to a bounded below complex* I_X *of objects in* $\mathcal{I}$ *with* $I^i_X = 0$ *for all* $i < n$*.*
(2) *Suppose that for all* $I, J \in \mathcal{I}$ *one also has* $I \oplus J \in \mathcal{I}$ *and that* $A \mapsto I^0(A)$ *is an additive functor* $\mathcal{A} \to \mathcal{I}$*. Then* $X \mapsto I_X$ *can in addition be chosen to be functorial in* X*, a monomorphism in each degree, and such that* $I_{X[k]} = I_X[k]$ *for all* $k \in \mathbb{Z}$*.*

We will only prove (and use) the second assertion. For the first one we refer to [Sta] 05T6.

Proof. By hypothesis, the quotient map $X \to \tau^{\geq n} X$ is a quasi-isomorphism. Replacing X by $\tau^{\geq n} X$, we may assume that $X^i = 0$ for $i < n$.

Let us first assume that $X = A$ is concentrated in degree 0. By hypothesis we are given a functorial monomorphism $u \colon A \hookrightarrow I^0(A)$ with $I^0(A)$ in $\mathcal{I}$. We define an exact sequence

$$0 \to A \to I^0(A) \to I^1(A) \to I^2(A) \to \cdots$$

inductively by $I^1(A) := I^0(\mathrm{Coker}(u))$ and $I^i(A) := I^0(\mathrm{Coker}(I^{i-2}(A) \to I^{i-1}(A)))$ and hence obtain a quasi-isomorphism of complexes $A \to I_A$ with $I_A^i := I^i(A)$ for $i \geq 0$ and $I_A^i := 0$ for $i < 0$. As $A \mapsto I^0(A)$ is functorial, $A \mapsto I_A$ is functorial in A.

Now let X be an arbitrary complex with $X^i = 0$ for $i < n$. By the functoriality of $A \mapsto I_A$ we obtain a complex of complexes

$$\cdots \to 0 \to I_{X^n} \to I_{X^{n+1}} \to \cdots,$$

i.e., a double complex $\tilde{I}_X := (I_{X^i}^j)_{i,j}$. As X is bounded below and $I_A^j = 0$ for all $j < 0$, the total complex I_X attached to this double complex exists.

If we consider X as a double complex by $X^{i,0} := X^i$ and $X^{i,j} := 0$ for $j \neq 0$, the morphisms $X^i \to I_{X^i}^0$ yield a morphism $\alpha_X \colon X \to \tilde{I}_X$ of double complexes which is a quasi-isomorphism in vertical direction. Hence it induces a quasi-isomorphism $X = \mathrm{Tot}(X) \to \mathrm{Tot}(\tilde{I}_X) = I_X$ by Lemma F.113. This quasi-isomorphism is functorial in X, a monomorphism in each degree and compatible with shift. $\square$

To construct resolutions of unbounded complexes one can combine Lemma F.194 with the following lemma.

Lemma F.195. ([Spa]$^\circ$ 3.7) *Let $\mathcal{I}$ be a class of bounded below complexes in $C(\mathcal{A})$ satisfying the following properties.*
(a) *If $X' \to X \to X''$ is a termwise split sequence of complexes in $C(\mathcal{A})$ such that X' and X'' are in $\mathcal{I}$, then X is in $\mathcal{I}$.*
(b) *For every bounded below complex Y in $C(\mathcal{A})$ there exists a quasi-isomorphism $Y \to I_Y$ with I_Y in $\mathcal{I}$.*
Let X be a complex. Then there exists a commutative diagram of complexes

(F.47.1)
$$\begin{array}{ccccccc} \tau^{\geq 0} X & \longleftarrow & \tau^{\geq -1} X & \longleftarrow & \tau^{\geq -2} X & \longleftarrow & \cdots \\ \downarrow {\scriptstyle f_0} & & \downarrow {\scriptstyle f_1} & & \downarrow {\scriptstyle f_2} & & \\ I_0 & \longleftarrow & I_1 & \longleftarrow & I_2 & \longleftarrow & \cdots \end{array}$$

such that
(1) *all complexes I_n are in $\mathcal{I}$,*
(2) *for all $n \geq 1$, the morphism $I_n \to I_{n-1}$ is a termwise split epimorphism whose kernel is a complex in $\mathcal{I}$,*
(3) *The vertical morphisms f_n are quasi-isomorphisms for all $n \geq 0$.*

Remark F.196. If I_Y depends functorially on Y, then we can obtain a diagram (F.47.1) satisfying (1) and (3) and depending functorially on X by simply setting $I_n := I_{\tau^{\geq -n} X}$.

From the commutative diagram (F.47.1) one obtains a morphism of complexes

(F.47.2)
$$X = \lim_n \tau^{\geq -n} X \longrightarrow \lim_n I_n$$

if these limits exist. As the morphisms f_n are quasi-isomorphisms, this morphism is a quasi-isomorphism if limits indexed by $\mathbb{N}$ are exact. Unfortunately this is almost never the case for those abelian categories that we are interested in, namely the abelian category of modules over a ring or a sheaf of rings. But the dual statement that colimits indexed by $\mathbb{N}$ (or any filtered colimits) are exact is true in these categories. Hence we deduce from the dual versions of Lemma F.194 and Remark F.196 the following corollary.

Corollary F.197. *Let $\mathcal{A}$ be an abelian category such that colimits indexed by the totally ordered set $\mathbb{N}$ exist and are exact (e.g., if $\mathcal{A}$ is a Grothendieck abelian category). Let $\mathcal{P}$ be a full additive subcategory of $\mathcal{A}$ such that for every object A in $\mathcal{A}$ there is an epimorphism $P_A \to A$ with P_A in $\mathcal{P}$ which is functorial in A.*

Then for every complex X in $C(\mathcal{A})$ there exists a quasi-isomorphism $P_X \to X$, functorial in X and compatible with shift, where P_X is a colimit indexed by $\mathbb{N}$ of bounded above complexes whose components are in $\mathcal{P}$.

(F.48) Derived functors on $D^+(\mathcal{A})$ and higher derived functors of left exact functors.

We will now translate the theory to a more "classical" language by linking derived functors on $D^+(\mathcal{A})$ to higher derived functors as (universal) δ-functors (see Definition F.201 below). We will also explain how to obtain quasi-isomorphisms to acyclic complexes via resolutions of complexes of acyclic objects, e.g., injective resolutions.

More precisely, let us consider an additive functor

$$F \colon \mathcal{A} \longrightarrow \mathcal{B}$$

between abelian categories $\mathcal{A}$ and $\mathcal{B}$. The induced triangulated functors $K^+(\mathcal{A}) \to D^+(\mathcal{B})$ and $K^-(\mathcal{A}) \to D^-(\mathcal{B})$ are again denoted by F. We are interested in right derived functors $R^+F \colon D^+(\mathcal{A}) \to D^+(\mathcal{B})$ and left derived functors $L^-F \colon D^-(\mathcal{A}) \to D^-(\mathcal{B})$. We will consider only the case R^+F. The case L^-F is entirely dual.

We will consider the following assumption on $\mathcal{A}$ and F that will be usually satisfied in our applications.

Condition (Ac) F.198. For every complex X in $K^+(\mathcal{A})$ there exists a quasi-isomorphism $X \to C_X$, where C_X is a right F-acyclic complex in $K^+(\mathcal{A})$. Moreover if $n \in \mathbb{Z}$ is an integer such that $X^i = 0$ for all $i < n$, then we may assume that $C_X^i = 0$ for all $i < n$.

Then the derived functor $R^+F \colon D^+(\mathcal{A}) \to D^+(\mathcal{B})$ exists and $R^+F(X) = F(C_X)$ by Theorem F.173.

Below we will give a criterion for (Ac) to hold via acyclic resolutions. A special case of this criterion and a principal example, where Condition (Ac) is satisfied, is the existence of enough injective objects in $\mathcal{A}$:

Example F.199. Assume that $\mathcal{A}$ has enough injective objects. We claim that Condition (Ac) is satisfied for every additive functor $F \colon \mathcal{A} \to \mathcal{B}$ to an abelian category $\mathcal{B}$.

Indeed, by Lemma F.194 every complex X in $K^+(\mathcal{A})$ admits a quasi-isomorphism to a bounded below complex I_X of injective objects with $I_X^i = 0$ for all $i < n$ if $X^i = 0$ for all $i < n$, some $n \in \mathbb{Z}$. As I_X is K-injective in $K^+(\mathcal{A})$ (Remark F.181 (2)), it is right acyclic for every functor (Proposition F.179) und in particular right F-acyclic.

Therefore the right derived functor $R^+F\colon D^+(\mathcal{A}) \to D^+(\mathcal{A}')$ exists, and $R^+F(X) \cong F(I_X)$ for every $X \in D^+(\mathcal{A})$.

We collect some properties of the functor $R^+F\colon D^+(\mathcal{A}) \to D^+(\mathcal{B})$ under the assumption that Condition (Ac) holds. Recall that in this case we have the right derived functors $R^iF := H^i \circ R^+F\colon D^+(\mathcal{A}) \to \mathcal{B}$ for $i \in \mathbb{Z}$, which we often restrict to functors $R^iF\colon \mathcal{A} \to \mathcal{B}$ and call them *higher derived functors of F*.

Remark F.200. Let $F\colon \mathcal{A} \to \mathcal{B}$ be an additive functor. Suppose that Condition (Ac) holds.
(1) Let X be in $D^+(\mathcal{A})$ and let $n \in \mathbb{Z}$ such that $H^i(X) = 0$ for all $i < n$. Then $R^iF(X) = 0$ for all $i < n$.

Indeed, by assumption $X \to \tau^{\geq n}X$ is a quasi-isomorphism and by (Ac) we can choose a quasi-isomorphism $\tau^{\geq n}X \to C_{\tau^{\geq n}X}$ to a right F-acyclic complex whose components are 0 in degrees $< n$. Then $RF(X) = RF(C_{\tau^{\geq n}X}) = F(C_{\tau^{\geq n}X})$ and hence $R^iF(X) = 0$ for all $i < n$.

(2) In particular, for every object A in $\mathcal{A}$, we find that $R^iF(A) = 0$ for all $i < 0$. In other words, $\dim^-(RF) \leq 0$ (and $= 0$ if $F \neq 0$). Moreover, from the long exact cohomology sequence attached to a short exact sequence in $\mathcal{A}$ (Remark F.167) it follows that R^0F is always left exact.

(3) The morphism $F \to RF$ of functors induces a morphism $F \to R^0F$ of functors $\mathcal{A} \to \mathcal{B}$, and the latter is an isomorphism if and only if F is left exact.

Indeed, by (2) the condition is clearly necessary. Conversely, if F is left exact and A is an object of $\mathcal{A}$, then the quasi-isomorphism $A \to C_A$ is the same as an exact sequence $0 \to A \to C_A^0 \to C_A^1 \to \dots$ and $0 \to F(A) \to F(C_A^0) \to F(C_A^1)$ is exact. Therefore $F(A) = H^0(F(C_A)) = R^0(A)$.

Considering R^iF as functors $\mathcal{A} \to \mathcal{B}$, we obtain by Remark F.167 a δ-functor $(R^iF)_{i\geq 0}$ in the following sense.

Definition F.201. *Let $\mathcal{A}$, $\mathcal{B}$ be abelian categories.*
(1) *A δ-functor from $\mathcal{A}$ to $\mathcal{B}$ is a collection $(T^i)_{i\geq 0}$ of functors $\mathcal{A} \to \mathcal{B}$ together with "connecting" morphisms*

$$\delta^i\colon T^i(A'') \to T^{i+1}(A') \text{ for each short exact sequence } 0 \to A' \to A \to A'' \to 0 \text{ in } \mathcal{A},$$

for all $i \geq 0$, such that
(a) *For each short exact sequence $0 \to A' \to A \to A'' \to 0$ in $\mathcal{A}$, the sequence*

$$0 \to T^0(A') \to T^0(A) \to T^0(A'') \to T^1(A') \to \cdots$$
$$\cdots \to T^i(A') \to T^i(A) \to T^i(A'') \to T^{i+1}(A') \to \cdots$$

is exact, and
(b) *for every commutative diagram with exact rows in $\mathcal{A}$*

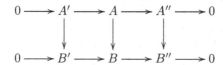

the connecting morphisms δ^i yield a commutative diagram

$$T^i(A'') \xrightarrow{\delta^i} T^{i+1}(A')$$
$$\downarrow \qquad\qquad\qquad \downarrow$$
$$T^i(B'') \xrightarrow{\delta^i} T^{i+1}(B').$$

(2) A δ-functor $T = (T^i)_i$ from $\mathcal{A}$ to $\mathcal{B}$ is called universal, if for every δ-functor $U = (U^i)_i$ from $\mathcal{A}$ to $\mathcal{B}$ and every morphism $f^0: T^0 \to U^0$ of functors, there exist unique morphisms $f^i: T^i \to U^i$ of functors, for all $i > 0$, such that for each short exact sequence $0 \to A' \to A \to A'' \to 0$ in $\mathcal{A}$, the diagram

$$T^i(A'') \xrightarrow{\delta^i} T^{i+1}(A')$$
$$\downarrow f^i(A'') \qquad\qquad \downarrow f^{i+1}(A')$$
$$U^i(A'') \xrightarrow{\delta^i} U^{i+1}(A').$$

is commutative.

There is the following criterion for a δ-functor to be universal.

Proposition F.202. ([Gro1] $^{\circ}$ 2.2.1) *Let $(T^i)_{i\geq 0}$ be a δ-functor from $\mathcal{A}$ to $\mathcal{B}$. Suppose that for every $i > 0$ and any object A of $\mathcal{A}$ there exists a monomorphism $u: A \to C$ such that $T^i(u) = 0$. Then $(T_i)_i$ is a universal δ-functor.*

One calls δ-functors satisfying the hypothesis of the proposition *effaceable*.

In order to apply the proposition to a δ-functor of the form $(R^i F)_{i\geq 0}$, one can often use monomorphisms to right F-acyclic objects C, where an object C of $\mathcal{A}$ is called *right F-acyclic* if it is right F-acyclic as complex concentrated in degree 0 for the induced functor $K^+(\mathcal{A}) \to D^+(\mathcal{B})$. Then $R^i F(C) = 0$ for $i > 0$, and a fortiori any morphism with target $R^i F(C)$ must vanish.

Example F.203. An injective object of $\mathcal{A}$ is K-injective as a complex concentrated in degree 0 (Remark F.181 (1)) and hence is right F-acyclic for every additive functor F (Proposition F.179 (1)).

The following lemma shows in particular that one obtains the "usual" definition of F-acyclic objects if F is a left exact functor $\mathcal{A} \to \mathcal{B}$ and $\mathcal{A}$ "has enough acyclic objects".

Proposition F.204. ([Lip2] $^{\circ}_{\times}$ (2.7.4)) *Suppose that $F: \mathcal{A} \to \mathcal{B}$ is left exact and suppose that every object of $\mathcal{A}$ admits a monomorphism into a right F-acyclic object of $\mathcal{A}$.*

Then every bounded below complex of right F-acyclic objects is a right F-acyclic complex, and for every X in $K^+(\mathcal{A})$ and every $n \in \mathbb{Z}$ such that $H^i(X) = 0$ for all $i < n$ there exists a quasi-isomorphism $X \to C_X$, where C_X is a complex with right F-acyclic components and $C_X^i = 0$ for all $i < n$. In particular, Condition (Ac) holds.

Moreover, the following conditions for an object A of $\mathcal{A}$ are equivalent.

(i) *A is right F-acyclic.*
(ii) *$R^i F(A) = 0$ for all $i > 0$.*
(iii) *The morphism $F(A) \to RF(A)$ is an isomorphism.*

Combining Proposition F.204, Remark F.200 (3) and Proposition F.202 we obtain:

Corollary F.205. *Suppose that $F \colon \mathcal{A} \to \mathcal{B}$ is left exact and that every object of $\mathcal{A}$ admits a monomorphism into a right F-acyclic object of $\mathcal{A}$. Then $F = R^0 F$ and $(R^i F)_{i \geq 0}$ is a universal δ-functor.*

Remark F.206. Suppose that in the situation of the corollary, $(T^i)_{\geq 0}$ is a δ-functor with $T^0 = F$ and such that $T^i(C) = 0$ for every right F-acyclic object C and all $i > 0$. Then the δ-functor $(T^i)_{\geq 0}$ is universal by Proposition F.202 and hence there is a unique isomorphisms of δ-functors $(T^i)_{i \geq 0} \xrightarrow{\sim} (R^i F)_{i \geq 0}$ that is the identity in degree 0.

A special case is if $\mathcal{A}$ has enough injective objects and $T^i(I) = 0$ for every injective object I of $\mathcal{A}$ and all $i > 0$.

The following proposition gives a recipe to simultaneously find acyclic objects and to check that Condition (Ac) holds.

Proposition F.207. ([Lip2] $\overset{O}{\underset{X}{}}$ (2.7.2) and its proof) *Let $F \colon \mathcal{A} \to \mathcal{B}$ be an additive functor of abelian categories. Let $\mathcal{I}$ be a class of objects of $\mathcal{A}$ such that*
(a) *For every object A of $\mathcal{A}$ there exists a monomorphism $A \to I$ with I in $\mathcal{I}$.*
(b) *For all I and J in $\mathcal{I}$ one has $I \oplus J$ in $\mathcal{I}$.*
(c) *If $0 \to A \to B \to C \to 0$ is an exact sequence in $\mathcal{A}$ with A and B in $\mathcal{I}$, then $C \in \mathcal{I}$ and the exact sequence $0 \to F(A) \to F(B) \to F(C) \to 0$ is exact.*
Then the following assertions hold.
(1) *The full subcategory $K^+(\mathcal{I})$ of $K^+(\mathcal{A})$ of complexes whose components are all in $\mathcal{I}$ is a triangulated subcategory of $K^+(\mathcal{A})$ and all complexes in $K^+(\mathcal{I})$ are right F-acyclic. In particular, every object of $\mathcal{I}$ is right F-acyclic.*
(2) *For every X in $K^+(\mathcal{A})$ and every $n \in \mathbb{Z}$ such that $H^i(X) = 0$ for all $i < n$ there exists a quasi-isomorphism $X \to I_X$ with $I_X \in K^+(\mathcal{I})$ and $I_X^i = 0$ for all $i < n$.*
In particular, Condition (Ac) holds, there exists a right derived functor $R^+ F$ of F, and $\dim^- R^+ F = 0$ if $F \neq 0$.

Note that (a) implies that $\mathcal{I}$ is non-empty and that (c) with $A = B \in \mathcal{I}$ shows that $0 \in \mathcal{I}$.

Remark F.206 above gives a criterion to identify the higher derived functors $R^i F = H^i \circ R^+ F$. This is of course less precise than identifying the derived functor $R^+ F$ itself. For this we can use Proposition F.193 and obtain:

Corollary F.208. *Let $F \colon \mathcal{A} \to \mathcal{B}$ be a left exact functor of abelian categories. Let $\mathcal{I}$ be a class of right F-acyclic objects in $\mathcal{A}$ such that every object of $\mathcal{A}$ admits a monomorphism into an object of $\mathcal{I}$ (e.g., if $\mathcal{A}$ has enough injective objects we can choose as $\mathcal{I}$ the class of all injective objects of $\mathcal{A}$). Consider F also as functor $K^+(\mathcal{A}) \to D(\mathcal{B})$. Let $Q \colon K^+(\mathcal{A}) \to D^+(\mathcal{A})$ be the localization functor.*

Let $G \colon D^+(\mathcal{A}) \to D(\mathcal{B})$ be a triangulated functor such that $G(D^{\geq 0}(\mathcal{A})) \subseteq D^{\geq 0}(\mathcal{B})$. Let $\xi \colon F \to G \circ Q$ be a morphism of triangulated functors such that $\xi(I) \colon F(I) \to G(I)$ is an isomorphism for every object I of $\mathcal{I}$. Then the morphism $RF \to G$ corresponding to ξ is an isomorphism.

(F.49) Hypercohomology spectral sequences.

The spectral sequence attached to a filtered complex, Section (F.24), can be generalized as follows. Let $\mathcal{T}$ be a triangulated category and let $R^0 \colon \mathcal{T} \to \mathcal{B}$ be a cohomological functor with values in some Grothendieck abelian category $\mathcal{B}$. For all $n \in \mathbb{Z}$ set $R^n := R^0 \circ S^n$, where S is the functor $X \mapsto X[1]$. Let

$$\cdots \longrightarrow X^{p+1} \xrightarrow{i^p} X^p \longrightarrow \cdots, \qquad p \in \mathbb{Z}$$

be a $\mathbb{Z}^{\mathrm{opp}}$-diagram in $\mathcal{T}$. We choose for all $p \in \mathbb{Z}$ a distinguished triangle

$$X^{p+1} \xrightarrow{i^p} X^p \longrightarrow X_p^{p+1} \xrightarrow{+1}.$$

As in Remark F.111, from the exact sequence

$$\cdots \to R^{p+q}(X^{p+1}) \to R^{p+q}(X^p) \to R^{p+q}(X_p^{p+1}) \to R^{p+q+1}(X^{p+1}) \to \cdots$$

we obtain a spectral sequence $(E_r, d_r)_{r \geq 1}$ with $E_1^{pq} = R^{p+q}(X_p^{p+1})$.

Now suppose that for all $n \in \mathbb{Z}$ one has $R^n(X^p) = 0$ for $p \gg 0$ (depending on n) and that i^p induces an isomorphism $R^n(X^{p+1}) \to R^n(X^p)$ for $p \ll 0$ (again depending on n). We set $X := \mathrm{colim}_{p \in \mathbb{Z}^{\mathrm{opp}}} X^p$, assuming that this colimit exists. Then the spectral sequence converges,

$$(F.49.1) \qquad\qquad E_1^{pq} = R^{p+q}(X_p^{p+1}) \Rightarrow R^n(X).$$

The filtration of the limit term is given by

$$F^p R^n(X) = \mathrm{Im}(R^n(X^p) \to R^n(X)).$$

In practice, one often has that i^p is an isomorphism for $p \ll 0$, and then the existence of X is automatic.

Example F.209. Let $T \colon \mathcal{A} \to \mathcal{B}$ be an additive functor of Grothendieck abelian categories. Let X be a complex with coefficients in $\mathcal{A}$ which is endowed with a descending filtration $(F^p X)_p$ such that for all $n \in \mathbb{Z}$ there exist p_0, p_1 with $F^p X^n = 0$ for $p \geq p_0$ and $F^p X = X$ for $p \leq p_1$. Applying the above construction to $R^0 := H^0 \circ RT = R^0 T$ we obtain a convergent spectral sequence

$$(F.49.2) \qquad\qquad E_1^{pq} = R^{p+q} T(\mathrm{gr}^p(X)) \Rightarrow R^n T(X).$$

We consider two special cases that are sometimes called *spectral sequences of hypercohomology*.

(1) First, we consider the "naive filtration" given by the subcomplexes $\sigma^{\geq p} X$ (F.14.3). In this case the spectral sequence takes the form

$$(F.49.3) \qquad\quad E_1^{pq} = R^{p+q} T(X^p[-p]) = R^q T(X^p) \Rightarrow R^n T(X).$$

(2) Consider the ascending filtration by the truncated complexes $\tau^{\leq n} X$ (F.14.1). To obtain a descending filtration we set $F^p X := \tau^{\leq -p} X$. Then $\mathrm{gr}^p X = H^{-p}(X)[p]$ and we obtain a convergent spectral sequence $E_1^{pq} = R^{2p+q} T(H^{-p} X) \Rightarrow R^{p+q} T(X)$. Replacing E_r^{pq} by $E_{r+1}^{-q, p+2q}$ we get the convergent spectral sequence

$$(F.49.4) \qquad\qquad E_2^{pq} = R^p T(H^q(X)) \Rightarrow R^{p+q} T(X).$$

As both truncations $\sigma^{\geq p}$ and $\tau^{\leq -p}$ are functorial, the spectral sequences (F.49.3) and (F.49.4) are both functorial in X.

Another special case is obtained from the filtrations induced by a double complex (Section (F.25)).

Example F.210. Let $T\colon \mathcal{A} \to \mathcal{B}$ be an additive functor of Grothendieck abelian categories and let X be a double complex with components in $\mathcal{A}$. Consider the two filtrations (F.25.1) yielding graded complexes

$$_I \operatorname{gr}^p(\operatorname{Tot}(X))\colon \qquad \cdots \longrightarrow X^{p,k-p} \xrightarrow{d_2^{p,k-p}} X^{p,k+1-p} \longrightarrow \cdots ,$$

$$_{II} \operatorname{gr}^p(\operatorname{Tot}(X))\colon \qquad \cdots \longrightarrow X^{k-p,p} \xrightarrow{d_1^{k-p,p}} X^{k+1-p,p} \longrightarrow \cdots ,$$

where in both cases the entries in degree k and $k+1$ are described. Now suppose that for all n there exist only finitely many non-zero $X^{i,j}$ with $i+j = n$. Then we obtain as special cases of (F.49.2) two convergent spectral sequences

(F.49.5)
$$_I E_1^{pq} = R^{p+q}T(X^{p,\bullet-p}) = R^q T(X^{p,\bullet}) \Rightarrow R^n T(\operatorname{Tot}(X)),$$
$$_{II} E_1^{pq} = R^{p+q}T(X^{\bullet-p,p}) = R^q T(X^{\bullet,p}) \Rightarrow R^n T(\operatorname{Tot}(X)).$$

As the differential in $_I E_1^{pq}$ (resp. $_{II} E_1^{pq}$) is given by d_2 (resp. d_1) we have

(F.49.6)
$$_I E_2^{pq} = R^q T(X^{p,\bullet}) \Rightarrow R^n T(\operatorname{Tot}(X)),$$
$$_{II} E_2^{pq} = R^{p+q}T(X^{\bullet-p,p}) = R^q T(X^{\bullet,p}) \Rightarrow R^n T(\operatorname{Tot}(X)).$$

(F.50) Grothendieck spectral sequence.

We apply Proposition F.176 to the setting considered in the previous section. Let $F\colon \mathcal{A} \to \mathcal{A}'$, $F'\colon \mathcal{A}' \to \mathcal{A}''$ be additive functors between abelian categories.

Proposition F.211. *Suppose that there exists a class $\mathcal{I}$ of objects in $\mathcal{A}$ containing 0 such that the following assumptions are satisfied.*
(a) *Every object of $\mathcal{I}$ is right F-acyclic, and every object of $\mathcal{A}$ admits a monomorphism to an object of $\mathcal{I}$.*
(b) *Every object of $\mathcal{A}'$ admits a monomorphism to a right F'-acyclic object of $\mathcal{A}'$.*
(c) *The functor F maps objects in $\mathcal{I}$ to right F'-acyclic objects of $\mathcal{A}'$.*
Then the canonical morphism

$$R^+(F' \circ F) \to R^+ F' \circ R^+ F$$

of functors $D^+(\mathcal{A}) \to D^+(\mathcal{A}'')$ is an isomorphism.

If $\mathcal{A}$ and $\mathcal{A}'$ have enough injective objects, then one can choose $\mathcal{I}$ as the class of all injective objects of $\mathcal{A}$ and (b) is satisfied for all functors F'.

Proof. By Lemma F.194 and Proposition F.204, Assumption (a) implies that Condition (Ac) is satisfied for $\mathcal{A}$ and F and that we can choose the quasi-isomorphisms $X \to C_X$ such that all components of C_X are in $\mathcal{I}$. By (c), $F(C_X)$ is a bounded below complex of right F'-acyclic objects and therefore $F(C_X)$ is right G'-acyclic objects by Assumption (b) via Proposition F.204. Therefore we can apply Proposition F.176. $\square$

Proposition F.212. (Grothendieck spectral sequence, [Sta] 015N) *Suppose that there exists a full additive subcategory $\mathcal{I}$ of $\mathcal{A}$ such that the assumptions (a), (b), and (c) of Proposition F.211 are satisfied. Then for all X in $D^+(\mathcal{A})$ there exists a converging spectral sequence*

$$E_2^{p,q} = R^p F'(R^q F(X)) \Rightarrow R^{p+q}(F' \circ F)(X),$$

which is functorial in X.

   Moreover, if $n \in \mathbb{Z}$ is such that $H^i(X) = 0$ for all $i < n$, then $E_2^{p,q} = 0$ whenever $q < n$ or $p < 0$.

(F.51) Derived bi-functors.

Let $\mathcal{K}_1$, $\mathcal{K}_2$, and $\mathcal{E}$ be triangulated categories with respective translation functors T_1, T_2, and T. We will define a triangulated bi-functor $\mathcal{K}_1 \times \mathcal{K}_2 \to \mathcal{E}$ to be a functor that is "triangulated in each variable". Here one has to give some care to signs and compatibility with translations functors.

   Given a functor $F \colon \mathcal{K}_1 \times \mathcal{K}_2 \to \mathcal{E}$, for each X in $\mathcal{K}_1$ and for each Y in $\mathcal{K}_2$ we have the partial functors

(F.51.1)
$$F_X \colon \mathcal{K}_2 \to \mathcal{E}, \qquad F_X(B) = F(X, B),$$
$$F_Y \colon \mathcal{K}_1 \to \mathcal{E}, \qquad F_Y(A) = F(A, Y).$$

Definition F.213. *A triangulated bi-functor $\mathcal{K}_1 \times \mathcal{K}_2 \to \mathcal{E}$ is a triple (F, θ_1, θ_2), where $F \colon \mathcal{K}_1 \times \mathcal{K}_2 \to \mathcal{E}$ is a functor and*

$$\theta_1 \colon F \circ (T_1 \times \mathrm{id}_{\mathcal{K}_2}) \xrightarrow{\sim} T \circ F, \qquad \theta_2 \colon F \circ (\mathrm{id}_{\mathcal{K}_1} \times T_2) \xrightarrow{\sim} T \circ F$$

are isomorphisms of functors such that
(a) For each Y in $\mathcal{K}_2$ and for each X in $\mathcal{K}_1$ the pairs (F_Y, θ_1) and (F_X, θ_2) are triangulated functors.
(b) The composed functorial isomorphisms

$$F(T_1 \times T_2) = F(T_1 \times \mathrm{id})(\mathrm{id} \times T_2) \xrightarrow{\ via\ \theta_1\ } TF(\mathrm{id} \times T_2) \xrightarrow{\ via\ \theta_2\ } TTF$$

$$F(T_1 \times T_2) = F(\mathrm{id} \times T_2)(T_1 \times \mathrm{id}) \xrightarrow{\ via\ \theta_2\ } TF(T_1 \times \mathrm{id}) \xrightarrow{\ via\ \theta_1\ } TTF$$

   are negatives of each other.
We also have the obvious notion of a *morphism of triangulated bi-functors.*

Example F.214. Let $\mathcal{A}$ be an additive category with a tensor product (see Section (F.19)). Suppose that $\mathcal{A}$ admits countable direct sums. Then forming the tensor complex yields a bi-functor $\otimes \colon C(\mathcal{A}) \times C(\mathcal{A}) \to C(\mathcal{A})$. It preserves homotopy in each variable and hence induces a bi-functor

(F.51.2)
$$\otimes \colon K(\mathcal{A}) \times K(\mathcal{A}) \longrightarrow K(\mathcal{A}).$$

It preserves termwise split sequences in each component. For a fixed complex Y in $K(\mathcal{A})$ we define $\theta_1 \colon X[1] \otimes Y \xrightarrow{\sim} (X \otimes Y)[1]$ to be the identity. For a fixed complex X in $K(\mathcal{A})$ we define $\theta_2 \colon X \otimes Y[1] \xrightarrow{\sim} (X \otimes Y)[1]$ to be the multiplication by $(-1)^i$ if restricted to $X^i \otimes Y^{j+1}$. Then $(- \otimes -, \theta_1, \theta_2)$ is a triangulated bi-functor.

   Now suppose that $\mathcal{A}_1$ and $\mathcal{A}_2$ are abelian categories and that $\mathcal{K}_i$ is a triangulated subcategory of $K(\mathcal{A}_i)$, $i = 1, 2$. Given a triangulated bi-functor $(F \colon \mathcal{K}_1 \times \mathcal{K}_2 \to \mathcal{E}, \theta_1, \theta_2)$ as above we consider the following two conditions.
(ACI) For all X in $\mathcal{K}_1$ there exists a quasi-isomorphism $X \to I_X$ such that

(a) the complex I_X is right F_Y-acyclic for all Y in $\mathcal{K}_2$ and
    (b) the partial functor $F_{I_X}\colon \mathcal{K}_2 \to \mathcal{E}$ sends quasi-isomorphisms to isomorphisms.
(ACII) For all Y in $\mathcal{K}_2$ there exists a quasi-isomorphism $Y \to I_Y$ such that
    (a) the complex I_Y is right F_X-acyclic for all X in $\mathcal{K}_1$ and
    (b) the partial functor $F_{I_Y}\colon \mathcal{K}_1 \to \mathcal{E}$ sends quasi-isomorphisms to isomorphisms.
Assume that Condition (ACI) is satisfied. Then for fixed Y in $\mathcal{K}_2$ the right derived functor $R(F_Y)$ exists. Moreover for X in $\mathcal{K}_1$ fixed, the functor $Y \mapsto R(F_Y)(X) = F_Y(I_X) = F_{I_X}(Y)$ sends quasi-isomorphisms to isomorphisms and hence factors through $\mathcal{K}_2 \to D^{\mathcal{K}_2}$. Hence we obtain a triangulated bi-functor

$$R_I F\colon D^{\mathcal{K}_1} \times D^{\mathcal{K}_2} \longrightarrow \mathcal{E}, \qquad R_I F(X,Y) = F(I_X, Y).$$

Similarly, if Condition (ACII) holds, one obtains a triangulated bi-functor $R_{II}F$. If both of the above conditions hold, then

$$R_I F(X,Y) = F(I_X, Y) \cong F(I_X, I_Y) = R_{II}F(I_X, Y) \cong R_{II}F(X,Y),$$

and thus $R_I F = R_{II} F$. Hence if at least one of the Conditions (ACI) or (ACII) hold, then we simply write $RF := R_I F$ or $RF := R_{II} F$ and call RF the *right derived bi-functor of* F.

(F.52) The derived Hom functor and Ext Groups.

Let $\mathcal{A}$ be an abelian category. We consider the bi-functor

$$\mathrm{Hom}_{\mathcal{A}}(-,-)\colon K(\mathcal{A})^{\mathrm{opp}} \times K(\mathcal{A}) \longrightarrow K(\mathrm{AbGrp}),$$

where for X and Y in $K(\mathcal{A})$ we denote by $\mathrm{Hom}_{\mathcal{A}}(X,Y)$ the Hom complex defined in (F.18). This is a triangulated bi-functor ([Lip2] $\overset{\mathrm{O}}{\underset{\mathrm{X}}{}}$ (2.4.3)).
    Let $\mathcal{K} \subseteq K(\mathcal{A})$ be a triangulated subcategory with enough K-injective objects. In practice, we will usually have $\mathcal{K} = K(\mathcal{A})$. Then the restriction of $\mathrm{Hom}_{\mathcal{A}}(-,-)$ to $K(\mathcal{A})^{\mathrm{opp}} \times \mathcal{K}$ satisfies Condition (ACII) of Section (F.51) and hence we obtain by Section (F.51) the derived bi-functor

(F.52.1) $$R\,\mathrm{Hom}_{\mathcal{A}}\colon D(\mathcal{A})^{\mathrm{opp}} \times D^{\mathcal{K}} \to D(\mathrm{AbGrp})$$

which is a triangulated bi-functor. It is called the *derived Hom functor*.
    Moreover, the choice of K-injective resolutions $Y \to I_Y$ for all Y in $\mathcal{K}$ induces a morphism of triangulated bi-functors

(F.52.2) $$\eta\colon \mathrm{Hom}_{\mathcal{A}}(X,Y) \longrightarrow \mathrm{Hom}_{\mathcal{A}}(X, I_Y) = R\,\mathrm{Hom}_{\mathcal{A}}(X,Y),$$

where as usual we omit the localization functors from the notation.
    If every X in $K(\mathcal{A})$ has a K-projective resolution $P_X \to X$ (i.e., a K-injective resolution $X^{\mathrm{opp}} \to (P_X)^{\mathrm{opp}}$ in $K(\mathcal{A})^{\mathrm{opp}}$), then we can also calculate $R\,\mathrm{Hom}_{\mathcal{A}}(X,Y) = \mathrm{Hom}_{\mathcal{A}}(P_X, Y)$. This is for instance the case if $\mathcal{A}$ is the category of modules over a ring by Theorem F.189.

Definition F.215. *For $i \in \mathbb{Z}$ and X and Y in $D(\mathcal{A})$ the i-th extension group (or: Ext group) of X by Y is the group*

$$\mathrm{Ext}^i_{\mathcal{A}}(X,Y) := \mathrm{Hom}_{D(\mathcal{A})}(X, Y[i]) = \mathrm{Hom}_{D(\mathcal{A})}(X[-i], Y).$$

In particular for all objects A and B in $\mathcal{A}$ the group $\operatorname{Ext}^i_{\mathcal{A}}(A,B)$ is defined by considering A and B as objects in $D(\mathcal{A})$ as usual.

If $\mathcal{A}$ is an R-linear category (Definition F.28) for some ring R, then the Ext groups are R-modules.

Remark F.216. Let X and Y be complexes in $D(\mathcal{A})$.

(1) Let $\mathcal{D}'$ be a triangulated subcategory of $D(\mathcal{A})$, such as $\mathcal{D}' = D^*(\mathcal{A})$ for $* \in \{b, +, -\}$ or, more generally, $\mathcal{D}' = D^*_{\mathcal{A}'}(\mathcal{A})$ for some plump subcategory $\mathcal{A}'$ (Definition F.161). If X and Y are both in $\mathcal{D}'$, then $\operatorname{Ext}^i_{\mathcal{A}}(X,Y) = \operatorname{Hom}_{\mathcal{D}'}(X, Y[i])$.

(2) As $\operatorname{Hom}_{D(\mathcal{A})}(\cdot, \cdot)$ is a cohomological functor in each variable (Example F.137), any two distinguished triangles $X' \to X \to X'' \to X[1]$ and $Y' \to Y \to Y'' \to Y[1]$ in $D(\mathcal{A})$ yield long exact sequences of abelian groups

(F.52.3)
$$\cdots \longrightarrow \operatorname{Ext}^{i-1}_{\mathcal{A}}(X, Y'') \longrightarrow \operatorname{Ext}^i_{\mathcal{A}}(X, Y') \longrightarrow \operatorname{Ext}^i_{\mathcal{A}}(X, Y) \longrightarrow \operatorname{Ext}^i_{\mathcal{A}}(X, Y'') \longrightarrow \cdots,$$
$$\cdots \longrightarrow \operatorname{Ext}^{i-1}_{\mathcal{A}}(X', Y) \longrightarrow \operatorname{Ext}^i_{\mathcal{A}}(X'', Y) \longrightarrow \operatorname{Ext}^i_{\mathcal{A}}(X, Y) \longrightarrow \operatorname{Ext}^i_{\mathcal{A}}(X', Y) \longrightarrow \cdots$$

(3) Every quasi-isomorphism $Y \to I_Y$ to some K-injective complex I by Proposition F.179 (7) for all i yields an isomorphism

(F.52.4)
$$\operatorname{Ext}^i_{\mathcal{A}}(X, Y) \xrightarrow{\sim} \operatorname{Hom}_{K(\mathcal{A})}(X, I_Y[i]).$$

Similarly for a quasi-isomorphism $P_X \to X$ with P_X a K-projective complex.

The relation between Ext groups and RHom is as follows.

Remark F.217. Suppose that $\mathcal{K}$ has enough K-injective objects (e.g., if $\mathcal{K} = K(\mathcal{A})$ for a Grothendieck abelian category $\mathcal{A}$, which will be the only case that is used by us). For all $i \in \mathbb{Z}$ there are isomorphisms, functorial in $X \in D(\mathcal{A})^{\mathrm{opp}}$ and $Y \in D^{\mathcal{K}}$,

(F.52.5)
$$H^i(R\operatorname{Hom}_{\mathcal{A}}(X, Y)) \xrightarrow{\sim} \operatorname{Ext}^i_{\mathcal{A}}(X, Y)$$

and in particular

(F.52.6)
$$H^0(R\operatorname{Hom}_{\mathcal{A}}(X, Y)) \xrightarrow{\sim} \operatorname{Hom}_{D(\mathcal{A})}(X, Y).$$

Indeed, for Y in $\mathcal{K}$ we can choose a quasi-isomorphism $Y \to I_Y$ with I_Y a K-injective complex. Then (F.52.5) is given as the composition

$$H^i(R\operatorname{Hom}_{\mathcal{A}}(X, Y)) \xrightarrow{\sim} H^0(R\operatorname{Hom}_{\mathcal{A}}(X, Y[i]))$$
$$\xrightarrow{\sim} H^0\operatorname{Hom}_{\mathcal{A}}(X, I_Y[i])$$
$$\xrightarrow{\sim} \operatorname{Hom}_{K(\mathcal{A})}(X, I_Y[i])$$
$$\xrightarrow{\sim} \operatorname{Ext}^i_{\mathcal{A}}(X, Y),$$

where the third isomorphism is (F.18.2) and the last isomorphism is (F.52.4).

Remark F.218. Let $b, c \in \mathbb{Z}$. Suppose $X \in D^{(-\infty, b]}(\mathcal{A})$ and $Y \in D^{[c, \infty)}(\mathcal{A})$. Then we can represent X (resp. Y) by a complex concentrated in degrees $\leq b$ (resp. in degrees $\geq c$) (Proposition F.154). For these representatives the complex $\operatorname{Hom}_{\mathcal{A}}(X, Y)$ is concentrated in degrees $[c - b, \infty)$.

Now suppose that $K(\mathcal{A})$ has enough K-injective objects. Then we can choose a quasi-isomorphism $Y \to I_Y$ with I_Y a K-injective complex concentrated in degrees $[c, \infty)$. Therefore we find $R\operatorname{Hom}_{\mathcal{A}}(X, Y) \in D^{[c-b, \infty)}(\mathcal{A})$, i.e.

$$\mathrm{Ext}^i_{\mathcal{A}}(X, Y) = 0 \qquad \text{for all } i < c - b.$$

Remark F.219. For objects A and B in the abelian category $\mathcal{A}$ we can describe $\mathrm{Ext}^i_{\mathcal{A}}(B, A)$ also as follows. An *extension of B by A of degree $i \geq 1$* is an exact sequence in $\mathcal{A}$ of the form

(F.52.7) $\qquad E: \qquad 0 \longrightarrow A \longrightarrow Z^{-i+1} \longrightarrow Z^{-i+2} \longrightarrow \ldots \longrightarrow Z^0 \to B \longrightarrow 0.$

An extension of B by A of degree 1 is just a short exact sequence $0 \to A \to Z^0 \to B \to 0$. Such an extension is called *split*, if this short exact sequence splits. Under the bijection constructed below, the split extension corresponds to the zero element of the group $\mathrm{Ext}^1_{\mathcal{A}}(B, A)$.

A *morphism $E \to \tilde{E}$* of extensions E and $\tilde{E}$ of B by A of the same degree i is a commutative diagram

$$
\begin{array}{ccccccccccc}
E: & 0 \longrightarrow & A & \longrightarrow & Z^{-i+1} & \longrightarrow & Z^{-i+2} & \longrightarrow \ldots \longrightarrow & Z^0 & \longrightarrow & B \longrightarrow 0 \\
& & \| & & \downarrow & & \downarrow & & \downarrow & & \| \\
\tilde{E}: & 0 \longrightarrow & A & \longrightarrow & \tilde{Z}^{-i+1} & \longrightarrow & \tilde{Z}^{-i+2} & \longrightarrow \ldots \longrightarrow & \tilde{Z}^0 & \longrightarrow & B \longrightarrow 0.
\end{array}
$$

Two extensions E and E' of B by A of the same degree i are called *equivalent* if there exists an extension $\tilde{E}$ of B by A of degree i and morphisms $\tilde{E} \to E$ and $\tilde{E} \to E'$. It will follow from Proposition F.220 below that this is indeed an equivalence relation.

For an extension E as in (F.52.7) let σE be the complex

$$\ldots \to 0 \to A \to Z^{-i+1} \to \ldots \to Z^0 \to 0 \to \ldots$$

with Z^j in degree j. We define $\delta(E) \in \mathrm{Ext}^i_{\mathcal{A}}(B, A)$ as the morphism $-fs^{-1}: B \to A[i]$ in $D(\mathcal{A})$, where s is the quasi-isomorphism $s: \sigma E \longrightarrow B = B[0]$ and $f: \sigma E \to A[i]$ the projection. In other words, $\delta(E)$ is the morphism in the distinguished triangle

$$A[i-1] \longrightarrow Z \longrightarrow B[0] \overset{\delta(E)}{\longrightarrow} A[i-1][1] = A[i]$$

associated to the exact sequence of complexes $0 \to A[i-1] \to Z \to B[0] \to 0$ (F.37.3).

Proposition F.220. ([Ver3] $^\circ$ III.3) *Two extensions E and E' are equivalent if and only if $\delta(E) = \delta(E')$ and one obtains a bijective map $E \mapsto \delta(E)$ between the set of equivalence classes of extensions of B by A of degree i and the set $\mathrm{Ext}^i_{\mathcal{A}}(B, A)$.*

(F.53) Injective dimension.

Let $\mathcal{A}$ be a Grothendieck abelian category.

Definition F.221. *Let $a \leq b$ be integers and set $d := b - a$. An object X in $D(\mathcal{A})$ is said to have* injective amplitude in $[a, b]$ *if the functor $R\mathrm{Hom}_{\mathcal{A}}(-, X)$ maps $\mathcal{A}$ into $D(\mathrm{AbGrp})^{[a,b]}$, i.e., $\mathrm{Ext}^i_{\mathcal{A}}(F, X) = 0$ for all $F \in \mathcal{A}$ and $i \notin [a, b]$. In this case, one also says that X has* injective dimension $\leq d$.

We say that X has finite injective dimension *if it is of injective dimension $\leq d$ for some d.*

Proposition F.222. *For an object X in $D(\mathcal{A})$ and integers $a \leq b$, the following are equivalent:*

(i) *There exists a complex $I^\bullet$ of injective objects I^i of $\mathcal{A}$ with $I^i = 0$ for $i \notin [a,b]$ such that X and $I^\bullet$ are isomorphic in $D(\mathcal{A})$.*

(ii) *The complex X has injective amplitude in $[a,b]$.*

Proof. *(i) $\Rightarrow$ (ii)*. We may assume that X is a complex of injective objects concentrated in degrees $[a,b]$. Then X is K-injective (Remark F.181) and hence $R\operatorname{Hom}(F,X) = \operatorname{Hom}(F,X)$ which shows $H^i(R\operatorname{Hom}(F,X)) = 0$ for $i \notin [a,b]$.

(ii) $\Rightarrow$ (i). We first show the following claim. Fix $i \in \mathbb{Z}$. If $H^i(R\operatorname{Hom}(F,X)) = 0$ for every $F \in \mathcal{A}$, then $H^i(X) = 0$. To prove the claim we may replace X by a K-injective complex. Then $R\operatorname{Hom}(F,X)$ is represented by the complex $\operatorname{Hom}_{\mathcal{A}}(F,X)$ and hence $H^i(\operatorname{Hom}_{\mathcal{A}}(F,X)) = 0$. We obtain a commutative diagram

$$
\begin{array}{ccc}
B^i(\operatorname{Hom}_{\mathcal{A}}(F,X)) & \xrightarrow{\ \sim\ } & Z^i(\operatorname{Hom}_{\mathcal{A}}(F,X)) \\
\downarrow & & \downarrow{\scriptstyle \cong} \\
\operatorname{Hom}_{\mathcal{A}}(F,B^i(X)) & \longrightarrow & \operatorname{Hom}_{\mathcal{A}}(F,Z^i(X)),
\end{array}
$$

where the right vertical arrow is an isomorphism since Hom is left exact. If $H^i(X) \neq 0$, then $B^i(X) \to Z^i(X)$ is a strict inclusion. Hence we find an object F in $\mathcal{A}$ (e.g., $F = Z^i(X)$) such that $\operatorname{Hom}_{\mathcal{A}}(F,B^i(X)) \to \operatorname{Hom}_{\mathcal{A}}(F,Z^i(X))$ is not surjective. This gives a contradiction.

The claim above implies that $X \in D^{[a,b]}(\mathcal{A})$. Therefore we find a quasi-isomorphism $X \to J$ for a complex I with J^i injective for all $i \in \mathbb{Z}$ and $J^i = 0$ for $i < a$. Moreover $I := \tau^{\leq b} J \to J$ is a quasi-isomorphism. Then I is concentrated in degrees $[a,b]$ and it is isomorphic to X in $D(\mathcal{A})$. Moreover $I^i = J^i$ is injective for all $i = a, \ldots, b-1$. It remains to show that I^b is injective.

Set $K := \sigma^{\leq b-1} I$ which is a complex of injective modules concentrated in degrees $[a, b-1]$. Then we obtain a distinguished triangle $I^b[-b] \longrightarrow I \longrightarrow K \xrightarrow{+1}$ in $D(\mathcal{A})$ and hence for every F in $\mathcal{A}$ an exact sequence

$$
\operatorname{Ext}_{\mathcal{A}}^b(F,K) \longrightarrow \operatorname{Ext}_{\mathcal{A}}^1(F,I^b) \longrightarrow \operatorname{Ext}_{\mathcal{A}}^{b+1}(F,I).
$$

By assumption, the term on the right vanishes and by the implication "(i) $\Rightarrow$ (ii)" the term on the left vanishes. Therefore $\operatorname{Ext}_{\mathcal{A}}^1(F,I^b) = 0$ for every F in $\mathcal{A}$. But this shows that $\operatorname{Hom}_{\mathcal{A}}(-,I^b)\colon \mathcal{A} \to (\mathrm{AbGrp})$ is exact and hence that I^b is injective in $\mathcal{A}$. $\qquad\square$

Example F.223. Let A be a ring and let $\mathcal{A}$ be the category of A-modules. Suppose that X is a complex in $D(A)$ such that $\operatorname{Ext}_A^i(A/\mathfrak{a}, X) = 0$ for all ideals $\mathfrak{a} \subseteq A$ and all $i \notin [a,b]$. Then X is of injective amplitude in $[a,b]$.

Indeed, one has $\operatorname{Ext}_A^i(A,X) = H^i(X)$ and hence $X \in D^{[a,b]}(A)$. Then the proof of "(ii) $\Rightarrow$ (i)" in Proposition F.222 shows that X is isomorphic in $D(A)$ to a complex of injective A-modules concentrated in degrees $[a,b]$, using Proposition G.21.

(F.54) Derived limits and homotopy limits.

Let $\mathcal{A}$ be a Grothendieck abelian category and let $\mathcal{I}$ be a small category. As $\mathcal{A}$ is complete (Remark F.58), we obtain a left exact functor $\lim_{\mathcal{I}}\colon \mathcal{A}^{\mathcal{I}} \to \mathcal{A}$. Moreover $\mathcal{A}^{\mathcal{I}}$ is again a Grothendieck category (Remark F.59). Hence the derived functor

$$
R\lim := R\lim_{\mathcal{I}}\colon D(\mathcal{A}^{\mathcal{I}}) \to D(\mathcal{A})
$$

exists (Corollary F.187). As usual we set $R^p \lim(X) := H^p(R\lim(X))$ for $X \in D(\mathcal{A}^{\mathcal{I}})$. Sometimes one writes $\lim^p(X)$ instead of $R^p \lim(X)$ but we will not use this notation.

Mostly we will be interested in the special case, where $\mathcal{I}$ is the category given by the partially ordered set $\mathbb{N}^{\mathrm{opp}}$, i.e., elements of $\mathcal{A}^{\mathcal{I}}$ are diagrams in $\mathcal{A}$ of the form

$$X_0 \leftarrow X_1 \leftarrow X_2 \leftarrow \dots.$$

Lemma F.224. ([Nee4]$^{\circ}$ A.3.6) *Suppose that in $\mathcal{A}$ countable products are exact. Let* $\dots \to E_2 \xrightarrow{u_2} E_1 \xrightarrow{u_1} E_0$ *be an $\mathbb{N}^{\mathrm{opp}}$-diagram in $\mathcal{A}$ and define*

$$\delta\colon \prod_n E_n \to \prod_n E_n, \qquad (e_n)_n \mapsto (e_n - u_{n+1}(e_{n+1}))_n.$$

Then the complex $\prod_n E^n \xrightarrow{\delta} \prod_n E_n$ (concentrated in degree 0 and 1) represents $R\lim E_n$. In particular one has

(F.54.1)
$$\lim_n E_n = \mathrm{Ker}(\delta),$$
$$R^1 \lim_n E_n = \mathrm{Coker}(\delta),$$
$$R^p \lim_n E_n = 0 \qquad \text{for } p \neq 0, 1.$$

Definition F.225. *Let $\mathcal{A}$ be an abelian category. A diagram $X\colon \mathbb{N}^{\mathrm{opp}} \to \mathcal{A}$ is said to satisfy the* Mittag-Leffler condition *or is* ML *if for every $n \in \mathbb{N}$ there exists $c = c(n) \geq n$ such that $\mathrm{Im}(X_k \to X_n) = \mathrm{Im}(X_c \to X_n)$ for all $k \geq c$.*

Clearly any $\mathbb{N}^{\mathrm{opp}}$-diagram X in which all transition maps $X_n \to X_{n-1}$ are epimorphisms satisfies the Mittag-Leffler condition.

Remark F.226. Let $\mathcal{A}$ be an abelian category. Then for every diagram $X\colon \mathbb{N}^{\mathrm{opp}} \to \mathcal{A}$ there exists a monomorphism $X \to M$ into an $\mathbb{N}^{\mathrm{opp}}$-diagram M that satisfies the Mittag-Leffler condition.

Indeed, for $n \in \mathbb{N}$ set $M_n := X_n \oplus X_{n-1} \oplus \dots \oplus X_0$ with transition maps given by the projections and define $X_n \to M_n$ by $(\mathrm{id}_{X_n}, u_n, \dots, u_1 \circ \dots \circ u_n)$, where $u_n\colon X_n \to X_{n-1}$ are the transition maps.

For modules over a ring the Mittag-Leffler condition ensures that the given diagram is right-lim-acyclic. This follows via Proposition F.207 from the following results.

Proposition F.227. ([Wei2]$^{\circ}$ 3.5) *Let R be a ring (not necessarily commutative). Let*

$$0 \longrightarrow (E_n)_n \longrightarrow (F_n)_n \longrightarrow (G_n)_n \longrightarrow 0$$

be an exact sequence of $\mathbb{N}^{\mathrm{opp}}$-diagrams of R-modules.
(1) If (F_n) is ML, then (G_n) is ML.
(2) If (E_n) is ML, then

$$0 \longrightarrow \lim_n E_n \longrightarrow \lim_n F_n \longrightarrow \lim_n G_n \longrightarrow 0$$

is exact.
(3) If (E_n) is ML, then $R^p \lim E_n = 0$ for all $p \neq 0$.

Analogous statements do not hold in arbitrary complete abelian categories, even if arbitrary products are exact. Neeman constructed in [Nee4] $^\bigcirc$ A.5 an abelian category with exact products and a ML system $(E_n)_n$ in it such that $R^1 \lim E_n \neq 0$. Roos gives in [Roo] $^\bigcirc$ 1.18 for all integers $m \geq 1$ an example of a quasi-affine noetherian regular scheme U (the complement of the special point in the spectrum of a regular local noetherian ring) and a projective system of coherent $\mathcal{O}_U$-modules $\mathscr{F}_n$ with surjective transition maps (in particular $(\mathscr{F}_n)_n$ is a ML system) such that $R^p \lim_n \mathscr{F}_n \neq 0$ if and only if $p \neq 0, m$.

For inverse systems in triangulated categories there is the notion of a homotopy limit which is closely related to the derived limit if the triangulated category is the derived category of a Grothendieck abelian category.

Definition F.228. *Let $(K_n, f_n \colon K_n \to K_{n-1})_{n \geq 0}$ be an inverse system in a triangulated category $\mathcal{D}$, i.e., it is an object of $\mathcal{D}^{\mathbb{N}^{\mathrm{opp}}}$. We call an object $K \in \mathcal{D}$ a homotopy limit of this system, and write $K = \operatorname{holim} K_n$, if the product $\prod_n K^n$ exists in $\mathcal{D}$, and there exists an exact triangle*

$$K \longrightarrow \prod_{n \in \mathbb{N}} K_n \longrightarrow \prod_{n \in \mathbb{N}} K_n \overset{+1}{\longrightarrow}$$

where the morphism $\prod K_n \to \prod K_n$ is given by "identity minus shift", i.e., "$(x_n)_n \mapsto (x_n - f_{n+1}(x_{n+1}))_n$".

Note that the homotopy limit, if it exists, is determined uniquely up to isomorphism, but *not up to unique isomorphism*. In particular, holim is not a functor. Nevertheless it has a weak functoriality property in the following sense.

Remark F.229. Let $\alpha \colon (K_n)_n \longrightarrow (L_n)_n$ be a map of inverse systems in a triangulated category $\mathcal{D}$, and assume that $\operatorname{holim} K_n$ and $\operatorname{holim} L_n$ exist. By a rotated version of (TR3) there exists a morphism $\eta \colon \operatorname{holim} K_n \to \operatorname{holim} L_n$ yielding a morphism of distinguished triangles

$$\begin{array}{ccccccc}
\operatorname{holim} K_n & \longrightarrow & \prod K_n & \longrightarrow & \prod K_n & \overset{+1}{\longrightarrow} & (\operatorname{holim} K_n)[1] \\
\downarrow{\scriptstyle \eta} & & \downarrow{\scriptstyle \prod \alpha_n} & & \downarrow{\scriptstyle \prod \alpha_n} & & \downarrow{\scriptstyle \eta[1]} \\
\operatorname{holim} L_n & \longrightarrow & \prod L_n & \longrightarrow & \prod L_n & \overset{+1}{\longrightarrow} & (\operatorname{holim} L_n)[1].
\end{array}$$

Moreover, if α is an isomorphism, then η is an isomorphism by the five lemma, Lemma F.119.

If the triangulated category $\mathcal{D}$ is of the form $D(\mathcal{A})$ for a Grothendieck abelian category $\mathcal{A}$, then arbitrary products and hence homotopy limits exist in $D(\mathcal{A})$ (Lemma F.188).

Lemma F.230. ([Sta] 08U1) *Let $F \colon \mathcal{A} \to \mathcal{B}$ be an additive functor from a Grothendieck abelian category $\mathcal{A}$ to an abelian category $\mathcal{B}$. By F.187, there exists a derived functor RF of F. Assume that RF commutes with countable products of objects of $D(\mathcal{A})$. Then for every inverse system $(E_n)_{n \geq 0}$, $E_n \in \mathcal{A}$ we have*

$$RF(\operatorname{holim} E_n) \cong \operatorname{holim}(RF E_n) \ \text{in } D(\mathcal{B}).$$

Note that the hypotheses on RF is satisfied in the following two cases.
(1) The abelian category $\mathcal{B}$ has exact countable products and F commutes with countable products (see the proof of [Sta] 08U1).

(2) The functor F has a left adjoint functor G, and G has a left derived functor LG. Then RF is right adjoint to LG (Proposition F.191) and in particular commutes with arbitrary products.

The next proposition relates the homotopy limit construction to the derived functors $R\lim$. We continue to denote by $\mathcal{A}$ a Grothendieck category. Then the category of inverse systems $\mathcal{A}^{\mathbb{N}^{\mathrm{opp}}}$ is also a Grothendieck category (Remark F.59). Every complex in $C(\mathcal{A}^{\mathbb{N}^{\mathrm{opp}}})$ can be considered as an inverse system $(X_n)_n$ of complexes in $C(\mathcal{A})$ and a morphism $(X_n)_n \to (Y_n)_n$ between such complexes is a quasi-isomorphism if and only if $X_n \to Y_n$ is a quasi-isomorphism for all n. Hence we obtain a functor

$$(\text{F.54.2}) \qquad\qquad D(\mathcal{A}^{\mathbb{N}^{\mathrm{opp}}}) \to D(\mathcal{A})^{\mathbb{N}^{\mathrm{opp}}}$$

which however usually is not faithful. To each object of the left hand side we can attach its derived limit in $D(\mathcal{A})$ and to each object of the right hand side we can attach a homotopy limit.

Proposition F.231. *Let $X = (X_n)_n$ be an object in $D(\mathcal{A}^{\mathbb{N}^{\mathrm{opp}}})$ viewed as an inverse system of complexes X_n. Then $R\lim X$ is a homotopy limit of $(X_n)_n$ in $D(\mathcal{A})^{\mathbb{N}^{\mathrm{opp}}}$.*

Proof. We start with three preliminary remarks.

(I). The functor $e_m\colon \mathcal{A}^{\mathbb{N}^{\mathrm{opp}}} \to \mathcal{A}$, $(Y_n)_n \mapsto Y_m$ has an exact left adjoint, namely the functor i_m that sends Z in $\mathcal{A}$ to the inverse system

$$\cdots \longrightarrow 0 \longrightarrow 0 \longrightarrow Z \xrightarrow{\mathrm{id}} Z \xrightarrow{\mathrm{id}} \cdots \xrightarrow{\mathrm{id}} Z,$$

where the first entry of Z sits in degree m. In particular, e_m preserves injective objects and the functor $K(\mathcal{A}^{\mathbb{N}^{\mathrm{opp}}}) \to K(\mathcal{A})$ induced by e_m preserves K-injective complexes (Corollary F.183).

(II). Let $J = (J_n)_n$ be an inverse system in $\mathcal{A}^{\mathbb{N}^{\mathrm{opp}}}$ that is injective as object of the abelian category $\mathcal{A}^{\mathbb{N}^{\mathrm{opp}}}$. We claim that $J_n \to J_{n-1}$ is a split epimorphism for all n. Indeed, for an object Z in $\mathcal{A}$ consider the monomorphism of inverse systems $i_{n-1}(Z) \to i_n(Z)$. As J is injective we obtain a surjective map

$$\mathrm{Hom}(Z, J_n) = \mathrm{Hom}(i_n(Z), J) \longrightarrow \mathrm{Hom}(i_{n-1}(Z), J) = \mathrm{Hom}(Z, J_{n-1}).$$

In particular, setting $Z = J_{n-1}$ we find a preimage of $\mathrm{id}_{J_{n-1}}$, which is a right inverse of $J_n \to J_{n-1}$.

(III). Let $(J_n)_n$ be an inverse system in $\mathcal{A}$ such that $J_n \to J_{n-1}$ is a split epimorphism for all n. Then

$$0 \to \lim_n J_n \longrightarrow \prod_n J_n \longrightarrow \prod_n J_n \longrightarrow 0$$

is a split exact sequence. Indeed, by the Yoneda lemma we may assume that $\mathcal{A}$ is the category of abelian groups. Then we can use Proposition F.224.

We now come to the proof of Proposition F.231. To calculate $R\lim X$ we choose a quasi-isomorphism $X \to I$, where $I = (I_n)_n$ in $K(\mathcal{A}^{\mathbb{N}^{\mathrm{opp}}})$ is K-injective and all terms are injective objects in $\mathcal{A}^{\mathbb{N}^{\mathrm{opp}}}$ (Theorem F.185). Then $R\lim X = \lim I_n$, where the limit is taken termwise. By Step (I), all complexes I_n are K-injective and consist of injective objects as well. Moreover, we have a termwise split exact sequence of complexes

$$0 \longrightarrow \lim I_n \longrightarrow \prod I_n \longrightarrow \prod I_n \longrightarrow 0$$

by Steps (II) and (III). By Lemma F.188, the products in the exact sequence represent the product in the derived category because I_n is K-injective. Therefore $R \lim X$ is a homotopy limit by Definition F.228.   □

Remark F.232. Let $\mathcal{A}$ be an abelian category. The functor $D(\mathcal{A}^{\mathbb{N}^{\mathrm{opp}}}) \to D(\mathcal{A})^{\mathbb{N}^{\mathrm{opp}}}$ (F.54.2) is essentially surjective (use [Sta] 0911 whose proof verbatim carries over to this situation). Hence for every inverse system $(X_n)_n$ in $D(\mathcal{A})$ we can choose an object $(\tilde{X}_n)_n$ in $D(\mathcal{A}^{\mathbb{N}^{\mathrm{opp}}})$ that gives rise to $(X_n)_n$, which however in general will not be unique up to isomorphism.

Now suppose that $\mathcal{A}$ is a Grothendieck category. Then Proposition F.231 shows that $R \lim \tilde{X}_n$ is a homotopy limit of $(X_n)_n$. Hence the isomorphism class of $\operatorname{holim} X_n$ does not depend on the isomorphism class of $(\tilde{X}_n)_n$.

In the category of modules over a ring many aspects of derived and homotopy limits become easier, for instance because arbitrary products are exact and Proposition F.227 holds. Another instance is the following result.

Lemma F.233. ([Sta] 07KZ) *Let A be a ring and let $(E_n)_{n \geq 0}$ be an inverse system of objects of $D(A\text{-Mod})$. For every $m \in \mathbb{Z}$, we have a short exact sequence of A-modules*

$$0 \to R^1 \lim H^{m-1}(E_n) \to H^m(\operatorname{holim} E_n) \to \lim H^m(E_n) \to 0.$$

(F.55) Homotopy colimits.

Definition F.234. *Let $\mathcal{D}$ be a triangulated category and let*

$$K_0 \xrightarrow{f_0} K_1 \xrightarrow{f_1} K_2 \longrightarrow \cdots$$

be an $\mathbb{N}$-diagram in $\mathcal{D}$. One says that an object K in $\mathcal{D}$ is a homotopy colimit *of the system $(K_n)_n$ if the direct sum $\bigoplus_n K_n$ exists and there is a distinguished triangle*

$$\bigoplus K_n \xrightarrow{\varepsilon} \bigoplus K_n \longrightarrow K \xrightarrow{+1},$$

where ε is the map given by $\operatorname{id}_{K_n} - f_n$. In this case we write $K = \operatorname{hocolim} K_n$.

If the homotopy colimit exists, then it is unique up to non-unique isomorphism. In particular hocolim is not a functor. As any map in $\mathcal{D}$ can be completed to a distinguished triangle, the homotopy colimit $\operatorname{hocolim} K_n$ exists if $\bigoplus K_n$ exists. This is for instance the case in the derived category of any Grothendieck abelian category (Lemma F.188).

If $\mathcal{A}$ is an abelian category that has exact countable direct sums (e.g., if $\mathcal{A}$ is a Grothendieck abelian category), then coproducts in $D(\mathcal{A})$ exist and can be computed by taking termwise direct sums of any representing complexes. This implies the following lemma, see [Sta] 093W.

Lemma F.235. *Let $\mathcal{A}$ be an abelian category that has exact countable direct sums. Let $(X_n)_n$ be an $\mathbb{N}$-diagram of complexes in $C(\mathcal{A})$. Then the termwise colimit $\operatorname{colim} X_n$ is a homotopy colimit in $D(\mathcal{A})$.*

Exercises

Exercise F.1. For a group G let BG be the category with a single object $*$ and with $\mathrm{End}_{BG}(*) = G$, where the composition is given by the multiplication in G. Let $\mathcal{G}$ be a groupoid, let x be an object in $\mathcal{G}$ and set $G := \mathrm{Aut}_{\mathcal{G}}(x)$. Show that the inclusion $BG \to \mathcal{G}$ is an equivalence if and only if $\mathcal{G}$ is connected.

Exercise F.2. Let A be a ring and let $S \subseteq A$ be a multiplicative subset. Consider S as the set of objects in a category, again denoted by S, where for $s, t \in S$ one sets $\mathrm{Hom}_S(s, t) := \{ u \in S \ ; \ us = t \}$. Composition in S is given by multiplication.
(1) Show that S is a filtered category.
(2) Let M be an A-module. Define a functor $M \colon S \to (A\text{-Mod})$ as follows. On objects, for $s \in S$ set $M(s) = M$, and for a morphism $u \colon s \to t$ let $M(u) \colon M \to M$ be the multiplication by u. Show that one has a functorial isomorphism $\mathrm{colim}_S M \overset{\sim}{\to} S^{-1}M$.

Exercise F.3. Let $\mathcal{C}$ be an abelian category. Let $u \colon X \to Y$ be an epimorphism in $\mathcal{C}$. Show that the following conditions are equivalent.
(i) For any morphism $g \colon Z \to X$ one has that $u \circ g$ is an epimorphism if and only if g is an epimorphism.
(ii) For all subobjects U of X with $U + \mathrm{Ker}(u) = X$ one has $U = X$.
If these conditions are satisfied, u is called an *essential epimorphism*. Show the following assertions.
(1) Let u and v be composable epimorphisms. Then $v \circ u$ is essential if and only if u and v are essential.
(2) If $u_i \colon X_i \to Y_i$ for $i = 1, \dots, r$ are essential epimorphisms, then $\bigoplus_{i=1}^r u_i$ is an essential epimorphism.

Exercise F.4. Let $\mathcal{C}$ be a category and let X be an object of $\mathcal{C}$. An object Y of $\mathcal{C}$ is called *retract of X* if there exist morphisms $Y \overset{i}{\to} X \overset{r}{\to} Y$ such that $r \circ i = \mathrm{id}_Y$. An endomorphism u of an object of $\mathcal{C}$ is called *idempotent* if $u^2 = u$. An idempotent endomorphism u of an object X is called *split* if there exists a retract $Y \overset{i}{\to} X \overset{r}{\to} Y$ of X such that $u = i \circ r$. If every idempotent in $\mathcal{C}$ admits a splitting, then $\mathcal{C}$ is said to be *idempotent complete*.
(1) Show that if the splitting of an idempotent u of an object X is given by a retract $Y \overset{i}{\to} X \overset{r}{\to} Y$, then there is a unique commutative diagram whose vertical maps are isomorphisms

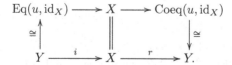

Therefore a splitting of u is unique up to unique isomorphism it it exists.
(2) Show that every small category $\mathcal{C}$ has an *idempotent completion*, i.e., there exists a fully faithful functor $i \colon \mathcal{C} \to \mathcal{C}'$, where $\mathcal{C}'$ is a small idempotent complete category such that every object in $\mathcal{C}'$ is the retract of an object of $\mathcal{C}$ under i.
Hint: Consider the Yoneda embedding $\mathcal{C} \to (\mathrm{PSh}(\mathcal{C}))$, where $(\mathrm{PSh}(\mathcal{C}))$ denotes the category of functors $\mathcal{C}^{\mathrm{opp}} \to (\mathrm{Sets})$, and define $\mathcal{C}'$ as the category of retracts in $(\mathrm{PSh}(\mathcal{C}))$ of the representable functors.

(3) Show that an idempotent completion $i \colon \mathcal{C} \to \mathcal{C}'$ has the following universal property and hence we can speak of *the* idempotent completion of $\mathcal{C}$. If $\mathcal{D}$ is a small idempotent complete category, then composition with i induces an equivalence of the category of functors $\mathcal{C}' \to \mathcal{D}$ with the category of functors $\mathcal{C} \to \mathcal{D}$.

(4) Show that the idempotent completion of an additive category is an additive category. *Remark*: If $\mathcal{C}$ is the category whose objects are the open sets of $\mathbb{R}^n$ for some $n \geq 0$ and whose morphisms are infinitely often differentiable maps between such open subsets, then its idempotent completion is the category of (Hausdorff, second countable) smooth manifolds, see `https://ncatlab.org/nlab/show/Karoubi+envelope`.

Exercise F.5. Let $\mathcal{C}$ be an additive category. Let X be an object of $\mathcal{C}$.

(1) Show that if the ring $\mathrm{Hom}_{\mathcal{C}}(X, X)$ is local, then X is indecomposable. Here we call a not necessarily commutative ring *local* if the sum of any two non-units is again a non-unit.

(2) Let $X = X_1 \oplus \cdots \oplus X_n = Y_1 \oplus \cdots \oplus Y_m$ be two decompositions into objects with local endomorphism ring. Show that $m = n$ and that there exists a permutation σ such that $X_i \cong Y_{\sigma(i)}$ for all $i = 1, \ldots, n$.

(3) The converse in (1) does not hold in general: Let $\mathcal{C}$ be the category of abelian groups. Show that $\mathbb{Z}$ is indecomposable in $\mathcal{C}$ but that the endomorphism ring of $\mathbb{Z}$ is not local.

Exercise F.6. An additive category is called *Krull-Schmidt category* if every object decomposes into a finite direct sum of objects having local endomorphism rings (see Exercise F.5). Show that every Krull-Schmidt category is idempotent complete (Exercise F.4).

Exercise F.7. Let $\mathcal{C}$ be an essentially small abelian category. A *bi-chain* in $\mathcal{C}$ is a sequence of triples $(X_n, i_n, p_n)_{n \geq 0}$, where X_n is an object in $\mathcal{C}$, $i_n \colon X_n \to X_{n-1}$ is a monomorphism, and $p_n \colon X_{n-1} \to X_n$ is an epimorphism. Such a bi-chain is said to *terminate* if there exists an integer $N \geq 1$ such that i_n and p_n are isomorphisms for all $n \geq N$. If every bi-chain in $\mathcal{C}$ terminates, one says that $\mathcal{C}$ satisfies the *bi-chain condition*.

(1) Show that an abelian category in which every object is of finite length satisfies the bi-chain condition.

(2) Show that a bi-chain $(X_n, i_n, p_n)_{n \geq 0}$ in $\mathcal{C}$ terminates if and only if the descending chain

$$\mathrm{End}(X_0) \supseteq \cdots \supseteq \mathrm{End}(X_{n-1}) \supseteq \mathrm{End}(X_n) \supseteq \cdots$$

becomes stationary. Here $\mathrm{End}(X_{n-1}) \to \mathrm{End}(X_n)$ is the injective map $u \mapsto p_n \circ u \circ i_n$. *Remark*: See Exercise 23.3 for an example of an abelian category that satisfies the bi-chain condition.

Exercise F.8. Let $\mathcal{C}$ be an abelian category that satisfies the bi-chain condition (Exercise F.7).

(1) Show that $\mathcal{C}$ is a Krull-Schmidt category (Exercise F.6).

(2) Show that an object of $\mathcal{C}$ is indecomposable if and only if its endomorphism ring is local and deduce that every object in $\mathcal{C}$ admits a decomposition into a finite direct sum of indecomposable objects having local endomorphism rings.

(3) Let u be an endomorphism of an object of X. Show that one has a *Fitting decomposition* with respect to u, i.e., that for sufficiently large integers r one has $X \cong \mathrm{Im}(u^r) \oplus \mathrm{Ker}(u^r)$.

Exercise F.9. Let $\mathcal{A}$ be an abelian category and fix $n \in \mathbb{Z}$.

(1) Show that the functor $\mathcal{A} \to C(\mathcal{A})$, $M \mapsto M[-n]$, is left adjoint to $X \mapsto \mathrm{Ker}(d^n \colon X^n \to X^{n+1})$ and right adjoint to $X \mapsto \mathrm{Coker}(d^{n-1})$.

(2) Let $D^n \colon \mathcal{A} \to C(\mathcal{A})$ be the functor that sends M to the complex

$$\cdots \longrightarrow 0 \longrightarrow M \xrightarrow{\mathrm{id}_M} M \longrightarrow 0 \longrightarrow \cdots$$

with M sitting in degrees n and $n+1$. Show that D^n is left adjoint to the functor $X \mapsto X^n$ and right adjoint to $X \mapsto X^{n+1}$.

Exercise F.10. Let $\mathcal{I}$ be a small category and for a category $\mathcal{A}$ consider the functor category $\mathcal{A}^{\mathcal{I}}$. Show that if $\mathcal{A}$ is additive (resp. abelian), then $\mathcal{A}^{\mathcal{I}}$ is additive (resp. abelian). *Remark:* Note that if $\mathcal{I}$ is the set $\mathbb{Z}$, considered as a discrete category, then $\mathcal{A}^{\mathbb{Z}}$ is the category of $\mathbb{Z}$-graded objects in $\mathcal{A}$ with morphisms of degree 0.

Exercise F.11. Let k be a field. Show that the category of filtered k-vector spaces is not abelian.

Exercise F.12. Let $\mathcal{A}$ be an abelian category and let $\mathcal{B} \subseteq \mathcal{A}$ be a full abelian subcategory. Suppose that for every epimorphism $X \to Y$ in $\mathcal{A}$ with Y in $\mathcal{B}$ there exists a morphism $Y' \to X$ with Y' in $\mathcal{B}$ such that the composition $Y' \to X \to Y$ is an epimorphism. Show that $\mathcal{B}$ is a plump subcategory of $\mathcal{A}$.

Exercise F.13. Let $F \colon \mathcal{A} \to \mathcal{C}$ be an exact functor of abelian categories. Then the full subcategory $\mathcal{B}$ of objects $X \in \mathcal{A}$ with $F(X) = 0$ is called the *kernel of F*. Show that $\mathcal{B}$ is a Serre subcategory of $\mathcal{A}$.

Exercise F.14. Let $\mathcal{A}$ be an abelian category that contains a small subcategory $\mathcal{A}_0$ such that the inclusion $\mathcal{A}_0 \to \mathcal{A}$ is an equivalence of categories. Let $\mathcal{B} \subseteq \mathcal{A}$ be a Serre subcategory. Show that there exists an abelian category $\mathcal{A}/\mathcal{B}$ and an exact functor

$$\pi \colon \mathcal{A} \to \mathcal{A}/\mathcal{B}$$

characterized by the following universal property. For any exact functor $F \colon \mathcal{A} \to \mathcal{C}$ of abelian categories such that $\mathcal{B}$ is contained in the kernel of F (Exercise F.13) there exists a factorization $F = \bar{F} \circ \pi$ for a unique exact functor $\bar{F} \colon \mathcal{A}/\mathcal{B} \to \mathcal{C}$.

Moreover, π is essentially surjective and its kernel is $\mathcal{B}$.
Hint: Show that the arrows u in $\mathcal{A}$ such that $\mathrm{Ker}(u), \mathrm{Coker}(u) \in \mathcal{B}$ form a saturated multiplicative system S and set $\mathcal{A}/\mathcal{B} = S^{-1}\mathcal{A}$.

Exercise F.15. Let $\mathcal{A}$ be an additive category. A short exact sequence in $\mathcal{A}$ is a sequence of the form

$$0 \longrightarrow X' \xrightarrow{i} X \xrightarrow{p} X'' \longrightarrow 0$$

such that i is a kernel of p and p is a cokernel of i. An *exact structure* on $\mathcal{A}$ is a class $\mathscr{E}$ of short exact sequences in $\mathcal{A}$, called *admissible*, that is closed under isomorphisms and that satisfies the axioms $(A)_0$ - $(C)^{\mathrm{opp}}$ below. Here we call a morphism i in $\mathcal{A}$ an *admissible monic* if there exists an admissible exact sequence $0 \to X' \xrightarrow{i} X \to X'' \longrightarrow 0$. Dually, one defines the notion of an *admissible epic*.

$(A)_0$ The identity of the zero object is an admissible epic.

(B) The composition of two admissible epics is an admissible epic.

(C) The pullback of an admissible epic along an arbitrary morphism exists and yields an admissible epic.

$(C)^{\mathrm{opp}}$ The pushout of an admissible monic along an arbitrary morphism exists and yields an admissible monic.

An *exact category* is a pair $(\mathcal{A}, \mathscr{E})$ consisting of an additive category and an exact structure.

(1) Show that $(A)_0$, (C), and $(C)^{\mathrm{opp}}$ imply the following property.
   (A) For all objects X in $\mathcal{A}$ the identity id_X is an admissible monic and an admissible epic.

(2) Show that in an exact category also the following property holds.
   $(B)^{\mathrm{opp}}$ The composition of two admissible monics is an admissible monic.
   Deduce that $\mathscr{E}$ is an exact structure for $\mathcal{A}$ if and only if $\mathscr{E}^{\mathrm{opp}}$ is an exact structure for $\mathcal{A}^{\mathrm{opp}}$.

(3) Show that all isomorphisms in $\mathcal{A}$ are admissible epics and admissible monics.

(4) Show that every split exact sequence $0 \to X \to X \oplus Y \to Y \to 0$ is admissible.

Remark: For (much) more on exact categories see [Büh] $\overset{\mathrm{O}}{\mathrm{X}}$.

Exercise F.16. Let $\mathcal{B}$ be an abelian category and let $\mathcal{A}$ be a full subcategory stable under extensions in $\mathcal{B}$. Show that $\mathcal{A}$ together with the class of sequences $0 \to X' \to X \to X'' \to 0$ that are exact in $\mathcal{B}$ is an exact category (Exercise F.15).

Remark: Conversely, one can show that if $(\mathcal{A}, \mathscr{E})$ is an exact category such that $\mathcal{A}$ is equivalent to a small category, then there exists a fully faithful functor $y \colon \mathcal{A} \to \mathcal{B}$ to an abelian category $\mathcal{B}$ such that the essential image of y is stable under extensions and such that a short sequence in $\mathcal{A}$ is exact if and only if its image under y is a short exact sequence in $\mathcal{B}$ ([Büh] $\overset{\mathrm{O}}{\mathrm{X}}$ App. A).

Exercise F.17. Let $\mathcal{A}$ be a Grothendieck abelian category and let $0 \to X' \to X \to X'' \to 0$ be a short exact sequence in $\mathcal{A}$.

(1) Show that if X is finitely generated, then X'' is finitely generated. Show that if X' and X'' are finitely generated, then X is finitely generated. (See Definition F.66.)

(2) Show that X is noetherian if and only if X' and X'' are noetherian.

Exercise F.18. Let A be a ring, let $E^{\bullet}$, $F^{\bullet}$ be complexes of A-modules and suppose that E is bounded above and that E^n and $H^n(E)$ are flat A-modules for all $n \in \mathbb{Z}$. Show that for all $n \in \mathbb{Z}$ there is a functorial isomorphism

$$\bigoplus_{i+j=n} H^i(E^{\bullet}) \otimes_A H^j(F^{\bullet}) \xrightarrow{\sim} H^n(E^{\bullet} \otimes_A F^{\bullet}).$$

Exercise F.19. Let $\mathcal{T}$ be a triangulated category and let $\mathcal{U}$ be a non-empty full additive subcategory of $\mathcal{T}$ such that if $E \to F \to G \to$ is a distinguished triangle in $\mathcal{T}$ and two of its terms are in $\mathcal{U}$, then the third is in $\mathcal{U}$. Show that $\mathcal{U}$ is stable under shifts and that if E is an object in $\mathcal{U}$, then every object of $\mathcal{T}$ isomorphic to E is also in $\mathcal{U}$. Deduce that $\mathcal{U}$ inherits from $\mathcal{T}$ the structure of a triangulated category.

Exercise F.20. Let $\mathcal{A}$ be a Grothendieck abelian category.

(1) Show that a complex I in $C(\mathcal{A})$ is K-injective if and only if for every exact complex E the Hom complex $\mathrm{Hom}_{\mathcal{A}}(E, I)$ is exact.

(2) Show that every K-injective complex is isomorphic in $K(\mathcal{A})$ to a K-injective complex K whose components K^n are injective objects of $\mathcal{A}$ for all $n \in \mathbb{Z}$.

(3) Show that a complex I in $C(\mathcal{A})$ is K-injective if and only if there exist complexes C_1, C_2 in $C(\mathcal{A})$ that are isomorphic to 0 in $K(\mathcal{A})$ and a K-injective complex with injective components K such that $I \oplus C_1 \cong K \oplus C_2$ in $C(\mathcal{A})$.

Remark: See [Gil] $\overset{\mathrm{O}}{\mathrm{X}}$ §6, and the Introduction.

Exercise F.21. Let $\mathcal{A}$ be a Grothendieck category. Let $Q\colon K(\mathcal{A}) \to D(\mathcal{A})$ be the canonical functor. Show that there exists a fully faithful right adjoint functor $I\colon D(\mathcal{A}) \to K(\mathcal{A})$ whose essential image consists of the full subcategory of K-injective complexes. Show that for each complex X in $K(\mathcal{A})$ the unit $X \to I(Q(X))$ of the adjunction is a K-injective resolution of X.

Exercise F.22. Let A be a not necessarily commutative ring and let $(E_i)_i$ be a family of left A-modules. Show that $\prod_i E_i$ is injective if and only if E_i is an injective A-module for all i.

Exercise F.23. Let A be a not necessarily commutative ring and let M be a left A-module. Show that M is an injective A-module if and only if for all left ideals $\mathfrak{a} \subseteq A$ every A-linear map $u\colon \mathfrak{a} \to A$ can be extended to an A-linear map $A \to E$.
Remark: See also Section (G.8).

Exercise F.24. Let A be an integral domain.
(1) Show that every injective A-module is *divisible* (i.e., for all $0 \neq a \in A$ scalar multiplication by a on E is surjective).
(2) Show that every torsion free and divisible module is injective. Deduce that the field of fractions of A is an injective A-module.
(3) Suppose that A is a principal ideal domain. Show that an A-module is injective if and only if it is divisible.
Hint: Exercise F.23

Exercise F.25. Let A be a not necessarily commutative ring.
(1) Show that the left A-module $E := \operatorname{Hom}_{\mathbb{Z}}(A, \mathbb{Q}/\mathbb{Z})$ is injective and that for every left A-module M and for all $0 \neq x \in M$ there exists an A-linear map $u\colon M \to E$ such that $u(x) \neq 0$.
(2) Let M be a left A-module. Show that

$$I(M) := E^{\operatorname{Hom}_A(M,E)}$$

is an injective A-module and that the map $e_M\colon M \to I(M)$ that sends $m \in M$ to $(u(m))_{u \in \operatorname{Hom}_A(M,E)}$ is A-linear, injective and functorial in M.
(3) Show that e_M is an injective hull of M.
Hint: Exercise F.24

Exercise F.26. Let A be a not necessarily commutative ring. Show that the following assertions are equivalent.
(i) A is left noetherian, i.e., every left ideal of A is finitely generated.
(ii) Every colimit of a small filtered diagram of injective left A-modules is again injective.
(iii) For all countable families $(E_n)_{n \in \mathbb{N}}$ of injective left A-modules the direct sum $\bigoplus_n E_n$ is an injective A-module.
Hint: To show that (i) implies (ii) use Exercise F.23, that every ideal $\mathfrak{a}$ of A is of finite presentation, and that hence $\operatorname{Hom}_A(\mathfrak{a}, -)$ commutes with filtered colimits. To show that (iii) implies (i) let $(\mathfrak{a}_n)_n$ be an ascending chain of ideals, let $\mathfrak{a} := \bigcup_n \mathfrak{a}_n$, and let $\mathfrak{a}/\mathfrak{a}_n \to I_n$ be an injective map of A-modules into an injective A-module. The induced map $\mathfrak{a} \to \bigoplus_n I_n$ can be extended to $u\colon A \to \bigoplus_n I_n$. Consider $u(1)$. (Here by ideal we mean left ideal, and similarly for modules.)

Exercise F.27. Let A be a noetherian ring and let $S \subseteq A$ be a multiplicative subset. Let E be an injective A-module. Show that $S^{-1}E$ is an injective A-module and an injective $S^{-1}A$-module.
Hint: Exercise F.26 and Exercise F.2.

Exercise F.28. Let $\mathcal{A}$ be an abelian category and let $p \geq 0$. Suppose that $\operatorname{Ext}_{\mathcal{A}}^{p}(X, Y) = 0$ for all objects $X, Y \in \mathcal{A}$. Show that $\operatorname{Ext}_{\mathcal{A}}^{i}(X, Y) = 0$ for all $i \geq p$ and all objects $X, Y \in \mathcal{A}$.

Exercise F.29. Let $\mathcal{A}$ be an abelian category. Show that the following assertions are equivalent.
(i) Every short exact sequence in $\mathcal{A}$ splits.
(ii) For all $X, Y \in \mathcal{A}$ one has $\operatorname{Ext}_{\mathcal{A}}^{1}(X, Y) = 0$.
(iii) For all $X, Y \in \mathcal{A}$ and for all $k \geq 1$ one has $\operatorname{Ext}_{\mathcal{A}}^{k}(X, Y) = 0$.
An abelian category satisfying these equivalent condition is called *semisimple*.
Hint: Exercise F.28.

Exercise F.30. Let $\mathcal{A}$ be an abelian category. Show that for an object X in $D^{b}(\mathcal{A})$ the following assertions are equivalent.
(i) X is isomorphic in $D^{b}(\mathcal{A})$ to a complex whose differentials are 0.
(ii) There exist morphisms $u^{p} \colon H^{p}(X)[-p] \to X$, $p \in \mathbb{Z}$, such that the resulting morphism $u \colon \bigoplus_{p} H^{p}(X)[-p] \to X$ satisfies $H^{p}(u) = \operatorname{id}_{H^{p}(X)}$ for all $p \in \mathbb{Z}$.
(iii) There exists an isomorphism $u \colon \bigoplus_{p} H^{p}(X)[-p] \to X$ in $D^{b}(\mathcal{A})$.
If X satisfies these equivalent conditions, X is called *decomposable*.

Exercise F.31. Let $\mathcal{A}$ be an abelian category.
(1) Let X be an object in $D^{b}(\mathcal{A})$ such that $\operatorname{Ext}^{p}(H^{i}(X), H^{j}(X)) = 0$ for all $p \geq 2$ and for all $i > j$. Show that X is decomposable (Exercise F.30).
(2) Suppose that $\operatorname{Ext}_{\mathcal{A}}^{2}(X, Y) = 0$ for all objects $X, Y \in \mathcal{A}$. Show that every complex in $D^{b}(\mathcal{A})$ is decomposable.
Hint: Exercise F.28
(3) Suppose that $\operatorname{Ext}_{\mathcal{A}}^{2}(X, Y) = 0$ for all objects $X, Y \in \mathcal{A}$ and that $\mathcal{A}$ has exact coproducts. Show that for any complex X in $C(\mathcal{A})$ there exists an isomorphism $\bigoplus_{i \in \mathbb{Z}} H^{i}(X)[-i] \xrightarrow{\sim} X$ in $D(\mathcal{A})$.

Exercise F.32. Let A be a Dedekind domain.
(1) Show that every A-module has injective dimension ≤ 1.
Hint: Use Proposition G.21 and that every ideal of A is a projective A-module.
(2) Show that every complex in $D^{b}(A)$ is decomposable (Exercise F.30).
Hint: Exercise F.31.

Exercise F.33. Let A be a ring, M an A-module and let $d \geq 0$ be an integer. Show that M has injective dimension $\leq d$ if and only if $\operatorname{Ext}_{A}^{d+1}(A/\mathfrak{a}, M) = 0$ for every ideal $\mathfrak{a}$ of A.
Hint: Proposition G.21.

Exercise F.34. Let A be a noetherian ring and let M be an A-module.
(1) Show that the injective dimension of the A-module M is the supremum of the injective dimensions of the $A_{\mathfrak{p}}$-modules $M_{\mathfrak{p}}$ where $\mathfrak{p}$ runs through all prime ideals (equivalently, all maximal ideals) of A.
Hint: Exercise F.33.
(2) Let $S \subseteq A$ be a multiplicative subset. Show that the injective dimension of the $S^{-1}A$-module $S^{-1}M$ is at most the injective dimension of the A-module M.
Deduce that if M is an injective A-module, then $S^{-1}M$ is an injective $S^{-1}A$-module.

Exercise F.35. Let k be a field and let $R := k[X]/(X^2)$. Show that R is an injective R-module. Show that the R-module k does not have finite injective dimension.

Exercise F.36. Let $\mathcal{A}$ be a Grothendieck abelian category and let $X \in D^b(\mathcal{A})$. Show that if $H^p(X)$ has finite injective dimension for all p, then X has finite injective dimension.

Exercise F.37. Let R be a ring and let E and F be bounded above complexes of projective R-modules. Show that every isomorphism $E \to F$ in $D(R)$ is induced by a homotopy equivalence $E \to F$ of complexes of R-modules.

Exercise F.38. Endow $\mathbb{N}$ with a topology by declaring a subset $U \subseteq \mathbb{N}$ to be open if $n \in U$ and $m < n$ imply $m \in U$ (i.e., it is the topological space attached to the partially ordered set $\mathbb{N}^{\mathrm{opp}}$). Let $\mathcal{A}$ be a Grothendieck abelian category.
(1) Show that there is an equivalence $\mathscr{F} \mapsto E(\mathscr{F})$ between the category of abelian sheaves $\mathscr{F}$ on $\mathbb{N}$ and the category of $\mathbb{N}^{\mathrm{opp}}$-diagrams in $\mathcal{A}$ such that $\Gamma(\mathbb{N}, \mathscr{F}) = \lim_n E(\mathscr{F})$.
(2) Deduce that $R^p \lim E(\mathscr{F}) = H^p(\mathbb{N}, \mathscr{F})$ for all $p \geq 0$.
(3) Generalize all the above assertions to the situation where one replaces $\mathbb{N}$ by the partially ordered set of ordinals that are smaller than some fixed ordinal.

Exercise F.39. Let $\mathcal{A}$ be an abelian category with countable products and enough injective objects and let X be a complex with components in $\mathcal{A}$. Let $(I_n)_n$ be a system of bounded below complexes with injective components that forms a resolution of $(\tau^{\geq -n} X)_n$ as in Lemma F.195.
(1) Show that $I := \lim I_n$ exists, is K-injective, and that $I = R\lim_n I_n$ in $D(\mathcal{A})$.
(2) Show that $X = \lim \tau^{\geq -n} X \to \lim_n I_n = I$ is an isomorphism in $D(\mathcal{A})$ if and only if $X \to R\lim \tau^{\geq -n} X$ is an isomorphism.
(3) Show that the equivalent conditions in (2) are satisfied if countable products are exact in $\mathcal{A}$.

G Commutative Algebra II

In this appendix, we collect some results from commutative algebra that were not stated in Appendix B.

On regular and Cohen-Macaulay rings

(G.1) More on flatness.

We first give some results on flat modules (also see Section (B.4)).

Proposition G.1. ([Mat2] Theorem 7.10) *Let A be a local ring and let M be a finitely generated flat A-module. Then M is a free A-module.*

Proposition G.2. ([Mat2] Theorem 22.5, [Sta] 046Y) *Let $\varphi\colon R \to A$ be a homomorphism of rings, and let M and N two A-modules of finite type. Let $u\colon M \to N$ be an A-linear homomorphism. Assume that N is flat over R and that one of the following conditions are satisfied.*
(a) *R and A are noetherian.*
(b) *A is an R-algebra of finite presentation, and M and N are A-modules of finite presentation.*
Then the following assertions are equivalent.
 (i) *u is injective and $\operatorname{Coker}(u)$ is a flat R-module.*
 (ii) *For every maximal ideal $\mathfrak{m}$ of A the morphism $u \otimes \operatorname{id}_{\kappa(\mathfrak{p})}\colon M \otimes_R \kappa(\mathfrak{p}) \to N \otimes_R \kappa(\mathfrak{p})$ is injective, where $\mathfrak{p} := \varphi^{-1}(\mathfrak{m}) \subset R$.*

Theorem G.3. (Lazard's theorem, [BouA10]^{O} §1.6, Théorème 1) *Let A be a ring and let M be an A-module. Then the following assertions are equivalent.*
 (i) *M is a flat A-module.*
 (ii) *M is a filtered colimit of finitely generated free A-modules.*
 (iii) *For every A-module N of finite presentation, the canonical map*

$$\operatorname{Hom}_A(N, A) \otimes_A M \longrightarrow \operatorname{Hom}_A(N, M)$$

is surjective.

© Springer Fachmedien Wiesbaden GmbH, ein Teil von Springer Nature 2023
U. Görtz und T. Wedhorn, *Algebraic Geometry II: Cohomology of Schemes*,
Springer Studium Mathematik – Master, https://doi.org/10.1007/978-3-658-43031-3

(G.2) Regularity and global dimension.

Proposition G.4. ([BouAC10] $^\circ$ §4 no. 2) *Let R be a local noetherian ring with residue field k. Then the following assertions are equivalent.*
(i) *R is regular.*
(ii) *There exists an integer $d \geq 0$ such that $\operatorname{Tor}_{d+1}^{R}(k, k) = 0$.*
(iii) *There exists an integer $e \geq 0$ such that $\operatorname{Tor}_{i}^{R}(M, N) = 0$ for all finitely generated R-modules M and N and for all $i > e$.*
If these equivalent conditions are satisfied, the minimal integers d and e satisfying (ii) resp. (iii) are both equal to $\dim R$.

Proposition G.5. ([BouAC10] $^\circ$ §4, no. 2.) *Let R be a regular noetherian ring. Then for every finitely generated R-module M there exists an exact sequence*

$$0 \longrightarrow F_{-n} \longrightarrow F_{-n+1} \longrightarrow \cdots \longrightarrow F_0 \longrightarrow M \longrightarrow 0$$

where the F_i are finitely generated free R-modules and where $n \leq \dim R$.

(G.3) Dimension of modules.

Definition G.6. *Let A be a ring and let M be a finitely generated A-module. Then $\dim_A(M) := \dim(\operatorname{Supp} M)$ is called the* dimension *of M.*

Recall that $\operatorname{Supp} M \subseteq \operatorname{Spec} A$ is a closed subset since M is finitely generated. Then $\dim_A(M) = -\infty$ if and only if $M = 0$.

Proposition G.7. ([BouAC89] $^\circ$ VIII 1.4, Prop. 9) *Let A be a ring and let $0 \to M' \to M \to M'' \to 0$ be a short exact sequence of finitely generated A-modules. Then $\dim(M) = \sup\{\dim(M'), \dim(M'')\}$.*

Proposition G.8. ([BouAC89] $^\circ$ VIII 3.1, Prop. 2) *Let A be a noetherian ring, let M be a finitely generated A-module, and let $\mathfrak{a} \subset A$ be an ideal that is contained in the Jacobson radical of A and which is generated by r elements. Then*

$$\dim M - r \leq \dim M/\mathfrak{a}M \leq \dim M.$$

Note that loc. cit. is formulated only for $M = A$ but the proof works verbatim for an arbitrary finitely generated A-module.

(G.4) Depth.

Let A be a local noetherian ring with maximal ideal $\mathfrak{m}$ and residue field κ. Recall that the *depth* of a finitely generated A-module $M \neq 0$ is defined as the maximal $r \geq 0$ such that there exists an M-regular sequence with r elements contained in $\mathfrak{m}$. It is denoted by $\operatorname{depth}_A(M)$.

More generally, let A be a noetherian ring (not necessarily local), let $I \subseteq A$ be an ideal and let M be a finitely generated A-module with $M \neq IM$. Then the length of a maximal M-regular sequence in I is well determined and it is called the *I-depth of M* and denoted by $\operatorname{depth}_A(I, M)$. If $M = IM$ we define the I-depth of M to be ∞.

Proposition G.9. ([BouAC10] ^{O} 1.4, Théorème 2) *Let A be a noetherian, $I \subseteq A$ an ideal and let M be a finitely generated A-module. Then*

$$\mathrm{depth}_A(I, M) = \inf\{\, i \geq 0 \; ; \; \mathrm{Ext}_A^i(A/I, M) \neq 0 \,\}.$$

Corollary G.10. ([BouAC10] ^{O} 1.4, Cor. 2 de Théorème 2) *Let A be a local noetherian ring and $M \neq 0$ be a finitely generated A-module. Then one has*

$$\mathrm{depth}_A(M) \leq \dim(\mathrm{Supp}(M)) < \infty.$$

Proposition G.11. ([BouAC10] ^{O} 1.7, Prop. 10, Cor. de Prop. 11, see also Lemma 14.23) *Let $A \to B$ be a local homomorphism of local noetherian rings and let k be the residue field of A.*
(1) *Let $\mathbf{f} = (f_1, \ldots, f_r)$ be a sequence of elements in the maximal ideal of B. Then $A \to B$ is flat and the image of $\mathbf{f}$ is $B \otimes_A k$ is regular if and only if $B/\mathbf{f}B$ is a flat A-algebra and $\mathbf{f}$ is regular.*

$$\mathrm{depth}(B) = \mathrm{depth}(A) + \mathrm{depth}(B \otimes_A k).$$

Dually to the notion of injective dimension, Section (F.53), we have the notion of *projective dimension*. For a ring A, an A-module M has *projective dimension d*, if d is the minimal natural number such that there exists a projective resolution

$$P_d \to \cdots \to P_1 \to M,$$

i.e., an exact sequence with all P_i projective A-modules.

Theorem G.12. (Auslander-Buchsbaum formula, [Mat2] Theorem 19.1) *Let A be a local noetherian ring and let $M \neq 0$ be a finitely generated A-module of finite projective dimension $\mathrm{projdim}(M)$. Then*

$$\mathrm{projdim}(M) + \mathrm{depth}(M) = \mathrm{depth}(A).$$

(G.5) Cohen-Macaulay modules.

Definition and Proposition G.13. ([BouAC10] ^{O} 2.1, Cor. de Prop. 1) *Let A be a noetherian ring. A finitely generated A-module M is called Cohen-Macaulay, if the following equivalent conditions are satisfied.*
(i) *For every maximal ideal $\mathfrak{m}$ of A with $M_\mathfrak{m} \neq 0$ one has $\mathrm{depth}_{A_\mathfrak{m}}(M_\mathfrak{m}) = \dim_{A_\mathfrak{m}}(M_\mathfrak{m})$.*
(ii) *For every prime ideal $\mathfrak{p}$ of A with $M_\mathfrak{p} \neq 0$ one has $\mathrm{depth}_{A_\mathfrak{p}}(M_\mathfrak{p}) = \dim_{A_\mathfrak{p}}(M_\mathfrak{p})$.*

Proposition G.14. ([BouAC10] ^{O} 2.2, Cor. de Prop. 2) *Let A be a local noetherian ring and let $M \neq 0$ be a finitely generated Cohen-Macaulay module. Then $\mathrm{Supp}\, M$ is equi-dimensional and for every ideal I of A one has*

$$\mathrm{depth}_A(I, M) = \dim M - \dim M/IM.$$

Definition G.15. *Let A be a noetherian ring, let M be a finitely generated A-module, and let $k \geq 0$ be an integer. Then M is said to have the property (S_k) if one has $\mathrm{depth}_{A_\mathfrak{p}}(M_\mathfrak{p}) \geq \inf\{k, \dim_{A_\mathfrak{p}}(M_\mathfrak{p})\}$ for every prime ideal $\mathfrak{p}$ of A.*

Every finitely generated A-module satisfies (S_0). A finitely generated A-module is Cohen-Macaulay if and only if it satisfies (S_k) for all $k \geq 0$.

(G.6) Reflexive modules.

Let A be a ring. For an A-module M, we denote by $M^\vee := \operatorname{Hom}_A(M, A)$ its A-dual. We then have the natural homomorphism $M \to M^{\vee\vee}$ from M into its double dual.

Definition G.16. *Let A be a ring. An A-module M is called* reflexive, *if the natural map $M \to M^{\vee\vee}$ is an isomorphism.*

Clearly locally free A-modules M of finite rank are reflexive (because this can be checked locally on $\operatorname{Spec} A$ and is clear for finite free modules). But in general, being reflexive is a weaker notion than being finite locally free. Every reflexive module (or, more generally, the dual of any module) over an integral domain is torsion-free.

Lemma G.17. ([Sta] 0AV3) *Let A be a noetherian domain and let M be a finitely generated A-module. Then the dual $M^\vee$ is reflexive.*

Reflexivity is a particularly useful notion for normal domains.

Proposition G.18. *Let A be a normal noetherian domain, $X = \operatorname{Spec} A$, and let M be a finite A-module. The following are equivalent.*
(i) *The module M is reflexive.*
(ii) *The module M is torsion-free and has the property (S_2) (Definition G.15).*
(iii) *The module M is torsion-free and $M = \bigcup_{\mathfrak{p}\in\operatorname{Spec} A,\ \operatorname{ht}(\mathfrak{p})=1} M_\mathfrak{p}$ (inside $M \otimes_A \operatorname{Frac}(A)$).*
(iv) *The module M is torsion-free, and whenever M' is a torsion-free A-module with $M \subseteq M'$ and $\operatorname{codim}_{\operatorname{Spec} A}(\operatorname{Supp}(M'/M)) \geq 2$, then $M = M'$.*

Proof. The equivalence of (i), (ii) and (iii) is proved in [Sta] 0AVB. Given (iii) and M' as in (iv), we have $M \subseteq M' \subseteq M \otimes K$ and $M_\mathfrak{p} = M'_\mathfrak{p}$ for all $\mathfrak{p}$ of height 1, and hence $M = M'$ using the description in (iii).

For implication (iv) $\Rightarrow$ (i) note that there exists an open $U \subseteq \operatorname{Spec}(A)$ whose complement has codimension ≥ 2 such that the restriction of M to U is locally free (the local rings $A_\mathfrak{p}$ for $\mathfrak{p}$ of height 1 are discrete valuation rings, and since M is torsion-free, $M_\mathfrak{p}$ is free). Therefore we can take $M' = M^{\vee\vee}$ in (iv) and obtain (i). $\square$

(G.7) Quotients by ideals generated by regular sequences.

Proposition G.19. ([BouAC89]$^\circ$ VIII, §5.3) *Let A be a local noetherian ring, $I \subsetneq A$ an ideal. Let A/I be regular. Then A is regular if and only if I is generated by a regular sequence.*

Proposition G.20. ([BouAC10]$^\circ$ §2.3, Prop. 4) *Let A be a local noetherian ring, $\mathbf{f} = (f_1, \ldots, f_c)$ a sequence in the maximal ideal of A, and let I be the ideal generated by $\mathbf{f}$. Assume that $\dim A/I = \dim A - c$. Then A is is Cohen-Macaulay if and only if $\mathbf{f}$ is a regular sequence and A/I is Cohen-Macaulay.*

On injective modules and Gorenstein rings

In the following sections, A will denote a ring.

(G.8) On injective modules.

Recall (Definition F.49) that an A-module M is *injective* if and only if the functor $\mathrm{Hom}_A(-, M)\colon (A\text{-Mod})^{\mathrm{opp}} \to (A\text{-Mod})$ is exact.

Proposition G.21. ([BouA10] ^{O} §1.7, Prop. 10, §5.5, Prop. 11) *Let A be a ring and let M be an A-module. Then the following assertions are equivalent.*
(i) *M is injective.*
(ii) *For every ideal $\mathfrak{a}$ of A every A-linear map $\mathfrak{a} \to M$ can be extended to A.*
(iii) *For every A-module N one has $\mathrm{Ext}^i_A(N, M) = 0$ for all $i > 0$.*
(iv) *For every ideal $\mathfrak{a}$ of A one has $\mathrm{Ext}^1_A(A/\mathfrak{a}, M) = 0$.*

Proposition G.22. ([BouA10] ^{O} §1.7, Cor. 2 de Prop. 10) *Let A be a principal ideal domain. Then an A-module M is an injective A-module if and only if it is divisible (i.e., for all $0 \neq a \in A$ the multiplication $M \to M$ by a is surjective).*

Proposition G.23. ([Mat2] §18, Lemma 5) *Let A be a noetherian ring, let $S \subseteq A$ be a multiplicative subset, and let I be an injective A-module. Then $S^{-1}I$ is an injective $S^{-1}A$-module.*

Proof. Let $\mathfrak{b} \subseteq S^{-1}A$ be an ideal. Then $\mathfrak{b} = S^{-1}\mathfrak{a}$ for some ideal $\mathfrak{a}$ of A. By Proposition G.21 it suffices to show that

$$\mathrm{Hom}_{S^{-1}A}(S^{-1}A, S^{-1}I) = S^{-1}\mathrm{Hom}_A(A, I)$$
$$\longrightarrow \mathrm{Hom}_{S^{-1}A}(S^{-1}\mathfrak{a}, S^{-1}I) = S^{-1}\mathrm{Hom}_A(\mathfrak{a}, I)$$

is surjective (for the first equality use that $\mathfrak{a}$ is of finite presentation since A is noetherian). This is clear as I is injective. $\square$

For different ideas of proofs see Exercise F.27 and Exercise F.34.

(G.9) Matlis duality.

In this section, A denotes a noetherian local ring with maximal ideal $\mathfrak{m}$ and residue class field κ. Every A-module M has an injective hull (Definition F.64). By definition, it is an embedding $M \hookrightarrow E$ of A-modules such that E is an injective A-module such that $N \cap M \neq 0$ for every non-zero submodule N of E. An injective hull is determined up to isomorphism (but not up to unique isomorphism), see Proposition F.65.

An injective hull of the A-module κ is called a *Matlis module for A*.

Proposition G.24. (Matlis duality, [BouAC10] ^{O} §8.3, Théorème 2) *Let E be a Matlis module for A. The functor*
$$D_A\colon N \mapsto \mathrm{Hom}_A(N, E)$$
is an anti-equivalence of the category of finite length A-modules with itself which is its own inverse. For every A-module M of finite length one has $\mathrm{lg}_A(D_A(M)) = \mathrm{lg}_A(M)$.

Denote by $(A\text{-Mod})_{\mathrm{fl}}$ be the full subcategory of the category of A-modules consisting of the A-modules of finite length, i.e., of finitely generated A-modules that are annihilated by some power of $\mathfrak{m}$. This is a plump subcategory (Definition F.43), in particular it is an abelian category.

Let $T \colon (A\text{-Mod})_{\mathrm{fl}} \to (A\text{-Mod})$ be an exact A-linear contravariant functor such that $T(\kappa)$ is an A-module of length 1. Then for every M in $(A\text{-Mod})_{\mathrm{fl}}$, $T(M)$ is also of finite length with $\lg_A T(M) = \lg_A M$.

For $n \geq 0$ set $E_n := T(A/\mathfrak{m}^n)$. Then for $m \geq n$, the canonical maps $A/\mathfrak{m}^m \to A/\mathfrak{m}^n$ induce injective A-linear maps $E_n \to E_m$. Set $E := \operatorname{colim}_n E_n$.

Let M be an A-module of finite length and choose an $n \geq 1$ such that $\mathfrak{m}^n M = 0$. For $x \in M$, the A-linear map $A/\mathfrak{m}^n \to M$ that sends 1 to x induces an A-linear map $\alpha_{x,M} \colon T(M) \to E_n \hookrightarrow E$ and the map

$$\theta_M \colon T(M) \longrightarrow \operatorname{Hom}_A(M, E), \qquad \lambda \mapsto \alpha_{x,M}(\lambda)$$

is independent of the choice of n and A-linear. It defines an A-linear map of A-linear functors $(A\text{-Mod})_{\mathrm{fl}}^{\mathrm{opp}} \to (A\text{-Mod})_{\mathrm{fl}}$

(G.9.1) $$\theta \colon T(-) \longrightarrow \operatorname{Hom}_A(-, E).$$

Proposition G.25. ([BouAC10] ^{O} §8.3, Théorème 3) *The A-module E is a Matlis module and θ is an isomorphism of functors.*

(G.10) Gorenstein rings.

Proposition/Definition G.26. (cf. [Mat2] Theorem 18.1) *Let A be a noetherian local ring of dimension n. Let κ be the residue class field of A. The following are equivalent.*

(i) *A has finite injective dimension as an A-module, i.e., A is isomorphic to a bounded complex of injective A-modules in $D(A)$.*

(ii) *$\operatorname{Ext}_A^i(\kappa, A) = 0$ for all $i \neq n$, and $\operatorname{Ext}_A^n(\kappa, A) \cong \kappa$.*

(iii) *There exists $i > n$, such that $\operatorname{Ext}_A^i(\kappa, A) = 0$.*

(iv) *The ring A is Cohen-Macaulay and $\operatorname{Ext}_A^n(\kappa, A) \cong \kappa$.*

We say that the local ring A is a Gorenstein ring, if A satisfies the above conditions. It has then injective amplitude contained in $[0, n]$.

As every local Artinian ring is zero-dimensional and hence Cohen-Macaulay, we have the following corollary.

Corollary G.27. *Let A be a local Artinian ring with residue field κ. Then the following assertions are equivalent.*

(i) *A is Gorenstein.*

(ii) *The A-module A is injective.*

(iii) *The κ-vector space $\operatorname{Hom}_A(\kappa, A)$ is one-dimensional.*

Proposition G.28. ([Mat2] Theorem 18.2) *Let A be a Gorenstein local ring. For every prime ideal $\mathfrak{p} \subset A$, the localization $A_{\mathfrak{p}}$ is Gorenstein.*

Because of this proposition, we can define the notion of Gorenstein ring as follows.

Definition G.29. *Let A be a noetherian ring. We say that A is a Gorenstein ring, if for every maximal ideal $\mathfrak{p} \subset A$ (equivalently: for every prime ideal $\mathfrak{p} \subset A$) the localization $A_{\mathfrak{p}}$ is a Gorenstein local ring in the sense of Definition G.26.*

Proposition G.30. ([BouAC10] ^{O} §3.8, Prop. 12) *Let $A \to B$ be a flat local homomorphism of local noetherian rings. Let κ_A be the residue field of A. Then the following assertions are equivalent.*
(i) *B is a Gorenstein ring.*
(ii) *A and $B \otimes_A \kappa_A$ are Gorenstein rings.*

Corollary G.31. ([BouAC10] ^{O} §3.8, Cor. 2 de Prop. 12) *Let A be a noetherian ring and let $I \subseteq A$ be an ideal that is contained in the Jacobson radical of A. Then A is Gorenstein if and only if the I-adic completion of A is Gorenstein.*

(G.11) Addenda on separable and inseparable field extensions.

Lemma G.32. ([BouAC] ^{O} V, §2.3, Lemma 4, Remark (1)) *Let A be a normal domain, K its field of fractions, let $p \geq 1$ be its characteristic exponent. Let K' be a purely inseparable extension of K and let A' be the integral closure of A in K'. Then*

$$A' = \{\, x' \in K' \; ; \; \exists m \geq 1 \colon x^{p^m} \in A \,\}$$

and for every prime ideal $\mathfrak{p}$ of A there exists a unique prime ideal $\mathfrak{p}'$ of A' lying over $\mathfrak{p}$. One has

$$\mathfrak{p}' = \{\, x' \in K' \; ; \; \exists m \geq 1 \colon x^{p^m} \in \mathfrak{p} \,\}.$$

We also need the following variant of Proposition B.97.

Proposition G.33. ([Sta] 030W; cf. [BouAII] ^{O} 15.4 Cor. 1) *Let K/k be a field extension of fields with positive characteristic p. The following are equivalent.*
(i) *The extension K/k is separable.*
(ii) *The ring $K \otimes_k k^{1/p}$ is reduced, where $k^{1/p} = k(\sqrt[p]{a},\ a \in k)$ is the extension of k obtained by adjoining all p-th roots of elements of k in some algebraic closure of k.*

Lemma G.34. ([Mat2] Theorem 26.5; [Sta] 07P2.) *Let k be a field of characteristic $p > 0$ and let K/k be a finitely generated field extension. Let $x_1, \dots, x_n \in K$. The following are equivalent.*
(i) *The elements $x_1, \dots, x_n$ are a differential basis of K over k, i.e., $dx_1, \dots, dx_n$ are a K-basis of $\Omega_{K/k}$.*
(ii) *The elements $x_1, \dots, x_n$ are a p-basis of K over k, i.e., the field K equals the compositum $kK^p(x_1, \dots, x_n)$ and $[K : kK^p] = p^n$.*
In particular, we have $p^{\dim_K \Omega_{K/k}} = [K : kK^p]$.

Bibliography

[ACGH] E. Arbarello, M. Cornalba, P. Griffiths, J. Harris, *Geometry of Algebraic Curves, Volume I*, Springer Grundlehren **267** (1985). DOI:10.1007/978-1-4757-5323-3

[ACG] E. Arbarello, M. Cornalba, P. Griffiths, *Geometry of Algebraic Curves, Volume II*, Springer Grundlehren **268** (2011). DOI:10.1007/978-3-540-69392-5

[AJL] L. Alonso Tarrío, A. Jeremías López, J. Lipman, *Bivariance, Grothendieck duality and Hochschild homology II: the fundamental class of a flat scheme-map*, Adv. math. **257** (2014), no. 1, 365–461.
DOI:10.1016/j.aim.2014.02.017   arXiv:1202.4367

[AlKl] A. Altman, S. Kleiman, *Introduction to Grothendieck duality*, Lecture Notes in Mathematics **146**, Springer (1970). DOI:10.1007/BFb0060932

[Art1] M. Artin, *Grothendieck topologies*, Harvard university (1962).

[Art2] M. Artin, *Algebraic approximation of structures over complete local rings*, Publ. math. IHES **36** (1969), 23–58. numdam:PMIHES_1969__36__23_0

[Art3] M. Artin, *Algebraization of Formal Moduli: I*, in: Global Analysis, Papers in Honor of K. Kodaira (D. Spencer, S. Iyanaga eds.), Princeton Mathematical Series **1586** (1970). DOI:10.1515/9781400871230-003

[Art4] M. Artin, *Algebraization of Formal Moduli: II – Existence of Modifications*, Annals of math. **91** (1970), 88–135. DOI:10.2307/1970602

[Art5] M. Artin, *Algebraic structure of power series rings*, Contemp. Math. **13** (1982), 223–227.

[Avr] L.L. Avramov, *Flat morphisms of complete intersections*, Soviet Math. Dokl. **16** (1975), 1413–1417. www.mathnet.ru/eng/dan39366

[Bad] L. Bădescu, *Algebraic Surfaces*, Springer Universitext (2001)
DOI:10.1007/978-1-4757-3512-3

[BBH] C. Bartocci, U. Bruzzo, D. Hernández-Ruipérez, *Fourier-Mukai and Nahm Transforms in Geometry and Mathematical Physics*, Progress in Mathematics **276**, Birkhäuser Boston (2009). DOI:10.1007/b11801

[BBM] P. Berthelot, L. Breen, W. Messing, *Théorie de Dieudonné Cristalline II*, Lecture Notes in Mathematics **930**, Springer (1982). DOI:10.1007/BFb0093025

[BDS] P. Balmer, I. Dell'Ambrogio, B. Sanders, *Grothendieck-Neeman duality and the Wirthmüller isomorphism*, Compos. Math. **152** (2016), no. 8, 1740–1776.
DOI:10.1112/S0010437X16007375   arXiv:1501.01999v2

[Beau] A. Beauville, *Complex Algebraic Surfaces*, 2nd edition, Cambridge University Press (2010). DOI:10.1017/CBO9780511623936

© Springer Fachmedien Wiesbaden GmbH, ein Teil von Springer Nature 2023
U. Görtz und T. Wedhorn, *Algebraic Geometry II: Cohomology of Schemes*,
Springer Studium Mathematik – Master, https://doi.org/10.1007/978-3-658-43031-3

[Bha] B. Bhatt, *Topics in Algebraic Geometry I – Abelian Varieties*, lecture notes (2017),
www.math.ias.edu/~bhatt/teaching/mat731f17/lectures.pdf

[BhSc] B. Bhatt, P. Scholze, *The pro-étale topology for schemes*, Astérisque **369** (2015),
99–201.
      smf.emath.fr/publications/la-topologie-pro-etale-sur-les-schemas arXiv:1309.1198v2

[BiLa] C. Birkenhake, H. Lange, *Complex abelian varieties*, 2nd ed., Springer (2004).
      DOI:10.1007/978-3-662-06307-1

[BLR] S. Bosch, W. Lütkebohmert, M. Raynaud, *Néron models*, Erg. Math., 3. Folge,
Bd. **21**, Springer (1990).   DOI:10.1007/978-3-642-51438-8

[BoBe] A. Bondal, M. van den Bergh, *Generators and representability of functors in
commutative and noncommutative geometry*, Mosc. Math. J. **3** (2003), no. 1, 1–36.
      DOI:10.17323/1609-4514-2003-3-1-1-36 arXiv:math/0204218v2

[Bom] E. Bombieri, *Counting points over finite fields*, Sém. Bourbaki **430** (1972-73),
234–241.   numdam:SB_1972-1973__15__234_0

[BoNe] M. Bökstedt, A. Neeman, *Homotopy limits in triangulated categories*, Compos. Math. **86** (1993), no. 2, 209–234.   numdam:CM_1993__86_2_209_0

[Bor] F. Borceux, *Handbook of Categorical Algebra 2*, Encyclopedia of Mathematics and
its Applications **51**, Cambridge University Press (1994).

[BouAI] N. Bourbaki, *Algebra*, Chapters 1–3, Springer (1989).
      link.springer.com/book/9783540642435

[BouAII] N. Bourbaki, *Algebra*, Chapters 4–7, Springer (1988).
      link.springer.com/book/10.1007/978-3-642-61698-3

[BouA8] N. Bourbaki, *Algèbre*, Chapitre 8, Hermann (1958). DOI:10.1007/978-3-540-35316-4

[BouA10] N. Bourbaki, *Algèbre*, Chapitre 10, Masson (1980), Springer (2007).
      DOI:10.1007/978-3-540-34493-3

[BouAC] N. Bourbaki, *Commutative algebra*, Chapters 1–7, 2nd printing, Springer (1989).
      link.springer.com/book/9783540642398

[BouAC89] N. Bourbaki, *Algèbre commutative*, Chapitres 8 et 9, Springer (2006)
      DOI:10.1007/3-540-33980-9

[BouAC10] N. Bourbaki, *Algèbre commutative*, Chapitre 10, Masson, Paris (1998).
      DOI:10.1007/978-3-540-34395-0

[BouGT] N. Bourbaki, *General topology*, Chapters 1–4, 2nd printing, Springer (1989).
      DOI:10.1007/978-3-642-61701-0

[BouLie23] N. Bourbaki, *Groupes et Algèbres de Lie*, Chapitres 2 et 3, Springer (2006).
      DOI:10.1007/978-3-540-33978-6

[BrHe] W. Bruns, J. Herzog, *Cohen-Macaulay rings*, Cambridge studies in adv. math. **39**,
revised ed., Cambridge Univ. Press (1998).   DOI:10.1017/CBO9780511608681

[Bri] M. Brion, *The Coherent Cohomology Ring of an Algebraic Group*, Algebras and Repr. Th. **16** (2013), 1449–1467.   DOI:10.1007/s10468-012-9364-0   arXiv:1204.2628v1

[Büh] T. Bühler, *Exact Categories*, Expositiones Mathematicae **28**, no. 1 (2010) 1–69.
DOI:10.1016/j.exmath.2009.04.004   arXiv:0811.1480v2

[CaEi] H. Cartan, S. Eilenberg, *Homological Algebra*, Princeton Landmarks in Math. **19**, Princeton Univ. Press (1999, originally 1956).
press.princeton.edu/books/ebook/9781400883844/homological-algebra-pms-19-volume-19

[Čes] K. Česnavičius, *Problems About Torsors over Regular Rings*, Acta math. Vietnam. **47** (1) (2022), 39–107.   DOI:10.1007/s40306-022-00477-y   arXiv:2201.06424v2

[ChRü] A. Chatzistamatiou, K. Rülling, *Higher direct images of the structure sheaf in positive characteristic*, Algebra Number Theory **5**, No. 6 (2011), 693–775.
DOI:10.2140/ant.2011.5.693   arXiv:0911.3599v2

[CJLO] R. Cheng, M. Larson, L. Ji, and N. Olander, *Theorem of the base*, to appear in SPEC: Stacks Project Expository Collection, Cambridge Univ. Press
www.math.columbia.edu/~rcheng/assets/theorem-of-the-base.pdf

[CLO] B. Conrad, M. Lieblich, M. Olsson, *Nagata compactification for algebraic spaces*, J. Inst. Math. Jussieu **12** (2012), 747–814.
DOI:10.1017/S1474748011000223   arXiv:0910.5008v2

[Con] B. Conrad, *Grothendieck duality and base change*, Lecture Notes in Mathematics **1750**, Springer (2000).   DOI:10.1007/b75857

[Coq] T. Coquand, *On seminormality*, preprint,
www.cse.chalmers.se/~coquand/min.pdf

[CoSi] G. Cornell, J. Silverman, *Arithmetic Geometry*, Springer (1986).
DOI:10.1007/978-1-4613-8655-1

[CSS] G. Cornell, J. Silverman, G. Stevens, *Modular Forms and Fermat's Last Theorem*, Springer (1997).   DOI:10.1007/978-1-4612-1974-3

[Deb] O. Debarre, *Complex Tori and Abelian Varieties*, SMF/AMS Texts and Monographs **11** (2005).   bookstore.ams.org/smfams-11

[Del] P. Deligne, *La conjecture de Weil: I*, Publ. Math. de l'IHÉS **43** (1974), 273–307.
numdam:PMIHES_1974__43__273_0

[DeMu] P. Deligne, D. Mumford, *The irreducibility of the space of curves of given genus*, Publ. Math. de l'IHÉS **36** (1969), 75–109.   numdam:PMIHES_1969__36__75_0

[DePa] P. Deligne, G. Pappas, *Singularités des espaces de modules de Hilbert, en les caractéristiques divisant le discriminant*, Compos. Math. **90** (1994), no. 1, 59–79.
numdam:CM_1994__90_1_59_0

[dJo] A.J. de Jong, *Smoothness, semi-stability and alterations*, Inst. Hautes Études Sci. Publ. Math. **83** (1996), 51–93.   numdam:PMIHES_1996__83__51_0

[dJOl] A. J. de Jong, N. Olander, *On weakly étale morphisms*, arXiv:2202.05875v1 (2022).

[EEGRO] E. Enochs, S. Estrada, J. R. García-Rozas, L. Oyonarte, *Flat cotorsion quasi-coherent sheaves. Applications*, Algebras and Representation Theory 7 (2004), no. 4, 441–456.   DOI:10.1023/B:ALGE.0000042145.72104.cc

[EGAI] A. Grothendieck, J.A. Dieudonné, *Eléments de Géométrie Algébrique I*, Publ. Math. IHÉS **4** (1960), 5–228.   numdam:PMIHES_1960__4__5_0

[EGAInew] A. Grothendieck, J.A. Dieudonné, *Eléments de Géométrie Algébrique I*, Springer (1971).

[EGAII] A. Grothendieck, J.A. Dieudonné, *Eléments de Géométrie Algébrique II*, Publ. Math. IHÉS **8** (1961), 5–222.   numdam:PMIHES_1961__8__5_0

[EGAIII] A. Grothendieck, J.A. Dieudonné, *Eléments de Géométrie Algébrique III*, Publ. Math. IHÉS **11** (1961), 5–167; **17** (1963), 5–91.   numdam:PMIHES_1961__11__5_0

[EGAIV] A. Grothendieck, J.A. Dieudonné, *Eléments de Géométrie Algébrique IV*, Publ. Math. IHÉS **20** (1964), 5–259; **24** (1965), 5–231; **28** (1966), 5–255; **32** (1967), 5–361.   numdam:PMIHES_1964__20__5_0

[EGM] S. Edixhoven, G. van der Geer, B. Moonen, *Abelian varieties*, in progress, www.math.ru.nl/~bmoonen/research.html#bookabvar

[EnEs] E. Enochs, S. Estrada, *Relative homological algebra in the category of quasi-coherent sheaves*, Adv. math. **194** (2005), 284–295.   DOI:10.1016/j.aim.2004.06.007

[FaCh] G. Faltings, C. Chai, *Degeneration of Abelian Varieties*, Erg. Math., 3. Folge, Bd. **22**, Springer (1990).   DOI:10.1007/978-3-662-02632-8

[FaKr] H. Farkas, I. Kra, *Riemann Surfaces*, Graduate Texts in Mathematic **71**, Springer (1980).   DOI:10.1007/978-1-4612-2034-3

[Fal] G. Faltings, *Endlichkeitssätze für abelsche Varietäten über Zahlkörpern*, Invent. math. **73** (1983), 349–366, see [CoSi]^O for an English translation.
                                                                eudml.org/doc/143051

[Fer] D. Ferrand, *Conducteur, descente et pincement*, Bull. Soc. Math. France **131** (2003) no. 4, pp. 553–585.   numdam:BSMF_2003__131_4_553_0

[FGA] A. Grothendieck, *Fondements de la géométrie algébrique*, Sém. Bourbaki, 1957–1962.   webusers.imj-prg.fr/~leila.schneps/grothendieckcircle/FGA.pdf

[FGAex] B. Fantechi et. al., *Algebraic Geometry: Grothendieck's FGA Explained*, Mathematical Surveys and Monographs **123**, AMS 2005. ncatlab.org/nlab/show/FGA+explained

[FrKi] E. Freitag, R. Kiehl, *Etale Cohomology and the Weil Conjecture*, Erg. Math., 3. Folge, Bd. **13**, Springer (1988).   DOI:10.1007/978-3-662-02541-3

[For] O. Forster, *Lectures on Riemann Surfaces*, Graduate Texts in Mathematics **81**, Springer (1977).   DOI:10.1007/978-1-4612-5961-9

[Fu] L. Fu, *Etale Cohomology Theory*, Nankai Tracts in Math. **13**, World Scientific (2011).
                                                                DOI:10.1142/9569

[FuKa] K. Fujiwara, F. Kato, *Foundations of Rigid Geometry I*, EMS Monographs in Mathematics (2018). DOI:10.4171/135 arXiv:1308.4734v5

[Ful1] W. Fulton, *Algebraic Curves, an introduction to algebraic geometry*, Benjamin 1969, revised version 2008, www.math.lsa.umich.edu/~wfulton/CurveBook.pdf

[Ful2] W. Fulton, *Intersection Theory*, 2nd ed., Springer (1998). DOI:10.1007/978-1-4612-1700-8

[GaRa] O. Gabber, L. Ramero, *Almost Ring Theory*, Lecture Notes in Mathematics **1800**, Springer (2003). DOI:10.1007/b10047 arXiv:math/0201175v3

[GaZi] P. Gabriel, M. Zisman, *Calculus of Fractions and Homotopy Theory*, Springer (1967). DOI:10.1007/978-3-642-85844-4

[Gil] J. Gillespie, *Exact model structures and recollements*, Journal of Algebra **458** (2016), 265–306. DOI:10.1016/j.jalgebra.2016.03.021 arXiv:1310.7530

[GiSa] P. Gille, T. Szamuely, *Central Simple Algebras and Galois Cohomology*, Cambridge studies in advanced mathematics **101**, Cambridge University Press (2006). DOI:10.1017/CBO9780511607219

[God] R. Godement, *Topologie algébrique et théorie des faisceaux*, Hermann (1958).

[Goo] J. Goodman, *Affine open subsets of algebraic varieties and ample divisors*, Ann. of math. **89** (1969), 160–183. DOI:10.2307/1970814

[GrHa] P. Griffiths, J. Harris, *Principles of algebraic geometry*, Wiley-Interscience (1978), reprinted 1994. DOI:10.1002/9781118032527

[GrK] E. Große-Klönne, *Das duale abelsche Schema*, Diplomarbeit Univ. Münster (1995).

[GrMay] J.P.C. Greenlees, J.P. May, *Derived Functors of I-adic Completion and Local Homology*, Journal of Algebra **149** (1992), 438–453. DOI:10.1016/0021-8693(92)90026-I

[Gro1] A. Grothendieck, *Sur quelques points d'algèbre homologique*, Tohoku Math. J. **9** (1957), 119–221. DOI:10.2748/tmj/1178244839

[Gro2] A. Grothendieck, *La théorie des classes de Chern*, Bull. Soc. Math. France **86** (1958), 137–154. DOI:10.24033/bsmf.1501

[Gro3] A. Grothendieck, *The Cohomology Theory of Abstract Algebraic Varieties*, in: Proc. ICM 1958, Edinburgh, 103–118. Cambridge University Press (1960). www.mathunion.org/fileadmin/ICM/Proceedings/ICM1958/ICM1958.ocr.pdf

[Gro4] A. Grothendieck, *Théorèmes de dualité pour les faisceaux algébriques cohérents*, Sém. Bourbaki **149** (1958), 169–193. numdam:SB_1956-1958_4__169_0

[Gro5] A. Grothendieck, *Techniques de construction en géométrie analytique. VI. Étude locale des morphismes: germes d'espaces analytiques, platitude, morphismes simples*, Séminaire Henri Cartan **13** (1960-1961), exp. 13, 1–13. numdam:SHC_1960-1961__13_1_A9_0

[Gross] B. Gross, *Hanoi lectures on the arithmetic of hyperelliptic curves*, 2012, people.math.harvard.edu/~gross/preprints/hanoi2.pdf

[GWI] U. Görtz, T. Wedhorn, *Algebraic Geometry I*, 2nd ed., Springer Spektrum (2020).
DOI:10.1007/978-3-658-30733-2

[Hal] J. Hall, *GAGA theorems*, arXiv:1804.01976 (2018).

[Har1] R. Hartshorne, *Residues and Duality*, Lecture Notes in Mathematics **20**, Springer (1966).
DOI:10.1007/BFb0080482

[Har2] R. Hartshorne, *Ample subvarieties of algebraic varieties*, Lecture Notes in Math. **156**, Springer (1970).
DOI:10.1007/BFb0067839

[Har3] R. Hartshorne, *Algebraic Geometry*, corrected 8th printing, Graduate Texts in Mathematics **52**, Springer (1997).
DOI:10.1007/978-1-4757-3849-0

[HeKu] J. Herzog, E. Kunz, *Die Wertehalbgruppe eines lokalen Rings der Dimension 1*, Ber. Heidelberger Akad. Wiss. **2** (1971).
DOI:10.1007/978-3-642-46267-2

[Her] J. Herzog, *Generators and relations of abelian semigroups and semigroup rings*, manuscripta math. **3** (1970), 175–193.
DOI:10.1007/BF01273309

[HiSi] M. Hindry, J. Silverman, *Diophantine Geometry*, Graduate Texts in Mathematics **201**, Springer (2000).
DOI:10.1007/978-1-4612-1210-2

[Hoc] M. Hochster, *Prime Ideal Structure in Commutative Rings*, Trans. AMS **142** (1969), 43–60.
DOI:10.2307/1995344

[Hu] R. Huber, *A generalization of formal schemes and rigid analytic varieties*, Math. Zeitschrift **217** (1994), 513–551.
DOI:10.1007/BF02571959

[Hub] A. Hubery, *Characterising the bounded derived category of an hereditary abelian category*, arXiv:1612.06674 (2016).

[Huy] D. Huybrechts, *Fourier-Mukai transforms in algebraic geometry*, Oxford University Press (2006).
DOI:10.1093/acprof:oso/9780199296866.001.0001

[ILN] S. Iyengar, J. Lipman, A. Neeman, *Relation between two twisted inverse image pseudofunctors in duality theory*, Compos. Math. **151** (2015), no. 4, 735–764.
DOI:10.1112/S0010437X14007672   arXiv:1307.7092

[Kah1] B. Kahn, *Fonctions zêta et L de variétés et de motifs*, Coll. NANO, Calvage Mounet (2018).
arXiv:1512.09250

[Kah2] B. Kahn, *Zeta and L-Functions of Varieties and Motives*, Cambridge Univ. Press (2020).
DOI:10.1017/9781108691536

[KaSh] M. Kashiwara, P. Schapira, *Categories and Sheaves*, Springer Grundlehren **332**, 2006.
DOI:10.1007/3-540-27950-4

[Kaw] T. Kawasaki, *On Arithmetic Macaulification of Noetherian Rings*, Trans. AMS **354** (2002), no. 1, 123–149.
DOI:10.1090/S0002-9947-01-02817-3

[Ked] K. S. Kedlaya, *Sheaves, Stacks, and Shtukas*, in *Perfectoid Spaces: Lectures from the 2017 Arizona Winter School*, ed. by B. Cais, AMS Mathematical Surveys and Monographs **242** (2019).
bookstore.ams.org/view?ProductCode=SURV/242
swc-math.github.io/aws/2017/2017KedlayaNotes.pdf

[Kem] G. Kempf, *On algebraic curves*, J. reine u. angew. Math. (Crelle) **295** (1977), 40–48.   eudml.org/doc/151915

[Kie] R. Kiehl, *Ein "Descente"-Lemma und Grothendiecks Projektionssatz für nicht-noethersche Schemata*, Math. Ann. **198** (1972), 287–316.   DOI:10.1007/BF01419561

[KiWe] R. Kiehl, R. Weissauer, *Weil Conjectures, Perverse Sheaves and l'adic Fourier Transform*, Erg. Math., 3. Folge, Bd. **42**, Springer (2001). DOI:10.1007/978-3-662-04576-3

[Kna] A. Knapp, *Elliptic Curves*, Math. Notes **40**, Princeton Univ. Press (1992).
press.princeton.edu/books/paperback/9780691085593/elliptic-curves-mn-40-volume-40

[KnMu] F. Knudsen, D. Mumford, *The projectivity of the moduli space of stable curves I: Preliminaries on "det" and "Div"*, Math. Scan. **39** (1976), 19–55.
DOI:10.7146/math.scand.a-11642

[KPR] H. Kurke, G. Pfister, M. Roczen, *Henselsche Ringe und algebraische Geometrie*, Math. Monogr., vol. **II**, VEB Deutscher Verlag der Wissenschaften (1975).

[Kun] E. Kunz, *Residues and Duality for Projective Algebraic Varieties*, with the assistance of and contributions by D. A. Cox and A. Dickenstein, AMS (2008).
bookstore.ams.org/view?ProductCode=ULECT/47

[KuPr] D. Kubrak, A. Prikhodko, *Hodge-to-de Rham degeneration for stacks*, IMRN (2021), rnab54.   DOI:10.1093/imrn/rnab054   arXiv:1910.12665v2

[Lan] K. W. Lan, *Arithmetic compactifications of PEL-type Shimura varieties*, London Math. Soc. Monographs **36**, Princeton Univ. Press (2013).   press.princeton.edu/...
www-users.cse.umn.edu/~kwlan/articles/cpt-PEL-type-thesis-revision.pdf

[Lang] S. Lang, *Algebra*, revised 3rd edition, Springer (2002). DOI:10.1007/978-1-4613-0041-0

[Lau] K. Lauter, *Geometric methods for improving the upper bounds on the number of rational points on algebraic curves over finite fields*, with an appendix by J.-P. Serre. J. Algebraic Geom. **10** (2001), no. 1, 19–36.   arXiv:math/0104247

[Law] T. Lawson, *In which I try to get the signs right for once*,
www-users.cse.umn.edu/~tlawson/papers/signs.pdf

[Lic] S. Lichtenbaum, *Curves over discrete valuation rings*, Amer. J. Math. **90** (1968), 380–405.   DOI:10.2307/2373535

[Lip1] J. Lipman, *Dualizing sheaves, differentials and residues on algebraic varieties*, Astérisque **117** (1984).   numdam:AST_1984__117__1_0

[Lip2] J. Lipman, *Notes on Derived Functors and Grothendieck duality*, in: Lipman, Hashimoto, Foundations of Grothendieck Duality for Diagrams of Schemes, Lecture Notes in Mathematics **1960**, Springer (2009).
DOI:10.1007/978-3-540-85420-3   www.math.purdue.edu/~lipman/Duality.pdf

[Lip3] J. Lipman, *Grothendieck Duality theories – abstract and concrete, I: pseudo-coherent finite maps*, arXiv:1908.09372v1 (2019).

[Liu] Q. Liu, *Algebraic geometry and arithmetic curves*, Oxford Graduate Texts in Math. **6**, Oxford University Press (2002). global.oup.com/academic/product/algebraic-geometry-and-arithmetic-curves-9780199202492

[LN] J. Lipman, A. Neeman, *Quasi-perfect scheme-maps and boundedness of the twisted inverse image functor*, Illinois J. Math. **51** (1), 209–236. DOI:10.1215/ijm/1258735333   arXiv:math/0611760

[LNS] J. Lipman, S. Nayak, P. Sastry, Pramathanath, *Variance and duality for Cousin complexes on formal schemes*, Cont. Math. **375**, AMS 2005.   DOI:10.1090/conm/375

[Loc] P. Lockhart, *On the discriminant of a hyperelliptic curve*, Trans. A.M.S. **342** (1994), 729–751.   DOI:10.1090/S0002-9947-1994-1195511-X

[LoKl] K. Lønsted, S. Kleiman, *Basics on families of hyperelliptic curves*, Compos. Math. **38** (1979), 83–111.   numdam:CM_1979__38_1_83_0

[Lor] D. Lorenzini, *An invitation to Arithmetic Algebraic Geometry*, Graduate Studies in Math. **9**, Amer. Math. Soc. (1996).   bookstore.ams.org/view?ProductCode=GSM/9

[Lu-DAG] J. Lurie, *Derived Algebraic Geometry*, (2011), www.math.ias.edu/~lurie/papers/DAG.pdf

[Lu-DAGXII] J. Lurie, *Derived Algebraic Geometry XII: Proper Morphisms, Completions, and the Grothendieck Existence Theorem*, (2011), www.math.ias.edu/~lurie/papers/DAG-XII.pdf

[Lu-HTT] J. Lurie, *Higher Topos Theory*, Annals of Mathematics Studies **170**, Princeton University Press (2009). press.princeton.edu/books/paperback/9780691140490/higher-topos-theory-am-170 www.math.ias.edu/~lurie/papers/HTT.pdf

[MaMo] S. Mac Lane, I. Moerdijk, *Sheaves in Geometry and Logic*, Universitext, Springer (1994).   DOI:10.1007/978-1-4612-0927-0

[Man] L. Mann, *A p-Adic 6-Functor Formalism in Rigid-Analytic Geometry*, arXiv:2206.02022v1 (2022).

[Mat1] H. Matsumura, *Commutative Algebra*, 2nd ed., Benjamin/Cummings (1981).

[Mat2] H. Matsumura, *Commutative ring theory*, Cambridge studies in advanced mathematics **8**, Cambridge University Press (1989).

[May] J.P. May, *A Concise Course in Algebraic Topology*, University of Chicago Press (1999).   press.uchicago.edu/ucp/books/book/chicago/C/bo3777031.html www.math.uchicago.edu/~may/CONCISE/ConciseRevised.pdf

[MFK] D. Mumford, J. Fogarty, F. Kirwan, *Geometric Invariant Theory*, Springer Erg. Math. **34**, 3rd ed., Springer (1994).   link.springer.com/book/9783540569633

[Mil1] J. Milne, *Étale Cohomology*, Princeton Univ. Press (1980). press.princeton.edu/books/hardcover/9780691082387/etale-cohomology-pms-33-volume-33

[Mil2] J. Milne, *Jacobian Varieties*, in [CoSi]$^{\circ}$.   DOI:10.1007/978-1-4613-8655-1

[Mil3] J. Milne, *The Riemann Hypothesis over Finite Fields*, in: The Legacy of Bernhard Riemann after One Hundred and Fifty Years (Lizhen Ji, Frans Oort, Shing-Tung Yau eds.), ALM **35**, 2015, 487–565.   jmilne.org/math/xnotes/pRH.pdf

[Mil4] J. Milne, *Abelian Varieties*, v2.0 (2008).   jmilne.org/math/CourseNotes/av.html

[MiMo] J. Milnor, J. Moore, *On the structure of Hopf algebras*, Annals of math. **82**, No. 2 (1965), 211–264.   DOI:10.2307/1970615

[Moc] S. Mochizuki, *Topics in absolute anabelian geometry I: generalities*, J. Math. Sci. Univ. Tokyo **19** (2012), no. 2, 139–242.
www.kurims.kyoto-u.ac.jp/~motizuki/Topics in Absolute Anabelian Geometry I.pdf

[Muk1] S. Mukai, *Duality between $D(X)$ and $D(\hat{X})$ and its application to Picard sheaves*, Nagoya Math. Journal **81** (1981), 153–175.   projecteuclid.org/...

[Muk2] S. Mukai, *Fourier Functor and its Application to the Moduli of Bundles on an Abelian Variety*, Adv. Studies in Pure Math. **10** (1987), 515–550.   DOI:10.2969/aspm/01010515

[Mum1] D. Mumford, *Abelian Varieties*, 2nd ed., Oxford Univ. Press (1974).

[Mum2] D. Mumford, *Curves and their Jacobians*, Univ. of Michigan (1975). Reprinted in D. Mumford, *The Red Book of Varieties and Schemes*, Lecture Notes in Mathematics **1358**, Springer (1999).

[Mur] J. Murre, *On contravariant functors from the category of preschemes over a field into the category of abelian groups*, Publ. math. IHES**23** (1964), 5–43.   numdam:PMIHES_1964__23__5_0

[Nag] M. Nagata, *Local rings*, Interscience Publishers (1962).

[Nav] A. Navarro, *On Grothendieck's Riemann-Roch theorem*, Expos. math. **35** (2017), no. 3, 326–342.   DOI:10.1016/j.exmath.2016.09.005   arXiv:1603.06740

[Nee1] A. Neeman, *The connection between the K-theory localization theorem of Thomason, Trobaugh and Yao and the smashing subcategories of Bousfield and Ravenel*, Ann. sci. E.N.S., 4e sér. **25** no. 5 (1992), 547–566.   DOI:10.24033/asens.1659

[Nee2] A. Neeman, *Stable homotopy as a triangulated functor*, Invent. math. **109**, Issue 1 (1992), 17–40.   DOI:10.1007/BF01232016

[Nee3] A. Neeman, *The Grothendieck duality theorem via Bousfield's techniques and Brown representability*, J. A. M. S. **9** (1996), no. 1, 205–236. DOI:10.1090/S0894-0347-96-00174-9

[Nee4] A. Neeman, *Triangulated Categories*, Ann. of Math. Studies **148**, Princeton Univ. Press (2001).   press.princeton.edu/...

[Nee5] A. Neeman, *Derived categories and Grothendieck duality*, in: Triangulated Categories, T. Holm, P. Jorgenson, R. Rouquier (eds.), LMS Lecture Note Series **375** (2010), 290–350.
DOI:10.1017/CBO9781139107075.007   core.ac.uk/download/pdf/13282696.pdf

[Nee6] A. Neeman, *Traces and residues*, Indiana Univ. Math. J. **64** (2015), no. 1, 217–229.
www.jstor.org/stable/26315456   openresearch-repository.anu.edu.au/handle/1885/13287

[Nee7] A. Neeman, *The Decomposition of* $\mathrm{Hom}_k(S,k)$ *into Indecomposable Injectives*, Acta Math Vietnam **40** (2015), 331–338.   DOI:10.1007/s40306-014-0110-z

[Neu] J. Neukirch, *Algebraic Number Theory*, Springer-Verlag (1999).
                                                                                    DOI:10.1007/978-3-662-03983-0

[NS1] S. Nayak, P. Sastry, *Grothendieck Duality and Transitivity I: Formal Schemes*, arXiv:1903.01779v3 (2019).

[NS2] S. Nayak, P. Sastry, *Grothendieck Duality and Transitivity II: Traces and Residues via Verdier's isomorphism*, arXiv:1903.01783v2 (2019).

[NSW] J. Neukirch, A. Schmidt, K. Wingberg, *Cohomology of Number Fields*, Grundl. math. Wiss. **323**, 2nd ed., Springer (2013).
              DOI:10.1007/978-3-540-37889-1   www.mathi.uni-heidelberg.de/~schmidt/NSW2e/

[Orl] D. Orlov, *Derived categories of coherent sheaves and equivalences between them*, Russian Math. Surveys **58**:3 (2003), 511–591.
                            DOI:10.1070/RM2003v058n03ABEH000629   arXiv:alg-geom/9712017

[Pan] I. Panin, *Riemann-Roch theorems for oriented cohomologies*, in: Axiomatic, enriched and motivic homotopy theory, NATO Sci. Ser. II Math. Phys. Chem. **131**, Kluwer (2004), 261–333.   DOI:10.1007/978-94-007-0948-5_8

[Pol] A. Polishchuck, *Abelian Varieties, Theta Functions and the Fourier Transform*, Cambridge University Press (2003).   DOI:10.1017/CBO9780511546532

[Pos] L. Positselski, *Remarks on derived complete modules and complexes*, arXiv:2002.12331 (2020).

[Ras] S. Raskin, *The Weil conjectures for curves*,
       math.uchicago.edu/~may/VIGRE/VIGRE2007/REUPapers/FINALFULL/Raskin.pdf (2007)

[Ray1] M. Raynaud, *Faisceaux amples sur les schémas en groupes et les espaces homogènes*, Lecture Notes in Mathematics **119**, Springer (1970).   DOI:10.1007/BFb0059504

[Ray2] M. Raynaud, *Anneaux Locaux Henséliens*, Lecture Notes in Mathematics **169**, Springer (1970).   DOI:10.1007/BFb0069571

[Roo] J.-E. Roos, *Derived Functors of inverse limits revisited*, J. London Math. Soc. **73** (2006), 65–83.   DOI:10.1112/S0024610705022416

[Rot] C. Rotthaus, *Rings with approximation theory*, Math. Ann. **287** (1990), 455–466.
                                                                                    eudml.org/doc/164697

[Rou] R. Rouquier, *Derived categories and algebraic geometry*, in: Triangulated Categories, T. Holm, P. Jorgenson, R. Rouquier (eds.), LMS Lecture Note Series **375** (2010), 351–370.   DOI:10.1017/CBO9781139107075.008

[Sam] P. Samuel, *Lectures On Old And New Results On Algebraic Curves*, Lecture notes Tata Inst. (1966).   www.math.tifr.res.in/~publ/ln/tifr36.pdf

[Sas1] P. Sastry, *Residues and duality for algebraic schemes*, Compos. Math. **101** (1996), no. 2, 133–178.   eudml.org/doc/90441

[Sas2] P. Sastry, *Base change and Grothendieck duality for Cohen-Macaulay maps*, Compos. Math. **140** (2004), no. 3, 729–777.
DOI:10.1112/S0010437X03000654 arXiv:math/0011138

[SaTo] P. Sastry, Y. Tong, *The Grothendieck trace and the de Rham integral*, Canad. Math. Bull. **46** (2003), no. 3, 429–440. DOI:10.4153/CMB-2003-043-3

[Sche] C. Scheiderer, *Quasi-augmented simplicial spaces, with an application to cohomological dimension*, J. Pure Appl. Algebra **81** (1992), 293–311.
DOI:10.1016/0022-4049(92)90062-K

[Scho] P. Scholze, *Lectures on Condensed Mathematics*, 2019,
www.math.uni-bonn.de/people/scholze/Condensed.pdf

[Ser1] J.-P. Serre, *Faisceaux Algébriques Cohérents*, Annals of math. **61** (1955), No. 2, 197–278. DOI:10.2307/1969915

[Ser2] J.-P. Serre, *Géométrie algébrique et géométrie analytique*, Ann. Inst. Fourier **6** (1956), 1–42. DOI:10.5802/aif.59

[Ser3] J. P. Serre, *Rigidité du foncteur de Jacobi d'échelon n ≥ 3*, App. à l'exposé 17 du séminaire Cartan 60/61. numdam:SHC_1960-1961__13_2_A4_0

[Ser4] J. P. Serre, *Prolongement de faisceaux analytiques cohérents*, Ann. Inst. Fourier **16** (1966), 363–374. DOI:10.5802/aif.234

[Ser5] J. P. Serre, *Cours d'Arithmétique*, Presses Univ. de France (1970). English translation: *A Course in Arithmetic*, Graduate Textes in Mathematics **7**, Springer (1973).

[Ser6] J. P. Serre, *Groupes algébriques et corps de classes*, Hermann (1975).

[SGA1] A. Grothendieck et al., *Revêtements Étales et Groupe Fondamental* (SGA 1), Lecture Notes in Mathematics **224**, Springer (1971), new ed. Doc. Math. **3**, Soc. Math. France (2003). DOI:10.1007/BFb0058656 arXiv:math/0206203

[SGA2] A. Grothendieck et al., *Cohomologie locale des faisceaux cohérents et théorèmes de Lefschetz locaux et globaux* (SGA 2), Adv. Studies in Pure Math. **2**, North-Holland (1968), new ed. Doc. Math. **4**, Soc. Math. France (2005).
smf.emath.fr/... arXiv:math/0511279

[SGA3] A. Grothendieck et al., *Schémas en groupes* (SGA 3), vol. I, Lecture Notes in Mathematics **151**, vol. II, Lecture Notes in Mathematics **152**, vol. III, Lecture Notes in Mathematics **153**, Springer (1970), new ed. vol. I, Doc. Math. **7**, vol. III, Doc. Math. **8**, Soc. Math. France (2011).
smf.emath.fr/... webusers.imj-prg.fr/~patrick.polo/SGA3/

[SGA4] M. Artin, A. Grothendieck, J. Verdier, *Théorie des Topos et Cohomologie Étale des Schémas* (SGA 4), Springer Lecture Notes in Mathematics **269, 270, 305** (1972–1973). DOI:10.1007/BFb0081551

[SGA4½] P. Deligne, *Cohomologie étale*, Springer Lecture Notes in mathematics **569**, Springer (1977). DOI:10.1007/BFb0091516

[SGA6] A. Grothendieck et al., *Théorie des intersections et théorème de Riemann-Roch*, Lecture Notes in Mathematics **225**, Springer (1971).   DOI:10.1007/BFb0066283

[Sharp] R. Sharp, *Necessary conditions for the existence of dualizing complexes in commutative algebra*, in: Sém. Algèbre P. Dubreil 1977/78, Springer Lecture Notes in Mathematics **740** (1979), 213–229.   DOI:10.1007/BFb0071048

[Shatz] S.S. Shatz, *The decomposition and specialization of algebraic families of vector bundles*, Comp. Math.**35** (1977), 163–187.   numdam:CM_1977__35_2_163_0

[Sil1] J. Silverman, *The Arithmetic of Elliptic Curves*, Graduate Texts in Mathematics **106**, 2nd ed., Springer (2009).   DOI:10.1007/978-0-387-09494-6

[Sil2] J. Silverman, *Advanced Topics in the Arithmetic of Elliptic Curves*, Graduate Texts in Mathematics **151**, Springer (1994).   DOI:10.1007/978-1-4612-0851-8

[Sou] J.-P. Soublin, *Anneaux et Modules Cohérents*, Journal of Algebra **15** (1970), 455–472.   DOI:10.1016/0021-8693(70)90050-5

[Spa] N. Spaltenstein, *Resolutions of unbounded complexes*, Compos. Math. **65**, no. 2 (1988), 121–154.   numdam:CM_1988__65_2_121_0

[Sta] Stacks project collaborators, *Stacks project*, stacks.math.columbia.edu

[Ste] B. Stenström, *Rings of Quotients*, Springer Verlag (1975).

[Sto] M. Stoll, *Arithmetic of hyperelliptic curves*, Lecture Notes Univ. Bayreuth (2014). mathe2.uni-bayreuth.de/stoll/teaching/ArithHypKurven-SS2014/ Skript-ArithHypCurves-pub-screen.pdf

[Tam] G. Tamme, *Introduction to étale cohomology*, Universitext, Springer (1994).   DOI:10.1007/978-3-642-78421-7

[tDi] T. tom Dieck, *Algebraic Topology*, 2nd printing, European Mathematical Society Textbooks in Mathematics (2010).   DOI:10.4171/048

[TeTy] M. Temkin, I. Tyomkin, *Ferrand pushouts for algebraic spaces*, European J. Math. **2** (2016), 960–983.   DOI:10.1007/s40879-016-0115-3   arXiv:1305.6014

[Tho] R. Thomason, *The classification of triangulated subcategories*, Compos. Math. **105** (1997), 1–27.   DOI:10.1023/A:1017932514274

[ThTr] R.W. Thomason, T. Trobaugh, *Higher Algebraic K-Theory of Schemes and of Derived Categories*, in *The Grothendieck Festschrift, Volume III* (ed. by P. Cartier et. al.), Progress in Mathematics **88**, Birkhäuser Boston (1990), 247–435.   DOI:10.1007/978-0-8176-4576-2_10

[To] B. Totaro, *The resolution property for schemes and stacks*, J. Reine Angew. Math. **577** (2004), 1–22.   DOI:10.1515/crll.2004.2004.577.1   arXiv:math/0207210

[Vak] R. Vakil, *The Rising Sea, Foundations of Algebraic Geometry* (2017). math.stanford.edu/~vakil/216blog/FOAGnov1817public.pdf

[Ver1] J.-L. Verdier, *Dualité dans la cohomologie des espaces localement compacts*, Sém. Bourbaki **300** (1965–66).   numdam:SB_1964-1966__9__337_0

[Ver2] J.-L. Verdier, *Base change for twisted inverse image of coherent sheaves*, in: Proc. Conf. Algebraic Geometry, Tata Inst. Fund. Res., Bombay, 1968, 393–408.
www.math.tifr.res.in/~publ/studies/SM_04-Algebraic-Geometry.pdf

[Ver3] J.-L. Verdier, *Des catégories dérivées des catégories abéliennes*, Asterisque **239** (1996). numdam:AST_1996__239__R1_0

[Was] L. Washington, *Elliptic Curves: Number Theory and Cryptography*, 2nd ed., Taylor and Francis (2008). DOI:10.1201/9781420071474

[Wed] T. Wedhorn, *Manifolds, Sheaves, and Cohomology*, Springer (2016).
DOI:10.1007/978-3-658-10633-1

[Wei1] C. Weibel, *Homotopy algebraic K-theory*, AMS Contemp. Math. **83** (1989), 461–488.
https://sites.math.rutgers.edu/~weibel/papers-dir/KH-theory.pdf

[Wei2] C. Weibel, *An introduction to homological algebra*, Cambridge Studies in Advanced Mathematics **38**, Cambridge Univ. Press (1994). DOI:10.1017/CBO9781139644136

[Wei3] C. Weibel, *The K-book: an introduction to algebraic K-theory"*, Grad. Studies in Math. **145**, AMS (2013).
bookstore.ams.org/view?ProductCode=GSM/145   sites.math.rutgers.edu/~weibel/Kbook.html

[Yas] T. Yasuda, *Non-adic Formal Schemes*, IMRN **2009**(13), 2417–2475.
DOI:10.1093/imrn/rnp021   arXiv:0711.0434

[Yek1] A. Yekutieli, *Dualizing complexes over noncommutative graded algebras*, J. Algebra **153** (1992), no. 1, 41–84. DOI:10.1016/0021-8693(92)90148-F

[Yek2] A. Yekutieli, *An explicit construction of the Grothendieck residue complex*, with an appendix by P. Sastry. Astérisque **208** (1992). numdam:AST_1992__208__3_0

[Yek3] A. Yekutieli, *Weak proregularity, derived completion, adic flatness, and prisms*, J. Algebra **583** (2021), 126–152. DOI:10.1016/j.jalgebra.2021.04.033   arXiv:2002.04901

[Zav] B. Zavyalov, *Almost coherent modules and almost coherent sheaves*, arXiv:2110.10773 (2021).

Detailed List of Contents

© Springer Fachmedien Wiesbaden GmbH, ein Teil von Springer Nature 2023
U. Görtz und T. Wedhorn, *Algebraic Geometry II: Cohomology of Schemes*,
Springer Studium Mathematik – Master, https://doi.org/10.1007/978-3-658-43031-3

Index of Symbols

$\bigwedge_R M$ exterior algebra, 24

$- \cup -$ cup product, 207

$- \otimes -$ tensor product of complexes, 762

$- \otimes^L -$ derived tensor product, 193

$- \amalg_- -$ pushout in a category, 743

$(\mathrm{Ab}(X))$ category of abelian sheaves on X, 153

$\mathscr{A}_{g,d,N}$ moduli space of principally polarized abelian varieties, 733

$\mathcal{C}/\mathcal{N}$ localization defined by null system, 785

$c_1(\mathscr{L})$ first Chern class of line bundle, 319

$c^A(\mathscr{E})$ Chern polynomial, 336

$c_i^A(\mathscr{E})$ i-th Chern class, 336

$\check{C}^n_{\mathrm{alt}}(\mathscr{U}, \mathscr{F})$ alternating Čech complex, 180

$\check{C}^n_{\mathrm{ord}}(\mathscr{U}, \mathscr{F})$ ordered Čech complex, 180

$C^*(R)$ bounded versions of $C(R)$, 153

$C^*(X)$ bounded versions of $C(X)$, 153

$\mathcal{C}_{/d}$ slice category over d, 739

$\mathcal{C}_{\backslash d}$ slice category under d, 739

$C_f^\bullet$ mapping cone of f, 757

ch^A Chern character, 340

$\chi(\mathscr{F})$ Euler characteristic, 320

$\mathscr{C}_i$ conormal sheaf, 11

Clopen(X)   set of open and closed subschemes of X, 96

$(\mathrm{Coh}(X))$ category of coherent $\mathscr{O}_X$-modules, 254

$\mathrm{colim}_{\mathcal{I}}$ colimit, 739

$\mathrm{Corr}_S(X, X')$ divisorial correspondences, 659

$\mathscr{C}\mathscr{O}_{X/S}$ scheme of open and closed subschemes, 96

$C(R)$ category of complexes in $(R\text{-Mod})$, 153

$\mathcal{C}_S$ localization of category $\mathcal{C}$ w.r.t. S, 782

$\check{C}(\mathscr{U}, \mathscr{F})$ Čech complex with respect to covering, 179

$C(X)$ category of complexes of $\mathscr{O}_X$-modules, 153

$\check{C}(X, \mathscr{F})$ Čech complex, 182

$\mathscr{C}_{Y/X}$ conormal sheaf, 11

$(D_1 \cdots D_t \cdot \mathscr{F})$ intersection number, 323

$D(\mathcal{A})$ derived category of abelian category $\mathcal{A}$, 785

$D^*_{\mathcal{A}'}(\mathcal{A})$ subcategory of $D^*(\mathcal{A})$, 791

$D^*(R)$ bounded versions of $D(R)$, 153

$D^*(X)$ bounded versions of $D(X)$, 153

$D_C(N)$ augmented extension by square zero ideal, 7

$D_{\mathrm{coh}}(X)$ derived category of complexes with coherent cohomology objects, 254

$D_{\mathrm{comp}}(A, I)$ derived I-complete complexes, 380

$D_{\mathrm{comp}}(X)$ derived complete complexes, 388

$D_{\mathrm{comp}}(X, Z)$ derived complete complexes, 388

$D_{\mathrm{comp}}(X, \mathscr{I})$ derived complete complexes, 388

$\deg(\mathscr{L})$ degree of line bundle, 327

$\deg_{\mathscr{L}}(Z)$ degree of closed subscheme w.r.t. line bundle, 327

$\deg(Z)$ degree of closed subscheme, 327

$\mathrm{Der}_R(A, M)$ R-derivations from A to M, 5

$\mathrm{Der}_S(\mathscr{O}_X, \mathscr{F})$ S-derivations from $\mathscr{O}_X$ to $\mathscr{F}$, 14

$D^{\mathcal{K}}$ localization of D w.r.t. $\mathcal{K}$, 789

$D_{\mathrm{qcoh}}(X)$ derived category of complexes with quasi-coherent cohomology objects, 243

$D(R)$ derived category of $(R\text{-Mod})$, 153

$D(X)$ derived category of $(\mathscr{O}_X\text{-Mod})$, 153

$\mathfrak{d}_{X/S}$ different of X over S, 121

$\mathscr{D}_{X/S}$ discriminant of X over S, 120

$\varepsilon_i(A)$ dimension of Koszul homology for local ring, 84

$\mathrm{Ex}_R(A, M)$ Extensions of A by M, 9

$\mathrm{Ext}_A^i(X, Y)$ i-th Ext group, 811

$\mathscr{E}xt^i(\mathscr{F}, \mathscr{G})$ Ext sheaf, 195

f^{an} analytification of f, 114

$\mathscr{F}_{\bar{s}}$ fiber functor, 124

$F^{\#}$ sheafification of presheaf F, 615

$f^!$ twisted inverse image functor, 464

f^t dual homomorphism of abelian schemes, 664

$f^{\times}$ right adjoint of Rf_*, 444

$\mathscr{F}^{\wedge}$ derived completion along closed subscheme, 388

$\mathscr{F}^{\wedge}_{/Z}$ derived completion along closed subscheme, 388

G^0 identity component of group scheme, 608

$\Gamma_Z(X, \mathscr{F})$ sections with support in Z, 177

$g(C)$ (arithmetic) genus of curve C, 527

© Springer Fachmedien Wiesbaden GmbH, ein Teil von Springer Nature 2023
U. Görtz und T. Wedhorn, *Algebraic Geometry II: Cohomology of Schemes*,
Springer Studium Mathematik – Master, https://doi.org/10.1007/978-3-658-43031-3

Index

© Springer Fachmedien Wiesbaden GmbH, ein Teil von Springer Nature 2023
U. Görtz und T. Wedhorn, *Algebraic Geometry II: Cohomology of Schemes*,
Springer Studium Mathematik – Master, https://doi.org/10.1007/978-3-658-43031-3